ATOMIC WEIGHTS OF THE ELEMENTS[†] (IUPAC 2007), Based on relative atomic mass of ^{12}C = 12 exactly

Name	Symbol	Atomic Number	Atomic Weight
Actinium*	Ac	89	(227)
Aluminum	Al	13	26.9815386(8)
Americium*	Am	95	(243)
Antimony	Sb	51	121.760(1)
Argon	Ar	18	39.948(1)
Arsenic	As	33	74.92160(2)
Astatine*	At	85	(210)
Barium	Ba	56	137.327(7)
Berkelium*	Bk	97	(247)
Beryllium	Be	4	9.012182(3)
Bismuth	Bi	83	208.98040(1)
Bohrium	Bh	107	(272)
Boron	B	5	10.811(7)
Bromine	Br	35	79.904(1)
Cadmium	Cd	48	112.411(8)
Cesium	Cs	55	132.9054519(2)
Calcium	Ca	20	40.078(4)
Californium*	Cf	98	(251)
Carbon	C	6	12.0107(8)
Cerium	Ce	58	140.116(1)
Chlorine	Cl	17	35.453(2)
Chromium	Cr	24	51.9961(6)
Cobalt	Co	27	58.933195(5)
Copernicium	Cn	112	(285)
Copper	Cu	29	63.546(3)
Curium*	Cm	96	(247)
Darmstadtium	Ds	110	(281)
Dubnium	Db	105	(268)
Dysprosium	Dy	66	162.500(1)
Einsteinium*	Es	99	(252)
Erbium	Er	68	167.259(3)
Europium	Eu	63	151.964(1)
Fermium*	Fm	100	(257)
Fluorine	F	9	18.9984032(5)
Francium*	Fr	87	(223)
Gadolinium	Gd	64	157.25(3)
Gallium	Ga	31	69.723(1)
Germanium	Ge	32	72.64(1)
Gold	Au	79	196.966569(4)
Hafnium	Hf	72	178.49(2)
Hassium	Hs	108	(270)
Helium	He	2	4.002602(2)
Holmium	Ho	67	164.93032(2)
Hydrogen	H	1	1.00794(7)
Indium	In	49	114.818(3)
Iodine	I	53	126.90447(3)
Iridium	Ir	77	192.217(3)
Iron	Fe	26	55.845(2)
Krypton	Kr	36	83.798(2)
Lanthanum	La	57	138.90547(7)
Lawrencium*	Lr	103	(262)
Lead	Pb	82	207.2(1)
Lithium	Li	3	6.941(2)
Lutetium	Lu	71	174.9668(1)
Magnesium	Mg	12	24.3050(6)
Manganese	Mn	25	54.938045(5)
Meitnerium	Mt	109	(276)
Mendelevium*	Md	101	(258)
Mercury	Hg	80	200.59(2)
Molybdenum	Mo	42	95.96(2)
Neodymium	Nd	60	144.242(3)
Neon	Ne	10	20.1797(6)
Neptunium*	Np	93	(237)
Nickel	Ni	28	58.6934(4)
Niobium	Nb	41	92.90638(2)
Nitrogen	N	7	14.0067(2)
Nobelium*	No	102	(259)
Osmium	Os	76	190.23(3)
Oxygen	O	8	15.9994(3)
Palladium	Pd	46	106.42(1)
Phosphorus	P	15	30.973762(2)
Platinum	Pt	78	195.084(9)
Plutonium*	Pu	94	(244)
Polonium*	Po	84	(209)
Potassium	K	19	39.0983(1)
Praseodymium	Pr	59	140.90765(2)
Promethium*	Pm	61	(145)
Protactinium*	Pa	91	231.03588(2)
Radium*	Ra	88	(226)
Radon*	Rn	86	(222)
Rhenium	Re	75	186.207(1)
Rhodium	Rh	45	102.90550(2)
Roentgenium	Rg	111	(280)
Rubidium	Rb	37	85.4678(3)
Ruthenium	Ru	44	101.07(2)
Rutherfordium	Rf	104	(267)
Samarium	Sm	62	150.36(2)
Scandium	Sc	21	44.955912(6)
Seaborgium	Sg	106	(271)
Selenium	Se	34	78.96(3)
Silicon	Si	14	28.0855(3)
Silver	Ag	47	107.8682(2)
Sodium	Na	11	22.98976928(2)
Strontium	Sr	38	87.62(1)
Sulfur	S	16	32.065(5)
Tantalum	Ta	73	180.94788(2)
Technetium*	Tc	43	(98)
Tellurium	Te	52	127.60(3)
Terbium	Tb	65	158.92535(2)
Thallium	Tl	81	204.3833(2)
Thorium*	Th	90	232.03806(2)
Thulium	Tm	69	168.93421(2)
Tin	Sn	50	118.710(7)
Titanium	Ti	22	47.867(1)
Tungsten	W	74	183.84(1)
Ununhexium	Uuh	116	(293)
Ununoctium	Uuo	118	(294)
Ununpentium	Uup	115	(288)
Ununquadium	Uuq	114	(289)
Ununtrium	Uut	113	(284)
Uranium*	U	92	238.02891(3)
Vanadium	V	23	50.9415(1)
Xenon	Xe	54	131.293(6)
Ytterbium	Yb	70	173.054(5)
Yttrium	Y	39	88.90585(2)
Zinc	Zn	30	65.38(2)
Zirconium	Zr	40	91.224(2)

[†]The atomic weights of many elements can vary depending on the origin and treatment of the sample. This is particularly true for Li; commercially available lithium-containing materials have Li atomic weights in the range of 6.939 and 6.996. The uncertainties in atomic weight values are given in parentheses following the last significant figure to which they are attributed.

*Elements with no stable nuclide; the value given in parentheses is the atomic mass number of the isotope of longest known half-life. However, three such elements (Th, Pa, and U) have a characteristic terrestrial isotopic composition, and the atomic weight is tabulated for these. **http://www.chem.qmul.ac.uk/iupac/AtWt/**

Online Web Learning

UMassAmherst

The results are in–and they prove that **OWL** will help you **study smarter and succeed** in chemistry!

OWL for Chemistry: Human Activity, Chemical Reactivity
The Chemist's Choice. The Student's Solution.

Strengthen your understanding with **OWL, the #1 online learning system for chemistry!**

Developed *by* chemistry instructors, **OWL** has already helped hundreds of thousands of students master chemistry through tutorials, interactive simulations, and algorithmically generated homework questions that provide instant answer-specific feedback.

OWL now features a modern, intuitive interface and is the only system specifically designed to support **mastery learning**, where you can work as long as you need to master each chemical concept and skill.

The newest version of **OWL for Chemistry: Human Activity, Chemical Reactivity** offers:

- **A wide range of assignment types**—tutorials, interactive simulations, short answer questions, and algorithmically generated homework questions that provide instant, answer-specific feedback, including *end-of-chapter questions from your textbook*

- **MarvinSketch**, an advanced molecular drawing program for drawing gradable structures

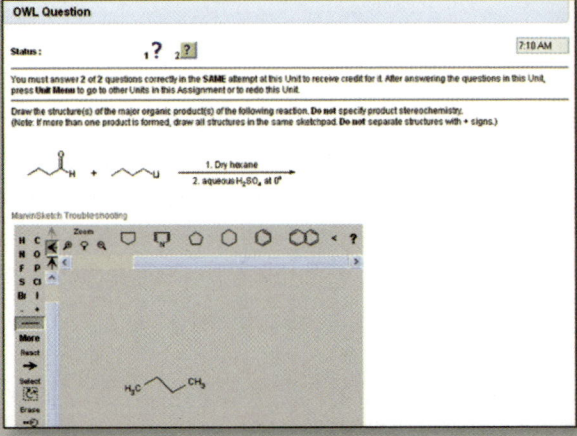

In addition, **OWL** also offers:

- **e-Books**, fully integrated electronic textbooks correlated to OWL questions
- **Jmol**, a molecular visualization program for rotating molecules and measuring bond distances and angles.

"I attribute my good grade in this course largely to **OWL**."

Student

"I liked the **step-by-step tutorials** and having all the charts (periodic table, etc.) a click away."

Student

CHEMISTRY

Human Activity, Chemical Reactivity

FIRST CANADIAN EDITION

Peter G. Mahaffy King's University College

Bob Bucat University of Western Australia

Roy Tasker University of Western Sydney

John C. Kotz State University of New York

Paul M. Treichel University of Wisconsin-Madison

Gabriela C. Weaver Purdue University

John McMurry Cornell University

NELSON / EDUCATION

NELSON / EDUCATION

Chemistry: Human Activity, Chemical Reactivity, First Canadian Edition
by Peter G. Mahaffy, Bob Bucat, Roy Tasker, John C. Kotz, Paul M. Treichel, Gabriela C. Weaver, and John McMurry

Vice President and Editorial Director:
Evelyn Veltch

Editor-in-Chief, Higher Education:
Anne Williams

Publisher:
Paul Fam

Senior Marketing Manager:
Sean Chamberland

Developmental Editor:
My Editor Inc.

Photo Researcher:
Joanne Tang

Permissions Coordinator:
Joanne Tang

Senior Content Production Manager:
Anne Nellis

Production Service:
Integra

Copy Editor:
Wendy Yano

Proofreader:
Integra

Indexer:
Integra

Manufacturing Coordinator:
Ferial Suleman

Design Director:
Ken Phipps

Managing Designer:
Franca Amore

Interior Design:
Liz Harasymczuk

Cover Design:
Olga Lavecchia

Cover Image:
Iain Cochrane/Photographer's Choice/Getty Images

Compositor:
Integra

Printer:
RR Donnelley

Library and Archives Canada Cataloguing in Publication

Chemistry : human activity, chemical reactivity /
Peter G. Mahaffy
... [et al.].—1st Canadian ed.

Includes index.
ISBN 978-0-17-610437-5

1. Chemistry—Textbooks. I. Mahaffy, Peter G.

QD31.3.C43 2010   540
C2010-900052-8

ISBN-13: 978-0-17-610437-5
ISBN-10: 0-17-610437-2

BRIEF CONTENTS

TABLE OF CONTENTS

Our Vision

Everything you smell, touch, and taste is composed of chemicals. So it is not surprising that a knowledge of chemistry is essential for any career involving the very small—like nanotechnology—to the everyday scale—like health care, engineering, and the environment—to the unimaginably big—like cosmology. People with an understanding of chemistry are needed to address the challenges posed by climate change; find new ways to harness energy; provide clean water and nutritious food for the world; contribute to the cure and prevention of disease; and apply our exponentially increasing knowledge of the brain and the genomes of living organisms.

Our vision for this project is to provide you with a broad range of learning resources that show chemistry as a current, creative, fascinating, and worthwhile human activity, with people like you building upon the experiments of imaginative scientists of the past. Chemistry is part of our culture, and we want to show that an understanding of chemistry is needed to understand our world, and that your knowledge can inform decision making.

We were each educated as chemists, and later specialized in research and development in chemistry education. This project gives us the opportunity to apply what we have discovered, and are still studying, about how students best learn chemistry. Our main aim is to communicate ideas clearly, using best practice pedagogy and the full range of media: text, graphics, animations, simulations, and interactive problem solving.

Our Approach

Chemistry: Human Activity, Chemical Reactivity presents chemistry as an engaging, developing human activity, as opposed to a body of facts, theories, and techniques handed down from the past.

The key features that differentiate our approach in these learning resources are given below, some followed by an example.

1.1 Case Study: Human Activity, Chemical Reactivity

Stressing that education in and about chemistry is critical to address challenges such as global climate change and sustainable supplies of clean water, food, energy, and medicine, the United Nations declared 2011 to be the International Year of Chemistry. You are about to enter your own year of studying that international and interwoven world of human activity and chemical reactivity. This first chapter tells the stories of two people who carry out research in chemistry to improve our world. You would not recognize them on the bus or in the supermarket as chemists. They are simply curious, hard-working people who use logic and creativity as they work with a research team to design and interpret experiments.

First we'll meet David Dolphin, a Canadian chemist who has designed and made new substances that have improved the quality of life for over a million people suffering from cancer or eye disease. Then we'll meet Gavin Flematti, an Australian chemist who, while he was a postgraduate student, identified a compound in smoke that causes plant seeds to germinate after a forest fire. He then found a way to make this compound in the laboratory.

You'll see with these two people that the work chemists carry out is challenging for many reasons, not least of which is that the particles they study—particles that make up everything in our world—are too small for us to see directly. So chemists develop and use instruments to extend their vision and to understand how the things we observe fit with molecular-level explanations.

As in so many areas of modern chemistry, in both of these stories, the interaction of electromagnetic radiation with matter plays a critical role. In the first case, visible light from a laser triggers the production of molecules that selectively destroy undesirable cells in body organs. In the second, selective absorption of electromagnetic radiation is the key to understanding the three-dimensional structures of newly discovered molecules in smoke.

Chemistry is presented as an exciting, developing human activity, rather than a body of facts, theories, and skills handed down from the past. Because modern chemistry is a global activity, examples of chemistry from around the world are included in case studies and elsewhere.

Each chapter opens with a case study that profiles people doing chemistry that matters.

Experimental facts are presented before the models used to explain them: that is, facts first, then theoretical modelling as an explanation second. In this way, an answer is always provided for the question "Why do we believe what we believe?" For example, in Chapter 3, we classify pure substances into categories based on their melting points, hardness, and electrical conductivity, and then develop the bonding models best able to explain them.

TABLE 3.3
Modelling the Nature of Metals

Observation	Inference
Metals are dense.	The particles comprising metals are closely packed.
Metals are good electrical conductors.	Metals contain charged particles that can move under the influence of an applied potential.
Metals are good conductors of heat.	Metals contain particles that can move from regions of high temperature to regions of low temperature, taking their energy with them.
Metals have high melting points.	There are strong forces of attraction holding the particles in fixed positions. Only when the energy of the particles is very high, achieved by high temperatures, can the particles move with respect to each other.
Metals are malleable and ductile.	Forcing particles in the solid metal to move past each other does not give rise to an unstable situation with high forces of repulsion.

FIGURE 3.9 A model for the structure of metals, portraying a lattice of positive ions (left) within a "sea" of electrons. For a simulation using this model to represent solid copper, see e3.8.

metallic ion

sea of mobile electrons

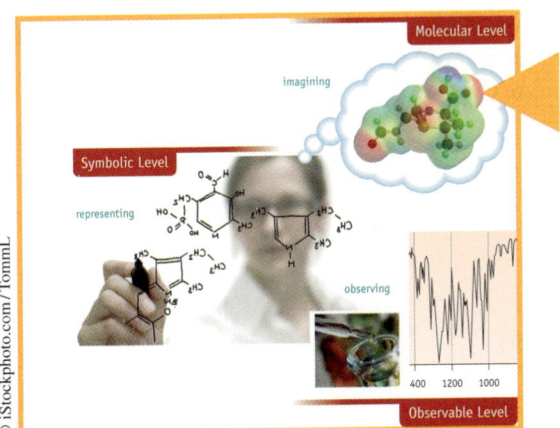

© iStockphoto.com/TommL

Learners can only understand chemistry fully if they can move seamlessly among the three thinking levels of chemistry: the *symbolic* level, the *observational* level, and the *molecular* level. The learning resource provides many opportunities to develop this skill for different substances, physical changes, and reactions. The word "ethanol," for example, may conjure up images of the symbolic level (the formula CH_3CH_2OH), the observational level (drops of clear liquid produced in a "still" or biofuel for a vehicle), and the molecular level (a physical or computer molecular model). In any given discussion, students will clearly understand which level they are "thinking at."

Examples from both inorganic and organic chemistry exemplify general chemistry concepts and emphasize that the same principles apply in all chemistry's sub-disciplines. The chemistry of carbon compounds is introduced early (see Chapter 4), due to both the importance of these compounds in controlling the delicate radiation balance of our planet and their central role in living chemical systems.

A significant challenge in learning chemistry is to develop accurate mental models of the dynamic, invisible molecular world. To develop your imagination and help you immerse yourself in this world, we use animations and simulations, and sometimes ask you to construct your own structures and simulations to answer a question.

Because modern chemistry is a global activity, examples of cutting edge chemistry from around the world are included in case studies and elsewhere. Some Canadian examples include:

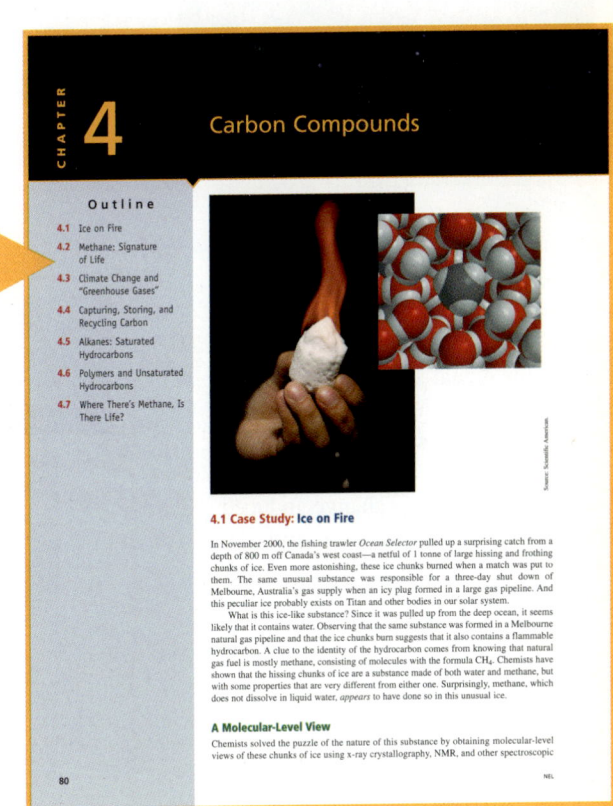

CHAPTER **4**   **Carbon Compounds**

Outline

4.1 Ice on Fire
4.2 Methane: Signature of Life
4.3 Climate Change and "Greenhouse Gases"
4.4 Capturing, Storing, and Recycling Carbon
4.5 Alkanes: Saturated Hydrocarbons
4.6 Polymers and Unsaturated Hydrocarbons
4.7 Where There's Methane, Is There Life?

4.1 Case Study: Ice on Fire

In November 2000, the fishing trawler *Ocean Selector* pulled up a surprising catch from a depth of 800 m off Canada's west coast—a netful of 1 tonne of large hissing and frothing chunks of ice. Even more astonishing, these ice chunks burned when a match was put to them. The same unusual substance was responsible for a three-day shut down of Melbourne, Australia's gas supply when an icy plug formed in a large gas pipeline. And this peculiar ice probably exists on Titan and other bodies in our solar system.

What is this ice-like substance? Since it was pulled up from the deep ocean, it seems likely that it contains water. Observing that the same substance was formed in a Melbourne natural gas pipeline and that the ice chunks burn suggests that it also contains a flammable hydrocarbon. A clue to the identity of the hydrocarbon comes from knowing that natural gas fuel is mostly methane, consisting of molecules with the formula CH_4. Chemists have shown that the hissing chunks of ice are a substance made of both water and methane, but with some properties that are very different from either one. Surprisingly, methane, which does not dissolve in liquid water, *appears* to have done so in this unusual ice.

A Molecular-Level View

Chemists solved the puzzle of the nature of this substance by obtaining molecular-level views of these chunks of ice using x-ray crystallography, NMR, and other spectroscopic

80   NEL

Courtesy of Dr. David Dolphin

David Dolphin with his research group who have come from around the world. In recognition of the impact of his work, Dolphin received Canada's top science prize in 2005. He has also been appointed an Officer to the Order of Canada and designated a "hero of chemistry" by the American Chemical Society.

David Dolphin and his research group in Vancouver have played a key role in the development of photodynamic therapy for the treatment of cancer and age-related macular degeneration.

Virginia Walker at Queen's University studies antifreeze proteins in fish, which may be useful in preventing methane clathrate plugs in pipelines.

The world-class Canadian Light Source in Saskatoon creates and stores a high-energy beam of electrons that produces synchrotron light that is one million times brighter than sunlight.

Chemists at the National Research Council laboratories in Ottawa use solid-state nuclear magnetic resonance (NMR) spectroscopy to confirm the structures of new crystal polymorphs. Canada's Ballard Power is a world leader in hydrogen-oxygen fuel cell technology.

Vaclav Smil in Manitoba has contributed to our understanding of the role of planetary nitrogen cycles, which has application both in the production of food and in our understanding of our atmosphere and oceans.

Rik Tykwinski's research group studies novel compounds containing alkyne functional groups.

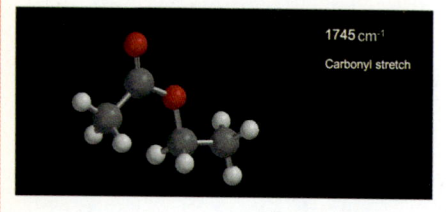

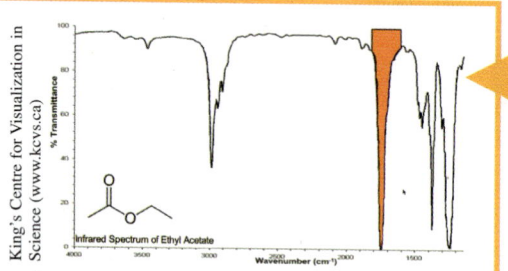

King's Centre for Visualization in Science (www.kcvs.ca)

The C=O bond in the model vibrates when the corresponding IR energy absorption peak in the spectrum is highlighted.

Spectroscopy provides convincing evidence to support our models of structure and bonding in molecules. Presented early, on a descriptive "need-to-know" basis, this evidence helps students to see the power of spectroscopic data to reveal aspects of molecular structure.

A central theme is the importance of chemical speciation: that is, that the chemistry of any system is the chemistry of the actual species present. For example, the pH-dependent speciation of carbonate is highlighted in Chapter 15 for its relevance to, for example, the climate-change induced changes in coral reefs around the world.

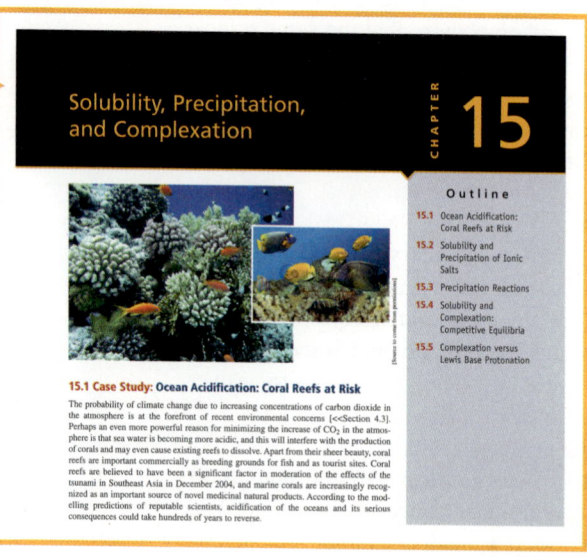

High-resolution mass spectrometry analysis of a pure molecular substance can give an accurate value of the relative mass of molecules with various isotopic compositions. From this, we can define the formula and composition of the compound.

Features known as "bottom-line statements" appear throughout each chapter. Students can work through a sequence of ideas and interactive ideas, then read the take-home message for review purposes.

A wide range of problem types helps students to apply their understanding in a variety of ways. In addition to the conventional drill-and-practice exercises are questions based on manipulating mathematical simulations, contextual questions, and group learning activities. Some activities require unusual—and often creative—solutions to facilitate new ways of thinking about concepts and to clarify relationships between often-confused terms.

4.13 Air can also be trapped inside clathrate hydrates. With reference to the figure below, what term(s) most correctly describe(s) the following forces of attraction:
(a) between nitrogen atoms in the N_2 molecules inside the clathrate hydrate
(b) between N_2 molecules and H_2O molecules of the clathrate hydrate cage
(c) between H_2O molecules that make up the clathrate hydrate cage
(d) between N_2 molecules and O_2 molecules inside the clathrate hydrate cage

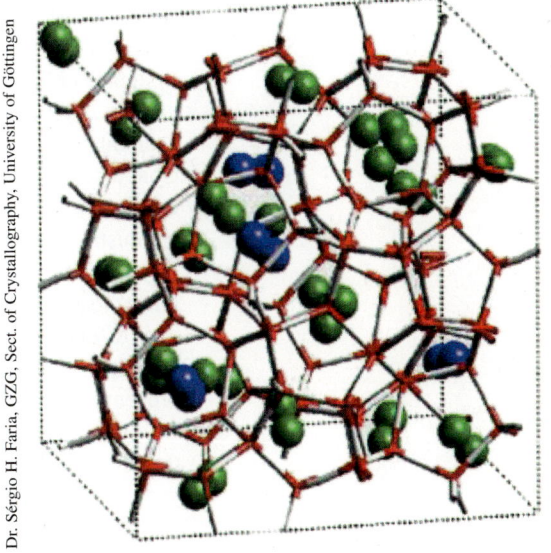

Dr. Sérgio H. Faria, GZG, Sect. of Crystallography, University of Göttingen

A significant challenge in learning chemistry is to develop accurate mental models of the dynamic, invisible molecular world. Both animations and simulations are used to do this, and at times students are asked to construct their own structures and simulations to answer a question. For example, in Chapter 21, the changes in electron density distribution in a substitution reaction are visualized using coloured electrostatic potential maps. With these visualizations students develop an intuitive feel for the reactivity of organic and inorganic functional groups. In Chapter 6, students are given print and electronic resources to imagine each ion in a solution as part of a dynamic system, surrounded by a cage of water molecules—there are so many situations where this model is essential for understanding reactivity.

Great care is taken with the use of language, terminology, chemical notation, and artwork to avoid confirming—or worse still, generating—misconceptions. For example, the resource avoids language that suggests that a chemical species can "attack" another molecule in some pre-destined way.

Organization of Learning Resources

The flow of ideas provides the experimental evidence for what we believe, followed by discussion of the models we use to make sense of that evidence. Chemistry is introduced through seven overarching themes, paralleling the order of coverage of most first-year chemistry courses, as follows.

Part 1: Chemistry: A Human Activity (Chapter 1)

Two "trigger" or concept-introducing case studies make up Chapter 1. Subsequent chapters begin with a case study to motivate learning and situate concepts in real-life, contemporary contexts. Following best practices in case studies, we introduce key concepts in a spiral fashion, explaining what is needed to make sense of the story and building on these concepts for deeper understanding as the need arises throughout the chapters.

Part 2: An Overview of Materials and Reactions (Chapters 2–7)

The building blocks of materials are introduced in Chapter 2, followed by the classification of substances based on their properties and the models that explain those properties (Chapter 3). Knowing the interest you are likely to have in making sense of living systems, carbon compounds are introduced in Chapter 4 in the context of planetary carbon cycles involving both inorganic and organic substances. Various methods of accounting for atoms in chemical reactions and reaction equations are presented in Chapter 5, using green chemistry and atom economy and atom efficiency, in addition to conventional stoichiometry. Chapter 6 gives an overview of the chemistry of water and chemistry in water, introducing in a spiral fashion key ideas of intermolecular forces. Energy flow in chemical reactions is introduced in Chapter 7, in the context of consideration of hydrogen as a potential major energy source/carrier. While the rather simple concept of enthalpy change as a "product" of reaction is considered here, the influence of enthalpy change of reaction on the direction of spontaneity is deferred until the presentation of chemical thermodynamics in Chapter 17.

To give you the evidence for "why we believe what we believe," we introduce in these early chapters (on a need-to-know basis, and at an appropriate level for your first year of chemistry) those tools that have revolutionized our understanding of structure and reactivity, such as IR (Chapter 3) and NMR (Chapter 9) spectroscopy and mass spectrometry (Chapter 3). For example, you will use animated IR spectra of different functional groups to see why chemists know they have different connectivity patterns (e3.19). Using an interactive digital learning resource, you will visualize the differences in symmetry of isomeric pentanone molecules that can be inferred from the number of peaks and chemical shift regions in their ^{13}C NMR spectra. (e9.5). A more detailed understanding of how NMR spectroscopy works is presented later in the resource package (Section 19.5).

Part 3: Relating the Structure and Behaviour of Substances (Chapters 8–12)

We begin with the question of what is known (and how it is known) about the structural arrangement of atoms of the component atoms in molecules. In Chapter 8, the atomic and substance properties of the elements are presented before we attempt to rationalize those properties. The chapter concludes with explanations of those properties emanating from the wave theory of electrons in atoms, with reference to the effective nuclear charge experienced by the valence electrons. Molecular recognition, the basis of modern biology, is used to introduce the shapes and structures of molecules and stereochemistry (Chapter 9), followed by development of the complementary models of structure and bonding that help explain how individual molecules differ from each other in shape and how they interact dynamically with each other (Chapter 10).

The interplay of gas, liquid, and solid states of atmospheric substances important to the radiation balance of our planet is used to introduce the states of matter (Chapter 11), followed by coverage of solutions and their behaviour (Chapter 12).

Part 4: Competing Influences on Chemical Reactions (Chapters 13–18)

A major theme throughout Chapters 13–16, which deal with equilibria, is that it's important to identify the species that are present in chemical reactions, and that every chemical process is a result of a competition between species. This theme is carried through in the treatment of acid-base reactions as competition for protons (Chapter 14); precipitation as competition between aquation and crystal lattice forces, and complexation as competition for Lewis bases (Chapter 15); and oxidation-reduction reactions as competition for electrons (Chapter 16). Speciation plots are a feature of some of these chapters. After you have some experience of these classes of reaction, with opportunities to develop powerful visualizations of reaction mixtures, the universally applicable field of thermodynamics is presented (Chapter 17). While on the one hand, we describe why spontaneous reactions proceed in the direction they do (including a probabilistic approach to the meaning of entropy), on the other, we give considerable attention to the interplay between thermodynamics and kinetics in what actually happens. This provides a segue into a formal treatment of reaction kinetics in Chapter 18, with a serious emphasis on multiple-particle visualization of reaction mixtures.

Part 5: Carbon Compounds: Patterns of Structure and Reactivity (Chapters 19–25)

The detailed chemistry of organic compounds is introduced, providing sufficient coverage for instructors wishing to incorporate organic chemistry anywhere from a few weeks to a full term in the first year. Underlying mechanistic threads are emphasized in the systematic introduction of the structure and reactions of alkenes and alkynes (Chapter 19), aromatic compounds (Chapter 20), alkyl halides (Chapter 21), alcohols (Chapter 22), carbonyl compounds (Chapters 23–24), and amines (Chapter 25). Digital Organic Flashware is available to assist you in visualizing the flow of electrons in reaction mechanisms, and attention is paid to replace static, deterministic descriptions of mechanistic processes with dynamic, probabilistic ones.

Part 6: Compounds of the Elements: Patterns of Structure and Reactivity (Chapters 26–27)

Patterns of structure and reactivity are highlighted in the detailed coverage of the descriptive chemistry of the main group (Chapter 26) and transition elements (Chapter 27). The unifying concepts of charge density of metal cations and polarization of electron clouds are used as rationalizing ideas.

Part 7: Chemistry of Materials, Life, and the Nucleus (Chapters 28-30)

Three complete chapters on the chemistry of modern materials (Chapter 28), an introduction to biochemistry (Chapter 29), and nuclear chemistry (Chapter 30) are available.

Visual Tour of e-learning Resources

In the margins of each chapter, distinct icons point to e-resources that offer a rich variety of electronic experiences students can access for learning either on their own or with other students.

Molecular Modelling

Students will be immersed in the molecular world through models and animations. Using their rich mental models, students are able to visualize and thereby interpret the subtlety and meaning of symbolic formulas, equations for reactions, and the mathematical relationships between quantities.

Molecular Modelling

e2.1 Represent molecular structures using models and structural formulas.

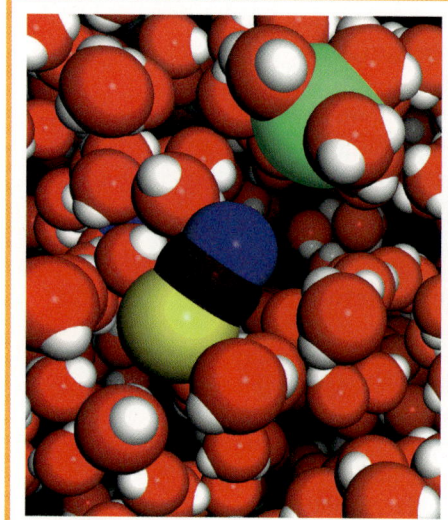

Molecular Modelling (Odyssey)

e2.4 Compare simulations of solid, liquid, and gaseous bromine.

Molecular Modelling (Odyssey)

Through building molecular models and constructing solution simulations of ions and molecules in Odyssey, students will acquire, at the molecular level, a "feel" for molecular flexibility and freedom of movement. Students can measure bond distances and approximate energies, change the temperature and pressure, and plot the results to discover mathematical relationships. In this way, students are able to experiment in a "molecular sandbox"!

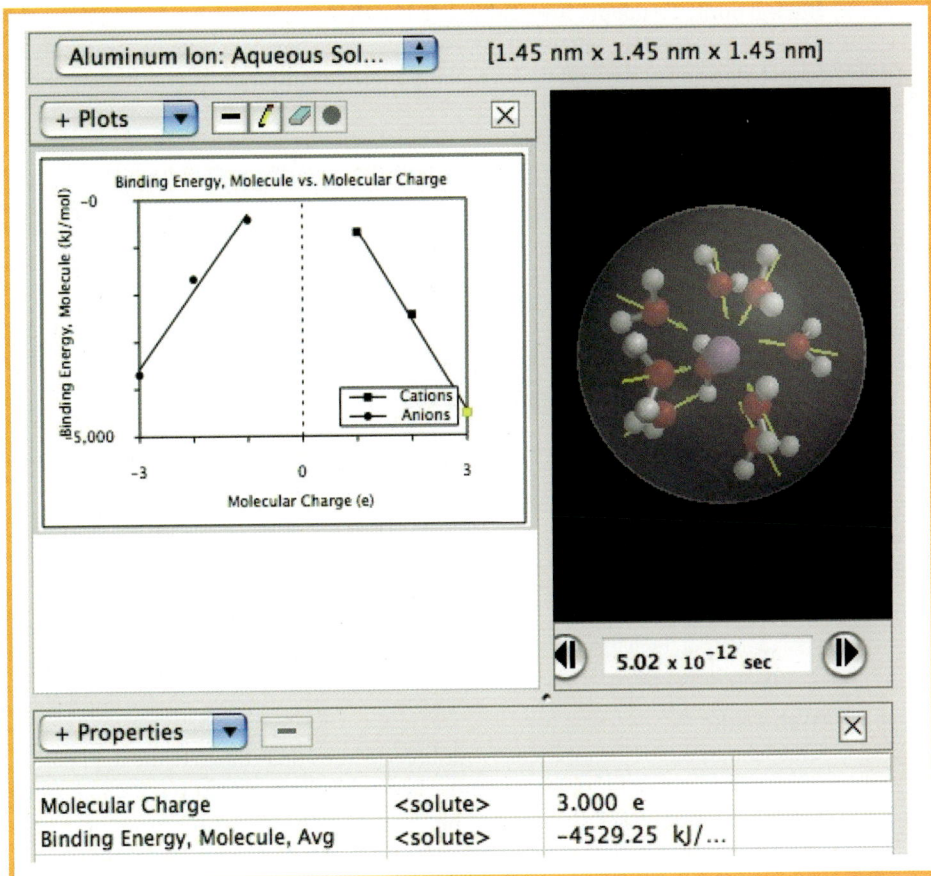

Think about It

These interactive resources involve active problem solving with immediate feedback. Some require interpretation of laboratory videos; others, manipulation of simulations.

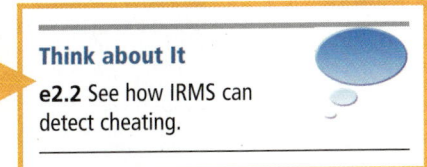

Think about It

e2.2 See how IRMS can detect cheating.

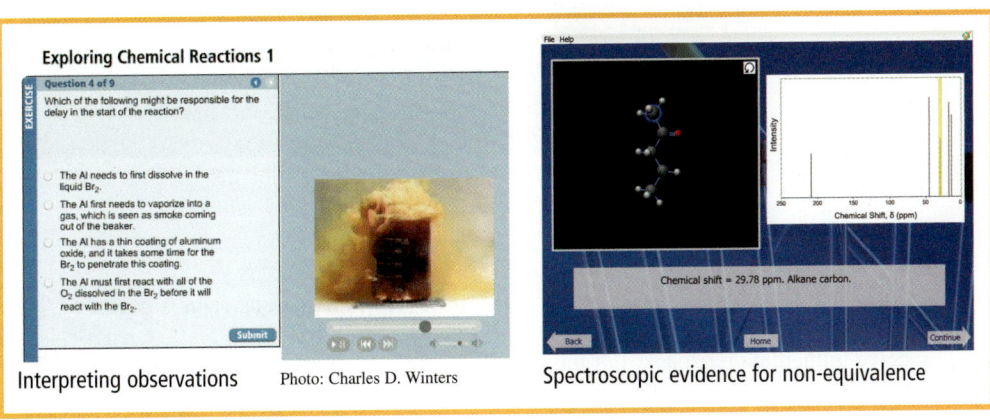

Interpreting observations

Photo: Charles D. Winters

Spectroscopic evidence for non-equivalence

Background Concepts and Taking It Further

These resources are intended for students who need to review prerequisite knowledge and skills and those wishing more detail on a topic.

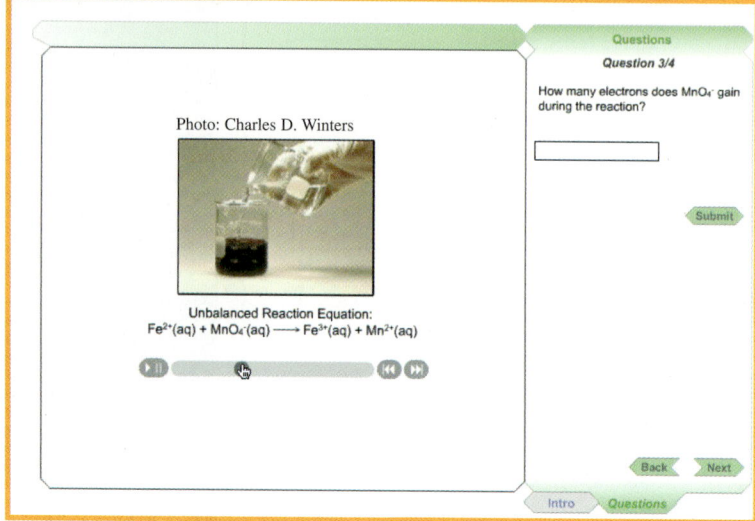

Web Link

This resource provides students with links to sites that illustrate the application of chemistry to real problems and the latest developments in research, such as the Protein Data Bank database of molecular structures.

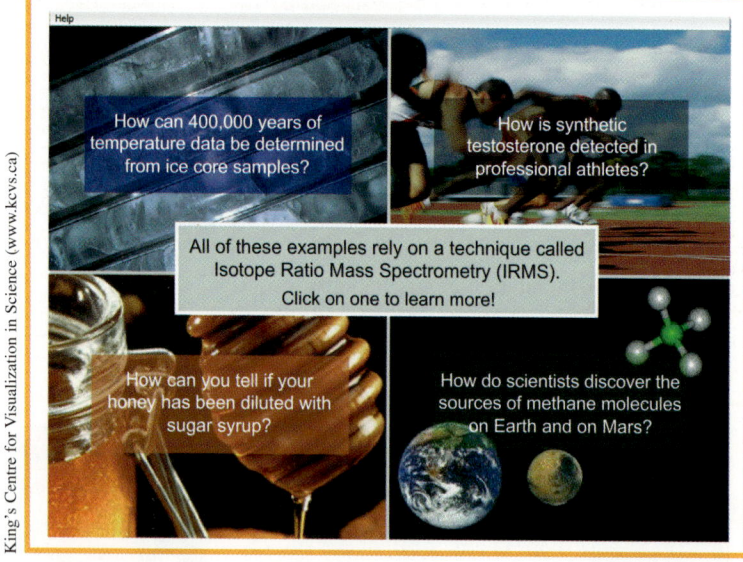

Interactive Exercises

Students can complete these exercises and receive immediate feedback. For the more challenging problems, students can access stepwise tutorial assistance for suggested strategies for solving the problems.

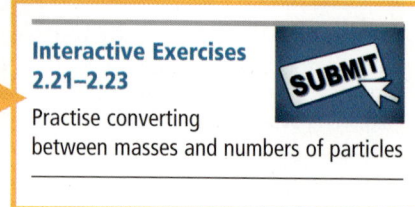

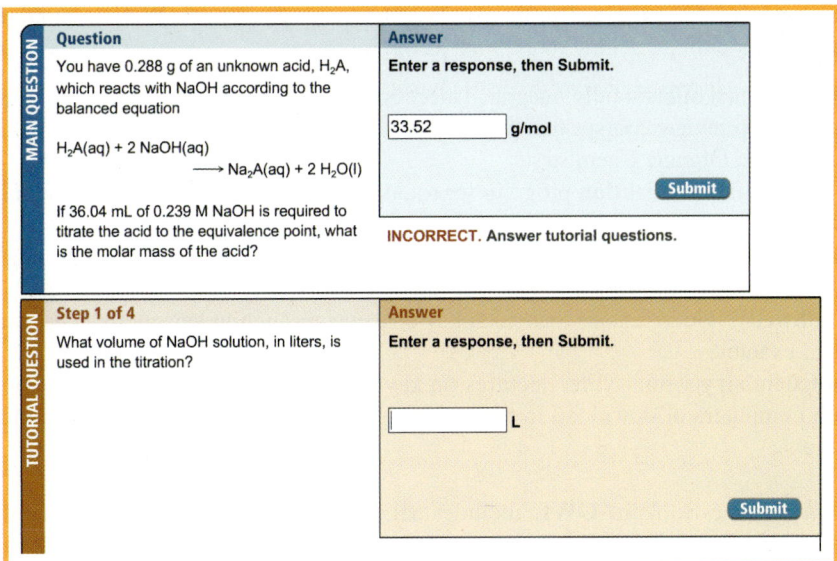

eBook

The interactive eBook effortlessly links students to the e-resources while they read the chapter. The eBook also includes full search capabilities and the ability to take notes, highlight meaningful text, and much more!

Chemistry: Human Activity, Chemical Reactivity uses the best of print and digital resources to bring chemistry into the 21st century. Visualization and interactivity are elevated to a new level by integrating text features that connect the print content to digital assets, such as an interactive and media-enabled virtual proving ground (VPG) eBook, Odyssey, Online Web Learning (OWL), Organic Chemistry Flashware, and more. Robust supplements are only a click away …

Instructor Teaching Resources

OWL®

OWL: Online Web Learning **The Chemist's Choice. The Student's Solution.**
OWL is a fully customizable and flexible web-based learning system. **OWL** supports mastery learning and offers numerical, chemical, and contextual parameterization to produce thousands of problems correlated to this learning resource. **OWL** to accompany *Chemistry: Human Activity, Chemical Reactivity,* First Canadian Edition, offers:

OWL Sign in to OWL at **www.cengage.com/owl** to view tutorials and simulations, develop problem-solving skills, and complete online homework assigned by your professor.

- **Simulations**—interactive simulations of chemical systems—are accompanied by guiding questions that lead students through an exploration of the simulation. These concept-building tools guide students to their own discovery of chemical concepts and relationships. Students can examine chemical systems by adjusting variables and observing how those changes affect the system.
- **Tutors**—interactive problem-solving tutors—ask questions and then give feedback that guides students toward solving problems. These problems are "parameterized" so that each time students use the material, they get a different problem of the same type.
- **Math Remediation Modules** provide math assistance to students with varying levels of math preparation and help students to master the basic math skills necessary for solving chemical problems.
- **Media-based Exercises**—interactive explorations of animations, movies, and graphic images—ask students to examine the chemical principles behind multimedia presentations of chemical events.
- **Parameterization**—numerical, chemical, and/or contextual—allows students to retry questions that cover the same concept, but with different numerical values, chemical systems, and question phrasing.

To address the various learning styles of today's students, **OWL** is continually enhanced with online learning tools, such as the following:

- **e-Book,** which offers a fully integrated electronic textbook correlated to OWL questions.
- **Quick Prep** review courses that help students learn essential skills to succeed in General and Organic Chemistry.
- A **molecular visualization** program for rotating molecules and measuring bond distances and angles.
- **Parameterized end-of-chapter questions** developed specifically for this learning resource.
- **Thinkwell™** video lessons that teach key concepts through video, audio, and whiteboard examples.
- **Go Chemistry®** mini-video lectures on key concepts that students can play on their computers or download to their video iPods, smart phones, or personal video players.

e-Book! The e-Book in **OWL** includes all chapters and sections in the textbook as assignable learning resources that are fully correlated to OWL homework content. The e-Book can be bundled with the text and/or ordered as a text replacement. Please consult your Nelson Education sales representative for pricing details (http://replocator.nelson.com/).

OWL with e-Book 4-Semester Access Students can purchase instant or printed access to **OWL with e-Book,** an electronic version of this textbook—fully integrated and linked to OWL questions. OWL includes end-of-chapter questions specific to this textbook, as well as more assignable, gradable content; more reliability; and more flexibility than any other system. Developed *by* chemistry instructors *for* teaching chemistry, OWL makes homework management a breeze and has already helped hundreds of thousands of students master chemistry through tutorials, interactive simulations, and algorithmically generated homework questions that provide instant, answer-specific feedback.

Nelson Education Teaching Advantage

The **Nelson Education Teaching Advantage (NETA) program** delivers research-based resources that promote student engagement and higher-order thinking and enable the success of Canadian students and educators.

The primary NETA components are a *Guide for Classroom Engagement* and our *Testing Advantage* resources.

Guide to Classroom Engagement The foundational principles underlying the *Guide to Classroom Engagement for Chemistry: Human Activity, Chemical Reactivity*, First Canadian Edition, are student-centred learning, deep learning, active learning, and creating positive classroom environments. Each NETA *Guide* includes a section outlining the research underlying these principles that will help you create engaging classrooms. The structure of the *Guide* was created by Dr. Roger Fisher, and validated by an interdisciplinary board of scholars of teaching and learning.

Editorial Advisory Board:
Norman Althouse, Haskayne School of Business, University of Calgary
Brenda Chant-Smith, Department of Psychology, Trent University
Scott Follows, Manning School of Business Administration, Acadia University
Glen Loppnow, Department of Chemistry, University of Alberta
Tanya Noel, Department of Biology, York University
Gary Poole, Director, Centre for Teaching and Academic Growth and School of Population and Public Health, University of British Columbia
Dan Pratt, Department of Educational Studies, University of British Columbia

Testing Advantage Resources Nelson Education Ltd. understands that the highest quality multiple-choice test bank provides the means to measure *higher-level thinking* skills as well as recall. In response to instructor concerns, and recognizing the importance of multiple-choice testing in today's classroom, we have created the Nelson Education Testing Advantage program (NETA) to ensure the value of our high quality test banks.

The testing component of our NETA program was created in partnership with David DiBattista, a 3M National Teaching Fellow, professor of psychology at Brock University, and researcher in the area of multiple-choice testing.

All NETA test banks include David DiBattista's guide for instructors, "Multiple Choice Tests: Getting Beyond Remembering." This guide has been designed to assist you in using Nelson test banks to achieve your desired outcomes in your course. See the Instructor's Resource DVD "NETA Guidelines" for this valuable resource.

Instructor's Resource DVD

ISBN 10: 0-17-647827-2
The **Instructor's Resource DVD for** *Chemistry: Human Activity, Chemical Reactivity* holds all instructor resources in one location, so there's no hassling with multiple media files or discs. The DVD contains:

- An ExamView® computerized test bank that is part of our Nelson Education Testing Advantage program (NETA), which ensures quality multiple-choice questions that provide the means to measure *higher-level thinking* skills as well as recall.
- Instructor's Guide to Classroom Engagement is an instructor's manual filled with chapter-specific lecture plans, instructor activities, and student activities.
- The full Solutions Manual with worked-out solutions for each exercise in the text.
- Microsoft® Word files of the test bank.
- Presentation lectures in PowerPoint for each chapter.
- An Image Bank of all text art and photos.

Turning Point

A tool for instant in-class quizzes and polls
Here's an easy way to increase students' participation in class! Nelson Education Ltd. is pleased to offer you book-specific **JoinIn™ on TurningPoint®** content for classroom response systems tailored to *Chemistry: Human Activity, Chemical Reactivity*. Our agreement to offer **TurningPoint®** software lets you pose book-specific questions and display students' answers seamlessly within the Microsoft® PowerPoint slides of your own lecture, in conjunction with the "clicker" hardware of your choice. It's a great tool for motivating students to come to class and pay attention. We provide the software and questions for each chapter of the text on ready-to-use Microsoft® PowerPoint slides.

Organic Chemistry Flashware

http://flashchem.nelson.com/
Organic Chemistry Flashware is a collection of interactive web-based multimedia courseware designed to assist students studying organic chemistry. This collection has been produced to enhance the traditional lecture experience and is optimized for both the individual computer user and classroom projection.

This innovative courseware consists of over 130 multimedia learning objects accessible online when you are ready to study. The home page allows users an entry point organized by common organic chemistry textbook topics lists and presentation styles.

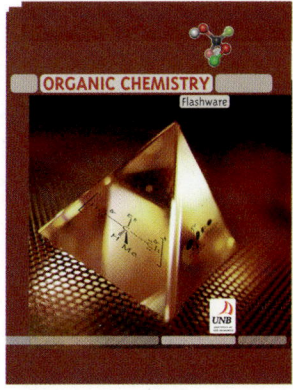

Chemistry e-Book

The interactive e-Book effortlessly links to the e-resources when reading the chapter. The e-Book also includes full search capabilities, as well as the facility to take notes, highlight, and much more! Instructors have access to all the e-resources noted by icons in the text, including animations; videos; Molecular Modelling, Background Concepts, Think about It, WebLink, and Taking It Further exercises; and interactive exercises. Use the e-resources as classroom tools or assign the interactive exercises as homework.

Student Learning Resources

Chemistry: Human Activity, Chemical Reactivity, First Canadian Edition with OWL will help you succeed in chemistry through its integrated, step-by-step approach to problem solving and its easy-to-understand presentation.

OWL (Online Web Learning)

The Chemist's Choice. The Student's Solution.
The results are in—and they prove that OWL will help you study smarter and succeed in chemistry!
This text is fully integrated with **OWL**, the number-one online learning system for chemistry. With **OWL** and the outstanding coverage in this text, you have the tools you need to succeed. **OWL** has already helped hundreds of thousands of students master chemistry through a wide range of assignment types, including tutorials, interactive simulations, and algorithmically generated homework questions that provide instant, answer-specific feedback, including end-of-chapter questions specific to this textbook.

ODYSSEY® is a unique teaching program for introductory and general chemistry classes in colleges and universities. Utilizing scientifically-based molecular simulations, *ODYSSEY* provides an interactive environment for learning and exploration. By using a unique symbol/link, the *Chemistry: Human Activity, Chemical Reactivity* integrated print/digital resource makes it easy for students and instructors to launch into the world of Odyssey to learn or teach a particular concept. If your instructor has chosen to bundle an Odyssey CD with your text, please install the Odyssey software onto your computer to access this interactive program.

Student Solutions Manual

ISBN-10: 0-17-647843-4
The **Student Solutions Manual** contains detailed solutions to all odd-numbered end-of-chapter exercises. Solutions match the problem-solving strategies used in the text. Prepared by Randy Dumont, McMaster University; technically checked by Rabin Bissessur, University of P.E.I.

Chemistry: Student Activity, Chemical Reactivity Workbook

ISBN-10: 0-17-647835-3
Study more effectively and improve your performance at exam time with this student workbook! The **Chemistry: Student Activity, Chemical Reactivity Workbook** focuses on the thinking processes required to succeed. Each chapter of this workbook contains chapter highlights/topic map, study strategies, exercises to strengthen visualization skills, a math skills primer, and suggestions for group study activities. Prepared by John Carran, Hugh Horton, and Nick Mosey of Queen's University, and Glen Loppnow of University of Alberta.

Integrated Media-Enabled e-Book

The integrated digital resources of *Chemistry: Human Activity, Chemical Reactivity* truly come to life within the interactive media enabled e-Book. The interactive e-Book effortlessly link to the e-resources noted by icons in the text, including animations; videos; Molecular Modelling, Background Concepts, Think About It, WebLink, and Taking It Further exercises; and interactive exercises. With active embedded multimedia, the media enabled e-Book truly provides you with resources that fit your learning styles. This e-Book is included with the purchase of a new text or may be purchased separately. Simply register once, using the sign-on card accompanying this text, and the full myriad of e-resources becomes available to you.

Acknowledgments

We have learned so much from each other as co-authors over the four years of working together. During that time, and over our entire careers, we have been continuously inspired and re-energized by our students, for whom we have created this learning resource. Colleagues at our universities, in our countries, and internationally, listed below, have contributed ideas, advice, critique, and peer review of different parts of this resource. We have been fortunate to draw on the experience, advice, and thoughtful review of Editorial Advisory Boards of leading university chemistry educators in Canada, Australia, and New Zealand.

We highlight the seminal contributions made by Jurgen Schnitker and his colleagues at Wavefunction, Inc. They have supported this project from its inception with the molecular dynamics *Odyssey* software, and advised over many discussions the best ways to build robust mental models of dynamic molecular systems for general chemistry students. We are also grateful for permission to use figures from Odyssey throughout the resource.

The resource package is published by Nelson Education Ltd. for distribution in Canada and internationally. Thank you to Scott Sinex who has graciously provided permission for the use of his Excel spreadsheets to enrich the e-resource material. It has been a true pleasure to share and co-develop a vision with Publisher Paul Fam for an integrated, contemporary package of text and electronic resources that incorporates important pedagogical innovations in teaching and learning chemistry. We are grateful to Evelyn Veitch and her amazing team at Nelson. Special thanks to Senior Marketing Manager, Sean Chamberland; Senior Content Production Manager, Anne Nellis; Art Buyer/Asset Coordinator, Sue Peden; Permissions Editor, Joanne Tang; and Copy Editor, Wendy Yano. Thank you to Katherine Goodes at My Editor Inc. for patiently and competently guiding us through the myriad of details needed to bring this to production. Also, without the foresight of Elizabeth Vella, then at Cengage Australia, who co-commissioned the project in 2005, the collaboration among authors and publishers would never have been brought forward.

Finally, we acknowledge individual contributions, as follows.

Peter Mahaffy

I have been sustained in my work by the daily, concrete support of my family, and offer my heartfelt gratitude for their role in making this vision become real. From writer Cheryl, I keep learning how to make words sing. Reuben, Naomi, and Miriam each inspire in different ways with their integrity, creativity, and love of learning. Brother and fellow chemist Paul has read several chapters and is always willing to talk shop. My love for the world of learning and teaching has been nourished by my mother and teacher, Arlena Mahaffy. My approach to learning and teaching has been shaped by many passionate and compassionate educators: If I single out only a few, they might be PhD mentor Mike Montgomery, Renaissance chemist Roald Hoffman, King's colleague Brian Martin, and University of Alberta colleague Margaret-Ann Armour.

Bob Bucat

For enriching me as an educator, Alex Johnstone (Scotland) and Peter Fensham (Australia) deserve my reverence. Most influential of all of my professional colleagues in my career and in this project is the late Professor Sir Noel Bayliss, my PhD supervisor and mentor of so many years ago, who demonstrated to me so often that it is possible to think like a molecule. How valuable this has been to me as a teacher! Above all, I dedicate my contribution to Lucinda, Sally, Jacqueline, Bob, Ben, and Michael, as well as their children (my 11 grandchildren) for the fact that I have spent so much less time with them than I would have liked. I will make it up.

Roy Tasker

I would like to dedicate my contribution to my family—Diane, Ken, and Skye—for their loving support and understanding for my all-too-often mental and physical absences from their lives during this project.

I would like to acknowledge three mentors who have influenced my approaches and priorities used in this project: Alex Johnstone (formerly at the University of Glasgow), Peter Atkins (formerly at Oxford University), and Loretta Jones (formerly at the University of Northern Colorado).

Mark Williams has been a valued colleague who has shared the load of testing so many of our interactive resources with our chemistry students.

Editorial Advisory Board Members and Reviewers

Canada
Bob Balahura, University of Guelph
Ghislain Deslongchamps, University of New Brunswick
Andy Dicks, University of Toronto
Randall Dumont, McMaster University
Noel George, Ryerson University
Glen Loppnow, University of Alberta
Michael Mombourquette, Queen's University
Susan Morante, Mount Royal College
Rashmi Venkateswaran, University of Ottawa
Peter Wassell, University of British Columbia
Mark Workentin, University of Western Ontario

Australia/New Zealand
Greg Dicinoski, University of Tasmania
Ian Gentle, University of Queensland
Richard Hartshorn, Canterbury University
Brynn Hibbert, University of New South Wales
Ian Rae, Melbourne University, Victoria
Richard Russell, University of Adelaide
Kevin Wainwright, Flinders University

Second Draft Reviewers

Tony Cusanelli, Capilano University
Hugh Horton, Queen's University
Lori Jones, University of Guelph
Ed Neeland, University of British Columbia
Shirley Wacowich-Sgarbi, Langara College
Todd Whitcombe, University of Northern British Columbia

Contributing Ideas and/or Manuscript Reviewers

Sushil Atreya, University of Michigan
Graham Chandler, The University of Western Australia
Mei-Hung Chiu, National Taiwan Normal University, Taipei
Tomislav Cvitas, University of Zagreb, Croatia
Emilio Ghisalberti, The University of Western Australia
William Gunter, Alberta Research Council
Jack Harrowfield, The University of Western Australia
Alastair Hay, University of Leeds
George Koutsantonis, The University of Western Australia
Frank Lincoln, The University of Western Australia
Marina Luetić (Croatia), University of Split, Croatia
Brian Martin, The King's University College
Cathy Middlecamp, University of Wisconsin-Madison
Ken Newman, The King's University College
Kris Ooms, The King's University College
Matthew Piggott, The University of Western Australia
Scott Stewart, The University of Western Australia
Grace Strom, The King's University College
Rik Tykwinski, University of Alberta
Roko Vladusić, University of Split, Croatia
Duncan Wild, The University of Western Australia

Courtesy of Dr. Peter Mahaffy

Peter Mahaffy's passion is for helping undergraduate students and others see the intricate web that connects chemistry to so many other aspects of life and to the health of our planet. After receiving his PhD in Physical Organic Chemistry from Indiana University, he moved to Canada where he is now Professor of Chemistry at the King's University College in Edmonton, Alberta, and Co-Director of the King's Centre for Visualization in Science. Peter collaborates regularly on research with undergraduate students in the areas of chemistry education, visualization in science, organic chemistry, and environmental chemistry. His recent focus is on the effective use of case-based and guided inquiry approaches to facilitate learning of chemistry. He chairs IUPAC's Committee on Chemistry Education (CCE) and led the team that obtained UN designation of 2011 as an International Year of Chemistry. Elected a fellow of the Chemical Institute of Canada (CIC) in 1999, Peter also received the CIC National Award for Chemistry Education in 2003. In 2008, he was awarded the 3M Teaching Fellowship, the highest honour in Canada for university teaching.

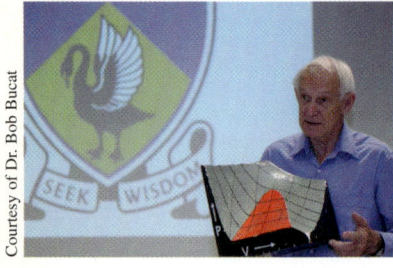

Courtesy of Dr. Bob Bucat

Bob Bucat is Professor of Chemistry at the University of Western Australia. After attaining a PhD in physical chemistry, his research interests have been related to the challenges of education in chemistry at the tertiary level. His research and teaching interests have centred on strategies to encourage thinking, the importance of language used (in word forms, symbolic forms, and graphics), and the challenges of visualization of molecular structures, and probabilistic and dynamic processes in reaction mixtures. He is a devotee of the notion of pedagogical content knowledge: both content knowledge and general pedagogical knowledge are necessary for good teaching, and each concept and topic has specific pedagogical challenges that demand specific teaching strategies. The chemical beliefs underlying his teaching revolve around recognition of the importance of the actual species present, and a conception that every chemical process is a result of a competition between species. Bob has been a titular member of the IUPAC Committee on Teaching Chemistry, and a member of the National Committee for Chemistry of the Australian Academy of Science. He is a winner of *awards for teaching excellence at UWA* and the *Medal of the Chemical Education Division of the Royal Australian Chemical Institute* for contributions to education in chemistry.

Courtesy of Dr. Roy Tasker

Roy Tasker is an Associate Professor of Chemistry, and has been teaching first-year university students at the University of Western Sydney (UWS) since 1985. With a PhD in inorganic chemistry from the University of Otago, and post-doctoral appointments at the Universities of Tasmania and Adelaide, he became interested in chemical education, and in particular, developing students' mental models of the molecular world. This led to the development of an integrated suite of molecular-level animations in the VisChem project in the early 1990s. These resources are still used at secondary and tertiary levels all over the World. During a three-year secondment with CADRE design—a multimedia production company—he gained experience in developing interactive multimedia resources to complement and supplement nine university-level chemistry and biochemistry textbooks. Since 2000, his research group has worked with students to study what and how students learn from molecular animations and simulations, and on this basis, developed the VisChem Learning Design. In 1999, he was awarded the inaugural UWS Award for Teaching Excellence; in 2002, the Royal Australian Chemical Institute Chemical Education Division Medal; and in 2008, an Australian Learning and Teaching Council Citation for Outstanding Contributions to Student Learning.

This learning resource adapts from and builds on the proven resources of two U.S. textbooks that have successfully served well over a million chemistry students: *Chemistry & Chemical Reactivity* by the author team of **John Kotz** (State University of New York), **Paul Treichel** (University of Wisconsin-Madison), and **Gabriela Weaver** (Purdue University); and *Fundamentals of Organic Chemistry* by **John McMurry** (Cornell University).

Human Activity, Chemical Reactivity

Courtesy of Dr. David Dolphin

Courtesy of Dr. Bob Bucat

1.1 Case Study: Human Activity, Chemical Reactivity

Stressing that education in and about chemistry is critical to address challenges such as global climate change and sustainable supplies of clean water, food, energy, and medicine, the United Nations declared 2011 to be the International Year of Chemistry. You are about to enter your own year of studying that international and interwoven world of human activity and chemical reactivity. This first chapter tells the stories of two people who carry out research in chemistry to improve our world. You would not recognize them on the bus or in the supermarket as chemists. They are simply curious, hard-working people who use logic and creativity as they work with a research team to design and interpret experiments.

First we'll meet David Dolphin, a Canadian chemist who has designed and made new substances that have improved the quality of life for over a million people suffering from cancer or eye disease. Then we'll meet Gavin Flematti, an Australian chemist who, while he was a postgraduate student, identified a compound in smoke that causes plant seeds to germinate after a forest fire. He then found a way to make this compound in the laboratory.

You'll see with these two people that the work chemists carry out is challenging for many reasons, not least of which is that the particles they study—particles that make up everything in our world—are too small for us to see directly. So chemists develop and use instruments to extend their vision and to understand how the things we observe fit with molecular-level explanations.

As in so many areas of modern chemistry, in both of these stories, the interaction of electromagnetic radiation with matter plays a critical role. In the first case, visible light from a laser triggers the production of molecules that selectively destroy undesirable cells in body organs. In the second, selective absorption of electromagnetic radiation is the key to understanding the three-dimensional structures of newly discovered molecules in smoke.

Spectroscopy is the study of characteristic patterns of absorption of electromagnetic radiation to identify substances and to provide information about the structures of their molecules. **Chromatography** is the science of separation of compounds in complex mixtures. These two techniques—part of an array of instrumental methods that have revolutionized the practice of chemistry—are at the heart of the stories about the success of these two people.

We emphasize that on your first reading, you should not expect to completely understand the chemistry related to these two stories. In Section 1.4 and later in the text, we return to the concepts introduced here. Instead, through these stories and the discussion questions that follow, we hope you will appreciate that advances in chemistry happen today, and that they can improve life for humans and the environment—the very reasons the United Nations declared an International Year of Chemistry. The advances described in this chapter and this book did not take place overnight, and they required sound knowledge of chemistry and other disciplines. They were made by logical thought processes and careful experiments, carried out by teams of creative people—people like those you meet every day. Perhaps most of all, we hope you appreciate that you could participate in significant research in chemistry or other sciences. You might want to read these stories again later for an even deeper appreciation.

1.2 Harnessing Light Energy and Exciting Oxygen

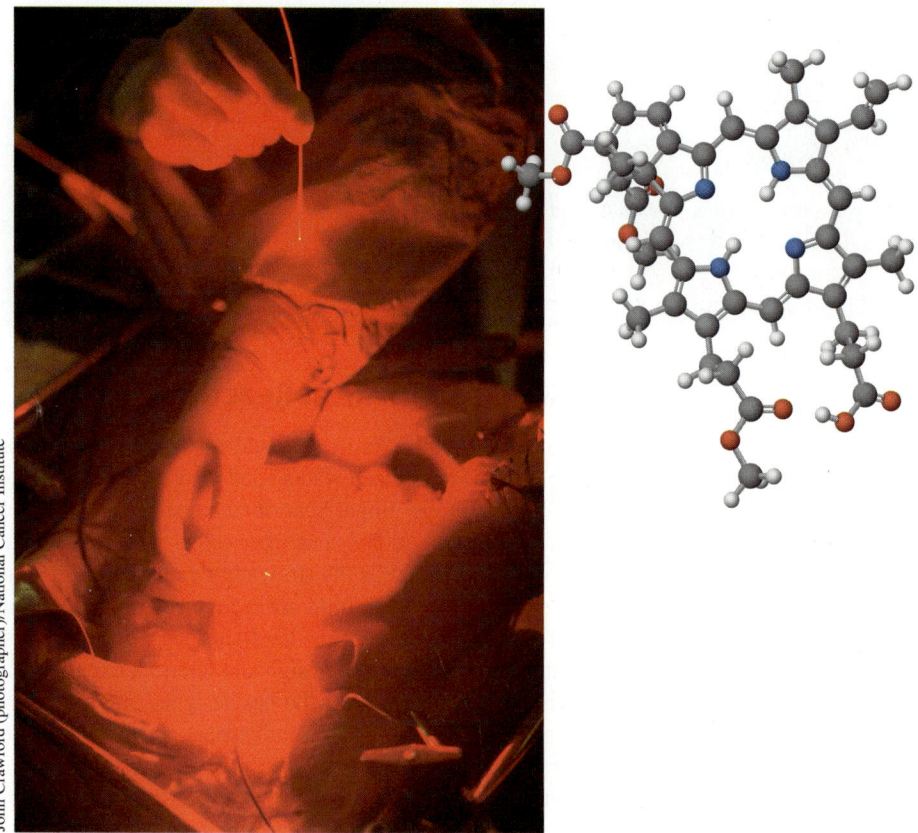

John Crawford (photographer)/National Cancer Institute

We live each day in a tropospheric ocean of oxygen and nitrogen molecules, and move around on the earth's crust, made mostly of oxygen atoms in combination with other elements such as silicon, nitrogen, carbon, and sulfur. Oxygen atoms are in the middle of H_2O molecules, which comprise the substance water, which in turn covers three-fourths of the surface of our planet. Without oxygen, life as we know it would be impossible. It makes up one-fifth of the air that we breathe. Our brains consume about 20% of the molecular oxygen (O_2) that we breathe in. This oxygen fuels numerous processes, including the electrical circuits that enable us to read and to think about our molecular world.

And yet, many forms of oxygen are highly toxic to human cells, including one form of molecular oxygen.

Two Energy States of Oxygen

How can molecular oxygen (O_2) be both essential to life and highly toxic? This riddle can be answered if we realize that O_2 can exist in different forms or energy states. This story is about two of those states: *singlet* oxygen and *triplet* oxygen. The terms *singlet* and *triplet* refer to differences in the energy levels of the electrons of O_2 molecules, explored further in Chapter 10. Those differences in the energies of the electrons of O_2 molecules cause a profound difference in chemical reactivity—the difference between life and death for cells.

In *triplet O_2*, the electrons in the O_2 molecules are in their lowest energy state (the *ground state,* [>>Section 8.5]). This is the oxygen we breathe. The electrons in singlet oxygen molecules are in higher energy (*excited*) states. *Singlet O_2* molecules can be formed when energy is transferred to triplet O_2 molecules as they collide with other "excited" molecules. Because singlet O_2 is exceptionally reactive, it doesn't last long.

While triplet O_2 in the air we breathe fuels life, singlet O_2 has very different physical properties and chemical reactivity. It bleaches coloured materials and polymers, and is highly toxic to dividing human cells. Singlet O_2 can be produced in laboratories as well as in nature. Respiration in animals and photosynthetic reactions in plants produce small amounts of singlet O_2, and chemists have learned a lot about how photosynthetic microbes and plants respond to the production of this toxic substance.

Chemotherapy and Photodynamic Therapy

Toxic or poisonous substances can sometimes be used to advantage in medicine. **Chemotherapy** is the use in medicine of substances that are selectively toxic to malignant cells or a disease-causing virus or bacterium. Because of its toxicity to dividing cells, singlet O_2 has found a role as a chemotherapeutic agent. An important chapter in the singlet O_2 story is the work of Professor David Dolphin from Canada's University of British Columbia. He and his research collaborators created Visudyne™, one of the world's most successful ophthalmic products. Since 2000, Visudyne™ has been used to treat more than a half-million people in 75 countries for age-related macular degeneration (AMD), the leading cause of vision loss for people over the age of 50 in the Western world.

Courtesy of Dr. David Dolphin

David Dolphin with his research group who have come from around the world. In recognition of the impact of his work, Dolphin received Canada's top science prize in 2005. He has also been appointed an Officer to the Order of Canada and designated a "hero of chemistry" by the American Chemical Society.

> *"I get my kicks out of solving problems, whether they're basic ones or more applied ... I know from my own graduate students that they're very excited about getting involved in research that might eventually benefit humanity."*
> **David Dolphin**

The original focus of Dolphin's research was treating skin and other cancers, rather than eye diseases. But, as often happens in chemistry research, a successful treatment for AMD was discovered unexpectedly. Dolphin's research group was studying *porphyrins,* a class of large cyclic molecules fundamentally important to the function of living systems. They absorb visible light strongly, and therefore are coloured. Examples of porphyrin or porphyrin-like substances are the green chlorophyll pigment in plants and the bright red hemoglobin in red blood cells (Figure 1.1). Their light-absorbing properties make them useful in technologies such as rewriteable CDs and DVDs. And this same property gives them a key role in **photodynamic therapy** (**PDT**)—the use of light in medical treatment.

Molecular Modelling (Odyssey)

e1.1 Examine a simulation of the heme group, separately and embedded in hemoglobin.

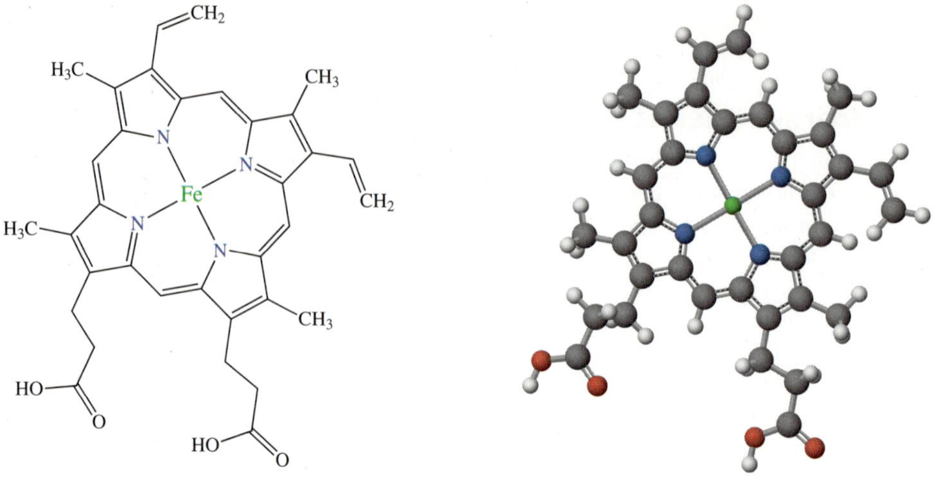

FIGURE 1.1 A heme group from a hemoglobin molecule. The iron atom (Fe, green) is in the centre of the large porphyrin ring, identified by its four nitrogen atoms (N, blue).

The light-absorbing properties of porphyrins also make them good **photosensitizers,** substances whose electrons can be excited by absorbing light of appropriate wavelength. They can then transfer their extra energy through molecular collisions to substances such as triplet O_2—producing singlet O_2. Dolphin had the imagination to see how porphyrin photosensitizers and light might work together with oxygen in medicine.

But Dolphin wasn't the first to think about photodynamic therapy. His breakthroughs built on many years of work by chemists in other countries. The basic technique was first developed in 1972 at the Roswell Park Cancer Institute in Buffalo, New York. PDT is a minimally invasive medical treatment with three components that work together: the administered drug (a photosensitizer), light to excite the photosensitizer molecules, and triplet O_2 in tissue (Figure 1.2). Excited photosensitizer molecules transfer energy to triplet O_2 in tissue, converting some of it to singlet O_2, which can kill rapidly growing cancer cells.

In Dolphin's PDT work, the photosensitizer is a porphyrin-containing compound, which is administered to the site of rapidly growing cells. The light source is usually a red light from light-emitting diodes or a laser diode.

Dolphin and his collaborators initially developed porphyrin PDT treatment for various cancers on or near the surface of the skin. The patient is given a dose of a photosensitizer or a substance that can be metabolized by the body into a photosensitizer. The drug might be injected at the site of the cancer or it might be designed to attach itself to *lipoprotein* molecules in the bloodstream so that it is transported to the affected area. After waiting an appropriate amount of time, the physician applies red light to the target site. The singlet O_2 that is formed lasts for only about a microsecond (1×10^{-6} s), but during that time it damages or destroys the rapidly growing cancer cells.

Lipoproteins are molecules that are combinations of proteins and fats, and are responsible for transporting fats and fat-like substances like cholesterol through the blood.

FIGURE 1.2 Excitation of a photosensitizer porphyrin molecule. Laser light energy (1) is absorbed by a photosensitizer molecule (2) in a cell, converting the molecule to its excited state (3). Energy is transferred (4) to a nearby triplet oxygen molecule to form a "lethal" singlet oxygen molecule (5).

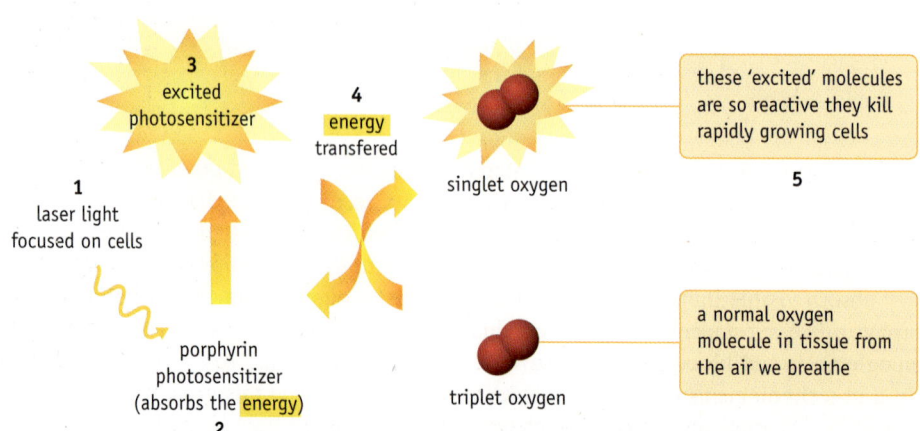

I   +   II

FIGURE 1.3 The two photosensitizer molecules that make up Visudyne™. The cyclic porphyrin rings are highlighted, and each bend in the structure in this representation symbolizes the presence of a carbon atom. Two molecules with the same formula but a different arrangement of atoms, such as these two, are called *isomers* [>>Section 4.5]. Can you spot the subtle difference between the two isomers?

It is important that, as far as possible, singlet O_2 is produced only at the site of the cancer, or otherwise healthy cells may be killed. A patient undergoing PDT must be careful about exposure to sunlight after treatment to avoid singlet O_2 production in other parts of the body where small amounts of the photosensitizer might still be present.

Unexpected Results: Effect on Vision

The development of Visudyne™ came from an unexpected turn in Dolphin's PDT cancer research. Patients undergoing PDT clinical trials for cancer reported beneficial effects on their vision. Researchers learned that some porphyrin photosensitizers had the unexpected effect of closing down abnormally growing blood vessels in the eyes of patients who also suffered from the most common "wet" form of age-related macular degeneration. Wet AMD is caused by the growth of an abnormal tangle of new blood vessels under the macula in the retina. These vessels then leak fluid and cause scar tissue, leading to the rapid loss of sight.

Visudyne™ is an equal mixture of two *isomeric* porphyrin-containing photosensitizers whose molecules have the structures shown in Figure 1.3.

The normal protocol for treating AMD with Visudyne™ can now be carried out in a doctor's office. The photosensitizer drug is administered intravenously, where it attaches to lipoprotein molecules and is carried selectively to the abnormal vessels in the eye. After five minutes, a red laser is shone into the patient's eye through a microscope for just over one minute. This excites the photosensitizers, which then transfer their excess energy to triplet O_2, producing singlet O_2 to destroy the abnormal blood vessel cells. New protocols, which combine Visudyne™ with other drugs, are also being tested.

Next Steps for PDT and Porphyrins

Chemists are now working to improve many aspects of PDT. One important area of focus is making new single-compound photosensitizers capable of strongly absorbing light of a specific wavelength that is not absorbed by blood. Other efforts focus on improved light sources that will penetrate more deeply into tissue without causing damage. Also needed are better methods of administering photosensitizers (or compounds that can be converted into photosensitizers) to patients. Computer modelling studies help chemists predict how to modify portions of photosensitizer molecules to improve their selective accumulation at cancer sites.

And just as research into PDT for skin cancer led to surprising results for an eye disease, understanding porphyrin chemistry may lead to unexpected applications. Research groups are now trying to use the light-absorbing ability of porphyrins to create inexpensive electrodes for fuel cells [>>Section 7.1], to make catalysts to produce ammonia gas from nitrogen gas [>>Section 13.1], and to convert water into hydrogen gas and oxygen gas.

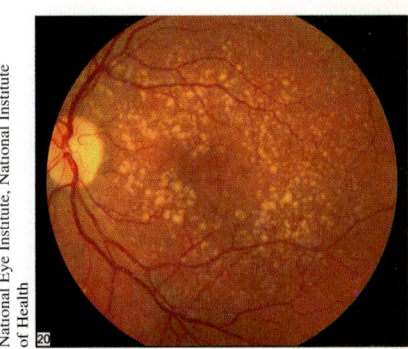

National Eye Institute, National Institute of Health

Advanced age-related macular degeneration.

National Eye Institute, National Institute of Health

National Eye Institute, National Institute of Health

Normal vision and age-related macular degeneration.

1.3 Where There's Smoke, There's Gavinone

Roger Rosentreter/Shutterstock

Throughout this textbook you'll read about reactions, compounds, and processes named after the people who discovered them. But can you imagine having a compound that might be worth millions of dollars named after you? That happened recently to an Australian chemistry student who discovered that trace amounts of a previously unknown compound in the smoke from wood fires causes plant seeds to start growing. The student's name is Gavin Flematti, and for a time the new compound was named Gavinone in his honour by the botanists on his research team.

The task wasn't easy. Gavinone is only one of thousands of compounds found in wood smoke, yet Flematti and his collaborators proved that very low concentrations (less than 1 part-per-billion) of this compound can cause seeds to germinate.

His discovery is important. Researchers knew that a compound causing seeds to germinate would be of great interest to agricultural and chemical industry. It could be used to control weeds, possibly saving farmers around the world millions of dollars. It might also help regenerate native plant species without the use of fire, or help plants grow in reclaimed soil on top of abandoned mine sites. Flematti and his research supervisors knew that they needed to start by understanding at the molecular level what causes germination after fires in nature. Eventually they might apply that knowledge to make substances in the laboratory that could be applied to seeds or to soil.

Let's try to understand the steps Flematti and his research team took in making this breakthrough. It took years of sleuthing as an honours and PhD chemistry student to isolate this single compound from smoke, prove that it causes seeds to germinate, identify the three-dimensional structure of its molecules, and then synthesize it in the laboratory.

How Do Seeds Germinate?

First, they needed to understand some plant biology. How do plant seeds survive in seed packets or on the ground and then germinate? Seeds can remain dormant until environmental conditions are suitable for the growth of the plant through the vulnerable seedling stage. Seed germination begins with the uptake of water by dry seeds. Then, a part of the embryo called the *radicle*, which develops into the primary root, penetrates the protective coating. This is a desirable characteristic for plants in nature, and can also be beneficial in agriculture. For example, wheat farmers often sow seed in dry land prior to seasonal rains, knowing that germination won't occur until the soil receives enough moisture. Seeds on the heads of grain also must remain dormant so that they do not germinate before harvesting.

On the other hand, the seeds of some crops remain dormant after planting, and not all of them germinate when the farmer or nursery owner would like. Crop growers would place great value on a substance that promotes germination of all seeds at the same time.

For example, it would be very useful if weed seeds in the ground could be made to germinate before the main crop is sown. The weed plants could then be sprayed with herbicide and eliminated as competitors for soil nutrients.

Where There Is Fire, There Is Smoke

Next, they needed information about fires. It is astonishing to see that, following a major forest fire, a blackened landscape can explode into colour the next spring. Seed germination was thought to be due to heat from the fire cracking open the seeds lying dormant in the ground. Heat is indeed important for some plants, but scientists in South Africa, Australia, and North America had shown that smoke alone, in the absence of heat, can also cause many seeds to germinate. Smoke is a very complex chemical soup, composed of many thousands of compounds. Flematti and his collaborators, along with competitors from several other countries, set out on a race to be the first to identify which bioactive compound or compounds in smoke promote germination.

Where There's Smoke, There's Chemistry

Smoke from a wide variety of plant materials earlier was shown to be effective in causing seeds to grow, but little was known about why this worked. Somehow, compounds in smoke must interact with others in seeds. Flematti did know that the compounds in smoke responsible for germination dissolve more easily in organic solvents than in water. This suggested that the active substances are probably one or more of the millions of *organic compounds* [>>Section 3.6] whose molecules contain carbon, hydrogen, and oxygen atoms. Although a number of components of smoke had been previously identified by other researchers, none of these promoted germination. The bioactive compound was like a "needle in the haystack," hiding among the thousands of compounds in smoke.

How to Find the "Needle in the Haystack"

The challenge Flematti faced is common in **natural products chemistry**, the branch of chemistry that studies compounds produced by living organisms. The general approach to finding the bioactive compound from so many possibilities involves several steps and a great deal of patience.

First, Flematti would need a reproducible source of the compounds in smoke from wood fires. Then, he would need to collect the complex mixture of compounds from that smoke, and determine if it helped seeds germinate. Since there are thousands of possible compounds that could be responsible, he would need to use repeated *chromatographic* separation techniques to divide the compounds into several groups, each of which contains fewer compounds. On each of those smaller groups of compounds, he could then use screening methods, called **bioassays**, to see if that group contained the active compound(s). This process of separation and bioassays would need to be repeated over and over for the "haystack" containing the "needle" to become smaller and smaller.

If a single compound could be isolated and shown to cause the bioactivity, the next challenge would be to determine its identity. Flematti would need to use powerful methods of structural determination, such as mass spectrometry [>>Chapter 3], nuclear magnetic resonance (NMR) spectroscopy, and x-ray crystallography [>>Chapter 9], to determine the composition and three-dimensional structure of the molecules of that compound. Finally, once the molecular structure of the substance was known, he might try to synthesize the same compound in the laboratory. This would help to prove that its structure was correct, and provide larger amounts for further testing and commercial use.

Chromatography is used to separate compounds from mixtures. Spectroscopic techniques, based on the way different molecules interact with electromagnetic radiation, are powerful tools for identifying compounds. A bioassay is a test for biological activity.

Mr. Brian Wiens, Head of Air Quality Science Unit, Environment Canada

Smoke from a forest fire in the Canadian Rockies.

Courtesy of Dr. Bob Bucat

Gavin Flematti (left) with his colleague and former PhD supervisor Professor Emilio Ghisalberti, both at the University of Western Australia, Perth.

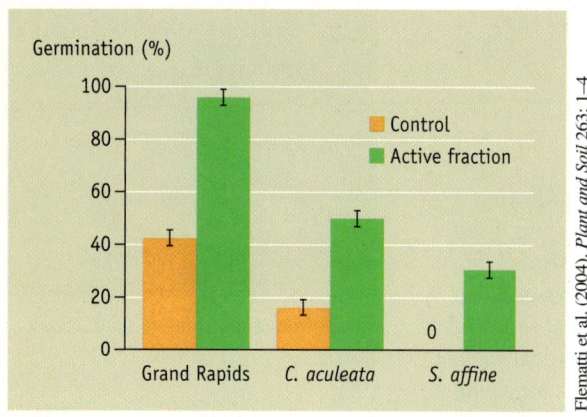

Flematti et al. (2004). *Plant and Soil* 263: 1–4

FIGURE 1.4 Bioassay effectiveness. Percent germination of Grand Rapids lettuce seeds and two Australian native plant seeds in response to smoke from burning filter paper (cellulose). Filtered water (with no smoke components) was used for the control experiments.

Producing and Testing Wood Smoke

Flematti and his collaborators burned laboratory filter paper as a source of smoke which is passed through water to make "smoke water". This seemed reasonable, since filter paper is mostly cellulose, the primary structural component of all green plants. They used a rapid bioassay technique developed at the University of Natal in South Africa to see if smoke caused germination. When smoke water was applied to the Grand Rapids variety of leaf lettuce seeds kept in the dark, germination took place within 24 hours. Only 45% of the seeds germinated in absence of smoke (called a "control" trial), while 90% germinated with smoke water application. Figure 1.4 shows their finding that these lettuce seeds and seeds of some Australian native plant species responded to smoke compounds.

They now knew that the active compound is found in burning cellulose, which is only one of many compounds in plants. This told them which "haystack" to start looking in.

Repeated Separations and Bioassays

Now the part of the task requiring the greatest patience—narrowing down the many possible compounds that might be causing the seed germination. Flematti used **extraction** techniques to separate the complex mixture of compounds in smoke into groups of compounds or fractions based on their acid/base and solubility properties. A bioassay was then used on each of those fractions to see if they contained the bioactive compound(s). Several types of *chromatography* were then used one after the other on the bioactive fractions, to create smaller and smaller groups of compounds. To give you an idea of the complexity of the problem, Figure 1.5 shows one of Gavin's gas chromatogram charts indicating the number of compounds found in just one fraction from plant-derived smoke water. Each "peak" or upward spike in the chromatogram represents a different chemical compound. One of these would eventually prove to be Gavinone. But which one?

FIGURE 1.5 Gas chromatogram of neutral fraction from plant-derived smoke water. Each peak in this chromatogram corresponds with a particular compound after it is pushed through the column by an unreactive gas. The horizontal axis is the time (minutes) before exit of each compound from the column. The vertical axis indicates the response of the detector at the exit, indicating the relative amount of substance reaching the detector at that particular time.

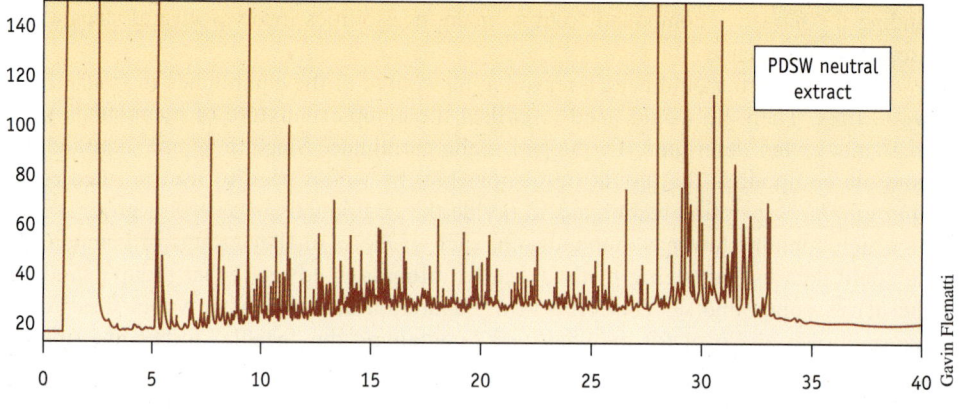

Gavin Flematti

Isolating the Bioactive Compound

After many rounds of chromatography separations, Flematti took 5 mg of the most active fraction and yet another solvent mixture. One of these fractions showed only about a dozen compounds. But he had less than 0.2 mg of material to work with in that fraction.

So as he had done many times before, back to the beginning again, burning an even larger amount of filter paper and repeating the many separations. This time he obtained a whopping amount of 1 mg (0.001 g) of each of three bioactive fractions, enough to work with using sophisticated instrumental techniques. From one of those fractions, Flematti separated out a single compound, and it was bioactive. The compound had a molecular weight of 150. For the first time ever, the compound in smoke causing seed germination had been isolated. The "needle in the haystack" had been found—but its identity, apart from its molecular weight, was still unknown.

What Is the Bioactive Compound?

As you will see in Chapter 3, a mass spectrometer can reveal the molecular weight of a compound, as well as how it breaks up into smaller pieces called fragments. Those fragments give information about the pattern of connectivity of atoms in each molecule. Large international databases list the fragment patterns in the mass spectra of every compound reported in the research literature. The *mass spectral fragmentation pattern* [>>Sections 3.9, 3.10] of the active component in smoke did not correspond with that of any other previously known compound. The active compound had never been reported in the chemistry literature before!

Mass spectra indicated a molecular weight of 150, and exact correspondence of the existence of certain isotopes by *high-resolution mass spectrometry* [>>Section 3.8] showed its formula to be $C_8H_6O_3$. From the molecular weight, the mass spectral fragmentation patterns, the wavelengths of ultraviolet light absorbed, its solubilities in different solvents and its acid-base properties, Flematti, his supervisors, and their colleagues gained insights into the structure of the molecules of the active compound. Flematti then synthesized in the laboratory several compounds that were possible fits with this information (Figure 1.6), but none showed germination activity.

The NMR spectrum [>>Section 9.4] of the active compound provided a map of the framework of hydrogen and carbon atoms in its molecules. This evidence finally convinced the research group that its molecules had the structure shown in Figure 1.7.

A search of the research literature indicated that this compound was not previously known. That meant that its properties could not be compared with reported properties of the compound observed by others. To prove that this was the correct structure for the active compound, he needed to synthesize in the laboratory the compound whose molecules have this structure. If the spectra and properties of the compound he made matched those of the bioactive ingredient of smoke, the chemical world would believe his evidence.

Synthesis of Gavinone In **synthetic chemistry**, researchers design a series of reactions, starting from one compound to produce another. But it is a gross oversimplification of Flematti's work to say that he began with some pyromeconic acid (Figure 1.8), converted this via a series of reactions (the product of one reaction being the starting point for the next reaction) to an intermediate compound, and then through another complicated sequence of reactions formed Gavinone. Unfortunately, at each step in a sequence, some of the starting material did not react in the way intended, and the yields become increasingly small. Flematti obtained only a very small amount of the target material, and invested much time in improving the synthetic method to increase its yield.

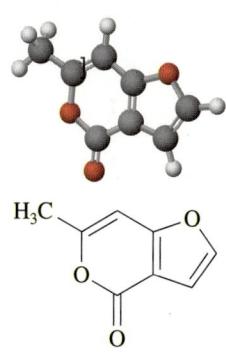

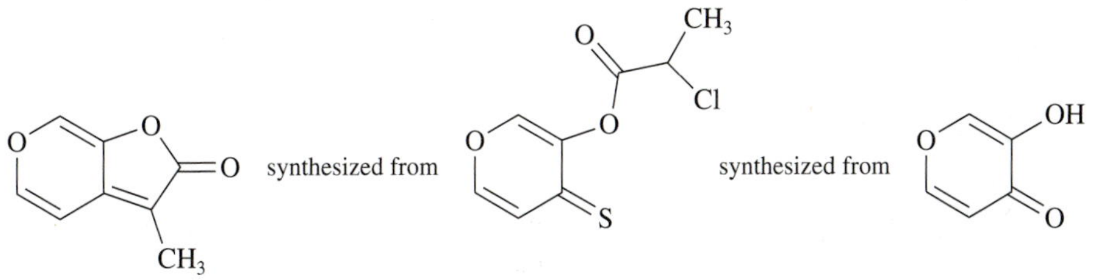

FIGURE 1.6 The molecular structure of one of the compounds made in the laboratory by Gavin Flematti. This molecule is very close in structure to the bioactive one, but is not quite the same—it's an isomer of Gavinone.

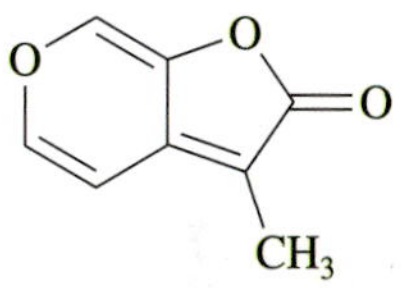

FIGURE 1.7 The molecular structure of the suspected active compound. How does this differ from the structure in Figure 1.6?

Molecular Modelling (Odyssey)

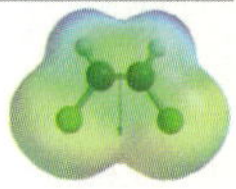

e1.2 Build a model of this molecule and simulate its flexibility.

3-methyl-2*H*-furo[2,3-*c*]pyran-2-one   synthesized from   intermediate compound   synthesized from   pyromecomic acid

FIGURE 1.8 The starting material and one of the intermediate compounds in Flematti's laboratory synthesis of Gavinone.

Comparing the Two Compounds

The NMR spectrum of the synthetic compound (Figure 1.8) corresponded exactly with that of the active ingredient of smoke (Figure 1.7), confirming that they are identical. Of course, if they are identical substances, the synthetic compound should also enhance seed germination. Bioassays showed that the synthetic Gavinone caused almost the identical level of germination as the compound from smoke.

Flematti's report of his work was published by the very prestigious journal *Science* (Volume 305, August 2004, page 977). The systematic name for his new compound, using the rules of the International Union of Pure and Applied Chemistry (IUPAC), is a mouthful: 3-methyl-2H-furo[2,3-c]pyran-2-one. Gavin Flematti's colleagues suggested the informal name *Gavinone* for this new compound as a tribute to his creativity, hard work, and success. He was also awarded the Cornforth Medal by the Royal Australian Chemical Institute for the most outstanding PhD thesis in any area of chemistry.

What's in a Name?

Names of compounds are taken very seriously in chemistry, so chemists can communicate clearly to each other about substances and their reactions. The informal name Gavinone was problematic for a couple of reasons. The "one" ending would imply to many chemists that the compound has a ketone functional group [>>Section 3.11], which is not consistent with the structure, and related compounds were soon discovered. After consulting with a linguist, Flematti proposed a new informal name for the class of compounds that trigger germination—*karrikins*, from the word "karrik," the word for "smoke" in the local aboriginal Nyungar dialect. Gavinone is now called *karrikenolide* and the research group is using the shortened abbreviation KAR_1—indicating that it is the first in an expected series of related compounds with similar biological activity.

The Future

KAR_1 has now been shown to promote germination of a range of vegetables and other plants in Australia, North America, and South Africa. Through Gavin Flematti's work, it is now possible to manufacture a seed germination promoter, rather than to extract it from smoke. Substantial benefits come from the ability to use a pure synthetic substance as a germinating agent rather than natural wood smoke, as smoke and smoke water contain many other compounds that are toxic and carcinogenic.

As with the PDT story, answers to research questions always open up new questions. What is there about the structure of KAR_1 that makes it a bioactive compound? What compounds in seeds interact with KAR_1, and why don't they interact with the thousands of other possibilities? Are there other, perhaps related compounds that cause germination in other seeds? Gavin Flematti has already started tackling these questions.

Courtesy of Australia Post

In subsequent work, Flematti identified and synthesized another compound that he thinks will cause germination of the wildflower, kangaroo paw, the floral emblem of Western Australia, shown here on an Australian stamp.

1.4 Chemical Reactivity, Your Activity

Some of the chemistry concepts in the PDT and Gavinone (KAR_1) stories may seem difficult at first reading. But relating them both to your own experience and to your prior understanding will help you make sense of them.

From the PDT story, the idea that oxygen exists in different forms or energy states that have strikingly different properties is not new. Nor does it only have uses in medicine. If you've ever been awed by the curtain of Northern or Southern Lights dancing thousands of kilometres across the sky, you've seen with your own eyes high energy forms of oxygen. High in the polar regions of the earth's atmosphere, electrons and other particles from the sun collide with oxygen and nitrogen atoms, exciting them to higher energy states. They give off this excess energy in the form of light. Excited oxygen between 100–300 km above the earth's surface emits the most commonly seen yellow-green colour of the Aurora.

Much closer to home, street lamps and "neon" signs both emit visible light from excited states of gases. The colour of the glow gives

Roman Krochuk/Shutterstock

Aurora Borealis over Kulusuk, Greenland.

information to a chemistry student about the gases in the lamp or sign. Sodium vapour street lights give off a characteristic yellow-orange colour, while mercury lamps are a bluish white. Mercury, neon, helium, and krypton gases in "neon" signs all emit light of different colours, as they move from their excited states to their ground states. Recognizing that the excited states of different elements emit light of different frequencies is key experimental evidence for our model of the electronic structure of atoms, discussed further in Chapter 8.

Perhaps you might think of the story of extraction, purification, and synthesis of Gavinone while next sipping on a morning cup of coffee. The first step in obtaining Gavinone was extracting it from thousands of compounds found in wood smoke. The aroma and flavour of the hot aqueous solution in a cup of java results from the interaction of several hundred compounds with receptors in a coffee drinker's mouth and nose. Compare the gas chromatogram of an air sample above roasted coffee beans in Figure 1.9 with the gas chromatogram of wood smoke from Flematti's lab in Figure 1.5. The compound in coffee that reaches the detector 21 minutes after injection (see x-axis) has been isolated, using many of the same chromatographic techniques Flematti used. Its name is 1,3,5-trimethylxanthine, and at doses above 250 mg, it causes restlessness, nervousness, insomnia, and tremors. You may recognize this addictive, toxic substance more readily by its common name, *caffeine*. Another compound in coffee, *trigonelline*, contributes to the bitter taste, and has been reported to have anti-tumour activity.

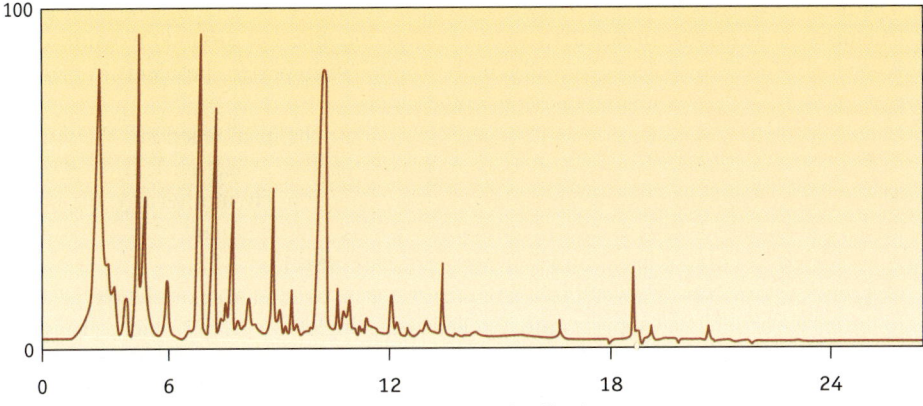

FIGURE 1.9 Gas chromatogram of air above roasted coffee beans.

We described Gavinone as an organic compound. What would it mean for a coffee package to be labelled "organic"? Most of the compounds in coffee from any source are organic—their molecules, just like those of Gavinone, are made of carbon, hydrogen, oxygen, and nitrogen atoms. But the "organic" label on food means something different, and usually refers to food that has been grown without the application of synthetic pesticides and herbicides.

Finally, your morning newscast or newspaper may well include a reference to "toxic chemicals." The PDT story shows that toxic chemicals, like singlet oxygen, can be used for beneficial purposes. In the Gavinone story, wood smoke, which you know to be toxic, is found to contain a compound that is highly beneficial for the environment. And even your morning cup of coffee contains both toxic and beneficial natural products. Many toxic substances have been produced in nature long before they were made by chemists. For over 5000 years, humans have sought natural products to destroy pathogens that cause disease. The World Health Organization estimates that 80% of the world's population still uses medicines from plants and herbs to meet most of their primary healthcare needs. Many of the pharmaceutical products in your medicine cabinet have been designed by slight modifications to the chemical structures of compounds first produced by plants or other organisms. And many are "toxic chemicals."

David Dolphin and Gavin Flematti share a passion for creative discovery and a disciplined approach to problem solving. In making sense of their stories and those in the chapters that follow, we hope you enjoy your own creative and disciplined activity of investigating, understanding, controlling, and using chemical reactivity.

References

Section 1.2 For further reading about photodynamic therapy, see QLT Inc. (n.d.). "Home Page." www.qltinc.com, accessed October 7, 2009; and R.R. Allison, et al. (2004). "Photosensitizers in Clinical PDT." *Photodiagnosis and Photodynamic Therapy* 1: 27–42.

Section 1.3 For more on Gavinone, see G. Flematti, et al. (2004). "A Compound from Smoke that Promotes Seed Germination." *Science* 305 (August 13): 977; and D. Nelson, et al. (2009). "Karrikins Discovered in Smoke Trigger Arabidopsis Seed Germination by a Mechanism Requiring Gibberellic Acid Synthesis and Light." *Plant Physiology* 149 (February): 863–73.

SUMMARY

Key Concepts

Key ideas in Chapter 1 are summarized below, along with several pointers to chapters where you will encounter them again in greater detail.

- Chemistry is a current, human activity, done by people like you around the world.
- Knowledge of chemistry and its skills can be applied to important problems in medicine, agriculture, environment, and many other areas.
- Electromagnetic radiation of the right energy can cause chemical reactions. We return to this idea in Chapter 21 on alkyl halides in the stratosphere. (Section 1.2)
- The toxicity of substances can be used to good advantage in **chemotherapy**, the medical use of chemical substances that are selectively toxic to malignant cells or to an agent such as a virus or bacterium that causes disease. (Section 1.2)
- Oxygen (O_2) can exist in several forms or energy states. Ground state, triplet oxygen, is in the air we breathe and supports life. Excited state, singlet oxygen, is toxic, causing severe damage to cells. Singlet oxygen can be produced in **photodynamic therapy (PDT)**, light-based medicine, and is used to treat cancers and eye diseases. (Section 1.2)
- **Chromatography**, a form of separation science, can be used to separate individual compounds from complex mixtures. Separation techniques, combined with spectroscopy, are important in identifying bioactive compounds such as those found in natural products. **Natural products**, compounds produced by living organisms, are central to the organic chemistry described in Chapters 4, 9, and 19–25. (Section 1.3)
- **Spectroscopy**, based on the absorption of electromagnetic radiation by molecules, can be a powerful tool to identify and characterize the three-dimensional structure of molecules. Evidence about connectivity of atoms in molecules can be obtained from mass spectrometry, infrared absorption spectroscopy, and nuclear magnetic resonance spectroscopy. We return to this idea in Chapter 3 (classifying substances), Chapter 9 (models of structure), Chapters 19–25 (organic chemistry), and Chapter 26 (transition metal chemistry). (Section 1.3)
- Modern medicine and the pharmaceutical sciences rely heavily on an understanding of chemistry applied to the synthesis of new substances and the study of how those new substances affect living organisms. (Section 1.2)
- The structures of porphyrins and related compounds such as hemoglobin are of central importance in biochemistry. We return to this in Chapter 16 (electron transfer reactions) and Chapter 29 (biochemistry). (Section 1.2)
- The two isomers of Visudyne™ in Figure 1.3 are very similar in structure, yet have important differences. Further details about regio- and stereoisomers are found in Chapter 9. (Section 1.2)

REVIEW QUESTIONS

Many of these questions can be answered by thinking about what you've just read in Chapter 1. In some cases, you will also find it helpful to browse the indicated sections of later chapters in the text, or online resources.

Section 1.2: Harnessing Light Energy and Exciting Oxygen

1.1 What three components are required for the use of PDT for treatment of skin cancer?

1.2 Could a green laser beam be substituted equally well for a red laser beam in PDT? Would any conditions need to change to make this substitution?

1.3 The location of a tumour is important in determining whether PDT may be an effective method of treatment. What factors related to tumour location are likely to be important in successful PDT?

1.4 Porphyrins are attractive choices for photosensitizers in medicine because this class of stable, coloured compounds is found so frequently in nature. When found in nature, the nitrogen atoms on the molecular framework of porphyrin molecules are often bound to metal atoms. Give an example of an important natural product where porphyrins molecules are bound to (a) magnesium atoms; (b) iron atoms; and (c) cobalt atoms.

1.5 Before porphyrins made in a laboratory can be used for PDT, they must be studied by toxicology. What is toxicology, and what information should be found about porphyrins from toxicological studies prior to their being approved for use in human medicine?

1.6 Light-based medicine has other applications besides the treatment of cancers and macular degeneration. List two such applications.

1.7 What is the definition of "chemotherapy"? Would you consider Louis Pasteur's vaccination of sheep against anthrax an example of chemotherapy? What about his use of garlic to prevent gangrene in World War I?

1.8 Would you consider garlic to be toxic? What are some known benefits and adverse effects of taking garlic for medicinal purposes?

1.9 Would you consider arsenic to be toxic? Consult Section 6.1 and indicate what considerations you might take into account in answering this question.

1.10 Small amounts of singlet oxygen can be produced under certain conditions in plants undergoing photosynthesis. What porphyrin-containing substance is found in plants that might act as a photosensitizer and absorb light, and then transfer energy from its excited state to ground state oxygen to produce singlet oxygen, just as Visudyne™ does in PDT?

1.11 David Dolphin's first scientific publications was in the prestigious journal *Nature* on the biological mechanism of Vitamin B-12. He later worked at Harvard University with U.S. Nobel laureate Robert Burns Woodward, whose research group synthesized Vitamin B-12 in the laboratory. The structure of a Vitamin B-12 molecule is shown below. Compare it to the structures of Visudyne™ in Figure 1.3. What is the connection between Vitamin B-12 and Dolphin's work, decades later, on photodynamic therapy?

Section 1.3: Where There's Smoke, There's Gavinone

1.12 Unlike Gavinone (KAR_1), not all compounds found in smoke are beneficial. Polycyclic aromatic hydrocarbons are one example of widespread pollutants produced in combustion reactions, including cigarette smoke. Consult Section 20.11 and list the elements that make up polycyclic aromatic hydrocarbons. Define each term in the name *polycyclic aromatic hydrocarbon*.

1.13 What is the definition of the term *natural product* given in Section 1.3?

1.14 Many other substances besides Gavinone have been named after people. What substance was named after each of the following people, and what is the significance of the substance?
(a) Bernard Belleau (Canada, 1925–1989)
(b) Richard Buckminster Fuller (USA, 1895–1983)

(c) Victor Francois Auguste Grignard (France, 1871–1935)

(d) Herbert H.M. Lindlar (Switzerland, 1909–)

(e) Marie Sklodowska Curie (France, 1867–1934)

(f) Giulio Natta (Italy, 1903–1979)

(g) Hugo Joseph Schiff (Germany, 1834–1915)

(h) Sir Geoffrey Wilkinson (Britain, 1921–1996)

(i) Georg Wittig (Germany, 1897–1987)

(j) Lise Meitner (Austria, 1878–1968)

1.15 Chemical processes, reactions, and theories have also been named after their discoverers. What importance do each of the following have in chemistry, and what people were they named after?

(a) Haber process

(b) Bohr model

(c) Erlenmeyer flask

(d) Van der Waals equation

(e) Gibbs free energy

(f) Lewisite

(g) Lewis base

(h) Schrödinger's equation

1.16 Sometimes discoverers in chemistry were not named or were less recognized for their contributions to chemistry. This was particularly true for early women pioneer scientists. What contributions did each of the following make to chemistry?

(a) Harriet Brooks-Pitcher

(b) Marie-Anne Pierette Paulze

(c) Rosalind Franklin

(d) May Sybil Leslie

(e) Ida Eva Tacke Noddack

(f) Marguerite Perey

(g) Margaret Hilda Thatcher

(h) Angela Merkle

1.17 From the simplified description in your text, create a flowchart that shows the processes Gavin Flematti used to determine which of the many possible compounds in wood smoke might be responsible for causing seeds to germinate.

Section 1.4: Chemical Reactivity, Your Activity

1.18 What is the definition of the term *organic* as used by chemists?

1.19 Using an Internet search engine, give one or more definitions most frequently found for the term *organic food*, and give two examples of food you might buy that are labelled "organic." Comment on whether you would expect those examples of "organic" food to contain any compounds that are not "organic" as understood by chemists.

1.20 The World Health Organization estimates that more than 80% of the world's population relies mostly on medicines obtained directly from herbs and plants for health care. List five examples of medicines produced by the pharmaceutical industry that are natural products first obtained from plants, and indicate their uses.

1.21 Pharmaceutical products are often first obtained from plants through a process of extraction, repeated cycles of separation and bioassays, and isolation and identification. The active compounds are then synthesized in laboratories before being placed on the market. How might synthetic drugs be the same and how might they be different from those extracted from plants?

1.22 What are the advantages and disadvantages of using pharmaceutical products from synthetic and from natural sources?

1.23 What are mycotoxins? Give one example. Is your example a natural product? Is it organic?

SUMMARY AND CONCEPTUAL QUESTIONS

1.24 Individuals with certain types of porphyria medical disorders are required to avoid bright sunlight and wear long-sleeve shirts, hats, and gloves. Patients undergoing PDT treatment for macular degeneration also need to take care about their exposure to sunlight for a few days after treatment. Look up what causes porphyria disorders, and explain what chemical substances these two groups of individuals have in common that cause photosensitivity.

1.25 A controversial suggestion has been made that porphyria provides a medical explanation for some historical legends of vampires and werewolves. Using what you learned in Question 1.24, what symptoms of porphyria seem consistent with reported behaviour of vampires?

1.26 T.M. Cox et al., in a research paper published by the respected medical journal *Lancet,* suggest that the madness suffered by British King George III (1738–1820) may have resulted from hereditary porphyria disease, and that his condition may have been made worse by arsenic poisoning. Arsenic interferes with the metabolism or breakdown of heme (a porphyrin in human blood) in humans. The researchers analyzed a sample of King George's hair and found levels of arsenic two orders of magnitude greater than samples from healthy controls. They suggest that the arsenic was likely an impurity in "emetic tartar" administered against the king's will by the royal physicians in the late 1780s. Emetic tartar is also known as *potassium antimony tartarate*. Refer to the periodic table [>>Section 2.12] to locate the metals antimony and arsenic, and suggest why emetic tartar obtained from natural sources in the eighteenth century was often contaminated with arsenic impurities. (Reference: T.M. Cox, N. Jack, S. Lofthouse, J. Watling, J. Haines, and M. Warren [2005]. "King George III and Porphyria: An Elemental Hypothesis and Investigation." *The Lancet* 366, 9482 [23–25 July]: 332–35.)

1.27 In spectroscopic techniques commonly used in chemistry to determine the structures of molecules, light of specific wavenumbers from certain regions of the electromagnetic spectrum interacts with molecules of a sample and is absorbed. That absorption of light energy has predictable effects on compounds, which gives valuable information about the structure of their molecules. Consult Section 3.8 and describe the effect that electromagnetic radiation from the infrared region of the spectrum has on molecular compounds. What information about the structure of a compound can be obtained from its infrared spectrum?

1.28 Organic compounds have common connectivity patterns of atoms, referred to as functional groups. The wavenumber of absorption by a molecular substance of infrared radiation is characteristic of the functional groups that are present in molecules of that substance. Consult the structure of a Gavinone molecule in Figure 1.7 and the discussion of functional groups in Section 3.8 and identify the functional groups present in Gavinone. At what wavenumbers would you predict Gavinone might absorb infrared radiation?

With the **eCHACR** single-sign-on access card bundled with your text, log on (http://login.cengage.com/sso) and access the e-book and click on any in-margin icons for dynamic molecular-level animations and simulations, videos of laboratory reactions, interactive exercises, reviews of background concepts, and other online supplementary materials as noted by the icons in the text margins.

Also go to www.chemistry.nelson.com <http://www.chemistry.nelson.com> for Answers to in-chapter exercises and selected Review Questions, Test Yourself questions, weblinks, crossword puzzles, flashcards, glossary of key terms, and other student resources.

Building Blocks of Materials

AP Photo/Marcio Jose Sanchez

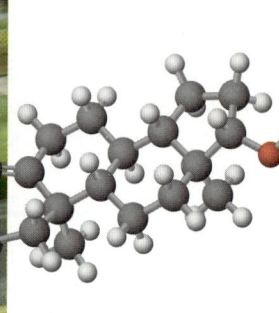

2.1 Case Study: Falsely Positive? The Chemistry of Drugs in Sport

In 2006, Floyd Landis, the winner of that year's prestigious Tour de France bicycle race, was accused of using performance-enhancing drugs. Urine samples collected after his dramatic victory in the mountainous 200 km Stage 17 of the race were alleged to have elevated levels of testosterone, a male sex hormone that builds muscle and improves stamina. Following an unsuccessful appeal to the Court of Arbitration for Sport, Landis was stripped of his winning yellow jersey and banned from racing for two years.

Cheating with the aid of chemistry is rife in many international sports. Chemistry can also be used to detect cheating, although it can sometimes be difficult to keep pace with new chemistry-assisted ways of enhancing performance. Let's examine the evidence used to decide whether an accused athlete is guilty of doping.

One way to classify performance-enhancing drugs is by origin—whether they are naturally produced in the human body through the transformation of substances in the diet (**endogenous substances**) or introduced to the body from an external source, such as from a laboratory synthesis (**exogenous substances**). Small but measurable differences in the masses of the individual atoms that make up these substances can reveal powerful insights into their origin.

When athletes take a known designer drug, such as THG (tetrahydrogestrinone, a.k.a. "the clear," Figure 2.1), that does not occur naturally in the body, the drug can be detected in their body fluid by chemical analysis. The method of choice is to test a urine or blood sample using a technique called **gas-chromatography** coupled with **mass spectrometry** (**GC-MS**), the same technique used by Gavin Flematti to identify his "needle in the haystack" in Section 1.3. First, chromatography is used to separate the compound(s) of interest from the hundreds of other compounds that are found in human body fluids. Then the compound is analyzed using a mass spectrometer, which provides information about the mass and structure of its molecules [>>Section 3.4]. This information is then fed into a computer database and compared with the mass spectrum of every known performance-enhancing drug. Detecting a drug such as THG can be challenging, however, due to the very low levels used, transformation of the drug in the body to other substances through metabolism, and the presence of other substances that can interfere with the analysis.

It is more complex to detect the illegal use of a substance that is both exogenous and endogenous, such as the steroid testosterone, which can be created in a laboratory and is also synthesized in the body from cholesterol. A logical test might be to look at how much of the substance appears in a sample using GC-MS measurements. However, normal levels of testosterone can differ from person to person and change during endurance competition and recovery. So even if a testing laboratory finds high levels of testosterone in an athlete's urine, how do we prove that the athlete has taken additional testosterone? It can be done. Let's look at the chemical analyses that were performed on Floyd Landis's samples.

Since testosterone is also endogenous and its levels can vary among athletes, doping agencies detect elevated levels by comparing the ratio of testosterone to another endogenous steroid found in all athletes, epitestosterone (Figure 2.1). It is generally agreed that a ratio of testosterone (T) to epitestosterone (E) greater than 6.0 indicates that the athlete has taken supplemental testosterone. Not all scientists agree: some claim that a ratio greater than 6.0 can occur by natural variation; others report that the ratio may not always exceed 6.0 after doping with testosterone, and athletes might even take exogenous epitestosterone along with testosterone to reduce the T/E ratio.

The procedure ultimately used in the Landis case was more definitive than GC-MS measurement of testosterone levels and analysis of the T/E ratio. Athletes who cheat by taking additional testosterone will have a mixture of exogenous and endogenous molecules of testosterone in their urine, so a highly specialized technique called **isotope ratio mass spectrometry (IRMS)** was used to tell the origin of molecules of the doping agent. IRMS, now called the "gold standard" of drug testing, uses a gas chromatograph to separate out a mixture of organic compounds found in human urine. The compounds that contain carbon atoms, such as testosterone, are next burned in a combustion chamber to produce carbon dioxide gas, which is then analyzed in the mass spectrometer. The technique determines the ratio of two different types of carbon atoms: carbon-12 and carbon-13. These two types of carbon atoms, called **isotopes** [>>Section 2.8], have the same number of protons (six) in their nuclei but differ in their number of neutrons: carbon-12 (^{12}C) has six neutrons, while carbon-13 (^{13}C) has seven. A carbon-13 isotope is therefore heavier and can be easily differentiated by the mass spectrometer.

Molecular Modelling

e2.1 Represent molecular structures using models and structural formulas.

The term *isotope* comes from Greek words meaning "same place," referring to two types of atoms that come from the same place in the periodic table (the same element) but with different numbers of neutrons in their nuclei.

(a) tetrahydrogestrinone

(b) testosterone

Key

bond in front
bond behind
carbon atom bonded to two hydrogen atoms and two carbon atoms
bond in plane

(c) epitestosterone

FIGURE 2.1 Chemical structures of steroid molecules. (a) Exogenous tetrahydrogestrinone (THG); (b) endogenous testosterone; and (c) endogenous epitestosterone. *Can you spot the differences?*

Chemistry can be used both to cheat and to detect cheating. What principles will guide your use of chemistry? One possibility might be to state (as Hippocrates did for medicine) that "as to chemistry, make a habit of two things—to help, or at least to do no harm."

What information about drug abuse can be obtained from the ratio of $^{13}C/^{12}C$ atoms in a compound from an athlete? In nature, only about 1% of carbon atoms are the ^{13}C isotope. Thus, the average $^{13}C/^{12}C$ ratio in the testosterone extracted from a normal body fluid sample is about 0.01. Exogenous testosterone has slightly lower ^{13}C content than endogenous testosterone, however, so the value of the $^{13}C/^{12}C$ ratio will vary by a few parts-per-thousand depending on the origin of the carbon atoms in the testosterone. The ^{13}C content is lower in exogenous testosterone because the starting materials used to synthesize testosterone are extracted from plants with a low ^{13}C content such as soy.

By contrast, endogenous testosterone is made from cholesterol, created when the body metabolizes plant and animal protein that is part of the human diet. As a result, endogenous testosterone is slightly higher in the relative number of ^{13}C atoms. So the ratio of $^{13}C/^{12}C$ atoms in testosterone from an athlete's sample gives compelling evidence of the origin of the carbon atoms—an athlete using synthetic testosterone will have a lower ratio. It was this evidence of a different isotope ratio in the sample from Floyd Landis that formed the basis of the case made against him by Tour de France officials.

Landis appealed his case. Because the different isotope ratios in endogenous and exogenous testosterone are so small, many of the questions in the appeal revolved around basic science and chemistry laboratory practice. These include whether any contamination of the samples occurred, and the expected range of errors in IRMS measurements. Landis maintains he is innocent, and he was back racing in 2009.

The Landis case shows the power of applying detailed knowledge of matter, the atoms that comprise it, and the slight differences in isotope ratios of those atoms to determine whether an athlete has cheated. e2.2 shows how IRMS measurements are made and interpreted. It also highlights other powerful ways to use isotope ratios, such as understanding Earth's climate several hundred thousand years ago, detecting the addition of sugar to honey, and learning whether conditions are favourable for life on Mars.

2.2 Classifying Matter

The label **matter** (or you might use the term *stuff*) is used for everything that we can see, as well as everything that we can't see, around us. We define matter as *anything that occupies space and has mass.*

We can classify matter in a number of ways (Figure 2.2). Is it solid, liquid, or gas? Is it homogeneous or heterogeneous? Is it pure or a mixture? Is it an element or a compound?

Think about It

e2.2 See how IRMS can detect cheating.

Web Link

e2.3 Why is ethics in chemistry so important?

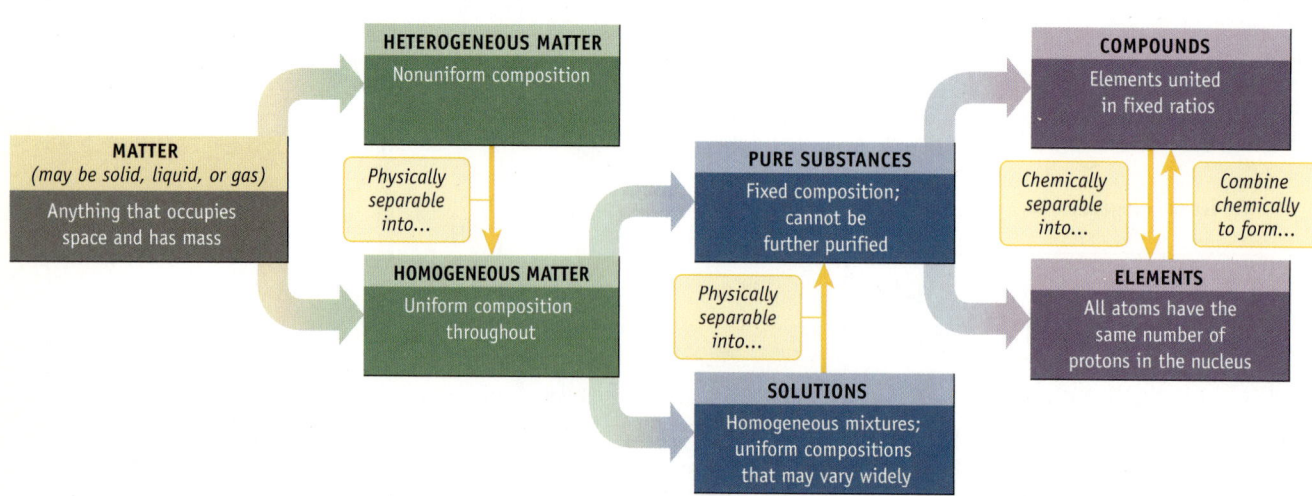

FIGURE 2.2 A flowchart for classifying matter. Solid, liquid, and gaseous states of matter.

Charles D. Winters

Bromine solid and liquid   Bromine gas and liquid

FIGURE 2.3 States of matter—solid, liquid, and gas. The photographs show the substance bromine in its solid, liquid, and gaseous states, and the insets show representations of how chemists imagine these forms at the molecular level. The tiny spheres represent atoms of bromine (Br). In all states, bromine consists of particles called molecules, each of which is composed of two bromine atoms (symbol Br_2), and these Br_2 molecules move as units in the liquid and gaseous states.

One way to classify matter, based on simple observation, is according to its **state** or **phase**—that is, whether it is solid, liquid, or gas (Figure 2.3). We recognize *solids* because they have a rigid shape and a fixed volume. *Liquids* also have fixed volume, but they are fluid: they are able to flow and take on the shape of a container up to its surface level. *Gases* are fluids that fill up the container they are in, whatever its volume and shape.

In making sense of solids, liquids, and gases, and the changes among them, chemists use the **kinetic-molecular model of matter** [>>Section 11.7]. Through the eyes of a chemist, all matter consists of extremely tiny particles (atoms, molecules, or ions) that are always moving, but experience attraction to other particles nearby:

- In solids, the particles are closely packed, usually in a regular array, and they vibrate rapidly about fixed positions.
- Liquids and gases are fluid because the particles are not fixed in specific locations and can move past one another.
- In the gas phase, molecules or atoms are far apart (compared with their size) and fly about very quickly, colliding with one another and with the container walls. The particles go to every part of their container, so the volume and shape of a gas sample are the volume and shape of the container.

An important idea of the kinetic-molecular model [>>Section 11.7] is that the higher the temperature, the faster the particles move. The energy of motion of the particles (their kinetic energy) acts to overcome forces of attraction between particles. In solids, the forces of attraction between particles dominate, and they are held in fixed positions, although they vibrate. A solid melts when its temperature is raised to the point at which the particles vibrate vigorously enough to push one another out of the way and move out of their positions. As the temperature increases even more, the particles in the liquid move still faster until they can escape the clutches of their neighbours (the forces of attraction between particles) and enter the gaseous state.

> Matter can be classified according to whether it is in the solid, liquid, or gaseous state. Chemists use the kinetic-molecular model to make sense of the nature of these states at the molecular level.

Homogeneous and Heterogeneous Mixtures

A mixture in which we cannot see the individual components, and which has uniform composition throughout (and, therefore, the same properties, such as colour and density, throughout) is called a **homogeneous mixture**. Homogeneous mixtures are called **solutions**. Common examples include air, gasoline, and a soft drink (sugar, flavouring agents, dissolved carbon dioxide, and water).

The *particles* referred to here (atoms, molecules, or ions) are not visible, even under conventional microscopes, and are much smaller than particles (or grains) of sand or sugar. In a grain of sugar, there may be as many as 10^{18} (1 000 000 000 000 000 000) molecules.

Molecular Modelling (Odyssey)

e2.4 Compare simulations of solid, liquid, and gaseous bromine.

In everyday usage, the term *solution* is usually applied to liquids, but in the scientific sense it also applies to gases mixed with other gases and even to some solid metal alloys, which contain two or more metals indistinguishably mixed over a range of compositions.

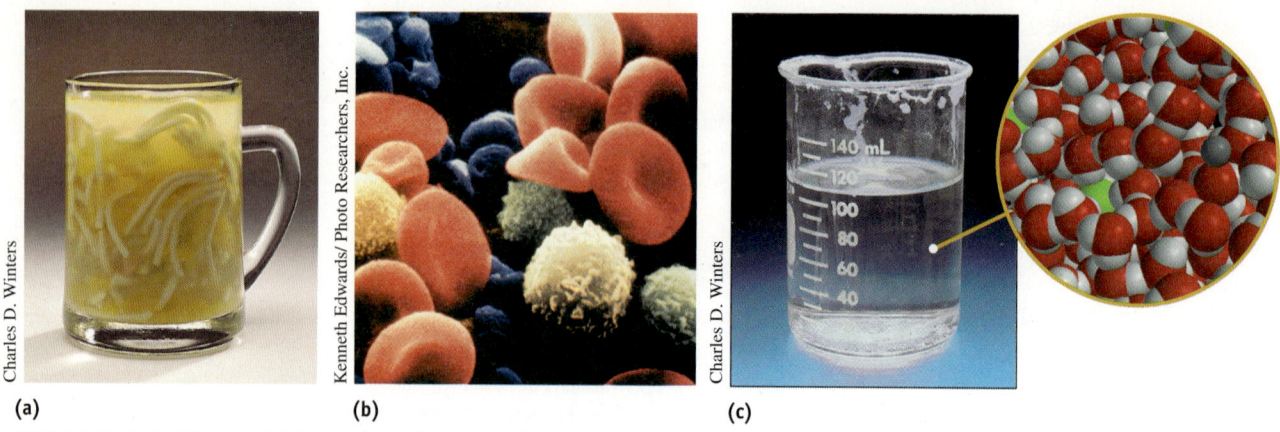

(a)   (b)   (c)

FIGURE 2.4 Mixtures. (a) A cup of noodle soup is a heterogeneous mixture. (b) Blood appears homogeneous to the unaided eye, but when viewed through an optical microscope it seems heterogeneous because we can distinguish both blood cells and a liquid. (c) A solution of salt in water is a homogeneous mixture. In the inset is a model of salt solution showing electrically charged particles (ions, here coloured green and grey) among the water molecules, but we cannot see the ions or the water molecules even under a powerful optical microscope.

Think about It

e2.5 Identify pure substances and heterogeneous or homogeneous mixtures.

A mixture in which the materials are not uniformly dispersed over the whole sample is called a **heterogeneous mixture**. The properties of such mixtures differ from region to region within the mixture. An example is a cup of noodle soup.

Some mixtures that seem to be homogeneous to the eye may better fit the heterogeneous description on closer examination. Blood, for example, may not look heterogeneous until you examine it under a microscope and red and white blood cells are revealed. Milk appears uniform to the unaided eye, but magnification reveals fat and protein globules. The classification as homogeneous or heterogeneous depends on the level of observation (Figure 2.4).

Matter can be classified according to whether it is homogeneous or heterogeneous.

Pure Substances and Mixtures

Matter that is composed of only a single substance is said to be *pure*—in contrast to *mixtures*, which contain more than one substance. By definition, pure substances are homogeneous. In chemistry, the term **substance** refers to a single, pure form of matter. So, water (with nothing dissolved in it), iron, and sugar (each with no impurities) are substances. On the other hand, and perhaps contrary to everyday use of the word *substance*, a salt solution in water, air (a mixture, mainly of nitrogen and oxygen gases), and gasoline (a mixture of hydrocarbon compounds) are not substances—they are mixtures of substances.

No sample of matter is absolutely pure: It always has at least a tiny fraction of other substances.

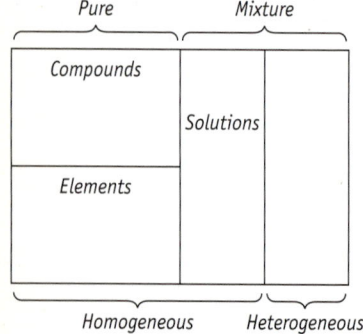

EXERCISE 2.1—CLASSIFICATION OF SUBSTANCES

The relational diagram (or Venn diagram) in the margin shows the relationships amongst some of the characteristics used to categorize substances.

Place each of the following substances into its appropriate box: (a) mud, (b) air, (c) vinegar, (d) gold, (e) milk, (f) sodium chloride, (g) an athlete's urine sample, and (h) testosterone.

2.3 Three Levels of Operation: Observable, Molecular, Symbolic

The characteristics of gases, liquids, and solids can be observed by the human senses when we see, measure, and handle samples of matter. We can also directly observe, for example, the colour and odour of a pure solid substance such as cholesterol or testosterone [>>Section 3.1], whether it dissolves in water, what its melting point is, and whether it conducts electricity or reacts with oxygen. This level of observations and experiments is called the **observable level of chemistry**.

When trying to make sense of observable chemical behaviour and events, chemists move to the level of atoms, molecules, and ions—a world of chemistry we understand with the help of mental models. Chemists believe that the observable behaviour of substances can be interpreted in terms of the arrangements, interactions, and movement of atoms, molecules, or ions.

These particles cannot be "seen" in the same way that one views the observable world, but they are no less real to chemists. This level of operation is referred to as the **molecular level**. The kinetic-molecular model of matter is a clear example of chemists modelling at the molecular level to rationalize phenomena at the observable level. The representation of how atoms connect to each other to make up a testosterone molecule (Figure 2.1) is another example of a molecular-level view.

Chemists communicate their observations and ideas with each other through words, labels, drawings, and symbols to refer to particular events and substances. Examples include $H_2O(\ell)$ to symbolize water in the liquid state, $C_{19}H_{28}O_2(s)$ for the solid state of testosterone, Br_2 for bromine molecules, $\rightarrow$ to mean chemical change, and colours to designate particular types of atoms or the distribution of electrons within molecules. These verbal and written representations are called the **symbolic level** of operation in chemistry.

To function properly in chemistry, we need to operate interchangeably and consciously at the observable, molecular, and symbolic levels (Figure 2.5).

The term *molecular level* refers to a scale of "zooming in" that can not be seen by optical microscopes, and should not be taken to mean that all substances consist of molecules. The sizes of molecules, atoms, and ions are of the same order.

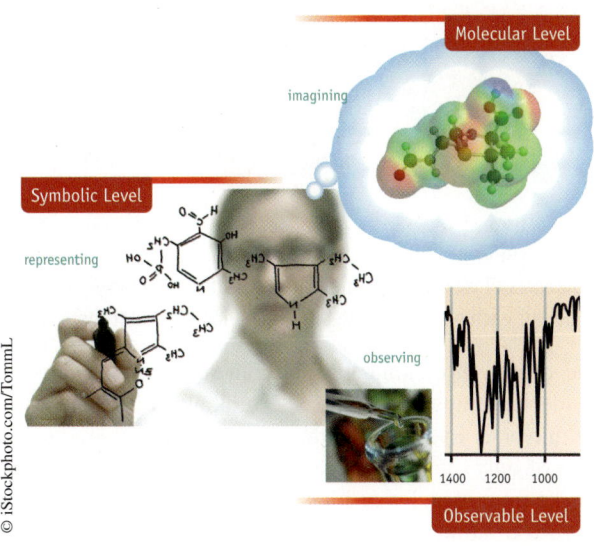

FIGURE 2.5 Levels of operation in chemistry. At the observable level, we use our senses and instruments to make observations and measurements. We develop mental models to imagine what is happening at the molecular level. At the symbolic level, we describe or represent structures, processes, and relationships by chemical notation and mathematical expressions.

EXERCISE 2.2—THINKING AT DIFFERENT LEVELS

At which level of operation (observable, molecular, or symbolic) is each of the following statements?

(a) The formation of water by the combustion of natural gas (methane) in air can be written

$$CH_4(g) + 2\,O_2(g) \longrightarrow CO_2(g) + 2\,H_2O(g)$$

(b) Water melts at 0°C.

(c) Water freezes when the temperature falls low enough that the energy of the water molecules is so low that they cannot escape from the forces of attraction between them, and they are unable to move with respect to each other.

2.4 Elements and Their Atoms

At the base of all understanding in chemistry is the view that tiny submicroscopic particles of different types, called **atoms**, are the building blocks for all matter.

There are relatively few different types of atom. Of all the substances known, a few—called **elements**—are such that all of the atoms in them are of only one type. The atoms of each element are different from those of every other element, and (except for nuclear transformations [>>Chapter 30]) atoms of one element cannot be changed into atoms of any other element.

There are 118 known elements, of which only about 90 are found in nature and the rest have been made—many of them fleetingly and in tiny amounts—in laboratories.

It is important to distinguish between the observable substances that we call elements and the atoms that are characteristic of each element. While all of the atoms in an element are of one type, it is not usually the case that the particles that comprise the substance are single atoms. For example, although the substance helium consists of helium atoms, the substance oxygen in the air is composed of molecules that have two oxygen atoms bound to each other (written as O_2 and referred to as molecular oxygen), just as bromine is composed of molecules that we would symbolize as Br_2. The most common form of phosphorus has P_4 molecules, while yellow sulfur powder is composed of S_8 molecules. Symbols such as $O_2(s)$, $Br_2(\ell)$, and $S_8(s)$ are the formulas of the **elemental substances**. Samples of elements are shown in Figure 2.6, and the element bromine is shown in its different states in Figure 2.3.

Mercury—liquid   Powdered sulfur—solid   Copper wire—solid   Iron chips—solid   Aluminum—solid

Charles D. Winters

FIGURE 2.6 Samples of elements. Each of these substances is composed of only one type of atom. The name and symbol for each element is listed in the periodic table at the front of this book.

Take in one of the most amazing and fundamental wonders of chemistry. Look around you and identify things you can see. Think also of other things you cannot see—anywhere in the universe. Every single thing or substance that you can see or think of is made up of atoms of only the 90 or so naturally occurring elements—most often in combinations that make up molecules or ions. The difference from one substance to another is in (a) which types of atom (of which elements); (b) the relative numbers of each type of atom; and (c) how the atoms are arranged in relation to each other.

The nature of atoms, the differences between them, their different arrangements in substances, and their influence on properties are the subject of more detailed discussion in Chapters 3, 8, 9, and 10.

EXERCISE 2.3—NAMES AND SYMBOLS OF ELEMENTS

Use the periodic table on the inside front cover of the book for the following questions:
(a) Find the names of the elements for the symbols Na, Cl, and Cr.
(b) Find the symbols for the elements zinc, nickel, and potassium.

All matter consists of atoms. Each element has a characteristic type of atom, different from those of every other element. Every substance is made up of some combination of atoms of the naturally occurring elements. It is important to distinguish between the kind of atom that is characteristic of an element and the elemental substance.

2.5 Compounds

A pure substance like testosterone, sugar, salt, or water, whose molecules or ions are composed of atoms of two or more different elements in fixed and definite proportions, is referred to as a **chemical compound**. Compounds are completely different substances from the elements of whose atoms they are composed. They have their own unique properties, such as colour, hardness, and melting point, which are different from those of the elements. Even though only 118 elements are known, there is no limit to the number of compounds that can be made from atoms of those elements. More than 20 million compounds are now known, and about a half million are added to the list each year.

Consider table salt (sodium chloride), which is composed of ions of the elements sodium and chlorine:

- Sodium is a shiny metal that reacts vigorously with water.
- Chlorine is a yellow-green gas that has a distinctive, suffocating odour and is a powerful irritant to our lungs and other tissues.
- Sodium chloride is a colourless, crystalline substance. Its properties are completely different from either of the substances sodium or chlorine (Figure 2.7).

A mixture of elements is different from a chemical compound. Iron (Fe), a shiny metal, and sulfur a yellow powder, can be mixed in any proportions, and the properties of both iron and sulfur remain observable (Figure 2.8a). However, the chemical compound iron pyrite, which is composed of atoms of iron and sulfur (Figure 2.8b), always has the same composition (46.55% Fe and 53.45% S by weight). Furthermore, iron pyrite has properties different from those of either iron or sulfur, or any mixture of these two elements.

EXERCISE 2.4—DIFFERENCE IN PROPERTIES BETWEEN A COMPOUND AND ITS ELEMENTS

Consult a reliable online chemical database to compare the properties of sucrose with the properties of the elemental substances carbon, hydrogen, and oxygen, of which sucrose is comprised.

Solid sodium, Na

+

Chlorine gas, Cl_2

Sodium chloride solid, NaCl

Charles D. Winters

FIGURE 2.7 A chemical compound. The compound sodium chloride (NaCl) bears no resemblance to either of the elemental substances sodium (Na) or chlorine (Cl_2). In sodium chloride, the elements sodium and chlorine have lost their identity as individual substances.

(a)   (b)

FIGURE 2.8 Mixtures and compounds. (a) In the dish is a mixture of iron chips and sulfur powder. The iron can be separated from the mixture with a magnet. (b) Iron pyrite, often found in nature as perfect golden cubes, is a chemical compound composed of atoms of iron and sulfur. The iron is not visible in the compound and cannot be removed from the mixture by physical means. Iron pyrite is a different substance from both iron and sulfur.

Think about It

e2.6 Watch a video on how to remove the iron in cereal.

Chemical Formulas of Compounds

The composition of any compound is represented by its **chemical formula**, which shows the relative numbers of the different types of atom of which it is composed. For example, carbon monoxide has the formula CO, which means that a sample of it contains the same number of carbon (C) atoms and oxygen (O) atoms—not as individual atoms, but combined in molecules with one C atom and one O atom in each. In a sample of carbon dioxide (CO_2), there are twice as many O atoms as C atoms; in water (H_2O), there are twice as many hydrogen (H) atoms as oxygen (O) atoms.

More specific interpretation of the chemical formula of a compound depends on whether we categorize the compound as a molecular compound, a covalent network compound, or an ionic compound [>>Chapter 3].

> **EXERCISE 2.5—MEANING OF SUBSCRIPTS IN A CHEMICAL FORMULA**
>
> What are the relative numbers of atoms of each element in (a) carbon monoxide, CO(g) (b) methane, CH_4(g) (c) ethyne, C_2H_2(g) and (d) glucose, $C_6H_{12}O_6$(s)?

The chemical formula of a compound indicates the relative numbers of atoms of each element in a sample.

2.6 Chemical Reactions, Chemical Change

When a mixture of the substances methane, CH_4(g), and oxygen, O_2(g), is heated, a rapid chemical reaction (called *burning* or *combustion*) happens. What does this mean? If we investigate what is in the reaction vessel after the reaction, we would find that there is almost none of either methane or oxygen left, but there are other substances—carbon dioxide and water—that were not present before the reaction. On the other hand, there are the same number of each type of atom—carbon, hydrogen, and oxygen—after the reaction as before. We can interpret this as a redistribution of atoms so that molecules of methane and oxygen that were present before the reaction no longer exist, and the atoms have re-arranged themselves into molecules of the substances carbon dioxide and water. We use the terms **reactants** for the substances that are present before reaction, and **products** for those new substances formed during the reaction.

Methane burning.

At the symbolic level, we can represent this chemical reaction by a **chemical equation**:

$$CH_4(g) + 2\ O_2(g) \longrightarrow CO_2(g) + 2\ H_2O(g)$$

This symbolic description not only tells us what the reactants are and what new substances are formed, but the symbol (g) also tells us that all of the reactants and products are in the gaseous state.

We can read this equation as follows: When methane and oxygen react with each other, and redistribution of atoms among molecules occurs, for every molecule of methane that reacts, two molecules of oxygen react, leading to formation of one molecule of carbon dioxide and two molecules of water.

When a lump of the solid element sodium, Na(s), is heated and put into a jar of the gaseous element chlorine, $Cl_2(g)$, a new substance, sodium chloride, NaCl(s), is formed and its properties are different from those of either sodium or chlorine. The reactants sodium and chlorine no longer exist—they have reacted and are "consumed" or "used up."

The concept of chemical reactions, and chemical equations to describe them, is the subject matter of Chapter 5.

Representations of molecules of reactants methane and oxygen.

Representations of molecules of products carbon dioxide and water.

EXERCISE 2.6—MEANING OF CHEMICAL EQUATIONS IN WORDS

Express in words the meaning of the following chemical equation for photosynthesis.

$$6\ CO_2(g) + 6\ H_2O(l) \longrightarrow C_6H_{12}O_6(aq) + 6\ O_2(g)$$

EXERCISE 2.7—MEANING OF CHEMICAL EQUATIONS IN WORDS

Express in words the meaning of the following chemical equation for a reaction important in formation of photochemical smog.

$$O_3(g) + NO(g) \longrightarrow O_2(g) + NO_2(g)$$

During a chemical reaction, there is a redistribution of the atoms so that new substances are formed. The products have completely different properties from the reactants.

Chemical and Physical Properties of Substances

Just as every person is different in looks and behaviour from every other person, so too every chemical substance differs from every other one. We describe the differences between substances in terms of both their chemical properties and physical properties.

The **chemical properties** of a substance describe its characteristic behaviour in reactions with other substances, and what happens if it does react. Does the substance react with oxygen in the air? Does it react with water? Does it react with chlorine gas? If so, what products are formed?

In contrast with chemical properties, the **physical properties** of a substance do not involve chemical transformation to new substances. Some physical properties commonly used to characterize substances include colour, state, melting point, boiling point, and density. Testosterone, for example, is an odourless, creamy, white solid with a melting point of 155 °C.

What are the physical and chemical properties of each elemental substance that makes it unique? Are there similarities among the properties of some of the elements? What is it about the atoms of each element that determines its characteristic properties? In what way are the atoms of element iron different from atoms of element chlorine? This chapter begins our exploration of the chemistry of the elements, which is developed in more detail in Chapter 8.

> Each substance can be described by its characteristic chemical and physical properties. Chemical properties refer to the chemical reactions that a substance undergoes, while physical properties do not involve chemical transformations.

2.7   Protons, Electrons, and Neutrons: Ideas about Atomic Structure

Around 1900, a series of experiments done by scientists such as John Joseph Thomson (1856–1940) and Ernest Rutherford (1871–1937) in Canada and then England established a model of the atom that is still the basis of modern ideas about the structure of atoms (Figure 2.9). According to this model:

- Atoms of every element are composed of the same three types of subatomic particles: **protons,** which have a positive electrical charge; **neutrons,** which are electrically neutral; and **electrons,** which have a negative electrical charge.
- The masses of protons and neutrons are much greater than the mass of electrons.
- The protons and neutrons together comprise a very small nucleus—which, therefore, contains all of the positive charge and nearly all of the mass of atoms.
- Electrons surround the nucleus and occupy most of the volume of atoms.
- The number of electrons around the nucleus is the same as the number of protons in the nucleus, so that atoms have zero net electrical charge.

Background Concepts

e2.7 Work through interactive tutorials on famous experiments that led to our model of the atom.

How small is an atom? The radii of atoms are between 30 and 300 pm (3×10^{-11} m to 3×10^{-10} m). To get a feeling for the incredible smallness of an atom, consider that one teaspoon of water (about 5 mL) contains about three times as many atoms as the Pacific Ocean contains teaspoons of water.

EXERCISE 2.8—RELATIVE SIZES OF THE NUCLEUS AND THE ATOM

We know now that the radius of the nucleus is about 0.001 pm, and the radius of an atom is approximately 100 pm. If an atom were an observable object with a radius of 100 m, it would approximately fill a small football stadium. What would be the radius of the nucleus of such an atom? Can you think of an object that is about that size?

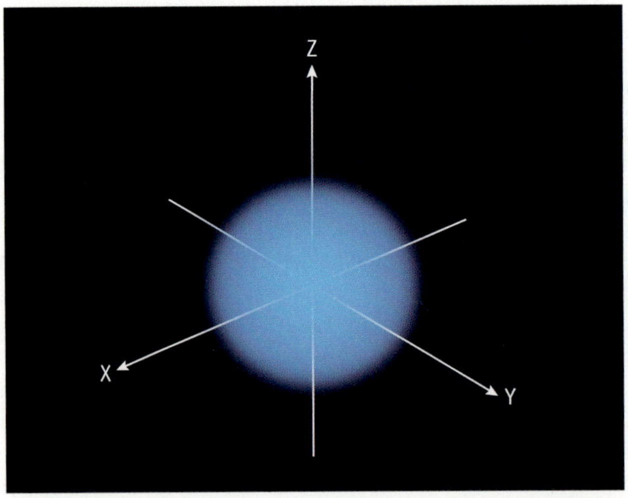

FIGURE 2.9 The structure of the atom. All atoms contain a nucleus with one or more protons (positive electric charge) and neutrons (no charge). A "cloud" of electrons (negative electric charge) is portrayed here, occupying the space around the nucleus. In an electrically neutral atom, the number of electrons equals the number of protons. The nucleus, occupying about 1/10 000 the size of the atom, is very dense matter.

Molecular Modelling

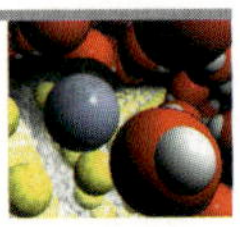

e2.8 Watch a cloud model of the atom become covered by a boundary surface.

Element Identity and Atomic Number

The defining feature of elements is the number of protons in the nuclei of their atoms. All atoms of an element have the same number of protons in the nucleus (and, therefore, the same number of electrons around the nucleus), and this is different from the number of protons in atoms of every other element. Hydrogen, the simplest element, has one proton in the nucleus of each of its atoms. All helium atoms have two protons, all lithium atoms have three protons, and all beryllium atoms have four protons. The number of protons in the nucleus of each atom of an element is its **atomic number**, symbol Z.

> Atoms are composed of a nucleus containing protons and neutrons, and electrons surrounding the nucleus. Protons have a positive electrical charge and electrons an equal negative charge, so there are equal numbers of them in an atom. Neutrons are uncharged. Every element has a different number of protons in its atoms, called the atomic number (Z).

Copper
29 ----- Atomic number
Cu ---- Symbol
63.546 --- Atomic weight

The entry in the periodic table for the element copper. Copper has $Z = 29$, meaning that in the nucleus of every atom of copper are 29 protons, and there are 29 electrons around the nucleus.

2.8 Isotopes of Elements

Although every atom of an element has the same number of protons in the nucleus, most elements have two or more types of atom with different numbers of neutrons. For example, while all carbon atoms ($Z = 6$) in a sample of an athlete's testosterone or other carbon-based molecules have 6 protons and 6 electrons, 98.89% of them have 6 neutrons in the nucleus, 1.11% of them have 7 neutrons, and 0.01% of them have 8 neutrons. Atoms of the same element with different numbers of neutrons are called *isotopes* of that element.

A fundamental idea in chemistry is that chemical properties of an element depend on the number of electrons in its atoms. So the chemical behaviour of different isotopes of an element is essentially the same. On the other hand, because the mass of an atom depends almost entirely on the numbers of protons and neutrons, atoms of different isotopes have different masses.

> Atoms of the same element that have different numbers of neutrons are called isotopes.

Isotope Identity and Mass Number

The sum of the number of protons and neutrons in atoms of an isotope is called its **mass number** and is given the symbol A.

$$\text{Mass number, } A = \text{number of protons} + \text{number of neutrons}$$
$$= Z + \text{number of neutrons}$$

Different isotopes of an element are defined by their mass number. For example, a boron atom ($Z = 5$) with 5 neutrons in its nucleus has mass number $A = 10$, while one with 6 neutrons has $A = 11$. The most common atom of uranium has 92 protons and 146 neutrons, so it has mass number $A = 238$. The convention used to refer to an atom of boron with $Z = 5$ and $A = 10$, for example, is $^{10}_{5}B$.

The subscript Z is optional because the element symbol tells us what the atomic number is. For example, the atoms described previously have the symbols $^{10}_{5}B$, $^{11}_{5}B$, and $^{238}_{92}U$, or simply ^{10}B, ^{11}B, and ^{238}U. In words, we say boron-10, boron-11, and uranium-238.

While scientists usually specify a particular isotope by its mass number (for example, ^{238}U), the isotopes of hydrogen are so important that they are given special names and symbols. All hydrogen atoms have one proton. An isotope with no neutrons in the nucleus, $^{1}_{1}H$, is called *protium,* or just "hydrogen." The isotope with one neutron, $^{2}_{1}H$, is called *deuterium* (symbol, D), while the nucleus of radioactive hydrogen-3, $^{3}_{1}H$, or *tritium* (symbol, T), has two neutrons.

The substitution of one isotope of an element for another can have an important effect on the physical properties of a compound. This is especially true when deuterium is

Think about It

e2.9 Practise using this notation for specifying an isotope.

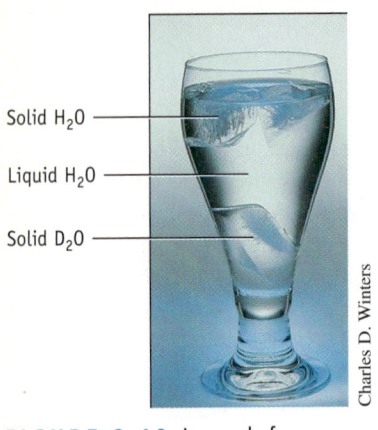

FIGURE 2.10 Ice made from "heavy water." Water containing ordinary hydrogen atoms ($^1_1$H, protium) forms a solid that is less dense at 0 °C *(d = 0.917 g mL^{-1})* than liquid H_2O *(d = 0.997 g mL^{-1})* and so it floats in the liquid. Similarly, "heavy ice" (D_2O, deuterium oxide) floats in "heavy water." However, D_2O ice is denser than liquid H_2O, so ice cubes made of D_2O sink in liquid H_2O.

Interactive Exercises 2.10–2.11

Practise using and interpreting the symbols for isotopes.

Think about It

e2.10 Use IRMS to show if maple syrup is genuine.

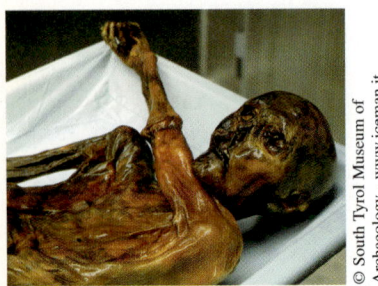

Ötzi the iceman.

Charles D. Winters

© South Tyrol Museum of Archaeology - www.iceman.it

substituted for hydrogen, because the mass of deuterium is double that of hydrogen (Figure 2.10). "Heavy water" used in some nuclear reactors is enriched in deuterium—most of the water molecules are D_2O.

WORKED EXAMPLE 2.1—ATOMIC COMPOSITION

What is the composition of an atom of phosphorus with 16 neutrons? What is its mass number? What is the symbol for atoms of this isotope?

Solution

For phosphorus, $Z = 15$, so a phosphorus atom has 15 protons.

$$\text{Mass number, } A = \text{number of protons} + \text{number of neutrons} = 15 + 16 = 31$$

The symbol for atoms of this isotope (^A_ZP) is $^{31}_{15}$P.

EXERCISE 2.9—ATOMIC COMPOSITION

(a) What is the mass number of an iron atom with 30 neutrons?
(b) How many protons, neutrons, and electrons are in an atom of the ^{64}Zn isotope?

Isotope Abundance

The story of Floyd Landis [<<Section 2.1] shows that tiny differences in the ^{13}C/^{12}C ratio of carbon atoms in testosterone from the average value of about 1% can be used to detect cheating by an athlete.

Scientists have used differences in the isotopic ratio of carbon and other atoms for many other important purposes. The natural abundance of isotopes is influenced not only by biochemical pathways, but also by other environmental conditions such as climate and soil. Analysis of isotope ratios of carbon, hydrogen, oxygen, and nitrogen has been used to determine the geological sources of oils, the authenticity of food products, and the origin of grapes used to make expensive wines. The dilution of genuine maple syrup with sugar water, for example, can be detected by IRMS, as can the replacement of natural vanillin with a synthetic product.

Differences in isotope ratios have also been used in forensic research to determine where humans were born and lived. Ötzi is a 5000-year-old mummy discovered in 1991 by two hikers in the melting ice of a glacier high in the Alps along the Austrian–Italian border. By comparing strontium (Sr), lead (Pb), and oxygen (O) isotopic compositions in the teeth, bones, and intestine of Ötzi with those of environmental sources such as water and rocks in the area, researchers were able to provide evidence that he spent most of his life in several valleys within 60 km of where the body was found.

What is the natural abundance of isotopes of some other elements? Let's look at the average **natural percent abundance** of different isotopes of hydrogen. A sample of water from a stream or lake consists almost entirely of H_2O, where the H atoms are the ^{1}H isotope. A few molecules, however, have one deuterium (^{2}H) atom. We could predict this because it is known that 99.985% of all hydrogen atoms on Earth are ^{1}H atoms. That is, the percent abundance of ^{1}H atoms in nature is 99.985%.

$$\text{Percent abundance of an isotope} = \frac{\text{number of atoms of that isotope}}{\text{total number of atoms of the element}} \times 100\%$$

The percent abundance of deuterium atoms is the remaining 0.015% of all hydrogen atoms. Tritium does not occur naturally.

In a naturally occurring sample of boron, the ^{10}B isotope has an abundance of 19.91% and ^{11}B has 80.09%. That means that if you could count out 10 000 boron atoms from an "average" natural sample, 1991 of them would be ^{10}B atoms and 8009 of them would be ^{11}B atoms.

WORKED EXAMPLE 2.2—ISOTOPES

Silver has two isotopes, one with 60 neutrons (percent abundance = 51.839%) and the other with 62 neutrons. What are the mass numbers and symbols of these isotopes? What is the percent abundance of the isotope with 62 neutrons?

Solution

Silver has $Z = 47$. $A = Z +$ number of neutrons.

For isotope 1, $A = 47 + 60 = 107$, and its symbol is $^{107}_{47}$Ag.

For isotope 2, $A = 47 + 62 = 109$, and its symbol is $^{109}_{47}$Ag.

The percent abundances of all isotopes must add up to 100%.

$\therefore$ percent abundance of ^{109}Ag $= 100.000\% - 51.839\% = 48.161\%$

EXERCISE 2.12—ISOTOPES

(a) Argon has three isotopes in which the number of neutrons in the atoms are 18 (0.337% abundant), 20 (0.063% abundant), and 22, respectively. What are the mass numbers and symbols of these three isotopes? What is the percent abundance of the isotope whose atoms have 22 neutrons?

(b) Gallium has two isotopes: ^{69}Ga and ^{71}Ga. How many protons and neutrons are in the nuclei of each of these isotopes? If the percent abundance of ^{69}Ga is 60.1%, what is the percent abundance of ^{71}Ga?

> For a discussion of the correct use of significant figures in calculations, see Appendix G

The percent abundance of an isotope of an element is the percentage of atoms of that isotope in relation to the total number of atoms of all of the isotopes of the element.

2.9 Relative Atomic Masses of Isotopes and Atomic Mass Units

What is the mass of an atom? Chemists have realized that careful experiments (see below) can give *relative* masses of atoms. For example, the mass of an atom of the oxygen ^{16}O isotope is estimated to be 1.33291 times the mass of an atom of the carbon ^{12}C isotope, and the mass of a ^{81}Br atom is 6.7430 times the mass of a ^{12}C atom.

Rather than use the absolute masses of atoms, which are very small (the mass of a ^{12}C atom is 1.9926×10^{-23} g), scientists find it convenient to use a relative scale of atomic masses. The standard on this relative scale is an atom of the ^{12}C isotope, assigned to be exactly 12. The **relative atomic mass (A_r)** of an atom of an isotope of any element is *its mass on a scale in which the mass of a ^{12}C atom is exactly 12.* An alternative definition is *the ratio of the mass of an atom to 1/12 of the mass of a ^{12}C atom.*

For example, experiment shows that $\dfrac{\text{mass of an } ^{16}\text{O atom}}{\text{mass of a } ^{12}\text{C atom}} = 1.33291$

If we assign a value of exactly 12 to mass of a ^{12}C atom, then

$$A_r(^{16}\text{O}) = 1.33291 \times 12 = 15.99492$$

The relative atomic masses of isotopes of all of the other elements are similarly assigned as ratios on this scale. Since the relative atomic mass of an isotope is a ratio, it has no units—it is just a number.

> An alternative name for the unit on the atomic mass scale is *Dalton*, symbol Da.

Sometimes the masses of atomic particles are expressed in **atomic mass units (u)**. On this scale, the unit of measurement, *the atomic mass unit, 1 u, is 1/12 of the mass of a ^{12}C atom.* In other words, a ^{12}C atom has a mass of exactly 12 u. The mass of a particle on the atomic mass unit scale is numerically equal to its relative atomic mass. So, the mass of an ^{16}O atom is 15.99492 u. The atomic masses on this scale of selected isotopes are listed in Table 2.2, and those of subatomic particles in Table 2.1, showing that the mass of a proton is over 1800 times the mass of an electron.

TABLE 2.1 Properties of Subatomic Particles*

| Particle | MASS | | Charge | Symbol |
	Grams	*Atomic Mass Units*		
Electron	9.109383×10^{-28}	0.0005485799	-1	$^{0}_{1}e$ or e^{-}
Proton	1.672622×10^{-24}	1.007276	$+1$	$^{1}_{1}p$ or p^{+}
Neutron	1.674927×10^{-24}	1.008665	0	$^{1}_{0}n$ or n^{0}

* These values and others in the book are taken from the National Institute of Standards and Technology website at http://physics.nist.gov/cuu/Constants/index.html

Measuring Atomic Mass and Isotope Abundance

How can we know the relative atomic masses and abundances of isotopes? These are determined experimentally using a mass spectrometer (Figure 2.11), which is also one technique used to determine the identity of molecules of substances such as gavinone [<<Section 1.3] and testosterone [<<Section 2.1]. A gaseous sample of an element is introduced into an evacuated chamber, and the molecules or atoms of the sample are converted to charged particles (ions) by the loss of one or more electrons. A beam of these ions is directed through a magnetic field, which causes the paths of the ions to be deflected. The degree of deflection depends on the ratio mass/charge of the particle. For ions all with a +1 charge, the lighter ions are deflected more. The separated ion beams are detected as an electric current at the end of the chamber. The amount of current measured is related to the number of ions of a particular mass and hence to the abundance of the ion. Knowing that most of the ions within the spectrometer have a +1 charge allows us to derive a value for mass.

Modern mass spectrometers can measure the relative atomic masses of isotopes to an accuracy of nine significant figures.

Except for carbon-12, whose relative atomic mass is defined to be exactly 12, relative atomic masses of isotopes do not have integer values, but they are very close to the mass number. For example, the relative atomic mass of boron-11 isotope (^{11}B, 5 protons and 6 neutrons) is 11.0093, and that of iron-58 isotope (^{58}Fe, 26 protons and 32 neutrons) is 57.9333.

Think about It

e2.11 See how a mass spectrometer focuses ions to produce a mass spectrum.

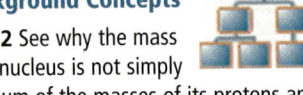

Background Concepts

e2.12 See why the mass of a nucleus is not simply the sum of the masses of its protons and neutrons.

EXERCISE 2.13—MASS RATIOS OF ISOTOPES

From the relative atomic masses of the isotopes of magnesium, listed in Table 2.2, calculate the following:
(a) the ratio of the mass of a ^{24}Mg atom to the mass of a ^{12}C atom
(b) the ratio of the mass of a ^{26}Mg atom to the mass of a ^{12}C atom
(c) the ratio of the mass of a ^{26}Mg atom to the mass of a ^{24}Mg atom

Relative atomic mass and relative abundance of an isotope can be measured by a mass spectrometer, which gives the mass/charge ratio of ions formed in an electron beam.

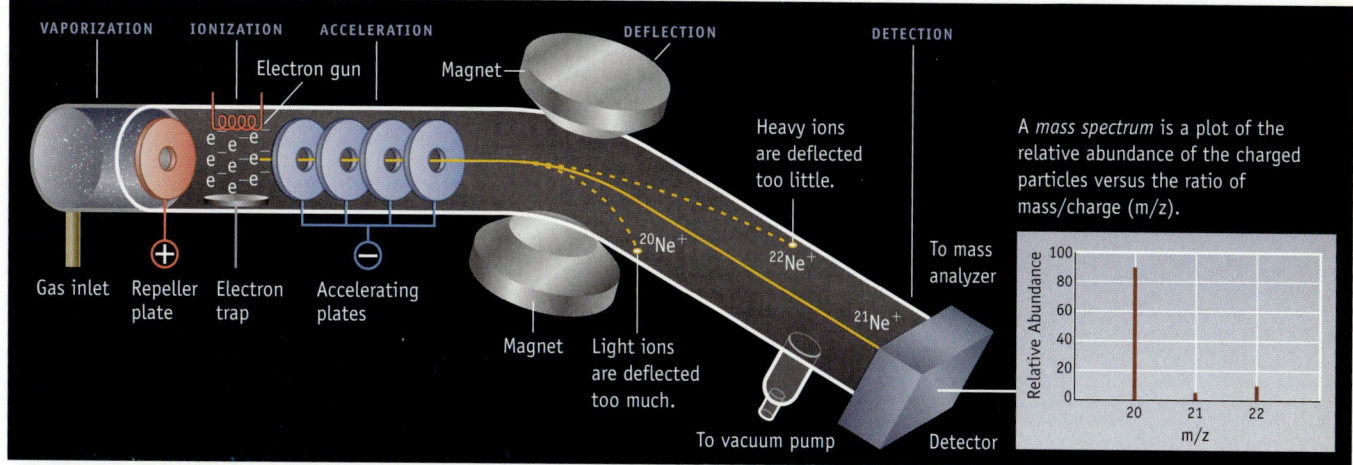

FIGURE 2.11 Mass spectrometer. A sample (in this case, neon) is introduced into the ionization chamber, where it is bombarded with high-energy electrons that strip electrons from the atoms or molecules of the substance. The resulting positively charge particles are accelerated by a series of negatively charged plates into a magnetic field perpendicular to the direction of the particles. The particles undergo deflection, with a radius of curvature that depends on both the mass and charge of the particles (as well as their speed and the magnetic field strength). In this diagram, $^{21}Ne^+$ particles are focused on the detector, while $^{22}Ne^+$ particles experience less deflections, and $^{20}Ne^+$ particles experience more. By changing the field strength, the beams of charged particles of different mass can be focused on the detector, and a spectrum of masses observed.

2.10 Atomic Weights of Elements

Because every sample of boron has some atoms with a relative atomic mass of 10.0129 and others with a relative atomic mass of 11.0093, the average relative mass of all atoms in a sample of boron is between these values. The **atomic weight** of an element is *the average relative atomic mass of a representative sample of its atoms, taking into account the relative abundances of its isotopes.* Since atomic weights are averaged values of ratios, they are unitless numbers.

In general, the atomic weight of an element can be calculated using the equation

$$\text{Atomic weight} = \left(\frac{\% \text{ abundance isotope 1}}{100}\right)(A_r \text{ of isotope 1})$$
$$+ \left(\frac{\% \text{ abundance isotope 2}}{100}\right)(A_r \text{ of isotope 2}) + \cdots$$

This equation gives an average, weighted by the abundance of each isotope of the element. For boron with two isotopes (^{10}B, 19.91% abundant; ^{11}B, 80.09% abundant), we find

$$\text{Atomic weight of boron} = \frac{19.91}{100} \times 10.0129 + \frac{80.09}{100} \times 11.0093 = 10.81$$

As illustrated by the data in Table 2.2, the atomic weight of an element is always closer to the mass of the most abundant isotope or isotopes.

The atomic weight of each stable element has been estimated experimentally, and the values are shown inside the front cover. The atomic weights of unstable (radioactive) elements are gradually changing because the percent abundances of the isotopes are changing. In these cases, the value given in parentheses is the atomic mass of the most stable isotope.

The atomic weight of an element is the weighted average of the relative atomic masses of its isotopes in a representative sample. Atomic weight values are not constants of nature, but can vary according to the fraction of an element's stable isotopes in any given sample of that element.

TABLE 2.2 Masses and Abundances of Isotopes, Atomic Weights of Elements

Element	Symbol	Atomic Weight	Mass Number	Isotopic Mass (u)	Natural Abundance (%)
Hydrogen	H	1.00794	1	1.0078	99.985
	D*		2	2.0141	0.015
	T†		3	3.0161	0
Boron	B	10.811	10	10.0129	19.91
			11	11.0093	80.09
Neon	Ne	20.1797	20	19.9924	90.48
			21	20.9938	0.27
			22	21.9914	9.25
Magnesium	Mg	24.305	24	23.9850	78.99
			25	24.9858	10.00
			26	25.9826	11.01

*D = deuterium; †T = tritium, radioactive.

Think about It

e2.13 Calculate an atomic weight from the atomic masses and abundance of its isotopes.

WORKED EXAMPLE 2.3—CALCULATING THE ATOMIC WEIGHT OF AN ELEMENT

Bromine has two naturally occurring isotopes. One has a relative atomic mass of 78.918338 and a percent abundance of 50.69%. The other, with relative atomic mass 80.916291, has a percent abundance of 49.31%. Calculate the atomic weight of bromine in a representative sample.

Solution

$$\text{Atomic weight of bromine} = \frac{50.69}{100} \times 78.918338 + \frac{49.31}{100} \times 80.916291 = 79.90$$

Interactive Exercises 2.15–2.17

Isotope abundance and atomic mass.

SUBMIT

EXERCISE 2.14—CALCULATING THE ATOMIC WEIGHT OF AN ELEMENT

Verify that the atomic weight of chlorine is 35.45, given the following information:
^{35}Cl: A_r = 34.96885; percent abundance = 75.77%
^{37}Cl: A_r = 36.96590; percent abundance = 24.23%

2.11 Amount of Substance and Its Unit of Measurement: The Mole

When we weigh a sample to determine the number of particles of the substance, we are making a connection between the observable world, the world we can see, with the molecular-level world of atoms, molecules, or ions.

Sometimes we need to know how much of one substance will react with a given amount of another, or how much of a product will be formed. The relative numbers of particles of reactants and products are given to us by the chemical equation that describes the reaction. So we need a practical method of counting atoms, molecules, and ions. The solution to this problem is to define a convenient amount of matter that contains a known number of particles. That unit of measurement for doing this is called the **mole** (abbreviation: mol). This is the unit of the quantity called **amount of substance**, for which we use the symbol *n*. The use of this concept allows us to weigh a sample of substance and know how many particles of it are in the sample.

In the particular case of elements composed of atoms, this unit is defined as follows: *A mole of atoms of an element is the amount of it that contains as many atoms as there are in exactly 12 grams of the carbon-12 isotope.* An alternative way to say this is: *A mole of atoms of an element is the amount of it whose mass in grams is numerically equal to its atomic weight.*

We read "*n*(Fe) = 0.10 mol" as "The amount of iron is 0.10 mol." Amount of substance is the quantity measured, and mole is the unit of measurement. In everyday language, the term *amount* is used to refer to any of mass (2 kg of oranges), volume (1 L of milk), or number (10 eggs). Because it has a specific meaning in chemistry, we will sometimes use the term *chemical amount*.

The key to understanding the concept of the amount of substance is recognizing that *one mole of particles of a substance contains the same number of particles, regardless of the nature of the particles*. For example, one 1 mol of sodium metal contains the same number of atoms as 1 mol of iron metal. The same principles apply to elements composed of molecules. One mole of an element composed of molecules (such as oxygen, O_2, and nitrogen, N_2), is the amount that contains the same number of molecules as there are atoms in exactly 12 g of ^{12}C. As a result, the number of molecules in 1 mol of oxygen (O_2) is the same as the number of atoms in 1 mol of iron (Fe).

How many specified particles are in 1 mol of a substance? The currently accepted best estimate of this number, based on many experiments, is 6.0221415×10^{23}. This number is referred to as the **Avogadro constant**, symbol N_A, named in honour of Amedeo Avogadro, an Italian physicist (1776–1856).

$$N_A = 6.0221415 \times 10^{23} \text{ mol}^{-1}$$

> Mole is the unit of amount of substance. One mole of specified particles of any substance has the same number of those particles as one mole of specified particles of any other substance.

The symbolic statement, "n(Fe) = 0.10 mol" corresponds with the more familiar "m(Cu) = 0.10 g" which we read as "The mass of copper is 0.10 grams." Mass is the quantity measured, and gram is the unit of its measurement.

Think about It

e2.14 Convert amounts to numbers of atoms, molecules, or ions.

Background Concepts

e2.15 Read about Amedeo Avogadro and the constant named in his honour.

EXERCISE 2.18—RATIOS OF AMOUNTS AND NUMBERS OF ATOMS

What is the ratio of the number of atoms in 2 mol of iron, Fe(s), to the number of atoms in 0.2 mol of rubidium, Rb(s)?

Molar Mass

The mass, in grams, of one mole of any element is called its **molar mass**—for which we use the symbol M. The unit of measurement of molar mass is g mol^{-1} (grams per mole). The molar mass of an element composed of individual atoms (as distinct from molecules) is the mass, in grams, that is numerically equal to its atomic weight.

Molar mass of sodium (Na) = 22.99 g mol^{-1} = mass of 6.022×10^{23} atoms

Molar mass of lead (Pb) = 207.2 g mol^{-1} = mass of 6.022×10^{23} atoms

The amount of sodium in a 45.98 g sample is 2.00 mol (n(Na) = 2.00 mol), and contains 12.044×10^{23} atoms. In a 621.6 g sample of lead, n(Pb) = 3.00 mol, and there are 18.066×10^{23} lead atoms.

It is often important to know the mass of an element that contains the same number of atoms as there are in a sample of another element. Figure 2.12 shows 1 mol of each of some common elemental substances. Although each of these "piles of atoms" has a different volume and different mass, each contains the same number (6.022×10^{23}) of atoms.

Amount of substance, and its unit the mole, is the cornerstone of quantitative chemistry. It is essential to be able to convert from amount (in mol) to mass and from mass to amount:

Amount to mass: mass (g) = amount (mol) $\times$ molar mass (g mol^{-1}) i.e., m = $n \times M$

$$\text{Mass to amount: amount (mol)} = \frac{\text{mass (g)}}{\text{molar mass (g mol}^{-1})} \quad \text{i.e., } n = \frac{m}{M}$$

For example, the mass of 0.35 mol of aluminium (M = 27.0 g mol^{-1}) is calculated as

$$\text{Mass} = 0.35 \text{ mol} \times 27.0 \text{ g mol}^{-1} = 9.5 \text{ g}$$

Think about It

e2.16–2.18 Practise converting among amounts, volumes, and masses of substances.

To cancel out units and arrive at an answer in the appropriate units, you need to recognize that mol^{-1} = 1/mol.

Charles D. Winters

Courtesy of Dr. Bob Bucat

(a)   (b)

FIGURE 2.12 One mole of some elements. (a) *(left to right)* Magnesium chips, tin, and silicon. *(above)* Copper beads. (b) one mole of nitrogen gas (N_2), which at 0 °C and 1 bar pressure occupies 24.8 L.

WORKED EXAMPLE 2.4—MASS AND AMOUNT

Consider the elements lead and tin, both in the same column of the periodic table.
(a) What is the mass of 2.50 mol of lead (Pb, atomic number = 82)?
(b) What is the amount (in mol) of 36.5 g of tin (Sn, atomic number = 50)?
(c) What is the ratio of the number of atoms in 518 g of lead to the number in 36.5 g of tin?

Solution

(a) M (Pb) = 207.2 g mol^{-1} (numerically equal to its atomic weight)

$$\text{Mass of 2.50 mol sample, } m = n \times M = 2.50 \text{ mol} \times 207.2 \text{ g mol}^{-1} = 518 \text{ g}$$

(b) M (Sn) = 118.7 g mol^{-1}

$$\text{Amount, } n(\text{Sn}) = \frac{m}{M} = \frac{36.5 \text{ g}}{118.7 \text{ g mol}^{-1}} = 0.308 \text{ mol}$$

(c) $$\frac{\text{Number of lead atoms}}{\text{Number of tin atoms}} = \frac{n(\text{Pb})}{n(\text{Sn})} = \frac{2.50 \text{ mol}}{0.308 \text{ mol}} = 8.12$$

WORKED EXAMPLE 2.5—CALCULATION OF AMOUNT

The density of mercury (Hg, a liquid metal) at 25 °C is 13.534 g mL^{-1}. What is the amount of mercury (in mol) in a 32.0 mL sample?

Solution

There is not a direct mathematical relationship between volume and amount of mercury. First we need to use the density of mercury to find the mass.

$$\text{Mass, } m(\text{Hg}) = \text{volume} \times \text{density} = 32.0 \text{ mL} \times 12.534 \text{ g mL}^{-1} = 401 \text{ g}$$

$$n(\text{Hg}) = \frac{m}{M} = \frac{401 \text{ g}}{200.6 \text{ g mol}^{-1}} = 2.00 \text{ mol}$$

(a) What is the mass (in g), of 1.5 mol of potassium (K)?
(b) What is the amount (in mol) in 200 g of copper (Cu)? How many atoms are there?

What mass of copper (Cu) has the same number of atoms as 1.00 g of zinc (Zn)?

Interactive Exercises 2.21–2.23 SUBMIT

Practise converting between masses and numbers of particles

In this section, we have dealt with the concept of molar mass as it applies to substances that are composed of atoms. In Chapter 3, this is broadened to situations where the substances are composed of molecules or collections of atoms or ions.

2.12 The Periodic Table of Elements

The periodic table of elements is one of the most useful tools in chemistry. It contains a wealth of information and is also used to organize many of the ideas of chemistry. We cannot sensibly talk about chemistry without some familiarity with its main features and terminology. In this section, we introduce some of the main features of the periodic table and the idea of periodicity. In Section 8.12, we will look systematically at the periodic variation of the properties of elements. Consult the inside cover of this textbook for a periodic table, showing the grid positions of each element.

Language of the Periodic Table

The main organizational features of the periodic table are the following:

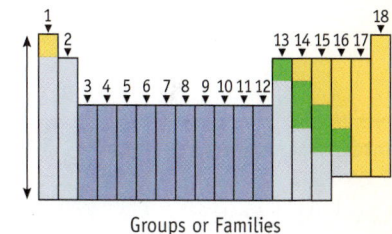

Groups or Families

- Elements are arranged so that those with similar chemical and physical properties lie in vertical columns called *groups* or families. The groups are numbered 1 to 18. The members of Groups 1, 2, and 13 to 18 are together referred to as the **main group elements;** the members of Groups 3 to 12 are the **transition elements**.
- The horizontal rows of the table are called *periods* or *rows*, and they are numbered from 1 for the period containing only hydrogen (H) and helium (He). For example, sodium (Na), in Group 1, is the first element in the third period. Mercury (Hg), in Group 12, is in the sixth period (or sixth row).

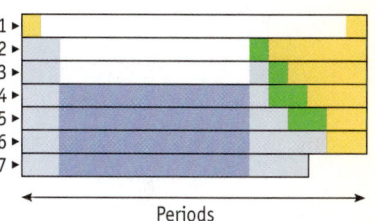

Periods

The periodic table can be divided into several regions according to the properties of the elements. On the periodic table inside the front cover, elements that we classify as *metals* [>>Section 3.5] are indicated in purple, those that we call *non-metals* are indicated in yellow, and elements that we refer to as *metalloids* or *semi-metals* appear in green. Across a given period, any element is more metallic than those further to the right side, and less metallic than those on their left. Within any group, any element has more metallic character than those above it—a trend which is most evident in Groups 13 to 16.

You are probably familiar with many properties of metals from everyday experience (Figure 2.13a). The *non-metallic elements*, on the right of a diagonal line in the periodic table from boron (B) to tellurium (Te), have a wide variety of properties. Some are solids (carbon, sulfur, phosphorus, and iodine). Those that are gases at room temperature include oxygen, nitrogen, fluorine, chlorine, and all of the Group 18 elements. One, bromine, is a liquid at room temperature (Figure 2.13b). With the exception of carbon in the form of graphite, non-metals do not conduct electricity, and this is one of the main features that distinguish them from metals.

Some of the elements on or near the diagonal line from boron (B) to tellurium (Te) have properties that make them difficult to classify clearly as either metals or non-metals. Chemists call them **metalloids** or, sometimes, *semi-metals* (Figure 2.13c). Chemists often

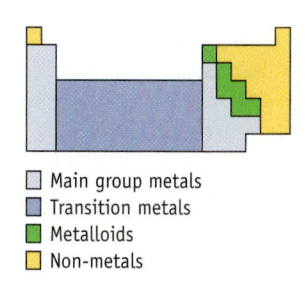

☐ Main group metals
☐ Transition metals
☐ Metalloids
☐ Non-metals

FIGURE 2.13 Representative elements. (a) Magnesium, aluminium, and copper are metals. (b) Only about 15 elements are classified as non-metals. Here are red-orange liquid bromine and purple solid iodine. (c) Six elements are generally classified as metalloids. Solid silicon is shown in various forms, including a thin wafer that holds printed electronic circuits.

disagree about which elements fit into this category. A *metalloid* can be defined as *an element that has some of the physical characteristics of a metal but some of the chemical characteristics of a non-metal;* in this textbook, boron, silicon, germanium, arsenic, antimony, and tellurium are included in this category. Antimony (Sb), for example, conducts electricity as well as many metals, while its chemistry resembles that of the non-metal phosphorus.

Developing the Periodic Table

Although the arrangement of elements in the periodic table can now be understood on the basis of atomic structure [>>Chapter 8], the table was originally developed from many experimental observations of the chemical and physical properties of elements and is the result of the ideas of a number of chemists in the 18th and 19th centuries.

In 1869, at the University of St. Petersburg in Russia, Dmitri Ivanovich Mendeleev (1834–1907) was pondering the properties of the elements as he wrote a textbook on chemistry. He realized that if the elements were arranged in a line in order of increasing atomic mass, he could recognize a repeating pattern of elements with similar properties. That is, he saw a **periodicity** or periodic occurrence of elements with similar properties.

Mendeleev organized the known elements into a table by lining them up in a horizontal row in order of increasing atomic mass. Every time he came to an element with properties similar to one already in the row, he placed the heavier element directly underneath the other, in a new row. For example, the elements lithium, beryllium, boron, carbon, nitrogen, oxygen, and fluorine were in a row. Sodium was the next element then known, and because its properties closely resembled those of lithium, Mendeleev started a new row with sodium under it. This process lead to the formation of columns, each of which contained elements (such as Li, Na, and K) with similar properties.

An important feature of Mendeleev's table—and a mark of his genius—was that he left empty spaces in his table for elements that were not known at the time, and even predicted quite accurately the properties of the elements that he believed "belonged" in those spaces and would one day be discovered. For example, a space was left between silicon (Si) and tin (Sn) in what is now Group 14. Based on the progression of properties in this group, Mendeleev was able to predict the properties of this missing element. His predictions were confirmed with the discovery of germanium (Ge) in 1886.

In Mendeleev's table, the elements were arranged generally in order of increasing mass. This puts tellurium (Te) after iodine (I), but on the grounds of their similarities to selenium (Se) and bromine (Br) respectively, Mendeleev reversed this order and put them in the same order as they are in the modern periodic table. Mendeleev assumed that the perceived anomaly was because the atomic masses known at that time were inaccurate—not a bad assumption based on the analytical methods of the day. His assumption that the properties of elements vary periodically with their masses was wrong. An apparent anomaly involving the position of chromium (Cr) disappeared when the accepted value of 43 for its atomic weight was re-evaluated to 52.

Interactive Exercises 2.24–2.26

Classifying elements in the periodic table

Near the beginning of the periodic table, each element has similar properties to the one eight places further along. So, $_4$Be, $_{12}$Mg, and $_{20}$Ca are examples of a periodic pattern of elements with similar properties. Further along, the repeating cycle is 18 elements—one such set is $_{20}$Ca, $_{38}$Sr, and $_{56}$Ba.

Think about It

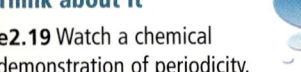

e2.19 Watch a chemical demonstration of periodicity.

In 1913, H.G.J. Moseley (1887–1915), a young English scientist working with Ernest Rutherford, corrected Mendeleev's assumption. Moseley was doing experiments in which he bombarded many different metals with electrons in a cathode-ray tube [>>e2.8] and examined the X-rays emitted in the process. In seeking some order in his data, he realized that the wavelength of the X-rays emitted by a given element were related in a precise manner to the *atomic number* of the element. Indeed, chemists quickly recognized that organizing the elements in a table by increasing atomic number validated the order in Mendeleev's table. We now have an improved version of the **law of chemical periodicity:** *The properties of the elements vary periodically with their atomic numbers.*

In the periodic table of the elements, the elements are arranged in families that recognize similar observable properties of the members of each family, although there are trends to more metallic nature as one goes from element to element toward the left of each row, and down each group.

Background Concepts

e2.20 Read about the attempts at portraying the periodic pattern of the elements.

WebLink

e2.21 Look at some of the wide range of periodic tables, and stories they have inspired.

SUMMARY

Key Concepts

- All matter consists of **atoms**. Every substance is made up of some combination of atoms of the naturally occurring **elements**. (Section 2.4)
- We can classify matter in several ways: by whether it is in the solid, liquid, or gaseous **state**; whether it is **homogeneous** or **heterogeneous;** pure or a mixture; and an element or a compound. The **kinetic-molecular model of matter** helps us make sense of the nature of solids, liquids, and gases. (Section 2.2)
- Learning chemistry requires operating consciously and interchangeably at three levels (Figure 2.5): the **observable** (what we can observe directly), the **molecular** (tiny particles, too small to see), and the **symbolic** (words, labels, drawings, and symbols) **levels**. (Section 2.3)
- More than 20 million **chemical compounds** are now known. Each is composed of atoms of two or more different elements in fixed and definite proportions. The composition of any compound is represented by its **chemical formula,** which indicates the relative numbers of atoms of each element. (Section 2.5)
- During a chemical reaction, there is a redistribution of the atoms so that new substances are formed. The **products** have completely different **chemical properties** from the **reactants**. By contrast, **physical properties** of substances do not involve chemical transformations. (Section 2.6)
- Atoms are composed of a nucleus containing **protons** and **neutrons**, and **electrons** surrounding the nucleus. By definition, all atoms of one element have the same number of protons. Atoms of the same element with different numbers of neutrons are called **isotopes**. **Isotope ratio mass spectrometry (IRMS)** measures small differences in the **relative atomic mass** and **natural percent abundance** of isotopes in substances. In the human activity of doing chemistry, IRMS has powerful applications to solve problems in forensics and environmental applications. (Section 2.1, Sections 2.7–2.9, e2.2)
- The **atomic weight** of an element is the weighted average of the relative atomic masses of its isotopes in a representative sample. (Section 2.10)
- The **mole** is the unit of **amount of substance** commonly used by chemists. One mole of specified particles of any substance has the same number of those particles as one mole of specified particles of any other substance. The best estimate of the number of particles in 1 mol of a substance is $N_A = 6.0221415 \times 10^{23} \text{mol}^{-1}$.
- N_A is referred to as the **Avogadro constant**. The mass in grams of one mole of any element is called its **molar mass**. (Section 2.11)
- The periodic table of elements is used to organize many ideas of chemistry. Elements are arranged so those with similar properties lie in vertical groups called groups, and the horizontal rows are called periods or rows. (Section 2.12)

Key Equations

$$\text{Percent abundance of an isotope} = \frac{\text{number of atoms of that isotope}}{\text{total number of atoms of the element}} \times 100\%$$

(Section 2.8)

$$\text{Atomic weight} = \left(\frac{\%\text{ abundance isotope 1}}{100}\right)(A_r \text{ of isotope 1}) +$$
$$\left(\frac{\%\text{ abundance isotope 2}}{100}\right)(A_r \text{ of isotope 2}) + \cdots \quad \text{(Section 2.10)}$$

Amount to mass: mass (g) = amount (mol) $\times$ molar mass (g mol^{-1}) i.e., $m = n \times M$

$$\text{Mass to amount: amount (mol)} = \frac{\text{mass (g)}}{\text{molar mass (g mol}^{-1})} \text{ i.e., } n = \frac{m}{M} \quad \text{(Section 2.11)}$$

REVIEW QUESTIONS

Section 2.3: Three Levels of Operation: Observable, Molecular, Symbolic

2.27 Figure 2.7 shows a piece of table salt and a representation of its internal structure. Which is the observable view and which is the molecular-level view? How are the observable and molecular-level views related?

Section 2.6: Chemical Reactions, Chemical Change

2.28 You have a sample of a white crystalline substance from your kitchen. You know that it is either salt or sugar. Although you could decide by taste, suggest another property that you can use to determine the sample's identity. (*Hint:* To find some pertinent information, you can use reliable chemical databases on the Internet or a handbook of chemistry in the library.)

2.29 In each case, decide whether the underlined property is a physical or chemical property.
(a) The normal colour of elemental bromine is <u>orange</u>.
(b) Iron <u>turns to rust</u> in the presence of air and water.
(c) Hydrogen can <u>explode</u> when ignited in air.
(d) The <u>density</u> of titanium metal is 4.5 g cm^{-3}.
(e) Tin metal <u>melts</u> at 232 °C.
(f) Chlorophyll, a plant pigment, is <u>green</u>.

2.30 In each case, decide whether the change is a chemical or physical change.
(a) A cup of household bleach changes the colour of your favourite T-shirt from purple to pink.
(b) Water vapour in your exhaled breath condenses in the air on a cold day.
(c) Plants use carbon dioxide from the air to make sugar.
(d) Butter melts when placed in the sun.

2.31 Which part of the description of a compound or element refers to its physical properties and which to its chemical properties?
(a) The colourless liquid ethanol burns in air.
(b) The shiny metal aluminium reacts readily with orange, liquid bromine.

2.32 Which part of the description of a compound or element refers to its physical properties and which to its chemical properties?
(a) Calcium carbonate is a white solid with a density of 2.71 g cm^{-3}. It reacts readily with an acid to produce gaseous carbon dioxide.
(b) Gray, powdered zinc metal reacts with purple iodine to give a white compound.

2.33 Give the mass number of each of the following atoms: (a) magnesium with 15 neutrons, (b) titanium with 26 neutrons, and (c) zinc with 32 neutrons.

2.34 Give the complete symbol (^A_ZX) for each of the following atoms: (a) potassium with 20 neutrons, (b) krypton with 48 neutrons, and (c) cobalt with 33 neutrons.

2.35 Cobalt has three radioactive isotopes used in medical studies. Atoms of these isotopes have 30, 31, and 33 neutrons, respectively. Give the symbol for each of these isotopes.

2.36 The three isotopes of magnesium are $^{24}_{12}$Mg (78.99% abundant), $^{25}_{12}$Mg (10.00% abundant), and $^{26}_{12}$Mg. What is the percent abundance of the $^{26}_{12}$Mg isotope?

2.37 Thallium has two stable isotopes, ^{203}Tl and ^{205}Tl. Knowing that the atomic weight of thallium is 204.4, which isotope is the more abundant of the two?

2.38 Calculate the atomic weight of lithium, given the following information:

^{6}Li: $A_r = 6.015121$, percent abundance = 7.50%
^{7}Li: $A_r = 7.016003$, percent abundance = 92.50%

2.39 Silver (Ag) has two stable isotopes, ^{107}Ag and ^{109}Ag. The isotopic mass of ^{107}Ag is 106.9051, and the isotopic mass of ^{109}Ag is 108.9047. The atomic weight of Ag is 107.868. Estimate the percentage of ^{107}Ag in a sample of the element.

(a) 0%   (b) 25%   (c) 50%   (d) 75%

2.40 Fill in the blanks in the table (one column per element).

Symbol	^{58}Ni	^{33}S		
Number of protons			10	
Number of neutrons			10	30
Number of electrons in the neutral atom				25
Name of element				

2.41 Fill in the blanks in the table (one column per element).

Symbol	^{65}Cu	^{86}Kr		
Number of protons			78	
Number of neutrons			117	46
Number of electrons in the neutral atom				35
Name of element				

2.42 Potassium has three naturally occurring isotopes (^{39}K, ^{40}K, and ^{41}K), but ^{40}K has a very low natural abundance. Which of the other two isotopes is the more abundant? Briefly explain your answer.

Section 2.11: Amount of Substance and Its Unit of Measurement: The Mole

2.43 What is the ratio of the number of molecules in 2 mol of hydrogen gas (H_2) to the number of molecules in 1 mol of oxygen gas (O_2)?

2.44 What is the ratio of the number of molecules in 100 mol of nitrogen gas (N_2) to the number of molecules in 300 mol of hydrogen gas (H_2)?

2.45 What is the ratio of the number of atoms in 0.1 mol of aluminium (Al) to the number of molecules in 0.1 mol of hydrogen gas (H_2)?

2.46 What mass of zinc has the same number of atoms as 6.355 g of copper?

2.47 What is the mass of aluminium that contains five times the number of atoms as 137.3 g of barium?

Section 2.12: The Periodic Table of Elements

2.48 Give the name and symbol of each of the Group 15 elements. State whether each is a metal, non-metal, or metalloid.

2.49 How many periods of the periodic table have 8 elements, how many have 18 elements, and how many have 32 elements?

2.50 Give the name and chemical symbol for the following:
(a) a non-metal in the second period
(b) a group 1 element
(c) the third-period Group 17 element
(d) an element that is a gas at 20 °C and 1 atmosphere pressure

2.51 Here are symbols for five of the seven elements whose names begin with the letter B: B, Ba, Bk, Bi, and Br. Match each symbol with one of the descriptions below:
(a) a radioactive element
(b) a liquid at room temperature
(c) a metalloid
(d) a Group 2 element
(e) a Group 15 element

2.52 Use the elements in the following list to answer the questions: sodium, silicon, sulfur, scandium, selenium, strontium, silver, and samarium. (Some elements will be entered in more than one category.)
(a) Identify those that are metals.
(b) Identify those that are main group elements
(c) Identify those that are transition metals.

2.53 Compare the elements silicon (Si) and phosphorus (P) using the following criteria:
(a) metal, metalloid, or non-metal
(b) possible conductor of electricity
(c) physical state at 25 °C (solid, liquid, or gas)

SUMMARY AND CONCEPTUAL QUESTIONS

2.54 *Crossword Puzzle:* In the 2 × 2 box shown here, each answer must be correct four ways: horizontally, vertically, diagonally, and by itself. Instead of words, use symbols of elements. When the puzzle is complete, the four spaces will contain the overlapping symbols of ten elements. There is only one correct solution.

1	2
3	4

Horizontal

1–2: Two-letter symbol for a metal used in ancient times

3–4: Two-letter symbol for a metal that burns in air and is found in Group 15

Vertical

1–3: Two-letter symbol for a metalloid

2–4: Two-letter symbol for a metal used in U.S. coins

Single squares: All one-letter symbols

1: A colourful non-metal

2: Colourless gaseous non-metal

3: An element that makes fireworks green

4: An element that has medicinal uses

Diagonal

1–4: Two-letter symbol for an element used in electronics

2–3: Two-letter symbol for a metal used with Zr to make wires for superconducting magnets

This puzzle first appeared in *Chemical & Engineering News*, p. 86, December 14, 1987 (submitted by S. J. Cyvin) and in *Chem Matters*, October 1988.

2.55 Review the periodic table.
(a) Name an element in Group 2.
(b) Name an element in the third period.
(c) Which element is in the second period in Group 14?
(d) Name the non-metal in Group 16 and the third period.
(e) Which Group 17 element is in the fifth period?
(f) Which Group 2 element is in the third period?
(g) Which Group 18 element is in the fourth period?
(h) Name a metalloid in the fourth period.

2.56 Review the periodic table.
(a) Name an element in Group 12.
(b) Name an element in the fifth period.
(c) Which element is in the sixth period in Group 14?
(d) Which element is in the third period in Group 16?
(e) Which Group 1 element is in the third period?
(f) Which Group 18 element is in the fifth period?
(g) Name the element in Group 16 and the fourth period. Is it a metal, non-metal, or metalloid?
(h) Name a metalloid in Group 15.

2.57 You are given 15 g each of yttrium, boron, and copper. Which sample represents the largest number of atoms?

2.58 Lithium has two stable isotopes: ^{6}Li and ^{7}Li. One of them has an abundance of 92.5%, and the other has an abundance of 7.5%. Knowing that the atomic mass of lithium is 6.941, which is the more abundant isotope?

2.59 Superman comes from the planet *Krypton*. If you have 0.00789 g of the gaseous element krypton, how many mol does this represent? How many atoms?

2.60 Put the following elements in order from smallest to largest mass:
(a) 3.79 10^{24} atoms Fe   (c)   8.576 mol C
(b) 19.921 mol H_2 (d)   7.4 mol Si

(e) 9.221 mol Na   (g) 9.2 mol Cl_2
(f) 4.07 10^{24} atoms Al

2.61 On the scale of relative atomic masses based on $A_r(^{12}C)$ = 12 exactly: $A_r(^1H)$ = 1.0078, $A_r(^{11}B)$ = 11.0093, $A_r(^{24}Mg)$ = 23.9850, and $A_r(^{63}Cu)$ = 62.9396.

What would the relative atomic masses of these isotopes be if the scale had been based upon the following:
(a) $A_r(^1H)$ = 1 exactly?
(b) $A_r(^{63}Cu)$ = 63 exactly?
(c) $A_r(^{24}Mg)$ = 24 exactly?

2.62 (a) On the scale of relative atomic masses based on $A_r(^1H)$ = 1 exactly, what would the relative atomic mass of ^{16}O be? What would be the value of the Avogadro constant under these circumstances?
(b) On the scale of relative atomic masses based on $A_r(^{16}O)$ = 16 exactly, what would the relative atomic mass of ^1H be? What would be the value of the Avogadro constant under these circumstances?

2.63 Consult e2.2 and follow the link to analysis of methane gas on Mars. For one of the methane samples, following combustion and purification, the carbon dioxide gas is analyzed by mass spectrometry and the relative intensities of the m/z = 44 peak is 20 000 002 and the m/z = 45 peak is 218 248. Calculate $\delta^{13}C$. Can you determine whether this methane sample is likely thermogenic or biogenic?

2.64 Consult e2.2 and follow the link to the detection of synthetic testosterone in professional athletes. From one athlete's urine samples, testosterone and other reference steroids were separated and combusted to produce carbon dioxide gas, which is analyzed by mass spectrometry. For the first place finisher, the relative intensities from CO_2 from his testosterone samples gave an m/z = 44 peak is 20 000 002 and the m/z = 45 peak is 218 228, and the relative intensities from CO_2 from reference steroids gave an m/z = 44 peak is 19 999 997 and the m/z = 45 peak is 219 135. Calculate $\delta^{13}C$ for his steroid sample and reference steroids. Can you determine whether this athlete has taken exogenous steroids to boost his performance?

2.65 Identify, from the list below, the information needed to calculate the number of atoms in 1 cm^3 of iron. Outline the procedure used in this calculation.
(a) the structure of solid iron
(b) the molar mass of iron
(c) the Avogadro constant
(d) the density of iron
(e) the temperature
(f) iron's atomic number
(g) the number of iron isotopes

2.66 The photo here depicts what happens when a coil of magnesium ribbon and a few calcium chips are placed in water.

Charles D. Winters

(a) Based on their relative reactivities, what might you expect to see when barium, another Group 2 element, is placed in water?

(b) Give the period in which each element (Mg, Ca, and Ba) is found. What correlation do you think you might find between the reactivity of these elements and their positions in the periodic table?

2.67 A jar contains some number of jelly beans. To find out precisely how many are in the jar you could dump them out and count them. How could you estimate their number without counting each one?

Charles D. Winters

With the **eCHACR** single-sign-on access card bundled with your text, log on (http://login.cengage.com/sso) and access the e-book and click on any in-margin icons for dynamic molecular-level animations and simulations, videos of laboratory reactions, interactive exercises, reviews of background concepts, and other online supplementary materials as noted by the icons in the text margins.

Also go to www.chemistry.nelson.com <http://www.chemistry.nelson.com> for Answers to in-chapter exercises and selected Review Questions, Test Yourself questions, weblinks, crossword puzzles, flashcards, glossary of key terms, and other student resources.

CHAPTER 3

Models of Structure to Explain Properties

Outline

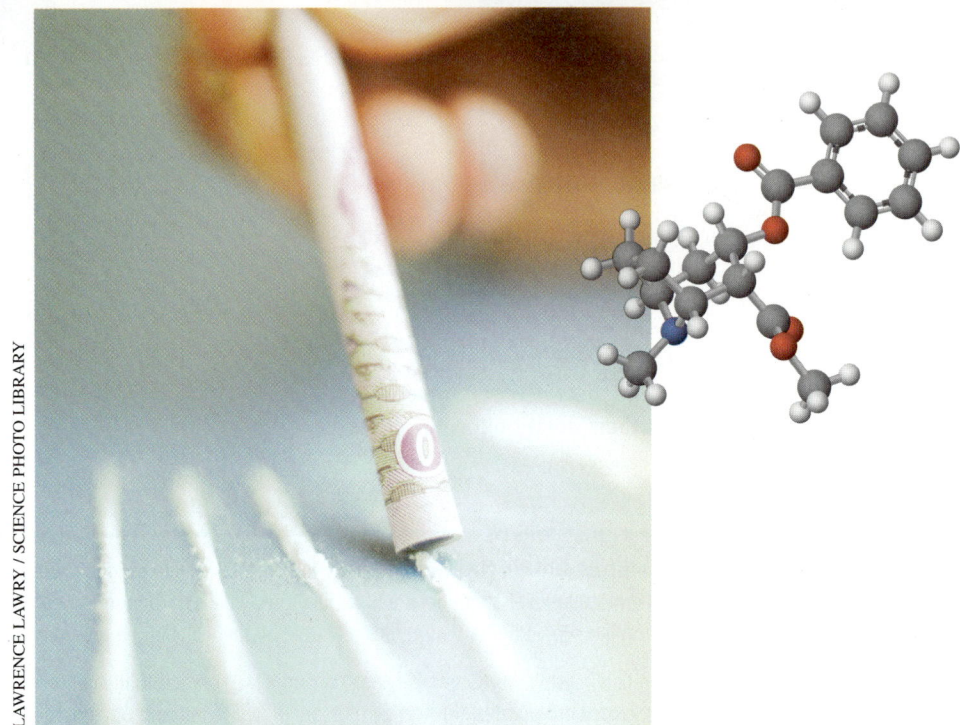

LAWRENCE LAWRY / SCIENCE PHOTO LIBRARY

3.1 Case Study: Is There a Stash on Your Cash?

Are you carrying cocaine in your wallet or purse? Studies done on paper currency from around the world suggest that it is highly likely that if forensic chemists relieved you of several $20 bills, they would be able to detect trace amounts of cocaine on the money using very sensitive analytical methods. Drugs are paid for with cash, handled by the same fingers that touch or use the drugs. Those paper banknotes are regularly put back into circulation, where they transfer tiny amounts of cocaine to other notes in money sorting machines.

How might you use equipment in a university chemistry laboratory to test paper money for very low levels of drug residues? You could start with a large bundle of notes, rinse them with a suitable organic solvent such as ethanol, and then evaporate the solvent. Suppose a barely detectable white powder is left behind—could it be cocaine? Or perhaps some other white stuff as innocuous as table salt? Since our perspiration contains salt, we often leave tiny deposits of sodium chloride on things we touch. How can you tell, for example, whether a trace amount of white powder on paper money is salt (sodium chloride), cocaine, table sugar (sucrose), or nicotine from the hands of smokers?

How can you tell whether a trace amount of white powder on paper money is salt (sodium chloride), cocaine, table sugar (sucrose), or nicotine from the hands of smokers?

These four white powders might look somewhat similar, but their properties show that they do not all belong to the same class of substances. Cocaine, sucrose, and nicotine are classified as *molecular substances*: on the basis of properties such as low melting and boiling points, low electrical conductivity, solubility in organic solvents, and softness, they are believed to consist of groupings of atoms, called *molecules*. The molecules of any substance are all the same—and different from those of every other substance. Salt is an *ionic substance*. The properties of many substances are so different that chemists believe that they have different structural arrangements at the atomic level. Depending on their properties they are categorized as *covalent network solids*, *ionic substances*, *metallic*, or *molecular substances*.

The observation that an unknown white substance on paper money could be dissolved from the surface of the paper by an organic solvent such as ethanol is one indication that it might be a molecular substance [>>Section 3.6]. So let's focus on how chemists distinguish among different molecular substances.

Chemists' knowledge of the molecular structures of substances has been revolutionized by powerful specialized instrumental techniques that give insights into what substances are like at the molecular level. In this chapter, we'll focus on two of the most useful and accessible of these techniques—mass spectrometry (MS) and infrared spectroscopy (IR)—each of which provides complementary information about the composition and structure of molecules. A third powerful technique, nuclear magnetic resonance spectroscopy (NMR) [>>Chapter 9], has transformed medical imaging and diagnosis. These same techniques were used to identify the structure of Gavinone from wood smoke in Chapter 1.

The sensitivity of modern chemical instruments can be very high. A mass spectrometer, which is the method of choice to measure levels of cocaine on paper money, can detect less than 1 μg (1×10^{-6} g) per note. Electronic "noses" are even more sensitive instruments, and are able to detect a few picograms of molecular compounds.

$1\ \mu\text{g (microgram)} = 1 \times 10^{-6}\ \text{g}$
$1\ \text{ng (nanogram)} = 1 \times 10^{-9}\ \text{g}$
$1\ \text{pg (picogram)} = 1 \times 10^{-12}\ \text{g}$

If a large enough sample of an unknown white powder was collected from paper money, you could compare its IR spectrum with the IR spectra of pure cocaine, nicotine, and sucrose, shown in Figure 3.1. The wavenumbers of radiation absorbed, and the percentage of energy absorbed (or transmitted) at each wavenumber are unique for each molecular substance. Essentially, each of these three white powders has its own distinct infrared spectral fingerprint, and forensics laboratories can identify unknown substances by comparing their IR spectra with many others stored in large computerized databanks.

In Section 3.11, you will see how IR spectra can be interpreted to tell which groups of atoms are attached to the "skeleton" of carbon-based molecules. When this information is supported with data from techniques such as NMR, MS, and x-ray crystallography, a chemist can decide which atoms are connected to each other in molecules of cocaine, nicotine, or sucrose; the shape of their molecules; and the distances between the atoms.

Using IR spectroscopy to identify cocaine on paper money requires washing a massive number of banknotes with a solvent. Mass spectrometers are much more sensitive instruments, and with specialized detectors they can identify a few nanograms of cocaine and other substances on a single bill. As you might expect, a mixture of compounds would be found on any given note, and amounts would vary widely depending on where the money came from. In a study of money from Canada, the U.S., and U.K., cocaine was found on almost every note, along with other compounds including the drug Ritalin, nicotine, the cosmetic ingredient triethanolamine, the mosquito repellent DEET, and dioctylphthalate, a plasticizer.

Typical single Canadian banknotes were found to contain 0.13–0.49 ng of cocaine, on currency taken from general circulation. Banknotes associated with criminal cases contained amounts 50–1000 times greater. You probably are carrying minute amounts of cocaine in your wallet or purse. It's the size of the stash on your cash that may be incriminating!

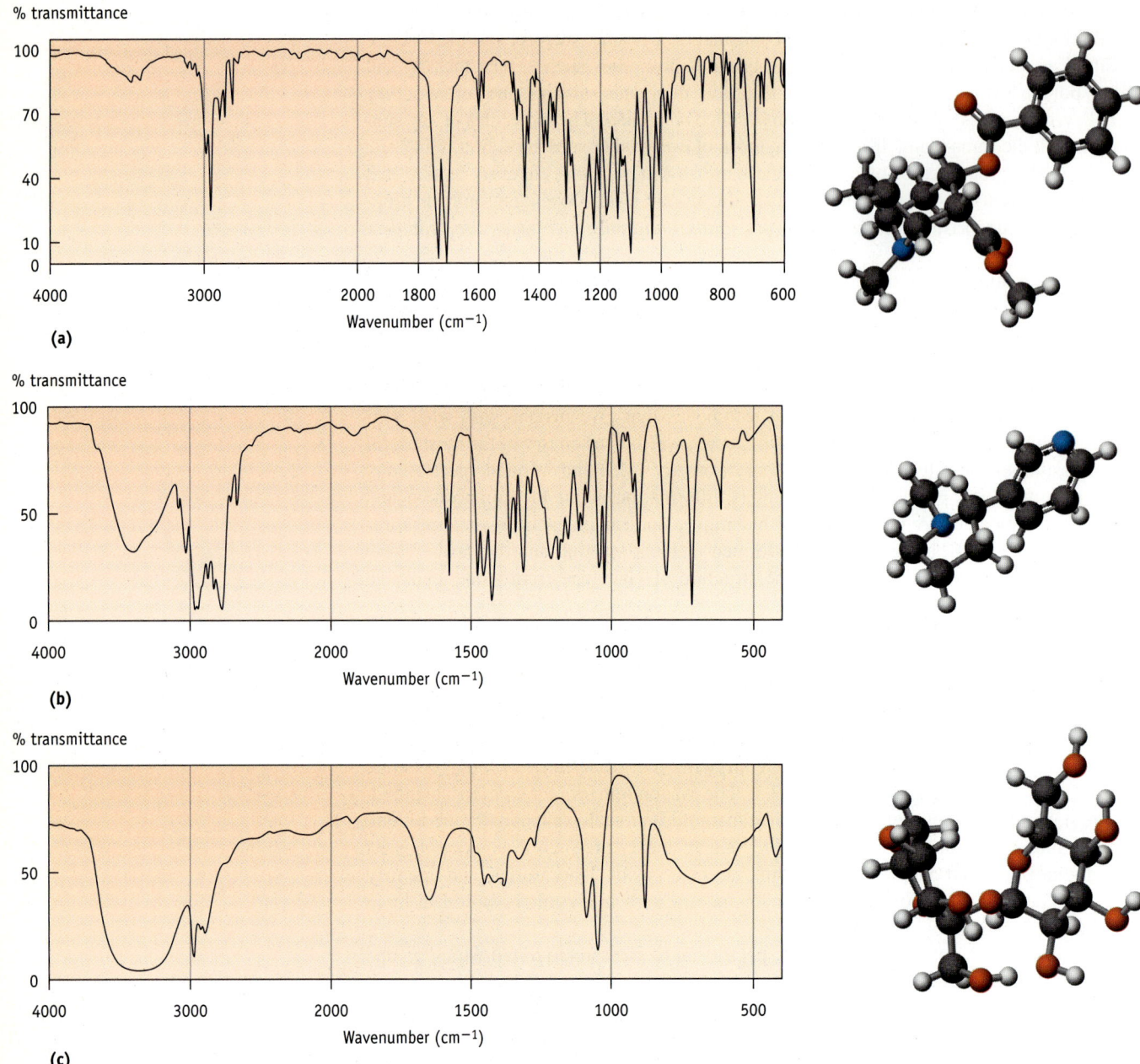

FIGURE 3.1 The infrared spectra of (a) cocaine, (b) nicotine, and (c) sucrose. The positions of the downward spikes tell us the wavenumbers of infrared radiation that are absorbed by each substance, expressed on the y-axis as the percentage of light transmitted through the sample. Every molecular substance has a characteristic absorption spectrum different from that of every other substance; the IR spectrum gives a "fingerprint" of a molecular substance. (IR spectra source: Georgia State Crime Lab Library).

3.2 Classifying Substances by Properties: An Overview

As seen in our case study in Section 3.1, some substances, such as cocaine and sucrose, absorb infrared radiation, giving us information about their arrangements of atoms. Suppose, however, that our white powder had been table salt, sodium chloride, which is an ionic substance. It would not absorb electromagnetic radiation in the infrared region where molecular substances do. This observation suggests that there are fundamental differences among the structures of substances. Many substances that absorb IR radiation are comprised of *molecules*, identical particles that have various atoms in a particular arrangement. Their IR spectra are attributed to selective increase of the energy of vibration of particular atoms or groups of atoms in the molecules.

Section 2.2 showed that chemists classify matter in various ways: (a) solids, liquids, or gases; (b) pure substances or mixtures; (c) homogeneous or heterogeneous; and (d) elements or compounds. The differences among the substances in each class are interpreted through models of the nature of matter at the molecular level. In this chapter, we see another way in which chemists categorize substances, according to differences in their *physical properties* and in the *different ways they interact with electromagnetic radiation*. Again, chemists use mental models to make sense of why different substances behave differently.

On the basis of general patterns in their characteristic properties, most substances are placed into one of four categories, as shown by the examples in Table 3.1:

- covalent network substances
- ionic substances
- metallic substances
- molecular substances

Every classification scheme has "fuzzy" edges, and some substances are not easy to assign unambiguously to one of these four categories. As discussed in Section 3.12, many such examples are found in the new materials being created at the borders between chemistry, biology, and technology.

Scientists use observations of properties of substances as evidence to draw inferences about the nature of the substances. They try to imagine what it is about those substances at the level of atoms and ions that gives rise to the observed properties. This is called **modelling**, and the imagined arrangement of atoms or ions in each type of substance is an example of a **model**.

We now examine each category more closely and describe the models that explain properties of each type of substance. We then introduce spectroscopic tools that have revolutionized our understanding of the structures of one of the most important categories—molecular substances.

Most substances can be placed into one of four categories (covalent network, ionic, metallic, or molecular) based on their characteristic properties. Chemists explain those properties with different models of molecular structure.

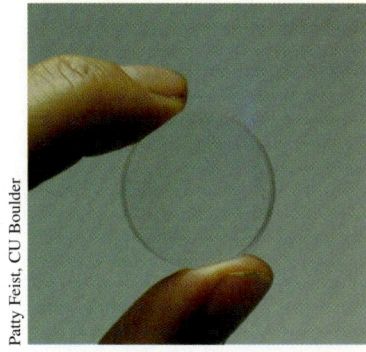

Since crystalline sodium chloride does not absorb electromagnetic radiation in the infrared region where molecular substances do, polished salt crystals are often used as sample holders for molecular compounds when recording their infrared spectra.

Think about It

e3.1 Watch video demonstrations of some tests of properties of substances.

TABLE 3.1 Categories of Substances Based on Physical Properties

	Covalent Network	Ionic	Metallic	Molecular
Examples	Carbon (C) as diamond, silicon (Si), silicon dioxide (SiO$_2$)	Sodium chloride (NaCl), magnesium fluoride (MgF$_2$)	Copper (Cu), gold (Au), silver (Ag), sodium (Na)	Oxygen (O$_2$), water (H$_2$O), methane (CH$_4$), cocaine (C$_{17}$H$_{21}$NO$_4$)
Composition	A few non-metallic elements, and a few compounds of non-metals	Many are compounds of metals with non-metals	Metallic elements and alloys	The non-metallic elements, compounds containing only non-metals
Melting point	Very high e.g., diamond 3550 °C, Si 1414 °C, SiO$_2$ 1650 °C	High e.g., NaCl 801 °C, MgF$_2$ 1260 °C	Mostly high, but variable e.g., Cu 1085 °C, Au 1064 °C, Ag 962 °C, Na 98 °C	Low e.g., O$_2$ −219 °C, CH$_4$ −182 °C, C$_{17}$H$_{21}$NO$_4$ 98 °C
Electrical conductivity				
- solid	Nil*	Nil	High	Nil
- molten	Nil	High	High	Nil

* An exception is solid carbon in the form of graphite.

3.3 Covalent Network Substances

If you are wearing a diamond on your finger, you are displaying a **covalent network substance**. Since these substances have very high melting points, we are only familiar with them as solids. They are generally very hard, and except for graphite, do not conduct electricity. Examples include the elements carbon (in the form of diamond), silicon, and boron, and some compounds of these elements, of which silicon dioxide (SiO_2), silicon carbide (SiC), and boron nitride (BN) are examples. Diamond is one of the hardest substances known, and in addition to its use in diamond rings, it is commonly used on the ends of large drills in mining and to cut stone, glass, and concrete.

A Model

The model used to account for the properties of substances of this type is a three-dimensional network of atoms, each atom bound to a number of other atoms by covalent bonds [>>Section 3.7]. A regular arrangement of the atoms in crystals of these substances (Figure 3.2) is confirmed by X-ray diffraction studies.

The lack of electrical conductivity is explained on the grounds that there are no mobile charged particles: the atoms are uncharged, and their electrons are "tied up" either on the atoms or in the bonds. The hardness of these materials can be attributed to the strong covalent bonds in several directions from each atom, so that atoms are not easily displaced. An explanation for their high melting points is that a large amount of energy must be supplied to break all of the covalent bonds in the lattice so that the atoms can move with respect to each other as they would in the liquid state.

Covalent network solids are modelled as three-dimensional networks of atoms, in which each atom is bound to other atoms, forming a network of covalent bonds. This model can rationalize their high melting points, lack of electrical conductivity, and hardness.

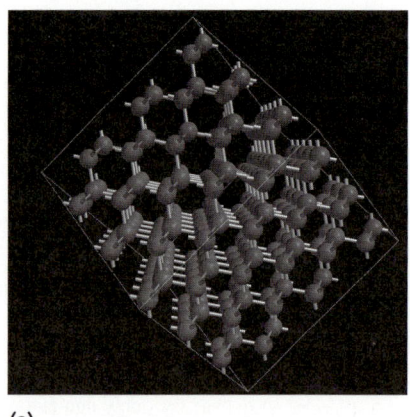
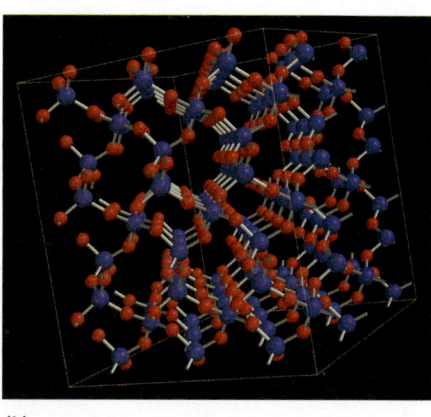

(a) (b)

FIGURE 3.2 Models of covalent network solids. (a) Diamond, in which each carbon atom is covalently bound to four other carbon atoms. (b) Silicon dioxide, in which each silicon atom is bound to four oxygen atoms, and each oxygen atom is bound to two silicon atoms. With every atom joined by covalent bonds to other atoms, this type of structure is sometimes called a "giant molecule." Nuclei positions are confirmed by x-ray diffraction, from which the covalent bonds are inferred.

Formulas of Covalent Network Solids

For an element that has a covalent network structure, the formula is simply the element's symbol—for example, C for diamond, and Si for silicon.

The *formula of a covalent network compound tells us the relative numbers of atoms of each element in the lattice*, rather than the absolute number of atoms in each molecule, as is the case for molecular compounds. For example, the formula of silicon carbide (SiC) means that there is an equal number (a 1:1 ratio) of Si atoms and C atoms in a crystal. The formula SiO_2 tells us that there are twice as many O atoms as Si atoms (the ratio of Si atoms to O atoms is 1:2) in a crystal of silicon dioxide.

WORKED EXAMPLE 3.1—FORMULAS OF COVALENT NETWORK SUBSTANCES

The oxide of germanium is a typical covalent network substance with the formula GeO_2. What does its formula tell us?

Solution

Whatever the arrangement of Ge atoms and O atoms in the lattice of this compound (and the formula gives no information about that), in the lattice there are twice as many O atoms as Ge atoms. In other words, the ratio of Ge atoms to O atoms is 1:2.

EXERCISE 3.1—FORMULAS OF COVALENT NETWORK SUBSTANCES

Boron nitride is a covalent network solid with an equal number of boron atoms and nitrogen atoms. What is its formula?

3.4 Ionic Substances

The salt at your dinner table and left on your skin (and paper money) when you perspire, like all ionic substances, is a compound [<<Section 2.5], composed of more than one element. A substance is classified as an **ionic compound** if it has a high melting temperature, does not conduct electricity when solid but does when molten or in solution (Table 3.1), and is hard, yet brittle (it shatters, rather than deforms). From these properties, chemists infer that (a) the constituent particles are held together very tightly, accounting for both hardness and high melting points; (b) any charged particles are held in fixed positions in the solid state, but that these become mobile when the compound is melted; and (c) distortion of the solid form brings about forces of repulsion between the particles, causing the crystal to shatter.

A Model

The currently accepted model of ionic compounds below their melting point is that they are composed of very many charged particles called **ions**, some with positive charge and some with negative charge. The ions, arranged regularly in a three-dimensional network called an **ionic lattice**, are held together by the *electrical force of attraction* that exists between each charged particle and surrounding oppositely charged particles (Figure 3.3).

The models of both covalent network substances and ionic substances propose three-dimensional lattices of atoms. The difference is that the model of covalent network substances has uncharged atoms at the lattice sites, while the lattice of ionic solids has both negatively charged and positively charged ions.

This model accounts for the physical properties of ionic substances:

- *High melting points.* Because each ion is surrounded by oppositely charged nearest neighbours, it is held tightly in one position. At room temperature, each ion can only vibrate about this position, and is not able to move past the other ions. Only when the temperature is high do the ions have enough vibrational energy that they can escape the attraction of neighbouring ions, and move through the lattice. Then, the lattice structure collapses and the substance becomes a liquid.
- *Electrical conductivity.* Solid ionic compounds do not conduct electricity when a potential is applied because there are no mobile charged particles—there are no "delocalized" electrons and the charged ions are firmly held in one place. When molten or in solution, the ions are able to move with respect to each other and conduction occurs by movement of ions.
- *Hardness.* Because the ions are firmly bound by attractions to the oppositely charged surrounding ions, they are not easily displaced by physical pressure.

Charles D. Winters

FIGURE 3.3 An ionic crystal. A crystal of sodium chloride can be visualized at the atomic level as a lattice of positively charged sodium ions (sodium atoms from which one electron has been removed) and negatively charged chloride ions (chlorine atoms that have gained one extra electron). A sodium ion does not "belong" to any one chloride ion, and vice versa.

Molecular Modelling (Odyssey)

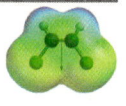

e3.3 Compare and contrast simulations of ionic solids.

• *Brittleness.* A severe hammer blow displaces layers of ions so that ions of like charge become adjacent. The repulsion between like charges then forces the lattice apart. A fine, carefully directed blow (in the direction of the planes of ions) can cause the crystal to cleave along a flat surface (Figure 3.4). For the same reason, however, random blows cutting across the planes of many layers of ions will cause the crystal to shatter.

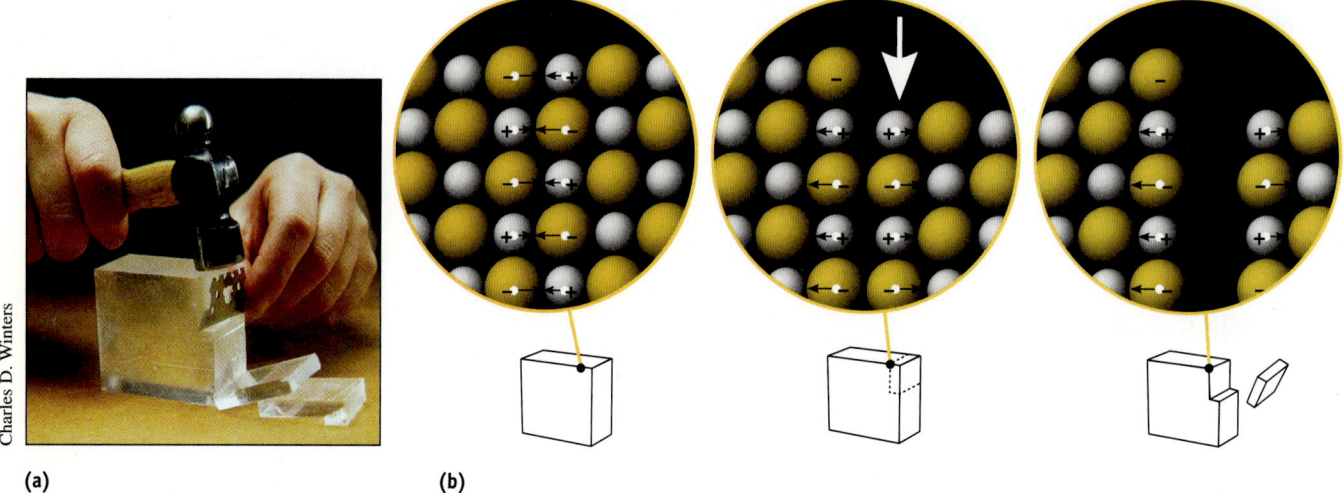

(a)   **(b)**

FIGURE 3.4 Cleaving an ionic crystal. (a) A craftsman uses a sharp, thin cutting instrument, and knows which direction to place it before striking with a hammer. (b) An explanation at the atomic level is that a layer of ions is displaced slightly with respect to the next layer, and ions of like charge become adjacent to each other. Forces of repulsion between the ions of like charge cause the crystal to cleave.

Think about It

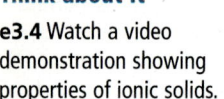

e3.4 Watch a video demonstration showing properties of ionic solids.

Many naturally occurring minerals are ionic compounds (Figure 3.5).

EXERCISE 3.2—IONIC COMPOUNDS

The melting points of sodium chloride (NaCl) and silicon carbide (SiC) are both high. Compare and contrast the origins of the high melting points of these two compounds.

The ionic lattice model of ionic compounds accounts for their high melting points, brittleness, hardness, lack of electrical conductivity when solid, and their ability to conduct electricity when molten or in solution.

Common Name	Name	Formula	Ions Involved
Calcite	Calcium carbonate	$CaCO_3$	Ca^{2+}, CO_3^{2-}
Fluorite	Calcium fluoride	CaF_2	Ca^{2+}, F^-
Gypsum	calcium sulfate dihydrate	$CaSO_4 \cdot 2\,H_2O$	Ca^{2+}, SO_4^{2-}
Hematite	Iorn(III) oxide	Fe_2O_3	Fe^{3+}, O^{2-}
Orpiment	Arsenic sulfide	As_2S_3	As^{3+}, S^{2-}

FIGURE 3.5 Some minerals that are ionic compounds. Note the different ions which comprise the lattices of these ionic substances.

What Are Ions? How Are They Formed? What Are the Charges on Them?

Ions are atoms or groups of atoms that as a result of processes during chemical reactions have different numbers of protons and electrons. If there are more protons than electrons, the ion has a positive charge and is called a **cation** (pronounced *cat-ion*). An ion with more electrons than protons has a negative charge and is called an **anion** (pronounced *ann-ion*).

A simple view of cation formation is to imagine an electron is removed from, say, a lithium-6 atom (Figure 3.6), represented as follows:

$$\text{Li atom} \longrightarrow \text{Li}^+ \text{ cation} + e^-$$

3 protons　　　　3 protons
3 electrons　　　2 electrons

Correspondingly, anion formation can be imagined to result from a fluorine-19 atom gaining an electron (Figure 3.6):

$$\text{F atom} + e^- \longrightarrow \text{F}^+ \text{ anion}$$

9 protons　　　　9 protons
9 electrons　　　10 electrons

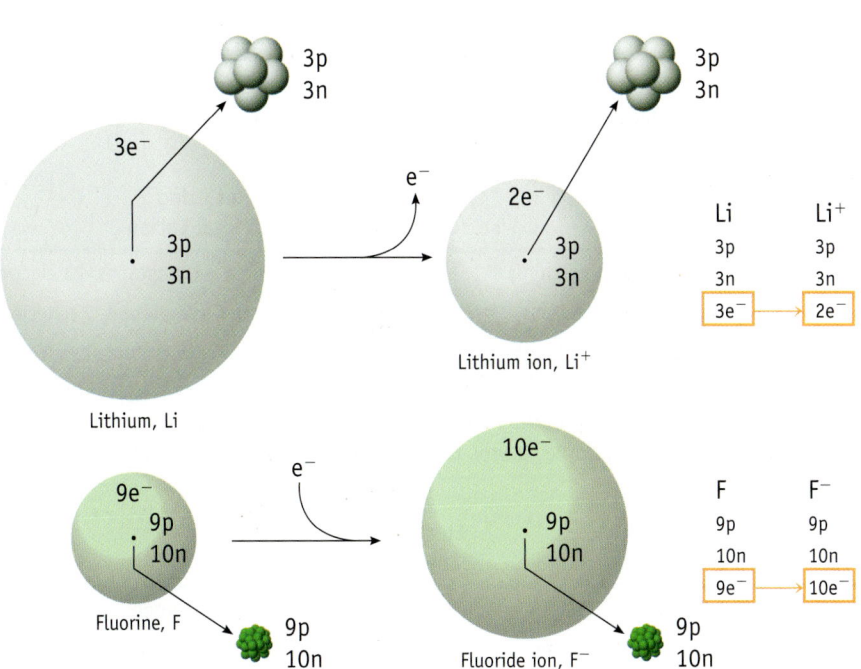

Lithium, Li

Lithium ion, Li$^+$

Li　　Li$^+$
3p　　3p
3n　　3n
3e$^-$ ⟶ 2e$^-$

Fluorine, F

Fluoride ion, F$^-$

F　　F$^-$
9p　　9p
10n　　10n
9e$^-$ ⟶ 10e$^-$

FIGURE 3.6 Ions. A lithium-6 atom is electrically neutral because it has 3 protons and 3 electrons. If an electron is removed from a Li atom, the resulting ion has 3 protons but only 2 electrons, and so has a net electrical charge of +1. We symbolize this lithium cation as Li$^+$. A fluorine-19 atom, with 9 protons and 9 electrons is electrically neutral, but if it acquires an extra electron, a fluorine anion with net charge −1 is formed. This anion is called a fluoride ion, and we use the symbol F$^-$.

How can we know which elements form cations and which form anions? The simple answer to this is that it requires experience, but the following guidelines are helpful:

- Cations are generally formed in chemical reactions by removal of one or more electrons from atoms of metallic elements (e.g., to form K$^+$ ions from K atoms, Mg^{2+} ions from Mg atoms, Al^{3+} ions from Al atoms).
- Anions are usually formed as the result of gaining one or more electrons by atoms of non-metallic elements (e.g., Cl$^-$ ions from Cl atoms, O^{2-} ions from O atoms, N^{3-} ions from N atoms).

How can we know what the charge is on different cations or anions? Common *monatomic ions* (those containing a single atom) are shown in the periodic table of Figure 3.7.

We observe some common patterns, although many ions don't fit these generalizations. The metals of Groups 1 and 2 form cations with positive charge equal to the group number, and Group 13 metals form ions with charge 3+.

FIGURE 3.7 Charges on some common monatomic cations and anions. The metallic elements form cations; the non-metallic elements form anions. The boxes group together ions with identical charge. The noble gases (Group 18 elements) form neither cations nor anions. Hydrogen forms a 1+ cation like the Group 1 elements, as well as a 1− anion like the Group 17 elements.

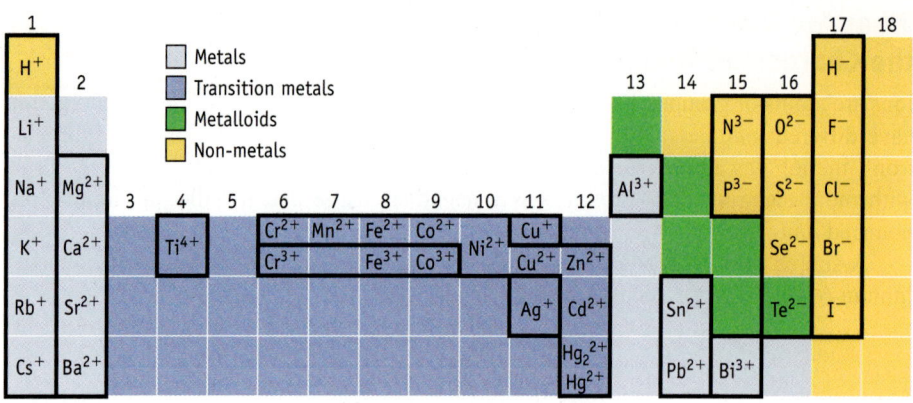

Transition metals (Groups 3–12) have cations with no easily predictable pattern of charges. Many have two types of ion with different charges. Iron-containing compounds, for example, may contain either Fe^{2+} or Fe^{3+} cations. Indeed, 2+ and 3+ ions are typical of many transition metals (see Figure 3.7).

Group	Metal Atom	Electron Change		Resulting Metal Cation
1	Na (11 protons, 11 electrons)	−1	⟶	Na^+ (11 protons, 10 electrons)
2	Ca (20 protons, 20 electrons)	−2	⟶	Ca^{2+} (20 protons, 18 electrons)
13	Al (13 protons, 13 electrons)	−3	⟶	Al^{3+} (13 protons, 10 electrons)

Transition metals (Groups 3–12) have cations with no easily predictable pattern of charges. Many have two types of ion with different charges. Iron-containing compounds, for example, may contain either Fe^{2+} or Fe^{3+} cations. Indeed, 2+ and 3+ ions are typical of many transition metals (see Figure 3.7).

Group	Metal Atom	Electron Change		Resulting Metal Cation
7	Mn (25 protons, 25 electrons)	−2	⟶	Mn^{2+} (25 protons, 23 electrons)
8	Fe (26 protons, 26 electrons)	−2	⟶	Fe^{2+} (26 protons, 24 electrons)
8	Fe (26 protons, 26 electrons)	−3	⟶	Fe^{3+} (26 protons, 23 electrons)

Non-metallic elements commonly form anions with negative charge equal to 18 minus their group number. For example, atoms of nitrogen (Group 15) gain three electrons and form anions with charge −3.

Group	Non-Metal Atom	Electron Change		Resulting Anion
15	N (7 protons, 7 electrons)	+3	⟶	N^{3-} (7 protons, 10 electrons)
16	S (16 protons, 16 electrons)	+2	⟶	S^{2-} (16 protons, 18 electrons)
17	Br (35 protons, 35 electrons)	+1	⟶	Br^- (35 protons, 36 electrons)

One of the most important ions is the H^+ cation, simply a proton, which can be thought of as an H atom that has lost an electron. The H^- anion can also be formed, and has one more electron than an H atom.

Numbers of Electrons on Monatomic Ions and on Noble Gas Atoms

The fact that atoms in different groups of the periodic table form ions with different charges makes more sense if we recognize that *cations formed from atoms of Groups 1, 2, and 13 have the same number of electrons as atoms of the noble gas that precedes it in the periodic table*. For example, the Mg^{2+} cation has 10 electrons (12 − 2)—the same as an atom of neon (Ne).

Ions of the non-metallic elements do not follow suit because to do so a chlorine atom would need to have 17 electrons removed, forming a Cl^{17+} cation! Instead, *atoms of the*

Think about It

e3.5 Work through tutorials on the charges on ions.

non-metallic elements commonly gain one or more electrons, forming anions with the same number of electrons as atoms of the next noble gas. For example, oxygen forms O^{2-} anions with 10 electrons $(8 + 2)$—the same as an atom of neon.

Ions Have Identities

Ions, whether cations or anions, have very different characteristics from the atoms from which they are derived. For example, the Fe^{2+} cation is a different species from an Fe atom—not just a somewhat modified Fe atom—and the Cl^- anion is a different species from a Cl atom. We use the term **chemical species** to refer to any set of identical particles with characteristic chemical behaviour [>>Section 6.1].

To reinforce our recognition of the identity of ions, we should be careful in our use of language. It is incorrect to refer to the Mg^{2+} ion as "magnesium." It is not magnesium—rather, it is the magnesium cation or, more specifically, the "magnesium-two-plus cation." The Cl^- ion should not be called "chlorine"; it is the chloride anion (see naming of ions below).

EXERCISE 3.3—IONS

Compare the number of electrons with the number of protons in the following:
(a) a sulfur atom and a sulfide ion (S^{2-})
(b) an aluminium atom and an aluminium ion (Al^{3+})
(c) a hydrogen atom and a hydrogen ion (H^+)

EXERCISE 3.4—IONS

List the numbers of protons in each of these common ions: N^{3-}, O^{2-}, F^-, Na^+, Mg^{2+}, and Al^{3+}. Also list the number of electrons in each ion. Compare the number of electrons in these ions with the number of electrons in a neon atom.

> The term *species* in chemistry refers to any atom, ion, molecule, or group of these that can exist as an independent entity and has characteristic behaviour. Ions are different species, with very different characteristics from the atoms from which they are derived.

Names of Monatomic Ions

The name of a monatomic cation is simply the name of the metal plus the word "cation." For example, the Al^{3+} ion is called the aluminium cation.

Some metals, especially in the transition series, have two or more cations with different charges. In these cases, each ion is named by putting its charge in parentheses after the metal name. For example, Co^{2+} is the cobalt(II) cation, and Co^{3+} is the cobalt(III) cation.

A monatomic anion is named by adding -ide to the name of the non-metal element from which is it derived (Figure 3.8).

Polyatomic Ions

Polyatomic ions are made of two or more atoms (Table 3.2). For example, the carbonate ion (CO_3^{2-}) is an anion that consists of one C atom and three O atoms, and has two units of negative charge because it has 32 electrons—two more than the number of protons (a total of 30) in the nuclei. Covalent bonds [>>Section 3.7] hold together the carbon and oxygen atoms within the carbonate anion.

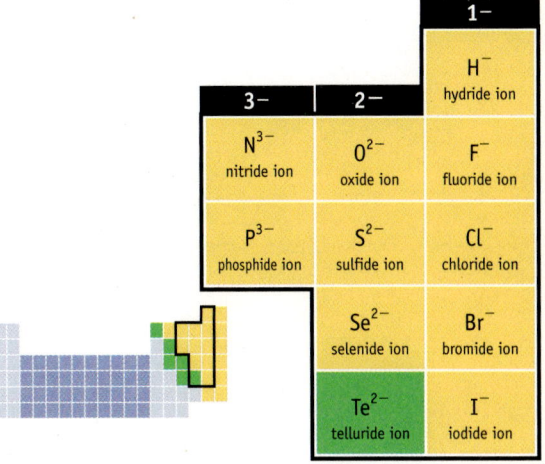

FIGURE 3.8 Names and symbols of some common monatomic anions.

TABLE 3.2 Formulas and Names of Some Common Polyatomic Ions

Formula	Name	Formula	Name
CATION: Positive Ion			
NH_4^+	ammonium ion		
ANIONS: Negative Ions			
Based on a Group 14 element		**Based on a Group 17 element**	
CN^-	cyanide ion	ClO^-	hypochlorite ion
CH_3COO^-	acetate ion	ClO_2^-	chlorite ion
CO_3^{2-}	carbonate ion	ClO_3^-	chlorate ion
HCO_3^-	hydrogencarbonate ion (or bicarbonate ion)	ClO_4^-	perchlorate ion
Based on a Group 15 element		**Based on a transition metal**	
NO_2^-	nitrite ion	CrO_4^{2-}	chromate ion
NO_3^-	nitrate ion	$Cr_2O_7^{2-}$	dichromate ion
PO_4^{3-}	phosphate ion	MnO_4^-	permanganate ion
HPO_4^{2-}	hydrogenphosphate ion		
$H_2PO_4^-$	dihydrogenphosphate ion		
Based on a Group 16 element			
OH^-	hydroxide ion		
SO_3^{2-}	sulfite ion		
SO_4^{2-}	sulfate ion		
HSO_4^-	hydrogensulfate ion (or bisulfate ion)		

Interactive Exercise 3.5

Practise remembering charges on ions.

Formulas of Ionic Compounds

Like covalent network substances, ionic compounds are not composed of molecules, and their formulas express relative numbers. *The formula of an ionic compound refers to the relative numbers of cations and anions in its lattice.* For example, the formula NaCl means that there is a 1:1 ratio of Na^+ cations and Cl^- anions—*not* separate molecules, each of which contains only one Na atom and one Cl atom. The formula $MgBr_2$ means that there are twice as many Br^- anions as Mg^{2+} cations in the lattice of magnesium bromide.

Why is the ratio of cations to anions different from compound to compound? This is because charges on the cations and anions are different from compound to compound, and the relative numbers of them must be such that the crystal has zero net electrical charge. The total of positive charges on all the cations must be equal to the total of negative charges on all of the anions.

In sodium chloride, the sodium ion (Na^+) has a +1 charge and the chloride ion (Cl^-) has a −1 charge. These ions must be present in a 1:1 ratio, and the formula is NaCl. The gem ruby is largely the compound composed of aluminium ions (Al^{3+}) and oxide ions (O^{2-}). These ions are in a 2:3 ratio: for every two Al^{3+} ions (total charge $2 \times +3 = +6$) in a crystal of this substance there are three O^{2-} ions (total charge $= 3 \times -2 = -6$), resulting in overall zero charge, and so the formula that we use is Al_2O_3.

Calcium, a Group 2 metal, forms a cation with a +2 charge. Examples of ionic compounds it forms when combined with different anions are shown in the following table:

Formula	Ions Present	Overall Charge on Formula
$CaCl_2$	$Ca^{2+} + 2\ Cl^-$	$(+2) + 2 \times (-1) = 0$
$CaCO_3$	$Ca^{2+} + CO_3^{2-}$	$(+2) + (-2) = 0$
$Ca_3(PO_4)_2$	$3\ Ca^{2+} + 2\ PO_4^{3-}$	$3 \times (+2) + 2 \times (-3) = 0$

In writing formulas, the convention is to give the symbol of the cation before that of the anion. When the formula has more than one polyatomic ion, parentheses are used for clarity.

Think about It

e3.6 Work through tutorials on naming polyatomic ions and ionic compounds.

WORKED EXAMPLE 3.2—FORMULAS OF IONIC COMPOUNDS

For each of the ionic compounds whose formulas are given, write the symbols for the ions present and give the relative number of each type of ion in its lattice: (a) $MgBr_2$, (b) Li_2CO_3, and (c) $Fe_2(SO_4)_3$.

Solution

The symbols for the ions, including their charges, are listed in Table 3.2.
(a) The ions are Mg^{2+} and Br^-. From the formula we see that there are twice as many Br^- ions as Mg^{2+} ions in the lattice—consistent with the compound having zero charge.
(b) The ions are Li^+ and CO_3^{2-}. The formula Li_2CO_3 tells us that there are twice as many Li^+ ions as there are CO_3^{2-} ions in the lattice. This achieves electrical neutrality.
(c) The ions are Fe^{3+} cations and SO_4^{2-} anions, in the ratio 2:3. We know that this must have Fe^{3+}, rather than Fe^{2+} cations, or the formula would be $FeSO_4$.

EXERCISE 3.6—FORMULAS OF IONIC COMPOUNDS

(a) Give the relative numbers and identity of the constituent ions in each of the following ionic compounds: $NaF(s)$, $Cu(NO_3)_2(s)$, and $NaCOOCH_3(s)$.
(b) Iron, a transition metal, forms ions having at least two different charges. Write the formulas of the compounds formed between two different iron cations and chloride ions.
(c) Write the formulas of all neutral ionic compounds that can be formed by combining the cations Na^+ and Ba^{2+} with the anions S^{2-} and PO_4^{3-}.

Interactive Exercises 3.7–3.9

Test yourself on charges on ions, naming ionic compounds, and writing ionic formulas.

The formulas of ionic compounds indicate the relative numbers of cations and anions in the lattice. The relative numbers of cations and anions in an ionic compound is such that the compound has no net electrical charge.

Molar Masses of Ionic Compounds

One mole of the substance sodium chloride, $NaCl(s)$, is the amount that contains 1 mol of Na^+ ions and 1 mol of Cl^- ions. One mole of $Al_2O_3(s)$ is the amount that contains 2 mol of Al^{3+} ions and 3 mol of O^{2-} ions. *The molar mass of an ionic compound is the sum of the molar masses of the ions in its formula, taking into account the number of each ion.*

WORKED EXAMPLE 3.3—MOLAR MASSES OF IONIC COMPOUNDS

What is the molar mass of calcium fluoride (CaF_2)?

Solution

$$M(CaF_2) = M(Ca^{2+}) + 2 \times M(F^-)$$

$$= 40.1 \text{ g mol}^{-1} + 2 \times 19.0 \text{ g mol}^{-1} = 78.1 \text{ g mol}^{-1}$$

Think about It

e3.7 Work through an experiment to determine the formula of a hydrated compound.

EXERCISE 3.10—MOLAR MASSES OF IONIC COMPOUNDS

Calculate the molar mass of the following:
(a) potassium bromide, $KBr(s)$
(b) ammonium chloride, $NH_4Cl(s)$
(c) aluminium nitrate, $Al(NO_3)_3(s)$
(d) magnesium hydrogenphosphate, $Mg(HPO_4)(s)$

Interactive Exercises 3.12–3.13

Calculate molar masses of compounds.

EXERCISE 3.11—MOLAR MASSES OF IONIC COMPOUNDS

Calculate the amount (in mol) of the following:
(a) 1.00 g of sodium hydroxide, $NaOH(s)$
(b) 0.0125 g of iron(II) nitrate, $Fe(NO_3)_2(s)$
(c) 0.500 t (tonnes) of calcium carbonate (limestone), $CaCO_3(s)$

3.5 Metals and Metallic Substances

The shiny gold band in which a diamond ring is set is a familiar metallic substance, as is the platinum, palladium, or rhodium catalyst that removes nitrogen oxide, carbon monoxide, and hydrocarbon combustion products from your auto exhaust. Some observations and inferences that form a model of the structure of metals are illustrated in Table 3.3.

Almost all **metals** have all the properties listed in Table 3.3. Mercury, rubidium, and cesium are exceptions. Mercury, for example, has a low melting point, and is a liquid at room temperatures, but its other properties are more consistent with those of metals than of any of the other classes of substance.

A Model

How do we make sense of these properties? Chemists imagine what we would see if we could inspect a lump of a metal closely enough to see atoms, ions and even electrons. The chemists' model of the structure of metals can be summarized as follows:

- Positively charged ions are arranged in a regular, three-dimensional pattern called a lattice (Figure 3.9).
- To form the ions, each atom sets free one or more electrons, which do not belong to any particular atom. The free electrons are said to be **delocalized**. The positive ions are arranged regularly within a continuous "sea" of delocalized electrons.
- Although the ions occupy fixed positions in the lattice, the delocalized electrons can move throughout the piece of metal.
- The forces binding the atoms together in the lattice are the electrical attractions between the positive ions and the sea of delocalized electrons. This is called **metallic bonding**.

Conduction of electricity is a flow of charged particles—either positive or negative. In the specific case of flow of electricity through metals (see above), the moving particles are electrons. This contrasts with the conduction of electricity through a molten ionic solid, in which the mobile particles are the anions and cations.

TABLE 3.3
Modelling the Nature of Metals

Observation	Inference
Metals are dense.	The particles comprising metals are closely packed.
Metals are good electrical conductors.	Metals contain charged particles that can move under the influence of an applied potential.
Metals are good conductors of heat.	Metals contain particles that can move from regions of high temperature to regions of low temperature, taking their energy with them.
Metals have high melting points.	There are strong forces of attraction holding the particles in fixed positions. Only when the energy of the particles is very high, achieved by high temperatures, can the particles move with respect to each other.
Metals are malleable and ductile.	Forcing particles in the solid metal to move past each other does not give rise to an unstable situation with high forces of repulsion.
Metals are lustrous.	Most of the light striking the surface of metals is reflected.

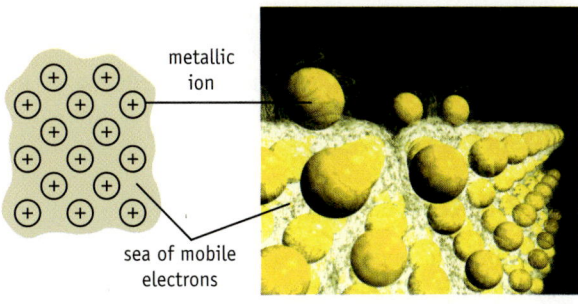

metallic ion

sea of mobile electrons

FIGURE 3.9 A model for the structure of metals, portraying a lattice of positive ions (left) within a "sea" of electrons. For a simulation using this model to represent solid copper, see e3.8.

At a simple level, this model can account for the characteristic properties of metallic substances as follows:

- Most metals are dense because the metal ions are closely packed in the lattice.
- Metals are good conductors of electricity because the delocalized electrons in the lattice are mobile. When a potential is applied across a piece of metal, electrons flow into one end and the same number of electrons flow out at the other end, with little resistance.
- Metals are good thermal conductors because the delocalized electrons transmit the energy of vibration of each positive ion to its neighbours. When one end of a block of metal is heated, the positive ions at that end vibrate rapidly. The delocalized electrons transmit this vibrational energy rapidly through the metal.
- Melting is achieved when the ions become mobile and move with respect to each other. Most metals have high melting points because a large amount of thermal energy is required to overcome the strong electrical forces between the positive ions and the sea of delocalized electrons.
- Metals are malleable and ductile because metallic bonding is retained even if regions of ions are displaced with respect to each other by distortion of the metal (Figure 3.10).
- The presence of free electrons causes most metals to be excellent reflectors of visible light and they are lustrous. This is not an easily explained property.

Molecular Modelling (Odyssey)

e3.8 Play a molecular-level simulation of several metals.

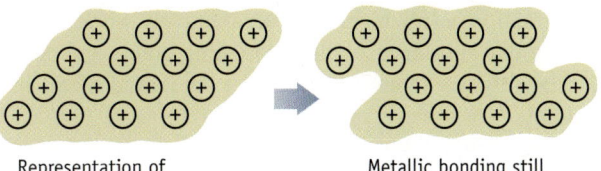

Representation of metallic bonding

Metallic bonding still operates after distortion

FIGURE 3.10 Malleability and ductility of metals. The conditions for metallic bonding are retained even if sections of the metal are forced to move past each other.

EXERCISE 3.14—METALLIC SUBSTANCES

What particles carry the charge when electricity flows in the following substances?
(a) solid magnesium
(b) molten magnesium
(c) molten magnesium chloride
(d) an aqueous solution of magnesium chloride

Metals are modelled as a lattice of positive ions within a mobile "sea" of electrons that acts as a "glue." This model can make sense of metals' high density, their ability to conduct electricity and heat, their malleability and ductility, and their relatively high melting points.

Symbols, Atomic Weights, and Molar Masses of Metals

The symbol Fe(s) can be used at two levels—to represent a piece of iron that we can see and hold, or to represent a model of the structure of iron, as described above.

The atomic weights and molar masses of metallic elements were explained in Chapter 2. For example, the atomic weight of iron is 55.845, and its molar mass is 55.845 g mol^{-1}—that is, 1 mol of iron, containing the number of atoms equal to the Avogadro constant, has mass 55.845 g.

3.6 Molecular Substances

Three of the candidates for our unidentified white powder in Section 3.1 are **molecular substances**—nicotine, cocaine, and sucrose (table sugar). Other molecular substances are listed in Table 3.1. Nitrogen (the main component of air), polyethylene food wrap, and methane (the main ingredient in natural gas used for cooking and heating), are also classified as molecular substances. The molecules of most molecular substances have only non-metal atoms. By far the most common of these are **organic compounds**, substances composed of molecules with carbon atoms joined to each other to form a framework to which other atoms or groups of atoms are joined.

These substances have in common the following characteristic properties:

- They have relatively low melting points and boiling points (many are gases or liquids at ordinary temperatures).
- They do not conduct electricity when an electric potential is placed across a solid or molten state sample.
- They are relatively soft; the surface of a solid sample is more easily scratched than samples of other types of substance.

From the above characteristic properties, chemists have made inferences about the nature of these substances. These inferences, based on the premise that all substances are composed of particles, are the following:

- Since conduction of electricity is the movement of electrically charged particles under the influence of an electric field, these substances do not contain charged particles that can move.
- The forces of attraction between particles are relatively weak. This is consistent with the understanding that substances with low melting points and boiling points need only to be at relatively low temperatures before the particles can move with respect to each other (and melt) or to have enough energy to separate from each other (and change into the gaseous state).

A Model

According to the currently accepted model, molecular substances are composed of very many particles called molecules, each of which has the same numbers of atoms of different elements. For example, all molecules of cocaine have 17 carbon atoms, 21 hydrogen atoms, 1 nitrogen atom, and 4 oxygen atoms connected together in a certain sequence (Figure 3.11a). All molecules of any substance are identical (except for isotopologues, [>>Section 3.8]), and they are different from those of every other substance either in (a) the numbers of atoms of each element in each molecule; (b) the **connectivity**, which is the sequence by which the atoms are joined to each other; and/or (c) the three-dimensional spatial arrangement of the atoms.

The number of atoms of each element in a molecule is expressed by its **molecular formula**. For example, the formula of water, H_2O, tells us that each water molecule contains 2 hydrogen atoms and 1 oxygen atom. Each molecule of sucrose, $C_{12}H_{22}O_{11}$, contains 12 carbon atoms, 22 hydrogen atoms, and 11 oxygen atoms. The formula of cocaine is $C_{17}H_{21}NO_4$.

Relatively strong forces of attraction must exist to maintain the arrangement of atoms in molecules. A simple view is that adjacent atoms have a force of attraction between them, and this force holding them together is called a **covalent bond**. For example, in a water molecule (Figure 3.11c), there are two covalent bonds—between the oxygen atom and each of the hydrogen atoms. These are referred to as O—H bonds. In a molecule of ethanol (Figure 3.11b), there is one C—C covalent bond, five C—H bonds, one C—O bond, and one O—H bond. Since these bonds are between atoms *within* molecules, they are labelled **intramolecular forces**.

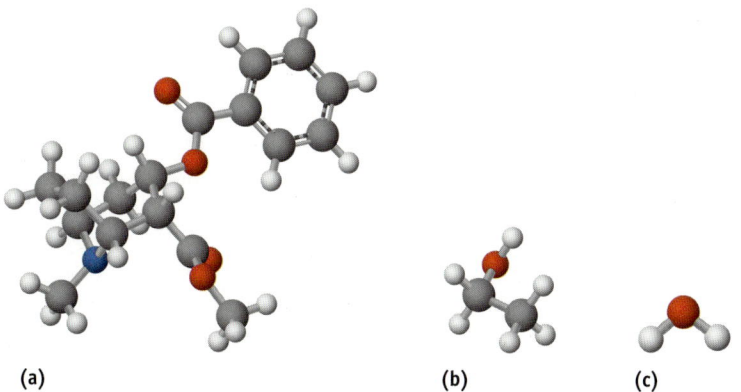

(a)   (b)   (c)

FIGURE 3.11 Representations of molecules of various substances. (a) Cocaine is composed of identical molecules composed of 17 carbon atoms, 21 hydrogen atoms, 1 nitrogen atom, and 4 oxygen atoms joined together in the arrangement shown here. (b) A representation of any one of the molecules of ethanol (commonly called "alcohol"). (c) A representation of a molecule of water.

The prefix *intra-* means "within" (e.g., intracellular), while *inter-* means "between," external to the unit of interest (e.g., international).

Besides intramolecular forces of attraction that hold atoms together in molecules (covalent bonds), there must also be **intermolecular forces** of attraction *between* molecules—otherwise, how could substances exist in the solid state? These intermolecular forces are much weaker than the covalent bonds within molecules. When the temperature of a solid molecular substance is raised, eventually (at the melting point) the molecules have enough kinetic energy to sufficiently overcome the intermolecular forces to move with respect to each other, and become liquid, *while the intramolecular covalent bonds remain unbroken*. At an even higher temperature (the boiling point), molecules have just enough kinetic energy to overcome their intermolecular forces and separate from each other to form a gas, while again the covalent bonds remain intact. The relatively low melting points and boiling points of molecular substances are consistent with the proposition that the intermolecular forces are relatively weak.

The power of our model for molecular substances can be shown by drawing a connection between the macroscopic and molecular-level views of an ice crystal. Why should there be such regularity of shape in ice and crystals of other molecular substances? The beautiful six-sided symmetrical shape of a snowflake (crystalline H_2O) that you can see with your eyes can be understood through a molecular-level view of water molecules, which are held together in six-sided rings by a remarkable interplay of intermolecular and intramolecular forces. Each six-sided ring has six water molecules. Each H_2O molecule is held together by O—H covalent bonds, and intramolecular forces between the O atom of one water molecule and the H atom of another hold water molecules together. This results in the shape seen in Figure 3.12.

Molecular Modelling (Odyssey)

e3.9 Simulate ice and its melting to form liquid water.

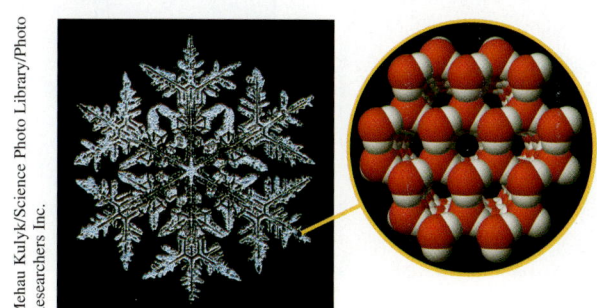

Mehau Kulyk/Science Photo Library/Photo Researchers Inc.

FIGURE 3.12 Snowflakes are six-sided structures, reflecting the arrangement of molecules in ice. The molecular structure of ice consists of six-sided rings formed by water molecules, in which each side of the rings has the sequence O —— H —— O, formed from an O —— H group of one water molecule, and an O atom of another.

Another unique property of water is that it occupies a larger volume when solid than when it is liquid. The lower density of ice than of liquid water has enormous consequences for the earth's climate. This, too, has a molecular-level explanation. H_2O molecules in ice have an open packing arrangement. When ice melts, the symmetrical structure breaks apart, and water molecules move closer together. This process is simulated in e3.9.

EXERCISE 3.15—MOLECULAR SUBSTANCES

Which of the following are composed of molecules?
(a) magnesium oxide, $MgO(s)$
(b) nitric oxide, $NO(g)$
(c) magnesium, $Mg(s)$
(d) ammonia, $NH_3(g)$
(e) sulfur, $S_8(s)$
(f) molten sodium chloride, $NaCl(l)$

EXERCISE 3.16—MOLECULAR SUBSTANCES

In sulfur, there are S—S covalent bonds; in silicon, there are Si—Si covalent bonds. The melting point of sulfur is 119 °C, while the melting point of silicon is 1400 °C. Explain why the melting points are so different.

Molecular substances can be identified by their relatively low melting points and boiling points; they do not conduct electricity; and their solids are relatively soft. Our structural model to explain these observable properties shows these substances made of many molecules. Intramolecular forces, called covalent bonds, hold atoms together within molecules. Molecules are attracted to each other through intermolecular forces.

3.7 Covalent Bonding

The commonly accepted model for molecular substances is that interaction between the electrons of atoms within each molecule leads to covalent bonding between atoms. The inability of molecular substances to conduct electricity is evidence that the molecules are electrically neutral (uncharged), meaning that the number of electrons in each molecule is the same as the number of protons in the nuclei of the atoms making up that molecule. Furthermore, the electrons are held tightly within each molecule and are not free to move from one molecule to another along a sample of the substance if an electric potential is applied.

The electrons around each atom within a molecule are best imagined as a disperse "cloud" of negatively charged matter around the nucleus [>>Chapter 8], attracted to the positively charged nucleus. In a covalent bond, the bound atoms share electron matter, and the atoms are held together by attraction of each nucleus for the shared electron matter (Figure 3.13).

Think about It

e3.10 Watch a portrayal of the merging of electron clouds when a bond forms.

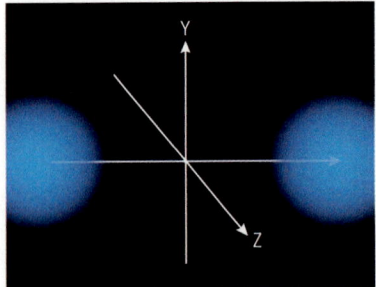 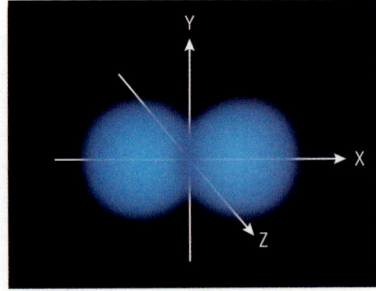

FIGURE 3.13 Depiction of shared electrons, representing a covalent bond in a hydrogen molecule (H_2). Both H nuclei (left) are attracted to the shared electron cloud, and so are held together in a bond. The existence of the molecule confirms that the attractions of the nuclei to the shared electrons dominate any repulsion between the two electrons and between the two nuclei.

In a covalent bond, the bound atoms share electron matter, and the atoms are held together by attraction of each nucleus for the shared electron matter.

When two electrons are shared, the covalent bond is called a **single bond**. An example is the H_2 molecule, which can be symbolically represented as H—H. Molecules of chlorine (Cl_2, represented as Cl—Cl) can similarly be described: each chlorine atom contributes one electron to a shared region of electron matter between the two Cl nuclei, and the mutual attraction of both nuclei to the shared electrons constitutes a single covalent bond (Figure 3.14). Since the shared electrons in Cl_2 molecules can be regarded as belonging to both atoms, each atom can be thought of as having one more electron than in an unbound atom.

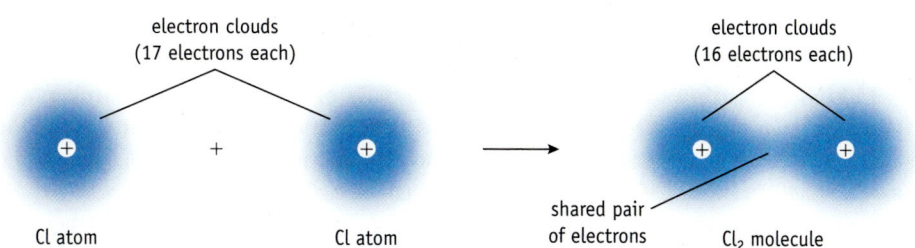

electron clouds (17 electrons each)

electron clouds (16 electrons each)

Cl atom   Cl atom   shared pair of electrons Cl_2 molecule

FIGURE 3.14 Depiction of covalent bond formation in a chlorine molecule (Cl_2).

In some molecules, the distance between the nuclei (the bond length) is shorter, and the energy needed to break the bond is greater than for single covalent bonds. This is due to the contribution of two electrons by each participating atom to a shared region with four electrons, called a **double bond**. Examples of substances whose molecules are best represented with double bonds are carbon dioxide and ethene.

There are also many compounds whose molecules have an oxygen atom bound by a single bond to each of two other atoms.

Infrared spectroscopy [>>Section 3.11] provides experimental evidence that the number of shared electrons alters the strength of bonds. A C=O double bond requires higher energy infrared radiation to cause it to vibrate than a C—O single bond.

O=C=O

Depiction of double bonds in molecules of carbon dioxide and ethene.

Structural formulas of molecules of water and ethanol. In each, an O atom is bound by single covalent bonds to two other atoms. The O atom in both cases can be considered to have eight valence electrons, including the two shared pairs of bonding electrons.

3.8 Composition and Formula by Mass Spectrometry

Let's return to the example in Section 3.1 of an unknown white powder extracted by washing large bundles of paper money with a solvent. It would not have been possible to identify cocaine on individual banknotes 50 years ago. Since then, spectroscopic techniques have become sensitive enough to identify and characterize trace amounts of molecular substances much too small to see. These techniques have revolutionized our understanding of chemistry. Any new compound synthesized in a laboratory is now routinely analyzed by IR, MS, and NMR spectroscopy to determine its identity.

How are these techniques used to tell us, for example, that the molecular substances cocaine, ethanol, and water have the structures shown in Figure 3.11? We need to find out how many atoms of each type a compound contains (composition and molecular formula), the pattern of connectivity between the atoms within a molecule (covalent bonds), and the arrangement of atoms in three-dimensional space.

We explore first how high-resolution mass spectrometry can be used to determine the composition and formula, and then how mass spectrometry and infrared spectroscopy can help determine the connectivity sequence of atoms within molecules. In the toolbox of a chemist, NMR spectroscopy and X-ray crystallography [>>Chapter 9] are other powerful techniques to give complementary information about molecular structure.

High-Resolution Mass Spectrometry

In Section 2.9, we described how mass spectrometry can distinguish different isotopes of elements. This same technique can be used to determine the composition (and, therefore, the formula) of a molecular compound. In one variation of mass spectrometry called

electron impact ionization mass spectrometry (EI), a substance is heated to produce a vapour, which is then passed into an evacuated chamber. There its molecules, denoted M, are bombarded by a high-energy electron beam, causing some of the molecules to eject an electron and form positive ions.

$$M(g) \longrightarrow M^+(g) + e^-$$

The M^+ ions are directed through a magnetic field where their paths are deflected by an angle that depends on the mass/charge (m/z) ratio. A detector records the relative numbers of ions arriving at each value of m/z. In the most common case of an ion like M^+ with $z = 1$, formed by ejection of just one electron, the value corresponds to the mass of the entire molecule. Such an M^+ ion is referred to as the **molecular ion** of the substance.

In **high-resolution mass spectrometry**, the mass of the molecular ion (relative to ^{12}C atoms) can be measured to seven significant figures. This method measures the mass so accurately that there is usually only one composition that corresponds with that mass.

To illustrate this method, suppose that we had a sample of a gaseous substance that we knew to be pure, whose molecules' relative mass was measured by a low accuracy technique to be 28.0. This measurement gives rise to two possibilities—carbon monoxide (CO) or nitrogen (N_2)—and does not allow us to decide which it is. The relative abundances and exact masses of the most common isotopes of several elements, including carbon, oxygen, and nitrogen are listed in Table 3.4.

A sample of carbon monoxide contains several **isotopologues**, molecules which are identical except that their atoms have different isotopic compositions—for example, $^{12}C^{16}O$, $^{12}C^{18}O$, $^{13}C^{16}O$, and $^{13}C^{18}O$. A mass spectrometer does not measure the average mass of carbon monoxide molecules; rather, it measures the **relative molecular mass (M_r)** of molecules of each isotopologue. *The relative molecular mass of an isotopologue is its mass on a scale in which the carbon-12 isotope is assigned a value of exactly 12.* By far the most abundant molecules are $^{12}C^{16}O$, whose relative molecular mass is 27.99491.

A sample of N_2 gas also contains $^{14}N^{14}N$, $^{14}N^{15}N$, and $^{15}N^{15}N$ isotopologues. Nearly all are $^{14}N^{14}N$ molecules, whose relative molecular mass is 28.00614. High-resolution mass spectrometry can measure accurately enough to distinguish between relative molecular masses of 27.99491 and 28.00614, telling us whether the sample of gas is CO or N_2. In addition, it can identify peaks that correspond with the masses and relative abundances of the less abundant isotopologues to confirm the identity of the compound.

Let's look at a somewhat more complex example. A sample of a liquid is injected into a high-resolution mass spectrometer, and the accurate mass of the most abundant molecular ions is measured to be 46.042706. What is the molecular formula of molecules of this substance?

In this example, the higher measured molecular ion mass makes it tedious to eliminate all possible combinations of common isotopes that might add up to the experimental mass of 46.042706. Simple computer algorithms that come with mass spectrometers can be used to automate this comparison. After taking into account experimental error in the measurement, the only reasonable combination of atoms that would give the correct relative mass for molecules of this substance is $^{12}C_2{}^1H_6{}^{16}O$ (relative molecular mass = 46.041865). Other peaks in the mass spectrum correspond with less

TABLE 3.4 Relative Abundances and Exact Masses of Isotopes of Several Elements

Element	Isotope	Relative Abundance	Exact Mass	Isotope	Relative Abundance	Exact Mass
carbon	^{12}C	98.90%	12.00000	^{13}C	1.10%	13.00335
oxygen	^{16}O	99.76%	15.99491	^{18}O	0.20%	17.99916
nitrogen	^{14}N	99.63%	14.00307	^{15}N	0.37%	15.00011
hydrogen	^1H	99.99%	1.00783	^2H	0.01%	2.01410
chlorine	^{35}Cl	75.78%	34.968852	^{37}Cl	24.20%	36.965902

abundant isotopologues with different isotopic composition, confirming that the elemental composition is C_2H_6O.

EXERCISE 3.17—RELATIVE MOLECULAR MASS FROM HIGH-RESOLUTION MASS SPECTROMETRY DATA

Identify the most common isotopes of carbon, oxygen, and hydrogen, and then using data from Table 3.4, calculate the mass of the molecular ion of the most abundant isotopologue of C_2H_6O. How does your value compare with the experimental value of 46.042706?

What mass would you obtain for C_2H_6O by adding up the atomic weights of two C atoms, six H atoms, and one O atom? The answer is 46.068. This value represents the mixture of various isotopologues that would be found in a representative sample of C_2H_6O molecules in nature due to the different isomers of carbon, hydrogen, and oxygen atoms. The value 46.041865 obtained from high-resolution mass spectrometry is different, because it represents the mass of the particular isotopologue $^{12}C_2{}^{1}H_6{}^{16}O$. *The value obtained from adding up atomic weights in the periodic table is referred to as the* **molecular weight** *of* C_2H_6O. The molecular weight of a molecular compound is the *average mass* of its molecules on a scale in which the carbon-12 isotope has a value of exactly 12. It is the sum of the atomic weights of the atoms in its molecules. For C_2H_6O, the numerical value of the molecular weight is 46.069.

More commonly, we will refer to the **molar mass (*M*)**, the mass, in grams, of one mole of a molecular compound. The molar mass is the sum of the molar masses of its constituent atoms [<<Section 2.11], and is numerically equal to the molecular weight. The unit of measurement is g mol^{-1}.

High-resolution mass spectrometry is used to determine the composition and formula of a molecular compound. The relative molecular mass of the most abundant isotopologues of a substance can be measured. The molecular weight is a weighted average mass of a substance's molecules, taking into account the different natural abundances of the isotopes of the elements in a particular sample.

> Molar mass (g mol^{-1}) is the mass of one mole of a substance, and is the sum of the molar masses of its constituent atoms.

Molecular weight is technically a mass, rather than a weight. The common use of the term *molecular weight* by chemists is a continuation of many decades of prior usage.

Think about It

e3.11 Work through tutorials on the relationships among amount, mass, and molar mass.

WORKED EXAMPLE 3.4—MOLAR MASS

What is the molar mass of cocaine?

Solution

The molecular formula of cocaine is $C_{17}H_{21}NO_4$. Its molar mass is the sum of the molar masses of each atom in its molecules, taking into account the number of each. The molar masses of carbon, hydrogen, nitrogen, and oxygen are obtained from the atomic weights on the periodic table. Taking atomic weights to three decimal places:

$$M(C_{17}H_{21}NO_4) = 17 \times M(C) + 21 \times M(H) + M(N) + 4 \times M(O)$$
$$= 17 \times 12.011 \text{ g mol}^{-1} + 21 \times 1.008 \text{ g mol}^{-1} + 14.007 \text{ g mol}^{-1} + 4 \times 15.999 \text{ g mol}^{-1}$$
$$= 303.358 \text{ g mol}^{-1}$$

EXERCISE 3.18—MOLAR MASS

Calculate the molar masses of the molecular substances:
(a) methane, $CH_4(g)$
(b) acetone, $C_3H_6O(l)$
(c) aniline, $C_6H_5NH_2(l)$

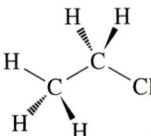
EXERCISE 3.19—MOLAR MASS

What is the mass of 0.0255 mol of each of the following compounds?
(a) $C_3H_7OH(l)$, isopropanol, rubbing alcohol
(b) $C_{11}H_{16}O_2(s)$, an antioxidant in foods, also known as BHA (butylated hydroxyanisole)
(c) $C_9H_8O_4(s)$, aspirin

WORKED EXAMPLE 3.5—MOLECULAR FORMULA FROM MASS SPECTRUM

The mass spectrum of a sample of chloroethane (CH_3CH_2Cl) in Figure 3.15 is recorded with an instrument that rounds off each ion's m/z value to the nearest integer. If the same sample was analyzed by a high-resolution mass spectrometer, the peaks at $m/z = 64$ and 66, which correspond to the molecular ion, would be expected to have values close to the exact masses of 64.007978 and 66.005028.
(a) What isotopologues for chloroethane cause the two molecular ion peaks at $m/z = 64$ and $m/z = 66$?
(b) What molecular weight would you obtain for chloroethane using values of the atomic weights of its constituent atoms? Why does this value not appear in the mass spectrum?

The structure of a chloroethane molecule.

FIGURE 3.15 Mass spectrum of chloroethane. The signal intensities (y-axis) are relative values compared to the most intense signal, which is assigned a value of 100%.

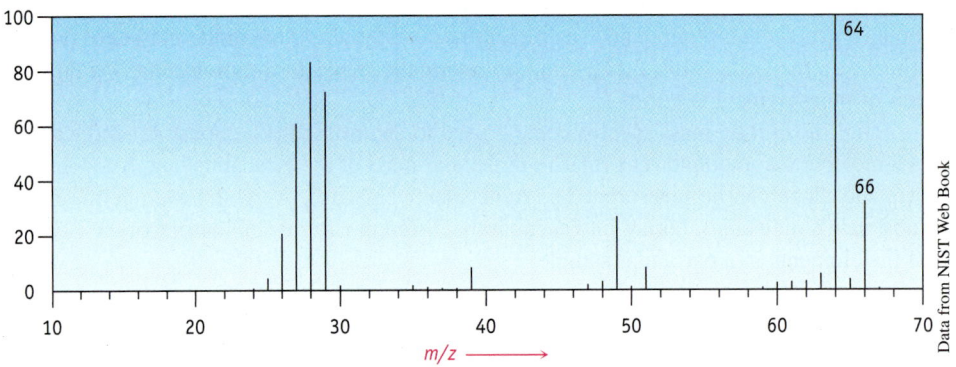

Solution

(a) The peak at a rounded m/z value of 64 corresponds to the ion $[^{12}C_2{}^1H_5{}^{35}Cl]^+$ with an exact mass of 64.007978. The peak at $m/z = 66$ is assigned to the isotopologue $[^{12}C_2{}^1H_5{}^{37}Cl]^+$, which has a rounded m/z value of 66 and an expected exact mass of 66.005028. Since 75.8% of all Cl atoms are ^{35}Cl and 24.2% are ^{37}Cl, the molecular ion peak at $m/z = 64$, $[^{12}C_2{}^1H_5{}^{35}Cl]^+$, has an intensity of three times (75.8/24.2) the molecular ion peak at $m/z = 66$, $[^{12}C_2{}^1H_5{}^{37}Cl]^+$.

(b) Adding the atomic weights for two C atoms, five H atoms, and one Cl atom gives 64.4927. This value is weighted by the relative abundance of each isotope and doesn't appear in the mass spectrum because a mass spectrometer records the spectrum for each isotopologue of C_2H_5Cl rather than an average value.

EXERCISE 3.23—MOLECULAR FORMULA FROM MASS SPECTRUM

The low-resolution mass spectrum of a compound containing only carbon, hydrogen, and chlorine atoms is shown below. How many chlorine atoms does a molecule of this compound contain? What is a reasonable molecular formula? Using Table 3.4, what would you predict the m/z value from a high-resolution mass spectrum to be for the peaks shown in the low-resolution spectrum at 112 and at 114?

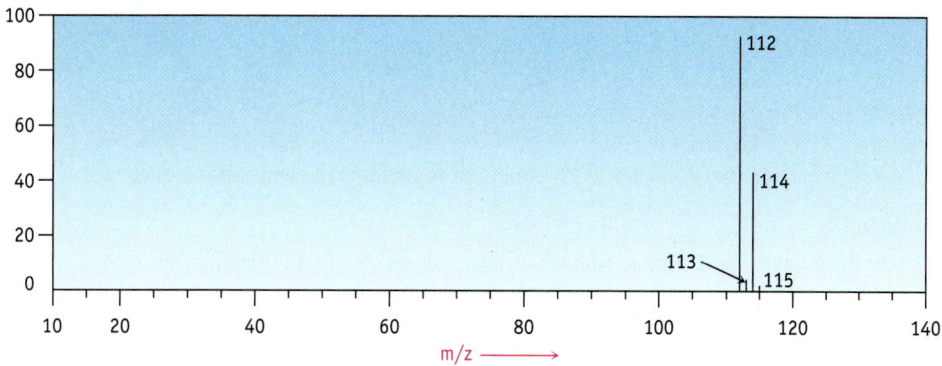

Relative abundance (%)

We've seen that high-resolution mass spectrometry analysis of an unknown liquid's molecular ion at 46.042706 corresponds with a molecular formula of C2H6O. If this instrumentation was not available, we could take a different approach to experimentally determining the formula—but the different technique would give us slightly different information. If we know that the liquid contains only carbon, hydrogen, and oxygen atoms, combustion analysis can be used to tell us the empirical formula, the simplest ratio of those atoms in each molecule. See e3.12, and later in Chapter 5 [>>e5.18], for details.

> High-resolution mass spectrometry analysis of a pure molecular substance can give an accurate value of the relative mass of molecules with various isotopic compositions. From this, we can define the formula and composition of the compound.

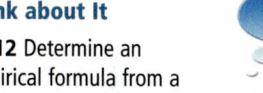

Think about It

e3.12 Determine an empirical formula from a combustion analysis.

3.9 Visualizing Connectivity in Molecules

From high-resolution mass spectrometry data, we know the composition and molecular formula of a substance, but not the connectivity sequence of its atoms. Let's return to the liquid in Section 3.8 with a molecular formula C_2H_6O, which could correspond to either ethanol or dimethyl ether. Molecules of these two substances have the same formula but different connectivity patterns—they are called **constitutional or structural isomers**. The experimental evidence for different connectivity patterns in ethanol and dimethyl ether comes from both IR spectroscopy and MS fragmentation patterns. Chapter 9 describes another powerful technique to tell them apart, NMR spectroscopy.

But first, how do we represent connectivity with visual symbols that others can interpret? Some sense of the different connectivities of the two C atoms, six H atoms, and one O atom in molecules of ethanol and dimethyl ether is given by **condensed formulas**, which show various groupings of atoms in molecules of the two substances:

$$CH_3CH_2OH \qquad\qquad H_3COCH_3$$
$$\text{ethanol} \qquad\qquad\qquad \text{dimethyl ether}$$

More detail is portrayed through the use of **structural formulas**, which indicate the connective sequence of all of the atoms as well as the covalent bonds between adjacent atoms and from ball and stick representations (Figure 3.16).

WORKED EXAMPLE 3.6—CONDENSED FORMULAS, STRUCTURAL FORMULAS, AND MOLECULAR FORMULAS

The acrylonitrile molecule is the building block for acrylic plastics (such as Orlon and Acrilan). Its condensed formula, structural formula, and molecular model are shown here. What is the molecular formula of acrylonitrile?

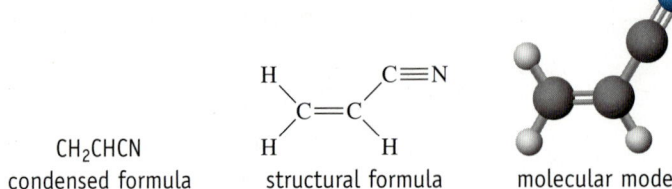

CH₂CHCN
condensed formula

structural formula

molecular model

Solution

Acrylonitrile has three C atoms, three H atoms, and one N atom. Therefore, its molecular formula is C_3H_3N. In writing molecular formulas of organic compounds, the convention is to write C first, then H, and finally other elements in alphabetical order.

FIGURE 3.16 Different representations of molecules. Ethanol and dimethyl ether have the same molecular formula C_2H_6O. On the top line are various representations of a molecule of ethanol; the bottom line shows corresponding ways of representing a molecule of dimethyl ether that clearly show the different connectivities. The lines between atoms represent the covalent bonds [<<Section 3.7].

EXERCISE 3.24—MOLECULAR FORMULAS

The styrene molecule (represented in various ways below) is the building block of molecules of polystyrene, a material used for insulated drinking cups and building insulation. What is the molecular formula of styrene?

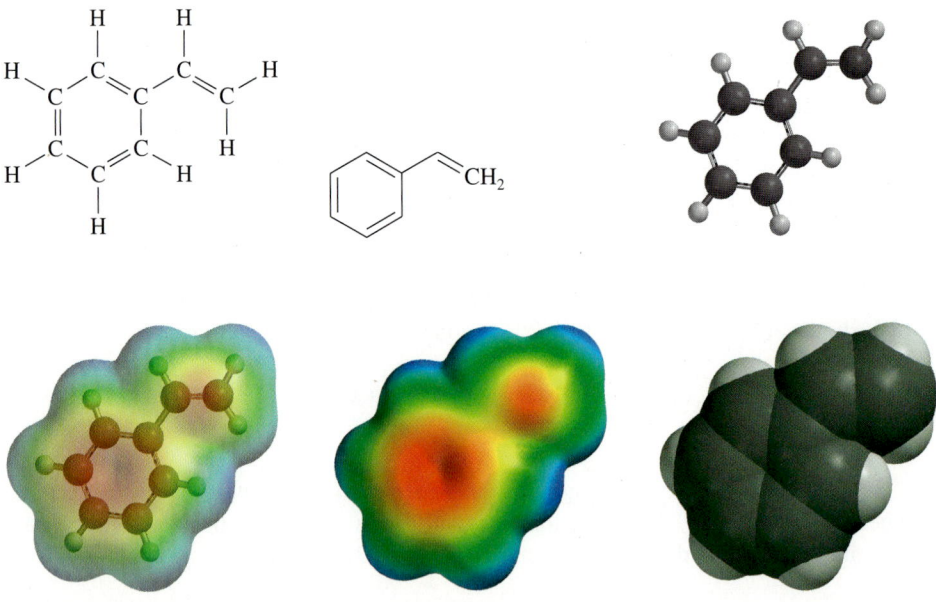

Because molecules are three-dimensional, it is difficult to represent their shapes on paper. Molecular models, which show the three-dimensional spatial arrangement of atoms, are often useful. Several kinds are used (Figure 3.17). For example, we can draw structural formulas with perspectives of depth indicated by wedges and hashes. In the ball-and-stick model, "balls" represent the atoms, and "sticks" represent the bonds. However, while they indicate the relative positions of the atoms in the molecule, ball-and-stick models do not attempt to portray the actual sizes of the atoms relative to the size of the molecule. Space-filling models better represent the actual sizes of atoms and their closeness to each other, from which we can predict the distance and orientation of approach of other molecules when they collide. A disadvantage of space-filling models is that some atoms may be hidden from view, and it can be difficult to appreciate where the nuclei are.

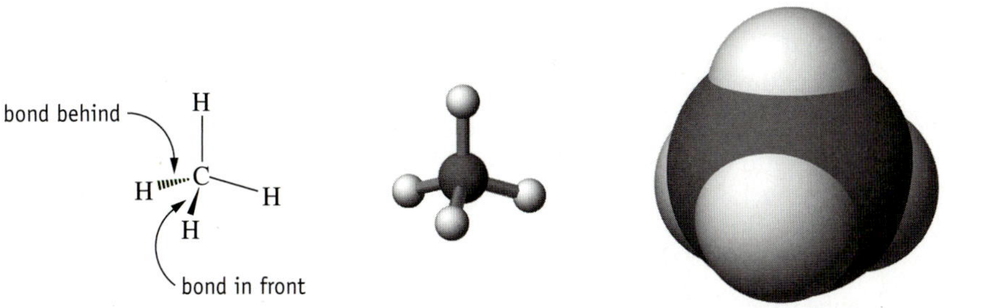

FIGURE 3.17 Ways of depicting a single methane (CH_4) molecule. Pseudo three-dimensional representations of structural, ball- and stick, and space-filling models on the computer screen can be rotated to allow viewing all parts of a molecule.

Even though the representations above tell us about the composition and connectivity in molecules, they are not pictures of molecules. No one has seen molecules, and no one knows exactly what they look like. Each representation is only a model, emphasizing certain features of a reality—whatever the reality is. Each model has deficiencies and there are no exact images of molecules.

> Molecules can be modelled by drawings, computer representations, or objects you can touch. The models indicate connectivity of atoms and relative orientations, but they should not be taken as exact images of the molecules that they represent.

3.10 Connectivity: Evidence from Mass Spectrometry

The high-resolution mass spectrum of chloroethane tells us that its formula is C_2H_5Cl. But how do we know that the atoms are connected together as CH_3CH_2Cl? Look more closely at the unit resolution mass spectrum for chloroethane in Figure 3.15. Note the presence of several peaks at lower mass than the molecular ion peaks at m/z 64 and 66 considered earlier. These are referred to as **fragment ions**. They are formed because the initial molecular ion formed in the electron beam, $[CH_3CH_2Cl]^+$, has so much energy that one or more of its covalent bonds break. The mass spectrometer records signals when the fragment ions arrive at the detector. *Thus, we can learn how atoms are connected in the molecule by studying the fragments formed when the molecule breaks apart.* Let's examine three fragmentation processes in chloroethane more closely (Figure 3.18): (1) the breaking of C—H bonds in the molecular ion leads to the loss of one H atom; (2) the breaking of a C—C bond leads to the loss of a —CH_3 group; and (3) the breaking of a C—Cl bond leads to the loss of a chlorine atom. In each case, fragment ions result that can be seen in the mass spectrum. From analysis of the difference in mass between the fragment ions and the molecular ion, we can learn important information about connectivity.

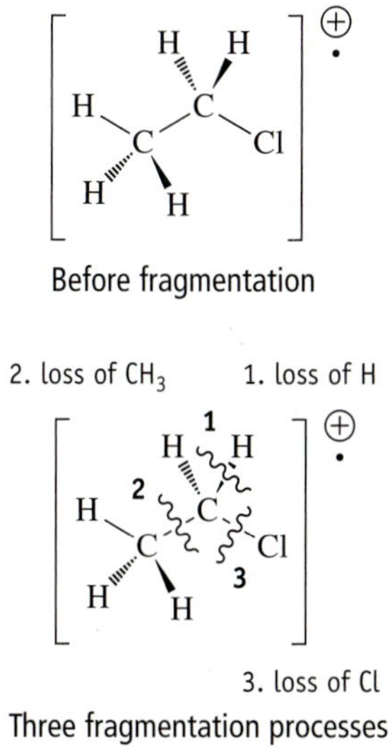

1. Fragmentation with loss of a hydrogen atom produces two peaks in the mass spectrum, at m/z ratios of 63 and 65, which corresponds to the two most abundant molecular ions, CH_3CH_2—$^{35}Cl^+$ ($m/z = 64$) and CH_3CH_2—$^{37}Cl^+$ ($m/z = 66$), each losing the mass of a hydrogen atom. One of two possible ways to lose a H atom would be:

$$[CH_3CH_2-Cl]^+ \longrightarrow H + [CH_3CH-Cl]^+$$

The positive charge is retained on the $[CH_3CH-Cl]^+$ fragment, so it shows up in the mass spectrum. Because the Cl atom is still a part of this fragment, two ions appear, one at $m/z = 63$, corresponding to $[CH_3CH-^{35}Cl]^+$, and one at $m/z = 65$, corresponding to the $[CH_3CH-^{37}Cl]^+$ isotopologue. Because of the relative isotopic abundances of ^{35}Cl and ^{37}Cl, the peak at 63 is about three times as intense as the peak at 65.

2. Fragmentation with loss of a —CH_3 group produces two peaks in the mass spectrum, at m/z ratios of 49 and 51, which corresponds to the two most abundant molecular

Before fragmentation

2. loss of CH_3 1. loss of H

3. loss of Cl

Three fragmentation processes

FIGURE 3.18 Breaking apart of molecular ions of chloroethane give MS fragment ions.

ions, $CH_3CH_2-^{35}Cl^+$ and $[CH_3CH_2-^{37}Cl]^+$, losing 15—the mass of a $-CH_3$ group.

$$[CH_3-CH_2-Cl]^+ \longrightarrow CH_3 + [CH_2-Cl]^+$$

Again, the positive charge is retained on the $[CH_2-Cl]^+$ fragment, so it shows up in the mass spectrum. Because the Cl atom is still a part of this fragment, two ions appear, one at $m/z = 49$, corresponding to $[CH_2-^{35}Cl]^+$, and one at $m/z = 51$, corresponding to $[CH_2-^{37}Cl]^+$. Due to the isotopic abundances of ^{35}Cl and ^{37}Cl, the peak at 49 is about three times as intense as the peak at 51.

3. Fragmentation with loss of a chlorine atom produces only one peak in the mass spectrum, at m/z ratio of 29, which corresponds to the two most abundant molecular ions, $[CH_3CH_2-^{35}Cl]^+$ and $[CH_3CH_2-^{37}Cl]^+$, each losing the mass of a chlorine atom.

$$[CH_3-CH_2-Cl]^+ \longrightarrow Cl + [CH_3CH_2]^+$$

Because the Cl atom (with two isotopes) is now gone from the $[CH_3CH_2Cl]^+$ molecular ion, only one fragment ion peak appears in the spectrum. The ion $[CH_3CH_2-^{35}Cl]^+$ loses its ^{35}Cl atom and the ion $[CH_3CH_2-^{37}Cl]^+$ loses its ^{37}Cl atom; thus, the two possible $[CH_3CH_2]^+$ fragment ions are now identical and have the same m/z value.

It is important to note that only charged particles are detected by the mass spectrometer. Uncharged fragments like H, CH_3, and Cl are not deflected by the magnets, and are not intercepted by the detector.

What about the example of the C_2H_6O sample discussed earlier? Ethanol and dimethyl ether cannot be distinguished by their molecular ion peaks in high-resolution mass spectrometry because they have the same number of C, H, and O atoms. However, the fragment ion patterns in Figure 3.19 have several subtle but important differences.

FIGURE 3.19 Mass spectra, indicating fragment ions of (a) ethanol and (b) dimethyl ether.

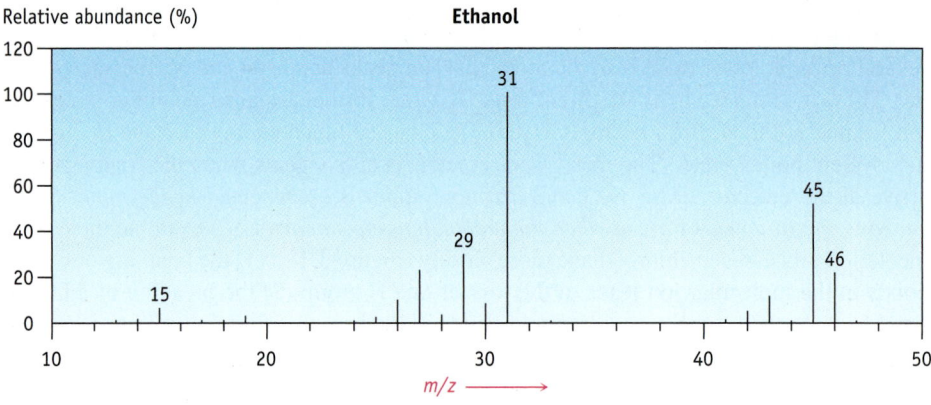

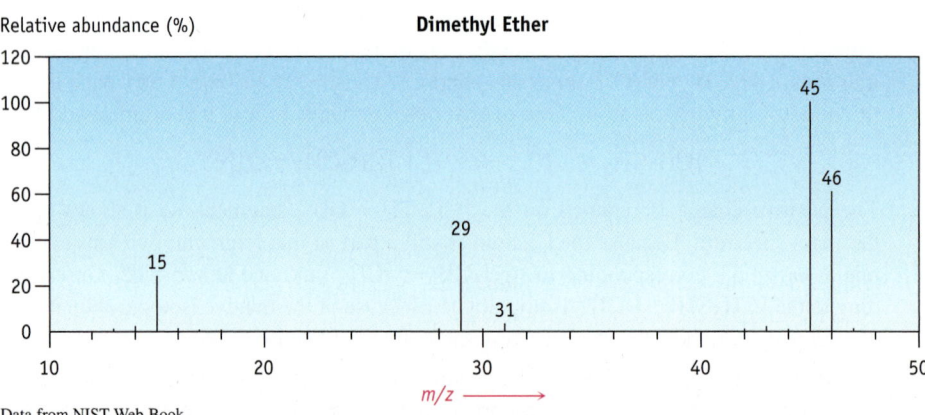

Data from NIST Web Book

For a detailed explanation of the fragmentation patterns in the mass spectra of ethanol and dimethyl ether, see e3.13.

> Fragment ions, formed in a mass spectrometer by the breaking apart of the molecular ion, give information about connectivity of molecular compounds.

3.11 Connectivity: Evidence from IR Spectroscopy

Spectroscopy and the Electromagnetic Spectrum

Spectroscopy is a method of finding out about the structure of molecules that depends on the interaction of molecular compounds with radiation. When a molecular compound is exposed to electromagnetic radiation with a range of frequencies, it absorbs some or all of certain frequencies, but transmits radiations of other frequencies. A plot of radiation energy versus the percentage of radiation transmitted is called an **absorption spectrum**.

The absorption of electromagnetic radiation causes different effects on molecular compounds, depending on the energy of the radiation and the molecular structure of the substance. As seen in e3.14, applying microwave radiation of the right energy to a molecular substance such as CF_2Cl_2 will cause tumbling of its molecules, the process that heats up the water molecules in your soup in a microwave oven. Ultraviolet radiation is high enough in energy to cause some covalent bonds to rupture, the process that leads to stratospheric ozone formation and depletion [>>Section 22.1]. Applying infrared radiation at certain energies will cause CF_2Cl_2 molecules to vibrate in ways that are characteristic of the connectivity pattern of its atoms; C—F and C—Cl covalent bonds stretch and bond angles bend.

How do microwave, ultraviolet, and infrared radiation differ? Along with visible light, X-rays, microwaves, and radio waves, they are all forms of electromagnetic radiation that differ only in their wavelengths and frequencies of vibration. Collectively, they make up the electromagnetic spectrum (Figure 3.20).

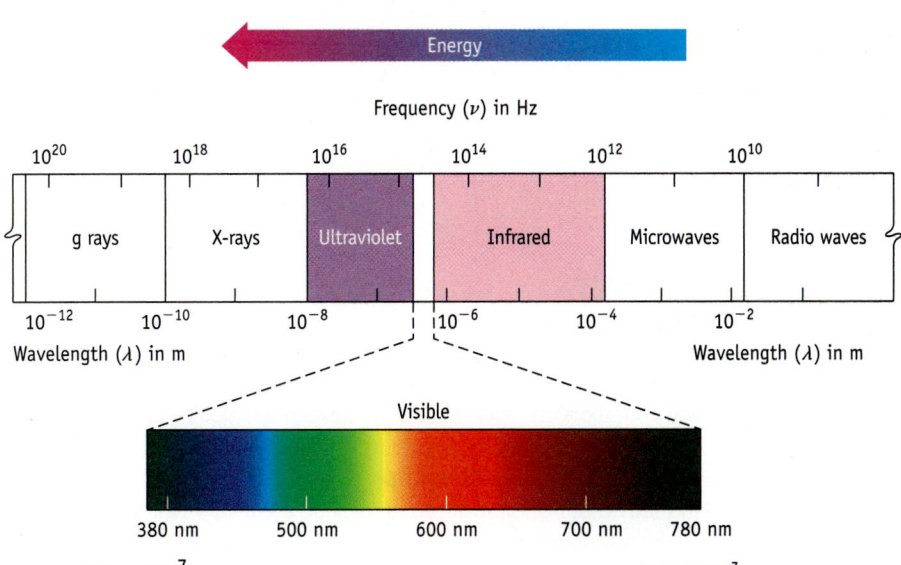

FIGURE 3.20 The electromagnetic spectrum. Radiation at the radio wave end (low frequency, ν, and long wavelength, λ) has the lowest energy, and that at the γ-ray end (high frequency and short wavelength) is the most energetic. Visible light (enlarged portion) is a very small part of the entire spectrum of radiation, and includes radiation whose wavelengths range from about 3.8×10^{-7} to 7.8×10^{-7} m (380 nm to 780 nm). Infrared radiation has slightly longer wavelengths (lower energies) than that of visible light.

The x-axis on an absorption spectrum measures the energy of the electromagnetic radiation and the y-axis the percentage of radiation transmitted or absorbed at each energy. It is important to distinguish among the terms *frequency, wavelength,* and *amplitude* (Figure 3.21) of any electromagnetic radiation.

The **frequency, ν** (Greek nu), is the number of wave peaks that pass by a fixed point per unit time, for which the unit of measurement is the *hertz* (symbol Hz) or reciprocal seconds (s^{-1}). For waveforms with smaller wavelengths, more wave peaks will pass by a fixed point in a given time period—the frequency will be higher.

FIGURE 3.21 Waveform characteristics. (a) Wavelength (λ, Greek letter lambda) is the distance between two successive wave maxima. Amplitude is the height of the waveform, measured from the centre. (b) and (c) What we perceive as different kinds of electromagnetic radiation are simply waveforms with different wavelengths and frequencies.

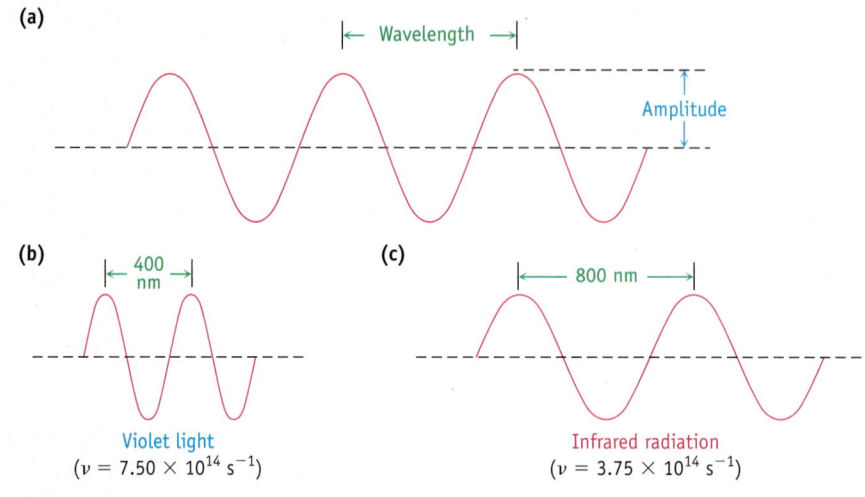

Think about It

e3.15 Work through tutorials to relate wavelength with frequency and energy.

In an **infrared absorption spectrum**, the property of electromagnetic waves most commonly expressed on the x-axis is the **wavenumber, $\tilde{\nu}$**. The wavenumber is the number of repeating units of a wave in a unit of space (rather than time)—and is the reciprocal of the wavelength expressed in centimetres. The units of wavenumber used in IR are reciprocal centimetres (cm^{-1}):

$$\text{Wavenumber } \tilde{\nu}\left(cm^{-1}\right) = \frac{1}{\lambda\left(cm\right)}$$

The wavenumber boundaries of a typical infrared spectrum (Figure 3.22) extend from 4000 to 400 cm^{-1}, which corresponds to a wavelength range of 2.5×10^{-4} to 2.5×10^{-3} cm.

FIGURE 3.22 The electromagnetic spectrum.

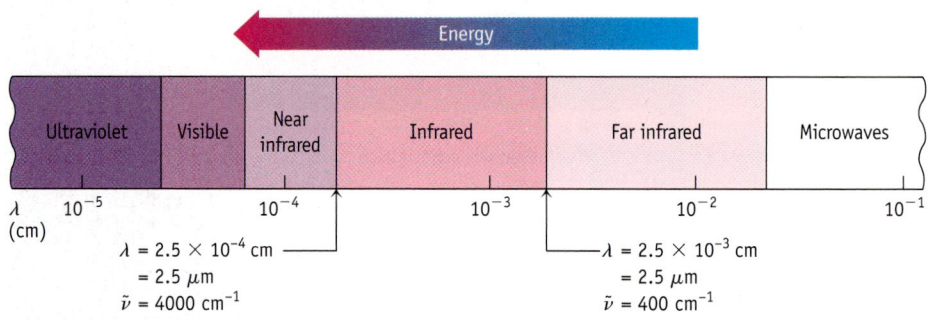

Functional Groups

What information does the infrared absorption spectrum of the molecular compound cocaine in Figure 3.1a reveal about connectivity of atoms within cocaine molecules?

The radiation wavenumbers at which infrared absorption peaks appear give us an experimental basis to identify commonly recurring connectivity patterns in portions of molecular compounds, called **functional groups**. The C=O groups in cocaine, for example, are found in many other organic compounds, and are called **carbonyl functional groups**. Any molecular compound with a carbonyl group will show an IR absorption at approximately the same energy as cocaine's carbonyl group.

While there are 18 million known carbon-based molecular compounds, each with its own physical and chemical properties, years of experience have shown us that these compounds can be classified into families according to characteristics of their functional groups. Compounds with the same functional groups often have similar chemical reactivity. Instead of needing to consider 18 million compounds and 18 million examples of reactivity, chemists generalize to a few dozen families of compounds based on their functional groups. The chemistry of compounds in each family are similar, and therefore reasonably predictable. Common examples from organic chemistry are given in Table 3.5.

Identification of Functional Groups by IR Spectroscopy

Why does a molecule with certain functional groups absorb some wavenumbers of infrared energy but not others? All molecules have a certain amount of energy, and this causes atoms to vibrate. Vibrations of atoms can be visualized as stretching or twisting of bonds, of which some common types, called *vibrational modes*, are shown in Figure 3.23.

The energy of a vibrational mode is not continuously variable, it is *quantized*; that is, the atoms can vibrate only at certain frequencies. For example, in the antisymmetric C—H bond stretching mode in Figure 3.23, the atoms vibrate at a particular frequency. Some C—H bonds stretch while others compress at the same time as if there were springs connecting the carbon atom to the hydrogen atoms.

When a molecular substance is irradiated with electromagnetic radiation, *energy is absorbed if the energy of the radiation matches the energy of a vibrational mode of its molecules*. The result of energy absorption is increased amplitude for the vibrational mode; in other words, the "spring" connecting the two atoms stretches and compresses a bit further.

> A car can be driven at any speed that we decide. If car speed were quantized, it might be able to travel only at, say, 20, 40, 60, 80, or 100 km/h, and not at any of the speeds in between these values.

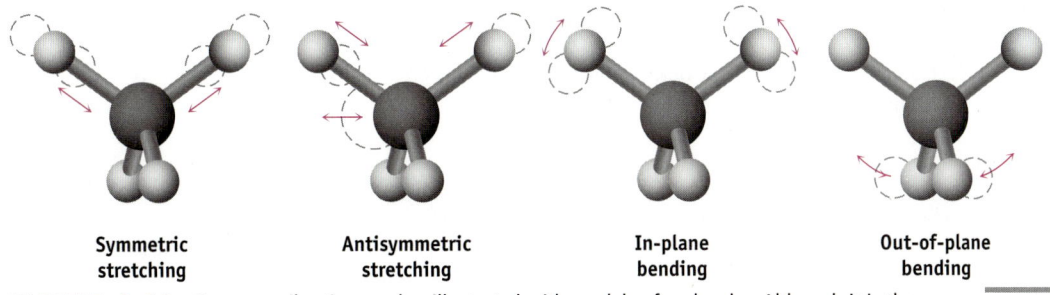

| Symmetric stretching | Antisymmetric stretching | In-plane bending | Out-of-plane bending |

FIGURE 3.23 Common vibration modes, illustrated with models of molecules. Although it is the atoms that vibrate, the movement is usually classified according to an imagined stretching or bending of bonds that results from the atoms vibrating.

Molecular Modelling

e3.16 Look at the stretching and bending vibrational modes of the CH_2Cl_2 molecule.

Every type of functional group in a molecule has vibrational modes with frequencies more or less unaffected by the rest of the molecule. These energies of vibration are different from the vibrational energies in other functional groups. So, the absorption peaks in the infrared absorption spectrum of a substance can tell us (with experience) which functional groups are present in its molecules.

In e3.17, we interpret the infrared spectrum of cocaine with animations showing how cocaine molecules are vibrationally excited by absorbing infrared radiation at different wavenumbers. You can see, for example, that the two intense absorbances at IR radiation of 1700 and 1736 cm^{-1} correspond to symmetric and antisymmetric stretching vibrations of the two C=O double bonds in ester functional groups. The absorbances around 3000 cm^{-1} correspond to stretching vibrations of the C—H single bonds. One functional group, the six-membered benzene ring, can be identified by a characteristic stretching vibration of the C—H bonds on the ring at 3005 cm^{-1} along with the out-of-plane bending vibration of the C—H bonds on the ring at 704 cm^{-1}.

Below 1500 cm^{-1}, most molecules show a complex pattern of absorption bands that correspond to bending and twisting vibrations in the compound. Most absorption peaks in this region are unique characteristics of molecules of each substance, rather than a particular functional group, and are often referred to as the *"fingerprint region"* of the infrared spectrum. The pattern of peaks in this region arise from the unique connectivity of each molecule and can be used to identify a particular substance or distinguish it from another one whose molecules have the same functional group.

The full interpretation of an IR spectrum is difficult because most organic molecules are so large that they have dozens of different bond stretching and bending motions, and so an IR spectrum usually contains dozens of absorption peaks. Fortunately, we don't need to interpret an IR spectrum fully to get useful information because *functional groups have*

> The ester functional group always contains a carbon atom connected to two oxygen atoms, one with a single bond and one with a double bond. The oxygen atom connected with a single bond is connected to a second carbon atom. Many esters show a strong C=O stretching infrared absorbance at 1735 cm^{-1}.

Ester

Think about It

e3.17 Examine the interactive IR spectrum of cocaine.

characteristic IR absorptions that don't change much from one compound to another. For example, many experimental observations show that absorption due to the C═O bond stretching vibration of a **ketone** is almost always in the range 1680 to 1750 cm^{-1}, and that due to the O—H stretching vibration of an **alcohol** is almost always in the range 3400 to 3650 cm^{-1} (Figure 3.24). The C═C stretching vibration of an **alkene** almost always occurs in the range 1640 to 1680 cm^{-1}. By learning to recognize where characteristic absorptions by functional groups occur, it is possible to get information about the connectivity patterns within molecules from IR spectra.

This can be illustrated through the IR spectra of cyclohexanol and cyclohexanone (Figure 3.24 and e3.18). Although both spectra contain many peaks, the characteristic absorptions of the different functional groups allow the compounds to be easily distinguished.

Of course it is also possible to obtain structural information from an IR spectrum by noticing which absorptions are *not* present. For example, if the spectrum of an unknown compound does *not* have an absorption near 3400 cm^{-1}, the unknown is not an alcohol; and if the spectrum does not have an absorption near 1715 cm^{-1}, the unknown is not a typical ketone.

Think about It

e3.18 Examine interactive IR spectra of other ketones and alcohols.

Some important functional groups are listed in Table 3.5, along with their characteristic infrared vibrational energies. e3.19 shows you how you can identify common functional groups with the help of animated IR spectra. With some practice, you can soon identify common functional groups in unknown compounds by looking for patterns in their IR spectra.

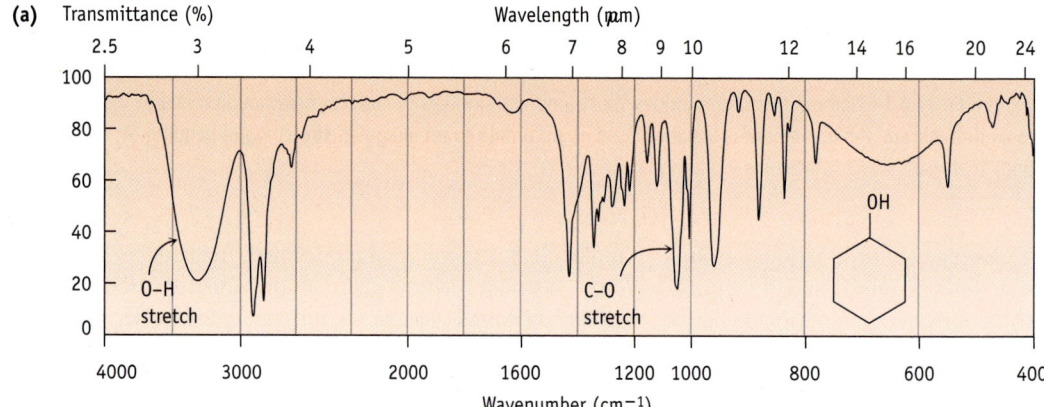

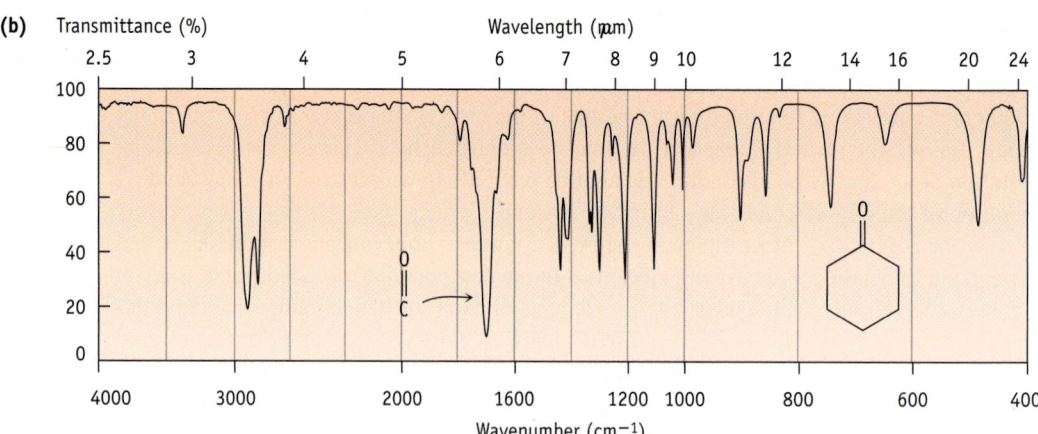

FIGURE 3.24 Infrared spectra of (a) cyclohexanol and (b) cyclohexanone. Cyclohexanol shows a characteristic alcohol O—H bond stretching absorption at 3300 cm^{-1} and a C—O stretching absorption at 1060 cm^{-1}; cyclohexanone shows a characteristic ketone C═O stretching absorption at 1715 cm^{-1}.

TABLE 3.5 Common Functional Group Names and Their Characteristic IR Absorptions

Functional Group	Structure	Type of Absorption	Wavenumber (cm^{-1})	Intensity
Hydrocarbons				
alkane		C—H stretch	2850–2980	medium–strong
alkene		=C—H stretch	3020–3100	medium
		C=C stretch	1640–1670	weak–medium, sharp
aromatic ring		C—H stretch	3000–3100	medium
		ring vibrations	1585–1600, 1400–1500	strong, sharp
		out of plane C—H bending	675–900	strong
alkyne	—C≡C—H	≡C—H stretch	3270–3340	strong, sharp
		C≡C stretch	2100–2260	medium, sharp
Carbonyl compounds				
carbonyl group (general class of compounds)		C=O stretch (note 1)	1540–1870	strong, sharp
aldehyde		C=O stretch	1720–1740	strong
		C—H stretch (attached to C=O)	2700–2850	medium, 2 peaks
ketone		C=O stretch	1705–1725	strong
carboxylic acid		C=O stretch	1700–1725	strong
		O—H stretch	2500–3300	strong, very broad
ester		C=O stretch	1730–1750	strong
		C—O stretch	1000–1300	strong, 2 bands
amide		C=O stretch	1640–1700	strong
		N—H stretch	3400–3500	strong
		C—N—H bend	1590–1655	medium
acid chloride		C=O stretch	1785–1815	strong
Other oxygen, nitrogen, and halogen compounds				
alcohol		O—H stretch	3200–3550	strong, broad
		C—O stretch	1000–1260	strong, sharp
ether		C—O stretch	1085–1150	medium

(Continued)

Functional Group	Structure	Type of Absorption	Wavenumber (cm⁻¹)	Intensity
amine		N—H stretch	3250–3400	medium
		C—N stretch	1020–1250	weak–medium
		C—N—H bend	1515–1650	medium–strong, sharp
nitrile	—C≡N	C≡N stretch	2210–2260	medium, very sharp
alkyl halide		C—F stretch (alkyl fluoride)	1000–1400	strong
		C—Cl stretch (alkyl chloride)	550–850	strong
	(X = F, Cl, Br, or I)	C—Br stretch (alkyl bromide)	515–690	strong
		C—I stretch (alkyl iodide)	500–600	strong
thiol		S—H stretch	2550–2600	weak
sulfide		C—S stretch	600–700	weak, variable
phosphate		P=O stretch	1250–1300	strong

Note 1: If the C=O group is conjugated with a C=C (i.e., C=C—C=O), the observed C=O absorption is shifted to lower wavenumbers by about 30 cm⁻¹

Think about It

e3.19 Examine interactive IR spectra of compounds with the functional groups in Table 3.5.

The symbol R is commonly used as shorthand notation to designate an unspecified hydrocarbon or alkyl group

You will find it easier to remember the positions of various key IR absorptions by dividing the infrared range from 4000 to 200 cm⁻¹ into four parts (Figure 3.25):

- The region from 4000 to 2500 cm⁻¹ corresponds to N—H, C—H, and O—H bond stretching motions. Both N—H and O—H bonds absorb in the 3300 to 3600 cm⁻¹ range, whereas C—H bond stretching occurs near 3000 cm⁻¹. Since almost all organic compounds have C—H bonds, almost all IR spectra have an intense absorption in this region. The exact wavenumber of absorbed IR radiation causing the C—H stretching vibration is important to note, as alkanes absorb just below 3000 cm⁻¹, while both alkene and alkyne stretching vibrations come above 3000 cm⁻¹.
- The region from 2500 to 2000 cm⁻¹ is where triple-bond stretching occurs. Both nitriles (RC≡N) and alkynes (RC≡CR′) absorb here.

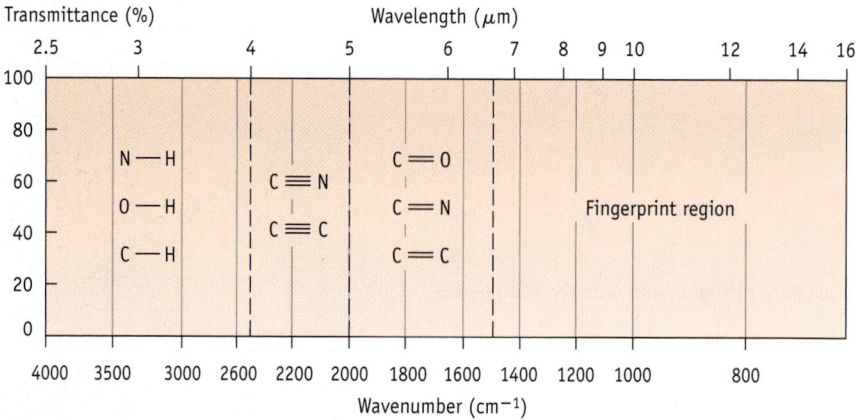

FIGURE 3.25 Single-bond, triple bond, double bond, and fingerprint absorption regions in the infrared spectrum.

- The region from 2000 to 1500 cm^{-1} is where $C=O$, $C=N$, and $C=C$ bonds absorb. Carbonyl groups generally absorb from 1670 to 1780 cm^{-1}, and alkene stretching normally occurs in the narrow range from 1640 to 1680 cm^{-1}. The exact position of a $C=O$ absorption is often diagnostic of the exact kind of carbonyl group in the molecule. Esters usually absorb at 1735 cm^{-1}, aldehydes at 1725 cm^{-1}, and open-chain ketones at 1715 cm^{-1}.
- The region below 1500 cm^{-1} is the so-called *fingerprint region*. A large number of absorptions due to various $C-O$, $C-C$, and $C-N$ single-bond vibrations occur here, forming a unique pattern that acts as an identifying "fingerprint" of each organic molecule.

Forensic scientists make good use of the unique IR "fingerprint" for each compound. Not only are libraries of drugs such as cocaine available for police authorities, but in some countries auto manufacturers must provide the IR spectra of the different layers of auto paints in their vehicles. Chemistry can then help police identify the model and age of a vehicle responsible for a hit-and-run accident. This is done by comparing the IR spectra from layers of a tiny paint fragment retrieved from the clothing of a pedestrian victim with spectra on file.

WORKED EXAMPLE 3.7—INTERPRETING IR SPECTRA

Refer to Table 3.5 and e3.19 and make educated guesses about the functional groups that cause the following IR absorptions:
(a) 1735 cm^{-1}
(b) 3500 cm^{-1}

Solution

(a) An absorption at 1735 cm^{-1} is in the carbonyl-group region of the IR spectrum, probably an ester.
(b) An absorption at 3500 cm^{-1} is in the $-OH$ (alcohol) region.

WORKED EXAMPLE 3.8—INTERPRETING IR SPECTRA

Acetone (CH_3COCH_3) and prop-2-en-1-ol ($H_2C=CHCH_2OH$) are isomers. How could you distinguish them by IR spectroscopy?

Solution

Identify the functional groups in each molecule, and refer to Table 3.5, which shows that acetone has a strong $C=O$ absorption at 1715 cm^{-1}, while prop-2-en-1-ol has an $-OH$ absorption at 3500 cm^{-1} and a $C=C$ absorption at 1660 cm^{-1}.

EXERCISE 3.25—INTERPRETING IR SPECTRA

What functional groups might molecules contain if they show IR absorptions at the following wavenumbers?
(a) 1715 cm^{-1}
(b) 1540 cm^{-1}
(c) 2210 cm^{-1}
(d) 1720 and 2500–3100 cm^{-1}
(e) 3500 and 1735 cm^{-1}

When the energy of infrared radiation matches the energy of a vibrational mode of molecules, absorption of energy takes place. Functional groups have characteristic IR absorptions that don't change much from one compound to another.

3.12 New Materials: Chemistry beyond the Molecule

Not all substances fit neatly with the four models described in this chapter (molecular substances, covalent network substances, ionic substances, and metallic substances). The blurring and extension of the boundaries that describe substances is evident in new materials that cross the borders between chemistry, biology, and technology. In studying new materials, we observe fascinating behaviour, such as molecules assembling themselves into large complexes called *supramolecular assemblies*. The intramolecular forces that hold together and organize molecules in these supramolecular assemblies are weak relative to covalent bonds. Understanding these "non-covalent bonding interactions" is the key to making sense of these new materials.

New language is needed to describe this area of research, including terms such as *supra*molecular chemistry, "chemistry *beyond* the molecule," and host–guest chemistry. A **host–guest complex** is a unique structural relationship between a large molecule or cluster of molecules (the host) that has a suitable overall shape and site to bind through non-covalent interactions to another molecule (the guest). A good example is methane clathrate hydrates in Section 4.1. The interplay of non-covalent forces that hold together molecules or parts of molecules in host–guest complexes is described with terms such as ion-ion, ion-dipole, dipole-dipole, π-π stacking, and hydrogen bonding interactions. We consider these interactions more fully in Chapters 6 and 20.

New interdisciplinary fields have sprung up to create and study these new materials. Edmonton is home to Canada's National Institute of Nanotechnology, Brisbane to the Australian Research Council Centre for Functional Nanomaterials, and Pretoria to the South African National Centre for Nano-structured Materials. At these and other centres, interdisciplinary teams use new materials to create artificial membranes, build electronic "noses" to detect odours, miniaturize electronic devices, build machines and motors that consist of single molecules or clusters of molecules, and engineer crystals with desired properties.

SUMMARY

Key Concepts

- Most substances can be placed into one of four categories (covalent network, ionic, metallic, or molecular), based on their characteristic properties. Chemists explain those properties with different **models** of molecular structure (Section 3.2). Some materials, however, don't neatly fit into these categories. (Section 3.12)
- **Covalent network substances** are modelled as three-dimensional networks of atoms, in which each atom is bound to other atoms, forming a network of covalent bonds. This model can rationalize their high melting points, lack of electrical conductivity and hardness. (Section 3.3)
- **Ionic compounds** are pictured as made of an **ionic lattice** of positively charged ions, called **cations**, and negatively charged ions, called **anions**. The ionic lattice model accounts for their high melting points, brittleness, hardness, lack of electrical conductivity when solid, and their ability to conduct electricity when molten or in solution. Ions are different **chemical species**, with very different characteristics from the atoms from which they are derived. (Section 3.4)
- **Metals** are modelled as a lattice of positive ions within a mobile "sea" of electrons that acts as a "glue." This model can make sense of metals' high density, their ability to conduct electricity and heat, their malleability and ductility, and their relatively high melting points. (Section 3.5)

- **Molecular substances** can be identified by their relatively low melting points and boiling points; they do not conduct electricity, and their solids are relatively soft. Our structural model to explain these observable properties shows these substances made of many molecules. Strong **intramolecular forces**, called **covalent bonds** hold atoms together within molecules. Molecules are attracted to each other through **intermolecular forces**. (Section 3.6)
- In a covalent bond, the bound atoms share electron matter, and the atoms are held together by attraction of each nucleus for the shared electron matter. The sharing of two electrons between atoms in a molecule is called a **single covalent bond**; the sharing of four electrons is called a **double bond**. (Section 3.7)
- Powerful experimental probes of molecular structure have revolutionized our understanding of modern chemistry. Complementary techniques are used to determine how many atoms of each type a molecular compound contains (composition and molecular formula), the pattern of connectivity between the atoms within a molecule (covalent bonds), and the arrangement of atoms in three-dimensional space. **High-resolution mass spectrometry** analysis of a pure molecular substance can give an accurate value of the relative molecular mass of **isotopologues**—molecules with various isotopic compositions. From this, we can define the formula and composition of the compound. (Section 3.8)
- Information about the connectivity sequence of atoms in molecules is obtained by two important techniques:
  - Examination of the pattern of **fragment ions** formed in a mass spectrometer by the breaking apart of the **molecular ion**. (Section 3.10)
  - The wavenumbers of infrared radiation absorbed by a substance are characteristic of the presence of certain **functional groups** in molecules of that substance. Energy is absorbed by a molecular substance at the energy of infrared radiation that matches the energy of a vibrational mode of its molecules. Functional groups have characteristic IR absorptions that don't change much from one compound to another. (Section 3.11)

REVIEW QUESTIONS

Section 3.2: Classifying Substances by Properties: An Overview

3.26 Classify each of the following as either ionic, metallic, or covalent:
  (a) A white solid that melts at 772 °C to form a colourless liquid. The solid does not conduct electricity, but it does when melted.
  (b) A yellow solid that melts at 119 °C to form a clear yellow liquid. Both solid and liquid are poor conductors of electricity.
  (c) A dark, shiny solid that sublimes to form a purple vapour. It is a poor conductor of electricity and heat.
  (d) A silvery solid that melts at 98 °C to form a silvery liquid. Both solid and liquid are good conductors of electricity.

3.27 Classify the following as either covalent network, ionic, metallic, or molecular:
  (a) iodine, $I_2(s)$
  (b) silicon dioxide, $SiO_2(s)$
  (c) sulfur dioxide, $SO_2(g)$
  (d) aluminium oxide, $Al_2O_3(s)$

Section 3.3: Covalent Network Substances

3.28 Compare the number of silicon atoms with the number of carbon atoms in the crystal lattice of silicon carbide (SiC).

Section 3.4: Ionic Substances

3.29 What charges are most commonly observed for monatomic ions of the following elements?
  (a) selenium   (c) iron
  (b) fluorine   (d) nitrogen

3.30 When a potassium atom becomes a monatomic ion, how many electrons does it lose or gain? What noble gas atom has the same number of electrons as a potassium ion?

3.31 Compare the total number of electrons with the total number of protons in the following:
  (a) an ammonium ion
  (b) a phosphate ion
  (c) a dihydrogenphosphate ion

3.32 Write formulas for ionic compounds composed of an aluminum cation and each of the following anions: (a) fluoride ion, (b) sulfide ion, and (c) nitrate ion.

3.33 Write the formulas for the four ionic compounds that can be made by combining each of the cations Na^+ and Ba^{2+} with the anions CO_3^{2-} and I^-. Name each of the compounds.

3.34 What is the ratio of the number of cations to the number of anions in the ionic lattice composed of the following:
(a) iron(II) ions and nitrate ions
(b) iron(III) ions and nitrate ions
(c) sodium ions and sulfate ions
(d) magnesium ions and phosphate ions
(e) ammonium ions and chloride ions

3.35 Name each of the following compounds, and state which ones are best described as ionic:
(a) ClF_3 (f) OF_2
(b) NCl_3 (g) KI
(c) $SrSO_4$ (h) Al_2S_3
(d) $Ca(NO_3)_2$ (i) PCl_3
(e) XeF_4 (j) K_3PO_4

3.36 Write the formula for each of the following compounds, and state which ones are best described as ionic:
(a) sodium hypochlorite
(b) boron triiodide
(c) aluminium perchlorate
(d) calcium acetate
(e) potassium permanganate
(f) ammonium sulfite
(g) potassium dihydrogen phosphate
(h) disulfur dichloride
(i) chlorine trifluoride
(j) phosphorus trifluoride

Section 3.5: Metals and Metallic Substances

3.37 Which of the following is the best description of metallic bonding?
(a) Electrons are shared between two atoms.
(b) Molecules attract each other.
(c) Electrons are set free.
(d) Electrons are transferred from one atom to another.

Section 3.6: Molecular Substances

3.38 Explain why the physical properties of the noble gases (Group 18) are similar to those of molecular substances, rather than to those of metallic, ionic, or covalent network substances.

3.39 Explain why the chemical reactivity of oxygen depends partly on the strength of covalent bonds, but its boiling point does not.

Section 3.8: Composition and Formula by Mass Spectrometry

3.40 Using Table 3.4, sketch the molecular ion peaks in the mass spectrum of the fumigant methyl bromide (CH_3Br). Which isotopologues would you expect to show most prominently in the mass spectrum? Relate your answer to the relative heights of peaks.

3.41 The most prominent molecular ion formed when a sample is analyzed by a high-resolution mass spectrometer experiment is observed at $m/z = 28.03125$. Assuming the compound contains only atoms found in Table 3.4, determine its molecular formula. If this sample were analyzed by a unit resolution mass spectrometer, the molecular ion would be observed at $m/z = 28$. What other molecular formulas would you have needed to consider as reasonable?

3.42 Hair analysis by mass spectrometry can provide evidence of drug use. Calculate the expected m/z ratio for the molecular ion peaks of the most important isotopologue of each of the following drugs. Could you distinguish them by mass spectrometry? Start by filling in any hydrogen atoms implied by the representations, but not explicitly drawn.

Methamphetamine

Cocaine

Phencyclidine

Tetrahydrocannabinol (THC)

Ecstasy

Section 3.11: Connectivity: Evidence from Infrared Spectroscopy

You may find it helpful to consult e3.19 for Questions 3.43–3.45.

3.43   At what wavenumbers in the infrared spectrum would you expect each of the following compounds to absorb?

(a)

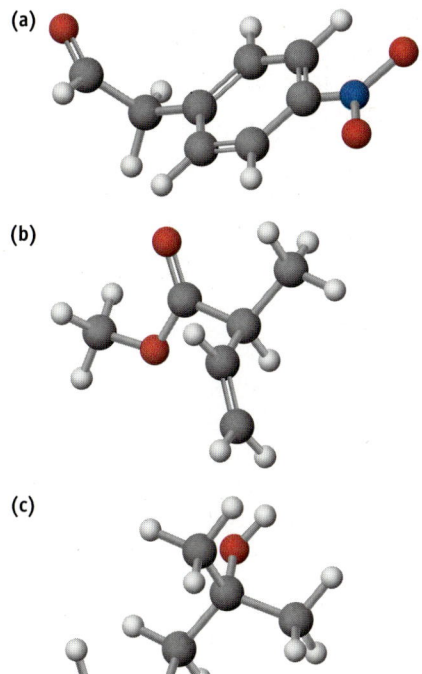

(b)

(c)

3.44   What kinds of functional groups might compounds contain if they show the following IR absorptions?
(a) 1670 cm^{-1}
(b) 1735 cm^{-1}
(c) 1540 cm^{-1}
(d) 1715 cm^{-1} and 2500–3100 cm^{-1} (broad)

3.45   Identify the functional groups in each of the following compounds, and then determine the approximate wavenumbers where they would show IR absorptions.

(a)   CO$_2$H

(b)   CO$_2$CH$_3$

(c)   C≡N   HO

(d)   O

(e) $$CH_3CCH_2CH_2COCH_3$$
           (with two C=O groups shown)

3.46   If C—O single-bond stretching occurs at 1000 cm^{-1} and C=O double-bond stretching occurs at 1700 cm^{-1}, which of the two requires more energy? How does your answer correlate with the relative strengths of single and double bonds?

3.47   Calculate the molar mass of each of the following compounds:
(a) Fe$_2$O$_3$, iron(III) oxide
(b) BCl$_3$, boron trichloride
(c) C$_6$H$_8$O$_6$, ascorbic acid (Vitamin C)

3.48   Assume you have 0.123 mol of each of the following compounds. What mass of each is present?
(a) C$_{14}$H$_{10}$O$_4$, benzoyl peroxide, used in acne medications
(b) Pt(NH$_3$)$_2$Cl$_2$, cisplatin, a cancer chemotherapy agent

3.49   Acetone, (CH$_3$)$_2$CO, is an important industrial solvent. If 1260 million kg of this organic compound is produced annually, what amount (in mol) is produced?

3.50   An Alka-Seltzer tablet contains 324 mg of aspirin (C$_9$H$_8$O$_4$), 1904 mg of NaHCO$_3$, and 1000 mg of citric acid (C$_6$H$_8$O$_7$). (The last two compounds react with each other to provide the "fizz," bubbles of CO$_2$, when the tablet is put into water.)
(a) Calculate the amount (in mol) of each substance in the tablet.
(b) If you take one tablet, how many molecules of aspirin are you consuming?

3.51   Which of the following samples has the largest number of ions?
(a) 1.0 g of BeCl$_2$
(b) 1.0 g of MgCl$_2$
(c) 1.0 g of CaS
(d) 1.0 g of SrCO$_3$
(e) 1.0 g of BaSO$_4$

3.52   A drop of water has a volume of about 0.05 mL. How many molecules of water are in a drop of water? (Assume water has a density of 1.00 g cm^{-3}.)

3.53   Capsaicin, the compound that gives the hot taste to chilli peppers, has the formula C$_{18}$H$_{27}$NO$_3$.
(a) Calculate its molar mass.
(b) If you eat 55 mg of capsaicin, what amount (in mol) have you consumed?
(c) What mass of carbon (in mg) is there in 55 mg of capsaicin?

3.54   Iron pyrite, often called "fool's gold," has the formula FeS$_2$. If you could convert 15.8 kg of iron pyrite to iron metal, what mass of the metal would you obtain?

3.55   A piece of nickel foil, 0.550 mm thick and 1.25 cm square, is allowed to react with fluorine (F$_2$) to give a nickel fluoride.
(a) How many moles of nickel foil were used? (The density of nickel is 8.902 g cm^{-3}.)
(b) If you isolate 1.261 g of the nickel fluoride, what is its formula?
(c) What is its complete name?

SUMMARY AND CONCEPTUAL QUESTIONS

3.56 You suspect that a solid is best classified as an ionic substance. What experimental tests would you do to confirm, or otherwise, your suspicion?

3.57 Why is the melting point of chlorine much lower than that of sodium chloride?

3.58 Explain why:
(a) sodium, Na(s), conducts electricity
(b) iodine, I_2(s), does not conduct electricity
(c) sodium iodide, NaI(s), does not conduct electricity
(d) molten sodium iodide and aqueous solutions of sodium iodide conduct electricity

3.59 Account for the differences in the melting points of the following:
(a) silicon dioxide (SiO_2) and carbon dioxide (CO_2)
(b) sodium sulfide (Na_2S) and hydrogen sulfide (H_2S)

3.60 Describe the kinds of attractive forces that must be overcome for each of the following to occur:
(a) sugar dissolves in water
(b) diamond is crushed
(c) water melts
(d) octane boils
(e) sodium chloride dissolves in water
(f) copper wire is stretched till it breaks
(g) the reaction represented by $Br_2(g) \longrightarrow 2\ Br(g)$

3.61 What information is provided by the following formulas? You will need to decide which category of compound each of the substances is.
(a) $CaCl_2$ (d) SiC
(b) $CaCO_3$ (e) HCl
(c) NH_3

3.62 Which of the following pairs of elements are likely to form ionic compounds when allowed to react with each other? Write appropriate formulas for the ionic compounds you expect to form, and give the name of each.
(a) chlorine and bromine
(b) phosphorus and bromine
(c) lithium and sulfur
(d) indium and oxygen
(e) sodium and argon
(f) sulfur and bromine
(g) calcium and fluorine

3.63 An ionic compound can dissolve in water because the cations and anions are attracted to water molecules. Which of the following cations should be most strongly attracted to water: Na^+, Mg^{2+}, or Al^{3+}? Explain briefly.

3.64 Pentan-2-one and pentan-3-one are isomers—both are molecules with five carbon atoms in a row and contain ketone carbonyl functional groups. In pentan-2-one, the carbonyl group is on the second carbon atom in the chain, and in pentan-3-one, the carbonyl group is on the middle carbon (carbon atom number 3).
(a) Draw structures of the two isomers and label them.
(b) The IR spectra of these two isomers are quite similar. What absorption peaks would you look for to identify a ketone?
(c) What would you expect the most abundant isotopologue of each of these isomers to be? Calculate the relative mass of that isotopologue, using data in Table 3.4.
(d) The low resolution mass spectra of the two ketones are shown below. Organic compounds containing the carbonyl functional group often show fragment ions that result from loss of a group attached to the carbonyl carbon atom. Which spectrum corresponds to pentan-2-one? To pentan-3-one?

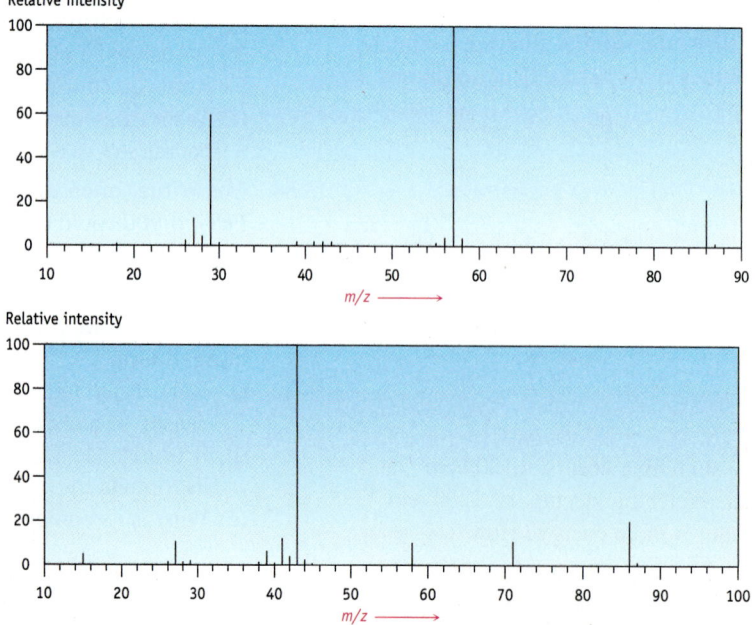

3.65 Analysis of a mixture of steroids by an expert revealed that the sample contained, among other things, testosterone, but not the banned drug clostebol.
(a) Redraw a molecule of clostebol and circle and name all the functional groups.
(b) Are the data in the following mass spectrum consistent with the expert's conclusion?

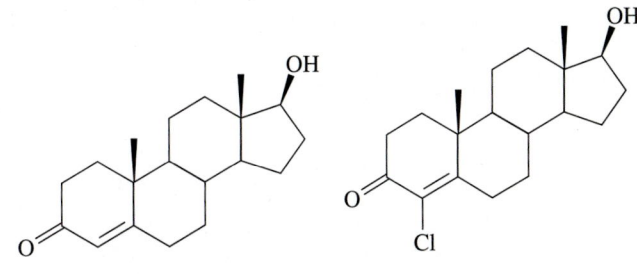

Testosterone **Clostebol**

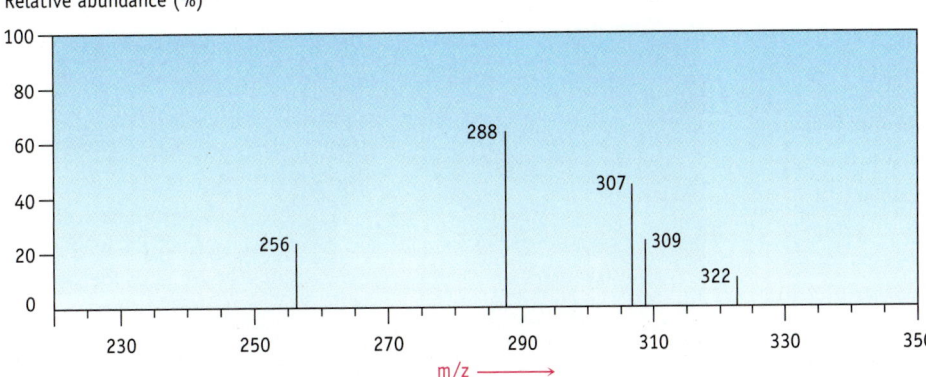

You may find it helpful to consult e3.19 for Questions 3.66–3.69.

3.66 How might you use IR spectroscopy to help distinguish between the following pairs of isomers?
(a) CH_3CH_2OH and CH_3OCH_3
(b)
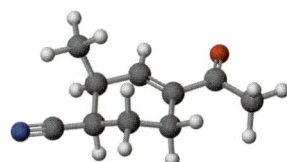 and $CH_3CH_2CH_2CH_2CH=CH_2$
(c)
$$CH_3CH_2\overset{O}{\overset{\|}{C}}OH \quad \text{and} \quad HOCH_2CH_2\overset{O}{\overset{\|}{C}}H$$

3.67 Which radiation wavenumbers in the infrared region would you expect to be absorbed by the molecule?

3.68 Propose structures for compounds that meet the following descriptions:
(a) C_5H_8, with IR absorptions at 3300 and 2150 cm^{-1}
(b) C_4H_8O, with a strong IR absorption at 3400 cm^{-1}
(c) C_4H_8O, with a strong IR absorption at 1715 cm^{-1}
(d) C_8H_{10}, with IR absorptions at 1600 and 1500 cm^{-1}

3.69 How would you use IR spectroscopy to distinguish between the following pairs of isomers?
(a) $(CH_3)_3N$ and $CH_3CH_2NHCH_3$
(b) CH_3COCH_3 and $CH_2=CHCH_2OH$
(c) CH_3COCH_3 and CH_3CH_2CHO

3.70 Would you expect the atomic weights of O atoms to be fixed constants? Should samples of H_2O collected from around the world and at different times of the year all have the same molecular weights? Explain.

Carbon Compounds

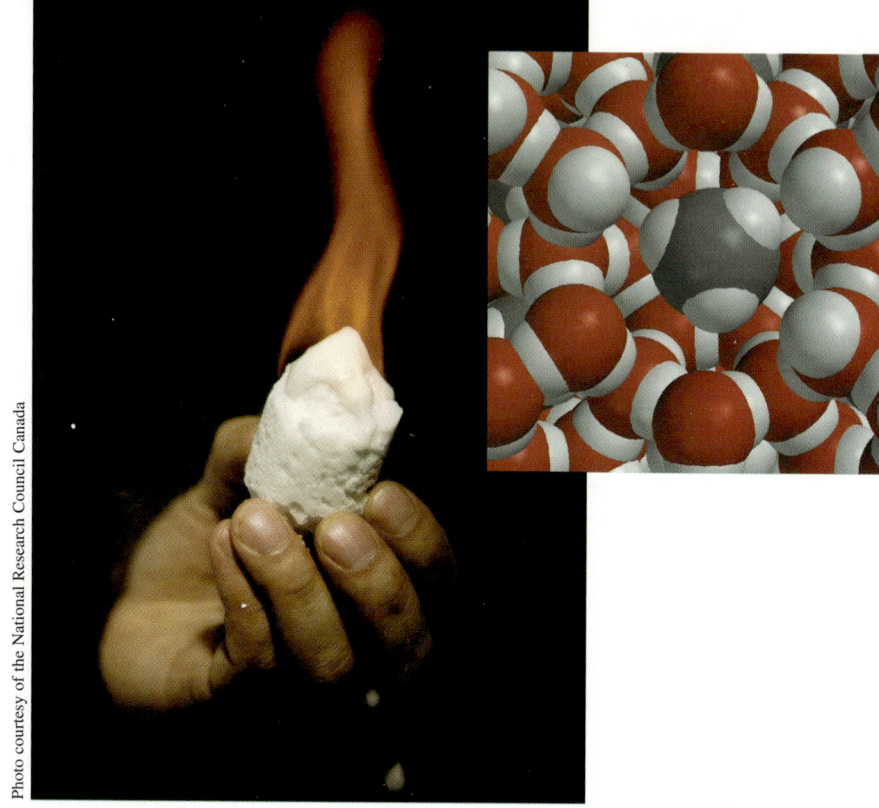

Photo courtesy of the National Research Council Canada

Wavefunction, Inc. (www.wavefun.com)

4.1 Case Study: Ice on Fire

In November 2000, the fishing trawler *Ocean Selector* pulled up a surprising catch from a depth of 800 m off Canada's west coast—a netful of 1 tonne of large hissing and frothing chunks of ice. Even more astonishing, these ice chunks burned when a match was put to them. The same unusual substance was responsible for a three-day shut down of Melbourne, Australia's gas supply when an icy plug formed in a large gas pipeline. And this peculiar ice probably exists on Titan and other bodies in our solar system.

What is this ice-like substance? Since it was pulled up from the deep ocean, it seems likely that it contains water. Observing that the same substance was formed in a Melbourne natural gas pipeline and that the ice chunks burn suggests that it also contains a flammable hydrocarbon. A clue to the identity of the hydrocarbon comes from knowing that natural gas fuel is mostly methane, consisting of molecules with the formula CH_4. Chemists have shown that the hissing chunks of ice are a substance made of both water and methane, but with some properties that are very different from either one. Surprisingly, methane, which does not dissolve in liquid water, *appears* to have done so in this unusual ice.

A Molecular-Level View

Chemists solved the puzzle of the nature of this substance by obtaining molecular-level views of these chunks of ice using x-ray crystallography, NMR, and other spectroscopic techniques. Analysis shows both methane and water molecules, with the water molecules forming cages that trap methane molecules inside them. But this new substance doesn't fit neatly into one of the four categories discussed in Section 3.2. The forces holding together the groups of water molecules and the forces between the water cage and the imprisoned methane molecules are neither covalent nor ionic bonds. The methane molecules aren't covalently bonded to the water, but rather are attracted through weak intermolecular forces of attraction, called **dispersion forces**. As a result, they have considerable freedom to rotate and move around inside the cage. This newly studied substance is an excellent example of a *supramolecular* or *host–guest complex*, described in Section 3.12. It is called a *methane clathrate hydrate*, in which guest methane molecules are trapped inside a cage made of host water molecules. The word **clathrate** comes from a Greek word meaning "I close" or "I confine." In chemistry, this term is used when one substance forms a crystalline cage structure ($H_2O(s)$ in this case) within which molecules of a second substance (CH_4 molecules in this case) are trapped. As shown in Figure 4.1, the bars on the host water cage are comprised of both covalent bonds within each water molecule and hydrogen bonds that hold water molecules together to form a symmetrical cavity.

Methane is not the only type of guest found in nature's water cages. Others include carbon dioxide and small carbon compounds. Gas hydrate or clathrate complexes are formed in nature when the host water and the guest are present at the right concentrations, and at relatively low temperature and high pressure. In oceans, methane hydrates are found at depths below 600 m and within the upper 700 m of sediment in temperate regions. Where temperatures are lower, they are also found in shallower water and in permafrost at high latitudes. Massive deposits have been found in the Gulf of Mexico and off the West Coast of Canada, where the fishing trawler made its surprising discovery. In oceans, we think methane clathrates are formed when methane from fossil fuels or methane-producing organisms is carried upward to regions where the temperature is low enough and the pressure high enough for these cage complexes to be stable.

Vast amounts of methane clathrate hydrates are being discovered. But as warming of the ocean and Arctic tundra frees methane from clathrate hydrates, more than 150 times their own volume of methane, a potent "greenhouse gas," is released.

A surge of interest in studying clathrates is motivated both by their promise and by concerns. We may be able to harvest the methane from clathrates as an abundant and relatively clean source of energy. Massive deposits of methane are stored in clathrates globally, with estimates ranging from 500 to 2500 Gt of stored carbon. This represents about 10–50% of the carbon thought to exist in all other fossil fuels. Much of this methane, however, would be very difficult to recover, as it is so widely dispersed off continental shelves around the world. Other promising applications for clathrates are as long-term molecular storage for carbon dioxide that would otherwise end up in the atmosphere; for temporary storage and transportation of natural gas; and in refrigeration and cold storage.

One challenge associated with methane clathrates is that they frequently cause giant ice plugs at the low temperature and high pressure in gas pipelines. About $1 billion is spent annually by the oil and gas industry to prevent and remove pipeline hydrate plugs. A second concern is strong geological evidence that slides and slumps under the ocean can be caused by destabilization of hydrate formations, leading in some cases to tsunamis. Finally, there are concerns about the effect of methane released from clathrates as a contributor to climate change. When methane hydrates are warmed to free methane from its ice cages, they release more than 150 times their own volume of methane gas, a potent "greenhouse gas."

Swapping Guests

If we know that both methane and other guests are trapped in hydrate cages, can we learn how to swap one type of guest for another through the icy cage bars? Think how useful it

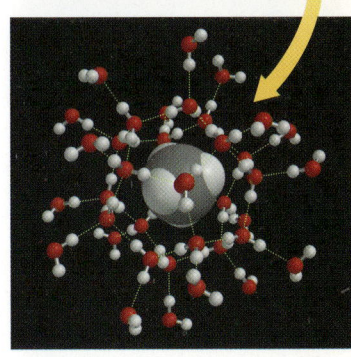

FIGURE 4.1 A single molecule of methane in a clathrate cage of water molecules (bottom). Cages of different sizes occupied by methane molecules are linked together to form a methane clathrate hydrate (top).

Molecular Modelling (Odyssey)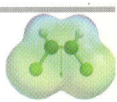

e4.1 Simulate methane molecules trapped in cages of water molecules.

John Ripmeester, Program Leader, Natural Research Council Canada.

might be to develop ways of locking up the "greenhouse gas" carbon dioxide formed in combustion processes, while at the same time obtaining methane fuel from abundant methane clathrates in the ocean! Chemist John Ripmeester's group at the National Research Council (NRC) in Ottawa has used a sophisticated set of tools, including NMR spectroscopy, to understand the structures of clathrate cages. And the NRC team and a research group led by Huen Lee at the Korea Advanced Institute of Science & Technology have independently studied conditions under which carbon dioxide molecules might replace methane molecules as guests inside clathrates of the type found on the ocean floor.

Figure 4.2 shows results of a promising experiment done by Lee's group in a high pressure reactor, where clathrates with methane and ethane (both constituents of natural gas deposits) were formed, and then carbon dioxide gas introduced. A molecular-level view of the results shows CO_2 molecules (represented simplistically as blue balls), replacing almost all of the CH_4 (red balls) and C_2H_6 (ethane, purple balls) molecules within the water clathrate cage (grey structures, with each vertex representing a water molecule). The structure of the clathrate cage changes when guest molecules of different size are introduced. The experimental evidence for this molecular-level view is found in the spectroscopy results in the bottom of the figure. On the left-hand side, an NMR spectrum shows the signals for the methane and ethane molecules in the cage disappearing almost completely over a 24-hour period. On the right-hand side, a Raman spectrum (a technique that, like infrared spectroscopy, detects vibrations in covalently bonded molecules) shows the size of the vibrational signals for carbon dioxide molecules in the cage increasing as more and more CO_2 molecules occupy the hydrate cages.

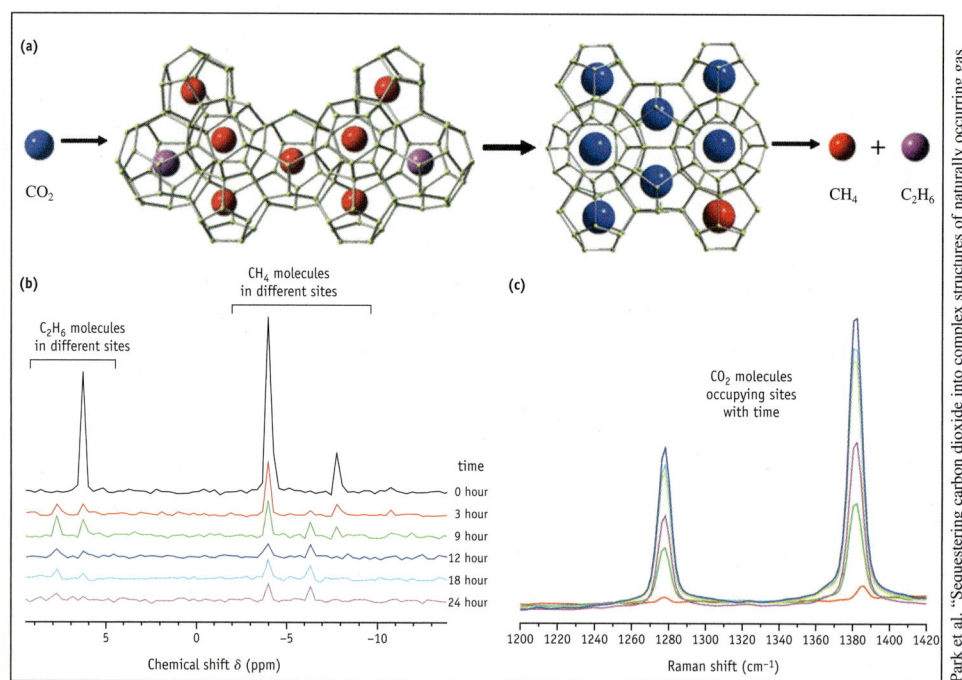

FIGURE 4.2 Replacement of methane and ethane molecules in a clathrate hydrate with carbon dioxide. Simplified representations of each of the gas molecules as single coloured spheres are shown.

Fishing for Solutions to Pipeline Clathrate Plugs

Ripmeester has teamed with molecular biologist Virginia Walker at Queen's University in Ontario to learn from chemistry in nature new ways to prevent hydrocarbon clathrate hydrates from forming in pipelines such as the one in Melbourne. Insects and polar fish have a variety of mechanisms to avoid freezing in the winter and in icy water. A collaboration between these chemistry and biology research groups has shown that an antifreeze protein produced by the winter flounder is quite effective at preventing both the formation and growth of methane hydrate crystals in laboratory tests. This antifreeze protein also appears to eliminate the poorly understood "memory effect" in hydrates, where once a hydrate has formed and melted, it seems to form a second time more easily.

This surprising discovery of methane captured in clathrate hydrates in the ocean and Arctic tundra is one of many examples of the location and uses of methane on our planet. We examine next the vital role it and other carbon compounds play in supporting human activity and life on Earth.

WORKED EXAMPLE 4.1—FORCES OF ATTRACTION IN SUPRAMOLECULAR COMPOUNDS

Another example of a supramolecular compound is a helium atom trapped inside a 60-carbon atom buckminsterfullerene cage.
(a) What term(s) most correctly describe(s) the forces of attraction between carbon atoms in the C_{60} cage?
(b) What term(s) most correctly describe(s) the forces of attraction between the helium atom and the carbon atoms of the C_{60} cage?

Solution

Supramolecular compounds are composed of guest molecules, atoms or ions held by weak non-covalent interactions to a host molecule.
(a) Covalent bonds hold together the carbon atoms in the C_{60} cage.
(b) Weak, non-covalent dispersion forces attract the guest helium atom to the network of carbon atoms (host) of the C_{60} cage.

Methane, carbon dioxide, and other small organic compounds can form clathrate hydrates, in which guest molecules are held inside "cages" of water molecules. Massive amounts of methane are stored in this way in nature, potentially useful as a source of energy and potentially dangerous to our climate if released into the atmosphere. The bonding in methane clathrate hydrates is an example of "supramolecular" or "host–guest" chemistry.

4.2 Methane: Signature of Life

While it is unlikely that you've set ice on fire, you may have used a Bunsen burner in the chemistry laboratory, lit your gas stove to cook a meal, or taken a shower with water heated by a natural gas water heater. Ice on fire and natural gas fuel have in common methane—a tasteless, odourless, colourless gas. It is the principal component in the fossil fuel *natural gas* (see below), and is also produced by micro-organisms and geological processes. Trace quantities of methane end up in our atmosphere where it plays an important role in regulating our planet's climate.

Methane is only one of more than 18 million compounds made of atoms of carbon combined with some combination of atoms of hydrogen, oxygen, nitrogen, halogens, and other elements called *organic compounds*. Methane is also called a **hydrocarbon**, since its molecules consist of carbon and hydrogen atoms joined together. The four

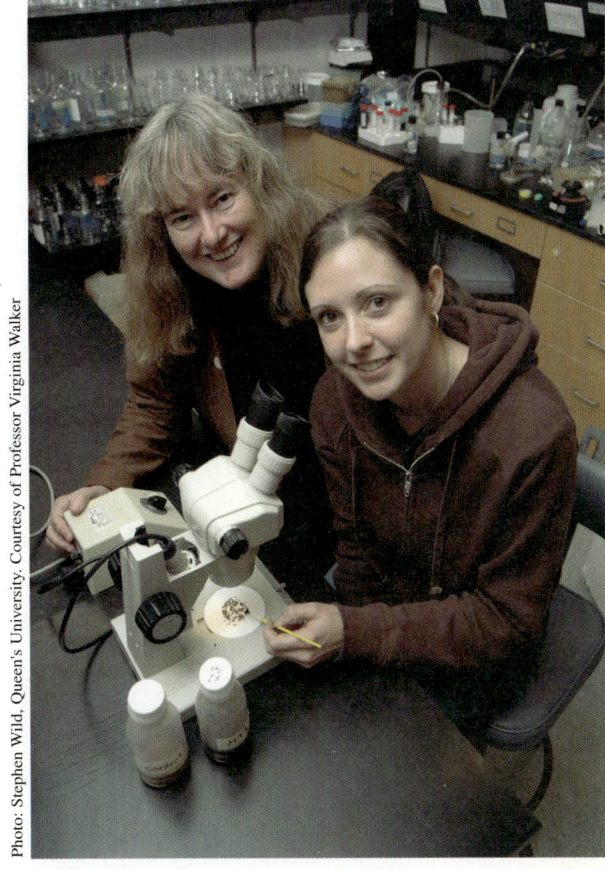

Photo: Stephen Wild, Queen's University. Courtesy of Professor Virginia Walker

Virginia Walker, Queen's University Canada.

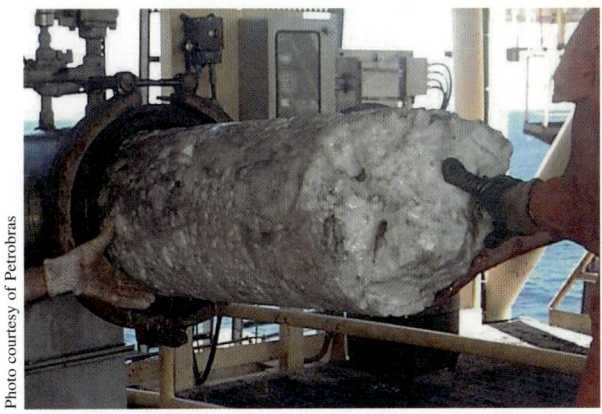

Photo courtesy of Petrobras

Methane clathrate hydrate plug removed from a pipeline.

Because methane is odourless and colourless, a leak of methane is not easily detected and could be very dangerous. For this reason, trace amounts of pungent sulfur compounds—with the same functional groups as those found in skunk spray—are added to natural gas so we can detect leaks.

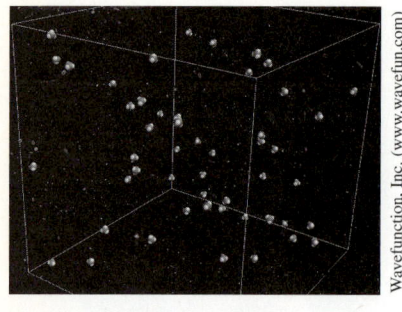

Wavefunction, Inc. (www.wavefun.com)

Methane, CH₄

FIGURE 4.3 Molecular-level representations of a sample of methane gas at 25 °C and 1 bar (top) and zooming in on a single methane molecule (bottom).

hydrogen atoms attached to the carbon atom in a CH_4 molecule do not lie in one plane, but they sit at the corners of a tetrahedron, as shown in Figure 4.3. The angle between any two C—H bonds is 109°.

EXERCISE 4.1—FORCES OF ATTRACTION IN METHANE

What term(s) most correctly describe(s) the forces of attraction *between carbon and hydrogen atoms within CH₄ molecules* in a sample of methane gas?

EXERCISE 4.2—FORCES OF ATTRACTION IN METHANE

What term(s) most correctly describe(s) the forces of attraction *between CH₄ molecules* in a sample of methane gas?

Methane Fuelling Human Activity

The burning of methane to produce energy is a combustion reaction, in which methane gas reacts with molecular oxygen in air to produce carbon dioxide gas and water vapour:

$$CH_4(g) + 2\,O_2(g) \longrightarrow CO_2(g) + H_2O(g)$$

The methane we burn is obtained from *natural gas*, which is a mixture of 80–95% methane and other hydrocarbons found in natural underground storage areas. Like other *fossil fuels*, natural gas can be formed from the remains of micro-organisms, plants, and animals. Buried under mud and sediment, these decaying organisms were exposed to heat and pressure in the earth's crust over millions of years. The complex mixture of organic compounds they provide is thought to produce fossil fuels such as coal, oil, and natural gas. Methane produced in this way is called *thermogenic methane*, because it results from the breakdown of organic matter through the application of heat from the interior of the earth.

While we are most familiar with using the combustion of methane to produce energy, methane is valuable as a feedstock in industrial chemical processes to produce molecular hydrogen, $H_2(g)$, and carbon monoxide, $CO(g)$. A process called *steam reforming* is used, where water vapour (steam) reacts with methane at high temperatures (~700–1100 °C) to produce carbon monoxide and hydrogen:

$$CH_4(g) + H_2O(g) \longrightarrow CO(g) + 3\,H_2(g)$$

To increase the yield of $H_2(g)$, the CO produced can be reacted with more water:

$$CO(g) + H_2O(g) \longrightarrow CO_2(g) + H_2(g)$$

The industrially important mixture of hydrogen and carbon monoxide that is formed is known as **synthesis gas** or **syngas**, because it can be used to synthesize other organic compounds such as methanol (CH_3OH), ethanol (CH_3CH_2OH), acetic acid (CH_3COOH), and acetic anhydride ($CH_3COOCOCH_3$), which in turn are the feedstocks (building blocks) for other organic compounds, including synthetic petroleum for fuels. South Africa produces much of its automotive fuel from syngas.

Methane has many other uses besides being burned by humans to produce energy, and it is found in many places on Earth—in underground fossil fuel reservoirs, the oceans, the Arctic tundra, and our atmosphere. This simple hydrocarbon is so important to our planet that it is sometimes called the "signature of life." Chemists have developed sophisticated techniques to detect methane in the earth's atmosphere and to tell how particular samples of atmospheric methane were likely made. They can use what's learned to look for evidence of methane and other organic compounds on other planets, telling us whether conditions favourable for life exist.

Feedstocks are substances used as starting materials to create other products in an industrial process.

Methanogens

Natural gas can also be formed by tiny micro-organisms called *methanogens* (methane creators). Some of these fascinating organisms live in the sediment of the ocean floor and consume the organic matter that falls down through the ocean, "passing" methane gas as a metabolic product. Since this methane results from biological activity (life processes), it is referred to as *biogenic methane,* or sometimes called **biogas**. These methanogenic organisms are typically found in environments that are very low in oxygen, such as in ocean sediments, wetlands, water-logged rice paddy soils, and landfills. Another example of biogenic methane is the methane produced and released to the atmosphere by methanogens that live in the digestive system of ruminant animals, such as cows, goats, and sheep. A similar symbiotic relationship between termites and methanogens leads to the production of methane when termites digest wood.

Abiogenic Methane

Although most of the methane on Earth has a biological origin, it can also be formed from non-biological processes, such as volcanic eruptions. Methane produced in this way is called *abiogenic* (without life) *methane.* In the ocean, abiogenic methane is known to form from molecular hydrogen and carbon dioxide in the reaction equation:

$$4\ H_2(g)\ +\ CO_2(g)\ \longrightarrow\ CH_4(g)\ +\ 2\ H_2O(g)$$

This reaction depends on temperature, pressure, and the presence of catalysts. On ocean ridges, in the presence of catalysts such as iron and magnesium in rocks, water heated by the earth's molten rocks (magma) produces hydrogen gas which can react with carbon dioxide to create abiogenic methane. Evidence of these "serpentinization" reactions is left in the form of vast deposits of the mineral serpentine.

Black smoker hydrothermal vents off Canada's West Coast.

Fissures in the ocean floor called "black smokers," first discovered near the Galapagos Islands, are an example of places where such chemical reactions produce methane. These towering chimneys are produced when seawater enters chambers containing magma, forming superheated water that reacts with rocks in the ocean crust. When the hot water finds its way out of a vent into the cold ocean water, salts and minerals precipitate out, creating chimneys. The temperature of water in these vents can approach 400 °C, but the water doesn't boil due to the very high pressure at these ocean depths. Near these vents, the water teems with "extremophile" microbes, able to survive under extreme conditions of temperature and pressure. Some are methanogens that produce methane from hydrogen gas. Others are *methanotrophs*, organisms that consume methane. Metal sulfides are responsible for the black colour of the fissures, as discussed in Section 6.7.

Previously unknown forms of life in the deep ocean are supported by methane. Exotic pink ice worms have been observed moving around methane clathrate hydrate surfaces at depths greater than 500 metres—living, perhaps, on bacteria that have synthesized food from methane.

Methane in Our Atmosphere

Some of the methane produced by natural processes and human activity finds its way into our atmosphere and doesn't "rain out," since methane gas is not soluble in water. Methane is the predominant hydrocarbon found in the earth's atmosphere, currently at levels of 0.0018% or 1800 ppb, almost three times that of pre-industrial levels (Figure 4.4). Most of that methane is biogenic, coming directly or indirectly from living organisms; about 70% comes from human activity, including energy and livestock production.

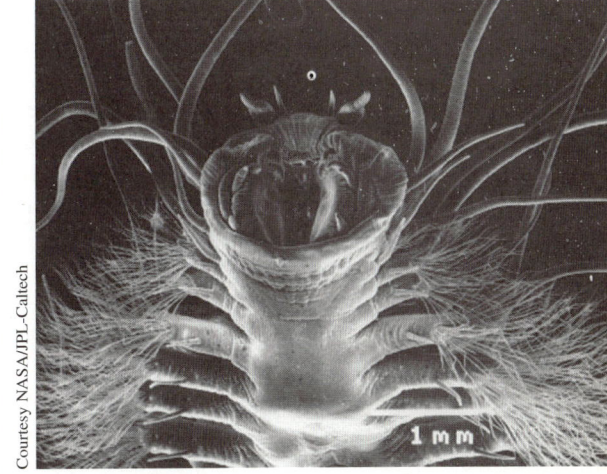

Close-up view of 2–3 cm long pink "ice worms" found burrowing in mounds of methane clathrate hydrates deep in the Gulf of Mexico.

B. Murton / Southampton Oceaography Centre/ Science Photo Library

Courtesy NASA/JPL–Caltech

FIGURE 4.4 Atmospheric concentration of CH_4 (ppb), CO_2 (ppm), and N_2O (ppb) over the past 2000 years.

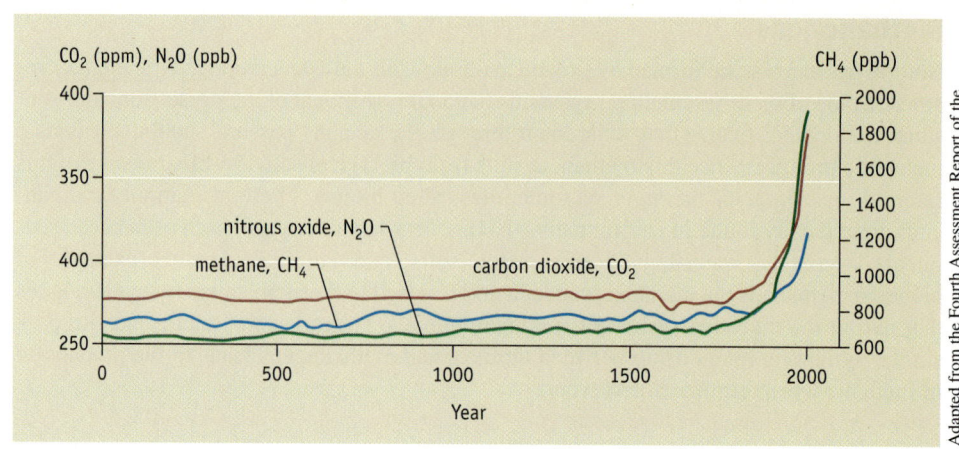

Adapted from the Fourth Assessment Report of the Intergovernmental Panel on Climate Change, 2007

A concentration of 1ppm of an atmospheric gas is 1 molecule of the gas in a million molecules of air, or 1 μmol of gas in 1 mol of air. One ppb is 1 part in a billion, and 1 ppt is 1 part in a trillion. These are called "trace" gases for a reason! An analogy for 1 ppb is 1 pinch of salt in 10 tonnes of potato chips.

Hydroxyl radicals ($\bullet$OH, with 9p$^+$, 10n, **9e$^-$**) in our atmosphere are very different in reactivity from the more familiar hydroxide ion in solution (OH$^-$, with 9p$^+$, 10n, **10e$^-$**), which has one more electron and a negative charge.

The average lifetime of a methane molecule in our atmosphere is about 10 years, and the primary way methane is destroyed is by colliding with a very reactive free radical species, hydroxyl ($\bullet$OH). Hydroxyl radicals are present in very low concentrations in our atmosphere (typically less than 1 ppt), and are formed by reaction of molecular oxygen (O_2) with ultraviolet light from the sun. A hydroxyl radical lasts only about a second before undergoing reaction—such as a successful collision to remove a H atom from a CH_4 molecule. In a series of subsequent reaction steps, methane is ultimately oxidized to carbon dioxide. The reaction of hydroxyl radicals with trace organic hydrocarbons like methane in our atmosphere, leading to their removal, is so important, that $\bullet$OH is sometimes referred to as the "tropospheric vacuum cleaner" or "the detergent of the atmosphere." Other organic compounds containing hydrogen atoms or double bonds also react with hydroxyl radical. The hydroxyl radical reaction with methane is only one example of the oxidizing environment on our planet—most compounds in the earth's atmosphere and crust are oxides of elements.

EXERCISE 4.3—REACTION OF CARBON COMPOUNDS WITH HYDROXYL RADICAL

While hydrocarbons such as CH_4 react readily with atmospheric hydroxyl radicals, chlorofluorocarbons (CFCs) such as CF_2Cl_2 do not. What difference in molecular structure between hydrocarbons and CFCs might be responsible for this difference in reactivity?

Where Did a Methane Sample Come From?

How can a chemist tell whether a sample of methane taken from our atmosphere or in a deposit under an ocean or lake is biogenic, thermogenic, or abiogenic? This is an important question for several reasons. Thermogenic methane is more likely to indicate the presence of a fossil fuel deposit. And as we'll see in Section 4.3, atmospheric methane is a powerful contributor to climate change. It's important to know whether that gas is coming from human or natural sources, so the right steps can be taken to control its release into the atmosphere.

Isotope Ratio Mass Spectrometry, the technique described in Section 2.1 to tell whether athletes cheat by taking synthetic testosterone, can help us identify the origin of a sample of methane gas. Biogenic methane contains a larger isotopic abundance of ^{12}C atoms (depleted in ^{13}C) than does abiogenic methane, since the processes used by organisms to create the methane take up less of the heavier ^{13}C isotope from CO_2 molecules. Different stable isotope ratios of carbon atoms in methane molecules can tell us whether a sample of methane was likely formed from ruminant animals, landfills, natural gas, or biomass burning. Additional evidence about the origin of a methane sample can be found from measuring the ratio of isotopes deuterium to protium (^{2}H/^{1}H) in methane molecules and identifying any other hydrocarbons such as ethane present in the sample.

Think about It

e4.2 Use isotope ratio mass spectrometry to determine the origin of methane.

Methane, the simplest hydrocarbon, is purified from natural gas and used in combustion reactions as a major source of energy. It is also an important feedstock for other organic compounds. Methane is formed in nature through thermogenic, abiogenic, and biogenic processes, and its origin can be determined by isotope ratio mass spectrometry. When released into the atmosphere, methane is eventually removed by reaction with hydroxyl radicals.

4.3 Climate Change and "Greenhouse Gases"

We are only beginning to understand the characters, plot lines, and our own roles as actors in the "great carbon mystery" of the chemistry of our atmosphere and its role in regulating climate. We know that methane and other carbon compounds play a crucial role in controlling Earth's climate, despite their very low concentration in Earth's atmosphere. *Climate is the weather patterns of an area, such as temperature, precipitation, and wind, measured over several decades.* The absorption of infrared radiation by methane, carbon dioxide, water vapour, and several other trace gases in our atmosphere warms the planet sufficiently to make Earth hospitable to life. Without these atmospheric "greenhouse gases," the average temperature on Earth's surface would be −18 °C, 33 degrees cooler than it is. At those temperatures, the oceans would be frozen and life as we know it would not exist. But the amount of these gases is critical. Mars, 228 million km from the sun, with an atmospheric mass less than 1% that of Earth's, has an average surface temperature of −47 °C. Venus, 108 million km from the sun, with an atmospheric mass 90 times that of Earth's, has an average surface temperature of 477 °C.

Our concern in this section will be with the disruptive effect on Earth's climate of substantial increases in these "greenhouse gases" over the past 250 years, due in part to the massive human activity of processing and burning natural gas and petroleum. But first let's take a chemist's view of how trace levels of gases and particles in our atmosphere can have such a great effect on Earth's climate.

Earth's Radiation Balance

When you look up into a clear sky, it's hard to visualize the complex chemical interactions constantly taking place, because they involve colourless gases such as carbon dioxide, methane, and water vapour in the troposphere absorbing or emitting radiation that falls outside of the visible region of the spectrum. The two images of Earth in Figure 4.5, taken by

Photo by Creative Services, University of Alberta

To a patient scientist, the unfolding greenhouse mystery is far more exciting than the plot of the best mystery novel. But it is slow reading, with new clues sometimes not appearing for several years. Impatience increases when one realizes that it is not the fate of some fictional character, but of our planet and species, which hangs in the balance as the great carbon mystery unfolds at a seemingly glacial pace.

David Schindler, University of Alberta scientist, *Nature,* **1999.**

FIGURE 4.5 (a) Long wave radiation and (b) short wave radiation scanning radiometer images of Earth, collected by the CERES instrument during the period March–May 2001. Courtesy of the Atmospheric Sciences Data Center and the CERES Science Team at NASA Langley Research Center.

Data courtesy of the Atmospheric Sciences Data Center and the CERES Science Team at NASA Langley Research Center. Images courtesy of Tom Bridgman, NASA Goddard Space Flight Center Scientific Visualization Studio

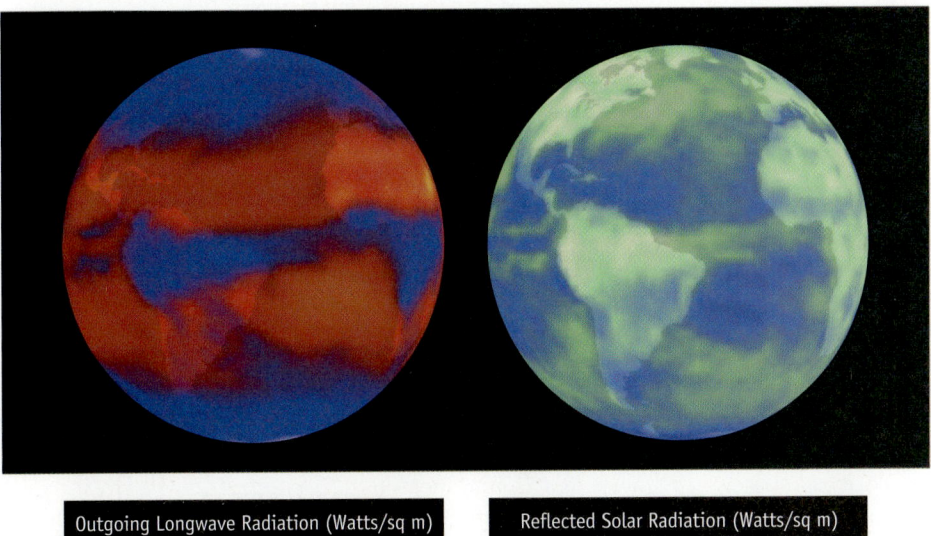

Outgoing Longwave Radiation (Watts/sq m)
100   150   200   250   300

Reflected Solar Radiation (Watts/sq m)
0   50   100   150   200

(a)   (b)

a scanning radiometer in space, help visualize two of the key components contributing to *Earth's radiation balance*—the delicate balance between incoming energy from the sun and energy radiated back into space that controls Earth's climate. A radiometer records energy from a source, in this case, Earth. The difference between the two images is the energy range in the electromagnetic spectrum of the recorded radiation. The left sphere shows the amount of infrared radiation (heat) emitted to space from Earth's surface and atmosphere; the right sphere shows the amount of sunlight reflected back to space by the ocean, land, ice, aerosols, and clouds. In each case, the amount of energy recorded by the radiometer is shown by a colour change on an arbitrary colour scale. Neither of these images represents the colour that you would see with your eyes from space.

Earth's radiation balance is controlled by many variables, including the wavelength and intensity of solar radiation; the tilt of the planet; the reflection of incoming solar radiation by Earth's surface and atmospheric aerosols and clouds; the emission of radiation from Earth, and the absorption of radiation by Earth's surfaces and gases in the atmosphere. Over geological time, this radiation balance has frequently shifted. At regular intervals (most recently every 100 000 years) the North American and Eurasian continents have been largely covered in sheets of ice several kilometres thick. Our knowledge of past temperature comes indirectly, from using isotope ratio mass spectrometry to measure very accurately the $^{18}O/^{16}O$ isotope ratio in water molecules. The way this information is translated into temperature is shown in e4.2.

Figure 4.6 shows a significant correlation between the surface temperature and atmospheric levels of CO_2. Note the time that it has taken over the past half-million years for each occurrence of a surface temperature change of several degrees. Only 250 years have passed since the Industrial Revolution, yet in that time Earth's radiation balance is shifting again. While many questions remain unanswered, a scientific consensus has emerged that, for the first time, this most recent shift in Earth's climate is due in significant part to human activity. Evidence of that change in climate comes

FIGURE 4.6 Historical trends in temperature and CO_2 concentrations, on a geological and recent time scale.

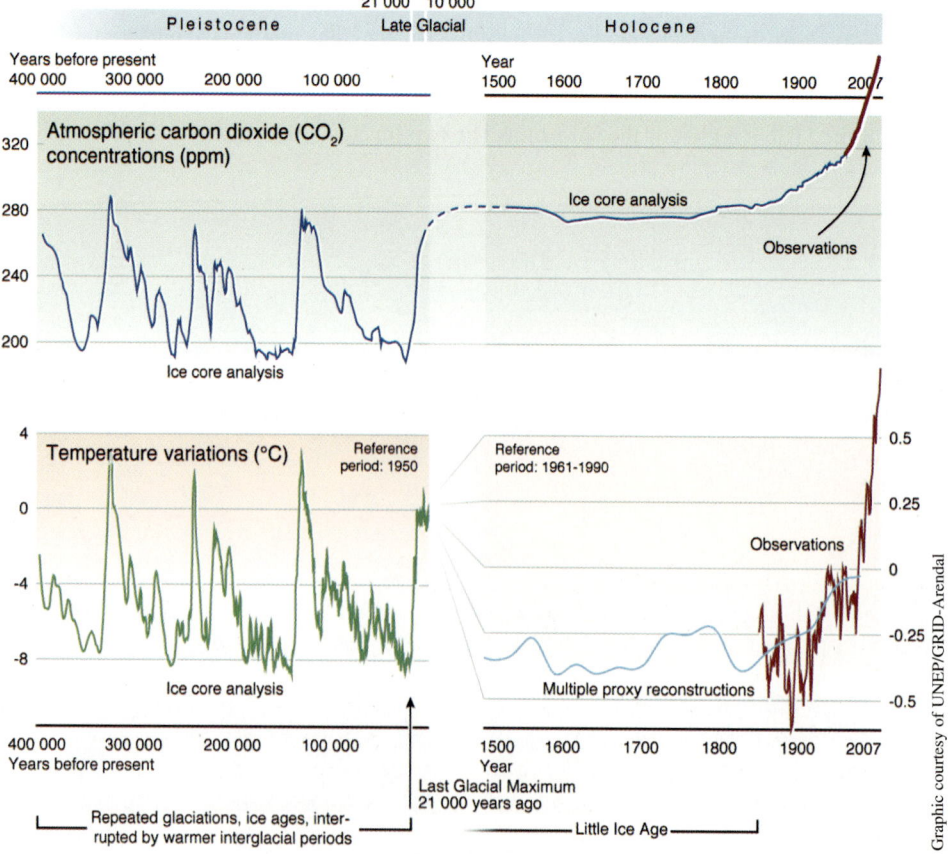

Graphic courtesy of UNEP/GRID-Arendal

from warmer average surface temperatures (Figure 4.7), melting of polar ice-packs, changes in precipitation and soil moisture in various regions, increased incidence of severe weather events, and the migration of destructive or disease-bearing insects like the mountain pine beetle [>>Section 23.1] or malaria-bearing mosquitoes into new habitats.

Adapted from the Fourth Assessment Report of the Intergovernmental Panel on Climate Change, 2007

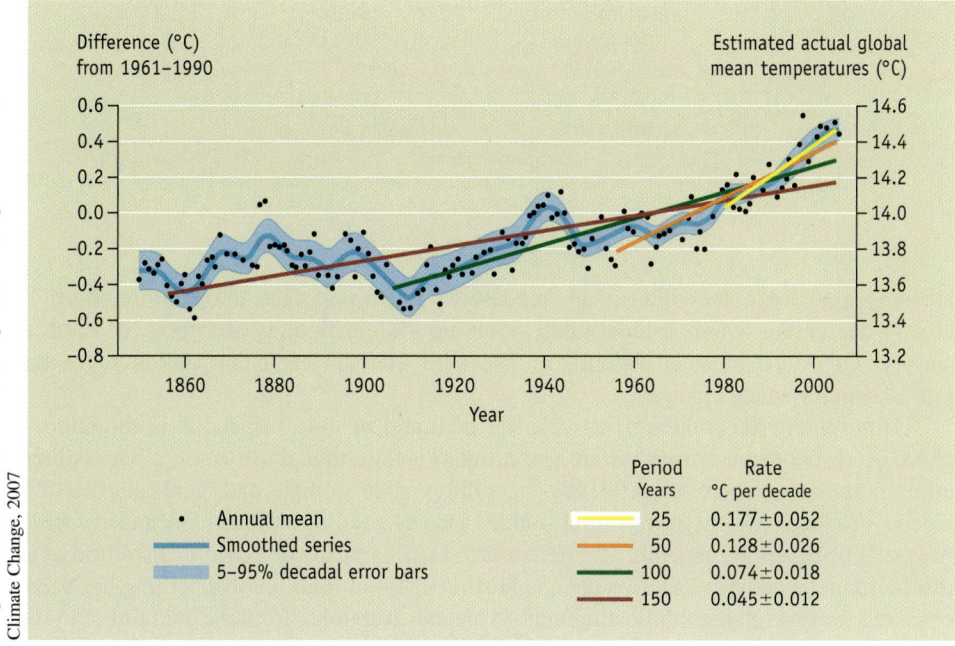

FIGURE 4.7 Global mean temperature changes since 1850, relative to the 1961–1990 normals.

Chemists can contribute particularly well in three ways to understanding and addressing this most recent climate change:

- Understanding the formation and behaviour of *aerosols,* fine particles suspended in the atmosphere that reflect visible light from the sun.
- Understanding the absorption of infrared radiation from the earth by "greenhouse gases" such as carbon dioxide, methane, nitrous oxide, water vapour, and chlorofluoro-carbons.
- Developing methods to capture and store carbon compounds responsible for climate change and advancing the science and technology needed to implement alternative forms of energy.

> Our knowledge of the earth's temperature over geological time scales comes from measuring the $^{18}O/^{16}O$ isotope ratio of water molecules in ice cores. Carbon dioxide levels are found by chemical analysis of the bubbles trapped in those same ice cores. A strong correlation exists between CO_2 levels and temperature over the past half-million years. Evidence suggests that human activity over the past 250 years is driving new changes in the earth's climate.

Reflection of Visible Light by Earth's Atmosphere: Clouds, Ice, and Aerosols

The type of electromagnetic radiation emitted by an object depends on its temperature. The sun is a much hotter object (surface temperature 5800 °C) than Earth (15 °C). The peak of the radiation emitted by the sun and reaching Earth's atmosphere is in the visible region of the spectrum. Some of this shorter wavelength radiation reaches Earth's surface and is absorbed (Figure 4.8). But as seen in the right-hand radiometer data from Earth in Figure 4.5, some of the visible sunlight is not absorbed by Earth's surface because it is reflected back into space by clouds, deserts, snow and ice, and aerosols. The fraction of sunlight reflected by Earth is called its **albedo**. As snow and

FIGURE 4.8 Energy distribution of radiation arriving from the sun and emitted by Earth.

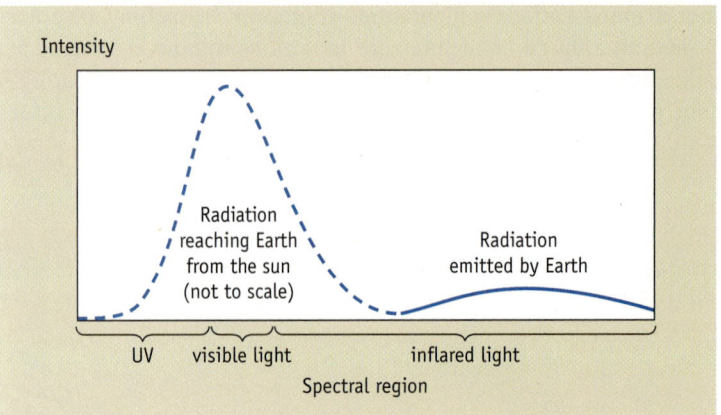

ice coverage diminishes due to an increase in Earth's surface temperature, Earth's albedo decreases, which means more incoming visible light is absorbed by Earth's surface. This causes further warming in a *positive feedback cycle* that may increase the rate at which climate changes.

Atmospheric **aerosols** are fine droplets of liquid or dust suspended in the atmosphere. Aerosols formed from sulfate and nitrate ions are thought to cause a net cooling effect in the atmosphere, by reflecting incoming visible sunlight and increasing Earth's albedo. Produced mostly from natural sources such as wind-blown dust, volcanoes, forest fires, and ocean spray, aerosols of different size and chemical composition also form as a result of human activities such as the combustion of fossil fuels. Global cooling has been observed for several years following major volcanic eruptions, from the creation of massive quantities of atmospheric aerosols. On the other hand, aerosols formed from elemental carbon (also known as *black carbon*), absorb visible light and can cause net atmospheric warming.

The lifetime of aerosols in the atmosphere depends on particle size. Particles smaller than 1 μm (referred to as PM_1) often stay in the atmosphere for relatively long periods of time and can be transported over long distances, while larger particles settle out more quickly close to the particle source.

Higher surface temperatures also cause an increase in evaporation of water from oceans and lakes, which may result in the formation of greater cloud cover. Since clouds reflect incoming visible sunlight, this increases Earth's albedo, causing a cooling effect. This is an example of a *negative feedback cycle* that may decrease the rate at which climate changes. This can be seen in Figure 4.5b, where clouds in the middle of South America substantially increase the amount of reflected short wavelength solar light.

Atmospheric aerosols, made of fine droplets of liquid or dust suspended in the atmosphere, cause cooling by reflecting incoming visible light from the sun. The atmospheric lifetime of particles depends on their size.

Absorption of Infrared Radiation by "Greenhouse Gases" The image of the globe in Figure 4.5a draws our attention to the loss of infrared or heat energy from the earth's surface, which cools our planet. We can't see infrared radiation—so what do these colours mean? The scale shows the infrared radiation emitted from different regions of the earth increasing on an arbitrary colour scheme from blue to red to yellow. Remember that the peak of the radiation leaving the surface of the earth is in the infrared region of the spectrum, as shown in Figure 4.8. What is the effect on the earth's climate of the presence of gases in our atmosphere that are able to absorb some of that infrared radiation? Recall from Section 3.8 that the absorption of infrared radiation by a molecular substance causes molecular vibrations—bonds stretch and twist. We use the term **radiative forcing** to describe the change in the radiation balance of our climate system due to absorption of some of that infrared radiation by naturally occurring or anthropogenic substances in our atmosphere. Besides water vapour, the most important atmospheric gases causing radiative forcing are carbon compounds.

That might be surprising, since much less than 1% of the gaseous substances in our atmosphere are made of carbon compounds. Most (99%) of the earth's atmosphere consists of only two gases: nitrogen (N_2: 78%) and oxygen (O_2: 21%). Most of these gas molecules are found very close to the surface of the earth, as shown by the decrease in the pressure of the atmosphere with increasing elevation. The region of our atmosphere closest to the surface of the earth, containing the highest density of gases, is called the *troposphere*. Trace gases found in unpolluted air include Argon (Ar: 0.97%), carbon dioxide (CO_2: 0.039%, or 390 ppm), methane (CH_4: 0.0018%, 1.8 ppm, or 1800 ppb), water vapour (H_2O: concentration varies widely), and others in even lower concentrations.

None of the three gases present in highest concentration—N_2, O_2, and Ar—are able to absorb infrared radiation emitted by the earth. This is because of a condition that must be met for any vibrating molecule to absorb infrared radiation—there must be a change in the relative positions of the centres of positive and negative charge in the molecule during its vibrations. This is referred to as a change in the **dipole moment** of the molecule, and will be discussed further in Section 6.3. For now it will be sufficient to know that this condition cannot be met by argon, nitrogen, or oxygen, since the centres of positive and negative charge are always in the same place in an isolated atom (Ar) or a molecule made of two identical atoms (N_2 and O_2). If a sample of argon, nitrogen, or oxygen gas were placed in an infrared spectrometer, no absorption peaks would be recorded.

Most of the trace carbon compounds in the atmosphere absorb heat energy emitted by the earth, evident from strong absorption peaks in their infrared spectra. These gases are called *"greenhouse gases"* because they are capable of absorbing some of the infrared radiation that would otherwise escape into space and cool the planet. Most attention has focused on carbon dioxide, since it is a product of all combustion reactions of hydrocarbons, and is produced every time fossil fuels are burned.

Past levels of carbon dioxide in our atmosphere can be measured by collecting and analysing pockets of air trapped in core samples taken from glaciers in Greenland and Antarctica. Figure 4.4 shows that over the past 1000 years, these levels were fairly constant at 280 ppm until about 1800, when they began a rise which appears to approach exponential growth. Human activity leading to the Industrial Revolution in 1750 is responsible for this large scale combustion of fossil fuels to produce very rapid increases in carbon dioxide levels in the atmosphere (Figure 4.9). Accurate direct measurements of

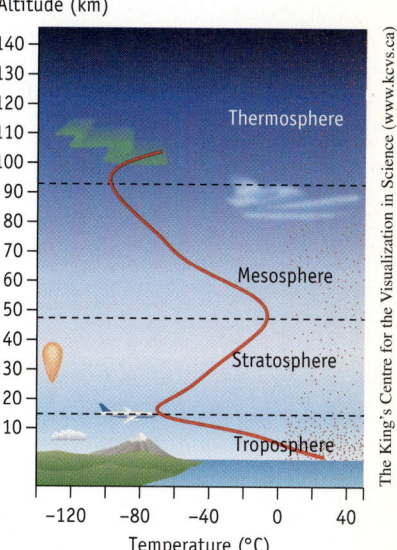

Temperature profile of the earth's atmosphere, with density of gases shown with red dots on the right. Go to e4.3 to study the profile further.

Think about It

e4.3 Explore regions of the earth's atmosphere and its density and temperature profiles.

The term *"greenhouse gas"* can be misleading, since the mechanism for warming of our atmosphere is completely different than that in a greenhouse. Greenhouses work mostly by preventing loss of heat by convection.

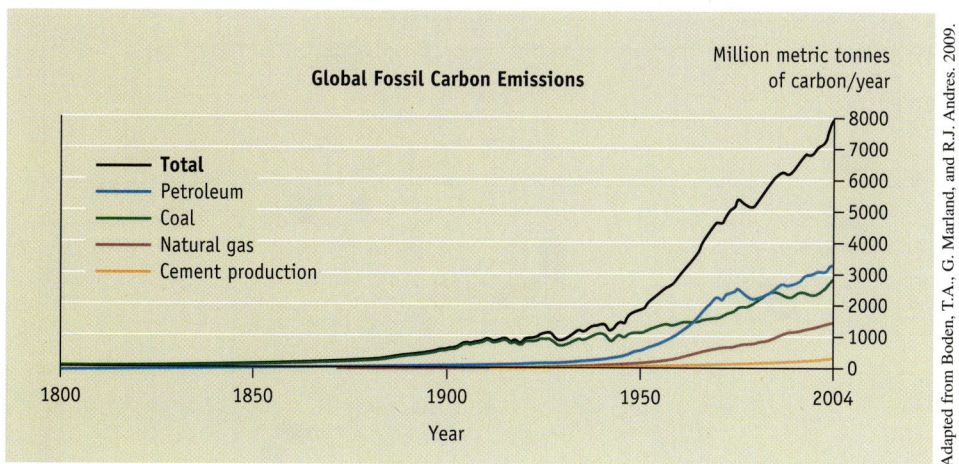

FIGURE 4.9 The increase in global emissions of carbon from fossil fuels since the Industrial Revolution.

carbon dioxide levels in the atmosphere have been recorded since 1958, after Roger Revelle, one of the first scientists to study global warming, warned that increased levels of carbon dioxide produced by humans was creating a global scale "geophysical experiment" with the earth's climate.

WORKED EXAMPLE 4.2—TEMPERATURE PROFILE OF THE ATMOSPHERE

With reference to e4.3, what happens to the temperature of the earth's atmosphere as you move from the troposphere into the stratosphere? What atmospheric chemistry causes this change in temperature profile? Is this chemistry mostly caused by natural processes or human activity?

Solution

This is the atmospheric region where the ozone "layer" is found. Large amounts of ozone (O_3) are produced each day from the reaction of ultraviolet light with molecular oxygen (O_2). O_2, in turn, is converted back to O_3. The net effect of these natural processes is the production of heat from ultraviolet light, causing the temperature to increase with altitude in this region.

EXERCISE 4.4—TEMPERATURE PROFILE OF THE ATMOSPHERE

With reference to e4.3, where in the earth's atmosphere does IR absorption by "greenhouse gases" mostly take place? Why does most IR absorption take place in this region of the atmosphere?

> The principal gases making up 99% of the earth's atmosphere (N_2 and O_2) do not directly absorb IR radiation, and are not "greenhouse gases."

Molecular-Level View of How "Greenhouse Gases" Cause Warming

How do carbon compounds such as carbon dioxide cause tropospheric warming? This is difficult to picture without the help of a dynamic model.

Think about It

e4.4 See a molecular-level picture of how "greenhouse gases" cause warming.

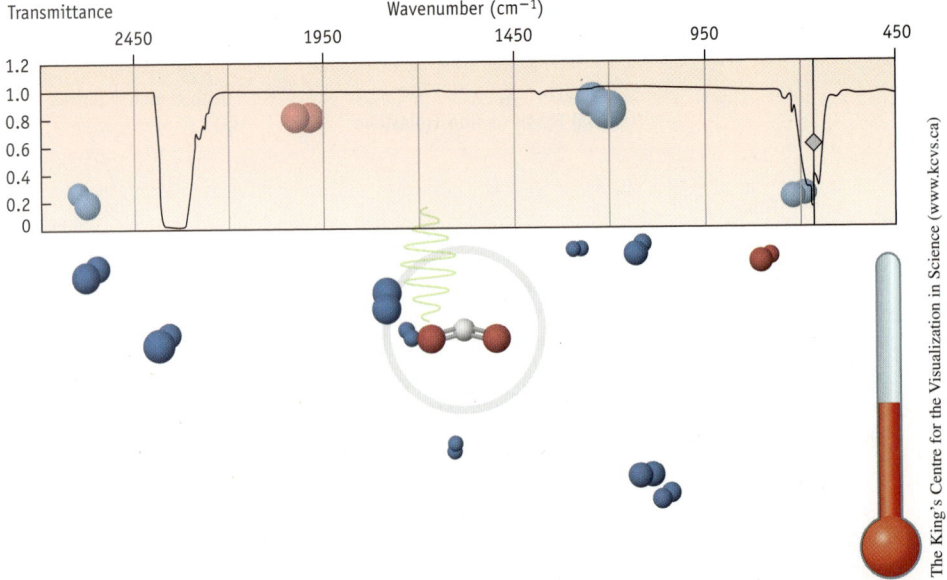

Open e4.4 and move the slider to the infrared absorption peak at 2350 cm^{-1} and then to the second peak at 670 cm^{-1}. What type of molecular vibration takes place when an infrared photon is absorbed by carbon dioxide in each case? Then, keeping the slider at 670 cm^{-1}, introduce (by clicking on the boxes) gas molecules (N_2—blue balls, and O_2—red balls) and a thermometer to monitor temperature. What happens to the velocity of surrounding N_2 and O_2 molecules when they collide with the vibrationally excited CO_2 molecule? Can you see, at the molecular level, why the temperature of the air goes up?

This molecular-level visualization shows one of the most important reasons why increased levels of "greenhouse gases" such as carbon dioxide cause our tropospheric temperature to rise. Carbon dioxide absorbs photons of infrared radiation if they are of exactly the right energy match to cause stretching or bending vibrations. In itself, that absorption of energy does not cause warming. Atmospheric warming takes place when that vibrational energy is transferred to other molecules in the atmosphere that are unable themselves to absorb infrared radiation. This process is called *collisional de-excitation* of carbon dioxide.

WORKED EXAMPLE 4.3—MOLECULAR-LEVEL VIEW OF ATMOSPHERIC WARMING

With reference to e4.4, what effect do photons at (a) 2350 cm^{-1} and (b) 665 cm^{-1} have on carbon dioxide molecules in the atmosphere? How does the absorption of infrared radiation at these wavenumbers cause tropospheric warming? Which of these absorption bands contributes more to tropospheric warming (at which wavenumber is more IR emitted by the earth)?

Solution

Absorption of IR photons at 2350 cm^{-1} causes the two C=O bonds in CO_2 molecules to stretch in an antisymmetric fashion, and absorption at 665 cm^{-1} causes the O—C—O bond angle to increase and decrease. Tropospheric warming happens when this vibrational energy is transferred to N_2 and O_2 molecules in the air through collisions. The earth emits much more IR at 665 cm^{-1}, as e4.5 shows.

Global Warming Potential and Infrared "Windows"

If an infrared photon in the region between 800 and 2200 cm^{-1} interacts with carbon dioxide, no absorption of energy takes place (e4.4). As seen in Section 3.11, molecular compounds can only absorb IR photons that have the right packet of energy to excite vibrational modes of that substance. Infrared radiation of other frequencies passes right through the substance without being absorbed. This has important implications for planetary cooling, as infrared radiation at these other frequencies emitted from the earth's surface will pass through our atmosphere into space, cooling the surface. Water vapour, with different covalent bonds than carbon dioxide, absorbs infrared radiation of different energies than carbon dioxide. Since carbon dioxide and water vapour are the two most abundant "greenhouse gases," combining their infrared spectra leaves only several regions or "windows" where infrared radiation can escape into space. The most significant "windows" are in spectral regions where the earth emits a great deal of IR radiation.

e4.5 shows that "greenhouse gases" such as methane, chlorofluorocarbons, and nitrous oxide can have a substantial global warming effect despite their very low atmospheric concentrations, because they strongly absorb infrared radiation in regions of the infrared spectrum where the earth emits IR and CO_2 and H_2O vapour do not absorb it.

This knowledge of the infrared spectrum of a "greenhouse gas" can be put together with information about its concentration, and the lifetime of that gas in the atmosphere to calculate its effect on Earth's radiation balance. The effect of a "greenhouse gas" on climate depends on *(a) its concentration, (b) how strongly and where in the infrared region it absorbs energy, and (c) its atmospheric lifetime.*

Think about It

e4.5 See how the radiative forcing of atmospheric gases relates to their IR spectra and the earth's emission.

One standard measure used to compare the ability of a gas to cause changes to the earth's climate is its **global warming potential**, always given relative to CO_2, which is arbitrarily assigned a value of 1. As shown in Table 4.1, 1 kg of methane has a global warming potential of 72 times that of 1 kg of carbon dioxide, over a 20-year period.

TABLE 4.1 Atmospheric Lifetime and Global Warming Potential of "Greenhouse Gases"

GAS	ATMOSPHERIC LIFETIME	GLOBAL WARMING POTENTIAL		
	(Years)	20-Year Horizon	100-Year Horizon	500-Year Horizon
Carbon dioxide(CO_2)	variable	1	1	1
Methane (CH_4)	12	72	25	7.6
CFC-12 (CF_2Cl_2)	100	11 000	10 900	5200
Nitrous oxide (N_2O)	114	289	298	153
Perfluoropropane (C_3F_8)	2600	6310	8830	12 530

Certain carbon compounds that have been introduced in the past 100 years have been found to have very long atmospheric lifetimes and global warming potentials. This is particularly true of *halocarbons,* in which carbon atoms are bonded to halogen atoms. A prime example is a class of compounds with no H atoms, called *chlorofluorocarbons* (CFCs), which are used for cooling in refrigerators and air conditioners. Even though CFCs are present in extremely low concentrations in our troposphere, their global warming potential is so great that we must pay careful attention to their introduction into the atmosphere. One kg of a chlorofluorocarbon (CF_2Cl_2) has a global warming potential of 10 900 times that of the same mass of carbon dioxide over a century! And as shown in Chapter 21, CFCs play a completely different, but also destructive role higher up in the stratosphere, where they interact with ultraviolet light from the sun to release chlorine free radicals that diminish our ozone layer.

Because of this concern, in most countries, CFCs have been replaced with HFCs (compounds with only **h**ydrogen, **f**luorine, and **c**arbon atoms) and HCFCs (compounds with only **h**ydrogen, **c**hlorine, **f**luorine, and **c**arbon atoms). The hydrogen atoms are bonded to carbon to make these new compounds reactive to hydroxyl radicals in the troposphere, in the same reaction we saw in Section 4.2 for $CH_4(g)$. In this way, HFCs and HCFCs can be destroyed in the troposphere before they reach the stratosphere, making them much less destructive to the ozone layer than CFCs, which have no C—H bonds. Unfortunately, however, CFCs, HFCs, and HCFCs all have large global warming potentials due to their long lifetimes in the troposphere and the strong absorption of infrared radiation by their C—F and C—Cl bonds. So they are considered serious contributors to climate change.

The Intergovernmental Panel on Climate Change assessment of overall changes in radiative forcing of anthropogenic gases since 1750, along with solar irradiance and other important contributors is shown in Figure 4.10. While most public attention is drawn to carbon dioxide, other "greenhouse gases" and contributors are also important. The relative contributions of changes in solar irradiance and anthropogenic contributions to climate change are also shown.

WORKED EXAMPLE 4.4—"GREENHOUSE GASES"

What factors contribute to CFC-12, CF_2Cl_2 (g), having a global warming potential (over a 20-year horizon) that is 11 000 times that of CO_2? Hint: Consult e4.5 for the IR spectrum of CFC-12 and data on atmospheric lifetimes.

Solution

Two factors contribute to its high global warming potential: its infrared absorption spectrum and atmospheric lifetime. CFC-12 has an atmospheric lifetime of 100 years, and the C—F bonds in CFC molecules absorb IR strongly in the IR "window" region where the earth emits radiation and CO_2 and H_2O vapour do not absorb. Fortunately, the atmospheric concentration of CFC-12 is much lower than that of CO_2, since 1 kg of CFC-12(g) has 11 000 times the global warming potential of 1 kg of $CO_2(g)$.

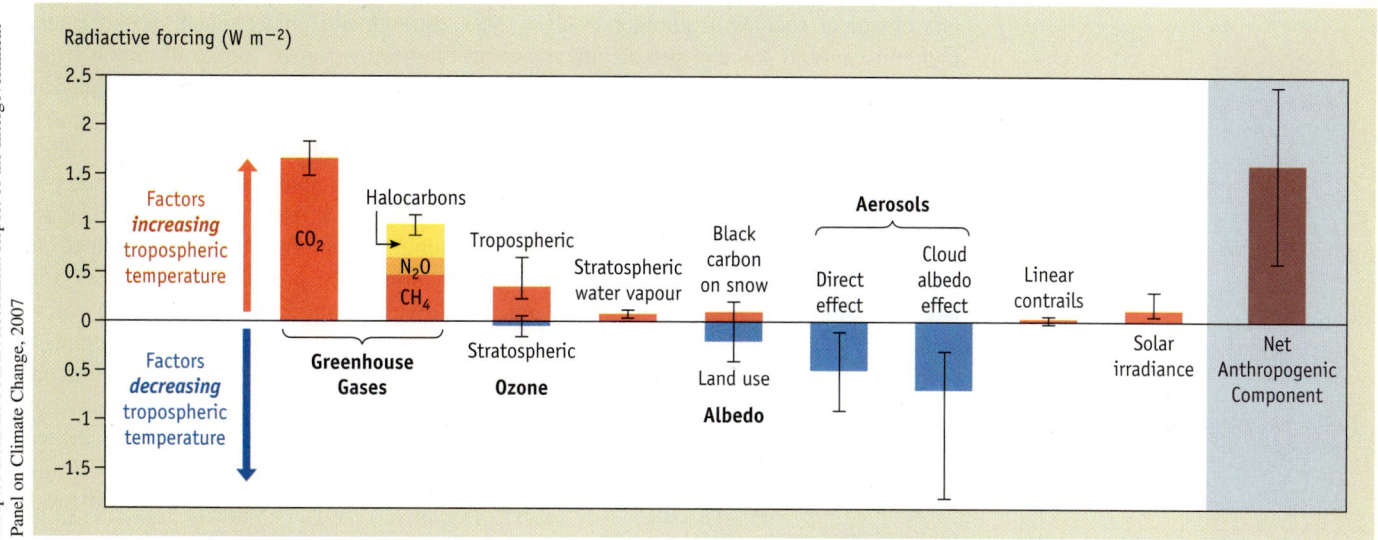

FIGURE 4.10 Global average estimates for the change in radiative forcing of anthropogenic "greenhouse gases" and other important contributors between 1750 and 2005.

EXERCISE 4.5—"GREENHOUSE GASES"

What factors contribute to nitrous oxide (N_2O) having a global warming potential (over a 20-year horizon) that is 289 times that of CO_2? You may wish to consult e4.5.

Trace carbon and other compounds in our atmosphere cause warming of our planet by absorbing infrared radiation that would otherwise escape into space. The molecular-level mechanism for warming involves the transfer of vibrational energy from those compounds to N_2 and O_2 gas molecules, increasing the temperature of the air.

Controlling Methane Sources

We end this section where we started, with methane (CH_4). Because its global warming potential is 72 times that of carbon dioxide (over a 20-year horizon), research is directed toward better understanding of present and future sources of methane in the atmosphere. Atmospheric scientists are concerned that modest changes to the earth's climate, such as an increase in temperature of several degrees in the Arctic, might release substantial quantities of methane from methane clathrate hydrates in the permafrost and in oceans, causing further increases in temperature that may lead to the release of more methane (a positive feedback cycle). So reducing thermogenic and biogenic methane has become a priority. Research areas include better maintenance of natural gas pipelines, controlling methane emissions in coal and petroleum production, modifying rice production techniques, and developing vaccines and changing feed to reduce methane production by methanogenic organisms in sheep and cattle. Methane emissions from livestock contribute up to 20% of global methane emissions. In New Zealand, the numbers are much higher—almost 90% of anthropogenic methane comes from ruminant livestock.

4.4 Capturing, Storing, and Recycling Carbon Compounds

This chapter began with the story of the *Ocean Selector* fishing trawler pulling up a massive catch of methane in the form of methane clathrate hydrate. We then examined the role methane plays on our planet and some ways carbon compounds are converted from one chemical species to another as they cycle through our environment. Fossil fuels, including natural gas and petroleum, represent long-term storage forms of carbon from plant and animal biomass, and chemists have taken these carbon compounds out of storage to convert them into synthetic polymers that have transformed almost every aspect of modern life. Yet we burn most fossil fuel resources to produce energy rather than using them to create molecular building blocks. And the main carbon compound produced in those hydrocarbon combustion reactions is carbon dioxide, whose concentration is building in our atmosphere, causing concern about its radiative forcing and effect on the earth's climate. Those climate change concerns now provide opportunities for chemists to play a lead role in finding innovative ways to reduce carbon dioxide emissions and recycle carbon compounds. One example is the research to study the replacement of methane and ethane by carbon dioxide in clathrate hydrates of the type found on the ocean floor [<<Section 4.1].

Imagine now another vessel trawling the ocean, but this time returning carbon dioxide to the deep ocean, perhaps releasing large blocks of dry ice (solid CO_2) or dragging a pipe that releases liquefied CO_2 captured from the emission stacks of a coal-fired powerplant.

This scenario is only one of many being explored for the storage of carbon dioxide to reduce atmospheric levels. In some countries, geological storage is being considered, where CO_2 is captured and injected into underground reservoirs. Ideally the injection of CO_2 gas into reservoirs can also help recover methane from coal beds (enhanced coal bed methane recovery) or oil from petroleum deposits (enhanced oil recovery). Large underground saline aquifers, such as the one under Barrow Island in Western Australia or the Viking Aquifer in Western Canada, are perhaps the ultimate choice for possible storage of large amounts of carbon dioxide for very long periods of time. The Viking Aquifer could store over 100 Gt of carbon, considering only the amount that would dissolve in the aquifer water.

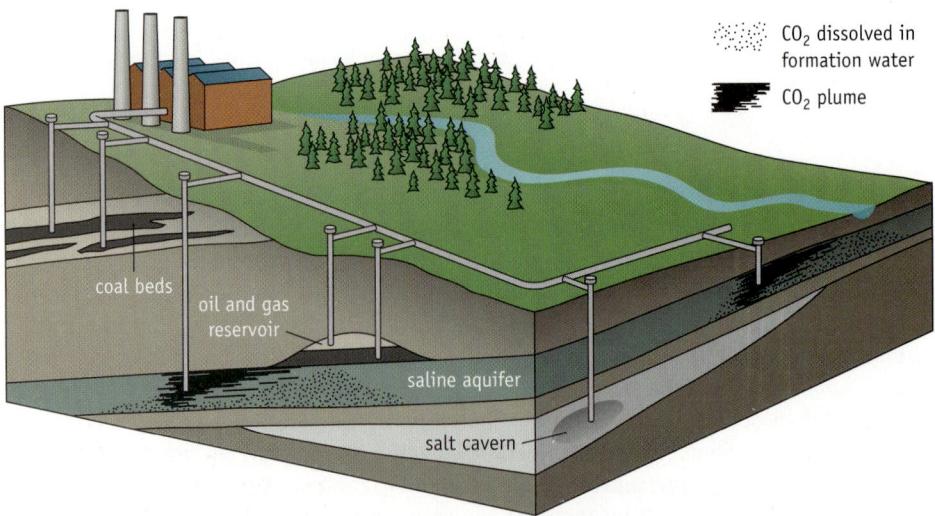

Geological CO_2 storage in aquifers and salt caverns.

Chemistry of Carbon Capture and Storage

In some examples of carbon compound storage, the molecular structure of carbon dioxide is unchanged. But chemical reactions can also be carried out to convert carbon dioxide into other compounds of carbon that will be immobilized for longer periods of time or used to synthesize other useful carbon compounds in natural or industrial plants. Methanogen organisms can work on $CO_2(g)$ deposited in coal beds, converting it into $CH_4(g)$. This process can be encouraged by the addition of methanogen growth media when the $CO_2(g)$ is sequestered.

What would happen to CO_2 deposited into the ocean by our vessel? Depending on the pressure and temperature at introduction into the ocean, it might form clathrate compounds, or more likely would react with water to produce the weak acid, carbonic acid (H_2CO_3). Several simultaneous chemical reactions then take place, where carbonic acid ionizes in water to produce hydrogencarbonate (bicarbonate) ions (HCO_3^-) and hydronium ions (H_3O^+), and some of the hydrogencarbonate ions can dissociate further to produce carbonate ions (CO_3^{2-}), and more hydronium ions.

$$CO_2(g) + 2\,H_2O(\ell) \rightleftharpoons HCO_3^-(aq) + H_3O^+(aq)$$

$$HCO_3^-(aq) + H_2O(\ell) \rightleftharpoons CO_3^{2-}(aq) + H_3O^+(aq)$$

These equilibrium reactions, which will be studied further in Chapters 15 and 16, do not go to completion, but can nevertheless convert significant quantities of CO_2 to CO_3^{2-}. If certain cations, such as Mg^{2+} or Ca^{2+} are present, insoluble salts such as $MgCO_3$ and $CaCO_3$ form precipitates, which can serve as much longer term storage for carbon compounds. Carbon is stored in the form of carbonates in sea shells, egg shells, rocks such as limestone and chalk, coral reefs, and minerals such as calcite. Learning from the chemistry taking place in the ocean, chemists can store carbon dioxide as carbonates in industrial processes. Serpentine or olivine rocks that are rich in magnesium can be crushed and reacted with carbon dioxide to make insoluble magnesium carbonate. But each step in this process requires energy, which presently comes from burning more fossil fuels and produces more $CO_2(g)$. And, as described in Section 16.1, the acidification of oceans by hydronium ions produced in these two reactions places coral reefs at risk.

Carbon Dioxide as a Feedstock and Solvent

Increasingly, chemists are working on the challenge to recycle carbon dioxide by using it as a feedstock to synthesize other carbon compounds directly rather than simply storing it. Human activity adds 1.5×10^{13} tonnes of $CO_2(g)$ annually to the atmosphere. Less than 1% of this amount, 100 million tonnes of CO_2, is presently used as a starting material in the synthesis of other carbon compounds, such as polycarbonate plastics, salicylic acid for pharmaceuticals, and urea, $(NH_2)_2CO(s)$—the world's most important nitrogen fertilizer. Carbon dioxide can also be turned back into a fuel by reaction with $H_2(g)$ to produce methanol, $CH_3OH(\ell)$, or $CH_4(g)$. For this overall process to make sense, the energy needed for the reaction to convert CO_2 into fuel must come from a renewable or plentiful source—such as solar or nuclear energy. While these are effective ways of recycling carbon, they are short-term solutions to removing carbon from the atmosphere—urea fertilizer, for instance, gets turned back into carbon dioxide soon after application to soil.

Finally, above 31 °C and 73 atm, CO_2 becomes a **supercritical fluid** with very different properties than either the gas or the liquid [>>Section 11.14]. Supercritical CO_2 now replaces many other organic solvents in processes such as removing the caffeine from coffee, and it can be recaptured and reused.

Technologies for storing even modest amounts of atmospheric carbon dioxide rely on being able to collect concentrated forms of carbon dioxide at the source of combustion—for example, in the stack of a coal-fired powerplant that produces electricity. This requires the separation of gases using chemical solvents or membranes. A much greater challenge is posed by trapping carbon dioxide produced by the myriad of mobile sources of combustion reactions, such as automobiles and airplanes. Since combustion reactions combine every carbon atom in the fuel with two oxygen atoms from O_2 to form CO_2, the product of combustion is about three times as heavy as the fuel that is burned.

WORKED EXAMPLE 4.5—CONCENTRATION AND TOTAL MASS

The concentration of $CO_2(g)$ in the earth's atmosphere as measured at the Mauna Loa observatory was 390 ppm in May 2009. If the total mass of air in the atmosphere is 5.2×10^{18} kg, and the average molar mass of air is 29.0 g mol^{-1}, calculate the total mass of CO_2 in the earth's atmosphere on that date.

Strategy

390 ppm of $CO_2(g)$ is the same as 390 μmol CO_2 mol^{-1} air. So we first need to convert the mass of the atmosphere into moles, using the average molar mass of air. Then we can determine the number of moles of CO_2, and use the molar mass of CO_2 to determine the mass of CO_2.

Solution

5.2×10^{18} kg air is 5.2×10^{21} g air. From this mass, we can deduce the amount (in mol) of air, using air's average molar mass of 29.0 g mol^{-1}.

$$n(\text{air}) = \frac{5.2 \times 10^{21} \text{ g}}{29.0 \text{ g mol}^{-1}} = 1.79 \times 10^{20} \text{ mol}$$

The amount of carbon dioxide then is

$$n(CO_2) = (1.79 \times 10^{20} \text{ mol}) \left(\frac{390}{1 \times 10^6} \right) = 7.0 \times 10^{16} \text{ mol}$$

This amount can be converted to mass, using the molar mass of CO_2,

$$m(CO_2) = (7.0 \times 10^{16} \text{ mol}) (44.01 \text{ g mol}^{-1}) = 3.1 \times 10^{18} \text{ g}$$

EXERCISE 4.6—CONCENTRATION AND TOTAL MASS

The amount of $CO_2(g)$ in the earth's atmosphere is often expressed in terms of only the total mass of carbon in the CO_2. Using your answer from Worked Example 4.5, what is the total mass of atmospheric carbon in the CO_2 in the earth's atmosphere in May 2009 ?

> Carbon dioxide (CO_2) levels in the atmosphere can be reduced by finding ways to capture it and store it in the ocean, in geological formations, and by conversion to other chemical species. A great deal of research is also directed toward reduction of emissions of methane and other "greenhouse gases" into the atmosphere.

Biopolymers: Carbon Dioxide Storage and Reactions in Nature

What other stories are told by the concentrations of trace atmospheric carbon compounds? Look closely in Figure 4.11 at how CO_2 levels have increased since the Mauna Loa Observatory started recording data in 1957. The overall trend has increased from 315 to 390 ppm over this time period. But examine the oscillatory behaviour superimposed on this upward moving graph. What causes CO_2 levels to rise and fall so regularly on an annual basis?

Plants and other organisms are much more efficient than humans at extracting carbon dioxide in its dilute form in the atmosphere and using it to synthesize useful carbon compounds. The global scale on which this biological **carbon sequestration**

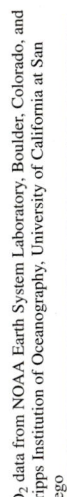

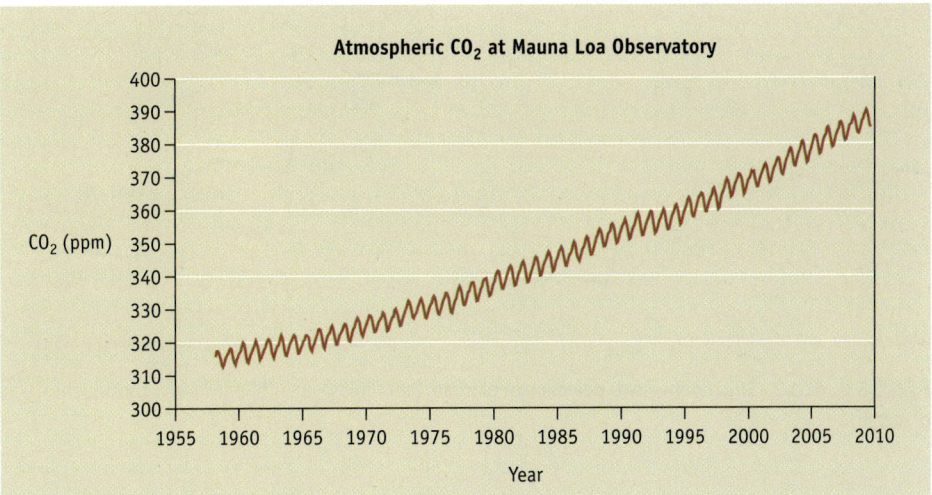

FIGURE 4.11 Measurements of monthly average atmospheric CO_2 levels from the observatory in Mauna Loa, Hawaii, since 1957.

happens is evident from the variation each year in the Mauna Loa $CO_2(g)$ measurements. Respiration, by which plants and animals use $O_2(g)$ and produce $CO_2(g)$, takes place throughout the year. In the process of photosynthesis, green plants use light from the sun to provide the energy needed to convert CO_2 into glucose, for which a simplistic chemical equation is shown below. Levels of CO_2 rise each winter, when green plants shed their leaves and total hemispheric photosynthesis is reduced, and fall again in the growing season when the leaves come out again.

$$6\,CO_2(g) + 6\,H_2O(\ell) \longrightarrow C_6H_{12}O_6(s) + 6\,O_2(g)$$

Glucose, the product of this photosynthetic process, is the *monomer* building block used by both plants and animals to synthesize bio*polymers* for structural and energy storage purposes. The polymers produced by glucose belong to a family of compounds called carbohydrates, known to be the most abundant carbon compounds found in nature. In plants, several thousand glucose monomers polymerize to form *cellulose,* the common structural material in plant cell walls. The filter paper burned by Flematti in the case study in Section 1.3 and the paper in this textbook are made of cellulose. Excess glucose is stored by plants in the form of a closely related polymer, *amylose,* one of the main constituents of starch.

Note the very subtle structural difference in the two polymers in Figure 4.12. The only difference is the three-dimensional connectivity of the glucose monomers to produce the polymer. In amylose, the oxygen atoms connecting each two glucose monomers are oriented differently in space than those in cellulose [>>Chapter 29].

Similar to the role played by starch in plants, *glycogen* is an energy storage polymer in animals—it releases glucose, which serves as an energy source for most organisms.

A *β*-glucose molecule.

EXERCISE 4.7—CARBON DIOXIDE STORAGE

Primo Levi (1975), in his book the *Periodic Table,* states: "Man [sic] has not tried until now ... to draw from the carbon dioxide in the air the carbon necessary to nourish him, clothe him, warm him, and for the hundred other more sophisticated needs of modern life. He has not done it because he hasn't needed to: he has found, and is still finding gigantic reserves of carbon, already organicized, or at least reduced." List five things you have used today to meet your needs that come from reserves of "organicized carbon," and indicate in what form you think the carbon was found.

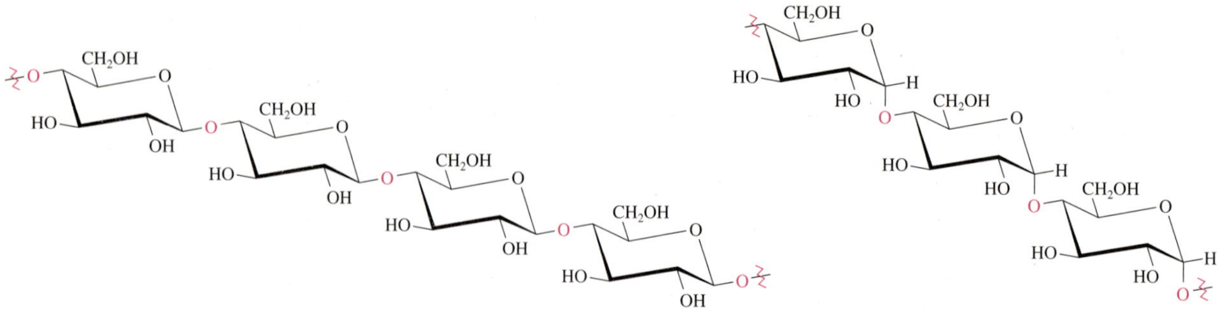

(a) Cellulose, a 1,4'-O-(β-D-glucopyranoside) polymer

(b) Amylose, a 1,4'-O-(α-D-glucopyranoside) polymer

FIGURE 4.12 Structures of two polymers of glucose: (a) cellulose, and (b) amylose, a constituent of starch.

In the future, that vessel depositing CO_2 back into the ocean will likely be propelled by energy from the combustion of **biofuel** derived from plants or algae rather than diesel fuel from a petroleum refinery. The dual realities of changing climate and dwindling fossil fuel resources has catalyzed a great deal of urgent research into *renewable* sources of carbon compound–based fuels. These can be produced rapidly by plants rather than over millennia as in petroleum, coal, oil, or natural gas. Photosynthetic processes in corn, soybean, sugar cane, palm oil, or switchgrass plants and some small algae use the sun's energy to convert CO_2 gas into *biomass*, recently living biological material that can be used for fuel or industrial purposes. Of particular interest is conversion of underutilized or waste products such as stalks, chaff, and manure to alcohols, oils, and gases that can be used in home cooking and heating, operation of motor vehicles, and direct electricity production. Anaerobic (low oxygen) digestion of biomass such as manure, sewage sludge, and municipal solid waste can be used to produce *biogas* [<<Section 4.2], a mixture of methane, carbon dioxide, and other gases. The chemical composition of biogas that has been cleaned up is very close to natural gas, and is sometimes called *renewable natural gas*. Much of the research is directed toward the energy balance in these processes. The production of biomass and conversion into fuels requires large amounts of energy, some of which presently comes from fossil fuel combustion.

Atmospheric CO_2 can also be captured very efficiently by plants and some other organisms, and stored in the form of glucose, through the process called photosynthesis. Glucose monomers are converted into polymers by plants and animals for structural purposes and energy storage.

4.5 Alkanes: Saturated Hydrocarbons

In Section 4.2, we took a close look at methane, the principal component of natural gas. The other components of natural gas vary widely depending on its source: A typical mixture is 70–90% methane (CH_4), 0–8% carbon dioxide (CO_2), and 0–20% of other simple hydrocarbons, called *alkanes*. Most important of these are ethane (C_2H_6), propane (C_3H_6), and butane (C_4H_{10}). In molecules of ethane, propane, and butane, two, three, and four carbon atoms, respectively, are joined by C—C single bonds. Molecules of these, and other substances with even longer chains of carbon atoms, are called **alkanes**, and they form the carbon skeletons for many important natural and synthetic materials.

Much of the world's current technology relies on *petroleum*, a complex mixture of molecules of much larger hydrocarbons than those found in natural gas. Burning fuels derived from petroleum provide by far the largest amount of energy in the industrial world. Petroleum and natural gas are also the chemical raw materials used to produce the carbon building blocks (feedstocks) for the manufacture of plastics, rubber, pharmaceuticals, and a vast array of other compounds.

Like natural gas, the composition of the petroleum that is pumped out of the ground varies greatly depending on its source. The primary components of petroleum are always alkanes, but, to varying degrees, nitrogen- and sulfur-containing compounds are also present. Aromatic compounds [<<Section 3.11] are usually present, but not alkenes and alkynes.

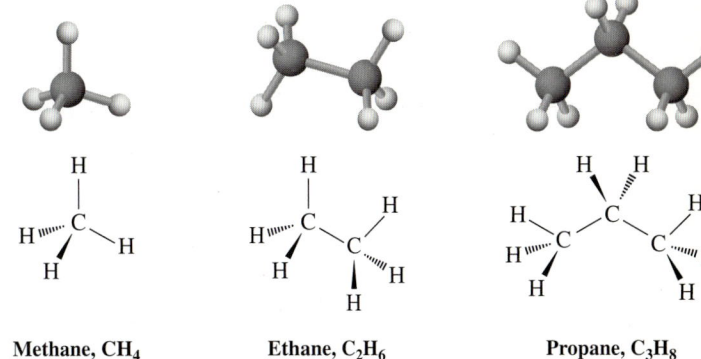

Methane, CH$_4$ **Ethane, C$_2$H$_6$**   **Propane, C$_3$H$_8$**

An early step in the petroleum-refining process required to produce useful materials is *distillation*, in which the crude petroleum is separated on the basis of boiling point into a series of simpler mixtures, called *fractions* (Figure 4.13). At the top of a distillation compound, a gaseous fraction is collected (mostly alkanes with one to four carbon atoms; this fraction is often burned off). Then fractions consisting of mixtures of larger-molecule hydrocarbons with higher boiling points, such as gasoline (bp 20–200 °C), kerosene (bp 175–275 °C), and heating oil or diesel fuel (bp 250–400 °C) are obtained. Finally, distillation under reduced pressure yields even higher boiling mixtures used as lubricating oils and waxes. Left over at the bottom of the distillation tower is an undistillable tarry residue of asphalt.

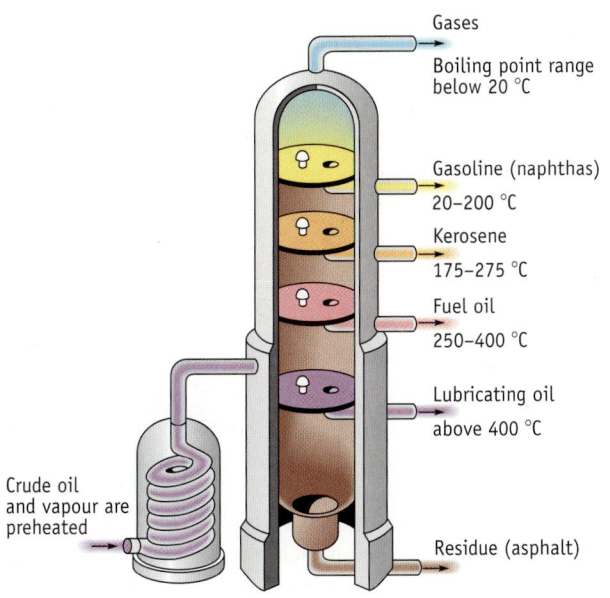

FIGURE 4.13 Fractional distillation separates petroleum into mixtures of alkanes based on boiling points.

The alkane molecules making up natural gas and petroleum are **saturated hydrocarbons**: "hydrocarbons" because they contain only carbon and hydrogen atoms; "saturated" because they have only single bonds (and no double or triple bonds, or rings) and so contain the maximum possible number of hydrogen atoms per carbon atom. They have the general formula C_nH_{2n+2}, where *n* is any integer.

A series of alkanes, such as that produced in petroleum refining, shows regular increases in both boiling point and melting point as molecular weight increases. As shown in Table 4.2, alkanes with smaller molecules are gases at room temperature, those with intermediate ones are liquids, and those with the larger ones are solids. In Chapters 5 and 11, these differences in properties are explained in terms of intermolecular forces between molecules of the substance.

Regardless of the number of carbon atoms in the chain of alkane molecules, experimental determinations show the average C—C and C—H bond lengths (C—C bond lengths of 154 ± 1 pm and C—H bond lengths of 109 ± 1 pm) [>>Section 9.3] and bond energies [>>Section 7.9] in molecules of different alkanes are nearly the same, and the geometry of the atoms connected to each carbon atom is tetrahedral, just as it is for CH_4. Other than in combustion reactions, alkanes also exhibit relatively low chemical reactivity.

Think about the different ways that carbon and hydrogen atoms can be connected in alkane molecules. Only one structure is possible connecting one carbon atom with four hydrogen atoms—methane (CH_4). Similarly, there is only one possible way two carbon atoms can be connected with six hydrogen atoms (ethane, CH_3CH_3) and only one possible combination of three carbons with eight hydrogen atoms (propane, $CH_3CH_2CH_3$). If larger numbers of carbon and hydrogen atoms are found in alkanes, however, different patterns of connectivity are observed. For example, there are molecules with *two* different connectivity patterns with the formula C_4H_{10}: the four carbon atoms can be connected in a straight-chain (butane), or they can have a branch (isobutane or 2-methylpropane, Figure 4.14). **Isomers** have the same numbers and kinds of atoms but differ in the way the atoms are arranged. Compounds like butane and 2-methylpropane, whose atoms are connected differently, are called **constitutional isomers**. Similarly, there are three connectivity patterns for molecules with the formula C_5H_{12}. As the number of carbon atoms in a series of alkanes increases, the number of possible constitutional isomers greatly increases; there are 5 isomers possible for C_6H_{14}, 9 isomers for C_7H_{16}, 18 for C_8H_{18} (which is found in gasoline), 75 for $C_{10}H_{22}$, and 366 319 for $C_{20}H_{42}$!

The term *straight-chain* means that all carbon atoms are linked together in a chain, without some carbon atoms branching off the chain. The chains of alkane molecules do not form a straight line—rotation occurs rapidly about C—C bonds, giving many different overall shapes for each molecule.

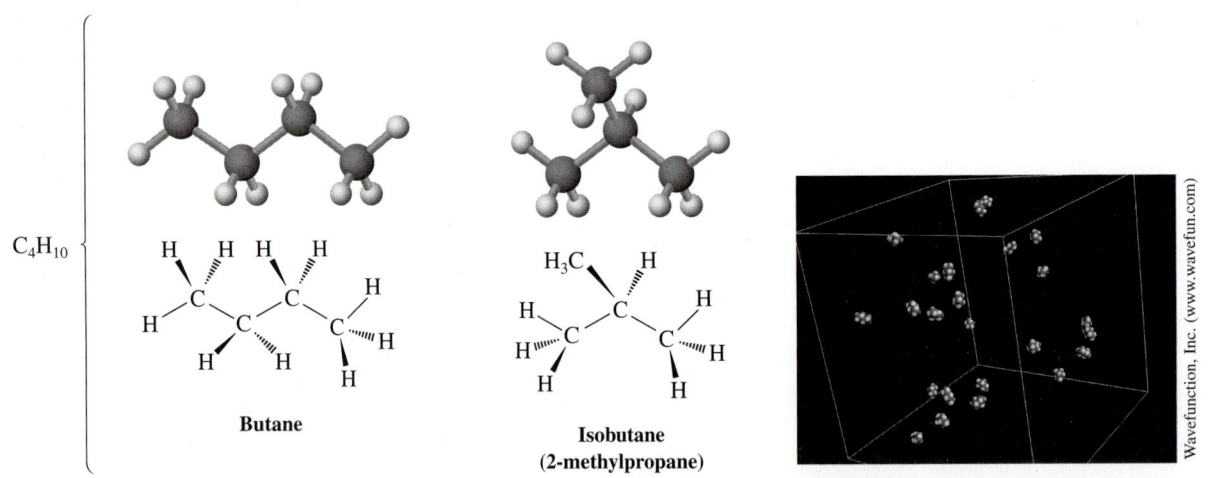

FIGURE 4.14 Constitutional isomers of butane (C_4H_{10}) and a representation of butane molecules in a sample of butane gas (right). e4.6 depicts the molecular motion of these molecules.

A given alkane molecule can be arbitrarily shown in many ways. For example, the straight-chain, four-carbon butane can be represented by any of the structures shown below:

TABLE 4.2 Names and States of Selected Straight-Chain Alkanes

Name	Molecular Formula	Structure	Molecular Weight	State at Room Temperature
Methane	CH_4		16.04	Gas
Ethane	C_2H_6		30.07	
Propane	C_3H_8		44.10	
Butane	C_4H_{10}		58.12	
Pentane	C_5H_{12} (pent- = 5)		72.15	Liquid
Hexane	C_6H_{14} (hex- = 6)		86.18	
Heptane	C_7H_{16} (hept- = 7)		100.20	
Octane	C_8H_{18} (oct- = 8)		114.23	
Nonane	C_9H_{20} (non- = 9)		128.26	
Decane	$C_{10}H_{22}$ (dec- = 10)		142.28	
Octadecane	$C_{18}H_{38}$ (octadec- = 18)		254.50	Solid
Eicosane	$C_{20}H_{42}$ (eicos- = 20)		282.55	

These structures are not intended to imply any particular three-dimensional geometry for butane; they only indicate the connections among atoms. In practice, chemists rarely draw all the bonds in a butane molecule, and usually refer to butane by the *condensed structure*, $CH_3CH_2CH_2CH_3$ or $CH_3(CH_2)_2CH_3$. In such representations, the C—C and C—H bonds are "understood" rather than shown. If a carbon atom has three hydrogen atoms bonded to it, we write CH_3; if a carbon has two hydrogen atoms bonded to it, we write CH_2, and so on. Still more simply, butane can even be represented as n-C_4H_{10}, where n signifies *normal*, straight-chain butane.

To recognize the isomers corresponding to a given formula, keep in mind the following points:

- Each alkane molecule has a framework of tetrahedrally oriented carbon atoms, and each C atom has four single bonds.
- An effective approach to drawing structural representations is to create a framework of carbon atoms and then fill the remaining positions around carbon with H atoms so that each C atom has four bonds.
- Free rotation occurs around C—C single bonds. Therefore, when atoms are drawn to form the skeleton of an alkane molecule, the emphasis is on how carbon atoms are connected to one another and not on how they might lie relative to one another in the plane of the paper. A physical or computer model is needed to create a more realistic view of the arrangement of atoms in three dimensions in space.

As you become proficient at drawing representations of individual alkane molecules, keep in mind that an accurate mental picture of the molecules in a sample of butane gas couldn't be drawn on paper at all—it would need to show many molecules tumbling in space, each with its atoms vibrating. Different molecules would also have many different overall shapes, due to rapid rotation occurring about C—C single bonds. Each butane molecule is weakly attracted to others by intermolecular forces. The optional "Odyssey" electronic resources accompanying this learning resource portray well dynamic pictures of multiple molecules interacting with each other.

Molecular Modelling (Odyssey)

e4.6 Simulate bulk matter samples of alkanes at different temperatures.

EXERCISE 4.8—STRUCTURAL REPRESENTATIONS

Draw structural representations of the five isomers of C_6H_{14} molecules.

Nomenclature (Names) of Alkanes

As you learn more about alkanes, it will be important to understand the systematic rules for naming them and other carbon compounds that chemists use to communicate with each other. In the International Union of Pure and Applied Chemistry (IUPAC, usually spoken as *eye-you-pac*) system of nomenclature, a chemical name has three parts: prefix, parent, and suffix. The parent name selects a main part of the molecule and tells how many carbon atoms are in that part; the suffix identifies the functional-group family that the molecule belongs to; and the prefix specifies the location(s) of various substituents on the main part:

Prefix—Parent—Suffix

Where are the substituents?   How many carbons?   What family?

A *substituent* is an atom or group of atoms that substitutes for a hydrogen atom on a hydrocarbon. A substituent might be an alkyl group, a halogen, an alcohol, or another functional group [<<Section 3.11].

Straight-chain alkanes are named according to the number of carbon atoms they contain, as shown in Table 4.3. With the exception of the first four compounds—methane, ethane, propane, and butane—whose names have historical origins, alkane molecules are named based on Greek numbers, according to the number of carbon atoms. The suffix *–ane* is added to the end of each name to identify the molecule as an alkane.

The partial structure that would remain if you imagine a hydrogen atom removed from an alkane molecule is called an **alkyl group**. Alkyl groups are named by replacing the *–ane* ending of the starting (parent) alkane with a *–yl* ending. For example, removal of a hydrogen atom from a methane molecule (CH_4) would give a *methyl group*, —CH_3, and removal of a hydrogen atom from ethane (CH_3CH_3), an *ethyl group*, —CH_2CH_3. Similarly, removal of a hydrogen atom from the end carbon of any *n*-alkane gives the series of *n*-alkyl groups shown in Table 4.3.

Think about It

e4.7 Name alkanes using the IUPAC system.

TABLE 4.3 Names of Common Alkanes and Alkyl Groups

Alkane	Name	Alkyl Group	Name (Abbreviation)
CH_4	Methane	—CH_3	Methyl (Me)
CH_3CH_3	Ethane	—CH_2CH_3	Ethyl (Et)
$CH_3CH_2CH_3$	Propane	—$CH_2CH_2CH_3$	Propyl (Pr)
$CH_3CH_2CH_2CH_3$	Butane	—$CH_2CH_2CH_2CH_3$	Butyl (Bu)
$CH_3CH_2CH_2CH_2CH_3$	Pentane	—$CH_2CH_2CH_2CH_2CH_3$	Pentyl

Interactive Exercises 4.9–4.10

SUBMIT

Draw structural formulas for alkanes and their isomers.

e4.7 outlines the IUPAC system for naming alkanes and Interactive Exercises 4.9 and 4.10 provides practice in applying these names. As we cover new functional groups in later chapters, the applicable IUPAC rules of nomenclature will be given.

Carbon atoms can be connected to form chains called alkanes. Different connectivity patterns lead to isomers. Alkanes and other organic compounds are given unique, systematic names using the IUPAC system of nomenclature.

4.6 Polymers and Unsaturated Hydrocarbons

The **polymer** making up the strong synthetic fishing net that hauled up the tonne of methane clathrate from the Pacific Ocean floor was synthesized from *unsaturated carbon compounds* derived from petroleum. Most fishing nets are made of nylon or polyethylene, both consisting of very large molecules (*macromolecules*) that make up polymeric substances produced by the petrochemical industry. To make a polymer such as polyethylene, chemists first treat petroleum in a process known as *cracking*. At very high temperatures, bond breaking or "cracking" can occur, and longer-chain hydrocarbon molecules in petroleum break into smaller molecular units. These reactions are carried out in the presence of *catalysts*, substances that speed up reactions and direct them toward specific products. Among the important products of cracking are ethylene (ethene) and other *alkenes*, which are used by chemists as the **monomers** (small molecule building blocks) in polymerization reactions to produce materials such as polyethylene.

While the hydrocarbons we've considered in Sections 4.1–4.5 are saturated and contain only C—C single bonds, others such as ethene are *unsaturated*. This means they have double bonds, triple bonds, or rings, including aromatic rings [<<Section 3.11] (Table 4.4). Organic compounds containing C=C double bonds are called *alkenes*, and their IUPAC names have an *–ene* suffix. Compounds with a C≡C triple bond are called *alkynes*, and their IUPAC names end in *–yne*. Aromatic compounds are made of cyclic molecules with electrons delocalized around the ring. Their reactivity is very different than that of alkenes, and is discussed in detail in Chapter 20.

The word *polymer* means "many parts" (from the Greek, *poly* and *meros*). Many *monomer* units react together to form a polymer.

TABLE 4.4 Classification of Hydrocarbons by Structural Features and Formula

Type of Hydrocarbon	Characteristic Features	General Formula	Example
Alkanes	C—C single bonds and all C atoms have four single bonds	C_nH_{2n+2}	CH_4, methane C_2H_6, ethane
Cyclic alkanes	C—C single bonds and all C atoms have four single bonds	C_nH_{2n}	C_6H_{12}, cyclohexane
Alkenes	C=C double bond	C_nH_{2n}	$H_2C=CH_2$, ethene (ethylene)
Alkynes	C≡C triple bond	C_nH_{2n-2}	HC≡CH, ethyne (acetylene)
Aromatics	Electrons delocalized around the ring	Structure dependent	Benzene, C_6H_6

Ethylene (also ethene, C_2H_4), the unsaturated hydrocarbon monomer from which polyethylene is made, is a product of petroleum refining and is the carbon compound synthesized on the largest scale globally, with over 75 000 000 tonnes produced per year. When ethylene is heated to between 100 and 250 °C at a pressure of 1000–3000 bar in the presence of a catalyst, polymers with molecular weights up to several million are formed.

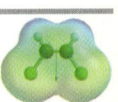

Molecular Modelling (Odyssey)

e4.8 Simulate bulk matter samples of alkenes.

Many $H_2C=CH_2$ ⟶

Ethylene   **A section of polyethylene**

Extensive use of synthetic polymers (sometimes called plastics) is a fairly recent development. A few synthetic polymers (Bakelite, rayon, and celluloid) were made early in the 20th century, but most of the products with which you are familiar originated in the last 50 years. More than 120 million tonnes of synthetic polymers are now produced annually in the world, about 20 kg per person each year. This huge growth is due to the properties of synthetic polymers—they are lightweight, act as thermal and electrical insulators, and can be formulated into a wide variety of materials ranging from soft packing materials to fibres that are stronger than steel.

We examine the classification and structures of polymers in detail in Chapters 19 and 24. Here, we draw attention to their importance in every aspect of our lives, as part of the cycle of carbon compounds derived from petroleum and fossil fuels. A chemist's perspective on fossil fuels is needed to inform choices society makes about burning hydrocarbons for energy while also preserving them as feedstocks for the production of polymers for future generations.

Think about It

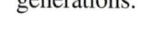

e4.9 Learn how to classify alkanes and alkenes.

WORKED EXAMPLE 4.6—UNITS OF UNSATURATION

A compound of formula C_9H_{14} contains no rings or triple bonds. How many double bonds does it have?

Solution

The formula of a saturated hydrocarbon molecule is C_nH_{2n+2}, where n is the number of carbon atoms. Every multiple bond or ring (called *units of unsaturation*) is evident by the compound having two H atoms fewer than a saturated compound. In this case, a saturated compound with nine C atoms would have twenty H atoms. C_9H_{14} is missing six H atoms relative to a completely saturated compound. Therefore, it has three units of unsaturation. Since we know that no triple bonds or rings are present, all of these must be double bonds. C_9H_{14} has three double bonds.

EXERCISE 4.11—UNITS OF UNSATURATION

How many units of unsaturation does a molecule of a compound with the formula C_6H_6 contain? Draw a reasonable structure for a molecule of this compound, if you know that it contains one ring.

EXERCISE 4.12—POLYMERS

Teflon is a polymer of the monomer tetrafluoroethene ($F_2C=CF_2$). Show the structure of Teflon by drawing several repeating units.

Compounds whose molecules contain double bonds, triple bonds, or rings are called unsaturated compounds. Alkenes such as ethene, which contain carbon–carbon double bonds, can react with many other alkenes in polymerization reactions to form polymers. An important polymer made from ethene monomer units is a macromolecular substance called polyethylene.

4.7 Where There Is Methane, Is There Life?

The cycle of carbon compounds, from methane and carbon dioxide to large polymers of glucose, is so fundamentally important on Earth that it's impossible to imagine life without them. Awareness of their role has led chemists to search for these compounds elsewhere in our solar system, to learn whether conditions might also be favourable for life on other planets. The recent and forthcoming exploration of Mars is a good example. Do other planets, too, have their own "great carbon mysteries" [<<Section 4.3]?

Methane on Mars

Knowing the importance of methane on Earth, and that most methane in our atmosphere is biogenic, scientists were very excited in 2003 to detect what could be the signature of methane in the atmosphere of Mars. Its infrared spectrum was observed by both Earth-based telescopes and a European Space Agency Mars orbiter spectrometer. Average levels of methane were about 10 ppb (compared to 1800 ppb on Earth), but data from high-resolution telescopes suggest much higher local concentrations in certain places. Since the lifetime of methane molecules in the Martian atmosphere is expected to be several hundred years, it is intriguing to imagine what sources might be replenishing those methane levels. Martian methane might be produced from abiogenic processes such as volcanoes or hydrogeochemical (serpentinization) processes. It might be coming from a source outside of Mars, such as a comet, meteorite, or interplanetary dust particles. But most intriguing is the speculation that it might be biogenic, produced by microbes such as methanogens, able to convert carbon monoxide and molecular hydrogen into methane. Other experimental findings in 2008 suggest past and recent liquid water on Mars, another molecule essential to life on Earth.

Each of these possibilities is being carefully examined by chemists. Volcanic methane on Earth is accompanied by sulfur dioxide gas (SO_2), which has not been found on Mars. Evidence for large cometary impacts and conditions suitable for serpentinization reactions has also been sought. If microbial life has recently produced the methane in the Martian atmosphere, we would expect to see other organic compounds they would likely have made. So far these haven't been detected, leading scientists to wonder whether a strong oxidizing agent might be present in the atmosphere, causing the destruction of both methane in the atmosphere and other organic compounds on the surface. After searching for almost 30 years, the spectral signature of the very strong oxidizing agent hydrogen peroxide (H_2O_2) was detected in the Martian atmosphere in 2003. Hydrogen peroxide might be created by photochemical reactions or in large Martian dust and electrical storms. Perhaps organic molecules have not been found on the surface because they are rapidly destroyed through oxidation reactions.

Some of the big questions about the origin of methane on Mars may be answered in 2012, when NASA's Mars Science Laboratory rover is scheduled to arrive on the planet. Among other experiments, a very high-resolution *tunable laser spectrometer* will be used to determine the ratio of $^{13}C/^{12}C$ isotopes, giving important clues as to whether the methane is biogenic or abiogenic. But both the chemical analysis and solving the puzzle will be challenging! The level of methane on Mars is about one half of 1% of that on Earth, so far fewer methane molecules are available to detect. Approximately 1% of each of those methane molecules contain ^{13}C atoms, and chemists will need to detect a *change* in the parts per thousand range in the ratio of $^{13}C/^{12}C$ isotopes to indicate whether living organisms have produced the methane. So a highly accurate and precise measurement of this isotope ratio will be required by a tiny instrument that can measure methane levels at 0.01 ppb. Furthermore, the interpretation of those levels will be difficult. To assess the signature of molecules indicating life on Earth requires comparing their $^{13}C/^{12}C$ isotopic ratio to that of a standard reference for carbon from inorganic mineral sources. Until recently, that standard source came from carbonate in the Pee Dee Belemnite (PDB) formation in South Carolina, U.S. High precision sampling from an inorganic reference source *on Mars* will be needed, requiring samples from carbonates in the surface, subsurface, and rocks, and this should be compared to the $^{13}C/^{12}C$ ratio in carbon dioxide and methane in the atmosphere, along with any organic carbon on the surface.

Other organic compounds found on the surface of Mars will also be tested for *chirality* [>>Section 9.8] or handedness. Many of the molecules of life on Earth are chiral, being identical in every respect to others except for the three-dimensional arrangement of the atoms. The name *chiral* comes from the Greek word *chiros*, meaning "hand," as the

NASA

Martian atmosphere.

subtle differences in shape of such molecules can be compared to the difference between your right and left hand. Since most of the building blocks of molecules of life on Earth, such as amino acids, carbohydrates, and nucleic acids, are chiral, it will be of great interest to see whether any other compounds of carbon exist on Mars and, if so, whether they are chiral.

While we don't have the evidence on other planets of fishing vessels pulling up hissing chunks of ice, it seems probable that methane and carbon dioxide clathrate hydrates exist under the permafrost of Mars and elsewhere in our solar system. Evidence consistent with their presence has been found on the gas giant planets and many of their moons. Models developed after analyzing data from the Cassini-Huygens mission to Titan, Saturn's largest moon, suggest that the large amounts of methane found there may be stored in the form of clathrates.

Finally, scientists have studied possible ways to make the atmosphere of Mars more like that of Earth and more hospitable to life. Recently published calculations suggest that the introduction of $\sim$0.4 Pa of the potent "greenhouse gas" perfluoropropane (C_3F_8) into the Martian atmosphere might be sufficient to create a "runaway" greenhouse effect that would release large amounts of CO_2 gas from the Martian poles. An intriguing finding, and one that should provide a note of caution when we think of the global experiment humans are presently carrying out by introducing halocarbons and other gases with large global warming potentials into Earth's atmosphere.

We've read only a few chapters in "the great carbon mystery novel." Many others remain to be written!

> Because of the importance of methane as a substance that signifies life on Earth, chemists are designing experiments to verify the detection of methane on other planets and determine whether it has been produced by recent forms of life.

References

For further reading about methane clathrate hydrates, consult I. Chatti, et al., (2005), "Benefits and Drawbacks of Clathrate Hydrates: A review of Their Areas of Interest," *Energy Conversion and Management,* 46, 1333–43; for climate change science, see *Climate Change 2007: Synthesis Report: An Assessment of the Intergovernmental Panel on Climate Change,* IPCC Fourth Assessment Report, Valencia, Spain: IPCC (www.ipcc.ch); and for methane on other planets, see the cover story in the May 2007 *Scientific American* by S.K. Atreya, "The Mystery of Methane on Mars and Titan," pp. 42–51.

SUMMARY

Key Concepts

- Methane (CH_4), the simplest **alkane**, is the primary hydrocarbon found in the fossil fuel natural gas. Methane is used in combustion reactions as a major source of energy, and it is an important feedstock for other organic compounds. Formed in nature through *thermogenic, abiogenic,* and *biogenic* processes, its origin can be determined by isotope ratio mass spectrometry. When released into the atmosphere, methane is eventually removed by reaction with hydroxyl radicals (Section 4.2). Because of the importance of methane as a substance that signifies life on Earth, chemists have designed experiments to detect methane on other planets such as Mars as an indicator of conditions that might be hospitable to supporting life. (Section 4.7)

- The earth's climate is regulated by the delicate *radiation balance* between incoming energy from the sun and energy radiated back into space. Methane, carbon dioxide, nitrous oxide, ozone, and halogenated organic compounds, while present in only trace levels as atmospheric gases, are important "greenhouse gases." They cause warming of our planet by absorbing infrared radiation that would otherwise escape into space. The mechanism for warming involves the transfer of vibrational energy from these compounds by *collisional de-excitation* to N_2 and O_2, the principal gases making up 99% of the earth's atmosphere. N_2 and O_2 do not absorb IR radiation directly, and are therefore not considered "greenhouse gases." (Section 4.3)

- The relative ability of a "greenhouse gas" to trap heat in the Earth's atmosphere, compared to CO_2, is referred to as its **global warming potential**. The effect of a "greenhouse gas" on climate depends on (a) its concentration, (b) how strongly and where in the infrared region it absorbs energy, and (c) its atmospheric lifetime. (Section 4.3)

- Atmospheric **aerosols**, made of fine droplets of liquid or dust suspended in the atmosphere, cause cooling by reflecting incoming visible light from the sun. (Section 4.3)

- Our knowledge of the earth's temperature through history comes from measuring the $^{18}O/^{16}O$ isotope ratio of water molecules in ice cores. Carbon dioxide levels are found by chemical analysis of the bubbles trapped in those same ice cores. A strong correlation exists between CO_2 gas levels and temperature over the past 400 000 years. Changes in the earth's climate in the past 250 years result in large part from human activity. (Section 4.3)

- Numerous strategies are used to address increasing levels of atmospheric "greenhouse gases" in the atmosphere. These include reducing emissions, capturing and storing "greenhouse gases," and converting carbon compounds such as CO_2 into useful materials through reactions to form other chemical species. (Section 4.4)

- Atmospheric CO_2 can also be captured very efficiently by plants and some other organisms, and stored in the form of glucose, through the process called photosynthesis. Glucose **monomers** are converted into **polymers** by plants and animals for structural purposes and energy storage (Section 4.4). Over long periods of time, these carbon polymers are converted into fossil fuels, including natural gas and petroleum. Chemists have taken fossil fuel–derived carbon compounds out of storage to convert them into synthetic polymers that have transformed almost every aspect of modern life (Section 4.6). But most fossil fuels are burned in combustion reactions to produce energy, along with carbon dioxide and water vapour.

- One strategy for reducing dependence on fossil fuels is to create and burn renewable **biofuels**, based on carbon compounds from recently living biomass. Much of the biofuel research is directed toward the energy balance in these processes, including trying to reduce the fossil fuel energy inputs presently required to produce the biomass and create and use biofuels. (Section 4.4)

- Massive amounts of methane, carbon dioxide, and other small organic compounds (referred to as "guests") are stored in nature as **clathrate** hydrates, in which guest molecules are held inside cages of "host" water molecules. Methane clathrate hydrates are potentially useful as a vast source of energy and potentially dangerous to our climate if released into the atmosphere. When warmed, methane clathrate hydrates release more than 150 times their own volume of methane, a potent "greenhouse gas." The bonding in methane clathrate hydrates is an example of the bonding found in "supramolecular" or "host-guest" compounds. (Section 4.1)

- Complex **hydrocarbons** are formed when carbon atoms are connected to form chains called **alkanes**. The primary components of the fossil fuel *petroleum* are much larger alkanes than those found in natural gas. (Section 4.5)

- The different connectivity patterns of alkanes lead to **isomers**, compounds with the same numbers and kinds of atoms that differ in the way the atoms are arranged. Alkanes and other organic compounds are given unique, systematic names using the IUPAC system of nomenclature. (Section 4.5)

- Compounds whose molecules contain double bonds, triple bonds, or rings, are called *unsaturated* compounds. Alkenes such as ethene, which contain carbon–carbon double bonds, can react with many other identical or different alkenes in polymerization reactions, to form polymers. An important polymer made from ethene monomer units is the macromolecular substance, polyethylene. (Section 4.6)

REVIEW QUESTIONS

Section 4.1: Ice on Fire

4.13 Air can also be trapped inside clathrate hydrates. With reference to the figure below, what term(s) most correctly describe(s) the following forces of attraction:
(a) between nitrogen atoms in the N_2 molecules inside the clathrate hydrate
(b) between N_2 molecules and H_2O molecules of the clathrate hydrate cage
(c) between H_2O molecules that make up the clathrate hydrate cage
(d) between N_2 molecules and O_2 molecules inside the clathrate hydrate cage

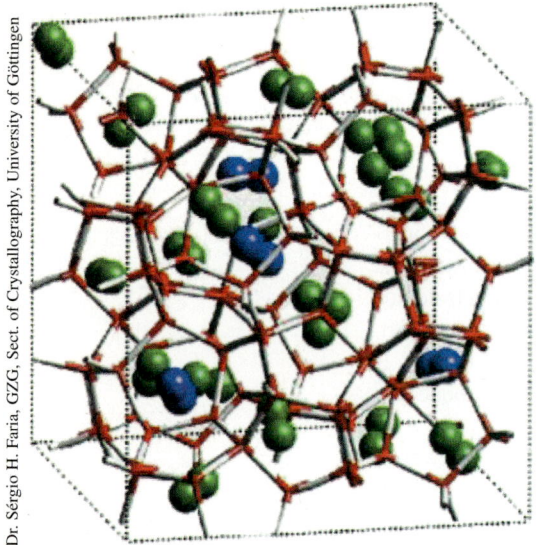

Section 4.2: Methane: Signature of Life

4.14 Because of concerns about their role in stratospheric ozone depletion, chlorofluorocarbons (CFCs) have been replaced as refrigerants with hydrochlorofluorocarbons (HCFCs) and hydrofluorocarbons (HFCs), such as HFC-134a (1,1,1,2-tetrafluoroethane). What differences would you expect in the reactions of hydroxyl radicals in the *troposphere* with CFCs and HFCs?

Section 4.3: Climate Change and "Greenhouse Gases"

4.15 It has been suggested that the term *greenhouse effect* should not be used to describe the warming of the earth's atmosphere by gases such as CO_2, CH_4, H_2O, etc. What phenomena are mainly responsible for the warming of air inside a greenhouse? How does this mechanism differ from the warming of our atmosphere by "greenhouse gases"?

4.16 With reference to e4.3, what atmospheric chemistry causes the temperature profile of the earth's atmosphere to switch directions when moving from the troposphere to the stratosphere? Is this chemistry mostly caused by natural processes or human activity?

4.17 Which of the following compounds would you expect to be "greenhouse gases" if found in our atmosphere? Helium, $He(g)$; ethane, $CH_3CH_3(g)$; ozone, $O_3(g)$; chlorine, $Cl_2(g)$; chloroform, $CHCl_3(g)$. Explain.

4.18 Consulting other sources as needed, what are the main atmospheric sources of nitrous oxide, N_2O? What is the ultimate fate of most atmospheric nitrous oxide?

Questions 4.19–4.29 relate to Figure 4.10, which summarizes scientific understanding of how radiative forcing of our climate has changed between 1750 and 2005.

4.19 What is the significance of IPCC selection of the date 1750 to compare to the present day?

4.20 Rank from least important to most important, the radiative forcing changes due to methane, carbon dioxide, halocarbons (i.e., CFCs and related compounds), and tropospheric ozone.

4.21 Do clouds cause warming or cooling? Explain.

4.22 A newspaper column states that the global surface air temperature increases since 1750 can be correlated with changes in solar irradiance rather than increased levels of CO_2 gas. Do you agree? Explain.

4.23 What is the radiative forcing effect of increased levels of tropospheric ozone?

4.24 What is the radiative forcing effect of increased levels of stratospheric ozone?

4.25 In which region of our atmosphere would you expect CF_2Cl_2 to exert its greatest direct radiative forcing effect?

4.26 In which region of our atmosphere would you expect CF_2Cl_2 to undergo the most significant chemical reactivity?

4.27 What is the difference between the radiative forcing and the global warming potential of methane?

4.28 What factors contribute to CH_4 having a global warming potential (over a 100-year horizon) that is 25 times that of CO_2?

4.29 Consult Figure 2.20 in the summary report of the 2007 Intergovernmental Panel on Climate Change on the companion website. Rank the following contributors with respect to our level of scientific understanding about their radiative forcing, from high understanding to low understanding: stratospheric ozone; tropospheric nitrous oxide; solar irradiance; jet contrails; tropospheric carbon dioxide.

Section 4.4: Capturing, Storing, and Recycling Carbon

4.30 What three factors are used to predict the effect a trace atmospheric "greenhouse gas" will have on the earth's climate?

4.31 List three of the most significant advantages and disadvantages of a strategy to provide energy by combustion of biofuels made from corn and soybean oil to replace fossil fuel combustion. What type of biofuels might minimize the disadvantages?

4.32 Use Figure 4.11 to determine what atmospheric levels of CO_2 (expressed in ppm) were in the year you were born. Compare them to present day values. What percent increase in atmospheric CO_2 levels has taken place in your lifetime?

4.33 Monthly atmospheric levels of carbon dioxide have been carefully measured at various stations around the world over the past 50 years. Data from the Mauna Loa station in Hawaii and from the South Pole are shown below. How do you account for the similarities and differences in the two data sets?

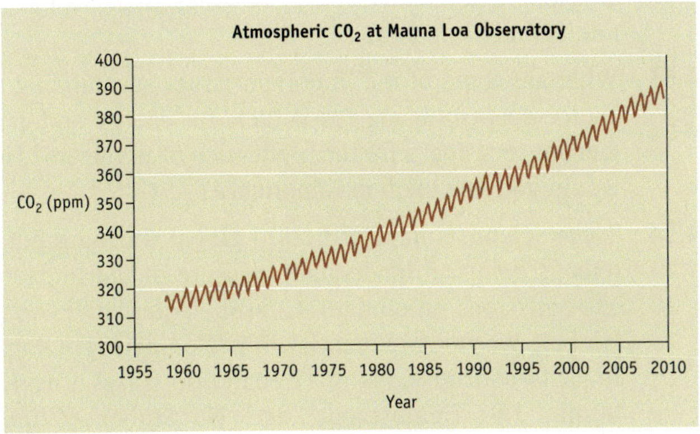

Data from Carbon Dioxide Information Analysis Center, Oak Ridge National Laboratory, U.S. Department of Energy

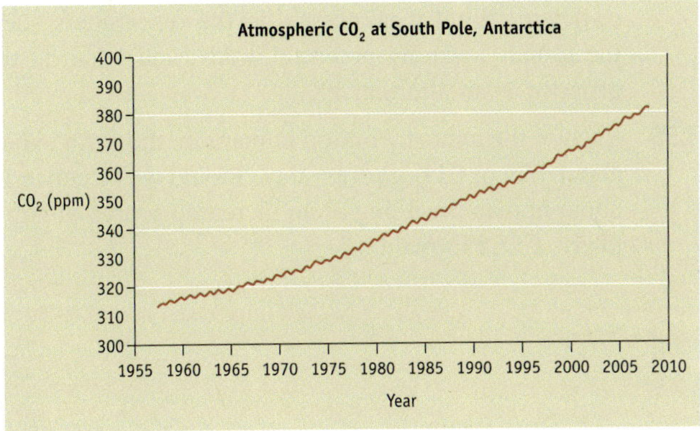

Data from Carbon Dioxide Information Analysis Center, Oak Ridge National Laboratory, U.S. Department of Energy

Section 4.6: Polymers and Unsaturated Hydrocarbons

4.34 How many units of unsaturation do each of the following compounds contain?

(a) 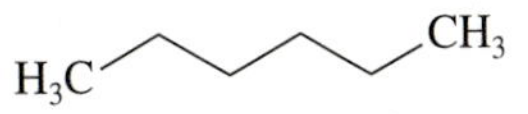

(b) 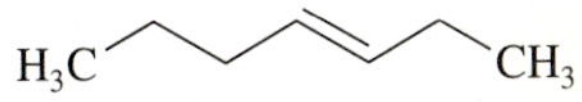

(c) 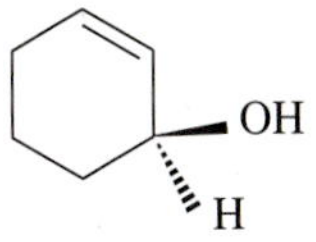

(d) 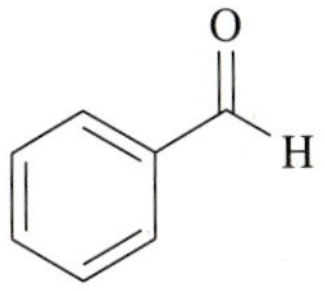

4.35 Draw four possible structures for each of the following formulas:

(a) C_6H_{10} (b) C_8H_8O (c) $C_7H_{10}Cl_2$

4.36 How many units of unsaturation does a compound with the formula C_4H_8O contain? Draw reasonable structures for molecules of three substances with this formula.

4.37 Identify the functional groups present in each of the molecules in Question 4.34.

Consult e4.7 on alkane nomenclature for explanation and worked examples, and then answer *Questions 4.38–4.40.*

4.38 Give IUPAC names for the following alkanes:

(a) The three isomers of C_5H_{12}

(b) 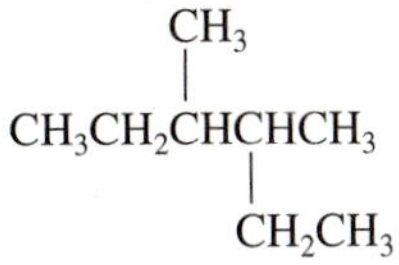

(c) 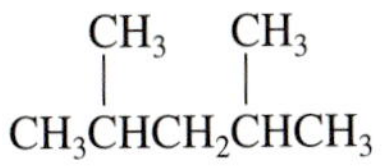

(d) 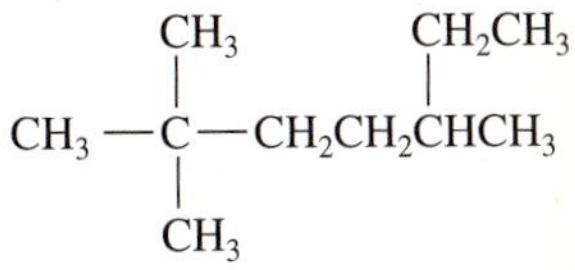

4.39 Draw structures corresponding to the following IUPAC names:

(a) 3,4-Dimethylnonane
(b) 3-Ethyl-4,4-dimethylheptane
(c) 2,2-Dimethyl-4-propyloctane
(d) 2,2,4-Trimethylpentane

4.40 Give a systematic IUPAC name for the following alkane:

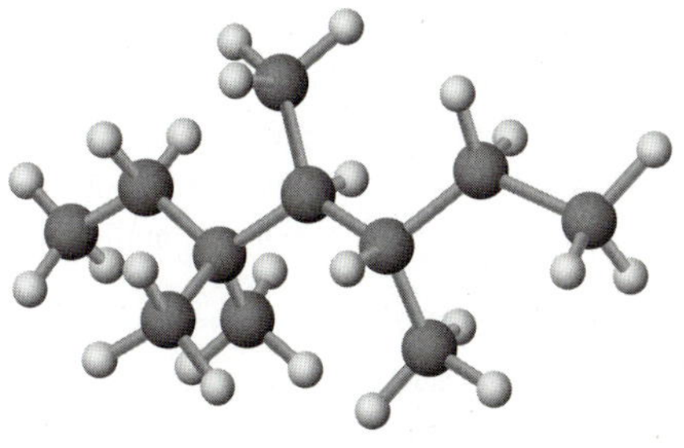

4.41 Saran is a copolymer of 1,1-dichloroethene and chloroethene (vinyl chloride). Show a possible structure for this polymer by drawing several repeating units.

4.42 What monomer unit might be used to prepare the polymer, polyvinyl chloride?

Vinyl chloride   **Poly(vinyl chloride)**

4.43 The spectroscopic signature for the strong oxidizing agent hydrogen peroxide (H_2O_2) has been recently detected on Mars. Explain the significance of this finding for helping to answer the question of whether conditions on Mars are favourable to support life.

SUMMARY AND CONCEPTUAL QUESTIONS

4.44 With reference to methane clathrate hydrate structures in Figure 4.1, and Exercises 4.1–4.2, rank in order of increasing strength the forces of attraction between carbon and hydrogen atoms in CH_4 molecules, between H_2O molecules, and between H_2O and CH_4 molecules in methane clathrate hydrate. Explain.

4.45 Exercise 4.3 showed that hydrocarbons such as CH_4 react readily with atmospheric hydroxyl radicals, but chlorofluorocarbons (CFCs), such as CF_2Cl_2 do not. Using the data in Table 7.3, compare the relative strengths of C—H, C—F, and C—Cl bonds, and suggest a reason for the difference in reactivity of hydrocarbons and CFCs with hydroxyl radicals.

4.46 With reference to e4.5, explain why perfluoropropane (C_3F_8) has been used in modelling studies as a gas that might be such a significant "greenhouse gas" that it could cause a "runaway greenhouse effect" on the planet Mars.

4.47 SF_6 has a global warming potential of 22 800 relative to CO_2 over a 100-year horizon. Consult e4.5 for the IR spectrum of CO_2. What features of the IR spectrum of SF_6 might you expect to be different than for CO_2, and how would those differences affect its global warming potential?

4.48 Consult the animated infrared spectrum of CH_2Cl_2 in e3.16. Compare its infrared spectrum to that of CO_2 and H_2O vapour in e4.5. Are there strong absorption peaks in the infrared spectrum of CH_2Cl_2 at wavenumbers where CO_2 and H_2O do not absorb IR? Would you expect CH_2Cl_2 to be an important "greenhouse gas"?

4.49 Methane, $CH_4(g)$, is one of four atmospheric gases most strongly linked to the presence of life. If you were responsible for designing a probe to explore another planet to determine whether conditions are favourable to support life as we know it on Earth, what three other gases would be highest on your list for analysis, and why?

4.50 Water vapour, $H_2O(g)$, is often left out of discussions of "greenhouse gases" in our atmosphere. Is water vapour a "greenhouse gas"? What might be some reasons for not including water vapour as a "greenhouse gas" affected by human activity?

4.51 In Worked Example 4.5, you estimated the total mass of CO_2 in the earth's atmosphere. Estimate the mass of CO_2 put into the atmosphere in one year by your vehicle. List your assumptions.

4.52 What are some of the important issues surrounding the choice between using fossil fuels for energy and developing alternative energy schemes? How would you rate them in importance?

4.53 What are some of the important issues surrounding the choice between using fossil fuels for energy and using them as feedstocks for the production of polymers? How would you rate them in importance?

4.54 Some scientists have criticized global warming potentials as being an inadequate measure of the impact of gas emissions on climate. Consult "Alternatives to the Global Warming Potential for Comparing Climate Impacts of Emissions of Greenhouse Gases," by K.P. Shine, J.S. Fuglestvedt, and N. Stuber (*Climate Change* 68, 3[February 2005]: 281–302), for one such criticism. What weaknesses do the researchers see in the global warming potential index, and what do they propose as an alternative?

4.55 Each of the almost 7 billion humans on the earth exhales about 1 kg of $CO_2(g)$ every day. Should we be concerned about human breathing as an increased source of atmospheric CO_2? Explain.

With the **eCHACR** single-sign-on access card bundled with your text, log on (http://login.cengage.com/sso) and access the e-book and click on any in-margin icons for dynamic molecular-level animations and simulations, videos of laboratory reactions, interactive exercises, reviews of background concepts, and other online supplementary materials as noted by the icons in the text margins.

Also go to www.chemistry.nelson.com <http://www.chemistry.nelson.com> for Answers to in-chapter exercises and selected Review Questions, Test Yourself questions, weblinks, crossword puzzles, flashcards, glossary of key terms, and other student resources.

CHAPTER 5

Chemical Reaction, Chemical Equations

Courtesy of Dr. Peter Mahaffy

GetStock/Alamy Images/Jim Pickerell

5.1 Case Study: Don't Waste a Single Atom!

Industrial chemical plants and "green chemistry" make front page news these days. But these new stories don't refer to highly efficient chemical "factories" such as the canola plants in the Northern Alberta field shown above. Each of those plants is literally a green chemical factory—converting chemicals in the air, soil, and water into nutritious oils that help to feed the world.

Over 100 years ago, it was realized that the crop growth needed to avoid starvation of an exploding world population was limited by the amount of nitrogen in forms that plants could use. Enter fertilizers such as ammonia (NH_3), which are synthesized in large human-built industrial plants from the $N_2(g)$ in the air. Plants such as canola can easily take up nitrogen atoms in the form of NH_3, using them as building blocks to produce proteins, nucleic acids, and many other organic nitrogen-containing compounds. Those compounds, in turn, are converted into other life-sustaining compounds by livestock that provide meat to feed the world. The production of natural products by plants and animals is highly efficient, and chemists often try to mimic these processes in laboratories and industrial plants >>Section 13.1.

> *To feed a hungry world while taking care of our environment, farmers need to work with agricultural, soil, and water chemists to find ways to account for and avoid wasting nitrogen atoms.*

Yet in the complex sequence of reactions and processes required to convert nitrogen atoms from ammonia into nitrogen atoms in forms suitable for human use, atoms are "wasted" at every step. It is estimated that of every 100 nitrogen atoms produced in the synthesis of NH_3 fertilizer, only 14 end up in the mouth of a vegetarian and 4 in the mouth of a meat-eater. The remainder are returned to the environment as many different compounds during fertilizer production, storage, transport, and application; uptake of fertilizer by crops; crop harvesting; uptake by livestock; and the formulation of food products from plants and animals for humans. To make matters worse, many of the "waste" compounds cause environmental problems. To feed a hungry world while taking care of our environment, farmers need to work with agricultural, soil, and water chemists to find ways to account for and avoid wasting nitrogen atoms.

Chemists also work with manufacturers to avoid waste in large-scale industrial plants that produce the building blocks for many materials important to life. Almost every process carried out in modern chemical industry is now scrutinized to minimize the waste of atoms, improve energy efficiency, and to use and produce safe materials that don't damage the environment or human health. This new emphasis by chemical industry is referred to as "green chemistry" or "sustainable chemistry."

One central feature of green chemistry approaches is to keep track of the atoms that go into and come out of industrial processes. Chemical industry uses the concept of *atom economy* or *atom efficiency*: processes should be designed that are efficient in the use of all of the atoms in the starting material. This means that as far as possible, all of the atoms in the starting materials should be in the desired products.

Here is an example. Many people predict that in the future, the fuel for our industrial society will be based on methanol, rather than fossil fuels like coal, oil, and natural gas. Methanol, $CH_3OH(\ell)$, is a clean-burning fuel; it emits very low levels of nitrogen- or sulfur-containing compounds into the atmosphere, although carbon dioxide is an inevitable combustion product. How is methanol currently made, and can chemists think of new ways that minimize waste (improve the atom efficiency) of the process?

One important current method for producing methanol [<<Section 4.2] is a two-step process starting with methane and steam. The mixture of carbon monoxide and hydrogen produced in the first step is called *syngas,* which reacts in the second step with hydrogen to produce methanol (as a gas at the high temperature of the process).

$$CH_4(g) + H_2O(g) \longrightarrow CO(g) + 3\ H_2(g)$$

$$CO(g) + 2\ H_2(g) \xrightarrow[\text{catalyst}]{\text{copper/zinc}} CH_3OH(g)$$

The overall reaction, which is the sum of these two steps, can be expressed by the equation

$$CH_4(g) + H_2O(g) \longrightarrow CH_3OH(g) + H_2(g)$$

This method poses two challenges. First, for every molecule of methane and steam that react, two of the hydrogen atoms do not end up in the CH_3OH product. Second, the energy required to produce the high temperatures for the first step account for 70% of the cost of methanol manufacture.

Professor George Olah and his research team at the University of Southern California are looking for ways to make methanol that are more atom-efficient, and more energy-efficient. One idea is to consider a one-step process by direct reaction of methane and oxygen, with a catalyst that redistributes the atoms differently than in combustion of methane. This reaction would incorporate all of the carbon, hydrogen, and oxygen atoms in starting materials into the product:

$$2\ CH_4(g) + O_2(g) \xrightarrow{\text{catalyst}} 2\ CH_3OH(g)$$

Olah is reported to have said, "You take methane and stick in just one oxygen atom [per molecule]." Easier said than done! Just because we can write a chemical equation does not mean that the corresponding reaction will proceed. A catalyst that brings this reaction about at low temperatures has not yet been developed.

A different approach is to develop catalysts that allow methane and gaseous bromine to react at moderate temperatures to form bromomethane (CH_3Br), which then can be reacted with water to form methanol.

$$CH_4(g) + Br_2(g) \xrightarrow{\text{catalyst}} CH_3Br(g) + HBr(g)$$
$$CH_3Br(g) + H_2O(\ell) \longrightarrow CH_3OH(\ell) + HBr(g)$$

The net result of these reactions can be expressed in one equation:

$$CH_4(g) + Br_2(g) + H_2O(\ell) \longrightarrow CH_3OH(\ell) + 2\ HBr(g)$$

The atom efficiency of this overall process does not appear to be high: none of the Br atoms in the reactants finish up in the product, and for each 1 mol of CH_3OH produced (32 g), 2 mol of HBr by-product (162 g) is also formed. However, if the HBr is allowed to react with oxygen in the air, it regenerates bromine (Br_2), which can be recycled to react with more methane. So, taking bromine out of the equation if it is recycled, the net reaction is 100% atom-efficient: all of the atoms in the reactant molecules finish up in the products. Progress is being made on developing catalysts for this methane-bromine reaction.

Perhaps the best solution would be to produce methanol by direct reaction of carbon dioxide and hydrogen:

$$CO_2(g) + 3\ H_2(g) \xrightarrow{\text{catalyst}} CH_3OH(\ell) + H_2O(\ell)$$

Apart from the production of water, which is a benign waste product, this reaction is highly atom-efficient. It's beauty and elegance lies in the fact that $CO_2(g)$, which is the inevitable waste product of combustion of any hydrocarbon, is being used here as a valuable starting material to produce more methanol fuel. If it could be implemented properly, the net result of cycles of producing and burning methanol fuel would not lead to releasing $CO_2(g)$ into the atmosphere.

This ingenious line of thinking is consistent with the guiding principles of the Green Chemistry Institute of the American Chemical Society. Six of those principles are listed here:

1. *Prevention*: It is better to prevent waste than to treat or clean up waste after it has been created.

2. *Atom Economy*: Synthetic methods should be designed to maximize the incorporation of all materials used in the process into the final product.

3. *Less Hazardous Chemical Syntheses*: Wherever practicable, synthetic methods should be designed to use and generate substances that possess little or no toxicity to human health and the environment.

4. *Designing Safer Chemicals*: Chemical products should be designed to affect their desired function while minimizing their toxicity.

5. *Safer Solvents and Auxiliaries*: The use of auxiliary substances (e.g., solvents, separation agents, etc.) should be made unnecessary wherever possible and innocuous when used.

6. *Design for Energy Efficiency*: Energy requirements of chemical processes should be recognized for their environmental and economic impacts and should be minimized. If possible, synthetic methods should be conducted at ambient temperature and pressure.

5.2 Chemical Reaction, Chemical Change

The terms *chemical reaction* and *chemical change* are interchangeable. Here, we begin with an operating definition that is useful to understand the basic idea of chemical reaction, before proceeding to a more sophisticated view.

Chemical reaction has been described in Section 2.6 as a process in which one or more new substances are formed as a result of redistribution of atoms. This basic idea of chemical change is most easily appreciated when all of the reactants and products are molecular substances. One such example is when a stream of chlorine gas, $Cl_2(g)$, is directed onto solid white phosphorus, $P_4(s)$, to form molten phosphorus trichloride, $PCl_3(\ell)$ (Figure 5.1). This is described by a balanced chemical equation:

$$P_4(s) + 6\ Cl_2(g) \xrightarrow[\text{of P and Cl atoms}]{\text{redistribution}} 4\ PCl_3(\ell)$$

Before chemical reaction, all of the P atoms are in P_4 molecules, and all of the Cl atoms are in Cl_2 molecules. We can imagine that at any moment during the reaction, millions of collisions are happening between Cl_2 molecules in the gas phase with P_4

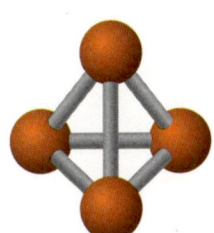

White phosphorus consists of P_4 molecules.

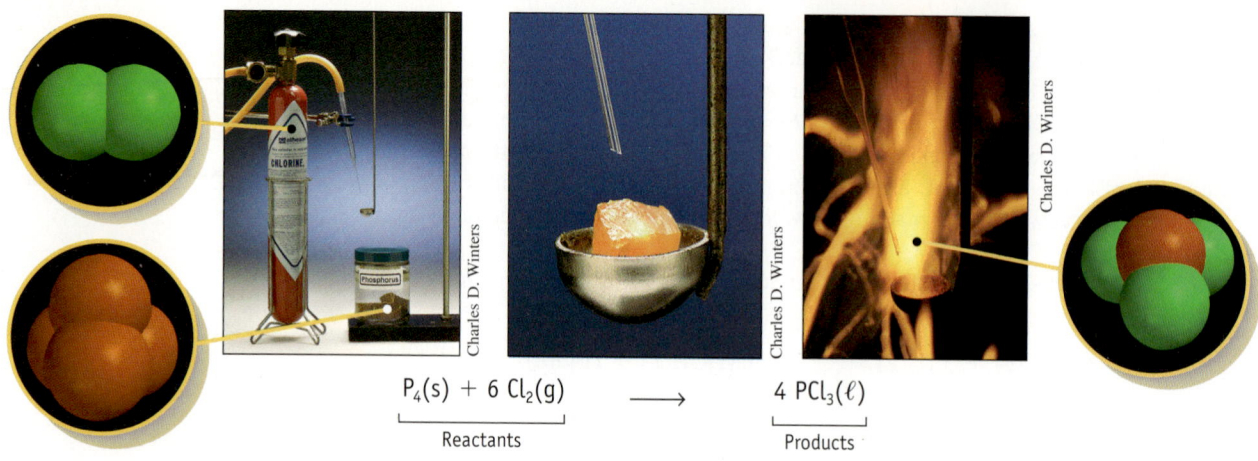

$$P_4(s) + 6\ Cl_2(g) \longrightarrow 4\ PCl_3(\ell)$$

Reactants   Products

FIGURE 5.1 Reaction of white phosphorus with chlorine gas. Chemical reaction is the redistribution of atoms in the reactant molecules to form molecules of a new substance. The phosphorus trichloride product is a different substance from both of the reactants, phosphorus and chlorine. The molecular building blocks of the substances are shown in the inserts. Go to the *Odyssey Stockroom* to see molecular-level simulations of these substances.

Think about It

e5.1 Watch this dramatic reaction and answer questions.

Think about It

e5.2 Watch a laboratory demonstration of the conservation of mass.

molecules on the surface of the phosphorus, and that as a consequence of the electrostatic interactions between them, some bonds (P—P and Cl—Cl) are weakened and break, while others (P—Cl) form. The result is a redistribution of atoms into different molecules.

All chemical reactions satisfy the **law of conservation of matter**: matter is neither created nor destroyed. This fundamental principle of chemical change, proposed by the great French scientist Antoine Lavoiser, can also be called the **law of conservation of atoms** because during chemical reaction, atoms are neither created nor destroyed—they are just redistributed so that the total number of atoms of each element in the reactants is the same as in the products. A consequence of this, called the **law of conservation of mass**, is that the total mass of substances that react is the same as the total mass of substances that are formed. If a total of 10 g of substances react, then 10 g of product substances are formed.

> Chemical reactions of molecular substances give rise to new substances as a result of redistribution of atoms. During reactions there is conservation of matter, conservation of mass, and conservation of atoms.

A Refined Definition of Chemical Reaction

Consideration of some types of chemical reaction, such as the following, forces us to refine our definition:

1. Some chemical reactions involve redistribution of ions. This is the case, for example, when we add a sodium hydroxide solution in water to a solution of acetic acid (CH_3COOH). Aquated hydroxide ions, $OH^-(aq)$, remove hydrogen ions, H^+, from acetic acid molecules, forming aquated acetate ions and water molecules:

In this and later equations, (aq) is the symbol for *aquated* (or *hydrated*), which means surrounded by water molecules [>>Section 6.5].

$$OH^-(aq) + CH_3COOH(aq) \xrightarrow[\text{to } OH^-\text{(aq) ions}]{\text{transfer of } H^+\text{ions}} H_2O(\ell) + CH_3COO^-(aq)$$

This is an example of an *acid-base reaction* [>>Chapter 15].

2. Some reactions, called oxidation-reduction reactions, involve redistribution of electrons [>>Sections 6.7 and 16.2]. If we put a piece of copper metal into a solution of silver nitrate, silver metal deposits on the copper surface, and the solution gradually turns blue (Figure 5.2); blue aquated copper ions are formed as a result

All photos: Charles D Winters

Pure copper
wire

Copper wire in dilute AgNO$_3$
solution; after several hours

Blue color due to
Cu^{2+} ions formed
in redox reaction

Silver crystals
formed after
several weeks

FIGURE 5.2 Reaction between copper metal and aquated silver ions. The reaction can be interpreted as due to transfer of electrons from copper atoms to aquated silver ions.

of transfer of electrons from copper atoms to silver ions that approach the metal surface:

$$\text{Cu(s)} + 2\,\text{Ag}^+(\text{aq}) \xrightarrow[\text{from Cu(s) to Ag}^+(\text{aq})]{\text{transfer of electrons}} \text{Cu}^{2+}(\text{aq}) + 2\,\text{Ag(s)}$$

The two processes described above involve OH$^-$(aq) ions, Ag$^+$(aq) ions, and Cu^{2+}(aq) ions as reactants or products. These ions cannot be called "substances" in the sense that we can have a jar of a substance. Although we can have a jar of water or sodium chloride, we cannot have a jar of OH$^-$(aq) ions alone, nor of Ag$^+$(aq) ions, nor Cu^{2+}(aq) ions. Chemists use the word **chemical species** to refer to any particles, all identical, that have characteristic chemical behaviour. The particles that comprise a chemical species may be atoms, molecules, or ions. For example, H$_2$O molecules, Na$^+$ ions in solid sodium chloride, and aquated Cu^{2+} ions in solution are all chemical species.

Chemical species have their own identity. For example, Cu^{2+}(aq) ions have chemical behaviour that is not at all like the chemical behaviour of solid copper, Cu(s). Cu^{2+}(aq) ions and Cu atoms in solid copper are different species.

Now we can refine our definition of **chemical reaction** (or **chemical change**) as *any process in which one or more new species form as a result of redistribution of atoms, ions, or electrons.*

The previous definition of chemical reaction based on formation of new substances applies at the level we can observe. This refined definition that refers to formation of different species is one that takes us into the molecular-level world of imagination.

A chemical reaction is a process in which new species form as a result of redistribution of atoms, ions, or electrons.

Molecular Modelling

e5.3 Watch an animation portraying this reaction between copper metal and aquated silver ions.

5.3 Chemical Equations: Chemical Accounting

A chemical equation is a statement, expressed in symbolic form, that chemists use to state information about a chemical reaction [<<Section 2.6]. The chemical equations in Section 5.2 are examples.

Balanced Chemical Equations

A **balanced chemical equation** is one that satisfies two requirements:

1. The total electrical charge on the species that react is equal to the total charge on the species that are formed. During a reaction, the reaction mixture will not accumulate positive or negative electrical charge.

In the case of the electron-transfer reaction between solid copper and aquated silver ions, the equation

$$Cu(s) + 2\,Ag^+(aq) \longrightarrow Cu^{2+}(aq) + 2\,Ag(s)$$

is balanced from the point of view of electrical charge, because it tells us that if one Cu atom reacts, then two units of positive charge (one on each of two silver ions) are removed, and two units of positive charge (on one copper ion) are produced.

2. The chemical equation must be consistent with the law of conservation of atoms. In this sense, the balanced chemical equation is the chemist's form of accountancy.

    For example, the equation

$$P_4(s) + 6\,Cl_2(g) \longrightarrow 4\,PCl_3(\ell)$$

is balanced with respect to conservation of atoms because, for each P_4 molecule that reacts, there are four P atoms (all in the P_4 molecule) and twelve Cl atoms (two in each of six Cl_2 molecules) in reactant species. In the species that are formed, there are also four P atoms (one in each of the four PCl_3 molecules) and twelve Cl atoms (three in each of the four PCl_3 molecules).

$$
\begin{array}{ccc}
6 \times 2 = & & 4 \times 3 = \\
\text{12 Cl atoms} & & \text{12 Cl atoms}
\end{array}
$$

$$P_4 \;+\; 6\,Cl_2 \longrightarrow 4\,PCl_3$$

$$\text{4 P atoms} \qquad\qquad \text{4 P atoms}$$

We can extend this logic to any level we choose. If 500 P_4 molecules react, then there are 2000 P atoms and 6000 Cl atoms in the reactant molecules, and the same numbers of each in the product molecules. If we consider the reaction of 1 mol of solid white phosphorus with chlorine, then there are 4 mol of P atoms and 12 mol of Cl atoms in the reactant molecules, and also 4 mol of P atoms and 12 mol of Cl atoms in the product molecules.

Interactive Exercise 5.1

SUBMIT

Work through the steps to balance a chemical equation.

WORKED EXAMPLE 5.1—BALANCED CHEMICAL EQUATIONS

Decide if the following chemical equations are balanced for both electrical charge and atoms:

(a) $N_2O_5(g) \longrightarrow 2\,NO_2(g) + \frac{1}{2}O_2(g)$

(b) $Mg(OH)_2(s) + H_3O^+(aq) \longrightarrow Mg^{2+}(aq) + 4\,H_2O(\ell)$

Solution

(a) No charged species react or are formed, so the equation is balanced for charge. It does not make sense to imagine the reaction of 1 molecule of N_2O_5, because this would lead to formation of half a molecule of O_2—clearly nonsensical. We need to consider reaction of 1 mol (6.02×10^{23} molecules) of N_2O_5. In that case, 2 mol of N atoms (in N_2O_5) are "consumed" and 2 mol of N atoms are in the products (in NO_2 molecules). And there are 5 mol of O atoms in the reactant molecules as well as 5 mol in the products (4 mol in NO_2 molecules and 1 mol in ½ mol of O_2). So the equation is balanced for atoms.

(b) If the reaction were to proceed according to this equation, for every 1 mol of solid $Mg(OH)_2$ that reacts, 1 mol of positive charge would be replaced by 2 mol, and the reaction vessel would accumulate positive charge. The equation is not balanced for charge. Inspecting the numbers of atoms of each element on each side of the equation, if 1 mol of $Mg(OH)_2$ reacts, 1 mol of Mg atoms in the reactants would be replaced by 1 mol in the products (as Mg^{2+} ions). However, a total of 3 mol of O atoms in the reactants would be replaced by 4 mol in the products, and 5 mol of H atoms would be replaced by 8 mol. So the equation is neither balanced for charge nor for atoms.

Comment

The equation in (b) would be balanced if there were 2 mol of $H_3O^+(aq)$ on the left side.

EXERCISE 5.2—BALANCED CHEMICAL EQUATIONS

Decide if the following chemical equations are balanced for both electrical charge and atoms:

(a) $Fe_2O_3(s) + 3\ CO(g) \longrightarrow 2\ Fe(s) + 3\ CO_2(g)$

(b) $CH_3COOH(aq) + C_6H_5CH_2OH(aq) \longrightarrow CH_3COOCH_2C_6H_5(aq) + H_2O(\ell)$

(c) $CaCO_3(s) + H_3O^+(aq) \longrightarrow Ca^{2+}(aq) + CO_2(g) + H_2O(\ell)$

Interactive Exercise 5.3

Practise the steps involved in balancing a chemical equation.

> In a balanced chemical equation, the total electrical charge on the reactant species is equal to the total charge on the product species, and the numbers of atoms of each element in the products is the same as in the reactants.

What Balanced Chemical Equations Can Tell Us

Qualitatively, a chemical equation tells us what the reactants are, what products are formed, and their physical states: the symbol (s) indicates a solid, (ℓ) a liquid, and (g) a gas.

Balanced chemical equations also give us quantitative information by means of the coefficients (the numbers in front of the formula of each product and reactant), although the coefficient 1 is usually not shown. In the balanced equation

$$P_4(s) + 6\ Cl_2(g) \longrightarrow 4\ PCl_3(\ell)$$

the coefficients are 1, 6, and 4. These coefficients indicate the relative amounts (in mol) of species that react and of species that are formed: for each 1 mol of solid P_4 that reacts with chlorine, it does so with 6 mol of Cl_2 gas, and 4 mol of liquid PCl_3 is formed. This information allows us to calculate the relative masses of $P_4(s)$ and $Cl_2(g)$ that react, and of $PCl_3(\ell)$ that is formed. This is called *stoichiometry*—the subject matter of Section 5.6.

As we shall see in Section 5.10, balanced equations are essential to the evaluation of the *atom economy* of chemical processes: they can tell us what percentage of the atoms of each sort in the reactants finish up in the desirable products—rather than in waste materials.

EXERCISE 5.4—RELATIVE AMOUNTS OF REACTANTS AND PRODUCTS

Aluminum reacts with oxygen to give aluminum oxide:

$$4\ Al(s) + 3\ O_2(g) \longrightarrow 2\ Al_2O_3(s)$$

What amount of $O_2(g)$, in mol, is needed for complete reaction with 6.0 mol of $Al(s)$? What amount of $Al_2O_3(s)$ can be produced?

> Balanced chemical equations identify the reactants and products and the relative amounts (in mol) of species that react and are formed.

What Balanced Chemical Equations Cannot Tell Us

It is easy to wrongly read into chemical equations information that they do not provide. Be aware of the following:

1. A balanced equation does not tell us the amounts of the species that react—only the *relative* amounts of them. Let's use, for example, a balanced chemical equation for combustion of methane:

$$CH_4(g) + 2\ O_2(g) \longrightarrow CO_2(g) + 2\ H_2O(g)$$

We would use this same equation to tell us about the relative amounts of reactants and products, regardless of whether 1 mol of methane reacts, or 0.001 mol, or 2000 mol, or 3.62 mol. For example, if 3.62 mol of methane reacts, the equation tells us that it does so with twice the amount (7.24 mol) of oxygen, to produce an equal amount (3.62 mol) of carbon dioxide and twice the amount (7.24 mol) of water vapour.

2. A chemical equation does not tell us the amounts of substances initially in a reaction mixture. The above equation for methane combustion does not imply that we are talking about a reaction mixture containing 1 mol of methane and 2 mol of oxygen. Indeed our reaction mixture may contain 0.01 mol of methane and 26 mol of oxygen, or 3 mol of methane and 0.0004 mol of oxygen, or any other amounts. It might also refer to a stream of methane flowing into a combustion zone (like a Bunsen burner or kitchen stove) where it reacts with oxygen available in the air. The amounts that actually react are determined by the amount of the *limiting reactant*, defined in Section 5.7.

3. A chemical equation does not tell us anything about the "natural tendency" of a reaction to happen—that is, the direction of reaction for a reaction mixture to come to *chemical equilibrium*, a condition described in Section 5.5, and in more detail in Chapter 14. Although we can write a balanced chemical equation such as

$$CO_2(g) + 2\,H_2O(g) \longrightarrow CH_4(g) + 2\,O_2(g)$$

this reaction does not have a "natural tendency" to happen—the opposite reaction does. Neither can a chemical equation tell us the amounts of species that will react before the reaction mixture reaches chemical equilibrium.

4. A chemical equation does not tell us whether the reaction is accompanied by the release of energy or by intake of energy, nor how much energy, when the reaction occurs [>>Chapter 7].

5. A chemical equation does not tell us how fast a reaction proceeds. Hydrogen ion transfer from acetic acid molecules to hydroxide ions in aqueous solution, represented by the chemical equation

$$OH^-(aq) + CH_3COOH(aq) \longrightarrow H_2O(\ell) + CH_3COO^-(aq)$$

is an extremely fast process, estimated to take about 10^{-10} seconds. On the other hand, even though a mixture of methane gas and oxygen gas at ambient temperatures has a natural tendency to react to form carbon dioxide and water, such a mixture may sit for years and years with no evidence of products, so slow is it to react.

Experimental observation is the only way of knowing about how fast a reaction goes. The measurement and sense-making of the different rates of reactions is the topic called *chemical kinetics* [>>Chapter 18].

6. A chemical equation does not tell us anything about the reaction mechanism—that is, how the reaction happens at the molecular level [>>Section 18.7]. For example, the reaction represented by

$$P_4(s) + 6\,Cl_2(g) \longrightarrow 4\,PCl_3(\ell)$$

almost certainly does not happen by simultaneous collision of six Cl_2 molecules in the gas phase with one P_4 molecule on the surface of solid phosphorus: this would be an extremely unlikely event.

Consider the reactions of hydrogen with iodine and bromine at temperatures above their boiling points. The balanced chemical equations for the reactions are similar:

$$H_2(g) + I_2(g) \longrightarrow 2\,HI(g)$$
$$H_2(g) + Br_2(g) \longrightarrow 2\,HBr(g)$$

Based on experimental evidence, chemists believe that the reaction of $H_2(g)$ with $I_2(g)$ occurs by direct collision and reaction between H_2 molecules and I_2 molecules, in one step corresponding with the chemical equation for the reaction. However, the reaction between $H_2(g)$ with $Br_2(g)$ is much more complicated, happening by a chain of events beginning with breaking of the Br—Br bond in Br_2 molecules. The net result of this chain of reactions is given by the balanced chemical equation for the reaction. We cannot predict these different mechanisms from the chemical equations.

Despite all the things that chemical reactions cannot tell us, they remain among the most important tools of chemistry for the information they do provide.

A chemical equation cannot tell us how much of the starting materials were present, how much of the substances reacted, the natural tendency of the reaction to happen, whether heat is evolved or absorbed, how fast the reaction happens, nor the mechanism of reaction.

5.4 Spontaneous Direction of Reaction

There is a directionality about chemical reactions: if a reaction "wants" to go one way, it does not "want" to go in the opposite way. That shouldn't be too surprising: water "wants" to run downhill, and does not "want" to run uphill, and no one has ever seen it go uphill without work being done to carry it or to pump it. For example, if we put sodium into chlorine gas, sodium chloride is formed

$$2\,Na(s) + Cl_2(g) \longrightarrow 2\,NaCl(s)$$

so we can say that this reaction has a "natural tendency" to occur. The opposite of this reaction, however, does not have a natural tendency to proceed:

$$2\,NaCl(s) \not\rightleftharpoons 2\,Na(s) + Cl_2(g)$$

We will see in Section 5.5 that the direction of "natural tendency" of a reaction is the direction that takes a reaction mixture toward the condition of *chemical equilibrium*, regardless of whether this reaction is fast or infinitely slow. If we attributed emotions to substances (which we don't), we would say it is the direction that they "want" to react.

When we know, from observation, in which direction a reaction has a "natural tendency" to occur, chemists use the following terms, all of which have a common meaning (Figure 5.3):

- The direction of "natural tendency" is the **spontaneous reaction** direction.
- The reactant species in a spontaneous reaction have higher **chemical potential** than the products.
- The product species in a spontaneous reaction are more **stable** than the reactant species.

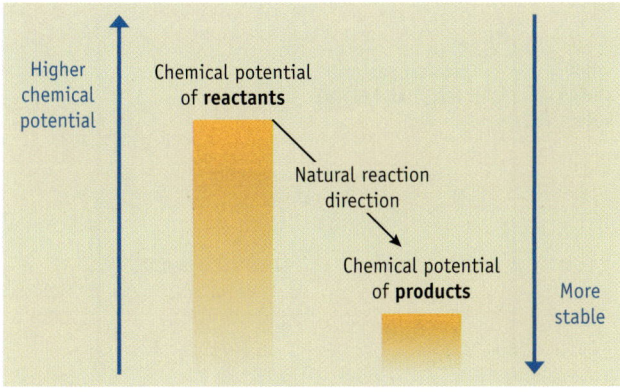

FIGURE 5.3 Terminology used in relation to the direction of "natural tendency" of a reaction.

Stability is relative. It is an inadequate statement to say that a mixture of sodium and chlorine is unstable: this mixture is unstable compared with solid sodium chloride (NaCl) but it is more stable than $NaCl_2$ (because we have no evidence of any tendency of this substance to be formed from sodium and chlorine). So, every statement of stability should be a comparative one, including words such as "than" or "compared with".

These ideas are central to the topic of *thermodynamics*, and are developed further in Chapter 17.

WORKED EXAMPLE 5.2—SPONTANEOUS REACTION DIRECTION AND RELATIVE STABILITY

We know from experience that a mixture of sodium and chlorine react to form sodium chloride.

(a) Which of the following is the spontaneous direction of reaction?

$$2\ Na(s) + Cl_2(g) \longrightarrow 2\ NaCl(s) \quad \text{or} \quad 2\ NaCl(s) \longrightarrow 2\ Na(s) + Cl_2(g)$$

(b) Which has higher chemical potential: a mixture of sodium and chlorine or solid sodium chloride?
(c) Which is more stable: a mixture of sodium and chlorine or solid sodium chloride?

Strategy

The answers to these questions can be derived from the definitions of the terms above.

Solution

(a) Our experience is that the direction of "natural tendency" is given by

$$2\ Na(s) + Cl_2(g) \longrightarrow 2\ NaCl(s)$$

So this is the spontaneous direction of reaction.
(b) The reactant species in the spontaneous reaction direction (the mixture of sodium and chlorine) have higher chemical potential than the product (sodium chloride).
(c) The product of the spontaneous reaction direction (sodium chloride) is more stable than the reactant species (the mixture of sodium and chlorine).

Comment

We can use a diagram specific to this reaction (Figure 5.4) that corresponds with the generalized diagram of Figure 5.3.

FIGURE 5.4 Terminology used in relation to the reaction between sodium and chlorine to form sodium chloride.

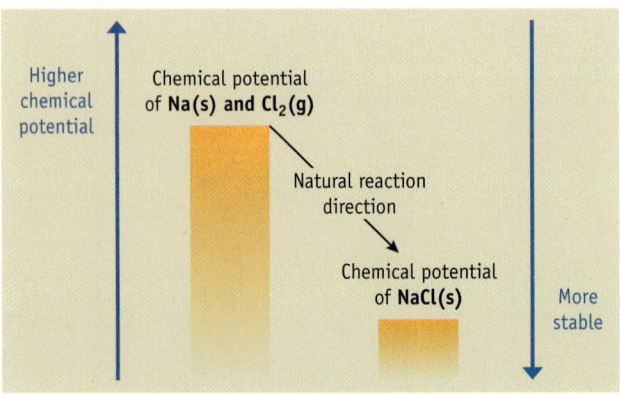

EXERCISE 5.5—SPONTANEOUS REACTION DIRECTION AND RELATIVE STABILITY

When we put a spoonful of sodium chloride into 1 L of water, we have observational evidence that there is a natural tendency for the salt to dissolve:

$$NaCl(s) \xrightarrow{H_2O} Na^+(aq) + Cl^-(aq)$$

For small amounts of sodium chloride, this is the spontaneous direction of reaction. Nobody has ever seen a dilute solution of salt in water separate out into salt and water without the expenditure of energy (such as electrical energy or solar energy to evaporate the water, or reverse osmosis to "push" only the water through a membrane).

(a) Which has higher chemical potential: a small amount of solid sodium chloride in water or a dilute solution of sodium chloride?

(b) Which is more stable: a small amount of solid sodium chloride in water or a dilute solution of sodium chloride?

If a reaction has a natural tendency to occur, this is the spontaneous reaction direction, and we say that the products are more stable than the reactants, and the reactants have higher chemical potential than the products.

5.5 The Condition of Dynamic Chemical Equilibrium

Most of the chemical reactions referred to so far "go to completion"—that is, reaction proceeds until there is essentially none of one or other (or both) reactant species remaining. There are, however, many reactions that proceed only to a situation where the reaction mixture contains significant quantities of reactant species as well as product species, although their concentrations do not change any further. While the reaction appears to have stopped, there is evidence that the reaction in one direction is counterbalanced by reaction also happening in the opposite direction, and at the same rate: while change is not observable at the macroscopic level, reactions in both directions occur at the molecular level. This situation is called **dynamic chemical equilibrium**, and in chemical equations the symbol $\rightleftharpoons$ is used to represent this condition.

For example, if we stir powdered limestone (calcium carbonate, $CaCO_3$) in water, a little of it dissolves with formation of aquated calcium ions and aquated carbonate ions. After a while, no more dissolves because it has reached the condition of dynamic chemical equilibrium: the rate at which $Ca^{2+}(aq)$ ions and $CO_3^{2-}(aq)$ ions come together to form solid $CaCO_3$ matches the rate at which solid $CaCO_3$ dissolves with formation of $Ca^{2+}(aq)$ ions and $CO_3^{2-}(aq)$ ions. This is represented by the chemical equation

In the case of dissolving substances, the condition of chemical equilibrium is called a *saturated solution* [>>Section 15.2].

$$CaCO_3(s) \xrightarrow{H_2O} Ca^{2+}(aq) + CO_3^{2-}(aq)$$

How far a reaction goes before it reaches chemical equilibrium is different from reaction to reaction. For example, at 25 °C, the dissolution of calcium carbonate in water achieves equilibrium when only a tiny amount has dissolved and the concentrations of $Ca^{2+}(aq)$ ions and $CO_3^{2-}(aq)$ ions are less than 1×10^{-4} mol L^{-1}. On the other hand, if we put a piece of copper into a 1 mol L^{-1} aqueous silver nitrate solution, the reaction represented by

$$Cu(s) + 2 Ag^+(aq) \rightleftharpoons Cu^{2+}(aq) + 2 Ag(s)$$

will occur until only about 1×10^{-10} mol L^{-1} of $Ag^+(aq)$ ions remain.

In principle, we can regard all reactions as ones that achieve the condition of chemical equilibrium, even if this happens when there are infinitesimal amounts of reactants remaining, or only infinitesimal amounts of products formed.

In a reaction mixture at equilibrium, that mixture of reactants and products is more stable than either a reaction mixture containing only reactants, or a reaction mixture containing only products. Any reaction that takes a reaction mixture toward the condition of chemical equilibrium is the *spontaneous direction of reaction*.

We will come across many cases of reactions in the condition of dynamic chemical equilibrium in the course of discussing everyday chemistry in this learning resource. The principles of systems at equilibrium are the focus of Chapters 13–17.

EXERCISE 5.6—EQUILIBRIUM, SPONTANEITY, AND STABILITY

When solid calcium carbonate is shaken up in water at 25 °C, a little of it dissolves and the mixture comes to equilibrium (called a *saturated solution*) with $Ca^{2+}(aq)$ ions and $CO_3^{2-}(aq)$ both at a concentration of 6×10^{-5} mol L^{-1}.

(a) Which of the following is the direction of spontaneous reaction?

$$CaCO_3(s) \longrightarrow \underset{6 \times 10^{-5} \text{ mol L}^{-1}}{Ca^{2+}(aq)} + \underset{6 \times 10^{-5} \text{ mol L}^{-1}}{CO_3^{2-}(aq)}$$

or

$$\underset{6 \times 10^{-5} \text{ mol L}^{-1}}{Ca^{2+}(aq)} + \underset{6 \times 10^{-5} \text{ mol L}^{-1}}{CO_3^{2-}(aq)} \longrightarrow CaCO_3(s)$$

(b) Which has higher chemical potential: a mixture of solid calcium carbonate in pure water or a mixture containing solid calcium carbonate and $Ca^{2+}(aq)$ ions and $CO_3^{2-}(aq)$, both at a concentration of 6×10^{-5} mol L^{-1}?

(c) Which is more stable: a mixture of solid calcium carbonate in pure water or a mixture containing solid calcium carbonate and $Ca^{2+}(aq)$ ions and $CO_3^{2-}(aq)$, both at a concentration of 6×10^{-5} mol L^{-1}?

The spontaneous direction of reaction takes the reaction mixture toward a condition of dynamic chemical equilibrium, when the reaction mixture is more stable than at any previous time.

5.6 Masses of Reactants and Products: Stoichiometry

The calculation of amounts and masses of reactants that react, and of products that are formed, in chemical reactions is called **stoichiometry** (derived from the Greek word *stoikheion* = element). The relative masses of starting materials that react and products formed can be calculated from the relative amounts (in mol) of the reactants and products, and their molar masses. Let's demonstrate this by reference to the reaction between phosphorus and chlorine, represented by the equation:

$$P_4(s) + 6 Cl_2(g) \longrightarrow 4 PCl_3(\ell)$$

Imagine a reaction vessel in which 1.00 mol of phosphorus (P_4, molar mass 123.9 g mol^{-1}) and 6.00 mol of chlorine gas (Cl_2, molar mass 70.9 g mol^{-1}) react. From the equation, we deduce that 4.00 mol of phosphorus trichloride (PCl_3, molar mass 137.3 g mol^{-1}) are formed. It is sometimes useful to show the amounts of the reactants and products in an *amounts table*. In the following amounts table, the masses of substances are also shown, in red.

The information in the amounts table can be shown graphically (Figure 5.5).

Equation	$P_4(s)$	$+$	$6 Cl_2(g)$	$\longrightarrow$	$4 PCl_3(\ell)$
Initial amount	1.00 mol		6.00 mol		0 mol
(Initial mass)	(123.9 g)		(6.00 mol × 70.9 g mol^{-1} = 425.4 g)		(0 g)
Change in amount	−1.00 mol		−6.00 mol		+4.00 mol
(Change in mass)	(123.9 g)		(425.4 g)		(4.00 mol × 137.3 g mol^{-1} = 549.2 g)
Final amount	0 mol		0 mol		4.00 mol
(Final mass)	(0 g)		(0 g)		(549.2 g)

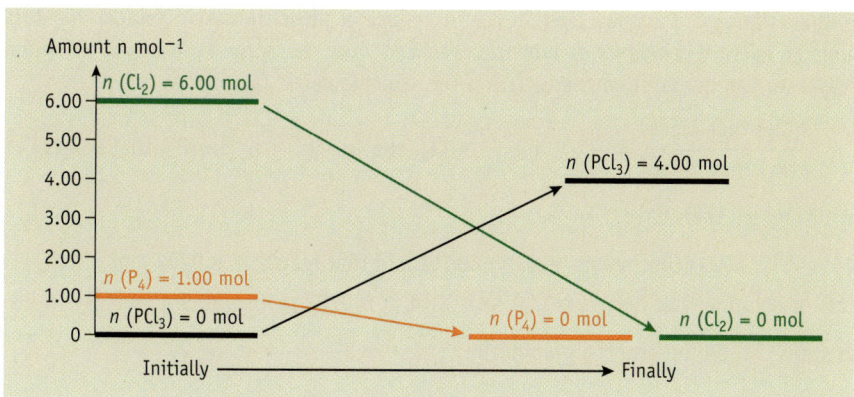

FIGURE 5.5 Schematic representation of amounts of reactants and products before and after 1.00 mol P_4 and 6.00 mol Cl_2 react.

The amounts table shows clearly that, as in every chemical reaction, mass is conserved: although the atoms are redistributed to form different molecules, the numbers of P atoms (four) and Cl atoms (twelve) remain unchanged, as does the total mass:

$$\text{Total mass of reactants} = 123.9 \text{ g} + 425.4 \text{ g} = 549.3 \text{ g}$$

$$= \text{mass of product, allowing for rounding off errors in the molar masses}$$

This does not mean, however, that the total amount (mol) of reactants is the same as the total amount of the products. In this reaction, if a total of 7 mol of substances react, only 4 mol of product is formed.

Think about It

e5.4 Watch two chemical reactions and calculate the reacting masses involved.

WORKED EXAMPLE 5.3—RELATIVE MASSES OF REACTANTS AND PRODUCTS

In a reaction between white phosphorus, $P_4(g)$, and chorine, $Cl_2(g)$, to form liquid phosphorus trichloride, $PCl_3(\ell)$,
(a) what mass of $Cl_2(g)$ will react completely with 1.45 g of $P_4(s)$?
(b) what mass of $PCl_3(\ell)$ is produced?
Set up an amounts table, presuming that there is just enough chlorine gas available to react with the phosphorus, showing the initial amounts and masses of reacting substances, the changes in the amounts and masses of reactants and products, and the amounts and masses of all substances after reaction.

Strategy

The first step in stoichiometric calculations of reacting amounts is always to write the appropriate balanced chemical equation. From the mass of $P_4(s)$ that reacts, the amount of it (in mol) can be calculated. From the amount of $P_4(s)$, we can use the balanced equation to deduce the amounts (and then the masses) of other reactants and of products.

Solution

The balanced chemical equation is

$$P_4(s) + 6 \; Cl_2(g) \longrightarrow 4 \; PCl_3(\ell)$$

From the mass of $P_4(s)$, calculate the amount of it:

$$\text{Amount of phosphorus, } n(P_4) = \frac{m}{M} = \frac{1.45 \text{ g}}{123.9 \text{ g mol}^{-1}} = 0.0117 \text{ mol}$$

(a) From the balanced chemical equation, we can see that the amount of $Cl_2(g)$ that reacts is six times the amount of $P_4(s)$ that reacts.

$$n(Cl_2) \text{ that reacts} = 6 \times n(P_4) \text{ that reacts} = 6 \times 0.0117 \text{ mol} = 0.0702 \text{ mol}$$

Alternative strategy: You may find it useful to use a **stoichiometric factor**: *the ratio of amounts of relevant reactants or products, deduced from the balanced chemical equation.* In this case, we can derive from the equation the stoichiometric factor:

$$\frac{n(\text{Cl}_2) \text{ that reacts}}{n(\text{P}_4) \text{ that reacts}} = \frac{6}{1} \qquad \text{or} \qquad n(\text{Cl}_2) \text{ that reacts} = 6 \times n(\text{P}_4) \text{ that reacts}$$

Substituting, we then have

$$n(\text{Cl}_2) \text{ that reacts} = 6 \times 0.0017 \text{ mol that reacts} = 0.0702 \text{ mol}$$

Mass of chlorine that reacts, $m(\text{Cl}_2) = n \times M = 0.0702 \text{ mol} \times 70.9 \text{ g mol}^{-1} = 4.98 \text{ g}$

These steps can be represented graphically:

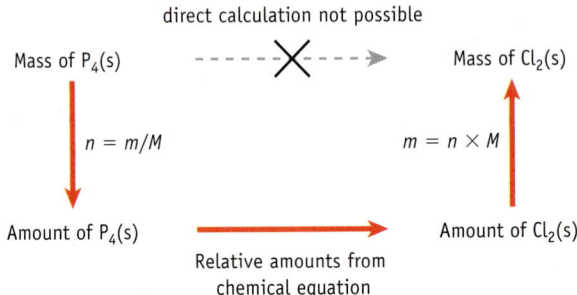

direct calculation not possible

Mass of $P_4(s)$ $\cdots\cdots\times\cdots\cdots\rightarrow$ Mass of $Cl_2(s)$

$n = m/M$ $m = n \times M$

Amount of $P_4(s)$ $\longrightarrow$ Amount of $Cl_2(s)$

Relative amounts from
chemical equation

(b) Because matter is conserved, the mass of $PCl_3(\ell)$ can be obtained by adding the masses of $P_4(s)$ and $Cl_2(g)$ that react

$$\text{Mass of PCl}_3 \text{ formed, } m(\text{PCl}_3) = 1.45 \text{ g} + 4.98 \text{ g} = 6.43 \text{ g}$$

Alternatively, from the balanced chemical equation we can deduce that the amount of $PCl_3(\ell)$ produced is four times the amount of $P_4(s)$ that reacts.

$$\text{Amount of PCl}_3 \text{ formed, } n(\text{PCl}_3) = 4 \times 0.0117 \text{ mol} = 0.0468 \text{ mol}$$

Mass of PCl_3 produced, $m(\text{PCl}_3) = n \times M = 0.0468 \text{ mol} \times 137.3 \text{ g mol}^{-1} = 6.43 \text{ g}$

Supposing that there was just enough chlorine gas to react with the phosphorus, the amounts table for this situation is the following:

Equation	$P_4(s)$	+	6 Cl_2 (g)	→	4 PCl_3 (ℓ)
Initial amount	0.0117 mol		$6 \times 0.0117 = 0.0702$ mol		0 mol
(Initial mass)	(1.45 g)		(4.98 g)		(0 g)
Change in amount	−0.0117 mol		−0.0702 mol		$+4 \times 0.0117 = 0.0468$ mol
(Change in mass)	(−1.45 g)		(−4.98 g)		(+6.43 g)
Final amount	0 mol		0 mol		0.0468 mol
(Final mass)	(0 g)		(0 g)		(6.43 g)

EXERCISE 5.7—RELATIVE MASSES OF REACTANTS AND PRODUCTS

Respiration of plants can be approximated to reaction of glucose with oxygen:

$$\text{C}_6\text{H}_{12}\text{O}_6(s) + 6 \text{ O}_2(g) \longrightarrow 6 \text{ CO}_2(g) + 6 \text{ H}_2\text{O}(\ell)$$

What mass of oxygen reacts with 25.0 g of glucose? What masses of carbon dioxide and water are formed? Set up an amounts table for the reaction.

**Interactive Exercises
5.8–5.10**

SUBMIT

Write a balanced
equation and calculate the
masses of reactants and products.

If we know the mass of a substance that has reacted or been produced, then the masses of the other substances that react or are produced can be deduced from the relative amounts (in mol) in the balanced chemical equation and their molar masses.

5.7 Reactions Limited by the Amount of One Reactant

Reaction mixtures seldom contain exactly the amounts of substances that react with each other. Reactions are often carried out with an excess of one reactant over the amount needed to react with the other. This is usually done to ensure that all of the other reactant is consumed. If we know the amount of the completely consumed reactant in the reaction mixture, this is the amount that reacts, and this can be used to calculate the amounts, and masses, of products.

Suppose you burn a "sparkler," a wire coated with magnesium (Figure 5.6) The magnesium burns in air, consuming oxygen and producing magnesium oxide, MgO(s).

$$2 \, Mg(s) + O_2(g) \longrightarrow 2 \, MgO(s)$$

The sparkler burns until the all of the magnesium has reacted. What about the oxygen? One mol of magnesium, Mg(s), will react with 0.5 mol of oxygen, O_2(g), but there is much more oxygen available in the air. The amount of MgO(s) produced depends on the amount of Mg(s) in the sparkler, and not on the amount of O_2(g) in the atmosphere. In this example, magnesium is called the **limiting reactant** (or *limiting reagent*) because the amount of MgO(s) formed is limited by the amount of Mg(s) present.

Let us look at an example of a limiting reactant situation using the reaction of oxygen with carbon monoxide to form carbon dioxide. A balanced chemical equation for the reaction is

$$2 \, CO(g) + O_2(g) \longrightarrow 2 \, CO_2(g)$$

Imagine a (very unlikely) mixture of four CO molecules and three O_2 molecules.

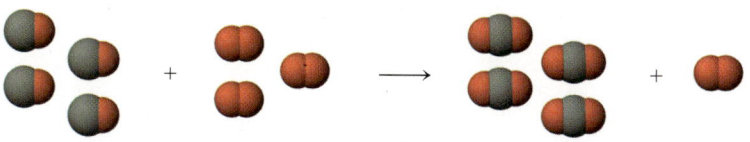

The four CO molecules require only two O_2 molecules for reaction, so one O_2 molecule would remain after reaction. Because more O_2 molecules are available than are required, the number of CO_2 molecules produced is limited by the number of CO molecules present. Even if there were four, ten, or one hundred O_2 molecules, the number of CO_2 molecules produced could not exceed four. Carbon monoxide is therefore the limiting reactant.

The same logic can be applied to reaction mixtures containing many trillions of molecules. The simplest way to do this is to scale up to amounts of substances, the unit of which is the mole. Carbon monoxide is also the limiting reactant in a mixture of 4 mol of CO gas with 3 mol of O_2 gas.

Stoichiometric Calculations in Limiting Reactant Situations

When there is a limiting reactant, as is usually the case, calculations of relative amounts, and relative masses, of reactants and products are as in Section 5.6, with all amounts based on the amount of the limiting reactant.

> **WORKED EXAMPLE 5.4—LIMITING REACTANT SITUATIONS**

The first step in the manufacture of nitric acid is the oxidation of ammonia to nitric oxide in the presence of a platinum-wire gauze catalyst (Figure 5.7).

$$4 \, NH_3(g) + 5 \, O_2(g) \longrightarrow 4 \, NO(g) + 6 \, H_2O(\ell)$$

FIGURE 5.6 Limiting reactant. When a sparkler burns, the amount of MgO(s) formed is limited by the amount of magnesium, and not by the amount of oxygen available for reaction.

Charles D. Winters

Think about It

e5.5 Watch two reactions and identify the limiting reactant in each one.

Charles D. Winters

Courtesy of Johnson Matthey

FIGURE 5.7 Oxidation of ammonia.
(a) Laboratory oxidation of ammonia on the surface of a platinum wire produces so much heat that the wire glows bright red. (b) Millions of tonnes of nitric acid are made annually by the oxidation of ammonia over a rhodium-platinum alloy gauze.

(a)

(b)

Suppose that 750 g of each of $NH_3(g)$ and of $O_2(g)$ are mixed. Are these reactants mixed in the correct stoichiometric ratio or is one of them a limiting reactant? That is, will one of them limit the quantity of $NO(g)$ that can be produced? How much $NO(g)$ can be formed? How much of the excess reactant is left over?

Strategy

From the masses of reactants, calculate the amounts (in mol). By comparing the relative amounts of the reactants present with the relative amounts needed according to the balanced chemical equation, decide which, if either, of the reactants is in short supply. Base your calculations of the amount of product on the amount of this limiting reactant.

Solution

First, find the amount of each reactant:

$$n(NH_3) = \frac{m}{M} = \frac{750 \text{ g}}{17.03 \text{ g mol}^{-1}} = 44.0 \text{ mol}$$

$$n(O_2) = \frac{m}{M} = \frac{750 \text{ g}}{32.00 \text{ g mol}^{-1}} = 23.4 \text{ mol}$$

To see if there is a limiting reactant, examine the ratio of amounts of reactants:

$$\text{Ratio of amounts of reactants present} = \frac{n(O_2)}{n(NH_3)} = \frac{23.4 \text{ mol } O_2}{44.0 \text{ mol } NH_3}$$

From the balanced chemical equation, we can tell the relative amounts of substances that will react together:

$$\text{Ratio of amounts that will react} = \frac{5 \text{ mol } O_2}{4 \text{ mol } NH_3} = \frac{55.0 \text{ mol } O_2}{44.0 \text{ mol } NH_3}$$

The chemical equation tells us that the 44.0 mol of ammonia requires 55.0 mol of oxygen to react with all of it, but the reaction mixture contains only 23.4 mol of oxygen. So, the amount of oxygen limits the amount of reaction, and some ammonia will remain unreacted. Calculations of the amount of ammonia that reacts, and of the amounts of nitric oxide and water formed, must be based on the amount of oxygen present (i.e., the amount of oxygen that reacts). Oxygen is the limiting reactant.

Now we can use the relative amounts of O_2 that reacts and NO that is produced, indicated by the balanced chemical equation, to calculate the amount of NO produced, and then its mass.

$$\text{Amount of nitric oxide produced, } n(NO) = \frac{4}{5} \times n(O_2) = \frac{4}{5} \times 23.4 \text{ mol} = 18.7 \text{ mol}$$

$$\text{Mass of nitric oxide produced, } m(NO) = N \times M = 18.7 \text{ mol} \times 30.01 \text{ g mol}^{-1} = 562 \text{ g}$$

Finally, using the balanced equation, we can calculate the amount, and then the mass, of unreacted ammonia left over. Again, the calculation needs to be based on the amount of limiting reactant present.

$$\text{Amount of ammonia that reacts, } n(NH_3) = \frac{4}{5} \times n(O_2) = \frac{4}{5} \times 23.4 \text{ mol} = 18.7 \text{ mol}$$

And then

$$\text{Amount of excess ammonia, } n(NH_3) = \text{ initial amount} - \text{amount reacted}$$
$$= 44.0 \text{ mol} - 18.7 \text{ mol} = 25.3 \text{ mol}$$
$$\therefore \text{ Mass of excess ammonia, } m(NH_3) = n \times M = 25.3 \text{ mol} \times 17.03 \text{ g mol}^{-1} = 431 \text{ g}$$

Comment

Because 431 g of NH_3 remains unreacted, we can say that 319 g (750 g − 431 g) reacted. Check that you can calculate that 28.1 mol (6/5 × 23.4 mol) of water is also produced, the mass of which is 506 g (28.1 mol × 18.02 g mol^{-1}).

An amounts table that summarizes this limiting reactant situation is shown here. As always (allowing for a slight discrepancy due to rounding-off errors) the total mass of reacted substances (750 g + 319 g = 1069 g) is the same as the total mass of products formed (562 g + 506 g = 1068 g). And the total mass of substances before reaction takes place (1500 g) is the same as the total mass after reaction. During chemical reaction, atoms are neither created nor destroyed—just redistributed to form different molecules. The number of molecules does not stay the same.

Equation	4 NH_3(g)	+	5 O_2(g)	→	4 NO(g)	+	6 H_2O(ℓ)
Initial amount	44.0 mol		23.4 mol		0 mol		0 mol
(Initial mass)	(750 g)		(750 g)		(0 g)		(0 g)
Change in amount	−18.7 mol		−23.4 mol		+18.7 mol		+28.1 mol
(Change in mass)	(−319 g)		(−750 g)		(+562 g)		(+506 g)
Final amount	25.3 mol		0 mol		18.7 mol		28.1 mol
(Final mass)	(431 g)		(0 g)		(562 g)		(506 g)

EXERCISE 5.11—LIMITING REACTANT SITUATIONS

Methanol (CH_3OH), which is used as a fuel, can be made by the reaction of carbon monoxide and hydrogen ("synthetic gas" or "syngas;" see Section 4.2):

$$CO(g) + 2 H_2(g) \longrightarrow CH_3OH(\ell)$$

Suppose 356 g of CO(g) and 65.0 g of H_2(g) are mixed and they react (in the presence of a catalyst).
(a) Which is the limiting reactant?
(b) What mass of methanol can be produced?
(c) What mass of the excess reactant remains after the reaction?
Draw up an amounts table for this situation.

EXERCISE 5.12—LIMITING REACTANT SITUATIONS

The "thermite" reaction produces iron metal and aluminum oxide from a mixture of powdered aluminum metal and iron(III) oxide.

$$Fe_2O_3(s) + 2 Al(s) \longrightarrow 2 Fe(s) + Al_2O_3(s)$$

Charles D. Winters

Thermite reaction. Iron(III) oxide reacts with aluminum metal to produce aluminum oxide and iron metal. The reaction produces so much heat that the iron melts and spews out of the reaction vessel.

**Interactive Exercises
5.13–5.16**

Identify the limiting
reactant and calculate masses of
products.

A mixture of 50.0 g each of $Fe_2O_3(s)$ and $Al(s)$ is used.
(a) Which is the limiting reactant?
(b) What mass of iron metal can be produced?
Complete an amounts table for this situation.

The amounts of substances that can react in a reaction mixture are limited by the
amount of the one that will be totally consumed. Stoichiometric calculations must
be based on the amount of this limiting reactant.

5.8 Theoretical Yield and Percent Yield

When a compound is made, we can use the masses of the starting materials and the balanced chemical equation to calculate the maximum mass of the product that could be
obtained. This is called the **theoretical yield**.

Usually, however, the mass of product obtained (the actual yield) is less than the theoretical yield. Some loss of product often occurs during the isolation and purification
steps. In addition, some reactions do not go completely to products, and some reaction
mixtures produce more than one set of products.

To provide information to other chemists who might want to carry out the same reaction, it is customary to report a **percent yield**, which is the actual yield expressed as a percentage of the theoretical yield (Figure 5.8).

$$\text{Percent yield} = \frac{\text{actual yield}}{\text{theoretical yield}} \times 100\%$$

(a)

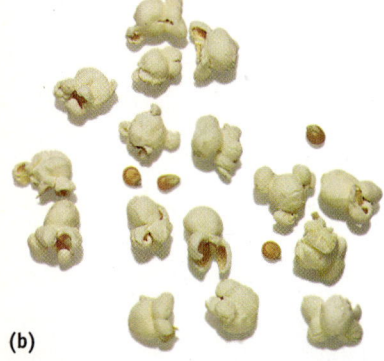

(b)

FIGURE 5.8 Percent yield analogy. If
we begin with 20 popcorn kernels, the
theoretical yield is 20 popped corns. Here,
only 16 of them popped, so the percent
yield is 80%.

Charles D. Winters

WORKED EXAMPLE 5.5—PERCENT YIELD AND THEORETICAL YIELD

Suppose that you made aspirin (acetylsalicylic acid, $M = 180.2$ g mol^{-1}) in the laboratory by
a reaction for which the chemical equation is

$$C_6H_4(OH)COOH(s) \quad + \quad (CH_3CO)_2O(\ell) \quad \longrightarrow \quad C_6H_4(OCOCH_3)COOH(s) \quad + \quad CH_3COOH(\ell)$$

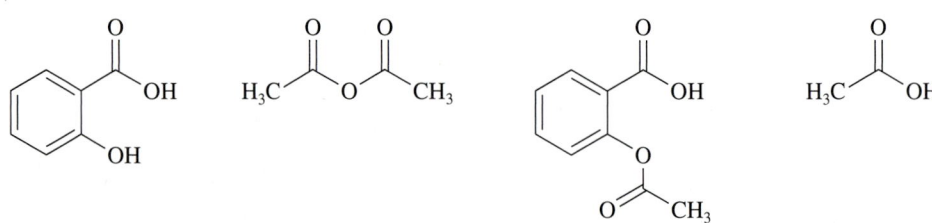

Salicylic acid   Acetic anhydride   Aspirin   Acetic acid

Suppose you started with 14.4 g of salicylic acid ($M = 138.1$ g mol^{-1}) and an excess of acetic
anhydride: salicylic acid is the limiting reactant. You obtain 6.26 g of aspirin. Calculate the
theoretical yield, and the percent yield of aspirin.

Strategy

From the amount of the limiting reactant, calculate the amount (in mol) of it, and then use the
balanced chemical equation to calculate the amount, and mass, of aspirin that could be produced
(the theoretical yield). The actual yield as a percentage of the theoretical yield is the percent yield.

Solution

$$\text{Amount of salicylic acid} = \frac{14.4 \text{ g}}{138.1 \text{ g mol}^{-1}} = 0.104 \text{ mol}$$

Since salicylic acid is the limiting reactant and the balanced equation tells us that 1 mol of aspirin will be produced for every 1 mol of salicylic acid that reacts, we should expect 0.104 mol of aspirin to be formed. Then the theoretical yield of aspirin (usually expressed as a mass) is

$$\text{Theoretical yield of aspirin} = 0.104 \; \cancel{\text{mol}} \times 180.2 \; \text{g} \; \cancel{\text{mol}^{-1}} = 18.7 \; \text{g}$$

and

$$\text{Percent yield of aspirin} = \frac{6.26 \; \cancel{\text{g}}}{18.7 \; \cancel{\text{g}}} \times 100\% = 33.5\%$$

EXERCISE 5.17—PERCENT YIELD AND THEORETICAL YIELD

Methanol (CH_3OH) can be burned in oxygen to provide energy, or it can react in the presence of a catalyst to form hydrogen gas, which can then be used as a fuel.

$$CH_3OH(\ell) \longrightarrow 2 \; H_2(g) + CO(g)$$

If 125 g of methanol reacts, what is the theoretical yield of $H_2(g)$? If 13.6 g of $H_2(g)$ is obtained, what is the percent yield?

Interactive Exercise 5.18

Calculate the percent yield in a reaction.

> The theoretical yield is the maximum possible mass of a product that could be obtained from a reaction mixture. The actual yield calculated as a percentage of the theoretical yield is the percent yield.

5.9 Stoichiometry and Chemical Analysis

Chemists use a variety of approaches to identify substances as well as to measure the amounts of components in mixtures. Most often they use spectroscopic and chromatographic techniques, but analysis based on the stoichiometry of chemical reactions still plays a central role.

Quantitative chemical analysis depends on knowing that the stoichiometry of a given reaction is strictly as indicated by the balanced equation—there are no side-reactions, and the reaction goes "to completion." That is, the percent yield is 100%. Then the amount of a component in the material being analyzed can be estimated from either (a) the amount of another substance that it reacts with or (b) the amount of substances formed from it.

An example of the first type is the analysis of a sample of vinegar containing an unknown amount of its essential ingredient, acetic acid. The acid reacts rapidly and completely with the hydroxide ions in a solution of sodium hydroxide, according to the equation

$$CH_3COOH(aq) + OH^-(aq) \longrightarrow CH_3COO^-(aq) + H_2O(\ell)$$

If the exact amount of hydroxide ions that react with the acetic acid can be measured, the amount of acetic acid present can be calculated. This type of analysis, called a *titration*, is discussed more fully in Section 14.10.

The second type is exemplified by the analysis of a sample of the mineral thenardite, which is largely sodium sulfate (Na_2SO_4). Sodium sulfate is soluble in water, forming aquated sodium ions and sulfate ions [>>Section 6.7]:

$$Na_2SO_4(s) \xrightarrow{H_2O} 2 \; Na^+(aq) + SO_4^{2-}(aq)$$

To find the quantity of Na_2SO_4 in an impure mineral sample, we can crush the rock and then stir it in water to dissolve the sodium sulfate. Then to this solution we can add an excess of aquated barium ions (in an aqueous barium chloride solution) with formation of barium sulfate, which is almost insoluble in water.

$$SO_4^{2-}(aq) + Ba^{2+}(aq) \longrightarrow BaSO_4(s)$$

Thenardite is sodium sulfate (Na_2SO_4). It is named after the French chemist Louis Thenard (1777–1857), a co-discoverer (with Gay-Lussac and Davy) of boron. Sodium sulfate is used in making detergent, glass, and paper.

(a) Na$_2$SO$_4$ solution   BaCl$_2$ solution

(b) BaSO$_4$, white solid   Solution with Na$^+$(aq), Cl$^-$(aq), and excess Ba^{2+}(aq) ions, and very few SO$_4^{2-}$ (aq) ions

(c) BaSO$_4$, white solid caught in filter

(d) Filter paper weighed

All photos: Charles D. Winters

FIGURE 5.9 Analysis for the sulfate content of a sample. The sulfate ions in a Na$_2$SO$_4$ solution react with barium ions in a BaCl$_2$ solution to form insoluble BaSO$_4$(s). The solid precipitate is collected on a filter paper of known mass and weighed. From the amount of BaSO$_4$(s) obtained and the balanced chemical equation, the amount of Na$_2$SO$_4$ in the sample can be calculated.

The method of analysis depends on the barium sulfate having known composition, and the experimental evidence that it precipitates as a pure compound. The solid barium sulfate is collected by filtration, dried in an oven, and weighed (Figure 5.9). We can find the amount of sulfate in the mineral sample because it is directly related to the amount of BaSO$_4$(s) that precipitates:

$$1 \text{ mol Na}_2\text{SO}_4 \text{ in mineral} \longrightarrow 1 \text{ mol BaSO}_4 \text{ in precipitate}$$

Methods of analysis such as this, which are based on the masses of reactants or products, are called *gravimetric analysis*. The following Worked Example and Exercises illustrate the method.

WORKED EXAMPLE 5.6—QUANTITATIVE ANALYSIS BASED ON STOICHIOMETRY

The mineral cerussite is mostly lead carbonate (PbCO$_3$). To analyze for the PbCO$_3$ content, a sample of the mineral is treated with nitric acid to dissolve the lead carbonate:

$$PbCO_3(s) + 2 H^+(aq) \longrightarrow Pb^{2+}(aq) + H_2O(\ell) + CO_2(g)$$

from HNO$_3$ solution

Dilute sulfuric acid solution is added to the resulting solution, and lead sulfate precipitates.

$$Pb^{2+}(aq) + SO_4^{2-}(aq) \longrightarrow PbSO_4(s)$$

from H$_2$SO$_4$ solution

Solid lead sulfate is filtered off and weighed. Suppose that in analysis of a 0.583 g sample of the mineral we obtained 0.628 g of PbSO$_4$(s). What is the mass percent of PbCO$_3$ in the mineral sample?

Strategy

The key is to recognize that 1 mol of PbCO$_3$ in the mineral will ultimately yield 1 mol of PbSO$_4$ precipitate.

Solution

First, calculate the amount of $PbSO_4$ (M = 303.3 g mol^{-1}) in the precipitate:

$$n(PbSO_4) = \frac{0.628 \ \cancel{g}}{303.3 \ \cancel{g} \ mol^{-1}} = 0.00207 \ mol$$

From the balanced equations, we can see that the amount of $PbCO_3$ in the mineral sample is also 0.00207 mol.

Now we can calculate the mass of $PbCO_3$ (M = 267.2 g mol^{-1}) in the 0.583 g sample of the mineral:

$$m(PbCO_3) = 0.00207 \ \cancel{mol} \times 267.2 \ g \ \cancel{mol^{-1}} = 0.5 \times 53 \ g$$

Finally, the mass percent of $PbCO_3$ in the mineral sample is the calculated mass of $PbCO_3$ as a percentage of the mass of the sample:

$$Mass \ percent \ of \ PbCO_3 = \frac{0.553 \ \cancel{g}}{0.583 \ \cancel{g}} \times 100\% = 94.9\%$$

> **Think about It**
>
> **e5.6** Work through a gravimetric analysis of an ionic compound.

> **Taking It Further**
>
> **e5.7** Learn how to determine the formula of a compound using combustion analysis.

EXERCISE 5.19—QUANTITATIVE ANALYSIS BASED ON STOICHIOMETRY

Nickel(II) sulfide (NiS) occurs naturally as the relatively rare mineral millerite. One of its occurrences is in meteorites. To analyze a mineral sample for the quantity of NiS, the sample is heated in nitric acid to form a solution containing all of the nickel as aquated nickel ions, $Ni^{2+}(aq)$:

$$NiS(s) + 2 \ H^+(aq) \longrightarrow Ni^{2+}(aq) + H_2S(g)$$

To the aqueous solution containing $Ni^{2+}(aq)$ ions is added a solution of the organic compound dimethylglyoxime (DMG, $C_4H_8N_2O_2$) to give the red precipitate of nickel dimethylglyoxime, $Ni(C_4H_7N_2O_2)_2$:

$$Ni^{2+}(aq) + 2 \ C_4H_8N_2O_2(aq) \longrightarrow Ni(C_4H_7N_2O_2)_2(s) + 2 \ H^+(aq)$$

Suppose that from a 0.468 g sample of millerite we obtain 0.206 g of $Ni(C_4H_7N_2O_2)_2$ precipitate. What is the mass percent of NiS in the mineral sample?

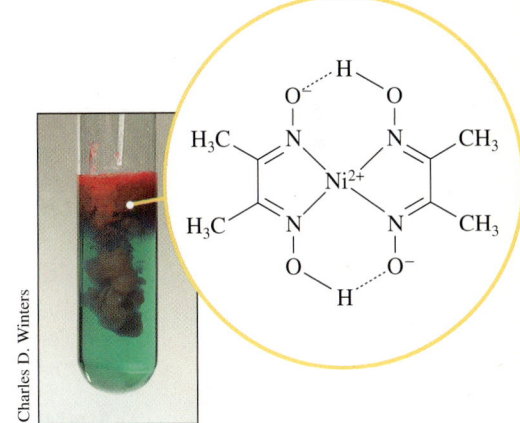

Charles D. Winters

A red precipitate of nickel dimethylglyoxime, formed when dimethylglyoxime is added to an aqueous solution with $Ni^{2+}(aq)$ ions.

> If you know that the reaction stoichiometry matches that of a balanced equation, the amount of a substance can be deduced from either the amount of another substance that it reacts with or the amount of a substance formed.

5.10 Atom Economy, Atom Efficiency

As we saw in the opening section of this chapter, large amounts of waste materials may be produced in the industrial manufacture of chemicals, even though the target product may be formed with 100% yield. Industry is being urged to follow the principle of **atom economy**: synthetic methods should be designed to maximize the incorporation of all materials used in the process into the final product. That is, as far as possible, all of the atoms in the starting materials should be in the desired products.

The **atom efficiency** of a reaction, calculated separately for atoms of each element, is a quantitative measure that can be calculated from the balanced equation. It expresses the percentage of atoms of each element that end up in the desired product. For example:

$$Carbon \ atom \ efficiency = \frac{number \ of \ C \ atoms \ in \ desired \ product}{number \ of \ C \ atoms \ in \ reactants} \times 100\%$$

The aim is to achieve 100% atom efficiency for atoms of each of the elements involved.

> The concepts of *atom economy* and *atom efficiency* are relatively new. You may find differences in the way that these are defined and used.

Sometimes chemists use the concept of **overall atom efficiency (OAE)**, which expresses the mass of the desired product as a percentage of the total mass of products:

$$\text{Overall atom efficiency} = \frac{\text{mass of desired product}}{\text{total mass of all products}} \times 100\%$$

An overall atom efficiency of 100% is desirable because that means that there are no waste materials amongst the products.

Another way of comparing the production of wastes between different processes is to use the **E-factor**, for which the objective is a value of 0.

$$\text{E-factor} = \frac{\text{mass of waste materials}}{\text{mass of desired product}}$$

WORKED EXAMPLE 5.7—ATOM ECONOMY

(a) Consider a now-outdated method of production of the sodium salt of phenol (C_6H_5ONa) by a method summarized by the following balanced chemical equation:

| sodium benzene sulfonate | sodium phenolate | sodium sulfite |

Calculate (i) the atom efficiency of the reaction with respect to each of the elements C, H, O, Na, and S; (ii) the overall atom efficiency; and (iii) the E-factor.

(b) Compare your answers with those for the modern process for phenol manufacture, in which the acetone also formed is a valuable product, rather than a waste.

Strategy

For part (a), consider the reaction of 1 mol of sodium benzene sulfonate and 2 mol of sodium hydroxide, and count the number of atoms of each element in the reactants and in the resultant products.

Solution

(a) (i)   Looking at the atom efficiencies of the elements, one by one:
- There are 6 mol of carbon atoms in the starting materials, and 6 mol in the desired product (sodium phenolate).

$$\text{Carbon atom efficiency} = \frac{\text{number of C atoms in } C_6H_5ONa}{\text{number of C atoms in all reactants}} \times 100\%$$

$$= \frac{6 \text{ mol}}{6 \text{ mol}} \times 100\% = 100\%$$

- Of the 7 mol of H atoms in the starting materials (5 mol in sodium benzene sulfonate and 2 mol in sodium hydroxide), only 5 mol finishes up in the sodium phenolate—a hydrogen atom efficiency of 71%.
- Of the 5 mol of oxygen atoms in the starting materials (3 mol in sodium benzene sulfonate and 2 mol in sodium hydroxide), only 1 mol finishes up in the sodium phenolate—an oxygen atom efficiency of 20%.
- Of the 3 mol of sodium atoms in the starting materials (1 mol in sodium benzene sulfonate and 2 mol in sodium hydroxide), only 1 mol finishes up in the sodium phenolate—a sodium atom efficiency of 33%.
- Of the 1 mol of sulfur atoms in the starting materials, none finishes up in the sodium phenolate—a sulfur atom efficiency of 0%.

(ii)   The mass of each substance can be calculated as the product of the amount (in mol) and the molar mass. Then the overall atom efficiency can be calculated as follows:

$$OAE = \frac{n \times M(\text{sodium phenolate})}{n \times M(\text{sodium phenolate}) + n \times M(\textit{sodium sulfite}) + n \times M(\text{water})} \times 100\%$$

$$= \frac{1 \text{ mol} \times 116 \text{ g mol}^{-1}}{(1 \text{ mol} \times 116 \text{ g mol}^{-1}) + (1 \text{ mol} \times 126 \text{ g mol}^{-1}) + (1 \text{ mol} \times 18 \text{ g mol}^{-1})} \times 100\%$$

$$= \frac{116 \text{ g}}{260 \text{ g}} \times 100\% = 44.6\%$$

This value tells us that, apart from the solvent, of the final reaction mixture only 44.6% is the desired product (at best, even if there were no losses or side-reactions).

(iii) $$E - \text{factor} = \frac{(1 \text{ mol} \times 126 \text{ g mol}^{-1}) + (1 \text{ mol} \times 18 \text{ g mol}^{-1})}{1 \text{ mol} \times 116 \text{ g mol}^{-1}} = \frac{144 \text{ g}}{116 \text{ g}} = 1.24$$

If we decide that water should not be regarded as a waste product, and remove it from our calculations, the E-factor improves to 1.08.

(b) Every atom in the starting materials finishes up in molecules of valuable products. So, the atom efficiencies of each of the elements is 100%, the OAE is 100%, and the E-factor is 0.

Comment

The modern method wastes less resources that the older way, and cleanup costs are reduced. Of course, a thorough analysis of the atom efficiency of this method needs to also consider the method of production of the cumene hydroperoxide starting material.

EXERCISE 5.20—ATOM ECONOMY

For the production of maleic anhydride by oxidation of (a) benzene and (b) butene (equations below), compare (i) the atom efficiency for atoms of each element, (ii) the overall atom efficiency, and (iii) the E-factor.

(a)

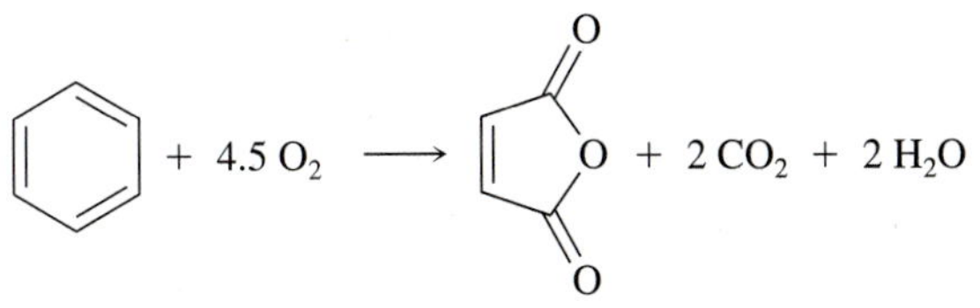

(b)

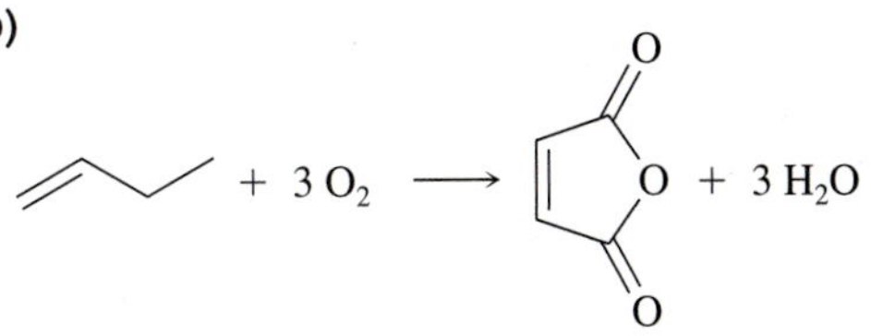

Atom Economy in Context

Of course, calculation of the atom efficiency of a process is only one factor to be taken into account when deciding how to make something on the industrial scale. Others include the energy requirements of the process, the cost of the starting materials, competing uses for the starting materials, the toxicity or hazardous nature of the waste materials, how easily the waste materials can be properly disposed of, how pure the desired product can be obtained, and the cost and nature of the solvent in which the reactions are carried out.

Another consideration is whether the by-products may be useful for some other purpose. For example, in the production of sodium phenolate described in Worked Example 5.7, sodium sulfite was considered a waste material, when in fact it may have been in demand for use as a reducing agent, preservative, and bleaching agent.

> Green chemistry policies recommend that industrial production of chemicals should be designed so that atom efficiencies for each element, and overall atom efficiencies, are maximized, while the E-factor is minimized.

SUMMARY

Key Concepts

- A **chemical reaction** is a process in which one or more new **chemical species** form as a result of redistribution of atoms, ions, or electrons. During a reaction, the **law of conservation of atoms** and the **law of conservation of mass** are satisfied. In reactions of molecular substances, the number of molecules may not be conserved. (Section 5.2)
- In **balanced chemical equations**, neither the charge nor the number of atoms of each element change as reactant species are converted to product species. The relative amounts (in mol) of species that react and are formed are indicated by the coefficients. (Section 5.3)
- A chemical equation cannot tell us how much of the starting materials were present, how much of the substances reacted, the spontaneous direction of reaction, whether heat is evolved or absorbed, how fast the reaction happens, or the mechanism of reaction. (Section 5.3)
- If we know the **spontaneous reaction** direction, the reactants are said to be higher in **chemical potential** than the products, and the products are said to be more **stable** than the reactants. (Section 5.4)
- A reaction mixture in a condition of **dynamic chemical equilibrium**, when there is no visible change at the observable level, is more stable than any other that was present as equilibrium was approached. Reaction toward equilibrium is the spontaneous direction of reaction. (Section 5.5)
- From the relative amounts of substances in the balanced chemical equation for a reaction we can deduce the relative masses of substances that react or are produced. This is called **stoichiometry**. (Section 5.6)
- The amount of reaction that can happen in a reaction vessel is governed by the amount of the **limiting reactant**: the reactant that is totally "consumed." Stoichiometric calculations must be based on the amount of the limiting reactant present. (Section 5.7)
- The maximum mass of a product that could be obtained from a specified reaction mixture is the **theoretical yield**. The actual yield calculated as a percentage of the theoretical yield is the **percent yield**. (Section 5.8)
- Quantitative chemical analysis depends on knowing that the stoichiometry of a reaction is strictly as indicated by the balanced equation. (Section 5.9)
- One of the principles of the green chemistry movement is that industrial production of chemicals should be designed so that **atom efficiencies** for each element and **overall atom efficiencies** are maximized, while the **E-factor** is minimized. (Section 5.10)

Key Equations

$$\text{Percent yield} = \frac{\text{actual yield}}{\text{theoretical yield}} \times 100\%$$ (Section 5.10)

$$\text{Carbon atom efficiency} = \frac{\text{number of C atoms in desired product}}{\text{number of C atoms in reactants}} \times 100\%$$

(Section 5.10)

$$\text{Overall atom efficiency} = \frac{\text{mass of desired product}}{\text{total mass of all products}} \times 100\%$$ (Section 5.10)

$$\text{E-factor} = \frac{\text{mass of waste materials}}{\text{mass of desired product}}$$ (Section 5.10)

REVIEW QUESTIONS

Section 5.3: Chemical Equations: Chemical Accounting

5.21 Decide if the following chemical equations are balanced for both electrical charge and atoms:

(a) $C_6H_6(\ell) + 3\,H_2(g) \longrightarrow C_6H_{12}(\ell)$

(b) $MnO_4^-(aq) + 5\,Fe^{2+}(aq) + 8\,H^+(aq)$
$\longrightarrow Mn^{2+}(aq) + 5\,Fe^{3+}(aq) + 4\,H_2O(\ell)$

(c) $Cu(OH_2)_4^{2+}(aq) + 4\,NH_3(aq)$
$\longrightarrow Cu(NH_3)_4^{2+}(aq) + 6\,H_2O(\ell)$

5.22 A major source of air pollution years ago was the metals industry. One common process involved "roasting" metal sulfides in the air:

$$2\,PbS(s) + 3\,O_2(g) \longrightarrow 2\,PbO(s) + 2\,SO_2(g)$$

If you heat 2.5 mol of PbS(s) in the air, what amount of $O_2(g)$ is required for complete reaction? What amounts of PbO(s) and $SO_2(g)$ are expected?

Section 5.4: Spontaneous Direction of Reaction

5.23 At 25 °C, liquid water is more stable than water vapour at 1 bar pressure.

(a) Which of the following is the spontaneous direction of reaction at 25 °C, if the pressure of $H_2O(g)$ is 1 bar?

$$H_2O(\ell) \longrightarrow H_2O(g) \quad \text{or} \quad H_2O(g) \longrightarrow H_2O(\ell)$$

(b) Which has higher chemical potential: liquid water or water vapour at 1 bar pressure?

5.24 At 25 °C, a reaction vessel with hydrogen and oxygen, each at 1 bar pressure, has higher chemical potential than liquid water.

(a) Which of the following is the spontaneous direction of reaction?

$$2\,H_2(g) + O_2(g) \longrightarrow 2\,H_2O(\ell)$$
$$\text{or}$$
$$2\,H_2O(\ell) \longrightarrow 2\,H_2(g) + O_2(g)$$

(b) Which is more stable: a mixture of hydrogen and oxygen, each at 1 bar pressure, or liquid water?

Section 5.5: The Condition of Dynamic Chemical Equilibrium

5.25 If we put a sufficiently large piece of copper into a 1 mol L^{-1} aqueous silver nitrate solution, reaction happens according to the chemical equation

$$Cu(s) + 2\,Ag^+(aq) \rightleftharpoons Cu^{2+}(aq) + 2\,Ag(s)$$

until, at equilibrium, the concentration of $Cu^{2+}(aq)$ ions is about 0.5 mol L^{-1}, and the concentration of $Ag^+(aq)$ ions is about 1×10^{-10} mol L^{-1}.

(a) Which has higher chemical potential: a mixture containing a piece of copper in a 1 mol L^{-1} aqueous silver nitrate solution or a mixture containing some solid silver in an aqueous solution in which the concentration of $Cu^{2+}(aq)$ ions is about 0.5 mol L^{-1} and the concentration of $Ag^+(aq)$ ions is about 1×10^{-10} mol L^{-1}?

(b) Which is more stable: a mixture containing a piece of copper in a 1 mol L^{-1} aqueous silver nitrate solution or a mixture containing some solid silver in an aqueous solution in which the concentration of $Cu^{2+}(aq)$ ions is about 0.5 mol L^{-1} and the concentration of $Ag^+(aq)$ ions is about 1×10^{-10} mol L^{-1}?

Section 5.6: Masses of Reactants and Products: Stoichiometry

5.26 Suppose 16.04 g of benzene (C_6H_6), is burned in oxygen:

$$C_6H_6(\ell) + \tfrac{15}{2}O_2(g) \longrightarrow 6\,CO_2(g) + 3\,H_2O(g)$$

(a) What mass of $O_2(g)$ is required for complete combustion of benzene?

(b) What is the total mass of products expected from 16.04 g of benzene?

5.27 If 10.0 g of carbon is combined with an exact, stoichiometric amount of oxygen (26.6 g) to produce carbon dioxide, what is the theoretical yield of CO_2, in grams?

5.28 The metabolic disorder diabetes causes a build-up of acetone (CH_3COCH_3) in the blood. Acetone, a volatile compound, is exhaled, giving the breath of untreated diabetics a distinctive odour. The acetone is produced by a breakdown of fats in a series of reactions. The equation for the last step is

$$CH_3COCH_2CO_2H \longrightarrow CH_3COCH_3 + CO_2$$

What mass of acetone can be produced from 125 mg of acetoacetic acid ($CH_3COCH_2CO_2H$)?

5.29 Your body deals with excess nitrogen by excreting it in the form of urea (NH_2CONH_2). The reaction producing it is the combination of arginine ($C_6H_{14}N_4O_2$) with water to give urea and ornithine ($C_5H_{12}N_2O_2$):

$$C_6H_{14}N_4O_2 + H_2O \longrightarrow NH_2CONH_2 + C_5H_{12}N_2O_2$$
$$\text{arginine} \qquad\qquad \text{urea} \qquad \text{ornithine}$$

If you excrete 95 mg of urea, what mass of arginine must have been used? What mass of ornithine must have been produced?

5.30 Iron metal and chlorine gas react to form iron(III) chloride:

$$2\,Fe(s) + 3\,Cl_2(g) \longrightarrow 2\,FeCl_3(s)$$

(a) Beginning with 10.0 g of iron, what mass of $Cl_2(g)$ is required for complete reaction? What mass of $FeCl_3(s)$ can be produced?

(b) If only 18.5 g of $FeCl_3(s)$ is obtained from 10.0 g of iron and excess $Cl_2(g)$, what is the percent yield?

(c) If 10.0 g of each of iron and chlorine are combined, what is the theoretical yield of iron(III) chloride?

5.31 Two beakers sit on a balance; one contains a solution of KI; the other contains a solution of $Pb(NO_3)_2$. The total mass is 167.170 g.

Solutions of KI and $Pb(NO_3)_2$ before reaction.

When the solution in one beaker is poured completely into the other, the following precipitate-forming reaction occurs:

$$Pb^{2+}(aq) + 2\,I^-(aq) \longrightarrow PbI_2(s)$$

What is the total mass of the beakers and solutions after reaction? Explain.

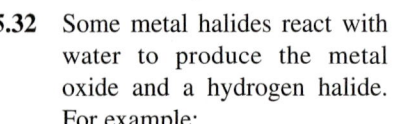

Solutions after reaction.

5.32 Some metal halides react with water to produce the metal oxide and a hydrogen halide. For example:

$$TiCl_4(\ell) + 2\,H_2O(\ell) \longrightarrow TiO_2(s) + 4\,HCl(g)$$

(a) If you begin with 14.0 mL of $TiCl_4(\ell)$, whose density is 1.73 g mL^{-1}, what mass of water is required for complete reaction?

(b) What mass of each product is expected?

5.33 The reaction of 750 g each of $NH_3(g)$ and $O_2(g)$ was found to produce 562 g of $NO(g)$:

$$4\,NH_3(g) + 5\,O_2(g) \longrightarrow 4\,NO(g) + 6\,H_2O(g)$$

(a) What mass of $O_2(g)$ is required to react with all of the 750 g of $NH_3(g)$?

(b) What mass of water is produced by this reaction?

Section 5.9: Stoichiometry and Chemical Analysis

5.34 Saccharin, an artificial sweetener, has the formula $C_7H_5NO_3S$. Suppose you have a sample of a saccharin-containing sweetener with a mass of 0.2140 g. After decomposition to free the sulfur and convert it to the SO_4^{2-} ion, the sulfate ion is trapped as water-insoluble $BaSO_4(s)$. The mass of $BaSO_4$ precipitate obtained is 0.2070 g. What is the mass percent of saccharin in the sweetener?

5.35 Sulfuric acid can be prepared starting with the sulfide ore, cuprite (Cu_2S). If each S atom in $Cu_2S(g)$ leads to one molecule of H_2SO_4, what mass of $H_2SO_4(\ell)$ can be produced from 1.00 tonne of $Cu_2S(g)$?

5.36 An unknown metal reacts with oxygen to give the metal oxide, MO_2. Identify the metal based on the following information:

Mass of metal = 0.356 g

Mass of sample after converting metal completely to oxide = 0.452 g

5.37 Titanium(IV) oxide, $TiO_2(s)$, is heated in hydrogen gas to give water and a new titanium oxide, $Ti_xO_y(s)$. If 1.598 g of $TiO_2(s)$ produces 1.438 g of $Ti_xO_y(s)$ what is the formula of the new oxide?

5.38 Thioridazine ($C_{21}H_{26}N_2S_2$) is a pharmaceutical used to regulate dopamine. (Dopamine, a neurotransmitter, affects brain processes that control movement, emotional response, and ability to experience pleasure and pain.) A chemist can analyze a sample of the pharmaceutical for the thioridazine content by decomposing it to convert the sulfur in the compound to sulfate ion. This is then "trapped" as water-insoluble barium sulfate:

$$SO_4^{2-}(aq, \text{from thioridazine}) + Ba^{2+}(aq) \longrightarrow BaSO_4(s)$$

Suppose a 12-tablet sample of the drug yielded 0.301 g of $BaSO_4(s)$. What is the thioridazine content, in milligrams, of each tablet?

5.39 A herbicide contains 2,4-D (2,4-dichlorophenoxyacetic acid), $C_8H_6Cl_2O_3$.

A 1.236 g sample of the herbicide was decomposed to liberate the chlorine as Cl^- ion. This was precipitated as $AgCl(s)$, with a mass of 0.1840 g. What is the mass percent of 2,4-D in the sample?

5.40 Potassium perchlorate is prepared by the following sequence of reactions:

$$Cl_2(g) + 2\ OH^-(aq) \longrightarrow Cl^-(aq) + ClO^-(aq) + H_2O(\ell)$$
$$3\ ClO^-(aq) \longrightarrow 2\ Cl^-(aq) + ClO_3^-(aq)$$
$$4\ ClO_3^-(aq) \longrightarrow 3\ ClO_4^-(aq) + Cl^-(aq)$$

If the $OH^-(aq)$ ions are added as KOH, in the end, evaporation leads to formation of crystals of $KClO_4(s)$. What mass of $Cl_2(g)$ is required to produce 234 kg of $KClO_4(s)$?

5.41 What mass of lime, $CaO(s)$, can be obtained by heating 125 kg of limestone that is 95.0% by mass $CaCO_3(s)$?

$$CaCO_3(s) \longrightarrow CaO(s) + CO_2(g)$$

5.42 Sulfuric acid can be produced from a sulfide ore such as iron pyrite by the following sequence of reactions:

$$4\ FeS_2(s) + 11\ O_2(g) \longrightarrow 2\ Fe_2O_3(s) + 8\ SO_2(g)$$
$$2\ SO_2(g) + O_2(g) \longrightarrow 2\ SO_3(g)$$
$$SO_3(g) + H_2O(\ell) \longrightarrow H_2SO_4(\ell)$$

Starting with 525 kg of $FeS_2(s)$ (and an excess of other reactants), what mass of pure H_2SO_4 can be prepared?

5.43 Copper metal can be obtained by roasting copper ore, which can contain cuprite (Cu_2S) and copper(II) sulfide:

$$Cu_2S(s) + O_2(g) \longrightarrow 2\ Cu(s) + SO_2(g)$$
$$CuS(s) + O_2(g) \longrightarrow Cu(s) + SO_2(g)$$

Suppose an ore sample contains 11.0% impurity in addition to a mixture of CuS and Cu_2S. Heating 100.0 g of the mixture produces 75.4 g of copper metal with a purity of 89.5%. What is the mass percent of CuS in the ore? What is the mass percent of Cu_2S?

Section 5.10: Atom Economy, Atom Efficiency

5.44 The following equations represent (a) an addition reaction, (b) a substitution reaction, (c) an elimination reaction, and (d) a rearrangement. Calculate for each, and compare amongst the various reactions (i) the atom efficiencies for atoms of each element, (ii) the overall atom efficiency, and (iii) the E-factor.

(a)

(b)

(c)

(d)

SUMMARY AND CONCEPTUAL QUESTIONS

5.45 A weighed sample of iron, $Fe(s)$, is added to liquid bromine, $Br_2(\ell)$, and allowed to react completely. The reaction produces a single product, which can be isolated and weighed. The experiment was repeated a number of times with different masses of iron but with the same mass of bromine. (See the graph on the next page.)

(a) What mass of $Br_2(\ell)$ is used when the reaction consumes 2.0 g of $Fe(s)$?
(b) What is the ratio of the amounts (in mol) of $Br_2(\ell)$ to $Fe(s)$ that react?
(c) What is the formula of the product?

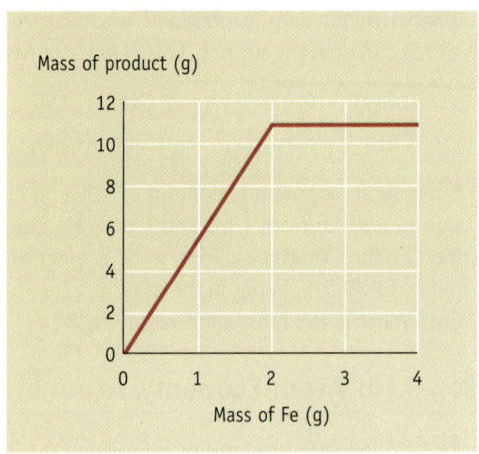

(d) Write the balanced chemical equation for the reaction of iron and bromine.

(e) What is the name of the reaction product?

(f) Which statement or statements best describe the experiments summarized by the graph?

   (i) When 1.00 g of Fe(s) is added to the $Br_2(\ell)$, Fe(s) is the limiting reagent.

   (ii) When 3.50 g of Fe(s) is added to the $Br_2(\ell)$, there is an excess of $Br_2(\ell)$.

   (iii) When 2.50 g of Fe(s) is added to the $Br_2(\ell)$, both reactants are used up completely.

   (iv) When 2.00 g of Fe(s) is added to the $Br_2(\ell)$, 10.0 g of product is formed. The percent yield must therefore be 20.0%.

5.46 Chlorine and iodine react according to the balanced equation

$$I_2(g) + 3\,Cl_2(g) \longrightarrow 2\,ICl_3(g)$$

Suppose that you mix $I_2(g)$ and $Cl_2(g)$ in a flask and that the mixture is represented by the diagram below:

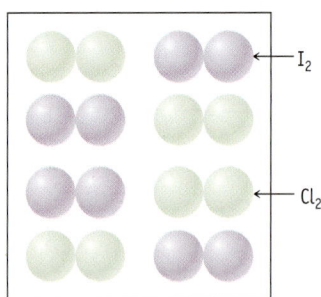

When the reaction between the $Cl_2(g)$ and $I_2(g)$ is complete, which panel below represents the outcome? Which compound is the limiting reactant?

(a)

(b)

(c)

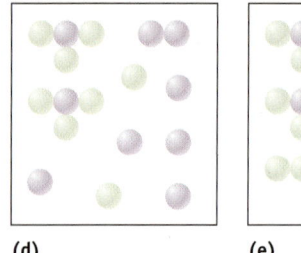

(d)

(e)

5.47 Cisplatin, $Pt(NH_3)_2Cl_2$, is a cancer chemotherapy agent. Notice that it contains NH_3 groups attached to platinum.

(a) What is the mass percent of Pt, N, and Cl in the cisplatin?

(b) Cisplatin is made by reacting ammonia in aqueous solution with $PtCl_4^{2-}(aq)$ ions (formed by ionization of K_2PtCl_4 dissolved in water):

$$K_2PtCl_4(aq) + 2\,NH_3(aq) \longrightarrow Pt(NH_3)_2Cl_2(aq) \\ + 2\,KCl(aq)$$

If you begin with 16.0 g of $K_2PtCl_4(g)$, what mass of ammonia should be used to completely react with it? What mass of cisplatin will be produced?

5.48 Iron(III) chloride, $FeCl_3(s)$, is produced by the reaction of iron and chlorine.

(a) If you place 1.54 g of iron metal gauze in chlorine gas, what mass of chlorine is required for complete reaction? What mass of iron(III) chloride is produced?

(b) Aquated iron(III) ions from dissolved iron(III) chloride reacts with $OH^-(aq)$ ions to produce iron(III) hydroxide. If you dissolve 2.0 g of iron(III) chloride with 4.0 g of NaOH(g) in water, what mass of iron(III) hydroxide is produced?

5.49 When zinc metal is added to an aqueous HCl solution, H_2 gas is a product:

$$Zn(s) + 2\,H^+(aq) \longrightarrow Zn^{2+}(aq) + H_2(g)$$

The three flasks shown below each contain 0.100 mol of HCl in solution. Zinc is added to each flask in the following quantities: Flask 1: 7.00 g; Flask 2: 3.27 g; Flask 3: 1.31 g.

Charles D. Winters

When the reactants are mixed, the $H_2(g)$ produced inflates the balloon attached to the flask. The results are as follows:

Flask 1: Balloon inflates completely but some Zn(s) remains when inflation ceases.

Flask 2: Balloon inflates completely. No Zn(s) remains.

Flask 3: Balloon does not inflate completely. No Zn(s) remains.

Explain these results completely. Perform calculations that support your explanation.

5.50 The reaction of aluminum and bromine is pictured below.

The white solid on the lip of the beaker at the end of the reaction is $Al_2Br_6(s)$. Which was the limiting reactant, $Al(s)$ or $Br_2(\ell)$?

Charles D. Winters

Before reaction

Charles D. Winters

After reaction

With the **eCHACR** single-sign-on access card bundled with your text, log on (http://login.cengage.com/sso) and access the e-book and click on any in-margin icons for dynamic molecular-level animations and simulations, videos of laboratory reactions, interactive exercises, reviews of background concepts, and other online supplementary materials as noted by the icons in the text margins.

Also go to www.chemistry.nelson.com <http://www.chemistry.nelson.com> for Answers to in-chapter exercises and selected Review Questions, Test Yourself questions, weblinks, crossword puzzles, flashcards, glossary of key terms, and other student resources.

Both photos: Bulletin of the World Health Organization, Volume 78 (9) 2000

6.1 Case Study: Arsenic Ain't Arsenic

Everyone knows that arsenic is poisonous. Or is it? People interested in crime fiction are aware of the use of "arsenic" for dastardly purposes. Of more profound significance, "arsenic" is common in groundwater, threatening the lives of many millions of people in Asia who drink water from wells. Long-term exposure to "arsenic" in water is known to lead to darkening and thickening of the skin, skin lesions, diseases of the blood vessels, and eventually to cancers. It has been estimated that more than 250 000 people die annually in Bangladesh as a result of "arsenic" intake.

> *Everyone knows that arsenic is poisonous. Or is it?*

Why is it then, that people can eat arsenic-containing foods? For example, lobster flesh contains significant levels of "arsenic," and yet we eat lobster with no ill effects. How can this be? The answer lies in differences in the chemical combination of the element arsenic with other elements. The following examples of *speciation* of arsenic and other elements illustrate the importance of this idea.

If we dissolve some salt, NaCl(s), in water, does the solution contain salt? That might seem a silly question, but we could ask more carefully if there are entities in sea water whose composition is given by the formula NaCl? Chemists believe there are not: there are sodium ions and chloride ions, more or less independent of each other, each surrounded by a "shell" of water molecules. These ions are said to be *aquated*, or *hydrated*, and are denoted by the symbols $Na^+(aq)$ and $Cl^-(aq)$. Each such aggregate of atoms, charged or uncharged, is called a *chemical species* [<<Section 3.4].

When we dissolve blue copper(II) sulfate, $CuSO_4(s)$, in water, we get a blue solution. We might think that the colour is due to the copper sulfate, but a solution made by dissolving copper(II) nitrate, $Cu(NO_3)_2(s)$, looks the same and has an identical absorption spectrum. If we presume that the properties of a solution (including colour) are due to the species in it, we have evidence that there is the same species in each of these solutions.

Since ionic solids dissociate into cations and anions when they dissolve in water [<<Section 3.4], we might speculate that Cu^{2+} ions are the species that absorb red light in aqueous solutions of both of these salts, so that we see the same blue colour. However, if we heat blue copper(II) sulfate pentahydrate, $CuSO_4.5H_2O(s)$, in an oven at 250 °C, the water in the crystals is driven off and they change to white anhydrous copper(II) sulfate, $CuSO_4(s)$. The colour change is reversed if water is added (Figure 6.1). Presumably there are Cu^{2+} ions in the anhydrous solid, so why isn't it also blue?

Perhaps the answer has something to do with the absence of water in the anhydrous salt. X-ray diffraction study of $CuSO_4.5H_2O(s)$ crystals indicates the molecular-level structure shown in Figure 6.2.

Photo by Martyn f. Chillmaid

FIGURE 6.1 Conversion of white $CuSO_4(s)$ to blue $CuSO_4.5H_2O(s)$, by addition of water.

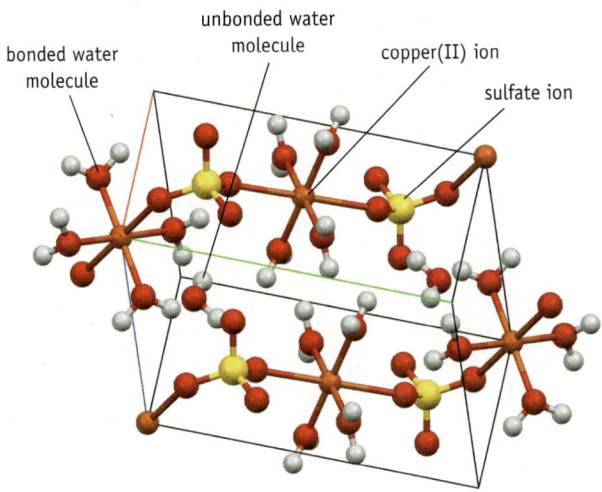

bonded water molecule — unbonded water molecule — copper(II) ion — sulfate ion

FIGURE 6.2 $CuSO_4.5H_2O(s)$ crystal structure. A portrayal of the spatial arrangement of the nuclei in a crystal of $CuSO_4.5H_2O(s)$, from data obtained by x-ray diffraction [>>Section 9.3]. Each Cu^{2+} ion is bound to the O atoms of four H_2O molecules, as well as to an O atom of each of two SO_4^{2-} ions. Other water molecules (one per Cu^{2+} ion, not bonded to the Cu^{2+} ion) are held between the bound water molecules by hydrogen bonding.

Anhydrous means "without water."

Chemists attribute the blue colour of both copper sulfate solutions and solid copper sulfate pentahydrate to absorption of red light by copper(II) ions surrounded by water molecules—called *aquated* copper(II) ions, $Cu^{2+}(aq)$ ions. Only this particular copper species has this property.

Although small amounts of copper compounds are necessary for healthy growth of plants, high concentrations in soils, near mine sites, for example, can be toxic to plants. Chemical species cannot be harmful to plants unless the species is *bioavailable*—that is, unless the plants take them up into their cells. $Cu^{2+}(aq)$ ions are a bioavailable species, toxic to plants at moderate levels. However, plants can thrive in many places where the soil has high concentrations of copper compounds. For example, when soils contain anions of naturally occurring oxalic acid, citric acid, tartaric acid, salicylic acid, lactic acid, or humic acid, these anions form bonds with Cu^{2+} ions to form *complex ions* that are not bioavailable.

As a consequence, it is difficult to specify safe levels of copper compounds in soils, because levels of toxicity depend on which copper species are present. A regulation that states the maximum allowed concentration of total copper in a soil is not very useful, since some copper species are not bioavailable, and therefore not toxic.

Let's return to the common understanding that "arsenic is poisonous." This statement makes no distinction among the different degrees of toxicity of various arsenic species. Certainly when we talk about the toxicity of arsenic, we do not mean the metallic element.

Arsenic compounds are categorized as either *inorganic arsenic* or *organic arsenic*. Inorganic arsenic includes arsenic trioxide (As_2O_3) and arsenic pentoxide (As_2O_5), as well as the arsenite ion (AsO_3^{3-}) and the arsenate ion (AsO_4^{3-}). All of these are poisonous as a result of reaction with sulfur-containing groups on certain enzymes. Arsenic trioxide has often been the poison of choice in real or fictional murders. Inorganic arsenic species in well waters have been identified as the cause of skin lesions and cancers referred to earlier.

The flesh of lobsters (and many other marine animals) contains organic arsenic species in their flesh. The predominant one is arsenobetaine, which, after eating, is excreted rather than metabolized. Many of us are thankful for that!

Don't be tempted to generalize that organic compounds of the metals are not poisonous—during the 1950s, the cause of a fatal neurological illness known then as Minamata disease was organic mercury compounds in fish and shellfish. People living in the area of Minamata Bay in Japan were affected by eating seafoods that were harvested close to where a fertilizer plant was dumping mercury compounds, which bioaccumulated up the marine food chain. The species mainly responsible for the tragedy was the methyl mercury ion (CH_3Hg^+).

We shall see in this and later chapters that understanding the chemistry of solutions and substances often requires understanding of the specific properties of particular species.

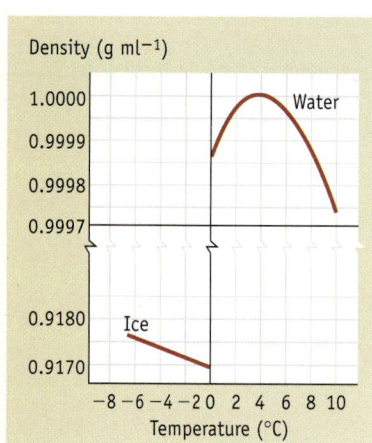

The structure of an arsenobetaine molecule.

6.2 The Remarkable Properties of Water

Before we talk about chemical reactions that happen in water, it is useful to consider the properties of water itself, and the way that chemists make sense of those properties. A comparison of the properties of water with those of other liquid substances will provide insights into the nature of liquid substances in general. You should keep in mind the basic features of the *kinetic-molecular model* of matter as it applies to liquids [<<Section 2.2].

We first present the properties of water without explanation, but you might notice that these properties, when compared with those of other molecular substances, suggest that the molecules of water are rather strongly attracted to each other. In Section 6.4, we rationalize these properties in terms of the structure of water molecules. Occasionally in this chapter we compare the properties of other substances with those of water, in order to make broad generalizations about the relationships between molecular structure and properties.

Change of Density with Temperature

When ice melts at 0 °C, the liquid water formed is 10% more dense than the ice (Figure 6.3), so ice floats in water. In contrast, the solid phase of nearly every other substance sinks in its liquid.

EXERCISE 6.1—TEMPERATURE DEPENDENCE OF WATER DENSITY

By estimating densities from the graph in Figure 6.3, calculate the volume of 100.0 g of water at the following temperatures:
(a) −5 °C (ice)
(b) 0 °C (ice)
(c) 0 °C (liquid water)
(d) 4 °C (liquid water)
(e) 10 °C (liquid water)
Sketch a graph of volume of 100 g H_2O *vs.* temperature from −5 °C to 10 °C.

FIGURE 6.3 The temperature dependence of the densities of ice and water.

Because of the way water's density changes near its freezing point, lakes in cold countries do not freeze solidly from the bottom up in the winter. When water near the surface of a lake cools, its density increases and it sinks, being displaced by warmer water from below. This "turnover" process continues until all of the water reaches 4 °C, the maximum density. This is how oxygen-rich water moves to the lake bottom, restoring the oxygen used during the summer, and nutrients are brought to the top layers. As more heat is lost from the surface, the colder water stays on top, because water cooler than 4 °C is less dense than at 4 °C. With further heat loss, ice forms on the surface, protecting the underlying water and aquatic life from further temperature decrease.

Specific Heat Capacity

The **specific heat capacity** (*c*) of a substance—the amount of heat required to raise the temperature of 1 g of it by 1 K—varies from liquid to liquid. In comparison with that of other liquids, the specific heat capacity of water is high (Table 6.1).

Molecular modelling (Odyssey)

e6.1 Examine the freedom of movement, and distance between molecules, in simulations of liquid water and ice.

Substance	c (J K^{-1} g^{-1})
Water, $H_2O(\ell)$	4.18
Ethanol, $C_2H_5OH(\ell)$	2.44
Diethyl ether, $C_2H_5OC_2H_5(\ell)$	2.37
Hexane, $C_6H_{14}(\ell)$	2.27
Acetone, $CH_3COCH_3(\ell)$	2.17
Carbon disulfide, $CS_2(\ell)$	1.00
Bromine, $Br_2(\ell)$	0.47

TABLE 6.1 Specific Heat Capacities of Liquids at 25 °C

EXERCISE 6.2—SPECIFIC HEAT CAPACITIES OF LIQUIDS

If it takes 60 s for a kettle to heat 1 kg water from 25 °C to 100 °C, how long would it take for the same kettle to heat 1 kg of ethanol through the same temperature range? Make the approximation that specific heat capacities do not vary with temperature.

The high specific heat capacity of water explains, in large part, why oceans and lakes exert a significant effect on weather. In autumn, when the temperature of the air is lower than the temperature of the ocean or lake, thermal energy is transferred from the water to the atmosphere, moderating the drop in air temperature. So much energy is made available for each 1 K drop in temperature that the decline in water temperature is gradual. For this reason, the temperature of the ocean or of a large lake generally remains higher than the average air temperature until late in the autumn.

Enthalpy Change of Vaporization

Vaporization or *evaporation* is the process in which a substance in the liquid state changes to the gaseous state. At the molecular level, we visualize molecules escaping from the liquid surface into the gaseous state (Figure 6.4). The molecules retain their identity: the O—H bonds in water are not broken, and both liquid and vapour are composed of H_2O molecules.

We know from common experience that we need to keep on providing heat to water while it is boiling and vaporizing (Figure 6.5). Since the temperature of the water remains constant, the additional energy does not increase the temperature of the water, but rather is used to separate the molecules from each other.

Energy changes under constant pressure conditions are called *enthalpy changes* [>>Section 7.4]. The symbol for enthalpy is *H*, and for enthalpy changes ΔH. The amount of energy required to bring about vaporization of one mole of a substance, called the *molar enthalpy change of vaporization* ($\Delta_{vap}H$) [>>Section 7.5] varies from substance to

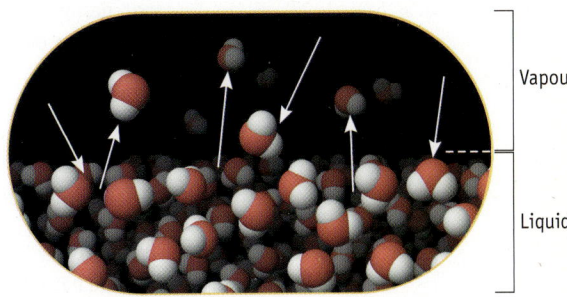

FIGURE 6.4 Vaporization. At any moment, some molecules at the surface of a liquid have enough energy to escape the attractions of their neighbours and enter the gas phase. We can imagine that often some molecules just fail to overcome the attractions of their neighbours and remain in the liquid state. All the while, some molecules in the gas phase can re-enter the liquid phase.

FIGURE 6.5 Vaporization of water during boiling, at the observable level and the imagined, molecular level. Energy must be supplied to separate the molecules in the liquid state against intermolecular forces of attraction. During condensation of the vapour to liquid, the same amount of energy is released.

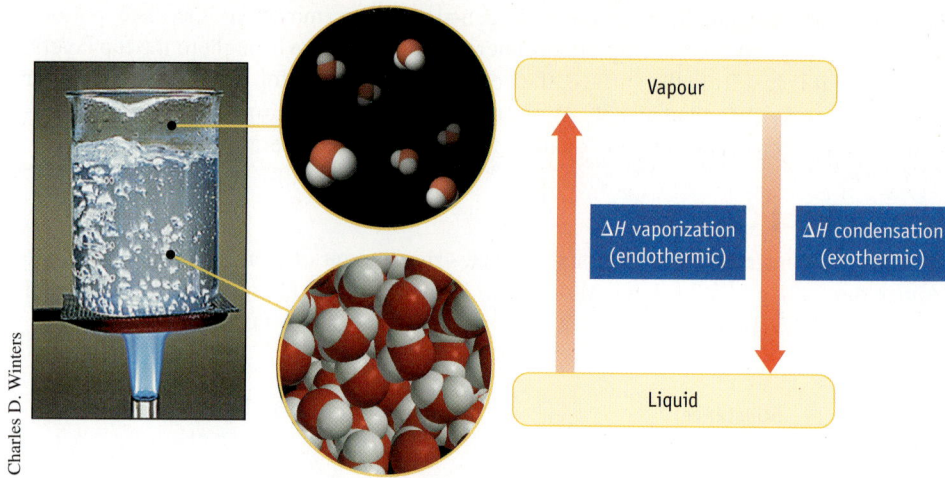

Charles D. Winters

Molecular modelling

e6.2 Watch an animation of the difficulty for water molecules to leave the liquid state

substance (Table 6.2). One mole of water requires much more energy to be vaporized than 1 mol of other common substances with small molecules.

TABLE 6.2 Molar Enthalpy Changes of Vaporization

Substance	$\Delta_{vap}H$ (kJ mol^{-1})
Water (H_2O)	40.7
Ammonia (NH_3)	23.3
Hydrogen chloride (HCl)	16.2
Methane (CH_4)	8.2
Nitrogen (N_2)	5.6

This large enthalpy change accompanying vaporization of water is useful for our own physical well-being. When we exercise vigorously, our body responds by sweating to rid itself of excess heat. Heat from our body is absorbed by water as it evaporates, and our body is cooled.

WORKED EXAMPLE 6.1—ENTHALPY CHANGE OF VAPORIZATION

1.00 L of water is put in a pan and maintained at 100 °C, at which its density is 0.958 g mL^{-1}. Over some time, the water evaporates. How much energy did the water absorb from its surroundings during vaporization? Compare this with the energy absorbed by vaporization of the same amount (in mol) of liquid N_2 at its boiling point (-195.8 °C).

Strategy

From the volume and density of water, we can calculate its mass. Using the molar mass of water (18.02 g mol^{-1}), the amount (in mol) can be calculated. We then use $\Delta_{vap}H$ for water at 100 °C (Table 6.2) to calculate the energy absorbed by evaporation of this amount.

Solution

We calculate the mass of water, and then the amount (in mol):

$$\text{Mass of water, } m(H_2O) = V \times d = 1000 \text{ mL} \times 0.958 \text{ g mL}^{-1} = 958 \text{ g}$$

$$\text{Amount of water, } n(H_2O) = \frac{m}{M} = \frac{958 \text{ g}}{18.02 \text{ g mol}^{-1}} = 53.2 \text{ mol}$$

Now we can calculate the amount of energy absorbed:

$$\text{Energy absorbed} = n \times \Delta_{vap}H = 53.2 \text{ mol} \times 40.7 \text{ kJ mol}^{-1} = 2.16 \times 10^3 \text{ kJ}$$

And the energy absorbed by evaporation of 53.2 mol of N_2 is

$$\text{Energy absorbed} = n \times \Delta_{vap}H = 53.2 \text{ mol} \times 5.6 \text{ kJ mol}^{-1} = 298 \text{ kJ}$$

Comment

Water absorbs about eight times as much energy as the same amount (mol) of nitrogen.

EXERCISE 6.3—ENTHALPY CHANGE OF VAPORIZATION

(a) The molar enthalpy of vaporization of methanol (CH_3OH) is 35.2 kJ mol^{-1} at its boiling point (64.6 °C). How much energy is absorbed by 1.00 kg of this alcohol during vaporization at 64.6 °C?

(b) How much energy is absorbed by the same amount (in mol) of ammonia during vaporization at its boiling point (see Table 6.2)?

Equilibrium Vapour Pressure

In an open vessel, liquids will eventually evaporate completely. In a closed flask (Figure 6.6), however, the pressure of a liquid eventually comes to a constant value, called the **equilibrium vapour pressure**, after which the amount of liquid remains constant. The higher the temperature is, the higher the equilibrium vapour pressure [>>Sections 11.14 and 17.9].

In common usage, the term *vapour pressure* is often used instead of *equilibrium vapour pressure*.

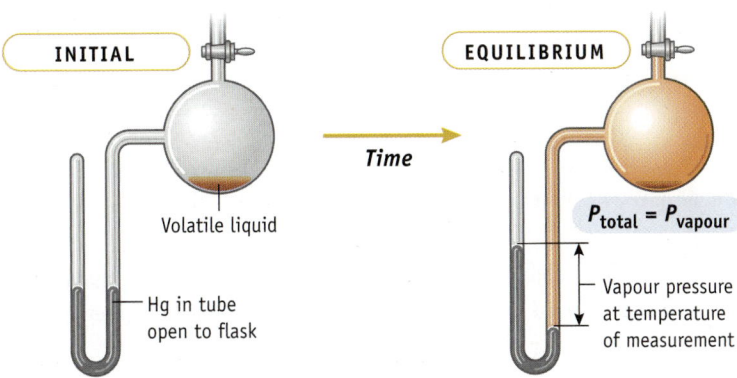

FIGURE 6.6 Equilibrium vapour pressure. Red-brown liquid bromine is placed in an evacuated flask (left), with no molecules in the vapour phase. Over time, some liquid evaporates and the increasing number of molecules in the vapour phase exert an increasing pressure. Eventually, the vapour pressure increases no further. This is the equilibrium vapour pressure of the liquid at the temperature of measurement (right).

At a given temperature, water has a lower equilibrium vapour pressure than other substances whose molecules are of similar size (Table 6.3).

The achievement of equilibrium vapour pressure can be rationalized if we visualize that some molecules are always "escaping" from the liquid into the vapour phase, and simultaneously molecules in the vapour phase collide with, and enter, the liquid phase. As more and more molecules occupy the vapour phase, the frequency of those returning to the liquid increases, and in time the frequency of molecules passing into the liquid becomes the same as the frequency of those passing across the surface in the other direction. At this point, there is no further increase in the vapour pressure, and no net change in the masses of the substance in each phase. The situation is an example of what chemists call a *dynamic equilibrium* [>>Chapter 13]. In the particular case of water, this equilibrium is represented as

$$H_2O(\ell) \rightleftharpoons H_2O(g)$$

At any specified temperature, each liquid achieves a particular equilibrium vapour pressure. The equilibrium vapour pressure of a liquid is a measure of the tendency of its molecules to escape from the liquid phase and enter the vapour phase. This tendency is referred to qualitatively as the *volatility* of the compound: the higher the equilibrium vapour pressure at a given temperature, the more volatile the compound.

TABLE 6.3 Equilibrium Vapour Pressures at 25 °C of Various Liquids

Substance	Equilibrium Vapour Pressure (kPa)
Water, $H_2O(\ell)$	3.17
Ethanol, $C_2H_5OH(\ell)$	7.87
Hexane, $C_6H_{14}(\ell)$	20.2
Bromine, $Br_2(\ell)$	28.7
Acetone, $CH_3COCH_3(\ell)$	30.8
Carbon disulfide, $CS_2(\ell)$	48.2
Diethyl ether, $C_2H_5OC_2H_5(\ell)$	71.7

Think about It

e6.3 See how vapour pressure ideas are important in a butane lighter.

FIGURE 6.7 Vapour pressure and boiling. When the vapour pressure of the liquid equals the atmospheric pressure, bubbles of vapour begin to form within the body of liquid, and the liquid boils.

Charles D. Winters

Molecular modelling

e6.4 Watch an animation portraying the inside of a bubble of boiling water.

FIGURE 6.8 Surface tension of water. (a) A series of photographs showing a falling water droplet, illuminated by a strobe light over a total duration of 0.05 s. The water droplets take on approximately the shape that minimizes its surface area. (b) The insect walking across the surface of water does not exert enough pressure to break through the cohesive force in the water surface.

Although pressures are now measured with units of kPa or bar, the normal boiling point of water was defined to be 100 °C at a time when the unit used was *atmosphere*. The boiling point at exactly 1 bar, called the *standard boiling point*, is 99.6 °C.

EXERCISE 6.4—EQUILIBRIUM VAPOUR PRESSURE AND VOLATILITY

List the substances in Table 6.3 in order of decreasing volatility.

Boiling Point

If we heat a liquid in a beaker, eventually a temperature is reached at which the vapour pressure of the liquid is the same as the pressure of the atmosphere acting down on the liquid surface. Bubbles of vapour form in the liquid and rise to the surface. We say that the liquid boils (Figure 6.7), and the temperature at which this happens is the **boiling point**.

Since the boiling point of a liquid is the temperature at which its vapour pressure is equal to the external pressure, its boiling point depends on the external pressure [>>Section 11.14]. When we use the term *boiling point*, we usually mean the particular boiling point when the external pressure is exactly 1 atm, called the *normal boiling point*.

Since water has a lower equilibrium vapour pressure than other liquids with similar-sized molecules, we should expect that we would have to heat it to a higher temperature before its vapour pressure is 1 atm (Table 6.4).

TABLE 6.4 Normal Boiling Points

Substance	T_b
Water (H_2O)	100 °C
Ammonia (NH_3)	−33.3 °C
Hydrogen chloride (HCl)	−84.8 °C
Methane (CH_4)	−161.5 °C
Nitrogen (N_2)	−195.8 °C

EXERCISE 6.5—BOILING POINT AND MOLAR ENTHALPY CHANGE OF VAPORIZATION

Compare the rank order of normal boiling points with the order of molar enthalpy changes of vaporization in Table 6.2. Is the comparison what you would expect if enthalpy of vaporization and boiling point both depend on the strengths of intermolecular forces between molecules?

Surface Tension

Why do drops of liquids take on a nearly spherical shape (Figure 6.8a), and not, say, the shape of a cube? How is it that insects can walk across the surface of water (Figure 6.8b)?

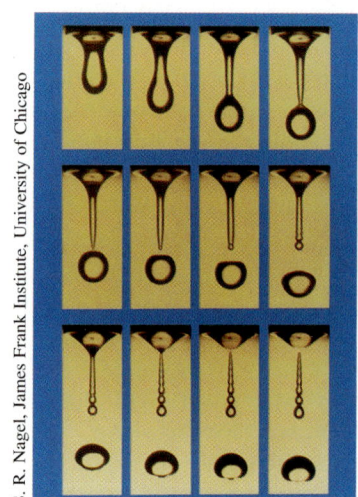

(a)

S. R. Nagel, James Frank Institute, University of Chicago

(b)

basel101658/Shutterstock

Liquids, and especially water, behave as though they have a "skin" stretched across their surface that minimizes the area of surface: energy is required to increase the surface area of a sample of any liquid. The **surface tension** of a liquid is the amount of energy required to increase its surface area, per square metre. Its units are $J\ m^{-2}$ (said as "joules per square metre"). Water has a very high surface tension compared with that of other liquids (Table 6.5).

Substance	Surface Tension at 20 °C ($J\ m^{-2}$)
Water (H_2O)	7.3×10^{-2}
Butanol (C_4H_9OH)	2.5×10^{-2}
Ethanol (C_2H_5OH)	2.3×10^{-2}
Octane (C_8H_{18})	2.2×10^{-2}
Diethyl ether ($C_2H_5OC_2H_5$)	1.7×10^{-2}

TABLE 6.5 Surface Tensions of Some Liquids with Small Molecules

EXERCISE 6.6—SURFACE TENSION OF LIQUIDS

Water drops are not spherical because the tendency for minimization of surface area is counteracted by the downward pull of gravitation (although they are spherical in the zero-gravity situation of an orbiting space shuttle). Would you expect drops of diethyl ether to be more or less spherical than drops of water?

The change of density of water with temperature is different from that of other common substances. Among substances with similarly sized molecules, water has remarkably high specific heat capacity, molar enthalpy change of vaporization, boiling point, and surface tension; and its equilibrium vapour pressure is relatively low.

6.3 Intermolecular Forces

Common sense tells us that in all molecular substances there are forces of attraction between the molecules—if there were not, the molecules would fly apart from each other and these substances would exist only in the gaseous state. Forces of attraction between molecules are called **intermolecular forces**—these are different from covalent bonds, which are *intramolecular forces*. While chemical changes are accompanied by breaking and formation of covalent bonds, the physical properties discussed in Section 6.2 are governed by intermolecular forces.

As we shall see, we can account for the remarkable properties of water on the grounds of stronger intermolecular forces than in other substances with molecules of about the same size. First, we need to investigate the origins of intermolecular forces.

EXERCISE 6.7—INTERMOLECULAR FORCES AND INTRAMOLECULAR FORCES

Are the following changes associated with intermolecular forces or intramolecular forces?
(a) dew forms on a cold night
(b) a piece of paper burns
(c) wet clothes are hung out to dry
(d) water changes to ice in a freezer

FIGURE 6.9 An electric dipole. The diagram shows the charges (left), or an alternative representation as an arrow (right) using the convention that the arrowhead is at the negatively charged end.

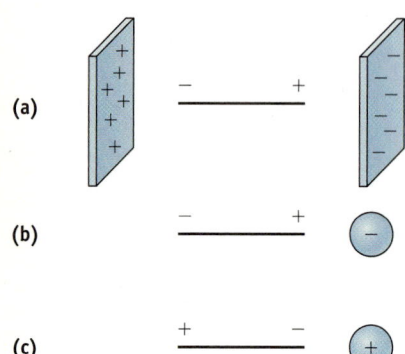

FIGURE 6.10 Orientation of a dipole free to rotate (a) in an electric field, (b) near a negative charge, and (c) near a positive charge.

Bond Polarity

Why should water have relatively strong intermolecular forces? A clue to the answer is provided by the experimental evidence, most easily measured in the vapour phase, that water molecules have a **dipole moment**: a force that aligns them toward a nearby electric charge, or toward the poles of an electrostatic field. A water molecule behaves in the same way as a linear object with a positive charge at one end, and a negative charge at the other. This is called an electric **dipole** (Figure 6.9).

If this dipole is free to rotate and is placed in an electric field, it will align itself with the field, the end with the (+) charge pointing toward the negative plate of the field (Figure 6.10). If the dipole is near a positive charge, it will rotate so that its (−) end is pointing toward it. And in the vicinity of a negative charge, it will rotate so that its (+) end is pointing to it.

Why should a water molecule be a dipole? The explanation for this is related to the electron distribution within the molecule. If we accept that a covalent bond is a shared pair of electrons between the bound atoms, it is logical to ask if the atoms share the pair of electrons equally. We should expect the two electrons in the bond of an H_2 molecule to be equally shared because the two nuclei are identical, but this would not be the case when the two atoms are different—as in HCl molecules, for example.

The ability of an atom to attract the bonding electrons that bind it to another atom in a molecule is called its **electronegativity**, symbol χ. Estimates of the electronegativities of the various elements are shown in Figure 6.11.

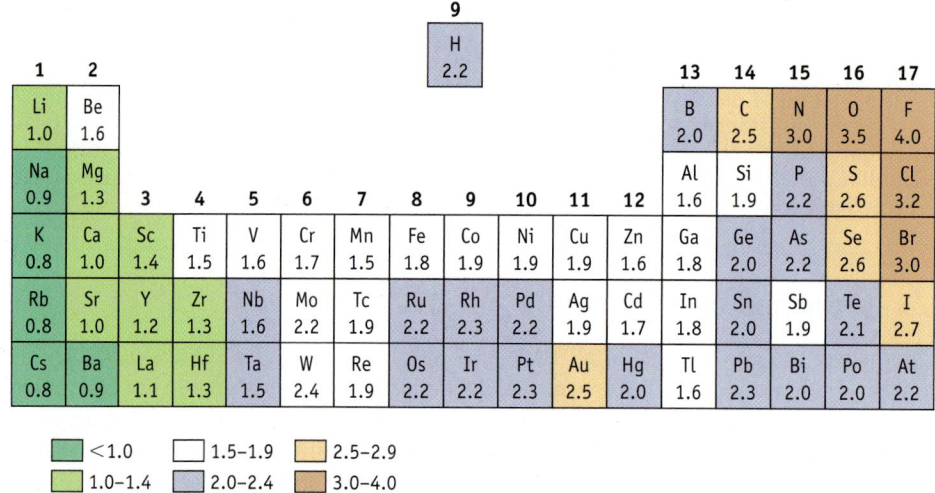

FIGURE 6.11 Estimates of the electronegativities of the elements.

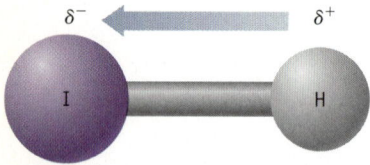

FIGURE 6.12 Polar covalent bond in an HI molecule.

In molecules of hydrogen iodide, for example, the I atom is more electronegative than the H atom, and the molecules have a charge separation as indicated in Figure 6.12. When atoms bound to each other have different charges, the bond is said to be a **polar bond**, and the magnitude of the charges is a measure of **bond polarity**.

Hydrogen fluoride molecules have one polar covalent bond, water molecules have two, and ammonia molecules have three (Figure 6.13).

The greater the difference in electronegativities of two atoms bound to each other, the greater the charge separation is, and the more polar the bond.

WORKED EXAMPLE 6.2—BOND POLARITIES

Using values of electronegativity listed in Figure 6.11, decide the direction of the dipole in the bonds B—F and B—Cl, and decide which is the more polar.

Strategy

In each pair of bound atoms, the more electronegative atom has the net partial negative charge, $\delta-$. Comparing the two bonds, the one with the greater difference of electronegativities ($\Delta\chi$) has the bigger charge separation, and is the more polar.

Solution

$\chi(F) > \chi(B)$, so in B–F bonds, the F atom has $\delta-$ charge and the B atom has $\delta+$ charge. $\chi(Cl) > \chi(B)$, so in B—Cl bonds, the Cl atom has $\delta-$ charge and the B atom has $\delta+$ charge.

In B—F bonds, $\Delta\chi = 4.0 - 2.0 = 2.0$, while in B—Cl bonds, $\Delta\chi = 3.2 - 2.0 = 1.2$. So B—F bonds are more polar than B—Cl bonds—in other words, the magnitudes of $\delta+$ and $\delta-$ are bigger in B—F bonds than in B—Cl bonds.

EXERCISE 6.9—BOND POLARITIES

Use electronegativity values in Figure 6.11 to decide the direction of the dipoles in each of the bonds in the following pairs, and decide which bond in each pair is more polar:
(a) H—F and H—I
(b) B—C and B—F
(c) C—Si and C—S

EXERCISE 6.10—BOND POLARITIES

Acrolein (C_3H_4O) is the starting material for manufacture of several plastics.

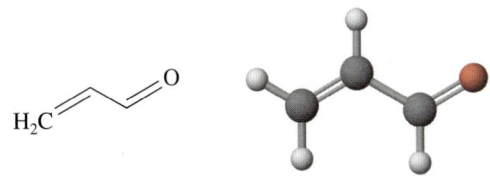

Using electronegativity values in Figure 6.11, decide the following:
(a) Which bonds in the molecule are polar and which are non-polar?
(b) Which is the most polar bond in the molecule? Which is the more negative atom of this bond?

If the two atoms bound by a covalent bond are not identical, the bonding electrons are unequally distributed, and the bond is a polar bond. The ability of an atom to attract bonding electrons is called its electronegativity. Net negative charge resides on the atom with greater electronegativity. The greater the difference between the electronegativity of bound atoms, the greater the charge separation.

Molecular Polarity and Dipole–Dipole Forces

In diatomic molecules with different atoms, such as HI, the bond is polar and so the molecule is a dipole. In the case of a substance whose molecules have more than one polar bond, whether or not the molecule is a dipole depends on its three-dimensional shape: if the arrangement of the bonds is symmetrical, the forces experienced in an electric field cancel out. For example, carbon dioxide molecules have two polar bonds, but because they point in opposite directions their effects cancel (Figure 6.14), and the molecules have zero dipole moment: they are *non-polar molecules* and we call such substances **non-polar substances**. By contrast, a water molecule has two O—H bonds at an angle of 104° to each

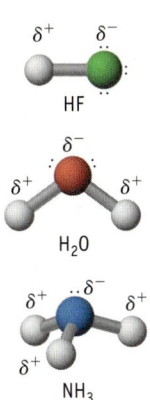

FIGURE 6.13 Molecules with polar covalent bonds. F, O, and N atoms are more electronegative than H atoms.

Molecular Modelling (Odyssey)

e6.5 Compare bonds between atoms with different electronegativities using electrostatic potential maps.

Interactive Exercise (Odyssey) 6.8

Compare the polarities of a range of different bonds.

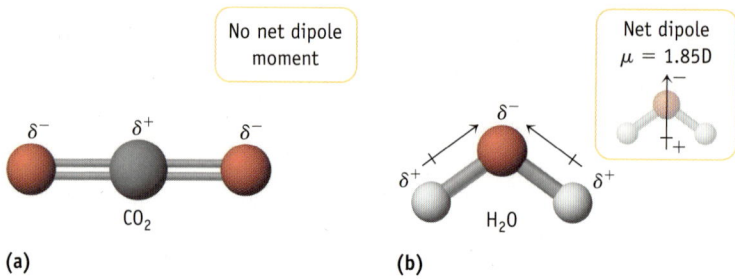

FIGURE 6.14 Bond polarities and dipole moments of molecules of carbon dioxide and water. Ball-and-stick representations show the two polar bonds in molecules of each substance. (a) Carbon dioxide molecule. In an electric field, the effects of the directly opposite bond dipoles cancel, so these molecules have zero dipole moment. (b) Water molecules. The net result of the bond polarities and the angle of the bonds to each other is that the molecule behaves as a dipole acting in the direction that bisects the H—O—H angle. The *debye*, abbreviation D, is the unit of measurement of *dipole moment* (symbol μ), the force experienced by a dipole to align itself with an electric field.

other. Since these bonds are not at 180° to each other, the molecule is a dipole, and is said to be a *polar molecule*, and we call water a **polar substance**.

The influence of symmetry on the polarity of molecules with three bonds is shown in Figure 6.15.

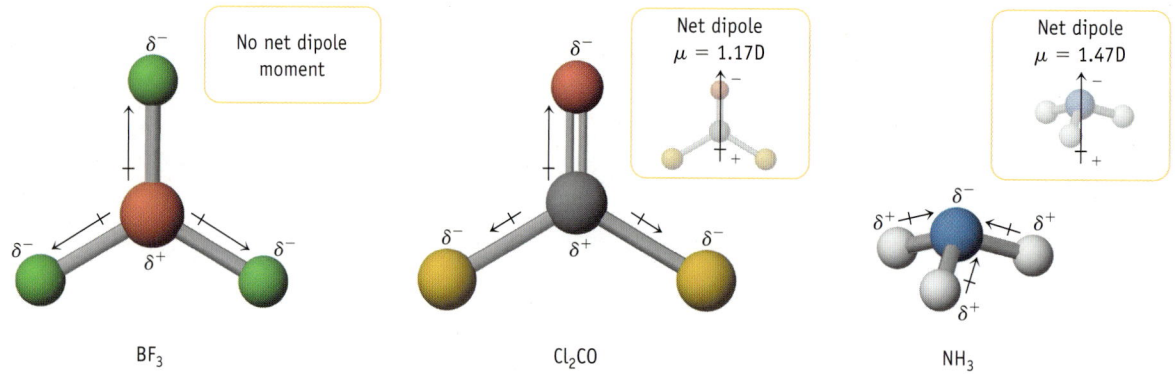

FIGURE 6.15 Dipole moments of molecules with three bonds. The symmetry of the trigonal-planar BF_3 molecule results in zero dipole moment. On the other hand, unsymmetrical planar Cl_2CO molecules and pyramidal NH_3 molecules have net dipole moments.

The dipole moments of various molecules with four atoms tetrahedrally arranged about a carbon atom are shown in Figure 6.16. The symmetrical molecules have zero dipole moment.

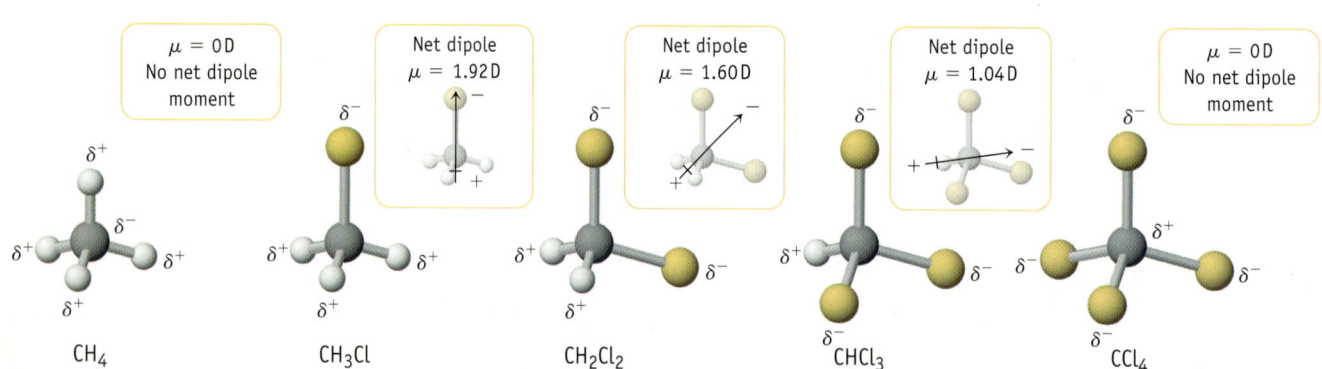

FIGURE 6.16 Dipole moments of tetrahedral molecules. Only in CH_4 and CCl_4 molecules are the bond dipoles exactly symmetrical, cancelling out to give zero dipole moment. The other molecules have dipole moments.

The principles described above are illustrated by the magnitudes of dipole moments of molecules of a few substances in Table 6.6.

TABLE 6.6 Dipole Moments of Selected Molecules

Molecule (AB)	Moment, μ (D)	Geometry	Molecule (AB_2)	Moment, μ (D)	Geometry
HF	1.78	linear	H_2O	1.85	bent
HCl	1.07	linear	H_2S	0.95	bent
HBr	0.79	linear	SO_2	1.62	bent
HI	0.38	linear	CO_2	0	linear
H_2	0	linear			

Molecule (AB_3)	Moment, μ (D)	Geometry	Molecule (AB_4)	Moment, μ (D)	Geometry
NH_3	1.47	trigonal-pyramidal	CH_4	0	tetrahedral
NF_3	0.23	trigonal-pyramidal	CH_3Cl	1.92	tetrahedral
BF_3	0	trigonal-planar	CH_2Cl_2	1.60	tetrahedral
			$CHCl_3$	1.04	tetrahedral
			CCl_4	0	tetrahedral

WORKED EXAMPLE 6.3—BOND POLARITY AND MOLECULAR POLARITY

Molecules of nitrogen trifluoride (NF_3) have a pyramidal shape like those of NH_3. Describe the polarity of NF_3 molecules.

Strategy

Decide on the direction of polarity of the bonds, with reference to the electronegativities listed in Figure 6.11. From the orientation of the bonds to each other, use the examples of Figure 6.15 to decide if the bond dipoles will cancel out.

Solution

Because F is more electronegative than N, each N—F bond is polar with a net partial negative charge on the three F atoms. Because the molecules are pyramidal (and not planar like BF_3), the bond dipoles do not cancel, and the molecules have a dipole moment. The positive end of the dipole is at the N atom.

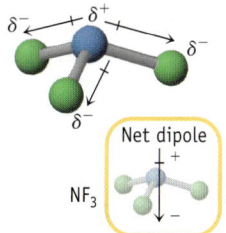

Molecular Modelling (Odyssey)

e6.6 Compare the shapes of molecules and their dipole moments.

WORKED EXAMPLE 6.4—BOND POLARITY AND MOLECULAR POLARITY

Because rotation around C═C double bonds is not possible, there is *cis*-1,2-dichloroethylene (A) and *trans*-1,2-dichloroethylene (B).

(a)   (b)

All bond angles are close to 120°. Is either of these polar? Estimates of electronegativities (Figure 6.11) are C: 2.5; H: 2.2; Cl: 3.2.

Strategy

Using electronegativity values, we decide on the bond polarities. Then, by reference to the molecular structures, we decide whether or not the effect of the bond polarities will cancel out in an electric field.

Solution

The electronegativity differences across the C—H and C—Cl bonds indicate that these are polar bonds, with a net displacement of electron density away from the H atoms [$H^{\delta+}$—$C^{\delta-}$] and toward the Cl atoms [$C^{\delta+}$—$Cl^{\delta-}$].

In molecule A, the Cl atoms are on the same side of the C≡C bond, and all four bond dipoles are in approximately the same direction. *cis*-Dichloroethylene (A) is polar. In molecule B, the two C—H bond dipoles cancel each other, and so do the two C—Cl bond dipoles. The movement of electron density toward the Cl atom on one end of the molecule is counterbalanced by an opposing movement on the other end. *trans*-Dichloroethylene (B) is non-polar.

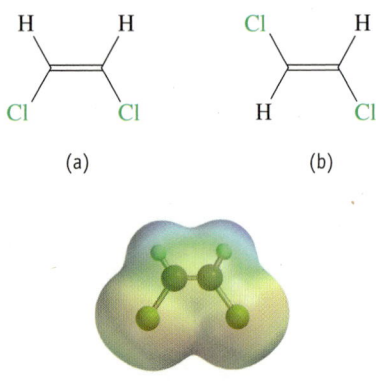

(a)   (b)

EXERCISE 6.12—BOND POLARITY AND MOLECULAR POLARITY

Molecules of carbon disulfide (CS_2) are linear, corresponding with the shape of CO_2 molecules, while molecules of SO_2 are bent, like water molecules. Are these molecules polar?

EXERCISE 6.13—MOLECULAR POLARITY

Use values of electronegativity listed in Figure 6.11. For each of the following molecules, decide whether the molecule is polar and, if it is, the direction of the molecular dipole:
(a) $BFCl_2$, whose shape is trigonal-planar, like that of BF_3
(b) NH_2Cl, whose shape is trigonal-pyramidal, like that of NH_3
(c) SCl_2, which is a bent molecule, like H_2O.

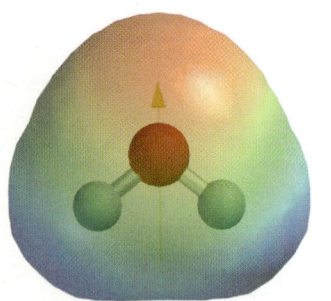

FIGURE 6.17 Electrostatic potential map of a water molecule. Although the molecule is neutral, because the electrons in the O—H bonds are unequally shared between the O atom and the H atoms, there is net negative charge at the O atom, and net positive charge at the H atoms.

Another useful way to represent a molecule is an *electrostatic potential map*, shown in Figure 6.17 for a water molecule. This is a space-filling model of the molecule with colour shading: red to indicate regions of the molecule where the negative charge of electrons exceeds the positive charge on the nuclei, and blue in regions where there is a deficiency of electrons and net positive charge. On a continuum from red to blue, the more red the colour the greater is the net negative charge, and the more blue the colour the greater the net positive charge. Intermediate colours (orange, yellow and green) indicate net charge closer to zero.

We can represent the behaviour of water molecules in an electric field as in Figure 6.18.

In water, we imagine many molecules moving with respect to each other, and making collisions with each other as they move around. As the polar molecules move past each other, they align themselves so that the positive end of one polar molecule is oriented toward the negative end of an adjacent molecule, and the attraction between unlike charges results in stronger intermolecular forces than if the molecules were non-polar. This type of intermolecular force, called a **dipole–dipole force**, is present in any substance, like water, whose molecules are dipoles. An instantaneous "snapshot" at the molecular level is represented in Figure 6.19.

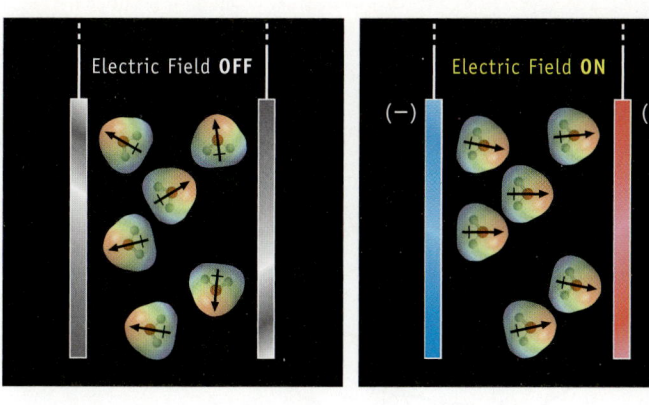

FIGURE 6.18 Alignment of water molecules in an electric field. When placed in an electric field (between charged plates), polar water molecules experience a force that tends to align them with the field. The negative ends of the molecules are drawn toward the positive plate, and vice versa. The orientation of the polar molecules affects the electrical capacitance of the plates (their ability to hold a charge), which provides a way to measure experimentally the magnitude of their dipole moment. This is a stylized representation—any reasonable sample of a substance between the plates of an applied electric field would consist of many millions of molecules, and not just six of them.

WORKED EXAMPLE 6.5—DIPOLE–DIPOLE FORCES

Would you expect there to be dipole–dipole forces operating in a sample of chloromethane (CH_3Cl)?

Strategy

If molecules of the substance are polar, its molecules are dipoles, and there will be dipole–dipole forces between molecules.

Solution

Chloromethane molecules are polar (See Table 6.6). There are dipole–dipole intermolecular forces.

EXERCISE 6.15—DIPOLE–DIPOLE FORCES

In which of the following substances would you expect there to be dipole–dipole forces of attraction acting between molecules?
(a) Sulfur dioxide, $SO_2(g)$
(b) Carbon dioxide, $CO_2(g)$
(c) Hydrogen chloride, $HCl(g)$

Substances whose molecules have a dipole moment are said to be polar. Whether or not molecules have a dipole moment depends on their three-dimensional shape. Polar molecules have polar bonds that are not symmetrically distributed. Forces of attraction between polar molecules are called dipole–dipole forces.

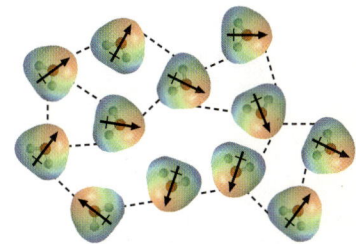

FIGURE 6.19 Non-random orientation of water molecules due to dipole–dipole forces. The molecules are in constant motion, changing "neighbours" often and quickly, so the environment of any molecule is ever-changing. Averaged over time, the orientation of the water molecule is not random, because there is a force of attraction between the positive end of any one molecule and the negative end of nearby molecules. To show this clearly, the spaces between the molecules (compared to the size of the molecules) are greatly exaggerated in this diagrammatic representation.

Hydrogen Bonding

Water's properties are even more extraordinary than those of other polar substances, and to account for this we need to explore a further characteristic of intermolecular forces in water (and a few other substances)—hydrogen bonding.

Many compounds, including water, that have H—F, H—O, or H—N bonds in their molecules have exceptional properties. Illustrations of this are the unexpectedly high boiling points of H_2O, HF, and NH_3 compared with related compounds (Figure 6.20). Because the temperature at which a substance boils depends on the forces of attraction between molecules, there is a clear indication of strong intermolecular attractions in H_2O, HF, and NH_3.

The electronegativities of F, O, and N are the highest of all the elements, and the large difference in electronegativity in N—H, O—H, and F—H bonds means that they are very polar, with a partial positive charge on the H atom in each case. So in these compounds there is an extreme form of intermolecular dipole–dipole interactions, called **hydrogen bonding**, between an H atom on one molecule and an O, N, or F atom on nearby molecules.

FIGURE 6.20 The boiling points of some simple compounds of hydrogen. The boiling point of water is compared with those of corresponding compounds of Group 16 elements (H_2S, H_2Se, and H_2Te). Also plotted are boiling points of hydrogen compounds of Group 17 elements (HF, HCl, HBr, and HI), Group 15 elements (NH_3, PH_3, AsH_3, and SbH_3), and Group 14 elements (CH_4, SiH_4, GeH_4, and SnH_4). The boiling points of H_2O, HF, and NH_3, in all of which hydrogen bonding operates, are considerably higher than expected by extension of the trend of decreasing boiling points with decreasing molar mass.

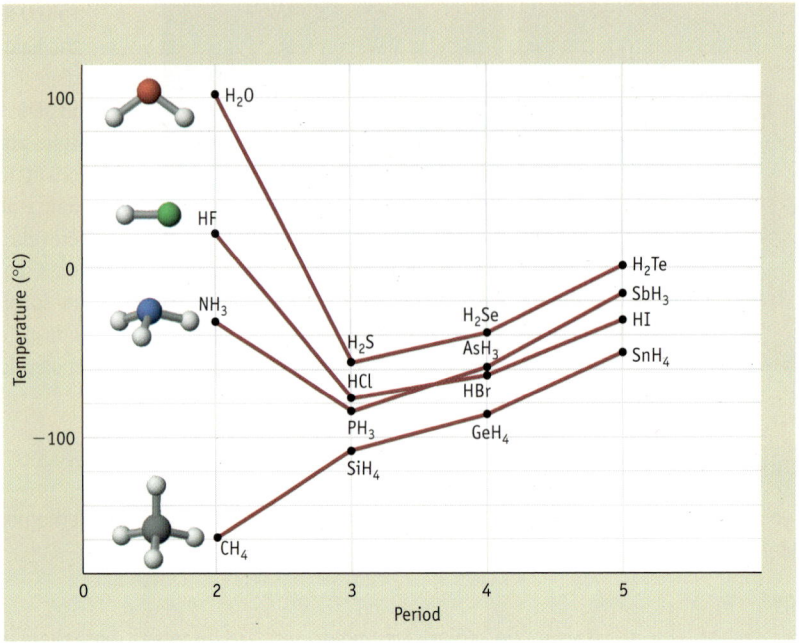

Molecular Modelling (Odyssey)

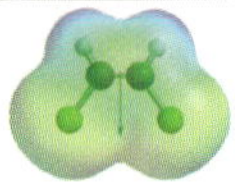

e6.8 Simulate the approach of two water molecules to optimize hydrogen bonding.

Molecular Modelling (Odyssey)

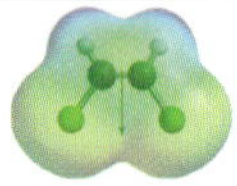

e6.9 Simulate water molecules in the liquid state showing hydrogen bonds forming and breaking.

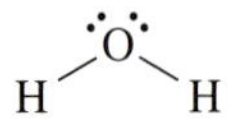

A Lewis "'dot structure" of a water molecule, showing two non-bonding pairs of electrons ("lone pairs") on the O atom.

In the case of water, there is a rapidly changing network of hydrogen bonding between the O atom on each molecule and one or both of the H atoms on nearby molecules (Figure 6.21).

While the difference between the electronegativities of O and H gives rise to polarity of the H—O bond, there are other factors that contribute to the strength of hydrogen bonding between water molecules:

- A hydrogen atom only has one electron, and when H is covalently bound to another atom, the electron is largely localized in the bond. Consequently, the positively charged hydrogen nucleus is rather exposed to adjacent molecules.
- The H atom is the smallest of all atoms. As a result, the H atom in a molecule can get very close to the O atom on other molecules. The force of attraction between opposite charges is inversely proportional to the square of the distance between them.
- A simple picture of the electron distribution in a water molecule is that some of the O atom's electrons are in the covalent bonds to H atoms, and two pairs of electrons not used in bonding (*non-bonding electrons*, or *lone pairs*) are localized on the O atom [>>Section 10.3]. These provide two localized centres of negative charge on the O atom, each of which can engage in hydrogen bonding to H atoms in nearby water molecules.

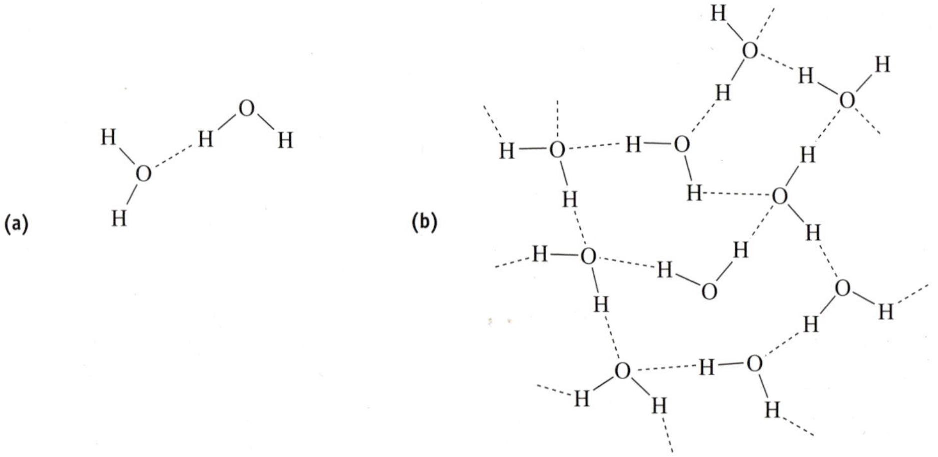

FIGURE 6.21 Hydrogen bonding in water. (a) A dotted line represents a hydrogen bond between the O atom on one H_2O molecule and an H atom on another H_2O molecule. (b) An imagined "'snapshot" of a rapidly changing network of hydrogen bonding in water. Hydrogen bonds are forces between atoms—not the O—H covalent bonds.

(a)

(b)

In general, extending our description to pure substances other than water, and to mixtures of substances, the attraction due to a hydrogen bond can be represented as

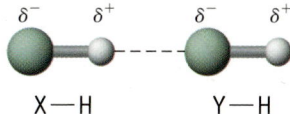

where the H atom is a "bridge" between the two electronegative atoms X and Y, and the dashed line represents the hydrogen bond. X and Y may be the same or different (Table 6.7). In water, for example, both X and Y are O atoms.

The energy of breaking hydrogen bonds is in the range 5–30 kJ mol^{-1}, which is considerably less than the 100–500 kJ mol^{-1} of single covalent bonds. In the particular case of water, the energy of the hydrogen bonds between water molecules is about 22 kJ mol^{-1}, whereas the bond energy of the O—H covalent bonds within the molecules is 463 kJ mol^{-1}.

Apart from the importance of hydrogen bonding in determining the properties of water and other liquids, these strong intermolecular forces play important roles in the structures of natural polymers such as proteins [>>Chapter 29] and DNA [>>Section 25.4], as well as synthetic polymers like nylon [<<Section 4.6].

Molecular Modelling (Odyssey)

e6.10 Simulate solutions of different substances where hydrogen bonding is present.

Interactive Exercise (Odyssey) 6.16

SUBMIT

Investigate hydrogen bonding in simulations of mixtures.

N—H - - - :N—	O—H - - - :N—	F—H - - - :N—
N—H - - - :O—	O—H - - - :O—	F—H - - - :O—
N—H - - - :F—	O—H - - - :F—	F—H - - - :F—

TABLE 6.7 The Most Common Circumstances of Hydrogen Bonding

WORKED EXAMPLE 6.6—MOLECULAR STRUCTURE AND HYDROGEN BONDING

Ethanol and dimethyl ether have the same composition, but their molecules have a different arrangement of atoms (they are *isomers*).

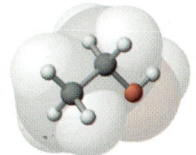   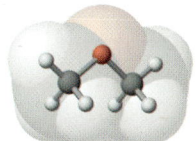

Ethanol, CH_3CH_2OH Dimethyl ether, CH_3OCH_3

In their liquid states, there is hydrogen bonding between the molecules of only one of these. Which one? Draw a diagram to represent of hydrogen bonding between molecules in this compound.

Strategy

Look for the atom sequences listed in Table 6.7.

Solution

Molecules of ethanol have an –OH group, so hydrogen bonding can contribute to the intermolecular forces of attraction.

In contrast, although molecules of dimethyl ether are polar, there is no opportunity for hydrogen bonding because they have no O—H covalent bond.

Comment

We will see in Section 6.4 that higher boiling points can be rationalized by stronger intermolecular forces. Indeed, ethanol boils at 78.3 °C, whereas dimethyl ether boils at −24.8 °C, which is why ethanol is a liquid and dimethyl ether is a gas at ambient conditions.

Interactive Exercise 6.18

SUBMIT

Predict whether hydrogen bonding would be involved in a liquid.

EXERCISE 6.17—MOLECULAR STRUCTURE AND HYDROGEN BONDING

Using structural formulas, describe with a diagram hydrogen bonding between molecules of methanol (CH_3OH).

Hydrogen bonding is a relatively strong form of dipole–dipole intermolecular attraction, most commonly formed between an H atom (joined to an F, O, or N atom) in one molecule, and an F, O, or N atom in another molecule.

Dispersion Forces in All Molecular Substances

Even in non-polar substances, there must be forces of attraction between the molecules, or these substances would only exist in the gas phase. A comparison of the boiling points and enthalpy changes of vaporization of substances with similar molar masses indicates that the intermolecular forces in polar substances are stronger than those in non-polar substances (Table 6.8). Comparisons are made between substances with similar molar masses because, as perusal of the data indicates, there is a correlation between the magnitude of intermolecular forces and molar masses.

TABLE 6.8 Boiling Points and $\Delta_{vap}H$ Values of Non-Polar and Polar Substances

NON-POLAR SUBSTANCES				POLAR SUBSTANCES		
	M (g mol^{-1})	T_b (°C)	$\Delta_{vap}H$ (kJ mol^{-1})		M (g mol^{-1})	T_b (°C)	$\Delta_{vap}H$ (kJ mol^{-1})
N_2	28	−196	5.57	CO	28	−192	6.04
SiH_4	32	−112	12.10	PH_3	34	−88	14.06
GeH_4	77	−90	14.06	AsH_3	78	−62	16.69
Br_2	160	59	29.96	ICl	162	97	53.1

Intermolecular forces of attraction in non-polar substances are called **dispersion forces** or *London forces*. The term *van der Waals forces* is also sometimes used. While dispersion forces are the only intermolecular forces operating in non-polar substances, these forces must also be in action in polar substances—along with dipole–dipole forces.

How can we explain these forces that operate between non-polar molecules? Averaged over time, the electrons in a molecule (other than those in covalent bonds) are symmetrically distributed. Chemists believe, however, that the electrons around a molecule (the *electron cloud*) are rather mobile and may fluctuate, so that at any instant the distribution over space may be distorted. In this case, we would have an **instantaneous dipole**—the electron cloud is not exactly symmetrically distributed, and there is a region with a slight excess of electron matter (and a transitory negative charge) and another region with a deficiency of electron matter (and a transitory positive charge). Imagine that the electron cloud can be pulled and pushed around by electrical charges in their immediate environment.

Now we can visualize that as two non-polar molecules approach each other, an instantaneous dipole in one of them can lead to distortions in the electron cloud of the other. That is, instantaneous dipoles can induce dipoles momentarily in neighbouring molecules, giving rise to attractions between them (Figure 6.22). So, the dispersion forces in non-polar substances are **instantaneous dipole–induced dipole forces**.

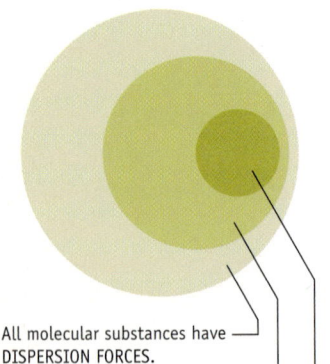

All molecular substances have DISPERSION FORCES.

Some molecular substances have DIPOLE–DIPOLE FORCES as well as dispersion forces.

Some molecular substances have very strong dipole–dipole forces called HYDROGEN BONDS as well as dispersion forces.

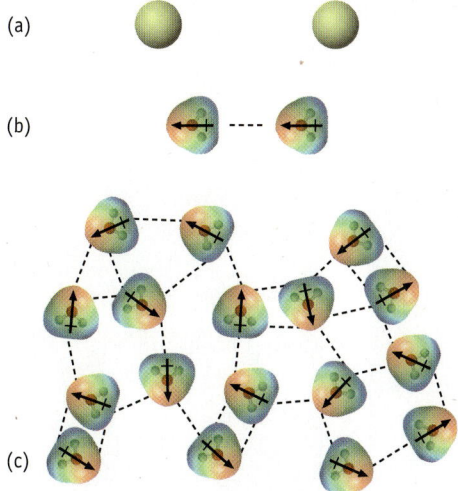

FIGURE 6.22 Dispersion forces: instantaneous dipole—induced dipole forces. (a) Non-polar molecules, or group 18 atoms, with uniform distribution of electron clouds. (b) An instantaneous dipole in one molecule induces an instantaneous dipole in another nearby. The momentary opposite charges give rise to attractions between the molecules. (c) Instantaneous dipole–induced dipole forces operate throughout the substance.

The distortion of the electron cloud of an atom or a molecule is called **polarization**, and the ease with which it can be distorted is called **polarizability**. This property is difficult to measure experimentally. It makes sense, however, that the electron cloud of a molecule with a large electron cloud, such as that of an I_2 molecule, can be polarized more readily than the electron cloud in a much smaller atom or molecule, such as He or H_2, since in the I_2 molecule the electrons are further from the nucleus and less tightly held. As a consequence, the dispersion forces in iodine are sufficiently strong that it is a solid at room temperatures.

We should not dismiss dispersion forces as weak and insignificant, especially in the case of large molecules. Since iodine is a solid at ambient conditions, the dispersion forces between I_2 molecules must be stronger than the total of all intermolecular forces (including hydrogen bonds) in water.

In general, within a family of compounds, such as the halogens—$F_2(g)$, $Cl_2(g)$, $Br_2(\ell)$, $I_2(s)$—or the alkanes—$CH_4(g)$, $C_2H_6(g)$, $C_3H_8(g)$, and so on—the higher the molar mass, the higher the melting and boiling points (Table 6.9), and, we can conclude, the stronger the dispersion forces are. Interpreting one step further, we can postulate that, within a family of compounds, the more electrons in the electron cloud of a molecule, generally the greater the polarizability of the electron cloud is.

Molecular Modelling (Odyssey)

e6.11 Simulate substances with different types of intermolecular bonding.

Iodine *sublimes* at 113.5 °C—that is, it changes directly from the solid to the vapour, rather than melt like most substances. Solid carbon dioxide ("dry ice") also sublimes [>>Section 11.13].

TABLE 6.9 Melting Points and Boiling Points of Non-Polar Halogens and Alkanes

T_f is the symbol for freezing point, which for any substance is the same as its melting point.

HALOGENS (GROUP 17 ELEMENTS)			ALKANES (C_nH_{2n})		
	T_f (°C)	T_b (°C)		T_f (°C)	T_b (°C)
$F_2(g)$	−219.6	−188.1	$CH_4(g)$	−182.4	−161.5
$Cl_2(g)$	−101.0	−34.6	$C_2H_6(g)$	−182.8	−88.6
$Br_2(\ell)$	−7.2	58.8	$C_3H_8(g)$	−187.6	−42.1
$I_2(s)$	113.5 (sublimes)		$C_4H_{10}(g)$	−138.2	−0.5

EXERCISE 6.19—DISPERSION FORCES

With reference to Table 6.9, decide which substance in each of the following pairs has the stronger dispersion forces, and explain why:
(a) fluorine or bromine
(b) ethane or butane

Dispersion forces, which operate between molecules of all molecular substances, can be attributed to instantaneous dipole–induced dipole forces. The ease with which the "electron cloud" of an atom is distorted is called its polarizability. The more easily polarizable the electron cloud, the greater the dispersion forces are.

Interactive Exercise 6.20

Predict what types of intermolecular forces are broken during boiling.

Interactive Exercise 6.21

Rank intermolecular forces in increasing strength.

6.4 Explaining the Properties of Water

Why does the density of water change with temperature? The lower density of ice at 0 °C compared to that of liquid water at 0 °C is attributed to a "locking in" of the H_2O molecules into a very open structure in ice (Figure 6.23).

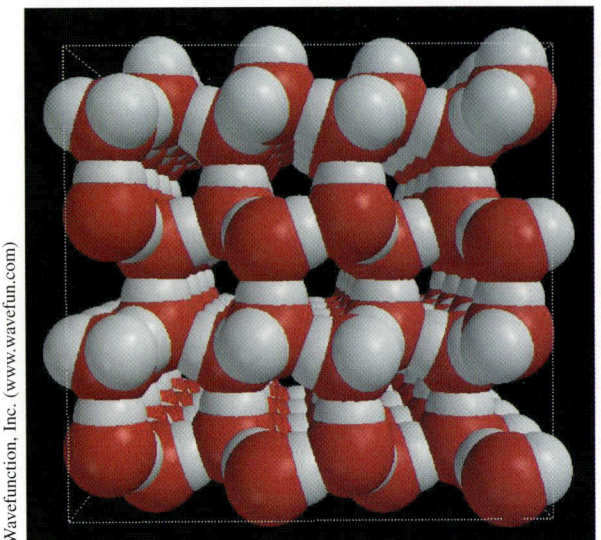
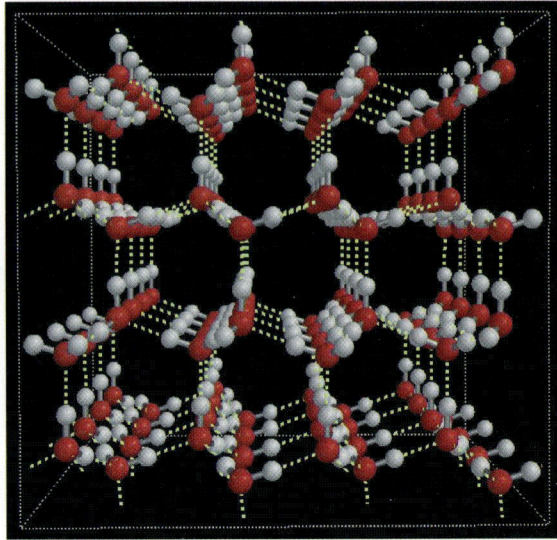

Wavefunction, Inc. (www.wavefun.com)

FIGURE 6.23 Open arrangement of water molecules in ice. A space-filling model is shown on the left, and the corresponding ball-and-spoke model, showing hydrogen bonds (yellow), is on the right. Notice the empty channels, a consequence of which is that the number of water molecules per unit volume (and the density) is relatively low. The space-filling model uses van der Waals radii, and the overlap indicates hydrogen bonding.

Molecular Modelling (Odyssey)

e6.12 Simulate ice and its hydrogen bonding between water molecules.

How can we explain this open structure? Each hydrogen atom on any molecule in ice can be attracted, through a hydrogen bond, to a lone pair of electrons on the oxygen atom of an adjacent molecule. And because its oxygen atom has two lone pairs of electrons, it can form two more hydrogen bonds with hydrogen atoms in adjacent molecules. The result is a tetrahedral arrangement of four H atoms around each O atom—two joined through covalent bonds, and two joined through hydrogen bonding (Figure 6.24).

FIGURE 6.24 The tetrahedral arrangement of four water molecules around each water molecule in ice. Each oxygen atom is covalently bonded to two hydrogen atoms in the same molecule, and hydrogen-bonded to hydrogen atoms in two other molecules. The hydrogen bonds are longer and weaker than the covalent bonds. This structural unit is repeated throughout the ice structure shown in Figure 6.23.

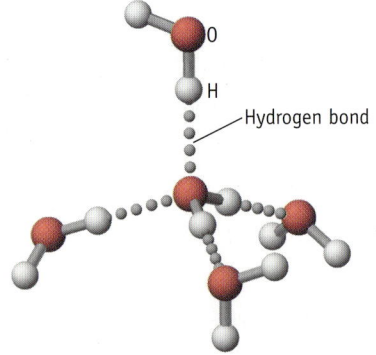

Molecular Modelling (Odyssey)

e6.13 Simulate ice melting and the corresponding change in molecular freedom.

Each of the four molecules surrounding the central one in Figure 6.24 is also surrounded by four molecules (one of which is the central one in the figure). Extension of this arrangement throughout the whole structure of ice results in a highly ordered, open-cage structure with lots of empty space.

When ice melts at 0 °C, the relatively large increase in density (Figure 6.3) can be explained by partial breakdown of the regular ice structure. The increase in density from 0 °C to 4 °C can be attributed to further gradual breakdown of the hydrogen-bonded structure. Just above the melting point, some of the water molecules remain in open ice-like structures, and as the temperature is raised, the ice-like structures collapse further, and the volume contracts. This increase in density can be regarded as a continuation of what happens when ice melts.

Water's density reaches a maximum at about 4 °C. With increasing temperature from 4 °C, the density decreases like every other substance because the increasing kinetic energy of the molecules results in occupation of more space per molecule, on average.

The high heat capacity of water can also be rationalized by reference to hydrogen bonding. As the temperature of water is increased, the average kinetic energy of the molecules increases. Energy must be provided to the molecules to overcome the forces of attraction between them. Gradual disruption of the hydrogen bonding network in water as the temperature increases requires more energy than for substances with weaker intermolecular forces.

Vaporization involves molecules in close proximity being separated from each other. The stronger are the intermolecular forces, the more energy is needed to separate the molecules. Consequently, the enthalpy change of vaporization of water is large because in the liquid phase the molecules are attracted to each other by hydrogen bonding.

The same reasoning can explain the relatively low equilibrium vapour pressure of water. If we compare water with a non-polar liquid at the same temperature, their molecules have the same average speeds, but the water molecules are held by stronger forces of attraction. Water molecules cannot escape so easily into the vapour phase.

The equilibrium vapour pressure of water at 25 °C is less than that of non-polar hexane, for example. So the temperature at which water's vapour pressure reaches atmospheric pressure, and at which it boils, is higher than is the case for hexane.

The high surface tension of water can similarly be attributed to hydrogen bonding. Molecules in the interior of a liquid interact with molecules all around them, while molecules in the surface layer are attracted only to those below (Figure 6.25). This leads to a net inward force of attraction on the surface molecules, resisting increase of surface area. So water behaves as though it has a thin skin that resists stretching or penetration. The surface tension of water is relatively high because the net inward force of hydrogen bonding to other molecules within the bulk of the liquid is high.

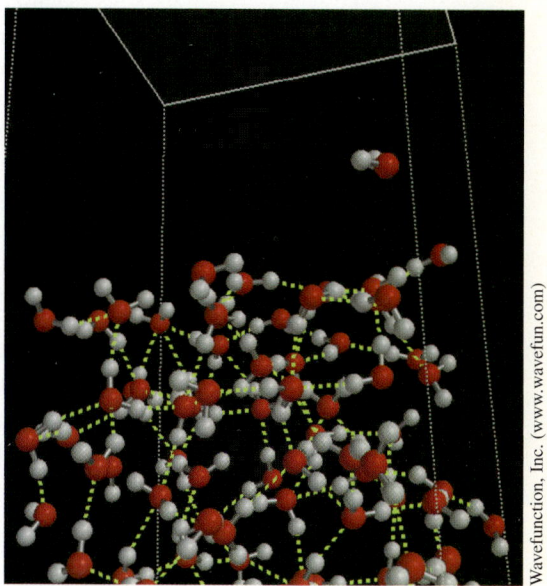

FIGURE 6.25 Intermolecular forces at the surface of a liquid. Water molecules in the bulk of the liquid are attracted by water molecules in every direction. Water molecules at the surface are attracted only by molecules below them.

EXERCISE 6.23—INTERMOLECULAR FORCES AND PROPERTIES

Select the substance in each of the following pairs that has the higher boiling point:
(a) Br_2 or ICl
(b) neon or krypton
(c) CH_3CH_2OH (ethanol) or C_2H_4O (ethylene oxide, structure to the right)

$H_2C - CH_2$ with O below

EXERCISE 6.24—INTERMOLECULAR FORCES AND PROPERTIES

Which type(s) of intermolecular force influence(s) the magnitude of the equilibrium vapour pressure of each of the following?
(a) hexane
(b) water
(c) carbon tetrachloride

Molecular Modelling (Odyssey)

e6.14 Examine a molecular-level simulation of the surface of liquid water.

Interactive Exercise 6.22

Select the compound with the higher boiling point.

The physical properties of molecular compounds depend on the strengths of intermolecular forces. The ways in which water is remarkable, compared with other substances with similarly sized molecules, can be attributed to strong hydrogen bonding between water molecules.

6.5 Water as a Solvent

Although water is a powerful solvent, and the most common, there are many other substances that can be used as solvents, such as ethanol, hexane, and acetone. The chemistry that occurs in one solvent is different from that in others.

One of the substances in a solution is generally considered to be the **solvent**, the medium in which one or more other substances, the **solutes**, are dissolved. Solutions with water as the solvent are called **aqueous solutions**. The *solubility* of a substance in water refers to the concentration of its solution when no more of the substance will dissolve.

Scarcity of water is fast becoming one of the world's great problems. But of course the total amount of water on the earth is constant, and is essentially the same now as it was when Julius Caesar ruled Rome. The real problem is the growing scarcity of "clean" water—that is, water with sufficiently low concentrations of substances dissolved in it. This is because water is capable of dissolving, to various extents, so many of the ionic substances to which it is exposed.

Groundwater dissolves salts from minerals, and water in the atmosphere dissolves gases such as carbon dioxide and sulfur dioxide, as well as materials from particles in the air. The water in seas, lakes, and rivers can dissolve a whole range of substances while in contact with the ground and the atmosphere over very long times. The water in seas and lakes generally contains enough dissolved oxygen to support marine life, and if decomposition of large biomasses in the water deplete the oxygen levels, fish and other organisms die. In dry regions, population growth has been such that the demand for water often exceeds that which is available in reservoirs. Increasingly, authorities are turning to groundwater supplies, but these are usually high in levels of salts, because of water's solvent ability, and they cannot be used without treatment.

As discussed in Section 6.1, in some parts of the world, water from wells contains dissolved arsenic compounds, and consumption over prolonged periods leads to health problems. In many mining areas, there is a problem of "acid mine drainage": acids and compounds of heavy metals (such as lead, copper, and cadmium) dissolved in the water used in the extraction of metals have rendered toxic the ground around tailings dumps. Mine rehabilitation is now a large industry.

Dissolving Ionic Salts

The water we drink, the oceans, and the aqueous solutions that are our body fluids contain many ions, most of which result from dissolving solid materials present in the environment.

The dissolution of an ionic substance in water involves the separation of the cations and anions, and the dispersal of these ions amongst the water molecules. Why should the ions separate and disperse throughout water? After all, the melting point of sodium chloride is very high (801 °C), suggesting that the forces of attraction between cations and anions in the crystal [<<Section 3.4] are quite strong.

The chemists' explanation is that each cation at the surface of the crystal is surrounded by water molecules oriented with their negative ends pointing toward the ion, and each anion is surrounded by the positive ends of several water molecules. The attraction of the water molecules for the cations and anions can be envisaged as a competition: if the attraction of the water molecules for the ions is strong enough, the salt dissolves, but if the strength of attractions between cations and anions in the crystal dominate, dissolution will not occur. The dissolution of an ionic compound is represented in Figure 6.26.

Molecular Modelling

e6.15 Watch animations of sodium chloride dissolving.

Molecular Modelling (Odyssey)

e6.16 Compare molecular-level simulations of aquated cations and anions and their energy factors.

FIGURE 6.26 Water as a solvent for ionic substances. Water molecules can experience attraction to both cations and anions by rotation so that the end with opposite charge to the ion is pointing toward it. When the ionic substance dissolves in water, each ion is surrounded by water molecules. The ions are said to be aquated.

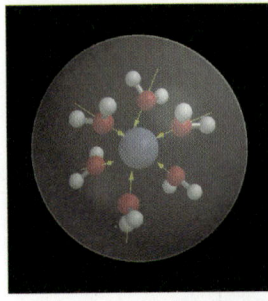

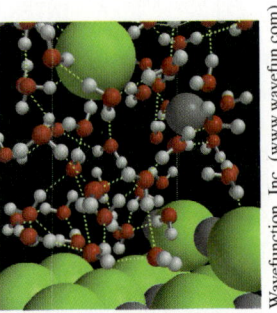

Wavefunction, Inc. (www.wavefun.com)

The attraction between either a cation or an anion and the oppositely charged end of a polar molecule is called an **ion-dipole force**. The surrounding of an ion or molecule of a solute by solvent molecules is called *solvation*. In the particular case of water as the solvent, a more particular term is **aquation** or *hydration*. The species comprising an ion surrounded by water molecules is called an **aquated ion** or *hydrated ion*. Aquated ions are represented by symbols that we have already used, such as $Na^+(aq)$ and $Cl^-(aq)$, where the suffix (aq) indicates aquation.

The aquated ions in an aqueous solution of an ionic compound are free to move in the solution. Under normal conditions, the movement of the aquated ions is random, and they are dispersed throughout the solution. However, if two electrodes (conductors of electricity, such as copper wire) are placed in the solution and connected to a battery, ion movement is no longer random: aquated cations move through the solution to the negative electrode, and aquated anions move to the positive electrode (Figure 6.27). If a light bulb is inserted into the electrical circuit, the bulb lights, demonstrating an electrical current made up of charges (the aquated ions) moving through the solution.

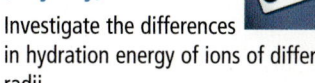

Interactive Exercise (Odyssey) 6.25
Investigate the differences in hydration energy of ions of different radii.

Interactive Exercise (Odyssey) 6.26
Investigate the differences in hydration energy of ions of different charges.

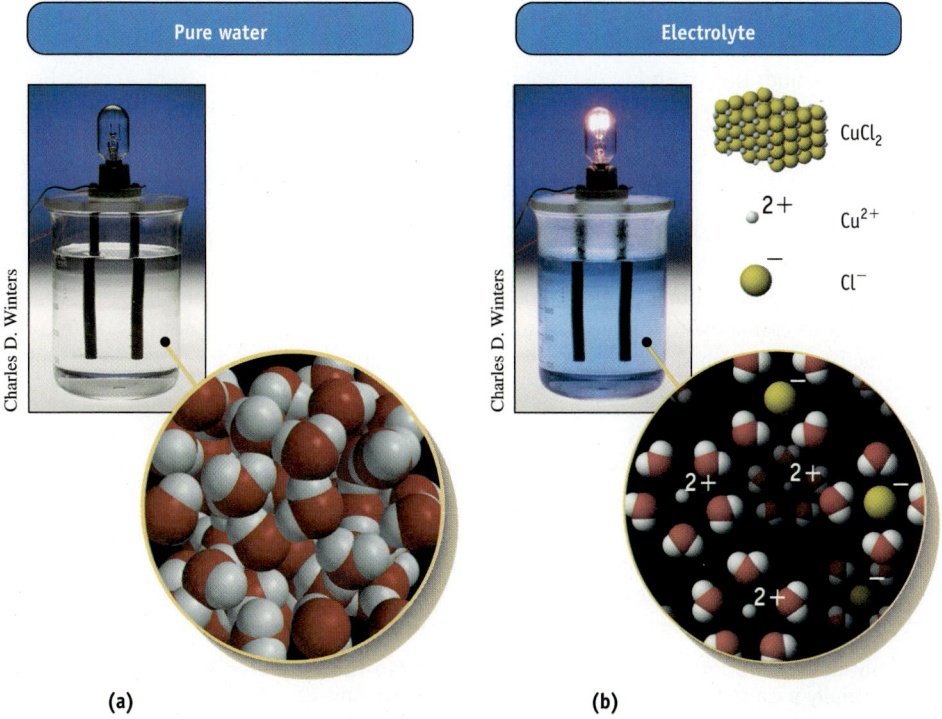

| Pure water | Electrolyte |

(a)   (b)

FIGURE 6.27 Electrical conductivity of water and of electrolyte solutions. (a) With an applied potential between two electrodes in water, there is no conduction of electricity because the concentration of ions is insignificant. (b) In a solution of an electrolyte such as $CuCl_2$, movement of the charged aquated ions between the electrodes (in opposite directions) constitutes an electric current.

Compounds whose aqueous solutions conduct electricity, because of the presence of ions in solution, are called **electrolytes**. When an ionic solid dissolves in water, the lattice separates into its ions (called **dissociation**) simultaneously with aquation. For example, the dissolution of sodium chloride in water is represented as

$$NaCl(s) \xrightarrow{H_2O} Na^+(aq) + Cl^-(aq)$$

Some ionic substances are only slightly soluble, but all that dissolves dissociates to form aquated ions. In that sense, they are strong electrolytes.

Molecular Modelling

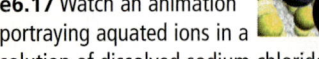

e6.17 Watch an animation portraying aquated ions in a solution of dissolved sodium chloride.

Substances that dissociate completely into ions when they dissolve in water are called **strong electrolytes**. All ionic solutes are strong electrolytes, unlike many molecular substances.

If we dissolve 0.01 mol of sodium chloride, NaCl(s), in 1 L of water, on the bottle we might put a label 0.01 mol L^{-1} sodium chloride solution. But although we put sodium chloride into the water, since the sodium chloride dissociated into Na^+(aq) and Cl^-(aq) ions, the solution does not contain a species "sodium chloride"—it contains only aquated sodium ions and aquated chloride ions. So it would be incorrect if we use the formula NaCl(aq) on a label or in a chemical equation because this implies NaCl particles surrounded by water molecules—and no such species exists.

The image that the only species in a solution of dissolved sodium chloride are aquated sodium ions and aquated chloride ions might be better appreciated if we think about how many water molecules separate these ions from each other, on average. For example, in a 0.01 mol L^{-1} solution of dissolved sodium chloride, on average there is only one Na^+(aq) ion and one Cl^-(aq) ion among every 5500 water molecules. We can conclude that in a dilute aqueous solution of a salt, the aquated ions are quite independent of each other.

A consequence of this is that chemical reactions of solutions are not reactions of the added solute, but of the chemical species that are in solution—aquated cations and anions—and this is how our chemical equations should represent them. For example, if we make a dilute solution of barium chloride, the dissolution process can be written:

$$BaCl_2(s) \xrightarrow{H_2O} Ba^{2+}(aq) + 2\,Cl^-(aq)$$

Now if we add sodium sulfate solution to this solution, a white precipitate forms, which analysis can show to be barium sulfate, BaSO₄. This is produced not by a reaction with BaCl₂(aq)—which does not exist—but by reaction of aquated barium ions with aquated sulfate ions:

$$Ba^{2+}(aq) + SO_4^{2-}(aq) \longrightarrow BaSO_4(s)$$

The Cl^-(aq) ions originally in solution, and the Na^+(aq) ions from the sodium sulfate solution, which are referred to as *spectator ions*, do not participate in this reaction, so they are not shown in the chemical equation.

In aqueous solutions, ionic compounds dissociate to form aquated ions, which are essentially independent of each other in dilute solutions.

EXERCISE 6.27—AQUATED IONS

What are the main species present in solution when some magnesium bromide, MgBr₂(s), is dissolved in water? Draw diagrams of these species.

Solubilities of Ionic Compounds Not all ionic compounds are very soluble in water. Many dissolve only to a small extent, and still others are essentially insoluble. Figure 6.28 lists broad guidelines that help predict whether a particular ionic compound is soluble in water. For example, sodium nitrate (NaNO₃) contains both an alkali metal cation (Na^+) and the nitrate anion (NO_3^-). Most compounds containing either of these ions are soluble in water. By contrast, calcium hydroxide is only slightly soluble in water.

SILVER COMPOUNDS

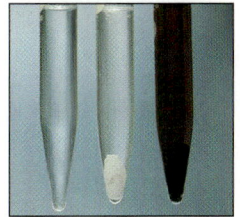

AgNO₃   AgCl   AgOH

(a) Nitrates are generally soluble, as are chlorides - except AgCl. Hydroxides are generally not soluble.

SULFIDES

(NH₄)₂S CdS Sb₂S₃ PbS

(b) Sulfides are generally not soluble. Exceptions include salts with NH₄⁺ and Na⁺ cations.

HYDROXIDES

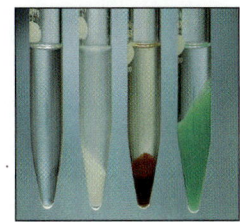

NaOH Ca(OH)₂ Fe(OH)₃ Ni(OH)₂

(c) Hydroxides are generally not soluble, except when the cation is a Group 1 metal.

All photos: Charles D. Winters

SOLUBLE COMPOUNDS	
Almost all salts of Na⁺, K⁺, NH₄⁺	
Salts of nitrate, NO₃⁻ chlorate, ClO₃⁻ perchlorate, ClO₄⁻ acetate, CH₃CO₂⁻	

EXCEPTIONS

Almost all salts of Cl⁻, Br⁻, I⁻	Halides of Ag⁺, Hg₂²⁺, Pb²⁺
Compounds containing F⁻	Fluorides of Mg²⁺, Ca²⁺, Sr²⁺, Ba²⁺, Pb²⁺
Salts of sulfate, SO₄²⁻	Sulfates of Ca²⁺, Sr²⁺, Ba²⁺, Pb²⁺

INSOLUBLE COMPOUNDS	EXCEPTIONS
Most salts of carbonate, CO₃²⁻ phosphate, PO₄³⁻ oxalate, C₂O₄²⁻ chromate, CrO₄²⁻	Salts of NH₄⁺ and the alkali metal cations
Most metal sulfides, S²⁻	
Most metal hydroxides and oxides	Ba(OH)₂ is soluble

FIGURE 6.28 Guidelines to predict the solubility of ionic compounds. If a compound contains one of the ions in the column to the left in the top chart, it is probably at least moderately soluble in water. There are a few exceptions, which are noted at the right. Most ionic compounds formed by the anions listed at the bottom of the chart are poorly soluble (with exceptions such as compounds with NH₄⁺ and the alkali metal cations).

WORKED EXAMPLE 6.7—SOLUBILITY OF IONIC COMPOUNDS

Predict whether the following ionic compounds are likely to be water-soluble. For the soluble compounds, write a chemical equation for dissociation into aquated ions.
(a) KCl(s)
(b) MgCO₃(s)
(c) Fe₂O₃(s)
(d) Cu(NO₃)₂(s)

Strategy

First recognize the cation and anion involved, and then decide the probable solubility based on the guidelines in Figure 6.28.

Solution

(a) Potassium chloride is composed of K⁺ and Cl⁻ ions. The presence of *either* of these ions means that the compound is probably soluble in water. The solution contains aquated K⁺ and Cl⁻ ions:

$$KCl(s) \xrightarrow{H_2O} K^+(aq) + Cl^-(aq)$$

NEL

(b) Magnesium carbonate is composed of Mg^{2+} and $CO_3{}^{2-}$ ions. Most carbonates, except those with Na^+ or $NH_4{}^+$ cations, are insoluble, so we would predict that $MgCO_3$ does not dissolve in water. (The experimental solubility of $MgCO_3$ is less than 0.2 g per 100 mL of water.)

(c) Iron(III) oxide is composed of Fe^{3+} and O^{2-} ions. Only Group 1 oxides are soluble, so we would predict that Fe_2O_3 is insoluble.

(d) Copper(II) nitrate is composed of Cu^{2+} and $NO_3{}^-$ ions. Almost all nitrates are soluble in water, so $Cu(NO_3)_2$ is water-soluble and produces aquated Cu^{2+} cations and $NO_3{}^-$ anions:

$$Cu(NO_3)_2(s) \xrightarrow{H_2O} Cu^{2+}(aq) + 2\ NO_3{}^-(aq)$$

EXERCISE 6.28—SOLUBILITY OF IONIC COMPOUNDS

Predict whether each of the following ionic compounds is soluble in water. For the soluble compounds, write a chemical equation for dissociation into aquated ions.

(a) $LiNO_3(s)$

(b) $CaCl_2(s)$

(c) $CuO(s)$

(d) $NaCH_3COO(s)$

The degree of solubility of ionic substances is the result of competition between anion-cation attractions in the lattice and ion-dipole attractions to water molecules. The species present in aqueous solutions are aquated cations and anions, essentially independent of each other at low concentrations. The chemistry of a solution is the sum of the chemistries of the aquated ions.

Dissolving Molecular Substances

Ethanol, $C_2H_5OH(\ell)$, is soluble in water in any proportion—that is, any mixture from those with a tiny fraction of ethanol to those with just a tiny fraction of water. On the other hand, gasoline, which is a mixture of compounds similar to octane, $C_8H_{18}(\ell)$, has negligible solubility in water. The difference between these two situations is that ethanol has polar molecules, while octane's molecules are non-polar.

Ethanol   Octane

The polar substances sucrose, $C_{12}H_{22}O_{11}(s)$, acetic acid, $CH_3COOH(\ell)$, and hydrogen chloride gas, $HCl(g)$, are all quite soluble in water. The non-polar substances benzene, $C_6H_6(\ell)$, carbon tetrachloride, $CCl_4(\ell)$, and hexane, $C_6H_{14}(\ell)$, are only slightly soluble in water, but they do dissolve in octane, which is non-polar. From observations such as these we can see a "rule of thumb" that "like dissolves like." Being careful not to take this guideline to an extreme, this means that most polar substances dissolve in polar solvents, and non-polar substances dissolve in non-polar solvents (Figure 6.29). The converse is also true; most polar substances do not dissolve in non-polar solvents, and most non-polar substances do not dissolve in polar solvents.

When liquid substances dissolve in each other, they are said to be *miscible* (similar to *mixable*). Liquids that do not dissolve in each other are *immiscible*.

What is the molecular basis for the "like dissolves like" guideline? A simple view is to recognize a natural tendency for things to mix, and to presume that they will mix unless there is an energy barrier preventing it. An observation that two liquids are miscible tells us that any energy barrier to mixing is insufficient to prevent the molecules from intermingling (and remaining intermingled). A more formal, mathematical treatment is to be found

Molecular Modelling (Odyssey)

e6.18 Compare molecular-level simulations of miscible and immiscible liquids.

(a) Ethylene glycol (HOCH₂CH₂OH), a polar compound used as antifreeze in automobiles, dissolves in water.

(b) Nonpolar motor oil (a hydrocarbon) dissolves in nonpolar solvents such as gasoline or CCl₄. It will not dissolve in a polar solvent such as water, however. Commercial spot removers use nonpolar solvents to dissolve oil and grease from fabrics.

FIGURE 6.29 "Like dissolves like." (a) Ethylene glycol, HOCH₂CH₂OH(ℓ), a polar compound used as an antifreeze in cars, dissolves in water. (b) Motor oil (consisting of non-polar hydrocarbons) does not dissolve in water. The two liquids remain in separate layers, with the less dense substance in the upper layer. In reality, molecules are closer together than is shown above.

in thermodynamics [>>Chapter 17]. A natural tendency for change to more probable mixed distributions is measured by a function called *entropy* (Figure 6.30).

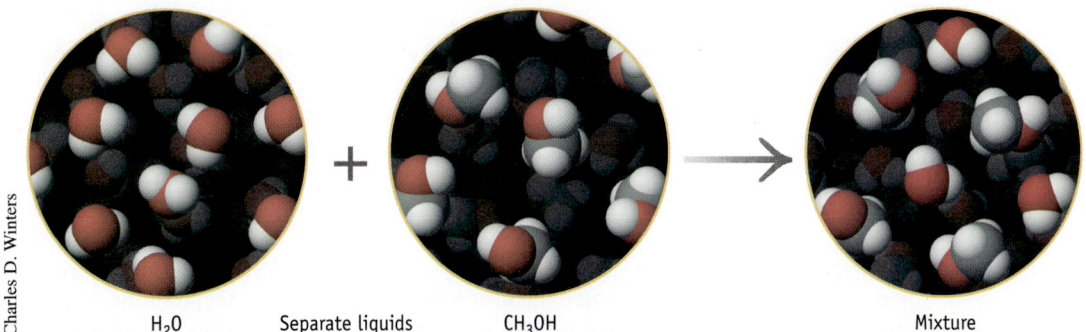

| H₂O | Separate liquids | CH₃OH | Mixture |

FIGURE 6.30 Driving the solution process—entropy. When two similar liquids—here, water and methanol—are mixed, the molecules remain intermingled (that is, water and methanol are miscible). The mixture has a more probable distribution of molecules than if the liquids remain separate, and this will happen if there is not too great an energy barrier resisting it. In reality, molecules are closer together than is shown above.

Whether or not there is too great an energy barrier to prevent the natural tendency of molecules of a solute to intermingle with water molecules depends on the interplay of intermolecular forces—between water molecules, between solute molecules, and between water molecules and solute molecules. To consider the process of dissolution, we imagine molecules of the solute being separated from each other (and dispersed amongst the solvent molecules), molecules of the solvent being separated from each other (because solute molecules have pushed in amongst them), and the interactions between solvent molecules and solute molecules in the mixture.

Why is ethanol soluble in water? Molecules of both of these liquids can engage in hydrogen bonding. On mixing, separation of the molecules from each other requires considerable energy—the system has gone "uphill" energetically. The observed miscibility can be rationalized on the grounds that hydrogen bonding is also possible between ethanol

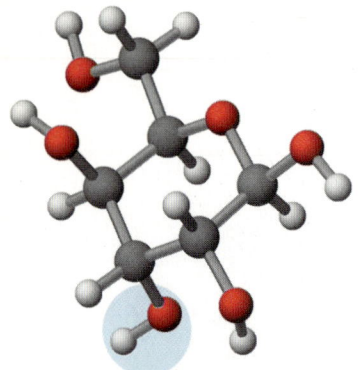

OH group

FIGURE 6.31 Representation of a glucose molecule. Glucose molecules have five –OH groups, allowing them to form hydrogen bonds with water molecules. As a consequence, the energy requirement for separation of water molecules and for separation of glucose molecules is sufficiently compensated that the tendency for glucose molecules and water molecules to mix is not prevented.

Molecular Modelling (Odyssey)

e6.19 Simulate hydrogen bonding sites in glucose and other similar compounds in water.

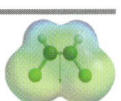

Molecular Modelling (Odyssey)

e6.20 Compare molecular-level simulations of a range of alcohols.

TABLE 6.10 Changing Length of Hydrocarbon Portion of Alcohol Molecules with One Terminal –OH Group

molecules and water molecules. So when the molecules of the two substances intermingle, there is sufficient compensation of the energy requirement of molecular separation that the natural tendency to mix is not prevented.

A similar argument can be applied to the considerable solubility of sucrose (table sugar) in water—a fact that we know well because of its use to sweeten drinks. The presence of –OH groups in molecules of sucrose and glucose (Figure 6.31) allows them to form strong hydrogen bonds to water molecules.

A different case of like substances that dissolve in each other is that of carbon tetrachloride, $CCl_4(\ell)$, and octane, $C_8H_{18}(\ell)$—both of which are non-polar. When these substances are mixed, the moderate energy required to overcome the dispersion forces between CCl_4 molecules, as well as the dispersion forces between C_8H_{18} molecules, is similar in magnitude to the energy regained due to dispersion forces between CCl_4 molecules and C_8H_{18} molecules. So the mixing of molecules is not prevented.

The "like dissolves like" guideline should not be taken too far: some substances with small molecules that have a hydrocarbon part and a polar part, such as ethanol (CH_3CH_2OH) and acetone (CH_3COCH_3), are soluble in both water and hydrocarbon solvents. It is also as well to realize that substances usually described as insoluble are seldom, if ever, entirely insoluble—usually at least a tiny amount dissolves.

EXERCISE 6.29—SOLUBILITIES OF MOLECULAR SUBSTANCES

Predict the solubilities of the following in water, and rationalize your prediction:
(a) ammonia, $NH_3(g)$
(b) hydrogen chloride, $HCl(g)$
(c) iodine, $I_2(s)$
(d) octane, $C_8H_{18}(\ell)$

EXERCISE 6.30—SOLUBILITIES OF MOLECULAR SUBSTANCES

Predict the solubilities of the following in octane, and rationalize your prediction:
(a) benzene, $C_6H_6(\ell)$
(b) water
(c) iodine, $I_2(s)$

Generally, polar solutes dissolve in polar solvents, and non-polar solutes dissolve in non-polar solvents, although this "like dissolves like" guideline has limitations. The "driving force" for dissolution of molecular substances is entropy—a natural tendency for the molecules to achieve a more probable distribution, unless there is an energy barrier that prevents them from doing so.

Polar and Non-Polar Parts of Solute Molecules We can gain further insights into factors affecting solubility from the solubilities of various alcohols in water and in hexane. Alcohol molecules have both polar and non-polar parts. Those listed in Table 6.10 have one alcohol –OH group attached to different lengths of hydrocarbon chain. The solubilities of this series of compounds, which are all polar (but have different degrees of polarity), show that there are limitations to the "like dissolves like" guideline.

Alcohol	Formula
Methanol	CH_3OH
Ethanol	CH_3CH_2OH
Propan-1-ol	$CH_3CH_2CH_2OH$
Butan-1-ol	$CH_3CH_2CH_2CH_2OH$
Pentan-1-ol	$CH_3CH_2CH_2CH_2CH_2OH$
Hexan-1-ol	$CH_3CH_2CH_2CH_2CH_2CH_2OH$

The solubility of methanol in water is due to the –OH portion of its molecules. The hydrogen bonding between the –OH group on methanol molecules and water molecules can compensate sufficiently the energy required to break the water–water and methanol–methanol hydrogen bonding. If hexan-1-ol molecules intermingle with water molecules, the quite large non-polar hydrocarbon portion of the molecules interrupts the hydrogen bonding between quite a few water molecules, with little ability for energy compensation—and so the hexan-1-ol molecules are rejected into a separate layer. As the length of the hydrocarbon portion increases, the solubility in water decreases: methanol and ethanol are miscible in water; propan-1-ol is moderately water-soluble; butan-1-ol is only slightly soluble; and pentan-1-ol and hexan-1-ol are essentially insoluble.

The opposite trend is seen for solubilities in hexane: methanol is only slightly soluble, while all of the others are soluble. In hexane and in hexan-1-ol, the main forces operating are dispersion forces. The energy requirement of separating hexane molecules and separating hexan-1-ol molecules is not great, and is compensated by dispersion forces between hexane molecules and hexan-1-ol molecules. In the case of methanol, however, the energy of separation of hydrogen-bonded methanol molecules is not sufficiently compensated by the formation of dispersion forces between hexane and methanol molecules.

We can also rationalize in this way the solubilities of alcohols with the same-sized hydrocarbon portion, but different numbers of –OH groups: although propan-1-ol and propan-2-ol are not soluble in water, propan-1,2-diol (propylene glycol) is.

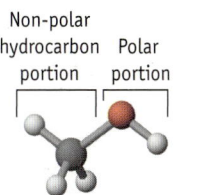

Non-polar hydrocarbon portion / Polar portion

Methanol molecule

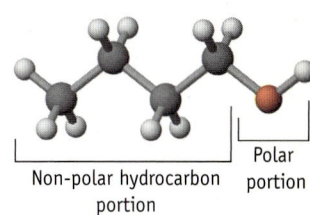

Non-polar hydrocarbon portion / Polar portion

Butan-1-ol molecule

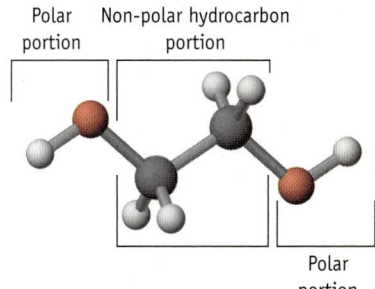

Polar portion / Non-polar hydrocarbon portion / Polar portion

Propan-1,2-diol

WORKED EXAMPLE 6.8—POLAR AND NON-POLAR PARTS OF MOLECULES

Which is more soluble in water: butan-1-ol, $CH_3CH_2CH_2CH_2OH(\ell)$, or butan-1,4-diol, $HOCH_2CH_2CH_2CH_2OH(\ell)$? Explain why.

Solution

Molecules of these two substances have the same-sized hydrocarbon portion, but butan-1,4-diol has two –OH groups while butan-1-ol has only one. When intermingled with water molecules, there is more opportunity for hydrogen bonding between butan-1,4-diol molecules and water molecules than between butan-1-ol molecules and water molecules. So, in the case of butan-1,4-diol, there is more compensation of the energy used if the solute molecules force apart water molecules.

EXERCISE 6.31—POLAR AND NON-POLAR PARTS OF MOLECULES

Which would you expect to be more soluble in hexane: butan-1-ol, $CH_3CH_2CH_2CH_2OH(\ell)$, or butan-1,4-diol, $HOCH_2CH_2CH_2CH_2OH(\ell)$? Explain why.

> If the molecules of a substance have a non-polar part and a polar part, the larger the non-polar part, the less likely that it will dissolve in water; and the more polar groups it has, the more likely that it will dissolve in water.

Ionization of Molecular Solutes Aqueous solutions of some molecular substances conduct electricity (Figure 6.32), indicating that the solution contains charged species. This is attributed to the breaking of a bond in the molecule, called **ionization**, as a result of interactions with water molecules:

$$A - B \xrightarrow{\text{H}_2\text{O}} A^+(aq) + B^-(aq)$$

Like ionic solutes, such molecular solutes are called *electrolytes*. Some of these are *strong electrolytes*: all of their molecules ionize in solution, and their aqueous solutions are good conductors of electricity. An example is hydrogen chloride, HCl(g), solutions of which are called hydrochloric acid.

$$HCl(g) \xrightarrow[\text{all ionized}]{\text{H}_2\text{O}} H^+(aq) + Cl^-(aq)$$

To indicate the involvement of water in the breaking of the H—Cl bond, ionization of HCl molecules is alternatively represented by the equation:

$$HCl(g) + H_2O(\ell) \longrightarrow H_3O^+(aq) + Cl^-(aq)$$

Only some of the molecules (usually a small fraction) of **weak electrolytes** ionize on dissolution in water, and so these conduct electricity poorly. An example is acetic acid:

$$CH_3COOH(aq) + H_2O(\ell) \xrightleftharpoons{\text{only some ionized}} CH_3COO^-(aq) + H_3O^+(aq)$$

The double arrows $\rightleftharpoons$ in the equation for ionization of a weak electrolyte indicate that a condition of dynamic chemical equilibrium is reached [<<Section 5.5].

Strong and weak electrolyte solutes are important in the chemistry of acids and bases in water [>>Chapter 14].

Many other molecular substances that dissolve in water do not ionize. They are called **non-electrolytes** because their solutions do not conduct electricity. Examples of non-electrolytes include sucrose, $C_{12}H_{22}O_{11}(s)$; ethanol, $CH_3CH_2OH(\ell)$; and the antifreeze ethylene glycol, $HOCH_2CH_2OH(\ell)$.

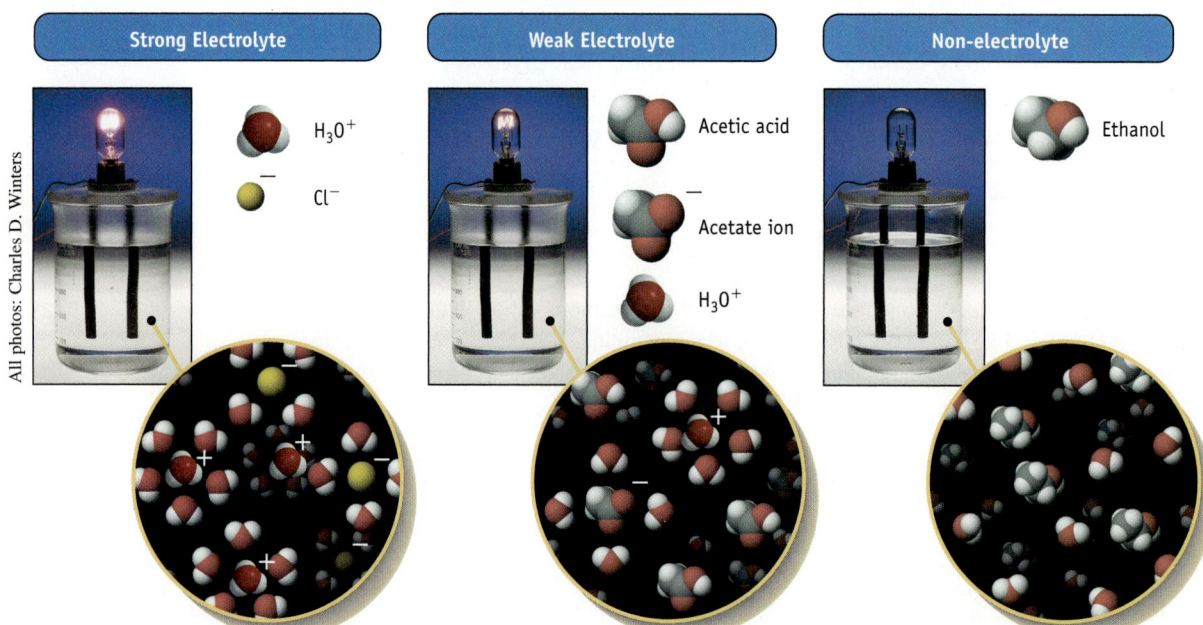

FIGURE 6.32 Classifying substances by the ability of their aqueous solutions to conduct electricity. (a) Solutions of a strong electrolyte conduct electricity well. All of the HCl molecules are ionized to form $H_3O^+(aq)$ ions and $Cl^-(aq)$ ions. (b) A solution of a weak electrolyte conducts electricity poorly because so few ions are present. (c) A solution of a non-electrolyte does not conduct electricity because no ions are present in solution. The molecular-level insets do not show the water molecules between the aquated ions for clarity.

EXERCISE 6.32—IONIZATION OF MOLECULAR SOLUTES

Write chemical equations to describe the dissolution of the following substances in water:
(a) hydrogen bromide (HBr), a strong electrolyte
(b) hydrogen fluoride (HF), a weak electrolyte
(c) formic acid (HCOOH), a weak electrolyte
(d) sucrose ($C_{12}H_{22}O_{11}$), a non-electrolyte

Some molecular substances that dissolve in water are strong electrolytes, some are weak electrolytes, and some are non-electrolytes.

6.6 Self-Ionization of Water

In a sample of pure water, there is an equilibrium condition between water molecules and tiny amounts of aquated hydrogen ions and hydroxide ions, the formation of which can be represented by the following equation:

$$H_2O(\ell) \rightleftharpoons OH^-(aq) + H^+(aq)$$

This reaction should not be interpreted simply as the breaking of an O—H bond in the water molecules. It requires energy to pull apart atoms joined by a bond, so why should the O—H bond break? The most useful view of this is recognition that the water molecules don't just "hang around" watching the ionization happen—rather, they participate in the process. A better interpretation of this process is that some water molecules "grab" hydrogen ions from others, represented by the following equation:

$$H_2O(\ell) + H_2O(\ell) \rightleftharpoons OH^-(aq) + H_3O^+(aq)$$

This proton-transfer reaction, called **self-ionization of water**, is fundamental in the chemistry of acids and bases [>>Chapter 14]. At 21 °C, the extent of this reaction when equilibrium is achieved is such that the concentration of both $H^+(aq)$ ions and $OH^-(aq)$ ions is 1×10^{-7} mol L^{-1}.

The species represented by H_3O^+, called a **hydronium ion**, can be regarded as an H^+ ion attached to a water molecule.

Hydronium ion (H_3O^+).

However, it seems possible that an H^+ ion might be surrounded by two water molecules, forming a species with formula $[H(H_2O)_2]^+$ or $H_5O_2^+$. In fact, it is generally accepted now that there is transitory formation in water of species that have a hydrogen ion surrounded by three water molecules ($H_7O_3^+$) or four water molecules ($H_9O_4^+$).

The actual species present can be visualized as continuously and rapidly changing, with water molecules falling off and jumping on, so that its identity is always changing. We use the formula of the aquated hydronium ion, $H_3O^+(aq)$, as an approximation. Alternatively, we use $H^+(aq)$ as an even simpler symbol of the rapidly changing species that are aquated hydrogen ions or "hydrated protons" (Figure 6.33).

The extent of water self-ionization depends on the temperature: at 90 °C, the concentration of the ions is 6×10^{-7} mol L^{-1}, and at 5 °C, the concentration is 0.4×10^{-7} mol L^{-1}.

Molecular Modelling (Odyssey)

e6.21 Watch an animation portraying self-ionization of water molecules, and investigate the nature of hydrated protons.

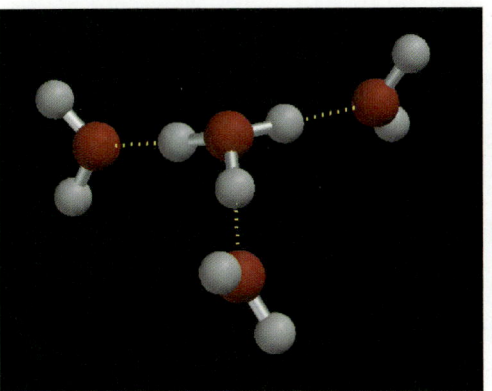 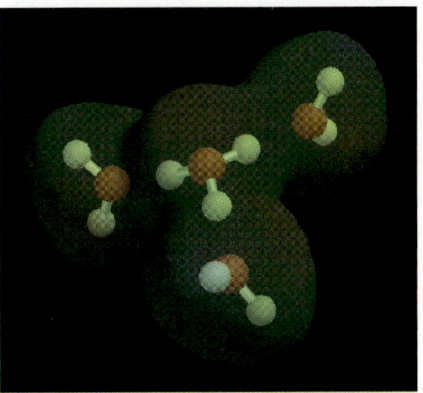

FIGURE 6.33 A model of the $H_9O_4^+$ ion. One of the transitory species comprising aquated hydrogen ions, or hydrated protons, $H^+(aq)$. The electrostatic potential map model indicates no localized positive charge.

Self-ionization of water leads to an equilibrium condition with small concentrations of $H^+(aq)$ ions and $OH^-(aq)$ ions. $H^+(aq)$ is continuously changing among several transitory species—one of which is the hydronium ion (H_3O^+).

6.7 Categories of Chemical Reaction in Water

In this section, we deal with reactions in water of the so-called inorganic compounds—the compounds generally of elements other than carbon. However, the dividing line between inorganic and organic compounds is not sharp.

The reactions of inorganic compounds can be categorized into four main types: precipitation reactions, oxidation-reduction reactions, acid-base reactions, and complexation reactions. In each of these categories, every reaction can be regarded as a competition between different species, and whether a reaction proceeds can be viewed as the result of a "driving force" provided by the winner of the competition. These ideas are summarized in Table 6.11.

TABLE 6.11 Types of Reaction in Water

Type of Reaction	Object of Competition	"Driving Force"
Precipitation	Cations and anions	Insolubility of precipitated salt
Oxidation-reduction	Electrons	Reduction potential of species reduced
Acid-base	Protons (H^+ ions)	Formation of weak electrolyte
Complexation	Ligands	Stability of complex formed

For a discussion of categories of "organic" reactions, see Chapter 19.

Precipitation Reactions

A **precipitation reaction** produces a solid deposit, known as a *precipitate*, of a substance that is insoluble in water, from cations and anions in solution. An example, shown in Figure 6.34, is the precipitation of silver chloride:

$$Ag^+(aq) + Cl^-(aq) \longrightarrow AgCl(s)$$

FIGURE 6.34 Precipitation of silver chloride. (a) Mixing aqueous solutions of silver nitrate and potassium chloride produces white, insoluble silver chloride (AgCl). The process, modelled at the molecular level in (b), (c), and (d), is essentially the opposite of dissolution.

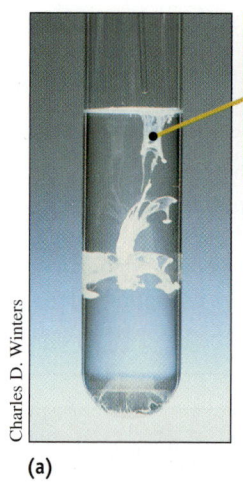

Charles D. Winters

(a)

(b) Initially the Ag^+ ions (silver colour) and Cl^- ions (green) are widely separated.

(c) Ag^+ and Cl^- ions approach and form ion pairs.

(d) As more and more Ag^+ and Cl^- ions come together, a precipitate of solid AgCl forms.

This reaction can be viewed as the result of a competition: attractions between the cations and the anions to form an ionic lattice competes with attraction of water molecules to the ions of both sorts (ion-dipole forces) to keep the ions aquated and in solution. Our observation of a precipitate tells us that in this case, the attractions between the Ag^+ cations and the Cl^- anions "win" the competition. This shouldn't surprise us, because when we shake up some silver chloride in water, infinitesimal amounts dissolve. So, we can say that the "driving force" for this reaction is the insolubility of silver chloride.

Potassium nitrate does not precipitate. In this case, the aquation of the ions outweighs the attractions between $K^+(aq)$ ions and $NO_3^-(aq)$ ions. This is consistent with the high solubility of potassium nitrate in water. Since the $K^+(aq)$ ions and $NO_3^-(aq)$ ions do not participate in the reaction, they are described as **spectator ions**. If the precipitate were filtered off and the solution then evaporated to dryness, a residue of potassium nitrate, $KNO_3(s)$, would be obtained.

If we were to mix solutions of silver acetate ($AgCH_3COO$) and sodium chloride, a silver chloride precipitate would again form, consistent with the same equation as that shown above—but the spectator ions would be different.

Many combinations of positive and negative ions form precipitates of insoluble substances. For example, lead chromate precipitates when an aqueous lead nitrate solution is added to a potassium chromate solution (Figure 6.35a):

$$Pb^{2+}(aq) + CrO_4{}^{2-}(aq) \longrightarrow PbCrO_4(s)$$

Almost all metal sulfides are insoluble in water. In nature, if a solution of a soluble metal salt comes in contact with a solution containing sulfide ions, the metal sulfide precipitates (Figure 6.35b):

$$Pb^{2+}(aq) + S^{2-}(aq) \longrightarrow PbS(s)$$

This is how many sulfide minerals such as iron pyrite are believed to have been formed. The black "smoke'" from undersea volcanoes [<<Section 4.2] consists of precipitated metal sulfides arising from sulfide anions and metal cations in the volcanic emissions.

With the exception of ions of the Group 1 elements, metal cations form slightly soluble hydroxides. For example, when solutions of iron(III) chloride ($FeCl_3$) and sodium hydroxide are mixed, iron(III) hydroxide precipitates (Figures 6.35c):

$$Fe^{3+}(aq) + 3\,OH^-(aq) \longrightarrow Fe(OH)_3(s)$$

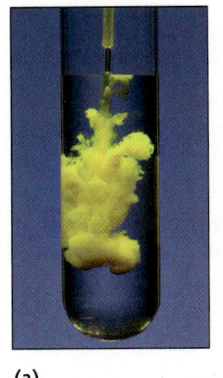  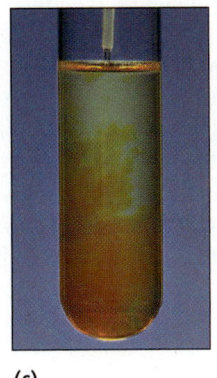 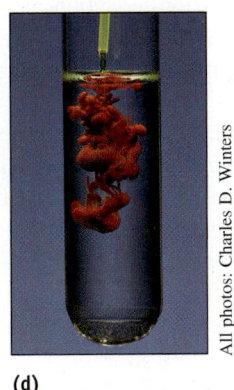

(a) (b) (c) (d)

All photos: Charles D. Winters

FIGURE 6.35 Precipitation reactions. Many ionic compounds are water-insoluble, and precipitate from solution: (a) A $Pb(NO_3)_2$ solution and K_2CrO_4 solution produce a yellow $PbCrO_4(s)$ precipitate. (b) A $Pb(NO_3)_2$ solution and a $(NH_4)_2S$ solution produce a black $PbS(s)$ precipitate. (c) A $FeCl_3$ solution and a NaOH solution produce a reddish $Fe(OH)_3(s)$ precipitate. (d) A $AgNO_3$ solution and a K_2CrO_4 solution produce a red $Ag_2CrO_4(s)$ precipitate.

Molecular Modelling

e6.22 Watch animations portraying what happens when sodium chloride solution is added to silver nitrate solution.

Interactive Exercise 6.33

Examine this reaction in detail.

WORKED EXAMPLE 6.9—DECIDING IF A PRECIPITATE FORMS

Is an insoluble product formed when aqueous solutions of silver nitrate and potassium chromate are mixed? If so, write the balanced equation.

Strategy

First, decide which ions are in the solutions. Then use information in Figure 6.28 to decide whether a cation from one reagent solution will combine with an anion from the other reagent solution to form an insoluble compound.

Solution

Both reagents, $AgNO_3(s)$ and $K_2CrO_4(s)$, are water-soluble. The ions Ag^+, $NO_3{}^-$, K^+, and $CrO_4{}^{2-}$, are released into solution and aquated when these compounds dissolve.

$$AgNO_3(s) \xrightarrow{H_2O} Ag^+(aq) + NO_3^-(aq)$$

$$K_2CrO_4(s) \xrightarrow{H_2O} 2\,K^+(aq) + CrO_4{}^{2-}(aq)$$

We can consider the possibilities that $Ag^+(aq)$ ions might combine with $CrO_4{}^{2-}(aq)$ ions, or that $K^+(aq)$ ions might combine with $NO_3{}^-(aq)$ ions. Ag_2CrO_4 is insoluble in water, whereas KNO_3 is soluble. So, silver chromate would be formed:

$$2\,Ag^+(aq) + CrO_4{}^{2-}(aq) \longrightarrow Ag_2CrO_4(s)$$

Interactive Exercise 6.34

Write equations for precipitation reactions.

Comment

This reaction is illustrated in Figure 6.35d. The $K^+(aq)$ ions and $NO_3^-(aq)$ ions are spectator ions.

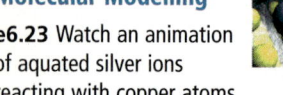
EXERCISE 6.36—PRECIPITATION REACTIONS

In each of the following cases, predict whether precipitation occurs when solutions of the two water-soluble reagents are mixed. Give the formula of any precipitate that forms, write a balanced chemical equation for the precipitation reactions that occur, and list the spectator ions.
(a) A sodium carbonate solution is mixed with a copper(II) chloride solution.
(b) A potassium carbonate solution is mixed with a sodium nitrate solution.
(c) A nickel(II) chloride solution is mixed with a potassium hydroxide solution.

> Whether precipitation happens on mixing solutions can be viewed as the result of a competition between cation-anion attractions to form an ionic lattice and ion-dipole attractions to water molecules to keep the ions aquated and in solution.

Oxidation-Reduction Reactions: Electron Transfer

The defining feature of **oxidation-reduction reactions**, sometimes called *redox* reactions, is the transfer of electrons from one species (atom, ion, or molecule) to another. Any oxidation-reduction reaction can be regarded as a competition between species for electrons: the reactant species that most strongly attracts electrons takes one or more electrons from a less competitive species.

The reactant species that takes the electrons is said to be *reduced*, and the accompanying change is **reduction**. The species from which electrons are taken is said to be *oxidized*, and the change accompanying the removal of electrons is **oxidation**. Because the species that is reduced brings about oxidation of the other, it is called the **oxidizing agent**, and the one that is oxidized is called the **reducing agent**. Obviously, neither oxidation nor reduction can occur without the other—they are necessarily simultaneous processes.

An example of an electron-transfer reaction is the reaction of copper metal with aquated silver ions in an aqueous silver nitrate solution (Figure 6.36).

FIGURE 6.36 The oxidation of copper metal by aquated silver ions. A piece of copper wire is placed in an $AgNO_3$ solution. Over time, the $Cu(s)$ reduces $Ag^+(aq)$ ions, forming $Ag(s)$ crystals, and the $Cu(s)$ is oxidized to blue aquated copper ions, $Cu^{2+}(aq)$.

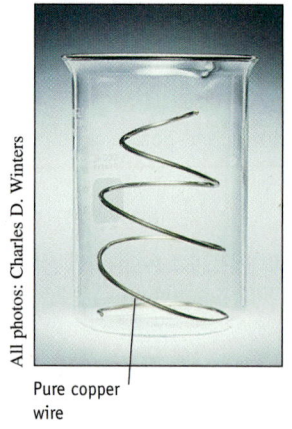

Pure copper wire

Copper wire in dilute $AgNO_3$ solution; after several hours

Blue color due to Cu^{2+} ions formed in redox reaction

Silver crystals formed after several weeks

All photos: Charles D. Winters

$Ag^+(aq)$ ions take electrons from $Cu(s)$ and are reduced to $Ag(s)$. The $Ag^+(aq)$ ion is the oxidizing agent.

$$Ag^+(aq) + e^- \longrightarrow Ag(s)$$

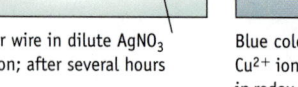

$$2\, Ag^+(aq) + Cu(s) \longrightarrow 2\, Ag(s) + Cu^{2+}(aq)$$

$Cu(s)$ provides electrons to $Ag^+(aq)$ ions, and is oxidized to $Cu^{2+}(aq)$ ions. Cu metal is the reducing agent.

$$Cu(s) \longrightarrow Cu^{2+}(aq) + 2\, e^-$$

In this case, the $Ag^+(aq)$ ion is the oxidizing agent (which is reduced), and copper metal is the reducing agent.

On the other hand, if we put silver metal into a solution with aquated Cu^{2+} ions, no reaction happens. We can conclude that $Ag^+(aq)$ ions are more strongly competitive for electrons than $Cu^{2+}(aq)$ ions, and chemists say that $Ag^+(aq)$ ions have a higher potential to be reduced—a higher *reduction potential*—than $Cu^{2+}(aq)$ ions. A ranking of reduction potentials of species is useful in predicting which species can be oxidized or reduced by a given species, and is fundamental to the chemistry described in Chapter 16.

Oxidation was originally defined as any process in which a reagent combines with oxygen, such as when magnesium burns in air (Figure 6.37):

$$2\,Mg(s) + O_2(g) \longrightarrow 2\,MgO(s)$$

Mg(s) releases two electrons per atom, and is oxidized to Mg^{2+} ions. Mg(s) is the reducing agent.

$$2\,Mg(s) + O_2(g) \longrightarrow 2\,MgO(s)$$

Each molecule of $O_2(g)$ gains four electrons and is reduced to two O^{2-} ions. $O_2(g)$ is the oxidizing agent.

If we recognize that MgO is an ionic compound comprising Mg^{2+} ions and O^{2-} ions ($Mg^{2+}O^{2-}$), then we can see that this reaction is also an oxidation-reduction reaction from the point of view of the electron-transfer definition.

We can analyze oxidation-reduction reactions, and the terminology we use, by dividing the general equation $X + Y \longrightarrow X^{n+} + Y^{n-}$ into two parts or *half-equations*, one for oxidation and one for reduction:

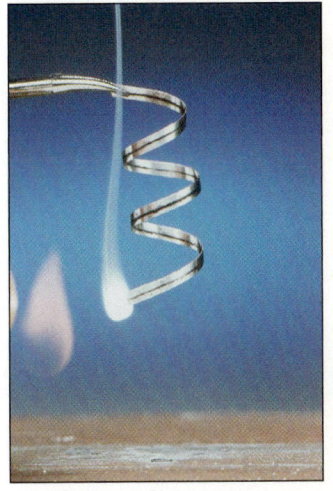

FIGURE 6.37 Burning magnesium in air.

Charles D. Winters

Half-Equation	Electron Transfer	Result
$X \longrightarrow X^{n+} + ne^-$	X transfers electrons to Y.	X is **oxidized** to X^{n+}.
		X is the **reducing agent**.
$Y + ne^- \longrightarrow Y^{n-}$	Y takes electrons from X.	Y is **reduced** to Y^{n-}.
		Y is the **oxidizing agent**.

Oxidation-reduction reactions can usually be recognized if you are familiar with the common oxidizing and reducing agents (Table 6.12).

Oxidizing Agent	Reaction Product	Reducing Agent	Reaction Product
$O_2(g)$, oxygen	Oxide ions, O^{2-}, or water, $H_2O(\ell)$	Hydrogen gas, $H_2(g)$	Hydrated proton, $H^+(aq)$, or water, $H_2O(\ell)$
Halogens: $F_2(g)$, $Cl_2(g)$, $Br_2(g)$, and $I_2(g)$	Aquated halide ions, $F^-(aq)$, $Cl^-(aq)$, $Br^-(aq)$, and $I^-(aq)$	Reactive metals such as Na(s), K(s), Fe(s), Al(s)	Aquated metal ions such as $Na^+(aq)$, $K^+(aq)$, $Fe^{2+}(aq)$ or $Fe^{3+}(aq)$, $Al^{3+}(aq)$
Hydrated protons, $H^+(aq)$, in dilute aqueous solutions of HCl, H_2SO_4	Hydrogen, $H_2(g)$	Carbon, C(s) as coke or charcoal	Carbon dioxide, $CO_2(g)$
Aquated nitrate ions, $NO_3^-(aq)$, in aqueous solutions of nitric acid, HNO_3	Nitrogen oxides* such as NO(g) and $NO_2(g)$	Carbon monoxide, CO(g)	Carbon dioxide, $CO_2(g)$

TABLE 6.12 Common Oxidizing and Reducing Agents

(continued)

TABLE 6.12 *(continued)*

Oxidizing Agent	Reaction Product	Reducing Agent	Reaction Product
Aquated dichromate ions, $Cr_2O_7^{2-}$ (aq)	Aquated chromium (III) ions, Cr^{3+}(aq), in acidic solution	Aquated hydrogensulfite ions, HSO_3^-(aq)	Aquated hydrogensulfate ions, HSO_4^-(aq)
Aquated permanganate ion, MnO_4^-(aq)	Aquated manganese (II) ions, Mn^{2+}(aq), in acid solution		
Aquated ions of relatively unreactive metals; Au^{3+}(aq), Ag^+(aq), Fe^{3+}(aq)	Metals Au(s), Ag(s), and Fe^{2+}(aq) ions, respectively		

* $NO(g)$ is mainly produced with dilute HNO_3 solutions, whereas $NO_2(g)$ is the main product in concentrated solutions.

WORKED EXAMPLE 6.10—OXIDATION-REDUCTION REACTIONS

The reaction of zinc metal in dilute hydrochloric acid solution can be represented by the equation

$$Zn(s) + 2\,H^+(aq) \longrightarrow Zn^{2+}(aq) + H_2(g)$$

This reaction equation can be separated into two half-equations:

$$Zn(s) \longrightarrow Zn^{2+}(aq) + 2\,e^-$$

$$2\,H^+(aq) + 2\,e^- \longrightarrow H_2(g)$$

Which species is oxidized? To what? Which species is the oxidizing agent? Which species is reduced? To what? Which species is the reducing agent?

Strategy

To answer this question, you need to recognize which species has electrons taken from it (and is therefore oxidized), and which species takes the electrons from it (and is therefore reduced).

Solution

In the first half-equation, we see that electrons are taken from metallic zinc, so this is the species oxidized. The product of its oxidation are Zn^{2+}(aq) ions. The second half-equation shows that the species that "grabbed" the electrons from zinc is H^+(aq) ions, which are therefore the oxidizing agent. Since H^+(aq) ions accept the electrons, they are reduced (to hydrogen gas), and the species that brought about their reduction as the source of electrons (i.e., the reducing agent) is zinc metal.

EXERCISE 6.37—OXIDATION-REDUCTION REACTIONS

The equation for the reaction between aquated iron(II) ions and aquated permanganate ions in solution is

$$5\,Fe^{2+}(aq) + MnO_4^-(aq) + 8\,H^+(aq) \longrightarrow 5\,Fe^{3+}(aq) + Mn^{2+}(aq) + 4\,H_2O(\ell)$$

This can be separated into two half-equations:

$$5\,Fe^{2+}(aq) \longrightarrow 5\,Fe^{3+}(aq) + 5\,e^-$$

$$MnO_4^-(aq) + 8\,H^+(aq) + 5\,e^- \longrightarrow Mn^{2+}(aq) + 4\,H_2O(\ell)$$

Which species is oxidized? To what? Which species is the oxidizing agent? Which species is reduced? To what? Which species is the reducing agent? (The species H^+(aq) and $H_2O(\ell)$ are not oxidized nor reduced.)

Oxidation-reduction reactions involve the transfer of electrons from one species to another, and can be regarded as competition between species for electrons. Oxidation and reduction are necessarily simultaneous processes. The species that takes the electrons is reduced, and the species from which electrons are taken is oxidized. The species that is reduced brings about oxidation of the other, so it is the oxidizing agent, and the one that is oxidized is the reducing agent. Oxidation-reduction equations can be separated into two half-equations—one for oxidation and one for reduction.

Acid-Base Reactions: Proton Transfer

Acids and bases are important compounds that are in some ways opposites: a base can neutralize the effect of an acid, and an acid can neutralize the effect of a base. For example, an acid will change the colour of litmus (a pigment obtained from lichens) from blue to red, while adding a base reverses the colour change. Acids and bases react with each other in aqueous solution by transfer of a proton (H^+) from the acid to the base. Acid-base chemistry is discussed in detail in Chapter 14.

Acids in an Aqueous Solution **Acids** have characteristic properties. They produce bubbles of $CO_2(g)$ when added to a metal carbonate such as $CaCO_3(s)$, and they react with the more reactive metals to produce hydrogen, $H_2(g)$ (Figure 6.38). Although tasting substances is *never* done in a chemistry laboratory, you have probably experienced the sour taste of acids such as acetic acid (in vinegar) or citric acid (in citrus fruit, and added to soft drinks).

(a) The juice of a red cabbage is normally blue-purple. On adding acid, the juice becomes more red. Adding base produces a yellow color.

(b) A piece of coral (mostly $CaCO_3$) dissolves in acid to give CO_2 gas.

(c) Zinc reacts with hydrochloric acid to produce zinc chloride and hydrogen gas.

All photos: Charles D. Winters

FIGURE 6.38 Some properties of acids and bases. (a) The colours of natural dyes are affected by acids and bases. The juice of a cabbage is blue-purple. On adding acid, the juice becomes redder. Adding base changes its colour to yellow. (b) Acids react readily with coral $CaCO_3(s)$ and other metal carbonates to produce $CO_2(g)$. (c) Acids react with many metals to produce $H_2(g)$. Here, HCl solution reacts with zinc metal to produce $H_2(g)$, leaving $Zn^{2+}(aq)$ ions and $Cl^-(aq)$ ions in solution.

Acids are electrolytes that in their ionization reaction with water form hydronium ions, described in Section 6.6. In the case of hydrogen chloride, this ionization can be represented by the equation

$$HCl(g) + H_2O(\ell) \longrightarrow H_3O^+(aq) + Cl^-(aq)$$

The view that the acid properties of a substance result from transfer of an H^+ ion from the acid to water molecules is called the *Brønsted-Lowry model*: the proton donor is called an *acid*, and the proton acceptor is called a *base*.

Acids, like hydrogen chloride, that are strong electrolytes are called **strong acids**. Acids that are weak electrolytes are called **weak acids**. An example is acetic acid:

$$CH_3COOH(aq) + H_2O(\ell) \rightleftharpoons CH_3COO^-(aq) + H_3O^+(aq)$$

In moderately dilute solutions, only about 1 to 5% of acetic acid molecules are ionized, depending on the concentration [>>Section 14.6]. That is, most of the acetic acid in solution is in the form of the species $CH_3COOH(aq)$. Different weak acids ionize to different extents.

Many weak acids, such as oxalic acid and acetic acid, are naturally occurring substances. Molecules of these "organic" acids contain the –COOH carboxylic acid group [>>Chapter 24]. Some common acids are listed in Table 6.13.

TABLE 6.13 Common Acids and Bases

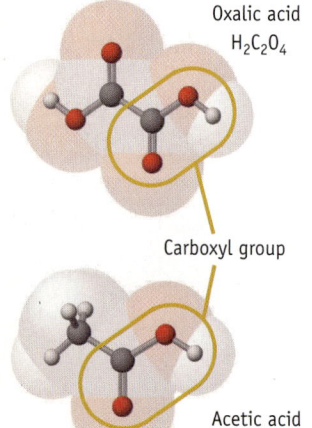

Oxalic acid
$H_2C_2O_4$

Carboxyl group

Acetic acid
CH3COOH

Strong Acids (Strong electrolytes)		Strong Bases (Strong electrolytes)	
HCl	Hydrochloric acid	LiOH	Lithium hydroxide
HBr	Hydrobromic acid	NaOH	Sodium hydroxide
HI	Hydroiodic acid	KOH	Potassium hydroxide
HNO_3	Nitric acid		
$HClO_4$	Perchloric acid		
H_2SO_4	Sulfuric acid		
Weak Acids (Weak electrolytes) *		**Weak Bases (Weak electrolytes) ***	
H_3PO_4	Phosphoric acid	NH_3	Ammonia
H_2CO_3	Carbonic acid	$CH_3CH_2NH_2$	Ethylamine
CH_3COOH	Acetic acid		
$H_2C_2O_4$	Oxalic acid		
$H_2C_4H_4O_6$	Tartaric acid		
$H_3C_6H_5O_7$	Citric acid		
$HC_9H_8O_4$	Aspirin		

* These are just a few common examples of the very many weak acids and weak bases.

Bases in Aqueous Solution   A characteristic of solutions of **bases** is that they have a higher concentration of $OH^-(aq)$ ions than does water, as a consequence of their ionization reaction with water. Bases that are strong electrolytes, such as sodium hydroxide and other water-soluble ionic hydroxides, are called **strong bases.**

$$NaOH(s) \xrightarrow{H_2O(\ell)} Na^+(aq) + OH^-(aq)$$

This equation does not portray NaOH as a proton acceptor, but the products of its ionization, $OH^-(aq)$ ions, are very powerful proton "grabbers."

Bases that are weak electrolytes are called **weak bases**. An example is ammonia—its ionization in aqueous solution is the result of transfer of H^+ ions from H_2O molecules to NH_3 molecules, to achieve an equilibrium condition:

$$NH_3(aq) + H_2O(\ell) \overset{\text{not all ionized}}{\rightleftharpoons} NH_4^+(aq) + OH^-(aq)$$

Some bases are listed in Table 6.13.

Neutralization: Reactions of Acids with Bases   If the self-ionization of water goes to only a tiny extent, we shouldn't be surprised if there is a considerable tendency for the opposite reaction to proceed:

$$OH^-(aq) + H_3O^+(aq) \longrightarrow 2\,H_2O(\ell)$$

That is indeed the case. If we mix an aqueous solution of a strong acid (such as hydrochloric acid, HCl) with a solution of a strong base (such as sodium hydroxide,

"Acid plus base gives salt plus water" is a common description of neutralization reactions, but the salt is only obtained if the water is evaporated from the solution.

NaOH), then the reaction represented by the above equation proceeds until there is essentially none of either $OH^-(aq)$ ions or $H_3O^+(aq)$ ions left. This reaction is called **acid-base neutralization**. Such reactions can be regarded as the result of competition for protons (H^+ ions), and the "driving force" is always the formation of water. The base, $OH^-(aq)$, "wins" the competition.

The neutralization reaction that happens when a solution of acetic acid, a weak acid, is added to a sodium hydroxide solution is best represented as

$$CH_3COOH(aq) + OH^-(aq) \longrightarrow CH_3COO^-(aq) + H_2O(\ell)$$

Since acetic acid is a weak acid, and exists mostly as aquated CH_3COOH molecules, it is preferable to represent the acid reactant species as $CH_3COOH(aq)$, rather than $H_3O^+(aq)$. In this case, the competition can be regarded as that between $OH^-(aq)$ ions and aquated $CH_3COOH(aq)$ molecules for H^+ ions. We can view this competition through the question, "Which weak electrolyte attracts its protons more—aquated CH_3COOH molecules or water?" Water is the weaker electrolyte, and $OH^-(aq)$ ions "win" the competition.

| The weaker the acid, the more it resists removal of protons from its molecules.

The neutralization reaction that occurs on mixing a hydrochloric acid solution and a solution of ammonia, a weak base, can be represented as

$$H_3O^+(aq) + NH_3(aq) \longrightarrow NH_4^+(aq) + H_2O(\ell)$$

The dominant species in ammonia solutions are aquated ammonia molecules, so it is preferable that the chemical equation shows this as the reactant base species, rather than $OH^-(aq)$ ions.

The competition for protons in this case is between $H_3O^+(aq)$ ions and NH_3 molecules. NH_3 molecules "win" the competition because $NH_4^+(aq)$ ions are a weaker acid than $H_3O^+(aq)$ ions.

EXERCISE 6.38—ACIDS AND BASES

Identify the acid reactant species and the base reactant species in each of the following reaction equations:

(a) $HCOOH(aq) + OH^-(aq) \longrightarrow HCOO^-(aq) + H_2O(\ell)$

(b) $NH_3(aq) + H_2CO_3(aq) \longrightarrow NH_4^+(aq) + HCO_3^-(aq)$

(c) $H_2C_4H_4O_6(aq) + HCO_3^-(aq) \longrightarrow HC_4H_4O_6^-(aq) + H_2CO_3(aq)$

(d) $CH_3NH_2(aq) + H_3O^+(aq) \longrightarrow CH_3NH_3^+(aq) + H_2O(\ell)$

EXERCISE 6.39—ACIDS AND BASES

In each case, write a balanced chemical equation for the neutralization reaction that happens, and list the spectator ions:
(a) A hydrogen bromide solution is mixed with a lithium hydroxide solution.
(b) A carbonic acid solution is mixed with a potassium hydroxide solution. (Carbonic acid has two protons that can be removed.)
(c) A methylamine solution is mixed with a nitric acid solution.
(d) A citric acid solution is mixed with a sodium hydroxide solution. (Citric acid has three protons that can be removed.)

Acid-base reactions can be viewed as competition between species for H^+ ions. From the perspective of the Brønsted-Lowry model, an acid is a species from which an H^+ ion is taken, and a base is the acceptor of an H^+ ion. Strong acids and strong bases are strong electrolytes; weak acids and weak bases are weak electrolytes.

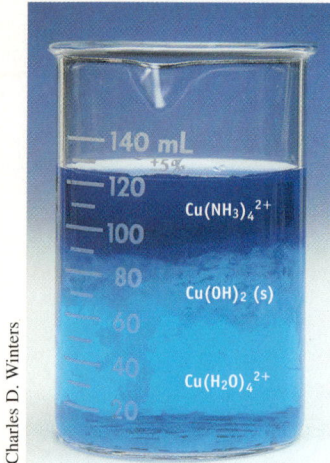

FIGURE 6.39 Formation of the complex ion $[Cu(NH_3)_4]^{2+}$. Ammonia was added to aqueous $CuSO_4$ solution (the light blue solution at the bottom). With OH^- ions in the ammonia solution, a pale blue $Cu(OH)_2$ precipitate is first formed (in the middle of the beaker). With more NH_3 solution, the deep blue $[Cu(NH_3)_4]^{2+}$ complex ion is formed (at the top of the solution).

Complexation Reactions, Lewis Acid-Base Reactions

When ammonia is added to a light blue copper (II) sulfate solution, the solution eventually turns deep blue. This is because of the formation of a species with the formula $[Cu(NH_3)_4]^{2+}$. In each such ion, four NH_3 molecules are joined by covalent bonds to a Cu^{2+} ion (Figure 6.39).

$$Cu^{2+}(aq) + 4\,NH_3(aq) \longrightarrow [Cu(NH_3)_4]^{2+}(aq)$$

This is an example of a class of reaction called **complexation**, and the product ion $[Cu(NH_3)_4]^{2+}$ is called a **complex ion**. The key to understanding this reaction is to realize that on each ammonia molecule, there are two electrons localized on the N atom that are not engaged in bonding to any other atom (Figure 6.40). This is called a **non-bonding pair** of electrons, or a *lone pair*. All complexation reactions involve bonding to species with lone pairs.

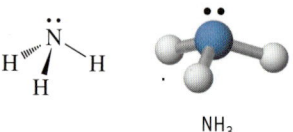

FIGURE 6.40 Representations of the ammonia molecule. There is a non-bonding pair of electrons on the N atom.

Each Cu–N covalent bond in the $[Cu(NH_3)_4]^{2+}$ complex ion can be imagined to have formed by sharing of an ammonia molecule's lone pair of electrons (Figure 6.41).

FIGURE 6.41 The complex ion formed by reaction of $Cu^{2+}(aq)$ and $NH_3(aq)$. A covalent bond is a shared pair of electrons. In formation of this complex ion, we imagine that both electrons of the Cu—N bond are from the lone pair on the N atom of the NH_3 molecule. Although this bond is referred to as a *coordinate bond*, it is no different from any other covalent bond in which one electron is considered to originate from each of the bonded atoms.

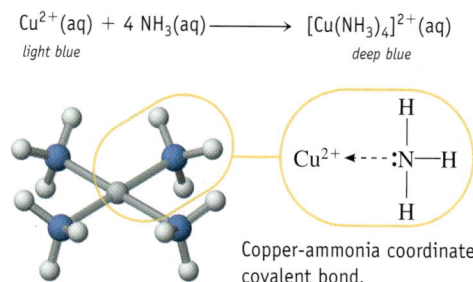

Copper-ammonia coordinate covalent bond.

Examples of this type of reaction include reaction between an $H^+(aq)$ ion and a water molecule to form a hydronium ion, and between an $H^+(aq)$ ion and an ammonia molecule to form an ammonium ion (Figure 6.42).

FIGURE 6.42 Formation of complex ions by sharing lone pairs with an H^+ ion. Each of these reactions can be regarded as both complexation reactions and acid-base reactions. In the top reaction, the H_2O molecule is a base (proton acceptor), and in the lower reaction an NH_3 molecule is a base.

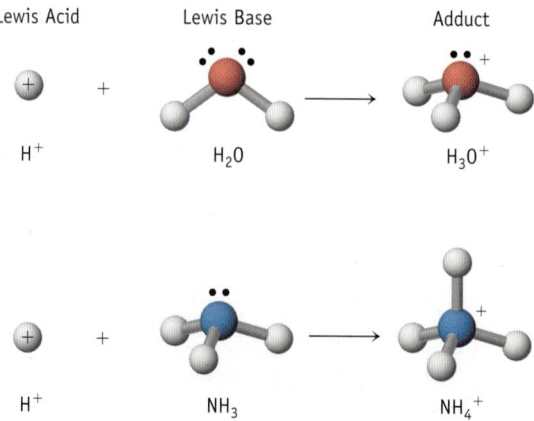

Each of these is an acid-base reaction involving proton transfer. The reaction between NH_3 molecules and an H^+ ions parallels that between NH_3 molecules and a Cu^{2+} ions (Figure 6.41). Since the NH_3 molecule is a base when it bonds to an H^+ ion using its own lone pair of electrons, then it is sensible to call it a base in the similar reaction with a Cu^{2+} ion—even though there is no transfer of protons.

This leads us to a more generalized view of acids and bases than the Brønsted-Lowry model, called the *Lewis acid-base model*. From the perspective of this model, a **Lewis acid** is a substance that accepts a pair of electrons from another atom, with formation of a new bond; a **Lewis base** is a substance that can donate a pair of electrons to another atom to form a new bond. A Lewis base is also called a *ligand* (pronounced *lie-gand*, derived from the Latin verb *ligare*, meaning "to bind"). The product of an acid-base reaction, in the Lewis model, is called an *acid-base adduct*.

$$B: + A \longrightarrow B - A$$

Base   Acid   Adduct

In the reaction between ammonia molecules and H^+ ions, NH_3 molecules are the Lewis base, and H^+ ions are the Lewis acid. In the reaction between ammonia molecules and Cu^{2+} ions, NH_3 molecules are the Lewis base, and Cu^{2+} ions are the Lewis acid.

WORKED EXAMPLE 6.11—COMPLEXATION REACTIONS

Write the formula, showing the overall charge, on the complex ion in which four Lewis base cyanide ions are bound to a Ni^{2+} ion.

Strategy

The charge on the ion is the net result of the charge on the metal ion and the sum of the charges on each Lewis base. The formula shows the Lewis acid metal ion and the number of Lewis base molecules or ions within square brackets, and the charge is written outside of the square brackets.

Solution

Each cyanide ion (CN^-) has -1 charge, so the four CN^- ions will contribute a total of -4 charge to the complex ion. Since the Ni^{2+} ion has $+2$ charge, the complex ion has a net charge of -2, and its formula is written as $[Ni(CN)_4]^{2-}$.

EXERCISE 6.40—COMPLEXATION REACTIONS

In each of the complexation reactions described by the following equations, identify the Lewis acid and the Lewis base(s).

(a) $Co^{2+}(aq) + 4\ Cl^-(aq) \longrightarrow [CoCl_4]^{2-}(aq)$

(b) $Fe^{2+}(aq) + 6\ CN^-(aq) \longrightarrow [Fe(CN)_6]^{4-}(aq)$

(c) $Ni^{2+}(aq) + 2\ CH_3NH_2(aq) \longrightarrow [Ni(CH_3NH_2)_2]^{2+}(aq)$

(d) $Cu^{2+}(aq) + 4\ Cl^-(aq) + 2\ NH_3(aq) \longrightarrow [Cu(NH_3)_2Cl_4]^{2-}(aq)$

> In complexation reactions, a bond is imagined to be formed from a lone pair of electrons on one of the reactant species. The lone-pair donor is called a Lewis base or ligand; and the recipient of the electron pair is called a Lewis acid.

Aquation of Metal Ions as Complexation In Section 6.5, we visualized aquated metal cations as metal ions surrounded, because of ion-dipole attractions, by water molecules. However, in the same way that we can think of complexation between NH_3 molecules and a metal ion as a result of donation of the lone pair of electrons on ammonia, we can postulate complexation of H_2O molecules (which have two lone pairs on the O atom) to metal ions.

Molecular Modelling

e6.26 Watch an animation of the formation of the $[Cu(NH_3)_4]^{2+}$ complex ion in solution.

For example, in aquated iron(II) ions, Fe^{2+}(aq), the Fe^{2+} ion is generally considered to be bound to six water molecules in the complex ion represented as $[Fe(OH_2)_6]^{2+}$.

$$Fe^{2+} + 6\ H_2O(\ell) \longrightarrow [Fe(OH_2)_6]^{2+}(aq)$$

The formula of the complex ion is written as $[Fe(OH_2)_6]^{2+}$ rather than $[Fe(H_2O)_6]^{2+}$ to emphasize that the O atom of the water molecule is directly bound to the metal ion. Corresponding complex ions are formed with other transition metal cations, and they are generally coloured (Figure 6.43). The general characteristics of complex ions and their compounds are described further in Chapter 27.

Molecular Modelling

e6.27 Watch an animation showing exchange of water molecules around a copper(II) ion in solution.

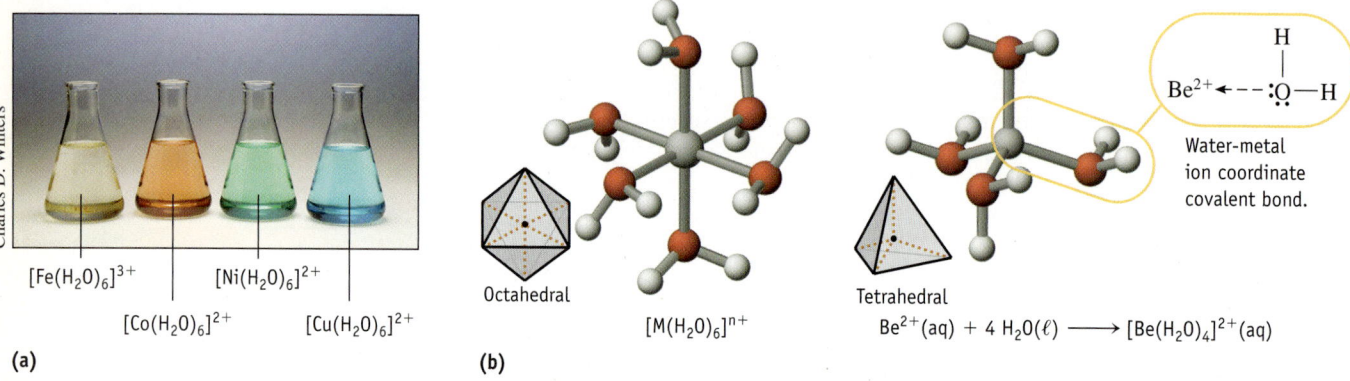

Charles D. Winters

$[Fe(H_2O)_6]^{3+}$ $[Ni(H_2O)_6]^{2+}$

$[Co(H_2O)_6]^{2+}$ $[Cu(H_2O)_6]^{2+}$

(a)

Octahedral

$[M(H_2O)_6]^{n+}$

(b)

Tetrahedral

Be^{2+}(aq) $+ 4\ H_2O(\ell) \longrightarrow [Be(H_2O)_4]^{2+}$(aq)

Water-metal ion coordinate covalent bond.

FIGURE 6.43 Metal cations in water. (a) Solutions of the nitrate salts of iron(III), cobalt(II), nickel(II), and copper(II) ions. All have characteristic colours of the complex ions formed between the metal ion and water molecules. (b) Models of the aquated metal ions. Most of these complex ions have six water molecules arranged octahedrally around the metal cation.

EXERCISE 6.41—AQUATION AS COMPLEXATION

Rewrite each of the following as complex ions with water molecules as Lewis bases. In each case, presume six water molecules in the complex ion.
(a) Fe^{2+}(aq)
(b) Mn^{2+}(aq)
(c) Al^{3+}(aq)
(d) Cr^{3+}(aq)

Aquated metal ions can be regarded as complex ions, with the water molecules acting as Lewis bases that donate a lone pair of electrons to the bond between metal ion and water molecule.

Complexation Reactions as Competition between Lewis Bases The formation of the $[Cu(NH_3)_4]^{2+}$ complex ion can be thought of as "addition" of ammonia molecules to a Cu^{2+} ion—to form an adduct. However, if we consider the aquated copper ion to be the $[Cu(OH_2)_6]^{2+}$ complex ion, then the reaction with ammonia is better regarded as the replacement of one type of Lewis base molecule (H_2O) by another (NH_3):

$$[Cu(OH_2)_6]^{2+} + 4\ NH_3(aq) \longrightarrow [Cu(NH_3)_4]^{2+}(aq) + 6\ H_2O(\ell)$$

This reaction can be seen as a competition between the Lewis bases H_2O and NH_3 to bond to the Cu^{2+} ion. From the evidence of what happens, we can conclude that the NH_3 "wins" the competition.

Although this equation shows that in the overall reaction six H_2O molecules are replaced by four NH_3 molecules, the reaction happens in a sequence of overlapping steps, with replacement of one NH_3 molecule at a time [>>Section 27.5].

When a small amount of sodium hydroxide solution is added to solutions of salts of the Al^{3+} ion, or to salts of Zn^{2+}, Sn^{4+}, Pb^{2+}, and Cr^{3+} ions, a metal hydroxide precipitates. These hydroxides are said to be *amphoteric* [>>Section 14.2], because on addition of more sodium hydroxide solution (as well as on addition of acidic solution), they dissolve. These reactions can be represented, in the case of aluminium ions, by the equations

$$Al^{3+}(aq) + 3\ OH^-(aq) \longrightarrow Al(OH)_3(s)$$

$$Al(OH)_3(s) + OH^-(aq) \longrightarrow [Al(OH)_4]^-(aq)$$

This description begs the question of why each aluminium(III) ion should bond with three OH^- ions first, but with four of them when more NaOH solution is added. A more powerful description is achieved if we consider the aquated Al^{3+} ion to be the $Al(OH_2)_6{}^{2+}$ complex ion. The hydroxide ion, OH^-, can act as a Lewis base because it has three lone pairs of electrons on the O atom. So the reactions with OH^- ions can be viewed as the result of competition between the H_2O and OH^- Lewis bases to bond to Al^{3+} ions, with sequential replacement of H_2O molecules by OH^- ions as the concentration of OH^- ions is increased:

$$[Al(OH_2)_6]^{3+} + OH^-(aq) \longrightarrow [Al(OH_2)_5(OH)]^{2+}(aq) + H_2O(\ell)$$

$$[Al(OH_2)_5(OH)]^{2+}(aq) + OH^-(aq) \longrightarrow [Al(OH_2)_4(OH)_2]^+(aq) + H_2O(\ell)$$

$$[Al(OH_2)_4(OH)_2]^+(aq) + OH^-(aq) \longrightarrow [Al(OH_2)_3(OH)_3](s) + 3\ H_2O(\ell)$$

$$[Al(OH_2)_3(OH)_3](s) + OH^-(aq) \longrightarrow [Al(OH_2)_2(OH)_4]^-(aq) + H_2O(\ell)$$

Now we have the more sensible representation that the Al^{3+} ions always have six Lewis base species bound to them, and as the OH^- ions become more competitive as their concentration increases, they are able to displace more H_2O molecules. In this sequence of reactions, the species $[Al(OH_2)_3(OH)_3](s)$ is aluminium hydroxide precipitate, previously described as $Al(OH)_3(s)$, and the species $[Al(OH_2)_2(OH)_4]^-(aq)$ is the aluminate ion, previously written as $[Al(OH)_4]^-(aq)$. As the concentration of $OH^-(aq)$ increases (and pH increases), the species present at highest concentration changes sequentially [>>Section 27.5].

Representation of the OH^- ion. The $-$ sign outside of the brackets indicates that the overall charge on the ion is -1, without specifying the location of the charge.

WORKED EXAMPLE 6.12—AQUATION AS COMPLEXATION

Rewrite the following equation in a way that shows complexation in aqueous solution as a competition between Lewis bases:

$$Fe^{3+}(aq) + 6\ CN^-(aq) \longrightarrow [Fe(CN)_6]^{3-}(aq)$$

Solution

The $Fe^{3+}(aq)$ ion can be represented as the complex ion $[Fe(OH_2)_6]^{3+}$. So in the complexation reaction, six H_2O Lewis base molecules are replaced by six CN^- Lewis base ions. The six displaced H_2O molecules are released into the bulk water.

$$[Fe(OH_2)_6]^{3+}(aq) + 6\ CN^-(aq) \longrightarrow [Fe(CN)_6]^{3-}(aq) + 6\ H_2O(\ell)$$

Comment

Check that the equation is balanced for both charge and numbers of each type of atom. Note that the charge on the complex ion changes when six neutral H_2O molecules are replaced by six CN^- ions.

EXERCISE 6.42—AQUATION AS COMPLEXATION

Rewrite each of the equations in Exercise 6.40 in a way that shows complexation in aqueous solution as a competition between Lewis bases.

Complexation reactions of aquated metal ions can be regarded as the result of competition between water and other Lewis bases to bond with the metal ion.

6.8 Solution Concentration

If we know the amount (in mol) of a species that reacts, we can use the balanced chemical equation to calculate the amount of other species that it will react with, or that are formed. Sometimes it is convenient to add reactant species in solution. If we measure the volume of solution added, we need to be able to calculate the amount of a species in that volume, and to do that we need to know its concentration.

The *concentration* of a solute expresses the amount of solute in a specified quantity of solution. There are a number of ways of expressing concentration, each with its own units. Here we will focus on the **amount concentration** (*c*), often called *molarity*, and which is defined as the amount of solute per litre of solution:

$$\text{amount concentration, or molarity, } c = \frac{\text{amount of solute (mol)}}{\text{volume of solution (L)}}$$

The unit for expressing amount concentration is mol L^{-1} (said as *moles per litre*), which is sometimes abbreviated to M.

For example, if 180.2 g, or 1.00 mol, of glucose is dissolved in enough water to give a total solution volume of 1.00 L, the concentration of glucose (*c*) is 1.00 mol L^{-1}. If 0.0100 mol of glucose were used, the concentration would be 0.0100 mol L^{-1}.

Of these two solutions, the one with $c = 1.00$ mol L^{-1} is said to be more *concentrated* than the other: it contains more molecules per litre than the one with $c = 0.0010$ mol L^{-1}. The 0.0010 mol L^{-1} solution is said to be the more *dilute*. Although chemists sometimes talk about concentrated or dilute solutions in an absolute (rather than relative) sense, there is no defined cut-off between which solutions are described as concentrated or dilute.

Concentration is an *intensive* property because it does not depend on the volume of sample. A 1 L bottle of a solution has the same concentration as a 10 mL sample of it—although the 1 L sample contains 100 times more of the solute (dissolved in 100 times more of the solution).

The amount concentration refers to the amount of solute per litre of *solution* and not per litre of *solvent*. If 1 L of water is added to 1 mol of a solid compound, the final volume is usually not exactly 1 L, so the concentration is not exactly 1 mol L^{-1}. When making aqueous solutions of a given amount concentration, we dissolve the solute in less than the desired volume of water, then add water until the desired solution volume is reached (Figure 6.44).

Molecular Modelling (Odyssey)

e6.28 Simulate a solution with a given concentration.

Think about It

e6.29 Watch the preparation of a solution of known concentration in the laboratory.

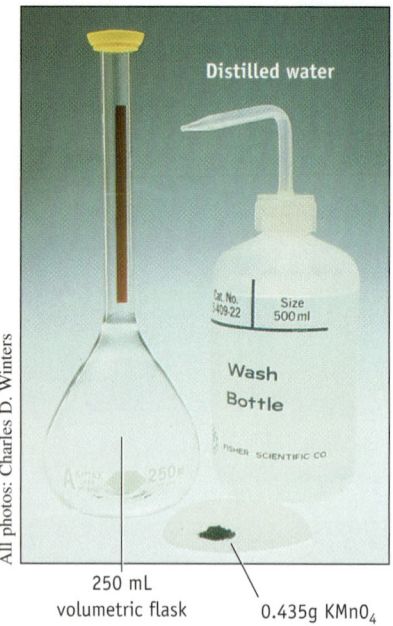

250 mL volumetric flask   0.435g KMnO₄

The KMnO₄ is first dissolved in a small amount of water.

Distilled water is added to fill the flask with solution just to the mark on the flask.

A mark on the neck of a volumetric flask indicates a volume of exactly 250 mL at 25 °C.

FIGURE 6.44 Making a solution of specified concentration. A 0.0110 mol L^{-1} KMnO₄ solution is made by adding some water to 0.435 g of KMnO₄(s), shaking to dissolve the solid and, after all of it has dissolved, making the total volume of solution 0.250 L. A Pasteur pipette is used to add the final drops to water so that the bottom of the meniscus is just level with the mark on the neck of the volumetric flask.

In our $KMnO_4$ solution, the concentration of the species $KMnO_4$ is zero. Because the balanced equation tells us that each mol of $KMnO_4$ that dissociates produces 1 mol of $K^+(aq)$ ions and 1 mol of $MnO_4^-(aq)$ ions, we can say:

$$[KMnO_4] = 0 \text{ mol L}^{-1}$$
$$[K^+] = 0.0110 \text{ mol L}^{-1}$$
$$[MnO_4^-] = 0.0110 \text{ mol L}^{-1}$$

Suppose that we dissolve glucose, $C_6H_{12}O_6(s)$, in water to make a 0.1 mol L^{-1} solution. Because glucose is a non-electrolyte, all of the dissolved glucose remains in solution as the molecular species $C_6H_{12}O_6$. So, $c = 0.10$ mol L^{-1}, and $[C_6H_{12}O_6] = 0.10$ mol L^{-1}.

WORKED EXAMPLE 6.14—CONCENTRATIONS OF SPECIES

What are the concentrations of the main species in a 0.025 mol L^{-1} $Fe(NO_3)_2$ solution?

Strategy

Recognize that ionic compounds are strong electrolytes. Use the equation for dissociation of the solute to see the concentrations (in moles per litre) of the anions and cations relative to the solute concentration.

Solution

Iron(II) nitrate dissociates in solution according to the equation

$$Fe(NO_3)_2 \xrightarrow{\;\;H_2O(\ell)\;\;} Fe^{2+}(aq) + 2\, NO_3^-(aq)$$

From this equation, we see that the amount of $Fe^{2+}(aq)$ ions is the same as the amount of dissolved solute in a given volume of solution.

$$\text{So } [Fe^{2+}] = c(Fe(NO_3)_2) = 0.025 \text{ mol L}^{-1}$$

We can also deduce from the dissociation equation that in a given volume of solution the amount of $NO_3^-(aq)$ ions is twice the amount of dissolved solute.

$$\text{So } [NO_3^-] = 2 \times c(Fe(NO_3)_2) = 0.050 \text{ mol L}^{-1}$$

Interactive Exercise 6.48

Calculate the concentrations of ions in a solution.

EXERCISE 6.49—CONCENTRATIONS OF SPECIES

For each solution, identify the ions that exist in aqueous solution, and specify the concentration of each ion:
(a) 0.25 mol L^{-1} $(NH_4)_2SO_4$ solution
(b) 0.123 mol L^{-1} Na_2CO_3 solution
(c) 0.056 mol L^{-1} HNO_3 solution

It is a little more complex to calculate the species concentrations in solutions of weak electrolytes such as acetic acid:

$$CH_3COOH(aq) + H_2O(\ell) \rightleftharpoons CH_3COO^-(aq) + H_3O^+(aq)$$

In a solution correctly labelled "1.0 mol L^{-1} acetic acid solution," about 4% of the acetic acid molecules ionize—and so the other 96% (0.96 mol L^{-1}) remains as aquated CH_3COOH molecules. Ionization of CH_3COOH molecules results in equal amounts (and equal concentrations) of $CH_3COO^-(aq)$ ions and $H_3O^+(aq)$ ions. So now we have a full description of the solution:

$$c = 1.0 \text{ mol L}^{-1}$$
$$[CH_3COOH] = 0.96 \text{ mol L}^{-1}$$
$$[CH_3COO^-] = 0.04 \text{ mol L}^{-1}$$
$$[H_3O^+] = 0.04 \text{ mol L}^{-1}$$

WORKED EXAMPLE 6.13—SOLUTION CONCENTRATION

Potassium permanganate, $KMnO_4(s)$, a common oxidizing agent that was used at one time as a germicide in the treatment of wounds and burns, is a shiny, purple-black solid that dissolves readily in water to give a deep purple solution. Suppose 0.435 g of $KMnO_4$ has been dissolved in water and the volume of solution then made up to 250 mL (Figure 6.44). What is the amount concentration of the $KMnO_4$ solution?

Strategy

First, from the mass of solute, and its molar mass, calculate the amount of it [<<Section 3.4]. From the amount of solute and volume of solution, the concentration can be calculated.

Solution

$$\text{Amount of } KMnO_4, \; n = \frac{m}{M} = \frac{0.435 \text{ g}}{158.0 \text{ g mol}^{-1}} = 0.00275 \text{ mol}$$

Now we can calculate the concentration (c) of the solution:

$$c = \frac{\text{amount of solute (mol)}}{\text{volume (L)}} = \frac{0.00275 \text{ mol}}{0.250 \text{ L}} = 0.0110 \text{ mol L}^{-1}$$

Comment

The procedure to make this solution is illustrated in Figure 6.44.

EXERCISE 6.43—SOLUTION CONCENTRATION

If 25.3 g of sodium carbonate, $Na_2CO_3(s)$, is dissolved in enough water to make 500 mL of solution, what is the solute concentration?

EXERCISE 6.47—SOLUTION CONCENTRATION

Which one of each of the following pairs is the more concentrated sucrose solution?
(a) (A) 0.20 mol sucrose in 200 mL solution or (B) 0.05 mol sucrose in 200 mL solution
(b) (A) 0.20 mol sucrose in 200 mL solution or (B) 0.50 mol sucrose in 1.00 L solution
(c) (A) 0.20 mol sucrose in 200 mL solution or (B) 0.20 mol sucrose in 100 mL solution
(d) (A) 0.020 mol sucrose in 20 mL solution or (B) 0.20 mol sucrose in 500 mL solution
(e) (A) 0.20 mol sucrose in 200 mL solution or (B) a 10 mL sample of solution A

Solute Concentration versus Concentration of Species

The $KMnO_4$ solution described above has a solute concentration 0.0110 mol L^{-1}. If we were to label this solution for storage and identification purposes, we perhaps might write:

$$0.0110 \text{ mol L}^{-1} KMnO_4 \text{ solution}$$

$$\text{or} \qquad KMnO_4 \text{ solution}, \; c = 0.0110 \text{ mol L}^{-1}$$

We call this value the "solute concentration," or the "labelled concentration," and it refers to how much of the $KMnO_4(s)$ was dissolved in solution (per litre). However, the solution does not contain any of the species with formula $KMnO_4$ because potassium permanganate is a strong electrolyte:

$$KMnO_4(s) \xrightarrow{H_2O} K^+(aq) + MnO_4^-(aq)$$

We often need to refer to the concentration of particular species in solution, and the symbol for this is the formula of the species in square brackets. For example, $[MnO_4^-(aq)]$ denotes the concentration of the $MnO_4^-(aq)$ ions in solution—although we usually write simply $[MnO_4^-]$.

Think about It

e6.30 Watch the preparation of a solution of known concentration using dilution in the laboratory.

Interactive Exercise 6.44

SUBMIT

See how to calculate the concentration of solute given the mass of solute and volume of solution.

Interactive Exercise 6.45

SUBMIT

See how to calculate the concentration of each ion in a solution given the mass of ionic solute and volume of solution.

Interactive Exercise 6.46

SUBMIT

See how to calculate the mass of solute required to make a solution of known concentration.

The amount concentration of a solute in solution (c) is the amount of solute per litre of solution:

$$\text{concentration, } c = \frac{\text{amount of solute (mol)}}{\text{volume of solution (L)}}$$

The concentration of any species in solution, which may be different from the solute concentration, is indicated by the symbol [species].

SUMMARY

Key Concepts

- The chemical behaviour exhibited by a solution is due to the particular species (aggregate of atoms, charge or uncharged) in solution. (Section 6.1)
- Compared with other substances whose molecules are of similar size, water is remarkable in its high **specific heat capacity**, molar enthalpy change of **vaporization**, **boiling point**, and **surface tension**, its low **equilibrium vapour pressure**, and the way its density changes with temperature. (Section 6.2)
- The remarkable physical properties of water can be attributed to its strong **intermolecular forces**. (Section 6.3)
- In all molecular substances there are **dispersion forces**, or **instantaneous dipole–induced dipole forces**, which result from transitory lack of symmetry of the molecules' electron clouds and induced **polarization** of the electron clouds of nearby molecules. Dispersion forces are generally weak compared to the other types of intermolecular force. (Section 6.3)
- If the molecules of a substance have bonds between dissimilar atoms, their different attractions for the bonding electrons (**electronegativity**) brings about a charge separation, and a **polar bond**. (Section 6.3)
- If the polar bonds in a molecule are symmetrically directed and the orienting effects in an electric field cancel out, the molecules (and the substance) are said to be **nonpolar**. (Section 6.3)
- In an electric field, if the forces on the polar bonds do not cancel out because of symmetry, a molecule is as a **dipole**: it experiences an orienting effect, and is said to be a polar molecule, and the substance of which it is composed, a **polar substance**. (Section 6.3)
- Between polar molecules there are **dipole–dipole forces** due to attractions between the positive ends of molecular dipoles with the negative ends of the dipoles of nearby molecules. (Section 6.3)
- In some polar substances there is a relatively strong form of dipole–dipole intermolecular attraction, called **hydrogen bonding**—most commonly formed between an H atom (joined to an F, O, or N atom) in one molecule and an F, O, or N atom in another molecule. (Section 6.3)
- The physical properties of molecular compounds depend on the strengths of intermolecular forces. The ways in which water is remarkable, compared with other substances with similarly sized molecules, can be attributed to the strength and direction of hydrogen bonding between water molecules. (Section 6.4)
- In **aqueous solutions,** ionic compounds undergo **dissociation** to form **aquated ions,** surrounded by water molecules as a result of **ion-dipole forces**. (Section 6.5)
- The degree of solubility of ionic substances is the result of competition between anion-cation attractions in the lattice and ion-dipole attractions to water molecules. The chemistry of a solution is the sum of the chemistries of the aquated ions. (Section 6.5)
- In cases of molecular substances, generally (but not in every case) polar solutes dissolve in polar solvents, and non-polar solutes dissolve in non-polar solvents. This can be seen as the consequence of a natural tendency for the molecules to achieve a more probable distribution by mixing, unless there is an energy barrier that prevents them from doing so. (Section 6.5)

- If the molecules of a substance have a non-polar part and a polar part, the larger the non-polar part the less likely that it will dissolve in water, and the more polar groups it has the more likely that it will dissolve in water. (Section 6.5)
- Some molecular substances that dissolve in water are **strong electrolytes**, some are **weak electrolytes**, and some are **non-electrolytes**. (Section 6.5)
- **Self-ionization of water** leads to an equilibrium condition with small, equal concentrations of $H^+(aq)$ ions and $OH^-(aq)$ ions. $H^+(aq)$ is continuously changing among several transitory species—one of which is the **hydronium ion** (H_3O^+). (Section 6.6)
- Common categories of reaction in aqueous solution are **precipitation**, **oxidation-reduction**, **acid-base neutralization**, and **complexation**. (Section 6.7)
- Whether or not precipitation happens on mixing solutions can be viewed as the result of a competition between cation-anion attractions an ionic lattice, and ion-dipole attractions to water molecules in solution. (Section 6.7)
- Oxidation-reduction reactions can be regarded as the result of competition between species for electrons. The species that "wins" the electrons undergoes **reduction**, and the species from which electrons are taken undergoes **oxidation**. (Section 6.7)
- In the Brønsted-Lowry model, acid-base reactions are the result of competition between species for H^+ ions: the species that "wins" the H^+ ion is a **base**, and the species from which an H^+ ion is removed is an **acid**. (Section 6.7)
- **Strong acids** and **strong bases** are strong electrolytes, **weak acids** and **weak bases** are weak electrolytes. (Section 6.7)
- In complexation reactions, a bond is imagined to be formed from a lone pair of electrons on one of the reactant species: the lone-pair donor is called a **Lewis base**, and the recipient is called a **Lewis acid**. (Section 6.7)
- Aquated metal ions can be regarded as **complex ions**, with the surrounding water molecules acting as Lewis bases. Complexation reactions of aquated metal ions are the result of competition between water and other Lewis bases to bond with the metal ion. (Section 6.7)
- The **amount concentration**, or molarity, of a solute in solution (c) is the amount of solute per litre of solution. The concentration of a particular species in solution, indicated by the symbol [species], may be different from the **solute concentration**. (Section 6.8)

Key Equations

$$\text{Amount concentration of solution, } c = \frac{\text{amount of solute (mol)}}{\text{volume of solution (L)}}$$

(Section 6.8)

REVIEW QUESTIONS

Section 6.1: Arsenic Ain't Arsenic

6.50 Give the formulas of (a) two carbonate species, and (b) two iron species, in aqueous solution.

Section 6.2: The Remarkable Properties of Water

6.51 From the equilibrium vapour pressures listed in Table 6.3, predict which has the higher boiling point: water or hexane.

6.52 If placed gently, a paper clip will float on the surface of water, even though it is more dense than water. Explain why this is possible. Is it more or less likely that a paper clip can float on the surface of octane? Explain.

Section 6.3: Intermolecular Forces

6.53 Are the following changes associated with intermolecular forces or intramolecular forces?
(a) Iron rusts.
(b) A rubber band is stretched.
(c) In the stratosphere, ultraviolet light causes O_2 molecules to form O atoms.
(d) Mothballs gradually "disappear."

6.54 Using electronegativity values in Figure 6.11, compare the direction and the magnitudes of the polarities of the following:
(a) Si—O and P—P bonds
(b) C=O and C=S bonds

6.55 Using electronegativity values in Figure 6.11, for each pair of bonds, indicate the more polar bond and use an arrow to show the direction of polarity in each bond:
(a) C—O and C—N
(b) P—Br and P—Cl
(c) B—O and B—S
(d) B—F and B—I

6.56 Urea, $(NH_2)_2CO$, is used in plastics and fertilizers. It is also the primary nitrogen-containing substance excreted by humans.

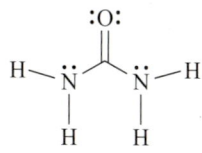

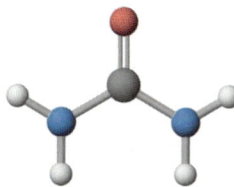

Using electronegativity values in Figure 6.11, decide the following:
(a) Which bonds in the molecule are polar and which are non-polar?
(b) Which is the most polar bond in the molecule? Which atom is the negative end of the bond dipole?

6.57 From the dipole moments of HF, HCl, HBr, and HI listed in Table 6.6, list the Group 17 atoms in decreasing order of electronegativity.

6.58 With reference to the values of electronegativity in Figure 6.11, consider the following molecules:
- H_2O, bent
- NH_3, trigonal-pyramidal
- CO_2, linear
- ClF
- CCl_4, tetrahedral

(a) In which molecules are the bonds most polar?
(b) Which of the molecules are *not* polar?
(c) Which atom in ClF is more negatively charged?

6.59 Which of the following molecules are polar? For each polar molecule, use values of electronegativity in Figure 6.11 to decide the direction of polarity:
(a) $BeCl_2$, linear
(b) HBF_2, trigonal-planar
(c) CH_3Cl, tetrahedral
(d) SO_3, trigonal-planar

6.60 Which of the following molecules are not polar? Use values of electronegativity in Figure 6.11 to decide in which molecule the bonds are most polar:
(a) CO
(b) BCl_3, trigonal-planar

(c) CF_4, tetrahedral
(d) PCl_3, trigonal-pyramidal
(e) GeH_4, tetrahedral

6.61 In which of the following substances would you expect there to be dipole–dipole forces of attraction acting between molecules?
(a) hexane, $C_6H_{14}(\ell)$
(b) hydrogen sulfide, $H_2S(g)$
(c) methane, $CH_4(g)$

6.62 In which of the following compounds would intermolecular hydrogen bonds be likely in the liquid state?
(a) diethyl ether $(C_2H_5OC_2H_5)$
(b) ammonia (NH_3)
(c) methane (CH_4)
(d) hydrogen fluoride (HF)
(e) acetic acid (CH_3COOH)
(f) bromine (Br_2)
(g) ethylene glycol $(HOCH_2CH_2OH)$
(h) methylamine (CH_3NH_2)

6.63 In which of the following compounds would intermolecular hydrogen bonding operate?
(a) hydrogen selenide (H_2Se)
(b) formic acid (HCOOH)
(c) hydrogen iodide (HI)
(d) acetone (CH_3COCH_3)

6.64 Molecules of acetic acid have the formula CH_3COOH. In the pure liquid, acetic acid consists of dimers (pairs of molecules joined together). Suggest, with a sketch, how this can be explained by hydrogen bonding between each pair of molecules.

6.65 Using the periodic table, predict which atoms have the more polarizable electron clouds: neon (Ne) or krypton (Kr). Explain.

6.66 What intermolecular force(s) must be overcome to do the following?
(a) melt ice
(b) sublime solid I_2
(c) convert liquid NH_3 to NH_3 vapour

6.67 What type of intermolecular force must be overcome in converting each of the following from a liquid to a gas?
(a) liquid O_2
(b) mercury
(c) methyl iodide (CH_3I)
(d) ethanol (CH_3CH_2OH)

6.68 Decide which type of intermolecular force operates in the following:
(a) liquid O_2
(b) methanol, $CH_3OH(\ell)$
(c) liquid sulfur dioxide, SO_2, whose molecules are V-shaped
(d) carbon dioxide, $CO_2(s)$

6.69 Rank the following substances in order of increasing strength of forces of attraction amongst the molecules or atoms. Which exist as gases at 25 °C and 1 atm?

(a) neon (Ne)

(b) methane (CH_4)

(c) carbon monoxide (CO)

(d) carbon tetrachloride (CCl_4)

6.70   What is the difference between *polarity* and *polarizability*? How are physical properties like boiling point affected by these?

Section 6.4: Explaining the Properties of Water

6.71   Insert one of *higher*, *smaller*, or *unchanged* to make correct sentences.

(a) The stronger the intermolecular forces in a liquid, its normal boiling point is _____.

(b) The weaker the intermolecular forces in a liquid, its equilibrium vapour pressure at a specified temperature is _____.

(c) The smaller the volume of a water in a sealed flask, its equilibrium vapour pressure is _____.

(d) The higher the temperature of a liquid, its equilibrium vapour pressure is _____.

(e) The more volatile a liquid, its boiling point is _____.

(f) The higher the boiling point of a liquid, its enthalpy change of vaporization is _____.

6.72   Oxygen and selenium are both members of Group 16. Account for the higher boiling point of water (H_2O) than that of hydrogen selenide (H_2S), as shown in Figure 6.20.

6.73   Rationalize the observation that propan-1-ol ($CH_3CH_2CH_2OH$) has a boiling point of 97.4 °C, whereas a compound with the same empirical formula, methyl ethyl ether ($CH_3CH_2OCH_3$), boils at 7.4 °C.

6.74   Ethanol, $CH_3CH_2OH(\ell)$, and dimethyl ether, $CH_3OCH_3(\ell)$, have the same compositional formula C_2H_6O. By reference to forces between the molecules in samples of each, predict which has the higher boiling point.

6.75   Which member of each of the following pairs of compounds has the higher boiling point?

(a) O_2 or N_2

(b) SO_2 or CO_2

(c) HF or HI

(d) SiH_4 or GeH_4

6.76   Which type(s) of intermolecular force influence(s) the magnitude of the enthalpy change of vaporization ($\Delta_{vap}H$) of each of the following?

(a) methanol

(b) octane

(c) acetone (CH_3OCH_3)

6.77   Using structural formulas, describe how hydrogen bonding can operate between molecules of methanol. What physical properties are likely to be affected by hydrogen bonding in methanol? Find values for these properties to check if they are consistent with your predictions.

6.78   With reference to Worked Example 6.4, which type(s) of intermolecular force influence(s) the magnitude of the equilibrium vapour pressure of the following?

(a) *cis*-1,2-dichloroethylene

(b) *trans*-1,2-dichloroethylene

Which of these compounds would you expect to have the higher equilibrium vapour pressure at a given temperature?

6.79   As well as the properties discussed in this section, water is relatively viscous (resistant to flow) compared with most other substances whose molecules are of a similar size.

(a) Account for the higher viscosity of water than hexane.

(b) Glycerol (propan-1,2,3-triol, $HOCH_2CHOHCH_2OH$) is even more viscous than water. Explain this.

6.80   Liquid ethylene glycol ($HOCH_2CH_2OH$) is one of the main ingredients in commercial antifreeze. Do you predict its viscosity to be greater or less than that of ethanol (CH_3CH_2OH)?

6.81   Maleic acid and fumaric acid have the same composition ($C_4H_4O_4$) and molecules of both can be described as an ethylene molecule with a carboxylic acid group (–COOH) replacing a H atom on each C atom. Maleic acid has both of the carboxylic acid groups on the same side of the double bond, while those of fumaric acid are on opposite sides. Rotation of one C atom with respect to the other in a double bond does not happen at ambient temperatures, so maleic acid and fumaric acid do not convert to each other.

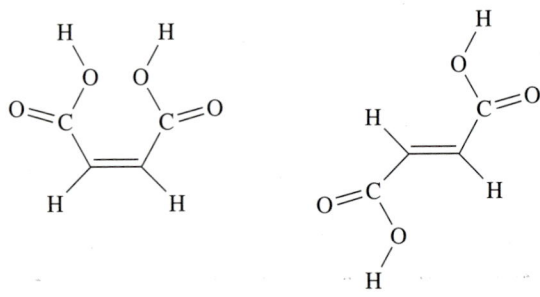

Maleic acid   Fumaric acid

Explain the fact that the melting point of fumaric acid (287 °C) is much higher than that of maleic acid (131 °C). (Hint: Hydrogen bonding can occur between different groups in the same molecule, as well as between different molecules.)

Interactive Exercise (Odyssey) 6.82

SUBMIT

Investigate general differences in hydration of different ions.

Section 6.5: Water as a Solvent

6.83 Which side of the water molecule dipole (the side with an O atom or the side with the H atoms) points at the following ions?
(a) $Ca^{2+}(aq)$ ions
(b) $Br^-(aq)$ ions
(c) $Cr_2O_7{}^{2-}(aq)$ ions
(d) $NH_4{}^+(aq)$ ions

6.84 What are the main species present in solution when some potassium sulfate, $K_2SO_4(s)$, is dissolved in water? Draw diagrams of these species.

6.85 What are the main possible reactant species in solution when some silver nitrate, $AgNO_3(s)$, and some potassium chloride, $KCl(s)$, are dissolved in the same sample of water?

6.86 Acetone (CH_3COCH_3) is a liquid commonly used as a laboratory solvent. It is easily contaminated with water absorbed from the air. Why does acetone absorb water so readily? Draw molecular structures showing how water and acetone can interact. What intermolecular force(s) is(are) involved in the interaction?

6.87 Cooking oil is not miscible with water. From this observation, what conclusions can you draw regarding the polarity or hydrogen-bonding ability of the molecules in cooking oil? Predict the miscibility of cooking oil in hexane.

6.88 (a) About equal volumes of the liquids water and carbon tetrachloride (CCl_4) are added to a test tube and shaken. What would you expect to see? (Carbon tetrachloride is colourless and is more dense than water.)
(b) With reference to the types of intermolecular force between each pair of compounds, describe what would happen if some hexane were now added and the test tube shaken.

Section 6.6: Self-Ionization of Water

6.89 Why are the concentrations of hydronium ion and hydroxide ion equal in pure water?

6.90 What ions are formed in the self-ionization of ammonia in liquid ammonia? What would be the relative concentrations of these ions?

Section 6.7: Categories of Chemical Reaction in Water

6.91 When a barium chloride solution and a potassium sulfate solution are mixed, a white precipitate of barium sulfate, $BaSO_4(s)$, forms. When an iron(II) chloride solution and a potassium sulfate solution are mixed, iron(II) sulfate, $FeSO_4(s)$, does not precipitate. Compare the solubilities of $BaSO_4(s)$ and $FeSO_4(s)$.

6.92 For each of the following cases, (i) write a balanced equation; and (ii) list the spectator ions.
(a) A precipitate of cadmium hydroxide, $Cd(OH)_2(s)$, forms when a cadmium chloride ($CdCl_2$) solution is mixed with a sodium hydroxide solution.
(b) A precipitate of nickel carbonate, $NiCO_3(s)$, forms when a nickel nitrate, $Ni(NO_3)_2$, solution is mixed with a potassium carbonate solution.
(c) A precipitate of copper(II) sulfide, $CuS(s)$, forms when a copper sulfate, $CuSO_4$, solution is mixed with an ammonium sulfide, $(NH_4)_2S$, solution.
(d) A precipitate of calcium oxalate, $CaC_2O_4(s)$, forms when a calcium nitrate, $Ca(NO_3)_2$, solution is mixed with a potassium oxalate, $K_2C_2O_4$, solution.

6.93 The reaction between aquated dichromate ions and aquated ethanol can be represented by the two half-equations

$$2\ Cr_2O_7{}^{2-}(aq) + 28\ H^+(aq) + 12\ e^-$$
$$\longrightarrow 4\ Cr^{3+}(aq) + 14\ H_2O(\ell)$$
$$3\ C_2H_5OH(aq) + 3\ H_2O(\ell)$$
$$\longrightarrow 3\ CH_3COOH(aq) + 12\ H^+(aq) + 12\ e^-$$

Which species is oxidized? To what? Which species is the oxidizing agent? Which species is reduced? To what? Which species is the reducing agent? (The species $H^+(aq)$ and $H_2O(\ell)$ are neither oxidized nor reduced.)

6.94 The reaction between hot molten sodium and chlorine gas (Figure 16.3) is an oxidation-reduction reaction that can be represented by the equation

$$2\ Na(\ell) + Cl_2(g) \longrightarrow 2\ Na^+Cl^-(g)$$

This reaction equation can be separated into two half-equations:

$$2\ Na(\ell) \longrightarrow 2\ Na^+ + 2\ e^-$$

$$Cl_2(g) + 2\ e^- \longrightarrow 2\ Cl^-$$

Which species is oxidized? To what? Which species is the oxidizing agent? Which species is reduced? To what? Which species is the reducing agent?

6.95 What ions are produced when nitric acid dissolves in water? The base barium hydroxide is moderately soluble in water. What ions are produced when it dissolves in water?

6.96 (a) Write balanced chemical equations to represent the neutralization of (i) a nitric acid solution with a potassium hydroxide solution, and (ii) a hydrochloric acid solution with a sodium hydroxide solution.
(b) Compare the two equations and comment.
(c) List the spectator ions in each case.
(d) Give the name and formula of the solid residue that would be left if each reaction mixture were evaporated to dryness.

6.97 Write the formulas, showing the overall charge, on the following complexes:
(a) An iron(II) ion is bound to six ammonia molecules.
(b) A zinc ion is bound to four cyanide ions.
(c) A manganese(II) ion is bound to six fluoride ions.
(d) An iron(III) ion is bound to six cyanide ions.
(e) A cobalt(II) ion is bound to four chloride ions.
(f) A nickel(II) ion is bound to four ammonia molecules and two water molecules.

6.98 Which of the following molecules can be a Lewis base?

(a)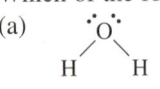

water (H_2O)

(b)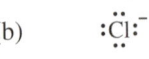

chloride ion (Cl^-)

(c)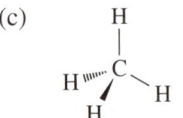

methane (CH_4)

(d)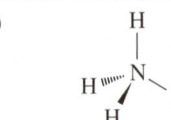

ammonium ion (NH_4^+)

(e)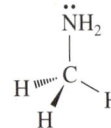

methylamine (CH_3NH_2)

(f) $:C{\equiv}N:^-$

cyanide ion (CN^-)

Section 6.8: Solution Concentration

6.99 To carry out an experiment, you need 250 mL of a 0.0200 mol L^{-1} $AgNO_3$ solution. You are given some solid $AgNO_3$, de-ionized water, and a 250 mL volumetric flask. Describe how you will make up the solution.

6.100 Which one of each of the pairs described in Exercise 6.47 has the largest amount of sucrose?

6.101 Which contains the greater mass of solute: 1.0 L of 0.1 mol L^{-1} NaCl solution or 1250 mL of 0.060 mol L^{-1} Na_2CO_3 solution?

Interactive Exercise 6.102

Calculate mass of solute in a solution.

6.103 What is the mass of $KMnO_4$ dissolved in 250 mL of a 0.0125 mol L^{-1} $KMnO_4$ solution?

Interactive Exercise 6.104

Calculate volume of solution containing a required mass of solute.

6.105 What volume of 0.123 mol L^{-1} NaOH solution has 25.0 g of dissolved NaOH?

Interactive Exercise 6.106

Calculate required mass of solute to prepare a solution of fixed concentration.

6.107 What mass of oxalic acid ($H_2C_2O_4$) is required to make 250 mL of a solution with concentration 0.15 mol L^{-1}?

6.108 For each solution, identify the ions that exist in aqueous solution, and specify the concentration of each ion:
(a) 0.12 mol L^{-1} $BaCl_2$ solution
(b) 0.0125 mol L^{-1} $CuSO_4$ solution
(c) 0.500 mol L^{-1} $K_2Cr_2O_7$ solution

6.109 If 6.73 g of Na_2CO_3 is dissolved in enough water to make 250 mL of solution, what is the solution concentration (c)? What are the concentrations of the Na^+(aq) ions and CO_3^{2-}(aq) ions?

SUMMARY AND CONCEPTUAL QUESTIONS

6.110 The figure below is a frame from the animation in e6.2 depicting the evaporation of water on the surface.

(a) Explain why the water molecule shown is pulled back into the liquid phase.
(b) Describe the forces involved.

6.111 The figure on the next page is a frame from the animation in e6.4 depicting the inside of a bubble in boiling water.

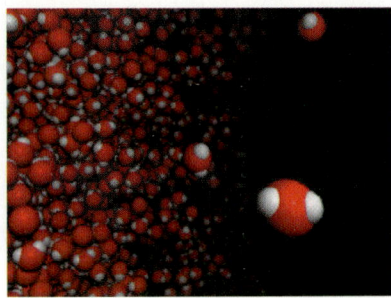

(a) What prevents the bubble from collapsing in on itself?

(b) Comment on the relative pressure of water vapour inside the bubble, with the atmospheric pressure on the surface.

6.112 The figure below is a frame from the animation in e6.15 depicting a chloride ion being hydrated and leaving the surface of the NaCl lattice.

(a) Why are the water molecules oriented toward the ion in a particular way?

(b) What are the competing forces involved in dissolving?

6.113 The figure below is a frame from the animation in e6.21 depicting a hydronium ion transferring a proton to another water molecule.

(a) Why do hydronium ions appear to diffuse so quickly through an aqueous solution?

(b) Describe what happens to the nuclei and electrons in an O—H bond during proton transfer to another water molecule.

6.114 The figure below is a frame from the animation in e6.21 depicting a hydroxide ion accepting a proton from another water molecule.

(a) Why do hydroxide ions appear to diffuse so quickly through an aqueous solution?

(b) Describe what happens to the nuclei and electrons in an O—H bond during proton transfer to the hydroxide ion.

6.115 The figure below is a frame from the animation in e6.22 depicting the formation of a AgCl precipitate.

(a) What are the competing forces in a precipitation reaction?

(b) Are there distinct AgCl molecules in the lattice?

6.116 The figures below are frames from the animation in e6.23 depicting (i) the reduction of the silver ion, and (ii) the oxidation of a copper atom.

(a) Explain what is happening in each case in terms of the atoms and ions involved.

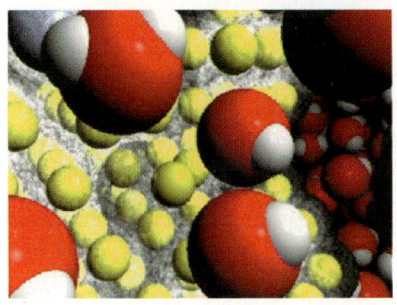

(b) What are the competing forces involved in these two processes?

6.117 The figure below is a frame from the animation in e6.24 depicting proton transfer from an acetic acid molecule to a water molecule.

(a) What happens to the electrons and the nuclei in the O–H bond of the CH_3COOH molecule in this process?

(b) What are the competing forces involved in this transfer?

6.118 The figure below is a frame from the animation in e6.25 depicting proton transfer from a water molecule to an an ammonia molecule.

(a) What happens to the electrons and the nuclei in the O–H bond of the water molecule in this process?

(b) What are the competing forces involved in this transfer?

6.119 The figure below is a frame from the animation in e6.26 depicting the formation of a tetraamminecopper(II) complex ion, $[Cu(NH_3)_4]^{2+}$.

(a) Identify the Lewis acid and Lewis base species involved.

(b) Why do the ammonia molecules replace the water molecules bonded to the copper(II) ion?

6.120 Classify each of the following as (i) a precipitation reaction, (ii) an oxidation-reduction reaction, (iii) an acid-base reaction, or (iv) a complexation reaction.

(a) $Ba^{2+}(aq) + CO_3^{2-}(aq) \longrightarrow BaCO_3(s)$

(b) $[Zn(OH_2)_6]^{2+} + 4\,CN^-(aq)$
$\longrightarrow [Zn(CN)_4]^{2-} + 6\,H_2O(\ell)$

(c) $Zn(s) + Fe^{2+}(aq) \longrightarrow Zn^{2+}(aq) + Fe(s)$

(d) $OH^-(aq) + CH_3CH_2COOH(aq)$
$\longrightarrow H_2O(\ell) + CH_3CH_2COO^-(aq)$

6.121 Classify each of the following as (i) a precipitation reaction, (ii) an oxidation-reduction reaction, (iii) an acid-base reaction, or (iv) a complexation reaction.

(a) $Ca(OH)_2(s) + 2\,H_3O^+(aq)$
$\longrightarrow Ca^{2+}(aq) + 4\,H_2O(\ell)$

(b) $CH_3COO(aq) + Ag^+(aq) + H_2O(\ell)$
$\longrightarrow AgCH_3COO(s) + H_3O^+(aq)$

(c) $Fe(OH)_3(s) + 3\,H_2C_2O_4(aq)$
$\longrightarrow [Fe(C_2O_4)_3]^{3-}(aq) + 3\,H_2O(\ell)$

(d) $5\,Fe^{2+}(aq) + MnO_4^-(aq) + 8\,H^+(aq)$
$\longrightarrow 5\,Fe^{3+}(aq) + Mn^{2+}(aq) + 4\,H_2O(\ell)$

6.122 How does the bond between the Lewis acid and a Lewis base in a complex ion differ from an "ordinary" covalent bond?

Interactive Exercise 6.123

Use Odyssey to build a simulation of 0.40 mol L^{-1} $FeCl_3$ solution containing 110 water molecules.

With the **eCHACR** single-sign-on access card bundled with your text, log on (http://login.cengage.com/sso) and access the e-book and click on any in-margin icons for dynamic molecular-level animations and simulations, videos of laboratory reactions, interactive exercises, reviews of background concepts, and other online supplementary materials as noted by the icons in the text margins.

Also go to www.chemistry.nelson.com <http://www.chemistry.nelson.com> for Answers to in-chapter exercises and selected Review Questions, Test Yourself questions, weblinks, crossword puzzles, flashcards, glossary of key terms, and other student resources.

Chemical Reactions and Energy Flows

Photolibrary/Superstock

GetStock/Alamy Images/Alex Segre

7.1 Case Study: Energy from Hydrogen?

Images of the burning airship Hindenburg have created a fear of hydrogen gas, $H_2(g)$, which has been enhanced by classroom demonstrations of explosions of mixtures of hydrogen and oxygen. And yet hydrogen-powered buses are being used on the streets of several cities, and Japan expects to have two million hydrogen-fuelled cars by 2020.

Why do we look to hydrogen as a key source of energy? Is it safe to use? How can chemists contribute to solving some of the challenges of a hydrogen-based future?

We use the term *source of energy* loosely. Hydrogen gas is not a primary source of energy like the sun. In fact, its production from other materials is usually *endothermic*, using up energy, and we obtain energy from hydrogen through its exothermic reaction with oxygen. So really hydrogen is just a carrier of energy; we can transport hydrogen to give us energy where it is needed.

Burning Energy Questions

Human activity and industrialization over the past 200 years has been fuelled by combustion of carbon-based fuels such as wood, coal, oil, and natural gas. As a result, the concentration of carbon dioxide in our atmosphere has steadily increased [<<Section 4.3]. This has contributed to global climate change [<<Section 4.3] and acidification of the oceans [>>Section 15.1]. One way to check this trend, motivated further by the recognition that fossil fuel reserves are limited, is to use alternative sources of energy that contribute less to carbon dioxide build-up.

Taking It Further
e7.1 The chemistry of fuels and energy sources.

Most alternative energy technologies are in developmental stages, requiring research in chemistry, physics, and engineering, as well as economics. Some depend on chemical processes, and some on physical processes. Some will have widespread application, and some will have niche uses. Let's look more closely at the potential and challenges for use of hydrogen.

Is Hydrogen Safe?

Gaseous fuels such as natural gas and hydrogen all pose safety challenges in their transportation and use. Yet experience with natural gas suggests that these challenges can be addressed. Above-ground pipelines carry natural gas across many countries. In the same way that gasoline vapour burns in a controlled manner in cars and natural gas burns steadily in a gas stove, the rate of burning of hydrogen can be limited to the rate at which it is provided to the combustion zone. Even if a pipeline bursts, hydrogen cannot burn more quickly than it is emitted from the fracture into the air.

Although a balloon containing both hydrogen and oxygen explodes violently, a balloon that contains only hydrogen burns, rather than explodes, because the rate of reaction is limited by the rate at which oxygen diffuses through air to the reaction site. The Hindenburg airship did not explode: it burned over some time, exhausting the surrounding air of its oxygen. Hydrogen's low density gives it an advantage over other fuels because when it escapes, it rises rapidly. For this reason, the gas in the Hindenburg burned at the top of the airship. Of the 97 persons on board, 35 died—most because they jumped. The survivors stayed in the cabin below the airship, which came to the ground relatively gently as the hydrogen was gradually replaced by oxygen-depleted air.

It is a simple matter to direct hydrogen and oxygen separately in a controlled manner to a reaction zone, whether this is direct combustion or reaction in a fuel cell.

Why Hydrogen?

Over the years, power generation has depended upon fuels that are increasingly hydrogen-rich (Figure 7.1).

For combustion of a given weight of fuel, the higher the percentage of atoms that are H atoms, the more heat is obtained (Table 7.1). Fuels with high *energy density*, on a weight basis, have obvious advantages for distribution and portability.

On the other hand, as Table 7.1 shows, 1 L of hydrogen gas can provide less energy than 1 L of other common fuels, and this poses challenges.

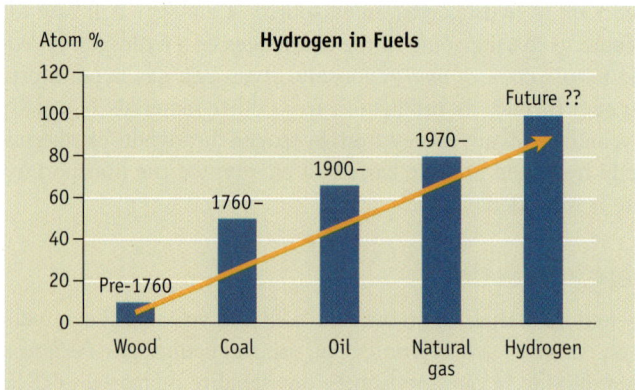

FIGURE 7.1 Hydrogen atom content of the main fuels used for power generation. As fuels have successively replaced each other, the H atom content has increased from 1 in 10 (wood) to 4 in 5 (methane). The hydrogen molecules in hydrogen gas contain nothing but H atoms.

	H₂ Gas (1 bar)	Liquid H₂	CH₄ Gas (1 bar)	Liquid Propane	Liquid Octane
Per gram	143	142	55	50	48
Per litre	12	9 970	36	25 600	34 000

TABLE 7.1 Energy Density of Various Fuels (kJ g^{-1} and kJ L^{-1})

In the context of current concerns about global climate change, an attractive feature of hydrogen as a fuel is that its combustion produces no carbon dioxide; essentially the only reaction is

$$2\ H_2(g) + O_2(g) \longrightarrow 2\ H_2O(g)$$

> *In the context of current concerns about global climate change, an attractive feature of hydrogen as a fuel is that its combustion produces no carbon dioxide.*

At the high temperatures of combustion, as in combustion of other fuels, small amounts of $N_2(g)$ and $O_2(g)$ from the air react to form oxides of nitrogen, $NO(g)$ and $NO_2(g)$. Accumulation of these in high-density traffic can contribute to photochemical smog formation [>>Sections 17.1, 17.10]. However, if the hydrogen-oxygen reaction is made to occur in a fuel cell at near ambient temperatures, oxides of nitrogen are not formed.

One of the attractions of using hydrogen is that it is easy to produce by applying about 1.5 volts of a DC electrical supply across two carbon rods in dilute salty water. The water "splits" into hydrogen and oxygen gases. During this *electrolysis* [>>Section 16.7], $H_2(g)$ forms at one of the carbon rods, and $O_2(g)$ at the other. The net reaction is the opposite to that of combustion of hydrogen. So all we need is some slightly salty water—and we have five oceans full of that!

In principle, electrolysis of water can easily be done at a local service station or even at home. Unfortunately, the reality is not so simple, because of improvements needed in the technology and because the infrastructure for hydrogen service stations is not in place. But research is bringing a *hydrogen economy* closer to reality.

Sources of Hydrogen

Hydrogen gas does not occur naturally on the earth—we have to make it from other materials. Most hydrogen is now created by high-temperature reaction between natural gas and steam, called *steam-methane reforming* (SMR), in a variation of the reaction used for production of syngas [<<Section 4.2], using a different catalyst:

$$CH_4(g) + 2\ H_2O(g) \longrightarrow CO_2(g) + 4\ H_2(g)$$

Unfortunately, like the direct combustion of methane, this process contributes to the increase of CO_2 concentration in our atmosphere. It makes sense to consider ways that do not produce $CO_2(g)$ at all.

In the electrolytic "splitting" of water, the amount of $H_2(g)$ formed is proportional to the amount of electricity (that is, the number of electrons) passed through the cell. So we need large amounts of electricity. Where from? We could simply plug into our wall sockets (using a transformer), but our energy source is perhaps a coal-burning, CO_2-producing power station—and that would defeat one of the purposes of using hydrogen.

The dream solution might be to use solar photovoltaic cells that produce DC electricity from sunlight (free) to electrolyze sea water (free). This is possible, but not yet commercially viable, due to the current cost of solar cells. But chemistry-based research into new materials brings cheaper solar cells closer to reality.

Taking a different tack, Kanan and Nocera have recently found a way to reduce the voltage necessary to electrolyze water. Adding cobalt ions to the aqueous solution catalyzes the production of oxygen at the electrode surface. And if a catalyst can be

See "*In Situ* Formation of an Oxygen-Evolving Catalyst in Neutral Water Containing Phosphate and Co$_2^+$," (M.W. Kanan and D.G. Nocera, *Science*, 321 (2008) 1072–75) and "Personalized Energy: The Home as a Solar Power Station and Solar Gas Station" (D. G. Nocera, *ChemSusChem* 2 (2009) 387 – 390).

found for the reduction process at the other electrode, perhaps the required voltage may be reduced further. This would lower the requirements of solar cells used to "split" water.

Storage of Hydrogen

If we store hydrogen gas at atmospheric pressure we would need a massive fuel tank to drive even a few kilometres. We can squeeze more gas into a container by pumping it in at high pressure. This increases the energy density of the hydrogen on a volume basis, but two disadvantages result: energy is used for compression, and to contain the gas at higher pressures, stronger, heavier containers must be used. Research into the design of light materials that can contain hydrogen gas at very high pressures will make a hydrogen economy more attainable.

Storage of hydrogen as a liquid comes into consideration because of its the relatively high energy density per litre (Table 7.1). However, to decrease the temperature of hydrogen to −253 °C to liquefy it, we would need to insulate our fuel tanks well to maintain this temperature and minimize boiling off. Furthermore, the energy needed to liquefy hydrogen is 40% of that available from its combustion.

Research into hydrogen storage for portable purposes focuses on its sponge-like absorption by metal alloys, such as Mg-Ni alloys, to form metal hydrides. These hydrides are not well understood, and they do not have fixed chemical formulas, but a container full of Mg-Ni alloy will absorb considerably more hydrogen than it would hold if it were empty (under 1 atm pressure of $H_2(g)$ in both cases).

The ideal alloy is one that reaches equilibrium with a moderate pressure of hydrogen gas. Then we can refuel by pumping in $H_2(g)$ at a pressure higher than the equilibrium pressure. As hydrogen is taken off to the combustion chamber, reducing the pressure of hydrogen gas, more is released. The challenge for chemical researchers is that an alloy that can contain lots of hydrogen is not suitable if it holds the hydrogen so tightly that its equilibrium pressure is too low to maintain a good supply. Also, the alloy must release hydrogen quickly enough for the demand. The design of appropriate alloys is in the hands of chemists.

Obtaining Energy from Hydrogen

Combustion, either to give mechanical energy directly (as in our cars) or to heat water to steam which can drive a turbine to create electricity, is an option. But the efficiency of these methods is only about 15%—about 85% of the energy from combustion is wasted.

However, if a combustion reaction is made to happen in a *fuel cell*, the efficiency rises to over 50%. A fuel cell is a battery that, instead of storing reactive chemicals inside it, has the reactants (in this case, H_2 gas and O_2 gas) continuously passed into it from external containers. Canada's Ballard Power is a world leader in hydrogen-oxygen fuel cell technology. Their most modern fuel cells use proton exchange membrane (PEM) technology (Figure 7.2), which allows the passage of hydrogen ions but not electrons. Different reactions occur on the catalytic surfaces of each of two electrodes: oxidation of hydrogen at the anode, and reduction of oxygen at the cathode [>>e16.9].

$$\textit{Anode}: H_2(g) \longrightarrow 2\,H^+(aq) + 2\,e^-$$
$$\textit{Cathode}: \tfrac{1}{2}O_2(g) + 2\,H^+(aq) + 2\,e^- \longrightarrow H_2O(\ell)$$

Fuel cells for mass use are a relatively new technology, and major improvements through research can be expected. The aims of current PEM cell developmental research are lower production costs, longer lifetime of the cells, and quick winter starting. Imagine the economic benefits arising from a 5% improvement in the energy efficiency of PEM fuel cells used in millions of vehicles.

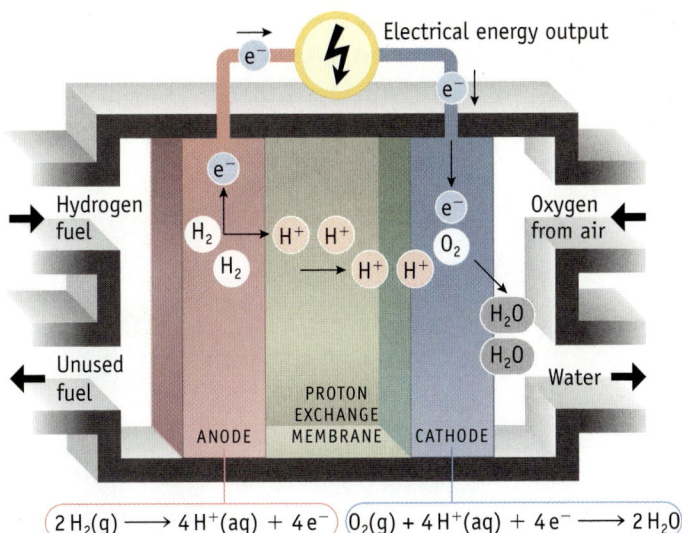

FIGURE 7.2 PEM fuel cell. The cell is maintained at 70 °C. At the anode, incoming $H_2(g)$ is oxidized to $H^+(aq)$ ions with the release of electrons. The $H^+(aq)$ ions can diffuse through the membrane to the cathode, where they are used up in the reduction of incoming $O_2(g)$ to form the only product, water. Electrons released at the anode cannot traverse the membrane. A flow of electrons (electricity) is directed through copper wires from the anode to cathode via electric motors which drive the vehicle.

In Conclusion . . .

The story of hydrogen as an energy source shows the many factors needed to obtain energy from chemical reactions at a commercial level. In a few years, some of you may play a part in the achievement of a hydrogen economy, in that way improving the quality of life of Earth's inhabitants. Human activity and chemical reactivity are also intertwined in the rest of this chapter, as we consider the amount of heat released or absorbed from many other reactions.

7.2   Chemical Changes and Energy Redistribution

Chemical changes are accompanied by redistributions of energy, as well of matter. There are many examples in our experience that demonstrate this. Most obvious of all, combustion reactions (of wood, paper, natural gas, auto fuel, coal) release energy as heat. When chemical reactions occur in batteries, electrical energy is supplied. We can often feel the heat released when a heap of vegetable matter decomposes. If we mix an acid solution and a base solution, the mixture feels hot, showing that heat has been released. On the other hand, if we dissolve ammonium nitrate in water, the solution feels cold.

Most importantly, but less obvious to our senses, we obtain energy for our bodies through the respiration reaction that is the net result of metabolism of foods, represented approximately by the equation

$$C_6H_{12}O_6(s) + 6\,O_2(g) \longrightarrow 6\,CO_2(g) + 6\,H_2O(g) + energy$$

Does the energy released during a chemical reaction depend on how much reacts? For a given amount of reaction, is the same amount of energy always released? Which chemical reactions have most potential use as energy sources? Why do some reactions release heat energy, while others cause absorption of heat? Of those that release heat, why do some release much more than others? These are the questions that define the subject of *thermochemistry*. To answer them, we need to make experimental measurements, such as the amount of energy made available when 1 mol of glucose reacts with oxygen. Thermochemistry is a quantitative subject.

By contrast, *thermodynamics* [>>Chapter 17] is the study of how energy considerations govern which chemical reactions can happen, and how far they go before reaching chemical equilibrium.

Exothermic and Endothermic Reactions

How can we account for the observations that heat is released during some reactions and is absorbed during others? If we accept that substances somehow store energy, there is a simple explanation. In those cases when the energy stored in the reactants is greater than that in the products, energy is left over, and this excess flows to the surroundings (Figure 7.3).

Energy Transfer in a Chemical Reaction

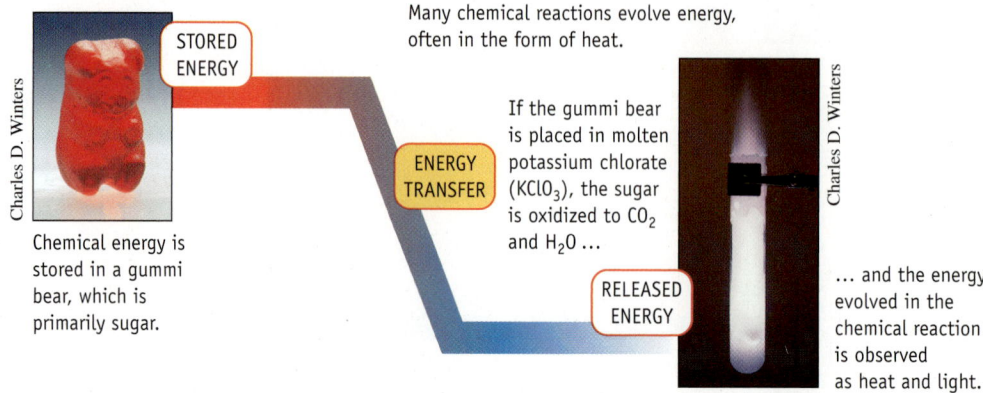

FIGURE 7.3 Energy released during chemical reaction. The energy stored in the reactants (sugar in the gummi bear and potassium chlorate) is more than that stored in the products (carbon dioxide and water), and the excess is released to the surroundings—in this case as both heat energy and light energy.

Think about It

e7.2 Watch a video of this reaction and the energy released when you consume a gummi bear.

FIGURE 7.4 Exothermic and endothermic processes. (a) In exothermic processes, the reactants have more energy than the products. (b) The opposite is the case in endothermic processes.

Processes that release energy as heat are called **exothermic** *processes*. If you are holding the reaction vessel, heat flows into your hand and it feels hot. **Endothermic** *processes are those in which the products have more energy than the reactants*, and the energy needed for the mixture to return to ambient temperature is taken as heat from the surroundings (Figure 7.4)—so the reaction vessel feels cold.

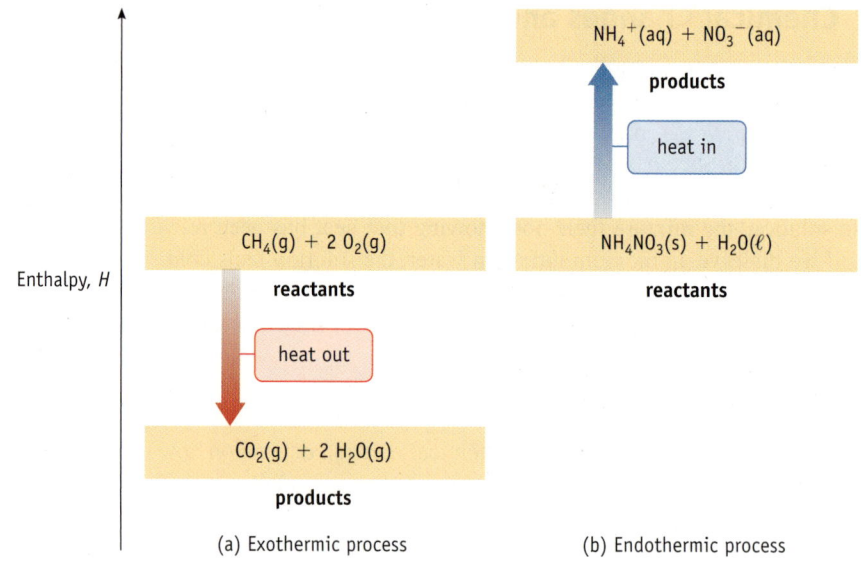

(a) Exothermic process   (b) Endothermic process

EXERCISE 7.1—EXOTHERMIC AND ENDOTHERMIC PROCESSES

Reaction of a nitric acid solution with a sodium hydroxide solution (called neutralization, and represented by the equation below) is an exothermic process.

$$H^+(aq) + OH^-(aq) \longrightarrow H_2O(\ell)$$

(a) Which has more energy: 1 mol of water or 1 mol each of $H^+(aq)$ ions and $OH^-(aq)$ ions that have not reacted?

(b) If you hold a test tube in which this reaction occurs, will your fingers feel hot or cold? Explain why.

In an exothermic process, the initial state has more energy than the final state and the excess energy is released as heat. In endothermic processes, the final state has more energy than the initial state.

7.3 Energy: Its Forms and Transformations

The concept of energy is not easy to visualize or to comprehend. We need to understand the nature of storage of energy in substances to make sense of the thermochemistry of reactions. When we talk about the energy of a sample of substance, we mean the energy contained within a sample and somehow related to movements or interactions of its atoms, molecules, or ions—and not to its energy as an object if, for example, the sample is thrown or dropped.

Energy Storage, Energy Interconversion

While we all have an everyday sense of what it means, **energy** can be defined as *the capacity to do work*. You do work when hiking up a mountain. This work is possible because of the provision of energy by metabolic reactions. Energy can be classified as *kinetic energy* or *potential energy*.

Kinetic energy *is energy due to motion*. In the context of thermochemistry, we are concerned with the movement of the atoms, molecules, or ions of which a substance is composed, including motion of translation, vibration, and rotation.

Potential energy *results from the position of objects*. We are concerned particularly with potential energy at the molecular level due to forces of attraction or repulsion between atoms, molecules, or ions, dependent on electrical charge distribution within them and their positions relative to each other.

If we heat some water in a beaker over a Bunsen burner (Figure 7.5), we can presume that the energy of the water increases. But where does the heat energy go? Part of the mystery of energy is that if we transfer some heat to water, it no longer exists as heat—unlike $1 coins, which if put into a jar, retain their identity. Certainly there are not small "packets" of heat floating amongst the water molecules! For comparison, if we shine white light into an aqueous copper sulfate solution, some red light is absorbed, but it seems obvious that the solution does not now contain light. Similarly, a solution that has absorbed heat does not contain heat.

FIGURE 7.5 Energy storage. Combustion of natural gas provides energy to the water in the beaker. Where does the energy go?

Charles D. Winters

Our current understanding is that the heat (thermal energy) is transformed to other forms of energy when absorbed by the water:

- The average kinetic energy of the molecules increases: they move on their trajectories more quickly, they vibrate more vigorously, and they tumble (rotate) faster.
- The potential energy of the collection of water molecules increases: the molecules become further apart, on average, against the attractions due to hydrogen bonds operating between the molecules [<<Section 6.3]. This is an increase in potential energy because work has to be done to increase the separation of the molecules, and they can do work in coming closer together again if the water is cooled.

The increase of temperature of water above a Bunsen burner can be described in terms of the energy stored in substances and conversion from one form of energy to another. The reactants in the burner flame are methane (in the case of natural gas) and oxygen in the air, and the products

Molecular Modelling (Odyssey)

e7.3 Examine simulations showing the relationship between kinetic and potential energy of vibrating molecules.

Molecular Modelling (Odyssey)

e7.4 Examine simulations illustrating components of kinetic energy.

Molecular Modelling (Odyssey)

e7.5 Use a simulation to see how kinetic energy depends on temperature.

Interactive Exercise (Odyssey) 7.2

Compare the kinetic energies of different substances.

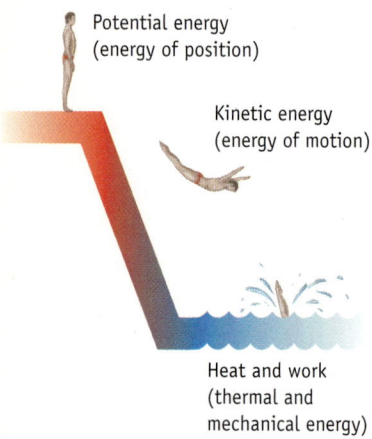

Potential energy
(energy of position)

Kinetic energy
(energy of motion)

Heat and work
(thermal and
mechanical energy)

FIGURE 7.6 Interconversion among different forms of energy. During the dive, potential energy decreases and kinetic energy increases. On entering the water, kinetic energy is converted to mechanical energy of waves, and eventually to slightly increased kinetic energy of the water molecules.

are mostly carbon dioxide and water vapour. The sum of the energies (both kinetic and potential) of $CH_4(g)$ and $O_2(g)$ is greater than the sum of the energies of the $CO_2(g)$ and $H_2O(g)$ formed, and the excess is converted to heat energy. Some of the heat energy released is absorbed by the water, where it is converted to kinetic and potential energy of the water molecules.

These energy conversions are just some of many examples of conversion from one form of energy to other forms. Energy conversions in a common physical situation are illustrated in Figure 7.6.

A mixture of methane and oxygen is said to have *chemical potential energy* because it has the capacity to do work by chemical reaction to form carbon dioxide and water. The heat released by reaction can be used to create steam in a turbine, which can then do work—such as generation of electrical energy. There are many examples of the energy of reacting substances (in excess of the energy of the products) being converted into other forms of energy—either in the laboratory (Figure 7.3) or outside of the laboratory (Figure 7.7).

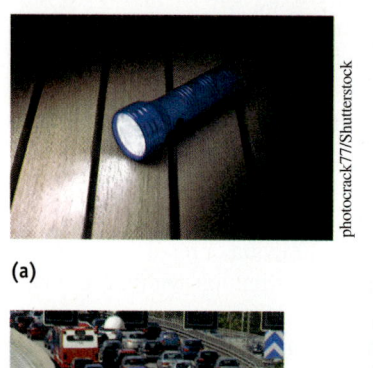

(a)

(b)

(c)

FIGURE 7.7 Conversion of chemical energy into other forms. (a) Excess energy of reactants (compared with that of products) in dry cells is converted to electrical energy, and then to light energy. (b) Excess energy of fuel/oxygen mixtures is converted to mechanical energy. (c) Excess energy of rocket fuel is converted to mechanical energy.

EXERCISE 7.3—ENERGY CONVERSION

A battery stores chemical potential energy. Into what types of energy can this potential energy be converted?

Energy is stored in substances as the kinetic energy and potential energy of its particles (atoms, ions, or molecules). Energy can be converted from one form to another.

Conservation of Energy, the First Law of Thermodynamics

The words *conserve* and *conservation* are used in relation to conserving food and conservation of the environment. In science, *conserve* means to maintain at a constant value.

An extraordinarily important principle is that although energy may be converted among various forms, the total amount of it neither lessens nor increases. It cannot appear, nor disappear, it is neither created nor destroyed. As a consequence, *the total energy of the universe is constant*. This is called the **law of conservation of energy** or the **first law of thermodynamics**.

In a car, the available chemical energy from the gasoline and oxygen reaction in the engine exactly matches the sum of the mechanical energy of motion of the car and other ineffective forms of energy, such as heat energy transferred to the surroundings and sound energy. The efficiency of the car engine, measured as the percentage of the chemical energy that is converted to the desirable form of energy, is much less than 100%. This is not because of disappearance of energy, but because some of the available energy is converted into forms that do not contribute to the operation of the car.

> Although energy can be converted from one form into another, the total amount of energy is constant.

Units of Energy Measurement

The unit for measuring energy is the joule (symbol J), defined as the energy used in lifting 1 metre against a force of 1 Newton. A Newton is the force that will give a 1 kg object an acceleration of 1 m s^{-2}. So $1 \text{ J} = 1 \text{ kg m}^2 \text{ s}^{-2}$. A kilojoule (kJ) is 1000 J.

To get some feeling for the magnitude of the joule, it is approximately the energy you would use to lift an object with mass 102 g (say, a small apple) by 1 m. If you drop a six-pack of full soft drink cans on your foot, the kinetic energy at the moment of impact is between 4 J and 10 J. Alternatively, a 50 kg person consumes 100 J of energy to go up a 20 cm step.

An older unit for measuring energy is the calorie (cal), which is the energy required to raise the temperature of 1.00 g of water by 1 Celsius degree. The conversion factor is 1 cal = 4.184 J.

The dietary Calorie (capital C) is often used to represent the energy content of foods when reading the nutritional information on labels. The dietary Calorie (Cal) is equivalent to 1000 calories (1 kcal). So, a breakfast cereal serving that provides 100 Cal of energy gives you 100 kcal or 418.4 kJ. To "burn off" a 100 Cal breakfast, a 50 kg person would need to climb about 840 m—1.5 times the height of the CN tower in Toronto. Of course, most of the energy obtained from our food goes into keeping us warm and maintaining body functions such as breathing and pumping blood, rather than into moving us around.

> The unit of measurement of energy is the joule (J).

Think about It

e7.6 Examine how the different energy units relate to each other.

7.4 Energy Flows between System and Surroundings

Chemical processes that occur in living systems, and most experiments in chemical laboratories carried out in beakers or flasks, are open to the atmosphere so that the pressure acting on them is essentially constant. The main focus of this chapter is a consideration of quantitative aspects of thermochemistry when changes occur under constant pressure conditions. First, some common terms need to be understood.

Temperature and Heat

The **temperature** *of an object refers to how hot it is*, and is the sole determinant of whether or not energy is transferred to another object in contact with it. Temperature is a measure of the average kinetic energy of its atoms, ions, or molecules. Temperature is an intensive property: it is independent of the amount of material. For example, a 5 mL sample and a 500 mL sample of water that have the same average kinetic energy of their molecules are at the same temperature.

Temperature can be estimated by measurement of any property of substances that is temperature dependent. In the mercury column thermometer, the volume of enclosed mercury increases as its temperature increases. The level of mercury in a thin tube of uniform diameter is calibrated to read temperature directly. Thermistors are devices whose electrical resistance varies with temperature. Measurement of the resistance, or of a property of a circuit dependent on the resistance, of a thermistor can be recorded by a digital device that is calibrated to display temperature directly.

Measurement scales of temperature are arbitrary: we can decide on any scale we wish. On the Celsius scale, 0 °C is set as the freezing point of water, 100 °C is set as the boiling point of water under 1 atm pressure, and the range of temperatures in between is divided into 100 equal intervals. On the Kelvin scale, the zero (0 K, called *absolute zero*) is the minimum possible temperature, and is the temperature at which atoms or molecules have no kinetic energy of translation. This corresponds with −273.16 °C. Units on the Kelvin scale are the same size as those on the Celsius scale.

The term **heat** strictly refers to *energy flowing from one object to another*, or from one part of an object to another. Although an object does not contain heat, it is common to say that it contains *thermal energy*. For example, it is often said that coal is a store of large amounts of thermal energy. This does not necessarily mean that the substance coal has lots of internal energy; rather this should be interpreted to mean that combustion of coal can supply large amounts of heat. We will see that this simply means that the sum of the energies of reacting amounts of coal and oxygen are far greater than the sum of the energies of the products of combustion (mainly carbon dioxide and water), and the excess energy is transferred to the surroundings as heat.

> The temperature of a substance refers to its level of hotness, and is a measure of the average kinetic energy of its particles. Heat is energy in transit.

Direction of Heat Transfer: Thermal Equilibrium

When two objects at different temperatures are brought into contact, *energy flows from the object with higher temperature to that with lower temperature*. In Figure 7.8, for example, the beaker of water and the piece of metal being heated have different temperatures. When the hot metal is plunged into the cold water, heat flows from the metal to the water until the two objects reach the same temperature. Then, the metal and water have reached *thermal equilibrium*. The distinguishing feature of thermal equilibrium is that no further temperature change occurs and the temperature throughout the entire system (metal and water) is the same.

FIGURE 7.8 Achieving thermal equilibrium. When the metal is totally submerged, energy is transferred as heat from the hot piece of metal to the cooler water until both are at the same temperature.

Charles D. Winters

Consistent with the law of conservation of energy, the amount of energy lost by a hotter object is the same as that gained by a cooler object in contact with it.

Although no change is evident at the macroscopic level when thermal equilibrium is reached, on the molecular level transfer of energy between individual molecules will continue to occur. This feature—no visible change on a macroscopic level, but processes still occurring at the molecular level—is a general feature of equilibrium situations [>>Chapter 14].

Think about It

e7.7 Watch a demonstration of heat transferred between substances.

EXERCISE 7.4—TEMPERATURE AND DIRECTION OF HEAT FLOW

The energy of the air in a classroom at 25 °C is much greater than the energy in a 50 mL beaker of water at 35 °C on a table in that classroom. In which direction does heat flow: from the air to the water, or from the water to the air?

Heat flows in the direction from an object at higher temperature to an object at lower temperature, until the objects are at the same temperature.

The System, the Surroundings, the Universe

Scientists use the term *system* to refer to the object, or collection of objects, being studied; that is, whatever we are focusing on (Figure 7.9). The *surroundings* include everything outside the system that can exchange energy with the system.

If we are studying the heat evolved during a chemical reaction, for example, the system might be defined as the reaction mixture. The reaction vessel, the air in the room, and anything else in contact with the vessel are the surroundings. At the molecular level, we might define a single atom or molecule as our system, while the surroundings are the atoms or molecules in its vicinity. On any occasion, we define the system and its surroundings to suit our purposes.

In principle, the surroundings include all of the universe that is not the system; that is,

$$system + surroundings = universe$$

For practical purposes, we usually need only consider those parts of the surroundings near and relevant to the system.

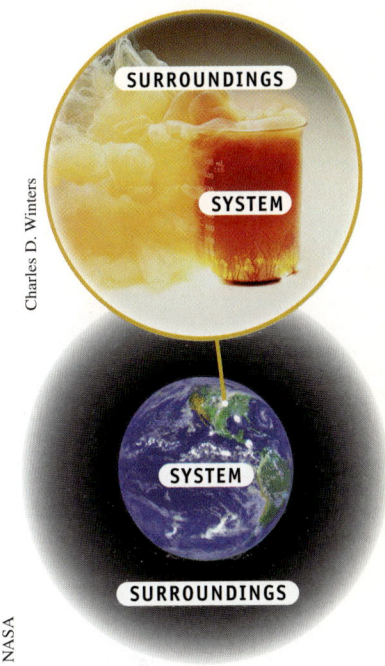

Charles D. Winters

NASA

FIGURE 7.9 Systems and their surroundings. Earth can be our system, with the rest of the universe as its surroundings. A chemical reaction occurring in a laboratory can also be our system, with the laboratory and the rest of the universe as its surroundings.

EXERCISE 7.5—SYSTEM AND SURROUNDINGS

For each of the following, define a system and its surroundings and give the direction of heat transfer between system and surroundings.
(a) Methane is burning in a gas oven in your home.
(b) Water drops, sitting on your skin after a dip in a swimming pool, evaporate.
(c) Water, at 25 °C, is placed in the freezing compartment of a refrigerator, where it cools and eventually solidifies.
(d) Aluminum and iron(III) oxide, $Fe_2O_3(s)$, are mixed in a flask sitting on a laboratory bench. A reaction occurs, and a large quantity of heat is evolved.

The object of focus, such as a reaction mixture, is called the system. The rest of the universe comprises the surroundings.

Internal Energy

The "energy content" of a system, *the sum of the kinetic energy and potential energy of the collection of atoms, ions, and molecules in the system*, is called the **internal energy** (*U*). Internal energy is an extensive property: the internal energy of a substance at a given temperature depends on the amount of it. So, 100 mL of water at 20 °C has ten times the internal energy of 10 mL of water at 20 °C. On the other hand, a cup of hot water at 70 °C has less internal energy than a bathtub of warm water at 40 °C—even though a 10 mL sample of water from the cup has more internal energy than a 10 mL sample from the bath.

Internal energy is said to be a **state function**: *for a given substance, its magnitude depends only on the amount of the substance and its conditions (pressure and temperature), and is independent of how the conditions were achieved or the previous history of the substance.* So, the internal energy of a 1 L sample of water at 1 bar pressure and 20 °C in Toronto is exactly the same as the internal energy of a 1 L sample of water at 1 bar pressure and 20 °C in Sydney, regardless of how the samples were made, and what the pressure and temperature of the samples have been at any time in the past. As we will see, this has important consequences.

We cannot measure exactly the internal energy of a system. This matters little, however, because it is more important in thermochemistry to know the change of internal energy of a system (ΔU), when a change (such as a chemical reaction) occurs. Conceptually, the internal energy change is defined as

$$\Delta U(system) = U(system)_{final} - U(system)_{initial}$$

Even though the internal energy of the system before and after the change is not known, it is relatively easy to measure the change ΔU [>>e7.9].

During exothermic processes, the internal energy of the system decreases, $U_{final} < U_{initial}$, so $\Delta U_{system} < 0$, and the numerical value of ΔH is negative. Conversely, in endothermic processes, $U_{final} > U_{initial}$, $\Delta U > 0$.

EXERCISE 7.6—STATE FUNCTIONS

Which of the following are state functions?
(a) the volume of a balloon
(b) the time it takes to drive from your home to your college or university
(c) the temperature of the water in a coffee cup
(d) the potential energy of a ball held in your hand

> The internal energy (U) of a system is the sum of the kinetic energy and potential energy of the collection of atoms, ions, and molecules in the system. It is a state function. Although the internal energy of a system cannot be measured, in thermochemistry we are more interested in the change of internal energy accompanying changes to the system.

Heat and Work: Different Forms of Energy Transfer

During a change such as a chemical reaction, the internal energy of a system may decrease in either, or both, of two ways:

- Energy is transferred as heat to the surroundings, bringing about an increase of temperature of the surroundings. The symbol q is used for the amount of heat transferred.
- Energy is transferred to the surroundings that brings about an increase of potential energy of the surroundings by doing *work* (w) on the surroundings—for example, the mechanical work of driving a car forward (if the system is the combustion chamber) or the electrical work used in operating a portable DVD player (if the system is a battery). If a system does work on its surroundings, energy must be expended and the internal energy of the system decreases. Conversely, if work is done by the surroundings on a system, the internal energy of the system increases.

The first law of thermodynamics requires that the change in the internal energy of a system is the sum of the energy transferred as heat and work. This is expressed by the general equation

$$\Delta U_{system} = q + w$$

This equation tells us that when an exothermic reaction occurs, the more of the excess internal energy that is used to do work on the surroundings, the less heat is transferred to the surroundings.

If we are referring only to an amount of heat, we do not need to give a sign to q. However, if we are doing "accounting" on the internal energy of a system, keeping records of energy in and energy out, there is a convention for use of signs: q and w are given positive values (> 0) if they increase the internal energy of the system.

Sign Convention for Energy Transfer as Heat (q) and Work (w)

Direction of Energy Transfer	Sign Convention	Effect on U_{system}
Heat flow from surroundings to system	$q > 0$ (+)	U increases, $\Delta U > 0$
Heat flow from system to surroundings	$q < 0$ (−)	U decreases, $\Delta U < 0$
Work done on system by surroundings	$w > 0$ (+)	U increases, $\Delta U > 0$
Work done on surroundings by system	$w < 0$ (−)	U decreases, $\Delta U < 0$

In cases where no work is done by or on the surroundings, $w = 0$, and so $\Delta U = q$. This provides the basis for measurement of ΔU_{system} accompanying a change: if no work is done, the change of internal energy is equal to the amount of heat transferred. This is the situation when the volume of the system is kept constant, and no external appliances are driven by the change that occurs [>>e7.9].

> The internal energy of a system may decrease either by transfer of heat to the surroundings or by work done on the surroundings. $\Delta U = q + w$. When $w = 0$, $\Delta U = q$.

Enthalpy and Enthalpy Change

In this chapter, our main focus is on changes that occur under conditions of constant pressure, as is the case for processes open to the atmosphere. If the pressure is kept constant while a change occurs, usually the volume of the system changes. For example, if a gas is produced by a chemical reaction, to maintain constant pressure, the volume increases. In such a case, work is done by the system in order to expand against the confining constant pressure. Supposing that the reaction is endothermic, the increase in the internal energy of the system is less than the amount of heat energy, q, transferred to it because some of that heat energy is used up to do the work of expansion.

Instead of having to make a correction for the work of expansion, it is convenient to introduce a new property of the system called **enthalpy** (**H**). The **enthalpy change** (ΔH) of a system *is equal to the amount of heat transferred between the system and surroundings during a process that occurs at constant pressure* (q_p), *if no work other than that due to expansion of the system occurs*. Under these conditions,

$$\Delta H_{system} = q_p$$

The difference between ΔH and ΔU of a system undergoing change at constant pressure is the amount of expansion work done on the surroundings (see e7.9). From now on we will only use enthalpy and enthalpy changes in our consideration of processes occurring at constant pressure.

Enthalpy is a state function that is an extensive property. For a process occurring at constant pressure, we can conceptualize the enthalpy change by

$$\Delta H(system) = H(system)_{final} - H(system)_{initial}$$

If we are clear that we are referring to the enthalpy of the *system* (and not of the surroundings or the universe), we can simplify this:

$$\Delta H = H_{final} - H_{initial}$$

For exothermic processes at constant pressure, $H_{final} < H_{initial}$, so $\Delta H < 0$. In endothermic processes, $H_{final} > H_{initial}$, and $\Delta H > 0$.

We cannot know nor measure the enthalpy of a system before and after a change in order to calculate ΔH, but we will see in Section 7.6 that it is relatively easy to experimentally measure enthalpy changes accompanying processes such as chemical reactions.

> The enthalpy (H) of a system changes by the amount of heat transferred between the system and surroundings during a process that occurs at constant pressure (q_p), if no work other than that due to expansion of the system occurs ($\Delta H = q_p$). For exothermic processes at constant pressure, $\Delta H < 0$; for endothermic processes, $\Delta H > 0$.

Taking It Further

e7.8 Compare energy flows at constant pressure and constant volume.

7.5 Enthalpy Changes Accompanying Changes of State

Heat must be provided to a solid at its melting point while melting occurs. This can be attributed to an increase in the potential energy of the atoms, molecules, or ions as they break free of the constraints of attractions to neighbouring atoms, molecules, or ions in the solid state. *The change of enthalpy of 1 mol of a solid substance during its conversion to a liquid at its melting point is called the* **molar enthalpy change of fusion** ($\Delta_{fus}H$). In the case of water, melting is represented by the equation

$$H_2O(s) \longrightarrow H_2O(\ell) \qquad \Delta_{fus}H = 6.00 \text{ kJ mol}^{-1}$$

This is an endothermic process: if 1 mol of ice melts at 0 °C, 6.00 kJ of energy is absorbed from the surroundings. If 1 mol of water freezes at 0 °C ($\Delta_{fus}H = -6.00 \text{ kJ mol}^{-1}$), 6.00 kJ of energy is released to the surroundings.

At the melting point, the average kinetic energy of molecules in the solid state is the same as in the liquid, although the form of the motion may largely change from vibrations to translation.

Fusion is another word for melting.

WORKED EXAMPLE 7.1—ENTHALPY CHANGE AND CHANGE OF STATE

How much heat is absorbed when 500 g of ice melts at 0 °C?

Strategy

Multiply the molar enthalpy change of fusion by the amount (in mol) that melt.

Solution

$$\text{Energy absorbed, } q = 6.00 \text{ kJ mol}^{-1} \times \frac{500 \text{ g}}{18.02 \text{ g mol}^{-1}} = 167 \text{ kJ}$$

EXERCISE 7.7—ENTHALPY CHANGE AND CHANGE OF STATE

How much heat is evolved when 1.0 L of water at 0 °C freezes to form ice at 0 °C?

The heat taken in when some liquid is converted to its vapour phase at its boiling point is called the enthalpy change of vaporization [<<Section 6.2]. For 1 mol of substance, this is the **molar enthalpy change of vaporization** ($\Delta_{vap}H$). For water, $\Delta_{vap}H$ at 100 °C is 40.65 kJ mol^{-1}. The heat taken in when 500 g of water is converted at 100 °C to water vapour is

$$q = 40.65 \text{ kJ mol}^{-1} \times \frac{500 \text{ g}}{18.02 \text{ g mol}^{-1}} = 1128 \text{ kJ}$$

Figure 7.10 graphs the cumulative heat absorbed as the temperature of 500 g of water is raised from −50 °C (ice) to 200 °C (water vapour). First, the temperature of the ice increases as heat is added. At 0 °C, the temperature remains constant as sufficient heat (167 kJ) is absorbed to melt the ice. When all the ice has melted, the liquid absorbs heat and the temperature rises to 100 °C. The temperature again remains constant as enough heat is absorbed (1128 kJ) to convert the liquid completely to vapour. Further heat absorption raises the temperature of the water vapour.

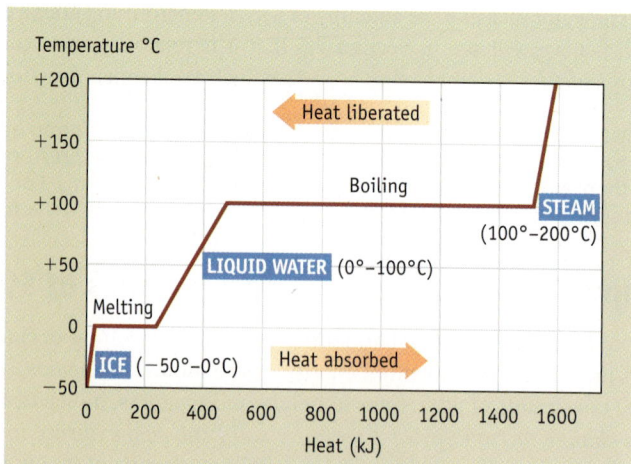

FIGURE 7.10 Heat absorption and temperature change of water. This graph shows the quantity of heat absorbed when 500 g of water is warmed from −50 °C (ice) to 200 °C. During each change of state, the added energy is used to overcome the forces attracting molecule to each other, not to increase the temperature of the substance. The different lengths of the constant-temperature sections reflect the difference between $\Delta_{fus}H$ and $\Delta_{vap}H$. The different gradients of the sloping portions of the graph are due to the different specific heats of ice, liquid water, and water vapour.

WORKED EXAMPLE 7.2—ENTHALPY CHANGE AND CHANGE OF STATE

Calculate the quantity of heat involved in each step shown in Figure 7.10 and the total quantity of heat required to convert 500 g of ice at –50 °C to vapour at 200 °C. $\Delta_{fus}H$ at 0 °C is 6.00 kJ mol^{-1}, and $\Delta_{vap}H$ at 100 °C is 40.65 kJ mol^{-1}. The specific heat capacity of ice is 2.06 J K^{-1} g^{-1}, that of liquid water is 4.184 J K^{-1} g^{-1}, and that of water vapour is 1.92 J K^{-1} g^{-1}.

Strategy

The problem is broken down into a series of steps: (1) warm the ice from −50 °C to 0 °C; (2) melt the ice at 0 °C; (3) raise the temperature of the liquid water from 0 °C to 100 °C; (4) evaporate the water at 100 °C; (5) raise the temperature of the vapour from 100 °C to 200 °C. For steps 2 and 4, use the values of $\Delta_{fus}H$ and $\Delta_{vap}H$ respectively (see Worked Example 7.1). Remembering that a Celsius degree is the same size as a Kelvin, for steps 1, 3, and 5, calculate the quantity of heat absorbed (q) by the formula based on the definition of specific heat (c) [<<Section 6.2]:

$$q = m \times c \times \Delta T$$

Solution

Step 1

q_1(ice through $\Delta T = 50$ K) $= (500 \text{ g})(2.06 \text{ J K}^{-1} \text{ g}^{-1})(50 \text{ K}) = 51\,500 \text{ J} = 51.5 \text{ kJ}$

Step 2 q_2(to melt ice at 0 °C) $= 6.00 \text{ kJ mol}^{-1} \times \dfrac{500 \text{ g}}{18.02 \text{ g mol}^{-1}} = 167 \text{ kJ}$

Step 3

q_3(water through $\Delta T = 100$ K) $= (500 \text{ g})(4.184 \text{ J K}^{-1} \text{ g}^{-1})(100 \text{ K}) = 209\,000 \text{ J} = 209 \text{ kJ}$

Step 4 q_4(to evaporate water at 100 °C) $= 40.65 \text{ kJ mol}^{-1} \times \dfrac{500 \text{ g}}{18.02 \text{ g mol}^{-1}} = 1128 \text{ kJ}$

Step 5

q_5(vapour through $\Delta T = 100$ K) $= (500 \text{ g})(1.92 \text{ J K}^{-1} \text{ g}^{-1})(100 \text{ K}) = 96\,000 \text{ J} = 96 \text{ kJ}$

Total heat absorbed, $q_{total} = q_1 + q_2 + q_3 + q_4 + q_5 = 1652 \text{ kJ}$

Comment

The conversion of liquid to vapour is by far the largest increment of energy. You may have noticed that the time taken to heat water to boiling on a stove is much less than the time to boil off the water.

EXERCISE 7.10—ENTHALPY CHANGE AND CHANGE OF STATE

How much heat must be absorbed to warm 25.0 g of liquid methanol (CH_3OH, M = 32.04 g mol^{-1}) from 25.0 °C to its boiling point (64.6 °C) and then to evaporate the methanol completely at that temperature? The specific heat capacity of liquid methanol is 2.53 J K^{-1} g^{-1}. For methanol, $\Delta_{vap}H$ is 37.4 kJ mol^{-1}.

Interactive Exercises 7.8–7.9

SUBMIT

Calculate the heat absorbed during vaporization and lost during cooling.

EXERCISE 7.11—ENTHALPY CHANGE AND CHANGE OF STATE

How much heat is released to the surroundings if 1.00 mL of mercury at 23.0 °C is transformed to solid at its freezing point, −38.8 °C? The density of liquid mercury is 13.6 g mL^{-1}, its specific heat capacity is 0.140 J K^{-1} g^{-1}, and $\Delta_{vap}H$ is 2.3 kJ mol^{-1}.

The amount of energy absorbed when 1 mol of a solid melts at its melting point is called the molar enthalpy change of fusion ($\Delta_{fus}H$). The energy absorbed when 1 mol of a liquid boils at its boiling point is called the molar enthalpy change of vaporization ($\Delta_{vap}H$).

7.6 Enthalpy Change of Reaction ($\Delta_r H$)

In cases where our system is a reaction mixture, the chemical reaction that occurs is accompanied by either an increase or a decrease in the enthalpy of the system. This change, called the **enthalpy change of reaction ($\Delta_r H$)**, *is the difference between the sum of the enthalpies of the products and the sum of the enthalpies of the reactants.* This is defined according to convention as

$$\Delta_r H = \sum H_{products} - \sum H_{reactants}$$

For example, the combustion of 1 mol of methane can be represented as

$$CH_4(g) + 2\,O_2(g) \longrightarrow CO_2(g) + 2\,H_2O(g)$$

If we define H_m as the molar enthalpy of a substance (the enthalpy of 1 mol of it) then *conceptually* $\Delta_r H$ is given by

$$\Delta_r H = [H_m(CO_2,g) + 2 \times H_m(H_2O,g)] - [H_m(CH_4,g) + 2 \times H_m(O_2,g)]$$

While this relationship is valid in principle, and can help us understand the meaning of $\Delta_r H$, in fact it cannot be used to estimate $\Delta_r H$ because values of H_m of reactants and products cannot be measured. The magnitude of $\Delta_r H$ can be estimated experimentally in a calorimeter, as described later in this section, or from values of *enthalpies of formation* of the reactants and products [>>Section 7.8].

The relationships for exothermic and endothermic chemical reactions are displayed in the following table and portrayed in Figure 7.11.

Exothermic reactions	$\sum H_{products} < \sum H_{reactants}$	$\Delta_r H < 0$	− value
Endothermic reactions	$\sum H_{products} > \sum H_{reactants}$	$\Delta_r H > 0$	+ value

FIGURE 7.11 Enthalpy diagrams for exothermic and endothermic reactions. The diagrams qualitatively show the relative magnitudes of the total enthalpies of reactants and products in each case.

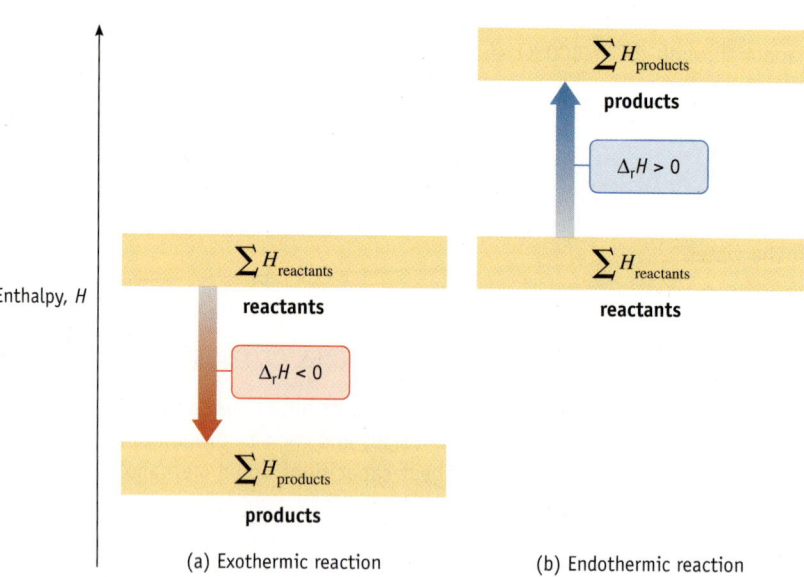

(a) Exothermic reaction　　(b) Endothermic reaction

The enthalpy change of systems in which chemical reactions occur is generally the net result of enthalpy increases due to breaking of both covalent bonds and intermolecular forces in the reactants, and decreases due to formation of new covalent bonds and intermolecular forces in the products. In Section 7.9, enthalpy changes of reaction are estimated from average values of bond energies between atoms.

The enthalpy change of a reaction mixture can be conceptualized as $\Delta_r H = \sum H_{products} - \sum H_{reactants}$. However, $\Delta_r H$ cannot be estimated in this way because the enthalpies of substances cannot be measured.

Quantitative Relationships

A value of the enthalpy change of reaction must specify the amounts of substances that react to bring about that change. A convention is to use a *thermochemical equation*, which links the value of $\Delta_r H$ to the amounts of reactants and products implied by a balanced chemical equation. For example, we can write

$$CH_4(g) + 2 O_2(g) \longrightarrow CO_2(g) + 2 H_2O(g) \qquad \Delta_r H = -890.3 \, kJ \, mol^{-1}$$

890.3 kJ per mole of what? This thermochemical equation is taken to mean that 890.3 kJ of energy is released when 1 mol of methane and 2 mol of oxygen react to form 1 mol of carbon dioxide and 2 mol of water vapour. Technically, this can be thought of as the enthalpy change for reaction of 1 mol of 'packets' of molecules, each packet containing one CH_4 molecule and two O_2 molecules.

There are some important principles concerning the signs and magnitudes of enthalpy changes of reactions:

1. The sign and magnitude of enthalpy change of reaction differs from reaction to reaction. This is obvious from comparison of $\Delta_r H$ for combustion of methane (above) with that for decomposition of limestone, CaCO3(s):

$$CaCO_3(s) \longrightarrow CaO(s) + CO_2(g) \qquad \Delta_r H = +179.0 \, kJ \, mol^{-1}$$

2. For a given reaction carried out under the same conditions (pressure, temperature, states of the reactants and products), the enthalpy change of reaction is always the same. For example, the enthalpy change accompanying combustion of methane at 1 bar pressure and at 25 °C is always $-890.3 \, kJ \, mol^{-1}$, regardless of where and when the reaction occurs. This is a necessary consequence of the definition of enthalpy change of reaction, and the fact that the enthalpies of the reactants and products are state functions. Indeed, this means that for any reaction occurring under a specified set of conditions, $\Delta_r H$ is a state function.

3. The magnitude of the enthalpy change of a reaction is proportional to the amounts of substances that react. $\Delta_r H$ for combustion of 2 mol of methane gas is $(2 \times -890.3 \, kJ \, mol^{-1}) = -1780.6 \, kJ \, mol^{-1}$.

4. For chemical reactions that are the reverse of each other, the magnitude of the enthalpy change is the same, although the sign is reversed. That is, if a reaction that proceeds in a given direction is exothermic, in the opposite direction it is endothermic to the same degree:

$$CO_2(g) + 2 H_2O(g) \longrightarrow CH_4(g) + 2 O_2(g) \qquad \Delta_r H = +890.3 \, kJ \, mol^{-1}$$

Molecular Modelling (Odyssey)

e7.9 Examine how energy changes in reactions can be modelled by computer.

WORKED EXAMPLE 7.3—ENTHALPY CHANGE OF REACTION

Combustion of propane, $C_3H_8(g)$, is represented by the following equation:

$$C_3H_8(g) + 5 O_2(g) \longrightarrow 3 CO_2(g) + 4 H_2O(\ell) \qquad \Delta_r H = -2220 \, kJ \, mol^{-1}$$

What is the change of enthalpy of the system (the reaction mixture) if 454 g of propane burns? Assuming that the system does no work, how much heat is transferred between the system and surroundings? In which direction?

Strategy

First, calculate the amount (in mol) of propane that burns. Then calculate by simple proportion the enthalpy change when 454 g mol of propane burns, from the enthalpy change when 1 mol burns.

Solution

$M(C_3H_8) = 44.10 \, g \, mol^{-1}$, so the amount of propane that burns is

$$n(C_3H_8) = \frac{m}{M} = \frac{454 \, \cancel{g}}{44.10 \, \cancel{g} \, mol^{-1}} = 10.3 \, mol$$

Now we multiply the enthalpy change per mol of propane burned by the amount of propane that burns:

$$\Delta_r H \text{ for } 10.3 \text{ mol} = 10.3 \cancel{\text{ mol}} \times -2220 \text{ kJ } \cancel{\text{mol}^{-1}} = -22\ 900 \text{ kJ}$$

Since no work is done, the amount of heat transferred is the same as the change of enthalpy of the system—that is, 22 900 kJ.

Since the enthalpy of the system decreases ($\Delta_r H$ is negative), the excess enthalpy is transferred as heat from the system to the surroundings—that is, the reaction is exothermic.

EXERCISE 7.12—ENTHALPY CHANGE OF REACTION

Nitrogen monoxide (NO), a gas recently found to be involved in a wide range of biological processes, reacts with oxygen gas to give brown NO_2 gas:

$$2 \text{ NO(g)} + O_2\text{(g)} \longrightarrow 2 \text{ NO}_2\text{(g)} \qquad \Delta_r H = -114.1 \text{ kJ mol}^{-1}$$

Is this reaction endothermic or exothermic? If 1.25 g of NO is converted completely to NO_2, how much heat is absorbed or evolved?

> For a given reaction, enthalpy change depends on the conditions and states of the reactants and products, and is proportional to the amounts that react. Reactions that are the reverse of each other have the same magnitude, but opposite sign, of $\Delta_r H$.

Measurement of Enthalpy Change of Reaction: Calorimetry

Although it is not possible to measure the exact enthalpy of substances, it is quite easy to experimentally measure the enthalpy change accompanying a chemical reaction using a device called a *calorimeter*. The measurement technique is called **calorimetry**.

The simplest example of a calorimeter for measuring enthalpy change of reaction under constant atmospheric pressure, the "coffee cup calorimeter" (Figure 7.12), consists of two nested Styrofoam coffee cups with a loose-fitting lid and a thermometer. Two solutions, each containing one of the reactants at the same temperature, are poured into the inner cup and stirred. The initial temperature of the solutions, the amounts (in mol) of the reactants, and the mass and specific heat capacity of the final solution must be known. If the reaction is exothermic, heat is evolved and the increase in temperature of the solution can be measured. In the case of endothermic reactions, the decrease in temperature is measured.

From the temperature change (ΔT), as well as the mass (m) and specific heat capacity of the solution (c), the amount of heat transferred (q) to or from the solution can be calculated [<<Section 6.2]:

$$q = c \times m \times \Delta T$$

From the known amounts (in mol) of species that react, it is then possible to calculate the amount of heat transferred per mol of specified reactant or product.

The accuracy of a measurement by calorimeter depends on the accuracy of measuring quantities such as temperature, mass, specific heat capacity, and amounts of reactants. In addition, it depends on how well the nested coffee cups prevent heat transfer between the calorimeter and its surroundings. A coffee cup calorimeter is an unsophisticated apparatus and the results obtained with it are not highly accurate, largely because some heat is transferred to or from the surroundings. In research laboratories, scientists utilize calorimeters that more effectively limit the heat transfer between

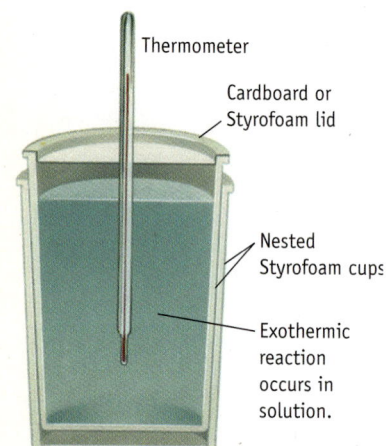

Thermometer

Cardboard or Styrofoam lid

Nested Styrofoam cups

Exothermic reaction occurs in solution.

FIGURE 7.12 A coffee cup calorimeter. This can measure enthalpy change of reaction under constant (atmospheric) pressure. The Styrofoam container is quite effective at preventing transfer of heat to or from the surrounding air.

Think about It

e7.10 Watch a demonstration of measuring a change in enthalpy using calorimetry.

Think about It

e7.11 Watch a demonstration of measuring a change in enthalpy using calorimetry.

calorimeter and surroundings, and they may also estimate and correct for any minimal heat transfer that does occur.

WORKED EXAMPLE 7.4—USING A CALORIMETER TO MEASURE Δ_rH

Suppose you place 0.500 g of magnesium pieces in a coffee cup calorimeter and add 100.0 mL of 1.00 mol L^{-1} HCl solution. The reaction that occurs is

$$Mg(s) + 2\,H^+(aq) \longrightarrow Mg^{2+}(aq) + H_2(g)$$

The temperature of the solution increases from 22.2 °C (295.4 K) to 44.8 °C (318.0 K). What is the enthalpy change for the reaction per mol of Mg that reacts? (Assume that the specific heat capacity of the solution is 4.20 J K^{-1} g^{-1} and the density of the HCl solution is 1.00 g mL^{-1}. Assume also that no energy is absorbed by the calorimeter.)

Strategy

We need to assume, or to demonstrate by calculation, that the magnesium is the limiting reactant [<<Section 5.7]. The amount of Mg(s) is 0.0206 mol and the amount of H$^+$(aq) ions is 0.100 mol—more than the 0.0412 mol needed to react with all of the Mg(s). Step 1 uses the mass of solution, the temperature increase, and its specific heat capacity [<<Section 6.2] to calculate the quantity of heat transferred to the solution (and evolved by reaction of 0.500 mol of Mg(s)). Step 2 calculates the quantity of heat that would be evolved if 1 mol of Mg(s) reacted—that is, Δ_rH.

Solution

Step 1   Calculate the quantity of heat evolved when 0.500 g of magnesium reacts.

$$\text{Mass of solution, } m = \text{mass of HCl solution} + \text{mass of magnesium}$$
$$= 100.0\text{ g} + 0.500\text{ g} = 100.5\text{ g}$$

We will use q = quantity of heat evolved by reaction = heat absorbed by solution

$$q \text{ when } 0.500\text{ g Mg(s) reacts} = c \times m \times \Delta T$$
$$= (4.20\text{ J K}^{-1}\text{ g}^{-1})(100.5\text{ g})(22.6\text{ K}) = 9.54 \times 10^3\text{ J}$$

Step 2   Calculate Δ_rH, the heat evolved if 1 mol of magnesium were to react. The molar mass of magnesium is 24.31 g mol^{-1}. That is, 1 mol has mass 24.31 g. So the heat evolved when 1 mol of Mg(s) reacts is given by

$$q \text{ when 1 mol of Mg(s) reacts} = (9.54 \times 10^3\text{ J}) \times \frac{24.31\text{ g}}{0.500\text{ g}} = 4.64 \times 10^5\text{ J (or 464 kJ)}$$

Since the reaction is obviously exothermic (and $\Delta_rH < 0$), we can now say

$$\Delta_rH = -464\text{ kJ mol}^{-1} \text{ (of Mg that reacts)}$$

EXERCISE 7.15—USING A CALORIMETER TO MEASURE Δ_rH

Assume that you mix 200 mL of 0.400 mol L^{-1} HCl solution with 200 mL of 0.400 mol L^{-1} NaOH solution in a coffee cup calorimeter. The temperature of both solutions before mixing was 25.10 °C. After mixing and reaction, the temperature becomes 27.78 °C. What is the molar enthalpy change of neutralization of the acid? (Assume that the densities of all solutions are 1.00 g mL^{-1} and their specific heat capacities are 4.20 J K^{-1} g^{-1}.)

Calorimetry is the measurement of enthalpy change of reaction by measurement of the quantity of heat evolved when a known amount of substances react.

Standard States, Standard Enthalpy Change of Reaction

At a given temperature, the magnitude of the enthalpy change of a reaction ($\Delta_r H$) depends on the pressure, the concentrations of the reactants and products in the reaction mixture, the physical state (solid, liquid, or gas) of each, as well as the form each of the substances is in (where more than one form exists). For example, the magnitude of $\Delta_r H$ for the combustion reaction

$$CH_4(g) + 2\,O_2(g) \longrightarrow CO_2(g) + 2\,H_2O(\ell \text{ or } g)$$

depends upon whether the water formed is in the liquid state ($\Delta_r H = -890.3$ kJ mol^{-1}) or the vapour state ($\Delta_r H = -802.2$ kJ mol^{-1}). Also, the magnitude of $\Delta_r H$ for the reaction

$$C(s) + O_2(g) \longrightarrow CO_2(g)$$

depends upon whether the carbon is in the form of diamond, graphite, or buckminsterfullerene (C_{60}).

To be able to specify the enthalpy change of a reaction with certainty, and to be able to compare $\Delta_r H$ values of different reactions, we need to specify the conditions. This is done by defining the **standard states** of the reactants and products:

- For pure substances, the standard state is its most stable form and state at 1 bar pressure at the temperature of interest. For example, the standard state of carbon at 25 °C is solid graphite (not diamond nor C_{60}), and not carbon vapour.
- For any gas, its standard state is when it is present at a pressure of 1 bar.
- For a species in aqueous solution, its standard state is when its concentration is exactly 1 mol L^{-1}, at a pressure of 1 bar.

For reaction of a specified amount of reactant at any specified temperature, the **standard enthalpy change of reaction** ($\Delta_r H°$) *is the enthalpy change of reaction when all the reactants and products are in their standard states.* For example, let's consider the decomposition of calcium carbonate to calcium oxide and carbon dioxide, for which $\Delta_r H°$ at 25 °C is +179.0 kJ mol^{-1}. At 25 °C and 1 bar pressure, the most stable form of calcium carbonate is the solid crystalline form calcite, the most stable form of calcium oxide is the solid, and the most stable form of carbon dioxide is the gas. So the thermochemical equation defining the conditions under which $\Delta_r H$ at 25 °C is the standard enthalpy change of reaction ($\Delta_r H°$) is

$$CaCO_3(s, \text{calcite}) \xrightarrow{25\,°C} CaO(s) + CO_2(g, 1\text{ bar}) \qquad \Delta_r H = \Delta_r H° = +179.0 \text{ kJ mol}^{-1}$$

Although *the enthalpy change of reaction depends on the temperature at which it takes place*, the standard enthalpy change of reaction does not define a temperature. Rather, there is a different value of $\Delta_r H°$ at each temperature. In this learning resource, we will consistently refer to reactions taking place at 25 °C to illustrate by example the principles of thermochemistry.

EXERCISE 7.16—STANDARD STATES OF SUBSTANCES

What are the standard states of the following elements, compounds, or species at 25 °C?
(a) bromine
(b) mercury
(c) sodium sulfate
(d) ethanol
(e) aquated chloride ions in aqueous sodium chloride solution

EXERCISE 7.17—STANDARD ENTHALPY CHANGE OF REACTION

Define the conditions under which the enthalpy change at 25 °C of the following reactions is the particular value that is called the standard enthalpy change of reaction ($\Delta_r H°$).

(a) $CO(g) + \frac{1}{2} O_2(g) \longrightarrow CO_2(g)$

(b) $Mg(s) + 2 H^+(aq) \longrightarrow Mg^{2+}(aq) + H_2(g)$

(c) $H^+(aq) + OH^-(aq) \longrightarrow H_2O(\ell)$

> The standard enthalpy change of reaction ($\Delta_r H°$) at a defined temperature is the enthalpy change of reaction when all the reactants and products are in their standard states.

7.7 Hess's Law

For many chemical reactions, measurement of the enthalpy change of reaction using a calorimeter is not possible. An example is the oxidation of carbon to carbon monoxide by oxygen:

$$C(s) + \frac{1}{2} O_2(g) \longrightarrow CO(g)$$

In practice, even if the reaction mixture initially contains exactly half as many moles of oxygen as there are of carbon, some carbon dioxide will be formed and some carbon left unreacted.

Fortunately, the enthalpy change of reaction for oxidation to CO can be calculated from measured enthalpy changes of reaction of other reactions. The calculation is based on **Hess's law**: *If a reaction can be written as the sum of two or more steps, its enthalpy change of reaction is the sum of the enthalpy changes of reaction of the steps*. It doesn't matter that this reaction does not actually occur via these two steps. Hess's law can be applied to any hypothetically proposed steps (as long as their sum is the same as the overall reaction).

The oxidation of C(s) to $CO_2(g)$ can be thought of as occurring in two steps: first the oxidation of C(s) to CO(g) (reaction 1), and then the oxidation of CO(g) to $CO_2(g)$ (reaction 2). Adding the equations for these two reactions gives the equation for the oxidation of C(s) to $CO_2(g)$ (reaction 3). The standard enthalpy changes of reaction at 25 °C (per mol of product, in each case) are listed for reactions 2 and 3:

Reaction 1: $C(s) + \frac{1}{2} O_2(g) \longrightarrow \cancel{CO(g)}$ $\Delta_r H°_1 = ?$

Reaction 2: $\cancel{CO(g)} + \frac{1}{2} O_2(g) \longrightarrow CO_2(g)$ $\Delta_r H°_2 = -283.0 \text{ kJ mol}^{-1}$

Reaction 3: $C(s) + O_2(g) \longrightarrow CO_2(g)$ $\Delta_r H°_3 = -393.5 \text{ kJ mol}^{-1}$

According to Hess's law, $\Delta_r H°_3 = \Delta_r H°_1 + \Delta_r H°_2$. Both $\Delta_r H°_2$ and $\Delta_r H°_3$ can be measured, so we can calculate $\Delta_r H°_1$:

$$-393.5 \text{ kJ mol}^{-1} = \Delta_r H°_1 + (-283.0 \text{ kJ mol}^{-1})$$
$$\Delta_r H°_1 = -110.5 \text{ kJ mol}^{-1}$$

To use another example, the standard enthalpy change of reaction for formation of 1 mol of liquid H_2O from $H_2(g)$ and $O_2(g)$ is different from that for formation of 1 mol of H_2O vapour. The difference is the *standard molar enthalpy change of vaporization* of water:

Reaction 1: $H_2(g) + \frac{1}{2} O_2(g) \longrightarrow \cancel{H_2O(\ell)}$ $\Delta_r H°_1$ at 25 °C $= -285.8 \text{ kJ mol}^{-1}$

Reaction 2: $\cancel{H_2O(\ell)} \longrightarrow H_2O(g)$ $\Delta_r H°_2$ at 25 °C $= ?$

Reaction 3: $H_2(g) + \frac{1}{2} O_2(g) \longrightarrow H_2O(g)$ $\Delta_r H°_3$ at 25 °C $= -241.8 \text{ kJ mol}^{-1}$

The relationship $\Delta_r H°_3 = \Delta_r H°_1 + \Delta_r H°_2$ makes it possible to calculate that the value of $\Delta_r H°_2$, the standard molar enthalpy change of vaporization of water, $\Delta_{vap} H°$, at 25 °C is 44.0 kJ mol^{-1}.

The Hess's law relationships discussed above can be visualized through *enthalpy level diagrams* (Figure 7.13). These represent the magnitude of the enthalpy of substances in their standard states on the vertical axis. Absolute values of enthalpy are not known, but differences between them are.

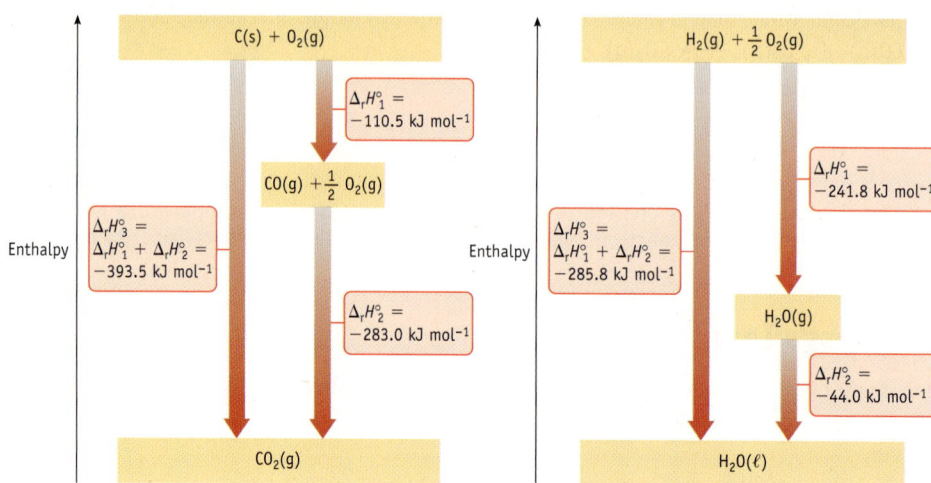

FIGURE 7.13 Enthalpy level diagrams. Enthalpies of substances, or combinations of substances, in their standard states are represented by the height on the vertical axis. The differences between these enthalpies, shown by arrows, represent standard enthalpy changes of reaction ($\Delta_r H°$). (a) The formation of 1 mol of $CO_2(g)$ from graphite, C(s), and $O_2(g)$ can occur in a single reaction, or by two stepwise reactions. By either pathway, the enthalpy change of the reaction mixture is −393.5 kJ mol⁻¹. (b) The formation of 1 mol of liquid water from $H_2(g)$ and $O_2(g)$ can occur either in a single reaction or by two hypothetical stepwise reactions. By either pathway, the total enthalpy change of the reaction mixture is −285.8 kJ mol⁻¹.

Enthalpy level diagrams can help us to see that Hess's law is a necessary consequence of the law of conservation of energy. In the example of formation of carbon dioxide, $\Delta_r H°_3 = -393.5$ kJ mol⁻¹. If it were the case that $\Delta_r H°_1 + \Delta_r H°_2 = -383.5$ kJ mol⁻¹ only, then we could make 1 mol of $CO_2(g)$ by reaction 3 with the release of 393.5 kJ, then convert it back to graphite and oxygen, using up only 383.5 kJ. We would have created 10 kJ of energy! This is not possible.

WORKED EXAMPLE 7.5—USING HESS'S LAW

Suppose you want to know the standard enthalpy change at 25 °C for the formation of methane from solid carbon (as graphite) and hydrogen gas:

$$C(s) + 2\,H_2(g) \longrightarrow CH_4(g) \quad \Delta_r H° = ?$$

The enthalpy change for this reaction cannot be measured in the laboratory because the reaction is infinitely slow. We can, however, measure the standard enthalpy changes for the combustion of carbon, hydrogen, and methane, separately.

Reaction 1: $C(s) + O_2(g) \longrightarrow CO_2(g)$ $\Delta_r H°_1$ at 25 °C = −393.5 kJ mol⁻¹

Reaction 2: $H_2(g) + \frac{1}{2}O_2(g) \longrightarrow H_2O(\ell)$ $\Delta_r H°_2$ at 25 °C = −285.8 kJ mol⁻¹

Reaction 3: $CH_4(g) + 2\,O_2(g) \longrightarrow CO_2(g) + 2\,H_2O(\ell)$ $\Delta_r H°_3$ at 25 °C = −890.3 kJ mol⁻¹

Use these standard enthalpy changes to obtain the standard enthalpy change at 25 °C for the formation of methane from graphite and hydrogen gas.

Strategy

The equations for the three reactions, as they are written, cannot be added together to obtain the equation for the formation of $CH_4(g)$ from its elements. To use Hess's law to solve this problem, we will have to manipulate the equations and manipulate the enthalpy changes accordingly. Variations of equations 2 and 3 will produce new equations that, along with equation 1, can be combined to give the desired net reaction.

Think about It

e7.13 Watch a demonstration of Hess's Law.

Think about It

e7.14 Work through a tutorial on enthalpy level diagrams.

Interactive Exercise 7.18

Use Hess's law to calculate a standard enthalpy change of reaction.

Look at the equation for the reaction whose standard enthalpy change we want to calculate, identify the reactants and products, and locate those substances in equations 1, 2, and 3. The reactants, C(s) and $H_2(g)$, are reactants in equations 1 and 2, but the product, $CH_4(g)$, is a reactant in equation 3. Equation 3 will need to be reversed to get $CH_4(g)$ on the product side.

We need to get the correct amounts of the reagents on each side. There is only 1 mol of $H_2(g)$ on the left side in equation 2, but we need 2 mol of $H_2(g)$ in the overall equation; this requires doubling the quantities in equation 2.

The other reagents in the equations will cancel when the equations are added. Equal amounts of $O_2(g)$ and $H_2O(\ell)$ appear on the left and right sides in the three equations, so they will cancel when the equations are added together.

Each manipulation requires corresponding manipulation of the standard enthalpy changes. Summing the adjusted equations and the adjusted enthalpy changes gives the desired equation and its standard enthalpy change.

Solution

To make $CH_4(g)$ a product in the overall reaction, we can reverse equation 3 while changing the sign of $\Delta_r H°_3$ (if a reaction is exothermic in one direction, its reverse must be equally endothermic):

Equation 3′: $CO_2(g) + 2 H_2O(\ell) \longrightarrow CH_4(g) + 2 O_2(g)$

$$\Delta_r H° = -\Delta_r H°_3 = +890.3 \text{ kJ mol}^{-1}$$

Next, since there are 2 mol of $H_2(g)$ on the reactant side in our desired equation, while equation 2 has only 1 mol of $H_2(g)$ as a reactant, we need to multiply the stoichiometric coefficients in equation 2 by 2, and multiply the value of $\Delta_r H°_2$ by 2.

Equation 2′: $2 H_2(g) + O_2(g) \longrightarrow 2 H_2O(\ell)$ $\Delta_r H° = 2 \times \Delta_r H°_2 = 2(-285.8 \text{ kJ mol}^{-1})$
$$= -571.6 \text{ kJ mol}^{-1}$$

Now we can rewrite the modified equations, and the modified standard enthalpy changes. When added together, $O_2(g)$, $H_2O(\ell)$, and $CO_2(g)$ all cancel to give the equation for the formation of methane from its elements.

Equation 1: $C(s) + \cancel{O_2(g)} \longrightarrow \cancel{CO_2(g)}$ $\Delta_r H° = \Delta_r H°_1 = -393.5 \text{ kJ mol}^{-1}$

Equation 2′: $2 H_2(g) + \cancel{O_2(g)} \longrightarrow \cancel{2 H_2O(\ell)}$ $\Delta_r H° = 2 \times \Delta_r H° = 2(-285.8 \text{ KJ})$
$$= -571.6 \text{ kJ mol}^{-1}$$

Equation 3′: $\cancel{CO_2(g)} + \cancel{2 H_2O(\ell)} \longrightarrow CH_4(g) + \cancel{O_2(g)}$
$$\Delta_r H° = -\Delta_r H°_3 = +890.3 \text{ kJ mol}^{-1}$$

Net equation: $C(s) + 2 H_2(g) \longrightarrow CH_4(g)$ $\Delta_r H°_{net} = \Delta_r H°_1 + 2 \times \Delta_r H°_2 + (-\Delta_r H°_3)$
$$= -74.8 \text{ kJ mol}^{-1}$$

Comment

This application of Hess's law can be summarized in an enthalpy level diagram:

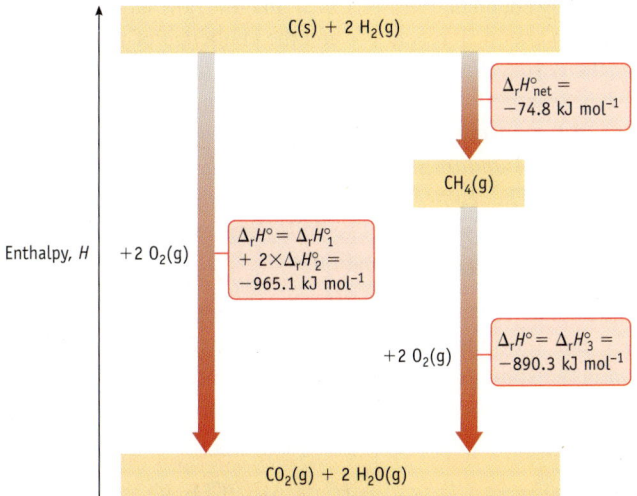

This diagram shows there are two ways to go from $[C(g) + 2\ H_2(g)]$ to $[CO_2(g) + 2\ H_2O(g)]$. The standard enthalpy changes along these two paths are $[\Delta_r H°_1 + 2 \times \Delta_r H°_2]$ and $[\Delta_r H°_{net} + \Delta_r H°_3]$. According to Hess's law,

$$\Delta_r H°_1 + 2 \times \Delta_r H°_2 = \Delta_r H°_{net} + \Delta_r H°_3$$

so
$$\Delta_r H°_{net} = \Delta_r H°_1 + 2 \times \Delta_r H°_2 + (-\Delta_r H°_3)$$

EXERCISE 7.19—USING HESS'S LAW

Graphite and diamond are two allotropes (different forms) of carbon. The standard enthalpy change for the process

$$C(\text{graphite}) \longrightarrow C(\text{diamond})$$

cannot be measured directly, but it can be evaluated using Hess's law.

(a) Calculate the standard enthalpy change at 25 °C, using experimentally measured standard enthalpy changes of combustion of graphite (-393.5 kJ mol^{-1}) and diamond (-395.4 kJ mol^{-1}).

(b) Draw an enthalpy level diagram that illustrates the use of Hess's law in this case.

EXERCISE 7.20—USING HESS'S LAW

Use Hess's law to calculate the standard enthalpy change at 25 °C for the formation of 1 mol of $CS_2(\ell)$ from C(s) and S(s) from the following standard enthalpy change values at 25 °C:

$C(s) + O_2(g) \longrightarrow CO_2(g)$	$\Delta_r H° = -393.5$ kJ mol^{-1}
$S(s) + O_2(g) \longrightarrow SO_2(g)$	$\Delta_r H° = -296.8$ kJ mol^{-1}
$CS_2(\ell) + 3\ O_2(g) \longrightarrow CO_2(g) + 2\ SO_2(g)$	$\Delta_r H° = -1103.9$ kJ mol^{-1}
$C(s) + 2\ S(s) \longrightarrow CS_2$	$\Delta_r H° = ?$

> Hess's law: If a reaction can be written as the sum of two or more steps, its enthalpy change of reaction is the sum of the enthalpy changes of reaction of the steps.

7.8 Standard Molar Enthalpy Change of Formation

Calorimetry and the application of Hess's law have made possible estimation of the enthalpy change for a great many chemical reactions. Just imagine how large would be a table of $\Delta_r H°$ values of every known chemical reaction—and the challenge of finding the value you need.

Fortunately, the $\Delta_r H°$ values of one particular type of reaction allow us to calculate the $\Delta_r H°$ values of any reaction—real or imagined. For any substance, the **standard molar enthalpy change of formation** ($\Delta_f H°$) is *the enthalpy change accompanying the reaction in which 1 mol of it in its standard state is formed from its component elemental substances in their standard states. The temperature must be specified.* For illustrative purposes, $\Delta_f H°$ values at 25 °C are used in this book. The following examples illustrate the meaning of standard molar enthalpy change of formation.

- $\Delta_f H°$ **for CO$_2$(g):** The component elements in carbon dioxide are carbon and oxygen. At 25 °C and 1 bar, the standard state of carbon dioxide is the gaseous state, $CO_2(g)$; that of carbon is the solid graphite form, C(s, graphite); and that of oxygen is the gas, $O_2(g)$. So, the standard molar enthalpy change of formation of $CO_2(g)$ is the enthalpy change accompanying formation of 1 mol of $CO_2(g)$ by the following reaction:

$$C(s, \text{graphite}) + O_2(g) \longrightarrow CO_2(g) \quad \Delta_r H° = \Delta_f H°(CO_2, g) = -393.5 \text{ kJ mol}^{-1} \text{ at } 25\ °C$$

- $\Delta_f H°$ **for NaCl(s):** At 25 °C and 1 bar, the standard states of the component elements are solid sodium, Na(s), and gaseous chlorine, $Cl_2(g)$. The standard molar enthalpy change of formation of NaCl(s) is the enthalpy change when 1 mol of NaCl(s) is formed by the following reaction:

$$Na(s) + \tfrac{1}{2}Cl_2(g) \longrightarrow NaCl(s) \quad \Delta_r H° = \Delta_f H°(NaCl, s) = -411.12 \text{ kJ mol}^{-1} \text{ at } 25 °C$$

- $\Delta_f H°$ **for $C_2H_5OH(\ell)$:** At 25 °C and 1 bar, the standard states of the component elements are C(s, graphite), $H_2(g)$, and $O_2(g)$. The standard molar enthalpy change of formation of $C_2H_5OH(\ell)$ is the enthalpy change accompanying formation of 1 mol of $C_2H_5OH(\ell)$ from 2 mol of C(s), 3 mol of $H_2(g)$, and $\tfrac{1}{2}$ mol of $O_2(g)$:

$$2 \text{ C(s)} + 3 \text{ } H_2(g) + \tfrac{1}{2}O_2(g) \longrightarrow C_2H_5OH(\ell)$$
$$\Delta_r H° = \Delta_f H°(C_2H_5OH, \ell) = -277.0 \text{ kJ mol}^{-1} \text{ at } 25 °C$$

The reaction defining the standard molar enthalpy change of formation need not be (and most often is not) a reaction that a chemist is likely to carry out in the laboratory. Ethanol, for example, is not made by reacting the elements together.

Table 7.2 (and Appendix C) list values of $\Delta_f H°$ at 25 °C for some common substances.

SUBSTANCE	NAME	$\Delta_f H°$ (kJ mol^{-1})
C(graphite)	graphite	0
C(diamond)	diamond	+1.8
$CH_4(g)$	methane	−74.87
$C_2H_6(g)$	ethane	−83.85
$C_3H_8(g)$	propane	−104.7
$C_2H_4(g)$	ethene (ethylene)	+52.47
$CH_3OH(\ell)$	methanol	−238.4
$C_2H_5OH(\ell)$	ethanol	−277.0
$C_{12}H_{22}O_{11}(s)$	sucrose	−2221.2
CO(g)	carbon monoxide	−110.53
$CO_2(g)$	carbon dioxide	−393.51
$CaCO_3(s)$*	calcium carbonate	−1207.6
CaO(s)	calcium oxide	−635.1
$H_2(g)$	hydrogen	0
$H_2O(\ell)$	liquid water	−285.83
$H_2O(g)$	water vapour	−241.83
$N_2(g)$	nitrogen	0
$NH_3(g)$	ammonia	−45.90
$NH_4Cl(s)$	ammonium chloride	−314.55
NO(g)	nitrogen monoxide	+90.29
$NO_2(g)$	nitrogen dioxide	+33.10
NaCl(s)	sodium chloride	−411.12
$S_8(s)$	sulfur	0
$SO_2(g)$	sulfur dioxide	−296.81
$SO_3(g)$	sulfur trioxide	−395.77

TABLE 7.2 Standard Molar Enthalpy Changes of Formation at 298 K of Selected Substances. Appendix C has a more comprehensive listing.

Data from the NIST Webbook (http://webbook.nist.gov)

* Data not in NIST database. Value is from J. Dean (ed.), *Lange's Handbook of Chemistry*, 14th ed. New York: McGraw-Hill, 1992.

Note that the elemental substances in Table 7.2 all have $\Delta_f H° = 0$ kJ mol^{-1}—as they must, from the definition of $\Delta_f H°$. The $\Delta_f H°$ values of some substances are positive, and some are negative. This simply means that the reactions forming some of these substances from their elements are exothermic ($\Delta_f H° < 0$), and some are endothermic ($\Delta_f H° > 0$). Most are exothermic reactions.

EXERCISE 7.21—STANDARD MOLAR ENTHALPY CHANGES OF FORMATION

Write equations for the reactions that define the standard molar enthalpy change of formation of the following:
(a) bromine, $Br_2(\ell)$
(b) solid iron (III) chloride, $FeCl_3(s)$
(c) sucrose, $C_{12}H_{22}O_{11}(s)$
Specify the standard states of the reactants in each equation.

EXERCISE 7.22—STANDARD MOLAR ENTHALPY CHANGES OF FORMATION

Write the balanced chemical equation for the reaction for which the standard enthalpy change ($\Delta_r H°$) is the standard molar enthalpy change of formation ($\Delta_f H°$) at 25 °C for methanol, $CH_3OH(\ell)$. Find the value of $\Delta_f H°$ for $CH_3OH(\ell)$ at 25 °C in Appendix C.

> The standard molar enthalpy change of formation of a compound ($\Delta_f H°$) is the enthalpy change accompanying the reaction in which 1 mol of it in its standard state is formed from its component elemental substances in their standard states.

Calculation of $\Delta_r H°$ from $\Delta_f H°$ Values

The standard enthalpy change of a reaction at any specified temperature can defined by

$$\Delta_r H° = \sum H°(\text{products}) - \sum H°(\text{reactants})$$

where $H°$ values are the *standard molar enthalpies* of reactants and products (molar enthalpies when in their standard states at the specified temperature). If we had lists of values of $H°$ of substances at a specified temperature, we could calculate $\Delta_r H°$ of any reaction at that temperature. However, since we cannot know exactly $H°$ of any substance, this is only an in-principle relationship.

The usefulness of tabulated values of $\Delta_f H°$ of substances is that they allow us to calculate the standard enthalpy change of reaction at a specified temperature by the following relationship:

$$\Delta_r H° = \sum n_i \Delta_f H°(\text{products}) - \sum n_i \Delta_f H°(\text{reactants})$$

This is provided that the $\Delta_r H°$ values of reactants and products are at the specified temperature (which we will take to be 25 °C).

To find $\Delta_r H°$ of a reaction, add up the $\Delta_f H°$ values of the product substances and subtract the sum of the $\Delta_f H°$ values of the reactant substances. We need to take into account the number of moles of each reactant and product in the chemical equation that we use to represent the reaction. This equation is a logical consequence of the definition of $\Delta_f H°$ and Hess's law.

Suppose we want to know the standard enthalpy change ($\Delta_r H°$) for decomposition of 1 mol of calcium carbonate (limestone) to calcium oxide (lime) and gaseous carbon dioxide under standard conditions at 25 °C:

$$CaCO_3(s) \longrightarrow CaO(s) + CO_2(g) \qquad \Delta_r H° \text{ at 25 °C} = ?$$

To do so, we can use the standard enthalpy changes of formation ($\Delta_f H°$) of the reactant and the products, listed in Table 7.2 or Appendix C:

Taking It Further

e7.14 Calculation of $\Delta_r H°$ from $\Delta_f H°$ values of substances—why does it work?

$$\Delta_r H° = \sum n_i \, \Delta_f H°(\text{products}) - \sum n_i \, \Delta_f H°(\text{reactants})$$
$$= [1 \times \Delta_f H°(\text{CaO, s}) + 1 \times \Delta_f H°(\text{CO}_2, \text{g})] - [1 \times \Delta_f H°(\text{CaCO}_3, \text{s})]$$
$$= [1 \times (-635.1 \text{ kJ mol}^{-1}) + 1 \times (-393.5 \text{ kJ mol}^{-1})]$$
$$- [1 \times (-1207.6 \text{ kJ mol}^{-1})]$$
$$= +179.0 \text{ kJ mol}^{-1}$$

The positive value of $\Delta_r H°$ shows that the decomposition of limestone to lime and carbon dioxide is endothermic: there is transfer of 179.0 kJ of heat for each mol of $CaCO_3$ decomposed from the surroundings into the system.

Think about It

e7.15 Work through a tutorial on using enthalpy changes of formation.

WORKED EXAMPLE 7.6—CALCULATION OF $\Delta_r H°$ FROM $\Delta_f H°$ VALUES

Nitroglycerin is a powerful explosive that forms four different gases when detonated (which is why it has explosive power):

$$2 \text{ C}_3\text{H}_5(\text{NO}_3)_3(\ell) \longrightarrow 3 \text{ N}_2(\text{g}) + \tfrac{1}{2} \text{O}_2(\text{g}) + 6 \text{ CO}_2(\text{g}) + 5 \text{ H}_2\text{O}(\text{g})$$

Calculate the standard enthalpy change at 25 °C when 10.0 g of nitroglycerin is detonated. For nitroglycerin, $\Delta_f H°$ at 25 °C is -364 kJ mol^{-1}. Use Table 7.2 or Appendix C to find other $\Delta_f H°$ values needed.

Interactive Exercise (Odyssey) 7.23

Calculate the energy released when nitroglycerin explodes.

Strategy

The equation given is for reaction of 2 mol of nitroglycerin. Use values of $\Delta_f H°$ for each of the reactants and products to calculate $\Delta_r H°$ for the reaction. From Table 7.2, $\Delta_f H°(\text{CO}_2(\text{g}))$ $= -393.5 \text{ kJ mol}^{-1}$, $\Delta_f H°(\text{H}_2\text{O (g)}) = -241.8 \text{ kJ mol}^{-1}$, and $\Delta_f H°(\text{N}_2(\text{g})) = \Delta_f H°(\text{O}_2(\text{g})) = 0 \text{ kJ mol}^{-1}$ for both $N_2(g)$ and $O_2(g)$. We need to take into account that when 2 mol of nitroglycerin reacts, it produces 3 mol of $N_2(g)$, 0.5 mol of $O_2(g)$, 6 mol of $CO_2(g)$, and 5 mol of $H_2O(g)$. From the mass of 2 mol of nitroglycerin, calculate by simple proportion the enthalpy change when 10.0 g of nitroglycerin explodes.

Interactive Exercise 7.24–7.25

Calculate a standard enthalpy change using standard molar enthalpy changes of formation.

Solution

For the balanced equation given (reaction of 2 mol of nitroglycerin):

$$\Delta_r H° = [3 \times \Delta_f H°(\text{N}_2, \text{g}) + 0.5 \times \Delta_f H°(\text{O}_2, \text{g}) + 6 \times \Delta_f H°(\text{CO}_2, \text{g})$$
$$+ 5 \times \Delta_f H°(\text{H}_2\text{O}, \text{g})] - [2 \times \Delta_f H°(\text{C}_3\text{H}_5(\text{NO}_3)_3, \ell)]$$
$$= [(3 \times 0 \text{ kJ mol}^{-1}) + (0.5 \times 0 \text{ kJ mol}^{-1}) + (6 \times -393.5 \text{ kJ mol}^{-1})$$
$$+ (5 \times -241.8 \text{ kJ mol}^{-1})] - [2 \times -364 \text{ kJ mol}^{-1}]$$
$$= -2842 \text{ kJ mol}^{-1} \text{ (for reaction of 2 mol nitroglycerin)}$$

For reaction of 1 mol of nitroglycerin, we can then write:

$$\Delta_r H° = \frac{-2842 \text{ kJ}}{2 \text{ mol}} = -1421 \text{ kJ mol}^{-1}$$

Now we can multiply $\Delta_r H°$ per mol that reacts by the amount (in mol) that reacts in this case. The molar mass of nitroglycerin is 227.1 g mol^{-1}. Then the enthalpy change for the reaction of 10.0 g of nitroglycerin is

$$\Delta_r H° = -1421 \text{ kJ mol}^{-1} \times \frac{10.0 \text{ g}}{227.1 \text{ g mol}^{-1}} = -62.6 \text{ kJ}$$

Comment

You might ask about the meaning of the standard enthalpy change of this reaction at 25 °C when it does not happen at 25 °C since explosions are accompanied by release of considerable heat. A simplified explanation is that the reactants are brought to 25 °C in a calorimeter, then the reaction made to happen, and then the whole system cooled to 25 °C again before the temperature change of a known volume of water surrounding the calorimeter is measured.

EXERCISE 7.26—CALCULATION OF $\Delta_rH°$ FROM $\Delta_fH°$ VALUES

Calculate the standard molar enthalpy change of combustion of liquid benzene (C_6H_6) at 25 °C.

$$C_6H_6(\ell) + 7.5\ O_2(g) \longrightarrow 6\ CO_2(g) + 3\ H_2O(\ell) \qquad \Delta_rH° \text{ at } 25\ °C = ?$$

$\Delta_fH°(C_6H_6(\ell)) = +49.0$ kJ mol^{-1}. Other $\Delta_fH°$ values can be found in Table 7.2 and Appendix C. Draw an enthalpy level diagram that shows the relationship between $\Delta_rH°$ for this reaction and the $\Delta_fH°$ values you use.

The standard enthalpy change of reaction at a given temperature can be calculated from values of the standard molar enthalpy changes of formation of the reactants and products at the same temperature:

$$\Delta_rH° = \sum n_i\Delta_fH°(\text{products}) - \sum n_i\Delta_fH°(\text{reactants})$$

7.9 Enthalpy Change of Reaction from Bond Energies

When a reaction involving only molecular substances occurs, the sign and magnitude of the enthalpy change can be approximated as the net result of bond-breaking processes (that require energy) and bond-forming processes (that release energy). If we know how much energy is associated with breaking or forming particular types of bonds, we can estimate the enthalpy changes of reactions.

The *molar enthalpy change of bond dissociation*, usually called **bond energy** (*D*) *is the enthalpy change for breaking a particular bond in the molecules of 1 mol of substance, with the reactants and products in the gas phase.* For example, for dissociation of the C—C bond in ethane:

$$H_3C—CH_3(g) \longrightarrow H_3C(g) + CH_3(g) \qquad \Delta_rH = -D = +346 \text{ kJ mol}^{-1}$$

Bond energy is the enthalpy change of *homolytic* dissociation [>>Section 7.9] such as that represented by the above equation, with one electron of the covalent bond remaining with each of the resulting atoms, ions, or molecules. In the case of bond dissociation in a neutral molecule, the resulting fragments are electrically neutral and because each has an "unpaired electron," they are *free radicals* [>>Section 10.4]. This differs from *heterolytic* dissociation in which both electrons of the bond remain with one of the dissociation fragments—in the case of ethane dissociation, with formation of CH_3^- and CH_3^+ ions.

Because energy must be supplied to break bonds, the process of breaking bonds in the gas phase is endothermic; so *D* has a positive value. On the other hand, the formation of bonds in the gas phase is always exothermic. The increase of enthalpy of a system when bonds form must be the same as the decrease of enthalpy when they are broken.

$$H_3C(g) + CH_3(g) \longrightarrow H_3C—CH_3(g) \qquad \Delta_rH = -D = -346 \text{ kJ mol}^{-1}$$

For a given type of bond (an O—H bond, for example), the bond energy is approximately the same in different compounds. Estimates of *average bond energies* are listed in Table 7.3.

We can see from the values in Table 7.3 that—not surprisingly—the average bond dissociation energy for the C—C single bond (346 kJ mol^{-1}) is less than that for the C=C double bond (610 kJ mol^{-1}), which in turn is less than that for the C≡C triple bond (835 kJ mol^{-1}).

A chemical reaction may involve the breaking of some bonds and formation of others. If the total energy released when new bonds form exceeds the energy required to break bonds in the reactant molecules, the reaction is exothermic. If the opposite is true, then the reaction is endothermic. Enthalpy changes of reactions can be approximately estimated from values of the average bond energies, using the following equation:

$$\Delta_rH = \sum D(\text{bonds broken}) - \sum D(\text{bonds formed})$$

The CH_3 species formed is sometimes represented as •CH_3 to indicate that it has an unpaired electron. Similarly, in other situations we can write •H, •Cl, and •NO_2.

Think about It

e7.16 Work through a tutorial on calculating enthalpy changes from bond energies.

Single Bonds

TABLE 7.3 Average Values of Some Single- and Multiple-Bond Energies (kJ mol^{-1})

	H	C	N	O	F	Si	P	S	Cl	Br	I
H	436	413	391	463	565	328	322	347	432	366	299
C		346	305	358	485	–	–	272	339	285	213
N			163	201	283	–	–	–	192	–	–
O				146	–	452	335	–	218	201	201
F					155	565	490	284	253	249	278
Si						222	–	293	381	310	234
P							201	–	326	–	184
S								226	255	–	–
Cl									242	216	208
Br										193	175
I											151

Multiple Bonds

N=N	418	C=C	610
N≡N	945	C≡C	835
C=N	615	C=O	745
C≡N	887	C≡O	1046
O=O (in O$_2$)	498		

I. Klotz and R. M. Rosenberg, *Chemical Thermodynamics*, 4th ed., p. 55, New York: John Wiley, 1994; and J. E. Huheey, E. A. Keiter, and R. L. Keiter, *Inorganic Chemistry*, 4th ed., Table E.1, New York: HarperCollins, 1993

WORKED EXAMPLE 7.7—USING BOND ENERGIES TO ESTIMATE ENTHALPY CHANGE OF REACTION

Reactions of hydrogen with unsaturated organic compounds to form saturated compounds are called *hydrogenation* [>>Section 19.12]. These are commonly used to convert vegetable oils, whose molecules contain C=C double bonds, to solid or semi-solid fats. Use average bond energies in Table 7.3 to estimate the enthalpy change for the hydrogenation of propene to propane:

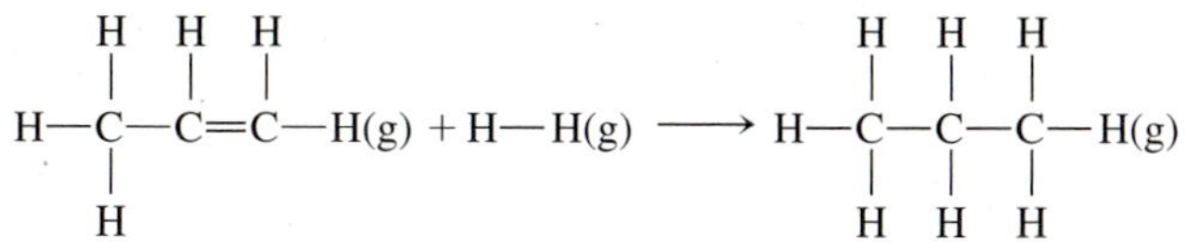

Strategy

Examine the reactants and products to determine which bonds are broken and which are formed. Add up the energies required to break bonds in the reactants and the energy evolved in forming bonds in the product. The difference in these energies is an estimate of the enthalpy change of the reaction.

Solution

In this case, the C=C bond in propene molecules and the H—H bond in hydrogen molecules are broken. For each molecule of propane produced, a C—C bond and two C—H bonds are formed. The following analysis applies for hydrogenation of 1 mol of propene.

Bonds broken: 1 mol of C=C bonds and 1 mol of H—H bonds

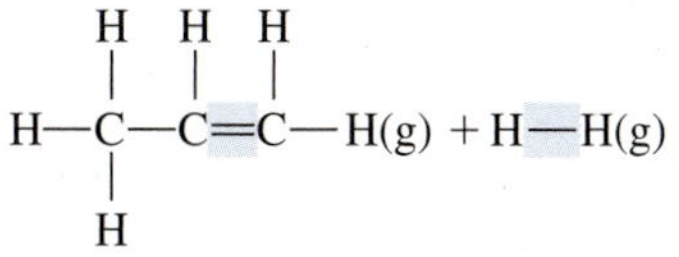

Energy required $= \sum D(\text{bonds broken}) = 610 \text{ kJ mol}^{-1} + 436 \text{ kJ mol}^{-1} = 1046 \text{ kJ mol}^{-1}$

Bonds formed: 1 mol of C—C bonds and 2 mol of C—H bonds

$$\begin{array}{ccccc} & H & H & H & \\ & | & | & | & \\ H- & C- & C- & C- & H(g) \\ & | & | & | & \\ & H & H & H & \end{array}$$

Energy evolved $= \sum D(\text{bonds formed}) = 346 \text{ kJ mol}^{-1} + 2(413 \text{ kJ mol}^{-1}) = 1172 \text{ kJ mol}^{-1}$

Now the energy change of reaction can be calculated.

$\Delta_r H = \sum D(\text{bonds broken}) - \sum D(\text{bonds formed}) = 1046 \text{ kJ mol}^{-1} - 1172 \text{ kJ mol}^{-1} = -126 \text{ kJ mol}^{-1}$

Comment

The calculated value of $\Delta_r H$ shows that the reaction is exothermic.

EXERCISE 7.27—USING BOND ENERGIES TO ESTIMATE ENTHALPY CHANGE OF REACTION

Acetone, a common industrial solvent, can be converted to isopropanol, rubbing alcohol, by hydrogenation. Calculate the approximate enthalpy change for this reaction using average bond energies.

$$\begin{array}{ccc} O & & H \\ \| & & \diagup \\ H_3C-C-CH_3(g) + H-H(g) & \longrightarrow & \begin{array}{c} O \\ | \\ H_3C-C-CH_3(g) \\ | \\ H \end{array} \end{array}$$

Acetone   **Isopropanol**

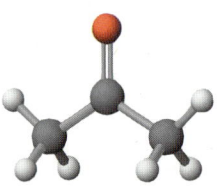 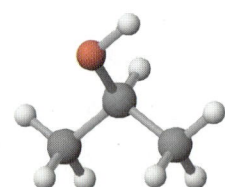

EXERCISE 7.28—USING BOND ENERGIES TO ESTIMATE ENTHALPY CHANGE OF REACTION

Using the average bond energies in Table 7.3, estimate the molar enthalpy change of combustion of gaseous methane, $CH_4(g)$. That is, estimate $\Delta_r H$ for the reaction of methane with O_2 to produce water vapour and carbon dioxide gas.

> Average bond energies (D) for different types of bond are tabulated, and enthalpy changes of reactions can be approximately calculated from $\Delta_r H = \sum D(\text{bonds broken}) - \sum D(\text{bonds formed})$.

7.10 Energy from Food

Why do we eat? Of course, we have a natural desire to do so, and we find it enjoyable. Some components of our food, such as water, are used directly in our bodies. Other food compounds are broken down by chemical reactions to obtain the molecular building blocks we need to make the many compounds that make up our bodies. Oxidation of compounds in foods also provides the energy we need to perform the activities of life. The many different chemical reactions that foods undergo in our bodies to provide energy and chemical building blocks fall into the area of biochemistry called *metabolism*.

Some substances in food, such as *carbohydrates*, are oxidized in the metabolic process. These oxidation processes are exothermic, releasing large quantities of energy. For example, the thermochemical equation for the oxidation of the sugar glucose ($C_6H_{12}O_6$) to form carbon dioxide and water is

$$C_6H_{12}O_6(s) + 6\,O_2(g) \longrightarrow 6\,CO_2(g) + 6\,H_2O(\ell) \qquad \Delta_rH° = -2803 \text{ kJ mol}^{-1}$$

If our bodies obtained energy directly from the oxidation reaction, we would have to nibble at food continuously, eating faster when we are involved in exercise, and more slowly when we are just sitting—a little like controlling the rate of supply of fuel to our cars by controlling the accelerator pedal. Fortunately, the body carries out oxidation by different pathways, in a more controlled way that allows it to obtain energy in small increments as needed. The cell carries out the oxidation of compounds such as glucose in one location and stores the energy in a small set of compounds that can be used almost anywhere in the cell.

The compound mainly used to store energy is adenosine-5′-triphosphate (ATP), whose structure is shown in Figure 7.14. In aerobic respiration (the oxidation of glucose and other compounds in the presence of air), about 30–32 mol of ATP is produced for each 1 mol of glucose that is metabolized.

As energy is required by the body, ATP undergoes a reaction with water (hydrolysis) to produce adenosine 5′-diphosphate (ADP), splitting off a hydrogenphosphate group in an exothermic process (Figure 7.14).

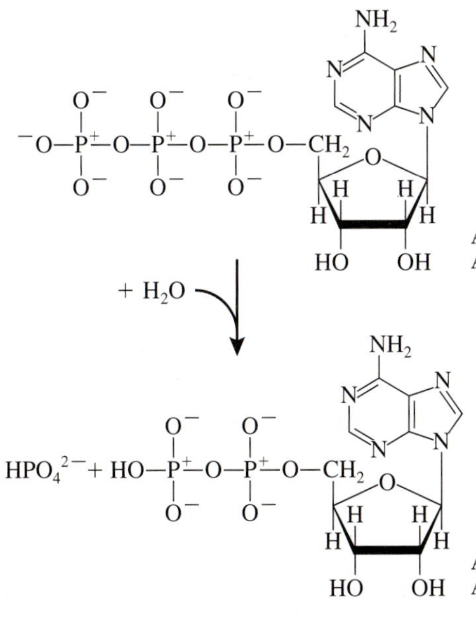

FIGURE 7.14 The exothermic hydrolysis of ATP to ADP as energy is required. During metabolic processes, the store of ATP is replenished by the opposite process (endothermic formation of ATP from ADP).

This exothermic reaction can be represented by the following unbalanced equation:

$$ATP(aq) + H_2O(\ell) \longrightarrow ADP(aq) + HPO_4{}^{2-}(aq) \qquad \Delta H \approx -24 \text{ kJ mol}^{-1}$$

Popular science articles sometimes say that "ATP has high energy bonds" because an oversimplified view is that the conversion from ATP to ADP involves just the breaking of a bond to a hydrogenphosphate group, with release of energy. This is not chemically sensible: to break any bond requires energy. More correctly, the total enthalpy of the reactants ATP and water is greater than that of the products ADP and a hydrogenphosphate group—resulting in excess energy as a result of reaction.

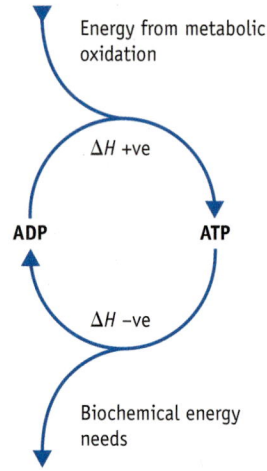

FIGURE 7.15 Cyclic interconversion between ADP and ATP is a means of energy storage and provision.

Metabolic processes result in formation of ATP from ADP by the reverse endothermic reaction:

$$\text{ADP(aq)} + \text{HPO}_4{}^{2-}\text{(aq)} \longrightarrow \text{ATP(aq)} + \text{H}_2\text{O(l)} \qquad \Delta H \approx +24\,\text{kJ}$$

In this way, ATP is cyclically cleaved and regenerated according to energy demands (Figure 7.15).

In cells, many chemical processes that would be endothermic on their own are linked with the hydrolysis of ATP. The combination of an energetically unfavourable process with the energetically favourable hydrolysis of ATP can yield a process that is energetically favourable. For example, most cells have a greater concentration of potassium ions and a smaller concentration of sodium ions inside the cells than outside them. The natural tendency, therefore, is for sodium ions to flow into the cell and for potassium ions to flow out. To maintain the correct concentrations, the cell must counteract this movement and actively pump sodium ions out of the cell and potassium ions into the cell. This activity goes in the direction that is not favoured, so it requires energy. To accomplish this feat, the cell links this pumping process to the hydrolysis of ATP to ADP. The energy released from the hydrolysis reaction provides the energy to run a molecular pump (an enzyme) that moves the ions in the direction the cell needs.

EXERCISE 7.31—ENERGY FROM FOOD

Which of the following statements are true?
(a) Breaking the P—O bond in ATP is an exothermic process.
(b) Making a new bond between the phosphorus atom in the hydrogenphosphate group being cleaved off ATP and the OH group of water is an exothermic process.
(c) Breaking bonds is an endothermic process.
(d) The energy released in the hydrolysis of ATP may be used to run endothermic reactions in a cell.

During metabolic processes, energy is stored by endothermic conversion of ADP to ATP. As energy is required by our bodies, energy is released by conversion of ATP to ADP.

SUMMARY

Key Concepts

- Although **energy** can be converted from one form into another, the total amount of energy is constant. (Section 7.3)
- The **temperature** of a substance refers to its level of hotness, and is a measure of the average kinetic energy of its particles. (Section 7.4)
- **Heat** is energy in transit. Heat flows in the direction from an object at higher temperature to an object at lower temperature, until the objects have come to thermal equilibrium. (Section 7.4)
- The **internal energy** (**U**) of a system is the sum of the **kinetic energy** and **potential energy** of the collection of atoms, ions, and molecules in the system. It is a **state function**. (Section 7.4)
- The internal energy of a system may decrease either by transfer of heat to the surroundings or by work done on the surroundings. $\Delta U = q + w$. When $w = 0$, $\Delta U = q$. (Section 7.4)
- The **enthalpy** (**H**) of a system changes by the amount of heat transferred between the system and surroundings during a process that occurs at constant pressure (q_p), if no work other than that due to expansion of the system occurs. Then $\Delta H = q_p$. (Section 7.4)

- The amount of energy absorbed when 1 mol of a solid melts at its melting point is called the **molar enthalpy change of fusion ($\Delta_{fus}H$)**. The energy absorbed when 1 mol of a liquid boils at its boiling point is called the **molar enthalpy change of vaporization ($\Delta_{vap}H$)**. (Section 7.5)
- The **enthalpy change of reaction ($\Delta_r H$)** can be conceptualized as $\Delta_r H = \sum H_{products} - \sum H_{reactants}$. However, $\Delta_r H$ cannot be estimated in this way because the enthalpies of substances cannot be measured. (Section 7.6)
- In **exothermic** reactions, $\sum H_{products} < \sum H_{reactants}$ and $\Delta_r H < 0$, while for **endothermic** reactions $\sum H_{products} > \sum H_{reactants}$ and $\Delta_r H > 0$. (Sections 7.4 and 7.6)
- **Calorimetry** is the measurement of enthalpy change of reaction by measurement of the quantity of heat evolved when a known amount of substances react. (Section 7.6)
- For a given reaction, the enthalpy change depends on the conditions and states of the reactants and products, and is proportional to the amounts that react. Reactions that are the reverse of each other have the same magnitude, but opposite sign, of $\Delta_r H$. (Section 7.6)
- The **standard enthalpy change of reaction ($\Delta_r H°$)** at a defined temperature is the enthalpy change of reaction when all the reactants and products are in their **standard states**. (Section 7.6)
- According to **Hess's law**, if a reaction can be written as the sum of two or more steps, its enthalpy change of reaction is the sum of the enthalpy changes of reaction of the steps. (Section 7.7)
- The **standard molar enthalpy change of formation ($\Delta_f H°$)** of a compound is the enthalpy change accompanying the reaction in which 1 mol of it in its standard state is formed from its component elements in their standard states. The temperature must be specified. (Section 7.8)
- $\Delta_f H°$ values at a given temperature can be used to calculate the standard enthalpy change of reactions at that temperature, with the equation $\Delta_r H° = \sum n_i \Delta_f H°(products) - \sum n_i \Delta_f H°(reactants)$ (Section 7.8)
- Average **bond energies (D)** for different types of bond have been estimated, and enthalpy changes of reactions can be approximately calculated from $\Delta_r H = \sum D(bonds\ broken) - \sum D(bonds\ formed)$. (Section 7.9)
- During metabolic processes, energy is stored by endothermic conversion of ADP to ATP. As energy is required by our bodies, energy is released by conversion of ATP to ADP. (Section 7.10)

REVIEW QUESTIONS

Section 7.1: Energy from Hydrogen?

7.32 (a) Write balanced chemical equations for the combustion in oxygen of (i) octane, (ii) methane gas, and (iii) hydrogen gas.
(b) Calculate and compare the mass of carbon dioxide produced by combustion of 1 mol of each of the fuels in (a).
(c) Calculate and compare the mass of carbon dioxide produced by combustion of 1 tonne of each of the fuels in (a).
(d) By reference to the energy density of these fuels given in Table 7.1, calculate and compare the mass of carbon dioxide produced by combustion of enough of each of the fuels in (a) to provide 1 GJ (1×10^9 J) of heat.

Section 7.2: Chemical Changes and Energy Redistribution

7.33 Melting of ice is an endothermic process. Which has more energy: a block of ice or the water that results from melting it?

7.34 For each of the following, decide whether the process is exothermic or endothermic:
(a) $H_2O(\ell) \longrightarrow H_2O(s)$
(b) $2\ H_2(g) + O_2(g) \longrightarrow 2\ H_2O(g)$

(c) $H_2O(\ell, 25°C) \longrightarrow H_2O(\ell, 15°C)$

(d) $H_2O(\ell) \longrightarrow H_2O(g)$

Section 7.5: Enthalpy Changes Accompanying Changes of State

7.35 What quantity of heat is required to vaporize 125 g of benzene (C_6H_6) at its boiling point, 80.1 °C? For benzene, $\Delta_{vap}H$ is 30.8 kJ mol^{-1}.

7.36 How much heat is required is required to raise the temperature of 1.00 kg of ethanol from 20.0 °C to the boiling point (78.29 °C) and then to change the liquid to vapour at that temperature? The specific heat capacity of liquid ethanol is 2.44 J K^{-1} g^{-1}, and $\Delta_{vap}H$ is 38.6 kJ mol^{-1}.

7.37 Calculate the quantity of heat required to convert 60.1 g of $H_2O(s)$ at 0.0 °C to $H_2O(g)$ at 100.0 °C. The enthalpy change of fusion of ice at 0 °C is 333 J g^{-1}; the enthalpy change of vaporization of liquid water at 100 °C is 2260 J g^{-1}.

Section 7.6: Enthalpy Change of Reaction ($\Delta_r H$)

7.38 Sucrose (table sugar, $C_{12}H_{22}O_{11}$) can oxidize to $CO_2(g)$ and $H_2O(\ell)$:

$$C_{12}H_{22}O_{11}(s) + 12\ O_2(g) \longrightarrow 12\ CO_2(g) + 11\ H_2O(\ell)$$
$$\Delta_r H = -5645 \text{ kJ mol}^{-1}$$

What is the enthalpy change for the oxidation of 5.00 g (1 teaspoonful) of sugar?

7.39 Reaction of 1 mol of $H_2(g)$ with $O_2(g)$ to form liquid water is represented by the equation

$$H_2(g) + \tfrac{1}{2}O_2(g) \longrightarrow H_2O(\ell) \quad \Delta_r H = -285.8 \text{ kJ mol}^{-1}$$

What is the enthalpy change accompanying decomposition of 12.6 g of liquid water to hydrogen and oxygen gases?

7.40 Calcium carbide (CaC2) is manufactured by the reaction of CaO with carbon at a high temperature (Calcium carbide is then used to make acetylene):

$$CaO(s) + 3\ C(s) \longrightarrow CaC_2(s) + CO(g)$$
$$\Delta_r H = +464.8 \text{ kJ mol}^{-1}$$

Is this reaction endothermic or exothermic? If 10.0 g of CaO is allowed to react with an excess of carbon, how much heat is absorbed or evolved by the reaction?

7.41 Acetic acid (CH_3COOH) is made industrially by the reaction of methanol and carbon monoxide:

$$CH_3OH(\ell) + CO(g) \longrightarrow CH_3COOH(\ell)$$
$$\Delta_r H = -355.9 \text{ kJ mol}^{-1}$$

If you produce 1.00 L of acetic acid (density = 1.044 g mL^{-1}) by this reaction, what quantity of heat is evolved?

7.42 Assume you mix 100.0 mL of 0.200 mol L^{-1} CsOH solution with 50.0 mL of 0.400 mol L^{-1} HCl solution in a coffee cup calorimeter. The following reaction occurs:

$$OH^-(aq) + H_3O^+(aq) \longrightarrow 2\ H_2O(\ell)$$

The temperature of both solutions before mixing was 22.50 °C, and it rises to 24.28 °C after the acid-base reaction. What is the enthalpy change of reaction per mol of $H_3O^+(aq)$ reacted? Assume the densities of the solutions are all 1.00 g mL^{-1} and the specific heat capacities of the solutions are 4.2 J K^{-1} g^{-1}.

7.43 Some cold packs take advantage of the fact that dissolving ammonium nitrate in water is an endothermic process. Adding 5.44 g of $NH_4NO_3(s)$ to 150.0 g of water in a coffee cup calorimeter (with stirring to dissolve the salt) resulted in a decrease in temperature from 18.6 °C to 16.2 °C. Calculate the enthalpy change per mol of $NH_4NO_3(s)$ dissolved. Assume that the solution (whose mass is 155.4 g) has a specific heat capacity of 4.2 J K^{-1} g^{-1}.

7.44 Almost insoluble AgCl(s) precipitates when aqueous $AgNO_3$ solution is mixed with NaCl solution:

$$Ag^+(aq) + Cl^-(aq) \longrightarrow AgCl(s) \quad \Delta_r H = ?$$

When 250 mL of 0.16 mol L^{-1} $AgNO_3$ solution and 125 mL of 0.32 mol L^{-1} NaCl solution are mixed in a coffee cup calorimeter, the temperature of the mixture rises from 21.15 °C to 22.90 °C. Calculate how much heat is evolved, and then calculate the enthalpy change for the precipitation of 1 mol of AgCl(s). Assume the density of the solution is 1.0 g mL^{-1} and its specific heat capacity is 4.2 J K^{-1} g^{-1}.

7.45 What are the standard states of the following at 25 °C?

(a) H_2O

(b) NaCl

(c) Hg

(d) CH_4

(e) $Na^+(aq)$ ions in an aqueous Na_2CO_3 solution

Section 7.7: Hess's Law

7.46 The standard enthalpy changes for the following reactions can be measured:

$$CH_4(g) + 2\ O_2(g) \longrightarrow CO_2(g) + 2\ H_2O(g)$$
$$\Delta_r H° = -802.4 \text{ kJ mol}^{-1}$$

$$CH_3OH(g) + \tfrac{3}{2}O_2(g) \longrightarrow CO_2(g) + H_2O(g)$$
$$\Delta_r H° = -676 \text{ kJ mol}^{-1}$$

(a) Use these values and Hess's law to determine the standard enthalpy change for the following reaction:

$$CH_4(g) + \tfrac{1}{2}O_2(g) \longrightarrow CH_3OH(g)$$

(b) Draw an enthalpy level diagram that shows the relationship between the $\Delta_r H°$ values.

7.47 The standard enthalpy change for the reaction

$$\tfrac{1}{2} N_2(g) + \tfrac{1}{2} O_2(g) \longrightarrow NO(g) \qquad \Delta_r H^\circ = ?$$

cannot be measured in a calorimeter because it is reactant-favoured. Use the experimentally measured standard enthalpy changes for the following reactions to calculate the standard enthalpy change for the formation of NO(g) from its elements:

$$N_2(g) + 3 H_2(g) \longrightarrow 2 NH_3(g)$$
$$\Delta_r H^\circ = -91.8 \text{ kJ mol}^{-1}$$

$$4 NH_3(g) + 5 O_2(g) \longrightarrow 4 NO(g) + 6 H_2O(g)$$
$$\Delta_r H^\circ = -906.2 \text{ kJ mol}^{-1}$$

$$H_2(g) + \tfrac{1}{2} O_2(g) \longrightarrow H_2O(g)$$
$$\Delta_r H^\circ = -241.8 \text{ kJ mol}^{-1}$$

Section 7.8: Standard Molar Enthalpy Change of Formation

7.48 Write the balanced chemical equation for the reaction for which the standard enthalpy change ($\Delta_r H^\circ$) is the standard molar enthalpy change of formation ($\Delta_f H^\circ$) at 25 °C for calcium carbonate, $CaCO_3(s)$. Find the value of $\Delta_f H^\circ$ for $CaCO_3(s)$ at 25 °C in Appendix C.

7.49 (a) Write the balanced chemical equation for the reaction for which the standard enthalpy change ($\Delta_r H^\circ$) is the standard molar enthalpy change of formation ($\Delta_f H^\circ$) at 25 °C for magnesium oxide, MgO(s). Find the value of $\Delta_f H^\circ$ for MgO(s) at 25 °C in Appendix C.

(b) What is the standard enthalpy change at 25 °C if 2.5 mol of magnesium with oxygen?

7.50 Use $\Delta_f H^\circ$ values in Appendix C to calculate standard enthalpy changes for the following:

(a) 1.0 g of white phosphorus burns, forming $P_4O_{10}(s)$

(b) 0.20 mol of NO(g) decomposes to $N_2(g)$ and $O_2(g)$

(c) 2.40 g of NaCl is formed from Na(s) and excess $Cl_2(g)$

(d) 250 g of iron is oxidized with oxygen to $Fe_2O_3(s)$

7.51 The first step in the production of nitric acid from ammonia involves the oxidation of $NH_3(g)$:

$$4 NH_3(g) + 5 O_2(g) \longrightarrow 4 NO(g) + 6 H_2O(g)$$

Use standard molar enthalpy change of formation values at 25 °C of the reactants and products at 25 °C to calculate the standard enthalpy change for this reaction at 25 °C. Draw an enthalpy level diagram that shows the relationship between $\Delta_r H^\circ$ for this reaction and the $\Delta_f H^\circ$ values you use.

7.52 For solid barium oxide, BaO(s), $\Delta_f H^\circ$ at 25 °C is $-553.5 \text{ kJ mol}^{-1}$, and for barium peroxide, $BaO_2(s)$, is $-634.3 \text{ kJ mol}^{-1}$.

(a) Calculate the standard enthalpy change at 25 °C for the following reaction. Is the reaction exothermic or endothermic?

$$BaO_2(s) \longrightarrow BaO(s) + O_2(g)$$

(b) Draw an enthalpy level diagram that shows the relationship between $\Delta_r H^\circ$ for this reaction and the $\Delta_f H^\circ$ values you use.

7.53 The standard enthalpy change at 25 °C for the oxidation of naphthalene, $C_{10}H_8(s)$, is measured by calorimetry:

$$C_{10}H_8(s) + 12 O_2(g) \longrightarrow 10 CO_2(g) + 4 H_2O(\ell)$$
$$\Delta_r H^\circ = -5156.1 \text{ kJ mol}^{-1}$$

Use this value, along with the $\Delta_f H^\circ$ values of $CO_2(g)$ and H_2O (ℓ) at 25 °C, to calculate the standard molar enthalpy change of formation of naphthalene at 25 °C.

7.54 Use Appendix C to find the standard molar enthalpy change of formation at 25 °C of oxygen atoms, oxygen molecules (O_2), and ozone (O_3). What is the standard state of O_2 at 25 °C? Is the formation of oxygen atoms from O_2 molecules exothermic? What is the standard enthalpy change for the formation of 1 mol of O_3 from O_2 at 25 °C?

7.55 Ready-to-eat meals used by military personnel and campers can be heated on a flameless heater. The source of energy in the heater is the reaction represented by

$$Mg(s) + 2 H_2O(\ell) \longrightarrow Mg(OH)_2(s) + H_2(g)$$

Calculate the standard enthalpy change at 25 °C for this reaction. What quantity of magnesium is needed to supply the heat required to warm 25 mL of water (d = 1.00 g mL^{-1}) from 25 °C to 85 °C?

7.56 Camping stoves are fuelled by propane, butane, gasoline, or ethanol. For butane, $\Delta_f H^\circ = -127.1 \text{ kJ mol}^{-1}$. Assume that gasoline is isooctane, C_8H_{18} (ℓ), with $\Delta_f H^\circ = -259.2 \text{ kJ mol}^{-1}$. Calculate the enthalpy change of combustion *per gram* of each of these fuels. Do you notice any great differences among these fuels? Are these differences related to their composition?

7.57 Hydrazine and 1,1-dimethylhydrazine both react spontaneously with O_2 and can be used as rocket fuels:

$$N_2H_4(\ell) + O_2(g) \longrightarrow N_2(g) + 2 H_2O(g)$$
hydrazine

$$N_2H_2(CH_3)_2(\ell) + 4 O_2(g) \longrightarrow 2 CO_2(g) + 4 H_2O(g)$$
1, 1 − dimethylhydrazine $\qquad\qquad + N_2(g)$

The standard molar enthalpy change of formation of $N_2H_4(\ell)$ at 25 °C is $+50.6 \text{ kJ mol}^{-1}$, and that of $N_2H_2(CH_3)_2(\ell)$ is $+48.9 \text{ kJ mol}^{-1}$. Use these values, with other $\Delta_f H^\circ$ values, to decide whether the reaction of hydrazine or 1,1-dimethylhydrazine with oxygen evolves more heat per gram of fuel.

7.58 Chloroform ($CHCl_3$) is formed from methane and chlorine in the following reaction:

$$CH_4(g) + 3 Cl_2(g) \longrightarrow 3 HCl(g) + CHCl_3(g)$$

Calculate $\Delta_r H^\circ$, the enthalpy change for this reaction, using the standard molar enthalpy change of formation of $CHCl_3(g)$, $\Delta_f H° = -103.1$ kJ mol^{-1}, and the standard enthalpy changes for the following reactions:

$$CH_4(g) + 2\,O_2(g) \longrightarrow 2\,H_2O(\ell) + CO_2(g)$$
$$\Delta_r H° = -890.4 \text{ kJ mol}^{-1}$$

$$2\,HCl(g) \longrightarrow H_2(g) + Cl_2(g)$$
$$\Delta_r H° = +184.6 \text{ kJ mol}^{-1}$$

$$C(graphite) + O_2(g) \longrightarrow CO_2(g)$$
$$\Delta_r H° = -393.5 \text{ kJ mol}^{-1}$$

$$H_2(g) + \tfrac{1}{2} O_2(g) \longrightarrow H_2O(\ell)$$
$$\Delta_r H° = -285.8 \text{ kJ mol}^{-1}$$

Section 7.9: Enthalpy Change of Reaction from Bond Energies

7.59 An important reaction in the stratosphere is the combination of oxygen atoms with ozone molecules to form oxygen molecules:

$$O_3 + O \longrightarrow 2\,O_2 \quad \Delta_r H° = -394 \text{ kJ mol}^{-1}$$

Using $\Delta_r H°$ and the bond energy data in Table 7.3, estimate the bond energy for the oxygen-oxygen bond in ozone molecules. Assume that the two oxygen-oxygen bonds in O_3 molecules are identical (but do not assume that they are either single or double bonds). How does your estimate compare with the listed bond energies of O—O single bonds and O=O double bonds?

7.60 Dinitrogen monoxide (or nitrous oxide), $N_2O(g)$, can decompose to nitrogen and oxygen gases:

$$2\,N_2O(g) \longrightarrow 2\,N_2(g) + O_2(g)$$

Use average bond energies to estimate the enthalpy change for this reaction.

7.61 The equation for the combustion of gaseous methanol is

$$2\,CH_3OH(g) + 3\,O_2(g) \longrightarrow 2\,CO_2(g) + 4\,H_2O(g)$$

(a) Using the average bond energies in Table 7.3 estimate the enthalpy change for this reaction. How much heat is transferred to the surroundings during combustion of 1 mol of gaseous methanol?

(b) Compare your answer in part (a) with a calculation of $\Delta_r H°$ from values of $\Delta_f H°$ of the reactants and products.

7.62 Dihydroxyacetone ($HOCH_2COCH_2OH$) is a component of some quick-tanning lotions. It reacts with the amino acids in the upper layer of skin and colours them brown in

a reaction similar to that occurring when food is browned as it cooks. Suppose that you can make this compound by reaction of acetone with oxygen in the gas phase.

Use average bond energies to estimate the enthalpy change for the reaction. Is the reaction exothermic or endothermic?

Section 7.10: Energy from Food

7.63 The exothermic oxidation of glucose has been represented above as

$$C_6H_{12}O_6(s) + 6\,O_2(g) \longrightarrow 6\,CO_2(g) + 6\,H_2O(\ell)$$
$$\Delta_r H° = -2803 \text{ kJ mol}^{-1}$$

Use $\Delta_f H°$ values at 25 °C to verify the value for $\Delta_r H°$. $\Delta_f H°(C_6H_{12}O_6(s)) = -1273.3$ kJ mol^{-1}.

SUMMARY AND CONCEPTUAL QUESTIONS

7.64 Reference is sometimes made to "zero emission" vehicles. In what ways are such vehicles different from conventional vehicles? To what extent is the label valid if we consider not only reactions of the fuel in the vehicle, but also all the processes by which the fuel was obtained from raw materials? Give examples.

7.65 The following terms are used extensively in thermodynamics. Define each and give an example.
(a) exothermic and endothermic reactions
(b) system and surroundings
(c) specific heat capacity of substance
(d) state function
(e) standard state of a substance
(f) enthalpy change of reaction ($\Delta_r H$)

(g) standard enthalpy change of reaction ($\Delta_r H°$)
(h) standard molar enthalpy change of formation ($\Delta_f H°$)

7.66 The first law of thermodynamics is often described as another way of stating the law of conservation of energy. Discuss whether this is a sensible portrayal.

7.67 Many people have tried to make a perpetual motion machine, but none have been successful although some have claimed success. Use the law of conservation of energy to explain why such a device is impossible.

7.68 (a) Using $\Delta_f H°$ values, calculate the total amount of heat released when 1 mol of methane is used to produce hydrogen and carbon dioxide by the steam-reforming reaction (Section 7.1), and the hydrogen is completely burned. (Add the $\Delta_r H°$ values of the two reactions.)

(b) Calculate the amount of heat released by complete combustion of 1 mol of methane in oxygen, and compare this with your answer in (a).

(c) Compare the amount of carbon dioxide produced from 1 mol of methane by the steam-reforming reaction (Section 7.1) with that produced by complete combustion of 1 mol of methane.

(d) Compare the use as a fuel of hydrogen produced by the steam-reforming process with methane from the perspectives of (i) energy obtainable (per mol and per kg), and (ii) carbon dioxide production.

7.69 You want to determine the value for the standard molar enthalpy of formation ($\Delta_f H°$) at 25 °C of $CaSO_4(s)$, for which the appropriate equation is

$$Ca(s) + \tfrac{1}{8}S_8(s) + 2\,O_2(g) \longrightarrow CaSO_4(s) \quad \Delta_f H° = ?$$

This reaction cannot be done directly. You know, however, that both calcium and sulfur react with oxygen to produce oxides in reactions that can be studied calorimetrically. You also know that basic $CaO(g)$ reacts with acidic $SO_3(g)$ to produce $CaSO_4(s)$ with $\Delta_r H° = -402.7$ kJ (for 1 mol of $CaSO_4(s)$, at 25 °C). Outline a method for determining $\Delta_f H°$ for $CaSO_4(s)$ and identify the information that must be collected by experiment. Using information in Table 7.2, confirm that $\Delta_f H°$ for $CaSO_4(s) = -1433.5$ kJ mol^{-1}.

7.70 Water can be decomposed to its elements, $H_2(g)$ and $O_2(g)$, using electrical energy or in a series of chemical reactions. The following sequence of reactions is one possibility:

$$CaBr_2(s) + H_2O(g) \longrightarrow CaO(s) + 2\,HBr(g)$$
$$Hg(\ell) + 2\,HBr(g) \longrightarrow HgBr_2(s) + H_2(g)$$
$$HgBr_2(s) + CaO(s) \longrightarrow HgO(s) + CaBr_2(s)$$
$$HgO(s) \longrightarrow Hg(\ell) + \tfrac{1}{2}O_2(g)$$

(a) Show that the net result of this series of reactions is the decomposition of water to its elements.

(b) If you use 1000 kg of water, what mass of $H_2(g)$ can be produced?

(c) Using the data in Table 7.2 and below, calculate the value of $\Delta_r H°$ at 25 °C for each step in the series. Are the reactions exothermic or endothermic?

$\Delta_f H°[CaBr_2(s)]$ at 25 °C $= -683.2$ kJ mol^{-1}

$\Delta_f H°[HgBr_2(s)]$ at 25 °C $= -169.5$ kJ mol^{-1}

(d) Comment on the commercial feasibility of using this series of reactions to produce $H_2(g)$ from water.

7.71 Explain the meaning of the term *bond dissociation energy*. Do enthalpy changes for a reactions that involve only bond-breaking processes (e.g., $C—H(g) \rightarrow C(g) + H(g)$) all have a positive sign, always have a negative sign, or vary? Explain briefly.

7.72 Bromine species play an important role in chemical reactions in the upper atmosphere.

(a) Use values of average bond energies (Table 7.3) to estimate the enthalpy changes accompanying the following three reactions of bromine species:

$$Br_2(g) \longrightarrow 2\,Br(g)$$
$$2\,Br(g) + O_2(g) \longrightarrow 2\,BrO(g)$$
$$BrO(g) + H_2O(g) \longrightarrow HOBr(g) + OH(g)$$

(b) Using average bond energies, estimate the standard molar enthalpy change of formation of $HOBr(g)$.

(c) Decide whether each of the reactions in (a) and (b) are exothermic or endothermic.

7.73 Suppose that 25 mm of rain falls over an area of 1 km^2. (Assume the density of rainwater is 1.0 g cm^{-3}.) The enthalpy change of vaporization of water at 25 °C is 44.0 kJ mol^{-1}. Calculate the amount of heat transferred to the surroundings from the condensation of water vapour in forming this quantity of liquid water. (The huge number tells you how much energy is "stored" in water vapour and why we think of storms as such great forces of energy in nature. It is interesting to compare this result with the energy given off, 4.2×10^6 kJ, when a tonne of dynamite explodes.)

7.74 Peanuts and peanut oil are organic materials that burn in air. How many burning peanuts does it take to provide the energy to boil a cup of water (250 mL of water)? To solve this problem, we assume each peanut, with an average mass of 0.73 g, is 49% peanut oil and 21% starch and the remainder is non-combustible. We further assume peanut oil is palmitic acid ($C_{16}H_{32}O_2$), with a standard enthalpy change of formation of -848.4 kJ mol^{-1}. Starch is a long chain of $C_6H_{10}O_5$ units, each unit having an enthalpy change of formation of -960 kJ.

8.1 Case Study: Horseflies, Elephants, and Electrons

"The Quantum Safari" from THE NEW WORLD OF MR. TOMPKINS. Written by George Gamow and illustrated by Russell Stannard. Copyright 2001 Cambridge University Press. Reprinted with the permission of Cambridge University Press

SURROUNDED BY AN INSECT PROBABILITY CLOUD

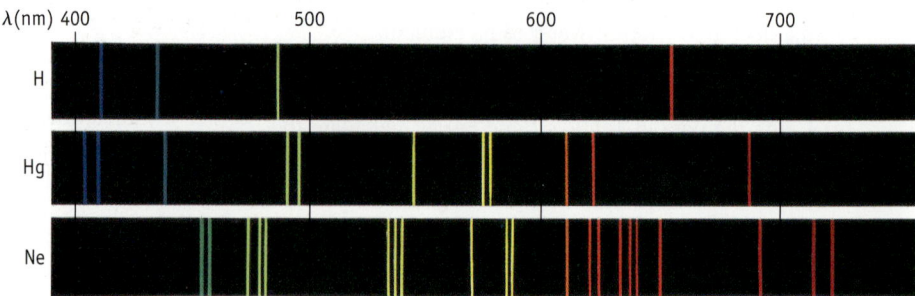

Switch on your imagination . . .

It's a warm summer afternoon in a mythical "quantum land." A boy is travelling through the jungle astride an elephant. He hears the drone of a large, aggressive horsefly buzzing around his head. Grabbing a fly swatter, he thrashes out at the insect. But as he swats, the horsefly becomes a blur—an indistinct buzzing cloud. He keeps swatting vigorously in all directions, especially where the horsefly blur is thickest. He can't locate the fly with the swatter. It seems to be everywhere, yet nowhere.

Thwack! He finally makes contact. The insect cloud vanishes, and the lifeless body of the troublesome horsefly arcs through the air and falls to the ground.

This scene is imagined in British physicist Russell Stannard's update to George Gamow's legendary popular science book about Mr. Tompkins—"a little clerk in a big city bank." The elephant and its rider are travelling through an imaginary land where small macroscopic objects such as horseflies follow the same mathematically predictable behaviour as electrons and atoms in the molecular-level world. This is because Planck's constant (*h*) [>>Section 8.3] has a value many times greater than it is in our world. As the traveller attempts to locate the horsefly with the swatter, the position of the horsefly becomes more and more uncertain, until all he can see is a blurry, buzzing insect "probability cloud," and most of his attempts to hit it miss the mark.

Stannard compares the quantum land insect that can't be clearly located until it is lifeless with the electrons that surround the nuclei of atoms in our world:

"With atomic electrons you have no more certainty of being able to hit one with a photon, than the lad had of hitting the insect. It's all down to probabilities—playing the odds. You shine a beam of light on the atom and most of the photons will miss; they pass through without having any effect at all. You just hope one of the photons will score a bull's-eye. (Gamow and Stannard, page 118)"

A barrage of swats in quantum land is analogous to a barrage of photons in our world. Understanding the quantum mechanical models of the behaviour of electrons in atoms is not easy. Many of the features of the mathematical models that revolutionized 20th century chemistry and physics are impossible to visualize, and some clash with our experiences of the observable world. In fact, mental images from our everyday experiences can be a barrier to understanding the atomic-scale world, although we can describe atomic-level behaviour accurately with mathematical language.

An example of a misleading image is the familiar representation of electrons around nuclei in atoms shown in Figure 8.1. These show pictures of particles moving in well-defined orbits, somewhat similar to planets moving around the sun. As you will see in this chapter, this is no longer the scientists' accepted view of electrons in atoms.

George Gamow and Russell Stannard, *The New World of Mr. Tompkins*, Cambridge: Cambridge University Press, 1999.

FIGURE 8.1 Unacceptable representations of electrons in atoms. These images are from a US scout merit badge, a Russian postage stamp, and the International Atomic Energy Agency. Images such as these are common, but seriously misleading.

The idea that a horsefly in our imaginary "quantum land" might be everywhere at once in a blurry cloud is not part of our experience and goes against our intuition. But it may help you understand the important idea that electrons in atoms cannot be precisely located in space. Their very small mass puts them in a grey area between what we accept to be waves (such as radio waves and ultraviolet light) and what we accept to be particles (such as protons, protein molecules, and cars). Whatever the nature of electrons in atoms, they have some properties of both waves and particles, so sometimes we need to resort to our experience of waves to think about them, and sometimes we resort to our experience of particles to describe them.

What we observe is not nature itself, but nature exposed to our method of questioning.
Werner Heisenberg, 1963

In the words of Werner Heisenberg, who contributed significantly to scientists' development of understanding of electrons in atoms:

"Light and matter are both single entities, and the apparent duality arises in the limitations of our language. It is not surprising that our language should be incapable of describing the processes occurring within the atoms, for, as has been remarked, it was invented to describe the experiences of daily life, and these consist only of processes involving exceedingly large numbers of atoms. (Heisenberg, The Physical Principles of the Quantum Theory, trans. by Carl Eckhart and Frank C. Hoyt, p 10. Chicago: University of Chicago Press, 1930)"

In Mr. Tompkin's quantum jungle, the horsefly shows the full weirdness of having both particle and wave attributes, but the elephant and its rider do not: they follow the laws of "Newtonian" physics that apply to objects. This is because the uncertainty of position of a particle depends on its wavelength, and very large objects like people and elephants have wavelengths so small as to be meaningless. Insects in this mythical world, however, are light enough that their wave-like properties become significant and observable. In our own world, the grey area between classical and quantum behaviour begins with objects very much smaller than horseflies—such as large molecules and small viruses. A beam of C_{60} molecules ("bucky balls"), for example, have recently been shown to undergo diffraction, a property characteristic of waves.

Atoms absorb photons of light, and this is attributed to increase of the energies of their electrons. How does this absorption of light compare with a successful swat of the horsefly in our story? Scientists in the 19th century knew that when white light is passed through a prism, it can be shown to consist of all the wavelengths of the visible spectrum (all the colours of the rainbow). However, it was observed that when a high voltage is applied to gas samples composed of hydrogen atoms, light is emitted that, when passed through a prism, is shown to consist of only a few particular wavelengths. Try to imagine the anguish of the scientific community at the realization that this could only be explained by the inference that the electron in a hydrogen atom can only have particular energies. Such an idea was difficult to accept, much less to explain.

Elements such as mercury and neon were subsequently found to emit their own characteristic set of wavelengths. It took another half-century to develop quantum mechanical models to explain these observations, concluding that the energy of electrons in atoms is *quantized*: they can only have particular energies, characteristic of each element. In this model, the energies of emitted photons of light correspond with the differences between energies that the electrons can have.

As an analogy, imagine a car that can only travel at the speeds 3 km h^{-1}, 8 km h^{-1}, 11 km h^{-1}, or 13 km h^{-1}—and at none of the intermediate speeds. If the car slowed from 8 km h^{-1}, its only possible speed would be 3 km h^{-1}. And in slowing down, energy is emitted that corresponds with the difference between the energy of the car travelling at 8 km h^{-1} and its energy when moving at 3 km h^{-1}. Different sorts of cars, corresponding with atoms of different elements, have different characteristic speeds, and therefore different-sized gaps between their energies.

Line spectra of atomic substances result when particular-sized parcels, or *quanta*, of energy are absorbed, raising electrons to higher energy levels. Then we have atoms in "excited states," just as singlet oxygen molecules have electrons in excited states [<<Section 1.2]. Returning to our previous analogy, each atomic substance requires a different-sized flyswatter, or different-sized photons, to excite its electrons. The reverse process can happen: energy is emitted by excited-state atoms as the energies of their electrons decrease to lower levels. Only particular wavelengths are emitted. If this emitted energy (radiation) is passed through a prism, we obtain a line spectrum.

Now we are in the rather strange situation that we know the exact energies of electrons in atoms, deduced from experimental emission spectra as well as from fundamental mathematical calculation, but we cannot define where they are in the atom. Indeed we cannot even describe *what* they are very well.

Applying images from our macroscopic world to electrons in atoms can be counterproductive to understanding. The behaviour of these electrons is best predicted and described by wrestling with some rather sophisticated mathematics. But the challenge has worthwhile payoffs of understanding and predictive power. Chemistry at the molecular level is all about what happens to electrons when different substances react. In a flurry of activity during collisions between molecules that give rise to reaction, electrons are transferred from one atom to another or shared. *Electrostatic potential maps* [<<Section 6.3; >>Section 14.5], used to predict reactivity, show the distribution of electronic charge in molecules.

Information about the atoms of which substances are composed can be obtained from the characteristic absorption or emission of energy by their electrons. *Atomic spectroscopy* can be thought of as using the appropriate molecular-level fly swatter to provide the exact amounts of energy to "excite" electrons to higher energy levels—and the magnitudes of these is different from element to element.

To illustrate the power of atomic spectroscopy, we turn to another image of a rider travelling on an elephant. This time, the observers are forensic chemists, and the elephant is a real one taken from history.

Instead of watching flies buzz around the head of the traveller, the forensic chemists' focus is the hair on the traveller's head, and modern atomic spectroscopy, rather than a fly swatter. The rider astride the elephant is Napoleon Bonaparte, the powerful ruler of France at the beginning of the 19th century. While waging war in Egypt, he imagined an empire in the Middle East, and him "riding into Asia on an elephant." His dreams were eventually shattered: Napoleon was imprisoned by the British and died in exile on the island of Saint Helena in 1821.

Design for the Elephant Fountain at the Place de la Bastille, 1813-14 (coloured engraving), Alavoine, Jean Antoine (1776-1834) / Musee de la Ville de Paris, Musee Carnavalet, Paris, France / Lauros / Giraudon / The Bridgeman Art Library International

Giant statue of an elephant, placed by Napoleon in the Place de la Bastille.

In 1955, the diaries of Napoleon's valet were published, and the descriptions of Napoleon's health before his death led to theories that arsenic poisoning was the cause of his death. Samples of his hair have been tested by chemists, using atomic spectroscopy. Specialized techniques can identify not only the element arsenic, but also its concentration and the particular arsenic-containing species. Several of these measurements have been reproduced, and lend support to the theory that Napoleon was poisoned. Arsenic levels in his hair were 40 times higher than normal values.

But the story of Napoleon's death is as full of contradictions as was his life. Very recent studies, while confirming high levels of arsenic, raise questions about whether he was intentionally poisoned. One study shows that the wallpaper used in the house in which he spent his final years contained high levels of an arsenic compound used to colour it green. In the humidity of Saint Helena, moulds may have reacted with that compound to produce the poisonous gas arsine. In 2008, an Italian group of 10 scientists reported results of analysis of Napoleon's hair at four different ages, as well as from his wife and son. All showed very high levels of arsenic, suggesting exposure to arsenic compounds in their everyday environment. About that time, small amounts of arsenic were sometimes given as a fashionable medication, thought to stimulate metabolism. There remains more work for the forensic fly swatters.

We look next at experimental evidence about the properties of the elements, and then go on to use a model of electrons in atoms as waves to make sense of the periodic variation of elements' properties.

8.2 Periodic Variation of Properties of the Elements

The periodic table [<<Section 2.12] reflects the **law of chemical periodicity**: *the properties of the elements vary periodically with their atomic numbers*. In this section, we discuss some of the properties that exhibit periodic variation—with some commonality observed in every 8th element as far as $Z = 20$, in every 18th element up to $Z = 56$, and then every 32nd element. Elements with similar properties are arranged in vertical groups in the periodic table. As a result of this periodicity, lithium, sodium, and potassium undergo some reactions that we regard as characteristic of Group 1 elements, while some common chemical behaviours of chlorine, bromine, and iodine are taken to be characteristic of Group 17 chemistry (Figure 8.2).

Perhaps you can appreciate the motivation that drove scientists to ask what it is about the nature of atoms that gives rise to this observed periodicity of behaviour and the number of elements in each cycle.

EXERCISE 8.1—CHEMICAL PERIODICITY

(a) Fluorine has atomic number 9. Which element has an atomic number 8 greater than that of fluorine? And which has an atomic number 18 greater than that? And 18 greater than that? And 32 greater than that? In which group(s) of the periodic table are these elements?

(b) What are the elements of Group 15? What is the difference between the atomic numbers of each successive pair of members?

In Mendeleev's day, the Group 18 elements were not known, so this group was missing from early periodic tables.

Some of the periodically varying properties are those of the elemental substances, such as solid aluminium crystals, liquid bromine, and chlorine gas. These properties, including melting and boiling points, metallic/non-metallic character, and ability to oxidize or reduce other substances, were generally known at the time of the work of Mendeleev (and others) leading to the conception of the periodic table. Some other so-called "atomic properties, such as atomic size, ion size, electronegativity, ionization energy, and electron affinity, are properties of isolated atoms of the elements rather than of the substances. These were not known in Mendeleev's time.

This chapter is concerned with application of the wave model to account for the periodically varying properties of the *main group elements*: members of the *s* block (Groups 1 and 2) and the *p* block (Groups 13–18). The model is extended to the *transition elements* (Groups 3–12) in Chapter 27.

Think about It

e8.1 See the pattern of periodic table trends in melting points and boiling points.

Melting Points and Boiling Points

The melting points and boiling points of the main block elemental substances show a marked periodic variation (Table 8.1). We can imagine the chemists of the 1860s wondering why the melting points of Cl_2, Br_2, and I_2 are low, and why the melting points of C

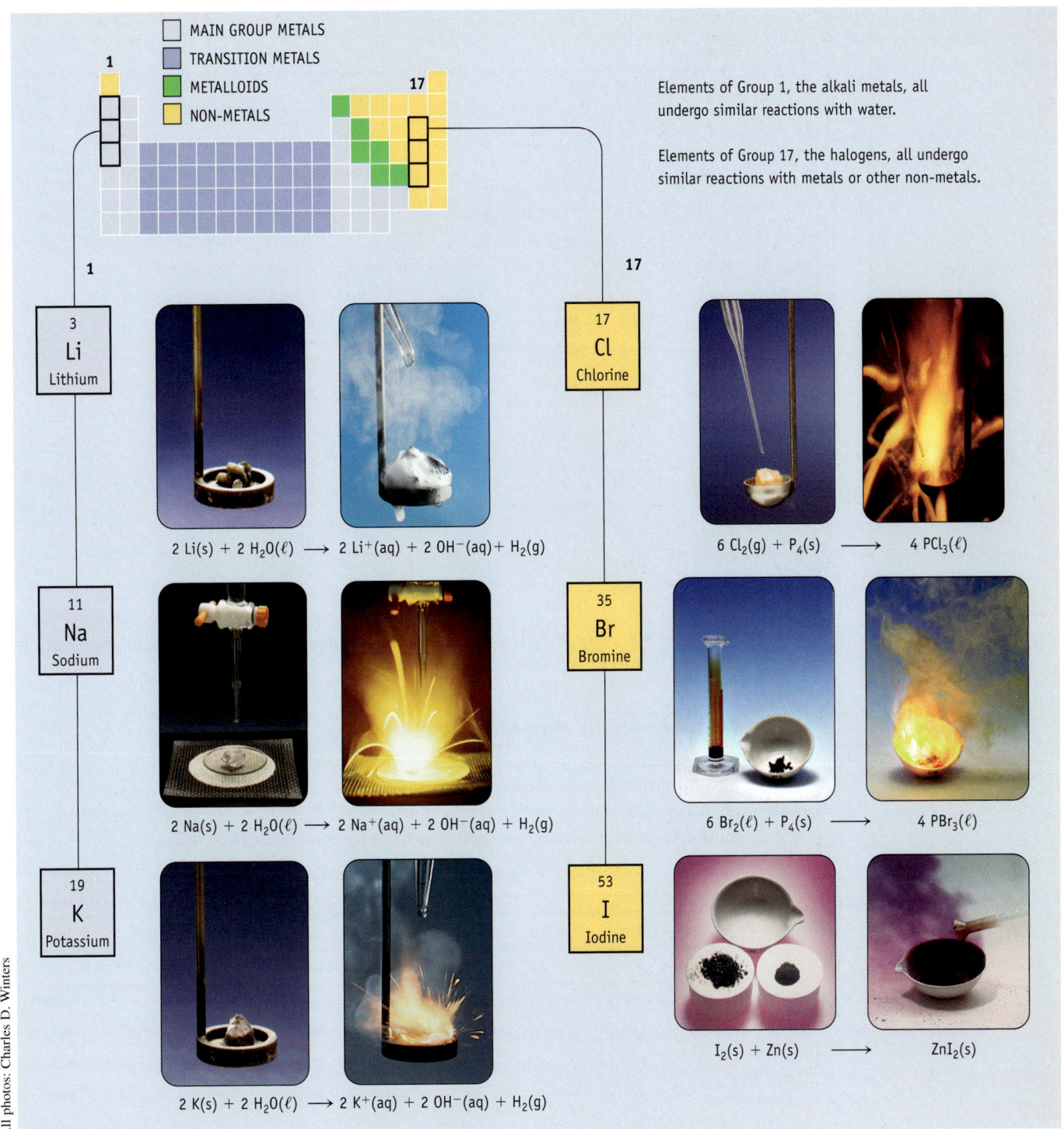

FIGURE 8.2 Examples of the periodicity of Group 1 and Group 17 elements. Dmitri Mendeleev developed the first periodic table by listing elements in order of increasing atomic weight. Elements at regular intervals in the sequence were recognized to have some common observable properties, such as their reaction with water (above), and these elements were placed in vertical columns. We now recognize that the periodic occurrence of similar properties is more meaningful if the elements are arranged in order of atomic number—which was not a known concept in Mendeleev's time.

All photos: Charles D. Winters

TABLE 8.1 Melting Points and Boiling Points of the Elemental Substances

Group	1	2	13	14	15	16	17	18
Period								
1	H₂(g)							He(g)
	−259 °C							←−272 °C
	−253 °C							−269 °C
2	Li(s)	Be(s)	B(s)	C(s)	N₂(g)	O₂(g)	F₂(g)	Ne(g)
	180 °C	1287 °C	2450 °C	3825 °C	−210 °C	−219 °C	−220 °C	−249 °C
	1347 °C	2500 °C	3931 °C	Graphite sublimes near 3800 °C	−196 °C	−183 °C	−188 °C	−246 °C
3	Na(s)	Mg(s)	Al(s)	Si(s)	P₄(s)	S₈(s)	Cl₂(g)	Ar(g)
	98 °C	650 °C	660 °C	1410 °C	44 °C	119 °C	−101 °C	−189 °C
	889 °C	1105 °C	2327 °C	2477 °C	280 °C	445 °C	−34 °C	−186 °C
4	K(s)	Ca(s)	Ga(s)	Ge(s)	As(s)	Se(s)	Br₂(ℓ)	Kr(g)
	63 °C	843 °C	30 °C	947 °C	Sublimes	220 °C	−7 °C	−157 °C
	757 °C	1483 °C	2250 °C	2830 °C	at 615 °C	685 °C	59 °C	−153 °C
5	Rb(s)	Sr(s)	In(s)	Sn(s)	Sb(s)	Te(s)	I₂(g)	Xe(g)
	39 °C	770 °C	156 °C	232 °C	630 °C	450 °C	114 °C	−112 °C
	679 °C	1350 °C	2070 °C	2687 °C	1635 °C	1390 °C	184 °C	−108 °C

and Si are both very high. In this chapter is presented a model of the energies of electrons in atoms that can make sense of these trends.

Metallic versus Non-Metallic Character

The physical properties that characterize metals, and which differentiate them from non-metals, have been discussed in Sections 2.12, 3.2, and 3.6. We cannot measure metallic character, so it is not possible to plot numerical values of metallic character versus atomic number. However, general periodic trends are recognizable (Table 8.2).

TABLE 8.2 Variation of Metallic Character of the Elemental Substances

→ **LESS METALLIC**

Group	1	2	13	14	15	16	17	18
Period								
1	H₂(g)							He(g)
2	Li(s)	Be(s)	B(s)	C(s)	N₂(g)	O₂(g)	F₂(g)	Ne(g)
3	Na(s)	Mg(s)	Al(s)	Si(s)	P₄(s)	S₈(s)	Cl₂(g)	Ar(g)
4	K(s)	Ca(s)	Ga(s)	Ge(s)	As(s)	Se(s)	Br₂(ℓ)	Kr(g)
5	Rb(s)	Sr(s)	In(s)	Sn(s)	Sb(s)	Te(s)	I₂(g)	Xe(g)
6	Cs(s)	Ba(s)	Tl(s)	Pb(s)	Bi(s)	Po(s)	At(s)	Rn(g)

MORE METALLIC ←

Metallic character is characterized by some chemical properties, as well as the more commonly recognized physical properties:

• The more metallic the element, the more ionic in character are its compounds with non-metals. This contrasts with the covalent molecular character of compounds formed between two different non-metals. For example, magnesium chloride,

$MgCl_2(s)$, is typical of ionic compounds [<<Sections 3.2, 3.5], while carbon tetra-chloride, $CCl_4(\ell)$, is a typical molecular compound [<<Sections 3.2, 3.3].

- As a general rule, the more metallic the element, the more basic are its oxides and hydroxides; the more non-metallic the element, the more acidic are its oxides and hydroxides. For example, metal oxides such as potassium oxide (K_2O) are basic because they produce aquated hydroxide ions in solution:

$$K_2O(s) + H_2O(\ell) \longrightarrow 2\,K^+(aq) + 2\,OH^-(aq)$$

On the other hand, sulfur dioxide is acidic, as demonstrated by the formation in water of the weak acid sulfurous acid:

$$SO_2(g) + H_2O(\ell) \longrightarrow H_2SO_3(aq)$$

These observation beg questions such as why the elements with atomic numbers 3, 11, 19, 37, 55, and 87 are more metallic than those whose atomic numbers are one less (2, 10, 18, 36, 54, and 86) or one more (4, 12, 20, 38, 56, and 88). Answers will emerge from applications of wave theory to electrons in atoms in this chapter, along with considerations of bonding between atoms in Chapter 10.

A note of caution: the degree of metallic character exhibited by elements depends on their oxidation state—the higher the oxidation state of a metal in its compounds, the more the compounds are like those of non-metallic elements. For example, lead(IV) chloride ($PbCl_4$) is more like a molecular compound than lead(II) chloride ($PbCl_2$). Also, chromium(VI) oxide (CrO_3) is more strongly acidic then chromium(III) oxide (Cr_2O_3).

EXERCISE 8.2—PERIODIC VARIATION OF METALLIC/NON-METALLIC CHARACTER

From their relative positions in the periodic table, which elemental substance in each of the following pairs do you think is more metallic?
(a) magnesium, Mg(s), or strontium, Sr(s)
(b) calcium, Ca(s), or selenium, Se(s)
(c) rubidium, Rb(s), or arsenic, As(s)

Reactivity as Oxidizing Agents and Reducing Agents

Oxidation-reduction reactions as electron-transfer reactions are summarized in Section 6.10 and discussed more deeply in Chapter 16. A periodic variation in oxidizing abilities and reducing abilities of the elemental substances is observed (Table 8.3).

TABLE 8.3 Periodic Trends in Oxidizing and Reducing Ability of Elemental Substances*

$\longrightarrow$ **MOST POWERFUL OXIDIZING AGENTS**

Group	1	2	13	14	15	16	17
Period							
2	Li(s)	Be(s)	B(s)	C(s)	$N_2(g)$	$O_2(g)$	$F_2(g)$
3	Na(s)	Mg(s)	Al(s)	Si(s)	$P_4(s)$	$S_8(s)$	$Cl_2(g)$
4	K(s)	Ca(s)	Ga(s)	Ge(s)	As(s)	Se(s)	$Br_2(\ell)$
5	Rb(s)	Sr(s)	In(s)	Sn(s)	Sb(s)	Te(s)	$I_2(g)$
6	Cs(s)	Ba(s)	Tl(s)	Pb(s)	Bi(s)	Po(s)	At(s)

MOST POWERFUL REDUCING AGENTS $\longleftarrow$

* Hydrogen and the Group 18 elements have been omitted.

Attempts to develop models of the nature of atoms need to explain why, for example, the elemental substances with atomic numbers 9 (fluorine), 17 (chlorine), and 35 (bromine) are powerful oxidizing agents, while the elemental substances with atomic numbers 3 (lithium), 11 (sodium), 19 (potassium), 37 (rubidium), and 55 (cesium) are powerful reducing agents. We will see that the quantum mechanical model of electrons in atoms helps us to rationalize the differences between how strongly atoms of various elements attract electrons—their own, as well as those of other substances.

EXERCISE 8.3—PERIODIC TRENDS OF OXIDIZING AND REDUCING ABILITIES

(a) Which is the more powerful oxidizing agent: bromine, $Br_2(g)$, or arsenic, As(s)?

(b) Which is the more powerful reducing agent: sodium, Na(s), or silicon, Si(s)?

(c) Which would you be most likely to use to bring about oxidation of a substance that is difficult to oxidize: chlorine, $Cl_2(g)$, or aluminium, Al(s)?

(d) What are the elements with atomic numbers 34, 35, 36, 37, and 38? Which of these elemental substances is the most powerful reducing agent?

Sizes of Atoms

Which do you think is bigger: sodium atoms ($Z = 11$) or polonium atoms ($Z = 84$)? Why? Keep your prediction in mind as you read on.

We cannot directly measure the size of an atom. There are, however, various ways that the sizes of atoms can be indirectly estimated, and these depend on measurements of distances between nuclei of adjacent atoms by means such as X-ray crystallography [>>Section 9.3].

In cases where atoms of the same element are bonded to each other in a molecule (such as Cl_2, O_2, P_4, and S_8) *an estimate of atomic size, called the* **covalent radius**, *is half the experimentally determined distance between the nuclei of the atoms.* For example, in the Cl_2 molecule the internuclear distance is 198 pm, so the covalent radius of a Cl atom is 99 pm (Figure 8.3). Similarly, the distance between the nuclei of adjacent C atoms in diamond is 154 pm, so a covalent radius of 77 pm is assigned to carbon. To test these estimates, we can add them together to predict the distance between C and Cl atoms in CCl_4: the predicted distance of 176 pm agrees with the experimentally measured C—Cl distance.

FIGURE 8.3 Estimating atomic radii. (a) The sum of the covalent radii of C and Cl is a good estimate of the C—Cl distance in molecules. (b) Each sphere in this diagram represents an aluminum atom in a crystal. Measuring the distance shown allows estimation of the metallic radius of aluminum.

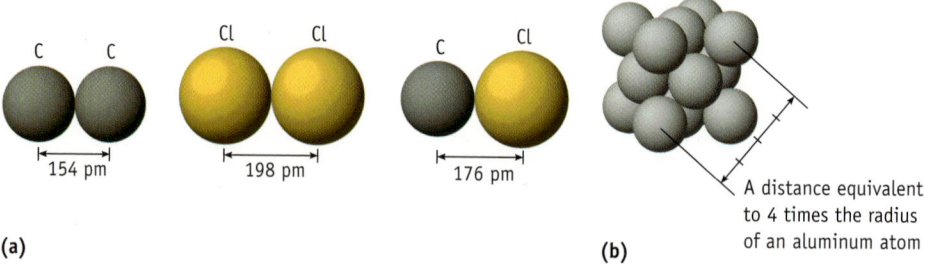

(a)

(b)

A distance equivalent to 4 times the radius of an aluminum atom

In the case of a metallic element, an estimate of atomic radius, called the **metallic radius**, is *half the experimentally determined distance between nuclei of adjacent atoms in a crystal* (Figure 8.3).

Estimates of the atomic radii of atoms of the elements have been compiled, and the periodic nature of these (with increasing atomic number) is evident in Figure 8.4. For the main group elements, atomic radii generally decrease going left-to-right across each period, and going up each group.

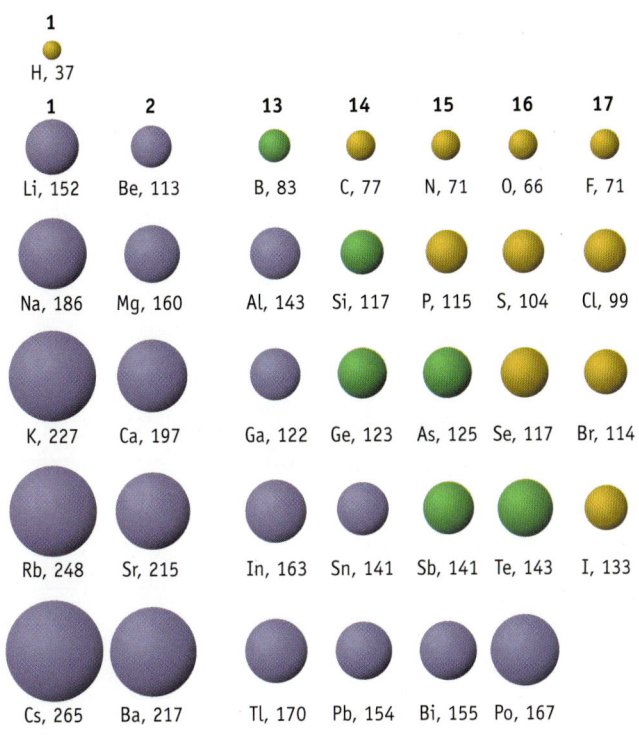

FIGURE 8.4 Atomic radii of main group elements. Values are in picometres (1 pm = 1 × 10⁻¹² m). Going from element to element, there are peaks at atomic numbers 3, 11, 19, 37, and 55 (in Group 1). Each of these is much bigger than the element immediately before it, and the elements after it are progressively smaller. The elements of Group 18 are not shown.

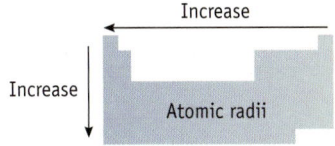

General trends in atomic radii of main block elements with position in the periodic table.

EXERCISE 8.5—PERIODIC TRENDS OF ATOMIC SIZE

Place the three elements Al, C, and Si in order of increasing atomic radius.

Think about It

e8.2 Examine the periodic trend in atomic radius. Is it counterintuitive?

Ionization Energies

For any element, the **first ionization energy** (*IE₁*) *is the minimum energy required to eject an electron from an atom in its lowest energy (ground) state.* For an element with symbol X (which must be in the gas phase), the first ionization energy is the energy required to bring about the process

$$X(g) \longrightarrow X^+(g) + e^-$$

To remove an electron, energy must be supplied to overcome the attraction of the positive charge of the nucleus, so all ionization energies have positive values.

Atoms other than hydrogen have a series of ionization energies, because more than one electron can be removed, in successive steps. For example, the first three ionization energies of magnesium are

Mg (g) $\longrightarrow$ Mg⁺(g) + e⁻   First ionization energy, $IE_1 = 738$ kJ mol⁻¹
Mg⁺(g) $\longrightarrow$ Mg²⁺(g) + e⁻   Second ionization energy, $IE_2 = 1451$ kJ mol⁻¹
Mg²⁺(g) $\longrightarrow$ Mg³⁺(g) + e⁻   Third ionization energy, $IE_3 = 7733$ kJ mol⁻¹

The first ionization energies of the main group elements show periodic variation, with the largest in each row at the Group 18 elements (Figure 8.5). The magnitudes of first ionization energies generally increase from element to element across each period (although there are small "dips'" for Groups 13 and 16 elements), and decrease as we go down each group.

Interactive Exercise 8.4

Arrange the elements in order of increasing atomic radius.

FIGURE 8.5 First ionization energies of the main group elements of the first four periods.

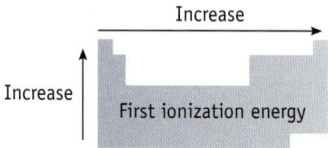

General trends in first ionization energy of main block elements with position in the periodic table.

Think about It

e8.3 Compare the periodic trend in ionization energies with that of atomic radii.

Interactive Exercise 8.6

Arrange the elements in order of increasing ionization energy.

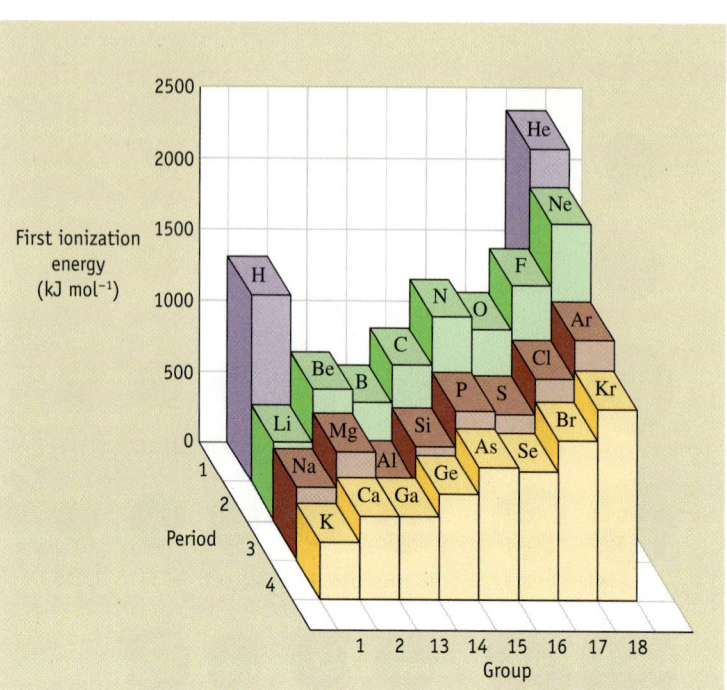

EXERCISE 8.7—TRENDS OF FIRST IONIZATION ENERGIES

Which of the following groups of elements is arranged correctly in order of increasing first ionization energy?
(a) $C < Si < Li < Ne$ (c) $Li < Si < C < Ne$
(b) $Ne < Si < C < Li$ (d) $Ne < C < Si < Li$

For any element, each successive ionization energy is larger than the previous one—that is, $IE_1 < IE_2 < IE_3 < IE_4$, and so on (Table 8.4). A rather remarkable observation, displayed for elements with $Z = 3 - 11$, is that for Group 1 elements IE_2 is very much greater than IE_1, for Group 2 elements IE_3 is very much greater than IE_2, for Group 13 elements IE_4 is very much greater than IE_3, and so on—with the sudden jump in magnitude occurring at the stepped line.

TABLE 8.4 Successive ionization energies of the elements with $Z = 3-11$

Z	Element	Number of Valence Electrons	Ionization energy (MJ mol⁻¹)*									
			IE_1	IE_2	IE_3	IE_4	IE_5	IE_6	IE_7	IE_8	IE_9	IE_{10}
3	Li	1	0.52	7.30	11.81							
4	Be	2	0.90	1.76	14.85	21.01						
5	B	3	0.80	2.43	3.66	25.02	32.82					
6	C	4	1.09	2.35	4.62	6.22	37.83	47.28				
7	N	5	1.40	2.86	4.58	7.48	9.44	53.27	64.36			
8	O	6	1.31	3.39	5.30	7.47	10.98	13.33	71.33	84.08		
9	F	7	1.68	3.37	6.05	8.41	11.02	15.16	17.87	92.04	106.43	
10	Ne	8	2.08	3.95	6.12	9.37	12.18	15.24	20.00	23.07	115.38	131.43
11	Na	1	0.50	4.56	6.91	9.54	13.35	16.61	20.11	25.49	28.93	141.37

*MJ mol⁻¹ = 1 × 10³ kJ mol⁻¹.

EXERCISE 8.8—SUCCESSIVE IONIZATION ENERGIES OF THE ELEMENTS

(a) For which of the elements of the third row of the periodic table is the fourth ionization energy much greater than the third (in comparison to the differences between other successive ionization energies)?

(b) Listed here are successive ionization energies of which third row element?

$IE_1 = 738$ kJ mol^{-1}
$IE_2 = 1\ 451$ kJ mol^{-1}
$IE_3 = 7\ 733$ kJ mol^{-1}
$IE_4 = 10\ 543$ kJ mol^{-1}

Charge on the Monatomic Ions

With some exceptions, the monatomic ions of elements in Groups 1, 2, 13, and 14 are cations, with increasing charge as we go from element to element: the ions of Group 1 have charge +1, those of Group 2 have charge +2, those of Group 13 have +3, and those of Group 14 have +4 (Table 8.5). The monatomic ions of elements of Groups 15, 16, and 17 are anions, with successively smaller charges equal in magnitude to 18 minus the group number (although there are exceptions, particularly with elements lower in the groups).

> The term *monatomic ion* refers to an ion formed from just one atom, such as the Al^{3+} ion or the Cl^- ion—in contrast to diatomic ions, such as the ClO^- ion, or the triatomic ions, CNO^- and AlO_2^-.

Group	1	2	13	14	15	16	17
Period 2	Li^+	Be^{2+}	B^{3+}	C^{4+}	N^{3-}	O^{2-}	F^-
Period 3	Na^+	Mg^{2+}	Al^{3+}	Si^{4+}	P^{3-}	S^{2-}	Cl^-

TABLE 8.5 Charges on Monatomic Ions of the Elements*

* The Group 18 elements have been omitted.

We will see that this periodic variation of charges of ions can be related to the magnitudes of ionization energies, which, in turn, can be rationalized by a powerful theory about the electrons in atoms.

Think about It

e8.4 Examine the trend in ionic charge across each period.

EXERCISE 8.9—CHARGES ON MONATOMIC IONS OF THE ELEMENTS

(a) Barium (Ba) is in Group 2 of the periodic table. What is the charge (the value of n) on Ba^{n+} ions?

(b) Selenium (Se) is in Group 16 of the periodic table. What is the charge (the value of n) on Se^{n-} ions?

Sizes of Ions

The **ionic radius** is *an estimate of the size of an ion in its crystalline compounds.* X-ray diffraction allows us to estimate the distance between nuclei of adjacent cations and anions, which is the sum of their radii (Figure 8.6). The absolute size of each ion cannot be determined from the internuclear distance in just one substance, but by comparing the distances in many compounds, chemists have built up a set of estimates of ion sizes.

Figure 8.7 shows the sizes of ions of some of the main group elements, as well as the sizes of their neutral "parent" atoms. Periodic trends are evident—as with atomic radii, as we go from element to element down a group, ions with the same charge are progressively bigger. Going across any period, the cations of elements of Groups 1, 2, and 13 decrease in size, as do the anions of elements of Groups 15, 16 and 17.

FIGURE 8.6 Ionic radius. The distance between the nuclei of a cation and adjacent anion is the sum of their radii r_+ and r_-.

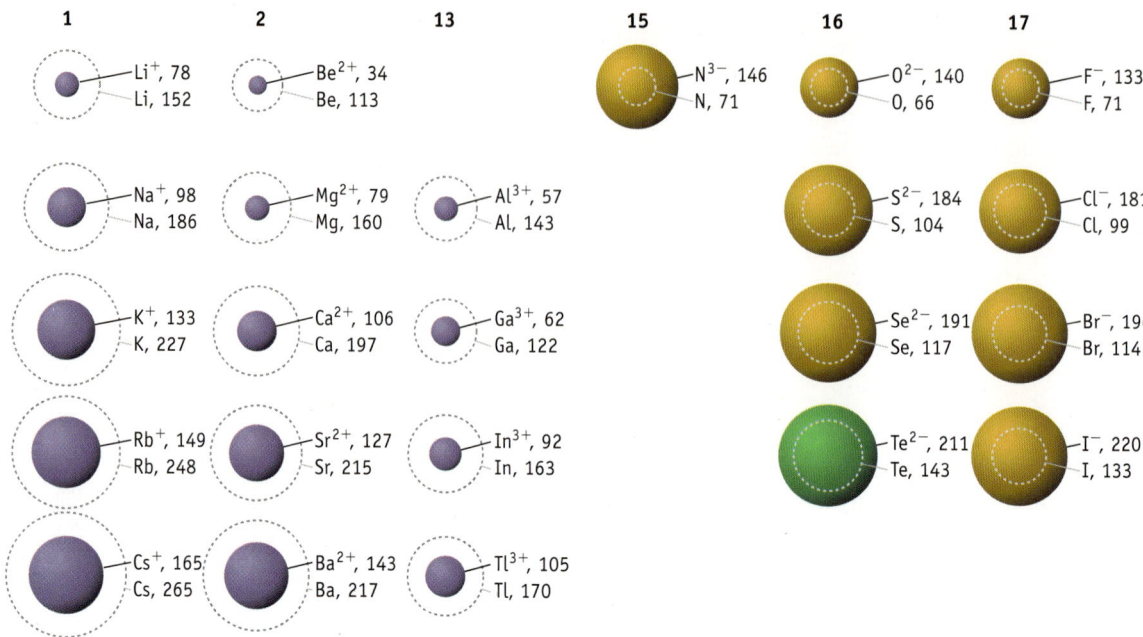

FIGURE 8.7 Estimates of sizes of some common ions, and the neutral atoms. Radii are in picometres (1 pm = 1×10^{-12} m).

It is useful to compare the sizes of *isoelectronic ions,* ions of different elements with the same number of electrons—such as O^{2-}, F^-, Na^+, and Mg^{2+} ions, all of which have 10 electrons—but different numbers of protons. Table 8.6 shows that in a series of isoelectronic ions, the more protons in an ion, the smaller it is.

TABLE 8.6 Sizes of Isoelectronic Ions

Ion	O^{2-}	F^-	Na^+	Mg^{2+}
Number of electrons	10	10	10	10
Number of protons in nucleus	8	9	11	12
Ionic radius (pm)	140	133	98	79

Two important generalizations can be made about the sizes of ions in comparison to the atoms from which they are derived:

- *Cations are smaller than their "parent" atoms.* Compare, for example, the radii of a Li^+ cation and a Li atom:

152 pm

78 pm

- *Anions are larger than their "parent" atoms.* Compare, for example, the radii of a F atom and a F^- ion:

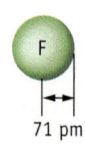

71 pm

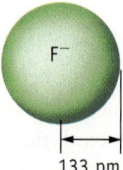

133 pm

Molecular Modelling (Odyssey)

e8.5 Compare the sizes of atoms and their ions. What do we mean by "size" at this level?

Interactive Exercise (Odyssey) 8.10

Compare the sizes of isoelectronic ions.

Interactive Exercise (Odyssey) 8.11

Compare the sizes of atoms and their ions.

In the case of elements that have more than one simple ion, the higher the charge on the ion, the smaller its radius. So, for example, while the radius of an iron atom is 126 pm, that of a Fe^{2+} ion is 78 pm, and that of a Fe^{3+} ion is 65 pm. Cations with high charge and small radius, such as the Al^{3+} ion, the Cr^{3+} ion, and the Pb^{4+} ion, which are said to have a high *charge density*, have salts with considerable covalent character, acidic oxides, high enthalpy of hydration, acidic solutions, and high lattice energies [>>Section 26.3].

Molecular Modelling (Odyssey)

e8.6 Examine the trend in ionic size with nuclear charge.

EXERCISE 8.13—RELATIVE SIZES OF ATOMS AND IONS

Select the atom or ion in each pair that has the larger radius:
(a) Cl atom or Cl^- ion
(b) Mg^{2+} ion or Ba^{2+} ion
(c) K atom or K^+ ion
(d) O^{2-} ion or Se^{2-} ion
(e) Cl^- ion or K^+ ion
(f) Pb^{2+} ion or Pb^{4+} ion

Interactive Exercise 8.12

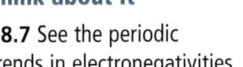

Compare the radii of different atoms and ions.

Electronegativities

The *electronegativity* (χ) of an element is the ability of its atoms to attract the electrons in its bonds with other atoms in a molecule [<<Section 6.3]. Differences between electronegativities of atoms govern the distribution of electrons over the molecule. This gives rise to molecular polarity which, in turn, influences melting points, boiling points, and solubilities [<<Section 6.3]. The distribution of electrons in molecules also influences reactivity, a particular case being the difference of acidity of H atoms in molecules of different acids [>>Section 14.5 and e14.15].

There is a periodicity of variation of electronegativities, with peaks at the Group 17 elements and troughs at the Group 1 elements (Figure 6.11). The consequence of this periodicity is that the trends of electronegativity across the periods, and down the groups, of the periodic table are similar to those of first ionization energy (Figure 8.5).

Hydrogen must be excluded from the generalization that electronegativities are smallest at the top of the groups and to the left of the periods; its electronegativity is about the same as that of phosphorus.

Think about It

e8.7 See the periodic trends in electronegativities.

Electron Affinity

For most elements, there is a force of attraction between an atom and electrons, and energy is released if an electron attaches to an atom to form a negative ion:

$$X + e^- \longrightarrow X^- \qquad \Delta_r H < 0$$

The magnitude of the attraction between an atom and electron is measured as the **electron affinity** (E_{ea}) defined as *the enthalpy change accompanying removal of one electron from the negative ion:*

$$X^- \longrightarrow X + e^- \qquad \Delta_r H = \text{electron affinity}, E_{ea}$$

The higher the electron affinity, the more energy must be used to remove the added electron, and the more positive its numerical value. Estimates of the magnitudes of the electron affinities of the main group elements are shown in Table 8.7.

The general trends of magnitudes of electron affinities of the elements are similar to those for ionization energies—increasing from element to element as we go across a period and up a group. There are some "blips" in the general trend across the periods: the magnitudes of electron affinities of the Group 2 and Group 15 elements are less than the general trend and the values for the Group 18 elements are negative (so energy must be used to attach an electron to an atom).

The IUPAC definition of electron affinity has been used here. It may be more intuitive to think of the magnitude of electron affinity as the amount of energy released when an atom and an electron come together. A negative value is the amount of work required to force an atom and an electron together.

TABLE 8.7 Electron Affinities of Main Group Elements (kJ mol^{-1})

Group	1	2	13	14	15	16	17	18
Period								
1	H							He
	72.8							−21
2	Li	Be	B	C	N	O	F	Ne
	59.8	*	23	122	−7	141	322	−29
3	Na	Mg	Al	Si	P	S	Cl	Ar
	52.9	*	44	134	72	200	349	−35
4	K	Ca	Ga	Ge	As	Se	Br	Kr
	48.3	2.4	36	116	77	195	325	−39
5	Rb	Sr	In	Sn	Sb	Te	I	Xe
	46.9	5.0	34	121	101	190	295	−41
6	Cs	Ba	Tl	Pb	Bi	Po	At	Rn
	45.5	14.0	30	35	101	186	270	−41

* Stable anions of these elements do not exist in the gas phase.

Think about It

e8.8–e8.9 See the periodic trends in electron affinity. Examine the trend in electron affinity down groups and across periods.

EXERCISE 8.14—PERIODIC VARIATION OF ELECTRON AFFINITIES

For each of the following pairs of ions predict which one requires more energy to remove an electron:
(a) N$^-$ ion or O$^-$ ion
(b) Cl$^-$ ion or Br$^-$ ion

EXERCISE 8.15—PERIODIC TRENDS OF PROPERTIES

Compare the three elements C, O, and Si:
(a) Place them in order of increasing atomic radius.
(b) Which has the largest first ionization energy?
(c) Which has the greatest electron affinity: O or C?

As we go from element to element in order of their atomic numbers, there is periodic variation of the properties of the elemental substances, such as melting and boiling points, degree of metallic character, and oxidizing or reducing properties. There is also periodic variation of "atomic" properties, such as atom size, ionization energy, charge on simple ions, ion size, electronegativity, and electron affinity. A model of the nature of atoms needs to be able to explain this periodicity.

8.3 Experimental Evidence about Electrons in Atoms

Why should there be similarities of chemical properties of the elements with, for example, atomic numbers 2, 10, and 18? The discovery of the electron, proton, and neutron [<<Section 2.1] prompted scientists to look for relationships between atomic structure and chemical behaviour. Since chemical properties were taken to be associated with the electrons in atoms, it seemed reasonable to propose that the periodic variation of properties is associated with periodic variation of arrangement or energy of the electrons. From about 1900, experimental and theoretical studies have given rise to understandings about the structure of atoms that correspond with nothing in our

everyday experiences, and yet which have enormous predictive power. The experimental findings revolutionized the way that scientists think about the nature of electrons in atoms.

Perhaps the two most significant experimental observations of that time were the line emission spectra of atoms and the diffraction of electron beams. Interpretation of these phenomena were at the heart of the development of the wave (or quantum mechanical) model of electrons in atoms.

Line Emission Spectra of Excited Atoms

When white light is directed through a prism and diffracted, we see a *continuous spectrum* that consists of all of the colours of the rainbow—that is, all of the wavelengths that make up visible light (Figure 8.8).

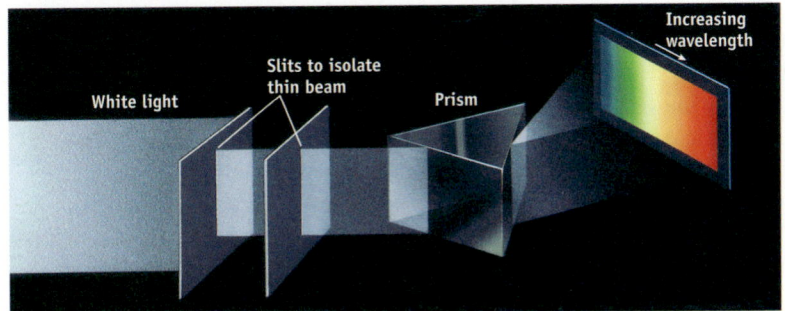

FIGURE 8.8 A spectrum of white light, produced by refraction in a prism. The light is first passed through a narrow slit to obtain a thin beam of light. The beam is then passed through a prism (or, in modern instruments, a diffraction grating). All of the colours, corresponding with a continuously varying range of wavelengths, are visible.

Think about It

e8.10 Review the electromagnetic spectrum and the relationships between energy, wavelength, and frequency.

A familiar example of atomic line emission is the light from a neon advertising sign, in which excited neon atoms emit orange-red light.

If a high voltage is applied to a gaseous sample of atoms of an element (obtained by high temperature vaporization at low pressure), the atoms absorb energy and are said to be "excited." The excited atoms can lose their extra energy by emission of light. However, as discussed in Section 8.1, unlike the continuous spectrum of white light, these *excited atoms emit only particular wavelengths of light—when this light is passed through a prism, only a few coloured lines are seen.* This is called a **line emission spectrum** (Figure 8.9).

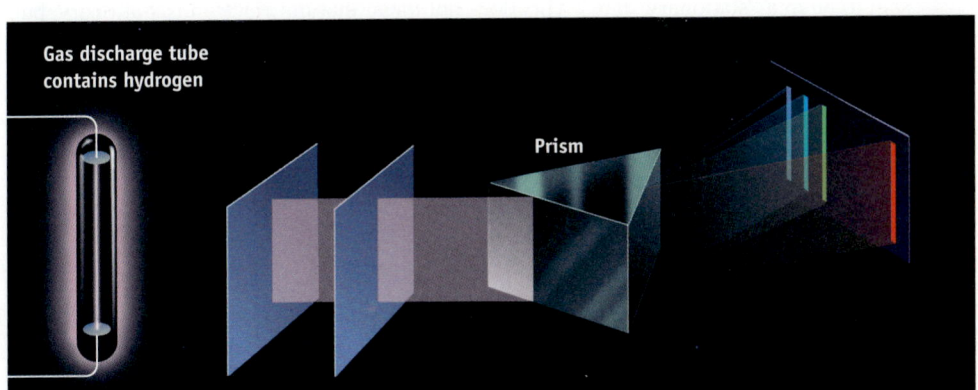

FIGURE 8.9 The line emission spectrum of hydrogen atoms. The emitted light is passed through a series of slits to create a narrow beam of light, which is then separated into its component wavelengths by a prism. A photographic plate or photocell can be used to detect the separate wavelengths as individual lines.

Every element has a unique line spectrum. Those of hydrogen, mercury, and neon are shown in Figure 8.10.

FIGURE 8.10 Line emission spectra from atoms of hydrogen, mercury, and neon. Excited gaseous atoms of different elements produce characteristic spectra that can be used to identify the elements as well as to determine how much of each element is present in a sample.

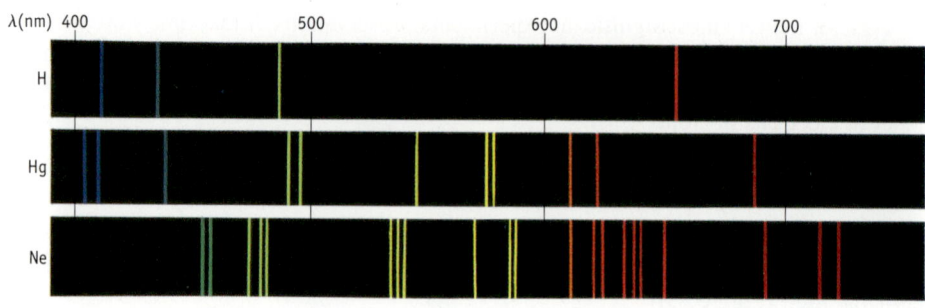

This was an amazing finding. Scientists of the time were puzzled as to what this meant. They came to the rather astonishing conclusion that the energy of electrons in atoms is **quantized**—*they can only have particular amounts of energy*, and not values between the "allowed" levels. This makes sense if we see that the energy of any emission line corresponds with the difference between two "allowed" energies, because as the energy of the electron changes from one "allowed" energy level to a lower "allowed" level, it emits radiation corresponding to the difference between the levels.

By analysis of the line emission spectrum of hydrogen atoms, Johann Balmer (1825–1898) and later Johannes Rydberg (1854–1919) realized that the spectral lines fit patterns. For example, the energies of photons of the red, green, and blue lines in the emission spectrum fit the following relationship, called the *Rydberg equation*:

$$E = R\left(\frac{1}{2^2} - \frac{1}{n^2}\right) \text{ when } n > 2$$

In this equation, n is an integer, and R, now called the *Rydberg constant*, has the value 2.179×10^{-18} J. If we substitute $n = 3$, the calculated energy corresponds with a wavelength (656.3 nm) that matches that of the red line in the hydrogen atom line spectrum. With $n = 4$, the wavelength obtained is that of the green line, and the value $n = 5$ gives the wavelength of the blue line.

For a long time, the question of why the spectral lines should fit this mathematical pattern remained unanswered. Niels Bohr (1885–1962), a Danish physicist, developed a model about electrons in atoms that explained, to some degree, why line emission spectra are observed, and predicted accurately the energies of the lines in the hydrogen atom spectrum. The following postulates of his model of the electron in a hydrogen atom are still useful:

- The energy of the electron can only have particular "allowed" values, and then it is said to be in a "stationary state." This does not mean that the electron is stationary, but that its energy in this state remains constant.
- While the electron is in any of its stationary states, it does not radiate energy.
- The energy of the electron can increase from that of one stationary state to another by absorption of a photon of radiation whose energy corresponds exactly with the difference of energies of the two states. A decrease of energy is accompanied by emission of a photon whose energy is equal to the difference between the energies of the higher and lower stationary states.

Bohr's images of the electron's stationary states were defined orbits of a particle revolving around the nucleus—similar to that of the planets circling the sun. This is no longer an acceptable model (Figure 8.1): while Bohr's initial model was useful as a first approximation or an initial understanding, it was replaced over the next decade by a much better understanding of the nature of electrons in atoms. Nevertheless, Bohr did deduce an equation for the energies of the stationary states of the electron in an H atom that was consistent with the experimentally derived Rydberg equation, and which is useful to understand the origin of the line emission spectra. Applying the mathematics of electrostatic

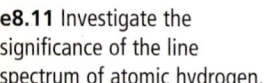

Think about It

e8.11 Investigate the significance of the line spectrum of atomic hydrogen.

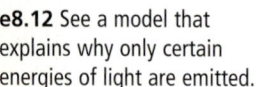

Think about It

e8.12 See a model that explains why only certain energies of light are emitted.

attraction between oppositely charged objects (the nucleus and the electron), and the mathematics of outward forces acting on revolving objects (the electron), Bohr deduced the following equation for the energy levels of electrons in atoms:

$$E_{electron} = -2.179 \times 10^{-18} \left(\frac{Z^2}{n^2} \right) \text{ J}$$

in which Z is the charge on the nucleus of the atom, and n, called the *principal quantum number*, is any integer 1, 2, 3, . . .

In the particular case of a hydrogen atom, $Z = 1$ and this equation becomes:

$$E_{electron} = -\frac{2.179 \times 10^{-18}}{n^2} \text{ J}$$

The ground state of an atom is when its electrons are in the lowest energy condition. In the case of hydrogen atoms, this is the stationary state defined by $n = 1$, the energy of which is

$$E_1 = -2.179 \times 10^{-18} \text{ J}$$

The energy of the electron in the next level ($n = 2$) is

$$E_2 = -\frac{2.179 \times 10^{-18}}{2^2} \text{ J} = -5.45 \times 10^{-19} \text{ J}$$

The energies of some of the stationary states of an H atom electron are shown in Figure 8.11.

We can calculate the energy of a photon absorbed or emitted when the electron undergoes a transition from one allowed level to another as the difference between their energies, ΔE. For example, when an electron in the $n = 1$ level of a hydrogen atom undergoes "excitation" to the $n = 2$ level:

$$\Delta E = E_2 - E_1 = -5.45 \times 10^{-19} \text{ J} - (-2.179 \times 10^{-18} \text{ J}) = 1.634 \times 10^{-18} \text{ J}$$

This is the energy of a photon absorbed due to transition of the energy of one electron. For 1 mol of atoms whose electrons undergo this transition:

$$\Delta E = (6.022 \times 10^{23})(1.634 \times 10^{-18}) \text{ J} = 9.84 \times 10^5 \text{ J} = 984 \text{ kJ}$$

Excitation and de-excitation between the $n = 1$ and $n = 2$ levels are represented in Figure 8.12.

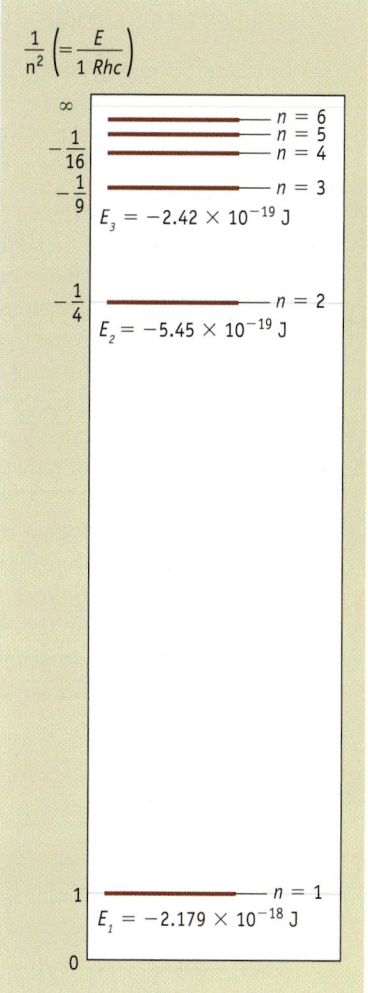

$$\frac{1}{n^2} \left(= \frac{E}{1\,Rhc} \right)$$

$n = 6$
$n = 5$
$n = 4$

$n = 3$
$E_3 = -2.42 \times 10^{-19}$ J

$n = 2$
$E_2 = -5.45 \times 10^{-19}$ J

$n = 1$
$E_1 = -2.179 \times 10^{-18}$ J

FIGURE 8.11 Energies of stationary states of the electron in an H atom, deduced from the Bohr model. The larger the value of n, the larger (less negative) the value of the energy. The negative numerical values result from the convention that the stronger the electrostatic attraction between oppositely charged objects, the more negative the value of the potential energy. If the electron were to obtain sufficient energy to become independent of the nucleus (that is, to be ionized), its energy is zero. The difference between successive energy levels becomes smaller as n becomes larger.

Think about It

e8.13 See why the Bohr model was so successful for interpreting the hydrogen spectrum.

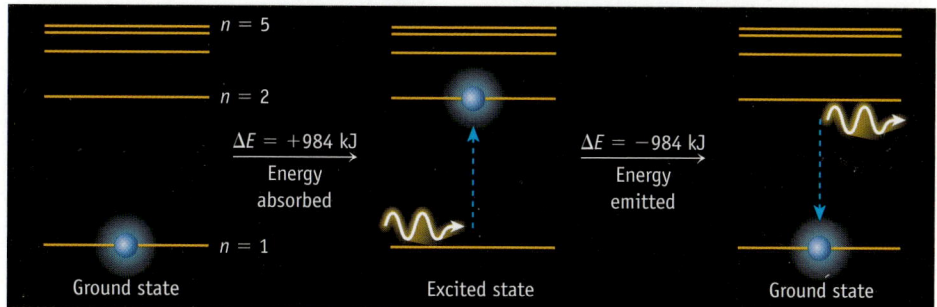

FIGURE 8.12 Absorption and emission of energy by 1 mol of H atoms due to transitions of the electron between the $n = 1$ and $n = 2$ levels. Energy is absorbed when the electron is excited from the $n = 1$ state to the $n = 2$ state. When the electron "relaxes" to the $n = 1$ state from $n = 2$, energy is emitted. Photons absorbed during this excitation, or emitted during relaxation to the ground state, have energy exactly 984 kJ mol^{-1}.

EXERCISE 8.16—ENERGIES OF ELECTRONS AND TRANSITIONS

(a) Calculate the energy of the $n = 3$ state of the H atom in (a) J atom^{-1}, and (b) kJ mol^{-1}.

(b) Using the energy of the $n = 3$ level calculated above, calculate the energy emitted per mol of electrons making the downward transition from the $n = 3$ level to the $n = 2$ level.

Charles D. Winters

The colours observed when the white solids sodium chloride (left), strontium chloride (middle), and boric acid (right) are soaked in methanol and ignited.

Bohr's model helps us to make sense of the observed line emission spectra. We can imagine that an electric potential can "excite" ground-state H atoms by raising the energy of their electrons—some to the $n = 2$ level, some to the $n = 3$ level, and so on. The electrons in the excited atoms naturally "relax" to lower levels (either directly or in a series of steps) and release the exact amount of energy corresponding with the difference between levels as quanta of radiation.

A series of emission lines having energies in the ultraviolet region (called the *Lyman series*, Figure 8.13) can be attributed to transition of electrons to the $n = 1$ state from states with $n > 1$. The *Balmer series* of lines in the visible region arises from transitions to the $n = 2$ state. Compounds of other elements with line emissions in the visible region are used to provide the colours of fireworks.

Think about It

e8.14 Apply these ideas to the colours of fireworks and flame tests for metals.

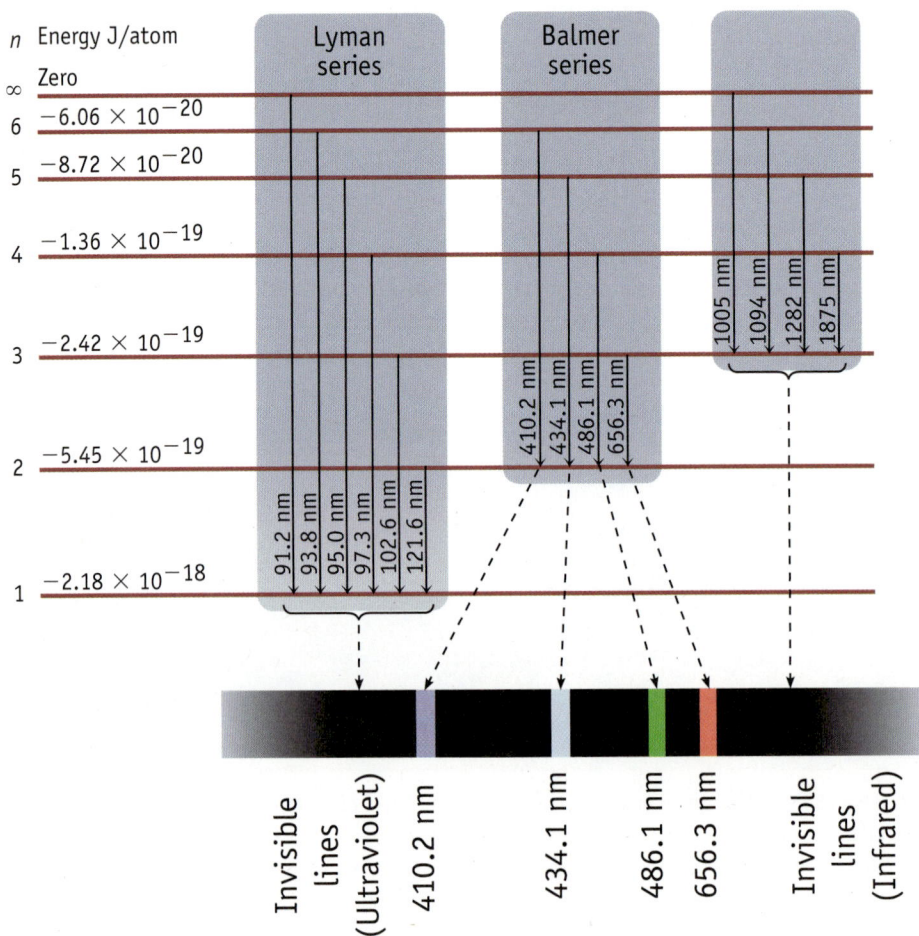

FIGURE 8.13 Some of the electronic transitions that can occur in an excited H atom. The Lyman series of lines in the ultraviolet region results from transitions to the $n = 1$ level. Transitions to the $n = 2$ level occur in the visible region (Balmer series). Lines in the infrared region result from transitions to the $n = 3$ or $n = 4$ levels. Transitions to the $n = 1$ level emit photons with more energy (shorter wavelength radiation) than those emitted by transitions to higher levels.

WORKED EXAMPLE 8.1—ENERGIES OF EMISSION LINES FROM EXCITED ATOMS

Calculate the wavelength of the green line in the visible spectrum of excited H atoms.

Strategy

First locate the green line in Figure 8.13 and determine which energy levels are involved. Then calculate the energy of the photon that would be emitted by calculating the difference between the energies of the two levels. Then, from the energy of the photon, calculate its wavelength.

Solution

The green line is the second most energetic line in the visible spectrum of hydrogen and arises from electrons moving from $n = 4$ to $n = 2$:

$$E_{photon} = E_4 - E_2 = \frac{-2.179 \times 10^{-18}\text{ J}}{4^2} - \frac{-2.179 \times 10^{-18}\text{ J}}{2^2}$$

$$= \left(-\frac{1}{16} + \frac{1}{4}\right)(2.179 \times 10^{-18}\text{ J}) = 4.086 \times 10^{-19}\text{ J}$$

Now we can apply Planck's equation that relates the energy, frequency, and wavelength of light:

$$E_{photon} = h\nu = \frac{hc}{\lambda} \quad \text{so } \lambda = \frac{hc}{E_{photon}}$$

in which h is Planck's constant (6.626×10^{-34} J s) and c is the speed of light (2.998×10^8 m s^{-1}):

$$\lambda = \frac{hc}{E_{photon}} = \frac{(6.626 \times 10^{-34}\text{ J s})(2.998 \times 10^8\text{ m s}^{-1})}{4.086 \times 10^{-19}\text{ J}} = 4.862 \times 10^{-7}\text{ m} = 486.2\text{ nm}$$

The experimental value is 486.1 nm (Figure 8.13). This represents excellent agreement between experiment and theory.

Interactive Exercise 8.17

Investigate the line emission spectrum of neon.

EXERCISE 8.18—ENERGY OF A SPECTRAL LINE OF ATOMIC HYDROGEN

The Lyman series of spectral lines for the H atom occurs in the ultraviolet region. They arise from transitions from higher levels to $n = 1$. Calculate the frequency and wavelength of the light emitted by de-excitation from the $n = 4$ level in this series.

Although Bohr's model of atoms was conceived through extraordinary insight and provides a useful rationalization of atomic line emission spectra, it has serious deficiencies:

- Scientists no longer accept that electrons are particles circling the nucleus. Based on this model, we should expect destruction of atoms because the electrons would spiral into the nucleus.
- Bohr's allocation of the values $n = 1, 2, 3$, etc. had no theoretical basis. These were used arbitrarily, because they gave results that fitted the experimental observations of line emission spectra of atomic gases.
- Bohr had no explanation for quantization of electron energies; the idea seemed necessary to explain the line emission spectra, but he had no sense why this should be.
- While it gave accurate correspondence with the energies of spectral lines for hydrogen (as well as for other one-electron species such as the He$^+$ ions and Li^{2+} ions), it failed rather badly in its prediction of the spectra of atoms of other elements, because it could not allow for inclusion of electron-electron repulsions in the calculations.

This was a model moulded to fit the experimental observations. Scientists of the time still sought understanding about atomic structure that was based on fundamental principles, could predict the experimental data, and was applicable to all atoms and ions. The observation described next, and its profound consequences, provided the means to achieve this.

The line emission spectra of excited atoms are interpreted to mean that the energies of electrons in atoms are quantized: the energies of the spectral lines correspond with differences between energy levels of the electrons, consistent with the equation

$$E = R\left(\frac{1}{2^2} - \frac{1}{n^2}\right) \qquad \text{when } n > 2$$

Wave Properties of the Electron: Wave-Particle Duality

During the early 1900s, scientists including Albert Einstein (1879–1955) were trying to explain the phenomenon known as the *photoelectric effect*: the ejection of electrons when light strikes the surface of a metal. This effect is the basis of photoelectric cells, which are now commonly used in automatic door openers in stores and elevators.

An electric potential is applied to the cell portrayed in Figure 8.14. When light strikes the cathode of the cell, electrons are ejected from the cathode surface and move to a positively charged anode. A stream of electrons—a current—flows through the cell. Thus, the cell can act as a light-activated switch in an electric circuit.

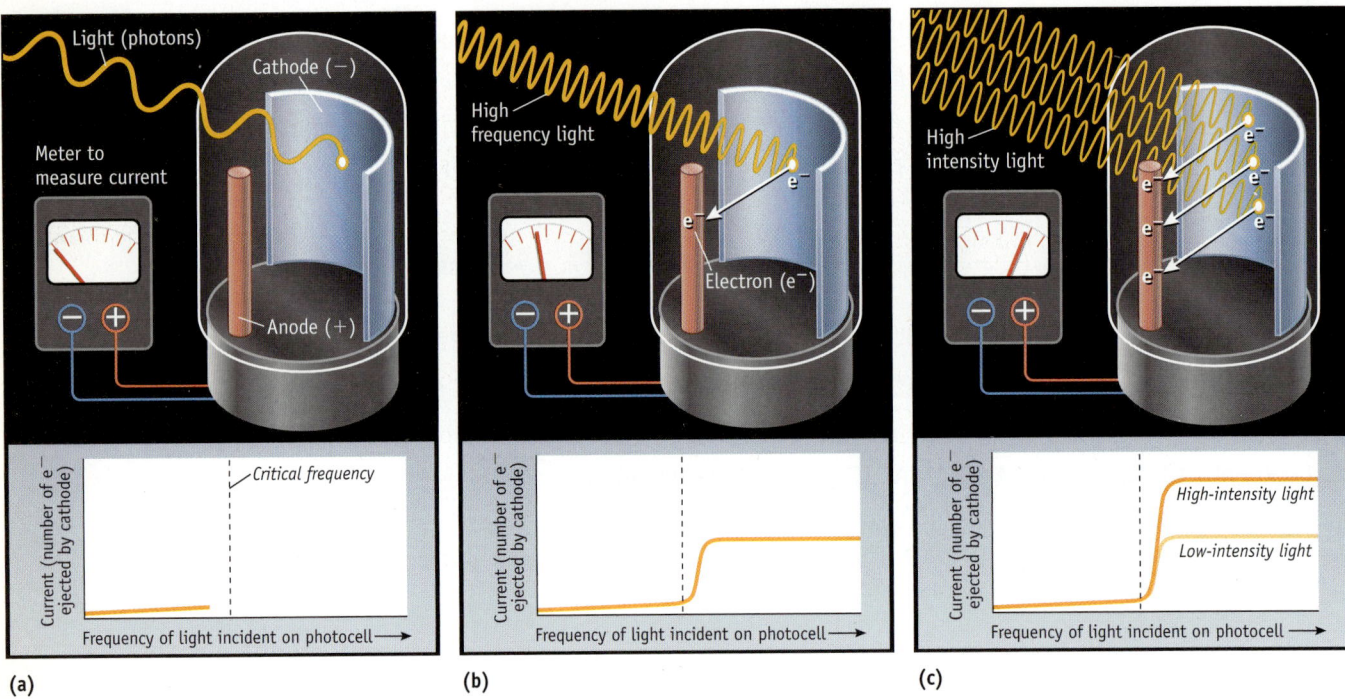

(a)   (b)   (c)

FIGURE 8.14 A photoelectric cell. (a) A light-sensitive cathode, usually a metal, ejects electrons if struck by light of sufficient energy (sufficiently high frequency). No current flows until the threshold frequency is reached. (b) With light of higher frequency than the threshold value, the excess energy allows electrons to escape the atom with greater speed. Current flow consists of ejected electrons moving to the anode. (c) With higher intensity light, more electrons are released from the surface. The onset of current is observed at the same frequency as with lower intensity light.

Think about It

e8.15 Simulate the photoelectric effect to study the relationship between the intensity and energy of light, the choice of metal, and the current produced.

Experiments show that electrons are ejected from the surface only if the frequency of the light is higher than a certain threshold value, dependent on the identity of the metal. If lower-frequency light is used, no effect is observed, regardless of the light's intensity (brightness). In contrast, if the frequency is above the minimum, increasing the light intensity causes a higher current (more electrons) to flow.

Einstein decided the experimental observations could be explained by the startling assumption that light has particle-like properties. He proposed that light, previously thought of only as waves, could be conceived as massless "particles," now called *photons*. The energy of each photon is proportional to the frequency of the radiation, as given by Planck's equation ($E = h\nu$).

Einstein's proposal helps us understand the photoelectric effect. It is reasonable to suppose that a high-energy particle bumping into an atom could cause the atom to lose an electron. It is also reasonable to suppose that an electron can be torn away from the atom only if the photon has enough energy. If electromagnetic radiation is described as a stream of photons, then the greater the intensity of light, the more photons are available to strike a surface per unit of time. However, the atoms of a metal surface will not lose electrons when the metal is bombarded by millions of photons if none of them is energetic enough.

Following Einstein's convincing argument that light can have the properties of particles, Louis Victor de Broglie (1892–1987) pondered whether a tiny object such as an electron, normally considered a particle, might also exhibit wave properties. In 1925, he proposed that a free electron of mass m moving with a velocity v should have an associated wavelength λ, related through Planck's constant h by the equation

$$\lambda = \frac{h}{mv}$$

This idea was revolutionary because it linked the particle properties of the electron (m and v) with a wave property (λ). Experimental proof was soon produced. In 1927, C.J. Davisson (1881–1958) and L.H. Germer (1896–1971), working at Bell Telephone Laboratories in New Jersey, found that a beam of electrons was diffracted like light waves by the atoms of a thin sheet of crystalline material acting as a diffraction grating (Figure 8.15) and that de Broglie's relation was followed quantitatively.

De Broglie's equation suggests that any moving particle can be thought of as a wave of particular wavelength—the higher the mass, the smaller the wavelength. Because h is very small, unless the product of mv is very small, the wavelength λ is insignificant in relation to the size of the object. For example, a 114 g baseball travelling at 180 km h^{-1} (25 m s^{-1}) can be calculated to have a wavelength of 1.8×10^{-34} m—immeasurable, and insignificant with respect to the size of the ball, and so we never think of it as a wave. Of course, if we were in a world with a much larger value of Planck's constant (as in Section 8.1), the batter would have trouble striking at an approaching wave!

For particles of extremely small mass, such as protons, electrons, and neutrons, the wavelength of the corresponding wave may be quite significant, as Worked Example 8.2 demonstrates.

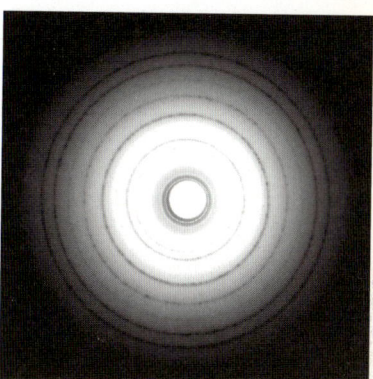

R.K. Bohn, Department of Chemistry, University of Connecticut

FIGURE 8.15 Diffraction patterns. This pattern is obtained by passing X-ray radiation through a thin film of magnesium oxide (MgO). To explain diffraction of an electron beam we need to think of moving electrons as a form of radiation.

WORKED EXAMPLE 8.2—USING DE BROGLIE'S EQUATION

Calculate the wavelength of radiation that corresponds with an electron of mass $m = 9.109 \times 10^{-28}$ g that travels at 40.0% of the speed of light.

Strategy

Use de Broglie's equation, taking particular care with the units. 1 J = 1 kg m^{-2} s^{-2}. Therefore, the electron mass must be in kg and speed in m s^{-1}.

Solution

Electron mass = 9.109×10^{-31} kg
Electron speed (40.0% of light speed) = $(0.400)(2.998 \times 10^8$ m s$^{-1}) = 1.20 \times 10^8$ m s^{-1}
Substituting these values into de Broglie's equation, we have

$$\lambda = \frac{h}{mv} = \frac{6.626 \times 10^{-34} \text{ J s}}{(9.109 \times 10^{-31} \text{ kg})(1.20 \times 10^8 \text{ m s}^{-1})} = 6.07 \times 10^{-12} \text{ m}$$

Check that the units cancel appropriately if you realize that 1 J = 1 kg m^2 s^{-2}

Comment

This wavelength can be compared with the atomic radius of a H atom of 37×10^{-12} m.

EXERCISE 8.19—USING DE BROGLIE'S EQUATION

Calculate the wavelength associated with a 1.0×10^2 g golf ball moving at 30. m s^{-1}. How fast must the ball travel to have a wavelength of 5.6×10^{-3} nm?

Diffraction experiments show that electrons can be described as having wave properties, yet the deviation of electrons moving in a magnetic field is a phenomenon characteristic of charged particles. Scientists now accept the notion of **wave-particle duality**: *the idea that electrons have properties of both particles and waves.* It would be an error to think that electrons have double identities, changing from one to the other as the occasion demands. Rather, duality is in our minds: the nature of the electron (whatever it is) is difficult, even mysterious, for us to understand, and sometimes we have to think about them as though they are particles and sometimes as waves.

The notion that electrons in atoms might be thought of as waves created the opportunity to formulate a new and extremely powerful model of electrons in atoms.

Electrons exhibit properties of waves, as well as of particles, with the relationship between the mass of the particle and the wavelength of the corresponding wave given by

$$\lambda = \frac{h}{mv}$$

This phenomenon is called wave-particle duality, where the duality is in our minds rather than some mysterious ability of electrons to switch their characteristics.

8.4 The Quantum Mechanical Model of Electrons in Atoms

During the 1920s, Austrian scientist Erwin Schrödinger (1887–1961) worked toward a theory of the behaviour of electrons in atoms, based on the hypothesis that they could be described by equations for wave motion. He developed the model that has come to be called *quantum mechanics* or *wave mechanics*. A simplified view of Schrödinger's objective was to mathematically formulate the conditions for three-dimensional "standing waves" that correspond with the stationary states of electrons in atoms.

The concept of standing waves is perhaps most easily understood by considering one-dimensional waves. If you tie down a string of length ℓ at both ends, as you would the string of a guitar, and pluck it, the string can vibrate as any of several standing waves (Figure 8.16). Importantly, the standing waves of the string are *quantized*: only certain wavelengths are possible. The condition for a standing wave in one dimension is given by the mathematical expression

$$\lambda = \frac{2\ell}{n} \qquad n = 1, 2, 3, 4, \ldots$$

This introduces the idea of a **quantum number**: *the parameter n in the general form of the equation for a standing wave, any "allowed" value of which gives rise to a solution for the equation.* We will see that similar, but more complicated, mathematical equations describe the stationary states of electrons (as three-dimensional waves) in atoms—and that solutions are possible only for "allowed" values of three quantum numbers.

We can also see from Figure 8.16 that standing waves in one dimension have at least two points of zero amplitude, called *nodes*. The distance between nodes of any standing wave is $\lambda/2$.

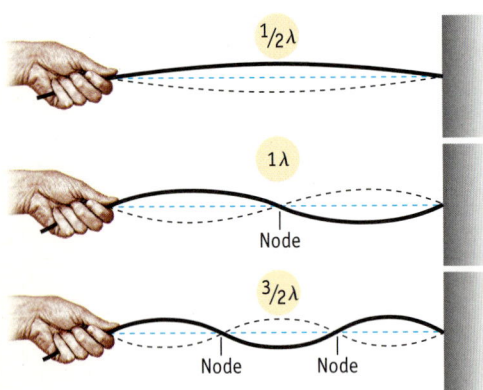

FIGURE 8.16 Standing waves. Only certain wavelengths of vibration are possible for a standing wave to exist. If the length of the string is ℓ, the first standing wave has $\lambda = 2\ell$ ($\ell = \lambda/2$), the second has $\lambda = \ell$ ($\ell = 2\lambda/2$), and the third has $\lambda = 2\ell/3$ ($\ell = 3\lambda/2$). The generalized equation for standing waves in one dimension is $\lambda = 2\ell/n$, where $n = 1, 2, 3, \ldots$. Vibrations of the string with, for example, $\lambda = 2\ell/3.5$, or $\lambda = 2.178\ell/3$ are not standing waves. In fact, vibrations with these wavelengths would be destructive waveforms.

EXERCISE 8.20—STANDING WAVES

Draw the stationary waves of the piece of string portrayed in Figure 8.16 when (a) $n = 4$; (b) $n = 5$.

The **Schrödinger equation** is *a generalized mathematical equation for three-dimensional standing "electron waves" around an atom's nucleus*. Its complexity is hidden in the form

$$H\psi = E\psi$$

in which E is the energy of an electron in a particular standing wave, and ψ, called the *wave function* or *atomic orbital*, is a function that varies over three-dimensional space in a unique way for each standing wave. The "Hamiltonian operator," H, represents a set of mathematical operations that, when performed on ψ for a standing electron wave, leads to a value of its energy, E. As for one-dimensional standing waves, solutions are obtained only under certain conditions.

A simplified (but still not very informative) version of the Schrödinger equation as it applies to the electron in a hydrogen atom is

$$H\psi = \frac{h^2}{2m}\left\{\frac{\partial^2\psi}{\partial x^2} + \frac{\partial^2\psi}{\partial y^2} + \frac{\partial^2\psi}{\partial z^2}\right\} - \frac{e^2}{r} = E\psi$$

The Schrödinger equation describes standing waves only when the values of three quantum numbers (not visible in the forms of the equations shown above) simultaneously have particular "allowed" values:

- The principal quantum number (n) must have an integer value 1, or 2, or 3, . . . up to ∞.
- For any value of n, the angular momentum quantum number (l) must have an integer value 0, or 1, or 2, . . . up to ($n-1$).
- For any value of l, the magnetic momentum quantum number (m_l) can have any integer value from $-l$, including 0, to $+l$.

Each value of n limits the range of "allowed" values of l, and each value of l limits the values of m_l. When any set of allowed values of n, l, and m_l is inserted into the Schrödinger equation, the solution to the equation corresponds with a standing electron wave. Each standing wave solution to the Schrödinger equation comprises two parts: an exact value of the energy of the electron wave (E) and a wave function (ψ), the significance of which we will discuss shortly. For example, if we insert into the Schrödinger equation the values $n=2$, $l=1$, and $m_l=0$, then the equation describes a standing electron wave, which we can solve to obtain the exact value of $E_{2,1,0}$ and the expression for $\psi_{2,1,0}$.

Since each stationary wave solution to the Schrödinger equation leads to a unique value of energy (E), we see that quantization of electron energies flows naturally from application of the wave model.

Chemists have historically used labels for various solutions to the wave equation, dependent on the value of quantum number l used in the wave equation:

Value of l Used	Label
$l = 0$	s orbital
$l = 1$	p orbital
$l = 2$	d orbital
$l = 3$	f orbital

As examples, a solution derived from the Schrödinger equation by using $n=2$ and $l=1$ (whatever value of m_l is used) is called a $2p$ orbital, and the wave function $\psi_{4,0,0}$ is that of a $4s$ orbital.

EXERCISE 8.22—ATOMIC ORBITALS

Will insertion of the following values in the Schrödinger equation satisfy the conditions for a standing electron wave?
(a) $n=4$, $l=2$, $m_l=-2$ (b) $n=1$, $l=2$, $m_l=-2$ (c) $n=3$, $l=1$, $m_l=-2$

Interactive Exercise 8.21

List allowed quantum numbers for a standing wave in an orbital.

Think about It

e8.16 Interpret the quantum numbers for orbitals.

Early studies of the emission spectra of elements classified lines into four groups on the basis of their appearance. These groups were labelled *sharp*, *principal*, *diffuse*, and *fundamental*. From these names came the labels we now apply to orbitals: *s*, *p*, *d*, and *f*.

Schrödinger and others applied the mathematics of waves to electrons in atoms. The conditions for the electrons to be standing waves include particular values of the quantum numbers n, l, and m_l.

Language Issues

There are some common terms that are important to understand:

- Technically, the term **orbital** refers to *a mathematical function (symbol ψ), which is one part (the other is the energy, E) of the solution to the wave equation for each stationary wave. For practical purposes, it is useful to think of orbitals as the non-uniform distribution of the electron matter over space around the nucleus.* The term *orbital* should not be confused with *orbit*, which refers to the path of an object revolving around some point; it is counterproductive to think of electrons travelling in orbits around the nucleus.

- *Orbitals of an atom that have been derived by use of the same value of n in the Schrödinger equation* are said to be in the same **shell**. For example, the 2s orbital ($n = 2$, $l = 0$) and the three 2p orbitals ($n = 2$, $l = 1$) are all part of the same $n = 2$ shell.

- In a given shell, *each set of orbitals derived by use of a particular value of l in the Schrödinger equation* are said to be in the same **sub-shell**. For example, in the $n = 3$ shell, the 3s sub-shell is comprised of the 3s orbital ($l = 0$, $m_l = 0$), the 3p sub-shell is comprised of three p orbitals ($l = 1$ and $m_l = -1$, $m_l = 0$, $m_l = 1$), while the 3d sub-shell is comprised of five d orbitals ($l = 2$ and $m_l = -2$, $m_l = -1$, $m_l = 0$, $m_l = 1$, $m_l = 2$).

- If a solution for a standing electron wave is derived from the Schrödinger equation by insertion of the quantum numbers $n = 2$ and $l = 1$, it is common for chemists to say that the electron "occupies" one of the (three) 2p orbitals, or to say that it is a 2p electron. It is important to realize that these are just shorthand forms of language: there is not a region of space, called an orbital, that an electron can enter and exit.

- It is common to say that an orbital has a certain energy, when we mean that an electron in a particular standing waveform has that energy.

- If just one electron in an atom has a particular standing waveform (defined by a particular set of n, l, and m_l), it is called an *unpaired electron*. If two electrons have the same standing waveform (they "occupy" the same orbital) they are said to be *paired electrons*. We will see that no more than two electrons can have the same standing waveform.

> The term *orbital* refers to a mathematical description of a standing electron waveform, and is different from an orbit. Orbitals of an atom derived with the same value of n are said to make up a shell. Orbitals in a shell with the same value of l together form a sub-shell. If only one electron has a particular waveform, it is said to be unpaired.

How Many Standing Waves Are Possible?

In a hydrogen atom, there is only one electron, but it may take the form of any of the many possible standing waves, each of different energy—although at any instant it can exist as only one of these. There are an infinite number of possible standing waveforms that electrons can have in atoms, one for every combination of allowed quantum numbers in the Schrödinger equation. We will be interested only in the first 30 or so. From the definition of sub-shell above:

Number of orbitals in a sub-shell = number of values of $m_l = 2l + 1$

Number of Orbitals in Types of Sub-shell	Type of Sub-shell	Number of Orbitals in Sub-shell
	s ($l = 0$)	1
	p ($l = 1$)	3
	d ($l = 2$)	5
	f ($l = 3$)	7

For each shell, the sub-shells and orbitals are summarized in Table 8.8.

Principal Quantum Number	Angular Momentum Quantum Number	Magnetic Quantum Number	Number and Type of Orbitals in the Sub-shell	
Symbol = n Values = 1, 2, 3 . . . n = number of sub-shells	Symbol = l Values = 0 . . . $n - 1$	Symbol = m_l Values = $-l$ . . . 0 . . . $+l$	Number of orbitals in shell = n^2 and number of orbitals in sub-shell = $2l + 1$	TABLE 8.8 Summary of the Quantum Numbers, Their Interrelationships, and the Orbital Information Conveyed
1	0	0	one 1s orbital (one orbital of one type in the $n = 1$ shell)	
2	0 1	0 +1, 0, −1	one 2s orbital three 2p orbitals (four orbitals of two types in the $n = 2$ shell)	
3	0 1 2	0 +1, 0, −1 +2, +1, 0, −1, −2	one 3s orbital three 3p orbitals five 3d orbitals (nine orbitals of three types in the $n = 3$ shell)	
4	0 1 2 3	0 +1, 0, −1 +2, +1, 0, −1, −2 +3, +2, +1, 0, −1, −2, −3	one 4s orbital three 4p orbitals five 4d orbitals seven 4f orbitals (sixteen orbitals of four types in the $n = 4$ shell)	

EXERCISE 8.24—STANDING WAVEFORMS FROM THE WAVE EQUATION

Complete the following statements regarding standing wave solutions to the wave equation:
(a) When $n = 2$, the value of l must be _____ or _____.
(b) When $l = 1$, the value of m_l must be _____ , _____ , or _____ , and the sub-shell is labelled _____.
(c) When $l = 2$, the sub-shell is called a _____ sub-shell.
(d) When a sub-shell is labelled s, the value of l is _____ and m_l has the value _____.
(e) When a sub-shell is labelled p, it has _____ orbitals.
(f) When a sub-shell is labelled f, there are _____ values of m_l and it has _____ orbitals.

Interactive Exercise 8.23

Describe orbitals using quantum numbers.

For a given shell, there can be one orbital with $l = 0$ (s orbital), three with $l = 1$ (p orbitals), five with $l = 2$ (d orbitals), and seven with $l = 3$ (f orbitals). Consequently, there can only be one orbital with $n = 1$, four with $n = 2$, nine with $n = 3$, and sixteen with $n = 4$.

Electron Spin

To complete our explanatory model of the nature of electrons in atoms, we need to refer to **electron spin,** *a property of electrons that results in them generating a small magnetic field*. If a beam of hydrogen atoms is directed through a vertical non-uniform magnetic field (Figure 8.17), half of the atoms are deflected upward, and the other half downward.

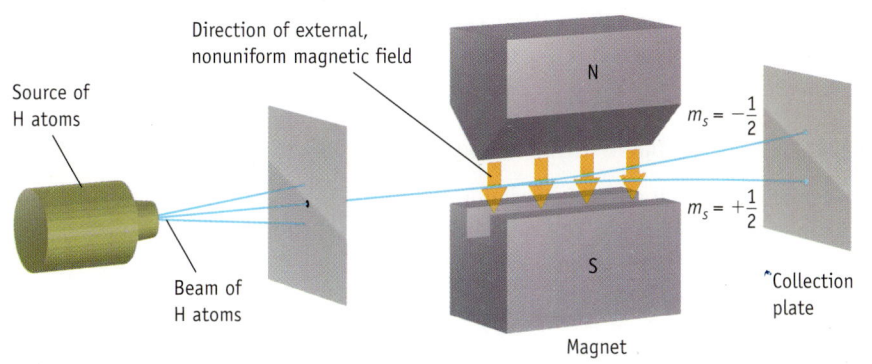

FIGURE 8.17 Splitting a beam of hydrogen atoms passing through a magnetic field.

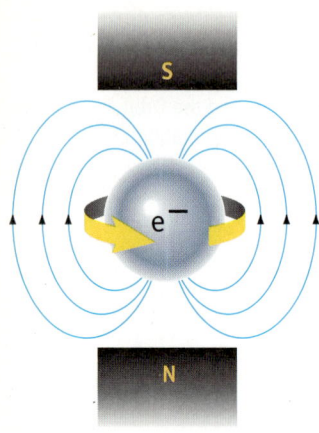

FIGURE 8.18 Electron spin and magnetism. The electron behaves as a "micro-magnet." Considering electrons as waves, the inherent property called spin cannot be explained. We need to resort to the "wave-particle duality" of electrons. Their magnetic field is as though they are rapidly spinning charged particles. Relative to an external magnetic field, only two spin directions are possible: clockwise or counterclockwise.

It seems that the electron in each H atom can generate either one of only two possible orientations of its magnetic field with respect to the external magnetic field, one of which causes attraction into the external field, and the other causes equal repulsion (Figure 8.18). In other words, electron spin is quantized. One orientation of electron spin is given a spin quantum number $m_s = +\frac{1}{2}$ and the other is given $m_s = -\frac{1}{2}$.

So now we need to recognize that the properties of electrons in atoms are governed by the "allowed" values of four quantum numbers—three of which (n, l, and m_l) emanate directly and naturally from stationary-wave solutions of the Schrödinger equation, and one (m_s) associated with the property of spin independently of the Schrödinger equation.

Electron spins govern whether or not substances are magnetic. Most substances are *slightly repelled by a magnet*, and are said to be **diamagnetic**. In contrast, some metals and compounds, said to be **paramagnetic**, *are attracted to a magnetic field*. The magnitude of the effect can be determined with an apparatus such as that illustrated in Figure 8.19a. The magnetism of most paramagnetic materials is so weak that you can observe the effect only in the presence of a strong magnetic field. For example, the oxygen we breathe is paramagnetic; it sticks to the poles of a strong magnet (Figure 8.19b).

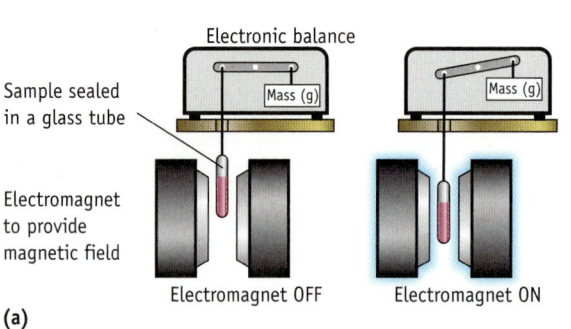

Charles D. Winters

(a) Electronic balance · Sample sealed in a glass tube · Mass (g) · Mass (g) · Electromagnet to provide magnetic field · Electromagnet OFF · Electromagnet ON · **(b)**

FIGURE 8.19 Observing and measuring paramagnetism. (a) A magnetic balance is used to measure the magnetism of a sample. The sample is weighed with the electromagnet off, and then with it on. If the substance is paramagnetic, the sample is drawn into the magnetic field and its apparent weight increases. (b) Liquid oxygen (boiling point −183 °C) clings to the poles of a strong magnet. Elemental oxygen is a paramagnetic substance.

Think about It

e8.17 See the meaning of magnetic spin and a video of the magnetic property of molecular oxygen.

The experimental phenomenon of paramagnetism is attributed to the spins of unpaired electrons. We observe experimentally that hydrogen atoms, each of which has a single electron, are paramagnetic. Helium, with two electrons, is diamagnetic. To account for this observation, we assume that the two electrons in the same orbital have opposite spin orientations. We say their *spins are paired*, and the magnetic field of one is cancelled out by the opposite magnetic field of the other. This explanation opens the way to understanding the arrangement of electrons in atoms with more than one electron.

EXERCISE 8.25—QUANTUM NUMBERS OF ELECTRONS IN ATOMS

Explain briefly why each of the following is not a valid set of quantum numbers for an electron in an atom. In each case, change a value (or values) to make the set valid.
(a) $n = 4$, $l = 2$, $m_l = 0$, $m_s = 0$
(b) $n = 3$, $l = 1$, $m_l = -3$, $m_s = -\frac{1}{2}$
(c) $n = 3$, $l = 3$, $m_l = -1$, $m_s = +\frac{1}{2}$

Electrons have a property called spin, which generates a small magnetic field in either of two opposite orientations, distinguished by spin quantum number $m_s = +\frac{1}{2}$ and $-\frac{1}{2}$. Paramagnetic substances have unpaired electrons and are attracted into magnetic fields.

Orbital "Shape"

We will see that as well as the energy of stationary electron waves, their distribution over space is important. Each solution to the Schrödinger equation (for each set of n, l, and m_l values) defines the energy of that standing wave exactly, but it does not define the position of the electron exactly. How can it, if the basis of the model is that the electron is a wave? Rather, one component of each solution is a mathematical expression that describes how ψ (the *wave function* or *orbital*) varies over the space around the nucleus.

For example, for the electron in an H atom, the wave function derived from the Schrödinger equation with quantum numbers $n = 1$, $l = 0$, and $m_l = 0$ (a $1s$ orbital) is

$$\psi_{1,0,0} = -\frac{1}{\sqrt{\pi}}\left(\frac{1}{a_o}\right)^{3/2} e^{-(r/a_0)}$$

For a $2p$ orbital defined by $n = 2$, $l = 1$ the wave function is

$$\psi_{2,1,0} = -\frac{1}{4\sqrt{2\pi}}\left(\frac{1}{a_o}\right)^{5/2} r\cos\theta\, e^{-(r/2a_0)}$$

In these equations, the variable r is the distance away from the nucleus, and a_o is a fixed distance—which happens to be the most probable distance from the nucleus of the electron in the $1s$ orbital of the H atom.

What does this mean? It turns out that ψ^2 (rather than ψ) is a measure of the probability of finding the electron at any point in space (if you think about the electron as a particle—even though the model is based on the assumption that it is not!). ψ^2 may be thought of as the superimposed sum of many snapshots of an atom, and is called the **electron density** distribution or an *electron cloud* (Figure 8.20a). One way of thinking about electron density is to imagine the electron as neither a wave nor a dynamic

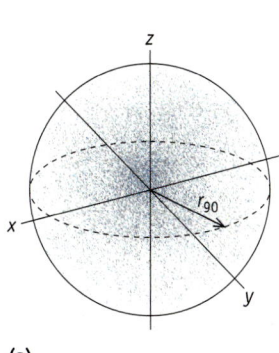

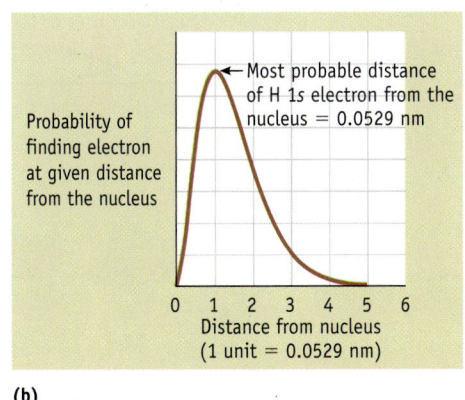

Probability of finding electron at given distance from the nucleus

← Most probable distance of H $1s$ electron from the nucleus = 0.0529 nm

0 1 2 3 4 5 6
Distance from nucleus
(1 unit = 0.0529 nm)

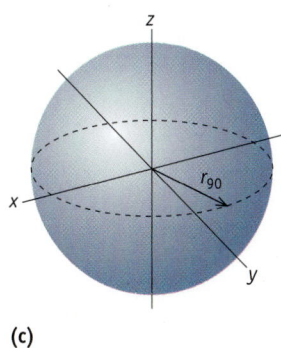

(a)   (b)   (c)

FIGURE 8.20 Different views of a $1s$ ($n = 1$ and $l = 0$) orbital. (a) Electron density picture of an electron in a $1s$ orbital. High density of dots corresponds with high probability of the location of the electron. Along all straight lines away from the nucleus, electron density decreases at the same rate, so that at the same distance from the nucleus, electron density is the same along each line. The distance r_{90} is the radius of a sphere with 90% of the electron density between the nucleus and the surface of the sphere. (b) A plot of the surface density ($4\pi r^2\psi^2$) with distance r from the nucleus, for an electron in the $1s$ orbital of a hydrogen atom. This gives the probability of the electron being on a sphere of radius r from the nucleus. (c) An isodensity surface for an electron in a $1s$ orbital. In the case of s orbitals, the contour surface of equal electron densities in all directions is a sphere. The isodensity surface shown here has been arbitrarily chosen to be the one that has 90% of the electron density between the nucleus and the surface of the sphere. The 50% isodensity surface (not shown) is a sphere with smaller radius.

Molecular Modelling (Odyssey)

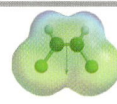

e8.18 Compare different ways to visualize the electron cloud in an atom.

particle, but as a smear of "electron matter" that is not uniformly distributed over space; the density of electron matter (or of dots in our picture) varies over the region around the nucleus.

s Orbitals

For any s orbital, the wave function (ψ) is similar to that shown for the H atom 1s orbital (above) in the sense that it has no angular term (i.e., no factor involving angles). This means that the value of ψ^2, and the electron density, depends only on distance from the nucleus, regardless of which direction we go. This electron density distribution in space is portrayed in the electron density picture (Figure 8.20a). Because of this, s orbitals are said to be *spherically symmetrical*.

For an electron in an s orbital, ψ^2 decreases with increasing radius (r): that is, along any straight line from the nucleus, the electron density probability decreases. However, Figure 8.20b (called a *surface density plot*) shows that the probability of finding the electron on the surface of a sphere has a maximum at some distance r from the nucleus. This is because the surface area of a sphere of radius r is $4\pi r^2$, so the probability of finding the electron on the surface of a sphere of radius r is $4\pi r^2\psi^2$. With increasing r, ψ^2 decreases and $4\pi r^2$ increases, so at some r, $4\pi r^2\psi^2$ is a maximum.

The "shape" of an electron probability distribution is most easily understood in terms of *a contour surface of equal probability (or equal density) of the electron*. Because ψ^2 for an s orbital has no angular dependence, such a contour surface is a sphere. This is called an **isodensity surface**, or simply, an *isosurface*. There are an infinite number of isosurfaces, depending on which electron density is chosen. A common choice is the isosurface that has 90% of the electron density between it and the nucleus (Figure 8.20c).

In a given atom, the larger the quantum number n is, the larger the radius of the spherical 90% isosurface of electrons in s orbitals (Figures 8.21 and 8.22). In other words, the

FIGURE 8.21 Surface density plots for electrons in the 1s, 2s, and 3s orbitals of an atom. The distance from the nucleus of maximum probability, and of the 90% isosurface, increase in the order $1s < 2s < 3s$. In loose shorthand language, it is common to say that the 2s orbital is bigger than the 1s orbital. Nodes are the distances at which the electron density = 0. The probability of finding an electron as we go away from the nucleus approaches but never quite reaches zero, even at very large distances.

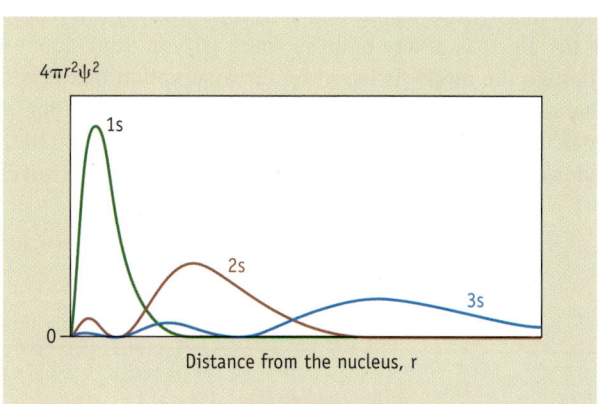

FIGURE 8.22 Atomic orbitals. Isodensity surfaces of electron densities in 1s, 2s, 2p, 3s, 3p, and 3d orbitals of a H atom.

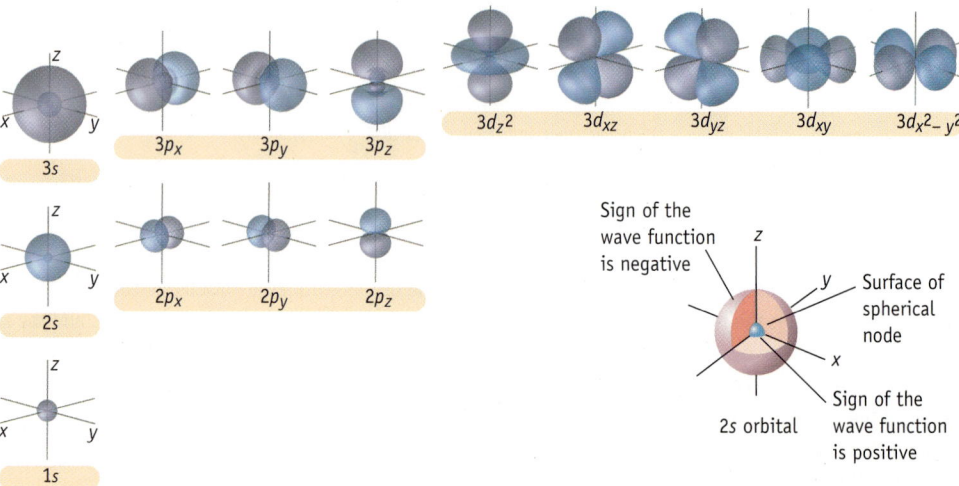

distribution of electrons in the 1*s* orbital is more compact than that in the 2*s* orbital, which, in turn, is more compact than that in the 3*s* orbital.

Just as standing waves in one dimension have nodes (Figure 8.16), the Schrödinger equation predicts nodes in the surface densities of electrons in *s* orbitals, other than the 1*s* orbital (Figure 8.21). The 2*s* orbital has one node, and the 3*s* orbital has two. In general, the number of nodes is given by $n - l - 1$.

p Orbitals

When the Schrödinger equation is solved for a standing electron wave with $l = 1$ (a *p* orbital), the resultant ψ has angular dependence. Consequently, the way that ψ^2 and the electron density vary over space depends on the direction away from the nucleus. This means that the isosurfaces for *p* orbitals, loosely referred to as the orbital "shape," are not spheres. Instead, the isosurfaces are approximately as shown in Figure 8.22. The three *p* orbitals in a sub-shell have orientations in mutually perpendicular directions in space (*x*, *y*, and *z*). After an axis direction is arbitrarily defined, the orbitals are labelled p_x, p_y, and p_z, according to the axis along which they lie.

The electron distributions in *p* orbitals have an imaginary plane passing through the nucleus that divides the electron density in half (Figures 8.22 and 8.23). On this imaginary plane, called a *nodal surface,* there is zero probability of finding the electron.

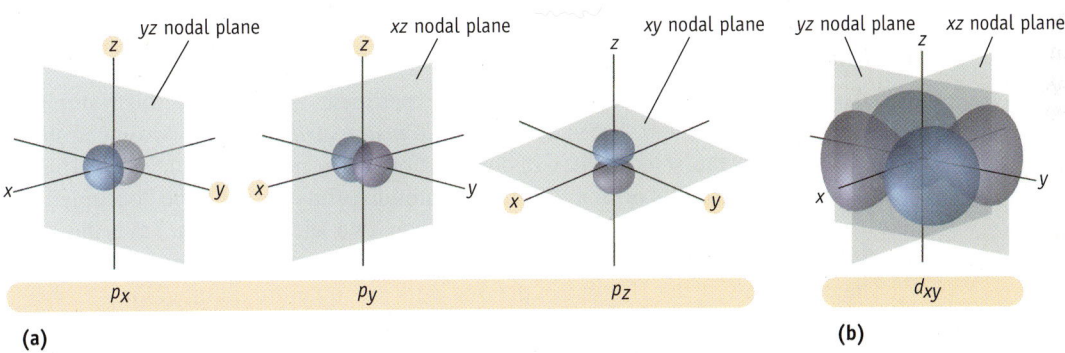

FIGURE 8.23 Nodal surfaces in *p* and *d* orbitals. A plane of zero electron density passing through the nucleus, perpendicular to the orbital axis, is called a *nodal surface*. The number of nodal surfaces is equal to the value of *l*. (a) Each *p* orbital ($l = 1$) has one nodal surface. (b) The d_{xy} orbital. Each of the five *d* orbitals ($l = 2$) have two nodal surfaces. Here the nodal surfaces are the *xz*- and *yz*-planes, so the regions of electron density lie in the *xy*-plane and between the *x*- and *y*-axes.

d Orbitals

Each of the five *d* orbitals has four regions of electron density and two nodal surfaces. The d_{xy} orbital lies in the *xy*-plane and the two nodal surfaces are the *xz*- and *yz*-planes (see Figure 8.23). The d_{xz} and d_{yz} orbitals lie in planes defined by the *xz*- and *yz*-axes. Of the two remaining, the $d_{x^2-y^2}$ orbital is easier to visualize: it results from two vertical planes slicing the electron density into quarters. The planes bisect the *x*- and *y*-axes, so the regions of electron density lie along the *x*- and *y*-axes. The d_{z^2} orbital has two main regions of electron density along the *z*-axis, but a "doughnut" of electron density also occurs in the *xy*-plane. This orbital has two nodal surfaces, but these are not flat.

EXERCISE 8.27—ORBITAL SHAPES

Draw circles representing cross-sections (passing through the nucleus) of isosurfaces to show the relative sizes of the following:
(a) a 50% isosurface and a 90% isosurface of the same *s* orbital
(b) 90% isosurfaces of a 1*s* orbital and a 2*s* orbital of the same atom

EXERCISE 8.28—ORBITAL SHAPES

Sketch a picture of the 90% isodensity surface of an s orbital and the p_x orbital. Be sure the latter drawing shows why the p orbital is labelled p_x and not p_y, for example.

Part of the solution of the Schrödinger equation for any stationary wave is the wave function (ψ) and ψ^2 defines the variation of electron density over the region around the nucleus. The "shape" of an orbital is the shape of a surface of points of equal (arbitrarily decided) electron density. In the case of s orbitals only, this isosurface is spherical.

8.5 Electron Configurations in Atoms

The distribution of electrons among the possible orbitals (that is, which standing wave-forms the electrons are in) is called the **electron configuration**. The **ground-state configuration** is *the distribution of electrons such that the atom has its lowest possible energy.* If *the electron configuration is such that an atom has more energy than the ground state*, it is said to be in an **excited state**.

Before we can apply our knowledge emanating from application of the wave theory of atoms, we also need to take into account the Pauli exclusion principle.

The Pauli Exclusion Principle

To make the quantum theory consistent with experiment, in 1925, the Austrian physicist Wolfgang Pauli (1900–1958) stated his *exclusion principle*: no two electrons in an atom can have the same values of the four quantum numbers n, l, m_l, and m_s. A direct consequence of this principle is that an atomic orbital may not be "occupied" by more than two electrons (that is, no more than two electrons can have the same waveform).

The $1s$ orbital of the H atom is derived with the quantum numbers $n = 1$, $l = 0$, and $m_l = 0$. To fully describe an electron with this waveform, its direction of electron spin must also be specified. Let us represent an orbital by a box ☐ and the electron by an arrow ↑ or ↓. Below is an *orbital box* representation of the electron configuration in a ground-state H atom, in which the electron has spin quantum number $m_s = +\frac{1}{2}$:

$$\boxed{\uparrow}$$
$$1s$$

In a ground-state helium atom, both electrons "occupy" the $1s$ orbital. The Pauli exclusion principle tells us that that each electron must have a different set of quantum numbers. For one of the electrons, $n = 1$, $l = 0$, $m_l = 0$, $m_s = +\frac{1}{2}$; for the other, $n = 1$, $l = 0$, $m_l = 0$, $m_s = -\frac{1}{2}$. Below is an orbital box diagram representing this configuration:

$$\boxed{\uparrow\downarrow}$$
$$1s$$

Because both of the electrons in this atom are paired, necessarily with opposite spins, we would predict that helium is diamagnetic, as experimentally observed.

Our understanding of the number of possible orbitals (Table 8.8), and the knowledge that an orbital can accommodate no more than two electrons, tells us the maximum number of electrons that can occupy each shell or sub-shell. Because a p sub-shell consists of three orbitals, it can accommodate up to six electrons. The five orbitals of a d sub-shell can accommodate up to ten electrons. The relationships among the quantum numbers and the maximum numbers of electrons in shells and sub-shells are shown in Table 8.9.

The direction of the spin arrow is arbitrary: it may point in either direction. Here we associate $m_s = +\frac{1}{2}$ with an up arrow $\boxed{\uparrow}$, but it could equally well be depicted as $\boxed{\downarrow}$.

n	Shell	Sub-shells Orbitals $2l + 1$	Number of Electrons in Sub-shell $2(2l + 1)$	Maximum Number of Electrons in Shell $2n^2$
1	s	1	2	2
2	s	1	2	8
	p	3	6	
3	s	1	2	18
	p	3	6	
	d	5	10	
4	s	1	2	32
	p	3	6	
	d	5	10	
	f	7	14	
5	s	1	2	50
	p	3	6	
	d	5	10	
	f	7	14	
	g*	9	18	
6	s	1	2	72
	p	3	6	
	d	5	10	
	f	7	14	
	g*	9	18	
	h*	11	22	

TABLE 8.9 Maximum Number of Electrons in Shells and Sub-shells

Interactive Exercise 8.29

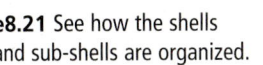 

What is the maximum capacity for electrons in shells and subshells defined by particular quantum numbers?

Think about It

e8.21 See how the shells and sub-shells are organized.

Think about It

e8.22 See how the energies of the orbitals vary for multi-electron atoms.

*These orbitals are not occupied in the ground state of any known element.

EXERCISE 8.30—ELECTRONS, ORBITALS, SUB-SHELLS, AND SHELLS

What is the maximum number of electrons that can be identified with each of the following sets of quantum numbers? In one case, the answer is "none". Explain why this is true.
(a) $n = 3$
(b) $n = 3$, $l = 2$
(c) $n = 4$, $l = 1$, $m_l = -1$, $m_s = -\frac{1}{2}$
(d) $n = 5$, $l = 0$, $m_l = +1$

> No two electrons in an atom can have the same values of n, l, m_l, and m_s. A direct consequence of this principle is that no more than two electrons can have the same orbital. As a result, the maximum number of electrons in the $n = 1$ shell is 2, in the $n = 2$ shell is 8, $n = 3$ shell is 18, and in the $n = 4$ shell is 32.

Assignment of Electrons to Orbitals

We are nearly at our goal: to be able to predict how many electrons are in each stationary waveform in ground-state atoms. We use an imagined process of creating atoms called the **aufbau principle** and go from element to element assigning one electron to the orbital that results in the lowest-energy atom (as well as adding one proton and the appropriate number of neutrons to the nucleus). The question is the order of orbital filling that achieves this outcome.

Aufbau is a German word that means "building up."

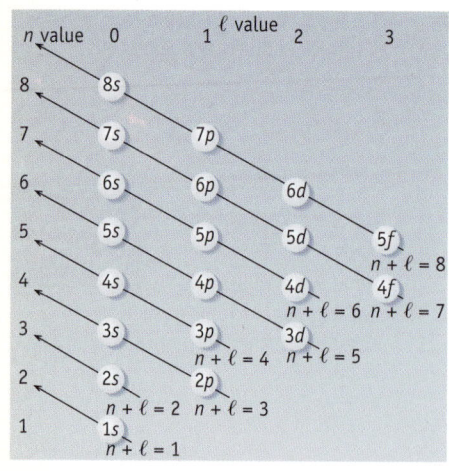

FIGURE 8.24 Order of assigning electrons to orbitals to achieve ground-state configuration. Electrons are assigned to orbitals in the order 1s, 2s, 2p, 3s, 3p, 4s, 3d, and so on.

Orbitals such as the three 2p orbitals are *degenerate* orbitals only in the absence of a magnetic field. In the presence of a powerful magnetic field, they "split" into orbitals of different energies.

Think about It

e8.23 Use the aufbau principle to hypothetically "create" electron configurations for atoms.

Interactive Exercise 8.31

SUBMIT

Deduce the electron configuration for a ground-state atom.

To apply the aufbau principle to achieve the correct electron configuration for ground-state isolated atoms, we follow a set of guidelines:

1. Hypothetically feed one more electron (and one more proton) into atoms of successive elements.

2. Use the order of assigning electrons to orbitals indicated in Figure 8.24.

3. In accordance with the Pauli exclusion principle, assign a maximum of two electrons (with opposite spins) per orbital.

4. Orbitals with the same energy, such as the three 2p orbitals, are called *degenerate orbitals*. Hund's rule says that paired electrons in one of the degenerate orbitals is a higher energy configuration (an excited state) than if these orbitals are singly occupied. Operationally speaking, we assign one electron to each of the degenerate orbitals before any pairing of electrons.

Using the aufbau principle, we can now hypothetically "create" some ground-state atoms.

WORKED EXAMPLE 8.3—ELECTRON CONFIGURATIONS OF GROUND-STATE ATOMS

(a) Deduce the electron configuration of ground-state atoms of silicon.

(b) Write one set of "allowed" values of the quantum numbers for each of the valence electrons (those with $n = 3$).

Solution

(a) Atoms of $_{14}$Si have 14 electrons. Using the aufbau principle, we assign these in order as follows:

- two electrons to the 1s orbital, with opposite spin directions
- two electrons to the 2s orbital
- two electrons to each of the 2p orbitals—a total of six
- two electrons to the 3s orbital

  Now there are only two electrons remaining to assign, and the next orbitals are the three 3p orbitals. We could consider putting both electrons into any one of the 3p orbitals, but this pairing would contravene Hund's rule—because of repulsion between electrons in the same orbital, paired electrons would result in a higher energy condition than if the electrons are unpaired in different orbitals. So we can write the orbital box representation for $_{14}$Si atoms:

  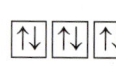 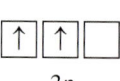
1s   2s   2p   3s   3p

(b) For both of the 3s electrons, the 3s label informs us that $n = 3$ and $l = 0$. Necessarily then, $m_l = 0$. These paired electrons (in the same orbital) must have opposite spins, with $m_s = +\frac{1}{2}$ and $-\frac{1}{2}$ (or the Pauli exclusion principle would be invalidated). So their configurations are as follows:

$$n = 3, \ l = 0, \ m_l = 0, \ m_s = +\frac{1}{2} \quad \text{and} \quad n = 3, \ l = 0, \ m_l = 0, \ m_s = -\frac{1}{2}$$

For both of the 3p electrons, $n = 3$ and $l = 1$. Each of the following is an "allowed" set of quantum numbers for one of these electrons:

$$n = 3, \ l = 1, \ m_l = 1, \ m_s = +\frac{1}{2} \qquad n = 3, \ l = 1, \ m_l = 1, \ m_s = -\frac{1}{2}$$
$$n = 3, \ l = 1, \ m_l = 0, \ m_s = +\frac{1}{2} \qquad n = 3, \ l = 1, \ m_l = 0, \ m_s = -\frac{1}{2}$$
$$n = 3, \ l = 1, \ m_l = -1, \ m_s = +\frac{1}{2} \qquad n = 3, \ l = 1, \ m_l = -1, \ m_s = -\frac{1}{2}$$

EXERCISE 8.32—ELECTRON CONFIGURATIONS OF GROUND-STATE ATOMS

Depict with an orbital box diagram the electron configuration of ground-state phosphorus atoms. Give one possible set of four quantum numbers for each of the electrons beyond those of the preceding noble gas.

The aufbau principle is used to predict the ground-state configuration of electrons. Electrons are assigned one at a time (consistent with the atomic number) into orbitals in the order indicated in Figure 8.24, while obeying the Pauli exclusion principle and Hund's rule.

Periodicity of Electron Configurations

Keep in mind that our goal is to connect the electron configurations of atoms of the elements with their positions in the periodic table. This will allow us ultimately to relate electron configurations to some chemically observed phenomena.

Also keep in mind the assumption of chemists that similarities and differences among chemical properties of the elements are attributable to similarities and differences among some features of the electrons in the atoms. In particular, characteristic chemical properties are attributed to the *"outermost" (highest-energy) electrons—those most likely to interact when atoms come close to each other*—called **valence electrons**. So, we shall particularly look for features of valence electrons, which in the case of the main group elements are those in the outermost shell. The remainder of this chapter is concerned mainly with the valence electrons of atoms, ions, and molecules.

The electron configurations of all of the elements up to $Z = 109$ are displayed in Table 8.10. The ground-state electron configuration of $_1$H atoms can be represented by an orbital box diagram, or by what is called *spdf* notation, as follows:

↑   or $1s^1$ (which signifies one electron in the $1s$ orbital)
$1s$

The ground-state electron configuration of atoms of helium ($_2$He) can be represented by

↑↓   or $1s^2$
$1s$

For $_{10}$Ne atoms, the predicted ground-state electron configuration is

↑↓   ↑↓   ↑↓ ↑↓ ↑↓   or $1s^2 2s^2 2p^6$
$1s$ $2s$ $2p$

The *spdf* notation can be abbreviated to $[\text{He}]2s^2 2p^6$, where [He] represents the electron configuration of the previous Group 18 element—in this case, $1s^2$ of He. The valence electron configuration ($2s^2 2p^6$) remains explicit. *The electrons corresponding with the configuration of the previous noble gas are called* **core electrons**.

For atoms of argon ($_{18}$Ar), the aufbau guidelines lead us to the electron configuration:

↑↓   ↑↓   ↑↓ ↑↓ ↑↓   ↑↓   ↑↓ ↑↓ ↑↓
$1s$ $2s$ $2p$ $3s$ $3p$

That is, in *spdf* notation: $1s^2 2s^2 2p^6 3s^2 3p^6$ or $[\text{Ne}]3s^2 3p^6$

For atoms of krypton ($_{36}$Kr) and xenon ($_{54}$Xe), we derive the following electron configurations:

$_{36}$Kr: $[\text{Ar}]4s^2 4p^6$
$_{54}$Xe: $[\text{Kr}]5s^2 5p^6$

The remarkable thing about these derived electron configurations is the commonality of the valence configurations—all $ns^2 np^6$, where n is the quantum number of the outermost occupied orbital. So application of the wave equation to atoms of elements with similar chemical properties has given us similar valence electron configurations. We will explore the significance of this later.

For atoms of lithium ($_3$Li), application of the aufbau principle leads us to the ground-state configuration:

↓↑   ↑   □ □ □   or $1s^2 2s^1$
$1s$ $2s$ $2p$

The derived ground-state configurations for atoms of the other Group 1 elements are

$_{11}$Na: $1s^2 2s^2 2p^6 3s^1$ or $[\text{Ne}]3s^1$ $_{37}$Rb: $[\text{Kr}]5s^1$
$_{19}$K: $1s^2 2s^2 2p^6 3s^2 3p^6 4s^1$ or $[\text{Ar}]4s^1$ $_{55}$Cs: $[\text{Xe}]6s^1$

Once again, we have the extraordinary outcome from application of the wave theory that all of the elements of Group 1 have the same valence electron configuration—in this case, ns^1.

$1s^1$ is read as "$1s$ – one."

TABLE 8.10   Electron Configurations of Atoms in the Ground State

Z	Element	Configuration	Z	Element	Configuration	Z	Element	Configuration
1	H	$1s^1$	38	Sr	$[Kr]5s^2$	74	W	$[Xe]4f^{14}5d^46s^2$
2	He	$1s^2$	39	Y	$[Kr]4d^15s^2$	75	Re	$[Xe]4f^{14}5d^56s^2$
3	Li	$[He]2s^1$	40	Zr	$[Kr]4d^25s^2$	76	Os	$[Xe]4f^{14}5d^66s^2$
4	Be	$[He]2s^2$	41	Nb	$[Kr]4d^45s^1$	77	Ir	$[Xe]4f^{14}5d^76s^2$
5	B	$[He]2s^22p^1$	42	Mo	$[Kr]4d^55s^1$	78	Pt	$[Xe]4f^{14}5d^96s^1$
6	C	$[He]2s^22p^2$	43	Tc	$[Kr]4d^55s^2$	79	Au	$[Xe]4f^{14}5d^{10}6s^1$
7	N	$[He]2s^22p^3$	44	Ru	$[Kr]4d^75s^1$	80	Hg	$[Xe]4f^{14}5d^{10}6s^2$
8	O	$[He]2s^22p^4$	45	Rh	$[Kr]4d^85s^1$	81	Tl	$[Xe]4f^{14}5d^{10}6s^26p^1$
9	F	$[He]2s^22p^5$	46	Pd	$[Kr]4d^{10}$	82	Pb	$[Xe]4f^{14}5d^{10}6s^26p^2$
10	Ne	$[He]2s^22p^6$	47	Ag	$[Kr]4d^{10}5s^1$	83	Bi	$[Xe]4f^{14}5d^{10}6s^26p^3$
11	Na	$[Ne]3s^1$	48	Cd	$[Kr]4d^{10}5s^2$	84	Po	$[Xe]4f^{14}5d^{10}6s^26p^4$
12	Mg	$[Ne]3s^2$	49	In	$[Kr]4d^{10}5s^25p^1$	85	At	$[Xe]4f^{14}5d^{10}6s^26p^5$
13	Al	$[Ne]3s^23p^1$	50	Sn	$[Kr]4d^{10}5s^25p^2$	86	Rn	$[Xe]4f^{14}5d^{10}6s^26p^6$
14	Si	$[Ne]3s^23p^2$	51	Sb	$[Kr]4d^{10}5s^25p^3$	87	Fr	$[Rn]7s^1$
15	P	$[Ne]3s^23p^3$	52	Te	$[Kr]4d^{10}5s^25p^4$	88	Ra	$[Rn]7s^2$
16	Si	$[Ne]3s^23p^4$	53	In	$[Kr]4d^{10}5s^25p^5$	89	Ac	$[Rn]6d^17s^2$
17	Cl	$[Ne]3s^23p^5$	54	Xe	$[Kr]4d^{10}5s^25p^6$	90	Th	$[Rn]6d^27s^2$
18	Ar	$[Ne]3s^23p^6$	55	Cs	$[Xe]6s^1$	91	Pa	$[Rn]5f^26d^17s^2$
19	K	$[Ar]4s^1$	56	Ba	$[Xe]6s^2$	92	U	$[Rn]5f^36d^17s^2$
20	Ca	$[Ar]4s^2$	57	La	$[Xe]5d^16s^2$	93	Np	$[Rn]5f^46d^17s^2$
21	Sc	$[Ar]3d^14s^2$	58	Ce	$[Xe]4f^15d^16s^2$	94	Pu	$[Rn]5f^67s^2$
22	Ti	$[Ar]3d^24s^2$	59	Pr	$[Xe]4f^36s^2$	95	Am	$[Rn]5f^77s^2$
23	V	$[Ar]3d^34s^2$	60	Nd	$[Xe]4f^46s^2$	96	Cm	$[Rn]5f^76d^17s^2$
24	Cr	$[Ar]3d^54s^1$	61	Pm	$[Xe]4f^56s^2$	97	Bk	$[Rn]5f^97s^2$
25	Mn	$[Ar]3d^54s^2$	62	Sm	$[Xe]4f^66s^2$	98	Cf	$[Rn]5f^{10}7s^2$
26	Fe	$[Ar]3d^64s^2$	63	Eu	$[Xe]4f^76s^2$	99	Es	$[Rn]5f^{11}7s^2$
27	Co	$[Ar]3d^74s^2$	64	Gd	$[Xe]4f^75d^16s^2$	100	Fm	$[Rn]5f^{12}7s^2$
28	Ni	$[Ar]3d^84s^2$	65	Tb	$[Xe]4f^96s^2$	101	Md	$[Rn]5f^{13}7s^2$
29	Cu	$[Ar]3d^{10}4s^1$	66	Dy	$[Xe]4f^{10}6s^2$	102	No	$[Rn]5f^{14}7s^2$
30	Zn	$[Ar]3d^{10}4s^2$	67	Ho	$[Xe]4f^{11}6s^2$	103	Lr	$[Rn]5f^{14}6d^17s^2$
31	Ga	$[Ar]3d^{10}4s^24p^1$	68	Er	$[Xe]4f^{12}6s^2$	104	Rf	$[Rn]5f^{14}6d^27s^2$
32	Ge	$[Ar]3d^{10}4s^24p^2$	69	Tm	$[Xe]4f^{13}6s^2$	105	Db	$[Rn]5f^{14}6d^37s^2$
33	As	$[Ar]3d^{10}4s^24p^3$	70	Yb	$[Xe]4f^{14}6s^2$	106	Sg	$[Rn]5f^{14}6d^47s^2$
34	Se	$[Ar]3d^{10}4s^24p^4$	71	Lu	$[Xe]4f^{14}5d^16s^2$	107	Bh	$[Rn]5f^{14}6d^57s^2$
35	Br	$[Ar]3d^{10}4s^24p^5$	72	Hf	$[Xe]4f^{14}5d^26s^2$	108	Hs	$[Rn]5f^{14}6d^67s^2$
36	Kr	$[Ar]3d^{10}4s^24p^6$	73	Ta	$[Xe]4f^{14}5d^36s^2$	109	Mt	$[Rn]5f^{14}6d^77s^2$
37	Rb	$[Kr]5s^1$						

Moving from group to group across the periodic table, for $_4$Be atoms, the ground-state electron configuration is $1s^22s^2$ or $[He]2s^2$. If we apply the aufbau principle to the remainder of the elements in this group, we see that the atoms all have the ns^2 valence electron configuration.

Because the valence electrons of atoms of *all the elements of Groups 1 and 2 occupy only s orbitals,* these are called **s-block elements**. We will see in what follows that *atoms of* *the elements in Groups 13–17 have partially filled p valence orbitals, and these, along with the elements of Group 18, which have filled s and p orbitals,* are called the ***p*-block elements**.

The ground-state electron configuration predicted for atoms of boron ($_5$B) is

| ↓↑ | | ↑↓ | | ↑ | | | | or | | $1s^2 2s^2 2p^1$ | | or | | [He]$2s^2 2p^1$ |
| 1s | | 2s | | 2p | | | |

Atoms of all of the Group 13 elements have 3 valence electrons, with configurations that can be denoted $ns^2 np^1$.

In Group 14, Carbon ($_6$C) is the first element for which we need to apply Hund's rule to predict the electron configuration. We find an $ns^2 np^2$ valence electron configuration in common with the atoms of all elements in this group:

| ↓↑ | | ↑↓ | | ↑ | ↑ | | | or | | $1s^2 2s^2 2p^2$ | | or | | [He]$2s^2 2p^2$ |
| 1s | | 2s | | 2p | | | |

Application of the aufbau principle to any of the Group 15 elements (the first of which is $_7$N) leads to the prediction that their atoms have 5 valence electrons, whose configuration can be denoted $ns^2 np^3$.

Oxygen is the first element for which we need to pair electrons (with opposite spins) in one of a set of degenerate orbitals. Once again, remarkably, the outcome from application of wave equation to electrons in atoms is that the valence electron configurations of the similarly behaving Group 16 elements is the same—in this case, $ns^2 np^4$. In the case of ground-state $_8$O atoms:

| ↓↑ | | ↑↓ | | ↑↓ | ↑ | ↑ | | or | | $1s^2 2s^2 2p^4$ | | or | | [He]$2s^2 2p^4$ |
| 1s | | 2s | | 2p | | | |

These patterns of electron configurations of atoms of elements in the same group are an extraordinary outcome of the application of mathematics of standing waveforms to electrons in atoms! Electrons were, after all, unheard of when elements were first allocated their places in the periodic table.

> When the aufbau principle is applied, it is found that atoms of elements in the same group have similar ground-state valence electron configurations—all s^1 for Group 1 elements, s^2 for Group 2 elements, $s^2 p^1$ for Group 13 elements, and so on.

8.6 Shielding and Effective Nuclear Charge

Using the outcomes of the wave model of electrons in atoms to explain why the elements have periodically varying properties assumes that the properties are dependent on characteristics of the "outermost" valence electrons. In particular, it requires understanding of how strongly the valence electrons are attracted toward the nucleus.

Periodic Variation of Effective Nuclear Charge in Atoms

You might think that the valence electron of potassium (nuclear charge +19) experiences a greater force of attraction to the nucleus than that of sodium (nuclear charge +11). Such a prediction assumes that the force of attraction between the valence electron and the nucleus is the only interaction operating. However the valence electron also experiences repulsions from the other electrons, and these repulsions have the effect of reducing the force that the valence electron experiences. *The effect of repulsions from other electrons in reducing the charge that the valence electrons "feel" at the nucleus* is called **shielding** or **screening**. This line of thinking gives rise to the concept of **effective nuclear charge (Z*),** *the nuclear charge experienced by any one valence electron due to reduction of the effect of the nuclear charge by repulsion from other electrons.* We will see that although Z increases continuously from element to element, Z^* varies with a periodically varying pattern.

A numerical value can be assigned to the effective nuclear charge experienced by a valence electron:

$$Z^* = Z - s$$

where s is a measure of the effect due to shielding. How much should we subtract for shielding?

Shielding, or screening, is repulsion by other electrons. It is inappropriate to think of it as "getting in the way" of interactions between the nucleus and the valence electron, as perhaps the names suggest. It is also incorrect to think that the force due to the charge on the nucleus is shared out amongst the electrons.

Think about It

e8.24 See how effective nuclear charge varies across the periodic table.

As a gross approximation, we could say that the nuclear charge experienced by a valence electron is "neutralized" on a 1:1 basis by repulsion from all of the "inner" core electrons:

$$Z* \approx Z - \text{number of core electrons}$$

If we apply this approximation to Na and K atoms, we find that the effective nuclear charge experienced by the valence electron in K atoms is similar to that for the valence electron of Na atoms:

$$Z*(_{11}\text{Na}) \approx +11 - 10 = +1$$
$$Z*(_{19}\text{K}) \approx +19 - 18 = +1$$

Similar calculations show that the effective nuclear charges for $_9$F, $_{17}$Cl, and $_{35}$Br atoms are nearly the same ($Z* \approx +7$). We can see that there is a periodic variation of $Z*$ values. To take our model through to its full explanatory power, however, we need to make more accurate calculations of effective nuclear charges.

Slater's rules for calculation of $Z*$ include a judgment of the average degree to which nuclear charge is lessened by shielding, depending on which shell the shielding electrons are in. Average estimates of the amount by which the nuclear charge is reduced are as follows:

- Shielding due to each of the other valence electrons $= 0.35$
- Shielding due to each electron in the penultimate $(n - 1)$ shell $= 0.85$
- Shielding due to each core electron (in $n - 2$, $n - 3$, . . . shells) $= 1.0$

WORKED EXAMPLE 8.4—EFFECTIVE NUCLEAR CHARGE

Use Slater's rules to calculate the effective nuclear charge experienced by a valence electron in ground-state atoms of (a) $_{10}$Ne, and (b) $_{11}$Na. In which of these atoms do the valence electrons experience the greater force of attraction to the nucleus?

Solution

(a) The electron configuration of ground-state $_{10}$Ne atoms is $1s^2 2s^2 2p^6$. So for any one of the eight valence electrons (in the $2s$ and $2p$ orbitals) there is shielding due to seven other valence electrons and two electrons in the penultimate shell (in the $1s$ orbital). Applying Slater's rules, we can calculate the effective nuclear charge experienced by any one of the valence electrons:

$$Z*(_{10}\text{Ne}) = +10 - (7 \times 0.35) - (2 \times 0.85) = +10 - 4.15 = +5.9$$

(b) For ground-state sodium atoms, the electron configuration is $1s^2 2s^2 2p^6 3s^1$. Shielding of the $1s$ valence electron is from eight electrons in the penultimate shell, and two core electrons. There is no repulsion by other valence electrons.

$$Z*(_{11}\text{Na}) = +11 - \left[(0 \times 0.35) - (8 \times 0.85) + (2 \times 1.0)\right] = +11 - 8.8 = +2.2$$

The relative magnitudes of $Z*$ tell us that each of the valence electrons in a ground-state $_{10}$Ne atom feel a much stronger pull toward the nucleus that the valence electron in a ground-state $_{11}$Na atom.

Comment

This comparison of the "pull" that valence electrons experience in atoms of different elements is at the heart of our ability to rationalize the periodic variation of properties of the elements in Section 8.7.

EXERCISE 8.34—EFFECTIVE NUCLEAR CHARGE

Calculate and compare the effective nuclear charges experienced by the valence electrons in ground-state atoms of Mg, P, and Ar. Are your values consistent with the trend of first ionization energies of atoms of these elements?

When we apply Slater's rules to valence electrons of the elements of the second, third, and fourth periods of the main block of the periodic table, we obtain the values of effective nuclear charge listed in Table 8.11. A periodic pattern is obvious.

TABLE 8.11 Values of Z^* in Ground-State Atoms

$_3$Li	$_4$Be	$_5$B	$_6$C	$_7$N	$_8$O	$_9$F	$_{10}$Ne
+1.3	+2.0	+2.6	+3.3	+3.9	+4.6	+5.2	+5.9
$_{11}$Na	$_{12}$Mg	$_{13}$Al	$_{14}$Si	$_{15}$P	$_{16}$S	$_{17}$Cl	$_{18}$Ar
+2.2	+2.9	+3.5	+4.2	+4.8	+5.5	+6.1	+6.8
$_{19}$K	$_{20}$Ca	$_{31}$Ga	$_{32}$Ge	$_{33}$As	$_{34}$Se	$_{35}$Br	$_{36}$Kr
+2.2	+2.8	+3.8	+4.5	+5.1	+5.8	+6.4	+7.1

Generalizations about the degree of shielding according to shell are necessarily approximate because the shielding depends to some extent on which orbitals the shielding electrons "occupy." For example, valence electrons in a $3s$ orbital are better shielded by electrons in a ("mostly inside") $2s$ orbital than by electrons in an interpenetrating $2p$ orbital, which have more of their electron density "outside" the valence electrons along one axis.

> Effective nuclear charge (Z^*) is the nuclear charge experienced by any valence electron, taking into account a reduction of the effect of the nuclear charge by repulsion from other electrons (shielding). Estimates of Z^* are made by applying Slater's rules, and we find that Z^* values vary periodically, consistent with the periodic variation of properties of the elements.

Effective Nuclear Charge for Valence Electrons in Ions

The effective nuclear charge experienced by the valence electrons in ions can be calculated in the same way as for atoms. Remember that both cations and anions have the same number of protons in the nucleus (Z) as their parent atoms, but they have different numbers of electrons.

WORKED EXAMPLE 8.5—EFFECTIVE NUCLEAR CHARGE IN IONS

Compare the effective nuclear charge experienced by the valence electrons in ground-state K atoms with that in K$^+$ ions. In which of these species does a valence electron experience the stronger "pull" toward the nucleus?

Solution

First we compare the electron configurations of the atom and ion, deduced as for atoms, but remembering that K$^+$ ions have one less electron than K atoms.

$$_{19}\text{K atom: } 1s^2 2s^2 2p^6 3s^2 3p^6 4s^1 \qquad _{19}\text{K}^+ \text{ ion: } 1s^2 2s^2 2p^6 3s^2 3p^6$$

Now we use Slater's rules to calculate Z^* in each species. In a K atom, the $4s$ valence electron is shielded by no other valence electrons, by 8 electrons in the penultimate ($n = 3$) shell, and by 10 core electrons:

$$Z^*(_{19}\text{K atom}) = +19 - \left[(0 \times 0.35) + (8 \times 0.85) + (10 \times 1.0)\right] = +18 - 16.8 = +2.2$$

In a K$^+$ ion, a $3p$ valence electron is shielded by 7 others, as well as by 8 electrons in the penultimate ($n = 2$) shell, and by 2 core electrons:

$$Z^*(_{19}\text{K}^+ \text{ ion}) = +19 - \left[(7 \times 0.35) + (8 \times 0.85) + (2 \times 1.0)\right] = +18 - 11.3 = +7.7$$

Even though there is the same positive charge on the nucleus of both species, the effective nuclear charges indicate that the valence electrons in a K$^+$ ion experience a much stronger "pull" toward the nucleus than the valence electron in a K atom. This is because there is much more repulsion from other electrons (shielding) in a K atom than in a K$^+$ ion.

Comment

By comparing the effective nuclear charges, it should not be surprising that the size of the K$^+$ cation is smaller than its "parent" K atom.

EXERCISE 8.35—EFFECTIVE NUCLEAR CHARGE IN IONS

Calculate and compare the effective nuclear charge for valence electrons in O atoms with that in O^{2-} ions. Are your answers consistent with the relative sizes of these two species?

8.7 Rationalizing the Periodic Variation of Properties

Now we can bring together the outcomes of experimental evidence and theoretical considerations to make sense, in a rigorous way, of the observed periodic trends of properties of the main group elements. In the following discussions, values of effective nuclear charge of atoms are taken from Table 8.11.

Sizes of Atoms

The two obvious trends in the sizes of atoms (Figure 8.4) are the following:

1. *From element to element across a row, atoms are smaller.* Effective nuclear charge increases as we go from elements in Group 1 to those in Group 18. Compare, for example, $_{12}Mg$ atoms ($Z^* = +2.9$, radius = 160 pm) and $_{16}S$ atoms ($Z^* = +5.5$, radius = 104 pm), which have valence electrons in the same shell. It is reasonable that electrons experiencing a higher effective nuclear charge are pulled inward more strongly, and so the atoms are smaller.

2. *From element to element down a group, atoms are bigger.* The effective nuclear charges on valence electrons become a little higher as we go to elements further down a group, but it seems that "occupation" of shells with higher quantum number n, with electron density dispersed further from the nucleus on average, more than compensates for this.

EXERCISE 8.36—EFFECTIVE NUCLEAR CHARGE AND ATOM SIZES

Use the concept of effective nuclear charge to rationalize the data that $r(_{84}Po) < r(_{11}Na)$.

Ionization Energies

Evidence about three important relationships concerning ionization energies was presented in Section 8.2:

1. *First ionization energies get larger from element to element across a row, and smaller down a group.* We can explain these trends using similar arguments to those used for the trends of atom and ion sizes.

2. *For any element, $IE_1 < IE_2 < IE_3$, and so on.* Let's use $_{13}Al$ to illustrate how these observations can be rationalized. We need to recognize that the magnitude of IE_1 is dependent on Z^* of Al atoms, IE_2 is dependent on Z^* of Al^+ ions, IE_3 is dependent on Z^* of Al^{2+} ions, and IE_4 is dependent on Z^* of Al^{3+} ions. This is because, for example, IE_2 is the energy required to remove a second valence electron from an Al atom—that is, a valence electron of an Al^+ ion. For the first three ionization steps, listed below are the appropriate electron configurations, and Z^* values.

$_{13}Al$ atom: $1s^2 2s^2 2p^6 3s^2 3p^1$ $Z^* = +13 - (2 \times 0.35) - (8 \times 0.85) - (2 \times 1.0) = +3.5$

$_{13}Al^+$ ion: $1s^2 2s^2 2p^6 3s^2$ $Z^* = +13 - (1 \times 0.35) - (8 \times 0.85) - (2 \times 1.0) = +3.9$

$_{13}Al^{2+}$ ion: $1s^2 2s^2 2p^6 3s^1$ $Z^* = +13 - (0 \times 0.35) - (8 \times 0.85) - (2 \times 1.0) = +4.2$

These estimates suggest that valence electron in Al^{2+} ions are held more tightly than those in Al^+ ions, and these more tightly than those in the Al atom. This is consistent with the experimental evidence that to remove each electron requires more energy than to remove the previous one.

3. *For any Group 1 element, IE_2 is very much greater than IE_1, and for any Group 2 elements, $IE_3 \gg IE_2$. Group 13 elements have $IE_4 \gg IE_3$, Group 14 elements have $IE_5 \gg IE_4$, and so on.* Again using aluminum, as an example to make sense of the evidence that $IE_4 \gg IE_3$ for Group 13 elements, we need to compare Z* of Al^{2+} ions with Z* of Al^{3+} ions:

$_{13}Al^{2+}$ ion: $1s^2 2s^2 2p^6 3s^1$ $Z^* = +13 - (0 \times 0.35) - (8 \times 0.85) - (2 \times 1.0) = +4.2$

$_{13}Al^{3+}$ ion: $1s^2 2s^2 2p^6$ $Z^* = +13 - (7 \times 0.35) - (0 \times 0.85) - (2 \times 1.0) = +8.6$

The very much greater effective nuclear charge experienced by the valence electrons in Al^{3+} ions compared with that in Al^{2+} ions is a convincing illustration of the power of the wave model of electrons in atoms to account for chemical behaviour of the elements.

Interactive Exercises 8.37–8.38

SUBMIT

Use effective nuclear charge to explain relative ionization energies.

EXERCISE 8.39—EFFECTIVE NUCLEAR CHARGE AND IONIZATION ENERGIES

Calculate the effective nuclear charges experienced by valence electrons in ground-state atoms of Na, Si, and Ar. Use these to rationalize the relative magnitudes of the first ionization energies of atoms of these elements.

Charges on the Monatomic Ions of the Elements

The evidence, reported in Section 8.2, that needs to be accounted for here, includes the following:

1. *Monatomic ions of Group 1 have charge +1, those of Group 2 have charge +2, those of Group 13 are +3, and those of Group 14 have +4.*

   An explanation for these facts is rather obvious from the consideration of successive ionization energies of the elements, discussed above. We can use the elements of the third row to illustrate. For sodium atoms, IE_1 is low, so one electron is relatively easily removed by other species (oxidizing agents) to form Na^+ ions. However, for this Group 1 element, IE_2 is very large, and removal of a second electron requires so much energy that this is not achievable by common oxidizing agents. Using corresponding arguments, we can see that while it may be possible for common oxidizing agents to remove two electrons from magnesium atoms (IE_1 and IE_2 are low), it is unlikely that they are able to remove the third electron (IE_3 is very large). In the case of aluminium in Group 13, it seems that removal of three electrons is possible (to form Al^{3+} ions) but removal of a fourth electron is usually not because IE_4 is so large.

2. *The monatomic ions of Groups 15, 16, and 17 are anions with negative charge = 18 − group number.*

   It seems reasonable that the elements of Groups 15, 16, and 17 do not form cations, because their first ionization energies are large (due to the high effective nuclear charge experienced by their valence electrons). Another way for an atom to form an ion is to accept one or more electrons, forming an anion. Our wave model of electronic structure allows us to evaluate whether an anion formed in this way is stable.

   Let's use chlorine ($_{17}Cl$) in Group 17 to develop a basis for rationalization. We can calculate the effective nuclear charge experienced by the seven valence electrons in chlorine atoms:

   $_{17}Cl$ atom: $1s^2 2s^2 2p^6 3s^2 3p^5$ $Z^* = +17 - (6 \times 0.35) - (8 \times 0.85) - (2 \times 1.0) = +6.1$

   The magnitude of Z* suggests that the valence electrons in a Cl atom are strongly held, but what if the atom accepts another electron—bearing in mind that another electron will introduce repulsions. To evaluate this, we need to calculate Z* of a Cl^- ion:

   $_{17}Cl^-$ ion: $1s^2 2s^2 2p^6 3s^2 3p^6$ $Z^* = +17 - (7 \times 0.35) - (8 \times 0.85) - (2 \times 1.0) = +5.8$

An explanation commonly given is that cations are stable when they have the electron configuration of the previous noble gas. This is not an explanation in itself, but noble gas configurations are relatively stable situations with high Z*.

This value suggests that the valence electrons are strongly held, and reconciles with the observation that Cl⁻ ions are stable. And if the ion accepts another electron, forming a Cl^{2-} ion? Let's calculate Z^*. Note that this ion has only one valence electron—in a $4s$ shell:

$$_{17}Cl^{2-} \text{ ion: } 1s^2 2s^2 2p^6 3s^2 3p^6 4s^1 \qquad Z^* = +17 - (0 \times 0.35) - (8 \times 0.85) - (10 \times 1.0) = +1.2$$

The magnitude of Z^* of Cl^{2-} ions suggests that the valence electrons experience negligible force of attraction to the atoms, and so it is perhaps not surprising that these ions are unstable, and do not exist.

EXERCISE 8.40—EFFECTIVE NUCLEAR CHARGE AND CHARGES ON SIMPLE IONS

Use the concept of effective nuclear charge to make sense of the fact that (a) nitrogen forms N^{3-} ions, and (b) sulfur commonly forms S^{2-} ions.

Sizes of Ions

Several relationships involving ion sizes [<<Section 8.2] can be explained by the effective nuclear charges experienced by valence electrons:

1. *Cations are smaller than their parent atoms.* Let's consider, for example, the radius of a K^+ ion (133 pm) compared with that of a K atom (227 pm). In Worked Example 8.5, we see that Z^* in K^+ ions is +7.7, while in K atoms Z^* is only +2.2. So it is not surprising that a K^+ ion is smaller than a K atom.

2. *Anions are larger than their parent atoms.* To make sense of this generalization, let's compare Z^* in S atoms with that in S^{2-} ions. The electron configurations of these two species are as follows:

$$_{16}S \text{ atom: } 1s^2 2s^2 2p^6 3s^2 3p^4 \qquad _{16}S^{2-} \text{ ion: } 1s^2 2s^2 2p^6 3s^2 3p^6$$

From these, we deduce

$$Z^*(_{16}S \text{ atom}) = +16 - \left[(5 \times 0.35) + (8 \times 0.85) + (2 \times 1.0)\right] = +16 - 10.5 = +5.5$$
$$Z^*(_{16}S^{2-} \text{ ion}) = +16 - \left[(7 \times 0.35) + (8 \times 0.85) + (2 \times 1.0)\right] = +16 - 11.3 = +4.7$$

Since the valence electrons on S atoms experience a stronger inward attraction than those in S^{2-} ions, it makes sense that S^{2-} ions are bigger than S atoms.

3. *Ions get smaller across a row, and larger as we go down a group.* These trends can be rationalized using similar arguments to those used for the trends in sizes of atoms.

EXERCISE 8.41—EFFECTIVE NUCLEAR CHARGE AND ION SIZES

Calculate and compare the effective nuclear charge for valence electrons in Na^+ ions, Mg^{2+} ions, and Al^{3+} ions. Are your answers consistent with the trend in sizes of these ions (Figure 8.7)?

Electronegativities

The general periodic trends in experimental estimates of electronegativities can be accounted for by comparison of effective nuclear charges, in the same way as for rationalization of trends of atomic sizes, first ionization energies, and ionic sizes.

EXERCISE 8.42—EFFECTIVE NUCLEAR CHARGE AND ELECTRONEGATIVITIES

Use the concept of effective nuclear charge to rationalize the relative values of the electronegativities of Mg and Cl.

Electron Affinities

The general trends of electron affinities (Table 8.7) can be explained by the effective nuclear charges experienced by valence electrons in the anion formed by attachment of an electron to an atom. For example, we have seen above that Z^* in a Cl^- ion is high (+5.8), indicating that Cl atoms strongly hold onto an extra electron—consistent with the high electron affinity.

Group 18 elements, on the other hand, have very low electron affinity, and this can be attributed to the very low Z^* in anions such as the Ne^- ion:

$$_{10}Ne^- \text{ ion: } 1s^2 2s^2 2p^6 3s^1 \quad Z^* = +10 - (0 \times 0.35) - (8 \times 0.85) - (2 \times 1.0) = +1.2$$

EXERCISE 8.43—EFFECTIVE NUCLEAR CHARGE AND ELECTRON AFFINITIES

Use the concept of effective nuclear charge to rationalize the differences in electronegativities of Mg, Cl, and Ar atoms in molecules.

Properties of the Elemental Substances

Atoms of the elements in Groups 16 and 17 have high ionization energies and electron affinities. Their valence electrons are not easily removed, but they attract electrons from other species strongly. These properties, which can be rationalized by consideration of the magnitude of Z^* in their atoms and ions, account for their reactions as oxidizing agents and the formation of anions.

Atoms of the *s*-block elements have low ionization energies and small electron affinities, so electrons are rather easily removed from them, but they have only very weak attractions for the electrons of other species. Consequently, they have a tendency to be powerful reducing agents, with formation of cations—again, consistent with the calculations of Z^* in their atoms and ions.

The very low melting points and boiling points of the Group 18 elements (Table 8.1) can be attributed to their existence as single atoms. Reasons for this include (a) their high ionization energy, which means that electrons are not easily removed from their atoms to form cations, or even to be shared in covalent bonds; and (b) their low electron affinity, which means that they have little ability to remove electrons from other species and form anions, or even to share the electrons from other species in covalent bonds. Furthermore, in atoms of these elements, the valence electrons experience relatively high effective nuclear charges (Table 8.11), and the strong inward force acting on the electrons causes their electron clouds to have low polarizability [<<Section 6.3]. This, in turn, implies weak dispersion forces between the atoms, giving rise to their very low melting points and boiling points.

What a powerful model we have been using! All of these observable properties, which strongly influence the chemical behaviour of substances, can now be rationalized by a model of electrons in atoms that considers and computes them as waves. This is an extraordinary tale of interaction between experiment observation and theoretical development, which required scientists to objectively accept what they could hardly believe. The experiments that put them on the path to solving the riddle of the nature of atoms required them to think "outside of the box" in ways to which everyday experiences could not have exposed them [<<Section 8.1].

> The periodical variation of properties of the elements can be explained by consideration of the inward pull on valence electrons in atoms, as estimated by relative values of effective nuclear charge, Z^*.

Every man's world picture is and always remains a construct of his mind and cannot be proved to have any other existence.

Erwin Schrödinger, *Mind and Matter*

WebLink

e8.25 Go to suggested websites to see periodic tables from a variety of perspectives.

SUMMARY

Key Concepts

- According to the **law of chemical periodicity,** as we go from element to element in order of their atomic numbers, there is periodic variation of the properties of the elemental substances as well as of the "atomic" properties of the elements. (Section 8.2)
- The observation of **line emission spectra** of excited atoms was interpreted to mean that the energies of electrons in atoms are **quantized**. The energy of each spectral line corresponds with the difference between two levels of electron energy. (Section 8.3)
- In some circumstances, electrons exhibit properties of waves as well as of particles. Recognition of this **wave-particle duality** (of the human mind, not of electrons) provided the impetus to model electrons in atoms as waves. (Section 8.3)
- The **Schrödinger equation** formulates the conditions for three-dimensional "standing waves" that correspond with the quantized energy levels of electrons in atoms. This wave equation corresponds with the description of standing waves only when it includes particular values of the **quantum numbers** n, l, and m_l. (Section 8.4)
- Standing wave solutions to the Schrödinger equation include the exact energy of the electron (E) and a wave function (ψ), whose magnitude varies over the space around the nucleus. At any point in space, the value of ψ^2 is a measure of the **electron density**, or probability. The "shape" of an **orbital** is characterized by a surface of points of equal electron density. Only in the case of s orbitals is this **isodensity surface** spherical. (Section 8.4)
- The term **shell** refers to all orbitals derived with the same value of quantum number n. A shell defined by $n = 1$ has one orbital, the $n = 2$ shell has four orbitals, the $n = 3$ shell has nine, and the $n = 4$ shell has sixteen. (Section 8.4)
- Electrons have a property called **spin**, which generates a small magnetic field in either of two opposite orientations, distinguished by spin quantum number $m_s = +\frac{1}{2}$ and $-\frac{1}{2}$. **Paramagnetic** substances have unpaired electrons. (Section 8.4)
- The distribution of electrons amongst the possible orbitals (that is, which standing waveforms the electrons can have) is called the **electron configuration**. The **ground-state configuration** is the distribution of electrons such that the atom has its lowest possible energy. If the electron configuration is such that an atom has more energy than the ground state, it is called an **excited state**. (Section 8.5)
- According to the Pauli exclusion principle, no two electrons in an atom can have the same values of n, l, m_l, and m_s. As a consequence, an orbital can have no more than two electrons, with opposite spins. (Section 8.5)
- Ground-state electron configurations can be predicted using the **aufbau principle**: electrons are assigned into orbitals in the order indicated in Figure 8.24, while obeying the Pauli exclusion principle and Hund's rule. (Section 8.5)
- When the aufbau principle is applied, it is found that atoms of elements in the same group of the periodic table have similar ground-state **valence electron** configurations: s^1 for Group 1 elements, s^2 for Group 2 elements, s^2p^1 for Group 13 elements, and so on. (Section 8.5)
- **Effective nuclear charge** (Z^*) is the nuclear charge experienced by valence electrons, taking into account reduction of the effect of the nuclear charge by repulsion from other electrons (**shielding**). Estimates of Z^* are made by applying Slater's rules. Z^* values vary periodically. (Section 8.6)
- The periodic variation of properties of the elements can be explained by consideration of the inward pull on valence electrons in atoms, as estimated by relative values of effective nuclear charge (Z^*). (Section 8.7)

Key Equations

$$E = R\left(\frac{1}{2^2} - \frac{1}{n^2}\right) \text{ when } n > 2 \qquad \text{(Section 8.3)}$$

$$\lambda = \frac{h}{mv} \text{ (Section 8.3)}$$

REVIEW QUESTIONS

Section 8.2: Periodic Variation of Properties of the Elements

8.44 (a) Which of calcium or arsenic is the better electrical conductor?

(b) Which of sodium or sulfur is the better thermal conductor?

(c) Which of carbon or lead would you expect to be the more malleable?

8.45 Of the two compounds magnesium chloride ($MgCl_2$) and phosphorus chloride (PCl_3), one melts at $-112\ °C$ and does not conduct electricity, either in the solid state or when molten. The other melts at $714\ °C$ and although it does not conduct electricity when solid, it does when molten. Which is which?

8.46 Fluorine, $F_2(g)$, is a powerful oxidizing agent. The element fluorine has atomic number 9. Give the names, symbols, and atomic numbers of the next three elements in the sequence of atomic numbers that occur as diatomic molecules with powerful oxidizing properties. What are the differences between the atomic numbers of successive pairs of these elements?

8.47 By reference to the periodic table (rather than Figure 8.4), arrange the following elements in order of increasing size of the atoms: Al, B, C, K, and Na.

8.48 (a) Using Figure 8.4, estimate the H–O and H–S distances in H_2O and H_2S, respectively.

(b) If the interatomic distance in Br_2 is 228 pm, what is the covalent radius of Br? Using this value, and that for Cl (99 pm), estimate the distance between atoms in BrCl.

8.49 Arrange the following atoms in the order of increasing first ionization energy: Si, K, P, and Ca.

Interactive Exercise 8.50

Compare the radii of different atoms and ions.

8.51 For each of the following pairs of ions predict which one requires more energy to remove an electron:

(a) Te^- ion or S^- ion

(b) Cl^- ion or P^- ion

8.52 Are the following processes endothermic or exothermic?

(a) $O^-(g) \longrightarrow O(g) + e^-(g)$

(b) $O(g) + e^-(g) \longrightarrow O^-(g)$

(c) $Ne^-(g) \longrightarrow Ne(g) + e^-(g)$

8.53 Compare the three elements B, Al, and C:

(a) Place the three elements in order of increasing atomic radius.

(b) Rank the elements in order of increasing ionization energy. (Do this without looking at Figure 8.4; then check your prediction against values in the graph.)

(c) Which element would you expect to have the highest electron affinity?

8.54 Compare the elements B, Al, C, and Si:

(a) Which has the most metallic character?

(b) Which has the largest atomic radius?

(c) Which has the greatest electron affinity?

(d) Place the three elements B, Al, and C in order of increasing first ionization energy.

8.55 Answer each of the following concerning periodic variation of properties:

(a) Place the following elements in order of increasing ionization energy: F, O, and S.

(b) Which has the largest first ionization energy: O, S, or Se?

(c) Which has the greatest electron affinity: Se, Cl, or Br?

(d) Which has the largest radius: O^{2-}, F^-, or F?

8.56 Name the element corresponding to each characteristic below:

(a) the element with the electron configuration $1s^2 2s^2 2p^6 3s^2 3p^3$

(b) the Group 2 element with the smallest atomic radius

(c) the element with the largest ionization energy in Group 15

(d) the element whose 2+ ion has the configuration $[Kr]4d^5$

(e) the element with the greatest electron affinity in Group 17

(f) the element whose electron configuration is $[Ar]3d^{10}4s^2$

Interactive Exercise 8.57

Classify elements based on their electron configurations.

8.58 Answer each of the following questions:
  (a) Of the elements S, Se, and Cl, which has the largest atoms?
  (b) Which has the larger radius, Br atoms or Br⁻ ions?
  (c) Which has the largest difference between the first and second ionization energy: Si, Na, P, or Mg?
  (d) Which has the largest ionization energy: N, P, or As?
  (e) Which of the following has the largest radius: O^{2-} ions, N^{3-} ions, or F^- ions?

Section 8.3: Experimental Evidence about Electrons in Atoms

8.59 Calculate the ionization energy of a ground state hydrogen atom; that is, the energy required to cause excitation of the electron from the $n = 1$ level to the $n = \infty$ level ($E = 0$).

8.60 The most prominent line in the spectrum of mercury is at 253.652 nm. Other lines are located at 365.015 nm, 404.656 nm, 435.833 nm, and 1013.975 nm.
  (a) Which of these lines represents the most energetic light?
  (b) What is the frequency of the most prominent line? What is the energy of one photon with this wavelength?
  (c) Are any of these lines found in the spectrum of mercury shown in Figure 8.10? What colour or colours are these lines?

8.61 A line in the Balmer series of emission lines of excited H atoms has a wavelength of 410.2 nm (Figure 8.13). What colour is the light emitted in this transition? Which transition of electron energy (defined by values of n_{initial} and n_{final}) gives rise to this emission line?

8.62 What are the wavelength and frequency of the radiation involved in the least energetic emission line in the Lyman series? Which transition of electron energy (defined by values of n_{initial} and n_{final}) gives rise to this emission line?

8.63 Consider only transitions involving the $n = 1$ through $n = 5$ energy levels for the H atom (where the energy level spacings below are not to scale).

  _____ $n = 5$

  _____ $n = 4$

  _____ $n = 3$

  _____ $n = 2$

  _____ $n = 1$

  (a) How many emission lines are possible, considering only the five quantum levels?
  (b) Photons of the highest frequency are emitted in a transition from the level with $n =$ ____ to a level with $n =$ ____.
  (c) The emission line having the longest wavelength corresponds to a transition from the level with $n =$ ____ to the level with $n =$ ____.

Interactive Exercise 8.64

Investigate the line emission spectrum of hydrogen.

8.65 The energy emitted when an electron moves from a higher energy state to a lower energy state in any atom can be observed as electromagnetic radiation.
  (a) Which involves the emission of less energy in the H atom, an electron moving from $n = 4$ to $n = 2$ or an electron moving from $n = 3$ to $n = 2$?
  (b) Which involves the emission of more energy in the H atom, an electron moving from $n = 4$ to $n = 1$ or an electron moving from $n = 5$ to $n = 2$? Explain fully.

Interactive Exercise 8.66

Predict the relative energies of transitions.

8.67 Calculate the wavelength associated with a neutron having a mass of 1.675×10^{-24} g and a kinetic energy of 6.21×10^{-21} J. (Recall that the kinetic energy of a moving particle is $E = \frac{1}{2}mv^2$.)

8.68 An electron moves with a velocity of 2.5×10^8 cm s^{-1}. What is its wavelength?

Section 8.4: The Quantum Mechanical Model of Electrons in Atoms

8.69 Assume that the line shown here is 10 cm long.

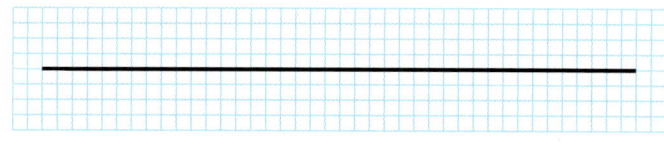

Using this line, do the following:
  (a) Draw a standing wave with one node between the ends. What is the wavelength of this wave?
  (b) Draw a standing wave with three evenly spaced nodes between the ends. What is its wavelength?
  (c) If the wavelength of the standing wave is 2.5 cm, how many waves fit within the boundaries? How many nodes are there between the ends?

8.70 (a) When a solution for the wave equation is derived using $n = 4$, what are the allowed values of l?
  (b) When $l = 2$ is used, what are the allowed values of m_l for standing electron waves?
  (c) For a 4s orbital, what are the allowed values of n, l, and m_l?
  (d) For a 4f orbital, what are the allowed values of n, l, and m_l?

8.71 An excited state of the H atom has the electron in a 4p orbital. List all allowed sets of quantum numbers n, l, and m_l for standing electron waves in this orbital.

8.72 Give the names of the orbitals derived by solution of the Schrödinger equation when the following values of quantum numbers have been used:
(a) $n = 4$, $l = 2$, $m_l = -2$
(b) $n = 4$, $l = 2$, $m_l = 0$
(c) $n = 3$, $l = 0$, $m_l = 0$
(d) $n = 2$, $l = 1$, $m_l = -1$
(e) $n = 5$, $l = 3$, $m_l = 2$

8.73 Explain briefly why each of the following is not a possible set of quantum numbers for a standing electron wave in an atom.
(a) $n = 2$, $l = 2$, $m_l = 0$
(b) $n = 3$, $l = 0$, $m_l = -2$
(c) $n = 6$, $l = 0$, $m_l = 1$

8.74 State which of the following atomic orbitals cannot exist according to the quantum theory: (a) 2*s*, (b) 2*d*, (c) 3*p*, (d) 3*f*, (e) 4*f*, and (f) 5*s*. Briefly explain your answers.

Interactive Exercises 8.75–8.76

Identify incorrect labelling of atomic orbitals.

8.77 How many sub-shells are in the shell defined by principal quantum number $n = 4$?

8.78 What is the maximum number of orbitals defined by each of the following sets of quantum numbers? When "none" is the correct answer, explain your reasoning.
(a) $n = 3$, $l = 0$, and $m_l = +1$
(b) $n = 5$, $l = 1$
(c) $n = 7$, $l = 5$
(d) $n = 4$, $l = 2$, and $m_l = -2$

Interactive Exercise 8.79

Identify incorrect labelling of atomic orbitals.

8.80 Explain briefly why each of the following is not a valid set of quantum numbers for an electron in an atom. In each case, change a value (or values) to make the set valid.
(a) $n = 2$, $l = 2$, $m_l = 0$, $m_s = +\frac{1}{2}$
(b) $n = 2$, $l = 1$, $m_l = -1$, $m_s = 0$
(c) $n = 3$, $l = 1$, $m_l = +2$, $m_s = +\frac{1}{2}$

8.81 (a) What are the *n* and *l* values for each of the following orbitals: 6*s*, 4*p*, 5*d*, and 4*f*?
(b) How many nodal planes exist for a 4*p* orbital? For a 6*d* orbital?

8.82 Complete the following table:

Orbital Type	Number of Orbitals in a Given Sub-shell	Number of Nodal Surfaces
s	_____	_____
p	_____	_____
d	_____	_____
f	_____	_____

8.83 Complete the following sentences:
(a) The quantum number *n* describes the _____ of an atomic orbital.
(b) The shape of an atomic orbital is given by the quantum number _____.
(c) A photon of green light has _____ (less or more) energy than a photon of orange light.
(d) The maximum number of orbitals that may be associated with the set of quantum numbers $n = 4$ and $l = 3$ is _____.
(e) The maximum number of orbitals that may be associated with the quantum number set $n = 3$, $l = 2$, and $m_l = -2$ is _____.
(f) When $n = 5$, the allowed values of *l* are _____.
(g) The number of orbitals in the $n = 4$ shell is _____.

Section 8.5: Electron Configurations in Atoms

8.84 Depict with an orbital box diagram the electron configuration of ground-state gallium atoms. Give a set of quantum numbers for the highest-energy electrons.

Interactive Exercise 8.85

Deduce the electron configuration for ground-state atoms.

8.86 (a) Which element has atoms with the ground-state configuration $1s^2 2s^2 2p^6 3s^2 3p^5$?
(b) Deduce the electron configuration of ground-state phosphorus atoms, and display it with both an orbital box diagram and the *spdf* notation.
(c) Write one possible set of quantum numbers for the valence electrons of calcium atoms in the ground state.

8.87 Deduce the electron configurations of ground-state atoms of (a) sulfur, and (b) aluminium.

8.88 Write the electron configurations, using both *spdf* notation and orbital box diagrams, of ground-state atoms of (a) P, and (b) Cl. Describe the relationship between each atom's electron configuration and its position in the periodic table.

8.89 Depict the electron configuration of ground-state atoms of each of the following elements, using *spdf* and noble gas notations:
(a) Arsenic (As). A deficiency of As can impair growth in animals even though larger amounts are poisonous.
(b) Krypton (Kr). It ranks seventh in abundance of the gases in the earth's atmosphere.

Interactive Exercises 8.90–8.91

Use effective nuclear charge to explain relative ionization energies.

Section 8.6: Shielding and Effective Nuclear Charge

8.92 (a) Calculate and compare the effective nuclear charge for valence electrons in ground-state atoms of elements with atomic numbers 17, 18, and 19. What do these values imply about the relative reactivity of atoms of Group 18 elements compared with atoms of the adjacent Group 17 and Group 1 elements, in terms of the ability of other species to remove electrons from them?

(b) Calculate the effective nuclear charge for valence electrons in (i) Ne^- ions, (ii) Ar^- ions, and (iii) Kr^- ions. What do these values imply about the relative reactivity of atoms of Group 18 elements, in terms of removing electrons from other species?

(c) What do the values of Z^* for Ne, Ar, and Kr atoms (Table 8.11) imply about the strength of the attraction between the valence electrons and the nuclei in Group 18 atoms? What might this mean in terms of the polarizability of the electron cloud in Group 18

atoms? How would you expect this to translate to the strengths of dispersion forces between these atoms, and the boiling points of the substances?

Section 8.7: Rationalizing the Periodic Variation of Properties

8.93 Calculate and compare the effective nuclear charge for valence electrons in (a) Mg atoms, (b) Mg^+ ions, and (c) Mg^{2+} ions. How do these values relate to the relative sizes of the successive ionization energies of magnesium atoms?

8.94 Successive ionization energies of an element are 590, 1145, 4912, 6491, and 8153 kJ mol^{-1}.
(a) In which group of the periodic table is the element?
(b) The element is in the fourth period. Which element is it?
(c) Use the concept of effective nuclear charge to rationalize the relative values.

8.95 Calculate and compare the effective nuclear charge for valence electrons in N^{3-} ions, O^{2-} ions, and F^- ions. Use these values to account for the relative sizes of these ions.

SUMMARY AND CONCEPTUAL QUESTIONS

8.96 Bohr pictured the electrons of the atom as being located in definite orbits about the nucleus, just as the planets orbit the sun. Criticize this model.

8.97 Light is given off by a sodium- or mercury-containing streetlight when the atoms are excited. The light you see arises for which of the following reasons?
(a) Electrons are moving from a given energy level to one of higher n.
(b) Electrons are being removed from the atom, thereby creating a metal cation.
(c) Electrons are moving from a given energy level to one of lower n.

8.98 How do we interpret the physical meaning of the square of the wave function (ψ)?

8.99 What does "wave-particle duality" mean? What are its implications in our modern view of atomic structure?

8.100 Suppose you live in a different universe where a different set of quantum numbers is required to obtain

solutions for electron standing waves in atoms of that universe. These quantum numbers have the following "allowed" values:

$N = 1, 2, 3, \ldots, \infty$

$L = N$

$M = -1, 0, +1$

How many orbitals are there all together in the first three electron shells?

8.101 Why is the radius of Li^+ so much smaller than the radius of Li? Why is the radius of F^- so much larger than the radius of F?

8.102 Which ions in the following list are not likely to be found in chemical compounds: K^{2+}, Cs^+, Al^{4+}, F^{2-}, and Se^{2-}? Explain briefly.

8.103 Explain how the ionization energy of atoms changes and why the change occurs when proceeding down a group of the periodic table.

CHAPTER 9

Molecular Shapes and Structures

Volker Steger / Science Photo Library

Volker Steger / Science Photo Library

Steve Gschmeissner / Science Photo Library

Steve Gschmeissner / Science Photo Library

9.1 Case Study: Molecular Handshakes and Recognition

Blockbuster movies like *Spider-Man* feature crime-busting humans with the ability to scale tall buildings to make the world a better place. But for ordinary people, the force of gravity far exceeds the force of attraction between our hands and smooth vertical surfaces. This isn't true, however, for all creatures—the gecko can defy gravity by walking up smooth walls and across ceilings. What is happening at the molecular level that gives gecko feet Spider-Man-like abilities?

> *What is happening at the molecular level that gives gecko feet Spider-Man-like abilities?*

As biomimetic chemists who try to create materials to mimic a gecko foot will attest, there is nothing simple about a gecko's walk up a smooth vertical surface. We do know that the non-covalent *intermolecular interactions* between molecules at the ends of gecko feet and the surfaces of a wall make this remarkable accomplishment possible. Scanning electron microscopy, has shown that gecko feet contain thousands of hair-like structures, each of which has bumps that end in a rounded hydrophobic tip (Figure 9.1a). The size, shape, and polarity of the molecules in these millions of tips allows a gecko to make use of non-bonded interactions to "stick" to the molecules in the dry smooth surface of a wall. Because the forces of interaction are weak, they can be broken relatively easily. A molecular-level image of the interaction as a gecko moves up a wall might be of sequential "hand shaking" and releasing between molecules in the hair-like tips at the end of its feet and molecules on the wall surface.

A biomimetic process is a human-made process that imitates nature.

The scanning electron microscope (SEM) image in Figure 9.1a shows a microscopic view of the hair-like tips at the end of gecko feet. But how do we know what the molecules at those tips look like? Because atoms and molecules are too small to directly observe with our senses, we need to "see" at another level with sophisticated instrumentation to recognize molecular-level differences between substances. We identify molecules of different substances based on absorption peaks and signal patterns in spectrometers and other instruments. For example, we've used infrared spectroscopy to tell apart the white cocaine, nicotine, and sucrose powders [<<Section 3.1], and gas chromatography–mass spectrometry to differentiate between and identify the compounds isolated from wood smoke that cause germination of seeds [<<Section 1.3].

FIGURE 9.1 (a) Scanning electron microscope (SEM) image of the hair-like tips at the end of the feet of geckos. Trying to mimic the properties of these feet, chemists at the University of Manchester, U.K., have synthesized gecko tape. (b) An SEM image of a tape made of a strong polyimide polymer using lithography, a method for fabricating materials on a micrometre scale. Each projection in the synthetic material is 2 μm long.

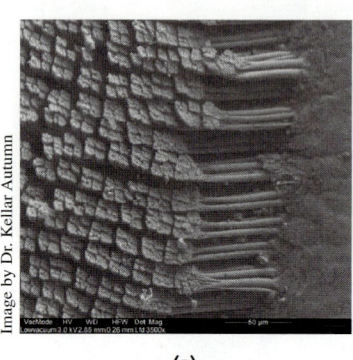
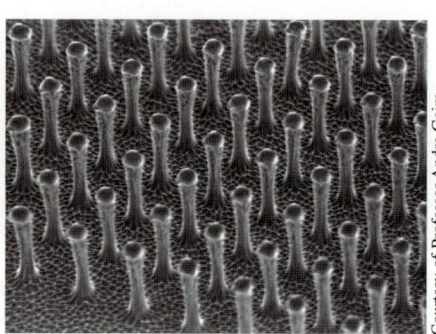

(a)   (b)

In this chapter, we introduce several new instrumental tools used to discern the overall shapes and three-dimensional structures of covalently bonded molecules, like those at the end of gecko feet, or those responsible for seed germination [<<Section 1.3]. Before narrowing our attention in Chapter 10 to models that explain how atoms are bonded to each other within molecules, we examine in this chapter what we know experimentally about the overall shapes of molecules. Those shapes aren't static—they change as a result of rotation about bonds, much as a human hand can be bent into different shapes by movement of fingers. And while we must understand the structures of individual molecules, the properties of substances are determined by dynamic molecules interacting with countless other molecules of the same and different kinds.

It can be useful to extend the analogy of a human hand further and describe molecules as "recognizing" each other's characteristic shapes and polarities by "shaking hands" with many other molecules. Countless weak forces of interaction between molecules lead to effects that we can observe directly with our senses, such as observing a gecko scamper up a wall.

Similar countless **molecular recognition** events are thought to be responsible for the communication of bacteria with each other. Single-celled organisms, bacteria can do little in isolation from each other, but chemical communication among bacteria that are close to each other allows them to turn on genes to coordinate their behaviour. This communication helps them participate in activities such as obtaining nutrient sources, exchanging DNA, and creating a protective environment for the resistance of antibiotics.

In nature, molecules never exist in isolation from other molecules, and recognition between molecules for others is the basis of modern biology. Ligands bind to proteins, nucleic acid strands bind to each other, and lipid bi-layers form spontaneously in all biological membranes, including cell membranes. And ion channel proteins show remarkable selectivity as they regulate the flow of ions across cell membranes.

Sometimes we can detect those molecular recognition events with our macroscopic senses. An example we explore more deeply throughout this chapter relates to the odours of spearmint and caraway oil. In both cases, they smell the way they do in large part because they contain carvone molecules. Carvone molecules in these two substances have the same molecular formula, $C_{10}H_{14}O$, and even exactly the same connectivity. But the different "type" of carvone molecules must have important, differences in structure, as the substances they make up don't smell the same. The subtle, but important, difference is in their orientation in three-dimensional space—their stereochemistry. These two stereoisomers of carvone have the same relationship to each other as your right hand has to your left—they are referred to as "handed" or *chiral* molecules. Just as initiating a handshake with your right hand will immediately recognize the difference between a friend's right and left hand, the "right" and "left" handed stereoisomers of carvone molecules have different non-covalent interactions with chiral receptor molecules in your nose. Your brain processes these different molecular handshakes as different odours.

Our understanding of molecular recognition and odour is in its infancy, and there is large individual variation in human senses. For example, the male pig sex pheromone, androstenone (described in Section 23.13), has an undetectable odour to some people; to others, it smells pleasant; and to a third group, it is disgusting.

The astounding scale and importance of molecular recognition was recognized over 50 years ago, in a study of the tobacco mosaic virus, the first plant virus to be discovered. X-ray crystallographer Rosalind Franklin showed the virus to be a hollow, rod-like supramolecular assembly in which a single strand of RNA is covered by a coat made of about 2130 protein monomer molecules. The remarkable role for molecular recognition became evident from the 1955 discovery that after the virus is separated into its constituent molecules by treatment with acetic acid, the RNA and protein molecules are able to automatically recognize, order, and assemble themselves in a test tube to form a functioning virus that can infect plants. This process is called **self-assembly**, and it is one of the most remarkable examples of the ability of molecules to recognize and interact with each other.

As one example of exploiting self-assembly, chemists are designing bioactive molecules called *peptide amphiphiles,* with the ability to assemble themselves into a putty-like material that may help regenerate bone growth to repair severe fractures.

In the rest of this chapter, we examine more carefully the three-dimensional shapes and dynamic features of molecules so important to molecular recognition, starting with experimental tools that tell us what we know about their structures.

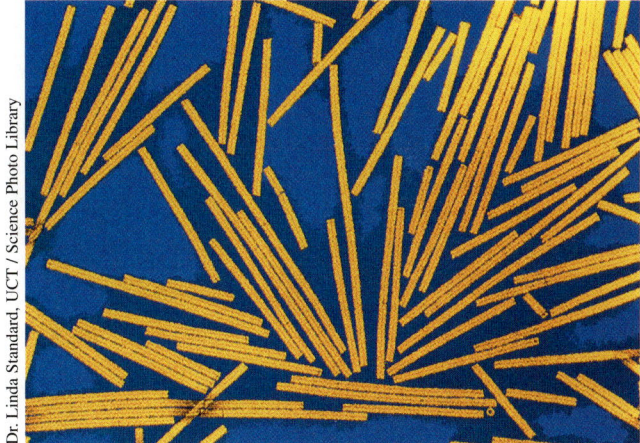

Coloured electron micrograph of a cluster of rod-shaped red virus particles of tobacco mosaic virus. Each particle is about 16 nm in diameter.

9.2 Experimental Tools for Molecular Structures and Shapes

We return to carvone. A *high-resolution mass spectrum* [<<Section 3.3] of carvone that has been extracted from spearmint oil tells us that the exact molar mass of its most abundant **isotopologue** is 150.104465, which is consistent with a molecular formula of $C_{10}H_{14}O$. So we know that each molecule of carvone has ten C atoms, fourteen H atoms, and one O atom. But in what connectivity patterns are the atoms joined together? Are there double bonds or rings? What is the spatial arrangement of atoms within the molecule that gives rise to its overall shape? These are the factors that cause carvone to have its characteristic behaviour.

The decomposition of carvone molecules into fragment ions within the *mass spectrometer* provides some limited information about how the atoms are connected to each other.

The *infrared spectrum* [<<Sections 3.1 and 3.3] of a substance tells us what groups of atoms (functional groups) are attached to the carbon "skeleton" of molecules. This provides additional information about connectivity. The infrared spectrum of carvone shows that it contains a carbonyl (C=O) functional group and an alkene (C=C).

X-ray crystallography shows the spatial relationship of atoms within crystalline substances. The location in space of each atom relative to the other atoms shows the overall shape of a molecule and gives detailed information about bond distances and bond angles. This structural information for a substance like carvone can be used in conjunction with powerful computer visualization software to draw accurate three-dimensional pictures showing how atoms in small or large molecules are arranged in

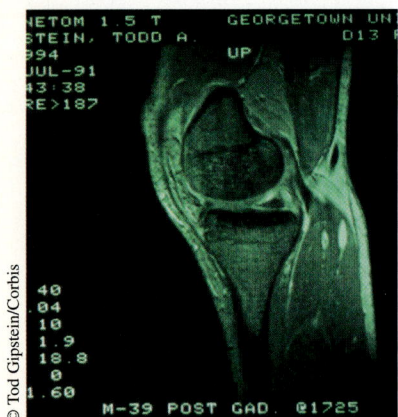

FIGURE 9.2 An MRI scan of a human knee, showing the presence of a ganglion cyst. Magnetic resonance imaging (MRI) uses an NMR instrument large enough to place a human body between the magnet poles instead of a thin glass tube. MRI determines the three-dimensional location of magnetic nuclei (usually ^1H) in the body. ^1H nuclei are present in abundance in water or fat found in soft tissues, allowing diagnosis of brain tumours, strokes, and damage to knees and other joints.

space. A significant limitation, however, is that this information is obtained for crystalline materials. Most chemistry takes place between molecules in solution, and often the overall shape of molecules changes significantly when a substance undergoes a phase change or dissolves in a solvent. This is particularly true for large molecules such as proteins.

Fortunately, a complementary technique exists to give us information about connectivity and molecular structure for molecules in solution, which have continually changing overall shapes. The technique is called *nuclear magnetic resonance (NMR) spectroscopy*, and may be more familiar to you for its applications in medicine, where it is the basis for magnetic resonance imaging (MRI, Figure 9.2). From an NMR spectrum, a map of the carbon-hydrogen framework of a molecule can be obtained. We return to the NMR spectrum of carvone and learn how to interpret NMR spectra in Section 9.4.

Finally, *polarimetry* can be used to tell apart certain isomeric substances whose molecules have the same connectivity, but different **configurations**, or three-dimensional orientation in space. Carvone molecules extracted from spearmint and caraway oils have the same connectivity, and so they appear identical by mass spectrometry, infrared spectroscopy, X-ray crystallography, and NMR spectroscopy. But a simple polarimetry experiment that measures the way substances rotate plane-polarized light shows that carvone molecules from these two sources have different configurations at one carbon atom. Molecules from these two sources have a relationship to each other that is similar to that of your right hand to your left hand. They are non-superimposable mirror images of each other [>>Section 9.8].

We now look more closely at X-ray crystallography and one type of NMR spectroscopy that is based on the magnetic properties of the small number of ^{13}C isotopes of the carbon atoms found in molecules. In Section 9.9, we'll see examples of how polarimetry is used.

> Instrumental techniques provide key experimental tools used by chemists to determine the molecular formulas, functional groups, three-dimensional arrangement of atoms, a map of the carbon framework, and the stereochemistry of molecules.

9.3 X-ray Crystallography

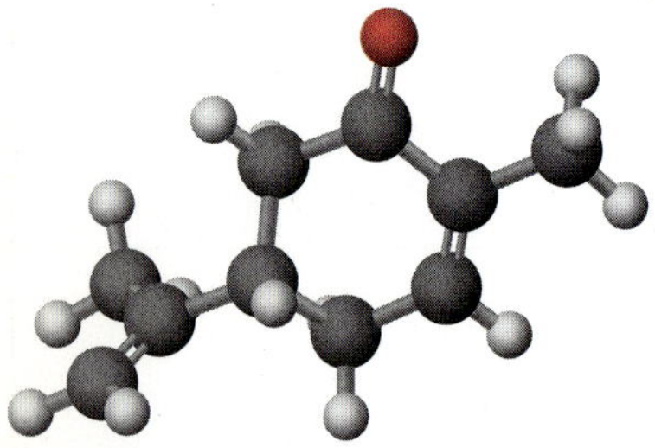

The overall shape of a molecule is determined by the lengths of each of the bonds that join atoms, the angles between the bonds, and the different ways in which parts of a molecule rotate in space about single bonds. From X-ray crystal structure data and complementary techniques such as electron diffraction, neutron diffraction, and microwave spectroscopy, an accurate picture of the position of each atom in three-dimensional space in molecules can be determined. From this, bond lengths and bond angles can be obtained. Examples of the information obtained from X-ray crystal structures are shown in Figures 9.3 and 9.4. Figure 9.3 is a representation of the three-dimensional structure of a carvone molecule from a sample extracted from caraway oil. The pure carvone was crystallized and its structure studied using X-ray crystallography.

FIGURE 9.3 Structure of a molecule of carvone extracted from caraway oil, based on X-ray crystal data. Molecular modelling software is used to visualize the spatial coordinates of each atom, and bonds are drawn wherever the distance between two adjacent atoms are close enough for them to reasonably share electrons.

Molecular Modelling (Odyssey)

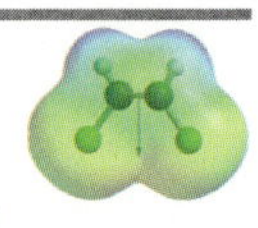

e9.1 Use X-ray crystal data to deduce a molecular structure.

Bond Lengths and Bond Angles

Bond length is the distance between the nuclei of two bonded atoms. Bond lengths are therefore related to the sizes of the atoms [<<Section 8.2]. But, for a given pair of atoms, an important trend is that single bonds are weaker and longer than double bonds, which are weaker and longer than triple bonds. Table 9.1 lists average bond lengths for a number of common chemical bonds. It is important to recognize that these are *average* values. Neighbouring parts of a molecule can affect the length of a particular bond. For example, Table 9.1 specifies that the average C—H bond has a length of 110 pm. In methane (CH_4), the measured bond length is 109.4 pm, whereas in acetylene (H—C≡C—H), the C—H bond is only 105.9 pm long. In some molecules, variations are as great as 10% from the average values listed in Table 9.1.

SINGLE BOND LENGTHS

TABLE 9.1 Some Average Single- and Multiple-Bond Lengths (pm)

Group	1	14	15	16	17	14	15	16	17	17	17
	H	C	N	O	F	Si	P	S	Cl	Br	I
H	74	110	98	94	92	145	138	132	127	142	161
C		154	147	143	141	194	187	181	176	191	210
N			140	136	134	187	180	174	169	184	203
O				132	130	183	176	170	165	180	199
F					128	181	174	168	163	178	197
Si						234	227	221	216	231	250
P							220	214	209	224	243
S								208	203	218	237
Cl									200	213	232
Br										228	247
I											266

Multiple bond lengths

C=C	134	C≡C	121
C=N	127	C≡N	115
C=O	122	C≡O	113
N=O	115	N≡O	108

Because atom sizes vary in a regular fashion with the position of the element in the periodic table (Figure 8.2), predictions of trends in bond length can be made. For example, the H—X distance in the hydrogen halides increases in the order predicted by the relative sizes of the halogens: H—F < H—Cl < H—Br < H—I. Likewise, bonds between carbon and another element in a given period decrease going from left to right, in a predictable fashion; for example, C—C > C—N > C—O > C—F. Trends for multiple bonds are similar. A C=O bond is shorter than a C=S bond, and a C=N bond is shorter than a C=C bond.

A bond angle is defined as the angle between the two straight lines connecting two atoms to a common central atom. For example, experimental determinations show that methane (CH_4), has an H—C—H bond angle of 109.5°. This is very close to the experimentally determined H—N—H bond angle in NH_3 of 107.5°, and the H—O—H angle in H_2O of 104.5°. By contrast, the corresponding H—C—H bond angle in ethene ($H_2C=CH_2$) is 120°, and in ethyne, the H—C≡C bond angle is 180°. In Section 10.7, we introduce a simple model used to rationalize the experimental bond angles that have been measured in many molecules.

Protein Crystal Structures

X-ray crystallography can also help determine the structures of much larger molecules than carvone, such as proteins. Figure 9.4 shows a representation of a much larger complex (so called, because it is made of several molecules)—in this case, two molecules of arachidonic acid ($C_{20}H_{32}O_2$) bound to the dimer of cyclooxygenase (blue), the enzyme that processes arachidonic acid in cells. Arachidonic acid is a signalling molecule that is important in the response to pain and inflammation.

The exact connectivity of atoms, and thus the structure of this large complex, can be determined from this data. While the size at which this structure is printed does not allow us to easily see the atom-to-atom connectivity, the data can be seen in the expanded portion. Twenty years ago, such detail was just a dream, and most proteins were shown as coloured circles in textbooks. Now the three-dimensional structures of

For further discussion of molecules important in the response to pain and inflammation, read about *non-steroidal anti-inflammatory* drugs (NSAIDs) in Section 20.1.

Molecular Modelling (Odyssey)

e9.2 Find the structures of proteins from Internet databases.

FIGURE 9.4 The structure of arachidonic acid and cyclooxygenase as determined from X-ray crystallography.

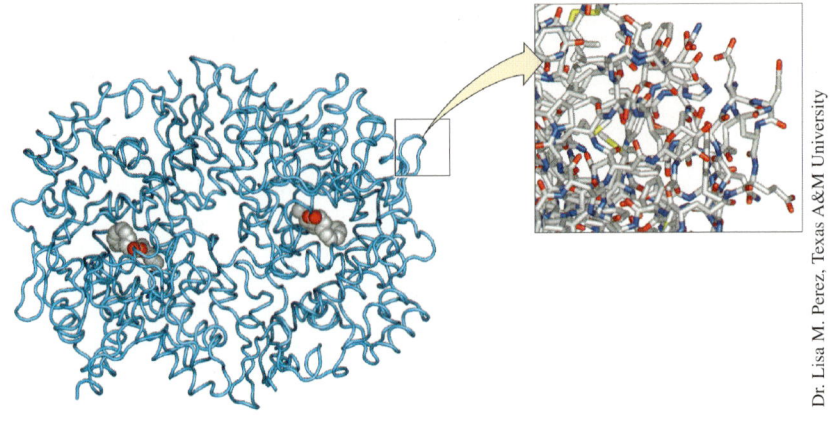

Dr. Lisa M. Perez, Texas A&M University

The Canadian Light Source in Saskatoon, which opened in 2004, creates and stores a high energy beam of electrons that produces synchrotron light a million times brighter than sunlight. The light is sent down separate beam-lines as sources for a wide range of instruments to solve protein crystal structures and many other lines of research.

Image courtesy of Erica Lukiwski, Canadian Light Source

FIGURE 9.5 Analyzing protein structures. To understand the mechanisms by which proteins function, we must study their three-dimensional structures. This is made possible by growing protein crystals, like these bovine lysozyme crystals, and analyzing their X-ray diffraction pattern using synchrotron light.

Taking It Further

e9.3 See how to produce an X-ray crystal structure.

Think about It

e9.4 Examine an interactive NMR spectrum of carvone.

thousands of proteins are known, and you can access them free on the Protein Data Bank (PDB) site [>>e9.2].

There is an art to growing crystals for X-ray analysis. While some small molecules form perfect crystals quickly, others can take months or years to produce. Large molecules like proteins are often very difficult to crystallize, and their diffraction patterns give less well-resolved pictures of the atomic positions. Protein crystal structures therefore are determined at a small number of specialized laboratories around the world. Despite these hurdles, it is now likely that the three-dimensional structures of most human proteins will be determined in your lifetime. This process has been greatly facilitated by the provision of powerful X-ray sources from synchrotron facilities, which are now used to solve most protein crystal structures (Figure 9.5). Often proteins are co-crystallized with small molecules, including potential drugs, to aid in the drug-discovery process or to understand how the protein works when it does chemistry, as in the case of the complex between arachidonic acid and cyclooxygenase.

> X-ray crystallography analysis of molecules of a solid can give the position of each atom in three-dimensional space. From this information, bond lengths and bond angles can be determined. Recent advances in X-ray source and computer-processing technologies have made it possible to determine three-dimensional structures of many large protein molecules.

9.4 ^{13}C NMR—Mapping the Carbon Framework of Molecules

The Number of Carbon Atoms in Unique Environments

How can ^{13}C NMR spectroscopy help map the carbon framework of molecules? Before seeing how it works, let's start with correlating NMR spectral signals from several substances with representations of their molecular structures. The ^{13}C NMR spectrum of carvone (Figure 9.6 and e9.4) contains 10 signals (also called *resonances*), one corresponding to each carbon atom in carvone molecules. The position of those peaks on the x-axis, labelled *chemical shift*, gives further information about the electronic environment around each carbon atom in the molecule. The carbonyl carbon atom gives a signal at 200 δ (ppm), and the four alkene carbon atoms come between 100 and 150 δ, to the left of the five alkane carbons that are found between 10 and 60 δ. The carbonyl and alkene carbon atoms correspond to lower magnetic fields in the NMR spectrum, and the alkane carbons to higher magnetic fields. All of these chemical shift values along the x-axis are relative to a standard reference signal at 0 δ.

Compare this spectrum to that of pentan-3-one in Figure 9.7a. Pentanone molecules have chains of five carbon atoms, but only three peaks appear in the ^{13}C NMR spectrum of pentan-3-one. By contrast, pentan-2-one, an isomer of pentan-3-one, has five peaks in its spectrum (Figure 9.7b). Can you spot the differences in the structures of molecules of the two isomeric pentanones that might lead to one having five peaks and one only three?

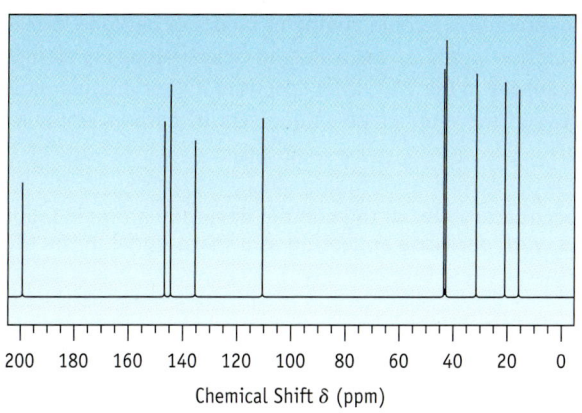

FIGURE 9.6 ¹³C NMR spectrum and structure of carvone. Each of the 10 peaks can be assigned to one of the carbon atoms in molecules of carvone. The left side of the chart is the low-field or downfield side and the right side is the high-field or upfield side. The standard reference signal (not shown in this spectrum) comes at a chemical shift value of 0 δ on the x-axis. The two signals just above 40 δ are so close together they almost merge at this resolution.

Examine computer models of these two molecules in e9.5, or build them with a molecular model kit or the model builder tool in Odyssey. You'll see that the pentan-3-one molecule is more symmetrical than its isomer. You could draw an imaginary line right through the middle of the C=O group, and if the bonds are rotated correctly, the halves on each side of that line will be identical. In pentan-3-one, both C2 and C4 are attached to a —CH₃ group and the carbonyl group. In pentan-2-one, C2 is the carbonyl carbon atom, while C4 is attached to a —CH₂ group and a —CH₃ group. Because of this symmetry, pentan-3-one has only three carbon atoms in unique environments, while no such symmetry exists with pentan-2-one and all five carbon atoms are in unique environments.

Think about It

e9.5 Interpret interactive NMR spectra of pentan-2-one and pentan-3-one.

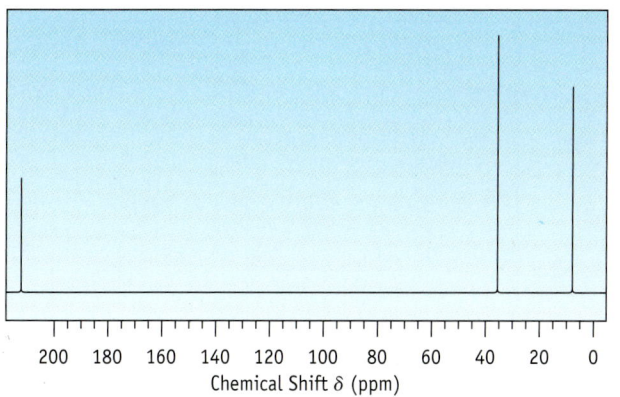

Pentan-3-one   Pentan-2-one

We refer to pentan-2-one and pentan-3-one as **isomers** of each other, because they have the same molecular formulas. They are also called **constitutional isomers** [<<Sections 3.9, 4.5], because the two molecules have different *connectivity* among the atoms. In one case, the carbonyl functional group (C=O) is at the second carbon atom (C2), and in the other case, at the third (C3). By contrast, other pairs of isomers, described in Section 9.6 are referred to as **stereoisomers**; because they have the same connectivity among atoms, but differ only in their arrangement in space.

Constitutional isomers
different connectivity among atoms

Different carbon skeletons:

$$CH_3CH_2CH_2CH_3 \text{ and } CH_3CHCH_3$$
$$\overset{|}{CH_3}$$

Different functional groups:
$$CH_3CH_2OH \text{ and } CH_3OCH_3$$

Different position of functional groups:
$$CH_3CHCH_3 \text{ and } CH_3CH_2CH_2NH_2$$
$$\overset{NH_2}{|}$$

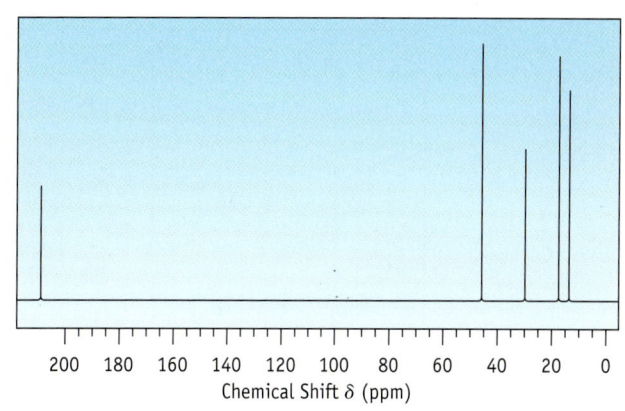

FIGURE 9.7 ¹³C NMR spectrum of (a) pentan-3-one, showing three resonances due to unique carbon atoms due to the symmetry in the molecule, and (b) pentan-2-one, showing five resonances due to all five carbon atoms being in unique environments.

Now look at the rotatable three-dimensional computer model of carvone in e9.4, and you'll see that none of the ten carbon atoms are equivalent to others—they are all in unique environments; and ten signals appear in the ^{13}C NMR spectrum. Other organic molecules show similar correlations between the number of unique sets of carbon atoms and the number of ^{13}C NMR signals.

At its simplest, ^{13}C NMR spectroscopy reveals the number of carbon atoms in unique environments in molecules, with each one appearing as a sharp signal in the NMR spectrum.

Chemical Shift As seen in these three examples, besides revealing how many carbon atoms in unique environments are found in a molecule, additional information can be obtained by looking at the position of each NMR signal along the x-axis, referred to as the **chemical shift**. Most ^{13}C resonances are between 0 and 220 δ downfield from the reference line, which usually results from a small amount of added tetramethylsilane (TMS), $(CH_3)_4Si$, arbitrarily assigned a value of 0 δ. TMS is used as a reference for both ^{13}C and ^{1}H NMR spectra [>>Section 19.6], because in both kinds of spectra it produces a single peak that occurs upfield (Figure 9.8, farther right on the chart) of other absorptions normally found in organic molecules. The carbon atoms in carbonyl groups are usually easily distinguishable on the basis of chemical shift, as are the carbon atoms in alkenes, aromatic rings, and alkynes. Complementary information from the position of stretching and bending frequencies in infrared spectroscopy can be used to confirm the presence of these functional groups.

^{13}C NMR spectroscopy can be used to differentiate several types of isomers, based on the number of signals in the spectrum and the chemical shifts of those signals.

Shown below are two **functional group isomers**, butan-1-ol and diethyl ether. Butan-1-ol has an alcohol functional group (R—OH), diethyl ether has an ether functional group (R—O—R). The number of signals in their ^{13}C NMR can be predicted by looking at structural representations of the two molecules and seeing how many carbon atoms are in unique environments in each case.

Functional group isomers are *constitutional isomers* (different connectivity among atoms), that have different functional groups.

Butan-1-ol   **Diethyl ether**

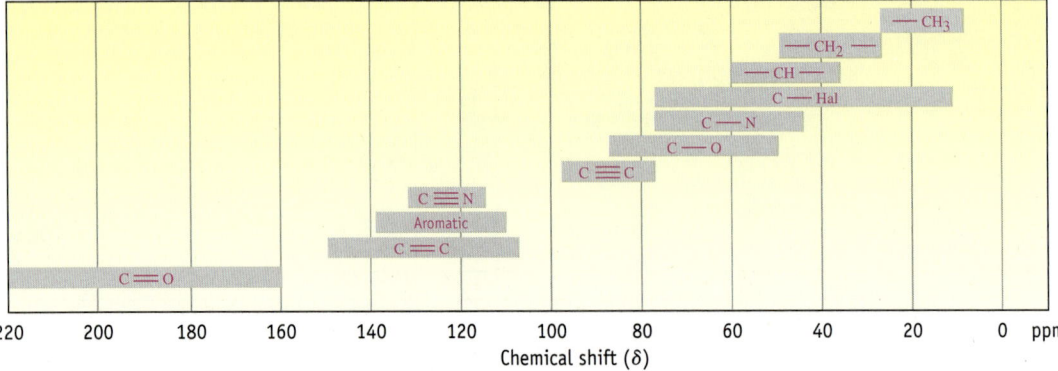

FIGURE 9.8 Chemical shift correlations for ^{13}C NMR resonances, showing how the electronic environment around a carbon atom in a molecule leads to predictable chemical shifts.

EXERCISE 9.1—DIFFERENTIATE ISOMERS BY ^{13}C NMR SPECTROSCOPY

The ^{13}C NMR spectra of butan-1-ol and diethyl ether are given below. Assign the correct spectrum to each compound. Peaks for CDCl$_3$, the solvent, are shown and identified in these spectra. Other spectra have been simplified for interpretation by removal of these peaks.

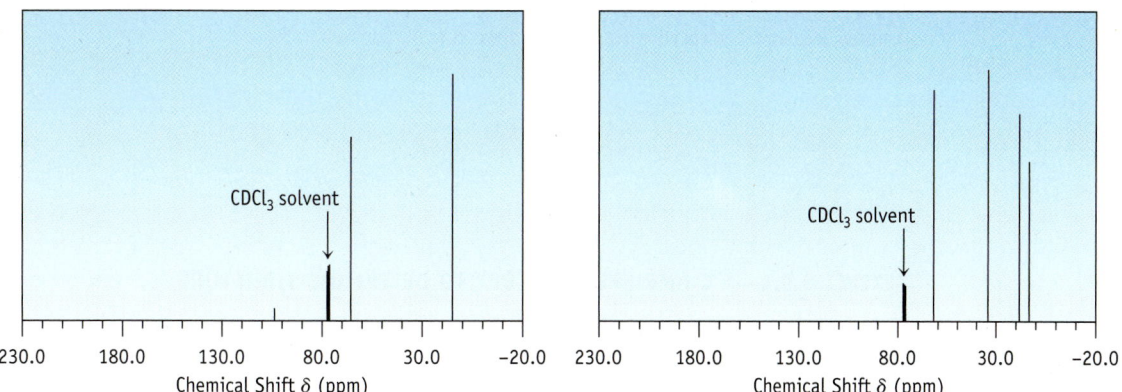

EXERCISE 9.2—DIFFERENTIATE ISOMERS BY IR SPECTROSCOPY

Referring back to the animated infrared spectra in e3.19, how would you tell apart the functional groups in butan-1-ol and in diethyl ether, based on their infrared spectra?

Exercise 9.3 shows structural representations of two positional isomers, propan-1-amine (n-propyl amine) and propan-2-amine (isopropylamine). Each molecule has an amine ($-NH_2$) functional group, but attached at a different place on the propane skeleton.

Propan-1-amine   **Propan-2-amine**

EXERCISE 9.3—DIFFERENTIATE POSITIONAL ISOMERS BY ^{13}C NMR SPECTROSCOPY

Predict the number of signals in the ^{13}C NMR spectra of propan-1-amine and propan-2-amine, and then assign the correct structure to each spectrum.

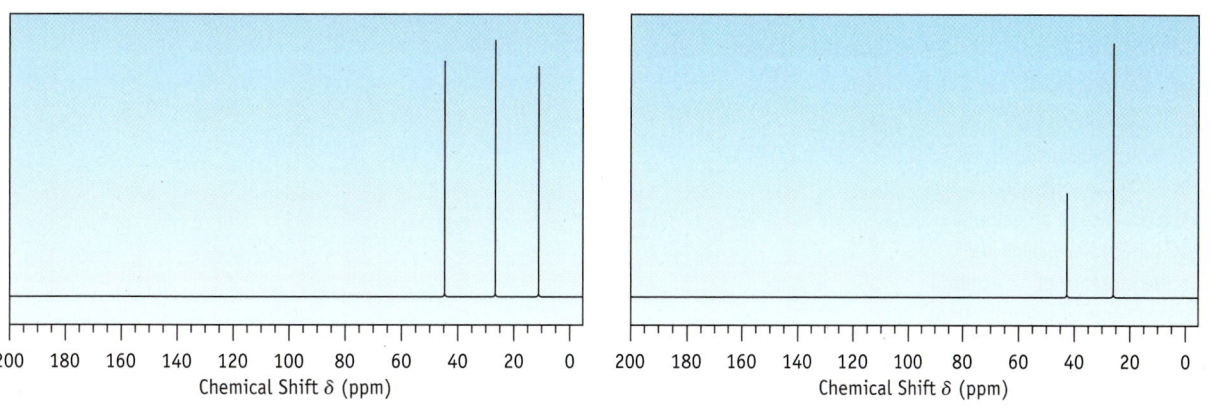

WORKED EXAMPLE 9.1—^{13}C NMR SPECTROSCOPY TO DETERMINE STRUCTURE

How many peaks would you expect in the ^{13}C NMR spectrum of methylcyclopentane?

Solution

Methylcyclopentane has a symmetry plane. As a result, it has only four kinds of carbon atoms in unique environments and four peaks in its ^{13}C NMR spectrum. Consult e9.4 for additional examples relating a structure to an NMR spectrum.

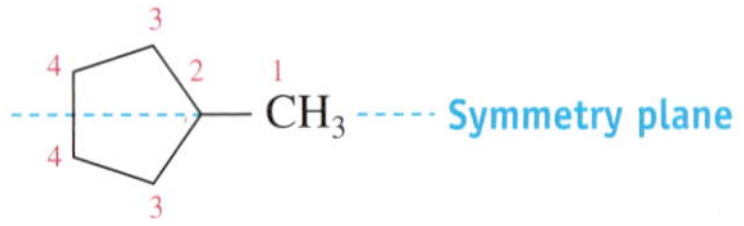

EXERCISE 9.4—^{13}C NMR SPECTROSCOPY TO DETERMINE STRUCTURE

How many resonances (peaks) would you expect in the ^{13}C NMR spectra of the following compounds?

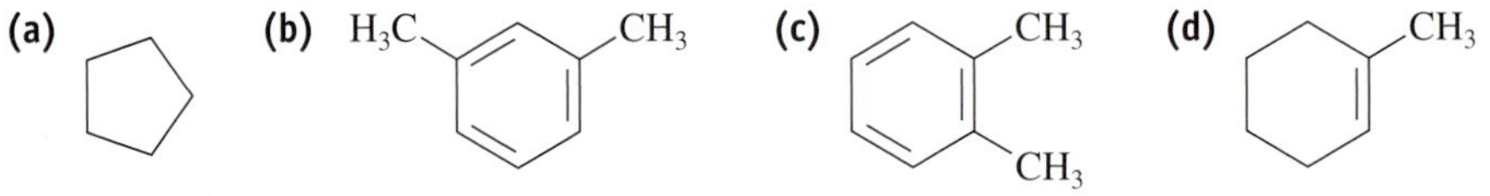

(a)   (b) H_3C ... CH_3 (c) CH_3 ... CH_3 (d) CH_3

How Does NMR Spectroscopy Work?

The nuclei of atoms in molecules that give NMR resonance signals must have a magnetic moment and therefore behave like a child's top spinning about an axis. In organic compounds, both ^{1}H and ^{13}C nuclei have this property, and so we will focus on these two nuclei in our explanation. ^{12}C, the most abundant carbon isotope, has no nuclear spin and is not observable by NMR. The only naturally occurring carbon isotope with a magnetic moment is ^{13}C, but its natural abundance is only 1.1%. Thus, only about 1 of every 100 carbon atoms in a sample of an organic compound is observable by NMR. Fortunately, the technical problems caused by this low abundance have been overcome by computer techniques, and ^{13}C NMR is a routine structural tool.

Since they're positively charged, these spinning ^{1}H and ^{13}C nuclei act like tiny magnets and interact with an external magnetic field (denoted B_0). In the absence of an external magnetic field, the nuclear spins of magnetic nuclei are oriented randomly. When a sample containing these nuclei is placed between the poles of a strong magnet, however, the nuclei adopt specific orientations, much as a compass needle orients itself in the earth's magnetic field.

A spinning ^{1}H or ^{13}C nucleus can orient so that its own tiny magnetic field is aligned either with (parallel to) or against (anti-parallel to) the external field. The two orientations don't have the same energy and therefore aren't equally likely. The parallel orientation is slightly lower in energy, making this spin state slightly favoured over the anti-parallel orientation (Figure 9.9).

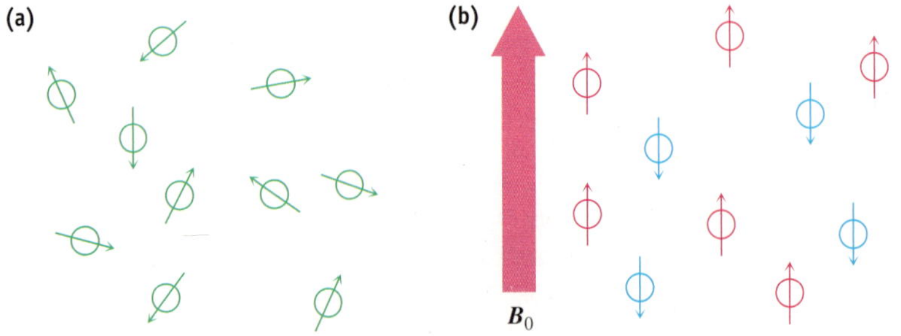

FIGURE 9.9 (a) Nuclear spins such as those of ^{1}H or ^{13}C are oriented randomly in the absence of an external magnetic field, but (b) have a specific orientation in the presence of an external field B_0. Note that some of the spins (red) are aligned parallel to the external field and others (blue) are anti-parallel. The parallel spin state is lower in energy.

If the oriented nuclei are now irradiated with electromagnetic radiation of the right frequency, energy absorption occurs and the lower-energy state "spin-flips" to the higher-energy state. When this spin-flip occurs, the nuclei are said to be "in resonance" with the applied radiation—hence, the name *nuclear magnetic resonance.*

The exact frequency necessary for resonance depends on the strength of the external magnetic field, the identity of the nuclei, and on their electronic environment. If a very strong external field is applied to a sample with particular nuclei in a given electronic environment, the energy difference between the two spin states is large, and higher-energy (higher-frequency) radiation is required (Figure 9.10). If a weaker magnetic field is applied, less energy is required to effect the transition between nuclear spin states.

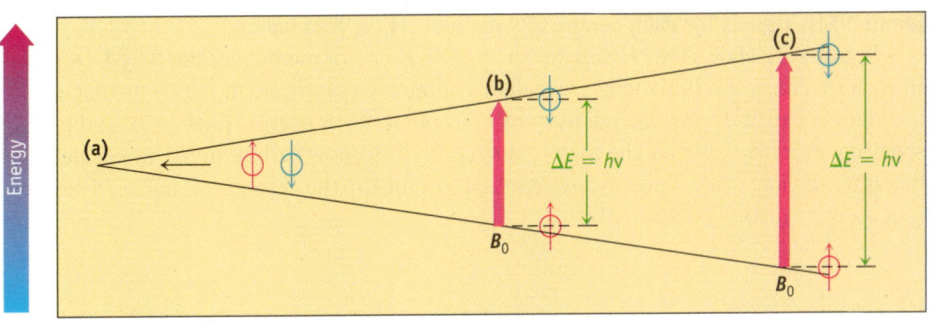

Strength of applied field, B_0 ⟶

FIGURE 9.10 The energy difference (ΔE) between nuclear spin states depends on the strength of the applied magnetic field. Absorption of energy of frequency ν converts a nucleus from a lower spin state to a higher spin state. (a) Spin states have equal energies in the absence of an applied magnetic field, but (b) have unequal energies in the presence of a magnetic field. At $\nu = 60$ MHz, $\Delta E = 2.4 \times 10^{-5}$ kJ mol^{-1}. (c) The energy difference between spin states is greater at larger applied fields. At $\nu = 500$ MHz, $\Delta E = 2.0 \times 10^{-4}$ kJ mol^{-1}.

A ^{13}C NMR experiment, then, starts with placing a sample, dissolved in a suitable solvent (usually deuterochloroform, CDCl$_3$) in a thin glass tube between the poles of a magnet (Figure 9.11). The strong magnetic field causes the ^{13}C nuclei in the molecule to align in one of the two possible orientations, and the sample is irradiated with radiofrequency (RF) energy. If the frequency of the RF irradiation is held constant and the strength of the magnetic field changed very slightly, each nucleus comes into resonance at a slightly different magnetic field strength.

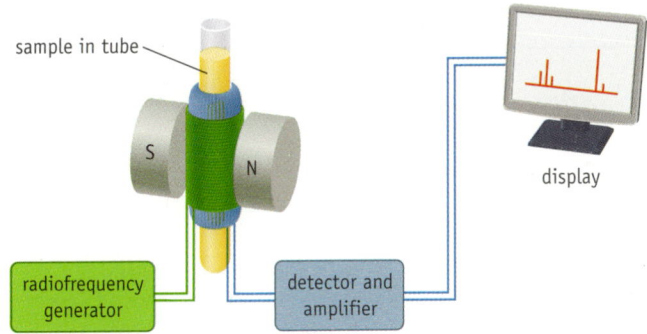

FIGURE 9.11 Schematic drawing of an early NMR spectrometer. A thin glass tube containing the sample solution is placed between the poles of a strong magnet and irradiated with radiofrequency (RF) energy.

In most labs, NMR spectra are no longer collected on fixed magnetic field instruments, but use Fourier transform spectroscopy techniques, where short pulses of radiofrequency excite the entire NMR spectrum. In addition, superconducting magnets that produce powerful fields in the range 2.62 to 4.7 tesla (T) are commonly used. At a magnetic field strength of 2.62 T, radiofrequency energy in the 100 MHz range (1 MHz = 10^6 Hz) is required to bring a ^{1}H nucleus into resonance, and RF energy of 25 MHz is required to bring a ^{13}C nucleus into resonance.

Chemical Shift From the description thus far, you might expect all ^{1}H nuclei in a molecule to absorb energy at the same frequency and all ^{13}C nuclei to absorb at the same frequency. If this were true, we would observe only a single NMR absorption in the ^{1}H or ^{13}C spectrum of a molecule, a situation that would be of little use for structure determination. In fact, the absorption frequency is not the same for all ^{1}H or all ^{13}C nuclei.

All nuclei are surrounded by circulating electrons. When an external magnetic field is applied to a molecule, the moving electrons around nuclei set up tiny local magnetic fields of their own. These local fields act in opposition to the applied field so that the *effective* field actually felt by the nucleus is a bit weaker than the applied field:

$$B_{effective} = B_{applied} - B_{local}$$

In describing this effect of local fields, we say that the carbon and hydrogen nuclei are shielded from the full effect of the applied field by their surrounding electrons. Since each specific ^1H or ^{13}C nucleus in a molecule is in a slightly different electronic environment, each specific nucleus is shielded to a slightly different extent and the effective magnetic field felt by each is not the same. These slight differences can be detected, and we see different NMR signals for each chemically distinct ^1H or ^{13}C nucleus.

Figure 9.12 shows the ^1H and the ^{13}C NMR spectra of methyl acetate (CH_3COOCH_3). In each spectrum, the horizontal axis tells the effective field strength felt by the nuclei, and the vertical axis indicates the intensity of absorption of RF energy. Each peak in the NMR spectrum corresponds to a chemically distinct hydrogen or carbon in the molecule. Note that different amounts of energy are required to spin-flip the ^1H and ^{13}C nuclei. These two spectra were recorded separately.

FIGURE 9.12 (a) The ^1H NMR spectrum, and (b) the ^{13}C NMR spectrum of methyl acetate, CH_3COOCH_3. The peak at the far right of each spectrum at 0 δ marked TMS is a calibration peak.

(a) Intensity

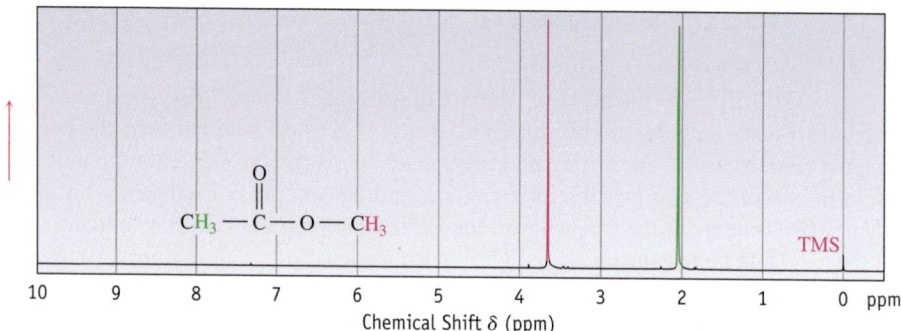

(b) Intensity

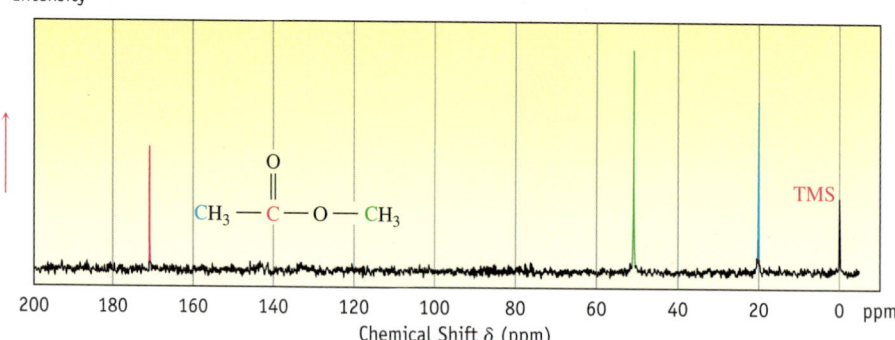

The ^{13}C spectrum of methyl acetate shown in Figure 9.12b has three peaks, one for each of the three chemically distinct carbons in the molecule. The ^1H spectrum shows only *two* peaks, however, even though methyl acetate has *six* hydrogen atoms. One peak is due to the $CH_3C=O$ hydrogen atoms and the other to the $—OCH_3$ hydrogen atoms. Because the three hydrogen atoms in each methyl group have the same chemical (and magnetic) environment, they are shielded to the same extent and are said to be *equivalent*. *Chemically equivalent nuclei show a single absorption.* The two methyl groups themselves, however, are non-equivalent, so the two sets of three hydrogen atoms absorb at different positions.

Remember that NMR spectra are displayed on charts that show the applied field strength increasing from left to right. The left side of the chart for the NMR spectra of methyl acetate in Figure 9.12 is the low-field (or *downfield*) side, and the right side is the high-field (or *upfield*) side. Nuclei that absorb on the downfield side of the chart require a lower field strength for resonance, implying that they have relatively little shielding by electrons. In some cases, this chemical shift is caused by the proximity of electronegative substituents that remove electron density; in others, it is due to magnetic fields created by π electrons. Nuclei that absorb on the upfield side require a higher field strength for resonance, implying that they are strongly shielded.

A small amount of TMS, $(CH_3)_4Si$, is used as a reference for both ^1H and ^{13}C spectra, because in both kinds of spectra it produces a single peak that occurs upfield (farther right on the chart) of other absorptions normally found in organic molecules. Almost all ^1H NMR absorptions occur 0–10 δ downfield from the ^1H absorption of TMS, and almost all ^{13}C NMR absorptions occur 0–220 δ downfield from the ^{13}C absorption of TMS.

By convention, the chemical shift of TMS is set as the zero point, and other peaks normally occur downfield, to the left on the chart. NMR charts are calibrated in units of frequency using an arbitrary scale called the *delta (δ) scale*, where 1 delta unit is equal to 1 part-per-million (ppm) of the spectrometer operating frequency. For example, if we were using a 100 MHz instrument to measure the ^1H NMR spectrum of a substance, 1 δ would be 1 ppm of 100,000,000 Hz, or 100 Hz. If we were measuring the spectrum with a 300 MHz instrument, however, then 1 δ = 300 Hz.

Although this method of calibrating NMR charts may seem complex, there's a good reason for it. As we saw earlier, the RF frequency required to bring a given nucleus into resonance depends on the spectrometer's magnetic field strength. But because there are many different kinds of spectrometers with many different magnetic field strengths available, chemical shifts given in frequency units (Hz) vary from one instrument to another. A resonance that occurs at 120 Hz downfield from TMS on one spectrometer might occur at 600 Hz downfield from TMS on another spectrometer with a more powerful magnet. A higher magnetic field (from a much more expensive instrument) allows better sensitivity, better resolution between peaks that are close together in chemical shift, and also leads to simplification of complex signals in a ^1H NMR spectrum.

By using a system of measurement in which NMR absorptions are expressed in *relative* terms (parts-per-million relative to spectrometer frequency) rather than in absolute terms (Hz), comparisons of spectra obtained on different instruments are possible. *The chemical shift of an NMR absorption in δ units is constant, regardless of the operating frequency of the instrument.* A ^{13}C nucleus that absorbs at 196.0 δ on a 100 MHz instrument also absorbs at 196.0 δ on an 800 MHz instrument.

WORKED EXAMPLE 9.2—^{13}C NMR SPECTROSCOPY

Cyclohexane shows an absorption at 27.1 δ in its ^{13}C NMR spectrum. How many hertz away from TMS is this on a spectrometer operating at 100 MHz? On a spectrometer operating at 400 MHz?

Solution

On a 100 MHz spectrometer, 1 δ = 100 Hz. Thus, 27.1 δ = 2,710 Hz away from the TMS reference peak. On a 400 MHz spectrometer, 1 δ = 400 Hz and 27.1 δ = 10 840 Hz.

WORKED EXAMPLE 9.3—^{1}H AND ^{13}C NMR SPECTROSCOPY TO DETERMINE STRUCTURE

How many signals would you expect *p*-dimethylbenzene to show in its ^1H and ^{13}C NMR spectra?

Solution

Look at the structure of the molecule, and count the number of kinds of chemically distinct ^1H and ^{13}C nuclei. Because of the molecule's symmetry, the two methyl groups in *p*-dimethylbenzene

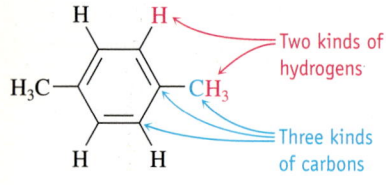

p-Dimethylbenzene

are equivalent and all four ring hydrogen atoms are equivalent. There are only two absorptions in the ^1H NMR spectrum. Also because of symmetry, there are only three absorptions in the ^{13}C NMR spectrum: one for the two equivalent methyl-group carbon atoms, one for the four equivalent C—H ring carbon atoms, and one for the two equivalent ring carbons bonded to the methyl groups.

For additional examples, consult e9.4.

EXERCISE 9.5—^1H AND ^{13}C NMR SPECTROSCOPY TO DETERMINE STRUCTURE

How many signals would you expect each molecule to show in its ^1H and ^{13}C NMR spectra?
(a) methane
(b) ethane
(c) propane
(d) cyclohexane
(e) dimethyl ether
(f) benzene
(g) $(CH_3)_3COH$
(h) chloroethane
(i) $(CH_3)_2C=C(CH_3)_2$

As we'll see in Section 22.4 the interpretation of ^1H NMR has an additional complication relative to ^{13}C NMR in that signals of ^1H nuclei are divided or split into multiple signals due to the presence of neighbouring ^1H nuclei. The number of split signals depends on the number of neighbours. While this complicates interpretation, it also yields additional invaluable information about the map of the C–H framework of a molecule.

Nuclear magnetic resonance (NMR) spectroscopy helps map the framework of carbon and hydrogen atoms of molecules in solution. We focus here on the following: (a) number of peaks—each non-equivalent kind of ^1H or ^{13}C nucleus in a molecule gives rise to a different peak; (b) chemical shift—the exact position of each peak is called its chemical shift and is correlated to the chemical environment of each ^1H or ^{13}C nucleus.

9.5 Conformations of Alkanes—Rotation about Single Bonds

Molecular Modelling (Odyssey)

e9.6 How many types of carbon atoms are in a butane molecule?

n-Butane (C_4H_{10}) has four carbon atoms, but the ^{13}C NMR spectrum shows only two distinct signals. Why? This suggests that the two —CH_2 (methylene) carbon atoms (C2 and C3) must on average experience the same influence of surrounding electrons (be in identical environments), as must the two —CH_3 (methyl) carbon atoms (C1 and C4). The methyl C atoms are in a different environment from that of the methylene C atoms, each giving a distinct signal.

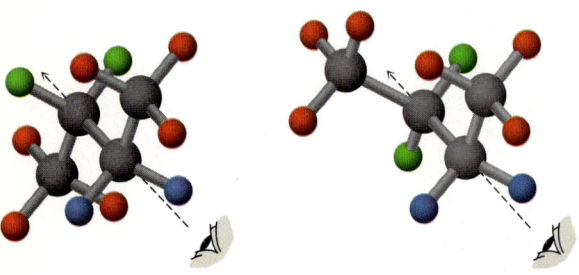

Observer   Observer

The evidence of only one signal for the methylene and one for the methyl carbon atoms gives us insights into the dynamic structure of butane molecules. Let's look more closely at two different minimum energy forms, called *conformers,* of an n-butane molecule, which result from rotation about the C2—C3 single bond. What would the ^{13}C NMR spectrum of n-butane look like if all molecules existed only as the conformer on the left (called the *anti conformer*)? The C1 and C4 methyl carbon atoms are in the same environment as each other, as are the C2 and C3 methylene atoms, but the methyl and methylene carbon atom environments are different from each other. We would expect to see two peaks in the ^{13}C NMR spectrum, one for the methyl and one for the methylene carbon atoms. Similarly, in the conformer on the right (called the *gauche conformer*), the two methyl and two methylene carbon atoms would be different from each other, giving us two peaks in the NMR spectrum if all butane molecules existed only in this form.

But are the methyl carbon atoms in the *two different conformers* in identical environments to each other in the molecule? Are the methylene carbon atoms? This is most easily seen by looking down the C2—C3 bond. In the anti conformer on the left, C1 and C4 have an angle of rotation (torsional angle) of 180°, while this angle is 60° in the gauche conformer on the right. Similarly, the C2 and C3 atoms have different environments in the two conformers relative to the other atoms. If these were the only two conformers present, we might expect four ^{13}C NMR signals for n-butane: one for C2/C3 in each conformer, and one for C1/C4 in each. However, only two ^{13}C NMR signals are observed. This can be explained if rotation about the C—C single bond is very fast with respect to the NMR timescale of about 10^{-3} sec, and the two rapidly interconverting conformers are not distinguishable. Consequently, one signal is recorded that corresponds with the average environment of the equivalent C2 and C3 atoms, and one with the average environment of the equivalent C1 and C4 atoms. You might think of a macroscopic analogy, such as the performer in Figure 9.13.

Conformations and Conformers

The different arrangements of atoms in molecules that result from rotation around a single bond are called **conformations**, which normally interconvert too rapidly to be isolated at room temperature. Rotation about single bonds occurs in such a way as to favour the energetically most stable conformations of a molecule. In many simple cases, it is evident that the energetically most stable conformation will be one in which substituents (with their electrons) within molecules are as far apart from each other as possible. Molecular recognition between two molecules is based not only on the individual molecule's bond lengths and bond angles, but also on their overall shapes, determined by their conformations. Dramatic changes to a protein's or nucleic acid's shape can be caused by applying heat or strong acid or base. This process is called *denaturing*. When a protein becomes denatured, it loses its ability to recognize other molecules and to function as it should in a cell. As shown in Figure 9.14, an egg changes shape, colour and texture upon heating, because its denatured protein has an overall shape very different from the native protein. Enzymes that become denatured lose their ability to act as catalysts.

FIGURE 9.13 Average positions of an object. When an object changes too fast for the device measuring those changes to record it, we see an average picture, rather than all the instantaneous positions of an object. This image of a performer skipping with a burning rope shows an average position for the flames, with the greatest intensity in regions where they are found most frequently during the time it takes to photograph the subject.

FIGURE 9.14 Dramatic changes in the shape, colour, and texture of an egg resulting from a denaturing of the proteins under the high temperature used for cooking. Egg white proteins are coiled up in the same pattern, stabilized by intramolecular bonds. During heating, these unfold into a disorganized mesh of protein chains.

Let's consider ethane, the simplest alkane that can undergo rotation about a C—C single bond to produce different conformations (Figure 9.15).

Chemists represent different conformations in two ways, as shown in Figure 9.16. *Sawhorse representations* view the molecule from an oblique angle and indicate spatial relationships by showing all the C—H bonds and the C—C bond. *Newman projections*

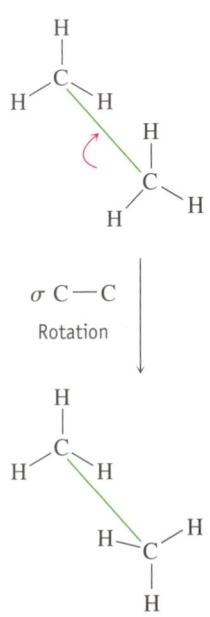

FIGURE 9.15 Two of the infinite number of conformations of ethane molecules. Rotation about the C—C single bond (green) changes the molecule from one conformation to the other.

view the molecule end-on, focusing on and sighting down a particular C—C bond and representing each of the carbon atoms in that bond by a circle. Bonds attached to the front carbon atom are represented by lines to the centre of the circle, and bonds attached to the rear carbon atom are represented by lines to the edge of the circle.

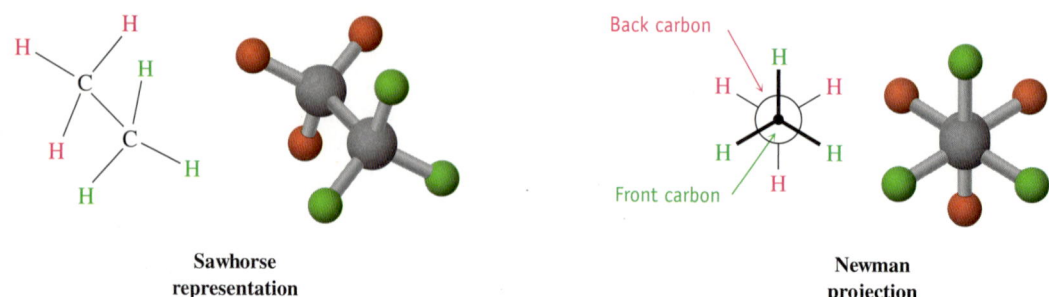

Sawhorse representation

Newman projection

FIGURE 9.16 A sawhorse representation and a Newman projection of an ethane molecule. The sawhorse projection views the molecule from an oblique angle, while the Newman projection views the molecule end-on, down the C—C bond.

Despite what we've just said, C—C rotation in ethane is not completely free. Experiments show that there is a small (12 kJ mol^{-1}) barrier to rotation and that the stability of the molecule changes as rotation takes it through different conformations. The conformation in which the molecule is lowest in energy and most stable is the one in which the electron pairs in all six C—H bonds are as far away from one another as possible (*staggered* when viewed end-on in a Newman projection). Conformations such as this one that are at energy minima are referred to as **conformers**. The highest energy, least stable conformation is the one in which the six C—H bonds are as close as possible (*eclipsed* in a Newman projection). This eclipsed conformation can be thought of as a higher energy transition state between two more stable staggered conformers (Figures 9.17 and 9.18).

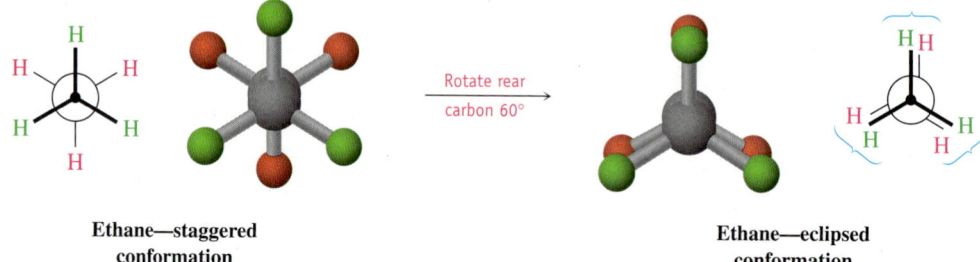

Ethane—staggered conformation

Ethane—eclipsed conformation

FIGURE 9.17 One of the three equivalent, staggered conformers (left) and one of the other infinite number of conformations (right) of ethane molecules. The staggered conformers are at energy minima for rotation about the C—C bond, and are lower in energy by 12.0 kJ mol^{-1} than the eclipsed conformations, which are at energy maxima for rotation about the C—C bond.

FIGURE 9.18 A graph of potential energy versus bond rotation in ethane. The energy minima are called conformers—in this case, the staggered conformers are 12 kJ mol^{-1} lower in energy than the eclipsed conformations.

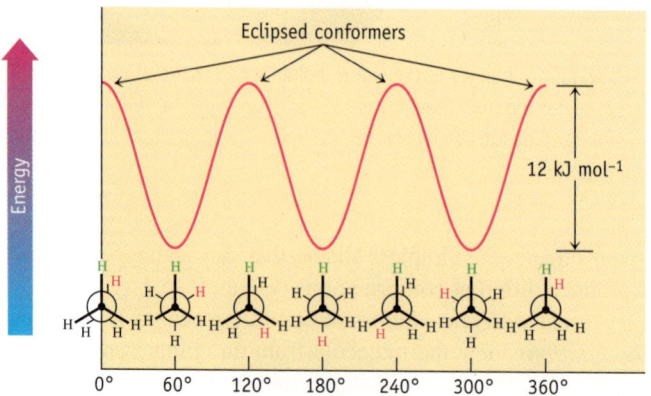

The same examination of the various conformers and conformations caused by rotation about C—C single bonds can be carried out for propane, butane, and all higher alkanes. The most favoured conformations for any alkane are ones in which all bonds have staggered arrangements, represent energy minima, and are referred to as conformers (Figure 9.19).

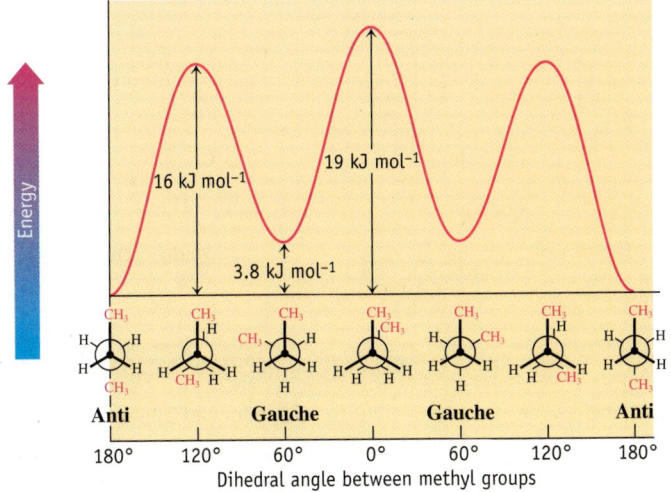

FIGURE 9.19 The most stable conformations of any alkane are conformers in which the bonds on adjacent carbons are staggered. Rotation about C—C bonds in the chain leads to many different structures, including one in which the carbon chain is fully extended as shown in this structure of decane. A sample of liquid decane would contain an equilibrium mixture of many such staggered conformers of decane molecules.

Molecular Modelling (Odyssey)

e9.8 Simulate octane in the liquid state and see the conformers.

EXERCISE 9.6—NEWMAN PROJECTIONS TO SHOW CONFORMATIONS

Sight along a C—C bond of propane and draw a Newman projection of the most stable conformer. Draw a Newman projection of the least stable conformation.

EXERCISE 9.7—NEWMAN PROJECTIONS TO SHOW CONFORMATIONS

Looking along the C2—C3 bond of butane, there are three staggered conformers (two of which have different energies) and three eclipsed conformations (two of which have different energies). Draw Newman projections of them.

WORKED EXAMPLE 9.4—NEWMAN PROJECTIONS TO SHOW CONFORMATIONS

Which of the butane conformers that you drew in Exercise 9.7 do you think is the most stable? Explain.

Solution

The two conformers are at rotational angles of 60 and 180°. The 180° conformer is most stable, as in this conformation the —CH3 groups on one C atom are further from those on the other C atom than in any other conformation.

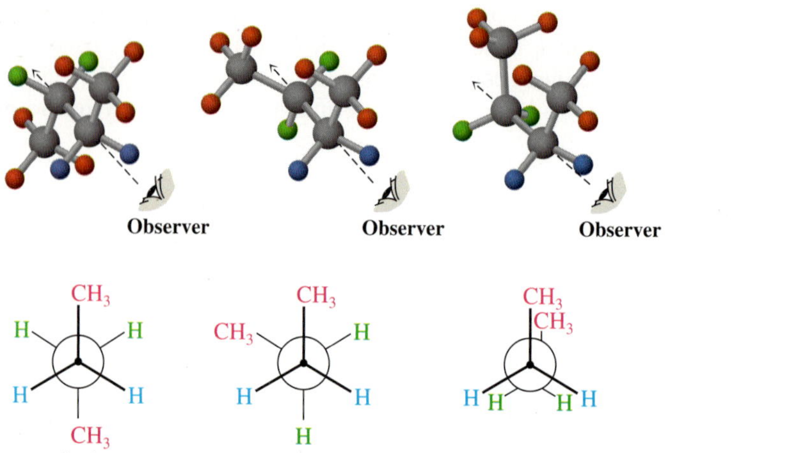

Observer   Observer   Observer

Anti   Gauche   Least stable eclipsed

EXERCISE 9.8—NEWMAN PROJECTIONS TO SHOW CONFORMATIONS

Looking along the C1—C2 bond of propane, draw the two different staggered conformers and the two different eclipsed conformations. Then, with reference to those drawings, sketch a graph of relative potential energy versus angle of rotation, as for ethane and butane.

Skeletal Structures Revisited

In the structures we've been using, a line between atoms represents the two electrons in a covalent bond. Since drawing every bond and every atom is tedious, chemists have devised a shorthand way of drawing *skeletal structures*, first introduced in Chapter 4. There are several conventions for drawing skeletal structures:

- Carbon atoms usually aren't shown. Instead, a carbon atom is assumed to be at the intersection of two lines (bonds) and at the end of each line. Occasionally, a carbon atom might be indicated for emphasis or clarity.
- Hydrogen atoms bonded to carbon atoms aren't shown. Because carbon always has a valence of four, we mentally supply the correct number of hydrogen atoms for each carbon.
- All atoms other than carbon and hydrogen *are* shown.

    The following structures give some examples.

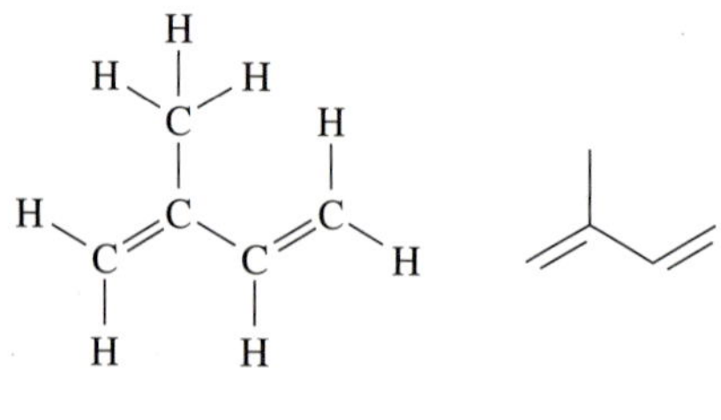

Isoprene, C_5H_8 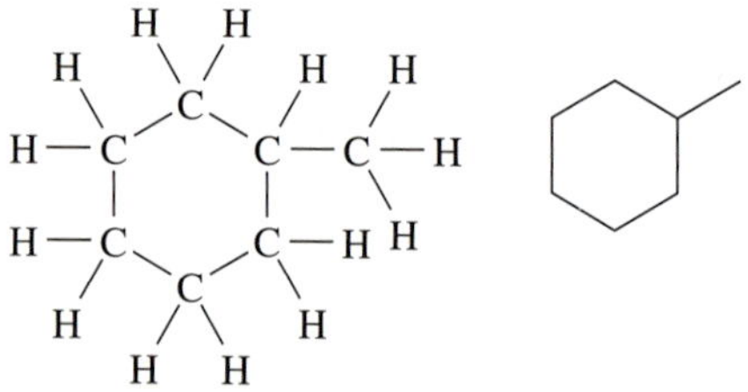   Methylcyclohexane, C_7H_{14}

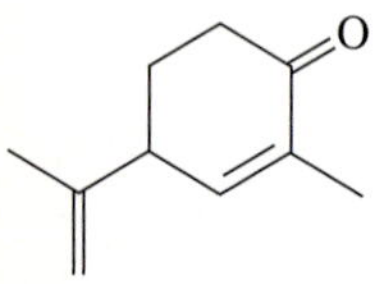

Carvone

WORKED EXAMPLE 9.5—SKELETAL STRUCTURE REPRESENTATIONS

Molecules of carvone, which are responsible for the odour of spearmint, have the following line structure. State how many hydrogen atoms are bonded to each carbon atom, and give the molecular formula of carvone.

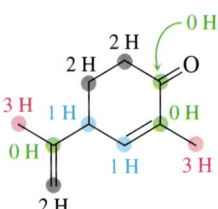

Carvone, $C_{10}H_{14}O$

Solution

Remember that the end of a line represents a carbon atom with three hydrogen atoms, CH_3; a two-line intersection is a carbon atom with two hydrogen atoms, CH_2; a three-line intersection is a carbon atom with one hydrogen atom, CH; and a four-line intersection is a carbon atom with no attached hydrogen atoms.

EXERCISE 9.10—SKELETAL STRUCTURE REPRESENTATIONS

Convert the following skeletal structures into molecular formulas, and tell how many hydrogen atoms are bonded to each carbon atom:

(a)   (b)   (c)

Pyridine   Cyclohexanone   Indole

Rapid rotation through many possible conformations is possible about C—C single bonds in most alkane molecules. Conformations that correspond to energy minima for a molecule are referred to as conformers. A staggered conformer for alkane molecules is more stable than an eclipsed conformation, minimizing the repulsion between electron pairs.

9.6 Restricted Rotation about Bonds

You have seen (Figure 9.18 and Worked example 9.4) that more energy is required for rotation of the larger methyl groups of butane past each other in eclipsed conformations than for the much smaller hydrogen atoms to rotate past each other in ethane. In many organic molecules, rotation about carbon–carbon bonds is even more restricted, and this plays an important role in determining the overall shape of a molecule. One example of such restricted rotation is found in compounds like carvone, whose molecules have a *ring* of carbon atoms. Such compounds are called cycloalkanes, or *alicyclic* (*aliphatic cyclic*) compounds [>>Section 9.7]. Since cycloalkane molecules consist of rings of —CH_2— units, those with only carbon and hydrogen atoms have the general formula $(CH_2)_n$, or C_nH_{2n}, and are represented by polygons in skeletal drawings:

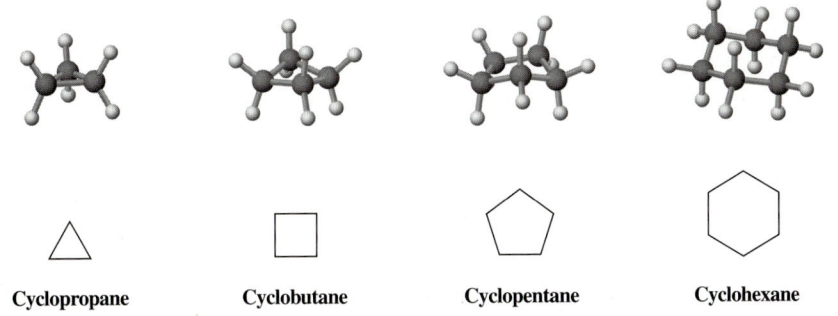

Cyclopropane   Cyclobutane   Cyclopentane   Cyclohexane

In many respects, the behaviour of cycloalkanes is similar to that of open-chain, acyclic alkanes. Both are non-polar and chemically inert to most reagents. There are, however, some important differences. One difference is that molecules of cycloalkanes are less flexible than their open-chain counterparts. Although open-chain alkane molecules have nearly free rotation around their C—C single bonds, cycloalkanes are more constrained.

Interactive Exercise 9.9

Convert molecular formulas into skeletal structures.

Interactive Exercise (Odyssey) 9.11

Model the molecular behaviour of straight-chain alkanes.

Molecular Modelling (Odyssey)

e9.9 Compare molecular-level simulations of linear and cyclic alkanes.

Interactive Exercise (Odyssey) 9.12

Model the molecular behaviour of cyclic alkanes.

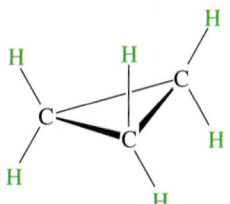

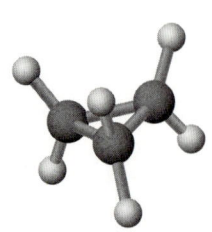

FIGURE 9.20 The structure of a cyclopropane molecule. No rotation is possible around the C—C bonds without breaking C—C bonds.

A cyclopropane molecule, for example, is geometrically constrained to be rigid and planar. No rotation around a C—C bond is possible in cyclopropane molecules without breaking open the ring (Figure 9.20).

Cis and Trans Stereoisomers

Because of their cyclic structures, hydrogen atoms or other substituents can be found on either the "top" or "bottom" side of a cycloalkane molecule, *relative to* other substituents. As a result, isomerism is possible in substituted cycloalkanes. For example, there are two 1,2-dimethylcyclopropane isomers: one with the two methyl groups on the same side of the ring and one with the methyl groups on opposite sides. Both isomers are stable compounds, and neither can be converted into the other without breaking bonds (Figure 9.21).

Unlike the *constitutional isomers* butane and isobutane [<<Section 4.5], which have different connectivity among atoms, the two 1,2-dimethylcyclopropanes have the *same* connections but differ in the spatial orientation of their atoms. They are therefore referred to as a pair of *stereoisomers* [<<Section 9.4]. The 1,2-dimethylcyclopropanes are special kinds of stereoisomers called *cis–trans* isomers. The prefixes *cis-* (Latin, "on the same side") and *trans-* (Latin, "across") are used to distinguish them.

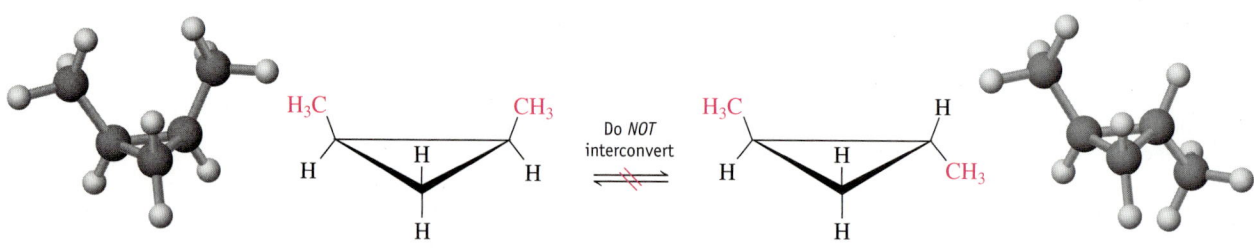

cis-1,2-Dimethylcyclopropane   trans-1,2-Dimethylcyclopropane

FIGURE 9.21 The two different 1,2-dimethylcyclopropane isomers. One has the methyl groups on the same side of the ring (*cis*) and the other on opposite sides of the ring (*trans*). The two molecules have the same connectivity, but differ in spatial orientation, and form a pair of *stereoisomers*. They cannot be interconverted by rotation about C—C single bonds—bonds must be broken and re-made.

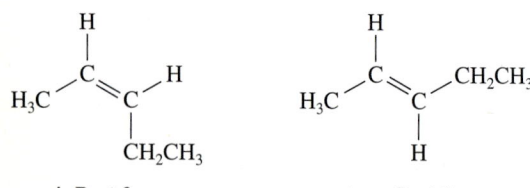

cis-Pent-2-ene   trans-Pent-2-ene

Think about It

e9.10 Examine interactive NMR and IR spectra of alkenes.

Molecular Modelling (Odyssey)

e9.11 Simulate isomerization about the double bond of alkene molecules.

Groups bonded directly to alkene carbon atoms have much higher barriers to rotation than those on C—C single bonds of non-cyclic (acyclic) alkanes. Since rotation about C=C bonds is very restricted at normal temperatures, pairs of stable stereoisomeric alkene molecules are also designated as *cis-* or *trans-*, depending on whether the two groups are on the same or opposite sides of a reference plane of the alkene.

Spectroscopic techniques identify *cis* and *trans* isomers of alkenes. The infrared spectra of isomeric alkenes show out-of-plane C—H bending vibrations at different characteristic frequencies (Figure 9.22), and ^{13}C NMR spectra of isomers are also different (Figure 9.23).

The differences in stereochemistry at a particular C=C bond that create stereoisomeric alkenes often have striking implications for molecular recognition. As you will see in Section 19.5, in the rod cells of the eye, the isomerization of the *cis*-alkene retinal (which is bound to the protein opsin) to the *trans*-alkene retinal changes its molecular shape to such an extent that it pushes against the protein and causes a conformational change. This leads to a nerve impulse being sent to the brain, which is interpreted as vision. In the absence of light, the *cis–trans* isomerization takes approximately 1100 years; in the presence of light of the right frequency, it occurs within 2 picoseconds (2×10^{-12} s).

Some inorganic compounds also have *cis-* and *trans-* stereoisomers, leading to important differences in reactivity. One remarkable example (Figure 9.24) is the two isomeric complexes with the formula $Pt(NH_3)_2Cl_2$, in which the N atoms of the NH_3 groups and the bonded Cl^- ions all lie in the same plane as the Pt(II) ion (called a *square planar* complex). In the *cis* isomer of this complex, the two bonded Cl^- ions are adjacent to each other; in

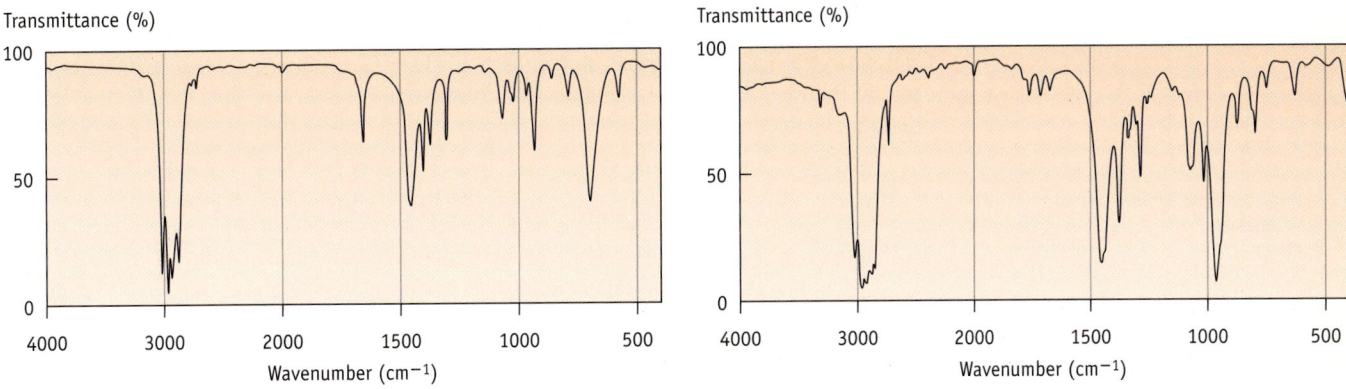

FIGURE 9.22 The infrared spectra of (a) *cis*-pent-2-ene and (b) *trans*-pent-2-ene show out-of-plane C—H bending vibrations of isomeric alkenes at different characteristic wavenumbers. *Cis*-alkenes typically absorb between 665 and 730 cm^{-1} and trans-alkenes between 960 and 980 cm^{-1}. In this case, the *cis*-pentene absorbance is at 698 cm^{-1} and the trans-pentene at 966 cm^{-1}.

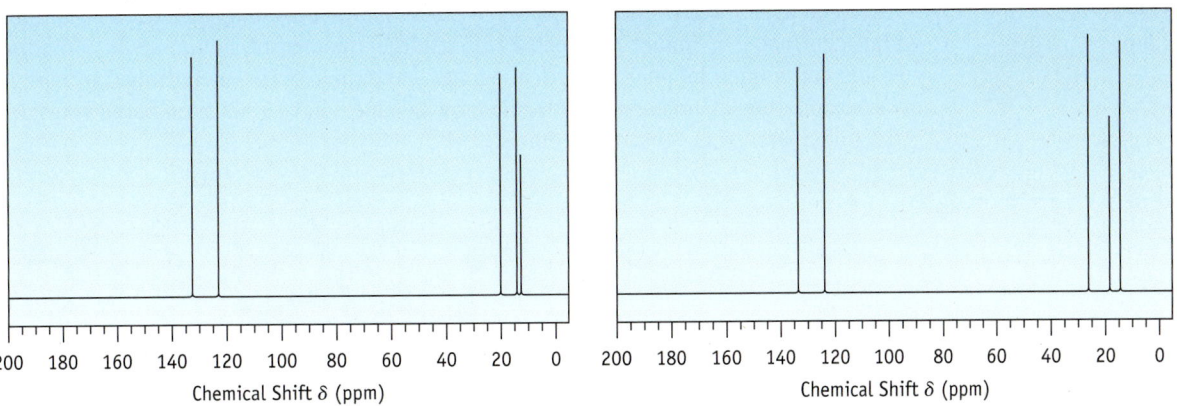

FIGURE 9.23 The ^{13}C NMR spectra of (a) *cis*-pent-2-ene and (b) *trans*-pent-2-ene. Each has the same number of signals, but slightly different chemical shifts show that the carbon frameworks have subtle differences.

the other (*trans*), they are on opposite sides of the metal ion. The *cis* isomer, called cisplatin, is effective in the treatment of testicular, bladder, and osteogenic sarcoma cancers, but the *trans* isomer has no effect on these diseases. In this case, as in the cyclopropane example, rotation about single bonds will not interconvert the two stereoisomers.

> Restricted rotation about C—C single bonds in cycloalkane molecules and about double bonds in alkene molecules contributes to the overall shape of molecules, and in some cases creates *cis–trans* stereoisomers. In a *cis* isomer, both substituents are on the same side of a cycloalkane ring or the plane of a C═C bond, whereas in a *trans* isomer, the substituents are on opposite sides of the ring or the plane of a C═C bond.

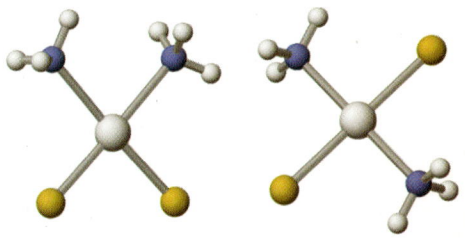

Cis isomer   *Trans* isomer

FIGURE 9.24 The square planar complex $Pt(NH_3)_2Cl_2$ exists in two geometries, *cis* and *trans*. The *cis* isomer, called cisplatin, is a potent chemotherapeutic agent.

9.7 **Cyclic Molecules**

In the early days of organic chemistry, the smaller cycloalkanes provoked a good deal of consternation among chemists. As we'll see in Chapter 10, in most alkane molecules, carbon atoms have bond angles carbon has bond angles of 109.5°, the angle formed within the geometrical shape of a tetrahedron, as this permits the bonding electron pairs around a carbon atom to be as far apart as possible. This bond angle is clearly not possible for cyclopropane and cyclobutane molecules. Cyclopropane must have a triangular shape with C—C—C bond angles near 60°, and cyclobutane molecules must have a square or rectangular shape with C—C—C bond angles near 90°. Nonetheless, these compounds *do* exist: that is, they are stable.

Let's look at the most common cycloalkanes, starting with the smallest.

Interactive Exercise (Odyssey) 9.13

Simulate isomerization of a double bond.

Cyclopropane, Cyclobutane, and Cyclopentane

Cyclopropane has symmetrical molecules with C—C—C bond angles of 60°, as indicated in Figure 9.25a. This large deviation from the normal 109.5° tetrahedral angle brings bonding electron pairs closer together in most alkanes, raising the energy of the molecule and making it more reactive than other alkanes. All six of the C—H bonds also have an eclipsed, rather than staggered, conformation.

Make molecular models of cyclobutane and cyclopentane molecules. You will see that they are slightly puckered rather than flat, as indicated in Figure 9.25b and c. This puckering makes the C—C—C bond angles a bit smaller than they would otherwise be and increases the angle strain. At the same time, though, the puckering relieves the eclipsing interactions of adjacent C—H bonds that would occur if the carbon atoms of the rings were all in the same plane.

(a)   (b)   (c)

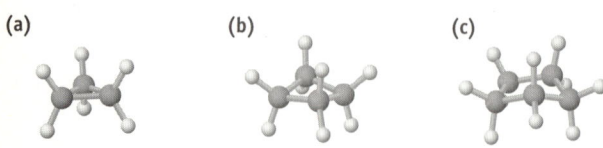

FIGURE 9.25 The structures of (a) cyclopropane, (b) cyclobutane, and (c) cyclopentane molecules. Cyclopropane's carbon skeleton is planar, but cyclobutane and cyclopentane molecules are slightly puckered.

Cyclohexane

Substituted cyclohexane molecules are the most common cycloalkanes because of their wide occurrence in nature. Molecules of a large number of compounds, including steroids and numerous pharmaceutical agents, have their shapes determined in part by cyclohexane rings. Cholesterol molecules, for instance, have 3 six-membered rings and 1 five-membered ring "fused" together.

Molecular Modelling (Odyssey)

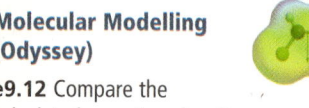

e9.12 Compare the calculated energies of cyclic and acyclic alkanes.

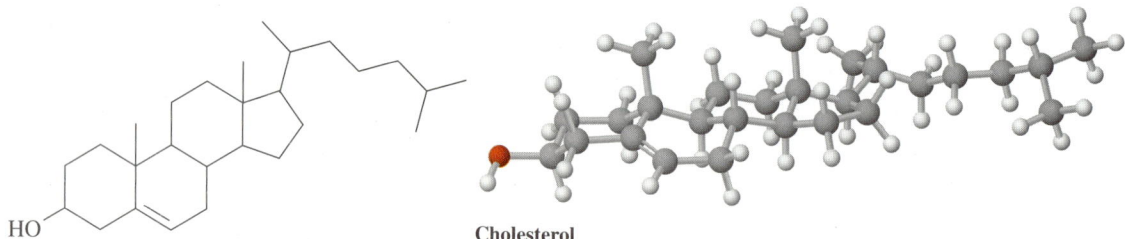

HO

Cholesterol

Testosterone, discussed in Section 2.1, is also made of molecules with the same four-fused ring skeleton as cholesterol, as are all chemical substances classified as steroids.

While the drawing of the steroid ring structure of a cholesterol molecule in the margin might imply that the carbon atoms all lie in a plane, it is evident from X-ray crystallography data used to produce the model that this is not the case. The carbon atoms in the four cyclohexane and cyclopentane rings that are fused together are puckered out of the plane.

Experimental evidence shows that individual cyclohexane rings are also not planar. If the carbon atoms in a cyclohexane ring were all at the corners of a planar hexagon, the C—C—C bond angles would be 120°. We've noted that almost all carbon atoms with four single bonds to other atoms have experimental bond angles close to 109.5°, the tetrahedral bond angle. Bond angles much higher or lower than this value would introduce bond angle strain and make the molecule less stable, or higher in energy.

As seen in Figure 9.26, cyclohexane molecules are puckered into a strain-free, three-dimensional shape called a *chair conformer*, in which the C—C—C bond angles are close to the typical 109.5° tetrahedral value. This conformer not only frees the molecule of angle strain but also from all C—H eclipsing interactions; neighbouring C—H bonds are staggered.

A chair conformer is drawn in three steps:

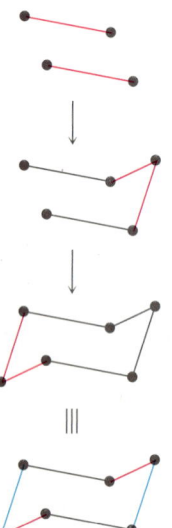

Step 1.   Draw two parallel lines, slanted downward and slightly offset from each other. This means that four of the cyclohexane carbons lie in a plane.

Step 2.   Place the top carbon atom above and to the right of the plane of the other four, and connect the bonds.

Step 3.   Place the bottom carbon atom below and to the left of the plane of the middle four, and connect the bonds. Note that the bonds to the bottom carbon atom are parallel to the bonds to the top carbon atom, as shown in the last structure.

(a)   **(b)**

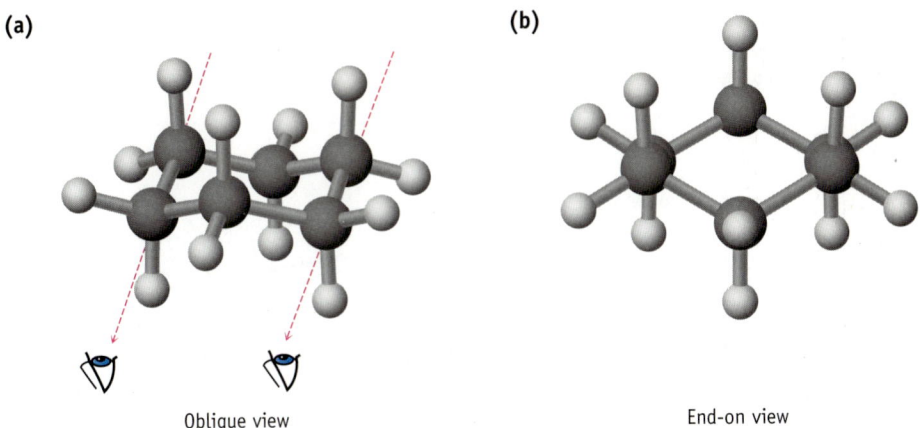

Oblique view   End-on view

FIGURE 9.26 The strain-free chair conformer of a cyclohexane molecule. (a) All C—C—C bond angles are close to 109° and all neighbouring C—H bonds are staggered, as evident in the end-on view in (b).

When viewing cyclohexane molecules, remember that the lower bond is in front and the upper bond is in back. If this convention is not defined, an optical illusion can make it appear that the reverse is true. For clarity, the cyclohexane rings shown here have the front (lower) bond heavily shaded to indicate its nearness to the reader.

This bond is in back.

This bond is in front.

Axial and Equatorial Bonds in Cyclohexane

The stable chair conformer of cyclohexane molecules has many chemical consequences. One is that there are two kinds of position for hydrogen atoms on the ring—axial positions and equatorial positions (Figure 9.27). Chair cyclohexane has six axial hydrogen atoms with C—H bonds that are perpendicular to the ring (parallel to the ring *axis*) and six equatorial hydrogen atoms that are approximately in the plane of the ring (around the ring *equator*).

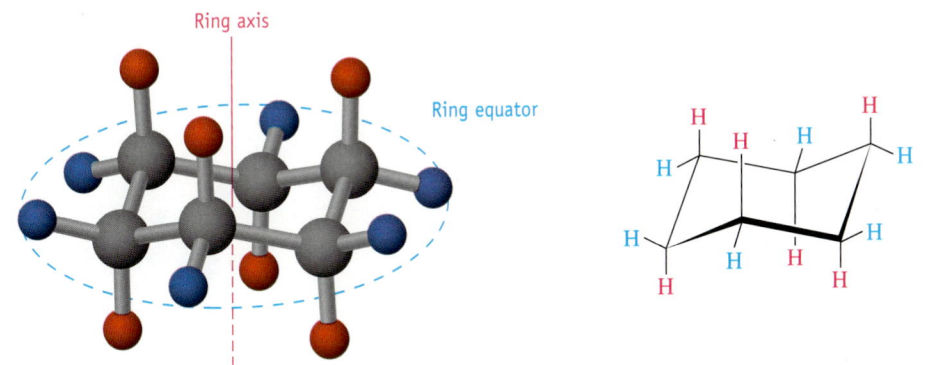

Ring axis

Ring equator

FIGURE 9.27 Axial (red) and equatorial (blue) hydrogen atoms in a molecule of cyclohexane. The six axial C—H bonds are parallel to the ring axis, and the six equatorial C—H bonds are in a band around the ring equator.

Each carbon atom in a cyclohexane molecule is bonded to one axial and one equatorial hydrogen atom, and each side of the ring has three axial and three equatorial hydrogen atoms in an alternating arrangement. For example, if the top side of the ring has axial hydrogen atoms on carbons 1, 3, and 5, then it has equatorial hydrogen atoms on carbon atoms 2, 4, and 6. Exactly the reverse is true for the bottom side: Carbon atoms 1, 3, and 5 have equatorial hydrogen atoms, but carbon atoms 2, 4, and 6 have axial hydrogen atoms.

Note that we haven't yet used the words *cis* and *trans* in this discussion of cyclohexane geometry. If two hydrogen atoms or other groups are both on the top or bottom side of a ring, they are *cis*, regardless of whether they are axial or equatorial and regardless of whether they are adjacent. Similarly, molecules with one group on the top side and one on the bottom side of the ring are *trans*.

Axial and equatorial bonds can be drawn following the procedure in Figure 9.28.

Axial bonds: The six axial bonds, one on each carbon, are parallel and alternate up–down.

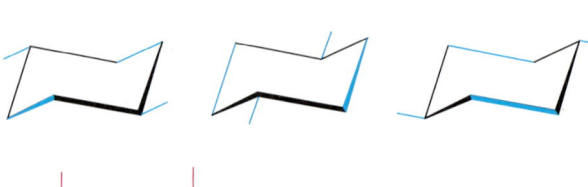

Equatorial bonds: The six equatorial bonds, one on each carbon, come in three sets of two parallel lines. Each set is also parallel to two ring bonds. Equatorial bonds alternate between sides around the ring.

Completed cyclohexane

FIGURE 9.28 A procedure for drawing axial and equatorial bonds in cyclohexane molecules.

EXERCISE 9.14—CHAIR CONFORMERS OF CYCLOHEXANE MOLECULES

Draw two chair structures for a methylcyclohexane molecule, one with the methyl group axial and one with the methyl group equatorial.

Conformational Mobility of Cyclohexane

Molecular Modelling

e9.13 See how the energy varies for different conformers of cyclohexane.

Because chair cyclohexane has two kinds of positions of substituents, axial and equatorial, we might expect to find two isomeric forms of a monosubstituted cyclohexane. In fact, though, there is only *one* methylcyclohexane, *one* bromocyclohexane, and so forth, because cyclohexane ring molecules are *conformationally mobile* at room temperature. Different chair cyclohexane conformers readily interconvert, resulting in the exchange of axial and equatorial positions. This interconversion of chair conformers (Figure 9.29) is simply the result of simultaneous rotation about carbon-carbon bonds within the ring, and is usually referred to as a *ring-flip*.

FIGURE 9.29 A ring-flip in a molecule of cyclohexane results from rotation about C—C single bonds in the chair form of the ring and interconverts axial and equatorial substituents.

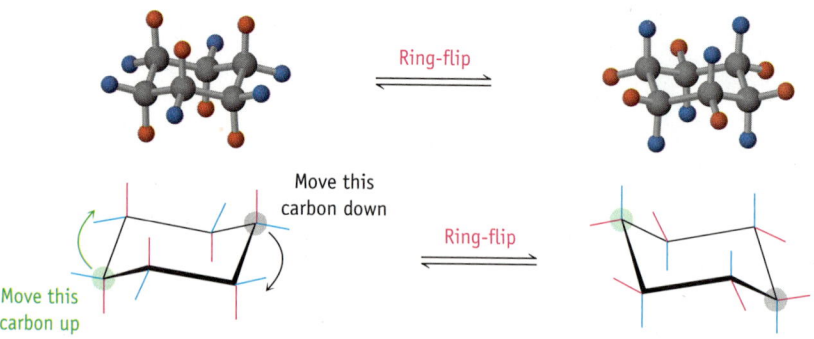

Conceptually, a chair cyclohexane molecule can be ring-flipped by keeping the middle four carbon atoms in place while folding the two ends in opposite directions. An axial substituent in one chair form becomes an equatorial substituent in the ring-flipped chair form and vice versa. For example, the axial methyl group in methylcyclohexane becomes equatorial after the molecule ring-flips. Because this interconversion occurs rapidly at room temperature, we can isolate only an interconverting mixture rather than distinct axial and equatorial isomers.

Although axial and equatorial methylcyclohexane molecules interconvert rapidly, they aren't equally stable. The equatorial conformer is more stable than the axial conformer by 7.6 kJ mol^{-1}, meaning that about 95% of methylcyclohexane molecules have their methyl group equatorial at any given instant. The energy difference is due to an unfavourable *steric* (spatial) interaction that occurs in the axial conformer between the electron cloud of the methyl group on carbon atom 1 and the electrons of the axial C—H bonds on carbon atoms 3 and 5. This so-called *1,3-diaxial interaction* introduces 7.6 kJ mol^{-1} of steric strain into the molecule because the axial methyl group and the nearby axial hydrogen atom are very close together (Figure 9.30).

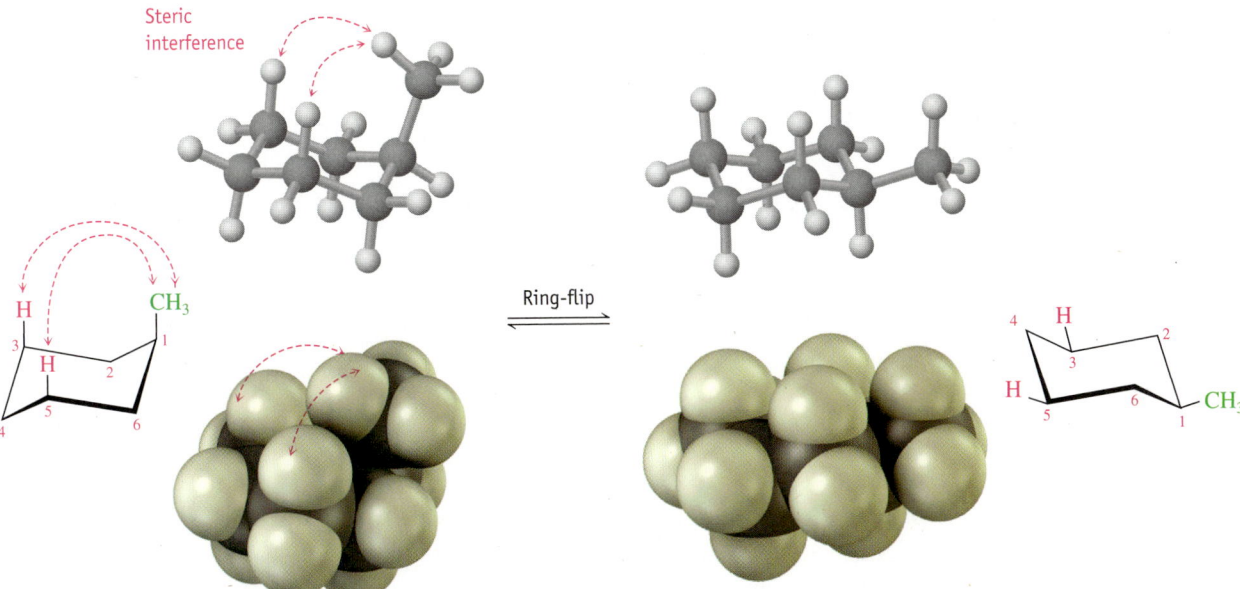

FIGURE 9.30 Axial versus equatorial methylcyclohexane molecules. The 1,3-diaxial steric interactions in axial methylcyclohexane (easier to see in space-filling models) make the equatorial conformer more stable by 7.6 kJ mol^{-1}.

The same analysis can be applied to other monosubstituted cyclohexane molecules. Most cyclohexane molecules are more stable with a substituent group in an equatorial position than in an axial position. As you might expect, the amount of steric strain increases as the size of the axial substituent group increases.

WORKED EXAMPLE 9.6—CHAIR CONFORMERS OF CYCLOHEXANE MOLECULES

Draw a 1,1-dimethylcyclohexane molecule, indicating the axial and equatorial methyl groups.

Solution

Draw a chair cyclohexane ring, and then put two methyl groups on the same carbon atom. The methyl group in the approximate plane of the ring is equatorial, and the other (above or below the ring) is axial.

EXERCISE 9.15—CHAIR CONFORMERS OF CYCLOHEXANE MOLECULES

Draw two different chair conformers of a bromocyclohexane molecule showing all hydrogen atoms. Label all positions as axial or equatorial. In which conformer do you think the molecule will be more stable?

> Cyclohexanes are the most common of all carbon rings in nature. Cyclohexane exists in a puckered, strain-free chair conformer in which all bond angles are near 109° and all neighbouring C—H bonds are staggered. Chair cyclohexane molecules can undergo a ring-flip that interconverts axial and equatorial bonds. Substituents on the ring are more stable in the equatorial than in the axial position.

Heterocyclic Compounds

Cyclic organic compounds that contain atoms of two or more elements in their rings are called **heterocyclic** compounds. Molecules with five- and six-membered heterocyclic rings containing one or more oxygen, nitrogen, or sulfur atoms play vital roles in many biological processes. Figure 9.31 shows examples, such as 2'-deoxyribose, the sugar component in DNA; adenine, a base in DNA and ATP; nicotine; and Viagra.

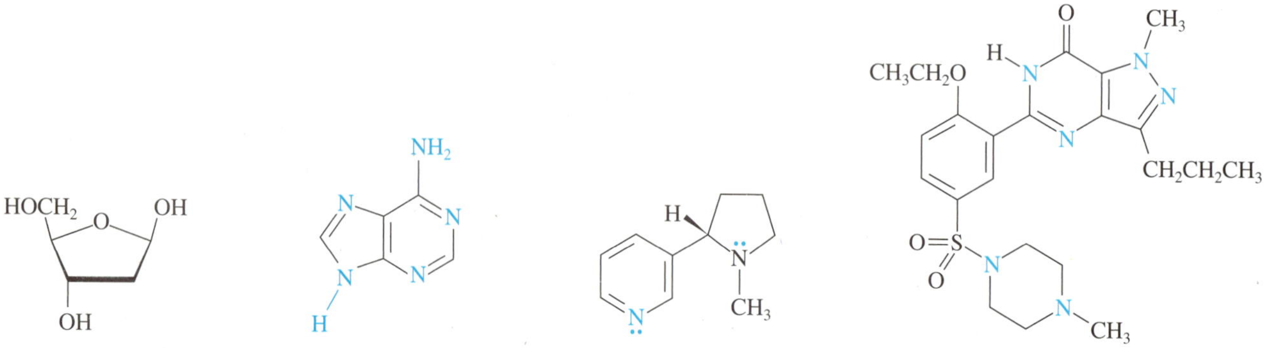

2-Deoxyribose   **Adenine (A)**   **Nicotine**   **Sildenafil**
                      A component of DNA, RNA   **(Viagra)**

FIGURE 9.31 Heterocyclic molecules with important biological functions.

Metal Coordination Complexes Containing Chelates

Metal *coordination complexes* with chelating ligands (molecules or ions that bind with more than one donor atom) often form five- and six-membered rings [>>Section 27.4]. The photodynamic therapy story in Chapter 1 introduced one of the most important metal complexes in the body, the heme unit in hemoglobin. Other examples include the pyrophosphate complexes formed with Mg^{2+} and Ca^{2+} ions in the process of softening water, and ethylenediaminetetraacetic acid (EDTA) complexes formed with lead(II) when a patient is treated for lead poisoning (Figure 9.32).

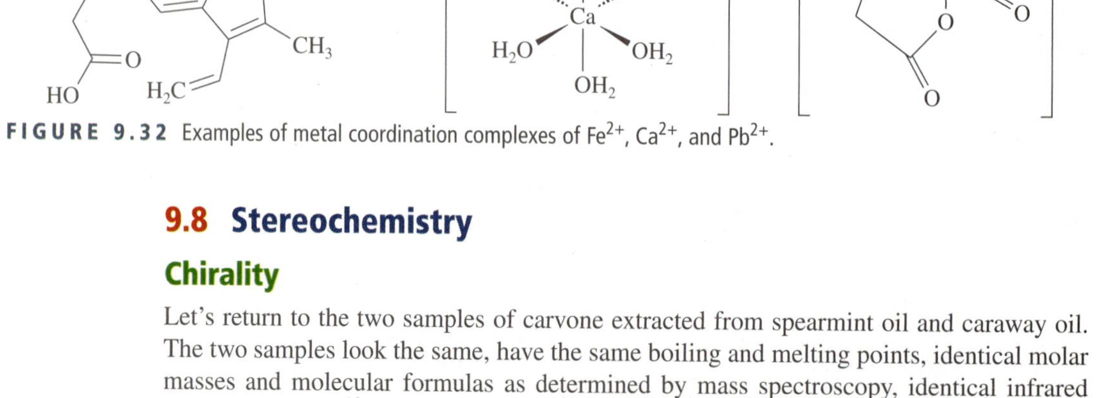

FIGURE 9.32 Examples of metal coordination complexes of Fe^{2+}, Ca^{2+}, and Pb^{2+}.

9.8 Stereochemistry

Chirality

Let's return to the two samples of carvone extracted from spearmint oil and caraway oil. The two samples look the same, have the same boiling and melting points, identical molar masses and molecular formulas as determined by mass spectroscopy, identical infrared spectra, identical ^{13}C NMR spectra, and X-ray crystallography shows molecules with the same atom connectivity, bond lengths, and bond angles. Yet receptors in your nose immediately recognize that carvone from these two sources are different. The subtle, but important difference between these two carvones can also be detected by placing them in a simple instrument called a *polarimeter*, which measures their *optical activity*. We'll return to optical activity in Section 9.9.

Look carefully at representations of molecules of the two different carvones (Figure 9.33).

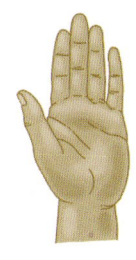

Molecular Modelling (Odyssey)

e9.14 See mirror-image isomers interact differently with a molecular receptor.

(a)

CH$_3$

(R)-Carvone
Spearmint Oil

(b)

CH$_3$

(S)-Carvone
Caraway Oil

Mirror

(R)-Carvone
Spearmint Oil

(S)-Carvone
Caraway Oil

FIGURE 9.33 (a) An (R)-carvone molecule extracted from spearmint oil, and (b) an (S)-carvone molecule extracted from caraway oil. The R/S designation is explained in Section 9.10. Ball and stick representations are shown on the right.

In both carvone molecules, the molecular formula and atom connectivity is the same. Only the spatial arrangement of the atoms is different. **Stereochemistry** *is the branch of chemistry concerned with the three-dimensional structures of molecules.* We've seen in Section 9.6 that *cis–trans* isomers of cycloalkanes differ only in their spatial arrangement. The same is true for *cis* and *trans* isomers of alkenes (Chapter 19). In this section, we focus on differences in spatial arrangement between molecules like carvone that are *handed* or **chiral** (ky-ral: Greek *cheir*, meaning "hand"). Our right and left hands are mirror images of each other. The result is that our hands, while similar, are not identical. When you hold your *left* hand up to a mirror, the reflection looks like a *right* hand. Try it.

Handedness (**chirality**) is important in organic chemistry and inorganic chemistry, and crucial in biochemistry, where carbohydrates, amino acids, nucleic acids, and many other naturally occurring molecules are chiral. To see how molecular handedness arises, look at the molecules shown in Figure 9.34. On the left of Figure 9.34 are three molecules, and on

Left hand   Right hand

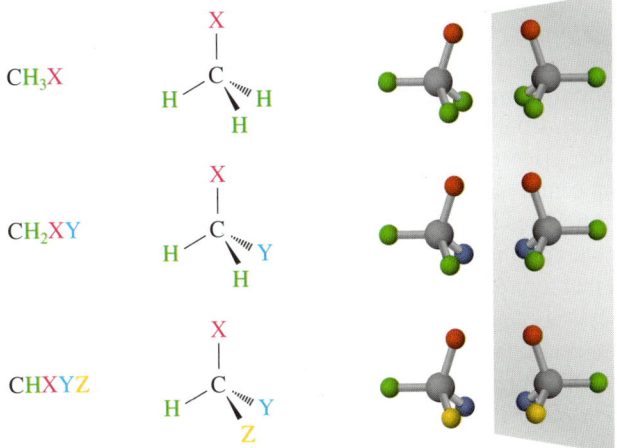

CH$_3$X

CH$_2$XY

CHXYZ

FIGURE 9.34 Three tetrahedral carbon atoms and their mirror images. Molecules of the type CH$_3$X and CH$_2$XY are identical to their mirror images, but a molecule of the type CHXYZ is not. A CHXYZ molecule is related to its mirror image in the same way that a right hand is related to a left hand.

the right are their images reflected in a mirror. The CH_3X and CH_2XY molecules are identical to their mirror images and thus are **achiral** (not handed). If you make molecular models of each molecule and of its mirror image, you find that you can superimpose one on the other. By contrast, the CHXYZ molecule is *not* identical to its mirror image. You can't superimpose a model of the molecule on a model of its mirror image for the same reason that you can't superimpose a left hand on a right hand: they simply aren't the same.

Two molecules, such as CHXYZ and its mirror image, are one category of stereoisomers, and the relationship between the two is named by calling them a pair of **enantiomers** (e-nan-tee-o-mer; Greek *enantio,* meaning "opposite"). Enantiomers are related to each other as a right hand is related to a left hand. The most common examples are found when a tetrahedral carbon atom is bonded to four different substituents (one need not be H). For example, lactic acid (2-hydroxypropanoic acid) exists as a pair of enantiomers because there are four different groups (—H, —OH, —CH_3, —COOH) bonded to the central carbon atom:

Lactic acid: a molecule of general formula CHXYZ

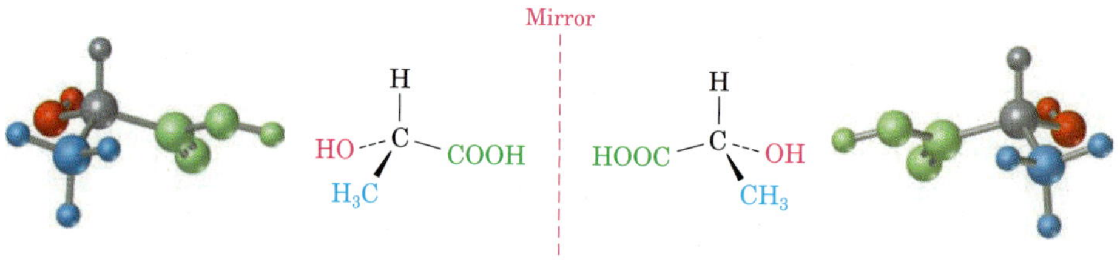

(+)-Lactic acid   **(−)-Lactic acid**

No matter how hard you try, you can't superimpose a molecule of "right-handed" lactic acid on top of a molecule of "left-handed" lactic acid. If any two groups match up, say —H and —COOH, the other two groups don't match (Figure 9.35).

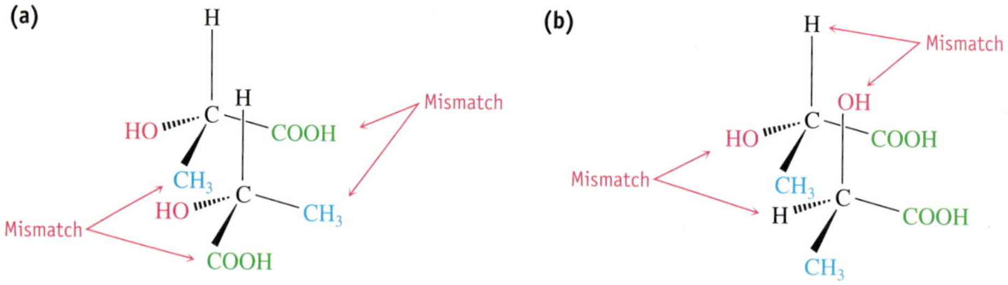

FIGURE 9.35 Attempts to superimpose the mirror-image forms of lactic acid. (a) When the —H and —OH substituents match up, the —COOH and —CH_3 substituents don't. (b) When the —COOH and —CH_3 substituents match up, —H and —OH don't. Regardless of how the molecules are oriented, they aren't identical.

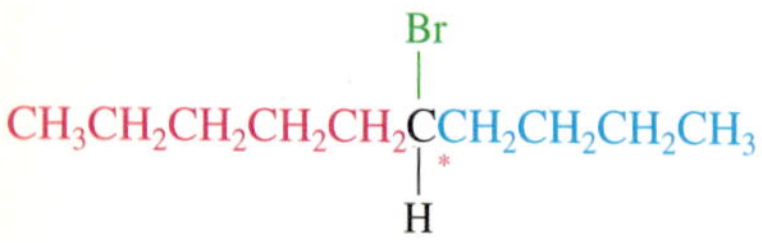

5-Bromodecane (chiral)

Substituents on carbon 5

—H

—Br

— $CH_2CH_2CH_2CH_3$ (butyl)

— $CH_2CH_2CH_2CH_2CH_3$ (pentyl)

How can you predict whether a given molecule is chiral or not? Throughout this chapter and the learning resource, we will focus on the most common (although not the only) cause of chirality in molecules: the presence of an atom (most commonly a carbon atom) bonded to four different groups. A good example is the central carbon atom in lactic acid. Such a carbon is referred to as a *chirality centre,* or **stereocentre**, because the stereochemistry at that centre is important. Note that *chirality* is a property of the entire molecule, whereas a difference in configuration at the stereocentre is the *cause* of chirality.

Identifying stereocentres in a complex molecule takes practice, because it's not always apparent that four different groups are bonded to a given carbon atom. The differences don't necessarily appear right next to the stereocentre. For example, 5-bromodecane is a chiral molecule because four different groups are bonded to C5 (marked by an asterisk). A butyl substituent is very similar to a pentyl substituent, but it isn't identical. The difference isn't apparent until four carbon atoms away from the stereocentre, but there's still a difference.

Carvone is an excellent example of such a chiral molecule, as is nootkatone, extracted from grapefruits; build a physical or computer model and check for yourself that the labelled atoms are indeed stereocentres. (It's helpful to note that carbon atoms in CH_2, CH_3, $C=C$, $C\equiv C$, and $C=O$ groups cannot be stereocentres because they have at least two identical bonds.)

A simple way to identify chiral molecules is to look for a plane of symmetry. *A molecule is not chiral if it contains a plane of symmetry.* A plane of symmetry is a plane that cuts through the middle of a molecule or other object so that one half of the object is an exact mirror image of the other half. A laboratory flask, for example, has a plane of symmetry. If you were to cut the flask in half, one half would be an exact mirror image of the other half. A hand, however, does not have a plane of symmetry. One "half" of a hand is not a mirror image of the other "half" (Figure 9.36).

Carvone (spearmint oil)

Nootkatone (grapefruit oil)

(a)   **(b)**

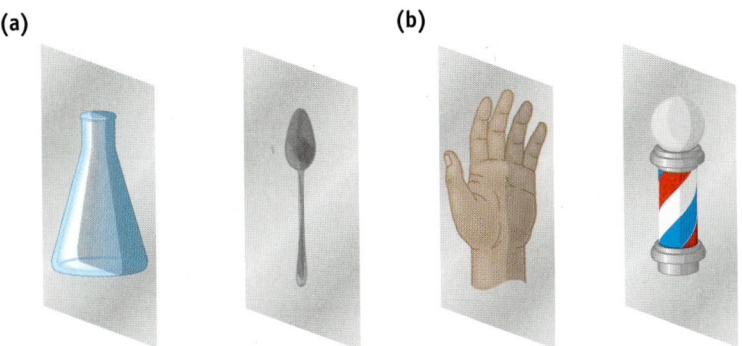

FIGURE 9.36 The meaning of a symmetry plane. (a) Objects like this flask or spoon have planes of symmetry passing through them, making the right and left halves mirror images. (b) Objects like this hand or barber's pole have no symmetry plane; the right "half" is not a mirror image of the "left" half.

A molecule that has a plane of symmetry in any of its possible conformations must be identical to its mirror image and hence must be *achiral* (without chirality). Thus, propanoic acid, CH_3CH_2COOH, contains a plane of symmetry when lined up as shown in Figure 9.37 and is therefore achiral. Lactic acid, $CH_3CH(OH)COOH$, however, has no plane of symmetry in any conformation and is chiral.

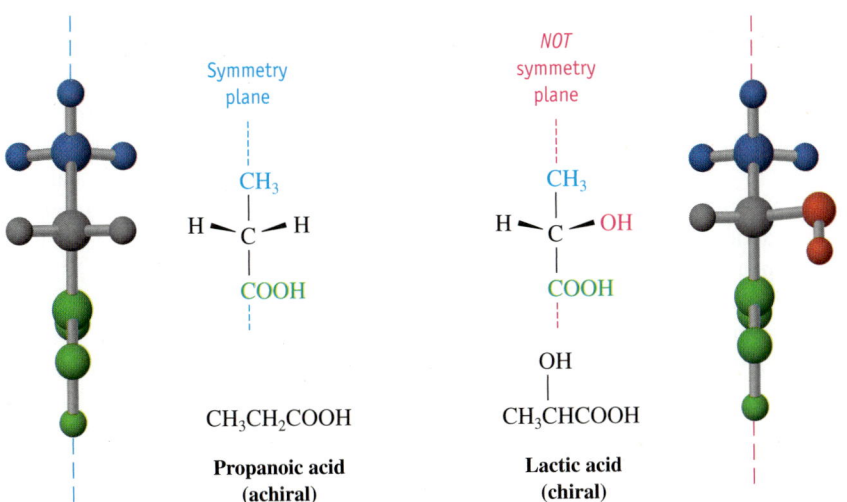

FIGURE 9.37 The achiral propanoic acid molecule versus the chiral lactic acid molecule. Propanoic acid has a plane of symmetry that makes one side of the molecule a mirror image of the other side. Lactic acid has no such symmetry plane.

$$CH_3CH_2 - \overset{\displaystyle OH}{\underset{\displaystyle H}{C}} - CH_3 \qquad \textbf{Butan-2-ol}$$

$$CH_3CH_2CH_2 - \overset{\displaystyle CH_3}{\underset{\displaystyle H}{\overset{*}{C}}} - CH_2CH_3$$

3-Methylhexane (chiral)

2-Methylcyclohexanone

CH₃

Molecular Modelling (Odyssey)

e9.15 See examples of chiral molecules.

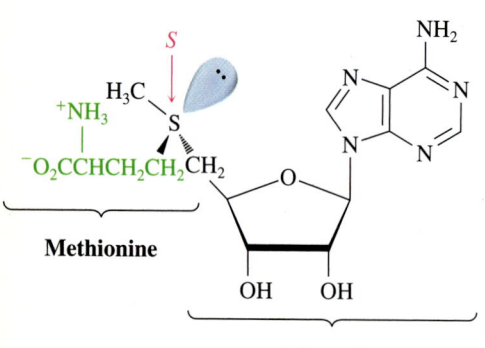

Methionine

OH   OH

Adrenaline

(S)-S-Adenosylmethionine

Molecular Modelling (Odyssey)

e9.16 Compare models of chiral and achiral metal complexes.

WORKED EXAMPLE 9.7—CHIRAL MOLECULES

Draw the structure of a molecule of a chiral alcohol.

Solution

An alcohol is a compound that contains the —OH functional group. To make an alcohol chiral, we need to have four different groups bonded to a single carbon atom, such as —H, —OH, —CH_3, and —CH_2CH_3.

WORKED EXAMPLE 9.8—CHIRAL MOLECULES

Is a 3-methylhexane molecule chiral?

Solution

Draw the structure of 3-methylhexane, and cross out all the CH_2 and CH_3 carbon atoms because they can't be stereocentres. Then look closely at any carbon atom that remains to see if it is bonded to four different groups. Carbon atom 3 is bonded to —H, —CH_3, —CH_2CH_3, and —$CH_2CH_2CH_3$, so the molecule is chiral.

WORKED EXAMPLE 9.9—CHIRAL MOLECULES

Is a 2-methylcyclohexanone molecule chiral?

Solution

Ignore the CH_3 carbon, the four CH_2 carbon atoms in the ring, and the C$=$O carbon atom because they can't be stereocentres. Then look carefully at C2, the only carbon atom that remains. Carbon 2 is bonded to four different groups: a —CH_3 group, an —H atom, a —C$=$O carbon atom in the ring, and a —CH_2— ring carbon atom, so 2-methylcyclohexanone is chiral.

Chirality at Non-Carbon Centres

While the simplest and most common examples of stereocentres involve tetrahedral carbon atoms with four different groups attached, chirality is also centrally important in compounds with stereocentres other than carbon atoms. Nitrogen, phosphorous, and sulfur atoms can all be chirality centres, as can certain transition metal atoms such as cobalt or zinc, which have six groups attached. In some cases where enantiomers can't rapidly interconvert, a pair of electrons can even act as a fourth "substituent," creating a chirality centre. A good example is the co-enzyme (S)-S-adenosylmethionine, which is known as a methyl group donor because in many metabolic pathways it is a source of transferring CH_3 groups. The second "S" in the name refers to the sulfur atom of methionine. The biologically active enantiomer has the (S) configuration at the sulfur atom (the first "S"), and the (R) enantiomer is known, but not biologically active. This R/S nomenclature is explained in Section 9.10.

A direct parallel can be drawn between a carbon stereocentre and a tetrahedral transition metal atom with four different Lewis base groups (called *ligands* in inorganic chemistry) attached to it, but such chiral tetrahedral metal complexes are rare. More commonly, certain metal complexes have octahedral shapes, with six groups attached to a metal ion pointing toward the vertices of an octahedron. Recall that the criterion for chirality is that a molecule must not be superimposable on its mirror image, which can often be easily seen by noting the *absence* of a plane of symmetry in the molecule. If there is a plane of symmetry, an octahedral complex is achiral. But chiral metal complexes are known in cases where the metal ion is bonded to three *bidentate* ligands (each ligand molecule bonded to the metal through two different atoms) or two bidentate ligands and two monodentate ligands that are *cis* to each other. Examples are two chiral complexes of cobalt in Figure 9.38.

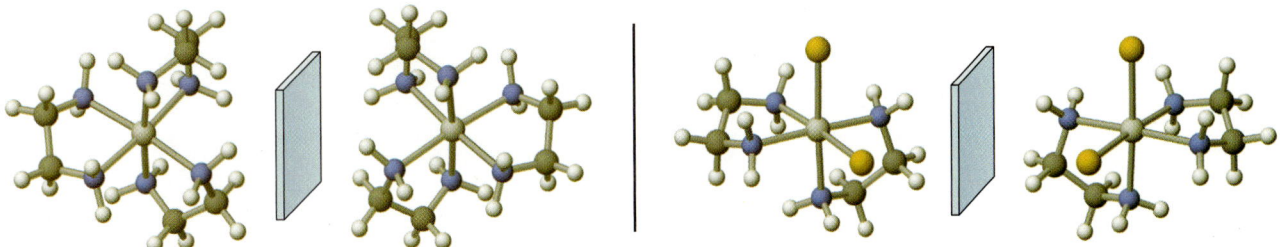

FIGURE 9.38 Chiral metal complexes. Both $[Co(en)_3]^{3+}$ and *cis*-$[Co(en)_2Cl_2]^+$ are chiral, because they are not superimposable on their mirror images. The abbreviation "en" refers to an ethylenediamine ligand, $NH_2CH_2CH_2NH_2$, which is called a *bidentate* ligand because each of its two nitrogen atoms can form a bond to the central cobalt ion. A plane of symmetry can not be found in either of these two complexes.

Much research is currently directed toward synthesizing new chiral transition metal complexes for use as catalysts in organic reactions. Since molecular recognition depends so strongly on chirality, chiral catalysts used in synthesis can selectively lead to the formation of one stereoisomer of many possible ones that could be formed in a reaction.

> A molecule that is not identical to its mirror image is said to be chiral, meaning "handed." A chiral molecule will not have a plane of symmetry. The usual source of chirality in a molecule is a stereocentre, a tetrahedral atom bonded to four different groups. Chiral compounds can exist as a pair of mirror-image stereoisomers called enantiomers.

Interactive Exercise (Odyssey) 9.16

Identify chiral metal complexes.

9.9 Optical Activity

Polarimetry Measurements

The systematic experimental study of stereochemistry started in the early 19th century with investigations by the French physicist Jean Baptiste Biot into the nature of *plane-polarized light.* A beam of ordinary light consists of electromagnetic waves that oscillate in an infinite number of planes at right angles to the direction of light travel. When a beam of ordinary light passes through a device called a *polarizer,* though, only the light waves oscillating in a *single* plane pass through and the light is said to be plane-polarized.

Biot made the remarkable observation that when a beam of plane-polarized light passes through a solution containing certain organic molecules, such as sugar or carvone, the plane of polarization is *rotated.* Not all substances exhibit this property, but those that do are said to be **optically active**.

The amount of rotation can be measured with an instrument called a polarimeter, represented in Figure 9.39. An optically active solution is placed in a sample tube, plane-polarized light is passed through the tube, and rotation of the plane occurs. The light then goes through a second polarizer called the *analyzer.* By rotating the analyzer until light passes through *it,* we can find the new plane of polarization and can tell to what extent rotation has occurred. The amount of rotation observed is denoted by α (Greek alpha) and is expressed in degrees.

In addition to determining the extent of rotation, we can also find the direction. From the vantage point of the observer looking at the analyzer, some optically active substances rotate plane-polarized light to the left (counterclockwise) and are said to be *levorotatory*; other substances rotate light to the right (clockwise) and are said to be *dextrorotatory.* By convention, rotation to the left is given a minus sign (−), and rotation to the right is given a plus sign (+). For example, the substance (−)-carvone is levorotatory and the substance (+)-sucrose is dextrorotatory.

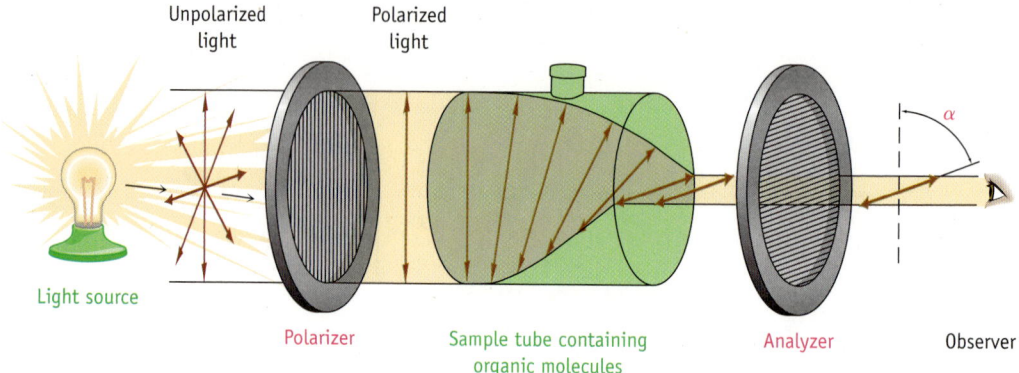

FIGURE 9.39 Schematic representation of a polarimeter. Plane-polarized light passes through an optically active solution, which rotates the plane of polarization.

Specific Rotation

The degree of rotation observed in a polarimetry experiment depends on the structure of the sample molecules and on the number of molecules encountered by the light beam. The number of molecules encountered depends, in turn, on sample concentration and sample path length. If the concentration of the sample in a tube is doubled, the observed rotation of the sample is doubled. If the concentration is kept constant but the length of the sample tube is doubled, the observed rotation is doubled.

To express optical rotation data so that comparisons can be made, we have to choose standard conditions. The *specific rotation, $[\alpha]_D$,* of a compound is defined as the observed rotation when light of 589.6 nanometre (nm; $1 \text{ nm} = 10^{-9}$ m) wavelength is used with a sample path length l of 1 decimetre (dm; 1 dm = 10 cm) and a sample concentration c of 1 g mL^{-1}. (Light of 589.6 nm, the so-called sodium D line, is the yellow light emitted from common sodium lamps.)

$$[\alpha]_D = \frac{\text{observed rotation (degrees)}}{\text{path length, } l \text{ (dm)} \times \text{concentration, } c \text{ (g mL}^{-1}\text{)}} = \frac{\alpha}{l \times c}$$

When optical rotation data are expressed in this standard way, the specific rotation $[\alpha]_D$ is a physical constant characteristic of a given optically active compound. Some examples are listed in Table 9.2.

TABLE 9.2 Specific Rotations of Some Compounds

Compound	$[\alpha]_D$ (Degrees)	Compound	$[\alpha]_D$ (Degrees)
Penicillin V	+233	Cholesterol	−31.5
Sucrose	+66.47	Morphine	−132
Camphor	+44.26	Acetic acid	0
Monosodium glutamate	+25.5	Benzene	0

WORKED EXAMPLE 9.10—OPTICAL ACTIVITY

A 1.20 g sample of cocaine, $[\alpha]_D = -16°$, was dissolved in 7.50 mL of chloroform and placed in a sample tube having a path length of 5.00 cm. What was the observed rotation?

Solution

Observed rotation (α) is equal to specific rotation, $[\alpha]_D$, multiplied by sample concentration (c), multiplied by path length (l):

$$\alpha = [\alpha]_D \times c \times l$$

where $[\alpha]_D = -16°$, $l = 5.00$ cm $= 0.500$ dm, and $c = 1.20$ g in 7.50 mL $= 0.160$ g mL^{-1}

$$\alpha = -16° \times 0.500 \times 0.160 = -1.3°$$

EXERCISE 9.17—OPTICAL ACTIVITY

Is cocaine (Worked Example 9.10) dextrorotatory or levorotatory?

EXERCISE 9.18—OPTICAL ACTIVITY

A 1.50 g sample of coniine, the toxic extract of poison hemlock, was dissolved in 10.0 mL of ethanol and placed in a sample tube with a path length of 5.00 cm. The observed rotation at the sodium D line was −121°. Calculate the specific rotation $[\alpha]_D$ for coniine.

Pasteur's Discovery of Enantiomers

Little was done after Biot's discovery of optical activity until 1848, when Louis Pasteur began work on a study of crystalline tartaric acid salts derived from wine. On recrystallizing a concentrated solution of sodium ammonium tartrate below 28°C, Pasteur made the surprising observation that two distinct kinds of crystals precipitated. Furthermore, the two kinds of crystals were mirror images and were related in the same way that a right hand is related to a left hand.

Working carefully with a pair of tweezers, Pasteur was able to separate the crystals into two piles, one of "right-handed" crystals and one of "left-handed" crystals, like those shown in Figure 9.40. Although the original sample (containing a 50:50 mixture of right- and left-handed crystals) was optically inactive, *solutions of crystals from each of the sorted piles were optically active* and their specific rotations were equal in amount but opposite in sign.

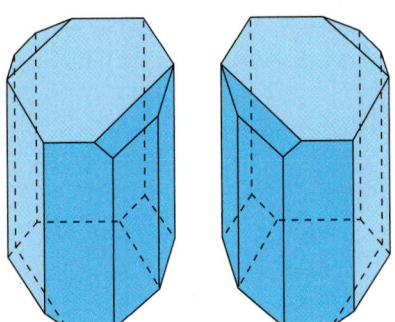

Sodium ammonium tartrate

FIGURE 9.40 Crystals of sodium ammonium tartrate, taken from Pasteur's original sketches. One of the crystals is dextrotatory in solution; the other is levorotatory.

We call this original 50:50 mixture of (+)- and (−)-substances a **racemate**. Racemates are often denoted by the symbol (±) to indicate that they contain equal amounts of dextrorotatory and levorotatory substances, each of which contains enantiomeric molecules of opposite chirality. Such mixtures show no optical activity because the (+) rotation caused by molecules of one enantiomer exactly cancels the (−) rotation from molecules of its mirror image, the other enantiomer. Substances containing only these enantiomers have identical chemical properties in the absence of a chiral environment.

Pasteur was far ahead of his time in seeing a connection between this experimental laboratory measurement and the molecular-level explanation behind it. It was not until 25 years later that his theories regarding the asymmetry of chiral molecules were confirmed.

Today, we would describe Pasteur's work by saying that he had discovered substances made of chiral molecules that are enantiomers of each other. *Two substances comprised of enantiomeric molecules* (also called *optical isomers*) have identical physical properties, such as melting points and boiling points and spectral features, but *differ in only two observable ways: Solutions of equal concentrations of these substances will rotate plane-polarized light the same amount, but in the opposite direction, and the substances will interact differently when placed in a chiral environment.*

Optical activity, the rotation of plane-polarized light by a substance, can be observed experimentally using a polarimeter, and quantified by the specific rotation of a substance, $[\alpha]_D$.

9.10 Sequence Rules for Specifying Configuration

Drawings provide visual representations of stereochemistry, but a systematic method for specifying the three-dimensional arrangement, or configuration, of substituents around a stereocentre in a single molecule is also necessary. The method approved by IUPAC identifies the relative priority of the four substituents around a stereocentre, using the same sequence rules we will see again in Section 19.5 for specifying *E* and *Z* alkene stereochemistry.

Rule 1   Look at the four atoms directly attached to the stereocentre, and assign priorities in order of decreasing atomic number. The atom with the highest atomic number is ranked first; the atom with the lowest atomic number is ranked fourth.

Rule 2   If a decision can't be reached by ranking the first atoms in the substituents, look at the second, third, or fourth atoms outward until the first difference is found.

Rule 3   Multiple-bonded atoms are equivalent to the same number of single-bonded atoms.

Having assigned priorities to the four groups attached to a stereocentre, we describe the stereochemical configuration around the stereocentre by orienting the molecule so that the group of lowest priority (4) is pointing directly back, away from us. We then look at the three remaining substituents, which now appear to radiate toward us like the spokes on a steering wheel (Figure 9.41). If a curved arrow drawn from the highest- to second-highest- to third-highest-priority substituent ($1 \rightarrow 2 \rightarrow 3$) is clockwise, we say that the stereocentre has an *R* configuration (Latin *rectus,* meaning "right"). If an arrow from $1 \rightarrow 2 \rightarrow 3$ is counterclockwise, the stereocentre has an *S* configuration (Latin *sinister,* meaning "left"). To remember these assignments, think of a car's steering wheel when making a *R*ight (clockwise) turn.

FIGURE 9.41 Assignment of configuration to a stereocentre. When the molecule is oriented so that the group of lowest priority (4) is toward the rear, the remaining three groups radiate toward the viewer like the spokes of a steering wheel. If the direction of travel $1 \rightarrow 2 \rightarrow 3$ is clockwise (right turn), the centre has the *R* configuration. If the direction of travel $1 \rightarrow 2 \rightarrow 3$ is counterclockwise (left turn), the centre is *S*.

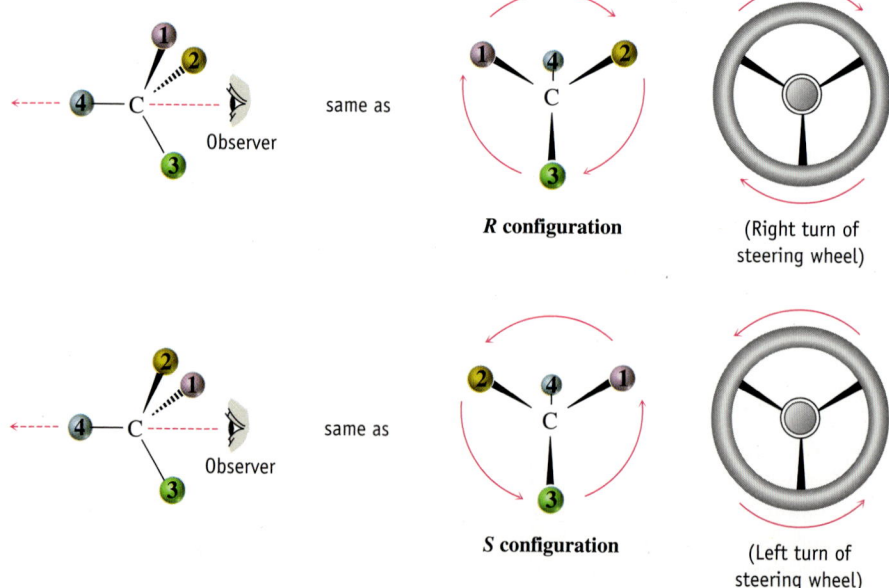

Look at a molecule of (−)-lactic acid in Figure 9.42 to see an example of how configuration is assigned. Sequence rule 1 says that —OH has priority 1 and —H has priority 4, but it doesn't allow us to distinguish between —CH₃ and —COOH because both groups have carbon as their first atom. Sequence rule 2, however, says that —COOH has higher priority than —CH₃ because O outranks H (the second atom in each group). Now, turn the molecule so that the fourth-priority group (–H) is oriented toward the rear, away from the observer. Since a curved arrow from 1 (—OH) to 2 (—COOH) to 3 (—CH₃) is clockwise (right turn of the steering wheel), a molecule of (−)-lactic acid has the *R* configuration. Applying the same procedure to a molecule of (+)-lactic acid leads to the opposite assignment.

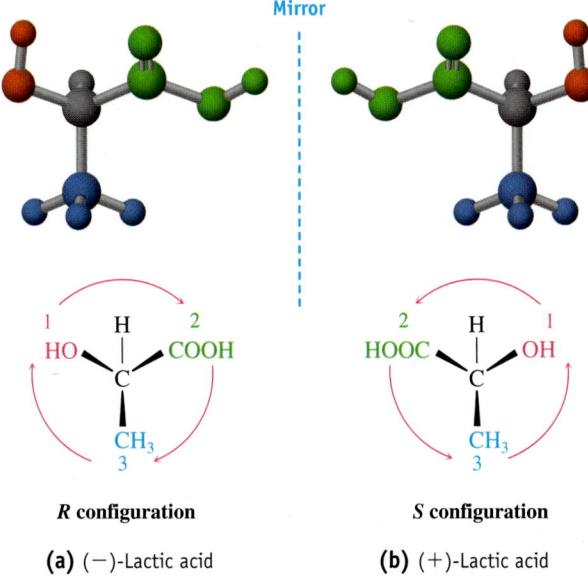

Mirror

R configuration

S configuration

(a) (−)-Lactic acid

(b) (+)-Lactic acid

FIGURE 9.42 Assignment of configuration of molecules of (a) (*R*)-(−)-lactic acid and (b) (*S*)-(+)-lactic acid.

Further examples are provided by naturally occurring (−)-glyceraldehyde and (+)-alanine, whose molecules have the *S* configurations shown in Figure 9.43. *Note that the sign of optical rotation, (+) or (−), which comes from an experimental polarimetry measurement of a substance, is not related to the R/S designation, which comes from assigning*

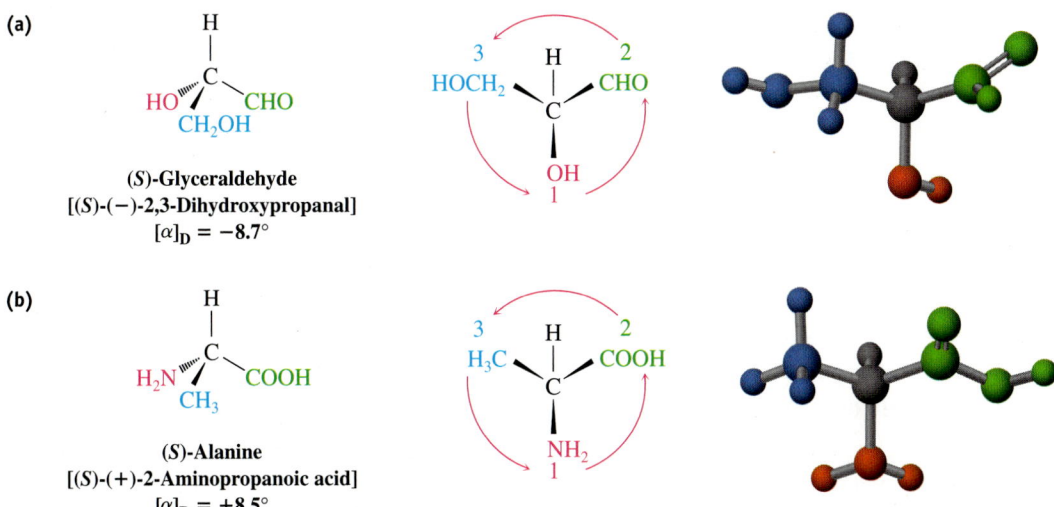

(a)

(S)-Glyceraldehyde
[(S)-(−)-2,3-Dihydroxypropanal]
[α]_D = −8.7°

(b)

(S)-Alanine
[(S)-(+)-2-Aminopropanoic acid]
[α]_D = +8.5°

FIGURE 9.43 Assignment of configuration to molecules of (a) (−)-glyceraldehyde, and (b) (+)-alanine. Both happen to have the *S* configuration, although the one substance is experimentally found to be levorotatory (−) and the other dextrorotatory (+).

the rules for priority to drawn structures of a molecule of that substance. Molecules of (*S*)-alanine make up a substance that when placed in a polarimeter is found experimentally to rotate plane-polarized light to the right and is designated dextrorotatory (+). Molecules of (*S*)-glyceraldehyde make up a substance that when placed in a polarimeter rotate plane-polarized light to the left and is designated levorotatory (−), but *there is no correlation between assignment of R/S configuration of molecules and the experimental determination of the direction of optical rotation (+) or (−) of a substance.*

WORKED EXAMPLE 9.11—*R/S* DESIGNATION

Orient each of the following drawings so that the lowest-priority group is toward the rear, and then assign *R* or *S* configuration:

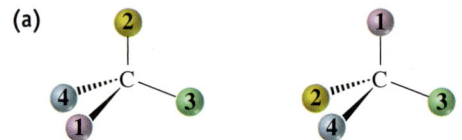

(a)

Solution

It takes practice to be able to visualize and orient a stereocentre in a chiral molecule in three dimensions. It is helpful to use a set of hand-held or computer molecular models to do so. You might start by indicating where the observer must be located—180° opposite the lowest-priority group. Then imagine yourself in the position of the observer, and redraw what you would see. In (a), you would be located in front of the page toward the top right of the molecule, and you would see group 2 to your left, group 3 to your right, and group 1 below you. This corresponds to an *R* configuration.

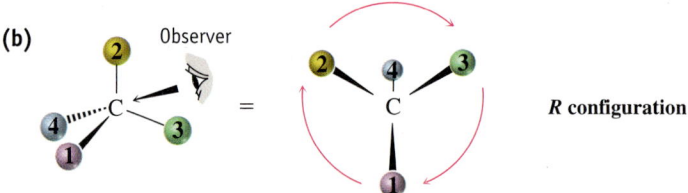

(b)   Observer   =   *R* configuration

In (b), you would be located behind the page toward the top *left* of the molecule from your point of view, and you would see group 3 to your left, group 1 to your right, and group 2 below you. This also corresponds to an *R* configuration.

(c)   Observer   =   *R* configuration

WORKED EXAMPLE 9.12—*R/S* DESIGNATION

Draw a three-dimensional representation of the tetrahedral geometry of a molecule of (*R*)-2-chlorobutane.

Solution

Identify the four substituents bonded to the stereocentre, and assign the priorities: (1) —Cl, (2) —CH$_2$CH$_3$, (3) —CH$_3$, (4) —H. To draw a tetrahedral representation of the molecule, orient the low-priority —H group away from you and imagine that the other three groups are coming out of the page toward you. Then place the remaining three substituents such that the direction of travel 1 → 2 → 3 is clockwise (right turn), and tilt the molecule toward you by 90° to bring the rear hydrogen into view. Again, using hand-held or computer molecular models is a great help in working problems of this sort.

Interactive Exercise 9.19

Draw structural formulas for stereoisomers.

Interactive Exercise (Odyssey) 9.20

Assign *R/S* configurations to molecular models.

(*R*)-2-Chlorobutane

EXERCISE 9.21—*R/S* DESIGNATION

Assign priorities to the substituents in each of the following sets:
(a) —H, —Br, —CH$_2$CH$_3$, —CH$_2$CH$_2$OH
(b) —CO$_2$H, —COOCH$_3$, —CH$_2$OH, —OH
(c) —Br, —CH$_2$Br, —Cl, —CH$_2$Cl

EXERCISE 9.22—*R/S* DESIGNATION

Assign *R/S* configurations to the following molecules:

(a)   (b)   (c)

The three-dimensional configuration of a stereocentre in a molecule is specified as either *R* or *S*, using the IUPAC-approved convention. There is no correlation between the *R/S* designation of a molecule and the (+) or (−) rotation of the plane polarized light by a substance made of those molecules.

9.11 Enantiomers, Diastereomers, and Meso Stereoisomers

Molecules like those of lactic acid and glyceraldehyde are relatively simple to deal with because each has only one stereocentre and only two enantiomeric forms. The situation becomes more complex, however, for molecules that have more than one stereocentre. Consider the amino acid threonine (2-amino-3-hydroxybutanoic acid). Since molecules of threonine have two stereocentres (C2 and C3), there are four possible stereoisomers, as shown in Figure 9.44. Check for yourself that the *R/S* configurations are correct.

The four stereoisomers of 2-amino-3-hydroxybutanoic acid can be grouped into two pairs of mirror-image enantiomers. The 2*S*,3*S* stereoisomer is the mirror image of 2*R*,3*R*, and the 2*S*,3*R* stereoisomer is the mirror image of 2*R*,3*S*. But what is the relationship between any two molecules that are not mirror images? What, for example, is the relationship between the 2*R*,3*R* isomer and the 2*R*,3*S* isomer? They are stereoisomers, yet they aren't enantiomers. Their relationship is described by a new term—**diastereomer**.

Diastereomers are a pair of stereoisomeric molecules that are not mirror images. Since we used the right-hand/left-hand analogy to describe the relationship between two enantiomers, we might extend the analogy by saying that the relationship between two diastereomers is that of hands from two different people. Your hand and your friend's hand look *similar*, but they aren't identical and they aren't mirror images. The

Molecular Modelling (Odyssey)

e9.17 Recognize diastereomers and enantiomers.

FIGURE 9.44 The four stereoisomers of 2-amino-3-hydroxybutanoic acid.

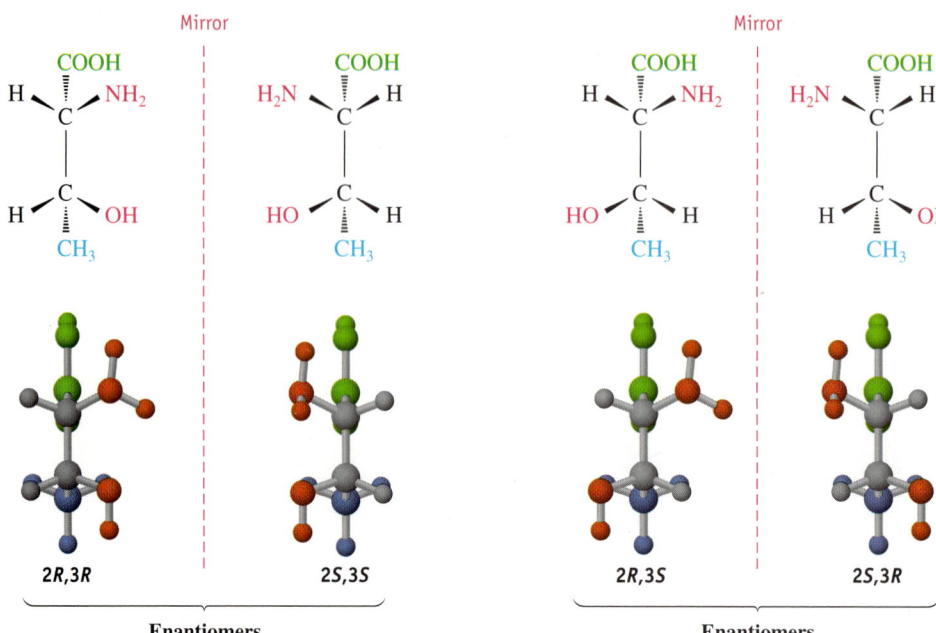

same is true of a pair of diastereomers—they're similar, but they aren't identical and they aren't mirror images.

Note carefully the difference between a pair of molecules that are enantiomers and a pair that are diastereomers: enantiomers have opposite configurations at *all* stereocentres; diastereomers have opposite configurations at *some* (one or more) stereocentres but the same configuration at others. A full description of the four threonine stereoisomers is given in Table 9.3.

Figure 9.47 summarizes these isomeric relationships.

TABLE 9.3 Relationships among the Four Stereoisomers of Threonine Molecules

Stereoisomer	Enantiomeric with	Diastereomeric with
2R,3R	2S,3S	2R,3S and 2S,3R
2S,3S	2R,3R	2R,3S and 2S,3R
2R,3S	2S,3R	2R,3R and 2S,3S
2S,3R	2R,3S	2R,3R and 2S,3S

Of the four stereoisomers of threonine molecules, only the 2S,3R isomer, $[\alpha]_D = -28.3°$, occurs naturally and is an essential human nutrient. Most biologically important organic molecules are chiral and usually only one stereoisomer is found in nature.

EXERCISE 9.23—R/S DESIGNATION

Assign *R* or *S* configuration to each stereocentre in the following molecules:

(a) H⋯C(Br)⋯CH₃ ; H⋯C(OH)⋯CH₃

(b) H⋯C(CH₃)⋯Br ; H₃C⋯C(OH)⋯H

(c) Br⋯C(CH₃)⋯H ; H⋯C(OH)⋯CH₃

EXERCISE 9.24—R/S DESIGNATION

Chloramphenicol is a powerful antibiotic isolated from the *Streptomyces venezuelae* bacterium. It is active against a broad spectrum of bacterial infections and is particularly valuable against typhoid fever. Assign *R* or *S* configuration to the stereocentres in chloramphenicol molecules:

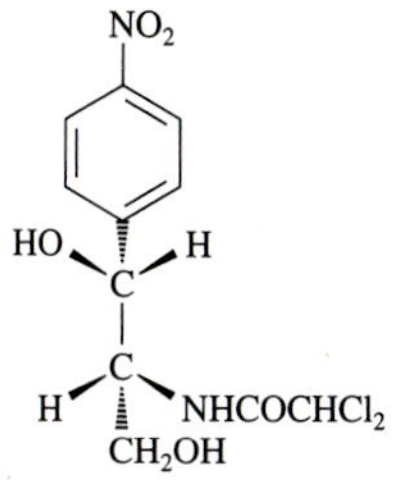

Chloramphenicol
$[\alpha]_D = +18.6°$

Meso Stereoisomers

Let's look at one more example at the related compounds made of molecules with two stereocentres: the tartaric acids used by Pasteur. The four possible stereoisomers can be drawn as follows:

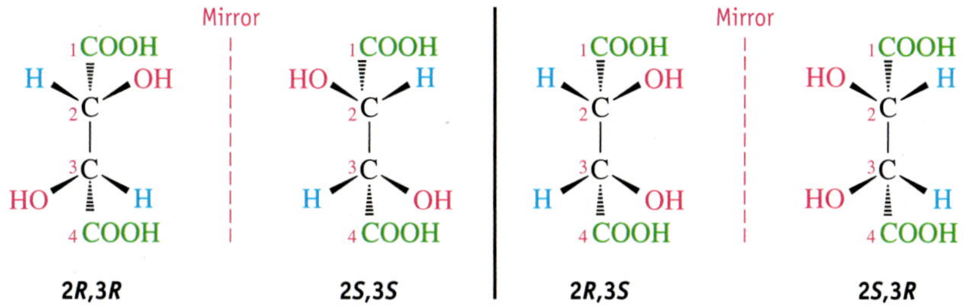

2R,3R   2S,3S   2R,3S   2S,3R

The mirror-image 2R,3R and 2S,3S structures are not identical and therefore represent an enantiomeric pair. A careful look, however, shows that the 2R,3S and 2S,3R structures *are* identical, as can be seen by rotating one structure 180°:

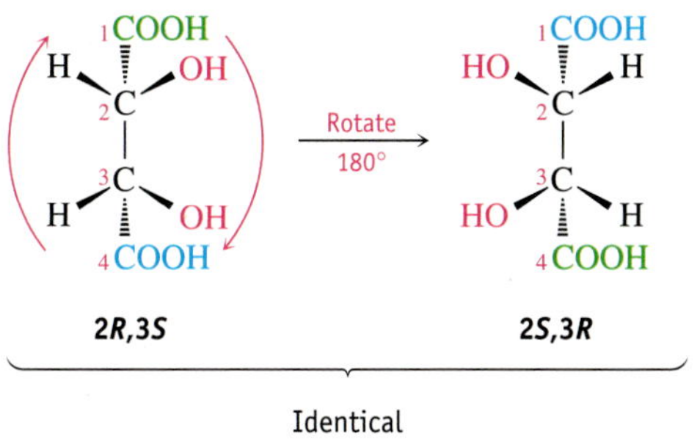

Rotate 180°

2R,3S   2S,3R

Identical

These two molecules are not isomers of any sort, but the same molecule, drawn with two different orientations. The 2R,3S and 2S,3R structures are identical because the molecule has a plane of symmetry and is therefore achiral. The symmetry plane cuts through the C2—C3 bond, making one half of the molecule a mirror image of the other half (Figure 9.45).

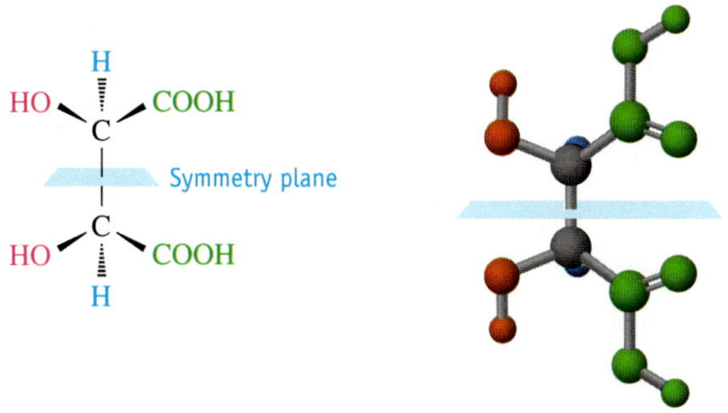

Symmetry plane

FIGURE 9.45 A symmetry plane cutting through the C2—C3 bond of a *meso*-tartaric acid molecule shows that it is achiral.

Because of the plane of symmetry, the tartaric acid stereoisomer shown in Figure 9.45 *is* achiral, despite the fact that it has two stereocentres. *Such molecules that are achiral, yet contain stereocentres, are called meso* (me-zo) *stereoisomers.* Thus, tartaric acid molecules exist in three stereoisomeric forms: a pair of enantiomers and one meso form.

Since tartaric acid molecules exist in different stereoisomeric configurations, the question arises whether substances made of these different stereoisomers have different physical properties. Some physical properties of substances comprised of the three stereoisomers of tartaric acid and of the racemate are shown in Table 9.4. The (+) and (−) substances, each made of molecules of only one of the enantiomers, have identical melting points, solubilities, and densities. They differ only in the sign of their rotation of plane-polarized light and in their interactions in chiral environments. The meso isomer, by contrast, is diastereomeric with the 2*R*,3*R* and 2*S*,3*S* forms. It is therefore a different molecule and will form a substance with different physical properties. The racemate is different in another way. Although made of a 50:50 mixture of enantiomers, racemates act as though they were pure substances, different from substances made of either enantiomer or from the meso form.

TABLE 9.4 Some Properties of Substances of the Stereoisomers of Tartaric Acid

Stereoisomer	Melting Point (°C)	$[\alpha]_D$ (Degrees)	Density (g cm^{-3})	Solubility at 20°C (g 100 mL^{-1} H$_2$O)
(+)	168–170	+12	1.7598	139.0
(−)	168–170	−12	1.7598	139.0
Meso	146–148	0	1.6660	125.0
(±)	206	0	1.7880	20.6

Our analogy of human hands can help again here in thinking about the differences between these substances at the molecular level. The interactions within a stack of 10 right hands (like a substance made of many molecules of one enantiomer) would be the same as the interactions within a stack of 10 left hands (like a substance made of many molecules of the other enantiomer). But those interactions would be different than those within a stack of 10 alternating right and left hands (like a racemate).

WORKED EXAMPLE 9.13—CHIRAL MOLECULES

Does a *cis*-1,2-dimethylcyclobutane molecule have any stereocentres? Is it a chiral molecule?

Strategy

To see whether a stereocentre is present, look for a carbon atom bonded to four different groups. To see whether the molecule is chiral, look for a symmetry plane. Not all molecules with stereocentres are chiral—meso molecules are an exception.

Symmetry plane

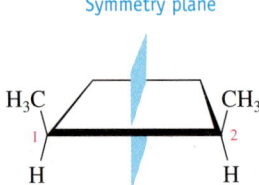

Solution

Looking at the structure of a *cis*-1,2-dimethylcyclobutane molecule, we see that both of the methyl-bearing ring carbon atoms (C1 and C2) are stereocentres. Overall, though, the molecule is achiral because there is a symmetry plane bisecting the ring between C1 and C2. Thus, *cis*-1,2-dimethylcyclobutane is a meso stereoisomer.

EXERCISE 9.25—MESO STEREOISOMERS

Which of the following molecules have meso stereoisomers?
(a) 2,3-dibromobutane   (b) 2,3-dibromopentane   (c) 2,4-dibromopentane

EXERCISE 9.26—MESO STEREOISOMERS

Which of the following structures represent meso molecules?

(a) 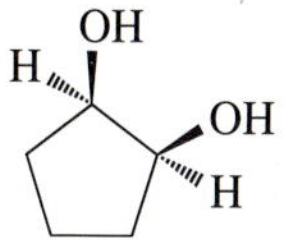

(b) 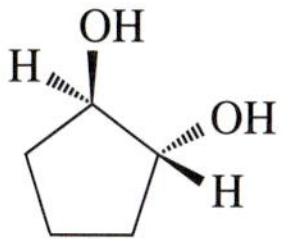

(c) 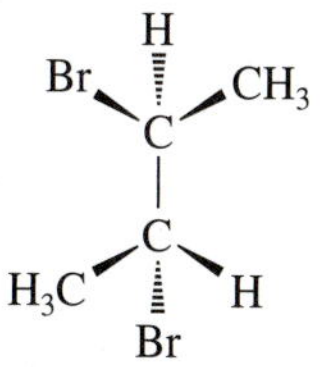

Enantiomer pairs have opposite configurations at all stereocentres, whereas diastereomer pairs have the same configuration for at least one centre but opposite configurations at the others. Meso stereoisomers contain stereocentres but are achiral molecules because they contain a plane of symmetry. Racemates are 50:50 mixtures of (+) and (−) enantiomers. Racemates and substances made of individual diastereomers differ in both their physical and biological properties.

The Use of ^{13}C NMR to Distinguish Diastereomers and Enantiomers

Since (+)-tartaric acid and (−)-tartaric acid are comprised of enantiomeric molecules, samples of each would give identical NMR spectra. Meso-tartaric acid molecules, however, have a diastereomeric relationship with either (+)-tartaric acid or (−)-tartaric acid molecules, and the ^{13}C nuclei in substances of those molecules would therefore be expected to give slightly different chemical shifts than for either of the enantiomeric pair. The chemical shifts for (+)-, (−)-, and meso-tartaric acid are summarized in Table 9.5.

TABLE 9.5 ^{13}C NMR Chemical Shifts of Tartaric Acid Stereoisomers

Stereoisomer	^{13}C NMR Chemical Shifts
(+)-tartaric acid	175.27 δ, 72.68 δ
(−)-tartaric acid	175.27 δ, 72.68 δ
meso-tartaric acid	174.80 δ, 73.31 δ

9.12 Molecules with More Than Two Stereocentres

One stereocentre gives rise to two stereoisomers (one pair of enantiomers), and two stereocentres give rise to a maximum of four stereoisomers (two pairs of enantiomers). In some cases, meso compounds can be formed which reduce the actual number of stereoisomers. *In general, a molecule with n stereocentres has a maximum of 2^n stereoisomers (2^{n-1} pairs of enantiomers).* For example, cholesterol has eight stereocentres. Thus, $2^8 = 256$ stereoisomers of cholesterol, or 128 pairs of enantiomers, are possible in principle, although many would be too strained to exist. Only one, however, is produced in nature—a remarkable demonstration of the sophistication of molecular recognition processes by chiral molecules in living systems.

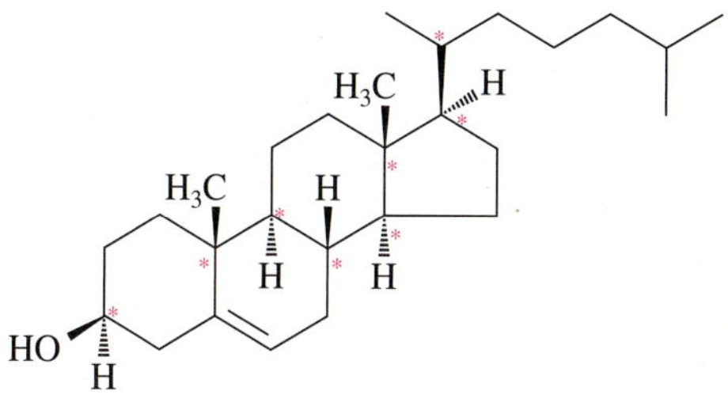

Cholesterol (eight stereocentres)

EXERCISE 9.27—STEREOISOMERS

Nandrolone is an anabolic steroid used by some athletes to build muscle mass. How many stereocentres do nandrolone molecules have? What is the maximum number of stereoisomers of nandrolone that are possible in principle?

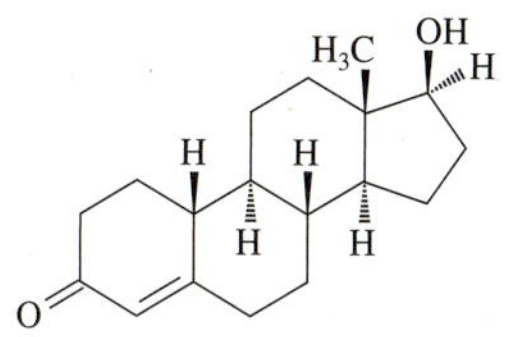

Nandrolone

9.13 Chiral Environments in Laboratories and Living Systems

Taking It Further

e9.18 See how you can separate enantiomeric substances.

Perhaps the single most important idea that should be distilled from this chapter is the importance of molecular recognition, and how this is determined by the overall shape and chirality of molecules. We now know that enantiomeric substances have identical physical properties, such as melting and boiling points, and common spectral features. Two compounds made of different diastereomers, by contrast, behave as completely different substances, and can be distinguished by NMR. Similarly, in the presence of another source of chirality (a chiral environment), compounds made of different enantiomers exhibit very different properties. Chiral environments can now routinely be generated in the laboratory through the use of chiral materials in chromatography columns, or through careful selection of chemical reactions involving chiral reagents.

The human body is, of course, a highly chiral environment. The enantiomers giving rise to the (+) compound of limonene give oranges their odour, but the other enantiomer that forms the (−) compound of limonene is found in lemons. The differences in odour result from different interactions with the chiral environment inside living cells in your nose.

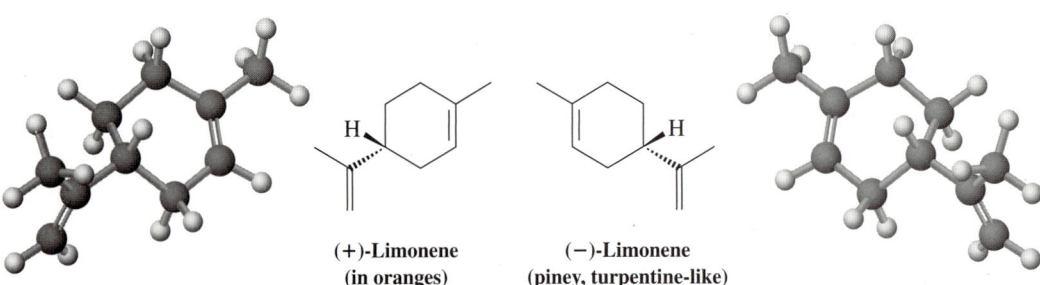

(+)-Limonene
(in oranges)

(−)-Limonene
(piney, turpentine-like)

More dramatic examples of how a change in chirality can affect molecular recognition and biological activity are found in many drugs, such as fluoxetine, a heavily prescribed medication sold under the trade name Prozac. Racemic fluoxetine is an extraordinarily effective antidepressant, but it has no activity against migraine headaches. The compound made of the pure *S* enantiomer, however, works remarkably well in preventing migraines and is now undergoing clinical evaluation.

Approved pharmaceutical agents come from many sources. Some are isolated directly from plants or bacteria, others are made by chemical modification of naturally occurring compounds, and still others are made entirely in the laboratory and have no relatives in nature.

Those drugs that come from natural sources, both directly or after chemical modification, are usually chiral and are generally found only as a single enantiomer rather than as a racemate. Penicillin V, for example, an antibiotic isolated from the *Penicillium* mould, has molecules with a 2S,5R,6R configuration. The compound of its enantiomer, which does not occur naturally but can be made in the laboratory, has essentially no biological activity.

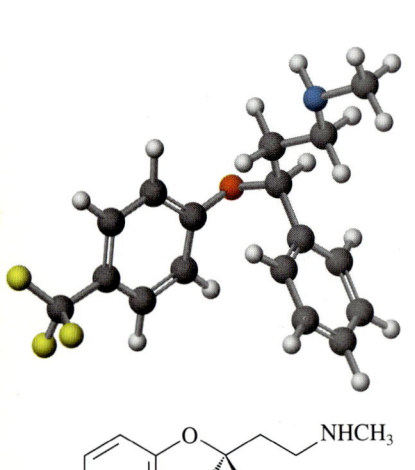

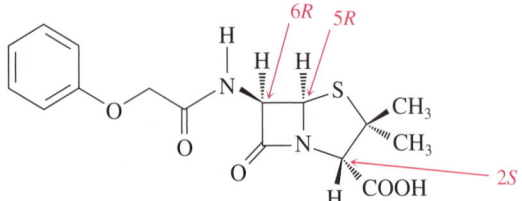

**(S)-Fluoxetine
(prevents migraine)**

Penicillin V (2S,5R,6R configuration)

In contrast to drugs from natural sources, in the past, drugs made entirely in the laboratory, if chiral, have usually been produced and sold as racemates. Ibuprofen, for example, contains one stereocentre, but only the compound from the *S* enantiomer is an analgesic/anti-inflammatory agent useful in treating aches and pains. While the *R* enantiomer of ibuprofen is inactive, an isomerase enzyme is able to slowly convert the *R* to the active *S* form. For this reason, the substance marketed under such trade names as Advil, Nuprin, and Motrin is a racemate of *R* and *S*.

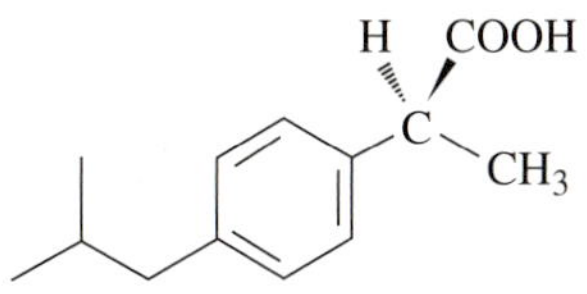

(S)-Ibuprofen

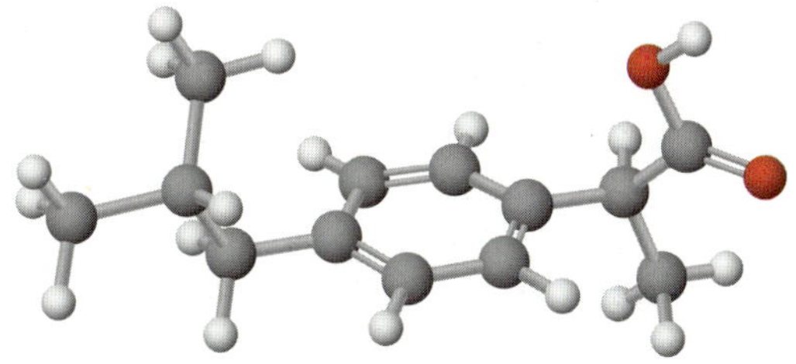

Not only is it wasteful to synthesize and administer a physiologically inactive enantiomer, many examples are now known where the presence of molecules of the "wrong" enantiomer in a racemate either affects the body's ability to utilize the "right" enantiomer or has unintended effects of its own. The presence of (*R*)-ibuprofen in the racemate, for instance, seems to slow substantially the rate at which the *S* enantiomer takes effect. Tragic lessons were learned from the synthesis and marketing of a racemate, thalidomide [>>Question 9.75], in the early 1960s as a sedative. When used by pregnant women, the racemate caused severe neurological damage in about 8000 newborns, including the deformity called phocomelia (from the Greek *phoke* = "seal," *melos* = "limb"), in which hands are attached directly to the shoulders and feet to the hips. Subsequent studies on rat embryos suggest that the birth defects may have resulted from only the (+) enantiomer.

To fully utilize the chiral environment of the human body to optimize the uptake and use of pharmaceutical products, chemists have now devised many methods of *enantioselective synthesis,* which allows them to prepare compounds containing only a single enantiomer rather than a racemate. Other chiral drugs can be resolved from a racemate using chiral chromatography or chemical reactions with chiral reagents. The production of single enantiomer drugs has now become a huge business, with annual sales estimated to exceed $175 billion dollars globally. Creative marketing emphasizes the different molecular recognition caused by the mirror-image relationship of enantiomers. Darvon, the 2*R*,3*S*(+) enantiomer of propoxyphene, is marketed by Eli Lily in an analgesic formulation. The 2*S*,3*R*(−) enantiomer, in a cough suppressant formulation, has been given the mirror-image trade name Novrad.

EXERCISE 9.28—STEREOISOMERS

A skeletal structure for a propoxyphene molecule is shown below. C2 and C3 are labelled, and the dashed lines mean that the stereochemistry is not shown at these two atoms. Draw three-dimensional representations of Darvon and Novrad molecules.

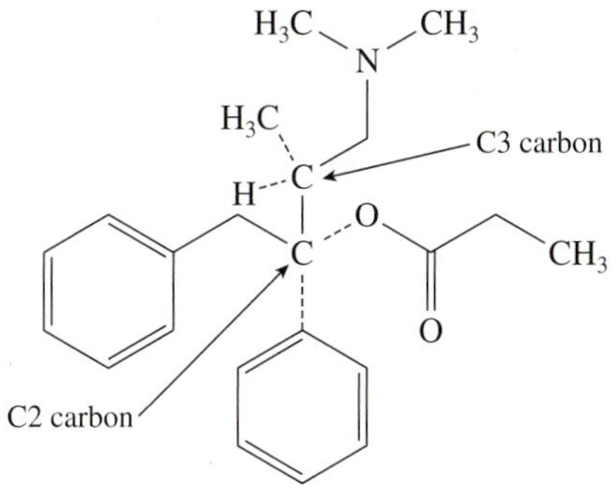

Propoxyphene

Why do compounds made of different stereoisomers like (+)- and (−)-carvone have different biological properties, such as odour? To exert its biological action, a chiral molecule must be recognized by a chiral receptor at a target site, much as a hand fits into a glove. But just as a right hand can fit only into a right-hand glove, so a particular stereoisomer is recognized only by a receptor having the proper complementary shape. Any other stereoisomer will be a misfit, like a right hand in a left-handed glove. A greatly simplified, "cartoon" representation of the interaction between a chiral molecule and a chiral biological receptor is shown in Figure 9.46. One enantiomer fits the receptor perfectly, but the other does not.

FIGURE 9.46 In this "cartoon" representation, (a) one enantiomer easily recognizes a chiral receptor site, and is able to exert its biological effect, but (b) the other enantiomer has the wrong stereochemistry for molecular recognition by the same receptor.

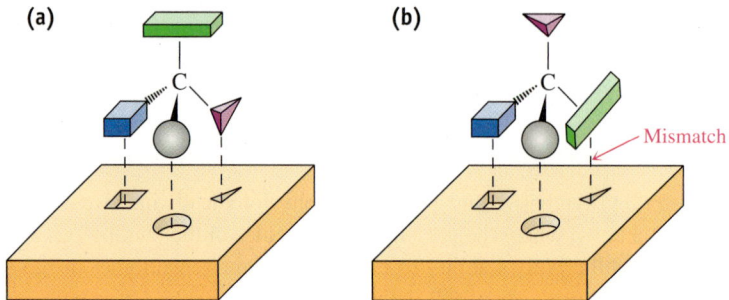

It would be of great human and commercial interest to the food, wine, perfume, and other industries if receptors for odour (olfactory receptors) such as those represented simplistically above could be systematically identified and their molecular structures completely characterized. We could then begin to understand better how molecular shape and polarity affect the odour of molecules. And significant steps have been taken toward achieving this goal. But the complexity of the chiral environment in living cells in the human body, including the nose, continues to defy easy characterization and understanding. For example, studies show that one human olfactory receptor is also found in sperm as well as in the nose, and a strong response to this receptor comes when it interacts with an aldehyde that is not naturally found in either the nose or in sperm. We do know that each of the thousand or more odour receptors found at the top of the nasal passage respond to a range of odourant molecules, and each odourant molecule can bind to a range of receptor types. Molecular recognition is certain to remain a fascinating and elusive area of study throughout your lifetime.

References

To learn more about reversible adhesion in nature and the laboratory, consult K. Autumn, "Gecko Adhesion: Structure, Function and Applications," *Materials Research Bulletin* 32 (June 2007): 473–78, and the rest of this special bulletin issue. For more on chiral drugs, see E. Thall, "When Drug Molecules Look in the Mirror," *Journal of Chemical Education* 73 (1996): 481–84.

SUMMARY

Key Concepts

- **Molecular recognition** is fundamental to a myriad of processes in living systems, and is controlled by the overall shapes and polarity of molecules as they interact with each other. The **chirality**, or handedness, of molecules is also of fundamental importance in molecular recognition. (Sections 9.1, 9.8)
- Instrumental techniques, including mass spectrometry, infrared spectroscopy, X-ray crystallography, nuclear magnetic resonance spectroscopy, and polarimetry, provide key experimental tools that are used to determine the molecular formulas, functional groups, three-dimensional arrangement of atoms, a map of the carbon framework, and the **stereochemistry** of molecules. (Section 9.2)
- X-ray crystallography can provide an accurate picture of the position of each atom in three-dimensional space in molecules of substances in the solid state. From this

information, bond lengths and bond angles can be determined. Recent advances in X-ray source and computer-processing technologies have made it possible to determine three-dimensional structures of many large protein molecules. (Section 9.3)

- Nuclear magnetic resonance (NMR) spectroscopy is the most valuable of the common spectroscopic techniques, as it provides a map of the framework of carbon and hydrogen atoms of molecules in solution. When ^{1}H and ^{13}C nuclei are placed in a magnetic field, their spins orient either with or against the field. On irradiation with radiofrequency (RF) waves, energy is absorbed and the nuclear spins flip from the lower energy state to the higher energy state. This absorption of energy is detected, amplified, and displayed as a NMR spectrum. In this chapter, we focus on two features of the information from NMR spectra:
  - *Number of peaks*. Each non-equivalent kind of ^{1}H or ^{13}C nucleus in a molecule gives rise to a different peak.
  - *Chemical shift*. The exact position of each peak is called its **chemical shift** and is correlated to the chemical environment of each ^{1}H or ^{13}C nucleus. Most ^{13}C absorptions fall in the range 0 to 220 δ downfield from the TMS reference signal, and most ^{1}H absorptions come between 0 and 10 δ. (Section 9.4)
- Rapid rotation is possible about C—C single bonds in most alkane molecules, which can rotate through many possible **conformations**. Conformations that correspond to energy minima for a molecule are referred to as **conformers**. A staggered conformer for alkane molecules is more stable than an eclipsed conformation, minimizing the repulsion between electron pairs. (Section 9.5)
- Restricted rotation about C—C single bonds in cycloalkane molecules and double bonds in alkene molecules helps to control the overall shape of molecules, and in some cases creates *cis–trans* stereoisomers. In a *cis* isomer, both substituents are on the same side of a cycloalkane ring or the plane of a C=C bond, whereas in a *trans* isomer, the substituents are on opposite sides of the ring or the plane of a C=C bond. (Section 9.6)
- A molecule that is not identical to its mirror image is said to be **chiral**, meaning "handed." A chiral molecule is one that does not contain a plane of symmetry. The usual cause of chirality is the presence of a tetrahedral atom bonded to four different groups—a so-called **stereocentre**. This can also be a nitrogen or sulfur atom, or a transition metal. Chiral compounds can exist as a pair of mirror-image stereoisomers called enantiomers, which are related to each other as a right hand is related to a left hand. When a beam of plane-polarized light is passed through a solution of a pure enantiomer, the plane of polarization is rotated and the compound is said to be **optically active**. (Sections 9.7–9.9)
- The three-dimensional **configuration** of a stereocentre in a molecule is specified as either *R* or *S*. Sequence rules are used to assign priorities to the four substituents on the chiral carbon, and the molecule is then oriented so that the lowest-priority group points directly away from the viewer. If an arrow drawn in the direction of decreasing priority for the remaining three groups is clockwise, the stereocentre has the *R* configuration. If the direction is counterclockwise, the stereocentre has the *S* configuration. There is no correlation between the *R/S* designation of a molecule and the (+) or (−) rotation of the plane-polarized light by a substance made of those molecules. (Section 9.10)
- The term **isomer** always refers to a relationship between molecular entities. It doesn't make sense to describe a single molecule as an isomer—it will have an isomeric relationship to one or more other molecules. (Figure 9.47)
- **Constitutional isomers** are pairs of molecules whose atoms are connected differently. Among the sub-categories of constitutional isomers are skeletal, **functional group**, and positional isomers. (Section 3.9, 9.4, Figure 9.47)
- **Stereoisomers** are pairs of molecules whose atoms are connected in the same way but with different arrangements in space. In many molecules, this results from the presence of one or more stereocentres. Among the sub-categories of stereoisomers are **enantiomers** (non-superimposable mirror-image stereoisomers) and **diastereomers** (non-mirror-image stereoisomers). Enantiomers have opposite configurations at all stereocentres, whereas diastereomers with stereocentres have the same configuration at one or more stereocentres but opposite configurations at the others. Meso stereoiso-

mers contain stereocentres but are **achiral** overall because they contain a plane of symmetry. **Racemates** are 50:50 mixtures of (+) and (−) enantiomers. Racemates and substances made of individual diastereomers differ in both their physical and biological properties. *Cis-trans* isomers are just a special kind of diastereomers, because they are non-mirror-image stereoisomers. (Sections 9.6, 9.8, 9.11)

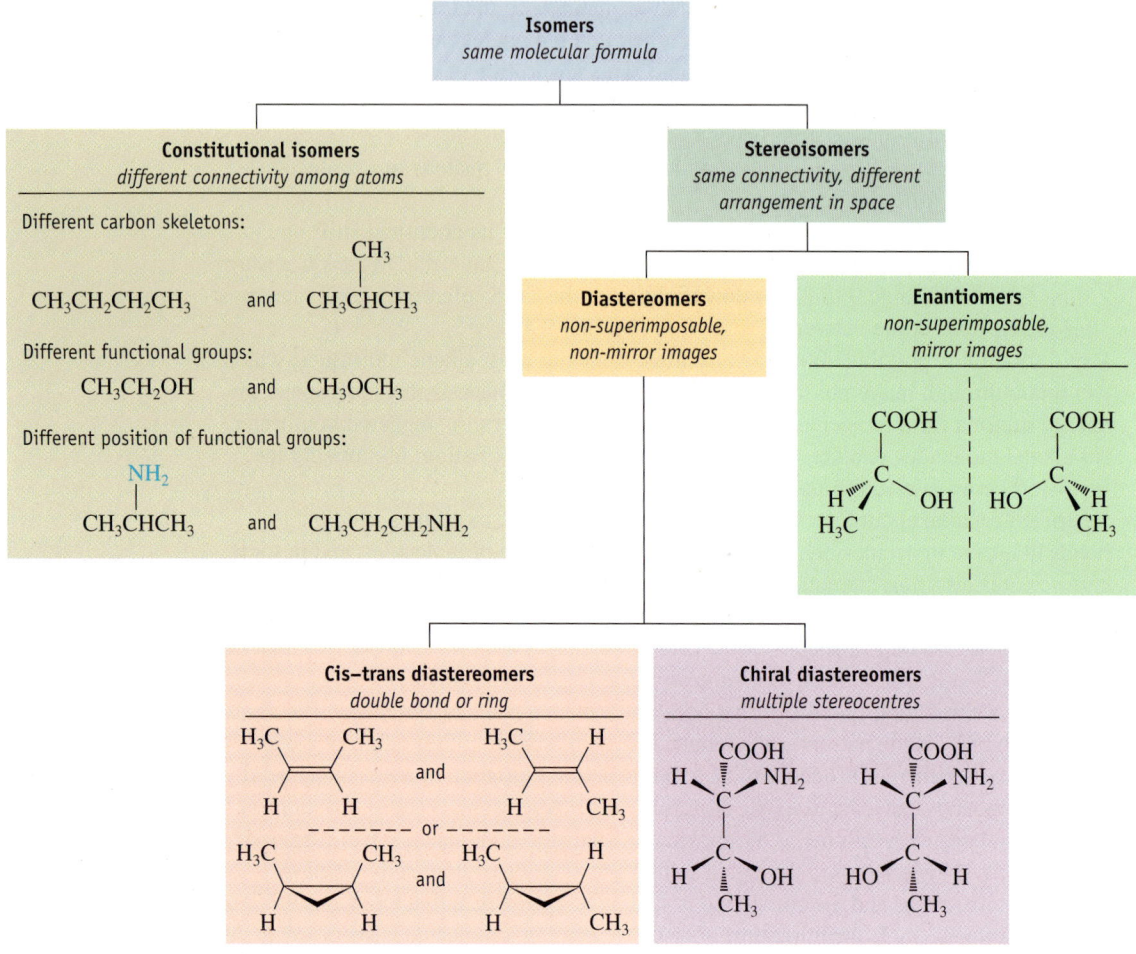

FIGURE 9.47 A summary of the different isomeric relationships.

REVIEW QUESTIONS

Section 9.4: ^{13}C NMR—Mapping the Carbon Framework of Molecules

9.29 Propose structures for compounds whose ^{13}C NMR spectra fit the following descriptions:
(a) a hydrocarbon with seven peaks in its spectrum
(b) a six-carbon compound with only five peaks in its spectrum
(c) a four-carbon compound with three peaks in its spectrum and a carbonyl functional group

9.30 How many peaks would you expect the compound with the following structure to have in its ^{13}C NMR spectrum?

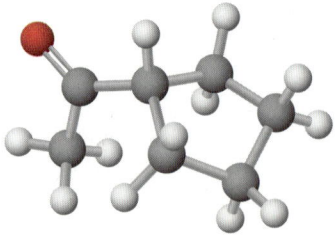

9.31 A common commercial NMR spectrometer uses electromagnetic radiation with a frequency of about 400 MHz. Is this a greater or lesser amount of energy than that used by IR spectroscopy [<< Section 3.11]?

9.32 How many signals would you expect the compound with the following structure to show in its ^{1}H and ^{13}C NMR spectra?

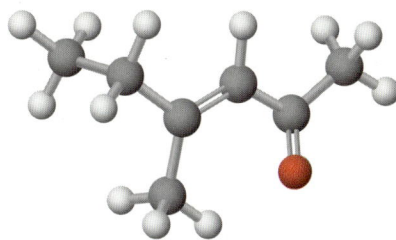

9.33 Sketch what you might expect the ^{13}C NMR spectra of the compound with the following structure to look like (yellow-green = Cl):

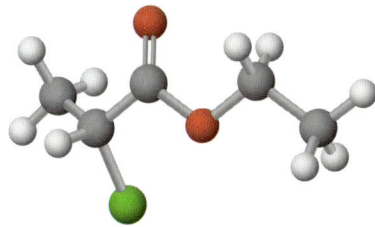

9.34 The following ^{13}C NMR absorptions, given in δ units, were obtained on a spectrometer operating at 400 MHz. Convert the chemical shifts from δ units into Hz downfield from TMS.
(a) 182.1 δ   (b) 23.5 δ   (c) 110.0 δ

9.35 When measured on a spectrometer operating at 400 MHz, chloroform ($CHCl_3$) shows an absorption in its ^{13}C NMR spectrum at 77.23 δ.
(a) How many ppm downfield from TMS does chloroform absorb?
(b) How many Hz downfield from TMS does chloroform absorb if the measurement is carried out on a spectrometer operating at 600 MHz?
(c) What is the position of the chloroform absorption in δ units measured on a 600 MHz spectrometer?

9.36 Describe the ^{13}C NMR spectra you would expect for the compounds with the following formulas:
(a) CH_3CHCl_2 (b) $CH_3CO_2CH_2CH_3$
(c) $(CH_3)_3CCH_2CH_3$

9.37 Compounds with the following formulas all show a single peak in their ^{13}C NMR spectra. List them in order of expected increasing chemical shift: CH_4, CH_2Cl_2, $HC(=O)C(=O)H$, $HC≡CH$, benzene.

9.38 How could you use ^{13}C NMR spectroscopy to distinguish between the following pairs of isomers?
(a) $(CH_3)_3N$ and $CH_3CH_2NHCH_3$
(b) CH_3COCH_3 and $CH_2=CHCH_2OH$
(c) CH_3COCH_3 and CH_3CH_2CHO

Section 9.5: Conformations of Alkanes—Rotation about Single Bonds

9.39 Propose skeletal structures for the following molecular formulas:
(a) C_4H_8 (b) C_3H_6O (c) C_4H_9Cl

9.40 The following molecular model is a representation of *para*-aminobenzoic acid (PABA), the active ingredient in many sunscreens. Indicate the positions of the multiple bonds and draw a skeletal structure (grey = C, red = O, blue = N, ivory = H).

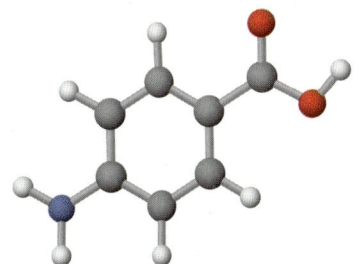

9.41 Sighting along the C2—C3 bond of 2-methylbutane, there are two different staggered conformers. Draw them both in Newman projections, decide which is more stable, and explain your choice.

9.42 Sighting along the C2—C3 bond of 2-methylbutane (see Question 9.41), there are also two possible eclipsed conformations. Draw them both in Newman projections, decide which you think is lower in energy, and explain.

9.43 The barrier to rotation about the C—C bond in bromoethane is 15.0 kJ mol^{-1}. If each hydrogen-hydrogen interaction in the eclipsed conformation is responsible for 3.8 kJ mol^{-1}, how much is the hydrogen-bromine eclipsing interaction responsible for?

Section 9.6: Restricted Rotation about Bonds

9.44 Neglecting *cis–trans* isomers, there are five substances with the formula C_4H_8. Draw their structures.

9.45 Which of the molecules you drew in Question 9.44 show *cis–trans* isomerism? Draw and name their *cis–trans* isomers.

Section 9.7: Cyclic Molecules

9.46 Draw *cis*-1,2-dichlorocyclohexane in a chair conformer, and explain why one group must be axial and one equatorial.

9.47 Draw *trans*-1,2-dichlorocyclohexane in a chair conformer, and explain why both groups must be axial or both equatorial.

9.48 On the following molecule, identify each substituent as axial or equatorial, and decide whether the conformer shown is the more stable or less stable chair form (grey = C, yellow-green = Cl, ivory = H).

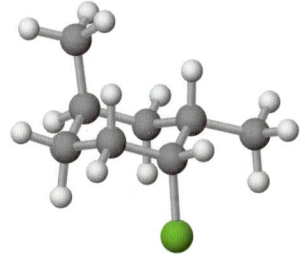

9.49 A trisubstituted cyclohexane with three substituents—red, green, and blue—undergoes a ring-flip to its alternative chair conformer. Identify each substituent as axial or equatorial, and show the positions occupied by the three substituents in the ring-flipped form.

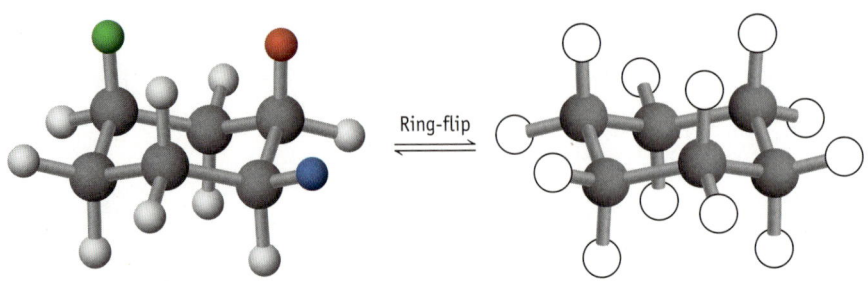

Ring-flip

9.50 *cis*-1-*tert*-Butyl-4-methylcyclohexane exists almost exclusively in the conformer shown. What does this tell you about the relative sizes of a *tert*-butyl substituent and a methyl substituent?

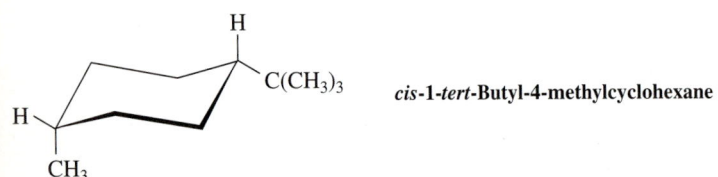

cis-**1-*tert*-Butyl-4-methylcyclohexane**

9.51 Draw *trans*-1,2-dimethylcyclohexane in its more stable chair conformer. Are the methyl groups axial or equatorial?

9.52 Draw *cis*-1,2-dimethylcyclohexane in its more stable chair conformer. Are the methyl groups axial or equatorial? Which is more stable: *cis*-1,2-dimethylcyclohexane or *trans*-1,2-dimethylcyclohexane (Question 9.51)? Explain.

9.53 *N*-Methylpiperidine has the conformer shown below. What does this tell you about the relative steric requirements of a methyl group versus an electron lone pair?

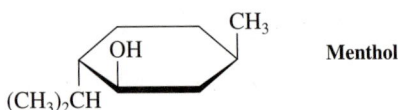

N-**Methylpiperidine**

9.54 One of the two chair conformers of *cis*-1-chloro-3-methylcyclohexane is more stable than the other by 15.5 kJ mol^{-1} (3.7 kcal mol^{-1}). Which is it?

9.55 Draw the three *cis*–*trans* isomers of menthol.

Menthol

$(CH_3)_2CH$

Section 9.8: Stereochemistry

9.56 Which of the following structures are identical? (Red = O, yellow-green = Cl)

(a)   **(b)**

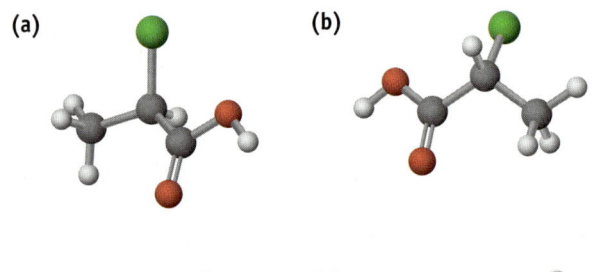

(c)   **(d)**

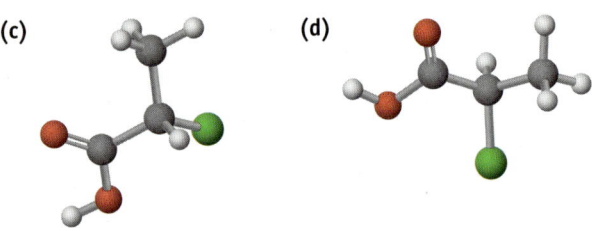

9.57 Which of the following molecules are chiral? Label all stereocentres.

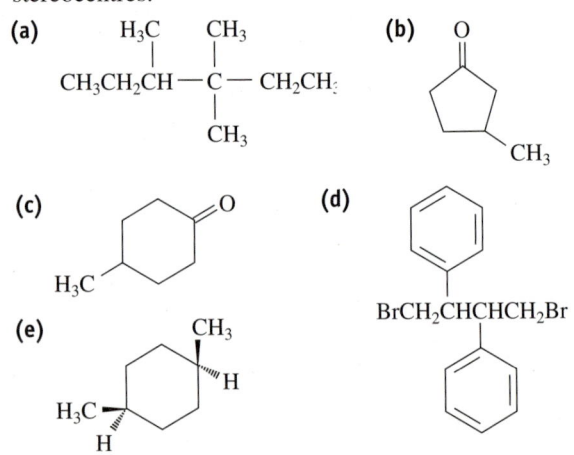

9.58 There are eight alcohols with the formula $C_5H_{12}O$. Draw their structures, and indicate which are chiral.

Section 9.9: Optical Activity

9.59 Cholic acid, the major steroid found in bile, was found to have a rotation of +2.22° when a 3.00 g sample was dissolved in 5.00 mL of alcohol in a sample tube with a 1.00 cm path length. Calculate $[\alpha]_D$ for cholic acid.

9.60 Polarimeters are so sensitive that they can measure rotations to the thousandth of a degree, an important advantage when only small amounts of a sample are available. For example, when 7.00 mg of ecdysone, an insect hormone that controls moulting in the silkworm moth, was dissolved in 1.00 mL of chloroform in a cell with a 2.00 cm path length, an observed rotation of +0.087° was found. Calculate $[\alpha]_D$ for ecdysone.

Section 9.10: Sequence Rules for Specifying Configuration

9.61 Assign *R* or *S* configuration to each stereocentre in pseudoephedrine, an over-the-counter decongestant found in cold remedies (red = O, blue = N).

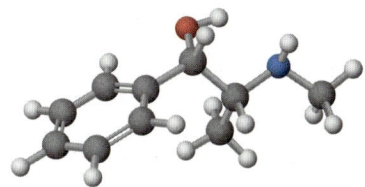

9.62 Assign priorities to the substituents in each of the following sets:
(a) —H, —OH, —OCH₃, —CH₃
(b) —Br, —CH₃, —CH₂Br, —Cl
(c) —CH=CH₂,   —CH(CH₃)₂,   —C(CH₃)₃,
    —CH₂CH₃
(d) —COOCH₃, —COCH₃, —CH₂OCH₃, —OCH₃

9.63 Assign priorities to the substituents in each of the following sets:

(a)

$$\text{—CH}_2\overset{\overset{\displaystyle CH_3}{|}}{\underset{\underset{\displaystyle CH_3}{|}}{C}}CH_3 \qquad \text{—}\bigcirc \qquad \text{—CH}_2\overset{\overset{\displaystyle CH_3}{|}}{C}HCH_2CH_3 \qquad \text{—CH}_2CH_2CH_2CH_2CH_3$$

(b) —SH   —NH₂   —SO₃H   —OCH₂CH₂OH

9.64 Draw tetrahedral representations of both enantiomers of the amino acid serine. State which of your structures is *S* and which is *R*.

$$\underset{\underset{\displaystyle NH_2}{|}}{\text{HOCH}_2\text{CH}}\overset{\overset{\displaystyle O}{\|}}{\text{C}}\text{OH} \qquad \textbf{Serine}$$

9.65 Assign *R* or *S* configuration to the stereocentres in the following molecules:

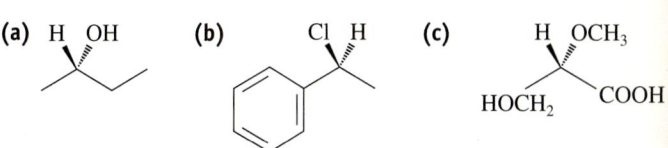

Section 9.11: Enantiomers, Diastereomers, and Meso Stereoisomers

9.66 Which, if any, of the following structures represent meso stereoisomers (red = O, blue = N, yellow-green = Cl)?

(a)

(b)

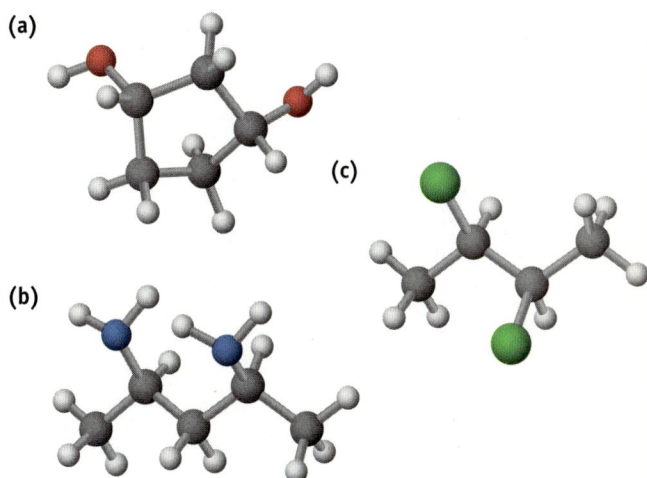

(c)

9.67 What is the relationship between the specific rotations of (2*R*,3*R*)-dihydroxypentane and (2*S*,3*S*)-dihydroxypentane? Between (2*R*,3*S*)-dihydroxypentane and (2*R*,3*R*)-dihydroxypentane?

9.68 What is the stereochemical configuration of the enantiomer of (2*S*,4*R*)-dibromooctane?

9.69 What are the stereochemical configurations of the two diastereomers of (2*S*,4*R*)-dibromooctane?

9.70 Draw examples of the following:
(a) a meso molecule with the formula C₈H₁₈
(b) a compound with two stereocentres, one *R* and the other *S*

Section 9.12: Molecules with More Than Two Stereocentres

9.71 Ribose, an essential part of ribonucleic acid (RNA), has the following structure:

Ribose

How many stereocentres does ribose have? Identify them with asterisks. How many stereoisomers of ribose are there?

9.72 Draw the structure of the enantiomer of ribose (see Question 9.71).

9.73 Draw the structure of a diastereomer of ribose (see Question 9.71).

9.74 On catalytic hydrogenation over a platinum catalyst, ribose (see Question 9.71) is converted into ribitol. Is ribitol optically active or inactive? Explain.

Ribitol

Section 9.13: Chiral Environments in Laboratories and Living Systems

9.75 One stereoisomer of thalidomide causes birth defects. (a) Identify the stereocentres in the two enantiomers of thalidomide. (b) Assign the stereocentres in each molecule as *R* or *S*.

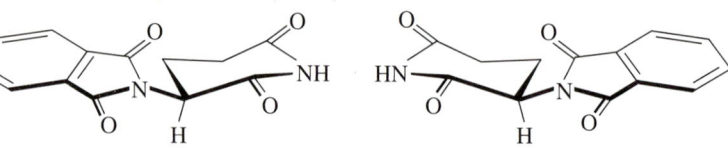

Thalidomide

9.76 The cancer drug Taxol is rich in stereochemistry:
(a) Identify all the stereocentres in Taxol.
(b) Calculate the maximum number of stereoisomers that could exist for Taxol.
(c) Nature produces only one isomer! How is this possible?

Taxol

SUMMARY AND CONCEPTUAL QUESTIONS

9.77 Amantadine is an anti-viral agent that is active against influenza A infection. Draw a three-dimensional representation of amantadine showing the chair cyclohexane rings.

Amantadine

9.78 NMR spectroscopy was one of the key experimental tools used by Gavin Flematti to identify Gavinone [<< Section 1.3]. Sketch the expected ^{13}C NMR spectrum of Gavinone.

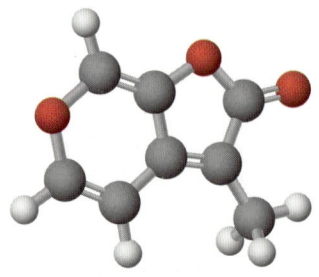

Gavinone

9.79 Draw two constitutional isomers of *cis*-1,2-dibromocyclopentane.

9.80 Draw a stereoisomer of *trans*-1,3-dimethylcyclobutane.

9.81 There are two different substances named *trans*-1,2-dimethylcyclopentane. What is the relationship between them?

9.82 Which of the following objects are chiral?
(a) basketball
(b) wine glass
(c) ear
(d) snowflake
(e) coin
(f) scissors

9.83 Penicillin V is a broad-spectrum antibiotic that contains three stereocentres. Identify them with asterisks.

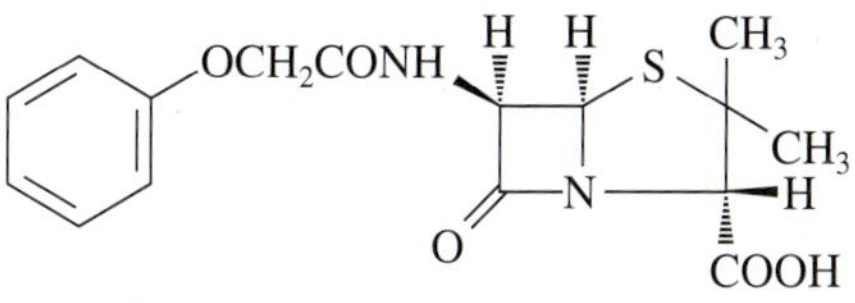

Penicillin V (antibiotic)

9.84 Draw chiral molecules that meet the following descriptions:
(a) a chloroalkane, $C_5H_{11}Cl$
(b) an alcohol, $C_6H_{14}O$
(c) an alkene, C_6H_{12}
(d) an alkane, C_8H_{18}

9.85 Naturally occurring (*S*)-serine has $[\alpha]_D = -6.83°$. What specific rotation do you expect for (*R*)-serine?

9.86 Draw a tetrahedral representation of a (*R*)-3-chloro-1-pentene molecule.

9.87 Draw all the stereoisomers of 1,2-dimethylcyclopentane. Assign *R*/*S* configurations to the stereocentres in all isomers, and indicate which stereoisomers are chiral and which, if any, are meso.

9.88 Using 3D representations, draw the meso form of each of the following molecules, and indicate the plane of symmetry in each:

(a) $CH_3CHCH_2CH_2CHCH_3$ with OH OH

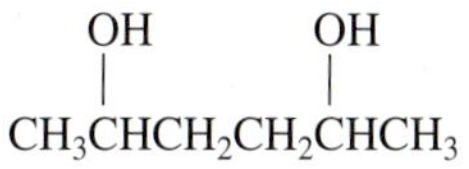

(b)

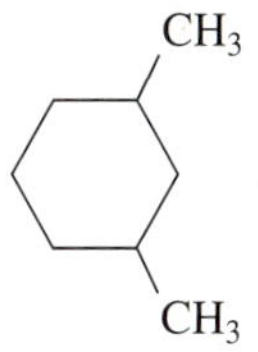

(c)

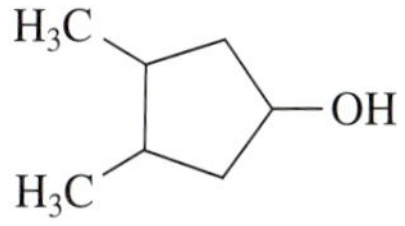

9.89 Assign *R* or *S* configuration to each stereocentre in the following molecules:

(a)

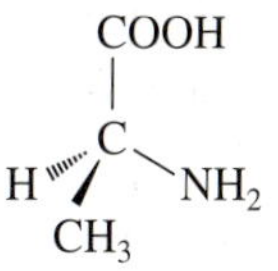

(b)

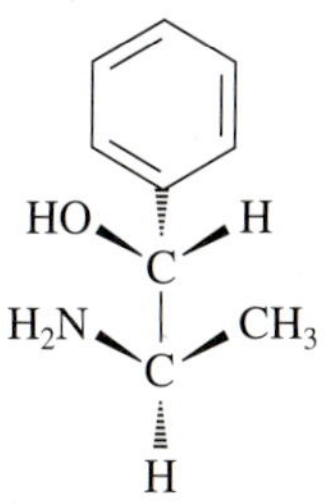

9.90 How many stereoisomers of 2,4-dibromo-3-chloropentane are there? Draw them and indicate which would make up an optically active sample.

9.91 Being as specific as possible, what kinds of isomers are the following pairs?
(a) (*S*)-5-chlorohex-2-ene and chlorocyclohexane
(b) (2*R*,3*R*)-dibromopentane and (2*S*,3*R*)-dibromopentane

9.92 State whether the following pairs of compounds are identical, constitutional isomers, or stereoisomers.
(a) *cis*-1,3-dibromocyclohexane and *trans*-1,4-dibromocyclohexane
(b) 2,3-dimethylhexane and 2,5,5-trimethylpentane

(c)

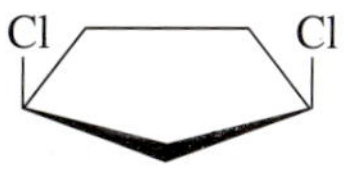

and

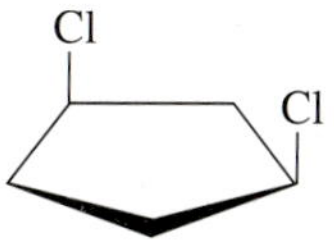

9.93 State whether the following Newman projection of 2-chlorobutane is *R* or *S*.

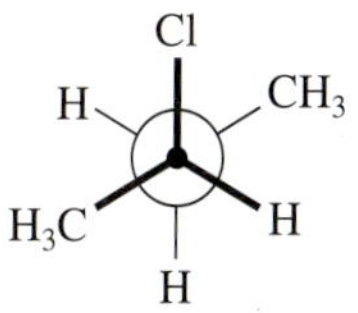

9.94 Draw a Newman projection that is enantiomeric with the one shown in Question 9.93.

10.1 Case Study: Breathing the Life of Imagination into Chemistry's Facts

Martin Shields/Getstock.com

What do you picture when you think of the substance methane? In your everyday experience with natural gas stoves or furnaces, methane is an odourless and colourless gas that reacts exothermically with oxygen in air to keep you warm and cook your food. In our economy and environment, natural gas pipelines carry methane, cows belch it into the atmosphere, and bacteria produce bubbles of it in pools of tarry asphalt, such as in the photo above from the LaBrea tar pits in Los Angeles. When methane is released into the atmosphere it warms the troposphere [<<Section 4.3]. In the laboratory, when we double the pressure of a sample of methane (keeping temperature constant), the volume of the sample is halved. If we cool a sample to −162 °C, it becomes a liquid. We can experience, observe and sometimes measure all of these facts of methane's chemistry.

> *"Because we are human, we are not satisfied with facts alone; and so there is added to our science the sustained effort to correlate them and breathe into them the life of imagination."*
>
> **Charles Coulson**

But in the human activity of doing chemistry, we aren't satisfied with facts alone. What mental models do you have of the methane gas inside the bubble in the asphalt pool? Can you 'breathe the life of imagination' into one or more **molecular-level** pictures in your mind's eye to help you explain the *observable* properties and reactivity of methane? Perhaps, through the walls of the bubble, you 'see' a sample of methane gas as composed of CH_4 molecules quite far apart from each other, but darting around, all the time vibrating and tumbling, and colliding with each other and the walls of the bubble. That mental model can help you picture the molecules being pushed closer together when we compress a gas sample and rapidly diffusing into the atmosphere when the bubble bursts.

You might get a different molecular-level view by imagining yourself sitting on one of the molecules and looking around at the others. From this perspective you could 'see' that in each molecule, a central carbon atom is joined in an equivalent way to four hydrogen atoms, which are tetrahedrally positioned around it. This model is informed by data obtained with instruments that give evidence about the three dimensional structure of methane molecules. This can be done by placing a sample of methane gas in an electron beam and observing the pattern of diffraction that results, or passing microwave or infrared radiation through a sample to detect which frequencies are absorbed. This molecular-level mental model can help you picture how methane molecules contribute to atmospheric warming, and how they might fit inside the cages of a clathrate hydrate [<<Section 4.1]

Now what might you see in your mind's eye if you could zoom in to take even a closer look at a single methane molecule? In this chapter we develop various mental models that come from the perspective of sitting on a methane molecule, but turning our attention from the other molecules to looking *within* the methane molecule: an **intramolecular level** view. While we accept that each methane molecule has one carbon nucleus and four hydrogen nuclei, each with their 'core' electrons, at this intramolecular level we focus on the distribution of the valence electrons between and around the nuclei. Here, too, we require our powers of imagination to develop models that help explain certain observed facts. Chemistry is a human activity, and an essential part of doing chemistry is the creative process of proposing how the electrons might be distributed within a molecule, and what happens when a molecule interacts with a molecule of another substance. What we understand about electron distribution within molecules of a substance is based on models; no one can see electrons to know where they are, nor whether they are localized in a particular region or free to roam over parts (or all) of the molecule. Our models of electron distribution need to make sense of experimental evidence. Chemists' curiosity demands that they try to answer questions like the following:

- Why are the nuclei in a methane molecule in their particular spatial positions with respect to each other?
- Why are the nuclei held together as one cluster called a molecule? Put another way, what is the origin of covalent bonding?
- Why is the H—C—H bond angle in methane molecules 109.5°, while the H—C—H bond angle in ethene molecules is 120°?
- Why is the distance between carbon atoms in ethane molecules greater than that between carbon atoms in ethene molecules?
- Why is ammonia a Lewis base, while ammonium ions are not?
- Why is oxygen gas (O_2) paramagnetic (attracted into a magnetic field) while nitrogen gas (N_2) is not?

There are many such questions, and chemists use many mental models to give sensible answers. No single mental model is adequate to explain the many different features of structure and reactivity. Complementary 'good enough' models are needed—very simple models such as Lewis structures rationalize patterns about connectivity, while much more sophisticated models such as molecular orbital theory are required to explain features of photoelectron spectroscopy.

We will return to the example of methane molecules at different times throughout this chapter. Conjure up your own picture of a molecule inside the methane gas bubble. By the end of this chapter, you should be able to see that methane molecule in several different, but complementary ways. Each of those models will be 'good enough' for explaining some observable properties of methane. But are any of the ways of 'seeing' a methane molecule real?

In his 1951 Tilden Lecture, Oxford University chemist Charles A. Coulson, whose work played an important role in developing our current theories of covalent bonding, described a simple chemical bond, such as the C—H bond in methane, as follows.

Charles Alfred Coulson (1910–1974), Oxford University's first Professor of Theoretical Chemistry.

"Sometimes it seems to me that a bond between two atoms has become so real, so tangible, (and) so friendly that I can almost see it. And then I awake with a little shock: for a chemical bond is not a real thing; it does not exist; no-one has ever seen it, no-one ever can. . . . Hydrogen I know, for it is a gas and we keep it in large cylinders; benzene I know, for it is a liquid and we keep it in bottles. The tangible, the real, the solid, is explained by the intangible, the unreal, (and) the purely mental. Yet that is what chemists are always doing . . ."

Imagining *scientific models* that fit patterns of existing experimental data and allow useful predictions is at the heart of doing chemistry. The rest of this chapter explains several complementary models for covalent bonding that are commonly used to explain observations about molecular structures. The simplest possible models are imagined to explain experimental observations about bonding. Then, when new facts of structure and bonding are uncovered, more sophisticated models must often be developed to explain them, but the simpler models are still used where they provide 'good enough' explanations for other observations.

> A bond is a force of attraction holding two atoms together. It is commonly said that a pair of electrons between two nuclei constitutes a covalent bond. We could philosophize forever about whether the bond is the pair of electrons or whether the pair of electrons gives rise to the force that is the bond.

Molecular Modelling (Odyssey)

e10.1 See how electron density isosurfaces can be used to indicate covalent bonds in a molecule.

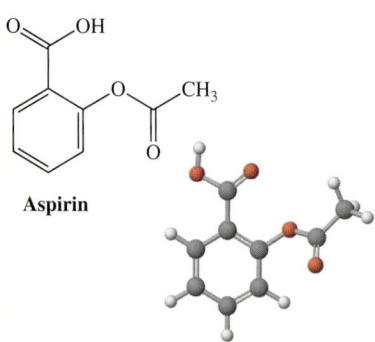

Aspirin

Molecular Modelling (Odyssey)

e10.2 Watch a simulation of flexibility within an aspirin molecule.

10.2 Covalent Bonding in Molecules

It is commonly accepted among chemists that only the valence electrons of atoms are involved in bonding them to other atoms in the same molecule. This chapter is concerned with covalent bonds, which are thought to result from sharing pairs of electrons between bonded atoms. The bond is the net force of attraction resulting from the combination of nucleus–electron attractions, nucleus–nucleus repulsions, and electron–electron repulsions. An oversimplified view is that two nuclei attracted to the same electron pair are bound together, so that the electron pair is acting as "glue." Not all models of bonding [>>Section 10.9], however, assume that valence electrons are localized between bonded atoms.

It is helpful to think of a continuum of bonding between a "pure" covalent bond (in which the pair of electrons is equally shared by the bonded atoms) and an ionic bond (in which the pair of electrons is unequally shared to the extent that it resides completely on the more electronegative atom). Along this continuum are *polar covalent bonds* (or bonds with fractional ionic character); the more the electron pair is unequally shared between atoms, the more polar the bond. The strength of attraction of atoms for the electron pair is measured on a scale of electronegativity [<<Section 6.3].

In this chapter, we focus on covalent bonding between atoms in molecules and ions, as distinct from network structures. Examples of compounds comprised of molecules with covalent bonds between the atoms include gases in our atmosphere (O_2, N_2, H_2O, and CO_2), common fuels (CH_4), and most of the compounds in our bodies. Covalent bonding is also responsible for the atom-to-atom connections in common ions such as CO_3^{2-}, CN^-, NH_4^+, NO_3^-, and PO_4^{3-}. In many, but not all cases, the molecules or ions are made up of only atoms of non-metals.

Although molecules and ions made up of only a few atoms are used here as examples to develop the basic principles of structure and bonding, the same principles and models of bonding apply to larger molecules, from aspirin to proteins and DNA with thousands of atoms.

> Covalent bonds can be thought of as the net force of attraction from sharing of electron pairs between bonded atoms. Polar bonds lie along a continuum between "pure" covalent bonds and ionic bonds: the more the electron pair is unequally shared between atoms, the more polar the bond.

10.3 Lewis Structures

The simplest model of accounting for the distribution of valence electrons in molecules are representations called **Lewis structures**, named in honour of American chemist Gilbert Newton Lewis (1875–1946). Covalent bonds are shown as a pair of dots, or a line, between bonded atoms.

A useful model of electron distribution in molecules must explain what we know from experiments. High-resolution mass spectrometry data tells us that the molar mass of methane is consistent with each molecule having one C atom and four H atoms (formula CH_4). We can deduce from IR spectroscopy and gas phase electron diffraction that the single C atom in each molecule is bound to four H atoms, and a zero dipole moment tells us that the molecule is symmetrical. From ^1H NMR spectroscopy [>>Section 19.6] we know that all four H atoms are in identical electronic environments. The following are alternative Lewis structures of methane molecules:

If we regard the bonding electrons as common to each of the bound atoms, the C atom has eight valence electrons (an octet), which is consistent with a "full" $n = 2$ valence shell. And each H atom has two valence electrons, corresponding with a "full" $n = 1$ valence shell.

The purpose of Lewis structures is to indicate how the valence electrons are distributed and not to portray the shape of the molecules. Spectroscopic evidence shows that methane molecules are highly symmetrical. A structural representation that includes what we know about the geometry of methane molecules shows the C atom in the centre of a tetrahedron with H atoms at each vertex.

In Lewis structures, a single pair of electrons in a bond is called a *single bond*, and a methane molecule has four C—H single bonds. In the F_2 molecule, only two of the fourteen valence electrons are assigned to the bond.

The three pairs of electrons that are placed on each of the F atoms are called *lone pairs* or *non-bonding pairs*. Each atom in this representation has a "full'" $n = 2$ valence shell of eight electrons.

Carbon dioxide (CO_2) and dinitrogen (N_2) are examples of molecules in which two atoms are bonded by sharing more than one electron pair.

In carbon dioxide, the carbon atom shares two pairs of electrons with each oxygen atom and so they are linked by a *double bond*. The valence shell of each oxygen atom in CO_2 has two bonding pairs and two lone pairs—a total of eight. In dinitrogen, the two nitrogen atoms share three pairs of electrons, so they are linked by a *triple bond*. Each N atom also has a lone pair of electrons—giving a total of eight electrons on each atom.

> Lewis structures are drawings that represent electron-pair bonds as a pair of dots or a line, between bonded atoms. Double bonds are shown as the sharing of two electron pairs, and triple bonds as sharing of three electron pairs. Some electrons do not participate in bonding and are represented as lone pairs of electrons on atoms.

The Octet Rule

In the Lewis structures we have seen so far, each atom except H shares four pairs (an octet) of electrons, so each has a noble gas configuration, corresponding with a "full" valence shell. *The tendency of molecules and polyatomic ions to have structures in which eight electrons are in the valence shell of each atom* is known as the **octet rule**. For example, in methane molecules, the C atom shares four bonding pairs; the C atom and each O atom in CO_2 molecules achieve the octet by forming double bonds; and in dinitrogen molecules, a triple bond gives an octet around each N atom.

Lone pairs of electrons are important features of a molecule's structure. They influence molecular shape [>>Section 10.7], and are the defining feature of many Lewis bases [<<Section 6.7].

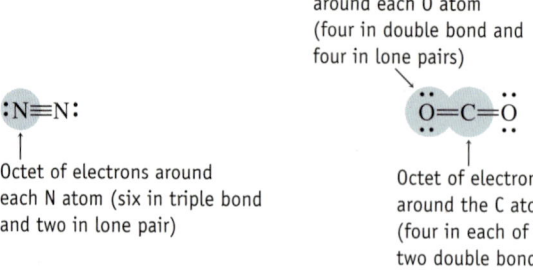

Octet of electrons
around each O atom
(four in double bond and
four in lone pairs)

:N≡N:

Octet of electrons around
each N atom (six in triple bond
and two in lone pair)

Ö=C=Ö

Octet of electrons
around the C atom
(four in each of
two double bonds)

Although the octet rule is widely applicable, exceptions do exist. Fortunately, many are obvious, such as when there are more than four bonds to an element or an odd number of valence electrons.

Hydrogen atoms typically form a bond to only one other atom, resulting in two electrons in its $n = 1$ valence shell—which is "full."

A measure of stability is associated with atoms having a "full" valence shell, making the octet rule useful. But keep in mind that it is more a *guideline* than a rule. In most cases, you should try to draw a Lewis structure in which each atom has eight electrons in its valence shell (or two in the case of hydrogen). A Lewis structure in which each atom achieves an octet is likely to be the most sensible, particularly for the second-period elements C, N, O, and F.

The octet rule refers to the stability of molecules and polyatomic ions whose structures have eight electrons in the valence shell of each atom. This is not an absolute rule; there are many stable molecules with structures having more than, or less than, eight electrons in the valence shell of atoms.

Drawing Lewis Structures

Lewis structures are not used to determine the structures of molecules: chemists have very powerful techniques to do that. Rather, given the structures of molecules (that is, the spatial arrangement of nuclei), Lewis structures are a useful visual model to make sense of molecular shape and reactivity.

We can't allocate valence electrons into a representation of a molecule unless we know its molecular structure: what atoms of each element are present, and which ones are sufficiently close to each other to be considered bound to each other (the *connectivity*). Today's diagnostic chemical tools (high-resolution mass spectrometry, NMR spectroscopy, infrared spectroscopy, and X-ray diffraction) provide this information.

Nevertheless, there are guidelines for the observed patterns of structures of molecules in which one atom is surrounded by others of different elements. For simple compounds, the first atom in a formula is often the central atom (e.g., SO_2, NH_4^+, NO_3^-). This is not always a reliable predictor, however. Notable exceptions include water (H_2O) and common acids (HNO_3, H_2SO_4), in which the acidic hydrogen is usually written first but N or S is the central atom. Alternative formulas for HNO_3 and H_2SO_4, which better indicate the connectivity, are $NO_2(OH)$ and $SO_2(OH)_2$.

While there are many exceptions, the central atom of a group of atoms comprising a molecule, or a part of a molecule, is usually the least electronegative. H atoms normally form only one covalent bond.

A systematic procedure is used to construct Lewis structures of molecules and ions whose spatial distribution of nuclei is known. Let's use formaldehyde (HCHO) as an example.

1. *Determine the total number of valence electrons in the molecule or ion.* In a neutral molecule this is the sum of the number of valence electrons on each atom, based on identifying in which group that atom is found on the periodic table. For an anion *add* a number of electrons equal to the negative charge, and for a cation *subtract* the number of electrons equal to the positive charge. For HCHO,

Number of valence electrons = [4 (from C)] + [2 × 1 (from H)] + [6 (from O)]

= 12 electrons (or 6 electron pairs)

2. Now we begin to distribute the valence electrons sensibly within the molecule. *Place a pair of electrons between each pair of bonded atoms to "make" a single bond.*

$$\text{Single bond} \longrightarrow \begin{array}{c} H \\ \backslash \\ C-O \\ / \\ H \end{array}$$

Here three electron pairs are used to make three single bonds (represented by single lines). Three pairs of electrons remain unallocated.

3. *Use any remaining pairs as lone pairs around each terminal atom (except H) so that each is surrounded by eight electrons.* If there are electrons left over after this step, assign them to the central atom. If the central atom is an element in the third or higher period, it can have more than eight electrons.

$$\text{Single bond} \longrightarrow \begin{array}{c} H \\ \backslash \\ C-\overset{..}{\underset{..}{O}}: \\ / \\ H \end{array} \quad \text{Lone pair}$$

Here all six pairs have been assigned, but notice that the C atom has a share of only six electrons. In molecules of all common compounds, C atoms share four electron pairs.

4. *If the central atom has fewer than eight electrons at this point, move one or more of the lone pairs on the terminal atoms to the region between the central atom and the terminal atom, forming multiple bonds.*

$$\begin{array}{c} H \\ \backslash \\ C-\overset{..}{\underset{..}{O}}: \\ / \\ H \end{array} \xrightarrow[\substack{\text{Move lone pair to create} \\ \text{double bond and satisfy} \\ \text{octet for C}}]{\text{Single bond}} \begin{array}{c} H \\ \backslash \\ C=\overset{..}{\underset{..}{O}} \\ / \\ H \end{array} \quad \substack{\text{Lone pair} \\ \\ \text{Two shared pairs;} \\ \text{double bond}}$$

As a general rule, double or triple bonds are formed *only* when both atoms are from the following list: C, N, O, S, or P. As examples, bonds such as $C=C$, $C\equiv N$, $C=O$, $C=S$, and $P=O$ will be incorporated.

Think about It

e10.3–10.5 Work through three tutorials on drawing Lewis structures.

WORKED EXAMPLE 10.1—DRAWING LEWIS STRUCTURES

Using the four steps outlined above, draw Lewis structures for an ammonia molecule (NH_3 with N the central atom), a hypochlorite ion (ClO^-), and a nitronium ion (NO_2^+ with N the central atom).

Solution for an NH_3 Molecule

1. Count the number of valence electrons.

$$\text{Number of valence electrons} = [5 \text{ (from N)}] + [3 \times 1 \text{ (from H)}]$$
$$= 8 \text{ electrons (4 electron pairs)}$$

2. Allocate a pair of electrons between the N atom and each H atom (using three of the four electron pairs).

$$\begin{array}{c} H-N-H \\ | \\ H \end{array}$$

3. Place the remaining pair of electrons on the central N atom.

$$\begin{array}{c} \overset{..}{N} \\ H-N-H \\ | \\ H \end{array}$$

Now the central N atom has a "full" valence shell (eight electrons), and each H atom has a share of one pair of electrons. No additional steps are required; this is a sensible Lewis structure.

Solution for a ClO⁻ Ion

1. With only two atoms, there is no central atom.

    Number of valence electrons = [7 (from Cl)] + [6 (from O)] + [1 (for negative charge)]

    $$= 14 \text{ electrons (7 electron pairs)}$$

2. Allocate one electron pair to the bonding region.

    $$Cl—O$$

3. Distribute the six remaining electron pairs around the C and O atoms: three pairs around each.

    $$\left[:\overset{..}{\underset{..}{Cl}}—\overset{..}{\underset{..}{O}}:\right]^{-}$$

As no electrons remain to be assigned and both atoms have an octet of electrons in their valence shells, this is a sensible Lewis structure.

Solution for an NO₂⁺ Ion

1. Count the total number of valence electrons in the ion.

    Number of valence electrons = [5 (from N)] + [12 (6 from each O)] − [1 (for positive charge)]

    $$= 16 \text{ electrons (8 electron pairs)}$$

2. Allocate one electron pair into each N—O internuclear region.

    $$O—N—O$$

3. Distribute the remaining six pairs of electrons on the terminal O atoms.

    $$\left[:\overset{..}{\underset{..}{O}}—N—\overset{..}{\underset{..}{O}}:\right]^{+}$$

4. The central N atom only has four electrons in its valence shell. Convert a lone pair of electrons on each O atom into a bonding electron pair to form two N=O double bonds.

    Move lone pairs to create double bonds and satisfy the octet for N.

    $$\left[:\overset{..}{O}\frown N\frown\overset{..}{O}:\right]^{+} \longrightarrow \left[\overset{..}{O}=N=\overset{..}{O}\right]^{+}$$

Each atom in the ion now has "full" valence shells: the N atom has four bonding pairs, and each O atom has two lone pairs and shares two bonding pairs.

Interactive Exercises 10.1–10.2

SUBMIT

Draw Lewis structures for molecules and ions from their formulas.

EXERCISE 10.3—DRAWING LEWIS STRUCTURES

Draw Lewis structures for an NH_4^+ ion, a CO molecule, an NO^+ ion, and a SO_4^{2-} ion.

Patterns of Molecular Structure

Lewis structures are useful to help understand the structure and reactivity of a molecule or ion. Chemists often rely on connectivity patterns in related molecules to draw Lewis structures.

Hydrogen Compounds The structures of molecules and ions of some common compounds of second-period non-metal elements with hydrogen are shown in Table 10.1.

TABLE 10.1 Common Hydrogen-Containing Molecules and Ions of the Second-Period Elements

Group 14		Group 15		Group 16		Group 17	
CH_4 methane	H | H—C—H | H	NH_3 ammonia	H—N̈—H | H	H_2O water	H—Ö—H	HF hydrogen fluoride	H—F̈:
C_2H_6 ethane	H H | | H—C—C—H | | H H	N_2H_4 hydrazine	H—N̈—N̈—H | | H H	H_2O_2 hydrogen peroxide	H—Ö—Ö—H		
C_2H_4 ethylene	H—C=C—H | | H H	NH_4^+ ammonium ion	$\left[\begin{array}{c} H \\ \| \\ H-N-H \\ \| \\ H \end{array}\right]^+$	H_3O^+ hydronium ion	$\left[\begin{array}{c} H-\ddot{O}-H \\ \| \\ H \end{array}\right]^+$		
C_2H_2 acetylene	H—C≡C—H	NH_2^- amide ion	$\left[H-\ddot{N}-H \right]^-$	OH^- hydroxide ion	$\left[:\ddot{O}-H \right]^-$		

These Lewis structures indicate the connectivity of atoms, and are models of the distribution of valence electrons within molecules, but they are not intended to indicate the shape of the molecules—that is, the three-dimensional spatial orientation of atoms. See Section 10.7.

These Lewis structures show patterns of electron distribution in molecules of some elements. For example, in uncharged molecules, N atoms commonly have three electron pairs in bonds. This is consistent with attainment of an octet in the valence electron shell: nitrogen atoms have five valence electrons (two in a lone pair, and three unpaired electrons), and each of the unpaired electrons can be paired with an electron from another atom to form a covalent bond. Similarly, carbon atoms commonly have four electron pairs in bonds, oxygen atoms have two, and fluorine atoms have one.

Group 14	Group 15	Group 16	Group 17
| —C— |	·· —N— |	·· —O— ··	·· :F— ··

The first two members of the series called the *alkanes* are CH_4 and C_2H_6 (Table 10.1). Suppose you draw a Lewis structure of the third member, propane (C_3H_8). We can rely on the idea that the atoms in propane molecules bond in predictable ways: carbon atoms can be expected to have four bonding electron pairs, and hydrogen atoms only one. The only arrangement of atoms that meets these criteria has three carbon atoms linked together by C—C single bonds. The remaining unpaired electrons on the C atoms pair off with the unpaired electron on H atoms to form C—H single bonds.

$$\begin{array}{ccc} H & H & H \\ \| & \| & \| \\ H-C-C-C-H \\ \| & \| & \| \\ H & H & H \end{array}$$

Propane, C_3H_8

WORKED EXAMPLE 10.2—PREDICTING MOLECULAR STRUCTURES

Draw Lewis structures for (a) CCl_4, and (b) NF_3.

Strategy

One way to solve this problem is to recognize that CCl_4 and NF_3 are similar to CH_4 and NH_3, respectively, except that the terminal atoms are halogens, rather than H atoms.

Solution

Recall that C atoms can be expected to form four bonds and N atoms three bonds. In addition, Cl and F atoms in the −1 oxidation state usually are terminal, forming just one covalent bond. So we can presume that the connectivity in these molecules is as displayed below:

$$\begin{array}{c} \text{Cl} \\ | \\ \text{Cl}-\text{C}-\text{Cl} \\ | \\ \text{Cl} \end{array} \qquad \begin{array}{c} \text{F}-\text{N}-\text{F} \\ | \\ \text{F} \end{array}$$

(a) CCl_4

Number of valence electrons = [4 (from C)] + [4 × 7 (from Cl)]

= 32 electrons (16 electron pairs)

The 32 valence electrons can be accommodated by placing two in each of the C—Cl bonds, and six lone pairs on each Cl atom; all atoms have "full" valence electron shells.

(b) NF_3

Number of valence electrons = [5 (from N)] + [3 × 7 (from F)]

= 26 electrons (13 electron pairs)

Put two electrons into each of three N—F single bonds, one lone pair on the N atoms, and three lone pairs on each terminal F atom. Our structure then has 26 valence electrons, and all atoms will have octets.

$$\begin{array}{c} :\ddot{\text{Cl}}: \\ | \\ :\ddot{\text{Cl}}-\text{C}-\ddot{\text{Cl}}: \\ | \\ :\ddot{\text{Cl}}: \end{array} \qquad \begin{array}{c} :\ddot{\text{F}}-\ddot{\text{N}}-\ddot{\text{F}}: \\ | \\ :\ddot{\text{F}}: \end{array}$$

Carbon tetrachloride   **Nitrogen trifluoride**

EXERCISE 10.4—PREDICTING MOLECULAR STRUCTURES

Predict Lewis structures for methanol (CH_3OH) and hydroxylamine (NH_2OH). (*Hint:* The formulas of these compounds are written to guide you in choosing the correct connectivity of atoms.)

In molecules of the second-period elements, carbon atoms usually have four electron pairs in bonds; nitrogen atoms have three (and a lone pair of electrons); oxygen atoms have two (and two lone pairs); and fluorine have one (and three lone pairs).

Oxoacids and Their Anions **Oxoacids** have oxygen atoms in their acidic groups. Lewis structures of common oxoacids and their anions are illustrated in Table 10.2. When pure (not dissolved in water), these acids are covalently bonded molecular compounds of non-metals. Nitric acid, for example, has properties that we associate with covalent molecular substances: it is a colourless liquid with a boiling point of 83°C.

In aqueous solution, HNO_3, H_2SO_4, and $HClO_4$ are ionized to form aquated anions and hydronium ions. A Lewis structure for the nitrate ion, for example, can be drawn with two N—O single bonds and one N=O double bond. To draw a Lewis structure for nitric acid, we can attach an H^+ ion to the O atom on one of the N—O single bonds.

$$\left[\begin{array}{c} :\ddot{\text{O}}-\text{N}=\ddot{\text{O}}: \\ | \\ :\ddot{\text{O}}: \end{array} \right]^{-} \qquad \begin{array}{c} \text{H}-\ddot{\text{O}}-\text{N}=\ddot{\text{O}}: \\ | \\ :\ddot{\text{O}}: \end{array}$$

Nitrate ion   **Nitric acid**

TABLE 10.2 Lewis Structures of Common Oxoacids and Their Anions

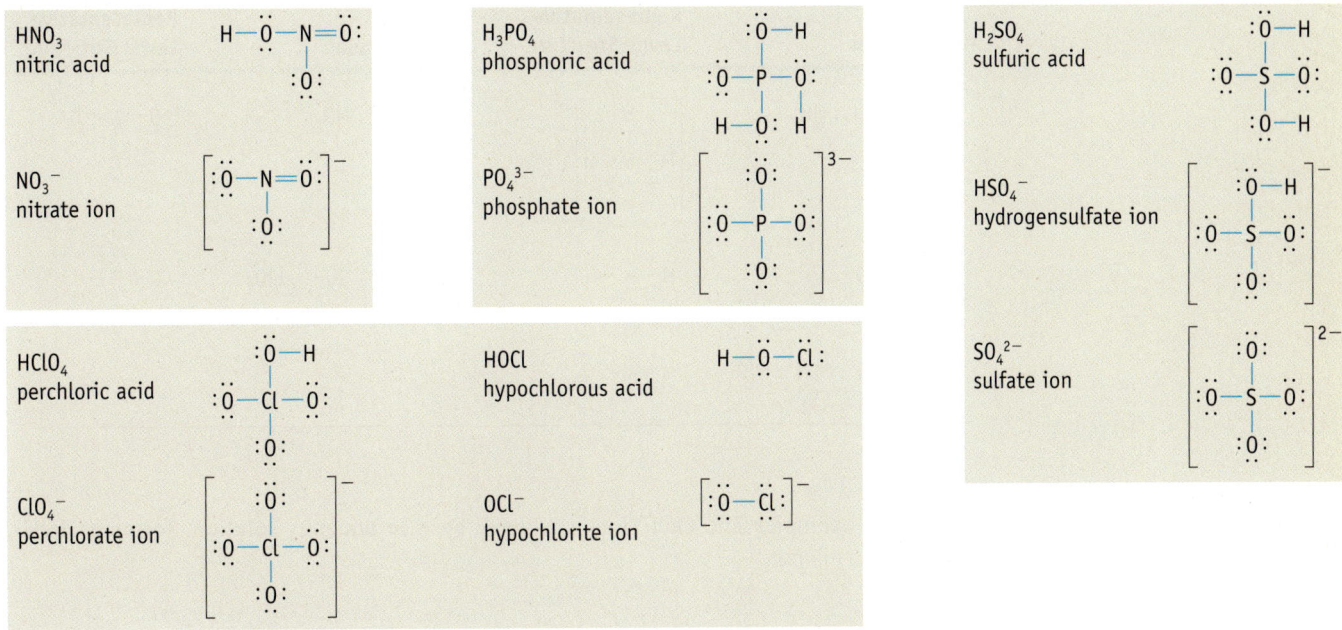

When a nitric acid molecule ionizes, the O—H bond is broken, the electrons of the bond staying with the O atom and are shown as a lone pair of electrons. As a result, HNO_3 and NO_3^- have the same number of electrons, 24, and their structures are closely related. As seen in Section 10.5, however, the structure of the nitrate ion is not adequately represented by a single Lewis structure.

EXERCISE 10.5—LEWIS STRUCTURES OF OXOACIDS AND THEIR ANIONS

Draw a Lewis structure for the dihydrogenphosphate anion ($H_2PO_4^-$) which is derived by ionization of phosphoric acid.

Isoelectronic Species The molecules and ions NO^+, N_2, CO, and CN^- are similar in that each has two atoms and the same total number of valence electrons, 10. We can write analogous Lewis structure for each in which the two atoms are linked with a triple bond. With three bonding pairs and one lone pair, each atom has an octet of electrons.

$$[:N\equiv O:]^+ \qquad :N\equiv N: \qquad :C\equiv O: \qquad [:C\equiv N:]^-$$

Molecules and ions with the same number of valence electrons and the same number and connectivity of atoms but different in some of the elements involved are said to be **isoelectronic species** (Table 10.3). You will find it helpful to think in terms of isoelectronic molecules and ions because this perspective offers another way to see relationships in bonding among common chemical substances.

Isoelectronic species have both similarities and important differences in their chemical properties. For example, both carbon monoxide (CO) and cyanide ion (CN^-) are very toxic, which results from the fact that they can bind to the iron of hemoglobin in blood and block the uptake of oxygen. They differ, though, in their acid-base chemistry: in aqueous solution, cyanide ion is a base and accepts H^+ to form HCN, whereas CO does not readily accept H^+. The isoelectronic species Cl_2 and OCl^- provide a

TABLE 10.3 Common Isoelectronic Molecules and Ions

Formulas	Representative Lewis Structure	Formulas	Representative Lewis Structure
BH_4^-, CH_4, NH_4^+	$\begin{bmatrix} H \\ \| \\ H-N-H \\ \| \\ H \end{bmatrix}^+$	CO_3^{2-}, NO_3^-	$\begin{bmatrix} :\ddot{O}-N=\ddot{O}: \\ \| \\ :\ddot{O}: \end{bmatrix}^-$
NH_3, H_3O^+	$\begin{matrix} H-\ddot{N}-H \\ \| \\ H \end{matrix}$	PO_4^{3-}, SO_4^{2-}, ClO_4^-	$\begin{bmatrix} :\ddot{O}: \\ \| \\ :\ddot{O}-P-\ddot{O}: \\ \| \\ :\ddot{O}: \end{bmatrix}^{3-}$
CO_2, OCN^-, SCN^-, N_2O NO_2^+, OCS, CS_2	$\ddot{O}=C=\ddot{O}$		

similar example. The OCl⁻ ion is a weak base in aqueous solution, forming HOCl, while Cl_2 is not.

EXERCISE 10.6—IDENTIFYING ISOELECTRONIC SPECIES

(a) Is the acetylide dianion (C_2^{2-}) isoelectronic with the N_2 molecule?
(b) Identify a common molecular (uncharged) species that is isoelectronic with the nitrite ion (NO_2^-).
(c) Identify a common ion that is isoelectronic with the HF molecule.

Isoelectronic molecules and ions have the same number of valence electrons and the same number and connectivity of atoms, but differ in some of the elements involved.

10.4 Exceptions to the Octet Rule

Although most molecular compounds and ions obey the octet rule, many do not. These include molecules and ions that have an atom with fewer than four pairs of electrons, those with more than four pairs, and those with an odd number of electrons.

Molecules in which an Atom Has Fewer Than Eight Valence Electrons

Atoms of boron, a non-metal in Group 13, have three valence electrons. Boron-containing molecules commonly have three covalent bonds to atoms of other elements. In these molecules, boron atoms have only six electrons in the valence shell—two short of an octet. Many boron compounds of this type are known, including such common compounds as boric acid, $B(OH)_3$ or H_3BO_3, borax ($Na_2B_4O_5(OH)_4 \cdot 8\ H_2O$. Figure 10.1), and the boron trihalides (BF_3, BCl_3, BBr_3, and BI_3).

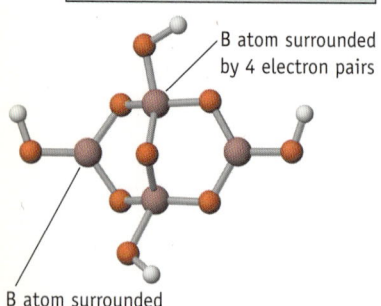

B atom surrounded by 4 electron pairs

B atom surrounded by 3 electron pairs

FIGURE 10.1 Borax. This common material, used in soaps, contains an interesting anion, $B_4O_5(OH)_4^{2-}$. Two B atoms are surrounded by four electron pairs and the other two B atoms are surrounded by only three electron pairs.

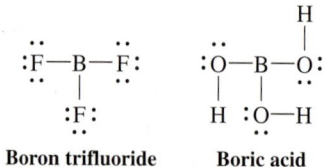

Boron trifluoride   **Boric acid**

Some boron compounds, such as BF_3, that are two electrons short of an octet are quite reactive. The boron atom can accommodate a fourth electron pair, acting as a Lewis acid [<<Section 6.7] when a lone pair is available on an atom of another compound (a Lewis base). For example, boron trifluoride reacts with ammonia to form a Lewis adduct with formula $H_3N—BF_3$.

The reaction of F^- ions with BF_3 atoms to form BF_4^- ions is another example.

Some compounds of boron have only six valence electrons. These usually react as Lewis acids.

Think about It

e10.6 Work a tutorial on Lewis structures for electron-deficient compounds.

Molecules in which an Atom Has More Than Eight Valence Electrons

Many compounds of elements in the third and higher periods have molecules or ions with a central atom surrounded by more than four valence electron pairs (Table 10.4). In most of these cases, the central atom is bonded to fluorine, chlorine, or oxygen atoms. For example, in sulfur hexafluoride (SF_6), a central S atom is bound by covalent bonds to six F atoms, so there are six electron pairs (12 electrons) in the valence shell of the S atoms.

TABLE 10.4 Lewis Structures in which an Atom Has More Than Eight Valence Electrons

If more than four groups are bound to a central atom, we can be sure that there are more than eight electrons around that atom. But there may be more than eight electrons when the central atom is bound to four or fewer atoms: the central atom in molecules of each of SF_4, ClF_3, and XeF_2 (Table 10.4) has five electron pairs in its valence shell.

A useful observation is that *only elements of the third and higher periods in the periodic table form compounds and ions in which an atom has more than eight valence electrons.* Compounds of second-period elements (B, C, N, O, and F) have a maximum of eight electrons in the valence shell of atoms of those elements. For example, nitrogen forms compounds and ions such as NH_3, NH_4^+, and NF_3, but NF_5 is unknown. Phosphorus, the third-period element just below nitrogen in Group 15, forms many compounds similar to nitrogen (PH_3, PH_4^+, PF_3), but it also readily accommodates five or six valence electron pairs in compounds such as PF_5 or in ions such as PF_6^-. Arsenic, antimony, and bismuth—the elements below phosphorus in Group 15—resemble phosphorus in their behaviour.

The usual explanation for the contrasting behaviour of second- and third-period elements centres on the differences between the "allowed" energies of electrons in atoms [<<Section 8.4]. The valence electrons in the second-period elements are in the $n = 2$ shell ($2s$ and $2p$ orbitals), which has a maximum "capacity" of eight electrons. For an atom to have more than eight valence electrons, the additional ones would need to have principal quantum number $n = 3$. The energies of electrons in the $n = 3$ shell are much higher than of those in the $n = 2$ shell—so much so that it is generally believed that for electrons to be in this high energy condition would render the compound unstable.

On the other hand, for elements in the third and higher periods, the energies of electrons in the orbitals of $n = 3$ shell ($3s$, $3p$, and $3d$) are relatively close, so that electrons in all of these orbitals can participate in bonding (that is, are valence electrons). For example, in compounds of phosphorus, the energies of electrons in the $3d$ orbitals differ little from those in the $3s$ and $3p$ orbitals. Consequently, occupation of $3d$ orbitals by electrons does not present too great an energy barrier.

Compounds of xenon are among the more interesting entries in Table 10.4. Noble gas compounds were not discovered until the early 1960s—largely because they were presumed to be entirely inert. Xenon difluoride can be synthesized by simply placing a flask containing xenon gas and fluorine gas in sunlight. After several weeks, crystals of colourless XeF_2 can be seen.

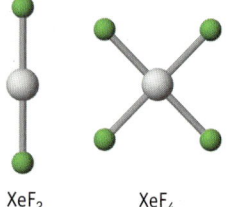

XeF_2 XeF_4

Many compounds of elements in the third and higher periods have molecules or ions with a central atom surrounded by more than four valence electron pairs. This is not the case for second-period elements.

Molecules with Odd Numbers of Valence Electrons

Two oxides of nitrogen—NO, with a total of 11 valence electrons, and NO_2, with 17—are among a small group of stable molecules in which an atom has an odd number of valence electrons. Because these compounds have an odd number of valence electrons, it is impossible to draw a Lewis structure in which all atoms obey the octet rule; at least one electron must be unpaired. For NO_2, several Lewis structures can be drawn; the commonly used one puts an unpaired electron on the N atom:

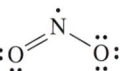

Below are two reasonable Lewis structures for the NO molecule:

Experimental evidence indicates that the bonding between N and O in NO molecules is intermediate between a double bond and a triple bond. It is not possible to write a Lewis structure for NO that adequately represents the properties of this substance. More sophisticated models are needed to understand bonding in molecules with odd numbers of valence electrons, and we return to this when molecular orbital theory is discussed in Section 10.9.

The nitrogen oxides, NO and NO_2, are members of a class of chemical substances called **free radicals**: chemical species (both atomic and molecular) with one or more unpaired electrons. Free radicals are generally quite reactive. Cl atom free radicals, for example, readily react with other compounds to produce molecules such as Cl_2 and HCl.

Free radicals are involved in many reactions in the atmosphere. For example, small amounts of NO are produced in vehicle exhaust. The NO rapidly forms NO_2, which decomposes in the presence of sunlight and oxygen to give more NO as well as ozone (O_3), an air pollutant that affects the respiratory system [>>Section 17.1, 17.10].

$$NO_2(g) + O_2(g) \xrightarrow{\text{uv radiation}} NO(g) + O_3(g)$$

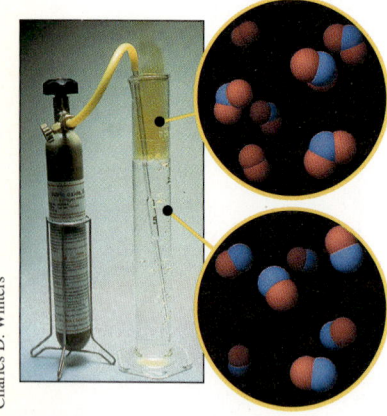

$$2\ NO(g) + O_2(g) \longrightarrow 2NO_2(g)$$

Colourless NO gas is bubbled from a high-pressure tank through water. When it emerges into the air, the NO reacts rapidly with O_2 to give brown NO_2 gas.

Neither NO nor NO_2 has the extreme reactivity of most free radicals, and they are sufficiently stable that they can be isolated. When NO_2 gas is cooled, however, two NO_2 molecules combine or "dimerize" to form colourless N_2O_4. This can be attributed to formation of a single N—N covalent bond by sharing of the unpaired electron on each N atom. Even though this bond is weak, the reaction is easily observed in the laboratory (Figure 10.2).

Think about It

e10.7 Watch a demonstration of this reaction between free radicals.

When cooled, NO_2 free radicals couple to form N_2O_4 molecules. $\longrightarrow$ N_2O_4 gas is colourless.

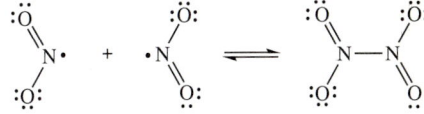

A flask containing an NO_2/N_2O_4 mixture (mainly brown NO_2) in warm water.

A flask containing an NO_2/N_2O_4 mixture (mainly colourless N_2O_4) in ice water.

FIGURE 10.2 Free radical chemistry. When cooled, the brown gas NO_2, a free radical, forms colourless N_2O_4, whose molecules have a N—N single bond.

Compounds whose molecules have an odd numbers of electrons must have at least one unpaired valence electron. These compounds are called free radicals, and they are usually quite reactive.

10.5 Resonance and Delocalized Electron Models

Ozone (O_3) a reactive, blue, diamagnetic gas with a characteristic pungent odour, is present at low concentrations in Earth's lower stratosphere, where it protects life from the sun's ultraviolet radiation [>>Section 21.1]. Using the guidelines of Section 10.3, a reasonable Lewis structure for the O_3 molecule has one single bond and one double bond, and all atoms have eight valence electrons.

If this were the structure of ozone molecules, one bond (O=O) would be shorter than the other bond (O—O). However, experimental data tells us that the two O—O bonds are the same length, suggesting an equal number of shared electrons in each bond. We conclude that this Lewis structure does not accurately represent the distribution of electrons in ozone molecules. A more sophisticated model of distribution of valence electrons is needed.

Imagine that you are on the central O atom of an ozone molecule, with one O atom on your right and the other on your left. We can imagine the electrons distributed in this molecule in two ways:

$$:O \overset{\ddot{O}}{=} \ddot{O}: \quad \text{and} \quad :\ddot{O} \overset{\ddot{O}}{=} O:$$

Is the double bond on your right or on your left? Perhaps a silly question, because we have already seen that neither of these models is consistent with equal-length bonds. Nobel laureate chemist Linus Pauling (1901–1994) proposed a theory of **resonance**: *If the valence electrons in a molecule or ion can be distributed in more than one sensible way, then neither is an accurate representation, and the actual structure is intermediate between them.* Structures with the different distributions of valence electrons in a molecule (such as the two shown above for ozone) are called *resonance structures*, and the "real" intermediate structure is referred to as a *resonance hybrid*.

It is known that ozone molecules are bent, or V-shaped. Experimental data shows that the O—O—O bond angle is not 180°, but 117°.

Resonance is portrayed in the following way:

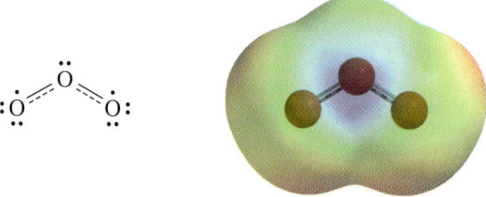

The double-headed arrow has a particular meaning in chemistry, so this diagram is to be interpreted as "The actual distribution of valence electrons in an O_3 molecule is neither of the structures shown, but is intermediate between them."

The double-headed arrow does not *imply that the distribution of valence electrons switches between one resonance structure and the other, and it does* not *mean that there is an equilibrium condition between molecules with the different structures.* Molecules with these distributions do not exist. It may seem a little odd that this representation does not attempt to describe what the actual structure is; rather, it tells us what the actual structure is not.

So what is the distribution of valence electrons in reality, represented here by the intermediate resonance hybrid? For one possible answer, we turn to a complementary bonding **electron delocalization model**. Each of the resonance structures of an O_3 molecule has four bonding electrons "localized" between two O atoms, and two "localized" between another pair of O atoms. Using the delocalization model, we imagine that the six bonding electrons in the molecule are spread uniformly over the two bonding regions—three electrons in each. Since we traditionally think of a bond as due a shared pair of electrons, we can say that each O—O region has 1.5 "classical" bonds—a single bond and a *partial bond*.

This is more formally expressed through a concept called **bond order**, the value of which is 1 for a single bond, 2 for a double bond, and 3 for a triple bond.

$$\text{Bond order} = \frac{1}{2} \times \frac{\text{Number of delocalized electrons}}{\text{Number of bonds over which they are delocalized}}$$

In the case of ozone molecules with six electrons delocalized over two oxygen–oxygen bonds:

$$\text{Bond order} = \frac{1}{2} \times \frac{6}{2} = 1.5$$

This model is consistent with the experimental knowledge that the oxygen–oxygen bond lengths in O_3 molecules are equivalent and 127.8 pm—intermediate between the typical length of an O=O double bond (121 pm) and an O—O single bond (132 pm).

The benzene molecule (C_6H_6) is a six-membered ring of carbon atoms with six equivalent carbon–carbon bonds, and a hydrogen atom attached to each C atom. The carbon–carbon bonds are 139 pm long—intermediate between the average length of a C=C double bond (134 pm) and a C—C single bond (154 pm). The resonance model suggests that the distribution of valence electrons in each benzene molecule is intermediate between the two resonance structures, with alternate double and single carbon–carbon bonds.

Resonance structures of benzene, C_6H_6

A representation of the experimental structure, according to the electron delocalization model, has the 18 bonding electrons in the six carbon–carbon bonds spread equally over the bonds. This is commonly portrayed in several ways:

For the six carbon–carbon bonds over which the 18 valence electrons are delocalized:

$$\text{Bond order} = \frac{1}{2} \times \frac{18}{6} = 1.5$$

Let us apply the concept of resonance to describe bonding in a carbonate ion (CO_3^{2-}) which has 24 valence electrons (12 pairs). Three resonance structures can be drawn using the guidelines for writing Lewis structures. These differ in our arbitrary choice of which O atom to connect to the C atom by a double bond.

The structure of a polyatomic ion is usually displayed in square brackets with the overall charge on the ion shown outside the bracket. In the case of the CO_3^{2-} ion, the −2 charge tells us that there are two more electrons in the species than there are protons in the nuclei—and does not attempt to indicate regions of negative (or positive) charge within the ion.

The resonance model is useful here, as none of the structures accurately describes the electron distribution in this ion. Rather, the actual resonance hybrid is intermediate between the three structures, in good agreement with experimental data that all three carbon–oxygen bond lengths are 129 pm—intermediate between typical C—O single bond (143 pm) and C=O double bond (122 pm) lengths.

A delocalized electron model of a carbonate ion has eight electrons spread over three bonds—2.67 electrons per bond.

$$\text{Bond order} = \frac{1}{2} \times \frac{8}{3} = 1.33$$

In aqueous solution, a hydrogen ion can bond to an O atom of a carbonate ion to give the hydrogencarbonate, or bicarbonate, ion. This ion can be described as a resonance hybrid of two Lewis structures.

A delocalized electron representation of the hydrogencarbonate ion is

and the bond order in the two bonds over which electrons are delocalized is 1.5.

Experimental data shows the structure of the ion to be intermediate between the three structures. The large brackets around the set of three structures shows visually that the real ion is one entity which is a composite of the three imaginary resonance structures, each of which shows a different distribution of electrons over the atoms of the ion.

In this delocalized electron picture of a carbonate ion, the lone pairs are not shown. It is difficult to do so in a way that has all O atoms equivalent. In such a case, we might consider delocalization of all 24 valence electrons (bonding and non-bonding) over the entire molecule. Then we could say that each O atom is surrounded by eight electrons—an octet—as is the C atom.

Two extreme cases of delocalization of valence electrons are (a) metals, for which our model is a lattice of cations within a "sea" of delocalized valence electrons, and (b) plasma, the so-called fourth state of matter [>>Section 11.1].

WORKED EXAMPLE 10.3—DRAWING RESONANCE STRUCTURES

Draw resonance structures for the nitrite ion (NO_2^-). Are the nitrogen–oxygen bonds single, double, or intermediate in value?

Strategy

Draw the dot structure in the usual manner. If multiple bonds are required, resonance structures may be needed. This will be the case if the octet of an atom can be completed by using an electron pair from more than one terminal atom to form a multiple bond. Bonds to the central atom cannot then be "pure" single or double bonds but are intermediate between the two.

Solution

Nitrogen is the central atom in the nitrite ion, which has a total of 18 valence electrons (9 pairs).

Valence electrons = 5 (for the N atom) + 12 (6 for each O atom) + 1 (for negative charge)

After forming N—O single bonds, and distributing lone pairs on the terminal O atoms, a pair remains, which is placed on the central N atom.

$$\left[:\overset{..}{\underset{..}{O}}-\overset{..}{N}-\overset{..}{\underset{..}{O}}: \right]^-$$

To complete the octet of electrons about the N atom, form an N=O double bond.

$$\left[:\overset{..}{O}=\overset{..}{N}-\overset{..}{\underset{..}{O}}: \quad\longleftrightarrow\quad :\overset{..}{\underset{..}{O}}-\overset{..}{N}=\overset{..}{O}: \right]^-$$

Two equivalent structures can be drawn, and the actual structure has delocalized electrons, which we can represent in this model as a resonance hybrid of these two structures. The nitrogen–oxygen bonds are neither single nor double bonds; rather, they are intermediate between single and double bonds.

Interactive Exercise 10.8

Practise drawing resonance structures.

EXERCISE 10.9—DRAWING RESONANCE STRUCTURES

Draw resonance structures for the nitrate ion (NO_3^-). Sketch a plausible Lewis dot structure for nitric acid (HNO_3).

If the valence electrons in a molecule or ion can be distributed in more than one sensible way, then neither is an accurate representation, and the actual structure is intermediate between them. The "real" intermediate structure is referred to as a resonance hybrid. Resonance is represented by a double-headed arrow. A resonance hybrid may also be represented by an electron delocalization model, with bonds intermediate between a single bond and a double bond.

10.6 Which Resonance Structures Are Most Important?

We've seen examples of molecules with several possible resonance structures that have the same number of single and double bonds. In other cases, several Lewis structures for the same molecule differ in the number of single, double, and triple bonds. For example, resonance structures for sulfur dioxide (SO_2, a V-shaped molecule with 18 valence electrons) include the following:

$$:\overset{..}{\underset{..}{O}}\diagdown\overset{\overset{..}{S}}{}\diagup\overset{..}{\underset{..}{O}}: \quad\longleftrightarrow\quad :O\diagup\diagup\overset{\overset{..}{S}}{}\diagdown\overset{..}{\underset{..}{O}}: \quad\longleftrightarrow\quad :\overset{..}{\underset{..}{O}}\diagdown\overset{\overset{..}{S}}{}\diagdown\diagdown O: \quad\longleftrightarrow\quad :O\diagup\diagup\overset{\overset{..}{S}}{}\diagdown\diagdown O:$$

$$\quad\quad A \quad\quad\quad\quad\quad\quad B \quad\quad\quad\quad\quad\quad C \quad\quad\quad\quad\quad\quad D$$

According to the resonance model, the actual (resonance hybrid) structure has a distribution of valence electrons that coincides exactly with none of these. But which of these does the actual structure most closely approximate? To decide this, chemists use the concept of **formal charge**: *the charge on each atom in a molecule or ion if we presume that bonding electrons are shared equally by the atoms that are directly bound to each other.* Electronegativity differences are ignored. Chemists use a general guideline that *the structure whose electron distribution most closely approximates that of the actual structure is the one with smallest values (positive or negative) of formal charge.*

Formal charge is calculated for *each atom* in a Lewis structure using the following relationship:

Formal charge on atom = [number of valence electrons] − [number of electrons in lone pairs]

$$-\tfrac{1}{2} \text{[number of electrons in bonds]} \qquad (10.1)$$

For the S atom in structure A:

$$\text{Formal charge} = +6 - 2 - \tfrac{1}{2}(4) = +2$$

For either O atom in structure A:

$$\text{Formal charge} = +6 - 6 - \tfrac{1}{2}(2) = -1$$

We can calculate and display the formal charges on each atom in all the SO_2 resonance structures:

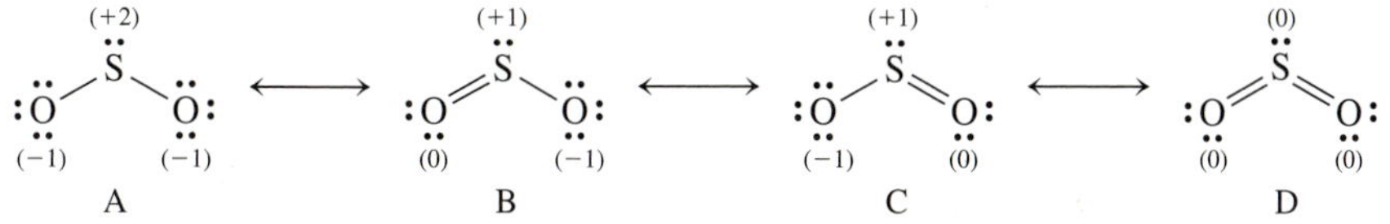

Since in structure D the formal charge on each atom is 0, chemists generally accept that the resonance hybrid structure is closer to that of D than it is to those of A, B, or C. While the actual structure is not identical to D, this is the one that used if a single structure represents an SO_2 molecule. This guideline works on the supposition that molecules are more stable when their atoms have the same number of valence electrons as they do when not combined.

This example highlights the need to be careful when applying the octet rule. We might have dismissed structure A because the S atom has only six valence electrons, and structure D because it has ten.

The sum of the formal charges on the atoms of a neutral molecule must be 0, and on the atoms of an ion must equal the charge on the ion. For example, the sum of formal charges on the atoms in a nitrate ion must be −1.

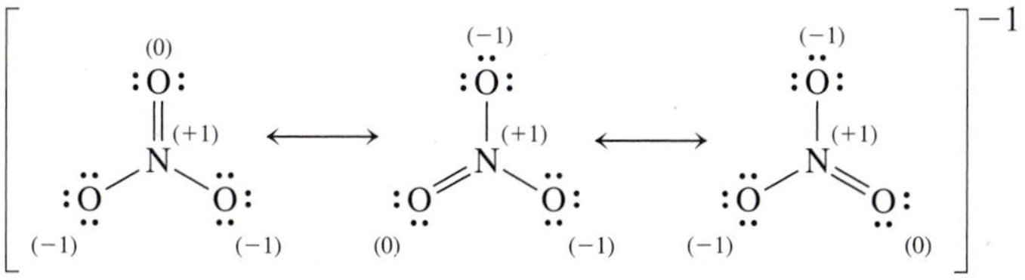

In this case, each of the resonance structures has the same combination of formal charges and the actual structure approximates each of them equally—consistent with the delocalized valence electron portrayal:

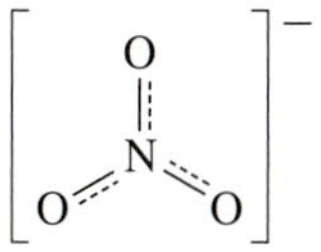

Experimental evidence tells us that the three bonds are the same length. We can presume that in this representation the formal charge on each O atom is not −1, −1, and 0, but the average of these, $-\tfrac{2}{3}$.

WORKED EXAMPLE 10.4—DECIDING WHICH RESONANCE STRUCTURES ARE MOST IMPORTANT

(a) Draw a resonance representation for a tetrahedral sulfate ion (SO_4^{2-}), using all possible Lewis structures with 12 or less electrons in the valence shell of the central S atom.

(b) Do any of the structures satisfy the octet rule?

(c) Calculate the formal charges on the atoms in each, and decide which most closely approximate the actual structure.

Strategy

Imagine that you are sitting on the S atom, contemplating how the 32 valence electrons are distributed in bonds and lone pairs.

Solution

(a) The number of valence electrons in the ion is $6 + (4 \times 6) + 2 = 32$. These can be distributed over a sulfate ion in 11 different ways (shown below) and the actual structure (a resonance hybrid) is assumed to be a "weighted average" of these, but not identical to any of them.

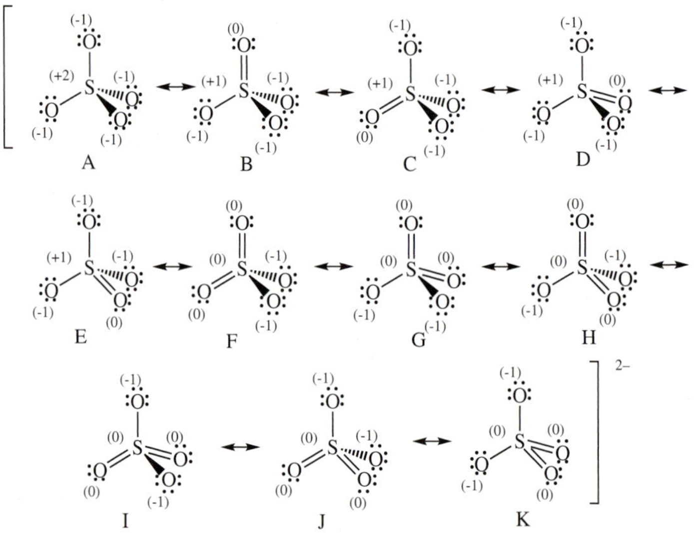

(b) In structure A, the octet rule is satisfied on all atoms.

(c) The formal charges on each atom in each of the electron distributions, calculated using the formal charge equation (10.1) above, are shown. The actual structure is not likely to be similar to A because it has large formal charges. (Furthermore, it suggests that the S atom has a negative charge and the O atoms have positive charges, contrary to what one would expect based on relative electronegativities.) The electron distributions in structures B, C, D, and E give rise to the same set of formal charges. And structures F, G, H, I, J, and K have the same set of formal charges. The formal charges in F, G, H, I, J, and K are lower than in any of the other resonance forms, so we presume that the actual structure most closely approximates these.

Comment

The structures most closely approximating the actual structure have 12 valence electrons on the S atom (and not an octet). If resonance distributions F, G, H, I, J, and K were the only resonance forms, we could say that electron delocalization picture of the actual structure would have a bond order of 1.5. The other resonance forms A, B, C, D, and E suggest less electron density in the bonding regions and more residing as non-bonding electrons—so a bond order less than 1.5. When chemists wish to portray the structure of a sulfate ion, they do not normally draw all 11 resonance forms, nor even the six equivalent forms closest to the resonance hybrid: rather, they draw any one of F, G, H, I, J, or K.

EXERCISE 10.11—DECIDING WHICH RESONANCE STRUCTURES ARE MOST IMPORTANT

Show that you can draw four reasonable Lewis structures that contribute to the actual electron distribution in a sulfuric acid molecule (H_2SO_4), and by calculation of formal charges on the atoms, decide which most closely approximate the actual distribution.

EXERCISE 10.12—DECIDING WHICH RESONANCE STRUCTURES ARE MOST IMPORTANT

Draw reasonable Lewis structures for a molecule of ozone, calculate formal charges on the atoms, and decide which most closely approximate the actual distribution.

> Formal charge is the charge on an atom in a molecule or ion if we presume that bonding electrons are shared equally by the atoms directly bound to each other. The resonance structure with smallest formal charge values (positive or negative) has an electron distribution most closely approximating that of the actual structure.

Interactive Exercise 10.10

Practise determining formal charges on atoms in a structure to predict the most important resonance structure.

Taking It Further

e10.9 Read about the difference between oxidation numbers and formal charges.

10.7 Spatial Arrangement of Atoms within Molecules

The physical and chemical properties of covalent compounds depend strongly on the structures of their molecules. In particular, most biochemical processes are controlled, in part, by the shapes and polarities of interacting molecules. The structures of molecules are determined by experimental techniques [<<Section 9.2], but it is useful to understand why molecules have their particular shapes, and to be able to predict the spatial arrangement of atoms with respect to each other in molecules.

Experimental evidence tells us that molecules of methane (CH_4) are tetrahedral, while molecules of xenon tetrafluoride (XeF_4) are flat and square planar. Molecules of carbon dioxide (CO_2) are linear and have no dipole moment, while molecules of sulfur dioxide (SO_2) are bent and have a dipole moment.

Chemists use a simple model, developed by Canadian chemist Ron Gillespie and Australian chemist Sir Ronald Nyholm, to rationalize the spatial distribution of bonds around atoms in molecules or polyatomic ions, such as the single bonds around the C atom in a methane molecule. This spatial distribution of bonds determines the shapes of small molecules with a central atom, and is a large factor in the shapes of large molecules such as proteins [<<Section 9.3]. The basis of the **valence shell electron-pair repulsion (VSEPR) model** is that *localized regions of high electron density in the valence shell of an atom (single and multiple bonds, non-bonding electrons, or a single electron) are oriented as far as possible from each other around that atom, in order to minimize repulsions between them*. The orientation of least repulsion energy depends on the number of pairs of electrons.

To apply this model, we need to refer to some three-dimensional geometry. The orientations about an atom of regions of high electron density that result in minimum repulsion energy (Figure 10.3) depend on the number of such regions (Table 10.5). The most common situations have two, three, or four regions of electron density.

Portrayal of the shapes of molecules with the formulas CH_4, XeF_4, CO_2, and SO_2. Multiple bonds and lone pairs are not shown.

Ronald Gillespie, from McMaster University, developed the VSEPR model with Sir Ronald Nyholm from Australia.

TABLE 10.5 Orientation of Regions of High Electron Density in the Valence Shell of an Atom

NUMBER OF REGIONS OF HIGH ELECTRON DENSITY	ORIENTATION OF ELECTRON REGIONS TO ATTAIN MINIMUM REPULSION ENERGY
2	Linear, and opposite
3	Trigonal-planar
4	Tetrahedral
5	Trigonal-bipyramidal
6	Octahedral

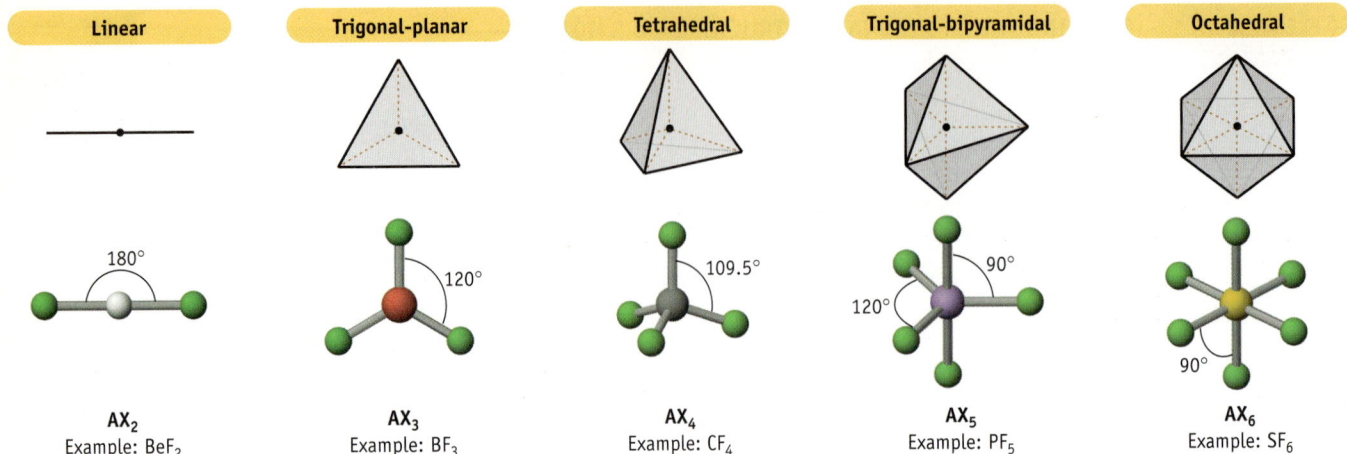

FIGURE 10.3 Orientations of localized regions of high electron density (bonds and lone pairs) in the valence shell of an atom. The orientations that achieve minimum repulsion energy depend on the number of regions.

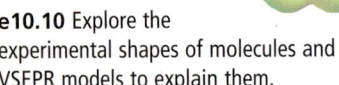

Molecular Modelling (Odyssey)

e10.10 Explore the experimental shapes of molecules and VSEPR models to explain them.

Think about It

e10.11 See why the ideal shapes in the VSEPR model are so intuitive.

Let's return to our molecule of focus in this chapter—methane. From gas phase electron diffraction, we know that in molecules of methane the H atoms are located at the corners of a tetrahedron, with the C atom in the centre. All four H atoms are in identical chemical environments. How can we explain this shape?

The Lewis structure of methane [<<Section 10.3] shows four regions of high electron density—all single bonds around the C atom in methane molecules. Using the VSEPR model, these will be oriented as far from each other as possible, and geometry tells us that this occurs when these bonds point toward the vertices of a tetrahedral shape with a C atom in the middle. This is consistent with the experimental evidence that the C—H bonds in methane molecules are at an angle of 109.5° to each other.

In a molecule of most alkanes, each C atom has four single bonds to other atoms, and these bonds are also directed tetrahedrally, regardless of the conformation [<<Section 9.5] of the molecule (Figure 10.4).

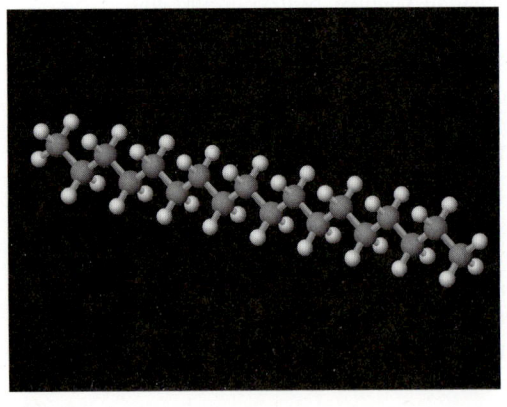

(a)

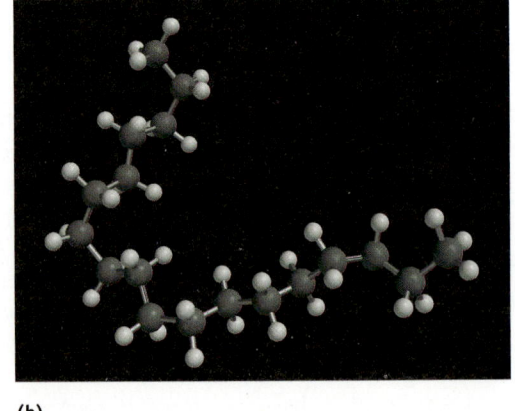

(b)

FIGURE 10.4 Ball-and-stick models of a molecule of octadecane (a) "stretched out,"' and (b) folded as a result of rotations about C—C single bonds after a 50 ps simulation in Odyssey. In both conformations, the spatial orientation of bonds around each C atom is tetrahedral, consistent with prediction using the VSEPR model.

Two Localized Regions of High Electron Density in a Valence Shell

Our model tells us that when there are two localized regions of high electron density in the valence shell of an atom, these regions will be arranged linear with each other, on opposite sides of the atom. This is the case for both beryllium chloride, $BeCl_2$ (four valence electrons in Be valence shell in two single bonds, no lone pairs), and carbon dioxide, CO_2 (eight valence electrons in the C atom in two double bonds, no lone pairs).

Only electron regions in the valence shell of the central atom affect the shape: the lone pairs on the Cl atoms in $BeCl_2$ molecules, and on the O atoms in CO_2 molecules, have no influence on spatial orientation of the bonds.

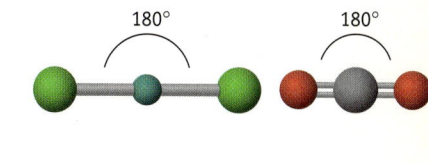

Three Localized Regions of High Electron Density in a Valence Shell

The simplest cases of molecules or ions that have three regions of high electron density in the valence shell of an atom are when all of the electron regions are bonds—whether single bonds or multiple bonds. Examples are boron trifluoride (BF_3, with six electrons in the valence shell of the B atom), and nitrate ions (NO_3^-, with eight electrons in the valence shell of the N atom).

If we had used any of the resonance forms of a nitrate ion, rather than the delocalized electron picture, we would have come to the same conclusion, because each of the resonance forms has three regions of electron matter (a double bond and two single bonds). Carbonate ions [<<Section 10.5] are isoelectronic with nitrate ions, and have the same geometry.

In the cases mentioned so far, the bonds around the central atom are equivalent, and the bond angles (120°) are those of a regular trigonal-planar arrangement. In the case of a formaldehyde molecule (right), however, there are two single bonds and one double bond around the carbon atom. The shape of the molecule is a slightly distorted trigonal-planar arrangement. Experiment evidence is that the H—C—H angle is less than 120°. This is generally interpreted to mean that double bond–single bond repulsions are greater than single bond–single bond repulsions. This is consistent with the greater electron density in double bonds, compared to that in single bonds.

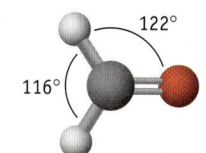

In those parts of molecules around the C atoms of a double bond, the bonds have an approximate trigonal-planar orientation. The simplest example is in ethene molecules, in which both sets of trigonal-planar bonds are in the same plane. Once again, the enhanced repulsion by the double bond is evident in the bond angles.

The same applies to both *cis*-but-2-ene and *trans*-but-2-ene.

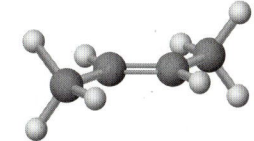

cis-**But-2-ene**   *trans*-**But-2-ene**

Although there is a trigonal-planar orientation of regions of high electron density around the two C atoms joined by the double bond, there is a tetrahedral arrangement of the four bonds (with no lone pairs) around the terminal C atoms.

The trigonal-planar arrangements of bonds around the C atoms of double bonds, in conjunction with the inability of double-bonded C atoms to rotate with respect to each other, is the reason that the melting points of polyunsaturated fatty acids have lower melting points than corresponding saturated fatty acids.

Stearic acid
(a saturated fatty acid)

Linolenic acid
(a polyunsaturated fatty acid)

Molecular Modelling (Odyssey)

e10.12 Compare molecular flexibilities in simulations of stearic acid and linolenic acid.

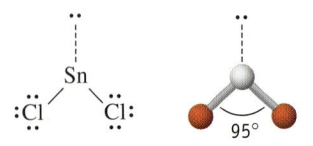

When saturated fatty acids, such as stearic acid, solidify, the regularly repeating structure at each C atom allows the molecules to fit closely together, maximizing the effect of dispersion forces. On the other hand, the bends and kinks in the carbon chain of unsaturated fatty acids prevent these molecules from packing together so snugly.

In some molecules, one of the three regions of electron density in the valence shell of an atom is a lone pair. In molecules of tin(II) chloride in the gas phase, although the VSEPR model tells us that the two bonding regions and the lone pair will assume a trigonal-planar arrangement, the shape of the molecule (defined by the positions of the nuclei) is described as bent, or V-shaped.

Why is the Cl—Sn—Cl bond angle much less than 120? This can be explained on the grounds that lone pair–single bond repulsion is generally stronger than single bond–single bond repulsion. This might be expected since the electrons of the lone pair are not drawn out toward another atom, so they occupy a larger area.

We can use the same argument to make sense of the shape of sulfur dioxide molecules (SO_2) and nitrite ions (NO_2^-). Both have three regions of high electron density, one of which is a lone pair. In each case, because the electron regions adopt a trigonal-planar arrangement, the species is V-shaped.

In the case of an SO_2 molecule, the enhanced repulsion by the lone pair is approximately cancelled by the enhanced repulsion between the double bonds—so the O—S—O bond angle is almost 120°. The N—O bonds in a NO_2^- ion have less electron density than the S—O bond in a SO_2 molecule, so it is no surprise that the O—N—O bond angle is only 115°.

Four Localized Regions of High Electron Density in a Valence Shell

We have already discussed molecules of methane and other alkanes in which the C atoms have four regions of high electron density around them—all bonding. In these cases, the distribution of electron regions is tetrahedral, as is the orientation of atoms around the C atoms.

Molecules of ammonia (NH_3) also have four electron regions around the N atom, and these are presumed to adopt a tetrahedral orientation, but since one of these is a lone pair, the shape of the molecule is a trigonal pyramid (Figure 10.5). A molecule of water also has four tetrahedrally oriented electron regions of which two are lone pairs, so the molecule is bent or V-shaped (Figure 10.5).

FOUR LOCALIZED REGIONS OF HIGH ELECTRON DENSITY

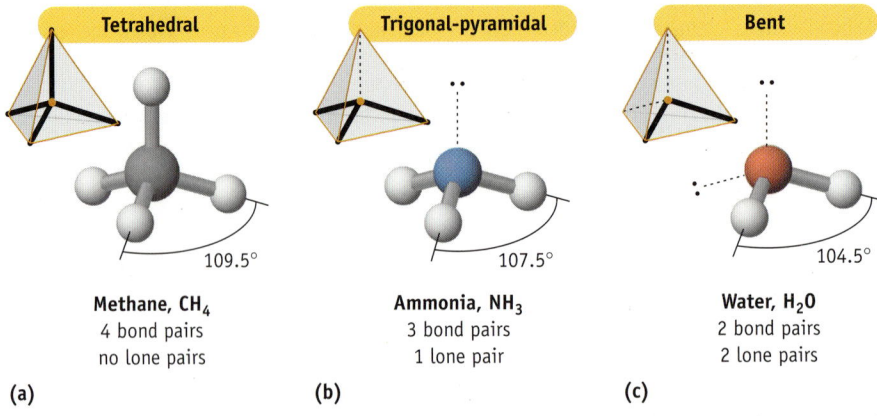

| Tetrahedral | Trigonal-pyramidal | Bent |

109.5°   107.5°   104.5°

Methane, CH₄   **Ammonia, NH₃**   **Water, H₂O**
4 bond pairs   3 bond pairs   2 bond pairs
no lone pairs   1 lone pair   2 lone pairs

(a)   (b)   (c)

FIGURE 10.5 The shapes of molecules of methane, ammonia, and water. All have four regions of high electron density around the central atom, so all have a tetrahedral distribution of these electron regions. (a) The C atom in a methane molecule has four bonding regions and no lone pairs, so the molecule has a tetrahedral shape. (b) The N atom in an ammonia molecule has three bonds and one lone pair, so the molecule has the shape of a trigonal pyramid. (c) In a molecule of water, the O atom has two bonds and two lone pairs, so the molecular shape is bent or V-shaped. The decrease in bond angles in the series can be explained by the fact that the lone pairs have a larger spatial requirement than the bonding pairs.

Earlier we asked why molecules of methane (CH$_4$) are tetrahedral, while those of xenon tetrafluoride (XeF$_4$) are planar with F—Xe—F bond angles 90°. A simple answer is that XeF$_4$ molecules have six localized electron regions in the valence shell of the Xe atom—four single bonds and two lone pairs—and our model tells us (Figure 10.3) that these will be oriented toward the corners of an octahedron, at 90° to each other. In which of the six directions are the lone pairs located? There are only two possibilities: they are at 90° to each other, or at 180° to each other. The experimental evidence is that the shape of the molecule is "square planar," from which we deduce that the most stable (least repulsion) condition is when the lone pairs are on opposite sides of the molecule.

Distribution of electron regions.

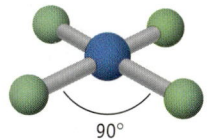

Position of atoms, giving shape of molecule.

Shapes of Small Molecules: A Summary

According to the VSEPR model, the orientation of localized electron regions in the valence shell of an atom minimizes repulsions among them. A physical model may help you remember the expected shapes for different numbers of electron regions. Blow up several balloons to a similar size, and imagine that each balloon represents an electron cloud. A repulsive force prevents other balloons from occupying the same space. When two, three, four, five, or six balloons are tied together at a central point (representing the nucleus and core electrons of a central atom), the balloons naturally form the shapes shown in Figure 10.6. These geometric arrangements minimize interactions between the balloons.

Taking It Further
e10.14 Read how shapes for central atoms with more than four regions of electron matter can be understood using the VSEPR model.

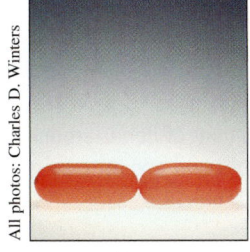
Linear

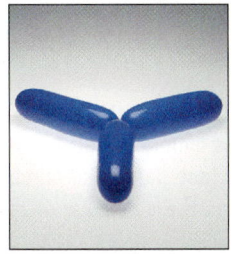
Trigonal planar

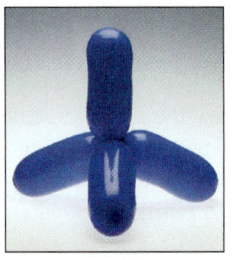
Tetrahedral

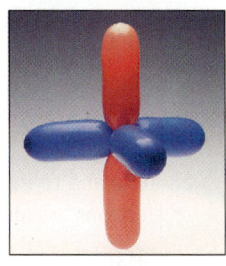
Trigonal bipyramidal

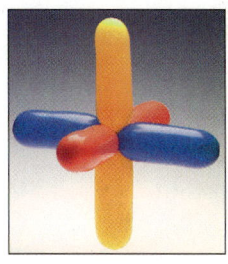
Octahedral

All photos: Charles D. Winters

FIGURE 10.6 A balloon model to help visualize the VSEPR model of localized regions of high electron density. If balloons of similar size and shape are tied together, they assume the arrangements shown.

The shapes of small molecules, defined by the positions of the atoms, depend on how many of these electron regions in a central atom's valence shell are bonding regions, and how many are lone pairs. This is summarized in Table 10.6. e10.11 and e10.15 extend this concept to atoms with five and six electron regions.

TABLE 10.6 Shapes of Small Molecules

NUMBER OF ELECTRON REGIONS IN VALENCE SHELL	ORIENTATION OF ELECTRON REGIONS	NUMBER OF ELECTRON REGIONS THAT ARE LONE PAIRS	EXAMPLE	SHAPE OF MOLECULE
2	Linear, opposite	0	BeCl$_2$	Linear
3	Trigonal-planar	0	BF$_3$	Trigonal-planar
		1	SO$_2$	V-shaped
4	Tetrahedral	0	CH$_4$	Tetrahedral
		1	NH$_3$	Trigonal-pyramidal
		2	H$_2$O	V-shaped

Shapes of Small Molecules and Molecular Polarity

One important implication of knowing the shape of a small molecule is that you can put this information together with what you know about the polarity of covalent bonds in the molecule to determine whether the molecule has a dipole moment. Polar molecules result from polar bonds which are not symmetrically distributed. Using the shapes in Figure 10.5, CH$_4$ should have no dipole moment, but both NH$_3$ and H$_2$O do, since electron pairs occupy at least one vertex of the tetrahedral central atom. See Section 6.3 for a full discussion.

Spatial Orientation of Atoms in Parts of Large Molecules

Lewis structures and the VSEPR model can also be used to find the relative positions of atoms in parts of more complicated molecules. Consider cysteine, a natural amino acid:

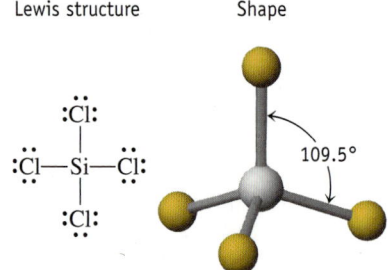

Cysteine, HSCH₂CH(NH₂)CO₂H

In a molecule of cysteine, four regions of electron matter occur around each of the S, N, C_2, C_3, and O atoms, so the arrangement around each of these atoms is tetrahedral. So, we would predict that the H—S—C, S—C—C, C—C—C, C—C—H, C—C—N, H—N—H, and C—O—H bond angles are approximately 109°. Finally, the O—C—O bond angle is 120° because the arrangement of electron regions around the C_1 atom is trigonal-planar.

WORKED EXAMPLE 10.5—PREDICTING THE SHAPES OF MOLECULES

Predict the shape of silicon tetrachloride ($SiCl_4$).

Strategy

The first step is to draw the Lewis structure. The Lewis structure does not need to be drawn with any particular shape because its purpose is merely to decide the number of bonds around an atom and to indicate whether there are any lone pairs. The number of bond and lone pairs of electrons around the central atom determines the molecular shape (Table 10.6).

Solution

The Lewis structure of $SiCl_4$ has four electron regions, all bond pairs, around the central Si atom, so we predict a tetrahedral structure for the $SiCl_4$ molecule, with Cl—Si—Cl bond angles of 109.5°. This agrees with the experimental structure for $SiCl_4$.

Lewis structure   Shape

109.5°

EXERCISE 10.13—PREDICTING THE SHAPES OF MOLECULES

What is the shape of a dichloromethane (CH_2Cl_2) molecule? Predict the Cl—C—Cl bond angle.

WORKED EXAMPLE 10.6—PREDICTING THE SHAPES OF MOLECULES AND IONS

What are the shapes of the ions (a) H_3O^+ and (b) ClF_2^+?

Strategy

In each case, draw the Lewis structures for each ion. Count the number of regions of electron matter (bonding electrons and lone pairs) around the central atom. Use Table 10.6 to decide on the arrangement of electron regions. Decide the shape of the ion, as defined by the relative positions of the atoms.

Solution

(a) The Lewis structure of the hydronium ion (H_3O^+) shows that the oxygen atom is surrounded by four electron regions, so we presume these to be oriented tetrahedrally.

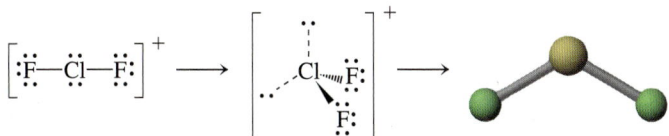

| Lewis structure | Orientation of electron regions, tetrahedral | Shape, trigonal pyramid |

Because one of the four regions is a lone pair, the central O atom and the three H atoms form a trigonal-pyramidal shape, corresponding with that of NH_3.

(b) Chlorine is the central atom in a ClF_2^+ ion. It is surrounded by four regions of high electron density, so these are distributed tetrahedrally. Because two of the four regions are lone pairs, the shape of the ion defined by the positions of the Cl and two F atoms is bent, or V-shaped.

EXERCISE 10.16—PREDICTING THE SHAPES OF MOLECULES AND IONS

Decide the three-dimensional orientation of electron regions around the B atom in a BF_3 molecule and a BF_4^- ion. Deduce the shape of these two species. What is the effect on the molecular geometry of BF_3 reacting with an F^- ion to form a BF_4^- ion?

EXERCISE 10.17—PREDICTING THE SHAPES OF MOLECULES AND IONS

Use Lewis structures and the VSEPR model to determine the distribution of regions of electron matter around atoms, and deduce the shapes of the following:

(a) a phosphate ion, PO_4^{3-}
(b) a phosphoric acid molecule, H_3PO_4
(c) a sulfate ion, SO_4^{2-}
(d) a sulfite ion, SO_3^{2-}
(e) an ethanol molecule
(f) acetone, $CH_3C(O)CH_3$, containing a $C{=}O$ double bond

10.8 The Valence Bond Model of Covalent Bonding

We know from experimental evidence that methane molecules exist as a unit comprising one C atom and four H atoms and that the H atoms are at the corners of a tetrahedron surrounding a C atom. The C—H bond distances are much shorter than the H—H distances, so we think of the molecule as being held together by bonds between the C atom and the H atoms. But what is a bond? When drawing a Lewis structure of a methane molecule, we portray a single bond as a shared pair of electrons. Which electrons? From which orbitals? If the electron configuration in a carbon atom is $1s^2 2s^2 2p^2$, which of these electrons are involved in the C—H bonds? Since the energies and distributions over space of electrons in the $2s$ orbital are different from those in the $2p$ orbital, how is it that the bonds and bond angles in CH_4 are identical?

Valence bond model

Bond due to electrons in overlapping orbitals

Molecule AB

Atom A   Atom B

Molecule AB

Bond due to electrons in a molecular orbital

Molecular orbital model

Comparison of the *valence bond* and *molecular orbital* models of covalent bonding.

In the following discussion, we think of electrons as waves, as a smear of negatively charged matter (an "electron cloud") or as a probabilistic summation over time of snapshots of the position of a particle [<<Chapter 8]. When we talk about electrons "occupying" an orbital, we mean they have a particular energy, and we can define regions of space in which the electron density is highest.

Molecular Modelling (Odyssey)

e10.16 Examine a calculated potential energy curve for the Cl_2 molecule.

The quantum mechanical model for the atom, which is the most successful way to explain the properties of atoms, describes electrons in atoms as waves. An atomic orbital has a specific energy related to electrostatic forces: an attractive force due to the positively charged atomic nucleus acting on an electron in that orbital, and a repulsive force acting on the electron due to the other electrons in the atom. If the orbital model of electrons in atoms is so powerful [<<Chapter 8], it seems reasonable that an orbital model could also be used to describe electrons in molecules.

Two common approaches to rationalizing chemical bonding based on orbitals are the **valence bond (VB) model** (or *valence bond theory*) and the **molecular orbital (MO) model** (often called *molecular orbital theory*). The former was developed largely by Linus Pauling and the latter by Robert S. Mulliken, another American chemist. The valence bond approach is closely tied to Lewis's idea of bonding electron pairs localized between atoms, and lone pairs localized on particular atoms: the electrons are considered to belong to the atoms from which the molecule is formed, although they are shared between atoms. By contrast, Mulliken's approach was to imagine orbitals that belong to the molecule as a whole, with electrons that are "spread out," or *delocalized*, over the molecule.

The VB model is generally used to provide a qualitative, visual picture of molecular structure and bonding, and is particularly useful for molecules made up of many atoms. It provides a good description of bonding for molecules in their ground, or lowest, energy state. In contrast, MO theory is essential if we want to know about the energies of electrons in molecules and, therefore, spectra, both in the ground state and in excited states. Among other things, it helps explain the colours of compounds, and for a few molecules such as NO and O_2, the MO theory is the only model to describe their bonding satisfactorily.

The valence bond model considers bonding electron pairs to be localized between atoms, and lone pairs to be localized on particular atoms. This contrasts with the molecular orbital approach that considers orbitals to belong to the molecule as a whole, with electrons that are "spread out," or delocalized, over the molecule.

The Valence Bond Model as Orbital Overlap

The essence of the valence bond model is that a covalent bond is the consequence of two electrons with opposite spins occupying an orbital formed by partial overlap of atomic orbitals on the atoms that are bonded. To take the simplest case possible, let's imagine formation of an H_2 molecule as two H atoms approach each other from at an infinite distance apart (even though H_2 is not usually formed in this way). Initially, when the two atoms are widely separated, they do not interact and the system (comprising these two atoms) has zero potential energy (Figure 10.7). As the atoms move closer together, the electrons on each atom begin to experience an attraction to the positive charge of the nucleus of the other atom, and the potential energy of the system is lowered—work must be done to separate them. At some point, repulsions between the nuclei, as well as between the electron clouds, "kick in." Calculations show that when the distance between the H atoms is 74 pm, the potential energy reaches a minimum, and the H_2 molecule is most stable. Significantly, 74 pm corresponds to the experimentally measured bond distance in the H_2 molecule.

Each H atom has a single electron. In the H_2 molecule, the two electrons are paired, mainly in the region of overlap of the $1s$ atomic orbitals, to form the bond. There is a net stabilization, measured by the extent to which the energy of the molecules is less than the energy of the separate atoms. The net stabilization can be calculated, and the calculated value approximates the experimentally determined bond energy. The agreement between theory and experiment on both bond distance and energy is evidence that this model has merit.

Bond formation is depicted in Figures 10.7 and 10.8 as occurring when the electron clouds on the two atoms overlap. This **overlap of atomic orbitals** distorts the electron cloud of each atom in such a way as to increase the probability of finding the bonding electrons in the region of space between the two nuclei. This outcome makes sense, because the distortion results in the electrons being situated so that they can be attracted equally to the two positively charged nuclei. Placing the electrons between the nuclei also matches the Lewis electron dot model.

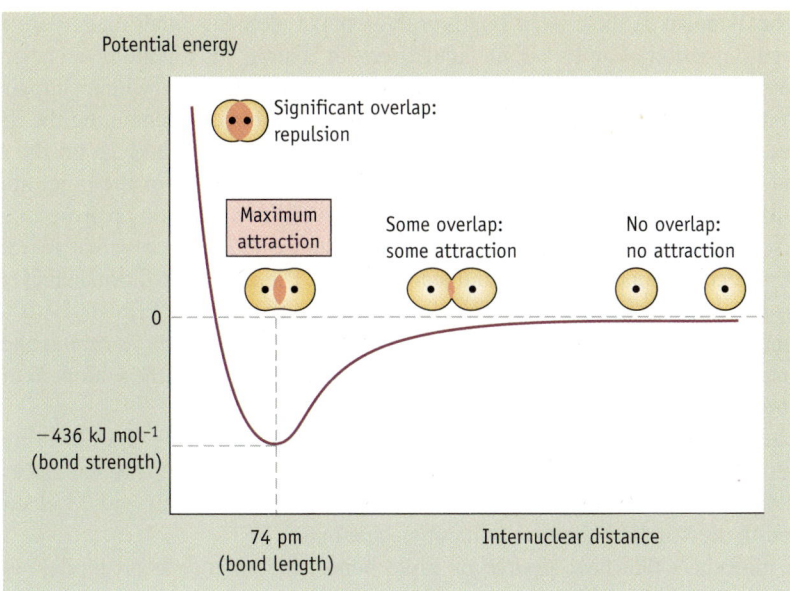

FIGURE 10.7 Potential energy change during H—H bond formation from isolated hydrogen atoms. The lowest energy situation is at an H—H separation of 74 pm, where there is overlap of 1s orbitals. At greater distances the overlap is less, and the bond is weaker. At H—H distances less than 74 pm, repulsions between the nuclei and between the electrons of the two atoms increase rapidly, and the potential energy curve rises steeply.

The covalent bond that arises from the overlap of two *s* orbitals, one from each of two atoms as in H_2, is called a **sigma (σ) bond**—a term designating bonds that are symmetrical as one rotates about the axis between the atoms (*cylindrical symmetry*).

The main points of the valence bond model of bonding are as follows:

* Orbitals overlap to form a bond between two atoms (see Figure 10.8).
* Two electrons, of opposite spin, can be accommodated in the overlapping orbitals. We usually imagine that one electron is supplied by each of the two bonded atoms.
* Because of orbital overlap, the bonding electrons have a higher probability of being within a region of space influenced by both nuclei. Both electrons are simultaneously attracted to both nuclei.
* The greater the degree of overlap of the orbitals, the stronger the bond is.

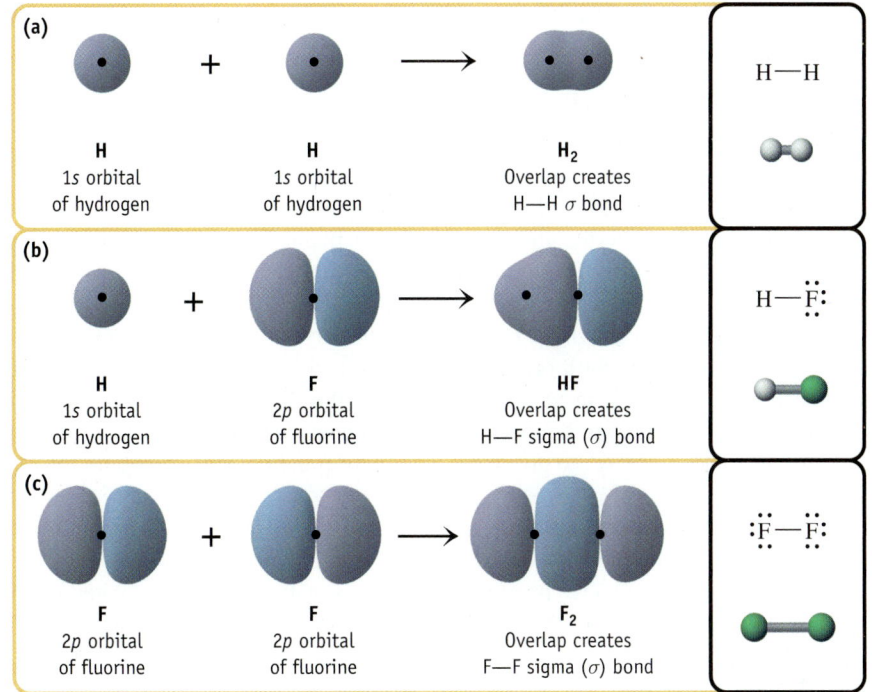

FIGURE 10.8 Covalent bond formation in H_2, HF, and F_2 (a) Overlap of hydrogen 1s atomic orbitals to form an H—H σ bond. (b) Overlap of hydrogen 1s and fluorine 2p atomic orbitals to form a σ bond in an HF molecule. (c) Head-on overlap of 2p atomic orbitals on fluorine atoms to form a σ bond in F_2.

Think about It

e10.17 Answer questions on the formation of a sigma bond.

What happens in the case of bonds with atoms of elements other than hydrogen? The imagined formation of an H—F molecule from an H atom and a F atom can be compared with formation of an H—H bond. We can imagine that an H atom (with an unpaired electron in its $1s$ orbital) approaches a fluorine atom along the axis containing the $2p$ orbital that has an unpaired electron. The orbitals ($1s$ on the H atom, and $2p$ on the F atom) become distorted as each atomic nucleus influences the electron of the other atom. Still closer together, the $1s$ and $2p$ orbitals overlap, and the two electrons pair up to give a σ bond (see Figure 10.8b). There is an optimal distance (92 pm) at which the energy is lowest; this corresponds to the bond distance in HF. The net stabilization achieved in this process is the H—F bond energy. The electron configuration of fluorine is $1s^2 2s^2 2p^5$. The electrons on the F atom that are not involved in bonding (two in the $2s$ orbital and four in the other two $2p$ orbitals) correspond with the lone pairs shown on the F atom in the Lewis structure.

Extension of this model gives a description of bonding in F_2. The $2p$ orbitals with unpaired electrons on the two atoms overlap in a head-on fashion, and the electron from each atom is paired in the resulting σ bond (Figure 10.8c). The $2s$ and $2p$ electrons not involved in the bond are the lone pairs on each atom.

In molecules that have double or triple bonds, the valence bond model suggests a second way in which p orbitals can overlap. Sideways overlap of p orbitals is called a **pi (π) bond**. A π bond has two regions of high electron density—one on either side of the internuclear axis. *A σ bond and a π bond together make a double bond, and a σ bond and two π bonds make up a triple bond.*

In the valence bond model, a covalent bond results from two electrons with opposite spins occupying an orbital formed by partial overlap of atomic orbitals. Sigma (σ) bonds may be formed by overlap of s orbitals on different atoms, an s orbital with a p orbital, or end-on overlap of two p orbitals. Pi (π) bonds have two regions of high electron density—one on either side of the internuclear axis.

The Valence Bond Model and Hybridization of Atomic Orbitals

Now that we have models for molecules of H_2, HF, and F_2, you might think we can create a model of molecules such as methane by overlap of orbitals on the C atom with the $1s$ orbital on each of the H atoms. However, if we apply the orbital overlap model used for H_2 and F_2 without modification to describe the bonding in CH_4, a problem arises. The three orbitals for the $2p$ valence electrons of a C atom are at right angles, 90° (Figure 10.9), and do not match the tetrahedral angle of 109.5°. Furthermore, a C atom in its ground state ($1s^2 2s^2 2p^2$) has only two unpaired electrons (in the $2p$ orbitals), not the four that are needed to allow formation of four bonds.

Chemists are resourceful, and the valence bond model can be "tweaked" to fit the experimental data. To describe the bonding in methane and other molecules, Linus Pauling proposed the theory of *orbital hybridization*. Imagine that the regions of highest electron density associated with s and p orbitals on atoms can be redistributed to form new sets of orbitals, called **hybrid orbitals**, which can bond to other atoms by orbital overlap. Different sets of hybrid orbitals can be invoked to rationalize different bonding situations (Figure 10.6). We need to know whether the distribution of electron regions around an atom is tetrahedral, trigonal-planar, or linear before we decide which set of hybrid orbitals to "conjure up" to give a description of bonding consistent with the experimental data.

For example, an isolated C atom has, according to the quantum model of atomic structure [<<Section 8.5], the electron configuration $1s^2 2s^2 2p_x^1 2p_y^1 2p_z^0$. To account for the shape of a methane molecule, we imagine that in this situation the $2s$ orbital and the three $2p$ orbitals are hybridized, and the result is a convenient set of four equivalent hybrid orbitals (called sp^3 orbitals) whose regions of highest electron density are directed out from the C atom in the tetrahedral directions. Because the hybrid orbitals have the same

"In 1931 . . . Professor Noyes . . . said that I probably would get the Nobel Prize someday. Well, I thought, that's nice of the old guy to say that, but I'm a little skeptical myself. And as the years went by, I thought, I don't do the sort of work for which Nobel Prizes are given."

Linus Pauling, 1954 Nobel Prize in Chemistry and 1962 Nobel Peace Prize.

Molecular Modelling

e10.18 Watch animations portraying the imaginary hybridization process and the changes in orbital energies.

energy, we can distribute the four valence electrons one to each orbital in accordance with Hund's rule.

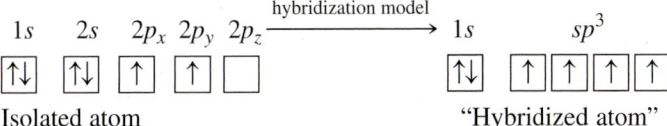

Now we can complete our picture of the bonding in methane according to this imagined situation: we overlap of each of the four equivalent sp^3 hybrid orbitals with the $1s$ orbitals of an H atom, resulting in four C—H single bonds oriented at 109.5° to each other (Figure 10.9).

Hybridization reconciles the known distribution of electron regions with the orbital overlap portrayal of bonding. A statement such as "the molecule is tetrahedral because the central atom is sp^3 hybridized" is backward. The tetrahedral distribution of electron regions around the C atom in methane is known, and a hybridization model of the C atom is one way to rationalize that knowledge.

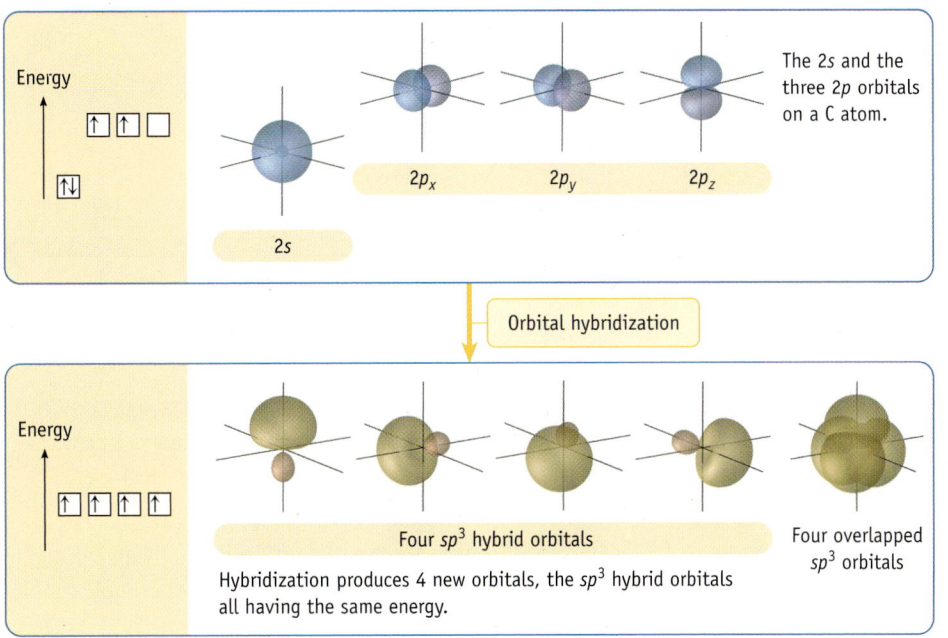

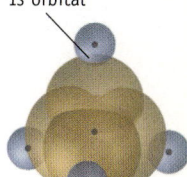

Each C–H bond uses one C atom sp^3 hybrid orbital and a H atom $1s$ orbital

Molecular model, CH_4 Orbital representation

FIGURE 10.9 A model of bonding in the CH_4 molecule, based on imagined hybrid orbitals.

Two important principles govern the use of imagined hybrid orbitals. First, *the number of hybrid orbitals must be the same as the number of atomic orbitals that are hybridized.* Second, *the hybrid orbitals are more directed from the central atom toward the terminal atoms than are the unhybridized atomic orbitals, leading to better orbital overlap and stronger bonds.*

The sets of hybrid orbitals that we can imagine to be formed from s and p atomic orbitals are illustrated in Figure 10.10. The following features are important:

- The most useful set of hybrid orbitals in a particular case will depend on how many localized regions of high electron density surround the atom in question.
- When four regions of electron matter (single bonds, multiple bonds, or lone pairs) are around an atom, use four hybrid orbitals imagined to have been formed by hybridization of an s orbital and three p orbitals in the valence shell of the atom. These sp^3 hybrid orbitals are directed at 109.5° to each other.
- When three regions of electron matter are around an atom, use three hybrid orbitals imagined to have been formed by hybridization of an s orbital and two p orbitals. These sp^2 hybrid orbitals are directed at 120° to each other, all in the same plane.
- When two regions of electron matter are around an atom, use two hybrid orbitals imagined to have been formed by hybridization of an s orbital and a p orbital. These sp hybrid orbitals are directed at 180° to each other, on opposite sides of the atom.

Other sets of hybrid orbitals (sp^3d^2 and sp^3d), formed from s, p, and d orbitals, are used to model the bonding in molecules that have an atom with either octahedral or trigonal-bipyramidal distribution of electron regions around it.

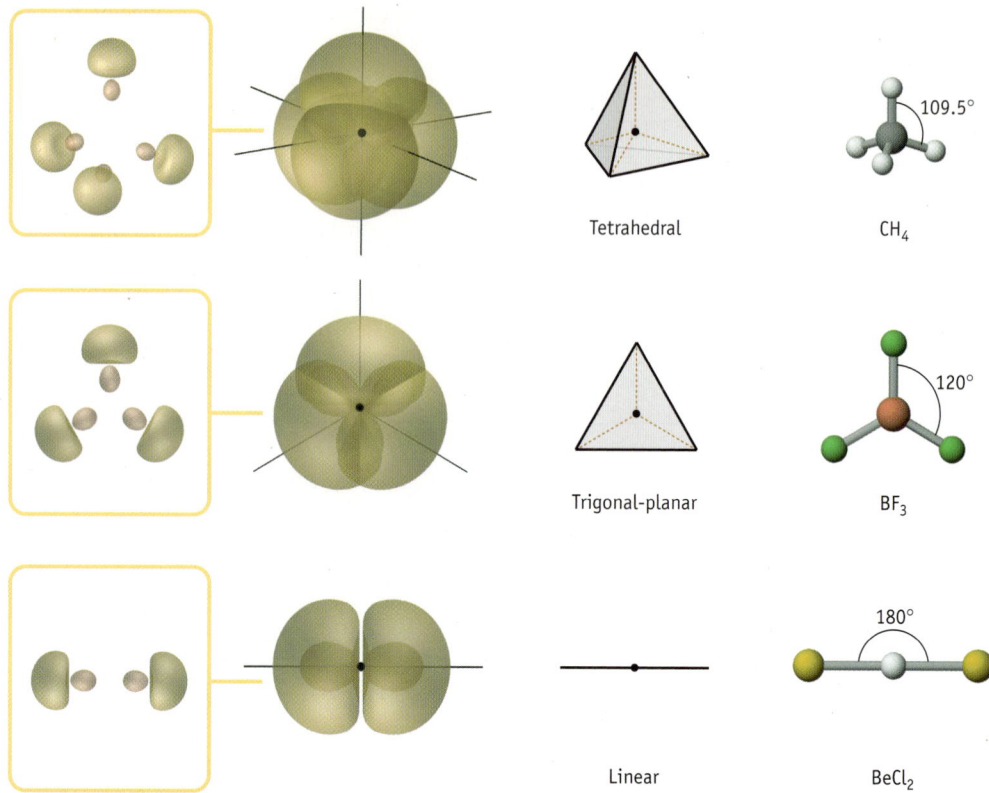

FIGURE 10.10 Hybrid orbitals invoked for 2, 3, or 4 electron regions around an atom. The spatial orientations of the hybrid orbital sets are shown in the right column. We cannot suggest the use of a particular hybrid orbital set unless we know how many localized regions of electron matter surround an atom.

Let us now examine cases of each type of hybridization in simple molecules. Keep in mind, however, that this model can be applied to atoms in even the most complex molecules, such as DNA.

> In a model of orbital hybridization, we imagine the regions of highest electron density associated with *s* and *p* orbitals are redistributed to form new sets of hybrid orbitals, which can overlap to form bonds to other atoms. Different sets of hybrid orbitals (sp^3, sp^2, and sp) are used for different bonding situations to rationalize experimental geometries.

Atoms with Four Regions of High Electron Density

We have already considered the bonding in methane, using the hybrid orbital model (Figure 10.10). An ammonia molecule has four localized electron regions (three single bonds and a lone pair) around the N atom. We imagine overlap of three of the sp^3 hybrid orbitals on N with the $1s$ orbitals on three H atoms to from single bonds, leaving one sp^3 hybrid orbital not involved in bonding. Of the total eight valence electrons, we can allocate two to each single bond and two as a lone pair in the orbital not involved in bonding (Figure 10.11). *Remember that hybridization models result from a thought exercise—we don't actually form NH_3 molecules by reacting a N atom with three H atoms.*

Water also has four localized electron regions (two single bonds and two lone pairs) in the valence shell of the O atom, so again we invoke an sp^3 hybridization model. Overlap of two of the sp^3 hybrid orbitals with $1s$ orbitals on H atoms account for two single bonds, each with two electrons, and non-bonding pairs of electrons in each of the other hybrid orbitals accounts for all eight valence electrons in the molecule (Figure 10.11).

Think about It

e10.19 Combine valence orbitals in a model to rationalize bonding in molecules.

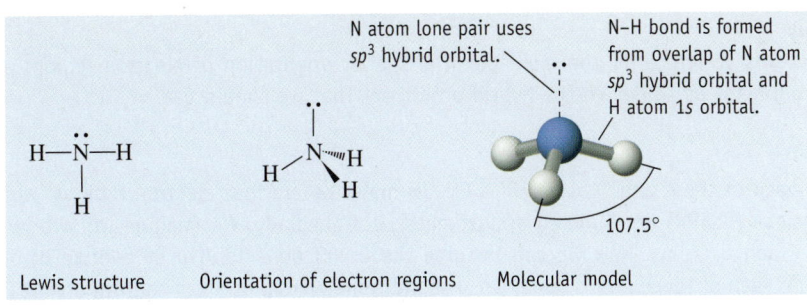

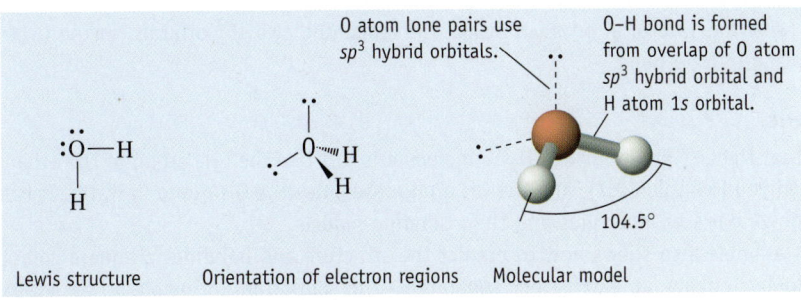

FIGURE 10.11 Models of bonding in an ammonia molecule and a water molecule. These models are based on imaginary formation of four sp^3 hybrid orbitals on the central atom, and overlap of some of these with 1s orbitals on the H atoms to from single bonds. The other hybrid orbitals are the location of lone (non-bonding) pairs of electrons.

WORKED EXAMPLE 10.7—A VALENCE BOND MODEL OF BONDING IN ETHANE

Describe the bonding in a molecule of ethane (C_2H_6) using the valence bond model of covalent bonding.

Strategy

First, draw the Lewis structure and predict the distribution of electron regions around both carbon atoms. Next, decide which hybridization model gives rise to hybrid orbitals with electron density in the appropriate directions (Figure 10.10). Finally, describe covalent bonds as regions of orbital overlap and distribute electron pairs among bonding and non-bonding localized regions.

Solution

Around each C atom are four electron regions (three single bonds to H atoms, and one to the other C atom), so we can presume a tetrahedral distribution of these. Presume hybridization to form four sp^3 orbitals. Now we can imagine the C—C bond formed by overlap of an sp^3 orbital on one C atom with an sp^3 orbital on the other, and each of the C—H bonds formed by overlap of an sp^3 orbital on the C atom with a hydrogen 1s orbital.

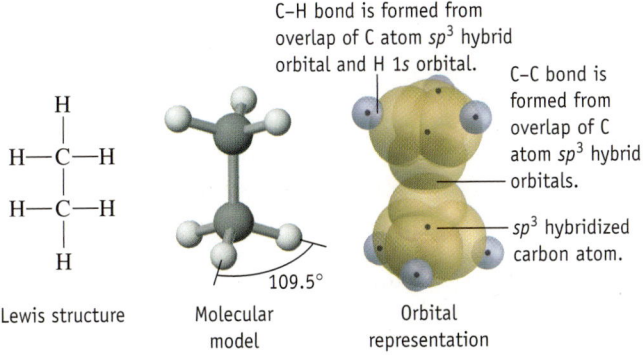

WORKED EXAMPLE 10.8—A VALENCE BOND MODEL OF BONDING IN METHANOL

Describe the bonding in a methanol molecule (CH_3OH) using the valence bond model of covalent bonding.

Strategy

Draw a Lewis structure for the molecule. The spatial orientation of electron regions around each atom helps us to decide the hybrid orbital set that we should use.

Solution

Around each of the C and O atoms in a CH_3OH molecule are four electron regions, which we can presume (VSEPR) are directed approximately tetrahedrally. We imagine sp^3 hybridization on the C and O atoms. Now we can imagine the C—O bond formed by overlap of one sp^3 orbital on each of these atoms, each C—H bond from overlap of an sp^3 orbital on the C atom with the $1s$ orbital on an H atom, and the O—H bond from overlap of an sp^3 orbital on the O atom with the $1s$ orbital on an H atom. The remaining two sp^3 orbitals on the O atom are the location of lone pairs.

Comment

Notice that the CH_3 group in the CH_3OH molecule is just like the CH_3 group in the ethane molecule (Worked Example 10.7), and the OH group resembles the OH group in water. It is helpful to recognize parts of molecules and their bonding models.

This example also shows how to predict the structure and bonding in a more complicated molecule by looking at each atom separately. This important principle is essential when dealing with molecules made up of many atoms.

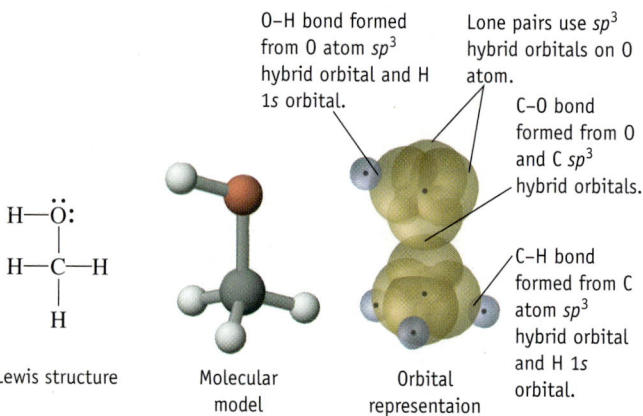

O–H bond formed from O atom sp^3 hybrid orbital and H $1s$ orbital.

Lone pairs use sp^3 hybrid orbitals on O atom.

C–O bond formed from O and C sp^3 hybrid orbitals.

C–H bond formed from C atom sp^3 hybrid orbital and H $1s$ orbital.

Lewis structure   Molecular model   Orbital representaion

EXERCISE 10.18—VALENCE BOND DESCRIPTION OF BONDING

Use the valence bond model to portray the bonding in (a) the hydronium ion (H_3O^+), and (b) methylamine (CH_3NH_2).

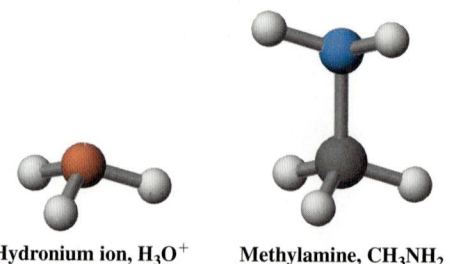

Hydronium ion, H_3O^+ Methylamine, CH_3NH_2

When molecules or parts of molecules have an atom whose valence shell has four bonding or non-bonding regions of high electron density, we imagine four sp^3 hybrid orbitals (from an s orbital and three p orbitals) that overlap with orbitals on atoms bonded to it.

Atoms with Three Localized Regions of High Electron Density

Molecules and ions in which an atom has a trigonal-planar distribution of electron density regions are common. For example, molecules of BF_3 and other boron halides are trigonal-planar, as are a number of other species, such as NO_3^- and CO_3^{2-} ions. The arrangement of electron regions around the C atoms in molecules of ethene ($CH_2=CH_2$), the central O atom in O_3 molecules, and N atom in NO_2^- ions are also trigonal-planar.

Our model accounts for trigonal-planar distribution of localized electron regions around an atom, by hybridizing an s orbital and two p orbitals to form three sp^2 hybrid orbitals in a plane 120° apart (Figure 10.12). If p_x and p_y orbitals are used in hybrid orbital formation, the three sp^2 hybrid orbitals will lie in the x_y-plane. The p_z orbital not used to form these hybrid orbitals is perpendicular to the plane containing the three sp^2 orbitals (Figure 10.12).

Three Localized Regions of Electron Density and Only Single Bonds A boron trifluoride molecule has trigonal-planar distribution of localized electron regions—all of them single bonds. Each B—F bond in the molecule can be imagined to result from overlap of an sp^2 orbital on the B atom with a p orbital on a F atom. Notice that the p_z orbital on the B atom, which is not included in hybridization, is not occupied by electrons.

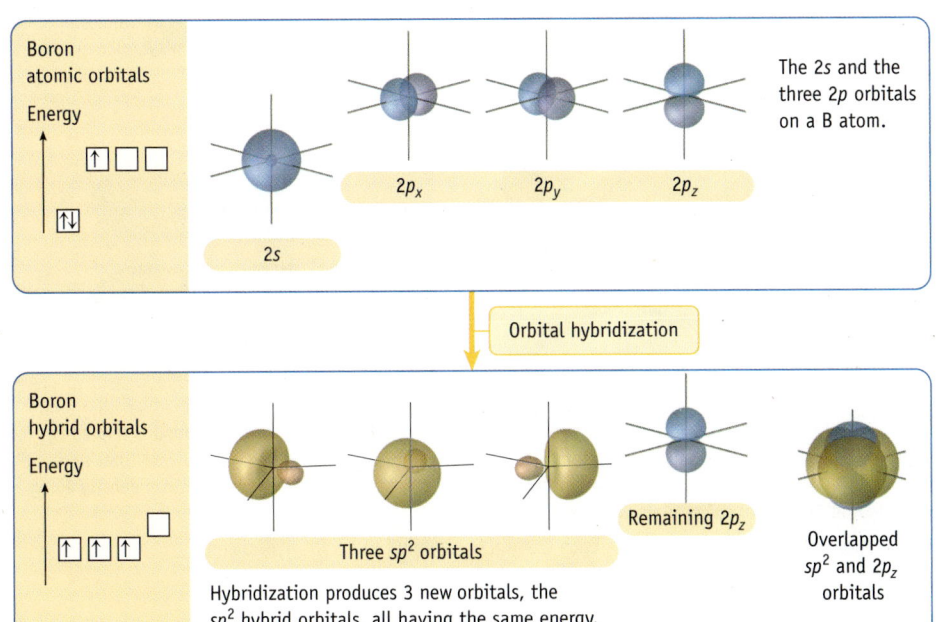

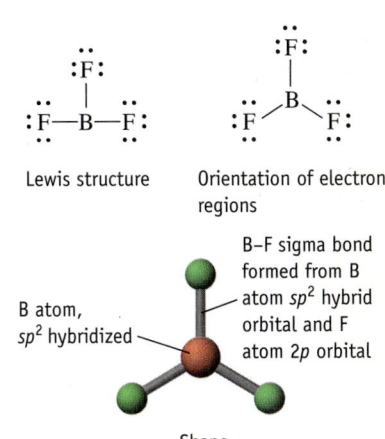

FIGURE 10.12 A valence bond model of bonding in a BF_3 molecule. This model is based on sp^2 hybridization of orbitals on the B atom, and the overlap of three of these singly occupied hybrid orbitals with a singly occupied p orbital on each F atom to form single bonds.

Three Localized Regions of Electron Density and Both Single and Double Bonds In the valence bond model, bond formation results from overlap of two orbitals on adjacent atoms, so a double bond requires *two* sets of overlapping orbitals and *two* electron pairs.

A molecule of ethene ($H_2C=CH_2$) has all six atoms in a plane, with H—C—H and H—C—C angles of approximately 120°. Around each C atom are three localized regions of high electron density—two single bonds and one double bond.

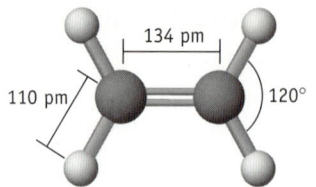

134 pm

110 pm

120°

A ball-and-stick model of an ethene molecule, C_2H_4

With a trigonal-planar arrangement of electron regions, we imagine sp^2 hybridization of orbitals on each C atom, so that each has three sp^2 hybrid orbitals in the plane of the molecule and an unhybridized p orbital perpendicular to that plane (Figure 10.13). Because each C atom is involved in four bonds, we put one electron in each of these orbitals.

Now we can visualize a C—C σ bond resulting from overlap of an sp^2 hybrid orbital on one C atom with an sp^2 hybrid orbital on the other, and two C—H bonds on each C atom from overlap of sp^2 orbitals with the $1s$ orbital on H atoms. This leaves an unhybridized p orbital on each C atom. Sideways overlap of these singly occupied p orbitals gives us a second C—C bond—a π bond with the electron pair occupying an orbital with electron density above and below the plane of the six atoms (Figure 10.13).

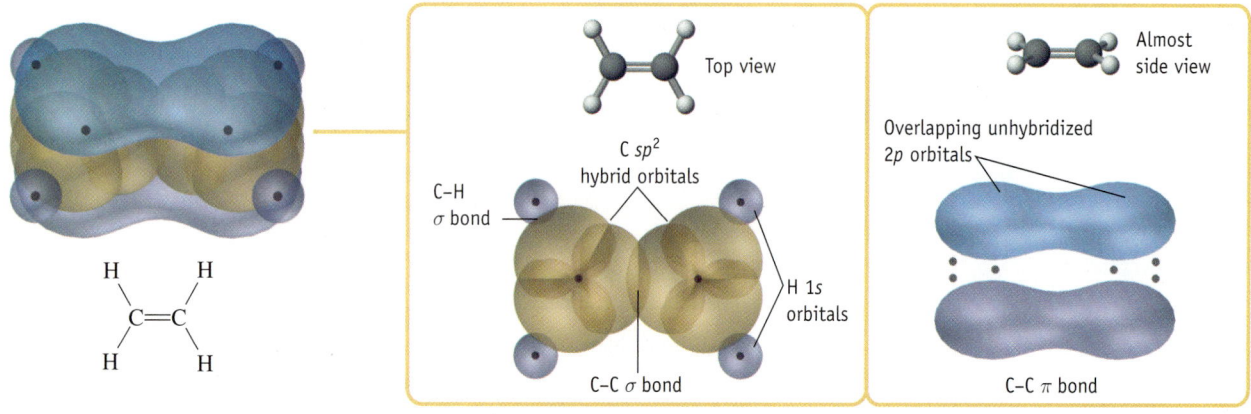

(a) Lewis structure and bonding of ethene, C_2H_4.

(b) The C–H σ bonds are formed by overlap of C atom sp^2 hybrid orbitals with H atom $1s$ orbitals. The σ bond between C atoms arises from overlap of sp^2 orbitals.

(c) The carbon–carbon π bond is formed by overlap of an unhybridized $2p$ orbital on each atom. Note the lack of electron density along the C–C bond axis.

FIGURE 10.13 The valence bond model of bonding in ethene (C_2H_4). Each C atom is assumed to be sp^2 hybridized.

A π bond can form *only* if (1) there are unhybridized p orbitals on adjacent atoms, and (2) the p orbitals are perpendicular to the plane of the molecule and parallel to one another. This happens only if the sp^2 orbitals of both carbon atoms are in the same plane. *The π bonding requires that all six atoms of the molecule (or portion of the molecule) lie in one plane.*

C=O, C=S, and C=N double bonds are quite common. Consider a molecule of formaldehyde (HCHO), with a C=O double bond (Figure 10.14). Localized electron regions directed in a trigonal-planar arrangement around the C atom invites use of sp^2 hybridization for the C atom. We picture C—O and C—H σ bonds formed by overlap of sp^2 hybrid orbitals with singly occupied orbitals on the O and H atoms. An unhybridized p orbital on the C atom is perpendicular to the molecular plane and is available for π bonding with an orbital on the O atom.

What orbitals on oxygen are used in this model? The approach in Figure 10.14 assumes sp^2 hybridization on the O atom. This uses one O atom sp^2 orbital in σ bond

Think about It

e10.20 Examine the valence bond description of bonding in ethene and formaldehyde.

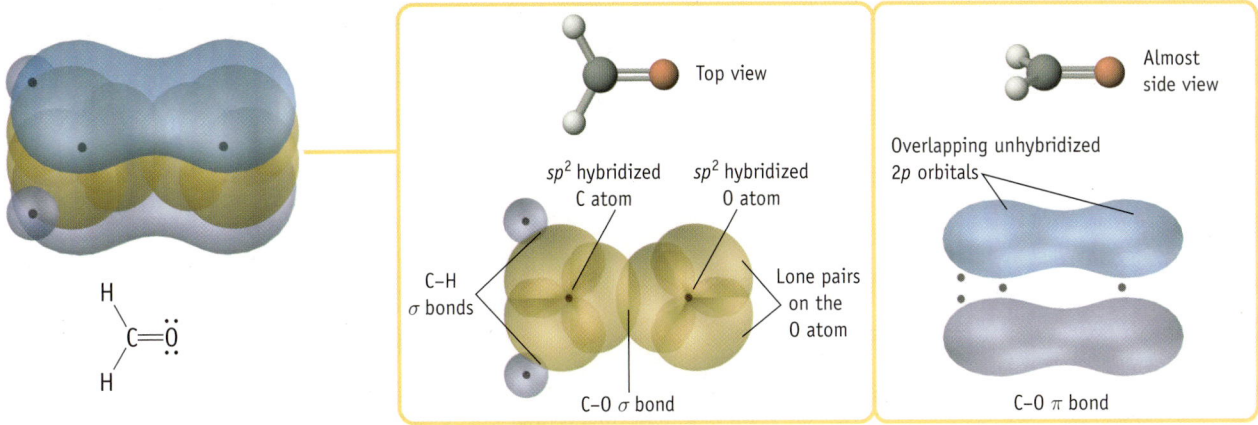

(a) Lewis structure and bonding of formaldehyde, CH_2O.

(b) The C–H σ bonds are formed by overlap of C atom sp^2 hybrid orbitals with H atom $1s$ orbitals. The σ bond between C and O atoms arises from overlap of sp^2 orbitals.

(c) The C–O π bond comes from the sideways overlap of p orbitals on the two atoms.

FIGURE 10.14 Valence bond description of bonding in a molecule of formaldehyde (CH_2O).

formation, leaving two sp^2 orbitals to accommodate lone pairs. The remaining p orbital on the O atom participates in the π bond.

WORKED EXAMPLE 10.9—BONDING IN A MOLECULE OF ACETIC ACID

Using the valence bond model, describe the bonding in a molecule of acetic acid (CH_3COOH), the important ingredient in vinegar.

Strategy

Write a Lewis structure and deduce the spatial orientation of localized electron regions around each atom using VSEPR. Use this geometry to decide which hybrid orbitals to use in σ bonding. If unhybridized p orbitals are available, π bonding between the C and O atoms can occur.

Solution

The C atom of the CH_3 group has tetrahedral distribution of electron regions, so our model tells us to use sp^3 hybridization: three sp^3 orbitals overlap with s orbitals on H atoms to form the C—H bonds, and the fourth is used to bond to the carbonyl C atom. This C atom has a trigonal-planar orientation of electron regions, so we use sp^2 hybridization. The C—C bond is pictured as formed using one of these sp^2 orbitals, and the other two are used to form the σ bonds to the two O atoms. The oxygen of the O—H group has four electron regions, which are tetrahedrally distributed and so we use sp^3 hybridization. This O atom uses two sp^3 orbitals to bond to the adjacent C and the H atoms, and two sp^3 orbitals accommodate the two lone pairs.

Finally, the C=O double bond can be described just as in the HCHO molecule (Figure 10.14). Guided by our model, we assume that both the C and O atoms are sp^2 hybridized, and the unhybridized p orbital remaining on each atom is used to form the C—O π bond.

EXERCISE 10.19—BONDING IN A MOLECULE OF ACETONE

EXERCISE 10.19—BONDING IN A MOLECULE OF ACETONE

Use the valence bond model to describe the bonding in a molecule of acetone (CH_3COCH_3).

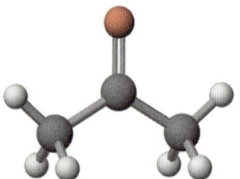

When an atom in a molecule, ion, or part of a molecule or ion has three regions of high electron density in its valence shell, and all are single bonds, we imagine overlap of three singly occupied sp^2 hybrid orbitals with singly occupied orbitals on atoms to which it is bonded, forming three σ bonds. When a double bond is present, the model shows an additional sideways overlap of two unhybridized p orbitals to form a π bond.

Atoms with Two Localized Regions of High Electron Density

For molecules in which an atom has two localized regions of high electron density, linear and on opposite sides of the atom, a valence bond model imagines one s and one p orbital hybridized to form two sp hybrid orbitals (Figure 10.15). If we choose to use the p_x orbital in hybridization, then the sp hybrid orbitals are oriented along the x-axis. The p_y and p_z orbitals are perpendicular to this axis.

Two Localized Regions of Electron Density and Only Single Bonds Beryllium dichloride ($BeCl_2$) is a solid at ambient temperatures and pressures. When it is heated to more than 520°C, however, it forms a vapour comprising linear $BeCl_2$ molecules.

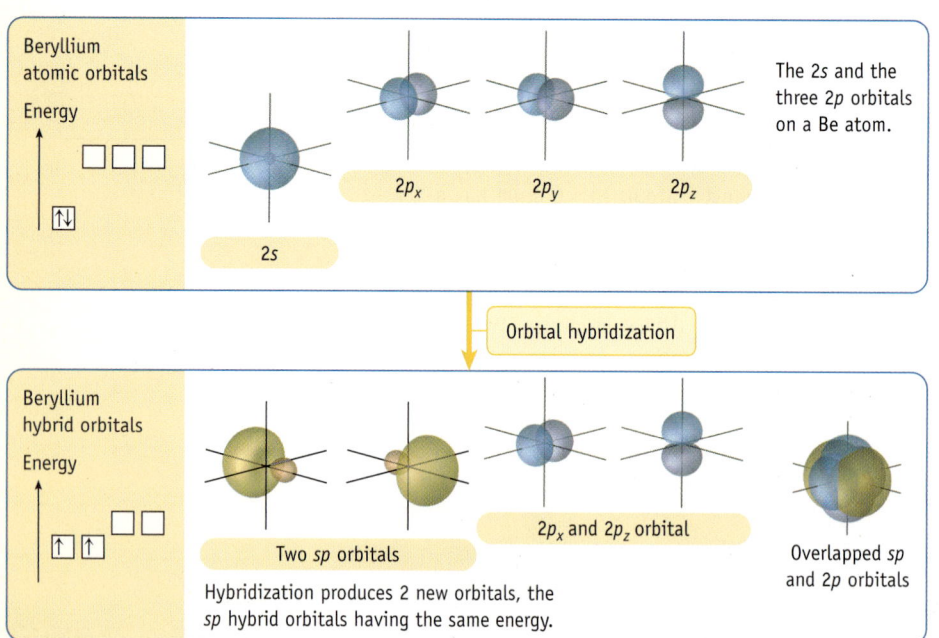

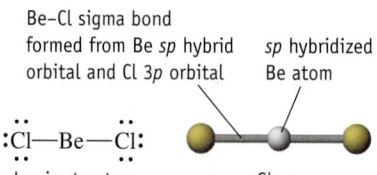

FIGURE 10.15 A valence bond model, with hybridization, of bonding in a linear molecule. Because only one p orbital on each atom is involved in hybridization, two p orbitals remain. These p orbitals are perpendicular to each other and to the axis along which the two sp hybrid orbitals lie.

In the gas phase, $BeCl_2$ is a linear molecule, so we visualize sp hybridization for the Be atom. Hybridization of the $2s$ and $2p_x$ orbitals gives the two sp hybrid orbitals that lie along the x-axis. Each Be—Cl bond can be considered to be due to overlap of an sp hybrid orbital on the Be atom with a $3p$ orbital on a Cl atom. In this molecule, there are only two electron pairs around the Be atom, so the p_y and p_z orbitals are not occupied (Figure 10.15).

Two Localized Regions of Electron Density Including Triple Bonds Molecules or parts of molecules that have a triple bond are linear. To explain triple bonds with the valence bond model, we consider three sets of overlapping atomic orbitals, with a pair of electrons in each set.

Acetylene (H—C≡C—H) has linear molecules with a triple bond. The four atoms lie in a straight line. If we invoke sp hybridization to account for this bonding, on each C atom, there are two sp orbitals at 180° to each other. We can imagine that an sp orbital on one C atom overlaps with an sp orbital on the other to form a C—C σ bond, and that the other sp orbital on each C atom overlaps with the $1s$ orbital on an H atom to form a C—H σ bond. Two unhybridized p orbitals remain on each carbon, and they are oriented so that it is possible for sideways side-on overlap to result in two C—C π bonds (Figure 10.16). So, in the C—C region, there are three bonds—one σ bond and two π bonds. The ten valence electrons in the molecule are paired up in each of the three σ bonds and the two π bonds.

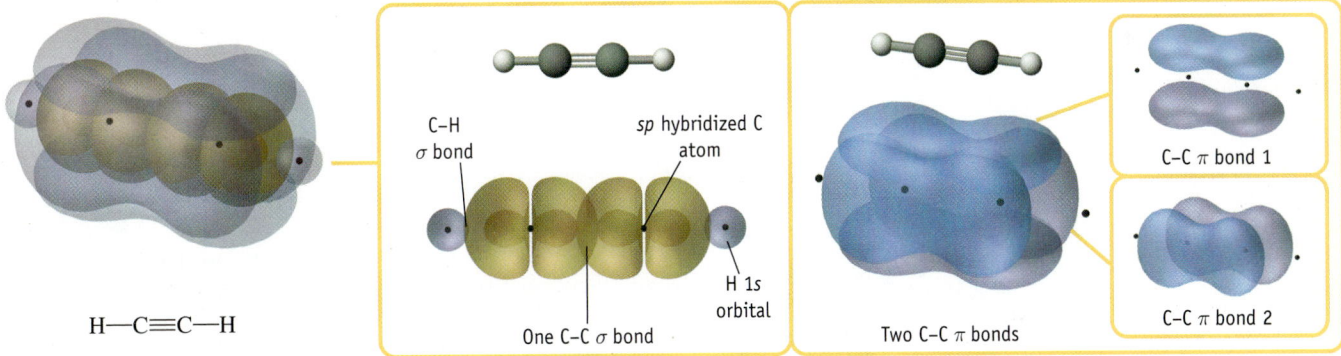

FIGURE 10.16 A valence bond model of the bonding in acetylene.

EXERCISE 10.20—MOLECULES WITH TRIPLE BONDS

Use the valence bond model to describe the bonding in a nitrogen molecule (N_2).

EXERCISE 10.21—BONDING AND HYBRIDIZATION

Estimate the H—C—H, H—C—C, and C—C—N bond angles in acetonitrile ($CH_3C≡N$). Indicate which hybridization scheme (of both C atoms and the N atom) you would use to account for bonding with the valence bond model, and describe the bonding.

When an atom in a molecule, ion, or part of a molecule or ion has two regions of high electron density in its valence shell, and both are single bonds, we imagine overlap of two singly occupied sp hybrid orbitals with singly occupied orbitals on atoms to which it is bonded, forming two σ bonds. When a triple bond is present, the model shows two additional π bonds, each resulting from sideways overlap of two unhybridized p orbitals.

The Valence Bond Model, Resonance, and Electron Delocalization

The valence bond model of bonding, incorporating hybridization of atomic orbitals, provides a helpful picture of bonding in molecules with multiple bonds that need the concepts of resonance or electron delocalization.

Resonance structures

OR

Delocalized electron model

Benzene (C_6H_6) is a classic example of a molecule to which the concept of resonance is applied. As seen in Section 10.5, the ring is flat, and all of the C—C bonds are the same length (139 pm). The C—C bond order is 1.5, the average of a single bond and a double bond.

To describe a benzene molecule from the valence bond approach, we recognize that the orientation of electron regions around each C atom is trigonal-planar, and assume sp^2 hybridization on each of them. Now we can imagine that each C—H bond is formed by overlap of an sp^2 orbital of a C atom with a $1s$ orbital of a H atom, and the C—C σ bonds from end-on overlap of sp^2 orbitals on adjacent C atoms. After accounting for the σ bonding, an unhybridized p orbital, with its electron density above and below the plane of the ring, remains on each C atom, and each is occupied by a single electron. If we imagine sideways overlap of these p orbitals on pairs of C atoms, we arrive at a picture of the benzene molecule with three π bonds—corresponding with either of the resonance forms.

However, we might ask why a p orbital on a C atom would sideways overlap with another on one adjacent C atom, but not with the p orbital on the other adjacent C atom. If we imagine sideways overlap of the p orbitals with those on both adjacent C atoms, our resulting picture is one of unbroken π bonding (above and below the plane of the ring) all the way around the ring of C atoms (Figure 10.17), consistent with the delocalized electron picture.

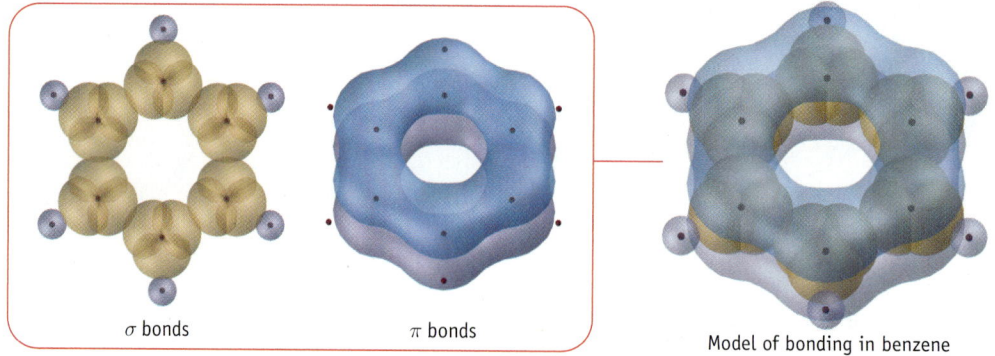

σ bonds π bonds

Model of bonding in benzene

σ and π bonding in benzene

FIGURE 10.17 Valence bond model of bonding with π electron delocalization in a molecule of benzene. (*Left*) The C atoms of the ring are bonded to each other through σ bonds imagined to be formed by end-on overlap of sp^2 hybrid orbitals on the C atoms. We also use C atom sp^2 hybrid orbitals (overlapping with the $1s$ orbitals on H atoms) to account for the C—H bonds. The π bonding framework of the molecule arises from sideways overlap of C atom unhybridized p orbitals in both directions. Since these orbitals are perpendicular to the ring, there are regions of high electron density of π bonding above and below the plane of the ring. (*Right*) A composite of σ and π bonding in benzene.

This delocalized electron picture of benzene is consistent with the C—C bond order 1.5, because there are three electrons on average between each pair of adjacent C atoms—two in the σ bond, and one in the π bonding system. Since a bond had traditionally been represented as two shared electrons, the C atoms are sometimes said to be joined by $1\frac{1}{2}$ 'classical two-electron bonds'.

The ozone molecule has two equivalent O—O bonds for which we can draw resonance structures, or a delocalized-electron picture with bond order 1.5 [<<Section 10.5]. We can make sense of the delocalized structure using the valence bond model if we specify a hybridization scheme and overlapping orbitals consistent with experimental findings. Since there is trigonal planar distribution of electron regions around the central O atom, we presume sp^2 hybridization of the O atoms. We can imagine σ bond formation by end-on overlap between sp^2 orbitals on adjacent O atoms. Now there is a p orbital on each of the O atoms, all perpendicular to the plane of the ion. There is no logical reason

why sideways overlap between the *p* orbitals on the O atoms should be preferred in any one of the directions, so we can presume electron density dispersed as π bonding over both O—O bonds.

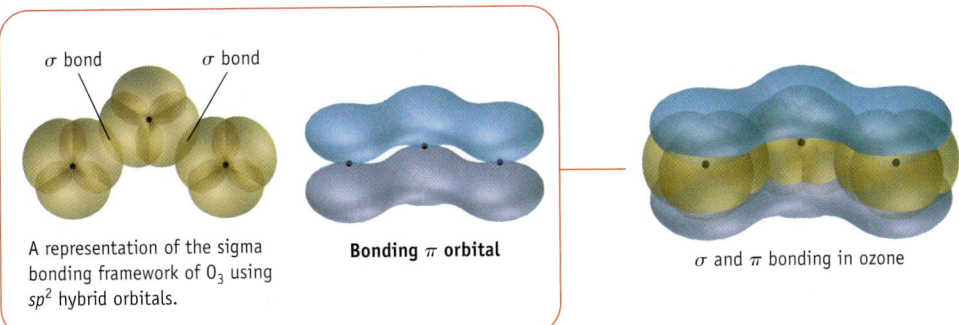

σ bond   σ bond

A representation of the sigma bonding framework of O_3 using sp^2 hybrid orbitals.

Bonding π orbital

σ and π bonding in ozone

When we apply valence bond modeling to molecules of substances, we often find limitations and difficulties when allocating electrons to particular orbitals on the various atoms. Neither is it a simple matter to justify the bond orders in terms of the numbers of electrons in each bond, as is done for benzene molecules, above. There is no escaping the fact that while any model gives us useful ways to visualize some things that we observe, each model has limitations. The delocalized electron model foreshadows the **molecular orbital model** of bonding [>>Section 10.9], in which the electrons are considered to belong to the molecule as a whole.

EXERCISE 10.22—VALENCE BOND MODEL AND ELECTRON DELOCALIZATION

Use the valence bond model, including the concept of π electron delocalization, to portray the bonding in acetate ion, with two equivalent C—O bonds.

When resonance structures are required in valence bond representations, we can explain equivalence of bonds by delocalization of electrons through sideways overlap of *p* orbitals on adjacent atoms.

Interactive Exercises 10.23–10.24

SUBMIT

Predict the hybridization of orbitals on selected atoms in a molecular structure.

10.9   Molecular Orbital Theory of Bonding

We introduce one more model of bonding in molecules and ions. We call it a complementary model, because it explains some experimental data that the previous models cannot. This doesn't mean that we throw away the other models—scientists usually use the simplest model available to make predictions and give satisfactory explanations of the phenomena under study.

Why Do We Need Another Model?

Experimental evidence that cannot be explained by Lewis structures, the VSEPR model, or the valence bond model and therefore requires a more sophisticated model includes the following:

"If the model is right, the rest is easy."

Sir Arthur Eddington

- Photoelectron spectroscopy informs us that the bonding electrons in methane do not have equal energies.
- Paramagnetism displayed by dioxygen (O_2) indicates that it has unpaired electrons.

The technique of **photoelectron spectroscopy (PES)** is used to estimate the energies of electrons in their various orbitals in molecules. High-energy light photons, with a particular and known energy, are directed at a substance, causing electrons to be ejected from the molecules. The kinetic energies of the ejected electrons can be measured, and it is

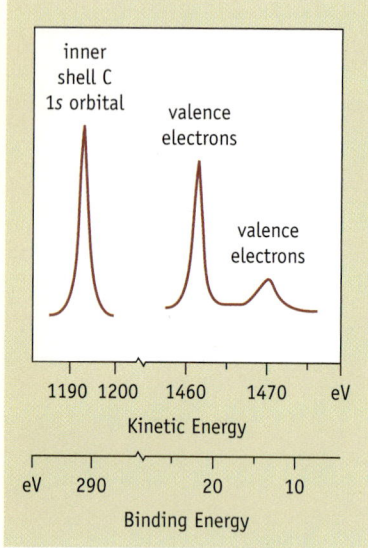

FIGURE 10.18 Photoelectron spectrum of methane. Core electrons have higher binding energy to the nucleus than valence electrons, so they require more energy to eject them, and they are ejected with less kinetic energy (for a given photon energy).

Think about It

e10.21 Watch a video demonstration of this paramagnetic liquid.

found that the ejected electrons have only certain values of kinetic energy. The technique assumes that electrons ejected with different kinetic energies have come from orbitals with different energies. The orbital energies can be calculated using a relationship based on conservation of energy principles:

Energy of photons = energy required to eject electrons from orbitals + kinetic energy of ejected electrons

Higher kinetic energy of ejected electrons indicates lower ionization energy of the electrons in the orbitals in the molecules from which they were ejected.

In the case of methane, three peaks of kinetic energy of ejected electrons are detected (Figure 10.18).

The peak corresponding with the lowest orbital energy is attributed to electrons ejected from the core $1s$ orbital of the carbon atoms. This leaves us with the conclusion that the valence electrons in methane are in two different energy levels—a finding inconsistent with the model of four equivalent single bonds produced by overlap of four equivalent sp^3 orbitals on each C atom with four equivalent $1s$ orbitals on the H atoms.

Dioxygen (O_2) is paramagnetic: it is attracted to the poles of a powerful magnet (Figure 10.19). This property is attributed to the presence of unpaired electrons in orbitals [<<Sections 8.4, 8.5], which is not predicted by any of the models of bonding in dioxygen discussed so far.

FIGURE 10.19 Liquid oxygen. Oxygen gas condenses (*left*) to a pale blue liquid at −183°C (*middle*). Oxygen is paramagnetic and clings to the poles of a magnet (*right*). In still conditions, a bubble containing pure oxygen gas, and sitting on the surface of a bowl of water, can be attracted toward a powerful magnet.

Some experimental data from photoelectron spectroscopy and magnetic measurements cannot be explained by the valence bond model of bonding, and require a more sophisticated model of bonding.

Principles of Molecular Orbital Theory

Molecular orbital (MO) theory is an alternative model used to consider the distribution of electrons that holds molecules together. In contrast to the localized bond and lone-pair electrons of the valence bond model, the MO model assumes that in any molecule the electrons occupy orbitals that are spread out, or delocalized, over the entire molecule. These orbitals are called *molecular orbitals.* As is the case with atomic orbitals, each molecular orbital has the capacity for two electrons, which have a particular amount of energy. The electron densities of electrons in different orbitals have different, and characteristic, distributions over space in relation to the positions of the nuclei. In other words, the contribution to the electron cloud due to electrons in a given molecular orbital has a particular "shape" (just as the electron density distributions in $2s$, $2p$, and $3d$ orbitals in atoms have characteristic "shapes").

Recall [<<Chapter 8] that the energies of electrons and their distributions over space around the nucleus in atoms is best achieved by regarding the electrons as waves, and then applying the wave equation to determine the characteristics of stable waveforms. Each waveform (or orbital) is defined by a solution to the wave equation called a wave function (ψ) whose value varies over xyz space. The two electrons "allowed" in each orbital have a characteristic energy, and the distribution of their density over space is defined by ψ^2. In

All photos: Charles D. Winters

this way, chemists have calculated the energies and "shapes" of electrons in the various orbitals of atoms.

Corresponding principles apply to molecular orbitals: electrons in a particular orbital have an exact energy, and for each orbital there is a characteristic distribution of electron density over the space around the various atoms. We assign electrons to the molecular orbitals of a ground-state molecule in order of energies from the lowest, according to the Pauli exclusion principle and Hund's rule [<<Section 8.5]. A problem is that chemists must use various approximate methods to obtain solutions to the wave equations for molecules and calculate the energies and wave functions for molecular orbitals. We will focus on one of these.

In MO theory, chemists begin with a given arrangement of atoms in the molecule at the known bond distances, and attempt to calculate the various solutions (the orbitals) to the wave equation. The approximate method that we will use here is called *Linear Combination of Atomic Orbitals (LCAO)*, and involves a mathematical operation on orbitals of the atoms to deduce orbitals of the molecule. We demonstrate by applying the method to homonuclear diatomic molecules, starting with H_2. **Homonuclear diatomic molecules** are composed of only two atoms of the same element, such as H_2, Li_2, and N_2.

In our H_2 molecule, we label the H atoms A and B. If these atoms were in their ground state and far apart, the $1s$ electron in each would be characterized by the wave functions $\psi_{1s}(A)$ and $\psi_{1s}(B)$. Recall that the three-dimensional distribution of electron density is related to ψ^2 [<<Section 8.4]—that is, mathematically described in each case by $[\psi_{1s}(A)]^2$ and $[\psi_{1s}(B)]^2$. Now imagine bringing the atoms together to form an H_2 molecule. According to the LCAO method, the wave functions for two molecular orbitals (σ_{1s} and σ^*_{1s}) can be calculated from the wave functions of the $1s$ atomic orbital on each of the bonded atoms using the following equations:

$$\sigma_{1s} = \psi_{1s}(A) + \psi_{1s}(B) \qquad \text{and} \qquad \sigma^*_{1s} = \psi_{1s}(A) - \psi_{1s}(B)$$

Then, we define the distribution of electron density for the first of the molecular orbitals:

$$(\sigma_{1s})^2 = [\psi_{1s}(A) + \psi_{1s}(B)]^2 = [\psi_{1s}(A)]^2 + [\psi_{1s}(B)]^2 + 2[\psi_{1s}(A) \times \psi_{1s}(B)]$$

Recall some algebra:
$(a + b)^2 = a^2 + b^2 + 2ab$

The electron density of an electron in this σ_{1s} molecular orbital is greater in the internuclear region (by the amount $2\,\psi_{1s}(A) \times \psi_{1s}(B)$) than the sum of the electron densities of the separate atoms if no bonding interaction took place, $[\psi_{1s}(A)]^2 + [\psi_{1s}(B)]^2$. The distribution of electron density of any electrons in this orbital is such that their effect is to pull the atoms together, so it is called a **bonding molecular orbital**. If the only electrons in the molecule were in this orbital, work would need to be done to separate the atoms, so we can say that the energy of electrons in the σ_{1s} bonding orbital are lower in energy than those in the $1s$ atomic orbitals. The approximate "shape" of the resulting σ_{1s} electron density distribution is shown in Figure 10.20. The symbol σ in the orbital name σ_{1s} is because it is cylindrically symmetrical (consistent with the symbols used for bonds in the valence bond model).

The distribution of electron density in the σ^*_{1s} molecular orbital is given by

$$(\sigma^*_{1s})^2 = [\psi_{1s}(A) - \psi_{1s}(B)]^2 = [\psi_{1s}(A)]^2 + [\psi_{1s}(B)]^2 - 2[\psi_{1s}(A) \times \psi_{1s}(B)]$$

In this case, the electron density in the internuclear region is less than the sum of the densities of the electrons in $1s$ orbitals if there were no bonding interactions between the atoms. The net effect is that repulsions between the nuclei dominate nucleus–electron attractions, and if the only electrons in the molecule were in this orbital, the atoms would experience a repulsion pushing them apart. Such molecular orbitals are called **antibonding molecular orbitals**. The asterisk is used to signify that a molecular orbital is antibonding. *Antibonding orbitals have no counterpart in valence bond theory.* If the only electrons in the molecule were in this orbital, work would need to be done to push the atoms together, so we can say that the electrons in the σ^*_{1s} antibonding orbital are higher in energy than those in the $1s$ atomic orbitals (Figure 10.20). The approximate "shape" of the electron density distribution in the σ^*_{1s} orbital is also cylindrically symmetrical, so we use the symbol σ.

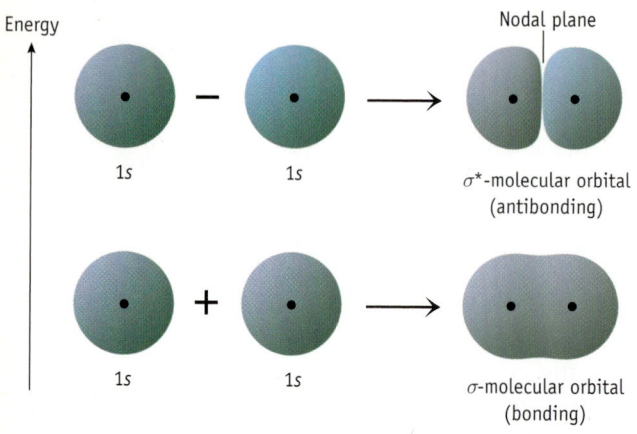

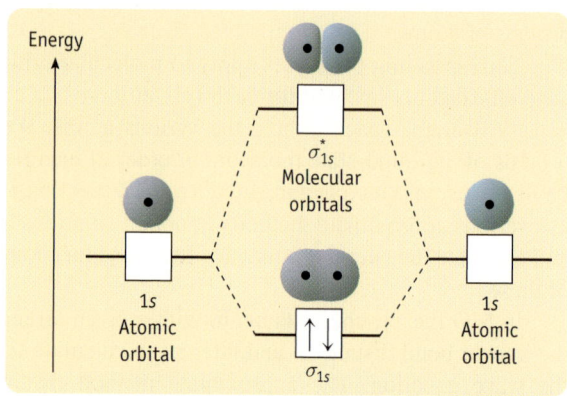

FIGURE 10.20 Molecular orbitals. (a) Bonding and antibonding σ molecular orbitals are derived by mathematical manipulation on two $1s$ atomic orbitals on adjacent atoms. Notice the presence of a node (a plane on which there is zero probability of finding an electron in the antibonding orbital. (b) A molecular orbital diagram for a ground-state H_2 molecule. The two electrons are placed in the σ_{1s} orbital, the lowest-energy molecular orbital.

Think about It

e10.22 Examine the differences between the bonding and antibonding molecular orbitals in the H_2 molecule.

The symbol σ_{1s}, read as "sigma $1s$," refers to a cylindrically symmetrical molecular bonding orbital whose energy and electron density distribution is deduced by mathematical derivation from $1s$ atomic orbitals. The symbol σ^*_{1s}, read as "sigma star $1s$" or "antibonding sigma $1s$" refers to a cylindrically symmetrical molecular antibonding orbital whose energy and electron density distribution is deduced by mathematical derivation from $1s$ atomic orbitals.

Orbitals are characterized as electron waves; therefore, a way to view molecular orbital formation is to assume that two electron waves, one from each atom, interfere with each other. The interference can be constructive, giving a bonding MO, or destructive, giving an antibonding MO.

The molecular orbital model assumes that the electrons occupy molecular orbitals that are spread over the entire molecule, rather than localized between bonded atoms. Bonding orbitals have the net effect of holding atoms together; antibonding orbitals have the net effect of repulsion between the atoms of the molecule. The two lowest-energy MOs are the σ_{1s} and σ^*_{1s} orbitals.

MO Electron Configuration in Ground-State H_2, He_2, and Some Ions

We now have the relative energies of electrons in lowest-energy molecular orbitals, and so we can deduce the electron configuration in ground-state molecules with just a few electrons, by assigning electrons to orbitals of successively higher energy, and application of the Pauli exclusion principle and Hund's rule [<<Section 8.5]. For example, the H_2 molecule has two electrons, and we can place both of these in the σ_{1s} orbital. We write the ground-state electron configuration of H_2 molecules as $(\sigma_{1s})^2$.

Both electrons in a ground-state H_2 molecule are in bonding orbitals—this rationalizes our knowledge that the molecule is more stable than the separate atoms—that is, that the H atoms are bonded together. To make some measure of the strength of bonding, we refer again to the concept of **bond order**. In this context, it has a similar meaning to its meaning in relation to the valence bond model [<<Section 10.5], but in the context of the molecular orbital model, it is defined as

$$\text{Bond order} = \tfrac{1}{2}(\text{number of electrons in bonding MOs} - \text{number of electrons in antibonding MOs}) \quad (10.2)$$

For ground-state H_2 molecules, the bond order is consistent with what we deduce in the valence bond model (that in the H_2 molecule there is a single bond):

$$\text{Bond order} = \tfrac{1}{2}(2 - 0) = 1$$

Let's contemplate a dihelium molecule (He$_2$). We assign the four electrons in the ground-state molecule as shown in Figure 10.21, deducing the electron configuration $(\sigma_{1s})^2(\sigma*_{1s})^2$.

In ground-state He$_2$ molecules,

$$\text{Bond order} = \tfrac{1}{2}(2 - 2) = 0$$

This suggests that there is negligible bonding interaction in ground-state He$_2$ molecules: the stabilizing effect of the two electrons in the σ_{1s} molecular orbital is offset by the destabilizing effect of the two electrons in the $\sigma*_{1s}$ orbital. This is consistent with what we already know—elemental helium exists in the form of single atoms and not as a diatomic molecule.

Some species have fractional bond orders. Consider the ion He$_2{}^+$. Its molecular orbital electron configuration is $(\sigma_{1s})^2(\sigma*_{1s})^1$. In this ion, there are two electrons in a bonding molecular orbital, but only one in an antibonding orbital. MO theory predicts that He$_2{}^+$ should have a bond order of 0.5; that is, a weak bond should exist between He atoms in such a species. This ion has been identified in the gas phase using special experimental techniques.

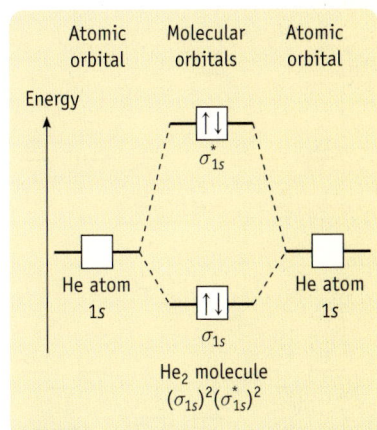

FIGURE 10.21 A molecular orbital energy-level diagram for a ground state dihelium molecule (He$_2$). This diagram provides a rationalization for the non-existence of the molecule. Both the bonding (σ_{1s}) and antibonding orbital ($\sigma*_{1s}$) have two electrons. The anti-bonding effect of the electrons in the $\sigma*_{1s}$ orbital cancels the bonding effect of the electrons in the σ_{1s} orbital.

WORKED EXAMPLE 10.10—DEDUCING ELECTRON CONFIGURATION AND BOND ORDER

Write the electron configuration of the H$_2{}^-$ ion in molecular orbital terms. What is the bond order of the ion?

Strategy

Count the number of valence electrons in the ion and then assign them in the MO energy-level diagram. Find the bond order using Equation 10.2.

Solution

This ion has three electrons (one each from the H atoms plus one for the negative charge). We deduce an electronic configuration $(\sigma_{1s})^2(\sigma*_{1s})^1$ and a bond order of 0.5. This suggests that the H$_2{}^-$ ion is stable, but weakly bound.

Think about It

e10.23 Examine the molecular orbital models for the dinitrogen and dioxygen molecules.

EXERCISE 10.25—DEDUCING ELECTRON CONFIGURATION AND BOND ORDER

What is the electron configuration of the H$_2{}^+$ ion? Compare the bond order of this ion with those of He$_2{}^+$ and H$_2{}^-$. Do you expect H$_2{}^+$ to exist?

Bond order (Equation 10.2) is defined as

Bond order = $\tfrac{1}{2}$(number of electrons in bonding MOs − number of electrons in antibonding MOs)

By allocating electrons to MOs from the lowest-energy orbitals, using the Pauli exclusion principle and Hund's rule, we can calculate bond orders to explain the stability of H$_2$ molecules and the instability of He$_2$ molecules.

MO Electron Configuration in Ground-State Li$_2$ and Be$_2$ Molecules

A principle of the LCAO method of calculating molecular orbital wave functions is that the atomic orbitals used should be of similar energy. This principle becomes important when we move past He$_2$ to Li$_2$ (dilithium) and molecules with even more electrons.

A lithium atom has electrons in two orbitals of the s type (1s and 2s), so $\psi_{1s} \pm \psi_{2s}$ combinations are theoretically possible. Because the 1s and 2s orbitals are quite different in energy, however, this interaction can be disregarded. So the molecular orbitals

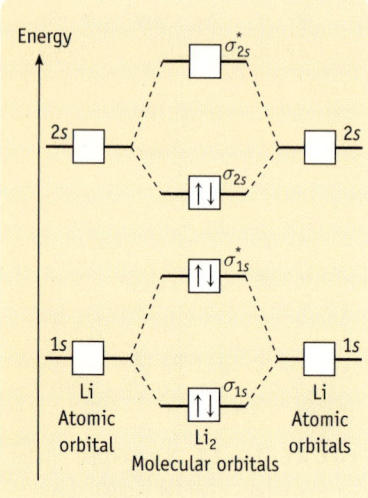

FIGURE 10.22 Energy-level diagram for the Li_2 molecule, derived by combination of atomic orbitals on Li atoms. The molecular orbitals are created by mathematically combining atomic orbitals of similar energies. The electron configuration is shown for a ground-state Li_2 molecule. According to the MO model, the electron configuration in a ground-state Li_2 molecule is $(\sigma_{1s})^2(\sigma*_{1s})^2(\sigma_{2s})^2$.

come only from $\psi_{1s} \pm \psi_{1s}$ and $\psi_{2s} \pm \psi_{2s}$ combinations, giving us molecular orbitals σ_{1s}, $\sigma*_{1s}$, σ_{2s}, and $\sigma*_{2s}$ (Figure 10.22), and the six electrons can be assigned in the usual way.

$$\text{Bond order} = \tfrac{1}{2}(4 - 2) = 1$$

The bonding effect of the σ_{1s} electrons is cancelled by the antibonding effect of the $\sigma*_{1s}$ electrons, so these pairs make no net contribution to bonding in Li_2. Bonding in Li_2 is due to the electron pair in the σ_{2s} orbital. The fact that the σ_{1s} and $\sigma*_{1s}$ electron pairs of Li_2 make no net contribution to bonding is exactly what we assume in drawing Lewis electron dot structures: core electrons make no contribution to bonding, and are ignored. In molecular orbital terms, core electrons are assigned to bonding and antibonding molecular orbitals that offset one another.

A diberyllium molecule (Be_2) is not expected to exist. Its electron configuration is

$$Be_2: \text{[core electrons]}(\sigma_{2s})^2(\sigma*_{2s})^2$$

The effects of σ_{2s} and $\sigma*_{2s}$ electrons cancel, and there is no net bonding. The bond order is 0, so the molecule would be unstable. Indeed, it does not exist in the ground state.

WORKED EXAMPLE 10.11—ELECTRON CONFIGURATION, BOND ORDER, AND STABILITY OF MOLECULES

Be_2 molecules are unstable. But what about the Be_2^+ ion? Deduce its electron configuration and calculate the bond order. Do you expect the ion to be stable?

Strategy

Count the number of electrons in the ion and place them in the MO energy-level diagram in Figure 10.22. Write the electron configuration and calculate the bond order using Equation 10.2.

Solution

The Be_2^+ ion has seven electrons (Be_2 has eight), of which four are core electrons (in the σ_{1s} and $\sigma*_{1s}$ MOs). The remaining three electrons are assigned to the σ_{2s} and $\sigma*_{2s}$ molecular orbitals, so the MO electron configuration is [core electron] $(\sigma_{2s})^2(\sigma*_{2s})^1$. This means the net bond order is 0.5, and so we would predict that Be_2^+ ions are stable (although the bond is weak).

EXERCISE 10.26—ELECTRON CONFIGURATION, BOND ORDER, AND STABILITY OF MOLECULES

Could the Li_2^- anion exist? What is the ion's bond order?

When there are more than four electrons in a diatomic molecule, bond order can be calculated from the deduced electron configuration without consideration of the electrons in the "core" σ_{1s} and $\sigma*_{1s}$ molecular orbitals because these make no net contribution to bonding.

Molecular Orbitals from *p* Atomic Orbitals

To use the LCAO method to calculate molecular orbitals for such important homonuclear diatomic molecules as N_2, O_2, and F_2, we need to use both *s* and *p* atomic orbitals because the molecules have more than eight electrons.

Mathematical combination of 1*s* and 2*s* atomic orbitals gives us σ molecular orbitals (both bonding and antibonding) as illustrated in Figure 10.20. Similarly, combination of the 2p atomic orbitals directed along the internuclear axis of the molecule gives rise to a σ_{2p} bonding MO and a σ^*_{2p} antibonding MO (Figure 10.23).

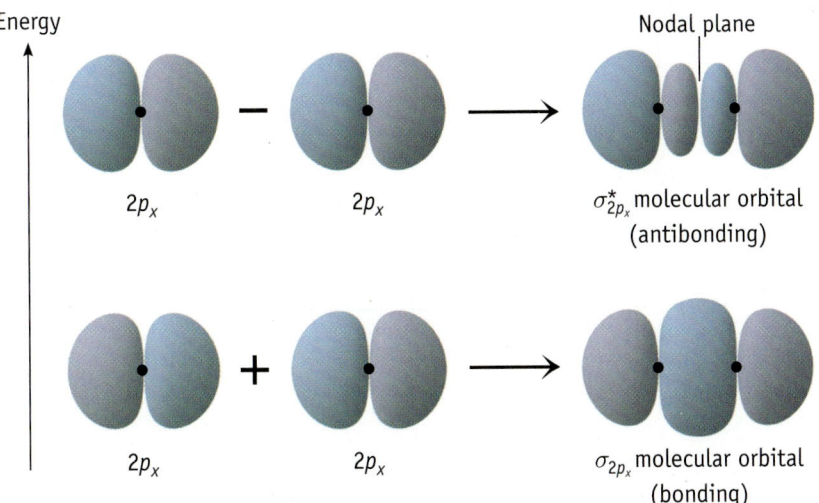

FIGURE 10.23 σ molecular orbitals from p atomic orbitals. A σ_{2p} bonding MO and a σ^*_{2p} antibonding MO are derived by mathematical combination of a 2p atomic orbital on each atom. Each MO can accommodate two electrons. The p orbitals in electron shells of higher n give molecular orbitals of the same basic shape.

In addition, each atom has two 2p atomic orbitals in planes perpendicular to the axis of the σ bonds in the molecule. If we define the axis of the σ bonds as the x direction, these are the 2p_y and 2p_z orbitals. Combination of the 2p_y orbitals on each atom gives rise to a π bonding molecular orbital (π_{2p_y}) and a π antibonding molecular orbital ($\pi^*_{2p_y}$). Similarly, combination of the 2p_z orbitals on each atom gives rise to π_{2p_z} and $\pi^*_{2p_z}$ molecular orbitals (Figure 10.24).

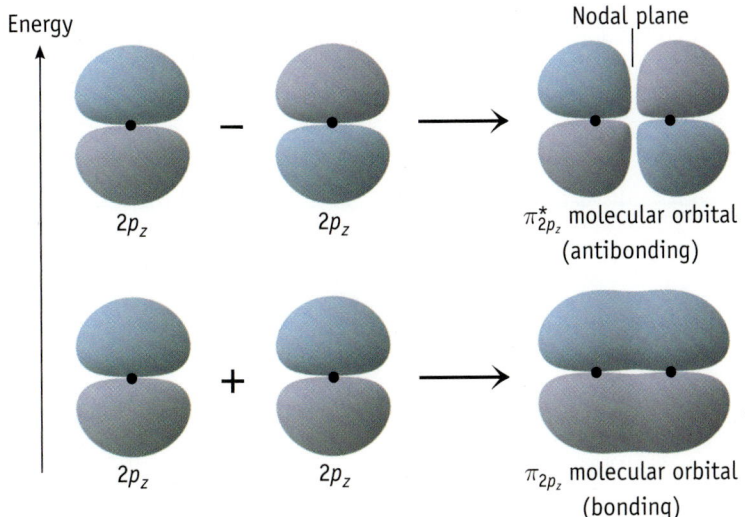

FIGURE 10.24 π molecular orbitals. Combination of 2p atomic orbitals that are parallel with each other and perpendicular to the σ bond axis produces a π_{2p} bonding MO and a π^*_{2p} antibonding MO. The figure portrays this for the 2p_z atomic orbitals. There is an equivalent pair of MOs produced by combination of the 2p_y atomic orbitals on each atom. The p orbitals in shells with higher n give molecular orbitals with the same basic shapes.

Since electrons in the 2p_y and 2p_z orbitals on each isolated atom are equivalent in energy and shape (except for direction), electrons in the bonding π_{2p_y} and π_{2p_z} molecular orbitals have equal energy. Similarly, electrons in the antibonding $\pi^*_{2p_y}$ and $\pi^*_{2p_z}$ molecular orbitals have the same energy. This is portrayed in the energy-level diagram (Figure 10.25) for molecular orbitals of most homonuclear diatomic molecules of the second-period elements.

FIGURE 10.25 Molecular orbital energy-level diagram for most homonuclear diatomic molecules of second-period elements. Although the diagram leads to the correct conclusions regarding bond order and magnetic behaviour for O_2, and F_2, the energy ordering of the MOs is correct only for B_2, C_2 and N_2. For O_2 and F_2 the σ_{2p} MO is lower in energy than the π_{2p} MOs. Table 10.7 takes this into account.

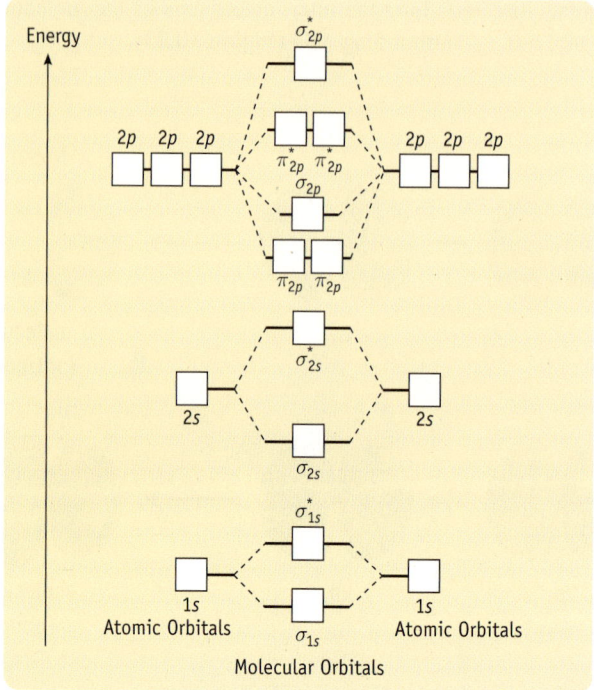

Think about It

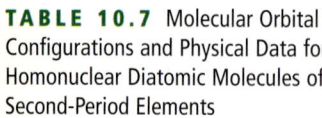

e10.24 Complete molecular orbital diagrams for homonuclear diatomic molecules to derive bond orders.

For diatomic molecules with 10 electrons or more, molecular orbital theory leads to π bonding orbitals and π^* antibonding orbitals (as well as σ_{1s} and σ^*_{1s} orbitals). A generalized order of energies of molecular orbitals is given in Figure 10.25.

Homonuclear Diatomic Molecules with 10–20 Electrons

Electron configurations can be predicted using the energy-level diagram in Figure 10.25, and these are tabulated in Table 10.7 for diatomic molecules from B_2 to F_2, along with bond orders, bond energies, bond lengths, and magnetic behaviour.

Table 10.7 has two noteworthy features:

1. There is a correlation between the electron configurations and the bond orders, bond lengths, and bond energies. The greater the bond order between a pair of atoms, the greater the energy required to break the bond, and the shorter the bond length. Dinitrogen (N_2), with a bond order of 3, has the largest bond energy and the shortest bond distance.

TABLE 10.7 Molecular Orbital Configurations and Physical Data for Homonuclear Diatomic Molecules of Second-Period Elements

	B_2	C_2	N_2	O_2	F_2
σ^*_{2p}	☐	☐	☐	☐	☐
π^*_{2p}	☐☐	☐☐	☐☐	↑ ↑	↑↓ ↑↓
σ_{2p}	☐	☐	↑↓	↑↓	↑↓
π_{2p}	↑ ↑	↑↓ ↑↓	↑↓ ↑↓	↑↓ ↑↓	↑↓ ↑↓
σ^*_{2s}	↑↓	↑↓	↑↓	↑↓	↑↓
σ_{2s}	↑↓	↑↓	↑↓	↑↓	↑↓
Bond order	1	2	3	2	1
Bond-dissociation energy (kJ mol^{-1})	290	620	945	498	155
Bond distance (pm)	159	131	110	121	143
Observed magnetic behaviour (paramagnetic or diamagnetic)	Para	Dia	Dia	Para	Dia

2. Ground-state dioxygen (O_2), which is paramagnetic, has two unpaired electrons. Dioxygen has 12 valence electrons (6 from each atom). If we arbitrarily decide that that σ bonds are in the x direction, ground-state O_2 has the molecular orbital configuration

$$[\text{core electrons}] \ (\sigma_{2s})^2(\sigma*_{2s})^2(\pi_{2p_y})^2(\pi_{2p_z})^2(\sigma_{2p_x})^2(\pi*_{2p_y})^1(\pi*_{2p_z})^1$$

This configuration leads to a bond order of 2, in agreement with experiment, and it specifies two unpaired electrons (in $\pi*_{2p}$ molecular orbitals). Molecular orbital theory succeeds here where the valence bond model fails. MO theory explains both the observed bond order and the paramagnetic behaviour of ground-state O_2 molecules.

WORKED EXAMPLE 10.12—MO ELECTRON CONFIGURATION FOR HOMONUCLEAR DIATOMIC IONS

When potassium reacts with O_2, potassium superoxide (KO_2) is one of the products. This is an ionic compound, in which the anion is the superoxide ion (O_2^-). Write the molecular orbital electron configuration for the ion in its ground state. Predict its bond order and magnetic behaviour.

Strategy

Use the energy-level diagram of Figure 10.25 to generate the electron configuration of this ion. Use Equation 10.2 to determine the bond order.

Solution

If we suppose that the σ bonds are in the x direction, and we assign the 13 electrons in the O_2^- ion, the ground state MO configuration is

$$[\text{core electrons}](\sigma_{2s})^2(\sigma*_{2s})^2(\pi_{2p_y})^2(\pi_{2p_z})^2(\sigma_{2p_x})^2(\pi*_{2p_y})^2(\pi*_{2p_z})^1$$

Based on this electron configuration, we would predict that the ground-state ion is paramagnetic to the extent of one unpaired electron—a prediction confirmed by experiment. The bond order is 1.5, because there are eight bonding electrons and five antibonding electrons. The bond order for O_2^- is lower than that for O_2, so we predict that the O—O bond length in an O_2^- ion will be longer than that in an O_2 molecule. In fact, the superoxide ion has an O—O bond length of 134 pm, whereas the bond length in O_2 molecules is 121 pm.

Comment

The superoxide ion (O_2^-) contains an odd number of electrons. It is one of several diatomic species (including NO and O_2) for which it is not possible to write a Lewis structure that accurately represents the bonding.

EXERCISE 10.27—MO ELECTRON CONFIGURATION FOR HOMONUCLEAR DIATOMIC IONS

The cations O_2^+ and N_2^+ are important components of the earth's upper atmosphere. Write the electron configuration of an O_2^+ ion in its ground state. Predict its bond order and magnetic behaviour.

Interactive Exercises 10.28–10.29

SUBMIT

Complete molecular orbital diagrams for some homonuclear diatomic molecules and ions to derive the bond orders.

Electrons can be assigned to molecular orbitals, starting with the lowest-energy orbital, to obtain electron configuration, and to deduce bond orders. A correlation is found between calculated bond orders and experimentally estimated bond lengths and bond energies. The electron configuration of ground-state dioxygen (O_2), has two unpaired electrons—consistent with the evidence of its paramagnetic behaviour.

Heteronuclear Diatomic Molecules

The compounds NO, CO, and ClF—all with molecules containing two atoms of different elements—are **heteronuclear diatomic molecules**. Although the energy levels of the atomic orbitals in the bonded atoms are not exactly the same, MO descriptions for heteronuclear diatomic molecules generally resemble those for homonuclear diatomic molecules. As a consequence, an energy-level diagram like that in Figure 10.25 can be used to evaluate the bond order and magnetic behaviour for most heteronuclear diatomic molecules.

For example, a molecule of nitrogen monoxide (NO) has 11 valence electrons. If they are assigned to the MOs using the energy-level diagram of Figure 10.25, we deduce the electron configuration

$$[\text{core electrons}]\ (\sigma_{2s})^2(\sigma^*_{2s})^2(\pi_{2p})^4(\sigma_{2p})^2(\pi^*_{2p})^2$$

The bond order is 2.5. The single unpaired electron is assigned to the π^*_{2p} molecular orbital. The molecule is paramagnetic, as predicted for a molecule with an odd number of electrons.

Polyatomic Molecules and Ions

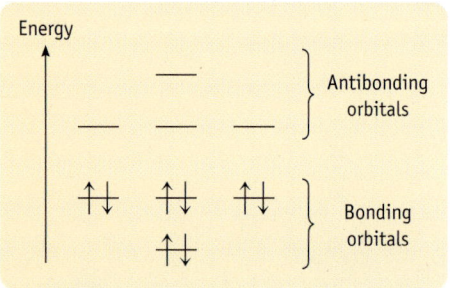

FIGURE 10.26 Molecular orbital energy-level diagram for a methane molecule. The eight valence electrons occupy four bonding MOs—not all equivalent. The two electrons in one of the bonding MOs have less energy than the six in the other three (equal energy) MOs.

Valence bond theory suggests that since the four bonds in methane are identical, all eight electrons have the same energy. However, as was pointed out in the introduction to this section, photoelectron spectroscopy data informs us that this is not the case. Molecular orbital theory can account for this. For methane, eight molecular orbitals are calculated by mathematical combination of the four valence orbitals on a C atom and the $1s$ orbitals on four H atoms. The outcome is necessarily eight molecular orbitals. Of these, four are bonding MOs and four are antibonding MOs, whose relative energies are portrayed in Figure 10.26.

The computer-calculated distribution of densities of electrons in the four bonding molecular orbitals is shown in Figure 10.27.

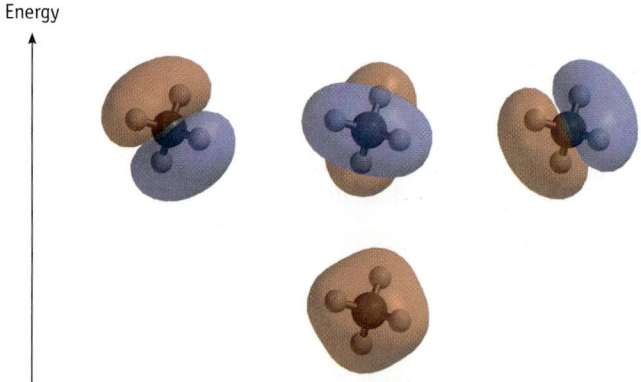

FIGURE 10.27 Bonding molecular orbitals of a methane molecule. The regions of high electron density are not localized along the C — H axes, but spread over the whole molecule.

The MO energy-level diagram of Figure 10.26 helps us to make sense of the photoelectron spectrum of methane presented in Figure 10.18. Electrons in three of the molecular orbitals of methane have the same energy, leading to one of the PES peaks, and this is different from the energy of electrons in the other molecular orbital, which gives rise to the other peak. The third peak at highest binding energy originates from core electrons on the C atom.

Here we have a fundamentally different perspective from the valence bond model: the energies of three pairs of electrons are different from another pair, and yet all four bonds are identical. The difference is that where the valence bond approach considers each electron pair to be localized in bonds in a methane molecule, the molecular orbital model considers the electron density of each orbital to dispersed over the whole molecule.

If MO theory shows that electron density is not localized along the C — H axes, does it even make sense to think of methane molecules as having C — H bonds? From the perspective of the MO model, we would answer, "No. There is bonding, but not localized bonds between particular atoms. The distribution of density of the four electron pairs in the four bonding MOs over the whole molecule suggests that there is bonding *among* the C atom and the four H atoms, but not *between* particular pairs of atoms."

To further illustrate the MO approach, let's look again at benzene molecules. On page 368 we discussed how the valence bond model can account for the π electrons being spread out over all six C atoms, by sideways overlap of *p* orbitals in both directions. We can now see how the same case can be made with MO theory. Six *p* orbitals contribute to the π system. Based on the premise that the number of molecular orbitals must equal the number of atomic orbitals, there must be six π molecular orbitals in benzene. Of these, three are bonding orbitals, and three are antibonding orbitals. An energy-level diagram for a benzene molecule (Figure 10.28) shows that the six electrons not engaged in σ bonding are in the three lowest-energy molecular orbitals, which are bonding orbitals.

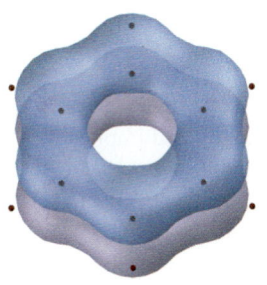 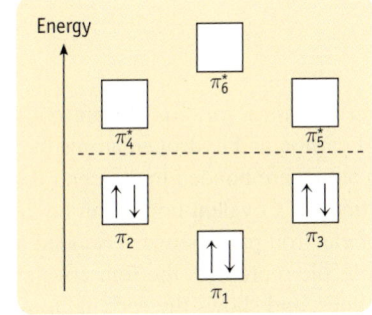

FIGURE 10.28 Molecular orbital energy-level diagram for benzene. Mathematical combination of the six unhybridized *p* orbitals on C atoms gives rise to six π molecular orbitals—three bonding and three antibonding. Because they are the lowest energy states, the three bonding molecular orbitals accommodate the six π electrons.

> Using the molecular orbital model for CH_4, electrons in three of the occupied (bonding) MOs have the same energy, but this is different from the energy of the electrons in the other MO. This is consistent with the experimental photoelectron spectrum of methane, and is not explainable using the valence bond model.

HOMOs and LUMOs

Chemists find it useful to consider the highest-energy MO that has electrons, called the **highest occupied molecular orbital (HOMO)**, and the lowest-energy MO that does not have electrons, called the **lowest unoccupied molecular orbital (LUMO)**. These are called the frontier orbitals. For example, in dioxygen the HOMO is the π^*_{2p} orbital, and the LUMO is the σ^*_{2p} orbital (Table 10.7).

These levels are important in the interpretation of spectra of molecular substances because the most probable electronic transition is from the HOMO to the LUMO, and the difference between the energies of these orbitals is identifiable as a peak in absorption spectra in the visible and ultraviolet regions. In addition, some chemical reactions can usefully be thought of as involving interactions between the HOMO of molecules of one reactant with the LUMO of molecules of another reactant. For example, the HOMO is the source of electrons in an electron donor, since it is the highest-energy orbital containing (the most loosely held) electrons. The LUMO is the key orbital in molecules of an electron acceptor, since it is the lowest-energy orbital that has capacity to accept electrons. In some cases, the HOMO–LUMO energy separation is an important factor in determining the chemical reactivity at the most reactive site of a molecule. You will see this powerful idea applied in later chapters [>>Section 19.2].

> The molecular orbitals that are important in the electronic spectra and reactivity of substances are called the *frontier orbitals*—the highest occupied molecular orbital (HOMO), and the lowest-energy MO that does not have electrons, the lowest unoccupied molecular orbital (LUMO).

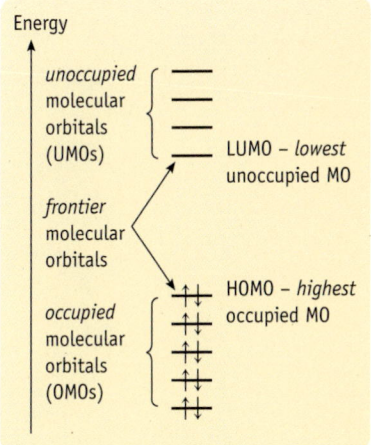

This diagram portrays the HOMO and LUMO frontier orbitals in a molecule.

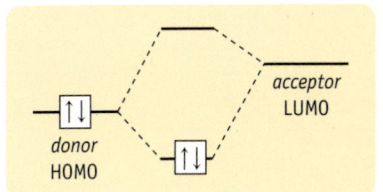

This diagram portrays the HOMO in one reactant molecule, and the LUMO in another reactant molecule, combining to form a new bonding MO.

SUMMARY

Key Concepts

- Several, complementary models at the **intramolecular level** are commonly used to explain observations about the forces holding together molecular structures. The simplest possible models are imagined to explain experimental observations about covalent bonding. Then, when new facts of structure and bonding are uncovered, more sophisticated models are developed to explain them (Section 10.1). A methane molecule (CH_4) is used in the summary below to illustrate the complementary models developed in this chapter.

- *Condensed structural formula:* Representing a methane molecule as CH_4 shows the number of carbon and hydrogen atoms in each molecule, and implies that all four hydrogen atoms are bonded to the central carbon atom. (Section 3.8)
- *Lewis structure:* Covalent bonds can be thought of as the net force of attraction from sharing of electron pairs between bonded atoms (Section 10.2). A Lewis structure for a CH_4 molecule represents the four covalent bonds between carbon and hydrogen atoms as lines, and shows the carbon atom as following the **octet rule**. (Section 10.3)
- *Resonance and delocalization models:* With only single bonds in its Lewis structure, these models are not needed for CH_4. **Resonance** representations are helpful when the valence electrons in a molecule or ion can be distributed in more than one sensible way, making any single Lewis structure inadequate to represent experimental observations about structure. An alternative to showing resonance hybrids is the **electron delocalization model,** showing dashed lines on a single Lewis structure to represent "partial bonds," intermediate between a single bond and a double bond. (Section 10.5)
- *Valence shell electron pair repulsion (VSEPR) model:* A tetrahedral geometry in CH_4 minimizes repulsion among the four regions of high electron density in the valence of the central carbon atom. (Section 10.7)
- *Valence bond (VB) model:* Considers the four bonding electron pairs in CH_4 to be **sigma (σ) bonds,** localized between the carbon and hydrogen atoms, and resulting from the partial overlap of singly occupied atomic orbitals on carbon and hydrogen atoms. Using the $2p$ orbitals on the carbon atom to overlap with s orbitals on each hydrogen atom, however, would give the wrong experimental geometry. Other molecules that have multiple bonds will have both sigma (σ) and pi (π) bonds. A **π bond** between two atoms is formed by sideways overlap of p orbitals, and has two regions of high electron density—one on either side of the internuclear axis. A σ bond and a π bond together make a double bond, and a σ bond and two π bonds make up a triple bond. (Section 10.8)
- *Orbital hybridization model:* Required to account for the experimentally observed tetrahedral geometry about the C atom in CH_4. Each of the four singly occupied sp^3 **hybrid orbitals** on the carbon atom overlaps with a singly occupied s orbital on a hydrogen atom to form a bond. Other hybrid orbitals (sp^2 and sp, respectively) are used to explain the observed geometry in trigonal-planar and linear molecules or portions of molecules. (Section 10.8)
- *Molecular orbital (MO) model:* Required to account for the observation from the photoelectron spectrum for methane, that its bonding electrons do not all have the same energies, as implied by the hybridization model. An MO model considers the valence electrons to be spread out over the molecule, rather than localized between bonded atoms. Electrons are assigned to MOs, starting with the lowest-energy orbital, to obtain electron configuration and to deduce **bond orders**. For CH_4, electrons in three of the occupied (bonding) molecular orbitals have the same energy, which is

different from the energy of the electrons in the other molecular orbital. The **highest occupied molecular orbital (HOMO)** and the **lowest unoccupied molecular orbital (LUMO)** can be used to understand the electronic spectra and reactivity of CH_4. (Section 10.9)

"I believe the chemical bond is not so simple as some people seem to think."

Robert Mulliken

Key Equations

Formal charge on atom = [number of valence electrons]
$$- \text{[number of electrons in lone pairs]} - \tfrac{1}{2} \text{[number of electrons in bonds]}$$
(Equation 10.1, Section 10.6)

Calculating bond order from the MO electron configuration:

Bond order = $\tfrac{1}{2}$(number of electrons in bonding MOs
$$- \text{number of electrons in antibonding MOs})$$
(Equation 10.2, Section 10.9)

REVIEW QUESTIONS

Section 10.3: Lewis Structures

10.30 Give the periodic group number and number of valence electrons for each of the following atoms:
(a) O   (d) Mg
(b) B   (e) F
(c) Na   (f) S

10.31 For elements in Groups 13–17 of the periodic table, give the number of bonds an element is expected to form if it follows the octet rule.

10.32 Which of the following elements are capable of forming compounds in which the indicated atom has more than four valence electron pairs?
(a) C   (d) F   (g) Se
(b) P   (e) Cl   (h) Sn
(c) O   (f) B

10.33 Draw a Lewis structure for each of the following molecules or ions:
(a) NF_3 (c) HOBr
(b) ClO_3^- (d) SO_3^{2-}

10.34 Draw a Lewis structure for each of the following molecules:
(a) chlorodifluoromethane, $CHClF_2$ (C is the central atom)
(b) acetic acid, CH_3COOH
(c) acetonitrile, CH_3CN (the framework is H_3C-C-N)
(d) allene, H_2CCCH_2

Section 10.4: Exceptions to the Octet Rule

10.35 In boron compounds, the B atom often is not surrounded by four valence electron pairs. Illustrate this with BCl_3. Show how the molecule can achieve an octet configuration by forming a coordinate covalent bond with ammonia (NH_3).

10.36 Which of the following compounds or ions do *not* have an octet of electrons surrounding the central atom: BF_4^-, SiF_4, SeF_4, BrF_4^-, XeF_4?

Section 10.5: Resonance and Delocalized Electron Models

10.37 Show all possible resonance structures for each of the following molecules or ions:
(a) SO_2
(b) NO_2^-
(c) SCN^-

10.38 Give the bond order for each bond in the following molecules or ions:
(a) CH_2O (c) NO_2^+
(b) SO_3^{2-} (d) NOCl

10.39 Draw resonance structures for the formate ion ($HCOO^-$), and then determine the $C-O$ bond order in the ion.

Section 10.6: Which Resonance Structures Are Most Important?

10.40 Determine the formal charge on each atom in the following molecules or ions:
(a) SCO
(b) $HCOO^-$ (formate ion)
(c) O_3
(d) HCOOH (formic acid)

10.41 Determine the formal charge on each atom in the following molecules and ions:
(a) NO_2^+ (c) NF_3
(b) NO_2^- (d) HNO_3

10.42 Three resonance structures are possible for nitrous oxide (N_2O), laughing gas.
(a) Draw the three resonance structures.
(b) Calculate the formal charge on each atom in each resonance structure.
(c) Based on formal charges and electronegativity, predict which resonance structure is the most reasonable.

10.43 Two resonance structures are possible for NO_2^-. Draw these structures and then find the formal charge on each atom in each resonance structure. If an H^+ ion is attached to NO_2^- (to form the acid HNO_2), does it attach to O or N?

Section 10.7: Spatial Arrangement of Atoms within Molecules

10.44 Draw a Lewis structure for each of the following molecules or ions. Describe the three-dimensional geometry of electron regions and the shape around the central atom for each of the following:
(a) NH_2Cl
(b) Cl_2O (O is the central atom)
(c) SCN^-
(d) HOF

10.45 The following molecules or ions all have two oxygen atoms attached to a central atom. Draw a Lewis structure for each one and then describe the three-dimensional geometry of electron regions and the shape around the central atom. Comment on similarities and differences in the series.
(a) CO_2 (c) O_3
(b) NO_2^- (d) ClO_2^-

10.46 Give approximate values for the indicated bond angles:
(a) O—S—O in SO_2
(b) F—B—F angle in BF_3
(c) Cl—C—Cl angle in Cl_2CO
(d) H—C—H (angle 1) and C—C≡N (angle 2) in acetonitrile

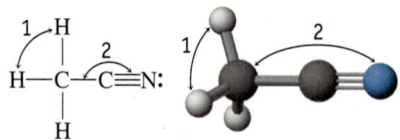

10.47 Phenylalanine is one of the natural amino acids and is a "breakdown" product of aspartame. Estimate the values of the indicated angles in the amino acid. Explain why the —CH_2—$CH(NH_2)$—COOH chain is not linear.

10.48 Acetylacetone has the structure shown here. Estimate the values of the indicated angles.

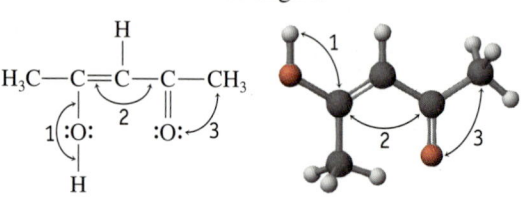

Section 10.8: The Valence Bond Model of Covalent Bonding

10.49 Draw the Lewis structure for NF_3. What is the three-dimensional geometry of electron regions and the shape around the central atom? Using the valence bond model, what hybrid orbitals on N and F overlap to form bonds between these elements?

10.50 What is the three-dimensional geometry of electron regions and the shape around the central atom for each of the following? Using the valence bond model, describe the hybrid orbital set used by the underlined atom in each molecule or ion:
(a) $\underline{C}Se_2$ (b) $\underline{S}O_2$ (c) $\underline{C}H_2O$ (d) $\underline{N}H_4^+$

10.51 Using the valence bond model, describe the hybrid orbital set used by each of the indicated atoms in the molecules below:
(a) the carbon atoms and the oxygen atom in dimethyl ether (H_3COCH_3)
(b) each carbon atom in propene

$$H_3C—\overset{\overset{\displaystyle H}{|}}{C}{=}CH_2$$

(c) the two carbon atoms and the nitrogen atom in the amino acid glycine

10.52 Draw the Lewis structure for AlF_4^-. What is the three-dimensional geometry of electron regions and the shape around the central Al atom? Using the valence bond model, what orbitals on Al and F overlap to form bonds between these elements?

10.53 Describe the O—S—O angle and the hybrid orbital set used by sulfur in each of the following molecules or ions:
(a) SO_2 (c) SO_3^{2-}
(b) SO_3 (d) SO_4^{2-}

Do all have the same value for the O—S—O angle? Does the S atom in all these species use the same hybrid orbitals?

10.54 Sketch the resonance structures for the nitrate ion (NO_3^-). Using the valence bond model, is the hybridization of the N atom the same or different in each structure? Describe the orbitals involved in bond formation by the central N atom.

10.55 Compare the structure and bonding in CO_2 and CO_3^{2-} with regard to the O—C—O bond angles, the C—O bond order, and the C atom hybridization used by the valence bond model.

10.56 Numerous molecules are detected in deep space. Three of them are illustrated here.

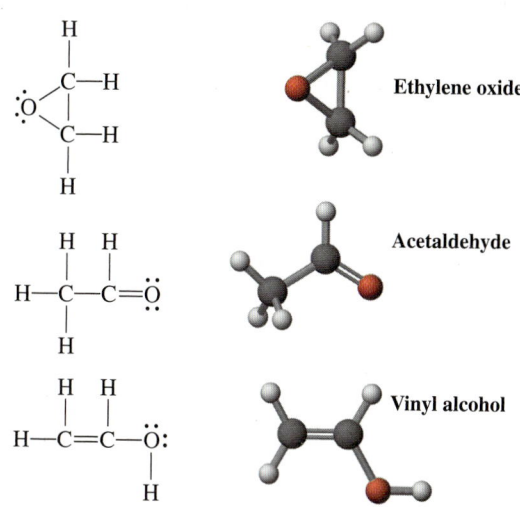

Ethylene oxide

Acetaldehyde

Vinyl alcohol

(a) Comment on the similarities or differences in the formulas of these compounds. Are they isomers?
(b) Using the valence bond model, indicate the hybridization of each C atom in each molecule.
(c) Indicate the value of the H—C—H angle in each of the three molecules.
(d) Are any of these molecules polar?
(e) Which molecule should have the strongest carbon–carbon bond? The strongest carbon–oxygen bond?

10.57 The compound sketched below is acetylsalicylic acid, commonly known as ASA or aspirin:

(a) What are the approximate values of the angles marked *A*, *B*, *C*, and *D*?
(b) Using the valence bond model, what hybrid orbitals are used by carbon atoms 1, 2, and 3?

Section 10.9: Molecular Orbital Theory of Bonding

10.58 How do valence bond and molecular orbital models differ in their explanation of the bond order of 1.5 for ozone?

10.59 The simple valence bond picture of O_2 does not agree with the molecular orbital view. Compare these two theories with regard to the peroxide ion (O_2^{2-}).
(a) Draw an electron dot structure for O_2^{2-}. What is the bond order of the ion?
(b) Write the molecular orbital electron configuration for O_2^{2-}. What is the bond order based on this approach?
(c) Do the two theories of bonding lead to the same magnetic character and bond order for O_2^{2-}?

10.60 Which of the homonuclear diatomic molecules of the second-period elements (from Li_2 to Ne_2) are paramagnetic? Which have a bond order of 1? Which have a bond order of 2? Which diatomic molecule has the highest bond order?

10.61 Which of the following molecules or molecule ions should be paramagnetic? What is the highest occupied molecular orbital (HOMO) in each one? Assume the molecular orbital diagram in Figure 10.25 applies to all of them.
(a) NO   (c) O_2^{2-} (e) CN
(b) OF^- (d) Ne_2^+

10.62 The CN molecule has been found in interstellar space. Assuming the electronic structure of the molecule can be described using the molecular orbital energy-level diagram in Figure 10.25, answer the following questions.
(a) What is the highest energy occupied molecular orbital (HOMO) to which an electron(s) is(are) assigned?
(b) What is the bond order of the molecule?
(c) How many net bonds are there? How many net π bonds?
(d) Is the molecule paramagnetic or diamagnetic?

SUMMARY AND CONCEPTUAL QUESTIONS

10.63 Many molecules have four electron pairs around a central atom. Give an example of one that has a pyramidal shape. Give an example of one with a bent structure. What bond angles are predicted in each case?

10.64 What is the difference between the three-dimensional geometry of electron regions and the shape defined by nuclei around a central atom of a molecule? Use the water molecule as an example in your discussion.

10.65 Compare the carbon–oxygen bond lengths in the formate ion ($HCOO^-$); in methanol (CH_3OH); and in the carbonate ion (CO_3^{2-}). In which species is the carbon–oxygen bond predicted to be longest? In which is it predicted to be shortest? Explain briefly.

10.66 What are the orders of the N—O bonds in NO_2^- and NO_2^+? The nitrogen–oxygen bond length in one of these ions is 110 pm and 124 pm in the other. Which bond length corresponds to which ion? Explain briefly.

10.67 Which has the greater O—N—O bond angle, NO_2^- or NO_2^+? Explain briefly.

10.68 Draw the Lewis structure for the sulfite ion (SO_3^{2-}). In aqueous solution, the ion interacts with H^+. Does H^+ attach itself to the S atom or the O atom of SO_3^{2-}?

10.69 Acrylonitrile (C_3H_3N) is the building block of the synthetic fibre Orlon.

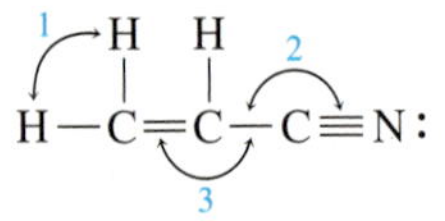

(a) Give the approximate values of angles 1, 2, and 3.
(b) Which is the shorter carbon–carbon bond?
(c) Which is the stronger carbon–carbon bond?
(d) Which is the most polar bond?

10.70 The cyanate ion (NCO^-) has the least electronegative atom, C, in the centre. The very unstable fulminate ion (CNO^-) has the same formula, but the N atom is in the centre.
(a) Draw the three possible resonance structures of CNO^-.
(b) On the basis of formal charges, decide on the resonance structure with the most reasonable distribution of charge.
(c) Mercury fulminate is so unstable it is used in blasting caps. Can you offer an explanation for this instability? (*Hint:* Are the formal charges in any resonance structure reasonable in view of the relative electronegativities of the atoms?)

10.71 Vanillin is the flavouring agent in vanilla extract and in vanilla ice cream.

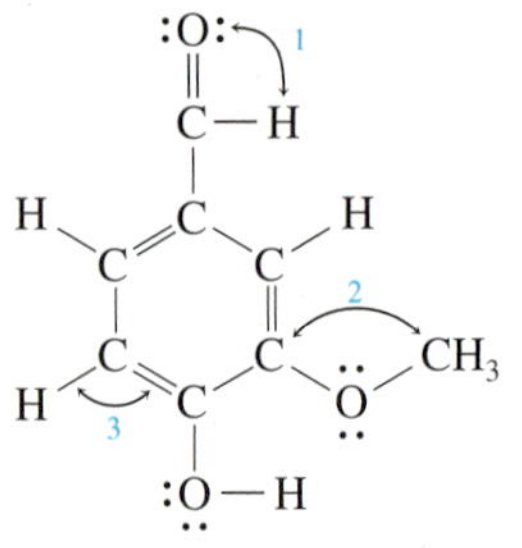

(a) Give values for the three bond angles indicated.
(b) Indicate the shortest carbon–oxygen bond in the molecule.
(c) Indicate the most polar bond in the molecule.

10.72 Nitric acid (HNO_3) has three resonance structures. One of them, however, contributes much less to the resonance hybrid than the other two. Sketch the three resonance structures and assign a formal charge to each atom. Which one of your structures is the least important?

10.73 Boron trifluoride (BF_3) can accept a pair of electrons from another molecule to form a coordinate covalent bond, as in the following reaction with ammonia:

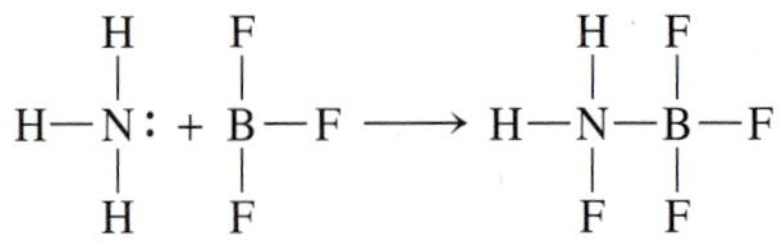

(a) What is the geometry about the boron atom in BF_3? In $H_3N \rightarrow BF_3$?
(b) Using the valence bond model, what is the hybridization of the boron atom in the two compounds?
(c) Does the valence bond model suggest that boron atom's hybridization will change on formation of the coordinate covalent bond?

With the **eCHACR** single-sign-on access card bundled with your text, log on (http://login.cengage.com/sso) and access the e-book and click on any in-margin icons for dynamic molecular-level animations and simulations, videos of laboratory reactions, interactive exercises, reviews of background concepts, and other online supplementary materials as noted by the icons in the text margins.

Also go to www.chemistry.nelson.com <http://www.chemistry.nelson.com> for Answers to in-chapter exercises and selected Review Questions, Test Yourself questions, weblinks, crossword puzzles, flashcards, glossary of key terms, and other student resources.

States of Matter

NASA

11.1 Case Study: Understanding Gases: Understanding Our World

You roll out of bed and look at the bright pinkish-tan sky. The forecast promises a quick move from the morning low of −87 °C to the afternoon high of −25 °C. At least you don't have to wait long for your morning cup of coffee. Water (if you could find some) boils at 8 °C today. But it's a bit hard to catch your breath—only 0.13% of the air is oxygen, and the average surface atmospheric pressure is 0.7 kPa.

What planet did you wake up on?

Our guide to the lonelier planets suggests you won't find much "atmosphere" on Mars and you won't have much company. What makes Earth (opening photo) so much more attractive to live on than Mars or other planets in our solar system? One important feature is the composition and pressure of the atmosphere surrounding Earth's outer crust, which has a pressure 100 times that on Mars. Literally, *atmosphere* means "ball of vapour" (Greek *atmos*—"vapour," and *sphaira*—"ball"). A good starting point to understand the almost ideal life-supporting features of our world is learning more about the properties of the substances that make up that "ball of gases" surrounding Earth's surface.

Earth's atmospheric composition fluctuates over time, and much of our world was not at all ideal for human life when its atmospheric composition was quite different. During the Pleistocene period 20 000 years ago, the concentration of CO_2 in the atmosphere was about half of its present value of 390 ppm, and the earth's average surface temperature was about 5 °C colder. The northern third of the globe, including large parts of North America, Europe, and Asia, lay under several kilometres of ice. So much of the earth's water was in the solid state that sea levels were more than 100 m lower than at present. About 10 000 years ago, the surface temperature of the earth increased by about 3–4 °C, allowing the emergence of agriculture. In contrast, in a very short period of time—the past 150 years—the unprecedented scale of human activity, coupled with our planet's own evolution, has significantly altered the chemical composition of our atmosphere, leading to signs of much more rapid climate change [<<Section 4.3].

How do we make rational decisions about the urgency with which we try to control the rate of addition of "greenhouse gases" and aerosols to the earth's atmosphere? Our starting point is with observations—current and historical data that show what substances we put into the atmosphere, how atmospheric composition has changed over time, and how this correlates with changes in our planet's climate. Informed action also requires that we use that understanding to glimpse into the future with as much certainty as possible. We do so with quantitative research *models* that predict how projected increases in "greenhouse gases" will affect climate variables such as temperature, precipitation, and snow and ice cover. The 2007 Nobel Peace Prize–winning Intergovermental Panel on Climate Change has produced research models that predict global mean surface temperatures will increase by about 3 °C by the end of the 21st century, depending on the level of activity humans undertake to check the rapid increase of atmospheric greenhouse gases.

Understanding the properties of gases and their role in our atmosphere enables chemistry learners to play a more informed role in decision making about our changing global climate. Applying that understanding will help us work to preserve those features that make ours an almost ideal world for life.

Earth's Atmosphere: A Ball of Gases?

How accurate is a mental model of the atmosphere as a "sphere" of "atmos," made up of a uniform mixture of gases surrounding the earth's surface? This picture helps us make sense of some features of the lower atmosphere. But striking changes take place in temperature and pressure at increasing altitudes. In fact, the regions of our atmosphere are defined by whether the temperature increases or decreases with increasing altitude, moving up from the troposphere to the stratosphere, then to the mesosphere and thermosphere. To illustrate, consult e11.1, a visualization of how the temperature, pressure, and composition of the atmosphere change with altitude.

Think about It

e11.1 Visualize how atmospheric temperature, pressure, and composition change with altitude.

> *The interplay among gas, liquid, and solid states of atmospheric substances plays an often unrecognized role in the complex and dynamic chemistry of our atmosphere.*

The average composition of Earth's troposphere to a height of 25 km (Table 11.1) is quite uniform, and consists mostly of N_2 and O_2 gas, with trace amounts of CO_2, H_2O, and gases such as Ar, Ne, and He. But above 100 km, in the thermosphere,

TABLE 11.1 Average Composition of Earth's Atmosphere to a Height of 25 km

Gas	Amount %	Source
N_2	78.08	Earth's interior, biologic
O_2	20.95	Biologic
Ar	0.93	Radioactivity
Ne	0.0018	Earth's interior
He	0.0005	Radioactivity
H_2O	0 to 4	Evaporation
CO_2	0.0390	Biologic, industrial
CH_4	0.00017	Biologic
N_2O	0.00003	Biologic, industrial
O_3	0.000004	Photochemical

the atmospheric composition changes significantly with altitude. Here, the relative rate at which gases *diffuse* determines their concentration at different altitudes. Above 400 km, very few particles are present and low-mass atoms such as helium undergo collisions infrequently with other particles. Those with sufficient kinetic energy can exceed the earth's escape velocity and exit the atmosphere. This is the origin of the term *exosphere* for the outermost region of our atmosphere.

We've focused so far on gases in Earth's atmosphere, but liquids and solids also play crucial roles in important atmospheric phenomena, such as the formation of a large Antarctic ozone "hole" each September and the cooling of our planet through scattering of visible light from the sun by aerosols and clouds. In the dark Antarctic winter, temperatures fall below −80 °C. This is so cold that water vapour and other gases condense to form liquids and crystalline solids, called polar stratospheric clouds (PSCs). Key chemical reactions that release chlorine atoms occur in thin films of liquid water on the surface of crystals of ice and nitric acid that make up PSCs. These chlorine atoms deplete large amounts of ozone [>>Section 21.1] over the South Pole during the Antarctic spring.

Scattering of the sun's visible light by atmospheric aerosols from sources such as volcanic eruptions and fossil fuel combustion increases the earth's reflectivity (albedo) and contributes to atmospheric cooling. Earth's atmosphere is much more than a "ball of vapour." The interplay between gas, liquid, and solid states of atmospheric substances plays an often unrecognized role in the complex and dynamic chemistry of our atmosphere.

We focus first in this chapter on the thermodynamic variables of pressure, temperature, volume, and amount (mol) that define the behaviour of gases such as those in our atmosphere. We then consider properties of liquids and solids and the phase transitions among these three states or phases.

The escape velocity of He atoms is defined as the speed at which their kinetic energy exceeds the potential binding energy due to the force of gravity, at which point they "break free" from gravitational force and escape the atmosphere.

Beyond Earth's Atmosphere: A Fourth State of Matter

We often limit our imagination to thinking of matter as existing in gas, liquid, and solid states. But the twinkle we see in the night sky coming through our atmosphere reminds us of the fourth and most common state of matter in our universe—ionized gases known as plasmas. All of the stars are made of plasma, a medium with charged particles so close together that each one influences many other neighbouring particles, affecting the electrical properties and behaviour of the medium.

The sun, our nearest and most familiar star and source of most of the energy on the earth, shines because massive amounts of energy are produced in nuclear fusion reactions within the plasmas of its core and radiated out into space. Temperatures in the core of the sun approach 15 million Kelvin. Those privileged to live in our planet's far north or south experience coloured Aurora light displays [<<Section 1.4], which receive their energy from the plasmas that make up solar winds encountering magnetic fields at the earth's poles.

Scientists create artificial plasmas for many purposes. Large-scale international research teams work to obtain energy from controlled *fusion* in plasmas made from deuterium and tritium nuclei at temperatures of about 100 million degrees Kelvin [>>Chapter 30]. On much smaller scales, plasmas of argon gas in **inductively coupled plasma (ICP) spectrometers** are used to identify the elements found in chemical compounds as a result of their characteristic spectral emissions. For example, forensic scientists used ICP spectrometry experiments to determine the particular species of arsenic found in samples of Napoleon's hair [<<Section 8.1] to evaluate claims that he was poisoned with inorganic

Lmsal/Stanford Univ/NASA/Science Photo Library

Loops of plasma arc above the sun's surface.

arsenic–containing compounds. Finally, you might use a plasma television screen to watch a program about any of these topics—the vibrant colours on the screen produced by plasma made of xenon and neon atoms.

11.2 Relationships among Gas Properties

Gas Pressure

Pressure is one of the key properties used to describe the state of any gas. From consulting weather forecasts, you will already be familiar with the idea that atmospheric gases exert pressure. Rising atmospheric pressure often indicates fair weather, and falling pressure suggests foul weather might be coming. A **barometer** is the device used to measure atmospheric pressure, which is given in SI-derived units of kilopascal or hectopascal.

The earliest unit of pressure, mm of mercury, comes directly from barometric measurements. A barometer can be made by filling a tube with a liquid, often mercury, and inverting the tube in a dish containing the same liquid (Figure 11.1). If the air has been removed completely from the vertical tube, the liquid in the tube assumes a level such that the pressure exerted by the mass of the column of liquid in the tube is balanced by the pressure of the atmosphere pressing down on the surface of the liquid in the dish.

At sea level, the mercury in a mercury-filled barometer will rise about 760 mm above the surface of the mercury in the dish. This is the historical reason why pressure was often reported in the unit of millimetres of mercury (mm Hg). Pressures were also reported as standard atmospheres (atm), a unit defined as follows:

$$1 \text{ standard atmosphere (1 atm)} = 760 \text{ mm Hg (exactly)}$$

The SI unit of pressure is the *pascal (Pa)*, named for the French mathematician and philosopher Blaise Pascal (1623–1662). **Pressure** is defined as the force exerted on an object divided by the area over which it is exerted.

$$1 \text{ pascal (Pa)} = 1 \text{ newton per square metre} = 1 \text{ N m}^{-2}$$

(The newton is the SI unit of force.) Because 1 Pa is a very small unit compared with ordinary pressures, the unit kilopascal (kPa) is used more frequently.

The pascal has a simple relationship to another unit of pressure called the *bar*, where 1 bar = 100 000 Pa. The thermodynamic data in Chapter 17 and in Appendix C are appropriate for conditions where atmospheric pressure is 1 bar. To summarize, the units used for pressure are

$$1 \text{ bar} = 1 \times 10^5 \text{ Pa (exactly)} = 1 \times 10^2 \text{ kPa} = 0.9872 \text{ atm}$$

or

$$1 \text{ atm} = 760 \text{ mm Hg (exactly)} = 101.3 \text{ kilopascal (kPa)} = 1.013 \text{ bar}$$

Gas Volume, Temperature, and Amount

Besides pressure (p), the other three thermodynamic properties of a gas are its temperature (T), volume (V), and amount (n).

Over a period of more than 100 years in the early history of chemistry, careful measurements were carried out to find quantitative relationships among these four variables. Table 11.2 summarizes the best known formulations of these relationships: Robert Boyle's (1627–1691) law, Jacques Charles's (1746–1823) law, and Amadeo Avogadro's (1776–1856) hypothesis.

The relationship determined by Charles has particular significance, as it leads to our definition of absolute zero. Figure 11.2 illustrates that the volumes of two different gas samples are directly proportional to temperature (at a constant pressure). When the plots of volume versus temperature are extended to lower temperatures, they extrapolate to zero

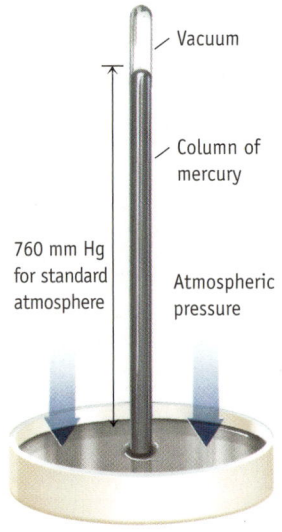

FIGURE 11.1 A barometer. The pressure of the atmosphere on the surface of the mercury in the dish is balanced by the downward pressure exerted by the column of mercury. The barometer was invented in 1643 by Evangelista Torricelli (1608–1647). A unit of pressure called the *torr* in his honour is equivalent to the pressure exerted by a vertical column of Hg, 1 mm high.

Labels in figure: Vacuum; Column of mercury; 760 mm Hg for standard atmosphere; Atmospheric pressure

Molecular Modelling (Odyssey)

e11.2 Simulate a gas to understand pressure in terms of collisions.

Background Concepts

e11.3 Read about the historical development of relationships between the pressure, volume, temperature, and amount of a gas.

Boyle's Law	Charles's Law	Avogadro's Hypothesis
pV = constant, or $V \propto (1/p)$	$V \propto T$	$V \propto n$
(at constant T, n)	(at constant p, n)	(at constant T, p)

TABLE 11.2 Historical Relationships among the Pressure, Volume, Temperature, and Amount of a Gas

Molecular Modelling (Odyssey)

e11.4 Explore quantitative relationships using simulations of different gases.

volume at the same temperature, -273.15 °C. (Of course, gases will not actually reach zero volume; they liquefy above that temperature.) This temperature is very significant. William Thomson (1824–1907), also known as Lord Kelvin, proposed a temperature scale—now known as the **Kelvin temperature scale**—for which the zero point is 0 K, which corresponds with -273.15 °C.

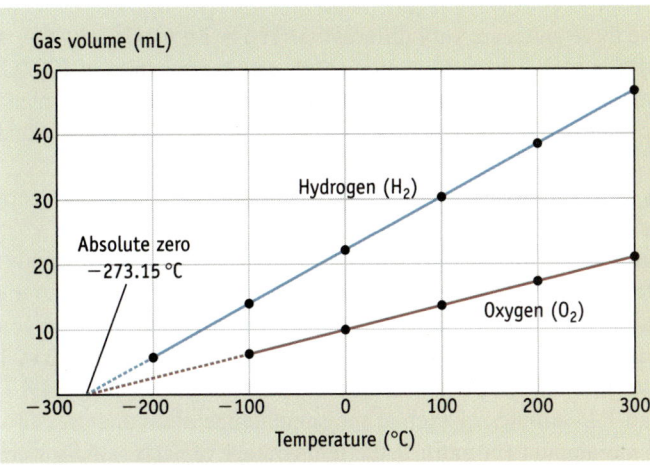

T (°C)	T (K)	Vol. H_2 (mL)	Vol. O_2 (mL)
300	573	47.0	21.1
200	473	38.8	17.5
100	373	30.6	13.8
0	273	22.4	10.1
−100	173	14.2	6.39
−200	73	6.00	—

FIGURE 11.2 Charles's law. The solid lines represent the volumes of samples of H_2 gas (0.00200 g) and O_2 gas (0.0200 g) at a pressure of 1.00 atm and different temperatures. The volumes decrease as the temperature is lowered at constant pressure. These lines, if extended, intersect the temperature axis at approximately -273 °C.

11.3 Do Different Gases Have Any Similar Properties?

Imagine isolating three trace atmospheric gases—hydrogen, carbon dioxide, and sulfur hexafluoride—and obtaining pure samples of each. How similar and how different would they be?

Gases present in very low concentrations are called "trace" gases.

Their natural abundance, uses, and chemical reactivity are strikingly different. Hydrogen gas, made of diatomic H_2 molecules, is a highly flammable gas used as a fuel. It reacts so rapidly with soil microbes and atmospheric hydroxyl free radicals that only trace amounts can be detected in our atmosphere, typically 1 molecule in more than 2 million particles of air, or 0.55 ppm (0.000055%). As carbon-based fuels needed to support human activity are phased out in response to concerns about our changing climate, hydrogen gas used in fuel cells will increase in importance [<<Section 7.1].

By contrast, carbon dioxide (CO_2), is so non-flammable that it is used in Class B and C fire extinguishers. After being released into our atmosphere through combustion and natural processes, carbon dioxide gas is taken up by plants in photosynthesis and also dissolves in ocean water where it becomes part of complex equilibria that acidify the oceans [>>Section 15.1]. Some of it ultimately ends up in calcium carbonate deposits that form the shells of marine organisms. Its atmospheric concentration [<<Section 4.3] has steadily increased over the past 150 years to its present value of 390 parts per million (390 ppm, 0.0390%).

Sulfur hexafluoride (SF_6) in our atmosphere comes entirely from human activity, released from uses such as an insulating gas in high-voltage electrical switches. Present atmospheric concentrations are only 3 parts-per-trillion (3 ppt, 0.0000000003%). SF_6 is so unreactive and insoluble in rain water that its atmospheric lifetime is over 3000 years.

These three gases interact differently with infrared and other wavelengths of electromagnetic radiation. As is the case for other gases made of diatomic molecules consisting of two identical atoms such as N_2 and O_2, hydrogen gas (H_2) does not absorb infrared radiation. It is therefore not a "greenhouse gas." CO_2 gas absorbs strongly in the infrared region. SF_6 is one of the most potent "greenhouse gases," with a global warming potential (GWP) [<<Section 4.3 and e4.5] of over 22 000. So even though atmospheric amounts of SF_6 and other gases whose molecules have central atoms attached to many halogen atoms are miniscule, they have the potential to play a significant role in the changing radiation balance of our planet. A good example is C_3F_8 gas, with an atmospheric lifetime of 2600 years and a GWP of 8600. Recent modelling studies designed to explore whether Mars could be made habitable show that the introduction of 0.4 Pa of C_3F_8 into the Martian atmosphere might melt the frozen CO_2 at polar caps and trigger a runaway greenhouse effect.

The molar masses of the three gases are very different: $M(H_2) = 2$ g mol^{-1}; $M(CO_2) = 44$ g mol^{-1}; and $M(SF_6) = 146$ g mol^{-1}. In a simulation at constant temperature of 120 °C, the differences in their dynamic behaviour at the molecular level are evident—the mean velocities of their molecules are $u_{mean}(H_2) = 2052$ m s^{-1}; $u_{mean}(CO_2) = 433$ m s^{-1}; and $u_{mean}(SF_6) = 242$ m s^{-1}.

Since these three gases are chemically so very different, it might surprise you to learn (see e11.5) that some of their thermodynamic properties are very similar.

As described in Section 11.2, the state of any gas can be described by four thermodynamic variables: pressure, temperature, volume, and amount. Their similarity refers to the relationship that can be established among these four variables. For example, suppose you have samples of the pure gases $N_2(g)$, $O_2(g)$, $H_2(g)$, $CO_2(g)$, and $SF_6(g)$. The volume of 1 mol of any of these gases at ambient temperature and pressure is approximately 24.8 L (Figure 11.3). Or, if you had 1 L samples of each at the same temperature and pressure, each contains almost the same amount (in mol). If the temperature of each sample were increased while holding the volume constant, the pressure of each gas would increase by the same amount. This uniform behaviour applies well under ambient conditions ($\sim$1 bar, $\sim$25 °C) for many gases.

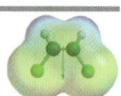
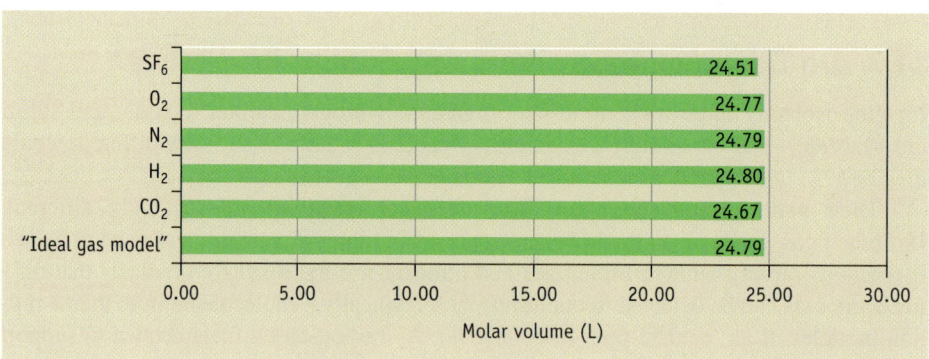

FIGURE 11.3 Molar volume of different atmospheric gases at $p = 1$ bar and $T = 298.15$ K. The ideal gas model is described quantitatively in Section 11.4.

We can make sense of this remarkably similar behaviour of different gases by imagining a molecular-level view of a sample of any one of these atmospheric gases. The chemist's molecular-level mental model, called the *kinetic–molecular theory*, shows each gas to consist of a vast number of particles that are constantly colliding with each other and with the walls of their containers. The number of molecule–molecule collisions in a gas sample is staggering. For example, we believe there are 1×10^{34} collisions per second involving molecules of N_2 and O_2 in the air of a typical automobile tire! Another important feature of our mental model of gases is that the separation between gas particles is much greater than the diameter of the particles. As a result of the relatively large distances between gas particles, the forces between them are very weak. Gas molecules move

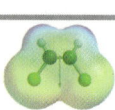

relatively independently, and different gases exhibit similar behaviour, which depends on pressure, temperature, volume, and the amount of the gas, rather than the identity of their constituent molecules.

In the absence of forces of attraction or repulsion between independent gas particles, we would predict that each particle moves in a straight line. The only time the trajectory of a particle changes is following collision with other particles or with the walls of a container.

> While different gases, such as those found in the atmosphere, have different uses, natural abundance, chemical reactivity, and interaction with electromagnetic radiation, they show almost universal thermodynamic behaviour under conditions near ambient pressure and temperature.

11.4 The Ideal Gas Equation

The almost universal thermodynamic behaviour of gases under certain limited conditions can be expressed mathematically by combining Boyle's law, Charles's law, and Avogadro's hypothesis to obtain $V\alpha\dfrac{nT}{p}$. If a proportionality constant, R, called the **gas constant**, is introduced, the mathematical equation is called the *ideal gas equation:*

$$V = R\left(\frac{nT}{p}\right)$$

or

$$pV = nRT \tag{11.1}$$

R is *the universal gas constant*—a number that you can use to interrelate the thermodynamic properties of any gas.

We call the equation $pV = nRT$ the **ideal gas equation**, because it mathematically expresses the behaviour of the chemists' molecular-level model of the gaseous state, called an "ideal gas," which assumes that the molecules of the gas have zero volume and that there are no interactions between gas molecules [>>Section 11.7]. It is quite remarkable that the equation derived from such a simple model describes so well the behaviour of many different real gases—regardless of their identity—under conditions where the pressure is about 1 bar or less and temperatures are at or above 25 °C.

To use the equation $pV = nRT$, we need a value for R, which can be determined experimentally from measurements of p, V, n, and T for many samples of gases.

$$R = 8.314 \text{ L kPa K}^{-1} \text{ mol}^{-1}$$

For example, under conditions of **standard ambient temperature and pressure (SATP)**, a gas temperature of 25 °C or 298.15 K and a pressure of 1×10^5 Pa (1×10^2 kPa), we can calculate that 1 mol of an "ideal" gas occupies 24.789 L.

$$V = \frac{RnT}{p} = \frac{(8.314 \text{ L kPa K}^{-1} \text{ mol}^{-1})(1.0000 \text{ mol})(298.15 \text{ K})}{(1.0 \times 10^2 \text{ kPa})} = 24.789 \text{ L}$$

It's sometimes necessary to work with other values for R, when V and p are given in other units, such as

$$R = 8.314 \text{ L kPa K}^{-1} \text{ mol}^{-1} = 8.314 \text{ m}^3 \text{ Pa K}^{-1} \text{ mol}^{-1} = 0.08206 \text{ L atm K}^{-1} \text{ mol}^{-1}$$

R can also be expressed in units that include energy. Since pressure is force per unit area:

$$1 \text{ Pa} = 1 \text{ N m}^{-2} = 1 \text{ kg m}^{-1}\text{s}^{-2} = 1 \text{ J m}^{-3}$$

therefore

$$R = 8.314 \text{ m}^3 \text{ Pa K}^{-1} \text{ mol}^{-1} = 8.314 \text{ J K}^{-1} \text{ mol}^{-1}$$

The ideal gas equation is a mathematical expression of a model, which describes the behaviour of many real gases under certain conditions. There is no such thing as an "ideal" gas, nor is anything "wrong" with a gas whose behaviour doesn't exactly fit this pattern. Like any model, it gives helpful explanations but has limitations.

Think about It

e11.7 Work a tutorial on the ideal gas equation.

Think about It

e11.8 Model the ideal gas equation with spreadsheets.

WORKED EXAMPLE 11.1—IDEAL GAS EQUATION

The nitrogen gas in an automobile airbag, with a volume of 65 L, exerts a pressure of 110.5 kPa at 25 °C. What amount of N_2 gas (in mol) is in the airbag?

Solution

To use the ideal gas equation with R having units of (L kPa K^{-1} mol^{-1}), the pressure must be measured in kPa and the temperature in K. Therefore,

$$p = 110.5 \text{ kPa}$$

$$T = 25 + 273 = 298 \text{ K}$$

Now substitute the values of p, V, T, and R into the ideal gas equation and solve for the amount of gas, n:

$$n \text{ (}N_2\text{)} = \frac{pV}{RT} = \frac{(110.5 \text{ kPa})(65 \text{ L})}{(8.314 \text{ L kPa K}^{-1} \text{ mol}^{-1})(298 \text{ K})} = 2.9 \text{ mol}$$

Notice that the units kPa, L, and K cancel to leave the answer in mol.

Interactive Exercises 11.1–11.2

Practise using the ideal gas equation.

EXERCISE 11.4—IDEAL GAS EQUATION

The balloon used by Jacques Charles in his historic flight in 1783 was filled with about 1300 mol of $H_2(g)$. If the temperature of the gas was 23 °C and its pressure was 1 bar, what was the volume of the balloon?

In using the ideal gas equation, we often encounter situations where we have a sample of a certain amount of gas (n is constant), but two of the three other variables (p, V, and T) change. For example, what would happen to the pressure of a sample of nitrogen in an automobile airbag if the same amount of gas were placed in a smaller bag and heated to a higher temperature? R is always a constant, and in this case n is also constant.

$$\frac{p_1V_1}{T_1} = nR = \frac{p_2V_2}{T_2} \quad \text{for a constant amount of gas, } n \qquad (11.2)$$

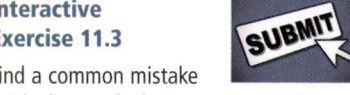

Interactive Exercise 11.3

Find a common mistake in ideal gas calculations.

WORKED EXAMPLE 11.2—IDEAL GAS EQUATION

Helium-filled balloons are used to carry scientific instruments high into the atmosphere to measure temperature and atmospheric composition. Suppose a balloon is launched when the temperature is 22.5 °C and the barometric pressure is 100.5 kPa. If the balloon's volume is 4.19×10^3 L (and no helium escapes from the balloon), what will the volume be at a height of 32.2 km, where the pressure is 10.1 kPa and the temperature is −33.0 °C?

Solution

Here we know the initial volume, temperature, and pressure of the gas. We want to know the volume of the same amount of gas at a new pressure and temperature. It is most convenient to use Equation 11.2 as a form of the ideal gas equation. Begin by setting out the information given in a table.

Initial Conditions	Final Conditions
$V_1 = 4.19 \times 10^3$ L	$V_2 = ?$ L
$p_1 = 100.5$ kPa	$p_2 = 10.1$ kPa
$T_1 = 22.5$ °C (295.7 K)	$T_2 = -33.0$ °C (240.2 K)

We can rearrange the ideal gas equation to calculate the new volume, V_2:

$$V_2 = \left(\frac{T_2}{p_2}\right) \times \left(\frac{p_1 V_1}{T_1}\right) = V_1 \times \frac{p_1}{p_2} \times \frac{T_2}{T_1}$$

$$= 4.19 \times 10^3 \text{ L} \times \frac{100.5 \text{ kPa}}{10.1 \text{ kPa}} \times \frac{240.2 \text{ K}}{295.7 \text{ K}}$$

$$= 3.38 \times 10^4 \text{ L}$$

In this case the pressure decreased by almost a factor of 10, which should lead to about a 10-fold volume increase. This increase is partly offset by a drop in temperature, which leads to a volume decrease. On balance, the volume increases because the pressure has dropped so substantially. Notice that the solution was to multiply the original volume (V_1) by a pressure factor (larger than 1 because the volume increases with a lower pressure) and a temperature factor (smaller than 1 because volume decreases with a decrease in temperature).

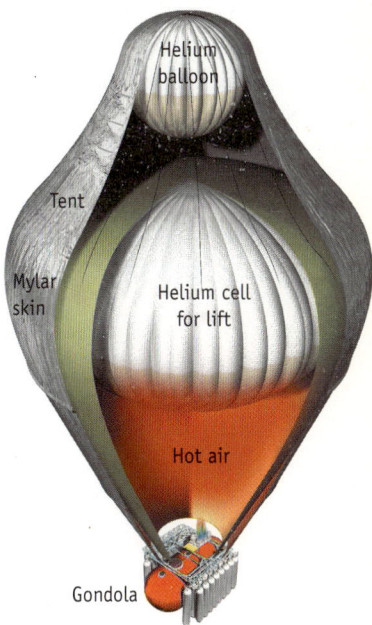

Tent

Mylar skin

Hot air

Gondola

A weather balloon is filled with helium. As it ascends into the troposphere, does the volume increase or decrease?

EXERCISE 11.5—IDEAL GAS EQUATION

You have a 22 L cylinder of helium at a pressure of 152 bar and a temperature of 31 °C. How many balloons can you fill, each with a volume of 5.0 L, on a day when the atmospheric pressure is 1.00 bar and the temperature is 22 °C?

EXERCISE 11.6—IDEAL GAS EQUATION

Gaseous ammonia is synthesized by the Haber-Bosch reaction:

$$N_2(g) + 3 H_2(g) \xrightarrow[\text{500 °C}]{\text{iron catalyst}} 2 NH_3(g)$$

Assume that 355 L of H_2 gas at 25.0 C and 72.2 kPa is combined with excess N_2 gas. What amount (in mol) of NH_3 gas can be produced? If this amount of NH_3 gas is stored in a 125 L tank at 25.0 °C, what is the pressure of the gas?

The ideal gas equation, $pV = nRT$, describes remarkably well the behaviour of many different real gases, regardless of their identity, under conditions where the pressure is about 1 bar or less and temperatures are at or above 25 °C.

11.5 The Density of Gases

At a given temperature and pressure, density tells us the mass of gas of a given volume.

Let us see how density relates to pressure and to the ideal gas equation. Because the amount (n) of any compound is given by its mass (m) divided by its molar mass (M), we can substitute m/M for n in the ideal gas equation.

$$pV = \left(\frac{m}{M}\right)RT$$

Density (ρ) is defined as mass divided by volume (m/V). We can rearrange this form of the gas equation to give the following equation:

$$\rho = \frac{m}{V} = \frac{pM}{RT} \qquad (11.3)$$

Density is directly proportional to the pressure and molar mass, and inversely proportional to the temperature of a gas. Equation 11.3 is useful because gas density can be calculated from the molar mass, or the molar mass can be found from a measurement of gas density at a given pressure and temperature.

Molecular Modelling (Odyssey)

e11.9 Simulate the difference in densities between gases and liquids.

Gas density. (a) The balloons are filled with nearly equal amounts of gas at the same temperature and pressure. One yellow balloon contains helium, a low-density gas ($\rho = 0.090$ g L^{-1}) relative to air. The other balloons contain air, a higher-density gas ($\rho = 1.2$ g L^{-1}).
(b) A hot-air balloon rises because the heated air has a lower density than the surrounding air.

Charles D. Winters

(a)

jacomstephens/iStock

(b)

Think about It

e11.10 Determine the molar mass of a gas from its density.

WORKED EXAMPLE 11.3—GAS DENSITY

Calculate the density of CO_2 at standard ambient pressure and temperature. Is CO_2 more or less dense than air (1.17 g L^{-1})?

Strategy

Use Equation 11.3 to relate gas density and molar mass. Here we know the molar mass (44.0 g mol^{-1}), pressure ($\rho = 100$ kPa), temperature ($T = 298.15$ K), and the gas constant (R). Only the density (ρ) is unknown.

Solution

The known values are substituted into Equation 11.3, which is then solved for density:

$$\rho = \frac{pM}{RT} = \frac{(100\ \text{kPa})(44.0\ \text{g}\ \text{mol}^{-1})}{(8.314\ \text{L}\ \text{kPa}\ \text{K}^{-1}\ \text{mol}^{-1})(298.15\ \text{K})} = 1.78\ \text{g}\ \text{L}^{-1}$$

The density of CO_2 is considerably greater than that of dry air at SATP (1.17 g L^{-1}).

Interactive Exercises 11.7–11.8

Practise calculations involving molar mass and gas density.

Interactive Exercise 11.9

Find a common mistake in gas density calculations.

EXERCISE 11.10—MOLAR MASS FROM DENSITY

The density of an unknown gas is 5.02 g L^{-1} at 15.0 °C and 99.3 kPa. Calculate its molar mass.

Gas density has practical implications. From the equation $\rho = pM/RT$ we recognize that the density of a gas is directly proportional to its molar mass at a given T and p. Dry air, which has an average molar mass of about 29 g mol^{-1}, has a density of about 1.2 g L^{-1} at 100 kPa and 25 °C. Gases with molar masses greater than 29 g mol^{-1} have densities larger than 1.2 g L^{-1} under these same conditions. As a consequence, gases such as CO_2, SO_2, and gasoline vapour can concentrate along the ground if released into the atmosphere (Figure 11.4). Conversely, gases such as H_2, He, CO, CH_4 (methane), and NH_3 rise if released into the atmosphere.

The significance of gas density has been tragically revealed in several events. One occurred in the African country of Cameroon in 1986, when Lake Nyos expelled a huge bubble of CO_2 into the atmosphere. Because CO_2 is denser than air, the CO_2 cloud hugged the ground, killing 1700 people living in a nearby village [>>Section 12.1].

11.6 Gas Mixtures and Partial Pressures

The air you breathe in (if it were dried to remove water vapour) has the approximate composition given in Table 11.3. What's your molecular-level picture of this ordinary air? Molecules of N_2 gas comprise 78% of the entities found in air. Are those molecules equally distributed around the room? Does the small amount of CO_2, which has a higher density than N_2, pool around your toes? In the lower troposphere, while the effect of gravity on different gases is not negligible, we can observe that atmospheric gases rapidly mix well, and samples of air are quite homogeneous. With some work, you could separate a sample of ordinary dry air into its constituent gases nitrogen, oxygen, carbon dioxide, argon, and trace amounts of other gases. If each of these samples of pure gases were then placed in previously evacuated containers of the same size as the sample of air, you could measure the pressure of each. Can you predict what those pressures would be?

FIGURE 11.4 Gas density. Because carbon dioxide from fire extinguishers is denser than air, it settles on top of a fire and smothers it. When CO_2 gas is released from the tank, it expands and cools significantly. The white cloud is solid CO_2 and condensed moisture from the air.

TABLE 11.3 Partial Pressures of Components of Dry Atmospheric Air**

Constituent	Molar Mass*	Amount Percent	Partial Pressure at Total Pressure of 1 bar
N_2	28.01	78.08	0.7912
O_2	32.00	20.95	0.2123
CO_2	44.01	0.033	0.00033
Ar	39.95	0.934	0.00946

*The average molar mass of dry air = 28.960 g mol^{-1}.

**If total pressure is 1 bar.

Working from the ideal gas equation, you know that the pressure of a gas (if temperature and volume are constant) is directly proportional to n, the amount of substance in the sample. Since the amount of N_2 is about four times the amount of O_2, its pressure is about four times as great:

$$p = n \times \frac{RT}{V} \quad \text{(the ideal gas equation)}$$

The pressure each gas in a mixture would exert if it were the only gas in a container of the same size at the same temperature is called its **partial pressure**. John Dalton (1766–1844) was the first to observe that the total pressure of a mixture of gases is the sum of their partial pressures. This observation is now known as **Dalton's law of partial pressures.** Mathematically, we can write Dalton's law of partial pressures as

$$p_{\text{total}} = p_A + p_B + p_C + \cdots \tag{11.4}$$

where p_A, p_B, and p_C are the partial pressures of the individual gases that make up the mixture and p_{total} is the total pressure.

In a mixture of gases, molecules of each gas behave independently of molecules of all others in the mixture. Therefore, we can consider the behaviour of each gas in a mixture separately. As an example, let us take a mixture of three gases, A, B, and C, and assume they obey the ideal gas equation. There are n_A mol of A, n_B mol of B, and n_C mol of C. Assume that the mixture ($n_{\text{total}} = n_A + n_B + n_C$) is contained in a given volume (V) at a given temperature (T). We can calculate the pressure exerted by each gas from the ideal gas equation

$$p_A V = n_A RT \quad p_B V = n_B RT \quad p_C V = n_C RT$$

where each gas (A, B, and C) is in the same volume V and is at the same temperature T. According to Dalton's law, the total pressure exerted by the mixture is the sum of the pressures exerted by each component:

$$p_{total} = p_A + p_B + p_C = n_A\left(\frac{RT}{V}\right) + n_B\left(\frac{RT}{V}\right) + n_C\left(\frac{RT}{V}\right)$$

$$p_{total} = (n_A + n_B + n_C)\left(\frac{RT}{V}\right)$$

$$p_{total} = n_{total}\left(\frac{RT}{V}\right) \tag{11.5}$$

For mixtures of gases, it is convenient to introduce a quantity called the **amount fraction** or **mole fraction (x)**, which is defined as the amount of a particular chemical species in a mixture divided by the total amount of all chemical species present. Mathematically, the mole fraction of a species A in a mixture with B and C is expressed as

$$x_A = \frac{n_A}{n_A + n_B + n_C} = \frac{n_A}{n_{total}}$$

Now we can combine this equation (written as $n_{total} = n_A/x_A$) with the equations for p_A and p_{total}, and derive Equation 11.6:

$$p_A = x_A p_{total} \tag{11.6}$$

This equation is useful because it tells us that the partial pressure of a gas in a mixture of gases is the product of its mole fraction and the total pressure of the mixture. For example, the mole fraction of N_2 in air is 0.78, so in a sample of air whose total pressure is 1×10^5 Pa (1 bar), its partial pressure is 7.8×10^4 Pa (0.78 bar).

Molecular Modeling (Odyssey)

e11.11 Use simulations of gas mixtures to calculate partial pressures.

Think about It

e11.12 Work through a tutorial on partial pressures of gases.

WORKED EXAMPLE 11.4—PARTIAL PRESSURES

Halothane ($CF_3CHBrCl$) is a non-flammable, non-explosive, and non-irritating gas that is commonly used as an inhalation anaesthetic. Suppose you mix 15.0 g of halothane vapour with 23.5 g of oxygen gas. If the total pressure of the mixture is 114 kPa, what is the partial pressure of each gas?

Solution

One way to solve this problem is to recognize that the partial pressure of a gas is given by the total pressure of the mixture multiplied by the mole fraction of the gas. Let us first calculate the mole fractions of halothane and of O_2.

Step 1. Calculate mole fractions:

$$n(CF_3CHBrCl) = \frac{m}{M} = \frac{15.0 \text{ g}}{197.4 \text{ g mol}^{-1}} = 0.0760 \text{ mol}$$

$$n(O_2) = \frac{m}{M} = \frac{23.5 \text{ g}}{32.00 \text{ g mol}^{-1}} = 0.734 \text{ mol}$$

$$x(CF_3CHBrCl) = \frac{n(CF_3CHBrCl)}{n(CF_3CHBrCl) + n(O_2)} = \frac{0.0760 \text{ mol}}{0.810 \text{ mol}} = 0.0938$$

Because the sum of the mole fractions of halothane and of O_2 must equal 1.000, the mole fraction of oxygen is 0.906.

$$x_{halothane} + x_{oxygen} = 1.000$$

$$0.0938 + x_{oxygen} = 1.000$$

$$x_{oxygen} = 0.906$$

Step 2. Calculate partial pressures:

$$\text{Partial pressure of halothane} = p_{\text{halothane}} = x_{\text{halothane}} \times P_{\text{total}}$$

$$p_{\text{halothane}} = 0.0938 \times P_{\text{total}} = 0.0938 \times 114 \text{ kPa}$$

$$p_{\text{halothane}} = 10.7 \text{ kPa}$$

The total pressure of the mixture is the sum of the partial pressures of the gases in the mixture.

$$p_{\text{halothane}} + p_{\text{oxygen}} = 114 \text{ kPa}$$

and so

$$p_{\text{oxygen}} = 114 \text{ kPa} - p_{\text{halothane}}$$

$$p_{\text{oxygen}} = 114 \text{ kPa} - 10.7 \text{ kPa} = 103 \text{ kPa}$$

Interactive Exercises 11.11–11.12

Practise calculations involving partial pressures.

EXERCISE 11.13—PARTIAL PRESSURES

The halothane-oxygen mixture described in Worked Example 11.4 is placed in a 5.00 L tank at 25.0 °C. What is the total pressure (in kPa) of the gas mixture in the tank? What are the partial pressures (in kPa) of each gas?

> The pressure each gas in a mixture would exert if it were in a container of the same size at the same temperature is called its partial pressure. Dalton's law of partial pressures states that the total pressure of a mixture of gases is the sum of their partial pressures.

11.7 The Kinetic-Molecular Theory of Gases

So far we have considered the observable properties of gases, properties such as pressure and volume that result from the behaviour of a system with a large number of particles. Now we look more closely at the **kinetic-molecular theory** as a *model* that describes the behaviour of matter at the *molecular level* [<<Section 2.2]. Thousands of experimental observations have led to the following postulates regarding the behaviour of gases:

- Gases consist of particles (molecules or atoms), whose separation is much greater than the size of the particles themselves (Figure 11.5) at reasonably low pressures and high temperatures. Thus, forces between the particles are negligible.
- The particles of a gas are in continual, random, and rapid motion. They move in straight lines unless they collide with each other or the walls of their container. These collisions take place without loss of the total energy of the system.
- The average kinetic energy of gas particles is proportional to the gas temperature. All gases, regardless of their molecular mass, have the same average kinetic energy at the same temperature.

Let's discuss the observed behaviour of gases from the point of view of this theoretical model.

Charles D. Winters

FIGURE 11.5 A molecular view of gases and liquids. The fact that a large volume of N_2 gas can be condensed to a small volume of liquid indicates that the distances between molecules in the gas phase are very large as compared with the distances between molecules in liquids. (Liquid N_2 boils at −196 °C.)

Molecular Speed and Kinetic Energy

How does your nose detect that your friend has just carried a pizza into your room? In scientific terms, we know that the odour-causing molecules of food enter the gas phase. After colliding in a very short period of time with many other molecules in the air, they travel to the cells of your nose, creating interactions that trigger a complex sequence of events ultimately perceived by your brain as odours. The same sort of molecular movement happens in the laboratory when bottles containing aqueous ammonia (NH_3) solutions and aqueous hydrochloric acid (HCl) solutions are placed side by side and opened (Figure 11.6). Molecules of the two compounds enter the gas phase and undergo

> In the "ideal gas" equation, we take the kinetic-molecular model to an extreme, by assuming that the particles of a gas have zero volume, and there is no force of interaction between particles. Under conditions where these assumptions do not hold, the ideal gas equation no longer fits observed behaviour of real gases very well.

Charles D. Winters

FIGURE 11.6 The movement of gas molecules. Open dishes of aqueous ammonia and hydrochloric acid were placed side by side. When molecules of NH_3 and HCl escape from solution to the atmosphere and encounter one another, we observe a cloud of solid ammonium chloride, NH_4Cl.

FIGURE 11.7 The distribution of molecular speeds in O_2 gas. A graph of the number of molecules with a given speed versus that speed shows the distribution of molecular speeds. The red curve shows the effect of increased temperature. Even though the curve for the higher temperature is "flatter" and broader than the curve for the lower temperature, the areas under the curves are the same because the number of molecules in the sample is fixed [>>e11.13].

Plots showing the relationship between the number of molecules and their speed or energy (Figure 11.7) are called Maxwell-Boltzmann distribution curves. The distribution of speeds (and kinetic energies) of molecules (Figures 11.7 and 11.8) is often used when explaining chemical phenomena.

Molecular Modelling (Odyssey)

e11.13 Simulate changes in speed distribution with temperature, molar mass, and gas composition.

Think about It

e11.14 Simulate the Maxwell-Botzmann distribution for different gases.

numerous collisions with other molecules before they encounter one another, at which time they react and form a cloud of tiny particles of solid ammonium chloride (NH_4Cl).

If you change the temperature of the environment of the containers in Figure 11.6 and measure the time needed for the cloud of ammonium chloride to form, you would find that this time is longer at lower temperatures. The reason is that the speed at which molecules move depends on the temperature. Let us expand on this idea.

The molecules in a gas sample do not all move at the same speed. Rather, as illustrated in Figure 11.7 for O_2 molecules, there is a distribution of speeds. Figure 11.7 shows the number of particles in a gas sample that are moving at certain speeds at a given temperature. We can make two important observations. First, at a given temperature, some molecules have high speeds and others have low speeds. Most of the molecules, however, have some intermediate speed, and their most probable speed corresponds to the maximum in the curve. For oxygen gas at 25 °C, for example, most molecules have speeds in the range of 200 m s^{-1} to 700 m s^{-1}, and their most probable speed is about 400 m s^{-1}. (These are very high speeds, indeed. A speed of 400 m s^{-1} corresponds to about 1440 kilometres per hour!) Collisions change the speed of individual molecules, but the distribution of speeds remains the same.

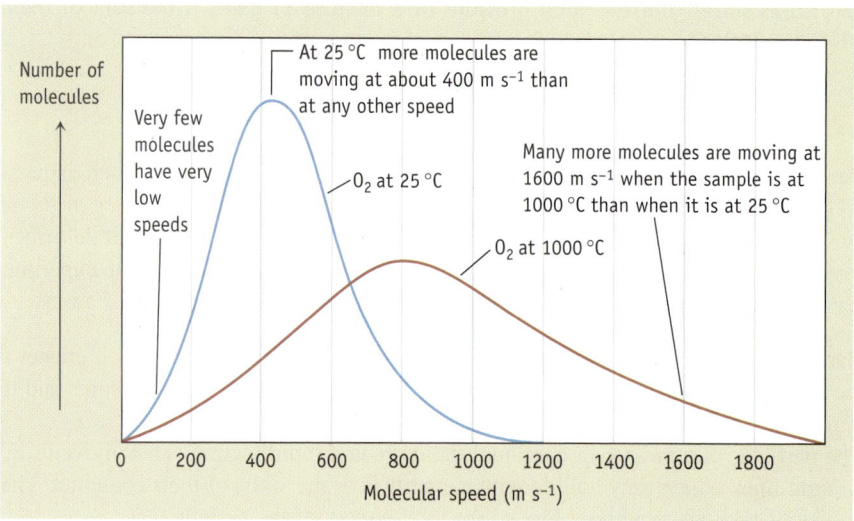

A second observation regarding the distribution of speeds is that as the temperature increases, the most probable speed increases, and the number of molecules travelling at very high speeds increases greatly.

The kinetic energy of a single molecule of mass m in a gas sample is given by the equation

$$E_k = \tfrac{1}{2}(\text{mass})(\text{speed})^2 = \tfrac{1}{2}mu^2$$

where u is the speed of that molecule. We can calculate the kinetic energy of a single molecule from this equation but not the kinetic energy of a collection of molecules, because the molecules in a gas sample move at different speeds. However, we can calculate the *average* kinetic energy of a collection of molecules by relating it to other averaged quantities of the system. In particular, the average kinetic energy is related to the average speed.

$$\overline{E_k} = \tfrac{1}{2}m\overline{u^2}$$

(The horizontal bar over the symbols E_k and u indicate an average value.) This equation states that the average kinetic energy of the molecules in a gas sample ($\overline{E_k}$) is related to the average of the squares of their speeds ($\overline{u^2}$) (called the "mean square speed").

Experiments also show that $\overline{E_k}$ is directly proportional to temperature with a proportionality constant of $\frac{3}{2}R$:

$$\overline{E_k} = \frac{3}{2}RT$$

where R is the gas constant expressed in SI units (8.314472 J K^{-1} mol^{-1}).

Because $\overline{E_k}$ is proportional to both $\frac{1}{2}m\overline{u^2}$ and T, temperature and $\frac{1}{2}m\overline{u^2}$ must also be proportional; that is, $\frac{1}{2}m\overline{u^2} \alpha\, T$. This relationship among mass, average speed, and temperature is expressed in Equation 11.7. Here the square root of the mean square speed ($\sqrt{\overline{u^2}}$, called the *root-mean-square* or *rms speed*), the temperature (T, in K), and the molar mass (M) are related.

$$\sqrt{\overline{u^2}} = \sqrt{\frac{3RT}{M}} \qquad (11.7)$$

This equation, sometimes called *Maxwell's equation* after James Clerk Maxwell, shows that the rms speeds of gas molecules are indeed related directly to the temperature (Figure 11.8). The rms speed is a useful quantity because it is directly related to the average kinetic energy and very close to the true average speed for a sample. (The average speed is 92% of the rms speed.)

All gases at the same temperature have the same average kinetic energy. However, if you compare a sample of one gas with another—say, compare O_2 and N_2—the molecules do not necessarily have the same average speed (Figure 11.8). Instead, Maxwell's equation shows that the smaller the molar mass of the gas, the greater the rms speed.

Think about It

e11.15 Model the speed distributions of gases using spreadsheet plots.

Think about It

e11.16 Work through a tutorial on calculating rms speeds.

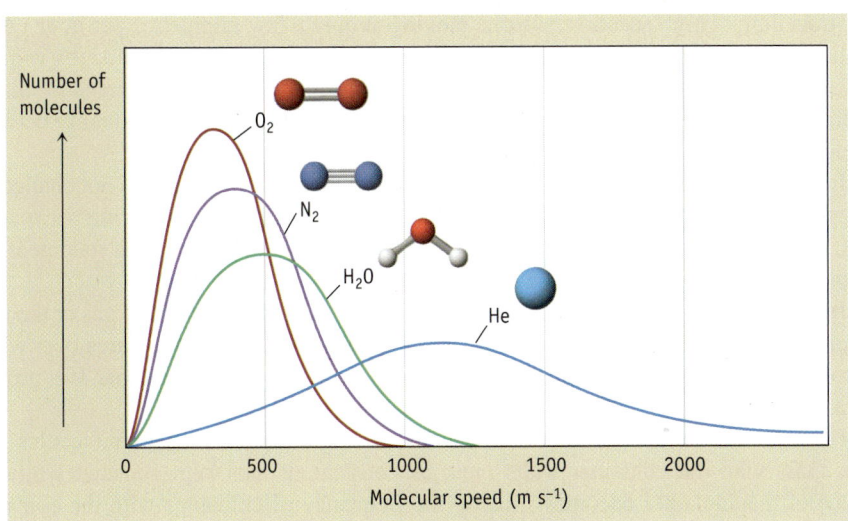

FIGURE 11.8 The effect of molar mass on the distribution of speeds. At a given temperature, molecules with higher molar masses have lower speeds.

WORKED EXAMPLE 11.5—RMS SPEED

Calculate the rms speed of oxygen molecules at 25 °C.

Solution

We use Equation 11.7 with M in kg mol^{-1} since R is in units of J K^{-1} mol^{-1}, and 1 J = 1 kg m^2 s^{-2}. The molar mass of O_2 is 32.0×10^{-3} kg mol^{-1}.

$$\sqrt{\overline{u^2}} = \sqrt{\frac{3(8.3145\text{ J K}^{-1}\text{ mol}^{-1})(298\text{ K})}{32.0 \times 10^{-3}\text{ kg mol}^{-1}}} = \sqrt{2.32 \times 10^5\text{ J kg}^{-1}}$$

To obtain the answer in metres per second, we use the relation $1 \text{ J} = 1 \text{ kg·m}^2 \text{ s}^{-2}$. This means we have

$$\sqrt{\overline{u^2}} = \sqrt{(2.32 \times 10^5 \, \cancel{J} \, \cancel{kg^{-1}}) \times \frac{(1 \, \cancel{kg} \, m^2 \, s^{-2})}{1 \, \cancel{J}}} = \sqrt{2.32 \times 10^5 \, m^2 \, s^{-2}} = 482 \text{ m s}^{-1}$$

This speed is equivalent to about 1735 kilometres per hour, although molecules do not move in a straight line because of collisions!

Interactive Exercises 11.14–11.15

Practise calculations involving rms speeds.

EXERCISE 11.16—RMS SPEED

Calculate the ratio of rms speeds of CO_2 and N_2 molecules at 25 °C.

Kinetic-Molecular Theory and the Ideal Gas Equation

The gas laws, which come from experiment, can be explained by the kinetic-molecular theory. Your mental model should envision pressure arising from collisions of gas molecules with the walls of its container (Figure 11.9). Pressure is related to the frequency and total force due to collisions.

$$\text{Gas pressure from collisions} = \frac{\text{total force due to collisions}}{\text{area}}$$

The total force on the container walls due to collisions depends on the number of collisions and the average force per collision. When the temperature of a gas increases, the average kinetic energy of the molecules increases, and so the average force of the collisions with the walls increases. (This is analogous to the difference in the force exerted by a car travelling at high speed versus one moving at only a few kilometres per hour.) Also, because the speed of gas molecules increases with temperature, more collisions *with the walls* occur per second. Thus, the collective force per square centimetre is greater—that is, the pressure increases. Mathematically, this is related to the direct proportionality between p and T when n and V are fixed; that is, $p = (nRT/V)$.

Increasing the amount of a gas at a fixed temperature and volume does not change the average collision force, but it does increase the number of collisions occurring per second. Thus, the pressure increases, which is consistent with the ideal gas equation statement that p is proportional to n when V and T are constant; that is, $p = n(RT/V)$.

If the pressure is held constant when either the number of molecules of gas or the temperature of the gas is increased, then the volume of the container (and the area over which the collisions can take place) must increase. This is expressed by stating that V is proportional to nT when p is constant: $V = nT(R/p)$.

Finally, if the temperature is constant, the average impact force of molecules of a given mass with the container walls must be constant. If n is kept constant while the volume of the container becomes smaller, the frequency of collisions with the container walls must increase. This means the pressure increases, consistent with $p = (1/V)(nRT)$.

> The observable gas laws can be explained by the kinetic-molecular theory, a model that describes the behaviour of matter at the molecular or atomic level. In the "ideal gas" equation, chemists take the kinetic molecular model to an extreme, by assuming that the particles of a gas have zero volume, and there is no force of interaction between particles.

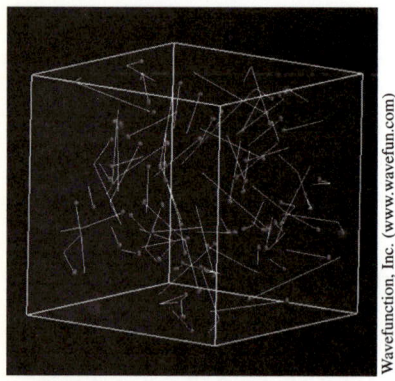

FIGURE 11.9 Gas pressure. According to the kinetic-molecular theory, gas pressure is caused by gas molecules bombarding the container walls. The Odyssey molecular dynamics simulation of a gas sample leaves "trails" to help you imagine the path molecules might take as they collide with each other and the walls of a container.

Wavefunction, Inc. (www.wavefun.com)

11.8 Diffusion and Effusion

When a pizza is brought into a room, even if there were no movement of the air caused by fans or people moving about, the aroma would reach everywhere in the room. The volatile aroma-causing molecules vaporize into the atmosphere, where they mix with molecules of oxygen, nitrogen, carbon dioxide, water vapour, and other entities present. This mixing of molecules of two or more gases due to their random molecular motions is called **diffusion**. Given time, the molecules of each component will completely mix with all other components of the mixture (Figure 11.10).

time →

(a)　　　　　　(b)

FIGURE 11.10 Diffusion. (a) Liquid bromine (Br_2) was placed in a small flask inside a larger container. (b) The cork was removed from the flask and, with time, bromine vapour diffused through the air in the larger container.

Think about It

e11.17 Watch HCl and NH_3 gases diffuse toward each other and react.

Diffusion is also illustrated by the experiment shown in Figure 11.6. The experiment is repeated with a different setup in Figure 11.11. Here cotton moistened with aqueous hydrochloric acid solution is placed at one end of a U-tube, and cotton moistened with aqueous ammonia solution is placed at the other end. Molecules of HCl and NH_3 diffuse through the air in the tube. When they meet, they produce white, solid NH_4Cl:

$$HCl(g) + NH_3(g) \longrightarrow NH_4Cl(s)$$

Closely related to diffusion is **effusion**, which is the movement of gas through a tiny opening in a container into another container where the pressure is very low (Figure 11.12). Thomas Graham (1805–1869), a Scottish chemist, studied the effusion of gases and found that the rates of effusion of gases—the amount of gas moving from one place to another in a given amount of time—are inversely proportional to the square root of their molar masses.

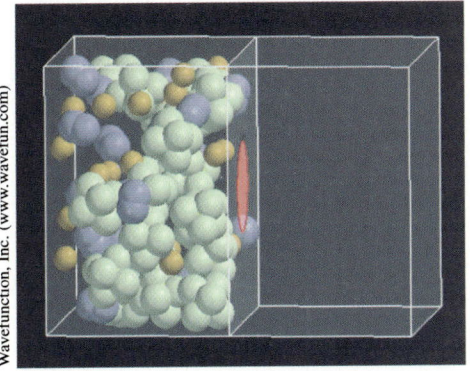

(a)　　　　　　(b)

FIGURE 11.12 Molecular dynamics simulation of the effusion of a mixture of SF_6, N_2, and He through a small hole, from the left to the evacuated right side of the simulation cell. Note that the lighter He atoms pass most frequently through the hole, followed by N_2 and then SF_6 molecules. (a) Before opening the hole, and (b) after a short period of effusion. See e11.18 for the simulation.

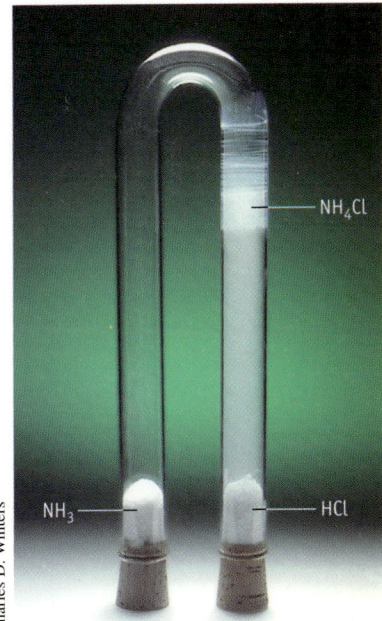

NH_4Cl

NH_3　　　HCl

FIGURE 11.11 Gaseous diffusion. HCl gas (from hydrochloric acid solution) and NH_3 gas (from aqueous ammonia solution) diffuse from opposite ends of a glass U-tube. When they meet, they react to form white, solid NH_4Cl. The NH_4Cl is formed closer to the end from which the HCl gas begins because HCl molecules are heavier than NH_3 molecules and diffuse slower. See also Figure 11.6.

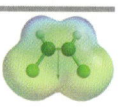

Molecular Modelling (Odyssey)

e11.18 Simulate the effect of molar mass on effusion and diffusion in Figure 11.12.

Think about It

e11.19 Work out how the rate of effusion relates to molar mass.

Based on many experimental results, the rates of effusion of two gases—that is, the amount of each gas passing through the hole per second—can be compared:

$$\frac{\text{Rate of effusion of gas 1}}{\text{Rate of effusion of gas 2}} = \sqrt{\frac{\text{molar mass of gas 2}}{\text{molar mass of gas 1}}} = \sqrt{\frac{M(\text{gas 2})}{M(\text{gas 1})}} \qquad (11.8)$$

The relationship in Equation 11.8—now known as *Graham's law*—is readily derived from Maxwell's equation by recognizing that the rate of effusion depends on the speed of the molecules. The ratio of the rms speeds is the same as the ratio of the effusion rates:

$$\frac{\text{Rate of effusion of gas 1}}{\text{Rate of effusion of gas 2}} = \frac{\sqrt{\overline{u^2} \text{ of gas 1}}}{\sqrt{\overline{u^2} \text{ of gas 2}}} = \frac{\sqrt{3RT/M(\text{gas 1})}}{\sqrt{3RT/M(\text{gas 2})}}$$

Cancelling out like terms gives the expression in Equation 11.8.

EXERCISE 11.17—GRAHAM'S LAW OF EFFUSION

Figure 11.12 shows a screenshot of the Odyssey simulation in e11.18 of a mixture of SF_6, N_2, and He effusing through the pores of a porous barrier. Using Graham's law, calculate the relative rates of effusion of SF_6, N_2, and He molecules.

WORKED EXAMPLE 11.6—MOLAR MASS USING GRAHAM'S LAW OF EFFUSION

Tetrafluoroethylene (C_2F_4) effuses through a barrier at a rate of 4.6×10^6 mol h^{-1}. An unknown gas, consisting of only boron and hydrogen, effuses at a rate of 5.8×10^6 mol h^{-1} under the same conditions. What is the molar mass of the unknown gas?

Solution

From Graham's law, we know that a sample made of molecules of low mass will effuse more rapidly than one made of higher mass molecules. Because the unknown gas effuses more rapidly than C_2F_4 ($M = 100.0$ g mol^{-1}), the unknown must have a molar mass less than 100 g mol^{-1}. Substitute the experimental data into Graham's law (Equation 11.8):

$$\frac{\text{Rate of effusion of unknown}}{\text{Rate of effusion of } C_2F_4} = \frac{5.8 \times 10^{-6} \text{ mol h}^{-1}}{4.6 \times 10^{-6} \text{ mol h}^{-1}} = 1.3$$

$$= \sqrt{\frac{M(C_2F_4)}{M(\text{unknown})}} = \sqrt{\frac{100.0 \text{ g mol}^{-1}}{M(\text{unknown})}}$$

To solve for the unknown molar mass, square both sides of the equation and rearrange:

$$1.6 = \frac{100.0 \text{ g mol}^{-1}}{M(\text{unknown})} \quad \text{so} \quad M(\text{unknown}) = \frac{100.0 \text{ g mol}^{-1}}{1.6} = 63 \text{ g mol}^{-1}$$

Note

A boron-hydrogen compound corresponding to this molar mass is B_5H_9, called pentaborane.

Interactive Exercise 11.18

Practise calculations using Graham's law.

EXERCISE 11.19—MOLAR MASS USING GRAHAM'S LAW OF EFFUSION

A sample of pure methane (CH_4) is found to effuse through a porous barrier in 1.50 min. Under the same conditions, an equal number of molecules of an unknown gas effuse through the barrier in 4.73 min. What is the molar mass of the unknown gas?

The mixing of molecules of two or more gases due to their random molecular motions is called diffusion. Effusion is the movement of gas through a tiny opening in a container into another container where the pressure is very low. Both diffusion and effusion depend on molar mass.

11.9 The Ideal Gas Model and Real Gas Behaviour

For gases near room temperature and a pressure of 1 bar, the ideal gas equation is remarkably successful in correlating the amount of gas and its pressure, volume, and temperature. For many gases under these conditions, these variables can be modelled with an error less than a few percent. But if we look at data for real gases at higher pressures (Figure 11.13) or lower temperatures (Figure 11.14), we find that the ideal gas model no longer predicts the behaviour of real gases very well.

As shown in Figure 11.13, at intermediate pressures, for $n = 1$ mol, and fixed V and T, pV/RT is less than the value of 1 predicted by the ideal gas equation. At very high pressures, it is greater than expected. How can we explain this? We can understand why the ideal gas model is deficient by recalling that in the ideal gas equation, we make two extreme assumptions about the kinetic-molecular model—that the particles of a gas have zero volume, and there are no forces of interaction between particles [<<Section 11.7].

At 1 bar and 25 °C, the volume occupied by a single molecule is *very* small relative to its share of the total gas volume. A helium atom with a radius of 31 pm, for example, has roughly the same relative volume as a pea has inside a basketball. Now suppose the pressure is increased significantly, to 1000 bar. The volume available to each molecule is a sphere with a radius of only about 200 pm, which means the situation is now like that of a pea inside a sphere a bit larger than a ping-pong ball.

The kinetic-molecular theory and the ideal gas equation are concerned with the volume available to the molecules to move about, not the volume of the molecules themselves. It is clear that the volume occupied by gas molecules is not negligible at higher pressures. For example, suppose you have a flask marked with a volume of 500 mL. This does not mean the space available to molecules is 500 mL. In reality, the available volume is less than 500 mL, especially at high gas pressures, because the molecules themselves occupy some of the volume. Therefore, at high pressures, real gases can't compress quite as much and their volume is somewhat greater than predicted by the ideal gas model.

Another assumption underlying the ideal gas equation is that collisions between molecules are elastic—that is, that the atoms or molecules of the gas bounce off each other without experiencing intermolecular forces and without loss of total energy in the system. We know these assumptions can't be true, since we observe that gases can be compressed to form liquids, although some gases require a very low temperature to do so (Figure 11.14). Liquid and solid states would not exist in the absence of intermolecular forces.

At high pressures, instead of the ideal gas picture of completely independent molecules striking the wall of a container, it would be more accurate to think of other molecules in their vicinity exerting a slight attraction for them. As a result of these intermolecular forces, molecules strike the wall with less force than they would in the absence of

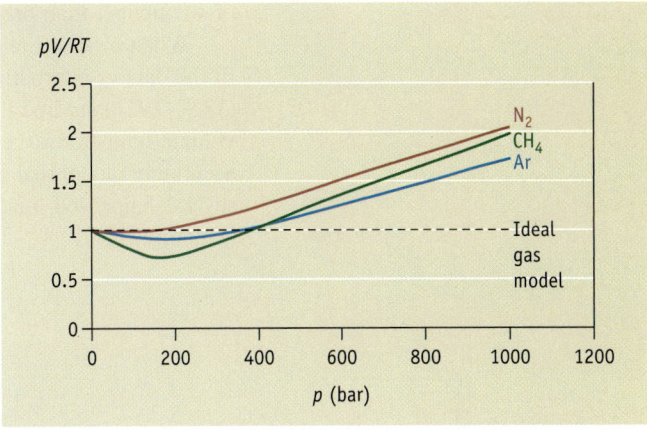

FIGURE 11.13 Effect of high pressure on the value of pV/RT for 1 mol of argon, nitrogen, and methane gases, all at 0 °C. The ideal gas model predicts that for 1 mol of any gas, $pV = RT$, so $pV/RT = 1$.

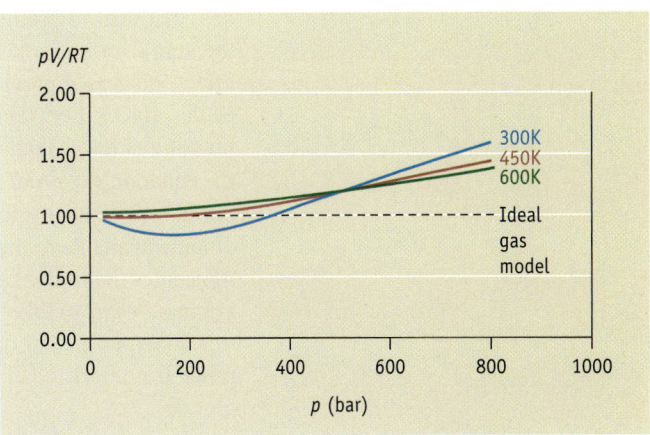

FIGURE 11.14 Effect of temperature and pressure on the on the value of pV/RT for 1 mol of methane gas, along with a prediction from the ideal gas model.

intermolecular attractive forces. Thus, because collisions between molecules in a real gas and the wall are "softer" than predicted by the model, the observed gas pressure is less than that predicted by the ideal gas equation. We might expect this effect to be particularly pronounced when the temperature is low.

An additional explanation might be that at very high pressures some gas molecules can form clusters of molecules, rather than discrete molecules. This would also result in a lower observed gas pressure, as fewer particles would be present to collide with the walls of a container than predicted by the ideal gas equation.

When models are limited in their ability to explain data, they need to be modified to fit the data or alternative models developed. The Dutch physicist Johannes van der Waals (1837–1923) studied the inability of the ideal gas equation to accurately model pressure-volume-temperature relationships at low T and high p, and developed an equation to more accurately model real gas properties at non-ambient conditions. This equation, only one of several improved mathematical models, is known as the **van der Waals equation** (Equation 11.9),

$$\left(p + a\left[\frac{n}{V}\right]^2\right)(V - bn) = nRT \tag{11.9}$$

Observed pressure · Container V · Correction for intermolecular forces · Correction for molecular volume

where a and b are experimentally determined constants (Table 11.4), different for each gas, that lead to prediction of best fit. Although Equation 11.9 might seem complicated, the terms in parentheses are those of the ideal gas equation, each corrected for the effects discussed previously. The pressure correction term, $a(n/V)^2$, accounts for intermolecular forces between particles. Because of intermolecular forces, the observed gas pressure is lower than the ideal pressure ($p_{observed} < p_{ideal}$, where p_{ideal} is calculated using the equation $pV = nRT$). Therefore, the term $a(n/V)^2$ is *added* to the observed pressure. The constant a typically has values in the range 1×10^{-4} to 0.1 kPa L^2 mol^{-2}. The actual volume available to the molecules is smaller than the volume of the container because the molecules themselves take up space. Therefore, we *subtract* an amount from the container volume ($= bn$) to take this factor into account. Here n is the number of moles of gas, and b is an experimental quantity that corrects for the molecular volume. Typical values of b range from 0.01 to 0.1 L mol^{-1}, roughly increasing with increasing molecular size.

TABLE 11.4 van der Waals Constants

Gas	a Values 10^{-2} kPa L^2 mol^{-2}	b Values L mol^{-1}
He	0.0346	0.0237
Ar	1.36	0.0322
H_2	0.247	0.0266
N_2	1.41	0.0391
O_2	1.38	0.0318
CO_2	3.64	0.0427
Cl_2	6.58	0.0562
H_2O	5.53	0.0305

A central part of doing chemistry involves trying to develop better conceptual and mathematical models to explain observations. The van der Waals equation makes similar predictions as the ideal gas equation for most gases at pressures near 1 bar. It is useful, however, because it fits the experimental pressure for real gases better at moderately high pressures. At even higher pressures, where the molar volume becomes almost as small as the b value, the $(V - bn)$ term approaches zero, and the equation gives meaningless predictions.

At high pressure and low temperature, two extreme assumptions of the ideal gas model fail—that the particles of a gas have zero volume, and there are no forces of interaction between particles. More sophisticated models than the ideal gas equation, such as the van der Waals equation, are needed to explain the properties of many real gases under such conditions.

Molecular Modelling (Odyssey)

e11.20 Simulate a gas at high pressures to see its real behaviour.

11.10 Liquid and Solid States—Stronger Intermolecular Forces

We turn now from gases to consider properties of liquids and solids. How differently do we picture the states of matter at the particulate level? Let's start with observations of the volumes occupied by an equal number of molecules of a substance in different states. Figure 11.15a shows a flask containing about 300 mL of liquid nitrogen. If all of the liquid were allowed to evaporate, the gaseous nitrogen at 1 bar and 25 °C would occupy a volume of 200 L. This is consistent with the model we have developed that shows a large amount of space exists between molecules in a gas, while in liquids the molecules are close together.

The large increase in volume when converting liquids to gases is striking. In contrast, the change in volume when a solid is converted to a liquid is less dramatic. Figure 11.15b shows the same amount of liquid and solid benzene side by side. They are not appreciably different in volume. This means that the atoms in the liquid are packed together about as tightly as the atoms in the solid state.

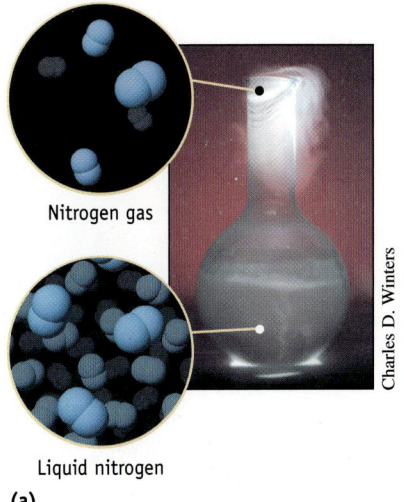

Nitrogen gas

Liquid nitrogen

(a)

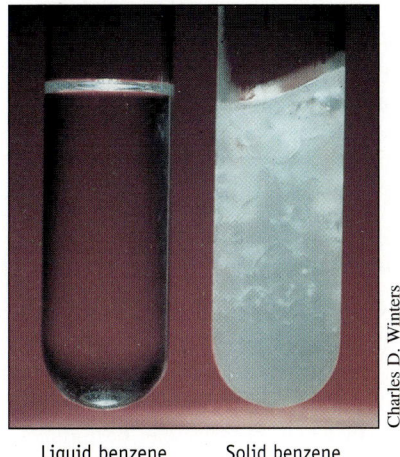

Liquid benzene   Solid benzene

(b)

FIGURE 11.15 Contrasting gases, liquids, and solids. (a) When a 300 mL sample of liquid nitrogen evaporates, it will produce more than 200 L of gas at 25 °C and 1.0 bar. (b) The same volume of liquid benzene (C_6H_6) is placed in two test tubes, and one tube *(right)* is cooled, freezing the liquid.

We also observe that gases can be compressed easily, a process that involves forcing the gas molecules closer together. The air-fuel mixture in a car's engine, for example, is compressed by a factor of about 10 before it is ignited. In contrast, the volume of liquid water changes only by 0.005% per bar of pressure applied to it. Molecules, ions, or atoms in liquid or solid states strongly resist forces that would push them closer together. Stopping your car also makes use of this principle, as auto hydraulic brake systems are based on the inability of oils to be compressed. If air gets in the brake line, the force required to operate the brakes can not be transferred (because the air is easily compressed instead), with catastrophic results.

In liquids and solids, strong forces pull the particles together and limit their motion. In the gas phase, the molecules have enough kinetic energy to escape these forces of attraction between molecules. In solid ionic compounds, the positively and negatively charged ions are held together by electrostatic attraction [<<Section 3.4].

In molecular solids and liquids, the *intermolecular* forces *between* molecules are based on various electrostatic attractions that are weaker than the forces between oppositely charged ions. By comparison, the binding energy due to forces of attraction between the ions in ionic compounds is usually in the range of 700 to 1100 kJ mol^{-1}, and bond energies for most covalent bonds are in the range of 100 to 400 kJ mol^{-1}. As a rough guideline, intermolecular forces are generally 15% (or less) of the values of bond energies (Table 11.5).

TABLE 11.5 Summary of Intermolecular Forces

Type of Interaction	Factors Responsible for Interaction	Approximate Energy (kJ mol^{-1})	Example
Ion–dipole	Ion change, magnitude of dipole	40–600	Na^+ ... H_2O
Dipole–dipole	Dipole moment (depends on atom electronegativities and molecular structure)	20–30	H_2O, HCl
Hydrogen bonding, X—H ... :Y	Very polar X—H bond (where X = F, N, O) and atom Y with lone pair of electrons (An extreme form of dipole interaction)	5–30	H_2O ... H_2O
Dipole–induced dipole	Dipole moment of polar molecule and polarizability of nonpolar molecule	2–10	H_2O ... I_2
Induced dipole–induced dipole (London dispersion forces)	Polarizability	0.05–40	I_2 ... I_2

Review of Types of Intermolecular Forces

We can understand the magnitude of intermolecular forces at work in liquids and solids first by picturing the structures of the individual entities that make up a substance, and then making an educated guess about the types of interactions between those entities. As discussed in Sections 6.4–6.7, intermolecular forces involve entities that are polar or those in which polarity can be induced (Table 11.5), and the particles need to be close enough to each other for these forces to have an effect. London dispersion forces are found between all molecules, both non-polar and polar, but dispersion forces are the only intermolecular forces we envision at work when non-polar molecules interact. Furthermore, several types of intermolecular forces can be simultaneously at work (Figure 11.16). A very large molecule, for example, can have polar or non-polar regions that interact with other molecules. In general, the strength of intermolecular forces is in the following order:

ion-dipole > dipole–dipole (including H-bonding) > dipole–induced dipole > induced dipole–induced dipole

Examples of the relative magnitude of these intermolecular forces are given in Figure 11.16.

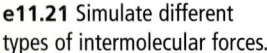

Molecular Modelling (Odyssey)

e11.21 Simulate different types of intermolecular forces.

Interactive Exercise 11.20

Represent the relationships among the types of intermolecular forces.

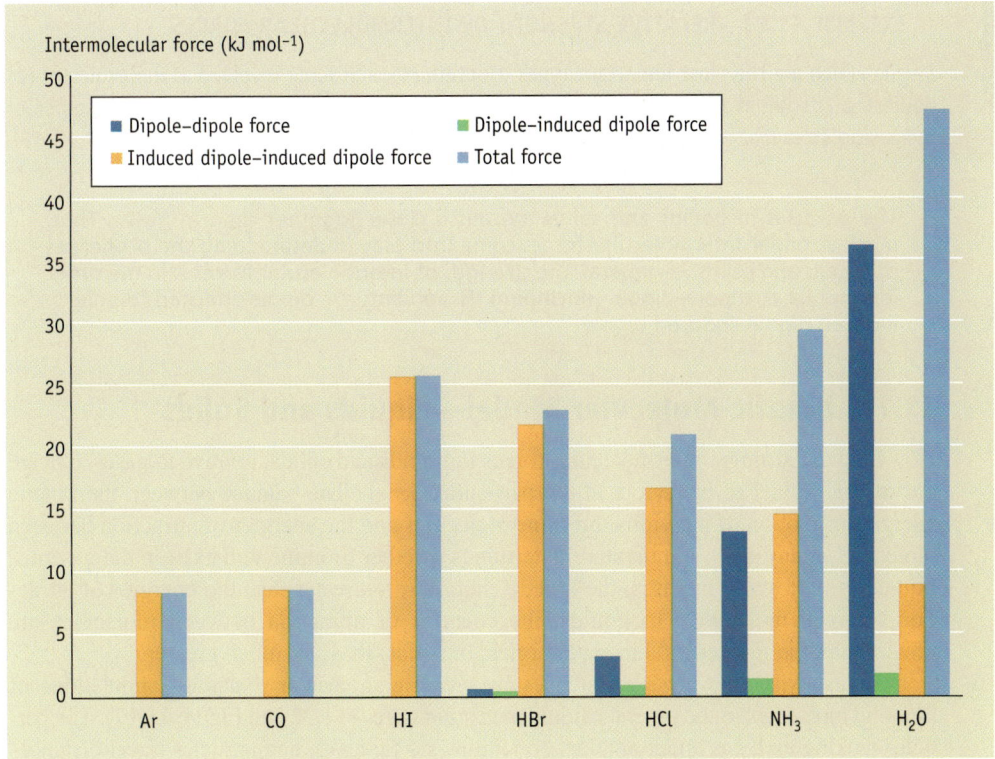

FIGURE 11.16 Intermolecular forces between like molecules of various substances. (Forces are reported as energies in kJ mol^{-1}). The total intermolecular force for weakly polar CO is small (8–9 kJ mol^{-1}) and consists entirely of dispersion forces. The polar molecules HI and HCl have larger intermolecular forces (21–26 kJ mol^{-1}). For HCl, dipole–dipole forces contribute 3.3 kJ mol^{-1} to the total force of 21.1 kJ mol^{-1}. Highly polar water molecules have the largest intermolecular forces (47.2 kJ mol^{-1}). Water molecules interact primarily through dipole forces, including hydrogen bonding, but induced forces are also present.

WORKED EXAMPLE 11.7—RELATIVE STRENGTH OF INTERMOLECULAR FORCES

Decide which are the most important intermolecular forces involved in each of the following and place them in order of increasing strength of interaction: (a) liquid methane (CH$_4$); (b) a mixture of water and methanol (CH$_3$OH); and (c) a solution of bromine in water.

Solution

For each example we consider the structure of the molecules present, and then decide whether they are polar. If polar, consider the possibility of hydrogen bonding.

(a) Methane is made of covalently bonded molecules. Based on the Lewis structure of those molecules, we can conclude that they are tetrahedral and non-polar. The only way methane molecules can interact with one another is through induced dipole/induced dipole forces.

(b) Both water and methanol are covalently bonded molecules, both are polar, and both have an O — H bond. They therefore interact through the special dipole–dipole force called hydrogen bonding.

(c) Non-polar molecules of bromine (Br$_2$) interact by induced dipole forces, whereas water is a polar molecule. Therefore, dipole/induced dipole forces are involved when Br$_2$ molecules interact with water.

The order of increasing strength of interactions is

$$\text{liquid CH}_4 < \text{H}_2\text{O and Br}_2 < \text{H}_2\text{O and CH}_3\text{OH}$$

EXERCISE 11.23—RELATIVE STRENGTH OF INTERMOLECULAR FORCES

Decide which are the most important types of intermolecular force involved in molecules of (a) liquid O_2; (b) liquid CH_3OH; and (c) O_2 dissolved in H_2O. Place the interactions in order of increasing strength.

The particles in liquids and solids are much closer together than in gases. Thus, much stronger intermolecular forces come into play in determining the properties of liquids and solids. In general, the strength of intermolecular forces is in the order ion-dipole > dipole–dipole (including H-bonding) > dipole–induced dipole > induced dipole–induced dipole.

11.11 Kinetic-Molecular Model—Liquids and Solids

How do these stronger intermolecular forces in liquids and solids, relative to gases, change our mental model of behaviour at the particulate level? The balance between the kinetic energies of motion of individual bromine molecules and the energies of attraction between molecules at any given temperature determines whether bromine will exist in the gaseous, liquid, or solid state. In gases, the kinetic energies are greater than the energies of attraction between molecules. In liquids, the energies of attraction between molecules are greater than the molecules' kinetic energies; in solids, they are much greater.

We can visualize these differences by creating molecular dynamics simulations of bromine molecules in the gaseous, liquid, and solid state (e11.22 and Figure 11.17). The particles making up a gas under ambient conditions are far apart, and particles travel distances of many molecular diameters before undergoing a collision. A molecular-level simulation of a liquid shows that particles also diffuse across the simulation cell, but only very slowly. If the liquid consists of molecules, rather than atoms, we imagine them rotating as well as translating. Our kinetic-molecular model of the structure of solids pictures the particles that make up solids—atoms, molecules, or ions—as close together and in an orderly arrangement with a high degree of symmetry [<<Section 3.4]. The particles of a solid move around in their lattice positions, but don't migrate very far under ordinary conditions.

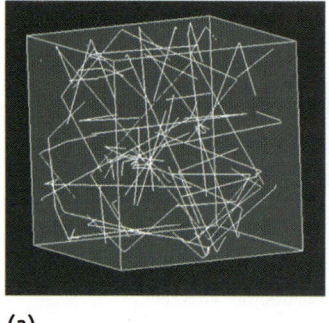
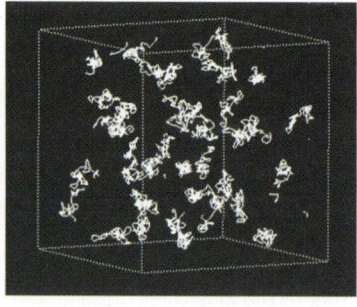
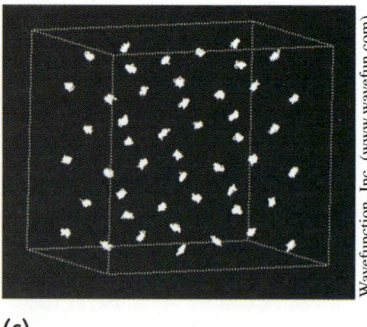

(a)   (b)   (c)

Wavefunction, Inc. (www.wavefun.com)

FIGURE 11.17 Simulation of the molecular motion in bromine (a) gas, (b) liquid, and (c) solid. The Br_2 molecules themselves are hidden in these representations, and you see only the molecular "trails" showing how the molecules moved over the same period of time in each phase. To emphasize the molecular motion, in each case the same number of molecules is shown in a cube of the same size. Since samples of $Br_2(s)$ or $Br_2(\ell)$ have many more molecules in a particular volume than for $Br_2(g)$, and the particles would be much closer together; the densities of the liquid and solid states are much greater than the gas. See e11.22 for an interactive Odyssey simulation.

11.12 Liquids: Properties and Phase Changes

Having seen how the states of liquids and gases differ, we now consider properties related to phase changes involving liquids and gases: vaporization and the vapour pressure of liquids. In this section, we review the concepts of vaporization and boiling point introduced in Sections 6.3 and 6.4 in the context of water, and develop further the idea of vapour pressure.

Vaporization

Vaporization or *evaporation* is the process in which a substance in the liquid state becomes a vapour. In this process, molecules escape from the liquid surface and enter the gaseous state.

Molecules in a liquid have a range of energies that closely resembles the distribution of energies for molecules of a gas. As with gases, the average energy for molecules in a liquid depends only on temperature: the higher the temperature, the higher the average energy and the greater the relative number of molecules with high kinetic energy.

In a sample of a liquid, at least a few molecules have more kinetic energy than the potential energy of the intermolecular attractive forces holding the liquid molecules to one another. If these fast-moving molecules find themselves at the surface of the liquid, and if they are moving in the right direction, they may be able break free of their neighbours and enter the gas phase (Figure 11.18). Intermolecular forces are discussed more fully in Section 6.3.

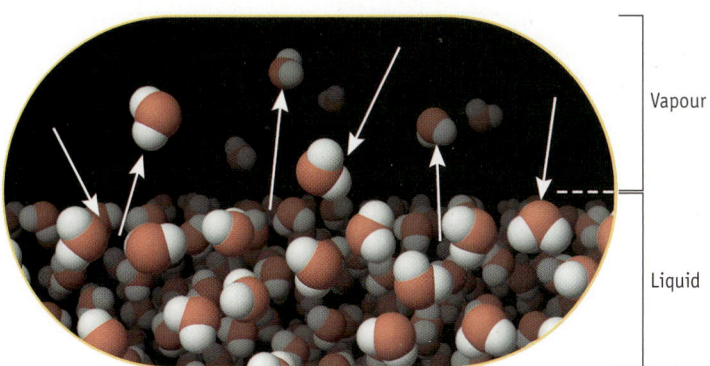

Vapour

Liquid

FIGURE 11.18 Vaporization. A few of the molecules at the surface of a liquid have enough energy and are moving in the right direction to escape the attractions of their neighbours and enter the gaseous state. At the same time, some molecules in the gaseous state can re-enter the liquid.

A few of the molecules in the gas phase will eventually transfer some kinetic energy by colliding with slower gaseous molecules and solid objects. If these molecules come in contact with the surface of the liquid again, they can re-enter the liquid phase in the process called **condensation**, which is the opposite of vaporization.

Molecular Modelling
e11.23 Watch an animation portraying water molecules at the liquid surface.

Vapour Pressure

If a liquid is placed in a closed flask, vaporization and condensation both take place until the pressure of the vapour comes to a constant value. At this point a liquid–vapour equilibrium has been established, and the *equilibrium vapour pressure* (often just called the **vapour pressure**) can be measured. The equilibrium vapour pressure of any substance is a measure of the tendency of its molecules to escape from the liquid phase and enter the vapour phase at a given temperature. This tendency is referred to qualitatively as the *volatility* of the compound. The higher the equilibrium vapour pressure is at a given temperature, the more volatile the compound [<<Section 6.2].

The distribution of molecular energies in the liquid phase is a function of temperature. At a higher temperature, more molecules have sufficient energy to escape the surface of the liquid. The equilibrium vapour pressure must, therefore, increase with temperature. All points along the vapour-pressure-versus-temperature curves in Figure 11.19 represent conditions of pressure and temperature at which liquid and vapour are in equilibrium. For example, at 60 °C the vapour pressure of water is 19.92 kPa. If water is placed in an evacuated flask that is maintained at 60 °C, liquid will evaporate until the pressure exerted by the vapour is 19.92 kPa (assuming enough water is in the flask so that some liquid remains when equilibrium is reached). If T and p locate a point not on the curve, the system is not at equilibrium.

FIGURE 11.19 Vapour pressure curves for diethyl ether [(C$_2$H$_5$)$_2$O], ethanol (C$_2$H$_5$OH), and water. Each curve represents conditions of *T* and *p* at which the two phases, liquid and vapour, are in equilibrium. These compounds exist as liquids for temperatures and pressures to the left of the curve and as gases under conditions to the right of the curve.

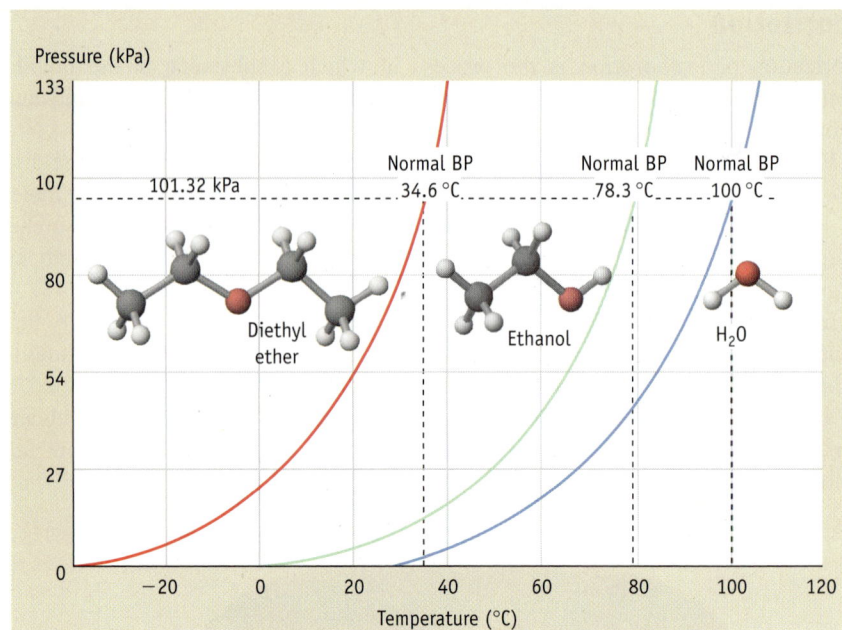

Think about It

e11.24 See how vapour pressure depends on temperature and molecular structure.

Interactive Exercise 11.24

Make predictions using vapour pressure curves.

EXERCISE 11.25—VAPOUR PRESSURE CURVES

Examine the vapour pressure curve for ethanol in Figure 11.19.
(a) What is the approximate vapour pressure of ethanol at 40 °C?
(b) Are liquid and vapour in equilibrium when the temperature is 60 °C and the pressure is 80.0 kPa? If not, does liquid evaporate to form more vapour, or does vapour condense to form more liquid?

EXERCISE 11.26—VAPOUR PRESSURE CURVES

If 0.50 g of pure water is sealed in an evacuated 5.0 L flask and the whole assembly is heated to 60 °C, will the pressure be equal to or less than the equilibrium vapour pressure of water at this temperature? What if you use 2.0 g of water? Under either set of conditions, is any liquid water left in the flask or does all of the water evaporate?

> The equilibrium vapour pressure of any substance is a measure of the tendency of its molecules to escape from the liquid phase and enter the vapour phase at a given temperature. A vapour pressure curve visualizes conditions of *T* and *p* at which the two phases, liquid and vapour, are in equilibrium.

11.13 Solids: Properties and Phase Changes

Crystalline Solids

Many kinds of solids exist in the world around us (Figure 11.20). Solid-state chemistry is one of the booming areas of science, especially because it relates to the development of interesting new materials [>>Chapter 28].

Solid-state chemistry can be organized by classifying solid materials as crystalline or amorphous (non-crystalline). Crystalline solids are further classified into a few types depending on their properties: metallic, ionic, molecular, and covalent network [<<Section 3.2]. Chemists use simple models of the structural arrangement of atoms, molecules, or ions at the particulate level to account for the bulk properties of the various categories of crystalline materials.

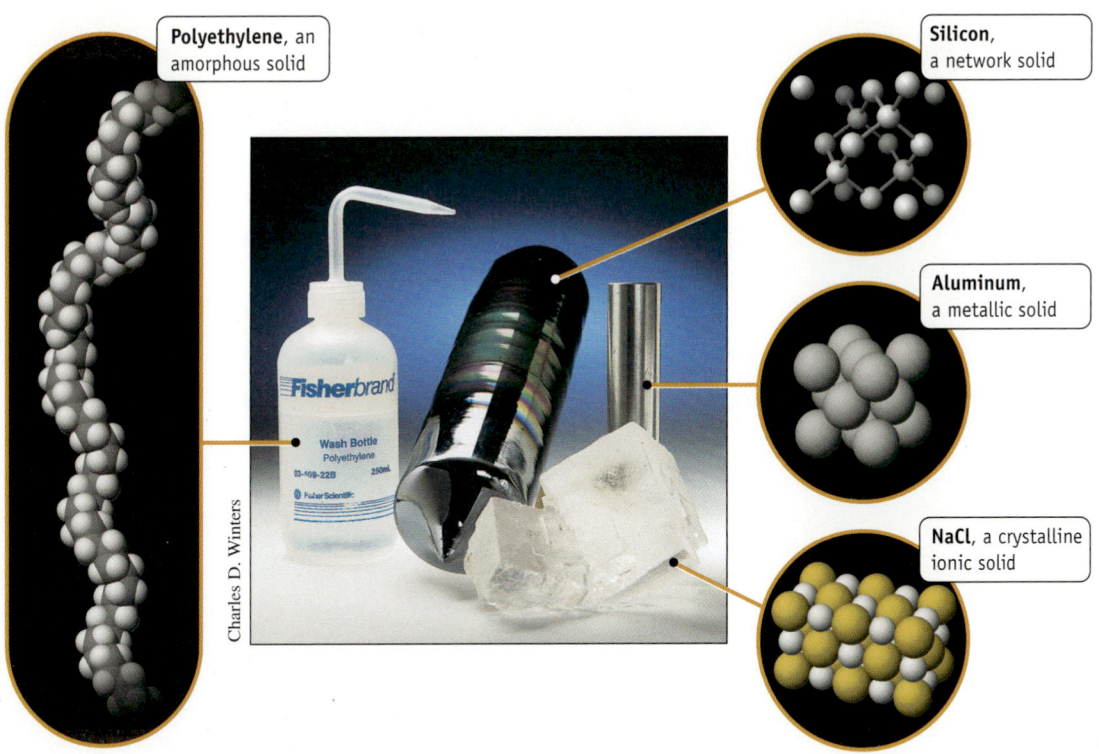

Charles D. Winters

Polyethylene, an amorphous solid

Silicon, a network solid

Aluminum, a metallic solid

NaCl, a crystalline ionic solid

FIGURE 11.20 The building blocks of some common solids, with models of their structures at the molecular level.

Crystals of metals (such as iron and copper), ionic compounds (such as salt and calcite, $CaCO_3$), molecular compounds (such as ice and sucrose), and covalent network solids (such as diamond and silicon) are all beautiful materials with flat faces that have particular angles between the faces—the same in all crystals of each material. This regularity at the visible level reflects a regularity of the arrangement of atoms, molecules, or ions at the molecular level that is known to exist through the experimental evidence of X-ray crystallography [<<Sections 9.2, 9.3]. Further information about the packing of atoms in solid metals is provided in e11.25.

We focus next on phase changes involving the interconversion of solids to liquids and to gases.

Taking It Further

e11.25 Explain physical properties of metals in terms of structural arrangements of their atoms.

Melting: Conversion of Solid to Liquid

The melting point of a solid is the temperature at which the crystal lattice collapses and the solid is converted to a liquid. Like the liquid-to-vapour transformation, melting requires energy, called the molar **enthalpy change of fusion** [<<Section 7.5].

Heat energy absorbed on melting = molar enthalpy change of fusion = $\Delta_{fus}H$ (kJ mol^{-1})

Heat energy evolved on freezing = molar enthalpy change of crystallization
$= -\Delta_{fus}H$ (kJ mol^{-1})

Enthalpy changes of fusion can range from just a few thousand joules per mole to many thousands of joules per mole (Table 11.6). A low melting temperature will certainly mean a low value for the enthalpy change of fusion, whereas high melting points are associated with high enthalpy changes of fusion. Figure 11.21 shows the enthalpy change of fusion for the metals of the fourth through the sixth periods.

Based on this figure and Table 11.6, we can make two statements: (1) metals that have notably low melting points, such as the alkali metals and mercury (mp = −39 °C), also have low enthalpy changes of fusion; and (2) transition metals have high enthalpy changes of fusion, with those of the third transition series being extraordinarily high. This trend parallels that seen with the melting points for these elements. Tungsten, which has the

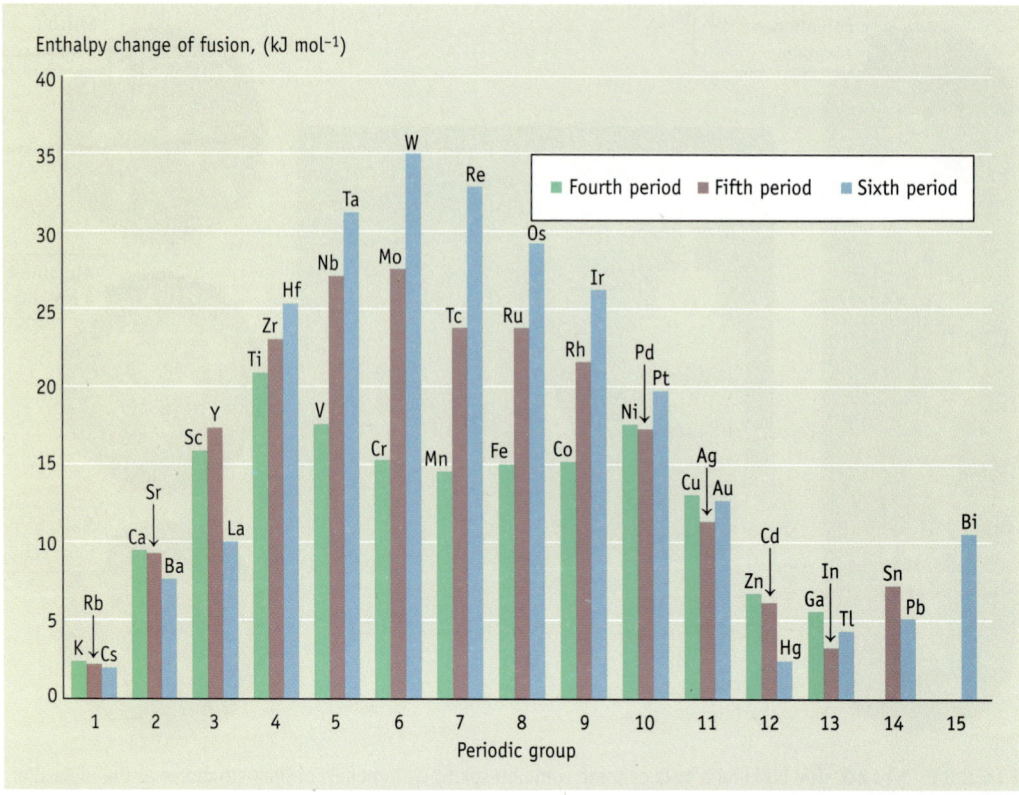

FIGURE 11.21 Enthalpy change of fusion of fourth-, fifth-, and sixth-period metals. Enthalpy changes of fusion range from 2–5 kJ mol^{-1} for Group 1 elements to 35.2 kJ mol^{-1} for tungsten. Notice that enthalpy changes of fusion generally increase for transition metals on descending the periodic table.

TABLE 11.6 Melting Points and Enthalpy Changes of Fusion of Some Substances

Substance	Melting Point(°C)	Enthalpy Change of Fusion(kJ mol^{-1})	Type of Interparticle Forces
Metals			
Hg	−39	2.29	Metal bonding [<<Section 3.5]
Na	98	2.60	
Al	660	10.7	
Ti	1668	20.9	
W	3422	35.2	
Molecular solids: Non-polar molecules			
O_2	−219	0.440	Dispersion forces only
F_2	−220	0.510	
Cl_2	−102	6.41	
Br_2	−7.2	10.8	
Molecular solids: Polar molecules			
HCl	−114	1.99	All three HX molecules have dipole–dipole forces.
HBr	−87	2.41	Dispersion forces increase with size and molar mass.
HI	−51	2.87	
H_2O	0	6.02	Hydrogen bonding
Ionic solids			
NaF	996	33.4	All ionic solids have extended ion-ion interactions.
NaCl	801	28.2	Note the general trend is the same as for lattice energies.
NaBr	747	26.1	
NaI	660	23.6	

highest melting point of all the known elements except for carbon, also has the highest enthalpy change of fusion among the transition metals. These properties affect the uses of this metal. For example, tungsten is used for the filaments incandescent light bulbs; no other material has been found to work better since the invention of the light bulb in 1908.

The melting temperature of a solid conveys a great deal of information. Table 11.6 provides data for several types of substances: metals, polar and non-polar molecular solids, and ionic solids. In general, non-polar molecular substances have low melting points. Melting points increase within a series of related molecular substances as the size and molar mass increase. This happens because dispersion forces are generally larger when the molar mass is larger. Increasing amounts of energy are required to break down the intermolecular forces in the solid, a principle that is reflected in an increasing enthalpy change of fusion.

The ionic compounds in Table 11.6 have higher melting points and higher enthalpy changes of fusion than do molecular solids. This is due to the strong ion-ion forces present in ionic solids, forces that are reflected in high lattice energies. Because ion-ion forces depend on ion size (as well as ion charge), there is a good correlation between lattice energy and the position of the metal or halogen in the periodic table. For example, the data show a decrease in melting point and enthalpy changes of fusion for sodium salts as the halide ion increases in size. This parallels the decrease in lattice energy seen with increasing ion size. Further information about the packing of ions in crystal lattices is provided in e11.26.

Taking It Further
e11.26 What factors determine the structural arrangements of ions in ionic solids?

Sublimation: Conversion of Solid to Vapour

Molecules can escape directly from the solid to the gas phase by sublimation (Figure 11.22).

$$\text{Solid} \longrightarrow \text{gas} \quad \text{heat energy required} = \Delta_{sub}H$$

Sublimation, like fusion and evaporation, is an endothermic process. The heat energy absorbed is called the **enthalpy change of sublimation, $\Delta_{sub}H$**. Water, which has a molar enthalpy change of sublimation of 51 kJ mol^{-1}, can be converted from solid ice to water vapour quite readily. A good example of this phenomenon is the sublimation of frost from grass and trees as night turns to day on a cold Canadian winter morning.

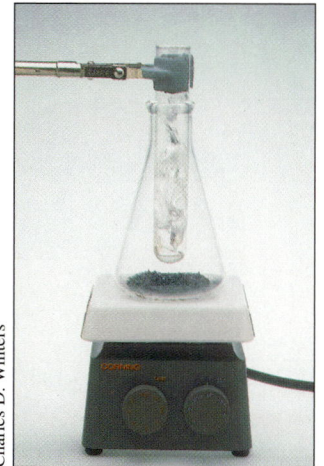

Iodine sublimes when heated →

FIGURE 11.22 Sublimation. Sublimation is the conversion of a solid directly to its vapour. Here, iodine (I_2) sublimes when heated in warm water. If an ice-filled test tube is inserted into the flask, the vapour condenses on the cold surface.

The enthalpy change of fusion is the heat energy absorbed when a solid melts. Another endothermic process is sublimation, the conversion of a solid directly to its vapour.

11.14 Phase Diagrams

Depending on the conditions of temperature and pressure, a substance can exist as a gas, liquid, or solid. In addition, under certain specific conditions, two (or even three) states can co-exist in equilibrium. It is possible to summarize this information for a substance in the form of a graph called a *phase diagram*. These are plots of pressure versus temperature for a substance, and they illustrate at what pressure and temperature the different solid, liquid, and gaseous states of matter for a substance are the most stable form. The location of different regions of a phase diagram can be predicted by remembering that our kinetic-molecular model suggests that low pressures and high temperatures favour the formation of gases. Chemists also use phase diagrams to visualize the conditions under which two phases will be found in equilibrium. On the phase diagram, two phases are in equilibrium at the temperature and pressure given by lines on the diagram separating phases.

Water

A phase diagram for water is shown in Figure 11.23. All points that do not fall on the lines in the figure represent conditions under which one state is the most stable. Line A-B represents conditions for solid–vapour equilibrium, and line A-C for liquid–solid equilibrium. The curve from point A to point D, representing the temperature and pressure combinations at which the liquid and vapour phases are in equilibrium, is the same curve plotted for water vapour pressure in Figure 11.19. Recall that the **normal boiling point**, 100 °C in the case of water, is the temperature at which the equilibrium vapour pressure is 1.00 atm or 101.325 kPa. The **standard boiling point** for water is 99.6 °C, the temperature at which the equilibrium vapour pressure is 1.00 bar.

FIGURE 11.23 Experimental phase diagram for water. The scale is intentionally exaggerated to be able to show the triple point and the negative slope of the line representing the liquid–solid equilibrium.

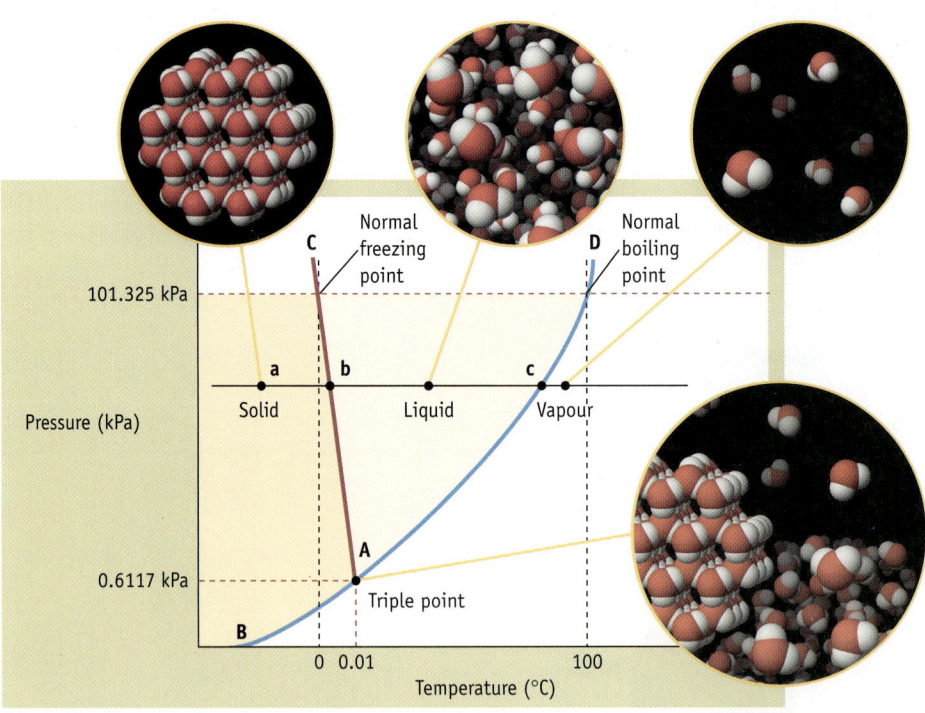

Molecular Modelling (Odyssey)

e11.27 Simulate phases of water and the melting transition.

Point A, appropriately called the **triple point**, occurs at a unique pressure and temperature for each substance, and indicates the conditions under which all three phases can co-exist in equilibrium. For water, the triple point is at $p = 0.61$ kPa and $T = 0.01$ °C.

The line A-C shows the conditions of pressure and temperature at which solid–liquid equilibrium exists. (Because no vapour pressure is involved here, the pressure referred to is the external pressure on the liquid.) *For water, this line has a negative slope;* the change for water is approximately −0.01 °C for each 100 kPa increase in pressure. That is, the higher the external pressure is, the lower the melting point.

The negative slope of the water solid–liquid equilibrium line can be explained from our knowledge of the structure of water and ice. Because ice is less dense than liquid water (due to the open lattice structure of ice), ice and water in equilibrium respond to increased pressure (at constant T) by melting ice to form more water because the same mass of water occupies less volume.

Ice Skating and the Solid–Liquid Equilibrium Ice is slippery stuff. It was long assumed that you can ski or skate on ice because the surface melts slightly from the pressure of a skate blade or ski, or that surface melting occurs because of frictional heating. This has always seemed an unsatisfying explanation, however, because it does not seem possible that just standing or sliding on a piece of ice could produce a pressure or temperature high enough to cause sufficient melting. Recently, surface chemists have studied ice surfaces and have come up with a better explanation. They have concluded that water molecules on the surface of ice are vibrating rapidly. The outermost layer or two of water molecules is almost liquid-like. This arrangement makes the surface slippery, explaining why we can ski on snow and skate on ice.

Think about It

e11.28 Watch the dramatic crushed can demonstration to see consequences of a phase transition.

Carbon Dioxide

While the basic features of the phase diagram for CO_2 (Figure 11.24) are the same as those for water, there are several important differences.

In contrast to water, the CO_2 solid–liquid equilibrium line has a *positive* slope. That is, increasing pressure on solid CO_2 in equilibrium with liquid CO_2 will cause at least some of the liquid to change to solid CO_2. Thus, increasing the pressure on liquid CO_2 will cause it to move to the denser phase, the solid phase. Because solid CO_2 is denser than the liquid, the newly formed solid CO_2 sinks to the bottom in a container of liquid CO_2.

Another feature of the CO_2 phase diagram is the triple point that occurs at a pressure of 518 kPa and 216.6 K (-56.6 °C). CO_2 cannot be a liquid at pressures lower than this. At a pressure of 100 kPa (1 bar), solid CO_2 is in equilibrium with the gas at a temperature of 197.5 K (-78.7 °C). As solid CO_2 warms to room temperature in an open container, it sublimes rather than melts. CO_2 is called *dry ice* for this reason; it looks like water ice, but it does not melt. Theatres and nightclubs place dry ice in warm water, where it rapidly sublimes to create a dense fog. At applied pressures greater than 518 kPa, however, solid CO_2 melts if the temperature is increased. As you could predict from Figure 11.24, carbon dioxide gas can be converted to a liquid at room temperature (20–25 °C) by exerting a moderately high pressure (between 518 and 7390 kPa) on the gas. In fact, CO_2 is regularly shipped in tanks as a liquid to laboratories and industrial companies.

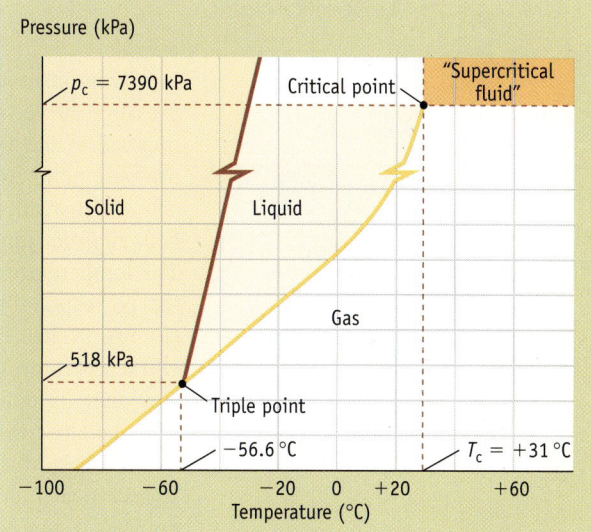

FIGURE 11.24 The phase diagram of CO_2. Notice in particular the positive slope of the solid–liquid equilibrium line.

EXERCISE 11.27—PHASE DIAGRAMS

Consider the phase diagram of CO_2 in Figure 11.24.
(a) Is the density of liquid CO_2 greater or less than that of solid CO_2?
(b) In what phase do you find CO_2 at 500 kPa and 0 °C?
(c) Can CO_2 be liquefied at 45 °C?

Interactive Exercise 11.28

Make predictions using a phase diagram.

A phase diagram is a plot of pressure versus temperature for a substance, and illustrates at what pressure and temperature the different solid, liquid, and gaseous phases of matter for that substance are the most stable form.

Critical Points

If pressurized CO_2 is placed in a high-pressure cell and the sample allowed to warm, a remarkable transformation can be observed. As the pressure and temperature increase, the clear boundary between liquid and gaseous CO_2 becomes difficult to distinguish, and above 7390 kPa and 31 °C it is no longer possible to observe two distinct phases. At this point, carbon dioxide has become a supercritical fluid, with properties that are very different from either the liquid or gas. A supercritical fluid is shown in Figure 11.25.

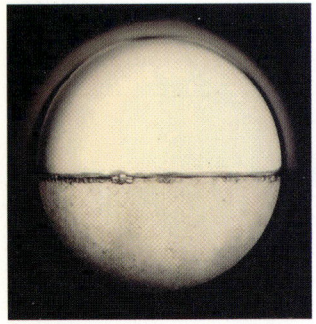

The separate phases of a substance are seen through the window in a high-pressure vessel.

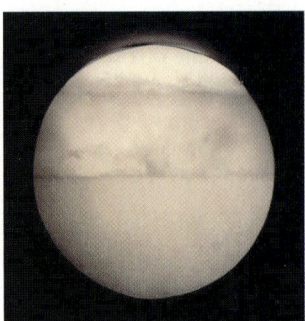

As the sample warms and the pressure increases, the mentscurs becomes less distinct.

As the temperature continues to increase, it is more difficult to distinguish the liquid and vapor phases.

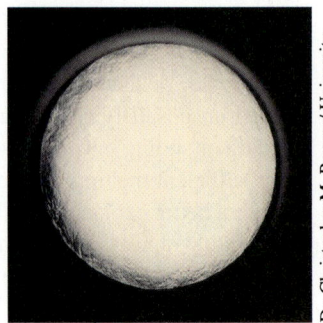

Once the critical T and P are reached, distinct liquid and vapor phases are no longer in evidence. This homogenous phase is called a *supercritical fluid*.

Dr. Christopher M. Rayner/ University of Leeds

FIGURE 11.25 A supercritical fluid.

Think about It

e11.29 Check your understanding of the different regions in a phase diagram.

This observation seems to contradict what we might first expect from a vapour pressure curve for a liquid. In general, the vapour pressure of a liquid continues to increase if the temperature is raised above the normal boiling point. But the vapour pressure–temperature curve (such as shown in Figure 11.19) doesn't continue without a limit. Instead, when a specific temperature and pressure are reached, the interface between the liquid and the vapour disappears. This point is called the **critical point**. The temperature at which this phenomenon occurs is the **critical temperature (T_c)** and the corresponding pressure is the **critical pressure (p_c)**, peculiar to each substance (Figure 11.26). The substance that exists under these conditions and at higher pressures and temperatures is called a **supercritical fluid**. It is like a gas under such a high pressure that its density resembles that of a liquid, while its viscosity (ability to flow) remains close to that of a gas.

A phase diagram shows us graphically the experimental conditions under which a supercritical fluid for any substance might be formed. For most substances, the critical point is at a very high temperature and pressure (Table 11.7). Water, for instance, has a critical temperature of 374 °C and a critical pressure of 2.2×10^4 kPa. Consider what the substance might look like at the molecular level under these conditions. At this high

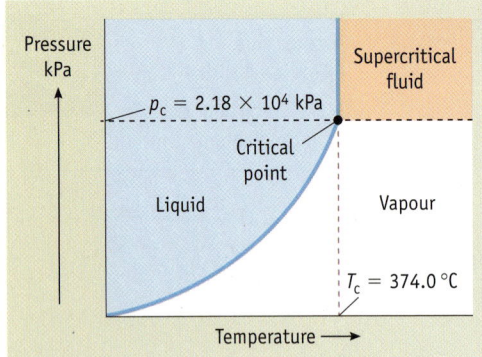

FIGURE 11.26 Critical temperature and pressure for water. The curve representing equilibrium conditions for liquid and gaseous water ends at the critical point; above that temperature and pressure, water becomes a supercritical fluid.

TABLE 11.7 Critical Temperatures and Pressures for Common Compounds

Compound	T_c (K)	p_c (MPa)
CH_4 (methane)	190.6	4.6
C_2H_6 (ethane)	305.3	4.9
C_3H_8 (propane)	369.9	4.3
CCl_2F_2	385.0	4.1
NH_3	405.6	11.4
H_2O	647.1	22.1
CO_2	304.1	7.4
SO_2	430.6	7.9

Source: Data taken from *88th CRC Handbook of Physics and Chemistry, 2007–2008*.

pressure, water molecules are forced closely together. The high temperature, however, means that each molecule has enough kinetic energy to exceed the forces holding molecules together. As a result, the supercritical fluid has a tightly packed molecular arrangement like a liquid, but the intermolecular forces of attraction that characterize the liquid state are less than the kinetic energy of the particles.

> When a specific temperature and pressure are reached, the interface between the liquid and the vapour for a substance disappears. This point is called the critical point, and the substance that exists under these conditions and at higher pressure and temperature is called a supercritical fluid.

Supercritical Fluids: Green Solutions for Solvent Extraction

Supercritical fluids can have unexpected properties, such as the ability to dissolve normally insoluble materials. Supercritical CO_2 is especially useful. Climate change concerns have added incentives to find new uses for carbon dioxide gas instead of disposing of it into the atmosphere. While it can be challenging to recover as a pure gas from combustion processes, CO_2 is widely available, essentially non-toxic, non-flammable, and inexpensive. It is relatively easy to reach carbon dioxide's critical temperature of 31.1 °C and critical pressure of 7390 kPa. The material is also easy to handle. CO_2 is highly useful because it does not dissolve water or highly polar compounds such as sugar, but it does dissolve relatively non-polar oils, which constitute many of the flavouring or odour-causing compounds in foods. As a result, food companies now use supercritical CO_2 to extract caffeine from coffee, for example, and essential oils from plants.

To decaffeinate coffee, the beans are treated with steam to bring the caffeine to the surface. The beans are then immersed in supercritical CO_2, which selectively dissolves the caffeine but leaves intact most of the compounds that give flavour to coffee. (Decaffeinated coffee contains less than 3% of the original caffeine.) The solution of caffeine in supercritical CO_2 is poured off, and the CO_2 is evaporated, trapped, and reused.

As chemists try to incorporate green chemistry principles into industrial processes, it is not surprising that other uses are being sought for this supercritical substance. For example, more than 10 billion kilograms of organic and halogenated solvents are used worldwide every year in cleaning applications. These cleaning agents can have deleterious effects on the environment, so it is hoped that many can be replaced by supercritical CO_2, which can then be recycled.

11.15 Polymorphic Forms of Solids

While we usually think of material as existing in one of four states [<<Section 11.1], solid material can often exist under different conditions in more than one form or crystal structure, called crystal **polymorphs** in the case of compounds, or **allotropes** for elements. A familiar example is the different allotropes of carbon, which include graphite, diamond, amorphous (coal), carbon nanotubes, and buckyballs. Ice has also been shown to exist in many different solid phases besides the hexagonal crystal form, Ice I (also called Ice I_h) that is produced by freezing liquid water under everyday conditions.

At extremely high pressures, above 2×10^8 kPa, the regular hydrogen-bonding linkages between water molecules break down, and more than a dozen other crystalline forms have been observed. The phase diagram for water has been extended to high pressures in Figure 11.27 to show the conditions under which a few of these additional polymorphs have been observed. With very different intermolecular interactions, we would expect these other forms of ice to show strikingly different physical properties—and they do.

The most familiar form of ice, seen here in the form of needle-like crystals growing at −20 °C over Jeffery Creek in the Canadian Rockies.

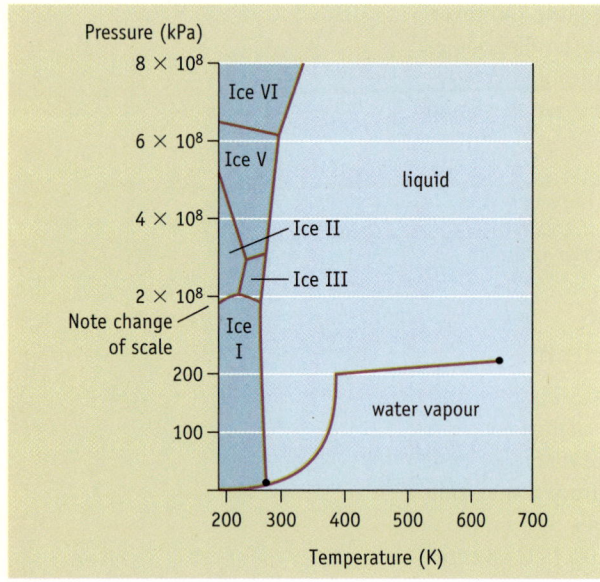

FIGURE 11.27 High-pressure region of the phase diagram for water, showing several other forms of ice, besides the familiar hexagonal form. None of these forms are stable below 2×10^8 kPa, even at −60 °C. Note the presence of additional triple points, where three phases are in equilibrium.

There are few places on the earth where we might find the requisite pressures to form these ice polymorphs, except in high pressure cells in laboratories. Ice VI, for example, melts at much higher temperatures than I_h. While an ice cube that melts at 80 °C might be an attractive addition to your tropical drink, a pressure of over 7×10^8 kPa is required! But analysis of the icy shells of satellites of Jupiter and Saturn suggests that other polymorphs of water might be formed under the extreme pressure of ice sheets up to 100 km thick. Scientists speculate that these other forms could lead to the explosive instability of ice that may have produced the observed giant trenches on Titan and Ganymede.

Polymorphs of crystalline materials other than water have become increasingly important in pharmaceutical chemistry and in the fabrication of new materials, but characterizing the crystalline structures of these new forms is very challenging as it relies on subtle interpretations of X-ray crystal structures.

Pharmaceutical ingredients often receive regulatory approval for only a single polymorph, and pharmaceutical companies have recently taken out patents on newly characterized polymorphs of existing antibiotics and other drugs. New crystalline forms of an active ingredient can have improved physical properties, such as water solubility that affect the drug's bioavailability. A new polymorph might also allow a pharmaceutical company access to compete in a significant market by producing a drug without violating an existing patent on a single polymorph held by another pharmaceutical company. Acetylsalicylic acid (aspirin) is an interesting example. First synthesized in 1853, aspirin soon became the world's best-selling drug. In the past five years, contradictory reports have been published with respect to whether a second form of acetylsalicylic acid with quite different crystal packing, called aspirin II, can be crystallized by using a different solvent. The X-ray crystal structure of aspirin II seemed to show a different arrangement of molecules, connected by hydrogen bonds. More careful analysis, however, shows that the evidence for a new form of aspirin is quite ambiguous, and that two distinct types of arrangement of molecules might occur within a single crystal. The controversy over the aspirin study has led to a re-examination of almost 15 000 crystal structures published in major prestigious academic journals, and the discovery that a worrying number of errors exist in these published data sets.

Canadian chemists have found that new spectroscopic tools can complement X-ray crystallography in confirming the structures of new polymorphs. In work that is still in progress, Gary Enright, Victor Tershkikh, Darren Brouwer, and John Ripmeester at the National Research Council in Ottawa have used ultra high field nuclear magnetic resonance spectroscopy to study the solid states of caffeine, the stimulant found in coffee and many soft drinks. While the most familiar applications of NMR spectroscopy is to study solutions of compounds, solid state NMR using high magnetic fields can provide unique insights into crystal structures that are difficult to determine by X-ray crystallography.

Using one of Canada's 900 MHz NMR spectrometers, the NRC research group has shown that when caffeine is crystallized in the solvent carbon disulfide in the absence of water, one crystalline form is produced. A second distinct polymorph of anhydrous caffeine can be created by heating those crystals to 430 K, followed by cooling them to 125 K, or by changing the solvent used to form the crystals. Using solid state NMR, they showed that differences in hydrogen bonding play a crucial role in determining how caffeine molecules pack together in the crystal lattices of these two polymorphs.

This NRC research shows how high field NMR spectroscopy can complement standard crystallographic techniques for determining molecular structures within crystals. This information is crucial to the emerging interdisciplinary field called crystal engineering, which uses detailed knowledge of crystallization processes and structures to design better ways of delivering and stabilizing active ingredients in pharmaceutical products.

Finally, while the NRC research may not lead to a better cup of java, getting cocoa butter to crystallize in the right polymorphic form to "melt in your mouth" is one of the keys to the sensual experience of eating high-quality chocolate. The most desired polymorphic form of cocoa butter (called polymorph V) melts 2–3 degrees below 37 °C, the temperature of the human body. Unfortunately, this is not the thermodynamically most stable polymorph, and when chocolate is stored at room temperature for several months, the cocoa butter can slowly be converted to the more stable polymorphic form, which makes the chocolate look cloudy or dusty, and melts just above, rather than below body temperature.

> Solid material can often exist under different conditions in more than one form or crystal structure, called polymorphs, which have become increasingly important in pharmaceutical chemistry and in the fabrication of new materials.

Taking It Further

e11.30 Read about Australian research on obtaining different phases within a single sample of solid silicon.

References

To learn more about crystal polymorphism, read the story on polymorphs of constituents of cocoa butter in Peter Fryer and Kerstin Pinschower, (2003), "The Materials Science of Chocolate," *Materials Research Society Bulletin,* (December): 25–29. The story of whether the defeat of Napoleon was due to a crystal polymorph of tin in buttons on soldier's uniforms crumbling in the Russian winter is told in Penny LeCouteur's book, (2003), *Napoleon's Buttons: How 17 Molecules Changed History*. New York: Putnam.

SUMMARY

Key Concepts

- Experimental observations of gases, carried out over several hundred years, established quantitative relationships between gas pressure and volume (Boyle's law), gas temperature and volume (Charles's law), gas temperature and pressure (Amonton's Law), and gas volume and amount (Avogadro's hypothesis). (Section 11.2, e11.3)
- While different gases, such as those found in the atmosphere, have different uses, natural abundance, chemical reactivity, and interaction with electromagnetic radiation, they show almost universal thermodynamic behaviour under certain conditions. **The ideal gas equation**, $pV = nRT$, describes remarkably well the behaviour of many different real gases, under conditions where the pressure is about 1 bar or less and temperatures at or above 25 °C. (Section 11.3, 11.4)
- The pressure each gas in a mixture would exert if it were alone in a container of the same size at the same temperature is called its **partial pressure**. **Dalton's law of partial pressures** states that the total pressure of a mixture of gases is simply the sum of their partial pressures. (Section 11.6)
- The observable gas laws can be explained by the **kinetic-molecular theory**, a model that describes the behaviour of matter at the molecular or atomic level. In the ideal gas equation, chemists take the kinetic-molecular model to an extreme by assuming that the particles of a gas have zero volume, and there is no force of interaction between particles. (Section 11.7)

- The mixing of molecules of two or more gases due to their random molecular motions is called **diffusion. Effusion** is the movement of gas through a tiny opening in a container into another container where the pressure is very low. Both diffusion and effusion depend on the molar mass of the gas. (Section 11.8)
- At high pressure and low temperature, two extreme assumptions of the ideal gas model fail—that the particles of a gas have zero volume, and there are no forces of interaction between particles. More sophisticated models than the ideal gas equation, such as the **van der Waals equation**, are needed to explain the properties of many real gases under such conditions. (Section 11.9)
- The particles in liquids and solids are much closer together than in gases. Thus, much stronger intermolecular forces come into play in determining the properties of liquids and solids. In general, the strength of intermolecular forces is in the order ion-dipole > dipole–dipole (including H-bonding) > dipole–induced dipole > induced dipole–induced dipole. (Section 11.10)
- The particles making up a gas under ambient conditions are far apart, and particles travel distances of many molecular diameters before undergoing a collision. A molecular-level simulation of a liquid shows that particles also diffuse across the simulation cell, but only very slowly. If the liquid consists of molecules, rather than atoms, we imagine them rotating as well as translating. Our kinetic-molecular model of the structure of solids pictures the particles that make up solids—atoms, molecules, or ions—as being close together and often in an orderly arrangement with a high degree of symmetry. (Section 11.11)
- The equilibrium **vapour pressure** of any substance is a measure of the tendency of its molecules to escape from the liquid phase and enter the vapour phase at a given temperature. A vapour pressure curve helps visualize conditions of T and p at which the two phases, liquid and vapour, are in equilibrium. (Section 11.12)
- The molar **enthalpy change of fusion** is the heat energy absorbed when a mole of a solid melts. Another endothermic process is sublimation, the conversion of a solid directly to its vapour. (Section 11.13)
- A phase diagram is a plot of pressure versus temperature for a substance, and illustrates at what pressure and temperature the different solid, liquid, and gaseous phases of matter for that substance are the most stable form. When a certain temperature and pressure are reached, the interface between the liquid and the vapour for a substance disappears. This point is called the **critical point**, and the substance that exists under these conditions and at higher pressure and temperature is called a **supercritical fluid**. (Section 11.14)
- Solid material can often exist under different conditions in more than one form or crystal structure, called **polymorphs**, which have become increasingly important in pharmaceutical chemistry and in the fabrication of new materials. (Section 11.15)

Key Equations

Ideal gas equation: $pV = nRT$ (11.1, Section 11.4)

If the amount of gas is constant, $\dfrac{p_1V_1}{T_1} = nR = \dfrac{p_2V_2}{T_2}$ for a constant amount of gas, n

(11.2, Section 11.4)

R = 8.314 L kPa K^{-1} mol^{-1} = 8.314 m^3 Pa K^{-1} mol^{-1} = 8.314 J K^{-1} mol^{-1}
= 0.08206 L atm K^{-1} mol^{-1}

Density of gases (where ρ is the gas density in g L^{-1}): $\rho = \dfrac{m}{V} = \dfrac{pM}{RT}$ (11.3, Section 11.5)

Dalton's law of partial pressures: $p_{total} = p_A + p_B + p_C + \ldots$ (11.4, Section 11.6)

The total pressure of a gas mixture: $p_{total} = n_{total}\left(\dfrac{RT}{V}\right)$ (11.5, Section 11.6)

The partial pressure of a gas (A) in a mixture $p_A = x_A p_{total}$ (11.6, Section 11.6)

Maxwell's equation: $\sqrt{\overline{u^2}} = \sqrt{\dfrac{3RT}{M}}$ (11.7, Section 11.7)

Graham's law: $\dfrac{\text{Rate of effusion of gas 1}}{\text{Rate of effusion of gas 2}} = \sqrt{\dfrac{\text{molar mass of gas 2}}{\text{molar mass of gas 1}}} = \sqrt{\dfrac{M\,(\text{gas 2})}{M\,(\text{gas 1})}}$

(11.8, Section 11.7)

van der Waals equation: (11.9, Section 11.9)

Observed pressure   Container V

$$\left(p + a\left[\dfrac{n}{V}\right]^2\right)(V - bn) = nRT$$

Correction for intermolecular forces   Correction for molecular volume

REVIEW QUESTIONS

Section 11.1: Understanding Gases: Understanding Our World

Consult e11.1 and use the visualization of the earth's atmosphere to answer *Review questions 11.29–11.31*.

11.29 Rank the following altitudes during the daytime at the equator in order of increasing temperature: 5 km, 25 km, 80 km, and 110 km.

11.30 How would you know if you were moving from the stratosphere into the mesosphere?

11.31 What happens to the temperature profile of the atmosphere when moving from the troposphere into the stratosphere? What chemical reactions are responsible for this change? Why don't these same reactions occur 1 km above the earth's surface in the troposphere, where the density of atmospheric gases is much greater?

11.32 Consult additional sources and describe in a couple sentences how a plasma television works. What gases are involved, and what interaction of electromagnetic radiation from the plasmas with other substances is responsible for the visible light you see when watching television?

11.33 With reference to Figure 4.6, what was different, relative to the present, about (a) the earth's atmospheric composition, and (b) the earth's surface temperature 30 000 years ago? 125 000 years ago? 260 000 years ago?

11.34 With reference to Figure 4.6, how do we know what the composition of the earth's atmosphere was 260 000 years ago? (Note: You may wish to consult the ice-core analysis visualization at www.kcvs.ca.)

Section 11.2: Relationships among Gas Properties

11.35 The average barometric pressure at an altitude of 10 km is 210 mm Hg. Express this pressure in atm, bar, and kPa.

11.36 You have 3.6 L of H_2 gas at 50.6 kPa and 25.0 °C. What is the pressure of this gas if it is transferred to a 5.0 L flask at 0.0 °C?

11.37 You have a sample of CO_2 in a flask A with a volume of 25.0 mL. At 20.5 °C, the pressure of the gas is 58.2 kPa. To find the volume of another flask B, you move the CO_2 to that flask and find that its pressure is now 12.6 kPa at 24.5 °C. What is the volume of flask B?

11.38 A sample of gas occupies 135 mL at 22.5 °C; the pressure is 22 kPa. What is the pressure of the gas sample when it is placed in a 252 mL flask at a temperature of 0.0 °C?

11.39 One of the cylinders of an automobile engine has a volume of 0.40 L. The engine takes in air at a pressure of 1.00 atm and a temperature of 15 °C and compresses the air to a volume of 0.050 L at 77 °C. What is the final pressure of the gas in the cylinder? (The ratio of before and after volumes—in this case, 0.40/0.050 or 8.0—is called the compression ratio.)

Section 11.3 Do Different Gases Have Any Similar Properties?

11.40 The mean velocities of molecules of $H_2(g)$, $CO_2(g)$, and SF_6 (g) at 120 °C are $u_{mean}(H_2) = 2052$ m s^{-1}; $u_{mean}(CO_2) = 433$ m s^{-1}; and $u_{mean}(SF_6) = 242$ m s^{-1}. Yet at this temperature and 1.00 bar pressure, the volume occupied by 1 mol of each of the gases is $(H_2) = 32.704$ L mol^{-1}; $(CO_2) = 32.625$ L mol^{-1}; and $(SF_6) = 32.542$ L mol^{-1}. Explain how their molar volumes can be so similar, when their mean velocities are so different.

11.41 Figure 11.3 gives values for the molar volume of H_2, CO_2, and SF_6 at 1.00 bar pressure and ambient temperature (298.15 K). Consult the online NIST database at http://webbook.nist.gov/chemistry/fluid/ and select isobaric data for these three gases to determine their molar volumes at a constant pressure of 1.00 bar and 448.15 K. At this higher temperature, would you still consider the molar volumes of these gases to be approximately the same, as they were at ambient temperature? Explain.

Section 11.4: The Ideal Gas Equation

11.42 A 1.25 g sample of CO_2 is contained in a 750 mL flask at 22.5 °C. What is the pressure of the gas?

11.43 A steel cylinder holds 1.50 g of ethanol (C_2H_5OH). What is the pressure of the ethanol vapour if the cylinder has a volume of 0.251 L and the temperature is 250 °C? (Assume all of the ethanol is in the vapour phase at this temperature.)

11.44 A balloon for long-distance flying contains 1.2×10^7 L of helium. If the helium pressure is 98.3 kPa at 25 °C, what mass of helium (in g) does the balloon contain?

11.45 A 1.007 g sample of an unknown gas exerts a pressure of 95.3 kPa in a 0.452 L container at 23 °C. What is the molar mass of the gas?

11.46 A 0.0125 g sample of a gas with an empirical formula of CHF_2 is placed in a 0.165 L flask. It has a pressure of 1.83 kPa at 22.5 °C. What is the molecular formula of the compound?

11.47 A helium-filled balloon of the type used in long-distance flying contains 1.2×10^4 m^3 of helium. Suppose you fill the balloon with helium on the ground, where the pressure is 98.3 kPa and the temperature is 16.0 °C. When the balloon ascends to a height of 3.2 km, where the pressure is only 80.0 kPa and the temperature is −33.0 °C, what volume is occupied by the helium gas? Assume the pressure inside the balloon matches the external pressure. Comment on the result.

Section 11.5: The Density of Gases

11.48 At 60 km above the earth's surface, the temperature is 250 K and the pressure is only 29.7 Pa. What region of the atmosphere is this? What is the density of air (in g L^{-1}) at this altitude? (Assume the molar mass of air is 28.96 g mol^{-1}.)

11.49 A gaseous organofluorine compound has a density of 0.355 g L^{-1} at 17 °C and 25.2 kPa. What is the molar mass of the compound?

11.50 Iron reacts with hydrochloric acid to produce iron(II) chloride and hydrogen gas:

$$Fe(s) + 2\ HCl(aq) \longrightarrow FeCl_2(aq) + H_2(g)$$

The H_2 gas from the reaction of 2.2 g of iron with excess acid is collected in a 10.0 L flask at 25 °C. What is the pressure of the H_2 gas in this flask?

11.51 Sodium azide, the explosive compound in automobile airbags, decomposes according to the following equation:

$$2\ NaN_3(s) \longrightarrow 2\ Na(s) + 3\ N_2(g)$$

What mass of sodium azide is required to provide the nitrogen needed to inflate a 75.0 L bag to a pressure of 1.32 bar at 25 °C?

Interactive Exercise 11.52

Calculate gas pressure using the gas laws and stoichiometry.

11.53 Hydrazine reacts with O_2 according to the following equation:

$$N_2H_4(g) + O_2(g) \longrightarrow N_2(g) + 2\ H_2O(\ell)$$

Assume the O_2 needed for the reaction is in a 450 L tank at 23 °C. What must the oxygen pressure be in the tank to have enough oxygen to consume 1.00 kg of hydrazine completely?

11.54 A self-contained breathing apparatus uses canisters containing potassium superoxide. The superoxide consumes the CO_2 exhaled by a person and replaces it with oxygen:

$$4\ KO_2(s) + 2\ CO_2(g) \longrightarrow 2\ K_2CO_3(s) + 3\ O_2(g)$$

What mass of KO_2 (in g) is required to react with 8.90 L of CO_2 at 22.0 °C and 102.3 kPa?

Section 11.6: Gas Mixtures and Partial Pressures

11.55 What is the total pressure in atmospheres of a gas mixture that contains 1.0 g of H_2 and 8.0 g of Ar in a 3.0 L container at 27 °C? What are the partial pressures of the two gases?

11.56 A cylinder of compressed gas is labelled "Composition (mol %): 4.5% H_2S, 3.0% CO_2, balance N_2." The pressure gauge attached to the cylinder reads 46.6 bar. Calculate the partial pressure of each gas, in atmospheres, in the cylinder.

Interactive Exercise 11.57

Calculate the mass of an anaesthetic in a tank.

**Interactive
Exercise 11.58**

Calculate the masses and
partial pressures in a balloon.

Section 11.7: The Kinetic-Molecular Model of Gases

**Interactive
Exercise 11.59**

Compare two gases using
kinetic-molecular theory.

11.60 Equal masses of gaseous N_2 and Ar are placed in separate flasks of equal volume at the same temperature. State whether each of the following statements is true or false. Briefly explain your answer in each case.
(a) There are more molecules of N_2 present than atoms of Ar.
(b) The pressure is greater in the Ar flask.
(c) The Ar atoms have a greater average speed than the N_2 molecules.
(d) The N_2 molecules collide more frequently with the walls of the flask than do the Ar atoms.

11.61 If the average speed of oxygen molecules is 4.28104 cm s^{-1} at 25 °C, what is the average speed of CO_2 molecules at the same temperature?

**Interactive
Exercise 11.62**

Rank gases in order of
increasing average molecular speed.

11.63 Calculate the rms speed for CO molecules at 25 °C. What is the ratio of this speed to that of Ar atoms at the same temperature?

11.64 The reaction of SO_2 with Cl_2 gives dichlorine oxide, which is used to bleach wood pulp and to treat wastewater:

$$SO_2(g) + 2\ Cl_2(g) \longrightarrow OSCl_2(g) + Cl_2O(g)$$

All of the compounds involved in the reaction are gases. List them in order of increasing average speed.

Section 11.8: Diffusion and Effusion

**Interactive
Exercise 11.65**

Compare the rates of
effusion of two gases.

11.66 Argon gas is 10 times denser than helium gas at the same temperature and pressure. Which gas is predicted to effuse faster? How much faster?

11.67 A gas whose molar mass you wish to know effuses through an opening at a rate one-third as fast as that of helium gas. What is the molar mass of the unknown gas?

11.68 A sample of uranium fluoride is found to effuse at the rate of 17.7 mg h^{-1}. Under comparable conditions, gaseous I_2 effuses at the rate of 15.0 mg h^{-1}. What is the molar mass of the uranium fluoride? (*Hint*: Rates must be converted to units of moles per time.)

Section 11.9: The Ideal Gas Model and Real Gas Behaviour

11.69 It has been stated that the pressure of 8.00 mol of Cl_2 in a 4.00 L tank at 27.0 °C should be 2990 kPa if calculated using the van der Waals' equation. Verify this result, using data from Table 11.4 and compare it with the pressure predicted by the ideal gas law.

11.70 You want to store 165 g of CO_2 gas in a 12.5 L tank at room temperature (25 °C). Calculate the pressure the gas would have using (a) the ideal gas law, and (b) the van der Waals equation. (For CO_2, $a = 3.64\ 10^{-2}$ kPa L^2 mol^{-2} and $b = 0.0427$ L mol^{-1}.)

Section 11.10: Liquid and Solid States— Stronger Intermolecular Forces

11.71 What type of forces must be overcome within the solid I_2 when I_2 dissolves in methanol (CH_3OH)? What type of forces must be disrupted between CH_3OH molecules when I_2 dissolves? What types of forces exist between I_2 and CH_3OH molecules in solution?

11.72 What intermolecular force(s) must be overcome to do the following?
(a) melt ice
(b) sublime solid I_2
(c) convert liquid NH_3 to NH_3 vapour

**Interactive
Exercise 11.73**

What intermolecular
forces are overcome during
evaporation?

11.74 What type of intermolecular force must be overcome in converting each of the following from a liquid to a gas?
(a) O_2 (c) CH_3I (methyl iodide)
(b) Hg   (d) CH_3CH_2OH (ethanol)

Section 11.12: Liquids: Properties and Phase Changes

11.75 Answer the following questions using Figure 11.19.
  (a) What is the approximate equilibrium vapour pressure of water at 60 °C?
  (b) At what temperature does water have an equilibrium vapour pressure of 80 kPa?
  (c) Compare the equilibrium vapour pressures of water and ethanol at 70 °C. Which is higher?

Interactive Exercise 11.76

Predict intermolecular forces based on vapour pressure curves.

11.77 Assume you seal 1.0 g of diethyl ether (see Figure 11.19) in an evacuated 0.10 L flask. If the flask is held at 30 °C, what is the approximate gas pressure in the flask? If the flask is placed in an ice bath, does additional liquid ether evaporate or does some ether condense to a liquid?

11.78 Refer to Figure 11.19 as an aid in answering these questions:
  (a) You put some water at 60 °C in a milk jug and seal the top very tightly so that gas cannot enter or leave the jug. What happens when the water cools?
  (b) If you put a few drops of liquid diethyl ether on your hand, does it evaporate completely or remain a liquid?

Interactive Exercise 11.79

Make predictions based on vapour pressure curves.

11.80 Answer each of the following questions with *increases, decreases,* or *does not change.*
  (a) If the intermolecular forces in a liquid increase, the normal boiling point of the liquid _____.
  (b) If the intermolecular forces in a liquid decrease, the vapour pressure of the liquid _____.
  (c) If the surface area of a liquid decreases, the vapour pressure _____.
  (d) If the temperature of a liquid increases, the equilibrium vapour pressure _____.

11.81 Construct a phase diagram for O_2 from the following information: normal boiling point, 90.18 K; normal melting point, 54.8 K; and triple point, 54.34 K at a pressure of 267 Pa. Very roughly estimate the vapour pressure of liquid O_2 at −196 °C, the lowest temperature easily reached in the laboratory. Is the density of liquid O_2 greater or less than that of solid O_2?

Section 11.13: Solids: Properties and Phase Changes

11.82 Benzene (C_6H_6) is an organic liquid that freezes at 5.5 °C (see Figure 11.15) to form beautiful, feather-like crystals. How much heat is evolved when 15.5 g of benzene freezes at 5.5 °C? (The enthalpy change of fusion of benzene is 9.95 kJ mol^{-1}.) If the 15.5 g sample is remelted, again at 5.5 °C, what quantity of heat is required to convert it to a liquid?

11.83 The specific heat capacity of silver is 0.235 J K^{-1} g^{-1}. Its melting point is 962 °C, and its enthalpy change of fusion is 11.3 kJ mol^{-1}. What quantity of heat (in J) is required to change 5.00 g of silver from a solid at 25 °C to a liquid at 962 °C?

Section 11.14: Phase Diagrams

11.84 Consider the phase diagram of CO_2 in Figure 11.24.
  (a) Is the density of liquid CO_2 greater or less than that of solid CO_2?
  (b) In what phase do you find CO_2 at 500 kPa and 0 °C?
  (c) Can CO_2 be liquefied at 45 °C?

Interactive Exercise 11.85

Make predictions based on a phase diagram.

11.86 Air conditioners used to use the chlorofluorocarbon CCl_2F_2 as the heat transfer fluid. The normal boiling point of CCl_2F_2 is −9.8 °C, and the enthalpy change of vaporization is 20.11 kJ mol^{-1}. The gas and the liquid have heat capacities of 117.2 J K^{-1} mol^{-1} and 72.3 J K^{-1} mol^{-1}, respectively. How much heat is evolved when 20.0 g of CCl_2F_2 is cooled from 40 °C to −40 °C?

11.87 The critical temperature and pressure of chloromethane (CH_3Cl) are 416 K and 6697 kPa, respectively. (Chloromethane's triple point is at 175.4 K and 8.7 kPa.) Can CH_3Cl be liquefied at or above room temperature? Explain briefly.

11.88 The critical point of H_2O is at a temperature 250 K higher than for H_2S, H_2Se, or H_2Te. At the molecular level, what needs to happen to the interactions between water molecules to reach the critical point? Why does water require so much higher a temperature to reach its critical point relative to the other Group 16 hydrides?

Section 11.15: Polymorphic Forms of Solids

11.89 With reference to Figure 11.27, why is Ice V not commonly found in your freezer when you make a tray of ice cubes?

11.90 Why might a pharmaceutical company producing drugs to treat Hepatitis C be interested in crystal polymorphism?

SUMMARY AND CONCEPTUAL QUESTIONS

11.91 To what temperature (in °C) must a 0.0255 L sample of oxygen at 90 °C be cooled for its volume to decrease to 0.0215 L? Assume the pressure and mass of the gas are constant.

11.92 You have a sample of helium gas at 33 °C, and you want to increase the average speed of helium atoms by 10.0%. To what temperature should the gas be heated to accomplish this?

11.93 A 1.0 L flask contains 10.0 g each of O_2 and CO_2 at 25 °C.
(a) Which gas has the greater partial pressure, O_2 or CO_2, or are they the same?
(b) Which molecules have the greater average speed, or are they the same?
(c) Which molecules have the greater average kinetic energy, or are they the same?

11.94 If equal masses of O_2 and N_2 are placed in separate containers of equal volume at the same temperature, which of the following statements is true? If false, explain why it is false.
(a) The pressure in the flask containing N_2 is greater than that in the flask containing O_2.
(b) There are more molecules in the flask containing O_2 than in the flask containing N_2.

11.95 State whether each of the following samples of matter is a gas. If there is not enough information for you to decide, write "insufficient information."
(a) A material is in a steel tank at 100 bar pressure. When the tank is opened to the atmosphere, the material suddenly expands, increasing its volume by 10%.
(b) A 1.0 mL sample of material weighs 8.2 g.
(c) The material is transparent and pale green in colour.
(d) One 1.0 m^3 of material contains as many molecules as 1.0 m^3 of air at the same temperature and pressure.

11.96 Each of the four tires of a car is filled with a different gas. Each tire has the same volume, and each is filled to the same pressure, 304 kPa, at 25 °C. One tire contains 116.0 g of air, another tire has 80.7 g of neon, another tire has 16.0 g of helium, and the fourth tire has 160.0 g of an unknown gas.
(a) Do all four tires contain the same number of gas molecules? If not, which one has the greatest number of molecules?
(b) How many times heavier is a molecule of the unknown gas than an atom of helium?
(c) In which tire do the molecules have the largest kinetic energy? The highest average speed?

11.97 The sodium azide required for automobile airbags is made by the reaction of sodium metal with dinitrogen oxide in liquid ammonia:

$$3 N_2O(g) + 4 Na(s) + NH_3(\ell) \longrightarrow$$
$$NaN_3(s) + 3 NaOH(s) + 2 N_2(g)$$

(a) You have 65.0 g of sodium and a 35.0 L flask containing N_2O gas with a pressure of 215 kPa at 23 °C. What is the theoretical yield (in grams) of NaN_3?
(b) Draw a Lewis structure for the azide ion. Include all possible resonance structures. Which resonance structure is most likely?
(c) What is the shape of the azide ion?

11.98 If 12.0 g of O_2 is required to inflate a balloon to a certain size at 27 °C, what mass of O_2 is required to inflate it to the same size (and pressure) at 5.0 °C?

11.99 It snows on Mars, but the snow is dry ice, solid CO_2. Consult the phase diagram for CO_2 in Figure 11.24. If the average atmospheric pressure on Mars is 0.61 kPa, what minimum temperature does it need to be for dry ice to form at the Martian polar caps?

11.100 Scientists have been looking for signs of water on Mars. Evidence has been found of water vapour in the atmosphere and substantial amounts of water ice at the Martian poles. What would happen to a cup of liquid water from Earth if you could transport it to the surface of Mars during the daytime? Consult the phase diagram for water in Figure 11.23 and assume that the surface pressure on Mars is 0.61 kPa and the surface temperature is at its daytime high of −17 °C. What would happen to that same cup of water if it landed at night, when the temperature is −60 °C?

11.101 The gas B_2H_6 burns in air to give H_2O and B_2O_3.

$$B_2H_6(g) + 3 O_2(g) \longrightarrow B_2O_3(s) + 3 H_2O(g)$$

(a) Three gases are involved in this reaction. Place them in order of increasing molecular speed. (Assume all are at the same temperature.)
(b) A 3.26 L flask contains B_2H_6 at a pressure of 34.1 kPa and a temperature of 25 °C. Suppose O_2 gas is added to the flask until B_2H_6 and O_2 are in the correct stoichiometric ratio for the combustion reaction. At this point, what is the partial pressure of O_2?

Interactive Exercise 11.102
Make predictions about gases based on physical quantities.

11.103 An automobile tire has a volume of 17 L. What mass of air is contained in the tire at 25 °C and a pressure of 324 kPa? (Molar mass of air = 28.96 g mol^{-1}.)

Interactive Exercise 11.104
Predict a theoretical yield from a reaction with a gas.

11.105 The density of air 20 km above the earth's surface is 92 g m^{-3}. The pressure of the atmosphere is 5.6 kPa and the temperature is 63 °C.

(a) What is the average molar mass of the atmosphere at this altitude?

(b) If the atmosphere at this altitude consists of only O_2 and N_2, what is the mole fraction of each gas?

Interactive Exercise 11.106

Calculate partial pressures of gases.

11.107 A study of climbers who reached the summit of Mount Everest without supplemental oxygen showed that the partial pressures of O_2 and CO_2 in their lungs were 4.7 kPa and 1.0 kPa, respectively. The barometric pressure at the summit was 33.7 kPa. Assume the lung gases are saturated with moisture at a body temperature of 37 °C (which means the partial pressure of water vapour in the lungs is $p(H_2O) = 6.3$ kPa). If you assume the lung gases consists of only O_2, N_2, CO_2, and H_2O, what is the partial pressure of N_2?

11.108 Ammonia gas is synthesized by combining hydrogen and nitrogen [>>Section 12.1]:

$$3 H_2(g) + N_2(g) \longrightarrow 2 NH_3(g)$$

(a) If you want to produce 562 g of NH_3, what volume of H_2 gas, at 56 °C and 99.3 kPa, is required?

(b) To produce 562 g of NH_3, what volume of air (the source of N_2) is required if the air is introduced at 29 °C and 99.3 kPa? (Assume the air sample has 78.1 mole % N_2.)

11.109 Relative humidity is the ratio of the partial pressure of water in air at a given temperature to the vapour pressure of water at that temperature. Calculate the mass of water per litre of air under the following conditions:

(a) at 20 °C and 45% relative humidity

(b) at 0 °C and 95% relative humidity

Under which circumstances is the mass of H_2O per litre greater? (Consult on-line sources for the vapour pressure of water.)

Interactive Exercise 11.110

Make predictions based on vapour pressure curves.

11.111 Mercury and many of its compounds are dangerous poisons if breathed, swallowed, or even absorbed through the skin. The liquid metal has a vapour pressure of 0.225 Pa at 24 °C. If the air in a small room is saturated with mercury vapour, how many atoms of mercury vapour occur per cubic metre?

11.112 Acetone (CH_3COCH_3) is a common laboratory solvent. It is usually contaminated with water, however. Why does acetone absorb water so readily? Draw molecular structures showing how water and acetone can interact. What intermolecular force(s) is(are) involved in the interaction?

11.113 Cooking oil is not miscible with water. From this observation, what conclusions can you draw regarding the polarity or hydrogen-bonding ability of molecules found in cooking oil?

11.114 The photos illustrate an experiment you can do yourself. Place 10 mL of water in an empty soft drink can and heat the water to boiling. Using tongs or pliers, quickly turn the can over in a pan of cold water, making sure the opening in the can is below the water level in the pan.

Both photos: Charles D. Winters

(a) (b)

(a) Describe what happens and explain it in terms of the subject of this chapter.

(b) Prepare a molecular-level sketch of the situation inside the can before heating and after heating (but prior to inverting the can).

11.115 The following figure is a plot of vapour pressure versus temperature for dichlorodifluoromethane (CCl_2F_2). The enthalpy change of vaporization of the liquid is 165 kJ g^{-1}, and the specific heat capacity of the liquid is about 1.0 J K^{-1} g^{-1}.

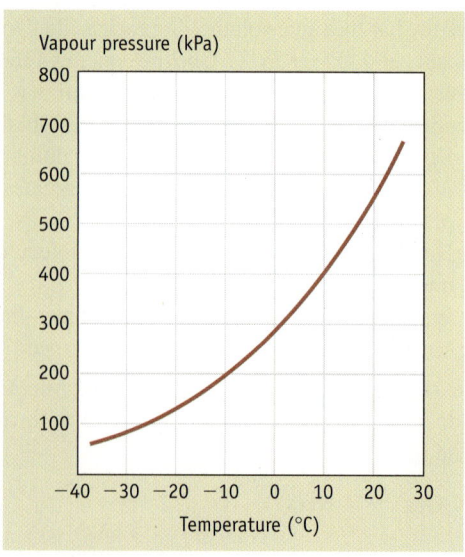

(a) What is the normal boiling point of CCl_2F_2?

(b) A steel cylinder containing 25 kg of CCl_2F_2 in the form of liquid and vapour is set outdoors on a warm day (25 °C). What is the approximate pressure of the vapour in the cylinder?

(c) The cylinder valve is opened, and CCl_2F_2 vapour gushes out of the cylinder in a rapid flow. Soon, however, the flow becomes much slower, and the outside of the cylinder is coated with ice frost. When the valve is closed and the cylinder is reweighed, it is found that 20 kg of CCl_2F_2 is still in the cylinder. Why is the flow fast at first? Why does it slow down long before the cylinder is empty? Why does the outside become icy?

(d) Which of the following procedures would be effective in emptying the cylinder rapidly (and safely)?

  (1) Turn the cylinder upside down and open the valve.

  (2) Cool the cylinder to −78 °C in dry ice and open the valve.

  (3) Knock off the top of the cylinder, valve and all, with a sledge hammer.

11.116 Figure 11.25 is a series of photos of propane (C_3H_8) as it changes from a mixture of liquid and vapour at equilibrium to the supercritical fluid. Draw a representation of the situation at the molecular level of the liquid–vapour equilibrium in the photo at the left and of the supercritical fluid at the right.

11.117 Ethanol (CH_3CH_2OH) has a vapour pressure of 7.9 kPa at 25 °C. What quantity of heat energy is required to evaporate 0.125 L of the alcohol at 25 °C? The enthalpy change of vaporization of the alcohol at 25 °C is 42.32 kJ mol^{-1}. The density of the liquid is 0.7849 g mL^{-1}.

11.118 The enthalpy change of vaporization of liquid mercury is 59.11 kJ mol^{-1}. What quantity of heat is required to vaporize 0.500 mL of mercury at 357 °C, its normal boiling point? The density of mercury is 13.6 g mL^{-1}.

11.119 Aspartame is an artificial sweetener. A model is available in Odyssey.

(a) Draw the structure of aspartame.

(b) Is the molecule capable of hydrogen bonding? If so, what are the sites of hydrogen bonding?

Thierry Orban/CORBIS Sygma

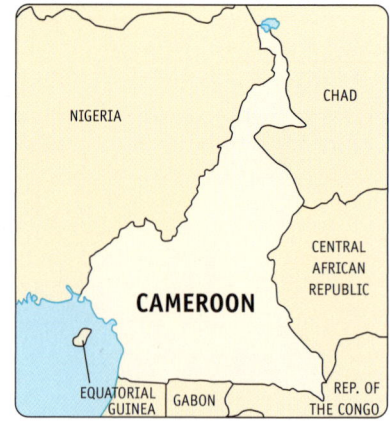

Courtesy of George Kling

12.1 Case Study: The Killer Lakes of Cameroon

It was the evening of Thursday, August 21, 1986. Suddenly people and animals around Lake Nyos in Cameroon collapsed and died. By the next morning, 1700 people, thousands of cattle, and unknown numbers of other animals were dead. The calamity had no apparent cause—no fire, no earthquake, no storm. Two years previously, 37 people had died in a similar way around nearby Lake Monoun. What had brought on these disasters?

Some weeks later, the mystery was solved. Lake Nyos and Lake Monoun are crater lakes, formed when extinct volcanic craters filled with water. Lake Nyos was lethal because it contains a very high concentration of dissolved carbon dioxide. The CO_2 is generated as a result of volcanic activity deep in the earth. Under the high pressures at the bottom of the lake, a large amount of CO_2 dissolves in the water.

On that fateful evening in 1986, something happened to disturb the lake, and the CO_2-saturated water at the bottom of the lake was carried to the surface. Here, at much lower pressure, the CO_2 is much less soluble and the water temporarily became supersaturated. Approximately 1 km^3 (1×10^{12} L) of carbon dioxide was released into the atmosphere, for the same reason that CO_2 in solution under high pressure in a bottle of carbonated beverage is released when we remove the cap. A fountain of gas and water shot up about 80 m.

Then, because it is more dense than air, it formed a layer at ground level that spread with the prevailing breeze at about 75 km per hour. Because the air near the ground was displaced by CO_2, many people and animals died from suffocation in towns as far as 26 km away.

> *On that fateful evening . . . the CO_2-saturated water at the bottom of the lake was carried to the surface. Here, at much lower pressure, the CO_2 is much less soluble and the water temporarily became supersaturated.*

The CO_2-laden water at the bottom of the lake, formed when CO_2 enters through vents under high pressure, is more dense than fresh water, so mixing of water layers is usually minimal. In many other lakes, there is a gradual seasonal "turning over" of the layers, and CO_2 is released from the rising water because its solubility decreases at lower pressures [>>Section 12.4]. It is presumed that a disturbance of Lake Nyos—perhaps a small earthquake, a strong wind, or an underwater landslide—caused the lake water to turn over suddenly, leading to a rapid release of CO_2.

Without intervention, no doubt the amount of CO_2 in the lake would begin to build up again, perhaps with a similar result in due course. A team of geologists from France and the U.S. worked to resolve this potential threat. In early 2001, scientists lowered a pipe, about 200 m long, into the lake. Now the pressure of escaping CO_2 causes a jet of water to rise as high as 50 m into the air. Over the course of a year, about 20 million m^3 of gas is released. While this has been a successful first step, more gas must be removed to make the lake entirely safe, so additional vents are planned.

Geologists estimate that Lake Nyos contains 300–400 million m^3 of CO_2, fed from the bottom of the lake at high pressure. This is about 16 000 times the amount found in similarly sized lakes, which dissolve CO_2 at the surface where it is in contact with low partial pressures of the gas.

The solubility of CO_2 in water, spectacularly and tragically demonstrated in this incident, is crucial to the well-being of our atmospheric environment. Seawater is an enormous "sink" for carbon dioxide emitted into the atmosphere from any source. The dissolved carbon dioxide forms hydrogencarbonate ions (HCO_3^-) and carbonate ions (CO_3^{2-}), and marine life is wonderfully able to use these to form insoluble calcium carbonate for their skeletons or exoskeletons [>>Section 15.1]. It is amazing to contemplate that some of the carbon atoms in a sea shell or carbonaceous sand that we are holding may once have been breathed out by our ancestors or produced by a fire that they lit.

The increasing CO_2 concentration of our atmosphere during modern industrial times [<<Section 4.3] is not because the seawater is saturated in this gas. The Lake Nyos story tells us that water can hold very much more CO_2 than seawater currently has dissolved in it. The main problem is the slow rate at which CO_2 is dissolved in seawater, mainly because of the lack of contact between the gas and the water, except at the interface. So, we have a lag time in water's ability to counteract our increasing emissions, but there is potential for "sequestering" vast amounts of CO_2 in the sea.

A soft drink is saturated with CO_2 gas under high pressure. When the cap is removed and the pressure reduced, CO_2 is much less soluble, and the excess is ejected from solution.

12.2 Solutions and Solubility

We come into contact with solutions every day: seawater and even our tap water (aqueous solutions containing mostly ionic salts), gasoline with additives to improve its properties, and household cleaners such as ammonia in water. We purposely make solutions. Adding sugar, flavouring, and sometimes CO_2 to water produces a palatable soft drink. Athletes drink beverages with dissolved salts to precisely match salt concentrations in body fluids. In medicine, saline solutions are infused into the body to replace lost fluids.

Although we tend to think of solutions as those with a liquid as solvent, in some cases the solvent is a gas or a solid—examples include air (a solution of nitrogen, oxygen, carbon dioxide, water vapour, and other gases) and solid solutions such as 18-carat gold, brass, bronze, and pewter. Nevertheless, the main objective in this chapter is to develop an understanding of solutes dissolved in liquid solvents.

The nature of solutions and the dissolution process is discussed in Section 6.5, and solution concentrations are the subject of Section 6.8.

A **saturated solution** *is one in which the concentration of solute will not increase further* (at a particular temperature), no matter how much of the solute is added and stirred; extra solute added will settle to the bottom (if a solid), form another layer (if a liquid), or escape into the air (if a gas). Although no change is observable at the macroscopic level, at the invisible molecular level, the processes of dissolving and separating continue—at the same rate, so that there is no net change. This is the condition called *dynamic equilibrium* [>>Chapter 13].

The extent to which various substances dissolve in a particular solvent differ widely. The **solubility** of a substance, in a particular solvent at a specified temperature, *is the concentration of the substance in a saturated solution.* Calculations concerning the concentrations in saturated solutions of sparingly soluble ionic salts are the subject of Chapter 15.

Although it may seem a contradiction, it is possible to have a solution in which the concentration of one or more dissolved solutes is higher than that of a saturated solution. Such solutions are called **supersaturated solutions**. These are unstable, and the excess solute eventually crystallizes from the solution until its equilibrium concentration is reached—although in some cases the rate of coming to equilibrium is slow. The oceans are supersaturated with respect to calcium carbonate [>>Section 15.1], but crystallization is very slow.

The solubility of most ionic salts in water decreases as the temperature is lowered. Supersaturated solutions are usually made by making a saturated solution and then gradually cooling it. If the rate of crystallization is slow, the solid may not precipitate even though the concentration has exceeded that of a saturated solution at the lower temperature. When disturbed in some manner (shaking, scratching the surface of the vessel, or adding a *seed crystal*), a supersaturated solution comes to equilibrium by precipitating solute (Figure 12.1).

Think about It

e12.1 Watch demonstrations involving saturated and unsaturated solutions.

Calcium sulfate is an exception among ionic salts in that its solubility decreases with increase of temperature—which is one reason that it forms a scale in kettles and industrial boilers.

FIGURE 12.2 Heat of crystallization. A heat pack relies on the heat evolved by the crystallization of sodium acetate from a supersaturated solution.

FIGURE 12.1 Supersaturated solutions. When a supersaturated solution is disturbed, the dissolved sodium salt (here sodium acetate, $NaCH_3COO$) crystallizes—usually rapidly. This situation can be compared with the waters of Lake Nyos when a physical disturbance caused water saturated with CO_2 at high pressure to be moved to a situation of low pressure where the equilibrium concentration is lower—so the heavily CO_2-laden water became temporarily supersaturated.

Supersaturated solutions are used in "heat packs" to apply heat to injured muscles. When exothermic crystallization of sodium acetate ($NaCH_3COO$) from a supersaturated solution is initiated, the temperature of a heat pack rises to about 50 °C (Figure 12.2).

Saturated solutions of ionic compounds are the subject matter of Chapter 15. In this chapter, we also deal with solutions that are not saturated.

Think about It

e12.2 Examine this heat pack reaction in more detail.

A saturated solution is one in a condition of dynamic equilibrium between dissolved and solid solute, and in which no more solute will dissolve (at a particular temperature). The solubility of a substance, in a particular solvent at a specified temperature, is the concentration of the substance in a saturated solution. A supersaturated solution is an unstable condition in which the concentration of dissolved solute is higher than the equilibrium concentration.

12.3 Enthalpy Change of Solution: Ionic Solutes

Among ionic compounds that are very soluble in water, the dissolution of some (such as sodium hydroxide) are exothermic processes, while others (such as ammonium nitrate) dissolve endothermically (Figure 12.3). The **molar enthalpy change of solution** ($\Delta_{sol}H$), *the enthalpy change accompanying dissolution of 1 mol of solute in a large volume of the solvent*, can be measured using a calorimeter [<<Section 7.6]. The example of the slightly exothermic dissolution of potassium fluoride in water is represented by the thermochemical equation

$$KF(s) \xrightarrow{H_2O} K^+(aq) + F^-(aq) \qquad \Delta_rH = \Delta_{sol}H = -16 \text{ kJ mol}^{-1}$$

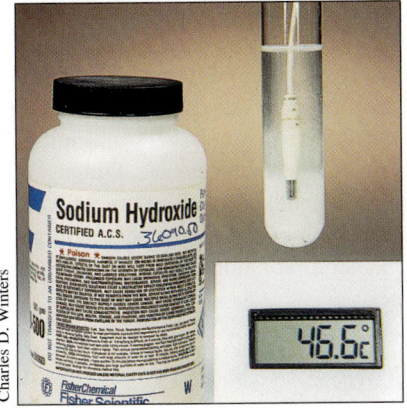
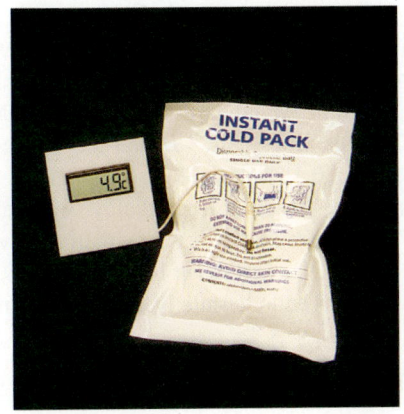

Charles D. Winters

Charles D. Winters

FIGURE 12.3 Dissolving ionic solids and enthalpy change of solution. Dissolving NaOH(s) in water is a strongly exothermic process. A "cold pack" contains solid ammonium nitrate, $NH_4NO_3(s)$, and a package of water. When the water and $NH_4NO_3(s)$ are mixed and the salt dissolves, the temperature of the system drops because of the endothermic dissolution of ammonium nitrate ($\Delta_{sol}H = +25.7 \text{ kJ mol}^{-1}$).

To understand the energetics of dissolution, let us analyze at the molecular level the process of dissolving 1 mol of potassium fluoride in water, with reference to the enthalpy-level diagram in Figure 12.4.

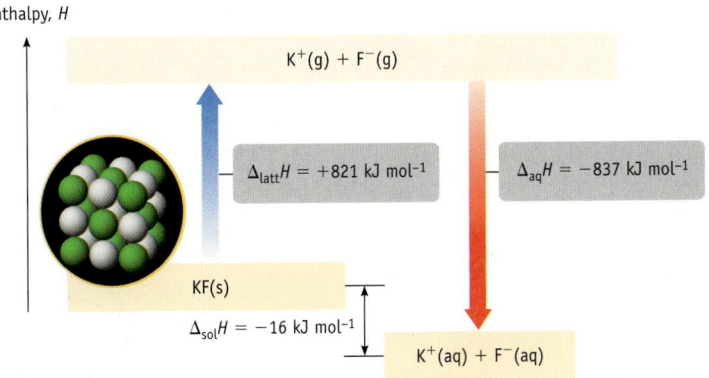

Enthalpy, *H*

$K^+(g) + F^-(g)$

$\Delta_{latt}H = +821 \text{ kJ mol}^{-1}$

$\Delta_{aq}H = -837 \text{ kJ mol}^{-1}$

KF(s)

$\Delta_{sol}H = -16 \text{ kJ mol}^{-1}$

$K^+(aq) + F^-(aq)$

FIGURE 12.4 Thermochemical analysis of enthalpy changes on dissolving KF in water. The dissolution of 1 mol of an ionic compound in water can be imagined to occur in two steps at the particulate level. Here, KF is first separated into isolated K^+ ions and F^- ions with an expenditure of 821 kJ mol^{-1}. These ions are then aquated, with $\Delta_{aq}H = -837$ kJ mol^{-1}. The net enthalpy change is -16 kJ mol^{-1}, a slightly exothermic process.

Solid potassium fluoride has an ionic lattice [<<Section 3.4]: it has alternating K^+ and F^- ions held in place by attractive forces due to their opposite charges. Bearing in mind the interpretation of the dissolution process [<<Section 6.9], the energy change to go from the reactants, KF(s) and $H_2O(\ell)$, to the products, $K^+(aq)$ and $F^-(aq)$, can be artificially broken down into two components:

1. The energy required to separate the ions against their mutual attractions is the *lattice enthalpy* $\Delta_{lattice}H$ [>>Section 26.3]. This process is highly endothermic ($\Delta_{lattice}H$ is positive and large) because the forces of attraction between these ions are strong.

2. The energy released when the separated ions are aquated is the **enthalpy of aquation, $\Delta_{aq}H$.** Again, strong forces of attraction (ion–dipole forces) are involved, so this process is highly exothermic. The enthalpy of aquation is actually the sum of the enthalpies of aquation of the $K^+(g)$ and $F^-(g)$ ions, but we can't separate these out.

These two artificial components of the dissolution process can be represented by chemical equations:

$$\text{Step 1} \qquad KF(s) \longrightarrow K^+(g) + F^-(g) \qquad\qquad \Delta_{lattice}H$$

$$\text{Step 2} \qquad K^+(g) + F^-(g) \xrightarrow{H_2O} K^+(aq) + F^-(aq) \qquad \Delta_{aq}H$$

Think about It

e12.3 Work through these steps with questions and feedback.

The overall dissolution process is the sum of these two steps. So, in accordance with Hess's law [<<Section 7.7], the enthalpy change of solution is the sum of the enthalpy changes of the two imagined steps:

$$\text{Overall:} \qquad KF(s) \xrightarrow{H_2O} K^+(aq) + F^-(aq)$$

$$\Delta_{sol}H = \Delta_{lattice}H + \Delta_{aq}H$$

Molecular Modelling (Odyssey)

e12.4 See how the charge and size of ions affects the strength of aquation.

For potassium fluoride and many other ionic compounds, the enthalpy change of solution is a relatively small difference between two very large enthalpy changes. Small differences of either lattice enthalpy or enthalpy of hydration of different ionic compounds can determine whether they dissolve endothermically or exothermically. These two enthalpy quantities, $\Delta_{lattice}H$ and $\Delta_{aq}H$, are both affected by ion sizes and ion charges [>>Section 26.4]. Salts composed of smaller ions with high charges (i.e., with high *charge density*) have larger lattice enthalpy because of strong mutual forces of attraction. On the other hand, such ions will interact strongly with water molecules, resulting in large aquation enthalpy. The net result is that to predict whether dissolution of an ionic compound is exothermic or endothermic is not a simple matter.

Interactive Exercise 12.1

Indicate whether ions are strongly or weakly aquated.

EXERCISE 12.2—ENTHALPY CHANGE OF SOLUTION

For lithium chloride, LiCl(s), and potassium chloride, KCl(s), predict the relative values of (a) lattice enthalpy, and (b) enthalpy of aquation of the ions. What can you predict about the comparative enthalpies of solution of these two salts?

The molar enthalpy change of solution, $\Delta_{sol}H$, is the enthalpy change accompanying dissolution of 1 mol of solute in a large volume of the solvent. In the case of an ionic solute, this can be regarded as the net result of energy expended as the lattice enthalpy, and energy gained as enthalpy of aquation.

$$\Delta_{sol}H = \Delta_{lattice}H + \Delta_{aq}H$$

12.4 Factors Affecting Solubility: Pressure and Temperature

Biochemists and physicians, among others, are interested in the solubility of gases such as CO_2 and O_2 in water or body fluids, and scientists and engineers need to know about the solubility of solids in various solvents. Pressure and temperature are two external factors that influence solubility. Both affect the solubility of gases in liquids, whereas only temperature is an important factor in the solubility of solids in liquids.

Pressure Effects on Solubility of Gases in Liquids

The solubility of a gas in a liquid is directly proportional to the partial pressure of the gas in contact with the solution (or to the total pressure if it is the only gas). This is a statement of **Henry's law:**

$$s = k_H \times p$$

where s is the gas solubility (the concentration of a saturated solution, in $mol\ L^{-1}$), p is the partial pressure of the gas above the solution (in kPa), and k_H is the Henry's law constant (Table 12.1). The value of this constant depends on the identity of both the solute and the solvent, as well as on the temperature. The magnitude of the constant for O_2 solubility at 25 °C tells us that for each increase of the partial pressure of $O_2(g)$ by 1 kPa, its solubility increases by $1.24 \times 10^{-5}\ mol\ L^{-1}$.

Carbonated soft drinks illustrate how Henry's law works. These beverages are packed in a chamber filled with carbon dioxide gas at high pressure, some of which dissolves in the drink. When the bottle is opened, the partial pressure of CO_2 above the solution drops, causing the solubility of CO_2 to drop. The solution is then a supersaturated solution, so gas is ejected from the solution as bubbles (Figure 12.5).

Henry's law has important consequences in scuba diving (Figure 12.6). When you dive, the pressure of the air you breathe must be adjusted to the external pressure of the water, or your lungs will collapse. In deeper dives, the gases in the scuba gear must be maintained at pressures of several atmospheres. At such high pressures, as Henry's law predicts, more gas dissolves in the blood. This can lead to a problem: if you ascend too rapidly, you can experience a painful and potentially lethal condition referred to as "the bends," in which bubbles of nitrogen gas form in the blood as the solubility of nitrogen decreases with decreasing pressure. In an effort to prevent the bends, divers may use a helium-oxygen mixture (rather than nitrogen-oxygen) because helium is less soluble in aqueous media than nitrogen.

TABLE 12.1 Henry's Law Constants of Gases in Water (25 °C)*

Gas	k_H (mol L^{-1} kPa^{-1})
N_2	6.32×10^{-6}
O_2	1.24×10^{-5}
CO_2	3.36×10^{-4}

*Converted from data in W. Stumm and J. J. Morgan: *Aquatic Chemistry*, p. 109. New York: Wiley, 1996.

Scuba is an acronym for **s**elf-**c**ontained **u**nderwater **b**reathing **a**pparatus.

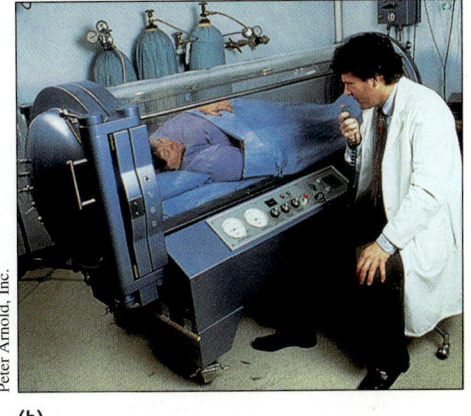

(a) (b)

FIGURE 12.6 Applications of Henry's law. (a) Scuba divers must be aware that the solubilities of gases in the blood increases as the pressure of the gases is increased. (b) A hyperbaric chamber. People who have problems breathing can be placed in a hyperbaric chamber where they are subjected to relatively high pressures of oxygen.

We can better understand the effect of pressure on solubility by consideration of the system at the molecular level. The solubility of a gas in a liquid is defined as the concentration of the dissolved gas when it is in equilibrium with the gas above the liquid. At equilibrium, the rate at which solute gas molecules escape the solution and enter the gaseous state equals the rate at which gas molecules re-enter the solution. An increase in gas pressure results in higher frequency of gas molecules striking the surface of the liquid and entering solution. The solution eventually reaches equilibrium when the concentration of gas dissolved in the solvent becomes high enough that the rate of gas molecules escaping is the same as the rate at which they are entering the solution.

Henry's law holds quantitatively only for gases that do not react chemically with the solvent. It does not work perfectly for ammonia, for example, because it reacts with water to form small concentrations of NH_4^+ ions and OH^- ions.

Provided that we know the values of the Henry's law constants for solubilities of particular gases in water at specified temperatures, Henry's law can be used to model the distribution of gaseous substances between the air and water bodies (if equilibrium is reached). Worked Example 12.1 is a simple example of this.

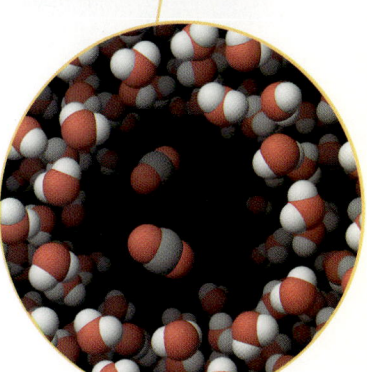

FIGURE 12.5 Gas solubility and pressure. Carbonated beverages are bottled under high CO_2 pressure. When the bottle is opened, the pressure is lowered and bubbles of CO_2 are ejected. After some time, an equilibrium between dissolved CO_2 and atmospheric CO_2 is reached, and at the very low levels of residual CO_2 concentration, the beverage is "flat."

WORKED EXAMPLE 12.1—USING HENRY'S LAW

What is the concentration of O_2 in a fresh water stream in equilibrium with air at 25 °C and 101.3 kPa (1 atm) pressure? The mole fraction of O_2 in air is 0.21. Express the answer in grams of O_2 per litre of solution.

Strategy

To use Henry's law to calculate the molar solubility of oxygen, the partial pressure of O_2 in air must first be calculated. The Henry's law constant for O_2 in water at 25 °C is listed in Table 12.1.

Solution

The mole fraction of O_2 in air is 0.21, so at a total air pressure of 101.3 kPa, the partial pressure of O_2 is $(0.21)(101.3 \text{ kPa}) = 21.3$ kPa. Using this pressure for p in Henry's law, we calculate the solubility s in mol L^{-1}:

$$s = k_H \times p = (1.24 \times 10^{-5} \text{ mol L}^{-1} \text{ kPa}^{-1})(21.3 \text{ kPa}) = 2.64 \times 10^{-4} \text{ mol L}^{-1}$$

This concentration can be expressed in g L^{-1}, using the molar mass of O_2:

$$\text{Solubility of } O_2 = (2.64 \times 10^{-4} \text{ mol L}^{-1})(32.0 \text{ g mol}^{-1}) = 8.5 \times 10^{-3} \text{ g L}^{-1}$$

Comment

This concentration of O_2 (0.0085 g L^{-1}) is quite low, but it is sufficient to provide the oxygen required by aquatic life.

EXERCISE 12.4—USING HENRY'S LAW

What is the concentration of CO_2 in water when the partial pressure of $CO_2(g)$ above the water is 33.4 kPa? (Although CO_2 reacts with water to give traces of $H^+(aq)$ ions and $HCO_3^-(aq)$ ions, the reaction occurs to such a small extent that Henry's law is obeyed at low partial pressures of $CO_2(g)$.)

Consistent with Henry's law, the solubility (s) of a gas in a liquid is directly proportional to the partial pressure of the gas in contact with the solution:

$$s = k_H \times p$$

The magnitude of k_H, the Henry's law constant, is characteristic of the solute-solvent combination and the temperature.

Temperature Effects on Solubility

The solubility of all gases in water decreases as the temperature is increased. The everyday observation of the appearance of bubbles as water is heated—even at temperatures well below the boiling point—is because air becomes less soluble in water as the temperature is increased.

This temperature dependence of solubility has significant environmental consequences. Fish in lakes and rivers often seek deeper water in summer because the warmer surface layers have lower oxygen concentrations. Thermal pollution, resulting from the use of surface water as an industrial coolant, can pose a special problem for marine life because warm effluent water returned to a natural water source is oxygen-deficient.

To understand the effect of temperature on the solubility of gases, it is useful to consider the enthalpy change of solution [<<Section 12.3]. Gases that dissolve to an appreciable extent in water usually do so in an exothermic process.

$$\text{Gas} + \text{water} \rightleftharpoons \text{saturated solution} + \text{heat} \qquad \Delta_{sol}H < 0$$

The reverse process, gas molecules exiting solution, is endothermic. These two processes can reach a condition of equilibrium, as implied by the $\rightleftharpoons$ symbol.

To predict how temperature affects solubility, we turn to **Le Chatelier's principle**: *A change in any of the factors that influence the condition of equilibrium brings about a change in the relative amounts of reactants and products in the direction that counteracts* *the applied change.* For example, if a solution of a gas in water is heated, a new condition of equilibrium will be achieved by evolution of some of the gas from solution, accompanied by absorption of heat. This adjustment results in less gas dissolved, or a lower solubility, at the higher temperature.

The solubility of solids in water also depends upon the temperature. In Figure 12.7 are plots of the solubilities of several salts over a range of temperatures. This graph shows that *the solubilities of many salts increase as the temperature is raised, but there are exceptions.*

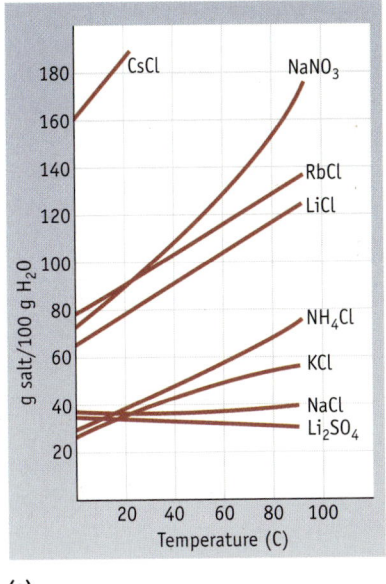
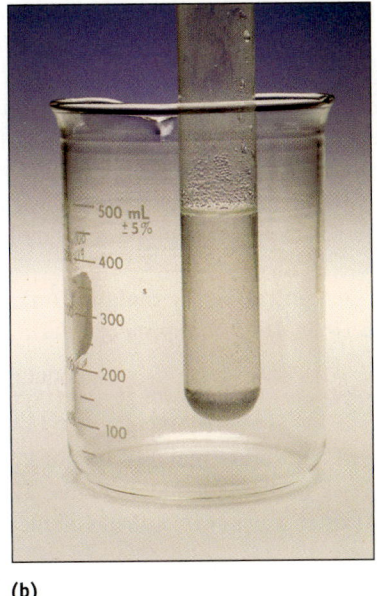

(a)   (b)   (c)

Photos: Charles D. Winters

FIGURE 12.7 The temperature dependence of the solubility of some ionic compounds in water. (a) The solubility of most ionic compounds increases with increasing temperature. (b) A solution of NH_4Cl in water. (c) NH_4Cl precipitates when the solution is cooled in ice.

Chemists take advantage of the variation of solubility with temperature to purify compounds, both in the chemistry laboratory and in industry. A solvent is chosen in which the solubility of the target substance increases dramatically as the temperature is raised. An impure sample of the compound is dissolved in the minimum volume of solvent at high temperature. The solution is filtered, and then cooled. As the temperature is lowered, crystals of the pure compound form (or the solution would be supersaturated). This process of purification by *recrystallization* depends for maximum effectiveness on differences between the solubilities of the desired compound and the impurities. Those impurities that are much less soluble at the higher temperature are separated by filtration before the solution is cooled. At the lower temperature, impurities that are more soluble than the target compound remain in solution while the crystals are growing. If the process is done slowly and carefully, it is sometimes possible to grow very large crystals (Figure 12.8).

EXERCISE 12.5—TEMPERATURE DEPENDENCE OF SOLUBILITY

Some lithium chloride, LiCl(s), is dissolved in 100 mL of water in a beaker, and some $Li_2SO_4(s)$ is dissolved in 100 mL of water in another. Both are at 10 °C and both are saturated solutions; some solid remains undissolved in each beaker. Describe what you would observe as the temperature is raised. Use the following data:

	SOLUBILITY (g per 100 mL)	
Compound	10 °C	40 °C
Li_2SO_4	35.5	33.7
LiCl	74.5	89.8

Charles D. Winters

FIGURE 12.8 Giant crystals of potassium dihydrogen phosphate. The crystal being measured by this researcher at Lawrence Livermore Laboratory in California weighs 318 kg and measures $66 \times 53 \times 58$ cm. The crystals were grown by suspending a thumbnail-sized seed crystal in a 2 m tank of saturated KH_2PO_4 solution. The temperature of the solution was gradually reduced from 65 °C over 50 days. The crystals are sliced into thin plates, which are used to convert light from a giant laser from infrared to ultraviolet.

The solubility of all gases in water decreases as the temperature is increased. A level of explanation is provided by Le Chatelier's principle: A change in any of the factors that influence the condition of equilibrium brings about a change in the relative amounts of reactants and products in the direction that counteracts the applied change. The solubilities of many salts increase as the temperature is raised, but there are exceptions.

12.5 More Units of Solute Concentration

Molar concentrations, or *molarity,* of solutions (and of species in solution) are defined in Section 6.8. Because the volume of a liquid sample changes with temperature (because its density changes), so too does the molar concentration; therefore, a statement of the molar concentration should specify the temperature at which it has been measured. Another problem with the use of molarity is that from the mass of a sample of solution, we cannot calculate the mass of dissolved solute unless we know the solution density.

Often it is useful to use mass-based ways of measuring solution concentrations, and four of these are molal concentration (or molality), mole fraction, mass percent, and parts-per-million.

The **molal concentration**, or **molality (m)** *of a dissolved solute is defined as the amount of solute (in mol) per kilogram of solvent.*

$$\text{Molality, } m = \frac{\text{amount of solute (mol)}}{\text{mass of solvent (kg)}}$$

The unit used to express molality is mol kg^{-1} (sometimes abbreviated as m). The solution in the flask on the left side of Figure 12.9 is made from 0.100 mol of K_2CrO_4 dissolved in 1.00 kg of water (with a final volume more than 1.00 L). We can calculate the molality as follows:

$$\text{Molality of } K_2CrO_4, m = \frac{0.100 \text{ mol}}{1.00 \text{ kg}} = 0.100 \text{ mol kg}^{-1}$$

Notice that different volumes of water were used to make the 0.100 mol L^{-1} and 0.100 mol kg^{-1} solutions of K_2CrO_4. This means that the molarity and the molality of a given solution are not the same (although the difference may be negligibly small when the solution is quite dilute).

The **mole fraction (x)**, also called *amount fraction,* of a solution component (either the solvent or any solute) *is the amount (in mol) of that component divided by the total amount of all of the components of the mixture.* We have previously used mole fraction in relation to mixtures of gases [<<Section 11.6]. Mole fraction has no units.

$$\text{Mole fraction of A, } x_A = \frac{n_A}{n_A + n_B + n_C + \cdots}$$

Consider a solution that contains 1.00 mol (46.1 g) of ethanol (C_2H_5OH) in 9.00 mol (162 g) of water. Here we have only two components:

$$x_{\text{ethanol}} = \frac{1.00 \text{ mol}}{1.00 \text{ mol} + 9.00 \text{ mol}} = 0.100$$

$$x_{\text{water}} = \frac{9.00 \text{ mol}}{1.00 \text{ mol} + 9.00 \text{ mol}} = 0.900$$

As for all mixtures, including solutions, the sum of the mole fractions of the components is 1.000:

$$x_{\text{water}} + x_{\text{ethanol}} = 0.100 + 0.900 = 1.000$$

The labels on common household products usually list the compositions of its various ingredients in terms of mass percent. **Mass percent** is *the mass of one component divided by the total mass of the mixture, multiplied by 100%:*

$$\text{Mass \% A} = \frac{m(A)}{m(A) + m(B) + m(C) + \cdots} \times 100\%$$

Charles D. Winters

$V_{\text{soln}} > 1.00$ L $V_{\text{soln}} = 1.00$ L
$V_{\text{H}_2\text{O}}$ added $= 1.00$ L $V_{\text{H}_2\text{O}}$ added < 1.00 L
0.100 molal solution   0.100 molar solution

FIGURE 12.9 Making 0.100 mol L^{-1} and 0.100 mol kg^{-1} solutions. In the flask on the right, 0.100 mol (19.4 g) of K_2CrO_4 was dissolved in enough water to make 1.00 L of solution. Slightly less than 1.00 L of water was added. In the flask on the left, exactly 1.00 kg of water was added to 0.100 mol of K_2CrO_4. Notice that the volume of solution is greater than 1.00 L. (The small pile of yellow solid in front of the flask is 0.100 mol of K_2CrO_4.)

Think about It

e12.5 Watch a video showing a potassium dichromate solution being made, and then calculate the solute concentration.

Molecular Modelling (Odyssey)

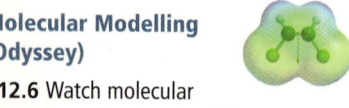

e12.6 Watch molecular simulations showing the differences between molarity and molality.

The ethanol-water solution described above has 46.1 g of ethanol and 162 g of water:

$$\text{Mass \% ethanol} = \frac{46.1 \text{ g}}{46.1 \text{ g} + 162 \text{ g}} \times 100\% = \frac{46.1 \text{ g}}{208.1 \text{ g}} \times 100\% = 22.2\%$$

If you know the mass percent of a solute, you can calculate its mole fraction or molality (or vice versa) because the masses of solute and solvent are known.

Naturally occurring solutions are often very dilute. Environmental chemists, biologists, geologists, oceanographers, toxicologists, and others frequently use **parts-per-million** **(ppm)** to express concentrations. The unit ppm *refers to relative masses of solute and solution*; a solution with 1.0 g of a solute in a sample with a total mass of 1 000 000 g, or one with 1.0 mg of solute in a total mass of 1 kg (1000 g), has a concentration of 1.0 ppm.

Because water has a density of 1.0 g mL^{-1}, a solute concentration of 1.0 mg L^{-1} of solution is nearly the same as 1.0 mg kg^{-1} (ppm) of water; that is, a concentration expressed in the units of ppm and mg L^{-1} are nearly the same.

> In measurement of gas concentrations, ppm can refer to the relative numbers of molecules of components rather than of masses [<<Section 4.2].

WORKED EXAMPLE 12.2—SOLUTION CONCENTRATION

Suppose that you add 1.2 kg of ethylene glycol ($HOCH_2CH_2OH$) as an antifreeze to 4.0 kg of water in the radiator of your car. What are the mole fraction, molality, and mass percent of the ethylene glycol?

Interactive
Exercise 12.6

Calculate the molality, mole fraction, and mass percent of a solute in solution.

Strategy

Calculate the amounts of ethylene glycol and water and then use the definitions of the various ways of expressing concentration.

Solution

1.2 kg of ethylene glycol ($M = 62.1$ g mol^{-1}) is 19 mol, and 4.0 kg of water is 220 mol.

$$\text{Mole fraction, } x_{\text{glycol}} = \frac{19 \text{ mol}}{19 \text{ mol} + 220 \text{ mol}} = 0.080$$

$$\text{Molal concentration} = \frac{19 \text{ mol}}{4.0 \text{ kg}} = 4.8 \text{ mol kg}^{-1}$$

$$\text{Mass \%} = \frac{1.2 \times 10^3 \text{ g}}{(1.2 \times 10^3 \text{ g}) + (4.0 \times 10^3 \text{ g})} \times 100\% = 23\%$$

WORKED EXAMPLE 12.3—SOLUTION CONCENTRATION

To adjust the pH of the water, you dissolve 560 g of $NaHSO_4$(s) in a swimming pool that contains 4.5×10^5 L of water at 25 °C. Assume that the density of the water is 1 kg L^{-1}. What is the sodium ion concentration in parts-per-million (if there are none from any other source)?

Strategy

First calculate the mass of sodium ions in 560 g of $NaHSO_4$(s). Then use this mass of sodium and the mass of water (calculated from its volume and density) to calculate how many grams in every 1,000,000 g of water. The molar masses of $NaHSO_4$ and Na^+ are 120 g mol^{-1} and 23.0 g mol^{-1}. The formula $NaHSO_4$ tell us that there is 1 mol of Na^+ ions in every 1 mol of $NaHSO_4$(s).

Solution

$$n(NaHSO_4) = \frac{m}{M} = \frac{560 \text{ g}}{120 \text{ g mol}^{-1}} = 4.67 \text{ mol} = n(Na^+) \text{ added}$$

$$m(Na^+ \text{ ions}) \text{ added} = n \times M = 4.67 \text{ mol} \times 23.0 \text{ g mol}^{-1} = 107 \text{ g}$$

$$m(Na^+ \text{ ions}) \text{ per 1000 kg } H_2O = 107 \text{ g} \left(\frac{1.0 \times 10^3 \text{ kg}}{4.5 \times 10^5 \text{ kg}} \right) = 0.24 \text{ g}$$

$$\therefore \text{Concentration of } Na^+ \text{ ions} = 0.24 \text{ ppm}$$

EXERCISE 12.7—SOLUTION CONCENTRATION

If you dissolve 10.0 g (about one heaped teaspoonful) of sugar (sucrose, $C_{12}H_{22}O_{11}$) in 250 g of water, what are the mole fraction, molality, and mass percent of sugar?

The molal concentration, or molality (m), of a dissolved solute is defined as the amount of solute (in mol) per kilogram of solvent. The mole fraction (x) of a solution component is the amount (in mol) of that component divided by the total amount of all of the components of the mixture. Mass percent is the mass of one component divided by the total mass of the mixture, multiplied by 100%. Parts-per-million (ppm) refers to relative masses of solute and solution.

12.6 Colligative Properties

When water contains any dissolved substance, the equilibrium vapour pressure of water over the solution is different from that over pure water. Similarly, the freezing point, boiling point, and osmotic pressure of any solution are different from pure water. *The magnitudes of these colligative properties depend only on the concentration of solute particles and, amazingly, are independent of the identity of the particles in solution.* With regard to these properties, it is as if water can count the number of particles dissolved in it, but cannot distinguish between them. This is rather surprising, given that so much of chemical behaviour depends on the ability of molecules to "recognize" each other [<<Section 9.1].

Lowering of Vapour Pressure by Non-Volatile Solutes

The equilibrium vapour pressure of a liquid substance at a particular temperature is the pressure of the vapour above it when the liquid and the vapour are in equilibrium (Figure 6.6). If we form a solution by dissolving in the liquid a non-volatile solute (insignificant vapour pressure of its own), the equilibrium vapour pressure of the solvent above the solution is lower than over the pure solvent, at the same temperature. For solutions said to be **ideal solutions**, *the vapour pressure of the solvent ($p_{solvent}$) is proportional to the mole fraction of solvent ($x_{solvent}$) in the solution.* At a specified temperature:

$$p_{solvent} = x_{solvent} \times p°_{solvent}$$

where $p°_{solvent}$ is the equilibrium vapour pressure above the pure solvent. This equation, called **Raoult's law,** tells us that, for example, if 95% of the molecules in a solution are solvent molecules ($x_{solvent} = 0.95$), then the vapour pressure of the solvent is 95% of that of the pure solvent.

Raoult's law is named after Francois M. Raoult (1830–1901), a professor of chemistry at the University of Grenoble in France, who did pioneering studies in this area.

The notion of an *ideal solution* corresponds with that of an *ideal gas*—a non-existent gas that obeys exactly $pV = nRT$ [<<Section 11.4]. In reality, there are no ideal solutions. Nevertheless, Raoult's law is a good approximation of solution behaviour in many instances, especially at low solute concentration.

WORKED EXAMPLE 12.4—VAPOUR PRESSURE OF SOLUTIONS. NON-VOLATILE SOLUTE

Ethylene glycol ($HOCH_2CH_2OH$) is used as an antifreeze additive to water. Suppose 651 g of ethylene glycol is dissolved in 1.50 kg of water. What is the vapour pressure of water over the solution at 90 °C? At this temperature, the equilibrium vapour pressure of pure water is 70.10 kPa. Assume ideal behaviour of the solution.

Strategy

To use Raoult's law, we must calculate the mole fraction of the solvent (water).

Solution

We first calculate the amounts of water and ethylene glycol, and then the mole fraction of water.

$$n(H_2O) = \frac{1.50 \times 10^3 \text{ g}}{18.02 \text{ g mol}^{-1}} = 83.2 \text{ mol}$$

$$n(HOCH_2CH_2OH) = \frac{651 \text{ g}}{62.07 \text{ g mol}^{-1}} = 10.5 \text{ mol}$$

$$\therefore x_{H_2O} = \frac{83.2 \text{ mol}}{83.2 \text{ mol} + 10.5 \text{ mol}} = 0.888$$

Next we apply Raoult's law:

$$p_{H_2O} = x_{H_2O} \times p^{\circ}_{H_2O} = (0.888)(70.10 \text{kPa}) = 62.2 \text{kPa}$$

EXERCISE 12.8—VAPOUR PRESSURE OF SOLUTIONS, NON-VOLATILE SOLUTE

Assume you dissolve 10.0 g of sucrose ($C_{12}H_{22}O_{11}$) in 225 g of water and warm the solution to 60 °C, at which temperature the equilibrium vapour pressure above pure water (p°) is 19.9 kPa. What is the vapour pressure of water over the solution?

We can derive from Raoult's law a direct relationship between the magnitude of the lowering of the vapour pressure, $\Delta p_{solvent}$, and the mole fraction of the *solute*. We begin with the meaning of $\Delta p_{solvent}$:

$$\Delta p_{solvent} = p^{\circ}_{solvent} - p_{solvent}$$

Substituting for $p_{solvent}$ in Raoult's law, we have

$$\Delta p_{solvent} = p^{\circ}_{solvent} - (x_{solvent} \times p^{\circ}_{solvent}) = (1 - x_{solvent})p^{\circ}_{solvent}$$

In any solution, the sum of the mole fractions of solvent and solutes is 1. If there is only one solute:

$$x_{solvent} + x_{solute} = 1 \qquad \text{so} \qquad 1 - x_{solvent} = x_{solute}$$

and the equation for $\Delta p_{solvent}$ can be rewritten as

$$\Delta p_{solvent} = x_{solute} \times p^{\circ}_{solvent}$$

This equation tells us that the amount by which the vapour pressure of the solvent above the solution is less than that of the pure solvent is *proportional to the mole fraction of solute*.

WORKED EXAMPLE 12.5—VAPOUR PRESSURE OF SOLUTIONS, NON-VOLATILE SOLUTE

Calculate directly the lowering of the vapour pressure of water in the solution described in Worked Example 12.4.

Strategy

We use the form of Raoult's law $\Delta p_{solvent} = x_{solute} \times p^{\circ}_{solvent}$

Solution

Since $x_{water} = 0.888$, we know that $x_{solute} = 1 - 0.888 = 0.112$
Substituting into the equation for $\Delta p_{solvent}$:

$$\Delta p_{solvent} = x_{solute} \times p^{\circ}_{solvent} = 0.112 \times 70.10 \text{ kPa} = 7.9 \text{ kPa}$$

Comment

This value reconciles with that calculated from Worked Example 12.4:

$$\Delta p_{solvent} = p^{\circ}_{solvent} - p_{solvent} = 70.10 \text{ kPa} - 62.2 \text{ kPa} = 7.9 \text{ kPa}$$

Interactive Exercise 12.9

Calculate the vapour pressure above a solution.

Taking It Further

e12.7 Read about vapour pressures of mixtures of volatile liquids.

Which solutions are most ideal? This question brings us to consider the forces of attraction between molecules of the solute and those of the solvent. For Raoult's law to be obeyed, the strength of the forces between solute and solvent molecules in the solution are similar to those between solvent molecules in the pure solvent and those between solute molecules in the pure solute. This is frequently the case when solvent and solute have molecules with similar structures. Solutions of one hydrocarbon in another (hexane, C_6H_{14}, dissolved in octane, C_8H_{18}, for example) usually follow Raoult's law quite closely. If solvent-solute interactions are stronger than solvent–solvent interactions, the vapour pressure is lower than calculated by Raoult's law. If the solvent-solute interactions are weaker than solvent-solvent interactions, the vapour pressure is higher than predicted.

For ideal solutions, the vapour pressure of the solvent is proportional to the mole fraction of solvent (Raoult's law). At a specified temperature,

$$p_{solvent} = x_{solvent} \times p°_{solvent} \qquad \text{or} \qquad \Delta p_{solvent} = x_{solute} \times p°_{solvent}$$

where $p°_{solvent}$ is the equilibrium vapour pressure above the pure solvent.

Freezing Point Depression

The freezing point of a solution is lower than that of the pure solvent (Figure 12.10a).

If an aqueous solution is cooled gradually, initially pure ice forms and the solutes remain in the residual solution (Figure 12.10b). As more ice forms, the concentration of the residual solution increases, and so its freezing point falls lower and lower. This is the essence of a method of separation of soluble impurities by a technique called *fractional crystallization*. Cooling is stopped before the whole sample has frozen, and the residual solution (containing most of the impurities) is run off. The sample may be melted and the process repeated a number of times until the desired degree of purity of the solvent has been attained.

FIGURE 12.10 Effect of a solute on the freezing point of water. (a) A jar of pure water *(left)* and a jar of water to which automobile antifreeze had been added *(right)* were kept overnight in the freezing compartment of a refrigerator. The solution did not reach its freezing point. (b) When a solution freezes, pure solvent solidifies. A purple solute was dissolved in water, and the temperature was gradually decreased. Pure ice formed along the walls of the tube, and the dye stayed in the residual solution.

Photos: Charles D. Winters

(a)

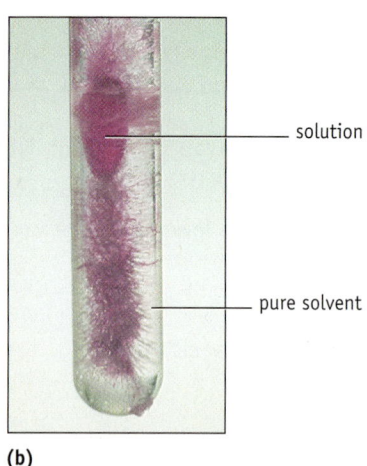

— solution

— pure solvent

(b)

At low solute concentrations, the difference between the freezing point of solvent and that of a solution (ΔT_f) is directly proportional to the molality (*m*) of the solute:

Freezing point depression, $\Delta T_f = K_f \times m_{solute}$

where K_f is the *molal freezing point depression constant*, characteristic of the solvent, for which the unit is K kg mol^{-1}. Values of K_f for a few common solvents are given in Table 12.2.

TABLE 12.2 Some Molal Freezing Point Depression Constants

Solvent	Normal freezing point of solvent (°C)	K_f(K kg mol^{-1})
Water	0.0	1.86
Benzene	5.50	5.12
Camphor	179.75	39.7

WORKED EXAMPLE 12.6—FREEZING POINT DEPRESSION

What mass of ethylene glycol ($HOCH_2CH_2OH$) must be added to 5.50 kg of water to lower the freezing point of the water from 0.0 °C to −10.0 °C?

Strategy

Select the value of K_f appropriate to water as the solvent from Table 12.2. Then calculate the molality of solution needed to bring about this freezing point depression, and finally the amount and mass of ethylene glycol required to make such a solution.

Solution

$$\text{Solution molality, } m = \frac{\Delta T_f}{K_f} = \frac{10.0 \text{ K}}{1.86 \text{ K kg mol}^{-1}} = 5.38 \text{ mol kg}^{-1}$$

Since $molality = \dfrac{n(\text{solute})}{m(\text{solvent}), \text{ in kg}}$, $n(\text{solute}) = (5.50 \text{ kg})(5.38 \text{ mol kg}^{-1}) = 29.6 \text{ mol}$

The molar mass of ethylene glycol is 62.07 g mol^{-1}, so we need

$$m(\text{solute}) = 29.6 \text{ mol} \times 62.07 \text{ g mol}^{-1} = 1840 \text{ g}$$

> Be careful with symbols here, because m is the symbol used for both mass and molal concentration.

EXERCISE 12.12—FREEZING POINT DEPRESSION

In cold climates, summer cottages are sometimes left unoccupied over winter. When doing so, the owners "winterize" the plumbing by putting antifreeze in the toilet tanks. Will the addition of 525 g of $HOCH_2CH_2OH$ to 3.00 kg of water ensure that the water will not freeze at −25 °C?

Interactive Exercise 12.11

SUBMIT

Calculate the mass of solute from the freezing point depression.

Molar Mass Determination from Freezing Point Depression

Measurement of freezing point of solutions, compared with the freezing point of pure solvent, was one of the first ways of determining the molar mass of compounds. In this technique, the freezing point of a solution made up from known masses of solute and solvent is measured. The freezing point depression allows calculation of the molality of the solution, and then of the molar mass that is consistent with this molality. The method is illustrated in Worked Example 12.7 for cases of non-electrolyte solutes.

WORKED EXAMPLE 12.7—MOLAR MASS FROM FREEZING POINT DEPRESSION

Phenylcarbinol is used in nasal sprays as a preservative. A solution of 0.52 g of the non-electrolyte in 25.0 g of water has a freezing point of −0.36 °C. What is the molar mass of phenylcarbinol?

Interactive Exercise 12.13

SUBMIT

Calculate the molar mass of a solute using freezing point depression.

Solution

Since the freezing point of pure water is 0 °C, $\Delta T_f = 0.36$ °C = 0.36 K. So we can calculate the molality of the solution, using the value of K_f appropriate for water as the solvent:

$$m = \frac{\Delta T_f}{K_f} = \frac{0.36 \text{ K}}{1.86 \text{ K kg mol}^{-1}} = 0.194 \text{ mol kg}^{-1}$$

The mass of solute per kg of solvent is $0.52 \text{ g} \times \dfrac{1000}{25.0} = 20.8 \text{ g}$

Then if the molar mass is M, $\dfrac{20.8 \text{ g}}{M} = 0.194 \text{ mol}$ and $M = \dfrac{20.8 \text{ g}}{0.194 \text{ mol}} = 107 \text{ g mol}^{-1}$

EXERCISE 12.14—MOLAR MASS FROM FREEZING POINT DEPRESSION

The organic compound called aluminon is used as a reagent to test for the presence of the aluminum ion in aqueous solution. A solution of 2.50 g of aluminon in 50.0 g of water freezes at −0.197 °C. What is the molar mass of aluminon?

The difference between the freezing point of a solvent and that of a solution with a non-electrolyte solute (ΔT_f) is directly proportional to the molality (m) of the solute:

$$\Delta T_f = K_f \times m_{solute}$$

where K_f is the molal freezing point depression constant, characteristic of the solvent.

Solutions of Electrolytes

Colligative properties depend not on the identity of what is dissolved but only on the concentration of solute particles. The significance of this in relation to the magnitude of colligative properties is illustrated here with specific reference to freezing point depression. When 1 mol of NaCl dissolves, 2 mol of ions form:

$$\text{NaCl(s)} \xrightarrow{\text{H}_2\text{O}} \text{Na}^+\text{(aq)} + \text{Cl}^-\text{(aq)}$$

which means that the effect on the freezing point of water is twice as large as that for 1 mol of sugar in the same volume of solution. This peculiarity was discovered by Raoult in 1884 and studied in detail by Jacobus Henrikus van't Hoff (1852–1911) in 1887. Later in that same year, Svante Arrhenius (1859–1927) provided the explanation for the behaviour of electrolytes based on the total concentration of ions in solution. A $0.100 \text{ mol kg}^{-1}$ solution of NaCl contains two solute species, $0.100 \text{ mol kg}^{-1}$ Na$^+$ ions and $0.100 \text{ mol kg}^{-1}$ Cl$^-$ ions. To estimate the freezing point depression, we need to use the total molal concentration of solute particles:

$$m_{\text{total}} = m(\text{Na}^+) + m(\text{Cl}^-) = (0.100 + 0.100) \text{ mol kg}^{-1} = 0.200 \text{ mol kg}^{-1}$$
$$\Delta T_f = (1.86 \text{ K kg mol}^{-1})(0.200 \text{ mol kg}^{-1}) = 0.372 \text{ K} = 0.372 \text{ °C}$$
$$\therefore \Delta T_f = 0.000 \text{ °C} - 0.372 \text{ °C} = -0.372 \text{ °C}$$

On the other hand, sugar (sucrose) is a non-electrolyte, and in a $0.100 \text{ mol kg}^{-1}$ sugar solution, the total molal concentration of particles (sucrose molecules) is $0.100 \text{ mol kg}^{-1}$, and the freezing point depression is 0.186 K

Any substance that is soluble in water can be spread on ice-covered roads to melt the ice, because it reduces the melting point of the ice. Salt is the most common substance used—it is cheap, dissolves readily in water, and has a relatively low molar mass so its effect per gram is large. In addition, salt is especially effective because it is an electrolyte: it dissolves to produce ions in solution.

To estimate the freezing point depression for a solution of an ionic compound, multiply the solution molality by the number of moles of ions produced from each 1 mol of the compound: 2 for NaCl, 3 for Na$_2$SO$_4$, 4 for LaCl$_3$, 5 for Al$_2$(SO$_4$)$_3$, and so on. This model gives a reasonable estimate of the effect of ionization of a solute on colligative properties, but it is not exact. Let us look at some experimental data (Table 12.3) of the freezing points of solutions of two ionic compounds, NaCl and Na$_2$SO$_4$. The measured freezing point depressions are larger than those calculated on the assumption that the salt is a non-electrolyte.

Salt is used on roads covered in snow or ice to lower the melting point, and so to help melt the ice.

Charles D. Winters

Mass %	m (mol kg^{-1})	ΔT_f (measured) (K)	ΔT_f (calculated) (K)	$\dfrac{\Delta T_f \text{ (measured)}}{\Delta T_f \text{ (calculated)}}$
NaCl				
0.00700	0.0120	0.0433	0.0223	1.94
0.500	0.0860	0.299	0.160	1.87
1.00	0.173	0.593	0.322	1.84
2.00	0.349	1.186	0.649	1.83
Na₂SO₄				
0.00700	0.00493	0.0257	0.00917	2.80
0.500	0.0354	0.165	0.0658	2.51
1.00	0.0711	0.320	0.132	2.42
2.00	0.144	0.606	0.268	2.26

TABLE 12.3 Freezing Point Depressions of Aqueous Solutions of NaCl and Na_2SO_4

We see that in the case of NaCl solute, experimental estimates of ΔT_f are less than twice the value calculated for a non-electrolyte solute. Similarly, measured values of ΔT_f of Na_2SO_4 solutions are less than three times that expected for a non-electrolyte solute. The ratio of the experimentally measured value of ΔT_f to the value calculated, assuming the solute to be a non-electrolyte, is called the *van't Hoff factor* (i).

$$i = \frac{\Delta T_f(\text{measured})}{\Delta T_f(\text{calculated})} = \frac{\Delta T_f(\text{measured})}{K_f \times m}$$

or

$$\Delta T_f = K_f \times m \times i$$

The numbers in the last column of Table 12.3 are values of the van't Hoff factor. They approach whole numbers (2, 3, and so on) only with very dilute solutions. In more concentrated solutions, the experimental freezing point depressions tell us that the solutions behave as if there are fewer ions in solution than expected. This behaviour, which is typical of all solutions of ionic compounds, is attributed to the attractions between ions in solution. The result is as if some of the positive and negative ions are paired, decreasing the total concentration of particles. Indeed, in more concentrated solutions, and especially in solvents less polar than water, ions are extensively associated in "ion pairs" and in even larger clusters. The consequence of ion pairing is that the concentration of particles is less than we might expect because an ion pair behaves, in relation to colligative properties, as though it were one particle. Consequently, the freezing point is less than we might otherwise expect. In such cases, the solute is said to have an *effective concentration*, or **activity** (a), different from its actual concentration. Its activity is the effective concentration deduced from the magnitude of physical properties, such as the freezing point depression. It is the value that makes the following equation satisfied:

$$\Delta T_f = K_f \times a$$

Estimation of the van't Hoff factor (i) was one of the first ways used to determine whether a solute was a strong electrolyte, weak electrolyte (and how weak), or a non-electrolyte.

All of the discussion above concerning freezing point depression of solutions of electrolytes applies correspondingly to boiling point elevation of solutions of electrolytes.

Think about It

e12.8 Use a graph of vapour pressure versus temperature to predict boiling points.

Think about It

e12.9 Use a graph of vapour pressure versus temperature to explain boiling point elevation.

Taking It Further

e12.10 Read about how solutes raise the boiling point of a solution.

WORKED EXAMPLE 12.8—FREEZING POINTS OF SOLUTIONS OF ELECTROLYTES

A 0.00200 mol kg^{-1} aqueous solution of an ionic compound, $Co(NH_3)_5(NO_2)Cl$, freezes at −0.00732 °C. How many mol of ions does 1.0 mol of the salt produce when it dissolves in water?

Strategy

First, calculate what ΔT_f would be if the solute were a non-electrolyte. Compare this value with the measured value of ΔT_f. The ratio reflects the number of ions produced.

Solution

Assuming that the solute is a non-electrolyte:

$$\Delta T_f = K_f \times m = (1.86 \text{ K kg mol}^{-1})(0.00200 \text{ mol kg}^{-1}) = 0.00372 \text{ K}$$

Bearing in mind that 0.00372 K = 0.00372 °C, we can now calculate the van't Hoff factor, i:

$$i = \frac{\Delta T_f(\text{measured})}{\Delta T_f(\text{calculated})} = \frac{0.00732 \text{ K}}{0.00372 \text{ K}} = 1.97$$

It appears that 1 mol of this compound produces 2 mol of ions when it dissolves.

Comment

In this case, the ions are $[Co(NH_3)_5(NO_2)]^+$ and Cl^-. The $[Co(NH_3)_5(NO_2)]^+$ ion is an example of a complex ion, with five NH_3 molecules and one NO_2^- ion acting as Lewis bases to bind to a Co^{2+} metal ion [<<Section 6.7, >>Section 27.4].

Interactive Exercise 12.15

Predict the freezing point of a solution.

EXERCISE 12.16—FREEZING POINTS OF SOLUTIONS OF ELECTROLYTES

Calculate the freezing point of 525 g of water that contains 25.0 g of NaCl dissolved in it. Assume that the van't Hoff factor (i) is 1.85.

In solutions of electrolytes

$$\Delta T_f = K_f \times m \times i$$

where $i = \frac{\Delta T_f(\text{measured})}{\Delta T_f(\text{calculated})}$ and $\Delta T_f(\text{calculated})$ is the freezing point depression predicted for a non-electrolyte solute. The van't Hoff factor i takes into account ionisation of the solute and that the activity (or effective concentration) of the solute is different from the labelled concentration.

Osmosis and Osmotic Pressure

Osmosis is *the movement of solvent molecules through a semipermeable membrane from a solution of lower solute concentration (or pure solvent) to a solution of higher solute concentration.* Transfer of water molecules by osmosis is extraordinarily important in biological systems, and in recent times "reverse osmosis" is used on a large scale for water purification. Solvent transfer by osmosis can be demonstrated with a simple experiment.

FIGURE 12.11 Osmosis. (a) The bag attached to the tube contains a solution that is 5% sugar and 95% water. The beaker contains pure water. The bag is made of a material that is semipermeable: it allows water molecules, but not sugar molecules, to pass through it. (b) Over time, water flows into the bag until the pressure exerted by the column of solution above the water level in the beaker is great enough to prevent further net flow into the bag. The pressure due to the height of the column is then the osmotic pressure (Π) of the solution in the bag.

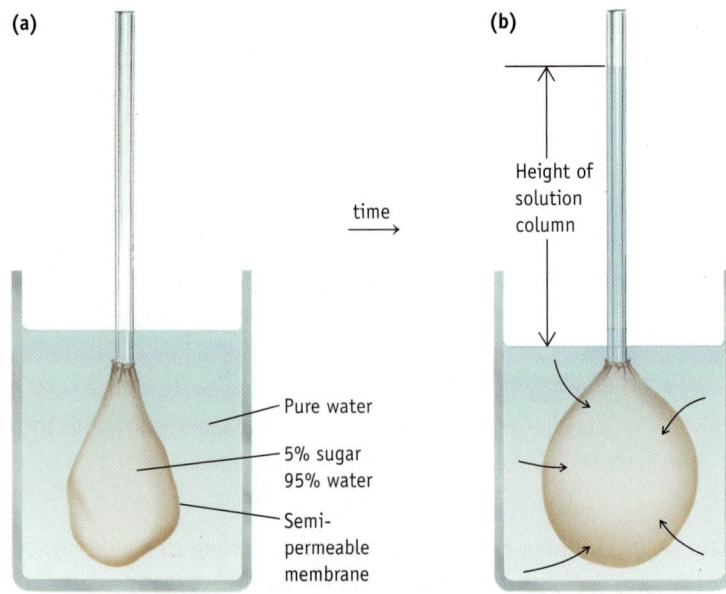

The beaker in Figure 12.11 contains pure water, and the bag and tube hold a sugar solution. The liquids are separated by a semipermeable membrane, a thin sheet of material (such as a vegetable tissue or cellophane) through which only certain types of molecules can pass. Here, water molecules can pass through the membrane but larger sugar molecules (or hydrated ions) cannot (Figure 12.12). When the experiment is begun, the liquid levels in the beaker and the tube are the same. Over time, however, the level of the sugar solution inside the tube rises, the level of pure water in the beaker falls, and the sugar solution becomes steadily more dilute. After a while, no further net change occurs; equilibrium is reached when the rate of solvent flow into the bag matches the rate of outward flow.

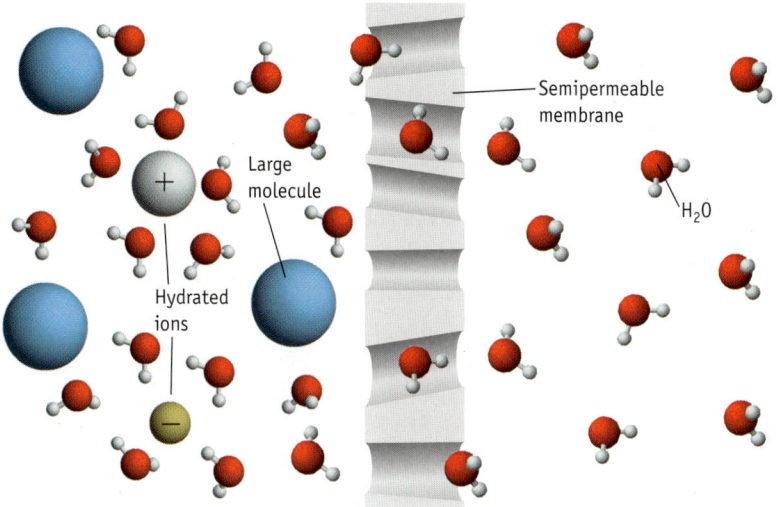

FIGURE 12.12 Osmosis at the molecular level. Osmotic flow through a membrane that is selectively permeable (semipermeable) to water. Dissolved substances such as aquated ions or large sugar molecules cannot diffuse through the membrane. The membrane, acting as a "molecular sieve," is represented here simply as a graphic, but it is actually a complex matrix of molecules.

Think about It

e12.11 Observe an animation of osmosis.

Osmosis brings about changes of the concentrations of the two solutions such that the more concentrated one becomes less concentrated, the less concentrated becomes more concentrated, and the concentrations of the two solutions come closer together. This is an example of a natural tendency in nature to even out differences.

From a molecular point of view, the semipermeable membrane allows the passage of water molecules, so they can move through the membrane in both directions. Initially, more water molecules pass through the membrane from the pure water side to the solution side than in the opposite direction, so there is net movement of water molecules from regions of low solute concentration to regions of high solute concentration.

Why does the system eventually reach equilibrium? Clearly, the solution in the bag in Figure 12.11 can never reach zero sugar concentration, when the concentrations inside and outside of the bag would be equal. However, the solution moves higher and higher in the tube as osmosis continues. Eventually the pressure exerted by this column of solution counterbalances the pressure exerted by the water moving through the membrane from the pure water side, and no further net movement of water occurs. The pressure created by the column of solution is called the **osmotic pressure (Π)** of the equilibrium system.

In fact, with the apparatus portrayed in Figure 12.11, the measured osmotic pressure would be that for the solution after some solvent has entered the bag and changed its concentration. To measure the osmotic pressure of a solution of known concentration, *a "back pressure" can be applied down the tube that is just sufficient to prevent osmosis occurring. This is the osmotic pressure of the solution.*

From experimental measurements on dilute solutions, it is found that osmotic pressure (Π) and molar concentration (c) are related by the equation

$$\Pi = cRT$$

where R is the gas constant, and T is the absolute temperature. Using a value for the gas constant of 8.314 L kPa mol^{-1} K^{-1} gives the osmotic pressure directly in kPa.

Alternatively, use of $R = 0.0821$ L atm K^{-1} mol^{-1} gives the osmotic pressure in atmospheres—in this case, 2.45 atm.

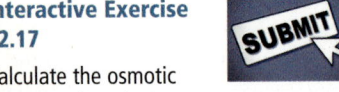

The osmotic pressure of a 0.100 mol L^{-1} solution of non-electrolyte at 25 °C is

$$\Pi = (0.100\ \text{mol L}^{-1})(8.314\ \text{L kPa K}^{-1}\ \text{mol}^{-1})(298\ K) = 248\ \text{kPa}$$

Because pressures of about 0.1 kPa (1×10^{-3} atm) are easily measured, concentrations as low as 1×10^{-4} mol L^{-1} can be accurately determined through measurements of osmotic pressure. This is an ideal method for measuring the molar masses of polymers and biologically important substances with very large molecules.

WORKED EXAMPLE 12.9—MOLAR MASS FROM OSMOTIC PRESSURE

β-carotene is the most important of the A vitamins [>>Section 19.1]. Its molar mass can be determined by measuring the osmotic pressure of a solution of known mass of β-carotene dissolved in chloroform. Calculate the molar mass of β-carotene if 10.0 mL of a solution containing 7.68 mg of β-carotene has an osmotic pressure of 3.54 kPa at 25.0 °C.

Strategy

Calculate the solution concentration from its osmotic pressure. Then, use the volume and concentration of the solution to calculate the amount of solute. Finally, find the molar mass of the solute from its mass and amount.

Solution

The osmotic pressure can be used to calculate the concentration of β-carotene:

$$\text{Molar concentration, } c = \frac{\Pi}{RT} = \frac{3.54\ \text{kPa}}{(8.314\ \text{L kPa K}^{-1}\ \text{mol}^{-1})(298\ K)} = 1.429 \times 10^{-3}\ \text{mol L}^{-1}$$

Now the amount of β-carotene dissolved in 10.0 mL (0.0100 L) of solvent can be calculated:

$$n(\beta\text{-carotene}) = (1.429 \times 10^{-3}\ \text{mol L}^{-1})(0.0100\ L) = 1.43 \times 10^{-5}\ \text{mol}$$

This amount of β-carotene (1.43×10^{-5} mol) is equivalent to 7.68 mg (7.68×10^{-3} g). This gives us a way to calculate the molar mass, using $n = m/M$:

$$\text{Molar mass, } M = \frac{m}{n} = \frac{7.68 \times 10^{-3}\ \text{g}}{1.43 \times 10^{-5}\ \text{mol}} = 538\ \text{g mol}^{-1}$$

Comment

β-carotene is a hydrocarbon with the formula $C_{40}H_{56}$. Does this formula correspond with the estimated molar mass?

EXERCISE 12.18—MOLAR MASS FROM OSMOTIC PRESSURE

A 1.40 g sample of polyethylene, a common plastic, is dissolved in enough benzene to give exactly 100 mL of solution. The measured osmotic pressure of the solution at 25 °C is 0.248 kPa. Calculate the average molar mass of the polymer.

Osmosis is of practical significance for people in the health professions. Patients who become dehydrated through illness often need to be given water and nutrients intravenously. Water cannot simply be dripped into a patient's vein, however. Rather, the intravenous solution must have the same overall solute concentration as the patient's blood: the solution must be iso-osmotic or *isotonic* with body cell fluid. If pure water were used, the inside of a blood cell would have a higher solute concentration, and water would flow into the cell. This *hypotonic* situation causes red blood cells to burst (*lyse*) (Figure 12.13). The opposite situation, *hypertonicity*, occurs if the intravenous solution is more concentrated than the contents of the blood cell. In this case the cell would lose water and shrivel up (*crenate*). The intravenous solution used for rehydration is a sterile 0.154 mol L^{-1} NaCl saline solution.

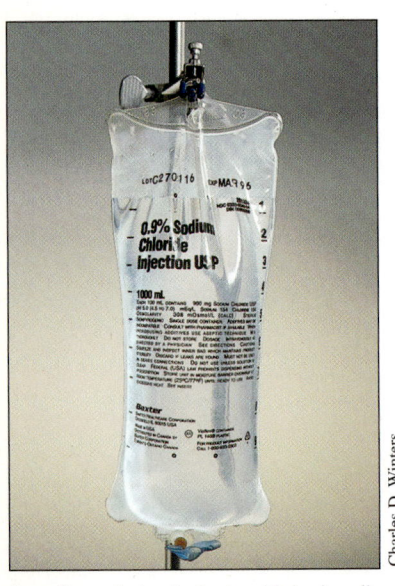

An saline solution isotonic with body cell fluid.

Charles D. Winters

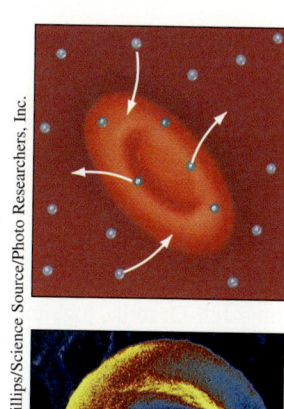
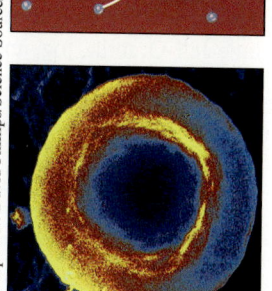
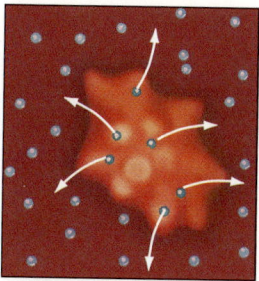
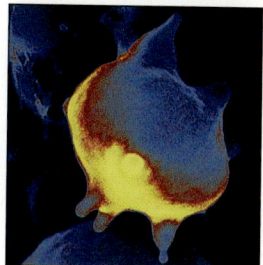
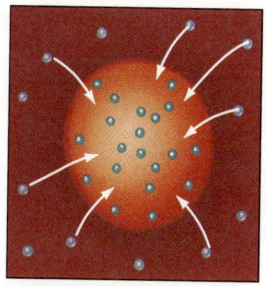
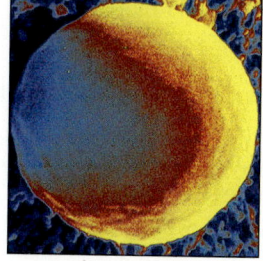

(a) Isotonic solution   **(b)** Hypertonic solution   **(c)** Hypotonic solution

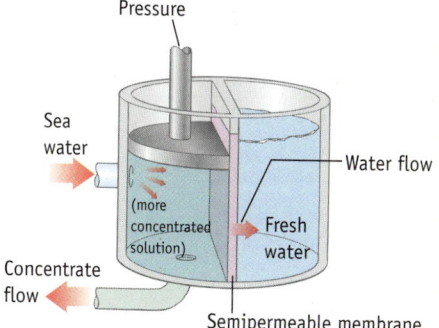

FIGURE 12.13 Osmosis and living cells. (a) A cell placed in an isotonic solution. The net movement of water into and out of the cell is zero because the concentration of solutes inside and outside the cell is the same. (b) In a hypertonic solution, the concentration of solutes outside the cell is greater than that inside. There is a net flow of water out of the cell, causing the cell to dehydrate, shrink, and perhaps die. (c) In a hypotonic solution, the concentration of solutes outside the cell is less than that inside. There is a net flow of water into the cell, causing the cell to swell and perhaps to burst (or lyse).

Finding sources of fresh water for human and agricultural use has been a challenge for centuries. Water has been the cause of numerous conflicts and a target in war. Although the earth has abundant water, 97% of it is too salty to drink or to irrigate crops. A large portion of the remaining 3% is in the form of ice in the polar regions.

One of the oldest ways to obtain fresh water from the oceans is evaporation. However, heating water requires large quantities of heat, and the salt left behind may not be suitable for consumption. A more modern way to extract fresh water from the oceans is desalination by *reverse osmosis*. A pressure is applied to sea water separated from fresh water by a semipermeable membrane (Figure 12.14). If the applied pressure is greater than the osmotic pressure of sea water, water flows in the non-spontaneous direction (from the sea water). In a continuous operation, concentrated sea water is returned to the ocean.

Osmosis is the movement of solvent molecules through a semipermeable membrane from a solution of lower solute concentration (or pure solvent) to a solution of higher solute concentration. The osmotic pressure (Π) of a solution is the pressure that must be applied against the spontaneous direction of osmosis to just prevent osmotic flow.

$$\Pi = cRT$$

FIGURE 12.14 Obtaining drinking water from sea water by reverse osmosis. The osmotic pressure of sea water is approximately 2.7 MPa. To obtain fresh water at a reasonable rate, a pressure of about 5 MPa is used. For comparison, the pressure in bicycle tires is usually about 0.3 MPa (300 kPa).

Osmotic Pressures of Solutions of Electrolytes

As is the case for freezing point depression of solutions of electrolytes, *osmotic pressures depend on the total activities of ions in solution, rather than the labelled solute concentration* (page 443):

$$\Pi = i \times cRT$$

where c is the solution labelled concentration and i is the van't Hoff factor.

EXERCISE 12.19—OSMOTIC PRESSURE OF ELECTROLYTE SOLUTIONS

Estimate the osmotic pressure of human blood at 37 °C. Assume blood is isotonic with a 0.154 mol L^{-1} NaCl solution, and assume the van't Hoff factor (i) is 1.9 for NaCl.

For solutions of electrolytes, osmotic pressures depend on the total activities of ions in solution, rather than the labelled solute concentration:

$$\Pi = i \times cRT$$

where c is the solution labelled concentration and i is the van't Hoff factor.

Gold colloidal dispersion. $AuCl_4^-(aq)$ ions are reduced to give colloidal gold metal—tiny particles of gold dispersed throughout the water.

12.7 Colloidal Dispersions

Earlier in this chapter [<<Section 12.2], a solution was broadly defined as a homogeneous mixture of two or more substances in a single phase. In a true solution (such as salt or sugar dissolved in water), no settling of the solute is observed and the solute particles are ions or relatively small molecules. You are also familiar with suspensions, which result, for example, if a handful of fine sand is added to water and shaken vigorously. Sand particles are still visible and gradually settle to the bottom. **Colloidal dispersions**, also called **colloids**, *are a state intermediate between a solution and a suspension*. Colloids include many of the foods you eat and the materials around you, and include Jell-O, milk, fog, and porcelain (Table 12.4).

TABLE 12.4 Types of Colloidal Dispersions

Type	Dispersing Medium	Dispersed Phase	Examples
Aerosol	Gas	Liquid	Fog, clouds, aerosol sprays
Aerosol	Gas	Solid	Smoke, airborne viruses, automobile exhaust
Foam	Liquid	Gas	Shaving cream, whipped cream
Foam	Solid	Gas	Styrofoam, marshmallow
Emulsion	Liquid	Liquid	Mayonnaise, milk, face cream
Gel	Solid	Liquid	Jelly, Jell-O, cheese, butter
Sol	Liquid	Solid	Gold in water, milk of magnesia, mud
Solid sol	Solid	Solid	Milk glass

Around 1860, the British chemist Thomas Graham (1805–1869) found that substances such as starch, gelatin, glue, and albumin from eggs diffused only very slowly when placed in water, compared with sugar or salt. In addition, these substances differ significantly in their ability to diffuse through a thin membrane: sugar molecules can diffuse through many membranes, but the very large molecules that make up starch, gelatin, glue, and albumin do not. Graham coined the word *colloid* (from the Greek, meaning "glue") to describe this class of substances that are distinctly different from true solutions and suspensions.

The particles in colloidal dispersions generally have high molar masses; this is true of proteins such as hemoglobin that have molar masses in the thousands. These particles are relatively large (say, 1000 nm diameter), and as a consequence they exhibit the *Tyndall effect*: they scatter visible light, making the dispersion appear cloudy (Figure 12.15). However, even though colloidal particles are relatively large compared with those in solutions, they are not so large that they settle out.

FIGURE 12.15 The Tyndall effect: colloidal dispersions scatter light. (a) Dust in the air scatters the light coming through the trees in a forest. (b) A narrow beam of light from a laser is passed through a NaCl solution (*left*) and then a colloidal mixture of gelatin and water (*right*).

(a)

(b)

Graham also gave us the words *sol* for a dispersion of a solid substance in a fluid medium, and *gel* for a dispersion that has a structure that prevents it from being mobile. Jell-O is a sol when the solid is first mixed with boiling water, but it becomes a gel when cooled. Other examples of gels (Figure 12.16) are the "gelatinous" precipitates of $Al(OH)_3$, $Fe(OH)_3$, and $Cu(OH)_2$.

Colloidal dispersions consist of finely divided particles that, as a result, have a very high surface area. For example, if you have one millionth of a mole of colloidal particles, each assumed to be a sphere with a diameter of 200 nm, the total surface area of the particles would be on the order of 200 000 000 cm^2, or the size of several football fields. It is not surprising, therefore, that many of the properties of colloids depend on the properties of surfaces.

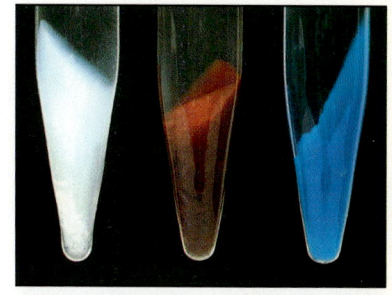

FIGURE 12.16 Gelatinous precipitates. (*Left*) $Al(OH)_3$; (*centre*) $Fe(OH)_3$; and (*right*) $Cu(OH)_2$.

EXERCISE 12.20—COLLOIDAL DISPERSIONS

A colloidal dispersion consists of spheres of diameter 100 nm in another medium.
(a) What is the volume ($V = \frac{4}{3}\pi r^3$) and surface area ($A = 4\pi r^2$) of each sphere?
(b) How many spheres are required to give a total volume of 1.0 mL? What is the total surface area of these spheres (in m^2)?

Colloidal dispersions are a state intermediate between a solution and a suspension. The particles in colloidal dispersions are larger than those of dissolved substances, but not so large that they settle out quickly. The dispersed particles have a high total surface area, and many of their properties are related to surface features.

Types of Colloids

Colloids are classified according to the state of the dispersed phase and the dispersing medium. Table 12.4 lists several types and gives examples of each.

Colloids with water as the dispersing medium can be classified as either *hydrophobic* (from the Greek, meaning "water-fearing") or *hydrophilic* ("water-loving"). A **hydrophobic colloid** *is one in which only weak attractive forces exist between the water and the surfaces of the colloidal particles.* Examples include dispersions of metals and of nearly insoluble salts in water. When compounds like AgCl precipitate, the result is often a colloidal dispersion. The precipitation reaction occurs too rapidly for ions to gather from long distances and make large crystals, so the ions aggregate to form small particles that remain suspended in the liquid.

The particles don't come together (coagulate) and form larger particles that settle out because they carry electric charges. An AgCl particle, for example, will adsorb onto its surface Ag^+ cations if they are present at substantial concentration in the surrounding solution. This is a result of attraction to Cl^- ions at the particle surface. So, the colloidal particles are positively charged, allowing them to attract a secondary layer of anions. The particles, now surrounded by layers of anions, repel one another and are prevented from coming together (Figure 12.17).

A stable hydrophobic colloid can be made to coagulate by introducing ions into the dispersing medium. Milk contains a colloidal suspension of hydrophobic clusters (called *micelles*) of casein protein molecules. When milk ferments, lactose (milk sugar) is converted to lactic acid, which ionizes to lactate ions and hydrogen ions. The charges on the surfaces of the colloidal particles are neutralized, and the milk coagulates; the milk solids come together in clumps called "curds." The same happens if acidic lemon juice is added to milk.

Soil particles are often carried by water in rivers and streams as hydrophobic colloids. When river water carrying large amounts of colloidal particles meets sea water with its high concentration of salts, the

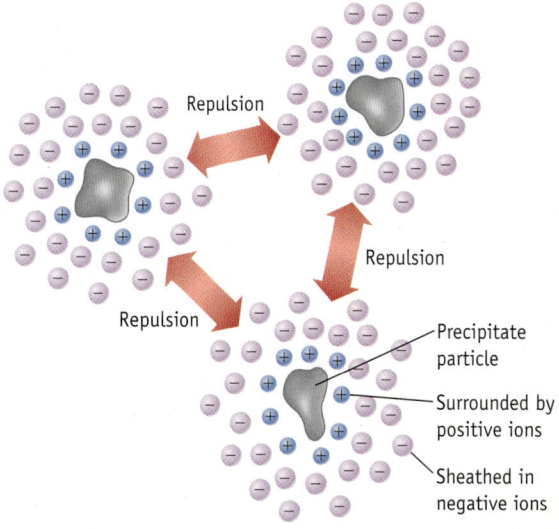

FIGURE 12.17 Hydrophobic colloids. A hydrophobic colloid is stabilized by positive ions absorbed onto each particle and a secondary layer of negative ions. Because the particles bear similar charges, they repel one another and precipitation is prevented.

FIGURE 12.18 Formation of silt. At the massive Ganges River delta in the Bay of Bengal, tributaries and distributaries of the Ganges and Brahmaputra Rivers deposit huge amounts of silt and clay. When hydrophobic colloidal particles come into contact with the salt water of the ocean, the charges on the particles are neutralized and they coagulate and settle out.

particles coagulate to form the silt seen at the mouth of the river (Figure 12.18). Municipal water treatment plants often add salts such as $Al_2(SO_4)_3$ to clarify water. In aqueous solution, aluminum ions exist as $[Al(H_2O)_6]^{3+}$ cations, which neutralize the charge on the hydrophobic colloidal soil particles, causing these particles to aggregate and settle out.

Hydrophilic colloids *are strongly attracted to water molecules.* They often have groups such as –OH and –NH₂ on their surfaces. These groups form strong hydrogen bonds to water, thereby stabilizing the colloid. Proteins and starch are important examples of hydrophilic colloids, and homogenized milk is the most common example.

Emulsions *are colloidal dispersions of one liquid in another,* such as oil or fat in water. Familiar examples include salad dressing, mayonnaise, and milk. If vegetable oil and vinegar are mixed to make a salad dressing, the mixture quickly separates into two layers because the non-polar oil molecules have relatively weak forces of attraction to the polar water and acetic acid (CH_3COOH) molecules. So why are milk and mayonnaise apparently homogeneous mixtures that do not separate into layers? The answer is that they contain an *emulsifying agent* such as soap or a protein. Lecithin is a phospholipid found in egg yolks, so mixing egg yolks with oil and vinegar stabilizes the colloidal dispersion known as mayonnaise. To understand this process further, let us look into the functioning of soaps and detergents, substances known as surfactants.

> Colloidal particles suspended in water may be either hydrophobic or hydrophilic. Colloidal particles of salts may adsorb ions from solution and repulsion between surface charges prevents coagulation. Charge neutralization occurs in the presence of high concentrations of ions, or of highly charged ions. Emulsions may be stable because of surfactants in solution.

Surfactants

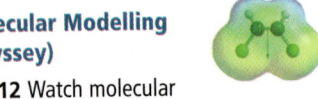

Hydrocarbon tail soluble in oil   Polar head soluble in water

Sodium stearate, a soap

Molecular Modelling (Odyssey)

e12.12 Watch molecular simulations of a soap molecule in a "cage" of water molecules, and the interface between 1-octanol and water.

Soaps and detergents are emulsifying agents. Soap is made by heating a fat with sodium or potassium hydroxide, which produces the anion of a fatty acid.

The fatty acid anion has a split personality: It has a non-polar, hydrophobic hydrocarbon tail that is soluble in other similar hydrocarbons and a polar, hydrophilic head that is soluble in water.

Oil cannot be readily washed away from dishes or clothing with water because oil is non-polar and insoluble in water. However, if we add soap, the polar heads of the soap anions interact with surrounding water molecules, leaving a cluster of non-polar hydrocarbon tails into which the oil particles can "dissolve" (Figure 12.19). If the oily material on a piece of clothing or a dish also contains some dirt particles, that dirt can now be washed away.

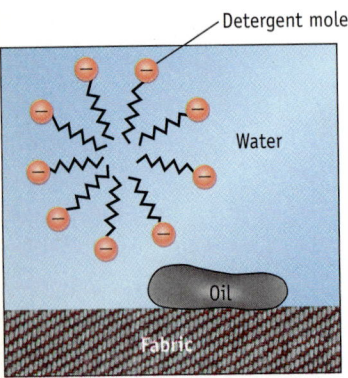
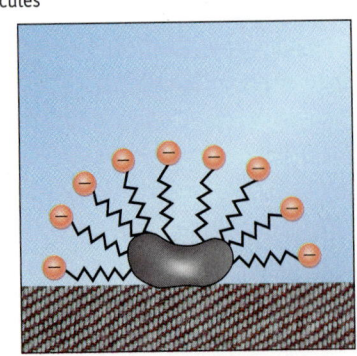
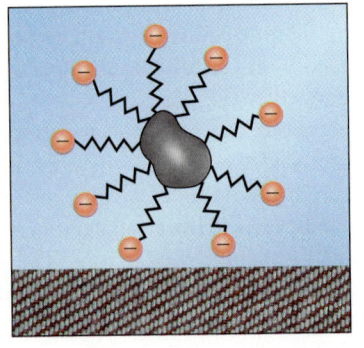
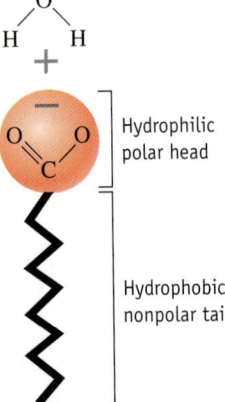

FIGURE 12.19 The cleaning action of soap. Soap molecules interact with water through the charged, hydrophilic end of the molecule. The long, hydrocarbon end of the molecule can bind through dispersion forces with hydrocarbons and other non-polar substances in grease and oil.

Substances such as soaps that affect the properties of surfaces, and therefore affect the interaction between two phases, are called surface-active agents, or **surfactants**. *A surfactant used for cleaning* is called a **detergent**.

Sodium laurylbenzenesulfonate

In general, synthetic detergents have the sulfonate group ($-SO_3^-$) as the polar head rather the carboxylate group ($-COO^-$). The carboxylate anions form an insoluble precipitate with $Ca^{2+}(aq)$ or $Mg^{2+}(aq)$ ions in water. Because many water sources have high concentrations of these ions, using soaps containing carboxylates produces bathtub rings and tattle-tale grey clothing. The synthetic sulfonate detergents have the advantage that they do not form such precipitates because their calcium salts are more soluble in water.

Think about It

e12.13 See everyday applications of these ideas.

> Soaps and detergents have molecules with a water-soluble polar "head" and a water-insoluble, oil-soluble, non-polar "tail." This structure affects the way that they interact with water, oily substances, and each other.

SUMMARY

Key Concepts

- The **solubility** of a substance, in a particular solvent at a specified temperature, is the concentration of the substance in a saturated solution. A **saturated solution** is in a condition of dynamic equilibrium between dissolved and solid solute, and a **supersaturated** solution is an unstable condition in which the concentration of solute is higher than that of a saturated solution. (Section 12.2)
- As well as molarity, concentration of solution may be expressed as **molality, mole fraction, mass percent** or **parts-per-million**. (Section 12.5)
- The **molar enthalpy change of solution** ($\Delta_{sol}H$) is the ΔH of dissolution of 1 mol of solute in a large volume of the solvent. For an ionic solute, this is the net result of energy expended as the lattice energy, and energy gained as enthalpy of aquation. (Section 12.3)
- **Henry's law**: At a given temperature, the solubility of a gas in a liquid is directly proportional to the partial pressure of the gas in contact with the solution. (Section 12.4)
- The solubilities of gases in water decrease as the temperature is increased. (Section 12.4)
- The solubilities of many salts in water increase as the temperature is raised, but there are exceptions. (Section 12.4)
- The magnitudes of the colligative properties (vapour pressure depression, boiling point elevation, freezing point depression, and osmotic pressure) depend only on the concentration of solute particles and are independent of the identity of the particles in solution. (Section 12.6)
- **Ideal solutions** are those in which the vapour pressure of the solvent is proportional to the mole fraction of solvent at a specified temperature (**Raoult's law**). (Section 12.6)
- Depression of freezing point of a solvent (ΔT_f) is directly proportional to the molality of the solute. (Section 12.6)
- When the solute is an electrolyte, the van't Hoff factor (i) takes into account (a) ionization of the solute, and (b) that the **activity** (or effective concentration) of the solute is different from the labelled concentration. (Section 12.6)
- **Osmosis** is the movement of solvent molecules through a semipermeable membrane from a solution of lower solute concentration (or pure solvent) to a solution of higher solute concentration. The **osmotic pressure** (Π) of a solution is the pressure that must

be applied against the spontaneous direction of osmosis to just prevent osmotic flow. (Section 12.6)

- The particles in **colloidal dispersions** are larger than those of dissolved substances, but not so large that they settle out quickly. The dispersed particles have a high total surface area, and many of their properties are related to surface features. (Section 12.7)
- Colloidal particles suspended in water may be either **hydrophobic** or **hydrophilic**. Colloidal particles of salts may adsorb ions from solution and repulsion between surface charges prevents coagulation. (Section 12.7)
- Soaps and **detergents** have molecules with a water-soluble polar "head" and a water-insoluble, oil-soluble, non-polar "tail." This structure affects the way that they interact with water, oily substances, and each other. (Section 12.7)

Key Equations

$\Delta_{sol}H = \Delta_{lattice}H + \Delta_{aq}H$	(Section 12.3)
$s = k_H \times p$	(Section 12.4)
Molality, $m = \dfrac{\text{amount of solute (mol)}}{\text{mass of solvent (kg)}}$	(Section 12.5)
Mole fraction of A, $x_A = \dfrac{n_A}{n_A + n_B + n_C + \cdots}$	(Section 12.5)
Mass % A $= \dfrac{m(A)}{m(A) + m(B) + m(C) + \cdots} \times 100\%$	(Section 12.5)
$p_{solvent} = x_{solvent} \times p°_{solvent}$	(Section 12.6)
$\Delta p_{solvent} = x_{solute} \times p°_{solvent}$	(Section 12.6)
For a non-electrolyte solute: $\Delta T_f = K_f \times m_{solute}$	(Section 12.6)
In solutions of electrolytes: $\Delta T_f = K_f \times m \times i$	(Section 12.6)
$i = \dfrac{\Delta T_f(\text{measured})}{\Delta T_f(\text{calculated for non-electrolyte solute})}$	(Section 12.6)
For a non-electrolyte solute: $\Pi = cRT$	(Section 12.6)
In solutions of electrolytes: $\Pi = i \times cRT$	(Section 12.6)

REVIEW QUESTIONS

Section 12.4: Factors Affecting Solubility: Pressure and Temperature

12.21 The Henry's law constant for O_2 in water at 25 °C is 1.24×10^{-5} mol L^{-1} kPa^{-1}. Which of the following is a reasonable prediction of the constant at 50 °C? Explain the reason for your choice.
(a) 6.20×10^{-6} mol L^{-1} kPa^{-1}
(b) 2.48×10^{-5} mol L^{-1} kPa^{-1}
(c) 1.24×10^{-5} mol L^{-1} kPa^{-1}
(d) 6.20×10^{-5} mol L^{-1} kPa^{-1}

12.22 An unopened soda can has an aqueous CO_2 concentration of 0.0506 mol L^{-1} at 25 °C. What is the pressure of CO_2 gas in the can?

12.23 You make a saturated NaCl solution at 25 °C. No solid is present in the beaker holding the solution. What can

be done to increase the amount of dissolved NaCl in this solution? (See Figure 12.7.)
(a) Add more solid NaCl.
(b) Raise the temperature of the solution.
(c) Raise the temperature of the solution and add some NaCl.
(d) Lower the temperature of the solution and add some NaCl.

Section 12.5: More Units of Solute Concentration

12.24 Sea water has a sodium ion concentration of 1.08×10^4 ppm. If the sodium is present in the form of dissolved sodium chloride, what mass of NaCl is in each litre of sea water? Sea water is denser than pure water because of dissolved salts. Its density is 1.05 g mL^{-1}.

12.25 Suppose you dissolve 2.56 g of succinic acid, $C_2H_4(CO_2H)_2(s)$, in 500.0 mL of water. Assuming that the density of water is 1.00 g mL^{-1}, calculate the molality, mole fraction, and mass percentage of acid in the solution.

12.26 Fill in the blanks in the table. Aqueous solutions are assumed.

Compound	Molality	Mass Percent	Mole Fraction
KNO_3	——	10.0	——
CH_3CO_2H	0.0183	——	——
$HOCH_2CH_2OH$	——	18.0	——

12.27 What mass of Na_2CO_3 must you add to 125 g of water to prepare 0.200 mol kg^{-1} Na_2CO_3 solution? What is the mole fraction of Na_2CO_3 in the solution?

Interactive Exercise 12.28

Practise calculations involving molality and mole fraction.

12.29 You wish to make an aqueous solution of glycerol, $C_3H_5(OH)_3$, in which the mole fraction of the solute is 0.093. What mass of glycerol must you add to 425 g of water to make this solution? What is the molality of the solution?

Interactive Exercise 12.30

Practise calculations comparing molality and molarity.

12.31 Hydrochloric acid is sold as a concentrated aqueous solution. If the molar concentration is 12.0 mol L^{-1} and its density is 1.18 g mL^{-1}, calculate (a) the molality of the solution, and (b) the mass percent of HCl in the solution.

12.32 The average lithium ion concentration in sea water is 0.18 ppm. What is the molal concentration of Li$^+$ ions in sea water?

Interactive Exercise 12.33

Practise calculations comparing molality and ppm.

Section 12.6: Colligative Properties

12.34 Pure ethylene glycol ($HOCH_2CH_2OH$) is added to 2.00 kg of water in the cooling system of a car. The vapour pressure of the water in the system when the

temperature is 90 °C is 60.93 kPa. What mass of glycol was added? (Assume the solution is ideal. At 90 °C, the vapour pressure of pure water is 70.10 kPa.)

12.35 Which has the higher equilibrium vapour pressure of water in contact with it: 0.30 mol kg^{-1} NH_4NO_3 solution or 0.15 mol kg^{-1} Na_2SO_4 solution?

12.36 105 g of iodine, $I_2(s)$, is dissolved in 325 g of carbon tetrachloride, $CCl_4(\ell)$, at 65 °C. Given that the vapour pressure of CCl_4 at this temperature is 70.80 kPa, what is the vapour pressure of the solution at 65 °C? (Assume that iodine does not contribute to the vapour pressure.)

12.37 A mixture of ethanol, $C_2H_5OH(\ell)$, and water has a freezing point of −16.0 °C.
(a) What is the molal concentration of the alcohol?
(b) What is the mass percent of alcohol in the solution?

12.38 You dissolve 15.0 g of sucrose ($C_{12}H_{22}O_{11}$) in a cup of water (225 g). What is the freezing point of the solution?

Interactive Exercise 12.39

Calculate the freezing point and boiling point of an aqueous solution.

12.40 The melting point of pure biphenyl ($C_{12}H_{10}$) is 70.03 °C. If 0.100 g of naphthalene is added to 10.0 g of biphenyl, the freezing point of the solution is 69.40 °C. K_f for biphenyl as the solvent is 8.00 K kg mol^{-1}. What is the molar mass of naphthalene?

12.41 Arrange the following aqueous solutions in order of decreasing freezing point. (The first is a non-electrolyte, the last three are strong electrolytes.)
(a) 0.20 mol kg^{-1} ethylene glycol solution
(b) 0.12 mol kg^{-1} K_2SO_4 solution
(c) 0.10 mol kg^{-1} $MgCl_2$ solution
(d) 0.12 mol kg^{-1} KBr solution

Interactive Exercise 12.42

Rank solutions in order of increasing melting point.

12.43 To make homemade ice cream, you cool the milk and cream by immersing the container in ice and a concentrated solution of rock salt (NaCl) in water. If you want to have a water-salt solution that freezes at −10.0 °C, what mass of NaCl must you add to 3.0 kg of water? (Assume the van't Hoff factor (i) for NaCl is 1.85.)

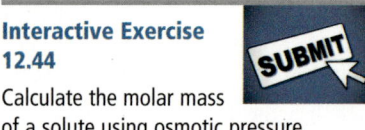

12.45 Calculate the osmotic pressure of a 0.0120 mol L^{-1} aqueous NaCl solution at 0 °C. Assume the van't Hoff factor (i) is 1.94 for this solution.

SUMMARY AND CONCEPTUAL QUESTIONS

12.46 If you dissolve equal amounts of NaCl and $CaCl_2$ in water, the dissolved $CaCl_2$ lowers the freezing point of the water almost 1.5 times as much as the dissolved NaCl. Why?

With the **eCHACR** single-sign-on access card bundled with your text, log on (http://login.cengage.com/sso) and access the e-book and click on any in-margin icons for dynamic molecular-level animations and simulations, videos of laboratory reactions, interactive exercises, reviews of background concepts, and other online supplementary materials as noted by the icons in the text margins.

Also go to www.chemistry.nelson.com <http://www.chemistry.nelson.com> for Answers to in-chapter exercises and selected Review Questions, Test Yourself questions, weblinks, crossword puzzles, flashcards, glossary of key terms, and other student resources.

Dynamic Chemical Equilibrium

Arthur C. Smith III, from Grant Heilman Photography

13.1 Case Study: Air into Bread

The term *equilibrium* has several meanings in everyday life, all of which have some association with the idea of balance. We talk of mental equilibrium, social equilibrium, and economic equilibrium. We might say that someone who stumbled, but did not fall over, regained their equilibrium. *Chemical equilibrium* is a condition existing in a reaction mixture in which there is no change in the amounts of substances, but—in contrast with the other meanings—is a dynamic situation: opposite reactions are taking place simultaneously and the same rate. Consequently, it is frequently referred to as *dynamic chemical equilibrium*.

Knowledge about chemical equilibrium has had profound significance in the control of a chemical reaction that has been claimed to be largely responsible for the survival of 40% of the world's population: the reaction of nitrogen, $N_2(g)$, with hydrogen, $H_2(g)$, to form ammonia. The ability to carry out this reaction in commercial quantities was sought by chemists during the 19th century because ammonia and some of its compounds are important agricultural fertilizers. This reaction allows transformation of "air into bread" in the sense that the main component of air is converted into ammonia and its compounds, which are needed to improve the yields of wheat and other grains—which, in turn, are used to make flour for bread.

Famous French chemist Henry Louis Le Chatelier studied this reaction, but he abandoned his research after the explosion of his experimental apparatus in 1901. Ultimate success came to German chemist Fritz Haber, who was awarded the 1918 Nobel Prize in

Chemistry for solving "the synthesis of ammonia from its elements," referred to as "the holy grail of synthetic inorganic chemistry."

Canadian scientist and writer Vaclav Smil has claimed that the Haber-Bosch process for making ammonia via this equilibrium reaction is the most important technical invention of the 20th century. Smil presents data showing that half of the nitrogen atoms in your body have at some time been involved in the Haber-Bosch reaction. Feeding the rapidly growing world population in the 20th century could not have been achieved without the extensive use of synthetic agricultural fertilizers to provide the nitrogen needed for crops. Smil claims that without having found a way to exploit this reaction, 40% of the current world population would not be alive.

The most important technical invention of the 20th century? Canadian scientist Vaclav Smil claims 40% of the world population would not be alive without the Haber-Bosch process for making ammonia.

It took a hundred years after the discovery of nitrogen as an element late in the 18th century to understand its importance in plant and animal growth. Justus von Liebig (1803–1873) wrote a book describing the critical importance of agriculture in the production of "digestible nitrogen." He showed that plant growth requires nitrogen atoms, among other things. We now know that molecules of amino acids, proteins, hemoglobin, and the nucleotides in DNA and RNA all contain N atoms, and that the human body requires about 2 kg of N atoms per year.

The Haber-Bosch synthesis of ammonia was spurred on by William Crookes, who warned in a now-famous address to the British Association of Science in September 1898, "All civilized nations stand in deadly peril of not having enough to eat," and "It is the chemist who must come to the rescue. . . . It is through the laboratory that starvation may ultimately be turned into plenty." Most agricultural reactive nitrogen in the world at this point came from two sources: guano (deposits of solidified bird excrements) from arid tropical islands, where low rainfall meant that water-soluble nitrogen compounds did not dissolve and leach out; and sodium nitrate from South America. But these were finite resources.

A "rescue" by chemists was needed because of the problem of speciation. Nitrogen atoms are plentiful in nature—78% of the mass of the earth's atmosphere is the stable molecular species dinitrogen (N_2), which plants cannot use. Although there are millions of N-containing compounds, dinitrogen is unreactive because of its strong $N\equiv N$ triple bond. To be available to organisms, N_2 molecules must be broken apart in chemical reactions to produce different chemical species with $N-C$, $N-O$, or $N-H$ bonds, referred to as *reactive nitrogen*. Processes that convert N in atmospheric N_2 into reactive nitrogen compounds that are bioavailable are called *fixation*, and *fixed nitrogen* is another name for reactive nitrogen.

Of course, food was produced before the advent of synthetic fertilizers, through the agency of micro-organisms and algae that have the complex molecular machinery needed to fix nitrogen. Lightning bolts use a brute force approach to provide the energy needed to split N_2 and produce nitrogen oxides. But at some point in the mid-20th century, the dietary requirements of an exploding human population outpaced the capacity of natural nitrogen fixation. The need for reactive nitrogen will increase further as the earth's population grows by another three or more billion people in the next 50 years.

Enter chemist Fritz Haber and his collaborator Carl Bosch, a metallurgist familiar with carrying out chemical reactions at high temperatures and pressures in the steel industry. Add the catalyst of wounded pride from Haber's scientific rivalry with physical chemist Walter Nernst, and a German political system that was far from social equilibrium just prior to World War I. Together, these formed the recipe for solving the large-scale synthesis of ammonia (Figure 13.1). As a result, in the past 20 years, human activity controlling chemical reactivity has now produced half of the synthetic nitrogen fertilizer ever used on the earth.

Oesper Collection in the History of Chemistry, University of Cincinnati

Archiv der Max-Planck-Gesellschaft, Berlin-Dahlem

FIGURE 13.1 Fritz Haber and his ammonia synthesis apparatus.

The balanced chemical equation for the synthesis reaction shows that 2 mol of ammonia gas are produced from a total of 4 mol of gaseous reactants. So, as we see in Section 13.6, we expect an increase in pressure to increase the yield of ammonia. And since a large amount of energy is needed to break the $N\equiv N$ bond, we can see why Haber focused on high temperature and pressure as the key to producing ammonia. Finding the right catalyst to increase the rate of reaction was even a bigger challenge. Many of us might have been discouraged by his first attempts. He showed that ammonia could be formed by using a finely divided iron catalyst heated to 1000 °C. The yield was a not-very-stunning 0.005–0.0125%!

Walter Nernst publicly ridiculed Haber's published values for the equilibrium yield of NH_3 in this reaction, severely wounding Haber's pride. To restore his reputation, Haber returned to the synthesis he had abandoned. And the rest is history.

Learning to control the yield and rate of this equilibrium reaction is of more than passing historical interest. Besides the benefits to agriculture, the result of this massive effort by chemists to remove the limits on nitrogen supply for food has some serious downsides. No chemical reaction proceeds in 100% yield. In the complex sequence of reactions and processes needed to convert N atoms from ammonia into N-containing species that can be taken up by humans, "wastage" occurs at every step. It is estimated that of every 100 N atoms produced by the Haber-Bosch process, only 14 end up in the mouths of vegetarians and 4 in the mouths of meat-eaters. The remainder are lost to the environment during production, storage, transport, application, uptake by and harvesting of crops, uptake by livestock, and the formulation of plant and animal food products (Figure 13.2).

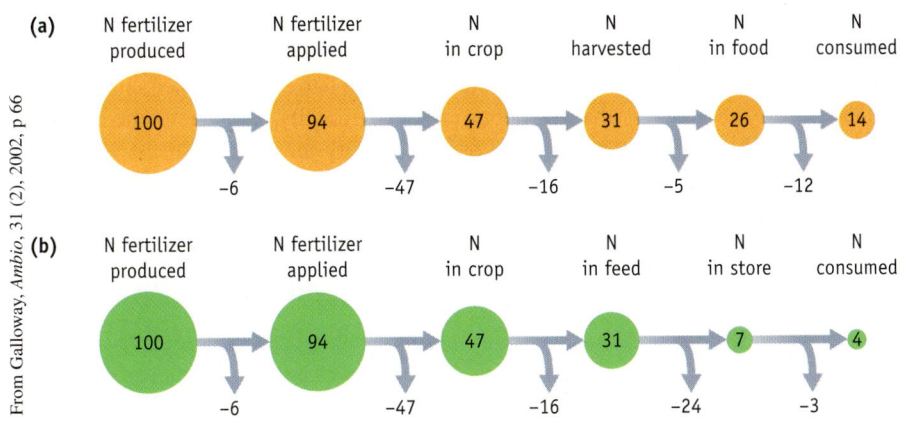

From Galloway, *Ambio*, 31 (2), 2002, p 66

FIGURE 13.2 Loss of N atoms from fixed nitrogen compounds. The fate of every 100 nitrogen atoms produced by the Haber-Bosch process from the factory to the mouth for (a) vegetarian diet and (b) carnivorous diet.

In Chapter 4, we focused on the important carbon cycle and the global changes in the earth's radiation balance that have resulted from substantial increases in the concentration of carbon dioxide in the atmosphere. But the cycles of transformation of an element into other species do not occur in isolation from cycles involving other elements. Nitrogen availability is essential for both terrestrial and marine plant growth, and plants remove carbon dioxide from the atmosphere. Nitrous oxide (N_2O) is a potent "greenhouse gas." During the period since the Industrial Revolution, as atmospheric concentrations of CO_2 have increased by 30%, the concentration of reactive nitrogen species has gone up by 300%. Human activity is now responsible for about half of the entire global cycle of nitrogen.

Other human activities besides ammonia synthesis produce large amounts of reactive nitrogen. The effects on health and the environment of nitrogen oxides (NO, NO_2, HNO_3, and NO_3^-) in the troposphere, produced during combustion of fossil fuel, are well known. Nitrous oxide (N_2O), which has a very long atmospheric lifetime, is released from complex reactions in soils [<<Section 4.3] and plays roles in both tropospheric warming and stratospheric ozone depletion. In some parts of our world, the production of food is severely nitrogen-limited; hunger and malnutrition result. Other places experience severe environmental degradation from excess reactive nitrogen produced in fossil fuel combustion and from nitrate run-off into ground and surface water as a result of intensive agriculture.

An integrated approach to understanding the beneficial and harmful effects of this escalation of N-containing species is just beginning, and chemistry has a central role to play. A series of international conferences has been established to bring together an interdisciplinary community to understand reactive nitrogen science.

You might think that the chemistry community would have quickly recognized Fritz Haber for his earth-shaking 1909 discovery of a way to commercially make ammonia from its elements. But the fragile political equilibrium in the world in the following decade was also earth-shaking, and here too Haber played a catalytic and controversial role. The successful scale-up of the reaction came just as Germany entered World War I, and reactive nitrogen immediately became important for the production of nitrates and TNT as explosives. Just as reactive nitrogen limits plant growth, it also became the limiting factor in the production of munitions for warfare.

With fears of naval blockades restricting access to nitrogen, Fritz Haber's professional work as a chemist was soon diverted to making nitric acid and other forms of reactive nitrogen for munitions. This was soon followed by his guiding the production of "chemical weapons"—poisonous gases such as chlorine, phosgene, and mustard gas—even though their use was forbidden by the Hague Conventions of 1899 and 1907. Haber returned to Berlin at the end of April 1915, after directing the first gas attack (chlorine) in military history in the trenches at Ypres, with 15 000 casualties. A few days later, his wife, Clara Immerwahr, a chemist and the first woman PhD at the University of Breslau, committed suicide. No autopsy report, letters, or newspaper announcements exist, but there is evidence that she was deeply troubled by his involvement in the production of poison gases.

Despite 1.3 million casualties from chemical weapons used by both sides during the war, Haber never expressed regrets about his role as head of Germany's Chemical Warfare Service. Due to this legacy, the scientific profession expressed such great ambivalence about the awarding of the Nobel Prize to Haber that the award ceremony for the 1918 prize was postponed until 1920. Following the war, Haber—who was part Jewish—was forced to resign his high profile position at the Kaiser-Wilhelm Institute and sent into exile by Hitler's rise to power.

Haber's story reminds us that for better and, occasionally, for worse, chemistry is a human activity. Numerous connections to human activity are evident in the rest of this chapter as we examine the factors that govern various dynamic chemical equilibria that are important to modern life and our environment.

13.2 Reaction Mixtures in Dynamic Chemical Equilibrium

The concept of dynamic chemical equilibrium is fundamental in chemistry. We will explore the consequences of the fact that in a closed system many reaction mixtures come to a state in which chemical reactions proceed simultaneously in opposite directions at the

Read about the challenges posed by the human activity of "fixing" too much nitrogen from the atmosphere in J. Rockström et al., (2009), "A Safe Operating Space for Humanity," *Nature*, 461 (Sept 24): 472–75. For a discussion of nitrogen and food production, see V. Smil, (2002), "Nitrogen and Food Production: Proteins for Human Diet," *Ambio*, 31(2), 126–31.

same rate, and that changing the concentrations or temperature can affect the relative amounts of reactants and products at equilibrium. Our goal will be to describe the chemical equilibrium condition for a variety of chemical reactions in quantitative terms.

Reversible Reactions

In some limestone caves, you can see remarkable stalactites and stalagmites, made mainly of calcium carbonate (Figure 13.3).

Photograph by Arthur N. Palmer, Professor Emeritus, State University of New York

FIGURE 13.3 Cave chemistry. Calcium carbonate stalactites cling to the roof of a cave, and stalagmites grow up from the cave floor. The chemistry producing these formations is a good example of the reversibility of chemical reactions.

The formation of stalactites and stalagmites depends on the **reversibility** of a chemical reaction—that is, *its ability to proceed in either direction, depending on the conditions*. Calcium carbonate exists in underground deposits in the form of limestone, a leftover from ancient oceans. If water seeping through the limestone contains a relatively high concentration of dissolved CO_2 (picked up when exposed to air), a reaction occurs that dissolves limestone to form an aqueous solution containing aquated Ca^{2+} and HCO_3^- ions.

$$CaCO_3(s) + CO_2(g) + H_2O(\ell) \xrightarrow{\text{reaction 1}} Ca^{2+}(aq) + 2\,HCO_3^-(aq)$$

When the mineral-laden water reaches a cave in which the air has only a very low concentration of CO_2, the opposite reaction occurs: CO_2 is released and solid $CaCO_3$ is deposited from solution.

$$Ca^{2+}(aq) + 2\,HCO_3^-(aq) \xrightarrow{\text{reaction 2}} CaCO_3(s) + CO_2(g) + H_2O(\ell)$$

Alternately dissolving and depositing $CaCO_3$ can be illustrated by a laboratory experiment with soluble salts (such as $CaCl_2$ and $NaHCO_3$) containing the $Ca^{2+}(aq)$ and $HCO_3^-(aq)$ ions. If you dissolve both of these salts in water, you will see bubbles of CO_2 gas and a precipitate of solid $CaCO_3$ forms (Figure 13.4) by reaction 2. If you then bubble CO_2 into the solution, the solid $CaCO_3$ dissolves by reaction 1. This experiment illustrates that many chemical reactions are reversible. Although all chemical reactions are in principle reversible, in practical terms, some reactions, such as cooking an egg, cannot be reversed.

What happens if the initial solution containing $Ca^{2+}(aq)$ and $HCO_3^-(aq)$ ions is in a closed container? As reaction 2 proceeds, the concentrations of the $Ca^{2+}(aq)$ and HCO_3^- (aq) ions become lower. Since the rates of reactions depend on the concentrations of reactants [>>Section 18.4], this reaction becomes slower and slower. At the same time, the concentrations of the products $CaCO_3(s)$ and $CO_2(g)$ increase, and they react (reaction 1) to re-form $Ca^{2+}(aq)$ and $HCO_3^-(aq)$ ions at an increasing rate. Eventually, the rate of formation of $CaCO_3$ by reaction 2 and the rate of dissolving $CaCO_3$ by reaction 1 become equal, and no further macroscopic change is observed: the amount of solid $CaCO_3$ and the concentrations of $Ca^{2+}(aq)$ and $HCO_3^-(aq)$ ions remain constant. This is not because nothing is happening, but because reaction is happening in both directions at the same rate.

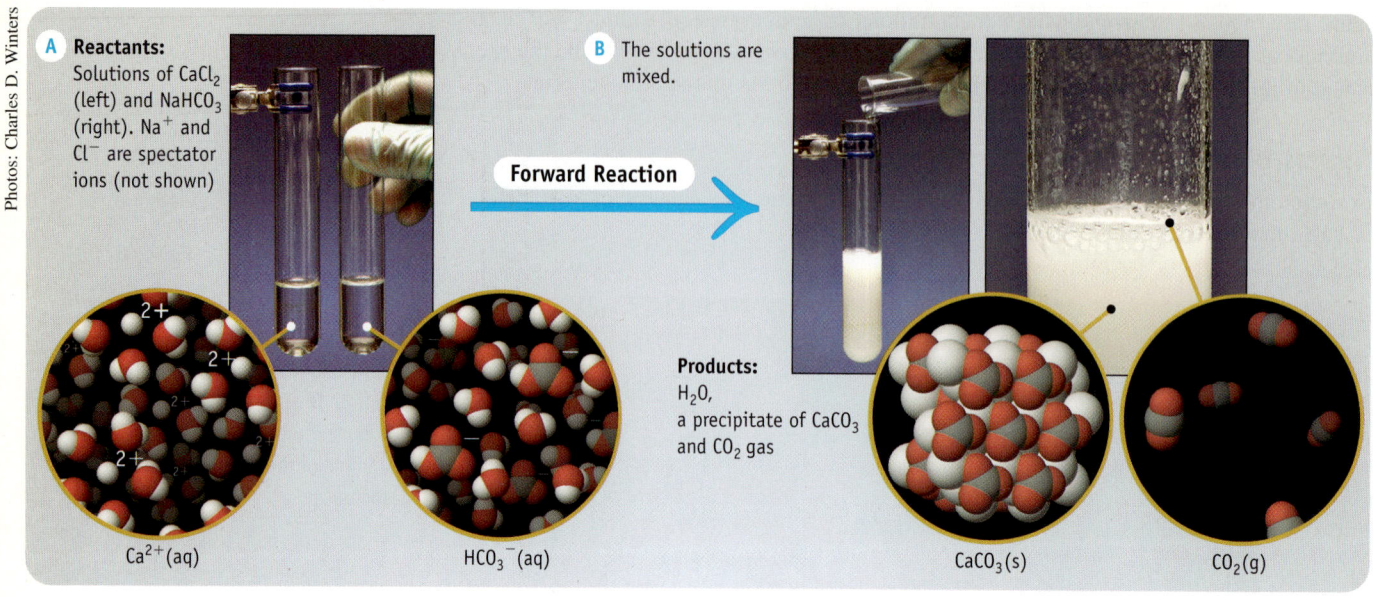

$$Ca^{2+}(aq) + 2\ HCO_3^-(aq) \longrightarrow CaCO_3(s) + CO_2(g) + H_2O(\ell)$$

The above reaction can be reversed:

$$Ca^{2+}(aq) + 2\ HCO_3^-(aq) \longleftarrow CaCO_3(s) + CO_2(g) + H_2O(\ell)$$

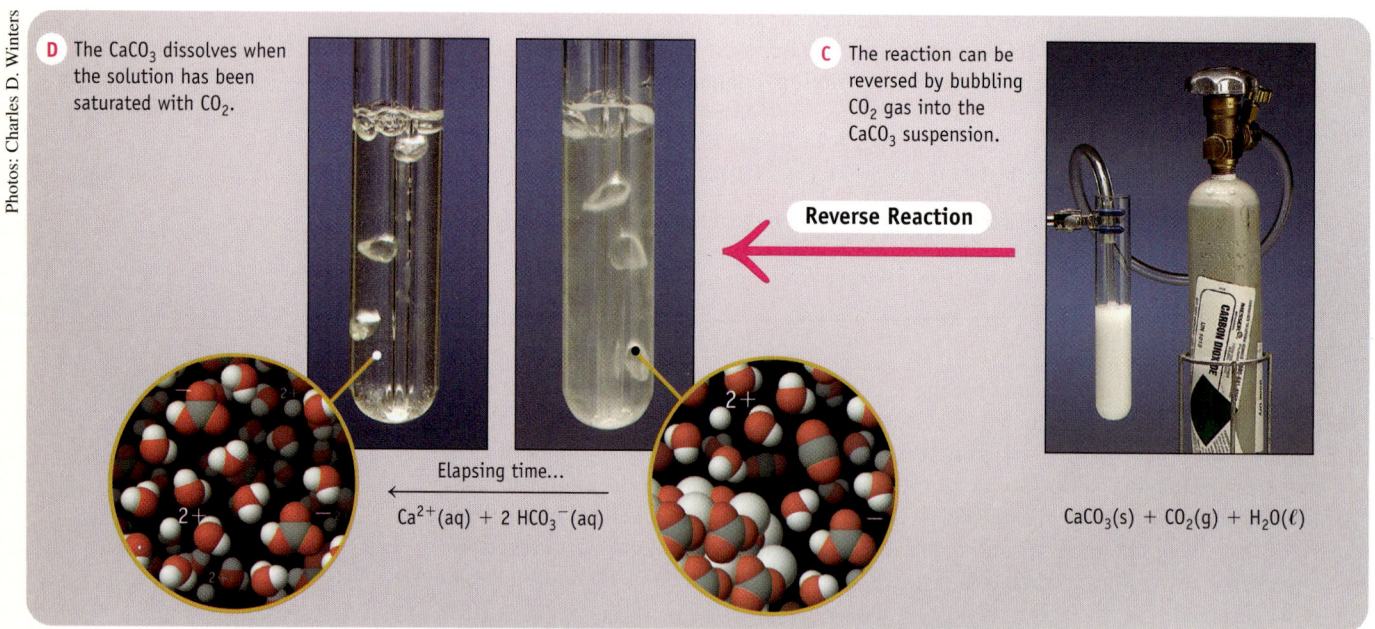

FIGURE 13.4 Reversible chemical reactions. The experiments pictured here demonstrate the reversibility of chemical reactions. All chemical reactions are in principle reversible, and, given enough time and the proper conditions, will achieve a state of dynamic chemical equilibrium.

The reaction mixture is said to be in a condition of **dynamic chemical equilibrium**, *a state in which the opposite reactions continue to occur at equal rates so that there is no change of concentrations of species.* We depict this situation by writing a balanced equation with the double arrow symbol of equilibrium, $\rightleftharpoons$.

$$Ca^{2+}(aq) + 2\ HCO_3^-(aq) \rightleftharpoons CaCO_3(s) + CO_2(g) + H_2O(\ell)$$

If you boil a solution into which CO_2 is bubbled (and $CaCO_3$ dissolved), CO_2 is driven off, and once more reaction 2 happens with precipitation of solid $CaCO_3$. This solid is the "scale" formed when water boils in a kettle or in industrial boilers. This chemistry is an example of acid-base chemistry [>>Chapter 14] and of competition between acid-base reactions and precipitation reactions [>>Section 15.2]. The chemistry described here is also relevant to the natural carbon cycle [<<Chapter 4], since the ultimate fate of much of the CO_2 emitted into the atmosphere is formation in the ocean of calcareous ooze and other forms of calcium carbonate.

Another example of a dynamic chemical equilibrium is the ionization of some of the molecules of acetic acid in aqueous solution to form aquated acetate ions and hydronium ions—the reaction responsible for the acidity of vinegar:

$$CH_3COOH(aq) + H_2O(\ell) \rightleftharpoons CH_3COO^-(aq) + H_3O^+(aq)$$

In a 1.0 mol L^{-1} aqueous solution of CH_3COOH at equilibrium at 25 °C, the concentrations of $CH_3COO^-(aq)$ and $H_3O^+(aq)$ ions are both 0.0042 mol L^{-1}. Even though the concentration of the acetic acid molecules is much higher than the concentrations of the ions, at equilibrium the rates of the reactions in the opposite directions are the same: just as many acetic acid molecules are formed per second as the number of them that ionize per second.

> Reversible chemical reactions can go in either direction, depending on the conditions. In a reaction mixture in dynamic chemical equilibrium, reactions in opposite directions occur at equal rates so that there is no change of concentrations of species.

Net Reaction

When a reaction mixture is at chemical equilibrium, reactions are occurring in opposite directions and at the same rate, so that the concentrations of reactants and products do not change over time. Even before the reaction mixture comes to equilibrium, when the concentrations of reactants are decreasing, and the concentrations of products are increasing, these reactions in opposite directions are simultaneously occurring—but not at equal rates. The reaction producing products happens at a faster rate than that producing reactants. In these circumstances, the term **net reaction** refers to *the changing concentrations of species due to unequal rates of the reactions in the opposite directions*. The difference between the rates of the reactions in the opposite directions, which is the apparent rate of change in the reaction mixture, is called the *net reaction rate*.

A study of the rates of reactions, how they change during the course of a reaction, and how they depend on concentrations and temperature, is called *reaction kinetics*, and is the subject of Chapter 18.

> Net reaction is the change of concentrations of species due to unequal rates of the reactions in opposite directions, when the system is not at equilibrium.

13.3 The Reaction Quotient and the Equilibrium Constant

The reaction between hydrogen and iodine gases at 425 °C to produce hydrogen iodide gas reaches a state of dynamic chemical equilibrium, represented by

$$H_2(g) + I_2(g) \rightleftharpoons 2\,HI(g)$$

Let's define, for reasons that will become apparent, a function of the concentrations of the reactants and products called the **reaction quotient** (Q) as follows:

$$Q = \frac{[HI]^2}{[H_2][I_2]}$$

Imagine a reaction vessel into which is put 0.0175 mol L^{-1} of both H_2 gas and I_2 gas at 425 °C, without any HI gas. We can calculate the numerical value of the reaction quotient Q in an imagined instant before any reaction takes place between the H_2 gas and I_2 gas:

$$Q = \frac{[HI]^2}{[H_2][I_2]} = \frac{(0)^2}{(0.0175)(0.0175)} = 0$$

Think about It

e13.1 Watch videos showing two other reversible reactions.

And now imagine that after an unspecified time, some reaction has happened so that the concentration of $H_2(g)$ has fallen to 0.0150 mol L^{-1}. From the stoichiometry of the reaction equation, we can say that the concentration of I_2 gas will also have fallen to 0.0150 mol L^{-1}, and the concentration of HI gas will have risen to 0.0050 mol L^{-1}. The numerical value of Q now can be calculated:

$$Q = \frac{[HI]^2}{[H_2][I_2]} = \frac{(0.0050)^2}{(0.0150)(0.0150)} = 0.111$$

Further imagine that at some later time, $[H_2]$ has decreased to 0.0100 mol L^{-1}. Then we can see that numerical value of Q has further increased while reaction is proceeding:

$$Q = \frac{[HI]^2}{[H_2][I_2]} = \frac{(0.0150)^2}{(0.0100)(0.0100)} = 2.25$$

After a relatively long time, experimental measurement would show that the gas concentrations are no longer changing, and that $[H_2] = [I_2] = 0.0037$ mol L^{-1} and $[HI] = 0.0276$ mol L^{-1}. Then

$$Q = \frac{[HI]^2}{[H_2][I_2]} = \frac{(0.0276)^2}{(0.0037)(0.0037)} = 56$$

This is evidence that the reaction mixture has reached a condition of dynamic chemical equilibrium (Figure 13.5).

FIGURE 13.5 The reaction between $H_2(g)$ and $I_2(g)$ reaches equilibrium. During reaction, the concentrations of the reactants $H_2(g)$ and $I_2(g)$ decrease, and the concentration of the product $HI(g)$ increases until dynamic chemical equilibrium is achieved, when the concentrations of all reactants and products remain constant. The constant value of the reaction quotient Q in the reaction mixture has the same value as any other equilibrium reaction mixture of these species at the same temperature.

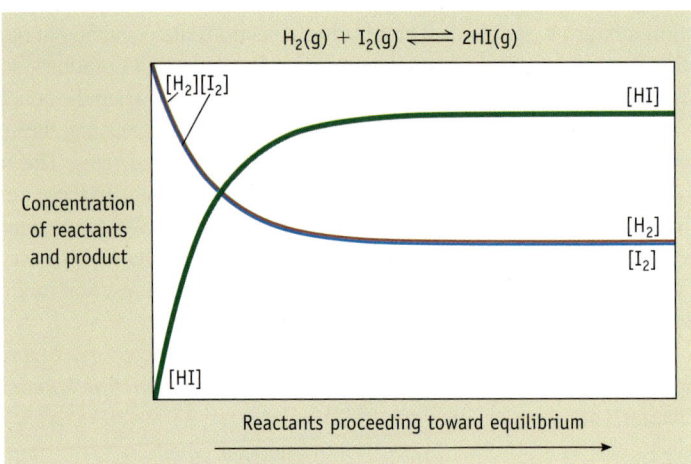

Think about It

e13.2 Simulate a reaction as it comes to equilibrium with different initial concentrations of reactants and products.

If we now imagine another reaction mixture, also at $425\ °C$, in which the starting concentrations of H_2 gas and I_2 gas are not equal: $[H_2] = 0.0200$ mol L^{-1} and $[I_2] = 0.0100$ mol L^{-1}, again we would find that the concentrations of H_2 gas and I_2 gas decrease, while that of HI gas increases—and the reaction quotient Q increases. Eventually, when the concentrations are no longer changing because chemical equilibrium has been reached, it would be found that $[H_2] = 0.0106$ mol L^{-1}, $[I_2] = 0.00060$ mol L^{-1}, and $[HI] = 0.0188$ mol L^{-1}. The numerical value of the reaction quotient Q at equilibrium can be calculated:

$$Q = \frac{[HI]^2}{[H_2][I_2]} = \frac{(0.0188)^2}{(0.0106)(0.00060)} = 56$$

The reaction quotient Q has the same value as in the previously equilibrium reaction mixture.

Finally, to illustrate the most significant point about quantitative treatment of chemical equilibrium situations, let's consider another hypothetical reaction mixture at $425\ °C$ in which there is some HI gas present at the start: $[H_2] = 0.0200$ mol L^{-1}, $[I_2] = 0.0150$ mol L^{-1}, and $[HI] = 0.0250$ mol L^{-1}. When the reaction mixture reaches equilibrium, it would be found that $[H_2] = 0.0092$ mol L^{-1}, $[I_2] = 0.0042$ mol L^{-1}, $[HI] = 0.0466$ mol L^{-1}, and $Q = 56$—the same as in the other equilibrium mixtures!

Taken together, the examples above illustrate the **law of equilibrium**: *For a given reaction at a specified temperature, all equilibrium mixtures have the same value of the reaction quotient (Q). This particular value of the reaction quotient is called the* **equilibrium constant**, denoted K.

To generalize, for a reaction which can be represented by the equation

$$a\,A + b\,B \rightleftharpoons c\,C + d\,D$$

there is a particular function of concentrations, called the reaction quotient (Q) defined by

$$Q = \frac{[C]^c[D]^d}{[A]^a[B]^b}$$

whose numerical value changes during the course of a reaction, until it becomes constant when chemical equilibrium is attained. For a given reaction at a specified temperature, all equilibrium mixtures have the same value of Q, called the equilibrium constant (K). Temperature must be specified because the value of Q at equilibrium depends upon the temperature. So, if we have a number of reaction mixtures, labelled 1, 2, 3, . . ., at which a particular reaction is at equilibrium at the same temperature, then even if the concentrations in the various reaction mixtures are different

$$\left\{\frac{[C]^c[D]^d}{[A]^a[B]^b}\right\}_{\text{Mixture 1}} = \left\{\frac{[C]^c[D]^d}{[A]^a[B]^b}\right\}_{\text{Mixture 2}} = \left\{\frac{[C]^c[D]^d}{[A]^a[B]^b}\right\}_{\text{Mixture 3}} = \ldots = K$$

$$\text{or, } Q_1 = Q_2 = Q_3 = \ldots = K$$

where the equilibrium constant (K) is a number, characteristic of each reaction at a specified temperature.

At equilibrium, because the concentrations of all species are constant, any function of the reactant and product concentrations is constant in any one reaction mixture. For example, for the reaction

$$H_2(g) + I_2(g) \rightleftharpoons 2\,HI(g)$$

all of the following functions come to a constant value in one reaction mixture:

$$[H_2][I_2] \quad \text{and} \quad \frac{[H_2]}{[HI]} \quad \text{and} \quad \frac{[HI]^7}{[I_2]^2}$$

However, only one particular function of concentrations (the one defined as Q) is the same in all vessels in which this reaction is at equilibrium at the same temperature.

> For a given reaction at a specified temperature, all equilibrium mixtures have the same value of the reaction quotient (Q). This numerical value of Q is called the equilibrium constant (K).

The form of Q and K

The following are generalizations that we can make about the function called reaction quotient Q, including the particular value of Q when the reaction mixture is at equilibrium, called the equilibrium constant K:

- The numerator (the top line) has concentrations of the species written on the right side of the chemical equation, and the denominator has the concentrations of the species on the left side of the equation. The concentration of each species is raised to the power of its stoichiometric coefficient in the balanced chemical equation. As a consequence, the expression for the reaction quotient (and the equilibrium constant) cannot be written without reference to a particular balanced chemical equation.
- For reactions involving solids, concentration terms for the solids are not included in expressions for Q or K. For example, for the decomposition of solid limestone (calcium carbonate, $CaCO_3$) into solid lime (calcium oxide, CaO) and gaseous carbon dioxide, we can write the chemical equation

$$CaCO_3(s) \rightleftharpoons CaO(s) + CO_2(g)$$

Think about It

e13.3 Simulate a reaction as it comes to equilibrium with different initial values of Q.

We could use the general definition of Q to write

$$Q = \frac{[CaO][CO_2]}{[CaCO_3]}$$

However, the concentrations (in mol L^{-1}) of solid $CaCO_3$ and solid CaO do not change during reaction. Since our concerns are about how Q changes during a reaction, and whether Q is greater than, less than, or equal to K, we can ignore these constant terms. In this case, we write

$$Q = [CO_2]$$

Another way to consider the non-inclusion of concentration terms for solids is that these constant terms can be transposed in the equation for Q, to give us a "pseudo Q" that we might call Q', which includes the constant solid concentrations in its value:

$$Q' = Q \times \frac{[CaCO_3]}{[CaO]} = [CO_2]$$

• For reactions taking place in aqueous solution, the molar concentration of water is usually very much higher than that of other reactants, and remains essentially constant during reaction. So, we do not include the concentration of water in the expression for Q or K. For example, the reaction of ammonia as a weak base in aqueous solution can be represented by the balanced chemical equation

$$NH_3(aq) + H_2O(\ell) \rightleftharpoons NH_4^+(aq) + OH^-(aq)$$

The reaction quotient for this reaction is written

$$Q = \frac{[NH_4^+][OH^-]}{[NH_3]}$$

WORKED EXAMPLE 13.1—EXPRESSING THE FORM OF THE REACTION QUOTIENT

Write the reaction quotient Q for the following reactions:
(a) The formation of ammonia from nitrogen and hydrogen

$$N_2(g) + 3 H_2(g) \rightleftharpoons 2 NH_3(g)$$

(b) The reaction of carbonic acid as a weak acid in aqueous solution

$$H_2CO_3(aq) + H_2O(\ell) \rightleftharpoons HCO_3^-(aq) + H_3O^+(aq)$$

Strategy

Remember that product concentrations are placed in the numerator, and reactant concentrations in the denominator. The concentration of each species should be raised to a power equal to the stoichiometric coefficient in the balanced equation. In reaction (b), the water concentration does not appear in the equilibrium constant expression.

Solution

(a) $Q = \dfrac{[NH_3]^2}{[N_2][H_2]^3}$

(b) $Q = \dfrac{[HCO_3^-][H_3O^+]}{[H_2CO_3]}$

Interactive Exercises 13.1–13.2

Practise writing expressions for Q.

EXERCISE 13.3—EXPRESSING THE FORM OF THE REACTION QUOTIENT

Write the reaction quotient Q appropriate for each of the following reactions:
(a) $PCl_5(g) \rightleftharpoons PCl_3(g) + Cl_2(g)$
(b) $CO_2(g) + C(s) \rightleftharpoons 2 CO(g)$

(c) $Cu(NH_3)_4^{2+}(aq) \rightleftharpoons Cu^{2+}(aq) + 4\ NH_3(aq)$

(d) $CH_3COOH(aq) + H_2O(\ell) \rightleftharpoons CH_3COO^-(aq) + H_3O^+(aq)$

Activity-based Equilibrium Constants

The ionization of formic acid in aqueous solution can be represented as

$$HCOOH(aq) + H_2O(\ell) \rightleftharpoons HCOO^-(aq) + H_3O^+(aq)$$

At 25 °C, the equilibrium constant for this reaction is 1.8×10^{-4}. Following what has been said above, we would interpret this to mean that we can write a mathematical function of species concentrations (Q) whose numerical value in all solutions in which this reaction is at equilibrium at 25 °C is 1.8×10^{-4}.

$$\text{At equilibrium at 25 °C, } Q = \frac{[HCOO^-][H_3O^+]}{[HCOOH]} = 1.8 \times 10^{-4} = K$$

In fact, the value of this function is only approximately the same from solution to solution at equilibrium and at the same temperature. The more different the concentrations of ions in the various solutions are, the more different the equilibrium values of Q. Rather than using the species concentrations in the expression for Q, it is more correct and more accurate to use the *activity* (a) of each species:

$$\text{At equilibrium at 25 °C, } Q = \frac{(a_{HCOO^-})(a_{H_3O^+})}{(a_{HCOOH})} = 1.8 \times 10^{-4} = K$$

The activity of a species differs from its concentration in two important ways:

- The activity is a measure of the effective concentration of the species [<<Section 12.6] rather than its actual concentration. Some properties of solutions, such as freezing point, depend on the concentrations of the species in solution, and activity refers to the apparent concentrations of species deduced from measurements of such properties. The activity of a species of specified concentration depends on the concentration of ions in the solution: the lower the concentration of the species, as well as of ions from other dissolved salts, the closer the numerical values of activity and concentration. One way to consider the activities of species is that they are the values of effective concentration that give rise to values of Q that are the same in all solutions at equilibrium at the same temperature.
- There are fundamental reasons why the value of Q must be unitless, but we can see that the value of the concentration-based Q for the above reaction has the units of mol L^{-1}. A unitless activity of any species is derived as a ratio of the effective concentration to 1 mol L^{-1}. So, for a species X that has effective concentration 0.025 mol L^{-1}, we have

$$a_X = \frac{0.025 \cancel{\text{ mol L}^{-1}}}{1 \cancel{\text{ mol L}^{-1}}} = 0.025$$

As the above calculation shows, the activity of a species has the same numerical value as its effective concentration. If all of the activities in the expression for Q are unitless, then Q is unitless.

At some time in the future, you may be carrying out some research in the areas of chemistry, biochemistry, environmental science, marine science, or agricultural science in saline waters. Then you will need to use estimates of activities of the various species in your calculations. In this text, we will almost always be concerned with solutions that are sufficiently dilute that there is negligible difference between the concentrations and activities of species. And for convenience, in consideration of equilibrium systems [>>Chapters 14 and 15] and other parts of thermodynamics [>>Chapter 17] we will refer to concentrations, and will continue to use the concentration symbol [...], even though the values we use will be those of unitless activities.

Reaction quotients are most correctly expressed as functions of activities of species rather than as concentrations. Activities are measures of effective concentration, and as a ratio compared with 1 mol L^{-1}, are unitless.

The Relationship between *Q* and *K* in Reaction Mixtures

When net reaction is occurring in a reaction mixture not yet at equilibrium, the numerical value of the reaction quotient (Q) changes as the concentrations of reactants and products change. Eventually chemical equilibrium is reached, and concentrations no longer change, and the value of Q is the equilibrium constant K for that reaction at the temperature of the reaction mixture. Only when the reaction mixture is at equilibrium does $Q = K$.

The use of the terms *reactants* and *products* can be ambiguous when we are talking about reversible reactions. If we start with a reaction mixture containing only $H_2(g)$ and $I_2(g)$, then these reagents are obviously the reactants. However, if our initial reaction mixture contains only $HI(g)$, then this substance can be considered to be the reactant. Sometimes the initial reaction might contain all of the reagents $H_2(g)$, $I_2(g)$, and $HI(g)$. They can all be called *reaction species*. Common usage is to refer to the species on the left side of the written reaction equation as *reactants*, and those on the right side of the equation as *products*, even though sometimes a reaction proceeds to equilibrium by conversion of species on the right side of the equation to those on the left. The direction in which we write an equation defines how Q is expressed, but has no bearing on what happens in the reaction mixture.

We have seen earlier in this chapter in relation to the reaction represented by

$$H_2(g) + I_2(g) \rightleftharpoons 2\,HI(g)$$

that if we start with some hydrogen gas and iodine gas at 425 °C, but no hydrogen iodide gas, then initially $Q = 0$. As net reaction occurs and the reaction mixture changes, with [HI] increasing and both [H_2] and [I_2] decreasing, Q increases. All the while $Q < K$, until the reaction mixture comes to equilibrium (when $Q = 56$).

If, however, we created a starting reaction mixture with high concentration of HI and small concentrations of $H_2(g)$ and $I_2(g)$, we may find that $Q > K$. For example, if we had a reaction mixture with [HI] = 0.100 mol L^{-1}, [H_2] = 0.0100 mol L^{-1}, and [I_2] = 0.0100 mol L^{-1}, we would have

$$Q = \frac{(0.100)^2}{(0.0100)(0.0100)} = 100$$

We can see that because $K = 56$, $Q > K$, and the reaction mixture is not at equilibrium. If net reaction were to happen in the direction to form more HI(g), the numerical value of Q would become even larger, and more different from K. Obviously, this would not happen: attainment of equilibrium would require net reaction in the opposite direction, and Q would decrease until its value is 56 (= K at 425 °C).

A comparison between the numerical value of Q with K can tell us if a reaction is at equilibrium, and if not, we can predict which direction of net reaction would bring the reaction mixture to equilibrium:

- $Q = K$. The reaction mixture is at equilibrium.
- $Q < K$. Some reactants must be converted to products for the reaction to reach equilibrium. This will decrease the reactant concentrations and increase the product concentrations, and Q will increase (until $Q = K$).
- $Q > K$. Some products must be converted to reactants for the reaction to reach equilibrium. This will increase the reactant concentrations and decrease the product concentrations, and Q will decrease (until $Q = K$).

Think about It

e13.4 Predict the direction of net reaction from the value of Q in relation to the value of K.

Think about It

e13.5 Watch an application of these ideas in the important technique of chromatography.

In a reaction mixture, if $Q = K$ the mixture is at equilibrium. If $Q < K$, equilibrium is attained if net reaction changes reactants into products. If $Q > K$, equilibrium is attained if net reaction changes products into reactants.

Spontaneous Reaction Direction, Stability, Gibbs Free Energy

If a reaction mixture is not at equilibrium, then *the direction in which net reaction would need to occur to bring a reaction mixture to the condition of chemical equilibrium is called a* **spontaneous reaction**.

In reaction mixtures with $Q < K$, the spontaneous direction of net reaction is conversion of reactants into products. In reaction mixtures with $Q > K$, the spontaneous direction of net reaction is conversion of products into reactants.

We will see in Chapter 17 that when a reaction comes to chemical equilibrium, the reaction mixture has a lower *Gibbs free energy* than at any point in time before equilibrium was attained. Indeed, minimization of Gibbs free energy of the reaction mixture is a criterion of being at chemical equilibrium.

The particular mixture of reactants and products present at equilibrium is said to be more *stable* [<<Section 5.4] than any of the other mixtures formed as the reaction proceeded toward equilibrium.

Following are some notes concerning these interrelated ideas:

1. The relationship between the numerical value of Q and K can tell us if a reaction mixture is at equilibrium, and if it is not, in which direction net reaction would need to occur for equilibrium to be attained. However, this relationship cannot tell us anything about how quickly equilibrium is achieved. Some reactions are fast; some are slow [>>Chapter 18].

2. In everyday language, the term *spontaneous* implies an immediate or instantaneous process. In chemistry, this term has a particular meaning about the direction of net reaction that would bring the value of Q toward K, and has nothing to do with how immediate or instantaneous that change occurs.

3. For a given reversible reaction, the spontaneous direction of reaction depends on the concentrations of reactants and products present in the reaction mixture. For every reaction (with a particular value of K at a given temperature), we can make some reaction mixtures with concentrations of species such that $Q < K$, and some others with $Q < K$. So, we cannot say that a reaction is spontaneous or non-spontaneous in some absolute sense; we can only say which is the spontaneous direction of reaction in a defined reaction mixture (if it is not at equilibrium).

> **Spontaneous Direction of Net Reaction**
>
> $Q < K$ reactants $\rightarrow$ products
> $Q = K$ reaction at equilibrium
> $Q > K$ reactants $\leftarrow$ products

WORKED EXAMPLE 13.2—DEDUCING THE DIRECTION OF SPONTANEOUS REACTION

Brown nitrogen dioxide, $NO_2(g)$, and colourless dinitrogen tetroxide, $N_2O_4(g)$, attain a condition of chemical equilibrium with $K = 170$ at 298 K.

$$2\ NO_2(g) \rightleftharpoons N_2O_4(g)$$

Suppose that in a reaction mixture, the concentration of $NO_2(g)$ is 0.015 mol L^{-1} and the concentration of $N_2O_4(g)$ is 0.025 mol L^{-1}. Is Q larger than, smaller than, or equal to K? If the system is not at equilibrium, which is the direction of spontaneous reaction: in the direction written or in the opposite direction?

Strategy

From the chemical equation, write the expression for Q and calculate its numerical value from the concentrations of reactant and product. Compare the numerical value of Q with K. The spontaneous direction of reaction is that direction of net reaction that would take the reaction mixture toward chemical equilibrium—that is, which would change Q closer to K.

Solution

$$Q = \frac{[N_2O_4]}{[NO_2]^2} = \frac{0.025}{(0.015)^2} = 110$$

The value of Q is less than K, so the reaction mixture is not at equilibrium. $Q < K$ (and K is a constant at a fixed temperature), so net reaction would happen in the direction that increases Q—that is, in the direction written: $[NO_2]$ decreases and $[N_2O_4]$ increases until $Q = K$.

Interactive Exercise 13.4

SUBMIT

Predict the spontaneous direction of a reaction from the relative values of Q and K.

EXERCISE 13.5—DEDUCING THE DIRECTION OF SPONTANEOUS REACTION

For the reversible transformation between butane and isobutane (2-methylpropane), represented below, $K = 2.50$ at 298 K—although net reaction to attain equilibrium is extremely slow.

Butane ⇌ Isobutane

$$CH_3CH_2CH_2CH_3 \rightleftharpoons CH_3CHCH_3 \quad (CH_3)$$

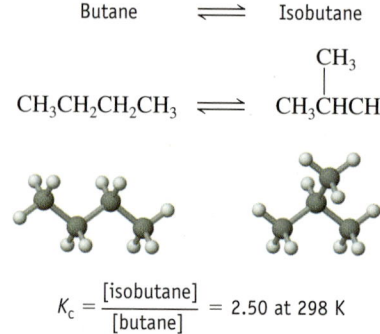

$$K_c = \frac{[\text{isobutane}]}{[\text{butane}]} = 2.50 \text{ at 298 K}$$

(a) Is a reaction mixture at 298 K at equilibrium if [butane] = 0.97 mol L^{-1} and [isobutane] = 2.18 mol L^{-1}? If not, in which direction would the reaction need to proceed to achieve equilibrium?

(b) Is a reaction mixture at 298 K at equilibrium when [butane] = 0.75 mol L^{-1} and [isobutane] = 2.60 mol L^{-1}? If not, in which direction would the reaction proceed to achieve equilibrium?

> The direction in which net reaction occurs to bring a reaction mixture to equilibrium is the spontaneous direction of reaction. As spontaneous reaction proceeds, the reaction mixture becomes more stable, reaching maximum stability at equilibrium. Equilibrium is the condition of minimum Gibbs free energy.

13.4 Quantitative Aspects of Equilibrium Constants

Every reaction has a unique value of its equilibrium constant at a specified temperature, and the value changes as temperature is changed. The magnitude of an equilibrium constant is indicative of how far a reaction will go before it comes to equilibrium.

Magnitude of K and Extent of Reaction

If a reaction has a large value of K, in a reaction mixture at equilibrium the concentrations of the products are greater than the concentrations of the reactants, and the reaction is said to be *product-favoured*.

An example is the reaction of nitric oxide and ozone, one of the reactions involved in depletion of stratospheric ozone:

$$NO(g) + O_3(g) \rightleftharpoons NO_2(g) + O_2(g)$$

At equilibrium $\quad Q = \dfrac{[NO_2][O_2]}{[NO][O_3]} = K = 6 \times 10^{34}$ at 25°C

The very large value of K indicates that, at equilibrium, $[NO_2][O_2] \gg [NO][O_3]$. If equal amounts of NO(g) and O_3(g) are mixed and the reaction allowed to come to equilibrium, virtually none of the reactants will remain (Figure 13.6a). Chemists say that the reaction has "gone to completion." On the other hand, if a reaction mixture is made with NO_2(g) and O_2(g) only, essentially no NO(g) and O_3(g) would be formed in attaining equilibrium.

Reactions with values of K < 1, are said to be *reactant-favoured*: at equilibrium, the concentrations of reactants are greater than concentrations of products (Figure 13.6b). This is the case, for example, with the formation of ozone from oxygen:

$$3\,O_2(g) \rightleftharpoons 2\,O_3(g)$$

At equilibrium at 25 °C $\quad \dfrac{[O_3]^2}{[O_2]^3} = K = 6 \times 10^{-58}$

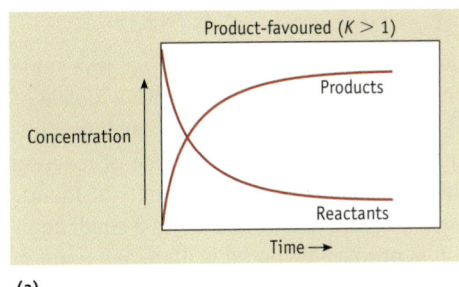

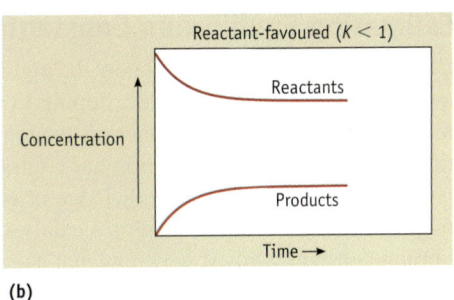

(a) (b)

FIGURE 13.6 Product-favoured and reactant-favoured reactions. (a) When $K \gg 1$, there is much more product than reactant at equilibrium. (b) When $K \ll 1$, there is much more reactant than product at equilibrium.

At equilibrium, $[O_3]^2 \ll [O_2]^3$. If pure $O_2(g)$ is placed in a flask, very little will have been converted to $O_3(g)$ when equilibrium is achieved.

When K is close to 1, it may not be immediately clear whether the reactant concentrations are larger than the product concentrations, or *vice versa* at equilibrium. It will depend on the coefficients of the concentration terms, and thus on the reaction stoichiometry. Table 13.1 lists some equilibrium constants for various reactions.

TABLE 13.1 Equilibrium Constants at 25 °C of Selected Reactions

Reaction	K	Product- or Reactant-Favoured
Combination reaction of non-metals		
$S(s) + O_2(g) \rightleftharpoons SO_2(g)$	4.2×10^{52}	$K > 1$; product-favoured
$2\,H_2(g) + O_2 \rightleftharpoons 2\,H_2O(g)$	3.2×10^{81}	$K > 1$; product-favoured
$N_2(g) + 3\,H_2(g) \rightleftharpoons 2\,NH_3(g)$	3.5×10^{8}	$K > 1$; product-favoured
$N_2(g) + O_2(g) \rightleftharpoons 2\,NO(g)$	1.7×10^{-3} (at 2300 K)	$K < 1$; reactant-favoured
Ionization of weak acids and bases		
$HCOOH(aq) + H_2O(\ell) \rightleftharpoons HCOO^-(aq) + H_3O^+(aq)$ formic acid	1.8×10^{-4}	$K < 1$; reactant-favoured
$CH_3COOH(aq) + H_2O(\ell) \rightleftharpoons CH_3COO^-(aq) + H_3O^+(aq)$ acetic acid	1.8×10^{-5}	$K < 1$; reactant-favoured
$H_2CO_3(aq) + H_2O(\ell) \rightleftharpoons HCO_3^-(aq) + H_3O^+(aq)$ carbonic acid	4.2×10^{-7}	$K < 1$; reactant-favoured
$NH_3(aq) + H_2O(\ell) \rightleftharpoons NH_4^+(aq) + H_3O^-(aq)$ ammonia	1.8×10^{-5}	$K < 1$; reactant-favoured
Dissolution of slightly soluble solids		
$CaCO_3(s) \rightleftharpoons Ca^{2+}(aq) + CO_3^{2-}(aq)$	3.8×10^{-9}	$K < 1$; reactant-favoured
$AgCl(s) \rightleftharpoons Ag^+(aq) + Cl^-(aq)$	1.8×10^{-10}	$K < 1$; reactant-favoured

---

EXERCISE 13.6—MAGNITUDE OF K AND EXTENT OF REACTION

Are the following reactions product-favoured or reactant-favoured?
(a) $Cu(NH_3)_4^{2+} \rightleftharpoons Cu^{2+}(aq) + 4\,NH_3(aq)$ $K = 1.5 \times 10^{-13}$
(b) $Cd(NH_3)_4^{2+} \rightleftharpoons Cd^{2+}(aq) + 4\,NH_3(aq)$ $K = 1.0 \times 10^{-7}$

If we have two reaction mixtures at equilibrium at the same temperature, one with $[Cu(NH_3)_4^{2+}] = [Cu^{2+}] = 1.0$ mol L^{-1}, and the other with $[Cd(NH_3)_4^{2+}] = [Cd^{2+}] = 1.0$ mol L^{-1}, in which mixture is the concentration of ammonia (NH_3) higher?

Think about It

e13.6 Watch videos of reactions and the significance of their equilibrium constants.

Think about It

e13.7 Simulate the approach to equilibrium under different starting conditions.

Reactions with $K \gg 1$ have higher concentrations of products than reactants at equilibrium and are said to be product-favoured. Reactions with $K \ll 1$ are reactant-favoured.

Estimating Equilibrium Constants

Chemists in industry need to know the equilibrium constants of reactions. If we mix some nitrogen and hydrogen at 25 °C, hoping to form ammonia, we would find no evidence of reaction. This could be for either of two reasons: (a) the reaction has a very small equilibrium constant, or (b) the reaction has a large equilibrium constant, but the rate of reaction is extremely slow. Without knowledge of equilibrium constants, large amounts of money might be wasted trying to develop a catalyst to speed up the reaction, if it has a small equilibrium constant. Only if we know that a reaction is product-favoured is there any point in trying to develop a catalyst. These are vital issues that faced Haber in the development of his process to "fix" nitrogen [<<Section 13.1 and >>Section 13.7]

Equilibrium constants for reactions can be calculated using knowledge of the field called chemical thermodynamics [>>Chapter 17], or they can be estimated by experiment. When the experimentally measured concentrations of all of the reactants and products are known *at equilibrium*, an equilibrium constant can be calculated by substituting the data into the reaction quotient, as in Worked Example 13.3.

Taking It Further

e13.8 Compare equilibrium constants based on gas pressures with those based on gas concentrations.

WORKED EXAMPLE 13.3—ESTIMATION OF EQUILIBRIUM CONSTANTS

Consider a reaction mixture at 852 K in which the reaction represented by the following equation has come to equilibrium.

$$2\ SO_2(g) + O_2(g) \rightleftharpoons 2\ SO_3(g)$$

The concentrations are measured to be $[SO_2] = 3.61 \times 10^{-3}$ mol L^{-1}, $[O_2] = 6.11 \times 10^{-4}$ mol L^{-1}, and $[SO_3] = 1.01 \times 10^{-2}$ mol L^{-1}. What is the equilibrium constant for this reaction at 852 K?

Strategy

We know the concentrations of all reacting species in the mixture, so we can calculate the value of Q. Since it is an equilibrium mixture, $Q = K$.

Solution

We substitute the species concentrations into the reaction quotient.

$$Q = \frac{[SO_3]^2}{[SO_2]^2[O_2]} = \frac{(1.01 \times 10^{-2})^2}{(3.61 \times 10^{-3})^2(6.11 \times 10^{-4})} = 1.28 \times 10^4 = K$$

More commonly, we measure the initial concentrations of reactants and the equilibrium concentration of one of the reactants or of one of the products—enabling us to deduce how much of the reactant species had reacted. The equilibrium concentrations of the rest of the reactants and products can then be inferred from the balanced chemical equation.

As an example, consider again the reaction of Worked Example 13.3. Suppose that 1.00 mol of $SO_2(g)$ and 1.00 mol of $O_2(g)$ are placed in a 1.00 L flask, this time at 1000 K. When equilibrium has been achieved, 0.925 mol of $SO_3(g)$ is present. The stoichiometric equation tells us that the amount of $SO_2(g)$ that reacted is the same as the amount of $SO_3(g)$ formed, and the amount of $O_2(g)$ that reacted is half the amount of $SO_3(g)$ formed.

It is useful to set up a table similar to the amounts table used in Section 5.4, except listing concentrations (in mol L^{-1} of substances, rather than amounts). We will call this an ICE table, for **i**nitial, **c**hange, and **e**quilibrium concentrations.

Think about it

e13.9 Determine the equilibrium constant for a reaction using an ICE table.

Equation	2 SO$_2$(g)	+	O$_2$(g)	$\rightleftharpoons$	2 SO$_3$(g)
Initial (mol L^{-1})	1.00		1.00		0
Change (mol L^{-1})	− 0.925		$-\frac{1}{2}(0.925)$		+ 0.925
Equilibrium (mol L^{-1})	1.00 − 0.925		1.00 $-\frac{1}{2}(0.925)$		0 + 0.925
	= 0.075		= 0.54		= 0.925

With the equilibrium concentrations now known, we can calculate K at 1000 K:

$$Q = \frac{[SO_3]^2}{[SO_2]^2[O_2]} = \frac{(0.925)^2}{(0.075)^2(0.54)} = 2.8 \times 10^2 = K$$

WORKED EXAMPLE 13.4—ESTIMATION OF EQUILIBRIUM CONSTANTS

An aqueous solution containing ethanol and acetic acid, both at an initial concentration of 0.810 mol L^{-1}, is heated to 100 °C. Reaction occurs between the alcohol and the carboxylic acid to form the ester, ethyl acetate:

$$C_2H_5OH(aq) + CH_3COOH(aq) \rightleftharpoons CH_3COOC_2H_5(aq) + H_2O(\ell)$$

At equilibrium, the acetic acid concentration is 0.748 mol L^{-1}. Calculate K for the reaction at 100 °C.

Strategy

Focus on deducing equilibrium concentrations. The balanced chemical equation informs us that the amount of ethanol that reacts is the same as the amount of acetic acid that reacts, and the amount of ethyl acetate that is formed. So the concentration of acetic acid must decrease by the same as the concentration of ethanol, as well as being the same as the increase in concentration of ethyl acetate.

Solution

The concentration of acetic acid decreases by $(0.810 - 0.748)$ mol L^{-1} = 0.062 mol L^{-1}. So the concentration of ethanol decreases by 0.062 mol L^{-1}, and the concentration of ethyl acetate increases by 0.062 mol L^{-1}. Construct an ICE table.

C_2H_5OH	+	CH_3COOH	$\rightleftharpoons$	$CH_3COOC_2H_5$	+	H_2O
Initial						
(mol L^{-1})		0.810		0.810		0
Change						
(mol L^{-1})		− 0.062		− 0.062		+ 0.062
Equilibrium						
(mol L^{-1})		0.748		0.748		0.062

We can substitute the equilibrium concentrations into the reaction quotient, and calculate K:

$$Q = \frac{[CH_3COOC_2H_5]}{[C_2H_5OH][CH_3COOH]} = \frac{0.062}{(0.748)(0.748)} = 0.11 = K$$

Interactive Exercise 13.8

Calculate the equilibrium constant using the equilibrium concentrations.

EXERCISE 13.7—ESTIMATION OF EQUILIBRIUM CONSTANTS

A solution is made by dissolving 0.050 mol of diiodocyclohexane ($C_6H_{10}I_2$) in the solvent carbon tetrachloride, $CCl_4(\ell)$. The total solution volume is 1.00 L. When the reaction

$$C_6H_{10}I_2(\text{in } CCl_4) \rightleftharpoons C_6H_{10}(\text{in } CCl_4) + I_2(\text{in } CCl_4)$$

has come to equilibrium at 35 °C, the concentration of iodine (I_2) is 0.035 mol L^{-1}.
(a) Calculate $[C_6H_{10}I_2]$ and $[C_6H_{10}]$ at equilibrium.
(b) Calculate the equilibrium constant for this reaction at 35 °C.

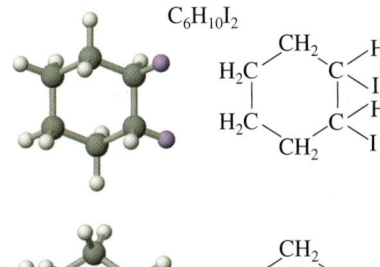

$C_6H_{10}I_2$

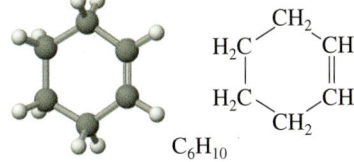

C_6H_{10}

If we know the concentrations of species in a reaction mixture at equilibrium, we can calculate the equilibrium constant at the temperature of measurement by substituting the concentrations into the reaction quotient.

Calculating Equilibrium Concentrations

Sometimes, the initial concentrations of reactants in a reaction mixture are known, as well as the value of K at the temperature of the mixture, and we want to know the concentrations present when equilibrium is attained. This can be important in an industrial situation to assess the economics of production of a chemical compound. Alternatively, during production by a slow chemical reaction, monitoring of the reaction mixture will allow chemists to assess if the reaction mixture is at, or nearly at, equilibrium. If not, they can evaluate the cost of further time in the reaction vessel against the value of the additional amount of product obtained by waiting for equilibrium to be attained. We look at some cases of this situation in Worked Examples 13.5 and 13.6. In the latter, arriving at an answer requires solution of a quadratic equation.

WORKED EXAMPLE 13.5—CALCULATING EQUILIBRIUM CONCENTRATIONS

For the reaction

$$H_2(g) + I_2(g) \rightleftharpoons 2\,HI(g)$$

the equilibrium constant at 425 °C is 55.64. If 1.00 mol each of $H_2(g)$ and $I_2(g)$ are placed in a 0.500 L flask at 425 °C, what are the concentrations of $H_2(g)$, $I_2(g)$, and $HI(g)$ when the reaction mixture comes to equilibrium?

Strategy

A common procedure is to assume that the concentration of one of the reactants changes by the amount x mol L^{-1} by the time that equilibrium is attained. The balanced chemical equation for the reaction then allows to calculate the changes in the concentrations of all reactants and products, and therefore the equilibrium concentrations, in terms of the parameter x. Substituting these equilibrium concentrations into the reaction quotient, with its known value at equilibrium, allows us to solve for the value of x.

Solution

Let's make the assumption that the decrease in concentration of $H_2(g)$ is x mol L^{-1}. From the balanced equation, we deduce that the concentration of $I_2(g)$ also decreases by x mol L^{-1}, and that the concentration of $HI(g)$ increases by $2x$ mol L^{-1}. It is useful to use an ICE concentrations table.

	$H_2(g)$	$+$	$I_2(g)$	$\rightleftharpoons$	$2\,HI(g)$
Initial concentration	(1 mol)/(0.500 L) = 2.00 mol L^{-1}		(1 mol)/(0.500 L) = 2.00 mol L^{-1}		0 mol L^{-1}
Concentration change	$- x$ mol L^{-1}		$- x$ mol L^{-1}		$+2x$ mol L^{-1}
Equilibrium concentration	$(2.00 - x)$ mol L^{-1}		$(2.00 - x)$ mol L^{-1}		$2x$ mol L^{-1}

Now we can substitute the equilibrium concentrations of the three species, expressed in terms of the unknown quantity x, into the reaction quotient, whose value we know:

$$Q = \frac{[HI]^2}{[H_2][I_2]} = \frac{(2x)^2}{(2.00 - x)(2.00 - x)} = \frac{(2x)^2}{(2.00 - x)^2} = K = 55.64$$

In this case, we have a simple path to finding the value of x by taking square roots:

$$\frac{2x}{2.00 - x} = \sqrt{K} = \sqrt{55.64} = 7.459$$

$$7.459(2.00 - x) = 2x$$

$$x = 1.58 \text{ mol } L^{-1}$$

Now we can calculate the equilibrium concentrations of all species:

$[H_2] = [I_2] = 2.00$ mol $L^{-1} - x$ mol $L^{-1} = 2.00$ mol $L^{-1} - 1.58$ mol $L^{-1} = 0.42$ mol L^{-1}

$[HI] = 2 \times x$ mol $L^{-1} = 2 \times 1.58$ mol $L^{-1} = 3.16$ mol L^{-1}

Comment

It is always wise to verify the answer by substituting the values back into the reaction quotient to see if your calculated values are consistent with the known value of K. In this case $(3.16)^2/(0.42)^2 = 57$. The small discrepancy with the given value of K (55.64) is because we know $[H_2]$ and $[I_2]$ to only two significant figures.

WORKED EXAMPLE 13.6—CALCULATING EQUILIBRIUM CONCENTRATIONS

The reaction

$$N_2(g) + O_2(g) \rightleftharpoons 2 NO(g)$$

contributes to air pollution whenever a fuel is burned in air at a high temperature. At 1500 K, $K = 1.0 \times 10^{-5}$. Suppose a sample of air has $[N_2] = 0.80$ mol L^{-1} and $[O_2] = 0.20$ mol L^{-1} before any reaction occurs. Calculate the equilibrium concentrations of reactants and products at 1500 K.

Solution

Define the decrease in $[N_2]$ to be x mol L^{-1}. From the balanced equation, we see that the decrease in $[O_2]$ will also be x mol L^{-1}, and the increase of $[NO]$ will be $2x$ mol L^{-1}. We can set up an ICE concentrations table.

	$N_2(g)$	+	$O_2(g)$	$\rightleftharpoons$	$2 NO(g)$
Initial (mol L^{-1})	0.80		0.20		0
Change (mol L^{-1})	$-x$		$-x$		$+2x$
Equilibrium (mol L^{-1})	$0.80 - x$		$0.20 - x$		$2x$

Now we can substitute these concentrations into Q (whose value is K at equilibrium):

$$Q = \frac{[NO]^2}{[N_2][O_2]} = \frac{(2x)^2}{(0.80 - x)(0.20 - x)} = K = 1.0 \times 10^{-5}$$

Multiplying out:

$$(1.0 \times 10^{-5})(0.80 - x)(0.20 - x) = 4x^2$$
$$(1.0 \times 10^{-5})(0.16 - 1.00x + x^2) = 4x^2$$
$$(4 - 1.0 \times 10^{-5})x^2 + (1.0 \times 10^{-5})x - 0.16 \times 10^{-5} = 0$$

This quadratic equation is of the generalized form $ax^2 + bx + c = 0$ and can be solved by the standard procedure for solving quadratic equations. There are two root solutions to this equation:

$$x = 6.3 \times 10^{-4} \qquad \text{or} \qquad x = -6.3 \times 10^{-4}$$

Only the first root is sensible. Now we can deduce the concentrations of reactants and products, to two significant figures:

$$[N_2] = 0.80 \text{ mol } L^{-1} - 6.3 \times 10^{-4} \text{ mol } L^{-1} = 0.80 \text{ mol } L^{-1}$$
$$[O_2] = 0.20 \text{ mol } L^{-1} - 6.3 \times 10^{-4} \text{ mol } L^{-1} = 0.20 \text{ mol } L^{-1}$$
$$[NO] = 2x = 2(6.3 \times 10^{-4} \text{ mol } L^{-1}) = 1.3 \times 10^{-3} \text{ mol } L^{-1}$$

> The solution to quadratic equations of the form
> $$ax^2 + bx + c = 0$$
> is given by
> $$x = \frac{-b \pm \sqrt{b^2 - 4ac}}{2a}$$
> Calculations of this sort can be made using a computer spreadsheet.

EXERCISE 13.10—CALCULATING EQUILIBRIUM CONCENTRATIONS

At some temperature, $K = 33$ for the reaction

$$H_2(g) + I_2(g) \rightleftharpoons 2 HI(g)$$

Assume the initial concentrations of both H_2 and I_2 are 6.00×10^{-3} mol L^{-1}. Find the concentration of each reactant and product at equilibrium at that temperature.

Interactive Exercise 13.9

Calculate equilibrium concentrations.

If we know the equilibrium constant of a reaction at the temperature of a reaction mixture and the starting concentrations of species, we can calculate the equilibrium concentrations of the species.

13.5 Reaction Equations and Equilibrium Constants

The form of the reaction quotient is derived directly from the chemical equation. Any chemical reaction can be represented by a variety of equations, each with their own form of the reaction quotient and value of the equilibrium constant. As a result, an equilibrium constant means nothing without specification of the chemical equation on which it is based. It doesn't matter that any chemical reaction can be described by more than one balanced chemical equation, and more than one equilibrium constant, as long as you use the K appropriate for the equation that you are using.

Doubling the Reaction Equation

The oxidation of carbon to give carbon monoxide can be described in equation 1:

$$\text{Equation 1: } C(s) + \tfrac{1}{2} O_2(g) \rightleftharpoons CO(g)$$

With the reaction defined in this way, for a reaction mixture at equilibrium at 25 °C, we can write

$$Q_1 = \frac{[CO]}{[O_2]^{1/2}} = 4.6 \times 10^{23} = K_1$$

The reaction can be equally well represented by equation 2, in which the coefficients are double those in equation 1:

$$\text{Equation 2: } 2\,C(s) + O_2(g) \rightleftharpoons 2\,CO(g)$$

Then the reaction quotient has a different form, and the equilibrium constant at 25 °C has a different value:

$$Q_2 = \frac{[CO]^2}{[O_2]} = 2.1 \times 10^{47} = K_2$$

Remembering that although we are using two different equations, the amounts of the various species in the reaction mixture do not change, we can show that $K_2 = (K_1)^2$:

$$K_2 = \frac{[CO]^2}{[O_2]} = \left\{ \frac{[CO]}{[O_2]^{1/2}} \right\}^2 = (K_1)^2$$

In summary, for a particular chemical reaction, if an equation (2) has coefficients twice that of another equation (1), K for equation (2) is the square of K for equation (1).

Similarly, if we multiply an equation by 3 (for example) the value of the equilibrium constant is cubed.

EXERCISE 13.11—REACTION EQUATIONS AND EQUILIBRIUM CONSTANTS

Which of the following correctly relates the equilibrium constants for the two reactions shown?

$$A + B \xrightleftharpoons{K_1} 2\,C$$
$$2A + 2B \xrightleftharpoons{K_2} 4\,C$$

(a) $K_2 = 2 \times K_1$ (b) $K_2 = K_1^2$ (c) $K_1 = 2 \times K_2$ (d) $K_1 = K_2^2$

If the coefficients of equation 1 are twice those of equation 2, K for equation 1 is the square of that for equation 2.

Reversing the Reaction Equation

What happens if a chemical equation is reversed? Let us compare the value of K for ionization of formic acid in aqueous solution written as follows:

Equation 1: $HCOOH(aq) + H_2O(\ell) \rightleftharpoons HCOO^-(aq) + H_3O^+(aq)$

$$Q_1 = \frac{[HCOO^-][H_3O^+]}{[HCOOH]} = K_1 = 1.8 \times 10^{-4} \text{ at } 25°C$$

with that of the reverse equation:

Equation 2: $HCOO^-(aq) + H_3O^+(aq) \rightleftharpoons HCOOH(aq) + H_3O^+(\ell)$

$$Q_2 = \frac{[HCOOH]}{[HCOO^-][H_3O^+]} = K_2 = 5.6 \times 10^3 \text{ at } 25°C$$

The second reaction quotient is the inverse of the first. The same relationship applies to the equilibrium constants:

$$K_2 = \frac{1}{K_1}$$

WORKED EXAMPLE 13.7—REACTION EQUATIONS AND EQUILIBRIUM CONSTANTS

A mixture of nitrogen, hydrogen, and ammonia is at equilibrium. When the chemical equation is written using whole-number coefficients, the value of K at 25 °C is 3.5×10^8:

Equation 1: $N_2(g) + 3 H_2(g) \rightleftharpoons 2 NH_3(g)$ $K_1 = 3.5 \times 10^8$

If the equation is written as in equation 2, what is the value of K_2, and how does the changed value of K affect the concentrations of the reactants and products?

Equation 2: $\frac{1}{2} N_2(g) + \frac{3}{2}H_2(g) \rightleftharpoons NH_3(g)$ $K_2 = ?$

Equation 3 for the decomposition of ammonia to the elements is the reverse of its formation as described by equation 1. What is the value of K_3?

Equation 3: $2 NH_3(g) \rightleftharpoons N_2(g) + 3 H_2(g)$ $K_3 = ?$

Solution

The coefficients in equation 1 are double those in equation 2. So $K_1 = (K_2)^2$.

$$\therefore K_2 = \sqrt{K_1} = \sqrt{3.5 \times 10^8} = 1.9 \times 10^4$$

Whichever chemical equation we write to represent an equilibrium reaction does not affect the concentrations of species in the reaction mixture. We find, within rounding-off errors, that one set of equilibrium concentrations is consistent with all values of K, as along as each reaction quotient used is appropriate for the way the chemical reaction is written. For example, an equilibrium reaction mixture at 25 °C has $[NH_3] = 0.098$ mol L^{-1}, $[N_2] = 0.028$ mol L^{-1}, and $[H_2] = 0.0010$ mol L^{-1}.

Using reaction equation 1, we can see that $Q = K$, and this is an equilibrium mixture:

$$Q_1 = \frac{[NH_3]^2}{[N_2][H_2]^3} = \frac{(0.098)^2}{(0.028)(0.0010)^3} = 3.5 \times 10^8 = K_1$$

The same concentrations can be seen to be at equilibrium if we use equation 2:

$$Q_2 = \frac{[NH_3]}{[N_2]^{1/2}[H_2]^{3/2}} = \frac{0.098}{(0.028)^{1/2}(0.0010)^{3/2}} = 1.9 \times 10^4 = K_2$$

Equation 3 is the reverse of equation 1, and so $K_3 = 1/K_1$:

$$K_3 = \frac{1}{K_1} = \frac{1}{3.5 \times 10^8} = 2.9 \times 10^{-9}$$

Comment

Although the reaction for formation of ammonia from nitrogen and hydrogen is product-favoured at 25 °C, the rate of achieving equilibrium is very slow indeed. This was the crux of the problem for Fritz Haber [>>Section 14.1]. The control of the reaction to achieve reasonable equilibrium yields in reasonable times is discussed in Section 14.7.

Interactive Exercise 13.12

SUBMIT

Compare the equilibrium constants for a reaction represented by different equations.

EXERCISE 13.13—REACTION EQUATIONS AND EQUILIBRIUM CONSTANTS

The conversion of oxygen to ozone has a very small equilibrium constant:

$$\tfrac{3}{2} O_2(g) \rightleftharpoons O_3(g) \qquad K = 2.5 \times 10^{-29} \qquad \text{at 25 °C}$$

(a) What is the value of K at 25 °C when the equation is written using whole-number coefficients?

$$3\,O_2(g) \rightleftharpoons 2\,O_3(g)$$

(b) What is the value of K for the conversion of ozone to oxygen?

$$2\,O_3(g) \rightleftharpoons 3\,O_2(g)$$

If the equation for an equilibrium reaction is the reverse of another, K for each reaction is the reciprocal of the other.

Deriving an Equilibrium Constant from Others

It is sometimes useful to add two balanced reaction equations to obtain the net equation for a process. This can be beneficial because if we know the equilibrium constants of the reactions whose equations we add, we can derive the equilibrium constant for the net process.

As an example, consider the reactions that take place when a solid silver chloride precipitate in water is in equilibrium with $Ag^+(aq)$ and $Cl^-(aq)$ ions, and ammonia is added to the solution. The ammonia reacts with silver ions to form a complex ion, $Ag(NH_3)_2^+$ [<<Section 6.8], and the silver chloride dissolves (Figure 13.7).

Photos: Charles D. Winters

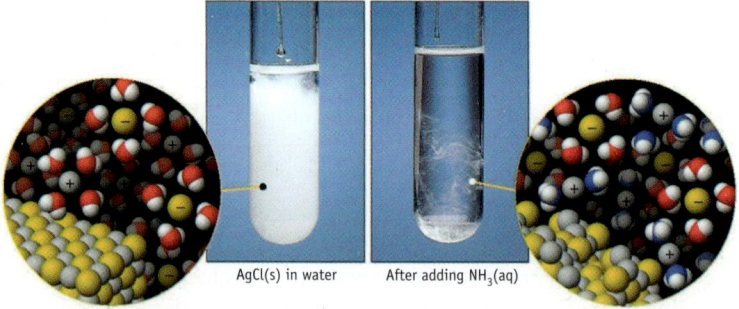

AgCl(s) in water   After adding NH₃(aq)

FIGURE 13.7 Dissolving solid silver chloride by adding ammonia. (Left) A precipitate of AgCl(s) is suspended in water. (Right) When a concentrated solution of ammonia is added, the ammonia reacts with the low level of $Ag^+(aq)$ ions in solution to form $[Ag(NH_3)_2]^+$ complex ions. Because the reaction between $Ag^+(aq)$ ions and ammonia has a large equilibrium constant, if enough ammonia is added, essentially all of the $Ag^+(aq)$ ions react to form the complex ions, and the AgCl precipitate dissolves. This is an example of competition between insolubility and complexation for metal ions [>>Section 15.4].

If we add the equation for dissolving solid AgCl and the equation for the reaction of $Ag^+(aq)$ ions with ammonia, we get the equation for the net reaction—dissolving solid AgCl in aqueous ammonia solution:

$$AgCl(s) \rightleftharpoons \cancel{Ag^+(aq)} + Cl^-(aq) \qquad K_1 = 1.8 \times 10^{-10} \text{ at 25°C}$$

$$\cancel{Ag^+(aq)} + 2\,NH_3(aq) \rightleftharpoons Ag(NH_3)_2^+(aq) \qquad K_2 = 1.6 \times 10^7 \text{ at 25°C}$$

Net: $AgCl(s) + 2\,NH_3(aq) \rightleftharpoons Ag(NH_3)_2^+(aq) + Cl^-(aq) \qquad K_{net} = ?$

We can see that the reaction quotient for the net reaction is the product of the reaction quotients for the other two reactions:

$$Q_1 \times Q_2 = [Ag^+][Cl^-] \times \frac{[Ag(NH_3)_2^+]}{[Ag^+][NH_3]^2} = \frac{[Ag(NH_3)_2^+][Cl^-]}{[NH_3]^2} = Q_{net}$$

When all three reactions are at equilibrium, $Q_1 = K_1$, $Q_2 = K_2$, and $Q_{net} = K_{net}$, and so

$$K_{net} = K_1 \times K_2 = (1.8 \times 10^{-10})(1.6 \times 10^7) = 2.9 \times 10^{-3} \text{ at } 25°C$$

EXERCISE 13.14—DERIVING AN EQUILIBRIUM CONSTANT FROM OTHERS

Calculate K for the reaction

$$SnO_2(s) + 2\ CO(g) \rightleftharpoons Sn(s) + 2\ CO_2(g)$$

at the same temperature as that at which the following information applies:

$$SnO_2(s) + 2\ H_2(g) \rightleftharpoons Sn(s) + 2\ H_2O(g) \qquad K_1 = 8.12$$
$$H_2(g) + CO_2(g) \rightleftharpoons H_2O(g) + CO(g) \qquad K_2 = 0.771$$

> If we can add the equations of two or more reactions to obtain a balanced chemical equation for another reaction, K for the net reaction is the product of the equilibrium constants of the other reactions at the same temperature.

13.6 Disturbing Reaction Mixtures at Equilibrium

If a reaction mixture is at equilibrium at a given temperature, $Q = K$. The condition of chemical equilibrium may be "disturbed" (that is, the mixture rendered no longer at equilibrium) if the conditions are changed such that $Q \neq K$. This can be brought about in either of the following ways, the first of which brings about a change in Q, and the second a change in K:

- *At a given temperature, if the concentrations of reactants or products (or both) are changed in such way that the value of Q is changed.* Then $Q \neq K$. There will be a tendency for net reaction in the direction that brings the value of Q closer to K, until $Q = K$ again.
- *If the temperature is changed, then the equilibrium constant K for the reaction changes* and so, if no net reaction occurs, $Q \neq K$. There will be a tendency for net reaction to occur in the direction that brings the value of Q toward K again.

A useful device for predicting the effect of disturbing a reaction mixture at equilibrium is **Le Chatelier's principle**: *If the condition of equilibrium of a reaction mixture is disturbed by either a change of concentrations of species, or by change of temperature, then the relative concentrations of reactants and products will change in such a way as to minimize the imposed change.* As we discuss below, the effects of different ways that equilibrium can be disturbed in terms of the relationship between Q and K, you can check that the outcomes are consistent with predictions made by Le Chatelier's principle.

Le Chatelier's principle, which has no theoretical basis, only allows us to make qualitative predictions about the direction of net reaction after equilibrium is disturbed. Predictions based on comparison of Q and K are theoretically based, and, as well as the direction of net reaction, can quantitatively predict the amounts that react before equilibrium is re-established.

Effect of Changing Concentrations

It is useful to consider two different ways that concentrations of species in a reaction mixture at equilibrium can be changed: (a) adding or removing reactants and products, and (b) in the case of gas-phase reaction mixtures, changing the volume.

Adding or Removing Reactants or Products Imagine that we have a reaction mixture with butane and methyl-2-propane (isobutane) in equilibrium at 25 °C. The equilibrium constant at 25 °C is 2.5.

$$CH_3CH_2CH_2CH_3(g) \rightleftharpoons CH_3CH(CH_3)CH_3(g)$$

At equilibrium $\quad Q = \dfrac{[CH_3CH(CH_3)CH_3]}{[CH_3CH_2CH_2CH_3]} = K = 2.5$

If we were to add some butane to the reaction vessel, keeping its volume constant, then $[CH_3CH_2CH_2CH_3]$ would be increased, and the value of Q would decrease. Then we would have $Q < K$, and the mixture no longer at equilibrium. For the mixture to come to equilibrium again, there would need to be net reaction in the direction that Q increases (decrease of butane concentration, consistent with Le Chatelier's principle) until $Q = K$ again.

For a specified increase of concentration of butane, we can calculate how much the concentration of butane needs to change in order that equilibrium is re-established (Worked Example 13.8).

If the concentration of isobutane were increased, Q would increase, and then $Q > K$. The direction of net reaction that would decrease Q is the direction that removes isobutane (consistent with Le Chatelier's principle).

Think about It

e13.10 Simulate the effect of adding a reagent to a system at equilibrium.

Think about It

e13.11 Calculate the effect of adding a reagent to a system at equilibrium.

Think about It

e13.12 See a graphical representation of the effect of adding a reagent to a system at equilibrium.

WORKED EXAMPLE 13.8—EFFECT OF CONCENTRATION CHANGES ON EQUILIBRIUM

Assume equilibrium has been established in a 1.00 L flask with [butane] = 0.500 mol L^{-1} and [isobutane] = 1.25 mol L^{-1}.

$$\text{butane} \rightleftharpoons \text{isobutane} \qquad K = 2.5$$

Then 1.50 mol of butane is added. What are the concentrations of butane and isobutane when equilibrium is re-established?

Strategy

After adding excess butane, $Q < K$. To re-establish equilibrium, the concentration of butane must decrease, and that of isobutane must increase. The decrease in butane concentration and the increase in isobutane concentration are both designated as x.

Solution

First organize the information in a modified ICE table.

Equation	Butane $\rightleftharpoons$	Isobutane
Initial (mol L^{-1})	0.500	1.25
Concentration immediately on adding butane (mol L^{-1})	0.500 + 1.50 = 2.00	1.25
Change in concentration to re-establish equilibrium (mol L^{-1})	$-x$	$+x$
Equilibrium (mol L^{-1})	2.00 − x	1.25 + x

Now we have, after equilibrium is re-established

$$Q = \frac{[\text{isobutane}]}{[\text{butane}]} = \frac{1.25 + x}{2.00 - x} = K = 2.5$$

which leads to $x = 1.07$ mol L^{-1}.

We can calculate the concentrations after equilibrium is re-established:

$$[\text{butane}] = 2.00 - x = 2.00 - 1.07 = 0.93 \text{ mol } L^{-1}$$
$$[\text{isobutane}] = 1.25 + x = 1.25 + 1.07 = 2.32 \text{ mol } L^{-1}$$

Comment

Check your answer to verify that $Q = K$ again.

EXERCISE 13.15—EFFECT OF CONCENTRATION CHANGES ON EQUILIBRIUM

In a reaction mixture at a certain temperature, butane and isobutane are in chemical equilibrium when [butane] = 0.20 mol L^{-1} and [isobutane] = 0.50 mol L^{-1}. Enough isobutane is added to the mixture to increase its concentration by 2.00 mol L^{-1}. What are the concentrations of butane and isobutane after equilibrium is re-established?

> If reagents are added to, or removed from, an equilibrium mixture so that $Q < K$, then equilibrium would be re-established by net reaction of reactants to products (increasing Q). This is consistent with application of Le Chatelier's principle.

Changing the Volume of a Gas-Phase Reaction Mixture

For a reaction that involves gases, what happens to the reaction quotient if the volume of the container is changed, changing the concentration of all species by the same ratio? Consider an equilibrium reaction mixture represented by the following:

$$2\ NO_2(g) \rightleftharpoons N_2O_4(g)$$

$$Q = \frac{[N_2O_4]}{[NO_2]^2} = K = 170 \text{ at } 298 \text{ K}$$

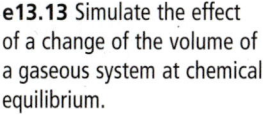

Think about It

e13.13 Simulate the effect of a change of the volume of a gaseous system at chemical equilibrium.

Imagine that the equilibrium mixture is in a cylinder with a movable piston. What happens to this system if the volume of the vessel is suddenly halved? The immediate result is that the concentrations of both gases will double. For example, assume that in the initial equilibrium mixture, $[N_2O_4]$ = 0.0279 mol L^{-1} and $[NO_2]$ = 0.0128 mol L^{-1}. $Q = (0.0279)/(0.0128)^2 = 170 = K$. When the volume is halved, $[N_2O_4]$ becomes 0.0558 mol L^{-1} and $[NO_2]$ is 0.0256 mol L^{-1}, and Q becomes $(0.0558)/(0.0256)^2 = 85.1$. If there were no net reaction, we would have $Q < K$. Re-establishment of equilibrium would require net reaction in the direction to increase Q (that is, in the direction that decreases $[NO_2]$ and increases $[N_2O_4]$).

Similar reasoning will lead to the following generalizations:

- A volume decrease of a gaseous reaction mixture at equilibrium will lead to a new equilibrium condition by net reaction that results in a decrease in the number of molecules in the system.
- A volume increase of a gaseous reaction mixture at equilibrium leads to a new equilibrium condition by net reaction that results in an increase in the number of molecules in the system.
- For a reaction in which there is no change in the number of molecules when reaction occurs (such as $H_2(g) + I_2(g) \rightleftharpoons 2\ HI(g)$), a volume change will not lead to a change in the relative amounts of reactants and products, because the equality $Q = K$ will not be disturbed.

To apply Le Chatelier's principle to situations in which the volume of a reaction system is changed, we need to consider the applied change as the change of concentration caused by the volume change (rather than by the volume change itself). According to this, if the total concentration of molecules is increased as a result of a volume decrease, then net reaction will happen in the direction that decreases the total concentration of molecules.

EXERCISE 13.16—EFFECT OF CHANGING THE VOLUME OF A GAS-PHASE REACTION

The formation of ammonia from its elements by the Haber process is an important industrial process.

$$3\ H_2(g) + N_2(g) \rightleftharpoons 2\ NH_3(g)$$

(a) How does the equilibrium composition change when extra H_2 is added? When extra NH_3 is added?

(b) What is the effect on the equilibrium when the volume of the system is increased? Does the equilibrium composition change or is the system unchanged?

If the volume of a gaseous reaction mixture at equilibrium is decreased, the direction of net reaction that would lead to a new equilibrium condition is that which results in a decrease in the number of molecules. This can be explained by the effect of compression on the value of the reaction quotient (Q).

Effect of Changing the Temperature

You can make a qualitative prediction about the effect of a temperature change on the equilibrium composition of a chemical reaction if you know whether the reaction is exothermic or endothermic. As an example, consider the endothermic reaction of $N_2(g)$ with $O_2(g)$ to give $NO(g)$:

$$N_2(g) + O_2(g) \rightleftharpoons 2\,NO(g) \qquad \Delta_rH^\circ = +180 \text{ kJ mol}^{-1}$$

$$\text{At equilibrium, } Q = \frac{[NO]^2}{[N_2][O_2]} = K$$

The equilibrium constant increases as temperature is raised, as is the case for all endothermic reactions.

K	Temperature
4.5×10^{-31}	298 K
6.7×10^{-10}	900 K
1.7×10^{-3}	2300 K

So if we have a reaction mixture at equilibrium ($Q = K$), and the temperature of the mixture is suddenly raised, K will increase, and so immediately $Q < K$. Equilibrium would be re-established by net reaction in the direction that Q increases; that is, in the endothermic direction written above.

This prediction is consistent with that of Le Chatelier's principle: if equilibrium is disturbed because the temperature is raised, then the direction of net reaction that re-establishes equilibrium is the one that would reduce the temperature of the system (the endothermic direction of reaction).

As another example, consider the combination of molecules of the brown gas NO_2 to form colourless $N_2O_4(g)$. An equilibrium between these compounds is readily achieved in a closed system (Figure 13.8).

$$2\,NO_2(g) \rightleftharpoons N_2O_4(g) \qquad \Delta_rH^\circ = -57.1 \text{ kJ mol}^{-1}$$

$$\text{At equilibrium } Q = \frac{[N_2O_4]}{[NO_2]^2} = K$$

As for other exothermic reactions, K decreases as temperature is raised.

K	Temperature
1300	273 K
170	298 K

So, if a reaction mixture of this sort is at equilibrium ($Q = K$) and the temperature is raised, K will decrease, and immediately $Q > K$. Re-establishment of equilibrium would need net reaction in the direction that decreases the magnitude of Q—that is, also in the endothermic direction, opposite to that written.

We can generalize: *when the temperature of a system at equilibrium is increased, equilibrium is disturbed because K changes and Q ≠ K. Re-establishment of equilibrium would be achieved by net reaction in the direction that is endothermic.* Conversely, if the temperature of a system at equilibrium is decreased, re-establishment of equilibrium would be achieved by net reaction in the direction that is exothermic.

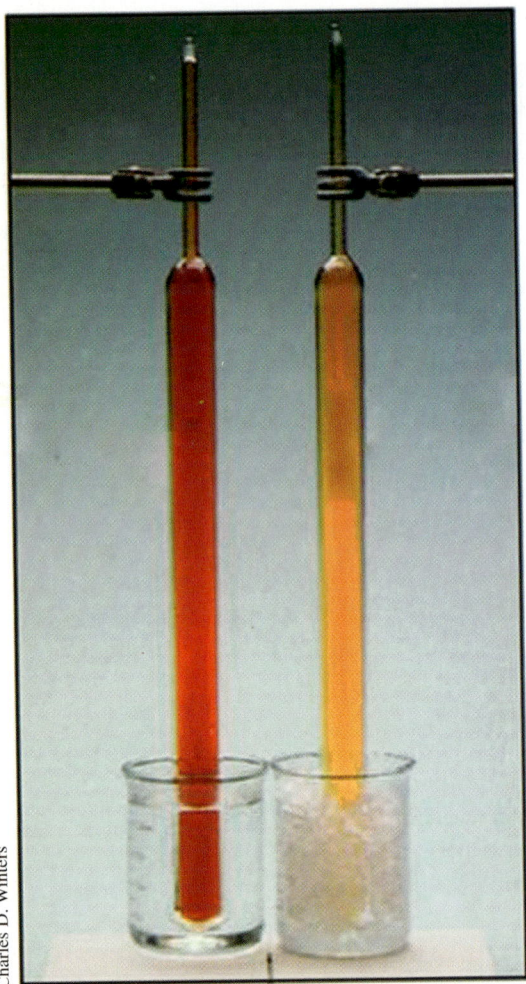

Charles D. Winters

FIGURE 13.8 Effect of temperature on an equilibrium mixture. The tubes in the photograph both contain gaseous NO_2 (brown) and N_2O_4 (colourless) at equilibrium. K is larger near 0 °C *(right)*, so the system contains relatively high concentration of the colourless $N_2O_4(g)$. At 50 °C *(left)*, K is smaller, and the reaction mixture contains a relatively high concentration of NO_2, as indicated by the darker brown colour.

Think about It

e13.14 Predict the effect of temperature changes on a system at equilibrium.

The dependence of K of reactions on temperature can be important. One useful outcome is in the control of ammonia production by the Haber-Bosch process, as discussed in Section 13.7. An undesirable environmental outcome is that although $N_2(g)$ and $O_2(g)$ in the atmosphere do not react noticeably, at the high temperatures in car engines and industrial furnaces, an equilibrium mixture contains a significant concentration of $NO(g)$. This is a reactant in reactions activated by sunlight that lead to "photochemical smogs" [>>Section 17.1].

EXERCISE 13.17—DEPENDENCE OF EQUILIBRIUM CONSTANT ON TEMPERATURE

Consider the disturbance of the following equilibrium systems, and predict what happens in terms of both (i) comparison of Q and K, and (ii) Le Chatelier's principle.

(a) Does the equilibrium concentration of $NOCl(g)$ increase or decrease if the temperature of the system is increased?

$$2\ NOCl(g) \rightleftharpoons 2\ NO(g) + Cl_2(g) \qquad \Delta_rH° = +77.1\ kJ\ mol^{-1}$$

(b) Does the equilibrium concentration of $SO_3(g)$ increase or decrease if the temperature is increased?

$$2\ SO_2(g) + O_2(g) \rightleftharpoons 2\ SO_3(g) \qquad \Delta_rH° = -198\ kJ\ mol^{-1}$$

For endothermic reactions, K increases as temperature is raised, so if the temperature of a system at equilibrium is raised, $Q < K$ and a new equilibrium condition would be attained by net reaction of reactants to products. The opposite is the case for exothermic reactions. This is consistent with application of Le Chatelier's principle.

13.7 Applying the Principles: The Haber-Bosch Process

As discussed in Section 13.1, one of the greatest advances in agriculture has been the manufacture of nitrogen-containing fertilizers, especially ammonia and ammonium compounds. This industry is based on the Haber-Bosch process: the synthesis of ammonia from its elements, nitrogen and hydrogen.

$$N_2(g) + 3\ H_2(g) \rightleftharpoons 2\ NH_3(g)$$

At 25°C, K (calculated value) $= 3.5 \times 10^8$ and $\Delta_rH° = -91.8\ kJ\ mol^{-1}$

At 450°C, K (experimental value) $= 0.16$ and $\Delta_rH° = -111.3\ kJ\ mol^{-1}$

This process (Figure 13.9) is a good example of the role that control of chemical equilibrium, as well as of reaction kinetics, can play in practical chemistry.

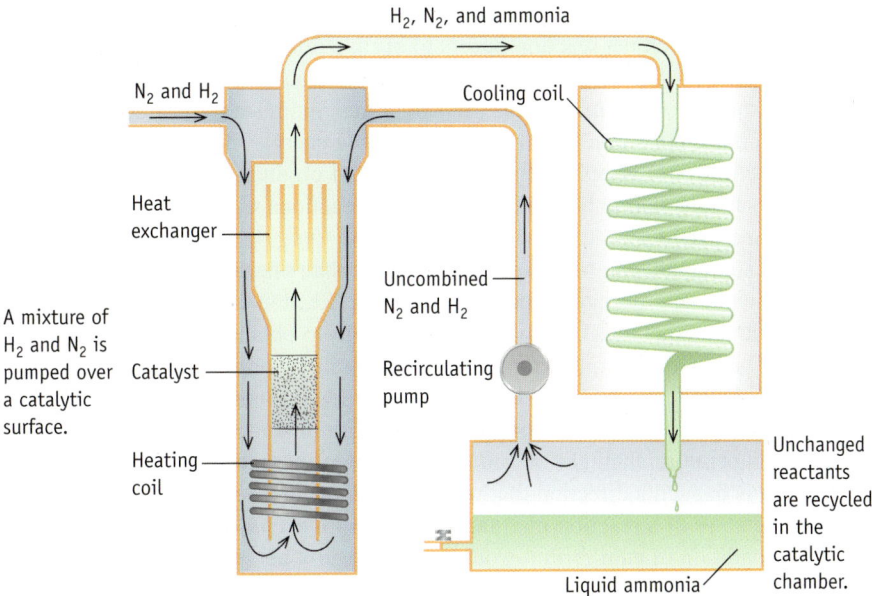

H_2, N_2, and ammonia

N_2 and H_2

Cooling coil

Heat exchanger

Uncombined N_2 and H_2

A mixture of H_2 and N_2 is pumped over a catalytic surface.

Catalyst

Recirculating pump

Heating coil

Liquid ammonia

Unchanged reactants are recycled in the catalytic chamber.

FIGURE 13.9 The Haber-Bosch process for ammonia synthesis. A mixture of $N_2(g)$ and $H_2(g)$ is pumped over a catalytic surface. The ammonia is collected as a liquid at −33 °C, and unreacted $N_2(g)$ and $H_2(g)$ are recycled in the catalytic chamber.

The chemical principles that are applied in this process are the following:

- At ambient temperatures, the reaction is exothermic and product-favoured ($K > 1$), but is slow. So it is carried out at a higher temperature to increase the reaction rate [<<Section 13.6].
- Although the reaction goes faster as the temperature is raised, the equilibrium constant decreases. So, for a given concentration of starting materials, the equilibrium concentration of ammonia is smaller at higher temperatures. In an industrial ammonia plant, it is necessary to balance reaction rate (faster at higher temperature) with product yield (greater at lower temperature).
- A catalyst is used to increase the reaction rate. An effective catalyst for the Haber-Bosch process is Fe_3O_4 mixed with KOH, SiO_2, and Al_2O_3 (all inexpensive chemicals). The catalyst is not effective at temperatures less than 400 °C, however, and the optimal temperature is about 450 °C.
- To increase the equilibrium concentration of ammonia, the reaction is carried out at a higher pressure.
- Ammonia is continually liquefied—by cooling in a chamber outside of the reaction vessel—and removed. The mixture of gases is then recycled into the reaction chamber.

SUMMARY

Key Concepts

- **Reversible** chemical reactions can go in either direction, depending on the conditions. In a reaction mixture in **dynamic chemical equilibrium**, the reactions in the opposite directions occur at equal rates so that there is no **net reaction**. (Section 13.2)
- For a particular reaction, all reaction mixtures have the same value of the **reaction quotient** (Q). This value of Q is called the **equilibrium constant** (K) at that temperature. (Section 13.3)
- Reaction quotients are most correctly expressed as functions of activities of species rather than concentrations. Activities are measures of effective concentration, and as a ratio compared with 1 mol L^{-1}, are unitless. (Section 13.3)
- In a reaction mixture, if $Q = K$, the mixture is at equilibrium. If $Q < K$, equilibrium is attained if net reaction changes reactants into products. If $Q > K$, equilibrium is attained if net reaction changes products into reactants. (Section 13.3)
- The direction in which net reaction occurs to bring a reaction mixture to equilibrium is the **spontaneous direction of reaction**. As spontaneous reaction proceeds, the reaction mixture becomes more stable, reaching maximum stability at equilibrium. Equilibrium is the condition of minimum Gibbs free energy. (Section 13.3)
- Reactions with $K \gg 1$ have higher concentrations of products than reactants at equilibrium and are said to be product-favoured. Reactions with $K \ll 1$ are reactant-favoured. (Section 13.4)
- Any chemical reaction can be represented by more than one equation. The value of K used must be that appropriate to the equation used. (Section 13.5)
- If we can add the equations of two or more reactions to obtain a balanced chemical equation for another reaction, K for the net reaction is the product of the equilibrium constants of the other reactions at the same temperature. (Section 13.5)
- If a change is made to a reaction mixture at equilibrium that changes Q, without changing temperature, then $Q \neq K$ and the direction of net reaction that would restore equilibrium is in such a direction that makes $Q = K$ again. (Section 13.6)
- If the temperature of a reaction mixture at equilibrium is changed, K changes and so $Q \neq K$. A condition of equilibrium would be restored by net reaction in the direction that makes Q the same as the new K. (Section 13.6)

Key Equations

For a reaction $a\,A + b\,B \rightleftharpoons c\,C + d\,D$, the reaction quotient (Q) is defined by

$$Q = \frac{[C]^c[D]^d}{[A]^a[B]^b}$$

For a system at equilibrium, $Q = K$. (Section 13.3)

REVIEW QUESTIONS

Section 13.3: The Reaction Quotient and the Equilibrium Constant

13.18 Write the reaction quotient (Q) appropriate for each of the following reactions:
(a) $2\,H_2O_2(g) \rightleftharpoons 2\,H_2O(g) + O_2(g)$
(b) $CO(g) + \frac{1}{2}O_2(g) \rightleftharpoons CO_2(g)$
(c) $C(s) + CO_2(g) \rightleftharpoons 2\,CO(g)$
(d) $NiO(s) + CO(g) \rightleftharpoons Ni(s) + CO_2(g)$

13.19 At 2000 K, the equilibrium constant (K) for the formation of nitric oxide, $NO(g)$, from nitrogen and oxygen is 4.0×10^{-4}:

$$N_2(g) + O_2(g) = \rightleftharpoons 2\,NO(g)$$

You have a flask in which, at 2000 K, the concentration of $N_2(g)$ is 0.50 mol L^{-1}, that of $O_2(g)$ is 0.25 mol L^{-1}, and that of $NO(g)$ is 4.2×10^{-3} mol L^{-1}. Is the system at equilibrium? If not, predict which way the reaction would proceed to achieve equilibrium.

13.20 At 500 K, $K = 5.6 \times 10^{12}$ for the dissociation of iodine molecules to iodine atoms:

$$I_2(g) \rightleftharpoons 2\,I(g)$$

A reaction mixture at 500 K has $[I_2] = 0.020$ mol L^{-1} and $[I] = 2.0 \times 10^{-8}$ mol L^{-1}. Is the mixture at equilibrium? If not, which way must net reaction proceed to reach equilibrium?

Section 13.4: Quantitative Aspects of Equilibrium Constants

13.21 The reaction

$$PCl_5(g) \rightleftharpoons PCl_3(g) + Cl_2(g)$$

is at equilibrium in a mixture at 250 °C. Measurements showed that $[PCl_5] = 4.2 \times 10^{-5}$ mol L^{-1}, $[PCl_3] = 1.3 \times 10^{-2}$ mol L^{-1}, and $[Cl_2] = 3.9 \times 10^{-3}$ mol L^{-1}. Calculate K for this reaction.

13.22 The reaction

$$C(s) + CO_2(g) \rightleftharpoons 2\,CO(g)$$

occurs at high temperatures. At 700 °C, a 2.0 L flask contains 0.10 mol of $CO(g)$, 0.20 mol of $CO_2(g)$, and 0.40 mol of $C(s)$ at equilibrium.
(a) Calculate K for the reaction at 700 °C.
(b) Calculate K for the reaction, also at 700 °C, if the amounts at equilibrium in the 2.0 L flask are 0.10 mol of $CO(g)$, 0.20 mol of $CO_2(g)$, and 0.80 mol of $C(s)$.
(c) Compare your answers in (a) and (b). Does the amount of carbon affect the value of K? Explain.

Interactive Exercise 13.23
Deduce the equilibrium constant from measured equilibrium concentrations.

13.24 You place 3.00 mol of pure SO_3 in an 8.00 L flask at 1150 K. At equilibrium, 0.58 mol of O_2 has been formed. Calculate K at 1150 K for the reaction

$$2\,SO_3(g) \rightleftharpoons 2\,SO_2(g) + O_2(g)$$

13.25 Suppose 0.086 mol of bromine (Br_2) is placed in a 1.26 L flask and heated to 1756 K, at which temperature some dissociation into atoms occurs:

$$Br_2(g) \rightleftharpoons 2\,Br(g)$$

If 3.7% of the $Br_2(g)$ is dissociated at this temperature, calculate K.

Interactive Exercises 13.26–13.27
Deduce the equilibrium constant from measured equilibrium concentrations.

13.28 Graphite and carbon dioxide are kept at constant volume at 1000 K until the reaction

$$C(graphite) + CO_2(g) \rightleftharpoons 2CO(g)$$

has come to equilibrium. At this temperature, $K = 0.021$. The initial concentration of $CO_2(g)$ is 0.012 mol L^{-1}. Calculate the equilibrium concentration of $CO(g)$.

13.29 Hydrogen and carbon dioxide gases react at a high temperature to give water and carbon monoxide:

$$H_2(g) + CO_2(g) \rightleftharpoons H_2O(g) + CO(g)$$

(a) Laboratory measurements at 986 °C show that there are 0.11 mol each of $CO(g)$ and $H_2O(g)$, and 0.087 mol each of $H_2(g)$ and $CO_2(g)$ at equilibrium in a 1.0 L container. Calculate the equilibrium constant for the reaction at 986 °C.

(b) Suppose 0.050 mol each of $H_2(g)$ and $CO_2(g)$ are placed in a 2.0 L container. When equilibrium is achieved at 986 °C, what amounts (in mol) of $CO(g)$ and $H_2O(g)$ would be present?

13.30 Carbonyl bromide decomposes to carbon monoxide and bromine:

$$COBr_2(g) \rightleftharpoons CO(g) + Br_2(g)$$

At 73 °C, K is 0.190. If you place 0.500 mol of $COBr_2(g)$ in a 2.00 L flask and heat it to 73 °C, what are the equilibrium concentrations of $COBr_2(g)$, $CO(g)$, and $Br_2(g)$? What percentage of the original $COBr_2(g)$ decomposed at this temperature?

13.31 Iodine dissolves in water, but its solubility in a non-polar solvent such as $CCl_4(\ell)$ is greater.

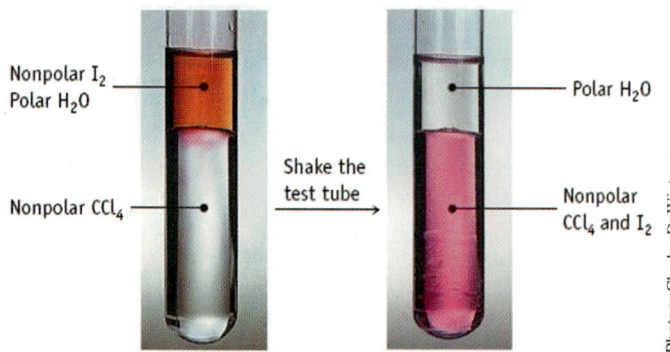

Extracting iodine (I_2) from water with the non-polar solvent CCl_4. $I_2(g)$ is more soluble in $CCl_4(\ell)$ and, after shaking a mixture of water and $CCl_4(\ell)$, the I_2 has accumulated in the more dense $CCl_4(\ell)$ layer.

The equilibrium constant at 25 °C is 85.0 for the reaction

$$I_2(aq) \rightleftharpoons I_2(CCl_4)$$

You place 0.0340 g of I_2 in 100.0 mL of water. After shaking it with 10.0 mL of $CCl_4(\ell)$, how much I_2 remains in the water layer?

Interactive Exercise 13.32

Calculate equilibrium concentrations.

13.33 Imagine that 3.60 mol of ammonia is placed in a 2.00 L vessel and allowed to decompose to the elements at 450 °C:

$$2 NH_3(g) \rightleftharpoons N_2(g) + 3 H_2(g)$$

If the experimental value of K is 6.3 for this reaction at 450 °C, calculate the equilibrium concentration of each reagent. What is the total pressure in the flask?

13.34 At 25 °C, K for the decomposition of ammonium hydrogensulfide is 1.8×10^{-4}

$$NH_4HS(s) \rightleftharpoons NH_3(g) + H_2S(g)$$

(a) When the pure salt decomposes at 25 °C in a flask, what are the equilibrium concentrations of $NH_3(g)$ and $H_2S(g)$?

(b) If $NH_4HS(s)$ is placed in a flask already containing 0.020 mol L^{-1} of $NH_3(g)$, and then the system is allowed to come to equilibrium, what are the equilibrium concentrations of $NH_3(g)$ and $H_2S(g)$?

Section 13.5: Reaction Equations and Equilibrium Constants

13.35 The equilibrium constant for the reaction

$$CO_2(g) \rightleftharpoons CO(g) + \tfrac{1}{2} O_2(g)$$

is 6.66×10^{-12} at 1000 K. Calculate K at 1000 K for the reaction

$$2 CO(g) + O_2(g) \rightleftharpoons 2 CO_2(g)$$

13.36 Which of the following correctly relates the equilibrium constants for the two reactions shown?

$$A + B \overset{K_1}{\rightleftharpoons} 2 C$$
$$C \overset{K_2}{\rightleftharpoons} \tfrac{1}{2} A + \tfrac{1}{2} B$$

(a) $K_2 = \dfrac{1}{(K_1)^{1/2}}$ (c) $K_2 = K_1^2$

(b) $K_2 = \dfrac{1}{K_1}$ (d) $K_2 = -(K_1)^{1/2}$

13.37 The equilibrium constant for the reaction represented as

$$N_2(g) + O_2 \rightleftharpoons 2 NO(g)$$

is 1.7×10^{-3} at 2300 K.

(a) What is K for the reaction when written as follows?

$$\tfrac{1}{2} N_2(g) + \tfrac{1}{2} O_2(g) \rightleftharpoons NO(g)$$

(b) What is K for the reaction represented as follows?

$$2 NO(g) \rightleftharpoons N_2(g) + O_2(g)$$

13.38 Calculate K for the reaction

$$Fe(s) + H_2O(g) \rightleftharpoons FeO(s) + H_2(g)$$

given the following information:

$$H_2O(g) + CO(g) \rightleftharpoons H_2(g) + CO_2(g) \quad K_1 = 1.6$$
$$FeO(s) + CO(g) \rightleftharpoons Fe(s) + CO_2(g) \quad K_2 = 0.67$$

Section 13.6: Disturbing Reaction Mixtures at Equilibrium

13.39 Consider the isomerization of butane at an unspecified temperature:

$$butane \rightleftharpoons isobutane$$

The system is originally at equilibrium with [butane] = 1.0 mol L^{-1} and [isobutane] = 2.5 mol L^{-1}.

(a) If isobutane is added suddenly, increasing its concentration by 0.50 mol L^{-1}, and the system shifts to a new equilibrium condition, what is the equilibrium concentration of each gas?

(b) If the concentration of butane is increased by 0.50 mol L^{-1}, and the system shifts to a new equilibrium condition, what is the equilibrium concentration of each gas?

13.40 The photo shows the result of heating and cooling a solution of Co^{2+} ions in water containing chloride ions (from HCl). The solution at the left is in a beaker of hot water, and the solution at the right is in a beaker of ice water.

Aqueous solutions containing Co^{2+} ions to which concentrated HCl solution has been added. The solution at the left is in a beaker of hot water, whereas the solution at the right is in a beaker of ice water.

The equation for the equilibrium existing in these solutions is

$$Co(H_2O)_6{}^{2+}(aq) + 4\ Cl^-(aq) \rightleftharpoons CoCl_4{}^{2-}(aq) + 6\ H_2O(\ell)$$

The Co(H$_2$O)$_6{}^{2+}$(aq) ion is pink, whereas the CoCl$_4{}^{2-}$(aq) ion is blue. Is the transformation of Co(H$_2$O)$_6{}^{2+}$(aq) to CoCl$_4{}^{2-}$(aq) exothermic or endothermic?

13.41 Dinitrogen trioxide, N$_2$O$_3$(g), decomposes to NO(g) and NO$_2$(g) in an endothermic process ($\Delta_r H° = 40.5$ kJ mol^{-1}) that comes to equilibrium:

$$N_2O_3(g) \rightleftharpoons NO(g) + NO_2(g)$$

Predict, by comparison of Q and K, the effect of the following changes on the relative concentrations of reactants and products as equilibrium is re-established:

(a) More N$_2$O$_3$(g) is added.

(b) More NO$_2$(g) is added.

(c) The volume of the reaction flask is increased.

(d) The temperature is lowered.

13.42 The decomposition of NH$_4$HS

$$NH_4HS(s) \rightleftharpoons NH_3(g) + H_2S(g)$$

is an endothermic process. By (a) comparison of Q and K, and (b) using Le Chatelier's principle, predict how increasing the temperature would affect the equilibrium. If more NH$_4$HS(s) is added to a flask in which this reaction is at equilibrium, how are the concentrations of the various substances affected? What if some additional NH$_3$(g) is put in the flask? What will happen to the partial pressure of NH$_3$(g) if some H$_2$S(g) is removed from the flask?

13.43 Heating a metal carbonate leads to endothermic decomposition:

$$BaCO_3(s) \rightleftharpoons BaO(s) + CO_2(g)$$

Predict the effect on the equilibrium of each change listed below. Answer by choosing (i) no net reaction, (ii) net reaction to form more BaCO$_3$(s), or (iii) net reaction to form more BaO(s) and CO$_2$(g).

(a) More BaCO$_3$(s) is added.

(b) More CO$_2$(g) is added.

(c) More BaO(s) is added.

(d) The temperature is raised.

(e) The volume of the reaction vessel is increased.

13.44 Carbonyl bromide decomposes to carbon monoxide and bromine:

$$COBr_2(g) \rightleftharpoons CO(g) + Br_2(g)$$

At 73 °C, K is 0.190. Suppose you placed 0.500 mol of COBr$_2$(g) in a 2.00 L flask and heated it to 73 °C. After equilibrium was achieved, you added an additional 2.00 mol of CO(g).

(a) How is the equilibrium mixture affected by adding more CO(g)?

(b) When equilibrium is re-established, what are the new equilibrium concentrations of COBr$_2$(g), CO(g), and Br$_2$(g)?

(c) How has the addition of CO(g) affected the percentage of COBr$_2$ that is decomposed?

13.45 Phosphorus pentachloride decomposes at higher temperatures:

$$PCl_5(g) \rightleftharpoons PCl_3(g) + Cl_2(g)$$

An equilibrium mixture at unspecified temperature consists of 3.120 g of PCl$_5$(g), 3.845 g of PCl$_3$(g), and 1.787 g of Cl$_2$(g) in a 1.00 L flask. If you add to this 1.418 g of Cl$_2$(g), how will the equilibrium be affected? What will the concentrations of PCl$_5$(g), PCl$_3$(g), and Cl$_2$(g) be when equilibrium is re-established?

SUMMARY AND CONCEPTUAL QUESTIONS

13.46 Neither solid $PbCl_2$ nor solid PbF_2 is appreciably soluble in water. If these solids are placed in equal amounts of water in separate beakers, in which beaker is the concentration of $Pb^{2+}(aq)$ greater? Equilibrium constants for these solids dissolving in water are as follows:

$$PbCl_2(s) \rightleftharpoons Pb^{2+}(aq) + 2\,Cl^-(aq) \quad K = 1.7 \times 10^{-5}$$
$$PbF_2(s) \rightleftharpoons Pb^{2+}(aq) + 2\,F^-(aq) \quad K = 3.7 \times 10^{-8}$$

13.47 In the gas phase, acetic acid exists as an equilibrium mixture of monomer and dimer molecules. (The dimer consists of two molecules linked through hydrogen bonds.)

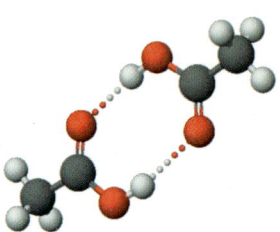

The equilibrium constant (K) at 25 °C has been determined to be 3.2×10^4. Imagine that acetic acid is present initially at a concentration of 5.4×10^{-4} mol L^{-1} at 25 °C and that no dimer is present initially.

(a) What percentage of the acetic acid is converted to dimer?

(b) If the temperature is increased, in which direction will net reaction occur? (Recall that hydrogen bond formation is an exothermic process.)

13.48 A reaction important in smog formation is

$$O_3(g) + NO(g) \rightleftharpoons O_2(g) + NO_2(g) \quad K = 6.0 \times 10^{34}$$

(a) If the initial concentrations are $[O_3] = 1.0 \times 10^{-6}$ mol L^{-1}, $[NO] = 1.0 \times 10^{-5}$ mol L^{-1}, $[NO_2] = 2.5 \times 10^{-4}$ mol L^{-1}, and $[O_2] = 8.2 \times 10^{-3}$ mol L^{-1}, is the system at equilibrium? If not, in which direction does net reaction proceed?

(b) If the temperature is increased, as on a very warm day, will the concentrations of the products increase or decrease? (*Hint:* You may have to calculate the enthalpy change for the reaction to find out if it is exothermic or endothermic.)

13.49 Sulfuryl chloride, $SO_2Cl_2(g)$, is a compound with very irritating vapours; it is used as a reagent in the synthesis of organic compounds. When heated to a sufficiently high temperature, it decomposes to $SO_2(g)$ and $Cl_2(g)$:

$$SO_2Cl_2(g) \rightleftharpoons SO_2(g) + Cl_2(g)$$
$$K = 0.045 \text{ at } 375\,°C$$

(a) Suppose 6.70 g of $SO_2Cl_2(g)$ is placed in a 1.00 L flask and then heated to 375 °C. What is the concentration of each of the compounds in the system when equilibrium is achieved? What fraction of $SO_2Cl_2(g)$ has dissociated?

(b) What are the concentrations of $SO_2Cl_2(g)$, $SO_2(g)$, and $Cl_2(g)$ at equilibrium in the 1.00 L flask at 375 °C if you begin with a mixture of 6.70 g of $SO_2Cl_2(g)$ and 1.00 atm $Cl_2(g)$? What fraction of $SO_2Cl_2(g)$ has dissociated?

(c) Compare the fractions of $SO_2Cl_2(g)$ dissociated in parts (a) and (b). Do they agree with expectations based on Le Chatelier's principle?

13.50 Hemoglobin (Hb) can form a complex with both O_2 and CO. For the reaction

$$HbO_2(aq) + CO(g) \rightleftharpoons HbCO(aq) + O_2(g)$$

at body temperature, K is about 200. If the ratio $[HbCO]/[HbO_2]$ comes close to 1, death is probable. What partial pressure of $CO(g)$ in the air is likely to be fatal? Assume the partial pressure of $O_2(g)$ is 0.20 atm.

Acid-Base Equilibria in Aqueous Solution

GetStock/Alamy Images/blickwinkel/Schmidbauer

Maria Dryfhout/Shutterstock

14.1 Case Study: Roses Are Red, Violets Are Blue, Hydrangeas Are Red or Blue

The colours of some flowers change according to their environment—especially the level of acidity of the soil. This is true for some varieties of the hydrangea shrub with their beautiful showy red or blue flowers.

Gardening books advise that the colour of red hydrangea flowers can be transformed to blue by adding aluminum sulfate to the soil. You might find this advice surprising since aquated Al^{3+} ions from aluminum sulfate are colourless! Why does flower petal colour depend on soil acidity? What role do aluminum ions play?

> *Red hydrangea flowers can be transformed to blue by adding aluminum sulfate to the soil. Why does flower petal colour depend on soil acidity? What role do aluminum ions play?*

The pigment in hydrangea petals belongs to the class of compounds called *anthocyanins*; more precisely, it is a *cyanidin*. Cyanidins are responsible for the red colour of roses, strawberries, raspberries, apple skins, rhubarb, and cherries, and for the purple colour of blueberries. The colour of cyanidins depends on the pH: red cabbage juice is only red in acidic solution and is purple in solutions with pH near 7. An extract of red rose petals is red in acidic solution but blue in a basic solution (Figure 14.1).

One hypothesis for this pH-dependent change in colour of hydrangeas is that the pigment is a weak acid [<<Section 6.7]. In acidic solutions, the pigment consists of large positively charged ions, but when its environment is made more basic, an H^+ ion is removed to form neutral molecules (the conjugate base of the weak acid, defined in

FIGURE 14.1 Cyanidin with added acid, base, and aluminum ions. *From left to right*: The pigment in red rose petals was extracted with ethanol; the extract was a faint red. After adding one drop of 6 mol L^{-1} HCl solution, the colour changed to a vivid red. Adding two drops of 6 mol L^{-1} basic NH$_3$ solution produced a green colour, and adding 1 drop each of HCl and NH$_3$ solutions gave a blue solution. Finally, adding a few milligrams of Al(NO$_3$)$_3$ turned the solution deep purple.

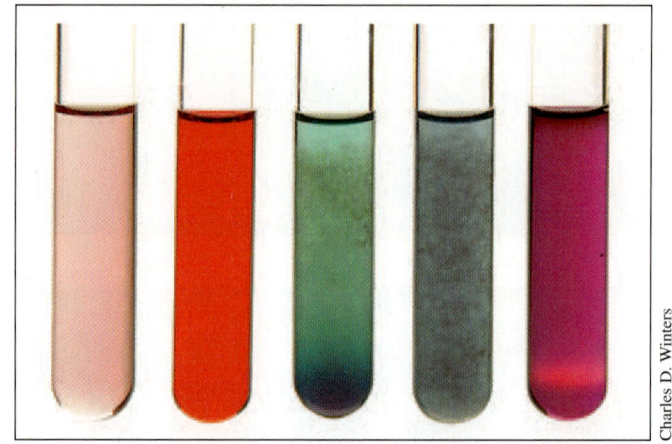

Charles D. Winters

Section 14.2) that absorb different wavelengths of light than the ions, and so are a different colour. If the pH can be controlled, then the colour can be controlled (Figure 14.2). This is the mode of colour change of most acid-base indicators, many of which are naturally occurring substances.

Cyanidin chloride in acidic solution

Cyanidin in basic solution

Low pH   High pH

FIGURE 14.2 Cyanidins at different pH. In basic solution (high pH) the cyanidin species has one less H$^+$ ion than that present in acidic solutions. The protonated species at low pH has a different colour from the deprotonated species at higher pH. A chloride ion shown at low pH is intended to indicate that the positive charges on the cations must be balanced by an equal number of negative charges on anions in the immediate environment.

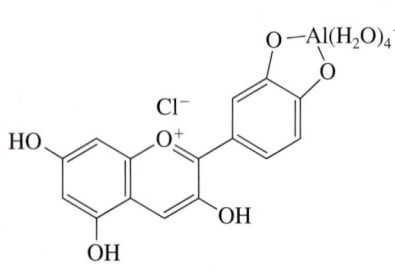

The low-pH species of cyanidin forms blue complex ions with Al^{3+}(aq) ions.

This seems consistent with the effect of adding aluminum sulfate to the soil because aquated aluminum ions, which can be represented as $[Al(OH_2)_6]^{3+}$ complex ions [<<Section 6.7] are weakly acidic:

$$[Al(OH_2)_6]^{3+}(aq) + H_2O(\ell) \rightleftharpoons [Al(OH_2)_5(OH)]^{2+}(aq) + H_3O^+(aq)$$

So any substance that provides $[Al(OH_2)_6]^{3+}$ ions might be expected to render the soil acidic enough to change the cyanidin to the protonated species, and if this species is blue, the explanation seems complete.

However, *in situ* experiments have shown that hydrangea petals do not turn blue if the soil is made acidic by reagents other than aluminum salts. Why does the blue colour of hydrangea blossoms in acidic soils depend on the presence of aluminum ions? One idea is that at lower soil pH values, cyanidin forms a complex ion with Al^{3+}(aq) ions, and the colour of this species is blue.

Now we need to reconsider why the colour is pH-dependent. As is so often the case, the answer lies in knowledge of speciation at the prevailing conditions [<<Section 6.1]. The $[Al(OH_2)_6]^{3+}$ ions can behave as polyprotic weak acids [>>Section 14.6], from which H$^+$ ions are successively removed as the solution they are in is gradually made more basic.

At around pH 7–9, after removal of three H^+ ions, the aluminum is mainly present as the uncharged, insoluble $[Al(OH_2)_3(OH)_3](s)$ species:

$$[Al(OH_2)_6]^{3+}(aq) + H_2O(\ell) \rightleftharpoons [Al(OH_2)_5(OH)]^{2+}(aq) + H_3O^+(aq)$$
$$[Al(OH_2)_5(OH)]^{2+}(aq) + H_2O(\ell) \rightleftharpoons [Al(OH_2)_4(OH)_2]^+(aq) + H_3O^+(aq)$$
$$[Al(OH_2)_4(OH)_2]^+(aq) + H_2O(\ell) \rightleftharpoons [Al(OH_2)_3(OH)_3](s) + H_3O^+(aq)$$

Consequently, if the soils are not sufficiently acidic, the aluminum is bound in the insoluble species and is not available to the plants. Making the soil solutions more acidic gives rise to formation of aluminum-containing ions that are soluble and bioavailable.

It is easier to turn red hydrangea blossoms blue than it is to turn blue blossoms red because it is easy to add aluminum salts to the soil but very difficult to remove them. To encourage the development of red hydrangeas, lime can be added to make the soil less acidic, converting the aluminum into the insoluble hydroxide complex, which is not bioavailable. However, if the pH is raised above about 6.4, $Fe^{3+}(aq)$ ions in the soil will be bound as the insoluble hydroxide $Fe(OH)_3$, and the plant can suffer from iron deficiency. Another strategy is to add high-phosphate fertilizers: phosphate ions ($PO_4{}^{3-}$) can complex with the $Al^{3+}(aq)$ ions to form an insoluble complex ion.

Once again, we see how speciation of an element affects its chemical behaviour [<<Section 6.1]. There are many examples in this chapter of dramatic difference of chemical properties brought about by the addition or removal of an H^+ ion to form a new species. A key to controlling the properties of a weak acid-base system is understanding of the pH range over which transformation to the different species occurs.

14.2 The Brønsted-Lowry Model of Acids and Bases

From the perspective of the **Brønsted-Lowry model** [<<Section 6.7], *acid-base reactions involve a competition between species for H^+ ions (protons): the species that "grabs" H^+ ions (the proton acceptor) is a base, and the species from which H^+ ions are removed (the proton donor) is an acid.* This is illustrated by the reaction in aqueous solution between acetic acid and hydroxide ions:

$$CH_3COOH(aq) + OH^-(aq) \longrightarrow CH_3COO^-(aq) + H_2O(\ell)$$
$$\text{acid} \qquad\qquad \text{base}$$

There are many naturally occurring acids (Figure 14.3).

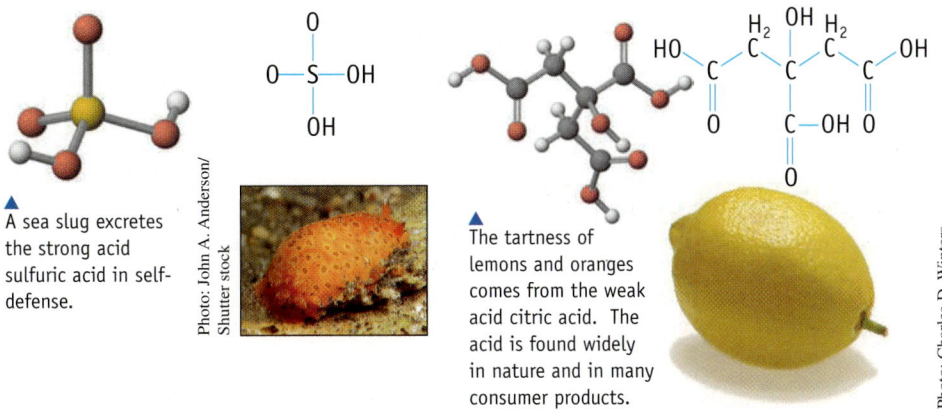

FIGURE 14.3 Natural acids. Hundreds of acids are found in nature. Our foods contain a wide variety, and many biochemically important molecules are acids.

▲ A sea slug excretes the strong acid sulfuric acid in self-defense.

Photo: John A. Anderson/Shutter stock

▲ The tartness of lemons and oranges comes from the weak acid citric acid. The acid is found widely in nature and in many consumer products.

Photo: Charles D. Winters

Weak bases also abound in nature. Ammonia, $NH_3(g)$, plays a part in the nitrogen cycle in the environment [<<Section 13.1]. Biological systems reduce nitrate ions to NH_3 and $NH_4{}^+$ ions and incorporate nitrogen into amino acids and proteins. Many natural bases are organic amines, which are derived from NH_3 by replacement of one or more of the H atoms with organic groups [>>Section 25.4].

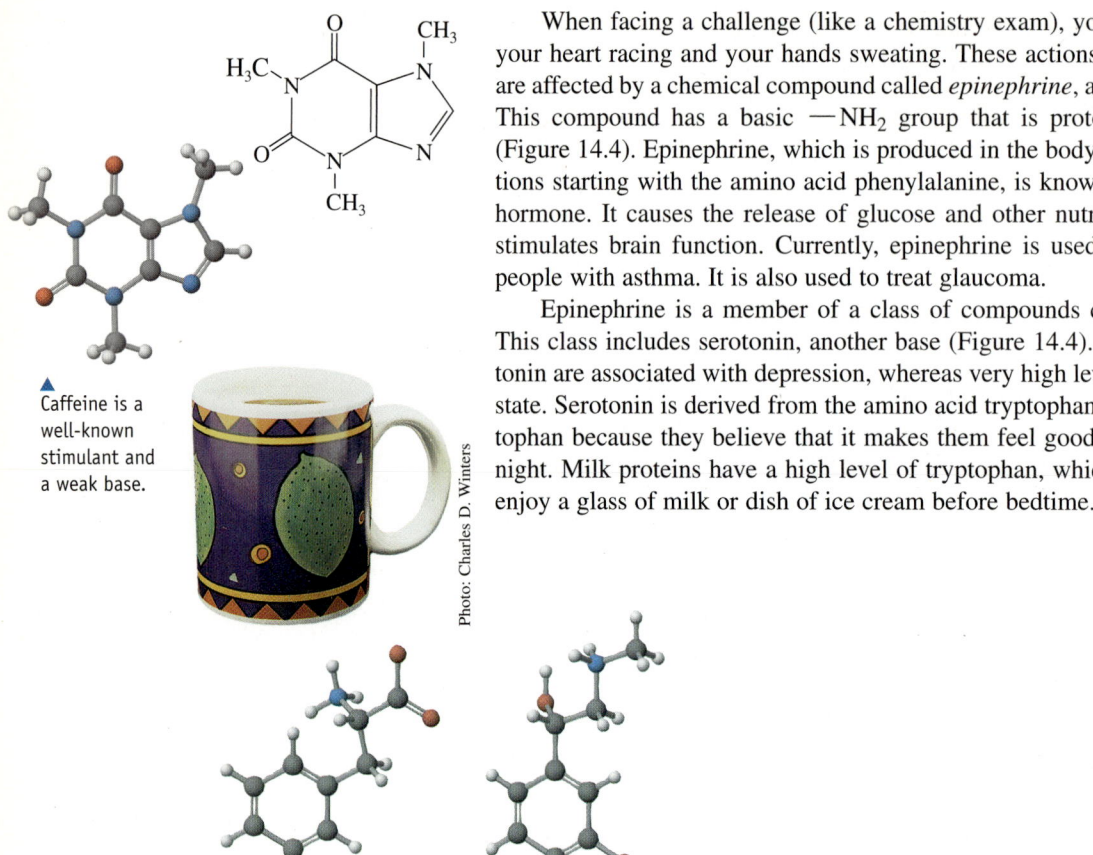

Caffeine is a well-known stimulant and a weak base.

Photo: Charles D. Winters

When facing a challenge (like a chemistry exam), you may have experienced your heart racing and your hands sweating. These actions of your nervous system are affected by a chemical compound called *epinephrine*, also known as *adrenaline*. This compound has a basic —NH₂ group that is protonated in acid solutions (Figure 14.4). Epinephrine, which is produced in the body through a chain of reactions starting with the amino acid phenylalanine, is known as the "flight or fight" hormone. It causes the release of glucose and other nutrients into the blood and stimulates brain function. Currently, epinephrine is used as a bronchodilator by people with asthma. It is also used to treat glaucoma.

Epinephrine is a member of a class of compounds called *neurotransmitters*. This class includes serotonin, another base (Figure 14.4). Very low levels of serotonin are associated with depression, whereas very high levels can produce a manic state. Serotonin is derived from the amino acid tryptophan. Some people take tryptophan because they believe that it makes them feel good and helps them sleep at night. Milk proteins have a high level of tryptophan, which may explain why you enjoy a glass of milk or dish of ice cream before bedtime.

Phenylalanine   **Epinephrine**   **Serotonin**

FIGURE 14.4 Biologically produced amine bases. Shown here are the structures of the protonated species of epinephrine and serotonin. In each case, the base is formed by removal of a proton from the —NH₃⁺ group, leaving an —NH₂ group. (See also J. Mann, *Murder, Magic, and Medicine*, New York, Oxford University Press, 1994.)

As well as their roles in our body chemistry, and in the chemistry of our aqueous environments, acids and bases are important in industry and in the home (Table 14.1).

> According to the Brønsted-Lowry model, acid-base reactions involve a competition between species for H^+ ions (protons): the proton acceptor is a base; the proton donor is an acid.

Characteristics of Acids, Bases, and Amphoteric Species

In aqueous solution, *strong acids* such as hydrogen chloride (whose solutions are called hydrochloric acid) are strong electrolytes: all of the molecules ionize as a result of an acid-base reaction with water molecules.

The continuously changing nature of the hydronium ion, H_3O^+(aq) or H^+(aq), is discussed in Section 6.8.

$$HCl(g) + H_2O(\ell) \longrightarrow H_3O^+(aq) + Cl^-(aq)$$

Acids	Use (Mostly in aqueous solution)	
Acetic acid, CH$_3$COOH(ℓ)	Flavouring, preservative	
Citric acid, H$_3$C$_6$H$_5$O$_7$(s)	Flavouring	
Phosphoric acid, H$_3$PO$_4$(ℓ)	Rust remover	
Boric acid, H$_3$BO$_3$(s)	Mild antiseptic, insecticide	
Hydrogen chloride, HCl(g)	Brick and ceramic tile cleaner	
Aluminum salts, NaAl(SO$_4$)$_2$.12H$_2$O	In baking powder, with sodium hydrogencarbonate	

BASES	USE (MOSTLY IN AQUEOUS SOLUTION)	
Sodium hydroxide, NaOH(s)	Oven and drain cleaners	
Ammonia, NH$_3$(g)	Household cleaner	
Sodium carbonate, Na$_2$CO$_3$(s)	Water softener, grease remover	
Sodium hydrogencarbonate, NaHCO$_3$(s)	Fire extinguishers, baking soda, antacid	
Sodium phosphate, Na$_3$PO$_4$(s)	Surface cleaner before painting or wallpapering	

TABLE 14.1 Some Common Acids and Bases and Their Household Uses

Courtesy of Miriam Mahaffy

Courtesy of Miriam Mahaffy

In this equation, the symbol ⟶ is used to indicate that the reaction "goes to completion." The formula HCl(aq) is not used in this equation because essentially none of this species exists in solution. In this text, (aq) refers to species that are in solution, and means "aquated" [<<Section 6.1]. The relative concentrations of species in a solution of strong acid HA are portrayed in Figure 14.5.

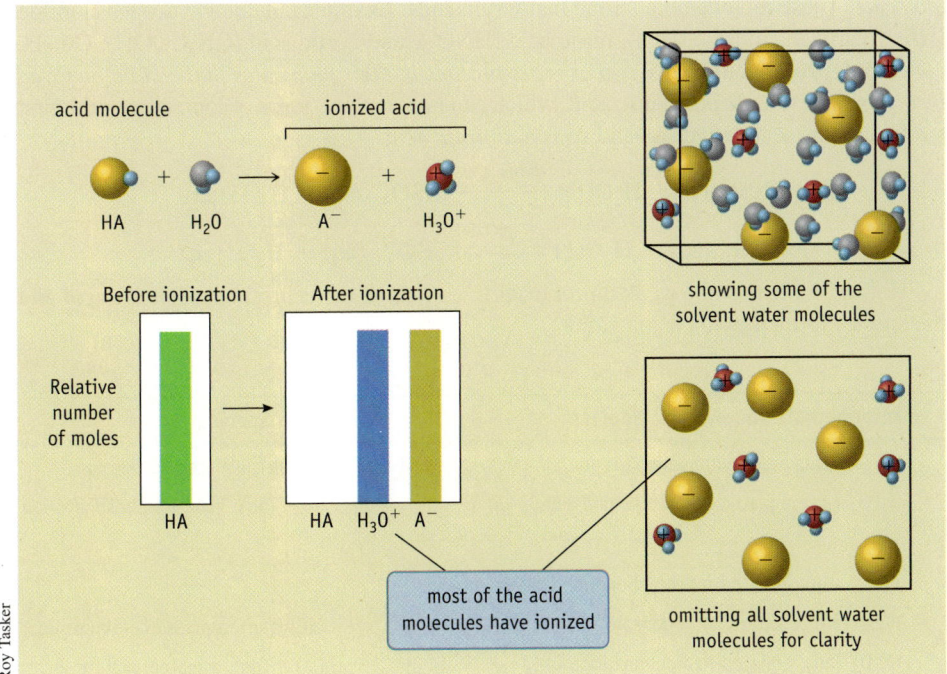

Roy Tasker

FIGURE 14.5 Ionization of a strong acid in aqueous solution. The bars indicate the relative amounts of species in an imagined solution when no ionization has occurred, and the amounts after ionization. In solutions of a strong acid, virtually no HA(aq) molecules remain. The diagrams shown here are intended to portray relative numbers of the various species, and give no sense of the enormous numbers of them, their close proximity to each other, nor their movement.

Think about It

e14.1 Distinguish between a strong acid and a concentrated acid solution.

In contrast, *weak acids* in aqueous solution are weak electrolytes: only some of the weak acid species ionize before a state of chemical equilibrium is reached (Figure 14.6).

FIGURE 14.6 Ionization of a weak acid in aqueous solution. At equilibrium, most of a weak acid species remains intact, and only a small concentration of $H_3O^+(aq)$ and $A^-(aq)$ ions are present.

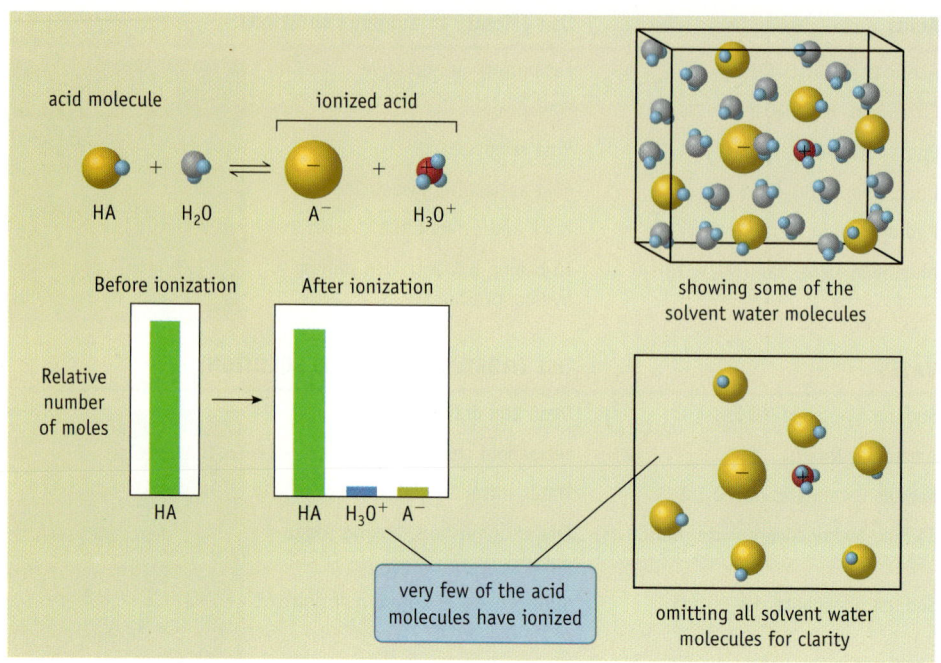

Think about It

e14.2 Identify the role of water molecules in acid-base reactions.

Weak acid species include aquated neutral molecules (such as acetic acid), aquated cations (such as $NH_4^+(aq)$ ions from dissolved ammonium nitrate), and aquated anions (such as $H_2PO_4^-(aq)$ ions from dissolved potassium dihydrogenphosphate).

$$CH_3COOH(aq) + H_2O(\ell) \rightleftharpoons CH_3COO^-(aq) + H_3O^+(aq)$$
$$NH_4^+(aq) + H_2O(\ell) \rightleftharpoons NH_3(aq) + H_3O^+(aq)$$
$$H_2PO_4^-(aq) + H_2O(\ell) \rightleftharpoons HPO_4^{2-}(aq) + H_3O^+(aq)$$

Some acids, called **monoprotic acids** *can have only one H^+ ion removed from each molecule.* These include both strong and weak acids. Examples include hydrogen fluoride (HF), hydrogen chloride (HCl), nitric acid (HNO_3), and acetic acid (CH_3COOH). Others, called **polyprotic acids**, *have two or more protons in their molecules that can be removed by bases.* An example is sulfuric acid, which can undergo two steps of ionization—the first strong and the second weak—and so is a *diprotic acid.*

$$H_2SO_4(\ell) + H_2O(\ell) \xrightarrow{\text{strong acid}} HSO_4^-(aq) + H_3O^+(aq)$$
$$HSO_4^-(aq) + H_2O(\ell) \underset{\text{weak acid}}{\rightleftharpoons} SO_4^{2-}(aq) + H_3O^+(aq)$$

Table 14.2 lists some weak diprotic acids, as well as the triprotic phosphoric acid and the species formed by their ionization.

TABLE 14.2 Weak Polyprotic Acids

Acid Species	Intermediate Amphoteric Species	Base Species
$H_2S(aq)$, aquated hydrogen sulfide	$HS^-(aq)$, aquated hydrogensulfide ion	$S^{2-}(aq)$, aquated sulfide ion
$H_3PO_4(aq)$, aquated phosphoric acid	$H_2PO_4^-(aq)$, aquated dihydrogenphosphate ion and $HPO_4^{2-}(aq)$, aquated monohydrogenphosphate ion	$PO_4^{3-}(aq)$, aquated phosphate ion
$H_2CO_3(aq)$, aquated carbonic acid	$HCO_3^-(aq)$, aquated hydrogencarbonate ion	$CO_3^{2-}(aq)$, aquated carbonate ion
$H_2C_2O_4(aq)$, aquated oxalic acid	$HC_2O_4^-(aq)$, aquated hydrogenoxalate ion	$C_2O_4^{2-}(aq)$, aquated oxalate ion

The terminology used for acids corresponds with that for bases. *Strong bases*, such as sodium hydroxide, are strong electrolytes in aqueous solution:

$$NaOH(s) \xrightarrow{H_2O} Na^+(aq) + OH^-(aq)$$

Weak bases, however, ionize in aqueous solution only to a condition of chemical equilibrium. The most common examples are aquated neutral molecules, such as ammonia, or aquated anions, such as acetate ions:

$$NH_3(aq) + H_2O(\ell) \rightleftharpoons NH_4^+(aq) + OH^-(aq)$$

$$CH_3COO^-(aq) + H_2O(\ell) \rightleftharpoons CH_3COOH(aq) + OH^-(aq)$$

Anions formed by removal of both of the protons of diprotic acids can accept two protons. Examples include $S^{2-}(aq)$, $CO_3^{2-}(aq)$, and $C_2O_4^{2-}(aq)$ ions. Successive acceptance of two H^+ ions by aquated carbonate ions is represented as

$$CO_3^{2-}(aq) + H_2O(\ell) \rightleftharpoons HCO_3^-(aq) + H_3O^+(aq)$$

$$HCO_3^-(aq) + H_2O(\ell) \rightleftharpoons H_2CO_3(aq) + H_3O^+(aq)$$

Aquated phosphate ions, $PO_4^{3-}(aq)$ can accept three protons.

The equation for ionization of $HCO_3^-(aq)$ ions illustrates a feature of all acid-base reactions. The $HCO_3^-(aq)$ and $CO_3^{2-}(aq)$ ions are related by the loss or gain of one H^+ ion, as are H_2O molecules and $H_3O^+(aq)$ ions. *Two species whose compositions differ by an H^+ ion are called a* **conjugate acid-base pair**.

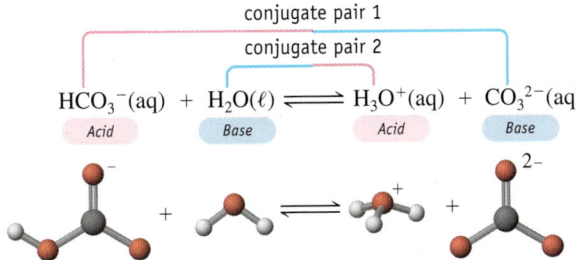

Think about It

e14.3 Identify conjugate acid-base pairs.

The $HCO_3^-(aq)$ ion is the conjugate acid of the base $CO_3^{2-}(aq)$ ion (which is the conjugate base of the acid $HCO_3^-(aq)$ ion); the $H_3O^+(aq)$ ion is the conjugate acid of the base H_2O. Every reaction considered an acid-base reaction in the Brønsted-Lowry sense involves H^+ transfer and has two conjugate acid-base pairs. Table 14.3 has a list of acid-base reactions with the conjugate pairs displayed.

Acid 1		Base 2		Base 1		Acid 2
HF(aq) aquated hydrogen fluoride molecules	+	H$_2$O(ℓ) water	$\rightleftharpoons$	F$^-$(aq) aquated fluoride ions	+	H$_3$O$^+$(aq) hydronium ions
HCOOH(aq) aquated formic acid molecules	+	CN$^-$(aq) aquated cyanide ions	$\rightleftharpoons$	HCOO$^-$(aq) aquated formate ions	+	HCN(aq) aquated hydrogen cyanide molecules
NH$_4^+$(aq) aquated ammonium ions	+	CO$_3^{2-}$(aq) aquated carbonate ions	$\rightleftharpoons$	NH$_3$(aq) aquated ammonia molecules	+	HCO$_3^-$(aq) aquated hydrogencarbonate ions
HPO$_4^{2-}$(aq) aquated hydrogenphosphate ions	+	SO$_3^{2-}$(aq) aquated sulfite ions	$\rightleftharpoons$	PO$_4^{3-}$(aq) aquated phosphate ions	+	HSO$_3^-$(aq) aquated hydrogensulfite ions
HC$_2$O$_4^-$(aq) aquated hydrogenoxalate ions	+	OH$^-$(aq) aquated hydroxide ions	$\rightleftharpoons$	C$_2$O$_4^{2-}$(aq) aquated oxalate ions	+	H$_2$O(ℓ) water
H$_3$O$^+$(aq) hydronium ions	+	HC$_2$O$_4^-$(aq) aquated hydrogenoxalate ions	$\rightleftharpoons$	H$_2$O(ℓ) water	+	H$_2$C$_2$O$_4$(aq) aquated oxalic acid

TABLE 14.3 Conjugate Acid-Base Pairs* in Some Reactions in Aqueous Solution

* In each case, acid 1 and base 1 are a conjugate pair, as are acid 2 and base 2.

Let's compare the equation for ionization in aqueous solution of an acid with that of a base:

$$CH_3COOH(aq) + H_2O(\ell) \rightleftharpoons CH_3COO^-(aq) + H_3O^+(aq)$$
$$NH_3(aq) + H_2O(\ell) \rightleftharpoons NH_4^+(aq) + OH^-(aq)$$

We can see that in the ionization of acids, water molecules are proton acceptors (bases), while in the ionization of bases they are proton donors (acids). *Such species that can behave as acids or bases, depending on their environment,* are said to be **amphoteric**. In the case of solvent species (such as water) that are amphoteric, the term *amphiprotic* is also used.

The aquated ions formed by removal of one H^+ ion from a diprotic acid are amphoteric (Table 14.2). An example is the aquated hydrogencarbonate ion:

$$HCO_3^-(aq) + H_3O^+(aq) \rightleftharpoons H_2CO_3(aq) + H_2O(\ell)$$
$$HCO_3^-(aq) + OH^-(aq) \rightleftharpoons CO_3^{2-}(aq) + H_2O(\ell)$$

Other examples are the aquated hydrogensulfide and hydrogenoxalate ions, $HS^-(aq)$, and $HC_2O_4^-(aq)$. Similarly, the $H_2PO_4^-(aq)$ and $HPO_4^{2-}(aq)$ ions formed by removal of either one or two H^+ ions from a triprotic acid are amphoteric.

EXERCISE 14.1—BRØNSTED-LOWRY MODEL OF ACIDS AND BASES

In each of the following equations, identify the acid on the left side and its conjugate base on the right. Similarly, identify the base on the left and its conjugate acid on the right.

(a) $HCOOH(aq) + H_2O(\ell) \rightleftharpoons HCOO^-(aq) + H_3O^+(aq)$

(b) $NH_3(aq) + H_2S(aq) \rightleftharpoons NH_4^+(aq) + HS^-(aq)$

(c) $HSO_4^-(aq) + OH^-(aq) \rightleftharpoons SO_4^{2-}(aq) + H_2O(\ell)$

Interactive Exercise 14.2

SUBMIT

Identify the conjugate acid-base pairs in chemical equations.

EXERCISE 14.3—BRØNSTED-LOWRY MODEL OF ACIDS AND BASES

Write balanced equations showing that the aquated hydrogenoxalate ion, $HC_2O_4^-(aq)$, can be both an acid and a base in the Brønsted-Lowry sense.

> Monoprotic acids have molecules with one removable H^+ ion, while polyprotic acids have more than one. Strong acids and bases are strong electrolytes; weak acids and bases are weak electrolytes. Species that can behave either as proton donors or proton acceptors are said to be amphoteric. Two species whose compositions differ by an H^+ ion are called a conjugate acid-base pair.

14.3 Water and the pH Scale

The properties of water are a recurring theme in this book [<<Section 3.3 and Chapter 6]. The properties of acids and bases in aqueous solution depend on the properties of water, and because acid-base reactions are so important in our bodies and in our external environment, we explore further the behaviour of water.

Water Self-Ionization and the Water Ionization Constant (K_w)

Long ago, Friedrich Kohlrausch (1840–1910) found that even after water is painstakingly purified, it can conduct a tiny electrical current, suggesting the presence of ions that can move through water. This is attributed to a proton transfer reaction between water molecules, referred to as *self-ionization of water* [<<Section 6.6], which comes to dynamic chemical equilibrium.

$$2\ H_2O(\ell) \rightleftharpoons H_3O^+(aq)\ +\ OH^-(aq)$$

We can apply the law of equilibrium [<<Section 13.3] to this situation.

At equilibrium $Q = [H_3O^+][OH^-] = K\ (or\ K_w)$

The equilibrium constant for self-ionization of water (symbol K_w) is known as the **ionization constant for water**. The balanced equation for water self-ionization tells us that concentrations of $H_3O^+(aq)$ ions and $OH^-(aq)$ ions in pure water must be the same, since there is no other source of these ions. Electrical conductivity measurements on pure water indicate that $[H_3O^+] = [OH^-] = 1.0 \times 10^{-7}$ mol L^{-1} at 25 °C, so K_w has a value of 1.0×10^{-14} at 25 °C.

Since proton transfer reactions are among the fastest reactions known, acid-base reactions in aqueous solution can generally be assumed to be at equilibrium. Then the following relationship, at the heart of all of the chemistry in this chapter, holds in all aqueous solutions (as well as in pure water):

$$[H_3O^+][OH^-] = K_w = 1.0 \times 10^{-14}\ at\ 25\ ^\circ C$$

Any aqueous solution in which $[H_3O^+] = [OH^-]$ is said to be a **neutral solution**. At 25 °C, the concentration of both of these ions in a neutral solution is 1.0×10^{-7} mol L^{-1}. This means that at any instant, at 25 °C only about two in every 10^9 water molecules are ionized. Pure water is a neutral solution, but so too are many solutions containing dissolved substances.

In solutions of acids in water at 25 °C, the concentration of $H_3O^+(aq)$ ions is greater than 1.0×10^{-7} mol L^{-1}. The concentration of $OH^-(aq)$ ions falls to a level less than 1.0×10^{-7} mol L^{-1}—by minimization of the self-ionization reaction until $Q = K_w$. *An aqueous solution in which $[H_3O^+] > [OH^-]$* is said to be an **acidic solution**.

Similarly, addition of a base to water increases the concentration of $OH^-(aq)$ ions, and the concentration of $H_3O^+(aq)$ ions decreases. *An aqueous solution in which $[OH^-] > [H_3O^+]$* is a **basic solution**.

Think about It

e14.4 Watch an animation of this ionization reaction at equilibrium.

The equation $Q = [H_3O^+][OH^-] = K_w$ is valid for any aqueous solution at any temperature. K_w is temperature dependent. The self-ionization reaction is endothermic, and K_w increases with temperature [<<Section 13.6].

TEMP.	K_w
10 °C	0.29×10^{-14}
15 °C	0.45×10^{-14}
20 °C	0.68×10^{-14}
25 °C	1.01×10^{-14}
30 °C	1.47×10^{-14}
50 °C	5.48×10^{-14}

So, the concentrations of $H_3O^+(aq)$ ions and $OH^-(aq)$ ions in pure water are not the same at all temperatures.

Think about It

e14.5 Look at the temperature dependence of K_w in more detail.

WORKED EXAMPLE 14.1—HYDRONIUM AND HYDROXIDE ION CONCENTRATIONS

What are the hydroxide and hydronium ion concentrations in a 0.0012 mol L^{-1} NaOH solution at 25 °C?

Strategy

NaOH is a strong base, so $[OH^-] = c(NaOH)$. We can then calculate $[H_3O^+]$ using the ionization constant for water.

Solution

$$c(NaOH) = 0.0012\ mol\ L^{-1},\ so\ [OH^-] = 0.0012\ mol\ L^{-1}$$

$$K_w = [H_3O^+][OH^-] = [H_3O^+](0.0012) = 1.0 \times 10^{-14}$$

$$So\ \ [H_3O^+] = \frac{1.0 \times 10^{-14}}{0.0012} = 8.3 \times 10^{-12}\ mol\ L^{-1}$$

In this and other calculations of equilibrium situations, we assume concentrations of species are their activities (<<Section 13.3), which are unitless.

Comment

The concentration of $OH^-(aq)$ ions was taken to be that from dissociation of NaOH, without any contribution from the self-ionization of water. This is because even if $[OH^-]$ were 1.0×10^{-7} mol L^{-1}, it would make an insignificant contribution since $0.0012 + 0.0000001$ is not different from 0.0012, to even five decimal places. In fact, in an acidic solution, $[OH^-] < 1.0 \times 10^{-7}$ mol L^{-1}.

Interactive Exercise 14.4

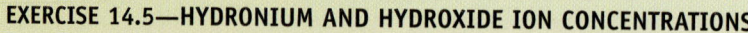

Challenge the statement that water at 95 °C is acidic.

EXERCISE 14.5—HYDRONIUM AND HYDROXIDE ION CONCENTRATIONS

In an HCl solution at 25 °C with $c(\text{HCl}) = 4.0 \times 10^{-3}$ mol L^{-1}, what are the concentrations of H_3O^+(aq) and OH^-(aq)?

Water undergoes a self-ionization reaction $2H_2O \ (\ell) \rightleftharpoons H_3O^+(aq) + OH^-(aq)$ for which the equilibrium constant, called the water ionization constant (K_w), is 1.0×10^{-14} at 25 °C. In neutral solutions $[H_3O^+] = [OH^-]$, so at 25 °C $[H_3O^+] = [OH^-] = 1.0 \times 10^{-7}$ mol L^{-1}. In acidic solutions $[H_3O^+] > [OH^-]$; in basic solutions $[H_3O^+] < [OH^-]$.

pH—A Logarithmic Scale of Hydronium Ion Concentrations

Concentrations of ions such as the H_3O^+(aq) ion can change by factors of as much as 10^{15}, so it is convenient to express them on a logarithmic scale. Values on this scale, called **pH**, are defined by

$$\text{pH} = -\log_{10} a_{H_3O^+}$$

where $a_{H_3O^+}$ is the activity of H_3O^+(aq) ions. For reasons expressed in Section 13.3, we will use the following approximation, which is quite accurate in dilute solutions:

$$\text{pH} = -\log_{10} [H_3O^+]$$

Similarly, we can define **pOH**, a logarithmic scale of concentrations of OH^-(aq) ions:

$$\text{pOH} = -\log_{10} [OH^-]$$

In neutral aqueous solutions, including pure water, at 25 °C:

$$\text{pH} = -\log_{10}(1.0 \times 10^{-7}) = 7.00 \quad \text{and} \quad \text{pOH} = -\log_{10}(1.0 \times 10^{-7}) = 7.00$$

An acidic solution with, for example, $[H_3O^+] = 1.0 \times 10^{-4}$ mol L^{-1}, and $[OH^-] = 1.0 \times 10^{-10}$ mol L^{-1} has

$$\text{pH} = -\log_{10}(1.0 \times 10^{-4}) = 4.00 \quad \text{and} \quad \text{pOH} = -\log_{10}(1.0 \times 10^{-10}) = 10.0$$

Similar calculations show that for any acidic solution at 25 °C, pH < 7.00 and pOH > 7.00. And for basic solutions at 25 °C, pH > 7.00 and pOH < 7.00. In summary (Figure 14.7):

- A neutral aqueous solution at any temperature has $[H_3O^+] = [OH^-]$
  At 25 °C, $[H_3O^+] = [OH^-] = 1.0 \times 10^{-7}$ mol L^{-1}, and pH = pOH = 7.00
- An acidic aqueous solution at any temperature has $[H_3O^+] > [OH^-]$
  At 25 °C, $[H_3O^+] > 1.0 \times 10^{-7}$ mol L^{-1}, pH < 7.00
  $[OH^-] < 1.0 \times 10^{-7}$ mol L^{-1}, pOH > 7.00
- A basic aqueous solution at any temperature has $[H_3O^+] < [OH^-]$
  At 25 °C, $[H_3O^+] < 1.0 \times 10^{-7}$ mol L^{-1}, pH > 7.00
  $[OH^-] > 1.0 \times 10^{-7}$ mol L^{-1}, pOH < 7.00

From the relationship between $[H_3O^+]$ and $[OH^-]$ in any solution, we can derive an expression between pH and pOH of the solution. In the following derivation, we define $pK_w = -\log K_w$, and we presume a temperature of 25 °C.

$$K_w = [H_3O^+][OH^-] = 1.0 \times 10^{-14}$$
$$-\log_{10}K_w = -\log_{10}[H_3O^+][OH^-] = -\log_{10}(1.0 \times 10^{-14})$$
$$pK_w = -\log_{10}[H_3O^+] + (-\log_{10}[OH^-]) = 14.00$$
$$pK_w = 14.00 = \text{pH} + \text{pOH}$$

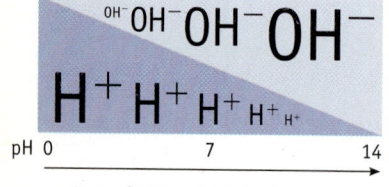

pH 0   7   14

As pH *increases*, [H⁺] *decreases*, and [OH⁻] *increases*

Think about It

e14.6, e14.7 Explore the relationships between pH, pOH, $[H_3O^+]$, and $[OH^-]$.

	[H₃O⁺]	pH	[OH⁻]	pOH
	1.0×10^{-15}	15.00	1.0×10^{1}	−1.00
	1.0×10^{-14}	14.00	1.0×10^{0}	0.00
	1.0×10^{-13}	13.00	1.0×10^{-1}	1.00
BASIC	1.0×10^{-12}	12.00	1.0×10^{-2}	2.00
	1.0×10^{-11}	11.00	1.0×10^{-3}	3.00
	1.0×10^{-10}	10.00	1.0×10^{-4}	4.00
	1.0×10^{-9}	9.00	1.0×10^{-5}	5.00
	1.0×10^{-8}	8.00	1.0×10^{-6}	6.00
NEUTRAL	1.0×10^{-7}	7.00	1.0×10^{-7}	7.00
	1.0×10^{-6}	6.00	1.0×10^{-8}	8.00
	1.0×10^{-5}	5.00	1.0×10^{-9}	9.00
	1.0×10^{-4}	4.00	1.0×10^{-10}	10.00
ACIDIC	1.0×10^{-3}	3.00	1.0×10^{-11}	11.00
	1.0×10^{-2}	2.00	1.0×10^{-12}	12.00
	1.0×10^{-1}	1.00	1.0×10^{-13}	13.00
	1.0×10^{0}	0.00	1.0×10^{-14}	14.00
	1.0×10^{1}	−1.00	1.0×10^{-15}	15.00

MORE BASIC ↑   MORE ACIDIC ↓

FIGURE 14.7 The relations among [H₃O⁺], pH, [OH⁻] and pOH at 25 °C. This figure shows data for selected solutions with various concentrations of $H_3O^+(aq)$ ions. Because K_w is constant, $[H_3O^+]$ and $[OH^-]$ are interdependent, as are pH and pOH. In each solution, the product of $[H_3O^+]$ and $[OH^-]$ is 1.0×10^{-14}, and the sum of pH and pOH is 14.00.

Think about It

e14.8 Examine trends in the pH of consumer products.

Think about It

e14.9 Challenge the idea that if you dilute an acidic solution enough it becomes basic.

WORKED EXAMPLE 14.2—pH AND pOH OF SOLUTIONS

In a solution at 25 °C, $[H_3O^+] = 8.0 \times 10^{-6}$ mol L⁻¹. Calculate pH and pOH of the solution.

Solution

We can calculate pH from the definition $pH = -\log_{10}[H_3O^+]$, and then pOH from the relationship $pH + pOH = 14.00$ at 25 °C.

$$pH = -\log_{10}[H_3O^+] = -\log_{10}(8.0 \times 10^{-6}) = -(-5.10) = 5.10$$
$$pOH = 14.00 - 5.10 = 8.90$$

Comment

An alternative way to calculate pOH is to first calculate $[OH^-]$:

$$[OH^-] = \frac{K_w}{[H_3O^+]} = \frac{1.0 \times 10^{-14}}{8.0 \times 10^{-6}} = 1.25 \times 10^{-9} \text{ mol L}^{-1}$$

$$pOH = -\log_{10}[OH^-] = -\log_{10}(1.25 \times 10^{-9}) = 8.90$$

EXERCISE 14.7—pH AND pOH OF SOLUTIONS

(a) What is the pH of a 0.0012 mol L⁻¹ NaOH solution at 25 °C?

(b) The pH of a diet soda is 4.32 at 25 °C. What are the concentrations of $H_3O^+(aq)$ and $OH^-(aq)$ in the soda?

(c) If the pH of a solution made by dissolving the strong base $Sr(OH)_2$ in water is 10.46 at 25 °C, what is the amount concentration (*c*) of $Sr(OH)_2$?

Interactive Exercise 14.6

Calculate ion concentrations from pH and pOH.

$pH = -\log_{10}[H_3O^+]$ and $pOH = -\log_{10}[OH^-]$. In aqueous solutions at 25 °C, $pH + pOH = 14.00$. At 25 °C, acidic solutions have pH < 7.0, and basic solutions have pH > 7.0.

14.4 Relative Strengths of Weak Acids and Bases

Weak acids differ in how strong they are—that is, in the extent to which they ionize in water. One way to compare their strengths is to measure the pH of solutions of different weak acids with the same concentration: the lower the pH is (the higher the $[H^+]$), the stronger the acid. For example, in 0.010 mol L^{-1} solutions of formic acid (HCOOH), acetic acid (CH$_3$COOH), and propanoic acid (CH$_3$CH$_2$COOH), the H_3O^+ ion concentrations are 1.3×10^{-3} mol L^{-1}, 4.2×10^{-4} mol L^{-1}, and 3.6×10^{-4} mol L^{-1} respectively. So we can say that formic acid is the strongest of these acids, and propanoic acid the weakest.

We will now move on to a more powerful quantitative measure of strengths of weak acids and weak bases.

Ionization Constants of Weak Acids and Bases

The equilibrium constant for ionization of a weak acid in aqueous solution is called its **acid ionization constant** (symbol K_a). If we use the general formula HA for a weak acid, ionization is represented as

$$HA(aq) + H_2O(\ell) \rightleftharpoons H_3O^+(aq) + A^-(aq)$$

$$\text{and at equilibrium} \quad Q = \frac{[H_3O^+][A^-]}{[HA]} = K_a$$

For weak acids, $K_a < 1$ because the product $[H_3O^+][A^-] < [HA]$. The stronger a weak acid is (the more easily the proton is removed), the larger its K_a. Table 14.4 displays K_a values at 25 °C for selected monoprotic weak acids. Appendix A has a more extensive list.

TABLE 14.4 K_a Values at 25 °C for Some Monoprotic Weak Acids in Aqueous Solution

NAME (FORMULA)	LEWIS STRUCTURE*	K_a
Chlorous acid (HClO$_2$)	H—Ö—Cl̈=Ö	1.1×10^{-2}
Nitrous acid (HNO$_2$)	H—Ö—N̈=Ö	7.0×10^{-4}
Hydrofluoric acid (HF)	H—F̈:	7.0×10^{-4}
Formic acid (HCOOH)	H—C(=O)—Ö—H	1.8×10^{-4}
Acetic acid (CH$_3$COOH)	H—C(H)(H)—C(=O)—Ö—H	1.8×10^{-5}
Propanoic acid (CH$_3$CH$_2$COOH)	H—C(H)(H)—C(H)(H)—C(=O)—Ö—H	1.3×10^{-5}
Hypochlorous acid (HClO)	H—Ö—Cl̈:	3.5×10^{-8}
Hydrocyanic acid (HCN)	H—C≡N:	6.2×10^{-10}

*Red type indicates the ionizable proton; all atoms have zero formal charge.

Interactive Exercise 14.8

Compare acid strengths on the basis of K_a values.

Taking It Further

e14.10 Read about how acid strength differs in different solvents.

Acid strength (arrow pointing up, increasing)

Chemists and biochemists use a logarithmic scale to report and compare acid strengths:

$$pK_a = -\log_{10} K_a$$

For example, at 25 °C, acetic acid has a pK_a value of 4.74:

$$K_a = 1.8 \times 10^{-5} \quad pK_a = -\log_{10}(1.8 \times 10^{-5}) = 4.74$$

The stronger a weak acid is, the larger K_a, and the smaller pK_a.

Propanoic acid   Acetic acid   Formic acid

Larger K_a, stronger acids, proton more easily removed

$\longrightarrow$

$K_a = 1.3 \times 10^{-5}$ $K_a = 1.8 \times 10^{-5}$ $K_a = 1.8 \times 10^{-4}$

$pK_a = 4.89$ $pK_a = 4.74$ $pK_a = 3.74$

$\longleftarrow$

Larger pK_a, weaker acids, protons more strongly held

Correspondingly, we can use *the equilibrium constant for ionization of a weak base in water* (called the **base ionization constant** [symbol K_b]) as a measure of its strength compared with other weak bases. We can use the symbol B as the general formula of weak bases:

$$B(aq) + H_2O(\ell) \rightleftharpoons BH^+(aq) + OH^-(aq)$$

At equilibrium $Q = \dfrac{[BH^+][OH^-]}{[B]} = K_b$

The relative values of K_b for ammonia, methylamine, and aniline indicate that of these, methylamine is the strongest weak base, and aniline is the weakest.

Ammonia   **Methylamine**   **Aniline**

$K_b = 1.8 \times 10^{-5}$ $K_b = 5.0 \times 10^{-4}$ $K_b = 4.0 \times 10^{-10}$

Some values of base ionization constants are listed on the right side of Table 14.5. It is also common to use pK_b values to specify base ionization constants on a logarithmic scale. The stronger the weak base is, the larger K_b, so the smaller pK_b:

$$pK_b = -\log_{10}K_b$$

EXERCISE 14.9—ACID AND BASE IONIZATION CONSTANTS

(a) At 25 °C, K_a for aquated benzoic acid, $C_6H_5COOH(aq)$, is 6.3×10^{-5}. Calculate pK_a.
(b) Is aquated chloroacetic acid, $ClCH_2COOH(aq)$, $pK_a = 2.87$, a stronger or weaker acid than aquated benzoic acid?

EXERCISE 14.10—ACID AND BASE IONIZATION CONSTANTS

Epinephrine hydrochloride has a $pK_a = 9.53$. What is the value of K_a? Where does the acid fit in Table 14.5?

Equilibrium constants for ionization of weak acids and weak bases in aqueous solution are called ionization constants, K_a and K_b. The larger the ionization constant is, the stronger the weak acid or base.

Relationship between K_a of an Acid and K_b of Its Conjugate Base

It makes sense that the stronger a weak acid is and the more easily it allows a proton to be removed from its molecules, the weaker its conjugate base (the less its ability to compete for a proton). This is confirmed by a quantitative expression linking K_a of a weak acid and K_b of its conjugate base, which we will derive for hydrocyanic acid, HCN(aq), and its conjugate base, CN^-(aq) ions.

Acid: $HCN(aq) + H_2O(\ell) \rightleftharpoons CN^-(aq) + H_3O^+(aq)$ K_a at 25 °C $= 4.0 \times 10^{-10}$

Base: $CN^-(aq) + H_2O(\ell) \rightleftharpoons HCN(aq) + OH^-(aq)$ K_b at 25 °C $= 2.5 \times 10^{-5}$

Presuming that both reactions are at equilibrium, we multiply the reaction quotients, and cancel out terms common to both the numerator and denominator:

$$K_a \times K_b = \left(\frac{[H_3O^+][CN^-]}{[HCN]}\right)\left(\frac{[HCN][OH^-]}{[CN^-]}\right) = [H_3O^+][OH^-] = K_w$$

That is, $K_a \times K_b = K_w \ (= 1.0 \times 10^{-14}$ at 25 °C)

This equation is useful because K_b for a weak base can be calculated if we know K_a of its conjugate acid, and vice versa. If we did not know K_b for aquated cyanide ions, for example, we can calculate

$$K_b(CN^-) \text{ at } 25\,°C = \frac{K_w}{K_a(HCN)} = \frac{1.0 \times 10^{-14}}{4.0 \times 10^{-10}} = 2.5 \times 10^{-5}$$

A logarithmic expression of this relationship is

$$pK_a + pK_b = pK_w \ \ (= 14.00 \text{ at } 25\,°C)$$

Again, using aquated cyanide ions for illustrative purposes:

$$pK_b(CN^-) = pK_w - pK_a(HCN) = 14.00 - 9.40 = 4.60$$

Think about It

e14.11 Simulate the relationship between the pK_a of an acid and the pK_b of its conjugate base.

Table 14.5 lists some weak acids in order of decreasing strength, and their conjugate bases in order of increasing strength. You can see from the K_a and K_b values of the conjugate acid-base pairs, that the stronger the acid is, the weaker its conjugate base.

Acids that are stronger than H_3O^+ are essentially completely ionized, so in this table their K_a values (in principle, infinitely large) are given as "large." Their conjugate bases have infinitesimally small concentrations of $OH^-(aq)$ ions, so their K_b values are given as "negligible." Similar arguments follow for strong bases and their conjugate acids.

TABLE 14.5 Ionization Constants for Some Acids and Their Conjugate Bases at 25 °C

Acid Name	Acid	K_a	Base	K_b	Base Name
Sulfuric acid	H_2SO_4	large	$HSO_4^-(aq)$	very small	Hydrogensulfate ion
Hydrochloric acid	HCl	large	$Cl^-(aq)$	very small	Chloride ion
Nitric acid	HNO_3	large	$NO_3^-(aq)$	very small	Nitrate ion
Hydronium ion	$H_3O^+(aq)$	1.0	$H_2O(\ell)$	1.0×10^{-14}	Water
Sulfurous acid	$H_2SO_3(aq)$	1.2×10^{-2}	$HSO_3^-(aq)$	8.3×10^{-13}	Hydrogensulfite ion
Hydrogensulfate ion	$HSO_4^-(aq)$	1.2×10^{-2}	$SO_4^{2-}(aq)$	8.3×10^{-13}	Sulfate ion
Phosphoric acid	$H_3PO_4(aq)$	7.5×10^{-3}	$H_2PO_4^-(aq)$	1.3×10^{-12}	Dihydrogenphosphate ion
Hydrofluoric acid	$HF(aq)$	7.0×10^{-4}	$F^-(aq)$	1.4×10^{-11}	Fluoride ion
Benzoic acid	$C_6H_5COOH(aq)$	6.3×10^{-5}	$C_6H_5COO^-(aq)$	1.6×10^{-10}	Benzoate ion
Acetic acid	$CH_3COOH(aq)$	1.8×10^{-5}	$CH_3COO^-(aq)$	5.6×10^{-10}	Acetate ion
Carbonic acid	$H_2CO_3(aq)$	4.2×10^{-7}	$HCO_3^-(aq)$	2.4×10^{-8}	Hydrogencarbonate ion
Hydrogen sulfide	$H_2S(aq)$	1.0×10^{-7}	$HS^-(aq)$	1.0×10^{-7}	Hydrogensulfide ion
Dihydrogenphosphate ion	$H_2PO_4^-(aq)$	6.2×10^{-8}	$HPO_4^{2-}(aq)$	1.6×10^{-7}	Hydrogenphosphate ion
Boric acid	$B(OH)_3(OH_2)$	7.3×10^{-10}	$B(OH)_4^-(aq)$	1.4×10^{-5}	Tetrahydroxoborate ion
Ammonium ion	$NH_4^+(aq)$	5.6×10^{-10}	$NH_3(aq)$	1.8×10^{-5}	Ammonia
Hydrogencarbonate ion	$HCO_3^-(aq)$	4.8×10^{-11}	$CO_3^{2-}(aq)$	2.1×10^{-4}	Carbonate ion
Hydrogensulfide ion	$HS^-(aq)$	1.0×10^{-12}	$S^{2-}(aq)$	1.0×10^{-2}	Sulfide ion
Hydrogenphosphate ion	$HPO_4^{2-}(aq)$	3.6×10^{-13}	$PO_4^{3-}(aq)$	2.8×10^{-2}	Phosphate ion
Water	$H_2O(\ell)$	1.0×10^{-14}	$OH^-(aq)$	1.0	Hydroxide ion
Ethanol	$C_2H_5OH(aq)$	1.0×10^{-16}	$C_2H_5O^-$	1.0×10^2	Ethoxide ion
Ammonia	$NH_3(aq)$	very small	NH_2^-	large	Amide ion
Hydroxide ion	$OH^-(aq)$	very small	O^{2-}	large	Oxide ion

Stronger weak acids ←

Stronger weak bases →

We can see from comparison of K_a and K_b values that, for example, the $NH_4^+(aq)$ ion is a much weaker acid than $HF(aq)$ since $5.6 \times 10^{-10} \ll 7.2 \times 10^{-4}$. Consistent with previous qualitative discussions, the conjugate base of the $NH_4^+(aq)$ ion is a stronger base than the conjugate base of $HF(aq)$ since $1.8 \times 10^{-5} \gg 1.4 \times 10^{-11}$. Note also, however, that even though the $NH_4^+(aq)$ ion is quite a weak acid, its conjugate base $NH_3(aq)$ is not a strong base. And even though the $F^-(aq)$ ion is a very weak base, its conjugate acid $HF(aq)$ is not a strong acid.

In the case of amphoteric species, we need to be careful in our interpretation of relative strengths as acids and bases. For example, we can see from Table 14.5 that water is a stronger weak acid than ammonia:

$$H_2O(\ell) + H_2O(\ell) \rightleftharpoons OH^-(aq) + H_3O^+(aq) \qquad K_a = 1.0 \times 10^{-14}$$
$$NH_3(aq) + H_2O(\ell) \rightleftharpoons NH_2^-(aq) + H_3O^+(aq) \qquad K_a \text{ very small}$$

We can deduce from these data that an $NH_2^-(aq)$ ion is a stronger base than an $OH^-(aq)$ ion, but they tell us nothing about the relative strengths of water and ammonia as bases. The base strengths of water and ammonia concern their ability to accept another proton, so we need to identify different reactions from Table 14.5 to see that ammonia is a stronger base than water:

$$NH_3(aq) + H_2O(\ell) \rightleftharpoons NH_4^+(aq) + OH^-(aq) \qquad K_b = 1.8 \times 10^{-5}$$
$$H_2O(\ell) + H_2O(\ell) \rightleftharpoons OH^-(aq) + H_3O^+(aq) \qquad K_b = 1.0 \times 10^{-14}$$

EXERCISE 14.12—IONIZATION CONSTANTS OF WEAK ACIDS AND THEIR CONJUGATE BASES

At 25 °C, K_a for lactic acid, $CH_3CHOHCOOH(aq)$, is 1.4×10^{-4}. What is K_b at 25 °C for its conjugate base, $CH_3CHOHCOO^-(aq)$? Where does this base fit in Table 14.5?

Interactive Exercise 14.11

Compare base strengths on the basis of the K_a values of their conjugate acids.

The stronger a weak acid is (the larger K_a is), the weaker its conjugate base (the smaller K_b is). For any conjugate acid-base pair $K_a \times K_b = K_w$ and $pK_a + pK_b = pK_w$.

Acid-Base Character of Aqueous Solutions of Salts

Every salt consists of a cation and an anion. Whether a solution of a salt is acidic, basic, or neutral depends on the interaction of both its cation and its anion with water. The influence of some ions on the pH of solution is listed in Table 14.6.

	Neutral		**Basic**			**Acidic**	
Cations	Li^+					NH_4^+	$CH_3NH_3^+$
	Na^+	Ca^{2+}				$[Al(OH_2)_6]^{3+}$	$[Fe(OH_2)_6]^{3+}$
	K^+	Ba^{2+}				$[Cr(OH_2)_6]^{3+}$	$[Cu(OH_2)_6]^{2+}$
Anions	Cl^-	NO_3^-	CH_3COO^-	CO_3^{2-}	HCO_3^-	HSO_4^-	
	Br^-	ClO_4^-	CN^-	S^{2-}	HS^-	$H_2PO_4^-$	
	I^-		F^-	PO_4^{3-}	HPO_4^{2-}	HSO_3^-	
			ClO^-	SO_3^{2-}			
			NO_2^-	SO_4^{2-}			
			$HCOO^-$				

TABLE 14.6 Influence of Some Aquated Ions on pH of Aqueous Solution

Aquated cations that are the conjugate acids of strong bases are such weak acids that they have no significant influence on the pH of solutions. These include the aquated ions of Groups 1 and 2 of the periodic table, such as Li^+, Na^+, K^+, Ca^{2+}, and Ba^{2+} ions.

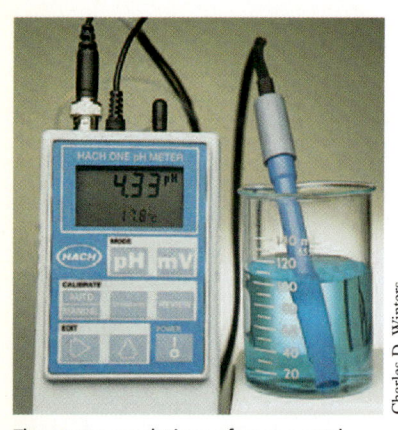

The aqueous solutions of many metal cations are acidic. A pH measurement shows that a solution of copper(II) sulfate is acidic. Among the common cations, Al^{3+} and transition metal ions Fe^{3+} and Cr^{3+} form the most acidic solutions. Salts of Al^{3+} and Fe^{3+} are used by gardeners to make soils more acidic.

Aquated ammonium ions, $NH_4^+(aq)$, lower the pH of solution:

$$NH_4^+(aq) + H_2O(\ell) \rightleftharpoons NH_3(aq) + H_3O^+(aq) \qquad K_a = 5.6 \times 10^{-10}$$

as do protonated amine species and the aquated ions of some metals with 2+ or 3+ charge:

$$[Al(OH_2)_6]^{3+} + H_2O(\ell) \rightleftharpoons [Al(OH_2)_5(OH)]^{2+} + H_3O^+(aq) \qquad K_a = 7.9 \times 10^{-6}$$

The aquated metal ions with most effect on pH are the $[Al(OH_2)_6]^{3+}$, $[Fe(OH_2)_6]^{3+}$, and $[Cr(OH_2)_6]^{3+}$ ions [>>Section 26.3], although hydrated ions of Cu^{2+}, Fe^{2+} and Mg^{2+} have measurable acidic reactions.

Aquated anions that are conjugate bases of strong acids, such as $Cl^-(aq)$ and NO_3^-(aq), are such weak bases that they have no effect on solution pH. Aquated anions that increase solution pH include the conjugate bases of weak monoprotic acids, such as $CH_3COO^-(aq)$ ions, as well as those formed by removal of all acidic protons from polyprotic acid molecules, such as $CO_3^{2-}(aq)$, $S^{2-}(aq)$, and $PO_4^{3-}(aq)$ ions. For example:

$$CO_3^{2-}(aq) + H_2O(\ell) \rightleftharpoons HCO_3^-(aq) + OH^-(aq) \qquad K_b = 2.1 \times 10^{-4}$$

Of the amphoteric anions formed by removal of just one proton from a diprotic acid, or either one or two protons from a triprotic acid, some, such as $HCO_3^-(aq)$ and $HPO_4^{2-}(aq)$, increase $[OH^-]$ of solution. On the other hand, there are others, such as $HSO_4^-(aq)$, $H_2PO_4^-(aq)$, and $HSO_3^-(aq)$, that increase $[H^+]$. The effect of any of these amphoteric anions on solution pH depends on how far its reaction as an acid proceeds compared with how far its reaction as a base proceeds. The aquated hydrogencarbonate ion, for example, is a stronger base than it is an acid (Table 14.5), and it increases the pH of solution:

$$HCO_3^-(aq) + H_2O(\ell) \underset{\text{as an acid}}{\overset{\text{reaction}}{\rightleftharpoons}} CO_3^{2-}(aq) + H_3O^+(aq) \qquad K_a = 4.8 \times 10^{-11}$$

$$HCO_3^-(aq) + H_2O(\ell) \underset{\text{as a base}}{\overset{\text{reaction}}{\rightleftharpoons}} H_2CO_3(aq) + OH^-(aq) \qquad K_b = 2.4 \times 10^{-8}$$

We can usually predict whether an aqueous solution of a salt is acidic, basic, or neutral by considering the influence of both its cation and its anion. However, to predict the pH of a solution of a salt like ammonium carbonate, which contains acidic $NH_4^+(aq)$ cations as well as basic $CO_3^{2-}(aq)$ anions, we need to compare K_a of the cation with K_b of the anion.

Think about It

e14.12 Explain the pH values of different salt solutions.

WORKED EXAMPLE 14.3—ACID-BASE CHARACTER OF SOLUTIONS OF SALTS

Predict whether aqueous solutions of each of the following are acidic, basic, or neutral:
(a) $NaNO_3$ (b) K_3PO_4 (c) $FeCl_2$ (d) NaH_2PO_4 (e) NH_4F

Strategy

First identify the cation and the anion, and then consider what acid-base interaction each has with water.

Solution

(a) *$NaNO_3$*: Aqueous solutions of this salt are neutral (pH = 7). The $Na^+(aq)$ ion does not react with water to a measurable extent. The $NO_3^-(aq)$ ion is the *very* weak conjugate base of a strong acid, so it has no influence on the solution pH.

(b) *K_3PO_4*: An aqueous K_3PO_4 solution is basic (pH > 7) because the $PO_4^{3-}(aq)$ ion is the conjugate base of the weak acid $HPO_4^{2-}(aq)$ ions. The $K^+(aq)$ ion does not react with water significantly.

(c) *$FeCl_2$*: An aqueous $FeCl_2$ solution is weakly acidic (pH < 7). The aquated Fe^{2+} ion, $Fe(H_2O)^{2+}$, is an acid. The $Cl^-(aq)$ ion is the *very* weak conjugate base of the strong acid HCl, so it does not react with water to form $OH^-(aq)$ ions.

(d) *NaH₂PO₄*: Amphoteric $H_2PO_4^-$(aq) ions have both an acidic reaction and a basic reaction with water:

$$H_2PO_4^-(aq) + H_2O(\ell) \underset{\text{as an acid}}{\overset{\text{reaction}}{\rightleftharpoons}} HPO_4^-(aq) + H_3O^+(aq)$$

$$H_2PO_4^-(aq) + H_2O(\ell) \underset{\text{as a base}}{\overset{\text{reaction}}{\rightleftharpoons}} H_3PO_4(aq) + OH^-(aq)$$

We can't predict whether a solution containing $H_2PO_4^-$(aq) ions is acidic or basic unless we know which of these reactions proceeds further. We need to know the equilibrium constants of the two reactions. In fact, $K_a = 6.2 \times 10^{-8}$, and $K_b = 1.3 \times 10^{-12}$, so the solution is acidic.

(e) *NH₄F*: This salt dissociates in water to form NH_4^+(aq) ions, which have an acidic reaction with water, and F^-(aq) ions, which have a basic reaction with water:

$$NH_4^+(aq) + H_2O(\ell) \rightleftharpoons NH_3(aq) + H_3O^+(aq) \qquad K_a(NH_4^+) = 5.6 \times 10^{-10}$$

$$F^-(aq) + H_2O(\ell) \rightleftharpoons HF(aq) + OH^-(aq) \qquad K_b(F^-) = 1.4 \times 10^{-11}$$

Because $K_a(NH_4^+) > K_b(F^-)$, we know that the NH_4^+(aq) ion is a stronger acid than the F^-(aq) ion is a base. On this evidence, we could predict (correctly) that solutions of NH_4F are mildly acidic.

EXERCISE 14.13—ACID-BASE CHARACTER OF SOLUTIONS OF SALTS

For each of the following salts, predict whether the pH of its aqueous solution is greater than, less than, or equal to 7:
(a) KBr   (b) NH_4NO_3 (c) $AlCl_3$ (d) Na_2HPO_4

The pH of solutions of salts depends on whether their cations and anions are acids or bases. Aquated anions that are conjugate bases of strong acids and aquated cations that are conjugate acids of strong bases have no effect on solution pH.

14.5 The Lewis Model of Acids and Bases

According to the Brønsted-Lowry model, ammonia is a base because its molecules can accept H^+ ions from other species:

$$H^+ + NH_3(aq) \rightleftharpoons H^+\!-\!NH_3 \text{ (i.e., } NH_4^+)$$

Since NH_3(aq) is a base in this reaction, it seems sensible that NH_3(aq) should be also regarded as a base in its reaction with Cu^{2+} ions to form $[Cu(NH_3)_4]^{2+}$ complex ions [<<Section 6.7]. A similarity with the above reaction is evident if we focus on formation of one of the $Cu^{2+}\!-\!NH_3$ bonds in the complex ion:

$$Cu^{2+} + NH_3(aq) \rightleftharpoons Cu^{2+}\!-\!NH_3$$

However, this reaction does not involve H^+ ion transfer. We now broaden the view of acid-base behaviour by introducing another model which encompasses the Brønsted-Lowry model but opens up chemical thinking about patterns of reactivity, especially in the fields of organic chemistry and metal ion complexation.

Lewis Acids and Bases—Electron Pair Transfer

In the 1930s, Gilbert N. Lewis developed a more general model of acids and bases that is not limited to substances that donate or accept protons. The *Lewis model* is based on the sharing of electron pairs between molecules (or ions) to form a covalent bond [<<Section 6.7]. A *Lewis base* is a molecule or ion that donates a pair of electrons to another atom to form a bond; a *Lewis acid* accepts a pair of electrons from another atom in the formation of a bond. The product of bond formation between a Lewis acid and a Lewis base is called an *adduct*.

$$\underset{\text{base}}{B\!:} + \underset{\text{acid}}{A} \longrightarrow \underset{\text{adduct}}{B\!-\!A}$$

A molecule or ion cannot be a Lewis base unless it has an electron pair available for bond formation with another molecule or ion (a Lewis acid). All molecules or ions that are bases in the Brønsted-Lowry model satisfy this condition, because without a "lone pair" of electrons they could not accept H^+ ions.

An H^+ ion is a Lewis acid because it accepts a pair of electrons when it forms a bond to a base. A compound such as $AlCl_3$ cannot be classified as an acid in the Brønsted-Lowry model since its molecules have no protons, but it is a Lewis acid because it accepts an electron pair from Lewis bases to form covalent bonds (Figure 14.8).

FIGURE 14.8 Lewis acid-base reactions. A Lewis base molecule donates a pair of electrons to a Lewis acid molecule, forming a covalent bond. The movement of an electron pair is represented by a curved arrow.

Classification of complexation reactions of metal cations as Lewis acid-base reactions, including the particular case of aquation of these ions, and a view of complexation as a competition between Lewis bases were discussed in Section 6.7. In these reactions, the terms *Lewis base* and *ligand* are used synonymously. There is further discussion in relation to the chemistry of transition metal ions in Chapter 27.

This more general Lewis model is also very helpful in finding patterns to the many reactions between electron pair donors and acceptors in organic chemistry. In fact, it is the basis for understanding and visualizing many of the molecular-level pathways of the reactions in Chapters 19–25.

> A Lewis base is a species that can provide a pair of electrons to a Lewis acid to form a covalent bond, forming a product called an adduct. The Lewis acid-base model includes species that are Brønsted-Lowry acids and bases.

Visualization of Reactive Sites of Organic Acids and Bases

Electrostatic potential maps [<<Section 6.4] are visual models that can help us to visualize possible reaction sites on Lewis acid and base species. These maps show space-filling models of molecules or ions with coloured surfaces. Regions of the surface are coloured red where the negative charge of electrons exceeds the positive charge on the nuclei (negative potential), and regions where there is excess positive charge (positive potential) are coloured blue. Gradations of positive potential are coloured in the order blue > green > yellow > orange > red.

Electrostatic potential maps for an acetic acid molecule reacting with a hydroxide ion to produce an acetate ion and a water molecule are shown in Figure 14.9.

Methanol molecules (CH_3OH) are amphoteric, in the sense that they can be either electron pair donors or acceptors, depending on the other reacting species in solution. The structural features giving rise to its dual reactivity can be seen from its electrostatic potential map, which shows excess negative charge (red) on the O atom, and net positive charge at the H atom in the –OH group.

Applying the Lewis model, we might predict that a methanol molecule can be an electron-pair donor because the O atom is electron rich. On the other hand, the most likely region of the molecule to accept an electron pair is the H atom in the —OH group.

Think about It

e14.13 Consider examples of Lewis acid-base reactions.

Interactive Exercise 14.14

Identify chemical species as Lewis acids or bases.

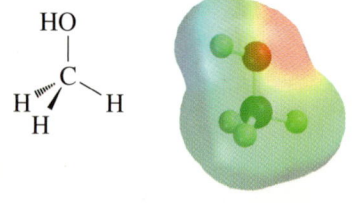

Molecular Modelling (Odyssey)

e14.14 Identify acidic and basic sites using electrostatic potential maps.

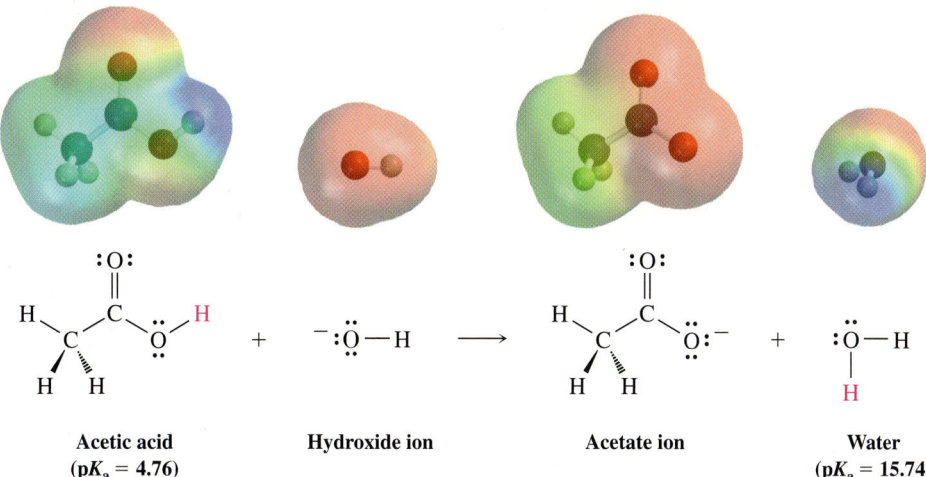

Acetic acid   **Hydroxide ion**   **Acetate ion**   **Water**
(pK$_a$ = 4.76)   **(pK$_a$ = 15.74)**

FIGURE 14.9 Electrostatic potential map portrayal of reactant and product species in reaction between an acetic acid molecule and a hydroxide ion. The red colour of the hydroxide ion shows that it is an excellent candidate to react as a Lewis base and donate an electron pair to acetic acid (the Lewis acid). The deepest blue part of the acetic acid molecule is the H atom that is bonded to an electronegative O atom, and the most likely region of the Lewis acid to receive an electron pair. With formation of a bond between this H atom and the OH$^-$ ion (resulting from donation of the electron pair by the OH$^-$ ion), the O—H bond in the acetic acid molecule breaks in such a way that both of the bond electrons remain on the O atom.

Methanol undergoes either of these reactions depending upon whether its molecules are in the presence of stronger acids or stronger bases.

Many of the reactions considered in later chapters involve *organic acids* and *organic bases* such as methanol. Organic acids are characterized by the presence of a hydrogen atom with net positive charge (blue in electrostatic potential maps) in their molecules. There are two main kinds: compounds like methanol and acetic acid, whose molecules contain an H atom bonded to an O atom (O—H), and compounds like acetone, whose molecules contain an H atom bonded to a C atom next to a C=O bond (O=C—C—H). Measures of their relative abilities to act as acids are their pK$_a$ values. Acetic acid is a stronger acid than either methanol or acetone. The different chemical behaviours of these types of organic acid are discussed in Chapter 24, and rationalized in terms of their electrostatic potential maps.

Molecules of some organic acids

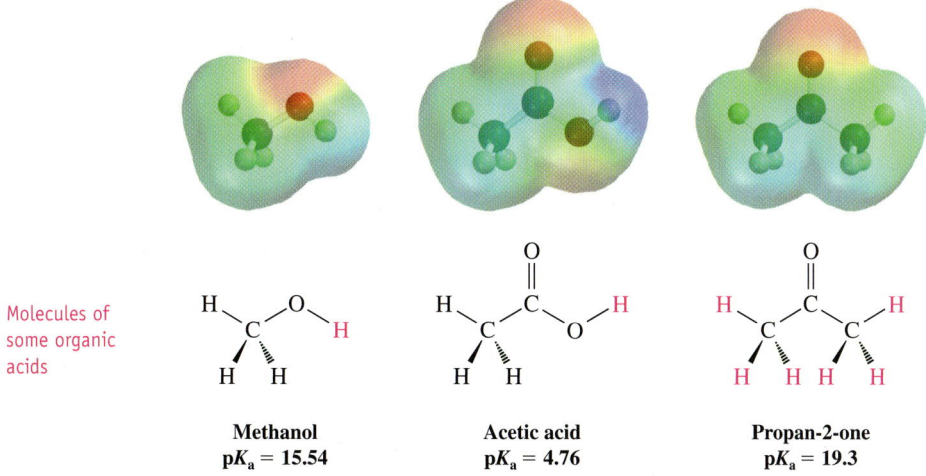

Methanol   **Acetic acid**   **Propan-2-one**
pK$_a$ = 15.54   **pK$_a$ = 4.76**   **pK$_a$ = 19.3**

Molecules of organic bases are characterized by the presence of a pair of electrons, either as a lone pair or a π bond, that can form bonds to H$^+$ ions or other Lewis acids. Nitrogen-containing compounds such as methylamine are the most common, but oxygen-containing compounds such as alcohols and ketones can also act as bases when reacting with a sufficiently strong acid.

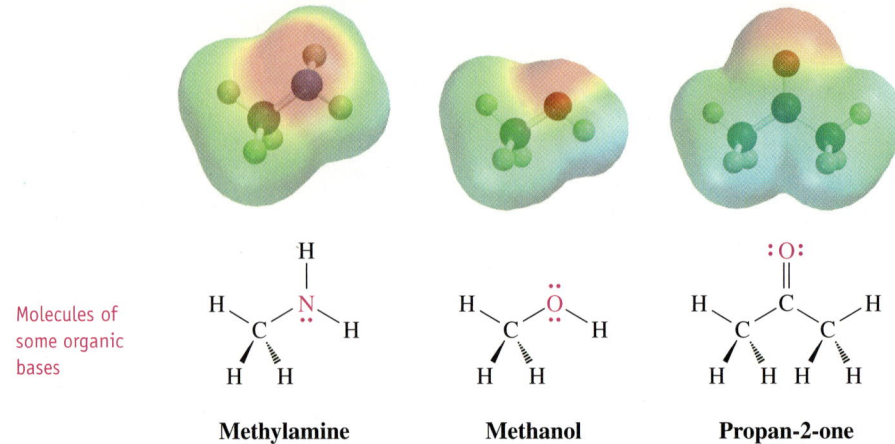

Molecules of some organic bases

Methylamine   **Methanol**   **Propan-2-one**

Taking It Further

e14.15 Read about how acid strength is related to molecular structure.

EXERCISE 14.15—LEWIS ACIDS AND BASES

The organic compound imidazole can act as both an acid and a base. Look at the following electrostatic potential map, and identify (a) the most acidic hydrogen atom, and (b) the most basic nitrogen atom in imidazole molecules.

Imidazole

WORKED EXAMPLE 14.4—LEWIS ACIDS AND BASES

Using curved arrows, show how acetaldehyde (CH_3CHO) can act as a Lewis base.

Strategy

A Lewis base donates an electron pair to a Lewis acid. We therefore need to locate the electron lone pairs on acetaldehyde and use a curved arrow to show the movement of one pair from the oxygen atom toward a Lewis acid such as the positively charged H atom on a molecule of a weak acid, HA, forming a new O—H bond.

Solution

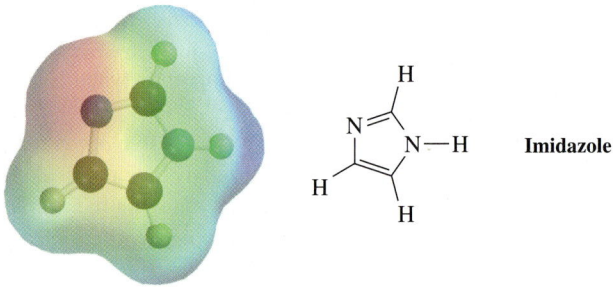

Acetaldehyde

EXERCISE 14.16—LEWIS ACIDS AND BASES

Carbon monoxide (CO) forms complexes with some metal ions. For example, $Ni(CO)_4$ and $Fe(CO)_5$ are well known. CO also forms a bond with the iron(II) ion in hemoglobin, which prevents the O_2 molecule bonding to the iron for transport to tissues. Is CO a Lewis acid or a Lewis base?

Which of the following are likely to act as Lewis acids, which as Lewis bases, and which as both? (Hint: First draw Lewis structures of each.)
(a) CH_3CH_2OH (b) $(CH_3)_2NH$ (c) Br^- ions
(d) $(CH_3)_3B$ (e) H_3CCl (f) $(CH_3)_3P$

Electrostatic potential map models have red regions indicating where the negative charge of electrons exceeds the positive charge on the nuclei (basic sites of molecules) and blue regions indicating where there is net positive charge (sites of acidic reactivity).

14.6 Equilibria in Aqueous Solutions of Weak Acids or Bases

In this section, we explore the power of quantitative calculations using equilibrium constants, first encountered in Chapter 13 but now applied to the context of equilibrium reactions involving weak acids and bases in aqueous solution.

Estimating K_a from Solute Concentration and Measured pH

The K_a and K_b values listed in Table 14.5 and in Appendix A have been determined by experiment. Several methods to do this are available, but one approach, illustrated by the following example, is to derive these values from the measured pH of aqueous solutions of acids and bases with known concentration.

A reminder about concentration definitions and symbols [<<Section 6.8]:

- c (solute) corresponds with the label that we would put on the solution container and refers to how much solute was put into each litre of solution.
- [species] refers to the concentration of a particular species in solution at equilibrium.

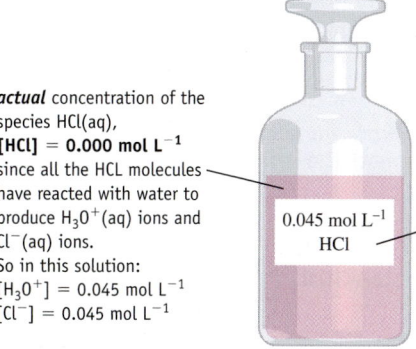

actual concentration of the species HCl(aq),
$[HCl] = 0.000$ mol L^{-1}
since all the HCL molecules have reacted with water to produce $H_3O^+(aq)$ ions and $Cl^-(aq)$ ions.
So in this solution:
$[H_3O^+] = 0.045$ mol L^{-1}
$[Cl^-] = 0.045$ mol L^{-1}

0.045 mol L^{-1}
HCl

label concentration
$c(HCl) = 0.045$ mol L^{-1}
describing how it was made
(for example by adding
0.045 mol HCl(g)
for every 1 L solution)

A 0.10 mol L^{-1} aqueous solution of lactic acid ($CH_3CHOHCOOH$) at 25 °C has a pH of 2.43. What is K_a for lactic acid, at 25 °C?

Strategy

We can calculate K_a if we put the equilibrium concentration of all reactant and product species into the reaction quotient. The pH of the solution tells us the equilibrium concentration of $H_3O^+(aq)$ ions, and we can then derive the equilibrium concentrations of the other species using principles of stoichiometry.

Solution

The equation for the ionization of lactic acid in aqueous solution is

$$CH_3CHOHCOOH(aq) + H_2O(\ell) \rightleftharpoons CH_3CHOHCOO^-(aq) + H_3O^+(aq)$$

From the solution pH, we can calculate the equilibrium concentration of $H_3O^+(aq)$ ions:

$$[H_3O^+] = 10^{-pH} = 10^{-2.43} = 3.7 \times 10^{-3} \text{ mol } L^{-1}$$

From the chemical equation for the reaction, we can deduce that if $[H_3O^+]$ has increased by 3.7×10^{-3} mol L^{-1}, then $[CH_3CHOHCOO^-]$ has increased by the same amount and $[CH_3CHOHCOOH]$ has decreased by the same amount. So, at equilibrium we have:

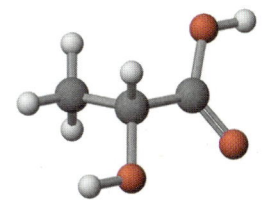

Lactic acid, $CH_3CHOHCO_2H$
Lactic acid is a weak monoprotic acid that occurs naturally in sour milk, apples, beer, and wine. It is also a product of the metabolism of glucose that provides energy when we exercise.

Think about It

e14.16 Calculate a value for K_a from a measured pH of an acid solution.

Interactive Exercise 14.18

Calculate K_a from a measured pH.

$$[CH_3CHOHCOOH] = (0.10 - 3.7 \times 10^{-3}) \text{ mol L}^{-1}$$
$$[H_3O^+] = 3.7 \times 10^{-3} \text{ mol L}^{-1}$$
$$[CH_3CHOHCOO^-] = 3.7 \times 10^{-3} \text{ mol L}^{-1}$$

These concentrations must satisfy the condition $Q = K$:

$$Q = \frac{[CH_3CHOHCOO^-][H_3O^+]}{[CH_3CHOHCOOH]} = \frac{(3.7 \times 10^{-3})(3.7 \times 10^{-3})}{[0.10 - (3.7 \times 10^{-3})]} = 1.4 \times 10^{-4}$$

i.e., K_a at 25 °C = 1.4×10^{-4}

Comment

In this calculation, it was assumed that all of the H_3O^+ ions have come from ionization of lactic acid, and ignored any contribution from self-ionization of water. This can be justified on the grounds that even if $[H_3O^+]$ from water self-ionization were 1.0×10^{-7} mol L^{-1}, this is insignificant compared to 3.7×10^{-3} mol L^{-1}. In fact, the water self-ionization is repressed in the presence of the acid (otherwise we would have $[H_3O^+][OH^-] > K_w$), and the contribution to $[H_3O^+]$ from water self-ionization even less than 1.0×10^{-7} mol L^{-1}.

EXERCISE 14.19—CALCULATING K_A FROM A MEASURED pH

A solution made from 0.055 mol of butanoic acid ($CH_3CH_2CH_2COOH$) dissolved in sufficient water to give 1.0 L of solution ($c = 0.055$ mol L^{-1}) has a pH of 2.72 at 25 °C. Determine K_a for butanoic acid at 25 °C.

From the solution concentration of a weak acid and the solution pH, we can estimate the ionization constant of the weak acid in aqueous solution.

Equilibrium Concentrations, pH, and Percentage Ionization from K_a

Knowing the equilibrium constants for weak acids and bases enables us to calculate the concentrations of all species at equilibrium, using the method described in Section 13.4 (*Calculating Equilibrium Concentrations*).

Think about It

e14.17 Calculate a value for solution pH from the K_a of an acid and its concentration.

Think about It

e14.18 Use a spreadsheet to simulate changes in pH for solutions of acids with different K_a values.

Interactive Exercise 14.20

Calculate equilibrium concentrations given K_a and pH.

WORKED EXAMPLE 14.6—EQUILIBRIUM CONCENTRATIONS, pH, AND PERCENT IONIZED FROM K_a

Calculate the pH of a 0.020 mol L^{-1} solution of benzoic acid (C_6H_5COOH) at 25 °C if $K_a = 6.3 \times 10^{-5}$ at 25 °C.

Strategy

Designate that the concentration of the weak acid decreases by x mol L^{-1} as equilibrium is reached. The stoichiometry of the balanced reaction equation allows us to determine the equilibrium concentrations of product species, in terms of x. It is useful to use an ICE table [<<Section 13.4].

Solution

The ionization equation is

$$C_6H_5COOH(aq) + H_2O(\ell) \rightleftharpoons C_6H_5COO^-(aq) + H_3O^+(aq)$$

If the concentration of benzoic acid decreases by x mol L^{-1}, the balanced chemical equation tells us that concentrations of benzoate ions and H_3O^+ ions each increase by x mol L^{-1}.

Concentrations (mol L^{-1})	C_6H_5COOH	+	H_2O	$\rightleftharpoons$	$C_6H_5COO^-$	+	H_3O^+
Initial	0.020				0		0
Change	$-x$				$+x$		$+x$
Equilibrium	$0.020 - x$				x		x

Interactive Exercise 14.21

Find the problem to match the solution.

SUBMIT

At equilibrium, the condition $Q = K$ must be satisfied:

$$Q = \frac{[C_6H_5COO^-][H_3O^+]}{[C_6H_5COOH]} = \frac{(x)(x)}{0.020 - x} = K_a = 6.3 \times 10^{-5}$$

From this we obtain the quadratic equation

$$x^2 + (6.3 \times 10^{-5})x - (1.26 \times 10^{-6}) = 0$$

Solving the quadratic equation, we derive $x = 0.0011$ mol L^{-1}.

So we have the equilibrium concentrations

$$[C_6H_5COO^-] = [H_3O^+] = x = 0.0011 \text{ mol L}^{-1}$$

$$[C_6H_5COOH] = 0.020 - x = 0.020 - 0.0011 = 0.019 \text{ mol L}^{-1}$$

$$\text{and pH} = -\log_{10}[H_3O^+] = -\log_{10}(0.0011) = 2.96$$

$$\text{and} \quad \text{\% of weak acid ionized} = \frac{x}{0.020} \times 100 = \frac{0.0011}{0.020} \times 100 = 5.5\%$$

Comment

As in Worked Example 14.5, the contribution to $[H_3O^+]$ from water self-ionization has been presumed insignificant in comparison to that due to ionization of the weak acid, and the final result justifies this approximation.

EXERCISE 14.22—EQUILIBRIUM CONCENTRATIONS, pH, AND PERCENT IONIZED FROM K_a

What are the equilibrium concentrations of $CH_3COOH(aq)$ molecules, $CH_3COO^-(aq)$ ions, and $H_3O^+(aq)$ ions in a 0.10 mol L^{-1} solution of acetic acid at 25 °C? What is the pH of the solution? What percentage of the aquated acetic acid is ionized at equilibrium?

The equilibrium constant can be used to estimate the percentage ionization of a weak acid in aqueous solution and the solution pH.

Dependence of Percentage Ionization on Magnitude of K_a

Using the method illustrated in Worked Example 14.6, we can use K_a to calculate $[H_3O^+]$, pH, and percentage ionization of any weak acid in aqueous solution. Table 14.7 displays the results of such calculations for 0.10 mol L^{-1} solutions of three weak acids with quite different K_a values.

Weak Acid	K_a	$[H_3O^+]$	pH	% Ionization
Benzoic acid, $C_6H_5COOH(aq)$	6.3×10^{-5}	2.5×10^{-3}	2.61	2.5%
Hypochlorous acid, $HOCl(aq)$	3.5×10^{-8}	5.9×10^{-5}	4.23	0.059%
Ammonium ion, $NH_4^+(aq)$	5.6×10^{-10}	7.5×10^{-6}	5.13	0.0075%

TABLE 14.7 Percentage Ionization of 0.1 mol L^{-1} Solutions of Some Weak Acids

The calculated values in Table 14.7 demonstrate that for solutions of weak acids with the same concentration, the larger K_a is, the more of it ionizes.

Interactive Exercise 14.23

SUBMIT

Find the problem to match the solution.

EXERCISE 14.24—MAGNITUDE OF IONIZATION CONSTANT AND PERCENTAGE IONIZATION

Verify the calculation results presented in Table 14.7.

For weak acid solutions of a given concentration and at a given temperature, the smaller K_a is, the lesser the percentage of the weak acid that ionizes.

Dependence of Percentage Ionization on Solution Concentration

The percentage ionization of a weak acid in aqueous solution depends on its concentration. Table 14.8 displays the results of calculations for three solutions of propanoic acid ($K_a = 1.3 \times 10^{-5}$ at 25 °C) with different concentrations.

TABLE 14.8 Percentage Ionization of Solutions of Propanoic Acid with Different Concentrations

c (CH$_3$CH$_2$COOH)	[H$_3$O$^+$]	pH	% Ionization
1.00 mol L^{-1}	3.6×10^{-3} mol L^{-1}	2.44	0.36
1.00×10^{-2} mol L^{-1}	3.5×10^{-4} mol L^{-1}	3.45	3.5
1.00×10^{-4} mol L^{-1}	3.0×10^{-5} mol L^{-1}	4.52	30

This is why in Table 14.7 it was essential to compare the percentage ionization of weak acids in solutions of the same concentration.

Table 14.8 shows that for propanoic acid, the more dilute its solutions are, the greater the percentage ionized (to CH$_3$CH$_2$COO$^-$(aq) and H$_3$O$^+$(aq) ions). The same is true of solutions of other weak acids. As a consequence, if we dilute 100-fold a solution of a weak acid, the concentration of H$_3$O$^+$(aq) ions changes by less than 100 times (and the pH increases by less than 2 units).

It is not a simple matter to rationalize this trend of percentage ionization with solution concentration, other than that this is a consequence of the requirement that at equilibrium in all of the three solutions, the reaction quotient has the same value ($= K_a$).

EXERCISE 14.25—DEPENDENCE OF PERCENTAGE IONIZATION ON SOLUTION CONCENTRATION

Verify the calculated values shown in Table 14.8.

At a given temperature, the more dilute the solution of a weak acid is, the greater the percentage of it that ionizes.

Effect of Common Ions on Percentage Ionization

At 25 °C, in a 0.10 mol L^{-1} solution of acetic acid ($K_a = 1.8 \times 10^{-5}$), the percentage of acid molecules ionized at equilibrium is 1.3%.

$$CH_3COOH(aq) + H_2O(\ell) \rightleftharpoons CH_3COO^-(aq) + H_3O^+(aq)$$

Imagine that we add to this solution a source of acetate ions (such as sodium acetate) or a source of hydronium ions (such as nitric acid solution). These ions are referred to as *common ions*: the acetate ion is "common" to both the acetic acid solution and sodium acetate, and the H$_3$O$^+$(aq) ion is "common" to both the acetic acid solution and the nitric acid solution.

Before adding a common ion, the equilibrium condition $Q = K_a$ must have been satisfied:

$$Q = \frac{[CH_3COO^-][H_3O^+]}{[CH_3COOH]} = K_a = 1.8 \times 10^{-5}$$

An instantaneous increase in concentration of either of CH$_3$COO$^-$(aq) ions or H$_3$O$^+$(aq) ions would bring about an out-of-equilibrium condition with $Q > K_a$.

Restoration of equilibrium would be achieved by net reaction in the direction that reduces Q—that is, by net reaction in the direction that reduces the concentrations of CH_3COO^- (aq) ions and H_3O^+(aq) ions. This is the **common ion effect**: *the presence of common ions reduces the extent of ionization of a weak acid.*

Worked Example 14.7 illustrates how we can calculate the percentage ionization in the presence of a common ion, and demonstrates the magnitude of the common ion effect.

WORKED EXAMPLE 14.7—THE COMMON ION EFFECT

Calculate the percentage ionization of a 0.10 mol L^{-1} acetic acid solution containing 0.05 mol L^{-1} sodium acetate.

Strategy

Recognize that sodium acetate is a salt that ionizes completely, so that the solution contains 0.05 mol L^{-1} of the common CH_3COO^-(aq) ions.

Solution

As usual, assume that, as a result of ionization, the concentration of acetic acid decreases by x mol L^{-1}. Then:

$$[CH_3COOH] = 0.10 - x \text{ mol } L^{-1}$$
$$[H_3O^+] = x \text{ mol } L^{-1}$$
$$[CH_3COO^-] = 0.05 + x \text{ mol } L^{-1}$$

At equilibrium, the $Q = K_a$ condition must be satisfied:

$$Q = \frac{[CH_3COO^-][H_3O^+]}{[CH_3COOH]} = \frac{(0.05 + x)(x)}{(0.1 - x)} = K_a = 1.8 \times 10^{-5}$$

This results in the quadratic equation $x^2 + (0.050)x - 1.8 \times 10^{-6} = 0$ for which the solution is $x = 3.6 \times 10^{-5}$ mol L^{-1}, and therefore

$$\text{Percentage ionization} = \frac{3.6 \times 10^{-5}}{0.10} \times 100 = 0.036\%$$

Comment

This is very much less than the 1.3% ionized in the absence of the external source of the common ion. In fact, in this situation, (100.000 – 0.036)% = 99.964% of the acetic acid remains un-ionized at equilibrium. We can also see from this calculation that $[H_3O^+] = x = 3.6 \times 10^{-5}$ mol L^{-1}, and so pH = 4.44. The situation where the common ion is the conjugate base of the weak acid is a *buffer solution*, used to control pH of solutions, and discussed in Section 14.9.

EXERCISE 14.27—THE COMMON ION EFFECT

(a) Calculate the pH and the percentage ionization in a 0.30 mol L^{-1} aqueous solution of formic acid, and compare this with the pH and the percentage ionization of formic acid if this solution also contained 0.10 mol L^{-1} of the salt sodium formate.

(b) Calculate the percentage ionization of formic acid in a 0.30 mol L^{-1} aqueous solution of formic acid (HCOOH) that also contains 0.10 mol L^{-1} hydrochloric acid (HCl).

The presence of common ions reduces the extent of ionization of a weak acid in aqueous solution.

Aqueous Solutions of Weak Bases

Just as acids can be molecular species or ions, so too can bases be molecular or ionic. Many molecular bases are based on nitrogen, with ammonia being the simplest, and many of these, such as caffeine and nicotine, occur naturally. The anionic conjugate bases of

Think about It

e14.19 Use a spreadsheet to simulate changes in pH for acid solutions involving common ions.

Think about It

e14.20 Simulate the common ion effect on ionization of acids.

Interactive Exercise 14.26

Calculate percentage ionization in the presence of a common ion.

weak acids, such as the aquated acetate ion, $CH_3COO^-(aq)$, make up another group of bases (Figure 14.10).

Ammonia, NH_3
$K_b = 1.8 \times 10^{-5}$

Caffeine, $C_8H_{10}N_4O_2$
$K_b = 2.5 \times 10^{-4}$

Benzoate ion, $C_6H_5CO_2^-$
$K_b = 1.6 \times 10^{-10}$

Phosphate ion, PO_4^{3-}
$K_b = 2.8 \times 10^{-2}$

Photo: Charles D. Winters

FIGURE 14.10 Examples of weak bases. Weak bases in water include substances whose molecules have one or more N atoms with lone pairs through which they can bond to an H^+ ion. Anions that are conjugate bases of weak acids are also weak bases.

For each type of calculation that has been applied earlier to weak acid solutions, there is a corresponding calculation applicable to solutions of weak bases. Worked Example 14.8 illustrates the calculation of the pH of a solution of a base.

WORKED EXAMPLE 14.8—pH OF A SOLUTION OF A WEAK BASE

What is the pH of a 0.015 mol L^{-1} aqueous sodium acetate solution of at 25 °C?

Strategy

Recognize that sodium acetate is a strong electrolyte in aqueous solution:

$$NaCH_3COO \xrightarrow{H_2O(\ell)} Na^+(aq) + CH_3COO^-(aq)$$

Aquated sodium ions do not affect solution pH, but the aquated acetate ion is a weak base:

$$CH_3COO^-(aq) + H_2O(\ell) \rightleftharpoons CH_3COOH(aq) + OH^-(aq)$$

We can calculate the concentration of $OH^-(aq)$ ions in a manner parallel to that in Worked Example 14.6.

Solution

Assume that $[CH_3COO^-]$ decreases by x mol L^{-1} as a result of ionization. We can use the stoichiometry of the reaction to estimate equilibrium concentrations of product species in terms of x.

Concentrations (mol L^{-1})	CH_3COO^-	$+$	H_2O	$\rightleftharpoons$	CH_3COOH	$+$	OH^-
Initial	0.015				0		0
Change	$-x$				$+x$		$+x$
Equilibrium	$0.015 - x$				x		x

Substituting equilibrium concentrations into the reaction quotient:

$$Q = \frac{[CH_3COOH][OH^-]}{[CH_3COO^-]} = \frac{(x)(x)}{(0.015 - x)} = K_b(CH_3COO^-) = 5.6 \times 10^{-10}$$

From this we obtain a quadratic equation: $x^2 + (5.6 \times 10^{-10})x - 8.4 \times 10^{-12} = 0$ for which the solution is $x = 2.9 \times 10^{-6}$ mol L^{-1}.

Now we can write the equilibrium concentrations:

$$[CH_3COOH] = [OH^-] = x = 2.9 \times 10^{-6} \text{ mol L}^{-1}$$

$$[CH_3COO^-] = 0.015 - x = 0.015 - 2.9 \times 10^{-6} = 0.015 \text{ mol L}^{-1}$$

$$[H_3O^+] = \frac{K_w}{[OH^-]} = \frac{1.0 \times 10^{-14}}{2.9 \times 10^{-6}} = 3.5 \times 10^{-9} \text{ mol L}^{-1}$$

And, finally, $pH = \log_{10}(3.5 \times 10^{-9}) = 8.46$

Comment

Alternatively, we could have first calculated pOH as $\log_{10}[OH^-]$, and then used the relationship $pH + pOH = pK_w$. Also, only a tiny percentage of the acetate ions ionize—as we might expect for such a small K_b.

EXERCISE 14.28—pH OF A SOLUTION OF A WEAK BASE

Sodium hypochlorite (NaOCl) is added as a disinfectant in swimming pools and water treatment plants, although the active agent is HOCl(aq) formed from the ClO$^-$(aq) ion weak base. What are the concentrations of HOCl(aq) and OH$^-$(aq), and the pH, of a 0.015 mol L^{-1} solution of NaOCl at 25 °C?

For weak bases in aqueous solution, there is a corresponding set of relationships to those that pertain to solutions of weak acids.

Solutions of Polyprotic Acids or Their Bases

More than one proton can be removed from molecules of polyprotic acids in acid-base reactions (Table 14.2). Many of these acids occur in nature, such as oxalic acid in rhubarb, citric acid in citrus fruit, malic acid in apples, and tartaric acid in grapes.

Phosphoric acid has three steps of ionization (K_a values are for 25 °C):

Step 1: $H_3PO_4(aq) + H_2O(\ell) \rightleftharpoons H_3O^+(aq) + H_2PO_4^-(aq)$ $K_{a1} = 7.5 \times 10^{-3}$

Step 2: $H_2PO_4^-(aq) + H_2O(\ell) \rightleftharpoons H_3O^+(aq) + HPO_4^{2-}(aq)$ $K_{a2} = 6.2 \times 10^{-8}$

Step 3: $HPO_4^{2-}(aq) + H_2O(\ell) \rightleftharpoons H_3O^+(aq) + PO_4^{3-}(aq)$ $K_{a3} = 3.6 \times 10^{-13}$

In any solution containing these phosphate species at equilibrium, the condition $Q = K$ is simultaneously obeyed for each step. That is, at the same time in any solution at equilibrium:

$$Q_{a1} = \frac{[H_3O^+][H_2PO_4^-]}{[H_3PO_4]} = K_{a1} = 7.5 \times 10^{-3}$$

$$Q_{a2} = \frac{[H_3O^+][HPO_4^{2-}]}{[H_2PO_4^-]} = K_{a2} = 6.2 \times 10^{-8}$$

$$Q_{a3} = \frac{[H_3O^+][PO_4^{3-}]}{[HPO_4^{2-}]} = K_{a3} = 3.6 \times 10^{-13}$$

The K_a value for each step is smaller than for the previous step: it is more difficult to remove an H$^+$ ion from a negatively charged H$_2$PO$_4^-$ ion than from a neutral H$_3$PO$_4$ molecule, and it is even more difficult to remove another from the doubly charged HPO$_4^{2-}$ ion.

A polyprotic acid. Malic acid is a diprotic acid occurring in apples. It is also classified as an alpha-hydroxy acid because it has an —OH group on the C atom next to the —COOH group (in the alpha position). It is one of a larger group of natural acids such as lactic acid, citric acid, and ascorbic acid. Alpha-hydroxy acids have been touted as ingredients in "anti-ageing" skin creams. They work by accelerating the natural process by which skin replaces the outer layer of cells with new cells.

For many inorganic polyprotic acids, such as phosphoric acid, carbonic acid, and hydrogen sulfide, K_a for the first ionization step is about 10^4 to 10^6 times bigger than that for the second step. As a consequence, the first ionization step produces about a million times more $H_3O^+(aq)$ ions than the second. For this reason, *if a polyprotic acid has $K_{a1} >> K_{a2}$, the concentration of hydronium ions resulting from the second and later ionization steps can be ignored in calculations of solution pH.* The same principle applies to the conjugate bases of polyprotic acids. This is illustrated by the calculation in Worked Example 14.9.

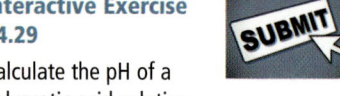

Interactive Exercise 14.29

Calculate the pH of a polyprotic acid solution.

WORKED EXAMPLE 14.9—pH OF A SOLUTION OF A POLYPROTIC BASE

The carbonate ion, $CO_3^{2-}(aq)$, reacts with water as a base in two stages, with the second stage much weaker than the first:

$$CO_3^{2-}(aq) + H_2O(\ell) \rightleftharpoons HCO_3^-(aq) + OH^-(aq) \qquad K_{b1} \text{ at } 25 \text{ °C} = 2.1 \times 10^{-4}$$
$$HCO_3^-(aq) + H_2O(\ell) \rightleftharpoons H_2CO_3(aq) + OH^-(aq) \qquad K_{b2} \text{ at } 25 \text{ °C} = 2.4 \times 10^{-8}$$

What is the pH of a 0.10 mol L^{-1} aqueous Na_2CO_3 solution of at 25 °C?

Strategy

Recognize that $Na^+(aq)$ ions do not affect solution pH, so the solution will be basic because of reactions of the $CO_3^{2-}(aq)$ ion. $K_{b2} << K_{b1}$, so for purposes of calculating pH, we can ignore the second ionization step.

Solution

Assume that reaction with water brings about a decrease of $[CO_3^{2-}]$ by x mol L^{-1}. Set up an ICE table.

ICE Table for Stage 1 of Reaction of $CO_3^{2-}(aq)$ as a Base

CONCENTRATIONS (mol L^{-1})	CO_3^{2-}	+	H_2O	$\rightleftharpoons$	HCO_3^-	+	OH^-
Initial	0.10				0		0
Change	$-x$				$+x$		$+x$
Equilibrium	$0.10 - x$				x		x

The equilibrium concentrations can be substituted into the reaction quotient for the first ionization step:

$$Q_{b1} = \frac{[HCO_3^-][OH^-]}{[CO_3^{2-}]} = \frac{(x)(x)}{0.10 - x} = K_{b1} = 2.1 \times 10^{-4}$$

Solving the resulting quadratic equation leads to the result:

$$[OH^-] = x = 4.6 \times 10^{-3} \text{ mol L}^{-1}$$

from which we derive pOH = $\log_{10}(4.6 \times 10^{-3})$ = 2.34, and pH = 14.00 − 2.34 = 11.66.

EXERCISE 14.30—pH OF A SOLUTION OF A POLYPROTIC ACID

What is the pH of a 0.10 mol L^{-1} solution of diprotic oxalic acid ($H_2C_2O_4$) at 25 °C? What are the concentrations of $H_3O^+(aq)$ and $HC_2O_4^-(aq)$?

In solutions of weak polyprotic acids with large differences in the successive K_a values, the pH depends primarily on the hydronium ions generated in the first ionization step. Similarly, in calculation of the pH of polyprotic base solutions, only the first ionization step needs to be considered.

Two Measures of Acidity of a Solution

We sometimes read statements about the *acidity* of natural waters. The term *acidity* has two meanings, and we need to be careful when using this term. To illustrate this point, let's compare 0.001 mol L^{-1} hydrochloric acid solution with 0.005 mol L^{-1} solution of propanoic acid ($K_a = 1.3 \times 10^{-5}$).

$$0.001 \text{ mol } L^{-1} \text{ HCl solution:} \quad [H_3O^+] = 0.001 \text{ mol } L^{-1}, \text{ pH} = 3.00$$
$$0.005 \text{ mol } L^{-1} \text{ CH}_3\text{CH}_2\text{COOH solution:} \quad [H_3O^+] = 0.00025 \text{ mol } L^{-1}, \text{ pH} = 3.60$$

Which of these solutions is more "acidic"? Sometimes, pH is an important criterion of acidity. For example, a significant problem of waterways contaminated by acid precipitation is that if the pH is low enough, the solubility of aluminum oxides and hydroxides is increased, forming $Al^{3+}(aq)$ ions that are highly toxic to marine life. Based on pH, we would say that the HCl solution is more acidic than the propanoic acid solution.

Sometimes, however, "acidity" can be used to mean how much base a solution can react with (its *neutralizing capacity*). From this perspective, the propanoic acid solution is more acidic because 1 L of it will react with five times as much base as 1 L of the hydrochloric acid solution. A base such as $OH^-(aq)$ ions will react not only with the equilibrium amount of $H_3O^+(aq)$ ions in a propanoic acid solution, but also with the ionizable hydrogen ions on all of the molecules:

$$CH_3CH_2COOH(aq) + OH^-(aq) \rightleftharpoons CH_3CH_2COO^-(aq) + H_2O(\ell)$$

How do we know that this reaction "goes to completion"? After all, propanoic acid is a weak acid, which means that it has some significant tendency to hold on to its protons. This question can be answered at two levels. Qualitatively, although the ionization constant of propanoic acid is small, that of water is much smaller still: water is a much weaker acid than propanoic acid. So we can regard this reaction as a competition between $OH^-(aq)$ ions and $CH_3CH_2COO^-(aq)$ ions for protons to form $H_2O(\ell)$ or $CH_3CH_2COOH(aq)$, respectively. The conjugate base of the weaker acid "wins."

This argument can be supported quantitatively. Applying the law of equilibrium to the neutralization of propanoic acid with $OH^-(aq)$ ions for a reaction mixture at equilibrium:

$$Q = \frac{[CH_3CH_2COO^-]}{[CH_3CH_2COOH][OH^-]} = K_{neut}$$

Now, multiplying numerator and denominator by $[H_3O^+]$, and separating terms to suit our purpose:

$$K_{neut} = \frac{[CH_3CH_2COO^-]}{[CH_3CH_2COOH][OH^-]} = \frac{[CH_3CH_2COO^-][H_3O^+]}{[CH_3CH_2COOH]} \times \frac{1}{[H_3O^+][OH^-]}$$

$$= \frac{K_a(CH_3CH_2COOH)}{K_w} = \frac{1.3 \times 10^{-5}}{1.0 \times 10^{-14}} = 1.3 \times 10^9$$

The equilibrium constant for the acid-base reaction is a massive number, indicating that the reaction is highly product-favoured.

Even in the case of reaction between a weaker acid and a weaker base (closer together in a list—see Table 14.5):

$$HOCl(aq) + NH_3(aq) \rightleftharpoons OCl^-(aq) + NH_4^+(aq)$$

the reaction is product-favoured:

$$K_{neut} = \frac{K_a(HOCl)}{K_a(NH_4^+)} = \frac{2.9 \times 10^{-8}}{5.6 \times 10^{-10}} = 52$$

Indeed, every reaction of an acid with the conjugate base of a weaker acid is product-favoured, and "goes to completion."

EXERCISE 14.31—ACIDITY OF SOLUTIONS

Rank the following solutions in order of increasing acidity in terms of (a) pH, and (b) amount (mol) of sodium hydroxide that will react with the same volume of each:
(a) 0.005 mol L^{-1} formic acid solution
(b) 0.001 mol L^{-1} hydrochloric acid solution
(c) 0.003 mol L^{-1} carbonic acid solution

> Reaction in aqueous solution between an acid and the conjugate base of any weaker acid is product-favoured and "goes to completion"—so a base can react with all of the removable protons of a weak acid (and not just the hydronium ions present) at equilibrium. The term *acidity* has two meanings: the pH of the solution, or the amount of base needed to neutralize the dissolved acid.

14.7 Speciation: Relative Concentrations of Species

Perhaps the most important aspect of acid-base chemistry is not about the pH of solutions in which the only solute is a weak acid or base, but the dependence of the relative concentrations of a weak acid and its conjugate base when strong acids or bases are added to control the pH.

In the gold mining industry, massive amounts of sodium cyanide are added to slurries of ore in water to dissolve the gold and separate it from "gangue" . This is to assist the oxidation (and dissolution) of gold metal by oxygen as a result of formation of a complex with cyanide ions:

$$4 \, Au(s) + O_2(g) + 8 \, CN^-(aq) + 2 \, H_2O(\ell) \longrightarrow 4 \, Au(CN)_2]^-(aq) + 4 \, OH^-(aq)$$

Cyanides are poisonous. Does this mean that workers in the industry are at grave risk? Certainly they would die if they drank enough of the slurry to consume more than about 1 g of sodium cyanide. However, the most lethal form of cyanides is the gas hydrogen cyanide (HCN). Workers would be at risk of breathing HCN gas if chemists did not understand the nature of the chemical equilibrium that exists in aqueous solution between HCN(aq) and the conjugate base CN$^-$(aq) ions.

$$HCN(aq) + H_2O(\ell) \rightleftharpoons CN^-(aq) + H_3O^+(aq) \qquad K_a \text{ at } 25 \, °C = 4.0 \times 10^{-10}$$

The more acidic the solution is (due to H$_3$O$^+$ ions from another source such as HCl), the more CN$^-$(aq) ions we would expect to "grab" a proton and exist as HCN(aq). So if the ore slurries were acidic, HCN gas could escape from solution into the air. The more basic the solution is, the more the cyanides would exist as the deprotonated base form CN$^-$(aq) ions, and the less as the protonated HCN(aq). By keeping the solution sufficiently basic, the concentration of HCN(aq) can be kept to a low enough level that the amount that escapes into the air is not dangerous.

But how basic does the solution need to be? While we need to be sure that the risk is negligible, the mine owners would not want the expense of adding more sodium hydroxide (to increase the pH) than necessary. Chemists can do quantitative equilibrium calculations to assist in this decision.

Distribution between Acid and Base Species as pH Is Changed

In any aqueous solution containing cyanide species, chemical equilibrium between HCN(aq) and the conjugate base CN$^-$(aq) ions is rapidly achieved:

$$HCN(aq) + H_2O(\ell) \rightleftharpoons CN^-(aq) + H_3O^+(aq) \qquad K_a \text{ at } 25 \, °C = 4.0 \times 10^{-10}$$

So the Q = K condition is obeyed:

$$Q = \frac{[CN^-][H_3O^+]}{[HCN]} = K_a$$

We can re-arrange this equation to produce another that gives us the ratio of concentrations of the weak acid and its conjugate base:

$$\frac{[HCN]}{[CN^-]} = \frac{[H_3O^+]}{K_a}$$

Since K_a is a constant at a given temperature, this equation tells us that the ratio of weak acid to its conjugate base is directly proportional to the concentration of H_3O^+ ions: if $[H_3O^+]$ in a solution is doubled, there is a net transfer of protons that decreases $[CN^-]$ and increases $[HCN]$ until the ratio $[HCN]/[CN^-]$ is doubled.

We can use this equation to calculate the ratio $[HCN]/[CN^-]$ at any pH to which we might want to adjust the solution. If $[H_3O^+] = K_a$ (i.e., pH $= pK_a$), then $[HCN]/[CN^-] = 1$, and $[HCN] = [CN^-]$. At this pH, the cyanide is equally distributed between the weak acid HCN(aq) and the conjugate base CN^-(aq) ions.

If we increase $[H_3O^+]$, we would expect to increase the concentration of HCN(aq) relative to the concentration of CN^-(aq) ions. For example, if $[H_3O^+] = 10 \times K_a$ (i.e., pH $= pK_a - 1$), then $[HCN]/[CN^-] = 10$, or $[HCN] = 10 \times [CN^-]$. Table 14.9 shows the ratio $[HCN]/[CN^-]$ at various pH values on either side of pH $= pK_a$.

$[H_3O^+]$	pH	$[HCN]/[CN^-]$
$[H_3O^+] = 1000 \times K_a = 4.0 \times 10^{-7}$ mol L^{-1}	pH $= pK_a - 3 = 6.40$	1000/1
$[H_3O^+] = 100 \times K_a = 4.0 \times 10^{-8}$ mol L^{-1}	pH $= pK_a - 2 = 7.40$	100/1
$[H_3O^+] = 10 \times K_a = 4.0 \times 10^{-9}$ mol L^{-1}	pH $= pK_a - 1 = 8.40$	10/1
$[H_3O^+] = K_a = 4.0 \times 10^{-10}$ mol L^{-1}	pH $= pK_a = 9.40$	1/1
$[H_3O^+] = 0.1 \times K_a = 4.0 \times 10^{-11}$ mol L^{-1}	pH $= pK_a + 1 = 10.40$	1/10
$[H_3O^+] = 0.01 \times K_a = 4.0 \times 10^{-12}$ mol L^{-1}	pH $= pK_a + 2 = 11.40$	1/100
$[H_3O^+] = 0.001 \times K_a = 4.0 \times 10^{-13}$ mol L^{-1}	pH $= pK_a + 3 = 12.40$	1/1000

TABLE 14.9 Distribution between HCN(aq) and CN^-(aq) at Various pH Values at 25 °C

This data can be graphed to form a *speciation plot* (Figure 14.11) that indicates the changing relative concentrations of the protonated species, HCN(aq), and the de-protonated species, CN^-(aq) ions, expressed as percentages of the total.

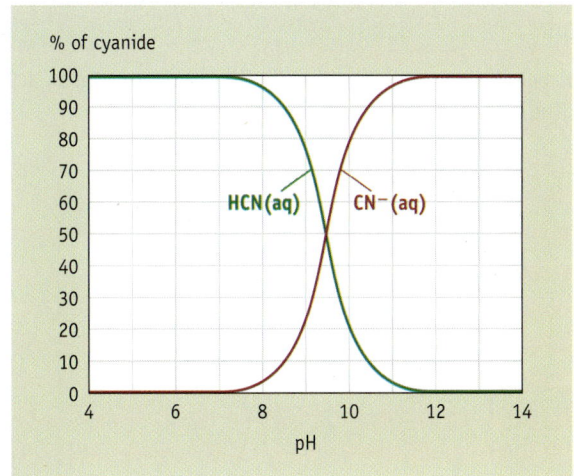

FIGURE 14.11 Speciation plot. The graphs show the changing distribution between the HCN(aq) and CN^-(aq) species as pH is changed. At a given pH, a point on the curve for either species is its percentage of the total cyanide concentration—in this case, $[HCN] + [CN^-] = 0.01$ mol L^{-1}. $[HCN] = [CN^-]$ at pH 9.40 ($= pK_a$).

We can see from Table 14.9 and the speciation graph that for every increase of 1 pH unit above pK_a, the ratio $[HCN]/[CN^-]$ decreases by a factor of 10. At pH $= pK_a + 2 = 11.40$, $[HCN]/[CN^-] = 1/100$: only 1% of the total cyanide is HCN.

Similar reasoning applies to the distribution between any weak acid–conjugate base pair as pH is changed. Using a generalized equation for weak acid–conjugate base equilibrium

$$HA(aq) + H_2O(\ell) \rightleftharpoons A^-(aq) + H_3O^+(aq)$$

in any aqueous solution, the distribution between the two species is given by

$$\frac{[HA]}{[A^-]} = \frac{[H_3O^+]}{K_a}$$

Regardless of the total concentration, the concentrations of a weak acid and its base are equal when $[H_3O^+] = K_a$ (that is, when $pH = pK_a$). In fact, the speciation plot for any monoprotic weak acid–conjugate base pair is identical to that for the HCN/CN⁻ pair, except that it is displaced so that the crossover of the lines (where $[HA] = [A^-]$) is at $pH = pK_a$. *The relative amounts of a weak acid and its conjugate base depend on the pH of solution in relation to pK_a of the acid*: at $pH < pK_a$, the concentration of the weak acid is greater than that of the base (and the lower the pH, the greater is the acid/base ratio), and at $pH > pK_a$, the concentration of the weak acid is less than that of the base (and the higher the pH is, the smaller the acid/base ratio).

A speciation plot for acetic acid-acetate ions ($pK_a = 4.74$) is shown in Figure 14.12.

FIGURE 14.12 $CH_3COOH(aq)$/ $CH_3COO^-(aq)$ speciation plot. Changing relative concentrations of acid and conjugate base with change of pH. At $pH = pK_a = 4.74$, $[CH_3COOH] = [CH_3COO^-]$. At $pH = pK_a - 1 = 3.74$, $[CH_3COOH] = 91\%$ of total, $[CH_3COO^-] = 9\%$ of total, and $[CH_3COOH]/[CH_3COO^-] = 10/1$.

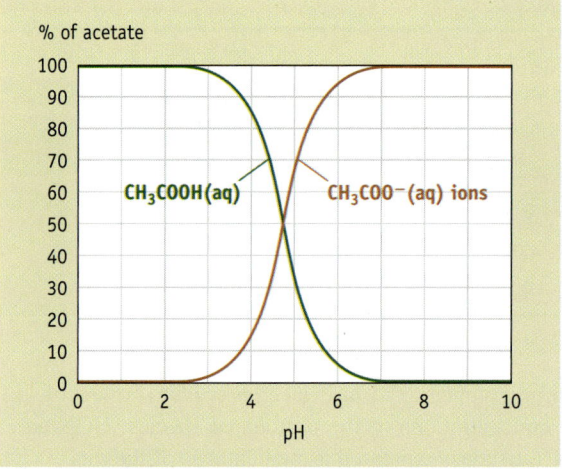

Think about It

e14.22 Use a spreadsheet to simulate differences in speciation plots in solutions of acids with different K_a values.

Interactive Exercise 14.32

SUBMIT

Interpret a speciation plot in terms of the species present at each point.

Interactive Exercise 14.33

SUBMIT

Interpret a speciation plot in terms of a titration graph.

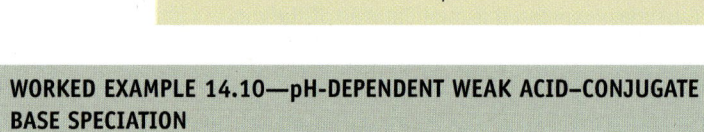

WORKED EXAMPLE 14.10—pH-DEPENDENT WEAK ACID–CONJUGATE BASE SPECIATION

The speciation plot below shows how the distribution between a weak acid, HA(aq), and its conjugate base, A⁻(aq), changes with pH.

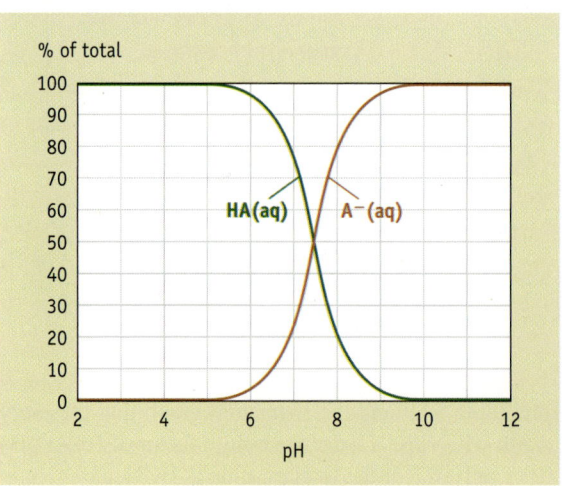

(a) Estimate pK_a and K_a of the weak acid.

(b) At what solution pH is $[HA]/[A^-] = 100/1$?

(c) At what solution pH is $[HA]/[A^-] = 1/1000$?

(d) With reference to Table 14.5 and Appendix A, suggest what the weak acid might be.

Strategy

We know that $[HA]$ will decrease as pH increases, so the green curve represents $[HA]$, and the red curve represents $[A^-]$. pH-dependent speciation is described by the equation $\dfrac{[HA]}{[A^-]} = \dfrac{[H_3O^+]}{K_a}$, which tells us that the $[HA]/[A^-]$ ratio depends on the relative magnitudes of $[H_3O^+]$ and K_a as pH changes. In particular, when $[H_3O^+] = K_a$ (i.e., pH $= pK_a$), then $[HA] = [A^-]$.

Solution

(a) We can estimate from the pH at which the curves intersect and $[HA] = [A^-]$, that pK_a is 7.4. From this, we derive $K_a = 10^{-7.4} = 4.0 \times 10^{-8}$.

(b) $\dfrac{[HA]}{[A^-]} = \dfrac{100}{1} = \dfrac{[H_3O^+]}{K_a} = \dfrac{[H_3O^+]}{4 \times 10^{-8}}$

So $[H_3O^+] = 4.0 \times 10^{-6}$ mol L^{-1}, and pH $= 5.4$.

(c) $\dfrac{[HA]}{[A^-]} = \dfrac{1}{1000} = \dfrac{[H_3O^+]}{K_a} = \dfrac{[H_3O^+]}{4 \times 10^{-8}}$

Then $[H_3O^+] = 4.0 \times 10^{-11}$ mol L^{-1}, and pH $= 10.4$.

(d) Allowing for error of reading the pH of the intersection of curves, the data is consistent with HA being hypochlorous acid, HOCl ($pK_a = 7.46$).

Comment

The answers in (b) and (c) are consistent with a change of the $[HA]/[A^-]$ ratio by a factor of 10 for each pH unit that the solution pH differs from pH $= pK_a$.

EXERCISE 14.34—pH-DEPENDENT WEAK ACID–CONJUGATE BASE SPECIATION

In a solution of the weak acid disinfectant hypochlorous acid (HOCl), calculate the ratio $[HOCl]/[OCl^-]$ in solution at 25 °C at (a) pH 4, (b) pH 7, and (c) pH 10.

The relative concentrations of a weak acid and its conjugate base in solution depend on the pH of the solution in relation to pK_a of the weak acid, and is quantitatively expressed by the relationship

$$\frac{[HA]}{[A^-]} = \frac{[H_3O^+]}{K_a}$$

If pH $= pK_a$, then $[HA] = [A^-]$.

Acid-Base Speciation and Complexation with Metal Ions

The change of relative amounts of weak acid and its conjugate base with change of pH is extraordinarily important environmentally. For example, while copper is a necessary trace element for plant growth, excessive concentrations of Cu^{2+}(aq) ions in soil solutions are toxic to plants. Many authorities specify maximum copper concentrations in water and soils around industrial and mining sites. However, natural waters usually contain substances that can act as Lewis bases and bind with Cu^{2+} ions to form stable complex ions. Many of these compounds are the conjugate bases of weak acids, such as citric acid, oxalic acid, tartaric acid, succinic acid, maleic acid, ascorbic acid, and a range of compounds called tannic acid.

Complexation with these Lewis bases renders the copper unavailable to the plants (it is not *bioavailable*), and so there can be quite high total concentrations of copper without deleterious effects on plants. On the other hand, at low pH (pH lower than the pK_a values of the various weak acids), H_3O^+ ions can successfully compete with the Cu^{2+} ions for the Lewis base

Taking It Further

e14.23 Read about applications of pH-dependent speciation plots for drugs.

molecules or ions: the corresponding weak acids are formed and Cu^{2+}(aq) ions are released. Cu^{2+} ions cannot be bound to molecules of the weak acids because the lone pairs of electrons at the bonding sites are preferentially "occupied" by H^+ ions. As a consequence, a particular total concentration of copper ions maybe toxic at low pH, but harmless at higher pH values (greater than the pK_a of the conjugate acids of the Lewis bases present in the environment).

In the same way, the ability of drug molecules that are weak acids to complex with metal ions in our bodies is pH-dependent: at pH less than pK_a, the drugs lose their ability as Lewis bases.

> Complexation to metal cations by Lewis bases decreases at low pH because of protonation of the bases, occupying the bonding sites on the molecules or ions of the bases.

Distribution among Species from Polyprotic Acids

In an aqueous solution of phosphoric acid at 25 °C, the following relationships are obeyed simultaneously:

$$Q_{a1} = \frac{[H_3O^+][H_2PO_4^-]}{[H_3PO_4]} = K_{a1} = 7.5 \times 10^{-3}$$

$$Q_{a2} = \frac{[H_3O^+][HPO_4^{2-}]}{[H_2PO_4^-]} = K_{a2} = 6.2 \times 10^{-8}$$

$$Q_{a3} = \frac{[H_3O^+][PO_4^{3-}]}{[HPO_4^{2-}]} = K_{a3} = 3.6 \times 10^{-13}$$

Re-arranging these equations, we obtain expressions showing how the relative concentrations of each weak acid–conjugate base pair depend on $[H_3O^+]$:

$$\frac{[H_3PO_4]}{[H_2PO_4^-]} = \frac{[H_3O^+]}{K_{a1}} = 7.5 \times 10^{-3}$$

$$\frac{[H_2PO_4^-]}{[HPO_4^{2-}]} = \frac{[H_3O^+]}{K_{a2}} = 6.2 \times 10^{-8}$$

$$\frac{[HPO_4^{2-}]}{[PO_4^{3-}]} = \frac{[H_3O^+]}{K_{a3}} = 3.6 \times 10^{-13}$$

These equations tell us that regardless of the total concentration of phosphate species in solution:

- at $[H_3O^+] = K_{a1} = 7.5 \times 10^{-3}$ mol L^{-1} (i.e., at pH = pK_{a1} = 2.12), $[H_3PO_4] = [H_2PO_4^-]$
- at $[H_3O^+] = K_{a2} = 6.2 \times 10^{-8}$ mol L^{-1} (i.e., at pH = pK_{a2} = 7.21), $[H_2PO_4^-] = [HPO_4^{2-}]$
- at $[H_3O^+] = K_{a3} = 3.6 \times 10^{-13}$ mol L^{-1} (i.e., at pH = pK_{a3} = 12.44), $[HPO_4^{2-}] = [PO_4^{3-}]$

Figure 14.13 shows relative concentrations of the various species as the pH of solution is changed.

FIGURE 14.13 Speciation plot of phosphate species with changing pH.

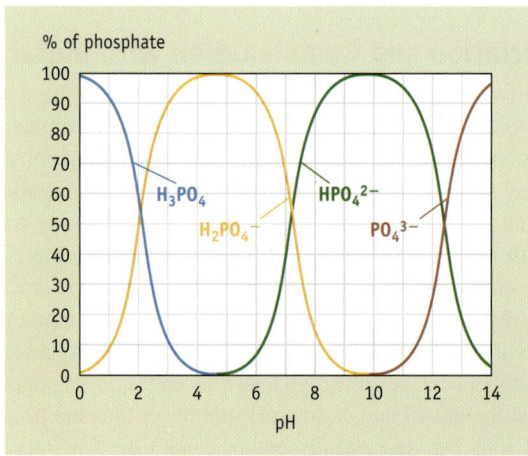

As was the case with monoprotic acids and their conjugate bases, for each phosphate acid-base pair a change of 1 pH unit brings about a tenfold change in the [acid]/[base] ratio. For example:

- at pH = 3.12, $[H_3PO_4]/[H_2PO_4^-] = 1/10$
- at pH 7.12, $[H_3PO_4]/[H_2PO_4^-] = 1/100\,000$. This is very near the pH at which $[H_2PO_4^-] = [HPO_4^{2-}]$
- at pH = 11.44, $[HPO_4^{2-}]/[PO_4^{3-}] = 10/1$

EXERCISE 14.35—pH-DEPENDENT SPECIATION OF POLYPROTIC ACID SPECIES

In an aqueous solutions containing phosphates at 25 °C at (a) pH 6.0, and (b) 9.0, calculate the ratios of species (i) $[H_3PO_4]/[H_2PO_4^-]$, (ii) $[H_2PO_4^-]/[HPO_4^{2-}]$, and (iii) $[HPO_4^{2-}]/[PO_4^{3-}]$. Which is the dominant species at each pH?

Speciation between the conjugate acid-base pair of each ionization step of a polyprotic acid simultaneously obeys a relationship of the form

$$\frac{[acid]}{[base]} = \frac{[H_3O^+]}{K_a}$$

14.8 Acid-Base Properties of Amino Acids and Proteins

All amino acids have a molecular structure that is commonly portrayed as in the diagrams below. Each amino acid has a different alkyl group (abbreviated –R) (Table 14.10).

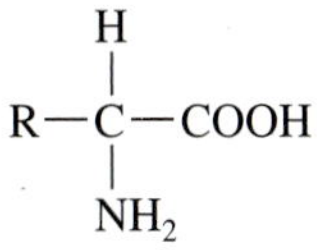

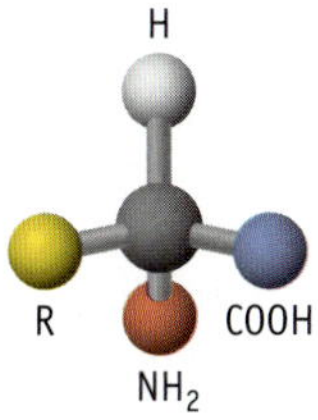

pH-Dependent Speciation of Amino Acids

Amino acids never exist as the uncharged molecules portrayed above. This can be illustrated with reference to the simplest amino acid, glycine, where R is a H atom. At low pH, the carboxylic and amine groups bonded to the same C atom (called the α-carbon atom) are both protonated, and glycine exists as ions with a single positive charge, $^+NH_3CH_2COOH$, abbreviated as H_2Gly^+.

This ion behaves like a typical weak diprotic acid, with two ionization steps: the proton on the –COOH group is removed in the first step (forming HGly), and that on the $-NH_3^+$ group in the second (forming Gly^-):

$$^+NH_3CH_2COOH(aq) + H_2O(\ell) \rightleftharpoons \,^+NH_3CH_2COO^-(aq) + H_3O^+(aq) \quad pK_{a1} = 2.34$$
$$^+NH_3CH_2COO^-(aq) + H_2O(\ell) \rightleftharpoons NH_2CH_2COO^-(aq) + H_3O^+(aq) \quad pK_{a2} = 9.60$$

All of the species have charged sites. Although the HGly species formed by removal of just one proton has net zero charge, this *zwitterion* has a positively charged $-NH_3^+$ group and a negatively charged $-COO^-$ group. The Gly^- species has a single negative charge.

In a manner exactly parallel to the case of carbonic acid (H_2CO_3), the relative concentrations of each acid-base conjugate pair are dependent on the relative values of $[H_3O^+]$ in solution and K_a of the acid species:

$$\frac{[H_2Gly^+]}{[HGly]} = \frac{[H_3O^+]}{K_{a1}} \quad \text{and} \quad \frac{[HGly]}{[Gly^-]} = \frac{[H_3O^+]}{K_{a2}}$$

Because the values of K_{a1} and K_{a2} are so different, if the pH of a solution containing glycine is gradually increased, essentially all of the –COOH protons are removed before any of the –NH_3^+ protons are, and this can be seen in a speciation plot (Figure 14.14).

FIGURE 14.14 pH-dependent glycine speciation. At pH = 2.34 = pK_{a1}, [H_2Gly^+] = [HGly]; at pH = 9.60 = pK_{a2}, [HGly] = [Gly^-].

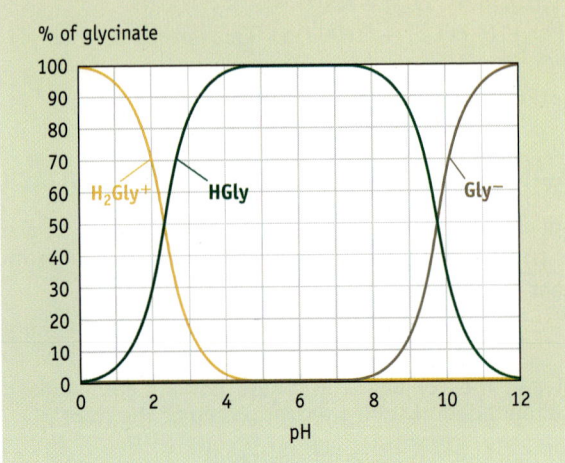

The pH midway between pK_{a1} and pK_{a2} (i.e., at pH = $\frac{1}{2}$ (pK_{a1} + pK_{a2}) = 5.97) is *when more of the glycine is in the form of zwitterions, with net zero charge, than at any other pH.* This is called the **isoelectric pH**. In the case of glycine, because pK_{a1} and pK_{a2} are very different, essentially 100% of it is present as the zwitterions at the isoelectric pH.

As can be seen in Table 14.10, in all of the amino acids, the –COOH group on the α-carbon atom is more easily deprotonated (pK_{a1} ~ 2) than the –NH_3^+ group (pK_{a2} ~ 9). The stronger acidity (lower pK_a) of the –COOH group in amino acids than in acids such as acetic acid and propanoic acid demonstrates the significant effect of other parts of the same molecule or ion—in this case, the electron-withdrawing effect of the amine group attached to the same C atom.

In both glutamic acid and aspartic acid, there is a carboxylic acid group on the side chain, so these are triprotic weak acids. The fully protonated species at low pH, abbreviated as $HGlu^+$, has a single positive charge. The three ionization steps are

$$H_3Glu^+(aq) + H_2O(\ell) \rightleftharpoons H_2Glu(aq) + H_3O^+(aq) \qquad pK_{a1} = 2.19$$
$$H_2Glu(aq) + H_2O(\ell) \rightleftharpoons HGlu^-(aq) + H_3O^+(aq) \qquad pK_{a2} = 4.25$$
$$HGlu^-(aq) + H_2O(\ell) \rightleftharpoons Glu^{2-}(aq) + H_3O^+(aq) \qquad pK_{a3} = 9.67$$

The changing speciation in solution as pH is increased is shown in Figure 14.15. Zwitterions (H_2Glu) are formed after removal of one H^+ ion, and have a —COOH, a —COO^- and a —NH_3^+ group. The isoelectric pH, at which zwitterions are the dominant species, is at pH = $\frac{1}{2}$ (2.19 + 4.25) = 3.22.

Molecular Modelling (Odyssey)

e14.24 Examine models of all the amino acids showing electrostatic potential maps.

FIGURE 14.15 pH-dependent speciation of glutamic acid. [H_3Glu^+] = [H_2Glu] at pH = pK_{a1} = 2.19, [H_2Glu] = [$HGlu^-$] at pH = pK_{a2} = 4.25, and [$HGlu^-$] = [Glu^{2-}] at pK_{a3} = 9.67. The isoelectric pH, at which the dominant species is the zwitterion H_2Glu with overall zero charge, is at pH = 3.2.

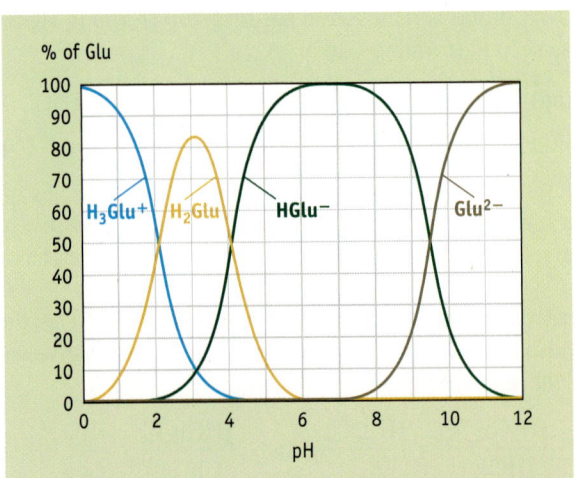

Taking It Further

e14.25 Read about how electrophoresis exploits the pH-dependent speciation of amino acids.

TABLE 14.10 Structures* of the 20 Amino Acids Found in Proteins (Names of the Amino Acids Essential to the Human Diet Are Shown in Red)

Name	Abbreviations	MW	Structure	pK_{a1} α-COOH	pK_{a2} α-NH$_3^+$	pK_a Side Chain	Isoelectric Point
Neutral amino acids							
Alanine	Ala (A)	89	CH$_3$CHCO$^-$ — NH$_3^+$ (with C=O)	2.34	9.69	—	6.01
Asparagine	Asn (N)	132	H$_2$NCCH$_2$CHCO$^-$ — NH$_3^+$ (two C=O)	2.02	8.80	—	5.41
Cysteine	Cys (C)	121	HSCH$_2$CHCO$^-$ — NH$_3^+$ (with C=O)	1.96	10.28	8.18	5.07
Glutamine	Gln (Q)	146	H$_2$NCCH$_2$CH$_2$CHCO$^-$ — NH$_3^+$ (two C=O)	2.17	9.13	—	5.65
Glycine	Gly (G)	75	CH$_2$CO$^-$ — NH$_3^+$ (with C=O)	2.34	9.60	—	5.97
Isoleucine	Ile (I)	131	CH$_3$CH$_2$CHCHCO$^-$ with CH$_3$, NH$_3^+$ (with C=O)	2.36	9.60	—	6.02
Leucine	Leu (L)	131	CH$_3$CHCH$_2$CHCO$^-$ with CH$_3$, NH$_3^+$ (with C=O)	2.36	9.60	—	5.98
Methionine	Met (M)	149	CH$_3$SCH$_2$CH$_2$CHCO$^-$ — NH$_3^+$ (with C=O)	2.28	9.21	—	5.74
Phenylalanine	Phe (F)	165	C$_6$H$_5$—CH$_2$CHCO$^-$ — NH$_3^+$ (with C=O)	1.83	9.13	—	5.48
Proline	Pro (P)	115	(ring structure) C—O$^-$ with N$^+$—H, H (with C=O)	1.99	10.60	—	6.30
Serine	Ser (S)	105	HOCH$_2$CHCO$^-$ — NH$_3^+$ (with C=O)	2.21	9.15	—	5.68

Name	Abbreviations	MW	Structure	pK_{a1} α-COOH	pK_{a2} α-NH$_3^+$	pK_a Side Chain	Isoelectric Point
Threonine	Thr (T)	119	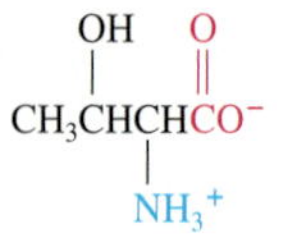	2.09	9.10	—	5.60
Tryptophan	Trp (W)	204	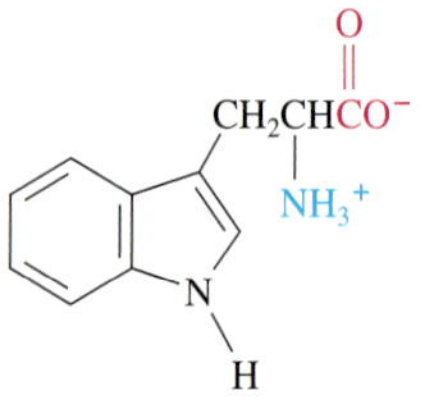	2.83	9.39	—	5.89
Tyrosine	Tyr (Y)	181	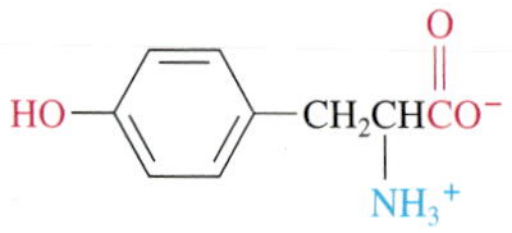	2.20	9.11	10.07	5.66
Valine	Val (V)	117	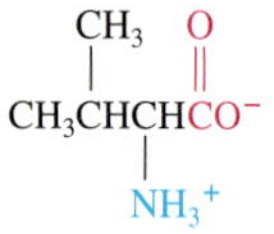	2.32	9.62	—	5.96
Acidic amino acids							
Aspartic acid	Asp (D)	133	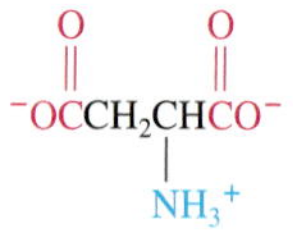	1.88	9.60	3.65	2.77
Glutamic acid	Glu (E)	147	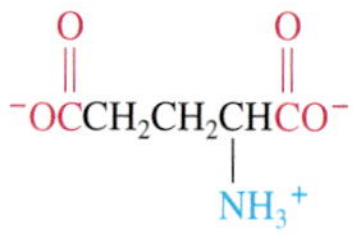	2.19	9.67	4.25	3.22
Basic amino acids							
Arginine	Arg (R)	174	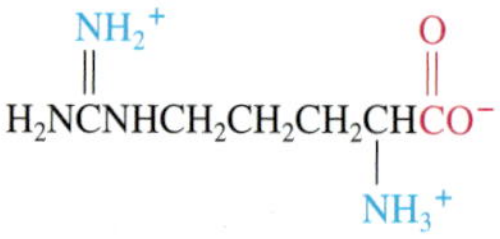	2.17	9.04	12.48	10.76
Histidine	His (H)	155	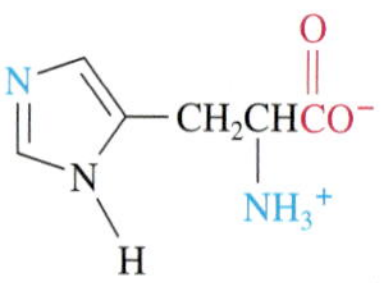	1.82	9.17	6.00	7.59
Lysine	Lys (K)	146	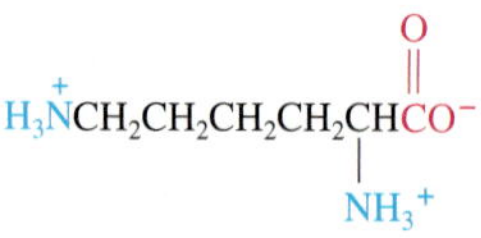	2.18	8.95	10.53	9.74

* The structures shown here are those of the dominant species at the physiological pH (7.4). Amino acids whose side chain has neither a —COOH group nor an —NH$_3^+$ group are called *neutral* amino acids. Those called *acidic* have a —COOH group in the side chain, and those with an amine group are classed as *basic*. In every case, the —COOH group on the α-carbon is the most acidic site (pK_{a1}). Although the values of pK_{a2} listed are pK_a values for the —NH$_3^+$ group on the α-carbon, in a few cases there is a more acidic site (lower pK_a) on the side chain.

Cysteine has an —SH group in its side chain, and tyrosine has an —OH group. These groups are much weaker acid sites than —COOH groups.

In each of arginine, histidine, and lysine, there is an amine group in the side chain. A consequence of this is that the fully protonated species at low pH have a 2+ charge—represented in the case of lysine as H_3Lys^{2+}. In these cases, a zwitterion with net zero charge is formed after removal of two protons—one from the —COOH group and one of the amine groups—and the isoelectric point is at pH $= \frac{1}{2}$ ($pK_{a2} + pK_{a3}$).

Fully protonated lysine ion, H_3Lys^{2+}

WORKED EXAMPLE 14.11—pH-DEPENDENT DISTRIBUTION AMONG AMINO ACID SPECIES IN SOLUTION

Use the pK_a values in Table 14.10 to deduce the ratios $[H_2Gly^+]/[HGly]$ and $[HGly]/[Gly^-]$ in aqueous solutions of glycine at pH 7.4 the pH of body fluids (blood pH). What is the dominant species at this pH, and what is its charge?

Strategy

We use the speciation equations in the same way as for speciation of polyprotic acid species in Section 14.7. $K_{a1} = 10^{-2.34} = 4.57 \times 10^{-3}$ and $K_{a2} = 10^{-9.60} = 2.51 \times 10^{-10}$. $[H_3O^+] = 10^{-7.4} = 3.98 \times 10^{-8}$ mol L^{-1}.

Solution

The speciation equations for glycine are

$$\frac{[H_2Gly^+]}{[HGly]} = \frac{[H_3O^+]}{K_{a1}} = \frac{3.98 \times 10^{-8}}{4.57 \times 10^{-3}} = 8.7 \times 10^{-6}$$

$$\frac{[HGly]}{[Gly^-]} = \frac{[H_3O^+]}{K_{a2}} = \frac{3.98 \times 10^{-8}}{2.51 \times 10^{-10}} = 160$$

Since, at this pH, $[H_2Gly^+] < [HGly]$ and $[HGly] > [Gly^-]$, we can say that HGly is the species at highest concentration. This species is the zwitterion with zero net charge (+1 on the protonated amine group, and −1 on the deprotonated carboxylate group).

Comment

The answer is qualitatively consistent with Figure 14.14 above.

EXERCISE 14.36—pH-DEPENDENT DISTRIBUTION AMONG AMINO ACID SPECIES IN SOLUTION

For each of (a) phenylalanine, (b) glutamic acid, and (c) lysine, calculate the [acid]/[base] ratio for each acid-base pair in solution at pH 7.4, the pH of body fluids. Write the structure of the dominant species at this pH, and give its net charge.

The pH-dependent speciation of amino acids is as for other polyprotic acids. The isoelectric pH for an amino acid is the pH at which the species has a site with positive charge and a site with a negative charge, but zero overall charge—called a zwitterion. Carboxylic acid protons are more acidic (more easily removed) than those on protonated amine groups.

14.9 Controlling pH: Buffer Solutions

The normal pH of human blood is 7.4. The addition of a small quantity of strong acid or base, say 0.01 mol, to 1 L of blood, leads to a change of only about 0.1 pH unit. By contrast, if you add 0.01 mol of HCl solution or NaOH solution to 1 L of water, the pH changes by 5 units. Blood, and many other body fluids, are said to be *buffered*: they contain reagents that minimize the change in pH when a relatively small amount of a strong acid or base is added (Figure 14.16).

FIGURE 14.16 Buffer solutions. (a) The pH meter indicates the pH of water in the beaker on the right that contains a trace of acid, as well as some bromophenol blue acid-base indicator. The solution at left is a buffer solution with pH about 7, also with the indicator. (b) When 5 mL of 0.1 mol L^{-1} HCl solution is added to each solution, the pH of the water drops by several units (as shown by the indicator colour change), whereas the unchanged indicator colour indicates that the pH of the buffer solution hardly changes.

Before

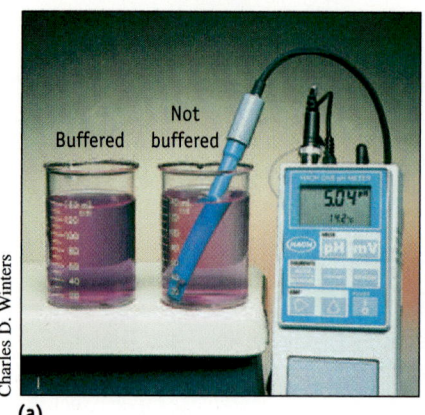

Buffered | Not buffered

After adding 0.10 M HCl

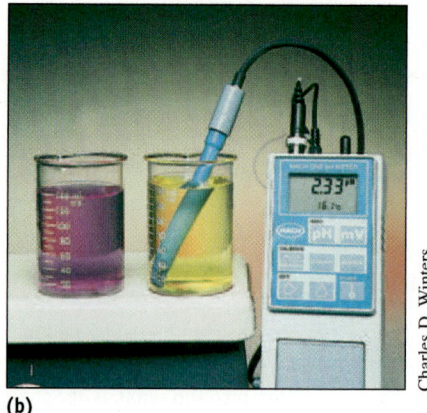

(a)   (b)

Think about It

e14.26 Simulate the addition of acid and base to buffered and unbuffered solutions.

TABLE 14.11 Some Commonly Used Buffering Reagents

Composition and Mode of Operation of Buffer Solutions

A **buffer solution** *minimizes the change of pH when some strong acid or base is added, and contains large amounts of both a weak acid and its conjugate base (relative to the amounts of acid or base that are added).* The optimal weak acid/base ratio is within the range 10:1 to 1:10. Some buffering agents commonly used in the laboratory are listed in Table 14.11.

Weak Acid	Conjugate Base	Acid K_a (pK_a)	Effective pH Range
Phthalic acid, $(HOOC)C_6H_4(COOH)$	Hydrogenphthalate ion $(HOOC)C_6H_4(COO^-)$	1.3×10^{-3} (2.89)	1.9–3.9
Acetic acid, CH_3COOH	Acetate ion, CH_3COO^-	1.8×10^{-5} (4.74)	3.7–5.8
Dihydrogenphosphate ion, $H_2PO_4^-$	Hydrogenphosphate ion, HPO_4^{2-}	6.2×10^{-8} (7.21)	6.2–8.2
Hydrogencarbonate ion, HCO_3^-	Carbonate ion, CO_3^{2-}	4.8×10^{-11} (10.32)	9.3–11.3
Hydrogenphosphate ion, HPO_4^{2-}	Phosphate ion, PO_4^{3-}	3.6×10^{-13} (12.44)	11.3–13.3

Buffer solutions function because the weak acid and its conjugate base are in chemical equilibrium. For example, in the case of a solution containing acetic acid and acetate ions there is an equilibrium, represented by the equation:

$$CH_3COOH(aq) + H_2O(\ell) \rightleftharpoons CH_3COO^-(aq) + H_3O^+(aq)$$

A qualitative way of considering the action of a buffer solution is that if some strong acid is added, most of the $H_3O^+(aq)$ ions added are "mopped up" by reaction with the weak base. And if a strong base is added, most of the $OH^-(aq)$ ions added are "mopped up" by the weak acid.

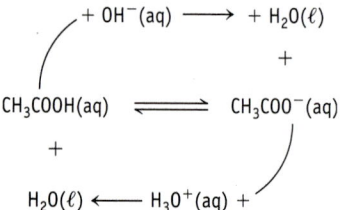

Think about It

e14.27 Watch a demonstration of a solution acting as a pH buffer.

Of course, $H_3O^+(aq)$ ions cannot be "mopped up" effectively if we add more acid than there is weak base in the solution, and $OH^-(aq)$ ions cannot be "mopped up" effectively if we add more base than the amount of HA (the "mopper-upper" of OH^- ions) in solution.

Although HCl and Cl^- are a conjugate acid-base pair, a solution containing these does not have buffering action because HCl is a strong acid, so there is no tendency for $Cl^-(aq)$ ions to react with $H_3O^+(aq)$ ions to form HCl. A solution containing carbonic acid (H_2CO_3) and carbonate ions (CO_3^{2-}) is not a buffer solution because these species are not a conjugate acid-base pair: they would react with proton transfer to form mainly hydrogencarbonate ions (HCO_3^-).

> Buffer solutions contain reagents that minimize the change in pH when strong acid or base is added. They have large amounts of both a weak acid and its conjugate base, relative to the amounts of acid or base that are added. Buffer solutions function because the weak acid and base are in equilibrium: added $H_3O^+(aq)$ ions are "mopped up" by the conjugate base, and added $OH^-(aq)$ ions are "mopped up"' by the weak acid.

Interactive Exercise 14.37

SUBMIT

Predict if specified solutions could act as buffers.

Quantitative Calculations of Buffer Solution pH

In the speciation calculations of Section 14.7, we were concerned with how the ratio of relatively small amounts of HA(aq) and A^-(aq) ions depend on the concentration of $H_3O^+(aq)$ ions from different sources. Buffer solution calculations use essentially the same relationship, but in this case the ratio of relatively large amounts of HA(aq) and A^-(aq) ions in solution control $[H_3O^+]$.

For any solution containing a weak acid HA(aq) and its base A^-(aq), we must have the relationship

$$Q = \frac{[H_3O^+][A^-]}{[HA]} = K_a$$

Transposing, we obtain the **buffer equation:**

$$[H_3O^+] = \frac{[HA]}{[A^-]} \times K_a \tag{14.1}$$

At a given temperature, K_a is constant, so this equation shows that the ratio $[HA]/[A^-]$ governs (and is proportional to) $[H_3O^+]$. When $H_3O^+(aq)$ or $OH^-(aq)$ ions are added, the redistribution resulting from "mopping up" will change the $[HA]/[A^-]$ ratio; it will increase if we add $H_3O^+(aq)$ ions, and decrease if we add $OH^-(aq)$ ions. If the amounts of HA(aq) and A^-(aq) ions are considerably greater than the amount of $H_3O^+(aq)$ or $OH^-(aq)$ ions added, the ratio $[HA]/[A^-]$ will change only a little—and so, according to Equation 14.1, the equilibrium concentration of $H_3O^+(aq)$ ions (and pH) will change only a little.

You may find it useful to use a logarithmic form of the buffer equation, called the *Henderson-Hasselbalch equation* (obtained by taking the negative logarithm of the terms in the Equation 14.1):

$$pH = pK_a + \log_{10}\frac{[A^-]}{[HA]} \tag{14.2}$$

Interactive Exercise 14.38

SUBMIT

Interpret the chemical significance of this equation from its graph.

L.J. Henderson (1878–1942) was a medical researcher at Harvard University, whose early studies included the buffering ability of blood. The main contribution of K.A. Hasselbalch at the University of Copenhagen was to express the buffer formula in logarithmic form in 1916, following the recently developed pH scale.

Worked Example 14.12, a rigorous calculation of the pH of a buffer solution using Equation 14.1, is used to illustrate easier pathways to the answer without sacrificing accuracy. You will see that because buffer solutions involve the common ion effect, we can presume that the species concentrations of the weak acid and the conjugate base can be taken to be the same as the "labelled concentrations."

WORKED EXAMPLE 14.12—pH OF A BUFFER SOLUTION

What is the pH of an acetic acid/sodium acetate buffer solution made with 0.350 mol of acetic acid and 0.300 mol of sodium acetate dissolved in 500.0 mL of solution?

Strategy

Calculate $c(CH_3COOH)$ and $c(CH_3COO^-)$. As usual, assume that $[CH_3COOH]$ decreases by x mol L^{-1} because of ionization.

Solution

We first calculate the labelled concentrations:

$$c(CH_3COOH) = \frac{n(CH_3COOH)}{V} = \frac{0.350 \text{ mol}}{0.500 \text{ L}} = 0.700 \text{ mol L}^{-1}$$

$$c(CH_3COO^-) = \frac{n(CH_3COO^-)}{V} = \frac{0.300 \text{ mol}}{0.500 \text{ L}} = 0.600 \text{ mol L}^{-1}$$

The equilibrium condition can be represented by the equation

$$CH_3COOH(aq) + H_2O(\ell) \rightleftharpoons CH_3COO^-(aq) + H_3O^+(aq)$$

Let's assume that in attaining equilibrium, the concentration of $CH_3COOH(aq)$ decreases by x mol L^{-1}. Then, at equilibrium, we have

$$[CH_3COOH] = (0.700 - x) \text{ mol L}^{-1}$$
$$[CH_3COO^-] = (0.600 + x) \text{ mol L}^{-1}$$
$$[H_3O^+] = x \text{ mol L}^{-1}$$

We can substitute these values into the $Q = K$ condition:

$$Q = \frac{[CH_3COO^-][H_3O^+]}{[CH_3COOH]} = \frac{(0.600 + x)(x)}{(0.700 - x)} = K_a = 1.8 \times 10^{-5}$$

Solving the resultant quadratic equation leads to $x = 2.1 \times 10^{-5}$.

$$\therefore [H_3O^+] = x = 2.1 \times 10^{-5} \text{ mol L}^{-1} \text{ and pH} = 4.68$$

Comment

We can see that, compared with $c(CH_3COOH)$ and $c(CH_3COO^-)$, the concentration change due to ionization of acetic acid is very small—as we should expect with the common ion effect in operation. In fact:

- $[CH_3COOH] = (0.700 - 2.1 \times 10^{-5}) \text{ mol L}^{-1} = 0.700 \text{ mol L}^{-1} = c(CH_3COOH)$ to three significant figures
- $[CH_3COO^-] = (0.600 + 2.1 \times 10^{-5}) \text{ mol L}^{-1} = 0.600 \text{ mol L}^{-1} = c(CH_3COO^-)$ to three significant figures

This is generally the case for buffer solutions, and so we can say, to a good approximation:

$$[CH_3COOH] = c(CH_3COOH) \quad \text{and} \quad [CH_3COO^-] = c(CH_3COO^-)$$

The conclusions from Worked Example 14.12 show that we can use a modified buffer equation:

$$[H_3O^+] = \frac{c(HA)}{c(A^-)} \times K_a \qquad (14.3)$$

WORKED EXAMPLE 14.13—BUFFER SOLUTION pH FROM LABELLED CONCENTRATIONS

What is the pH of an acetic acid/sodium acetate buffer solution made with 0.350 mol of acetic acid and 0.300 mol of sodium acetate dissolved in 500.0 mL of solution?

Strategy

Substitute directly into Equation 14.3.

Solution

We know that

$$c(CH_3COOH) = 0.700 \text{ mol L}^{-1} \quad \text{and} \quad c(CH_3COO^-) = 0.600 \text{ mol L}^{-1}$$

$$[H_3O^+] = \frac{c(CH_3COOH)}{c(CH_3COO^-)} \times K_a = \frac{0.700}{0.600} \times (1.8 \times 10^{-5}) = 2.1 \times 10^{-5} \text{ mol L}^{-1}$$

Comment

This is the same result as we obtained in Worked Example 14.12, and demonstrates that we do not need to account for ionization of the acid and base in buffer solutions. The appearance in the last line of the appropriate unit for $[H_3O^+]$ may be confusing. Recall that these equations actually apply to activities, rather than concentrations, and that activity in dilute solutions is the ratio of concentration to 1 mol L^{-1}. So, more rigorously we might have written:

$$a_{H_3O^+} = \frac{[H_3O^+]}{1 \text{ mol L}^{-1}} = 2.1 \times 10^{-5} \quad \text{so} \quad [H_3O^+] = 2.1 \times 10^{-5} \text{ mol L}^{-1}$$

Furthermore, inspection of Equation 14.3 shows that it is not necessary to use the actual concentrations of the weak acid and its base—only the ratio of them. This is sometimes convenient.

In Worked Example 14.13, we calculated $c(CH_3COOH)$ and $c(CH_3COO^-)$ from the amounts of each (n) and the solution volume (V). Since the acid and its base are in the same solution, we use the same value of V, and this cancels out when we substitute into Equation 14.3, leading us to another version of the buffer equation:

$$[H_3O^+] = \frac{c(CH_3COOH)}{c(CH_3COO^-)} \times K_a = \frac{n(CH_3COOH)}{\cancel{V}} \times \frac{\cancel{V}}{n(CH_3COO^-)} \times K_a$$
$$= \frac{n(CH_3COOH)}{n(CH_3COO^-)} \times K_a$$

Now we have an equation that is convenient to use if we have the amounts of the weak acid and its conjugate base (or even the ratio of acid to base), regardless of what the volume of solution is. Generalizing for the weak acid HA and its conjugate base A$^-$:

$$[H_3O^+] = \frac{n(HA)}{n(A^-)} \times K_a \tag{14.4}$$

WORKED EXAMPLE 14.14—BUFFER SOLUTION pH FROM THE AMOUNTS OF WEAK ACID AND BASE

What is the pH of an acetic acid/sodium acetate buffer solution made with 0.350 mol of acetic acid and 0.300 mol of sodium acetate dissolved in 500.0 mL of solution?

Strategy

Substitute the amounts of the reagents directly into Equation 14.4.

Solution

$$[H_3O^+] = \frac{n(CH_3COOH)}{n(CH_3COO^-)} \times K_a = \frac{0.350}{0.300} \times (1.8 \times 10^{-5}) = 2.1 \times 10^{-5} \text{ mol L}^{-1}$$

Comment

This is the same answer as we obtained in both of Worked Examples 14.12 and 14.13, and shows that if we have the amounts of the weak acid and its conjugate base, then regardless of what the volume of the buffer solution is, it is not necessary to convert these into concentrations to calculate $[H_3O^+]$ or pH.

The approximations $[HA] = c(HA)$ and $[A^-] = c(A^-)$, as well as $n(HA)/n(A^-) = [HA]/[A^-] = c(HA)/c(A^-)$ render the mathematics of deriving $[H_3O^+]$ and pH in buffer solutions straightforward. However, these approximations apply only to buffer solutions, because of the common ion effect, and you need to recognize whether a buffer solution is involved before you use them.

Think about It

e14.28 See how the pH of a buffer solution depends on the pK_a and acid-base concentration ratio.

Think about It

e14.29 Calculate the pH of
a buffer solution.

**Interactive
Exercise 14.39**

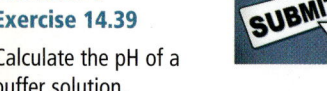

Calculate the pH of a
buffer solution.

The various forms of the buffer equation (Equations 14.1–14.4) show that the pH of a buffer solution is controlled by two factors: the strength of the weak acid (indicated by K_a) and the ratio of amounts of the weak acid and its conjugate base. The pH is established primarily by the value of K_a, and is fine-tuned by adjusting the acid/base ratio. When the concentrations of acid and its conjugate base are the same, the ratio [acid]/ [base] is 1, so pH = pK_a. If there is more of the conjugate base in the solution than of the weak acid, pH > pK_a. Conversely, if there is more weak acid than its base in solution, pH < pK_a.

EXERCISE 14.40—PH OF BUFFER SOLUTIONS

Calculate the pH of a solution in which 2.00 g of benzoic acid (C_6H_5COOH) and 2.00 g of sodium benzoate ($NaC_6H_5COO^-$) are dissolved in enough water to make 1.00 L of solution at 25 °C.

In buffer solutions, $[H_3O^+]$ is governed by, and is proportional to, the ratio $[HA]/[A^-]$—expressed by the buffer equation

$$[H_3O^+] = \frac{[HA]}{[A^-]} \times K_a \quad \text{or} \quad pH = pK_a - \log_{10}\frac{[HA]}{[A^-]}$$

The pH depends mainly on the value of K_a, and is fine-tuned by variation of the $[HA]/[A^-]$ ratio. When $[HA] = [A^-]$, pH = pK_a. Alternative forms of the buffer equation are

$$[H_3O^+] = \frac{c(HA)}{c(A^-)} \times K_a \quad \text{and} \quad [H_3O^+] = \frac{n(HA)}{n(A^-)} \times K_a$$

Design of a Buffer Solution of Specified pH

A buffer solution containing a particular weak acid–conjugate base pair can be made in any several ways. For example, if we were to make a buffer solution comprising acetic acid and acetate ions in aqueous solution, we could do so by any of the following ways:

- Put the some acetic acid and some sodium acetate (from their reagent jars) into the same solution.
- Put some acetic acid into water and neutralize *some* of it by adding sodium hydroxide solution. This will leave some acetic acid in solution and convert some of it to acetate ions.
- Dissolve some sodium acetate into water and neutralize *some* of it by adding hydrochloric acid solution. This will leave some acetate ions in solution and convert some of them into acetic acid.

The last two methods are related to what happens when (a) a weak acid is titrated with a strong base, or (b) a weak base is titrated with a strong acid, and are discussed in some detail in Section 14.10.

To design and make a buffer solution that can be effective at a particular pH, we need to do the following:

Think about It

e14.30 Simulate the making
of a buffer solution with a
specified pH.

Think about It

e14.31 Work through a
tutorial on making a buffer
solution with a specified pH.

Think about It

e14.32 Work through a
tutorial on the calculations
involved in making a buffer solution with
a specified pH.

- Choose a weak acid/conjugate base system with pK_a of the acid within 1 unit of the desired pH. If you choose an acid whose pK_a is more than 1 unit away from the target pH, the required [weak acid]/[conjugate base] ratio will be outside of the range 1/10 to 10/1—which are approximate boundaries of buffer action against the addition of both acid and base.
- Calculate the [weak acid]/[conjugate base] ratio that will fine-tune the solution pH to the target value.
- Make a solution containing both the weak acid and its conjugate base with the appropriate ratio, using one of the three methods above.
- Use amounts of the weak acid and its conjugate base well in excess of the amounts of strong acid or base that are likely to added to the solution, and which would otherwise cause dramatic changes of pH (see *buffer capacity*, below).

WORKED EXAMPLE 14.15—DESIGNING A BUFFER SOLUTION

Suppose that you want to make 1.0 L of a buffer solution with a pH of 4.30. Choose from the following weak acids and their conjugate bases:

Weak Acid	Conjugate Base	K_a	pK_a
HSO_4^-	SO_4^{2-}	1.2×10^{-2}	1.92
CH_3COOH	CH_3COO^-	1.8×10^{-5}	4.74
HCO_3^-	CO_3^{2-}	4.8×10^{-11}	10.32

Which combination should be selected, and what should be the ratio of amounts of acid and base?

Strategy

Choose an acid-base pair with an acid whose pK_a value is close to the pH needed (4.30). Then use Equation 14.1 to calculate the appropriate acid/base ratio.

Solution

Of the acids listed, only acetic acid (CH_3COOH) has a pK_a value close to the desired pH of 4.30. Apply Equation 14.1, using the target $[H^+] = 10^{-4.30}$ mol $L^{-1} = 5.0 \times 10^{-5}$ mol L^{-1}.

$$[H_3O^+] = 5.0 \times 10^{-5} = \frac{[CH_3COOH]}{[CH_3COO^-]} \times K_a = \frac{[CH_3COOH]}{[CH_3COO^-]} \times 1.8 \times 10^{-5}$$

This leads us, by re-arrangement, to

$$\frac{[CH_3COOH]}{[CH_3COO^-]} = \frac{5.0 \times 10^{-5}}{1.8 \times 10^{-5}} = \frac{2.8}{1}$$

If you add 0.28 mol of acetic acid and 0.10 mol of sodium acetate (or any other molar amounts in the ratio 2.8/1) to enough water to make 1.0 L of solution, the resulting buffer solution will have pH 4.30.

Alternative Solution

You may find it preferable to use Equation 14.2:

$$pH = 4.30 = pK_a - \log_{10}\frac{[CH_3COO^-]}{[CH_3COOH]} = 4.74 - \log_{10}\frac{[CH_3COOH]}{[CH_3COO^-]}$$

$$\log_{10}\frac{[CH_3COOH]}{[CH_3COO^-]} = 4.74 - 4.30 = 0.44$$

$$\frac{[CH_3COOH]}{[CH_3COO^-]} = \frac{2.8}{1}$$

Comment

Because we are calculating a ratio of concentrations, it is often useful if the answer is expressed as a ratio (2.8/1), rather than as a number (2.8).

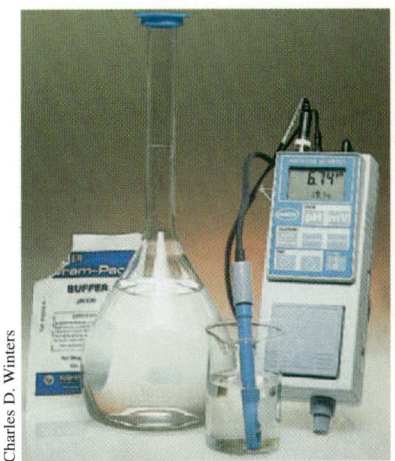

An off-the-shelf buffer solution. The solid weak acid and its base in the packet are dissolved in water.

Another method to make the buffer solution described in Worked Example 14.15 is to dissolve some sodium acetate in water, and then add HCl solution until the pH is the desired value, as indicated by a pH meter. Alternatively, gradually add NaOH solution into an acetic acid solution until the desired pH is reached.

Commercially available buffers are often sold as premixed, dry ingredients—different mixes for each solution pH. To use them, you simply dissolve the ingredients in water.

**Interactive
Exercise 14.41**

SUBMIT

Select the appropriate
acid-base pair to prepare a specified
buffer solution.

EXERCISE 14.42—DESIGNING BUFFER SOLUTIONS

Which of the following combinations of reagents would be the best to buffer a solution at a
pH near 9?
(a) HCl and NaCl
(b) NH_3 and NH_4Cl
(c) CH_3COOH and $NaCH_3COO$

EXERCISE 14.43—DESIGNING BUFFER SOLUTIONS

Describe how to make a buffer solution using $NaH_2PO_4(s)$ and $Na_2HPO_4(s)$ so that the pH is 7.5.

> To design a buffer solution for a particular pH, choose a weak acid with pK_a within
> 1 unit of the target pH, and calculate the [weak acid]/[conjugate base] ratio that
> will adjust the solution pH to the target value.

pH Change of Buffer Solutions

Although buffer solutions minimize the change of solution pH on addition of relatively
small amounts of strong acid or base, some change necessarily occurs. To calculate the pH
change, we need simply to calculate the weak acid/conjugate base ratio both before and
after addition of the strong acid or base, and use it in Equation 14.1.

To calculate by how much pH changes for addition of a specified amount of $H_3O^+(aq)$,
we work in amounts (in mol) of the weak acid HA(aq) and its conjugate base A^-(aq) rather
than concentrations:

- Calculate the amounts of HA(aq) and A^-(aq) in the sample of buffer solution.
- Calculate the amount of H_3O^+(aq) added.
- Use the stoichiometry of the equation for neutralization to calculate the amount of
  HA(aq) that reacts and the amount of A^-(aq) ions produced. Assume that this reaction
  "goes to completion."
- Calculate the new amounts $n(HA)$ and $n(A^-)$ in the solution.
- Calculate the new ratio $n(HA)/n(A^-)$.
- Calculate the new pH using the ratio $n(HA)/n(A^-)$ and K_a of the weak acid HA(aq).

Think about It

e14.33 Predict the effect on
the pH of a buffer solution when an acid
or base is added.

Think about It

e14.34 Work through a
tutorial on calculating the
change in pH of a buffer solution
when an acid or base is added.

WORKED EXAMPLE 14.16—pH CHANGE OF A BUFFER SOLUTION

What is the change in pH when 1.00 mL of 1.00 mol L^{-1} HCl solution is added to (a) an acetic
acid/sodium acetate buffer solution made with 0.350 mol of acetic acid and 0.300 mol of
sodium acetate dissolved in 500.0 mL of solution; and (b) 500 mL of pure water?

Solution

(a) Adding HCl to the Buffer Solution
   In Worked Example 14.12, we calculated that the pH of the buffer solution (before the
   addition of HCl solution) is 4.68.
   When HCl solution is added, the amount of H_3O^+(aq) ions added is

$$n(H_3O^+) = c \times V = (1.00 \text{ mol } L^{-1})(0.00100 \text{ } L) = 1.00 \times 10^{-3} \text{ mol}$$

   Buffering is achieved by the neutralization reaction

$$CH_3COO^-(aq) + H_3O^+(aq) \longrightarrow CH_3COOH(aq) + H_2O(\ell)$$

   Now we can calculate the amounts (in mol) of CH_3COOH(aq) and CH_3COO^-(aq) in the 0.500 L
   of buffer solution after it has "mopped up" the 1.00×10^{-3} mol of H_3O^+(aq) ions added:

$$n(CH_3COOH) = (0.350 + 0.001) \text{ mol} = 0.351 \text{ mol}$$

$$n(CH_3COO^-) = (0.300 - 0.001) \text{ mol} = 0.299 \text{ mol}$$

We can calculate the new pH by substituting the altered amounts of weak acid and its base into Equation 14.4:

$$[H_3O^+] = \frac{n(HA)}{n(A^-)} \times K_a = \frac{n(CH_3COOH)}{n(CH_3COO^-)} \times K_a = \frac{0.351}{0.299} \times (1.8 \times 10^{-5})$$
$$= 2.11 \times 10^{-5} \text{ mol L}^{-1}$$

from which we derive pH = 4.68 and ΔpH = 0.

(b) Adding HCl to Pure Water

Before the addition of HCl solution, pH of the water is 7.00.

Since we add 1.00×10^{-3} mol of strong acid to 0.500 L of water:

$$[H_3O^+] = \frac{n}{V} = \frac{1.00 \times 10^{-3} \text{ mol}}{0.500 \text{ L}} = 2.00 \times 10^{-3} \text{ mol L}^{-1} \quad \text{So pH = 2.70, and } \Delta\text{pH = 4.30.}$$

Comment

This calculation demonstrates the effectiveness of the buffer solution since, within the number of significant figures used, the pH of the buffer solution changes insignificantly on addition of the strong acid. On the other hand, the added acid solution changes the pH of the same volume of water 4.3 units.

If some strong base is added to the buffer solution, you can use a corresponding method to calculate the change of pH, except that buffering would occur as the result of the reaction represented by

$$CH_3COOH(aq) + OH - (aq) \rightleftharpoons CH_3COO^-(aq) + H_2O(\ell)$$

EXERCISE 14.45—pH CHANGE OF A BUFFER SOLUTION

Calculate the pH of 0.500 L of a buffer solution with 0.50 mol L^{-1} formic acid and 0.70 mol L^{-1} sodium formate (a) before, and (b) after adding 10.0 mL of 1.0 mol L^{-1} HCl solution.

Interactive
Exercise 14.44

Calculate the pH of a buffer solution before and after addition of base.

> To calculate how much pH changes on addition of a specified amount of H_3O^+(aq) to a buffer solution, calculate (a) the amounts of weak acid and conjugate base in the buffer solution, (b) the amount of acid that reacts and base that is formed, (c) the new ratio n(acid)/n(base), and (d) the new pH.

Buffer Capacity

We have seen earlier, in Worked Example 14.15, that acetic acid/acetate ion buffer solutions with a [CH$_3$COOH]/[CH$_3$COO$^-$] ratio 2.8/1 have pH 4.30. Each of the following solutions has pH 4.30, because all have the same value of the acid/base ratio:

A a solution that has c(CH$_3$COOH) = 2.80 mol L^{-1} and c(CH$_3$COO$^-$) = 1.00 mol L^{-1}

B a solution that has c(CH$_3$COOH) = 0.280 mol L^{-1} and c(CH$_3$COO$^-$) = 0.100 mol L^{-1}

C a solution that has c(CH$_3$COOH) = 0.0280 mol L^{-1} and c(CH$_3$COO$^-$) = 0.0100 mol L^{-1}

This shows an interesting characteristic of buffer solutions: the pH is dependent only on the ratio of the weak acid and its conjugate base; it is independent of the absolute concentrations of them. As a consequence, the pH of a buffer solution does not change when it is diluted.

Although the three solutions described above have the same pH, they are different in their abilities to minimize pH changes on addition of strong acid or strong base. **Buffer capacity** refers to *the ability of a buffer solution to minimize pH change on addition of strong acids or bases.* Solutions with high concentrations of the weak acid and its conjugate base have high buffer capacity. In comparison with the same volume of a buffer solution with lower concentrations, (a) for the addition of a given amount of strong acid or base, the change of pH is smaller; and (b) more strong acid or base can be added before buffering ability is lost. Both of these aspects of buffer capacity are illustrated in Worked Example 14.17.

WORKED EXAMPLE 14.17—BUFFER CAPACITY

Compare the buffering capacities of the three acetic acid/acetate ion buffer solutions A, B, and C above by considering how much the pH changes when 2.00 mL of 1.00 mol L^{-1} HCl solution is added to 100 mL of each of the buffer solutions.

Strategy

Use the method in Worked Example 14.16 for each buffer solution.

Solution

The amount of H$_3$O$^+$(aq) ions added in the HCl solution to each solution is

$$n(H_3O^+) = c \times V = (1.00 \text{ mol } L^{-1})(0.00200 \text{ L}) = 2.00 \times 10^{-3} \text{ mol}$$

Buffering is achieved by the reaction

$$CH_3COO^-(aq) + H_3O^+(aq) \longrightarrow CH_3COOH(aq) + H_2O(\ell)$$

Buffer solution A: In 100 mL of this solution, $n(CH_3COOH) = (2.80 \text{ mol } L^{-1})(0.100 \text{ L}) = 0.280$ mol, and $n(CH_3COO^-) = (1.00 \text{ mol } L^{-1})(0.100 \text{ L}) = 0.100$ mol

After "mopping up" the added H$_3$O$^+$(aq) ions, the amounts of CH$_3$COOH(aq) and CH$_3$COO$^-$(aq) ions are

$$n(CH_3COOH) = (0.280 + 0.002) \text{ mol} = 0.282 \text{ mol}$$

$$n(CH_3COO^-) = (0.100 - 0.002) \text{ mol} = 0.098 \text{ mol}$$

So, using Equation 14.4, we can see that [H$_3$O$^+$] will have changed to

$$[H_3O^+] = \frac{n(CH_3COOH)}{n(CH_3COO^-)} \times K_a = \frac{0.282}{0.098} \times (1.8 \times 10^{-5}) = 5.18 \times 10^{-5} \text{ mol } L^{-1}$$

and so pH = 4.29 and ΔpH = 0.01.

Buffer solution B: The solution contains $n(CH_3COOH) = 0.028$ mol, and $n(CH_3COO^-) = 0.010$ mol. After "mopping up" 0.002 mol of H$_3$O$^+$(aq) ions:

$$n(CH_3COOH) = (0.028 + 0.002) \text{ mol} = 0.030 \text{ mol}$$

$$n(CH_3COO^-) = (0.010 - 0.002) \text{ mol} = 0.008 \text{ mol}$$

and $\quad [H_3O^+] = \dfrac{0.030}{0.008}(1.8 \times 10^{-5}) = 6.75 \times 10^{-5} \text{ mol } L^{-1} \quad$ so pH = 4.17 and ΔpH = 0.13.

Buffer Solution C: The solution contains $n(CH_3COOH) = 0.0028$ mol, and $n(CH_3COO^-) = 0.0010$ mol. If we add 0.002 mol of H$_3$O$^+$(aq) ions, this is more than the amount of CH$_3$COO$^-$(aq) ions available to "mop it up," so this solution cannot buffer against that amount of added H$_3$O$^+$(aq) ions.

In fact, there would be an excess 0.001 mol of H$_3$O$^+$(aq) ions in the 100 mL solution, so we would have [H$_3$O$^+$] = (0.001 mol)/(0.100 L) = 0.01 mol L^{-1}, which corresponds with pH 2.0.

Comment

These calculations show that buffer solution A has higher buffer capacity than B, which in turn has higher buffer capacity than C. For the same amount of strong acid solution added, solution A controls the pH more tightly than solution B, and solution C is unable to buffer against the addition of the acid sample.

EXERCISE 14.46—BUFFER CAPACITY

Compare (i) the pH and (ii) the pH change on addition of 0.40 g of $NaOH(s)$ to the following buffer solutions:
(a) 100 mL of solution with $c(HCO_3^-) = 0.50$ mol L^{-1} and $c(CO_3^{2-}) = 0.25$ mol L^{-1}
(b) 100 mL of solution with $c(HCO_3^-) = 0.080$ mol L^{-1} and $c(CO_3^{2-}) = 0.40$ mol L^{-1}

The more concentrated the weak acid and base are in a solution, the higher the buffer capacity: (a) for a given amount of added strong acid or base, the less is the pH change; and (b) the more strong acid or base can be added before the solution loses its effectiveness.

14.10 Acid-Base Titrations

A **titration** *is a method of quantitative analysis* [<<Section 5.9] *based on a reaction of known stoichiometry, and depends on finding the volume of a solution containing the amount of one reagent that reacts exactly with a known amount of another.* Because the method is based on measurement of reacting volumes, it is called *volumetric analysis*. Titrations may be based on any type of reaction (acid-base, oxidation-reduction, precipitation, complexation), but our focus here is on those based on reactions between acids and bases, called *acid-base titrations*.

The Methodology of Acid-Base Titrations

Suppose you are asked to determine the mass of oxalic acid ($H_2C_2O_4$) in an impure sample. This acid reacts with bases such as sodium hydroxide with known stoichiometry corresponding with the following balanced equation:

$$H_2C_2O_4(aq) + 2\,OH^-(aq) \longrightarrow C_2O_4^{2-}(aq) + 2\,H_2O(\ell)$$

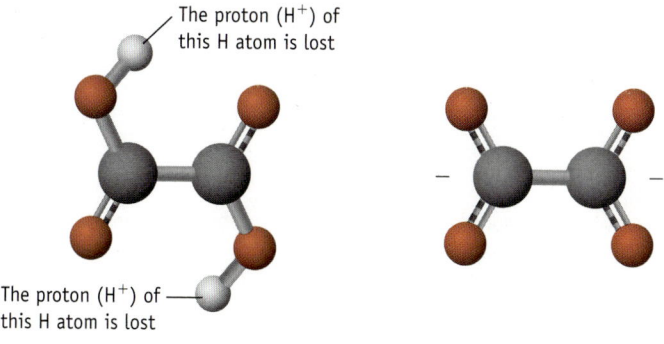

The proton (H^+) of this H atom is lost

The proton (H^+) of this H atom is lost

Oxalic acid $H_2C_2O_4$ **Oxalate anion $C_2O_4^{2-}$**

Oxalic acid. Oxalic acid has two carboxylic acid groups from which H^+ ions can be removed by reaction with bases. Hence, 1 mol of oxalic acid requires 2 mol of OH^- for complete reaction.

You can take an accurately measured mass of the impure sample, dissolve it in water, and add to it in small increments a solution of sodium hydroxide to react with the oxalic acid. You can determine the amount of oxalic acid present in the sample if the following essential elements of the titration procedure are met:

• You can determine when the amount of $OH^-(aq)$ ions in the added solution is just enough to react with all the oxalic acid present in solution (that is, when exactly twice as many mol of $OH^-(aq)$ ions have been added as there were of oxalic acid in the sample). This is called the *equivalence point*.
• You know the concentration of the sodium hydroxide solution and volume of it that has been added when the equivalence point is reached.

The solution containing oxalic acid is placed in a flask along with an *acid-base indicator*, a dye that changes colour when the solution reaches a certain pH. For each acid-base reaction,

it is possible to calculate the pH at the equivalence point (in this case, it is the pH of a
sodium oxalate solution) and choose an indicator that changes colour near that pH. The
NaOH solution of known concentration in a burette is added drop-wise to the flask. As long
as some oxalic acid is present in solution, all of the base added from the burette reacts. At
the equivalence point, the amount of OH⁻(aq) ions that have been added is exactly the
same as the amount of oxalic acid that was in the solution. From the volume of NaOH solution added and its concentration, the amount of OH⁻(aq) ions that have been added can be
calculated:

$$n(\text{OH}^- \text{ ions}) \text{ added} = c(\text{NaOH solution}) \times V(\text{solution added})$$

From the amount of OH⁻(aq) ions that react in the titration, we can use the balanced
equation to find the amount, and then the mass, of oxalic acid in the impure sample, as
illustrated in Worked Example 14.18.

WORKED EXAMPLE 14.18—ACID-BASE TITRATION

A 1.034 g sample of impure oxalic acid is dissolved in water and an appropriate acid-base
indicator added. The sample requires 34.47 mL of 0.485 M NaOH solution to reach the equivalence point. What is the mass of oxalic acid and what is its mass percent in the sample?

Solution

The amount of OH⁻(aq) ions reacted is given by

$$n(\text{OH}^-) = c \times V = 0.485 \text{ mol L}^{-1} \times 0.03447 \text{ L} = 0.0167 \text{ mol}$$

The balanced equation for the reaction is

$$\text{H}_2\text{C}_2\text{O}_4(\text{aq}) + 2 \text{ OH}^-(\text{aq}) \longrightarrow \text{C}_2\text{O}_4{}^{2-}(\text{aq}) + 2 \text{ H}_2\text{O}(\ell)$$

$$\text{So } n(\text{H}_2\text{C}_2\text{O}_4) = 0.5 \times n(\text{OH}^-) = 0.5 \times 0.0167 \text{ mol} = 0.00836 \text{ mol}$$

Then we can calculate the mass of oxalic acid in the sample:

$$m = n \times M = 0.00836 \text{ mol} \times 90.04 \text{ g mol}^{-1} = 0.753 \text{ g}$$

and then the mass percent of oxalic acid in the sample:

$$\text{Mass \%} = \frac{0.753 \text{ g}}{1.034 \text{ g}} \times 100 = 72.8\%$$

Comment

This method of analysis presumes that there are no substances in the impure oxalic acid
sample that react with the base.

EXERCISE 14.47—ACID-BASE TITRATION

A 25.0 mL sample of vinegar (which contains acetic acid) requires 28.33 mL of a
0.953 mol L⁻¹ solution of NaOH for titration to the equivalence point. What mass of acetic
acid is in the vinegar sample, and what is the molar concentration of acetic acid in the
vinegar?

In Worked Example 14.18, the concentration of the base used in the titration was known.
Usually, this has to be found by prior measurement. Titration to accurately determine the
concentration of a solution is called *standardization*. One approach to standardization is to
weigh accurately a sample of a pure, solid acid or base (known as a *primary standard*),
dissolve it water, and then titrate the solution with a solution of the base or acid to be
standardized (Worked Example 14.19). Alternatively, the solution to be standardized can be
titrated with another solution whose concentration is already known (Exercise 5.20). This is
often done with standard solutions purchased from chemical supply companies.

WORKED EXAMPLE 14.19—STANDARDIZING SOLUTION CONCENTRATION BY TITRATION

A 0.263 g sample of sodium carbonate (Na_2CO_3) dissolved in water requires 28.35 mL of HCl solution for exact neutralization. What is the molar concentration of the HCl solution (i.e., $c(HCl)$)?

In this example, Na_2CO_3 is a primary standard: it can be obtained pure, can be weighed out easily and accurately, and reacts completely with strong acids with known stoichiometry.

Strategy

The amount of $CO_3^{2-}(aq)$ ions is the same as the amount of sodium carbonate, which can be calculated from its mass and molar mass. The balanced equation for the reaction between $CO_3^{2-}(aq)$ ions and $H^+(aq)$ ions (from the HCl) is

$$CO_3^{2-}(aq) + 2\,H^+(aq) \longrightarrow H_2O(\ell) + CO_2(g)$$

From this, deduce the amount of $H^+(aq)$ ions in 28.35 mL of HCl solution. Then calculate $[H_3O^+]$, which is the same as $c(HCl)$ because HCl is a strong acid.

Solution

$$n(CO_3^{2-}) = n(Na_2\,CO_3) = \frac{m}{M} = \frac{0.263\ \text{g}}{106.0\ \text{g mol}^{-1}} = 0.00248\ \text{mol}$$

From the balanced equation, we can say that the amount of $H^+(aq)$ ions in 28.35 mL is

$$n(H^+) = 2 \times n(CO_3^{2-}) = 2 \times 0.00248\ \text{mol} = 0.00496\ \text{mol}$$

And then we can calculate the concentration of the HCl solution:

$$c(HCl) = [H^+] = \frac{n}{V} = \frac{0.00496\ \text{mol}}{0.02835\ \text{L}} = 0.175\ \text{mol L}^{-1}$$

EXERCISE 14.49—STANDARDIZING SOLUTION CONCENTRATION BY TITRATION

Hydrochloric acid solution can be purchased from chemical supply houses with an accurately known concentration of 0.100 mol L^{-1}, and this solution can be used to standardize a solution of a base. If titration of 25.00 mL of a sodium hydroxide solution to the equivalence point requires 29.67 mL of 0.100 mol L^{-1} HCl solution, what is $c(NaOH)$?

Interactive Exercise 14.48

Calculate the molar mass of an acid using titration data.

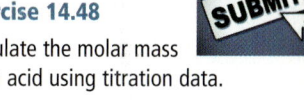

An acid-base titration is a method of quantitative analysis, based on a reaction of known stoichiometry, and depends on finding the volume of a solution containing the amount of acid (or base) that reacts exactly with a known amount of base (or acid).

Strong Acid–Strong Base Titrations

The reaction that occurs when a solution of a strong acid is added to a solution of a strong base is best represented as

$$H_3O^+(aq) + OH^-(aq) \longrightarrow H_2O(\ell)$$

Figure 14.17 illustrates how the pH changes as 0.100 mol L^{-1} NaOH solution is added drop-wise to 50.0 mL of 0.100 mol L^{-1} HCl solution.

Before any base is added, the 0.100 mol L^{-1} HCl solution has $[H^+] = 0.100$ mol L^{-1} and pH 1.00. As NaOH solution is added, $[H_3O^+]$ decreases, and the pH increases. The volume of the solution increases. For illustrative purposes, let us find the pH of the solution after 10.0 mL of 0.100 mol L^{-1} NaOH solution (0.00100 mol of OH$^-$ ions) has been added. The total volume of solution is now 60.0 mL.

$$\text{Initially, } n(H_3O^+) = c \times V = 0.100\ \text{mol L}^{-1} \times 0.0500\ \text{L} = 0.00500\ \text{mol}$$

$$n(H_3O^+) \text{ reacted} = n(OH^-) \text{ added} = 0.00100\ \text{mol}$$

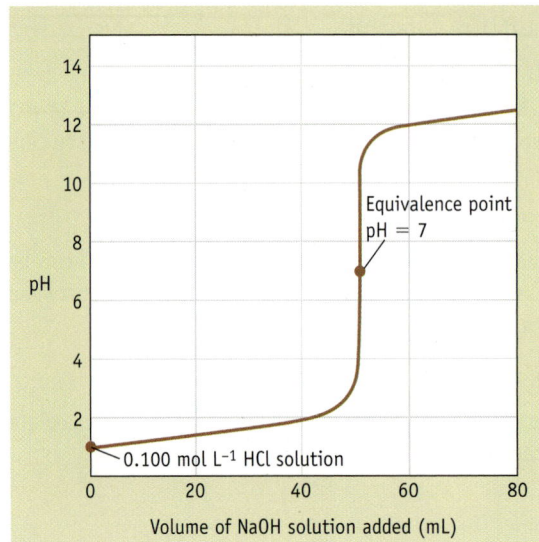

FIGURE 14.17 Change in pH as a solution of a strong acid is titrated with a solution of a strong base.

50.0 mL of 0.100 mol L⁻¹ HCl solution titrated with 0.100 mol L⁻¹ NaOH solution

Volume of base added	pH
0.0	1.00
10.0	1.18
20.0	1.37
40.0	1.95
45.0	2.28
48.0	2.69
49.0	3.00
50.0	7.00
51.0	11.00
55.0	11.68
60.0	11.96
80.0	12.36
100.0	12.52
very large amount	13.00 (maximum)

$$n(\text{H}_3\text{O}^+) \text{ remaining in } 60.0 \text{ mL} = (0.00500 - 0.00100) \text{ mol} = 0.00400 \text{ mol}$$

$$\text{So } [\text{H}_3\text{O}^+] = \frac{n}{V} = \frac{0.00400 \text{ mol}}{0.0600 \text{ L}} = 0.0667 \text{ mol L}^{-1}$$

$$\text{and pH} = -\log_{10} 0.0667 = 1.176$$

The pH increases very rapidly near the equivalence point. At the equivalence point, the solution is identical to a NaCl solution and the pH is 7.00 (at 25 °C). This is the case for all strong acid-strong base titrations.

After all of the $\text{H}_3\text{O}^+(\text{aq})$ ions have reacted, and the slightest excess of NaOH solution has been added, the solution is basic, and the pH increases further as more NaOH solution is added.

> At the equivalence point in titration of a solution of a strong base into a solution of a strong acid, pH = 7.0. The pH at any point prior to the equivalence point can be calculated from the amount of unreacted acid.

Weak Acid–Strong Base Titrations

We illustrate weak acid–strong base titrations with the titration of 100.0 mL of 0.100 mol L⁻¹ acetic acid solution with 0.100 mol L⁻¹ NaOH solution, represented by the equation

$$\text{CH}_3\text{COOH}(\text{aq}) + \text{OH}^-(\text{aq}) \longrightarrow \text{CH}_3\text{COO}^-(\text{aq}) + \text{H}_2\text{O}(\ell)$$

Figure 14.18 shows how the pH changes during this titration.

Let us focus on three important points on this curve:

1. *The pH before titration begins.* The pH of the solution before any base is added can be calculated from $c(\text{CH}_3\text{COOH})$ and K_a (Worked Example 14.6).

2. *The pH at the equivalence point.* At the equivalence point the solution is the same as a 0.050 mol L⁻¹ sodium acetate solution. The only species affecting the pH are the $\text{CH}_3\text{COO}^-(\text{aq})$ ions, and as for any base, the pH can be calculated from $c(\text{CH}_3\text{COO}^-)$ and K_b (Worked Example 14.8).

3. *The pH at halfway to the equivalence point.* In the particular case discussed here, 100 mL of base solution is needed to reach the equivalence point. When 50 mL has been added, the pH is equal to the pK_a of the weak acid—an observation that is discussed in more detail below.

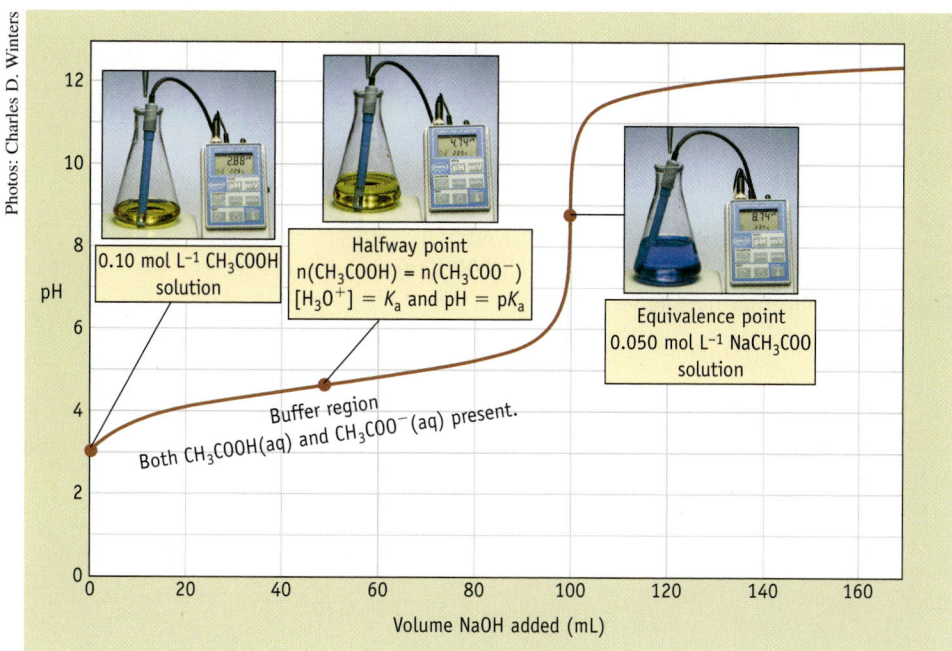

FIGURE 14.18 Change in pH during the titration of a solution of a weak acid with a solution of a strong base. Here 100.0 mL of 0.100 mol L^{-1} CH_3COOH solution is titrated with 0.100 mol L^{-1} NaOH solution. At the start, the pH is 2.87. When exactly half the acid has been reacted, pH = pK_a of the acid (4.74). At the equivalence point, the solution is basic because of the CH_3COO^-(aq) ions, and pH = 8.72.

Think about It

e**14.38** Simulate different acid-base titrations and plot their titration curves.

Think about It

e**14.39** Interpret a pH titration curve.

As NaOH solution is added, CH_3COOH(aq) reacts with the base and CH_3COO^-(aq) ions are formed. So, during the titration the solution contains both the weak acid CH_3COOH(aq) and its conjugate base CH_3COO^-(aq) ions. These are the components of a buffer solution, which explains why the pH does not change much with addition of base in the intermediate stages of the titration. To calculate $[H_3O^+]$ or pH, we can use Equation 14.1 or 14.2:

$$[H_3O^+] = \frac{[\text{weak acid remaining in solution}]}{[\text{conjugate base produced}]} \times K_a$$

$$\text{or} \quad pH = pK_a - \log_{10} \frac{[\text{weak acid remaining in solution}]}{[\text{conjugate base produced}]}$$

When *exactly* half of the amount of CH_3COOH(aq) has been reacted with base to form its conjugate base CH_3COO^-(aq) ions, $[CH_3COOH] = [CH_3COO^-]$. Then

$$[H_3O^+] = \frac{[CH_3COOH]}{[CH_3COO^-]} \times K_a = K_a \quad \text{or} \quad pH = pK_a - \log_{10} \frac{[CH_3COOH]}{[CH_3COO^-]} = pK_a$$

In the particular case of titration of acetic acid solution with solution of strong base, when exactly half the volume required to reach the equivalence point has been added, $[H_3O^+] = K_a = 1.8 \times 10^{-5}$, and pH = p$K_a$ = 4.74.

WORKED EXAMPLE 14.20—TITRATION OF CH_3COOH SOLUTION WITH NaOH SOLUTION

What is the pH of the solution when 90.0 mL of 0.100 mol L^{-1} NaOH solution has been added to 100.0 mL of 0.100 mol L^{-1} CH_3COOH solution?

Strategy

This problem involves two steps: (a) a stoichiometry calculation to find the amount of acetic acid remaining and amount of its conjugate base formed after adding the specified volume of NaOH solution, and (b) a calculation to find the pH of this buffer solution.

Solution

We can calculate the amount of acetic acid initially, and the amount of $OH^-(aq)$ ions in 90.0 mL of the NaOH solution:

$$\text{Initial } n(CH_3COOH) = c \times V = 0.100 \text{ mol } L^{-1} \times 0.100 \text{ L} = 0.0100 \text{ mol}$$

$$n(OH^-) \text{ added} = c \times V = 0.100 \text{ mol } L^{-1} \times 0.090 \text{ L} = 0.0090 \text{ mol}$$

From the balanced chemical equation for the reaction, we know that the amount of $OH^-(aq)$ ions added is the same as the amount of $CH_3COOH(aq)$ that has reacted and the amount of $CH_3COO^-(aq)$ ions that have formed. So

$$n(CH_3COOH) = 0.0100 \text{ mol} - 0.0090 \text{ mol} = 0.0010 \text{ mol}$$

$$n(CH_3COO^-) = 0.0090 \text{ mol}$$

Now we can calculate the pH of the buffer solution, recognizing that the ratio $n(CH_3COOH)/n(CH_3COO^-)$ is the same as the ratio $[CH_3COOH]/[CH_3COO^-]$.

$$[H_3O^+] = \frac{n(CH_3COOH)}{n(CH_3COO^-)} \times K_a = \frac{0.0010}{0.0090} \times (1.8 \times 10^{-5}) = 2.0 \times 10^{-6} \text{ mol } L^{-1}$$

$$pH = -\log_{10}(2.0 \times 10^{-6}) = 5.70$$

Comment

pH 5.70 is consistent with Figure 14.18 and is sensible for an added volume between the halfway point (pH = 4.74) and the equivalence point (pH = 8.72).

Interactive Exercise 14.50

Calculate the pH at the equivalence point.

EXERCISE 14.51—TITRATION OF A SOLUTION OF A WEAK ACID WITH A SOLUTION OF A STRONG BASE

What is the pH of the solution when (a) 35.0 mL or (b) 100.0 mL of 0.100 mol L^{-1} NaOH solution has been added to 100.0 mL of 0.100 mol L^{-1} acetic acid solution?

At the equivalence point in titration of a solution of a weak acid with a strong base, pH > 7, and can be calculated using K_b of the conjugate base. Prior to the equivalence point, the solution contains both residual weak acid and some of its conjugate base and is a buffer solution. When exactly half the amount of strong base solution needed to reach the equivalence point has been added, pH = pK_a of the weak acid.

Titration of Polyprotic Weak Acids with Strong Base Solution

For a diprotic acid such as oxalic acid ($H_2C_2O_4$) with K_{a1} (5.9×10^{-2}) much greater than K_{a2} (6.4×10^{-5}), neutralization happens in distinct steps, because the first H^+ ion is much more easily removed than the second:

$$H_2C_2O_4(aq) + OH^-(aq) \longrightarrow HC_2O_4^-(aq) + H_2O(\ell)$$

$$HC_2O_4^-(aq) + OH^-(aq) \longrightarrow C_2O_4^-(aq) + H_2O(\ell)$$

Figure 14.19 shows the curve for the titration of 100 mL of 0.100 mol L^{-1} oxalic acid solution with 0.100 mol L^{-1} NaOH mol L^{-1}. The first sharp rise in pH occurs when 100 mL of base has been added, indicating that the first proton of the acid has been titrated—the first equivalence point. When the second proton has been removed (the second equivalence point), the pH again rises significantly.

At the second equivalence point, the pH is due essentially to the weak base $C_2O_4^{2-}$ (aq) ions. From its concentration (0.033 mol L^{-1}) and K_b (1.6×10^{-10}), we can calculate the pH to be 8.4.

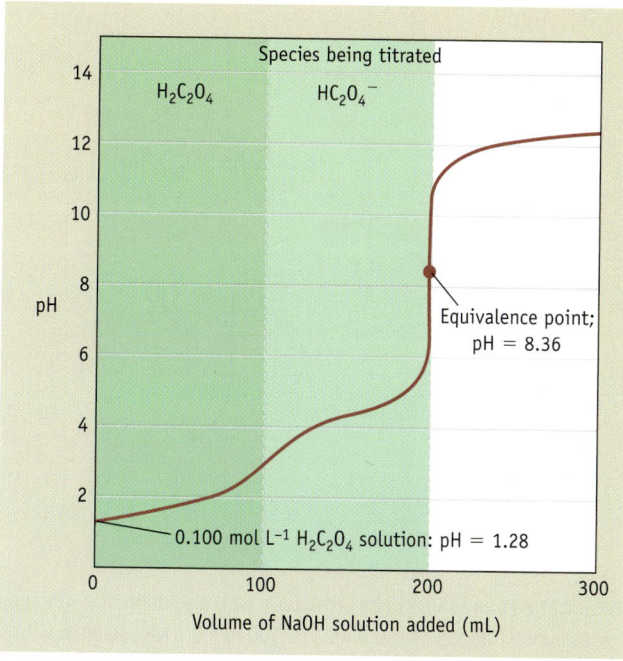

FIGURE 14.19 Titration curve for a diprotic acid. The curve for the titration of 100.0 mL of 0.100 mol L^{-1} solution of oxalic acid with 0.100 mol L^{-1} NaOH solution. The first equivalence point (at 100.0 mL) occurs when the most acidic H^+ ion of each $H_2C_2O_4$ molecule has been removed, and the second (at 200.0 mL) occurs after the second H^+ ion has been removed.

During the titration up to the first equivalence point, the reaction mixture is a buffer solution, containing $H_2C_2O_4(aq)$, and its conjugate base, hydrogenoxalate ions, $HC_2O_4^-$ (aq). When the added volume is exactly half that needed to reach the first equivalence point, $[H_2C_2O_4] = [HC_2O_4^-]$ and pH = pK_{a1} of oxalic acid. When the volume of NaOH solution added is between the first and second equivalence points, the reaction solution is a buffer solution consisting of aquated hydrogenoxalate ions, $HC_2O_4^-(aq)$, and its conjugate base, aquated oxalate ions, $C_2O_4^{2-}(aq)$. When the volume of NaOH solution added is exactly midway between the first and second equivalence points, $[HC_2O_4^-] = [C_2O_4^{2-}]$ and pH = pK_{a2} of oxalic acid. To a good approximation, at the first equivalence point the pH of the reaction mixture is given by

$$pH = \tfrac{1}{2}(pK_{a1} + pK_{a2})$$

EXERCISE 14.52—TITRATION OF A DIPROTIC WEAK ACID WITH A STRONG BASE

100.0 mL of 0.010 mol L^{-1} solution of carbonic acid (H_2CO_3) is titrated against 0.010 mol L^{-1} NaOH solution. What is the pH (a) at the beginning of the titration, (b) when the volume of NaOH solution added is half that needed to arrive at the first equivalence point, (c) at the first equivalence point, (d) when the volume of NaOH solution added is exactly midway between the first and second equivalence points, and (e) at the second equivalence point.

In titration of a diprotic weak acid with $K_{a1} \gg K_{a2}$, the most acidic proton is removed from nearly all molecules before the second one is removed from them. Prior to the first equivalence point, the solution is a buffer solution, and when the volume of NaOH solution is exactly half that needed to reach the first equivalence point, pH = pK_{a1}. Between the first and second equivalence points, the solution is a buffer solution, and when the volume added is midway between the first and second equivalence points, pH = pK_{a2}.

Weak Base–Strong Acid Titrations

Figure 14.20 shows the pH curve for titration of 100.0 mL of 0.100 mol L^{-1} NH_3 solution with 0.100 mol L^{-1} HCl solution:

$$NH_3(aq) + H_3O^+(aq) \longrightarrow NH_4^+(aq) + H_2O(\ell)$$

Taking It Further
e14.40 Read about how acid-base indicators can be used to indicate the pH of a solution.

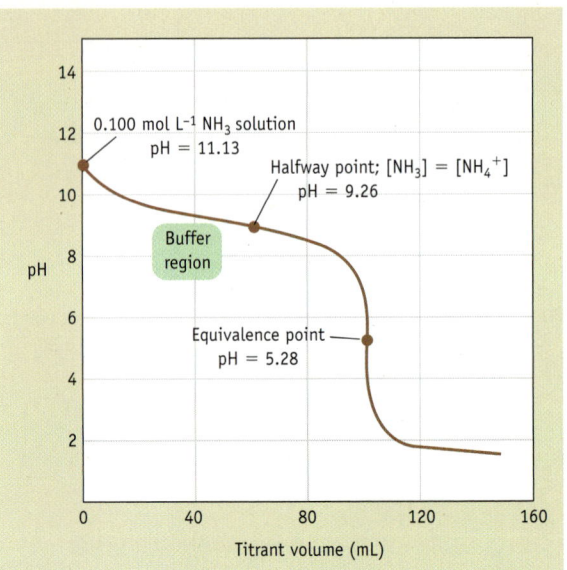

FIGURE 14.20 Titration of a solution of a weak base with a solution of a strong acid. The graph shows how pH changes during titration of 100.0 mL of 0.100 mol L^{-1} NH$_3$ solution with 0.100 mol L^{-1} HCl solution. The pH at the halfway point is equal to the pK_a for the weak acid NH$_4^+$(aq) ions (9.26). At the equivalence point, the pH of the solution is due to the weakly acidic NH$_4^+$(aq) ions, and the pH is about 5.

The initial pH (0.100 mol L^{-1} NH$_3$ solution) is 11.13, and can be calculated using K_b(NH$_3$) = 1.8 × 10^{-5}. As the titration progresses, the important species in solution are the unreacted conjugate base, NH$_3$(aq), and the weak acid NH$_4^+$(aq) ions formed. These comprise a buffer solution, for which the equilibrium can be represented as

$$NH_4^+(aq) + H_2O(\ell) \rightleftharpoons NH_3(aq) + H_3O^+(aq) \qquad K_a = 5.6 × 10^{-10}$$

At the halfway point, the concentrations of NH$_4^+$(aq) and NH$_3$(aq) are the same, so

$$[H_3O^+] = \frac{[NH_4^+]}{[NH_3]} × K_a = 5.6 × 10^{-10}$$

$$[H_3O^+] = K_a \qquad \text{and} \qquad pH = pK_a = -\log_{10}(5.6 × 10^{-10}) = 9.25$$

As the addition of HCl solution to NH$_3$ solution continues, the pH decrease is slow because of the buffering action. Near the equivalence point, the pH drops rapidly. At the equivalence point, the solution contains only NH$_4^+$(aq) ions and Cl$^-$(aq) ions. Because the NH$_4^+$(aq) ion is a weak acid, the pH of solution can be calculated as for any solution of a weak acid (Worked Example 14.20).

WORKED EXAMPLE 14.21—TITRATION OF A WEAK BASE WITH A STRONG ACID

What is the pH of the solution at the equivalence point in the titration of 100.0 mL of 0.100 mol L^{-1} ammonia solution with 0.100 mol L^{-1} HCl solution?

Strategy

Recognize that at the equivalence point, the reaction solution is the same as a solution of NH$_4$Cl. The Cl$^-$(aq) ions in such a solution do not influence the pH, but the NH$_4^+$(aq) ions are a weak acid. As for the solution of any weak acid, calculate its pH from its concentration and K_a. First do a stoichiometry calculation to find the concentration of NH$_4^+$(aq) ions at the equivalence point.

Solution

We are titrating 0.0100 mol of NH$_3$(aq) ($n = c × V$), so 0.0100 mol of H$_3$O$^+$(aq) ions (in 100.0 mL of 0.100 mol L^{-1} HCl solution) is needed for exact reaction with the ammonia. The amount of $NH_4^+(aq)$ ions produced will be 0.0100 mol, and this will be in 200.0 mL of reaction solution. So

$$c(NH_4^+) = \frac{n}{V} = \frac{0.0100\ mol}{0.200\ L} = 0.0500\ mol\ L^{-1}$$

The pH is determined just as it would be for a 0.0500 mol L^{-1} NH_4Cl solution, using the ionization reaction of the weak acid:

$$NH_4^+(aq) + H_2O(\ell) \rightleftharpoons NH_3(aq) + H_3O^+(aq)$$

Assuming that x mol L^{-1} of $NH_4^+(aq)$ ions ionize:

$$Q = \frac{[NH_3][H_3O^+]}{[NH_4^+]} = \frac{(x)(x)}{0.0500 - x} = K_a = 5.6 \times 10^{-10}$$

Solution gives us $x = [H_3O] = 5.3 \times 10^{-6}$ mol L^{-1}, and pH = 5.28.

EXERCISE 14.53—TITRATION OF A WEAK BASE WITH A STRONG ACID

Calculate the pH after 75.0 mL of 0.100 mol L^{-1} HCl solution has been added to 100.0 mL of 0.100 mol L^{-1} NH_3 solution.

During titration of a solution of a weak base with a solution of a strong acid, the conjugate acid of the weak base is gradually formed, resulting in formation of a buffer solution. When the volume of strong acid solution added is exactly half that required for exact neutralization, pH = pK_a of the conjugate weak acid.

SUMMARY

Key Concepts

- From the perspective of the **Brønsted-Lowry model**, acid-base reactions involve transfer of a H^+ ion (proton) from an acid to a base. When a proton is transferred, the acid is converted to its conjugate base, and the base is converted to its conjugate acid. (Section 14.2)
- **Monoprotic acids** have molecules with one removable H^+ ion, while **polyprotic acids** have more than one. (Section 14.2)
- Strong acids and bases are strong electrolytes; weak acids and bases are weak electrolytes. (Section 14.2)
- **Amphoteric species** can be proton donors or proton acceptors. (Section 14.2)
- Water undergoes a self-ionization reaction for which the equilibrium constant, called the **ionization constant for water (K_w)**, (= $[H_3O^+][OH^-]$), is 1.0×10^{-14} at 25 °C. At this temperature, **pH + pOH = 14.00**. (Section 14.3)
- In **neutral solutions** at 25 °C, $[H_3O^+] = [OH^-] = 1.0 \times 10^{-7}$ mol L^{-1}, and pH = pOH = 7.00. (Section 14.3)
- In **acidic solutions**, $[H_3O^+] > [OH^-]$. At 25 °C, $[H_3O^+] > 1.0 \times 10^{-7}$ mol L^{-1}, pH < 7, and pOH > 7. (Section 14.3)
- In **basic solutions**, $[H_3O^+] < [OH^-]$. At 25 °C, $[H_3O^+] < 1.0 \times 10^{-7}$ mol L^{-1}, pH > 7, and pOH < 7. (Section 14.3)
- Equilibrium constants for ionization of weak acids in aqueous solution are called **acid ionization constants (K_a)**, and those for ionization of weak bases are called **base**

ionization constants (K_b). The larger the ionization constant is, the stronger the weak acid or base. (Section 14.4)

- The stronger a weak acid is, the weaker its conjugate base. For a **conjugate acid-base pair,** $K_a \times K_b = K_w$ and $pK_a + pK_b = pK_w$. (Section 14.4)
- A Lewis base can provide a pair of electrons to a Lewis acid to form a covalent bond. The Lewis acid-base model includes species that are Brønsted-Lowry acids and bases. (Section 14.5)
- For solutions of a weak acid (or weak base) with the same concentration and temperature, the smaller K_a (or K_b) is, the less the percentage ionization. (Section 14.6)
- At a given temperature, the more dilute the solution of a weak acid or weak base is, the greater the percentage of it that ionizes. (Section 14.6)
- The presence of a common ion reduces the extent of ionization of a weak acid or weak base in aqueous solution. This is the **common ion effect**. (Section 14.6)
- In solutions of weak polyprotic acids with large differences in the successive K_a values, the pH depends primarily on the hydronium ions generated in the first ionization step. (Section 14.6)
- In reactions between a weak acid and the conjugate base of a weaker acid, the base reacts with all of the removable protons of the acid (and not just the hydronium ions present from ionization). (Section 14.6)
- The term *acidity* can refer to either the pH of a solution or the amount of base needed to neutralize the dissolved acid. (Section 14.6)
- The relative concentrations of a weak acid and its conjugate base in solution depends on the pH of the solution in relation to pK_a of the weak acid, according to the relationship $\frac{[HA]}{[A^-]} = \frac{[H_3O^+]}{K_a}$. If pH = pKa, then [HA] = [A$^-$]. The way that the relative concentrations change with change of pH can be shown in a speciation plot. This applies for each ionization step of a polyprotic acid. (Section 14.7)
- Complexation to metal cations by Lewis bases decreases at low pH because of protonation of the bases occupying the bonding sites on the molecules or ions of the bases. (Section 14.7)
- The pH-dependent speciation of amino acids is the same as for polyprotic acids. Carboxylic acid protons are more easily removed than those on protonated amine groups. The **isoelectric pH** is the pH at which the species has zero overall charge—called a zwitterion. (Section 14.8)
- A **buffer solution** minimizes the change in pH when a strong acid or base is added. It contains relatively large amounts of a weak acid and its conjugate base. Added H_3O^+(aq) ions are "mopped up" by the conjugate base, and added OH$^-$(aq) ions are "mopped up" by the weak acid. (Section 14.9)
- To design a buffer solution for a particular pH, choose a weak acid with pK_a within 1 unit of the target pH, and calculate the [weak acid]/[conjugate base] ratio that will adjust the solution pH to the target value. (Section 14.9)
- The more concentrated the weak acid and base are in a solution, the higher the **buffer capacity**: (a) for a given amount of added strong acid or base, the less the pH change, and (b) the more strong acid or base can be added before the solution loses its effectiveness. (Section 14.9)
- Analysis by acid-base **titration** is based on a reaction of known stoichiometry, and depends on finding the volume of a solution that contains the amount of acid (or base) reacting exactly with a known amount of base (or acid). (Section 14.10)
- In titration of a solution of a strong base into a solution of a strong acid, at the equivalence point, pH = 7.0. (Section 14.10)
- In titration of a solution of a weak acid with a strong base, at the equivalence point, pH > 7. Near the middle of the titration, the solution is a buffer solution. When exactly half the amount of strong base solution needed to reach the equivalence point has been added, pH = pK_a of the weak acid. (Section 14.10)
- In titration of a diprotic weak acid with $K_{a1} \gg K_{a2}$, the most acidic proton is removed from nearly all molecules before the second one is removed. When the volume of the

base solution added is exactly half that needed to reach the first equivalence point, the solution is a buffer solution with $pH = pK_{a1}$. When the volume added is exactly midway between the first and second equivalence points, $pH = pK_{a2}$. (Section 14.10)

- During titration of a solution of a weak base with a solution of a strong acid, when the volume of strong acid solution added is exactly half that required for exact neutralization, $pH = pK_a$ of the conjugate weak acid formed. (Section 14.10)

Key Equations

Water ionization constant: $K_w = [H_3O^+][OH^-] = 1.0 \times 10^{-14}$ at 25 °C   (Section 14.3)

$pH = -\log_{10}[H_3O^+]$ and $pOH = -\log_{10}[OH^-]$ (Section 14.3)

$pK_w = -\log_{10} K_w$ (Section 14.3)

At 25 °C, $pK_w = pH + pOH = 14.00$ (Section 14.3)

For ionization of a weak acid, HA: at equilibrium,

$$Q = \frac{[H_3O^+][A^-]}{[HA]} = K_a = \text{acid ionization constant}$$ (Section 14.4)

For ionization of a weak base, B: at equilibrium,

$$Q = \frac{[BH^+][OH^-]}{[B]} = K_b = \text{base ionization constant}$$ (Section 14.4)

$pK_a = -\log_{10} K_a$ and $pK_b = -\log_{10} K_b$ (Section 14.4)

For a conjugate acid-base pair, $K_a \times K_b = K_w$ and $pK_a + pK_b = pK_w$ (Section 14.4)
In aqueous solution, the relative concentrations of a weak acid and its conjugate base is

given by $\dfrac{[HA]}{[A^-]} = \dfrac{[H_3O^+]}{K_a}$ If $pH = pK_a$, $[HX] = [X^-]$ (Section 14.7)

In buffer solutions, $[H_3O^+] = \dfrac{[HA]}{[A^-]} \times K_a$ (14.1) or $pH = pK_a + \log_{10}\dfrac{[A^-]}{[HA]}$ (14.2)

Alternatively, $[H_3O^+] = \dfrac{c(HA)}{c(A^-)} \times K_a$ (14.3) and $[H_3O^+] = \dfrac{n(HA)}{n(A^-)} \times K_a$ (14.4)

(Section 14.9)

REVIEW EXERCISES

Section 14.2: The Brønsted-Lowry Model of Acids and Bases

14.54 Write balanced equations showing that the aquated hydrogenphosphate ion, $HPO_4^{2-}(aq)$, is an amphoteric species.

Section 14.3: Water and the pH Scale

14.55 State whether solutions at 25 °C with the following species concentrations are acidic, basic, or neutral.
(a) $[H_3O^+] = 5 \times 10^{-10}$ mol L^{-1}
(b) $[H_3O^+] = 7 \times 10^{-7}$ mol L^{-1}
(c) $[OH^-] = 3 \times 10^{-9}$ mol L^{-1}

14.56 If slightly acidic tap water with $[H_3O^+] = 5 \times 10^{-6}$ mol L^{-1} is diluted by a factor of 1000 with pure water, what is the concentration of hydronium ions?

14.57 An aqueous solution has a pH of 3.75. What is the hydronium ion concentration of the solution? Is the solution acidic or basic?

14.58 What is the pH of a 0.0075 mol L^{-1} HCl solution? What is $[OH^-]$ in this solution?

14.59 The aquated trimethylammonium ion, $(CH_3)_3NH^+(aq)$, is the conjugate acid of the weak base aquated trimethylamine, $(CH_3)_3N(aq)$. For $(CH_3)_3NH^+(aq)$, $pK_a = 9.80$. What is K_b for $(CH_3)_3N(aq)$?

Section 14.4: Relative Strengths of Weak Acids and Bases

14.61 Write balanced equations showing that the aquated hydrogenphosphate ion, $HPO_4^{2-}(aq)$, is an amphoteric species.

14.62 A weak acid has $K_a = 6.5 \times 10^{-5}$. What is pK_a for this acid?

14.63 Which is the stronger of the following two acids: (a) benzoic acid (C_6H_5COOH), $pK_a = 4.20$, or (b) 2-chlorobenzoic acid (ClC_6H_4COOH) $pK_a = 2.88$?

14.64 Use Table 14.5 to answer the following questions regarding acids and bases in aqueous solution.
(a) Which is the stronger acid, H_2SO_4 or H_2SO_3?
(b) Is benzoic acid (C_6H_5COOH) stronger or weaker than acetic acid?
(c) Which has the stronger conjugate base, acetic acid or boric acid?
(d) Which is the stronger base, ammonia or the acetate ion?
(e) Which has the stronger conjugate acid, ammonia or the acetate ion?

14.65 Several acids are listed here with their respective equilibrium constants at 25 °C:

$$C_6H_5OH(aq) + H_2O(\ell) \rightleftharpoons H_3O^+(aq) + C_6H_5O^-(aq)$$
$$K_a = 1.3 \times 10^{-10}$$

$$HCOOH(aq) + H_2O(\ell) \rightleftharpoons H_3O^+(aq) + HCOO^-(aq)$$
$$K_a = 1.8 \times 10^{-4}$$

$$HC_2O_4^-(aq) + H_2O(\ell) \rightleftharpoons H_3O^+(aq) + C_2O_4^{2-}(aq)$$
$$K_a = 6.4 \times 10^{-5}$$

(a) Which is the strongest acid? Which is the weakest acid?
(b) Which acid has the weakest conjugate base?
(c) Which acid has the strongest conjugate base?

14.66 State which of the following species has the strongest conjugate base and briefly explain your choice.
(a) $HSO_4^-(aq)$
(b) $CH_3COOH(aq)$
(c) $HClO(aq)$

14.68 An aqueous solution of $K_2CO_3(s)$ is basic. Which species is responsible for this? Write a balanced equation that demonstrates why.

14.69 If each of the salts listed here were dissolved in water to give a 0.10 mol L^{-1} solution, which solution would have the highest pH? Which would have the lowest pH?
(a) $Na_2S(s)$ (d) $NaF(s)$
(b) $Na_3PO_4(s)$ (e) $NaCH_3COO(s)$
(c) $NaH_2PO_4(s)$ (f) $AlCl_3(s)$

Section 14.5: The Lewis Model of Acids and Bases

14.70 Trimethylamine, $(CH_3)_3N$, is a common reagent. It interacts readily with diborane gas, $B_2H_6(g)$. The latter dissociates to BH_3, and this forms a complex with the amine, $(CH_3)_3N–BH_3$. Is BH_3 a Lewis acid or a Lewis base?

14.71 Show how the species in part (a) can act as Lewis bases in their reactions with HCl, and show how the species in part (b) can act as Lewis acids in their reaction with OH^- ions.
(a) CH_3CH_2OH, $(CH_3)_2NH$, $(CH_3)_3P$
(b) H_3C^+, $(CH_3)_3B$, Br^- ions

Section 14.6: Equilibria in Aqueous Solutions of Weak Acids or Bases

14.72 Methylamine (CH_3NH_2) is a weak base:

$$CH_3NH_2(aq) + H_2O(\ell) \rightleftharpoons CH_3NH_3^+(aq) + OH^-(aq)$$

If the pH of a 0.065 mol L^{-1} solution of methylamine is 11.70, what is the value of K_b?

14.73 A 2.5×10^{-3} mol L^{-1} aqueous solution of an acid has pH 3.80 at 25 °C.
(a) What is the hydronium ion concentration in the solution?
(b) Is the acid a strong acid, a moderately weak acid (K_a about 10^{-5}), or a very weak acid (K_a about 10^{-10})?

14.74 What are the equilibrium concentrations of HF(aq), F^-(aq) ions, and H_3O^+(aq) in a 0.10 mol L^{-1} solution of hydrofluoric acid (HF) at 25 °C? What is the pH of the solution? What percentage of the HF(aq) is ionized at equilibrium?

14.75 Does the pH of the solution increase, decrease, or stay the same when you
(a) dissolve solid ammonium chloride in a dilute aqueous ammonia solution?
(b) dissolve solid sodium acetate in a dilute aqueous acetic acid solution?
(c) dissolve solid sodium chloride in a dilute sodium hydroxide solution?

14.76 What is the pH of a solution that has $c(NH_4Cl) = 0.20$ mol L^{-1} and $c(NH_3) = 0.20$ mol L^{-1}?

14.77 What are the equilibrium concentrations of $NH_3(aq)$, $NH_4^+(aq)$, and $OH^-(aq)$ in a 0.15 mol L^{-1} aqueous ammonia solution at 25 °C? What is the pH of the solution? What percentage of the ammonia is ionized?

14.78 In aqueous solution, the weak base methylamine (CH_3NH_2) has $K_b = 4.2 \times 10^{-4}$ at 25 °C.

$$CH_3NH_2(aq) + H_2O(\ell) \rightleftharpoons CH_3NH_3^+(aq) + OH^-(aq)$$

Calculate the equilibrium hydroxide ion concentration in a 0.25 mol L^{-1} solution at 25 °C, and the percentage ionization. What are the pH and pOH of the solution?

14.79 Compare the percentage ionization of aquated hypochlorite ions, $ClO^-(aq)$, as a weak base in (a) 1.00 mol L^{-1} solution, (b) 0.100 mol L^{-1} solution, and (c) 0.0100 mol L^{-1} sodium hypochlorite solution. Is the trend consistent with that shown in Table 14.8 for a weak acid?

14.80 Calculate (a) the pH, and (b) [HSO$_3^-$] in a 0.45 mol L^{-1} H_2SO_3 solution.

14.81 In aqueous solution, hydrazine is a base that can accept two protons:

$$N_2H_4(aq) + H_2O(\ell) \rightleftharpoons N_2H_5^+(aq) + OH^-(aq)$$
$$K_{b_1} = 8.5 \times 10^{-7}$$

$$N_2H_5^+(aq) + H_2O(\ell) \rightleftharpoons N_2H_6^{2+}(aq) + OH^-(aq)$$
$$K_{b_2} = 8.9 \times 10^{-16}$$

(a) In a 0.010 mol L^{-1} aqueous hydrazine solution, what are [OH$^-$], [N$_2$H$_5^+$(aq)] and [N$_2$H$_6^{2+}$]?

(b) What is the pH of the solution?

14.82 Aquated ethylenediamine, $H_2NCH_2CH_2NH_2(aq)$, can interact with water in two steps, forming OH$^-$ in each step ($K_{b1} = 8.5 \times 10^{-5}$, $K_{b2} = 2.7 \times 10^{-8}$). If you have a 0.15 mol L^{-1} aqueous solution of the amine, calculate [H$_3$NCH$_2$CH$_2$NH$_3^{2+}$] and [OH$^-$], as well as pH.

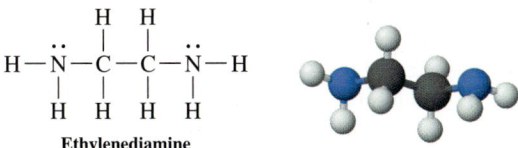

Ethylenediamine

Section 14.7: Speciation: Relative Concentrations of Species

14.83 In aqueous solution, aquated methylammonium ions, $CH_3NH_3^+(aq)$, is in equilibrium with its conjugate base, aquated methylamine, $CH_3NH_2(aq)$, and the distribution between the two species depends on pH:

$$CH_3NH_3^+(aq) + H_2O(\ell) \overset{pK_a=3.38}{\rightleftharpoons} CH_3NH_2(aq) + H_3O^+(aq)$$

At each of pH 3, 7 and 11, deduce which of the two species is present at greater concentration, and calculate the ratio [CH$_3$NH$_3^+$]/[CH$_3$NH$_2$]. At what pH is [CH$_3$NH$_3^+$] = [CH$_3$NH$_2$]?

14.84 Suppose that some benzoic acid and hydrocyanic acid are both in the same aqueous solution whose pH we gradually increase from 2 to 12. Deduce the pH range over which one of the acids is mostly in its conjugate base (deprotonated) form, while the other remains mostly protonated. Which of the acids is deprotonated first?

14.85 Consider the speciation plot of species $H_2A(aq)$, HA^-(aq), and A^{2-}(aq), where $H_2A(aq)$ is a weak diprotic acid.

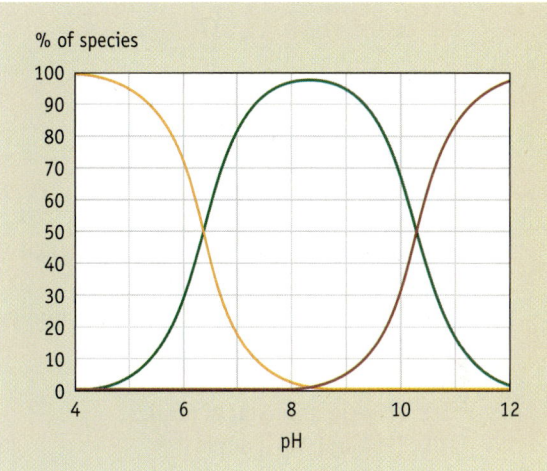

(a) Which species is at the highest concentration at (i) pH 6, (ii) pH 8, and (iii) pH 10?

(b) Deduce, as well as you can, the values of pK_{a1} and pK_{a2}.

(c) Which weak diprotic acid might this be?

14.86 Consider the speciation plot of species $H_2A(aq)$, HA^-(aq), and A^{2-}(aq), where $H_2A(aq)$ is a weak diprotic acid.

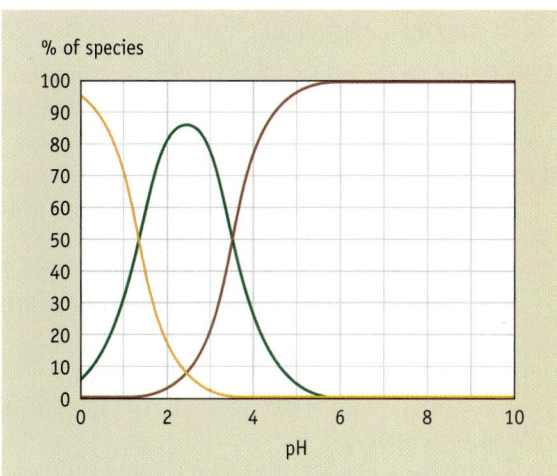

(a) Which species is at the highest concentration at (i) pH 4, (ii) pH 6, and (iii) pH 8?

(b) Deduce, as well as you can, the values of pK_{a1} and pK_{a2}.

(c) Which weak diprotic acid might this be?

(d) Compare the ability of this substance to be a Lewis base that forms complexes with metal ions at pH 1 with that at pH 7.

Section 14.8: Acid-Base Properties of Amino Acids and Proteins

14.87 Write the structure, and list the overall charge, of the dominant glutamic acid species at (a) pH 1.0, (b) pH 3.0, (c) pH 8.0, and (d) pH 12.0.

14.88 The protonated form of alanine can be drawn as follows:

$$H_3C - \underset{\overset{|}{\overset{+}{NH_3}}}{CH} - C\overset{\overset{O}{\|}}{} - OH$$

It can be represented as H_2Ala^+, with successive ionization constants $pK_{a1} = 2.3$ and $pK_{a2} = 9.7$.
(a) Calculate the ratio of the specified conjugate acid-base pairs at the pH values indicated:
  (i) $[H_2Ala^+]/[HAla]$ at pH 1.3
  (ii) $[H_2Ala^+]/[HAla]$ at pH 4.3
  (iii) $[HAla]/[Ala^-]$ at pH 4.3
  (iv) $[HAla]/[Ala^-]$ at pH 10.7
(b) Draw the structure of the alanine zwitterion.
(c) Estimate the isoelectric pH for alanine.

Section 14.9: Controlling pH: Buffer Solutions

14.89 What is the pH of a formic acid/formate ion buffer solution at 25 °C with $c(HCOOH) = 0.50$ mol L^{-1} and $c(HCOO^-) = 0.70$ mol L^{-1}?

14.90 Calculate the pH of 1.00 L of a hydrogencarbonate/carbonate buffer solution at 25 °C containing 15.0 g of $NaHCO_3$ and 18.0 g of Na_2CO_3.

14.91 A buffer solution is made by dissolving formic acid to be 0.050 mol L^{-1} and sodium formate to be 0.035 mol L^{-1}.
(a) What is the pH of this solution at 25 °C?
(b) To what value would the ratio $[HCOOH]/[HCOO^-]$ need to be changed to raise the pH by 0.5?

14.92 A buffer solution is made by dissolving 1.360 g of $KH_2PO_4(s)$ and 5.677 g of $Na_2HPO_4(s)$ in the same sample of water.
(a) What is the pH of the buffer solution?
(b) What mass of $KH_2PO_4(s)$ must be added to decrease the solution pH by 0.5?

14.93 Describe how to make a hydrogencarbonate ion/carbonate ion buffer solution with pH 10.00.

14.94 Design three buffer solutions, each with pH 4.60, using (a) benzoic acid and sodium benzoate, (b) acetic acid and sodium acetate, and (c) propanoic acid and sodium propanoate.

14.95 A buffer solution is made by dissolving 4.95 g of sodium acetate in 250 mL of 0.150 mol L^{-1} acetic acid solution.
(a) What is the pH of this buffer solution?
(b) What is the resultant pH if 82 mg of $NaOH(s)$ is dissolved in 100 mL of the buffer solution?

14.96 A buffer solution is made by dissolving 0.125 mol of ammonium chloride in 500 mL of 0.500 mol L^{-1} ammonia solution.
(a) What is the pH of the buffer solution?
(b) If 0.0100 mol of HCl gas is bubbled into the buffer solution, what is the new pH?

Interactive Exercise 14.97

Calculate the pH of a buffer before and after addition of base.

Section 14.10: Acid-Base Titrations

14.98 What volume of 0.812 mol L^{-1} HCl solution is required to exactly neutralize 1.45 g of NaOH (i.e., to titrate to the equivalence point)?

14.99 Solutions of potassium hydrogen phthalate, $KHC_8H_4O_4(s)$, are used to standardize solutions of bases.

$$HC_8H_4O_4^-(aq) + OH^-(aq) \longrightarrow C_8H_4O_4^{2-}(aq) + H_2O(\ell)$$

If 0.902 g of potassium hydrogen phthalate is dissolved in water and titrated to the equivalence point with 26.45 mL of NaOH solution, what is the molar concentration of the NaOH solution?

14.100 An unknown solid acid is either triprotic citric acid or diprotic tartaric acid. To determine which acid you have, you titrate a sample of the solid with NaOH. A 0.956 g sample requires 29.1 mL of 0.513 mol L^{-1} NaOH solution for titration to the equivalence point. What is the unknown acid?

14.101 What is the pH of the solution when we add the following to 100.0 mL of 0.200 mol L^{-1} solution of hypochlorous acid (HOCl): (a) 100.0 mL, and (b) 80.0 mL of 0.100 mol L^{-1} NaOH solution?

14.102 You require 36.78 mL of 0.0105 mol L^{-1} HCl solution to reach the equivalence point in the titration of 25.0 mL of aqueous ammonia solution.
(a) What was $c(NH_3)$ of the solution before any HCl solution was added?
(b) What are $[H_3O^+]$, $[OH^-]$, and $[NH_4^{2+}]$ at the equivalence point?
(c) What is the pH of the solution at the equivalence point?

14.103 A solution of the weak base aniline ($C_6H_5NH_2$) in 25.0 mL of water requires 25.67 mL of 0.175 mol L^{-1} HCl solution to reach the equivalence point.

$$C_6H_5NH_2(aq) + H_3O^+(aq) \longrightarrow C_6H_5NH_3^+(aq) + H_2O(\ell)$$

(a) What was the $c(C_6H_5NH_2)$ of the solution before any HCl solution was added?

(b) What are $[H_3O^+]$, $[OH^-]$, and $[C_6H_5NH_3^{?+}]$ at the equivalence point?

(c) What is the pH of the solution at the equivalence point?

14.104 You titrate 25.0 mL of 0.10 mol L^{-1} NH_3 solution with 0.10 mol L^{-1} HCl solution.

(a) What is the pH of the solution before any HCl solution is added?

(b) What is the pH at the equivalence point?

(c) What is the pH at the halfway point of the titration?

(d) What indicator would you use to correctly indicate that the equivalence point has just been reached (See e14.40 for indicator data)?

(e) Calculate the pH of the solution after adding 5.0, 15.0, 20.0, 22.0, and 30.0 mL of the acid. Combine this information with that in parts (a)–(c) and plot the titration curve.

14.105 Construct a rough plot of pH vs. volume of base for the titration of 25.0 mL of 0.050 mol L^{-1} HCN solution with 0.075 mol L^{-1} NaOH solution.

(a) What is the pH of the solution before any NaOH solution is added?

(b) What is the pH at the halfway point of the titration?

(c) What is the pH when 95% of the NaOH solution required for exact neutralization has been added?

(d) What volume of base solution is required to reach the equivalence point?

(e) What is the pH at the equivalence point?

(f) What indicator would be most suitable for this titration (See e14.40 for indicator data)?

(g) What is the pH when 105% of the required base has been added?

14.106 Suggest an indicator to use in each of the following titrations (See e14.40 for indicator data).

(a) A solution of the weak base pyridine is titrated with HCl solution.

(b) A formic acid solution is titrated with NaOH solution.

(c) A solution of ethylenediamine, a weak diprotic base, is titrated with HCl solution.

SUMMARY AND CONCEPTUAL PROBLEMS

14.107 Consider a cube containing about 55 000 water molecules:

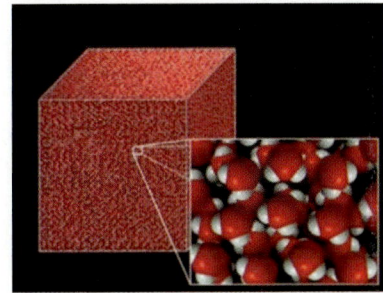

If this water included some acid, sufficient to raise the $[H_3O^+]$ to 0.10 mol L^{-1}, then the pH would be 1.0. In the cube of water molecules above, there would be about 100 H_3O^+ ions, portrayed below with all the water molecules faded back:

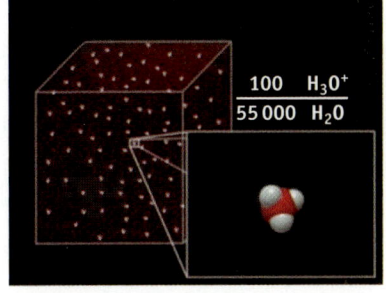

(a) If the pH was 2, how many hydronium ions would be present in the cube?

(b) If the pH was 3, how many hydronium ions would be present in the cube?

(c) If the pH was 4, how could you visualize the ratio of hydronium ions to water molecules?

(d) If the pH of water from a tap is around 6.0, what is the approximate ratio of H_3O^+ ions to water molecules?

14.108 25.00 mL of each of 0.0100 mol L^{-1} solutions of acids HA and HB, are titrated separately with a NaOH solution. The solutions have the following pH values at the equivalence point: from HA, pH = 9.5, and from HB, pH = 8.5.

(a) Which is the stronger acid, HA or HB?

(b) Which of the conjugate bases, A^- or B^-, is the stronger base?

(c) Which of the solutions (HA or HB) requires more NaOH solution for exact neutralization?

14.109 Describe how a buffer solution can minimize the pH of a solution when a strong base is added. Use a solution of containing both sodium carbonate and sodium hydrogencarbonate as an example, and include balanced chemical equations in your answer.

14.110 Use Equation 14.1 to explain how the pH of a buffer solution changes if (a) we use a conjugate acid-base pair with a larger ionization constant of the weak acid, and (b) we use a buffer solution with lower [acid]/base ratio.

14.111 Speciation plots are often used to visualize the species in a solution of an acid or base as the pH is varied. The diagram for 0.100 mol L^{-1} acetic acid is shown here:

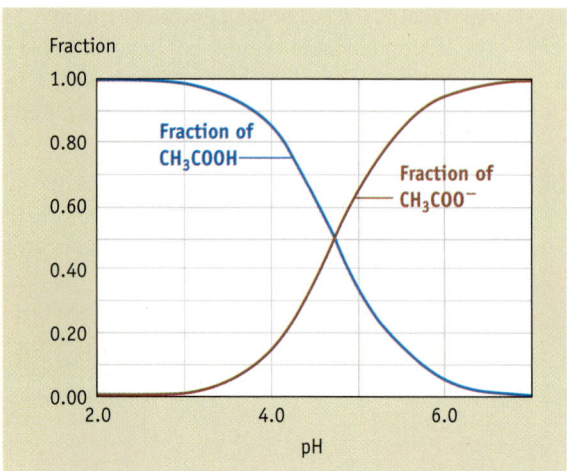

The plot shows how the fraction $\alpha = [CH_3COOH]/([CH_3COOH] + [CH_3COO^-])$, changes as the pH increases (blue curve). The red curve shows how the fraction of acetate ion, $CH_3COO^-(aq)$, changes as the pH increases. Speciation plots are another way of visualizing the relative concentrations of acetic acid and acetate ion when a strong base is added to a solution of acetic acid in the course of a titration.

(a) Explain why the fraction of acetic acid declines and that of acetate ion increases as the pH increases.

(b) Which species predominates at pH 4, acetic acid or acetate ion? What is the situation at pH 6?

(c) Consider the point where the two lines cross. The fraction of acetic acid in the solution is 0.5, and so is that of acetate ion. That is, their concentrations are equal. Explain why the pH at this point is 4.74.

14.112 The pH-dependent speciation for the important acid-base system of carbonic acid (H_2CO_3) is graphed here:

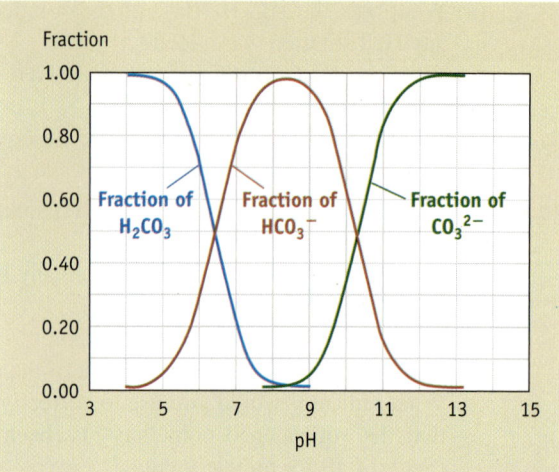

(a) Explain why the fraction of hydrogencarbonate ions, HCO_3^-, (as a fraction of the amounts of all carbonate species) rises and then falls as the pH increases.

(b) What can you deduce about the relative concentrations of species when the pH is 6.0? When the pH is 10.0?

(c) To make a buffer solution at pH 11.0, what should be the ratio $[HCO_3^-]/[CO_3^{2-}]$?

14.113 The chemical name for aspirin is acetylsalicylic acid. It is believed that the analgesic and other desirable properties of aspirin are due not to the aspirin itself but rather to the simpler compound salicylic acid, $C_6H_4(OH)CO_2H$, that results from the breakdown of aspirin in the stomach.

Salicylic acid

(a) Experiment shows that 1.00 g of salicylic acid dissolves in 460 mL of water. If the pH of this solution is 2.4, what is K_a for the acid?

(b) If you have salicylic acid in your stomach, and if the pH of gastric juice is 2.0, calculate the percentage of salicylic acid that will be present in the stomach in the form of the salicylate ion, $C_6H_4(OH)COO^-(aq)$.

(c) Assume you have 25.0 mL of a 0.014 mol L^{-1} solution of salicylic acid and titrate it with 0.010 mol L^{-1} NaOH. What is the pH at the halfway point of the titration? What is the pH at the equivalence point?

CHAPTER 15

Solubility, Precipitation, and Complexation

Manamana/Shutterstock

Ian Scott/Shutterstock

15.1 Case Study: Ocean Acidification: Coral Reefs at Risk

The probability of climate change due to increasing concentrations of carbon dioxide in the atmosphere is at the forefront of recent environmental concerns [<<Section 4.3]. Perhaps an even more powerful reason for minimizing the increase of CO_2 in the atmosphere is that sea water is becoming more acidic, and this will interfere with the production of corals and may even cause existing reefs to dissolve. Apart from their sheer beauty, coral reefs are important commercially as breeding grounds for fish and as tourist sites. Coral reefs are believed to have been a significant factor in moderation of the effects of the tsunami in Southeast Asia in December 2004, and marine corals are increasingly recognized as an important source of novel medicinal natural products. According to the modelling predictions of reputable scientists, acidification of the oceans and its serious consequences could take hundreds of years to reverse.

> *Perhaps an even more powerful reason for minimizing the increase of CO_2 in the atmosphere is that sea water is becoming more acidic, and this will interfere with the production of corals and may even cause existing reefs to dissolve.*

To understand this issue, we need to understand the chemistry underlying the processes that occur when CO_2 dissolves in sea water and makes it more acidic. In particular, we need to be able to model the distribution among species $CO_2(aq)$, $H_2CO_3(aq)$, $HCO_3^-(aq)$, $CO_3^{2-}(aq)$, and $CaCO_3(s)$, and how the distribution depends on pH. The following is a summary.

Dissolving CO_2 in water involves a series of reactions that come to equilibrium:

- Equilibrium between gaseous CO_2 in the atmosphere and CO_2 dissolved in the sea water, to an extent dependent on the partial pressure of CO_2 in the atmosphere [<<Henry's law, Section 12.4]:

$$CO_2(g) \underset{H_2O(\ell)}{\rightleftharpoons} CO_2(aq)$$

- Reaction between dissolved carbon dioxide and water to form aquated carbonic acid, $H_2CO_3(aq)$:

$$CO_2(aq) + H_2O(\ell) \rightleftharpoons H_2CO_3(aq)$$

- Two stages of ionization of aquated carbonic acid as a weak diprotic acid [<<Section 14.6]:

$$H_2CO_3(aq) + H_2O(\ell) \rightleftharpoons HCO_3^-(aq) + H_3O^+(aq)$$
$$HCO_3^-(aq) + H_2O(\ell) \rightleftharpoons CO_3^{2-}(aq) + H_3O^+(aq)$$

The pH of the oceans is due mainly to species other than these carbonate species, especially aquated borate ions, $B(OH)_4^-(aq)$. At present, the average pH of sea water near the surface of the oceans is about 8.07, and at this pH the distribution among the species derived from dissolved CO_2 is portrayed in the speciation plot [<<Section 14.7] of Figure 15.1. At this pH, the concentrations of the various carbonate species are $[H_2CO_3] = 1.3 \times 10^{-5}$ mol L^{-1}, $[HCO_3^-] = 1.923 \times 10^{-3}$ mol L^{-1}, and $[CO_3^{2-}] = 1.91 \times 10^{-4}$ mol L^{-1}.

FIGURE 15.1 pH-dependent distribution among carbonate species in aqueous solution. At pH 8.1, $HCO_3^-(aq)$ ions are dominant: the relative amounts are 0.6% $H_2CO_3(aq)$, 90.4% $HCO_3^-(aq)$, and 9.0% $CO_3^{2-}(aq)$.

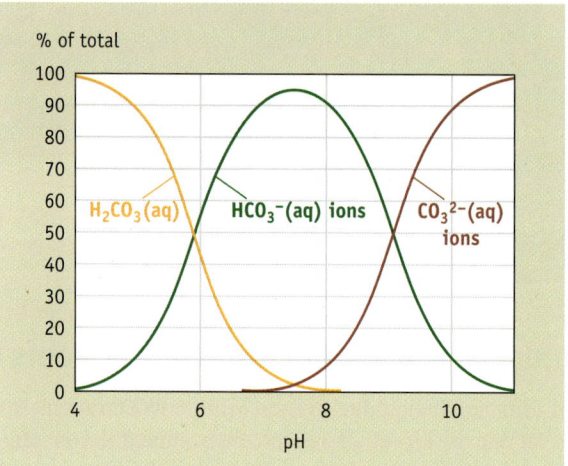

Recall [<<Section 14.7] that $[H_2CO_3] = [HCO_3^-]$ at pH = pK_{a1} and $[HCO_3^-] = [CO_3^{2-}]$ at pH = pK_{a2}, so we can deduce from this plot that $pK_{a1} = 5.9$ and $pK_{a2} = 9.1$. These values are different from those in Table 14.5 ($pK_{a1} = 6.38$ and $pK_{a2} = 10.32$), which are appropriate at 25 °C for very dilute solutions containing no other substances except the carbonate species. In sea water, the species are in the presence of much higher concentrations of other ions: for example, $[Cl^-]$ is about 230 times greater than $[HCO_3^-]$. Because of interactions between ions and formation of temporary ion-pairs, the carbonate species have an effective concentration or *activity* [<<Section 13.3] different from their actual concentration. The values $pK_{a1} = 5.9$ and $pK_{a2} = 9.1$ are values that are appropriate when using concentrations (rather than activities) in sea water.

Now if we consider the solubility in sea water of the mineral aragonite, a slightly soluble form of calcium carbonate, we can represent its equilibrium by the equation

$$CaCO_3(s) \xrightleftharpoons[]{H_2O(\ell)} Ca^{2+}(aq) + CO_3^{2-}(aq)$$

The equilibrium constant for this reaction, called the *solubility product* (K_{sp}), is 7.7×10^{-7} at 25 °C. When this reaction is at equilibrium, at the limit of solubility of aragonite, the solution is called a *saturated solution*. Applying the law of equilibrium, in a saturated solution of aragonite,

$$Q = [Ca^{2+}][CO_3^{2-}] = K_{sp} = 7.7 \times 10^{-7}$$

The concentration of calcium ions near the surface of the oceans is about 1.05×10^{-2} mol L^{-1}. We can use this to calculate the reaction quotient (Q) in ocean water near the surface:

$$Q = [Ca^{2+}][CO_3^{2-}] = (1.05 \times 10^{-2})(1.91 \times 10^{-4}) = 2.0 \times 10^{-6}$$

Since $Q (= 2.0 \times 10^{-6}) > K (= 7.7 \times 10^{-7})$, the water contains more of these ions than it would at equilibrium; it is *supersaturated*. For the ocean water to come to equilibrium, some solid aragonite would need to precipitate:

$$Ca^{2+}(aq) + CO_3^{2-}(aq) \longrightarrow CaCO_3(s)$$

However, the precipitation process is very slow because the $Ca^{2+}(aq)$ and $CO_3^{2-}(aq)$ ions are at very low concentrations and the frequency with which they collide with each other, remain together, and eventually form clusters of Ca^{2+} and CO_3^{2-} ions in a lattice arrangement is very small. Hardly any deposition of the solid occurs except through the action of marine organisms and animals, such as corals and molluscs, which create their shells and plates from aragonite—a process called *calcification*. These calcareous structures can only be made if the sea water is saturated with respect to aragonite.

If the concentration of CO_2 in the atmosphere increases further, the concentration of $H_2CO_3(aq)$ will increase, and its ionization will cause a lowering of the pH. From Figure 15.1, we can see that a lowering of pH would cause a redistribution among the carbonate species, decreasing the concentration of $CO_3^{2-}(aq)$ ions. This is consistent with the generalization that the salts with anions that are weak bases become more soluble if the pH of solution is decreased [>>Section 15.2].

Table 15.1 shows data from a report of the Royal Society, published in 2005, concerning concentrations of the carbon-containing species in the surface layer of the ocean, if the atmospheric CO_2 concentration rises to double what it was prior to the Industrial Revolution.

Some corals are composed of another form of calcium carbonate, called calcite, which has a different crystalline structure from aragonite and is slightly more soluble.

TABLE 15.1 Carbonate Species Concentrations in Surface Ocean Water*

	Pre-industrial	Today	2 × Pre-industrial
CO_2 in atmosphere	280 ppm	380 ppm	560 ppm
$[H_2CO_3]$ (μmol kg^{-1})	9	13	19
$[HCO_3^-]$ (μmol kg^{-1})	1768	1867	1976
$[CO_3^{2-}]$ (μmol kg^{-1})	225	185	141
Total dissolved carbonate species (μmol kg^{-1})	2003	2065	2136
Average pH near surface	8.18	8.07	7.92

* The data do not correspond exactly with values given in the text. This is to be expected because ocean water differs considerably, depending upon geographical location and depth. Note that the concentrations of species in solution are given in units of μmol kg^{-1}, rather than μmol L^{-1}. The values differ by a factor of 1.03 (the density of sea water).

1 μmol (1 micromole) $= 1 \times 10^{-6}$ mol.

Source: Report of the Royal Society, (2005). *Ocean Acidification Due to Increasing Atmospheric Carbon Dioxide* (June). Available at www.royalsoc.ac.uk.

A pH change from 8.18 to 7.92 may not seem much, but it represents an 80% change in $[H_3O^+]$—enough to bring about a 37% reduction in $[CO_3^{2-}]$. Such a change would bring sea water closer to the equilibrium situation for dissolved aragonite. The consequences of this could be that calcification is more difficult for the marine organisms, so more friable coral structures might be formed, and perhaps that existing corals might dissolve away.

Aragonite becomes more soluble with decreasing temperature and increasing pressure and, therefore, with increasing depth in the ocean. There is not much mixing between levels of the ocean, so going downward there is a rather sharp delineation, called the *saturation horizon*, below which the water is unsaturated with respect to aragonite ($Q < K_{sp}$), and above which the water is supersaturated ($Q > K_{sp}$). The depth of the saturation horizon varies geographically (between 0.5 and 2.5 km) and is closest to the surface at low latitudes. If the atmospheric CO_2 concentration increases further and $[CO_3^{2-}]$ decreases further, the saturation horizon will move closer to the surface. The fear is that surface waters in the Southern Ocean will approach unsaturation.

Apart from where there are "upwellings" of deep ocean water to the surface (equatorial Pacific Ocean and the Arabian Sea), mixing between layers of the oceans is very slow. Large changes of atmospheric CO_2 levels in the past happened over very long periods of time—slow enough for increases of acidification to be buffered by the beds of carbonate "ooze" on the ocean floor. A significant issue with the current increase of atmospheric CO_2

levels is the speed with which it has happened—too fast for the natural buffering processes to keep up. For the same reasons of slow inter-layer mixing, low pH values being brought about today may not return to conditions similar to those of pre-industrial times for tens of thousands of years, even if CO_2 were capped now.

We need to be cautious, and we need to understand the chemistry relevant to this issue. In this chapter, quantitative aspects of solubility of slightly soluble salts, their precipitation from supersaturated solutions, and the pH-dependence of solubility of salts like carbonates are discussed.

15.2 Solubility and Precipitation of Ionic Salts

Precipitation of almost insoluble ionic salts from aqueous solution is described in Section 6.7 as competition in which attraction of oppositely charged ions to form a solid lattice "wins" over attraction of water molecules to the ions (ion-dipole forces) to form aquated ions which remain in solution. For example, when a $0.1\ mol\ L^{-1}$ calcium chloride solution is mixed with a $0.1\ mol\ L^{-1}$ sodium carbonate solution, solid calcium carbonate forms— much more quickly than at the low ion concentrations in sea water [<<Section 15.1].

$$\underset{\text{from CaCl}_2}{Ca^{2+}(aq)} + \underset{\text{from Na}_2\text{CO}_3}{CO_3{}^{2-}(aq)} \rightleftharpoons CaCO_3(s)$$

There are very large differences among the solubilities of ionic salts [<<Section 6.5]. This chapter is about ionic salts, which are sometimes called "insoluble." No salt is absolutely insoluble, so we will use the term *slightly soluble* for these compounds such as calcium carbonate (Figure 15.2).

FIGURE 15.2 Precipitation of slightly soluble ionic substances in natural situations. (a) Metal sulfides (and hydroxides) in a black smoker [<<Section 4.1]. (b) Stalactites composed of $CaCO_3(s)$.

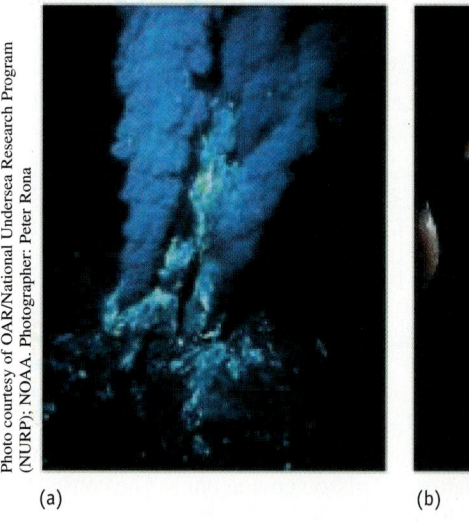
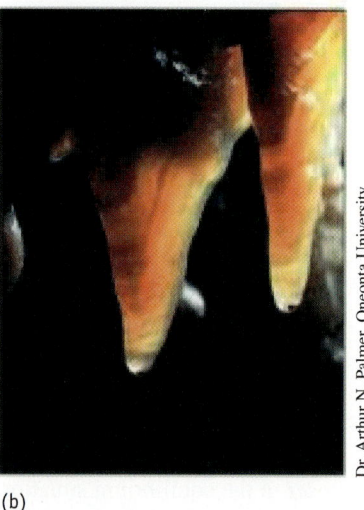

(a)   (b)

Photo courtesy of OAR/National Undersea Research Program (NURP); NOAA. Photographer: Peter Rona

Dr. Arthur N. Palmer, Oneonta University

Now we want to discuss solubility in a more quantitative way, and to explore conditions under which some compounds precipitate and others do not. We have seen in Section 15.1 how important this is in natural situations. In Section 15.3, we will see how quantitative knowledge about solubilities of sparingly soluble salts helps people in the mining industry to separate metals from each other by the process called *selective precipitation*.

> Precipitation of slightly soluble ionic salts from aqueous solution can be viewed as attraction of the cations and anions for each other to form a solid, "winning" a competition over aquation of the ions.

Solubility Equilibria: Saturated Solutions

If some silver bromide, $AgBr(s)$, is put into water and stirred, only a tiny amount of it dissolves, and a condition of equilibrium is reached:

$$AgBr(s) \rightleftharpoons Ag^+(aq) + Br^-(aq)$$

Think about It

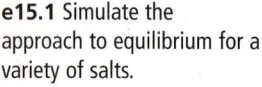

e15.1 Simulate the approach to equilibrium for a variety of salts.

When no more AgBr can dissolve, the solution is said to be a **saturated solution**. The concentrations of both $Ag^+(aq)$ and $Br^-(aq)$ ions in a saturated solution at 25 °C can be shown to be 7.35×10^{-7} mol L^{-1}. The extent to which a slightly soluble salt dissolves before becoming a saturated solution can be expressed through the *law of equilibrium* [<<Section 13.3], and in such cases, the equilibrium constant is called the **solubility product (K_{sp})**:

$$Q = [Ag^+][Br^-] = K_{sp}$$

From the experimentally measured concentrations of ions in saturated AgBr solutions, we can estimate the magnitude of the solubility product at 25 °C:

$$K_{sp} = [Ag^+][Br^-] = (7.35 \times 10^{-7})(7.35 \times 10^{-7}) = 5.40 \times 10^{-13}$$

The form of the reaction quotient for saturated aqueous solutions of ionic salts can be derived from the balanced chemical equation describing the equilibrium condition, generalized as follows:

$$A_xB_y(s) \rightleftharpoons x\, A^{y+}(aq) + y\, B^{x-}(aq) \qquad K_{sp} = [A^{y+}]^x[B^{x-}]^y$$

The following are two examples of different types:

$$CaF_2(s) \rightleftharpoons Ca^{2+}(aq) + 2\, F^-(aq) \qquad K_{sp} = [Ca^{2+}][F^-]^2 = 5.3 \times 10^{-11}$$
$$Ag_2SO_4(s) \rightleftharpoons 2Ag^+(aq) + SO_4{}^{2-}(aq) \qquad K_{sp} = [Ag^+]^2[SO_4{}^{2-}] = 1.2 \times 10^{-5}$$

Solubility products of a few salts are given in Table 15.2, and more are listed in Appendix B.

> Recall that we do not include the concentration of solid reactants (in this case, AgBr(s)) in the reaction quotient (Q) because their concentration does not change during reaction and we can regard their constant concentration as being already included in the value of Q [<<Section 13.3].

TABLE 15.2 Solubility Products of Some Common Slightly Soluble Salts*

Formula	Name	K_{sp} (25 °C)	Common Names/Uses
$CaCO_3$	Calcium carbonate	3.4×10^{-9}	Calcite, iceland spar
$MnCO_3$	Manganese(II) carbonate	2.3×10^{-11}	Rhodochrosite (forms rose-coloured crystals)
$FeCO_3$	Iron(II) carbonate	3.1×10^{-11}	Siderite
CaF_2	Calcium fluoride	5.3×10^{-11}	Fluorite (source of HF and other inorganic fluorides)
$AgCl$	Silver chloride	1.8×10^{-10}	Chlorargyrite
$AgBr$	Silver bromide	5.4×10^{-13}	Used in photographic film
$CaSO_4$	Calcium sulfate	4.9×10^{-5}	The hydrated form is commonly called gypsum
$BaSO_4$	Barium sulfate	1.1×10^{-10}	Barite (used in "drilling mud" and as a component of paints)
$SrSO_4$	Strontium sulfate	3.4×10^{-7}	Celestite
$Ca(OH)_2$	Calcium hydroxide	5.5×10^{-5}	Slaked lime

* The values in this table were taken from *Lange's Handbook of Chemistry*, 15th ed., New York, NY, McGraw-Hill Publishers, 1999. Additional K_{sp} values are given in Appendix B.

Do not confuse the *solubility* of a salt with its *solubility product*. *Solubility* here refers to the amount of a salt in a specified volume of a saturated solution, and may be expressed in moles per litre, grams per 100 mL, or other units. The *solubility product* is an equilibrium constant. Nonetheless, there is a connection between them: if one is known, the other can be calculated.

Interactive Exercise 15.2

SUBMIT

Identify some common misconceptions about the solubility of salts.

EXERCISE 15.1—REACTION QUOTIENTS FOR SLIGHTLY SOLUBLE SALTS IN AQUEOUS SOLUTION

Write a balanced equation for dissolution of each of the following slightly soluble salts in water. Also write the expression for the reaction quotient (Q), and find the value of Q at equilibrium (the solubility product, K_{sp}) at 25 °C (see Appendix B).
(a) AgI(s)   (b) BaF_2(s)   (c) Ag_2CO_3(s)

Think about It

e15.2 Work through a tutorial on writing expressions for solubility products.

A slightly soluble salt and its aquated ions are at equilibrium (a saturated solution) when the reaction quotient (Q) is the same as the equilibrium constant (called the solubility product) for that salt at the temperature of the solution—that is, when $Q = K_{sp}$.

Think about It

e15.3 See how ion concentrations are determined using atomic absorption spectroscopy.

Relating Solubility and Solubility Product

Solubility products of slightly soluble salts are determined by careful experimental measurements of the concentrations of ions in saturated solutions at the temperature of interest. We will generally use values appropriate to 25 °C, but the same principles, with different values, apply at all temperatures.

WORKED EXAMPLE 15.1—K_{SP} FROM ION CONCENTRATIONS IN SATURATED SOLUTION

Think about It

e15.4 Work through a tutorial on determining K_{sp} from ion concentrations.

Calcium fluoride, the main component of the mineral fluorite, dissolves to a slight extent in water:

$$CaF_2(s) \rightleftharpoons Ca^{2+}(aq) + 2\,F^-(aq)$$

Calculate the solubility product for CaF_2 if the concentration of $Ca^{2+}(aq)$ ions is measured to be 2.4×10^{-4} mol L^{-1}.

Strategy

Write the expression for the reaction quotient for CaF_2 solution equilibrium, and then substitute the numerical values of the equilibrium concentrations of the ions.

Solution

Interactive Exercise 15.3

Calculate K_{sp} from ion concentrations in saturated solution.

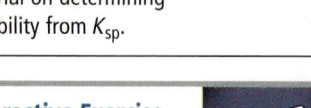

The balanced equation solution shows that the concentration of $F^-(aq)$ ions is double the concentration of $Ca^{2+}(aq)$ ions:

$$[F^-] = 2\,[Ca^{2+}] = 4.8 \times 10^{-4}\ \text{mol}\ L^{-1}$$

Now we can calculate the solubility product since at equilibrium $Q = K_{sp}$:

$$Q = [Ca^{2+}][F^-]^2 = K_{sp} = (2.4 \times 10^{-4})(4.8 \times 10^{-4})^2 = 5.5 \times 10^{-11}$$

EXERCISE 15.4—K_{SP} FROM ION CONCENTRATIONS IN SATURATED SOLUTION

The concentration of $Ba^{2+}(aq)$ ions in a saturated aqueous barium fluoride solution is 3.6×10^{-3} mol L^{-1}. Calculate $K_{sp}(BaF_2)$ at the temperature of measurement.

Solubility products can be used to estimate the solubilities of slightly soluble salts or to determine whether a salt will precipitate when solutions containing its cation are mixed with solutions containing its anion. Let's first look at an example of the estimation of the solubility of a salt from its K_{sp} value.

WORKED EXAMPLE 15.2—SOLUBILITY FROM K_{SP}

Think about It

e15.5 Work through a tutorial on determining solubility from K_{sp}.

K_{sp} for $BaSO_4$ is 1.1×10^{-10} at 25 °C. Calculate the solubility of barium sulfate in pure water at 25 °C in (a) moles per litre, and (b) grams per litre.

Strategy

First estimate the solubility in mol L^{-1}. This can be done using the law of equilibrium, noting from the balanced equation that the amount (in mol) of salt that dissolves per litre is the same as the concentrations of $Ba^{2+}(aq)$ ions and $SO_4^{2-}(aq)$ ions.

Solution

Interactive Exercise 15.5

Calculate solubility from K_{sp}.

(a) The equation for the equilibrium in a saturated $BaSO_4$ solution is

$$BaSO_4(s) \rightleftharpoons Ba^{2+}(aq) + SO_4^{2-}(aq) \quad Q = [Ba^{2+}][SO_4^{2-}] = K_{sp} = 1.1 \times 10^{-10}$$

Let us denote the solubility of $BaSO_4$ by x mol L^{-1}; that is, x moles of $BaSO_4$ dissolve per litre. Therefore, at equilibrium (i.e., in the saturated solution), $[Ba^{2+}] = [SO_4^{2-}] = x$ mol L^{-1}. We can substitute into the reaction quotient

$$Q = [Ba^{2+}][SO_4^{2-}] = K_{sp} = 1.1 \times 10^{-10} = (x)(x) = x^2$$

from which we can calculate the value of x:

$$\text{Solubility, } x = [Ba^{2+}] = [SO_4^{2-}] = \sqrt{1.1 \times 10^{-10}} = 1.0 \times 10^{-5} \text{ mol } L^{-1}$$

(b) To find its solubility in g L^{-1}, we need just to multiply by the molar mass of $BaSO_4$:

$$\text{Solubility (g } L^{-1}) = (1.0 \times 10^{-5} \text{ mol } L^{-1})(233 \text{ g mol}^{-1}) = 0.0023 \text{ g } L^{-1}$$

WORKED EXAMPLE 15.3—SOLUBILITY FROM K_{SP}

K_{sp} for MgF_2 in aqueous solution at 25 °C is 5.2×10^{-11}. Calculate the solubility of the salt at 25 °C in (a) moles per litre, and (b) grams per litre.

Strategy

From the balanced equation, we know that if 1 mol of MgF_2 dissolves, 1 mol of $Mg^{2+}(aq)$ ions and 2 mol of $F^-(aq)$ ions form in the solution. So the MgF_2 solubility (in mol L^{-1}) is the same as the concentration of $Mg^{2+}(aq)$ ions in solution. If the solubility of MgF_2 is x mol L^{-1}, then at equilibrium $[Mg^{2+}] = x$ mol L^{-1} and $[F^-] = 2x$ mol L^{-1}.

Solution

(a) Equilibrium in the saturated solution is defined by

$$MgF_2(s) \rightleftharpoons Mg^{2+}(aq) + 2 F^-(aq) \qquad Q = [Mg^{2+}][F^-]^2 = K_{sp} = 5.2 \times 10^{-11}$$

We can express the concentrations of ions in terms of x:

$$K_{sp} = [Mg^{2+}][F^-]^2 = (x)(2x)^2 = 4x^3 = 5.2 \times 10^{-11}$$

Solving the equation for x, the solubility of MgF_2 in water:

$$\text{Solubility of } MgF_2, x = \sqrt[3]{\frac{K_{sp}}{4}} = \sqrt[3]{\frac{5.2 \times 10^{-11}}{4}} = 2.4 \times 10^{-4} \text{ mol } L^{-1}$$

(b) Solubility in grams per litre:

$$\text{Solubility (g } L^{-1}) = (2.4 \times 10^{-4} \text{ mol } L^{-1})(62.3 \text{ g mol}^{-1}) = 0.015 \text{ g } L^{-1}$$

Comment

Problems like this might provoke you to ask, "Aren't you counting things twice when you multiply x by 2 and then square it as well?" in the expression $K_{sp} = (x)(2x)^2$. The answer is no—the 2 in the $2x$ term is based on the stoichiometry of the compound, while the power of 2 on $[F^-]$ arises from the rules for writing reaction quotients.

EXERCISE 15.6—SOLUBILITY FROM K_{SP}

Calculate the solubility of $Ca(OH)_2$ in (a) moles per litre, and (b) grams per litre, both at 25 °C, using the value of K_{sp} in Appendix B.

Interactive Exercise 15.7

SUBMIT

Thinking task: Identify errors in proposed solutions to questions.

In some instances, the *relative* solubilities of salts can be deduced by comparing values of solubility products—but you must be careful! For example, let's compare solubility products for silver chloride and silver chromate, in both cases at 25 °C:

$$AgCl(s) \rightleftharpoons Ag^+(aq) + Cl^-(aq) \qquad K_{sp} = 1.8 \times 10^{-10}$$
$$Ag_2CrO_4(s) \rightleftharpoons 2 Ag^+(aq) + CrO_4^{2-}(aq) \qquad K_{sp} = 1.1 \times 10^{-12}$$

The solubility product for Ag_2CrO_4 is smaller than that for AgCl. However, if you determine solubilities from K_{sp} values as in Worked Examples 15.2 and 15.3, you will find that the solubility of Ag_2CrO_4 (6.5×10^{-5} mol L^{-1}) is greater than that of AgCl (1.3×10^{-5} mol L^{-1}). From this example, and countless others, we conclude that comparisons of the solubilities of two salts by direct comparison of their K_{sp} values can only be correctly made if the salts have the same cation-to-anion ratio. This means, for example, that you can compare solubilities of 1:1 salts (such as the silver halides) by comparison of their K_{sp} values, and you can likewise for various 1:2 salts (such as the lead halides), but you cannot compare the solubility of a 1:1 salt with that of a 1:2 salt by inspection of K_{sp} values.

EXERCISE 15.8—COMPARING SOLUBILITIES FROM SOLUBILITY PRODUCTS

Using K_{sp} values from Appendix B, predict which salt in each pair is the more soluble in water at 25 °C.

   (a)   AgCl(s) or AgCN(s)
   (b) $Mg(OH)_2(s)$ or $Ca(OH)_2(s)$
   (c) $Ca(OH)_2(s)$ or $CaSO_4(s)$

> The solubility of a slightly soluble salt can be deduced from the value of its solubility product at the appropriate temperature. Relative solubilities of different salts can only be deduced from the relative values of the solubility products when they are salts of the same type—all 1:1 salts, or all 1:2 salts, for example.

Complexity Leading to Errors in Solubility Calculations

The K_{sp} value reported for lead(II) chloride ($PbCl_2$) is 1.7×10^{-5}. If we assume the only equilibrium reaction in a saturated solution is

$$PbCl_2(s) \rightleftharpoons Pb^{2+}(aq) + 2\,Cl^-(aq)$$

we calculate that the solubility of $PbCl_2$ is 0.016 mol L^{-1}. The experimental estimate of the solubility of the salt, however, is 0.036 mol L^{-1}—more than twice the calculated value! What is the problem? There are several, as summarized by the diagram below:

Although lead(II) chloride is regarded as a strong electrolyte, at any time there is a tiny concentration of the un-ionized salt, $PbCl_2(aq)$. This concentration can be estimated, to give us a value of the equilibrium constant for the equilibrium between solid and dissolved salt. This applies to other slightly soluble salts.

The main problem in the lead(II) chloride case, and in many others, is that the $Pb^{2+}(aq)$ ions and $Cl^-(aq)$ ions are not the only species formed in solution: at any instant, some of the cations and anions are temporarily joined as *ion pairs*. So, although the solubility product listed is valid for the chemical reaction described by the equation, the equation does not encompass everything that is happening in the saturated solution.

Other problems that lead to discrepancies between calculated and experimental solubilities are the reactions of ions (particularly anions) with water, and complex ion formation. An example of the former effect is the reaction of sulfide ions with water, a reaction that is strongly product-favoured because the sulfide ion is a strong base.

$$S^{2-}(aq) + H_2O(\ell) \rightleftharpoons HS^-(aq) + OH^-(aq) \qquad K_b = 1 \times 10^5$$

This means that the solubilities of metal sulfides such as nickel sulfide and lead sulfide are best described by chemical equations such as

$$NiS(s) + H_2O(\ell) \rightleftharpoons Ni^{2+}(aq) + HS^-(aq) + OH^-(aq)$$
$$PbS(s) + H_2O(\ell) \rightleftharpoons Pb^{2+}(aq) + HS^-(aq) + OH^-(aq)$$

As a consequence of such reactions, nickel sulfide, for example, dissolves to a greater extent than is predicted by consideration only of the equilibrium condition:

$$NiS(s) \rightleftharpoons Ni^{2+}(aq) + S^{2-}(aq) \qquad Q = [Ni^{2+}][S^{2-}] = K_{sp}(NiS)$$

Complex ion formation can explain why lead chloride becomes more soluble if we add excess chloride ions from another source—by adding sodium chloride, for example. This is contrary to what we might expect by application of Le Chatelier's principle, and can be attributed to the formation of the complex ion $PbCl_4^{2-}$:

$$PbCl_2(s) + 2\,Cl^-(aq) \rightleftharpoons PbCl_4^{2-}(aq)$$

Interactive Exercise 15.9

Predict solubilities taking into account other reactions in solution.

EXERCISE 15.10—PARTICIPATION OF IONS IN OTHER REACTIONS

Explain why the solubility of $Ag_3PO_4(s)$ in water may be greater in water than is calculated from K_{sp} of the salt.

> Calculations of solubilities from solubility products gives accurate estimates only if the cation and/or the anion are not involved in equilibrium reactions other than the one between the solid salt and its aquated ions in saturated solution.

Solubility of Salts and the Common Ion Effect

In a saturated aqueous silver acetate solution, aquated silver ions and acetate ions are in equilibrium with solid silver acetate:

$$AgCH_3COO(s) \rightleftharpoons Ag^+(aq) + CH_3COO^-(aq)$$

What would happen if the silver ion concentration were increased—for example, by adding a concentrated silver nitrate solution? Le Chatelier's principle [<<Section 13.6] suggests—and we observe—that more silver acetate precipitate will form.

The effect of adding silver ions to a saturated silver acetate solution is another example of the **common ion effect** [<<Section 14.6]. Adding a common ion to a saturated solution of a salt lowers the salt solubility (Figure 15.3).

 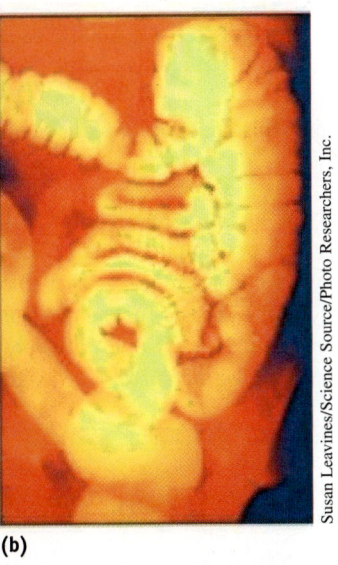

(a)   (b)

Charles D. Winters

Susan Leavines/Science Source/Photo Researchers, Inc.

Think about It

e15.6 Watch a lab demonstration of the common ion effect on the solubility of a salt.

Think about It

e15.7 Simulate the common ion effect on the solubility of different salts.

FIGURE 15.3 A medical application of barium sulfate's slight solubility and the common ion effect. Barium sulfate, a white solid, is sparingly soluble in water ($K_{sp} = 1.1 \times 10^{-10}$). (a) The mineral barite, which is mostly barium sulfate. (b) Barium sulfate is opaque to x-rays, so physicians use it to examine the digestive tract. A patient drinks a slurry containing $BaSO_4$, and the progress of the slurry through the digestive organs can be followed by x-ray tracking. This photo is an x-ray of a gastrointestinal tract after a person has ingested barium sulfate slurry. It is fortunate that $BaSO_4$ is only slightly soluble in water, because water-soluble barium salts are toxic. Just to be sure, some potassium sulfate is added to the slurry to reduce the solubility of barium sulfate through the common ion effect (see Exercise 15.11).

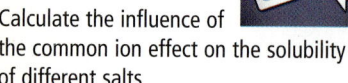
WORKED EXAMPLE 15.4—THE COMMON ION EFFECT AND SALT SOLUBILITY

What is the solubility (mol L^{-1}) of AgCl in (a) water, and (b) 0.55 mol L^{-1} NaCl solution?

Strategy

If we define the solubility to be x mol L^{-1} (different in the two solutions), in each saturated solution we can express $[Ag^+]$ and $[Cl^-]$ in terms of x. Substitute these values in the reaction quotient, and then knowing that in a saturated solution $Q = K$, we can solve for the value of x.

Solution

The equilibrium condition is represented by the equation

$$AgCl(s) \rightleftharpoons Ag^+(aq) + Cl^-(aq)$$

and the law of equilibrium must be satisfied

$$Q = [Ag^+][Cl^-] = K_{sp} = 1.8 \times 10^{-10}$$

(a) If the solubility is x mol L^{-1}, $[Ag^+] = [Cl^-] = x$ mol L^{-1}

$$\text{So } Q = [Ag^+][Cl^-] = (x)(x) = x^2 = 1.8 \times 10^{-10}$$

$$\text{Solubility, } x = [Ag^+] = \sqrt{K_{sp}} = 1.3 \times 10^{-5} \text{ mol } L^{-1}$$

(b) In this case, it may be useful to set up an ICE table [<<Section 13.4] that displays the initial concentrations of each ion, the changes in concentrations, and the equilibrium concentrations:

	$[Ag^+]$	**$[Cl^-]$**
Initial (before adding any AgCl)	0 mol L^{-1}	0.55 mol L^{-1}
Change in concentration	$+x$ mol L^{-1}	$+x$ mol L^{-1}
Equilibrium concentration	x mol L^{-1}	$(0.55 + x)$ mol L^{-1}

Then, the equilibrium condition is defined by

$$Q = [Ag^+][Cl^-] = (x)(0.55 + x) = K_{sp} = 1.8 \times 10^{-10}$$

which re-arranges to the quadratic equation:

$$x^2 + 0.55\,x - 1.8 \times 10^{-10} = 0$$

Solution of this equation gives us Solubility $= x = 3.3 \times 10^{-10}$ mol L^{-1}

Comment

The solubility of AgCl in the presence of Cl^-(aq) ions from another source (3.3×10^{-10} mol L^{-1}) is much less than that in pure water (1.3×10^{-5} mol L^{-1}). This is the common ion effect. With experience, you will come to realize that since the solubility in water is only 1.3×10^{-5} mol L^{-1}, the solubility in the presence of the common ion is even less than that. We can therefore make the approximation $0.55 + x = 0.55$ with very little error. Then, the equilibrium condition reduces to $0.55x = K_{sp}$ and we obtain $x = 3.3 \times 10^{-10}$ mol L^{-1} without the need to solve a quadratic equation.

WORKED EXAMPLE 15.5—THE COMMON ION EFFECT AND SALT SOLUBILITY

Calculate the solubility of silver chromate (Ag_2CrO_4) at 25 °C in 0.0050 mol L^{-1} K_2CrO_4 solution.

$$Ag_2CrO_4(s) \rightleftharpoons 2\,Ag^+(aq) + CrO_4^{2-}(aq) \qquad K_{sp} = 1.1 \times 10^{-12}$$

For comparison, the solubility of Ag_2CrO_4 in pure water is 1.3×10^{-4} mol L^{-1}.

Strategy

Assume the solubility of Ag_2CrO_4 is x mol L^{-1}, and set up an ICE table to deduce the concentrations of the cation and the anion in terms of x.

Solution

Realize from the chemical equation that the increase in $[Ag^+]$ is double that of $[CrO_4^{2-}]$, and set up an ICE table:

	$[Ag^+]$	$[CrO_4^{2-}]$
Initial (before adding any Ag_2CrO_4)	0 mol L^{-1}	0.0050 mol L^{-1}
Change in concentration	$+2x$ mol L^{-1}	$+x$ mol L^{-1}
Equilibrium concentration	$2x$ mol L^{-1}	$(0.0050 + x)$ mol L^{-1}

The equilibrium condition is specified by

$$Q = [Ag^+]^2[CrO_4^-] = (2x)^2(0.0050 + x) = K_{sp} = 1.1 \times 10^{-12}$$

This gives us a cubic equation: $4x^3 + 0.02x^2 - 1.1 \times 10^{-12} = 0$.

We can solve this equation using solver software, or we can simplify the problem by using our chemistry knowledge. You can make the approximation that x is very small with respect to 0.0050, so $(0.0050 + x) \approx 0.0050$. This assumption is reasonable because $[CrO_4^{2-}]$ is 0.00013 mol L^{-1} when Ag_2CrO_4 is dissolved in pure water, so it is certain that x is even smaller in the presence of extra chromate ion. Then, the equilibrium condition reduces to

$$(2x)^2(0.0050) = K_{sp} = 1.1 \times 10^{-12}$$

Solving, we find x, the solubility of silver chromate in the presence of excess chromate ion:

$$\text{Solubility of } Ag_2CrO_4 = x = 7.4 \times 10^{-6} \text{ mol } L^{-1}$$

Comment

The solubility of silver chromate in the presence of the common ion is indeed less than that in pure water. You can check the validity of the approximation made by substituting this value of x into the mathematical statement of the equilibrium condition to ensure that the equation is satisfied.

EXERCISE 15.13—THE COMMON ION EFFECT AND SALT SOLUBILITY

Calculate the solubility at 25 °C of $BaSO_4$ (a) in pure water, and (b) in the presence of 0.010 mol L^{-1} $Ba(NO_3)_2$. At 25 °C, K_{sp} for $BaSO_4$ is 1.1×10^{-10}.

The presence of a "common ion" (either the cation or the anion) reduces the solubility of a slightly soluble salt in water.

pH-Dependence of Solubility of Salts whose Anions Are Bases

Silver acetate is only slightly soluble in water:

$$AgCH_3COO(s) \rightleftharpoons Ag^+(aq) + CH_3COO^-(aq) \qquad K_{sp} \text{ at } 25 °C = 1.9 \times 10^{-3}$$

If this were the only reaction to occur when silver acetate dissolves in water, we could calculate that in a saturated solution $[Ag^+] = [CH_3COO^-] = 0.04$ mol L^{-1}. However, the aquated acetate ion is a weak base, and some reaction with water (hydrolysis) occurs:

$$CH_3COO^-(aq) + H_2O(\ell) \rightleftharpoons CH_3COOH(aq) + OH^-(aq) \qquad K_b = 5.6 \times 10^{-10}$$

Because of the decrease of concentration of $CH_3COO^-(aq)$ ions, more silver acetate dissolves than we would have predicted if we ignored this hydrolysis reaction.

Hydrolysis occurs to an even greater extent with slightly soluble phosphates, carbonates, cyanides, sulfides, and hydroxides because their aquated anions are stronger bases than $CH_3COO^-(aq)$ ions. In fact, the aquated sulfide ion is a strong base and the hydrolysis reaction is very much product-favoured. If, for example, we add some lead sulfide to water and stir, two reactions occur:

$$PbS(s) \rightleftharpoons Pb^{2+}(aq) + S^{2-}(aq) \qquad K_{sp} = 3 \times 10^{-33}$$

$$S^{2-}(aq) + H_2O(\ell) \rightleftharpoons HS^-(aq) + OH^-(aq) \qquad K_b = 1 \times 10^5$$

So the net result of lead sulfide dissolving (slightly) in water is better described by the overall equation deduced by adding the above equations:

$$PbS(s) + H_2O(\ell) \rightleftharpoons Pb^{2+}(aq) + HS^-(aq) + OH^-(aq) \quad K = (K_{sp})(K_b) = 3 \times 10^{-28}$$

You might think that since the S^{2-}(aq) ion is quite a strong base, lead sulfide might be very soluble, because "all" of the S^{2-}(aq) ions produced react with water. However, there will always be some S^{2-}(aq) ions in equilibrium with the HS^-(aq) ions formed by hydrolysis, and at even a tiny concentration we will have $Q = K_{sp}$ because of the extremely small K_{sp}.

If we add some solution of a strong acid to saturated solutions of these slightly soluble salts with basic anions, more of the solid will dissolve because the basic anion reacts with the added H_3O^+(aq) ions. If there are only small amounts of excess solid, and moderate amounts of strong acid are added, the solid may dissolve completely. However $CuS(s)$ is so slightly soluble ($K_{sp} = 6 \times 10^{-37}$) that it does not dissolve in even moderately high concentrations of H_3O^+(aq) ions.

Many metal sulfides dissolve in acidified solution:

$$FeS(s) + 2 H_3O^+(aq) \rightleftharpoons Fe^{2+}(aq) + H_2S(aq) + 2 H_2O(\ell)$$

as do metal phosphates:

$$Ag_3PO_4(s) + 3 H_3O^+(aq) \rightleftharpoons 3 Ag^+(aq) + H_3PO_4(aq) + 3 H_2O(\ell)$$

and metal hydroxides:

$$Mg(OH)_2(s) + 2 H_3O^+(aq) \rightleftharpoons Mg^{2+}(aq) + 4 H_2O(\ell)$$

The dissolving of solid silver acetate in acidic solution can be regarded as the result of two equilibrium reactions:

$$AgCH_3COO(s) \rightleftharpoons Ag^+(aq) + CH_3COO^-(aq) \qquad K_{sp} = 1.9 \times 10^{-3}$$

$$CH_3COO^-(aq) + H_3O^+(aq) \rightleftharpoons CH_3COOH(aq) + H_2O(\ell) \qquad K = \frac{1}{K_a} = \frac{1}{1.8 \times 10^{-5}} = 5.5 \times 10^4$$

The net result of these reactions is product-favoured:

$$AgCH_3COO(s) + H_3O^+(aq) \rightleftharpoons Ag^+(aq) + CH_3COOH(aq) + H_2O(\ell)$$

$$K_{net} = (K_{sp})\left(\frac{1}{K_a}\right) = (1.9 \times 10^{-3})(5.5 \times 10^4) = 1 \times 10^2$$

In such a case, we have another factor in the competition that governs extent of dissolution of a salt. Not only does aquation of the ions compete with attraction between the ions in the solid lattice, but the CH_3COO^-(aq) ions compete to "grab" protons to form CH_3COOH(aq). This latter process alters the balance of competition compared with what happens in neutral solution, and the silver acetate dissolves.

$$AgCH_3COO(s) \xrightarrow{K_{sp} = 1.9 \times 10^{-3}} Ag^+(aq) + CH_3COO^-(aq)$$

$$+$$

$$H_3O^+(aq)$$

$$\left\updownarrow K = 5.5 \times 10^{+4}\right.$$

$$CH_3COOH(aq) + H_2O(\ell)$$

Think about It

e15.9 Simulate the effect of pH on solubility of selected salts.

Think about It

e15.10 Calculate the effect of pH on solubility of selected salts.

This phenomenon has important ramifications if natural waterways become acidic, because metal ions can be 'leached' from minerals.

Interactive Exercise 15.14
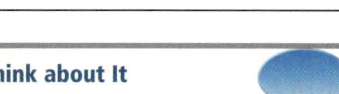

Thinking task: Identify errors in proposed solutions to questions.

The more soluble the salt and the more strongly basic its anion, the more likely that a slightly soluble salt will dissolve when the solution is acidified. Conversely, the more insoluble the salt and the more weakly basic the cation, the less effect acidification has on solubility. In the extreme case of a conjugate base of a strong acid (as is the case for AgCl and AgBr), solubility is not affected at all by acidification because the anions have zero tendency to "grab" protons.

If an acid is added to a water-insoluble metal carbonate such as $CaCO_3$, the salt dissolves [<<Section 13.2] with effervescence:

$$CaCO_3(s) + 2\,H_3O^+(aq) \longrightarrow Ca^{2+}(aq) + 3\,H_2O(\ell) + CO_2(g)$$

You can think of this as the net result of a series of equilibrium reactions—a solubility equilibrium reaction and some acid-base equilibria—for which the equations and their equilibrium constants at 25 °C are displayed:

$$CaCO_3(s) \rightleftharpoons Ca^{2+}(aq) + CO_3^{2-}(aq) \qquad K_{sp} = 3.4 \times 10^{-9}$$
$$CO_3^{2-}(aq) + H_2O(\ell) \rightleftharpoons HCO_3^-(aq) + OH^-(aq) \qquad K_{b1} = 2.1 \times 10^{-4}$$
$$HCO_3^-(aq) + H_2O(\ell) \rightleftharpoons H_2CO_3(aq) + OH^-(aq) \qquad K_{b2} = 2.4 \times 10^{-8}$$
$$2[OH^-(aq) + H_3O^+(aq) \rightleftharpoons 2\,H_2O(\ell)] \qquad K = \left(\frac{1}{K_w}\right)^2 = 1 \times 10^{28}$$

Net: $CaCO_3(s) + 2\,H_3O^+(aq) \rightleftharpoons Ca^{2+}(aq) + 2\,H_2O(\ell) + H_2CO_3(aq)$

$$K_{net} = (K_{sp})(K_{b1})(K_{b2})\left(\frac{1}{K_w}\right)^2 = 1.7 \times 10^8$$

The carbonic acid product is in equilibrium with carbon dioxide:

$$H_2CO_3(aq) \rightleftharpoons CO_2(g) + H_2O(\ell) \quad K \approx 10^5$$

Carbon dioxide has a limited solubility in water, and if the amount produced is greater than that needed to form a saturated solution, the excess gas bubbles out of the solution— and we see "effervescence."

A more delicate version of the chemistry described above is the increased solubility of aragonite coral, $CaCO_3(s)$, in ocean waters if the pH is lowered as a result of increasing CO_2 concentration in the atmosphere [<<Section 15.1]. The higher solubility of calcium carbonate corals in more acidic sea water is not the only situation in which calcium carbonate solubility is pH-dependent. Natural limestone deposits (once coral formations) are "weathered" by gradual dissolution by rain water that has been acidified by dissolving $CO_2(g)$ from the air; exposed rock formations are shaped by this. The run-off water carries increased calcium ion loads, and, in other situations where the air is CO_2-depleted, can form stalactites and stalagmites [<<Section 13.2]. In the artificial environment, statues created from marble (another form of calcium carbonate) are disfigured as a result of this same phenomenon (Figure 15.4).

Fletcher & Baylis / Photo Researchers, Inc.

FIGURE 15.4 The reaction between calcium carbonate, $CaCO_3(s)$, and acidified water has been the cause of deterioration of marble statues in many parts of the world, but especially in Europe. The marble, a form of calcium carbonate, has reacted over many years with $H_3O^+(aq)$ ions in "acid rain"—rain with dissolved acidic pollutants derived from oxidation in the atmosphere of SO_2 and NO_2.

EXERCISE 15.15—PH-DEPENDENCE OF SOLUBILITY OF SALTS

Which slightly soluble compound in each pair would you predict to be more soluble in nitric acid solution than in pure water?
(a) $PbCl_2(s)$ or $PbS(s)$
(b) $Ag_2CO_3(s)$ or $AgI(s)$
(c) $Al(OH)_3(s)$ or $AgCl(s)$

Think about It

e15.11 Watch a demonstration of the reaction of $Co^{2+}(aq)$ ions with hydroxide ions, and then with nitric acid.

Slightly soluble salts in which the anion is a weak base become more soluble as the pH of solution lowers. The weaker the anion is as a base, the more pH-dependent the solubility of such salts.

15.3 Precipitation Reactions

Charles D. Winters

FIGURE 15.5 Minerals are often insoluble salts. The minerals shown here are light purple fluorite (calcium fluoride), black hematite (iron(III) oxide), and rust brown goethite, a mixture of iron(III) oxide and iron(III) hydroxide.

Most metal-bearing ores contain a metal in the form of a nearly insoluble salt (Figure 15.5). Many industrial methods for extracting metals from their ores involve dissolving metal salts to obtain the metal ions in solution. The solution is then separated from insoluble materials, and a reagent is added to precipitate selectively a salt of only the desired metal. In the case of nickel, for example, the Ni^{2+} ion can be precipitated as nickel sulfide or nickel carbonate:

$$Ni^{2+}(aq) + S^{2-}(aq) \longrightarrow NiS(s)$$
$$Ni^{2+}(aq) + CO_3^{2-}(aq) \longrightarrow NiCO_3(s)$$

The final step in obtaining the metal is to reduce the cation to the metal either chemically or electrochemically [>>Chapter 16]. A challenge in extracting metals is that some ores contain salts of several different metals. Sometimes, however, differences in the solubilities of salts of the metals can be used to advantage in separating one from the others, as we shall see below in a discussion of *separation by selective precipitation*.

Our immediate goal is to determine the conditions under which a precipitate can form. For example, if $Ag^+(aq)$ and $Cl^-(aq)$ ions are present at specified concentrations, will solid AgCl precipitate from the solution? Precipitation of AgCl(s) is often the first step in recovery of silver from solutions of silver salts in industry.

Deciding if a Salt Will Precipitate: Q versus K_{sp}

Silver chloride, like silver bromide, is used in photographic film. It dissolves to a very small extent in water:

$$AgCl(s) \rightleftharpoons Ag^+(aq) + Cl^-(aq)$$

In a saturated solution at 25 °C:

$$Q = [Ag^+][Cl^-] = K_{sp} = 1.8 \times 10^{-10}$$

Let's look at this situation from the other direction: if a solution contains $Ag^+(aq)$ and $Cl^-(aq)$ ions at some specified concentrations, will solid AgCl precipitate from solution? We can answer this question by comparison of the value of Q with K_{sp} appropriate to the temperature of the reaction mixture [<<Section 13.3]:

- If $Q = K_{sp}$, the solution is saturated and at equilibrium.
- If $Q < K_{sp}$, the solution is unsaturated, precipitation does not happen, and more of the salt can dissolve, increasing the concentrations of cation and anion until $Q = K_{sp}$.
- If $Q > K_{sp}$, the solution is *supersaturated*: precipitation to reduce the concentrations of ions would bring the system to equilibrium (when $Q = K_{sp}$)—although this tells us nothing about how quickly this would happen.

We have seen in Section 15.1 that the $Q > K_{sp}$ condition applies in surface ocean waters in relation to calcium carbonate, even if precipitation is extremely slow unless achieved through the agency of marine organisms.

First, we will apply the Q vs. K_{sp} criteria to solutions made by stirring a sparingly soluble salt in water.

WORKED EXAMPLE 15.6—IS THE SOLUTION SATURATED?

Solid AgCl has been placed in a beaker of water at 25 °C, and stirred. After some time, the concentrations of $Ag^+(aq)$ and $Cl^-(aq)$ ions are each 1.2×10^{-5} mol L^{-1}. Has the system reached equilibrium? If not, how much more AgCl can dissolve?

Strategy

Calculate the reaction quotient (Q) and see if $Q = K_{sp}$. At 25 °C, $K_{sp} = 1.8 \times 10^{-10}$.

Solution

$$Q = [Ag^+][Cl^-] = (1.2 \times 10^{-5})(1.2 \times 10^{-5}) = 1.4 \times 10^{-10} \ < K_{sp}$$

The solution is not saturated, and AgCl(s) will continue to dissolve until $Q = K_{sp}$, which we can calculate to be when $[Ag^+] = [Cl^-] = 1.3 \times 10^{-5}$ mol L^{-1}. An additional 0.1×10^{-5} mol of AgCl will dissolve per litre of solution before it is saturated.

EXERCISE 15.16—IS THE SOLUTION SATURATED?

Solid PbI$_2$ ($K_{sp} = 9.8 \times 10^{-9}$) is placed in a beaker of water. After a period of time, the lead(II) concentration is measured and found to be 1.1×10^{-3} mol L^{-1}. Has the system reached equilibrium? That is, is the solution saturated? If not, can more PbI$_2$ dissolve?

In the case of a slightly soluble salt, if $Q = K_{sp}$, the solution is saturated. If $Q < K_{sp}$, the solution is unsaturated and more can dissolve, and if $Q > K_{sp}$, the solution is supersaturated and precipitation would bring the system to equilibrium. These statements say nothing about how fast dissolution or precipitation happens.

Precipitation when Reagent Solutions Are Mixed

If we mix two solutions (known volume and concentration of each), one of which contains the cation of a slightly soluble salt, and the other containing the anion, we can decide (a) whether some of the salt can precipitate as a solid, and if not, (b) to what concentration one of the ions must be raised to make precipitation possible.

Think about It

e15.12 A simulation: is precipitation possible?

WORKED EXAMPLE 15.7—MIXING SOLUTIONS: IS THE MIXTURE SATURATED?

Suppose the concentration of Mg^{2+}(aq) ions in a solution at 25 °C is 1.5×10^{-6} mol L^{-1}. If enough of a concentrated NaOH solution is added to make $[OH^-] = 1.0 \times 10^{-4}$ mol L^{-1}, can precipitation of Mg(OH)$_2$(s) occur? $K_{sp} = 5.6 \times 10^{-12}$ at 25 °C. If not, can precipitation occur if $[OH^-]$ is increased to 1.0×10^{-2} mol L^{-1}? Assume that volume change due to addition of the NaOH solution is small enough to be ignored.

Think about It

e15.13 Calculate if precipitation is possible.

Strategy

Compare Q with K_{sp} to decide whether the system is at equilibrium, unsaturated, or supersaturated. Although we are considering precipitation, values of K_{sp} are based on the equation for reaction in the opposite direction (dissolving).

Solution

$$Mg(OH)_2(s) \rightleftharpoons Mg^{2+}(aq) + 2OH^-(aq)$$
$$Q = [Mg^{2+}][OH^-]^2 = (1.5 \times 10^{-6})(1.0 \times 10^{-4})^2 = 1.5 \times 10^{-14} \ < K_{sp}$$

Since $Q < K_{sp}$, the solution is not saturated, and precipitation does not occur. When $[OH^-]$ is increased to 1.0×10^{-2} mol L^{-1}, we can calculate:

$$Q = (1.5 \times 10^{-6})(1.0 \times 10^{-2})^2 = 1.5 \times 10^{-10} \ > K_{sp}$$

$Q > K_{sp}$, so precipitation of Mg(OH)$_2$ is now possible, and can continue until $[Mg^{2+}]$ and $[OH^-]$ ion have decreased to the levels that $Q = K_{sp}$.

WORKED EXAMPLE 15.8—MIXING SOLUTIONS: IS THE MIXTURE SATURATED?

Suppose you mix 100.0 ml of 0.0200 mol L^{-1} BaCl$_2$ solution with 50.0 mL of 0.0300 mol L^{-1} Na$_2$SO$_4$ solution, and the mixture is at 25 °C. Can BaSO$_4$ precipitate? For BaSO$_4$, $K_{sp} = 1.1 \times 10^{-10}$ at 25 °C.

Interactive Exercise 15.17

Predict if a precipitate can form in a solution.

Interactive Exercise 15.18

Thinking task: Identify errors in proposed solutions to questions.

Strategy

Find the concentrations of $Ba^{2+}(aq)$ and $SO_4^{2-}(aq)$ ions in the mixture. Then calculate Q and compare it with K_{sp} for $BaSO_4$ at 25 °C.

Solution

The concentrations of the ions in the mixture can be calculated by simple proportion, recognizing that the volume occupied by $Ba^{2+}(aq)$ ions increases from 100 mL to 150 mL, and the volume occupied by $Cl^-(aq)$ ions increases from 50 mL to 150 mL.

$$[Ba^{2+}] = (0.0200 \text{ mol L}^{-1}) \times \frac{0.1000 \text{ L}}{0.1500 \text{ L}} = 0.0133 \text{ mol L}^{-1}$$

$$[SO_4^{2-}] = (0.0300 \text{ mol L}^{-1}) \times \frac{0.0500 \text{ L}}{0.150 \text{ L}} = 0.0100 \text{ mol L}^{-1}$$

Now the reaction quotient can be calculated:

$$Q = [Ba^{2+}][SO_4^{2-}] = (0.0133)(0.0100) = 1.33 \times 10^{-4} > K_{sp}$$

Since $Q > K_{sp}$, some $BaSO_4$ can precipitate (until $Q = K_{sp}$).

EXERCISE 15.19—MIXING SOLUTIONS: IS THE MIXTURE SATURATED?

Can $SrSO_4$ precipitate form in a solution at 25 °C containing aquated strontium ions, $Sr^{2+}(aq)$, at a concentration of 2.5×10^{-4} mol L^{-1} if enough of the soluble salt $Na_2SO_4(s)$ is added that $[SO_4^{2-}] = 2.5 \times 10^{-4}$ mol L^{-1}? K_{sp} at 25 °C for $SrSO_4$ is 3.4×10^{-7}.

If we mix a solution that contains the cation of a slightly soluble salt with another containing the anion, we can decide whether some of the salt can precipitate as a solid by comparing Q in the mixture with K_{sp}.

Adjusting the Concentration of One Ion

Now let's turn to the problem of deciding to what concentration one of the ions of a slightly soluble salt must be increased, in a solution of known concentration of the other, so that the solution becomes saturated (and on the verge of precipitating some of the salt). Again, this is a matter of application of the law of equilibrium: we need to adjust the concentration of the added ion until $Q = K_{sp}$.

WORKED EXAMPLE 15.9—ADJUSTING THE CONCENTRATION OF ONE ION

The concentration of aquated barium ions in an aqueous solution at 25 °C is 0.010 mol L^{-1}.
(a) What concentration of aquated sulfate ions is required to render the solution saturated (and on the verge of precipitation of some $BaSO_4(s)$)?
(b) If $[SO_4^{2-}]$ is increased to 0.015 mol L^{-1}, what is the concentration of $Ba^{2+}(aq)$ ions left in solution?
(c) When $[SO_4^{2-}] = 0.015$ mol L^{-1}, what percent of the $Ba^{2+}(aq)$ ions remain in solution?

Strategy

If the concentration of sulfate ions were gradually increased, at some level the situation becomes $Q = K_{sp}$, and the solution would be saturated. We need to calculate $[SO_4^{2-}]$ that achieves this condition in a solution with $[Ba^{2+}] = 0.010$ mol L^{-1}.

Solution

The equilibrium condition in any saturated solution of $BaSO_4$ at 25 °C is described by

$$BaSO_4(s) \rightleftharpoons Ba^{2+}(aq) + SO_4^{2-}(aq) \qquad Q = [Ba^{2+}][SO_4^{2-}] = K_{sp} = 1.1 \times 10^{-10}$$

(a) For a saturated solution at 25 °C with $[Ba^{2+}] = 0.010$ mol L^{-1}:

$$Q = [Ba^{2+}][SO_4^{2-}] = (0.010)[SO_4^{2-}] = K_{sp} = 1.1 \times 10^{-10}$$

$$\therefore [SO_4^{2-}] = \frac{K_{sp}}{[Ba^{2+}]} = \frac{1.1 \times 10^{-10}}{0.010} = 1.1 \times 10^{-8} \text{ mol L}^{-1}$$

(b) Imagine that we gradually add a solution containing sulfate ions. For each tiny addition, we would instantaneously have $Q > K_{sp}$, and some $BaSO_4$ would precipitate till once again $Q = K_{sp}$. Most, but not all, of the added sulfate would be removed in the precipitate. Gradually $[SO_4^{2-}]$ increases, and $[Ba^{2+}]$ decreases. We can calculate $[Ba^{2+}]$ when $[SO_4^{2-}]$ reaches 0.015 mol L^{-1}.

$$[Ba^{2+}] = \frac{K_{sp}}{[SO_4^{2-}]} = \frac{1.1 \times 10^{-10}}{0.015} = 7.3 \times 10^{-9} \text{ mol L}^{-1}$$

(c) Originally, the concentration of $Ba^{2+}(aq)$ ions was 0.010 mol L^{-1}, so the percentage remaining is

$$\% \text{ Ba}^{2+} \text{ ions still in solution} = \frac{7.3 \times 10^{-9} \text{ mol L}^{-1}}{0.010 \text{ mol L}^{-1}} \times 100 = 7.3 \times 10^{-5} \%$$

Comment

When the concentration of $SO_4^{2-}(aq)$ ions has been increased to 0.015 mol L^{-1}, essentially all of the $Ba^{2+}(aq)$ ions have been removed as $BaSO_4(s)$ precipitate.

EXERCISE 15.20—ADJUSTING THE CONCENTRATION OF ONE ION

What is the minimum concentration of $I^-(aq)$ ions that can cause precipitation of $PbI_2(s)$ from a 0.050 mol L^{-1} $(PbNO_3)_2$ solution at 25 °C? K_{sp} for PbI_2 at 25 °C is 9.8×10^{-9}. What concentration of $Pb^{2+}(aq)$ ions remains in solution when the concentration of $I^-(aq)$ ions is 0.0015 mol L^{-1}?

> At a known concentration of one of the ions of a slightly soluble salt, we can use the condition $Q = K_{sp}$ to calculate the concentration of the other ion that will bring about a saturated solution, on the verge of some precipitation.

Separation of Metal Cations by Selective Precipitation

Sometimes during mineral extraction processes, solutions containing cations of the desired metal also contain lower concentrations of cations of other metals, regarded as contaminants. The chemist's task is to design ways of obtaining the desired metal without contamination by the others. One approach is to consider exploiting differences in the solubilities of salts of the metals. The task is simple if, for example, the hydroxide of the wanted metal is soluble, while the hydroxides of the contaminant metals are insoluble: just add some base and filter off the insoluble hydroxides. Then, the ions of the desired metal can be reduced to the metal with relatively few impurities.

In some cases, the chemical behaviours of the cations of the target metal and another metal are similar, and the only opportunity for separation based on solubilities is to take advantage of different solubilities of salts of the same anion type (sulfides, for example). The concentration of the anion is gradually increased, and the less soluble salt precipitates out first. If the solubilities are sufficiently different, most of the less soluble salt is precipitated and removed before any of the more soluble salt begins to precipitate.

Considered at a more quantitative level, as the concentration of the precipitating anion is increased, the values of Q for both salts increase. At some concentration of the anion, $Q = K_{sp}$ for the salt with the smaller K_{sp}. Any further increase in the concentration of the anion brings about some precipitation of this salt—while no precipitation of the other salt occurs because still $Q < K_{sp}$. The metal cation of the less soluble salt is continuously removed from solution by precipitation until the concentration of the anion is such that $Q = K_{sp}$ for the more soluble salt. Any further increase in concentration of the anion would lead to precipitation of both salts. Armed with the knowledge of K_{sp} for each of the salts, you can predict the percentage of the less soluble salt that would precipitate (and be filtered off) before the more soluble salt begins to precipitate (Worked Example 15.10). This is called separation by **selective precipitation**.

Think about It

e15.14 Watch a demonstration separating metal ions using selective precipitation.

Interactive Exercise 15.21

SUBMIT

Devise a way to separate metal ions using selective precipitation.

Taking It Further

e15.15 Read about chemical control of concentration of anions in selective precipitation.

WORKED EXAMPLE 15.10—SEPARATION OF METALS BY SELECTIVE PRECIPITATION

In the nickel extraction industry, it is common, near the end of the process, to have aqueous solutions containing $Cu^{2+}(aq)$ ions as well as the desired $Ni^{2+}(aq)$ ions. It is important to separate the two metal cations before reducing the $Ni^{2+}(aq)$ ions to nickel metal, or the metal will be contaminated. This is done by selective precipitation of the sulfides, whose solubilities are quite different ($K_{sp} = 8 \times 10^{-34}$ for CuS, $K_{sp} = 3 \times 10^{-19}$ for NiS). Suppose an aqueous solution has $[Cu^{2+}] = [Ni^{2+}] = 0.0100$ mol L^{-1}. As the concentration of aquated sulfide ions, $S^{2-}(aq)$ is increased in small increments.
(a) Which will begin to precipitate first: CuS or NiS?
(b) At what $[S^{2-}]$ will the less soluble sulfide just begin to precipitate?
(c) At what $[S^{2-}]$ will the more soluble sulfide just begin to precipitate?
(d) What percent of the cation of the salt that precipitates first remains in solution just before the second sulfide begins to precipitate?

Strategy

Recognize that as we increase the anion concentration, the onset of precipitation of a salt occurs when the solution is a saturated solution of that salt. The $Q = K_{sp}$ condition is applied to each salt separately. The calculations are similar to those in Worked Example 15.9.

Solution

The solubility equilibria for the two metal sulfides are described by

$$NiS(s) \rightleftharpoons Ni^{2+}(aq) + S^{2-}(aq) \qquad K_{sp} = 3 \times 10^{-19}$$
$$CuS(s) \rightleftharpoons Cu^{2+}(aq) + S^{2-}(aq) \qquad K_{sp} = 8 \times 10^{-34}$$

(a) The solution becomes saturated with respect to NiS when

$$Q = [Ni^{2+}][S^{2-}] = (0.0100)[S^{2-}] = K_{sp} = 3 \times 10^{-19} \quad \text{i.e., when } [S^{2-}] = 3 \times 10^{-17} \text{ mol L}^{-1}$$

and saturated with respect to CuS when

$$Q = [Cu^{2+}][S^{2-}] = (0.0100)[S^{2-}] = K_{sp} = 8 \times 10^{-34} \quad \text{i.e., when } [S^{2-}] = 8 \times 10^{-32} \text{ mol L}^{-1}$$

Obviously, CuS(s) will begin to precipitate first—when $[S^{2-}]$ reaches 8×10^{-32} mol L^{-1}, which is well below the concentration at which the solution becomes saturated in NiS.
(b) CuS(s) will *just* begin to precipitate when the solution is saturated with respect to that salt—that is, when $[S^{2-}] = 8 \times 10^{-32}$ mol L^{-1}.
$[S^{2-}]$ at which the solution just becomes saturated with respect to CuS, and at which CuS just begins to precipitate, is calculated in (a) to be 8×10^{-32} mol L^{-1}.
(c) $[S^{2-}]$ at which the solution just becomes saturated with respect to NiS, and at which NiS just begins to precipitate, has been calculated in (a) to be 3×10^{-17} mol L^{-1}.
(d) When NiS just begins to precipitate, and $[S^{2-}] = 3 \times 10^{-17}$ mol L^{-1}, the solution is still a saturated solution in CuS, and so we apply the $Q = K_{sp}$ relationship again:

$$Q = [Cu^{2+}][S^{2-}] = [Cu^{2+}][3 \times 10^{-17}] = K_{sp} = 8 \times 10^{-34}$$
$$\therefore [Cu^{2+}] = 3 \times 10^{-17} \text{ mol L}^{-1}$$

So the percentage of the original amount of $Cu^{2+}(aq)$ remaining in solution is

$$\% \ Cu^{2+} \ \text{not precipitated} = \frac{3 \times 10^{-17}}{0.0100} \times 100 = 3 \times 10^{-13} \ \%$$

Comment

An insignificant fraction of the copper remains in solution as $Cu^{2+}(aq)$ ions when NiS is just about to precipitate. If all CuS precipitate is filtered off before any NiS precipitates, the NiS precipitate will be essentially pure. Then nickel metal can be obtained (by reduction of NiS) with negligible contamination by copper.

EXERCISE 15.22—SEPARATION OF METALS SELECTIVE PRECIPITATION

If a solution at 25 °C has $[Mg^{2+}] = 0.200$ mol L^{-1} and $[Cu^{2+}]$ 0.100 mol L^{-1}, at what concentration of $OH^-(aq)$ ions could we get maximum separation of the cations by precipitation of one of them as the slightly soluble hydroxide? At 25 °C, K_{sp} for $Mg(OH)_2$ is 5.6×10^{-12}, and K_{sp} for $Cu(OH)_2$ is 2.2×10^{-20}.

When two metals have salts of the same anion with sufficiently different solubilities, they may be effectively separated by selective precipitation. In a solution containing aquated cations of both metals, if the concentration of the aquated anion is raised incrementally, just as the solution becomes saturated with respect to the most soluble salt, the less soluble salt has essentially entirely precipitated from solution.

15.4 Solubility and Complexation: Competitive Equilibria

As we have seen in Section 15.2, solid calcium carbonate is almost insoluble in water, but dissolves in acidic solutions because the tendency for basic $CO_3^{2-}(aq)$ anions to undergo an acid-base reaction with $H_3O^+(aq)$ ions is greater than that to form a solid lattice with $Ca^{2+}(aq)$ ions. In contrast, sparingly soluble silver chloride does not dissolve in acidic solutions because $Cl^-(aq)$ ions do not react as bases. AgCl(s) does, however, dissolve if ammonia is added (Figure 15.6). In this case, the dissolution of the solid is not because the anion is a base, but because the Ag^+ cation is a Lewis acid that reacts with $NH_3(aq)$ Lewis base [<<Section 6.7] to form a complex ion:

$$AgCl(s) + 2 \ NH_3(aq) \rightleftharpoons [Ag(NH_3)_2]^+(aq) + Cl^-(aq)$$

Because the solid AgCl precipitate disappears, we usually say that it has dissolved. More correctly, it has undergone a chemical change to a species that is soluble.

We can view dissolving AgCl(s) by reaction with ammonia as a two-step process. First, AgCl(s) dissolves minimally in water to give very low equilibrium concentrations of $Ag^+(aq)$ and $Cl^-(aq)$ ions. Second, the $Ag^+(aq)$ ions combine with $NH_3(aq)$ to form the complex ion. Lowering the concentration of $Ag^+(aq)$ ions by complexation means that the first process is not at equilibrium ($Q < K$), so more AgCl(s) dissolves, and more $Ag^+(aq)$ ions are made available to react with $NH_3(aq)$. The interdependent reactions proceed until they are both at equilibrium:

$$AgCl(s) \rightleftharpoons Ag^+(aq) + Cl^-(aq) \qquad K_{sp} = 1.8 \times 10^{-10}$$

$$Ag^+(aq) + 2 \ NH_3(aq) \rightleftharpoons [Ag(NH_3)_2]^+(aq) \qquad K_f = 1.6 \times 10^7$$

The equilibrium constant for the formation of a complex ion such as $[Ag(NH_3)_2]^+$ is called the **formation constant**, K_f [>>Section 27.5] (and sometimes called the *stability constant*, K_{stab}). The large value of this equilibrium constant means that the reaction is highly product-favoured, and provides the "driving force" for dissolving AgCl(s) in the

Think about It

e15.16 Watch a video portraying a precipitate dissolving due to complexation.

Think about It

e15.17 Interpret a video portrayal of competition between precipitation and complexation.

All photos: Charles D. Winters

AgCl(s),
$K_{sp} = 1.8 \times 10^{-10}$

(a) AgCl precipitates on adding NaCl solution to AgNO₃ solution (see Figure 5.4).

$[Ag(NH_3)_2]^+(aq)$

(b) The precipitate of AgCl dissolves on adding aqueous NH₃ to give water-soluble $[Ag(NH_3)_2]^+$.

AgBr(s),
$K_{sp} = 5.4 \times 10^{-13}$

(c) The silver-ammonia complex ion is changed to insoluble AgBr on adding NaBr solution.

$[Ag(S_2O_3)_2]^{3-}(aq)$

(d) Solid AgBr dissolves on adding Na₂S₂O₃ solution. The product is the water-soluble complex ion $[Ag(S_2O_3)_2]^{3-}$.

FIGURE 15.6 Competition: precipitation vs. complexation. Beginning with a precipitate of AgCl(s), adding aqueous ammonia solution dissolves the precipitate to form the soluble complex ion $[Ag(NH_3)_2]^+(aq)$. Silver bromide is even more stable than $[Ag(NH_3)_2]^+(aq)$, so AgBr(s) forms on addition of bromide ions. If thiosulfate ions, $S_2O_3^{2-}(aq)$, are then added, AgBr(s) dissolves due to the formation of $[Ag(S_2O_3)_2]^{3-}(aq)$, a complex ion with a large formation constant.

$$
\begin{aligned}
K_{sp} &\times K_f \\
&= [Ag^+][Cl^-] \times \frac{[Ag(NH_3)_2^+]}{[Ag^+][NH_3]^2} \\
&= \frac{[Ag(NH_3)_2^+][Cl^-]}{[NH_3]^2} \\
&= K_{net}
\end{aligned}
$$

presence of NH₃(aq). The net equilibrium constant for reaction of AgCl(s) with ammonia in aqueous solution can be shown to be a function of K_{sp} and K_f:

$$K_{net} = \frac{[Ag(NH_3)_2^+][Cl^-]}{[NH_3]^2} = K_{sp} \times K_f = (1.8 \times 10^{-10})(1.6 \times 10^7) = 2.9 \times 10^{-3}$$

Even though the value of K_{net} is not large, at low concentrations of $Ag^+(aq)$ ions and high concentrations of NH₃(aq), the extent of conversion of AgCl(s) into $[Ag(NH_3)_2]^+(aq)$ ions is quite high.

The competition for $Ag^+(aq)$ ions between $Cl^-(aq)$ ions and NH₃(aq) molecules is portrayed diagrammatically below:

$$
AgCl(s) \xrightleftharpoons{K_{sp} = 1.8 \times 10^{-10}} Ag^+(aq) + Cl^-(aq)
$$

$$+$$

$$2\ NH_3(aq)$$

$$K_f = 1.6 \times 10^{+7}$$

$$[Ag(NH_3)_2]^+(aq)$$

There is another competitor for the $Ag^+(aq)$ ions "lurking": the cations are attracted to the water molecules by ion-dipole forces. However, the evidence that AgCl(s) is only slightly soluble in water tells us that these attractions are dominated by attractions between $Ag^+(aq)$ ions and $Cl^-(aq)$ ions.

In the above case, the NH₃(aq) Lewis base "wins" the competition. However, silver bromide ($K_{sp} = 5.4 \times 10^{-13}$) is more insoluble than silver chloride, meaning that $Br^-(aq)$ ions are better competitors for $Ag^+(aq)$ ions than are $Cl^-(aq)$ ions—sufficiently so that they "win" in competition with NH₃(aq) molecules, and ammonia does not bring about dissolution of solid AgBr in water. On the other hand, aquated thiosulfate ions, $S_2O_3^{2-}(aq)$, form complex ions with a sufficiently high formation constant that they can bring about dissolution of AgBr (Figure 15.6).

Formation constants have been measured for many complex ions [>>Section 27.5 and Appendix D]. By way of example, values of a few silver(I) ion complexes are given:

$$Ag^+(aq) + 2\,Cl^-(aq) \rightleftharpoons [AgCl_2]^-(aq) \qquad K_f = 1.1 \times 10^5$$
$$Ag^+(aq) + 2\,S_2O_3{}^{2-}(aq) \rightleftharpoons [Ag(S_2O_3)_2]^{3-}(aq) \qquad K_f = 2.0 \times 10^{13}$$
$$Ag^+(aq) + 2\,CN^-(aq) \rightleftharpoons [Ag(CN)_2]^-(aq) \qquad K_f = 1.0 \times 10^{22}$$

The formation of each silver(I) complex is strongly product-favoured. The cyanide complex ion $[Ag(CN)_2]^-(aq)$ is the most stable of the three adducts.

If the complex $[AgCl_2]^-(aq)$ ion did not form, we would expect (Le Chatelier's principle) that if we had a saturated solution of silver chloride

$$AgCl(s) \rightleftharpoons Ag^+(aq) + Cl^-(aq)$$

an increase in the concentration of $Cl^-(aq)$ ions would cause more $AgCl(s)$ to precipitate. Instead, if we add some concentrated hydrochloric acid solution or sodium chloride solution, the precipitate dissolves due to the formation of the complex ion.

WORKED EXAMPLE 15.11—COMPETITION: PRECIPITATION VERSUS COMPLEXATION

What is the value of the equilibrium constant (K_{net}) for dissolving $AgBr(s)$ in a solution containing the thiosulfate ion, $S_2O_3{}^{2-}(aq)$ (as shown in Figure 15.6)? Explain why $AgBr(s)$ dissolves readily on adding aqueous sodium thiosulfate solution.

Strategy

Summing several equilibrium processes gives the net chemical equation. K_{net} is the product of the values of K of the summed chemical equations.

Solution

The overall reaction for dissolution of $AgBr(s)$ in the presence of aquated thiosulfate ions is the sum of two equilibrium processes:

$$AgBr(s) \rightleftharpoons Ag^+(aq) + Br^-(aq) \qquad K_{sp} = 5.4 \times 10^{-13}$$
$$Ag^+(aq) + 2\,S_2O_3{}^{2-}(aq) \rightleftharpoons Ag(S_2O_3)_2{}^{3-}(aq) \qquad K_f = 2.0 \times 10^{13}$$

The equation for the net chemical reaction is

$$AgBr(aq) + 2\,S_2O_3{}^{2-}(aq) \rightleftharpoons Ag(S_2O_3)_2{}^{3-}(aq) + Br^-(aq) \qquad K_{net} = K_{sp} \times K_f = 10$$

The value of K_{net} is greater than 1, indicating a decidedly product-favoured reaction, and we would predict that $AgBr(s)$ dissolves in aqueous $Na_2S_2O_3$ solution, as observed.

EXERCISE 15.24—COMPETITION: PRECIPITATION VERSUS COMPLEXATION

Calculate the value of the equilibrium constant (K_{net}) for dissolving $Cu(OH)_2(s)$ in aqueous ammonia to form the complex $[Cu(NH_3)_4]^{2+}(aq)$ ion.

> Salts that are only slightly soluble in water can dissolve as a result of reaction between the metal cation and a Lewis base to form a complex ion. This can be regarded as the result of two competing reactions for the cation: formation of a solid lattice and formation of the complex ion. Equilibrium constants for formation of complex ions, called formation constants (K_f), are measures of stability of the complex ions.

15.5 Complexation versus Lewis Base Protonation

As discussed in Section 15.4, the large value of K_f for the $[Ag(NH_3)_2]^+(aq)$ complex ion tells us that it is very stable: in the presence of moderate concentrations of $NH_3(aq)$ in aqueous solution, Ag^+ ions will exist mainly as the $[Ag(NH_3)_2]^+(aq)$ ions because Lewis

Interactive Exercise 15.23

Calculate the equilibrium constant (K_{net}) for a competitive equilibrium.

Think about It

e15.18 Calculate the metal ion concentration in the presence of a Lewis base.

base NH_3 molecules displace Lewis base water molecules bound to the Ag^+ ions in the species that we write as $Ag^+(aq)$ or $[Ag(OH_2)_6]^+(aq)$.

$$Ag^+(aq) + 2\,NH_3(aq) \rightleftharpoons [Ag(NH_3)_2]^+(aq) \qquad K_f = 1.6 \times 10^7$$

Just as pH affects the extent of solubility of salts with anions that are weak bases (discussed in Section 15.2), so too does pH affect the extent of formation of complex ions with Lewis bases that are weak Brønsted-Lowry bases, such as ammonia:

$$NH_3(aq) + H_2O(\ell) \rightleftharpoons NH_4^+(aq) + OH^-(aq) \qquad K_b = 1.8 \times 10^{-5}$$

The speciation between $NH_4^+(aq)$ ions and $NH_3(aq)$ molecules in aqueous solution as pH is changed has been discussed previously [<<Section 14.7] and is displayed in Figure 15.7.

FIGURE 15.7 pH-dependent speciation between $NH_4^+(aq)$ ions and $NH_3(aq)$ molecules in aqueous solution. The concentrations of the two species are equal at $pH = pK_a(NH_4^+) = 9.25$.

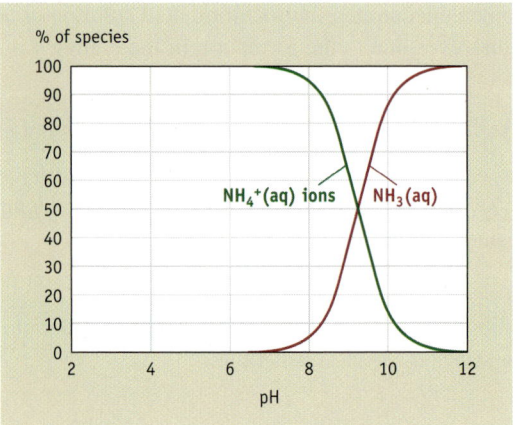

$NH_4^+(aq)$ ions have no ability to be Lewis bases because they do not have a lone pair of electrons. So, if the pH of solution is reduced below about pH 8, the concentration of $NH_3(aq)$ falls away sharply, limiting the amount of the $[Ag(NH_3)_2]^+$ complex ion that is formed, as shown in Figure 15.8.

FIGURE 15.8 Speciation plot showing the percentages present as $Ag^+(aq)$ ions and $[Ag(NH_3)_2]^+$ ions as pH is changed. The total concentration of silver-containing species is 0.005 mol L^{-1}, and the total concentration of ammonia in solution is 0.01 mol L^{-1}. Below about pH 7, the concentration of the $[Ag(NH_3)_2]^+$ ions is negligible because ammonia exists mainly as the non-Lewis base $NH_4^+(aq)$ ions.

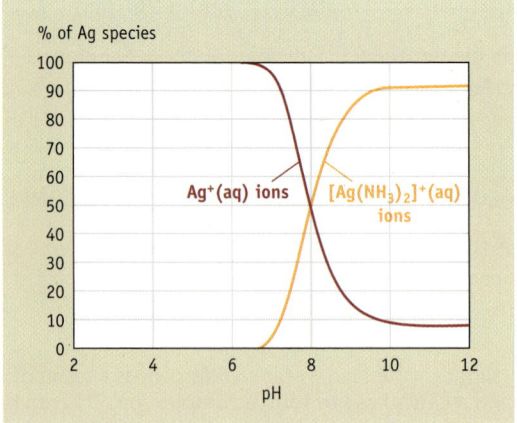

This situation can be regarded as competition between $H_3O^+(aq)$ ions and $Ag^+(aq)$ ions to react with $NH_3(aq)$ molecules. As the concentration of $H_3O^+(aq)$ ions increases (as pH decreases), it makes sense that they become more competitive, and "win" the competition.

One of the powerful techniques of science is the ability to model in response to "What if?" questions. Some fairly simple modelling allows us to compare the pH-dependent speciation of $Ag^+(aq)$ ions in the presence of $NH_3(aq)$ Lewis base, with that in the presence of Lewis base L, whose conjugate acid $HL^+(aq)$ has $pK_a = 6.00$ (Figure 15.9).

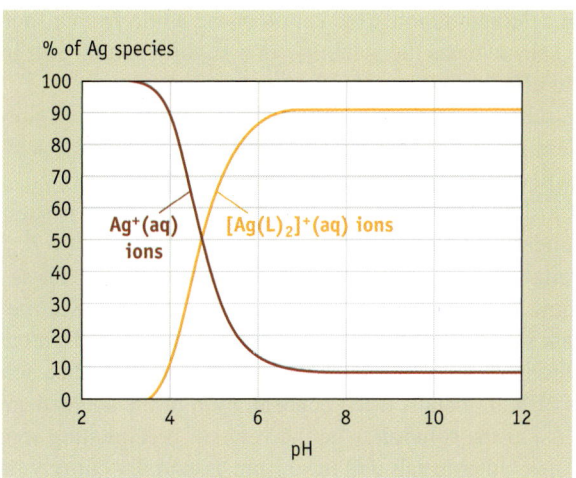

FIGURE 15.9 Modelled pH-dependent speciation between $Ag^+(aq)$ ions and $[Ag(L)_2]^+$ ions if pK_a for $HL^+(aq)$ ions is 6.00. The hypothetical $HL^+(aq)$ ion with $pK_a = 6.00$ is a stronger acid than $NH_4^+(aq)$ ions with $pK_a = 9.25$. The conjugate base, $L(aq)$, would be less competitive to "grab" H^+ ions than $NH_3(aq)$, so $[Ag(L)_2]^+$ complex ions would be the dominant species over a wider pH range (toward the low pH end) than is the case for $[Ag(NH_3)_2]^+$.

An important consequence of this competition between acid-base speciation and complexation with metal ions, concerning the bioavailability of copper ions, was discussed in Section 14.7.

EXERCISE 15.25—COMPLEXATION VERSUS LEWIS BASE PROTONATION

In aqueous solution, CN^- ions can form stable complex ions with $Ag^+(aq)$ ions. The weak acid HCN has $K_a = 3.5 \times 10^{-4}$. Discuss and explain how the relative amounts of $[Ag(CN)_2]^+(aq)$ and $Ag^+(aq)$ ions change as pH of solution is gradually increased from pH 1 to pH 13.

Metal ion complexes with a Lewis base that is also a Brønsted-Lowry weak base are less stable at lower pH because of competition of $H_3O^+(aq)$ ions for the base. In comparing Lewis bases, the stronger they are as Brønsted-Lowry bases (the higher the pK_a of the conjugate acid), the more strongly they retain the H^+ ion on the Lewis base binding site, and the higher the pH is before they release the H^+ ion and complexation occurs.

SUMMARY

Key Concepts

- Acidification of seawater causes re-distribution among carbonate species, reducing the concentration of $CO_3^{2-}(aq)$ ions, bringing seawater closer to saturation in $CaCO_3$, and making deposition of corals by marine organisms more difficult. (Section 15.1)
- Precipitation of slightly soluble ionic salts from aqueous solution can be viewed as the result of competition between attraction of the cations and anions for each other to form a solid and aquation of the ions. (Section 15.2)
- When a slightly soluble solid salt and its aquated ions are at equilibrium, the solution is said to be a **saturated solution**. In such systems, the equilibrium constant is called the **solubility product (K_{sp})**. A saturated solution is one in which $Q = K_{sp}$. (Section 15.2)
- The solubility of a slightly soluble salt at a given temperature can be deduced from its solubility product at that temperature, provided that neither the cation nor anion are involved in other reactions. Relative solubilities of different salts can be deduced from the relative values of the solubility products only when they are salts of the same type—all 1:1 salts or all 1:2 salts, for example. (Section 15.2)

- The presence of a "common ion" (the cation or the anion from a source other than the slightly soluble salt) reduces the solubility of a slightly soluble salt in water. This is the **common ion effect**. (Section 15.2)
- Slightly soluble salts whose anion is a weak base become more soluble the more acidic the solution is. The weaker the anion is as a base, the less acidic the solution in which the salt will dissolve. (Section 15.2)
- In a solution of a slightly soluble salt, if $Q = K_{sp}$, the solution is saturated. If $Q < K_{sp}$, the solution is under-saturated and more can dissolve, and if $Q > K_{sp}$, it is supersaturated and precipitation would bring the system to equilibrium. These statements say nothing about how fast dissolution or precipitation happens. (Section 15.3)
- When two metals salts with the same anion have sufficiently different solubilities, they may be effectively separated by **selective precipitation**. In a solution containing aquated cations of both metals, if the concentration of the aquated anion is raised incrementally, just as the solution is on the verge of precipitating some of the most soluble salt, the less soluble salt will have been essentially entirely removed from solution as precipitate that can be filtered off. (Section 15.3)
- Slightly soluble salts may dissolve as a result of reaction between the metal cation and a Lewis base to form a complex ion. This can be regarded as the result of two competing reactions for the cation: formation of a solid lattice and formation of the complex ion. Equilibrium constants for formation of complex ions from the metal ion and the Lewis base, called **formation constants** (K_f), are measures of stability of the complex ions. (Section 15.4)
- Metal ion complexes with a Lewis base that is also a Brønsted-Lowry weak base are less stable at lower pH because of competition of $H_3O^+(aq)$ ions for the base. In comparing Lewis bases, the stronger they are as Brønsted-Lowry bases (the higher the pK_a of the conjugate acid), the higher the pH is at which the Lewis base is protonated. (Section 15.5)

REVIEW QUESTIONS

Section 15.2: Solubility and Precipitation of Ionic Salts

15.25 Write a balanced equation for equilibrium of each of the following slightly soluble salts with their aquated ions in aqueous solution, write the expression for the reaction quotient (Q), and find in Appendix B the value of Q at equilibrium (the solubility product, K_{sp}) at 25 °C.
(a) PbSO(s)   (b) NiCO$_3$(s)   (c) Ag$_3$PO$_4$(s)

15.26 When some solid thallium(I) bromide is added to 1.00 L of water and stirred, only a little of the salt dissolves. Measurements show that the concentration of both Tl$^+$(aq) ions and Br$^-$(aq) ions in equilibrium with the solid salt at 25 °C is 1.9×10^{-3} mol L^{-1}. What is K_{sp} for TlBr at 25 °C?

15.27 Write a balanced equation for equilibrium between slightly soluble strontium fluoride, SrF$_2$(s), and its aquated ions in a saturated aqueous solution. In a saturated solution at 25 °C, $[Sr^{2+}] = 1.0 \times 10^{-3}$ mol L^{-1}. What is K_{sp} for SrF$_2$ at 25 °C?

15.28 At 25 °C, K_{sp} for radium sulfate (RaSO$_4$) is 4.2×10^{-11}. If 25 mg of radium sulfate is stirred in 100 mL of water at 25 °C, does all of it dissolve? If not, how much dissolves?

15.29 If 55 mg of lead(II) sulfate is stirred in 250 mL of water at 25 °C, does all of it dissolve? If not, how much dissolves?

Interactive Exercise 15.30

SUBMIT

Calculate K_{sp} from ion concentrations in saturated solution.

15.31 Use K_{sp} values to decide which compound in each of the following pairs is the more soluble:
(a) PbCl$_2$(s) or PbBr$_2$(s)
(b) HgS(s) or FeS(s)
(c) Fe(OH)$_2$(s) or Zn(OH)$_2$(s)

15.32 Calculate the solubility of Zn(CN)$_2$ at 25 °C (a) in pure water, and (b) in the presence of 0.10 mol L^{-1} Zn(NO$_3$)$_2$. At 25 °C, K_{sp} for Zn(CN)$_2$ is 8.0×10^{-12}.

Section 15.3: Precipitation Reactions

15.33 Can AgCl precipitate if you add 5.0 mL of 0.025 mol L^{-1} HCl solution to 100.0 mL of 0.0010 mol L^{-1} silver nitrate solution at 25 °C?

15.34 You have a solution at 25 °C that has a Pb^{2+}(aq) ion concentration of 0.0012 mol L^{-1}. If enough of a chloride salt is dissolved in the solution that the [Cl^-] is 0.010 mol L^{-1}, will $PbCl_2$(s) precipitate?

15.35 Sodium carbonate is dissolved in a solution in which the concentration of Ni^{2+}(aq) ions is 0.0024 mol L^{-1}. Can precipitation of $NiCO_3$(s) occur at 25 °C with the following concentrations of the carbonate ion?
(a) 1.0×10^{-6} mol L^{-1}
(b) [CO_3^{2-}] = 1.0×10^{-4} mol L^{-1}

15.36 The cations Ba^{2+} and Sr^{2+} can be precipitated as almost insoluble sulfates.
(a) If you incrementally add sodium sulfate to a solution containing these metal cations, each with a concentration of 0.10 mol L^{-1}, which is precipitated first, $BaSO_4$(s) or $SrSO_4$(s)?
(b) What will be the concentration of the first cation that precipitates (Ba^{2+} or Sr^{2+}) when the second, more soluble salt begins to precipitate?

15.37 Salts of Fe^{3+}, Pb^{2+}, and Al^{3+} ions are found in nature, and all are important economically. If you have a solution containing these three ions, each at a concentration of 0.10 mol L^{-1}, what is the order in which their hydroxides precipitate as aqueous NaOH solution is slowly added to the solution?

15.38 In principle, the ions Ba^{2+} and Ca^{2+} can be separated by the difference in solubility of their fluorides, BaF_2(s) and CaF_2(s). If you have a solution that is 0.10 mol L^{-1} in both Ba^{2+}(aq) and Ca^{2+}(aq) ions, and the concentration of F^-(aq) ions is gradually increased, CaF_2(s) will begin to precipitate first.
(a) What concentration of fluoride ions will precipitate the maximum amount of Ca^{2+} ion without precipitating BaF_2(s)?
(b) What concentration of Ca^{2+}(aq) ions remain in solution when BaF_2(s) just begins to precipitate?

Section 15.4: Solubility and Complexation: Competitive Equilibria

15.39 Solid gold(I) chloride, AuCl(s), dissolves in a solution containing an excess of cyanide ions as a result of formation of a water-soluble complex ion:

$$AuCl(s) + 2\ CN^-(aq) \rightleftharpoons [Au(CN)_2]^-(aq) + Cl^-(aq)$$

Show that this equation is the sum of two other equations, one for dissolving AuCl(s) and the other for the formation of the $[Au(CN)_2]^-$ ion from Au^+(aq) ions and CN^-(aq) ions. Calculate K_{net} for the overall reaction.

Section 15.5: Complexation versus Lewis Base Protonation

15.40 Ethylamine ($CH_3CH_3NH_2$) form complexes with Cu^{2+}(aq) ions. pK_a for the ethylammonium ion ($CH_3CH_3NH_2^+$) is 10.66. Discuss and explain how the relative amounts of $[Cu(CH_3CH_3NH_2)]^{2+}$(aq) and Ag^+(aq) ions change as pH of solution is gradually increased from pH 1 to pH 13.

SUMMARY AND CONCEPTUAL QUESTIONS

15.41 The pH of a saturated calcium hydroxide solution at 25 °C is measured to be 12.68. What is K_{sp} for $Ca(OH)_2$ at 25 °C?

15.42 Which of the following barium salts would you expect to dissolve in HCl solution: $Ba(OH)_2$(s), $BaSO_4$(s), or $BaCO_3$(s)?

15.43 A sample of hard water contains about 2.0×10^{-3} mol L^{-1} Ca^{2+}(aq) ions. A soluble fluoride-containing salt such as NaF is added to "fluoridate" the water (to aid in the prevention of dental caries). What is the maximum concentration of F^-(aq) ions that can be present without precipitating CaF_2(s)?

15.44 Suggest a method for separating a precipitate consisting of a mixture of CuS(s) and $Cu(OH)_2$(s).

With the **eCHACR** single-sign-on access card bundled with your text, log on (http://login.cengage.com/sso) and access the e-book and click on any in-margin icons for dynamic molecular-level animations and simulations, videos of laboratory reactions, interactive exercises, reviews of background concepts, and other online supplementary materials as noted by the icons in the text margins.

Also go to www.chemistry.nelson.com <http://www.chemistry.nelson.com> for Answers to in-chapter exercises and selected Review Questions, Test Yourself questions, weblinks, crossword puzzles, flashcards, glossary of key terms, and other student resources.

16 Electron Transfer Reactions and Electrochemistry

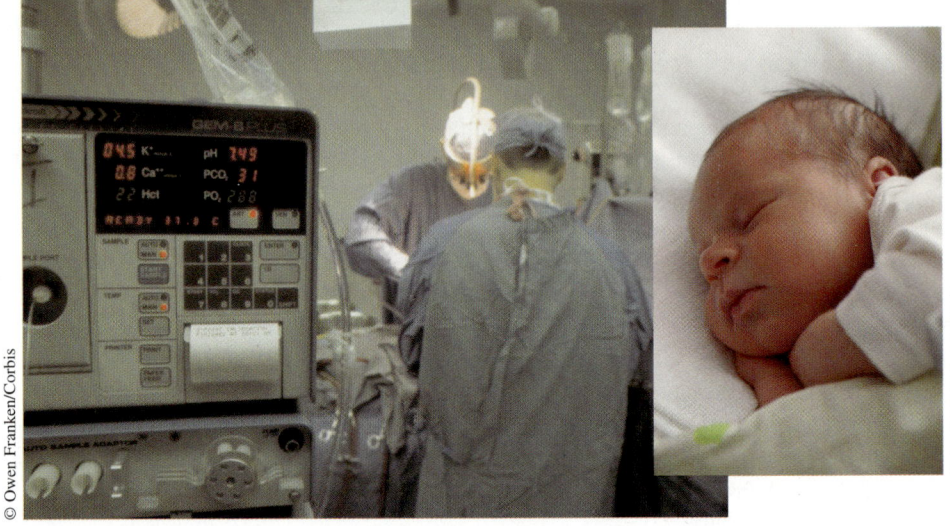

© Owen Franken/Corbis

Robert Milek/Shutterstock

16.1 Case Study: Blood Oxygen—How Much Is Too Much?

Blood exchanges oxygen and carbon dioxide in our lungs, so the blood in our arteries has more O_2 and less CO_2 than the blood in our veins. In response to certain symptoms, blood gas analysis (BGA) is often carried out for anesthetized patients undergoing surgery, or to monitor fetal blood condition. BGA usually provides estimates of, among other variables, blood pH and the partial pressures of O_2 and CO_2 exerted by the blood as it passes through our lungs. Analysis of arterial blood may indicate that not enough O_2 is being transported to the patient's lungs, that the patient is not getting rid of enough CO_2, or that there is a kidney malfunction.

The solubility of O_2 in water is quite low, and if it were the same in blood, we would not get enough O_2. Blood does provide us with enough O_2 because O_2 chemically combines with substances in it. O_2 transport to tissues is carried out mainly by hemoglobin in blood, and by myoglobin (Figure 16.1) in muscles. Both of these O_2 carriers are protein

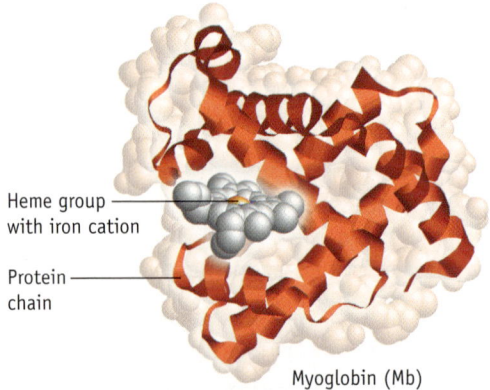

Heme group with iron cation

Protein chain

Myoglobin (Mb)

FIGURE 16.1 Myoglobin, the protein that binds O_2 in muscles. At the centre of the molecule is a heme group. An oxygen molecule binds to the iron ion in the centre of this group. When the protein is "oxygenated," it is red, giving muscle tissue its characteristic colour.

molecules that have a heme group, at the centre of which is an iron ion to which O_2 molecules bind.

Myoglobin has a stronger affinity than hemoglobin for O_2: most of the binding sites in myoglobin are taken up by O_2 at lower partial pressures than are necessary for "saturation" of the sites in hemoglobin (Figure 16.2). Under normal circumstances, hemoglobin becomes saturated with O_2 in the lungs, where the O_2 partial pressure is about 13 kPa. In tissue capillaries, however, the O_2 partial pressure drops, and hemoglobin releases O_2.

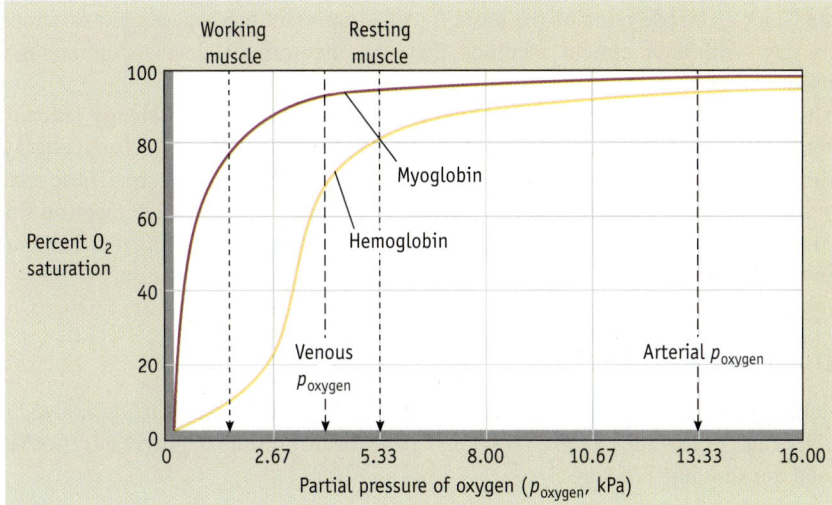

Molecular Modelling (Odyssey)

e16.1 Investigate the structure of a myoglobin molecule, with and without a bound O_2 molecule.

FIGURE 16.2 Binding of O_2 by myoglobin and hemoglobin at pH 7.4 in a normal adult. Interestingly, fetal blood has a low partial pressure of O_2—about 4 kPa, equivalent to living at an altitude of 8000 m.

Some of the released O_2 binds with myoglobin and is stored for use in times of severe oxygen deprivation, such as during strenuous exercise.

At high concentrations, oxygen is toxic; it can cause damage to the central nervous system, lungs, and eyes. For this reason, blood O_2 levels must be kept within certain limits, and monitoring blood O_2 levels is important, especially in newborn children.

Bio-sensor devices have been developed to measure blood O_2 levels. The most important of these are tiny *electrochemical cells*, which operate on the same principle as the batteries we use in electronic devices. They are composed of two compartments, one of which is a reference electrode and the other is an electrode in contact with the blood to be monitored.

> *... blood O_2 levels must be kept within certain limits, and monitoring blood O_2 levels is important, especially in newborn children. A number of devices, called bio-sensors, have been developed for measurement of blood O_2. The most important of these are tiny electrochemical cells ...*

In one type of electrochemical sensor, the cell voltage depends on the concentration of O_2 in the blood, according to a relationship called the *Nernst equation*, described in Section 16.5.

More common electrochemical sensors monitor a reaction of O_2 in a known volume of blood in the cell to estimate the amount (mol) of O_2 in the blood sample. An electric potential applied to the cell causes chemical change to occur, by an electron-forcing process called *electrolysis* [>>Section 16.7]. The applied potential brings about chemical reduction of dissolved oxygen in the blood sample:

$$O_2(aq) + 2\,H_2O\,(\ell) + 4\,e^- \longrightarrow 4\,OH^-(aq)$$

The device "counts" the number of electrons that move through the circuit and, from that value, determines the amount of dissolved O_2 in the sample. The equation for the reaction tells us that the amount of O_2 that reacts is one-quarter of the amount of electrons transferred. This instrument has been miniaturized to fit into the tip of a narrow catheter so that it can be inserted into an artery.

Measurement of O_2 concentration in aqueous solution using electrochemistry has been a well-established laboratory procedure for some time, but applying this technology to a mixture as complicated as blood did not work well because many other substances in blood interfered with the measurement. The key development that led to the new device was incredibly simple. Biochemist L.C. Clark, Jr., modified the laboratory device by covering the electrode with a polyethylene membrane. The membrane screens most of the components of blood from the electrode, but is permeable to oxygen, which passes through the membrane to the sensing electrode. Clark obtained a patent for his invention in 1956, and the Clark electrode remains the basis for this important medical diagnostic procedure to this day—although optical methods that can measure O_2 levels continuously are also used.

This chapter is about converting chemical energy into electrical energy (because we want portable and convenient sources of electricity), and converting electrical energy into chemical energy (because we want the substances produced in doing so). These transformations are possible with *oxidation-reduction reactions,* which involve electron transfer from one species to another. The transfer of electrons along a conducting medium (such as copper wire) is electricity.

16.2 Oxidation-Reduction Reactions

The defining characteristics of the class of chemical reaction called *oxidation-reduction reactions* (and sometimes *redox reactions*) have been presented previously [<<Section 6.7], and are summarized here:

- Reaction to form new substances is the result of transfer of electrons from one reactant species to another. The reactant species from which electrons are taken is said to be *oxidized*, and the species that gains the electrons is said to be *reduced*.
- Oxidation and reduction are necessarily simultaneous processes: one cannot happen without the other.
- The *oxidizing agent* (the species that brings about oxidation by taking electrons) is reduced, and the *reducing agent* (the species from which electrons are taken to bring about reduction of the other reactant) is oxidized.
- The amounts of oxidizing agent and reducing agent that react are such that the number of electrons taken from the reducing agent is the same as the number of electrons gained by the oxidizing agent.

These aspects of oxidation-reduction reactions are illustrated below for the reaction between copper metal and aquated silver ions (see also Figure 6.36):

Electrons are removed from Cu(s): it is oxidized to Cu^{2+} (aq) ions.
Cu(s) is the reducing agent.

$$Cu(s) + 2\,Ag^+(aq) \longrightarrow Cu^{2+}(aq) + 2\,Ag(s)$$

Ag^+(aq) ions gain electrons, and are reduced to Ag(s).
The Ag^+ (aq) ion is the oxidizing agent.

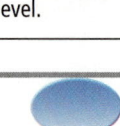

Molecular Modelling

e16.2 Watch an animation of oxidation and reduction processes at the molecular level.

Think about It

e16.3 Watch videos of other oxidation-reduction reactions.

The reaction can be imagined to consist of two components, called *half-reactions,* one for oxidation and one for reduction. These can be represented by two *half-equations,* the sum of which is the equation shown above:

Oxidation: $Cu(s) \longrightarrow Cu^{2+}(aq) + 2\,e^-$

Reduction: $2\,Ag^+(aq) + 2\,e^- \longrightarrow 2\,Ag(s)$

In this relatively straightforward example, we can easily deduce that oxidation and reduction have occurred, decide which species is oxidized and which is reduced, and balance the equation. This is not always the case, so chemists use the concept of *oxidation state* to help understand these reactions.

Oxidation and reduction occur simultaneously when electrons are transferred from one species to another. The species from which electrons are taken is oxidized (and is the reducing agent), and the species that takes the electrons is reduced (and is the oxidizing agent).

Oxidation State

How can you tell if a reaction is an oxidation-reduction reaction? How can you tell which substance has gained electrons (or had electrons taken from it) and so decide which substance is the oxidizing agent (or reducing agent)? Sometimes it is clear, but if not, then the best way is to look for a change in an element's **oxidation state**—which is *a measure of the degree of oxidation of an element in a compound, compared with when it is uncombined*. The higher the oxidation state of an element is, the greater its degree of oxidation. The oxidation state of an atom in a molecule or ion can be thought of as the charge that atom would have if all bonding electrons were imagined to belong to the more electronegative atom in each bond. A positive oxidation state means fewer electrons on the atom than in the uncombined element, and since it has "lost" electrons, it is in a higher oxidation state. *Oxidation is accompanied by an increase in the oxidation state of an element in the species, and reduction is accompanied by a decrease in the oxidation state of an element.*

The oxidation state of an element in any substance or species (molecule or ion) is assigned according to a set of guidelines:

1. *The oxidation state of each atom in a pure element is zero.* The oxidation state of Cu in metallic copper is 0, as it is for each atom in $O_2(g)$, $I_2(s)$, or $S_8(s)$.

2. *For a simple, monatomic ion, the oxidation state is equal to the charge on the ion.* For example, the oxidation state of magnesium atoms in Mg^{2+} ions is +2, consistent with removal of two electrons from an Mg atom to form an Mg^{2+} ion. In the oxide ion (O^{2-}), the oxygen atoms are in oxidation state −2.

3. *The algebraic sum of the oxidation states of the atoms in a neutral molecule is zero; in a polyatomic ion, the sum is equal to the charge on the ion.* To apply this rule, we need to take into account the number of atoms of each element in the molecule or ion. For example, for a water molecule (H_2O) and a dichromate ion ($Cr_2O_7^{2-}$) we would apply

   Charge on H_2O molecule = 0 = 2(oxidation state of H) + 1(oxidation state of O)

   Charge on $Cr_2O_7^{2-}$ ion = −2 = 2(oxidation state of Cr) + 7(oxidation state of O)

4. *In all of its compounds and ions, fluorine is in oxidation state −1.* Since F is the most electronegative element and only forms single bonds, attribution of bonding electrons to the most electronegative element always leaves F atoms with one more electron than in uncombined F atoms.

5. *In most compounds and ions, hydrogen is in oxidation state +1.* An exception is when hydrogen forms a binary compound with a metal; the metal forms a positive ion and each H atom becomes a hydride ion, H^-. So, in CaH_2 the oxidation state of Ca is +2 and that of hydrogen is −1.

6. *In most compounds, O is in oxidation state −2.* Exceptions are members of a class of compounds called peroxides, in which O is in oxidation state −1. For example, in hydrogen peroxide (H_2O_2), H is in oxidation state +1, and O is in oxidation state −1.

Bear in mind that oxidation states do not reflect the actual electric charge on an atom in a molecule or polyatomic ion. Oxidation states assume that the electrons of any bond in a molecule "belong" entirely to the more electronegative atom, so that the atoms have unit positive or negative charges, which is not accurate. For example, in H_2O, the H atoms are not H^+ ions, and the O atoms are not O^{2-} ions. This is not to say, however, that atoms in molecules do not bear an electric charge of any kind. In water, for example, dipole moments indicate the O atom has a charge of about −0.4 (or 40% of the charge of an electron) and the H atoms each have a charge of about +0.2. These are the charges that give rise to the polarity of water molecules [<<Section 6.3].

Think about It

e16.4 Work through a tutorial on assigning oxidation states.

So why use the concept of oxidation state? It is an artificial but useful way of dividing up the electrons among the atoms in a molecule or polyatomic ion. Because the division of electrons changes in an oxidation-reduction reaction, we use this method to decide whether this type of reaction has occurred, to distinguish the oxidizing and reducing agents, and to balance the reaction equation.

WORKED EXAMPLE 16.1—CALCULATING OXIDATION STATES

Determine the oxidation state of the indicated element in each of the following compounds or ions:
(a) aluminum in aluminum oxide (Al_2O_3)
(b) phosphorus in phosphoric acid (H_3PO_4)
(c) sulfur in the sulfate ion ($SO_4{}^{2-}$)
(d) each Cr atom in the dichromate ion ($Cr_2O_7{}^{2-}$)

Solution

(a) Al_2O_3 is a neutral compound (net charge = 0). Assume that O is in oxidation state −2, and use guideline 3 above.

$$\text{Charge on } Al_2O_3 = 0 = 2(\text{oxidation state of Al}) + 3(\text{oxidation state of O})$$
$$= 2(\text{oxidation state of Al}) + 3(-2)$$
$$\therefore \text{ Oxidation state of Al} = +3$$

(b) H_3PO_4 molecules have zero overall charge. Assume H and O are in oxidation states +1 and −2, respectively.

$$\text{Charge on } H_3PO_4 = 0 = 3(\text{oxidation state of H}) + (\text{oxidation state of P}) + 4(\text{oxidation state of O})$$
$$= 3(+1) + (\text{oxidation state of P}) + 4(-2)$$
$$\therefore \text{ Oxidation state of P} = +5$$

(c) The sulfate ion ($SO_4{}^{2-}$) has charge −2. Because this is not a peroxide, assume each O atom is in oxidation state −2.

$$\text{Charge on } SO_4{}^{2-} = -2 = (\text{oxidation state of S}) + 4(\text{oxidation state of O})$$
$$= (\text{oxidation state of S}) + 4(-2)$$
$$\therefore \text{ Oxidation state of S} = +6$$

(d) The net charge on $Cr_2O_7{}^{2-}$ ion is −2. Assume each O atom is in oxidation state −2.

$$\text{Charge on } Cr_2O_7{}^{2-} \text{ ion} = -2 = 2(\text{oxidation state of Cr}) + 7(\text{oxidation state of O})$$
$$= 2(\text{oxidation state of Cr}) + 7(-2)$$
$$\therefore \text{ Oxidation state of Cr} = +6$$

Interactive Exercise 16.1

SUBMIT

Assign the oxidation state to an element in a species from its chemical formula.

EXERCISE 16.2—CALCULATING OXIDATION STATES

Calculate the oxidation state of the bold atom in each ion or molecule:
(a) **Fe**$_2O_3$ (b) H$_2$**S**O$_4$ (c) **C**O$_3{}^{2-}$ (d) N**O**$_2{}^{+}$

The oxidation state of an element in a compound is a measure of the degree of oxidation, compared with compared with when it is uncombined. A set of guidelines can be used to calculate oxidation states.

Recognizing Oxidation and Reduction

You can tell whether a reaction involves oxidation and reduction by assessing the oxidation state of each element and noting whether any of them are different in the products from those in the reactants. In many cases, however, this analysis is not necessary. It is obvious that an oxidation-reduction reaction has occurred if an uncombined element is converted to a compound, or if the reaction involves a well-known oxidizing or reducing agent.

Like oxygen, $O_2(g)$, the halogens $F_2(g)$, $Cl_2(g)$, $Br_2(\ell)$, $I_2(s)$ are always oxidizing agents in their reactions with metals and non-metals. An example is the reaction of chlorine gas with sodium metal (Figure 16.3).

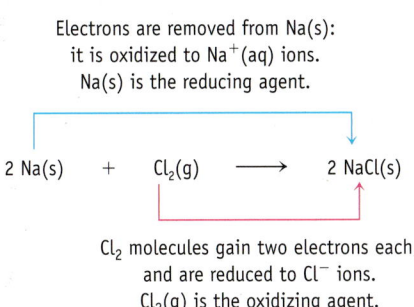

Electrons are removed from Na(s):
it is oxidized to $Na^+(aq)$ ions.
Na(s) is the reducing agent.

$$2\ Na(s)\ +\ Cl_2(g)\ \longrightarrow\ 2\ NaCl(s)$$

Cl_2 molecules gain two electrons each,
and are reduced to Cl^- ions.
$Cl_2(g)$ is the oxidizing agent.

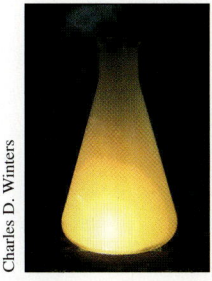

FIGURE 16.3 Sodium-chlorine reaction. Sodium metal reduces chlorine gas; chlorine gas oxidizes sodium metal.

Each chlorine molecule ends up as two Cl^- ions, having acquired two electrons (from two Na atoms). So, the oxidation state of each Cl atom has decreased from 0 to -1. This means $Cl_2(g)$ has been reduced and is the oxidizing agent. The oxidation state of the Na atoms increases from 0 to $+1$. Sodium metal has been oxidized and is the reducing agent.

Figure 16.4 shows the reaction of another excellent oxidizing agent—the aquated nitrate ions in concentrated nitric acid solution—with copper metal. The equation for the reaction is

NO_2 gas

Oxidation number of Cu changes from 0 to $+2$. Cu
is oxidized to Cu^{2+} and is the reducing agent.

$$Cu(s)\ +\ 2\ NO_3^-(aq)\ +\ 4\ H^+(aq)\ \longrightarrow\ Cu^{2+}(aq)\ +\ 2\ NO_2(g)\ +\ 2\ H_2O(\ell)$$

N in NO_3^- changes from $+5$ to $+4$ in NO_2. NO_3^-
is reduced to NO_2 and is the oxidizing agent.

Copper metal oxidized to blue-green
$Cu^{2+}(aq)$ ions

FIGURE 16.4 The reaction of copper with nitric acid solution. Copper (a reducing agent) reacts vigorously with $NO_3^-(aq)$ ions in concentrated nitric acid solution, an oxidizing agent. $NO_3^-(aq)$ ions are reduced to the brown gas $NO_2(g)$, and Cu(s) is oxidized to blue-green $Cu^{2+}(aq)$ ions.

The symbols $H^+(aq)$ and $H_3O^+(aq)$ are alternatives for the same continuously changing species [<<Section 6.6]. In Chapter 14, we commonly used $H_3O^+(aq)$ to emphasize the idea of acid-base reactions as proton transfer. In this chapter, the equations for oxidation-reduction reactions are simpler if we use $H^+(aq)$ rather than $H_3O^+(aq)$, and this is done without sacrificing meaning.

The oxidation state of N atoms has been decreased from $+5$ (in the $NO_3^-(aq)$ ion) to $+4$ (in $NO_2(g)$), so aquated nitrate ions are the oxidizing agent. Copper metal is the reducing agent: from each Cu atom two electrons are taken, forming a $Cu^{2+}(aq)$ ion.

Although the oxidation state of N atoms decreases during this reaction, it is not correct to say that nitrogen has been reduced. The aquated NO_3^- ion is the species reduced, and is the species that interacts with, and brings about oxidation of, the copper metal. The electrons from Cu metal are taken up by the entire $NO_3^-(aq)$ ion species, and not just by the N atoms in that species.

In the above reactions of metals with chlorine and nitric acid solution, the metals are oxidized and are reducing agents. This is typical behaviour of many metals, especially those of Group 1 and Group 2. An example is the reduction of water to hydrogen gas by potassium:

$$2\ K(s)\ +\ 2\ H_2O(\ell)\ \longrightarrow\ 2\ K^+(aq)\ +\ 2\ OH^-(aq)\ +\ H_2(g)$$

Think about It

e16.5 Compare the reactions of copper and zinc metals with nitric acid solutions.

Interactive Exercise 16.3

SUBMIT

Use oxidation states to identify oxidation and reduction processes.

FIGURE 16.5 The thermite reaction. So much heat is evolved that the iron produced is molten. This reaction has been used for welding section of railway track together, especially in remote locations. The molten iron is run between the sections, and on cooling holds them together.

Aluminum can reduce iron(III) oxide to iron metal in the "thermite reaction" (Figure 16.5):

$$Fe_2O_3(s) + 2\,Al(s) \longrightarrow 2\,Fe(\ell) + Al_2O_3(s)$$

WORKED EXAMPLE 16.2—RECOGNIZING OXIDATION AND REDUCTION

For the reaction of aquated iron(II) ions with permanganate ions in acidic solution, decide which atoms undergo a change in oxidation state and identify the oxidizing and reducing agents:

$$5\,Fe^{2+}(aq) + MnO_4^-(aq) + 8\,H^+(aq) \longrightarrow 5\,Fe^{3+}(aq) + Mn^{2+}(aq) + 4\,H_2O(\ell)$$

Strategy

Determine the oxidation states of the atoms in each molecule or ion. An increase of oxidation state of an element indicates oxidation, and a decrease indicates reduction.

Solution

The oxidation state of Mn in MnO_4^-(aq) ions is +7, and in Mn^{2+}(aq) ions is +2. So, the MnO_4^- (aq) ions have been reduced and are the oxidizing agent. The oxidation state of Fe increases from +2 in Fe^{2+}(aq) ions to +3 in Fe^{3+}(aq) ions. The Fe^{2+}(aq) ions are oxidized and are the reducing agent in this reaction.

EXERCISE 16.4—RECOGNIZING OXIDATION AND REDUCTION

Reaction between ethanol and aquated dichromate ions in acidified aqueous solution occurs in a device that has been used in breath tests for alcohol. Identify the oxidizing and reducing agents, and the substances oxidized and reduced.

$$3\,C_2H_5OH(aq) + 2\,Cr_2O_7^{2-}(aq) + 16\,H^+(aq) \longrightarrow 3\,CH_3COOH(aq) + 4\,Cr^{3+}(aq) + 11\,H_2O(\ell)$$

The reaction between aquated dichromate ions and ethanol is the basis of a breath test used in a Breathalyzer. The higher the alcohol level, the more green Cr^{3+}(aq) ions are formed. The bottom photo shows a home breath-test kit.

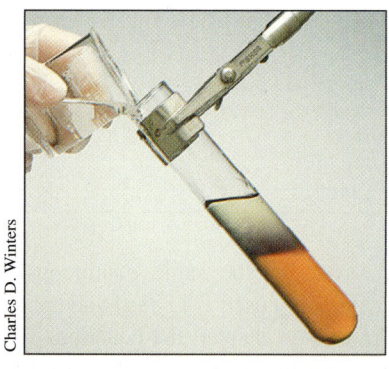

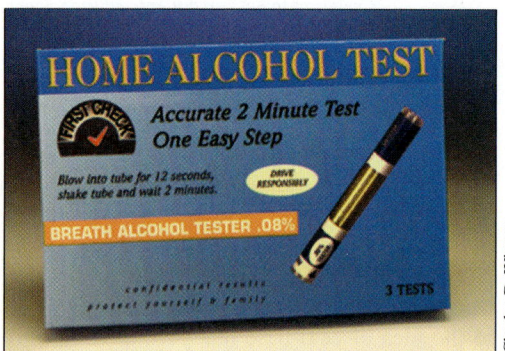

Background Concepts

e16.6 Review how to balance oxidation-reduction reactions.

The balanced chemical reaction for a process will not always be provided for you, so it is important to be able to balance equations when you know what the reactants and products are, as outlined in Background Concepts e16.6.

> Oxidation is accompanied by an increase in the oxidation state of an element in the species, and reduction by a decrease.

Oxidation-Reduction Reactions as Competition

Oxidation-reduction reactions can be seen as the result of competition between species to "grab" electrons: species differ in their ability to compete for electrons, and in a reaction mixture the more powerful competitor "wins" the competition, takes electrons from the other reactant (which is oxidized), and is reduced. You will see in Section 16.4 that species are ranked in their ability to compete for electrons on a scale of *reduction potentials*.

The use of the terms *grab* and *win* might suggest desire and intent on the part of atoms, molecules, and ions (to attribute such intent to inanimate objects is called *anthropomorphism*). They do not, of course, have such feelings of competitive spirit or intent, but all species resist removal of their own electrons (ionization energy) and most have an attraction for the electrons of other species in their immediate environment. The use of terms that imply competitiveness for electrons can induce powerful images at the molecular level of chemical reactivity.

In the reaction between copper metal and aquated silver ions (above), Ag^+(aq) ions compete more strongly for electrons than does Cu(s) metal. And because the reaction does not go in the opposite direction, we can conclude that Cu^{2+}(aq) ions cannot take electrons from Ag(s) metal.

Chemists say that the reaction in the direction represented by

$$Cu(s) + 2\,Ag^+(aq) \longrightarrow Cu^{2+}(aq) + 2\,Ag(s)$$

is the direction of *spontaneous reaction*—not because it is fast or instantaneous (which it is not), but because reaction has a natural tendency to occur in that direction. More correctly, this is the direction of reaction that takes the reaction mixture toward the equilibrium condition $Q = K$ [<<Sections 5.4, 13.3]. The equilibrium constant here is very large ($K \approx 10^{34}$), so this reaction goes essentially "to completion." How far a reaction goes before equilibrium is achieved is a measure of the difference between the abilities of the reactant species to compete for electrons. We will see in Section 16.5 that the reduction potentials of species (their abilities to compete for electrons) depend on their concentrations.

In voltaic cells, such as household batteries, spontaneous oxidation-reduction reactions occur and make energy available to us [>>Section 16.3]. Reactions in the non-spontaneous direction can happen only if we expend some energy on the reaction mixture—such as by applying an electric potential to drive an electric current in electrolysis [>>Section 16.7].

Chapter 17 discusses a fundamental theory, called thermodynamics, about the spontaneous direction of reaction in terms of change of Gibbs free energy accompanying a reaction at specified conditions. This theory applies to all reactions—not just those occurring in electrochemical cells.

> Oxidation-reduction reactions can be regarded as the result of competition between species for electrons: the "winner" is reduced and oxidizes the other species. The outcome of competition determines the direction of spontaneous reaction: the direction of reaction that takes the reaction mixture toward equilibrium.

In our everyday use of language, the word *spontaneous* means immediate, instantaneous, or very rapidly. In chemistry, this word has a special meaning: the direction of reaction that would take a reaction mixture closer to equilibrium, regardless of how quickly or slowly this happens [>>Section 17.2].

16.3 Voltaic Cells: Electricity from Chemical Change

As the name suggests, **electrochemistry** is *a field of study about the interaction between electricity and chemistry*—both spontaneous chemical reaction producing electricity and electricity forcing non-spontaneous reactions to occur. In laboratory and industrial situations, these transformations are usually made to occur in *electrochemical cells*, which are of two types:

- **Voltaic cells** or *galvanic cells,* which are *an arrangement that directs the transfer of electrons in a spontaneous oxidation-reduction reaction from one compartment of the cell, through a conductor, to another compartment.* The two compartments are called half-cells. In this way, substances (the materials oxidized and reduced) are sacrificed, and electrical energy is harnessed. Voltaic cells are the subject of this and following sections.
- **Electrolysis cells** (or electrolytic cells), in which *the application of an electrical potential, and the carrying of an electrical current (electrons), forces a non-spontaneous oxidation-reduction reaction to occur. This forced non-spontaneous process is called* **electrolysis**. During electrolysis, electrical energy is expended in order to make substances which may be otherwise difficult to make. Electrolysis is discussed in Section 16.7.

An example is the electrolysis of water (Figure 16.13), in which water is transformed into $H_2(g)$ and $O_2(g)$. Electrolysis is also used to electroplate one metal onto another (Figure 16.13b), to obtain reactive metals such as $Al(s)$ and $Cl_2(g)$ from their compounds, and to make important substances such as chlorine on an industrial scale [>>Section 26.11].

Electrolysis cells

Electrical energy → Chemical energy

Voltaic cells

The name *voltaic cell* recognizes the developmental research of Count Alessandro Volta (1745–1827). The alternative name *galvanic cell* honours Luigi Galvani (1737–1798).

Background Concepts

e16.7 Read about Galvani's experiments that led him to the idea of "animal electricity," and Volta's experiments to explain the phenomenon.

If we put a piece of copper metal into an aqueous silver nitrate solution, over time metallic silver is deposited and the solution takes on the blue-green colour typical of aquated Cu^{2+} ions (Figure 6.36 and e16.2):

$$Cu(s) + 2\ Ag^+(aq) \longrightarrow Cu^{2+}(aq) + 2\ Ag(s)$$

Carried out in this way, this spontaneous reaction releases energy as heat, which dissipates and is wasted. If, however, the reaction occurs in an apparatus (a voltaic cell) that allows electrons to be transferred from a reactant in one half-cell through an electrical circuit to the other reactant in another half-cell, we will have "harnessed" the available energy as electrical energy (Figure 16.6). The movement of electrons through the circuit constitutes an electric current that can be used to operate an LED, a computer, an MP3 player, or an electric motor.

FIGURE 16.6 A voltaic cell based on the reaction between copper metal and aquated silver ions. One half-cell contains copper metal dipping into a solution containing $Cu^{2+}(aq)$ ions, and the other has silver metal dipping into a solution containing $Ag^+(aq)$ ions. Electrons flow through the external circuit from the half-cell where $Cu(s)$ is oxidized (the anode) to the half-cell where $Ag^+(aq)$ ions are reduced (the cathode). In the salt bridge, which contains aqueous $NaNO_3$ solution, negative $NO_3^-(aq)$ ions migrate toward the anode, and positive $Na^+(aq)$ ions migrate toward the cathode. When $[Cu^{2+}] =$ 1.0 mol L^{-1} and $[Ag^+] = 1.0$ mol L^{-1} and the cell is at 25 °C, a cell potential of 0.46 V is generated. In battery-operated devices, the voltmeter is replaced by devices that operate by using electrical energy.

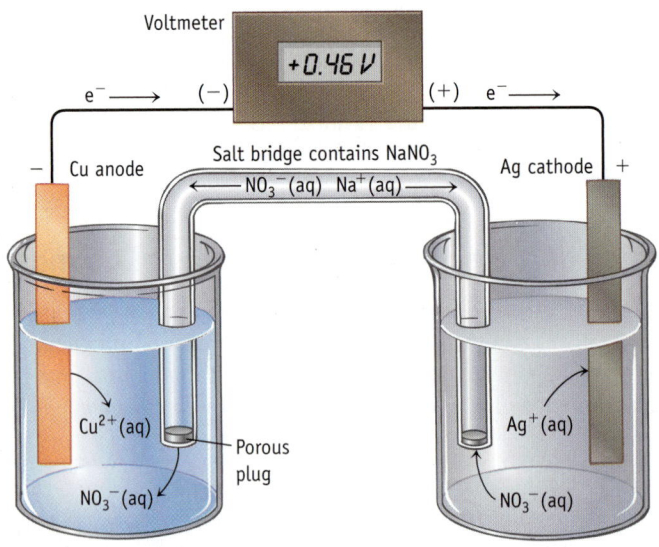

Net reaction: $Cu(s) + 2\ Ag^+(aq) \longrightarrow Cu^{2+}(aq) + 2\ Ag(s)$

The essential elements of a voltaic cell are the following:

- The species to be oxidized is in one compartment (called a *half-cell*), and the species to be reduced is the other.
- Strips or rods of conducting materials (called *electrodes*) dip into each half-cell to conduct electrons either out of or into the half-cell. The oxidation and reduction processes take place on the surfaces of the electrodes. The electrodes are joined by a

conducting material, such as copper wire (called the *external circuit*), through which electrons are transferred from the electrode where oxidation occurs to the one where reduction occurs.
- The electrode at which oxidation occurs is called the **anode**, and that at which reduction occurs is the **cathode**.
- Electrons flow in the external circuit from the anode to the cathode.
- According to international convention, the anode is labelled with a (−) sign and the cathode with a (+) sign.
- The two half-cells are connected with a *salt bridge* that allows cations and anions to move between the two half-cells (the *internal circuit*) so that there is no build-up of electrical charge in the half-cells. The electrolyte used in the salt bridge is chosen so that its ions do not react with the reagents in either half-cell.

In the cell shown in Figure 16.6, the copper metal strip that is oxidized also acts as an electrode (anode), providing a path for electrons taken from the copper to be transferred to the silver electrode (cathode), where electrons reduce aquated silver ions in solution near the surface of the electrode. To balance the number of electrons removed from Cu metal at the anode and gained by $Ag^+(aq)$ ions at the cathode, for every Cu atom oxidized at the anode, two $Ag^+(aq)$ ions are reduced at the cathode.

In the absence of a salt bridge, voltaic cell reactions would not proceed because there would be a build-up of positive charge (due to formation of $Cu^{2+}(aq)$ ions) in the anode half-cell and of negative charge (due to removal of $Ag^+(aq)$ ions) in the cathode half-cell. In the salt bridge, anions move toward the anode, and cations move toward the cathode (Figure 16.6). As $Cu^{2+}(aq)$ ions are formed at the anode, negative ions enter that half-cell from the salt bridge and positive ions leave, so that the overall charge in the half-cell remains zero. Similarly, as $Ag^+(aq)$ ions are reduced at the cathode, positive ions enter from the salt bridge and negative ions exit.

In a voltaic cell, the negatively charged electrons and anions move in a direction that completes a "circle": electrons move from anode to cathode in the external circuit, and anions move from cathode to anode in the internal circuit (the salt bridge). *Electrons do not travel through the internal circuit.*

A generalized representation of voltaic acid processes and terminology is shown in Figure 16.7.

A simple salt bridge can be made by adding gelatin to a solution of an electrolyte. Gelatin makes the contents semi-rigid (jelly-like) so that the solutions in the half-cells are unable to mix, and yet allows ions to move from one half-cell to the other through the salt bridge. Porous glass disks and permeable membranes are alternatives to a salt bridge.

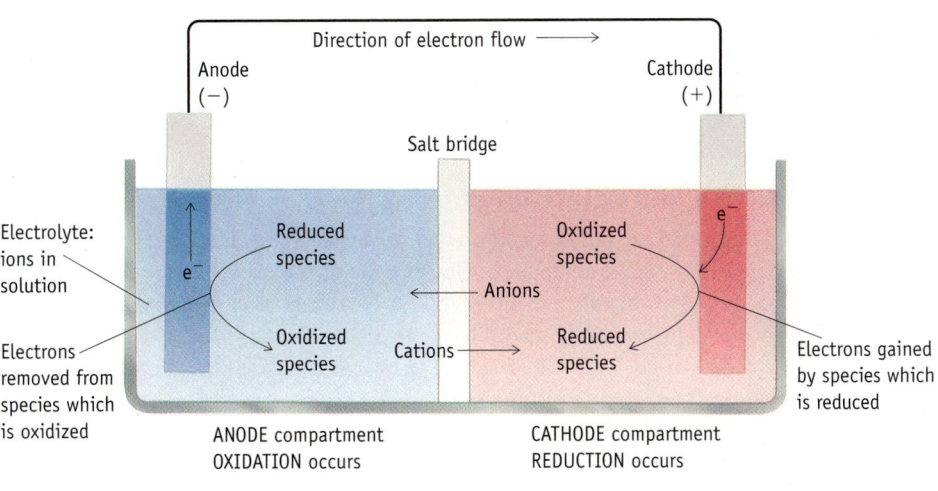

FIGURE 16.7 Summary of voltaic cell processes and terms. Electrons move from the anode, the site of oxidation, through the external circuit to the cathode, the site of reduction. Charge balance in each half-cell is achieved by migration of ions through the salt bridge. Anions move toward the anode, and cations move toward the cathode.

WORKED EXAMPLE 16.3—VOLTAIC CELLS

Describe how to set up a voltaic cell to generate an electric current using the reaction represented by

$$Fe(s) + Cu^{2+}(aq) \longrightarrow Cu(s) + Fe^{2+}(aq)$$

Think about It
e16.8 Examine the components of a voltaic cell.

Which electrode is the anode and which is the cathode? In which direction do electrons flow in the external circuit? In which direction do cations and anions flow in the salt bridge? Write equations for the half-reactions that occur at each electrode.

Strategy

First, identify the two half-cells that make up the cell. Then decide in which half-cell oxidation occurs and in which reduction occurs.

Solution

This voltaic cell is similar to the one portrayed in Figure 16.6. One half-cell contains an iron electrode in a solution of an iron(II) salt such as $Fe(NO_3)_2$. The other half-cell contains a copper electrode in a solution of a copper(II) salt such as $Cu(NO_3)_2$. The two half-cells are linked with a salt bridge containing an electrolyte such as KNO_3. The reaction equation tells us that iron is oxidized, so the iron electrode is the anode. Because copper(II) ions are reduced, the copper electrode is the cathode.

$$\text{Oxidation, anode:} \quad Fe(s) \longrightarrow Fe^{2+}(aq) + 2\,e^-$$
$$\text{Reduction, cathode:} \quad Cu^{2+}(aq) + 2\,e^- \longrightarrow Cu(s)$$

In the external circuit, electrons flow from the iron electrode (anode) to the copper electrode (cathode). In the salt bridge, anions flow toward the half-cell containing $Fe(s)$ and $Fe^{2+}(aq)$ ions and cations flow in the opposite direction.

Interactive Exercise 16.5

SUBMIT

Identify the processes at the anode and cathode in a voltaic cell.

EXERCISE 16.6—VOLTAIC CELLS

Describe how to set up a voltaic cell based on the following half-reactions:

$$\text{Reduction half-reaction:} \quad Ag^+(aq) + e^- \longrightarrow Ag(s)$$
$$\text{Oxidation half-reaction:} \quad Ni(s) \longrightarrow Ni^{2+}(aq) + 2\,e^-$$

Which is the anode and which is the cathode? What is the overall cell reaction? What is the direction of electron flow in an external wire connecting the two electrodes? Describe the ion flow in a salt bridge (with $NaNO_3$) connecting the cell compartments.

In voltaic cells, the electrons transferred during a spontaneous reaction are harnessed as an electrical current. The oxidation half-reaction occurs at the anode, and the reduction half-reaction occurs cathode. The electrodes in the half-cells are connected by an external circuit (the flow of electrons) and an internal circuit (a salt bridge for transfer of ions to maintain charge neutrality).

Voltaic Cells with Inert Electrodes

In the half-cells described so far, the metal used as an electrode is also a reactant or a product in the oxidation-reduction reaction. However, with the exception of carbon in the form of graphite, most non-metals are unsuitable as electrode materials because they do not conduct electricity. In half-cells where a metal is not a reactant or product, an *inert* (unreactive) electrode is used. These are made of materials that conduct an electric current but are neither oxidized nor reduced in the cell.

Consider a voltaic cell based on the following product-favoured reaction:

$$2\,Fe^{3+}(aq) + H_2(g) \longrightarrow 2\,Fe^{2+}(aq) + 2\,H^+(aq)$$

None of the reactants and products can be used as electrode materials, so inert electrodes are placed into the half-cells to conduct electrons in or out. Graphite is commonly used: it is a conductor of electricity, inexpensive, and does not oxidize under the conditions encountered in most cells. Platinum and gold metals are also used in laboratory experiments because both are chemically inert under most circumstances. They are, however, costly.

The *hydrogen electrode* (Figure 16.8) is particularly important in the field of electrochemistry because it is used as a reference in assigning cell voltages, as discussed in Section 16.4. Depending which half-cell the hydrogen electrode is connected to in a voltaic cell, it may be the anode compartment, where the reaction on the platinum surface is described by the half-equation

$$H_2(g) \longrightarrow 2\,H^+(aq) + 2\,e^-$$

and the electrons are conducted away to the cathode through the platinum electrode. Alternatively, it may be the cathode, in which case electrons are conducted into the half-cell through the platinum electrode, on the surface of which reduction takes place:

$$2\,H^+(aq) + 2\,e^- \longrightarrow H_2(g)$$

A half-cell involving transformations between $Fe^{3+}(aq)$ and $Fe^{2+}(aq)$ ions can be constructed with a platinum electrode dipping into a solution that contains iron ions in the two oxidation states. Electrons are transferred from the electrode surface to or from the reactant species, depending on whether the half-cell is the anode or cathode compartment in the voltaic cell.

A voltaic cell in which $Fe^{3+}(aq)$ ions are reduced to $Fe^{2+}(aq)$ ions by H_2 gas is depicted in Figure 16.9. In this cell, the hydrogen electrode is the anode, and the compartment containing $Fe^{3+}(aq)$ and $Fe^{2+}(aq)$ ions is the cathode.

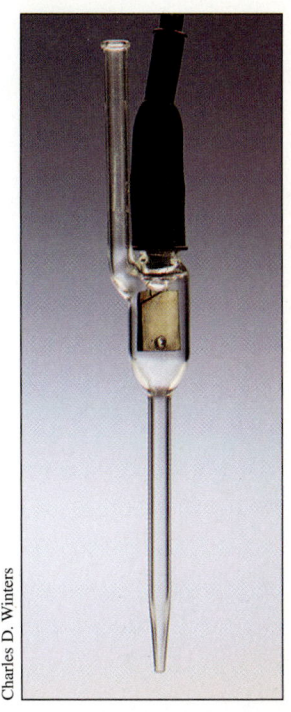

Charles D. Winters

FIGURE 16.8 Hydrogen electrode. Hydrogen gas is bubbled over a platinum electrode in an acidic aqueous solution. These electrodes function best if they have a large surface area to maximize contact between the gas and the electrode surface. Often platinum wires are woven into a gauze or the metal surface is roughened to increase the surface area.

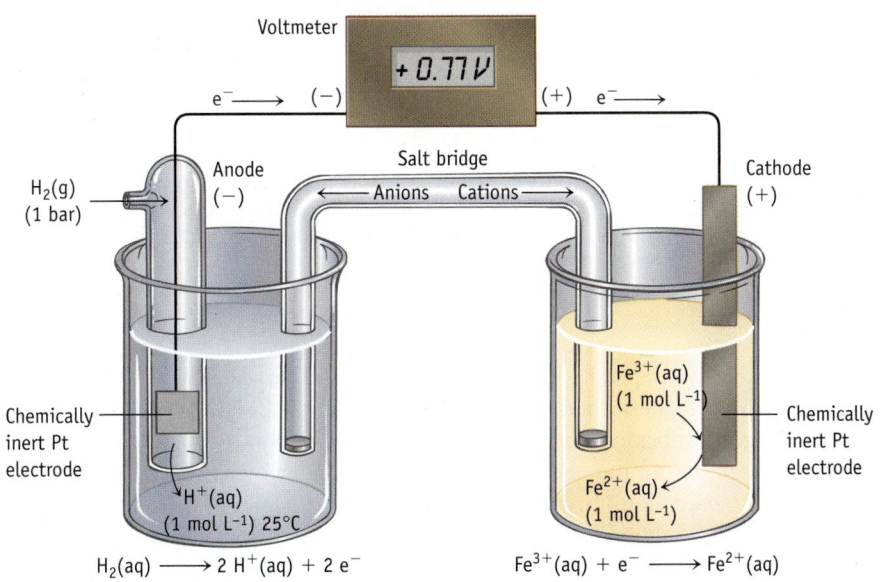

FIGURE 16.9 A voltaic cell with inert electrodes. This cell has $Fe^{3+}(aq)$ ions and $Fe^{2+}(aq)$ ions, both at 1.0 mol L^{-1} in the cathode compartment, and $H_2(g)$ at 1 bar pressure and $H^+(aq)$ ions at 1.0 mol L^{-1} in the anode compartment. At 25 °C, the electric potential between the cathode and electrode is +0.77 V.

> If the components of a half-cell are gases, liquids, or dissolved species, electrons can be transferred to or from the external circuit through inert electrodes—that is, electrodes made of conducting materials that are too unreactive to be oxidized or reduced.

Electrochemical Cell Conventions

Chemists use a shorthand notation to refer to half-cells and cells. When referring to a half-cell in isolation, the following notation, in which the single vertical line indicates a phase boundary or surface, is used:

$$\text{oxidized species} \mid \text{reduced species}$$

A half-cell containing copper metal dipping into a solution with $[Cu^{2+}] = 1.0$ mol L^{-1} is written

$$Cu^{2+}(aq, 1.0\ mol\ L^{-1})|Cu(s)$$

and the anode half-cell in Figure 16.9 is represented as

$$H^+(aq, 1.0\ mol\ L^{-1})|H_2(g, 1.0\ bar)|Pt$$

The notation used for cells, if we know which half-cell is the anode and which is the cathode, is to write the anode half-cell notation on the left and the cathode half-cell notation on the right, separated by the symbol ‖ to represent the salt bridge. The anode half-cell representation is reversed to imply oxidation. The voltaic cell shown in Figure 16.6 is written as

$$Cu(s)|Cu^{2+}(aq, 1.0\ mol\ L^{-1})‖Ag^+(aq, 1.0\ mol\ L^{-1})|Ag(s)$$

and the notation used to represent the cell portrayed in Figure 16.9 is

$$Pt|H_2(g, 1.0\ bar)|H^+(aq, 1.0\ mol\ L^{-1})‖Fe^{3+}(aq, 1.0\ mol\ L^{-1}), Fe^{2+}(aq, 1.0\ mol\ L^{-1})|Pt$$

EXERCISE 16.7—ELECTROCHEMICAL CELL CONVENTIONS

Draw the cells, showing electron movement and ion movements, represented by the following:
(a) $Zn(s)|Zn^{2+}(aq)‖Ni^{2+}(aq)|Ni(s)$
(b) $C(s)|Fe^{2+}(aq)|Fe^{3+}(aq)‖Ag^+(aq)|Ag(s)$
(c) $Mg(s)|Mg^{2+}(aq)‖Br_2(\ell)‖Br^-(aq)|Pt(s)$

The convention for representing isolated half-cells places the oxidized species on the left. A vertical line represents a boundary between phases. For cells, if the identity of the anode is known, it is written on the left with order of species reversed. Double vertical lines represent the salt bridge.

Batteries

The word *battery* technically refers to a collection of two or more voltaic cells. For example, the 12-volt battery used in cars is made up of six voltaic cells connected in series. Each voltaic cell develops a voltage of 2 volts.

The **batteries** that we depend on as portable energy sources in our everyday lives are voltaic cells. The cells described above have limited practical use as batteries: they are neither compact nor robust. In most situations, it is also important that the cell produces a constant voltage, but in the cells described so far, the voltage decreases as the reaction proceeds toward equilibrium [>>Section 16.6]. Also, the currents are small. Attempting to draw a large current results in a drop in voltage because the current depends on how fast ions in solution migrate to the electrode. Ion concentrations near the electrode become depleted if current is drawn rapidly, resulting in a decline in voltage.

The amount of electricity that can be drawn from a voltaic cell depends on the quantity of reagents available for reaction. A voltaic cell must have a large mass of reactants to produce current over a prolonged period. In addition, a voltaic cell that can be recharged is attractive. Recharging a cell means replenishing the reagents that were oxidized or reduced by electrolysis [>>Section 16.7]. In the cells described so far, the movement of ions in the cell during cell operation mixes the reagents and they cannot be "unmixed."

Batteries can be classified as primary and secondary. *Primary batteries* cannot be returned to their original state by recharging, so when the reactants are consumed, these batteries are "dead" and are discarded. In *secondary batteries*, also called *storage batteries* or *rechargeable batteries*, the reactions can be reversed, and the batteries can be re-used.

e16.9 gives some details about how fuel cells and nickel-metal hydride batteries work and describes other common voltaic cells, such as the alkaline and lead acid storage battery.

Taking It Further

e16.9 Read about commercial voltaic cells and the challenges involved.

16.4 Cell emf, and Half-Cell Reduction Potentials

In this section we discuss measurement of the voltage or "driving force" for electron flow, generated in voltaic cells, and consider this from the perspective of competition for electrons. A quantitative measure of "competitiveness" for electrons is called *half-cell reduction potential*.

Cell emf, Competition for Electrons

In a voltaic cell, electrons flow in the external circuit from the anode to the cathode. If we gradually increase an electric potential in the direction opposite to the electron flow, at some point the electron flow will be stopped. The applied potential required to stop electron flow is called the *cell electromotive force*, or **cell emf** (symbolized E_{cell}), for which the literal meaning is "force causing electrons to move." Cell emf has units of volts (V).

We can regard the cell emf as a measure of the difference between the abilities of the two half-cells to compete for electrons (that is, the difference between the *potentials* of the two half-cells to attract electrons). Each type of voltaic cell has a characteristic cell emf: 1.5 V for an alkaline cell, about 1.25 V for a Ni-Cd battery, and about 2.0 V for each of the cells in a lead storage battery.

For example, measurement shows that the cell

$$Zn(s)|Zn^{2+}(aq, 1.0 \text{ mol L}^{-1})\|Cu^{2+}(aq, 1.0 \text{ mol L}^{-1})|Cu(s)$$

has $E_{cell} = 1.10$ V. This means that the potential of the $Cu^{2+}(aq)|Cu(s)$ half-cell (the cathode) to attract electrons is 1.10 V greater than that of the $Zn^{2+}(aq)|Zn(s)$ half-cell (the anode). It is not difficult to show experimentally that, for a voltaic cell with particular reagents, the cell emf depends on the concentrations of species in solution, the pressure of any gaseous reagents, and the temperature. For purposes of comparison, we measure cell emf values under *standard conditions*, which means that all reactants and products are in their *standard states* [<< Section 7.6] defined as follows:

- Solutes in aqueous solution have a concentration of 1.0 mol L^{-1}.
- Gaseous substances have a pressure of 1.0 bar.
- Solids and liquids (not solutions) are pure.

A cell emf measured under these conditions is called the **standard cell emf** and is denoted by $E°_{cell}$.

Experimentally, it is found that $E°_{cell}$ values are additive—just like distances along a straight line, or amounts of money. To illustrate, listed here are typical experimental values of the standard cell emf values at 25 °C for two simple voltaic cells:

$$Zn(s)|Zn^{2+}(aq, 1.0 \text{ mol L}^{-1})\|Cu^{2+}(aq, 1.0 \text{ mol L}^{-1})|Cu(s) \qquad E°_{cell} = 1.10 \text{ V}$$
$$Cu(s)|Cu^{2+}(aq, 1.0 \text{ mol L}^{-1})\|Ag^{+}(aq, 1.0 \text{ mol L}^{-1})|Ag(s) \qquad E°_{cell} = 0.46 \text{ V}$$

These measurements lead to the following interpretations:

- The potential of a $Cu^{2+}(aq, 1.0 \text{ mol L}^{-1})|Cu(s)$ half-cell to attract electrons is 1.10 V greater than that of a $Zn^{2+}(aq, 1.0 \text{ mol L}^{-1})|Zn(s)$ half-cell.
- The potential of a $Ag^{+}(aq, 1.0 \text{ mol L}^{-1})|Ag(s)$ half-cell to attract electrons is 0.46 V greater than that of a $Cu^{2+}(aq, 1.0 \text{ mol L}^{-1})|Cu(s)$ half-cell.

This suggests that the potential of a $Ag^{+}(aq, 1.0 \text{ mol L}^{-1})|Ag(s)$ half-cell to attract electrons is 1.56 V (1.10 V + 0.46 V) greater than that of a $Zn^{2+}(aq, 1.0 \text{ mol L}^{-1})|Zn(s)$ half-cell. Indeed, experimentally we find

$$Zn(s)|Zn^{2+}(aq, 1.0 \text{ mol L}^{-1})\|Ag^{+}(aq, 1.0 \text{ mol L}^{-1})|Ag(s) \qquad E°_{cell} = 1.56 \text{ V}$$

This additivity of cell emf values gives us a means of quantitatively measuring and ranking the abilities of half-cells to compete for electrons—that is, for the species in the half-cells to act as oxidizing agents.

One volt is the potential needed to impart one joule of energy to an electric charge of one coulomb (1 J = 1 V × 1 C).

One coulomb is the amount of charge that passes a point in an electric circuit when a current of one ampere flows for one second (1 C = 1 A × 1 s).

It is redundant to write that the concentrations of the ions in these cells is 1.0 mol L^{-1}, because that is a necessary condition for specification of standard cell emf values. We will generally not include concentrations when listing $E°_{cell}$ values; 1.0 mol L^{-1} is implied. The symbol E_{cell} is used to refer to cells with concentrations of species not restricted to 1.0 mol L^{-1}.

Cell emf (E_{cell}) is a measure of the difference between the abilities of species in the half-cells to compete for electrons. When all of the species are in their standard states, the cell emf is called the standard cell emf ($E°_{cell}$) at the temperature of measurement.

Half-Cell Reduction Potentials

If we take the cell emf to be a quantitative measure of the difference in potentials of half-cells to attract electrons, since we know that the potential to attract electrons is greater for the cathode half-cell than the anode half-cell, we can write:

$$E_{cell} = E_{cathode} - E_{anode}$$

where $E_{cathode}$ and E_{anode} are examples of **half-cell reduction potential** ($E_{half-cell}$), also called *electrode potential*, which is a quantitative measure of the ability of a half-cell to attract electrons.

So, if we had a numerical value for the reduction potential of a half-cell from the emf of any cell containing that half-cell, we would be able to calculate the reduction potential of the other half-cell. The problem is that we are unable to measure the absolute value of half-cell reduction potentials: we can only measure differences between reduction potentials as cell emf values. To overcome this, chemists have arbitrarily assigned a value to the reduction potential of a particular half-cell as a reference point. *The reference half-cell against which we compare the reduction potentials of other half-cells is the **standard hydrogen electrode (SHE)**, arbitrarily assigned zero reduction potential.* The standard hydrogen electrode, portrayed in Figure 16.10, is described by the notation

$$H^+(aq, 1.0 \text{ mol L}^{-1})|H_2(g, 1.0 \text{ bar}) \qquad E = E°_{H^+/H_2} = 0.0 \text{ V}$$

$E°_{H^+/H_2} = 0.0\,V$ does not mean that the SHE has no electron-grabbing ability—simply that chemists chose this arbitrary value to fix all other values. They could have chosen to assign a reference potential of 1.0 V or −1.35 V to the SHE, or to assign 0.0 V to any other electrode. Whatever the reference point chosen, the differences on the scale remain the same.

Suppose you set up a number of half-cells and connect each in turn to a standard hydrogen electrode. Your apparatus might look like the voltaic cell in Figure 16.10.

FIGURE 16.10 A voltaic cell constructed from a standard hydrogen electrode and a $Zn^{2+}(aq)|Zn(s)$ half-cell. The reading on the voltmeter shows that the reduction potential of one half-cell is 0.76 V greater than that of the other. From experience of which reaction happens when the reagents are in direct contact, or from how the voltmeter is connected to give a reading with a positive value, we can deduce that the standard hydrogen electrode is the cathode (+), and is therefore the half-cell with the higher reduction potential.

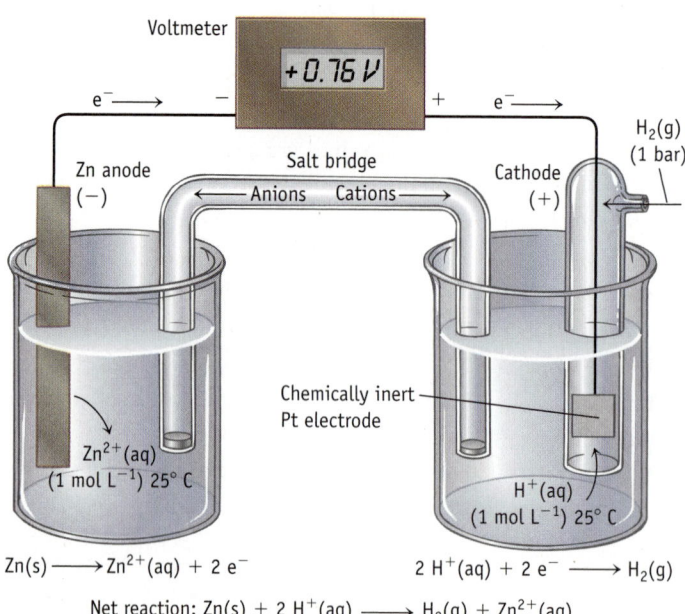

We can use the particular case of the cell shown in Figure 16.10 to consider some important issues:

1. *Direction of the oxidation-reduction reaction.* In the cell pictured in Figure 16.10, depending upon the relative abilities of species to compete for electrons, one of two

reactions (each the reverse of the other) will happen. Either $Zn^{2+}(aq)$ ions are reduced by H_2 gas, in which case the SHE would be the anode (−):

$$Zn^{2+}(aq) + H_2(g) \longrightarrow Zn(s) + 2\,H^+(aq)$$

or the $H^+(aq)$ ions are reduced by $Zn(s)$, and the SHE would be the cathode (+):

$$Zn(s) + 2\,H^+(aq) \longrightarrow Zn^{2+}(aq) + H_2(g)$$

The same reaction happens in a voltaic cell as would happen if the reagents were in direct contact: that is, $H^+(aq)$ ions are reduced, and $Zn(s)$ is oxidized.

2. *Direction of electron flow in the external circuit.* In a voltaic cell, electrons always flow from the anode (−) to the cathode (+)—in this case, from the $Zn^{2+}(aq)|Zn(s)$ half-cell to the $H^+(aq)|H_2(g)$ half-cell. The $H^+(aq)|H_2(g)$ half-cell has the greater attraction for electrons (higher reduction potential).

3. *Relative oxidizing and reducing abilities of cell reagents.* Since the cell reaction is

$$Zn(s) + 2\,H^+(aq) \longrightarrow Zn^{2+}(aq) + H_2(g)$$

and not the reverse, we can conclude that in a competition between $H^+(aq)$ ions and $Zn^{2+}(aq)$ ions to "grab" electrons, $H^+(aq)$ ions "win." In other words, the $H^+(aq)$ ion is a better oxidizing agent than the $Zn^{2+}(aq)$ ion. We can also deduce that $Zn(s)$ metal is a better reducing agent than H_2 gas.

4. *Quantitative estimates of half-cell reduction potentials.* Since the voltmeter reading tells us that the SHE has a reduction potential 0.76 V higher than the $Zn^{2+}(aq)|Zn(s)$ half-cell, and the SHE has been assigned a value of 0 V, we can calculate the reduction potential of the $Zn^{2+}(aq)|Zn(s)$ half-cell:

$$E_{cell} = 0.76\text{ V} = E_{cathode} - E_{anode} = 0\text{ V} - E_{Zn^{2+}/Zn}$$
$$\therefore E_{Zn^{2+}/Zn} = -0.76\text{ V}$$

If we construct a voltaic cell from a standard hydrogen electrode and a $Cu^{2+}(aq, 1.0\text{ mol L}^{-1})|Cu(s)$ half cell, we would measure a cell emf of 0.337 V with the SHE as the anode. We can deduce the reduction potential of the $Cu^{2+}(aq)|Cu(s)$ half cell:

$$0.34\text{ V} = E_{Cu^{2+}/Cu} - 0\text{ V} \qquad \therefore E_{Cu^{2+}/Cu} = +0.34\text{ V}$$

Hundreds of electrochemical cells like that shown in Figure 16.10 can be set up, allowing us to determine half-cell reduction potentials, and the relative oxidizing or reducing abilities of the chemical species in the half-cells. A few results of such measurements are shown in Figure 16.11, where all half-equations are written in the reduction direction.

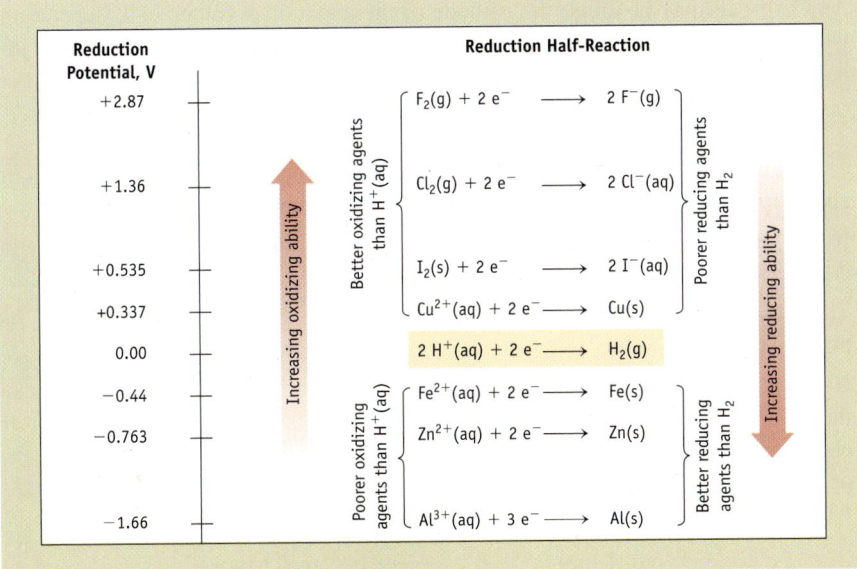

FIGURE 16.11 A potential ladder for reduction half-reactions. Half-cell potentials on a scale relative to $E°_{SHE}$ are listed. The higher the half-equation is, the more powerful an oxidizing agent the species on the left (the better it competes for electrons, and the more easily it is reduced). Conversely, the lower the half-equation is, the better the species on the right are as reducing agents.

The half-cell reduction potential of a half-cell ($E_{half-cell}$) is a quantitative measure of its ability to attract electrons. A cell emf is given by $E_{cell} = E_{cathode} - E_{anode}$. Half-cells can be placed in a list in order of reduction potentials, in which the standard hydrogen electrode (SHE) is arbitrarily assigned a value of 0 V.

Standard Half-Cell Reduction Potentials

Half-cell reduction potentials (and, therefore, cell emf values) change with changing concentrations of reactants and products. In order to sensibly compare half-cell reduction potentials, we do so when all reactants and products are in their standard states. Under these conditions, half-cell reduction potentials are called **standard half-cell reduction potentials** ($E°_{half-cell}$) (or *standard electrode potentials*). Standard potentials also depend on the temperature, but we will restrict ourselves to consideration of cells at 25 °C. The same principles apply at any other temperature.

When all of the reactant species in both half-cells are in their standard states, we have the relationship

$$E°_{cell} = E°_{cathode} - E°_{anode} \qquad (16.1)$$

In principle, by linking any standard half-cell to a standard hydrogen electrode to form a voltaic cell, we can estimate its standard reduction potential from (a) the cell emf (which is the standard cell emf, $E°_{cell}$), and (b) noting whether the SHE is the cathode or anode. By doing this for all half-cells, we can create a table of standard half-cell reduction potentials (Table 16.1, with a more complete list in Appendix E).

Note the following important points about the use of the table of standard half-cell reduction potentials:

1. The values of standard reduction potentials are appropriate to 25 °C. Values for different temperatures can be found in other databases.

2. Any half-cell may be either the cathode or the anode in a voltaic cell, depending on which other cell is joined to it. However, all half-cell equations are written in the reduction direction:

$$\text{oxidized species} + \text{electron(s)} \longrightarrow \text{reduced species}$$

In this way, the table compares the abilities of species to be reduced: that is, their ability to "grab" electrons and to bring about oxidation of other species.

3. If two half-cells are connected to form a voltaic cell, the half-cell higher in the table is the cathode (where reduction occurs).

4. The reaction that occurs in a voltaic cell is a spontaneous reaction, and $E°_{cell}$ has a positive numerical value.

5. Half-cell reduction potentials depend on the identity of the reactants and products and their concentrations, but not on the amount of material that reacts. Therefore, changing the stoichiometric coefficients in a half-equation does not change the value of $E°_{half-cell}$. For example, for the reduction of Fe^{3+}(aq) ions, $E°_{Fe^{3+}/Fe^{2+}} = 0.771V$ whether the half-equation is written as

$$Fe^{3+}(\text{aq, 1 mol } L^{-1}) + e^- \longrightarrow Fe^{2+}(\text{aq, 1 mol } L^{-1})$$

Reduction Half-Reaction		$E°$ (V)
$F_2(g) + 2\ e^-$	$\longrightarrow 2\ F^-$ (aq)	+2.87
$H_2O_2(aq) + 2\ H^+(aq) + 2\ e^-$	$\longrightarrow 2\ H_2O(\ell)$	+1.77
$PbO_2(s) + SO_4{}^{2-}(aq) + 4\ H^+(aq) + 2\ e^-$	$\longrightarrow PbSO_4(s) + 2\ H_2O(\ell)$	+1.685
$MnO_4{}^-(aq) + 8\ H^+(aq) + 5\ e^-$	$\longrightarrow Mn^{2+}(aq) + 4\ H_2O(\ell)$	+1.51
$Au^{3+}(aq) + 3\ e^-$	$\longrightarrow Au(s)$	+1.50
$Cl_2(g) + 2\ e^-$	$\longrightarrow 2\ Cl^-(aq)$	+1.36
$Cr_2O_7{}^{2-}(aq) + 14\ H^+ + 6\ e^-$	$\longrightarrow 2\ Cr^{3+}(aq) + 7\ H_2O(\ell)$	+1.33
$O_2(g) + 4\ H^+(aq) + 4\ e^-$	$\longrightarrow 2\ H_2O(\ell)$	+1.229
$Br_2(\ell) + 2\ e^-$	$\longrightarrow 2\ Br^-(aq)$	+1.08
$NO_3{}^-(aq) + 4\ H^+(aq) + 3\ e^-$	$\longrightarrow NO(g) + 2\ H_2O(\ell)$	+0.96
$OCl^-(aq) + H_2O(\ell) + 2\ e^-$	$\longrightarrow Cl^-(aq) + 2\ OH^-(aq)$	+0.89
$Hg^{2+}(aq) + 2\ e^-$	$\longrightarrow Hg(\ell)$	+0.855
$Ag^+(aq) + e^-$	$\longrightarrow Ag(s)$	+0.799
$Hg_2{}^{2+}(aq) + 2\ e^-$	$\longrightarrow 2\ Hg(\ell)$	+0.789
$Fe^{3+}(aq) + e^-$	$\longrightarrow Fe^{2+}(aq)$	+0.771
$I_2(s) + 2\ e^-$	$\longrightarrow 2\ I^-(aq)$	+0.535
$O_2(g) + 2\ H_2O(\ell) + 4\ e^-$	$\longrightarrow 4\ OH^-(aq)$	+0.40
$Cu^{2+}(aq) + 2\ e^-$	$\longrightarrow Cu(s)$	+0.337
$Sn^{4+}(aq) + 2\ e^-$	$\longrightarrow Sn^{2+}(aq)$	+0.15
$2\ H^+(aq) + 2\ e^-$	$\longrightarrow H_2(g)$	0.00
$Sn^{2+}(aq) + 2\ e^-$	$\longrightarrow Sn(s)$	−0.14
$Ni^{2+}(aq) + 2\ e^-$	$\longrightarrow Ni(s)$	−0.25
$V^{3+}(aq) + e^-$	$\longrightarrow V^{2+}(aq)$	−0.255
$PbSO_4(s) + 2\ e^-$	$\longrightarrow Pb(s) + SO_4{}^{2-}(aq)$	−0.356
$Cd^{2+}(aq) + 2\ e^-$	$\longrightarrow Cd(s)$	−0.40
$Fe^{2+}(aq) + 2\ e^-$	$\longrightarrow Fe(s)$	−0.44
$Zn^{2+}(aq) + 2\ e^-$	$\longrightarrow Zn(s)$	−0.763
$2\ H_2O(\ell) + 2\ e^-$	$\longrightarrow H_2(g) + 2\ OH^-(aq)$	−0.8277
$Al^{3+}(aq) + 3\ e^-$	$\longrightarrow Al(s)$	−1.66
$Mg^{2+}(aq) + e^-$	$\longrightarrow Mg(s)$	−2.37
$Na^+(aq) + e^-$	$\longrightarrow Na(s)$	−2.714
$K^+(aq) + e^-$	$\longrightarrow K(s)$	−2.925
$Li^+(aq) + e^-$	$\longrightarrow Li(s)$	−3.045

Increasing power of oxidizing agents

Increasing power of reducing agents

* In volts (V) versus the standard hydrogen reaction.

TABLE 16.1 Standard Half-Cell Reduction Potentials in Aqueous Solution at 25 °C*

Taking it Further

e16.10 Compare $E°$ values in acidic and basic conditions.

or as

$$2\ Fe^{3+}(aq,\ 1\ mol\ L^{-1}) + 2\ e^- \longrightarrow 2\ Fe^{2+}(aq,\ 1\ mol\ L^{-1})$$

A reduction potential measures the competitiveness of species to "grab" electrons and be reduced, irrespective of how much of it is reduced.

EXERCISE 16.9—STANDARD HALF-CELL REDUCTION POTENTIALS

A voltaic cell is formed from a standard half-cell represented by $Cl_2(g, 1.0\ bar)|Cl^-(aq, 1.0\ mol\ L^{-1})|Pt$ and a standard hydrogen electrode, both at 25 °C. The SHE is found to be the anode, and the resultant standard cell emf is measured to be 1.36 V. What is the standard reduction potential for the $Cl_2(g)|Cl^-(aq)|Pt$ half-cell at 25 °C. Be sure to assign the correct sign (+ or −) to your answer.

Half-cell reduction potentials depend on concentrations of reaction species. When all species are in their standard states, the half cell reduction potential is called the standard half-cell reduction potential ($E°_{half-cell}$).

Calculating Standard Cell emf

The standard half-cell reduction potentials were obtained by measuring the cell emf when standard half-cells are joined to the standard hydrogen electrode in voltaic cells. Since standard reduction potentials are additive, we can calculate standard cell emf values.

Think about It

e16.11 Use standard reduction potentials to calculate standard cell emf.

Interactive Exercises 16.10–16.11

Work through calculations involving the emf of voltaic cells.

Interactive Exercise 16.12

Spot the errors in these solutions to problems.

WORKED EXAMPLE 16.4—$E°_{cell}$ FROM STANDARD HALF-CELL REDUCTION POTENTIALS

A voltaic cell is constructed from the half-cells $Ag^+(aq, 1.0\ mol\ L^{-1})|Ag(s)$ and $Cu^{2+}(aq, 1.0\ mol\ L^{-1})|Cu(s)$.
(a) Which half-cell is the cathode and which is the anode?
(b) What is the overall cell reaction?
(c) Write the shorthand notation for the cell, according to convention.
(d) What is $E°_{cell}$ at 25 °C?

Strategy

Identify the two half-equations in Table 16.1 that are appropriate to the competition for electrons between these two half-cells. These are

$$Ag^+(aq) + e^- \longrightarrow Ag(s) \qquad E°_{Ag^+/Ag} = +0.799\ V$$
$$Cu^{2+}(aq) + 2e^- \longrightarrow Cu(s) \quad E°_{Cu^{2+}/Cu} = +0.337\ V$$

Solution

(a) Since the standard $Ag^+(aq)|Ag(s)$ half-cell has the higher reduction potential (higher ability to attract electrons), it is the cathode.

(b) The half-reactions in the two half-cells are

$$Cathode\ (reduction): \quad Ag^+(aq) + e^- \longrightarrow Ag(s)$$
$$Anode\ (oxidation): \quad Cu(s) \longrightarrow Cu^{2+}(aq) + 2e^-$$

To balance the number of electrons removed from $Cu(s)$ and taken by $Ag^+(aq)$ ions, for 1 mol of $Cu(s)$ oxidized, 2 mol of $Ag^+(aq)$ ions must be reduced. Multiply the cathode half-reaction by 2, and add the two half-equations to obtain the equation for the overall cell reaction:

$$2\ Ag^+(aq) + Cu(s) \longrightarrow 2\ Ag(s) + Cu^{2+}(aq)$$

(c) The convention for designating the cell is with the anode on the left.

$$Cu(s)|Cu^{2+}(aq, 1.0\ mol\ L^{-1})\|Ag^+(aq, 1.0\ mol\ L^{-1})|Ag(s)$$

(d) Since all the species are in their standard states, the cell emf is $E°_{cell}$. We can use equation 16.1:

$$E°_{cell} = E°_{cathode} - E°_{anode} = E°_{Ag^+/Ag} - E°_{Cu^{2+}/Cu} = +0.799\ V - 0.337\ V = +0.462\ V$$

Comment

If we were to consider that the overall reaction were

$$2\ Ag(s) + Cu^{2+}(aq) \longrightarrow 2\ Ag^+(aq) + Cu(s)$$

then the $Cu^{2+}(aq)|Cu(s)$ half-cell would be the cathode, and $Ag^+(aq)|Ag(s)$ the anode. Calculation of cell emf leads to:

$$E°_{cell} = E°_{cathode} - E°_{anode} = E°_{Cu^{2+}/Cu} - E°_{Ag^+/Ag} = +0.337\ V - (+0.799\ V) = -0.462\ V$$

The negative sign for $E°_{cell}$ indicates that when the ions are in their standard states (1.0 mol L^{-1}), this is not the spontaneous direction of reaction: the reaction would need to go in the opposite direction to come closer to chemical equilibrium. For the indicated reaction to occur, a voltage of at least 0.46 V would have to be applied to the system from an external source of electricity [see *Electrolysis*, >>Section 16.7].

EXERCISE 16.13—$E°_{cell}$ FROM STANDARD HALF-CELL REDUCTION POTENTIALS

Suppose that in a voltaic cell, the overall reaction is described by the equation

$$Zn(s) + 2\ Ag^+(aq) \longrightarrow Zn^{2+}(aq) + 2\ Ag$$

Assuming all species are in their standard states, identify the half-reactions that occur at the anode and the cathode, and calculate the standard cell emf at 25 °C.

EXERCISE 16.14—$E°_{cell}$ FROM STANDARD HALF-CELL REDUCTION POTENTIALS

Calculate the value of $E°_{cell}$ at 25 °C for each of the following voltaic cell reactions. Decide whether each is written in the direction of spontaneous reaction.

(a) $2\ I^-(aq) + Zn^{2+}(aq) \longrightarrow I_2(s) + Zn(s)$

(b) $Zn^{2+}(aq) + Ni(s) \longrightarrow Zn(s) + Ni^{2+}(aq)$

(c) $2\ Cl^-(aq) + Cu^{2+}(aq) \longrightarrow Cu(s) + Cl_2(g)$

(d) $Fe^{2+}(aq) + Ag^+(aq) \longrightarrow Fe^{3+}(aq) + Ag(s)$

> For any cell, the standard cell emf (at specified temperature) can be calculated from the standard half-cell reduction potentials (at the same temperature):
>
> $$E°_{cell} = E°_{cathode} - E°_{anode}$$

Relative Oxidizing and Reducing Abilities, Predicting Spontaneous Reactions

We can make the following interpretations from Table 16.1, although we should bear in mind that the numerical values of standard half-cell potentials apply only when all of the species in the reaction equation are in their standard states.

1. The more positive the value of $E°_{half-cell}$, the better the oxidizing ability of the ion or compound on the left side of the half-reaction. This means that fluorine, $F_2(g)$, is the best oxidizing agent in the table ($E°_{F_2/F^-} = +2.87\,V$). Other powerful oxidizing agents include $H_2O_2(aq)$, $MnO_4^-(aq)$, $Cl_2(g)$, $Cr_2O_7^{2-}(aq)$, and the most common of all, $O_2(g)$. These are all good competitors for electrons, and so are easily reduced. The oxidizing abilities of the reagents on the left of the half-equations decrease from the top to the bottom of the table, and the poorest oxidizing agent in Table 16.1 is the $Li^+(aq)$ ion ($E°_{Li^+/Li} = -3.045\,V$).

2. The more negative the value of $E°_{half-cell}$, the less likely the half-reaction will occur as a reduction, and the more likely the reverse half-reaction will occur (as an oxidation). So, $Li(s)$ is the best reducing agent in the table, and $F^-(aq)$ is the weakest. The species on the right of the half-equations increase in reducing ability from the top to the bottom. Those at the bottom are poor at retaining their electrons, so are easily oxidized and therefore powerful reducing agents. The best reducing agents are the reactive metals $Li(s)$, $K(s)$, $Mg(s)$, $Al(s)$, and $Zn(s)$.

3. The reaction between any species on the left of a half-equation in this table and any substance on the right of a lower half-equation is the direction of spontaneous reaction, if all reactants and products are present in their standard states. For example,

Zn(s) can reduce Fe^{2+}(aq) ions, H^+(aq) ions, Cu^{2+}(aq) ions, and I_2(s), while Cu(s) can reduce I_2(s)—but not H^+(aq) ions, Fe^{2+}(aq) ions, or Zn^{2+}(aq) ions.

Reduction Half-Reaction

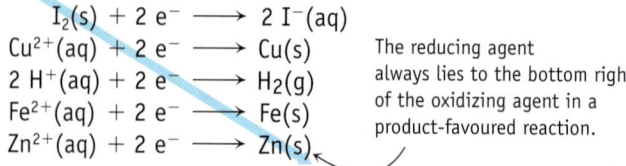

$$I_2(s) + 2\ e^- \longrightarrow 2\ I^-(aq)$$
$$Cu^{2+}(aq) + 2\ e^- \longrightarrow Cu(s)$$
$$2\ H^+(aq) + 2\ e^- \longrightarrow H_2(g)$$
$$Fe^{2+}(aq) + 2\ e^- \longrightarrow Fe(s)$$
$$Zn^{2+}(aq) + 2\ e^- \longrightarrow Zn(s)$$

The reducing agent always lies to the bottom right of the oxidizing agent in a product-favoured reaction.

Think about It

e16.12 Use standard reduction potentials to predict spontaneous reactions.

Interactive Exercise 16.15

Make predictions on the basis of standard reduction potentials.

Interactive Exercise 16.16

Make predictions about relative oxidizing and reducing abilities.

WORKED EXAMPLE 16.5—RELATIVE OXIDIZING AND REDUCING ABILITIES

Use the table of standard half-cell reduction potentials to do the following. Presume that all reagents are in their standard states and at 25 °C.
(a) Rank the halogens, F_2(g), Cl_2(g), Br_2(ℓ), and I_2(s), in decreasing order of their abilities as oxidizing agents.
(b) Decide whether hydrogen peroxide in acidic solution, H_2O_2(aq), is a stronger oxidizing agent than chlorine gas, Cl_2(g).
(c) Decide which of the halogens is capable of oxidizing Cr^{3+}(aq) ions to $Cr_2O_7^{2-}$(aq) ions.

Strategy

Use the fact that reagents on the left side of the half-equations are able to oxidize substances on the right side of half-equations lower down. So the higher the reduction potential of a reagent, the more substances it can oxidize, and the more powerful an oxidizing agent it is.

Solution

(a) The standard reduction potential of F_2(g) is higher than those of the other halogens: of the halogens, it is the most powerful oxidizing agent. For example, F_2(g) can oxidize Cl^-(aq) ions, H_2O(ℓ) and Fe^{2+}(aq) ions—none of which can be oxidized by I_2(s). Their abilities as oxidizing agents is in decreasing order of their reduction potentials; that is, $F_2(g) > Cl_2(g) > Br_2(\ell) > I_2(s)$.

(b) The relevant half-equations, and standard reduction potentials, are

$$H_2O_2(aq) + 2\ H^+(aq) + 2\ e^- \longrightarrow 2\ H_2O(\ell) \qquad E°_{H_2O_2/H_2O} = +1.77\ V$$
$$Cl_2(g) + 2\ e^- \longrightarrow 2\ Cl^-(aq) \qquad E°_{Cl_2/Cl^-} = +1.36\ V$$

The standard half-cell potential for reduction of H_2O_2(aq) is greater than that of Cl_2(g). When all of the species present are in their standard states, H_2O_2(aq) can oxidize some reagents that Cl_2(g) cannot—such as Mn^{2+}(aq) ions. H_2O_2(aq) is the more powerful oxidizer.

(c) The halogens that can oxidize Cr^{3+}(aq) ions to $Cr_2O_7^{2-}$(aq) ions at standard concentrations are those whose standard reduction potentials are higher than that for reduction of $Cr_2O_7^{2-}$(aq) ions:

$$Cr_2O_7^{2-}(aq) + 14\ H^+(aq) + 6\ e^- \longrightarrow 2\ Cr^{3+}(aq) + 7\ H_2O(\ell) \qquad E°_{Cr_2O_7^{2-}/Cr^{3+}} = 1.33\ V$$

Only F_2(g) and Cl_2(g) can oxidize Cr^{3+}(aq) ions at standard concentrations

---

EXERCISE 16.17—RELATIVE OXIDIZING AND REDUCING ABILITIES

Look up the standard reduction potentials at 25 °C for the Cu^{2+}(aq)|Cu(s), Sn^{2+}(aq)|Sn(s), Fe^{2+}(aq)|Fe(s), Zn^{2+}(aq)|Zn(s), and Al^{3+}(aq)|Al(s) half-cells.
(a) Which metal is most easily oxidized?
(b) Which metal(s) is(are) capable of reducing Fe^{2+}(aq) ions to Fe(s)?
(c) Write a balanced chemical equation for the reaction between Fe^{2+}(aq) ions and Sn(s). Is this the direction of spontaneous reaction?
(d) Write a balanced chemical equation for the reaction between Zn^{2+}(aq) ions and Sn(s). Is this the direction of spontaneous reaction?

If reagents are in their standard states, the spontaneous direction of reaction (toward equilibrium) is that between a species on the left side of a half-reaction in the table of standard half-cell reduction potentials and any species on the right side of a half-reaction with lower reduction potential.

16.5 Voltaic Cells under Non-standard Conditions

In practice, the reagents in voltaic cells are seldom in their standard states. Even if a cell is constructed with all dissolved species at 1 mol L^{-1} concentration, reactant concentrations decrease and product concentrations increase as the battery is used, and the cell reaction proceeds. Changing concentrations of reactants and products change the cell emf. In the following sections we consider how E_{cell} depends on concentrations of reactants and products.

Dependence of Cell emf on Concentrations

Based on both theory and experimental results, scientists have found experimentally that cell emf depends on concentrations of reactants and products, and on temperature, in a manner expressed by the **Nernst equation**:

$$E_{cell} = E^\circ_{cell} - \frac{RT}{nF} \ln Q$$

In this equation, R is the gas constant (8.3144 J K^{-1} mol^{-1}); T is the temperature (K); n is the amount (mol) of electrons transferred between oxidizing and reducing agents in the balanced equation for the cell reaction; F is the Faraday constant (9.6485338 × 10^4 C mol^{-1}); and Q is the reaction quotient calculated from the concentrations of all species in the equation for the cell reaction.

The Nernst equation is easily derived from the fundamental equation that specifies how the Gibbs free energy change of a reaction at non-standard compositions ($\Delta_r G$) differs from that when all reagents are in their standard states ($\Delta_r G^\circ$) [>>Section 17.6]:

$$\Delta_r G = \Delta_r G^\circ + RT \ln Q$$

The free energy terms are eliminated using the relationships between free energy change and cell emf [>>Section 17.8]:

$$\Delta_r G = -nFE_{cell} \quad \text{and} \quad \Delta_r G^\circ = -nFE^\circ_{cell}$$

The Nernst equation expresses how much E_{cell} (at non-standard conditions) differs from E°_{cell}. When all reagents are in their standard states, $Q = 1$ and $E_{cell} = E^\circ_{cell}$. The smaller Q is (smaller [products]/[reactants] ratio), the larger E_{cell} is. The larger Q is, the smaller E_{cell} is, and in a cell with Q sufficiently large, we can have $E_{cell} < 0$, and the spontaneous reaction direction is opposite to that in the standard cell.

Substituting values for the constants in the Nernst equation, and using $T = 298$ K, gives the form of the Nernst equation that we will use from now on:

$$E_{cell} = E^\circ_{cell} - \frac{0.0257 \text{ V}}{n} \ln Q \tag{16.2}$$

The values of Q and n are derived from the equation for the cell reaction, which must be written out to apply the Nernst equation correctly.

WORKED EXAMPLE 16.6—CELL EMF UNDER NON-STANDARD CONDITIONS AT 25 °C

A voltaic cell is set up at 25 °C with the following half-cells: Al^{3+}(aq, 0.0010 mol L^{-1}) | Al(s) and Ni^{2+}(aq, 0.50 mol L^{-1})|Ni(s). Write an equation for the reaction that occurs when the cell generates an electric current and calculate the cell emf.

Strategy

First, use Table 16.1 to decide which half-cell is the cathode and which is the anode. Add the oxidation and reduction half-equations, ensuring the same number of electrons are

Think about It

e16.13 Use graphs to visualize the relationships expressed by the Nernst Equation

Walter Nernst (1864–1941) was a German physicist and chemist known for his work in the field of thermodynamics. He was also a scientific rival of Fritz Haber [<<Section 13.1].

A Faraday is the quantity of electric charge (in coulombs) carried by 1 mol of electrons. The factor RT/F has units of volts (V) because 1 J = 1 C x V.

Background Concepts

e16.14 Read about Michael Faraday's contributions to electrochemistry.

Think about It

e16.15 Calculate the cell potential under non-standard conditions at 25 °C.

Interactive Exercise 16.18

Calculate E_{cell} under non-standard conditions at 25 °C.

Interactive Exercise 16.19

Spot the errors in these solutions to problems.

transferred, to get the equation for the overall reaction. Calculate the standard cell emf ($E°_{cell}$) as the difference between the half-cell reduction potentials. Finally, use equation 16.2 (the Nernst equation) to calculate the non-standard cell emf (E_{cell}).

Solution

The relevant half-equations in Table 16.1 are

$$Ni^{2+}(aq) + 2\ e^- \longrightarrow Ni(s) \qquad E°_{Ni^{2+}/Ni} = -0.25\ V$$
$$Al^{3+}(aq) + 3\ e^- \longrightarrow Al(s) \qquad E°_{Al^{3+}/Al} = -1.66\ V$$

We see that the $Ni^{2+}(aq)|Ni(s)$ half-cell is the cathode (higher reduction potential). We obtain a balanced equation for the overall cell reaction by multiplying the reduction half-equation by 3, and the oxidation half-equation by 2.

$$Cathode: 3\left\{Ni^{2+}(aq) + 2\ e^- \longrightarrow Ni(s)\right\}$$
$$Anode: 2\left\{Al(s) \longrightarrow Al^{3+}(aq) + 3\ e^-\right\}$$
$$Overall: 2\ Al(s) + 3\ Ni^{2+}(aq) \longrightarrow 2\ Al^{3+}(aq) + 3\ Ni(s)$$
$$E°_{cell} = E°_{cathode} - E°_{anode} = (-0.25\ V) - (-1.66\ V) = +1.41\ V$$

From the equation for the overall cell reaction, we can deduce $n = 6$, and

$$Q = \frac{[Al^{3+}]^2}{[Ni^{2+}]^3} = \frac{(0.0010)^2}{(0.50)^3} = 8.0 \times 10^{-6}$$

Now we substitute into the Nernst equation:

$$E_{cell} = E°_{cell} - \frac{0.0257\ V}{n} \times \ln\frac{[Al^{3+}]^2}{[Ni^{2+}]^3} = +1.41\ V - \frac{0.0257\ V}{6} \times \ln(8.0 \times 10^{-6})$$
$$= +1.41\ V - \frac{0.0257\ V}{6} \times (-11.7) = +1.46\ V$$

Comment

Cell emf depends on concentrations because half-cell reduction potentials depend on concentrations. (The Nernst equation can be applied to half-cell reduction potentials also.) The lower the concentrations of $Ni^{2+}(aq)$ and $Al^{3+}(aq)$ are, the lower the reduction potentials of the $Ni^{2+}(aq)|Ni(s)$ and $Al^{3+}(aq)|Al(s)$ half-cells. Because the $[Al^{3+}]$ in the anode half-cell is much further from 1 mol L^{-1} than is $[Ni^{2+}]$ in the cathode, the difference between the two non-standard half-cell potentials is greater than the difference between the standard half-cell potentials, and so $E_{cell} > E°_{cell}$.

EXERCISE 16.20—CELL EMF UNDER NON-STANDARD CONDITIONS AT 25 °C

A voltaic cell is set up with an aluminum electrode in a 0.025 mol L^{-1} $Al(NO_3)_3$ aqueous solution and an iron electrode in a 0.50 mol L^{-1} $Fe(NO_3)_2$ aqueous solution. Calculate the cell emf at 25 °C.

When the species in a voltaic cell are not in their standard states, cell emf can be calculated using the Nernst equation:

$$E_{cell} = E°_{cell} - \frac{RT}{nF}\ln Q = E°_{cell} - \frac{0.0257\ V}{n}\ln Q \quad \text{at 25 °C}$$

This calculates the difference between the cell emf and the standard cell emf. The values of Q and n are both derived from a written equation for the cell reaction.

pH Meters and Ion-Selective Electrodes

Worked Example 16.6 demonstrates the calculation of a cell emf if species concentrations are known. It is also useful to apply the Nernst equation in the opposite sense, using a measured cell potential to determine an unknown concentration. A device that does just this is the pH meter (Figure 16.12). In a voltaic cell in which $H^+(aq)$ is a reactant or product, the cell emf varies with changes of $[H^+]$, in accordance with the Nernst equation. The cell emf is measured, and this is used to calculate pH. Worked Example 16.7 illustrates how E_{cell} depends on the hydrogen ion concentration in a simple cell.

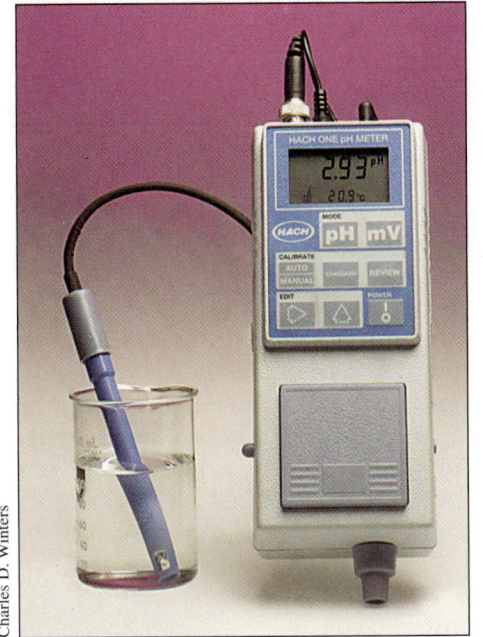

(a)

(b)

FIGURE 16.12 Measuring pH. (a) A portable pH meter that can be used in the field. (b) The tip of a glass electrode for measuring pH.

Charles D. Winters

WORKED EXAMPLE 16.7—PRINCIPLES OF THE PH METER

A voltaic cell is set up with the half-cells $Cu^{2+}(aq, 1.0 \text{ mol L}^{-1})|Cu(s)$ and $H^+(aq)|H_2(g, p = 1 \text{ bar})|Pt$. The concentration of $H^+(aq)$ ions in the $H^+(aq)|H_2(g)$ half-cell is unknown. At 298 K, $E_{cell} = 0.490$ V. Calculate the pH of the solution.

Strategy

Using the data of Table 16.1, E°_{cell} can be calculated. The only unknown quantity in equation 16.2 (the Nernst equation) is the concentration of hydrogen ions, from which we can calculate the solution pH.

Solution

The relevant half-equations in Table 16.1 are

$$Cu^{2+}(aq) + 2\,e^- \longrightarrow Cu(s) \qquad E^\circ_{Cu^{2+}/Cu} = +0.337 \text{ V}$$
$$2\,H^+(aq) + 2\,e^- \longrightarrow H_2(g) \qquad E^\circ_{H^+/H_2} = 0.00 \text{ V}$$

From these we see that the $Cu^{2+}(aq,)|Cu(s)$ half-cell is the cathode, and we can derive

$$\textit{Overall:}\ Cu^{2+}(aq) + H_2(g) \longrightarrow Cu(s) + 2\,H^+(aq)$$
$$E^\circ_{cell} = E^\circ_{cathode} - E^\circ_{anode} = +0.337\,V - 0.00\,V = +0.337 \text{ V}$$

Now we can substitute into the Nernst equation, using $n = 2$:

$$E_{cell} = E^\circ_{cell} - \frac{0.0257 \text{ V}}{n} \ln Q = E^\circ_{cell} - \frac{0.0257 \text{ V}}{2} \times \ln \frac{[H^+]^2}{[Cu^{2+}] \times p_{H_2}}$$

$$0.49\,V = +0.337 \text{ V} - \frac{0.0257 \text{ V}}{2} \times \ln \frac{[H^+]^2}{(1.00)(1.00)}$$

$$\ln[H^+]^2 = 2 \times \ln[H^+] = -11.9$$

$$\therefore\ [H^+] = 2.6 \times 10^{-3} \text{ mol L}^{-1} \quad \text{and} \quad pH = 2.59$$

Taking It Further

e16.16 Read about the sensitivity of half-cell reduction potentials to pH.

Comment

pH meters are voltaic cells that measure emf values, calibrated to indicate pH directly.

In practice, using a hydrogen electrode in a pH meter is not practical. The apparatus is bulky and delicate, and platinum (for the electrode) is costly. Modern pH meters use a *glass electrode,* so-called because it contains a thin glass membrane separating the cell from the solution whose pH is to be measured (Figure 16.12). Inside the glass electrode is a silver wire coated with AgCl and a solution of HCl; outside is the solution of unknown pH to be evaluated. A calomel electrode—a common reference electrode using a mercury(I)–mercury redox couple ($Hg_2Cl_2|Hg$)—serves as the second electrode of the cell. The emf across the glass membrane depends on [H^+].

Concentrations of ions other than H^+(aq) ions can also be measured using the Nernst equation dependence of cell emf on the concentration of the ion of interest. Collectively, the electrodes used to measure ion concentrations are known as **ion-selective electrodes**. Some houses are equipped with water softeners whose function is to remove Ca^{2+}(aq) and Mg^{2+}(aq) ions from household water and replace them with Na^+ ions. They work by passing the water over an *ion-exchange resin.* To test whether the resin is functioning, the water is periodically sampled for Ca^{2+} ions. This can be done with a built-in ion-selective electrode to estimate the concentration of this ion. When the electrode indicates that the [Ca^{2+}] has reached a specified level, it sends a signal to regenerate the ion-exchange resin.

Ion-selective electrodes are very important analytical tools in monitoring water quality in the natural environment, measuring concentrations of ions such as fluoride, bromide, and many metal ions. They are also used to measure the concentrations of dissolved gases in blood [<<Section 16.1].

Interactive Exercise 16.21

Calculate the ion concentration from cell potential data.

EXERCISE 16.22—ION CONCENTRATION FROM CELL EMF

One half-cell in a voltaic cell at 25 °C is constructed from a silver wire dipped into a $AgNO_3$ solution of unknown concentration. The other half-cell consists of a zinc electrode in a 1.0 mol L^{-1} $Zn(NO_3)_2$ solution. The cell emf is measured to be 1.48 V. What is the concentration of Ag^+(aq) ions?

pH meters and other ion-selective electrodes depend on measurement of the emf of a voltaic cell in which one half-cell contains the solution with unknown concentration of the ion of interest. The readout is calibrated, on the basis of the Nernst equation, to read the concentration of the ion directly.

pH-Dependence of Oxidizing Power of Oxoanions

A 1 mol L^{-1} nitric acid solution will oxidize and dissolve metallic copper (Figure 16.4), but a 1 mol L^{-1} copper nitrate solution will not—even though in both cases the concentration of the NO_3^-(aq) ion oxidizing agent is 1 mol L^{-1}. In acidic solutions, NO_3^-(aq) ions are a more powerful oxidizing agent than in neutral solutions.

In general, the more acidic the solution is, the more powerful aquated oxoanions such as NO_3^-(aq), MnO_4^-(aq), ClO_4^-(aq), and OCl^-(aq) are as oxidizing agents. This is a natural outcome of the dependence of half-cell reduction potentials on the concentrations of all species participating in the reaction—not just the oxidizing and reducing agents. H^+(aq) ions participate in the reduction of NO_3^-(aq) ions, as can be seen from the half-equation for reduction:

$$NO_3^-(aq) + 4\,H^+(aq) + 3\,e^- \longrightarrow NO(g) + 2\,H_2O(\ell)$$

When NO_3^-(aq) ions are reduced, the product may be either NO_2(g) or NO(g), depending on the H^+(aq) concentration. Usually a mixture of the gases is produced. We use the equation for NO formation here for illustrative purposes.

Let's calculate the half-cell potential for reduction of NO_3^-(aq) ions to NO(g) in 1 mol L^{-1} nitric acid solution, assuming that NO(g) is at a pressure of 1 bar. In this case, [NO_3^-] = 1.0 mol L^{-1}, [H^+] = 1.0 mol L^{-1}, and p_{NO} = 1.0 bar, so all species are in their standard states. In that case, using Table 16.1,

$$E_{NO_3^-/NO} = E°_{NO_3^-/NO} = +0.96\text{ V}$$

On the other hand, 1 mol L^{-1} $NaNO_3$ solution is neutral, so $[NO_3^-] = 1.0$ mol L^{-1} and $[H^+] = 1 \times 10^{-7}$ mol L^{-1}. If we again assume that $p_{NO} = 1$ bar, we can use equation 16.2 (the Nernst equation) to calculate the half-cell reduction potential in the non-standard conditions, remembering to derive the value of n and the expression for Q directly from the written half-equation above.

$$E_{NO_3^-/NO} = E^{\circ}_{NO_3^-/NO} - \frac{0.0257 \text{ V}}{3} \times \ln \frac{p_{NO}}{[NO_3^-][H^+]^4}$$

$$= +0.96 \text{ V} - \frac{0.0257 \text{V}}{3} \times \ln \frac{1.0}{(1.0)(1.0 \times 10^{-7})^4}$$

$$= +0.96 \text{ V} - \frac{0.0257 \text{V}}{3} \times 64 = +0.96 \text{ V} - 0.55 \text{ V} = +0.41 \text{ V}$$

Clearly, in the 1 mol L^{-1} HNO_3 solution, $NO_3^-(aq)$ ions have a higher reduction potential than in the 1 mol L^{-1} $NaNO_3$ solution, so they are capable of oxidizing a larger number of reagents in the acidic solution than in the neutral solution.

EXERCISE 16.23—PH-DEPENDENCE OF OXIDIZING POWER OF OXOANIONS

What is the reduction potential at 25 °C for reduction of $MnO_4^-(aq)$ ions, at 1 mol L^{-1} to $Mn^{2+}(aq)$ ions at 1 mol L^{-1} at (a) pH = 0.0, and (b) pH = 5.0? At which of these pH values are $MnO_4^-(aq)$ ions the more powerful oxidizing agent?

When the half-equation for a reduction process involves $H^+(aq)$ ions, the half-cell reduction potential depends on the concentration of these ions (and, therefore, pH), in accordance with the Nernst equation. A consequence is that the lower the pH of solution, the more powerful oxoanions are as oxidizing agents.

16.6 Standard Cell emf and Equilibrium Constant

Recall that the reaction that takes place in a voltaic cell is in the spontaneous direction; that is, toward equilibrium. When a voltaic cell provides an electric current, the reactant concentrations decrease and the product concentrations increase. The cell emf also changes; as reactants are converted to products, E_{cell} decreases. Eventually no further net reaction occurs because the cell reaction has reached chemical equilibrium, and the cell emf falls to zero.

This situation can be analyzed using equation 16.2 (the Nernst equation):

$$E_{cell} = E^{\circ}_{cell} - \frac{RT}{nF} \ln Q = E^{\circ}_{cell} - \frac{0.0257 \text{ V}}{n} \ln Q \text{ at } 25 \text{ °C}$$

When $E_{cell} = 0$, the reactants and products are at equilibrium and the reaction quotient (Q) is equal to the equilibrium constant (K). Substituting into the Nernst equation:

$$E_{cell} = 0 = E^{\circ}_{cell} - \frac{RT}{nF} \ln K = E^{\circ}_{cell} - \frac{0.0257 \text{ V}}{n} \ln K \text{ at } 25 \text{ °C}$$

Re-arrangement gives an equation that relates the standard cell emf and equilibrium constant for the cell reaction. We will generally use the version that is applicable specifically to 25 °C:

$$\ln K = \frac{nFE^{\circ}_{cell}}{RT} = \frac{nE^{\circ}_{cell}}{0.0257 \text{ V}} \quad \text{at } 25 \text{ °C}$$

This relationship can be used to determine equilibrium constants for cell reactions, using standard cell emf values calculated from standard reduction potentials in Table 16.1.

The farther apart half-reactions are on the table of standard reduction potentials, the larger $E°$ is, and the larger K is. The magnitude of $E°_{cell}$ is a measure of how far a cell with all reactant species in their standard states is from equilibrium.

WORKED EXAMPLE 16.8—$E°_{cell}$ AND K FOR THE CELL REACTION

Calculate the equilibrium constant at 25 °C for the reaction

$$Fe(s) + Cd^{2+}(aq) \rightleftharpoons Fe^{2+}(aq) + Cd(s)$$

Strategy

In Table 16.1, find the standard reduction potentials at 25 °C for the appropriate half-reactions and calculate $E°_{cell}$. Use the half-equations to deduce the balanced equation for the cell reaction and then the value of n.

Solution

The half-reactions, written in the reduction direction, are

$$Cathode,\ reduction: Cd^{2+}(aq) + 2\ e^- \longrightarrow Cd(s) \qquad E°_{Cd^{2+}/Cd} = -0.40\ V$$
$$Anode,\ oxidation: Fe^{2+}(aq) + 2\ e^- \longrightarrow Fe(s) \qquad E°_{Fe^{2+}/Fe} = -0.44\ V$$

From these we deduce

$$E°_{cell} = E°_{cathode} - E°_{anode} = -0.40\ V - (-0.44\ V) = +0.04\ V$$

If we reverse the anode half-equation and add it to the cathode half-equation, we get the balanced equation given in the problem, showing that $n = 2$. Now we can calculate K at 25 °C:

$$\ln K = \frac{nE°_{cell}}{0.0257\ V} = \frac{(2)(0.04\ V)}{0.0257\ V} = 3.1 \qquad \therefore K = 22$$

Comment

The relatively small standard cell emf (0.040 V) indicates that when all the reactant species in the cell are at standard concentrations (and $Q = 1$), the reaction is not far from equilibrium. The small equilibrium constant is in accord with this observation.

Interactive Exercise 16.24

Calculate K from E_{cell} for a given reaction.

EXERCISE 16.25—$E°_{cell}$ AND K FOR THE CELL REACTION

Calculate the equilibrium constant at 25 °C for the reaction

$$2\ Ag^+(aq) + Hg(\ell) \rightleftharpoons 2\ Ag(s) + Hg^{2+}(aq)$$

There is a direct relationship between the standard cell emf and the equilibrium constant for the overall cell reaction:

$$\ln K = \frac{nFE°_{cell}}{RT} = \frac{nE°_{cell}}{0.0257\ V} \qquad at\ 25\ °C$$

16.7 Electrolysis: Chemical Change Using Electrical Energy

During *electrolysis* (Figure 16.13a), *reactions are forced to go in the non-spontaneous direction by application of electrical energy* to reagents in an electrochemical cell called an *electrolysis cell*. Electroplating (Figure 16.13b) is another example of electrolysis. Here, electrons (as an electric current) are forced into a metal object which is dipping into a solution of a salt of the metal to be plated out. At the surface of the object (cathode), metal ions in solution are reduced and form a metallic deposit on the surface.

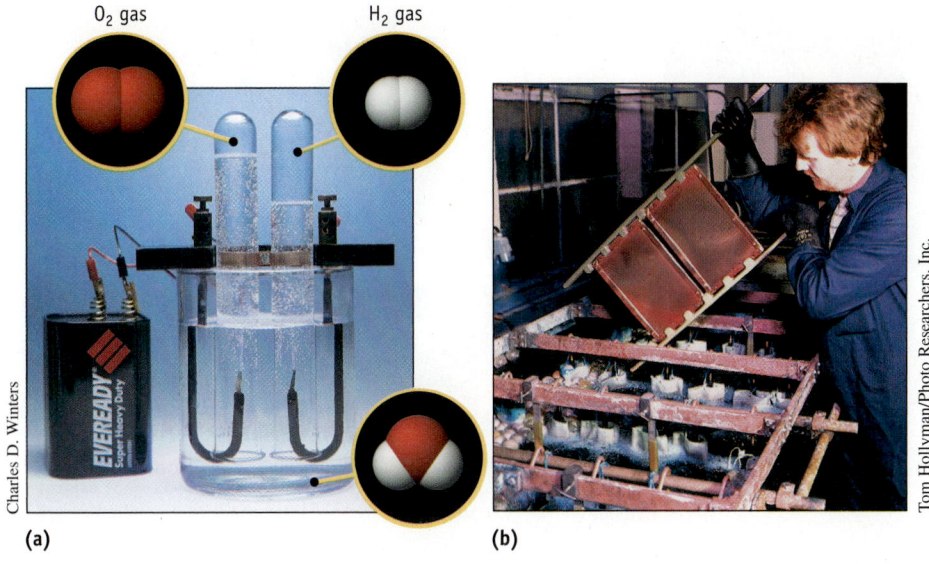

O₂ gas   H₂ gas

(a)   (b)

FIGURE 16.13 Electrolysis. (a) Electrolysis of water produces hydrogen and oxygen gas. (b) Electroplating adds a layer of metal to the surface of an object, either to protect the object from corrosion or to improve its physical appearance. The procedure uses an electrolysis cell, set up with the object to be plated as the cathode. The insets show only a single molecule of the very many molecules of the substances.

Electrolysis is an important procedure because it is widely used in the refining of metals such as aluminum and in the production of substances such as chlorine gas. These important topics are discussed in Chapter 26.

> During electrolysis, reactions are forced to go in the non-spontaneous direction by application of electrical energy to reagents in an electrochemical cell called an electrolysis cell.

Electrolysis of Molten Salts

All electrolysis cells have some features in common. The material to be electrolyzed is either a molten salt or a solution. As was the case with voltaic cells, the movement of ions in the liquid or solution constitutes the electric current within the cell. The cell has two electrodes that are connected to a source of DC (direct-current) potential. If a high enough voltage is applied, chemical reactions occur at the two electrodes. Reduction occurs at the surface of the electrode to which electrons are forced to flow (the cathode, labelled (−)): electrons are transferred to a chemical species in the cell. At the other electrode (the anode, labelled (+)), oxidation occurs as electrons are taken from a chemical species in the cell and transferred to the electrode.

Let's first focus on the chemical reactions that occur at each electrode in the simplest electrolysis cell—electrolysis of a molten salt. In molten NaCl, at temperatures higher than 800 °C, sodium ions (Na^+) and chloride ions (Cl^-) are freed from their rigid arrangement in the solid crystalline lattice and become mobile. If a sufficiently high external voltage is applied to the cell, at the cathode, sodium ions accept electrons and are reduced to sodium metal (a liquid at this temperature). Simultaneously, at the anode, electrons are taken from chloride ions, with formation of chlorine gas (Figure 16.14). The cell reaction is not in the spontaneous direction. The energy required to make this reaction occur is provided by the electric current.

Cathode, reduction: $2\,Na^+ + 2\,e^- \longrightarrow 2\,Na(\ell)$

Anode, oxidation: $2\,Cl^- \longrightarrow Cl_2(g) + 2\,e^-$

Net reaction: $2\,Na^+ + 2\,Cl^- \longrightarrow 2\,Na(\ell) + Cl_2(g)$

FIGURE 16.14 The production of sodium and chlorine by the electrolysis of molten NaCl. In the molten state, sodium ions migrate to the cathode, where they are reduced to sodium metal. Chloride ions migrate to the anode, where they are oxidized to chlorine gas.

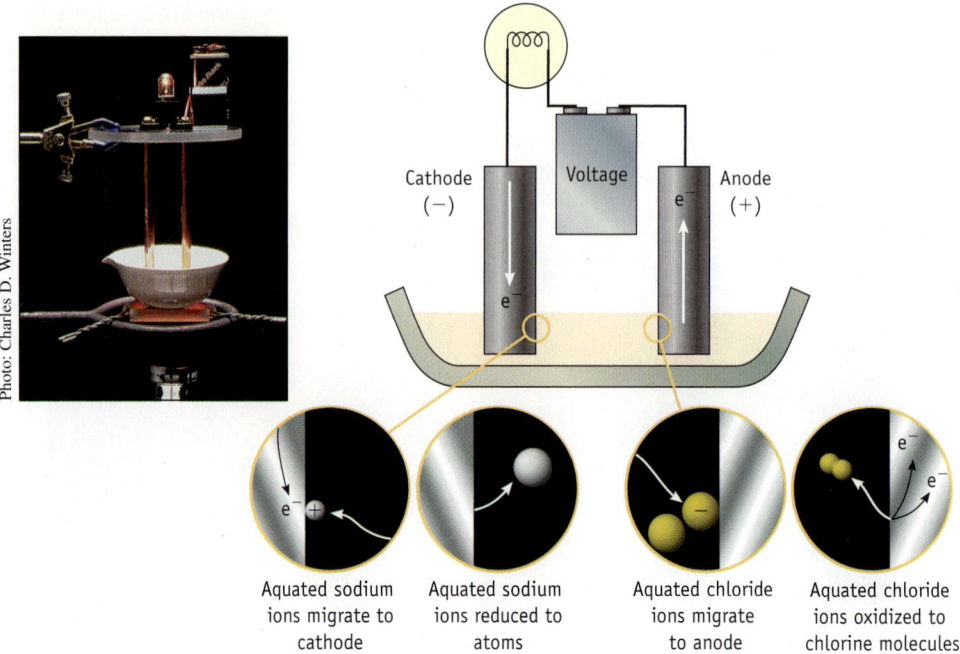

Photo: Charles D. Winters

Aquated sodium ions migrate to cathode

Aquated sodium ions reduced to atoms

Aquated chloride ions migrate to anode

Aquated chloride ions oxidized to chlorine molecules

Some clarification is needed about terms and sign conventions in electrochemistry. In both voltaic cells and electrolysis cells, the electrode where oxidation takes place is called the anode, and the electrode where reduction takes place is called the cathode. However, in voltaic cells the anode is labelled (−), and the cathode (+), while in electrolysis cells the anode is labelled (+), and the cathode (−).

> During the electrolysis of molten salts, when the applied emf is sufficiently high, the cation of the salt is reduced at the cathode, and the anion is oxidized at the anode. These are the only possible reactions.

Electrolysis of Aqueous Solutions

In molten NaCl, the only species present are sodium ions (Na^+) and chloride ions (Cl^-). Only one of these (Cl^-) can be oxidized, and only one (Na^+) can be reduced. Electrolysis of aqueous solutions is more complicated than electrolysis of molten salts, however, because of the presence of water, which can also be oxidized or reduced in electrochemical processes. We can regard the processes at each electrode as competitions between water molecules and aquated ions to "grab" electrons at the cathode or to have electrons removed at the anode. At the cathode, the species reduced is the one that competes most strongly for electrons, and at the anode, the species oxidized is the one that most easily gives up electrons.

The purification of copper from impure "blister copper" [>>Section 27.3] is an ingenious application of electrolysis in aqueous solution on the industrial scale that depends on these principles of competition.

Consider the electrolysis of aqueous sodium iodide solution (Figure 16.15), and let's imagine that all species are present in their standard states, except that the solution is neutral, so that $[H^+] = 1.0 \times 10^{-7}$ mol L^{-1}. In addition to water molecules, near the surfaces of the electrodes are $Na^+(aq)$ ions, $I^-(aq)$ ions, and $H^+(aq)$ ions. The only reduction reactions possible at the cathode are the following two, with reduction potential for formation of $H_2(g)$ adjusted for the non-standard $H^+(aq)$ concentration:

$$2\ H_3O^+(aq) + 2\ e^- \longrightarrow H_2(g) + 2\ H_2O(\ell) \qquad E_{H_3O^+/H_2} = -0.41\ V$$
$$Na^+(aq) + e^- \longrightarrow Na(s) \qquad E = E^\circ_{Na^+Na} = -2.71\ V$$

Experimental observations show that H_2 gas forms at the cathode is consistent with the much higher reduction potential of $H_3O^+(aq)$ ions than that of $Na^+(aq)$ ions. As the concen-

The half-equation for formation of $H_2(g)$ is listed in Table 16.1 as

$$2\ H^+(aq) + 2\ e^- \longrightarrow H_2(g)$$

for which $E^\circ = 0$ V when $[H^+] = 1$ mol L^{-1} and $p(H_2) = 1$ bar. The value of $E = -0.41$ V is derived by application of the Nernst equation with $[H^+] = 1 \times 10^{-7}$ mol L^{-1}. Correspondingly, the value $E = +0.82$ V for reduction of $O_2(g)$ is that for the non-standard concentration of $H^+(aq)$ ions.

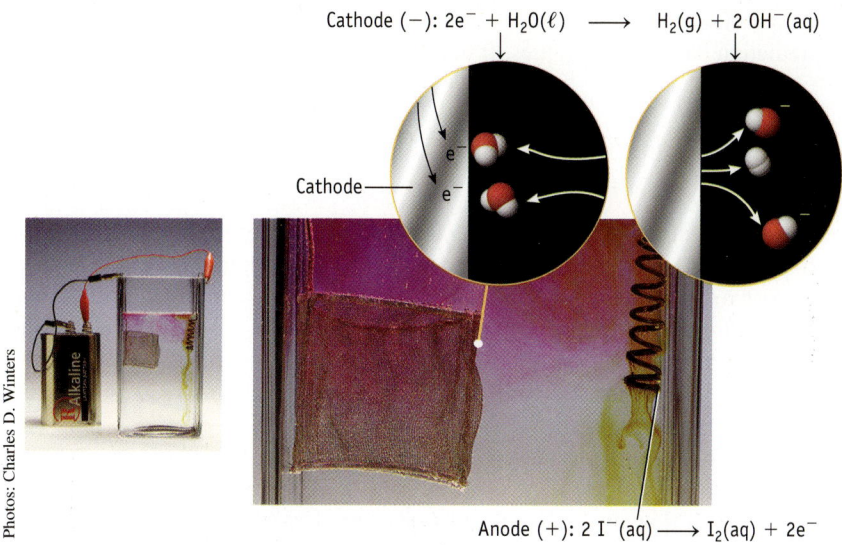

Cathode (−): 2e⁻ + H₂O(ℓ) ⟶ H₂(g) + 2 OH⁻(aq)

Cathode

Anode (+): 2 I⁻(aq) ⟶ I₂(aq) + 2e⁻

FIGURE 16.15 Electrolysis of aqueous NaI solution. An aqueous NaI solution is electrolyzed, with a potential applied using an external source of electricity. A drop of phenolphthalein has been added to the solution in this experiment so that the formation of OH⁻(aq) can be detected (by the red colour of the indicator in basic solution). Iodine forms at the anode, and H₂(g) and OH⁻(aq) ions form at the cathode.

tration of H_3O^+(aq) ions is decreased by reduction, the concentration of OH^-(aq) ions necessarily increases [<<Section 14.3]. The observation that iodine (I_2) is produced at the anode, rather than oxygen gas, is consistent with what we would predict from reduction potentials. The two relevant half-equations, written in the reduction direction, are the following:

$$O_2(g) + 4\,H^+(aq) + 4\,e^- \longrightarrow 2\,H_2O(\ell) \qquad E_{O_2/H_2O} = +0.82\ V$$

$$I_2(aq) + 2\,e^- \longrightarrow 2\,I^-(aq) \qquad E^\circ_{I_2/I^-} = +0.54\ V$$

We would predict that oxidation of water to form $O_2(g)$ would require the application of 1.23 V (i.e., $+0.82\ V - [-0.41\ V]$), whereas oxidation of I^-(aq) ions to form $I_2(aq)$ requires an external voltage of only 0.95 V (i.e., $+0.54\ V - [-0.41\ V]$). The reaction that requires the smaller applied voltage is the one that occurs.

What happens if an aqueous tin(II) chloride ($SnCl_2$) solution is electrolyzed? As before, consult Table 16.1 and consider all possible half-reactions. In this case, aquated Sn^{2+} ions are much more easily reduced ($E^\circ = -0.14\ V$) than water ($E^\circ = -0.83\ V$) at the cathode, so tin metal is produced (Figure 16.16). At the anode, two oxidation reactions are possible: Cl^-(aq) ions to $Cl_2(g)$ or H_2O to $O_2(g)$. Experiments show that chloride ion is preferentially oxidized, rather than water, so the reactions are

Cathode, reduction:　$Sn^{2+}(aq) + 2\,e^- \longrightarrow Sn(s)$

Anode, oxidation:　$\underline{2\,Cl^-(aq) \longrightarrow Cl_2(g) + 2\,e^-}$

Overall reaction:　$Sn^{2+}(aq) + 2\,Cl^-(aq) \longrightarrow Sn(s) + Cl_2(g)$

$E^\circ_{cell} = E^\circ_{cathode} - E^\circ_{anode} = (-0.14\ V) - (+1.36\ V) = -1.50\ V$

The formula I_2(aq) used here is an over-simplification, since iodine (I_2) is quite insoluble in water. In solutions containing iodide ions, however, iodine goes into solution as the aquated tri-iodide ion, I_3^-(aq), and this is the main species present.

$$I_2(s) + I^-(aq) \longrightarrow I_3^-(aq)$$

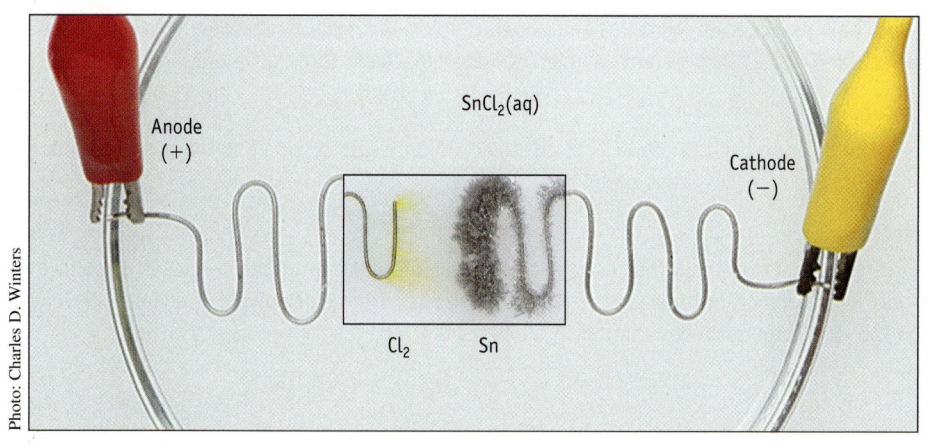

SnCl₂(aq)

Anode (+)

Cathode (−)

Cl₂　Sn

FIGURE 16.16 Electrolysis of aqueous tin(II) chloride solution. Tin metal is produced by reduction of Sn^{2+}(aq) ions at the cathode. Chlorine gas is formed at the anode by oxidation of Cl^-(aq) ions, even though the potential for the oxidation of Cl^-(aq) ions is less than that for the oxidation of water (that is, chlorine should be less easily oxidized than water). The result of chemical kinetics, this process illustrates the complexity of some aqueous electrochemistry.

Photos: Charles D. Winters

Photo: Charles D. Winters

Formation of $Cl_2(g)$ at the anode in the electrolysis of aqueous $SnCl_2$ solution is contrary to prediction based on $E°$ values. If the electrode reactions were

$$\textit{Cathode, reduction:} \quad Sn^{2+}(aq) + 2\,e^- \longrightarrow Sn(s)$$

$$\textit{Anode, oxidation:} \quad 2\,H_2O(\ell) \longrightarrow O_2(g) + 4\,H^+(aq) + 4\,e^-$$

$$E°_{cell} = (-0.14\ V) - (+1.23\ V) = -1.37\ V$$

we would expect that application of an external potential between 1.37 V and 1.50 V would produce $O_2(g)$ at the anode rather than $Cl_2(g)$. To explain the formation of chlorine instead of oxygen, we must take into account rates of reaction. In the commercially important production of $Cl_2(g)$ by electrolysis of aqueous NaCl solution [>>Section 26.11], a voltage is used that is high enough to oxidize both $Cl^-(aq)$ ions and H_2O. However, chloride ions are oxidized much faster than H_2O, with the result that $Cl_2(g)$ is the major product.

Differences from predictions made using Table 16.1 results from the fact that $E°$ values are measured under conditions of no electron flow: cell emf values are measures of the potential for electron flow. In industrial situations, very large currents are used in order to obtain substantial quantities of products over reasonable periods of time. Especially when gaseous products are formed, an electrically insulating layer of gas can build up around the electrode where the gas is forming, and the applied potential needs to be increased to achieve reaction. *The extra potential needed to make reaction occur, above that predicted from the table of standard reduction potentials*, is called **overvoltage**. The magnitude of the overvoltage cannot be theoretically calculated; it depends on the substances being produced, the magnitude of the current being passed, and the nature of the electrode surfaces.

Another instance in which rates are important is the industrial application of zinc metal onto the surface of steel (*galvanizing*) by electrolysis in acidic zinc sulfate solutions. This is one example of *electroplating*—electrolytically depositing a coating of one metal onto the surface of another. Consultation of Table 16.1 would suggest that water will be reduced to hydrogen in preference to the reduction of aquated zinc ions to zinc metal. However, at large currents, hydrogen formation develops a considerable overvoltage (> 1 V), so that if the applied voltage is gradually increased, an emf is reached at which zinc ions are reduced, but water is not. In this case, overvoltage works in our favour: without it, hydrogen would be produced rather than the coating of zinc.

One other factor—the concentration of species in solution—must be taken into account when discussing electrolysis processes. As shown in Section 16.5, the potential for a species in solution to be oxidized or reduced depends on concentration. Unless standard conditions are used, predictions based on $E°$ values are merely qualitative. In addition, the rate of a half-reaction depends on the concentration of the reactive species at the electrode surface. At a very low concentration, the rate of an oxidation-reduction reaction may depend on the rate at which an ion diffuses from the solution to the electrode surface.

Think about It

e16.17 Examine the processes involved in the electrolysis of water.

WORKED EXAMPLE 16.9—ELECTROLYSIS

Predict how products of electrolysis of aqueous solutions made by dissolving NaF(s), NaCl(s), NaBr(s), and NaI(s) in different samples of water are likely to be different.

Strategy

The main criterion used to predict the chemistry in an electrolysis cell is the ease of oxidation and reduction, an assessment based on $E°$ values.

Solution

The cathode reaction in all four examples solutions presents no problem—water is reduced to H_2 gas in preference to reduction of $Na^+(aq)$ ions, as in the electrolysis of aqueous NaI solution. So, the main cathode reaction in all cases is

$$2\,H_2O(\ell) + 2\,e^- \longrightarrow H_2(g) + 2\,OH^-(aq)$$

$$E°_{cathode} = -0.83\ V$$

At the anode, we need to compare the ease of oxidation of halide ions (to elemental halogens) and water (to oxygen gas). Based on $E°$ values, the ease of oxidation of halide ions is $I^-(aq) > Br^-(aq) > Cl^-(aq) \gg F^-(aq)$. Fluoride ion is much more difficult to oxidize than water, and electrolysis of an aqueous solution containing this ion results exclusively in formation of $O_2(g)$. The main anode reaction for NaF solutions is

$$2\,H_2O(\ell) \longrightarrow O_2(g) + 4\,H^+(aq) + 4\,e^-$$

Therefore, in this case,

$$E°_{cell} = (-0.83\,V) - (+1.23\,V) = 2.06\,V$$

Aquated bromide and iodide ions are considerably easier to oxidize than aquated chloride ions. Since $Cl_2(g)$ is the main product in the electrolysis of solutions of chloride salts, $Br_2(\ell)$ and $I_2(s)$ may be expected as the main products in the electrolysis of aqueous NaBr and NaI solutions. For NaBr solution, the main anode reaction is

$$2\,Br^-(aq) \longrightarrow Br_2(\ell) + 2\,e^-$$

so $E°_{cell}$ is

$$E°_{cell} = (-0.83\,V) - (+1.08\,V) = -1.91\,V$$

EXERCISE 16.27—ELECTROLYSIS

Predict the chemical reactions that will occur at the two electrodes in the electrolysis of an aqueous sodium hydroxide solution. What is the minimum voltage needed to cause this reaction to occur?

Interactive Exercise 16.26

Write equations for reactions occurring at each electrode.

SUBMIT

During electrolysis of aqueous solutions, there is competition between water molecules and aquated ions to "grab" electrons at the cathode and to have electrons removed at the anode. When large currents are applied in industrial situations, the production of gaseous products may involve application of an overvoltage (the amount by which the applied emf must be greater than predicted by the table of standard reduction potentials), and the substances produced may be different from those predicted by use of the table of standard reduction potentials.

16.8 Corrosion of Iron

We are all aware of corrosion of iron or steel, at least because of problems with rust on our automobiles (Figure 16.17). On a grander scale, it is estimated that 20% of the iron produced is used to replace iron that has rusted away.

FIGURE 16.17 Corrosion of iron. Rust, referred to as "structural cancer," results in major economic losses.

Charles D. Winters

Péter Gudella/Shutterstock

Corrosion is *the deterioration of metals as the result of oxidation in the presence of air and water.* The metal oxides formed do not have the strength, flexibility, or conductive ability of the metals, and so the functionality of the metal object decreases. In the case of iron corrosion, iron is converted to red-brown rust, which is a hydrated iron(II) oxide of variable composition, approximately represented as $Fe_2O_3.nH_2O(s)$, where n is about 3. The overall reaction of iron corrosion can be represented as

$$2\ Fe(s) + \tfrac{3}{2}\ O_2(g) + n\ H_2O(\ell) \longrightarrow Fe_2O_3.nH_2O(s)$$

Corrosion usually does not happen uniformly over the surface of a piece of iron, even of a flat sheet, and corrosion can happen even when the iron is not in direct contact with air or water. These are clues to the fact that corrosion is an electrochemical process, with oxidation and reduction occurring in different regions of the surface of a piece of iron. In a simple representation of iron corrosion (Figure 16.18), there are regions on the surface where oxidation of iron occurs (called *anodic regions*):

$$\textit{Anodic regions: } Fe(s) \longrightarrow Fe^{2+}(aq) + 2\ e^-$$

and regions where reduction of oxygen occurs (called *cathodic regions*):

$$\textit{Cathodic regions: } O_2(g) + 2\ H_2O(\ell) + 4\ e^- \longrightarrow 4\ OH^-(aq)$$

Electrons can transfer from an anodic region to a cathodic region through the metal, and electrical neutrality can be maintained by migration of ions through moisture between the regions. There is a direct correspondence between the process of iron corrosion and a voltaic cell, and a potential between a cathodic region and an anodic region drives the electron-transfer process.

FIGURE 16.18 Corrosion of iron as an electrochemical process. This is a spontaneous process corresponding with a voltaic cell.

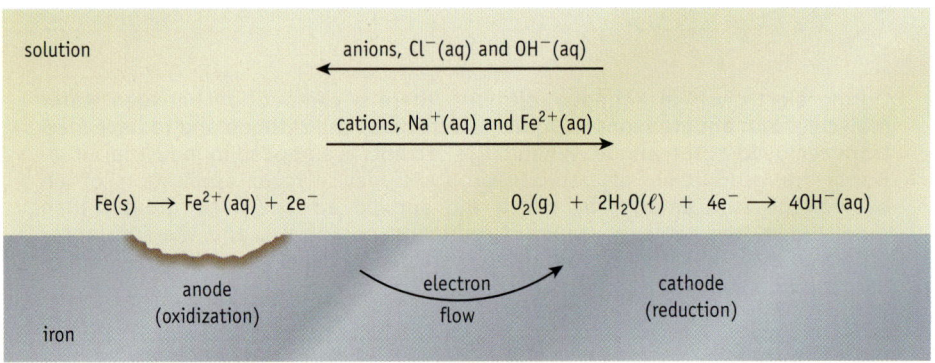

OH$^-$(aq) ions migrate from the cathodic region through moisture toward the anodic region, and Fe^{2+}(aq) ions migrate in the opposite direction. OH$^-$(aq) ions move through water much faster than Fe^{2+}(aq) ions, so close to the anodic region, iron(II) hydroxide precipitates out:

$$Fe^{2+}(aq) + 2\ OH^-(aq) \longrightarrow Fe(OH)_2(s)$$

This is quickly oxidized in air to iron(III) hydroxide, $Fe(OH)_3(s)$, which gradually changes in structure to the hydrated oxide $Fe_2O_3.nH_2O(s)$.

Acceleration of the Rate of Corrosion

There are a number of factors that accelerate the corrosion of iron. These are related to a decrease of resistance to the migration of electrons through the moisture, or to an increase of the electric potential between the cathodic and anodic regions, which can be thought of as an increase of "driving force" of the electrons from anode to cathode. Bear in mind that an increase of cell potential can be brought about by an increase of the reduction potential at the cathode or a decrease of the reduction potential at the anode. An awareness of the chemistry of corrosion can help us to avoid conditions that cause rapid rusting.

Think about It

e16.18 Examine the processes involved in the rusting of iron.

Contact with Salty Water It is well known that iron exposed to salty water is prone to rapid rusting. This is very evident in cars that are driven in Canadian cities where salt is spread onto the roads to melt snow. The presence of high concentrations of ions in the salty water lowers the resistance to electron flow through the metal by increasing the ability of the "salt bridge" (the salty water) to carry current. The rate of electron flow (as number of electrons transferred from anode to cathode per unit time) is limited by how quickly ions flow through the aqueous medium to maintain electrical neutrality overall.

Deformation of the Iron Some deformation of the shape of iron often occurs during production or in the fabrication of structures from different pieces of iron. In rural situations, deformation of fencing wire can occur if wire is tightly wrapped around itself to maintain tension. All of these circumstances can bring about high probability of corrosion at the site of deformation (Figure 16.19). This can be attributed to deformation of the iron crystal lattice, making it more easily oxidized. In electrochemical terms, the reduction potential ($E_{Fe^{2+}/Fe}$) is lower for iron with lattice deformation than it is for unstressed iron with regular packing of its ions. This means that deformed iron gives up its electrons more easily, and the cell emf is greater.

(a)

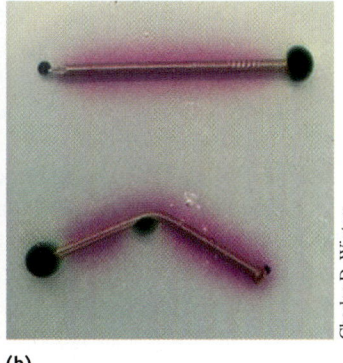
(b)

FIGURE 16.19 Iron corrosion at sites of deformation. (a) Corroded fencing wire in a rural situation. (b) In a laboratory, two iron nails were placed in a gel that contains dissolved phenolphthalein and potassium ferricyanide, $K_3[Fe(CN)_6]$. Over time, anodic regions of corrosion are indicated by blue coloration due to reaction between Fe^{2+}(aq) ions and $Fe(CN)_6^{3-}$(aq) ions to form blue $Fe_4[Fe(CN)_6]_3$. Cathodic regions are indicated by the red colour of phenolphthalein due to increased concentration of OH^-(aq) ions produced by the cathode reaction. Where the iron is deformed by bending the nail, corrosion occurs.

Acidic Conditions It is common knowledge that rusting occurs rather quickly in an acidic environment, and this can be explained using electrochemical principles and an increase of cell emf. We can apply equation 16.2 (the Nernst equation) to the dependence of the reduction potential of the cathode half-reaction:

$$O_2(g) + 2\,H_2O(\ell) + 4\,e^- \longrightarrow 4\,OH^-(aq)$$

$$E_{O_2/OH^-} = E°_{O_2/OH^-} - \frac{0.0257\text{ V}}{4} \times \ln \frac{[OH^-]^4}{p_{O_2}}$$

The more acidic the solution is at the cathodic region, the higher $[H^+]$ is and the lower $[OH^-]$. The Nernst equation tells us that the lower $[OH^-]$ is, the higher the cathode reduction potential (E_{O_2/OH^-}). Consequently, the cell emf ($= E_{cathode} - E_{anode}$) is greater and so the current is greater (more electrons are transferred per unit time).

Contact with a More Noble Metal It is not unusual for plumbers to connect copper water pipes with a steel fitting and then to find some years later that water flow is restricted because the steel has corroded so badly that the rust blocks the pathway (Figure 16.20).

Copper is a more noble metal than iron: copper is less easily oxidized than iron, because the Cu^{2+}(aq)/Cu(s) half-cell has a higher reduction potential than the Fe^{2+}(aq)/Fe(s) half-cell. Putting Cu(s) and Fe(s) in contact in a moist environment creates a voltaic cell in which iron is the anode (where oxidation of iron occurs) and the more

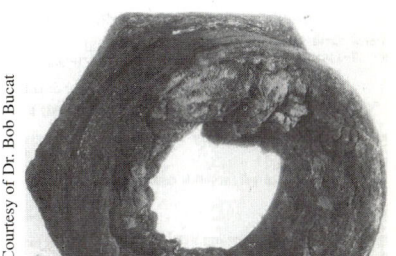

FIGURE 16.20 Corrosion of an iron fitting that has been joined to copper tubing.

"Concrete cancer" in a concrete bridge as a result of corrosion of iron reinforcing mesh.

noble metal is the cathode (where reduction of oxygen occurs). A potential difference to "drive" electrons has been created.

This phenomenon has given rise to a problem with car engines constructed of a steel engine block and an aluminum cylinder head. In this case, of the two metals in contact, steel is the more noble and the aluminum head is prone to corrosion.

Differential Aeration In some circumstances, there are regions on the surface of an iron object where the concentrations (or pressures) of oxygen are different—called *differential aeration*—and rusting is accelerated in those regions with lower $O_2(g)$ concentration. This might seem odd: since corrosion is a reaction with oxygen, you might think that corrosion would preferentially occur where $O_2(g)$ concentration is higher. We fall back on our chemical principles for an explanation.

We can see from the application of the Nernst equation above that the reduction potential of the cathode half-reaction depends on the $O_2(g)$ pressure: the higher the $O_2(g)$ pressure is, the higher E_{O_2/OH^-}. If there are two regions with different oxygen pressures, a potential difference (to "drive" electrons) is created. The region of higher reduction potential, where $p(O_2)$ is higher, will be the cathode (where O_2 is reduced), and regions where $p(O_2)$ is lower will, by default, be anodic regions (where iron is oxidized).

Unless precautions are taken, this phenomenon has enormously significant consequences. One example is the "pitting" of boiler plates: rather than rusting uniformly, rusting "drills" holes in the boiler plates and hot water leaks out. If there is depression on the inside surface, and $p(O_2)$ is lower in the depression than elsewhere, an electrochemical cell is set up in which the depression is the anode, and iron is oxidized. This creates a deeper depression, in which it is likely that there will be even lower $p(O_2)$, leading to greater degree of differential aeration, until the localized depression "eats" through the steel plate.

Another consequence is corrosion of steel piles of a pier or bridge in sea water. There is differential aeration between the surface of the piles just out of water and that under the water. Regions out of the water (higher $p(O_2)$) become cathodic, and corrosion occurs at the anodic regions just below the water line.

In construction of large concrete structures, the wet concrete is usually poured over steel "reinforcing" mesh. It is important that the concrete completely envelopes the mesh, so that it is not in contact with air or water. If the ends of the mesh are exposed to the air, we will have a situation of differential aeration in which the mesh that is enclosed by concrete is the anode region. This means that corrosion of the steel mesh will occur within the concrete. Rust occupies a greater space than the iron from which it came. This creates huge pressures, causing the concrete to crack, and even crumble.

EXERCISE 16.28—CORROSION OF IRON

Steel water pipes connected with brass fittings can corrode rapidly. Account for this phenomenon, and explain why the rate of iron corrosion can vary dramatically from city to city, depending upon the water quality.

EXERCISE 16.29—CORROSION OF IRON

The steel piles supporting a pier corrode most rapidly just below the water line. Why is corrosion worse in this region compared with parts of the pier out of the water, or well below the water line?

Corrosion of iron is an electrochemical process in which iron is oxidized an anodic region and water is reduced at a cathodic region. The product is rust, a hydrated iron oxide. The rate of corrosion of iron is accelerated by the presence of salty water (reducing the resistance to ion migration), deformation of the iron (making it more reactive), more acidic environments (increasing the reduction potential of the cathode half-reaction), contact with a more noble metal (forming a voltaic cell with iron as the anode), or circumstances of differential aeration (creating a cell, with anode in the region of lower oxygen pressure).

Protection against Corrosion of Iron

There are a number of measures that are commonly taken to prevent corrosion, or at least to minimize it. Most of these are based on electrochemical principles.

Sacrificial Anodes When two metals are in contact with each other and an electrolyte solution, a voltaic cell is formed with the more reactive metal as the anode. This can be used to our benefit. For example, if a lump of one of the reactive metals zinc, magnesium, or aluminum is attached to steel that is prone to rusting, the reactive metal becomes the anode and is oxidized away, while only reduction of oxygen occurs on the iron. This method is used on cars in coastal environments or that are driven on roads on which salt has been spread, as well as on steel-hulled ships and oil-drilling rigs (Figure 16.21). In hot water systems with stainless steel storage tanks, it is common to put an aluminum rod in the tank: the aluminum corrodes preferentially. In all of these cases, because the more reactive metal is oxidized away in order to preserve the iron structure, it is called a *sacrificial anode*. They need to be replaced periodically.

(a)

(b)

FIGURE 16.21 Sacrificial anodes. (a) Zinc anodes on a ship. More than one is necessary because the protective action is less effective the further away from the anode due to electrical resistance of the sea water electrolyte. (b) An aluminum anode used on an oil rig.

Because sacrificial anodes function by causing the steel structure to be the cathode of a voltaic cell, this method of corrosion prevention is also called *cathodic protection*—as is protection by an applied potential, described in the next section.

Applied Electric Potential Using their knowledge of corrosion as an electrochemical phenomenon, corrosion scientists often preserve a steel structure by application of a "back emf" between the structure and some expendable, cheap iron such as compressed car bodies. The potential is applied in the direction such that there is electron flow from the scrap iron toward the steel structure. In this way, the scrap iron is made the anode, and it corrodes away, while the structure becomes the cathode where only oxygen reduction occurs. This is a common technique used with jetties, piers, and underground pipelines (Figure 16.22).

In sea water situations, the anode is often made of a noble (unreactive) metal. In these cases, oxidation of water (to oxygen) occurs at the anode, rather than oxidation of the anode itself. Titanium plated with platinum is commonly used for this purpose. Although these anodes are expensive, they do not need to be replaced.

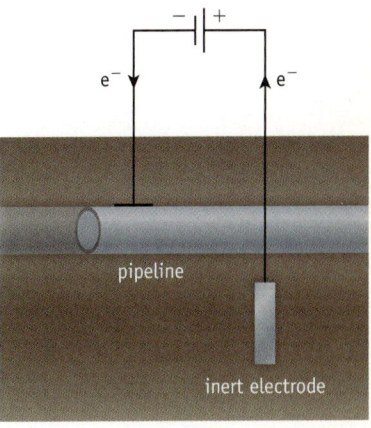

FIGURE 16.22 Cathodic protection of an underground pipeline by application of an electric potential.

Coatings Coating the iron to prevent any contact with air or moisture is a common method of corrosion protection. These may take the form of paints (Figure 16.23) or metallic coatings.

FIGURE 16.23 Corrosion protection by paint. Continuous repainting of the 53 000 tonnes of steel in the Sydney harbour bridge is the task of about 30 full-time workers.

Metallic coatings are of two types—those of metals more reactive than iron, and those of less reactive metals. The first category include coatings of zinc (called "galvanized iron") and aluminum, commonly used on the walls and roofs of factories, silos, and houses, as well as for piping, fencing wire, and nails. Because the metal coating is more reactive than iron, it is preferentially oxidized, and is called a *sacrificial coating*. Even though the coatings are thin, they are long-lasting because in the air they form surface layers of oxides (on aluminum) and carbonates (on zinc) that render them less reactive.

Galvanizing iron sheets is carried out either by dipping the sheets into molten zinc, or by electroplating (Figure 16.13b).

Coating with metals less reactive than iron includes tin plating and chrome plating. These function in the same way as paints, and are used for reasons of low cost or appearance. They suffer from the disadvantage that if the coating is damaged, even by scratching, an electrochemical cell is formed in which the iron is the anode, and oxidation of iron will occur faster than if not coated at all.

Structures fabricated with galvanized iron.

Corrosion Inhibitors It is common practice to add chemicals called *corrosion inhibitors* to boiler waters and the cooling water in automobile radiators. These usually work by forming a protective film over the iron surface. For example, those containing nitrite or chromate ions form an oxide layer.

Alloying Some alloys (intimate mixtures of metals) are more resistant to corrosion than the component metals. Stainless steels are an example, the most common of which contains 18% chromium and 8% nickel. The effectiveness of these stainless steels is due to formation of a chromium oxide layer (Cr_2O_3) similar to that formed on aluminium [>>Section 26.7].

EXERCISE 16.30—PROTECTION AGAINST CORROSION OF IRON

Explain how corrosion of iron is slowed by undamaged coatings of (a) tin, (b)zinc, and (c) paint. In each case, explain what will happen if the protective coating is damaged and the iron is exposed.

Protection of iron against corrosion can be achieved by use of sacrificial anodes (which are preferentially oxidized), applied electric potential that renders the iron the cathode, paint (which prevents contact with air), sacrificial metal coatings, coating with less reactive metals (preventing contact with air), use of corrosion inhibitors (which put a protective film on the surface of the iron), or by using corrosion-resistant alloys.

SUMMARY

Key Concepts

- Oxidation and reduction occur simultaneously when electrons are transferred from one species to another. The species from which electrons are taken is oxidized (and is the reducing agent), and the species which takes the electrons is reduced (and is the oxidizing agent). (Section 16.2)
- The **oxidation state** of an element in a compound is a measure of the degree of oxidation, compared with when it is the uncombined element. Oxidation is accompanied by an increase in the oxidation state of an element in the species, and reduction by a decrease. (Section 16.2)
- Oxidation-reduction reactions can be regarded as the result of competition between species for electrons: the "winner" is reduced, and oxidizes the other species. (Section 16.2)
- **Electrochemistry** concerns interchange of chemical energy into electrical energy by spontaneous processes in **voltaic cells**, and electrical energy into chemical energy by "forced," non-spontaneous processes in **electrolysis**. (Section 16.3)
- In voltaic cells, the oxidation half-reaction occurs at the **anode**, and the reduction half-reaction occurs at the **cathode**. Electrons flow from anode to cathode through an external circuit, and migration of ions through the salt bridge maintains charge neutrality. (Section 16.3)
- **Cell emf** (E_{cell}) is a measure of the difference between the abilities of species in the half-cells to compete for electrons. When all of the species are in their standard states, the cell emf is called the **standard cell emf** ($E°_{cell}$). Values of $E°_{cell}$ must be accompanied by specification of the temperature. (Section 16.4)
- The **half-cell reduction potential** ($E_{half-cell}$) is a quantitative measure of a half-cell's ability to attract electrons. A cell emf is the difference between half-cell reduction potentials: $E_{cell} = E_{cathode} - E_{anode}$. (Section 16.4)
- Half-cells can be placed in a list in order of reduction potentials, in which the **standard hydrogen electrode** (**SHE**) is arbitrarily assigned a value of 0 V. Half-cell reduction potentials depend on concentrations of reaction species. When all species are in their standard states, the half cell reduction potential is called the **standard half-cell reduction potential**, $E°_{half-cell}$. (Section 16.4)
- For any cell, the standard cell emf (at specified temperature) can be calculated from the standard half-cell reduction potentials (at the same temperature): $E°_{cell} = E°_{cathode} - E°_{anode}$. (Section 16.4)
- If reagents are in their standard states, the spontaneous direction of reaction (toward equilibrium) is that between a species on the left side of an half-reaction in the table of standard reduction potentials and any species on the right side of a half-reaction with lower standard reduction potential. (Section 16.4)
- When the species in a voltaic cell are not in their standard states, cell emf can be calculated using the **Nernst equation**. This calculates the difference between the cell emf and the standard cell emf. The values of Q and n are derived from a written equation for the cell reaction. (Section 16.5)
- pH meters and other ion-selective electrodes depend on the variation of the emf of a voltaic cell in which one half-cell contains the solution with unknown concentration of the ion of interest. (Section 16.5)
- When the half-equation for a reduction process involves $H^+(aq)$ ions, the reduction potential depends on the concentration of $H^+(aq)$ ions, in accordance with the Nernst equation. Oxoanions are more powerful oxidizing agents the lower the pH of solution. (Section 16.5)
- There is a direct relationship between the standard cell emf and the logarithm of the equilibrium constant for the overall cell reaction. (Section 16.6)

- During the electrolysis of molten salts, when the applied emf is sufficiently high, the cation of the salt is reduced at the cathode, and the anion is oxidized at the anode. These are the only possible reactions. (Section 16.7)
- During electrolysis of aqueous solutions, there is competition between water molecules and aquated ions to "grab" electrons at the cathode, and to have electrons removed at the anode. When large currents are applied in industrial situations, the production of gaseous products may involve application of an **overvoltage** (the amount by which the applied emf must be greater than predicted by the table of standard reduction potentials), and the substances produced may be different from those predicted by use of the table of standard reduction potentials. (Section 16.7)
- **Corrosion** of iron is an electrochemical process in which iron is oxidized at anodic region and water is reduced at a cathodic region. The product is rust, a hydrated iron oxide. The rate of corrosion of iron is accelerated by the presence of salty water (reducing the resistance to ion migration), deformation of the iron (making it more reactive), more acidic environments (increasing the reduction potential of the cathode half-reaction), contact with a more noble metal (forming a voltaic cell with iron as the anode), or circumstances of differential aeration (creating a cell, with anode in the region of lower oxygen pressure). (Section 16.8)
- Protection of iron against corrosion can be achieved by use of sacrificial anodes (which are preferentially oxidized), applied electric potential that renders the iron the cathode, paint (which prevents contact with air), sacrificial metal coatings, coating with less reactive metals (preventing contact with air), use of corrosion inhibitors (which put a protective film on the surface of the iron), or use of corrosion-resistant alloys. (Section 16.8)

Key Equations

$$E_{cell} = E_{cathode} - E_{anode} \qquad \text{(Section 16.5)}$$

$$E^{\circ}_{cell} = E^{\circ}_{cathode} - E^{\circ}_{anode} \qquad \text{(16.1, Section 16.5)}$$

$$E_{cell} = E^{\circ}_{cell} - \frac{RT}{nF} \ln Q = E^{\circ}_{cell} - \frac{0.0257 \text{ V}}{n} \ln Q \quad \text{at 25°C} \quad \text{(16.2, Section 16.6)}$$

$$\ln K = \frac{nFE^{\circ}_{cell}}{RT} = \frac{nE^{\circ}_{cell}}{0.0257 \text{ V}} \quad \text{at 25°C} \qquad \text{(Section 16.6)}$$

REVIEW QUESTIONS

Section 16.2: Oxidation-Reduction Reactions

16.31 What is the oxidation state of each element in the ions or compounds with the formulas given?
(a) PF_6^-
(b) $H_2AsO_4^-$
(c) UO^{2+}
(d) N_2O_5
(e) $POCl_3$
(f) XeO_4^{2-}

16.32 Calculate the oxidation state of carbon atoms (or the average, in the case of molecules with two C atoms) in each of the following compounds: (a) methane (CH_4), (b) methanol (CH_3OH), (c) formaldehyde (CH_3CHO), (d) acetic acid (CH_3COOH), (e) carbon monoxide (CO), and (f) carbon dioxide (CO_2). Is reaction in the sequence methane → methanol → formaldehyde → acetic acid → carbon monoxide → carbon dioxide oxidation or reduction?

16.33 Which of the equations below describe oxidation-reduction reactions?
(a) $OH^-(aq) + H^+(aq) \longrightarrow H_2O(\ell)$
(b) $Cu(s) + Cl_2(g) \longrightarrow CuCl_2(s)$
(c) $CO_3^{2-}(aq) + 2 H^+(aq) \longrightarrow CO_2(g) + H_2O(\ell)$
(d) $2 S_2O_3^{2-}(aq) + I_2(s) \longrightarrow S_4O_6^-(aq) + 2 I^-(aq)$

16.34 Which two of the following equations describe oxidation-reduction reactions? In each case, give reasons for your answer. What class of reaction is the remaining reaction?
(a) $Zn(s) + 2 NO_3^-(aq) + 4 H^+(aq) \longrightarrow Zn^{2+}(aq) + 2 NO_2(g) + 2 H_2O(\ell)$
(b) $Zn(OH)_2(s) + 2 H^+(aq) \longrightarrow Zn^{2+}(aq) + 2 H_2O(\ell)$
(c) $Ca(s) + 2 H_2O(\ell) \longrightarrow Ca(OH)_2(s) + H_2(g)$

16.35 Balance the following equations for oxidation-reduction reactions in acidic solution.
(a) $Sn(s) + H^+(aq) \longrightarrow Sn^{2+}(aq) + H_2(g)$
(b) $Cr_2O_7^{2-}(aq) + Fe^{2+}(aq) \longrightarrow Cr^{3+}(aq) + Fe^{3+}(aq)$
(c) $MnO_2(s) + Cl^-(aq) \longrightarrow Mn^{2+}(aq) + Cl_2(g)$
(d) $HCHO(aq) + Ag^+(aq) \longrightarrow HCOOH(aq) + Ag(s)$

16.36 Balance the following half-equations for atoms and charge.
(a) $UO_2^+(aq) \longrightarrow U^{4+}(aq)$ (in acidic solution)
(b) $ClO_3^-(aq) \longrightarrow Cl^-(aq)$ (in acidic solution)
(c) $N_2H_4(aq) \longrightarrow N_2(g)$ (in basic solution)
(d) $ClO^-(aq) \longrightarrow Cl^-(aq)$ (in basic solution)

Section 16.3: Voltaic Cells: Electricity from Chemical Change

16.37 A voltaic cell is constructed using the reaction of chromium metal and aquated iron(II) ions:

$$2\,Cr(s) + 3\,Fe^{2+}(aq) \longrightarrow 2\,Cr^{3+}(aq) + 3\,Fe(s)$$

Complete the following sentences: Electrons in the external circuit flow from the ___ electrode to the ___ electrode. Negative ions move in the salt bridge from the ___ half-cell to the ___ half-cell. The half-reaction at the anode is ___ and that at the cathode is ___.

16.38 The half-cells $Fe^{2+}(aq)|Fe(s)$ and $O_2(g)|H_2O$ (in acid solution) are linked to create a voltaic cell.
(a) Write equations for the oxidation and reduction half-reactions and for the overall cell reaction.
(b) Which half-reaction occurs in the anode compartment and which occurs in the cathode compartment?
(c) Complete the following sentences: Electrons in the external circuit flow from the ___ electrode to the ___ electrode. Negative ions move in the salt bridge from the ___ half-cell to the ___ half-cell.

16.39 Magnesium metal is oxidized and silver ions are reduced in a voltaic cell using $Mg^{2+}(aq, 1\ mol\ L^{-1})|Mg$ and $Ag^+(aq, 1\ mol\ L^{-1})|Ag$ half-cells:

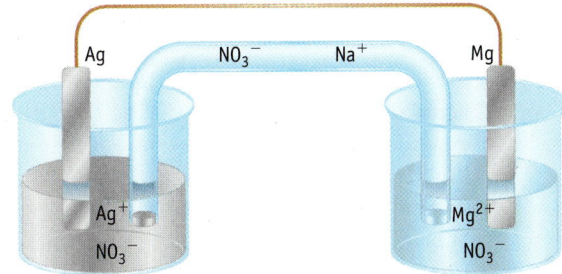

(a) Label each part of the cell.
(b) Write equations for the half-reactions occurring at the anode and the cathode as well as an equation for the net reaction in the cell.
(c) Trace the movement of electrons in the external circuit. Assuming the salt bridge contains a $NaNO_3$

solution, trace the movement of the $Na^+(aq)$ and $NO_3^-(aq)$ ions in the salt bridge that occurs when a voltaic cell produces current. Why is a salt bridge required in a cell?

Section 16.4: Cell emf, and Half-Cell Reduction Potentials

16.40 When a voltaic cell is constructed from a $Ag^+(aq, 1.0\ mol\ L^{-1})|Ag(s)$ half-cell and a standard hydrogen electrode, the cell emf is measured to be 0.80 V, and the $Ag^+(aq)|Ag(s)$ half-cell is found to be the cathode. When another cell is composed of a $Ag^+(aq), 1.0\ mol\ L^{-1}|Ag(s)$ half-cell and a $Zn^{2+}(aq, 1.0\ mol\ L^{-1})|Zn(s)$ half-cell, the cell emf is measured to be 1.56 V, and the $Zn^{2+}(aq)|Zn(s)$ half-cell is the anode. What is the reduction potential of the $Zn^{2+}(aq)|Zn(s)$ half-cell?

16.41 Figure 16.11 lists values of some half-cell reduction potentials in the scale in which the SHE is assigned a reduction potential of 0 V. If chemists had arbitrarily chosen the $Fe^{2+}(aq)|Fe(s)$ as the arbitrary zero of half-cell reduction potentials, what would have been the reduction potentials of (a) the $Cu^{2+}(aq)|Cu(s)$ half-cell, and (b) the $Zn^{2+}(aq)|Zn(s)$ half-cell?

16.42 The half-cell represented by $Cd^{2+}(aq, 1.0\ mol\ L^{-1})|Cd(s)$ is joined to a standard hydrogen electrode, using a salt bridge, to form a voltaic cell at 25 °C. The SHE is the cathode, and E_{cell} (= $E°_{cell}$ under these standard conditions) is 0.40 V. What is $E°_{Cd^{2+}/Cd}$ at 25 °C? Be sure to assign the correct sign (+ or −) to your answer.

16.43 (a) For the half-cell written as $NO_3^-(aq)|NO(g)|Pt$, the half-equation for the reduction half-reaction is

$$NO_3^-(aq) + 4\,H^+(aq) + 3\,e^- \longrightarrow NO(g) + 2\,H_2O(\ell)$$

Specify the conditions under which $E_{NO_3^-/NO} = E°_{NO_3^-/NO}$ at any given temperature.
(b) When a voltaic cell is formed from a standard $NO_3^-(aq)|NO(g)|Pt$ half-cell and a SHE, all at 25 °C, the SHE is the anode and $E°_{cell} = 0.96$ V. Calculate $E°_{NO_3^-/NO}$ at 25 °C.

16.44 Balance each of the following unbalanced equations for cell reactions in acidic conditions, then calculate the standard cell emf, and decide whether the equation is written in the direction of spontaneous reaction.
(a) $Sn^{2+}(aq) + Ag(s) \longrightarrow Sn(s) + Ag^+(aq)$
(b) $Al(s) + Sn^{4+}(aq) \longrightarrow Sn^{2+}(aq) + Al^{3+}(aq)$
(c) $ClO_3^-(aq) + Ce^{3+}(aq) \longrightarrow Cl^-(aq) + Ce^{4+}(aq)$
(d) $Cu(s) + NO_3^-(aq) \longrightarrow Cu^{2+}(aq) + NO(g)$

16.45 You want to set up a series of voltaic cells with specific cell voltages. A $Zn^{2+}(aq, 1.0\ mol\ L^{-1})|Zn(s)$ half-cell is in one compartment. Identify several half-cells that you could use so that the cell emf will be close to (a) 1.1 V, and (b) 0.5 V. Consider cells in which the $Zn^{2+}(aq)|Zn(s)$ half-cell is either the cathode or the anode.

16.46 The following half-cells are available: Ag^+ (aq, 1.0 mol L^{-1})|Ag(s), Zn^{2+}(aq, 1.0 mol L^{-1})|Zn(s), Cu^{2+}(aq, 1.0 mol L^{-1})|Cu(s), and Co^{2+}(aq, 1.0 mol L^{-1})|Co(s). Linking any two half-cells makes a voltaic cell. Given four different half-cells, six voltaic cells are possible. These are labelled, for simplicity, Ag-Zn, Ag-Cu, Ag-Co, Zn-Cu, Zn-Co, and Cu-Co.

(a) In which of the voltaic cells is the copper electrode the cathode? In which of the cells is the cobalt electrode the anode?

(b) Which cell generates the highest voltage? Which cell generates the lowest voltage?

16.47 In the table of standard reduction potentials (Table 16.1), locate the half-equations for the reduction of the non-metals F_2(g), Cl_2(g), Br_2(ℓ), and I_2(s) (to halide ions), and O_2(g), S(s), and Se(s) (to H_2X) in aqueous acidic solution.

(a) Among the species in these half-equations, which is the least powerful oxidizing agent?

(b) Which species is the least powerful reducing agent?

(c) Which species is(are) capable of oxidizing H_2O(ℓ) to O_2(g)?

(d) Which species is(are) capable of oxidizing H_2S(g) to S(s)?

(e) Is O_2(g) capable of oxidizing I^-(aq) ions to I_2(s) in acidic solution?

(f) Is S(s) capable of oxidizing I^-(aq) ions to I_2(s) in acidic solution?

(g) Is the equation H_2S(aq) + Se(s) $\longrightarrow$ H_2Se(aq) + S(s) written in the direction of spontaneous reaction, assuming all species are at standard concentrations at 25 °C?

(h) Is the equation H_2S(aq) + I_2(s) $\longrightarrow$ $2 H^+$(aq) + $2 I^-$(aq) + S(s) written in the direction of spontaneous reaction, if all species are at standard concentrations at 25 °C?

16.48 Which metal in the following list is easiest to oxidize: Fe, Ag, Zn, Mg, Au? Which metal is the most difficult to oxidize?

(a) With which of the metals Cu(s), Al(s), Ag(s), Fe(s), and Zn(s) is reaction with H^+(aq) ions to produce H_2(g) a spontaneous reaction?

(b) Select from Table 16.1 three oxidizing agents that are capable of oxidizing Cl^-(aq) ions to Cl_2(g).

16.49 Determine which of the following equations refer to spontaneous oxidation-reduction reactions. Assume that every species listed is in its standard state and the temperature is at 25 °C.

(a) Ni^{2+}(aq) + H_2(g) $\longrightarrow$ Ni(s) + $2 H^+$(aq)

(b) Fe^{3+}(aq) + $2 I^-$(aq) $\longrightarrow$ Fe^{2+}(aq) + I_2(s)

(c) Br_2(ℓ) + $2 Cl^-$(aq) $\longrightarrow$ $2 Br^-$(aq) + Cl_2(g)

(d) $Cr_2O_7^{2-}$(aq) + $6 Fe^{2+}$(aq) + $14 H^+$(aq) $\longrightarrow$
$2 Cr^{3+}$(aq) + $6 Fe^{3+}$(aq) + $7 H_2O$(ℓ)

16.50 Which of the following elements is the best reducing agent: Cu(s), Zn(s), Fe(s), Ag(s), or Cr(s)?

Interactive Exercises 16.51–16.52  Make predictions about ease of reduction on the basis of standard reduction potentials.

16.53 Which of the following equations are written in the direction of spontaneous reaction when all species are at standard concentrations at 25 °C?

(a) Zn(s) + I_2(s) $\longrightarrow$ Zn^{2+}(aq) + $2 I^-$(aq)

(b) $2 Cl^-$(aq) + I_2(s) $\longrightarrow$ Cl_2(g) + $2 I^-$(aq)

(c) $2 Na^+$(aq) + $2 Cl^-$(aq) $\longrightarrow$ 2 Na(s) + Cl_2(g)

(d) 2 K(s) + H_2O(ℓ) $\longrightarrow$ $2 K^+$(aq) + H_2(g)
$+ 2 OH^-$(aq)

16.54 (a) Which halogen is most easily reduced in acidic solution: F_2(g), Cl_2(g), Br_2(ℓ), or I_2(s)?

(b) Identify the halogens that are better oxidizing agents than MnO_2(s) in acidic solution.

16.55 A KI solution is added drop-wise to a pale blue solution $Cu(NO_3)_2$ solution. The solution changes to a brown colour and a precipitate forms. In contrast, no change is observed if a KCl solution or a KBr solution is added to aqueous $Cu(NO_3)_2$ solution. Consult Table 16.1 to explain the dissimilar results seen with the different halides. Write an equation for the reaction that occurs when a KI solution and a $Cu(NO_3)_2$ solution are mixed.

Section 16.5: Voltaic Cells under Non-standard Conditions

16.56 The half-cells Fe^{2+}(aq, 0.024 mol L^{-1})|Fe(s) and H^+ (aq, 0.056 mol L^{-1})|H_2(1.0 bar) are linked by a salt bridge to create a voltaic cell. Determine the cell emf (E_{cell}) at 298 K.

16.57 One half-cell in a voltaic cell at 25 °C is constructed from a silver wire dipped into a 0.25 mol L^{-1} $AgNO_3$ solution. The other half-cell consists of a zinc electrode in a 0.010 mol L^{-1} $Zn(NO_3)_2$ solution. Calculate the cell emf.

16.58 (a) Write the balanced half-equation for reduction of aquated dichromate ions, $Cr_2O_7^{2-}$(aq), to aquated chromium(III) ions, Cr^{3+}(aq), in acidic solution, and use Table 16.1 to decide the reduction potential of a half-cell containing these species, both at 1.0 mol L^{-1}, at pH = 0.0 and at 25 °C.

(b) Calculate the reduction potential of this half-cell at pH = 4.0 if all other conditions remain unchanged.

(c) List some species that 1.0 mol L^{-1} $Cr_2O_7^{2-}$(aq) ions can oxidize at pH = 0.0 but cannot at pH = 4.0—assume $[Cr^{3+}]$ = 1.0 mol L^{-1} and at 25 °C in both cases.

16.59 Consider an electrochemical cell at 25 °C based on the half-reactions

$$Ni^{2+}(aq) + 2\,e^- \longrightarrow Ni(s)$$
and $$Cd^{2+}(aq) + 2\,e^- \longrightarrow Cd(s)$$

in which $[Ni^{2+}]$ and $[Cd^{2+}]$ are both $1.0\ mol\ L^{-1}$.

(a) Draw the cell and label each of the components, including the anode, cathode, and salt bridge.

(b) Use the equations for the half-reactions to write a balanced, net equation for the overall cell reaction.

(c) What is the sign of each electrode?

(d) What is the value of $E°_{cell}$?

(e) In which direction do electrons flow in the external circuit?

(f) If a salt bridge containing $NaNO_3$ solution connects the two half-cells, in which direction do the $Na^+(aq)$ ions move? In which direction do the $NO_3^-(aq)$ ions move?

(g) Calculate the equilibrium constant for the cell reaction at 25 °C.

(h) If the concentration of $Cd^{2+}(aq)$ ions is reduced to $0.010\ mol\ L^{-1}$, keeping $[Ni^{2+}] = 1.0\ mol\ L^{-1}$, what is E_{cell}? Is the net reaction still that given in question (b)?

16.60 (a) Is it easier to reduce water in acidic or basic solution? To evaluate this, consider the half-reaction

$$2\,H_2O(\ell) + 2\,e^- \longrightarrow 2\,OH^-(aq) + H_2(g)$$
$$E°_{cell} = -0.83\ V\ at\ 25°C$$

(b) What is the reduction potential for water in solutions at pH 7 (neutral) and pH 1 (acidic)? Comment on the value of $E°$ at pH 1.

Section 16.6: Standard Cell emf and Equilibrium Constant

16.61 Calculate $E°_{cell}$ at 25 °C and then the equilibrium constant at 25 °C for the following reactions:

(a) $2\,Fe^{3+}(aq) + 2\,I^-(aq) \longrightarrow 2\,Fe^{2+}(aq) + I_2(s)$

(b) $I_2(s) + 2\,Br^-(aq) \longrightarrow 2\,I^-(aq) + Br_2(aq)$

16.62 Aquated iron(II) ions undergo a disproportionation reaction to give Fe(s) and aquated iron(III) ions. That is, iron(II) ions are both oxidized and reduced in the same reaction:

$$3\,Fe^{2+}(aq) \longrightarrow Fe(s) + 2\,Fe^{3+}(aq)$$

(a) Separate the equation for the disproportionation into two half-equations, one for oxidation and one for reduction.

(b) From Table 16.1, decide if the equation is written in the direction of spontaneous reaction (assuming $[Fe^{3+}]$ and $[Fe^{2+}]$ are both $1.0\ mol\ L^{-1}$ and the temperature is 25 °C).

(c) Calculate the equilibrium constant for this reaction at 25 °C.

Interactive Exercises
16.63–16.64
Calculate K_{sp} from E_{cell} for a given reaction.

16.65 Use the standard reduction potentials (Appendix E) for the half-reactions

$$[AuCl_4]^-(aq) + 3\,e^- \longrightarrow Au(s) + 4\,Cl^-(aq)$$
and $$Au^{3+}(aq) + 3\,e^- \longrightarrow Au(s)$$

to calculate the value of the formation constant (K_f) for the complex ion $[AuCl_4]^-(aq)$ at 25°C.

16.66 When a voltaic cell is constructed based on the following half-reactions, with all species in their standard states and at 25 °C, a cell emf of 0.146 V is measured.

Anode: $Ag(s) \longrightarrow Ag^+(aq) + e^-$
Cathode: $Ag_2SO_4(s) + 2\,e^- \longrightarrow 2\,Ag(s) + SO_4^{2-}(aq)$

(a) What is the standard reduction potential of the cathode at 25 °C?

(b) Calculate the solubility product (K_{sp}) for Ag_2SO_4 at 25 °C.

Section 16.7: Electrolysis: Chemical Change Using Electrical Energy

16.67 Draw a diagram of the apparatus used to electrolyze molten NaCl. Identify the anode and the cathode. Trace the movement of electrons through the external circuit and the movement of ions in the electrolysis cell.

16.68 Draw a diagram of the apparatus used to electrolyze aqueous $CuCl_2$ solution. Identify the reaction products, the anode, and the cathode. Trace the movement of electrons through the external circuit and the movement of ions in the electrolysis cell.

16.69 Which product, $O_2(g)$ or $F_2(g)$, would you expect to form at the anode in the electrolysis of an aqueous KF solution? Explain your reasoning.

16.70 An aqueous solution of sodium sulfide (Na_2S) is placed in a beaker with two inert platinum electrodes. When the cell is attached to an external battery, electrolysis occurs.

(a) Hydrogen gas and aquated hydroxide ions form at the cathode. Write an equation for the half-reaction that occurs at this electrode.

(b) Sulfur is the primary product at the anode. Write an equation for its formation.

Section 16.8: Corrosion of Iron

16.71 Given a choice between galvanized steel and chrome-plated steel for a bumper on your off-road vehicle, which would you choose? Why?

16.72 Make a 1:1 correspondence between the components of a voltaic cell and the processes in corrosion of iron.

16.73 Shown here is a diagrammatic representation of a steel wharf that is protected from corrosion by application of

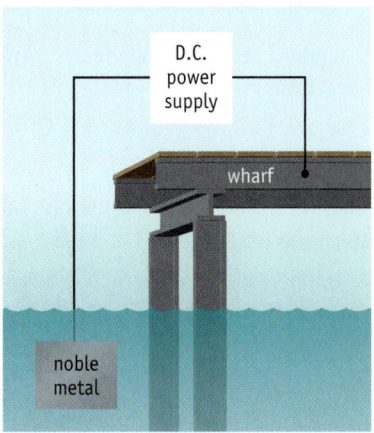

an electric potential between the wharf and a block of a noble metal:

Indicate:
(a) the path and direction of electron flow
(b) the direction of movement of $Na^+(aq)$ and $Ca^{2+}(aq)$ ions through the sea water
(c) the direction of movement of $Cl^-(aq)$ ions through the sea water
(d) the anode and an equation for the half-reaction at the anode
(e) the cathode and an equation for the half-reaction at the cathode
(f) the signs on the terminals of the DC power supply

SUMMARY AND CONCEPTUAL QUESTIONS

16.74 If you are going away on a holiday, you might clean the stainless steel kitchen sink and leave a sponge or cloth on the sink for three weeks. Explain why you come back to find evidence of rusting underneath the sponge where there is an oxygen deficiency, rather than at the surface exposed to air.

16.75 What advice would you have given to a yachtsman, ignorant of electrochemistry, who had trouble with two of his yachts:
(a) One was made of copper alloy sheets (toxic to barnacles) with steel screws, which began to leak.
(b) He carelessly left a steel wrench in the bottom of an aluminium yacht for a long time. When he came back, there was a wrench-shaped hole in the hull.

16.76 Why are food containers protected against corrosion by tin-plating, rather than galvanizing?

16.77 An electric potential is applied between the steel piles of a pier and a block of platinum-coated titanium, with the intention of corrosion by cathodic protection. A worker connects the applied emf in the wrong direction. What would be the consequences?

16.78 Four metals, A, B, C, and D, exhibit the following properties:
(a) Only A and C react with 1.0 mol L^{-1} HCl solution to form $H_2(g)$.
(b) When C is added to solutions of the ions of the other metals, metallic B, D, and A are formed.
(c) Metal D reduces $B^{n+}(aq)$ ions to form metallic B and $D^{n+}(aq)$ ions.

Based on this information, arrange the four metals in order of increasing ability to act as reducing agents.

Spontaneous Change: How Far?

CHAPTER 17

AFP/Getty Images

17.1 Case Study: Photochemical Smog and Chemical Equilibrium

Many urban areas experience smogs of different sorts, with different chemical and physical origins. For many decades, London was famous for its "pea-soupers"—a haze due a combination of damp, foggy atmospheric conditions and smoke particles and sulfur dioxide emitted by industries and coal-fired home heating systems. Hence the name *smog* (= *sm*oke + f*og*). Thousands, if not millions, of deaths have been attributed to such smogs, largely because of harmful effects of both particles and sulfur dioxide on the respiratory system. After the deaths of 4000 Londoners were attributed to a four-day smog in 1952, and the introduction of the Clean Air Act of 1956, London smogs are now almost a thing of the past.

Since the 1950s, another form of air pollution called **photochemical smog** has been drastically affecting the quality of people's lives. This type of smog results from dramatically increasing populations in larger cities and their demands for industrial products and dependence on motor vehicles for transport. The photograph above of Santiago, Chile, shows a brown photochemical smog in a geographical basin and under clear skies—the classical conditions for formation of such smogs [>>Section 17.10].

Around the world, many cities—Los Angeles, Mexico City, and Beijing are notorious—have photochemical smog problems. In Canada, the urban belt from Windsor to Quebec City has experienced severe photochemical smogs, as has the Lower Fraser Valley in British Columbia. The highest concentrations of the pollutant ozone, $O_3(g)$, are found quite a distance east of Vancouver, where the *primary pollutants* are emitted into the air. While these primary pollutants slowly react to form the real "nasties" due to the initiating effect of sunlight, as described in Section 17.10, the air mass containing them usually shifts eastward.

In December of 2005 in Tehran, Iran, public places were closed and many people were hospitalized due to the effects of a severe photochemical smog. In the coming decades, most of the world's population increase will occur in urban areas of poorer countries. The population of Dar-es-Salam in Tanzania doubles every 12 years. Socioeconomic development with increasing urbanization in countries of the western Pacific region brings with it serious environmental challenges. According to the World Health Organisation, urban air pollution (much of it attributable to vehicles, industry, and energy generation) kills about 800 000 people each year.

While the family economic circumstances for people of India benefit from the recent production of a very cheap, no-frills car, the prospect of a significant fraction of India's 1.2 billion people driving cars without pollution controls provides an environmental dilemma at the national level.

An understanding of the origins and chemistry of photochemical smogs can help policy makers make wise choices about regulations to minimize their formation. To understand these processes, and how we might intervene to minimize smog formation, we need to understand the distinction between thermodynamics (how far a reaction goes before equilibrium is reached, and how this varies with temperature) and kinetics (how fast a reaction mixture proceeds toward the equilibrium condition). The thermodynamics and kinetics of a reaction are independent factors: the question of how fast cannot be predicted from that of how far. In what follows, examples are used to make this important distinction.

> *To minimize smog formation, we need to understand the distinction between thermodynamics and kinetics. The thermodynamics and kinetics of a reaction are independent factors: the question of how fast cannot be predicted from that of how far.*

Photochemical smogs are the result of a complicated series of chemical reactions among substances including mainly $NO_2(g)$, formed from $NO(g)$, and *volatile organic compounds*, called *VOCs* (such as low molecular weight hydrocarbons), in an atmospheric "soup." These reactions are initiated by the action of ultraviolet radiation from the sun on NO_2 molecules in the polluted air of industrialized cities with heavy traffic [>>Section 17.10]. A simplified, unbalanced equation for the overall process, which gives no indication of the mechanism of how it happens, is the following:

$$NO + O_2 + VOCs \xrightarrow{\text{uv light}} O_3 + HNO_3 + \text{"organic"peroxides}$$

VOCs are emitted into the atmosphere from car exhaust as unburnt fuels. You might rightly inquire about the source of $NO(g)$ in the air, since our common experience is that $N_2(g)$ and $O_2(g)$ do not react with each other at room temperature. After all, the air we breathe consists mostly of N_2 and O_2 molecules. The concentration of $NO(g)$ in the air of crowded cities subject to photochemical smogs builds up during peak traffic periods, suggesting that the $NO(g)$ is emitted by cars. And yet the gasoline burned in cars contains almost no N-containing compounds that might give rise to $NO(g)$ as a combustion product (like CO and CO_2 are formed from reaction of oxygen with the carbon-containing molecules in fuels).

In fact, $NO(g)$ is formed from reaction between $N_2(g)$ and $O_2(g)$ at the high temperatures in the combustion chambers of cars and industrial furnaces by the reaction

$$\text{Reaction 1:} \quad N_2(g) + O_2(g) \rightleftharpoons 2\,NO(g)$$

The chemistry underlying this negligible formation of $NO(g)$ at low temperatures, but significant formation at high temperatures, contrasts with the chemistry of similar observations for the reaction

$$\text{Reaction 2:} \quad 2\,H_2(g) + O_2(g) \rightleftharpoons 2\,H_2O(g)$$

Let's begin with thermodynamic analysis of reaction 2. From data listed in Appendix C, we can calculate that at 25 °C the equilibrium constant for this reaction is $1 \times 10^{+80}$—an enormous equilibrium constant! This means that at equilibrium at 25 °C:

$$Q = \frac{(p_{H_2O})^2}{(p_{H_2})^2(p_{O_2})} = K = 1 \times 10^{+80}$$

So if we had a reaction mixture at 25°C in which $p_{H_2} = 1$ bar, and $p_{O_2} = 1$ bar, it would be at equilibrium only if $p_{H_2O} = 1 \times 10^{+40}$ bar. This is an impossibly large pressure, but it does tell us that a reaction mixture with, say, 1 bar of $H_2(g)$ and $O_2(g)$ and no $H_2O(g)$ is a long way from equilibrium. Other ways of saying this are that the reaction to form water is *thermodynamically spontaneous*, or that the reaction mixture of $H_2(g)$ and $O_2(g)$ is *thermodynamically unstable* with respect to one consisting mostly of water vapour (or, conversely, that a reaction mixture containing mostly water vapour with tiny amounts of $H_2(g)$ and $O_2(g)$ is *more stable* than one containing mostly $H_2(g)$ and $O_2(g)$ with a tiny amount of water). Colloquially, we sometimes say that the $H_2(g)$ and $O_2(g)$ "want" to react to form H_2O.

And yet we can watch a clear glass jar containing $H_2(g)$ and $O_2(g)$ at ambient temperatures for years without any signs that reaction takes place. This is not because the reaction mixture is at equilibrium but because it is so very slow at reacting to achieve equilibrium. We refer to the apparently stable reaction mixture as *kinetically inert* (or, sometimes, *kinetically stable*).

The term *inert* means unreactive. To say that a reaction mixture containing $H_2(g)$ and $O_2(g)$ is *kinetically inert* means that no reaction is visible because of the very slow rate of its reaction (rather than for the thermodynamic reason that the reaction has a very small equilibrium constant). A mixture of $H_2(g)$ and $O_2(g)$ cannot do what it "wants" to do.

This is not something that we could have predicted: the rates of reactions can only be found through measurement [>>Chapter 18]. If we raise the temperature of the reaction mixture (or even parts of it) to about 200 °C, the reaction happens very quickly indeed (often described as "explosively"). This is not because the equilibrium constant for the reaction is bigger at the higher temperature (indeed, like all exothermic reactions, the equilibrium constant is smaller at higher temperatures), but because, like all reactions, it proceeds to equilibrium faster at higher temperatures [>>Section 18.6].

This analysis is important because if we know that $H_2(g)$ and $O_2(g)$ do not react because the rate of reaction is so slow, although the equilibrium constant is large, it may be worth investing in research to design catalysts that can speed up the reaction. On the other hand, there is no point in trying to speed up a reaction that does not "want" to go— that is, if the mixture of reactants is more stable than the desired products. In fact, the design of catalysts for this reaction between $H_2(g)$ and $O_2(g)$ has been the key component in the development of hydrogen-oxygen fuel cells [<<Section 7.1, e16.9] and will be a major factor in the development of a "hydrogen economy" [<<Section 7.1] in the future.

On the other hand, for reaction 1, the starting point of photochemical smogs, the reason for no visible reaction at room temperature is very different indeed—as indicated by the thermodynamic analysis of Worked Example 17.13. At 25 °C, the equilibrium constant for this reaction is calculated to be 4×10^{-31}:

$$Q = \frac{(p_{NO})^2}{(p_{N_2})(p_{O_2})} = K = 4 \times 10^{-31}$$

In a reaction mixture at 25 °C in which $p_{N_2} = 0.8$ bar and $p_{O_2} = 0.2$ bar (their approximate partial pressures in air), equilibrium would be achieved when p_{NO} is approximately 1×10^{-15} bar. This is extremely small, and we can say that a mixture of $N_2(g)$ and $O_2(g)$ is very close to equilibrium: it is a stable mixture and does not "want" to react to any significant extent. The thermodynamic analysis tells us not to worry about doing research to find a catalyst to speed up the reaction, because a slow rate is not the reason for negligible amounts of product formation.

However, equilibrium constants change as the temperature of the system is changed, and one aspect of the incredible power of thermodynamics is the ability to calculate the equilibrium constants of a reaction at various temperatures [>>Section 17.9]. Worked Example 17.13 shows how we can calculate the equilibrium constant for reaction of $N_2(g)$ and $O_2(g)$ to form $NO(g)$ at 2000 K, which approximates the temperature of combustion in cars and furnaces. We see that the equilibrium constant for this endothermic reaction increases to 4×10^{-4}. As a consequence, in a reaction mixture at equilibrium at 2000 K, if $p_{N_2} = 0.8$ bar and $p_{O_2} = 0.2$ bar, the equilibrium pressure of $NO(g)$ would be a rather significant 0.025 bar. At this temperature, the rate of achievement of equilibrium is fast. If we have thousands of cars producing $NO(g)$ at this level with each piston stroke, large amounts of $NO(g)$ will be emitted into the atmosphere, contributing to the conditions necessary for photochemical smog formation, as described in Section 17.10.

Perhaps you have realized that when car exhaust gases cool down in the atmosphere, the opposite of reaction 1 must have an equilibrium constant of $1/(4 \times 10^{-31})$ or $2.5 \times 10^{+30}$, and that $NO(g)$ "wants" to decompose. This is correct: indeed, $NO(g)$ is thermodynamically unstable with respect to decomposition to $N_2(g)$ and $O_2(g)$ at ambient temperatures. However, the reaction is very slow, so $NO(g)$ is kinetically inert and persists for long periods, entering into other reactions before it can form $N_2(g)$ and $O_2(g)$.

In light of the above discussion, it should be apparent that the atmosphere is a non-equilibrium system. The concentrations of components are not what they would be if spontaneous reactions proceeded rapidly to equilibrium. The same is also true of living systems: our bodies and the trunks of trees, for examples, would change to a mixture of gases if kinetics allowed thermodynamics to express itself.

The interaction between thermodynamics and kinetics is incredibly important in what happens and in what chemists can make happen. In this chapter, we develop the powerful principles of thermodynamics, while chemical kinetics is the subject matter of Chapter 18.

> What chemical reactions happen, and don't happen, may be the result of either thermodynamic or kinetic factors. Nitric oxide pollutant is emitted from cars and industrial furnaces because the equilibrium constant for its formation from $N_2(g)$ and O_2 (g) increases to a significant value at the high temperatures of combustion, and not because of a faster spontaneous reaction at high temperatures.

17.2 Spontaneous Direction of Change and Equilibrium

Along with Chapter 7, this chapter provides an introduction to the subject of *chemical thermodynamics*. Chapter 7 is about the enthalpy of substances and the enthalpy changes of mixtures undergoing chemical reaction, with the discussion guided by the first law of thermodynamics. In this chapter, we explore two more laws of thermodynamics. These allow us to determine the direction of spontaneous chemical changes, to account for the "driving forces" for reaction mixtures to change in the direction that achieves chemical equilibrium, and to calculate the equilibrium constants of reactions at specified temperatures.

Chemists use the term *spontaneous* to refer to the direction of reaction that would take a reaction mixture closer to a state of chemical equilibrium [<<Sections 5.4, 13.3]. To say that a reaction is spontaneous does not say anything about the rate of the reaction: spontaneous reactions may be extremely fast or so slow that change is undetectable.

This chapter is about the factors that determine the direction of a spontaneous chemical change. A chemical system at equilibrium will not change in a (non-spontaneous) way that results in the system no longer being in equilibrium. A reaction mixture that is not at equilibrium will also not react in a way that takes it further from the equilibrium condition.

If a reaction proceeds, it must be a spontaneous reaction. If, however, the reaction does not happen, that does not mean that the reaction is not spontaneous—it may be that the reaction is spontaneous, but infinitely slow.

Enthalpy Change of Reaction—Insufficient Criterion of Spontaneity

In previous chapters, we have encountered many spontaneous chemical reactions: $H_2(g)$ and $O_2(g)$ react to form $H_2O(g)$; $CH_4(g)$ burns in air with formation of $CO_2(g)$ and $H_2O(g)$; $Na(\ell)$ and $Cl_2(g)$ react to form $NaCl(s)$; and $H^+(aq)$ and $OH^-(aq)$ react to form $H_2O(\ell)$. All of these spontaneous chemical reactions are exothermic. So it is perhaps tempting to conclude that a decrease of enthalpy of a chemical system is the criterion that determines whether a process is spontaneous. This is not the case, as is evident from consideration of some common spontaneous changes that are not exothermic:

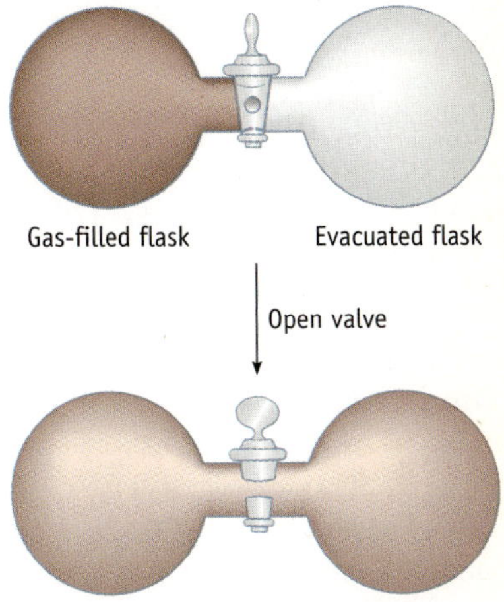

- *Dissolving $NH_4NO_3(s)$ in water.* This spontaneous process is endothermic, with $\Delta_r H°$ $= +25.7$ kJ mol^{-1}.
- *Expansion of a gas into a vacuum.* If we open the valve between one flask containing a gas and another that has been evacuated (Figure 17.1), there is a spontaneous flow of gas from one flask into the other until the pressure is the same in both. The expansion of a gas is endothermic and is the basis of refrigeration.
- *Phase changes.* The spontaneous melting of ice at a temperature above 0 °C is an endothermic process; to melt 1 mol of ice requires about 6 kJ.
- *Energy transfer.* The temperature of a cold soft drink placed in a warm environment will rise until it reaches the ambient temperature. There has been spontaneous transfer of energy from the surroundings. The energy of the soft drink has increased: this is an endothermic process.

We can gain further insight into spontaneity if we think about a specific chemical system— for example, the reaction $H_2(g) + I_2(g) \rightleftharpoons 2\ HI(g)$. Equilibrium in this system can be approached from either direction. Reaction of a mixture containing only $H_2(g)$ and $I_2(g)$ until there is an equilibrium mixture containing $H_2(g)$, $I_2(g)$, and $HI(g)$ is a spontaneous process that is endothermic. On the other hand, the decomposition of pure $HI(g)$ to form $H_2(g)$ and $I_2(g)$ until equilibrium is achieved is also spontaneous, but in this case the process is exothermic.

From these examples, and many others, we conclude that a decrease of enthalpy of the reaction mixture is not a sufficient criterion to determine whether a process is spontaneous.

> In chemistry, the term *spontaneous* refers to the direction of reaction that would take a reaction mixture closer to a state of chemical equilibrium, regardless of how fast or slow that reaction occurs. We cannot know if a reaction is spontaneous from knowledge of whether it is exothermic or endothermic.

Gas-filled flask   Evacuated flask

Open valve

When the valve is opened the gas expands irreversibly to fill both flasks.

FIGURE 17.1 Spontaneous equalization of pressure of a gas.

When discussing chemical reactions, the term *direction of reaction* does not refer to change of physical position as it does when we talk about direction of travel in a car. Rather, it refers to net conversion of species on the left side of the equation into those on the right side of the equation ("reaction to the right," with reaction quotient Q decreasing), or net change of species on the right to those on the left (with Q increasing).

17.3 Entropy: Dispersal of Energy and Matter

To predict whether a process is spontaneous, we can use a thermodynamic property called **entropy (S)**, and the **second law of thermodynamics**: *Any spontaneous process is accompanied by an increase in the entropy of the universe.* Ultimately, this law allows us to predict the conditions at equilibrium as well as the direction of spontaneous change toward equilibrium.

The concept of entropy is built around the idea that *spontaneous change results in the dispersal of energy.* Sometimes, the dispersal of energy is more easily recognized as dispersal of matter. Because the logic underlying these ideas is based on statistics, let's examine the statistical nature of entropy more closely. In the following over-simplified representations of chemical systems, we are concerned with identifying the most probable distribution of matter in a system.

> According to the second law of thermodynamics, every spontaneous process is accompanied by an increase in the entropy of the universe.

Recall from Section 7.4 that universe = system + surroundings

Maximization of Entropy as Most Probable Dispersal of Matter

At one level, achievement of maximum entropy can be thought of as achievement of the most probable distribution of atoms if there are no forces (such as chemical bonds)

preventing this condition. Let's examine a system qualitatively to gain insight into the most likely distribution of matter in a system that approximates the absence of force between particles—the noble gases of Group 18.

Consider a flask A containing neon atoms connected by a valve to an evacuated flask B with the same volume. Assume that when the valve connecting the flasks is opened, the atoms originally in flask A can move randomly throughout flasks A and B.

First, imagine the possible distributions of two neon atoms (Figure 17.2). At any given instant, the atoms will be in one of four possible configurations shown, along with their probabilities, in the figure.

FIGURE 17.2 Dispersal of matter. The expansion of neon gas over two flasks. There are four possible arrangements for two atoms in this apparatus. There is a 50% probability that there will be one atom in each flask at any instant of time.

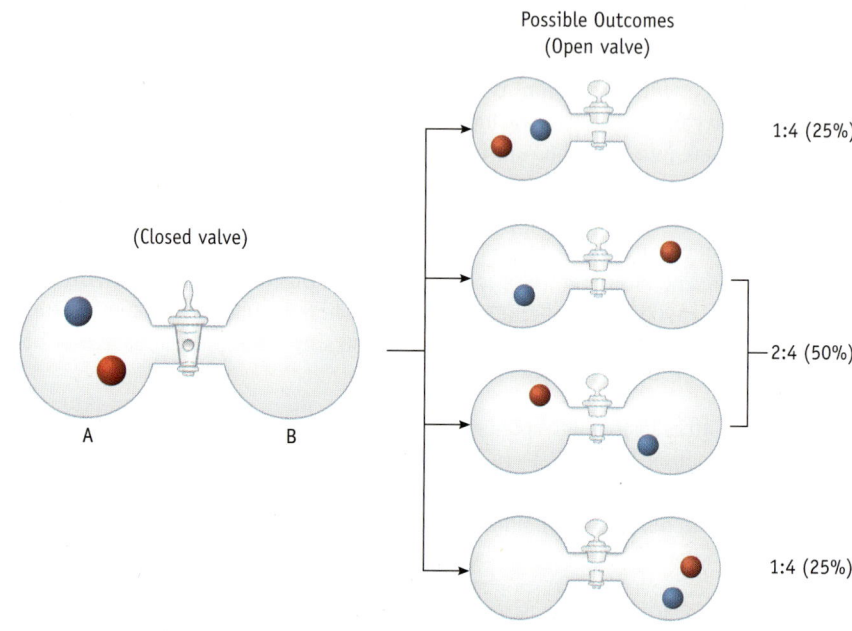

Molecular Modelling (Odyssey)

e17.1 Watch a simulation of the spontaneous dispersal of matter.

If we next consider a system that has three atoms originally in flask A, we would find there is only a one-in-eight chance that the three atoms will remain in that flask. With 10 atoms, there is only a 1-in-1024 chance of finding all the atoms in flask A. The probability of n atoms remaining in the initial flask in this two-flask system is $(1/2)^n$. If flask A contained 1 mol of neon, the probability of all the atoms being found in that flask when the connecting valve is opened is $(1/2)^N$, where N is the Avogadro constant—a probability too small to comprehend! If we calculate the probabilities for all the other possible arrangements of the 1 mol of neon in this scenario, we would find that the most probable arrangement, by a huge margin, is that in which the atoms are, on average, equally distributed between the two flasks.

This analysis shows that in a system such as is depicted in Figure 17.2, it is highly probable that neon atoms will flow from one flask into an evacuated flask until the pressures in the two flasks are equal. Conversely, the opposite process, in which all the neon atoms in the apparatus congregate in one of the two flasks, is highly improbable. What happens, and what we see, in the absence of constraints, is simply the condition that is statistically the most probable—that is, of highest entropy.

The logic applied to the expansion of a gas into a vacuum can be used to rationalize the mixing of two gases. If flasks containing $O_2(g)$ and $N_2(g)$ are connected (in an experimental set-up like that in Figure 17.2), the two gases will diffuse into each other, eventually leading to a mixture in which O_2 and N_2 molecules are evenly distributed. A mixture of O_2 and N_2 molecules will not separate into samples of pure $O_2(g)$ and $N_2(g)$. The important point is that what began as a relatively ordered system with N_2 and O_2 molecules in separate flasks spontaneously moves toward a system in which each gas is maximally dispersed. *For gases at room temperature, the entropy-driven dispersal of matter is equivalent to an increase in disorder of the system.*

A more disordered system is one in which its components are less ordered, or have fewer constraints on their arrangement of energy or of matter. By way of analogy, if 100 people are arranged into 10 lines of 10 people, the system is more ordered (has less disorder) than if the 100 people are milling around randomly. If the 100 people are confined to only one section of the available space, the system is more constrained than if they can go anywhere. If 50 of the people are required to walk at exactly 2 km h^{-1} and the other 50 at exactly 3 km h^{-1}, the system is more constrained (less disordered) than if the people can walk at any speed.

The equivalence of matter dispersal with disorder is true for some solutions as well. For example, when a water-soluble compound is placed in water, it is highly probable that the molecules or ions of the compound will ultimately become distributed evenly throughout the solution (Figure 17.3). The process leads to a mixture, a system in which the solute and solvent molecules are more widely dispersed.

It is also evident from these examples of matter dispersal that if we wanted to put all of the neon atoms in Figure 17.2 back into one flask, or recover the KMnO$_4$ crystals, we would have to do work on the system in some way. For example, we could use a pump to force all of the neon atoms from one side to the other.

Does the formation of a mixture always lead to greater disorder? It does when gases are mixed. With liquids and solids this is usually the case as well, but not always. Exceptions can be found, especially when considering aqueous solutions. For example, Li$^+$ and OH$^-$ ions are highly ordered in the solid LiOH crystal lattice [<<e11.26], but they become more disordered when they enter into solution. However, the solution process is accompanied by aquation, a process in which water molecules become tightly bound to the ions. As a result, the water molecules are constrained to a more ordered arrangement than in pure water (Figure 17.4). Thus, two opposing effects are at work here: a decrease in order for the Li$^+$ and OH$^-$ ions when dispersed in the water, and an increased order for water. The higher degree of ordering due to solvation dominates, and the result is a higher degree of order (lower entropy) overall in the system.

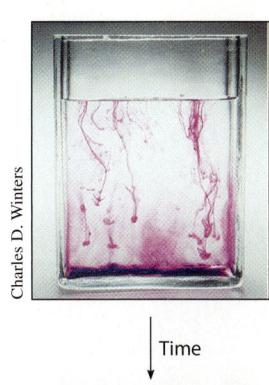

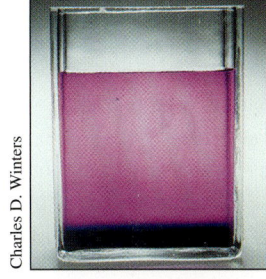

FIGURE 17.3 Matter dispersal. A small quantity of purple KMnO$_4$(s) is added to water (*top*). With time, the solid dissolves and the highly coloured MnO$_4^-$(aq) ions (as well as the K$^+$(aq) ions) become dispersed throughout the solution (*bottom*).

Molecular Modelling (Odyssey)

e17.2 Compare the ordering influence of the different alkali metal ions.

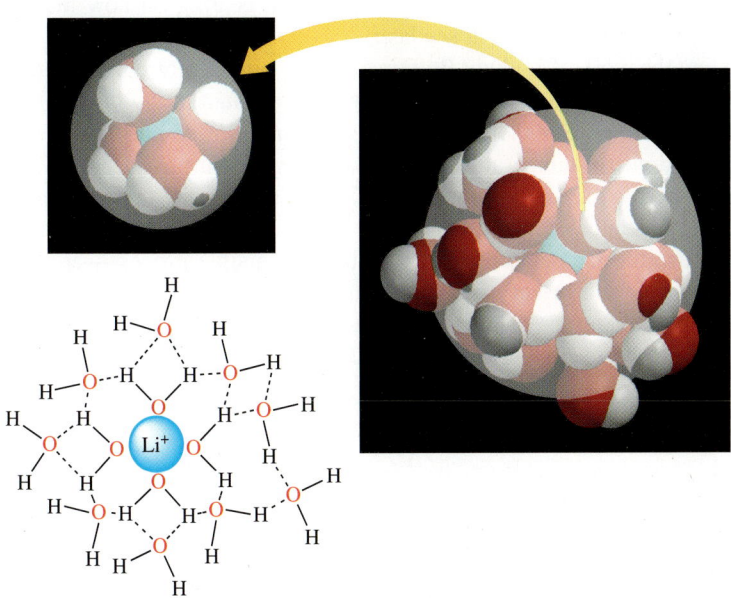

FIGURE 17.4 Order and disorder in solid and solution. Lithium hydroxide is a crystalline solid with an orderly arrangement of Li$^+$ and OH$^-$ ions. When it dissolves in water, the ions become less ordered. Water molecules in pure water have a highly disordered arrangement, but they become highly ordered around the Li$^+$ ions. This can be modelled in *Odyssey* as shown in e17.2. The result is a net increase of order of the system when LiOH(s) dissolves.

Taking It Further

e17.3 See how maximizing entropy is the most probable way energy can be dispersed.

When most other salts dissolve in water, the entropy increase due to dispersal of the ions exceeds the entropy decrease due to constraining of water molecules. Li$^+$ ions have a powerful ordering influence because they are so small, and the strength of electrostatic attractions decreases rapidly with distance between the charged species. This is one of the characteristic features of ions with high *charge density* [>>Section 26.3].

Ludwig Boltzmann (1844–1906). Engraved on his tombstone in Vienna, Austria, is his equation defining entropy: $S = k \log W$. The symbol k is a constant now known as the Boltzmann constant.

It is possible to measure the absolute entropy of a sample of substance. By contrast, only changes of enthalpy can be measured.

Maximization of entropy of a system corresponds with the maximum probability of distribution matter—the system of least order—if there are no energy constraints.

The Boltzmann Equation for Entropy

Ludwig Boltzmann (1844–1906) developed the idea of looking at the distribution of matter and energy as a way to calculate entropy. At the time of his death, the scientific world had not yet accepted his ideas. In spite of this, he was firmly committed to his theories and the *Boltzmann equation* for entropy was engraved on his tombstone in Vienna, Austria. The equation is

$$S = k \log W$$

where k is the Boltzmann constant, and W represents the number of different ways that the energy can be distributed over the available energy levels. The more ways that energy can be dispersed, the higher the entropy is. Boltzmann concluded that the state of maximum entropy, in which W has the maximum value, is when the system has come to equilibrium.

17.4 Entropy and the Second Law of Thermodynamics

The entropy (S) of a system is a measure of the extent of disorder of energy and matter: the further a system is from a condition of perfect order, the higher the entropy is. For any substance under a given set of conditions, a numerical value for entropy can be determined.

Like enthalpy (H), entropy is a state function: the entropy change for any process depends only on the initial and final states of the system, and not on the pathway by which the process occurs [<<Section 6.4]. This point has a bearing on how the entropy of a substance is determined, as we will see later.

Following the work of Ludwig Boltzmann, the point of reference for entropy values is established by the **third law of thermodynamics**: *There is no disorder in a perfect crystal at 0 K—that is, S = 0.* The entropy of a sample of an element or compound at specified conditions (of temperature and pressure) is the entropy increase if it were converted from a perfect crystalline form at 0 K to its state at the defined conditions. The entropy of a substance at any temperature can be obtained by measuring the heat required to raise the temperature from 0 K by a reversible process—approximated by adding heat slowly and in very small increments. The entropy increase accompanying each incremental change is calculated using the following equation, where q_{rev} is the heat absorbed in the reversible change and T is the temperature (Kelvin) at which the change occurs:

$$\Delta S = \frac{q_{rev}}{T} \tag{17.1}$$

Since the unit of measurement of heat transferred is the joule (J), and of temperature is Kelvin (K), the unit of measurement of entropy of a sample of substance is J K^{-1}.

Adding the entropy changes for the incremental steps gives the total entropy of a substance. Because it is necessary to add heat to raise the temperature, *all substances have positive entropy values at temperatures above 0 K.* When a substance undergoes a phase change as heat energy is supplied at its melting point or boiling point, its entropy increases sharply (Figure 17.6).

Taking It Further

e17.4 Read about the differences between reversible and irreversible processes.

Molecular Modelling (Odyssey)

e17.5 Watch simulations of phase changes and the corresponding changes in entropy.

WORKED EXAMPLE 17.1—ENTROPY CHANGE

Compare the increase of entropy of a large mass of water when 200 kJ of heat is transferred to it at (a) 80 °C, and (b) 10 °C. In each case, presume that the change is reversible and that the mass of water is sufficiently large that its temperature can be considered to be constant.

Strategy

Use equation 17.1. Change the temperature into kelvins.

Solution

(a)
$$\Delta S = \frac{q_{rev}}{T} = \frac{200 \times 1000 \text{ J}}{353.2 \text{ K}} = 566 \text{ J K}^{-1}$$

(b)
$$\Delta S = \frac{q_{rev}}{T} = \frac{200 \times 1000 \text{ J}}{283.2 \text{ K}} = 706 \text{ J K}^{-1}$$

Comment

For the same amount of heat transferred, the change of entropy is less for the system that is already most disordered (the water at the higher temperature).

EXERCISE 17.1—ENTROPY CHANGE

Compare the increase of entropy of (a) a large mass of water, and (b) a large mass of hexane, when 1000 kJ of heat is transferred to each at 50 °C. In each case, presume that the change is reversible and that the mass of water is sufficiently large that its temperature can be considered to be constant. Comment on your answers, remembering that the heat capacity of water is greater than that of hexane.

WORKED EXAMPLE 17.2—ENTROPY CHANGE

When liquid water at 100 °C changes to vapour at 100 °C, the heat absorbed is 40.7 kJ mol^{-1}. What is the molar entropy change of vaporization of water at 100 °C?

Strategy

Use equation 17.1.

Solution

$$\Delta S = \frac{q_{rev}}{T} = \frac{40.7 \times 1000 \text{ J mol}^{-1}}{373.2 \text{ K}} = 109 \text{ J K}^{-1} \text{ mol}^{-1}$$

Comment

Entropy is an extensive property, so the answer in Worked Example 17.1 has the units J K^{-1}. Molar entropy is an intensive property (does not depend on the size of the sample), and the answer here has the units J K^{-1} mol^{-1}.

EXERCISE 17.3—ENTROPY CHANGE

When ammonia changes from liquid to vapour at its normal boiling point (239.7 K), the heat absorbed is 29.1 kJ mol-1. Calculate the molar entropy change of vaporization of ammonia at its normal boiling point.

> Entropy (S) is a measure of the lack of order resulting from dispersal of energy and matter. Calculation of the entropy of a substance at a specified temperature is based on (a) S = 0 in a perfect crystal at 0 K, and (b) for each increment of change absorbing energy q$_{rev}$ reversibly up to that temperature, $\Delta S = \dfrac{q_{rev}}{T}$.

Standard Molar Entropy of Substances, S°

The **standard molar entropy** (S°) of a substance at a specified temperature is *the entropy of 1 mol of that substance when it is in its standard state* [<<Section 7.6] at that temperature. Molar entropy is the entropy of a sample divided by the amount of sample, so the units for molar entropy of a substance are J K^{-1} mol^{-1}. Molar entropy is an intensive

Interactive Exercise 17.2

Calculate molar entropy changes for phase changes.

Think about It

e17.6 Calculate the change in entropy accompanying a phase change.

property. Standard molar entropies of substances can be estimated and tabulated at any temperature. In this text, values at 298 K are used to illustrate the principles of their use. A few values of $S°$ of substances at 298 K are given in Table 17.1, and a more extensive list appears in Appendix C.

TABLE 17.1 Standard Molar Entropies of Some Substances at 298 K

Element	$S°$ (J K^{-1} mol^{-1})	Compound	$S°$ (J K^{-1} mol^{-1})
C(graphite)	5.6	$CH_4(g)$	186.3
C(diamond)	2.377	$C_2H_6(g)$	229.2
C(vapour)	158.1	$C_3H_8(g)$	270.3
Ca(s)	41.59	$CH_3OH(\ell)$	127.2
Ar(g)	154.9	CO(g)	197.7
$H_2(g)$	130.7	$CO_2(g)$	213.7
$O_2(g)$	205.1	$H_2O(g)$	188.84
$N_2(g)$	191.6	$H_2O(\ell)$	69.95
$F_2(g)$	202.8	HCl(g)	186.2
$Cl_2(g)$	223.1	NaCl(s)	72.11
$Br_2(\ell)$	152.2	MgO(s)	26.85
$I_2(s)$	116.1	$CaCO_3(s)$	91.7

Some interesting and useful generalizations can be drawn from the data given in Table 17.1 and Appendix C:

- *When comparing the same or similar substances, the molar entropies of gases are much larger than those for liquids, and the molar entropies of liquids are larger than those for solids.* Compare, for example, the standard molar entropies at 25 °C of liquid water and water vapour; of $I_2(s)$, $Br_2(\ell)$, and $Cl_2(g)$; and of C(s, graphite) and C(g). These differences can be explained on the grounds that in a solid, the particles are highly ordered in fixed positions in the lattice. When the solid melts, its particles have more freedom to change their relative positions, resulting in an increase in disorder (Figure 17.5). And when a liquid evaporates, restrictions due to forces between the particles nearly disappear, and another large entropy increase occurs because the particles have almost unrestricted, random motion over the whole of the available space.

Think about It

e17.7 See the properties that determine the entropy of a substance.

(a)

(b)

FIGURE 17.5 Entropy and states of matter. (a) The entropy of liquid bromine is 152.2 J K^{-1} mol^{-1}, and that for less ordered bromine vapour is 245.47 J K^{-1} mol^{-1}. (b) The entropy of ice, which has a highly ordered molecular arrangement, is smaller than that for liquid water.

- *Substances composed of large molecules have more entropy (per mole) than substances with smaller molecules, and those whose molecules have complex structures have more entropy than those with simpler molecules.* These generalizations apply particularly in series of related compounds. Values of $S°$ at 25 °C for methane, $CH_4(g)$, ethane, $C_2H_6(g)$, and propane, $C_3H_8(g)$, are 186.3, 229.2, and 270.3 J K^{-1} mol^{-1}, respectively. The dependence on molecular structure can be seen with substances whose particles have similar molar mass, but different numbers of atoms: Ar, CO_2, and C_3H_8 have standard molar entropies of 154.9, 213.7, and 270.3 J K^{-1} mol^{-1}, respectively. In more complex molecules, there are more possibilities for different arrangements of the atoms because there are more ways for the molecule to rotate, twist, and vibrate in space (referred to as degrees of freedom).

- *Comparing two systems with the same number of atoms of each element, and in the same phase, the one with more, smaller molecules has more entropy.* For example, at 25 °C, the entropy of 1 mol of $N_2O_4(g)$ at 1 bar is 304.38 J K^{-1}, while the entropy of 2 mol of $NO_2(g)$ at 1 bar is 480.08 J K^{-1} ($S° = 240.04$ J K^{-1} mol^{-1}). This can be explained on the grounds that 2 mol of N atoms and 4 mol of O atoms are more constrained when they are present as 1 mol of N_2O_4 molecules than as 2 mol of NO_2 molecules. In the absence of energy considerations, 2 mol of NO_2 molecules is a more probable arrangement of these atoms than is 1 mol of N_2O_4 molecules.

- *The entropy of any substance increases as the temperature is raised.* Large increases in entropy accompany changes of state (Figure 17.6).

$S°$ (J K^{-1} mol^{-1})

186.3

methane

229.2

ethane

270.3

propane

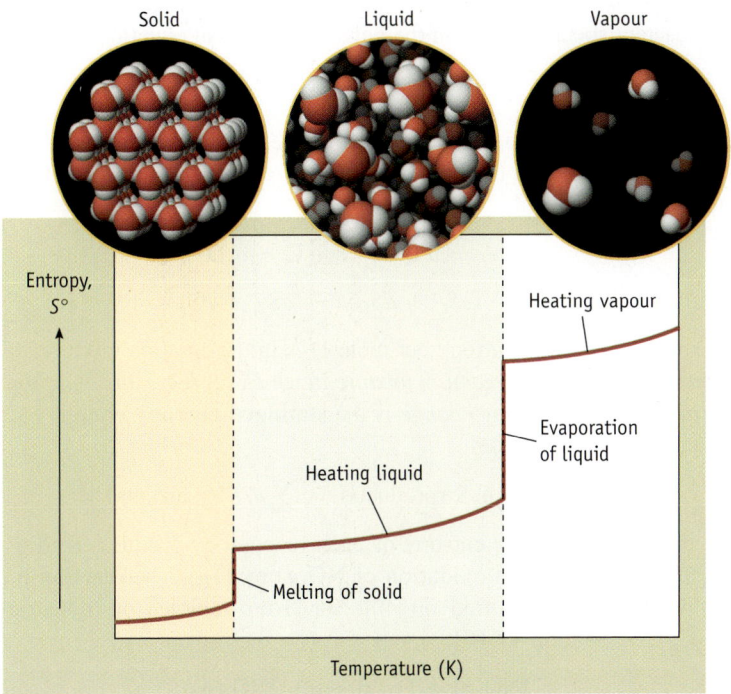

FIGURE 17.6 Entropy of a substance as temperature is raised. For each of the three states of matter, entropy increases with increasing temperature. A large increase in entropy accompanies a phase change from solid to liquid, and liquid to gas.

WORKED EXAMPLE 17.3—COMPARISON OF ENTROPIES OF SUBSTANCES

Predict which of the following has the higher entropy when the substances are in their standard states at a specified temperature. Explain your reasoning.

(a) 1 mol of $I_2(g)$ or 1 mol of $I_2(s)$

(b) 1 mol of benzene vapour, $C_6H_6(g)$, or 3 mol of acetylene vapour, $C_2H_2(g)$

Solution

(a) Substances have higher entropy in the vapour phase than in the liquid phase, so we should expect $S°$ of iodine vapour to be larger than $S°$ of solid iodine. ($S°$ for $I_2(g)$ is 260.69 J K^{-1} mol^{-1}; and $S°$ for $I_2(s)$ is 116.135 J K^{-1} mol^{-1}.)

Interactive Exercise 17.4

Compare entropies of substances.

Think about It

e17.8 See what happens to the entropy when substances are mixed.

(b) Both substances are in the vapour phase, and in both cases there are 6 mol of C atoms and 6 mol of H atoms. These atoms are in a more constrained arrangement in the larger benzene molecules, which we would therefore expect to have less entropy.

EXERCISE 17.5—COMPARISON OF ENTROPIES OF SUBSTANCES

For each of the pairs listed, predict which substance has the higher entropy and explain your reasoning.
(a) 1 mol of $O_2(g)$ or 1 mol of $O_3(g)$
(b) 1 mol of $SnCl_4(\ell)$ or 1 mol of $SnCl_4(g)$
(c) 3 mol of $O_2(g)$ or 2 mol of $O_3(g)$

> The standard molar entropy ($S°$) of a substance at a specified temperature is the entropy of 1 mol of that substance when it is in its standard state at that temperature. In general, $S°$ of a substance in the gas phase is much larger than the liquid, and this is larger than for the solid phase. Substances with large and complex molecules have higher $S°$ than those with small molecules or simple structures. Of two systems with the same number of elements of each type, and in the same state of matter, the one with more, smaller molecules has more entropy.

Standard Entropy Change of Reaction ($\Delta_r S°$)

Interactive Exercise 17.6

SUBMIT

Use a relational diagram to show relationships among concepts.

The **entropy change** for a chemical reaction ($\Delta_r S$) *is the difference between the entropies of the product species and the reactant species*, taking into account the amounts of the product and reactant species shown in the balanced chemical equation

$$\Delta_r S = \sum n_i\, S(\text{products}) - \sum n_i\, S(\text{reactants})$$

where S is the entropy per mole of each product and reactant, and n_i is the number of moles of each reactant or product in the balanced reaction equation. For a reaction represented by the generalized equation

$$a\,A + b\,B \longrightarrow c\,C + d\,D$$

$$\Delta_r S = (cS_C + dS_D) - (aS_A + bS_B)$$

where, for example, S_A is the entropy per mole of A in the reaction mixture.

In the particular case of a reaction mixture in which all reactants and products are in their standard states, the entropy change is the **standard entropy change ($\Delta_r S°$),** which we can evaluate from the equation

$$\Delta_r S° = \sum n_i\, S°(\text{products}) - \sum n_i\, S°(\text{reactants})$$

where $S°$ is the standard molar entropy of each reactant or product. To illustrate, let's calculate $\Delta_r S°$ at 25 °C for the oxidation of NO(g) by $O_2(g)$—a reaction important in urban areas where NO is emitted into the air from vehicles or industrial furnaces [<<Section 17.1, >>Section 17.10].

$$2NO(g) + O_2(g) \longrightarrow 2NO_2(g)$$

We subtract the total entropy of the reactants, 2 mol of NO(g) and 1 mol of $O_2(g)$, from the total entropy of the products, 2 mol of $NO_2(g)$, all at 25 °C:

In the statement $\Delta_r S° = -146.7$ J K^{-1} mol^{-1} we need to ask "per mol of what?" In this case, the statement applies to 1 mol of a collection of reactant particles, each collection comprising two NO molecules and one O_2 molecule. Alternatively, it applies to 1 mol of a collection of product particles, each collection comprising two NO_2 molecules.

$$\Delta_r S° = [2 \times S°(NO_2, g)] - [2 \times S°(NO, g) + 1 \times S°(O_2, g)]$$

$$= [2 \times 240.0\,\text{J K}^{-1}\,\text{mol}^{-1}] - [(2 \times 210.8\,\text{J K}^{-1}\,\text{mol}^{-1})$$
$$+ (1 \times 205.1\,\text{J K}^{-1}\,\text{mol}^{-1})]$$

$$= 480.0\,\text{J K}^{-1}\,\text{mol}^{-1} - 626.7\,\text{J K}^{-1}\,\text{mol}^{-1} = -146.7\,\text{J K}^{-1}\,\text{mol}^{-1}$$

The negative value of $\Delta_r S°$ means that the entropy of the system decreases, as we should expect when the atoms in 3 mol of gaseous substances are constrained into 2 mol of larger molecules.

$\Delta_r S°$ depends on the amounts that react and are formed. From the balanced chemical equation used, we see that this answer means that when 2 mol of NO(g) and 1 mol of O_2(g) react to form 2 mol of NO_2(g) at 25 °C, the entropy of the system (the reaction mixture) decreases by 146.7 J K^{-1}.

Consistent with the calculation of $\Delta_r H°$ values [<<Section 7.6], if we use the balanced equation

$$NO(g) + \tfrac{1}{2} O_2(g) \longrightarrow NO_2(g)$$

we calculate a value half of that deduced previously (within rounding off errors):

$$\Delta_r S° = [1 \times S°(NO_2, g)] - [1 \times S°(NO, g) + \tfrac{1}{2} \times S°(O_2, g)]$$

$$= [1 \times 240.0\,J\,K^{-1}\,mol^{-1}] - [(1 \times 210.8\,J\,K^{-1}\,mol^{-1})$$
$$+ (\tfrac{1}{2} \times 205.1\,J\,K^{-1}\,mol^{-1})]$$

$$= 240.0\,J\,K^{-1}\,mol^{-1} - 313.4\,J\,K^{-1}\,mol^{-1} = -73.4\,J\,K^{-1}\,mol^{-1}$$

WORKED EXAMPLE 17.4—STANDARD ENTROPY CHANGE OF REACTION

Calculate the standard entropy changes for the following processes at 298 K. Do the calculations match your prediction?

(a) Evaporation of 1 mol of liquid ethanol to ethanol vapour:

$$C_2H_5OH(\ell) \longrightarrow C_2H_5OH(g)$$

(b) Oxidation of 1 mol of ethanol vapour by oxygen gas:

$$C_2H_5OH(g) + 3\,O_2(g) \longrightarrow 2\,CO_2(g) + 3\,H_2O(g)$$

Think about It

e17.9 Calculate the change in entropy in a chemical reaction.

Interactive Exercise 17.7

Calculate the standard entropy change and interpret the sign.

Strategy

Subtract the entropy of the reactants from the entropy of the products, taking into account the amount (in mol) of each reactant and product in the balanced equation.

Solution

(a) We calculate $\Delta_r S°$ when 1 mol of $C_2H_5OH(\ell)$ reactant is converted to 1 mol of $C_2H_5OH(g)$ product, each in their standard states in the reaction mixture at 25 °C.

$$\Delta_r S° = \sum n_i\,S°(products) - \sum n_i\,S°(reactants)$$

$$= 1 \times S°(C_2H_5OH, g) - 1 \times S°(C_2H_5OH, \ell)$$

$$= (1 \times 282.70\,J\,K^{-1}\,mol^{-1}) - (1 \times 160.7\,J\,K^{-1}\,mol^{-1})$$

$$= (282.70 - 160.7)\,J\,K^{-1}\,mol^{-1} = +122.0\,J\,K^{-1}\,mol^{-1}$$

The large positive value for the entropy change (increase in entropy) is expected because the process converts ethanol from a more ordered state (liquid) to a less ordered state (vapour).

(b) From the equation, we presume conversion of 1 mol of $C_2H_5OH(g)$ and 3 mol of $O_2(g)$ into 2 mol of $CO_2(g)$ and 3 mol of $H_2O(g)$, all in their standard states at 25 °C.

$$\Delta_r S° = [2 \times S°(CO_2, g) + 3 \times S°(H_2O, g)] - [1 \times S°(C_2H_5OH, g) + 3 \times S°(O_2, g)]$$

$$= [(2 \times 213.74\,J\,K^{-1}\,mol^{-1}) + (3 \times 188.84\,mol^{-1})]$$

$$-[(1 \times 282.70\,J\,K^{-1}\,mol^{-1}) + (3 \times 205.07\,J\,K^{-1}\,mol^{-1})]$$

$$= +96.09\,J\,K^{-1}\,mol^{-1}$$

We would predict an increase in entropy for this reaction because the total amount of gases increases from 4 mol of reactants to 5 mol of products.

Calculate the standard entropy change ($\Delta_r S°$) at 298 K for each of the following processes, using the standard molar entropy values in Appendix C. Do the calculated values of $\Delta_r S°$ match your predictions?

(a) Decomposition of calcium carbonate: $CaCO_3(s) \longrightarrow CaO(s) + CO_2(g)$

(b) Reaction of $N_2(g)$ and $H_2(g)$ to form $NH_3(g)$: $N_2(g) + 3 H_2(g) \longrightarrow 2 NH_3(g)$

The standard entropy change of reaction ($\Delta_r S°$) at a specified temperature is the difference between the sum of the standard molar entropies of products and the standard molar entropies of reactants at that temperature, taking into account the amounts of the products and reactants shown in the balanced chemical equation:

$$\Delta_r S° = \sum n_i S°(\text{products}) - \sum n_i S°(\text{reactants})$$

17.5 Entropy Changes and Spontaneity

Entropy changes are the key to deciding if a process, such as a chemical reaction, is spontaneous. You might ask how this is possible when, as you have seen earlier, the entropy of a system may either decrease (oxidation of NO) or increase (evaporation and oxidation of ethanol) during a spontaneous reaction. An answer is provided by the *second law of thermodynamics*: *A spontaneous process is one that results in an increase of entropy in the universe.* This criterion of spontaneity requires us to assess entropy changes in both the system under study and the surroundings.

In thermodynamic considerations, the universe is considered to have two components: the system and its surroundings [<<Section 7.4]. The entropy change of the universe accompanying any process is the sum of the entropy changes of the system (the reaction mixture) and of the surroundings:

$$\Delta S_{\text{univ}} = \Delta S_{\text{sys}} + \Delta S_{\text{surr}}$$

According to the second law of thermodynamics:

- If $\Delta S_{\text{univ}} > 0$ for a reaction described by a chemical equation, that reaction is spontaneous: the reaction mixture would come closer to equilibrium if the reaction proceeds in the direction written.
- If $\Delta S_{\text{univ}} < 0$, the reaction as written is not spontaneous, but reaction in the opposite direction, with $\Delta S_{\text{univ}} > 0$, is spontaneous.
- If $\Delta S_{\text{univ}} = 0$, the system is at equilibrium.

We can calculate a standard entropy change at any specified temperature in a corresponding way:

$$\Delta S°_{\text{univ}} = \Delta S°_{\text{sys}} + \Delta S°_{\text{surr}}$$

The relationship applies at any temperature, but the values of the terms differ from temperature to temperature, so the temperature must always be specified. Again, we will generally apply the thermodynamic principles to systems at 25 °C. Worked Example 17.5 illustrates how we can calculate the value of $\Delta S°_{\text{univ}}$ to predict whether a chemical reaction is spontaneous when all reactants and products in the reaction mixture are in their standard states at 25 °C.

WORKED EXAMPLE 17.5—PREDICTING SPONTANEITY FROM $\Delta S°_{\text{univ}}$

Consider the reaction currently used to manufacture methanol:

$$CO(g) + 2 H_2(g) \longrightarrow CH_3OH(\ell)$$

In a reaction mixture at 25 °C with the partial pressures of $CO(g)$ and $H_2(g)$ both at 1 bar, and in contact with pure liquid methanol, will reaction in the direction indicated in the equation take the system toward chemical equilibrium?

Since spontaneous change happens continuously, the entropy of the universe is continuously increasing. This is in contrast to energy, which, according to the first law of thermodynamics, is constant.

Think about It

e17.10 Predict spontaneity from the change in entropy in the universe.

Strategy

Predict the direction of spontaneous reaction by calculating the value of $\Delta S°_{univ}$ accompanying the change in the direction indicated by the equation. $\Delta S°_{univ}$ is the sum of $\Delta S°_{sys}$ and $\Delta S°_{surr}$. For a chemical reaction, the system is the reaction mixture, and $\Delta S°_{sys}$ is $\Delta_r S°$. The value of $\Delta_r S°$ at a given temperature can be calculated from the standard molar entropies of reactants and products, as in Worked Example 17.4. $\Delta S°_{surr}$ can be calculated from $\Delta_r H°$ and the temperature of reaction.

Solution

First, calculate $\Delta S°_{sys}$ ($\Delta_r S°$) at 25 °C, using $S°$ values from Table 17.1:

$$\Delta S°_{sys} = \Delta_r S° = \sum n_i S°(\text{products}) - \sum n_i S°(\text{reactants})$$

$$= [1 \times S°(CH_3OH, \ell)] - [1 \times S°(CO, g) + 2 \times S°(H_2, g)]$$

$$= (1 \times 127.2 \text{ J K}^{-1} \text{mol}^{-1}) - [(1 \times 197.7 \text{ J K}^{-1} \text{mol}^{-1})$$

$$+(2 \times 130.7 \text{ J K}^{-1} \text{mol}^{-1})]$$

$$= -331.9 \text{ J K}^{-1} \text{mol}^{-1}$$

We should have expected a decrease of entropy of the system because the atoms in 3 mol of gases are rearranged in this reaction to form 1 mol of a liquid substance.

Now calculate $\Delta S°_{surr}$. The standard enthalpy change of reaction is $\Delta_r H°$, so the energy transferred to the surroundings (q_{surr}) if the reaction mixture is held at 25 °C is $-\Delta_r H°$. Then

$$\Delta S°_{surr} = \frac{q_{surr}}{T} = -\frac{\Delta_r H°}{T}$$

$\Delta_r H°$ is calculated as described in Section 7.8:

$$\Delta H°_{sys} = \Delta_r H° = \sum n_i \Delta_f H°(\text{products}) - \sum n_i \Delta_f H°(\text{reactants})$$

$$= [1 \times \Delta_f H°(CH_3OH, \ell) - [1 \times \Delta_f H°(CO, g) + 2 \times \Delta_f H°(H_2, g)]$$

$$= (1 \times -238.4 \text{ kJ mol}^{-1}) - [(1 \times -110.5 \text{ kJ mol}^{-1}) + (2 \times 0 \text{ kJ mol}^{-1})]$$

$$= -127.9 \text{ kJ mol}^{-1}$$

The negative value of $\Delta H°_{sys}$ means that the reaction is exothermic, and heat is transferred to the surroundings. As a result, the entropy of the surroundings increases.

If we make the simplifying assumption that the process is reversible and occurs at a constant temperature of 298 K, we can calculate the entropy change of the surroundings. We need to convert the value of $\Delta H°_{sys}$ from kJ mol^{-1} to J mol^{-1}.

$$\Delta S°_{surr} = -\frac{\Delta H°_{sys}}{T} = -\frac{-127.9 \times 10^3 \text{ J mol}^{-1}}{298 \text{ K}} = +429.2 \text{ J K}^{-1} \text{mol}^{-1}$$

The pieces are now in place to calculate the entropy change in the universe accompanying the formation of $CH_3OH(\ell)$ from $CO(g)$ and $H_2(g)$, all present in their standard states at 298 K:

$$\Delta S°_{univ} = \Delta S°_{sys} + \Delta S°_{surr} = -331.9 \text{ J K}^{-1} \text{mol}^{-1} + 429.2 \text{ J K}^{-1} \text{mol}^{-1}$$

$$= +97.3 \text{ J K}^{-1} \text{mol}^{-1}$$

The positive value indicates an increase in the entropy of the universe. It follows from the second law of thermodynamics that the reaction is spontaneous under these conditions.

Comment

We shortly consider another property of the reaction mixture, Gibbs free energy, which allows us to predict the direction of spontaneous reaction by consideration of the system alone, without consideration of the surroundings.

Consider the reaction represented as

$$Si(s) + 2\,Cl_2(g) \longrightarrow SiCl_4(g)$$

In a reaction vessel in which all of the reactants and products are in their standard states at 25 °C, is this reaction spontaneous? Answer this question by calculating $\Delta S°_{sys}$, $\Delta S°_{surr}$, and $\Delta S°_{univ}$.

> The criterion of spontaneity of a defined direction of reaction is given by the second law of thermodynamics: A spontaneous process is one that results in an increase of entropy in the universe.

Contributions of $\Delta_r S°$ and $\Delta_r H°$ to Spontaneity of Reaction

The criterion for a spontaneous reaction in a standard-state reaction mixture at a specified temperature is $\Delta S°_{univ} > 0$, or $\Delta S°_{sys} + \Delta S°_{surr} > 0$. In Worked Example 17.5, we found that reaction in the direction indicated by the equation is spontaneous, because $\Delta S°_{univ} = +97.3$ J K^{-1} mol^{-1}. This, in turn, is because an unfavourable $\Delta S°_{sys}$ (-331.9 kJ mol^{-1}) is more than counterbalanced by a favourable $\Delta S°_{surr}$ ($+429.2$ J K^{-1} mol^{-1}). It is evident that the value of $\Delta S°_{univ}$ depends on the relative values of $\Delta S°_{sys}$ and $\Delta S°_{surr}$, both of which may be either positive or negative.

There are four different categories of reaction, depending on the signs of $\Delta S°_{sys}$ and $\Delta S°_{surr}$, that govern what we can say about the spontaneity of a reaction in a standard state reaction mixture. These are displayed in Table 17.2, after taking into account that $\Delta S°_{sys} = \Delta_r S°$, and that the sign of $\Delta S°_{surr}$ is the opposite to that of $\Delta_r S°$.

TABLE 17.2 $\Delta_r S°$, $\Delta_r H°$ and Spontaneity in a Standard State Reaction Mixture

Type	$\Delta_r S°$	$\Delta_r H°$	Spontaneity of Reaction
1	$\Delta_r S° > 0$	$\Delta_r H° < 0$ ($\Delta S°_{surr} > 0$)	Spontaneous. $\Delta S°_{univ} > 0$
2	$\Delta_r S° < 0$	$\Delta_r H° < 0$ ($\Delta S°_{surr} > 0$)	Depends on relative magnitudes of $\Delta_r S°$, $\Delta_r H°$
3	$\Delta_r S° > 0$	$\Delta_r H° > 0$ ($\Delta S°_{surr} < 0$)	Depends on relative magnitudes of $\Delta_r S°$, $\Delta_r H°$
4	$\Delta_r S° < 0$	$\Delta_r H° > 0$ ($\Delta S°_{surr} < 0$)	Not spontaneous. $\Delta S°_{univ} < 0$

Interactive Exercise 17.11

Predict spontaneity of reaction.

Combustion of butane gas (like most combustion reactions) is an example of a Type 1 reaction:

$$2\,C_4H_{10}(g) + 13\,O_2(g) \longrightarrow 8\,CO_2(g) + 10\,H_2O(g)$$

In this reaction, entropy increases ($\Delta_r S°$ at 25 °C $= +312.4$ J K^{-1} mol^{-1}) because the atoms are distributed among more molecules in the products than in the reactants—so $\Delta S°_{sys} > 0$. The reaction is exothermic ($\Delta_r H°$ at 25 °C $= -5315$ kJ mol^{-1}), $\Delta S°_{surr} > 0$. Both factors favour reaction by contributing to a positive value of $\Delta S°_{univ}$.

On the other hand, consider synthesis of the high-energy rocket fuel hydrazine, $N_2H_4(\ell)$ from $N_2(g)$ and $H_2(g)$:

$$N_2(g) + 2\,H_2(g) \longrightarrow N_2H_4(\ell)$$

This would be a desirable way of making hydrazine, because both nitrogen and hydrogen are available cheaply in large supplies. However, we can calculate from thermodynamic data that the reaction is endothermic ($\Delta_r H°$ at 25 °C $= +50.63$ kJ mol^{-1}) and the entropy change of the system is negative ($\Delta_r S°$ at 25°C $= -331.4$ J K^{-1} mol^{-1}). This is a Type 4 reaction: at 25 °C the reaction is non-spontaneous, and so there is no point trying to make hydrazine this way. If all of the species are present in their standard states ($p(N_2) = p(H_2) = 1$ bar, and $N_2H_4(\ell)$ is pure) at 25 °C, reaction in the direction indicated would take the reaction mixture further from equilibrium.

In the case of reactions of Types 2 and 3, entropy and enthalpy changes oppose each other. A process may be favoured by the enthalpy change but disfavoured by the entropy change (Type 2), or vice versa (Type 3). In either instance, whether or not a process is spontaneous depends on which factor dominates. In fact, the spontaneity of these reactions depend on the temperature of the system, for reasons discussed in Section 17.6.

The formation of ammonia from nitrogen and hydrogen is a Type 2 reaction:

$$N_2(g) + 3 H_2(g) \longrightarrow 2 NH_3(g)$$

This reaction is exothermic ($\Delta_r H° < 0$, $\Delta S°_{surr} > 0$), and has $\Delta_r S° < 0$ (and $\Delta S°_{sys} < 0$). We cannot decide if this reaction is spontaneous in a reaction mixture with each reactant and the product all in the standard state of 1 bar pressure without calculating $\Delta S°_{univ}$ (or, as we shall see in Section 17.6, $\Delta_r G°$).

An example of a Type 3 reaction is the decomposition of solid ammonium chloride:

$$NH_4Cl(s) \longrightarrow NH_3(g) + HCl(g)$$

This reaction involves an entropy increase, but it is endothermic, so we cannot decide if the reaction is spontaneous in a standard state reaction mixture unless we calculate the relative magnitudes of $\Delta S°_{sys}$ and $\Delta S°_{surr}$.

EXERCISE 17.12—CONTRIBUTIONS OF $\Delta_r S°$ AND $\Delta_r H°$ TO SPONTANEITY OF REACTION

Classify each of the following reactions as one of the four types listed in Table 17.2:

Reaction	$\Delta_r H°$ (kJ mol^{-1})	$\Delta_r S°$ (J K^{-1} mol^{-1})
(a) $CH_4(g) + 2 O_2(g) \longrightarrow 2 H_2O(l) + CO_2(g)$	−890.6	−242.8
(b) $2 Fe_2O_3(s) + 3 C(graphite) \longrightarrow 4 Fe(s) + 3 CO_2(g)$	+467.9	+560.7
(c) $C(graphite) + O_2(g) \longrightarrow CO_2(g)$	−393.5	+3.1
(d) $N_2(g) + 3 F_2(g) \longrightarrow 2 NF_3(g)$	−264.2	−277.8

Decisions about spontaneity of reactions based on assessing both $\Delta_r S°$ and $\Delta S°_{surr}$ are cumbersome. In Section 17.6, we introduce an easier way (derived from the same principles) involving changes of the Gibbs free energy of the system.

> When all species in a reaction mixture are in their standard states, reaction in a specified direction is spontaneous if $\Delta_r S° > 0$ and $\Delta_r H° < 0$, and non-spontaneous if $\Delta_r S° < 0$ and $\Delta_r H° > 0$. In cases where $\Delta_r S° > 0$ and $\Delta_r H° > 0$, or $\Delta_r S° < 0$ and $\Delta_r H° < 0$, we cannot predict the direction of spontaneous reaction without calculations to see if $\Delta S°_{univ} > 0$.

Thermodynamics, Time, and Life

Chapters 7 and 17 bring together the three laws of thermodynamics. C.P. Snow is believed to have paraphrased these laws. The first law was transmuted into "You can't win!"—referring to the fact that energy is always conserved, so a process in which you get back more energy than you put in is impossible. The second law was paraphrased as "You can't break even!" We see in Section 17.6 that the Gibbs free energy provides a rationale for this interpretation. Only part of the energy from a chemical reaction can be converted to useful work; the rest will be committed to the redistribution of matter or energy.

The second law of thermodynamics requires that disorder increases with time. Because all spontaneous natural processes result in increased disorder as time progresses, it is evident that increasing entropy and time "point" in the same direction. A snowflake will melt in a warm room, but you won't see a glassful of water molecules reassemble themselves into snowflakes at any temperature above 0 °C. Molecules of perfume will diffuse throughout a room, but they won't collect again on your body. All spontaneous processes

First law: The total energy of the universe is a constant.
Second law: The total entropy of the universe is always increasing.
Third law: The entropy of a pure, perfectly formed crystalline substance at 0 K is zero.

result in energy becoming more dispersed throughout the universe. This is what scientists mean when they say that the second law is an expression of time in a physical—as opposed to psychological—form. For this reason, entropy has been called "time's arrow."

Neither the first law nor the second law of thermodynamics can be proven experimentally. But there has never been an example demonstrating that either is false. Albert Einstein once remarked that thermodynamic theory "is the only physical theory of the universe content [which], within the framework of applicability of its basic concepts, will never be overthrown." Einstein's statement does not mean that people have not tried (and are continuing to try) to disprove the laws of thermodynamics. Claims to have invented machines that perform useful work without expending energy—perpetual motion machines—are frequently made. However, no perpetual motion machine has ever been shown to work.

Roald Hoffmann, a chemist who shared the 1981 Nobel Prize in chemistry, has said, "One amusing way to describe synthetic chemistry, the making of molecules that is at the intellectual and economic center of chemistry, is that it is the local defeat of entropy" (*American Scientist* (Nov.–Dec. 1987): 619–21). "Local defeat" in this context refers to an unfavourable entropy change in a system because, of course, the entropy of the universe must increase if a process is to occur. Many chemical syntheses are entropy-disfavoured. Chemists find ways to accomplish them by balancing the unfavourable changes in the system with favourable changes in the surroundings to make these reactions occur.

"Local defeat" of entropy is apparent when we observe a pine tree with all of its symmetry at so many levels, when we think of the non-randomness of a human body in its form and its function, and when we see a model of the extraordinary structure of DNA. The formation of these low-entropy forms are accompanied by very large increases of entropy in their surroundings.

Finally, thermodynamics addresses one of the great mysteries: the origin of life. Life as we know it requires the creation of extremely complex molecules such as proteins and nucleic acids. Their formation must have occurred from atoms and small molecules. Assembling thousands of atoms into a highly ordered state in biochemical compounds clearly requires a local decrease in entropy. Some have said that life is a violation of the second law of thermodynamics. A more logical view, however, is that the local decrease is offset by an increase in entropy in the rest of the universe.

> Because the entropy of the universe is always increasing, it has been called "time's arrow." Living systems include many examples of localized decreases of entropy, but these are offset by larger increases of entropy in the rest of the universe.

17.6   Gibbs Free Energy

The method used so far to determine whether a process is spontaneous requires evaluation of two quantities, ΔS°_{sys} and ΔS°_{surr}. Wouldn't it be convenient to have a single thermodynamic function that serves the same purpose? A function associated with a system only and which does not require assessment of the surroundings—would be even better. Such a function exists! It is called **Gibbs free energy (G)**, often referred to simply as **free energy**, in honour of J. Willard Gibbs.

The free energy of a system at a specified temperature T, is defined mathematically as

$$G = H - TS$$

where H is the enthalpy of the system and S is the entropy of the system. Shortly, in a discussion of free energy change as a criterion of the spontaneous direction of reaction, we will see why free energy was defined in this way.

A sample of a substance has a particular amount of free energy. The amount of free energy in a sample of a substance is seldom known or even of interest. Rather, we are concerned with changes in free energy in the course of chemical reactions. We will see that the change in magnitude of free energy of a reaction mixture for a defined direction of reaction allows us to decide whether or not it is the direction that takes the mixture toward equilibrium.

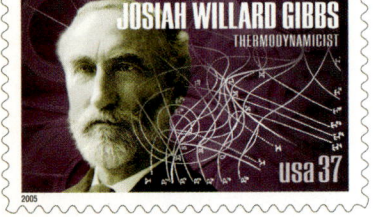

J. Willard Gibbs (1839–1903) made profound contributions to thermodynamics and received the first PhD in science awarded by an American university.

"In his later years he was a tall, dignified gentleman, with a healthy stride and ruddy complexion, performing his share of household chores, approachable and kind (if unintelligible) to students. Gibbs was highly esteemed by his friends, but American science was too preoccupied with practical questions to make much use of his profound theoretical work during his lifetime. He lived out his quiet life at Yale, deeply admired by a few able students but making no immediate impression on American science commensurate with his genius."

J.G. Crowther. (1969). Famous American Men of Science. Freeport, NY: Books for Libraries Press

We can define a property of substances called Gibbs free energy, G. The change of free energy for a specified direction of chemical reaction in a defined reaction mixture allows us to deduce if it is the spontaneous direction of reaction.

Free Energy Change of Reaction, $\Delta_r G$

The **free energy change of reaction** ($\Delta_r G$) at any stage of a reaction can be defined as

$$\Delta_r G = \sum n_i\, G(\text{products}) - \sum n_i\, G(\text{reactants})$$

where n_i is the number of moles of each product and reactant species in the balanced chemical equation, and G is the free energy per mole of each species under the conditions of the specified reaction mixture. If we represent a chemical reaction by the generalized equation

$$a\,\text{A} + b\,\text{B} \longrightarrow c\,\text{C} + d\,\text{D}$$

then

$$\Delta_r G = (c \times\ G_C + d \times\ G_D) - (a \times\ G_A + b \times\ G_B)$$

where, for example, G_A is the free energy per mole of A in the reaction mixture. $\Delta_r G$ changes during the course of a reaction as the amounts of reactants A and B decrease and the amounts of C and D increase, so a value for $\Delta_r G$ is applicable only to a reaction mixture with a particular composition at a particular time during the reaction.

We cannot know the exact magnitudes of the molar free energies of substances, nor do we need to know them to determine the value of $\Delta_r G$, as we will see in Worked Examples 17.6 and 17.7. In this sense, free energy and enthalpy are similar: substances possess defined amounts of enthalpy, but we do not have to know what the actual value of H is to obtain or use $\Delta_r H$ [<<Section 7.8].

Let's see first how to use free energy change as a way to determine whether a reaction is spontaneous. We can then ask further questions about the meaning of the term *free energy* and its use in calculating equilibrium constants to determine if reactions are product-favoured or reactant-favoured.

> In principle, the free energy change of reaction can be defined as $\Delta_r G = \sum n_i\, G(\text{products}) - \sum n_i\, G(\text{reactants})$ although this equation cannot be used to calculate $\Delta_r G$ since the molar free energies of substances is not known. $\Delta_r G$ changes as the amounts of reactants and products change during chemical reaction.

> This equation for $\Delta_r G$ is an approximation because (a) the free energy of 1 mol of a substance depends on its pressure (if it is a gas) or its concentration (if it is a solute in solution), and (b) there is another contribution to $\Delta_r G$ due to mixing of the reactants and products.

$\Delta_r G$ and Spontaneity of Reaction

According to the second law, a specified reaction direction is spontaneous if $\Delta S_{\text{univ}} > 0$. This applies whether or not the reactants and products are in their standard states. We can derive a subsidiary, but much more convenient, criterion in the case where the enthalpy of reaction is $\Delta_r H$. Although the following derivation is valid whether the reaction is exothermic or endothermic, perhaps it is easier to visualize an exothermic process during which, to maintain the system at constant temperature, the amount of heat (q_{surr}) transferred to the surroundings is equal in magnitude to $\Delta_r H$. This causes an increase in the entropy of the surroundings (q_{surr}/T), if we presume that the temperature change of the surroundings is insignificant:

$$\Delta S_{\text{univ}} = \Delta S_{\text{sys}} + \Delta S_{\text{surr}} = \Delta_r S + \frac{q_{\text{surr}}}{T} = \Delta_r S - \frac{\Delta_r H}{T}$$

Multiplying through this equation by $-T$, gives us

$$-T\Delta S_{\text{univ}} = \Delta_r H - T\Delta_r S$$

Gibbs defined the free energy as $G = H - TS$. Because enthalpy and entropy are state functions [<<Section 7.4], free energy is also a state function, and for a change occurring at constant temperature:

$$\Delta_r G = \Delta_r H - T\Delta_r S = -T\Delta S_{\text{surr}} - T\Delta S_{\text{sys}} = -T(\Delta S_{\text{surr}} + \Delta S_{\text{sys}}) = -T\Delta S_{\text{univ}}$$

and

$$\Delta S_{\text{univ}} = -\frac{\Delta_r G}{T}$$

Because of the minus sign in this relationship, if $\Delta S_{univ} > 0$, $\Delta_r G$ is < 0, and if $\Delta S_{univ} < 0$, $\Delta_r G$ is > 0. If $\Delta S_{univ} = 0$, $\Delta_r G = 0$.

Now we can translate our criteria for spontaneity based on the entropy change of the *universe* to criteria based on the free energy change of the *system*. Bearing in mind that the free energies of reactants and products depends on the amounts of them in a reaction mixture, for a defined reaction mixture we have the following:

- If $\Delta_r G < 0$, the reaction is spontaneous.
- If $\Delta_r G = 0$, the reaction is at equilibrium.
- If $\Delta_r G > 0$, the reaction is not spontaneous (but reaction in the opposite direction is).

These criteria of spontaneity based on the value of $\Delta_r G$ can be shown to reconcile with our previous criteria based on the relative values of reaction quotient Q and equilibrium constant K [<<Section 13.3]:

- If $Q < K$, the reaction is spontaneous in the direction written.
- If $Q = K$, the reaction is at equilibrium.
- If $Q > K$, the reaction is not spontaneous (but it is spontaneous in the opposite direction).

To demonstrate this, consider again our generalized reaction

$$a\,A + b\,B \longrightarrow c\,C + d\,D$$

and let's imagine a reaction vessel initially containing only A and B. After the tiniest amount of reaction, unless equilibrium is already reached, Q is very small and $Q < K$. At this time

$$(c \times G_C + d \times G_D) < (a \times G_A + b \times G_B) \quad \text{so} \quad \Delta_r G < 0$$

The spontaneous direction of reaction is such that free energy of the reaction mixture decreases ($\Delta_r G < 0$), which will happen if more of A and B react to form C and D. The free energy of a substance depends on the amount of it, so as reaction proceeds and Q increases, the free energies of A and B decrease and the free energies of C and D increase, and $\Delta_r G$ becomes less negative. More of A and B react, the total free energy of the reaction mixture decreases, Q increases, and $\Delta_r G$ becomes less negative until $Q = K$. In the reaction mixture at that time

$$(c \times G_C + d \times G_D) = (a \times G_A + b \times G_B) \quad \text{so} \quad \Delta_r G = 0$$

At this stage, the free energy of the reaction mixture is less than for any previous composition during the reaction, and the reaction is at equilibrium. How far the reaction goes before coming to equilibrium differs from reaction to reaction, and depends on the relative magnitudes of the free energies of each reactant and product.

If we imagine that some more of A and B were to react, then the total free energy of the reaction mixture would increase, and we would have $Q > K$ and $\Delta_r G > 0$. This change would not be spontaneous—rather, for the reaction mixture to come back to the point of equilibrium would be a spontaneous change. These changes are summarized in Figure 17.7.

The free energy change for a reaction mixture is given by $\Delta_r G = \Delta_r H - T\Delta_r S$. A criterion of spontaneity that focuses solely on the reaction mixture is given by $\Delta_r G < 0$. The condition of chemical equilibrium (when $Q = K$) is when $\Delta_r G = 0$, and G of the system is less than in any other reaction mixture derived by reaction from the starting mixture.

The Standard Free Energy Change of Reaction ($\Delta_r G°$)

Think about It

e17.11 See how the free energy change of reaction depends on entropy and enthalpy changes.

At any instant during a chemical reaction held at constant temperature and pressure $\Delta_r G = \Delta_r H - T\Delta_r S$. In the particular case of a reaction mixture in which all of the reactants and products are present in their standard states, the free energy change of reaction is the **standard free energy change of reaction ($\Delta_r G°$).** It is defined by

$$\Delta_r G° = \Delta_r H° - T\Delta_r S° \tag{17.2}$$

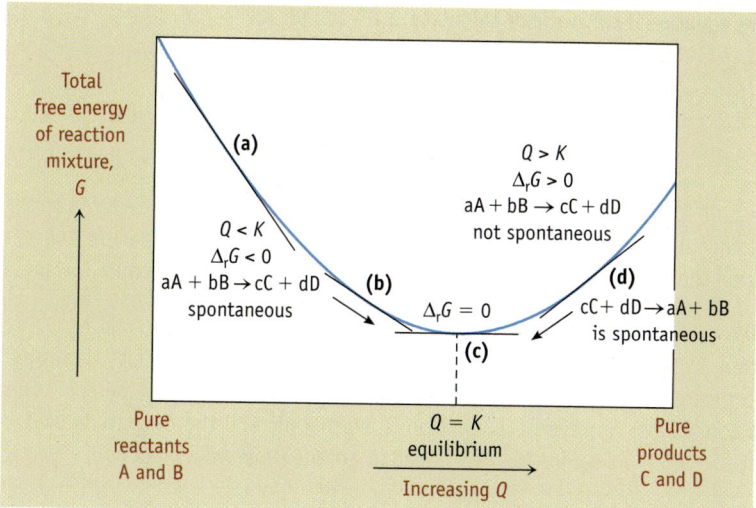

FIGURE 17.7 Change of free energy of reaction mixture during reaction. If reaction were to proceed from a mixture of reactants only entirely to a mixture of products only, Q would change from $-\infty$ to $+\infty$. In fact, net reaction ceases at some intermediate value of Q, when the mixture has attained chemical equilibrium. Prior to equilibrium, for example, at instants corresponding with points (a) and (b), conversion of reactants into products is spontaneous: Q is increasing, G is decreasing, and $\Delta_r G < 0$. For the reaction mixture instantaneously at (b), $\Delta_r G$ is less negative than at (a). $\Delta_r G$ at any instant during the reaction can be represented as the slope of the curve of G: prior to equilibrium, at (a) and (b), it is negative and becoming closer and closer to 0. At equilibrium (c), $\Delta_r G = 0$ when the free energies of products and reactants are equal, the slope of the curve is 0, and the total free energy of the system is less than for the reaction mixture at any previous time. For the reaction to go further, it would be a non-spontaneous event defying the second law of thermodynamics by changing to a system (such as (d)) of higher free energy ($\Delta_r G > 0$).

$\Delta_r G° < 0$ is used as a criterion of reaction spontaneity for standard state reaction mixtures, and, as we shall see, is directly related to the value of the equilibrium constant, and hence to whether reactions are reactant-favoured or product-favoured. If $\Delta_r G° < 0$, the spontaneous direction of reaction for a reaction mixture in which all of the reactants and products are present in their standard states is the direction indicated by the chemical equation.

For any reaction, $\Delta_r G°$ can be calculated from the values of $\Delta_r H°$ and $\Delta_r S°$, as illustrated in Worked Example 17.6.

Think about It

e17.12–e17.13 Explore the relationships among $\Delta_r H°$, $\Delta_r S°$, and $\Delta_r G°$.

WORKED EXAMPLE 17.6—CALCULATING $\Delta_r G°$ FROM $\Delta_r H°$ AND $\Delta_r S°$

Calculate the standard free energy change of reaction ($\Delta_r G°$) for the formation of methane at 298 K by the reaction

$$C(graphite) + 2\,H_2(g) \longrightarrow CH_4(g)$$

Strategy

$\Delta_r H°$ and $\Delta_r S°$ at 298 K can be calculated from values of $\Delta_f H°$ and $S°$ of reactants and products listed in Appendix C. Remember that $S°$ values are given in units of J K^{-1} mol^{-1}, whereas $\Delta H°$ values are given in units of kJ mol^{-1}.

Think about It

e17.14 Calculate how the free energy change of reaction depends on entropy and enthalpy changes.

Solution

First we calculate $\Delta_r H°$ and $\Delta_r S°$ at 298 K:

$\Delta_r H° = [1 \times \Delta_f H°(CH_4, g)] - [1 \times \Delta_f H°(C, graphite) + 2 \times \Delta_f H°(H_2, g)]$

$= [1 \times (-74.9\text{ kJ mol}^{-1})] - [(1 \times 0\text{ kJ mol}^{-1}) + (2 \times 0\text{ kJ mol}^{-1})] = -74.9\text{ kJ mol}^{-1}$

$\Delta_r S° = [1 \times S°(CH_4, g)] - [1 \times S°(C, graphite) + 2 \times S°(H_2, g)]$

$= [1 \times 186.3\text{ J K}^{-1}\text{ mol}^{-1}] - [(1 \times 5.6\text{ J K}^{-1}\text{ mol}^{-1}) + (2 \times 130.7\text{ J K}^{-1}\text{ mol}^{-1})])$

$= -80.7\text{ J K}^{-1}\text{ mol}^{-1} = -0.0807\text{ kJ K}^{-1}\text{ mol}^{-1}$

Interactive Exercise 17.13

Calculate the free energy change of reaction from entropy and enthalpy changes.

Now using equation 17.2, we can calculate $\Delta_r G°$ at 298 K:

$$\Delta_r G° = \Delta_r H° - T\Delta_r S°$$

$$= -74.9 \text{ kJ mol}^{-1} - (298 \text{ K})(-0.0807 \text{ kJ K}^{-1} \text{ mol}^{-1})$$

$$= -74.9 - (-24.1) \text{ kJ mol}^{-1} = -50.8 \text{ kJ mol}^{-1}$$

Since $\Delta_r G° < 0$, we can say that a for reaction mixture in which the reactants and products are in their standard states (pure solid graphite, $p(H_2) = 1$ bar, $p(CH_4) = 1$ bar) at 298 K, reaction in the direction shown in the chemical equation would take the reaction mixture toward equilibrium.

Comment

In this case, $\Delta_r S°$ is negative and so $-T\Delta_r S°$ has a positive value. If the entropy change of the system were the only contributor to $\Delta_r G°$, the reaction would not be spontaneous at the specified conditions. However, the reaction is exothermic and the magnitude of $\Delta_r H°$ is larger than that of $T\Delta_r S°$, so $\Delta_r G° < 0$. This is called an *enthalpy-driven reaction*.

EXERCISE 17.14—CALCULATING $\Delta_r G°$ FROM $\Delta_r H°$ AND $\Delta_r S°$

Using values of $\Delta_f H°$ and $S°$ to find $\Delta_r H°$ and $\Delta_r S°$, calculate the standard free energy change ($\Delta_r G°$) at 298 K, for the formation of 2 mol of $NH_3(g)$ from $N_2(g)$ and $H_2(g)$ when all of the substances are at 1 bar pressure (and 25 °C): $N_2(g) + 3 H_2(g) \longrightarrow 2 NH_3(g)$

The standard free energy change of reaction ($\Delta_r G°$) is the free energy change of reaction in the particular case of a reaction mixture in which all of the reactants and products are present in their standard states. The temperature must be specified.

$$\Delta_r G° = \Delta_r H° - T\Delta_r S°$$

Standard Molar Free Energy Change of Formation ($\Delta_f G°$)

The **standard molar free energy change of formation ($\Delta_f G°$)** of a compound is the free energy change when 1 mol of the compound in its standard state is formed from its component elements in their standard states. The temperature of measurement must be specified. The values for a few substances are listed in Table 17.3. By definition, *the standard molar free energy change of formation of an element in its standard state is zero*.

For example, $\Delta_f G°$ of $CH_4(g)$ is the free energy change when 1 mol of $CH_4(g)$ is produced by the following reaction, when all of the reactants and the product are in their standard states at the temperature of interest:

$$C(\text{graphite}) + 2 H_2(g) \longrightarrow CH_4(g) \qquad \Delta_r G° = \Delta_f G°(CH_4, g)$$

Put another way, $\Delta_f G°$ of $CH_4(g)$ is the standard free energy change when 1 mol of $CH_4(g)$ is produced by the this reaction, at the temperature of interest.

TABLE 17.3 Standard Molar Free Energy Change of Formation of Substances at 298 K

Substance	$\Delta_f G°$ (kJ mol^{-1})	Substance	$\Delta_f G°$ (kJ mol^{-1})
$H_2(g)$	0	$CO_2(g)$	−394.4
$O_2(g)$	0	$CH_4(g)$	−50.8
$N_2(g)$	0	$H_2O(g)$	−228.6
$C(s, \text{graphite})$	0	$H_2O(\ell)$	−237.2
$C(s, \text{diamond})$	2.900	$NH_3(g)$	−16.4
$CO(g)$	−137.2	$Fe_2O_3(s)$	−742.2

EXERCISE 17.15—STANDARD MOLAR FREE ENERGY CHANGE OF FORMATION ($\Delta_f G°$)

For each of the following substances, write the equation for the reaction whose standard free energy change of reaction ($\Delta_r G°$) is the standard molar free energy change of formation ($\Delta_f G°$) of the substance.

(a) $NH_3(g)$
(b) $Fe_2O_3(s)$
(c) $CH_3CH_2COOH(\ell)$
(d) $Ni(s)$

> The standard molar free energy change of formation ($\Delta_f G°$) of a compound is the standard free energy change of reaction ($\Delta_r G°$) when 1 mol of the compound is formed from its component elements at a specified temperature. For any element, $\Delta_f G°$ is zero.

$\Delta_r G°$ of Reaction from $\Delta_f G°$ of Reactants and Products

The standard free energy change of a reaction can be conceptualized as

$$\Delta_r G° = \sum n_i\, G°(\text{products}) - \sum n_i\, G°(\text{reactants})$$

where n_i are the stoichiometric coefficients of product and reactant species in the balanced chemical equation. Because we cannot measure molar free energies of substances, we cannot use this equation to calculate values of $\Delta_r G°$. However, Hess's law applies to free energy changes as well as to enthalpy changes. So just as the standard enthalpy change for a reaction ($\Delta_r H°$) can be calculated using values of $\Delta_f H°$ of the reactants and products at the same temperature [<<Section 7.8], there is a corresponding equation to obtain the standard free energy change of a reaction ($\Delta_r G°$) at a given temperature, from values of $\Delta_f G°$ of reactants and products at the same temperature:

$$\Delta_r G° = \sum n_i\, \Delta_f G°(\text{products}) - \sum n_i\, \Delta_f G°(\text{reactants})$$

WORKED EXAMPLE 17.7—$\Delta_r G°$ FROM $\Delta_f G°$ VALUES

Calculate the standard free energy change for the combustion of 1 mol of methane at 25 °C, using the standard molar free energy changes of formation of the products and reactants at 25 °C.

$$CH_4(g) + 2\, O_2(g) \longrightarrow 2\, H_2O(g) + CO_2(g)$$

Strategy

Because $\Delta_f G°$ values are for 1 mol of a substance, each value of $\Delta_f G°$ must be multiplied by the number of moles defined by the stoichiometric coefficient in the balanced chemical equation.

Solution

$\Delta_r G° = [2 \times \Delta_f G°(H_2O, g) + 1 \times \Delta_f G°(CO_2, g)] - [1 \times \Delta_f G°(CH_4, g) + 2 \times \Delta_f G°(O_2, g)]$

$= [2 \times (-228.6\ \text{kJ mol}^{-1}) + 1 \times (-394.4\ \text{kJ mol}^{-1})]$

$\quad -[1 \times (-50.8\ \text{kJ mol}^{-1}) + 2 \times (0\ \text{kJ mol}^{-1})]$

$= -801.0\ \text{kJ mol}^{-1}$

Comment

The large negative value of $\Delta_r G°$ indicates that in a standard state reaction mixture at 298 K, the reaction would need to proceed in the direction written to come to equilibrium.

EXERCISE 17.17—$\Delta_r G°$ FROM $\Delta_f G°$ VALUES

Calculate the standard free energy change at 25 °C for the oxidation of 1.00 mol of $SO_2(g)$ by $O_2(g)$ to form $SO_3(g)$.

> $\Delta_r G°$ at a specified temperature can be calculated from values of $\Delta_f G°$ of reactants and products at the same temperature:
>
> $$\Delta_r G° = \sum n_i\, \Delta_f G°(\text{products}) - \sum n_i\, \Delta_f G°(\text{reactants})$$

Think about It

e17.15 Work through a tutorial on calculating $\Delta_r G°$ from $\Delta_f G°$ values.

Interactive Exercise 17.16

Calculate $\Delta_r G°$ from $\Delta_f G°$ values.

Free Energy as Available Work

The term *free energy* was chosen for good reason. In any process, *the free energy change is the maximum energy available to do useful work*. Mathematically, this is expressed as $\Delta_r G = w_{max}$. In this sense, the word *free* means "available." To illustrate the logic behind this concept, consider the reaction discussed in Worked Example 17.6, which brings about a decrease in enthalpy of the reaction mixture ($\Delta_r H° < 0$) as well as a decrease in entropy ($\Delta_r S° < 0$).

$$C(graphite) + 2\, H_2(g) \longrightarrow CH_4(g) \quad \Delta_r H° = -74.9 \text{ kJ mol}^{-1}, \Delta_r S° = -80.7 \text{ JK}^{-1} \text{ mol}^{-1}$$

$$\text{At 298K,} \quad \Delta_r G° = \Delta_r H° - T\Delta_r S° = -50.8 \text{ kJ mol}^{-1}$$

Although it might seem that all the enthalpy available from the reaction (74.9 kJ per mol of graphite that reacts) could be transferred to the surroundings and be available to do work, this is not the case. The negative entropy change means that the reaction mixture becomes more ordered as a result of the reaction. A portion of the energy from the reaction is used to create this more ordered system, and is not available to do work. The energy left over "free," or available, to do work (the free energy change) is 50.8 kJ mol^{-1}. Of the 74.9 kJ mol^{-1} decrease in enthalpy of the reaction mixture, 24.1 kJ mol^{-1} ($= T\Delta_r S°$) is used by the system itself to increase the order of the arrangement of atoms.

A similar analysis of the enthalpy and entropy changes applies to any reaction, whatever the combination of $\Delta H°$ and $\Delta S°$. In a process that is favoured by both the enthalpy and entropy changes, the free energy change exceeds the enthalpy change of reaction.

> In any process, the free energy change is the maximum energy that becomes available to do useful work: $\Delta_r G = w_{max}$.

$\Delta_r G°$ and Reaction Spontaneity—A Qualitative Perspective

The standard free energy change of a reaction ($\Delta_r G°$) is the difference between the free energies of products and reactants when each of the reactants and products in the reaction mixture is in its standard state (reaction quotient $Q = 1$). The sign of $\Delta_r G°$ tells us which way reaction would need to go for this reaction mixture to become closer to equilibrium (the spontaneous direction of reaction), and the magnitude of $\Delta_r G°$ is an indication of how far this reaction mixture is from equilibrium.

For example, if $\Delta_r G° < 0$, we know that in a standard reaction mixture, reaction would need to go in the direction written to reach equilibrium. Since the standard mixture has $Q = 1$, K must be greater than 1, and we would describe such a reaction as product-favoured (Figure 17.8). If $\Delta_r G°$ is a small negative value, then reaction would not need to proceed far for equilibrium to be reached, so that K would not be a large number (although greater than 1). If $\Delta_r G°$ is a large negative value, Q would not need to change considerably before equilibrium is reached, so the magnitude of K would be large.

FIGURE 17.8 Free energy changes during a reaction for which $\Delta_r G° < 0$. In a standard reaction mixture ($Q = 1$), if $\Delta_r G° < 0$ there must be conversion of reactants into products to reach equilibrium. For such a reaction, $K > 1$. If a reaction mixture contains only reactants ($Q = -\infty$), as the mixture changes during conversion of reactants into products, at some point in time we would have $Q = 1$ prior to equilibrium being reached. The larger K is, the more a standard reaction mixture ($Q = 1$) would need to change before $Q = K$.

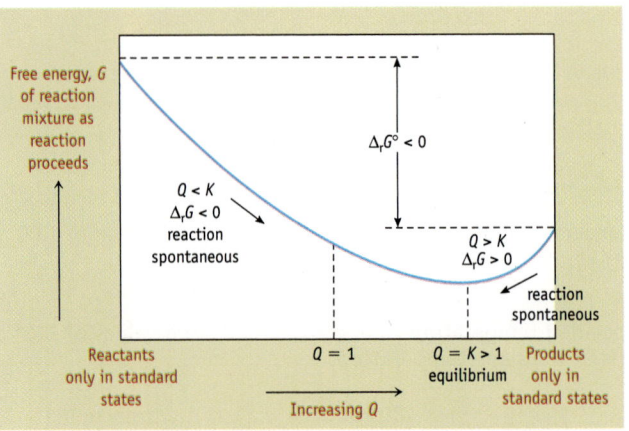

For a reaction with $\Delta_r G° > 0$, a standard reaction mixture ($Q = 1$) would need to react in the direction opposite to that written to reach equilibrium. In such cases, $K < 1$, and these reactions are reactant-favoured (Figure 17.9).

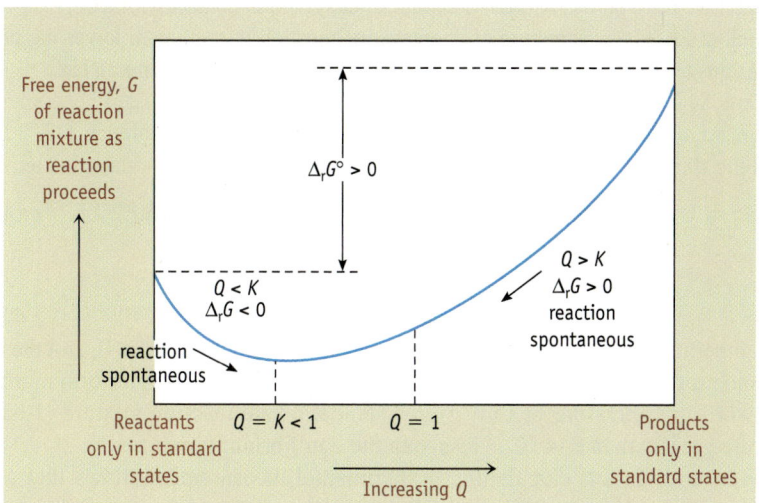

FIGURE 17.9 Free energy changes during a reaction for which $\Delta_r G° > 0$. In a standard reaction mixture ($Q = 1$) since $\Delta_r G° > 0$, products (as defined by the reaction equation) must change into reactants for equilibrium to be achieved. During reaction, Q decreases, so $K < 1$.

EXERCISE 17.18—$\Delta_r G°$ AND SPONTANEITY

If hydrogen gas can be produced cheaply, it could be burned directly or converted to another fuel such as methane $CH_4(g)$:

$$3\ H_2(g) + CO(g) \longrightarrow CH_4(g) + H_2O(g)$$

Calculate $\Delta_r G°$ at 298 K for the reaction. Is the written equation in the spontaneous reaction direction at 25 °C when all reactants and products are in their standard states?

For reactions with $\Delta_r G° < 0$, $K > 1$ and a reaction mixture with all reactants and products in their standard states would need to go in the direction written to reach equilibrium. Such reactions are product-favoured. Reactant-favoured reactions have $\Delta_r G° > 0$ and $K < 1$, so the spontaneous direction of reaction in a standard reaction mixture is the opposite of that written.

Dependence of $\Delta_r G°$ and Spontaneity of Reaction on Temperature

In Table 17.2, we saw that we cannot tell by inspection whether a reaction is spontaneous if either (a) $\Delta_r H° < 0$ and $\Delta_r S° < 0$, or (b) $\Delta_r H° > 0$ and $\Delta_r S° > 0$. This is because in the first case, $\Delta_r H°$ is negative and $-T\Delta_r S°$ is positive, while in the second case, $\Delta_r H°$ is positive and $-T\Delta_r S°$ is negative. In either case, we need to calculate the value of $\Delta_r G°$ to see which term is dominant, and if $\Delta_r G°$ is negative or positive.

$\Delta_r H°$ and $\Delta_r S°$ change very little over moderate changes of temperature. So, for reactions of the two types described above, there is some temperature at which $\Delta_r H° = -T\Delta_r S°$ and $\Delta_r G° = 0$, and the standard reaction mixture is at equilibrium.

The conversion of limestone to lime and carbon dioxide gas is represented by the equation

$$CaCO_3(s) \longrightarrow CaO(s) + CO_2(g)$$

Thermodynamic data relevant to these substances at 25 °C, obtained from Appendix C, are shown below:

	$CaCO_3(s)$	$CaO(s)$	$CO_2(g)$
$\Delta_f G°$ (kJ mol^{-1})	−1129.16	−603.42	−394.36
$\Delta_f H°$ (kJ mol^{-1})	−1207.6	−35.09	−393.51
$S°$ (J K^{-1} mol^{-1})	91.7	38.2	213.74

From these data we can calculate that $\Delta_r G° = +131.38$ kJ mol^{-1}, $\Delta_r H° = +179.0$ kJ mol^{-1}, and $\Delta_r S° = +160.2$ J K^{-1} mol^{-1}. Although the entropy change favours reaction, the large positive unfavourable enthalpy change dominates at this temperature. The positive value of $\Delta_r G°$ tells us that if we have pure $CaCO_3(s)$ and pure $CaO(s)$ both in contact with $CO_2(g)$ at 1 bar pressure, and all at 25 °C, the direction of spontaneous reaction is such as to lower the pressure of $CO_2(g)$. At this temperature, the equilibrium pressure of $CO_2(g)$ is less than 1 bar.

Since $\Delta_r S°$ is positive, if we raise the temperature of the system, $T\Delta_r S°$ becomes increasingly large. At some higher temperature, $T\Delta_r S°$ will exactly counterbalance $\Delta_r H°$, $\Delta_r G°$ will be 0, and the standard-state system will be at equilibrium. we can calculate this temperature:

$$\Delta_r G° = 0 = \Delta_r H° - T\Delta_r S° = 179 \times 10^3 \text{ J mol}^{-1} - T(160.2 \text{ J K}^{-1}\text{mol}^{-1})$$

$$T = \frac{179 \times 10^3 \text{ J mol}^{-1}}{160.2 \text{ J K}^{-1} \text{ mol}^{-1}} = 1117 \text{ K } (=844 \text{ °C})$$

At temperatures higher than 844 °C, we would expect that $\Delta_r G° < 0$, and the direction of spontaneous reaction in a standard-state mixture would be such as to form more $CO_2(g)$. In other words, the pressure of $CO_2(g)$ in a standard-state reaction mixture (1 bar) at any temperature greater than 844 °C is less than the equilibrium pressure.

How accurate is this calculation? Experimental measurement shows that a reaction mixture containing the two solids and $CO_2(g)$ at 1 bar pressure is at equilibrium at about 900 °C. The reasonably small difference between this and the calculated temperature is mainly due to the assumption that $\Delta_r H°$ and $\Delta_r S°$ do not change with temperature changes—which is not strictly true. Chemists and chemical engineers use more accurate mathematical relationships than those used here.

Worked Example 17.8 demonstrates how calculation of the temperature at which a standard reaction mixture is at equilibrium can give us the boiling point of a substance. This is possible because the standard free energy change for conversion between a liquid and its vapour applies when the vapour is at 1 bar (its standard state)—the same condition as at the boiling point.

WORKED EXAMPLE 17.8—FINDING THE TEMPERATURE AT WHICH $\Delta_r G° = 0$

At what temperature is the standard reaction mixture represented by the following equation at equilibrium?

$$H_2O(\text{pure liquid}) \rightleftharpoons H_2O(g, 1.0 \text{ bar})$$

Strategy

We know from experience (and we can show by calculation) that for this equation $\Delta_r H° > 0$ and $\Delta_r S° > 0$, so $-T\Delta_r S° > 0$. From the data in Appendix C, we can calculate $\Delta_r H°$ and $\Delta_r S°$ at 25 °C. If we assume that these values are constant over all temperatures of interest, there can be only one temperature at which $\Delta_r H° = -T\Delta_r S°$ and $\Delta_r G° = 0$.

Solution

Using the data from Appendix C to evaluate thermodynamic values at 25 °C:

$$\Delta_r H° = [1 \times \Delta_f H°(H_2O, g)] - [1 \times \Delta_f H°(H_2O, \ell)]$$

$$= [1 \times -241.83 \text{ kJ mol}^{-1}] - [1 \times -285.83 \text{ kJ mol}^{-1}] = +44.00 \text{ kJ mol}^{-1}$$

$$\Delta_r S° = [1 \times S°(H_2O, g)] - [1 \times S°(H_2O, \ell)]$$

$$= [1 \times 188.84 \text{ J K}^{-1} \text{ mol}^{-1}] - [1 \times 69.95 \text{ J K}^{-1} \text{ mol}^{-1}] = +118.89 \text{ J K}^{-1} \text{ mol}^{-1}$$

Converting the value of $\Delta_r S°$ into its value in kJ K^{-1} mol^{-1}, we can find the temperature at which $\Delta_r H° = -T\Delta_r S°$ (and $\Delta_r G° = 0$).

$$+44.00 \text{ kJ mol}^{-1} = T \times \frac{+118.89}{1000} \text{ kJ K}^{-1} \text{ mol}^{-1}$$

$$T = 370.1 \text{ K } (=96.9 \text{ °C})$$

Comment

The temperature at which water vapour at 1 bar pressure is in equilibrium with liquid water is, by definition, the boiling point of water at that pressure. So we have estimated here the standard boiling point of water (not the normal boiling point, which applies at 1 atm pressure), the experimental value of which is 99.6 °C. The closeness of our calculation to the experimental value indicates the power of thermodynamics, spoiled only slightly by the inexactness of our assumption that $\Delta_r H°$ and $\Delta_r S°$ have the same values at 298 K as at 370.1 K.

EXERCISE 17.19—FINDING THE TEMPERATURE AT WHICH $\Delta_r G° = 0$

Oxygen was first made by Joseph Priestley (1733–1804) by heating HgO(s):

$$2\ HgO(s) \longrightarrow 2\ Hg(\ell) + O_2(g)$$

Use the thermodynamic data of Appendix C that apply at 25 °C to estimate the temperature at which HgO(s) in a sealed vessel decomposes sufficiently to be in equilibrium with 1 bar pressure of $O_2(g)$.

For the standard reaction mixture defined in Worked Example 17.8, at any $T < 370.1$ K, $\Delta_r G° > 0$ and the spontaneous direction of reaction is the opposite to that written: the vapour pressure would need to decrease for the system to come to equilibrium at that temperature. At any temperature T < 370.1 K, $\Delta_r G° > 0$. So we can see that for a standard reaction mixture, reactions of this type are spontaneous below the equilibrium temperature, and non-spontaneous at higher temperatures.

For reactions with $\Delta_r H° < 0$ and $\Delta_r S° < 0$, we can also calculate a temperature at which a standard reaction mixture is at equilibrium. At temperatures below this, the written reaction is non-spontaneous, and at temperatures above the equilibrium temperature the reaction becomes spontaneous.

EXERCISE 17.21—SPONTANEITY AND TEMPERATURE

Determine whether at 25 °C, $\Delta_r S° < 0$ or $\Delta_r S° > 0$ for each of the reactions listed below. Predict how an increase in temperature will affect the value of $\Delta_r G°$.

(a) $N_2(g) + 2\ O_2(g) \longrightarrow 2\ NO_2(g)$

(b) $2\ C(s) + O_2(g) \longrightarrow 2\ CO(g)$

(c) $CaO(s) + CO_2 \longrightarrow CaCO_3(s)$

(d) $2\ NaCl(s) \longrightarrow 2\ Na(s) + Cl_2(g)$

For a reaction in which either (a) $\Delta_r H° < 0$ and $\Delta_r S° < 0$, or (b) $\Delta_r H° > 0$ and $\Delta_r S° > 0$, because $\Delta_r G° = \Delta_r H° - T\Delta_r S°$, we can calculate a temperature at which $\Delta_r G° = 0$, using the assumption that $\Delta_r H°$ and $\Delta_r S°$ are independent of temperature. At this temperature, the reaction in a standard-state reaction mixture is at equilibrium. The direction of spontaneous reaction is different above that equilibrium temperature than below it.

Free Energy Change of Reaction in Non-standard Reaction Mixtures

When reactants and products are not in their standard states, the free energy change of a reaction ($\Delta_r G$) is different from the standard free energy change ($\Delta_r G°$). A quantitative statement of the magnitude of the difference is the following:

$$\Delta_r G = \Delta_r G° + RT \ln Q \qquad (17.3)$$

Interactive Exercise 17.20

Predict how temperature will affect $\Delta_r G°$.

Think about It

e17.16 Simulate the relationship between $\Delta_r G°$ and temperature.

Think about It

e17.17 Predict whether the reaction is reactant- or product-favoured.

Perhaps you have recognized a correspondence between the equation for the difference between $\Delta_r G$ and $\Delta_r G°$ and the Nernst equation for the difference between E_{cell} and $E°_{cell}$ [<<Section 16.5]. The similarity should not be surprising since $\Delta_r G$ and E_{cell} are both measures of "driving force" of reaction toward equilibrium.

where R is the universal gas constant, 8.314 J K^{-1} mol^{-1}, and T is the temperature (in K).

From this equation, we can deduce that in a reaction mixture with $Q = 1$ (when all reactants and products have solution concentration = 1 mol L^{-1} or pressure = 1 bar), since $\ln 1 = 0$, $\Delta_r G = \Delta_r G°$.

The power of this relationship is indicated by Worked Example 17.9 and the following exercise.

WORKED EXAMPLE 17.9—CALCULATION OF $\Delta_r G$ FOR A NON-STANDARD REACTION MIXTURE

Equilibrium between liquid water and water vapour can be represented by

$$H_2O(\ell) \rightleftharpoons H_2O(g)$$

At 25 °C (298 K), the equilibrium vapour pressure of water is 3.17 kPa or 0.0317 bar.
(a) Use thermodynamic data from Appendix C to calculate $\Delta_r G°$ at 25 °C, and interpret the meaning of your answer.
(b) Calculate $\Delta_r G$ at 25 °C for a sealed vessel containing liquid water and water vapour with a pressure of (i) 2.0 bar, and (ii) 0.10 bar, and in each case make sense of your answer in terms of reaction spontaneity.

Strategy

For part (a), calculate $\Delta_r G°$ at 25 °C from values of $\Delta_f G°$ at 25 °C in Appendix C. For part (b), use equation 17.3.

Solution

From Appendix C, at 25 °C, $\Delta_f G°(H_2O, \ell) = -237.15$ kJ mol^{-1}, $\Delta_f G°(H_2O, g) = -228.59$ kJ mol^{-1}.

(a) At 25 °C, $\Delta_r G° = \Delta_f G°(H_2O, g) - \Delta_f G°(H_2O, \ell)$

$$= -228.59 \text{ kJ mol}^{-1} - (-237.15 \text{ kJ mol}^{-1}) = +8.56 \text{ kJ mol}^{-1}$$

Since we have calculated a value of the *standard* free energy change, our result is applicable to a reaction mixture containing all reactants and products in their standard states—pure liquid water and water vapour whose pressure is 1.0 bar. $\Delta_r G° > 0$, so the reaction is not spontaneous in the direction written. Rather, to reach equilibrium, there would need to be reaction in the opposite direction:

$$H_2O(g) \longrightarrow H_2O(\ell)$$

This is consistent with the knowledge that the vapour pressure in this reaction vessel (1.0 bar) is higher than the equilibrium vapour pressure (0.0317 bar) at 25 °C.

(b) (i) Since liquid water has a constant concentration, $Q = p_{H_2O}$. $\Delta_r G°$ has units of kJ mol^{-1}, while R is expressed in J K^{-1} mol^{-1} and a conversion needs to be made.

$$\Delta_r G = \Delta_r G° + RT \ln Q$$

$$= +8.56 \times 10^3 \text{ J mol}^{-1} + (8.314 \text{ J K}^{-1} \text{ mol}^{-1})(298 \text{ K})(\ln 2.0)$$

$$= +8.56 \times 10^3 \text{ J mol}^{-1} + 1.72 \times 10^3 \text{ J mol}^{-1} = +10.28 \text{ kJ mol}^{-1}$$

Not surprisingly, since the vapour pressure is even higher than 1.0 bar, and the reaction mixture further from equilibrium than a standard reaction mixture, $\Delta_r G$—the "driving force" toward equilibrium by forming more $H_2O(\ell)$—is higher than that in the standard reaction mixture in part(a).

(b) (ii) $\Delta_r G = \Delta_r G° + RT \ln Q$

$$= +8.56 \times 10^3 \text{ J mol}^{-1} + (8.314 \text{ J K}^{-1} \text{ mol}^{-1})(298 \text{ K})(\ln 0.10)$$

$$= +8.56 \times 10^3 \text{ J mol}^{-1} + (-5.70 \times 10^3 \text{ J mol}^{-1}) = +2.86 \text{ kJ mol}^{-1}$$

The positive value of $\Delta_r G$ tells us that net reaction in the direction indicated by the equation would not take the reaction mixture toward equilibrium; net reaction in the opposite direction would. This is consistent with the fact that the vapour pressure is greater than the equilibrium vapour pressure. The fact that $\Delta_r G$ here is less than for the standard reaction mixture ($\Delta_r G°$) is consistent with the fact that a vapour pressure of 0.10 bar is closer to the equilibrium vapour pressure than is 1.0 bar.

Comment

Perhaps you can see that if we choose smaller and smaller values of the vapour pressure, we will come to a pressure such that $\Delta_r G = 0$ and the reaction mixture is at equilibrium. Try $p = 0.0317$ bar. This leads us to Section 17.7.

EXERCISE 17.22—CALCULATION OF $\Delta_r G$ FOR A NON-STANDARD REACTION MIXTURE

Using the value of $\Delta_r G°$ from Worked Example 17.9, calculate $\Delta_r G$ for the reaction $H_2O(l) \rightleftharpoons H_2O(g)$ in a vessel containing water vapour with a pressure of 0.020 bar in contact with liquid water. Interpret your answer in terms of spontaneity of the reaction.

For non-standard reaction mixtures, $\Delta_r G = \Delta_r G° + RT \ln Q$.

17.7 The Relationship between $\Delta_r G°$ and K

The magnitude of the equilibrium constant K for a reaction, at a specified temperature, is a measure of how far the reaction would need to proceed before the reaction mixture comes to equilibrium. At the same temperature, the magnitude of $\Delta_r G°$ is a measure of how far a reaction mixture with reaction quotient $Q = 1$ is from equilibrium (when all reactants and products are in their standard states). So perhaps it is not surprising that there is a direct mathematical relationship between these two measures of "driving force" of reaction.

During a chemical reaction, the concentrations of reactants decrease and the concentrations of products increase until chemical equilibrium is reached. At this time, $Q = K$ and $\Delta_r G = 0$. We can substitute these values appropriate to the state of chemical equilibrium into equation 17.3 ($\Delta_r G = \Delta_r G° + RT \ln Q$), leading us to

$$0 = \Delta_r G° + RT \ln K$$

Re-arranging, we get a very useful relationship between the standard free energy change of a reaction and its equilibrium constant, both measured at the same temperature:

$$\Delta_r G° = -RT \ln K$$

This equation allows calculation of an equilibrium constant at the same temperature as thermochemical data in tables or obtained from an experiment. Alternatively, it provides a direct route to determine the standard free energy change of a reaction from experimentally determined equilibrium constants.

The logic used in deriving the equation relating $\Delta_r G°$ and K corresponds with that used to derive the relationship between $E°$ and K for electrochemical cell reactions from the Nernst equation [<<Section 16.5].

Interactive Exercise 17.23

SUBMIT

Use a concept map to show relationships between thermodynamic concepts.

Interactive Exercise 17.24

SUBMIT

Thinking task: Identify errors in proposed solutions to questions.

WORKED EXAMPLE 17.10—THE RELATIONSHIP BETWEEN $\Delta_r G°$ AND K

Determine the standard free energy change ($\Delta_r G°$) at 25 °C for the formation of 1.0 mol of ammonia gas from nitrogen gas and hydrogen gas, and use this value to calculate the equilibrium constant for this reaction at 25 °C.

Strategy

The free energy change to form 1.0 mol of $NH_3(g)$ from its elements is the free energy of formation of $NH_3(g)$. Calculate the equilibrium constant at 25 °C from $\Delta_r G°$ at 25 °C using $\Delta_r G° = -RT \ln K$.

Solution

Begin by specifying a balanced equation for the chemical reaction:

$$\tfrac{1}{2}\,N_2(g) + \tfrac{3}{2}\,H_2(g) \longrightarrow NH_3(g)$$

From Appendix C, we find that $\Delta_r G° = \Delta_f G°(NH_3, g) = -16.37$ kJ mol^{-1} at 25 °C.

$$\Delta_r G° = -RT \ln K$$

$$-16.37 \times 10^3 \ \cancel{J} \ \cancel{mol^{-1}} = -(8.314 \ \cancel{J} \ \cancel{K^{-1}} \ \cancel{mol^{-1}})(298 \ \cancel{K}) \ln K$$

$$\ln K = 6.607$$

$$K = 740$$

Comment

This is the core reaction of the industrial formation of ammonia by the Haber Process [<<Sections 13.1, 13.7]. Although the reaction is product-favoured at 25 °C, it is slow at that temperature. If we try to increase the rate of reaction by increasing the temperature, the equilibrium constant of this exothermic reaction decreases. The challenge has been to develop a catalyst that will allow economical rates of reaction at sufficiently low temperatures that the equilibrium constant is not too small.

WORKED EXAMPLE 17.11—THE RELATIONSHIP BETWEEN $\Delta_r G°$ AND K

The value of K_{sp} for AgCl(s) at 25 °C is 1.8×10^{-10}. Use this value to determine $\Delta_r G°$ at 25 °C for the process Ag$^+$(aq) + Cl$^-$(aq) $\longrightarrow$ AgCl(s).

Strategy

The chemical equation given is the opposite of the equation used to define K_{sp}, so its equilibrium constant is $1/K_{sp}$. This value is used to calculate $\Delta_r G°$ at the same temperature.

Solution

For Ag$^+$(aq) + Cl$^-$(aq) $\longrightarrow$ AgCl(s)

$$K = \frac{1}{K_{sp}} = \frac{1}{1.8 \times 10^{-10}} = 5.6 \times 10^9$$

$$\Delta_r G° = -RT \ln K = -(8.314 \ J \ \cancel{K^{-1}} \ mol^{-1})(298 \ \cancel{K}) \ln (5.6 \times 10^9)$$

$$= -56 \times 10^3 \ J \ mol^{-1} = -56 \ kJ \ mol^{-1}$$

Comment

The negative value of $\Delta_r G°$ indicates that if we have a reaction mixture with the reactant species in their standard states (a solution with both [Ag$^+$] and [Cl$^-$] at 1.0 mol L^{-1}), then the direction of spontaneous reaction is precipitation of solid AgCl. Earlier, we described the experimental determination of $\Delta_r G°$ from thermochemical measurements— that is, from $\Delta_r H°$ and $\Delta_r S°$ values or from $\Delta_f G°$ values. Here is another method to determine $\Delta_r G°$.

EXERCISE 17.25—THE RELATIONSHIP BETWEEN $\Delta_r G°$ AND K

From the thermodynamic data in Appendix C, calculate $\Delta_r G°$ at 25 °C for the reaction

$$C(s) + CO_2(g) \longrightarrow 2\,CO(g)$$

Use this to calculate the equilibrium constant at 25 °C.

For the reaction $Ag^+(aq) + 2 NH_3(aq) \longrightarrow [Ag(NH_3)_2]^+(aq)$ the equilibrium constant at 25 °C is 1.6×10^7. Calculate $\Delta_r G^\circ$ at 25 °C.

For a specified reaction, $\Delta_r G^\circ$ and K at the same temperature are mathematically related: $\Delta_r G^\circ = -RT \ln K$

17.8 $\Delta_r G^\circ$ and E°_{cell} for Voltaic Cell Reactions

We have seen that the standard cell emf (E°_{cell}) is a measure of the "driving force" of the overall reaction that takes place in a voltaic cell, and that there is a mathematical relationship between E°_{cell} and K of the reaction [<<Section 16.6]. Since there is also a mathematical relationship between $\Delta_r G^\circ$ and K, it follows that for voltaic cell reactions there must be a mathematical link between $\Delta_r G^\circ$ and E°_{cell} at the same temperature.

This mathematical relationship is easily derived from

$$\Delta_r G^\circ = -RT \ln K \quad \text{or} \quad \ln K = -\frac{\Delta_r G^\circ}{RT}$$

and, for voltaic cells

$$\ln K = +\frac{nFE^\circ_{cell}}{RT}$$

where n is the number of electrons transferred in the equation used to represent the overall cell reaction, F is the charge in coulombs per mol of electrons (96 450 C mol^{-1}), R is the gas constant, and T is the temperature. We can equate the terms on the right side of these two equations (at the same temperature), giving us the relationship we are looking for:

$$\Delta_r G^\circ = -nFE^\circ_{cell}$$

The negative sign in this equation is logical because a spontaneous reaction in a standard cell would be characterized by a positive value of E°_{cell} and a negative value of $\Delta_r G^\circ$.

WORKED EXAMPLE 17.12—$\Delta_r G^\circ$ AND E°_{cell} AT THE SAME TEMPERATURE

The standard cell emf for the reduction of silver ions with copper metal is 0.46 V at 25 °C. Calculate $\Delta_r G^\circ$ for this reaction at 25 °C.

Strategy

Use the relationship $\Delta_r G^\circ = -nFE^\circ_{cell}$. The value of n can be deduced from the half-equations used to derive a balanced equation for the cell reaction.

Solution

We use half-equations to obtain a balanced equation for the cell reaction:

$$Cathode\ half-reaction: 2 \times [Ag^+(aq) + e^- \longrightarrow Ag(s)]$$

$$Anode\ half-reaction: Cu(s) \longrightarrow Cu^{2+}(aq) + 2e^-$$

$$Overall\ reaction: 2\ Ag^+(aq) + Cu(s) \longrightarrow 2\ Ag(s) + Cu^{2+}(aq)$$

We see that $n = 2$. Substituting:

$$\Delta_r G^\circ = -nFE^\circ_{cell} = -(2)(96450\ C\ mol^{-1})(0.462\ V) = -89\,100\ C\ V\ mol^{-1}$$

Because 1 C V = 1 J

$$\Delta_r G^\circ = -89\,100\ J\ mol^{-1} = -89.1\ kJ\ mol^{-1}$$

Think about It

e17.18 Calculate $\Delta_r G^\circ$ from E°_{cell}.

Comment

This example demonstrates an effective method of obtaining thermodynamic values from simple electrochemical measurements. The calculated value of $\Delta_r G°$ now allows estimation of K for the cell reaction.

EXERCISE 17.27—$\Delta_r G°$ AND $E°_{cell}$ AT THE SAME TEMPERATURE

The electrochemical cell with the net reaction represented by the following equation has $E°_{cell} = -0.76$ V at 25 °C.

$$H_2(g) + Zn^{2+}(aq) \longrightarrow Zn(s) + 2\ H^+(aq)$$

Calculate $\Delta_r G°$ at 25 °C. What is the direction of spontaneous reaction in a cell at 25 °C in which all species are in their standard states?

For reactions that take place in a voltaic cell, $\Delta_r G° = -nFE°_{cell}$.

Related "Driving Forces" of Reaction: $\Delta_r G°$, K, and $E°_{cell}$

Recall the mathematical relationship between the standard cell emf ($E°_{cell}$) and the equilibrium constant (K) of the cell reaction [<<Section 16.6] at a given temperature. In Section 17.7, we see a simple relationship between the standard free energy change of a reaction ($\Delta_r G°$) and K, at the same temperature. And above we have a direct link between $\Delta_r G°$ and $E°_{cell}$, at a given temperature. The existence of relationships among these three properties of any chemical reaction should not surprise us, because they are all measures of the "driving force" of reaction: how far a reaction will go before reaching chemical equilibrium. Since each of the relationships is quantitative, we now have, for any reaction at a defined temperature, a set of mathematical relationships among $\Delta_r G°$, K, and $E°_{cell}$ (in the case of voltaic cell reactions). If we know the value of any one of these at a specified temperature, we can calculate the value of the other two at that temperature. The equations are displayed in Figure 17.10.

FIGURE 17.10 Relationships among the "driving forces" $\Delta_r G°$, K, and $E°_{cell}$ of a reaction at a given temperature. For spontaneous reaction in a standard-state mixture ($Q = 1$), $K > 1$, $E°_{cell} > 0$, and $\Delta_r G° < 0$.

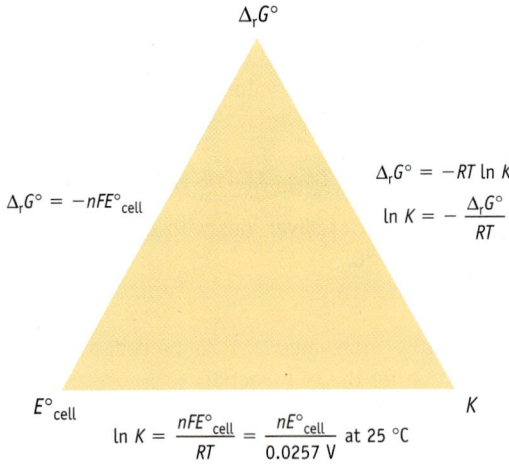

Interactive Exercise 17.29

Use a concept map to show relationships among concepts.

Think about It

e17.19 Calculate $\Delta_r G°$ and the equilibrium constant K from $E°_{cell}$.

EXERCISE 17.28—RELATIONSHIPS AMONG $\Delta_r G°$, K, AND $E°_{cell}$

From standard reduction potentials at 25 °C, calculate $E°_{cell}$ at 25 °C, and then $\Delta_r G°$ and K at 25 °C, for the following:

(a) $2\ Fe^{3+}(aq) + 2\ I^-(aq) \longrightarrow 2\ Fe^{2+}(aq) + I_2(s)$

(b) $I_2(s) + 2\ Br^-(aq) \longrightarrow 2\ I^-(aq) + Br_2(\ell)$

The three "driving forces" of reaction—$\Delta_r G°$, $E°_{cell}$, and K—are mathematically related at the same temperature.

Of two systems with the same number of elements of each type, and in the same state of matter, the one with more, smaller molecules has more entropy. (Section 17.4)

- The **standard entropy change of reaction** ($\Delta_r S°$) at a specified temperature is the difference between the sum of the standard molar entropies of products and the standard molar entropies of reactants at that temperature, taking into account the amounts of the products and reactants shown in the balanced chemical equation

$$\Delta_r S° = \sum n_i S°(\text{products}) - \sum n_i S°(\text{reactants}) \qquad \text{(Section 17.4)}$$

- When all species in a reaction mixture are in their standard states, reaction in a specified direction is spontaneous if $\Delta_r S° > 0$ and $\Delta_r H° < 0$, and non-spontaneous if $\Delta_r S° < 0$ and $\Delta_r H° > 0$. In cases where $\Delta_r S° > 0$ and $\Delta_r H° > 0$, or $\Delta_r S° < 0$ and $\Delta_r H° < 0$, we cannot predict the direction of spontaneous reaction without calculations to see if $\Delta S°_{univ} > 0$. (Section 17.5)
- Living systems include many examples of localized decreases of entropy, but these are offset by larger increases of entropy in the rest of the universe. (Section 17.5)
- **Gibbs free energy** is a property of substances defined by $G = H - TS$. (Section 17.6)
- The **free energy change of reaction** ($\Delta_r G$) changes as reaction proceeds and the amounts of reactants and products change. It can be defined in principle as

$$\Delta_r G = \sum n_i G(\text{products}) - \sum n_i G(\text{reactants}) \qquad \text{(Section 17.6)}$$

- The free energy change of reaction in a defined reaction mixture is given by $\Delta_r G = \Delta_r H - T\Delta_r S$. A criterion of spontaneity that focuses solely on the reaction mixture is given by $\Delta_r G < 0$. The condition of chemical equilibrium is when $\Delta_r G = 0$, and G of the system is less than in any other reaction mixture derived by reaction from the starting mixture. (Section 17.6)
- In any process, the free energy change is the maximum energy that becomes available to do useful work: $\Delta_r G = w_{max}$. (Section 17.6)
- The **standard free energy change of reaction** ($\Delta_r G°$) is the free energy change of reaction in the particular case of a reaction mixture in which all of the reactants and products are present in their standard states. The temperature must be specified:

$$\Delta_r G° = \Delta_r H° - T \Delta_r S°\qquad \text{(Section 17.6)}$$

- The **standard molar free energy change of formation** ($\Delta_f G°$) of a compound is the standard free energy change of reaction ($\Delta_r G°$) when 1 mol of the compound is formed from its component elements at a specified temperature. For any element, $\Delta_f G°$ is zero. (Section 17.6)
- $\Delta_r G°$ at a specified temperature can be calculated from values of $\Delta_f G°$ of reactants and products at the same temperature:

$$\Delta_r G° = \sum n_i \Delta_f G°(\text{products}) - \sum n_i \Delta_f G°(\text{reactants}) \quad \text{(Section 17.6)}$$

- For reactions with $\Delta_r G° < 0$, a reaction mixture with all reactants and products in their standard states would need to go in the direction written to reach equilibrium. For reactions with $\Delta_r G° > 0$, the spontaneous direction of reaction in a standard reaction mixture is the opposite of that written. (Section 17.6)
- For a reaction in which either (a) $\Delta_r H° < 0$ and $\Delta_r S° < 0$, or (b) $\Delta_r H° > 0$ and $\Delta_r S° > 0$, because $\Delta_r G° = \Delta_r H° - T\Delta_r S°$, and $\Delta_r H°$ and $\Delta_r S°$ are almost independent of temperature, we can calculate a temperature at which $\Delta_r G° = 0$ and in a standard reaction mixture the reaction is at equilibrium. The direction of spontaneous reaction is different above that equilibrium temperature than below it. (Section 17.6)
- For non-standard reaction mixtures, $\Delta_r G = \Delta_r G° + RT \ln Q$. (Section 17.6)
- For any reaction, there is a mathematical relationship between $\Delta_r G°$ and K at the same temperature: $\Delta_r G° = -RT \ln K$ (Section 17.7)
- For reactions that take place in a voltaic cell, $\Delta_r G° = -nFE°_{cell}$. (Section 17.8)
- The dependence of equilibrium constants, including equilibrium vapour pressures, on temperature is given by the mathematical relationship

$$\ln K = -\frac{\Delta_r H^\circ}{RT} + \frac{\Delta_r S^\circ}{R} \qquad \text{(Section 17.9)}$$

A graph of $\ln K$ vs. $1/T$ is a straight line with slope $-\Delta_r H^\circ/R$. Alternatively, if we know K at one temperature, we can calculate it at another through the relationship

$$\ln K_1 - \ln K_2 = -\frac{\Delta_r H^\circ}{R}\left(\frac{1}{T_1} - \frac{1}{T_2}\right) \qquad \text{(Section 17.9)}$$

- What happens chemically may be the result of either thermodynamic or kinetic factors. (Sections 17.1 and 17.10)
- Nitric oxide pollutant is emitted from cars and industrial furnaces because the equilibrium constant for its formation from $N_2(g)$ and O_2 (g) increases to a significant value at the high temperatures of combustion (and not because of a faster spontaneous reaction high temperatures). (Sections 17.1 and 17.10)
- The overall result of the multitude of reactions leading to photochemical smogs is the synergistic oxidation of NO and VOCs: NO(g) is formed in combustion chambers where the temperature is high enough that the equilibrium constant for its formation from $N_2(g)$ and $O_2(g)$ is significant. The oxidation processes are initiated by absorption of ultraviolet light by $NO_2(g)$ and catalyzed by OH radicals. Oxidation of neither NO nor VOCs would occur without the presence of the other. (Section 17.10)

Key Equations

$\Delta S = \dfrac{q_{rev}}{T}$	(17.1, Section 17.4)
$\Delta_r S^\circ = \sum n_i\, S^\circ(\text{products}) - \sum n_i\, S^\circ(\text{reactants})$	(Section 17.4)
$\Delta S^\circ_{univ} = \Delta S^\circ_{sys} + \Delta S^\circ_{surr}$	(Section 17.5)
$\Delta_r G = \Delta_r H - T\Delta_r S$	(Section 17.6)
$\Delta_r G^\circ = \Delta_r H^\circ - T\,\Delta_r S^\circ$	(17.2, Section 17.6)
$\Delta_r G^\circ = \sum n_i\, \Delta_f G^\circ(\text{products}) - \sum n_i\, \Delta_f G^\circ(\text{reactants})$	(Section 17.6)
$\Delta_r G = w_{max}$	(Section 17.6)
$\Delta_r G = \Delta_r G^\circ + RT \ln Q$	(17.3, Section 17.6)
$\Delta_r G^\circ = -RT \ln K$	(Section 17.7)
$\Delta_r G^\circ = -nFE^\circ_{cell}$	(Section 17.8)
$\ln K = -\dfrac{\Delta_r H^\circ}{RT} + \dfrac{\Delta_r S^\circ}{R}$	(Section 17.9)
$\ln K_1 - \ln K_2 = -\dfrac{\Delta_r H^\circ}{R}\left(\dfrac{1}{T_1} - \dfrac{1}{T_2}\right)$	(17.4, Section 17.9)

REVIEW QUESTIONS

Section 17.4: Entropy and the Second Law of Thermodynamics

17.33 When ice at 0 °C changes to liquid water at 0 °C, the heat absorbed is 6.01 kJ mol^{-1}. What is the molar entropy change of fusion (melting) of water at 0 °C?

17.34 Calculate the molar entropy change for the vaporization of ethanol at its normal boiling point, 78.0 °C. The enthalpy

change of vaporization of ethanol is 39.3 kJ mol^{-1}. Which substance has the higher entropy in each of the following pairs?

(a) dry ice (solid CO_2) at −78 °C or CO_2 (g) at 0 °C

(b) liquid water at 25 °C or liquid water at 50 °C

(c) pure alumina, $Al_2O_3(s)$, or ruby (ruby is Al_2O_3 in which some of the Al^{3+} ions in the crystalline lattice are replaced with Cr^{3+} ions)

(d) 1 mol of $N_2(g)$ at 1 bar pressure or 1 mol of $N_2(g)$ at 10 bar pressure (both at 298 K)

17.35 By comparing the formulas for each pair of compounds, decide which is expected to have the higher entropy. Assume both are at the same temperature. Check your answers using data in Appendix C.
(a) NaCl(s), NaCl(g), or NaCl(aq)
(b) $H_2O(g)$ or $H_2S(g)$
(c) $C_2H_4(g)$ or $N_2(g)$ (with the same molar mass)
(d) $H_2SO_4(\ell)$ or $H_2SO_4(aq)$

17.36 Use values of $S°$ of substances at 298 K to calculate the standard entropy change ($\Delta_r S°$) for each of the following processes and comment on the sign of the change:
(a) KOH(s) $\longrightarrow$ KOH(aq)
(b) Na(g) $\longrightarrow$ Na(s)
(c) $Br_2(\ell)$ $\longrightarrow$ $Br_2(g)$
(d) HCl(g) $\longrightarrow$ HCl(aq)

17.37 Calculate the standard entropy change for the formation of 1 mol of gaseous ethane (C_2H_6) at 25 °C:

$$2\ C(graphite) + 3\ H_2(g) \longrightarrow C_2H_6(g)$$

Interactive Exercise 17.38
Calculate the standard entropy change for the formation of substances.

Interactive Exercise 17.39
Calculate the standard entropy change and interpret the sign.

17.40 Calculate the standard entropy change for the formation of 1 mol of each of the following compounds from their elements at 25 °C:
(a) HCl(g)   (b) $Ca(OH)_2(s)$

17.41 Calculate the standard molar entropy change ($\Delta_r S°$) for each of the following reactions:

1: $C(s) + 2\ H_2(g) \longrightarrow CH_4(g)$

2: $CH_4(g) + \frac{1}{2}O_2(g) \longrightarrow CH_3OH(\ell)$

3: $C(s) + 2\ H_2(g) + \frac{1}{2}O_2(g) \longrightarrow CH_3OH(\ell)$

Verify that these values are related by the equation $\Delta_r S°_1 + \Delta_r S°_2 = \Delta_r S°_3$. What general principle is illustrated here?

Interactive Exercise 17.42
Find the error in thinking about entropy changes.

Section 17.5: Entropy Changes and Spontaneity

17.43 In a reaction mixture at 25 °C in which $p(H_2) = p(Cl_2) = p(HCl) = 1$ bar, does reaction in the direction

$$H_2(g) + Cl_2(g) \longrightarrow 2\ HCl(g)$$

take the mixture toward equilibrium? Calculate $\Delta S°_{sys}$, and $\Delta S°_{surr}$ to find $\Delta S°_{univ}$.

17.44 Production of iron in a blast furnace can be represented as reduction of iron(II) oxide by carbon:

$$2\ Fe_2O_3(s) + 3\ C(graphite) \longrightarrow 4\ Fe(s) + 3\ CO_2(g)$$

Calculations show that at 25 °C, $\Delta_r H° = +467.9$ kJ mol^{-1} and $\Delta_r S° = +560.7$ J K^{-1} mol^{-1}. Show that this reaction will not occur in a standard reaction mixture unless the temperature is raised.

17.45 Calculate the standard enthalpy change and the standard entropy change for the decomposition at 25 °C of liquid water to form gaseous hydrogen and oxygen. In a reaction mixture at 25 °C in which all of the reactants and products are in their standard states, is this reaction spontaneous? Explain your answer.

17.46 Classify each of the reactions according to one of the four reaction types summarized in Table 17.2.

$$Fe_2O_3(s) + 2\ Al(s) \longrightarrow 2\ Fe(s) + Al_2O_3(s)$$
$$\Delta_r H° = -851.5\ kJ\ mol^{-1},$$
$$\Delta_r S° = -375.2\ J\ K^{-1}\ mol^{-1}$$
(b) $N_2(g) + 2\ O_2(g) \longrightarrow 2\ NO_2(g)$
$$\Delta_r H° = +66.2\ kJ\ mol^{-1}, \Delta_r S° = -121.6\ J\ K^{-1}\ mol^{-1}$$

17.47 In a reaction mixture at 25 °C with all reactants and products in their standard states (1 bar pressure), the following reaction is not spontaneous:

$$COCl_2(g) \longrightarrow CO(g) + Cl_2(g)$$

Would you have to raise or lower the temperature to make it spontaneous?

Section 17.6: Gibbs Free Energy

17.48 Using values of $\Delta_f H°$ and $S°$ of the reactants and products at 25 °C, calculate $\Delta_r G°$ at 25 °C for each of the following reactions. In each case, decide if the reaction is spontaneous in a reaction mixture at 25 °C in which all of the reactants and products are in their standard states. Are the reactions enthalpy- or entropy-driven?

(a) $2\ Pb(s) + O_2(g) \longrightarrow 2\ PbO(s)$

(b) $CaO(s) + CO_2(g) \longrightarrow CaCO_3(s)$

17.49 Using values of $\Delta_f H°$ and $S°$ of the reactants and products at 25 °C, calculate $\Delta_r G°$ at 25 °C for each of the following reactions. In each case, decide if the reaction is spontaneous in a reaction mixture at 25 °C in which all of the reactants and products are in their standard states. Are the reactions enthalpy- or entropy-driven?

(a) $2 \, Al(s) + 3 \, Cl_2(g) \longrightarrow 2 \, AlCl_3(s)$

(b) $6 \, C(graphite) + 3 \, H_2(g) \longrightarrow C_6H_6(\ell)$

17.50 Almost 5 billion kg of benzene, $C_6H_6(\ell)$, are made each year. Benzene is used as a starting material for many other compounds and as a solvent, although it is also a carcinogen, and its use is restricted. One compound that can be made from benzene is cyclohexane, $C_6H_{12}(\ell)$:

$$C_6H_6(\ell) + 3 \, H_2(g) \longrightarrow C_6H_{12}(\ell)$$

$\Delta_rH° = -206.7 \, kJ \, mol^{-1}$, $\Delta_rS° = -316.5 \, J \, K^{-1} \, mol^{-1}$

Is this reaction spontaneous in a standard-state reaction mixture at 25 °C? Is the reaction enthalpy- or entropy-driven?

17.51 Consider the reaction represented by the equation

$$2 \, C(graphite) + 2 \, H_2(g) + O_2(g) \longrightarrow CH_3COOH(\ell)$$

From values of $\Delta_fH°$ and $S°$ of the reactants and products at 25 °C, calculate $\Delta_rH°$ and $\Delta_rS°$ at 25 °C and then the standard free energy change of reaction ($\Delta_rG°$) at 25 °C. How does this value compare with the value of the standard molar free energy change of formation ($\Delta_fG°$) of $CH_3COOH(\ell)$ at 25 °C? Explain.

17.52 Using values of $\Delta_fG°$ of reactants and products at 25 °C, calculate $\Delta_rG°$ at 25 °C for each of the following reactions. In each case, decide if the reaction is spontaneous in a reaction mixture at 25 °C in which all reactants and products are in their standard states.

(a) $2 \, K(s) + Cl_2(g) \longrightarrow 2 \, KCl(s)$

(b) $2 \, H_2S(g) + 3 \, O_2(g) \longrightarrow 2 \, H_2O(g) + 2 \, SO_2(g)$

(c) $4 \, NH_3(g) + 7 \, O_2(g) \longrightarrow 4 \, NO_2(g) + 6 \, H_2O(g)$

17.53 For the reaction

$$BaCO_3(s) \longrightarrow BaO(s) + CO_2(g)$$

$\Delta_rG° = +219.7 \, kJ \, mol^{-1}$ at 25 °C. Using this value and other data from Appendix C, calculate $\Delta_fG°$ for $BaCO_3(s)$ at 25 °C.

17.54 Using data in Appendix C, calculate the standard boiling point of methanol.

17.55 The decomposition of liquid nickel carbonyl, $Ni(CO)_4$, is represented as follows:

$$Ni(CO)_4(\ell) \longrightarrow Ni(s) + 4 \, CO(g)$$

$$\Delta_rG° \text{ at } 25 °C = 40 \, kJ \, mol^{-1}$$

Select from $S°$ ($320 \, J \, K^{-1} \, mol^{-1}$) and $\Delta_fH°$ ($-632 \, kJ \, mol^{-1}$) of $Ni(CO)_4(\ell)$ and data from Appendix C to estimate the temperature at which a reaction mixture with 1 bar pressure of carbon monoxide is at equilibrium.

17.56 Hydrogenation, the addition of hydrogen to an organic compound, is a reaction of industrial importance. Calculate $\Delta_rH°$, $\Delta_rS°$, and $\Delta_rG°$ at 25 °C for the hydrogenation of octene, $C_8H_{16}(g)$, to octane, $C_8H_{18}(g)$. Is the reaction product- or reactant-favoured (that is, is the reaction spontaneous) in a standard-state reaction mixture at 25 °C?

$$C_8H_{16}(g) + H_2(g) \longrightarrow C_8H_{18}(g)$$

In addition to data in Appendix C, the following information is required:

$C_8H_{16}(g)$: $\Delta_fH°$ at 25 °C $= -82.93 \, kJ \, mol^{-1}$, $S°$ at 25 °C $= 462.8 \, J \, K^{-1} \, mol^{-1}$

$C_8H_{18}(g)$: $\Delta_fH°$ at 25 °C $= -208.45 \, kJ \, mol^{-1}$, $S°$ at 25 °C $= 463.639 \, J \, K^{-1} \, mol^{-1}$

17.57 At 25 °C and with all reactants and products in their standard states, is the combustion of ethane a spontaneous reaction?

$$C_2H_6(g) + \tfrac{7}{2} \, O_2(g) \longrightarrow 2 \, CO_2(g) + 3 \, H_2O(g)$$

Does your calculated answer agree with your preconceived idea of this reaction?

17.58 Estimate the temperature at which decomposition of $HgS(s)$ into $Hg(\ell)$ and $S_8(g)$ is at equilibrium, if all reactant and product species are in their standard states at that temperature.

Section 17.7: The Relationship between $\Delta_rG°$ and K

17.59 Calculate K at 25 °C for the formation of $NO(g)$ from its elements:

$$\tfrac{1}{2} \, N_2(g) + \tfrac{1}{2} \, O_2(g) \longrightarrow NO(g)$$

$$\Delta_rG° \text{ at } 25 °C = +86.58 \, kJ \, mol^{-1}$$

Comment on the connection between the sign of $\Delta_rG°$ and the magnitude of K.

17.60 Calculate $\Delta_rG°$ and K at 25 °C for the reaction

$$C_2H_4(g) + H_2(g) \longrightarrow C_2H_6(g)$$

Comment on the sign of $\Delta_rG°$ and the magnitude of K.

17.61 The equilibrium constant (K) is 0.14 at 25 °C for the reaction represented by

$$N_2O_4(g) \rightleftharpoons 2 \, NO_2(g)$$

Calculate $\Delta_rG°$ from this information and compare your calculated value with that determined from the $\Delta_fG°$ values in Appendix C.

Section 17.8: $\Delta_r G°$ and $E°_{cell}$ for Voltaic Cell Reactions

17.62 From standard reduction potentials at 25 °C, calculate $E°$ at 25 °C, and then $\Delta_r G°$ and K at 25 °C for each of the following:

(a) $Zn^{2+}(aq) + Ni(s) \longrightarrow Zn(s) + Ni^{2+}(aq)$

(b) $Cu(s) + 2\,Ag^+(aq) \longrightarrow Cu^{2+}(aq) + 2\,Ag(s)$

Section 17.9: Dependence of Equilibrium Constants on Temperature

17.63 Methanol is now widely used as a fuel in race cars. Consider the following reaction as a possible way of making methanol:

$$C(graphite) + \tfrac{1}{2}O_2(g) + 2\,H_2(g) \longrightarrow CH_3OH(\ell)$$

Calculate K for the formation of methanol at 298 K using this reaction. Would a different temperature be preferable for making methanol by this reaction?

17.64 The equilibrium vapour pressure of ethanol at 25 °C is 0.0787 bar, and $\Delta_{vap}H°$ is 39.33 kJ mol^{-1}. Estimate (a) the equilibrium vapour pressure at 25 °C, and (b) the boiling point of ethanol at 1.0 bar pressure (the temperature at which the equilibrium vapour pressure is equal to 1.0 bar).

17.65 Estimate the boiling point of water in Denver, Colorado (where the altitude is 1.60 km and the atmospheric pressure is 0.829 atm).

SUMMARY AND CONCEPTUAL QUESTIONS

17.66 When calcium carbonate is heated strongly, CO_2 gas is evolved. The equilibrium pressure of the gas is 1.0 bar at 897°C, and $\Delta_r H°$ at 298 K is 179.0 kJ mol^{-1}.

$$CaCO_3(s) \longrightarrow CaO(s) + CO_2(g)$$

Estimate the value of $\Delta_r S°$ at 897°C for this reaction.

17.67 The reaction used by Joseph Priestley to make oxygen is

$$2\,HgO(s) \longrightarrow 2\,Hg(\ell) + O_2(g)$$

(a) For this reaction, predict whether the signs of $\Delta S°_{sys}$, $\Delta S°_{surr}$, $\Delta S°_{univ}$, $\Delta_r H°$, and $\Delta_r G°$ are greater than zero, equal to zero, or less than zero, and explain your prediction. Using data from Appendix C, calculate the value of each of these quantities to verify your prediction.

(b) Calculate K at 298 K. Is the reaction product-favoured?

17.68 A crucial reaction for the production of synthetic fuels is the conversion of coal to H_2 with steam:

$$C(s) + H_2O(g) \longrightarrow CO(g) + H_2(g)$$

(a) Calculate $\Delta_r G°$ for this reaction at 25°C assuming $C(s)$ is graphite.

(b) Calculate K for the reaction at 25 °C.

(c) Is the reaction spontaneous at 25 °C with all reactants and products in their standard states? If not, at what temperature will it become so?

17.69 Methanol is relatively inexpensive to produce. Much consideration has been given to using it as a precursor to other fuels such as methane, which could be obtained by the decomposition of the alcohol:

$$CH_3OH(\ell) \longrightarrow CH_4(g) + \tfrac{1}{2}O_2(g)$$

(a) What are the sign and magnitude of the standard entropy change for the reaction? Does the sign of $\Delta_r S°$ agree with your expectation? Explain briefly.

(b) Is the reaction spontaneous under standard conditions at 25°C? Use thermodynamic values to prove your answer.

(c) If it is not spontaneous at 25 °C, at what temperature does the reaction become spontaneous?

17.70 Wet limestone is used to scrub SO_2 gas from the exhaust gases of power plants. One possible reaction gives a hydrated calcium sulfite:

$$CaCO_3(s) + SO_2(g) + \tfrac{1}{2}H_2O(\ell)$$
$$\rightleftharpoons CaSO_3 \cdot \tfrac{1}{2}H_2O(s) + CO_2(g)$$

Another reaction gives a hydrated calcium sulfate:

$$CaCO_3(s) + SO_2(g) + \tfrac{1}{2}H_2O(\ell) + \tfrac{1}{2}O_2(g)$$
$$\rightleftharpoons CaSO_4 \cdot \tfrac{1}{2}H_2O(s) + CO_2(g)$$

Which is the more product-favoured reaction? Use the data below and in Appendix C.

$CaSO_3 \cdot \tfrac{1}{2}H_2O(s)$: At 25 °C, $\Delta_f H° = -1574.65$ kJ mol^{-1},
$$S° = 121.3 \text{ J K}^{-1} \text{ mol}^{-1}$$

$CaSO_4 \cdot \tfrac{1}{2}H_2O(s)$: At 25 °C, $\Delta_f H° = -1311.7$ kJ mol^{-1},
$$S° = 134.8 \text{ J K}^{-1} \text{ mol}^{-1}$$

17.71 Sulfur undergoes a phase transition between 80 and 100 °C:

$$S_8(s, \text{rhombic}) \longrightarrow S_8(s, \text{monoclinic})$$
$$\Delta_r H° = 3.213 \text{ kJ mol}^{-1}, \Delta_r S° = 8.7 \text{ J K}^{-1}$$

(a) Estimate $\Delta_r G°$ for the transition at 80.0 °C and 110.0 °C. What do these results tell you about the stability of the two forms of sulfur at each of these temperatures?

(b) Calculate the temperature at which $\Delta_r G° = 0$. What is the significance of this temperature?

17.72 If copper metal is heated in air, a black film of CuO(s) forms on the copper surface. In the presence of hydrogen gas, the copper oxide is reduced to the metal at high temperatures:

$$CuO(s) + H_2(g) \longrightarrow Cu(s) + H_2O(g)$$

Is this reaction product- or reactant-favoured under standard conditions at 298 K?

17.73 Silver(I) oxide can be formed by the reaction of silver metal and oxygen:

$$4 Ag(s) + O_2(g) \longrightarrow 2 Ag_2O(s)$$

(a) Calculate $\Delta_r H°$, $\Delta_r S°$, and $\Delta_r G°$ for the reaction.

(b) What is the pressure of $O_2(g)$ in equilibrium with Ag(s) and $Ag_2O(s)$ at 25 °C?

(c) At what temperature would the pressure of $O_2(g)$ in equilibrium with Ag(s) and $Ag_2O(s)$ be 1.0 bar?

17.74 Calculate $\Delta_f G°$ for HI(g) at 350 °C, given that the reaction

$$\tfrac{1}{2} H_2(g) + \tfrac{1}{2} I_2(g) \rightleftharpoons HI(g)$$

is at equilibrium at 350 °C when $p(H_2) = 0.132$ bar, $p(I_2) = 0.295$ bar, and $p(HI) = 1.61$ bar. At 350 °C, I_2 is a gas.

17.75 Mercury vapour is dangerous because it is toxic and can be breathed into the lungs. Estimate the vapour pressure of mercury at two different temperatures from the following data:

	$\Delta_f H°$ (kJ mol^{-1})	$S°$ (J K^{-1} mol^{-1})	$\Delta_f G°$ (kJ mol^{-1})
Hg(ℓ)	0	76.02	0
Hg(g)	61.38	174.97	31.88

Estimate the temperature at which K for the process Hg(ℓ) $\longrightarrow$ Hg(g) is equal to (a) 1.00, and (b) 1.33×10^{-3}. What is the vapour pressure at each of these temperatures? (Experimentally, it is found that the vapour pressure is 1.33×10^{-3} bar at 126.2 °C and 1.00 bar at 356.6 °C.) (The temperature at which $p = 1.00$ bar can be calculated from thermodynamic data. To find the other temperature, you will need to use the temperature at which $p = 1.00$ bar and the Clausius-Clapeyron equation (equation 17.4)).

17.76 Some metal oxides can be decomposed to the metal and oxygen under moderate conditions. Is the decomposition of silver(I) oxide product-favoured at 25 °C?

$$2 Ag_2O(s) \longrightarrow 4 Ag(s) + O_2(g)$$

If not, can it become so if the temperature is raised? At what temperature is the reaction product-favoured?

17.77 Explain why each of the following statements is incorrect:

(a) Entropy increases in all spontaneous reactions.

(b) Reactions with a negative free energy change ($\Delta_r G° < 0$) are product-favoured and occur with rapid transformation of reactants to products.

(c) All spontaneous processes are exothermic.

(d) Endothermic processes are never spontaneous.

17.78 Decide whether each of the following statements is true or false. If false, rewrite it to make it true.

(a) The entropy of a substance increases as it goes from the liquid to the vapour state at any temperature.

(b) An exothermic reaction will always be spontaneous.

(c) Reactions with a positive $\Delta_r H°$ and a positive $\Delta_r S°$ can never be product-favoured.

(d) If $\Delta_r G°$ for a reaction is negative, the reaction has $K > 1$.

17.79 Entropy, as well as enthalpy, plays a role in the solution process. If $\Delta_r H°$ for the solution process is zero, explain how the process can be driven by entropy.

17.80 Write a chemical equation for the oxidation of $C_2H_6(g)$ by $O_2(g)$ to form $CO_2(g)$ and $H_2O(g)$.

(a) Predict whether the signs of $\Delta S°_{sys}$, $\Delta S°_{surr}$, and $\Delta S°_{univ}$ will be greater than zero, equal to zero, or less than zero. Explain your prediction.

(b) Predict the signs of $\Delta_r H°$ and $\Delta_r G°$ Explain how you made this prediction.

(c) Will the value of K be very large, very small, or near 1? Will the equilibrium constant (K) for this system be larger or smaller at temperatures greater than 298 K than at 298 K? Explain how you made this prediction.

17.81 The normal melting point of benzene (C_6H_6) is 5.5 °C. For the process of melting, what is the sign of each of the following?

(a) $\Delta_r H°$ (c) $\Delta_r G°$ at 5.5 °C (e) $\Delta_r G°$ at 25.0 °C

(b) $\Delta_r S°$ (d) $\Delta_r G°$ at 0.0 °C

17.82 For each of the following processes, give the algebraic sign of $\Delta_r H°$, $\Delta_r S°$, and $\Delta_r G°$. No calculations are necessary; use your common sense.

(a) The decomposition of liquid water to give gaseous oxygen and hydrogen, a process that requires a considerable amount of energy.

(b) Dynamite is a mixture of nitroglycerin ($C_3H_5N_3O_9$) and diatomaceous earth. The explosive decomposition of nitroglycerin gives gaseous products such as water, CO_2, and others; much heat is evolved.

(c) The combustion of gasoline in the engine of your car, as exemplified by the combustion of octane:

$$2 C_8H_{18}(g) + 25 O_2(g) \longrightarrow 16 CO_2(g) + 18 H_2O(g)$$

17.83 Iodine, I_2(s), dissolves readily in carbon tetrachloride, $\Delta_r H° = 0$ kJ mol^{-1}:

$$I_2(s) \longrightarrow I_2(\text{in } CCl_4 \text{ solution})$$

What is the sign of $\Delta_rG°$? Is the dissolving process entropy-driven or enthalpy-driven? Explain.

17.84 The formation of diamond from graphite is a process of considerable importance.
  (a) Using data in Appendix C, calculate $\Delta_rS°$, $\Delta_rH°$, and $\Delta_rG°$ for this process.
  (b) The calculations will suggest that this process is not possible under any conditions. However, the synthesis of diamonds by this reaction is a commercial process. How can this contradiction be rationalized?

(Note: In the synthesis, high pressures and temperatures are used.)

17.85 Oxygen dissolved in water can cause corrosion in hot-water heating systems. To remove oxygen, hydrazine (N_2H_4) is often added. Hydrazine reacts with dissolved O_2 to form water and N_2.
  (a) Write a balanced chemical equation for the reaction of hydrazine and oxygen. Identify the oxidizing and reducing agents in this redox reaction.
  (b) Calculate $\Delta_rH°$, $\Delta_rS°$, and $\Delta_rG°$ for this reaction.

With the **eCHACR** single-sign-on access card bundled with your text, log on (http://login.cengage.com/sso) and access the e-book and click on any in-margin icons for dynamic molecular-level animations and simulations, videos of laboratory reactions, interactive exercises, reviews of background concepts, and other online supplementary materials as noted by the icons in the text margins.

Also go to www.chemistry.nelson.com <http://www.chemistry.nelson.com> for Answers to in-chapter exercises and selected Review Questions, Test Yourself questions, weblinks, crossword puzzles, flashcards, glossary of key terms, and other student resources.

Spontaneous Change: How Fast?

Charles D. Winters

18.1 Case Study: Winds of Change

After eating certain foods, such as beans, cabbage, and broccoli, you may have experienced the passing of what is known politely as flatulence (from *flatus,* the medical term for intestinal gas). There is, of course, chemistry involved in passing gas. And an enzyme produced by a fungus can help prevent it.

Foods that give rise to gas contain oligosaccharides—substances whose molecules are oligomers (polymers with just a few repeating units) of monosaccharides. Oligosaccharide molecules must be broken down by chemical reaction in an aqueous medium (hydrolyzed) in the gastrointestinal (GI) tract to form monosaccharides, which can be absorbed into the bloodstream. Oligosaccharides hydrolyze at very different rates in the GI tracts of different people. When the rate is too slow, residual unhydrolyzed oligosaccharides are fermented by anaerobic micro-organisms in the colon to produce gases such as CO_2, H_2, CH_4, small amounts of hydrogen sulfide (H_2S), and other smelly compounds that comprise flatus.

Chemists have produced a commercial product to increase the rate of hydrolysis of oligosaccharides and help people tackle the winds of digestion. Its maker's advertising material states that it "is a food enzyme from a natural source that breaks down the complex sugars in gassy foods, making them more digestible."

Enzymes are biological catalysts: they speed up chemical reactions. α-Galactosidase, one of the enzymes in products designed to reduce flatulence, speeds up the breakdown of oligosaccharides whose molecules contain a galactose unit, such as raffinose, stachyose, and verbascose, to the simple sugars galactose and sucrose. Sucrose is then broken down to glucose by another enzyme, sucrase. The overall process can be represented as follows:

$$\text{Oligosaccharide} + H_2O \xrightarrow[\text{sucrase}]{\text{galactosidase}} \text{galactose} + \text{glucose}$$

α-Galactosidase, one of the enzymes in products designed to reduce flatulence, speeds up the breakdown of oligosaccharides that would otherwise ferment in the colon . . .

Biotechnology companies purify the enzyme α-galactosidase from a genetically engineered form of the common fungus, *Aspergillus niger*.

The rate at which the uncatalyzed hydrolysis of oligosaccharides proceeds under the conditions of temperature and pH found in the GI tract is very slow: the reaction is thermodynamically spontaneous, but the oligosaccharides are kinetically inert [<<Section 17.1]. An enzyme catalyst speeds up the rate of the reaction so it can go to completion before the oligosaccharides pass through to the colon.

The chemistry of flatulence is of more than just passing interest. The Intergovernmental Panel on Climate Change has drawn attention to the contribution of methane gas in the atmosphere to climate change. Methane has 72 times the global warming potential of an equal mass of CO_2 over a 20-year time period, and 25 times that of CO_2 over a 100-year period. The agricultural sector plays a particularly important role in tackling methane emissions; the United Nations Food and Agricultural Organization estimates that one-third of the world's arable land is used to produce feed for livestock. While most people produce 0.5 to 2.0 L per day of gas, ruminant animals produce much larger amounts. Methanogen bacteria in the rumen of cows are estimated to produce from 100 to 500 L of methane gas per day. Globally, ruminant livestock produce about 80 million tonnes of methane annually, or 28% of global methane emissions related to human activity. Strategies to control this greenhouse gas must be a serious part of policies to address climate change.

Many different approaches are being tested by researchers to reduce the production of methane in the livestock industry. These include the use of enzymes to speed up the breakdown of oligosaccharides, changing animal diets to include grains and legumes that hydrolyze more readily, controlling the methanogen organisms in animal stomachs, and introducing new micro-organisms that metabolize oligosaccharides to produce products other than methane gas. One of the benefits of this research is improved feed efficiency. When less of the energy from animal feed is gone with the wind through methane production, more remains to be used to produce meat or milk.

Enzymes are naturally occurring substances, usually proteins, that increase reaction rates. There are many of them, and each one acts on a specific substance to increase the rate of a specific reaction. By understanding the mechanisms of reactions (the pathways by which they occur), chemists design and make catalysts to increase the rate of production of industrial substances, dramatically reducing the cost of manufacture. Even more powerfully, catalysts can be used to make different products from the same materials. It is often the case that particular starting materials can react in several ways simultaneously to produce different products, the relative amounts of each dependent on the rates of the different reaction paths. If different catalysts can be developed that increase the rate of formation of each product specifically, then each catalyst leads to production of mainly just one of the possible products. An example of this ability to "dial up" a product by use of specific catalysts is the production of different substances from syngas, described in Section 18.6. It is hard to imagine, however, that chemists will ever develop synthetic catalysts that match the naturally occurring enzymes in their efficiency and specificity.

18.2 The Concept of Reaction Rate

When carrying out a chemical reaction, chemists are concerned with two issues: (a) the direction of spontaneous reaction and the *extent* to which the reaction will go before reaching equilibrium, and (b) the *rate* at which the reaction proceeds toward equilibrium. Chapters 14 and 17 address the first question. In this chapter we turn to **chemical kinetics**, *the study of the rates of chemical reactions.*

Although we might see the sudden formation of a precipitate or a coloured reaction product, giving the appearance that all of the product molecules or ions are formed simultaneously, product molecules or ions are generated continuously over the course of time until equilibrium is attained. The duration of this product formation may be short or long, depending on the reaction and the conditions under which it occurs.

Our study of chemical kinetics is divided into two parts. The first part concerns experimental measurement at the observable level: what the reaction rate means, how to

Oligosaccharides literally means "a few" saccharides or simple sugars. Oligosaccharides are sugars whose molecules have a few (typically 3–10) simple sugar molecules linked together to form small polymers. Part of the digestion process involves breaking down the oligosaccharide molecules to molecules of simple sugars, which are ultimately broken down to CO_2 and H_2O, with the release of energy.

determine a reaction rate experimentally, and how factors such as temperature and the concentrations of reactants influence rates. The second part considers chemists' hypotheses about chemical reactions at the molecular level. Here, the focus is on the *reaction mechanism*, the detailed pathways of reactions. The goal is to reconcile experimental data from the observable world of chemistry with our best understandings of how and why chemical reactions occur at the molecular level—and then to apply this information to control important reactions.

An easy way to estimate the speed of a car is to measure how far it travels during a measured time period. Two measurements are made: distance travelled (Δd) and time elapsed (Δt). The average speed over the time elapsed is $\Delta d/\Delta t$. If a car travels 2.4 km in 4.5 min (0.075 h), its average speed is (2.4 km)/(0.075 h) = 32 km h^{-1}. Of course, during that 4.5 minutes, the instantaneous speed of the car may be quite different from its average speed.

The concept of *rate* is encountered in many everyday circumstances. Common examples are the speed of a train as a measure of the distance travelled per unit time (e.g., kilometres per hour, km h^{-1}) and the rate of flow of water from a tap as volume per unit time (e.g., litres per minute, L min^{-1}). In each case, a change is measured over an interval of time.

Similarly, we measure the **rate of a chemical reaction** as the *change in concentration of a substance per unit of time.*

$$\text{Rate of reaction} = \frac{\text{change in concentration}}{\text{change in time}} = \frac{\Delta c}{\Delta t}$$

The change in two quantities, concentration and time, must be measured. For a rate study, the concentration of a substance undergoing reaction can be determined by a variety of methods. Often, concentrations are obtained by measuring a property such as the absorbance of light that is related to concentration (Figure 18.1).

(a)   (b)   (c)

All photos: Charles D. Winters

FIGURE 18.1 An experiment to measure rate of reaction. (a) A few drops of blue food dye were added to water, followed by a solution of bleach. Initially, the concentration of dye was about 3.4×10^{-5} mol L^{-1} and the bleach, ClO$^-$(aq), concentration was about 0.034 mol L^{-1}. (b, c) The intensity of the dye colour faded as it reacted with the bleach. The absorbance of the solution can be measured at various times using a spectrophotometer, and these values can be used to determine the concentration of the dye.

During a chemical reaction, the amounts of reactants decrease with time and the amounts of products increase. It is possible to describe the rate of reaction based on either the decrease in concentration of a reactant or the increase in concentration of a product per unit of time. Consider the decomposition of N$_2$O$_5$ in a solvent:

$$2\,N_2O_5(\text{in solution}) \longrightarrow 4\,NO_2(\text{in solution}) + O_2(g)$$

The rate of this reaction for any interval of time can be expressed as the change in concentration of the reactant N$_2$O$_5$ divided by the change in time:

[N$_2$O$_5$] indicates the concentration of the species N$_2$O$_5$ in mol L^{-1}. Recall also that we calculate a change in a quantity X as $\Delta X = X_{\text{final}} - X_{\text{initial}}$, so ΔX is negative for the change of concentration of a reactant (and positive for change of concentration of a product).

$$\text{Rate of reaction} = -\frac{\text{change in [N}_2\text{O}_5]}{\text{change in time}} = -\frac{\Delta[N_2O_5]}{\Delta t}$$

The minus sign is required by convention—since the concentration of [N$_2$O$_5$] decreases with time (Δ[N$_2$O$_5$] has a negative value), and the rate of change is always expressed as a positive quantity.

Data for a typical experiment carried out at 30.0 °C are presented as a graph of $[N_2O_5]$ vs. time in Figure 18.2. For example, the average rate of disappearance of N_2O_5 between 40 and 55 minutes is given by

$$-\frac{\Delta[N_2O_5]}{\Delta t} = -\frac{(1.10 - 1.22)\ \text{mol L}^{-1}}{(55 - 40)\ \text{min}} = \frac{0.12\ \text{mol L}^{-1}}{15\ \text{min}} = 0.0080\ \text{mol L}^{-1}\ \text{min}^{-1}$$

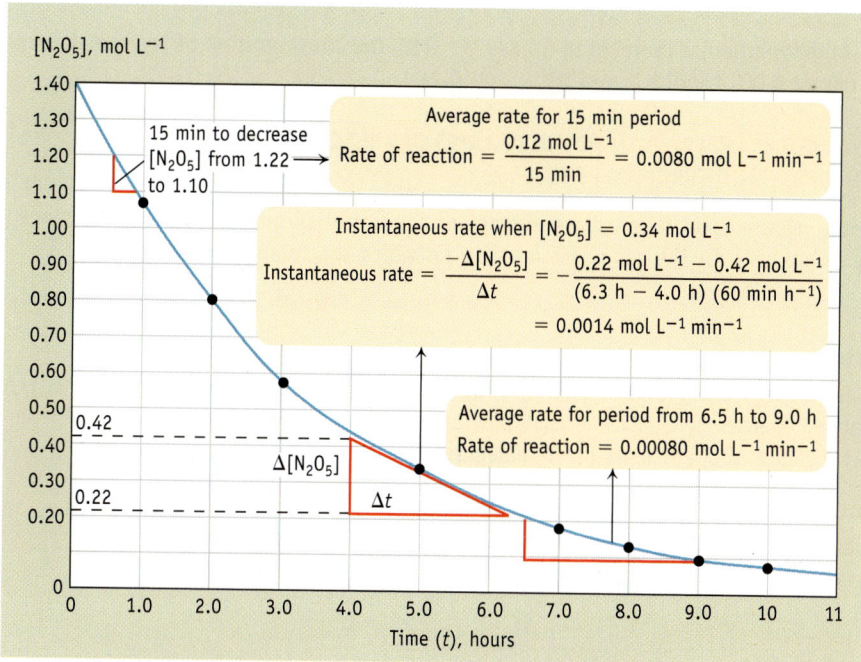

FIGURE 18.2 A plot of reactant concentration vs. time for the decomposition of N_2O_5. The average rate between 40 and 55 minutes is 0.0080 mol L^{-1} min^{-1}, while that between 6.5 and 9.0 hours is much slower—0.00080 mol L^{-1} min^{-1}. The instantaneous rate when $[N_2O_5] = 0.34$ mol L^{-1} is 0.0014 mol L^{-1} min^{-1}.

The rate of this reaction can also be expressed in terms of the rate of increase of either NO_2 or O_2. The balanced chemical equation tells us that the rate of increase of $[NO_2]$ is twice the rate of decrease of $[N_2O_5]$:

$$\frac{\Delta[NO_2]}{\Delta t} = 2 \times -\frac{\Delta[N_2O_5]}{\Delta t} = 2 \times 0.0080\ \text{mol L}^{-1}\ \text{min}^{-1} = 0.0160\ \text{mol L}^{-1}\ \text{min}^{-1}$$

The rate of increase of $[O_2]$ is half the rate of decrease of $[N_2O_5]$:

$$\frac{\Delta[O_2]}{\Delta t} = \tfrac{1}{2} \times -\frac{\Delta[N_2O_5]}{\Delta t} = \tfrac{1}{2} \times 0.0080\ \text{mol L}^{-1}\ \text{min}^{-1} = 0.0040\ \text{mol L}^{-1}\ \text{min}^{-1}$$

Each of these three statements of the rate of the reaction is valid and correct, but they are numerically different from each other. Therefore, it is important to specify which reagent is referred to when a rate of reaction is stated. To define a single rate of reaction, we can divide each value of $\Delta[\text{reagent}]/\Delta t$ by the appropriate coefficient in the balanced chemical equation:

$$\text{Rate} = \tfrac{1}{2} \times -\frac{\Delta[N_2O_5]}{\Delta t} = \tfrac{1}{4} \times \frac{\Delta[NO_2]}{\Delta t} = \frac{\Delta[O_2]}{\Delta t} = 0.0040\ \text{mol L}^{-1}\ \text{min}^{-1}$$

The graph of concentration vs. time (Figure 18.2) is not a straight line because the rate of the reaction changes over time. The concentration of N_2O_5 decreases rapidly at the beginning of the reaction but more slowly near the end. The average rate of reaction for the time interval from 6.5 to 9.0 hours is only one-tenth of the average rate of reaction between 40 and 55 minutes:

$$-\frac{\Delta[N_2O_5]}{\Delta t} = -\frac{(0.1 - 0.22)\ \text{mol L}^{-1}}{(540 - 390)\ \text{min}} = \frac{0.12\ \text{mol L}^{-1}}{150\ \text{min}} = 0.00080\ \text{mol L}^{-1}\ \text{min}^{-1}$$

The procedure we have used to calculate the reaction rate gives the *average rate of reaction* over the specified time interval. The rate of reaction at any instant, the *instantaneous rate of reaction*, is determined by drawing a tangent to the concentration vs. time curve at the specified instant (Figure 18.2) and obtaining the rate from the slope of this line. For example, when $t = 5.0$ h (and $[N_2O_5] = 0.34$ mol L^{-1}):

$$\text{Instantaneous rate} = -\frac{\Delta[N_2O_5]}{\Delta t} = -\frac{-0.20 \text{ mol } L^{-1}}{140 \text{ min}} = +1.4 \times 10^{-3} \text{ mol } L^{-1} \text{ min}^{-1}$$

At that particular moment in time ($t = 5.0$ h), the concentration of N_2O_5 is decreasing at a rate of 0.0014 mol L^{-1} min^{-1}.

WORKED EXAMPLE 18.1—RELATIVE RATES OF [REACTANT] DECREASE AND [PRODUCT] INCREASE

Give the relative rates for decrease of reactant concentration and the increases of concentrations of products for the gas-phase reaction for which the balanced equation is

$$4 \text{ PH}_3(g) \longrightarrow P_4(g) + 6 \text{ H}_2(g)$$

Strategy

Because $PH_3(g)$ is a reactant, and $P_4(g)$ and $H_2(g)$ are products, the value of $\Delta[PH_3]/\Delta t$ is negative, whereas $\Delta[P_4]/\Delta t$ and $\Delta[H_2]/\Delta t$ are positive. Relative rates are deduced from the relative values of coefficients in the balanced equation.

Solution

The amount of $P_4(g)$ formed is one-quarter the amount of $PH_3(g)$ that reacts, so

$$\frac{\Delta[P_4]}{\Delta t} = \tfrac{1}{4} \times -\frac{\Delta[PH_3]}{\Delta t}$$

The amount of $H_2(g)$ formed is 1.5 times the amount of $PH_3(g)$ that reacts, so

$$\frac{\Delta[H_2]}{\Delta t} = \tfrac{3}{2} \times -\frac{\Delta[PH_3]}{\Delta t}$$

A single statement of rate of reaction can be obtained by dividing each Δ[reagent]$/\Delta t$ by the appropriate coefficient in the balanced equation.

$$\tfrac{1}{4} \times -\frac{\Delta[PH_3]}{\Delta t} = +\frac{\Delta[P_4]}{\Delta t} = +\tfrac{1}{6} \times \frac{\Delta[H_2]}{\Delta t}$$

WORKED EXAMPLE 18.2—ESTIMATION OF RATE OF REACTION

Data collected on the concentration of a dye as a function of time (Figure 18.1) are shown in the graph. What is the average rate of change of the dye concentration over the first 2 min? What is the average rate of change during the fifth minute (from $t = 4$ min to $t = 5$ min)? Estimate the instantaneous rate at $t = 4.0$ min.

Strategy

To find the average rate over a given time period, calculate the difference in concentration at the beginning and end of a time period ($[dye]_{final} - [dye]_{initial}$) and divide by the time duration. To find the instantaneous rate, draw a line tangent to the graph at the given time. The slope of the line is the instantaneous rate.

Solution

(a) The concentration of dye decreases from 3.4×10^{-5} mol L^{-1} at $t = 0$ min to 1.7×10^{-5} mol L^{-1} at $t = 2.0$ min.

$$\text{Average rate} = -\frac{\Delta[dye]}{\Delta t} = \frac{(1.7 \times 10^{-5} - 3.4 \times 10^{-5}) \text{ mol } L^{-1}}{2.0 \text{ min}} = 8.5 \times 10^{-6} \text{ mol } L^{-1} \text{ min}^{-1}$$

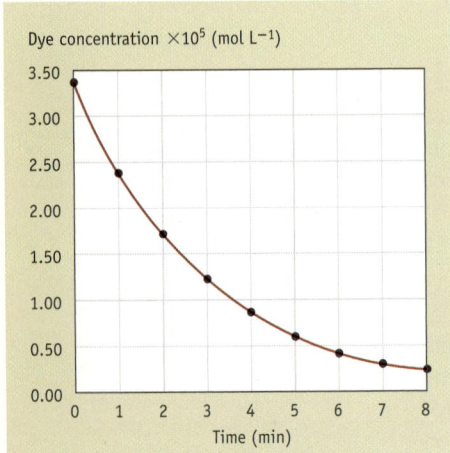

Dye concentration $\times 10^5$ (mol L^{-1})

Time (min)

(b) [Dye] decreases from 0.9×10^{-5} mol L^{-1} at $t = 4.0$ min to 0.6×10^{-5} mol L^{-1} at $t = 5.0$ min.

$$\text{Average rate} = -\frac{\Delta[\text{dye}]}{\Delta t} = \frac{(0.6 \times 10^{-5} - 0.9 \times 10^{-5}) \text{ mol L}^{-1}}{1.0 \text{ min}}$$

$$= 3.0 \times 10^{-6} \text{ mol L}^{-1} \text{ min}^{-1}$$

(c) From the slope of the tangent to the curve, the instantaneous rate at 4 min is 3.5×10^{-6} mol L^{-1} min^{-1}.

Comment

Notice that the average rate between $t = 4$ min and $t = 5$ min is less than half the average rate in the first minute.

EXERCISE 18.2—RELATIVE RATES OF [REACTANT] DECREASE AND [PRODUCT] INCREASE

What are the relative rates of appearance or disappearance of each product and reactant in the decomposition of nitrosyl chloride, $NOCl(g)$?

$$2 \text{ NOCl}(g) \longrightarrow 2 \text{ NO}(g) + \text{Cl}_2(g)$$

EXERCISE 18.3—ESTIMATION OF RATE OF REACTION

Sucrose decomposes to fructose and glucose in acidic solution. A plot of the concentration of sucrose as a function of time is shown. What is the rate of change of the sucrose concentration over the first 2 h? What is the rate of change over the last 2 h? Estimate the instantaneous rate at $t = 4.0$ h.

> The rate of a chemical reaction is measured as the change in concentration of a reactant or product (which must be specified) per unit of time. We distinguish between the average rate over a given time duration and the instantaneous rate at any moment.

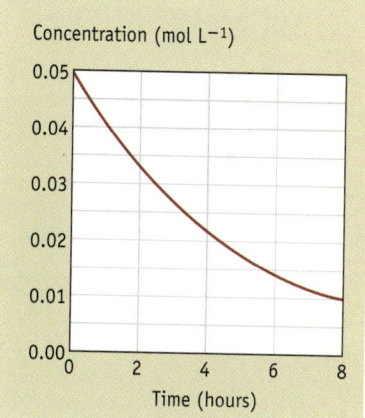

[Sucrose] vs. time during the decomposition of sucrose.

18.3 Conditions that Affect the Rate of a Reaction

A fundamental principle in chemistry is that for a reaction to occur, atoms, molecules, or ions of the reactants must collide with each other. For reactions that occur in the gas phase or in solution, in which the reactant particles are mobile, several factors—reactant concentrations, temperature, and presence of catalysts—affect the rate of a reaction. If the reactant is a solid, the surface area available for reaction will also affect the rate of reaction. This section makes sense of the dependence of reaction rates on these factors by imagining what happens at the molecular level.

The "iodine clock reaction" shown in Figure 18.3 illustrates the effect of concentration and temperature. Reaction mixtures contain different concentrations of dissolved hydrogen peroxide (H_2O_2) and iodide ions (I^-), a fixed amount of Vitamin C (ascorbic acid), and some starch (which is an indicator of the presence of iodine, I_2). The reaction of interest is the slow oxidation of $I^-(aq)$ ions to $I_2(s)$ by $H_2O_2(aq)$:

$$\textit{Reaction 1}: \text{H}_2\text{O}_2(aq) + 2 \text{ I}^-(aq) + 2 \text{ H}^+(aq) \longrightarrow 2 \text{ H}_2\text{O}(\ell) + \text{I}_2(s)$$

The fixed quantity of Vitamin C allows us to monitor the rate of reaction 1. As $I_2(s)$ is formed by reaction 1, Vitamin C rapidly reduces it back to $I^-(aq)$ ions

$$\textit{Reaction 2}: \text{I}_2(s) + \text{C}_6\text{H}_8\text{O}_6(aq) \longrightarrow \text{C}_6\text{H}_6\text{O}_6(aq) + 2 \text{ H}^+(aq) + 2 \text{ I}^-(aq)$$

When eventually all of the Vitamin C has just been used up in this way, any further $I_2(s)$ produced by reaction 1 forms a blue–black complex with starch. The time taken for the solution to turn blue-black is the time it takes for reaction 1 to produce the amount of I_2 that just reacts with the fixed amount of Vitamin C.

Reaction 1 indicates the correct stoichiometry, but is a simplification. Iodine, $I_2(s)$, is almost insoluble in water. However, in solutions containing $I^-(aq)$ ions, it dissolves by forming tri-iodide ions, $I_3^-(aq)$.

$$\text{I}_2(s) + \text{I}^-(aq) \longrightarrow \text{I}_3^-(aq)$$

As a consequence, the overall reaction is perhaps more accurately represented by the following equation, in which only two-thirds of the iodide ions are oxidized, and the other one-third combine with the I_2 formed to form tri-iodide ions:

$$\text{H}_2\text{O}_2(aq) + 3\text{I}^-(aq) + 2\text{H}^+(aq)$$
$$\longrightarrow 2\text{H}_2\text{O}(\ell) + \text{I}_3^-(aq)$$

For experiment A in Figure 18.3, the time required is 51 seconds. When the concentration of iodide ion is smaller (experiment B), the time required is 1 minute and 33 seconds. In experiment C, when the concentrations are the same as in experiment B but the reaction mixture is at a higher temperature, the reaction occurs more rapidly (56 seconds). These observations can be explained by consideration of the frequencies and energies of collisions between reactant molecules and ions [>>Section 18.6].

(a) Initial Experiment.
The blue color of the starch–iodine complex develops in 51 seconds.

(b) Change Concentration.
The blue color of the starch–iodine complex develops in 1 minute, 33 seconds when the solution is less concentrated than A.

(c) Change Temperature.
The blue color of the starch–iodine complex develops in 56 seconds when the solution is less concentrated than A but at a higher temperature.

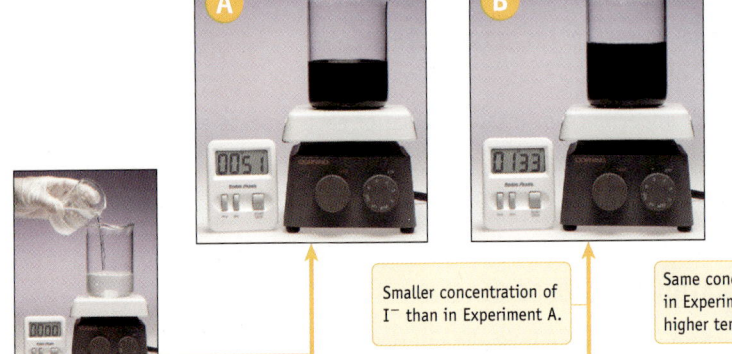

Hot bath

All photos: Charles D. Winters

Solutions containing vitamin C, H_2O_2, I^-, and starch are mixed.

Smaller concentration of I^- than in Experiment A.

Same concentrations as in Experiment B, but at a higher temperature.

FIGURE 18.3 The iodine clock reaction. This experiment demonstrates that for this reaction, the higher the concentration of I^-(aq) and the higher the temperature, the faster the reaction. (You can do these experiments yourself with reagents available in the supermarket. For details, see S.W. Wright, (2002), "The Vitamin C Clock Reaction," *Journal of Chemical Education* 79: 41.)

Catalysts are *substances that accelerate chemical reactions and are regenerated after performing their function in the reaction.* For example, the decomposition of hydrogen peroxide, H_2O_2(aq), to water and oxygen is thermodynamically spontaneous:

$$2\,H_2O_2(aq) \longrightarrow O_2(g) \;+\; 2\,H_2O(\ell)$$

but the rate of the decomposition reaction is extremely slow. Adding a manganese salt, an iodide-containing salt, or a biological substance called an *enzyme* causes this reaction to occur rapidly, as shown by the immediate release of heat, and vigorous bubbling as gaseous oxygen escapes from the solution (Figure 18.4). Catalysts and their mode of functioning are discussed more fully in Section 18.6.

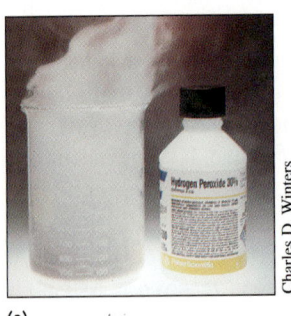
Charles D. Winters
(a)

Thomas Eisner with Daniel Aneshansley, Cornell University
(b)

Charles D. Winters
(c)

FIGURE 18.4 Catalyzed decomposition of H_2O_2(aq). (a) A 30% H_2O_2 solution, poured onto the black solid MnO_2 catalyst, rapidly decomposes to O_2(g) and $H_2O(\ell)$. Steam forms because of the high enthalpy change of reaction. (b) A bombardier beetle uses the enzyme-catalyzed decomposition of H_2O_2 as a defence mechanism. The enthalpy change of the reaction lets the insect eject hot water and other irritating chemicals with explosive force. (c) An enzyme in potato catalyzes the decomposition of hydrogen peroxide, producing bubbles of O_2(g).

The surface area of a solid reactant can also affect the reaction rate. Only molecules or ions at the surface of a solid can come in contact with other reactant particles. The smaller the particles of a solid are, the bigger the surface area of a fixed amount of a solid. When particle size is very small, the effect of large surface area on rate can be quite dramatic (Figure 18.5). Farmers know that explosions of fine dust particles (suspended in the air in an enclosed silo or at a feed mill) represent a major hazard because of the potential for very rapid reaction with oxygen in the air.

> The rate at which a given reaction occurs is increased by higher reactant concentrations, higher temperature, the presence of a catalyst, and (in the case of solid reactants) increase in surface area. Catalysts accelerate chemical reactions, but are regenerated after performing their function in the reaction.

(a)

18.4 Dependence of Rate on Reactant Concentration

One goal in studying kinetics is to determine how variation of the concentrations of reactants affects the rates of reactions. We will see in Section 18.7 that this knowledge can shed insights on the mechanisms of reactions—that is, the molecular events that lead to formation of the products.

We can investigate the effects of varying the concentration of any reactant by measuring the rate when there are different concentrations of the reactant (with the concentrations of the other reactants and the temperature held constant) in the reaction mixture. Figure 18.2 shows how the concentration of N_2O_5 in solution changes with time during its decomposition to NO_2 and O_2. When $[N_2O_5] = 0.34$ mol L^{-1}, the instantaneous rate of reaction of N_2O_5 is 0.0014 mol L^{-1} min^{-1}. When $[N_2O_5] = 0.68$ mol L^{-1}, a tangent to the curve shows that the instantaneous rate is 0.0028 mol L^{-1} min^{-1}. That is, doubling the concentration of N_2O_5 doubles the reaction rate. And if $[N_2O_5]$ is halved to 0.17 mol L^{-1}, the reaction rate is also halved. We can conclude that for this reaction, the reaction rate is directly proportional to the concentration of reactant N_2O_5:

$$\text{Rate of reaction} \propto [N_2O_5]$$

(b)

FIGURE 18.5 The combustion of lycopodium powder. (a) The spores of this common fern burn only with difficulty when piled in a dish. (b) If the spores are ground to a fine powder and sprayed into a flame, combustion is rapid.

In other reactions we find experimentally that there are different relationships between reaction rate and reactant concentration. For example, the reaction rate may be independent of concentration, or it may depend on the reactant concentration raised to some power (that is, rate $\propto$ [reactant]^n). If the reaction involves several reactants, the reaction rate may depend on the concentrations of only one of them, some, or all.

Think about It

e18.2 Watch a video of a reaction, and deduce the dependence of the rate on concentration.

Experimental Rate Equations

The experimentally observed relationship between reactant concentration and reaction rate is expressed by a mathematical relationship called a **rate equation**, or *rate law.* For the decomposition of N_2O_5 the rate equation is

$$\text{Rate of reaction} = k[N_2O_5]$$

where the proportionality constant (k) is called the **rate constant**.

For a reaction of the general form

$$a\text{A} + b\text{B} \longrightarrow x\text{X}$$

we can write a general form of the rate equation:

$$\text{Rate} = k[\text{A}]^m[\text{B}]^n \qquad (18.1)$$

The *exponents m and n are not necessarily the same as the stoichiometric coefficients (a and b) in the balanced chemical equation. The exponents must be determined by experiment.* They are often positive whole numbers, but they may also be fractions, or zero. The rate dependence can only be determined by experiment and cannot be deduced from the stoichiometric equation.

In the case of homogeneous catalysis, when the catalyst is in the same phase as the reactants, the concentration of the catalyst might also be included in the rate equation, even though the catalytic species does not appear in the overall equation for the reaction. Consider, for example, the decomposition of hydrogen peroxide in aqueous solution in the presence of catalytic aquated iodide ions:

$$2 \ H_2O_2(aq) \xrightarrow{\text{I}^-(aq) \ catalyst} 2 \ H_2O(\ell) + O_2(g)$$

Measurements show that the rate equation is

$$\text{Reaction rate} = k[H_2O_2][I^-]$$

Here the power of each concentration term is 1, even though the in the equation the coefficient of H_2O_2 is 2 and $I^-(aq)$ ions are not in the balanced equation.

> For a reaction of the general form $aA + bB \longrightarrow xX$, the dependence of reaction rate on the concentrations of reactants is expressed by a rate equation of the form Rate = $k[A]^m[B]^n$, in which the term k is called the rate constant. The values of exponents m and n can only be determined experimentally.

Think about It

e18.3 Using simulations of reactions, deduce the dependence of the rate on concentration.

The Order of a Reaction

The **order of a reaction** with respect to each reactant is the exponent of its concentration term in the rate equation (equation 18.1), and the **overall reaction order** is the sum of the exponents on all concentration terms. For example, the decomposition of $H_2O_2(aq)$ in the presence of aquated iodide ions (above) is first order with respect to $[H_2O_2]$, also first order with respect to $[I^-]$, and second order overall. This tells us that the rate doubles if either one of $[H_2O_2]$ or $[I^-]$ is doubled (and the other held constant), and increases four-fold if the concentrations of both reactants are doubled.

Consider another example, the reaction between $NO(g)$ and $Cl_2(g)$:

$$2 \ NO(g) + Cl_2(g) \longrightarrow 2 \ NOCl(g)$$

The experimentally determined rate equation for this reaction is

$$\text{Rate} = k[NO]^2[Cl_2]$$

This reaction is second order with respect to $[NO]$, first order in $[Cl_2]$, and third order overall. We can see how this rate equation is derived by examining some experimental data for the rate of disappearance of $NO(g)$, in the table below:

Experiment	[NO] (mol L^{-1})	[Cl$_2$] (mol L^{-1})	Rate (mol $L^{-1} \ s^{-1}$)
1	0.250	0.250	1.43×10^{-6}
2	0.500	0.250	5.72×10^{-6}
3	0.250	0.500	2.86×10^{-6}
4	0.500	0.500	11.4×10^{-6}

Comparing experiments 1 and 2, when $[Cl_2]$ is held constant and $[NO]$ is doubled from 0.250 mol L^{-1} to 0.500 mol L^{-1}, the reaction rate increases four-fold. So we have rate is proportional to $[NO]^2$. Comparing experiments 1 and 3, when $[NO]$ is held constant and $[Cl_2]$ is doubled, the rate doubles. So rate is proportional to $[Cl_2]$. From a comparison of experiments 1 and 4, we see that if both $[NO]$ and $[Cl_2]$ are doubled, the rate increases eight-fold—consistent with the rate equation.

The decomposition of ammonia on a platinum surface at 856 °C is interesting because it is zero order: the reaction rate is independent of $NH_3(g)$ concentration.

$$2 \ NH_3(g) \longrightarrow N_2(g) + 3 \ H_2(g) \qquad \text{Rate} = k[NH_3]^0 = k$$

Knowledge of reaction order is important because it gives us insight into the question of how a reaction occurs at the molecular level. This is discussed in Section 18.7.

EXERCISE 18.4—RATE EQUATIONS AND ORDER OF REACTION

The reaction between ozone and nitrogen dioxide at 231 K is first order in both $[NO_2]$ and $[O_3]$:

$$2\ NO_2(g) + O_3(g) \longrightarrow N_2O_5(g) + O_2(g)$$

(a) Write the rate equation for the reaction.
(b) If the concentration of $NO_2(g)$ is tripled, what is the change in the reaction rate?
(c) What is the effect on reaction rate if the concentration of $O_3(g)$ is halved?

> The dependence of rate of a reaction on reactant concentrations is given by a rate equation of the form Rate $= k[A]^m[B]^n$. The order of the reaction with respect to $[A]$ is m, the order with respect to $[B]$ is n, and the overall order is $m + n$.

The Rate Constant (k)

The rate constant (k) in a rate equation is a proportionality constant that applies at a particular temperature. It is an important quantity because it enables us to find the reaction rate for reaction mixtures with different reactant concentrations at the same temperature. To see how to use k, consider the substitution of a Cl^- ion by a water molecule in molecules of the cancer chemotherapy agent *cis*-platin, $Pt(NH_3)_2Cl_2$ [<<Section 9.6].

$$Pt(NH_3)_2Cl_2(aq) \quad + \quad H_2O(\ell) \quad \longrightarrow \quad [Pt(NH_3)_2(H_2O)Cl]^+(aq) \quad + \quad Cl^-(aq)$$

The experimentally derived rate equation for this reaction is

$$\text{Rate} = k[Pt(NH_3)_2Cl_2]$$

and at 25 °C the rate constant (k) is 0.090 h^{-1}. Knowing k allows us to calculate the rate at any concentration of *cis*-platin. For example, when $[Pt(NH_3)_2Cl_2] = 0.018$ mol L^{-1}:

$$\text{Rate} = k[Pt(NH_3)_2Cl_2] = (0.090\ h^{-1})(0.018\ \text{mol L}^{-1}) = 0.0016\ \text{mol L}^{-1}\ h^{-1}$$

Reaction rates ($\Delta[R]/\Delta t$) have units of mol L^{-1} time^{-1} (change of concentration per unit time). Rate constants must have units consistent with the units for the other terms in the rate equation.

- First-order reactions: the units of k are time^{-1}.
- Second-order reactions: the units of k are L mol^{-1} time^{-1}.
- Zero-order reactions: the units of k are mol L^{-1} time^{-1}.

> The rate constant (k) is the proportionality constant in the rate equation. Its value is different from reaction to reaction at the same temperature, and for a given reaction its value changes as temperature is changed.

Determining a Rate Equation: Method of Initial Rates

A rate equation can only be determined experimentally. One way to do so is by using the "method of initial rates." The *initial rate* is the instantaneous reaction rate at the start of the reaction (at $t = 0$). An approximate value of the initial rate can be obtained by mixing the reactants and determining $\Delta[\text{product}]/\Delta t$ or $\Delta[\text{reactant}]/\Delta t$ immediately. The method involves measuring the initial rates in several different experimental "runs," each with different initial concentrations, and seeing how the initial rates depend upon the reactant concentrations.

To illustrate this method, let's look at the reaction of aquated hydroxide ions with methyl acetate in aqueous solution to produce acetate ions and methanol:

$$CH_3COOCH_3(aq) \;+\; OH^-(aq) \;\longrightarrow\; CH_3COO^-(aq) \;+\; CH_3OH(aq)$$

Initial concentrations and reaction rates were measured for several experimental runs at 25 °C:

INITIAL CONCENTRATIONS (mol L^{-1})			INITIAL REACTION RATE
Run	[CH$_3$COOCH$_3$]	[OH$^-$]	(mol L^{-1} s^{-1})
1	0.050	0.050	0.00034
2	0.050	0.10	0.00069
3	0.10	0.10	0.00137

These rate data show that when the initial concentration of either reactant is doubled while the concentration of the other is held constant, the initial reaction rate doubles. So the reaction is first order with respect to each of these reactants, and the rate equation is

$$Rate = k[CH_3COOCH_3][OH^-]$$

From the rate equation, the value of the rate constant (k) can be found by substituting values for the rate and reactant concentrations in any run into the rate equation. Using the data from run 1, we have

$$Rate = 0.00034 \text{ mol L}^{-1}\text{ s}^{-1} = k(0.050 \text{ mol L}^{-1})(0.050 \text{ mol L}^{-1})$$

$$k = \frac{0.00034 \text{ mol L}^{-1}\text{s}^{-1}}{(0.050 \text{ mol L}^{-1})(0.050 \text{ mol L}^{-1})} = 0.14 \text{ L mol}^{-1}\text{ s}^{-1}$$

Interactive Exercise 18.5

Predict changes in initial rates from the rate law.

Think about It

e18.4 Deduce the rate equation from initial reaction rates at different initial concentrations.

Interactive Exercise 18.6

Deduce the rate equation and calculate the rate constant from initial rate data.

WORKED EXAMPLE 18.3—DETERMINING A RATE EQUATION FROM INITIAL RATES

The rate of the reaction between CO(g) and NO$_2$(g), represented by

$$CO(g) + NO_2(g) \longrightarrow CO_2(g) + NO(g)$$

was studied at 540 K with various starting concentrations of CO(g) and NO$_2$(g), and the experimental data are shown in the table. Determine the rate equation and the value of the rate constant at 540 K.

RUN	INITIAL CONCENTRATIONS (mol L^{-1})		INITIAL RATE
	[CO]	[NO$_2$]	(mol L^{-1} h^{-1})
1	5.10×10^{-4}	0.350×10^{-4}	3.4×10^{-8}
2	5.10×10^{-4}	0.700×10^{-4}	6.8×10^{-8}
3	5.10×10^{-4}	0.175×10^{-4}	1.7×10^{-8}
4	1.02×10^{-3}	0.350×10^{-4}	6.8×10^{-8}
5	1.53×10^{-3}	0.350×10^{-4}	10.2×10^{-8}

Strategy

We need to determine m and n in a rate equation of the form

$$\text{Rate} = k[CO]^m[NO_2]^n$$

We can compare experimental runs in which the concentration of one reactant is held the same, and then decide how the rate of reaction changes as the concentration of the other is varied.

Solution

In the first three runs, $[CO]$ is the same. In run 2, $[NO_2]$ is double that in run 1, and the initial rate is also doubled. So, $n = 1$ and the reaction is first order with respect to $[NO_2]$. This conclusion is confirmed by run 3: with $[NO_2]$ half that in run 1, the initial rate is half that in run 1.

Comparison of the data in run 1 and 4 (with constant $[NO_2]$) shows that doubling $[CO]$ doubles the initial rate, and the data from runs 1 and 5 show that tripling $[CO]$ triples the initial rate. So the reaction is also first order with respect to $[CO]$, and $m = 1$. We can now say that the rate equation for this reaction is

$$\text{Rate} = k[CO][NO_2]$$

To estimate the value of the rate constant (k), we can use the data from any run. From run 1, for example:

$$\text{Rate} = (3.4 \times 10^{-8} \text{ mol L}^{-1} \text{ h}^{-1}) = k(5.10 \times 10^{-4} \text{ mol L}^{-1})(0.350 \times 10^{-4} \text{ mol L}^{-1})$$

$$k = 1.9 \text{ L mol}^{-1} \text{ h}^{-1}$$

WORKED EXAMPLE 18.4—USING A RATE EQUATION TO ESTIMATE INITIAL RATES

Using the rate equation and rate constant determined for the reaction of $CO(g)$ and $NO_2(g)$ at 540 K in Worked Example 18.3, determine the initial rate of the reaction at 540 K when $[CO] = 3.8 \times 10^{-4} \text{ mol L}^{-1}$ and $[NO_2] = 0.650 \times 10^{-4} \text{ mol L}^{-1}$.

Strategy

Substitute the rate constant and the initial reactant concentrations into the rate equation.

Solution

$$\text{Rate} = k[CO][NO_2] = (1.9 \text{ L mol}^{-1} \text{ h}^{-1})(3.8 \times 10^{-4} \text{ mol L}^{-1})(0.650 \times 10^{-4} \text{ mol L}^{-1})$$

$$= 4.7 \times 10^{-8} \text{ mol L}^{-1} \text{ h}^{-1}$$

EXERCISE 18.7—DETERMINING A RATE EQUATION FROM INITIAL RATES

The initial rate of the reaction between nitrogen monoxide and oxygen gases

$$2 \text{ NO}(g) + O_2(g) \longrightarrow 2 \text{ NO}_2(g)$$

was measured in five experimental runs at 25 °C with various initial concentrations of $NO(g)$ and $O_2(g)$. Data are shown in the table. Determine the rate equation. What is the value of the rate constant (k) at 25 °C?

RUN	INITIAL CONCENTRATIONS (mol L^{-1})		INITIAL RATE
	[NO]	[O$_2$]	(mol L^{-1} s^{-1})
1	0.020	0.010	0.028
2	0.020	0.020	0.057
3	0.020	0.040	0.114
4	0.040	0.020	0.227
5	0.010	0.020	0.014

To determine the rate equation for a reaction by the method of initial rates, the reaction is carried out several times with different concentrations of the reactants. The initial rates in the various runs are then compared.

18.5 Concentration-Time Relationships: Integrated Rate Equations

In an alternative method of deducing the rate equation for a reaction, during a single experimental run, the concentrations of reactants or products are monitored at various times. The way that reaction rate changes with time for a zero-order reaction is different from that for a first-order reaction, which in turn is different from that for a second-order reaction.

It is often useful to predict what a reactant or product concentration will be after a particular time has elapsed, or at what time the concentration of a reactant or product will reach a particular value. One way to make these predictions is to use a mathematical equation that relates concentration and time during a reaction. That is, we would like to have an equation that will describe concentration vs. time curves like the one shown in Figure 18.2. The object of this method is to see whether the concentration vs. time dependence during an experimental run corresponds with that expected for a zero-order, first-order, or second-order reaction. To do so, we need a mathematical analysis of the concentration vs. time dependence of each type of reaction.

First-Order Reactions

By definition, for a reaction that is first order with respect to the concentration of reactant R,

$$-\frac{\Delta[R]}{\Delta t} = k[R]$$

Using calculus, this relationship can be transformed into a very useful equation called an **integrated rate equation** (because integral calculus is used in its derivation):

$$\ln \frac{[R]_t}{[R]_0} = -kt$$

Here $[R]_0$ and $[R]_t$ are concentrations of the reactant at time $t = 0$ and at a later time, t. Time $t = 0$ need not correspond with the start of the reaction. The timer may be started at any time—for example, when a measuring instrument is turned on. The ratio of concentrations, $[R]_t/[R]_0$, is the fraction of reactant R (that was present at $t = 0$) that is still present at time t.

The integrated rate equation is useful in three ways:

- If $[R]_t/[R]_0$ is measured after some amount of time t has elapsed, then k can be calculated.
- If $[R]_0$ and k are known, then the concentration of reactant R after time t has elapsed ($[R]_t$) can be calculated.
- If k is known, then the time that will elapse before a particular fraction ($[R]_t/[R]_0$) remains can be calculated.

Taking It Further

e18.5 Derive the integrated rate equations from first principles.

Cyclopropane   Propene

WORKED EXAMPLE 18.5—USING THE INTEGRATED RATE EQUATION FOR A FIRST-ORDER REACTION

Cyclopropane, $C_3H_6(g)$, has been used in the past as an anesthetic, in a mixture with oxygen, although this practice has almost ceased because it is flammable. When cyclopropane is heated, the highly strained three-membered ring can open, with formation of an alkene. The reaction has been experimentally determined to be a first-order process:

Rate $= k$[cyclopropane]   At 25 °C, $k = 5.4 \times 10^{-2}$ h^{-1}

If the initial concentration of cyclopropane is 0.050 mol L^{-1}, how much time (in hours) will elapse at 25 °C before its concentration has dropped to 0.010 mol L^{-1}?

Strategy

Apply the first-order integrated rate equation.

Solution

The first-order integrated rate equation applied to this reaction is

$$\ln \frac{[\text{cyclopropane}]_t}{[\text{cyclopropane}]_0} = -kt$$

Substituting into this equation

$$\ln \frac{0.010 \ \cancel{\text{mol L}^{-1}}}{0.050 \ \cancel{\text{mol L}^{-1}}} = -(5.4 \times 10^{-2} \ \text{h}^{-1})t$$

$$t = \frac{-\ln 0.20}{5.4 \times 10^{-2} \ \text{h}^{-1}} = \frac{-(-1.61)}{5.4 \times 10^{-2} \ \text{h}^{-1}} = 30 \ \text{h}$$

Think about It

e18.6 Use the integrated rate equation for a first-order reaction.

Interactive Exercise 18.8

Calculate a rate constant using the integrated rate law for a first-order reaction.

WORKED EXAMPLE 18.6—USING THE INTEGRATED RATE EQUATION FOR A FIRST-ORDER REACTION

Hydrogen peroxide decomposes in a dilute sodium hydroxide solution at 20 °C in a first-order reaction:

$$2 \ H_2O_2(aq) \longrightarrow 2 \ H_2O(\ell) + O_2(g) \qquad \text{Rate} = k[H_2O_2], \ k = 1.06 \times 10^{-3} \ \text{min}^{-1}$$

What is the fraction remaining after exactly 100 min if the initial concentration of H_2O_2 is 0.020 mol L^{-1}? What is the concentration of hydrogen peroxide after exactly 100 min?

Strategy

Substitute the known values of k and t into the integrated rate equation for a first-order reaction to calculate the ratio $[H_2O_2]_t/[H_2O_2]_0$.

Solution

$$\ln \frac{[H_2O_2]_t}{[H_2O_2]_0} = -kt = -(1.06 \times 10^{-3} \ \cancel{\text{min}^{-1}})(100 \ \cancel{\text{min}}) = -0.106$$

$$\text{Fraction remaining} = \frac{[H_2O_2]_t}{[H_2O_2]_0} = 0.90$$

Substituting $[H_2O_2]_0 = 0.020$ mol L^{-1} gives $[H_2O_2] = 0.018$ mol L^{-1}.

EXERCISE 18.9—USING THE INTEGRATED RATE EQUATION FOR A FIRST-ORDER REACTION

Sucrose, a sugar, decomposes in acid solution to give glucose and fructose. The reaction is first order with respect to [sucrose], and at 25 °C, $k = 0.21$ h^{-1}. If the initial concentration of sucrose is 0.010 mol L^{-1}, what is its concentration after 5.0 h?

The way that concentration changes with time during a first-order reaction is described by the integrated rate equation:

$$\ln \frac{[R]_t}{[R]_0} = -kt$$

Second-Order Reactions

For a reaction R → products that is second order with respect to [R], as well as second order overall, the rate equation is

$$-\frac{\Delta[R]}{\Delta t} = k[R]^2$$

Using the methods of calculus, this relationship can be transformed into the following integrated rate equation that relates reactant concentration and time:

$$\frac{1}{[R]_t} - \frac{1}{[R]_0} = kt$$

The same symbolism used with first-order reactions applies.

WORKED EXAMPLE 18.7—USING THE INTEGRATED RATE EQUATION FOR A SECOND-ORDER REACTION

For the gas-phase decomposition of HI(g):

$$HI(g) \longrightarrow \tfrac{1}{2}H_2(g) + \tfrac{1}{2}I_2(g)$$

the rate equation is experimentally determined to be

$$-\frac{\Delta[HI]}{\Delta t} = k[HI]^2$$

where $k = 30$ L mol^{-1} min^{-1} at 443 °C. How much time does it take for the concentration of HI to drop from 0.010 mol L^{-1} to 0.0050 mol L^{-1} at 443 °C?

Strategy

Substitute the values of [HI]$_0$, [HI]$_t$, and k into the second-order integrated rate equation.

Solution

Here [HI]$_0 = 0.010$ mol L^{-1} and [HI]$_t = 0.0050$ mol L^{-1}. We have

$$\frac{1}{[R]_t} - \frac{1}{[R]_0} = \frac{1}{0.0050 \text{ mol L}^{-1}} - \frac{1}{0.010 \text{ mol L}^{-1}} = (30 \text{ L mol}^{-1} \text{ min}^{-1})t$$

$$\{(2.0 \times 10^2) - (1.0 \times 10^2)\}\text{ L mol}^{-1} = (30 \text{ L mol}^{-1} \text{ min}^{-1})t$$

$$t = 3.3 \text{ min}$$

EXERCISE 18.11—USING THE INTEGRATED RATE EQUATION FOR A SECOND-ORDER REACTION

Using the rate constant for HI(g) decomposition given in Worked Example 18.7, calculate [HI] after 12 min if [HI]$_0 = 0.010$ mol L^{-1}.

The way that concentration changes with time during a second-order reaction is described by the integrated rate equation

$$\frac{1}{[R]_t} - \frac{1}{[R]_0} = kt$$

Zero-Order Reactions

For a reaction R → products that is zero order with respect to [R], the rate equation is

$$-\frac{\Delta[R]}{\Delta t} = k[R]^0 = k$$

This leads to the integrated rate equation

$$[R]_t = [R]_0 - kt$$

Graphical Methods for Determining Reaction Order

The integrated rate equations for zero-, first-, and second-order reactions provide a convenient way to determine the order of a reaction and its rate constant. Re-arranged, each of these equations has the form $y = mx + b$ as illustrated in the table ($x = t$ in each case). This is the equation for a straight line, where m is the slope of the line and b is the y-intercept (the value of y when x is zero).

Zero Order	First Order	Second Order
$[R]_t = -kt + [R]_0$	$\ln [R]_t = -kt + \ln [R]_0$	$\dfrac{1}{[R]_t} = +kt + \dfrac{1}{[R]_0}$
$y \quad mx \quad b$	$y \quad mx \quad b$	$y \quad mx \quad b$

To determine the reaction order, we can plot the experimental concentration time data in different ways until a straight line plot is achieved. The mathematical relationships for zero-, first-, and second-order reactions are summarized in Table 18.1.

Order	Rate Equation	Integrated Rate Equation	Straight Line Plot	Slope	Units of k
0	$-\Delta[R]/\Delta t = k[R]^0$	$[R]_0 - [R]_t = kt$	$[R]_t$ vs. t	$-k$	mol L^{-1} time^{-1}
1	$-\Delta[R]/\Delta t = k[R]^1$	$\ln ([R]_t/[R]_0) - kt$	$\ln [R]_t$ vs. t	$-k$	time^{-1}
2	$-\Delta[R]/\Delta t = k[R]^2$	$1/[R]_t - 1/[R]_0 = kt$	$1/[R]_t$ vs. t	k	L mol^{-1} time^{-1}

TABLE 18.1 Mathematical Characteristics of Zero-, First-, and Second-Order Reactions

As an example of how we can decide the order of a reaction from concentration vs. time data, consider the concentration-independent decomposition of ammonia on a platinum surface.

$$2 NH_3(g) \longrightarrow N_2(g) + 3 H_2(g) \qquad Rate = k [NH_3]^0 = k$$

From the experimental data, you can make graphs of $[R]_t$ vs. time, $\ln [R]_t$ vs. time, and $1/[R]_t$ vs. time to see which, if any, is a straight line. The straight line obtained when the concentration at time t, $[R]_t$, is plotted against time (Figure 18.6) is evidence that this reaction is zero order with respect to $[NH_3]$.

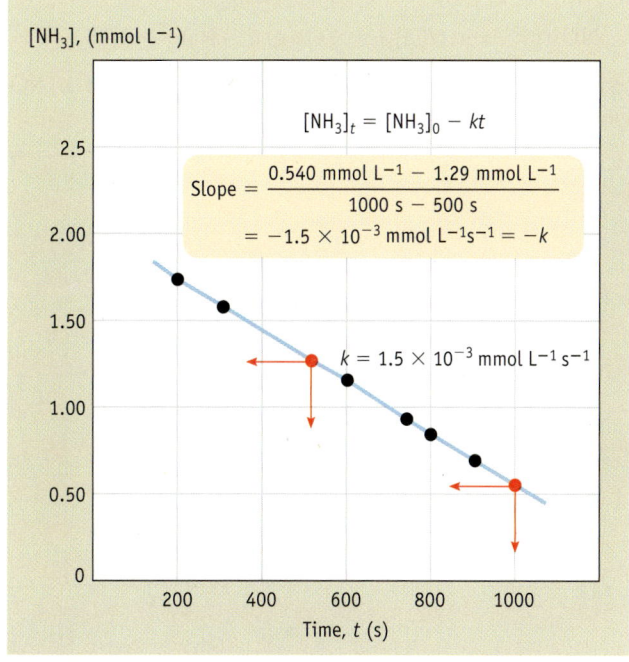

FIGURE 18.6 Straight line plot for a zero-order reaction. A graph of $[NH_3]$ vs. time for the decomposition of $NH_3(g)$ on a metal surface at 856 °C is a straight line, indicating a zero-order reaction. The rate constant (k) is found from the slope of the line. The points chosen for calculation of the slope are in red (1 mmol = 1×10^{-3} mol).

The rate constant (k) can be determined from the slope of the line:

$$\text{slope} = -k = -1.5 \times 10^{-3} \text{ mmol L}^{-1} \text{ s}^{-1} = -1.5 \times 10^{-6} \text{ mol L}^{-1} \text{ s}^{-1}$$
$$k = 1.5 \times 10^{-6} \text{ mol L}^{-1} \text{ s}^{-1}$$

As Figure 18.2 shows, a plot of [R] vs. time for a first-order reaction is a curve. A plot of ln [R] vs. time, however, is a straight line with a negative slope. Consider the decomposition of hydrogen peroxide, referred to in Worked Example 18.6:

$$2 \text{ H}_2\text{O}_2(\text{aq}) \longrightarrow 2 \text{ H}_2\text{O}(\ell) + \text{O}_2(\text{g}) \qquad \text{Rate} = k[\text{H}_2\text{O}_2]$$

Experimental data are listed in Figure 18.7. A graph of ln [H_2O_2] vs. time is a straight line, showing that the reaction is first order in [H_2O_2]. Slope $= -k = 1.05 \times 10^{-3}$ min^{-1}.

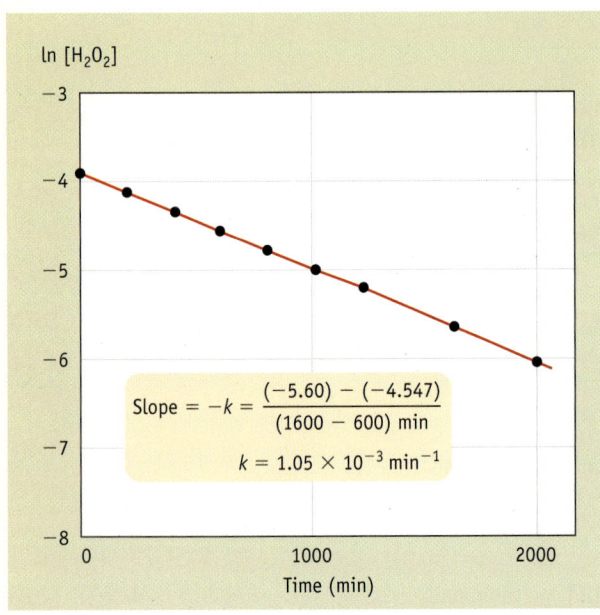

Time (min)	[H_2O_2] (mol L^{-1})	ln [H_2O_2]
0	0.0200	−3.912
200	0.0160	−4.135
400	0.0131	−4.335
600	0.0106	−4.547
800	0.0086	−4.76
1000	0.0069	−4.98
1200	0.0056	−5.18
1600	0.0037	−5.60
2000	0.0024	−6.03

FIGURE 18.7 Straight line plot for a first-order reaction. If data for the decomposition of hydrogen peroxide are plotted as ln [H_2O_2] vs. time, the result is a straight line with a negative slope. This indicates a first-order reaction. The rate constant $k = -$slope.

The decomposition of gaseous NO_2 is a second-order process:

$$\text{NO}_2(\text{g}) \longrightarrow \text{NO}(\text{g}) + \tfrac{1}{2}\text{O}_2(\text{g}) \qquad \text{Rate} = k[\text{NO}_2]^2$$

This rate dependence can be verified by showing that a plot of $1/[\text{NO}_2]$ vs. time is a straight line (Figure 18.8). Here, the slope $= k$.

FIGURE 18.8 Straight line plot for a second-order reaction. A plot of $1/[\text{NO}_2]$ vs. time for the decomposition of $\text{NO}_2(\text{g})$ is a straight line, confirming that this is a second-order reaction. The slope of the line equals the rate constant.

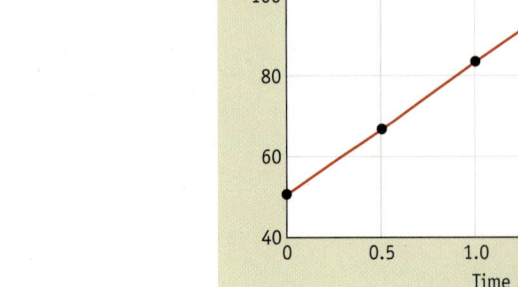

Think about It

e18.7 Plot different graphs of kinetic data to deduce the order of reaction.

EXERCISE 18.13—DETERMINING REACTION ORDER BY GRAPHICAL METHODS

Data for the decomposition of N_2O_5 in a particular solvent at 45 °C are as follows:

$[N_2O_5]$ (mol L^{-1})	t (min)
2.08	3.07
1.67	8.77
1.36	14.45
0.72	31.28

Separately plot $[N_2O_5]$, ln $[N_2O_5]$, and $1/[N_2O_5]$ vs. time (t). What is the order of the reaction? What is the rate constant for the reaction?

Interactive Exercise 18.12

Deduce the order of reaction using graphical methods.

The order of a reaction can be deduced from a single run by seeing which of the integrated rate equations are consistent with the experimental concentration vs. time data. For a reactant R, a straight line plot will be obtained as follows:

- Zero-order reaction: [R] vs. t (slope $= -k$)
- First-order reaction: ln [R] vs. t (slope $= -k$)
- Second-order reaction: 1/[R] vs. t (slope $= k$)

Half-Life and First-Order Reactions

The **half-life ($t_{1/2}$)** of a reaction is *the time required for the concentration of a reactant to decrease to one-half its initial value*: the longer the half-life, the slower the reaction. Half-life is used primarily when dealing with first-order processes.

Half-life is a term often encountered when dealing with radioactive elements. Radioactive decay is a first-order process, and half-life is used to describe how fast a radioactive element decays [>>Section 30.5; Worked Example 18.9].

The half-life ($t_{1/2}$) of a reactant R is the time from $t = 0$ at which

$$[R]_t = \tfrac{1}{2} [R]_0 \quad \text{or} \quad \frac{[R]_t}{[R]_0} = \tfrac{1}{2}$$

To evaluate $t_{1/2}$ for a first-order reaction, we substitute $[R]_t/[R]_0 = \tfrac{1}{2}$ and $t = t_{1/2}$ into the integrated first-order rate equation:

$$\ln \frac{[R]_t}{[R]_0} = \ln \tfrac{1}{2} = -\ln 2 = -0.693 = -kt_{1/2}$$

Re-arranging, we derive a relationship between half-life and the rate constant showing that the *half-life of a first-order reaction is independent of reactant concentration*: from any instant during the course of a reaction, the time taken for the concentration of reactant to be halved is the same.

$$t_{1/2} = \frac{0.693}{k}$$

To illustrate the concept of constant half-life, consider the first-order decomposition of H_2O_2(aq), referred to earlier in Worked Example 18.6 and Figure 18.7.

$$2\ H_2O_2(aq) \longrightarrow 2\ H_2O(\ell) + O_2(g)$$

At the temperature of measurement, the rate constant (k) was determined to be 1.05×10^{-3} min^{-1}. From this, we can calculate the half-life of H_2O_2(aq):

$$t_{1/2} = \frac{0.693}{k} = \frac{0.693}{1.05 \times 10^{-3} \text{min}^{-1}} = 660 \text{ min}$$

In Figure 18.9, the plot of $[H_2O_2]$ as a function of time shows that $[H_2O_2]$ falls by half over each 660 min period. The initial $[H_2O_2]$ is 0.020 mol L^{-1}, and it drops to 0.010 mol L^{-1} after 660 min. The concentration drops again by half (to 0.0050 mol L^{-1}) after another 660 min. That is, after two half-lives (1320 min), the concentration is $(\frac{1}{2}) \times (\frac{1}{2}) = (\frac{1}{2})^2 = \frac{1}{4}$, or 25% of the initial concentration. After three half-lives (1980 min), the concentration has dropped to $(\frac{1}{2}) \times (\frac{1}{2}) \times (\frac{1}{2}) = (\frac{1}{2})^3 = \frac{1}{8}$, or 12.5% of the initial value—that is, 0.0025 mol L^{-1}.

FIGURE 18.9 Constant half-life of a first-order reaction. The concentration vs. time curve shows that the concentration of H_2O_2(aq) is halved every 660 min. Every first-order reaction has a similar concentration vs. time plot.

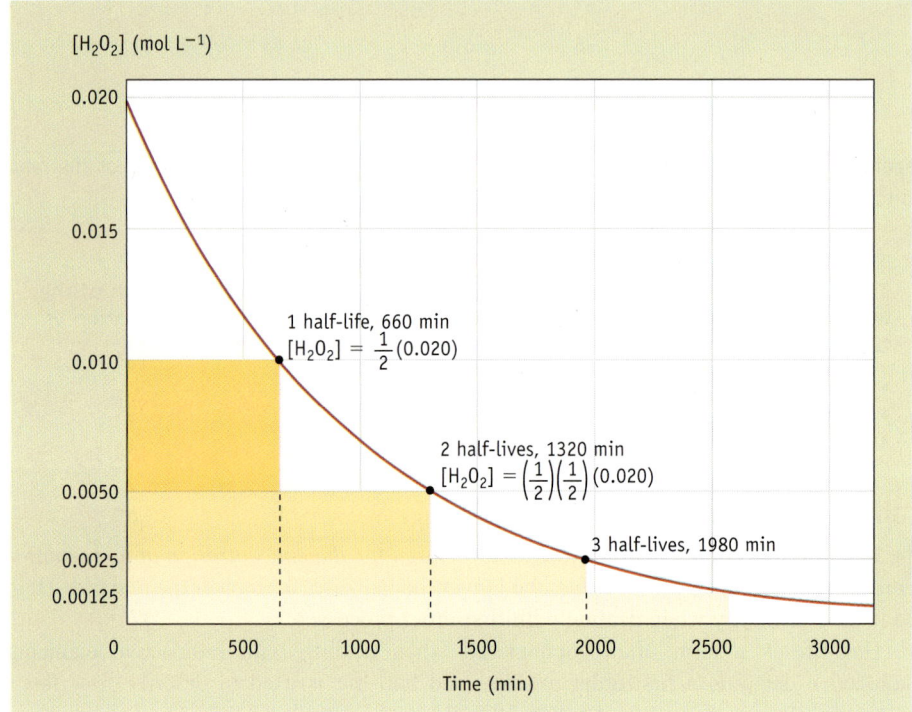

The half-life of first-order reactions is commonly used as an indicator of the relative rates of reaction. It is difficult to visualize whether a reaction is fast or slow from the value of the rate constant, but this is easily assessed from the half-life, 660 min or 11 h. We can immediately deduce that the concentration of reactant will fall to 25% over 22 h, to 12.5% over 44 h, and to about 6% over 88 h.

The half-lives of zero-order and second-order reactions depend on the initial concentration. You can show that for zero-order reactions

$$t_{1/2} = \frac{[R]_0}{2k}$$

And for second-order reactions

$$t_{1/2} = \frac{1}{k[R]_0}$$

WORKED EXAMPLE 18.8—HALF-LIFE OF A FIRST-ORDER REACTION

Sucrose ($C_{12}H_{22}O_{11}$) decomposes to fructose and glucose in acidic solution by a first-order process:

$$Rate = k[sucrose] \qquad k = 0.208 \text{ h}^{-1} \text{ at } 25 \text{ °C}$$

How long will it take for 87.5% of the initial concentration of sucrose to decompose?

Strategy

After 87.5% of the sucrose has decomposed, 12.5% remains. To reach this point, three half-lives are required (after one half-life, 50% remains; after two half-lives, 25%

Think about It

e18.8 Determine the half-life from a concentration vs. time graph.

remains; and after three half-lives, 12.5% remains). Calculate the half-life and multiply by 3.

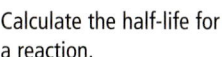
Interactive Exercise 18.14

Calculate the half-life for a reaction.

Solution

The half-life for the reaction is

$$t_{1/2} = \frac{0.693}{k} = \frac{0.693}{0.208 \text{ h}^{-1}} = 3.33 \text{ h}$$

It will take 10 h for 87.5% to decompose: 12.5% remains after three half-lives—i.e., at $t = 3 \times 3.33 \text{ h} = 10 \text{ h}$

WORKED EXAMPLE 18.9—HALF-LIFE OF A FIRST-ORDER PROCESS

Radioactive radon-222 gas (^{222}Rn) from natural sources seeps into the basement of a home. The half-life of ^{222}Rn is 3.8 d (3.8 days). If a basement has 4.0×10^{13} atoms of ^{222}Rn per litre of air, and the radon gas is trapped in the basement, how many atoms of ^{222}Rn will remain after 30 days?

Strategy

From the half-life, calculate the rate constant. Knowing $[^{222}\text{Rn}]_0$ and k, use the integrated rate equation for first-order reactions to calculate $[^{222}\text{Rn}]_t$ when $t = 30$ d.

Solution

The rate constant (k) is given by

$$k = \frac{0.693}{t_{1/2}} = \frac{0.693}{3.8\text{d}} = 0.182 \text{ d}^{-1}$$

In the integrated rate equation, we substitute $[^{222}\text{Rn}]_0 = 4.0 \times 10^{13}$ atoms L^{-1}, $k = 0.182$ d^{-1}, and $t_{1/2} = 3.8$ d.

$$\ln \frac{[R]_t}{[R]_0} = \ln \frac{[^{222}\text{Rn}]_{30}}{4.0 \times 10^{13} \text{ atoms L}^{-1}} = -kt = -(0.182 \text{ d}^{-1})(30 \text{ d}) = -5.5$$

$$\frac{[^{222}\text{Rn}]_{30}}{4.0 \times 10^{13} \text{ atoms L}^{-1}} = 0.0042$$

$$[^{222}\text{Rn}]_{30} = 1.7 \times 10^{11} \text{ atoms L}^{-1}$$

EXERCISE 18.15—HALF-LIFE OF A FIRST-ORDER PROCESS

Americium is used in smoke detectors and in medicine for the treatment of some malignancies. One isotope of americium, ^{241}Am, has a rate constant (k) for radioactive decay of 0.0016 y^{-1} (0.0016 years^{-1}). In contrast, radioactive iodine-125 (^{125}I), which is used for studies of thyroid functioning, has a rate constant for decay of 0.011 d^{-1}.
(a) What are the half-lives of these isotopes?
(b) Which element decays faster?
(c) If you are given a dose of iodine-125, and have 1.6×10^{15} atoms, how many atoms remain after 2.0 days?

The half-life ($t_{1/2}$) of a reaction is the time required for the concentration of a reactant to decrease to one-half its initial value. For first-order reactions, the half-life is independent of the reactant concentration and is related to the rate constant at a given temperature by

$$t_{1/2} = \frac{0.693}{k}$$

18.6   A Microscopic View of Reaction Rates: Collision Theory

Throughout this learning resource, we have turned to the molecular level of chemistry to understand chemical phenomena. In this chapter, we look at the way reactions occur at the molecular level to provide some insight into the various influences on rates of reactions.

We will consider reactions from the point of view of the kinetic-molecular model of matter (<<Section 2.2, 11.7) that assumes that molecules, atoms, or ions of reactants are in rapid and random motion within the reaction mixture. They frequently collide with other molecules and strike the walls of the vessel. For a reaction to occur, the **collision theory of reaction rates** gives us three conditions that must be met:

1. Reactant particles must collide with each another if they are to react.
2. The combined energies of colliding molecules must be at least a certain level, different from reaction to reaction.
3. The molecules must collide in an orientation that can lead to appropriate re-arrangement of the atoms.

Each of these conditions will be discussed within the context of the effects of concentration and temperature on reaction rate.

> According to the collision theory of reaction rates, reaction is possible only when molecules collide with sufficient total energy and in particular orientations.

Reactant Concentration and Reaction Rate

Consider the important atmospheric reaction of nitric oxide and ozone [<<Section 17.10], which has first-order dependence on the concentration of each reactant:

$$NO(g) + O_3(g) \longrightarrow NO_2(g) + O_2(g) \qquad Rate = k[NO][O_3]$$

Why does this reaction depend in this way upon the concentrations of these reactants? Let's consider the reaction in the gas phase at the molecular level. According to collision theory, if reaction is to take place, molecules must collide with one another. It is logical to suppose that the experimentally measurable rate of conversion (mol L^{-1} s^{-1}) of reactants into products is related to the frequency of collisions between the particles (although not all collisions result in reaction), and this is in turn related to their concentrations (Figure 18.10). Doubling the concentration of either reagent in the $NO(g)/O_3(g)$ reaction mixture, say NO, will lead to twice the frequency of collisions between NO and O_3 molecules. Figure 18.10a shows a single NO molecule moving randomly among sixteen O_3 molecules. In a given time period, it might collide with two O_3 molecules. The number of

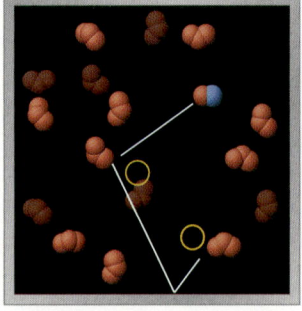 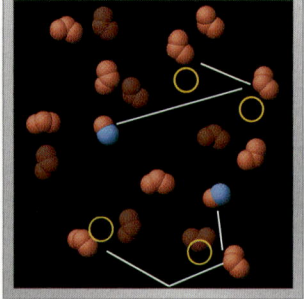 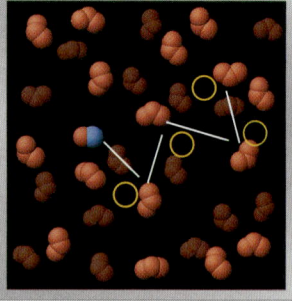

(a) 1 NO, 16 O₃: 2 hits s⁻¹   **(b) 2 NO, 16 O₃:** 4 hits s⁻¹   **(c) 1 NO, 32 O₃:** 4 hits s⁻¹

FIGURE 18.10 The effect of concentration on the frequency of molecular collisions. (a) A single NO molecule, moving among sixteen O_3 molecules, is shown colliding with two of them per second. (b) If two NO molecules move among sixteen O_3 molecules, we would predict that four NO–O_3 collisions would occur per second. (c) If the number of O_3 molecules is doubled (to 32), the frequency of NO–O_3 collisions is also doubled, to four per second. In fact, in real conditions at ambient pressures, each gas molecule will make millions of collisions per second.

collisions between NO and O_3 molecules in the same time period will double, however, if the number of NO molecules is doubled (to 2, as shown in Figure 18.10b) or if the number of O_3 molecules is doubled (to 32, as in Figure 18.10c). Since the frequency of collisions between the two reactant molecules is directly proportional to the concentration of each reactant, the rate of reaction is also proportional to the concentration of each reactant.

We will see [>>Section 18.7] that the reasoning used here necessarily applies only to the rate-determining step of reactions.

Since the frequency of collisions between reactant molecules is proportional to the concentration of each reactant, collision theory suggests that the rate of reaction may be dependent on the concentration of each reactant.

Temperature, Reaction Rate, and Activation Energy

In a laboratory or in chemical industry, a chemical reaction is often carried out at a high temperature so that the reaction occurs more quickly. Conversely, it is sometimes desirable to lower the temperature to slow down a chemical reaction (to avoid an uncontrollable reaction or a potentially dangerous explosion, or just to slow down the rate of deterioration of foods). But how and why does temperature influence reaction rate?

A discussion of the effect of temperature on reaction rate goes back to the distribution of energies for molecules in a sample of a gas or liquid. Recall that the molecules in a sample of a gas or liquid have a range of energies, described as a Maxwell Boltzmann distribution of energies [<<Section 11.12, Figure 11.8). That is, at any instant in a sample of a gas or liquid, there is a range of energies of the molecules. As the temperature is increased, the average energy of the molecules in the sample increases, as does the fraction having higher energies (Figure 18.11).

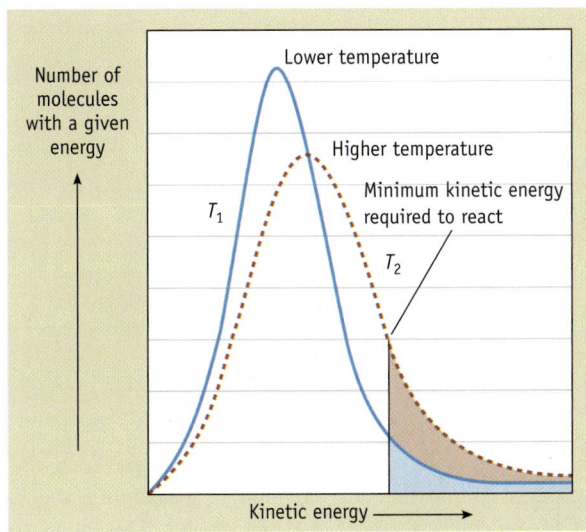

FIGURE 18.11 Distribution of kinetic energies of molecules. The vertical axis gives the relative number of molecules possessing the energy indicated on the horizontal axis, at two temperatures. The higher the temperature, the more the distribution is displaced to higher kinetic energies. The graph indicates, for an unspecified reaction, the minimum kinetic energy that the molecules must have if collisions are to result in re-arrangement to form product molecules. The higher the temperature, the larger the fraction of the molecules that have this minimum threshold of kinetic energy.

Chemists postulate that collisions can be "productive" (that is, reaction results) only if the colliding molecules have at least a certain level of kinetic energy, dependent on the reaction. In addition, the collisions must take place with certain orientations of the reacting molecules (see below). Otherwise, the colliding molecules will "bounce" off each other, and continue on their way.

The minimum combined kinetic energy that a pair of colliding particles must have in excess of the average for their collision to result in reaction is called the **activation energy** (**E_a**). For reactions with low activation energy, a large proportion of collisions may be sufficiently energetic to result in formation of product molecules. These reactions are fast—the number of product molecules form per unit time is large. In cases where the activation energy is high, only a small fraction of collisions may be productive, and formation of a given number of product molecules will take longer.

An analogy is that if two eggs are brought into contact at low energies, the eggs retain their identity. If, however, contact is made at speeds beyond a threshold level, the eggs may be "re-arranged."

As an illustration of the meaning of activation energy, consider the reaction between $NO_2(g)$ and $CO(g)$ to form $NO(g)$ and $CO_2(g)$:

$$NO_2(g) + CO(g) \rightleftharpoons NO(g) + CO_2(g)$$

At the molecular level, we imagine that in the reaction mixture there are an enormous number of collisions between NO_2 molecules and CO molecules happening per second, and in each productive collision there is transfer of an O atom from an NO_2 molecule to a CO molecule. Over time, there is an increase in the amounts (or number of molecules) of $NO(g)$ and $CO_2(g)$.

Let's consider a collision between one NO_2 molecule and one CO molecule as a system. The energy of this system changes as interaction and re-arrangement occur. We show this by using a **reaction energy diagram** (Figure 18.12). The horizontal axis (labelled "reaction progress") represents the passage of time during the re-arrangement of atoms between the colliding molecules. The vertical axis represents the changing potential energy of the tiny system as the re-arrangement progresses.

FIGURE 18.12 A reaction energy diagram portraying activation energy. The potential energy of a system comprising one NO_2 molecule and one CO molecule undergoing a collision leading to reaction changes during the course of the reaction event. The reaction $NO_2(g) + CO(g) \longrightarrow NO(g) + CO_2(g)$ has an activation energy of 132 kJ mol^{-1}. The activation energy of the reverse reaction $(NO(g) + CO_2(g) \longrightarrow NO_2(g) + CO(g))$ is 358 kJ mol^{-1}. The reaction between $NO_2(g)$ and $CO(g)$ is exothermic, with $\Delta_r H = 226$ kJ mol^{-1}.

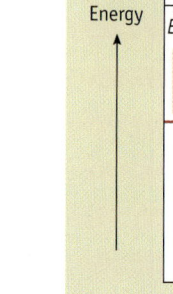

Think about It

e18.9 Manipulate a reaction energy diagram to see the relationships among variables.

As an N—O bond in the NO_2 molecule stretches and weakens, the potential energy of this system increases. While the N—O bond is weakening, the O atom begins to interact with the CO molecule, forming a C—O bond. This bond formation releases energy and the potential energy of the system decreases. At some point during the transfer of the O atom from the NO_2 molecule to the CO molecule, *the energy of the system reaches a maximum in an unstable arrangement of atoms* called the **transition state**.

In the unstable transition state, with the O atom "shared" between the two colliding particles, there is an equal probability that the O atom transfer will be completed with formation of a NO molecule and a CO_2 molecule (a productive collision), or that the O atom will return to where it came from, leaving an NO_2 molecule and a CO molecule to move away from each other (an unproductive collision).

The potential energy of the transition state exceeds the combined average potential energies of the two separated reactant molecules by $132/N_A$ kJ, where N_A is the Avogadro constant (6.02×10^{23}). This is an extremely small amount of energy, and it is easier for us to think in terms of 1 mol of the collisional events just described, in which case we say that the activation energy is 132 kJ mol^{-1}.

Although for purposes of considering convenient amounts of energy, we consider 1 mol of collisions, it is important to appreciate that reaction does not occur by 1 mol of simultaneous productive collisions. Rather, at any instant during the course of a reaction, some productive collisions will have already occurred, some may be in various phases of occurrence, and some reactant molecules will not yet have met their counterparts in a productive reaction.

The potential energy of the transition state exceeds the potential energy of separated product molecules (NO and CO_2) by $358/N_A$ kJ (358 kJ mol^{-1}). If, over the course of time, 1 mol of products are formed, there will be a release of 226 kJ (= 358 kJ − 132 kJ) of heat: the reaction is exothermic with enthalpy change of reaction $\Delta_r H = -226$ kJ. The sign and magnitude of $\Delta_r H$ tells us nothing about the magnitude of the activation energy of reaction.

A discussion of the way that chemists interpret the effect of increase of temperature on rate of reaction depends on two important ideas:

1. Reaction energy diagrams like Figure 18.12 portray the changing potential energy of the collection of species taking part in one productive collisional event. In a reaction mixture leading to formation of 0.1 mol of product, about 6×10^{22} such collisional events happen over the duration of the reaction—not all at the same time. While a reaction energy diagram refers to energy changes during one collisional event, energy distribution diagrams (Figure 18.11) show the range of energies of all of the molecules in a sample. These two vastly different scales need to be reconciled to interpret the effect of temperature on reaction rates.

2. *Rate of reaction* does not refer to how fast a collisional event like that portrayed in the reaction energy diagram of Figure 18.12 happens. That event only leads to the formation of one molecule or ion of product. Since rate of reaction can be measured in terms of the amount of product formed per unit of time, it is related to the frequency of productive collisions—that is, the number of productive collisions taking place in the reaction mixture per unit of time.

Molecular Modelling (Odyssey)

e18.10 Simulate a reaction between gases that illustrates these points.

The conversion of $NO_2(g)$ and $CO(g)$ to products at room temperature is slow because in only a small fraction of the collisions do the combined kinetic energies of the reactant molecules exceed the average by the activation energy (132 kJ mol^{-1})—so relatively few molecules of NO and CO_2 are formed per second. The reaction can be made to go faster (producing more product molecules per unit of time) by increasing the temperature of the reaction mixture. The higher the temperature, the higher the average kinetic energy of the molecules is [<<Section 11.7] and so the greater the frequency of productive collisions. This is partly because at higher speeds and at higher temperature, there will be more collisions per unit of time, but more significantly because a greater fraction of the collisions will be such that the reactant molecules have at least as much energy as the minimum threshold energy for the collisions to be productive (Figure 18.11).

> The potential energy of a pair of colliding molecules increases as they interact, reaching a maximum with the formation of an unstable arrangement called the transition state. The minimum combined kinetic energy that a pair of colliding particles must have in excess of the average for their collision to result in reaction is the activation energy (E_a). At any temperature, molecules of a substance have a range of kinetic energies, and the higher the temperature, the greater the fraction of colliding molecules that will have energy equal to or greater than the activation energy.

Orientation of Colliding Molecules

That reactant molecules must have a sufficiently high kinetic energy is a necessary but not sufficient condition for a productive collision. Not only must NO_2 and CO molecules collide with sufficient energy, but they must also collide in a particular orientation with respect to each other. In the case of the reaction between NO_2 and CO, we can imagine that the transition state structure has one of the O atoms of NO_2 molecule forming a partial bond to the C atom of the CO molecule (Figure 18.12). Orientation is an important factor determining the rate of reaction and affects the value of the rate constant (k). This situation is more likely to result when collision between NO_2 and CO molecules happens with one of the O atoms of the NO_2 molecule oriented toward the C atom of the CO molecule (Figure 18.13a). Collisions with other orientations, such as those portrayed in Figure 18.13b and c, have a lower probability of formation of products. The less probable are collisions of the type shown in Figure 18.13a, the smaller value of k, and the slower the reaction.

FIGURE 18.13 Importance of orientation of colliding molecules. Reaction between a NO_2 molecule and a CO molecule is more likely if they collide with orientation as in (a) than if they have either of the orientations (b) or (c). In the latter cases, electron shift from between N and O atoms in one molecule to between C and O atoms in the other is more difficult.

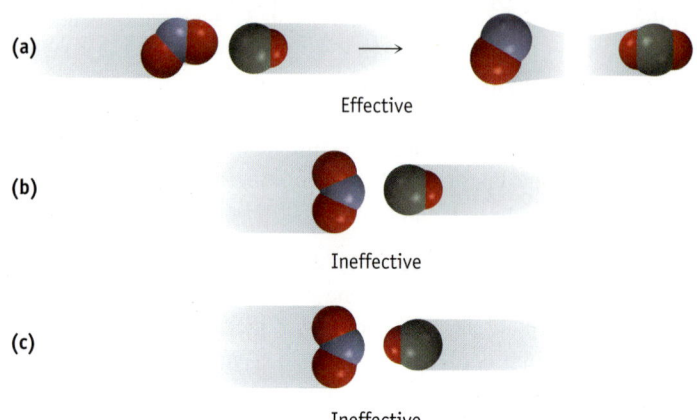

(a)

Effective

(b)

Ineffective

(c)

Ineffective

Molecular Modelling (Odyssey)

e18.11 Simulate the outcomes of different collision geometries in a reaction.

Imagine what happens during reactions of molecules with many more atoms than those of NO_2 and CO. Perhaps in only a small fraction of the collisions will their orientation at the instant of collision lead to re-arrangement into a transition state. No wonder some reactions are so slow. Conversely, it is amazing that so many are so fast!

For productive collisions, colliding molecules with sufficient combined energy must have a particular orientation with respect to each other.

The Arrhenius Equation

The observation that reaction rates depend on the energy and frequency of collisions between reacting molecules, on the temperature, and on whether the collisions have the correct geometry is summarized by the **Arrhenius equation**:

$$k = \text{rate constant} = A e^{-E_a/RT}$$

Frequency factor

Fraction of molecules with minimum energy for reaction

where R is the gas constant with value of 8.3145 J K^{-1} mol^{-1}. The parameter A is called the *frequency factor* (units L mol^{-1} s^{-1}), which combines the frequency of collisions and the probability that collisions have an appropriate orientation of particles. The value of A is specific to each reaction and is temperature-dependent. The factor $e^{-E_a/RT}$ is interpreted as the *fraction of molecules having the minimum energy required for reaction*; its value is always less than 1. This fraction changes significantly with temperature. Values of this fraction for reactions with a typical value of 40 kJ mol^{-1} activation energy are as follows:

Temperature (K)	$e^{-E_a/RT}$
298	9.7×10^{-8}
400	5.9×10^{-6}
600	3.3×10^{-4}

The Arrhenius equation is valuable because it can be used (a) to calculate the value of the activation energy from the temperature dependence of the rate constant, and (b) to calculate the rate constant for a given temperature if the activation energy and *A* are known. Taking the natural logarithm of each side of the Arrhenius equation gives us

$$\ln k = \ln A + \left(\frac{-E_a}{RT}\right)$$

This can be re-arranged slightly to give an equation for a straight line relationship between ln *k* and 1/*T*:

$$\ln k = -\frac{E_a}{R}\left(\frac{1}{T}\right) + \ln A$$

If ln *k* is plotted vs. 1/*T*, the result is a straight line with a slope of $(-E_a/R)$. Now we have a way to calculate E_a from experimental values of *k* at several temperatures, as illustrated in Worked Example 18.10.

This equation shows us that the degree of dependence of *k* on temperature (more strictly, of ln *k* on 1/*T*) depends on the value of E_a. The value of E_a of many reactions is such that an increase of 10 °C near ambient temperatures brings about a doubling of the reaction rate.

Think about It

e18.12 Simulate changes in variables in the Arrhenius equation.

WORKED EXAMPLE 18.10—DETERMINATION OF E_a

Using the experimental data of rate constants at various temperatures shown in the table, calculate the activation energy (E_a) for the reaction

$$2\,N_2O(g) \longrightarrow 2\,N_2(g) + O_2(g)$$

Think about It

e18.13 Plot the appropriate graph to determine the activation energy for a reaction.

Experiment	T (K)	k (L mol^{-1} s^{-1})
1	1125	11.59
2	1053	1.67
3	1001	0.380
4	838	0.0011

Strategy

Calculate ln *k* and 1/*T* for each data point, and then plot ln *k* vs. 1/*T*. E_a is calculated from the resulting straight line (slope $= -E_a/R$).

Solution

The data are transformed into 1/*T* and ln *k*, and plotting these data gives the graph shown in Figure 18.14.

Experiment	T (K)	1/T (K^{-1})	k (L mol^{-1} s^{-1})	ln k
1	1125	8.889 × 10^{-4}	11.59	2.4501
2	1053	9.497 × 10^{-4}	1.67	0.513
3	1001	9.990 × 10^{-4}	0.380	-0.968
4	838	11.9 × 10^{-4}	0.0011	-6.81

The slope can be obtained from the equation for the trendline through the data in a spreadsheet. Manually, we can estimate the slope between the large blue points on the line:

$$\text{Slope} = \frac{\Delta(\ln k)}{\Delta(1/T)} = \frac{2.0 - (-5.6)}{(9.0 - 11.5)(10^{-4})K^{-1}} = -3.0 \times 10^4\ K$$

The activation energy is evaluated from

$$\text{Slope} = -3.0 \times 10^4\ K = -\frac{E_a}{R} = -\frac{E_a}{8.31\ J\ K^{-1}\ mol^{-1}}$$

$$E_a = (-3.0 \times 10^4\ K)(8.31\ J\ K^{-1}\ mol^{-1}) = 250\ kJ\ mol^{-1}$$

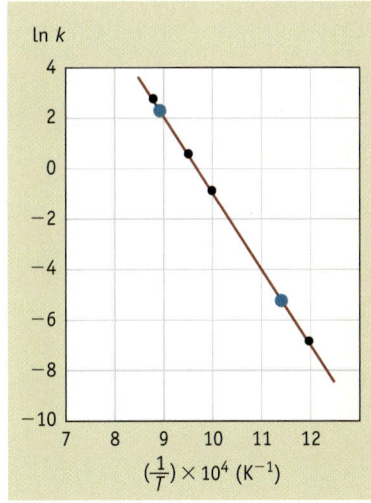

FIGURE 18.14 Arrhenius equation plot. A plot of ln *k* versus 1/*T* for the reaction $2\,N_2O(g) \longrightarrow 2\,N_2(g) + O_2(g)$. The slope of the line gives E_a.

As well as the graphical method for evaluating E_a used in Worked Example 18.10, the value of E_a can be obtained algebraically, though less accurately. Knowing k at two different temperatures, we can write an equation for each of these conditions:

$$\ln k_1 = -\frac{E_a}{RT_1} + \ln A \qquad \text{and} \qquad \ln k_2 = -\frac{E_a}{RT_2} + \ln A$$

If one of these equations is subtracted from the other, we have

$$\ln k_2 - \ln k_1 = \ln \frac{k_2}{k_1} = -\frac{E_a}{R}\left(\frac{1}{T_2} - \frac{1}{T_1}\right)$$

Worked Example 18.11 and e18.14 demonstrate the use of this equation.

Sometimes it is useful to use the relationship

$$\left(\frac{1}{T_2} - \frac{1}{T_1}\right) = \frac{T_1 - T_2}{T_1 T_2}$$

Think about It

e18.14 Use an algebraic method to determine the activation energy for a reaction.

Interactive Exercise 18.16

Calculate E_a from the temperature dependence of k.

Think about It

e18.15 Using spreadsheet simulations, explore how reaction rate varies with temperature.

WORKED EXAMPLE 18.11—CALCULATING E_a FROM THE TEMPERATURE DEPENDENCE OF K

Using values of k determined at two different temperatures, calculate the value of E_a for the decomposition of HI:

$$2\ HI(g) \longrightarrow H_2(g) + I_2(g)$$

At 650 K, $k = 2.15 \times 10^{-8}$ L mol^{-1} s^{-1} At 700 K, $k = 2.39 \times 10^{-7}$ L mol^{-1} s^{-1}

Strategy

Here, the values of k_1, T_1, k_2, and T_2 are given, so we can use the algebraic method, with E_a the only unknown.

Solution

$$\ln \frac{2.39 \times 10^{-7}\ \text{L mol}^{-1}\ \text{s}^{-1}}{2.15 \times 10^{-8}\ \text{L mol}^{-1}\ \text{s}^{-1}} = \frac{-E_a}{8.314\ \text{J K}^{-1}\ \text{mol}^{-1}}\left(\frac{1}{700\ \text{K}} - \frac{1}{650\ \text{K}}\right)$$

Solving this equation for E_a, we find $E_a = 182$ kJ mol^{-1}.

EXERCISE 18.17—CALCULATING E_a FROM THE TEMPERATURE DEPENDENCE OF K

The colourless gas N_2O_4 decomposes to the brown gas NO_2 in a first-order reaction:

$$N_2O_4(g) \longrightarrow 2\ NO_2(g)$$

The rate constant is 4.5×10^3 s^{-1} at 274 K and 1.00×10^4 s^{-1} at 283 K. What is the activation energy (E_a) for this reaction?

The effect of temperature on the rate constant of a reaction is given by the Arrhenius equation

$$k = Ae^{-E_a/RT}$$

The activation energy of reaction can be calculated from the straight line plot of $\ln k$ vs. $1/T$, for which slope = $-E_a/R$. A rate constant of a reaction at one temperature can be calculated algebraically from that at another temperature using the relationship

$$\ln k_2 - \ln k_1 = \ln \frac{k_2}{k_1} = -\frac{E_a}{R}\left(\frac{1}{T_2} - \frac{1}{T_1}\right)$$

Effect of Catalysts on Reaction Rate

Catalysts are substances, specific to each reaction, that speed up the rate of formation of products in a chemical reaction. We have seen examples of catalysts in earlier discussions in this chapter. In biological systems, catalysts called *enzymes* influence the rates of most reactions [>>Section 18.9]. For example, galactosidase catalyzes the hydrolysis of oligosaccharides in our GI tract, as described in Section 18.1.

Catalase is an enzyme that speeds up the decomposition of hydrogen peroxide, ensuring that this highly toxic compound does not accumulate in our bodies. The decomposition of hydrogen peroxide in aqueous solution is catalyzed by MnO_2 and certain enzymes (Figure 18.4), and by hydroxide ions (Worked Example 18.6).

The function of catalysts is to provide an alternative pathway of reaction with a lower activation energy. Although catalyst molecules and ions are involved in the mechanism of reactions that they speed up, they are regenerated over and over again, so that they can act on other (as yet unreacted) reactant molecules. In this way, very often only small amount of catalyst can bring about rapid reaction of much larger amounts of substances.

To illustrate the participation of the catalyst in **homogeneous catalysis**, in which *all reactants and the catalyst are in the same phase*, consider the first-order interconversion of *cis*-but-2-ene to the slightly more stable isomer, *trans*-but-2-ene.

cis-but-2-ene   **Transition state**   **trans-but-2-ene**

End rotates

π bond breaks

Think about It

e18.16 Examine the reaction energy diagram for this reaction.

The activation energy for the uncatalyzed conversion is relatively large (264 kJ mol^{-1}) because the C=C bond must be broken to allow one end of the molecule to rotate with respect to the other. As a result, the frequency of molecules undergoing this reaction is very low (the reaction is slow), and high temperatures are required for a reasonable rate of reaction.

The conversion of *cis*-but-2-ene to *trans*-but-2-ene is greatly accelerated by iodine, which acts a catalyst for this reaction. The presence of iodine allows the isomerization reaction to occur at a temperature several hundred degrees lower than the uncatalyzed reaction. Iodine is not "consumed," and it does not appear in the overall balanced chemical equation. It does, however, appear in the experimentally derived rate equation:

$$\text{Rate} = k[\text{cis-}C_4H_8][I_2]^{1/2}$$

The rate of the *cis–trans* conversion increases because the presence of iodine provides an alternative, low-activation energy pathway for the reaction. That is, it changes the *mechanism* of the reaction, described in more detail in Section 18.7. Chemists who have studied this reaction hypothesize that the reaction occurs in a sequence of steps, shown in Figure 18.15. Iodine molecules first dissociate to form iodine atoms (Step 1). During some of the collisions between I atoms and *cis*-but-2-ene molecules, there is rapid electron redistribution from the C=C double bond of the butene into the C—I region of space (Step 2). This converts the double bond between the carbon atoms to a single bond (the π bond is broken) and allows the ends of the molecule to rotate relative to each other (Step 3). The C—I bond then breaks, and the double bond can re-form—sometimes with the butene molecule in the *trans* configuration (Step 4).

Iodine atoms released in Step 4 are now free to interact with other molecules of *cis*-but-2-ene and catalyze their isomerization. In this way, molecule after molecule of *cis*-but-2-ene is converted to the *trans* isomer with the assistance of the same I atom, and there are many I atoms participating in this "catalytic cycle." The catalytic cycle of any I atom is

Think about It

e18.17 Watch a video of a reaction and identify the catalyst.

FIGURE 18.15 The mechanism of the iodine-catalyzed isomerization of *cis*-but-2-ene. *Cis*-but-2-ene is converted to *trans*-but-2-ene in the presence of a small amount of iodine. Catalyzed reactions are often portrayed in this way to emphasize what chemists refer to as a "catalytic cycle."

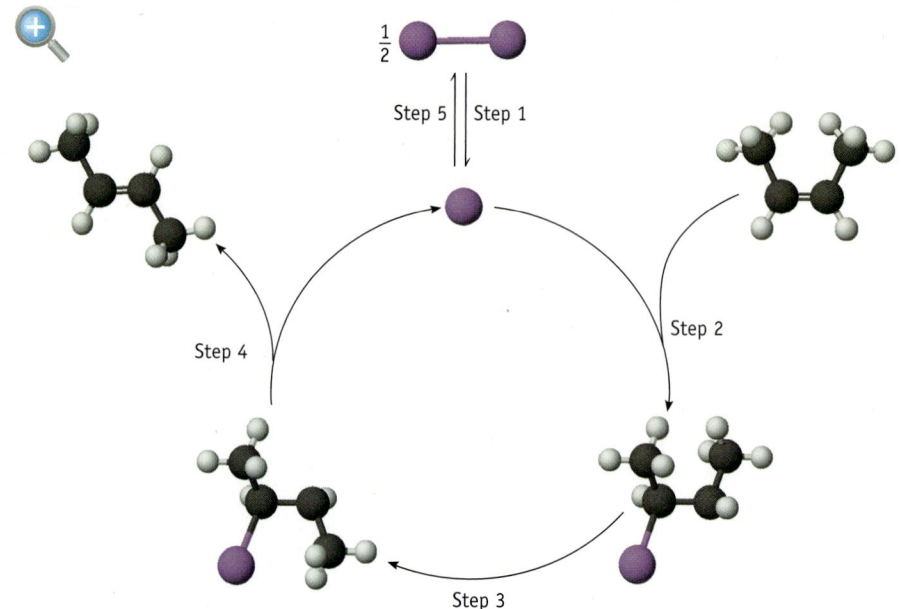

Representations of "reaction mechanisms" (how a reaction happens at the molecular level) are different from balanced equations for the overall reaction. They zoom in on just one of the very many collision events that lead to the formation of products each second, and ignore 'unproductive' collisions. The zoom icon 🔍 is intended to remind you that this is what we might observe if we had the means to see at the molecular level (we don't!). The zoom icon is used below, especially in Section 18.7, as well as in Section 19.8 and beyond.

broken if it collides with another I atom and electron redistribution leads to regeneration, perhaps temporarily, of molecular iodine, I_2 (Step 5). In the end, there is as much I_2 present as at the start of the reaction.

A reaction energy diagram for the catalyzed reaction (Figure 18.16) shows an overall activation energy barrier much lower than that in the uncatalyzed reaction. In addition, the energy diagram for the catalyzed reaction has five stages, each representing one of the steps in the reaction mechanism. This proposed mechanism includes a series of *reaction intermediates*: relatively high energy species formed in one step of the reaction and consumed in a later step. These include iodine atoms and the free radical species formed when an iodine atom adds to a *cis*-but-2-ene molecule.

FIGURE 18.16 Reaction energy diagram for the iodine-catalyzed isomerization of *cis*-but-2-ene. The energy diagram for the iodine-catalyzed reaction is represented by the red curve, and that for the uncatalyzed reaction is shown by the black curve. A catalyst accelerates a reaction by providing a reaction pathway (mechanism) with a lower activation energy. A larger fraction of reactant molecules have sufficient energy for collisions to be productive, so more molecules react per unit of time.

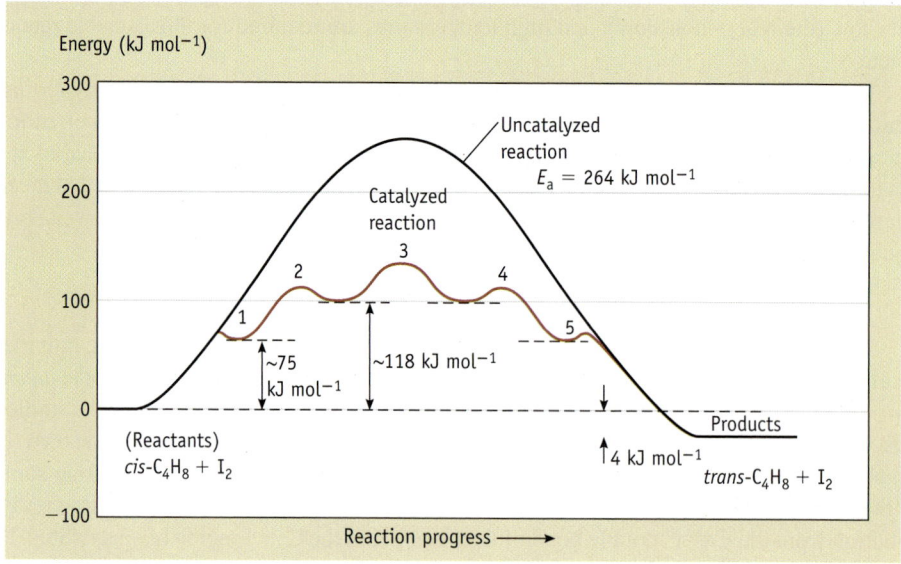

In **heterogeneous catalysis,** *the catalyst is in a different phase from the reactants.* An example is the catalytic hydrogenation of ethylene to ethane [>>Section 19.11].

$$C_2H_4(g) + H_2(g) \xrightarrow{\text{Pt(s)}} C_2H_6(g)$$

This reaction is very slow in the gas phase at room temperature, but is accelerated by the surface of platinum metal acting catalytically (Figure 18.17).

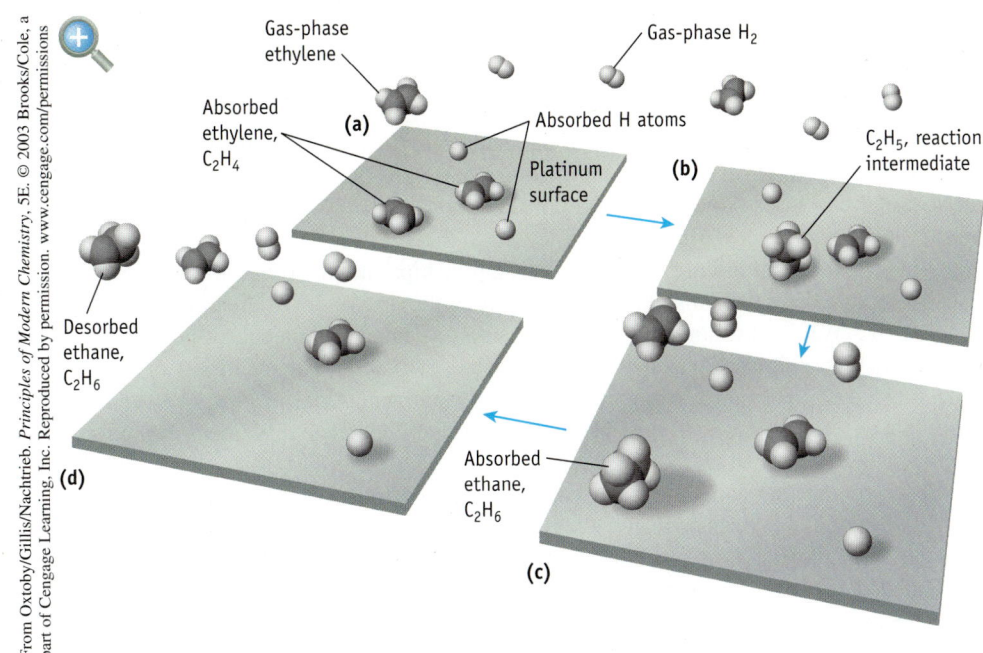

FIGURE 18.17 Heterogeneous catalysis of ethylene hydrogenation. (a) Some H_2 molecules that collide with the platinum surface form Pt-H bonds, weakening and breaking the H—H bonds, and leaving H atoms bound to Pt atoms on the surface. C_2H_4 molecules also attach to the surface, through bonds from each C atom to a Pt atom, with breaking of the π bond of the C=C bond in each molecule. (b) Reaction with adjacent H atoms first leads to formation of C_2H_5, and then (c) to formation of ethane, C_2H_6. (d) As new C—H bonds are formed, the Pt—C bonds are weakened and the C_2H_6 molecules leave the surface. In their place, other C_2H_4 and H_2 molecules can bond to the surface and undergo reaction.

We have seen [<<Section 17.1] in the context of formation of photochemical smog, that what happens in the "reaction mixture" called the atmosphere is not necessarily governed by the formation of the most stable products. Sometimes, the product that is formed is a consequence of relative rates of reaction rather than relative stabilities. This is also true of biological systems: if the substances comprising our own bodies quickly reached equilibrium with the air around them, we would be transformed mostly into carbon dioxide and water. The same is true of tree trunks and other vegetation.

Similarly, what happens in industry may be governed by kinetics, rather than by thermodynamics. For example, synthesis gas or syngas [<<Section 4.2], comprising a mixture of carbon monoxide and hydrogen, can be produced by reaction with steam and hydrocarbon materials such as methane (natural gas) or coke. Depending on the circumstances, including the presence of particular catalytic materials at high temperatures and pressures, the carbon monoxide and hydrogen in syngas can react in a number of ways, each forming a different product. For example, methanol, methane, and a liquid fuel of composition approximating C_8H_{18} are all possible products, and may even be formed simultaneously.

$$CO(g) + 2\,H_2(g) \longrightarrow CH_3OH(\ell)$$

$$CO(g) + 3\,H_2(g) \longrightarrow CH_4(g) + H_2O(\ell)$$

$$8\,CO(g) + 17\,H_2(g) \longrightarrow C_8H_{18}(g) + 8\,H_2O(\ell)$$

A copper/zinc catalyst specifically accelerates the rate of formation of methanol so that this important substance is the dominant product. By using this catalyst, methanol can be made from syngas, which has been derived from either coke or natural gas, depending on which is the most abundant, or least expensive, at the site of the plant. In places with abundant supplies of coal, but little natural gas to use as a fuel, a nickel catalyst is used with syngas produced from coke, and the production of methane is specifically accelerated. Similarly, while many countries import all or most of their petroleum, some countries with large coal deposits maintain their independence from crude oil suppliers by reaction of the syngas components from coke in the presence of silicon-based catalysts called zeolites. The dominant product is then a synthetic liquid fuel. World leaders in this technology, called the Sasol process, are South Africa and New Zealand.

Catalysts are substances that accelerate reactions. They are specific to the reaction accelerated. Catalyst molecules, atoms, or ions participate in the reaction, providing a mechanistic pathway with lower activation energy than that of the uncatalyzed reaction. Catalyst molecules, atoms, or ions are regenerated, becoming available to react again.

18.7 Reaction Mechanisms

One of the most important reasons to study reaction rates is that rate equations give us insights into **reaction mechanisms**, the sequence of bond-making and bond-breaking steps that occurs during the conversion of molecules of reactants to molecules of products. For example, the above discussion of how iodine catalyzes the isomerization of *cis*-but-2-ene to *trans*-but-2-ene was based on a consideration of the reaction mechanism in the absence and presence of iodine.

While rate equations are quantitative relationships about observable change, the study of reaction mechanisms places us squarely within the realm of the molecular level of chemistry. We want to analyze the changes that atoms and molecules undergo when they react. One primary reason to do this is to see patterns in reactivity of similar molecules, such as organic molecules with the same functional groups. We then can relate this description back to the observable world and experimental measurements of reaction rates.

Based on the rate equation for a reaction and the application of chemical intuition, chemists can often make an educated guess about the mechanism for the reaction. In some reactions, the conversion of reactant molecules to product molecules in a single step is envisioned. For example, chemists propose that nitrogen dioxide and carbon monoxide molecules react in a single-step reaction at each productive collision between them (Figure 18.12). The uncatalyzed isomerization of *cis*-but-2-ene to *trans*-but-2-ene is also believed to be a single-step reaction (Figure 18.16).

Most chemical reactions, however, occur in a sequence of steps. We saw an example with the five steps proposed for the iodine-catalyzed but-2-ene isomerization reaction. Another example of a reaction believed to occur in several steps is the reaction of $Br_2(g)$ and $NO(g)$ in the gas phase. The balanced chemical equation for this reaction indicates the overall result of the steps, but not the mechanism by which it happens:

$$Br_2(g) + 2\ NO(g) \longrightarrow 2\ BrNO(g)$$

A single-step reaction would require that three reactant molecules collide simultaneously with sufficient energy, and in just the right orientation, to be productive. Such an event has a low probability of occurring. It is reasonable to look for a mechanism that occurs in a series of steps, with each step involving only one or two molecules. For example, in one proposed mechanism (Figure 18.18), Br_2 molecules and NO molecules combine to produce molecules of an intermediate species, Br_2NO. Each of these then reacts with another NO molecule to form two molecules of the reaction product. The equation for the overall reaction is obtained by adding the equations for these two steps:

Step 1: $Br_2 + NO \rightleftharpoons Br_2NO$

Step 2: $\underline{Br_2NO + NO \longrightarrow 2\ BrNO}$

Overall reaction: $Br_2 + 2\ NO \longrightarrow 2\ BrNO$

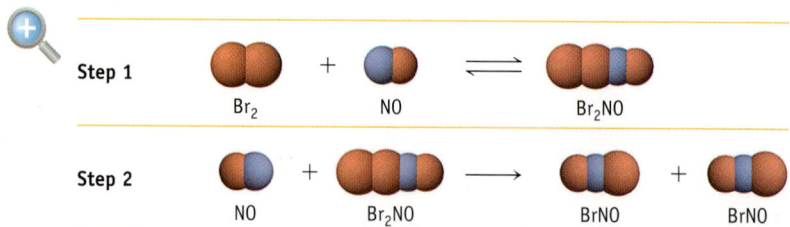

FIGURE 18.18 A reaction mechanism. A representation of the proposed two-step mechanism by which each molecule of Br_2 and a molecule of NO are converted to two molecules of NOBr. This is just one of a multitude of such two-step events that happen over the course of the reaction, each productive collision occurring when reactant molecules (of each step) collide with sufficient energy and appropriate orientation.

A reaction mechanism is a sequence of bond-making and bond-breaking steps that occurs during the conversion of molecules of reactant to molecules of products. Chemists can never prove what the mechanism is, but experimental evidence can help them to propose mechanisms that are consistent with the evidence.

Elementary Steps

Each step in a proposed multi-step reaction mechanism is called an **elementary step**: a single molecular event such as the formation or rupture of a chemical bond or the displacement of atoms as a result of a molecular collision. They can be thought of as the simplest possible events at the molecular level. For example, each of step 1 and step 2 in the reaction mechanism proposed for the reaction between Br_2 and NO in the gas phase is an elementary step.

Unlike the balanced overall equations for chemical reactions, the equation for an elementary step portrays productive collisional events: it specifies which particles collide with each other and which ones are formed if the collision is productive. Each step has its own activation energy barrier (E_a) and rate constant (k). The steps must add up to give the balanced equation for the overall reaction. A series of steps that satisfactorily explains the kinetic properties of a chemical reaction and is consistent with other evidence about the reaction constitutes a possible reaction mechanism.

While chemists can never prove what the mechanism is, they can use experimental evidence, along with models of structure and reactivity, to make educated guesses about how reactions proceed at the molecular level. To see how this is done, we first describe three types of elementary steps in terms of the concept of *molecularity*.

Elementary steps are the individual molecular events that comprise a reaction mechanism. Unlike overall chemical equations, the equation for an elementary step describes particular collisional events—that is, which particles collide, which bonds are broken or formed, and which particles are formed.

Molecularity of Elementary Steps

Elementary steps in a mechanism are classified by their **molecularity**: *the number of reactant molecules (or ions, atoms, or free radicals) that collide and undergo change.* When one molecule is the only reactant in an elementary step, the reaction is a *unimolecular* process. A *bimolecular* elementary process involves two molecules, which may be identical ($A + A \longrightarrow$ products) or different ($A + B \longrightarrow$ products). For example, in the mechanism proposed for the decomposition of ozone in the stratosphere [>>Section 21.2] a unimolecular step is followed by a bimolecular step:

Step 1: Unimolecular $O_3 \longrightarrow O_2 + O$

Step 2: Bimolecular $\underline{O_3 + O \longrightarrow 2\,O_2}$

Overall reaction: $2\,O_3 \longrightarrow 3\,O_2$

A *termolecular* elementary step involves three molecules. It could involve three molecules of the same or a different type ($3\,A \longrightarrow$ products; $2\,A + B \longrightarrow$ products; or $A + B + C \longrightarrow$ products). As you might suspect, the simultaneous collision of three molecules is not likely, unless one of the molecules involved is in high concentration, such as a solvent molecule. In fact, most termolecular processes involve the collision of two reactant molecules and a third, inert molecule. The function of the inert molecule is to absorb the excess energy produced when a new chemical bond is formed by the first two molecules. For example, while a N_2 molecule is a necessary participant in a termolecular elementary step between oxygen molecules and oxygen atoms that produces ozone in the stratosphere, it remains unreacted:

$$O + O_2 + N_2 \longrightarrow O_3 + N_2\,(\text{excited})$$

The molecularity of an elementary step is the number of reactant molecules, ions, or atoms that collide and undergo change. A unimolecular step involves reaction of only one particle, a bimolecular step involves two, and a termolecular step involves three.

Rate Equations for Elementary Steps

Since the stoichiometric equation of any proposed elementary step defines the particles that participate in the productive collisions (unlike the overall reaction equation), the rate equation of an elementary step is derivable directly from its stoichiometric equation. Examples are given in the following table.

Elementary Step	Molecularity	Rate Equation
A → product	Unimolecular	Rate = $k[A]$
A + B → product	Bimolecular	Rate = $k[A][B]$
A + A → product	Bimolecular	Rate = $k[A]^2$
2 A + B → product	Termolecular	Rate = $k[A]^2[B]$

For example, the rate equations for each of the two steps in the decomposition of ozone are

$$\text{Rate for unimolecular step } 1 = k[O_3]$$
$$\text{Rate for bimolecular step } 2 = k'[O_3][O]$$

When a reaction mechanism consists of two elementary steps, the two steps will probably occur at different rates. The two rate constants *(k* and *k'* in this example) are not expected to have the same value (nor the same units, because the two steps have different molecularities).

Although the rate equation for an elementary step can be written directly from its equation, the order of a chemical reaction cannot be deduced from the overall balanced equation, because the equation does not give us any information about the mechanism of the reaction. The mechanism can only be deduced experimentally (by seeing how the rate of reaction depends on the concentrations of each reactant).

Conversely, if you find experimentally that a reaction is first order, you cannot conclude that it occurs in a single, unimolecular elementary step. Similarly, a second-order rate equation does not imply that the reaction occurs in a single, bimolecular elementary step. An example illustrating this is the first-order decomposition of $N_2O_5(g)$ in the gas phase:

$$2\,N_2O_5(g) \longrightarrow 4\,NO_2(g) + O_2(g) \qquad \text{Rate} = k[N_2O_5]$$

Although it is a first-order reaction, chemists are certain that the mechanism involves a series of unimolecular and bimolecular steps.

Chemical intuition is required to see how the experimentally observed rate equation for the overall reaction is consistent with a possible mechanism involving a sequence of elementary steps. We will provide only a glimpse of the subject in the next section.

Think about It

e18.18 Propose the most likely mechanism for a reaction consistent with the rate equation.

Interactive Exercise 18.18

Write rate equations for elementary steps.

WORKED EXAMPLE 18.12—ELEMENTARY STEPS OF A REACTION

In aqueous solution, the aquated hypochlorite ion $OCl^-(aq)$ reacts to form aquated chlorate ions, $ClO_3^-(aq)$, and chloride ions, $Cl^-(aq)$:

$$3\,ClO^-(aq) \longrightarrow ClO_3^-(aq) + 2\,Cl^-(aq)$$

This reaction is thought to occur in two elementary steps:

Step 1: $ClO^-(aq) + ClO^-(aq) \longrightarrow ClO_2^-(aq) + Cl^-(aq)$

Step 2: $ClO_2^-(aq) + ClO^-(aq) \longrightarrow ClO_3^-(aq) + Cl^-(aq)$

What is the molecularity of each elementary step? Write the rate equation for each elementary step. Show that the sum of the elementary steps is the equation for the overall reaction.

Strategy

The molecularity is the number of ions or molecules involved in an elementary step. The rate equation for an elementary step can be deduced directly from the equation describing it.

Solution

Because each elementary step involves the collision of two ions, each step is bimolecular. The rate equations are

$$\text{Step 1: Rate} = k[ClO^-]^2$$

$$\text{Step 2: Rate} = k[ClO_2^-][ClO^-]$$

If we add the equations for the two elementary steps, we see that the $ClO_2^-(aq)$ ion is an intermediate, a product of the first step and a reactant in the second step. We can cancel it out, and we are left with the stoichiometric equation for the overall reaction:

$$\text{Step 1:} \quad ClO^-(aq) + ClO^-(aq) \longrightarrow \cancel{ClO_2^-(aq)} + Cl^-(aq)$$
$$\text{Step 2:} \quad \cancel{ClO_2^-(aq)} + ClO^-(aq) \longrightarrow ClO_3^-(aq) + Cl^-(aq)$$
$$\text{Sum of steps:} \quad 3\,ClO^-(aq) \longrightarrow ClO_3^- + 2\,Cl^-(aq)$$

EXERCISE 18.19—ELEMENTARY STEPS OF A REACTION

Nitrogen monoxide is reduced by hydrogen to give nitrogen and water:

$$2\,NO(g) + 2\,H_2(g) \longrightarrow N_2(g) + 2\,H_2O(g)$$

One possible mechanism for this reaction is

$$2\,NO \longrightarrow N_2O_2$$
$$N_2O_2 + H_2 \longrightarrow N_2O + H_2O$$
$$N_2O + H_2 \longrightarrow N_2 + H_2O$$

What is the molecularity of each of the three steps? What is the rate equation for the third step? Show that the sum of these elementary steps gives the net reaction.

Molecularity refers the number of particles that participate in an elementary step. Elementary steps may be unimolecular, bimolecular, or termolecular. Unlike overall reactions, the rate equation of an elementary step is derivable directly from its stoichiometric equation.

Reaction Mechanisms and Rate Equations

Experimental observations of how the rate of a reaction depends on concentrations of reactants can give us a rate equation for the reaction. We can use this experimental data about rate—along with our imagination, intuition, and good "chemical sense"—to make an educated guess about how a reaction occurs at the molecular level. Often several mechanisms can be proposed that correspond with the observed rate equation, although some (or all) of these postulated mechanisms may be incorrect.

A great deal of experience and chemical intuition is required to propose sensible reaction mechanisms. However, given a proposed mechanism for a reaction, you can more readily decide if it is consistent with the experimental rate equation. Of course, you will still not know if that mechanism is *the* mechanism.

One important principle of chemical kinetics is that *products of a reaction can never be produced at a rate faster than the rate of the slowest step.* If one step in a multi-step reaction mechanism is slower than the others, then the rate of the overall reaction is limited by this slowest step, called the **rate-determining step**, or *rate-limiting step*. Sometimes, the overall reaction rate and the rate of the slowest step are nearly the same.

Let's imagine that a reaction takes place with a mechanism consisting of two elementary steps, the first step is of which is slow and the second fast:

$$\text{Elementary step 1:} \qquad A + B \xrightarrow[\text{Slow, } E_a \text{ large}]{k_1} X + M$$

$$\text{Elementary step 2:} \qquad M + A \xrightarrow[\text{Fast, } E_a \text{ small}]{k_2} Y$$

$$\text{Overall reaction:} \qquad \overline{2\,A + B \longrightarrow X + Y}$$

You are already familiar with rate-determining steps. The rate at which a crowd exits a cinema or sporting arena is limited by the rate at which they can pass through the doors or gates. A colloquial term for such a rate-determining step is *bottleneck*.

In the first step, a molecule of A and a molecule of B collide. Only a small fraction of such collisions result in formation of molecules of one of the products (X) and a reaction intermediate (M), so the reaction is slow. Molecules of the intermediate collide with another A molecule, and a large fraction of these collisions are productive to form the second product Y. The products X and Y are formed as the result of two elementary steps. The rate-determining step is the first elementary step. That is, the rate of the overall reaction is the rate of the first step. This elementary step is bimolecular and so has the rate equation

$$\text{Rate} = k_1[A][B]$$

The overall reaction would be expected to follow this same second-order rate equation.

Let's apply these ideas to a real reaction. Experiment shows that the reaction of nitrogen dioxide with fluorine has a second-order rate equation:

$$2\,NO_2(g) + F_2(g) \longrightarrow 2\,FNO_2(g) \qquad \text{Rate} = k[NO_2][F_2]$$

We have some experimental evidence, and now we try to make sense of this evidence. The experimental rate equation immediately rules out the possibility that the reaction occurs in a single step. If the reaction takes place in a single termolecular elementary step, we would expect that the rate of reaction would have second-order dependence on $[NO_2]$. It follows that the mechanism must include at least two steps. We can also conclude from the rate equation that the rate-determining elementary step must involve NO_2 and F_2 in a 1:1 ratio. The simplest possible mechanism that we could propose is the following:

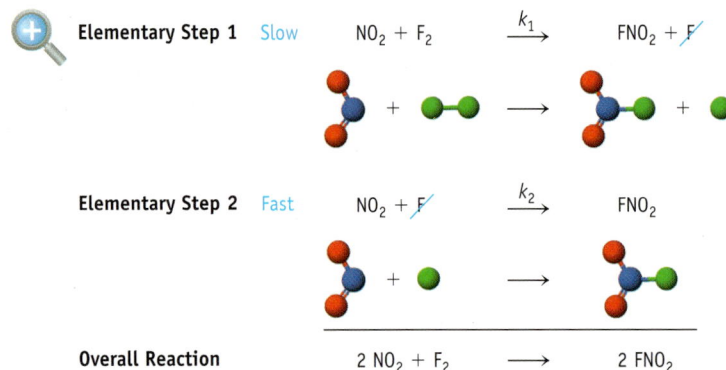

This proposed mechanism suggests that molecules of the product $FNO_2(g)$ are produced by a pathway that has as its first step a reaction between a molecule of $NO_2(g)$ and a molecule of $F_2(g)$ to produce one molecule of the product (FNO_2) plus one F atom. In a second step, the F atom produced in the first step reacts with another NO_2 molecule to form a second molecule of product. If we assume that the first, bimolecular elementary step is rate-determining, the overall rate equation would be expected to be the same as for this first step: rate $= k_1[NO_2][F_2]$. This is the same as the experimentally observed rate equation. This agreement between the rate equation derived from our proposed mechanism and the experimentally deduced rate equation tells us that the proposed mechanism is *perhaps* how the reaction happens.

The F atom formed in the first step of the reaction is a reaction intermediate. It does not appear in the equation describing the overall reaction. Reaction intermediates usually have only a fleeting existence, but occasionally they have long enough lifetimes to be observed. One test of a proposed mechanism is the detection, often by spectroscopic techniques, of an intermediate.

It is important to realize that the reactants of any elementary step of a multi-step reaction do not undergo reaction in unison. In the above case, different F_2 molecules react with NO_2 molecules at various times, and the resultant F atoms quickly react with other NO_2 molecules. As a result, some FNO_2 product molecules are formed before some of the NO_2 molecules have undergone reaction by either the first step or the second. In this way, FNO_2 molecules are formed over the course of time, in a way that follows second-order kinetics. A sense of this idea is seen in Figure 18.19, which is a snapshot of a simulation a of reaction mixture. The snapshot is taken from e18.20.

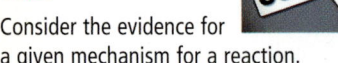

Interactive Exercise 18.20

Consider the evidence for a given mechanism for a reaction.

Think about It

e18.19 Apply these ideas to the depletion of ozone in the atmosphere.

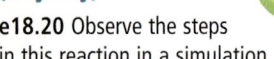

Molecular Modelling (Odyssey)

e18.20 Observe the steps in this reaction in a simulation.

one of the collisions between an
intermediate F atom and a
reactant molecule, NO$_2$

two *intermediate* atoms, F

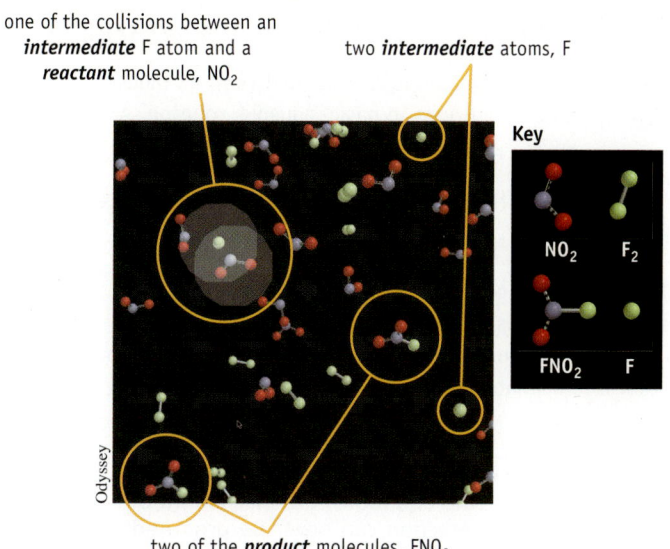

Key

NO$_2$ F$_2$

FNO$_2$ F

two of the *product* molecules, FNO$_2$

FIGURE 18.19 A "snapshot" of a reaction mixture during the second-order reaction between NO$_2$(g) and F$_2$(g). At this point in time, there are some unreacted NO$_2$ and F$_2$ molecules, and some transitory F atom intermediates formed by step 1. One of the collisions in step 2 is highlighted with halos. Although step 1 forms equal numbers of FNO$_2$ molecules and F atoms, there are more FNO$_2$ molecules than F atoms in the reaction mixture because some F atoms have reacted with NO$_2$ molecules to form more FNO$_2$ molecules.

WORKED EXAMPLE 18.13—ELEMENTARY STEPS AND REACTION MECHANISMS

Oxygen atom transfer from nitrogen dioxide to carbon monoxide in the gas phase produces nitrogen monoxide and carbon dioxide (page 688) by a second-order reaction at temperatures less than 500 K:

$$NO_2(g) + CO(g) \longrightarrow NO(g) + CO_2(g) \qquad Rate = k[NO_2]^2$$

Can this reaction occur in one bimolecular elementary step whose stoichiometry is the same as the overall reaction?

Strategy

Write the rate equation that you predict would be experimentally observed if the balanced overall reaction took place in the single elementary step described. If this rate equation corresponds with the observed rate equation, then you will have deduced a possible mechanism for the overall reaction.

Solution

If the reaction occurs by a multitude of events, each comprising the collision of one NO$_2$ molecule with one CO molecule, we would expect the rate equation to be

$$Rate = k[NO_2][CO]$$

This rate equation is not the same as that observed, so the mechanism must involve more than a single step. In one possible mechanism whose rate equation would agree with that observed, the reaction involves two bimolecular steps, the first one slow and the second one fast:

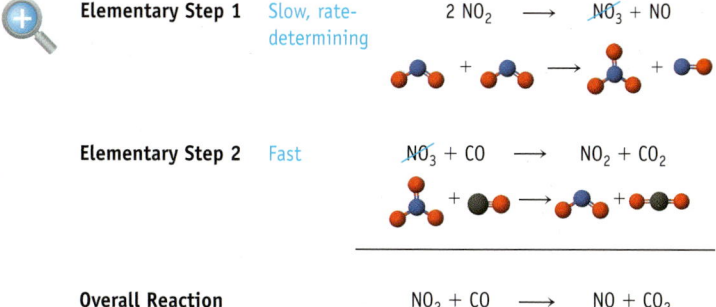

Elementary Step 1	Slow, rate-determining	2 NO$_2$ $\longrightarrow$ NO$_3$ + NO
Elementary Step 2	Fast	NO$_3$ + CO $\longrightarrow$ NO$_2$ + CO$_2$
Overall Reaction		NO$_2$ + CO $\longrightarrow$ NO + CO$_2$

The first (rate-determining) step has a rate equation that agrees with experiment, so this mechanism may be the way the reaction occurs. We have to bear in mind, however, that this is no proof, and there may be other possible mechanisms that obey this rate equation.

EXERCISE 18.21—ELEMENTARY STEPS AND REACTION MECHANISMS

The Raschig reaction produces aquated hydrazine, $N_2H_4(aq)$, an industrially important reducing agent, from $NH_3(aq)$ and $OCl^-(aq)$ ions in basic, aqueous solution. A proposed mechanism is

Step 1 (fast): $NH_3(aq) + OCl^-(aq) \longrightarrow NH_2Cl(aq) + OH^-(aq)$

Step 2 (slow): $NH_2Cl(aq) + NH_3(aq) \longrightarrow N_2H_5^+(aq) + Cl^-(aq)$

Step 3 (fast): $N_2H_5^+(aq) + OH^-(aq) \longrightarrow N_2H_4(aq) + H_2O(\ell)$

(a) What is the overall balanced stoichiometric equation?
(b) Which step of the three is rate-determining?
(c) Write the rate equation for the rate-determining elementary step.
(d) What reaction intermediates are involved?

> Although one or more mechanisms may be proposed that are consistent with the experimental evidence of rate equations, the mechanism of a reaction cannot be proven. The products of a reaction can never be produced at a rate faster than that of the rate-determining step—the slowest elementary step in the reaction mechanism.

18.8 Nucleophilic Substitution Reactions

An outstanding example of how chemists use experimental evidence to postulate reaction mechanisms is provided by the class of chemical reaction called **nucleophilic substitution reactions** [>>Section 21.5]. These involve the substitution of one electron-rich *nucleophile* (such as a chloride ion or a hydroxide ion) in molecules of a substance (referred to as the *substrate*) for an atom, group of atoms, or an ion, called the *leaving group*. A good nucleophile needs to be able to donate an electron pair to an electron-deficient (*electrophilic*) site on the substrate, and the leaving group must be able to readily accommodate an electron pair (that is, a weak base).

Two examples of reactions of this type are described by the following balanced equations:

Reaction 1: $OH^- + H_3C-Br \longrightarrow H_3C-OH + Br^-$

Reaction 2: $H_2O + (H_3C)_3C-Br \longrightarrow (H_3C)_3C-OH + Br^- + H_3O^+$

In both reactions, there is replacement of a –Br atom by an –OH group. We will see, however, that experimental evidence indicates that they occur by quite different mechanisms. Reaction 1 is an example of a subset of nucleophilic substitution reactions said to proceed by an S_N2 *mechanism*, and reaction 2 is an example of a subset that proceed by an S_N1 *mechanism*. In each case, the S_N part of the label stands for **s**ubstitution, **n**ucleophilic.

The S_N2 Mechanism of Nucleophilic Substitution Reactions

The rate of reaction 1 can be shown experimentally to have first-order dependence on the concentrations of each of the substrate CH_3Br and the substituting nucleophile OH^-. So reaction 1 is second order overall (hence the label S_N2), and the rate equation is

Rate of decrease of $CH_3Br = k[CH_3Br][OH^-]$

This evidence suggests that the rate-determining elementary step is due to bimolecular collisions between CH_3Br molecules and OH^- ions (in each of the multitude of events leading to formation of molecules of CH_3OH).

There is further experimental evidence that provides insight into the mechanism of nucleophilic substitution reactions that have second-order rate equations: if the substrate is chiral [<<Section 9.8], the product is of opposite configuration to the substrate. For

example, substitution of Br^- in (S)-2-bromobutane by OH^- results in formation of (R)-butan-2-ol with second-order kinetics:

(S)-2-bromobutane → **(R)-butan-2-ol**

A mechanism, now generally accepted, that has been proposed to account for both the second-order rate equation and the *inversion of configuration* involves single-step bimolecular collisions between the substrate molecules and the nucleophilic OH^- ions. It is proposed that, in the case of substitution into chiral $CH_3CH_2CH(Br)CH_3$, of all the collisions happening in a reaction mixture with sufficient energy, the only productive ones are those in which the OH^- ion nucleophile interacts with the substrate molecule on the side opposite to the $C—Br$ bond (Figure 18.20).

As the two particles react, the $C—Br$ bond weakens simultaneously with the gradual formation of a C–OH bond. In the maximum-energy transition state, the $C—H$, $C—CH_3$ and $C—CH_2CH_3$ bonds are planar. As the re-arrangement proceeds, the $C—OH$ bond formation is completed, resulting in a $CH_3CH_2CH(OH)CH_3$ molecule with the stereochemistry at the C atom inverted, and the release of the Br^- ion with the electron pair from the former $C—Br$ bond. The reaction energy diagram for this single-step mechanism with formation of a transition state is similar to that in Figure 18.12.

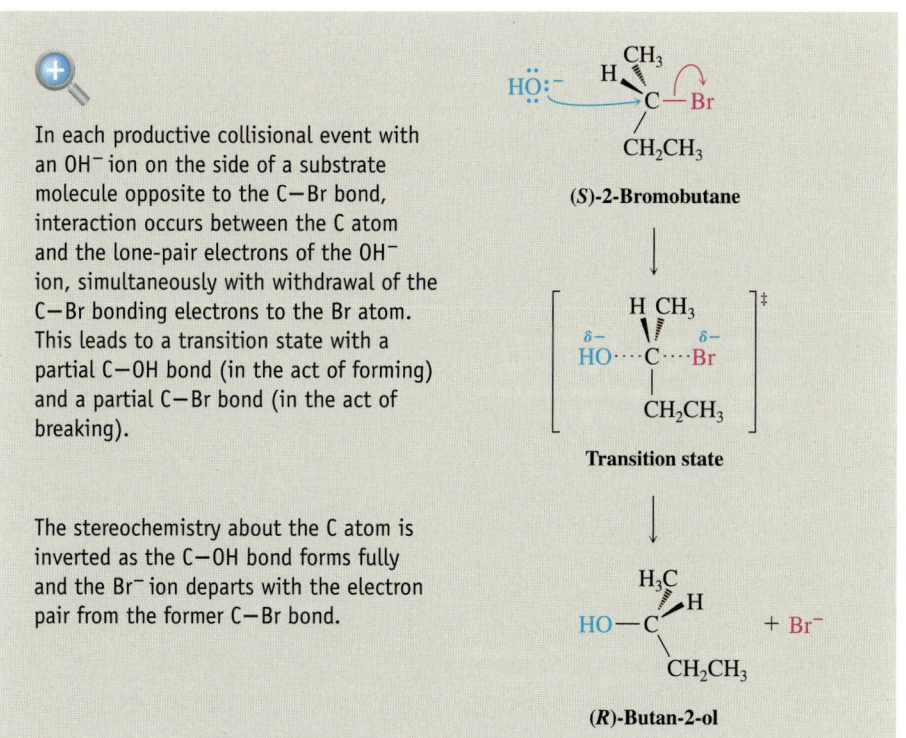

In each productive collisional event with an OH^- ion on the side of a substrate molecule opposite to the C—Br bond, interaction occurs between the C atom and the lone-pair electrons of the OH^- ion, simultaneously with withdrawal of the C—Br bonding electrons to the Br atom. This leads to a transition state with a partial C—OH bond (in the act of forming) and a partial C—Br bond (in the act of breaking).

The stereochemistry about the C atom is inverted as the C—OH bond forms fully and the Br^- ion departs with the electron pair from the former C—Br bond.

(S)-2-Bromobutane

Transition state

(R)-Butan-2-ol

FIGURE 18.20 Accepted reaction mechanism for S_N2 substitution of Br^- by OH^- in a molecule of (S)-2-bromobutane, showing inversion of configuration. The figure portrays one of the many productive collisional events that occur in the reaction mixture over time. The event is a bimolecular elementary step, so the frequency of such events is proportional to the concentration of each reactant.

Since CH_3Br and CH_3OH molecules are not chiral, the inversion of configuration cannot be distinguished in the case of substitution of Br^- by OH^- in methyl bromide (reaction 1 above).

The mechanism proposed for nucleophilic substitution reactions that proceed by the S_N2 mechanism is based on the experimental observation of second-order kinetics and stereochemical evidence: the stereochemistry of chiral substrates is inverted in the product. These reactions are proposed to proceed by a single bimolecular step during which the nucleophile interacts with the substrate on the side opposite to the leaving group.

The S$_N$1 Mechanism of Nucleophilic Substitution Reactions

Experimental studies on reaction 2 (above) show that its rate is first order with respect to the concentration of the substrate (CH$_3$)$_3$CBr (and hence the label S$_N$1), but independent of the concentration of the substituting nucleophile OH$^-$. The rate equation is

$$\text{Rate} = k[(\text{H}_3\text{C})_3\text{CBr}]$$

This suggests that the rate-determining elementary step of reaction 2 is a unimolecular step involving (H$_3$C)$_3$CBr molecules. Again, in the case of chiral substrate molecules there is stereochemical experimental evidence that gives chemists an insight as to how reactions of this type occur: the product of an S$_N$1 reaction is a racemate [<<Section 9.9]. For example, in the first-order substitution of the nucleophile CH$_3$OH for Br$^-$ in (R)-3-bromo-3-methylhexane, the product is a racemate of (R)- and (S)-3-methoxy-3-methylhexanes.

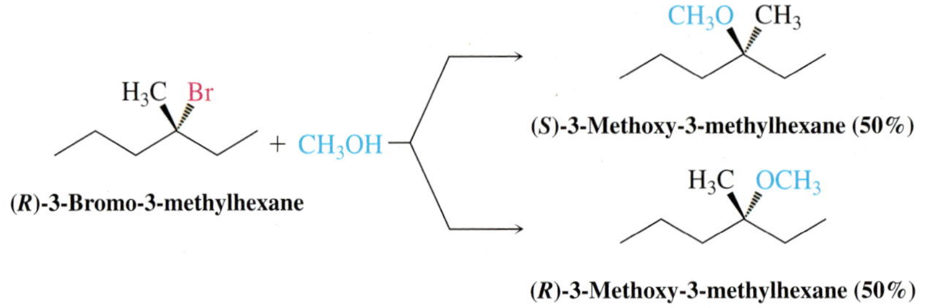

H$_3$C Br

+ CH$_3$OH

(R)-3-Bromo-3-methylhexane

CH$_3$O CH$_3$

(S)-3-Methoxy-3-methylhexane (50%)

H$_3$C OCH$_3$

(R)-3-Methoxy-3-methylhexane (50%)

The commonly accepted postulation for the mechanism of S$_N$1 reactions includes a rate-determining unimolecular first elementary step in which the leaving group separates from substrate molecules, in each case forming a planar *carbocation* intermediate. A second, faster, elementary step involves productive collisions between the carbocations and nucleophile molecules or ions. This S$_N$1 mechanism is illustrated in Figure 18.21 for reaction 2, in which case the final step involves transfer of a proton to the solvent.

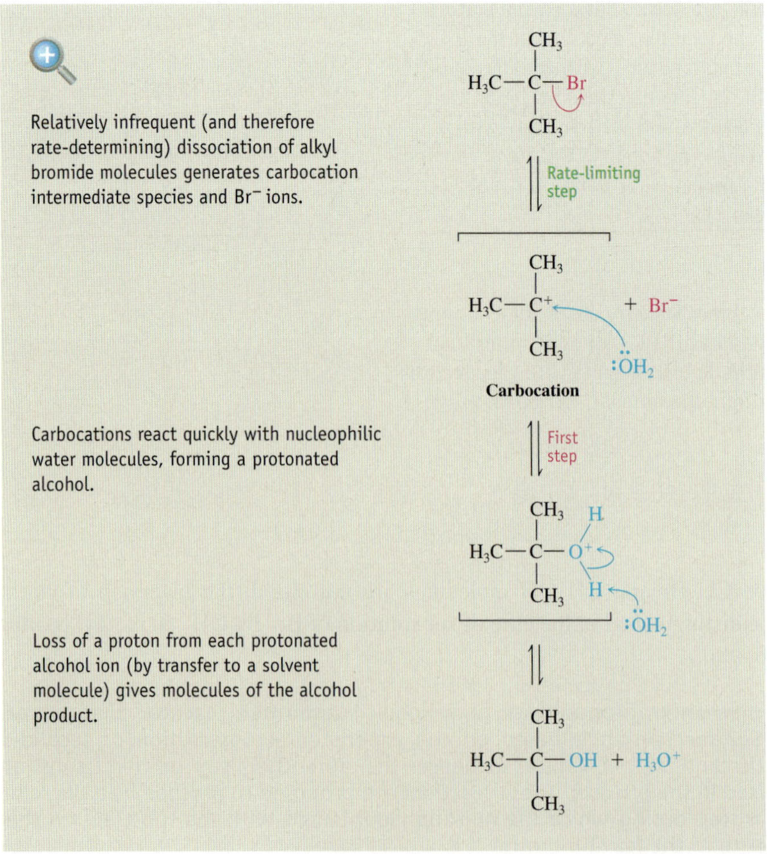

FIGURE 18.21 Accepted reaction mechanism for S$_N$1 substitution of H$_2$O for Br$^-$ in molecules of tertiary butyl bromide. Postulation of a rate-determining, bond-breaking first elementary step is consistent with the observed first-order kinetics. In this case, since the substrate molecule is not chiral, the stereochemical characteristics of the S$_N$1 mechanism cannot be observed.

Relatively infrequent (and therefore rate-determining) dissociation of alkyl bromide molecules generates carbocation intermediate species and Br$^-$ ions.

CH$_3$
H$_3$C—C—Br
CH$_3$

Rate-limiting step

CH$_3$
H$_3$C—C$^+$ + Br$^-$
CH$_3$
                  :ÖH$_2$
Carbocation

Carbocations react quickly with nucleophilic water molecules, forming a protonated alcohol.

First step

CH$_3$ H
H$_3$C—C—O$^+$
CH$_3$ H
              :ÖH$_2$

Loss of a proton from each protonated alcohol ion (by transfer to a solvent molecule) gives molecules of the alcohol product.

CH$_3$
H$_3$C—C—OH + H$_3$O$^+$
CH$_3$

Because the carbocations are planar, there is a 50% probability that nucleophiles will collide with them on either side of the plane of the ion. In the case of carbocations formed from a chiral substrate, the consequence of reaction with nucleophiles will be a 50:50 mixture of enantiomers of the product (Figure 18.22).

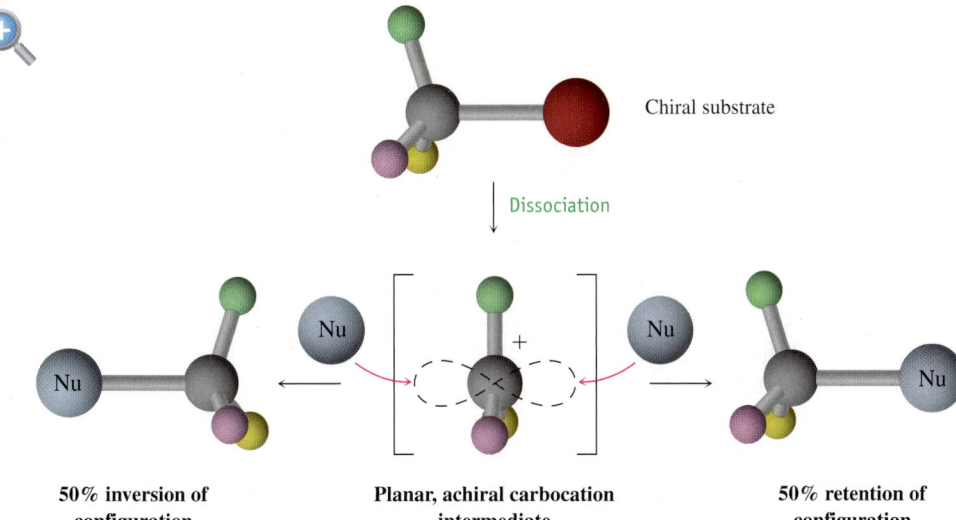

Chiral substrate

Dissociation

50% inversion of configuration

Planar, achiral carbocation intermediate

50% retention of configuration

FIGURE 18.22 Racemate formation in S_N1 reaction mechanism from a chiral substrate.

The mechanism proposed for nucleophilic substitution reactions that proceed by the S_N1 mechanism is based on the experimental observation of first-order kinetics and the stereochemical evidence that a racemate of the product is formed from a chiral substrate. Chemists propose that these reactions proceed by infrequent unimolecular dissociation of the substrate molecules to form planar carbocations, with which the nucleophiles can react on either side with 50% probability.

Factors Affecting the Mechanism by which S_N Reactions Proceed

The above discussion of nucleophilic substitution reactions has focused on the kinetic and stereochemical experimental evidence leading to postulation of two widely accepted mechanisms for these reactions. Sections 21.5–21.7 is a more detailed treatment of why these reactions occur, and of the factors that govern whether they occur by S_N1 or S_N2 mechanisms (or by both pathways simultaneously). One of these factors is the size of the groups attached to the carbon atom of the substrate at which substitution occurs: the larger the groups, the less likely it is an S_N2 mechanism. An illustration of this is provided by the mechanisms of reactions 1 and 2 at the beginning of this section, and a simple explanation is that large substituent groups hinder the nucleophile species in accessing and bonding to the carbon atom in a bimolecular step. Other factors include the identity of the leaving group and the nature of the solvent.

Technically, we can consider that nucleophilic substitution reactions of alkyl halides proceed by both S_N1 or S_N2 mechanisms simultaneously, and that various factors govern the proportion going by each pathway. Researchers have identified that substitution can also happen by pathways other than S_N1 and S_N2.

EXERCISE 18.22—NUCLEOPHILIC SUBSTITUTION REACTIONS

(*S*)-2-Bromooctane reacts with hydroxide ions in aqueous solution by an S_N2 mechanism.
(a) Write a rate law for this reaction.
(b) How will the rate of substitution change if the concentration of (*S*)-2-bromooctane is doubled, keeping the concentration of hydroxide ions constant?
(c) How will the rate of substitution change if the concentration of hydroxide ions is doubled, keeping the concentration of (*S*)-2-bromooctane constant?
(d) Draw the transition state formed during the reaction.
(e) Draw molecules of the product.

EXERCISE 18.23—NUCLEOPHILIC SUBSTITUTION REACTIONS

Iodide ions react with *tert*-butyl chloride to form *tert*-butyl iodide by an S_N1 mechanism.

(a) How will the rate of substitution change if the concentration of *tert*-butyl chloride is doubled, keeping the concentration of iodide ions constant?

(b) How will the rate of substitution change if the concentration of iodide ions is doubled, keeping the concentration of *tert*-butyl chloride constant?

(c) Draw the carbocation intermediate that is formed.

(d) Why does it not make sense to ask if inversion of configuration takes place in this case?

(e) Suggest why the proportion of this reaction that proceeds by the S_N2 mechanism is negligible.

18.9 Enzymes: Nature's Catalysts

In Section 18.1, we discussed the action of enzymes in the particular context of hard-to-digest foods. **Enzymes** are powerful catalysts, typically producing a reaction rate that is 10^7 to 10^{14} times faster than the uncatalyzed rate. Metal ions are often part of an enzyme. Carboxypeptidase, for example, contains Zn^{2+} ions at the active site.

In 1913, Leonor Michaelis and Maud L. Menten proposed a general theory of enzyme action based on kinetic observations. They assumed that the substrate (S, the reactant) and the enzyme (E) form a complex (ES). This complex then breaks down, releasing the enzyme and the product (P).

$$E + S \rightleftharpoons ES \rightleftharpoons E + P$$

Table 18.2 lists a few important enzymes.

TABLE 18.2 Biologically Important Reactions Catalyzed by Enzymes

Enzyme	Enzyme Function, or Reaction Catalyzed
Carbonic anhydrase	$CO_2(aq) + H_2O(\ell) \longrightarrow H_2CO_3(aq)$
Chymotrypsin	Cleavage of peptide linkages in proteins
Urease	$(H_2N)_2CO(aq) + 2\ H_2O(\ell) + H^+(aq) \longrightarrow 2\ NH_4^+(aq) + HCO_3^-(aq)$
Catalase	$2\ H_2O_2(aq) \longrightarrow 2\ H_2O(\ell) + O_2(g)$
Acetylcholinesterase	Regenerates acetylcholine, an important substance in the transmission of nerve impulses, from acetate and choline
Hexokinase and glucokinase	Both enzymes catalyze the formation of a phosphate ester linkage to an –OH group of a sugar. Glucokinase is a liver-specific enzyme, and the liver is the major organ for the storage of excess dietary sugar as glycogen.

Carbonic anhydrase is one of the many enzymes that play important roles in biological processes. Carbon dioxide dissolves in water to a small extent to produce carbonic acid, which ionizes to give $H^+(aq)$ and $HCO_3^-(aq)$ ions:

Reaction 1: $CO_2(g) \xrightarrow{\ H_2O(\ell)\ } CO_2(aq)$

Reaction 2: $CO_2(aq) + H_2O(\ell) \rightleftharpoons H_2CO_3(aq)$

Reaction 3: $H_2CO_3(aq) + H_2O(\ell) \rightleftharpoons HCO_3^-(aq) + H_3O^+(aq)$

Carbonic anhydrase speeds up reactions 1 and 2. Many of the $H^+(aq)$ ions produced by ionization of H_2CO_3 (reaction 3) are picked up by hemoglobin in the blood as hemoglobin loses O_2. The resulting $HCO_3^-(aq)$ ions are transported back to the lungs. When hemoglobin again takes on O_2, it releases $H^+(aq)$ ions. These ions and $HCO_3^-(aq)$ ions re-form $H_2CO_3(aq)$, from which $CO_2(g)$ is liberated and exhaled.

A simple experiment illustrates the effect of carbonic anhydrase (Figure 18.23). First, a small amount of NaOH solution is added to a cold, aqueous CO_2 solution. The solution becomes basic immediately because there is not enough $H_2CO_3(aq)$ in the solution to use up the added $OH^-(aq)$ ions, and reaction 2 only slowly produces more of it. After some seconds, however, dissolved $CO_2(aq)$ slowly produces more $H_2CO_3(aq)$, which reacts with $OH^-(aq)$ ions, and the solution again becomes acidic.

(a)　　$t = 0$　　**(b)**　　$t = 9$ sec　　**(c)**　　$t = 28$ sec　　**(d)**　　$t = 34$ sec　　**(e)**　　$t = 37$ sec

FIGURE 18.23 CO_2 equilibria in water. (a) A cold solution of CO_2 in water. (b) A few drops of a dye (bromothymol blue) are added to the cold solution. The yellow colour of the dye indicates an acidic solution. (c) A less-than-stoichiometric amount of sodium hydroxide is added, converting $H_2CO_3(aq)$ to $HCO_3^-(aq)$ ions (and some $CO_3^{2-}(aq)$ ions). (d) The blue colour of the dye indicates a basic solution. (e) The blue colour begins to fade after some seconds as $CO_2(aq)$ slowly forms more $H_2CO_3(aq)$. The amount of $H_2CO_3(aq)$ formed is sufficient to consume the added $OH^-(aq)$ ions and the solution again becomes acidic.

Now we try the experiment again, this time adding a few drops of blood to the solution (Figure 18.24). Carbonic anhydrase in blood speeds up reactions 1 and 2 by a factor of about 10^7, as evidenced by the more rapid reaction that occurs under these conditions.

(a)　　$t = 0$　　**(b)**　　$t = 3$ sec　　**(c)**　　$t = 15$ sec　　**(d)**　　$t = 17$ sec　　**(e)**　　$t = 21$ sec

FIGURE 18.24 Action of carbonic anhydrase. (a) A few drops of blood are added to a cold solution of CO_2 in water. (b) The dye indicates an acidic solution. (c) A less-than-stoichiometric amount of sodium hydroxide is added, converting $H_2CO_3(aq)$ to $HCO_3^-(aq)$ ions (and some $CO_3^{2-}(aq)$ ions). (d) The dye's blue colour indicates a basic solution. (e) The blue colour begins to fade after a few seconds as more $H_2CO_3(aq)$ forms, and the solution again becomes acidic. The formation of $H_2CO_3(aq)$ is more rapid in the presence of the enzyme.

Here is another experiment you can do with carbonic anhydrase. Take a sip of very cold carbonated beverage. The tingling sensation you feel on your tongue and in your mouth is not from the CO_2 bubbles. Rather, it comes from the protons released when carbonic anhydrase accelerates the formation of $H^+(aq)$ ions from dissolved H_2CO_3. Acidification of nerve endings creates the tingling feeling.

The enzymes trypsin, chymotrypsin, and elastase are digestive enzymes, catalyzing the hydrolysis of peptide bonds [>>Chapter 29]. They are synthesized in the pancreas and secreted into the digestive tract.

Acetylcholinesterase is involved in transmission of nerve impulses. Many pesticides interfere with this enzyme, so farm workers are often tested to be sure they have not been overexposed to agricultural toxins.

The liver has the primary role in maintaining blood glucose levels. This organ produces glucose with phosphate groups (PO_4^{3-}) attached. The enzyme glucose phosphatase in the liver has the function of removing the phosphate group before the glucose enters the blood.

Enzyme action. The tingling feeling you get when you drink a carbonated beverage comes from the $H^+(aq)$ ions released by $H_2CO_3(aq)$. The acid is formed rapidly from dissolved CO_2 in presence of the enzyme carbonic anhydrase.

SUMMARY

Key Concepts

- **Rate of a chemical reaction** is the change in concentration of a reactant or product (which must be specified) per unit of time. We distinguish between the average rate over a given time duration and the instantaneous rate at any moment. (Section 18.2)
- The rate of a given reaction is increased by higher reactant concentrations, higher temperature, the presence of a **catalyst**, and (in the case of solid reactants) larger surface area. (Section 18.3)
- The dependence of the rate of a reaction between reactants A and B on the concentrations of reactants is expressed by a **rate equation:** Rate $= k[A]^m[B]^n$, in which k is the **rate constant,** particular to that reaction and a defined temperature. The **order of the reaction** with respect to [A] is m, the order with respect to [B] is n, and the **overall reaction order** is $m + n$. The values of m and n can only be determined experimentally. One method to do so is the method of initial rates. (Section 18.4)
- **Integrated rate equations** express how the concentration of reactant R varies with time during a reaction. Using these, by trial and error we can see if experimental data are consistent with zero-order kinetics ([R] vs. t is a straight line), first-order kinetics (ln [R] vs. t is a straight line), or second-order kinetics (1/[R] vs. t is a straight line), as well as estimate the value of the rate constant at the temperature at which the data were collected. (Section 18.5)
- For a first-order reaction, the time required for the concentration of a reactant to decrease to one-half its initial value (the **half-life, $t_{1/2}$**) is independent of the reactant concentration, and so does not change during the course of a reaction. This is not the case for second-order reactions. (Section 18.5)
- According to the **collision theory of reaction rates,** reaction is possible only when molecules collide with sufficient total energy and in particular orientations. Since the frequency of collisions between reactant molecules is proportional to the concentration of each reactant, collision theory suggests why the rate of reaction may depend on the concentration of each reactant, although knowledge about reaction mechanisms accounts for why this does not necessarily apply. (Sections 18.6 and 18.7)
- The potential energy of a pair of colliding molecules increases as they interact, reaching a maximum with the formation of an unstable arrangement called the **transition state.** The changes of total potential energy of reacting particles is represented by a **reaction energy diagram**. The minimum combined kinetic energy that a pair of colliding particles must have in excess of the average for their collision to result in reaction is the **activation energy (E_a)**. At any temperature, molecules of a substance have a range of kinetic energies, and the higher the temperature, the greater the fraction of colliding molecules that will have energy equal to or greater than the activation energy. (Section 18.6)
- For productive collisions, colliding molecules with sufficient combined energy must have a particular orientation with respect to each other. (Section 18.6)
- The effect of temperature on the rate constant of a reaction is given by the **Arrhenius equation.** The activation energy of reaction can be calculated from the straight line plot of ln k vs. $1/T$, for which slope $= -E_a/R$. A rate constant at one temperature can be calculated algebraically from the value at another temperature, if E_a is known, by using the logarithmic form of the Arrhenius equation. (Section 18.6)
- Catalysts are substances that accelerate reactions. They are specific to the reaction accelerated. Catalyst molecules, atoms, or ions participate in the reaction, providing a mechanistic pathway with lower activation energy than that of the uncatalyzed reaction. Catalyst molecules, atoms, or ions are regenerated, becoming available to react again. (Section 18.6)
- A **reaction mechanism** is a sequence of bond-making and bond-breaking steps that occurs during the conversion of molecules of reactant to molecules of products.

Chemists can never prove what the mechanism is, but experimental evidence can help them to propose mechanisms that are consistent with the evidence. More than one proposed mechanism may be consistent with the evidence. (Section 18.7)

- **Elementary steps** are the individual molecular events that comprise a reaction mechanism. The **molecularity** of an elementary step is the number of reactant molecules, ions, or atoms that collide and undergo change. A unimolecular step involves reaction of only one particle, a bimolecular step involves two, and a termolecular step involves three particles. Unlike overall reactions, the rate equation of an elementary step is derivable directly from its stoichiometric equation. (Section 18.7)
- The products of a reaction can never be produced at a rate faster than that of the **rate-determining step**—the slowest elementary step in the reaction mechanism. (Section 18.7)
- The generally accepted mechanisms of S_N1 and S_N2 **nucleophilic substitution reactions** are based on the experimental collection of kinetics data and stereochemical evidence in those cases where the substrates are chiral. S_N1 reactions are proposed to proceed by a single bimolecular step during which the nucleophile interacts with the substrate on the side opposite to the leaving group. A two-step mechanism proposed for S_N1 reactions includes infrequent unimolecular dissociation of the substrate molecules to form planar carbocations, with which the nucleophiles can react on either side with 50% probability. (Section 18.8)
- **Enzymes** are naturally occurring substances that catalyze particular reactions. (Section 18.9)

Key Equations

Rate of reaction $= \dfrac{\text{change in concentration}}{\text{change in time}} = \dfrac{\Delta c}{\Delta t}$ (Section 18.2)

Rate equation: Rate $= k[A]^m[B]^n$ (18.1, Section 18.4)

Zero-order reaction: $-\dfrac{\Delta[R]}{\Delta t} = k[R]^0 = k$

First-order reaction: $-\dfrac{\Delta[R]}{\Delta t} = k[R]$

Second-order reaction: $-\dfrac{\Delta[R]}{\Delta t} = k[R]^2$

Integrated forms of the rate equations: (Section 18.5):

  Zero-order reaction: $[R]_t = [R]_0 - kt$

  First-order reaction: $\ln \dfrac{[R]_t}{[R]_0} = -kt$

  Second-order reaction: $\dfrac{1}{[R]_t} - \dfrac{1}{[R]_0} = kt$

Half-life of a first-order reaction: $t_{1/2} = \dfrac{0.693}{k}$ (Section 18.5)

Arrhenius equation: Rate constant, $k = Ae^{-E_a/RT}$ (Section 18.6)

Dependence of k on T: $\ln k = -\dfrac{E_a}{R}\left(\dfrac{1}{T}\right) + \ln A$ (Section 18.6)

Algebraic relationship between E_a values at two temperatures:

$\ln k_2 - \ln k_1 = \ln \dfrac{k_2}{k_1} = -\dfrac{E_a}{R}\left(\dfrac{1}{T_2} - \dfrac{1}{T_1}\right)$ (Section 18.6)

REVIEW QUESTIONS

Section 18.2: The Concept of Reaction Rate

18.24 Give the relative rates of disappearance of reactants and formation of products for each of the following reactions:
(a) $2 O_3(g) \longrightarrow 3 O_2(g)$
(b) $2 HOF(g) \longrightarrow 2 HF(g) + O_2(g)$

18.25 In the synthesis of ammonia if $-\Delta[H_2]/\Delta t = 4.5 \times 10^{-4}$ mol L^{-1} min^{-1}, what is $\Delta[NH_3]/\Delta t$?

$$N_2(g) + 3 H_2(g) \longrightarrow 2 NH_3(g)$$

18.26 Phenyl acetate, an ester, reacts with water according to the equation

$$\underset{\text{Phenyl acetate}}{CH_3\overset{\overset{\displaystyle O}{\|}}{C}OC_6H_5} + H_2O \longrightarrow \underset{\text{Acetic acid}}{CH_3\overset{\overset{\displaystyle O}{\|}}{C}OH} + \underset{\text{Phenol}}{C_6H_5OH}$$

The data in the table were collected for this reaction at 5 °C.

Time (s)	[Phenyl acetate] (mol L^{-1})
0	0.55
15.0	0.42
30.0	0.31
45.0	0.23
60.0	0.17
75.0	0.12
90.0	0.085

(a) Plot the phenyl acetate concentration vs. time and describe the shape of the curve observed.
(b) Calculate the rate of change of the phenyl acetate concentration during the period 15.0 s to 30.0 s and also during the period 75.0 s to 90.0 s. Compare the values and suggest a reason why one value is smaller than the other.
(c) What is the rate of change of the phenol concentration during the time period 60.0 s to 75.0 s?
(d) What is the instantaneous rate at 15.0 s?

Section 18.3: Conditions that Affect the Rate of a Reaction

18.27 Using the experimentally observed rate equation Rate $= k[A]^2[B]$, state the order of the reaction with respect to [A] and [B]. What is the total order of the reaction?

Interactive Exercise 18.28

Predict changes in initial rates from the rate law.

Section 18.4: Dependence of Rate on Reactant Concentration

18.29 At a certain temperature, the rate constant (k) is 0.090 h^{-1} for the reaction

$$Pt(NH_3)_2Cl_2(aq) + H_2O(\ell) \longrightarrow$$
$$Pt(NH_3)_2(H_2O)Cl^+(aq) + Cl^-(aq)$$

and the rate equation is Rate $= k[Pt(NH_3)_2Cl_2]$. Calculate the rate of reaction when the concentration of $Pt(NH_3)_2Cl_2$ is 0.020 mol L^{-1}. What is the rate of change of $[Cl^-]$?

Interactive Exercise 18.30

Deduce the rate law and calculate the rate constant from initial rate data.

18.31 The data in the table are for the reaction of NO(g) and O_2 (g) at 660 K:

$$2 NO(g) + O_2(g) \longrightarrow 2 NO_2(g)$$

REACTANT CONCENTRATION (mol L^{-1})		RATE OF DISAPPEARANCE OF NO (mol L^{-1} s^{-1})
[NO]	[O₂]	
0.010	0.010	2.5×10^{-5}
0.020	0.010	1.0×10^{-4}
0.010	0.020	5.0×10^{-5}

(a) Determine the order of the reaction for each reactant.
(b) Write the rate equation for the reaction.
(c) Calculate the rate constant.
(d) Calculate the rate (in mol L^{-1} s^{-1}) at the instant when $[NO] = 0.015$ mol L^{-1} and $[O_2] = 0.0050$ mol L^{-1}.
(e) At the instant when NO(g) is reacting at the rate 1.0×10^{-4} mol L^{-1} s^{-1}, what is the rate at which O_2 (g) reacts and NO_2 (g) forms?

18.32 A reaction has the experimental rate equation Rate $= k[A]^2$. How will the rate change if the concentration of A is tripled? If the concentration of A is halved?

18.33 The reaction

$$2 NO(g) + 2 H_2(g) \longrightarrow N_2(g) + 2 H_2O(g)$$

was studied at 904 °C, and the data in the table were collected.

REACTANT CONCENTRATION (mol L^{-1})		RATE OF APPEARANCE OF N$_2$ (mol L^{-1} s^{-1})
[NO]	**[H$_2$]**	
0.420	0.122	0.136
0.210	0.122	0.0339
0.210	0.244	0.0678
0.105	0.488	0.0339

(a) Determine the order of the reaction for each reactant.
(b) Write the rate equation for the reaction.
(c) Calculate the rate constant for the reaction.
(d) Find the rate of appearance of N$_2$(g) at the instant when [NO] = 0.350 mol L^{-1} and [H$_2$] = 0.205 mol L^{-1}.

18.34 Data for the following reaction are given in the table below.

$$CO(g) + NO_2(g) \longrightarrow CO_2(g) + NO(g)$$

	CONCENTRATION (mol L^{-1})		INITIAL RATE
Run	**[CO]**	**[NO$_2$]**	**(mol L^{-1} h^{-1})**
1	5.0×10^{-4}	0.36×10^{-4}	3.4×10^{-8}
2	5.0×10^{-4}	0.18×10^{-4}	1.7×10^{-8}
3	1.0×10^{-3}	0.36×10^{-4}	6.8×10^{-8}
4	1.5×10^{-3}	0.72×10^{-4}	?

(a) What is the rate equation for this reaction?
(b) What is the rate constant for the reaction?
(c) What is the initial rate of the reaction in run 4?

18.35 Carbon monoxide reacts with O$_2$ to form CO$_2$:

$$2 CO(g) + O_2(g) \longrightarrow 2 CO_2(g)$$

Information on this reaction is given in the table below.

[CO] (mol L^{-1})	[O$_2$] (mol L^{-1})	Rate (mol L^{-1} min^{-1})
0.02	0.02	3.68×10^{-5}
0.04	0.02	1.47×10^{-4}
0.02	0.04	7.36×10^{-5}

(a) What is the rate law for this reaction?
(b) What is the order of the reaction with respect to [CO]? What is the order with respect [O$_2$]? What is the overall order of the reaction?
(c) What is the value for the rate constant (k)?

18.36 Data for the reaction 2 NO(g) + O$_2$(g) $\longrightarrow$ 2 NO$_2$(g) are given in the table.

	CONCENTRATION (mol L^{-1})		INITIAL RATE
Run	**[NO]**	**[O$_2$]**	**(mol L^{-1} h^{-1})**
1	3.6×10^{-4}	5.2×10^{-3}	3.4×10^{-8}
2	3.6×10^{-4}	1.04×10^{-2}	6.8×10^{-8}
3	1.8×10^{-4}	1.04×10^{-2}	1.7×10^{-8}
4	1.8×10^{-4}	5.2×10^{-2}	?

(a) What is the rate equation for this reaction?
(b) What is the rate constant for the reaction?
(c) What is the initial rate of the reaction in run 4?

Section 18.5: Concentration-Time Relationships: Integrated Rate Equations

18.37 Gaseous NO$_2$ decomposes when heated:

$$2 NO_2(g) \longrightarrow 2 NO(g) + O_2(g)$$

The disappearance of NO$_2$(g) is a first-order reaction with $k = 3.6 \times 10^{-3}$ s^{-1} at 300 °C.
(a) If we have a sample of gas in a flask at 300 °C, what fraction of it will be present after another 150 s?
(b) How long must a sample be held at 300 °C so that 99% of the NO$_2$(g) has decomposed?

18.38 The rate equation for the hydrolysis of sucrose to fructose and glucose

$$C_{12}H_{22}O_{11}(aq) + H_2O(\ell) \longrightarrow 2 C_6H_{12}O_6(aq)$$

is Δ[sucrose]/$\Delta t = k$[sucrose]. After 2.57 h at 27 °C, the sucrose concentration decreased from 0.0146 mol L^{-1} to 0.0132 mol L^{-1}. Find the rate constant (k) at the temperature of the hydrolysis.

18.39 The decomposition of SO$_2$Cl$_2$(g) is a first-order reaction:

$$SO_2Cl_2(g) \longrightarrow SO_2(g) + Cl_2(g)$$

The rate constant for the reaction is 2.8×10^{-3} min^{-1} at 600 K. If the initial concentration of SO$_2$Cl$_2$ is 1.24×10^{-3} mol L^{-1}, how long will it take for the concentration to drop to 0.31×10^{-3} mol L^{-1}?

Interactive Exercise 18.40–18.41

Calculate reactant concentrations using the integrated rate law.

18.42 The thermal decomposition of formic acid (HCOOH) is a first-order reaction with a rate constant of 2.4×10^{-3} s^{-1} at a given temperature. How long will it take for three-fourths of a sample of HCOOH to decompose?

18.43 Ammonium cyanate (NH$_4$NCO) re-arranges in water to give urea, (NH$_2$)$_2$CO:

$$NH_4NCO(aq) \longrightarrow (NH_2)_2CO(aq)$$

The rate equation for this process is Rate = k [NH$_4$NCO]^2, where $k = 0.0113$ L mol^{-1} min^{-1}. If the original concentration of NH$_4$NCO in solution is 0.229 mol L^{-1}, how long will it take for the concentration to decrease to 0.180 mol L^{-1}?

18.44 Common sugar, sucrose, breaks down in dilute acid solution to form glucose and fructose. Both products have the same formula, C$_6$H$_{12}$O$_6$:

$$C_{12}H_{22}O_{11}(aq) + H_2O(\ell) \longrightarrow 2 C_6H_{12}O_6(aq)$$

The rate of this reaction has been studied in acid solution, and the data in the table were obtained.

Time (min)	$[C_{12}H_{22}O_{11}]$ (mol L^{-1})
0	0.316
39	0.274
80	0.238
140	0.190
210	0.146

(a) Plot ln[sucrose] vs. time and 1/[sucrose] vs. time. What is the order of the reaction?

(b) Write the rate equation for the reaction and calculate the rate constant (k).

(c) Estimate the concentration of sucrose after 175 min.

18.45 Data for the reaction of phenyl acetate with water are given in Question 18.26. Plot these data as ln [phenyl acetate] and 1/[phenyl acetate] vs. time. Based on the appearance of the two graphs, what can you conclude about the order of the reaction with respect to phenyl acetate? Working from the data and the rate equation, determine the rate constant for the reaction.

18.46 Data for the decomposition of dinitrogen oxide

$$2 N_2O(g) \longrightarrow 2 N_2(g) + O_2(g)$$

on a gold surface at 900 °C are given below. Verify that the reaction is first order by preparing a graph of ln $[N_2O]$ vs. time. Derive the rate constant from the slope of the line in this graph. Using the rate equation and value of k, determine the decomposition rate at 900 °C when $[N_2O] = 0.035$ mol L^{-1}.

Time (min)	$[N_2O]$ (mol L^{-1})
15.0	0.0835
30.0	0.0680
80.0	0.0350
120.0	0.0220

18.47 Gaseous NO_2 decomposes at 573 K.

$$2 NO_2(g) \longrightarrow 2 NO(g) + O_2(g)$$

The concentration of NO_2 was measured as a function of time. A graph of 1/$[NO_2]$ vs. time gives a straight line with a slope of 1.1 L mol^{-1} s^{-1}. What is the rate equation for this reaction? What is the rate constant?

Interactive Exercise 18.48

SUBMIT

Deduce the order of reaction using graphical methods.

18.49 The rate equation for the decomposition of $N_2O_5(g)$ (forming $NO_2(g)$ and $O_2(g)$) is $-\Delta[N_2O_5]/\Delta t = k[N_2O_5]$.

The value of k is 5.0×10^{-4} s^{-1} for the reaction at a particular temperature.

(a) Calculate the half-life of $N_2O_5(g)$.

(b) How long does it take for $[N_2O_5]$ to decrease to one-tenth of its original value?

18.50 Gaseous azomethane ($CH_3N{=}NCH_3$) decomposes in a first-order reaction when heated:

$$CH_3N{=}NCH_3(g) \longrightarrow N_2(g) + C_2H_6(g)$$

The rate constant for this reaction at 425 °C is 40.8 min^{-1}. If the initial quantity of azomethane in the flask is 2.00 g, how much remains after 0.0500 min? What quantity of $N_2(g)$ is formed in this time?

18.51 Formic acid decomposes at 550 °C according to the equation

$$HCOOH(g) \longrightarrow CO_2(g) + H_2(g)$$

The reaction follows first-order kinetics. In an experiment, it is determined that 75% of a sample of HCO_2H has decomposed in 72 seconds. Determine $t_{1/2}$ for this reaction.

Section 18.6: A Microscopic View of Reaction Rates: Collision Theory

18.52 Calculate the activation energy (E_a) for the reaction

$$N_2O_5(g) \longrightarrow 2 NO_2(g) + \frac{1}{2} O_2(g)$$

from the observed rate constants: k at 25 °C $= 3.46 \times 10^{-5}$ s^{-1} and k at 55 °C $= 1.5 \times 10^{-3}$ s^{-1}.

18.53 If the rate constant for a reaction triples when the temperature rises from 3.00×10^2 K to 3.10×10^2 K, what is the activation energy of the reaction?

18.54 When heated to a high temperature, cyclobutane (C_4H_8) decomposes to ethylene:

$$C_4H_8(g) \longrightarrow 2 C_2H_4(g)$$

The activation energy (E_a) for this reaction is 260 kJ mol^{-1}. At 800 K, the rate constant $k = 0.0315$ s^{-1}. Determine the value of k at 850 K.

18.55 The reaction of H_2 molecules with F atoms

$$H_2(g) + F(g) \longrightarrow HF(g) + H(g)$$

has an activation energy of 8 kJ mol^{-1} and an enthalpy change of reaction of 133 kJ mol^{-1}. Draw a reaction energy diagram for this process. Indicate the activation energy and enthalpy change of reaction on this diagram.

Section 18.7: Reaction Mechanisms

18.56 What is the rate law for each of the following *elementary* reactions?

(a) $NO(g) + NO_3(g) \longrightarrow 2 NO_2(g)$

(b) $Cl(g) + H_2(g) \longrightarrow HCl(g) + H(g)$

(c) $(CH_3)_3CBr(aq) \longrightarrow (CH_3)_3C^+(aq) + Br^-(aq)$

18.57 What is the rate law for each of the following *elementary* reactions?

(a) $Cl(g) + ICl(g) \longrightarrow I(g) + Cl_2(g)$
(b) $O(g) + O_3(g) \longrightarrow 2\ O_2(g)$
(c) $2\ NO_2(g) \longrightarrow N_2O_4(g)$

18.58 The reaction of $NO_2(g)$ and $CO(g)$ is thought to occur in two steps:

Step 1: Slow $NO_2(g) + NO_2(g) \longrightarrow NO(g) + NO_3(g)$

Step 2: Fast $NO_3(g) + CO(g) \longrightarrow NO_2(g) + CO_2(g)$

(a) Show that the elementary steps add up to give the overall stoichiometric equation.
(b) What is the molecularity of each step?
(c) For this mechanism to be consistent with kinetic data, what must be the experimental rate equation?
(d) Identify any intermediates in this reaction.

18.59 Ozone (O_3) in the earth's upper atmosphere decomposes according to the equation:

$$2\ O_3 \longrightarrow 3\ O_2$$

The mechanism of the reaction is thought to proceed through an initial fast, reversible step followed by a slow, second step.

Step 1: Fast, reversible $O_3 \rightleftharpoons O_2 + O$

Step 2: Slow $O_3 + O \longrightarrow 2\ O_2$

(a) Which of the steps is rate-determining?
(b) Write the rate equation for the rate-determining step.

18.60 The mechanism for the reaction of CH_3OH and HBr is believed to involve two steps. The overall reaction is exothermic.

Step 1: Fast, endothermic $CH_3OH + H^+ \rightleftharpoons CH_3OH_2^+$

Step 2: Slow $CH_3OH_2^+ + Br^- \longrightarrow CH_3Br + H_2O$

(a) Write an equation for the overall reaction.
(b) Draw a reaction energy diagram for this reaction.
(c) Show that the rate equation for this reaction is $\Delta[CH_3OH]/\Delta t = k[CH_3OH][H^+][Br^-]$.

18.61 Iodide ion is oxidized in acid solution by hydrogen peroxide (Figure 18.3).

$$H_2O_2(aq) + 2\ H^+(aq) + 2\ I^-(aq) \longrightarrow I_2(aq) + 2\ H_2O(\ell)$$

A proposed mechanism is

Step 1: Slow $H_2O_2(aq) + I^-(aq) \longrightarrow H_2O(\ell) + IO^-(aq)$

Step 2: Fast $H^+(aq) + IO^-(aq) \longrightarrow HOI(aq)$

Step 3: Fast $HOI(aq) + H^+(aq) + I^-(aq)$
$$\longrightarrow I_2(aq) + H_2O(\ell)$$

(a) Show that the three elementary steps add up to give the overall stoichiometric equation.
(b) What is the molecularity of each step?
(c) For this mechanism to be consistent with kinetic data, what must be the experimental rate equation?

Interactive Exercise 18.62

Given the reaction mechanism, predict the rate equation.

18.63 A proposed mechanism for the reaction of NO_2 and CO is

Step 1: Slow, endothermic $2\ NO_2(g) \longrightarrow NO(g) + NO_3(g)$

Step 2: Fast, exothermic $NO_3(g) + CO(g)$
$$\longrightarrow NO_2(g) + CO_2(g)$$

Overall reaction: Exothermic $NO_2(g) + CO(g)$
$$\longrightarrow NO(g) + CO_2(g)$$

(a) Identify each of the following as a reactant, product, or intermediate: $NO_2(g)$, $CO(g)$, $NO_3(g)$, $CO_2(g)$, and $NO(g)$.
(b) Draw a reaction energy diagram for this reaction. Indicate on this drawing the activation energy for each step and the overall reaction enthalpy change.

18.64 A three-step mechanism for the reaction of $(CH_3)_3CBr$ and H_2O is proposed:

Step 1: Slow $(CH_3)_3CBr \longrightarrow (CH_3)_3C^+ + Br^-$

Step 2: Fast $(CH_3)_3C^+ + H_2O \longrightarrow (CH_3)_3COH_2^+$

Step 3: Fast $(CH_3)_3COH_2^+ + Br^- \longrightarrow (CH_3)_3COH + HBr$

(a) Write an equation for the overall reaction.
(b) Which step is rate-determining?
(c) What rate equation is expected for this reaction?

SUMMARY AND CONCEPTUAL QUESTIONS

18.65 Run the Odyssey simulation e18.20, which portrays a reaction mixture during reaction of $NO_2(g)$ and $F_2(g)$, discussed in Section 18.7. Stop the clip at various times, and count the numbers of each species.

(a) The F_2 molecules react in step 1 of the mechanism. Do they all react simultaneously?
(b) What happens to the number of NO_2 molecules as time passes?
(c) What happens to the number of F_2 molecules as time passes?
(d) What are the relative rates of change of $[NO_2]$ and $[F_2]$?
(e) Trace the path of some NO_2 molecules. Identify some that take part in step 1 of the mechanism [<<Section 18.7], and some that take part in step 2.

(f) Try to identify and follow a F atom. What is its fate?

(g) What happens to the number of F atoms as time passes? Does this number change as quickly as do the numbers of NO_2 molecules and F_2 molecules? Explain.

18.66 Hydrogenation reactions, processes in which H_2 is added to a molecule, are usually catalyzed. An excellent catalyst is a very finely divided metal suspended in the reaction solvent. Explain why finely divided rhodium, for example, is a much more efficient catalyst than a small block of the metal.

18.67 The following statements relate to the reaction with the following experimentally determined rate law: Rate $= k[H_2][I_2]$.

$$H_2(g) + I_2(g) \longrightarrow 2\,HI(g)$$

Determine whether each of the following statements is true. If a statement is false, indicate why it is incorrect.

(a) The reaction must occur in a single step.

(b) This is a second-order reaction overall.

(c) Raising the temperature will cause the value of k to decrease.

(d) Raising the temperature lowers the activation energy for this reaction.

(e) If the concentrations of both reactants are doubled, the rate will double.

(f) Adding a catalyst in the reaction will cause the initial rate to increase.

18.68 One mechanism by which chlorine atoms contribute to the destruction of the earth's ozone layer is by the following sequence of reactions

$$Cl + O_3 \longrightarrow ClO + O_2$$
$$ClO + O \longrightarrow Cl + O_2$$

where the O atoms in the second step come from the decomposition of ozone by sunlight:

$$O_3 \longrightarrow O + O_2$$

What is the net equation on summing these three equations? Why does this lead to ozone loss in the stratosphere? What is the role played by Cl in this sequence of reactions? What name is given to species such as ClO?

18.69 Decide whether each of the following statements is true. If false, rewrite the sentence to make it correct.

(a) The rate-determining elementary step in a reaction is the slowest step in a mechanism.

(b) It is possible to change the rate constant by changing the temperature.

(c) As a reaction proceeds at constant temperature, the rate remains constant.

(d) A reaction that is third order overall must involve more than one step.

18.70 Decide whether each of the following statements is incorrect. If the statement is incorrect, rewrite it to be correct.

(a) Reactions are faster at a higher temperature because activation energies are lower.

(b) Rates increase with increasing concentration of reactants because there are more collisions between reactant molecules.

(c) At higher temperatures, a larger fraction of molecules have enough energy to get over the activation energy barrier.

(d) Catalyzed and uncatalyzed reactions occur by the same mechanism.

With the **eCHACR** single-sign-on access card bundled with your text, log on (http://login.cengage.com/sso) and access the e-book and click on any in-margin icons for dynamic molecular-level animations and simulations, videos of laboratory reactions, interactive exercises, reviews of background concepts, and other online supplementary materials as noted by the icons in the text margins.

Also go to www.chemistry.nelson.com <http://www.chemistry.nelson.com> for Answers to in-chapter exercises and selected Review Questions, Test Yourself questions, weblinks, crossword puzzles, flashcards, glossary of key terms, and other student resources.

Alkenes and Alkynes

JJ Morales / Shutterstock

19.1 Case Study: Pheasant Wattles, Photoprotection, and Photonics

Your grandmother may have urged you to eat carrots to improve your vision. Perhaps she was unknowingly spreading a myth introduced by British Intelligence during World War II that a diet rich in carrots (rather than secret airborne interception radar) was responsible for the success of Royal Air Force personnel in seeing and intercepting incoming bombers from long distances at night.

Or perhaps your grandmother was aware of the important link between the chemistry of alkenes and vision. The alkene β-carotene, responsible for giving carrots their characteristic orange colour, can be converted in your intestinal lining and liver into Vitamin A (retinol), which plays an important role in the chemistry of vision.

The structures of molecules of β-carotene and Vitamin A are represented in Figure 19.1. β-carotene is a highly **conjugated alkene**—that is, it has alternating double and single bonds. It is also a polymer constructed of eight isoprene monomer units (Figure 19.1c) and a dimer of Vitamin A (Figure 19.1b). Many other important compounds are polymers of the alkene isoprene, forming a class of compounds called *terpenes*. The carotenoids, including β-carotene, are terpenes whose molecules are derived from eight isoprene units. Carotenoids with no oxygen atoms are called *carotenes*. Other polymers of isoprene are the family of steroid compounds whose molecules are recognizable by their four fused rings, and are responsible for regulating many vital functions in plants, animals, and fungi.

The orange colour of a solution of β-carotene results from its absorption of blue visible light with wavelengths of about 480–490 nm. Not all of the white light that hits the solution is therefore transmitted through it—the transmitted light is deficient in blue light and appears to be orange, which is the complementary colour of blue. The colours of substances containing other conjugated alkenes can be correlated with the length of the chain of alternating single and double bonds. In general, as the conjugated chain length increases, so does the wavelength of visible light that substance will absorb to

FIGURE 19.1 (a) The structural formula for β-carotene, a terpene polymer. The lower structure uses heavy and dashed lines to emphasize it is made from eight isoprene monomer units (one shown separately in (b)). β-Carotene is a dimer of Vitamin A, with structure shown in (c).

excite an electron from the highest occupied molecular orbital of its molecules [<<Section 10.9]. This is discussed in Section 19.2. The orange colour of β-carotene has become one of its most marketable attributes, and β-carotene is produced commercially for use in colouring margarine, cheese, baked goods, and pet foods (Figure 19.2a). Young children sometimes develop orange faces or hands as a result of the deposition of serum carotenoids in outer layers of skin, and foods rich in β-carotene can change the colour of other visible features of animals. In some male birds such as pheasants (Figure 19.2b), the colour of ornaments such as spurs and feathers is important in mating behaviour. Males with redder, more saturated ornaments are more successful in attracting females or dominating other males, and this colouring is known to be made more brilliant by carotenoids in the diet.

The red-orange hue gives pheasants a colourful advantage. But why do plants produce β-carotene in the first place? The answer is tied in with another brightly coloured substance, chlorophyll.

FIGURE 19.2 (a) Cognis Nutrition and Health produces β-carotene from microscopic algae *Dunaliella salina*, on Australian algae farms, for use in improving the colour of margarine, cheese, baked goods, pet foods, and the fish, birds, and other pets that consume those pet foods. (b) Male common pheasant with an orange wattle.

The green colour of plants comes from chlorophyll (Figure 19.3), the pigment whose molecules contain light-absorbing porphyrin rings with a central magnesium ion that is complexed to the nitrogen atoms of the porphyrin ring. Porphyrin rings are also found in molecules of hemoglobin and Visudyne™ [<<Section 1.2], the photosensitizer used to treat age-related macular degeneration. Both Visudyne™ and chlorophyll have the powerful ability to absorb light and transfer energy to other molecules. In the case of Visudyne™, this energy is transferred to ground-state oxygen to produce singlet oxygen, used therapeutically as a chemotherapeutic agent. Chlorophyll uses its ability to absorb energy to drive photosynthesis, one of the most important chemical reactions of plants and some algae. The light energy absorbed by chlorophyll is transferred to other molecules, leading to a series of reactions that ultimately converts carbon dioxide and water into oxygen and glucose [<<Section 4.4].

Wherever chlorophyll is found in the protein complexes responsible for photosynthesis, carotenoids are found close by, and they are thought to help protect the molecular photosynthesis machinery. β-carotene and other carotenoids play at least two important roles in photosynthesis. β-carotene, which is orange, absorbs visible light of different wavelengths than chlorophyll, which is green. One of the roles of β-carotene is to increase the efficiency of photosynthesis by transferring to chlorophyll the energy it receives from the blue light. But β-carotene plays an even more important role. A plant can't adjust to too much sunlight by moving into the shade or lathering on sunscreen. In some cases, harvesting too much light by chlorophyll leads to the formation of free radicals such as singlet oxygen [<<Section 1.2], which can damage proteins and membranes involved in photosynthesis. If carotenoid molecules are physically close enough to the chlorophyll in the photosynthetic reaction centre, they can mop up (technically called "quench") the singlet oxygen produced by chlorophyll before it causes too much damage. Since singlet oxygen is a strong oxidizing agent, the carotenoids are referred to as **antioxidants**, and this role of β-carotene in plants is called a *photoprotective role*.

> *β-carotene and other carotenoids give male pheasants the bright orange-red colours important in mating. These same compounds function as important antioxidants in plants and humans, intercepting oxygen-containing free radicals before they cause cellular damage.*

Carotenoids such as β-carotene and lycopene, which is found in tomatoes, have also been shown to function as important antioxidants in humans, intercepting singlet oxygen and other oxygen containing free radicals before they cause cellular damage. Much research is in progress on the benefits and possible adverse effects of taking dietary supplements of carotenoids such as lycopene and β-carotene. Even more fundamentally, β-carotene is needed in the diet to produce Vitamin A. In places in the developing world where few items are available to supplement rice in the diet, Vitamin A deficiency is a major problem, often leading to blindness and an increase in the severity of certain diseases. One approach to ensuring a diet with sufficient β-carotene has been to genetically engineer a strain of rice rich in carotenoids, giving them a characteristic yellow-orange colour, and the name "golden rice." A gene from maize has been used to produce new strains of "golden rice" with as much as 37 μg carotene per g of rice, 25 times the amount in the first genetically modified strains (Figure 19.4).

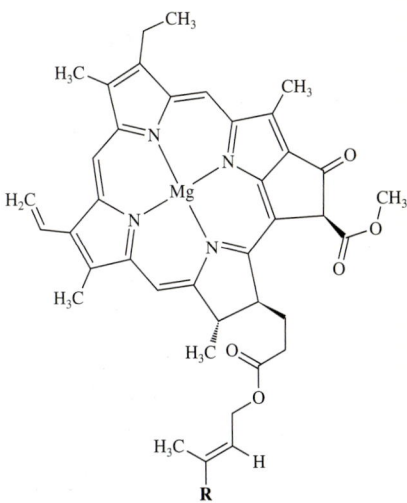

FIGURE 19.3 The structure of a chlorophyll *a* molecule. Compare this to the porphyrin structures in heme (Figure 1.1) and Visudyne™ (Figure 1.3).

Wild Type   Np *Psy/crtl*   Zm *Psy/crtl*

FIGURE 19.4 Two strains of "golden rice" (*centre* and *right*), genetically engineered to contain more carotene than wild rice (*left*).

Dr. Roy Tasker

Molecular Modelling (Odyssey)

e19.1 Simulate the flexibility of a carotene molecule and a carbon nanotube.

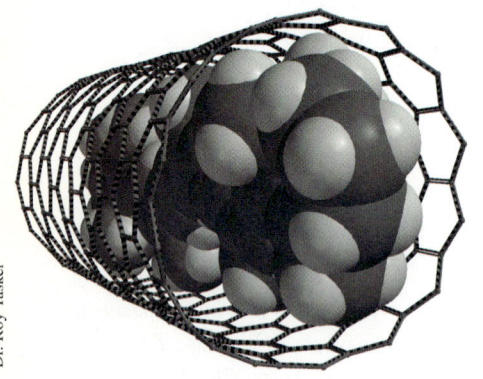

FIGURE 19.5 A schematic representation of β-carotene (space-filling model representation) in a single-walled carbon nanotube cage, which may protect the alkene from reacting with free radicals such as singlet oxygen.

Finally, the light-absorbing properties of conjugated alkenes such as β-carotene that make them so important in the chemistry of vision and the mating behaviour of pheasants may also play a role in the next generation of information processing devices. β-Carotene and other conjugated alkenes efficiently absorb light, making them prime candidates for use in photonics devices for processing data and creating optical-neural networks. Optical-neural networks are rapid information-processing devices modelled after neural networks in the human brain. But the ability of β-carotene to react so easily with free radicals like singlet oxygen limits their possible applications, as β-carotene is easily degraded under environmental conditions where photonics devices might be used. Kazuhiro Yanagi's research group at the Nanotechnology Research Institute in Higashi, Japan, has proposed a way to overcome this challenge by locking β-carotene molecules in a single-walled cage of carbon nanotubes, as shown in Figure 19.5. This is similar to the encapsulation of methane gas in water cages (methane clathrate hydrates), discussed in Chapter 4. Researchers think that the carbon cage might protect the β-carotene from reacting with free radicals and may permit further developments that use the light-absorbing properties of carotenes and substituted carotenes in optical devices.

While β-carotene plays an important role in protecting plants from too much sunlight, the simplest alkene, ethene (commonly called ethylene), is made by all parts of most plants, with functions ranging from ripening fruit to providing signals that help plants grow in the right way to avoid too much shade! It is also the organic compound produced on the largest scale industrially, with annual global production exceeding 120 million tonnes. What is all that ethylene used for? In the rest of this chapter you will encounter numerous reactions that readily convert ethylene and alkenes into compounds containing other functional groups and into polymers such as polyethylene.

Just a few of the many possible reactions of alkenes are shown above. You will need some tools to find patterns of reactivity of alkenes. We start by looking more closely at the electronic structure of the C=C bond in alkene molecules and of the reagents they react with to produce these many products.

Alkenes can be converted into numerous other functional groups.

An example of the importance of alkenes, β-carotene, a conjugated alkene, plays a key role in the chemistry of vision, the bright colouration of plants and animals, the provision of necessary vitamins, and in the creation of modern materials for optical-neural networks.

19.2 Electronic Structure of Alkenes

Because of their double bond, alkene molecules are referred to as *unsaturated*. They have fewer H atoms per C atom than related alkanes, which we describe as *saturated* [<<Section 4.5]. Ethylene (or ethene), for example, has the formula C_2H_4, whereas ethane has the formula C_2H_6.

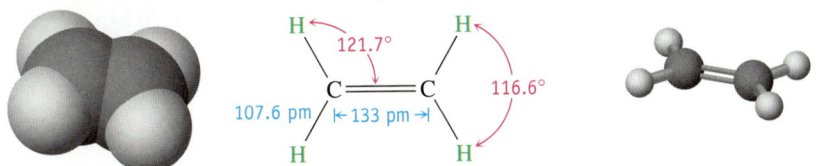

Ethylene: C_2H_4
(fewer hydrogen atoms—*unsaturated*)

Ethane: C_2H_6
(more hydrogen atoms—*saturated*)

Other structural features of alkenes are also important. The experimentally determined C=C bond length (133 pm) in ethylene molecules is much shorter and stronger (bond energy: 611 kJ mol^{-1}) than the C—C single bond (length: 154 pm; bond energy: 376 kJ mol^{-1}) in ethane. Furthermore, all six atoms lie in a plane. This is clear from the representations of ethylene molecules below. Those same features are found in the alkene portions of other molecules, such as Vitamin A (retinol) in Figure 19.1c. Evidence from X-ray crystallography shows that all eight C atoms making up the four conjugated alkenes attached to the cyclohexene ring lie in a plane.

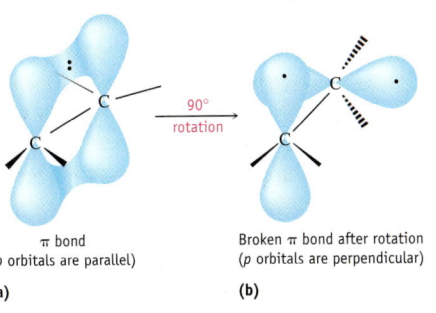

Three representations of a molecule of ethylene.

Section 9.5 discussed the rotation that occurs around C—C single bonds, giving open-chain alkanes like ethane and propane many rapidly interconverting conformers. The same is not true for double bonds [<<Section 9.6]. While the rotation barrier for a single bond is only about 12 kJ mol^{-1}, the energy barrier to rotation around a double bond is as great as the strength of the second bond, an estimated 268 kJ mol^{-1}.

Several complementary models of covalent bond formation help explain these experimental observations. The valence bond model of overlapping hybridized atomic orbitals, introduced in Section 10.8, helps us understand observations about organic structures such as bond length, planarity, and the high barrier to rotation about C=C bonds. Other observations such as the differing colours of substances containing conjugated alkene molecules with different chain lengths, require a molecular orbital model [<<Section 10.9] for a satisfactory explanation.

Applying the model of hybridization in Section 10.8 to alkenes, each C atom in double bonds has three equivalent sp^2 hybrid orbitals, which lie in a plane at angles of about 120° to one another. The fourth C atom orbital is an unhybridized p orbital perpendicular to the sp^2 plane. If you imagine two such C atoms approaching each other, they form a σ bond by head-on overlap of sp^2 orbitals, and a π bond by sideways overlap of p orbitals. The doubly bonded C atoms and the four attached atoms lie in a plane, with bond angles of approximately 120° (Figure 19.6).

Think about It

e19.2 Visualize the model for hybridization of atomic orbitals.

Think about It

e19.3 Watch a portrayal of the model showing overlap of orbitals to form a double bond.

Molecular Modelling (Odyssey)

e19.4 See how the energy changes as a result of rotation around the C=C bond.

FIGURE 19.6 (a) Our model suggests that each C atom in a C=C bond is sp^2 hybridized, with the π bond formed by sideways overlap of one $2p$ orbital on each C atom. (b) For rotation to take place around a double bond, the π bond would have to break.

π bond
(*p* orbitals are parallel)

Broken π bond after rotation
(*p* orbitals are perpendicular)

(a)

(b)

In the hybridization model, the head-on overlap of sp^2 hybrid orbitals on each C atom of the C=C bond of an alkene molecule forms a σ bond, and the sideways overlap of $2p$ orbitals forms a π bond. This explains why the C=C is shorter and stronger than a C—C single bond, and also why the alkene C and H atoms are all in a plane. For rotation to occur around a double bond, the π bond would have to break and re-form, explaining the high energy barrier for rotation in alkenes.

A molecular orbital (MO) model of C=C π bonds is helpful to explain the colours of different conjugated alkenes. As described for the hydrogen molecule in Section 10.9, a thought exercise can be carried out by bringing together the two p atomic orbitals on adjacent C atoms. This can be done in two ways, with an additive combination of the two p orbitals giving a π bonding molecular orbital, and a subtractive combination of the two p orbitals giving an antibonding molecular orbital (designated as a π^* orbital). In a ground-state alkene with only 2 π electrons, only the lower energy bonding molecular orbital would be occupied. This orbital is referred to as the **highest occupied π molecular orbital (HOMO)**. If electromagnetic radiation of the correct frequency is absorbed by a sample of the alkene, one of those electrons in the HOMO will move to the **lowest unoccupied π^* molecular orbital (LUMO)**, which is an antibonding molecular orbital.

Molecular orbital description of the C=C π bond.

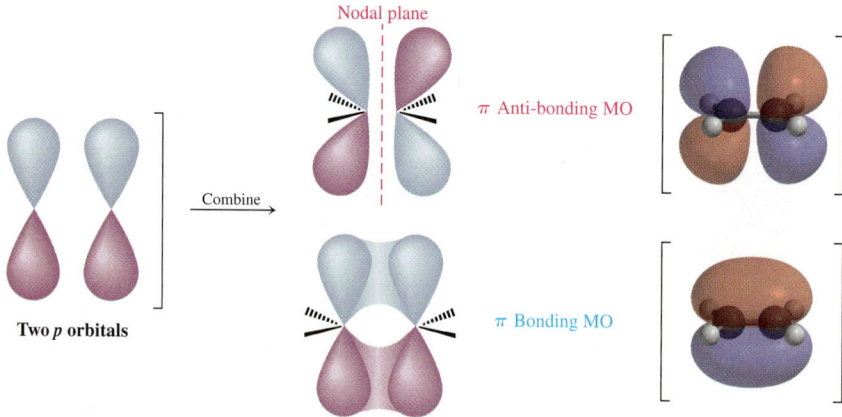

A solution of a conjugated alkene like β-carotene absorbs blue light (and therefore appears orange), because blue light from the visible region of the electromagnetic spectrum provides exactly the right quantum of energy to excite an electron from the HOMO to the LUMO. A solution of an isolated alkene such as cyclohexene is colourless, because the energy of a photon required to excite a π electron from the HOMO to the LUMO of cyclohexene molecules is in the ultraviolet, rather than the visible, region of the spectrum. For a series of conjugated alkenes, the energy gap between the HOMO and LUMO decreases as the number of conjugated C=C bonds increases. Since the wavelength of visible light absorbed changes in this series, so do the colours of substances made of these conjugated alkenes.

The hybridization model of bonding in alkenes explains some properties of organic molecules, but more satisfactory explanations of spectroscopic results and examples of unusual stability or reactivity come from applying molecular orbital (MO) models.

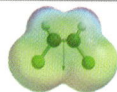

Molecular Modelling (Odyssey)

e19.5 See the HOMO/LUMO energy gap change as the number of conjugated double bonds increases.

19.3 Naming Alkenes

Alkenes are given IUPAC names with a series of rules similar to those used for alkanes, with the suffix *–ene* used in place of *–ane* to identify the family. There are three steps:

Step 1 *Name the parent hydrocarbon.* Find the longest carbon chain that contains the double bond, and name the compound using the suffix *-ene*.

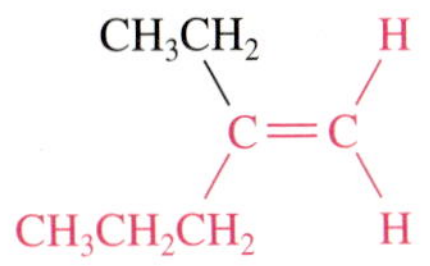

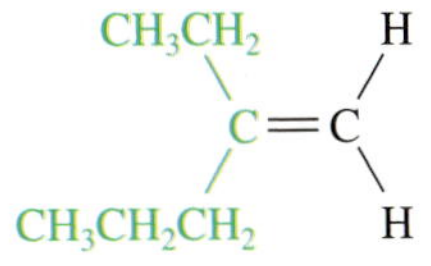

$$CH_3CH_2 \quad\quad H$$
$$\diagdown \quad\quad \diagup$$
$$C=C$$
$$\diagup \quad\quad \diagdown$$
$$CH_3CH_2CH_2 \quad\quad H$$

Named as a *pentene*   *NOT*   as a hexene, since the double bond is
not contained in the six-carbon chain

Step 2   *Number the C atoms in the chain.* Begin numbering at the end nearer the double bond. If the double bond is equidistant from the two ends, begin numbering at the end nearer the first branch point. This rule ensures that the double-bonded C atoms are assigned the lowest possible numbers.

Step 3   *Write the full name.* Number the substituents according to their position in the chain and list them alphabetically. Indicate the position of the double bond by giving the number of the *first* alkene C atom. If more than one double bond is present, give the position of each and use one of the suffixes *–diene, –triene*, and so on. Following IUPAC recommendations, the number locating the position of the double bond is placed before the *–ene* suffix, such as but-2-ene.

In other sources, you may encounter an older convention, in which but-2-ene is named 2-butene, but this is not recommended by IUPAC.

Cycloalkenes are named in a similar way, but because there is no chain end to begin from, we number the cycloalkene so that the double bond is between C1 and C2 and so that the first substituent has as low a number as possible. It is not necessary to specify the position of the double bond in the name because it's always between C1 and C2:

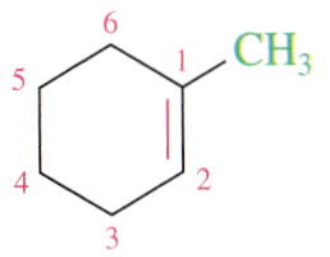

1-Methylcyclohexene 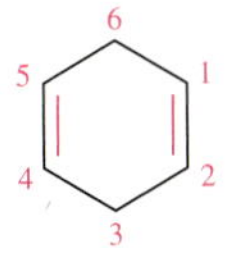   **Cyclohexa-1,4-diene** 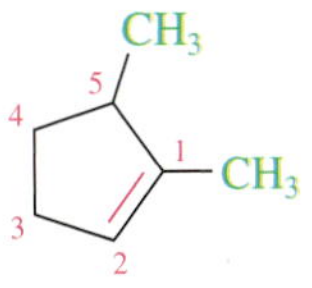   **1,5-Dimethylcyclopentene**

For historical reasons, there are a few alkenes whose names don't conform to the rules. For example, the alkene corresponding to ethane should be called *ethene*, but the name *ethylene* has been used for so long that it is accepted by IUPAC. The common (also called trivial) name for *propene* is *propylene*.

WORKED EXAMPLE 19.1—IUPAC NAMES

What is the IUPAC name of the following alkene?

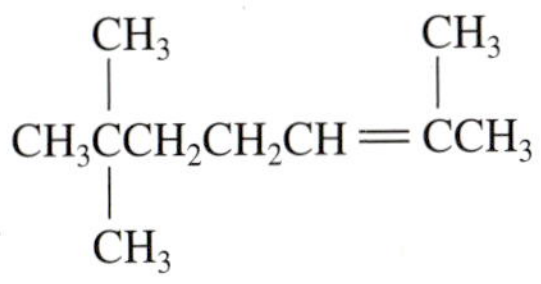

$$\begin{array}{cc} CH_3 & CH_3 \\ | & | \\ CH_3CCH_2CH_2CH=CCH_3 \\ | \\ CH_3 \end{array}$$

Strategy

First, find the longest chain containing the double bond—in this case, a heptene. Next, number the chain beginning at the end nearer the double bond, and identify the substituents at each position. In this case, there are methyl groups at C2 and C6 (two):

Solution

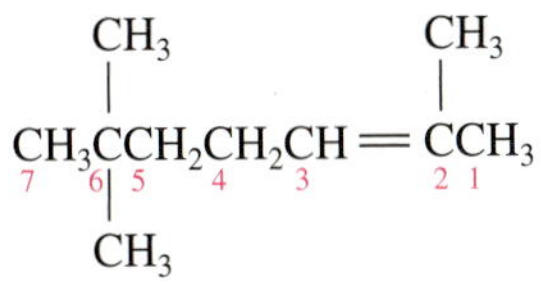

$$\begin{array}{cc} CH_3 & CH_3 \\ | & | \\ CH_3CCH_2CH_2CH=CCH_3 \\ {\scriptstyle 7 \ \ 6 \ \ 5 \ \ \ 4 \ \ \ 3 \ \ \ \ 2 \ 1} \\ | \\ CH_3 \end{array}$$

The full name is 2,6,6-trimethylhept-2-ene.

EXERCISE 19.2—IUPAC NAMES OF ALKENES

Draw structures of molecules of compounds with the following IUPAC names:
(a) 2-methylhex-1-ene
(b) 4,4-dimethylpent-2-ene
(c) 2-methylhexa-1,5-diene
(d) 3-ethyl-2,2-dimethylhept-3-ene

19.4 *E* and *Z* Isomers of Alkenes

The high barrier to rotation around C=C bonds is of more than just theoretical interest; it also has chemical consequences. Imagine the situation for a disubstituted alkene such as but-2-ene. (*Disubstituted* means that two substituents other than H are bonded to the double-bonded C atoms.) The two methyl groups in but-2-ene can be either on the same side of the double bond or on opposite sides, a situation reminiscent of substituted cycloalkanes [<<Section 9.6]. Figure 19.7 shows the two but-2-ene isomers.

Because bond rotation can't occur, the two but-2-enes do not interconvert and are diastereomers [<<Section 9.8], different chemical compounds. As with disubstituted cycloalkanes, we call such compounds *cis* and *trans isomers*. The isomer with both substituents on the same side is *cis*-but-2-ene, and the isomer with substituents on opposite sides is *trans*-but-2-ene.

cis-**But-2-ene**   *trans*-**But-2-ene**

FIGURE 19.7 *Cis* and *trans* isomers of but-2-ene. The *cis* isomer has both methyl groups on the same side of the double bond, and in the *trans* isomer they are on opposite sides.

Cis–trans isomerism is not limited to disubstituted alkenes. It occurs whenever both double-bonded C atoms are attached to two different groups (Figure 19.8). If one of the double-bonded C atoms is attached to two identical groups, however, then *cis–trans* isomers do not exist.

FIGURE 19.8 For *cis* and *trans* isomers to exist, each alkene C=C bond carbon atom must be bonded to two different groups.

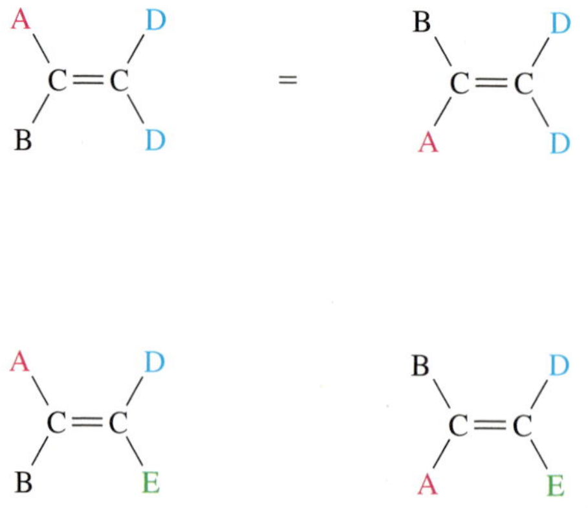

These two compounds are identical; they are not cis–trans isomers.

These two compounds are not identical; they are cis–trans isomers.

Although the interconversion of *cis* and *trans* alkene isomers doesn't normally occur, it can be brought about by treating an alkene with a strong acid catalyst. If we interconvert *cis*-but-2-ene to *trans*-but-2-ene and allow them to reach equilibrium, we find that they aren't of equal stability. The *trans* isomer is more favoured than the *cis* isomer by a ratio of 76:24.

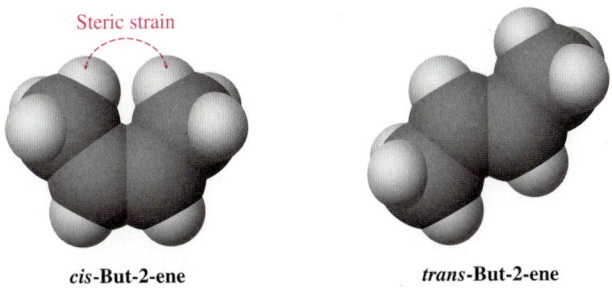

trans (76%)   cis (24%)

Cis alkenes are less stable than their *trans* isomers because of steric (spatial) interference between the large substituents on the same side of the double bond. This is similar to the interference, or *steric strain,* that we saw in axial methylcyclohexane [<<Section 9.7].

Steric strain

cis-But-2-ene   *trans*-But-2-ene

WORKED EXAMPLE 19.2—*CIS–TRANS* ISOMERS

Draw the *cis* and *trans* isomers of 5-chloropent-2-ene.

Strategy

First, draw the molecule without indicating isomers to see the overall structure: $ClCH_2CH_2CH=CHCH_3$. Then locate the two substituent groups on the same side of the double bond for the *cis* isomer and on opposite sides for the *trans* isomer.

Solution

cis-5-Chloropent-2-ene   *trans*-5-Chloropent-2-ene

**Interactive Exercise
(Odyssey) 19.3**

SUBMIT

Build models of *cis* and *trans* isomers of an alkene and calculate the difference in energy.

EXERCISE 19.4—*CIS–TRANS* ISOMERS

Which of the following compounds can exist as *cis–trans* isomers? Draw each *cis–trans* pair.
(a) $CH_3CH=CH_2$
(b) $(CH_3)_2C=CHCH_3$
(c) $ClCH=CHCl$
(d) $CH_3CH_2CH=CHCH_3$
(e) $CH_3CH_2CH=C(Br)CH_3$
(f) 3-methylhept-3-ene

The interconversion of *cis* and *trans* $C=C$ bonds occurs quite rapidly under certain circumstances in nature. An important example is the perception of light, leading to vision, which is triggered by a *cis* to *trans* isomerization of the $C=C$ bonds in molecules of retinal. *Cis*-retinal is produced from dietary β-carotene, which is converted into Vitamin A by enzymes in the small intestine and liver. The alcohol functional group is then oxidized to an aldehyde, along with an isomerization of the double bond between C11 and C12, giving 11-*cis*-retinal (Figure 19.9).

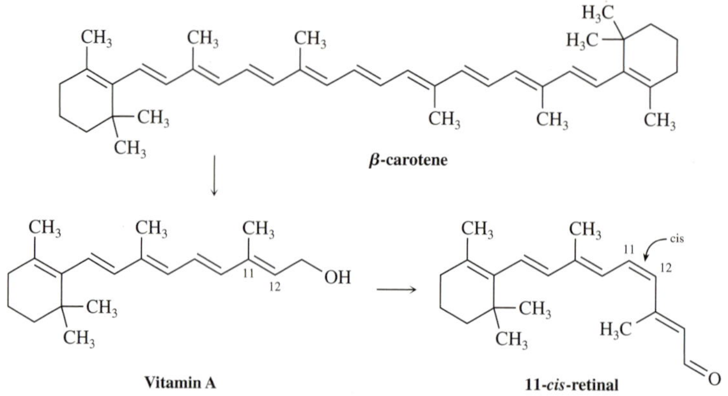

FIGURE 19.9 Molecules of β-carotene converted to Vitamin A and then to 11-*cis*-retinal, the first step in the chemistry of vision.

β-carotene

Vitamin A

11-*cis*-retinal

In the rod cells of the eye, 11-*cis*-retinal binds to the protein opsin, producing the substance rhodopsin. When visible light strikes the rod cells, the C11–C12 *cis* double bond isomerizes to a *trans* double bond (Figure 19.10), converting rhodopsin into *trans*-rhodopsin, also called metarhodopsin II. The molecular geometry of 11-*cis* and 11-*trans* retinal are sufficiently different (Figure 19.11) that the protein changes its overall shape, causing a nerve impulse to be sent to the brain where it is perceived as vision. In the absence of light, this *cis*–*trans* isomerization takes about 1100 years. In the presence of light, it occurs within 2×10^{-11} seconds.

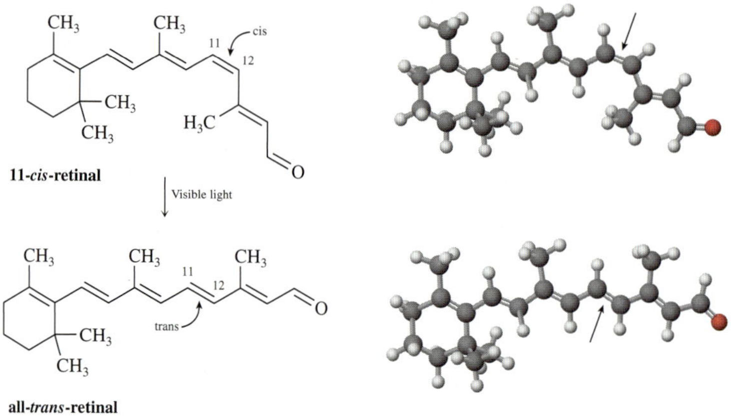

FIGURE 19.10 Isomerization of molecules of 11-*cis*-retinal to 11-*trans*-retinal leads to a striking change in the overall geometry of the molecule, evident in the space filling model representations on the right.

11-*cis*-retinal

Visible light

all-*trans*-retinal

FIGURE 19.11 Representations of the helical protein opsin bound to *cis*-retinal (*left*) and after the *cis*–*trans* isomerization (*right*). Because the all-*trans*-retinal isomer is such a different shape, it does not fit well into the protein, and so it causes a series of conformational changes within nanoseconds (10^{-9} s), eventually leading to retinal no longer binding to the protein. This change in molecular geometry causes a nerve impulse to be sent to the brain, which is perceived as vision.

Dr. Roy Tasker

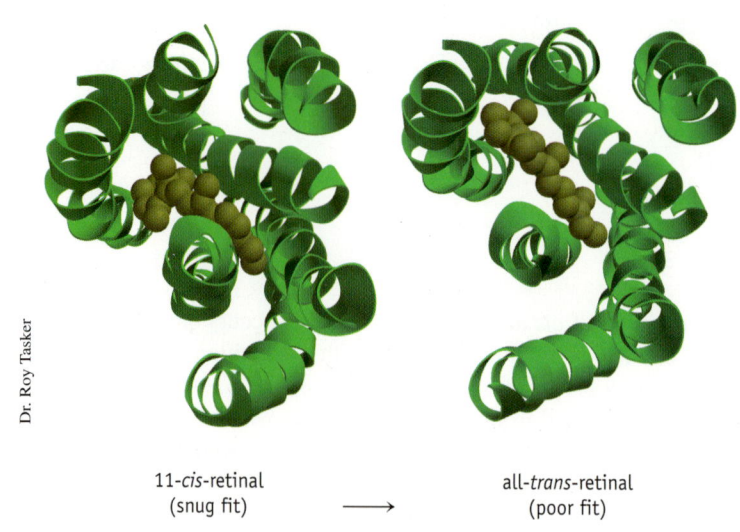

11-*cis*-retinal
(snug fit)

all-*trans*-retinal
(poor fit)

The *cis–trans* naming system used in the previous section works only with disubstituted alkenes. With trisubstituted and tetrasubstituted double bonds, a more general method is needed for describing double-bond geometry.

Applying the IUPAC approved ***E, Z* system of nomenclature** (Figure 19.12), the same set of *sequence rules* (introduced in Section 9.10) to specify configuration of substituents around a stereocentre is used to assign priorities to the substituent groups on the double-bonded C atoms. Considering the double-bonded C atoms separately, we decide which of the two attached groups on each is higher in priority. If the higher-priority groups on each C atom are on opposite sides of the double bond, the alkene is designated *E*, for the German *entgegen*, meaning "opposite." If the higher-priority groups are on the same side, the alkene is designated *Z*, for the German *zusammen*, meaning "together."

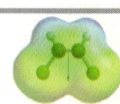
Trisubstituted means three substituents other than H on the double bond; *tetrasubstituted* means four substituents other than H.

FIGURE 19.12 *E, Z* system of nomenclature for alkenes.

Lower   Higher

$$\text{C} = \text{C}$$

Higher   Lower

E double bond (Higher-priority groups are on opposite sides.)

Higher   Higher

$$\text{C} = \text{C}$$

Lower   Lower

Z double bond (Higher-priority groups are on the same side.)

Called the *Cahn–Ingold–Prelog rules* after the chemists who proposed them, a summary of the sequence rules can be obtained in section 9.10.

WORKED EXAMPLE 19.3—ASSIGNING *E, Z* CONFIGURATION

Assign *E* or *Z* configuration to the double bond in molecules of the following compound:

$$\begin{array}{ccc} \text{H} & & \text{CH(CH}_3)_2 \\ & \text{C} = \text{C} & \\ \text{H}_3\text{C} & & \text{CH}_2\text{OH} \end{array}$$

Strategy

Look at each double-bonded C atom individually, and assign priorities. Then see whether the two high-priority groups are on the same or opposite sides of the double bond.

Solution

The left-hand C atom has two substituents, —H and —CH$_3$, of which —CH$_3$ receives higher priority by rule 1. The right-hand C atom also has two substituents, —CH(CH$_3$)$_2$ and —CH$_2$OH, which are equivalent by rule 1. By rule 2, however, —CH$_2$OH receives higher priority than —CH(CH$_3$)$_2$ because —CH$_2$OH has an O atom as its highest second atom, whereas —CH(CH$_3$)$_2$ has a C atom as its highest second atom. The two high-priority groups are on the same side of the double bond, so the compound has *Z* configuration.

Low   H   CH(CH$_3$)$_2$ Low

$$\text{C} = \text{C}$$

High   H$_3$C   CH$_2$OH   High

C, C, H bonded to this carbon

O, H, H bonded to this carbon

Z configuration

EXERCISE 19.6—ASSIGNING *E, Z* CONFIGURATIONS

Which member in each of the following sets is higher in priority?
(a) —H or —Br   (b) —Cl or —Br   (c) —CH$_3$ or —CH$_2$CH$_3$
(d) —NH$_2$ or —OH   (e) —CH$_2$OH or —CH$_3$ (f) —CH$_2$OH or —CH=O

EXERCISE 19.7—ASSIGNING *E, Z* CONFIGURATIONS

Assign *E* or *Z* configuration to molecules of the following compounds:

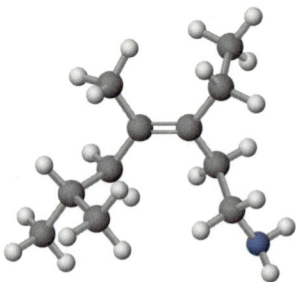

EXERCISE 19.8—ASSIGNING *E, Z* CONFIGURATIONS

Assign *E* or *Z* configuration to molecules of the following compound (blue = N):

When double-bonded C atoms are each attached to two different groups, *cis* and *trans* isomers of the compound are possible. The π bond restricts rotation about the C=C bond. Systematic IUPAC names for such alkenes make use of the *E* (*entgegen*), *Z* (*zusammen*) convention.

19.5 Spectroscopy of Alkenes

Infrared Spectroscopy

Recall from Section 3.11 that organic functional groups can be identified by their characteristic infrared absorption peaks, most of which occur in the region above 1450 cm^{-1} (Figure 19.13).

FIGURE 19.13 Characteristic absorption regions in the infrared spectra of molecules with single, double, and triple bonds.

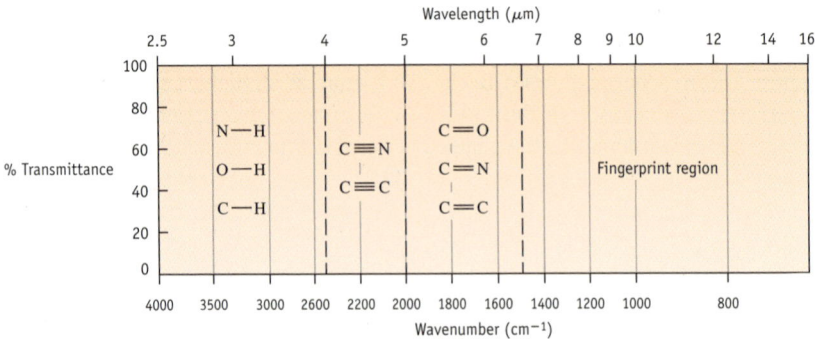

e19.7 uses animated infrared spectra of alkene molecules to show how they can be identified by several characteristic IR absorptions. A C—H bond attached to a C=C carbon atom is slightly shorter (106 pm, ethene) and stronger than a C—H bond attached to an alkane C atom (109 pm, ethane). The energy (wavenumber) of infrared radiation corresponding to the stretching vibration of an alkene C—H bond is greater than for an alkane C—H bond. Examine the region around 3000 cm⁻¹ in Figure 19.14 and the animated IR spectrum in e19.7 and note the exact absorption wavenumber for all C—H stretching vibrations. Absorption peaks just below 3000 cm⁻¹ correspond to symmetric and antisymmetric stretching vibrations of the alkane C—H bonds. Those just above 3000 cm⁻¹ correspond to symmetric and antisymmetric stretching vibrations of H atoms attached to C atoms of C═C bonds, indicating the presence of an alkene that has H atoms attached directly to the C═C. In the spectrum of hex-1-ene, both types of C—H bonds are present.

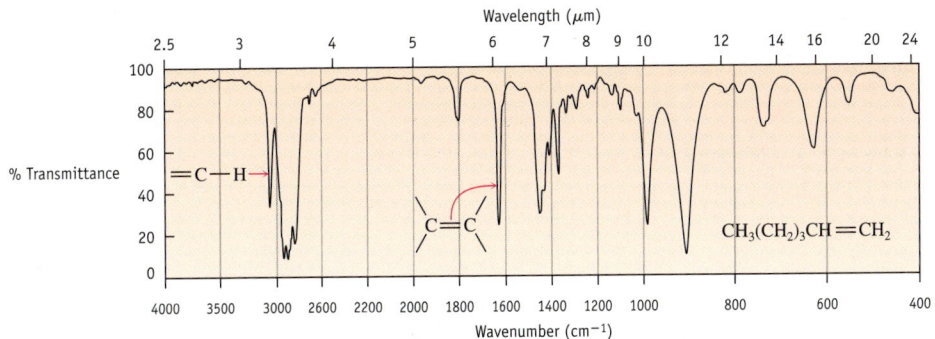

FIGURE 19.14 Infrared spectrum of hex-1-ene.

In alkenes whose molecules have unsymmetrically substituted C═C bonds, a C═C stretching vibration often is seen at 1640–1680 cm⁻¹, and if the alkene has H atoms attached directly to one of the C atoms of a C═C, out-of-plane bending vibrations are found at about 900 cm⁻¹. The exact wavenumber corresponding to those out-of-plane bending vibrations can be used to determine the pattern of substitution of H atoms on the C═C carbon atoms (number of H atoms and whether they are *cis* or *trans* to each other).

Think about It

e19.8 Examine IR spectra of three alkenes to confirm where alkene absorptions appear.

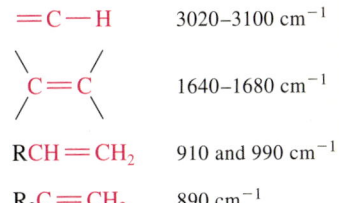

═C—H	3020–3100 cm⁻¹
C═C	1640–1680 cm⁻¹
RCH═CH₂	910 and 990 cm⁻¹
R₂C═CH₂	890 cm⁻¹

Summary of key infrared absorptions characteristic of alkenes.

¹³C NMR Spectroscopy

Recall from Section 9.4 that a ¹³C NMR spectrum of a compound identifies the number of C atoms in unique environments in a molecule. As illustrated in the ¹³C NMR spectrum of hex-1-ene in Figure 19.15, a sharp resonance line is observed for each of the six C atoms, all of which are in unique environments.

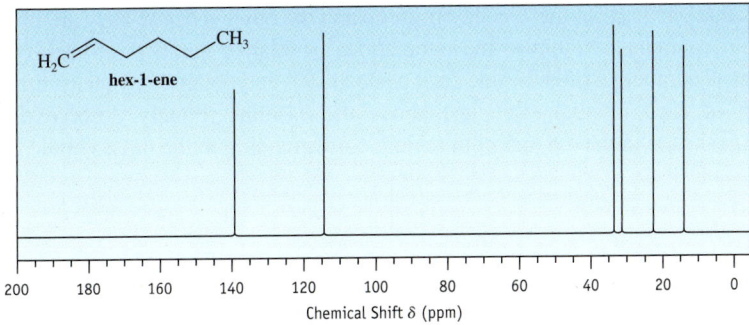

FIGURE 19.15 ¹³C NMR spectrum of hex-1-ene.

Carbon atoms in alkenes and aromatic rings appear downfield from C atoms that are part of single bonds. Typical chemical shifts for alkenes range from $\delta = 110$–150 ppm. e19.9 gives examples of correlations between the three-dimensional structures of alkene molecules and their ¹³C NMR spectra.

Think about It

e19.9 Work through an analysis of a ¹³C NMR spectrum of an alkene.

EXERCISE 19.9—¹³C NMR SPECTROSCOPY OF ALKENES

Which of the six absorbance lines in Figure 19.15 should be assigned to the two alkene C atoms?

^{1}H NMR Spectroscopy

When used with ^{13}C NMR, ^{1}H NMR provides a "map" of the carbon-hydrogen framework in an organic molecule. While appearing to be more complex than a ^{13}C NMR spectrum, the ^{1}H NMR spectrum for a compound can provide even more information—including the *number of unique "types" of H atoms* in a molecule, the *relative number of each type*, and how H atoms are *positioned relative to neighbouring H atoms*.

^{1}H chemical shifts correlate well with the electronic environment surrounding the H atoms, as seen in Table 19.1. Note that protons directly attached to alkene C atoms (referred to as vinylic H atoms), typically absorb downfield (ranging from about δ = 4.5–6.5 ppm) from protons attached to alkane C atoms (δ = 1–5 ppm). They often appear as clusters of peaks, rather than sharp single peaks, due to the influence of neighbouring H atoms.

TABLE 19.1 Correlation of ^{1}H NMR Chemical Shift with the Electronic Environment

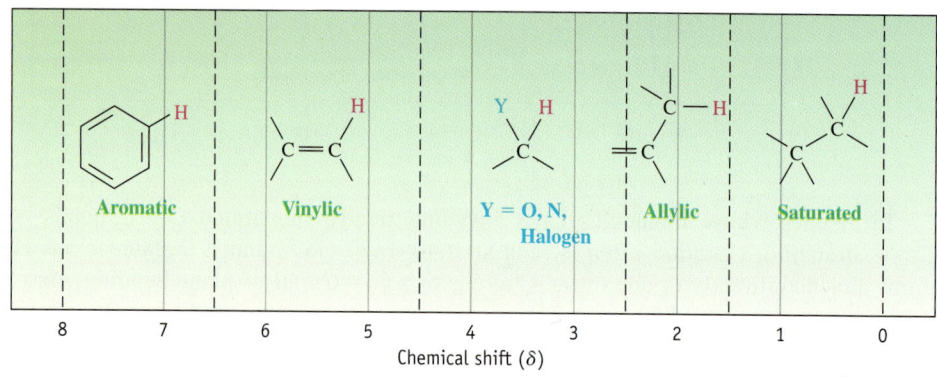

At this point in your work with ^{1}H NMR spectra, focus on the *number of peak clusters*, which will tell you how many unique H atoms are found in the molecule, paying particular attention to those appearing between 4.5 and 6.5 ppm. You should also note the *integration* of each of those peak clusters, which tells you the relative numbers of each type of unique H atoms.

For example, 3-bromopropene in Figure 19.16 has four unique types of H atoms, and four peak clusters appear in the ^{1}H NMR spectrum. While you might first think that the protons labelled "c" and "d" are equivalent to each other, the restricted rotation about the C=C bond means that one is *cis* to the —CH$_2$Br group and the other is *trans*. They are in different electronic environments. Note also the integrations or relative numbers of each type of H atom, shown as numbers written beside each peak cluster, and obtained by measuring the relative height of each integral trace just above the baseline. Finally observe that three absorbance peaks are in the 4.5–6.5 ppm region, corresponding to the three vinylic H atoms.

Think about It

e19.10 See why some peaks in a ^{1}H NMR spectrum like Figure 19.16 are clusters rather than sharp peaks.

Think about It

e19.11 Examine ^{1}H NMR spectra of five alkenes to confirm where vinylic and allylic resonances appear.

FIGURE 19.16 ^{1}H NMR of 3-bromopropene, with assignments. Numbers beside each of the four absorbance (peak) clusters indicates the integration, or relative number of H atoms responsible for each absorbance.

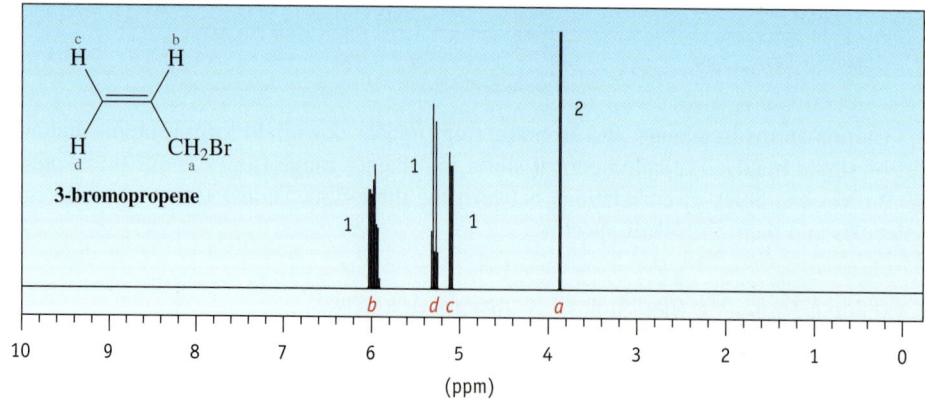

The IR and ^1H NMR spectra for a compound with formula C_6H_{10} are given below. The ^{13}C NMR shows three sharp resonances, at $\delta = 127$, 25, and 23 ppm. Analyze the two spectra and identify the compound.

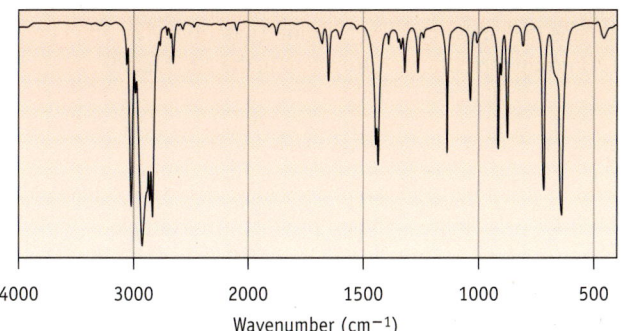

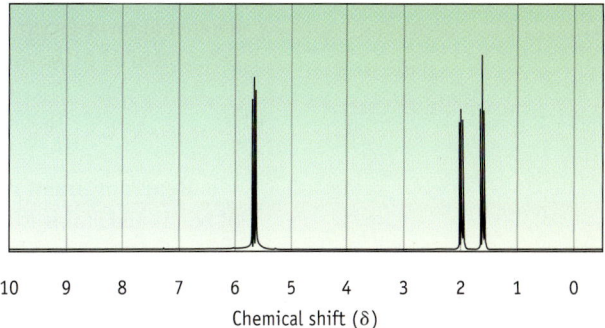

19.6 Classifying Organic Reactions by Type

Knowing something about the electronic structure of alkene molecules can help us understand the chemical reactivity of alkenes. Before looking at alkene reactions, we'll first outline some basic principles that underlie all organic reactions.

An organic chemical reaction can be classified either by *what kind* of reaction occurs or by the *reaction mechanism* by which it is believed to proceed—that is, *how* the reaction has occurred at the molecular level.

Addition, Elimination, Substitution, Rearrangement

We can tell what *kind of reaction* has taken place by observing how the products differ from the reactants. This is done using spectroscopic techniques like IR and NMR spectroscopy, and mass spectrometry to understand the structures of molecules of reactants and products, and see what overall changes have taken place during a reaction. There are four broad types of organic reactions: *addition, elimination, substitution,* and *rearrangement.*

Addition reactions occur when molecules of two reactants add together to form molecules of a single new product with no atoms "left over." An example that we'll be studying soon is the reaction of an alkene with HCl to yield an alkyl chloride:

These two reactants ...

H—Cl

+

Ethylene
(an alkene)

Chloroethane
(an alkyl halide)

... add to give
this product.

Elimination reactions are, in a sense, the opposite of addition reactions and occur when reactant molecules split into molecules of two products. An example is the reaction of an alkyl halide with base to yield an acid and an alkene:

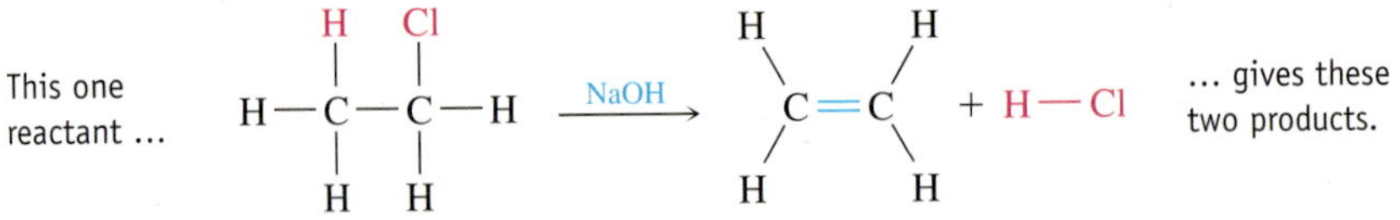

This one reactant ...

Chloroethane
(an alkyl halide)

Ethylene
(an alkene)

... gives these two products.

Substitution reactions occur when an atom or group in a molecule is replaced by another atom or group. An example is the nucleophilic substitution reaction of an alkyl halide [>>Section 22.5], where an —OH group replaces a —Br group in the alkyl halide molecule.

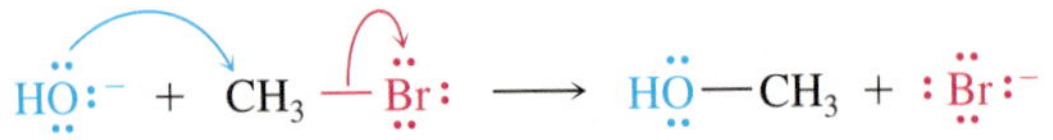

Rearrangement reactions occur when molecules of a reactant undergo a reorganization of bonds and atoms to yield molecules of a single isomeric product. An example is the conversion of 11-*cis*-retinal into its isomer 11-*trans*-retinal in Figure 19.9, important in vision.

One important way to classify organic reactions is based on spectroscopic analysis of reactants and products to see what kind of reaction has occurred. There are four broad types of organic reactions: addition, elimination, substitution, and rearrangement.

EXERCISE 19.11—CLASSIFYING ORGANIC REACTIONS BY TYPE

Classify the following reactions as addition, elimination, substitution, or rearrangement:
(a) $CH_3Br + KOH \longrightarrow CH_3OH + KBr$
(b) $CH_3CH_2OH \longrightarrow H_2C{=}CH_2 + H_2O$
(c) $H_2C{=}CH_2 + H_2 \longrightarrow CH_3CH_3$

19.7 Visualizing the Mechanisms of Organic Reactions

How do reactions occur? How do atoms rearrange? The answer lies in the valence electrons. Most reactions occur with a flurry of electron activity leading to a change of connectivity between atoms. But how can you picture what's happening at a molecular level? Imagine that you could record video footage with a molecular-level video cam—what would you see? From a distance, you would see an uncountable number of reactant molecules moving around rapidly, surrounded by even more solvent molecules. Some of them will be clustered to the solvent through various intermolecular forces. Reactant molecules will be colliding with each other much too frequently to even hope to track them. Now zoom in on a single pair of reactant molecules. Most probably they will bump into each other without enough energy or with the wrong orientation, and nothing will happen. But every so often, one of those collisions will lead to a transfer of electrons and perhaps a reorganization of atoms—many such successful collisions constitute a chemical reaction. If this impossible filming exercise could be carried out, and you could analyze each frame of the video, you would be the first chemist to ever witness a **reaction mechanism** [<<Section 18.7]. We need a way to show how a chemical reaction happens at the molecular level—in how many separate steps and which steps are faster than others. For each step in the reaction mechanism, we can write a chemical equation that shows which particles collide and interact; we can use curved arrows to indicate what electron shifts take place, what bonds are broken and formed, and in what order; and we can label which steps are faster than others.

Understanding reaction mechanisms takes time and patience, but the rewards are great. Mechanisms can form the basis for chemical "intuition," allowing you to understand patterns of reactivity when you encounter new reactions, reagents, and processes, with much less blind memorization. But it is easy to forget that you are zooming in only a single pair of successfully reacting molecules, with an imaginary camera capable of shooting this molecular-level film. Chemists at best can make educated guesses or models of reaction mechanisms, and as for any model, no description of a mechanism can be proven with complete certainty. Evidence at the observable level that our understanding of a particular reaction mechanism is a reasonable molecular-level explanation comes from consistency with experimental data on the rate law for the reaction, knowledge about changes of stereochemistry, and the presence or absence of reactive intermediates.

In practice, a reaction mechanism is described by a series of chronologically ordered steps, with each step using symbols such as curved arrows to represent clearly the motion of electrons. When these steps are added together, the result is the net reaction. A mechanism suggests when, in the sequence of events that make up the reaction, particular bonds in the reactant molecules are broken and particular bonds in the product molecules are formed. Remember that for each sequence of collisions leading to the products there are millions of unproductive collisions.

The electrons that initiate chemistry are usually not hidden between nuclei in a strong C—C σ bond. More often, these electrons are in accessible positions such as π bonds or lone pairs. The number of electrons that move in any mechanistic step depends greatly on the type of reaction and conditions. Most organic reactions that we will examine are considered **polar reactions**, in which the description of the mechanism shows two electrons (electron pairs) moving together when bonds are formed or broken.

> Electrons in molecules are extraordinarily mobile and a flurry of electrons will shift in the changing environment of a chemical reaction. Curved arrows are a way of keeping track of these overall changes.

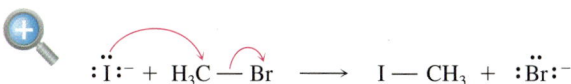

In polar reactions such as these, the electrons in lone pairs of ions or on oxygen or nitrogen as well as the π electrons of double bonds can initiate chemical reactions.

When electrons move in pairs, bonds break in a *heterolytic* manner and form in a *heterogenic* manner. That is, when a bond is broken, both bonding electrons remain with one of the two fragments remaining, and when a bond is formed, both bonding electrons required are contributed by one reaction partner. In **radical reactions** involving unpaired electrons, bond breaking and forming reactions are called *homolytic* and *homogenic*, respectively. Figure 19.17 shows three one-step mechanisms, including a step where electrons move in pairs (*polar reactions*), a reaction where single electrons move (*free radical reactions*), and a reaction where three pairs of electrons move simultaneously (*concerted reactions*, so called because electron pairs are moving so bonds break and form at the same time, or in concert).

Think about It

e19.12 Watch an animation explaining the meaning of curved arrow notation.

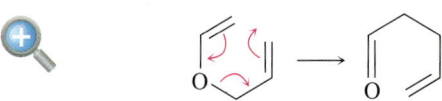

A polar substitution reaction showing heterolysis of the C–Br bond and heterogenesis of the C–I bond.

FIGURE 19.17 Examples of one-step reaction mechanisms for polar, free radical, and concerted reactions.

A radical elimination reaction featuring homolysis of the O–O bond.

A concerted rearrangement reaction showing three electron pairs moving simultaneously.

Notice how the electron movement in the reaction mechanisms in Figure 19.17 is indicated with curved arrows. The arrow indicates *movement of electrons* from where the tail is to where the head is. The number of electrons moving is indicated by the type of arrow: a double-barbed arrow denotes two electrons moving; a single-barbed (fish-hook) arrow represents movement of a single electron. *Note that curved arrows are never used to show atoms moving, but only electron movement.*

Understanding a reaction mechanism, how a reaction occurs at the molecular level, is one of the most important tasks of a chemist, as it helps find patterns of reactivity in the myriad of reactions. Most reactions in this book are classified as polar reactions based on their mechanisms—electrons move in pairs as bonds break and form.

Polar Reaction Mechanisms

To see how polar reactions occur, we first need to look more deeply into the effects of bond polarity on organic molecules. We saw in Section 6.3 that certain bonds in a molecule are polar. In bonds between carbon and a more electronegative atom such as chlorine or oxygen, the bond is polarized so that the carbon atom bears a partial positive charge ($\delta+$) and the electronegative atom bears a partial negative charge ($\delta-$). In bonds between carbon and a less electronegative atom such as a metal, the opposite polarity results. Electrostatic potential maps show electron-rich regions of a molecule in red and electron-deficient regions in blue.

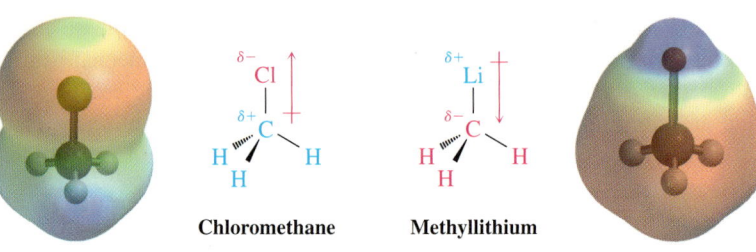

Chloromethane   **Methyllithium**

> **Molecular Modelling (Odyssey)**
>
> **e19.13** Examine electrostatic potential maps to identify electron-rich and electron-deficient regions in a molecular model.

What effect does bond polarity have on chemical reactions? *The fundamental characteristic of all polar reactions is that electron-rich sites in one molecule react with electron-poor sites in another molecule because the negatively charged region of one molecule and the positively charged region of a nearby molecule are attracted to each other.* Bonds form when there is a transfer of a pair of electrons from an electron-rich atom to an electron-poor atom, and bonds break when one atom leaves with both bonding electrons.

A generalized polar reaction

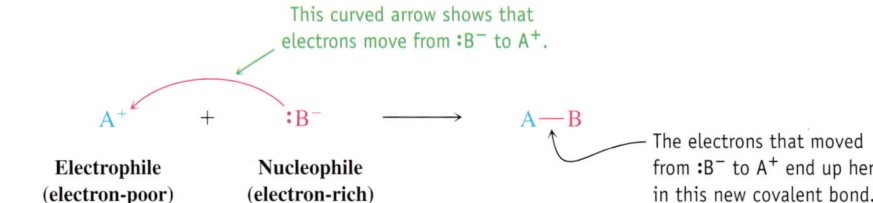

> **Think about It**
>
> **e19.14** Watch animations of electron movement portrayed with curved arrows involving nucleophiles, electrophiles, and leaving groups.

In referring to polar reactions, chemists use the words *nucleophile* and *electrophile* (Figure 19.18). A **nucleophile** is a molecule or ion that is "nucleus loving" and thus attracted to a positive charge. A nucleophile has an electron-rich atom and can form a bond to its reaction partner (the electrophile) by donating both bonding electrons. Nucleophiles often have lone pairs of electrons and are often negatively charged. Molecules of ammonia and water, and hydroxide and chloride ions are examples. Each of these could also be considered a *Lewis base* [<<Section 6.7]. An **electrophile**, by contrast, is "electron loving" and thus attracted to a negative charge. An electrophile has an electron-poor atom and can form a bond by accepting an electron pair from a nucleophile. Electrophiles are often, though not always, positively charged. Molecules or ions of acids (H^+ donors), alkyl halides, and carbonyl compounds are examples. Each of these could also be considered a *Lewis acid*.

WORKED EXAMPLE 19.4—BOND POLARITY

What is the direction of bond polarity in the amine functional group $C-NH_2$?

Solution

Look at the electronegativity values in Figure 6.11 to see which atoms withdraw electrons more strongly. Nitrogen atoms are more electronegative than carbon atoms, so an amine is polarized with C atoms $\delta+$ and N atoms $\delta-$.

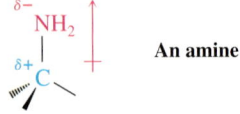

An amine

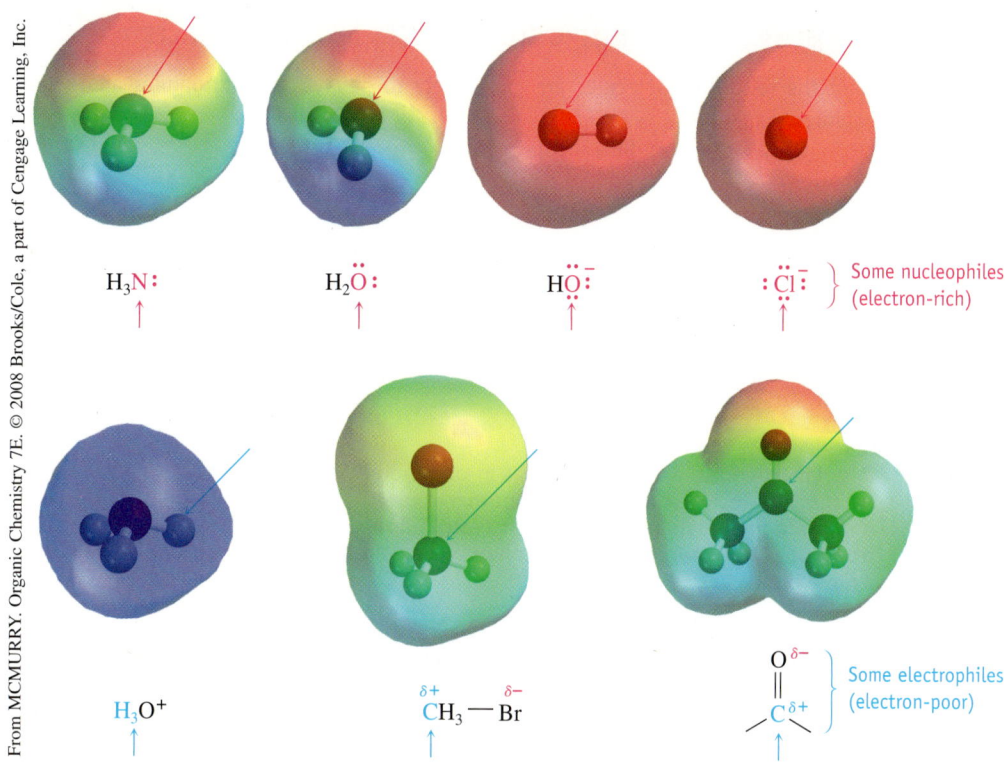

FIGURE 19.18 Examples of nucleophiles and electrophiles. Arrows indicate nucleophilic and electrophilic centres.

WORKED EXAMPLE 19.5—RECOGNIZING ELECTROPHILES AND NUCLEOPHILES

Determine whether each of the following species is likely to be an electrophile or a nucleophile:
(a) NO_2^+ ions   (b) CH_3OH molecules

Strategy

Electrophiles have an electron-poor site, either because they are positively charged or because they have a functional group containing an atom that is positively polarized. Nucleophiles have an electron-rich site, either because they are negatively charged or because they have a functional group containing an atom that has a lone pair of electrons.

Solution

(a) NO_2^+ (nitronium ion) is likely to be an electrophile because it is positively charged.
(b) CH_3OH (methanol) molecules can be either nucleophiles, because they have two lone pairs of electrons on oxygen, or electrophiles, because they have polar C—O and O—H bonds.

$$\overset{\delta+}{CH_3} - \overset{\delta-}{\underset{\cdot\cdot}{\overset{\cdot\cdot}{O}}} - \overset{\delta+}{H}$$

Electrophilic   Nucleophilic   Electrophilic

EXERCISE 19.13—BOND POLARITY OF FUNCTIONAL GROUPS

What is the direction of bond polarity in molecules with the following functional groups? (See Figure 6.11 for electronegativity values.)
(a) aldehyde   (b) ether
(c) ester   (d) alkylmagnesium bromide, R—MgBr

Interactive Exercise (Odyssey) 19.12

SUBMIT

Identify the most electrophilic and nucleophilic centres in molecules using electrostatic potential maps.

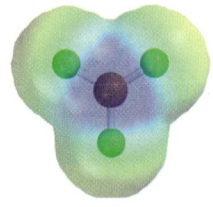

EXERCISE 19.14—RECOGNIZING ELECTROPHILES AND NUCLEOPHILES

Which of the following molecules are most likely to behave as electrophiles and which as nucleophiles? Explain.

(a) NH_4^+ (b) CN^- (c) Br^- (d) CH_3NH_2 (e) $H—C\equiv C—H$

EXERCISE 19.15—ELECTROSTATIC POTENTIAL MAPS

An electrostatic potential map of a molecule of boron trifluoride is shown. Is a BF_3 molecule likely to be an electrophile or a nucleophile? Draw a Lewis structure for a BF_3 molecule, and explain the result.

In polar reactions, electron-rich sites in one molecule (a Lewis base or nucleophile) react with electron-poor sites in another molecule (a Lewis acid or electrophile). Bonds are made when an electron-rich atom donates a pair of electrons to an electron-poor atom, and bonds are broken when one atom leaves with both electrons from a bond.

19.8 Electrophilic Addition of HX to Alkenes: Hydrohalogenation

Let's look in detail at a typical polar reaction of alkenes, the reaction of ethylene with HCl. Spectroscopic techniques have been used to show that when ethylene is reacted with HCl at room temperature, chloroethane is produced. Overall, the reaction can be represented by the chemical equation:

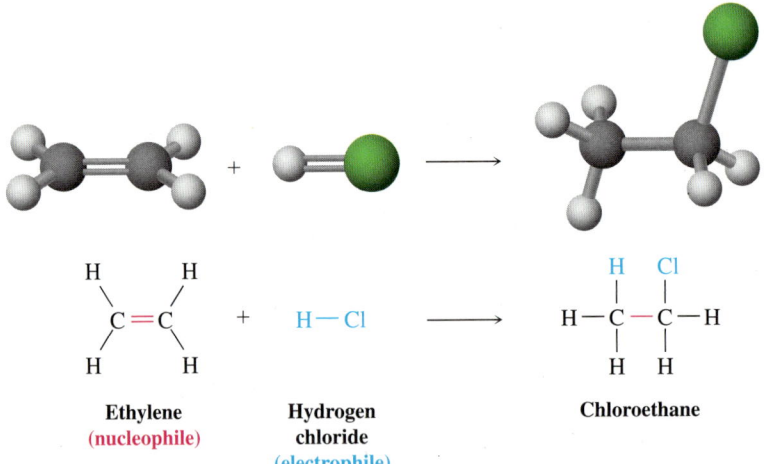

Ethylene	**Hydrogen**	**Chloroethane**
(nucleophile)	**chloride**	
	(electrophile)	

Can we make sense of why alkenes react in this way, using the general concepts discussed in Sections 19.6 and 19.7? First, the kind of reaction is classified as *addition*, as two molecules add together to form a molecule of the product. It is called a *hydrohalogenation reaction* since the Lewis acid or electrophile HCl adds across the C=C bond of ethylene molecules, with the H atom adding to one C atom and the halogen atom (Cl) to the other. Since HCl molecules are electrophiles, we classify this reaction as an *electrophilic addition to an alkene*. Now let's make an educated guess about the reaction mechanism—how the transfer of electrons occurs in each step as reactant molecules are transformed into product molecules.

What do we know about ethylene molecules? Remember that the atomic orbital hybridization model shows a C=C bond as resulting from orbital overlap of two sp^2 hybridized C atoms. The σ bond results from $sp^2—sp^2$ overlap, and the π bond results from sideways $p—p$ overlap.

How might we expect a substance containing molecules with a C=C bond to react? We know that alkanes are relatively unreactive because their molecules' valence electrons are tied up in strong, non-polar, C—C and C—H bonds. The bonding electrons in alkane molecules are considered relatively inaccessible to external reagents because they are sheltered in σ bonds between nuclei. The situation for *alkenes* is quite different, however. For one thing, double bonds, consisting of a π and σ bond, have greater electron density than single bonds—four electrons in a double bond versus only two electrons in a single bond. As well, the electrons in the π bond are accessible to external reagent molecules because they are located above and below the plane of the carbon and attached atoms rather than between the nuclei (Figure 19.19). Even though the total bond strength of C=C bonds in alkenes are greater than for C—C single bonds in alkanes, alkenes are more reactive than alkanes, as the weaker π bonds can break, transferring a pair of electrons.

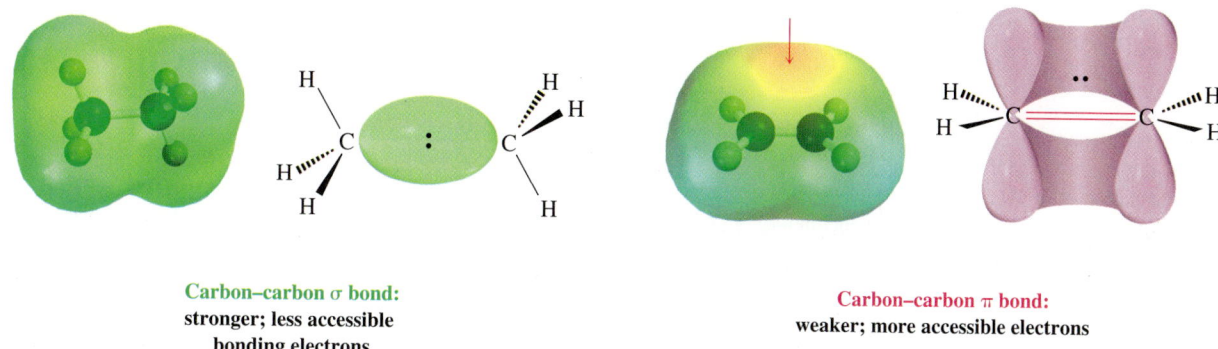

Carbon–carbon σ bond:
stronger; less accessible
bonding electrons

Carbon–carbon π bond:
weaker; more accessible electrons

FIGURE 19.19 A comparison of C—C single and double bonds. A double bond is both more accessible to approaching reactants than a single bond, and more electron-rich (more nucleophilic).

Both electron richness and electron accessibility lead to the prediction that C=C bonds should behave as *nucleophiles*. That is, the chemistry of alkenes should involve reaction of the electron-rich π bond with electron-poor reagents. This is exactly what we find: The most common reaction of alkenes is their reaction with electrophiles.

And what do we know about HCl? As a strong acid, HCl is a powerful proton (H$^+$) donor. Because a proton is positively charged and electron-poor, it is a good electrophile. So the reaction of H$^+$ ions with ethylene molecules is a typical electrophile–nucleophile combination, characteristic of all polar reactions.

On the basis of this understanding of electronic structures and patterns of reactivity, we can imagine the mechanism for electrophilic addition reaction between ethylene and HCl to proceed as represented in the steps shown in Figure 19.20a. Imagine one of the many collisions between an ethylene molecule and an HCl molecule (or an H$^+$ ion, since HCl is a strong acid) leading to a flurry of electron activity as the electrons respond to the environment in which the reagent species find themselves. In this case, two electrons from the ethylene π bond move toward the H$^+$ ion, forming a new σ bond between the H and one of the ethylene C atoms. This electron motion is shown by a double-headed arrow that starts with the electrons and points to the H$^+$. The other ethylene C atom, having lost its share of the π electrons, is now left with a vacant *p* orbital, has only six valence electrons, and carries a positive charge. It is called a *carbocation* (short for carbon cation) and is an electrophile. In the second step, this electrophilic carbocation accepts an electron pair from the nucleophilic Cl$^-$ anion to form a C—Cl bond and produce the neutral addition product. Once again, a curved arrow in Figure 19.20a shows the path of the electron-pair movement from Cl$^-$ ion to the electrophilic C atom. This mechanism can't be proven, but is consistent with many lines of evidence, including what the products are, the observed rate laws for these reactions, the effect of changing solvents, and stereochemical observations of products when substituted alkenes are used.

Remember that, for each of the productive collisions leading to products over time, such as the one shown in Figure 19.20, many collisions will occur with insufficient energy or the wrong orientation between reacting molecules. In writing steps in a postulated reaction mechanism, we show only the ones that lead to the products that we know are formed.

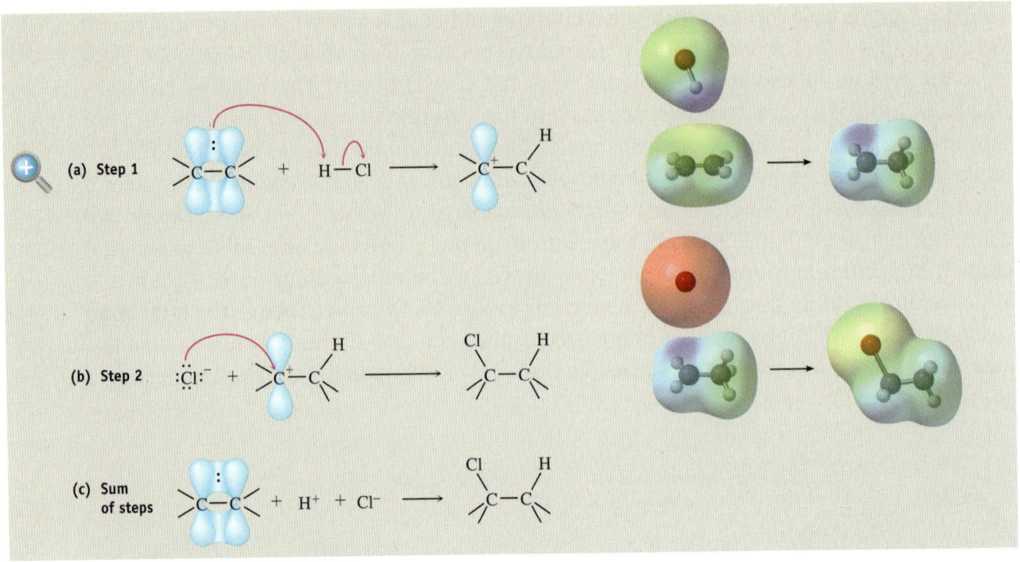

FIGURE 19.20 Two representations of the **accepted reaction mechanism** of the electrophilic addition of HCl to ethylene, in which we "zoom" in on an event showing a reacting ethylene molecule successfully colliding with an H^+ ion, leading to the formation of a product molecule. Formation of each product molecule takes place in two reaction steps. Step 1 involves a collision between the electrophile H^+ and the alkene nucleophile, leading to a transfer of the two π electrons and forming a new C—H σ bond. In step 2, Cl^- donates an electron pair to the positively charged C atom, forming a C—Cl σ bond and yielding the neutral alkyl halide product. Note that the sum of the two mechanistic steps gives the overall reaction, and both atoms and charge are conserved in each step. Electrostatic potential maps show the electron-rich and electron-poor regions in the reactants, the intermediate, and the product of the reaction.

When viewing the representation of a reaction mechanism, remind yourself that the equations you are looking at are different from the equations for overall reactions. They are an educated guess at what the accepted steps in a reaction mechanism might "look like" if you could see them at a molecular level (you can't!). Remember too that you are choosing not to look at the solvent or any of the unproductive collisions between species, but are zooming in on just one of the relatively few events at the individual molecule level that actually leads to the formation of individual molecules of products. The 'zoom icon' at the top reminds you of this.

Molecular Modelling (Odyssey)

e19.15 Examine a three-dimensional simulation of this mechanism.

Think about It

e19.16 Watch animations of the electron movement in this mechanism, portrayed with curved arrows and the orbitals involved.

WORKED EXAMPLE 19.6—ADDITION OF HX TO ALKENES

What product would you expect from reaction of HCl with cyclohexene?

Solution

HCl molecules add to the double-bonded functional group in cyclohexene molecules in exactly the same way as HCl adds to ethylene, yielding an addition product.

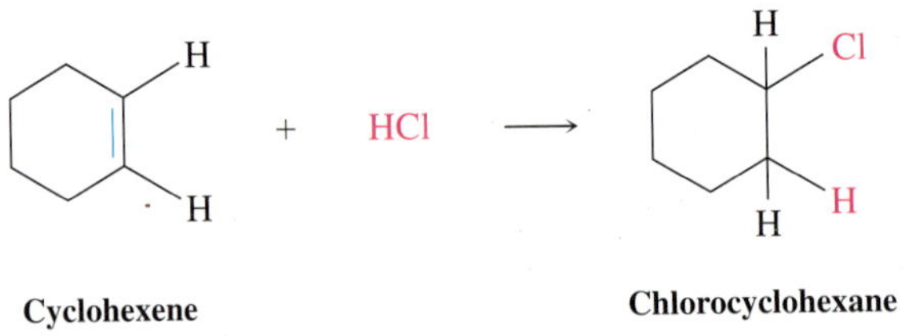

Cyclohexene   +   HCl   ⟶   Chlorocyclohexane

EXERCISE 19.16—ADDITION OF HX TO ALKENES

Reaction of HCl with methylprop-2-ene yields 2-chloro-2-methylpropane but not 1-chloro-2-methylpropane. What is the structure of the carbocation formed during the reaction? Show the detailed, step-wise mechanism of the reaction, using curved arrows to show movement of electron pairs.

$$(CH_3)_2C=CH_2 + HCl \longrightarrow (CH_3)_3C-Cl$$

The accepted mechanism for electrophilic addition of HCl to an alkene takes place in two steps, initiated by a collision between a molecule of the alkene nucleophile and an H^+ ion electrophile, leading to formation of a carbocation intermediate, which reacts rapidly with a Cl^- ion to form the observed alkyl halide product.

Visualizing Energy Changes in the Reaction Mechanism

A *reaction energy diagram* [<<Section 18.6] shown in Figure 19.21 can be very helpful in depicting the energy changes that occur during each step of a reaction mechanism. The vertical axis of the diagram represents the changing energy of the "system" comprising the reacting species that we are so closely examining in the mechanism, and the horizontal axis represents the progress of reaction events from beginning (left) to end (right). We can break this diagram into regions corresponding to each step of our mechanism.

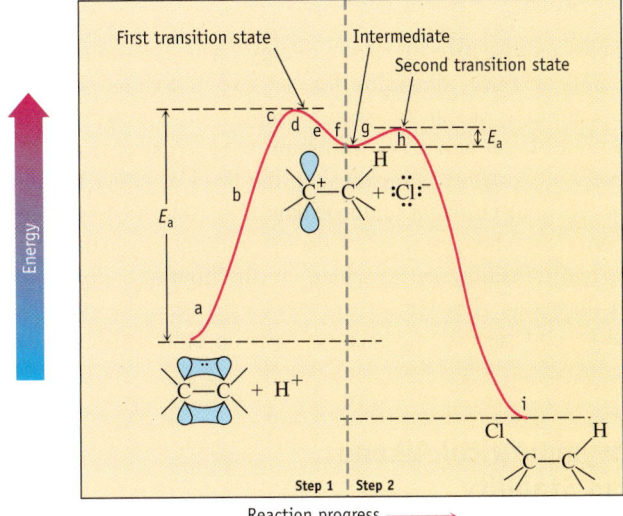

FIGURE 19.21 Visualizing energy changes for the reaction of one molecule of ethylene and an H^+ ion from HCl. Energy increases from the ground state (a) as the reactant molecules approach each other (b and c); this increase continues until a point of maximum energy, a transition state (d) is reached. Successful reaction releases energy (e) until a high energy reactive intermediate (f) is formed. Further distribution of energy takes place as the C—Cl bond forms (g) until a second transition state (h) is reached. The energy of the product molecules (i) is lower than that of the reactants.

We describe the accepted pathway chemists believe, but cannot definitely prove, by which this reaction event proceeds. At the beginning of the reaction, the ethylene and HCl molecules have the total amount of energy indicated by the reactant level on the left side of the diagram. As the two species bump into each other, their electron clouds repel each other, causing the energy level to rise. If the collision has occurred with sufficient force and suitable orientation, the reactants continue to approach each other despite the repulsion until the new C—H bond starts to form and the H—Cl bond starts to break. At some point, a structure of maximum energy, called the *transition state*, is reached. The transition state represents the highest-energy structure involved in this step of the reaction and can neither be isolated nor directly observed. Nevertheless, we can imagine it to have a partially broken C=C π bond and a partially formed new C—H bond.

As discussed in Section 18.6, experimental measurements at the macroscopic, observational level allow us to determine the magnitude of the energy difference between reactants and transition state, called the *activation energy* (E_a). A large activation energy results in a slow reaction (i.e., not many product molecules being formed per second) because in only a small fraction of the collisions do the reacting molecules have enough energy to reach the transition state. A small activation energy results in a rapid reaction because in most of the collisional events, rearrangement through to the transition state occurs. Most organic reactions have activation energies in the range 40 to 125 kJ mol^{-1}. Reactions with activation energies less than 80 kJ mol^{-1} usually can take place at or below room temperature, whereas reactions with higher activation energies normally require heating. At higher temperatures, the colliding molecules have enough energy to overcome the activation barrier.

Once the transition state has been reached, the reaction proceeds to yield the carbocation reaction intermediate. The combined energy of the participants decreases as the new C—H bond forms fully, leading to the carbocation intermediate. This minimum point represents the energy level of the carbocation product of the first step. The energy change for the first step is simply the difference between the total energy of the reactant species and the carbocation intermediates in the diagram. Since the carbocations are less stable than the alkene reactants, energy is absorbed in this first step.

Carbocations are good examples of **reaction intermediates**, described in Section 18.6. As soon as intermediates are formed in the first step, they react with Cl^- ions in a second, faster step to give the final product, chloroethane. This second step has its own activation energy (E_a), its own transition state, and its own energy change. We can view the second transition state as a complex between the electrophilic carbocation intermediate and nucleophilic Cl^- anion, in which the new C—Cl bond is just starting to form.

Figure 19.21 helps us visualize all of the energy changes implied by our proposed reaction mechanism. In essence, diagrams for the two individual steps are joined in the middle so that the *products* of step 1 (the carbocations) are the *reactants* for step 2. The carbocation intermediates lie at an energy minimum between steps 1 and 2. Because the energy level of these intermediates is higher than the level of either the initial reactants (ethylene + HCl) or the final products (chloroethane), the intermediates are reactive and can't be isolated. They are, however, at an energy minimum on this reaction energy diagram, unlike the two transition states, which are the highest energy points along the pathway to product formation. The barrier to reaction between the carbocations and the chloride ions in the second step is low because electron cloud repulsion between these species is greatly reduced.

Addition to Unsymmetrical Alkenes and Carbocation Stability

Other halogen acids such as HBr and HI (generically shown as HX) also add to alkenes, and the reaction proceeds even more rapidly with alkenes that have alkyl groups instead of one or more H atoms on the C atoms of the C=C bond. Three examples are shown below:

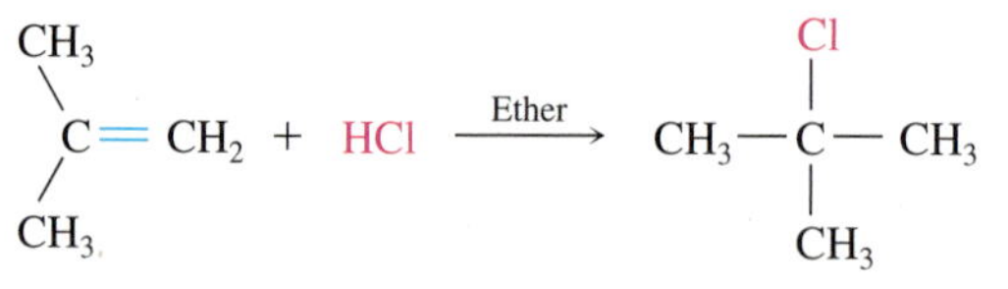

2-Methylpropene   **2-Chloro-2-methylpropane**

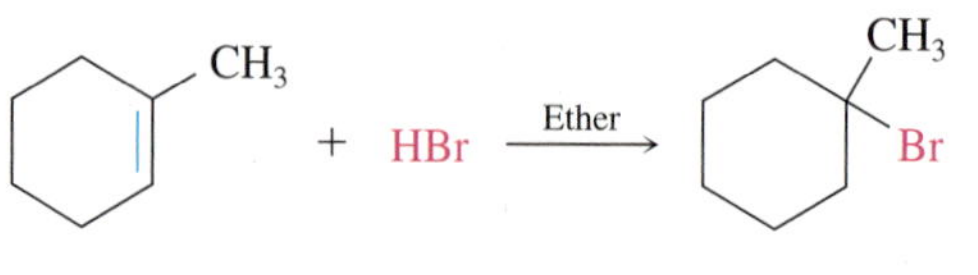

1-Methylcyclohexene   **1-Bromo-1-methylcyclohexane**

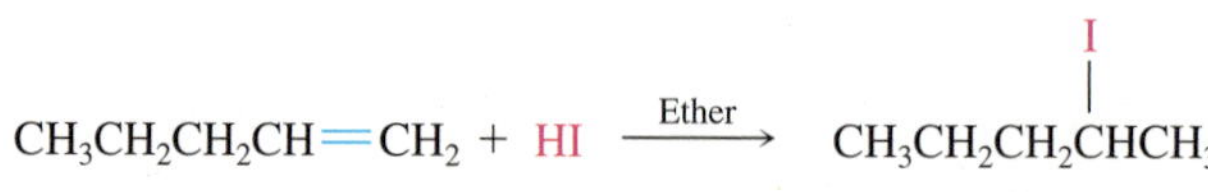

$CH_3CH_2CH_2CH$=CH_2 + HI $\xrightarrow{\text{Ether}}$ $CH_3CH_2CH_2CHCH_3$

Pent-1-ene   **2-Iodopentane**

In each case, an unsymmetrically substituted alkene forms a single addition product rather than the mixture that might have been expected. For example, 2-methylpropene *might* have reacted with HCl to give 1-chloro-2-methylpropane as well as 2-chloro-2-methylpropane, but it doesn't. We say that reactions are **regioselective** (*REE-jee-oh-selective*) when one of the two possible directions of addition is preferred.

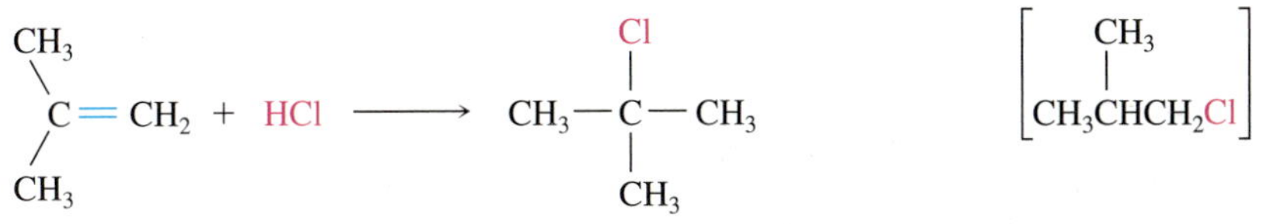

A regioselective reaction.

2-Methylpropene   **2-Chloro-2-methylpropane**   **1-Chloro-2-methylpropane**
(*sole product*)   (*NOT* formed)

In the 1860s, early in the history of organic chemistry, Russian chemist Vladimir Markovnikov looked at many electrophilic addition reactions of alkenes and noted that in the case of unsymmetrical alkenes, the proton adds to the C atom with the greater number of H atoms (and fewer alkyl groups) (Figure 19.22).

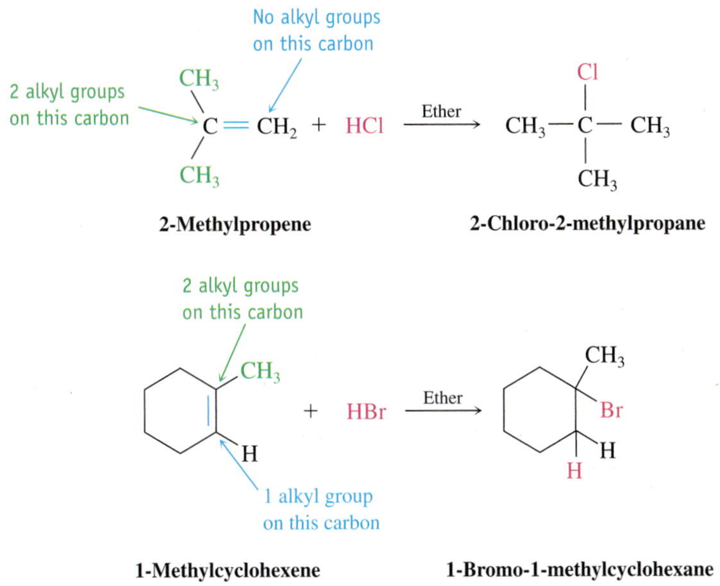

FIGURE 19.22 The regiochemistry of addition to these unsymmetrical alkenes is consistent with Markovnikov's observations.

2-Methylpropene   **2-Chloro-2-methylpropane**

1-Methylcyclohexene   **1-Bromo-1-methylcyclohexane**

When both double-bonded C atoms have the same degree of substitution, a mixture of two addition products results:

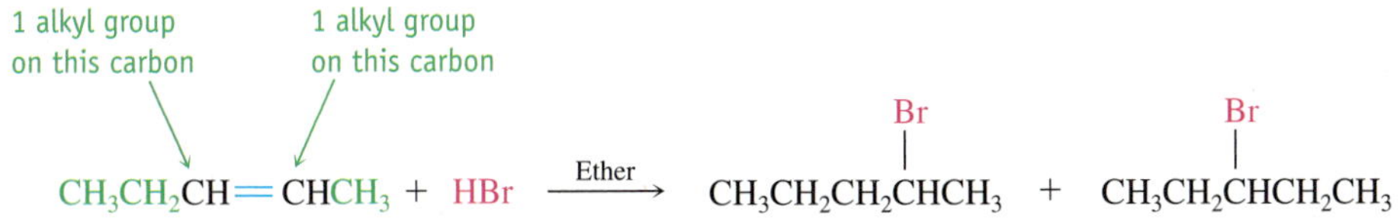

Pent-2-ene   **2-Bromopentane**   **3-Bromopentane**

Why are these patterns observed? Recall that carbocations are thought to be formed as reaction intermediates in the rate-determining first step of the reaction mechanism. Using 2-methylpropene as an example, we can see that different carbocation intermediates are formed along the reaction pathways leading to the two possible products. Since 2-chloro-2-methylpropane is the major product observed, there must be a preference for the more highly substituted carbocation (tertiary, with three alkyl groups attached), relative to a less highly substituted one (primary, with only one alkyl group attached).

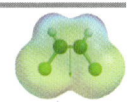

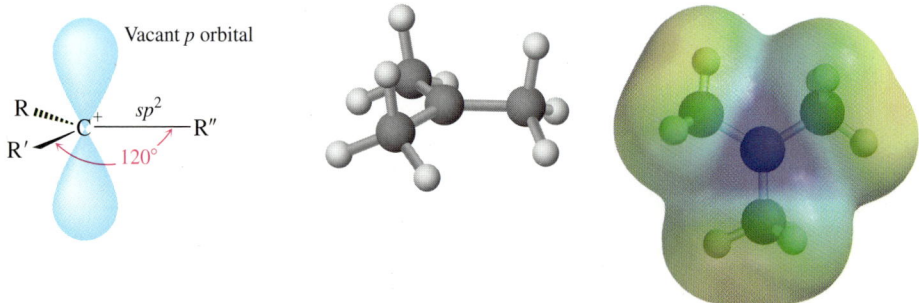

tert-Butyl carbocation
(tertiary; 3°)

2-Chloro-2-methylpropane

2-Methylpropene

Isobutyl carbocation
(primary; 1°)

1-Chloro-2-methylpropane
(*NOT* formed)

Why should increasing the number of alkyl groups attached to a carbocation lead to a more favourable pathway for reaction? To understand this pattern of reactivity, we need to learn more about the structure and stability of substituted carbocations. Regarding structure, evidence shows that carbocations are *planar*. The positively charged C atom can be thought of with the hybridization bonding model as being sp^2 hybridized, with the three substituents bonded to the carbocation C atom oriented to the corners of an equilateral triangle. Because there are only six electrons in the C atom valence shell and all six are used in the three σ bonds, the p orbital extending above and below the plane is vacant.

Vacant *p* orbital

sp^2

120°

Regarding stability, measurements show that carbocations are more stable and lower in energy on the reaction energy diagram with increasing alkyl substitution: *more highly substituted carbocations are more stable than less highly substituted ones because alkyl groups tend to donate electrons to the positively charged C atom. The more alkyl groups present, the more electron donation there is, and the more stable the carbocation.*

Methyl

Primary (1°)

Secondary (2°)

Tertiary (3°)

Less stable ←————— Stability —————→ More stable

This knowledge helps to explain Markovnikov's observations. In the reaction of 1-methylcyclohexene with HBr, for example, the intermediate carbocation might have either *three* alkyl substituents (a tertiary cation, 3°) or *two* alkyl substituents (a secondary cation, 2°). Because the tertiary cation is more stable than the secondary one, it's the tertiary cation that forms as the reaction intermediate, thus leading to the observed tertiary alkyl bromide product. The more highly substituted carbocation will be lower in energy (more stable) than the less substituted one (less stable). In reactions where two high energy intermediates are possible, our observation of what products are formed suggests that the more stable intermediate also has a lower activation energy of formation, and therefore is the preferred pathway.

(A tertiary carbocation)   1-Bromo-1-methylcyclohexane

1-Methylcyclo-hexene   + HBr

(A secondary carbocation)   1-Bromo-2-methylcyclohexane
(NOT formed)

Markovnikov, working in the 1860s, didn't know about carbocation intermediates. A modern explanation for his observations that incorporates what we now know about the underlying mechanism, is called **Markovnikov's rule**: *In addition reactions of HX to unsymmetrical alkenes, the proton adds preferentially to the C atom that will lead to the most stable carbocation intermediate.*

WORKED EXAMPLE 19.7—ADDITION OF HX TO ALKENES

What product would you expect from the reaction of HCl with 1-ethylcyclopentene?

Strategy

Begin by identifying the functional group(s) in the reactant molecules and deciding which is likely to be an electrophile and which a nucleophile. In this case, the reactant molecules are electron-rich alkenes that will probably undergo an electrophilic addition reaction with the electrophile, HCl. Use your knowledge of the reaction mechanism to predict the product. You know that electrophilic addition reactions take place to give the most stable carbocation intermediate, following the pattern observed by Markovnikov, so H^+ will add to the double-bonded C atom that has one alkyl group (C2 on the ring), and the Cl^- will add to the $C=C$ carbon atom that has two alkyl groups (C1 on the ring).

Solution

The expected product is 1-chloro-1-ethylcyclopentane.

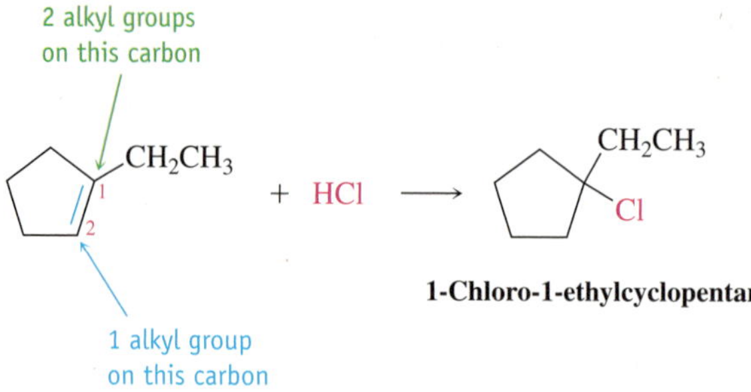

2 alkyl groups
on this carbon

CH_2CH_3

+ HCl ⟶

CH_2CH_3

Cl

1-Chloro-1-ethylcyclopentane

1 alkyl group
on this carbon

EXERCISE 19.17—ADDITION OF HX TO ALKENES

Predict the products of the following reactions:

(a) $CH_3CH_2CH=CH_2$ + HCl ⟶ ?

(b) CH_3
$|$
$CH_3C=CHCH_2CH_3$ + HI ⟶ ?

(c) 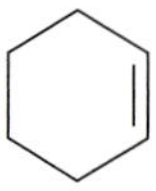 ⬡ + HCl ⟶ ?

EXERCISE 19.18—ADDITION OF HX TO ALKENES

What alkenes would you start with to prepare the following alkyl halides?

(a) Bromocyclopentane

(c) 1-iodo-1-isopropylcyclohexane

(b) Br
$|$
$CH_3CH_2CHCH_2CH_2CH_3$

(d) 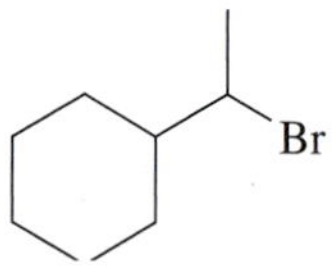 ⬡ Br

---

EXERCISE 19.19—ADDITION OF HX TO ALKENES

Show the structures of the carbocation intermediates you would expect in the following reactions:

(a) CH_3 CH_3
$|$ $|$
$CH_3CH_2C=CHCHCH_3$ + HBr ⟶ ?

(b) 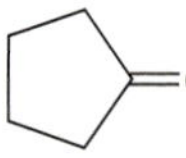 ⬠$=CHCH_3$ + HI ⟶ ?

Electrophilic Addition of HX to Conjugated Dienes: Allyl Cation Intermediates

Although much of the chemistry of alkenes and conjugated dienes is similar, a striking difference shows up in their electrophilic addition reactions with reagents like HX and X_2 [<<Section 19.10]. When HX adds to an alkene whose molecules have isolated double

bonds, the reaction usually occurs with predictable regiochemistry, giving the product derived from the most stable carbocation intermediate. When HX adds to a conjugated diene, however, mixtures of products are usually obtained. For example, reaction of HBr with buta-1,3-diene yields two products:

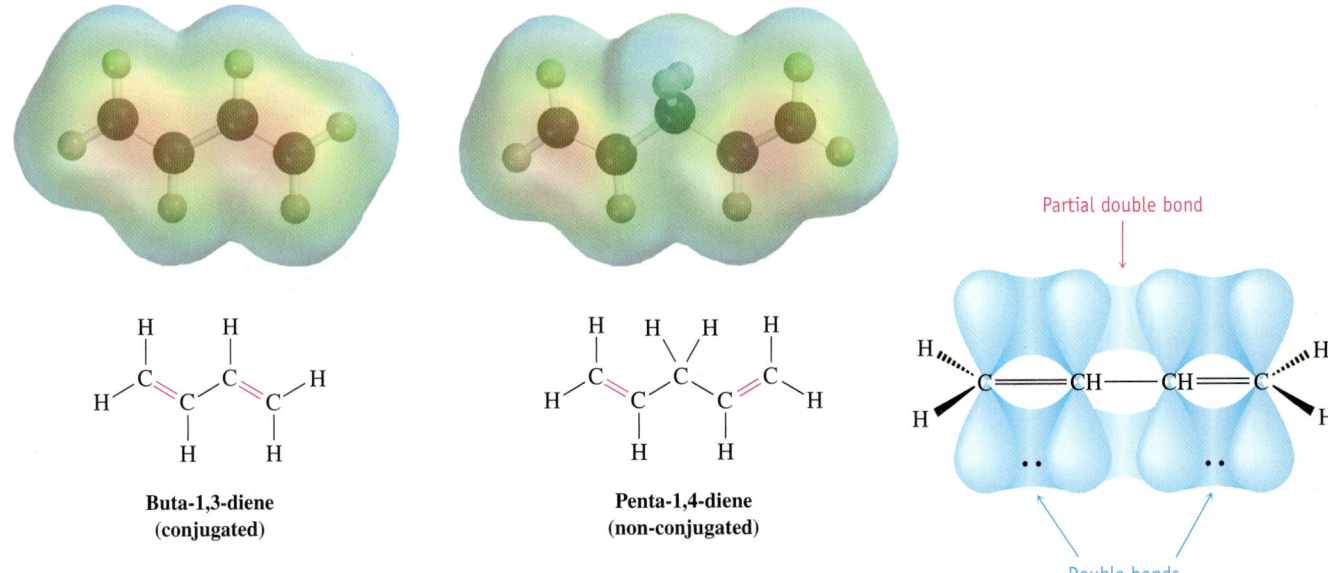

3-Bromobut-1-ene is the product with the expected regiochemistry, but 1-bromobut-2-ene is an unexpected product. Besides an addition reaction, rearrangement takes place, and the double bond has moved to a position between C2 and C3, while HBr has added to C1 and C4, a result described as *1,4-addition.*

Is the starting material, buta-1,3-diene, different from other alkenes? Like β-carotene, it is a conjugated diene, whereas another diene, penta-1,4-diene, is non-conjugated with isolated double bonds. An electrostatic potential map shows the difference clearly. Penta-1,4-diene molecules have two separate electron-rich sites that could react with a proton. In buta-1,3-diene molecules, the electrons are delocalized over all four C atoms.

Molecular Modelling (Odyssey)

e19.18 Compare electrostatic potential map models of these dienes.

The orbital view of buta-1,3-diene in Figure 19.23 shows this even more clearly. *There is an electronic interaction between the two double bonds of molecules of a conjugated diene* because of p orbital overlap across the central single bond. You might predict that the C2—C3 bond would be shorter than a normal C—C single bond, and it is.

FIGURE 19.23 An orbital view of a buta-1,3-diene molecule. Each of the four C atoms has a p orbital, allowing for an electronic interaction across the C2—C3 single bond.

Our understanding of the reaction mechanism helps account for the formation of the 1,4-addition product. An *allylic* carbocation is involved as an intermediate in the reaction, where the word *allylic* means "next to a double bond." When H adds to an electron-rich π bond of a buta-1,3-diene molecule, two carbocation intermediates are possible—a primary non-allylic carbocation and a secondary allylic carbocation. Allylic carbocations are more stable, and also have a lower activation barrier for formation, so they provide a preferred energy pathway for reacting molecules.

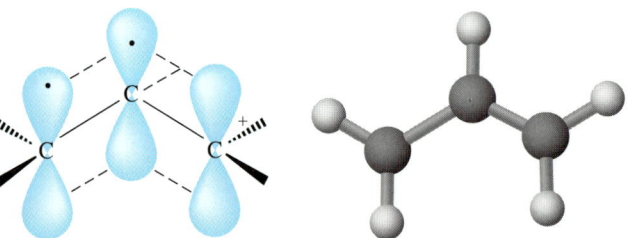

Why are allylic carbocations particularly stable? To see one explanation, look at the orbital picture of an allylic carbocation in Figure 19.24. Using the hybrid orbital model, the positively charged C atom has a vacant *p* orbital that can overlap with the *p* orbitals of the neighbouring double bond.

FIGURE 19.24   An orbital picture of an allylic carbocation. The vacant *p* orbital on the positively charged C atom can overlap with the *p* orbitals on the double-bonded C atoms, spreading out or *delocalizing* the positive charge.

From an electronic viewpoint, an allylic carbocation is symmetrical. Our bonding model shows all three C atoms to be sp^2 hybridized, and each has a *p* orbital. The *p* orbital on the central C atom can overlap equally well with *p* orbitals on either of the two neighbouring C atoms, and the two electrons are free to move about over the entire three-orbital array. This *electron delocalization* results in a more stable (lower energy) carbocation.

As discussed in Section 10.5, we can't adequately represent this delocalized cation by drawing one Lewis structure with the vacant *p* orbital on the right and the double bond on the left, or by a Lewis structure with the vacant *p* orbital on the left and the double bond on the right. *Neither representation of the true structure is correct by itself; the true structure of the allylic carbocation has a distribution of electrons halfway between the two.* The resonance model requires that we show a hybrid of two resonance structures to depict the charge delocalization. This can be pictured on a single structure by an electrostatic potential map, which shows an equal distribution of positive charge on the two end carbon atoms.

An allylic carbocation

An allylic carbocation is more stable and lower in energy than a typical non-allylic carbocation, due to the positive charge being spread out over a larger molecular volume, as it is shared over an extended π orbital network rather than remaining centred on only one atom.

The delocalization of charge in an allylic carbocation helps us understand patterns of reactivity. As is evident from examining the electrostatic potential map or the resonance structure model, when the allylic carbocations produced by protonation of buta-1,3-diene molecules react with Br$^-$ ions to complete the addition, reaction can occur at either C1 or C3, because both share the positive charge. Therefore both 1,2- and 1,4-addition products are formed when this reaction is carried out.

Molecular Modelling (Odyssey)

e19.19 Examine the electrostatic potential map and bond distances to visualize electron delocalization.

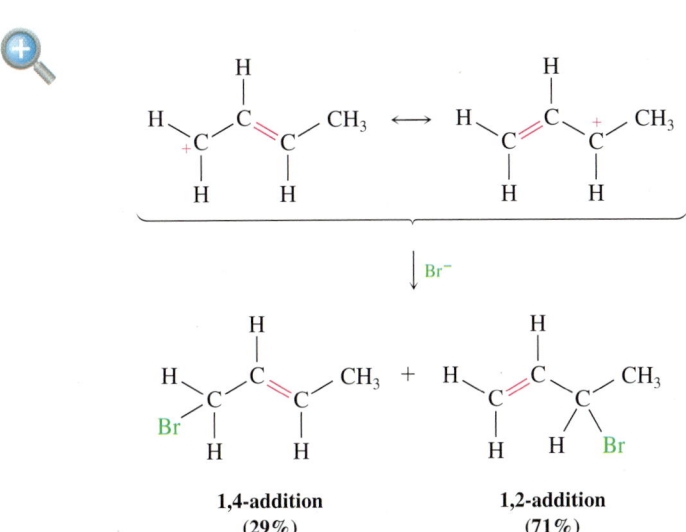

1,4-addition
(29%)

1,2-addition
(71%)

EXERCISE 19.20—ADDITION OF Br$_2$ TO DIENES

Buta-1,3-diene, H$_2$C=CH—CH=CH$_2$, reacts with Br$_2$ to yield a mixture of 1,2- and 1,4-addition products. Show the structure of each product.

19.9 Addition of H$_2$O to Alkenes: Hydration

Just as HX adds to alkenes to yield alkyl halides, water adds to yield alcohols (ROH), a process called *hydration*. Industrially, hundreds of thousands of tonnes of ethanol are produced globally each year by this method:

$$C=C + H_2O \xrightarrow[250\ °C]{H_3PO_4\ \text{catalyst}} CH_3CH_2OH$$

Ethylene

Ethanol

Hydration is thought to take place by reaction of an alkene with aqueous acid by a mechanism similar to that of HX addition. As pictured in the mechanism scheme below, in each successful collisional event, reaction of the alkene double-bonded electrons with H^+ yields a carbocation intermediate, which then reacts with a water molecule as a nucleophile to yield a protonated alcohol (ROH_2^+) product. Loss of H^+ from the protonated alcohol gives the neutral alcohol and regenerates the acid catalyst (Figure 19.25). The addition of water to an unsymmetrical alkene follows the same pattern as addition of HX, following Markovnikov's rule, and giving the more highly substituted alcohol as product.

FIGURE 19.25 Accepted reaction mechanism for the addition of H_2O to an alkene.

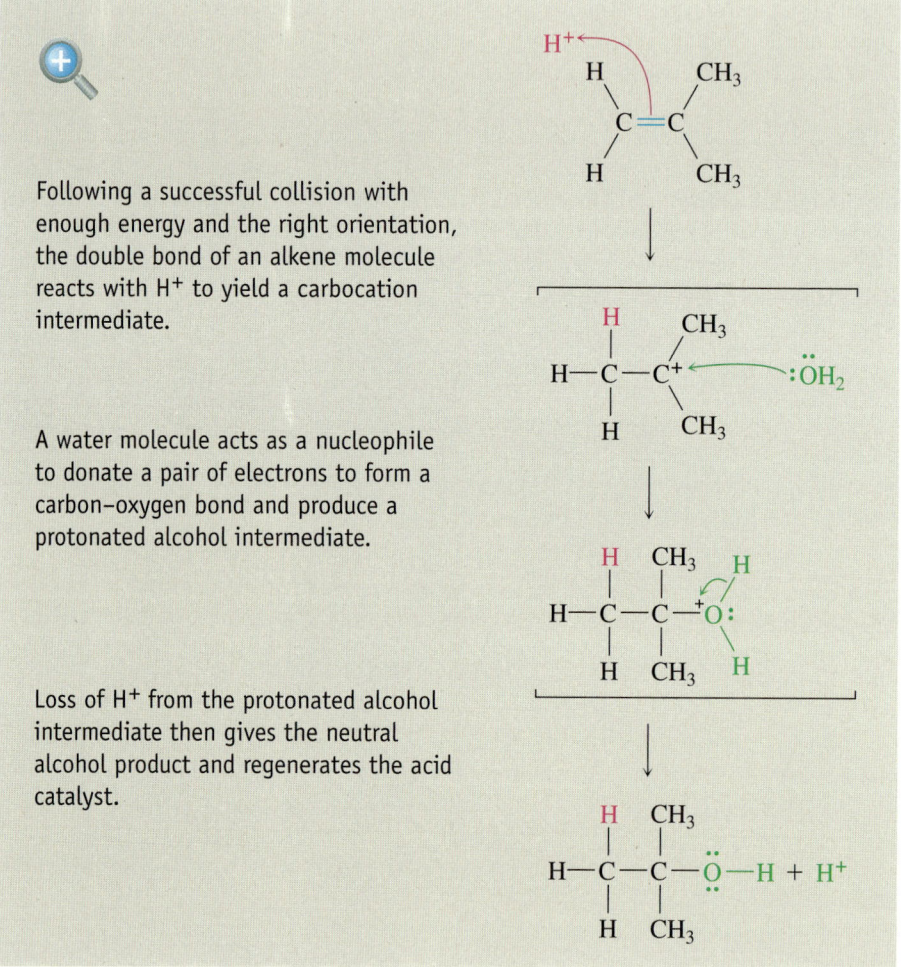

Following a successful collision with enough energy and the right orientation, the double bond of an alkene molecule reacts with H^+ to yield a carbocation intermediate.

A water molecule acts as a nucleophile to donate a pair of electrons to form a carbon–oxygen bond and produce a protonated alcohol intermediate.

Loss of H^+ from the protonated alcohol intermediate then gives the neutral alcohol product and regenerates the acid catalyst.

Unfortunately, the reaction conditions required for hydration are so severe that molecules are sometimes destroyed by the high temperatures and strongly acidic conditions. For example, the hydration of ethylene to produce ethanol requires a phosphoric acid catalyst and reaction temperatures of up to 250 °C.

As a result, chemists are constantly searching for better procedures that use new reactions and catalysts. Figure 19.26 shows three examples. The first two examples are new reactions. With an understanding of reaction mechanisms, chemists have developed addition reactions with the opposite regiochemistry from that observed by Markovnikov, referred to as *non-Markovnikov addition* (Figure 19.26b). The last example shows a remarkable catalyst, a protein isolated from cells. While fumarase catalyzes hydration of this alkene called fumaric acid, it does not catalyze the same reaction with the isomeric *cis*-alkene called maleic acid. Chemists have a difficult time creating catalysts that match the reactant selectivity that nature displays in its catalysts.

(a) Oxymercuration–demercuration (Markovnikov addition)

1. Hg(OAc)₂, H₂O
2. NaBH₄

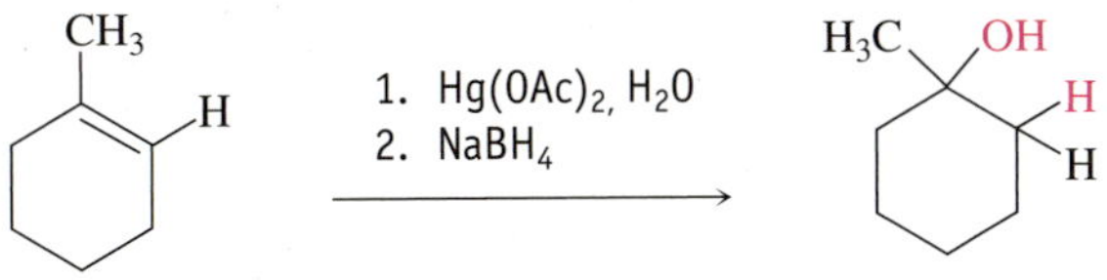

(b) Hydroboration (non-Markovnikov addition)

1. BH₃
2. H₂O₂, NaOH, H₂O

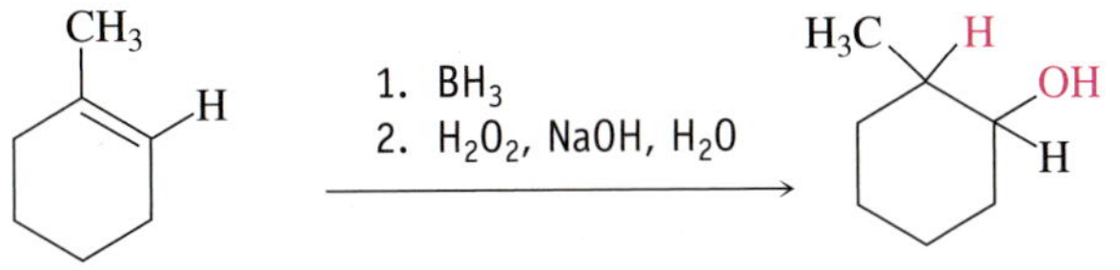

(c) Enzymatic hydration

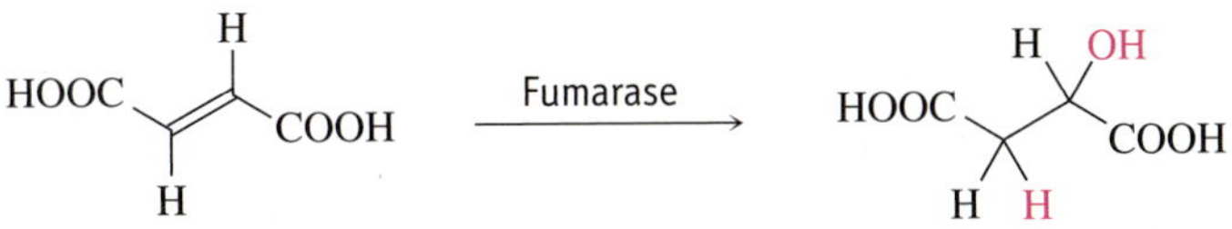

Fumarase

FIGURE 19.26 Newer methods for hydration include the use of mercury or boron reagents (that lead to non-Markovnikov products) and catalysts like the protein fumarase.

WORKED EXAMPLE 19.8—ADDITION OF H₂O TO ALKENES

What product would you expect from addition of water to methylenecyclopentane?

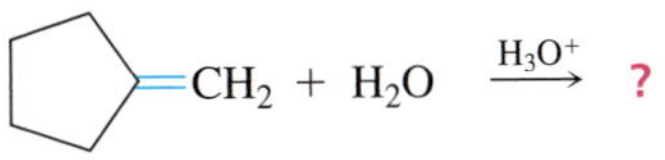

Methylenecyclopentane

Strategy

Following the normal pattern, the H⁺ should add to the C atom of the alkene molecule that will produce the most stable carbocation (a tertiary one), with the OH adding to the other C atom (the ring C atom). Thus, the product will be a tertiary alcohol. If the H⁺ had added to the ring C atom, the intermediate carbocation would have been a primary one.

Solution

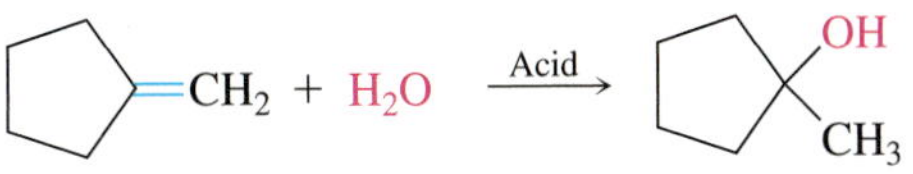

Acid

EXERCISE 19.21—ADDITION OF H₂O TO ALKENES

What product would you expect to obtain from addition of water to the following alkenes?

(a)
$$CH_3CH_2\overset{\overset{\displaystyle CH_3}{|}}{C}\!=\!CHCH_2CH_3$$

(c) 2,5-Dimethylhept-2-ene

(b) 1-Methylcyclopentene

What alkenes might the following alcohols be made from?

(a)

$$CH_3CH_2CHCH_3$$
OH

(b)

$$CH_3CH_2-\overset{\overset{\displaystyle OH}{|}}{\underset{\underset{\displaystyle CH_3}{|}}{C}}-CH_2CH_3$$

(c)

OH
CH₃
CH₃

19.10 Addition of X₂ to Alkenes: Halogenation

Many other substances besides HX and H₂O add to alkenes. Bromine and chlorine, for instance, add readily to yield 1,2-dihaloalkanes. Millions of tonnes of 1,2-dichloroethane (also called ethylene dichloride) are synthesized each year by addition of Cl_2 to ethylene. The product is used both as a solvent and as a starting material for the synthesis of poly(vinyl chloride), PVC.

$$\underset{H}{\overset{H}{\diagdown}}C=C\underset{H}{\overset{H}{\diagup}} + Cl_2 \longrightarrow H-\overset{\overset{\displaystyle Cl}{|}}{\underset{\underset{\displaystyle H}{|}}{C}}-\overset{\overset{\displaystyle Cl}{|}}{\underset{\underset{\displaystyle H}{|}}{C}}-H$$

Ethylene

1,2-Dichloroethane
(ethylene dichloride)

Addition of Br_2 also acts as a simple and rapid laboratory test for unsaturation, which can be shown dramatically with a reaction of bacon with Br_2 (Figure 19.27).

FIGURE 19.27 Bacon reacts with bromine. The fat in bacon is partially unsaturated. Here you see the colour of bromine vapour fade when a strip of bacon is introduced.

A few minutes →

Charles D. Winters

Charles D. Winters

For most spot-test analyses of this kind, a sample of unknown structure is dissolved in dichloromethane (CH_2Cl_2) and several drops of Br_2 are added. Immediate disappearance of the reddish Br_2 colour signals a positive test and indicates that the sample is an alkene.

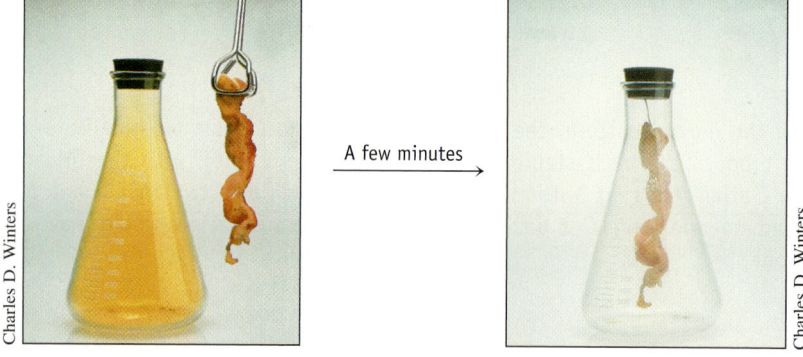

Cyclopentene

$\xrightarrow{\text{Br}_2 \text{ in CH}_2\text{Cl}_2}$

1,2-Dibromocyclopentane (95%)

By analogy with what we've seen thus far, a possible mechanism for the reaction of Br_2 or Cl_2 with an alkene might involve formation of a bond between a C atom and a Br^+ ion, displacing the other Br atom as a Br^- anion. The net result would be that an electrophilic Br⁺ ion adds to an alkene molecule in much the same way that H⁺ does, giving a carbocation intermediate that reacts further with Br⁻ to yield the dibromo addition product.

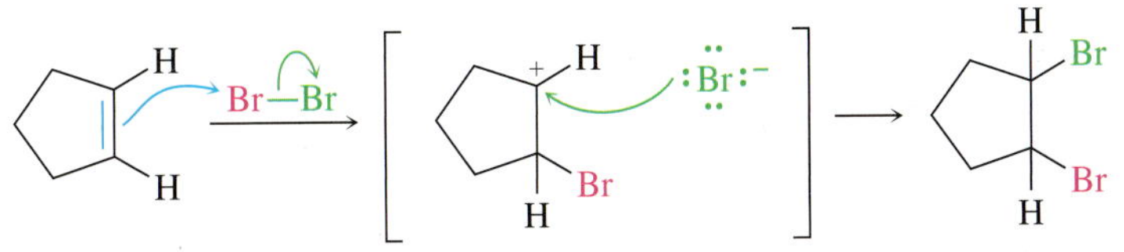

Although this mechanism looks reasonable, it's not completely consistent with known facts. In particular, the mechanism doesn't explain the *stereochemistry* of halogen addition. That is, the mechanism doesn't explain what product stereoisomers [<<Section 9.8] are formed in the reaction.

Let's look again at the mechanism for reaction of Br₂ with cyclopentene and assume that a Br⁺ ion adds from the bottom side of the molecule shown in Figure 19.28 to form the carbocation intermediate. (The addition could just as well occur with a Br⁺ ion adding from the top side, but we'll first consider only one possibility for simplicity.) Because the positively charged C atom is planar and *sp²* hybridized, it could react with a Br⁻ anion in the second step of the reaction from either side of the plane to give a *mixture* of products. One possible product has the two Br atoms on the same side of the ring (*cis*), and the other has them on opposite sides (*trans*). We find, however, that only *trans*-1,2-dibromocyclopentane is produced; none of the *cis* product is formed.

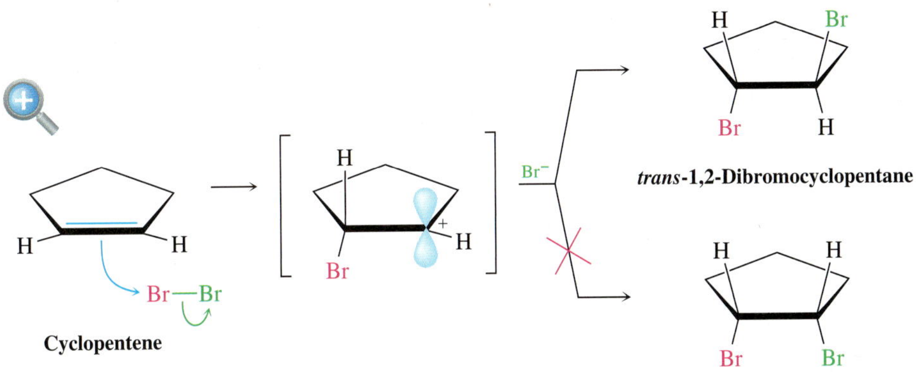

FIGURE 19.28 Accepted reaction mechanism for addition of Br₂ to cyclopentene.

Because the two Br atoms add to opposite faces of the cyclopentene double bond, we say that the reaction occurs with *anti stereochemistry*, meaning that the two bromine atoms have made C—Br bonds from directions approximately 180° apart. This result is best explained by imagining that the reaction intermediate is not a true carbocation but is instead a *bromonium ion* (R₂Br⁺), formed by overlap of bromine lone-pair electrons with the vacant *p* orbital of the neighbouring C atom (Figure 19.29). Since the bromine atom that is part of the bromonium ion effectively "shields" one side of the molecule, reaction with Br⁺ ion in the second step occurs from the opposite, more accessible side to give the anti product.

A bromonium ion intermediate is postulated in this mechanism, shielding one face of the double bond and resulting in addition with anti stereochemistry, in this case, giving a product where the two Br atoms are *trans* to each other. Addition of the bromide ion in the second step could equally well have been to the other C atom that is part of the bromonium ion.

Think about It

e19.21 Watch animations of the electron movement in the mechanism for bromination of an alkene.

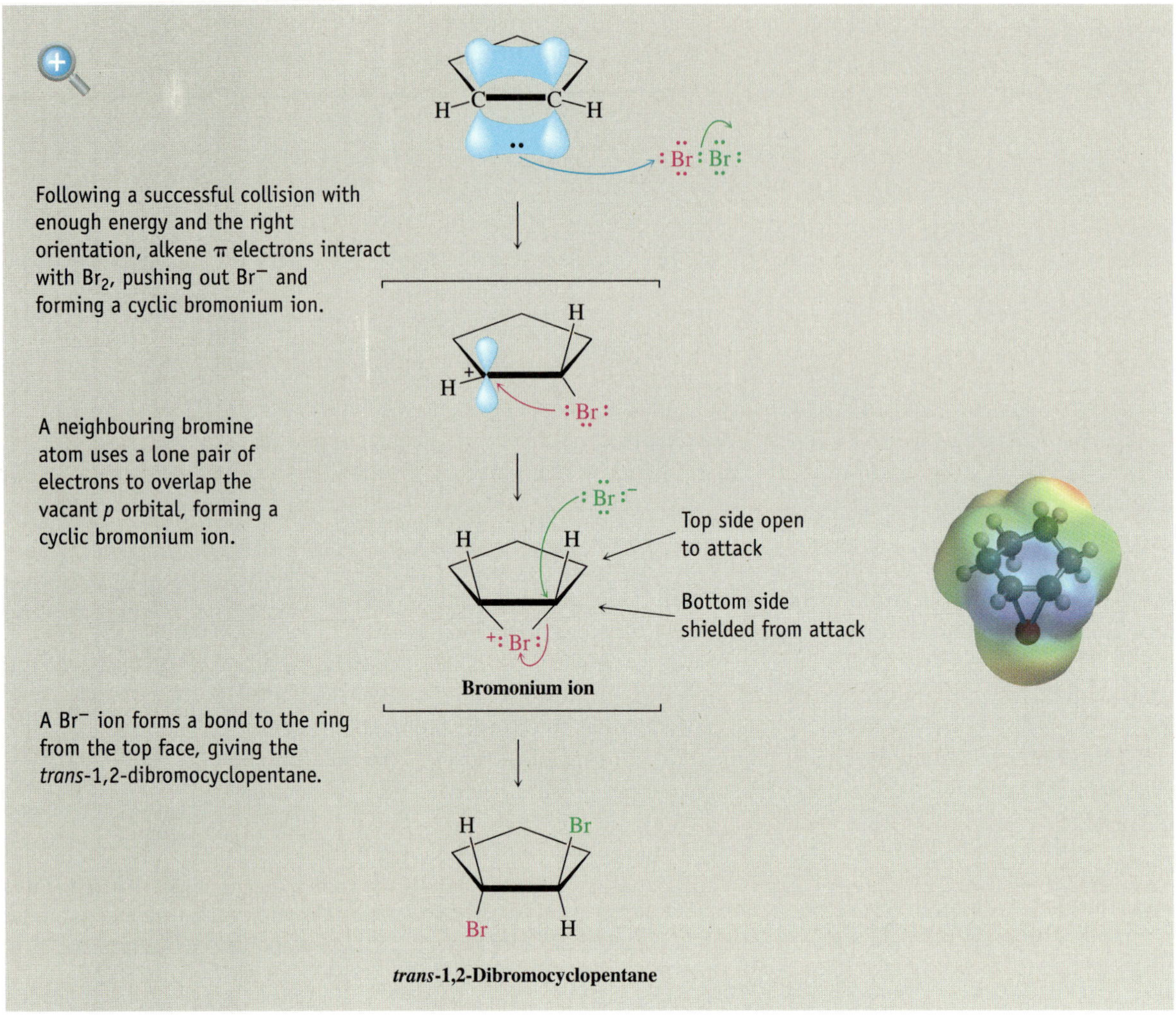

Following a successful collision with enough energy and the right orientation, alkene π electrons interact with Br₂, pushing out Br⁻ and forming a cyclic bromonium ion.

A neighbouring bromine atom uses a lone pair of electrons to overlap the vacant *p* orbital, forming a cyclic bromonium ion.

Top side open to attack

Bottom side shielded from attack

Bromonium ion

A Br⁻ ion forms a bond to the ring from the top face, giving the *trans*-1,2-dibromocyclopentane.

trans-**1,2-Dibromocyclopentane**

FIGURE 19.29 **Accepted reaction mechanism** for the addition of Br₂ to an alkene. The electrostatic potential map of the bromonium ion intermediate indicates the positive charge is shared over the Br and two C atoms.

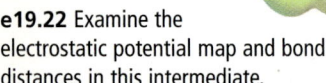

Molecular Modelling (Odyssey)

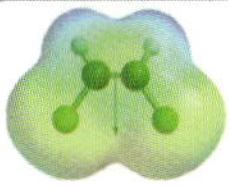

e19.22 Examine the electrostatic potential map and bond distances in this intermediate.

EXERCISE 19.23—ADDITION OF Br₂ TO ALKENES

Draw the structure of the second *trans* product molecule formed in the mechanism in Figure 19.29, and convince yourself that it would be an enantiomer of the product shown, with an expected 50/50 ratio of these two enantiomers.

EXERCISE 19.24—ADDITION OF Br₂ TO ALKENES

What product(s) would you expect to obtain from addition of Br₂ to 1,2-dimethylcyclohexene? Show the stereochemistry of the product molecules.

EXERCISE 19.25—ADDITION OF Br₂ TO ALKENES

Show the structure of the intermediate bromonium ion formed in Exercise 19.24.

19.11 Addition of H₂ to Alkenes: Hydrogenation

Addition of H_2 to the $C=C$ bond o1f alkene molecules occurs when alkenes are exposed to an atmosphere of hydrogen gas in the presence of a catalyst. We describe the result by saying that the alkene is *hydrogenated*, or *reduced*. For most alkene hydrogenations, either palladium metal or platinum (as PtO_2) is used as the catalyst.

Reduction in organic chemistry refers to an increase in electron density on carbon atoms by forming a C—H bond or breaking a C—O, C—X, or C—N bond.

An alkene   **An alkane**

Catalytic hydrogenation of alkenes, unlike most other organic reactions, is a *heterogeneous* process, rather than a homogeneous one. That is, the hydrogenation reaction occurs on the surface of solid catalyst particles rather than in solution. Unlike the bromination reaction in the last section, the hydrogenation reaction occurs with *syn stereochemistry* (the opposite of *anti*), meaning that both hydrogen atoms add to the double bond from the same side.

1,2-Dimethylcyclohexene   ***cis*-1,2-Dimethylcyclohexane**
                                        (82%)

In addition to its usefulness in the laboratory, alkene hydrogenation is also of great commercial value. In the food industry, unsaturated vegetable oils are catalytically hydrogenated on a vast scale to produce the saturated fats used in margarine. Converting some of the alkenes in unsaturated vegetable oils to alkanes raises the melting point of the material, producing a product that is a solid and therefore spreadable on your dinner buns. In the process, some *trans*-fats, which are not as healthy, are also produced. Next time you read a margarine container label, look for "hydrogenated" oils, "trans-fat free" labels, and β-carotene (which may simply be listed as a "colourant") to colour your wattle.

Think about It

e19.23 Watch an animation of the hydrogenation of an alkene on a catalytic metal surface.

EXERCISE 19.26—ADDITION OF H₂ TO ALKENES

What product would you expect to obtain from catalytic hydrogenation of the following alkenes?
(a) $(CH_3)_2C=CHCH_2CH_3$
(b) 3,3-dimethylcyclopentene

19.12 Oxidation of Alkenes: Hydroxylation and Cleavage

Hydroxylation of an alkene—the addition of an —OH group to each of the alkene C atoms in an alkene molecule—can be carried out by reaction of the alkene with potassium permanganate ($KMnO_4$) in basic solution. Since oxygen adds to the alkene during the reaction, we call this an *oxidation*. The reaction occurs with syn stereochemistry and yields a 1,2-dialcohol, or *1,2-diol*, product (also called a *glycol*). For example, cyclohexene gives *cis*-cyclohexane-1,2-diol in 37% yield.

Oxidation in organic chemistry refers to decreasing the electron density on C atoms by breaking a C—H bond or forming a C—O, C—X, or C—N bond.

Cyclohexene   ***cis*-Cyclohexane-1,2-diol**
                          (37%)

When oxidation of the alkene is carried out with $KMnO_4$ in acidic rather than basic solution, *cleavage* of the double bond in alkene molecules occurs and carbonyl-containing products are obtained. If the double-bonded C atoms are completely substituted with alkyl groups (tetra-substituted), the two carbonyl-containing products are ketones; if an H atom is present on the double bond, one of the carbonyl-containing products is a carboxylic acid; and if two H atoms are present on one C atom, CO_2 is formed:

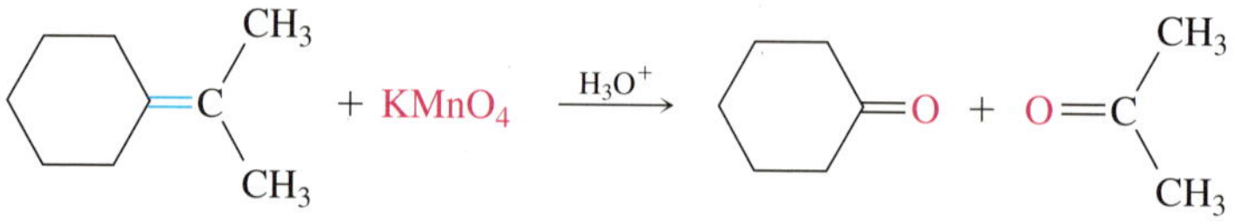

Isopropylidenecyclohexane   Cyclohexanone   Acetone
                                                  (two ketones)

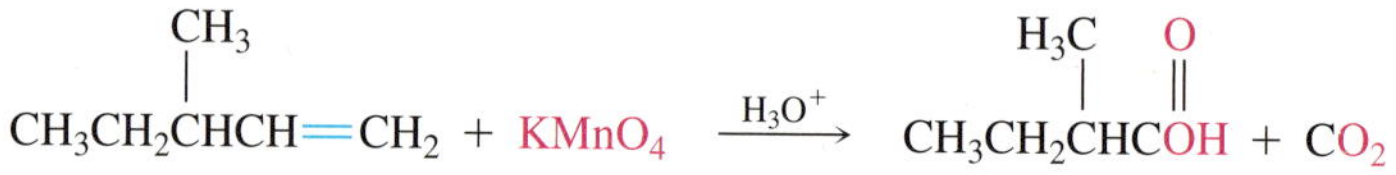

3-Methylpent-1-ene   2-Methylbutanoic acid
                                        (45%)

WORKED EXAMPLE 19.9—OXIDATION OF ALKENES

Predict the product of reaction of pent-2-ene with aqueous acidic $KMnO_4$ solution.

Solution

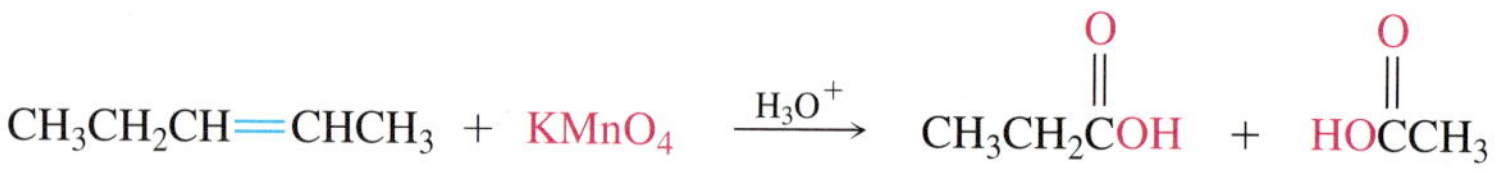

Pent-2-ene   Propanoic acid   Acetic acid

WORKED EXAMPLE 19.10—OXIDATION OF ALKENES

What alkene gives a mixture of acetone and propanoic acid on reaction with acidic $KMnO_4$ solution?

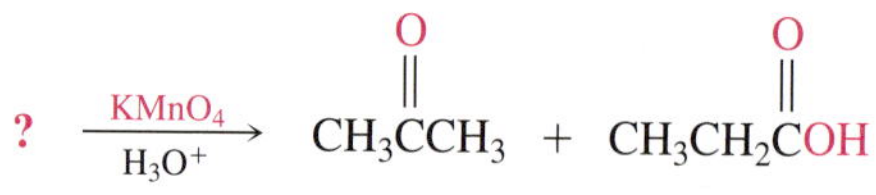

Acetone   Propanoic acid

Strategy

When solving a problem that asks how to synthesize a given product, *always work from the product back to the starting materials*. Look at the product molecule, identify the functional group(s) it contains, and ask yourself, "How can I synthesize that functional group?" In this case, the products are a ketone and a carboxylic acid, which can be prepared by reaction of an alkene with acidic $KMnO_4$. To find the starting alkene that gives the cleavage products shown, remove the oxygen atoms from the two products, join the fragments with a double bond, and replace the —OH by —H.

Solution

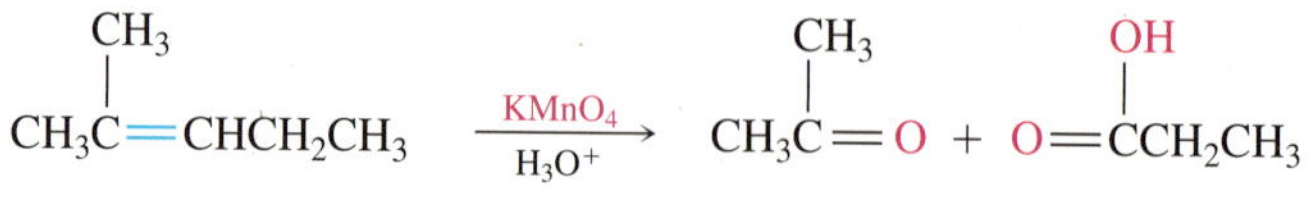

2-Methylpent-2-ene   Acetone   Propanoic acid

EXERCISE 19.27—OXIDATION OF ALKENES

Predict the product of the reaction of 1,2-dimethylcyclohexene with the following reagents:
(a) $KMnO_4$, H_3O^+
(b) $KMnO_4$, OH^-, H_2O

EXERCISE 19.28—OXIDATION OF ALKENES

Propose structures for alkenes that yield the following products on treatment with acidic $KMnO_4$:
(a) $(CH_3)_2C{=}O + CO_2$
(b) 2 mol $CH_3CH_2CO_2H$ for each mol of starting material

19.13 Free Radical Addition to Alkenes: Polymerization

In Section 4.6, you were introduced to polymers, substances made of large molecules built up by repetitive bonding together of many smaller molecules, called *monomers*. Examples of biological polymers include terpenes [<<Section 19.1], built of repeating isoprene units; cellulose, built of repeating sugar units; proteins, built of repeating amino acid units; and nucleic acids, built of repeating nucleotide units. Many of the most important synthetic polymers are addition polymers built of repeating alkenes, such as ethylene, propylene, vinyl chloride, styrene, and tetrafluoroethylene. These addition polymers are made by directly adding monomer units together, usually under conditions of high pressure, high temperature, and with a suitable catalyst.

Ethylene, for example, undergoes an addition polymerization reaction to yield polyethylene, whose molecules have up to several thousand monomer units incorporated into their long hydrocarbon chains. Ethylene polymerization is usually carried out at high pressure (1000–3000 atm) and high temperature (100–250 °C) with a catalyst such as benzoyl peroxide.

Many $H_2C{=}CH_2$ $\longrightarrow$

Ethylene   **A section of polyethylene**

This reaction to produce long polymer chains, along with most industrial addition polymerization reactions, has been found to follow free radical rather than polar reaction mechanisms. It is therefore called a *free radical chain reaction*. Benzoyl peroxide is selected as a catalyst because its molecules have an exceptionally weak $O{-}O$ single bond, which dissociates homolytically rather than heterolytically. This process *initiates* a chain reaction, which occurs hundreds or thousands of times as the polymer chain grows.

The mechanism for radical polymerization of an alkene generally involves three kinds of steps: *initiation, propagation,* and *termination*. At the molecular level, we picture the key propagation step as the addition of a *radical* to the ethylene double bond in a process similar to what takes place in the addition of an *electrophile* to an alkene. In writing the free radical mechanism, a curved half-arrow (or "fishhook") is used to show the movement of a single electron, as distinct from the full curved arrow used to show the movement of an electron pair in a polar reaction.

Step 1   *Initiation:* Reaction begins when a few radicals are generated by the catalyst. For example, when benzoyl peroxide is used as initiator, the $O{-}O$ bond is broken on heating to yield benzoyloxy radicals. A benzoyloxy radical can then add to the C=C bond of an ethylene molecule to generate a carbon atom radical. One electron from the C=C bond pairs up with the unpaired electron on the benzoyloxy radical to form a C—O bond, and the other electron remains on the C atom.

Benzoyl peroxide →(Heat) **Benzoyloxy radical** = BzO·

BzO· + H₂C=CH₂ ⟶ BzO—CH₂CH₂·

Step 2 *Propagation:* Polymerization occurs when a C atom radical formed in step 1 adds to another ethylene molecule. Repetition of this step for hundreds or thousands of times builds the polymer chain.

BzOCH₂CH₂· + H₂C=CH₂ ⟶ BzOCH₂CH₂CH₂CH₂· $\xrightarrow[\text{many times}]{\text{Repeat}}$ BzO(CH₂CH₂)ₙCH₂CH₂·

Step 3 *Termination:* At the molecular level, we picture polymerization eventually stopping when a reaction that consumes the radical occurs. For example, combination of two chains by chance meeting is a possible chain-terminating reaction:

2 R—CH₂CH₂· ⟶ R—CH₂CH₂CH₂CH₂—R

Many substituted ethylenes, called *vinyl monomers,* undergo radical-initiated polymerization, yielding polymers whose molecules have substituent groups regularly spaced along the polymer chain. Propylene, for example, yields polypropylene when polymerized (although a non-radical method of polymerization is used industrially).

H₂C=CHCH₃ ⟶ ⎛—CH₂CHCH₂CHCH₂CHCH₂CH—⎞ with CH₃ groups

Propylene   **Polypropylene**

Table 19.2 shows some commercially important alkene polymers, the alkene monomers from which they are derived, their common uses, and United States production figures.

WORKED EXAMPLE 19.11—POLYMERS

Show the structure of poly(vinyl chloride), a polymer made from H₂C=CHCl monomers, by drawing several repeating units.

Strategy

The electrons from the double bond in the monomer molecules are used to make the new C—C bonds in the polymer chain, so the polymer will contain an alkane chain rather than alkenes. The C atoms in the double bond of the monomer will be the ones forming the new single bonds in the polymer chain.

Solution

The general structure of poly(vinyl chloride) is

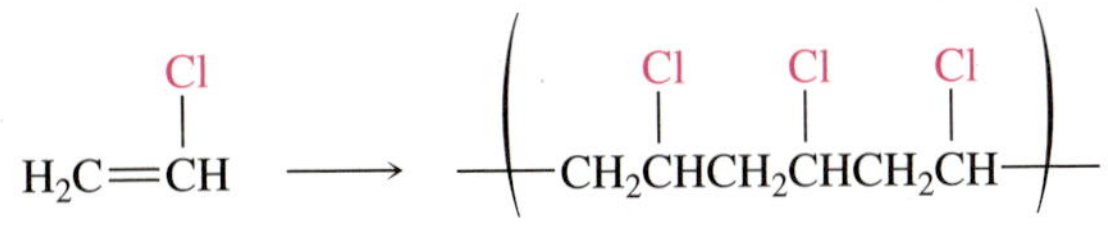

TABLE 19.2　Addition Polymers from Ethylene Derivatives

Formula	Monomer Common Name	Polymer Name (Trade Names)	Uses	U.S. Polymer Production (Metric tonnes year^{-1})
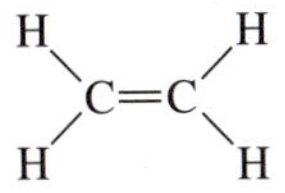 H₂C=CH₂	Ethylene	polyethylene (polythene)	squeeze bottles bags, films, toys and moulded objects, electric insulation	7 million
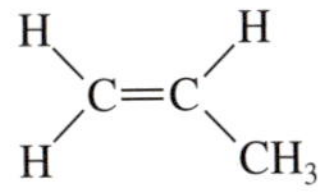 H₂C=CHCH₃	Propylene	polypropylene (Vectra, Herculon)	bottles, films, indoor-outdoor carpets	1.2 million
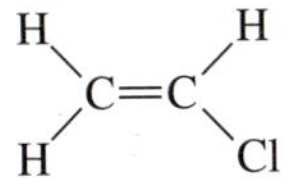 H₂C=CHCl	Vinyl chloride	polyvinyl chloride (PVC)	floor tile, raincoats, pipe	1.6 million
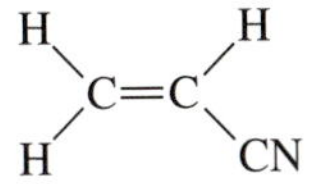 H₂C=CHCN	Acrylontrile	polyacrylontrile (Ortan, Acritan)	rugs, fabrics	0.5 million
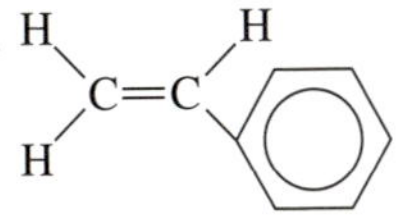 H₂C=CH(C₆H₅)	Styrene	polystyrene (Styrofoam, Styron)	food and drink coolers, building material insulation	0.9 million
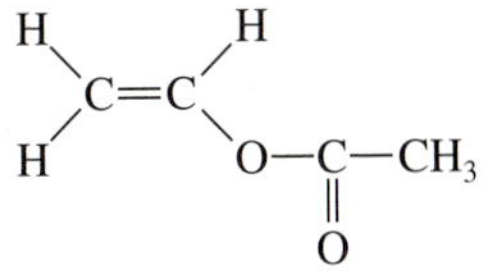 H₂C=CH-O-C(=O)-CH₃	Vinyl acetate	polyvinyl acetate (PVA)	latex paint, adhesives, textile coatings	200 000
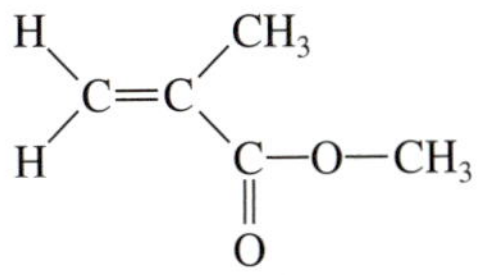 H₂C=C(CH₃)-C(=O)-O-CH₃	Methyl methacrylate	polymethyl methacrylate (Plexiglas, Lucite)	high-quality transparent objects, latex paints, contact lenses	200 000
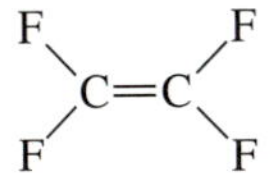 F₂C=CF₂	Tetrafluoroethylene	polytetrafluoroethylene (Teflon)	gaskets, insulation, bearings, pan coatings	6 000

19.14 Alkynes: Electronic Structure, Spectroscopy, and Reactions

Alkynes are hydrocarbons whose molecules contain a C≡C bond. The general formula for an alkyne with one triple bond is C_nH_{2n-2} with four H atoms fewer than a corresponding alkane, C_nH_{2n+2}. Because alkynes occur much less commonly than alkenes, we'll look at them only briefly.

Alkynes are named by general rules similar to those used for alkanes [<<Section 4.5] and alkenes. The suffix *–yne* is used in the parent hydrocarbon name to denote an alkyne, and the position of the triple bond is indicated by its number in the chain. Numbering begins at the chain end nearer the triple bond so that the triple bond receives as low a number as possible.

$$\overset{8}{C}H_3\overset{7}{C}H_2\overset{6}{C}H\overset{5}{C}H_2\overset{4}{C}\equiv\overset{3}{C}\overset{2}{C}H_2\overset{1}{C}H_3$$
$$\underset{CH_3}{|}$$

Begin numbering at the end nearer the triple bond.

6-Methyloct-3-yne

Compounds containing both double and triple bonds are called *enynes* (not ynenes). Numbering of the hydrocarbon chain starts from the end nearer the first multiple bond, whether double or triple. If there is a choice in numbering, double bonds receive lower numbers than triple bonds:

$$\underset{7\quad\;6\;5\quad4\quad3\quad2\quad1}{HC\equiv CCH_2CH_2CH_2CH=CH_2}$$

Hept-1-en-6-yne

$$\overset{CH_3}{\underset{1\quad2\;3\quad4\quad5\quad6\quad7\quad8\;9}{HC\equiv CCH_2CHCH_2CH_2CH=CHCH_3}}$$

4-Methylnon-7-en-1-yne

EXERCISE 19.29—IUPAC NOMENCLATURE

Give IUPAC names for the following compounds:

(a) $CH_3CH_2C\equiv CCH_2\overset{\overset{\displaystyle CH_3}{|}}{C}HCH_3$

(b) $HC\equiv C\overset{\overset{\displaystyle CH_3}{|}}{\underset{\underset{\displaystyle CH_3}{|}}{C}}CH_3$

(c) $CH_3\overset{\overset{\displaystyle CH_3}{|}}{C}HCH_2C\equiv CCH_3$

(d) $CH_3CH=CHCH_2C\equiv CCH_3$

Some properties and reactions of alkynes can be understood by using the hybrid orbital model, in which a C≡C bond is thought of as a σ bond formed by overlap of two *sp* hybridized C atoms and two π bonds, each formed by overlap of 2*p* orbitals on the two C atoms (Figure 19.30).

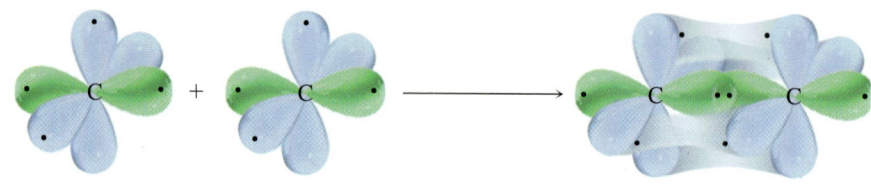

Spectroscopy of Alkynes

Terminal alkyne molecules, with an H atom directly attached to one of the C atoms of the triple bond, have even a shorter and stronger C—H bond than alkanes or alkenes. The IR wavenumber corresponding to C—H stretching is even higher, at 3300 cm^{-1}. If the alkyne is unsymmetrically substituted, a C≡C stretching absorbance is also seen at 2100–2260 cm^{-1}. The animated infrared spectrum of oct-1-yne in e19.25 is an example.

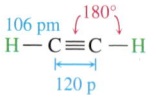

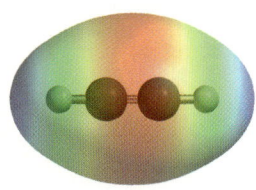

Three representations of molecules of ethyne.

Molecular Modelling

e19.24 Watch an animation portraying the use of the hybrid orbital model to form a triple bond.

FIGURE 19.30 Bonding in alkynes. Using the hybrid orbital model, two s and p need to be italicized - here and elsewhere throughout the chapter where s, p, or sp, sp2, sp3, etc. appear hybridized C atoms are brought together to form an alkyne, resulting in orbital overlap to give one σ and two π bonds.

Think about It

e19.25 Observe molecular vibrations responsible for specific absorptions in IR spectra of alkynes.

$$—C \equiv C— \quad 2100 - 2260 \ cm^{-1}$$

$$\equiv C—H \quad 3300 \ cm^{-1}$$

Characteristic IR stretching absorbances for alkynes.

In ^{13}C NMR spectra, alkyne C atoms have absorbances slightly upfield ($\delta = 75–100$ ppm) from alkene C atoms ($\delta = 100–119$ ppm). In ^{1}H NMR spectra, an H atom attached to a terminal C atom of an alkyne molecule absorbs very close in chemical shift to that of simple —CH_3 groups. So IR and ^{13}C NMR would be the spectroscopic methods of choice for identifying the presence of an alkyne.

EXERCISE 19.30—IR SPECTRA OF ALKYNES

Which absorbances would allow you to characterize the unknown compound whose infrared spectrum is shown below as an alkyne? Do molecules of this compound also contain any H atoms bonded to alkane C atoms? Why or why not?

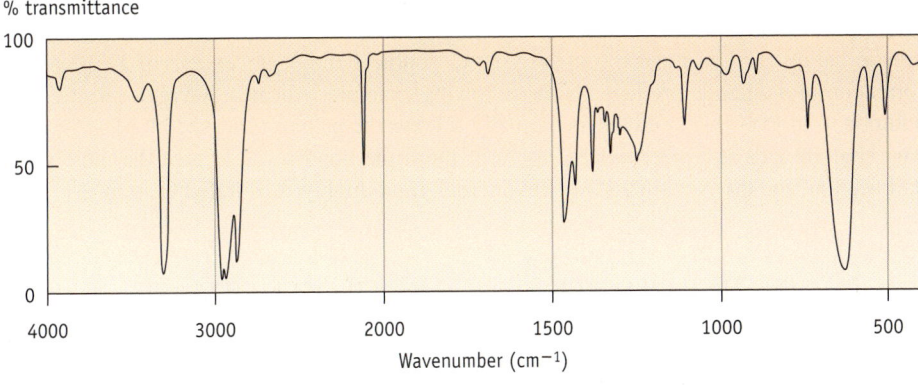

EXERCISE 19.31—NMR SPECTRA OF ALKYNES

Now consider the ^{13}C NMR spectrum of the same unknown compound in Exercise 19.30. How many unique kinds of C atoms does it have? Which ones confirm that it is an alkyne?

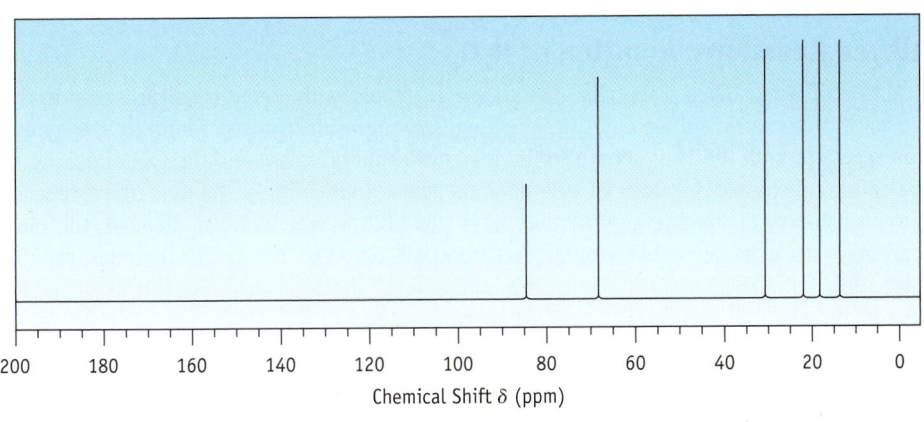

Identify the unknown compound whose IR and ^{13}C NMR spectra are given in Exercise 19.30–19.31, and whose mass spectrum below shows a weak molecular ion at $m/z = 82$.

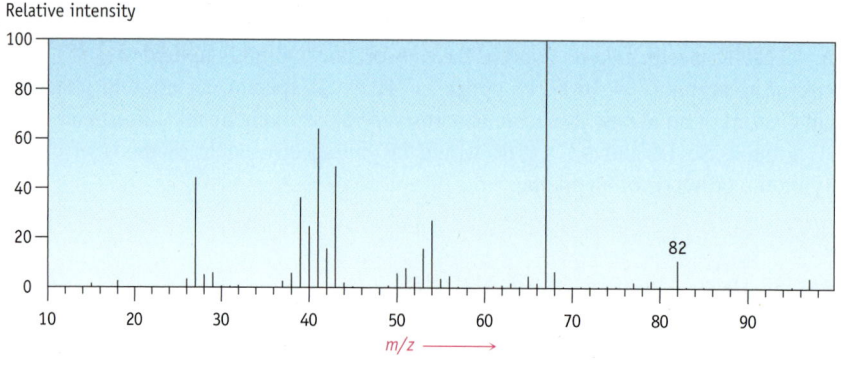

Alkyne Reactions: Addition of HX

Alkynes give addition products on reaction with HCl, HBr, and HI just as alkenes do. Although the reaction can usually be stopped after addition of 1 equivalent of HX to yield a *vinylic* halide (*vinylic* means "on the C=C double bond"), an excess of HX leads to formation of a dihalide product. As the following example indicates, the regioselectivity of addition to mono-substituted alkynes usually follows the pattern observed by Markovnikov. The H atom adds to the terminal C atom of the triple bond so as to give the most stable carbocation, and the X atom adds to the internal, more highly substituted C atom.

$$CH_3CH_2CH_2CH_2C{\equiv}CH + HBr \longrightarrow CH_3CH_2CH_2CH_2\overset{\overset{\displaystyle Br}{|}}{C}{=}CH_2$$

Hex-1-yne   **2-Bromohex-1-ene**

Alkyne Reactions: Addition of X₂

Bromine and chlorine add to alkynes to give dihalide addition products with anti stereochemistry:

$$CH_3CH_2CH_2CH_2C{\equiv}CH + Br_2 \xrightarrow{CCl_4}$$

$$\underset{Br\quad\quad H}{\overset{CH_3CH_2CH_2CH_2\quad\quad Br}{C{=}C}}$$

Hex-1-yne

(E)-1,2-Dibromohex-1-ene

Alkyne Reactions: Addition of H₂O

Addition of water takes place when an alkyne is treated with aqueous sulfuric acid in the presence of mercuric sulfate catalyst. Markovnikov regioselectivity is found for the hydration reaction, with the H attaching to the less substituted C atom and the OH attaching to the more substituted C atom of alkyne molecules. Interestingly, though, the expected vinylic alcohol, or *enol* (*ene* = alkene; *ol* = alcohol), is not isolated. Instead, the enol rearranges to a more stable isomer, a ketone ($R_2C{=}O$). Enols and ketones rapidly interconvert [>>Section 24.8]. With few exceptions, the keto–enol equilibrium heavily favours the ketone, and enols are almost never isolated.

$$CH_3CH_2CH_2C{\equiv}CH + H_2O \xrightarrow[HgSO_4]{H_2SO_4} \left[CH_3CH_2CH_2\overset{\overset{\displaystyle OH}{|}}{C}{=}CH_2 \right] \longrightarrow CH_3CH_2CH_2\overset{\overset{\displaystyle O}{||}}{C}CH_3$$

Pent-1-yne   **An enol**   **Pentan-2-one (78%)**

A mixture of two ketones results when an internal alkyne (R—C≡CR') is hydrated, but only a single product is formed from reaction of a terminal alkyne (R—C≡C—H).

$$CH_3CH_2C≡CCH_3 + H_2O \xrightarrow[HgSO_4]{H_2SO_4} CH_3CH_2\overset{\overset{O}{\|}}{C}CH_2CH_3 + CH_3CH_2CH_2\overset{\overset{O}{\|}}{C}CH_3$$

Pent-2-yne   **Pentan-3-one**   **Pentan-2-one**
(an internal alkyne)

$$CH_3CH_2CH_2C≡CH + H_2O \xrightarrow[HgSO_4]{H_2SO_4} CH_3CH_2CH_2\overset{\overset{O}{\|}}{C}CH_3$$

Pent-1-yne   **Pentan-2-one**
(a terminal alkyne)

WORKED EXAMPLE 19.12—ADDITION OF H₂O TO ALKYNES

What product would you obtain by hydration of 4-methylhex-1-yne?

Solution

Addition of water to 4-methylhex-1-yne according to the expected pattern of regioselectivity will yield a product with the OH group attached to C2 rather than C1. This enol then isomerizes to yield a ketone.

$$\underset{\textbf{4-Methylhex-1-yne}}{CH_3CH_2\overset{\overset{CH_3}{|}}{C}HCH_2C≡CH} + H_2O \xrightarrow[HgSO_4]{H_2SO_4} \left[CH_3CH_2\overset{\overset{CH_3}{|}}{C}HCH_2\overset{\overset{OH}{|}}{C}=CH_2 \right]$$

$$\longrightarrow \underset{\textbf{4-Methylhexan-2-one}}{CH_3CH_2\overset{\overset{CH_3}{|}}{C}HCH_2\overset{\overset{O}{\|}}{C}CH_3}$$

EXERCISE 19.33—ADDITION OF H₂O TO ALKYNES

What product would you obtain by hydration of oct-4-yne?

EXERCISE 19.34—ADDITION OF H₂O TO ALKYNES

What alkynes would you start with to prepare the following ketones by a hydration reaction?

(a) $CH_3CH_2CH_2\overset{\overset{O}{\|}}{C}CH_3$ (b) $CH_3CH_2CH_2\overset{\overset{O}{\|}}{C}CH_2CH_3$

Alkyne Reactions: Addition of H₂

Alkynes are easily converted into alkanes by reduction with 2 equivalents of H_2 over a palladium catalyst. The reaction proceeds through an alkene intermediate, and can be stopped at the alkene stage if the right catalyst is used. The catalyst most often used for this purpose is the Lindlar catalyst, a specially prepared form of palladium metal. Because hydrogenation occurs with *syn* stereochemistry, alkynes give *cis* alkenes when reduced:

The term '2 equivalents' means that 2 mol of H_2 are added for every 1 mol of alkyne.

$$CH_3(CH_2)_3C≡C(CH_2)_3CH_3$$
Dec-5-yne

$$\xrightarrow[Pd/C]{2\ H_2} CH_3(CH_2)_8CH_3$$
Decane (96%)

$$\xrightarrow[\substack{Lindlar\\catalyst}]{H_2} \underset{CH_3(CH_2)_3 \qquad (CH_2)_3CH_3}{\overset{H \qquad\qquad H}{C=C}}$$

***cis*-Dec-5-ene (96%)**

Alkyne Reactions: Formation of Acetylide Anions

The most striking difference between the chemistry of alkenes and alkynes is that terminal alkynes ($R—C\equiv C—H$) are weakly acidic, with $pK_a \approx 25$ [<<Section 14.4]. Alkenes, by contrast, are even weaker acids, with $pK_a \approx 44$. When a terminal alkyne is treated with a very strong base such as sodium amide, $NaNH_2$, the terminal H atom in the molecule is removed and an *acetylide anion* is formed:

$$R—C\equiv C—H + :NH_2\ Na^+ \longrightarrow R—C\equiv C:^-\ Na^+ + :NH_3$$

Acetylide anion

The presence of an unshared electron pair on the negatively charged alkyne C atom of acetylide anions makes them both basic and nucleophilic. As a result, acetylide anions react with alkyl halides such as bromomethane to substitute for the halogen and yield a new alkyne product. This is a very useful method for preparing larger alkynes from simpler precursors. Terminal alkynes can be prepared by reaction of acetylene itself, and internal alkynes can be prepared by further reaction of a terminal alkyne:

$$HC\equiv CH \xrightarrow{NaNH_2} HC\equiv C^-\ Na^+ \xrightarrow{RCH_2Br} HC\equiv CCH_2R$$

Acetylene   **A terminal alkyne**

$$RC\equiv CH \xrightarrow{NaNH_2} RC\equiv C^-\ Na^+ \xrightarrow{R'CH_2Br} RC\equiv CCH_2R'$$

A terminal alkyne   **An internal alkyne**

One limitation to the reaction of an acetylide anion with an alkyl halide is that only primary alkyl halides (RCH_2X) can be used, for reasons discussed when we consider the mechanism in Chapter 21.

WORKED EXAMPLE 19.13—SYNTHESIS OF ALKYNES

What alkyne and what alkyl halide would you use to prepare pent-1-yne?

Strategy

As always when synthesizing a compound, start from the product and work the problem backwards. Draw the structure of the target molecule, and identify the alkyl group(s) attached to the triple-bonded C atoms. In the present case, one of the alkyne C atoms has a propyl group attached to it, and the other has an H atom attached to it. Thus, pent-1-yne could be prepared by treatment of acetylene with $NaNH_2$ to yield sodium acetylide, followed by reaction with 1-bromopropane.

Solution

This propyl group comes from 1-bromopropane.

$$H—C\equiv C—CH_2CH_2CH_3 + Na^+Br^- \longleftarrow H—C\equiv C:^-\ Na^+ + CH_3CH_2CH_2Br$$

Pent-1-yne   **1-Bromopropane**

$$H—C\equiv C:^-\ Na^+ + :NH_3 \longleftarrow H—C\equiv C—H + :NH_2^-\ Na^+$$

Sodium acetylide   **Acetylene**

EXERCISE 19.35—SYNTHESIS OF ALKYNES

Show the alkyne and alkyl halide from which the following products can be obtained. Where two routes look feasible, list both.

(a) $\underset{\underset{CH_3}{|}}{CH_3CHCH_2CH_2C}\equiv CH$ 　　(b) $CH_3CH_2CH_2C\equiv CCH_3$ 　　(c) $\underset{\underset{CH_3}{|}}{CH_3CHC}\equiv CCH_3$

19.15　New Directions: Polyynes and Polyenynes

This chapter began with the example of β-carotene, a conjugated alkene extracted from carrots. Research attention is now paid to conjugated alkynes (polyynes) and compounds with alternating double and triple bonds (polyenynes). Naturally occurring compounds containing conjugated C≡C triple bonds have been isolated from plants, fungi, and marine sponges, and have been found to have powerful biological activity, such as antibacterial, pesticidal, and antifungal properties.

The compound containing an ene-triyne (one double bond conjugated with three triple bonds) shown below, has been isolated from a daisy plant, and found to be toxic to mosquito and black fly larvae, and adult nematodes. It is known as a phototoxic substance, which makes the skin susceptible to damage by light. As part of a global effort to synthesize new compounds with potential to check the spread of mosquito-borne diseases such as malaria and West Nile virus, a Canadian research group led by Rik Tykwinski has synthesized this and other related polyenyne compounds in the laboratory.

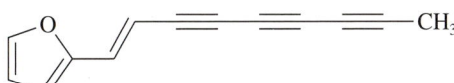

Ene-triyne synthesized by Tykwinski's research group.

Courtesy of Dr. Rik Tykwinski

Canadian chemist Rik Tykwinski.

Tykwinski's group and other research groups in other parts of the world have found that, like β-carotene, conjugated polyynes and polyenynes also show promise as electronically interesting new optical materials for data storage and other applications. The new landscape opening up through the exploration of polyalkynes has been imaginatively portrayed on the cover of an issue of the *Journal of Organic Chemistry* (Figure 19.31a), featuring Tykwinski's work (Figure 19.31b).

Reprinted with permission from *J. Org. Chem.*, January 25, 2002, 67. Copyright 2002 American Chemical Society

(a)

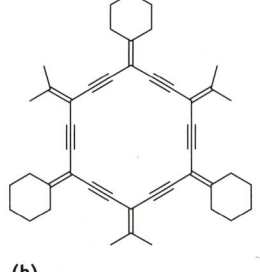

(b)

FIGURE 19.31 (a) The cover of the *Journal of Organic Chemistry* featuring new aspects of polyenynes and polyynes, and (b) a new polyalkyne synthesized by Tykwinski's group.

SUMMARY

Key Concepts

- Alkenes are hydrocarbons whose molecules contain one or more $C=C$ bonds. Because they contain fewer H atoms than related alkanes, alkenes are often referred to as unsaturated. An example of the importance of alkenes, β-carotene, a **conjugated alkene**, plays a key role in the chemistry of vision, the bright colouration of plants and animals, the provision of necessary vitamins, and in the creation of modern materials for optical-neural networks. (Section 19.1)

- Using the hybrid orbital model, a double bond can be thought of as consisting of a σ bond formed by head-on overlap of sp^2 hybrid orbitals on each C atom of the $C=C$ bond, and a π bond formed by sideways overlap of $2p$ orbitals on each C atom. Because rotation around the double bond is not possible, substituted alkenes can exist as *cis–trans stereoisomers*. The geometry of a double bond can be described using an ***E, Z system of nomenclature*** by applying a series of sequence rules. (Sections 19.2, and 19.4)

- A molecular orbital model is needed to give more satisfactory explanations of spectroscopic results and examples of unusual stability or reactivity of alkenes. (Section 19.2)

- The structures of alkene molecules can be characterized by their IR, ^{13}C NMR, and ^{1}H NMR spectra. (Section 19.5)

- One important way to classify organic reactions is based on spectroscopic analysis of reactants and products to see what kind of reaction has occurred. There are four broad types of organic reactions: **addition, elimination, substitution,** and **rearrangement**. (Section 19.6)

- A full description of how a reaction occurs at the molecular level is called its **reaction mechanism**. There are two kinds of organic mechanisms: polar and radical. In **polar reactions**, the most common kind, electron-rich sites in one molecule (a Lewis base or **nucleophile**) react with electron-poor sites in another molecule (a Lewis acid or **electrophile**). Bonds are made when an electron-rich atom donates a pair of electrons to an electron-poor atom, and bonds are broken when one atom leaves with both electrons from a bond. **Radical reactions** involve odd-electron species and occur when each reactant donates one electron in forming a new bond. (Section 19.7)

- The chemistry of alkenes is dominated by addition reactions of electrophiles. When HX reacts with an alkene, **Markovnikov's rule** predicts that the H atom will add to the C atom that has fewer alkyl substituents and the X group will add to the C atom that has more alkyl substituents. Our postulated mechanism for this reaction takes place in two steps, initiated by a collision between a molecule of the alkene nucleophile and an H^+ ion electrophile, leading to formation of a carbocation intermediate, which reacts rapidly with a Cl^- ion to form the observed alkyl halide product. More highly substituted carbocations are more stable than less highly substituted ones because alkyl groups tend to donate electrons to the positively charged C atom. The more alkyl groups present, the more electron donation there is and the more stable the carbocation. (Section 19.8)

- Many electrophiles besides HX add to alkenes. Br_2 and Cl_2 add to alkenes to give 1,2-dihalide addition products having *anti stereochemistry*. Addition of H_2O (*hydration*) takes place on reaction of the alkene with aqueous acid, and addition of H_2 (*hydrogenation*) occurs in the presence of a metal catalyst such as platinum or palladium. *Oxidation* of alkenes is carried out using potassium permanganate ($KMnO_4$). Under basic conditions, $KMnO_4$ reacts with alkenes to yield *cis*-1,2-diols. Under neutral or acidic conditions, $KMnO_4$ cleaves double bonds to yield carbonyl-containing products. (Sections 19.9–19.12)

- Conjugated alkenes, such as 1,3-butadiene, have molecules with alternating single and double bonds. The mechanism for the *1,4-addition* of electrophiles to conjugated dienes is thought to involve formation of a resonance-stabilized *allylic* carbocation intermediate. Allylic carbocation intermediates are very stable due to delocalization of the positive charge over several C atoms of the cation. (Section 19.8)

- Many simple alkenes undergo *polymerization* when treated with a radical catalyst. Polymers are large molecules formed by the repetitive bonding together of many small monomer units. (Section 19.13)
- **Alkynes** are hydrocarbons whose molecules contain C≡C triple bonds. Much of the chemistry of alkynes is similar to that of alkenes. For example, alkynes react with 1 equivalent of HBr and HCl to yield *vinylic* halides, and with 1 equivalent of Br₂ and Cl₂ to yield 1,2-dihalides. Alkynes can also be hydrated by reaction with aqueous sulfuric acid in the presence of mercuric sulfate catalyst. The reaction leads to an intermediate enol that immediately isomerizes to a ketone. Alkynes can be hydrogenated with the Lindlar catalyst to yield a *cis* alkene. Terminal alkynes are weakly acidic and can be converted into *acetylide anions* by treatment with a strong base. Reaction of an acetylide anion with a primary alkyl halide then gives an internal alkyne. This important reaction leads to formation of a new C—C bond. (Section 19.14)
- Naturally occurring compounds containing conjugated alkynes (polyynes) and alternating double and triple bonds (polyenynes) have been isolated from plants, fungi, and marine sponges, and have been found to have powerful biological activity, such as antibacterial, pesticidal, and antifungal properties. (Section 19.15)

Key Reactions

1. Reactions of alkenes
   (a) Addition of HX, where X = Cl, Br, or I (Section 19.8)

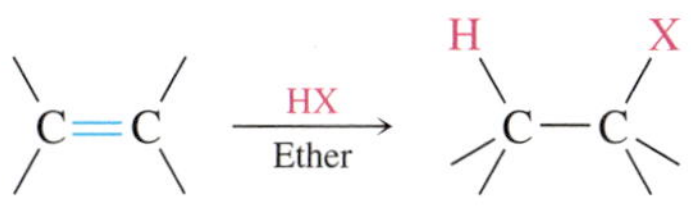

   Markovnikov's rule: The proton adds preferentially to the C atom that will lead to the most stable carbocation intermediate.

   (b) Addition of H₂O (Section 19.9)

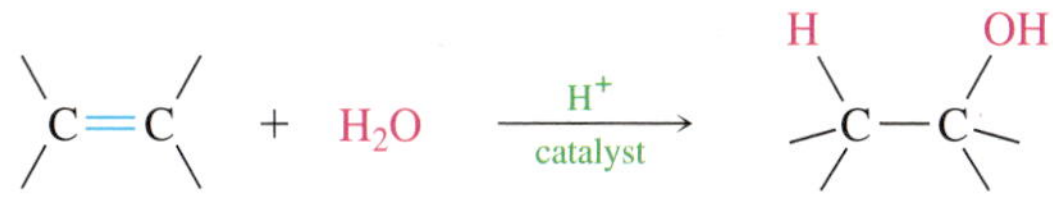

   Markovnikov's rule: The proton adds preferentially to the C atom that will lead to the most stable carbocation intermediate.

   (c) Addition of X₂, where X = Cl, Br (Section 19.10)

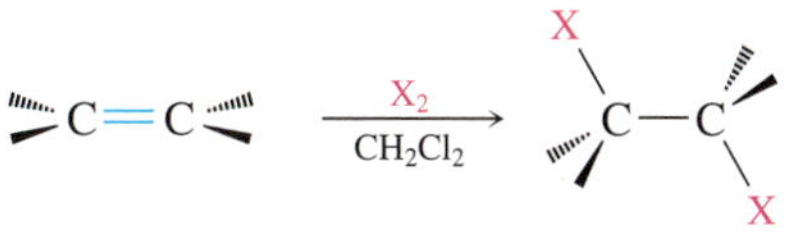

   Anti addition

   (d) Addition of H₂ (Section 19.11)

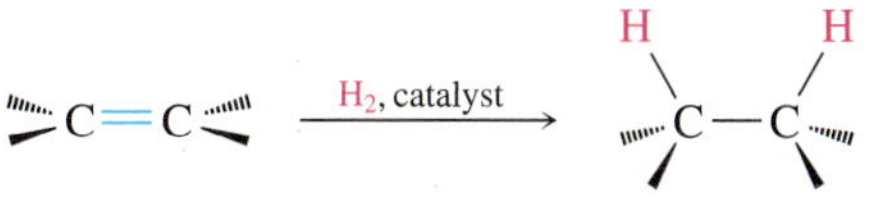

   Syn addition

   (e) Hydroxylation with KMnO₄ (Section 19.12)

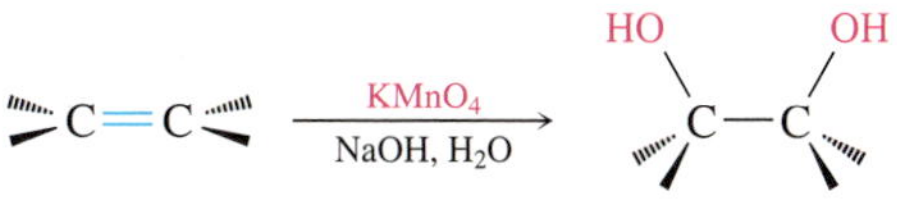

   Syn addition

   (f) Oxidative cleavage of alkenes with acidic KMnO₄ (Section 19.12)

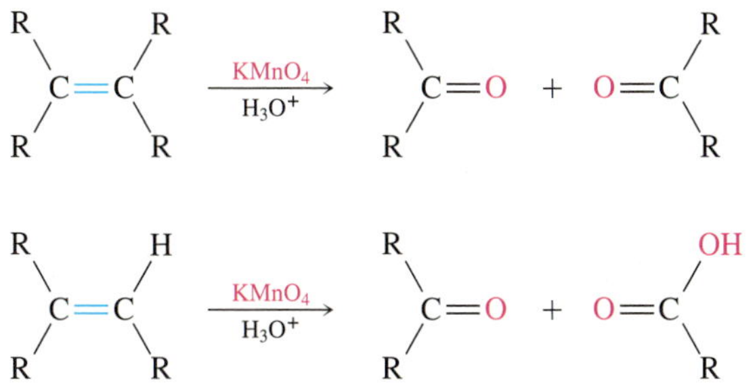

(g) Free radical polymerization of alkenes (Section 19.13)

$$n\ H_2C=CH_2 \xrightarrow[\text{initiator}]{\text{Radical}} (CH_2CH_2)_n$$

2. Reactions of alkynes
   (a) Addition of H_2 (Section 19.14)

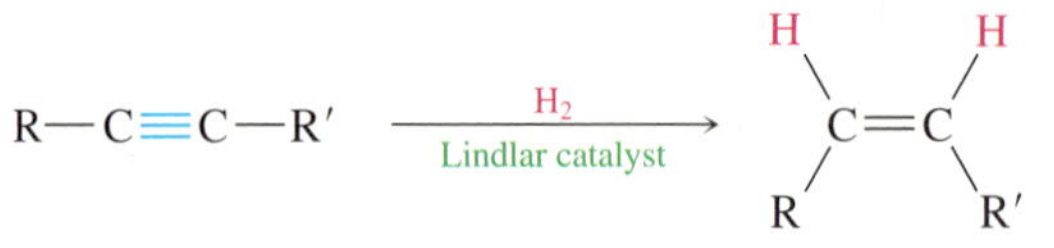

$$R—C\equiv C—R' \xrightarrow[\text{Lindlar catalyst}]{H_2} \begin{array}{c} H \quad\quad H \\ C=C \\ R \quad\quad R' \end{array} \quad \text{Syn addition}$$

A cis alkene

(b) Addition of HX, where X = Cl, Br, or I (Section 19.14)

$$—C\equiv C— + HX \longrightarrow \begin{array}{c} H \quad\quad X \\ C=C \end{array}$$

Markovnikov's rule: The proton adds preferentially to the C atom that will lead to the most stable carbocation intermediate.

(c) Addition of X_2, where X = Cl, Br (Section 19.14)

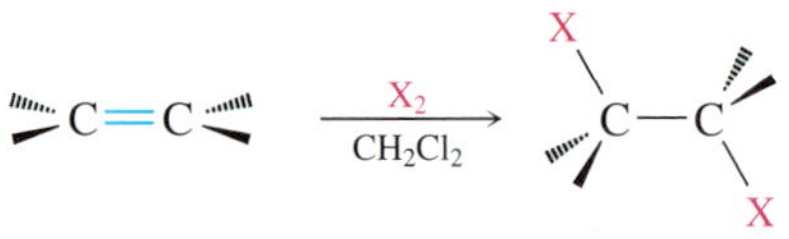

$$C=C \xrightarrow[\text{CH}_2\text{Cl}_2]{X_2} \begin{array}{c} X \\ C—C \\ X \end{array} \quad \text{Anti addition}$$

(d) Addition of H_2O (Section 19.14)

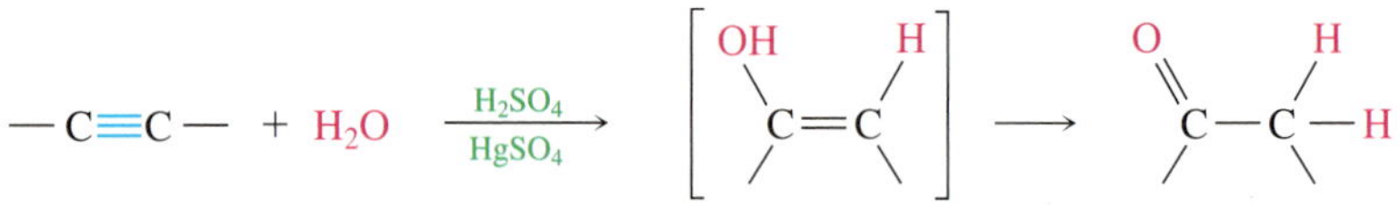

$$—C\equiv C— + H_2O \xrightarrow[\text{HgSO}_4]{H_2SO_4} \left[\begin{array}{c} OH \quad\quad H \\ C=C \end{array} \right] \longrightarrow \begin{array}{c} O \quad\quad H \\ C—C—H \end{array}$$

(e) Acetylide anion formation (Section 19.14)

$$R—C\equiv C—H \xrightarrow{\text{NaNH}_2} R—C\equiv C:^- \ Na^+ + NH_3$$

(f) Reaction of acetylide anions with alkyl halides (Section 19.14)

$$R—C\equiv C:^- \ Na^+ + R'CH_2X \longrightarrow R—C\equiv C—CH_2R' + NaX$$

(A particularly important reaction, since it leads to formation of a new C—C bond.)

REVIEW QUESTIONS

Section 19.1: Pheasant Wattles, Photoprotection, and Photonics

19.36 Circle the isoprene units in the molecule of Vitamin A. How does the number of isoprene units compare with that in β-carotene (Figure 19.1a)?

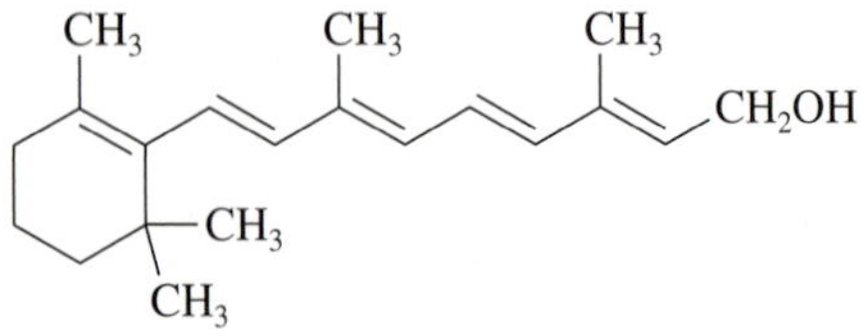

19.37 Why is β-carotene often added to commercial food for ornamental fish?

19.38 What role does β-carotene play in photosynthesis?

19.39 Identify the functional groups in each of the following molecules:

(a) $CH_3CH_2C\equiv N$

(b) 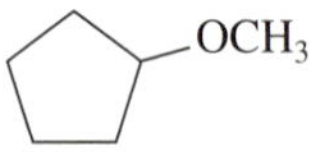 OCH_3

(c) 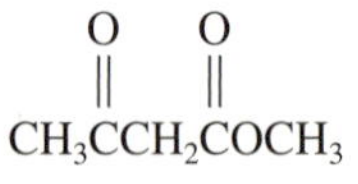 $CH_3CCH_2COCH_3$ (with two O double bonds)

(d)

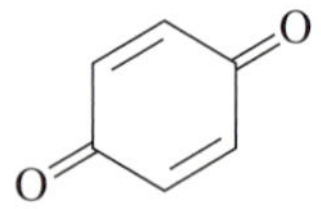

(e) 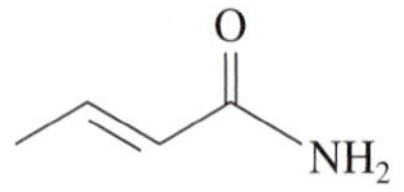 NH_2

(f)

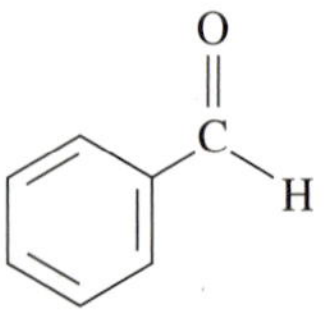

Section 19.2: Electronic Structure of Alkenes

19.40 Indicate the hybridization predicted for each C atom in the following molecules, using the hybrid orbital model [<<Section 10.8].

(a)

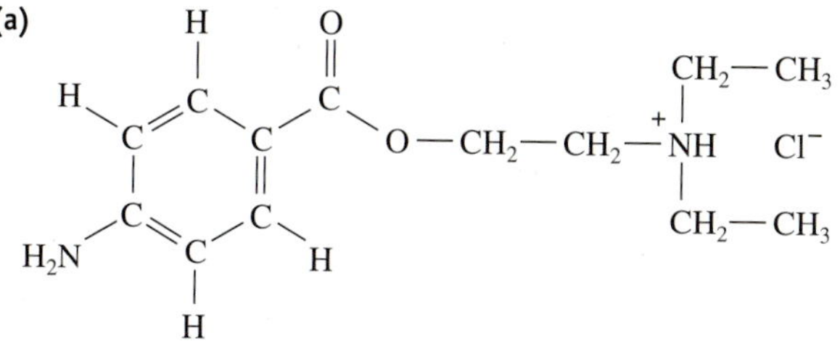

Procaine

(b)

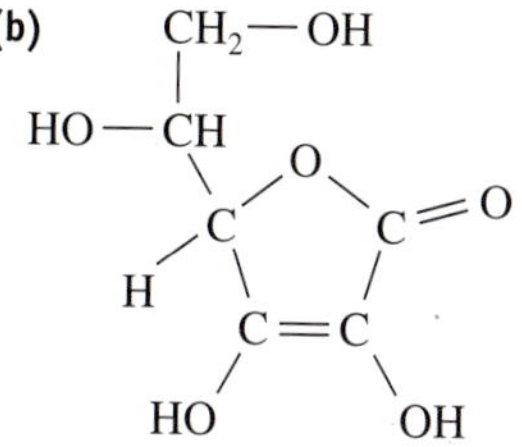

Vitamin C

19.41 The molecular structure of α-carotene is shown below. Identify any differences relative to the β-carotene structure shown in Figure 19.1.

The molecular structure of lycopene is also shown below. α-carotene is orange (absorbs blue light), and lycopene is the main red pigment (absorbs green light) in tomatoes. The difference in energy between the HOMO and LUMO in these two molecules are given below:

$$\alpha\text{-carotene: } \Delta E_{(LUMO - HOMO)} = 2.74 \text{ ev}$$
$$\text{lycopene: } \Delta E_{(LUMO - HOMO)} = 2.64 \text{ ev}$$

Explain why the colours of the substances are different and why one absorbs at a higher wavelength than the other.

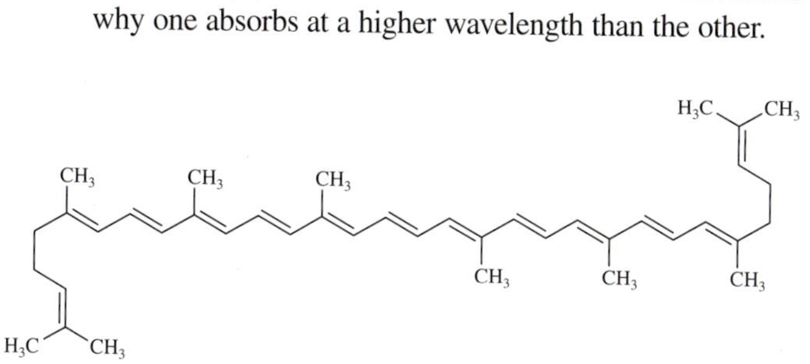

Lycopene

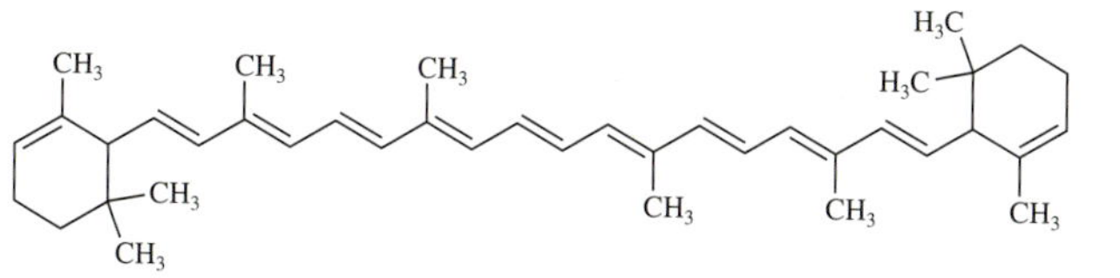

α-carotene

Section 19.3: Naming Alkenes

19.42 Draw molecular structures corresponding to the following IUPAC names:
(a) 3-propylhept-2-ene
(b) 2,4-dimethylhex-2-ene
(c) octa-1,5-diene
(d) 4-methylpenta-1,3-diene
(e) *cis*-4,4-dimethylhex-2-ene
(f) (*E*)-3-methyl-3-heptene

19.43 Draw the structures of molecules of the following cycloalkenes:
(a) *cis*-4,5-dimethylcyclohexene
(b) 3,3,4,4-tetramethylcyclobutene

19.44 A compound of formula C_9H_{14} contains no rings or triple bonds in its molecules. How many double bonds does it have?

19.45 Give IUPAC names for the compounds whose molecules are shown:

(a) $H_2C=CHCH_2\overset{\overset{\displaystyle CH_3}{|}}{C}HCH_3$

(b) $CH_3CH_2CH=CHCH_2CH_2CH_3$

(c) $H_2C=CHCH_2CH_2CH=CHCH_3$

(d) $CH_3CH_2CH=CHCH(CH_3)_2$

19.46 Name the cycloalkenes whose molecules are shown below:

(a)   (c)

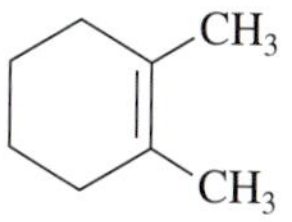

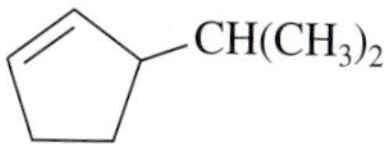

(b)

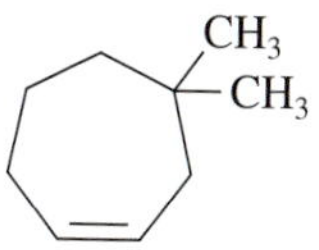

19.47 Which of the following molecules have *cis* and *trans* isomers?

(a) $CH_3\overset{\overset{\displaystyle CH_3}{|}}{C}=CHCH_2CH_3$

(b) $ClCH_2CH_2\overset{\overset{\displaystyle H_3C}{|}}{C}=\overset{\overset{\displaystyle CH_3}{|}}{C}CH_2CH_2Cl$

(c) HO

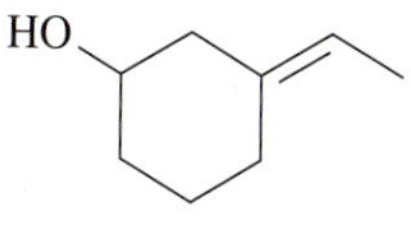

Section 19.4: *E* and *Z* Isomers of Alkenes

19.48 Name the following alkenes, including the *cis* or *trans* designation:

(a)

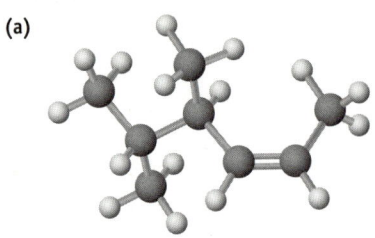

(b)

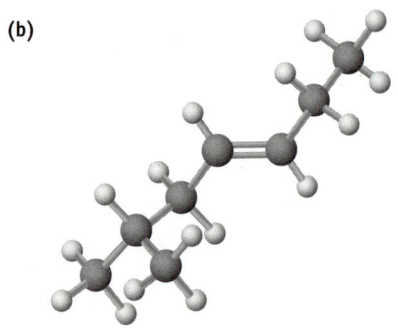

19.49 Which is higher in priority?

$$\underset{\text{—C—OH}}{\overset{O}{\|}} \quad \text{or} \quad \underset{\text{—C—OCH}_3}{\overset{O}{\|}}$$

19.50 Rank the following sets of substituents in order of priority according to the sequence rules:
(a) $-CH_3$, $-Br$, $-H$, $-I$
(b) $-OH$, $-OCH_3$, $-H$, $-CO_2H$
(c) $-CH_3$, $-CO_2H$, $-CH_2OH$, $-CHO$
(d) $-CH_3$, $-CH=CH_2$, $-CH_2CH_3$, $-CH(CH_3)_2$

19.51 Assign *E* or *Z* configuration to the following alkene molecules:

(a)
$$\underset{CH_3}{\overset{HOCH_2}{}}C=C\underset{H}{\overset{CH_3}{}}$$

(b)
$$\underset{Cl}{\overset{HO-C}{\overset{\|}{\overset{O}{}}}}C=C\underset{OCH_3}{\overset{H}{}}$$

Section 19.5: Spectroscopy of Alkenes

19.52 The ^{13}C NMR spectra for two alkenes are shown. One has the formula C_5H_{10} and the other C_5H_8. Which is which? Explain.

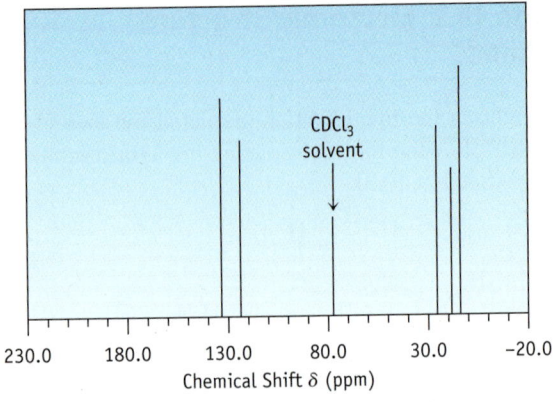

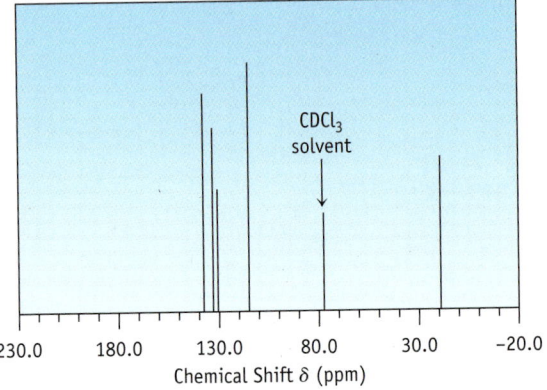

Section 19.6: Classifying Organic Reactions by Type

19.53 Identify the following reactions as addition, elimination, substitution, or rearrangement:

(a) $CH_3CH_2Br \;+\; NaCN \longrightarrow CH_3CH_2CN \;(+ NaBr)$

(b)

$$\text{(cyclohexanol)} \xrightarrow[\text{catalyst}]{\text{Acid}} \text{(cyclohexene)} \;(+ H_2O)$$

(c)

$$\text{(cyclopentadiene)} + \text{(methyl vinyl ketone)} \xrightarrow{\text{Heat}} \text{(bicyclic ketone)}$$

(d)

$$\text{(cyclohexane)} + O_2N-NO_2 \xrightarrow{\text{Light}} \text{(nitrocyclohexane)} + HNO_2$$

Section 19.7: Visualizing the Mechanisms of Organic Reactions

19.54 Which of the following ions or molecules are most likely to behave as electrophiles and which as nucleophiles?
(a) Cl^- (d) CN^-
(b) $N(CH_3)_3$ (e) CH_3^+
(c) Hg^{2+}

Section 19.8: Electrophilic Addition of HX to Alkenes: Hydrohalogenation

19.55 Draw a reaction energy diagram for a two-step reaction that releases energy and whose first step is faster than its second step. Label the parts of the diagram corresponding to reactants, products, transition states, reaction intermediate, activation energies, and overall energy change.

19.56 Consider the reaction energy diagram shown, and answer the following questions:

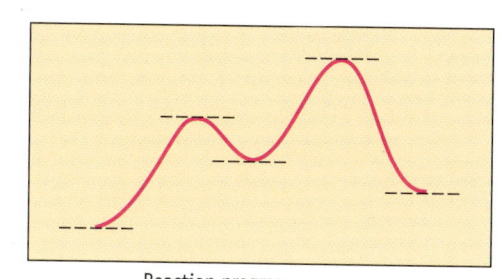

Reaction progress ⟶

(a) Indicate the overall energy change for the reaction. Is it positive or negative?
(b) How many steps are involved in the reaction?
(c) Which step is faster?
(d) How many transition states are there? Label them.

19.57 Reaction of HCl with pent-2-ene yields a mixture of two addition products. Write the reaction and show the two products. Then show the detailed, step-wise mechanism for the reaction, using curved arrows to show movement of electron pairs.

Sections 19.9–19.13: Reactions of Alkenes

19.58 When isopropylidenecyclohexane is treated with strong acid at room temperature, isomerization occurs by the mechanism shown below to yield 1-isopropylcyclohexene:

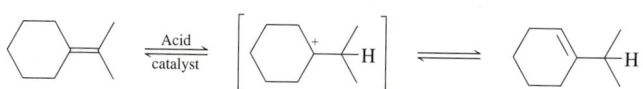

Isopropylidenecyclohexane   **1-Isopropylcyclohexene**

At equilibrium, the product mixture contains about 30% isopropylidenecyclohexane and about 70% 1-isopropyl-cyclohexene.
(a) What kind of reaction is occurring? Is the mechanism polar or radical?
(b) Draw curved arrows to indicate electron flow in each step of the mechanism.

19.59 We've seen that the stability of carbocations depends on the number of alkyl groups attached to the positively charged C atom—the more alkyl groups, the more stable the cation. Draw the two possible carbocation intermediates that might be formed in the reaction

HCl with 2-methylpropene, determine which is more stable, and predict which product will form.

19.60 What alkenes would give the following alcohols on hydration? (Red = O.)

(a)

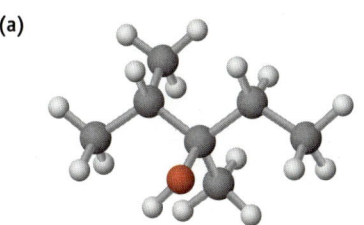

(b)

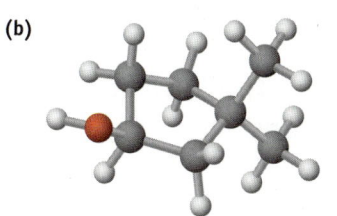

19.61 Predict the products of the following reactions of styrene. Indicate regioselectivity where relevant. (The aromatic ring is inert to all the indicated reagents.)

Styrene

(a) Styrene + H_2 $\xrightarrow{Pd}$?   (b) Styrene + Br_2 ⟶ ?
(c) Styrene + HBr ⟶ ?   (d) Styrene + $KMnO_4$ $\xrightarrow{NaOH, H_2O}$?

19.62 Suggest structures for alkenes that give the following reaction products. There may be more than one answer for some cases.

(a) ? $\xrightarrow{H_2/Pd\ catalyst}$ **2-Methylhexane**

(b) ? $\xrightarrow{Br_2\ in\ CH_2Cl_2}$ **2,3-Dibromo-5-methylhexane**

(c) ? $\xrightarrow{HBr}$ **2-Bromo-3-methylheptane**

(d) ? $\xrightarrow[H_2O]{KMnO_4, OH^-}$

$$CH_3CHCH_2CHCHCH_2CH_3$$
with CH$_3$, HO, OH groups

19.63 Using an oxidative cleavage reaction, explain how you would distinguish between the following two isomeric cyclohexadienes:

and

19.64 What products would you expect to obtain from reaction of cyclohexa-1,3-diene with each of the following?
(a) 1 mol Br_2 in CH_2Cl_2
(b) 1 mol HCl
(c) 1 mol DCl (D = deuterium, ^2H)
(d) H_2 over a Pd catalyst

19.65 Plexiglas, a clear plastic used to make many moulded articles, is made by polymerization of methyl methacrylate. Draw a molecular level picture of a representative segment of Plexiglas.

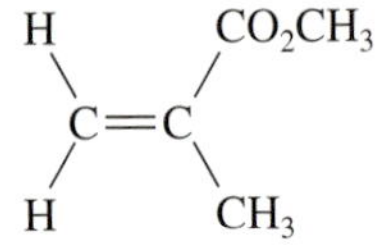

$$\begin{array}{c} H \\ \diagdown \\ C=C \\ \diagup \quad \diagdown \\ H \qquad CH_3 \end{array} \begin{array}{c} CO_2CH_3 \end{array}$$

Methyl methacrylate

19.66 What monomer unit might be used to prepare the following polymer?

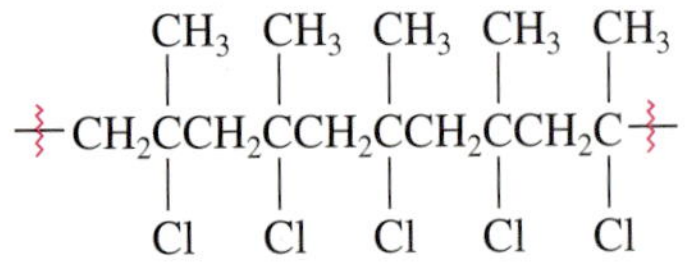

$$\begin{array}{ccccc} CH_3 & CH_3 & CH_3 & CH_3 & CH_3 \\ | & | & | & | & | \\ \xi CH_2CCH_2CCH_2CCH_2CCH_2C\xi \\ | & | & | & | & | \\ Cl & Cl & Cl & Cl & Cl \end{array}$$

Section 19.14: Alkynes: Electronic Structure, Spectroscopy, and Reactions

19.67 How many ^{13}C NMR absorptions would you expect for the following:
(a) hex-1-yne   (b) hex-2-yne   (c) hex-3-yne

19.68 How might you differentiate between hex-1-yne and hex-2-yne by IR spectroscopy?

19.69 Predict the products of the following reactions of hex-1-yne:

(a) $\xrightarrow{\text{1 equiv HBr}}$ **?** (c) $\xrightarrow{\text{H}_2,\ \text{Lindlar catalyst}}$ **?**

(b) $\xrightarrow{\text{1 equiv Cl}_2}$ **?**

19.70 Predict the products of the following reactions of dec-5-yne:

(a) $\xrightarrow{\text{H}_2,\ \text{Lindlar catalyst}}$ **?** (c) $\xrightarrow{\text{H}_2\text{O, H}_2\text{SO}_4,\ \text{HgSO}_4}$ **?**

(b) $\xrightarrow{\text{2 equiv Br}_2}$ **?**

19.71 How would you prepare *cis*-but-2-ene starting from prop-1-yne, an alkyl halide, and any other reagents needed? (This problem can't be worked in a single step. You'll have to carry out more than one reaction.)

19.72 What alkynes would you hydrate to obtain the following ketones?

(a)
$$\begin{array}{cc} CH_3 & O \\ | & \| \\ CH_3CHCH_2CCH_3 \end{array}$$

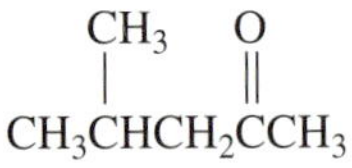

(b)
$$\begin{array}{c} O \\ \| \\ C-CH_3 \end{array}$$

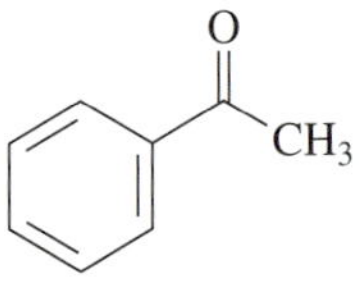

SUMMARY AND CONCEPTUAL QUESTIONS

19.73 Draw and name molecules that meet the following descriptions:
(a) an alkene, C_6H_{12}, that does not have *cis* and *trans* isomers
(b) the *E* isomer of a trisubstituted alkene, C_6H_{12}
(c) a cycloalkene, C_7H_{12}, with a tetrasubstituted double bond

19.74 α-Farnesene is a constituent of the natural waxy coating found on apples.
(a) Circle the isoprene units in the molecule.
(b) What is its IUPAC name?
(c) Indicate, where appropriate, *E* or *Z* configuration for each of the double bonds.

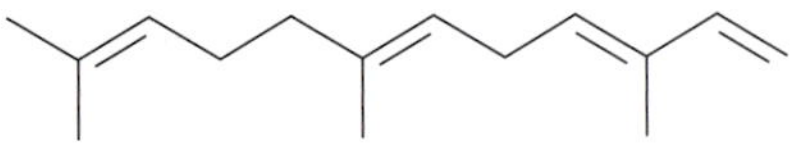

α-**Farnesene**

19.75 Methoxide ion (CH_3O^-) reacts with bromoethane in a single step according to the following equation for the mechanism:

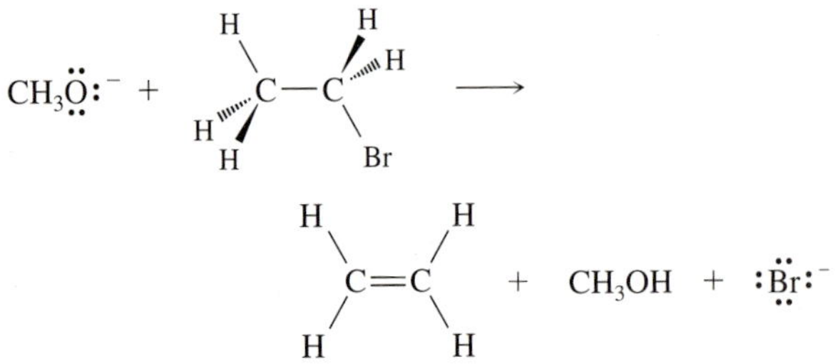

$$CH_3\ddot{O}\!:^- +$$

Identify the bonds broken and formed, and draw curved arrows to represent the flow of electrons during the reaction.

19.76 Follow the flow of electrons indicated by the curved arrows in the following reaction mechanism, and predict the products that result:

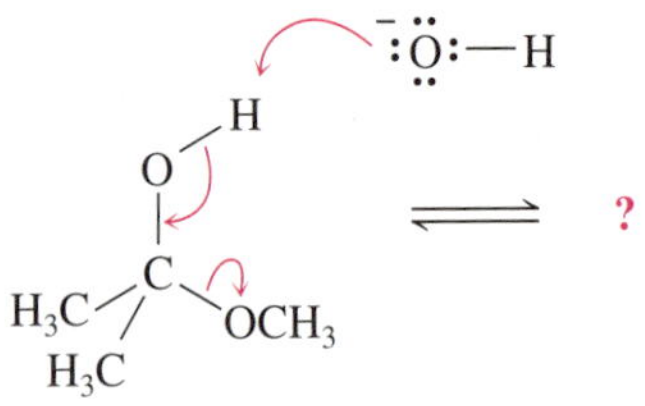

$\rightleftharpoons$ **?**

19.77 Draw three additional resonance structures for the benzyl cation, and comment on how electron delocalization affects its stability.

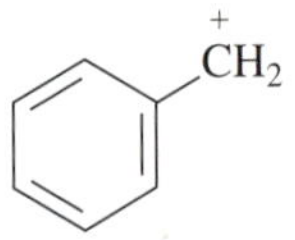

Benzyl cation

19.78 In light of your answer to Question 19.77, what product would you expect from the following reaction? Explain.

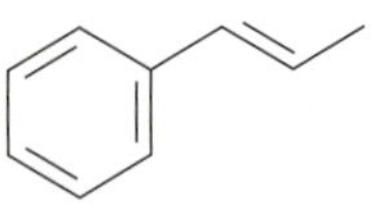

$+ HCl \longrightarrow$ **?**

19.79 Reaction of 2-methylpropene with CH_3OH in the presence of H_2SO_4 catalyst yields methyl *tert*-butyl ether, $CH_3OC(CH_3)_3$, by a mechanism analogous to that of acid-catalyzed alkene hydration. Write the detailed, step-wise mechanism.

19.80 Compound A has the formula C_8H_8. It reacts rapidly with acidic $KMnO_4$ but reacts with only 1 equivalent of H_2 over a palladium catalyst. On hydrogenation under conditions that reduce aromatic rings, A reacts with 4 equivalents of H_2 and hydrocarbon B, C_8H_{16}, is produced. The reaction of A with $KMnO_4$ gives CO_2 and a carboxylic acid C, $C_7H_6O_2$. What are the structures of A, B, and C? Write all the reactions.

19.81 Each mole of compound A, C_9H_{12}, absorbs 3 equivalents of H_2 on catalytic reduction over a palladium catalyst to give B, C_9H_{18}. On reaction with $KMnO_4$, compound A gives, among other things, a ketone that was identified as cyclohexanone. On treatment with $NaNH_2$ in NH_3, followed by addition of iodomethane, compound A gives a new hydrocarbon C, $C_{10}H_{14}$. What are the structures of A, B, and C?

19.82 The sex attractant of the common housefly is a hydrocarbon named *muscalure* ($C_{23}H_{46}$). On treatment of muscalure with aqueous acidic $KMnO_4$, two products are obtained: $CH_3(CH_2)_{12}COOH$ and $CH_3(CH_2)_7COOH$. Propose a structure for muscalure.

19.83 How would you synthesize muscalure (Question 19.82) starting from acetylene and any alkyl halides needed? (The double bond in muscalure is *cis*.)

19.84 Poly(vinyl pyrrolidone), prepared by from *N*-vinylpyrrolidone, is used both in cosmetics and as a synthetic blood substitute. Draw a representative segment of the polymer.

N-Vinylpyrrolidone

19.85 Draw a reaction energy diagram for the addition of HBr to pent-1-ene. Let one curve on your diagram show the formation of 1-bromopentane product and another curve on the same diagram show the formation of 2-bromopentane product. Label the positions for all reactants, intermediates, and products.

19.86 Make sketches of what you imagine the transition-state structures to look like in the reaction of HBr with pent-1-ene (Question 19.85).

19.87 Methylenecyclohexane, on treatment with strong acid, isomerizes to yield 1-methylcyclohexene. Propose a mechanism by which the reaction might occur.

Methylenecyclohexane   **1-Methylcyclohexene**

19.88 α-Terpinene ($C_{10}H_{16}$) is a pleasant-smelling hydrocarbon that has been isolated from oil of marjoram. On hydrogenation over a palladium catalyst, each mole of α-terpinene reacts with 2 mol of hydrogen to yield a new hydrocarbon, $C_{10}H_{20}$. On reaction with acidic $KMnO_4$, α-terpinene yields oxalic acid and 6-methyheptane-2,5-dione. Propose a structure for α-terpinene.

Oxalic acid   **6-Methylheptane-2,5-dione**

With the **eCHACR** single-sign-on access card bundled with your text, log on (http://login.cengage.com/sso) and access the e-book and click on any in-margin icons for dynamic molecular-level animations and simulations, videos of laboratory reactions, interactive exercises, reviews of background concepts, and other online supplementary materials as noted by the icons in the text margins.

Also go to www.chemistry.nelson.com <http://www.chemistry.nelson.com> for Answers to in-chapter exercises and selected Review Questions, Test Yourself questions, weblinks, crossword puzzles, flashcards, glossary of key terms, and other student resources.

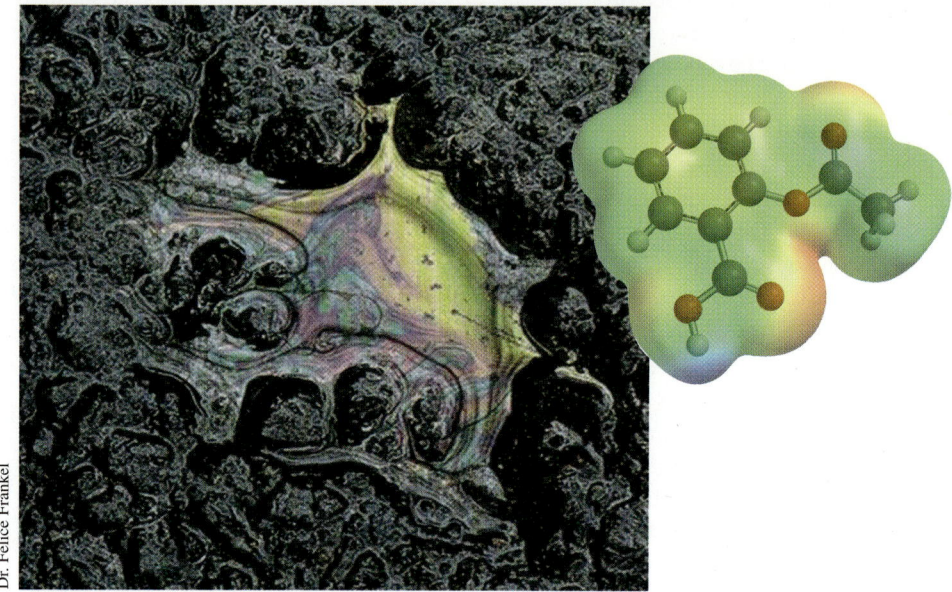

Dr. Felice Frankel

20.1 Case Study: Aromatic Roots of Chemistry: Curing, Death, and Dyeing

Dolor, rubor, calor, and *tumor.*

You probably don't recognize these Latin names for symptoms of inflammation from Celsus' *Encyclopedia*, written in about 25 BC. But you would recognize pain, redness, warmth, and swelling if you experienced a sprained ankle or bee sting. They are classic signs of inflammation, a local, usually non-specific response of the body's defence system to cellular or tissue injury. In mammals, we know that injured cells release chemical compounds that help eliminate toxic substances and dispose of damaged tissue so that healing can begin. Sometimes, the inflammatory response doesn't achieve its goal, and inflammatory disorders such as chronic arthritis become evident.

Attempts to relieve symptoms of inflammation with substances extracted from natural products are as old as the symptoms themselves. One of the oldest recorded natural products is willow bark extract, used as early as 400 BC by the Roman army for joint pain. Willow bark constituents were found to reduce both pain and fever in clinical trials in Northern Europe in the 1760s by Edward Stone, who was looking for a remedy for malaria. In 1827, the active agent in willow bark was identified as the compound *salicin,* which could be converted by reaction with water into salicyl alcohol and then oxidized to give salicylic acid.

Salicylic acid is even more effective than salicin for reducing fevers and also has both analgesic and anti-inflammatory properties. Unfortunately, salicylic acid is too corrosive to the walls of the stomach for everyday use. Conversion of the —OH functional group in salicylic acid into an ester, however, yields acetylsalicylic acid (ASA, aspirin), which proved just as potent as salicylic acid but less corrosive to the stomach (Figure 20.1). Molecules of each of these compounds contain an "aromatic ring," described in Section 20.2.

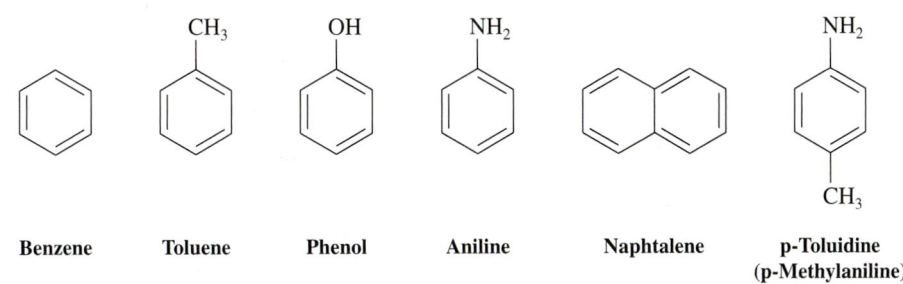

FIGURE 20.1 Production of ASA from salicin. Like many biologically active compounds extracted from plants, the rest of the salicin molecule is joined to a sugar, or carbohydrate molecule—in this case, glucose.

| Salicin | Salicylic alcohol | Salicylic acid | Acetylsalicylic acid (ASA) |

Dyes and Medicine—The Beginnings of Industrial Organic Chemistry

Herman Kolbe first synthesized salicylic acid from phenol in a laboratory in 1874, making it unnecessary to extract starting materials from plants. The larger scale conversion of salicylic acid into ASA was done in the Bayer dye factory on the shores of the Rhine River in Germany. This development foreshadowed what was to come in the evolution of industrial organic chemistry. Several ingredients important to building the modern pharmaceutical industry were present in Germany at the end of the 19th century. Skilled dye workers were available as a result of the emigration from France to Germany by many skilled Huguenot silk weavers and dyers following the revocation of the Edict of Nantes; starting materials were available from coal tar; and the water needed for chemical processes was available from the Rhine River.

The connection between dyes and medicine was established by several accidental discoveries. One was the attempt by William Perkin to synthesize the anti-malarial compound quinine from coal tar, the black sticky nuisance liquid material left over from heating coal in the absence of air to produce coal gas and coke for fuel for the steel industry. From coal tar, chemists had been able to isolate several important compounds, many of which had strong aromas, now known for different reasons as aromatic compounds. These included benzene, toluene, phenol, aniline, and naphthalene. Perkin treated aniline with the strong oxidizing agent potassium permanganate, but instead of isolating quinine, a beautiful lavender-lilac coloured solution appeared. He had made "mauveine," a new synthetic dye soon in high demand in the Paris fashion industry and used to colour Queen Victoria's silk dress in 1862. Ironically, when Perkin repeated the reaction with purified aniline, the dye was not produced—it had resulted from a toluidine impurity in the original solution.

William Henry Perkin with mauveine dye from coal tar.

| Benzene | Toluene | Phenol | Aniline | Naphtalene | p-Toluidine (p-Methylaniline) |

"But this same tar, which for a hundred years was thrown away ... turns out to be one of the most useful things in the world. It is one of the strategic points in war and commerce. It wounds and heals. It supplies munitions and medicines. It is like the magic purse of Fortunatus from which anything wished for could be drawn. The chemist puts his hand into the black mass and draws out all the colors of the rainbow. This evil-smelling substance beats the rose in the production of perfume and surpasses the honey-comb in sweetness."

Edwin Slosson, *Creative Chemistry*, 1919

A second accidental discovery from the coal tar industry took place when two young physicians in Strasbourg were asked to treat patients who had intestinal worms by giving them naphthalene. To their great surprise, the worms survived, but the patient's fevers went down. They learned that the pharmacy had incorrectly supplied another aromatic compound, acetanilide, as the medicine, instead of naphthalene. This led to the discovery that acetanilide is a pain-relieving compound (analgesic), and another dye factory began to produce it. Bayer soon found that a substituted acetanilide was just as effective, which was marketed as Phenacetin. Acetanilide was later used in the synthesis of sulfanilamide, a drug that ushered in the beginning of the antibiotics era and was credited with saving the lives of thousands of wounded during World War II. The success of these early discoveries was the beginning of the industrial development of the pharmaceutical and synthetic organic chemistry industries in Germany and elsewhere.

Acetanilide   **Phenacetin**   **Sulfanilamide**
                                                 (a sulfa drug)

Multiple Uses of Chemicals

Simple aromatic compounds soon became vitally important in treating pain and inflammation resulting from cellular or tissue injury. And chemists made great contributions to relieving pain and suffering by learning how to synthesize these aromatic compounds in laboratories. But at this time in history, nations also experienced large-scale inflammatory threats and injuries from other nations. And the chemists belonging to those human communities soon learned to use their synthetic expertise to convert some of these aromatic compounds into substances with the ability to cause pain, as well as cure it. The new synthetic organic chemistry industry was called into the service of war. Substances such as toluene and phenol, whose molecules contain six-membered "aromatic" rings were found to react readily with nitric acid, leading to substitution of three of the H atoms on the ring with nitro ($-NO_2$) groups. Phenol, which had revolutionized surgery as an antiseptic, could be turned into an explosive, trinitrophenol, also called picric acid. By the late 19th century, a yellow dye, trinitrotoluene (TNT), was found to be even more useful as an explosive, particularly since it was stable enough to melt and pour it into shell cases (Figure 20.2). TNT replaced picric acid, and became a primary explosive dyeing red the trenches of World Wars I and II.

FIGURE 20.2 On Dec 6, 1917, Halifax, Nova Scotia, was devastated by the world's largest human-caused accidental explosion. The explosion, which killed over 1600 people and injured 9000, triggered an 18 m-high tsunami and a massive pressure wave. It was caused by the detonation of 2000 tonnes of picric acid and 200 tonnes of TNT on a French cargo ship loaded with wartime explosives that collided with another ship.

2,4,6-Trinitrophenol   **2,4,5-Trinitrotoluene**
(picric acid)   **(TNT)**

Library and Archives Canada PA-166585

Today, your education in chemistry raises awareness of the ethical issues related to multiple uses of chemicals by humans. You may wish to familiarize yourself with international agreements such as the Chemical and Biological Weapons Conventions, which seek to guide human hands away from using chemistry for destructive purposes in response to inflammatory political situations.

Modern pharmaceutical chemistry continues to develop new drugs to treat inflammation, called non-steroidal anti-inflammatory drugs (NSAIDs). e20.1 discusses the chemistry of the next generation of NSAIDS, following ASA.

Taking It Further

e20.1 Read about the next generation of NSAIDS.

Aromatic compounds, first known for their aromas, played a vital role in the development of industrial organic chemistry. By the early 20th century, this industry produced dyes, pharmaceutical products, and explosives.

20.2 Benzene and Aromaticity: Reactivity, Structure, and Spectroscopy

Chemistry played a role in checking crime on the streets of early 19th century London by introducing brightly burning fuels for street lamps. To replace whale oil, "illuminating gas" was prepared by heating coal in the absence of oxygen. We now know illuminating gas to be a mixture of hydrogen, ethylene, and methane. Despite John Lee Comstock's optimistic view in his 1841 chemistry text that "… coal gas, by the brilliant light it shed on our streets, has worked, and is now working, a moral reformation," more than coal gas was needed to solve London's crime. The fuel for these lamps, however, played a key role in the development of organic chemistry. In 1825, English scientist Michael Faraday analyzed an oil that had separated out from cylinders of a lamp gas prepared from whale oil, and isolated a new "aromatic" compound, benzene (C_6H_6).

The chemistry of benzene was actively explored following Faraday's discovery, and surprising results were obtained. Unlike alkenes, benzene (C_6H_6) does not react with bromine (Br_2) to give products resulting from addition of two bromine atoms across its $C=C$ double bonds. A Lewis acid catalyst is needed to get any reaction of benzene with bromine to take place, and the *substitution* product with formula C_6H_5Br, rather than the *addition* product $C_6H_6Br_2$, is formed. Furthermore, only one mono-bromo substitution product is observed; no isomers are formed.

It was 40 years before German chemist August Kekule was able to account for these observations. If the carbon atoms in benzene molecules formed a straight or branched chain, more than one bromination product would be observed, as the bromine atom could substitute for one of several non-equivalent hydrogen atoms. In 1865, he proposed that benzene molecules had all six carbon atoms linked together in a ring with alternating double and single bonds, which he represented as shown in Figure 20.3. Note how carbon–carbon single and double bonds were represented in his drawings of the mid-19th century.

Despite the double bonds shown in these representations of a benzene molecule, the chemistry of benzene is strikingly different from the unsaturated alkenes and polyenes studied in Chapter 19. While cyclohexene reacts readily with electrophiles such as HCl and H_3O^+ to give addition products, and with $KMnO_4$ to give a 1,2-diol, benzene does none of these things.

Molecular Modelling (Odyssey)

e20.2 Compare the calculated energies of isomers of C_6H_6.

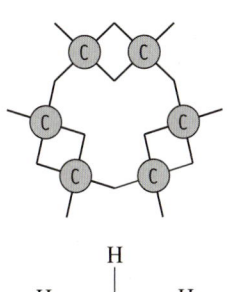

FIGURE 20.3 Kekule's representation of a benzene molecule (C_6H_6) along with a Lewis structure, showing alternating double and single bonds.

KMnO₄ / H₂O — OH, OH (Cyclohexene)

H₃O⁺ — OH

HCl / Ether — Cl

Cyclohexene

Benzene

KMnO₄ — No reaction

H₃O⁺ — No reaction

HCl / Ether — No reaction

Molecular Modelling (Odyssey)

e20.3 Compare calculated bond lengths and electron distributions in benzene and cyclohexa-1,3-triene.

Unlike alkenes, benzene does not undergo electrophilic addition reactions.

If a Lewis acid catalyst is present, benzene does react, but rather than addition reactions, it undergoes *substitution reactions*, where an H atom on the benzene ring can be *replaced* with a substituent such as a Br atom, a nitro ($-NO_2$) group, a sulfonyl ($-SO_3H$) group, an alkyl ($-R$) group, or an acyl ($-C=0$) group. These reactions and their mechanisms are discussed fully in Sections 20.5–20.7.

Not only is the *reactivity* of benzene unusual compared to alkenes, but other evidence suggests unusual electronic *structure* and *stability* for benzene. Most $C-C$ single bonds have lengths near 154 pm, and most $C=C$ double bonds are about 134 pm long, but all the carbon–carbon bonds in benzene molecules are found experimentally to be 139 pm long, intermediate between typical single and double bonds. So even though the Kekule and Lewis structures of benzene molecules show alternating double and single bonds, these representations are not consistent with experimental structural data.

The low reactivity of benzene to reagents that would normally give addition products suggests that benzene might be more stable than alkenes, and that reactions that would remove π bonds in the ring are not favoured. But the stability of a substance is very hard to quantify, because it must always be defined relative to other substances. For example, in Chapter 7, we compared the total amount of energy of a substance to that of its constituent elements in their standard states. How might we experimentally measure the stability of benzene, and what other compounds could we compare it to? One useful approach is to compare the enthalpy change of hydrogenation of benzene, whose Lewis structure shows three alternating double and single bonds, with that of cyclohexa-1,3-diene, whose molecules have two alternating carbon–carbon double bonds, and cyclohexene, with only one. If 3 mol of hydrogen gas are added to 1 mol of benzene, the product is cyclohexane. Two mol of hydrogen gas added to 1 mol of cyclohexa-1,3-diene produce the same cyclohexane product, as does the addition of 1 mol of hydrogen to 1 mol of cyclohexene. Since all three of these alkenes undergo hydrogenation to give the same final product, cyclohexane, their relative enthalpy changes of hydrogenation give a direct experimental measure of their relative energy (or relative stability). The numbers shown in Figure 20.4 are quite startling. Benzene is 150 kJ mol^{-1} more stable than expected for a model compound with three isolated double bonds.

FIGURE 20.4 Comparison of the enthalpy change of hydrogenation of cyclohexene, cyclohexa-1,3-diene, and benzene.

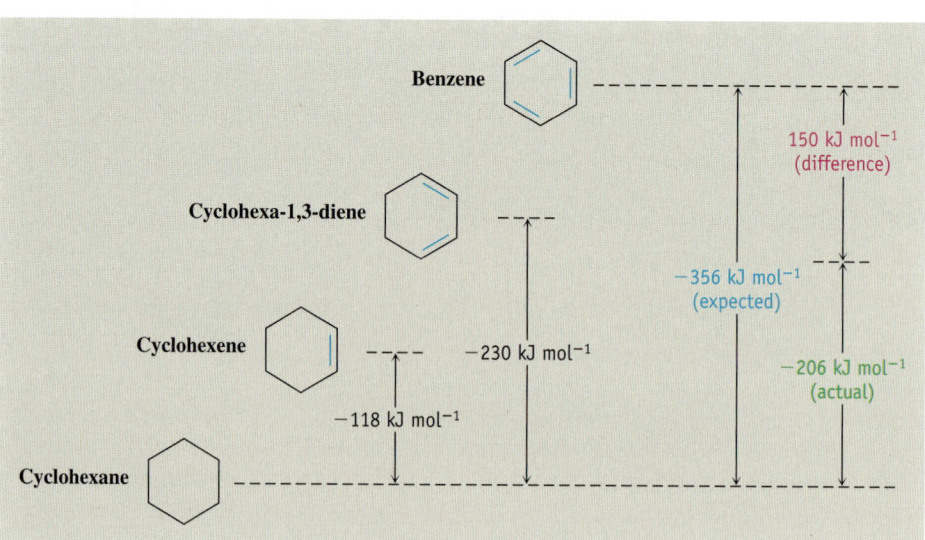

Cyclohexene, an isolated alkene, has $\Delta_r H° = -118$ kJ mol^{-1} for hydrogenation. For hydrogenation of cyclohexa-1,3-diene, a conjugated alkene, $\Delta_r H° = -230$ kJ mol^{-1}, almost twice the value of cyclohexene. The energy is lower than a similar molecule with two isolated double bonds. We might say that cyclohexa-1,3-diene is 6 kJ mol^{-1} more stable than expected, relative to a model compound that has two isolated alkene groups. Using the same logic, we would expect a molecule with six carbon atoms and three isolated alkene groups to have an enthalpy change of hydrogenation three times that of cyclohexene, -118 kJ mol$^{-1} \times 3$, or -356 kJ mol^{-1}. Surprisingly, we find that benzene has $\Delta_r H° = -206$ kJ mol^{-1}, releasing 150 kJ mol^{-1} *less* energy than expected in this hydrogenation reaction. Benzene has 150 kJ mol^{-1} less energy than expected, or we could say benzene is 150 kJ mol^{-1} more stable than expected. Our model of the electronic structure of benzene molecules will have to take this into account.

EXERCISE 20.1—RELATIVE STABILITY OF DIENES

What approximate value would you expect for the enthalpy change of hydrogenation of cyclohexa-1,4-diene relative to cyclohexa-1,3-diene, based on the values given in Figure 20.4? (Hint: start by asking yourself whether the π electrons can be delocalized in this case.)

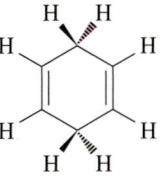

Cyclohexa-1,4-diene

> Since direct experimental measures of aromaticity are difficult, the unusual stability of benzene can be estimated by experiments that compare its energy to that of a selected model compound such as cyclohexene or cyclohexa-1,3-diene.

Finding Patterns: Hückel's 4n + 2 Rule for Aromaticity

As numerous other new cyclic conjugated compounds were synthesized, some showed similar low reactivity and high stability as benzene. Others proved to be so unstable they were difficult or impossible to synthesize. What seemed to differentiate among them was the number of π electrons in the conjugated ring. In 1931, over 100 years after Faraday's isolation of benzene, German physicist Erich Hückel looked at these patterns and formulated a helpful guideline for predicting which existing and yet unmade compounds would be expected to have similar properties as benzene, and considered **aromatic**. After considering the reactivity, experimental geometry, and stability of numerous cyclic conjugated compounds, Erich Hückel formulated a rule to predict the aromaticity of compounds:

Hückel's rule: *Substances whose molecules contain monocyclic, planar, completely conjugated rings with a total of 4n+2 π electrons, where n is an integer (n = 0, 1, 2...) should be exceptionally stable relative to model compounds and are therefore aromatic. Substances whose molecules contain monocyclic, planar, completely conjugated rings with 4n π electrons (n = 1, 2 ...) should be exceptionally unstable relative to model compounds and are described as anti-aromatic.*

Models for the Electronic Structure of Benzene

A hybrid orbital model of the electronic structure of conjugated molecules can be helpful in counting the number of electrons in the π system of a molecule to apply Hückel's rule. Using benzene as an example, each of the carbon atoms in the ring is known experimentally to be trigonal-planar, consistent with a model of sp^2 hybridization. As with an alkene molecule, one π electron from a p orbital on each of the carbon atoms is contributed to the conjugated system, in this case for a total of 6. This is a $4n + 2$ number, where n = 1.

Three different representations of a benzene molecule are shown in Figure 20.5. The electrostatic potential maps clearly show the red electron-rich regions distributed evenly above and below the plane of the ring. A single Lewis structure representation with double and single bonds cannot adequately represent a flat, symmetrical benzene molecule, which experiments show has the shape of a regular hexagon. All C—C—C bond angles are 120°, and each carbon–carbon bond length is identical, intermediate in length between a single and double bond. A resonance model can be used to better represent the distribution of electrons in the molecule as a resonance hybrid of two representations. Benzene doesn't oscillate back and forth between two forms; this model suggests its true structure is somewhere between the two. Each carbon–carbon connection has a bond order of 1.5, midway between that of a single bond and a double bond [<<Section 10.5]. Chemists also represent benzene

molecules with a single drawing, using a solid or dashed circle in the middle of the ring to represent the delocalized π electrons. If the circle represents one π electron between each pair of C atoms, then each bond has three electrons (two σ and one π), which is consistent with a bond order of 1.5.

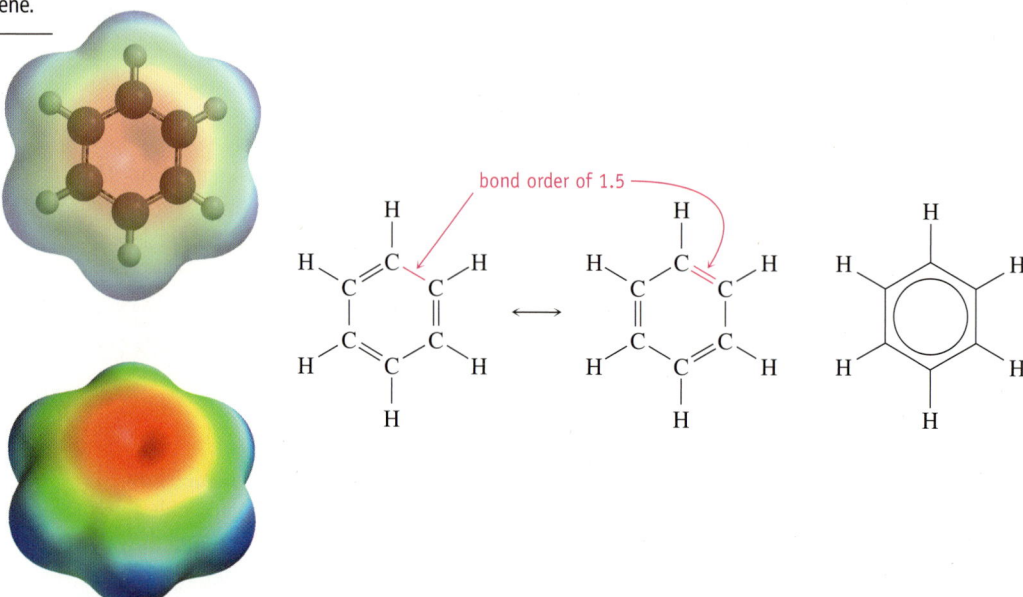

FIGURE 20.5 Alternative delocalized electron representations of a benzene molecule. Electrostatic potential maps, a pair of resonance structures, and a circle representation. It is helpful to work with multiple representations of aromatic compounds, each of which reveals important features of the molecular structure, but has significant limitations if used as the only representation.

A hybrid orbital model is consistent with the representations in Figure 20.5. Because experimental evidence shows all six carbon–carbon bonds to be equivalent, it is misleading to think of benzene molecules as having three localized alkene π bonds in which a given p orbital overlaps with only *one* adjacent p orbital. A more accurate picture would show each p orbital overlapping equally well with *both* neighbouring p orbitals, leading to a structure for benzene in which the π electrons are shared around the ring in two doughnut-shaped clouds (Figure 20.6).

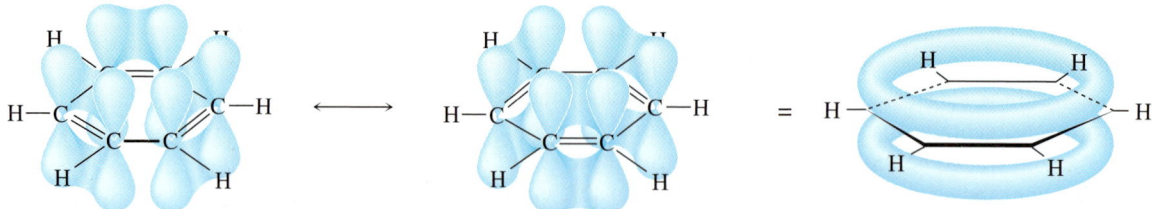

FIGURE 20.6 A hybrid orbital picture of a benzene molecule. Each of the six carbon atoms can be thought of as having a p orbital that can overlap equally well with neighbouring p orbitals on both sides. The π electrons are then shared around the ring in doughnut-like clouds above and below the plane of the ring. This delocalization of electrons helps explain the observed stability of benzene.

Evidence for Aromaticity: Spectroscopy

How consistent is the model in Figure 20.6 with experimental observations? Because the delocalized π electrons are spread around the ring over a larger volume of the molecule, the energy of benzene is lower than a hypothetical triene molecule that does not have this delocalization. We've already seen one experimental measure of aromaticity. If benzene were to react with electrophiles in addition reactions, this complete delocalization of π electrons would be disrupted, consistent with the observation that benzene undergoes substitution reactions that preserve the aromatic character of its molecules rings, rather than addition reactions.

In the early history of understanding aromaticity, chemists had to rely entirely on evidence such as a substance's unusual stability (hydrogenation data) and reactivity (undergoes substitution not addition reactions). After the commercialization of NMR spectroscopy in the 1950s, however, a new and simple means of probing the delocalization of π electrons characteristic of **aromatic compounds** became available. We can now observe that the hydrogen atoms directly attached to the aromatic ring of benzene molecules absorb at $\delta = 7.37$, compared with $\delta = 4.5$–6.5 ppm for most vinylic hydrogen atoms on alkene molecules. Since the hybrid orbital model predicts that hydrogen atoms are attached to sp^2 hybridized carbon atoms in the molecules of both aromatic compounds and in alkenes, this seems surprising at first. But when an aromatic ring is oriented perpendicular to a strong magnetic field, the delocalized π electrons circulate around the ring, producing a small local magnetic field. The resulting field lines in Figure 20.7 show that this induced field should reinforce the applied magnetic field outside the ring. Hydrogen atoms attached to the carbon atoms of aromatic rings experience an effective magnetic field greater than the applied magnetic field, and would be deshielded by the induced field, leading to a downfield chemical shift—exactly what is observed!

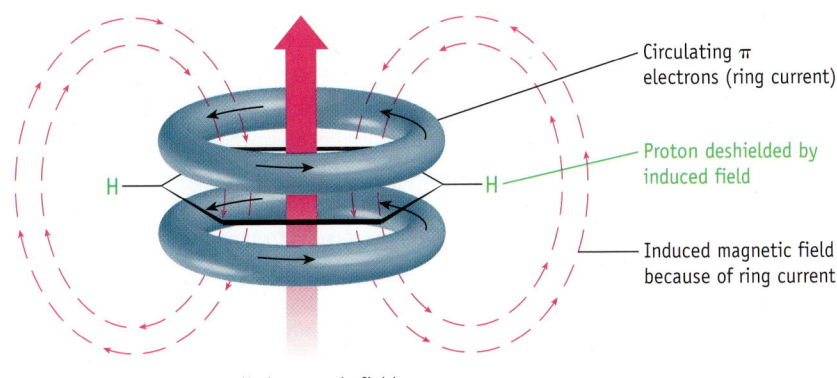

Circulating π electrons (ring current)

Proton deshielded by induced field

Induced magnetic field because of ring current

Applied magnetic field

FIGURE 20.7 Deshielding of hydrogen atoms in the plane of aromatic ring molecules. For discussion of shielding and chemical shifts, see Section 9.4.

To test this NMR approach to "measuring" aromaticity even further, chemists have synthesized substances that have H atoms located on the inside of a much larger ring predicted to be aromatic by Hückel's rule. The six hydrogen atoms on the inside of an [18] annulene molecule are strongly shielded, appearing at $\delta = -3.0$, even upfield from the usual reference compound, TMS. Again, exactly as predicted!

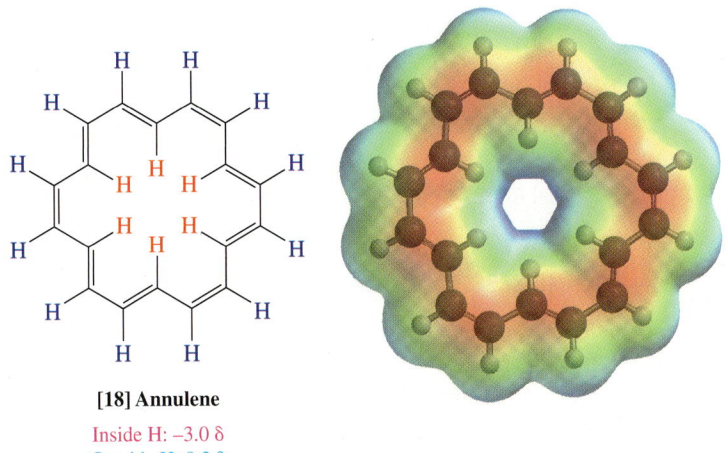

[18] Annulene
Inside H: $-3.0\ \delta$
Outside H: $9.3\ \delta$

A molecule of [18] annulene fits Hückel's rule, since it is monocyclic, planar, and completely conjugated, with 18 π electrons (4n + 2, where n = 4). Since Hückel's formulation of this rule, chemists have learned that it can be applied to polycyclic, heterocyclic, and charged compounds, considered further in Section 20.3 and 20.11. At times, the concept of aromaticity has been considered somewhat fuzzy, as experimental evidence to support claims that particular compounds were aromatic hinged on finding appropriate reference compounds for comparison of stability and reactivity. ^{1}H NMR spectroscopy has helped to resolve much of this controversy.

EXERCISE 20.2—HÜCKEL'S RULE

Draw Lewis structures for each of the following compounds. Then determine whether each is monocyclic, planar, and completely conjugated. Finally, use Hückel's rule to classify each as aromatic or anti-aromatic.

(a) cyclobutadiene
(b) cyclohepta-1,3,5-triene

(c) cyclopenta-1,3-diene
(d) cyclooctatetraene

> ^{1}H NMR spectroscopy provides a simple and elegant test of aromaticity. Look at the chemical shifts of hydrogen atoms directly bonded to ring carbon atoms in a molecule to see whether they fall into the $\delta = 6.5$–8.5 region.

^{13}C NMR Spectroscopy

Carbon atoms of aromatic rings absorb between $\delta = 110$–140. While these are at different chemical shifts and readily distinguished from alkane and carbonyl carbon atoms, they are not easy to distinguish from alkene carbon atoms on the basis of chemical shift alone. For benzene-like aromatic hydrocarbons, however, there will often be more than two absorptions in this region, depending on the symmetry of the molecule and the number of unique carbon atoms in the ring. Toluene molecules have four unique carbon atoms in the benzene ring, and four sharp resonances appear in the spectrum (Figure 20.8). Since no electronegative elements are present, the carbon atom signals are in similar electronic environments, and the chemical shifts are close together. The NMR spectrum needs to be expanded in this region to show all four signals. The methyl group attached to the benzene ring shows up as a fifth signal.

FIGURE 20.8 ^{13}C NMR spectrum of toluene, with the aromatic region enlarged for visibility. The ^{13}C atoms in the solvent, in this case CHCl$_3$, show an absorbance at 77 ppm.

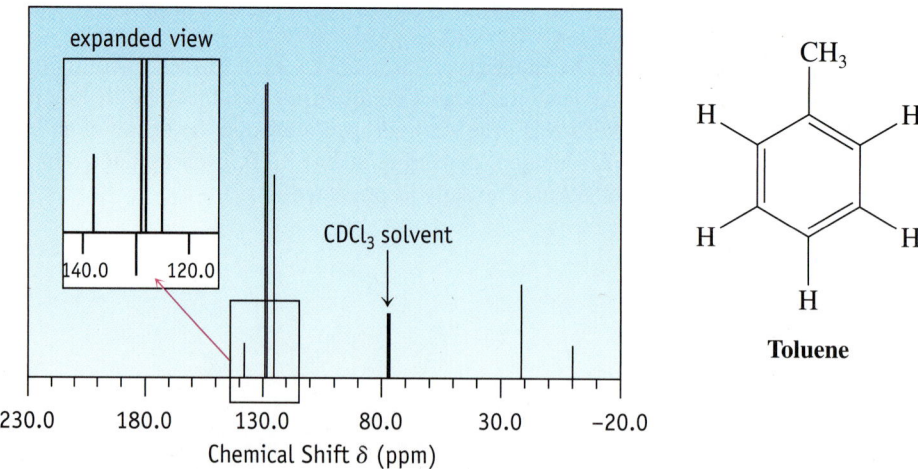

EXERCISE 20.3—^{13}C NMR SPECTRA OF AROMATIC COMPOUNDS

Draw structures for the following compounds, and determine how many ^{13}C signals you would expect to see in a ^{13}C NMR spectrum.

(a) benzene
(b) chlorobenzene

(c) naphthalene
(d) 1,3-dichlorobenzene

Infrared Spectroscopy

As is the case for alkenes, hydrogen atoms attached to aryl carbon atoms show stretching vibrations just above 3000 cm^{-1}, rather than just below 3000 cm^{-1} for alkyl C—H bonds. The IR spectrum of toluene in Figure 20.9 shows both of these stretching vibrations, and the region between 1450 and 1600 cm^{-1} shows characteristic vibrations of the aromatic ring itself. Out-of-plane bending vibrations of the C—H bonds in the 690–900 cm^{-1} region and weak absorptions in the 1660–2000 cm^{-1} region can be used to determine the pattern of substitution on an aromatic ring In e20.5, animations of key vibrational modes for benzene and toluene are shown.

Key spectroscopic data used to identify aromatic compounds are summarized in Table 20.1.

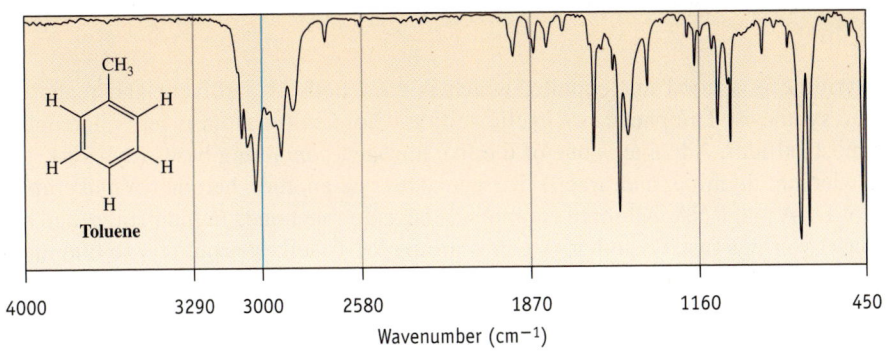

FIGURE 20.9 Infrared spectrum of toluene.

EXERCISE 20.4—IR SPECTRA OF AROMATIC COMPOUNDS

Draw the structure of a toluene molecule, and then consulting its infrared spectrum in Figure 20.9 and the animated spectra of benzene and toluene in e20.5, identify the following:
(a) the aryl C—H and alkyl C—H stretching frequencies
(b) the out-of-plane C—H bending vibrations
(c) any other characteristic vibrations of the aromatic ring

Think about It

e20.5 Observe molecular vibrations responsible for specific peaks in IR spectra of benzene and toluene.

TABLE 20.1 Summary of Key Spectroscopic Data Used to Identify Aromatic Compounds

	KIND OF ABSORPTION	
Spectroscopy	**Position**	**Interpretation**
Infrared (cm^1)	3030	Aryl C—H stretch
	1500 and 1600	Two absorptions due to ring motions
	690–900	Intense C—H out-of-plane bending
^{1}H NMR (δ)	2.3–3.0	Benzylic protons
	6.5–8.0	Aryl protons
^{13}C NMR (δ)	110–140	Aromatic ring carbons

20.3 Aromatic Heterocycles and Ions

Heterocycles

Certain **heterocyclic compounds**, whose molecules contain a *heteroatom* other than carbon, such as oxygen, nitrogen, phosphorous, or sulfur in the ring, also exhibit unusual stability and reactivity, similar to benzene-like aromatic compounds.

Consider the structures of heterocyclic molecules of the compounds pyridine, thiophene, and adenine.

Pyridine
^{1}H NMR shifts at
$\delta = 7.3, 7.7,$ and 8.6

Thiophene
^{1}H NMR shifts at
$\delta = 7.1, 7.3$

Adenine
^{1}H NMR shifts for C-H
$\delta = 8.11, 8.14$

Pyridine is a good nucleophile, which is used both as a solvent and reagent in organic synthesis. Thiophene is a cyclic sulfide whose sulfur atom is less nucleophilic than most sulfides. Adenine is one of the five nitrogen-containing bases (cytosine, guanine, adenine, thymine, and uracil) that pair with one another, helping to make up the code of DNA and RNA. All three compounds have double bonds in their rings, and adenine has two rings fused together. Are they aromatic? To tell, we could try to find model compounds with which to compare their stability and reactivity—but this is not easy in the case of heterocyclic compounds. We can, however, see whether these cyclic compounds demonstrate ring currents by looking at the chemical shifts of hydrogen atoms on the rings–they fall squarely into the aromatic region $\delta = 6.5$–8.5 of the ^{1}H NMR spectrum!

Could you predict this using Hückel's rule if you didn't have NMR data? The structure of pyridine looks very similar to benzene, with a N atom replacing a C—H in the ring. The hybrid orbital model can help us count the number of π electrons. If we regard the N atom as sp^2 hybridized, a single electron will be in a p orbital on N, overlapping with the p orbital on the neighbouring carbon atoms. The lone pair of electrons on N would be in an sp^2 orbital, in the plane of the benzene ring, and would not be part of the π system. So pyridine would have 6 π electrons, making it aromatic by applying the 4n + 2 rule, with n = 1. The aromatic behaviour of the substance is consistent with our prediction from looking at the electronic structure of its molecules.

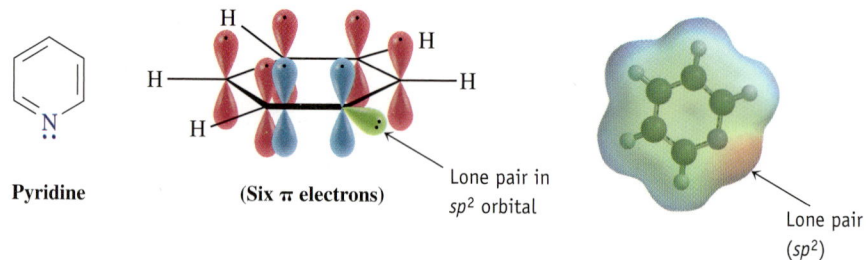

Pyridine

(Six π electrons)

Lone pair in
sp^2 orbital

Lone pair
(sp^2)

What about thiophene? Four electrons are contributed to the π system from the two alkene groups in the molecule. Again, the hybrid orbital model can be helpful in counting the number of electrons contributed to the π system by the S atom. VSEPR guidelines [<<Section 10.7] would suggest that the four pairs of electrons around the S atom should point toward the corners of a tetrahedron. But if this were so, the lone pairs of electrons on the S atom would not be in the same plane as the other p orbitals in the ring and could not be delocalized over the rest of the molecule. However, if the S atom were considered to be sp^2 hybridized, one lone pair of electrons would be in

the plane of the ring in an sp^2 orbital and the other would be in a p orbital perpendicular to the ring and able to overlap with the p orbitals on both adjacent carbon atoms. The two electrons in the p orbital on the S atom would then become part of the π system, giving a total of 6 π electrons in the ring, consistent with other aromatic compounds.

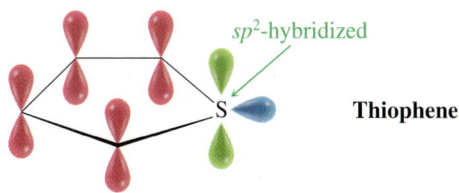

Thiophene

WORKED EXAMPLE 20.1—AROMATICITY OF HETEROCYCLIC COMPOUNDS

Count the number of electrons in the π system for an adenine molecule and use Hückel's rule to predict whether the compound to be aromatic.

Solution

First note that adenine has two rings fused together. Hückel's rule can be extended to molecules with multiple fused rings, as discussed in Section 20.11. Now count the number of π electrons, taking care to use the hybrid orbital model for each N atom in the ring. Three of the four N atoms in the rings contribute only a single electron to the π system, just as in pyridine. In the case of the N atom with an H atom attached to it, the three sp^2 orbitals might all be considered as being involved in bonding: two of them overlap with sp^2 orbitals on adjacent C atoms to form π bonds, and one overlaps with an s orbital on the H atom to form a third δ bond. This leaves the lone pair of electrons in the p orbital on that N atom and allows it to be delocalized into the π system. This gives adenine molecules a total of 10 π electrons, which fits the 4n + 2 rule, where n = 2. Adenine should be aromatic, and its ^{1}H NMR spectrum shows this to be the case.

EXERCISE 20.5—AROMATICITY OF HETEROCYCLIC COMPOUNDS

Furan

Furan is an oxygen-containing heterocyclic compound. Count the number of electrons in the π system of a furan molecule and use Hückel's rule to determine whether it is aromatic. What chemical shift in the ^{1}H NMR would you expect for the H atoms on the ring?

Ions

Unusual acid-base behaviour of some cyclic compounds can also be explained with the help of the concept of aromaticity. Compare the acidity of cyclopentane and cyclopenta-1,3-diene. Cyclopentane has a pK_a value of ~59—that is, the equilibrium constant for the reaction in which a proton is transferred from its molecule to water is 1×10^{-59}, so small that it is very difficult to measure. This is consistent with pK_a values for other hydrocarbons, which are very weak acids.

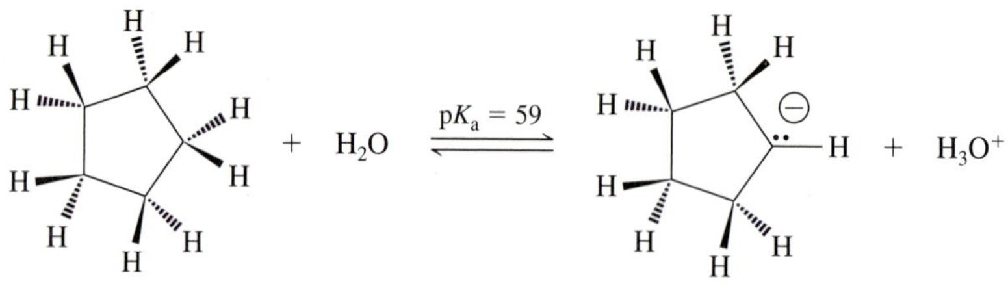

Cyclopentane

Cyclopentyl anion

Cyclopenta-1,3-diene, however, is surprisingly acidic by comparison. Its pK_a value is 18 ($K_a = 1 \times 10^{-18}$)—much more acidic than cyclopentane. While still a weak acid, much more cyclopenta-1,3-diene ionizes in water than is the case for cyclopentane. What could account for this remarkable difference?

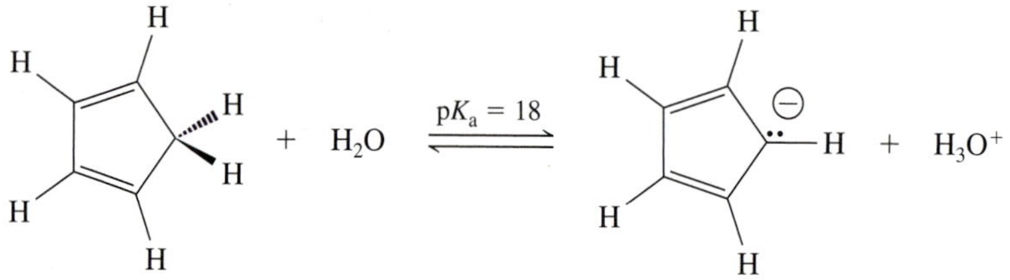

Cyclopenta-1,3-diene

Cyclopenta-1,3-dienyl anion

The strength of an acid can be measured by the extent to which it ionizes in water, measured by the magnitude of its K_a, or acid ionization constant [<<Section 14.4]. Recall also that the equilibrium constant K is related to $\Delta_r G°$, the free energy difference between the acid reactant and the product, its conjugate base, by the relationship [<<Section 17.7]

$$\Delta_r G° = -RT \ln K$$

Any structural factor that changes the free energy of either the acid or its conjugate base will therefore affect $\Delta_r G°$ and K_a.

So what does the concept of aromaticity have to do with acid ionization constants? Cyclopenta-1,3-dienyl anion, which is the conjugate base of cyclopenta-1,3-diene, has 6 electrons in the π system if the carbanion is considered to be sp^2 hybridized, as in the case of thiophene. Since the conjugate base of cyclopenta-1,3-diene is aromatic, it is much lower in energy relative to cyclopenta-1,3-diene than the cyclopentyl anion is relative to cyclopentane. Since ΔS in these two examples is comparable, $\Delta_r G°$ for the ionization of cyclopenta-1,3-diene in water is a much smaller positive number than for cyclopentane, and the K_a is correspondingly greater, consistent with the differences in experimental pK_a values.

EXERCISE 20.6—AROMATICITY OF CYCLIC IONS

Classify each of the following ions as aromatic or anti-aromatic.

(a) 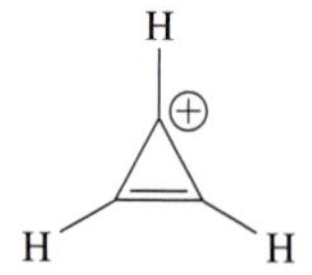 (b) 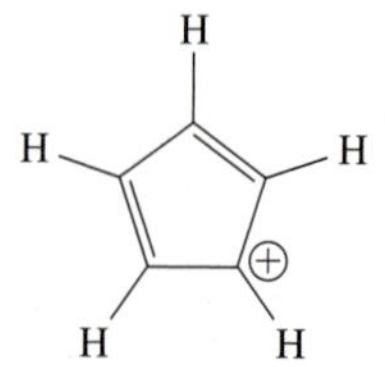 (c)

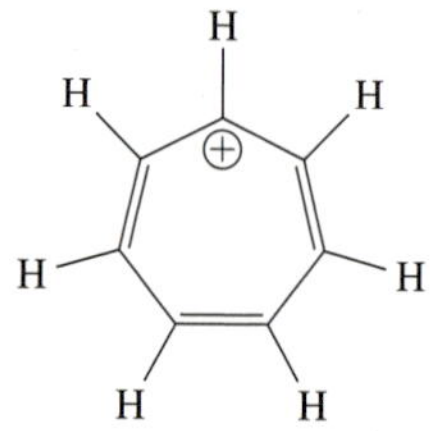

As was the case with compounds made of molecules with cyclic, completely conjugated carbon rings, aromaticity can be correlated using Hückel's 4n + 2 rule after counting the number of π electrons in the rings of molecules making up heterocyclic compounds and ions. Care must be taken in counting which electrons on heteroatoms are part of the π system.

20.4 **Naming Aromatic Compounds**

As a result of the extensive history of knowledge, aromatic substances have acquired a larger number of common names than any other class of organic compounds. IUPAC rules allow for some of those most commonly used to be retained. As seen in IUPAC rules, methylbenzene is commonly known as toluene, hydroxybenzene as phenol, and aminobenzene as aniline. Several other aromatic compounds whose common names are permitted are shown below:

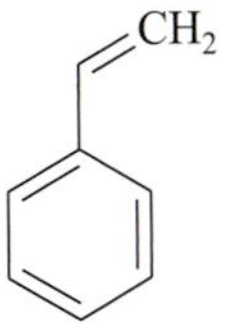

| Styrene | Xylene
(para-xylene shown) | Benzaldehyde | Benzoic acid | Benzonitrile |

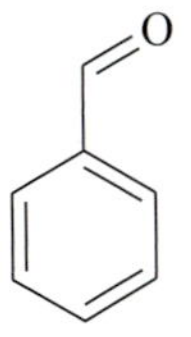

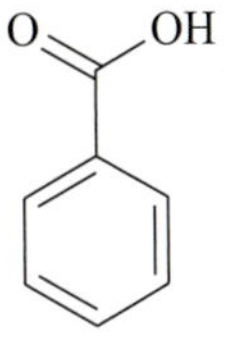

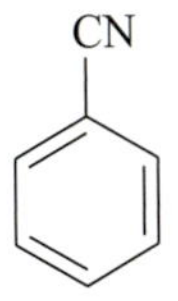

Monosubstituted benzene molecules are systematically named in the same manner as other hydrocarbons, with *–benzene* used as the parent name. $C_6H_5Br(\ell)$ is named bromobenzene, and $C_6H_5CH_2CH_3(\ell)$ is ethylbenzene. The name *phenyl* (pronounced "*FEN-nil*") is used for the $-C_6H_5$ unit when the benzene ring is considered as a substituent, and the name *benzyl* is used for the $-CH_2C_6H_5$ group.

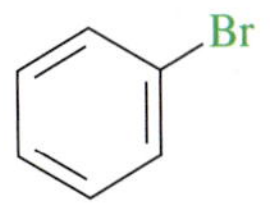

| **Bromo**benzene | **Ethyl**benzene | **A phenyl group** | **A benzyl group** |

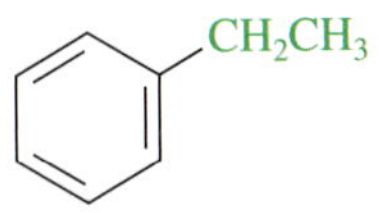

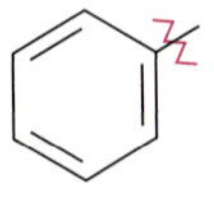

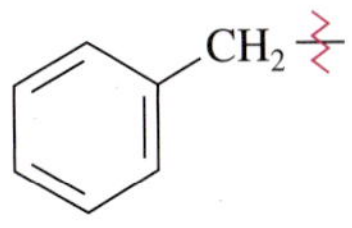

Disubstituted benzene molecules are named using one of the prefixes *ortho-* (*o*), *meta-* (*m*), or *para-* (*p*). An *ortho*-disubstituted benzene has its two substituents in a 1,2 relationship on the ring; a *meta*-disubstituted benzene has its two substituents in a 1,3 relationship; and a *para*-disubstituted benzene has its substituents in a 1,4 relationship:

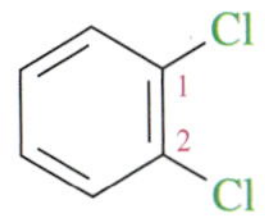

| *ortho*-**Dichlorobenzene**
1,2 disubstituted | *meta*-**Xylene**
1,3 disubstituted | *para*-**Chlorobenzaldehyde**
1,4 disubstituted |

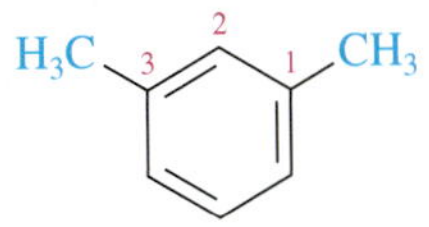

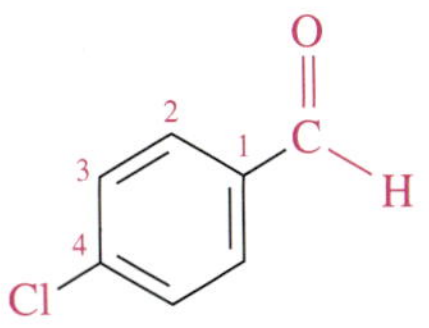

Benzene rings with more than two substituents are named by numbering the position of each substituent on the ring so that the lowest possible numbers are used. The substituents are listed alphabetically when writing the name.

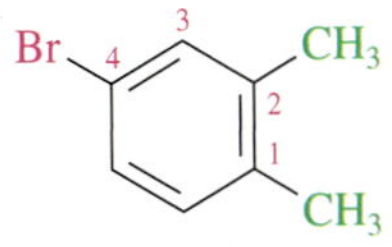

| **4-Bromo-1,2-dimethylbenzene** | **2-Chloro-1,4-dinitrobenzene** | **2,4,6-Trinitrotoluene (TNT)** |

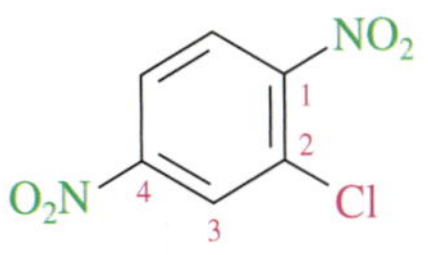

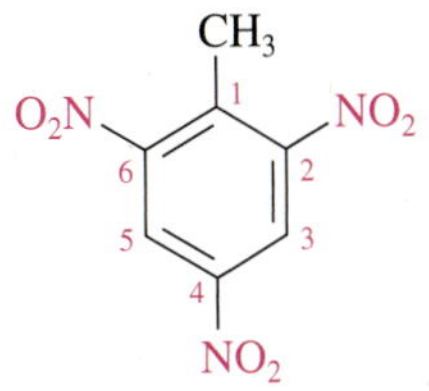

In the third example shown, note that *toluene* is used as the parent name rather than *benzene*. Any of the monosubstituted aromatic compounds whose common names have been accepted can serve as a parent name, with the principal substituent ($-CH_3$ in toluene, for example) assumed to be on carbon atom 1. The following two examples further illustrate this practice:

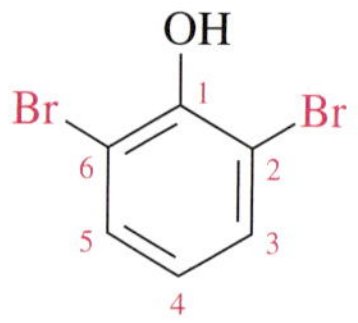

2,6-Dibromophenol   *m*-**Chloro**benzoic acid

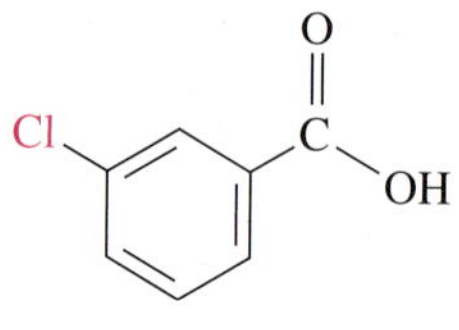

WORKED EXAMPLE 20.2—NOMENCLATURE OF AROMATIC COMPOUNDS

What is the IUPAC name of the following compound?

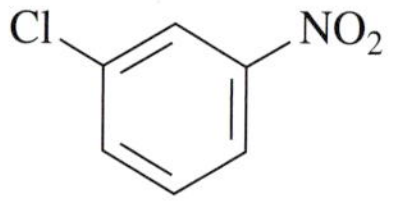

Solution

Because the nitro group ($-NO_2$) and chloro group are on carbon atoms 1 and 3, they have a *meta* relationship. Citing the two substituents in alphabetical order gives the IUPAC name 1-chloro-3-nitrobenzene, or *m*-chloronitrobenzene.

EXERCISE 20.8—NOMENCLATURE OF AROMATIC COMPOUNDS

Indicate whether the following compounds are *ortho-*, *meta-*, or *para*-substituted.

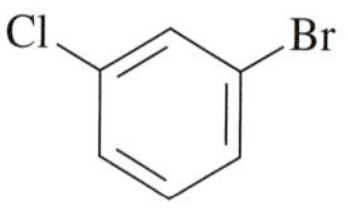

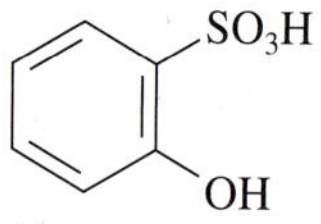

EXERCISE 20.9—NOMENCLATURE OF AROMATIC COMPOUNDS

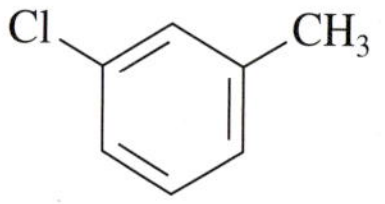

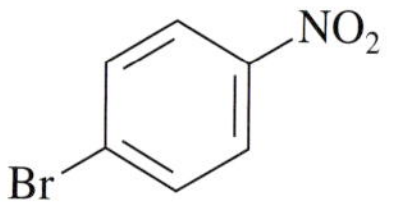

Give IUPAC names for the following compounds:

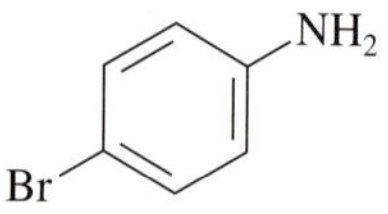

20.5 Electrophilic Aromatic Substitution: Bromination

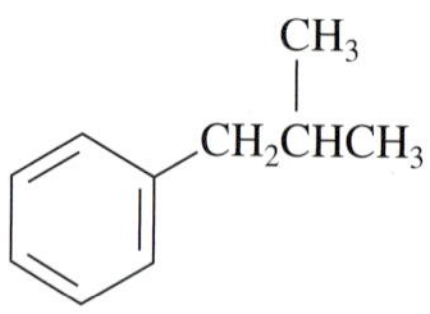

Let's look more closely at the most common reaction of aromatic compounds, **electrophilic aromatic substitution**. The name derives from observing that electron-poor regions of a Lewis acid or electrophile E^+ react with electron-rich regions of an aromatic Lewis base or nucleophile, leading to substitution of the electrophile for one of the ring hydrogen atoms. Addition reactions do not generally take place, as they would destroy the stable aromatic ring.

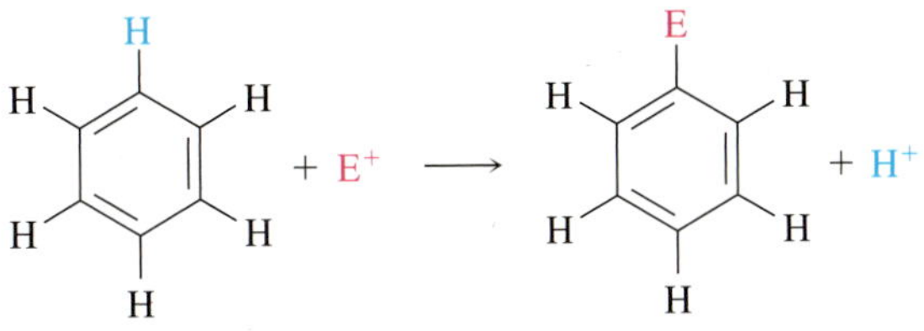

Many different substituents can be introduced onto an aromatic ring by electrophilic substitution reactions. It is possible to *halogenate* an aromatic ring (substitute a halogen: —F, —Cl, —Br, or —I for a hydrogen atom), *nitrate* it (substitute a nitro group: —NO$_2$), *sulfonate* it (substitute a sulfonic acid group: —SO$_3$H), *alkylate* it (substitute an alkyl group: —R), or *acylate* it (substitute an acyl group: —COR). Starting with only a few simple materials, we can prepare many different substituted aromatic compounds (Figure 20.10).

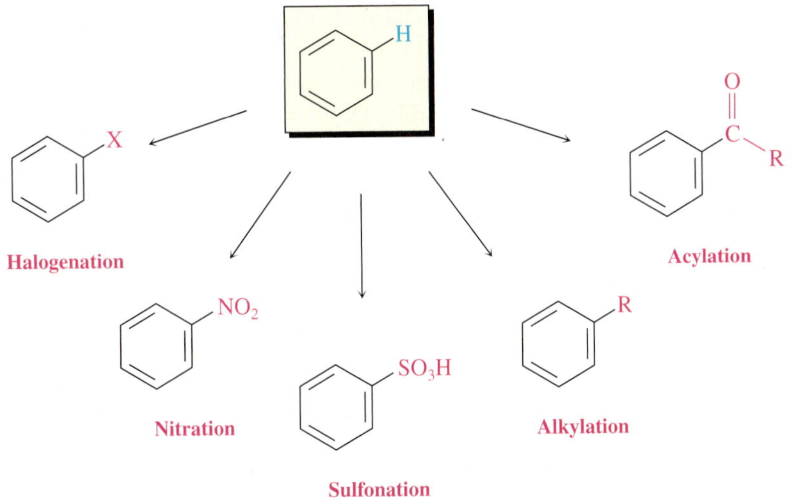

Halogenation

Nitration

Sulfonation

Alkylation

Acylation

FIGURE 20.10 Important electrophilic aromatic substitution reactions.

All five of the substitution reactions in Figure 20.10 have in common that the reagents needed to carry out the transformation include a Lewis acid catalyst, and in each case, substitution rather than addition takes place. Can we propose a reasonable reaction mechanism that is consistent with these observations? Let's examine the bromination of benzene. Benzene reacts with Br$_2$ in the presence of the Lewis acid catalyst FeBr$_3$ to yield the substitution product bromobenzene:

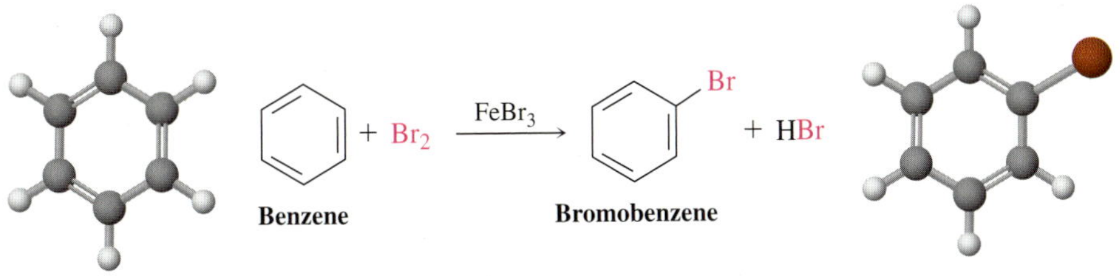

Benzene

Bromobenzene

As evident from the electrostatic potential map in Figure 20.5, the π electrons in the aromatic ring of a benzene molecule are the electron-rich, nucleophilic site, much as was the case for an alkene. Let's zoom down to the molecular level and visualize a reaction mechanism that is consistent with observations of this substitution reaction. Recall from Sections 19.8–19.10 the accepted reaction mechanism for electrophilic addition reactions of alkenes. When a reagent such as HCl adds to an alkene, we picture at the molecular level the successful events leading to products starting with one of the electrophilic H$^+$ ions colliding with one of the many alkene molecules, forming a bond to one carbon atom and leaving a positive charge on the other carbon atom. The carbocation intermediate then goes on to form the addition product when it successfully collides with a nucleophilic Cl⁻ ion.

Alkene   **Carbocation intermediate**   **Addition product**

It would be reasonable to picture an electrophilic aromatic substitution reaction mechanism beginning in a similar way, but we need to account for some important experimental differences. One is that aromatic rings are less reactive toward electrophiles than alkenes are. For example, Br_2 in CH_2Cl_2 solution reacts instantly with most alkenes but does not react with benzene. For bromination of benzene to take place, a catalyst such as $FeBr_3$ is needed. Our molecular-level view of the mechanism shows a catalyst making a Br_2 molecule more electrophilic by reacting with it to give a $Br^+FeBr_4^-$ species that reacts as if it were Br^+.

$$Br-Br + FeBr_3 \longrightarrow Br^+ \; Br-FeBr_3^- \longrightarrow Br^+ \; ^-FeBr_4$$

Molecular Modelling (Odyssey)

e20.6 Examine the evidence for the delocalized positive charge on this carbocation intermediate.

By analogy with the mechanism for alkene reactions, it would be reasonable to postulate that one of the few collisions between partners with the right energy and orientation, involves an electrophilic Br^+ ion reacting with one of the many electron-rich (nucleophilic) benzene rings to yield a non-aromatic carbocation intermediate. This carbocation is allylic [<<Section 19.9] and using Lewis structure representations requires three resonance forms to show the distribution of electrons and the positive charge on the intermediate.

Although more stable than a non-allylic carbocation, the reaction intermediate in electrophilic aromatic substitution is nevertheless much less stable than the starting aromatic reactant. Reaction of electrophiles with benzene has a relatively high activation energy and is rather slow. A reaction energy diagram compares the first steps of the mechanism for reaction of a single E^+ with an alkene molecule and with a benzene molecule. The benzene reaction is slower (has a higher E_a) because the aromatic benzene molecules in the starting material are so much lower in energy (more stable) than alkene molecules.

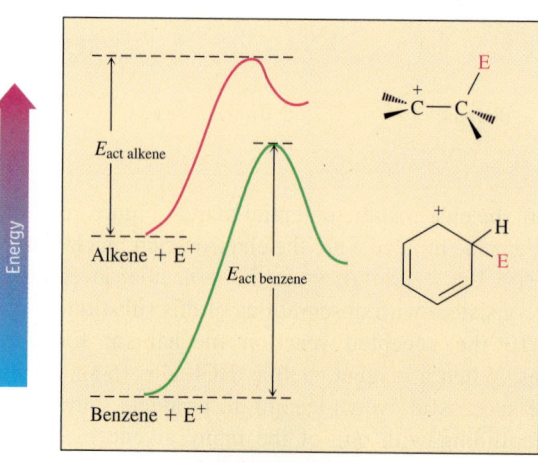

Reaction progress ⟶

A second point of difference is that substitution takes place rather than addition. In the case of alkenes, a carbocation intermediate is formed by the addition of an H^+ to the $C=C$ bond. Br^- ions then add to the carbocation to yield addition products. In the case of aromatic compounds, the carbocation intermediate is formed by an electrophilic Br^+ adding to the $C=C$. A base then removes an H^+ ion from the bromine-bearing carbon atom, providing the ring with the two electrons it needs to re-form the neutral aromatic ring of the substitution product. The net effect is the substitution of a Br atom for an H atom on the aromatic ring by the overall mechanism shown in Figure 20.11.

Think about It

e20.7 Watch animations of the electron-pair movement and orbitals involved in this mechanism.

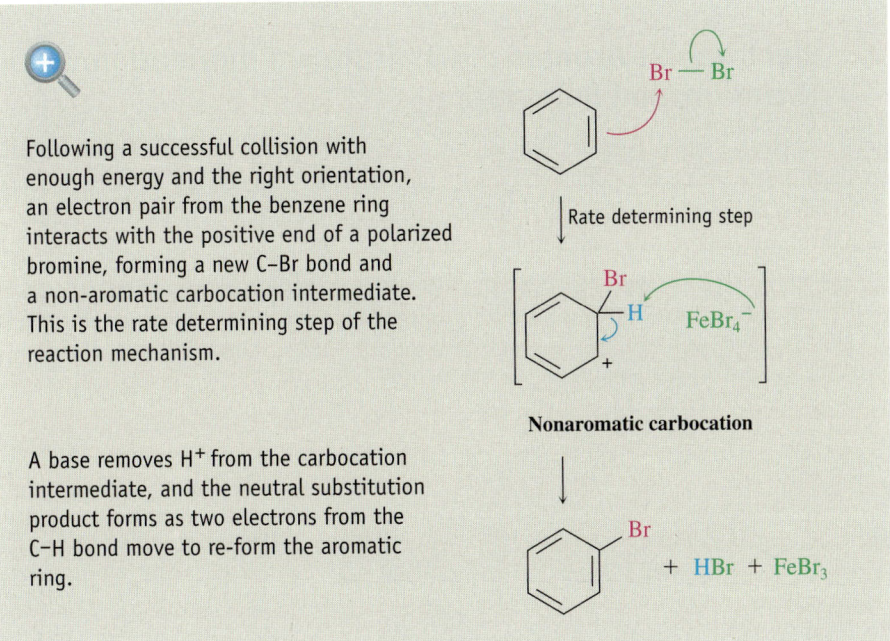

Following a successful collision with enough energy and the right orientation, an electron pair from the benzene ring interacts with the positive end of a polarized bromine, forming a new C–Br bond and a non-aromatic carbocation intermediate. This is the rate determining step of the reaction mechanism.

Rate determining step

Nonaromatic carbocation

A base removes H^+ from the carbocation intermediate, and the neutral substitution product forms as two electrons from the C–H bond move to re-form the aromatic ring.

FIGURE 20.11 Accepted reaction mechanism for the electrophilic aromatic bromination reaction. We zoom in on what happens when each of the reacting benzene molecules undergoes successful collisions that lead to formation of products.

Why does the reaction of Br_2 with benzene take a different course than its reaction with an alkene? The answer is straightforward: If *addition* occurred, the molecules would lose their stable aromatic rings, and the overall reaction would be energetically unfavourable. When *substitution* occurs, though, the stability of the aromatic ring is retained and the reaction is energetically favourable. A reaction energy diagram for the energy changes implied by this reaction mechanism is shown in Figure 20.12.

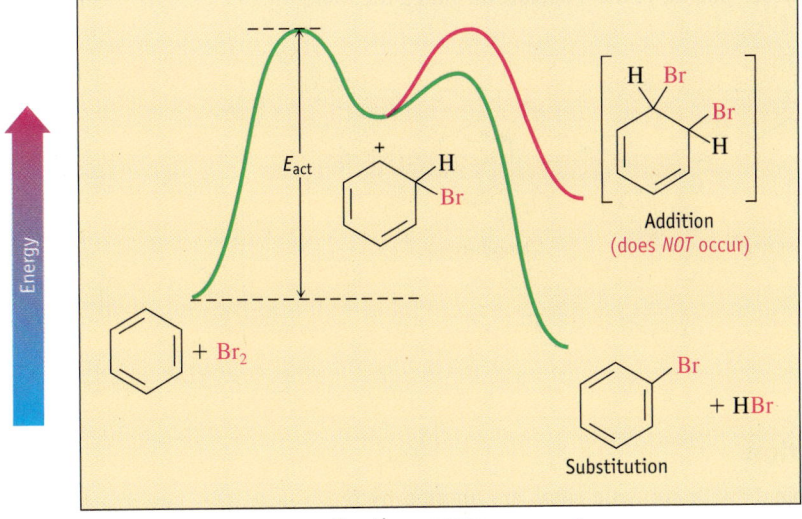

FIGURE 20.12 Visualizing energy changes in the accepted reaction mechanism for the successful reaction of one molecule of benzene with one molecule of Br_2. The reaction is postulated to occur in two steps, and the energy of the products is assumed to be lower than the energy of the reactants.

Three possible products might form on reaction of toluene with Br_2 and $FeBr_3$. Draw and name them.

The accepted mechanism for electrophilic aromatic substitution is thought to be a two-step polar reaction, in which the π electrons of the aromatic ring form a bond to the electrophile to yield a delocalized carbocation intermediate, which then loses an H^+ to give a substituted aromatic product.

20.6 Electrophilic Aromatic Substitution: Chlorination, Nitration, and Sulfonation

Thousands of research papers studying the mechanisms for electrophilic aromatic substitution reactions have been published, and, while some differences are evident, a similar pattern of reactivity is seen in the substitution reactions of aromatic compounds with many different electrophiles. It has therefore been suggested that molecular-level views of the mechanisms for most of these reactions are very similar, with the chief difference being the nature of the electrophile that reacts with the aromatic ring in the first step. To help you recognize the similarities in these patterns of reactivity, consider a simplified general mechanism that seems plausible for most electrophilic aromatic substitution reactions of an electrophile E^+ with molecules of an aromatic compound (Figure 20.13).

FIGURE 20.13 Accepted (generalized) reaction mechanism for the substitution of an electrophile E^+ for a hydrogen atom on an aromatic ring.

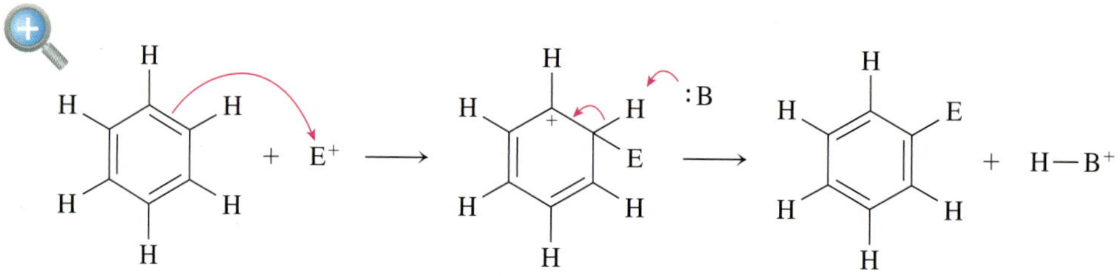

The most important variation in all these reactions is in the nature of E^+ and the manner in which it is generated. Let's look briefly at several of these other reactions.

Think about It

e20.8 Watch animations of the electron-pair movement and orbitals involved in the general mechanism.

Chlorination

Compounds with aromatic rings are chlorinated by reaction with Cl_2 in the presence of $FeCl_3$ catalyst. This kind of reaction, shown below for a benzene molecule, is used in the synthesis of numerous pharmaceutical agents, including the tranquilizer diazepam (Valium).

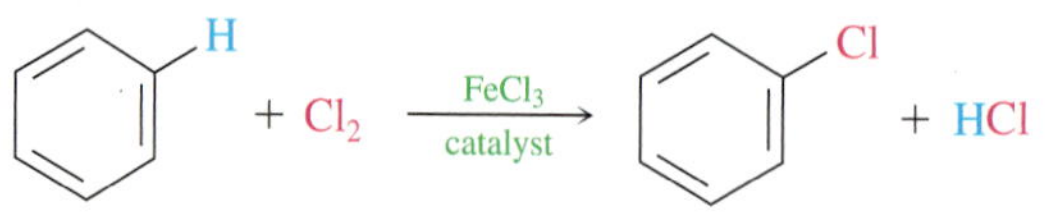

Benzene   Chlorobenzene (86%) 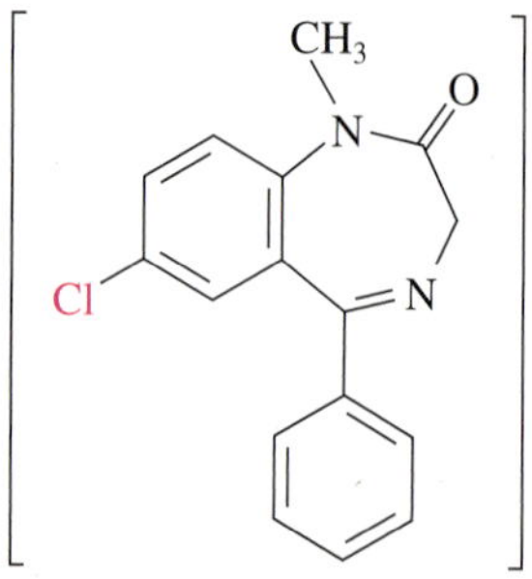   Diazepam

Nitration

Compounds with aromatic rings are nitrated by reaction with a mixture of concentrated nitric and sulfuric acids. Nitration of compounds with aromatic rings is a key step in the synthesis of explosives such as TNT (2,4,6-trinitrotoluene), picric acid (2,4,6-trinitro-phenol), dyes, and many pharmaceutical agents [<<Section 20.1].

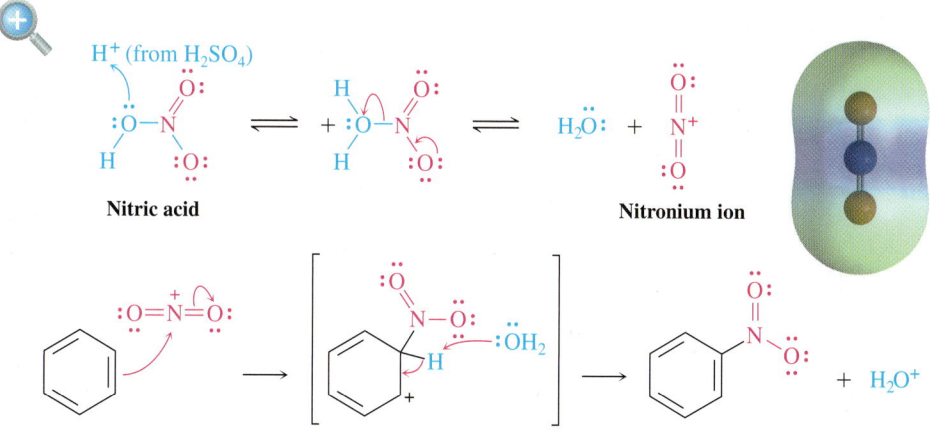

Benzene   **Nitrobenzene (85%)**   **Trinitrotoluene (TNT)**

Considering the accepted reaction mechanism, spectroscopic evidence shows that the electrophile E^+ in this case is the nitronium ion (NO_2^+), which can be formed by protonation of HNO_3 by H_2SO_4, followed by loss of water. NO_2^+ can then react with a benzene molecule in much the same way Br^+ does. Removal of an H^+ in the final step regenerates the aromatic ring, yielding the substitution product, nitrobenzene.

Nitric acid   **Nitronium ion**

Nitrobenzene

> **Think about It**
>
> **e20.9** Watch animations of the electron-pair movement and orbitals in the nitration mechanism.

> **Molecular Modelling (Odyssey)**
>
> **e20.10** Examine the electrostatic potential map of the nitronium ion to identify the electrophilic site.

Sulfonation

Compounds with aromatic rings are sulfonated by reaction with so-called *fuming sulfuric acid,* a mixture of SO_3 and H_2SO_4. Aromatic sulfonation is a key step in the synthesis of such compounds as the sulfa drug family of antibiotics.

Benzene   **Benzenesulfonic acid (95%)**   **Sulfanilimide (a sulfa drug)**

The reactive electrophile is thought to be either SO_3 or HSO_3^+, formed by protonation of SO_3 by H_2SO_4.

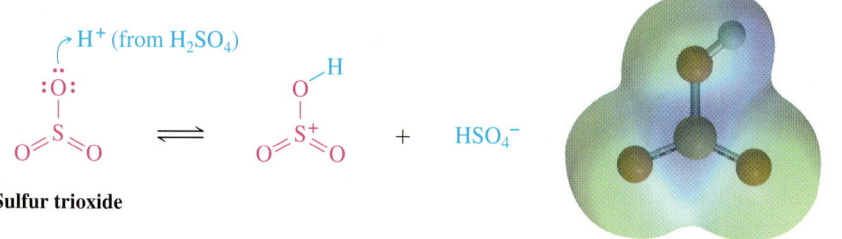

Sulfur trioxide

> **Molecular Modelling (Odyssey)**
>
> **e20.11** Examine the electrostatic potential map of the HSO_3^+ ion to identify the electrophilic site.

Sulfonation is thought to occur by the reactive electrophile in a similar mechanism as seen for bromination and nitration.

WORKED EXAMPLE 20.3—MECHANISM FOR SULFONATION

Show a reasonable detailed, step-wise mechanism for the reaction of a benzene molecule with fuming sulfuric acid to yield benzenesulfonic acid.

Solution

The mechanism would be expected to follow the same general pattern seen in Figure 20.13. We assume that the electrophile is HSO_3^+, formed as shown above, which then adds to an aromatic ring to give a charge-delocalized carbocation intermediate. HSO_4^-, the conjugate base of H_2SO_4, then pulls off an H^+ ion to regenerate the aromatic ring and yield the final substitution product.

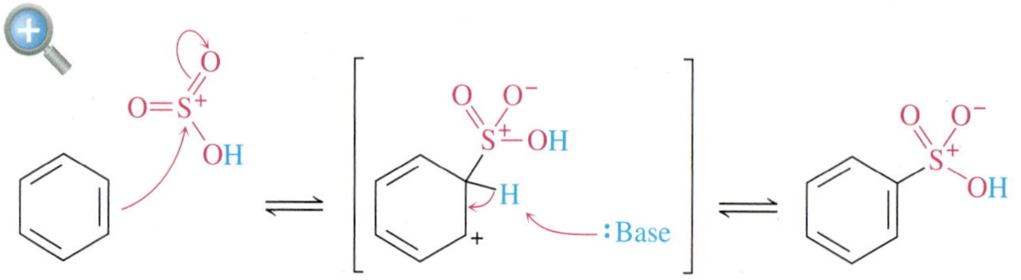

Benzenesulfonic acid

EXERCISE 20.11—MECHANISM FOR SULFONATION

With reference to the reaction mechanism above, how can you account for the observation that deuterium (^2H) atoms replace all six hydrogen atoms in the aromatic ring when benzene is treated with D_2SO_4?

A generalized mechanism (Figure 20.13) for electrophilic aromatic substitution reactions helps explain similar reactivity patterns of aromatic compounds with different electrophilic reagents, E^+.

20.7 Electrophilic Aromatic Substitution: Friedel-Crafts Alkylation and Acylation Reactions

In *Friedel-Crafts alkylation*, reaction between an aromatic compound and an alkyl chloride (RCl) in the presence of a Lewis acid catalyst $AlCl_3$ results in substitution of an alkyl group (R) for an H atom on the aromatic ring. For example, benzene reacts with 2-chloropropane in the presence of $AlCl_3$ to yield isopropylbenzene (also called cumene):

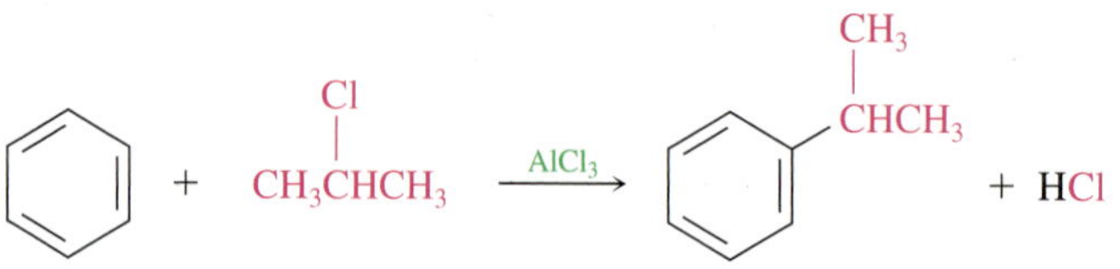

| Benzene | 2-Chloropropane | Cumene (85%) (Isopropylbenzene) |

This, and the closely related *Friedel-Crafts acylation reaction*—in which an *acyl* (pronounced "*A-sil*") *group*, —C(=O)R, from a carboxylic acid chloride molecule, Cl—C(=O)R, substitutes for an H atom—are two of the most important electrophilic aromatic substitution reactions. In both cases, a new carbon–carbon bond is formed. Carbon–carbon bond-forming reactions are necessary to synthesize larger molecules from smaller building blocks. An example of the Friedel-Crafts acylation reaction is the synthesis of acetophenone from benzene and acetyl chloride, in the presence of the catalyst $AlCl_3$, to produce acetophenone, a ketone, with formation of a new carbon–carbon bond.

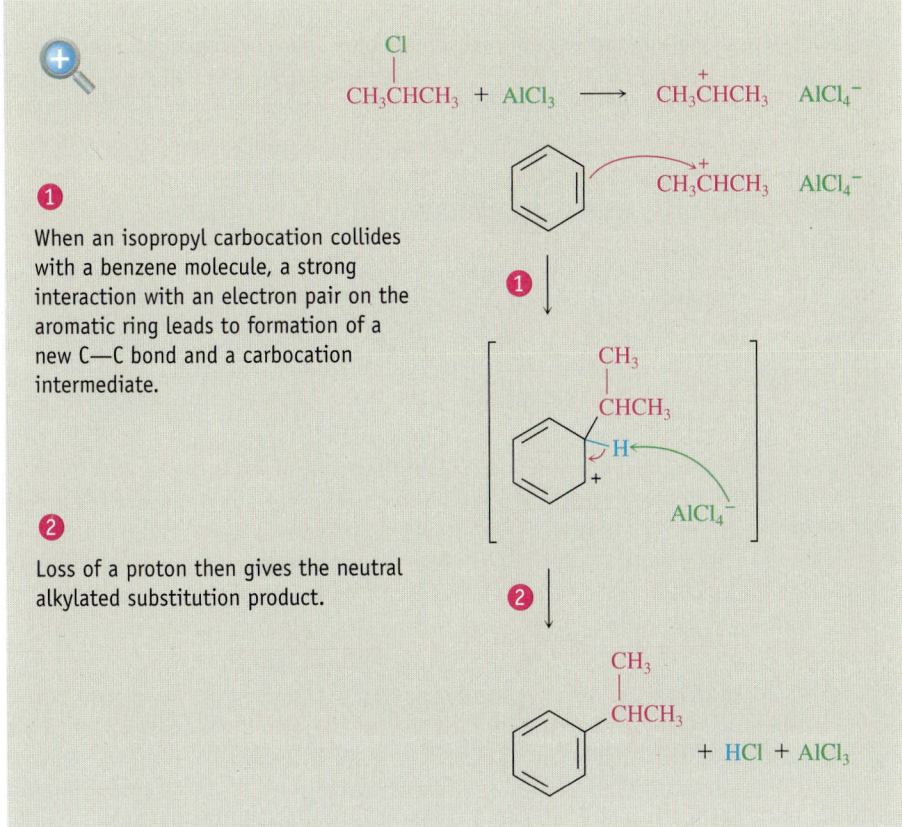

Benzene   **Acetyl chloride**   **Acetophenone (95%)**

Although very useful, the Friedel-Crafts alkylation reaction has several limitations. For one thing, only *alkyl* halides can be used; aryl halides such as chlorobenzene don't react. In addition, Friedel-Crafts reactions don't succeed on aromatic rings that are already substituted by the groups $-NO_2$, $-C\equiv N$, $-SO_3H$, or $-COR$. Such aromatic rings are much less reactive than benzene for reasons discussed in Section 20.8.

A plausible mechanism for Friedel-Crafts reactions shows aluminum chloride catalyzing the reaction by helping an alkyl chloride molecule ionize, in much the same way that $FeBr_3$ helps Br_2 ionize [<<Section 20.5]. A simplified representation of the electrophile E^+ would be a carbocation R^+, which then adds to the aromatic ring in the same general pattern seen in Figure 20.13. A representation of the overall Friedel-Crafts mechanism for the successful events leading to synthesis of isopropyl benzene is shown in Figure 20.14.

1 When an isopropyl carbocation collides with a benzene molecule, a strong interaction with an electron pair on the aromatic ring leads to formation of a new C—C bond and a carbocation intermediate.

2 Loss of a proton then gives the neutral alkylated substitution product.

FIGURE 20.14 Accepted reaction mechanism for the Friedel-Crafts alkylation reaction. The Friedel–Crafts acylation reaction is thought to proceed by a similar mechanism, with the electrophile being an acyl ion, generated by $AlCl_3$-assisted ionization of a carboxylic acid chloride (RCOCl). The positive charge on the new carbocation intermediate that is formed can be delocalized, as shown by resonance structures.

Think about It

e20.12 Watch animations of the electron-pair movement and orbitals involved in this alkylation mechanism.

WORKED EXAMPLE 20.4—MECHANISM FOR FRIEDEL-CRAFTS ACYLATION

Show a reasonable step-wise mechanism for the reaction of molecules of benzene with a carboxylic acid chloride (RCOCl), in the presence of $AlCl_3$, and then draw the structure of the expected product.

Solution

A reasonable mechanism for Friedel-Crafts acylation starts with the $AlCl_3$ catalyzing the reaction by helping the carboxylic acid chloride ionize, in much the same way that $FeBr_3$ helps

Br$_2$ ionize [<<Section 20.5]. A simplified representation of the electrophile E$^+$ would be an acyl cation [RCO]$^+$, whose electrostatic potential map shows the region around the carbon atom blue and electrophilic. The acyl cation then adds to the benzene ring giving a delocalized carbocation. AlCl$_4^-$ then pulls off an H$^+$ ion to regenerate the aromatic ring and yield the final substitution product. Note the formation of a new carbon–carbon bond as part of a new ketone functional group.

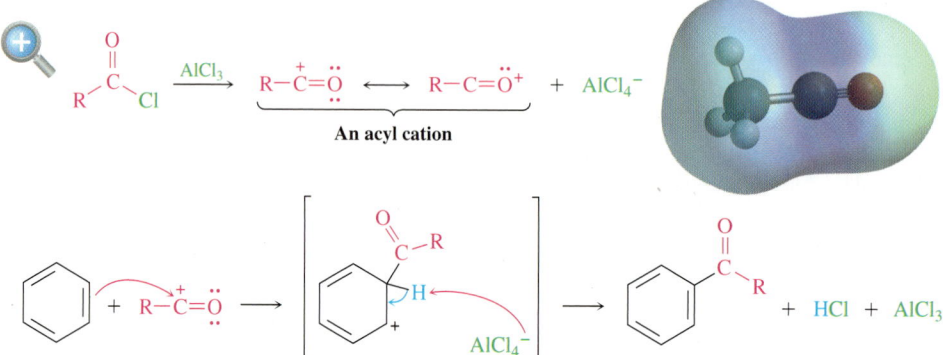

An acyl cation

Molecular Modelling (Odyssey)

e20.13 Examine the electrostatic potential map of the acyl cation to identify the electrophilic site.

Friedel-Crafts alkylation and acylation reactions are especially important synthetic reactions, as they lead to the formation of C—C bonds. They can be thought of as proceeding by similar mechanisms to other electrophilic aromatic substitution reactions.

20.8 Substituent Effects in Electrophilic Aromatic Substitution

Only one product can form when a single electrophilic substitution occurs on benzene, but what happens if we carry out an electrophilic substitution reaction on an aromatic compound with a ring that already has a substituent? Some puzzling patterns are seen. For example, phenol reacts with nitric acid in the presence of sulfuric acid to give a mixture of *ortho*- and *para*-substituted products. If excess nitric acid is used, substitution takes place in three positions and 2,4,6-trinitrophenol (picric acid) is produced.

Nitration reaction of phenol 1000 times faster than for benzene

Phenol

o-Nitrophenol
(50%)

p-Nitrophenol
(50%)

Nitrobenzene, however, reacts to give almost exclusively the *meta*-substituted product.

Nitration reaction of nitrobenzene 17 million times slower than for benzene

Nitrobenzene

m-dinitrobenzene
(major product)

In addition to the differences in orientation of phenol and nitrobenzene, a striking difference in reaction rate is experimentally seen. Phenol reacts 1000 times faster than benzene, the usual standard for comparison. Nitrobenzene reacts 17 million times slower than benzene.

When considering many other electrophilic aromatic substitution reactions on compounds whose rings already have a substituent, patterns are seen—namely that substituents already present on an aromatic ring can have two effects:

- Substituents change the *reactivity* of the aromatic ring. Some substituents (such as —OH, —NH$_2$ and —R) increase the rate of reaction of the compound toward further electrophilic substitution, relative to benzene. These substituents are called *activating groups*. Others (such as —NO$_2$, —CN, and —Cl) slow the rate of reaction relative to benzene. These substituents are called *deactivating groups*.

NO$_2$	Cl	H	OH

Relative rate of nitration 6×10^{-8} 0.033   1   1000

Reactivity →

- Substituents affect the *orientation* of the reaction. The three possible substituted products (*ortho*, *meta*, and *para*) are usually not formed in equal amounts. Instead, the nature of the substituent already present on the ring determines where the second substitution will take place on the ring. An —OH or —R group directs further substitution toward the *ortho* and *para* positions, for instance, while an —NO$_2$ or —CN group directs further substitution primarily toward the *meta* position.

Substituents can be classified into three groups, as shown in Figure 20.15: *ortho*- and *para*-directing activators, *ortho*- and *para*-directing deactivators, and *meta*-directing deactivators. No *meta*-directing activators are known. Note how the directing effects of the groups correlate with their reactivities. All *meta*-directing groups are deactivating, and most *ortho*- and *para*-directing groups are activating. The halogens are unique in being *ortho*- and *para*-directing deactivators.

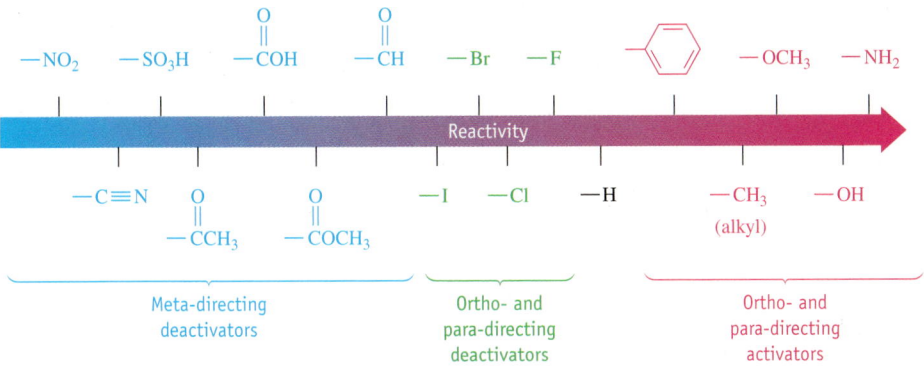

FIGURE 20.15 Substituent effects in electrophilic aromatic substitution reactions. All activating substituents are *ortho*- and *para*-directing toward subsequent reaction. All deactivating substituents other than halogens are *meta*-directing. Halogens are *ortho*- and *para*-directing and mildly deactivating.

WORKED EXAMPLE 20.5—SUBSTITUENT EFFECTS IN ELECTROPHILIC AROMATIC SUBSTITUTION

Which would you expect to react faster in an electrophilic aromatic substitution reaction: chlorobenzene or ethyl benzene?

Solution

From Figure 20.15, you can see that a chloro substituent is deactivating, whereas an alkyl substituent is activating. So ethyl benzene should react faster than chlorobenzene.

Use Figure 20.15 to rank the compounds in each of the following groups in order of increasing rate of reaction to electrophilic aromatic substitution.
(a) nitrobenzene, phenol (hydroxybenzene), toluene
(b) phenol, benzene, chlorobenzene, benzoic acid
(c) benzene, bromobenzene, benzaldehyde, aniline (aminobenzene)

For each of the aromatic compounds in Exercise 20.12, show the products you would expect from reaction with Br_2 in the presence of $FeBr_3$.

Using Models of Electronic Structure to Explain Substituent Effects

What makes a compound with a substituted aromatic ring activating or deactivating relative to benzene? Recall that the first step in the proposed mechanism for electrophilic aromatic substitution involves the π electrons on the aromatic ring forming a bond with an electrophile. This is usually the rate-determining step in the reaction. Compare the electrostatic potential maps of molecules of phenol (activated), chlorobenzene (weakly deactivated), and nitrobenzene (strongly deactivated). The ring with the —OH substituent is quite electron-rich (red), while the electron withdrawing —Cl makes the ring less electron-rich (yellow), and the strongly electron withdrawing —NO_2 group makes the ring even less electron-rich (green).

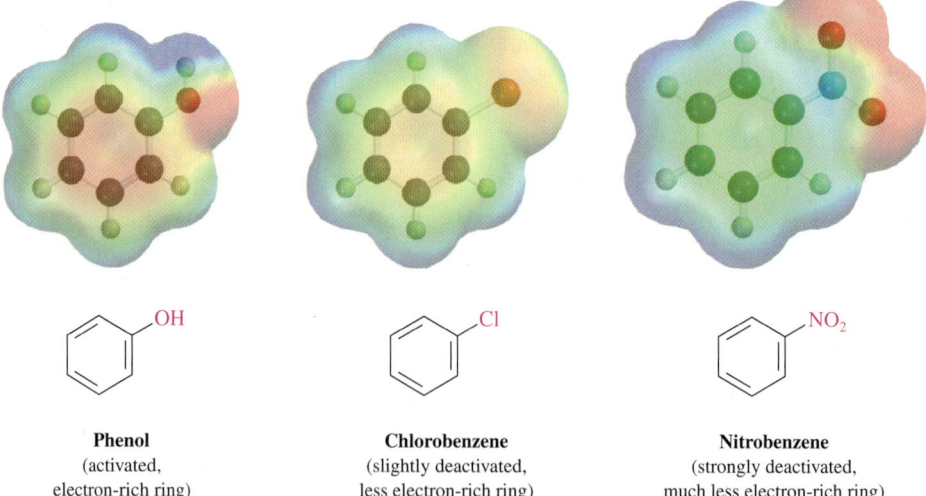

Phenol	Chlorobenzene	Nitrobenzene
(activated, electron-rich ring)	(slightly deactivated, less electron-rich ring)	(strongly deactivated, much less electron-rich ring)

Molecular Modelling (Odyssey)

e20.14 Use electrostatic potential maps to identify the effect of substituents on location of electron density in the ring.

The common characteristic of all activating groups is that they push electrons toward the ring, thereby making the ring more electron-rich. Activating substituents would be expected to delocalize the positive charge on the carbocation intermediate over a larger molecular volume, and lower the activation energy for its formation. *The common characteristic of all deactivating groups is that they withdraw electrons from the ring,* thereby making the ring less electron-rich. This would be expected to destabilize the positive charge on the carbocation intermediate, and raise the activation energy for its formation.

Orienting Effects in Substituted Aromatic Rings: *Ortho*- and *Para*-Directors

Let's look at the accepted mechanism for nitration of phenol as an example of how *ortho*- and *para*-directing substituents might be seen at the molecular level to change both the orientation and rate of substitution. In the first step of the mechanism, we could envision reaction with the electrophilic nitronium ion (NO_2^+) to occur either *ortho*, *meta*, or *para* to the —OH group, giving the carbocation intermediates shown in Figure 20.16. Yet only the *ortho* and *para* products are formed, suggesting that the *ortho* and *para* intermediates are more stable and have a lower activation energy for formation than the *meta* intermediate. The resonance model for drawing Lewis structures can help visualize this, because the carbocation intermediates that would result from *ortho* and *para* reaction have more resonance forms, including one that allows the positive charge to be stabilized by electron donation from the substituent oxygen atom. Since the *ortho* and *para* intermediates are more stable than the *meta* intermediate, they should be formed faster.

Molecular Modelling (Odyssey)

e20.15 Examine electrostatic potential maps to see the effect of the —OH substituent on delocalization of the positive charge in the carbocation intermediate.

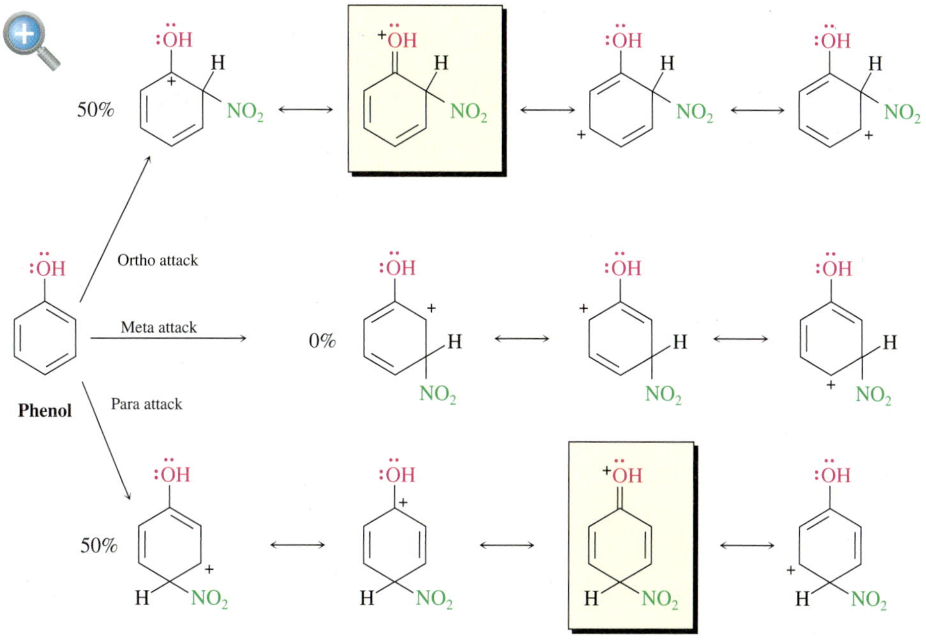

FIGURE 20.16 Reaction intermediates that would be expected in the reaction mechanism for three possible sites of nitration of phenol. The *ortho* and *para* intermediates should be lower in energy than the *meta* intermediate because they are able to delocalize the positive charge of the carbocation over a larger molecular volume. This is evident from the resonance model, where a fourth resonance structure shows electron donation from the oxygen atom contributes to charge delocalization, but only in the case of *ortho* and *para* intermediates.

In general, any substituent that has a lone pair of electrons on the atom directly bonded to the aromatic ring makes possible an electron-donating resonance interaction with the carbocation resulting from *ortho* or *para* substitution and thus acts as an *ortho*- and *para*-director:

Orienting Effects in Substituted Aromatic Rings: *Meta*-Directors

The influence of *meta*-directing substituents can be explained using the same kinds of arguments used for *ortho*- and *para*-directors. Look at the accepted reaction mechanism for chlorination of benzaldehyde, a *meta*-director (Figure 20.17). Considering possible substitution in *ortho*, *meta*, or *para* positions, none of the structures for the three possible carbocation intermediates look very favourable, as the partial positively charged carbon atom of the aldehyde group is attached to the ring of a positively charged carbocation intermediate. Substituents with an atom that has a partial positive charge ($\delta+$) attached to the ring (typical of those that are *meta* directing) deactivate the ring toward further substitution by raising the energy of the carbocation intermediate

Molecular Modelling (Odyssey)

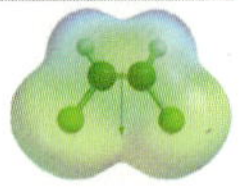

e20.16 Examine electrostatic potential maps of the intermediates to see how the —NO_2 substituent affects the carbocation intermediate.

relative to substitution of benzene. *But the best of three undesirable alternatives for the site of substitution on benzaldehyde is meta*, as the resonance structures for *ortho* and *para* attack place the positive charge directly on the ring carbon atom bonded to the aldehyde group, where it is disfavoured by a repulsive interaction with the partial positively charged carbon atom of the C=O group. As the best of three poor alternatives, the *meta* intermediate is favoured and is formed faster than the *ortho* and *para* intermediates.

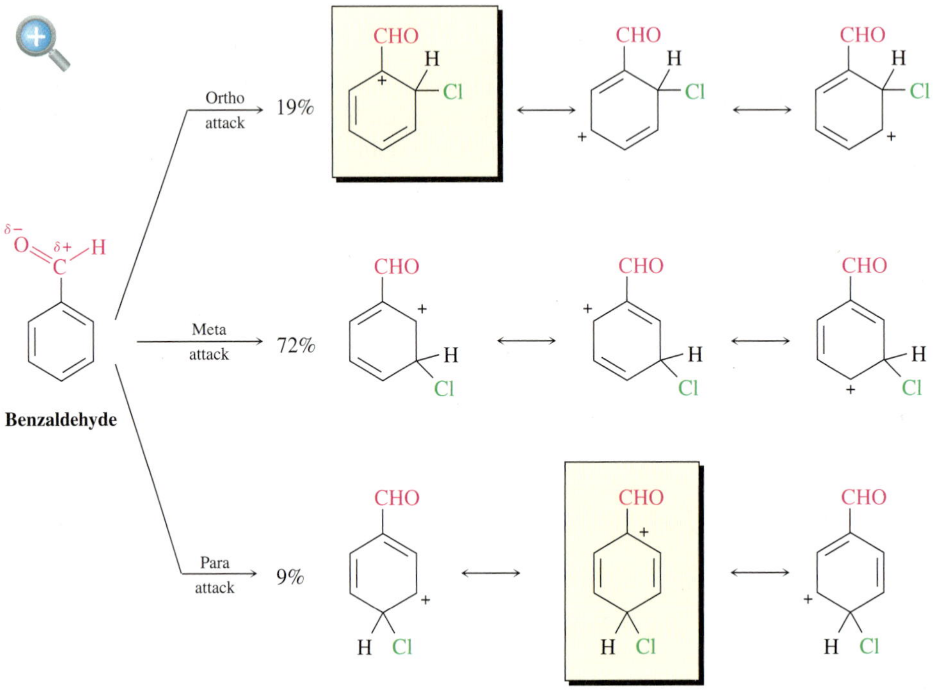

FIGURE 20.17 Expected reaction intermediates in the reaction mechanism for three possible sites of nitration of benzaldehyde. The *ortho* and *para* intermediates are expected to be higher in energy than the *meta* intermediate, because the positive charge of the carbocation is located in part on the carbon atom bonded directly to the partial positively charged carbon atom (δ+) of the C=O group. As a result, *meta* direction for substitution is a more favourable option.

WORKED EXAMPLE 20.6—ORIENTATION OF ELECTROPHILIC AROMATIC SUBSTITUTION

What product(s) would you expect from bromination of aniline ($C_6H_5NH_2$)?

Solution

From Figure 20.17, the —NH_2 substituent is predicted to be *ortho*- and *para*-directing. Since an amino group is *ortho*- and *para*-directing, we expect to obtain a mixture of *o*-bromoaniline and *p*-bromoaniline.

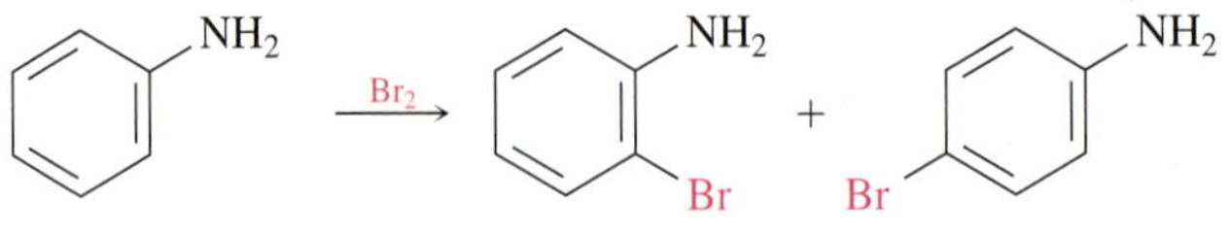

Aniline **o-Bromoaniline** **p-Bromoaniline**

EXERCISE 20.14—ORIENTATION OF ELECTROPHILIC AROMATIC SUBSTITUTION

What product(s) would you expect from sulfonation of the following compounds?

(a) nitrobenzene (b) bromobenzene (c) toluene
(d) benzoic acid (e) benzonitrile

In general, any substituent with a positively polarized atom ($\delta+$) directly attached to the ring destabilizes the most those carbocation intermediates resulting from *ortho* or *para* substitution. Reaction is, therefore, directed to the *meta* position.

Meta directors

20.9 Oxidation and Reduction of Aromatic Compounds

Consistent with the stability of aromatic rings, aromatic compounds do not usually react with strong oxidizing agents such as $KMnO_4$. (Recall from Section 19.12 that $KMnO_4$ cleaves alkene $C=C$ bonds.) Alkyl groups attached to the aromatic ring are readily attacked by oxidizing agents, however, and are converted into carboxyl groups ($-COOH$). For example, butylbenzene is oxidized by $KMnO_4$ to give benzoic acid. The mechanism of this reaction is complex and won't be considered here. It is thought to involve attack on the side-chain $C-H$ bonds at the position next to the aromatic ring (the *benzylic position*) to give radical intermediates.

$CH_2CH_2CH_2CH_3$ $\xrightarrow[\text{H}_2\text{O}]{\text{KMnO}_4}$ $COOH$

Butylbenzene   **Benzoic acid (85%)**

Just as aromatic rings in compounds are usually inert to oxidation, they are also inert to reduction under typical alkene hydrogenation conditions. Only if high temperatures and pressures are used does reduction of an aromatic ring occur. For example, *o*-dimethylbenzene (*o*-xylene) gives 1,2-dimethylcyclohexane if reduced at high pressure. But very high pressures of H_2 gas are needed:

CH_3 CH_3 $\xrightarrow[\text{2000 psi, 25 °C}]{\text{H}_2, \text{Pt; ethanol}}$ CH_3 CH_3

o-**Xylene**   **1,2-Dimethylcyclohexane (100%)**

EXERCISE 20.15—OXIDATION OF AROMATIC COMPOUNDS

What aromatic products do you expect to obtain from oxidation of the following substances with $KMnO_4$?

(a) *m*-Chloroethylbenzene   **(b)**   Tetralin

20.10 Organic Synthesis: Thinking Backward

The laboratory synthesis of organic molecules from simple precursors is carried out for many reasons. As seen in Section 20.1, in the pharmaceutical industry, new organic molecules are designed and synthesized for evaluation as medicines. In other areas of chemical industry, syntheses are undertaken to devise more efficient routes (both economic and atom efficiency) to known compounds, to make new compounds, or to use approaches more consistent with other green chemistry principles [<<Section 5.1]. The main goals of

introducing synthetic reactions in this learning resource are to learn how substances important to modern life can be created from smaller building blocks, and to help you learn
organic chemistry. Devising a route for the synthesis of an organic molecule requires that
you approach chemical problems in a logical way, draw on all your knowledge of how
carbon–carbon bonds can be made and functional groups modified, and organize that
knowledge into a workable plan.

One important strategy in devising a multi-step organic synthesis is to *work backward,*
a process called **retrosynthetic analysis**. After inspecting the starting materials available
and the desired product to see how many carbon atoms need to be added and what functional group modifications made, ask yourself: "What is the immediate precursor of that
product?" Having found an immediate precursor, work backward again, one step at a time,
until a suitable starting material is found (see Worked Example 20.7).

WORKED EXAMPLE 20.7—MULTI-STEP SYNTHESIS

Synthesize *m*-chloronitrobenzene starting from benzene.

Strategy

Start by inspecting the starting material and product. Note that both structures contain a benzene ring, but two new functional groups will need to replace H atoms on benzene molecules
to give the product. Then work backward by asking, "What is an immediate precursor of *m*-
chloronitrobenzene?"

m-Chloronitrobenzene

There are two substituents on the ring, a —Cl group, which is *ortho-* and *para*-directing,
and an —NO$_2$ group, which is *meta*-directing. We can't nitrate chlorobenzene because the
wrong isomers (*o-* and *p*-chloronitrobenzenes) would result, but chlorination of nitrobenzene
should give the desired product.

"What is an immediate precursor of nitrobenzene?" Benzene, which can be nitrated.

Solution

We've solved this problem in two steps, but they need to be done in the right order.

Starting with either benzene or toluene, how would you synthesize the following substances? Assume that *ortho* and *para* isomers can be separated.
(a) 2-bromo-4-nitrotoluene
(b) 1,3,5-trinitrobenzene
(c) 2,4,6-tribromoaniline

20.11 Polycyclic Aromatic Hydrocarbons

As seen in the example of adenine in Section 20.3, the concept of aromaticity—the unusual chemical stability that arises in compounds with cyclic conjugated molecules like benzene—can be extended beyond simple monocyclic compounds to include polycyclic aromatic compounds. Naphthalene, familiar for its use in mothballs and as the intended (but not supplied) deworming medicine in Section 20.1, has two benzene-like rings fused together and is the simplest and best-known **polycyclic aromatic hydrocarbon** (PAH). Hückel's 4n+2 rule, when first formulated, was restricted to monocyclic, completely conjugated systems. But compounds like naphthalene, with 10 π electrons (4n+2, n = 2) around the periphery of the ring skeleton, show the same patterns of stability and reactivity as benzene. A naphthalene molecule has chemical shifts of $\delta = 7.8$ and 7.45 in the ^{1}H NMR, consistent with the ring currents expected for a delocalized aromatic compound. An electrostatic potential map of a naphthalene molecule indicates that the 10 π electrons are completely delocalized around both rings (Figure 20.18).

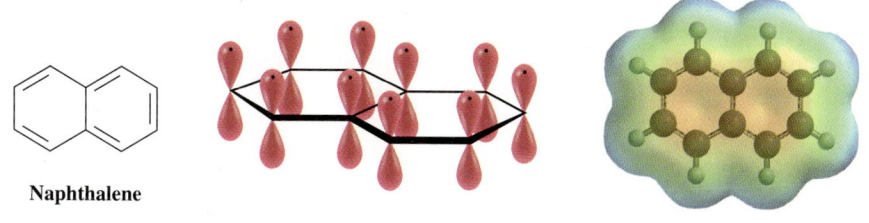

Naphthalene

FIGURE 20.18 A Lewis structure, an orbital representation, and electrostatic potential map of a naphthalene molecule. The 10 π electrons are completely delocalized around both rings.

Many polycyclic aromatic hydrocarbons are more complex than naphthalene. Perhaps the most notorious PAH is benz[*a*]pyrene, whose molecules have five benzene-like rings and is one of the carcinogenic (cancer-causing) substances found in chimney soot, cigarette smoke, and charcoal-grilled meat. Exposure to even a tiny amount of benz[*a*]pyrene is sufficient to induce a skin tumour in susceptible mice. The first description of chemical carcinogenesis may have been made by London surgeon Percivall Pott, who traced the occupational occurrence of scrotal cancer in 18th century chimney sweeps to their repeated exposure to chimney soot, which we now know contains trace levels of benz[*a*]pyrene.

Benz[*a*]pyrene   **A diol epoxide**

After benz[*a*]pyrene is taken into the body by eating or inhaling, the body attempts to rid itself of the foreign substance by converting it into a water-soluble metabolite called a *diol epoxide,* which can be excreted. Unfortunately, the diol epoxide metabolite reacts with and binds to cellular DNA, thereby altering the DNA and leading to mutations or cancer. On prolonged exposure, even benzene itself can cause certain types of cancer, so breathing the fumes of benzene and other volatile aromatic compounds in the laboratory should be avoided.

PAHs are formed and released into the atmosphere as a result of incomplete (usually oxygen-deficient) combustion of hydrocarbons, especially wood and coal. Those like benz[*a*]pyrene, with higher molar masses than simple aromatic compounds, do not last long in the gas phase, but rapidly condense on the surfaces of suspended atmospheric particles, often called particulate matter (PM) [<<Section 4.3]. The threat to human health from PAHs in the air depends on the nature of the PAH compounds and the diameter of the PM on which it is adsorbed. Of particular concern is *fine* or *respirable* particulate matter, given the symbol $PM_{2.5}$ since they contain only particles smaller than 2.5 μm. Respirable particles are those that are not easily trapped out by natural filtering mechanisms such as mucous membranes and cilia on the walls of bronchial tubes. As a result $PM_{2.5}$ can penetrate deep into lung tissue, where harmful adsorbed PAHs can cause severe damage over long periods of time.

> The concept of aromaticity can be extended to polycyclic aromatic hydrocarbons (PAHs). Some PAHs are known to be carcinogenic, and can find entry into humans by being absorbed on fine particulate matter (PM) that is inhaled.

20.12 Moving out of Flatland: Graphite, Nanotubes, and Fullerenes

As you read this chapter, you might be taking notes with a macromolecular aromatic substance—your pencil "lead," which is actually made of graphite. What does it look like at the molecular level? In a series of PAH molecules, moving from naphthalene through anthracene and pyrene to coronene, the increasing number of benzene rings fused together causes the ratio of hydrogen atoms to carbon atoms to decrease. In benzene, the ratio of H to C atoms is 1:1; in coronene, it is 1:2.

> Graphite is also classified as a network solid [<<Section 3.3].

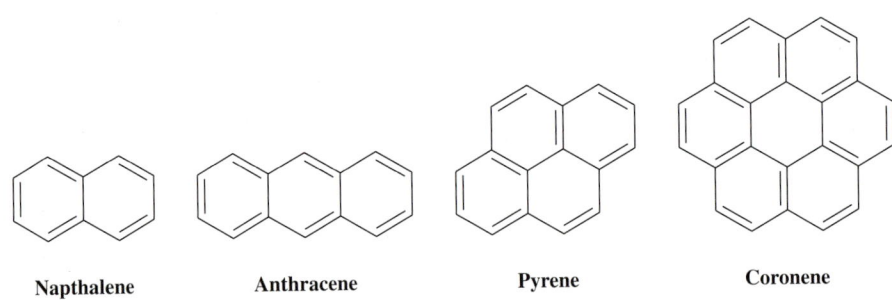

Napthalene **Anthracene** **Pyrene** **Coronene**

If this fused ring system extended further into space, the H/C ratio would eventually approach zero, producing *graphite,* an allotrope of carbon. Zoom in on graphite at the molecular level, and you would see an infinite two-dimensional sheet of fused aromatic rings, with very strong interactions within each graphite sheet and very weak intermolecular interactions between sheets. As you take notes, the sheets of graphite molecules slide past each other easily and get deposited on your paper.

You saw in Sections 19.1 and 19.15 that conjugated polyenes (alkenes) and polyynes (alkynes) respond rapidly to light, making them interesting candidates for use in photonic devices for processing data and creating optical-neural networks. Having now seen the delocalized π systems of aromatic hydrocarbons, you should not be surprised that large conjugated polymers of benzene (polyphenylenes) are also hot areas of research in the development of electronic materials.

One-Dimensional Wires

In the field of **nanotechnology**, chemists try to design, create, and control devices and systems on the nanometre, or 10^{-9} m, scale. Long polyphenylene chains are one class of compounds that have been studied for their potential as molecular wires, which are conjugated molecules that might serve as *one-dimensional electronic conductors*. You might imagine connecting a chain of benzene molecules together to create a polyphenylene chain, and attaching it somehow on each end to an electrode. Current could then be passed through the wire, with potential use in molecular-level electronic devices. Such polyphenylene chains have been synthesized and modified in various ways, including the use of alkynes as spacers between the benzene rings.

Two-Dimensional Sheets and Monolayers

Elegant synthetic work is also carried out to create and study new polycyclic aromatic hydrocarbons that can form thin *two-dimensional sheets*. One example is the dehydrogenation and fusion of benzene rings of the compound on the left below (a polyphenylene) to remove 108 hydrogen atoms and produce a rigid, planar polycyclic aromatic hydrocarbon. The two molecules have strikingly different structures and properties. Due to steric interactions, not all of the π systems of benzene rings in the molecule on the left are in the same plane. The molecule on the right is planar and rigid. Polar substituents have been added to the outside of similar rigid PAHs to dramatically change their properties, producing compounds that function as liquid crystals.

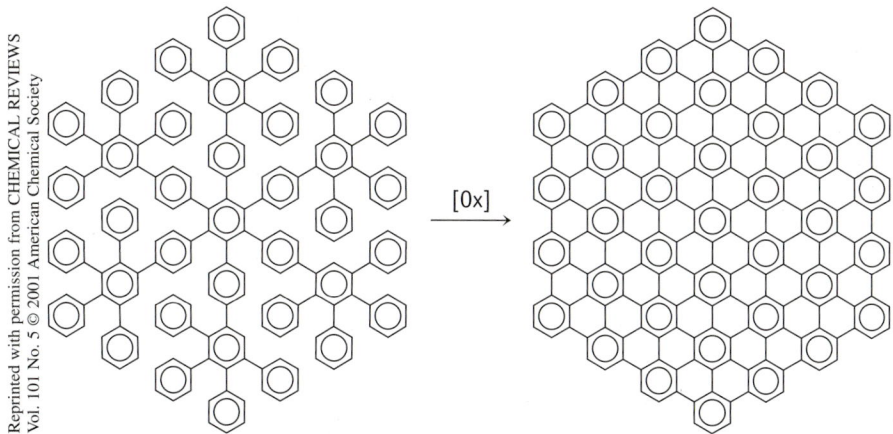

Reprinted with permission from CHEMICAL REVIEWS Vol. 101 No. 5 © 2001 American Chemical Society

Intriguingly, molecules of some PAHs can be made to aggregate together (a process called *self-assembly* [<<Section 9.1]) to create layers that are exactly one molecule thick (monolayers). Using techniques important in the field of nanotechnology, these *self-assembled monolayers* can be deposited on surfaces to create two-dimensional crystals that can be studied with techniques such as *scanning tunnelling microscopy* (STM). Studying the monolayers by STM is made easier due to the electron-rich π core of the PAH system, which shows up as bright spots.

Three-Dimensional Tubes and Spheres

You may have a feeling by now that the world of aromatic compounds is a true flatland, as all of the structures of aromatic compounds we have seen are planar. But a molecular-level view of a single-walled carbon nanotube, such as the one shown enclosing a

β-carotene molecule in Figure 19.5, opens up a new dimension. A carbon nanotube can be viewed at the molecular level as made of rolled-up graphite sheets to create a three-dimensional fibre. When coiled and joined together, the carbon framework forms some of the strongest fibres known, and has electrical properties that can be tuned by changes in structure to range from metal-like conductivity to that of semiconductors. Carbon nanotubes are made by passing a current between two graphite electrodes under an atmosphere of inert helium gas. Some of the graphite vaporizes and then condenses again, forming carbon nanotubes on the cathode.

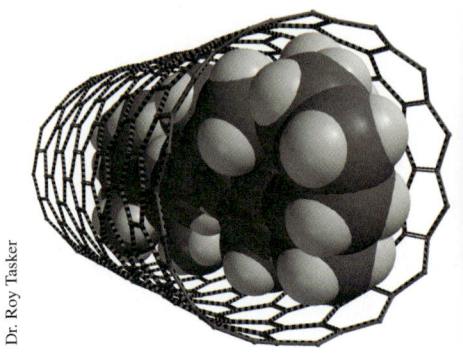 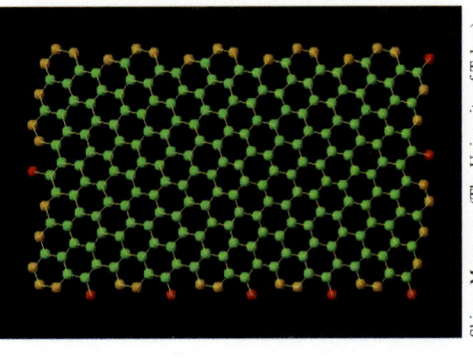

And so graphite, an allotrope of carbon used in your pencil, can be converted into three-dimensional tubes. Looking at the most recently discovered allotrope of carbon, buckminsterfullerene, will complete your emergence from the flatland of individual planar aromatic compounds, one-dimensional molecular wires, and two-dimensional polycyclic aromatic hydrocarbon sheets. Since the discovery of buckminsterfullerene (C_{60}) in 1985 by Robert Curl, Harry Kroto, and Richard Smalley, the three-dimensional image of a molecular-sized soccer ball has captured the imagination of chemists (Figure 20.19). Reminiscent of the discovery of mauveine by Perkin, the deep purple colour of C_{60} in solution was one of the keys to its discovery and isolation. Buckminsterfullerene was only the first of thousands of other related spherical carbon compounds synthesized in the past 30 years. Applications may be found as superconductors, carbon films, and as cages to trap and release other species, perhaps making possible the delivery of drugs.

When we introduced the concept of aromaticity, we described it as a bit "fuzzy," as the experimental measurements of aromaticity hinged on comparing the reactivity and

 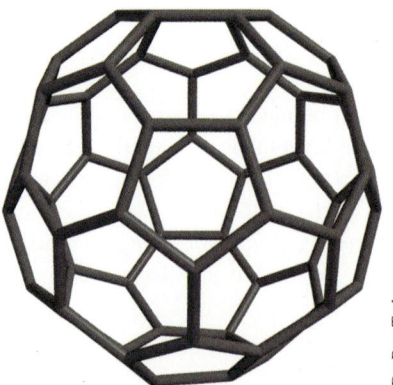 

FIGURE 20.19 Buckminster Fuller's dome, a molecular-level view of C_{60}, and crystals of buckminsterfullerene from the Leopold-Franzens-Universitat in Innsbruck. (Photograph by Gschnaller Jochen.) Note that a few five-membered rings are needed to create the spherical buckminsterfullerene, just as Buckminster Fuller used in his geodesic dome, displayed at the Montreal Expo 67 World Fair.

stability of simple cyclic aromatic compounds with appropriate reference compounds. The aromaticity of fullerenes was also controversial in the years after they were discovered, as these compounds are less stable than their planar counterparts, due to the strain introduced by changing the geometry of planar carbon atoms into pyramidal ones. But the same evidence from NMR spectroscopy that proved so helpful in establishing the aromaticity of simple planar compounds has also shown that fullerenes can sustain ring currents, and must have cyclic delocalization of the π electrons around the sphere. Particularly elegant work has been done by synthetic chemists who have placed substituents on the outside of fullerene rings to probe the expected ring currents and correlate them with chemical shifts.

New aromatic compounds continue to show promise in the development of materials unknown a few years ago.

SUMMARY

Key Concepts

- **Aromatic compounds**, such as benzene, played a vital role in the development of industrial organic chemistry. By the early 20th century, this industry produced dyes, pharmaceutical products, and explosives. At the beginning of the 21st century, two- and three-dimensional aromatic compounds have promising applications in the development of new materials such as superconductors, carbon films, and molecular cages. (Sections 20.1, 20.13)
- **Hückel's rule** states that substances whose molecules contain monocyclic, planar, completely conjugated rings with a total of 4n+2π electrons, where n is an integer (n = 0, 1, 2, …), are exceptionally stable relative to model compounds and are aromatic. Substances whose molecules contain monocyclic, planar, completely conjugated rings with 4n π electrons (n = 1, 2, . . .) are exceptionally unstable relative to model compounds and are *anti-aromatic*. While aromaticity is difficult to measure directly, the stability of a compound can be estimated by experiments that compare its energy to that of a carefully chosen model compound. But ^{1}H NMR spectroscopy provides a simple and elegant test of aromaticity. Look at the chemical shifts of hydrogen atoms directly bonded to ring carbons in a molecule to see whether they fall into the $\delta = 6.5$–8.5 region. (Section 20.2)
- Hückel's rule can also be extended to apply to **polycyclic aromatic hydrocarbons (PAH)**, **heterocyclic compounds**, and ions (Sections 20.3, 20.11). Some PAHs are known to be carcinogenic, and can find entry into humans by being absorbed on fine particulate matter (PM) that is inhaled. (Section 20.11)
- The most common reaction of aromatic compounds is **electrophilic aromatic substitution**. Bromination, chlorination, iodination, nitration, sulfonation, Friedel-Crafts alkylation, and Friedel-Crafts acylation can all be carried out. Friedel-Crafts alkylation is particularly useful for making C—C bonds in preparing a variety of alkylbenzenes but is limited because only alkyl halides can be used and strongly deactivated rings do not react. (Sections 20.5–20.7)
- The mechanism for electrophilic aromatic substitution is thought to be a two-step polar reaction, in which the π electrons of the aromatic ring first react with the electrophile to yield a delocalized carbocation intermediate, which then loses H$^+$ to give a substituted aromatic product. (Section 20.5)
- Substituents on the benzene ring affect both the reactivity of the ring toward further substitution and the orientation of that further substitution. Substituents can be classified either as *activators* (react faster than benzene) or *deactivators* (slower than benzene), and either as *ortho-* and *para*-directors or as *meta*-directors. In

general, substituents that have a lone pair of electrons on the atom directly bonded to the aromatic ring acts as *ortho*- and *para*-directors, and their reactions are faster than for benzene. Substituents with a positively polarized atom ($\delta+$) directly attached to the ring are *meta*-directors, and their reactions are slower than benzene. (Summarized in Figure 20.13, Section 20.8)

- The side chains of alkylbenzenes have unique reactivity because of the neighbouring aromatic ring. An alkyl group attached to the aromatic ring can be converted to a carboxyl group (—COOH) by oxidation with aqueous $KMnO_4$. Aromatic rings can be reduced to yield cyclohexanes on catalytic hydrogenation at high pressure. (Section 20.9)

- **Retrosynthetic analysis** (working backward) is helpful in rationalizing how to carry out a multi-step synthesis. (Section 20.10)

Key Reactions

1. Electrophilic aromatic substitution

    (a) Bromination (Section 20.5)

    (b) Chlorination (Section 20.6)

    (c) Nitration (Section 20.6)

    (d) Sulfonation (Section 20.6)

    (e) Friedel-Crafts alkylation (Section 20.7)

    (f) Friedel-Crafts acylation (Section 20.7)

2. Oxidation of aromatic side chains (Section 20.9)

3. Hydrogenation of aromatic rings (Section 20.9)

REVIEW QUESTIONS

Section 20.1: Aromatic Roots of Chemistry: Curing, Death, and Dyeing

20.17 In this text, you encounter many organic reactions that are named after German chemists. What conditions were present in the dye factories on the banks of the Rhine River in the late 19th century that helped establish such a major role for Germany in the development of industrial organic chemistry?

20.18 Section 20.1 discusses the multiple uses of chemicals such as toluene and phenol and the electrophilic aromatic nitration reactions that convert toluene into trinitrotoluene and phenol into picric acid. Swedish chemist Alfred Nobel built his wealth on the safe manufacture and use of nitroglycerine, another substance with multiple uses. Alarmed by the possible harmful consequences of his inventions, he established the Nobel prizes to advance civilization and promote peace. Consulting additional sources, list the uses of nitroglycerine in warfare, construction, and medicine.

20.19 What is the chemical connection between the reagents needed during World Wars I and II to nitrate toluene and phenol, the massive deposits of the excrement of seabirds, bats, and seals (guano) in Chile, and the Haber synthesis of ammonia, described in Section 13.1?

Section 20.2: Benzene and Aromaticity: Reactivity, Structure, and Spectroscopy

20.20 Draw a Lewis structure for a 1,3,5,7-tetramethylcyclooctatetraene molecule, and consider its geometry and number of π electrons. Is it aromatic? Explain.

20.21 With reference to your answer to Question 20.20, the experimental geometry of a 1,3,5,7-tetramethylcyclooctate-

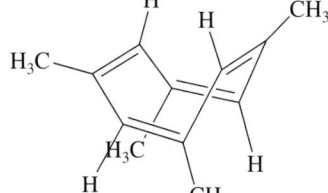

trene molecule shows that the ring is non-planar and tub-shaped, with alternating C—C single and double bonds, with bond lengths of 148 pm and 133 pm, respectively. Explain why you think the ring prefers a tub to a planar conformation.

20.22 Using the resonance model, consider the structure of a naphthalene molecule, shown with numbering in Question 20.62. Account for the experimental observation that not all of the C—C bonds are the same length. The C_1—C_2 bond is 136 pm long, while the C_2—C_3 bond is 139 pm long.

20.23 One resonance structure of an anthracene molecule is shown in Section 20.12. Draw the other three.

20.24 How many peaks would you expect to see in the ^{13}C NMR spectrum of anthracene (Question 20.23)? At approximately what chemical shifts?

20.25 How many clusters of peaks would you expect to see in the ^{1}H NMR spectrum of anthracene (Question 20.23)? At approximately what chemical shifts?

20.26 Identify the wavenumber of and describe the key vibrational modes in the IR spectrum of anthracene that characterize it as an aromatic hydrocarbon.

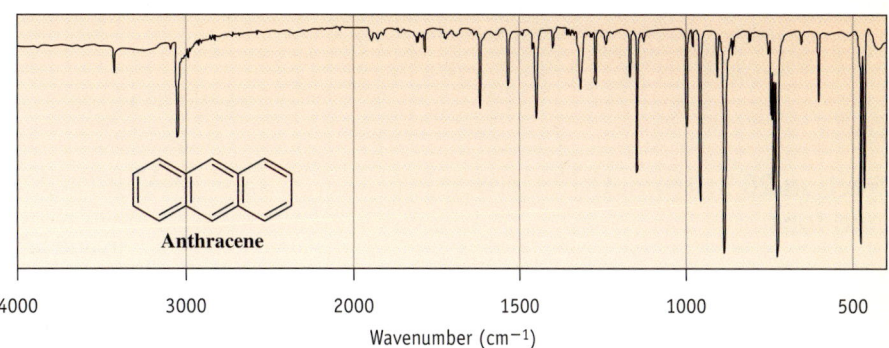

20.27 The structure of a molecule of phenanthrene is shown. How many π electrons can be delocalized around the perimeter of the molecule? Is phenanthrene aromatic?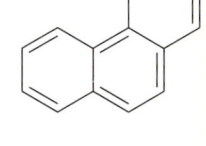

20.28 One structure of a phenanthrene molecule is shown in Question 20.27. Draw four additional resonance structures for phenanthrene.

20.29 Using the resonance structures drawn for a phenanthrene molecule in Question 20.28, predict which of the C—C bonds should be the shortest.

20.30 The structure of a molecule of benz[α]pyrene is shown in Section 20.11. Why is benz[α]pyrene considered aromatic?

20.31 The ^{13}C and ^{1}H NMR spectra for a compound with the molecular formula C_8H_9Br are shown below. Calculate the number of units of unsaturation present, and propose a reasonable structure that is consistent with this data. In the ^{1}H NMR spectrum, the peaks at

$\delta = 7.3, 7.0, 2.6,$ and 1.2 have integrations of 2, 2, 2, and 3, respectively.

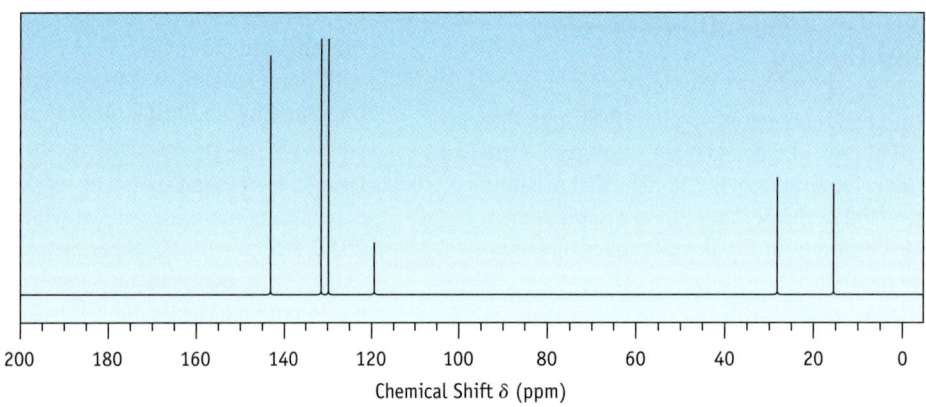

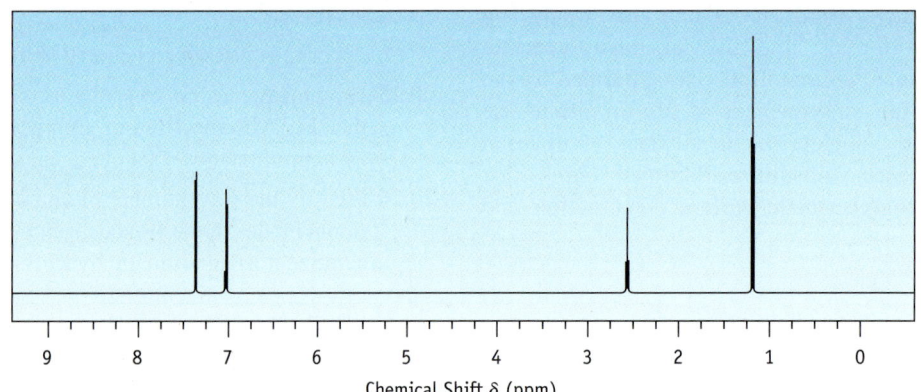

Section 20.3: Aromatic Heterocycles and Ions

20.32 When treated with alkali metals, two electrons can be added to the π system of molecules of 1,3,5,7-tetramethylcyclooctatetraene (Question 20.20), creating a dianion. Draw the structure of the 1,3,5,7-tetramethylcyclooctatetraene dianion and count the number of π electrons in the cyclic conjugated system. Is the dianion aromatic?

20.33 What would you predict the geometry of the 1,3,5,7-tetramethylcyclooctatetraene dianion (Question 20.32) be, and how would it compare with a 1,3,5,7-tetramethylcyclooctatetraene molecule?

20.34 Pyridine is a cyclic nitrogen-containing compound that shows many of the properties associated with aromaticity. For example, pyridine undergoes electrophilic substitution reactions. Draw an orbital picture of a pyridine molecule, and account for its aromatic properties.

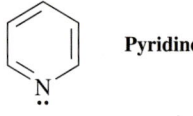

Pyridine

20.35 Indole is an aromatic heterocycle in which each molecule has a benzene ring fused to a pyrrole ring. Using the hybrid orbital model, draw an orbital picture of a molecule. How many π electrons does the molecule have? Is it aromatic?

Indole

20.36 Which would you expect to be more stable: cyclononatetraenyl radical, cation, or anion?

20.37 How might you convert 1,3,5,7-cyclononatetraene to an aromatic substance?

Section 20.4: Naming Aromatic Compounds

20.38 Draw structures of molecules of substances with the following IUPAC names:
(a) *p*-bromochlorobenzene
(b) *p*-bromotoluene
(c) *m*-chloroaniline
(d) 1-chloro-3,5-dimethylbenzene

20.39 Give IUPAC names for the substances whose molecular structures are shown (red = O, blue = N).

(a)

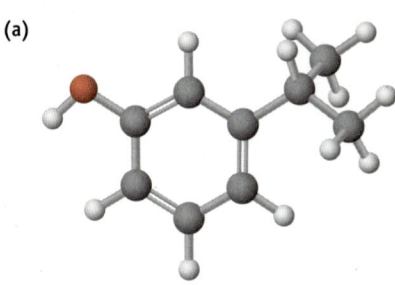

(b)

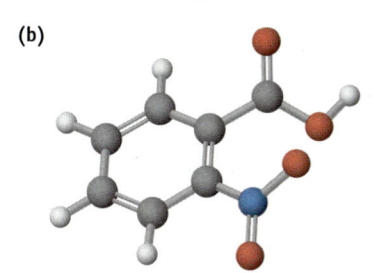

20.40 Draw molecules of all aromatic compounds with the formula C_7H_7Cl and name them.

20.41 Draw molecules of all isomeric (a) dinitrobenzenes and (b) bromodimethylbenzenes and name them.

20.42 Draw structures of molecules of the following substances:
(a) *m*-bromophenol
(b) 1,3,5-benzenetriol
(c) *p*-iodonitrobenzene
(d) 2,4,6-trinitrotoluene (TNT)
(e) *o*-aminobenzoic acid
(f) 3-methyl-2-phenylhexane

Sections 20.5–20.7: Electrophilic Aromatic Substitution

20.43 Write a detailed, step-wise mechanism, showing the steps involved in the Friedel-Crafts alkylation of benzene with CH_3Cl.

20.44 Write a detailed, step-wise mechanism that explains the fact that deuterium atoms (D, ^2H) slowly replace hydrogen atoms (^1H) on the aromatic ring when benzene is treated with D_2SO_4.

Interactive Exercise 20.45

SUBMIT

Use the calculated energies of the reactants and products of bromination reactions to explain why benzene does not behave like an ordinary alkene.

Section 20.8: Substituent Effects in Electrophilic Aromatic Substitution

20.46 Identify each of the following groups as an activator or deactivator and as an *o*-, *p*-director or *m*-director:

(a) $-N(CH_3)_2$ (d)

(b)

(c) $-OCH_2CH_3$

20.47 Draw structures of molecules of the products from reaction of each of the following substances with (i) Br_2, $FeBr_3$ and (ii) CH_3COCl, $AlCl_3$ (red = O), and name the products.

(a)   **(b)**

20.48 Draw all resonance structures of the three possible carbocation intermediates (for *ortho*, *meta* and *para* substitution) to show why a methoxy group ($-OCH_3$) on a benzene ring directs bromination toward *ortho* and *para* positions.

20.49 Draw all resonance structures of the three possible carbocation intermediates (for *ortho*, *meta* and *para* substitution) and show why an acetyl group ($CH_3C=O$) on a benzene ring directs bromination toward the *meta* position.

20.50 Draw structures of molecules of the expected product(s) for the reaction of methyl phenyl ether with Br_2 in the presence of $FeBr_3$. Explain why other product(s) are not formed.

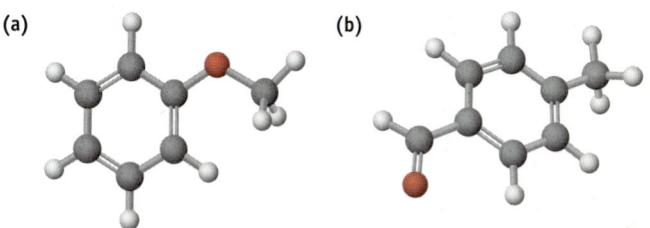

$$\xrightarrow{Br_2,\ FeBr_3}$$

Methyl phenyl ether

20.51 Draw structures of molecules of the expected product(s) for the reaction of benzophenone with Br_2 in the presence of $FeBr_3$? Explain why other product(s) are not formed.

$$\xrightarrow{Br_2,\ FeBr_3}$$

Benzophenone

20.52 Would you expect methyl phenyl ether (Question 20.50) or benzophenone (Question 20.51) to react more rapidly with Br_2 in the presence of $FeBr_3$? Explain.

20.53 Predict the major product(s) of the following reactions:

(a)

$$\xrightarrow[\text{AlCl}_3]{\text{CH}_3\text{CH}_2\text{Cl}} \text{ ?}$$

(b)

$$\xrightarrow[\text{AlCl}_3]{\text{CH}_3\text{CH}_2\text{COCl}} , \text{ ?}$$

(c)

$$\xrightarrow[\text{H}_2\text{SO}_4]{\text{HNO}_3} \text{ ?}$$

(d)

$$\xrightarrow[\text{H}_2\text{SO}_4]{\text{SO}_3} \text{ ?}$$

20.54 Predict the major product(s) of mononitration of the following substances:
(a) bromobenzene
(b) benzonitrile (cyanobenzene)
(c) benzoic acid
(d) nitrobenzene
(e) phenol
(f) benzaldehyde

20.55 Which of the substances listed in Question 20.54 react faster than benzene and which react slower?

20.56 Rank the compounds in each group according to their reactivity toward electrophilic substitution:
(a) chlorobenzene, *o*-dichlorobenzene, benzene
(b) *p*-bromonitrobenzene, nitrobenzene, phenol
(c) fluorobenzene, benzaldehyde, *o*-dimethylbenzene

Section 20.10: Organic Synthesis: Thinking Backward

20.57 How would you synthesize the following substances starting from benzene?
(a) *o*-bromotoluene
(b) 2-bromo-1,4-dimethylbenzene

20.58 How would you synthesize from benzene the compound whose molecules are shown? (green colour = Cl)

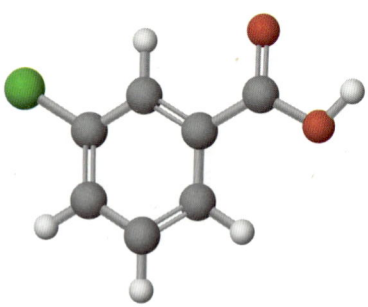

20.59 How would you synthesize from benzene the compound whose molecules are shown below? More than one step is needed.

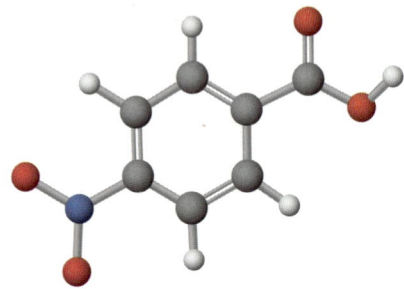

20.60 Starting with benzene, how would you synthesize the following substances? Assume that you can separate *ortho* and *para* isomers if necessary.
(a) *m*-bromobenzenesulfonic acid
(b) *o*-chlorobenzenesulfonic acid
(c) *p*-chlorotoluene

20.61 Starting from any aromatic hydrocarbon of your choice, how would you synthesize the following substances? *Ortho* and *para* isomers can be separated if necessary.
(a) *o*-nitrobenzoic acid   (b) *p*-*tert*-butylbenzoic acid

Section 20.11: Polycyclic Aromatic Hydrocarbons

20.62 Explain by drawing resonance structures of the intermediate carbocations why naphthalene undergoes electrophilic aromatic substitution at C1 rather than at C2.

$$+ \text{Br}_2 \xrightarrow{\text{FeBr}_3} + \text{HBr}$$

SUMMARY AND CONCEPTUAL QUESTIONS

20.63 Would you expect the trimethylammonium group to be an activating or deactivating substituent? Explain.

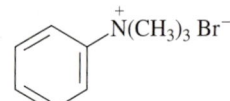

Phenyltrimethylammonium bromide

20.64 The orientation of electrophilic aromatic substitution on a disubstituted benzene ring is usually controlled by whichever of the two groups already on the ring is the more powerful activator. Name and draw the structure(s)

of the major product(s) of electrophilic chlorination of these substances:
(a) *m*-nitrophenol
(b) *o*-methylphenol
(c) *p*-chloronitrobenzene

20.65 Predict the major product(s) you would expect to obtain from sulfonation of the following substances:
(a) *o*-chlorotoluene
(b) *m*-bromophenol
(c) *p*-nitrotoluene

20.66 Rank the following aromatic compounds in the expected order of their reactivity toward Friedel-Crafts acylation. Which compounds are unreactive?
(a) bromobenzene
(b) toluene
(c) anisole ($C_6H_5OCH_3$)
(d) nitrobenzene
(e) *p*-bromotoluene

20.67 In some cases, the Friedel-Crafts acylation reaction can occur intramolecularly—that is, within the same molecule. Predict the product of the following reaction:

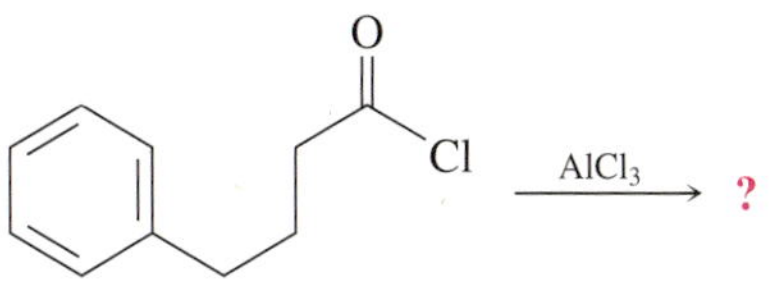

20.68 Explain why bromination of biphenyl occurs at the *ortho* and *para* positions rather than at the *meta* positions. Use the resonance model to show structures of the carbocation intermediates in your explanation.

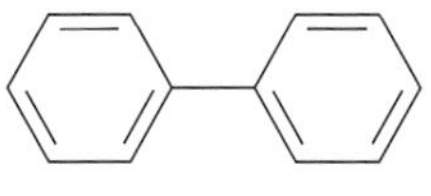

Biphenyl

20.69 In light of your answer to Question 20.68, at what position and on which ring would you expect nitration of 4-bromobiphenyl to occur?

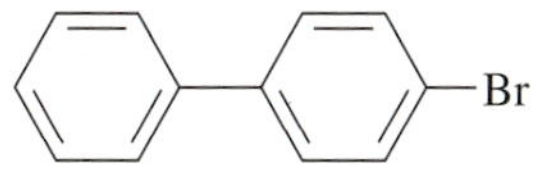

4-Bromobiphenyl

20.70 We've seen in Section 19.8 that allylic carbocations are stable as a result of the delocalization of positive charge over the molecule. Draw resonance structures to account for a similar stabilization of benzylic carbocations.

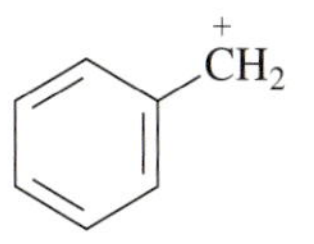

A benzylic carbocation

20.71 Addition of HBr to 1-phenylpropene yields (1-bromopropyl)benzene as the exclusive product. Propose a mechanism for the reaction, and explain why none of the other regioisomers is produced.

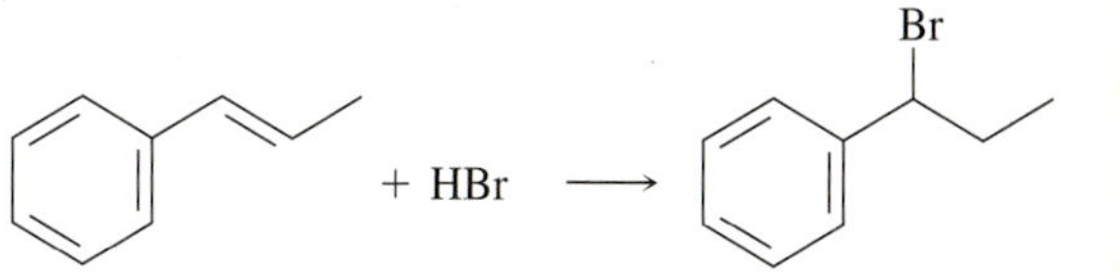

20.72 With reference to Question 20.71, why does HBr only add across the external $C=C$, and not across one of the double bonds in the benzene ring?

20.73 The following syntheses have flaws in them. What is wrong with each?

(a)

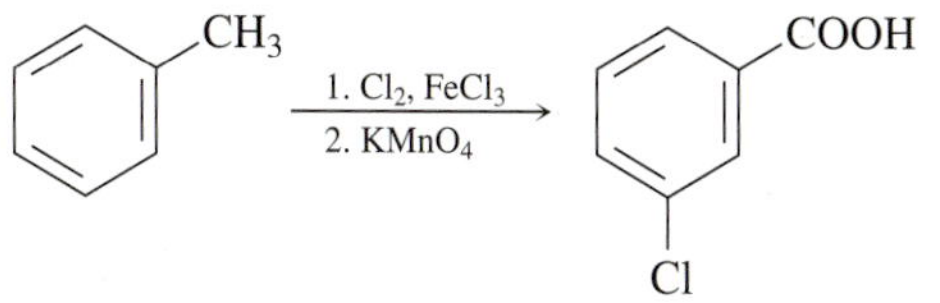

(b)

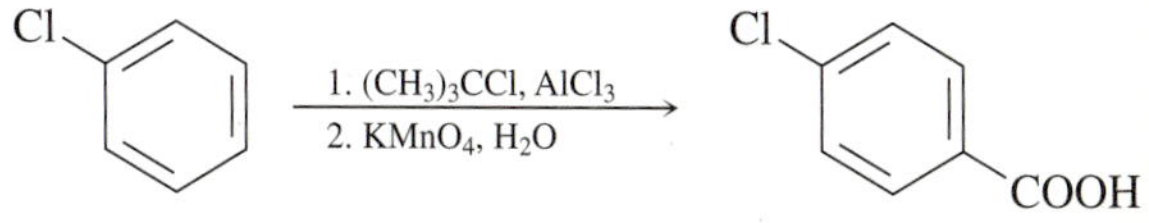

20.74 Carbocation intermediates thought to be formed in the reaction mechanism for an alkene with a strong acid catalyst can react with aromatic rings in a Friedel-Crafts reaction. Propose a mechanism to account for the industrial synthesis of the food preservative BHT from *p*-cresol and 2-methylpropene:

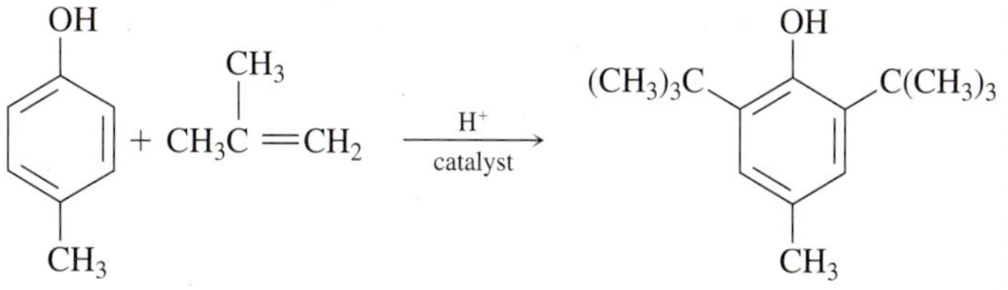

p-**Cresol**   **BHT**

20.75 You know the proposed mechanism of HBr addition to alkenes, and you know the effects of various substituent groups on aromatic substitution. Use this knowledge to predict which of the following two alkenes reacts faster with HBr. Explain your answer by drawing resonance structures of the relevant carbocation intermediates to show electron delocalization.

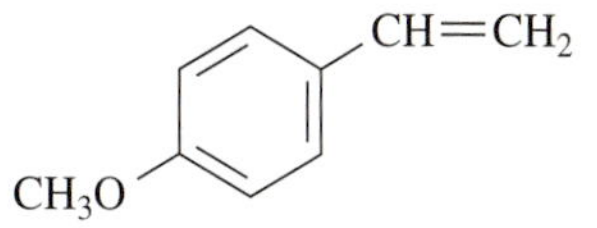

and

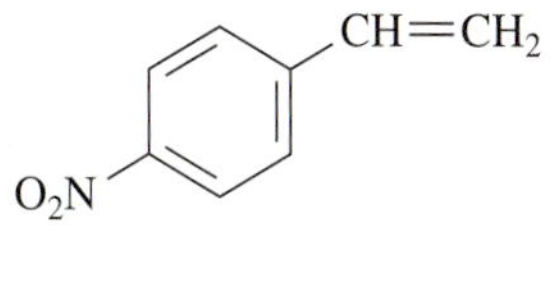

20.76 Acetaminophen is marketed under the name of Tylenol (among others) and is prepared in only three steps from phenol.

(a) What three functional groups are present in acetaminophen?

(b) What is the molecular formula of acetaminophen?

(c) Nitration of phenol can produce two products. What is the desired product A? Draw resonance structures to show why the delocalization of charge favours nitration occurring at the desired position.

(d) Of the two potential choices, *ortho* or *meta*, formation of which other isomer is supported by the resonance structures you drew?

(e) After reduction of the nitro group, the last step to install the acetyl (—COCH₃) group is accomplished with a reagent containing what functional group?

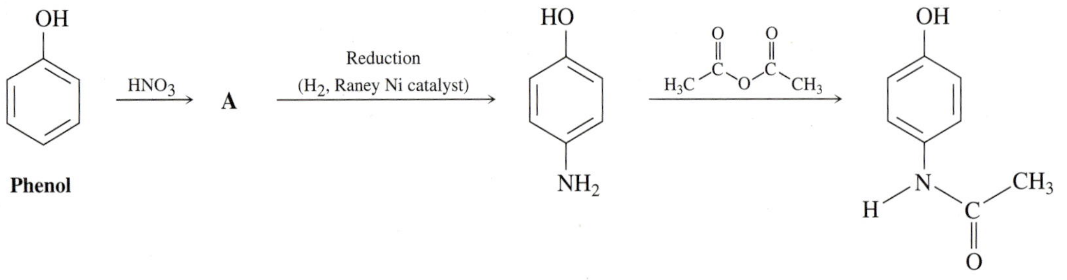

Phenol

**Acetaminophen
(Tylenol)**

20.77 Ibuprofen is categorized as a non-steroidal anti-inflammatory drug (NSAID). Draw the mechanism for the Friedel-Crafts acylation of isobutylbenzene to form intermediate A. Use resonance structures to show why delocalization of charge favours the formation of the desired product over the undesired isomer.

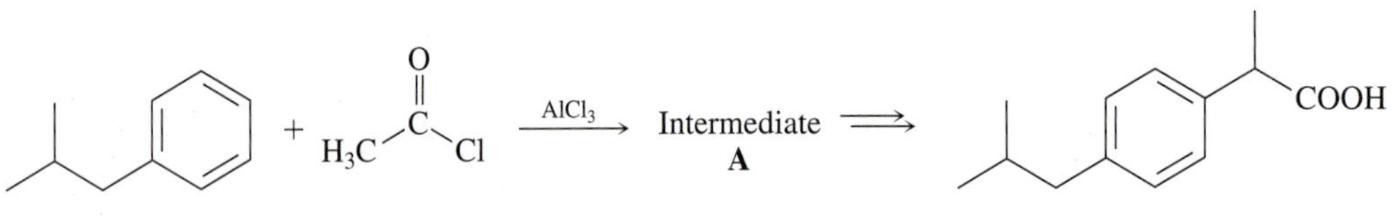

Isobutylbenzene

Ibuprofen

20.78 Recently, controversy has surrounded NSAIDs with more complex structures that target enzymes called cyclooxygenases. Several of these so-called COX-2 inhibitors have been withdrawn from the market. Identify the aromatic rings in the structures *o*-Celebrex and Vioxx.

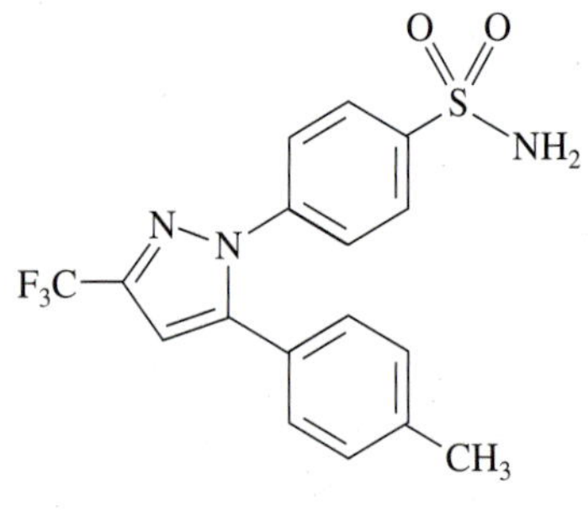

**Celecoxib
(Celebrex)**

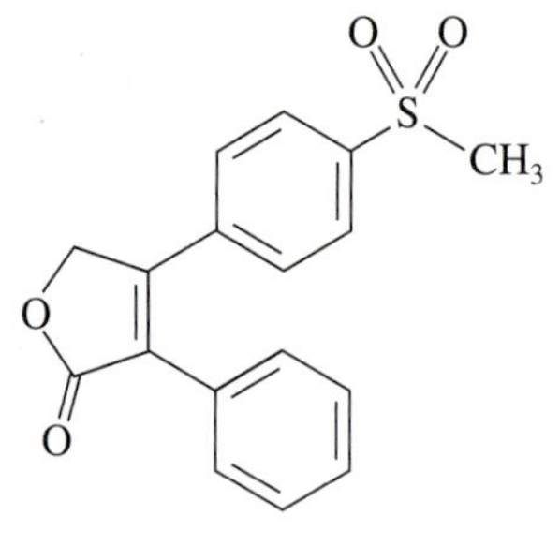

**Rofecoxib
(Vioxx)**

Alkyl Halides

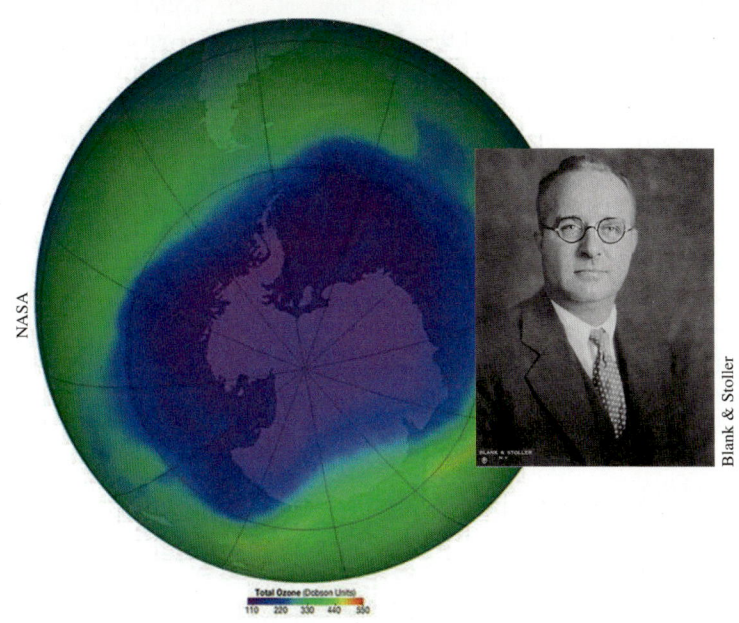

NASA

Blank & Stoller

21.1 Case Study: Chlorofluorocarbons and Stratospheric Ozone Depletion

In the early 20th century, toxic substances such as sulfur dioxide, chloromethane, and ammonia were used as refrigerants. Thomas Midgley, Jr. (photo above) set out to change that in 1928. He recognized that changing the number and type of substituents on a carbon atom in a molecule would significantly alter the physical properties of a substance. So Midgley carefully surveyed where the atoms making up molecules of known refrigerants were positioned on the periodic table. Selecting fluorine as a likely candidate, he synthesized in just three days the first **chlorofluorocarbon** (**CFC**)—"miracle compounds" containing only **c**arbon, **c**hlorine, and **f**luorine atoms [<<Section 4.3]. His tests showed these new CFCs to be odourless, colourless, non-flammable, non-corrosive, and non-toxic. And they had just the right thermodynamic properties (boiling points, specific heats, and heats of vaporization) to make them ideal refrigerants and propellants. Midgley dramatically demonstrated the unre-active, non-toxic properties of CFCs to an American Chemical Society meeting in 1930, where he introduced the new gases by inhaling them and then blowing out a candle. The first patent for CFCs was awarded to the Frigidaire Corporation in 1928, and within a few years, the compressors of millions of refrigerators were charged with CFCs.

The uses of CFCs grew dramatically, not only for air conditioning and refrigeration equipment, but also as inhalers for asthma sufferers, propellants for aerosol cans, and foaming agents in the production of expanded plastic foams. By 1987, over 1 million tonnes of CFCs were produced annually in the world.

Unfortunately, as we have often learned about other "miracle compounds," the very properties that made CFCs so useful led to completely unexpected effects on our environment—in this case, severe damage to the ozone layer in our stratosphere (photo above). Midgley and his collaborators didn't foresee that CFCs would undergo almost no reactions when released into our troposphere (lower atmosphere), and the only way they are removed is by their slow transport up to the stratosphere where they undergo photochemical reactions initiated by interaction with ultraviolet light from the sun. Three chemists were awarded the 1995 Nobel Prize in Chemistry for doing the detective work needed to find the connection between trace levels of CFCs in our atmosphere and the depletion of stratospheric ozone that shields life on the earth from the sun's ultraviolet light. U.S. chemists Mario Molina and F. Sherwood Roland and Dutch chemist Paul Crutzen shared this Nobel prize, the first awarded for research into the effects of human-made substances on our environment.

A low steady-state concentration of ozone (O_3) exists in our stratosphere, as every day it is formed from and broken down naturally by sunlight in a photochemical reaction to molecular oxygen (O_2). In the process of this natural cycle, both molecular oxygen and ozone absorb certain wavelengths of the sun's ultraviolet light, converting ultraviolet light to thermal energy, which heats the stratosphere. But CFCs have provided new catalytic cycles for depleting stratospheric ozone. Several lines of experimental evidence show one reasonable mechanism for this process. A CFC molecule is thought to absorb a photon of ultraviolet light of a particular energy in the stratosphere to produce a chlorine atom radical, which then initiates a radical chain reaction that destroys ozone, as shown in Figure 21.1. This process only takes place in the stratosphere, because ultraviolet photons of these energies are not available at lower altitudes in the troposphere. Since chlorine atoms serve as catalysts, thousands of ozone molecules can be destroyed by a single chlorine atom before it is removed from the stratosphere.

Molecular Modelling

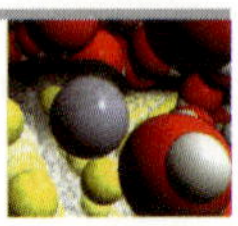

e21.1 Compare the effect of UV with other forms of radiation on a CF_2Cl_2 molecule.

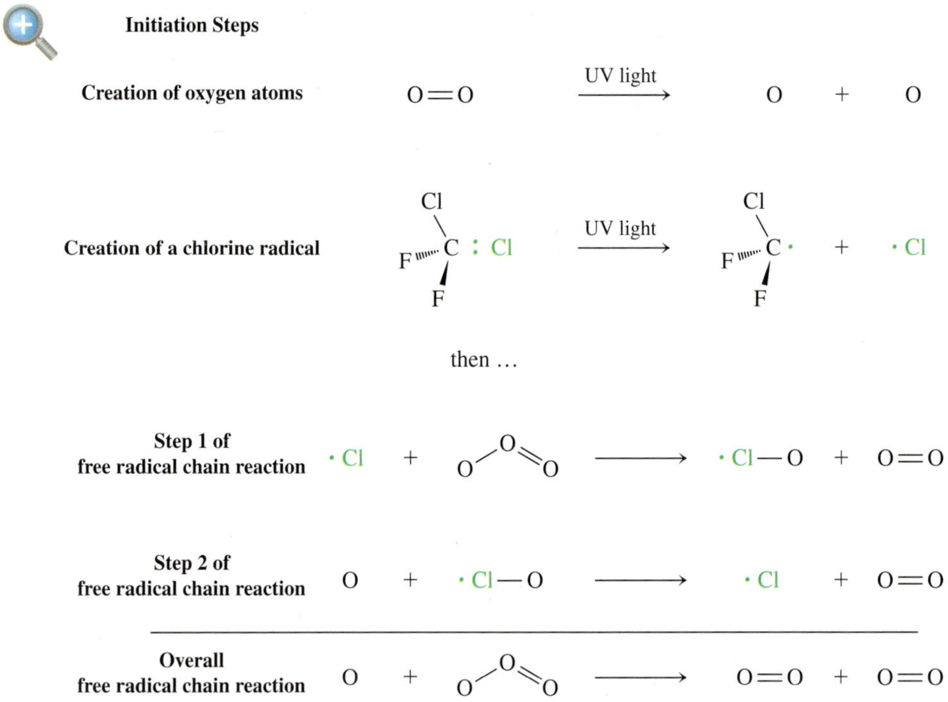

FIGURE 21.1 Simplified mechanism for the radical chain reaction thought to be responsible for the destruction of stratospheric ozone by chlorine atoms from CFCs.

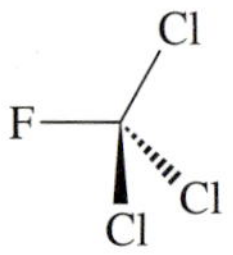

CFC-11 (trichlorofluoromethane)

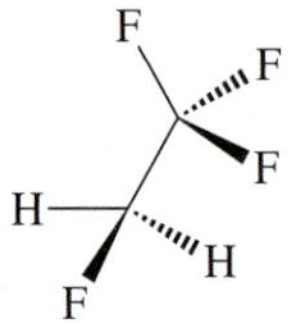

HFC-134a (1,1,1,2-tetrafluoroethane)

However, replacing one or more of the halogen atoms on CFC or fluorocarbon molecules with a hydrogen atom makes these compounds much more reactive in the troposphere. Trace amounts of reactive hydroxyl radicals (•OH) are able to remove hydrogen atoms from atmospheric molecules with C—H bonds, leading to other chemical reactions that oxidize and remove these partially halogenated organic compounds from the troposphere before they can travel up to the stratosphere. This reactivity has led to a short-term strategy of replacing CFCs as refrigerants with compounds such as **HCFCs** (containing **h**ydrogen, **c**hlorine, **f**luorine and **c**arbon) and **HFCs** (containing **h**ydrogen, **f**luorine, and **c**arbon). The average atmospheric lifetime of HCFCs and HFCs are much shorter than for CFCs. CFC-12 (CCl_2F_2), for example, has an atmospheric lifetime of 140 years. Its replacement compound, HFC-134a lasts on average for only 12 years when released into the atmosphere.

CFCs have ideal properties for use as refrigerants. But these gases are unreactive when released into the troposphere. The only way they break down in the atmosphere is by slow transport up into the stratosphere where they react with ultraviolet light, releasing chlorine atoms that act as catalysts to destroy ozone.

21.2 Alkyl Halides—Synthetic and Natural

Less well-known than CFC refrigerants are the many other **alkyl halides** synthesized by chemists for a vast array of uses in modern industrial processes. The substitution of one or more hydrogen atoms on an alkane molecule or aromatic ring with F, Cl, Br, or I atoms causes remarkable changes in a substance's physical properties, chemical reactivity, interaction with electromagnetic radiation, and toxicity.

Trichloroethylene
(a solvent)

Halothane
(an inhaled anesthetic)

Dichlorodifluoromethane
(a refrigerant)

Bromomethane
(a fumigant)

The number and type of halogen substituents determines these properties. Alkanes such as methane and propane burn (react with oxygen) very readily, making them useful as fuels. By contrast, completely halogenated carbon compounds such as carbon tetrachloride (tetrachloromethane, CCl_4), are non-flammable and quite unreactive. These and other properties of carbon tetrachloride led to its use as a dry cleaning solvent, refrigerant, and fire extinguisher. As we have seen in Section 4.3, the unreactivity of completely halogenated alkanes such as chlorofluorocarbons (CFCs) in the troposphere gives them a long atmospheric lifetime. When this property is coupled with their strong absorbance of infrared radiation, they become potent greenhouse gases. And those same CFCs react with ultraviolet light in the stratosphere, leading to the catalytic destruction of ozone, described in Section 21.1.

Some synthetic, completely fluorinated hydrocarbons are so unreactive that they are used to transport oxygen and carbon dioxide through the human body as artificial blood substitutes. And they have other uses. Tetrafluoroethene (C_2F_4) is the monomer used to make the very slippery and unreactive polymer polytetrafluoroethene, known by the tradename Teflon®. Teflon® is used in cooking products, and when formulated into porous membranes, gives the breathable, yet water-repelling properties desirable in fabrics such as Gore-Tex®.

Partially halogenated alkanes such as trichloromethane (chloroform, $CHCl_3$) and halothane ($C_2HBrClF_3$) have been widely used as inhaled anesthetic agents, replaced now with partially halogenated ethers. Dichloromethane (CH_2Cl_2) and most other simple alkyl chloride compounds are fat soluble (lipophilic) and water insoluble (hydrophobic).

These properties, along with a low boiling point, make dichloromethane suitable as a solvent for paint strippers and for extracting substances such as caffeine from coffee. However, due to increasing evidence of its toxicity and potential to cause cancer in mice and other animals, dichloromethane is now often replaced as an extraction solvent with non-chlorinated alternatives. Other examples of partially chlorinated organic compounds are solvents such as trichloroethylene (Cl_2C=$CHCl$), used in the modern electronics industry for cleaning semiconductor chips and other components.

Chemical industries in North America and Europe produced large quantities of higher molecular weight chlorinated organic compounds as insecticides and other pesticides for use in World War II and later for agricultural purposes. These alkyl or aromatic chlorides, such as DDT, show much greater toxicity to insects than to

Chloroform

Dichloromethane

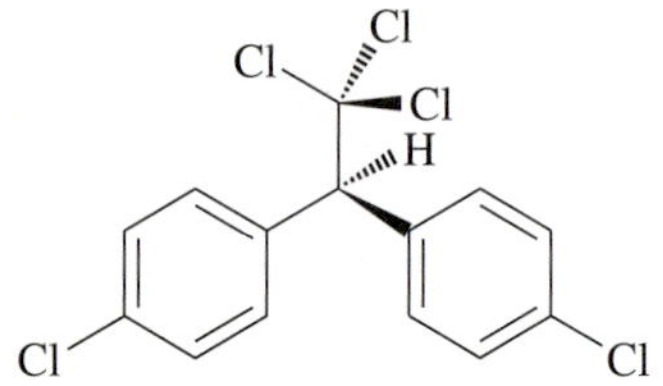

DDT

DDT is an abbreviation for the common name dichlorodiphenyltrichloroethane.

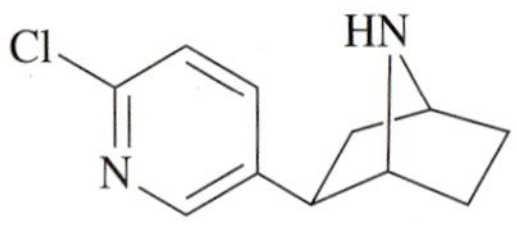

Epibatidine

A toxin (from the Greek word *toxikon*, meaning "arrow poison") is a poisonous substance produced by living cells or organisms. The term is often misused as referring to harmful synthetic substances.

humans, but they are highly lipophilic (fat soluble) and concentrate in fatty tissues of organisms.

Because of its fat solubility, DDT concentrations increase as they travel up the food chain from insects to birds to mammals. This phenomenon is called **biomagnification**. DDT and other chlorinated organic compounds oxidize and break down very slowly under environmental conditions. Despite these environmental concerns, DDT is still widely used to keep malaria-bearing mosquito populations in check in the developing world, as affordable alternatives are not available in many places. In some places where DDT has been banned, significant increases in malaria have been seen.

Not long ago, naturally occurring halogen-substituted organic compounds were thought to be relatively rare, except in salt-rich marine environments. Some of those that were known showed remarkable biological activity. Epibatidine (see structure), for example, is a compound isolated by medicinal chemists studying the toxins from the skin of frogs used traditionally in poison arrow tips.

Epibatidine is one of the most powerful analgesic (pain relieving) compounds known, but it has proven to be too toxic to use for human pain relief. Because the species of frog from which epibatidine was isolated is endangered, medicinal chemists have synthesized this compound in the laboratory and have modified its structure to produce new, less toxic compounds that still possess its powerful analgesic properties. Chemists have now identified several thousand alkyl halides produced by marine organisms, plants, fungi, lichens, bacteria, algae, insects, and even humans. About four million tonnes of chloromethane, for instance, are released annually from sources such as marine algae and phytoplankton, forest fires, and volcanoes. An intriguing recent finding is that Australian termites produce chloroform ($CHCl_3$) in their mounds. Termites worldwide may account for up to 15% of global chloroform emissions—which, as you can imagine, is a hard number to quantify!

Why do organisms produce organohalogen compounds, many of which are toxic? Toxic organohalogen compounds can be used for self-defence—either as feeding deterrents, as irritants to predators, or as natural pesticides. Marine sponges, coral, and sea hares, for example, release foul-tasting organohalogen compounds that deter fish, starfish, and other predators from eating them. More remarkably, even humans appear to produce halogenated compounds as part of their defence against infection. The human immune system contains a peroxidase enzyme capable of carrying out halogenation reactions on fungi and bacteria, thereby killing the pathogen.

Much remains to be learned—only a few hundred of the more than 500 000 known species of marine organisms have been examined—but it's already clear that alkyl halide compounds are an integral part of the world around us.

> Thousands of halogenated organic compounds have been identified in nature and synthesized in laboratories. Toxic naturally occurring alkyl halides are often synthesized by organisms as defence mechanisms against infections and predators, and their structures form the basis of many important antibiotics in medicine.

21.3 Naming Alkyl Halides

Alkyl halides are named in the same corresponding way as alkanes [<<Section 4.5], by considering the halogen as a substituent on the parent alkane chain. There are three steps:

Step 1   *Find the longest chain, and name it as the parent.* If a multiple bond is present, the parent chain must contain it.

Step 2   *Number the carbon atoms of the parent chain,* beginning at the end nearer the first substituent, regardless of whether it is alkyl or halo. Assign each

substituent a number according to its position on the chain. If there are substituents the same distance from both ends, begin numbering at the end nearer
the substituent with alphabetical priority.

CH₃   Br

CH₃CHCH₂CHCHCH₂CH₃
 1 2 3 4 5 6 7

CH₃

5-Bromo-2,4-dimethylheptane

Br   CH₃

CH₃CHCH₂CHCHCH₂CH₃
 1 2 3 4 5 6 7

CH₃

2-Bromo-4,5-dimethylheptane

Step 3   *Write the name.* List all substituents in alphabetical order and use one of the
prefixes di-, tri-, and so forth if more than one of the same substituent is
present.

Cl Cl

CH₃CHCHCHCH₂CH₃
 1 2 3 4 5 6

CH₃

2,3-Dichloro-4-methylhexane

In addition to their systematic names, many simple alkyl halides are also named by
identifying first the alkyl group and then the halogen. For example, CH₃I can be called
either iodomethane or the common name methyl iodide. Common names such as carbon
tetrachloride for CCl₄ are also frequently used.

CH₃I

Iodomethane
(or methyl iodide)

Cl

CH₃CHCH₃

2-Chloropropane
(or isopropyl chloride)

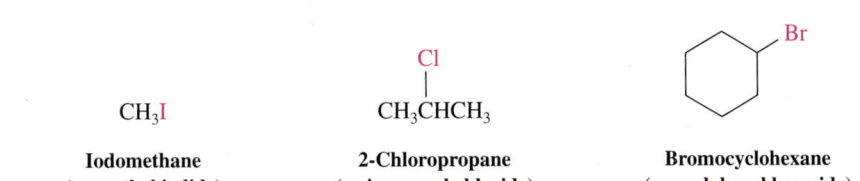

Bromocyclohexane
(or cyclohexyl bromide)

EXERCISE 21.1—NOMENCLATURE OF ALKYL HALIDES

Give the IUPAC names of the substances whose molecules are shown here:

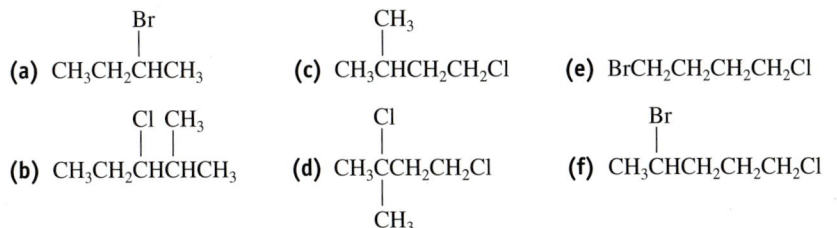

(a) CH₃CH₂CHCH₃ (Br)

(b) CH₃CH₂CHCHCH₃ (Cl CH₃)

(c) CH₃CHCH₂CH₂Cl (CH₃)

(d) CH₃CCH₂CH₂Cl (Cl, CH₃)

(e) BrCH₂CH₂CH₂CH₂Cl

(f) CH₃CHCH₂CH₂CH₂Cl (Br)

Interactive Exercise 21.2
Use the drawing tool to draw structural formulas for alkyl halides.

EXERCISE 21.3—NOMENCLATURE OF ALKYL HALIDES

Draw structures of molecules of the following substances:
(a) 2-chloro-3,3-dimethylhexane
(b) 3,3-dichloro-2-methylhexane
(c) 3-bromo-3-ethylpentane
(d) 2-bromo-5-chloro-3-methylhexane

21.4 Synthesis of Alkyl Halides

We often use the symbol X as a short hand designation for a halogen atom attached to a carbon chain. Sometimes X can also refer to other substituents—if so, this is indicated.

We've already seen several methods for making alkyl halides, including the addition reactions of HX and X_2 with alkenes [<<Sections 19.9–19.11].

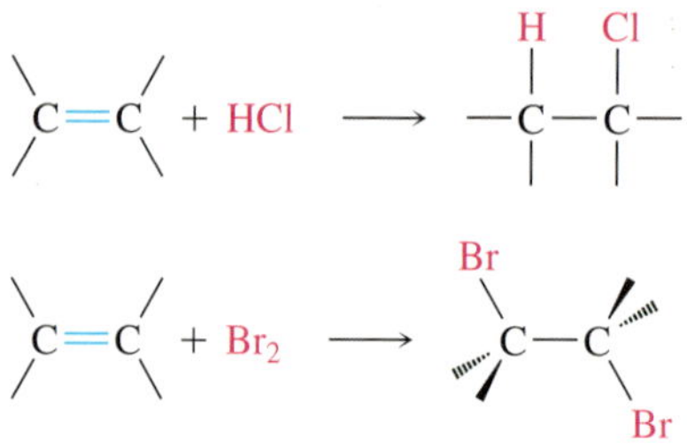

$$\backslash C=C / + HCl \longrightarrow -\overset{\overset{\displaystyle H}{|}}{C}-\overset{\overset{\displaystyle Cl}{|}}{C}-$$

$$\backslash C=C / + Br_2 \longrightarrow \overset{\displaystyle Br}{\underset{\displaystyle Br}{C-C}}$$

$$CH_4 + Cl_2 \xrightarrow{h\nu} CH_3Cl + HCl$$

Methane   **Chloromethane**

The most general method for making alkyl halides is to treat an alcohol with hydrogen halide (HX). 1-Methylcyclohexanol, for example, is converted into 1-chloro-1-methylcyclohexane by reaction with HCl:

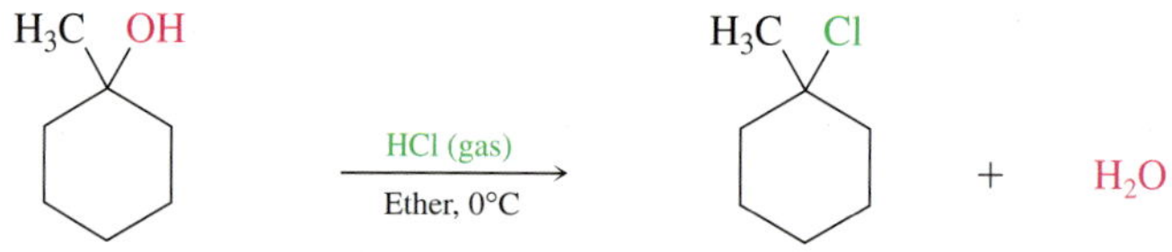

1-Methylcyclohexanol   **1-Chloro-1-methylcyclohexane**
                                          (90%)

For reasons that will be discussed in Section 21.7, the reaction works best with tertiary alcohols. Primary and secondary alcohols react much more slowly.

$$R-OH + HX \longrightarrow R-X + H_2O$$

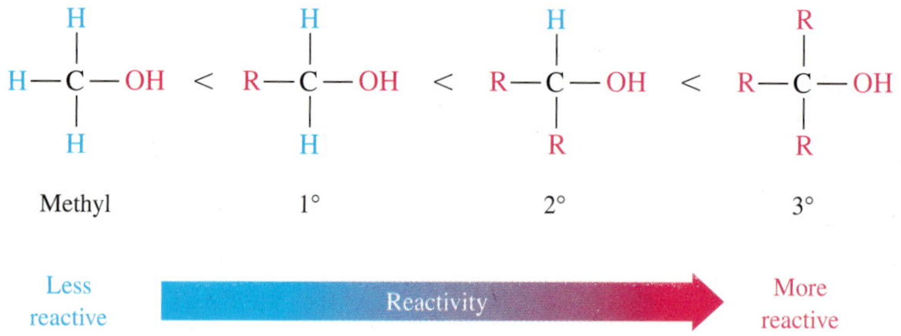

Methyl   1°   2°   3°

Less reactive   Reactivity   More reactive

Primary and secondary alcohols are best converted into alkyl halides by treatment with any of thionyl chloride ($SOCl_2$), phosphorous pentachloride (PCl_5, see Figure 21.2), or phosphorus tribromide (PBr_3). These reactions normally take place in high yield.

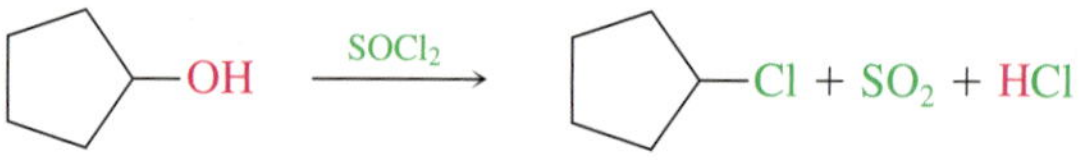

Cyclopentanol   **Chlorocyclopentane**

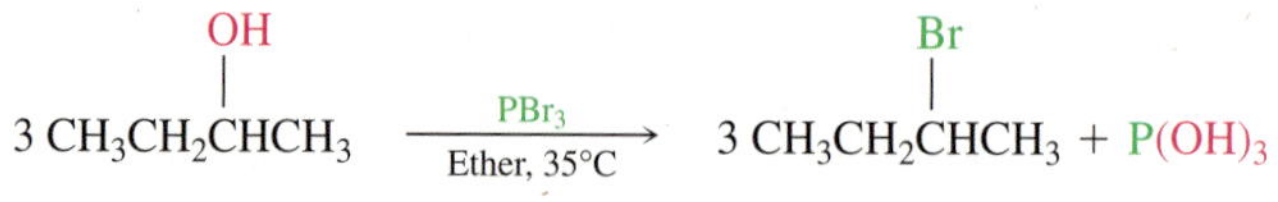

$$3\ CH_3CH_2\overset{\overset{\displaystyle OH}{|}}{C}HCH_3 \xrightarrow[\text{Ether, 35°C}]{PBr_3} 3\ CH_3CH_2\overset{\overset{\displaystyle Br}{|}}{C}HCH_3 + P(OH)_3$$

Butan-2-ol   **2-Bromobutane**
                                  (86%)

WORKED EXAMPLE 21.1—SYNTHESIS OF ALKYL HALIDES

Predict the product of the following reaction:

OH
|
[benzene ring]—CHCH$_3$ $\xrightarrow{\text{SOCl}_2}$

Strategy

This is one of many reactions in organic chemistry that convert one functional group into another. You may find it useful to use to keep track of these reactions that interconvert functional groups by making flash cards with the reactants on one side and the products on the reverse.

Solution

OH
|
[benzene ring]—CHCH$_3$ $\xrightarrow{\text{SOCl}_2}$ [benzene ring]—CHCH$_3$
|
Cl

EXERCISE 21.5—SYNTHESIS OF ALKYL HALIDES

How would you prepare the following alkyl halides from the appropriate alcohols?

(a) 2-Chloro-2-methylpropane

(b) 2-Bromo-4-methylpentane

CH$_3$
|
(c) BrCH$_2$CH$_2$CH$_2$CH$_2$CHCH$_3$

CH$_3$ Cl
| |
(d) CH$_3$CH$_2$CHCH$_2$CCH$_3$
|
CH$_3$

EXERCISE 21.6—SYNTHESIS OF ALKYL HALIDES

Predict the products of the following reactions:

OH   CH$_3$
| |
(a) CH$_3$CH$_2$CHCH$_2$CHCH$_3$ + PBr$_3$ $\longrightarrow$?

(b) [cyclohexane ring]—CH$_3$, OH + HCl $\longrightarrow$?

(c) H$_3$C, H$_3$C [cyclopentane ring] OH + SOCl$_2$ $\longrightarrow$?

Primary or secondary alkyl halides can be synthesized from alcohols by treatment with either thionyl chloride (SOCl$_2$) or phosphorus tribromide (PBr$_3$). Tertiary alkyl halides can be made from alcohols by reaction with HCl or HBr.

21.5 Nucleophilic Substitution Reactions

In 1896, the German chemist Paul Walden made the remarkable discovery that (+)- and (−)-malic acids could be interconverted by a series of simple substitution reactions. When Walden treated (−)-malic acid with PCl$_5$, he isolated (+)-chlorosuccinic acid. This, on reaction with wet Ag$_2$O, gave (+)-malic acid. Similarly, reaction of (+)-malic acid with PCl$_5$ gave (−)-chlorosuccinic acid, which was converted into (−)-malic acid when treated with wet Ag$_2$O. The full cycle of reactions reported by Walden is shown in Figure 21.2.

FIGURE 21.2 Walden cycle for converting (−)-malic acid into (+)-malic acid.

$$
\underset{\substack{\text{(−)-Malic acid} \\ [\alpha]_D = -2.3°}}{\text{HOCCH}_2\text{CHCOH}} \quad \xrightarrow[\text{Ether}]{\text{PCl}_5} \quad \underset{\text{(+)-Chlorosuccinic acid}}{\text{HOCCH}_2\text{CHCOH}}
$$

$$
\Big\uparrow \text{Ag}_2\text{O, H}_2\text{O} \qquad\qquad \Big\downarrow \text{Ag}_2\text{O, H}_2\text{O}
$$

$$
\underset{\text{(−)-Chlorosuccinic acid}}{\text{HOCCH}_2\text{CHCOH}} \quad \xleftarrow[\text{Ether}]{\text{PCl}_5} \quad \underset{\substack{\text{(+)-Malic acid} \\ [\alpha]_D = +2.3°}}{\text{HOCCH}_2\text{CHCOH}}
$$

Remember from Section 9.10 that you can't tell the configuration of a stereocentre of a molecule by looking at the sign of optical rotation of a substance.

At the time, the results were astonishing. Since (−)-malic acid was converted into (+)-malic acid, *some reactions in the cycle must have occurred with an inversion, or change, in the configuration of the stereocentre in the malic acid molecules.* But which ones, and how?

Today we refer to the transformations taking place in Walden's cycle as *nucleophilic substitution reactions* [<<Section 18.8] because each step involves the substitution of one nucleophile (chloride ion, Cl⁻, or hydroxide ion, OH⁻) by another. Nucleophilic substitution reactions are one of the most common and versatile reaction types in organic chemistry. After reviewing the mechanisms by which nucleophilic substitution reactions are thought to occur, we return to the Walden cycle in Section 21.6 to make sense of his observations about the interconversion of malic acid.

In nucleophilic substitution reactions a nucleophile (symbolized Nu: or Nu:⁻) reacts with a substrate R—X and substitutes for a leaving group X:⁻ to yield the product R–Nu. If the nucleophile is neutral (Nu:), then the product is positively charged to maintain charge conservation; if the nucleophile is negatively charged (Nu:⁻), the product is neutral. How can you identify a good nucleophile? It is a species that can form a bond by donating an electron pair to an electron-deficient (electrophilic) site on the substrate molecule, which is usually a carbon atom with a partial positive charge. You might predict that species such as hydroxyl ions, HO⁻, would be good nucleophiles (species with an affinity for electrophilic carbon atoms) because they are strong bases (species with an affinity for protons), and this is observed. But some weak bases, such as halide ions, are also good nucleophiles. A good leaving group must be able to readily accommodate an electron pair—that is, a weak base.

Think about It

e21.2 Watch an animation portraying the changes in bonding in a successful nucleophilic substitution.

Molecular Modelling (Odyssey)

e21.3 Examine an electrostatic potential surface map of a chloromethane molecule.

Neutral Nu: $\qquad$ Nu: + R—X $\longrightarrow$ Nu⁺—R + X:⁻

Negatively charged Nu:⁻ $\qquad$ Nu:⁻ + R—X $\longrightarrow$ Nu—R + X:⁻

Alkyl halides make excellent substrates for nucleophilic substitution reactions, as their C—X bonds are very polar, with an electrophilic site on the carbon atom bonded to the halogen atom. Note the colour of the carbon and halogen atoms in the electrostatic potential map of an alkyl halide molecule below.

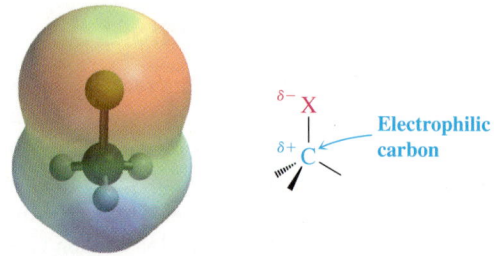

Substances whose molecules have many different functional groups can be prepared using nucleophilic substitution reactions. We've already seen examples in previous chapters. The reactions of acetylide anions with methyl or primary alkyl halides [<<Section 19.15], for instance, are nucleophilic substitution reactions in which acetylide ion nucleophiles replace halide ion leaving groups. Table 21.1 lists other examples.

$$R—C≡C:^- + CH_3Br \longrightarrow R—C≡C—CH_3 + Br:^-$$

An acetylide anion
(nucleophile)

TABLE 21.1 Some Common Nucleophiles and the Product of Nucleophilic Substitution Reactions with Bromomethane

Common Nucleophiles		Reaction Products with CH₃Br	
Formula	Name	Formula	Name
$H:^-$	Hydride	CH_4	Methane
$CH_3\ddot{S}:^-$	Methanethiolate	CH_3SCH_3	Dimethyl sulfide
$H\ddot{S}:^-$	Hydrosulfide	$HSCH_3$	Methanthiol
$N≡C:^-$	Cyanide	$N≡CCH_3$	Acetonitrile
$:\ddot{I}:^-$	Iodide	ICH_3	Iodomethane
$H\ddot{O}:^-$	Hydroxide	$HOCH_3$	Methanol
$CH_3\ddot{O}:^-$	Methoxide	CH_3OCH_3	Dimethyl ether
$^-:N=\overset{+}{N}=\ddot{N}:^-$	Azide	N_3CH_3	Azidomethane
$:\ddot{Cl}:^-$	Chloride	$ClCH_3$	Chloromethane
$CH_3CO_2:^-$	Acetate	$CH_3CO_2CH_3$	Methyl acetate
$H_3N:$	Ammonia	$H_3\overset{+}{N}CH_3Br^-$	Methylammonium bromide
$(CH_3)_3N:$	Trimethylamine	$(CH_3)\overset{+}{N}CH_3Br^-$	Tetramethylammonium bromide

WORKED EXAMPLE 21.2—NUCLEOPHILIC SUBSTITUTION REACTIONS

What is the substitution product from reaction of 1-chloropropane with NaOH?

Strategy

Write representations of the two reactant species, and identify the *nucleophile* (an electron pair donor, in this instance, OH⁻), the *leaving group* (a weak base, in this instance, Cl⁻), and the *substrate* (in particular, the carbon atom connected to the leaving group). Then replace the —Cl group attached to the carbon atom on a propane molecule with —OH and write the complete reaction equation.

Solution

$$CH_3CH_2CH_2Cl + Na^+ \, ^-OH \longrightarrow CH_3CH_2CH_2OH + Na^+ \, ^-Cl$$

1-Chloropropane   **Propan-1-ol**

WORKED EXAMPLE 21.3—NUCLEOPHILIC SUBSTITUTION REACTIONS

How would you prepare propane-1-thiol ($CH_3CH_2CH_2SH$) from an alkyl halide using a nucleophilic substitution reaction?

Strategy

Identify the group in the product molecule that is introduced by nucleophilic substitution. In this case, the product contains an —SH group, so it might be prepared by reaction of SH⁻ (hydrosulfide ion) with an alkyl halide such as 1-bromopropane.

Solution

$$CH_3CH_2CH_2Br + Na^+ {}^-SH \longrightarrow CH_3CH_2CH_2SH + Na^+ {}^-Br$$

1-Bromopropane   **Propane-1-thiol**

EXERCISE 21.8—NUCLEOPHILIC SUBSTITUTION REACTIONS

What substitution products would you expect to obtain from the following reactions?

(a) $CH_3CH_2CHCH_3$ (with Br substituent) $+ \; LiI \; \longrightarrow \; ?$

(b) CH_3CHCH_2Cl (with CH_3 substituent) $+ \; HS^- \; \longrightarrow \; ?$

(c) (phenyl)$-CH_2Br + NaCN \; \longrightarrow \; ?$

EXERCISE 21.9—NUCLEOPHILIC SUBSTITUTION REACTIONS

How might you make the following substances by using nucleophilic substitution reactions?
(a) $CH_3CH_2CH_2CH_2OH$
(b) $(CH_3)_2CHCH_2CH_2N_3$

Following Walden's observations on (+)-and (−)-malic acid, a series of investigations was carried out during the 1920s and 1930s to clarify the mechanism of nucleophilic substitution reactions and to find out how inversions of configuration occur. Experimental evidence of how the rates of substitution reactions depend on concentration and what happens to the stereochemistry of a chiral substrate have given us insight into how these reactions might take place at the molecular level.

This experimental evidence has led to suggestions that most nucleophilic reactions occur by one of two important pathways, named the S_N1 *mechanism* and the S_N2 *mechanism*. In both cases, the S_N part of the name stands for *substitution, nucleophilic*. The *1* and *2* designations refer to unimolecular and bimolecular rate-determining steps in the mechanisms. Building on the introduction given in Section 18.8 in the context of kinetics data, we will examine more closely in the next two sections the factors that make S_N1 mechanisms more probable in some cases and S_N2 more probable in others.

> In nucleophilic substitution reactions, a nucleophile (symbolized Nu: or Nu:$^-$) reacts with a substrate (symbolized R—X) and substitutes for a leaving group X:$^-$ to yield the product R—Nu. A good nucleophile needs to be able to make a bond by donating an electron pair to an electron-deficient (electrophilic) site on the substrate molecule, and the leaving group must be able to readily accommodate an electron pair (a weak base). Experimental evidence suggests two mechanisms for most nucleophilic substitution reactions, the S_N1 and the S_N2 mechanisms.

21.6 The S$_N$2 Mechanism for Substitution Reactions

Think about It

e21.4 Watch an animation portraying the orbitals involved in this reaction.

As introduced in Section 18.8, a well-studied nucleophilic substitution reaction takes place in solution between hydroxide ion (OH^-) and methyl bromide (CH_3Br). Experimental rate law data shows that it is first order in OH^- and first order in CH_3Br, and therefore second order overall.

Reaction rate = Rate of disappearance of $CH_3Br = k \, [OH^-]^1[CH_3Br]^1$

This experimental kinetics evidence is consistent with a molecular-level explanation that each event leading to reaction involves two reacting species (OH^- ions and CH_3Br molecules) colliding with each other in the rate-determining step of the reaction. We would expect that if the concentration of either or both reactant species is increased, there would be more effective collisions between them, leading to an increase in the rate of production of the products. This is referred to as a *bimolecular step* in the pathway.

$$HO:^- \ + \ CH_3\text{—}Br: \ \longrightarrow \ HO\text{—}CH_3 \ + \ :Br:^-$$

A second important experimental observation mentioned briefly in Section 18.8 comes from monitoring stereochemical changes between reactants and products in cases where the substrate is chiral. The reaction of OH^- with (*S*)-2-bromobutane gives an alcohol product, (*R*)-butan-2-ol, whose molecules have the opposite configuration of 2-bromobutane.

(*S*)-2-Bromobutane   (*R*)-2-Butanol

This stereochemical evidence suggests that an accurate molecular-level model for this reaction pictures a nucleophile forming a new bond on the opposite side of the substrate molecule from which the leaving group departed, giving *inversion of configuration* of the substrate. In this case, since the relative priorities of a —Br group and an —OH group are the same, inversion of configuration converts the (*S*)-alkyl halide into an (*R*)-alcohol.

In 1937, E.D. Hughes and C. Ingold explained both the kinetics and stereochemistry of similar reactions of substrates and nucleophiles by proposing a pathway called the *S_N2 reaction mechanism* (substitution, nucleophilic, bimolecular). While many collisions take place at any moment in solution between nucleophile and substrate molecules, only certain ones will have sufficient energy and the right orientation of the colliding molecules to lead to product formation. The S_N2 mechanism is proposed as a single-step process without intermediates when a productive collision takes place between a nucleophile and a substrate molecule from a direction 180° away from the leaving group. As the nucleophile comes in on one side of the molecule, interaction of its electron pair with the substrate causes repulsion with the other bonds to the substrate, causing the leaving group $X:^-$ to depart from the other side of the molecule, taking with it the electron pair from the C—X bond. In the transition state for the reaction, the new Nu—C bond is partially forming at the same time the old C—X bond is partially breaking, and the negative charge is shared by both the incoming nucleophile and the outgoing leaving group. The accepted mechanism is shown in Figure 18.18 for the reaction of OH^- with (*S*)-2-bromobutane.

Molecular Modelling (Odyssey)

e21.5 Examine a simulation of the inversion of stereochemistry at the carbon atom.

EXERCISE 21.10—KINETICS OF S_N2 REACTION

What effects would the following changes have on the rate of the S_N2 reaction between CH_3I and sodium acetate?
(a) The CH_3I concentration is tripled.
(b) Both CH_3I and $NaCH_3COO$ concentrations are doubled.

Rates of Reactions and the S_N2 Mechanism

The proposal that two molecules (OH^- and (*S*)-2-bromobutane in this case) collide in the rate-determining step of this bimolecular S_N2 mechanism in Figure 21.3 is consistent with the experimental observation of a second-order rate law.

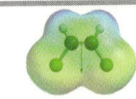
Stereochemistry of the S$_N$2 Mechanism

Look carefully at the stereochemistry of one of the productive collisions that lead to the S$_N$2 reaction mechanism shown in Figure 21.3.

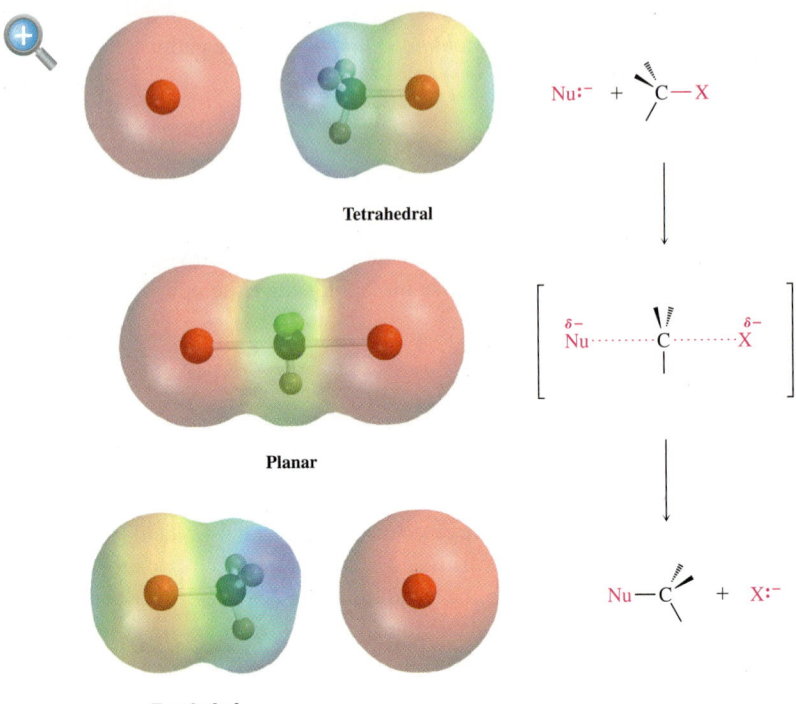

Tetrahedral

Planar

Tetrahedral

FIGURE 21.3 Stereochemistry of the S$_N$2 reaction mechanism.

As an incoming nucleophile collides with the substrate and electron redistribution occurs to cause the leaving group to depart from the opposite side, the configuration of the substrate molecule *inverts*. (S)-2-bromobutane gives (R)-butan-2-ol, for example, explained by an inversion of configuration that occurs through a planar transition state of the substrate molecules. This allows us to account for Walden's observations at the beginning of Section 21.5. While the reactions of chlorosuccinic acid with water and silver oxide occur with retention of configuration, the conversion of the —OH group in malic acid molecules to a —Cl group with PCl$_5$ is a good example of evidence for an S$_N$2 reaction mechanism that involves inversion of configuration (Figure 21.4). This reaction is the key to the inversion of configuration of molecules of (−)-malic acid to (+) malic acid.

FIGURE 21.4 Stereochemistry of reactions in the Walden cycle.

$$
\underset{\substack{\text{(−)-Malic acid} \\ [\alpha]_D = -2.3°}}{\underset{\overset{|}{\text{OH}}}{\text{HOCCH}_2\text{CHCOH}}}
\xrightarrow[\substack{\text{Ether} \\ \text{(inversion of} \\ \text{configuration)}}]{\text{PCl}_5}
\underset{\substack{\text{(+)-Chlorosuccinic acid}}}{\underset{\overset{|}{\text{Cl}}}{\text{HOCCH}_2\text{CHCOH}}}
$$

(retention of configuration) ↑ Ag$_2$O, H$_2$O

Ag$_2$O, H$_2$O ↓ (retention of configuration)

$$
\underset{\substack{\text{(−)-Chlorosuccinic acid}}}{\underset{\overset{|}{\text{Cl}}}{\text{HOCCH}_2\text{CHCOH}}}
\xleftarrow[\substack{\text{Ether} \\ \text{(inversion of} \\ \text{configuration)}}]{\text{PCl}_5}
\underset{\substack{\text{(+)-Malic acid} \\ [\alpha]_D = +2.3°}}{\underset{\overset{|}{\text{OH}}}{\text{HOCCH}_2\text{CHCOH}}}
$$

WORKED EXAMPLE 21.4—S$_N$2 REACTIONS

What product would you expect to obtain from the S$_N$2 reaction of (S)-2-iodooctane with a sodium cyanide solution (NaCN)?

Strategy

Identify the nucleophile (cyanide ion) and the leaving group (iodide ion). Then carry out the substitution, making sure to invert the configuration at the stereocentre. (S)-2-iodooctane reacts with CN$^-$ to yield (R)-2-methyloctanenitrile.

Solution

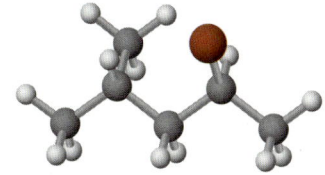

(S)-2-Iodooctane   (R)-2-Methyloctanenitrile

EXERCISE 21.11—S$_N$2 REACTIONS

What product would you expect to obtain from the S$_N$2 reaction of (S)-2-bromohexane with sodium acetate (NaCH$_3$COO)? Show the stereochemistry of both product and reactant molecules.

EXERCISE 21.12—S$_N$2 REACTIONS

Assign configuration to the following molecule, and draw the structure of molecules of the product that would result on nucleophilic substitution reaction with HS$^-$ (reddish-brown atom = Br):

Interactive Exercise 21.13

Draw structural formulas indicating the stereochemistry of S$_N$2 reactions.

SUBMIT

Steric Effects in the S$_N$2 Mechanism

The ease with which a nucleophile can interact with a substrate in an S$_N$2 reaction mechanism depends on spatial accessibility. Since the S$_N$2 mechanism involves passing through a transition state with five groups partially bonded to the substrate carbon atom, bulky substrates, in which the halogen-bearing carbon atom is difficult to access, would be expected to react much more slowly than those in which the carbon atom is more accessible (Figure 21.5). Fewer of the collisions between bulky reacting molecules are likely to be productive.

Molecular Modelling (Odyssey)

e21.8 Examine calculated three-dimensional electron distributions to understand the basis of steric hindrance.

FIGURE 21.5 Steric hindrance in the S$_N$2 reaction mechanism. These models show that as the substrate carbon atoms become more accessible, the rates of reaction with nucleophiles increase. Of these four substrates, bromomethane molecules have the most accessible carbon atom and therefore collisions with nucleophiles are more likely to lead to product formation.

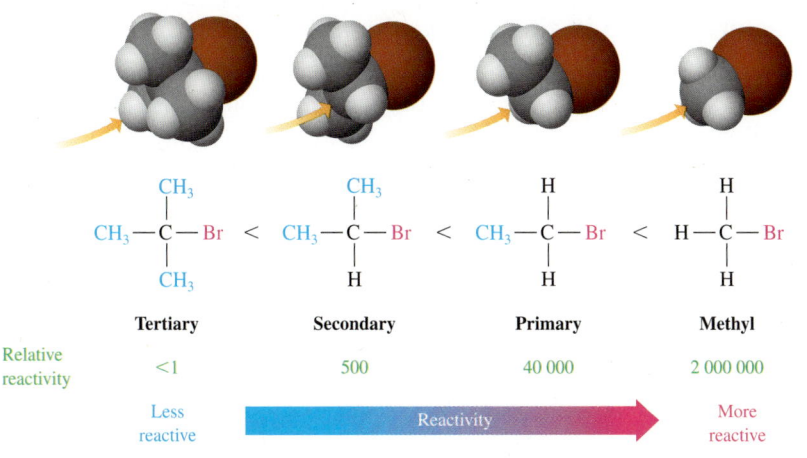

	Tertiary		Secondary		Primary		Methyl
Relative reactivity	<1		500		40 000		2 000 000

Less reactive ——— Reactivity ———> More reactive

Methyl halides (CH_3X) are the most reactive substrates in conditions favouring the S_N2 mechanism, followed by primary alkyl halides (RCH_2—X) such as ethyl and propyl. Alkyl branching next to the leaving group slows the reaction greatly for secondary halides (R_2CH—X), and further branching effectively halts the reaction for tertiary halides (R_3C—X).

Vinylic (R_2C=CRX) and aryl (Ar—X) halides are not shown on this reactivity list because they are completely unreactive toward S_N2 displacements. This lack of reactivity is due to steric hindrance. The electron-rich incoming nucleophile would experience a great deal of electron repulsion if it burrowed through part of the substrate molecule to carry out a displacement.

Vinylic halide

WORKED EXAMPLE 21.5—KINETICS OF S_N2 REACTIONS

Which would you expect to be faster: the S_N2 reaction of OH^- ion with 1-bromopentane or with 2-bromopentane?

Strategy

Decide which substrate is less hindered.

Solution

Since 1-bromopentane is a 1° halide and 2-bromopentane is a 2° halide, reaction with the less hindered 1-bromopentane is faster.

Primary
$CH_3CH_2CH_2CH_2CH_2Br$

1-Bromopentane

Br — Secondary
$CH_3CH_2CH_2CHCH_3$

2-Bromopentane

EXERCISE 21.14—KINETICS OF S_N2 REACTIONS

Which of the following S_N2 reactions would you expect to be faster?
(a) reaction of CN^- (cyanide ion) with $CH_3CH(Br)CH_3$ or with $CH_3CH_2CH_2Br$
(b) reaction of I^- with $(CH_3)_2CHCH_2Cl$ or with H_2C=CHCl

---

Molecular Modelling

e21.9 See how interaction of solvent with reactants, transition state, and leaving group affects the reaction rate.

The Leaving Group in the S_N2 Mechanism

Another variable that can affect substitution reactions is the identity of the leaving group displaced by the incoming nucleophile. Because the leaving group is expelled with a negative charge in most reactions that follow S_N2 mechanisms, the best leaving groups are those that give the most stable anions (conjugate bases of strong acids). A halide ion (I^-, Br^-, or Cl^-) is the most common leaving group, although others are also possible. Anions such as F^-, OH^-, OR^-, and NH_2^- are rarely leaving groups.

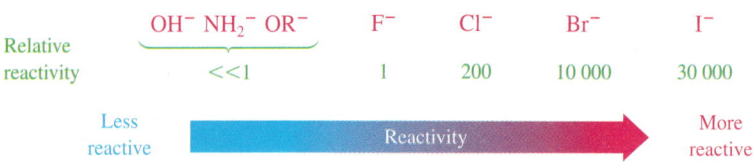

Rank the following compounds in order of their expected reactivity toward S$_N$2 reaction: CH$_3$I, CH$_3$F, and CH$_3$Br.

The S$_N$2 mechanism is proposed as a single-step process without intermediates when a productive collision takes place between a nucleophile and the substrate from a direction 180° away from the leaving group. As the nucleophile comes in on one side of the molecule, its electron pair interacts with the other electron pairs bonded to the substrate. This displaces the leaving group from the other side of the molecule, giving the substitution product.

21.7 The S$_N$1 Mechanism for Substitution Reactions

As introduced in Section 18.8, other nucleophilic substitution reactions have been found to give different experimental rate laws, stereochemical outcomes, and show a different order of substrate reactivity than in reactions best explained at the molecular level by S$_N$2 mechanisms. Let's look at some examples.

The substitution reaction between H$_2$O and (CH$_3$)$_3$CBr looks somewhat similar to the reaction between OH$^-$ and CH$_3$Br in Section 21.6.

The experimental rate of the first reaction above, however, does not depend on the concentration of the nucleophile, as it does in the second reaction, which follows an S$_N$2 mechanism. In the first reaction, the experimental rate depends only on the concentration of the substrate and is independent of the nucleophile concentration. It is first order in (CH$_3$)$_3$CBr alone and therefore first order overall.

$$\text{Reaction rate} = \text{rate of disappearance of } (CH_3)_3CBr = k\,[(CH_3)_3CBr]^1$$

This experimental evidence is inconsistent with an S$_N$2 molecular-level explanation involving a collision between H$_2$O and (CH$_3$)$_3$CBr in the rate-determining step of the reaction. A molecular-level explanation is needed that involves in each successful collisional event only one molecule in the rate-determining step, the (CH$_3$)$_3$CBr substrate. This is referred to as a *unimolecular step* in the pathway.

And as seen in Section 18.8, if we substitute a chiral substrate for the achiral (CH$_3$)$_3$CBr, we find a different stereochemical outcome than the inversion of configuration observed in reactions following S$_N$2 pathways. (*R*)-3-bromo-3-methylhexane reacts with CH$_3$OH to give a racemic mixture of (*R*)- and (*S*)-3-methoxy-3-methylhexanes.

In a hydroxylic or protic solvent, like CH$_3$OH, molecules have a proton bonded to an oxygen atom.

The observation of first-order rate laws and complete or partial racemization of configuration is often seen in substitution reactions involving *tertiary* substrate molecules, and

Think about It

e21.10 Watch an animation of the accepted mechanism for this reaction.

only under neutral or acidic conditions in a hydroxylic solvent such as water or alcohol. We saw in Section 21.4, for example, that alkyl halides can be prepared from alcohols by treatment with HCl or HBr. Tertiary alcohols react rapidly, but primary and secondary alcohols react more slowly.

$$R_3COH \gg R_2CHOH > R_2CH_2OH > CH_3OH$$

What's going on here? Clearly, nucleophilic substitution reactions are taking place in each of these examples. Yet the kinetics is first order instead of second order, the stereochemistry gives racemization instead of inversion, and the reactivity order of the substrate molecules $3° > 2° > 1°$ is opposite from the S_N2 order. These lines of evidence suggest a second reaction mechanism under these conditions, called the *S_N1 reaction mechanism* (substitution, nucleophilic, unimolecular).

Unlike what is postulated in an S_N2 mechanism, where the leaving group from a successfully reacting substrate molecule is displaced *at the same time* that the incoming nucleophile approaches, we have a mental model of the S_N1 reaction mechanism as occurring by loss of that leaving group *before* the incoming nucleophile approaches. Loss of the leaving group gives a carbocation intermediate, which then reacts with the nucleophile in a second step to yield the substitution product. This mechanism is illustrated in Figure 18.21 for the reaction between H_2O and $(CH_3)_3CBr$. In this case, after the nucleophile H_2O forms a bond to the carbocation, a proton is removed by the solvent in the third and final step to give the alcohol product.

Let's examine more closely how the S_N1 mechanism fits observations from kinetics and stereochemistry, and data on the order of reactivity of different substrates.

Rates of Reactions and the S$_N$1 Mechanism

The *unimolecular* S_N1 mechanism shows only one molecule involved in the rate-determining step of the reaction mechanism. To be consistent with this mechanism, we would expect the experimental rate law of the reaction to be first order, depending on the concentration of only one substance—the substrate. This is exactly what is observed.

EXERCISE 21.16—KINETICS OF S$_N$1 MECHANISM

What effect would the following changes have on the rate of the S_N1 reaction of tert-butyl alcohol with HBr?
(a) The HBr concentration is tripled.
(b) The HBr concentration is halved, and the *tert*-butyl alcohol concentration is doubled.

Stereochemistry of the S$_N$1 Reaction

If S_N1 reactions occur through carbocation intermediates, their stereochemistry should be different from that of S_N2 reactions. Because our model of bonding suggests carbocations are planar and sp^2 hybridized, they should be achiral. The positively charged carbon atoms can therefore react with a nucleophile equally well from either of the two faces, so that a racemate is formed—50% of them form one enantiomer, 50% the other. In an S_N1 reaction of molecules of a single enantiomer of a chiral substrate, the product should be formed as a racemate, as is seen in the reaction of (*R*)-3-bromo-3-methylhexane with CH_3OH (Figure 21.6).

Complete racemization, as shown in the simplified figure here, does not always occur. Under some conditions, the leaving group hasn't diffused completely away from the carbocation intermediate and partially blocks the entry of the nucleophile to form a new bond.

FIGURE 21.6

Stereochemistry of the S_N1 reaction mechanism. Racemization of chiral substrates is explained by the involvement of planar, achiral carbocation intermediates. Reaction on the opposite faces of the carbocations give enantiomeric products.

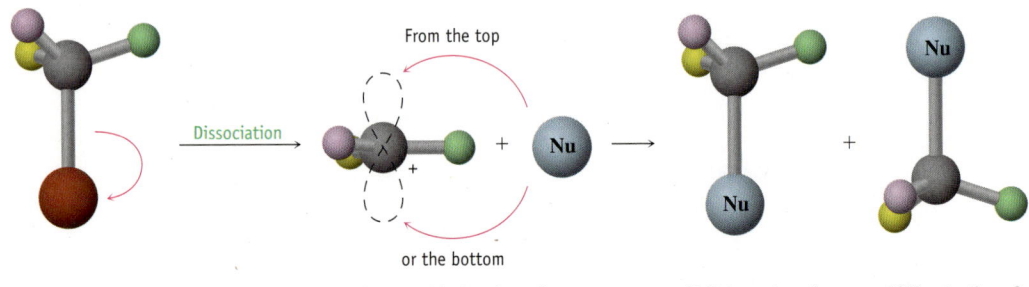

Chiral substrate   Planar, achiral carbocation   50% inversion of   50% retention of
                          intermediate   configuration   configuration

---

WORKED EXAMPLE 21.6—STEREOCHEMISTRY OF S$_N$1 REACTION

What stereochemistry would you expect for the S$_N$1 reaction of (R)-phenylbutan-1-ol with HCl?

Strategy

First draw the starting alcohol, showing its correct stereochemistry. Then protonate the —OH group to create a good leaving group, and after H$_2$O is lost to form a carbocation intermediate, Cl$^-$ reacts with the intermediate to give the racemate.

Solution

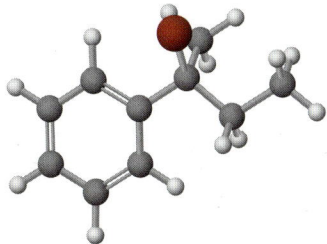

(R)-Phenylbutan-1-ol	**(R)-1-Phenyl-1-chlorobutane**	**(S)-1-Phenyl-1-chlorobutane**
	(50%, retention)	(50%, inversion)

EXERCISE 21.17—STEREOCHEMISTRY OF S$_N$1 REACTION

What product would you expect to obtain from the S$_N$1 reaction of (S)-3-methyloctan-3-ol with HBr? Show the stereochemistry of both starting material and product molecules.

EXERCISE 21.18—STEREOCHEMISTRY OF S$_N$1 REACTION

Assign configuration to the following substrate, and show the stereochemistry and identity of the product you would obtain by S$_N$1 reaction with H$_2$O (reddish-brown atom = Br).

Relative Reactivity of Substrates in the S$_N$1 Mechanism

This mechanism explains why tertiary substrates react so much more rapidly than primary or secondary ones: S$_N$1 reactions can occur only when stable carbocation intermediates are formed. The more stable the carbocation intermediates, the faster the S$_N$1 reaction, since reactions with more stable carbocations formed in the rate-determining step have lower activation energies. This also explains the reactivity order of alcohols with HBr (R$_3$COH $\gg$ R$_2$CHOH > R$_2$CH$_2$OH > CH$_3$OH)—it is the same as the stability order of carbocation intermediates (3° > 2° > 1°).

Leaving Groups in the S$_N$1 Mechanism

The best leaving groups in reactions that follow S$_N$1 mechanisms are those that give the most stable anions (conjugate bases of strong acids), just as in S$_N$2 reactions. If an S$_N$1 reaction is carried out under acidic conditions—as occurs when a tertiary alcohol reacts with HX to yield an alkyl halide—the first step of the mechanism will be protonation of the alcohol group by HX, so neutral water can be the leaving group in the rate-determining step. The observed S_N1 reactivity order of leaving groups is:

$$F^- << Cl^- = H_2O < Br^- < I^-$$

Worse leaving
group   Reactivity   Better leaving
                                                                             group

The S_N1 mechanism is believed to begin with loss of the leaving group from the substrate molecule to give a carbocation intermediate in the rate-determining step. The carbocation intermediate then reacts with the nucleophile in a second step to yield the substitution product.

21.8   Substitution Reactions in Living Organisms

All chemistry—whether carried out in flasks by chemists or in cells by living organisms—follows the same rules. Most biological and laboratory reactions can be explained by similar pathways. Among the most common biological substitution reactions is *methylation*, the transfer of a —CH$_3$ group from an electrophilic donor molecule to a nucleophile. A laboratory chemist might choose CH$_3$I for such a reaction, but living organisms use complex molecules such as *S*-adenosylmethionine as biological methyl-group donors. Since the sulfur atom in *S*-adenosylmethionine has a positive charge (a *sulfonium* ion, R$_3$S$^+$), it is an excellent leaving group for S_N2 displacements on the methyl carbon. An example of such a biological methylation takes place in the adrenal medulla during the biological synthesis of adrenaline from norepinephrine.

Can you identify the nucleophile, substrate, and leaving group in this biological S_N2 reaction?

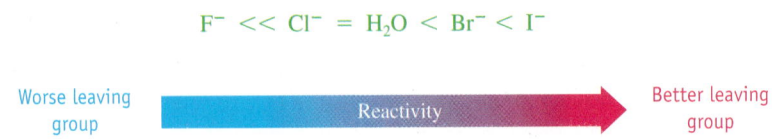

Norepinephrine

S-Adenosylmethionine

S_N2 reaction

Adrenaline

While *S*-adenosylmethionine is a much more complex molecule than CH$_3$I, both play exactly the same role—they transfer a methyl group by a process we visualize as an S_N2 reaction mechanism.

Another example of a biological S_N2 reaction is involved in the response of organisms to certain toxic substances. Many reactive S_N2 substrates with deceptively simple structures are quite toxic to living organisms. Bromomethane (methyl bromide), for example, has been widely used as a soil sterilizer and an effective fumigant to kill insects and fungi. Its toxicity derives from its ability to transfer a methyl group to a nucleophilic amino group

(—NH_2) or mercapto group (—SH) in enzymes, thus altering the enzyme's normal biological activity. However, despite its effectiveness as a fumigant, the production and use of bromomethane has been severely cut back, as it undergoes photochemical dissociation in the stratosphere to release bromine atoms that are even more damaging than chlorine atoms in the catalytic processes that destroy ozone (Figure 21.1).

21.9 Elimination Reactions

Another type of reaction can compete with substitution when a nucleophile/base reacts with an alkyl halide. 2-Chloro-2-methylpropane (a tertiary alkyl chloride) reacts with H_2O to give the expected alcohol as a substitution product—but a second product, an alkene, is also formed. The alkene molecules are formed by *elimination of two atoms from substrate molecules, the chloride ion leaving group, along with a proton from adjacent carbon atoms.*

2-Chloro-2-methylpropane   **2-Methylpropan-2-ol**   **2-Methylpropene**
                                              (64%)   **(36%)**

It is not unusual to find evidence for competition between these two types of reaction in the same round-bottom flask. A nucleophile/base can *substitute* for the leaving group in an S_N1 or S_N2 reaction, and the nucleophile/base can cause *elimination* of HX, leading to formation of an alkene.

The elimination of HX from alkyl halides is an extremely useful method for synthesizing alkenes, but it is not always easy to predict the regiochemistry of this reaction. Which proton will be removed from an unsymmetrical alkyl halide molecule, if more than one adjacent carbon atom contains a hydrogen atom? We observe that elimination reactions almost always give *mixtures* of alkene products, and the best we can usually do is to predict which will be the major one.

According to **Zaitsev's rule**, a predictive guideline formulated by the Russian chemist Alexander Zaitsev in 1875, *base-induced elimination reactions generally give the more highly substituted alkene product—that is, the alkene with the larger number of alkyl substituents on the double bond.* For example, treatment of 2-bromobutane with KOH in ethanol gives primarily but-2-ene (disubstituted molecules; two alkyl group substituents on the double-bonded carbon atoms) rather than but-1-ene (monosubstituted molecules; one alkyl group substituent on the double-bonded carbon atoms).

2-Bromobutane   **But-2-ene (81%)**   **But-1-ene (19%)**

21.10 E1 and E2 Elimination Reaction Mechanisms

The E2 Mechanism for Elimination Reactions

Some of the same experimental lines of evidence used to understand substitution reactions have been applied to elimination reactions. When an alkyl halide capable of undergoing elimination (such as 2-bromobutane above) is treated with strong base, such as hydroxide ion or alkoxide ion (RO^-), a second-order rate law is observed, just as in the case of the S_N2 reaction.

$$\text{Rate of reaction} = k \, [\text{base}]^1 [\text{substrate}]^1$$

This experimental rate law is consistent with a molecular-level view of this reaction called the **E2 reaction mechanism** (for *elimination, bimolecular*), as shown in Figure 21.7.

Think about It

e21.11 Watch an animation portraying the mechanism and orbitals involved in the E2 reaction.

FIGURE 21.7 Accepted E2 mechanism for one productive collision in reaction of an alkyl halide with strong base.

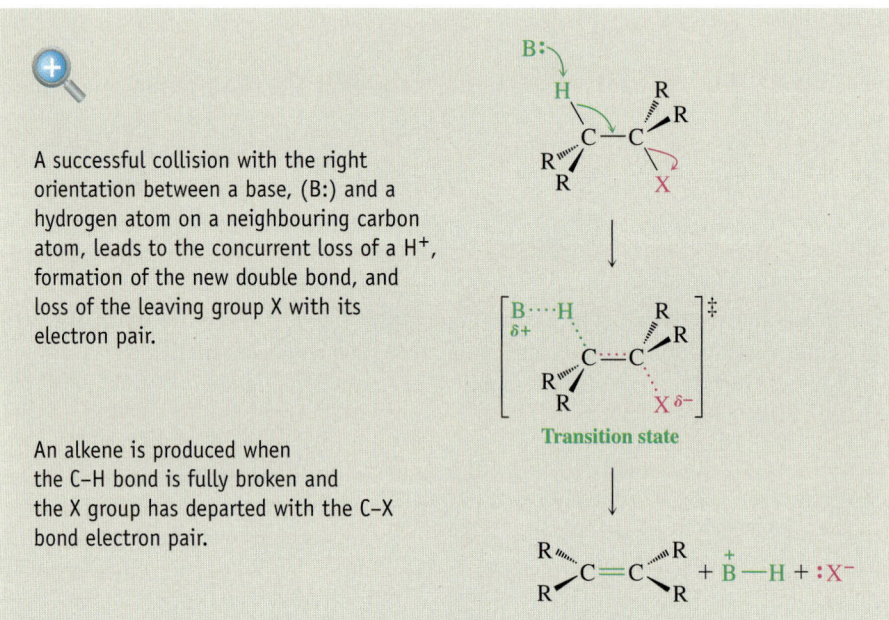

A successful collision with the right orientation between a base, (B:) and a hydrogen atom on a neighbouring carbon atom, leads to the concurrent loss of a H^+, formation of the new double bond, and loss of the leaving group X with its electron pair.

An alkene is produced when the C–H bond is fully broken and the X group has departed with the C–X bond electron pair.

Transition state

Like the S_N2 mechanism, the E2 pathway is postulated to take place in one step without intermediates. As a base with sufficient energy and the right orientation collides with an H atom attached to a C atom adjacent to the C atom bonded to the leaving group, the C—H and C—X bonds begin to break and the C═C double bond begins to form. When the leaving group departs, it takes with it the two electrons from the former C—X bond. Both the base and the substrate are involved in the rate-determining step (a bimolecular process), consistent with the observed second-order rate law. E2 elimination takes place with a particular stereochemical arrangement: *the base removes a proton that is anti to [<<Section 9.5] and in the same plane as the leaving group on the neighbouring carbon atom.*

WORKED EXAMPLE 21.7—E2 REACTION

What product would you expect from reaction of 1-chloro-1-methylcyclohexane with KOH in ethanol?

Strategy

We know that treatment of an alkyl halide with a strong base such as KOH yields an alkene. To find the products in a specific case, draw the structure of the starting material and locate the hydrogen atoms on each neighbouring carbon. Then generate the potential alkene products by removing HX in as many ways as possible. The major product will be the one that has the most highly substituted double bond—in this case, 1-methylcyclohexene.

Solution

1-Chloro-1-methylcyclohexane   1-Methylcyclohexene   Methylenecyclohexane
                                          (major)   (minor)

EXERCISE 21.19—E2 REACTION

Ignoring double-bond stereochemistry, what elimination products would you expect from the following alkyl halides.

(a) $CH_3CH_2CHCHCH_3$ (with Br and CH_3 substituents)

(b) $CH_3CHCH_2-C-CHCH_3$ (with CH_3, Cl, CH_3, and CH_3 substituents)

(c) cyclohexyl-$CHCH_3$ (with Br substituent)

EXERCISE 21.20—E2 REACTION

What alkyl halides might the following alkenes have been made from?

(a) $CH_3CHCH_2CH_2CHCH=CH_2$ (with two CH_3 substituents)

(b) cyclopentene with two CH_3 substituents

The E2 mechanism for elimination reactions is postulated to take place in one step without intermediates. A base removes an H^+ attached to a carbon atom next to the one bonded to the leaving group. At the same time, the C—X bonds begin to break and the C=C double bond begins to form, giving an alkene product.

The E1 Mechanism for Elimination Reactions

The reaction of 2-chloro-2-methylpropane with H_2O at the beginning of Section 21.9 does not show a second-order rate law as in the case of reactions explained by the E2 mechanism. Doubling the concentration of H_2O has no effect on the reaction rate, but doubling the concentration of the alkyl chloride causes the rate to double. The reaction is first order in the substrate only, and therefore first order overall.

$$\text{Rate of reaction} = k\,[\text{2-chloro-2-methylpropane}]^1$$

To explain this experimental rate law, an **E1 reaction mechanism** (*elimination, unimolecular*), which is analogous to and often competes with the S_N1 mechanism, has been proposed. The E1 mechanism can be formulated as shown in Figure 21.8 for the elimination of HCl from 2-chloro-2-methylpropane.

The mechanisms for E1 elimination reactions are proposed to begin with the same unimolecular dissociation we saw in the S_N1 mechanism, but the dissociation is followed by a base removing an H^+ from a carbon atom adjacent to the intermediate carbocation, rather than by substitution. In fact, the E1 and S_N1 reaction mechanisms normally occur in competition whenever an alkyl halide is treated in a hydroxylic solvent with a nucleophile that is not a strong base. The best E1 substrates are also the best S_N1 substrates, and mixtures of substitution and elimination products are usually obtained. 2-Chloro-2-methylpropane produces a 64:36 mixture of 2-methyl-2-propanol (S_N1) and 2-methylpropene (E1) when warmed to 65 °C in 80% aqueous ethanol.

Think about It

e21.12 Watch an animation portraying the mechanism and orbitals involved in the E1 reaction.

FIGURE 21.8 Accepted E1 mechanism for one productive collision in the elimination of HCl from 2-chloro-2-methylpropane.

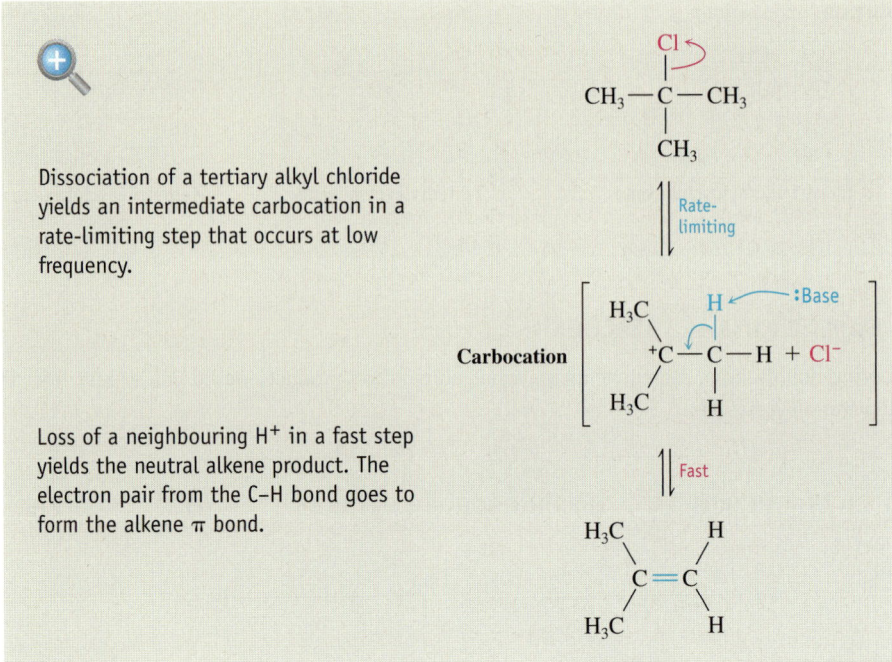

Dissociation of a tertiary alkyl chloride yields an intermediate carbocation in a rate-limiting step that occurs at low frequency.

Loss of a neighbouring H⁺ in a fast step yields the neutral alkene product. The electron pair from the C–H bond goes to form the alkene π bond.

EXERCISE 21.21—E1 REACTION MECHANISM

What effect on the rate of an E1 reaction of 2-chloro-2-methylpropane would you expect if the concentration of the alkyl halide were tripled?

The E1 mechanism for elimination reactions starts with the same unimolecular dissociation of the substrate as in the S_N1 mechanism, but the dissociation is followed by a base removing an H^+ from a carbon atom adjacent to the intermediate carbocation to give an alkene.

21.11 Which Mechanism Predominates?

Now that we've seen four different mechanisms to explain observations about nucleophilic substitution/elimination reactions, you may be wondering which gives the best explanation in any given case. Will substitution or elimination occur? Will the reaction be unimolecular or bimolecular? There are no rigid answers to these questions, and they depend on further considerations, such as the nature of the solvent and the temperature. But it is possible to make some broad generalizations.

TABLE 21.2 A Summary of Substitution and Elimination Reactions

Halide Type	S_N1	S_N2	E1	E2
RCH₂X (primary)	Does not occur	Highly favoured	Does not occur	Occurs when strong bases are used
R₂CHX (secondary)	Can occur with benzylic and allylic halides	Occurs in competition with E2 reaction	Can occur with benzylic and allylic halides	Favoured when strong bases are used
R₃CX (tertiary)	Favoured in hydroxylic solvents	Does not occur	Occurs in competition with S_N1 reaction	Favoured when bases are used

- *A primary substrate (RCH₂X)* is more likely to react by an S_N2 pathway if a good nucleophile such as I^-, Br^-, RS^-, NH_3, or CN^- is used, and by an E2 pathway if a strong base such as hydroxide ion or an alkoxide ion (RO^-) is used.

- *A secondary substrate (R₂CHX)* can usually react by both S_N2 and E2 pathways to give a mixture of substitution and elimination products. E2 is favoured when strong bases are present.
- *A tertiary substrate (R₃CX)* reacts by an E2 pathway if a strong base is used or by a mixture of S_N1 and E1 pathways under neutral or acidic conditions.

WORKED EXAMPLE 21.8—PREDICTING THE MECHANISM

Predict which mechanism(s) predominate(s): S_N1, S_N2, E1, or E2.

Strategy

Look to see whether the substrate is primary, secondary, or tertiary, and determine whether substitution or elimination has occurred. Then apply the generalizations summarized above.

Solution

The substrate is a secondary alkyl halide, and in the presence of a strong base an elimination reaction has occurred to give an alkene. This is an E2 reaction.

EXERCISE 21.22—PREDICTING THE MECHANISM

Which mechanism(s) should predominate: S_N1, S_N2, E1, or E2?

(a) 1-Bromobutane + NaN_3 ⟶ 1-Azidobutane

(b)
$$CH_3CH_2\overset{\overset{\displaystyle Cl}{|}}{C}HCH_2CH_3 + KOH \longrightarrow CH_3CH_2CH{=}CHCH_3$$

(c)

21.12 Grignard Reagents

Alkyl halides react with magnesium metal in ether solvent to yield organomagnesium halides, called **Grignard reagents** after their discoverer, Victor Grignard. Grignard reagents contain a carbon–metal bond and are examples of **organometallic compounds**.

$$R - X + Mg \xrightarrow{\text{ether}} R - Mg - X$$

where R = 1°, 2°, or 3° alkyl, aryl, or alkenyl, and X = Cl, Br, or I.

Chlorobenzene, for instance, reacts rapidly with magnesium metal in ether to give phenylmagnesium chloride.

Molecular Modelling (Odyssey)

e21.13 Examine the three-dimensional electrostatic potential map of methyl magnesium chloride.

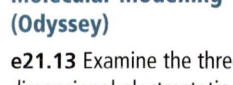

Chlorobenzene   Phenylmagnesium chloride

FIGURE 21.9 Grignard reagents. Note the difference in polarity of the C—Mg bond in phenylmagnesium chloride (centre electrostatic potential map) and the C—Cl bond in chloromethane (far *right*).

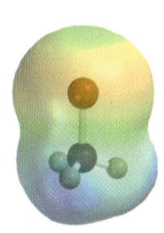

Compare the electrostatic potential map of a phenylmagnesium chloride molecule in the centre of Figure 21.9 with that of a chloromethane molecule on its right. The carbon atom bonded to the magnesium metal atom is electron-rich (negative electrostatic potential, red), while the carbon atom bonded directly to the chlorine atom in the molecule on the right is electron-deficient (positive electrostatic potential, blue). By the transformation of an alkyl halide into a Grignard reagent, we have switched the normal polarity of the carbon atom, creating both nucleophilic and basic carbon. This switch in polarity of a carbon atom from being electrophilic to nucleophilic is referred to with the expressive German word, *umpolung*, which means "polarity reversal."

Because of its electron-rich carbon atom, a Grignard reagent is both a base and a nucleophile; it therefore reacts with acids and electrophiles. For example, it reacts with HCl, H_2O, or an alcohol ROH by accepting H^+ and yielding a hydrocarbon. The overall sequence, $R-X \longrightarrow R-MgX \longrightarrow R-H$, is a useful method for converting an organic halide into a hydrocarbon:

$$CH_3CH_2CH_2CH_2CH_2CH_2Br \xrightarrow[\text{Ether}]{\text{Mg}} CH_3CH_2CH_2CH_2CH_2CH_2MgBr$$

1-Bromohexane　　　　　　　　　　　　　**1-Hexylmagnesium bromide**

$$\downarrow H_2O$$

$$CH_3CH_2CH_2CH_2CH_2CH_3$$

Hexane (85%)

The most important reactions of Grignard reagents are with carbonyl compounds to form alcohols in which a C—C bond has been formed to the alcohol carbon atom [>>Chapter 22].

WORKED EXAMPLE 21.9—SYNTHESIS USING GRIGNARD REAGENTS

By using several reactions in sequence, you can accomplish transformations that can't be done in a single step. How would you prepare the alkane methylcyclohexane from the alcohol 1-methylcyclohexanol?

1-Methylcyclohexanol　　　　**Methylcyclohexane**

Strategy

Working backward, we know that alkanes can be made from alkyl halides and that alkyl halides can be made from alcohols. Carrying out the two reactions sequentially converts 1-methylcyclohexanol into methylcyclohexane.

Solution

1-Methylcyclohexanol　　　**1-Bromo-1-methylcyclohexane**　　　**Methylcyclohexane**

EXERCISE 21.23—SYNTHESIS USING GRIGNARD REAGENTS

An advantage to preparing an alkane from a Grignard reagent is that deuterium (D; the ^2H isotope of hydrogen) can be placed at a specific site in a molecule. How might you convert 2-bromobutane into 2-deuteriobutane?

EXERCISE 21.24—SYNTHESIS USING GRIGNARD REAGENTS

How could you convert 4-methyl-pentan-1-ol into 2-methylpentane?

$$CH_3CHCH_2CH_2CH_2OH$$ with a CH_3 group on the second carbon   **4-Methylpentan-1-ol**

Alkyl halides react with magnesium metal to form organomagnesium halides, or Grignard reagents. These organometallic compounds react with acids to yield the corresponding alkanes.

SUMMARY

Key Concepts

- **Chlorofluorocarbons (CFCs)** are alkyl halides with ideal properties for use as refrigerants. But these gases are unreactive when released into the troposphere. The only way they break down in the atmosphere is by filtering up into the stratosphere where they react with ultraviolet light, releasing chlorine atoms that act as catalysts to destroy ozone. They are now replaced with **HCFCs** and **HFCs**, which are more reactive in the troposphere and have lower ozone-depleting potential. (Section 21.1)
- Thousands of halogenated organic compounds have been identified in nature and synthesized in laboratories. Toxic naturally occurring *alkyl halides* are often synthesized by organisms as defence mechanisms against infections and predators, and their structures form the basis of many important antibiotics in medicine. (Section 21.2)
- **Alkyl halides** are usually prepared from alcohols by treatment either with HX (for tertiary alcohols) or with $SOCl_2$ or PBr_3 (for primary and secondary alcohols). (Section 21.4)
- Treatment of an alkyl halide with a nucleophile/base results either in substitution or elimination reactions. Nucleophilic substitution reactions require a nucleophile (symbolized Nu: or Nu:⁻) and a substrate (symbolized R—X) that is bonded to a good leaving group (symbolized X: or X:⁻). Two mechanisms, S_N2 and S_N1, are used to explain experimental evidence from kinetics and stereochemistry. In the S_N2 reaction mechanism, a few of the many collisions causes the entering nucleophile to successfully react with the substrate molecule from a direction 180° away from the leaving group, resulting in an umbrella-like inversion of configuration at the carbon atom. S_N2 reactions are strongly inhibited by increasing steric bulk of the reagents and are favoured only for primary substrates. In the S_N1 reaction mechanism, the substrate first dissociates to a carbocation, which then reacts with a nucleophile in a second step. As a consequence, S_N1 reactions take place with racemization of configuration at the substrate carbon atom and are favoured only for tertiary or other substrates that give stable carbocation intermediates. (Sections 21.5–21.7)
- *Elimination reactions* are also thought to occur by two mechanisms: E2 and E1. In the **E2 reaction mechanism,** a base abstracts a proton from a neighbouring carbon atom at the same time that the leaving group departs from a substrate molecule. In the **E1 reaction mechanism**, the substrate first dissociates to form a carbocation, which can subsequently lose H⁺ from a neighbouring carbon atom to give an alkene. The reaction occurs on tertiary substrates in neutral or acidic hydroxylic solvents. The major product formed by an elimination reaction can be predicted by **Zaitsev's rule**. (Sections 21.9–21.10)
- Alkyl halides react with magnesium metal to form organomagnesium halides, or **Grignard reagents**. These **organometallic compounds** react with acids to yield the corresponding alkanes. (Section 21.12)

Key Reactions

1. Synthesis of alkyl halides from alcohols
   (a) Reaction of tertiary alcohol with HX, where X = Cl, Br (Section 21.4)

   (b) Reaction of primary and secondary alcohols with PBr₃ and SOCl₂ (Section 21.4)

$$ROH + PBr_3 \longrightarrow RBr$$
$$ROH + SOCl_2 \longrightarrow RCl$$

2. Reactions of alkyl halides
   (a) S_N2 reaction mechanism: reaction of nucleophile with alkyl halide, involving inversion of configuration of the substrate (Section 21.6)

   Substrate must be primary or secondary.

   (b) S_N1 reaction mechanism: carbocation intermediate is involved (Section 21.7)

   Substrate must be tertiary or (occasionally) secondary.

   (c) E2 reaction mechanism (Section 21.10)

   (d) E1 reaction mechanism (Section 21.10)

   Best for tertiary substrates in neutral or acidic solvents.
   Carbocation intermediate is involved.

   (e) Formation and protonation of Grignard reagents (Section 21.12)

$$RX + Mg \longrightarrow RMgX$$
$$RMgX + H_2O \longrightarrow RH$$

REVIEW QUESTIONS

Section 21.1: Chlorofluorocarbons and Stratospheric Ozone Depletion

21.25 A potent ozone depleting substance is methyl bromide (CH₃Br). Using additional sources, determine the major natural and human sources of this substance. For which of the major human sources are ready replacements available?

21.26 HCFC and HFC compounds have been developed to replace CFCs in refrigerants and air conditioners. Draw structures of representative HCFC and HFC molecules. Which of these is seen as short-term replacements, and which are considered longer-term solutions? What are the advantages and disadvantages of each?

Section 21.2: Alkyl Halides—Synthetic and Natural

21.27 Halothane is an inhalational general anesthetic that was first used in the U.K. in the mid-1950s to replace some of the ethers and other volatile compounds that were then in use. Using additional sources, give the IUPAC name and draw the structure of a halothane molecule. What functional groups are present in molecules of halothane?

21.28 In the 1970s and 1980s, enflurane and isoflurane replaced halothane (Question 21.27) as inhalational anesthetics. Using additional sources, draw structures of molecules of these two compounds. What functional groups are present in each?

21.29 Using additional sources, look up recent review papers by Gordon Gribble at Dartmouth College in the U.S. on naturally produced organohalogen compounds. How many organohalogen compounds does he think are known? What environments are the largest sources of naturally occurring organohalogen compounds?

21.30 Consulting the review paper in Question 21.29, what advantage might marine or terrestrial organisms gain by synthesizing organobromine or organochlorine compounds?

Sections 21.5–21.7: Nucleophilic Substitution Reactions

21.31 Describe the effects of the following variables on both S_N2 and S_N1 reactions:
(a) substrate structure   (b) leaving group

21.32 Draw molecules of the product you would expect from reaction of each of the following molecules with (i) SCH_3^- and (ii) OH^- (yellow-green atom = Cl).

(a)

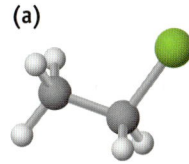

(c)

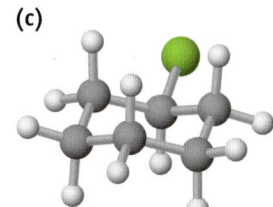

(b)

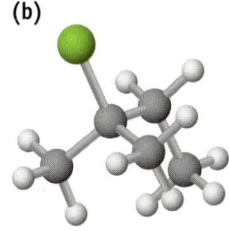

21.33 Which ion in each of the following pairs is a better leaving group?
(a) F^- or Br^- (b) Cl^- or NH_2^- (c) OH^- or I^-

21.34 Assign R or S configuration to the following molecule, write the product you would expect from S_N2 reaction with NaCN, and assign R or S configuration to the product (red = O, yellow-green = Cl).

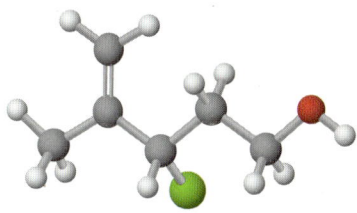

21.35 From what alkyl bromide was the following alkyl acetate made by S_N2 reaction? Write the reaction, showing all stereochemistry.

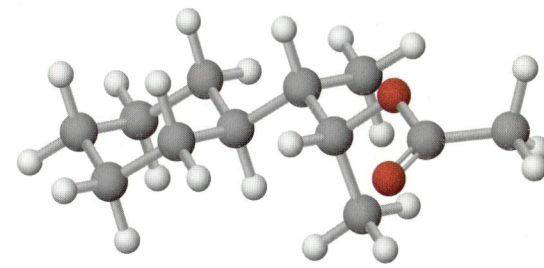

21.36 Which alkyl halide in each of the following pairs will react faster in an S_N2 reaction with OH^-?
(a) Bromobenzene or benzyl bromide, $C_6H_5CH_2Br$
(b) CH_3Cl or $(CH_3)_3CCl$
(c) $CH_3CH=CHBr$ or $H_2C=CHCH_2Br$

21.37 What effect would you expect the following changes to have on the S_N2 reaction of CH_3Br and CN^- to give CH_3CN?
(a) The concentration of CH_3Br is tripled and that of CN^- is halved.
(b) The concentration of CH_3Br is halved and that of CN^- is tripled.
(c) The concentration of CH_3Br is tripled and that of CN^- is doubled.
(d) The reaction temperature is raised.
(e) The volume of the reacting solution is doubled by addition of more solvent.

21.38 What effect would you expect the following changes to have on the S_N1 reaction of $(CH_3)_3CBr$ with CH_3OH to give $(CH_3)_3COCH_3$?
(a) The concentration of $(CH_3)_3CBr$ is doubled and that of CH_3OH is halved.
(b) The concentration of $(CH_3)_3CBr$ is halved and that of CH_3OH is doubled.
(c) The concentrations of both $(CH_3)_3CBr$ and CH_3OH are tripled.
(d) The reaction temperature is lowered.

21.39 Rank each set of compounds with respect to S_N2 reactivity:

(a)

$$CH_3CCl(CH_3) \quad CH_3CH_2CH_2Cl \quad CH_3CH_2CHCH_3(Cl)$$

with CH_3 groups on the first carbon

(b)

$$CH_3CHCHCH_3(Br) \quad CH_3CHCH_2Br(CH_3) \quad CH_3Br$$

21.40 Order the following compounds with respect to both S_N1 and S_N2 reactivity:

$$CH_3CCl(CH_3)(CH_3) \quad \text{(benzyl)}CH_2Cl \quad \text{(phenyl)}Cl$$

21.41 What is wrong with each of the following reactions?

(a) $CH_3CH_2CCH_2CH_3(Br)(CH_3) \xrightarrow{NaCN} CH_3CH_2CCH_2CH_3(CN)(CH_3)$

(b) $CH_3CHCH_2CH_2CH_2OH(CH_3) \xrightarrow{NaBr} CH_3CHCH_2CH_2CH_2Br(CH_3)$

(c) $CH_3CH_2CCH_3(OH)(CH_3) \xrightarrow{HBr} CH_3CH=CCH_3(CH_3)$

Sections 21.9–21.11: Elimination Reactions and the Elimination/Substitution Continuum

21.42 Draw the structure of molecules of the product you expect from E2 reaction of the following molecule with NaOH (yellow-green = Cl):

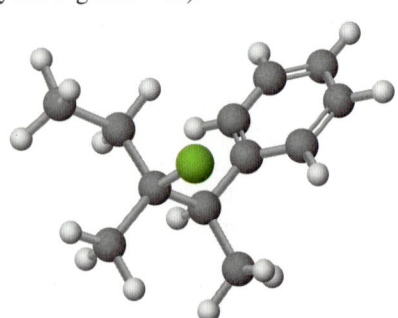

21.43 Propose a structure for an alkyl halide that can give a mixture of three alkenes on E2 reaction.

21.44 Heating either *tert*-butyl chloride or *tert*-butyl bromide with ethanol yields the same reaction mixture: approximately 80% *tert*-butyl ethyl ether $[(CH_3)_3COCH_2CH_3]$ and 20% 2-methylpropene. Explain why the identity of the leaving group has no effect on the product mixture.

21.45 What effect would you expect the following changes to have on the rate of the reaction of 1-iodo-2-methylbutane with CN^-?

$$CH_3CH_2CHCH_2I(CH_3) + CN^- \longrightarrow CH_3CH_2CHCH_2CN(CH_3)$$

1-Iodo-2-methylbutane

(a) CN^- concentration is halved and 1-iodo-2-methylbutane concentration is doubled.

(b) Both CN^- and 1-iodo-2-methylbutane concentrations are tripled.

21.46 What effect would you expect on the rate of reaction of ethyl alcohol with 2-iodo-2-methylbutane if the concentration of the alkyl halide is tripled?

21.47 Identify the following reactions as either S_N1, S_N2, E1, or E2:

(a)

$$\text{(phenyl)}CHCH_3(Br) \xrightarrow{KOH} \text{(phenyl)}CH=CH_2$$

(b)

$$\text{(phenyl)}CHCH_3(Br) \xrightarrow[\text{Heat}]{CH_3OH} \text{(phenyl)}CHCH_3(OCH_3)$$

21.48 Predict the major alkene product from the following eliminations:

(a)

$$\text{cyclohexane ring with } H_3C, H \text{ and } H, Br \xrightarrow{KOH} ?$$

(b)

$$CH_3CHCBr(H_3C)(CH_3)(CH_2CH_3) \xrightarrow[\text{Heat}]{CH_3CO_2H} ?$$

21.49 Treatment of an alkyl chloride (C_4H_9Cl) with strong base gives a mixture of three isomeric alkene products. What is the structure of molecules of the alkyl chloride, and what are the structures of the three products?

Section 21.12: Grignard Reactions

21.50 Why do you suppose it's not possible to prepare a Grignard reagent from a bromoalcohol such as 4-bromopentan-1-ol?

$$CH_3CHCH_2CH_2CH_2OH(Br) \xrightarrow[]{Mg} \!\!\!\!\!\times\!\!\!\! CH_3CHCH_2CH_2CH_2OH(MgBr)$$

4-Bromopentan-1-ol

SUMMARY AND CONCEPTUAL QUESTIONS

21.51 Many of the earliest known naturally occurring organohalogen compounds are found in marine environments. Why do you think this is the case?

21.52 How might you prepare the following compounds using a nucleophilic substitution reaction at some step?

(a) CH_3CH_2Br

(b) $CH_3CH_2CH_2CH_2CN$

(c) $CH_3\overset{\overset{\displaystyle CH_3}{|}}{\underset{\underset{\displaystyle CH_3}{|}}{C}}CH_3$ (as drawn: $CH_3OC(CH_3)_2CH_3$ with two CH_3 groups)

(d) $CH_3CH_2CH_2\overset{+}{N}=N=N^-$

(e) CH_3CH_2SH

(f) $CH_3\overset{\overset{\displaystyle O}{||}}{C}OCH_3$

21.53 How would you prepare the following compounds, starting with cyclopentene and any other reagents needed?
(a) chlorocyclopentane
(b) cyclopentanol
(c) cyclopentylmagnesium chloride
(d) cyclopentane

21.54 Predict the product(s) of the following reactions:

(a)

$\xrightarrow[\text{Ether}]{\text{HBr}}$?

(b) $CH_3CH_2CH_2CH_2OH \xrightarrow{SOCl_2}$?

(c)

$\xrightarrow[\text{Ether}]{PBr_3}$?

(d) $CH_3CH_2CH(Br)CH_3 \xrightarrow[\text{Ether}]{Mg} A \xrightarrow{H_2O} B$

21.55 What products do you expect from reaction of 1-bromopropane with the following reagents?
(a) NaI   (d) Mg, then H_2O
(b) NaCN   (e) $NaOCH_3$
(c) NaOH

21.56 Predict the product and describe the stereochemical changes during reactions of the following nucleophiles with (R)-2-bromooctane:
(a) CN^- (b) $CH_3CO_2^-$ (c) Br^-

21.57 Draw all isomers of C_4H_9Br, name them, and arrange them in order of decreasing reactivity in the S_N2 reaction.

21.58 Ethers can be prepared by S_N2 reaction of an alkoxide ion with an alkyl halide:

$$R-O^- + R'-Br \longrightarrow R-O-R' + Br^-$$

Suppose you wanted to prepare cyclohexyl methyl ether. Which route would be better: reaction of methoxide ion (CH_3O^-) with bromocyclohexane, or reaction of the alkoxide of cyclohexoxide with bromomethane? Explain.

Cyclohexyl methyl ether

21.59 How could you prepare diethyl ether ($CH_3CH_2OCH_2CH_3$) starting from ethyl alcohol and any inorganic reagents needed? More than one step is needed. (See Question 21.58.)

21.60 The S_N2 reaction can occur *intramolecularly*, meaning within the same molecule. What product would you expect from treatment of 4-bromo-1-butanol with base?

$$BrCH_2CH_2CH_2CH_2OH \xrightarrow{\text{Base}} BrCH_2CH_2CH_2CH_2O^- Na^+ \longrightarrow ?$$

21.61 *trans*-1-Bromo-2-methylcyclohexane yields the non-Zaitsev elimination product 3-methylcyclohexene on treatment with KOH. What does this result tell you about the stereochemistry of E2 reactions?

trans-1-Bromo-2-methylcyclohexane   3-Methylcyclohexene

21.62 How can you explain the fact that treatment of (R)-2-bromohexane with NaBr yields 2-bromohexane racemate?

21.63 Reaction of HBr with (R)-3-methylhexan-3-ol yields (±)-3-bromo-3-methylhexane. Explain.

$$CH_3CH_2CH_2\overset{\overset{\displaystyle OH}{|}}{\underset{\underset{\displaystyle CH_3}{|}}{C}}CH_2CH_3 \qquad \text{3-Methylhexan-3-ol}$$

21.64 (S)-Butan-2-ol slowly racemizes to give (±)-butan-2-ol on standing in dilute sulfuric acid. Propose a mechanism to account for this observation.

$$CH_3CH_2\overset{\overset{\displaystyle OH}{|}}{C}HCH_3 \qquad \text{Butan-2-ol}$$

21.65 Compound A is optically inactive and has the formula $C_{16}H_{16}Br_2$. On treatment with strong base, compound A gives hydrocarbon B ($C_{16}H_{14}$), which absorbs 2 equivalents of H_2 when reduced over a palladium catalyst. Hydrocarbon B also reacts with acidic $KMnO_4$ to give two carbonyl-containing products. One product, C, is a carboxylic acid with the formula $C_7H_6O_2$. The other product is oxalic acid (HO_2CCO_2H). Formulate the reactions involved, and suggest structures for A, B, and C.

Alcohols, Phenols, and Ethers

George Bailey/Shutterstock

22.1 Case Study: Cyclodextrins: A Spoonful of Sugar Helps the Medicine Go Down

A spoonful of sugar helps the medicine go down. Since long before the days when Mary Poppins first sang these lines, parents and doctors have faced the challenge of getting patients, especially very young ones, to swallow bitter medications. Some drugs with disgusting tastes need to be given under the tongue or as chewable tablets to be properly absorbed and act on the target site in the body. Some analgesics [<<Section 20.1], such as acetaminophen and ibuprofen, are bitter and irritating to the throat if taken by patients who can't easily swallow whole tablets. Another pharmaceutical product whose bitter taste can prevent its effective use is nicotine. While toxic, nicotine is taken in small doses, usually in the form of chewing gum, by smokers trying to quit. Besides its intolerable taste, other difficulties in preparing stable nicotine formulations come from its physical properties and chemical reactivity. Nicotine is a colourless oil that is volatile, absorbs water, and turns brown on exposure to air or light. In its pure form, it decomposes rapidly in air.

Chemists have seen research opportunities in the bitter taste and instability of some drugs, and have learned that a spoonful of sugar, indeed, can help the medicine go down. But a closer look at the sugar now added to formulations of some bitter medications reveals that it is not table sugar (sucrose) or ordinary glucose. Rather, it is a small polymer of glucose, called a *cyclodextrin*. And the mode of action of cyclodextrins is much more elegant and effective than simply masking the bitter taste with sucrose, as Mary Poppins had in mind. Cyclodextrins work by encapsulating the drug—think of them as providing a temporary molecular-level pill container that a patient can chew or swallow with the pill inside.

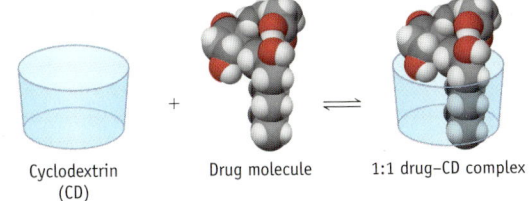

Cyclodextrin (CD) + Drug molecule ⇌ 1:1 drug–CD complex

A molecular-level "cartoon" showing a 1:1 drug-cyclodextrin complex.

Not a spoonful of ordinary sugar ... You might think of cyclodextrins as providing a temporary molecular-level pill container that a patient can chew or swallow with the pill inside it.

Molecular Modelling (Odyssey)

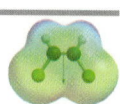

e22.1 See a simulation demonstrating the flexibility of a cylcodextrin molecule.

A small polymer molecule such as β-cyclodextrin is called an *oligomer* (from the Greek, *oligos*, meaning "few"), made by combining a *few* monomer units. The OH groups are coloured red to highlight them.

Taking it Further

e22.2 See different-sized structures of cyclodextrins used for different purposes.

Like glucose and sucrose, cyclodextrins contain both R—O—H (hydroxyl or alcohol) and R—O—R (ether) functional groups, classifying them as both alcohols and ethers. The structures, properties, and reactivities of simple alcohols and ethers are the focus of this chapter. In Chapter 29, you will see that in aqueous solution, some of the alcohol and ether functional groups present in the cyclic form of certain carbohydrates are converted into aldehydes and ketones when they are in equilibrium with open-chain forms.

All three sugars shown above have cyclic molecules. Glucose molecules have one six-membered ring; sucrose molecules have two rings, one containing five atoms and one containing six. A cyclodextrin molecule has a much larger ring, called a *macrocycle*, since it contains a ring with a large number of atoms. The molecule of β-cyclodextrin is a small polymer made of a ring of seven glucose rings.

Because this small polymer contains only a few glucose monomers, it is called an **oligomer**. Other well-known cyclodextrins contain six glucose rings (α-cyclodextrin) or eight glucose rings (γ-cyclodextrin). Cyclodextrins are produced on thousand-tonne scale by breaking down corn or potato starch, a polymer of glucose, with enzymes in the presence of water.

Let's look more closely at the structure and properties of cyclodextrins to see how they work at the molecular level to help people take bitter medicine and how they can also help drugs reach and act on their intended sites in the body.

Taste: A Molecular-Level View

At the molecular level, we envision taste as related to the interaction of water soluble substances with receptor molecules in the mouth. Sweet taste receptors are concentrated near the front of the tongue; sour taste receptors on both edges; and receptors interacting with bitter substances are at the back. Substances dissolved in saliva, an aqueous solution, come intimately in contact with these receptors and trigger interactions that our brains interpret as pleasant or unpleasant tastes. Many active components of drugs are not very soluble in water, which means they don't fully interact with taste receptors when taken orally. The bitter taste of a drug may be minimized if it has limited solubility, but this same property can limit the dissolving and absorption processes in the gastrointestinal tract and elsewhere needed to provide adequate drug levels at the target site in the body.

Glucose (β form)

Sucrose

β-Cyclodextrin

Cyclodextrin Host–Guest Complexes

How might a cyclodextrin molecule interact with molecules of the active component of a drug to affect its taste, solubility, and absorption? As far back as the 1950s, evidence was obtained that cyclodextrins can form *host–guest complexes* [<<Section 3.12] when

relatively non-polar compounds (guests) of the right size and shape are held in place by non-covalent forces in the cavity of a cyclodextrin host. Modern laboratories characterize these complexes using structural techniques including NMR spectroscopy, where chemists can observe changes in chemical shifts of nuclei of guest molecules after they enter the interior of a host molecule. The size of the cavity depends on the number of glucose monomer molecules making up the macrocyclic cyclodextrin. The overall molecular shape of the host is often simplistically represented as a cone (Figure 22.1), with an interior diameter of the cavity being about 0.50, 0.60, and 0.80 nm for α-, β-, and γ-cyclodextrins, respectively. The alcohol functional groups on the exterior of the cyclodextrin cavity make it very polar and **hydrophilic**, and the interior cavity is relatively non-polar.

Host–guest complexes fall under the sub-discipline of *supramolecular chemistry*, which focuses on *non-covalent interactions in molecules*. Examples seen earlier include methane clathrate hydrate [<<Section 4.1] and β-carotene in a carbon nanotube (Figure 19.5).

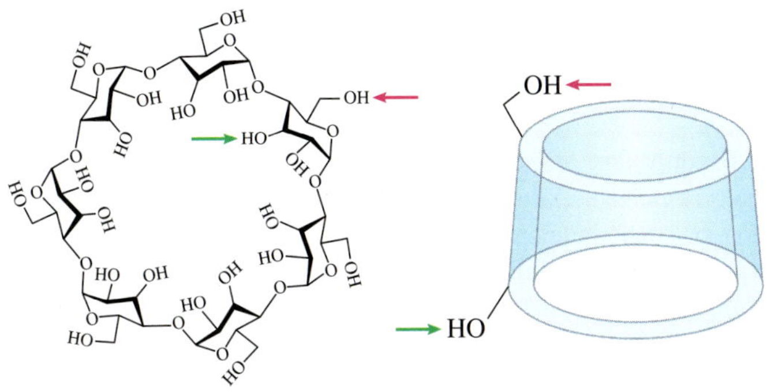

FIGURE 22.1 Representations of β-cyclodextrin. The alcohol functional groups, which are capable of hydrogen bonding, make the outside of the cyclodextrin very polar. The inside of the macrocycle is much less polar, and binds readily to non-polar guest molecules, such as the active ingredients in many pharmaceutical products.

Adapted by permission from Macmillan Publishers Ltd: NATURE REVIEWS DRUG DISCOVERY Vol. 3 No. 12 © 2004 www.nature.com/nrd

Cyclodextrins are quite soluble in water, as a result of hydrogen bonds forming between solvent water molecules and the alcohol groups on the outside of the molecule. The cavity is thought to be initially occupied by water molecules. When less-polar guest molecules of the right dimensions such as nicotine, an analgesic, or another drug is introduced, they replace water molecules in the cavity, releasing the water molecules into solution. Intermolecular dispersion forces between the non-polar guest and the relatively non-polar interior of the macrocycle hold the guest inside the host. Most often, one, two, or three cyclodextrin host molecules contain one or more trapped guest molecules (Figure 22.2).

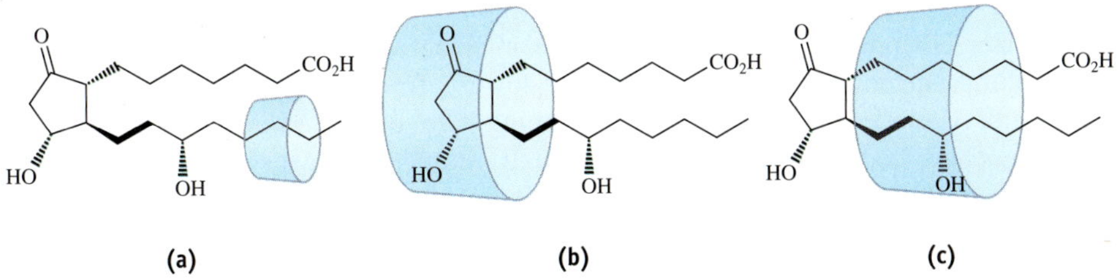

(a) (b) (c)

FIGURE 22.2 Molecular-level representation of the formation of a host–guest complex of molecules of prostaglandin E2 and (a) α-cyclodextrin, (b) β-cyclodextrin, and (c) γ-cyclodextrin.

Adapted by permission from Macmillan Publishers Ltd: NATURE REVIEWS DRUG DISCOVERY Vol. 3 No. 12 © 2004 www.nature.com/nrd

So how does a cyclodextrin "spoonful of sugar" help a bitter drug go down? A bitter-tasting drug is administered as a fine dispersion of the drug and cyclodextrin. In some cases, a host–guest complex is already formed in the drug formulation. In other cases, as soon as the two substances interact in an aqueous solution such as saliva, a host–guest complex is formed, with two results. The polar outside of the cyclodextrin complex often increases the solubility of the drug, and locking the bitter drug inside the non-polar cavity keeps it from interacting with the taste receptors for bitter compounds on the back of the tongue.

But even more important than the taste of drug is its *bioavailability*: a measure of what fraction of the administered drug reaches the circulatory system. Being able to change the rate at which a drug dissolves or how much will eventually dissolve before the drug breaks down (metabolizes) is an important factor in getting the right amount of medication to the right target. In many cases, the formation of cyclodextrin complexes can help. A wide range of drugs are now approved and marketed as cyclodextrin formulations, including nitroglycerin for chest pain (angina) and heart conditions, Prostaglandin E2 (PGE$_2$) for the induction of labour, Prostaglandin E1 (PGE$_1$) for arterial disease and erectile dysfunction, and the antibiotic cephalosporin. The polarity and hydrophilic nature of the outside of cyclodextrin host molecules can be fine-tuned by converting some or all of the alcohol (—OH) functional groups on the outside of cyclodextrins into ethers such as methoxy (—OCH$_3$) groups, longer chain alcohols (—CH$_2$CH$_2$OH), or alkyl sulfonyl groups (—CH$_2$CH$_2$CH$_2$CH$_2$SO$_3^-$).

The strategy of locking a bitter guest inside a cyclodextrin cage has also been applied to food chemistry. One important example is removing the harsh taste and unpleasant (to some) smell caused by trace alcohols and carbonyl compounds in products with soybean paste. Soy products are important sources of protein for people in many countries of the world, and adding cyclodextrins improves the taste and smell of soy-based food product formulations. Research groups in Korea and Japan have also shown that β-cyclodextrin can be used to remove cholesterol from milk used to produce cheddar cheese and other dairy products.

Desirable flavours and aromas can be enhanced through cyclodextrin complexation. Many compounds responsible for desirable flavours and aromas are lost easily through oxidation, evaporation, or other chemical reactions such as polymerization. Cyclodextrin complexation can help preserve desirable flavours and odours. Lemon peel oil in most formulations is oxidized by air in several days, but when complexed with β-cyclodextrin, it is stable for years. Dissolving the complex in an aqueous solution like saliva releases the flavour.

The textile industry invests in cyclodextrin research, with research directed at binding cyclodextrins to natural and synthetic fibres. Numerous possible applications can be envisioned. Complexation of substances such as insect repellents, fragrances, or drugs would make possible their release over time. Cyclodextrins bound to textiles might also be used to remove compounds responsible for undesirable odours from sweat, cigarette smoke, or other environmental pollutants. New dressings with cyclodextrins have also proven effective in eliminating odour from wounds.

Finally, applications of cyclodextrins bring us back to the photodynamic therapy story in Chapter 1. In PDT, porphyrin-containing photosensitizer compounds need to be delivered to the site of a tumour, after which a visible-light laser beam is focused on the site. This causes the photosensitizer to produce singlet oxygen at the site of the tumour, which selectively damages or destroys rapidly growing cancer cells. One of the major challenges faced by medicinal chemists working in this area is the need to deliver photosensitizer compounds to particular organs or receptors. This requires formulations with appropriate solubility that won't break down before they get to the target organ. Dimers of cyclodextrins [<<e22.2], containing two of the same or two different monomers have been found to form exceptionally stable complexes with porphyrin-like photosensitizers, and their solubility can in principle be changed by modifying the outside of the cyclodextrin host.

Very recent preliminary research in virology is generating excitement at the possibility that cyclodextrins may even play some role in blocking the replication of the HIV virus by preventing the build-up of cholesterol in cell compartments.

> Cyclodextrins are macrocyclic oligomers made of glucose monomers containing alcohol and ether functional groups. With hydrophilic exteriors and hydrophobic interiors, they form supramolecular complexes important in pharmaceutical chemistry and textile and other industries.

To learn more about cyclodextrins in pharmacy, read T. Loftsson and D. Duchêne, (2007), "Cyclodextrins and Their Pharmaceutical Applications," *International Journal of Pharmacy*, 329: 1–11. For overviews of cyclodextrins, see J. Szejtli, (2004), "Past, Present, and Future of Cyclodextrin Research," *Pure and Applied Chemistry*, 76, 10: 1825–45; and http://notendur.hi.is/thorstlo/cyclodextrins_eng.htm.

Other Compounds with Alcohol Functional Groups

Many other compounds besides carbohydrates like cyclodextrin have *alcohol* functional groups, with a hydroxyl group (—OH) bonded to a saturated carbon atom. Related compounds are **phenols**, with a hydroxyl group bonded directly to an aromatic ring. Alcohols, phenols, and ethers occur widely in nature and have many industrial, pharmaceutical, and biological applications. Ethanol, for instance, is a fuel additive, an industrial solvent, and a

Taking it Further

e22.3 Read about the many uses of ethanol as a biofuel, drug, and poison.

beverage. Menthol (an alcohol), is a flavouring agent; BHT (butylated hydroxytoluene, a phenol) is an antioxidant food additive; and diethyl ether (the familiar "ether" of medical use) was once popular as an anesthetic agent.

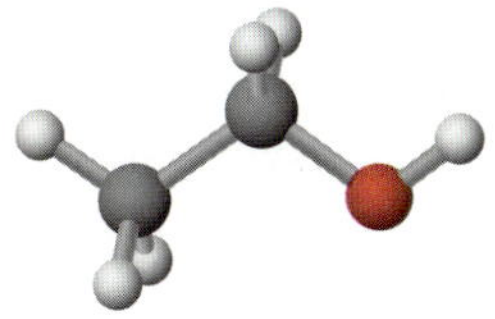

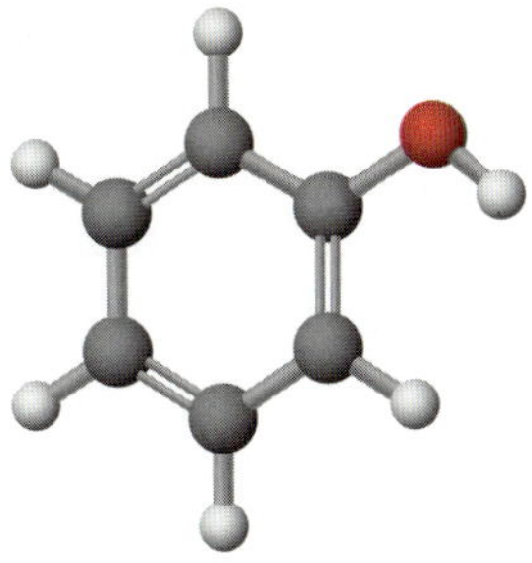

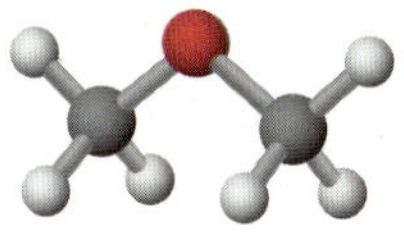

CH_3CH_2OH	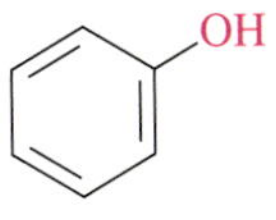	CH_3OCH_3
Ethanol	**Phenol**	**Dimethyl ether**

Lacking an —OH group, ethers are much less reactive than alcohols, and have extensive use as relatively inert, low-boiling solvents.

22.2 Naming Alcohols, Phenols, and Ethers

Alcohols

The **alcohols** on the larger opening of a cyclodextrin cone are classified as secondary alcohols, and those on the smaller opening considered primary. This classification of alcohols as primary (1°), secondary (2°), or tertiary (3°), depends on the number of carbon substituents bonded to the hydroxyl-bearing carbon atom:

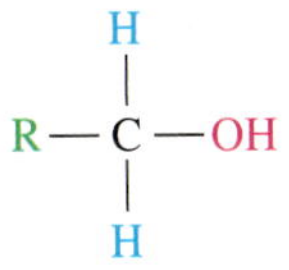

A primary alcohol (1°)	A secondary alcohol (2°)	A tertiary alcohol (3°)

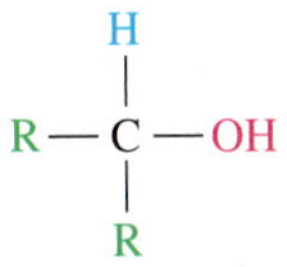

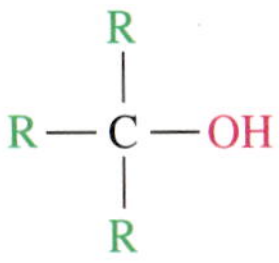

Simple alcohols are named in the IUPAC system as derivatives of the parent alkane, using the suffix *–ol*. The naming procedure is as follows:

Step 1   Select the longest carbon chain containing the hydroxyl group, and replace the *–e* ending of the corresponding alkane with *–ol*.

Step 2   Number the carbons of the parent chain beginning at the end nearer the hydroxyl group.

Step 3   Number all substituents according to their position on the chain, and write the name listing the substituents in alphabetical order.

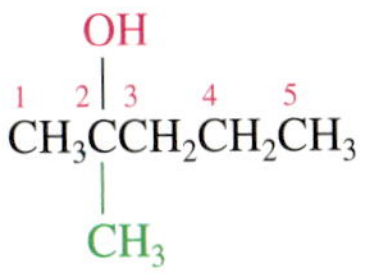

2-Methylpentan-**2-ol**	*cis*-**Cyclohexane-1,4-diol**	**3-Phenyl**butan-**2-ol**

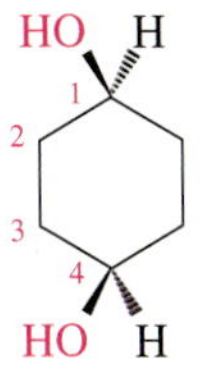

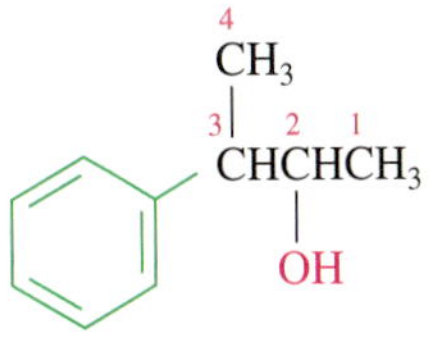

Some well-known alcohols also have common names. For example:

Benzyl alcohol
(phenylmethanol)

CH_2OH

tert-Butyl alcohol
(2-methylpropan-2-ol)

CH_3
CH_3COH
CH_3

Ethylene glycol
(ethane-1,2-diol)

$HOCH_2CH_2OH$

Glycerol
(propane-1,2,3-triol)

$HOCH_2CHCH_2OH$
OH

Phenols

The word *phenol* is used both as the name of a specific substance (hydroxybenzene) and as the family name for all hydroxy-substituted aromatic compounds. Phenols are named as substituted aromatic compounds according to the rules discussed in Section 20.4, with –*phenol* used as the parent name rather than –*benzene*.

m-**Methylphenol**
(*m*-**Cresol**)

H_3C — OH

2,4-Dinitrophenol

OH
O_2N — NO_2

Ethers

Simple ethers that contain no other functional groups are named by identifying the two organic groups attached to the oxygen atom and adding the word **ether**.

tert-**Butyl** methyl ether

H_3C — O — C — CH_3
H_3C CH_3

Ethyl phenyl ether

O — CH_2CH_3

If more than one ether linkage is present, or if other functional groups are present, the ether part is named as an *alkoxy* substituent on the parent compound.

p-**Dimethoxybenzene**

CH_3O — OCH_3

4-*tert*-Butoxycyclohex-1-ene

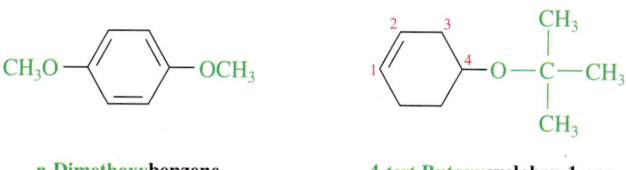

CH_3
O — C — CH_3
CH_3

EXERCISE 22.2—NAMING ALCOHOLS

Draw structures of molecules of compounds with the following IUPAC names
(a) 2-methylhexan-2-ol
(b) hexane-1,5-diol
(c) 2-ethylbut-2-en-1-ol
(d) cyclohex-3-en-1-ol
(e) 2-bromophenol
(f) 2,4,6-trinitrophenol (picric acid, from Section 20.1)

EXERCISE 22.3—IDENTIFYING ALCOHOLS

Identify the alcohols in Exercise 22.2 as primary, secondary, or tertiary.

EXERCISE 22.4—NAMING ETHERS

EXERCISE 22.4—NAMING ETHERS

Name, using IUPAC rules, the ethers whose structural formulas are shown

(a)   CH₃ CH₃
       │ │
       CH₃CHOCHCH₃

(b)   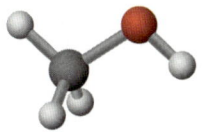 OCH₂CH₂CH₃

(c)   Br— 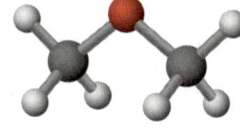 —OCH₃

(d)   (CH₃)₂CHCH₂OCH₂CH₃

22.3 Alcohols, Phenols, and Ethers: Structure and Reactivity

Alcohols, phenols, and ethers can be thought of as organic derivatives of water molecules in which one or both of the hydrogen atoms have been replaced by alkyl groups: H—O—H becomes R—O—H or R—O—R′. Molecules of all three classes of compounds have nearly the same geometry as water. The C—O—H or C—O—C bonds have an approximately tetrahedral angle—109° in methanol and 112° in dimethyl ether, for example—and the oxygen atom can be thought of using the hybridization model as sp^3 hybridized.

In Chapters 19–21, you've seen how electrostatic potential maps for organic molecules with several different functional groups can be used to help understand properties and reactivity. Consider the electrostatic potential map for a typical alcohol, ethanol:

Electron density isosurface of ethanol, coloured by electrostatic potential

Transparent version showing ball and spoke model within

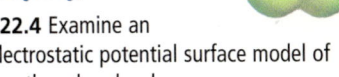

Molecular Modelling (Odyssey)

e22.4 Examine an electrostatic potential surface model of an ethanol molecule.

Can you use this representation of the electron distribution in ethanol and your understanding of how structure affects physical properties and reactivity to make sense of the following four experimental observations?

1. Cyclodextrins are quite soluble in water, due to the —OH groups on the outside of the "cone."

2. Alcohols and phenols differ significantly from the hydrocarbons and alkyl halides we've studied thus far. As shown in Figure 22.3, alcohols have higher boiling points than alkanes or haloalkanes of similar molecular weight. For example, the molecular weights of propan-1-ol (mol wt = 60 u), butane (mol wt = 58 u), and chloroethane (mol wt = 65 u) are similar, but propan-1-ol boils at 97.4 °C, compared with −0.5 °C

for the alkane and 12.3 °C for the chloroalkane. Similarly, phenols have higher boiling points than aromatic hydrocarbons. Phenol itself, for example, boils at 182°C, whereas toluene boils at 110.6 °C.

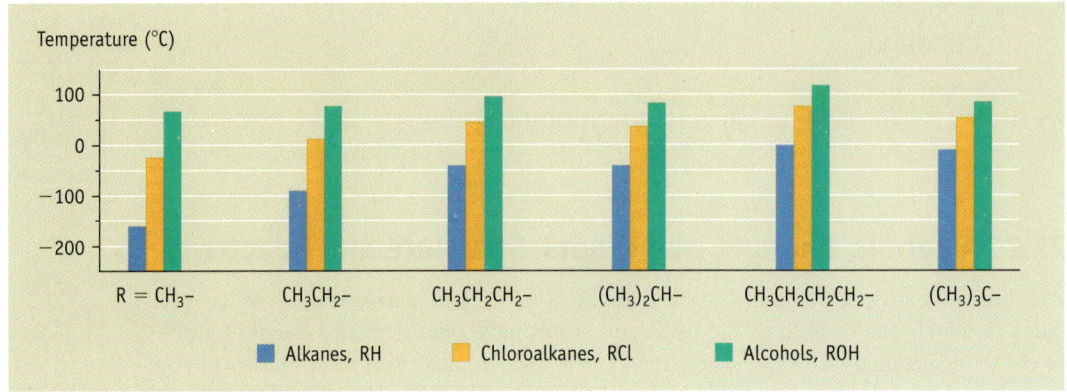

FIGURE 22.3 A comparison of boiling points for some alkanes, chloroalkanes, and alcohols.

3. Compare the O—H stretching region of the infrared spectra of liquid and vapour phase ethanol. When ethanol is in the liquid phase, the O—H stretching absorbance is broad, and extends from 3100 to 3600 cm^{-1}. The vapour phase sample of ethanol shows a sharp absorbance at much higher wavenumber, 3650 cm^{-1} (Figure 22.4).

FIGURE 22.4 Infrared spectra of the O—H stretching region of (a) a thin film of liquid ethanol, and (b) vapour phase infrared spectrum of ethanol.

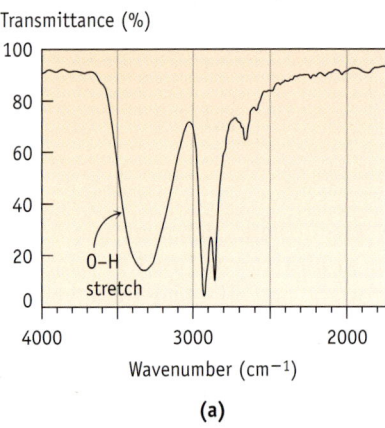

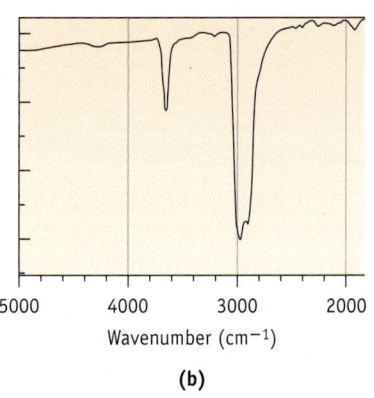

4. An alcohol is a weak acid, reacting like water with a strong base such as $NaNH_2$ to give up a proton and produce its conjugate base, an alkoxide ion.

All four of these observations can be explained with help from the electrostatic potential map of an alcohol, which shows that the O—H bond within each alcohol molecule is a very polar bond. As a result, just as in water, strong *intermolecular hydrogen bonding* [<<Section 6.3] takes place between alcohol or phenol molecules. The partial positive charge on the —OH hydrogen atom of one molecule is attracted to the partial negative charge of the oxygen atom of another molecule, resulting in a force of attraction between the two molecules (Figure 22.5) [<<Section 6.3]. This results in unusually high boiling points for alcohols because these forces must be overcome for a molecule to break free from the liquid and enter the vapour phase. Ethers, because they lack hydroxyl groups, can't form hydrogen bonds and therefore have much lower boiling points. This makes them useful as solvents, as they can be easily removed by evaporation after a reaction is completed, as the first step in isolating reaction products.

Hydrogen bonding can also occur between alcohol and water molecules, explaining the solubility of cyclodextrins in water [<<Section 6.3].

Molecular Modelling (Odyssey)

e22.5 Examine the hydrogen bonding in a liquid alcohol and an aqueous solution of alcohol.

Molecular Modelling (Odyssey)

e22.6 Examine hydrogen bonding in a dimer of methanol.

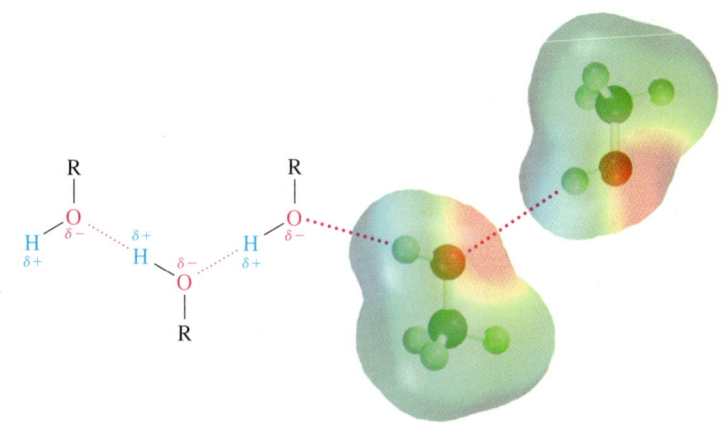

FIGURE 22.5 Hydrogen bonding in alcohols and phenols. The force of attraction between molecules as a result of a partial positive charge on an —OH hydrogen atom of one molecule and a partial negative charge on an —OH oxygen atom of another. The electrostatic potential map of methanol shows the polarity clearly, with the oxygen atoms appearing red and the —OH hydrogen atoms blue.

From MCMURRY. Organic Chemistry 6E. © 2004 Brooks/Cole, a part of Cengage Learning, Inc.

What about the differences in the position and sharpness of the O—H stretching vibration in the liquid and gas phase infrared spectra for ethanol (CH_3CH_2OH)? In a sample of a large number of alcohol molecules, such as are present in a liquid phase sample, we would expect some of the hydrogen bonds between molecules to be quite strong and others to be weaker, and they would constantly be changing. This is shown clearly in e22.5, an Odyssey simulation of molecules of CH_3OH in the liquid phase. A broad signal (3100–3600 cm^{-1}) is observed for the O—H stretching vibration, as the degree of hydrogen bonding affects the vibrational wavenumber corresponding to the O—H stretch. What happens to hydrogen-bonding interactions between alcohol molecules if a liquid phase sample of an alcohol such as ethanol is heated until it vaporizes? In the gas phase, molecules are much farther apart than in the liquid phase, and so hydrogen-bonding interactions become negligible. Elegant evidence that this is exactly what happens comes from seeing that the O—H stretching vibration for a gas phase sample of ethanol gives a very sharp peak at higher wavenumber, at 3650 cm^{-1}.

The polarity of the RO—H bond in alcohols and phenols is also evident in their reactivity as weak acids. Strong bases can pull off a proton from alcohols, producing their conjugate bases, called alkoxide ions (RO$^-$).

22.4 Spectroscopy of Alcohols and Phenols

Infrared Spectroscopy of Alcohols and Phenols

Since most simple alcohols are liquids and IR spectra are recorded under conditions where hydrogen bonding occurs, the O—H stretching band is usually seen as an intense broad peak, centred at 3300–3400 cm^{-1}, as seen in the animated infrared spectrum of cyclopentanol in Figure 22.6. Alcohols also show a strong C—O stretching absorption near 1050 cm^{-1}, whose exact position varies slightly for 1°, 2° and 3° alcohols. The predictable absorption peak for an alkane C—H stretching vibration occurs just below 3000 cm^{-1}.

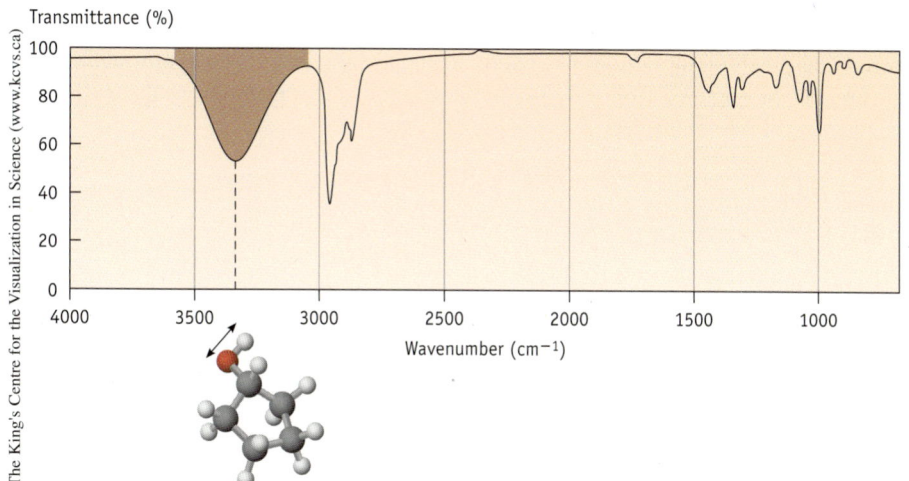

The King's Centre for the Visualization in Science (www.kcvs.ca)

FIGURE 22.6 Frame from animated spectrum (e22.7) of cyclopentanol showing O—H, C—O, and C—H stretching vibrations.

Think about It

e22.7 Compare the animated IR bond stretching modes and spectra of compounds with and without an –OH group.

Refer to the animated infrared spectra of benzene and toluene in e22.7 to remind yourself of the characteristic absorbance positions used to identify compounds with aromatic rings.

The infrared spectra of phenols are consistent with what we know about compounds with O—H functional groups and aromatic rings. In the spectrum of phenol itself (Figure 22.7), the mono-substituted aromatic ring can be characterized by the aromatic C—H stetching vibration just above 3000 cm^{-1}, by the aromatic bands at 1500 and 1600 cm^{-1}, and by the out-of-plane bending vibrations at 690 and 760 cm^{-1}.

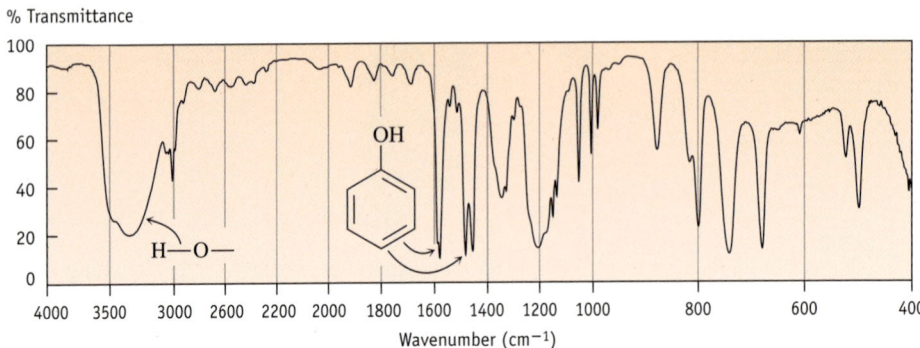

FIGURE 22.7 Infrared spectrum of phenol.

EXERCISE 22.5—IR SPECTROSCOPY OF ALCOHOLS

In the infrared spectrum of benzyl alcohol ($C_6H_5CH_2OH$), identify the peaks corresponding to the O—H stretch, the alkane C—H stretch, the aromatic C—H stretch, the C—O stretch, and the benzene ring. Is benzyl alcohol an example of a phenol?

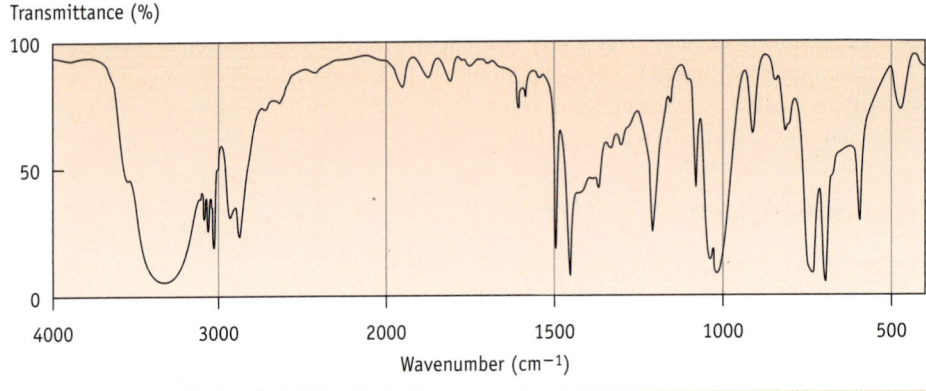

Nuclear Magnetic Resonance Spectroscopy

Alcohols show characteristic absorptions in the ^{1}H NMR spectrum. The ^{1}H NMR spectrum of propan-1-ol ($CH_3CH_2CH_2OH$) in Figure 22.8 will remind you of some features of spectral interpretation introduced in Chapter 9, and to explain what new information about structure can be obtained from observing that some of the absorption peaks in ^{1}H NMR spectra are split into multiplets.

FIGURE 22.8 ^{1}H NMR spectrum of propan-1-ol.

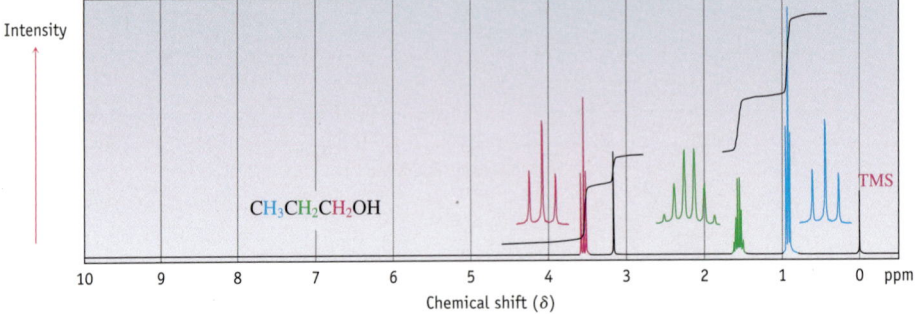

Identify the absorptions for the CH_3, CH_2, CH_2, and OH hydrogen atoms. Focus first on the chemical shift and integrations. The CH_3 methyl group (relative integration 3) absorbs upfield, close to TMS, as expected, since it is quite far away from the electron-withdrawing OH group. The two CH_2 and CH_2 methylene groups (each with relative integration 2) absorb downfield with the CH_2 group furthest downfield, since it is directly attached to the electronegative oxygen atom. The O—H hydrogen atom (relative integration 1) absorbs at 3.2 δ in the normal range for alcohols (3.0–4.5 δ).

But note that the signals for the CH_3, CH_2, and CH_2 hydrogen atoms groups are each split into multiple absorptions. Called **spin–spin splitting**, the phenomenon of multiple absorptions is caused by the interaction, or *coupling*, of the nuclear spins of neighbouring atoms. The tiny magnetic field of one nucleus affects the magnetic field felt by a neighbouring nucleus. As a general rule, called the *n + 1 rule*, protons that have *n* equivalent neighbouring protons (within three bonds on saturated systems) split into *n + 1* peaks in their NMR absorption. An explanation of this rule is given in e22.8.

Applying this rule, the CH_3 group has two neighbouring hydrogen atoms (on the CH_2 group) within three bonds and is therefore split into a triplet. The CH_2 group has five neighbouring hydrogen atoms (three on the neighbouring CH_3 and two on the neighbouring CH_2), and is split into a sextet. You might expect the CH_2 to be influenced by the magnetic field of both the O—H and CH_2 hydrogen atoms, since both are within three bonds. But the O—H hydrogen atom is a singlet, and the CH_2 is a triplet, affected only by the two neighbouring CH_2 hydrogen atoms.

Once again, the polarity of the O—H group, as evident from the electrostatic potential map, can help to explain this. Most samples contain small amounts of acidic impurities, which catalyze an exchange of the O—H proton on a time scale that is faster than the time it takes for the neighbouring CH_2 hydrogen atoms to undergo nuclear spin flipping. So spin–spin splitting between O—H and neighbouring C—H hydrogen atoms is not usually seen in the NMR spectra of alcohols of standard commercial purity. If an alcohol is rigorously purified to remove any traces of acid, it is possible to see splitting between these neighbouring groups.

Taking It Further

e22.8 See how extra information can be extracted from a ^{1}H NMR spectrum due to spin-spin coupling.

No NMR coupling observed

Additional evidence for rapid proton exchange in alcohols comes from observing that when D_2O (water whose molecules are isotopically substituted with deuterium rather than hydrogen atoms) is added to the NMR tube containing an alcohol sample, the O—H signal disappears. The O—H H atoms are replaced by deuterium atoms, which resonate at a much different frequency.

^{13}C nuclei bonded to electron-withdrawing OH groups are deshielded and absorb at a lower frequency in the ^{13}C NMR spectrum than do typical alkane carbons. Most alcohol carbon absorptions fall in the range $\delta = 50–80$ ppm, as illustrated for cyclohexanol below.

Cyclohexanol   **Phenol**

^{13}C absorptions for phenols come in the region expected for aromatic carbons, as shown for phenol. Again, the carbon directly bonded to the —OH group gives rise to a peak substantially downfield from the others.

Mass Spectrometry

Often molecular ion signals, from which we can obtain the molecular weight of a compound, are very weak in intensity or absent completely in the mass spectra of alcohols. This is because the molecular ion formed when an alcohol is ionized in a mass spectrometer is usually quite unstable and forms fragments [<<Section 3.10] by two characteristic pathways, alpha-cleavage and dehydration. In the alpha-cleavage pathway, a C—C bond nearest the hydroxyl group is broken, yielding a neutral radical plus a charged oxygen-containing fragment:

Alpha cleavage

In the dehydration pathway, water is eliminated, yielding an alkene radical cation:

Dehydration

Dehydration can be recognized by the loss of water with a mass of 18 mass units, giving a fragment radical cation that has the same parity as the molecular ion—that is, if the molecular ion has an even m/z value, the fragment formed by loss of a neutral water molecule will also have an even m/z value. In most other mass spectral fragmentation patterns, a free radical with an odd molar mass is lost, giving a fragment ion with different parity from the starting molecular ion.

Both of these characteristic fragmentation pathways are apparent in the mass spectrum of butan-1-ol (Figure 22.9). Only a very weak signal for a molecular ion is seen at $m/z = 74$. Loss of water from the molecular ion gives a peak at $m/z = 56$, and the peak at $m/z = 31$ results from alpha-cleavage.

FIGURE 22.9 Mass spectrum of butan-1-ol.

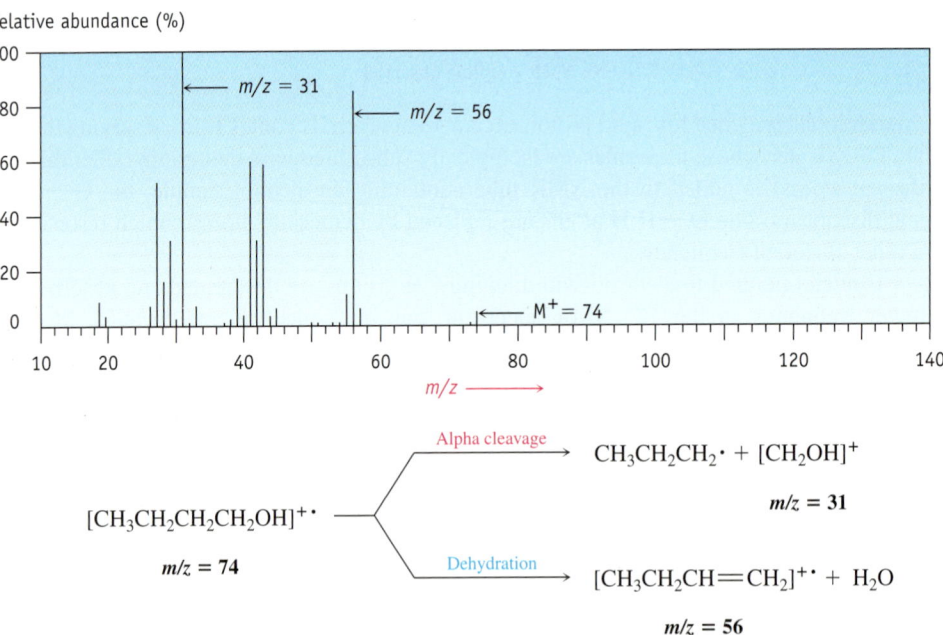

EXERCISE 22.6—SPECTROSCOPY OF ALCOHOLS

Assume that you need to make 5-cholestene-3-one from cholesterol. How could you use infrared spectroscopy to tell if the reaction occurred? What differences would you look for in the infrared spectra of the starting material and product? Refer back to the discussion in Section 3.11 and e3.19 as needed.

$$\text{Cholesterol} \xrightarrow[\text{H}_3\text{O}^+]{\text{CrO}_3} \text{5-Cholestene-3-one}$$

| **Cholesterol** | **5-Cholestene-3-one** |

22.5 Synthesis of Alcohols

Alcohols occupy a central position in organic chemistry. They can be synthesized from many other kinds of compounds (alkenes, alkyl halides, ketones, aldehydes, and esters, among others), and they can be transformed into an equally wide assortment of compounds.

We've already seen, for instance, that alcohols can be synthesized by hydration of alkenes [<<Section 19.10]. Treatment of an alkene with water and an acid catalyst leads to the product with regiochemistry predicted by Markovnikov, forming a tertiary alcohol in the example below.

1-Methylcyclohexene   **1-Methylcyclohexanol**

The most general method for making alcohols is by reduction of carbonyl compounds—the formal addition of H_2 across a $C={O}$ double bond. Let's consider why this is categorized as a reduction reaction and list the kinds of carbonyl reductions that can be carried out.

A carbonyl compound   **An alcohol**

where [H] is a generalized reducing agent

Oxidation and Reduction in Organic Chemistry

From examples in Chapter 16, you are familiar with oxidation and reduction reactions of inorganic compounds. In organic chemistry, where polar covalent bonds are common, redox reactions are harder to spot by inspecting a chemical equation. *It is often helpful to think of an oxidation reaction of organic compounds as one that results in a decrease of electron density by a carbon atom.* This loss is usually caused either by bond formation between carbon and a more electronegative atom (usually oxygen, nitrogen, or a halogen) or by bond breaking between carbon and a less electronegative atom (usually hydrogen).

Conversely, *a reduction of organic compounds results in a gain of electron density by a carbon atom.* This gain is usually caused either by bond formation between carbon and a less electronegative atom or by bond breaking between carbon and a more electronegative atom, summarized in Figure 22.10.

FIGURE 22.10 Oxidation levels of some common types of organic compounds.

CH_3CH_3	$H_2C{=}CH_2$	$HC{\equiv}CH$	
CH_3OH	$H_2C{=}O$	HCO_2H	CO_2
CH_3Cl	CH_2Cl_2	$CHCl_3$	CCl_4
CH_3NH_2	$H_2C{=}NH$	$HC{\equiv}N$	

Low oxidation level $\longrightarrow$ High oxidation level

The most important oxidation reactions of alcohols produce carbonyl compounds, which we say are at a higher *oxidation level* than alcohols. But these carbonyl compounds can also be reduced to produce alcohols.

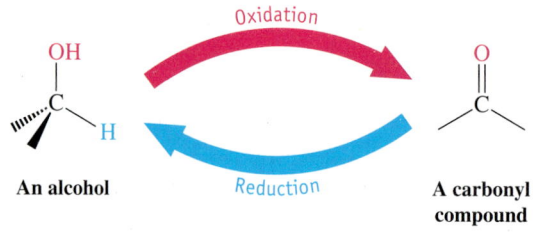

An alcohol — Oxidation / Reduction — A carbonyl compound

WORKED EXAMPLE 22.1—OXIDATION LEVEL OF COMPOUNDS

Rank the following series of compounds in order of increasing oxidation level:

(a) $CH_3CH{=}CH_2$
(b) $CH_3\overset{\displaystyle OH}{\underset{\displaystyle |}{C}}HCH_3$
(c) $CH_3\overset{\displaystyle O}{\overset{\displaystyle \|}{C}}CH_3$
(d) $CH_3CH_2CH_3$

Solution

Each of the four compounds in this series has three carbon atoms, so a direct comparison of the number of C—O and C—H bonds can be made. Comparing (a) and (b), the alcohol in (b) has one carbon atom that has formed a new bond to oxygen, and another carbon atom that has formed a new bond to hydrogen. Thus, the oxidation levels of molecules of (a) and (b) are quite similar. Comparing (c) and (b), the ketone in (c) has a higher oxidation level, since the central carbon atom in the ketone has two bonds to oxygen instead of one in (b), and it has lost a hydrogen atom. Compound (d) has two more C—H bonds than compound (a). So we would expect the order of increasing oxidation level to be (d) < (a) ≅ (b) < (c).

During oxidation of organic compounds, an oxygen, nitrogen, or halogen atom is often added to a carbon atom, while during reduction, hydrogen atoms are often added.

Reduction of Aldehydes and Ketones

Aldehydes and ketones are easily reduced to yield alcohols. Aldehydes are converted into primary alcohols, and ketones are converted into secondary alcohols. The reduction of the carbonyl compounds can be recognized by noting that hydrogen atoms are added to carbon atoms, and the number of carbon–oxygen bonds is reduced from two to one.

An aldehyde   **A primary alcohol**   **A ketone**   **A secondary alcohol**

Many reducing reagents are available, but sodium borohydride ($NaBH_4$) is usually chosen because of its safety and ease of handling.

Aldehyde reduction

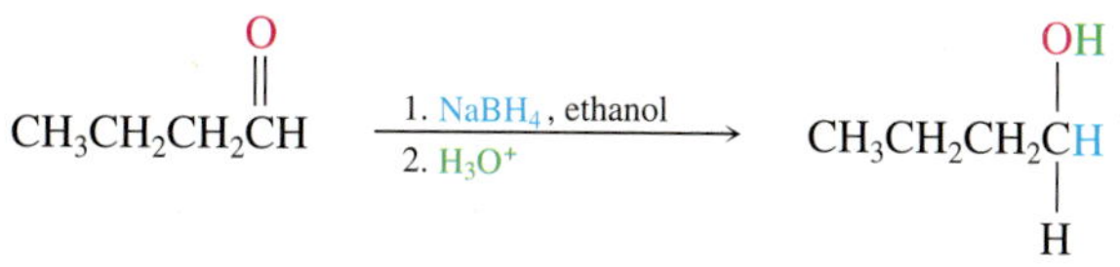

Butanal

Butan-1-ol (85%)
(a 1° alcohol)

Ketone reduction

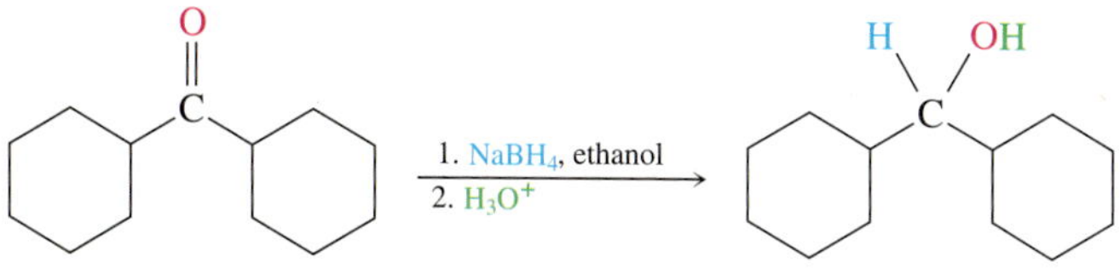

Dicyclohexyl ketone

Dicyclohexylmethanol (88%)
(a 2° alcohol)

Reduction of Esters and Carboxylic Acids

Esters and carboxylic acids are reduced to give primary alcohols:

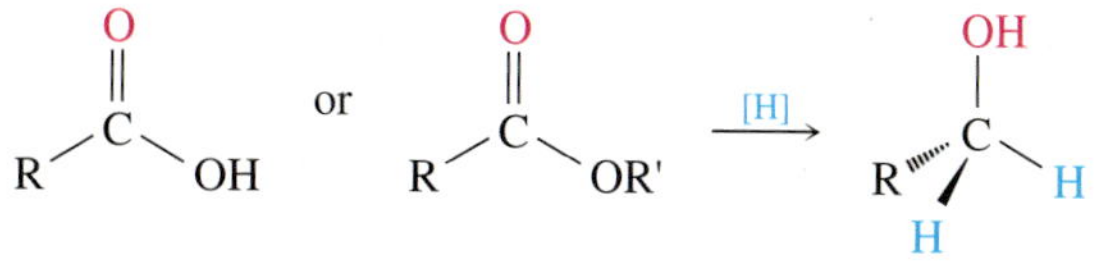

A carboxylic acid   **An ester**   **A primary alcohol**

These reactions proceed more slowly than the corresponding reductions of aldehydes and ketones, so the more powerful reducing agent $LiAlH_4$ is used rather than $NaBH_4$. ($LiAlH_4$ will also reduce aldehydes and ketones.) Note that only one hydrogen atom is added to the carbonyl carbon atom during the reduction of an aldehyde or ketone, but two hydrogen atoms are added to the carbonyl carbon atom during reduction of an ester or carboxylic acid.

Carboxylic acid reduction

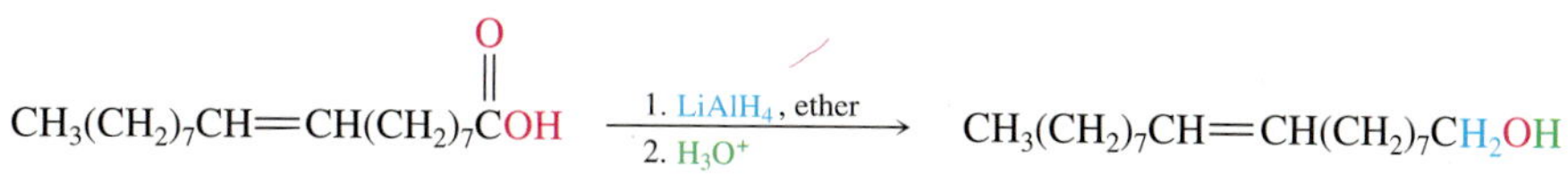

$$CH_3(CH_2)_7CH=CH(CH_2)_7COH \xrightarrow[\text{2. } H_3O^+]{\text{1. } LiAlH_4 \text{, ether}} CH_3(CH_2)_7CH=CH(CH_2)_7CH_2OH$$

Octadec-9-enoic acid
(oleic acid)

Octadec-9-en-1-ol (87%)

Ester reduction

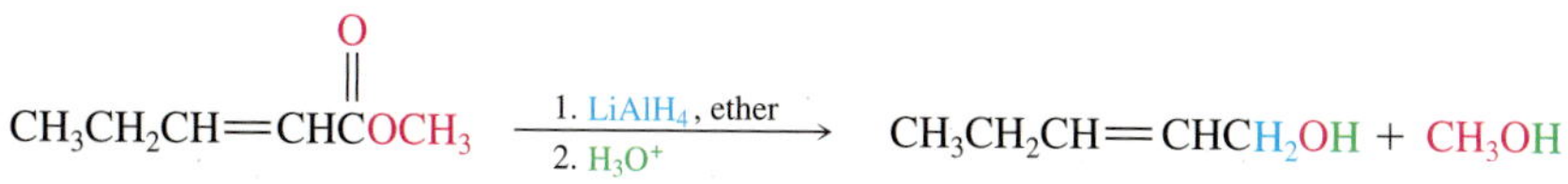

$$CH_3CH_2CH=CHCOCH_3 \xrightarrow[\text{2. } H_3O^+]{\text{1. } LiAlH_4 \text{, ether}} CH_3CH_2CH=CHCH_2OH + CH_3OH$$

Methyl pent-2-enoate

Pent-2-en-1-ol (91%)

WORKED EXAMPLE 22.2—REDUCTION OF CARBONYL COMPOUNDS

Predict the product of the following reaction:

$$CH_3CH_2CH_2\overset{\overset{\displaystyle O}{\|}}{C}CH_2CH_3 \xrightarrow[\text{2. H}_3\text{O}^+]{\text{1. NaBH}_4} ?$$

Solution

Ketones are reduced by treatment with NaBH₄ to yield secondary alcohols. Reduction of hexan-3-one yields hexan-3-ol.

$$CH_3CH_2CH_2\overset{\overset{\displaystyle O}{\|}}{C}CH_2CH_3 \xrightarrow[\text{2. H}_3\text{O}^+]{\text{1. NaBH}_4} CH_3CH_2CH_2\overset{\overset{\displaystyle OH}{|}}{C}HCH_2CH_3$$

Hexan-3-one   **Hexan-3-ol**

WORKED EXAMPLE 22.3—REDUCTION OF CARBONYL COMPOUNDS

What carbonyl compound(s) might be reduced to obtain the following alcohol?

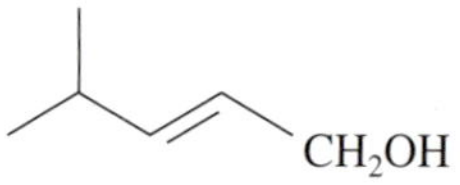

Solution

Identify the alcohol as primary, secondary, or tertiary. A primary alcohol can be prepared by reduction of an aldehyde, an ester, or a carboxylic acid; a secondary alcohol can be prepared by reduction of a ketone; and a tertiary alcohol cannot be prepared by reduction. In this case, the target molecule is a primary alcohol, which can be prepared by reduction of an aldehyde, an ester, or a carboxylic acid. LiAlH₄ is needed for the ester and carboxylic acid reductions.

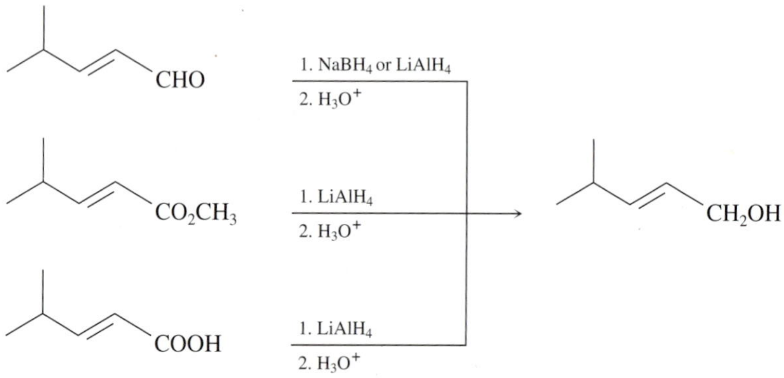

EXERCISE 22.7—REDUCTION OF CARBONYL COMPOUNDS

What reagents would you need to carry out the following reactions?

(a) $CH_3\overset{\overset{\displaystyle O}{\|}}{C}CH_2CH_2\overset{\overset{\displaystyle O}{\|}}{C}OCH_3 \xrightarrow{?} CH_3\overset{\overset{\displaystyle OH}{|}}{C}HCH_2CH_2\overset{\overset{\displaystyle O}{\|}}{C}OCH_3$

(b) $CH_3\overset{\overset{\displaystyle O}{\|}}{C}CH_2CH_2\overset{\overset{\displaystyle O}{\|}}{C}OCH_3 \xrightarrow{?} CH_3\overset{\overset{\displaystyle OH}{|}}{C}HCH_2CH_2CH_2OH$

The most general method of alcohol synthesis involves reduction of a carbonyl compound. Aldehydes, esters, and carboxylic acids yield primary alcohols on reduction; ketones yield secondary alcohols.

22.6 Reactions of Alcohols

Alcohol molecules undergo reactions at both their O—H and C—O bonds. As evident from the electrostatic potential map of ethanol, both of these bonds are exceptionally polar. Strong bases should be able to remove a proton from the —OH group and we would predict that nucleophiles should react at the electrophilic carbon atom of the C—O bond. The electron pairs on the alcohol oxygen atom can also be donated to other electrophiles.

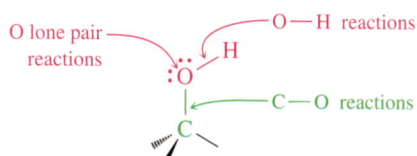

Alcohols as Weak Acids and Weak Bases

Water also undergoes reactions at its O—H bonds, and donation of its lone pair of electrons from the oxygen atom to form a bond with Lewis acids. Alcohols and phenols, like water, are both weakly acidic and weakly basic. As weak Lewis bases, alcohols and phenols are reversibly protonated by strong acids to yield *oxonium ions* (ROH_2^+):

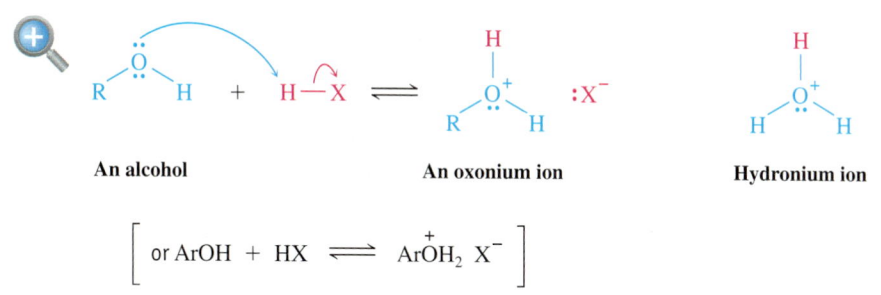

| An alcohol | An oxonium ion | Hydronium ion |

$$\left[\text{or ArOH} + \text{HX} \rightleftharpoons \text{ArOH}_2^+ \text{ X}^- \right]$$

Note the similarity between an oxonium ion and the hydronium ion, obtained by protonating water.

As weak acids, alcohols and phenols dissociate to a slight extent in dilute aqueous solution by donating a proton to water, generating H_3O^+ and an *alkoxide ion (RO$^-$)* or a *phenoxide ion (ArO$^-$)*.

$$\text{RO}-\text{H} + \text{H}_2\ddot{\text{O}}: \rightleftharpoons \text{R}\ddot{\text{O}}:^- + \text{H}_3\text{O}:^+$$

Table 22.1 gives the pK_a values of some common alcohols and phenols.

Alcohol or Phenol	pK_a	
$(CH_3)_3COH$	18.00	Weaker acid
CH_3CH_2OH	16.00	
[HOH, water]*	[15.74]	
CH_3OH	15.54	
p-Methylphenol	10.17	
Phenol	9.89	
p-Bromophenol	9.35	
p-Nitrophenol	7.15	Stronger acid

* Value for water is shown for reference. While K_w is 1×10^{-14}, the K_a for water (and from this the pK_a) is obtained by dividing the K_w by the concentration of water (55.49 M).

Molecular Modelling (Odyssey)

e22.9 Compare the relative acidity of alcohols using electrostatic potential maps.

Review Sections 14.4–14.6 for the definition of pK_a and implications for the relative reactivity of acids. A more complete table of pK_a values is found in Appendix A.

TABLE 22.1 pK_a Values of Some Alcohols and Phenols

Molecular Modelling (Odyssey)

e22.10 Use electrostatic potential maps to predict where an alcohol acts as an acid or a base.

The data in Table 22.1 show that alcohols are about as acidic as water. They react with alkali metals just as water does to yield alkoxide ions that are strong bases. These alkoxide ions are very useful in synthesis as strong bases and nucleophiles.

$$2 \, CH_3OH + 2 \, Na \longrightarrow 2 \, CH_3O^- \, Na^+ + H_2$$

Methanol   **Sodium methoxide**

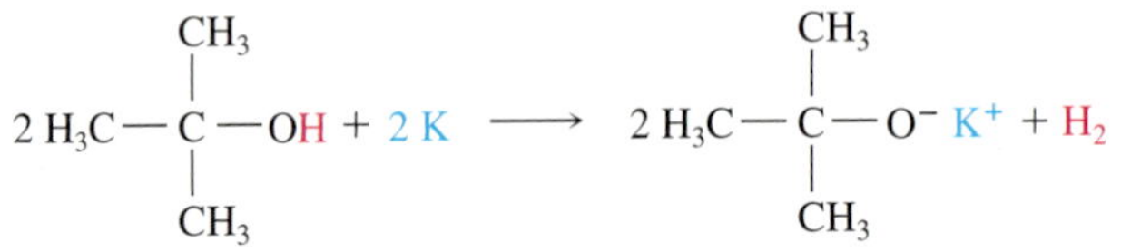

tert-**Butyl alcohol**   **Potassium *tert*-butoxide**

As measured by their differences in K_a values, phenols are about a million times more acidic than alcohols. In fact, some nitro-substituted phenols approach or surpass the acidity of carboxylic acids. One practical consequence of this acidity is that phenols are soluble in dilute NaOH solutions, but alcohols of about the same molecular weight are not.

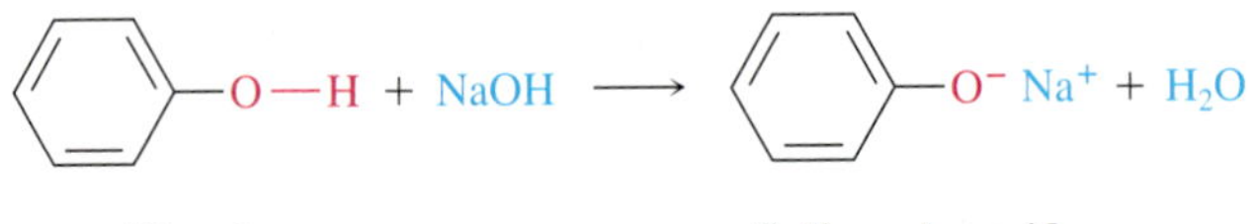

Phenol   **Sodium phenoxide**

This striking increase in acidity results from the ability of the phenoxide anion to delocalize its charge over the aromatic ring. This lowers the energy of the phenoxide anion and increases the tendency of the corresponding phenol to be deprotonated (Figure 22.11).

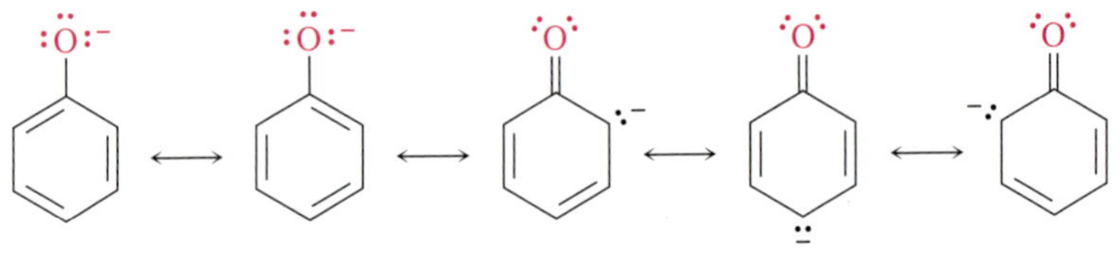

FIGURE 22.11 The negative charge is less localized on a phenoxide oxygen atom than on the methoxide oxygen atom. Distributing this negative charge over a larger volume of the molecule stabilizes the phenoxide ion (relative to phenol) more than an alkoxide ion is stabilized (relative to an alcohol). Thus, we observe that phenols are much more acidic than alcohols, and have a lower pK_a.

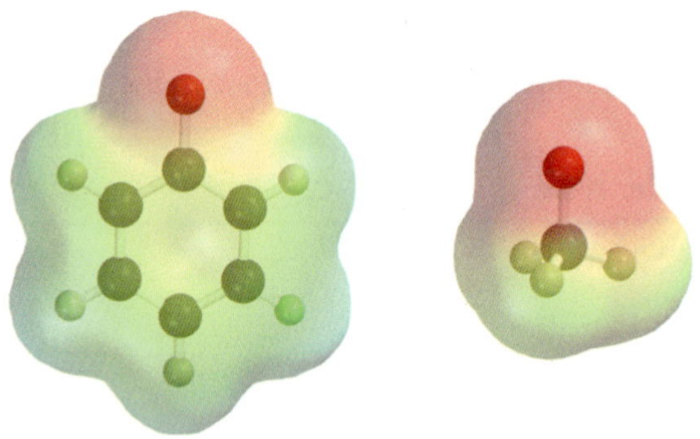

Phenoxide ion

Methoxide ion

Molecular Modelling (Odyssey)

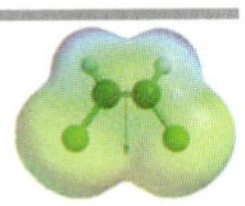

e22.11 Use electrostatic potential maps to see why phenols are much stronger acids than other alcohols.

Conversion into Ethers

Alcohols are converted into ethers by formation of the corresponding alkoxide ion followed by reaction with an alkyl halide, a reaction known as the *Williamson ether synthesis*. As noted in Section 22.3, the alkoxide ion needed in the reaction can be prepared by reaction of an alcohol with $NaNH_2$.

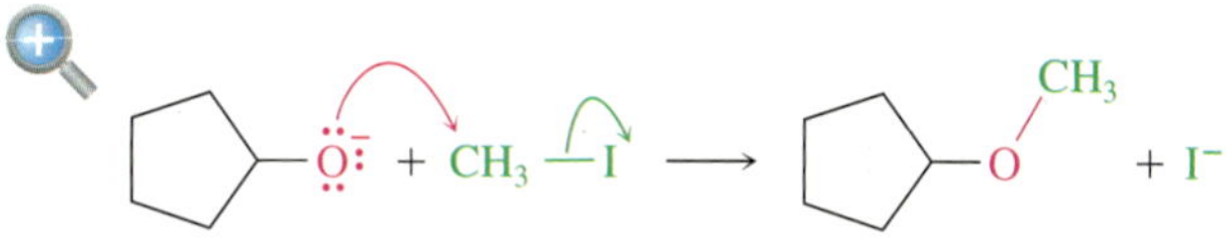

Cyclopentoxide ion   **Cyclopentyl methyl ether**
                                          (74%)

This reaction is classified as a nucleophilic substitution reaction, in which halide ion is replaced by the alkoxide ion. The mechanism of this reaction is most consistent with an S_N2 pathway [<<Section 21.6]. Alkyl halides that are not sterically hindered, such as CH_3I or a primary halide, work best because competitive E2 elimination of HX can occur with more hindered substrates. Unsymmetrical ethers are best prepared by reaction of the more hindered alkoxide partner with the less hindered alkyl halide partner, rather than vice versa. For example, *tert*-butyl methyl ether is best synthesized by reaction of *tert*-butoxide ion with iodomethane, rather than by reaction of methoxide ion with 2-chloro-2-methylpropane.

S_N2 reaction

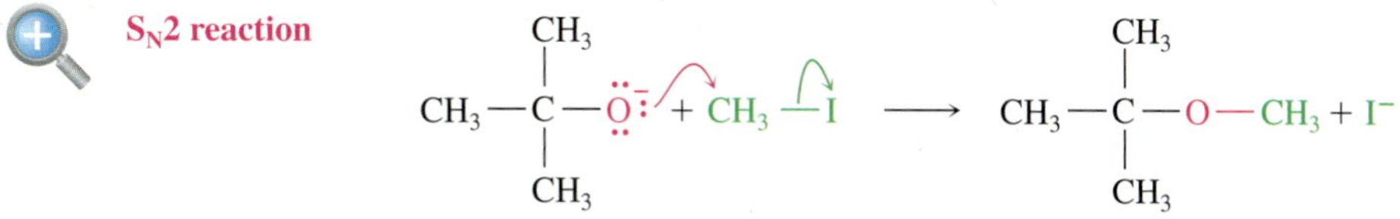

tert-Butoxide   **Iodomethane**   **_tert_-Butyl methyl ether**
ion

EXERCISE 22.8—WILLIAMSON SYNTHESIS OF ETHERS

Treatment of cyclohexanol with $NaNH_2$ gives an alkoxide ion that undergoes reaction with iodoethane to yield cyclohexyl ethyl ether. Write the reaction equation, showing all the steps. Could you equally well make cyclohexyl ethyl ether by treating ethanol with $NaNH_2$ and reacting the product with cyclohexyl iodide?

EXERCISE 22.9—WILLIAMSON SYNTHESIS OF ETHERS

Rank the following alkyl halides in order of their expected reactivity toward an alkoxide ion in the Williamson ether synthesis: bromoethane, 2-bromopropane, chloroethane, 2-chloro-2-methylpropane.

Dehydration of Alcohols

In Section 19.10, we saw that alkenes can undergo acid-catalyzed hydration (addition of H_2O) to produce alcohols, with regiochemistry following the pattern observed by Markovnikov. Under appropriate conditions, the reverse of this reaction can also be made to happen, where alcohols undergo *dehydration* (elimination of H_2O) to give alkenes. Although many ways of carrying out the reaction have been devised, a method that works particularly well for secondary and tertiary alcohols is treatment with a strong acid, often at elevated temperatures. For example, when 1-methylcyclohexanol is treated with aqueous sulfuric acid, dehydration occurs to yield 1-methylcyclohexene:

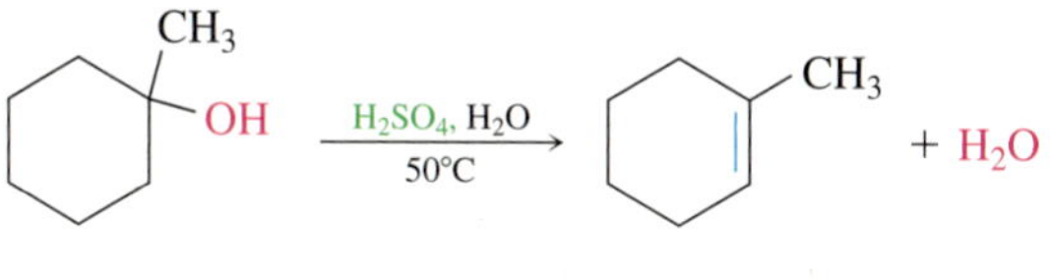

1-Methylcyclohexanol   **1-Methylcyclohexene (91%)**

Acid-catalyzed dehydrations usually follow Zaitsev's rule [<<Section 21.9] and yield the more highly substituted alkene as major product. For example, 2-methyl-butan-2-ol gives primarily 2-methylbut-2-ene (trisubstituted) rather than 2-methylbut-1-ene (disubstituted):

$$CH_3CH_2-\underset{\underset{\displaystyle CH_3}{|}}{\overset{\overset{\displaystyle OH}{|}}{C}}-CH_3 \xrightarrow[\text{25°C}]{H_2SO_4, H_2O} CH_3CH=\underset{\underset{\displaystyle CH_3}{|}}{C}CH_3 \ + \ CH_3CH_2\underset{\underset{\displaystyle CH_3}{|}}{C}=CH_2$$

2-Methylbutan-2-ol **2-Methylbut-2-ene** **2-Methylbut-1-ene**
(major) (minor)

As might be expected for a tertiary substrate under strongly acidic conditions, the mechanism of this acid-catalyzed dehydration is consistent with an E1 process [<<Section 21.10]. A molecular-level view might show that the strong acid protonates the alcohol oxygen, the protonated intermediate then loses water to generate a carbocation, and loss of H⁺ from a neighbouring carbon atom then yields the alkene product (Figure 22.12).

FIGURE 22.12 Accepted reaction mechanism for the acid-catalyzed dehydration of a tertiary alcohol.

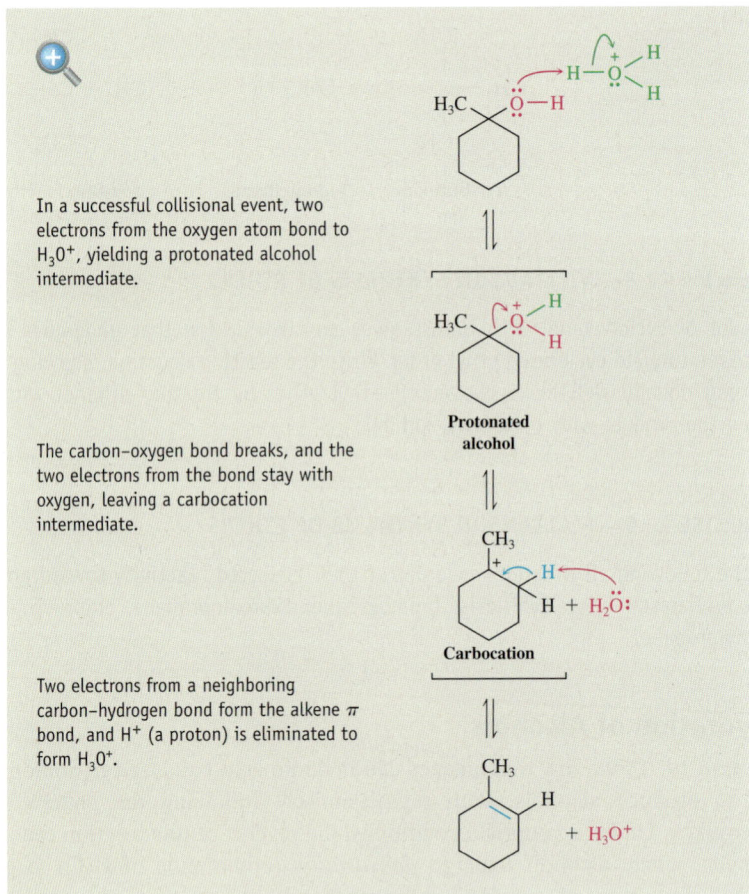

In a successful collisional event, two electrons from the oxygen atom bond to H_3O^+, yielding a protonated alcohol intermediate.

The carbon–oxygen bond breaks, and the two electrons from the bond stay with oxygen, leaving a carbocation intermediate.

Two electrons from a neighboring carbon–hydrogen bond form the alkene π bond, and H⁺ (a proton) is eliminated to form H_3O^+.

WORKED EXAMPLE 22.4—DEHYDRATION OF ALCOHOLS

Predict the major product of the following reaction:

$$CH_3CH_2\underset{\underset{\displaystyle H}{|}}{\overset{\overset{\displaystyle H_3C \ \ OH}{| \ \ \ |}}{C}}HCHCH_3 \xrightarrow{H_2SO_4, H_2O} \ ?$$

Solution

Treatment of an alcohol with H_2SO_4 leads to dehydration and formation of the more highly substituted alkene product (Zaitsev's rule). Thus, dehydration of 3-methylpentan-2-ol yields 3-methylpent-2-ene as the major product rather than 3-methylpent-1-ene.

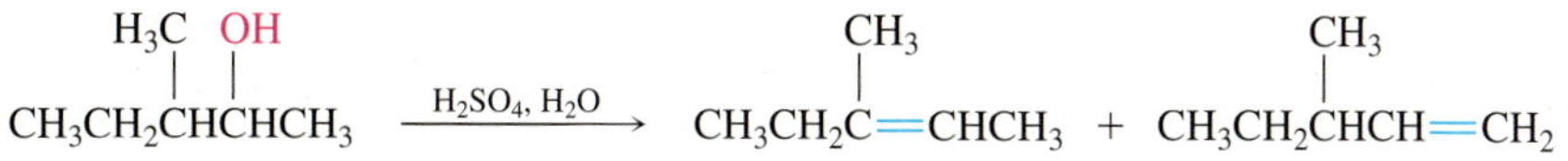

$$H_3C \quad OH$$
$$CH_3CH_2CHCHCH_3 \xrightarrow{H_2SO_4,\ H_2O} CH_3CH_2C=CHCH_3 \ + \ CH_3CH_2CHCH=CH_2$$

3-Methylpentan-2-ol   **3-Methylpent-2-ene**   **3-Methylpent-1-ene**
                                      (major)   (minor)

EXERCISE 22.10—DEHYDRATION OF ALCOHOLS

Predict the products you would expect from the following reactions. Indicate the major product in each case.

(a) $H_3C \quad OH$
$$CH_3CHCCH_2CH_3 \xrightarrow{H_2SO_4}$$
$$\qquad\quad CH_3$$

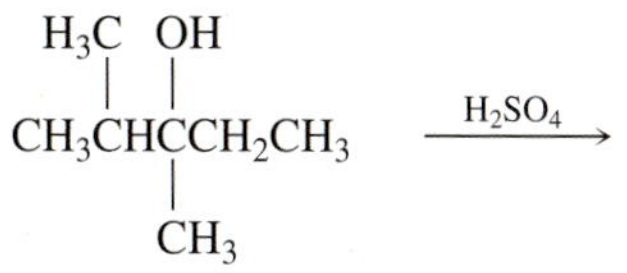

(b) $\qquad\qquad OH$
$$CH_3CH_2CH_2CCH_3 \xrightarrow{H_2SO_4}$$
$$\qquad\qquad\quad CH_3$$

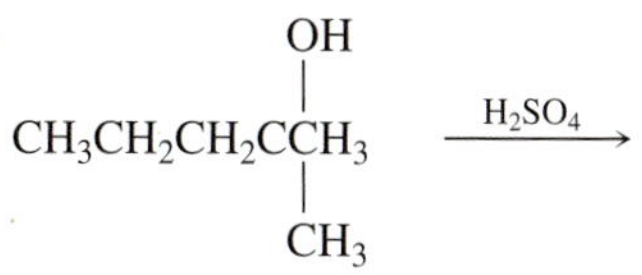

Oxidation of Alcohols

One of the most valuable reactions of alcohols is their oxidation to yield carbonyl compounds: $CH-OH \longrightarrow C=O$. Primary alcohols yield aldehydes or carboxylic acids, and secondary alcohols yield ketones, but tertiary alcohols don't normally react with oxidizing agents.

Primary alcohol

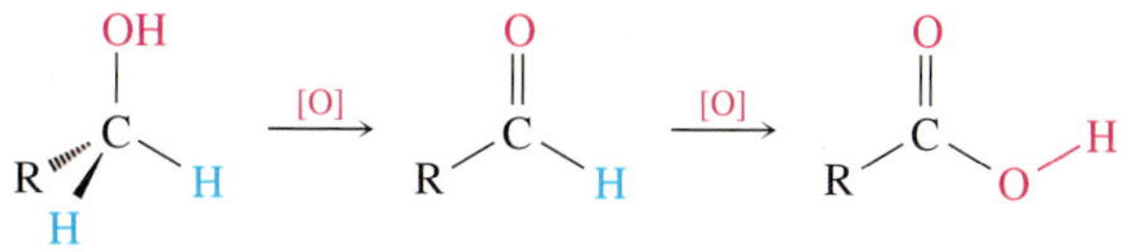

An aldehyde   **A carboxylic acid**

Secondary alcohol

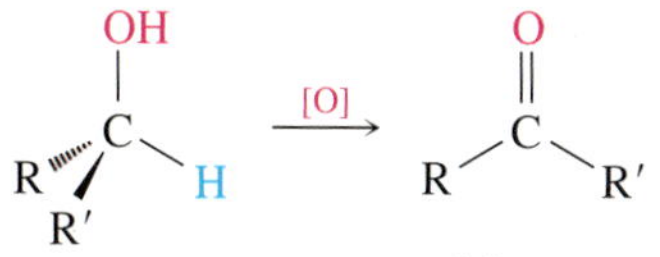

A ketone

Primary alcohols are oxidized either to aldehydes or to carboxylic acids, depending on the reagents chosen. Probably the best method for synthesizing an aldehyde from a primary alcohol on a laboratory scale (as opposed to an industrial scale) is by use of pyridinium chlorochromate (PCC, $C_5H_6NCrO_3Cl$) in dichloromethane solvent. This reagent is too expensive for large-scale use in industry, however.

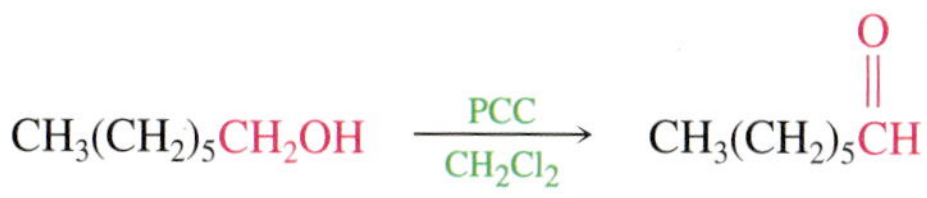

$$CH_3(CH_2)_5CH_2OH \xrightarrow[CH_2Cl_2]{PCC} CH_3(CH_2)_5CH$$

Heptan-1-ol   **Heptanal (78%)**

Many oxidizing agents, such as chromium trioxide (CrO_3) and sodium dichromate ($Na_2Cr_2O_7$) in aqueous acid solution oxidize primary alcohols to carboxylic acids. Although aldehydes are intermediates in these oxidations, they usually can't be isolated because further oxidation takes place too rapidly.

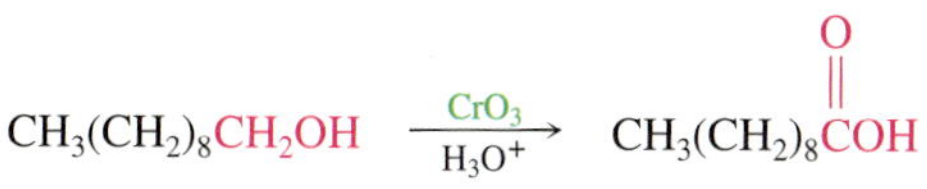

$$CH_3(CH_2)_8CH_2OH \xrightarrow[H_3O^+]{CrO_3} CH_3(CH_2)_8COH$$

Decan-1-ol   **Decanoic acid (93%)**

Secondary alcohols are oxidized easily to produce ketones. Sodium dichromate dissolved in aqueous acetic acid is often used as the oxidant, although pyridinium chlorochromate also works.

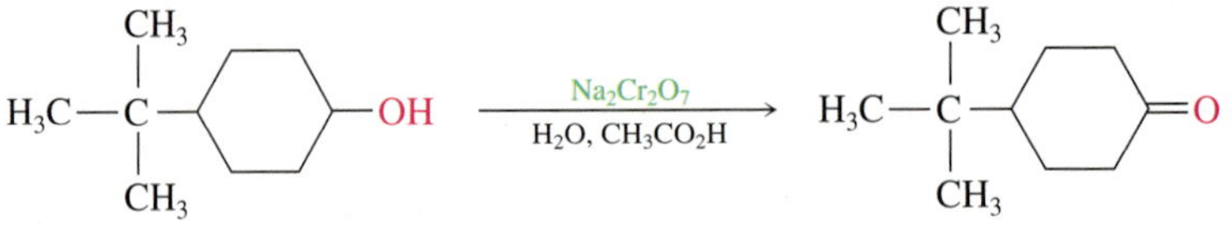

4-*tert*-Butylcyclohexanol → **4-*tert*-Butylcyclohexanone (91%)**

Due to the toxicity of the chromium by-products of this reaction in waste water, more environmentally friendly oxidizing agents such as household bleach (NaOCl) in acetic acid have been developed, and shown to work well for many secondary alcohols.

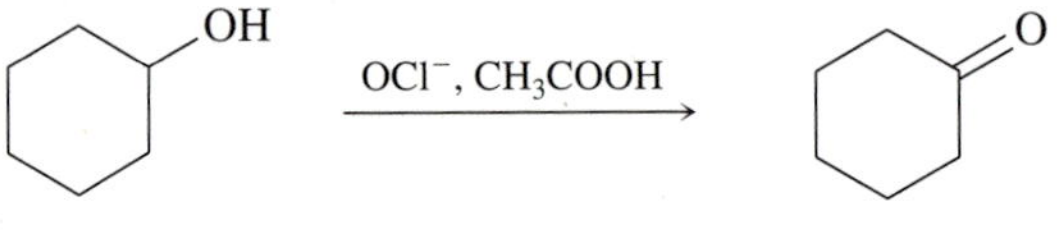

Cyclohexanol   **Cyclohexanone**

WORKED EXAMPLE 22.5—OXIDATION OF ALCOHOLS

What product would you expect from reaction of benzyl alcohol with CrO_3?

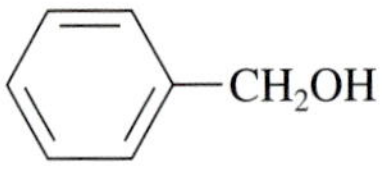

Benzyl alcohol

Solution

Treatment of a primary alcohol with CrO_3 yields a carboxylic acid. Oxidation of benzyl alcohol yields benzoic acid.

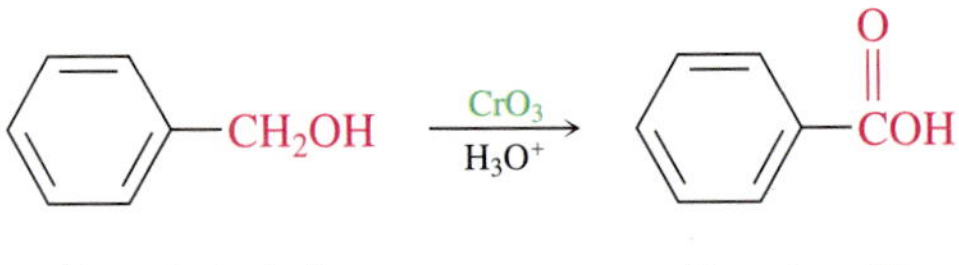

Benzyl alcohol   **Benzoic acid**

EXERCISE 22.11—OXIDATION OF ALCOHOLS

What alcohols would give the following products on oxidation?

(a)

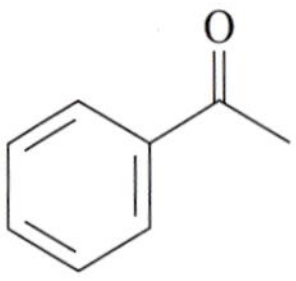

(b)

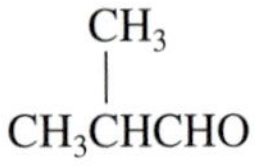

$$CH_3$$
$$CH_3CHCHO$$

(c)

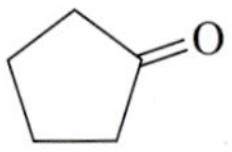

EXERCISE 22.12—OXIDATION OF ALCOHOLS

What products would you expect to obtain from oxidation of the following alcohols with CrO_3?
(a) cyclohexanol
(b) hexan-1-ol
(c) hexan-2-ol

What products would you expect to obtain from oxidation of the alcohols in Exercise 22.12 with pyridinium chlorochromate (PCC)?

Alcohols are weak acids that can be converted into their alkoxide anions on treatment with a strong base. Alcohols can also be dehydrated to yield alkenes, converted into ethers by reaction of their anions with alkyl halides, and oxidized to yield carbonyl compounds. Primary alcohols give either aldehydes or carboxylic acids when oxidized, secondary alcohols yield ketones, and tertiary alcohols are not oxidized.

22.7 Synthesis, Reactions, and Environmental Importance of Phenols

Synthesis of Phenols

Since the hydroxyl group in phenol molecules is attached directly to a benzene ring, the methods used to make alcohols can't generally be used to make phenols. Phenols can be synthesized from aromatic starting materials by a two-step sequence. The starting compound is first sulfonated by treatment with SO_3/H_2SO_4 [<<Section 20.6], and the sulfonic acid product is then converted into a phenol by high-temperature reaction with a NaOH solution.

Toluene   *p*-Toluenesulfonic   *p*-Methylphenol
                      acid   (72%)

p-Cresol (4-methylphenol) is used industrially both as an antiseptic and as a starting material to prepare the food additive BHT. Show how you could synthesize *p*-cresol from benzene.

Alcohol-like Reactions of Phenols

Phenols and alcohols react very differently despite the fact that both have —OH groups. Phenols can't be dehydrated by treatment with acid and can't be converted into halides by treatment with HX. Phenols can, however, be converted into ethers by conversion into phenoxide ions in base, followed by S_N2 reaction with alkyl halides. Williamson ether synthesis with phenols occurs easily because phenols are more acidic than alcohols and are more readily converted into their conjugate base phenoxide ions.

o-Nitrophenol   1-Bromobutane   Butyl *o*-nitrophenyl ether (80%)

Write a detailed, step-wise mechanism for the reaction of *o*-nitrophenol with K_2CO_3 and 1-bromobutane to produce butyl *o*-nitrophenyl ether.

Electrophilic Aromatic Substitution Reactions of Phenols

In addition to reactions of the —OH functional group, the aromatic rings of phenols can react as discussed in Section 20.8. The —OH group is an activating, *ortho*- and *para*-directing substituent in electrophilic aromatic substitution reactions. As a result, phenols are reactive substrates for electrophilic halogenation, nitration, and sulfonation.

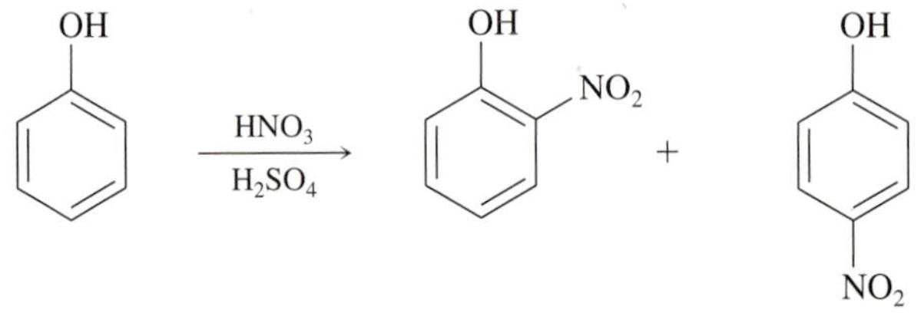

Phenol   **o-Nitrophenol** (50%)   **p-Nitrophenol** (50%)

The formation of picric acid by reacting phenol with nitric and sulfuric acid in Section 20.8 is a good example.

Oxidation of Phenols: Quinones

Treatment of a phenol with a strong oxidizing agent such as sodium dichromate yields a cyclohexadienedione, or *quinone*:

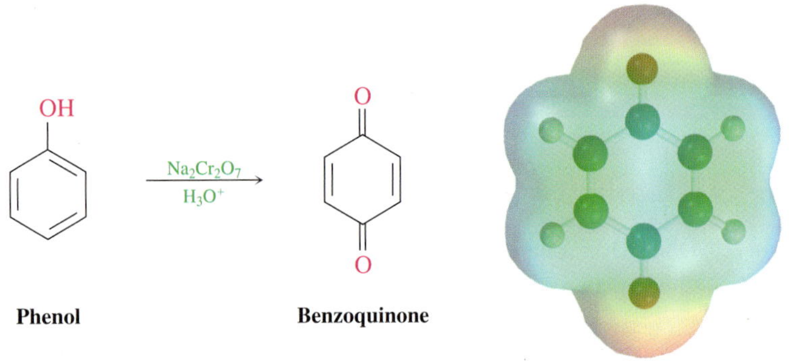

Phenol   **Benzoquinone**

Quinones are an interesting and valuable class of compounds because of their oxidation-reduction properties. They can be easily reduced to *hydroquinones* (*p*-dihydroxybenzenes) by $NaBH_4$ or $SnCl_2$, and hydroquinones can be easily oxidized back to quinones by $Na_2Cr_2O_7$. Hydroquinone is used, among other things, as a developer of photographic film because it reduces Ag^+ on film to metallic silver.

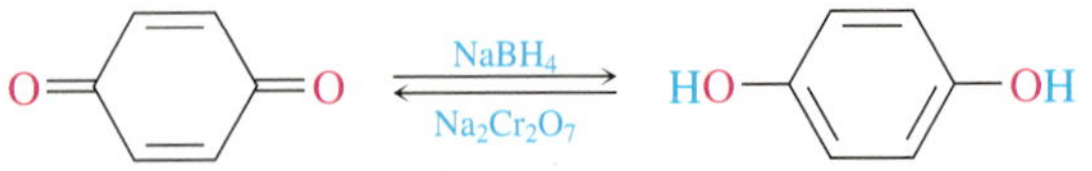

Benzoquinone   **Hydroquinone**

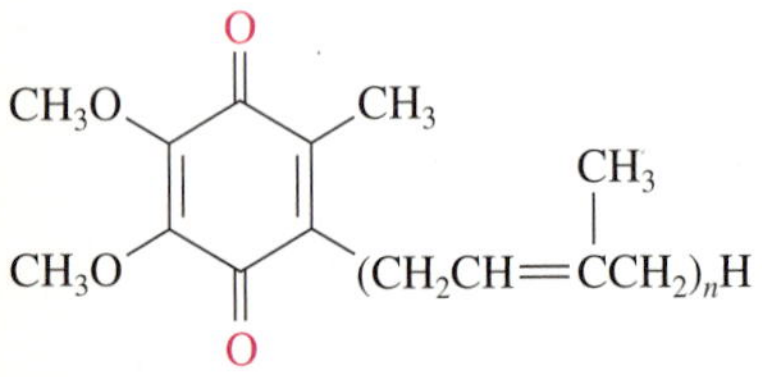

The oxidation-reduction properties of quinones are crucial to the functioning of living cells, where compounds called *ubiquinones* act as biochemical oxidizing agents to mediate the electron-transfer processes involved in energy production. Ubiquinones, also called *coenzymes Q*, are components of the cells of all aerobic organisms, from the simplest bacterium to humans. They are so named because of their *ubiquitous* occurrence (seeming to be everywhere) in nature.

Ubiquinones (*n* = 1–10)

Ubiquinones function within the mitochondria of cells to mediate the respiration process in which electrons are transported from the biological reducing agent NADH to molecular oxygen. Although a complex series of steps is involved in the overall process, the ultimate result is a cycle whereby NADH is oxidized to NAD^+, O_2 is reduced to water, and energy is produced. Ubiquinone acts only as an intermediary and is itself unchanged.

Bisphenol-A, Nonylphenol, and Endocrine Disruptors

17-β-estradiol is an alcohol of great importance to humans—one of three major endogenous (produced in the human body) steroidal sex hormones we refer to as estrogen. The endocrine system, which includes the pituitary, thyroid, and adrenal glands, releases hormones that circulate through the bloodstream to regulate sexual function and reproduction, as well as development, metabolism, and other processes. Recent scientific research has shown that some natural and synthetic compounds have the ability to disrupt the endocrine system in humans and wildlife. This can be done in several ways: by possessing key structural features in common with alcohols like 17-β-estradiol and mimicking their function, by preventing the synthesis of endogenous hormones, or by disrupting the function of estrogens in the body (*antagonistic* effect).

These compounds, known as possible **endocrine disruptors**, are released into the environment from numerous sources, including plasticizers, brominated fire retardants, and cosmetic agents. Whether they can effect reproduction in mammals, fish, birds, reptiles, amphibia, and aquatic invertebrates at the very low levels found in the environment is the subject of considerable controversy. Analytical capabilities have increased dramatically over the past decades, allowing the detection of lower and lower levels of these controversial compounds in environmental samples. But it's not yet clear in many cases what the effect of long-term exposure to trace levels of these substances might be. While many of these contain chlorine or bromine, several of the most active non-halogenated substances thought to be endocrine disruptors are phenols, including bisphenol-A and nonylphenol.

Bisphenol-A (BPA) was shown to be an estrogenic substance when distilled water heated in an autoclave to sterilize polycarbonate plastic flasks used for endocrinology research was found to increase the rate of growth of human mammary cancer cells. BPA is used on the billion kg scale as a monomer in the manufacture of polycarbonate plastics, and trace amounts of the monomer were identified in the autoclaved distilled water by NMR and mass spectrometry. The government of Canada has drafted legislation prohibiting the sale and advertising of baby bottles containing BPA.

17-β-Estradiol

4-Nonylphenol

Genistein

Bisphenol-A

Alkylated phenols, such as octyl and nonylphenol, are formed during sewage treatment as a result of the break down of larger polyethoxylates, and used in the synthesis of diverse products such as detergents, plastics, and spermicides.

Of course, not only synthetic substances affect the endocrine system. We ingest much higher quantities of many known plant-based estrogenic compounds called *phytoestrogens*. Foods that are particularly rich in these compounds are flax seed, soybeans, and red clover. Genistein is one example of a natural phytoestrogen found in clover that has been shown to cause infertility in sheep.

> Phenols are aromatic counterparts of alcohols. Although similar in some respects, phenols are more acidic than alcohols because the negative charge of phenoxide anions is spread over a larger molecular volume. Phenols undergo electrophilic aromatic substitution and can be oxidized to yield quinones. Alkylated phenols, formed during sewage treatment and industrial processes and other naturally occurring substances, can act as endocrine disruptors.

22.8 Synthesis and Reactions of Ethers

Synthesis of Ethers

As discussed in Section 22.6, ethers are easily prepared from alcohols by conversion to the alkoxide ion followed by S_N2 reaction with a primary alkyl halide (Williamson synthesis).

$$R-O-H \longrightarrow R-O^- \xrightarrow[S_N2]{R'X} R-O-R'$$

An alcohol   An alkoxide ion   An ether

Reactions of Ethers

Ethers are unreactive to most common reagents, a property that accounts for their frequent use as reaction solvents. Halogens, mild acids, bases, and nucleophiles have no effect on most ethers. In fact, ethers undergo only one general reaction—they are cleaved by strong acids such as aqueous HI or HBr.

Acidic ether cleavages are typical nucleophilic substitution reactions. They are thought to take place by either an S_N1 or S_N2 pathway, depending on the structure of the ether. Ethers with only primary and secondary alkyl groups react by an S_N2 pathway, in which nucleophilic halide ion reacts with the protonated ether at the less highly substituted site. The ether oxygen atom stays with the more hindered alkyl group, and the halide bonds to the less hindered group. For example, ethyl isopropyl ether yields isopropyl alcohol and iodoethane on cleavage by HI:

Ethyl isopropyl ether   Isopropyl alcohol   Iodoethane

Ethers with a tertiary alkyl group probably cleave by an S_N1 mechanism because they can produce stable intermediate carbocations. In such reactions, the ether oxygen atom stays with the less hindered alkyl group and the halide bonds to the tertiary group. Like most S_N1 reactions, the cleavage is fast and often takes place at or below room temperature.

tert-Butyl cyclohexyl ether   Cyclohexanol   2-Bromo-2-methylpropane

WORKED EXAMPLE 22.6—REACTIONS OF ETHERS

Predict the products of the reaction of *tert*-butyl propyl ether with HBr.

Solution

Identify the substitution pattern of the two groups attached to oxygen—in this case a tertiary alkyl group and a primary alkyl group. Then recall the guidelines for ether cleavages. An ether with only primary and secondary alkyl groups usually undergoes cleavage by S_N2 reaction of a nucleophile at the less hindered alkyl group, but an ether with a tertiary alkyl group usually undergoes cleavage by an S_N1 mechanism. In this case, an S_N1 cleavage of the tertiary C—O bond will occur, giving propan-1-ol and a tertiary alkyl bromide.

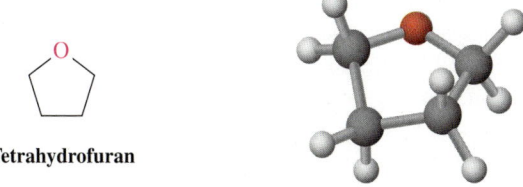

tert-**Butyl propyl ether**	**2-Bromo-2-methylpropane**	**Propan-1-ol**

EXERCISE 22.16—REACTIONS OF ETHERS

What products do you expect from the reaction of the following ethers with HI?

(a) $CH_3CH_2OCH_2CH_3$ **(b)** [structure] OCH_2CH_3 **(c)** [structure] $CH_3 \\ CH_3COCH_2CH_3 \\ CH_3$

Ethers have two organic groups bonded to the same oxygen atom. They are prepared by reaction of an alkoxide ion with a primary alkyl halide—the Williamson synthesis, thought to occur by an S_N2 mechanism. Ethers are inert to most reagents but are cleaved by the strong acids HBr and HI.

22.9 Cyclic Ethers: Crown Ethers and Epoxides

For the most part, cyclic ethers behave like acyclic ethers. The chemistry of the ether functional group is the same whether it's in an open chain or in a ring. For example, the cyclic ether tetrahydrofuran (THF) is often used as a solvent because it is so unreactive.

Tetrahydrofuran

Crown Ethers

An interesting parallel can be drawn between large ring **crown ethers** whose molecules form a "crown" of oxygen atoms linked by hydrocarbon spacers, such as —CH_2CH_2— and the large rings of macrocyclic cyclodextrins in Section 22.1. The oxygen atoms make the interior of the ring cavities of crown ethers hydrophilic and bind well to metal ions, forming host–guest complexes. The exterior of the cavities are hydrophobic; as a result, crown ether compounds dissolve in non-polar hydrocarbon solvents. Encapsulating a metal ion inside a crown ether gives chemists a way of transporting a metal ion to a place

Molecular Modelling (Odyssey)

e22.12 Consider the factors that determine whether a metal ion can bind to a crown ether.

where it would not otherwise go—into a non-polar organic solvent or through the membrane of a cell wall. The size of the cavity in the crown ether can be tailor-made to fit metal ions of different sizes (Figure 22.13).

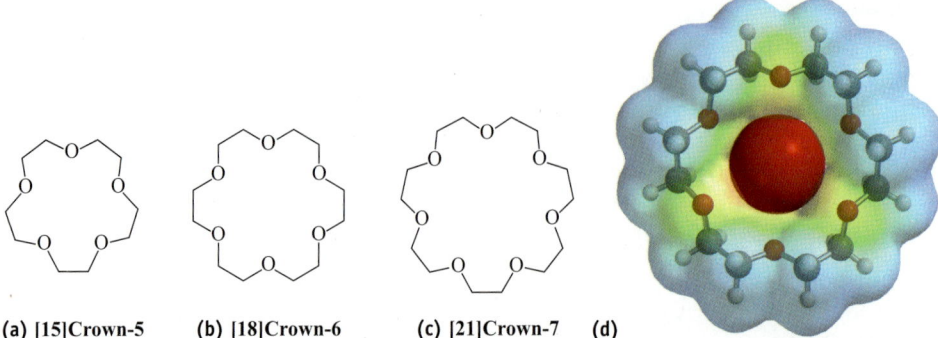

(a) [15]Crown-5   **(b) [18]Crown-6**   **(c) [21]Crown-7**   **(d)**

FIGURE 22.13 (a) [15]crown-5 has the right size cavity to readily solvate a sodium ion, (b) [18]crown-6, a potassium ion, and (c) [21]crown-7, a cesium ion. In (d), an electrostatic potential map of [18]crown-6 is shown with a K^+ ion inside the cavity, surrounded by electron-rich O atoms. The numbering in crown ether names refers to the total number of atoms in the ring in square brackets, and the number of oxygen atoms following the word "crown."

Crown ethers are macrocyclic ethers able to solvate metal ions and transport them into non-polar environments.

Three-Membered Cyclic Ethers: Epoxides

The one sub-group of cyclic ethers that is much more reactive than open-chain or larger ring ethers are the three-membered ring containing compounds called *epoxides*, or *oxiranes*. The strain of the three-membered ring gives epoxides unique chemical reactivity.

Epoxides are prepared by reaction of an alkene with a peroxyacid (RCO_3H). *m*-Chloroperoxybenzoic acid (MCPBA) is often used because it is more stable and more easily handled than other peroxyacids.

$$CH_2Cl_2, 25°C$$

Cycloheptene

1,2-Epoxycycloheptane
(78%)

Epoxide rings are cleaved by treatment with acid just like other ethers. The major difference is that epoxides react under much milder conditions because the strain of the three-membered ring is relieved if the ether cleaves. Dilute aqueous acid solution at room temperature converts an epoxide to a 1,2-diol (also called a *glycol*). Over 10 000 kt of ethylene glycol are produced annually, for the production of polyesters for fibre and polyethylene terephthalate (PET) and for antifreeze solutions in automobile radiators. One important industrial preparation of ethylene glycol uses the acid-catalyzed ring opening of ethylene oxide. Note that the name *ethylene glycol* refers to the glycol derived *from* ethylene. Similarly, *ethylene oxide* is the common name of the epoxide derived from ethylene.

$$H_3O^+$$

Ethylene oxide

Ethylene glycol
(ethane-1,2-diol)

The accepted mechanism for acid-catalyzed epoxide cleavage is described by S_N2 reaction of H_2O with the protonated epoxide.

1,2-Epoxycyclo-hexane

trans-Cyclo-hexane-1,2-diol (86%)

EXERCISE 22.17—REACTIONS OF EPOXIDES

Show the structure of the product you would obtain by treatment of the following epoxide with aqueous acid. What is the stereochemistry of the product if the ring opening takes place by a nucleophilic substitution reaction following an S_N2 mechanism with displacement by the nucleophile from the backside of the substrate?

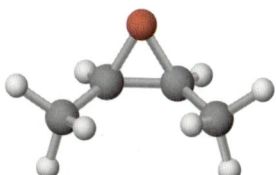

Nucleophiles other than water, such as alcohols and amines, can be used to open epoxides. Epoxide ring opening by an amine molecule is proposed to be a key step in the mechanism by which benzo[α]pyrene and other polycyclic aromatic hydrocarbons in chimney soot and cigarette smoke cause cancer [<<Section 20.11]. Benzo[α]pyrene is converted by metabolic oxidation into a diol epoxide, which then undergoes nucleophilic ring opening by reaction with an amino group in cellular DNA. With its DNA thus altered, the cell is unable to function normally.

Benzo[a]pyrene   **A diol epoxide**

Finally, the alcohol groups on the exterior of cyclodextrin molecules can be made to react with epoxides to form supramolecular polymers whose intriguing shapes and potential uses have captured the interest of chemists. Several steps are needed to create a supramolecular complex that has been described as a "molecular tube." First, between 15 and 100 α-cyclodextrin molecules are threaded onto a polyethylene glycol (polymer of ethylene glycol, PEG) polymer chain, and bulky substituted benzene rings are put on the end to serve as "stoppers" to prevent the cyclodextrin molecules from coming off the chain. This results in an intriguing supramolecular assembly called a "molecular necklace."

This work is described fully in A. Harada, et. al. (2006), "Cyclodextrin-Based Supramolecular Polymers," *Advances in Polymer Science*, 201: 1-43.

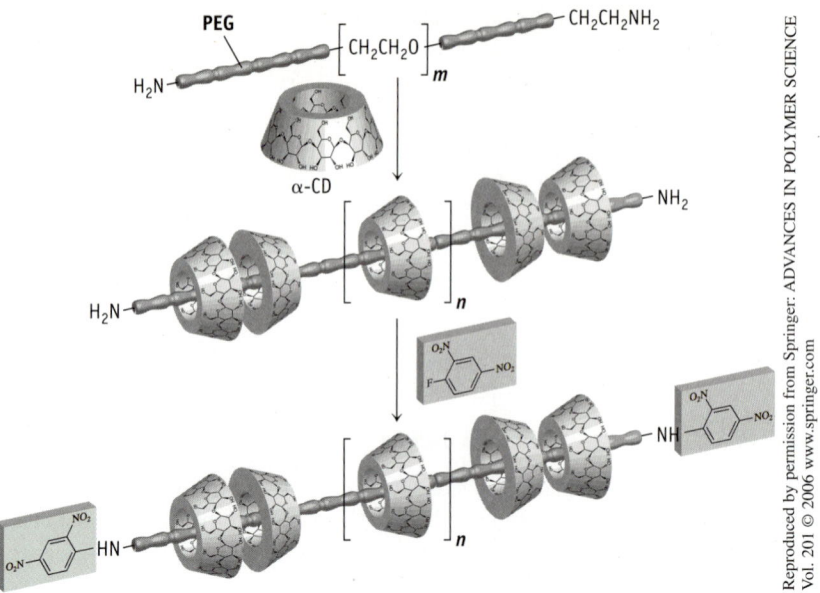

Adjacent α-cyclodextrin units in the necklace are then linked together by reacting the alcohol groups on the outside of each cyclodextrin cup with an epoxide, epichlorohydrin. The resulting bonds stitch together the cyclodextrin units. Finally, the bulky aromatic "stoppers" are removed by treatment with strong base, releasing the PEG chain and leaving a "molecular tube," with an average molar mass of 2×10^4 g mol^{-1}. This tube was shown to be able to efficiently serve as a host and encapsulate $I_3{}^-$ ions from a solution of KI and I_2, something α-cyclodextrin is unable to do.

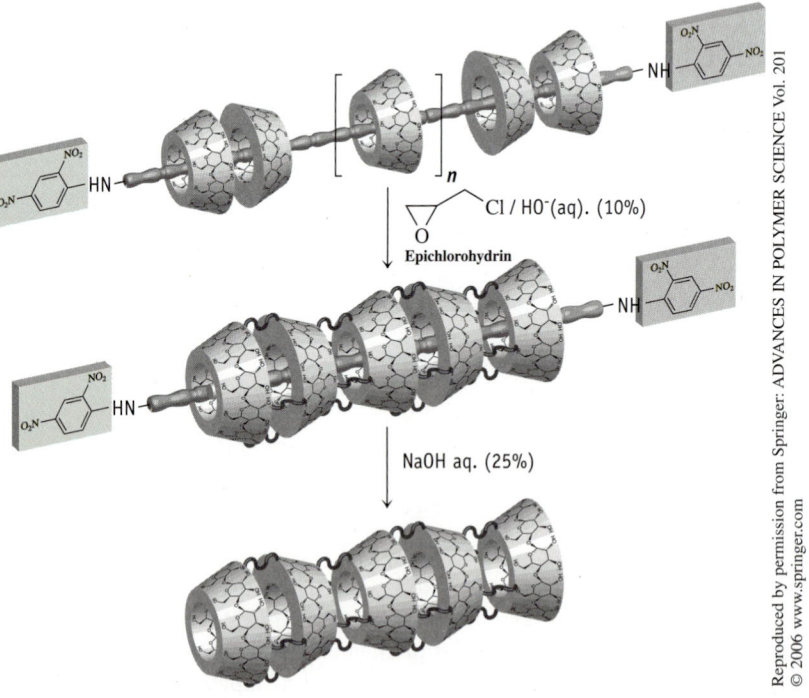

Epoxides—cyclic ethers whose molecules have an oxygen atom in a three-membered ring—differ from other ethers in their ease of cleavage. The high reactivity of the strained three-membered ether ring allows epoxides to react with aqueous acid, yielding diols (glycols).

22.10 Thiols and Sulfides

Sulfur is the element just below oxygen in the periodic table, and many oxygen-containing organic compounds have sulfur analogs. **Thiols (R—SH)**, are sulfur analogs of alcohols, and **sulfides (R—S—R′)** are sulfur analogs of ethers. Both classes of compounds are widespread in living organisms.

Thiols are named in the same way as alcohols, with the suffix *–thiol* used in place of *–ol*. The —SH group itself is referred to as a *mercapto group*.

CH_3CH_2SH

Ethanethiol   **Cyclohexanethiol** 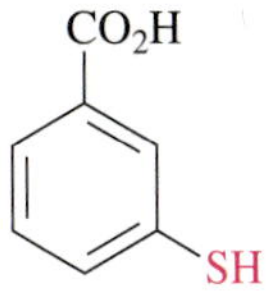   ***m*-Mercaptobenzoic acid**

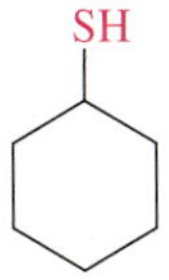

Sulfides are named in the same way as ethers, with *sulfide* used in place of *ether* for simple compounds and with *alkylthio* used in place of *alkoxy* for more complex substances.

CH_3-S-CH_3

Dimethyl sulfide   **Methyl phenyl sulfide** 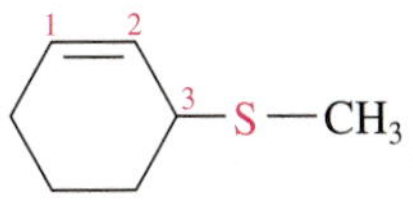   **3-(Methylthio)cyclohexene**

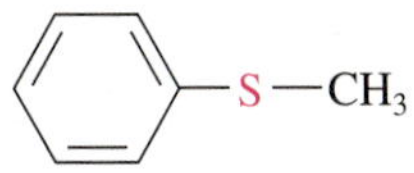

Thiols can be prepared from the corresponding alkyl halide by S_N2 displacement with a sulfur nucleophile such as hydrogensulfide anion, SH^-:

$$CH_3(CH_2)_6CH_2-Br + Na^+ \ ^-:\!\ddot{S}H \longrightarrow CH_3(CH_2)_6CH_2SH + NaBr$$

1-Bromooctane   **Sodium**   **Octane-1-thiol**
                         hydrosulfide

Sulfides are prepared by treating a primary or secondary alkyl halide with a *thiolate ion*, RS^-, the sulfur analog of an alkoxide ion. Reaction is thought to occur by an S_N2 mechanism analogous to the Williamson ether synthesis [<<Section 22.6]. Thiolate anions are among the best nucleophiles known, so these reactions usually work well.

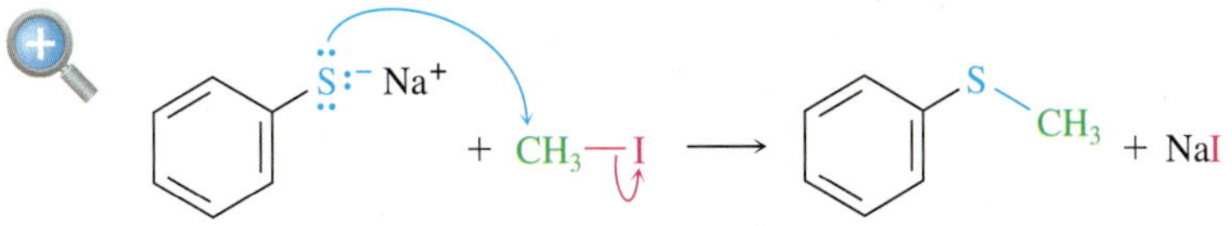

$$+ \ CH_3-I \longrightarrow \ + NaI$$

Sodium benzenethiolate   **Methyl phenyl sulfide**
                                        (96%)

The most unforgettable physical characteristic of thiols is their appalling odour. Skunk scent is due primarily to the simple thiols 3-methylbutane-1-thiol and 2-butene-1-thiol. Thiols and sulfides are also responsible both for the odour and the antibiotic properties of onion and garlic.

Thiols can be oxidized by mild reagents such as bromine to yield *disulfides* (R—S—S—R), and disulfides can be reduced back to thiols by treatment with zinc metal and acetic acid:

$$2 \ R-SH \ \underset{Zn, \ H^+}{\overset{Br_2}{\rightleftharpoons}} \ R-S-S-R + 2 \ HBr$$

A thiol   **A disulfide**

We'll see in Chapter 29 that this thiol–disulfide interconversion is extremely important in biochemistry because disulfide "bridges" form cross-links that help stabilize the three-dimensional structure of proteins.

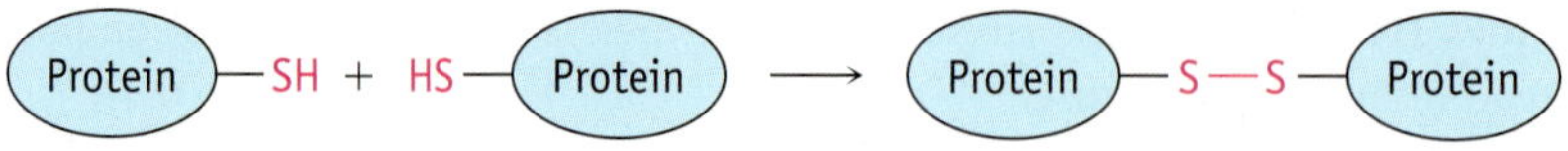

A cross-linked protein

EXERCISE 22.18—NAMING THIOLS

Name the following thiols by IUPAC rules:

(a)

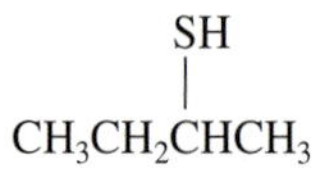

$$\underset{\text{CH}_3\text{CH}_2\text{CHCH}_3}{\overset{\overset{\text{SH}}{|}}{}}$$

(b)

$$\underset{\underset{\text{CH}_3}{|}}{\overset{\overset{\text{CH}_3}{|}\ \overset{\text{SH}}{|}\ \overset{\text{CH}_3}{|}}{\text{CH}_3\text{CCH}_2\text{CHCH}_2\text{CHCH}_3}}$$

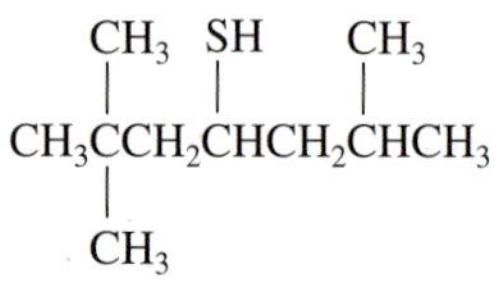

(c)

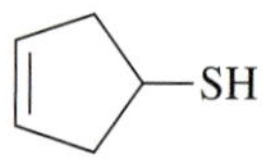

EXERCISE 22.19—NAMING THIOLS AND SULFIDES

Give IUPAC names to the compounds with the following structures:

(a) $CH_3CH_2SCH_3$

(b)

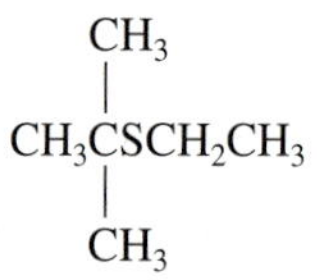

$$\underset{\underset{\text{CH}_3}{|}}{\overset{\overset{\text{CH}_3}{|}}{\text{CH}_3\text{CSCH}_2\text{CH}_3}}$$

(c)

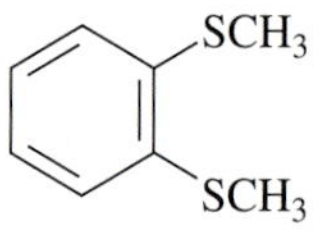

EXERCISE 22.20—SYNTHESIS OF THIOLS

2-Butene-1-thiol is a component of skunk spray. How would you synthesize this substance from 2-buten-1-ol? From methyl but-2-enoate ($CH_3CH=CHCO_2CH_3$)? More than one step is required in both instances.

Sulfides (R—S—R′) and thiols (R—SH) are sulfur analogs of ethers and alcohols. Thiols are prepared by S_N2 reaction of an alkyl halide with HS^-, and sulfides by further alkylation of thiols with an alkyl halide.

SUMMARY

Key Concepts

- **Alcohols, phenols**, and **ethers** are organic derivatives of water in which one or both of the water molecule hydrogen atoms have been replaced by organic groups. (Sections 22.2–22.3)
- Cyclodextrins are macrocyclic **oligomers** made of glucose monomers containing alcohol and ether functional groups. With hydrophilic exteriors and hydrophobic interiors, they form supramolecular complexes important in pharmaceutical chemistry and the textile and other industries. (Section 22.1)
- Alcohols are compounds whose molecules have an –OH bonded to an alkyl group. The structures of alcohol and phenol molecules can be characterized by their IR, MS, ^{13}C NMR and ^{1}H NMR spectra. (Section 22.4)

- Alcohols can be prepared in many ways, including hydration of alkenes. The most general method of alcohol synthesis involves reduction of a carbonyl compound. Aldehydes, esters, and carboxylic acids yield primary alcohols on reduction; ketones yield secondary alcohols. (Section 22.5)
- Alcohols are weak acids and can be converted into their alkoxide anions on treatment with a strong base or with an alkali metal. Alcohols can also be dehydrated to yield alkenes, converted into ethers by reaction of their anions with alkyl halides, and oxidized to yield carbonyl compounds. Oxidation of organic compounds often adds an oxygen, nitrogen, or halogen atom to a carbon atom, while reduction often adds hydrogen atoms. Primary alcohols give either aldehydes or carboxylic acids when oxidized, secondary alcohols yield ketones, and tertiary alcohols are not oxidized. (Section 22.6)
- Phenols are aromatic counterparts of alcohols. Although similar to alcohols in some respects, phenols are more acidic than alcohols because the negative charge of phenoxide anions is spread over a larger molecular volume. Phenols undergo electrophilic aromatic substitution and can be oxidized to yield quinones. Some alkylated phenols, formed during sewage treatment and other industrial process, as well as other naturally occurring compounds can act as **endocrine disruptors**. (Section 22.7)
- Ether molecules have two organic groups bonded to the same oxygen atom. They are prepared by reaction of an alkoxide ion with a primary alkyl halide—the Williamson synthesis, thought to occur by an S_N2 mechanism. Ethers are inert to most reagents but are cleaved by the strong acids HBr and HI. **Crown ethers** are macrocyclic ethers able to solvate metal ions and transport them into non-polar environments. Epoxides are cyclic ethers whose molecules have an oxygen atom in a three-membered ring—differ from other ethers in their ease of cleavage. The high reactivity of the strained three-membered ether ring allows epoxides to react with aqueous acid, yielding diols. (Sections 22.8–22.9)
- **Sulfides (R—S—R′)** and **thiols (R—SH)** are sulfur analogs of ethers and alcohols. Thiols are prepared by S_N2 reaction of an alkyl halide with HS^-, and sulfides are prepared by further alkylation of the thiol with an alkyl halide. (Section 22.10)

Key Reactions

1. Synthesis of alcohols
   (a) Reduction of aldehydes to yield primary alcohols (Section 22.5)

$$\underset{RCH}{\overset{O}{\parallel}} \quad \xrightarrow[\text{2. H}_3\text{O}^+]{\text{1. NaBH}_4} \quad RCH_2OH$$

   (b) Reduction of ketones to yield secondary alcohols (Section 22.5)

$$\underset{RCR'}{\overset{O}{\parallel}} \quad \xrightarrow[\text{2. H}_3\text{O}^+]{\text{1. NaBH}_4} \quad \underset{RCHR'}{\overset{OH}{\mid}}$$

   (c) Reduction of esters to yield primary alcohols (Section 22.5)

$$\underset{RCOR'}{\overset{O}{\parallel}} \quad \xrightarrow[\text{2. H}_3\text{O}^1]{\text{1. LiAlH}_4} \quad RCH_2OH + R'OH$$

   (d) Reduction of carboxylic acids to yield primary alcohols (Section 22.5)

$$\underset{RCOH}{\overset{O}{\parallel}} \quad \xrightarrow[\text{2. H}_3\text{O}^+]{\text{1. LiAlH}_4} \quad RCH_2OH$$

2. Reactions of alcohols
   (a) Conversion to conjugate base alkoxide ions (Section 22.3)

$$ROH + NaNH_2 \longrightarrow RO^- + H\text{-}NH_2$$

(b) Dehydration to yield alkenes (Section 22.6)

$$-\overset{\overset{\displaystyle H}{|}}{C}-\overset{\overset{\displaystyle OH}{|}}{C}-\xrightarrow{H_2SO_4} \quad \overset{\diagdown}{C}=\overset{\diagup}{C} + H_2O$$

(c) Oxidation to yield carbonyl compounds (Section 22.6)

$$RCH_2OH \xrightarrow[\text{chlorochromate}]{\text{Pyridinium}} \overset{\overset{\displaystyle O}{\|}}{R}CH \qquad \textbf{An aldehyde}$$

$$RCH_2OH \xrightarrow[\text{H}_3O^+]{CrO_3} \overset{\overset{\displaystyle O}{\|}}{R}COH \qquad \textbf{A carboxylic acid}$$

$$\overset{\overset{\displaystyle OH}{|}}{R}CHR' \xrightarrow[\text{Acetic acid}]{Na^+ OCl^-} \overset{\overset{\displaystyle O}{\|}}{R}CR' \qquad \textbf{A ketone}$$

(d) Conversion into ethers (Section 22.6)

$$2\,ROH + 2\,Na \longrightarrow 2\,RO^- Na^+ + H_2$$

$$RO^- Na^+ + R'Br \xrightarrow[\text{reaction}]{S_N2} ROR'$$

3. Synthesis of phenols (Section 22.7)

4. Synthesis of ethers
   (a) Williamson ether synthesis (Section 22.6)

$$RO^- Na^+ + R'Br \xrightarrow[\text{reaction}]{S_N2} ROR'$$

   (b) Epoxides (Section 22.9)

5. Reactions of ethers
   (a) Acidic cleavage with HBr or HI (Section 22.8)

$$ROR' + HI \longrightarrow ROH + R'I$$

   (b) Epoxide cleavage with aqueous acid (Section 22.9)

6. Synthesis of thiols (Section 22.10)

$$Na^+ SH^- + RBr \xrightarrow[\text{by } S_N2 \text{ mechanism}]{\text{Substitution}} RSH$$

7. Synthesis of sulfides (Section 22.10)

$$RS^- Na^+ + R'Br \xrightarrow[\text{mechanism}]{S_N2} RSR'$$

REVIEW QUESTIONS

Section 22.1: Cyclodextrins

22.21 How do the alcohol —OH functional groups on the outside of cyclodextrin molecules facilitate their use in drug delivery?

22.22 The interior cavity of cyclodextrin molecules is quite non-polar. Refer to Section 4.1, and describe the forces of attraction that hold prostaglandin E2 molecules inside that cavity, as shown in Figure 22.2.

22.23 Consult the review article "Cyclodextrins and Their Pharmaceutical Applications," by T. Loftsson and D. Duchêne (2007) (*International Journal of Pharmacy*, 329: 1–11) and list five pharmaceutical applications of cyclodextrins that have not been mentioned in Section 22.1. In light of the title for Section 22.1, what is meant by the term *sublingual tablet*, used to describe cyclodextrin-drug complexes?

22.24 The history of cyclodextrins reveals something of the timeline for the development of organic chemistry. With reference to Loftsson and Duchêne's article in Question 22.23, when was the first production of cyclodextrins by heating starch discovered? When was the structure of the first cyclodextrins determined? When was their ability to complex with organic compounds discovered?

22.25 In light of the diverse applications of cyclodextrins to form host–guest complexes that can remove undesirable odours and aromas, how might they be used in medicine for the treatment of wounds?

Section 22.2: Naming Alcohols, Phenols, and Ethers

22.26 Give IUPAC names for the following compounds:

(a)

$$\underset{\underset{CH_3}{|}}{CH_3}CHCH_2CHCHCH_3$$
with OH groups

(b)
benzene ring—CH_2CH_2CCH_3 with OH and CH_3

(c) HO—cyclohexane with CH_3, CH_3

(d) cyclopentane with H, Br, H, OH

22.27 Draw structures of molecules of the following compounds.
(a) ethyl isopropyl ether
(b) 3,4-dimethoxybenzoic acid
(c) 2-methyl-2,5-heptanediol
(d) *trans*-3-ethylcyclohexanol
(e) 4-allyl-2-methoxyphenol (eugenol, from oil of cloves)

22.28 Name the compounds that have the following molecular structures

(a)

$$HOCH_2CH_2\underset{\underset{}{|}}{C}HCH_2OH$$
with CH_3

(b) $CH_3CHCHCH_2CH_3$ with HO and CH_2CH_2CH_3

(c) cyclopentane with Ph, H and OH, H

(d)

$$(CH_3)_2CHCCH_2CH_2CH_3$$
with SH and CH_3

22.29 Draw and name structures of the eight isomeric alcohols with the formula $C_5H_{12}O$.

22.30 Which of the eight molecules of alcohols you identified in Question 22.29 are chiral?

22.31 Draw and name structures of the six ethers that are isomeric with the alcohols you drew in Question 22.29. Which are chiral?

Section 22.3: Alcohols, Phenols, and Ethers: Structure and Reactivity

22.32 Explain, at the molecular level, why the absorption peak for the O—H stretching vibration in the IR spectrum of liquid ethanol in Figure 22.4 is so much broader and at lower wavenumber than for ethanol vapour.

Interactive Exercise (Odyssey) 22.33

Predict the most stable structure for a diol molecule.

Interactive Exercise (Odyssey) 22.34

Build models of alcohol molecules, measure their dipole moments, and compare hydrogen bonding with other compounds.

Section 22.4: Spectroscopy of Alcohols and Phenols

22.35 When the ^{1}H NMR spectrum of an alcohol is run in deuterated dimethyl sulfoxide (DMSO) solvent rather than in deuterochloroform, exchange of the O—H proton is slow, and spin–spin splitting is seen between the O—H proton and C—H protons on the adjacent carbon. What spin multiplicities would you expect for the hydroxyl protons in the following alcohols in DMSO?
(a) 2-methylpropan-2-ol   (d) propan-2-ol
(b) cyclohexanol   (e) cholesterol
(c) ethanol   (f) 1-methylcyclohexanol

22.36 How would you use ^{1}H NMR, ^{13}C NMR, IR, and MS to differentiate between the two isomeric compounds benzyl alcohol and p-cresol?

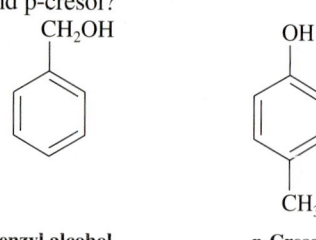

Benzyl alcohol   *p*-Cresol

22.37 Propose a structure consistent with the following spectral data for a compound $C_8H_{18}O_2$:

IR: 3350 cm^{-1}

^{1}H NMR: 1.24 δ (12 H, singlet); 1.56 δ (4 H, singlet); 1.95 δ (2 H, singlet)

22.38 The ^{1}H NMR spectrum shown is that of 3-methyl-but-3-en-1-ol. Assign all the observed absorption peaks to specific protons and account for the splitting patterns.

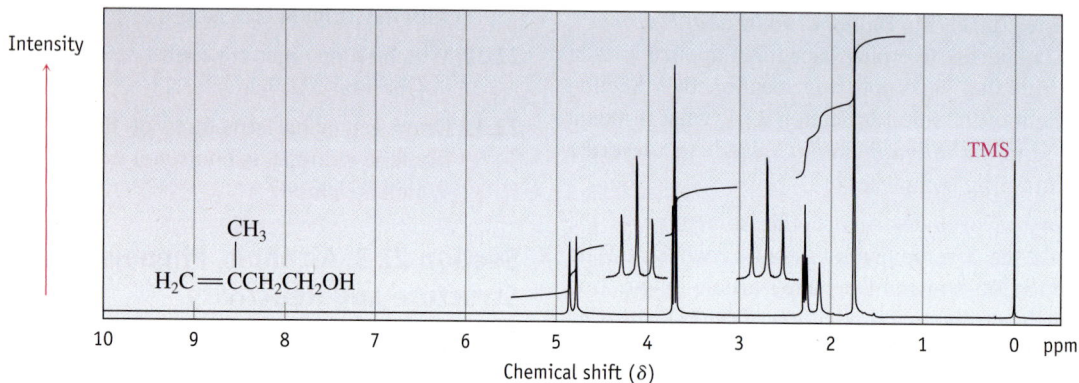

22.39 Propose structures for alcohols that have the following ^{1}H NMR spectra:
(a) $C_5H_{12}O$

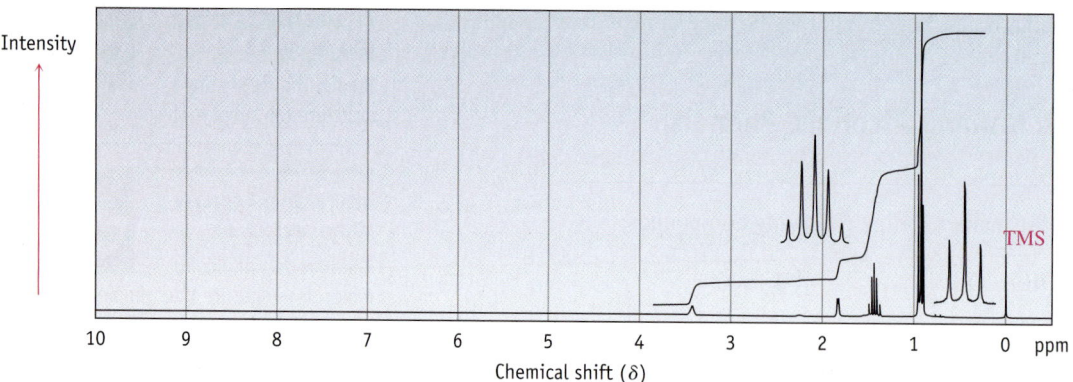

(b) $C_8H_{10}O$

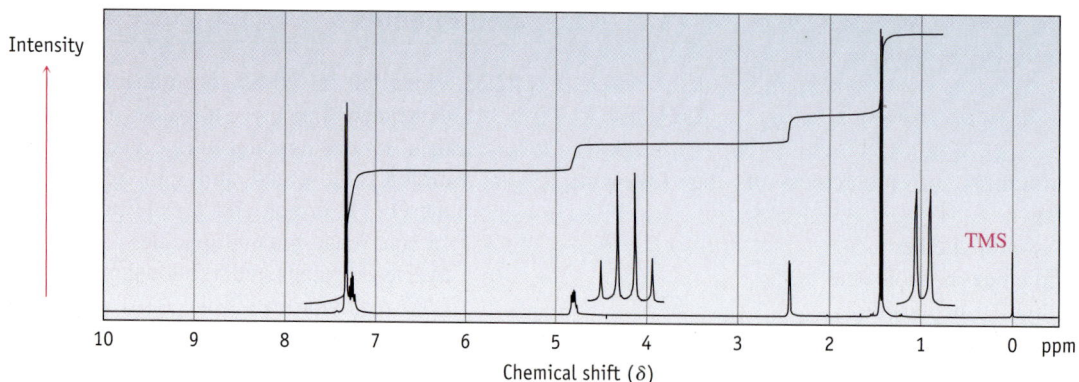

Sections 22.5–22.8: Synthesis and Reactions

22.40 What carbonyl compounds give the following alcohols on reduction with LiAlH$_4$? Show all possibilities.

(a)

CH$_2$OH

(c)

OH
H

(b)

OH
CHCH$_3$

22.41 Draw the structure of molecules of the carbonyl compound(s) from which the following alcohol might have been prepared, and show the products you would obtain by treatment of the alcohol with NaNH$_2$, followed by CH$_3$CH$_2$I.

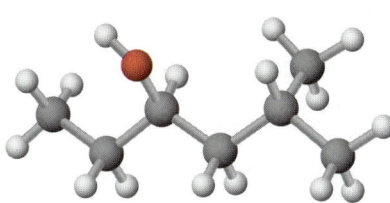

22.42 From what alcohols, and using what reagents, might the following alkenes have been made?

(a)

CH$_3$

CH$_3$

(b) CH$_3$CH$_2$CH=CHCH$_2$CH$_2$CH$_3$

22.43 What products would you expect from oxidation of the following alcohols with OCl$^-$ in acetic acid?
(a) cyclohexanol
(b) hexan-2-ol
(c) isopropyl alcohol

22.44 Which of the eight alcohols you identified in Question 22.29 would react with aqueous acidic CrO$_3$? Show the products you would expect from each reaction.

22.45 Show the HI cleavage products of the ethers you drew in Question 22.31.

22.46 How would you make the following ethers?

(a) CH$_3$OCH$_2$CH$_2$CH$_3$ **(c)**

CH$_3$
CH$_2$OCHCH$_3$

(b)

OCH$_3$

22.47 Predict the product(s) of the following transformations:

(a)

CH$_2$OH

$\xrightarrow{\text{PCC}}$ **?**

(b)

CH$_3$
OCH$_2$CHCH$_3$

$\xrightarrow{\text{HBr}}$ **?**

(c)

CH$_3$ O
H$_2$C=CHCHCH$_2$COCH$_3$

$\xrightarrow[\text{2. H}_2\text{O}]{\text{1. LiAlH}_4}$ **?**

(d)

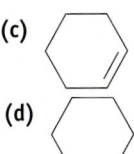

$\xrightarrow{\text{HBr}}$ **?**

22.48 Show how you could synthesize the following substances from cyclohexanol:

(a)

O

(c)

(b)

Br

(d)

22.49 Show how you could synthesize the following substances from propan-1-ol:

(a)

O
CH$_3$CH$_2$CH

(c) CH$_3$CH$_2$CH$_2$O$^-$Na$^+$

(b)

O
CH$_3$CH$_2$COH

(d) CH$_3$CH$_2$CH$_2$Cl

22.50 Predict the likely products of reaction of the following ethers with HI:

(a)

CH$_3$
CH$_3$CH$_2$OCHCH$_3$

(b)

CH$_3$
CH$_3$CCH$_2$OCH$_3$
CH$_3$

22.51 How would you make the following compounds from 2-phenylethanol?
(a) benzoic acid
(b) ethylbenzene
(c) 2-bromo-1-phenylethane
(d) phenylacetic acid (C$_6$H$_5$CH$_2$COOH)
(e) phenylacetaldehyde (C$_6$H$_5$CH$_2$CHO)

22.52 Give the structures of the major products you would obtain from reaction of phenol with the following reagents:
(a) Br$_2$ (1 mol)   (c) NaOH, then CH$_3$I
(b) Br$_2$ (3 mol)   (d) Na$_2$Cr$_2$O$_7$, H$_3$O$^+$

22.53 What products would you obtain from reaction of butan-1-ol with the following reagents?
(a) PBr$_3$
(b) CrO$_3$, H$_3$O$^+$
(c) NaNH$_2$
(d) pyridinium chlorochromate (PCC)

22.54 What products would you obtain from reaction of 1-methylcyclohexanol with the following reagents?
(a) HBr
(b) H$_2$SO$_4$
(c) CrO$_3$
(d) Na
(e) product of part (d), then CH$_3$I

22.55 What alcohols would you oxidize to obtain the following products?

(a) [cyclopentanone structure] =O
(b) [benzene ring]—CHO
(c) CH$_3$ | CH$_3$CHCOOH

22.56 Show the alcohols you would obtain by reduction of the following carbonyl compounds:

(a) CH$_3$ | CH$_3$CHCH$_2$CHO
(c) O CH$_3$ || | CH$_3$CH$_2$CCH$_2$CHCH$_3$
(b) [benzene ring with two COOH groups ortho]

22.57 Predict the product(s) of the following reactions:

(a) [benzene ring with OH and HO (hydroquinone)] →(Na$_2$Cr$_2$O$_7$) **?**

(b) CH$_3$ | CH$_3$CHCH$_2$CH$_2$CH$_2$Br →(Na$^+$ $^-$SH) **?**

(c) [cyclopentane]—SH →(Br$_2$) **?**

22.58 Rank the following substances in order of increasing acidity: (most acidic H atoms are designated)

O || CH$_3$CCH$_3$ O O || || CH$_3$CCH$_2$CCH$_3$ [benzene ring]—OH   O || CH$_3$COH

Acetone   **Pentane-2,4-dione**   **Phenol**   **Acetic acid**
pK$_a$ = 19   pK$_a$ = 9   pK$_a$ = 9.9   pK$_a$ = 4.7

22.59 Which, if any, of the substances in Question 22.58 are strong enough acids to react substantially with NaOH? (The pK$_a$ of H$_2$O is 15.7.)

22.60 Is *tert*-butoxide anion a strong enough base to react with water? In other words, does the following reaction take place as written? (The pK$_a$ of *tert*-butyl alcohol is 18.)

CH$_3$ | CH$_3$CO$^-$ Na$^+$ + H$_2$O →(**?**) CH$_3$COH + NaOH | CH$_3$ CH$_3$ | | CH$_3$

22.61 Sodium hydrogencarbonate (NaHCO$_3$) is the sodium salt of carbonic acid (H$_2$CO$_3$), pK$_a$ = 6.4. Which of the substances shown in Question 22.58 will react with sodium hydrogencarbonate?

22.62 Assume that you have two unlabelled bottles, one that contains phenol (pK$_a$ = 9.9) and one that contains acetic acid (pK$_a$ = 4.7). In light of your answer to Question 22.61, propose a simple way to tell what is in each bottle.

Section 22.9: Cyclic Ethers: Crown Ethers and Epoxides

22.63 The Nobel Prize in 1987 was co-awarded to Dr. Charles Pedersen for the discovery of crown ethers. These molecules are able to bind metal cations such as Na$^+$ or K$^+$. Where would you expect the cations to bind? Why? The molecule shown binds Na$^+$ effectively, but it does not bind Cs$^+$ because it is a much larger cation. How would you design a molecule that binds Cs$^+$?

[crown ether structure]   **A crown ether**

22.64 Nonoxynol 9 is a potent spermicide made by reacting ethylene oxide with *p*-nonylphenoxide. Write a detailed step-wise mechanism for this reaction.

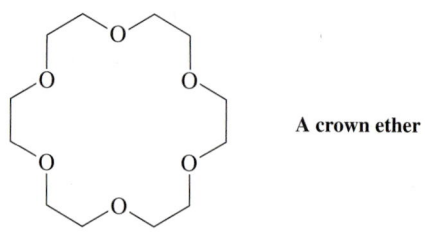

p-Nonylphenoxide   Ethylene oxide   Nonoxynol 9

SUMMARY AND CONCEPTUAL QUESTIONS

22.65 Named *bombykol,* the sex pheromone secreted by the female silkworm moth has the formula $C_{16}H_{28}O$ and the systematic name (10*E*,12*Z*)-10,12-hexadecadien-1-ol. Draw a molecule of bombykol showing correct geometry for the two double bonds.

22.66 When 4-chloro-1-butanethiol is treated with a strong base such as sodium hydride (NaH), tetrahydrothiophene is produced. Suggest a mechanism for this reaction.

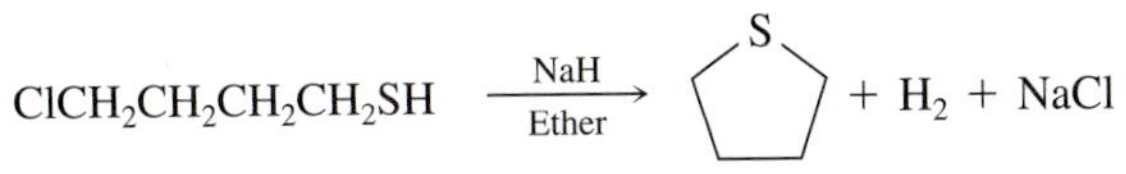

Tetrahydrothiophene

22.67 Why can't the Williamson ether synthesis be used to prepare diphenyl ether?

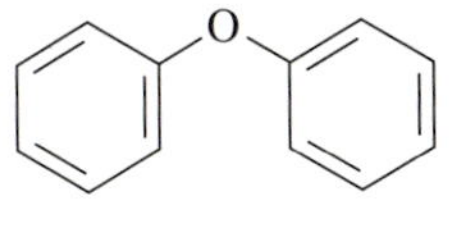

Diphenyl ether

22.68 Starting from benzene, how would you make benzyl phenyl ether? More than one step is required.

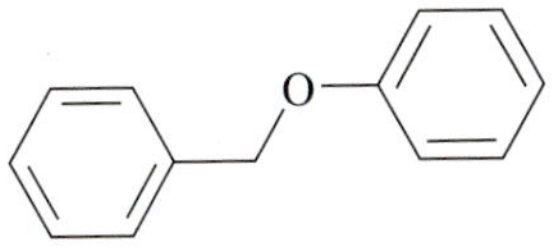

Benzyl phenyl ether

22.69 It is found experimentally that a substituted cyclohexanol with an axial –OH group reacts with CrO_3 more rapidly than its isomer with an equatorial –OH group. Draw both *cis*- and *trans*-4-*tert*-butylcyclohexanol, and predict which oxidizes faster. (The large *tert*-butyl group is equatorial in both.)

22.70 Since all hamsters look pretty much alike, pairing and mating is governed by chemical means of communication. Investigations have shown that dimethyl disulfide (CH_3SSCH_3) is secreted by female hamsters as a sex attractant for males. How would you synthesize dimethyl disulfide in the laboratory if you wanted to trick your hamster?

22.71 The herbicide 2,4,5-T (2,4,5-trichlorophenoxyacetic acid) can be synthesized by heating a mixture of 2,4,5-trichlorophenol and $ClCH_2CO_2H$ with NaOH. Show the mechanism of the reaction.

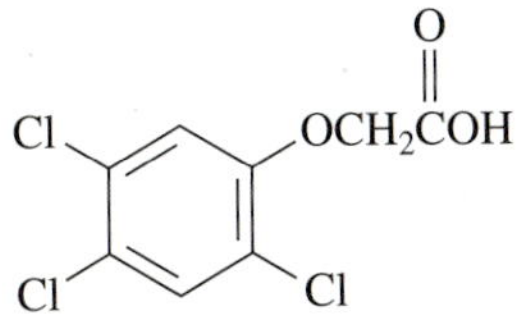

2,4,5-T

22.72 *tert*-Butyl ethers can be prepared by the reaction of an alcohol with 2-methylpropene in the presence of an acid catalyst. Propose a mechanism for this reaction.

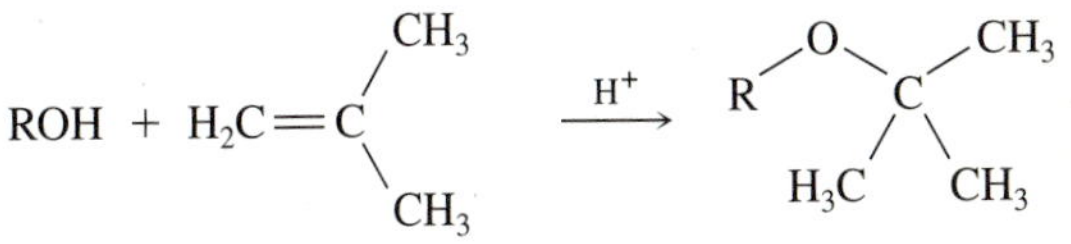

22.73 *tert*-Butyl ethers react with trifluoroacetic acid (CF_3CO_2H) to yield an alcohol and 2-methylpropene. For example:

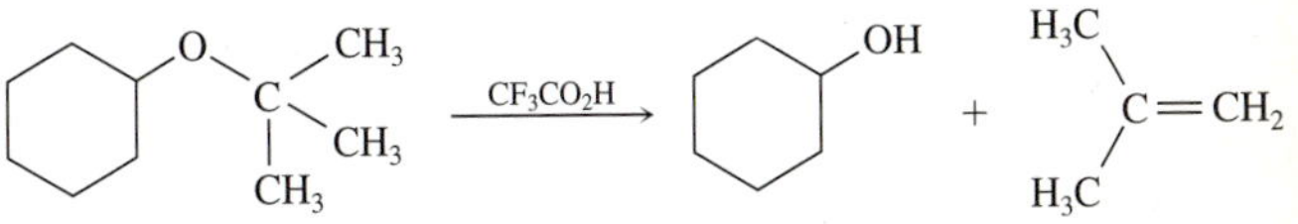

Tell what kind of reaction is occurring, and propose a mechanism.

22.74 How would you synthesize the following ethers?

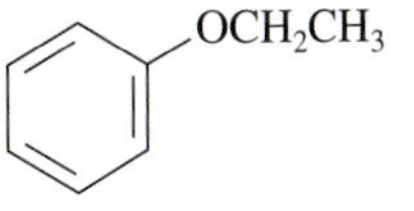

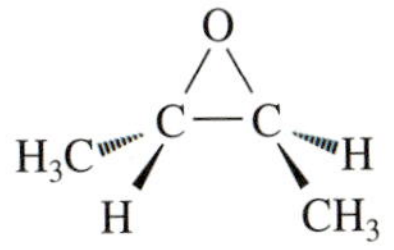

22.75 Identify the reagents a through d in the following scheme:

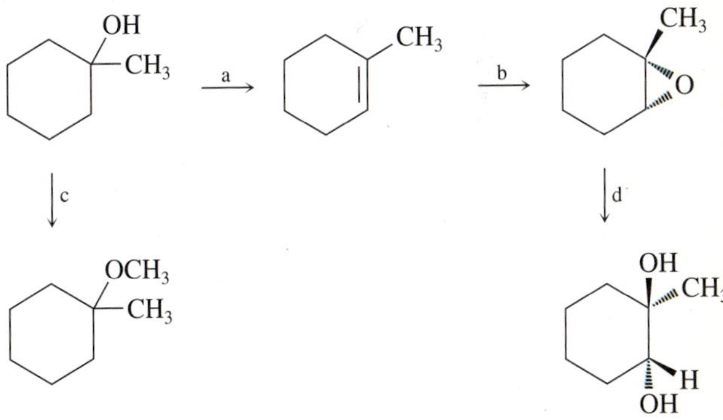

22.76 What cleavage product would you expect from reaction of tetrahydrofuran with hot aqueous HI?

Tetrahydrofuran

22.77 Methyl phenyl ether can be cleaved to yield iodomethane and lithium phenoxide when heated with LiI. Propose a mechanism for this reaction.

22.78 Reduction of butan-2-one with $NaBH_4$ yields butan-2-ol. Explain why the product is chiral but not optically active.

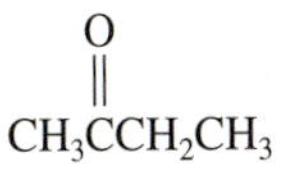

Butan-2-one

22.79 Compound A ($C_8H_{10}O$) has the IR and ^1H NMR spectra shown. Propose a structure consistent with the observed spectra, and assign each peak in the NMR spectrum. Note that the absorption at 5.5 δ disappears when D_2O is added.

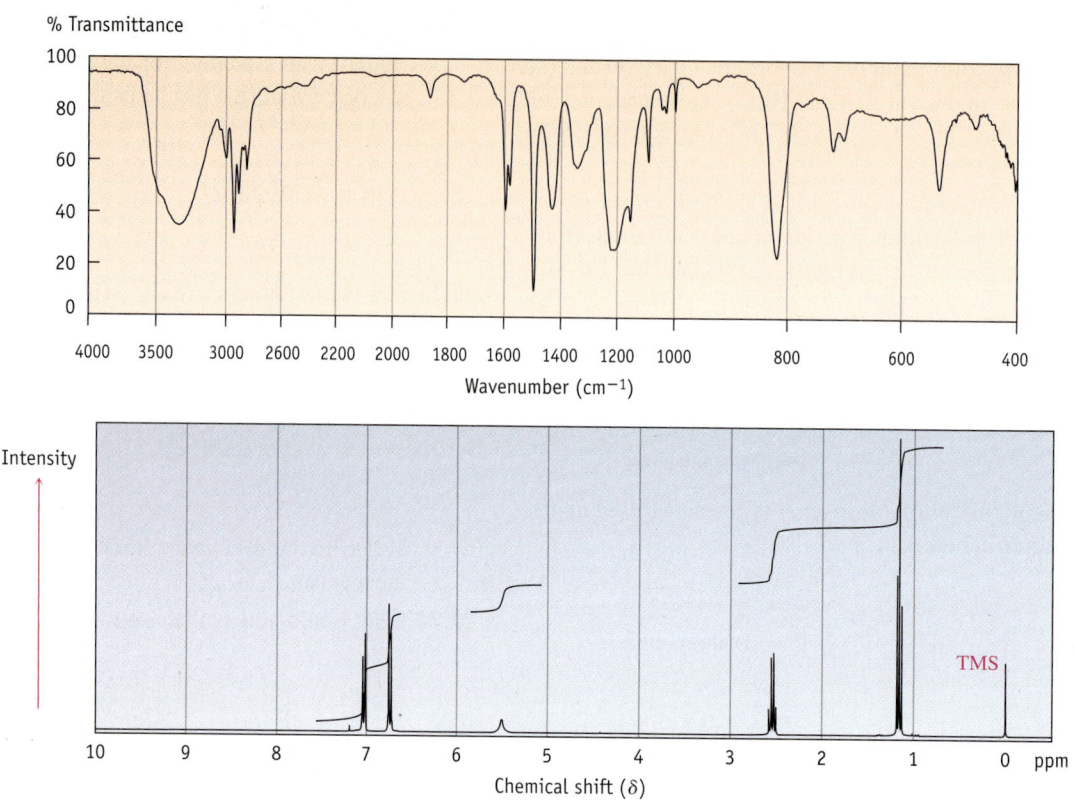

22.80 Propose a structure for a compound $C_{15}H_{24}O$ that has the following ^1H NMR spectrum. The peak marked by an asterisk disappears when D_2O is added to the sample.

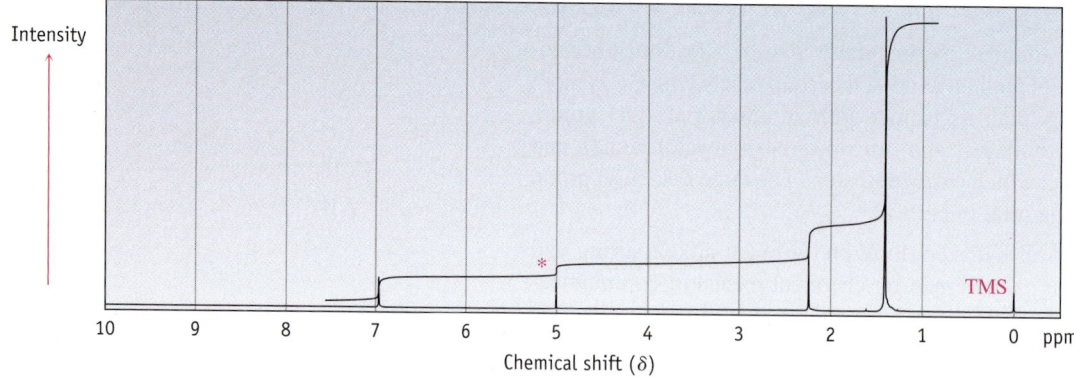

With the **eCHACR** single-sign-on access card bundled with your text, log on (http://login.cengage.com/sso) and access the e-book and click on any in-margin icons for dynamic molecular-level animations and simulations, videos of laboratory reactions, interactive exercises, reviews of background concepts, and other online supplementary materials as noted by the icons in the text margins.

Also go to www.chemistry.nelson.com <http://www.chemistry.nelson.com> for Answers to in-chapter exercises and selected Review Questions, Test Yourself questions, weblinks, crossword puzzles, flashcards, glossary of key terms, and other student resources.

Carbonyl Compounds: Part I

Gunter Marx/Getstock.com

© Canadian Forest Service

23.1 Case Study: Making Scents of the Mountain Pine Beetle

In our introduction to the chemistry of carbon compounds in Chapter 4, we saw that the absorption of infrared electromagnetic radiation by methane, carbon dioxide, water vapour, and other trace gases in our atmosphere is beneficial, providing climatic conditions that make our planet hospitable to life. But we also saw evidence suggesting that substantial increases in these "greenhouse gases" over the past 250 years, due mostly to the massive human activity of processing and burning fossil fuels, is significantly altering the earth's climate.

> *Studying and controlling surging populations of insects such as the mountain pine beetle is a necessary part of the global response to our rapidly changing climate—and here, too, chemistry has a role to play.*

What does this have to do with insects such as beetles? Besides fluctuations in temperature, precipitation, and increased frequency of severe weather events, much more subtle results of changing climate are also increasingly evident. Warmer average temperatures and changes in the availability of water play a crucial role in the life cycle of cold-blooded insects, and changes in the earth's climate have already led to substantial increases in the range of climatically suitable habitats for insects such as mosquitoes and beetles.

The implications of these changes in insect habitat are far-reaching. Australia recently created a malaria control group to tackle this insect-borne disease contracted by 500 million people a year. Over 1 million die from malaria per year—many are children under the age of five. Even slight increases in temperature and precipitation in heavily populated regions of Asia and Africa may create conditions for episodes of explosive reproduction of the anopheles mosquito, which transmits malaria.

Studying and controlling surging populations of insects that carry serious diseases has become a necessary part of the global response to our rapidly changing climate—and here, too, chemistry has a role to play.

The rice-grain sized mountain pine beetle (opening photo) is another insect whose populations have expanded over the past several decades into northern regions and higher elevations that previously had unsuitable climates for the species to survive. This beetle attacks pine trees, particularly the lodgepole pine, and is now one of the most significant causes of mortality in mature forests in Western North America (opening photo). In the past several years, populations have reached epidemic proportions in millions of hectares of British Columbia, with concerns that recent detection of beetles blown over the Continental Divide into Alberta may be the beginning of large-scale losses of the vast and economically important boreal forest. This raises concerns, as the boreal forest is one of the largest carbon storehouses on the earth, and losses of trees in this ecosystem may reinforce other changes to climate (a positive feedback cycle, see Section 4.3). Variables that are favourable for large-scale migration of pine beetles into these new habitats include a certain number of days with warm enough temperatures; winters without sustained periods of −40 °C temperatures, which control the population; and lower precipitation, which affects the ability of pine trees to resist attack.

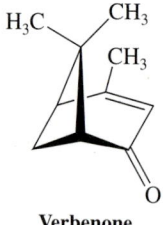

Verbenone

One of the tools available to forest ecologists to prevent the spread of mountain pine beetles is the ketone S-(−)-verbenone. This compound is called a **pheromone**, because it is an organic compound used to exchange information between individuals within a species. Many pheromones contain carbonyl groups—verbenone is a ketone. It is produced in trace quantities by pine beetles to deter other pine beetles from gathering on a tree that has already been attacked by large numbers. Sleuthing chemists and entomologists have proposed that the following steps might be involved. Attack on a pine tree is usually initiated by a female beetle, which bores into the tree. Healthy trees produce sticky sap as a defence mechanism, which traps a single beetle before it can reproduce. So the female beetle releases aggregant (gathering) pheromones to attract other beetles in sufficient numbers that some survive the bubbles of sap. As the females bore into the tree, they disrupt the vascular system and also transmit a mouth-borne blue stain fungus, which further slows down the flow of sap. As male beetles arrive and enter the tunnels created by females, they make a shrill creaking noise by rubbing together a scraper on their abdomen against a file on the inside of their wing. In response, the scent of the female beetles changes—instead of producing aggregant pheromones that are alcohols, they produce anti-aggregant ketone pheromones like verbenone to send a chemical message to other beetles that this tree is occupied and they should move on to another one. Perhaps the advantage gained by beetles by the use of pheromone is to reduce overcrowding of the new brood.

Applying laminated flakes containing trace amounts of synthesized verbenone to pine trees has been shown to trick other pine beetles and prevent them from attacking a stand of trees. This gives humans one possible tool to use with others to help adapt to ecosystem changes resulting from climate change. While it remains to be seen how significant a tool this one will prove to be, it has helped us learn new things about the fascinating world of insect communication through chemicals.

The Carbonyl Group

In this and the next chapter, we discuss compounds whose molecules, like verbenone, contain a **carbonyl functional group (C=O)**, the most important and widely occurring functional group in both organic and biological chemistry. Carbonyl compounds are everywhere in nature. Most biologically important molecules contain carbonyl groups, as do many pharmaceutical agents and many of the synthetic chemicals that touch our everyday lives.

Examples of substances whose molecules contain carbonyl groups as part of various functional groups include acetic acid (the chief component of vinegar), formaldehyde (a monomer for insulation and adhesive resins), acetaminophen (an over-the-counter pain reliever), acetone (an industrial solvent), polyethylene terephthalate (PET, the polyester material used in clothing, known by Dacron® and other trade names), and tetradec-11-enal (an aldehyde that is the sex pheromone of the spruce budworm).

Acetic acid (a carboxylic acid) **Formaldehyde** (an aldehyde) **Acetaminophen** (an amide) **Acetone** (a ketone)

Polyethylene terephthalate (a polyester) $CH_3CH_2CH=CH(CH_2)_9CH$ with O **Tetradec-11-enal** (an aldehyde)

Recalling the carbon cycle in Chapter 4, the oxidation of carbon compounds in combustion reactions produces the most important of the "greenhouse gases": carbon dioxide ($O=C=O$), which you might think of as containing two carbonyl groups in one triatomic molecule. Plants use CO_2 as a carbon source to produce glucose and other carbohydrates, which also contain carbonyl groups.

In this chapter, we focus on aldehydes and ketones.

23.2 Naming Aldehydes and Ketones

Aldehyde molecules have one carbon and one hydrogen atom attached to the carbonyl carbon atom. **Ketones** have two carbon atoms attached.

Aldehydes are named by replacing the terminal –*e* of the corresponding alkane name with –*al*. The parent chain must contain the —CHO group, and the —CHO carbon atom is always numbered as carbon 1. For example:

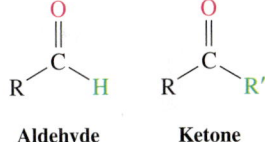

Aldehyde   **Ketone**

CH_3CH with O CH_3CH_2CH with O CH_3CHCH_2CHCH with CH_3 and CH_2CH_3

Ethanal (acetaldehyde) **Propanal** (propionaldehyde) **2-Ethyl-4-methylpentanal**

The longest chain in 2-ethyl-4-methylpentanal is a hexane, but this chain does not include the —CHO group and thus is not the parent.

For more complex aldehydes in which the —CHO group is attached to a ring, the suffix –*carbaldehyde* is used:

CHO CHO

Cyclohexanecarbaldehyde   **2-Naphthalenecarbaldehyde**

Some simple and well-known aldehydes also have common (trivial) names, as indicated in Table 23.1.

Interactive Exercise 23.1

Draw structural formulas of molecules of aldehydes and ketones.

TABLE 23.1 Names of Some Simple Aldehydes

Formula	Common Name	Systematic Name
HCHO	Formaldehyde	Methanal
CH_3CHO	Acetaldehyde	Ethanal
CH_3CH_2CHO	Propionaldehyde	Propanal
$CH_3CH_2CH_2CHO$	Butyraldehyde	Butanal
$CH_3CH_2CH_2CH_2CHO$	Valeraldehyde	Pentanal
$H_2C{=}CHCHO$	Acrolein	Prop-2-enal
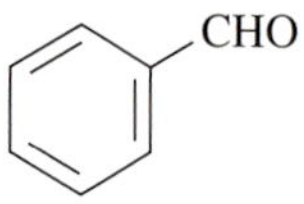 CHO	Benzaldehyde	Benzenecarbaldehyde

Ketones are named by replacing the terminal –*e* of the corresponding alkane name with –*one*. The parent chain is the longest one that contains the ketone group, and numbering begins at the end nearer the carbonyl carbon. For example:

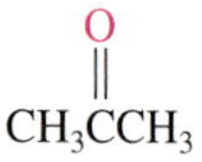

$$CH_3CCH_3$$

Propanone (acetone)

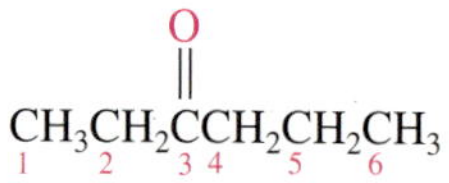

$$CH_3CH_2CCH_2CH_2CH_3$$

Hexan-3-one

$$CH_3CH{=}CHCH_2CCH_3$$

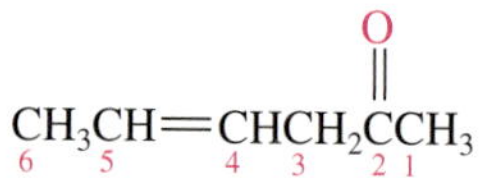

Hex-4-en-2-one

A few ketones also have common names:

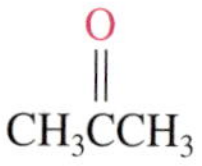

$$CH_3CCH_3$$

Acetone

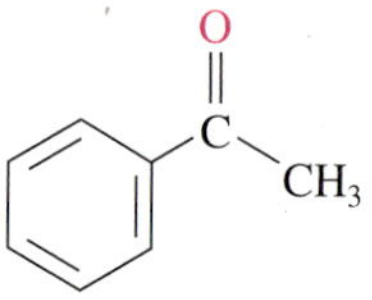

Acetophenone

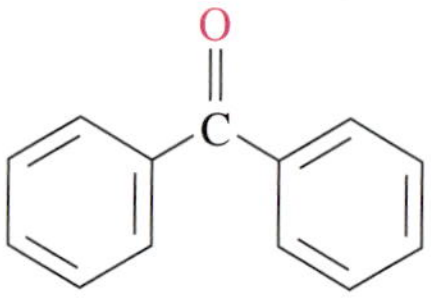

Benzophenone

When it's necessary to refer to the $-C({=}O)R$ group as a substituent, the general term *acyl* is used. Other common examples: $-C({=}O)CH_3$ is an acetyl group, $-C({=}O)H$ is a formyl group, $-C({=}O)Ar$ is an aroyl group, and $-C({=}O)C_6H_5$ is a benzoyl group.

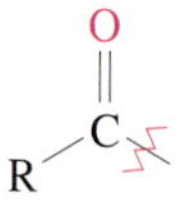

An acyl group (R = alkyl, alkenyl) 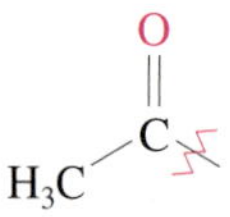   **Acetyl** 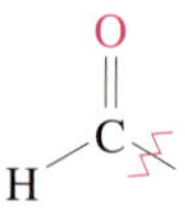   **Formyl** 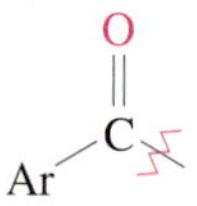   **Aroyl (Ar = aromatic)** 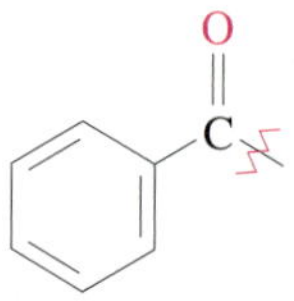   **Benzoyl**

When another group is present with higher priority for naming than a ketone, the doubly bonded oxygen atom is considered a substituent, and the prefix *oxo*– is used. For example:

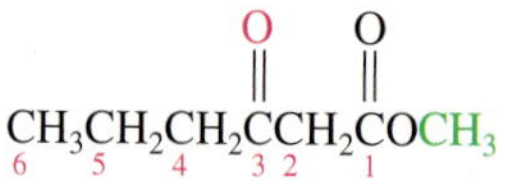

$$CH_3CH_2CH_2CCH_2COCH_3$$

Methyl 3-oxohexanoate

EXERCISE 23.2—NAMING ALDEHYDES AND KETONES

Name the following aldehydes and ketones:

(a) $CH_3CH_2CCH(CH_3)_2$ (with O double-bonded to the third carbon)

(b) (benzene ring) CH_2CH_2CHO

(c) $CH_3CCH_2CH_2CH_2CCH_2CH_3$ (with two O double bonds)

(d) (cyclohexane ring with) H.., CH₃, H, CHO

(e) $OHCCH_2CH_2CH_2CHO$

(f) (cyclohexanone ring with) O, H₃C.., H, H, CH₃

EXERCISE 23.3—NAMING ALDEHYDES AND KETONES

Draw structures of molecules corresponding to the following IUPAC names:
(a) 3-methylbutanal
(b) 3-methylbut-3-enal
(c) 4-chloropentan-2-one
(d) phenylacetaldehyde
(e) 2,2-dimethylcyclohexanecarbaldehyde
(f) cyclohexane-1,3-dione

23.3 Synthesis and Spectroscopy

A chemist wishing to synthesize the anti-aggregant pheromone verbenone to deter mountain pine beetles from attacking a stand of trees could start with the same compound beetles use when they make verbenone, and from the same source. α-Pinene is one of several *terpenes* found in the resins of coniferous trees such as the pine. Like the terpenes β-carotene and Vitamin A [<<Section 19.1], α-pinene is a natural product whose molecules are constructed of isoprene monomer units. α-Pinene can be extracted from the sticky hydrocarbon pine resin secretions given off by pine trees when they respond to a pine beetle attack. It is thought that beetles themselves, or perhaps micro-organisms in their intestinal tracts, can oxidize α-pinene to produce the aggregant pheromone verbenol, and more completely oxidize α-pinene to the anti-aggregant pheromone verbenone (Figure 23.1). The same oxidation reactions occur slowly in air.

FIGURE 23.1 α-Pinene produced by pine trees, and the beetle pheromones verbenol and verbenone synthesized from α-pinene. The wavy line connecting the —OH group to the hydrocarbon ring in verbenol means that two possible stereoisomers could be formed, one with the —OH group pointing up in the representation, and one with the —OH group pointing down.

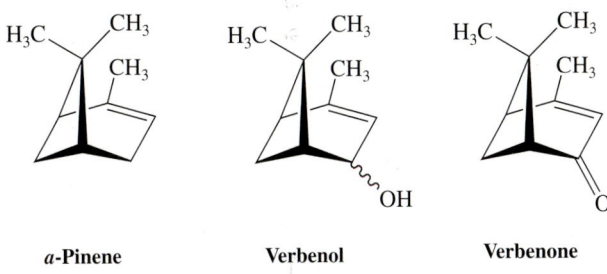

a-Pinene   Verbenol   Verbenone

Spectroscopy of Aldehydes and Ketones

How would synthetic chemists tell that the laboratory transformation in this series of two oxidation reactions starting from α-pinene has successfully produced alcohol and ketone products? Spectroscopic methods will quickly tell the story, and infrared spectroscopy would be the method of choice to recognize changes in functional groups from an alkene only to an alkene/alcohol and then to an alkene/ketone. If this synthesis hadn't been performed before, chemists would look up or record the spectra of simpler model compounds with the same functional groups, to compare with the new compounds they hope to create. Since α-pinene, verbenol, and verbenone all have a cyclohexene ring where the reactive sites are located, reasonable model compounds for spectral analysis might be cyclohexene, cyclohex-2-en-1-ol, and cyclohex-2-en-1-one, shown below. They can be easily distinguished by features in their IR and ^{13}C NMR spectra, as shown in Worked Example 23.1.

Think about It

e23.1 Compare the animated IR spectra of compounds with and without a carbonyl group.

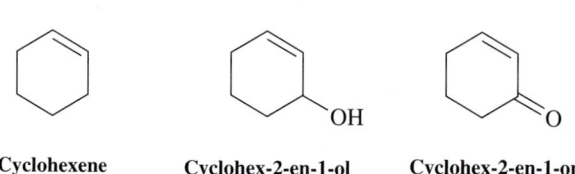

| Cyclohexene | Cyclohex-2-en-1-ol | Cyclohex-2-en-1-one |

WORKED EXAMPLE 23.1—IR SPECTROSCOPY

Infrared spectra for the three model compounds are shown below. Referring back to the animated IR spectra in e23.1, which IR spectrum belongs to which model compound?

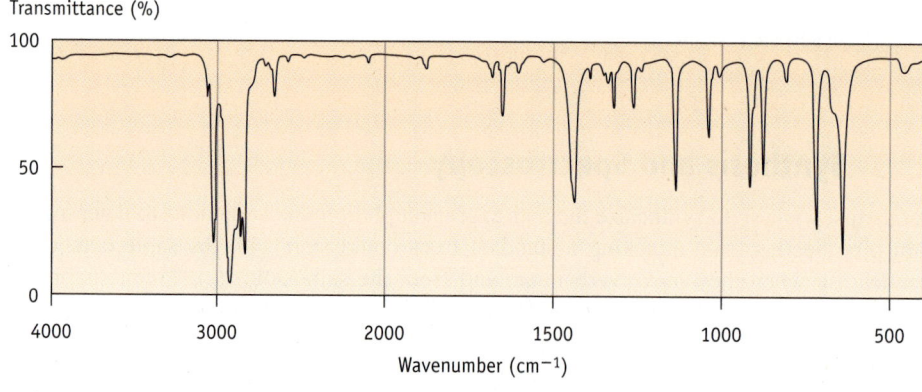

Transmittance (%)

IR spectrum A

Transmittance (%)

IR spectrum B

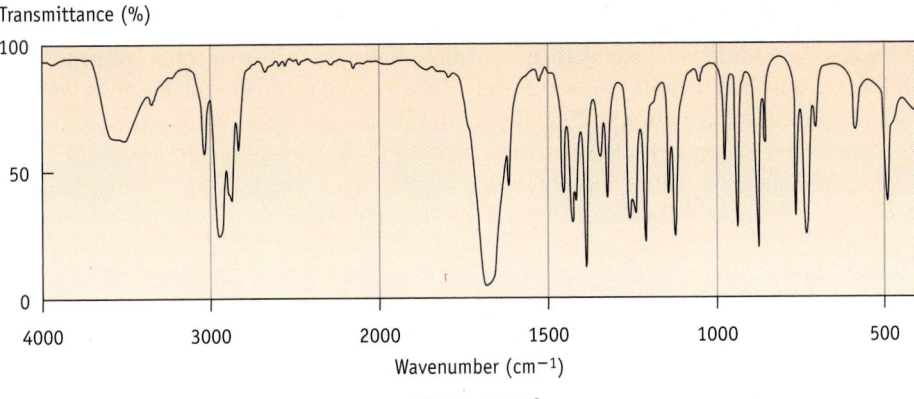

IR spectrum C

Solution

Molecules of all three compounds contain alkene functional groups, with characteristic absorptions just above 3000 cm^{-1} for the alkene C—H stretching vibrations. The alkene (spectrum A) and alcohol (spectrum B) show sharp absorptions at about 1650 cm^{-1} from C=C stretching. The alcohol is easily recognized by a broad, H-bonded O—H stretching band at 3400 cm^{-1} and a C—O stretch at 1060 cm^{-1}. The most intense band in the IR spectrum of cyclohex-2-ene-1-one (spectrum C) is due to stretching of the carbonyl C=O group at 1685 cm^{-1}. A strong absorption in the region between 1670 and 1780 cm^{-1} is characteristic of all carbonyl-containing compounds, and typically occurs between 1710 and 1730 cm^{-1} for open-chain aldehydes and ketones. In this case, the absorption is shifted to lower wavenumber due to the conjugation of the C=O bond with a C=C of the alkene, which leads to delocalization of electrons in the carbonyl group and a somewhat weaker bond. This conjugation also shifts the stretching wavenumber of the C=C group in cyclohex-2-ene-1-one. The weak O—H stretching absorption in the ketone spectrum is due to either a trace of water in the sample or the presence of a small amount of the enol form of the ketone [>>Section 24.8].

WORKED EXAMPLE 23.2—^{13}C NMR SPECTROSCOPY

^{13}C NMR spectroscopy provides complementary information to distinguish among the three model compounds in Worked Example 23.1. Match the ^{13}C NMR spectral assignments for the three compounds below with the correct structures, and identify any carbon atoms which come downfield from the others.

  Spectrum A: 6 sharp peaks at 200, 151, 130, 38, 26, and 23 δ.
  Spectrum B: 3 sharp peaks at 127, 25, and 23 δ.
  Spectrum C: 6 sharp peaks at 131, 120, 65, 32, 25, and 19 δ.

Solution

Cyclohexene (spectrum B) is a more symmetrical molecule than the other two, and has only three unique carbon atoms [<<Section 9.4], giving three signals. The two alkene carbon atoms appear at 127 δ and the other two carbon atoms at 25 and 23 δ. Each of the six carbon atoms are unique in both cyclohex-2-en-1-ol (spectrum C) and cyclohex-2en-1-one (spectrum A), but the carbonyl carbon atom of the ketone is far downfield at 200 δ.

Aldehydes and ketones can be identified by a very intense C=O infrared stretching vibration, typically between 1710 and 1730 cm^{-1}. ^{13}C NMR spectroscopy provides complementary information, with carbonyl carbon atoms appearing far downfield, between 190 and 220 δ.

Synthesis of Aldehydes and Ketones

The syntheses of the ketone verbenone from oxidation of an alcohol verbenol and cyclohex-2-en-1-one from cyclohex-2-en-1-ol are examples of one of the best ways to make aldehydes and ketones [<<Section 22.6]. Primary alcohols can be oxidized to give aldehydes, and secondary alcohols to ketones. Pyridinium chlorochromate (PCC) in dichloromethane is usually chosen as the oxidizing agent for making aldehydes, while PCC, CrO_3, and $Na_2Cr_2O_7$ are all effective for making ketones:

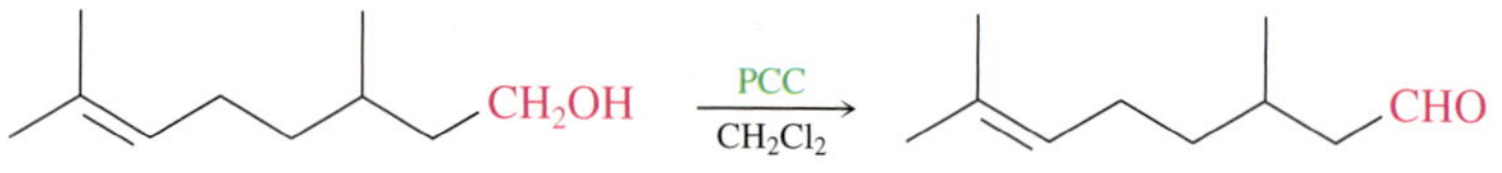

Citronellol → **Citronellal (82%)**

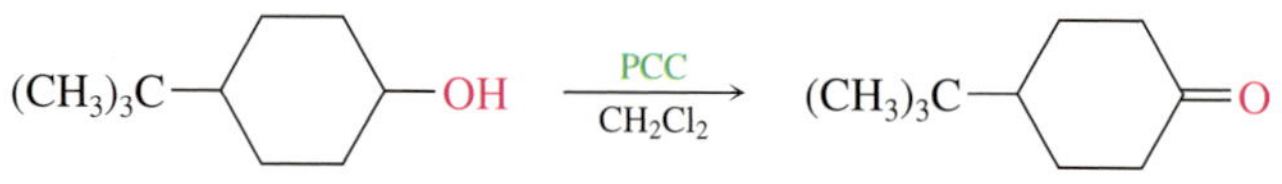

4-*tert*-Butylcyclohexanol → **4-*tert*-Butylcyclohexanone (90%)**

Other methods for preparing ketones include the hydration of a terminal alkyne to yield a methyl ketone [<<Section 19.14] and the Friedel-Crafts acylation of an aromatic ring to yield an alkyl aryl ketone [<<Section 20.7].

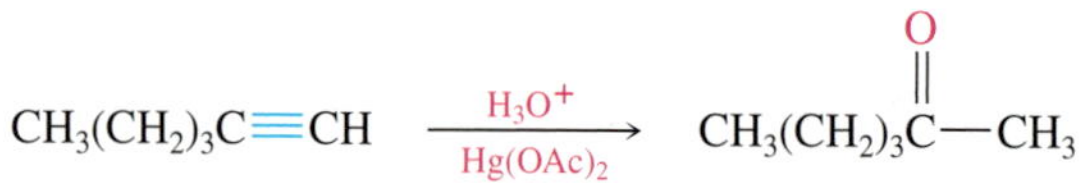

$$CH_3(CH_2)_3C\equiv CH \xrightarrow[Hg(OAc)_2]{H_3O^+} CH_3(CH_2)_3\overset{\overset{\displaystyle O}{\|}}{C}-CH_3$$

Hex-1-yne → **Hexan-2-one (78%)**

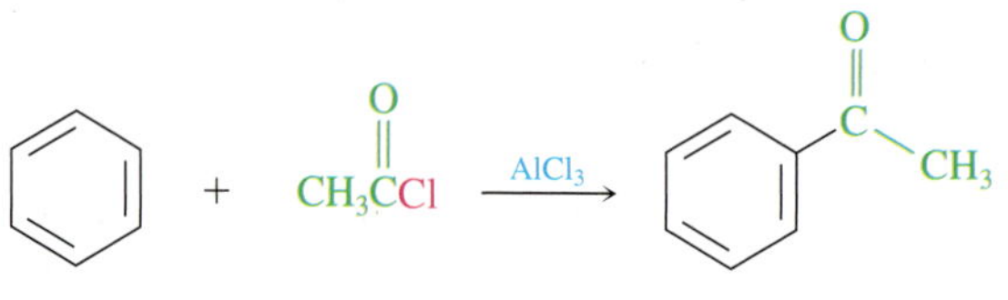

Benzene + **Acetyl chloride** → **Acetophenone (95%)**

EXERCISE 23.4—SYNTHESIS OF ALDEHYDES

How could you synthesize pentanal from the following starting materials?
(a) 1-pentanol   (b) $CH_3CH_2CH_2CH_2COOH$ (c) 5-decene

EXERCISE 23.5—SYNTHESIS OF KETONES

How could you synthesize hexan-2-one from the following starting materials?

(a) $CH_3CH_2CH_2CH_2\overset{\overset{\displaystyle OH}{|}}{C}HCH_3$ (b) $CH_3CH_2CH_2CH_2C\equiv CH$ (c) $CH_3CH_2CH_2CH_2\overset{\overset{\displaystyle CH_3}{|}}{C}=CH_2$

One of the most useful methods of synthesizing aldehydes and ketones is by oxidation of primary and secondary alcohols, respectively.

23.4 Structure and Reactivity of the Carbonyl Group

The carbonyl carbon–oxygen double bond is similar in some respects to the carbon–carbon double bond of alkene molecules. Like alkenes, molecules of carbonyl compounds are planar about the double bond and have bond angles of approximately 120°. Consistent with this experimental geometry, a hybrid orbital model of the bonding involving the carbonyl carbon atom would represent it as being sp^2 hybridized and forming three σ bonds. Using this model, carbon and oxygen atoms of the carbonyl group are thought of as being connected by one σ and one π bond. The relatively weak π bond, which results from side-to-side overlap of two p orbitals, is then the key to the reactivity of carbonyl-containing compounds. The oxygen atom of a carbonyl group also has two non-bonding pairs of electrons, which occupy its remaining two orbitals.

Electronic structure of the carbonyl group

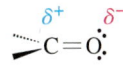

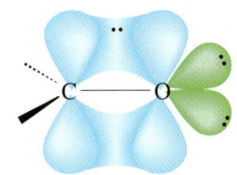

Carbonyl group

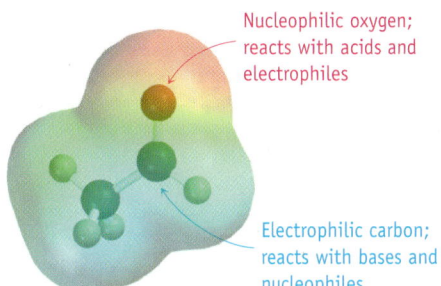

Nucleophilic oxygen; reacts with acids and electrophiles

Electrophilic carbon; reacts with bases and nucleophiles

But the two carbon atoms making up a $C=C$ bond of an alkene molecule can be similarly pictured as being connected by one σ and one π bond. Despite these similarities, the reactivity of carbonyl-containing compounds and alkenes is strikingly different. As seen in Chapter 19, electrophiles add to alkenes, while the reactions in this chapter show that nucleophiles add to the carbonyl carbon atom of aldehydes and ketones. What might cause such a significant difference in reactivity?

An explanation can be found by looking at the electrostatic potential map of ethanal above. Unlike the $C=C$ bonds of alkenes, $C=O$ double bonds are polar because of the high electronegativity difference between oxygen and carbon atoms. The carbonyl carbon atom has a partial positive charge, and therefore in reactions is electrophilic and a site for reaction with nucleophiles. Conversely, the carbonyl oxygen atom has a partial negative charge, is nucleophilic, and often reacts with a proton or other electrophiles. We'll see in this and the next two chapters that most carbonyl group reactions are the result of this bond polarization and involve breaking the π bond.

23.5 Nucleophilic Addition Reactions

For the reasons discussed in Section 23.4, the most common reaction of aldehydes and ketones is **nucleophilic addition**, in which a nucleophile *adds* to the electrophilic carbonyl carbon atom of a molecule. In this category we include reactions such as reduction with a source of hydride ion (H^-); hydration with H_2O; acetal or ketal formation from reaction with an alcohol ROH; Grignard addition with RMgX; and imine formation from reaction with RNH_2. The nucleophile can be either negatively charged ($:Nu^-$) or neutral ($:Nu$). Negatively charged nucleophiles add to aldehydes and ketones under basic conditions, and neutral nucleophiles are usually added under acidic conditions. A list of common nucleophiles for reaction with carbonyl-containing compounds is shown in Table 23.2. Neutral nucleophiles usually have a hydrogen atom that can be removed in a subsequent step as a proton (H^+).

We can use what we know of the polarity of the carbonyl group and the reactivity patterns of Lewis acids and bases to postulate a reasonable reaction mechanism that helps us make sense at the molecular level of why these different nucleophiles react with aldehydes and ketones, and why they give addition products. The mechanism changes slightly under acidic and under basic conditions, as summarized in Figure 23.2.

TABLE 23.2 Nucleophiles for Addition Reactions to Carbonyl Groups

Nu—H	Name	Reagent	Nu⁻	Name	Reagent
HO—H	Water	H_2O	HO^-	Hydroxide ion	LiOH
RO—H	An alcohol	ROH	$H:^-$	Hydride ion	$NaBH_4$
H_2N—H	Ammonia	NH_3	$R_3C:^-$	A carbanion	R_3CMgX
RHN—H	An amine	RNH_2	RO^-	An alkoxide ion	NaOR
			$N{\equiv}C:^-$	A cyanide ion	KCN

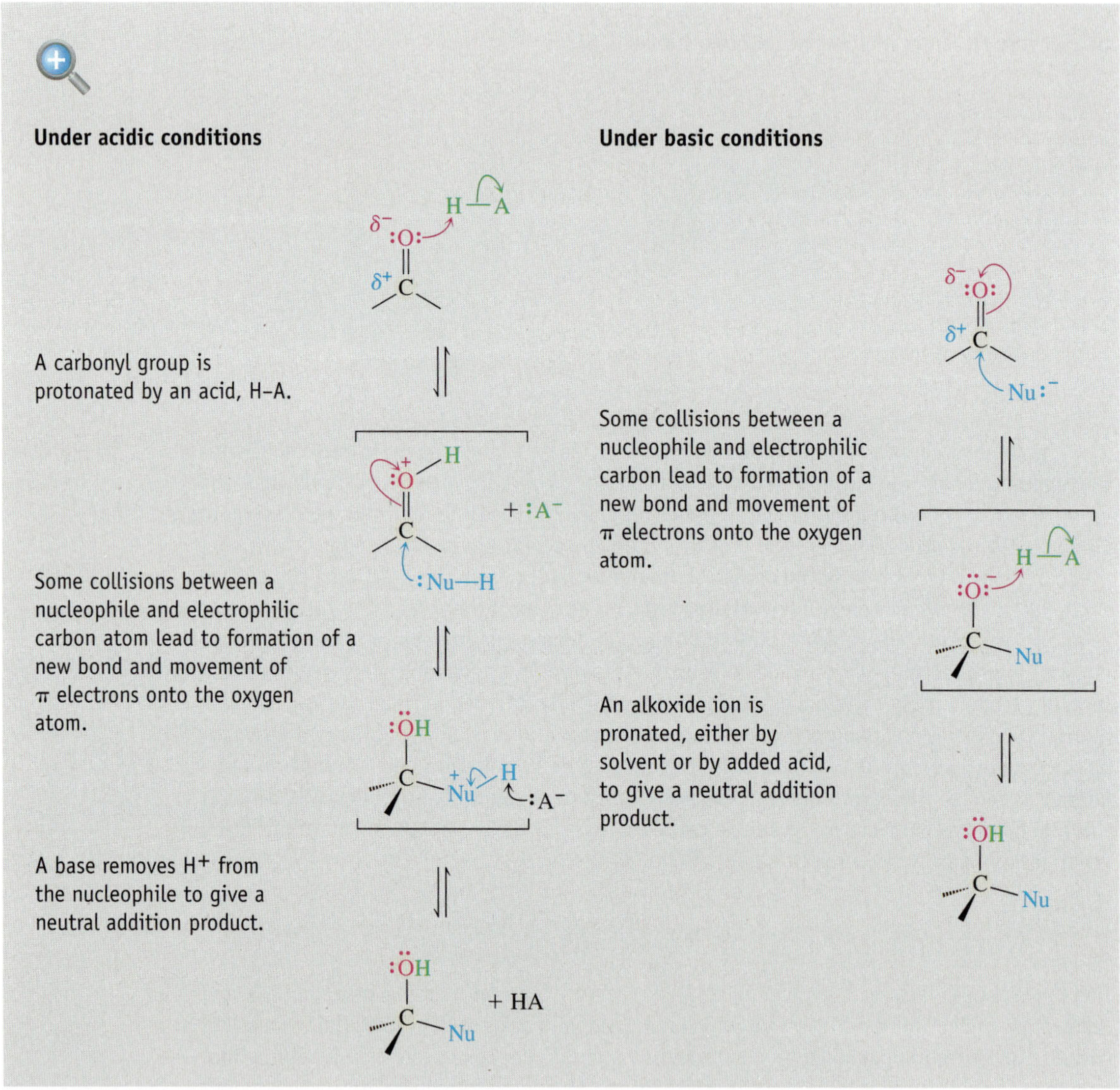

Under acidic conditions

A carbonyl group is protonated by an acid, H–A.

Some collisions between a nucleophile and electrophilic carbon atom lead to formation of a new bond and movement of π electrons onto the oxygen atom.

A base removes H⁺ from the nucleophile to give a neutral addition product.

Under basic conditions

Some collisions between a nucleophile and electrophilic carbon lead to formation of a new bond and movement of π electrons onto the oxygen atom.

An alkoxide ion is pronated, either by solvent or by added acid, to give a neutral addition product.

FIGURE 23.2 Accepted general reaction mechanism for nucleophilic addition to a carbonyl group under acidic (*left*) and basic (*right*) conditions. The fundamental difference between the two conditions is the timing for when the oxygen atom receives a proton.

Think about It

e23.6 Watch an animation to help understand these mechanisms.

Under the proposed mechanism for basic conditions, one of the few successful collisional events that leads to formation of product molecules starts with a pair of electrons on a nucleophile :Nu⁻ being shared with the electrophilic carbonyl carbon atom, causing cleavage of the C=O π bond, with the oxygen atom of the carbonyl group accepting a lone pair of electrons and becoming an alkoxide ion. Because the geometry of the carbonyl carbon atom changes from trigonal-planar to tetrahedral in this step, the intermediate is called a *tetrahedral intermediate*. The alkoxide ion is then protonated, either by solvent or added acid, giving a neutral addition product.

Under acidic conditions, the first step proposed for the reaction mechanism is protonation of the nucleophilic oxygen atom of the carbonyl group, followed by reaction of a nucleophile with the electrophilic carbonyl carbon atom. This causes cleavage of the C=O π bond, with the oxygen atom of the carbonyl group accepting a lone pair of electrons and becoming a tetrahedral intermediate. In the final step, solvent removes H^+ from the nucleophile to give a neutral addition product.

> Under basic conditions (in the presence of $^-$OH or $^-$OR), evidence suggests that negatively charged reaction intermediates are formed; positively charged intermediates are not usually seen. Under acidic conditions (the presence of H^+), evidence suggests that positively charged intermediates are formed; negatively charged intermediates are not usually observed.

WORKED EXAMPLE 23.3—NUCLEOPHILIC ADDITION REACTIONS

What product would you expect from nucleophilic addition of aqueous hydroxide ion to acetaldehyde?

Solution

A negatively charged hydroxide ion is a nucleophile, which can add to the C=O carbon atom and give an alkoxide ion intermediate. Protonation will then yield a 1,1-dialcohol.

Acetaldehyde

EXERCISE 23.6—NUCLEOPHILIC ADDITION REACTIONS

What product would you expect if the nucleophile cyanide ion (CN^-) was added to acetone and the intermediate was then protonated?

EXERCISE 23.7—NUCLEOPHILIC ADDITION REACTIONS

What product would you expect if the nucleophile methoxide ion (CH_3O^-) was added to benzaldehyde and the intermediate was then protonated?

Molecular Modelling (Odyssey)

e23.7 Use models to predict reactivity in nucleophilic addition reactions.

Why do the reaction steps shown in Figure 23.2 stop with the neutral addition product? We could imagine that a driving force might exist for the addition product to re-form the exceptionally stable C=O bond if it could lose a leaving group capable of accepting a pair of electrons. This question can be answered by noting a fundamental difference between reactions of aldehydes and ketones in this chapter and carboxylic acid derivatives in Chapter 24.

This reactivity difference is so important that compounds containing carbonyl groups are usually divided into two categories, based on what is attached to the carbonyl carbon atom, which affects the types of reactions they undergo. When a nucleophile reacts with the carbonyl carbon atom of an aldehyde or ketone, it typically undergoes an *addition* rather than a *substitution* reaction, since neither hydrogen nor carbon atoms can readily stabilize a negative charge, and so groups such as an $H:^-$ or $R:^-$ rarely act as leaving groups.

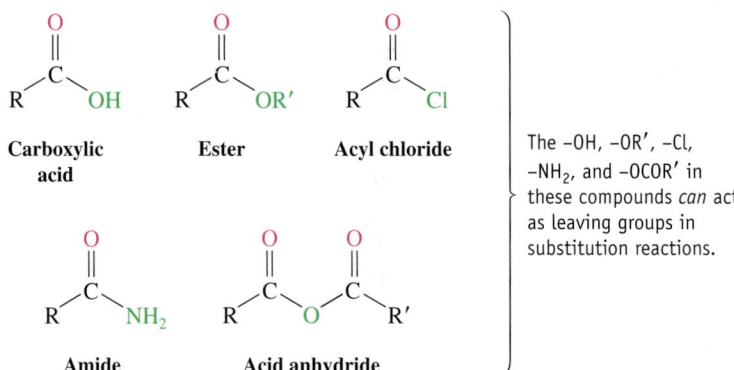

The –H and –R' in these compounds *can't* act as leaving groups in substitution reactions.

Aldehyde **Ketone**

In Chapter 24, we will see that the carbon atoms of carbonyl groups in carboxylic acid molecules and their derivatives are bonded to more electronegative atoms (i.e., oxygen, halogen, nitrogen) that *can* stabilize a negative charge. As a result, when a nucleophile reacts with the carbonyl carbon atom of a carboxylic acid derivative, the addition product is not the final product—these atoms capable of stabilizing a negative charge act as leaving groups, making possible *substitution* reactions.

Carboxylic acid **Ester** **Acyl chloride**

The –OH, –OR', –Cl, –NH$_2$, and –OCOR' in these compounds *can* act as leaving groups in substitution reactions.

Amide **Acid anhydride**

The C=O bond of aldehyde and ketone molecules is very polar, and nucleophiles react at the electrophilic carbonyl carbon atom. After a nucleophile has added to a carbonyl group, no suitable leaving groups are present, so the final product is usually the addition product. By contrast, in carboxylic acid derivatives [>>Chapter 24], atoms capable of accepting a negative charge and becoming leaving groups are present, leading to net substitution reactions.

23.6 Addition of Water: Hydration

Aldehydes and ketones undergo nucleophilic addition reaction with water to yield 1,1-diols, called *geminal (gem) diols.* The term *geminal* refers to the position of both –OH groups on the same carbon atom. The reaction is reversible, and the diol product can eliminate water to regenerate a ketone or aldehyde:

Acetone **Acetone hydrate (a gem diol)**

The equilibrium distribution between gem diols and aldehydes/ketones depends on the structure of the carbonyl compound. Although the equilibrium strongly favours the less sterically crowded carbonyl compound in most cases, the gem diol is favoured for a few simple aldehydes. For example, an aqueous solution of acetone consists of about 0.1% gem diol and 99.9% ketone, whereas an aqueous solution of formaldehyde (CH$_2$O) consists of 99.9% gem diol and 0.1% aldehyde.

Just as with many other nucleophilic addition reactions, the addition reaction of water to aldehydes and ketones is slow in pure water but is catalyzed by both acid and base. The catalyst never changes the equilibrium distribution of reactants and products; it affects only the rate at which the hydration reaction occurs [<<Section 18.6].

Zooming in to picture what evidence suggests is happening at the molecular level, we can postulate a reasonable reaction mechanism that puts together what we have just learned about the polarity of the C=O group with this new information that both base and acid catalyze this nucleophilic addition reaction of water.

The base-catalyzed addition reaction is thought to take place in several steps, as shown in Figure 23.3. The reacting nucleophiles are negatively charged hydroxide ions.

Molecular Modelling (Odyssey)

e23.8 Use electrostatic potential maps to understand base-catalyzed hydration.

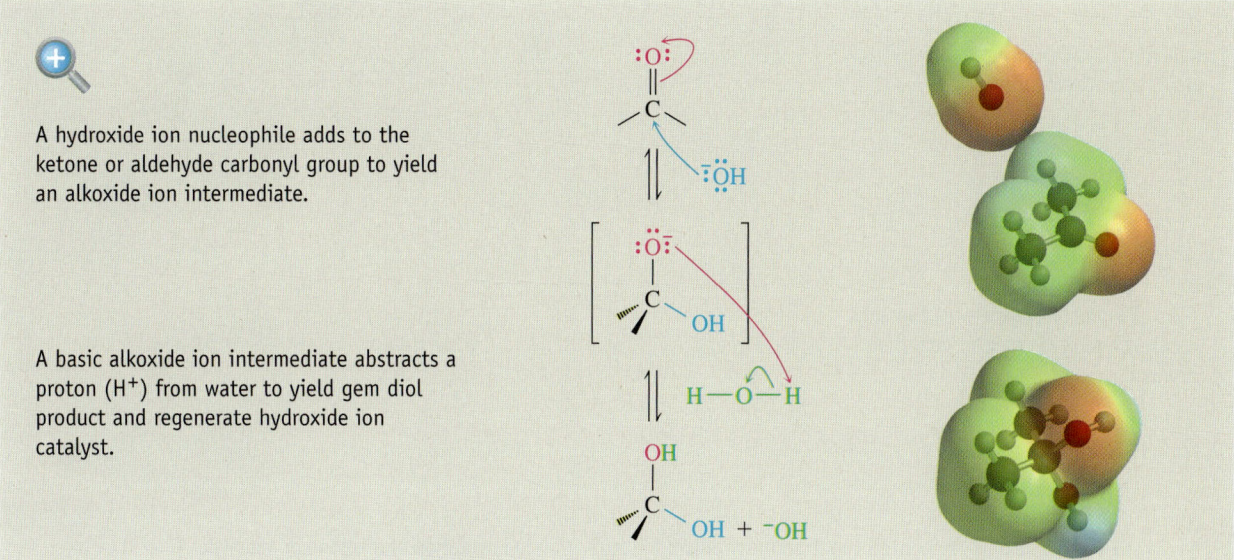

A hydroxide ion nucleophile adds to the ketone or aldehyde carbonyl group to yield an alkoxide ion intermediate.

A basic alkoxide ion intermediate abstracts a proton (H⁺) from water to yield gem diol product and regenerate hydroxide ion catalyst.

FIGURE 23.3 **Accepted reaction mechanism** for base-catalyzed hydration of a ketone or aldehyde. Hydroxide ion is a more reactive nucleophile than neutral water.

An acid-catalyzed hydration reaction is also thought to take place in several steps (Figure 23.4). An acid catalyst first protonates the basic oxygen atom of the carbonyl group, and subsequent nucleophilic addition of neutral water yields a protonated gem diol. Loss of a proton then gives the gem diol product.

These two reaction mechanisms for the addition of water have been shown to be consistent with important features of other nucleophilic addition reactions of aldehydes and ketones. Note the difference between the base-catalyzed and acid-catalyzed processes.

Molecular Modelling (Odyssey)

e23.9 Use electrostatic potential maps to understand acid-catalyzed hydration.

EXERCISE 23.8—ADDITION OF WATER

When dissolved in water, trichloroacetaldehyde (chloral, CCl_3CHO) exists primarily as the gem diol chloral hydrate (better known as "knockout drops"). Show the structure of a chloral hydrate molecule.

EXERCISE 23.9—ADDITION OF WATER

The oxygen atoms in water molecules are primarily (99.8%) ^{16}O, but water enriched with the heavy isotope ^{18}O is also available. When a ketone or aldehyde is dissolved in $H_2{}^{18}O$, the isotopic label becomes incorporated into the carbonyl group: $R_2C{=}O + H_2O^* \longrightarrow R_2C{=}O^* + H_2O$ (where $O^* = {}^{18}O$). Explain.

Think about It

e23.10 Watch an animation to help you understand hydration mechanisms.

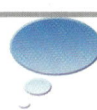

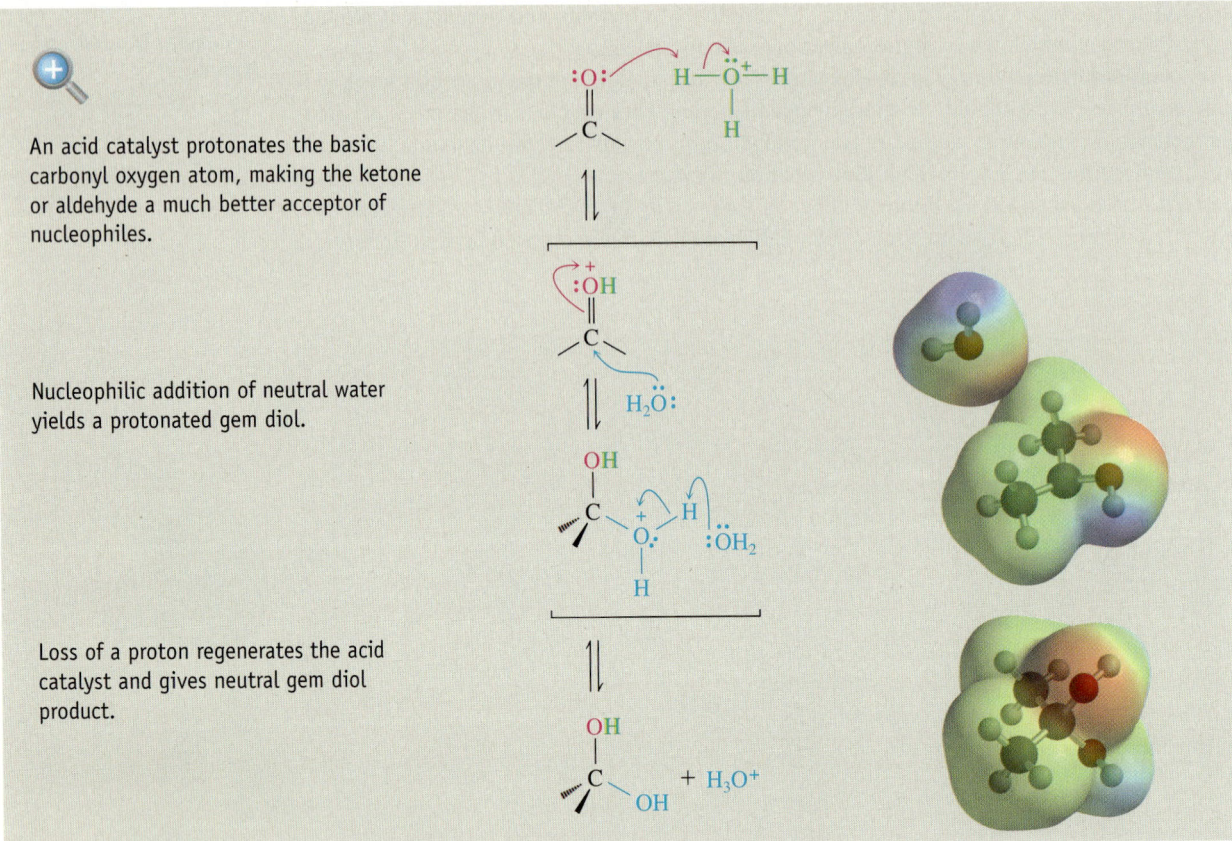

An acid catalyst protonates the basic carbonyl oxygen atom, making the ketone or aldehyde a much better acceptor of nucleophiles.

Nucleophilic addition of neutral water yields a protonated gem diol.

Loss of a proton regenerates the acid catalyst and gives neutral gem diol product.

FIGURE 23.4 Accepted reaction mechanism for acid-catalyzed hydration of a ketone or aldehyde. An acid catalyst protonates the carbonyl starting material, making it more electrophilic and reactive.

Base catalyzed addition of water to aldehydes and ketones takes place at a faster rate than the uncatalyzed reaction because hydroxide ion is a much better nucleophile than neutral water. The acid-catalyzed hydration reaction takes place at a faster rate because the carbonyl compound is converted by protonation into a much better electrophile.

23.7 Addition of Alcohols in Nature and the Laboratory

Recall the cyclodextrin pill "containers" in Section 22.1. Look closely again at the representations below of molecules of glucose and β-cyclodextrin. What functional groups do you recognize in each?

Glucose (β form)

β-**Cyclodextrin**

Perhaps you recognized alcohol and ether functional groups. But observe that one carbon atom in each six-membered ring contains two functional groups. When two groups are found together on a single carbon atom, their reactivity changes sufficiently that we sometimes name this as a new functional group. The **hemiacetal** carbon atom in cyclic glucose (orange) can be recognized as the only one that has two oxygen atoms attached to it—one is part of an alcohol group, the other part of an ether. As is the case with most sugars (carbohydrates), glucose molecules exist as an equilibrium mixture of two forms— *an open chain form containing an aldehyde group that reacts with a hydroxyl group on the same molecule to form a cyclic hemiacetal form.*

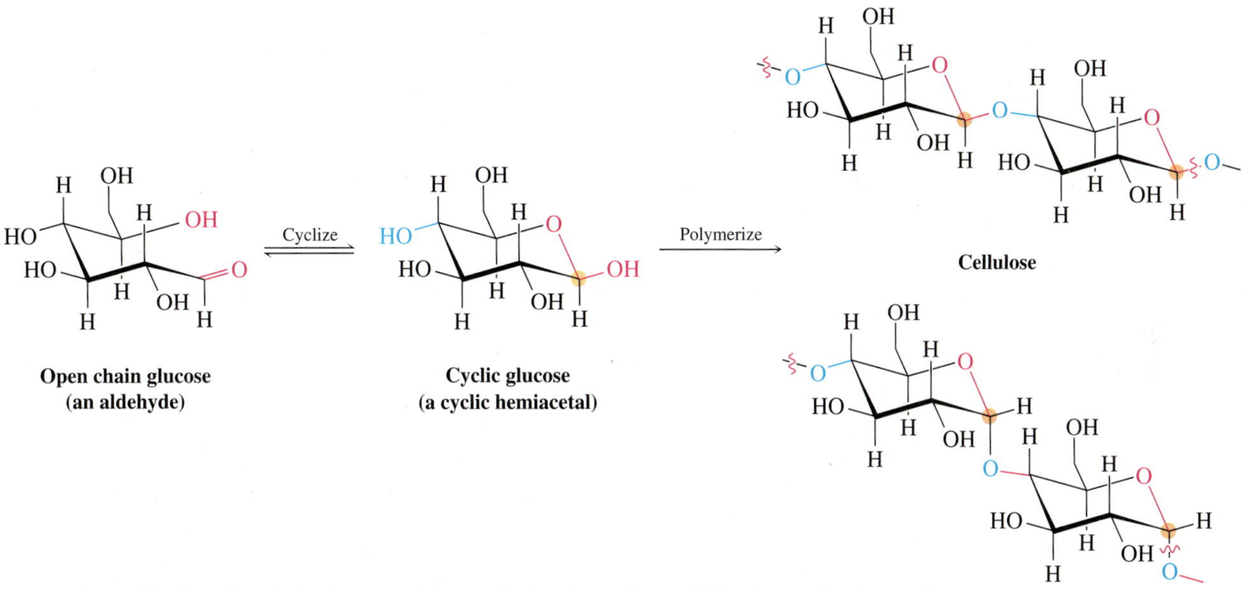

**Open chain glucose
(an aldehyde)**

**Cyclic glucose
(a cyclic hemiacetal)**

Cellulose

Starch

This equilibrium lies far in favour of the cyclic hemiacetal: no aldehyde can be detected spectroscopically in solutions of glucose. We've seen that cyclodextrins are low molecular weight cyclic polymers of glucose obtained by the enzymatic hydrolysis of starch. Nature is very efficient at polymerizing the hemiacetal groups in glucose molecules by making cellulose or starch, both of which contain **acetal** functional groups. Acetal functional groups have *two oxygen atoms attached to the same carbon atom, both of which are part of ether groups.* Glucose hemiacetal groups react with other glucose hydroxyl groups (blue) to give acetal linkages (orange). If this reaction occurs at the blue hydroxyl group, two diastereomers result. One is called cellulose; the other is starch. Because they are diastereomers, they have completely different properties. Due to the selectivity of enzymes in our digestive systems, we can process starches, such as that in potatoes, but we cannot digest the cellulose of the wood in trees.

This reaction is more general than in carbohydrates. Aldehydes react with two moles of alcohols in the presence of an acid catalyst to yield acetals ($RHC(OR')_2$), and ketones react under the same conditions to produce **ketals** ($R_2C(OR')_2$), compounds that have two ether –OR groups bonded to the same carbon atom:

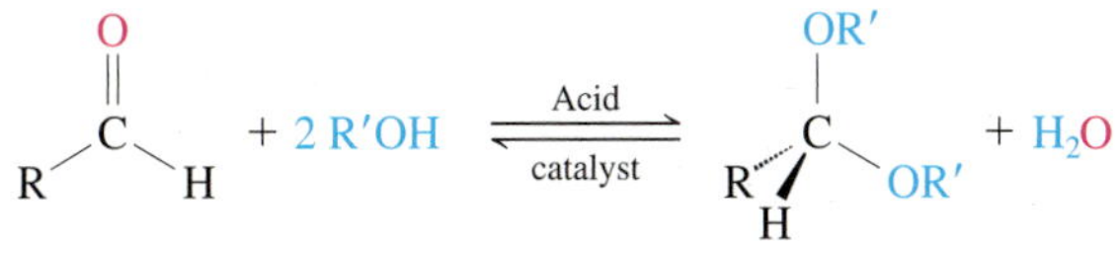

Aldehyde

An acetal

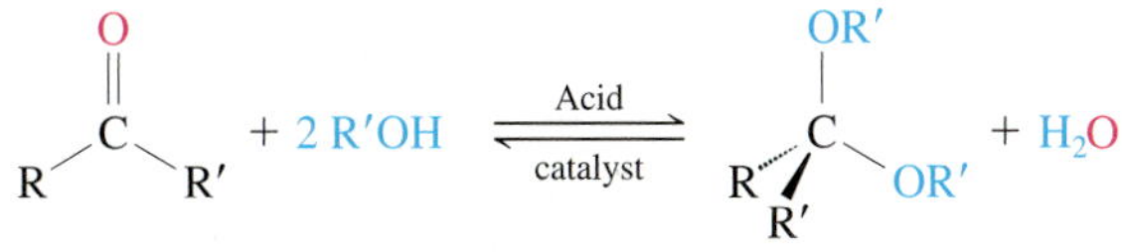

Ketone

A ketal

Given the similarity in acid-base properties between water and alcohols, it is not surprising that the accepted mechanism for acid-catalyzed nucleophilic addition of an alcohol to form an acetal or ketal is similar to that of acid-catalyzed hydration [<<Section 23.6]. A successful initial nucleophilic addition step to an aldehyde yields a hemiacetal, identical to the first step in the addition of HOH across the C=O bond. However, a difference is that this hemiacetal can go on to react with a second equivalent of alcohol to yield an acetal. The proposed mechanism involves formation of a relatively stable reactive intermediate called an **oxonium ion**. The oxonium ion resembles a protonated aldehyde or ketone, and it has similar reactions.

> An oxonium ion is related to H_3O^+, with three covalent bonds to an oxygen atom. Note that every atom has an octet, and this is an exceptionally stable reaction intermediate.

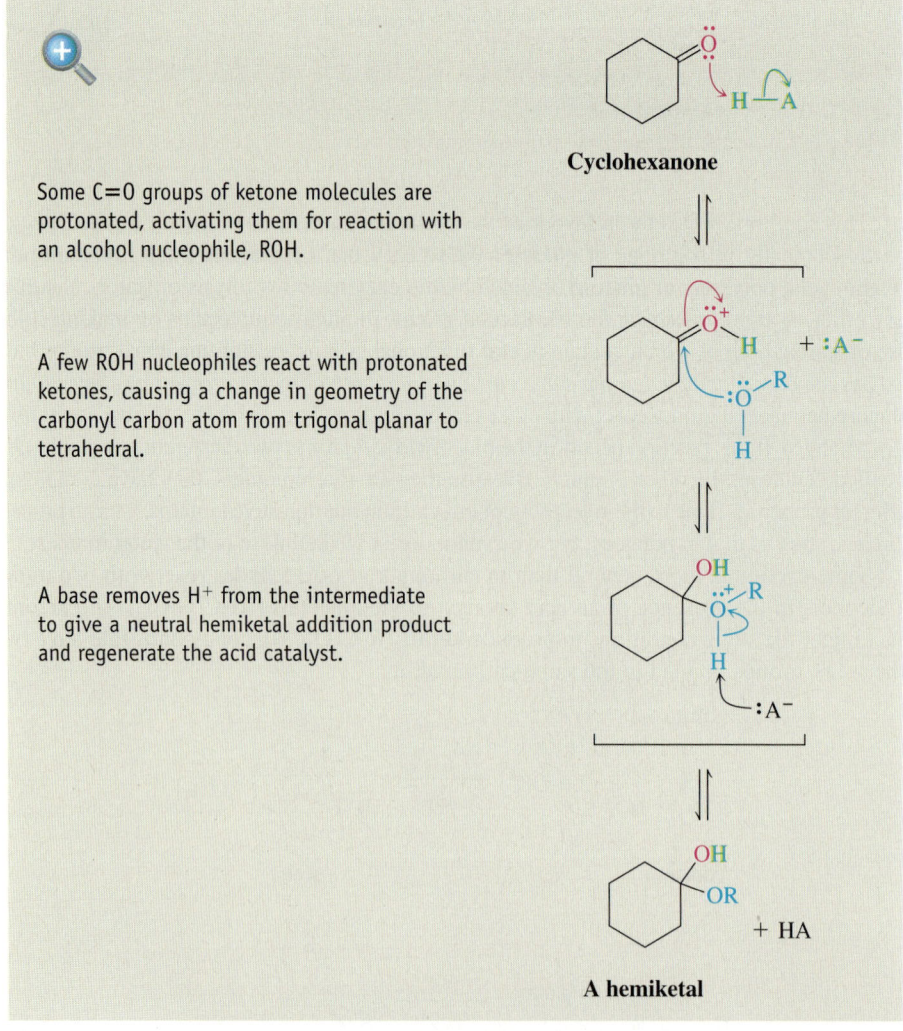

For example, reaction of cyclohexanone with an equal amount of methanol yields a **hemiketal**, and with twice the amount of methanol, the *dimethylketal*. A plausible mechanism is shown in Figure 23.5.

Some C=O groups of ketone molecules are protonated, activating them for reaction with an alcohol nucleophile, ROH.

A few ROH nucleophiles react with protonated ketones, causing a change in geometry of the carbonyl carbon atom from trigonal planar to tetrahedral.

A base removes H⁺ from the intermediate to give a neutral hemiketal addition product and regenerate the acid catalyst.

Think about It

e23.11 Watch an animation to understand this mechanism.

FIGURE 23.5   Accepted reaction mechanism for acid-catalyzed formation of a hemiketal. This mechanism is very similar to Figure 23.4, acid-catalyzed hydration.

Now let's look at the molecular level at how a ketal might be formed from the hemiketal. The overall reaction involves the substitution of an —OR group for an —OH group on the hemiketal carbon atom. In Chapter 21, two fundamental mechanisms for substitution reactions were discussed, S_N1 and S_N2. Ketal formation is more consistent with the pattern seen in S_N1 mechanisms, because the reactive intermediate is a very stable oxonium ion. We could postulate as the first step the protonation of the —OH group to make it a much weaker base and a better leaving group (Figure 23.6). Next, water leaves, giving a stable oxonium ion intermediate. The stability of this intermediate makes sense when considering that a lone pair of electrons from the —OR group can be delocalized onto the adjacent carbocation centre, producing a stabilized oxonium ion. This creates conditions that enable an alcohol molecule to react with the oxonium ion, in the same manner as when it first reacts with the protonated carbonyl group to produce the hemiketal (Figure 23.5).

As with hydration, all the steps in ketal formation are reversible, and the reaction can be made to go either from the carbonyl compound to a ketal or in the opposite direction, from the ketal back to a carbonyl compound, depending on reaction conditions. The reaction in the direction to produce ketal is favoured by conditions that remove water from the medium. The reverse reaction is favoured in the presence of a large excess of water, which drives the equilibrium toward formation of the carbonyl compound.

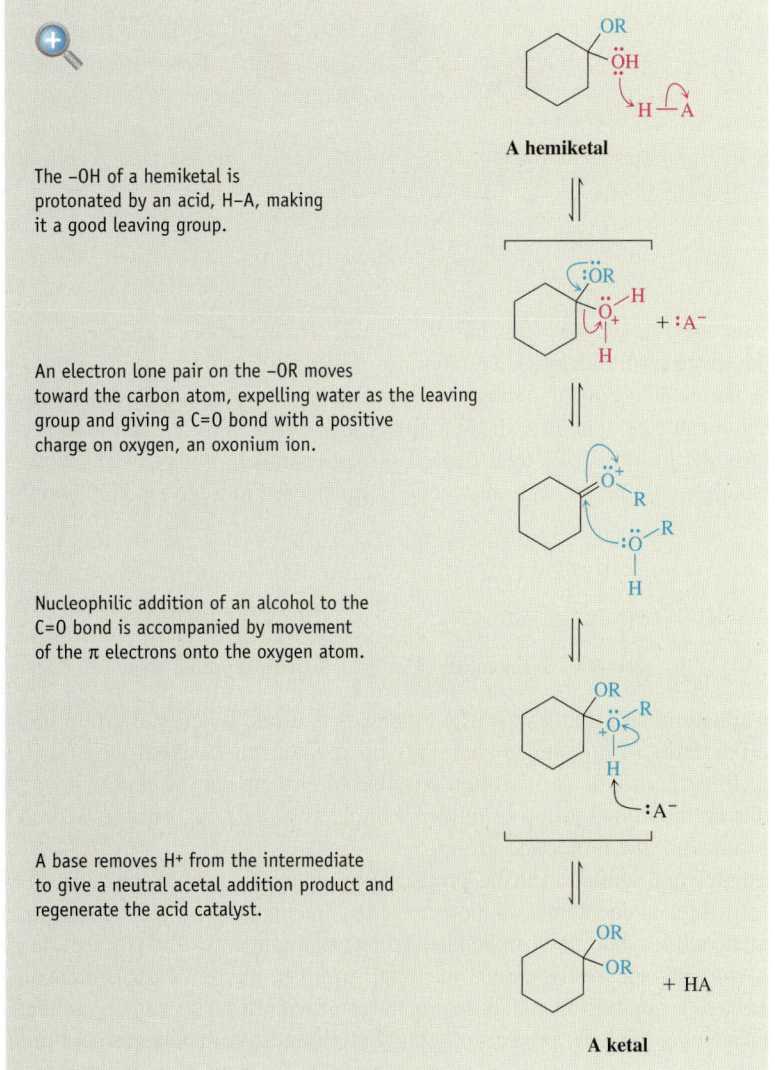

The –OH of a hemiketal is protonated by an acid, H–A, making it a good leaving group.

An electron lone pair on the –OR moves toward the carbon atom, expelling water as the leaving group and giving a C=O bond with a positive charge on oxygen, an oxonium ion.

Nucleophilic addition of an alcohol to the C=O bond is accompanied by movement of the π electrons onto the oxygen atom.

A base removes H+ from the intermediate to give a neutral acetal addition product and regenerate the acid catalyst.

FIGURE 23.6 Accepted reaction mechanism for the acid-catalyzed formation of a ketal from a hemiketal.

Think about It

e23.12 Watch an animation to understand the mechanism for acetal and ketal formation.

Many molecules cyclize to form hemiacetals and acetals or hemiketals and ketals. Polyether antibiotics, like monensin, contain carbonyl groups that react with hydroxyl groups to form hemiacetal and acetal rings. The result of these cyclizations is a complex structure that orients functional groups in specific places in three dimensions to accomplish tasks such as folding to enable more oxygen atoms to bind to metal ions. The ability of monensin to transport metal ions across membranes gives it potent antibacterial activity.

Monensin

Adding metal ion stabilizes cyclizations

A protecting group is a temporary, reversible conversion of a reactive functional group into an unreactive one, so as not to compete with reaction at a second site on a molecule possessing more than one reactive functional group.

Acetals and ketals are valuable to organic chemists because they can serve as **protecting groups** for aldehydes and ketones. To see what this means, suppose you wish to reduce the ester group of methyl 4-oxopentanoate to obtain 5-hydroxypentan-2-one. This reaction can't be done in a single step because of the presence in the molecule of the ketone carbonyl group. If we treat methyl 4-oxopentanoate with LiAlH$_4$, both ester and ketone groups would be reduced, in reactions considered in Sections 23.9 and 24.7.

Methyl 4-oxopentanoate   **5-Hydroxypentan-2-one**

This situation isn't unusual. It often happens that one functional group in a complex molecule interferes with intended chemistry on another functional group elsewhere in the molecule. In such situations, it's often possible to circumvent the problem by *protecting* the interfering functional group to render it unreactive, carrying out the desired reaction, and then removing the protecting group.

Aldehydes and ketones can be protected by converting them into acetals or ketals, respectively. Acetals and ketals, like other ethers, are stable to bases, reducing agents, and various nucleophiles, but they can be cleaved by treatment with acid [<<Section 22.8]. For example, you can selectively reduce the ester group in methyl 4-oxopentanoate by converting the keto group into a ketal, reducing the ester with LiAlH$_4$ in ether [>>Section 24.7], and then removing the ketal protecting group by treatment with aqueous acid to regenerate the ketone.

$$\underset{\substack{\text{Methyl 4-oxopentanoate}}}{CH_3CCH_2CH_2COCH_3} \xrightarrow[\text{H}^+ \text{ catalyst}]{2\ CH_3OH} \underset{\substack{OCH_3}}{CH_3CCH_2CH_2COCH_3}$$

1. LiAlH$_4$
2. H$_2$O

$$2\ CH_3OH\ +\ \underset{\substack{\text{5-Hydroxypentan-2-one}}}{CH_3CCH_2CH_2CH_2OH} \xleftarrow{H_3O^+} \underset{OCH_3}{CH_3CCH_2CH_2CH_2OH}$$

Hemiacetal groups are formed by nucleophilic addition reaction of equal amounts of an alcohol and an aldehyde, and can be recognized as having an —OH and an —OR group attached to the same carbon atom, which will also be bonded to an H atom and another C atom. Acetal groups are formed by reaction of twice the amount of an alcohol with an aldehyde and can be recognized as having two —OR groups attached to the same carbon atom that is also bonded to an H atom and another C atom. Similarly, hemiketals and ketals are formed by nucleophilic addition of one and two moles of an alcohol, respectively, to one mole of a ketone. Ketals and acetals are important protecting groups in organic synthesis.

WORKED EXAMPLE 23.4—ADDITION OF ALCOHOLS

What product would you obtain from the acid-catalyzed reaction of 1 mol of 2-methylcyclopentanone with 2 mol of methanol?

Solution

A ketone reacts with 2 mol of an alcohol in the presence of acid to yield a ketal. To draw the product, replace the oxygen atom of the ketone with two —OCH$_3$ groups from the alcohol.

EXERCISE 23.10—ADDITION OF ALCOHOLS

What product would you expect from the acid-catalyzed reaction of cyclohexanone and the same amount (in mol) of ethanol? Twice the amount of ethanol?

EXERCISE 23.11—ADDITION OF ALCOHOLS

When an aldehyde or ketone is treated with a diol such as ethylene glycol (1,2-ethanediol) and an acid catalyst, a *cyclic* acetal or ketal is formed. Draw the structure of the product you would obtain from benzaldehyde and ethylene glycol.

EXERCISE 23.12—ADDITION OF ALCOHOLS

Show how you might carry out the following transformation. (A protection step is needed.)

$$\underset{}{HCCH_2CH_2COCH_3} \longrightarrow \underset{}{HCCH_2CH_2CH_2OH}$$

23.8 Addition of Amines to Form Imines

Ammonia and primary amines (R′NH$_2$) add to aldehydes and ketones to yield molecules with the **imine functional group (R$_2$C=NR′)**. At the molecular level, we can think of imines as being formed in several steps—first by addition to the carbonyl group of the nucleophilic amine, followed by loss of water from the amino alcohol addition product.

Think about It

e23.13 Watch an animation to understand the mechanism for imine formation.

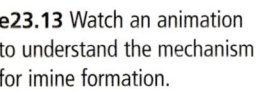

A ketone or aldehyde   Amino alcohol intermediate   An imine

Imines are common intermediates in numerous biological pathways and processes. In the context of biosynthesis, the bacterium *Bacillus subtillus* makes alanine from pyruvic acid and ammonia. The key step is the nucleophilic addition of ammonia to the ketone group of pyruvic acid. The resulting imine C=N bond is further reduced (analogous to the reduction of C=O bonds) to the amino acid.

$$CH_3CCOOH + :NH_3 \Longrightarrow \left[\begin{array}{c} NH \\ \| \\ CH_3CCOOH \end{array} \right] \xrightarrow[\text{enzyme}]{\text{Reducing}} \begin{array}{c} NH_2 \\ | \\ CH_3CHCOOH \end{array}$$

Pyruvic acid   An imine   Alanine

Imines are also important in the route by which amino acids are degraded in the body. The amino acid alanine, for instance, reacts with the aldehyde pyridoxal phosphate, a derivative of Vitamin B-6, to yield an imine that is then further degraded. We'll see further examples in Chapter 29.

Pyridoxal phosphate   Alanine   An imine

WORKED EXAMPLE 23.5—REACTION WITH AMINES

What product do you expect from the reaction of butan-2-one with hydroxylamine (NH$_2$OH)?

Solution

The overall reaction in formation of an imine from a ketone involves the C=O group replaced with a C=NR bond, with loss of the carbonyl oxygen atom and two amine hydrogen atoms as water.

$$\begin{array}{c} O \\ \| \\ CH_3CH_2CCH_3 \end{array} + H_2NOH \longrightarrow \begin{array}{c} NOH \\ \| \\ CH_3CH_2CCH_3 \end{array} + H_2O$$

EXERCISE 23.13—REACTION OF KETONES

Write the products you would obtain from treatment of cyclohexanone with the following:
(a) CH₃NH₂   (b) CH₃CH₂OH, H⁺   (c) NaBH₄

(a) CH_3NH_2 (b) CH_3CH_2OH, H^+ (c) $NaBH_4$

EXERCISE 23.14—REACTION WITH AMINES

Show the reagents needed to prepare the following molecule from a carbonyl compound and an amine (blue = N):

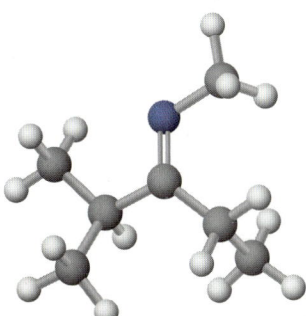

Imines ($R_2C{=}NR'$) are formed by addition of a nucleophilic amine to the carbonyl group of an aldehyde or ketone, followed by loss of water from the amino alcohol addition product. Imines are particularly important in the organic chemistry of living systems.

23.9 Reduction of Aldehydes and Ketones

As discussed in Section 22.5, reduction reactions in organic chemistry result in a gain of electron density by carbon atoms in molecules. This is usually caused by formation of a new bond between carbon and a less electronegative atom such as hydrogen or by bond breaking between carbon and a more electronegative atom. Aldehydes and ketones undergo reduction to produce alcohols by addition of hydride ion nucleophiles ($:H^-$) to the carbonyl carbon atom.

An aldehyde   **A primary alcohol**   **A ketone**   **A secondary alcohol**

Hydride ions ($H:^-$) are much too reactive to exist on their own as reagents. Instead, reducing agents such as $LiAlH_4$ (lithium aluminum hydride) and $NaBH_4$ (sodium borohydride) are used as sources to deliver hydride ions to a carbonyl carbon atom. Different functional groups require different reducing agents. $NaBH_4$ is powerful enough to reduce aldehydes and ketones but not to reduce carboxylic acids. For carboxylic acids, $LiAlH_4$ is required. By comparing the electronegativities of H to either B or Al atoms [<<Section 6.3], we can make sense of why $LiAlH_4$ is a stronger reducing agent than $NaBH_4$. Al is more electropositive than B, so the hydride of $LiAlH_4$ is more reactive than the hydride of $NaBH_4$.

We can oversimplify the representation of the reducing agent as $H:^-$ to focus on what we understand to happen at the molecular level in the reaction mechanism (Figure 23.7). Using acetone as an example of a reacting ketone, it has been proposed that a few of the many collisions of the nucleophile molecules with acetone molecules lead to sharing of the hydride electron pair to form a bond to the carbon atom of the C=O group. This would be expected to cause a change in the geometry of the carbonyl carbon atom to tetrahedral from trigonal-planar. Using a hybrid orbital model, this could be thought of as resulting in rehybridization of the C=O carbon atom from sp^2 to sp^3 with two electrons from the C=O bond moving to the oxygen atom, giving an alkoxide ion. The final step might reasonably involve addition of H^+ to the alkoxide ion to yield a neutral alcohol product. An example would be the reduction of a ketone or aldehyde to an alcohol by $NaBH_4$.

FIGURE 23.7 Accepted reaction mechanism for the reduction of a ketone by a hydride source.

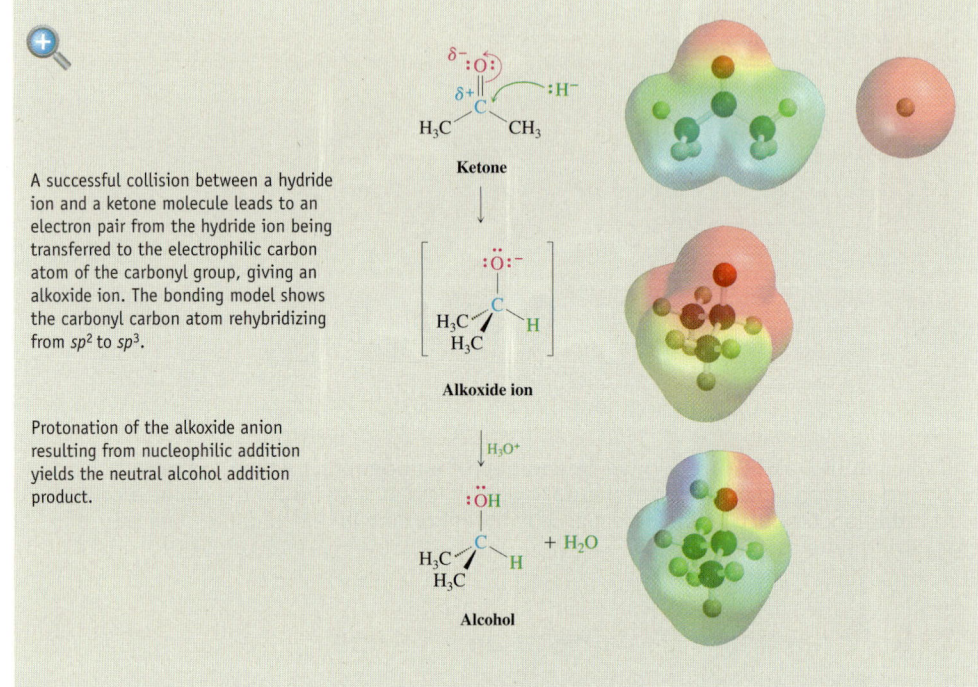

A successful collision between a hydride ion and a ketone molecule leads to an electron pair from the hydride ion being transferred to the electrophilic carbon atom of the carbonyl group, giving an alkoxide ion. The bonding model shows the carbonyl carbon atom rehybridizing from sp^2 to sp^3.

Protonation of the alkoxide anion resulting from nucleophilic addition yields the neutral alcohol addition product.

Think about It

e23.14 Watch an animation to understand the mechanism for $LiAlH_4$ reduction.

Think about It

e23.15 Watch an animation to understand the mechanism for $NaBH_4$ reduction.

Unlike these simple laboratory reducing agents, nature uses more complex molecules that undergo *reversible* oxidation and reduction reactions. NADH is a co-enzyme that plays an important role as a reducing agent in metabolic processes.

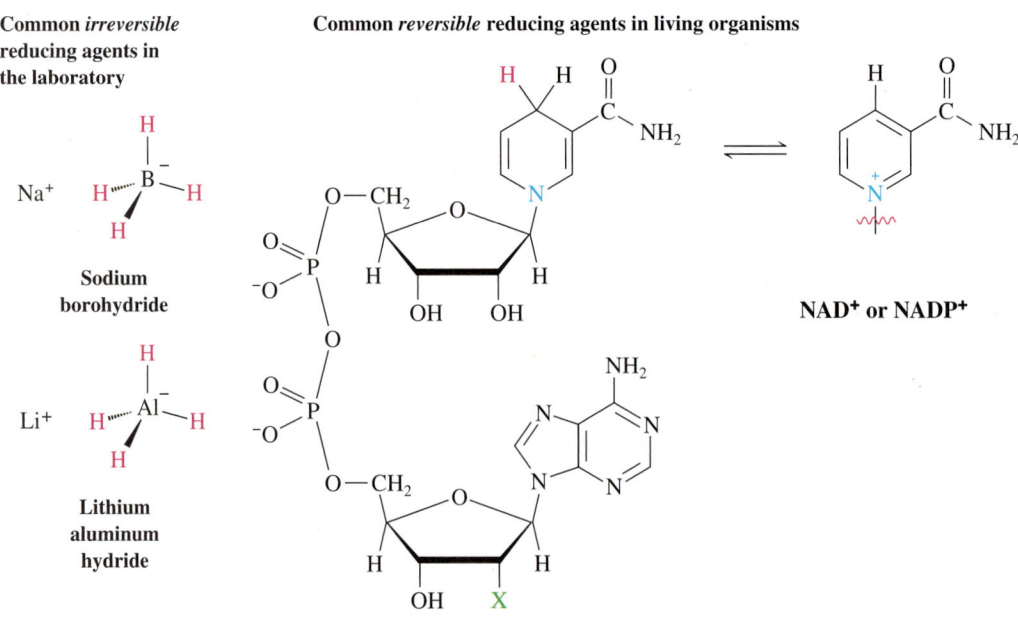

Common *irreversible* reducing agents in the laboratory

Common *reversible* reducing agents in living organisms

Sodium borohydride

Lithium aluminum hydride

NAD⁺ or NADP⁺

X = OH   Reduced nicotinamide adenine dinucleotide, (NAD**H**)
X = OPO_3^{2-} Reduced nicotinamide adenine dinucleotide phosphate, (NADP**H**)

What is the source of hydride in NADH or NADPH? (The "H" denotes the reduced form.) We know that the electronegativity difference between C and H atoms is very small. How does nature offset the cost making an H:⁻ when breaking an unpolarized C—H bond? The answer is *aromaticity*. Donation of hydride from NADH or NADPH produces a molecule with a stable aromatic ring [<<Chapter 20].

> Aldehydes and ketones can be reduced to alcohols using reagents that can deliver hydride ions, such as LiAlH₄ (lithium aluminum hydride) and NaBH₄ (sodium borohydride). Reversible oxidation and reduction reactions are common in biological systems.

23.10 Addition of Grignard Reagents: Alcohol Formation

We saw in Section 21.12 that organohalides react with magnesium metal in ether solution to give **Grignard reagents** (RMgX).

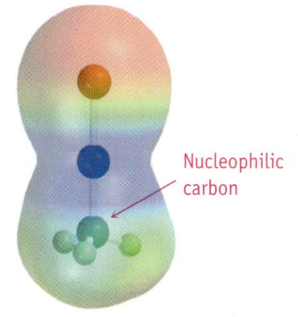

CH₃MgCl

Nucleophilic carbon

$$R\!-\!X \xrightarrow[\text{Ether}]{\text{Mg}} \overset{\delta-}{R}\!-\!\overset{\delta+}{MgX}$$

Alkyl halide   **Grignard reagent**

From the electrostatic potential map of CH₃MgCl, it is apparent that the carbon–magnesium bond of a Grignard reagent is polarized in a different way than in the more common cases where carbon atoms are bonded to more electronegative elements such as oxygen, nitrogen, or a halogen. We have referred to this reversing of polarity as *umpolung* [<<Section 21.12], and it happens because Mg atoms are less electronegative than C atoms, making the C atom in a Grignard reagent both nucleophilic and basic. Grignard reagents therefore react as though they were carbon anions, or *carbanions* (R:⁻), and they add to aldehydes and ketones just as water and alcohols do. The reaction is thought to first produce a tetrahedral magnesium alkoxide intermediate, which is then protonated to yield the neutral alcohol on treatment with aqueous acid. Unlike the addition of water and alcohols, though, the nucleophilic addition of a Grignard reagent is irreversible.

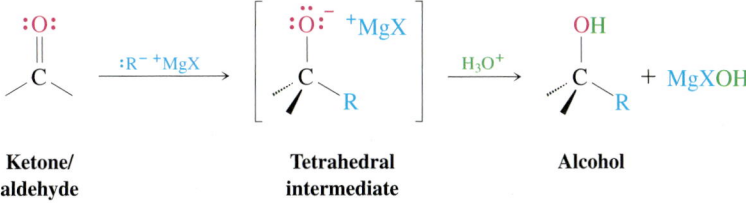

Ketone/ aldehyde   **Tetrahedral intermediate**   **Alcohol**

The Grignard reaction is a versatile way to synthesize alcohols while at the same time making a new C—C bond to the carbon atom of a carbonyl group. For example, formaldehyde (CH₂O) reacts with Grignard reagents to give primary alcohols (RCH₂OH); other aldehydes react with Grignard reagents to give secondary alcohols (R₂CHOH); and ketones give tertiary alcohols (R₃COH):

Formaldehyde and Grignard reagents give primary alcohols.

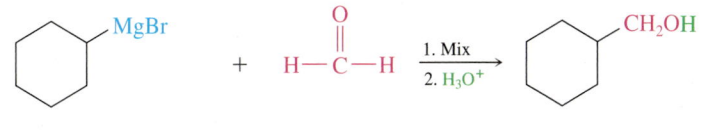

Cyclohexylmagnesium bromide   **Formaldehyde**   **Cyclohexylmethanol (65%) (a 1° alcohol)**

Other aldehydes and grignard reagents give secondary alcohols.

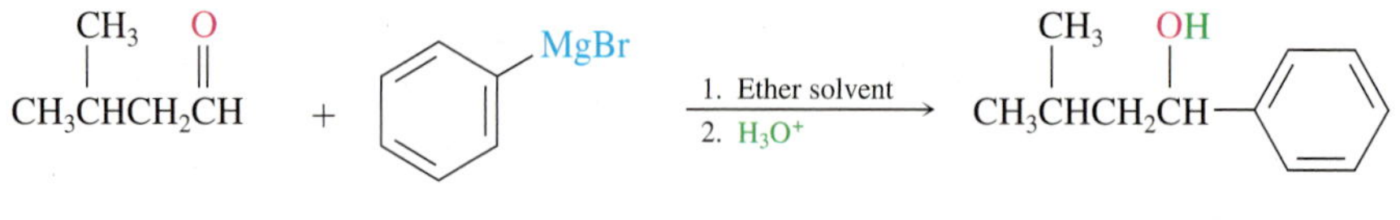

3-Methylbutanal

**Phenylmagnesium
bromide**

1. Ether solvent
2. H_3O^+

3-Methyl-1-phenylbutan-1-ol (73%)
(a 2° alcohol)

Ketones and Grignard reagents give tertiary alcohols.

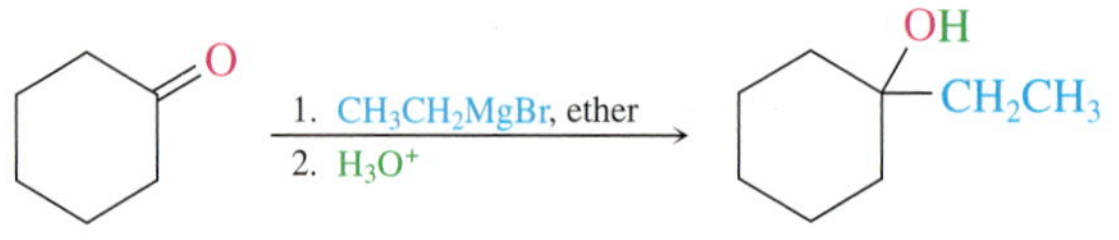

Cyclohexanone

1. CH_3CH_2MgBr, ether
2. H_3O^+

1-Ethylcyclohexanol (89%)
(a 3° alcohol)

How do the alcohols produced in Grignard reactions of aldehydes and ketones compare to those produced by reduction of aldehydes and ketones with $NaBH_3$ or $LiAlH_4$? The key difference is that simple reduction with hydride ion adds an H atom to the carbonyl carbon atom. Nucleophilic addition of a carbanion from a Grignard reagent forms a C—C bond to the carbonyl carbon atom. Since C—C bond formation reactions are essential to build larger organic molecules from smaller ones, this reaction is a most useful addition to our toolbox of reagents.

The Grignard reaction also has limitations. A Grignard reagent can't be prepared from an organohalide that has other reactive functional groups in the same molecule. Some of these groups cause the Grignard reagent to react with itself. Others, possessing acidic RCOO**H**, RO**H**, or RN**H**$_2$ protons in the molecule, destroy the strongly basic Grignard reagent as it's formed by protonation. In general, Grignard reagents can't be prepared from compounds that have the following functional groups in the molecule:

$$-CHO, -COR, -CONR_2, -C\equiv N, -NO_2, -SO_2R$$
Grignard reagent reacts with these groups.

$$-OH, -NH, -SH, -COOH$$
Grignard reagent is destroyed by the acidic hydrogen.

WORKED EXAMPLE 23.6—USING GRIGNARD REAGENTS

How can you use the addition of a Grignard reagent to a ketone to synthesize 2-phenylpropan-2-ol?

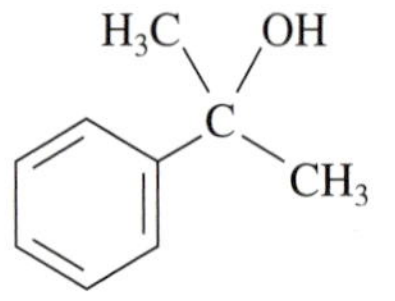

2-Phenylpropan-2-ol

Solution

Look at the product, and identify the groups bonded to the alcohol carbon atom. In this instance, there are two methyl groups (—CH$_3$) and one phenyl (—C$_6$H$_5$). One of the three must come from a Grignard reagent, and the remaining two must come from a ketone.

So the possibilities are addition of CH₃MgBr to acetophenone and addition of C₆H₅MgBr to acetone.

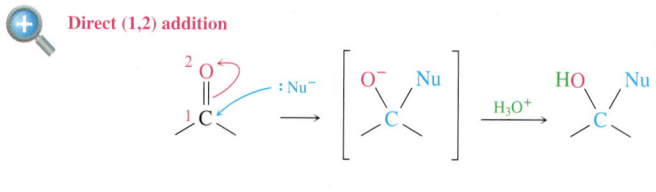

| Acetophenone | 2-Phenylpropan-2-ol | Acetone |

EXERCISE 23.15—USING GRIGNARD REAGENTS

Show the products obtained from addition of CH_3MgBr to the following compounds:

(a) **(b)** **(c)**

$$CH_3CH_2CH_2CCH_2CH_3$$

Grignard reactions between organomagnesium compounds and aldehydes and ketones produce alcohols with one or more carbon atoms added to the carbonyl carbon. Since C—C bonds are formed in the process, they are very important tools for making larger molecules from smaller ones.

23.11 Conjugate Addition Reactions

In the case of an α, β-unsaturated aldehyde or ketone, *conjugate (1,4) addition* of a nucleophile to the conjugated C=C bond can compete with direct (1,2) addition to the C=O bond. An α, β-unsaturated carbonyl compound is one whose molecules have a C=C double bond between the α carbon atom (the one next to the C=O group) and the β-carbon (the one two carbon atoms away from the C=O group). An α, β-unsaturated carbonyl compound has its C=C and C=O bonds conjugated, much as a conjugated diene does [<<Section 19.1].

Direct (1,2) addition

Conjugate (1,4) addition

α, β-Unsaturated aldehyde/ketone

Enolate ion

Saturated aldehyde/ketone

Think about It

e23.19 Watch an animation to understand conjugate addition.

The initial product of conjugate addition is thought to be a resonance-stabilized **enolate anion** [>>Section 24.9], which undergoes protonation on the α-carbon atom to give a saturated aldehyde or ketone product. For example, methylamine reacts with but-3-en-2-one to give an amino ketone addition product.

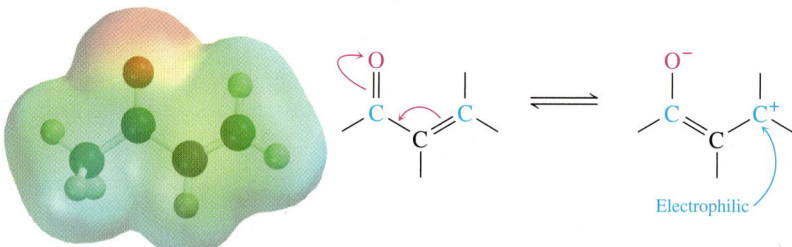

But-3-en-2-one   **Conjugate addition product**

Conjugate addition occurs because the electronegative oxygen atom of the α, β-unsaturated carbonyl-containing compound withdraws electrons from the β-carbon atom, thereby making it more electron-deficient and more electrophilic than a typical alkene C=C bond.

In the case of O, N, and S nucleophiles, the 1,4-addition reaction often leads to a thermodynamically more stable product than direct 1,2-addition to the C=O group.

Conjugate additions are common in many biological pathways. An example is the conversion of fumarate to malate by reaction with water, a step in the citric acid cycle by which acetate is metabolized to CO_2.

Fumarate   **Malate**

EXERCISE 23.16—CONJUGATE ADDITION

The following compound was prepared by a conjugate addition reaction between an α, β-unsaturated ketone and an alcohol. Draw the structures of the two reactants.

Molecules of α, β-unsaturated carbonyl compounds have a C=C bond conjugated with a C=O bond, so π electrons can be delocalized over both bonds. As a result of this delocalization, the β-carbon atom is electron-deficient and the site for nucleophilic attack, referred to as 1,4- or conjugate addition.

23.12 Oxidation of Aldehydes

Aldehydes are easily oxidized to yield carboxylic acids (RCHO $\longrightarrow$ RCOOH), but ketones are unreactive toward oxidation. This reactivity difference is a consequence of structure: aldehyde molecules have a hydrogen atom attached to the carbonyl carbon atom that can be removed during oxidation, but ketones do not.

An aldehyde   A ketone

One of the simplest methods for oxidizing an aldehyde is to use silver ion (Ag^+) in dilute aqueous ammonia, a mixture called *Tollens' reagent*. As the oxidation proceeds, a shiny mirror of silver metal is deposited on the walls of the reaction flask, forming the basis of a simple test to detect the presence of an aldehyde functional group in a sample of unknown structure. A small amount of the unknown is dissolved in ethanol in a test tube, and a few drops of Tollens' reagent are added. If the test tube becomes silvery, the unknown is an aldehyde.

Benzaldehyde   Benzoic acid

Oxygen, $O_2(g)$, in air can also convert aldehydes into carboxylic acids. Sunlight promotes this reaction. For both of these reasons, aldehydes such as vanilla (or vanillin) are stored in tightly sealed, dark bottles.

Vanillin   An oxidation product
(an aldehyde)   (a carboxylic acid)

The ethanol in wine can be oxidized into acetaldehyde (ethanal), which can be further oxidized to acetic acid by bacteria and yeasts, giving wine a tainted, vinegar flavour and aroma.

Ethanol   Acetaldehyde   Acetic acid

WORKED EXAMPLE 23.7—OXIDATION OF ALDEHYDES

What product would you obtain from the oxidation of 3-methylbutanal with Tollens' reagent?

Solution

Write the structure of the aldehyde molecule, and then replace the —H atom bonded to the carbonyl group by —OH.

$$CH_3CHCH_2C-H \xrightarrow[\text{reagent}]{\text{Tollens}} CH_3CHCH_2C-OH$$

3-Methylbutanal   **3-Methylbutanoic acid**

EXERCISE 23.17—OXIDATION REACTIONS

Predict the products of the reaction of the following substances with Tollens' reagent:

(a)

$$CH_3CH_2CH_2CH_2CH$$

(b)

$$CH_3CH_2CH_2CH_2CCHO$$
with CH_3 groups

(c)

cyclohexanone structure

Aldehydes (but not ketones) are easily oxidized to produce carboxylic acids, by reagents such as the Tollens' reagent and O_2 in the air.

23.13 Making Scents of Human Activity?

"We cannot expect humans to behave like moths flying up the concentration gradient toward the desirable source," says University of Chicago's neuroendocrinology researcher Martha McClintock. Yet scientists have learned that the type of chemical signalling through pheromones used by moths and the mountain pine beetle also plays a critical role in communication for organisms ranging from bacteria to plants, fungi, insects, fish, and mammals. Do humans transmit unseen chemical signals to each other as part of mate selection? Entrepreneurs would have us believe so, and we see humans flock like moths to thousands of websites that advertise the mate-attracting ability of purported human pheromones with ketone-containing steroidal structures such as androstenone and androstadienone.

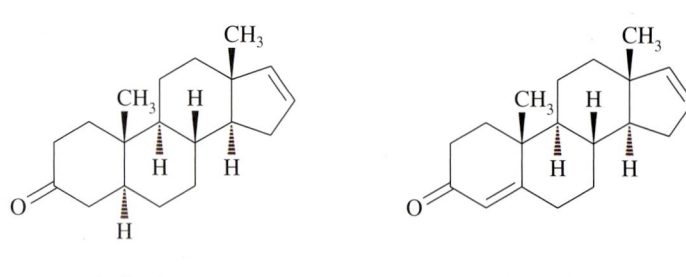

Androstenone   **Androstadienone**

As implied by McClintock, human behaviour takes place within a much more complex context of social and other factors than that of mountain pine beetles or moths, and receptors for pheromones are more elusive than is the case for the mountain pine beetle, where antennae are thought to serve this purpose.

Research into the role of odours in communication of humans and other primates is ongoing, and there are indications that humans may use pheromones. The ketone androstadienone in the air, for example, produced by scent glands, has been shown from positron

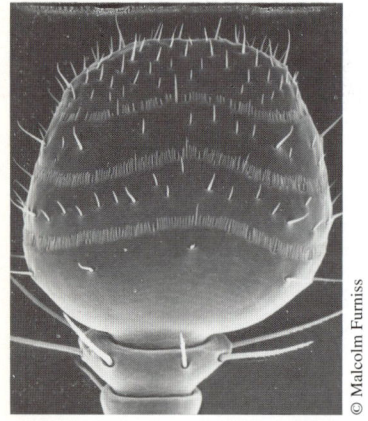

Scanning electron micrograph of structures thought to act as receptors for chemical signals from mates on the antennae of bark beetles.

emission tomography studies to increase activity of the limbic system (structures in the brain responsible for emotion, memory formation, and motivation) in women, and alter skin temperature and conductance in both men and women. It has also been reported that it is possible to alter the menstrual cycle of women by exposure to chemical signals of other women. And we know, too, that human male glands also produce androstenone, which has been shown to be a sex pheromone in pigs. The effect of this scent on human activity isn't yet clear!

Fundamental puzzles remain. Further research to establish that link between messages and chemicals may help make sense of the role of chemical communication in some of our most complex human activities.

SUMMARY

Key Concepts

- Compounds whose molecules containing the **carbonyl functional group** ($C=O$) are the most important and widely occurring compounds in both organic and biological chemistry. An example of their importance is found in the role carbonyl compounds play in chemical signalling (**pheromones**) in insects and other organisms, including humans. (Sections 23.1 and 23.13)
- Carbonyl compounds are classified into two general categories, based on differences in structure and reactivity. (Section 23.1)

RCHO R$_2$CO	**Aldehydes** and **ketones** are similar in their reactivity and are distinguished by the fact that the substituents on the acyl carbon atom of their molecules can't act as leaving groups.
RCOOH RCOCl RCOOCOR′ RCOOR′ RCONH$_2$	**Carboxylic acids** and their derivatives—**acyl chlorides**, **acid anhydrides**, **esters**, and **amides**—are distinguished by the fact that the substituents on the acyl carbon atom of their molecules can act as leaving groups.

- **Aldehydes** and **ketones** can be identified readily by a very intense $C=O$ stretching vibration in the infrared spectrum, typically between 1710 and 1730 cm^{-1}. ^{13}C NMR spectroscopy provides complementary information, with carbonyl carbon atoms appearing far downfield, near 200 ppm. (Section 23.3)
- One of the most useful methods of synthesizing aldehydes and ketones is by oxidation of primary and secondary alcohols, respectively. (Section 23.3)
- As evident from the electrostatic potential map, the carbonyl carbon atom has a partial positive charge, and therefore is electrophilic (a Lewis acid) and reacts with nucleophiles. Conversely, the carbonyl oxygen atom has a partial negative charge and is nucleophilic (a Lewis base). Most carbonyl group reactions are the result of this bond polarization and involve breaking the π bond. (Section 23.4)
- Aldehydes and ketones behave similarly in much of their chemistry. Both undergo nucleophilic addition reactions, and a variety of products can be prepared. Many of these reactions are catalyzed by either bases or acids, and the proposed mechanisms are different. For example, aldehydes and ketones undergo a reversible addition reaction with water to yield 1,1-dialcohols, or *gem diols*. Similarly, they react with alco-

hols to yield **acetals (RHC(OR′)₂)** and **ketals (R₂C(OR′)₂)**, which are valuable as carbonyl **protecting groups.** Primary amines add to aldehydes and ketones to give **imines (R₂C=NR′).** In addition, aldehydes and ketones are reduced by $NaBH_4$ to yield primary and secondary alcohols, respectively, and they react with **Grignard reagents** to give alcohols, accompanied by the formation of new C—C bonds. (Sections 23.6–23.10)

- Closely related to the direct (1,2) addition of nucleophiles to aldehydes and ketones is the conjugate (1,4) addition of nucleophiles to the C—C double bond of α, β-unsaturated aldehydes and ketones. Both direct and conjugate addition reactions are common in biological pathways. (Section 23.11)

Key Reactions

1. Reaction of aldehydes and ketones with alcohols to yield acetals and ketals (Section 23.7)

2. Reaction of aldehydes and ketones with amines to yield imines (Section 23.8)

3. Reaction of aldehydes and ketones with reducing agents to yield alcohols (Section 23.9)

An aldehyde   **A primary alcohol**   **A ketone**   **A secondary alcohol**

4. Reaction of aldehydes and ketones with Grignard reagents to yield alcohols, while creating new C—C bonds to the C=O carbon atom (Section 23.10)

5. Conjugate (1,4) nucleophilic addition reactions (Section 23.11)

6. Oxidation of aldehydes (Section 23.12)

An aldehyde

REVIEW QUESTIONS

Section 23.1: Making Scents of the Mountain Pine Beetle

23.18 A Canadian Forest Service study suggest that over a 21-year period, mountain pine beetle damage to the forest could lead to the release of almost a billion megatonnes of CO_2 into the atmosphere, roughly equivalent to five years of transportation sector emissions. Explain how pine beetle damage leads to CO_2 emissions into the atmosphere.

23.19 Consult additional sources to read about insect pheromones. List the functional groups for the ones you find. Do many of them contain carbonyl groups?

23.20 What strategies are commonly used to employ insect pheromones that have been identified and characterized to help reduce the population of insects?

23.21 Identify the kinds of carbonyl groups in the following molecules (red = O, blue = N):

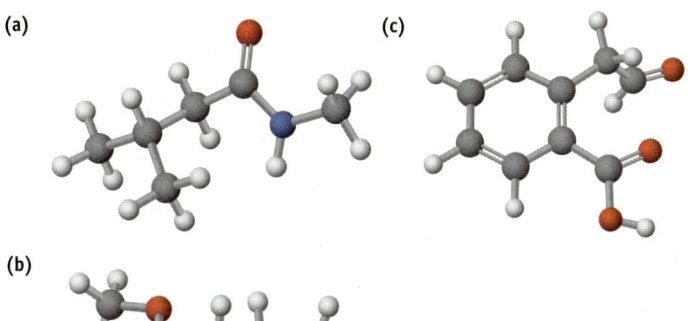

(a)

(b)

(c)

23.22 Identify the different kinds of carbonyl functional groups in the following molecules:

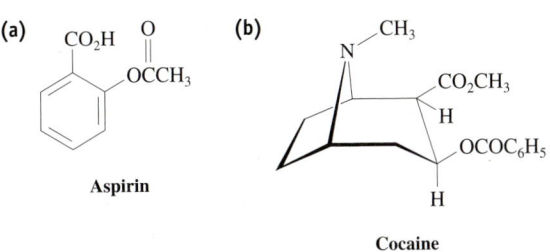

(a) Aspirin

(b) Cocaine

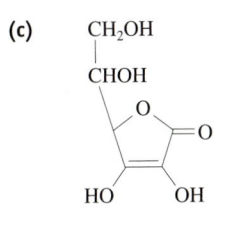

(c) Ascorbic acid (Vitamin C)

23.23 Draw structures of molecules that meet the following descriptions:
(a) a cyclic ketone, C_6H_8O
(b) a diketone, $C_6H_{10}O_2$
(c) an aryl ketone, $C_9H_{10}O$
(d) a 2-bromo aldehyde, C_5H_9BrO

Section 23.2: Naming Aldehydes and Ketones

23.24 Draw structures of molecules corresponding to the following names:
(a) bromoacetone
(b) 3-methyl-2-butanone
(c) 3,5-dinitrobenzaldehyde
(d) 3,5-dimethylcyclohexanone
(e) 2,2,4,4-tetramethyl-3-pentanone
(f) butanedial
(g) (S)-2-hydroxypropanal
(h) 3-phenyl-2-propenal

23.25 Draw and name the seven aldehydes and ketones with the formula $C_5H_{10}O$.

23.26 Which of the compounds you identified in Question 23.25 have chiral molecules?

23.27 Give IUPAC names for the compounds whose molecules are shown:

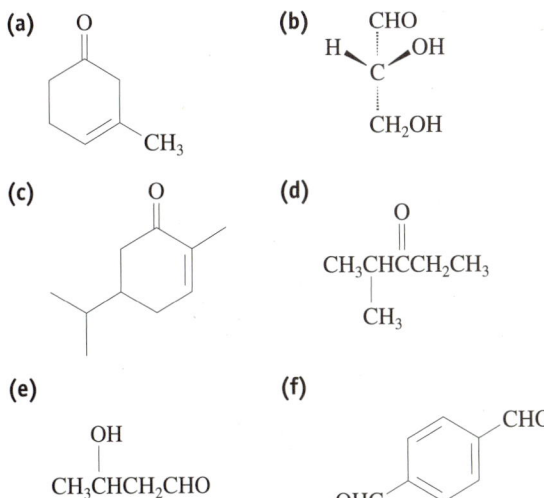

(a)

(b)

(c)

(d)

(e)

(f)

23.28 Give an example of each of the following:
(a) an acetal
(b) a gem diol
(c) an imine
(d) a hemiacetal

Section 23.3: Synthesis and Spectroscopy

23.29 How would you carry out the following transformations? More than one step may be required.

(a) hex-3-ene ⟶ hexan-3-one

(b) benzene ⟶ 1-phenylethanol

23.30 Compound A, MW = 86, shows an IR absorption at 1730 cm^{-1} and a very simple ^{1}H spectrum with peaks at 9.7 δ (1H, singlet) and 1.2 δ (9H, singlet). Propose a structure for A.

23.31 Compound B is isomeric with A (Question 23.30), and shows an IR absorption at 1715 cm^{-1}. The ^{1}H spectrum of B has peaks at 2.4 δ (1H, septet, J = 7 Hz), 2.1 δ (3H, singlet), and 1.2 δ (6H, doublet, J=7Hz). Propose a structure for B.

23.32 The ^{1}H NMR spectrum shown is that of a compound with formula $C_9H_{10}O$. How many double bonds and/or rings does this compound contain? If the unknown has an IR absorption at 1690 cm^{-1}, propose a likely structure.

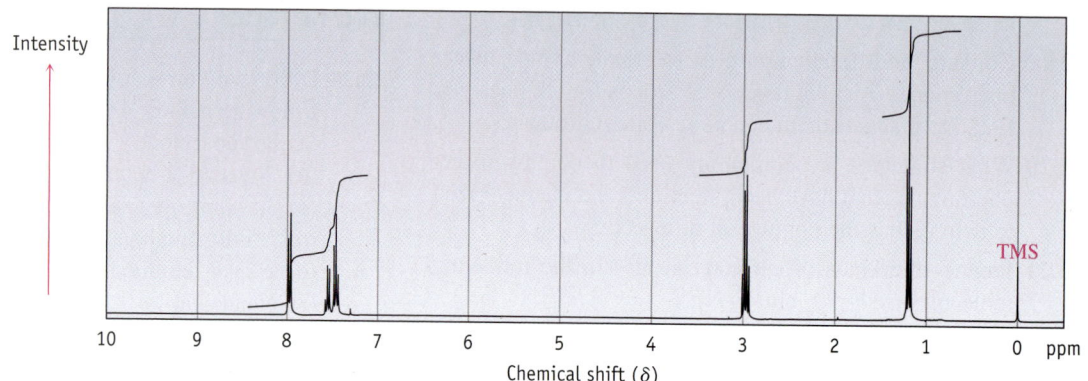

Sections 23.4–23.10: Reactions of Aldehydes and Ketones

23.33 Predict the products of the reaction of phenylacetaldehyde ($C_6H_5CH_2CHO$) with the following reagents:
(a) $NaBH_4$, then H_3O^+
(b) Tollens' reagent
(c) NH_2OH
(d) CH_3MgBr, then H_3O^+
(e) CH_3OH, H^+ catalyst

23.34 Answer Question 23.33 for the reaction of acetophenone ($C_6H_5COCH_3$) with the same reagents.

23.35 Identify the nucleophile that has added to acetone to give the following products:

(a) CH₃CHCH₃ with OH

(b) CH₃CCH₂CH₃ with OH and CH₃

(c) CH₃CCH₃ with NCH₃

(d) CH₃CCH₃ with OH and SCH₃

23.36 Reaction of butan-2-one with HCN yields a *cyanohydrin* product [$R_2C(OH)CN$], creating a new stereocentre. Explain why the product is not optically active.

23.37 In light of your answer to Question 23.36, what stereochemistry would you expect the product from the reaction of phenylmagnesium bromide with butan-2-one to have?

23.38 How can you explain the observation that the S_N2 reaction of (dibromomethyl)-benzene with NaOH yields benzaldehyde rather than (dihydroxymethyl) benzene?

(Dibromomethyl)benzene ⟶ (NaOH, H₂O) ⟶ Benzaldehyde

23.39 Show the structures of the intermediate hemiacetals/ ketals and the final acetals and ketals that result from the following reactions:

(a) [acetophenone] + CH₃CHCH₃ (with OH) ⟶ (H⁺ catalyst)

(b) CH₃CH₂CCH₂CH₃ + [cyclopentyl]—OH ⟶ (H⁺ catalyst)

23.40 Show the structures of the alcohols and aldehydes or ketones you would use to make the following acetals and ketals:

(a)

$$CH_3CH_2CHCH_2CHOCH_3$$

with CH₃ and OCH₃ groups

(b)

$$CH_3CH_2O \quad OCH_2CH_3$$
$$C6H5 \quad CH_3$$

(c)

spiro dioxolane structure

23.41 Treatment of a ketone or aldehyde with hydrazine (H_2NNH_2) yields an *azine* ($R_2C=N-N=CR_2$). Show how the reaction occurs.

23.42 When glucose is treated with $NaBH_4$, reaction occurs to yield *sorbitol*, a commonly used food additive. Show how this reduction occurs.

Glucose $\xrightarrow[\text{H}_2\text{O}]{\text{NaBH}_4}$ $HOCH_2CHCHCHCHCH_2OH$ (with OH, HO OH OH)

Glucose　　　　　**Sorbitol**

23.43 How might you use a Grignard addition reaction to prepare the following alcohols from an aldehyde or ketone?

(a)
$$CH_3CCH_3$$
with OH and CH₃

(b) cyclohexane with OH and CH₃

(c)
$$CH_3CH_2CCH_2CH$$
with OH and CH₃

23.44 How might you use a Grignard reaction to prepare the following alcohol (red = O)?

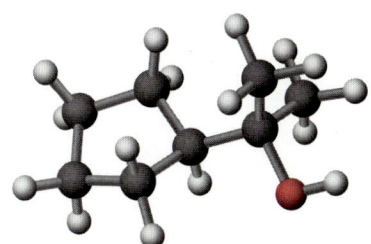

23.45 Show how you would use a Grignard reaction on an aldehyde or ketone to synthesize the following compounds from an aldehyde or ketone:
(a) pentan-2-ol
(b) 2-phenylbutan-2-ol
(c) 1-ethylcyclohexanol
(d) diphenylmethanol

23.46 Show the products that result from the reaction of phenylmagnesium bromide with the following reagents:
(a) CH_2O
(b) benzophenone ($C_6H_5COC_6H_5$)
(c) pentan-3-one

23.47 Show how you could make the following alcohols using a Grignard reaction of an aldehyde or ketone. Show all possibilities.

(a) $CH_3CHCH_2CH_2CH_2OH$ (with CH₃)

(b) cyclohexane with CH(OH)CH₃

(c) $CH_3CH_2CHCH=CHCH_3$ (with OH)

23.48 Show how you could make the following alcohols using a Grignard reaction of an aldehyde or ketone. Show all possibilities.

(a) benzene with CH_2OH

(b) benzene with C(HO)(CH₃)(CH₂CH₃)

(c) cyclohexane with C(HO)(CH₃)(CH₃)

(d) $CH_3CH_2CHCH_2CHCH_3$ (with CH₃ and OH)

(e) CH_3CH_2COH (with CH₂CH₃ and CH₂CH₃)

(f) $CH_3CH_2 \cdots C \cdots CH_2CH_3$ (with OH and H)

Section 23.11: Conjugate Addition Reactions

23.49 Draw the product(s) obtained by conjugate addition of the following reagents to cyclohex-2-enone:
(a) H_2O　　(c) CH_3OH
(b) NH_3　　(d) CH_3CH_2SH

23.50 Carvone is the major constituent of spearmint oil. What products would you expect from the reaction of carvone with the following reagents?
(a) $LiAlH_4$, then H_3O^+
(b) C_6H_5MgBr, then H_3O^+
(c) H_2, Pd catalyst
(d) CH_3OH, H^+

carvone structure

Carvone

SUMMARY AND CONCEPTUAL QUESTIONS

23.51 Starting from cyclohex-2-enone and any other reagents needed, how would you make the following substances? (More than one step may be required.)

(a)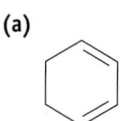

(b)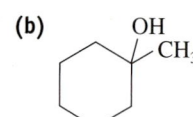

(c)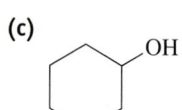

(d)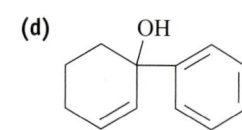

23.52 How could you convert bromobenzene into benzoic acid (C_6H_5COOH)? (More than one step is required.)

23.53 Show the products from the reaction of pentan-2-one with the following reagents:

(a) NH_2OH **(b)**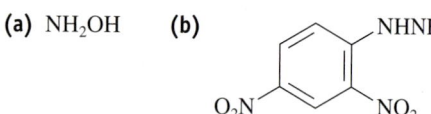

(c) CH_3CH_2OH, H^+

23.54 How would you synthesize the following compounds from cyclohexanone?

(a) (structure: methylcyclohexene) **(b)** (structure: trans-1,2-cyclohexanediol)

(c) (structure: 1-bromo-1-methylcyclohexane) **(d)** (structure: 1-cyclohexylcyclohexanol)

23.55 Treatment of a ketone or aldehyde with a thiol in the presence of an acid catalyst yields a *thioacetal* ($R_2C(SR')_2$). To what other reaction is this thioacetal formation analogous? Show how the reaction occurs.

23.56 How would you synthesize tamoxifen, a drug used in the treatment of breast cancer, from benzene, the following ketone, and any other reagents needed?

(reaction scheme: ketone with $C=O$ and $(CH_3)_2NCH_2CH_2O$ substituent converts to Tamoxifen with $C=C$, CH_2CH_3 and $(CH_3)_2NCH_2CH_2O$ substituents)

Tamoxifen

23.57 Identify the reagents a through d in the following scheme:

(reaction scheme from aldehyde with O and H, step a to alcohol with H OH, step b to ketone with O, step c to ketone with O and Br, step d to compound with CH_3O OCH_3 and Br)

23.58 Interfering with ecdysone, the moulting hormone for insects, is a common mechanism of action for some insecticides.

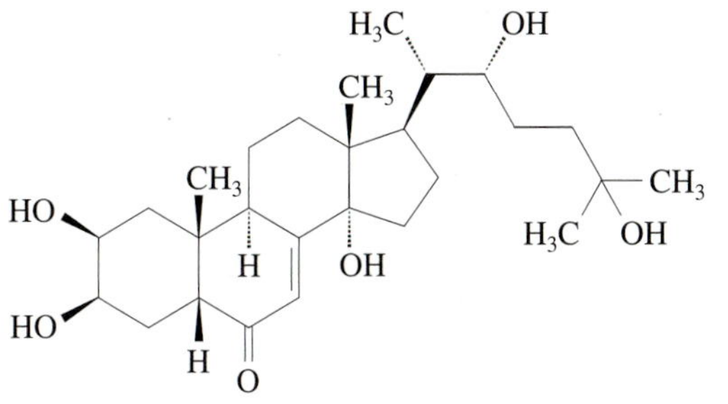

Ecdysone

(a) Categorize each of the hydroxyl groups as primary, secondary, or tertiary.
(b) Oxidation of which hydroxyl groups of ecdysone will produce aldehydes?
(c) Oxidation of which hydroxyl groups of ecdysone will produce ketones?
(d) Oxidation of which hydroxyl groups will be unsuccessful?
(e) How many chirality centres are in ecdysone?
(f) Addition of a reducing agent to ecdysone can produce two different classes of products. Draw the mechanisms for 1,4-addition of hydride to produce a ketone (after protonation) and a 1,2-addi-tion of hydride to produce an alcohol with an α-alkene.
(g) From where does the "one" of ecdys*one* originate?

23.59 Each of the 1,4- and 1,2-addition reactions of hydride to ecdysone (Question 23.58) produce two products depending on which side of the molecule the hydride attacks.
(a) Draw these stereoisomers.
(b) What term describes the relationship between the 1,4-addition products? Between the 1,2-addition products? (Hint: Refer back to Chapter 9 if needed.)
(c) Do you expect that the products of either 1,4- or 1,2-addition will be in equal 50:50 ratios? Why or why not?

23.60 The production of Vitamin C (ascorbic acid) highlights the importance of oxidation and reduction chemistry of oxygen-containing functional groups.
(a) The synthesis of Vitamin C starts with D-glucose, shown here in its open-chain form. How many chirality centres are in D-glucose?
(b) The reduction of D-glucose involves the conversion of what functional group into what kind of alcohol?
(c) The critical step in Vitamin C production is the oxidation step to intermediate A, which contains a single carbonyl group. Draw the eight possible oxidation products and identify them as aldehydes, ketones, or carboxylic acids.

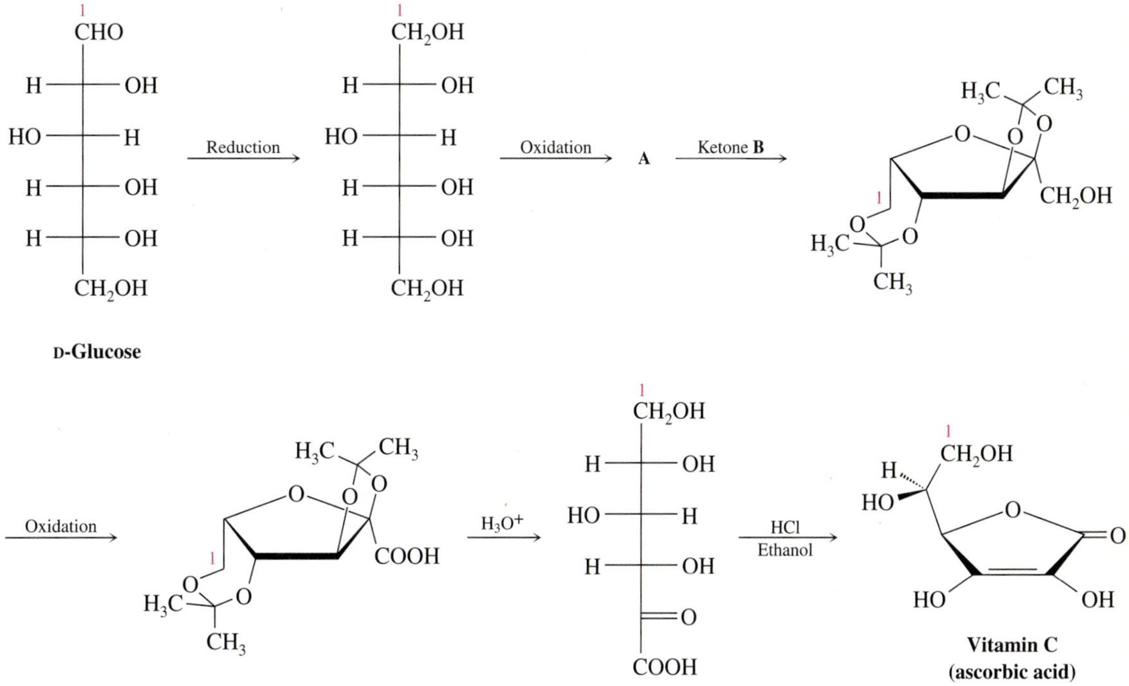

(d) The oxidation is carried out by an enzyme from a bacterium. Why would an enzyme be used instead of a chemical oxidizing agent such as $KMnO_4$?

(e) The subsequent steps also rely on oxidation chemistry; however, before the second oxidation is carried out, intermediate A is reacted with an excess of a ketone to make acetals. What ketone is used?

(f) These acetals are cleaved with H_3O^+ after the second oxidation step. What purpose do these acetals serve? What could these groups be called?

(g) Concentrated HCl promotes an esterification reaction [>>Chapter 24] to produce Vitamin C, which has a single chiral centre. Is the configuration R or S?

Carbonyl Compounds: Part II

© Andrew McClenaghan/Photo Researchers, Inc.

24.1 Case Study: Structure, Medicinal Activity, and Serendipity

Although you should never underestimate the value of hard work and logical thinking, serendipity often plays a role in scientific breakthroughs. A carboxylic acid derivative (a cyclic amide) played the key role in what has been called "the supreme example [of luck] in all scientific history." It occurred in the late summer of 1928 when Scottish bacteriologist Alexander Fleming went on vacation, leaving in his lab a culture plate recently inoculated with the bacterium *Staphylococcus aureus*.

While Fleming was away, an extraordinary chain of events occurred. First, a nine-day cold spell lowered the laboratory temperature to a point where the *Staphylococcus* on the plate could not grow. During this time, spores from a colony of the mould *Penicillium notatum* (opening photo), which were being grown on the floor below, wafted up into Fleming's lab and landed in the culture plate. The temperature then rose, and both *Staphylococcus* and *Penicillium* began to grow. On returning from holiday, Fleming discarded the plate into a tray of antiseptic, intending to sterilize it. Evidently, though, the plate did not sink deeply enough into the antiseptic, because when Fleming glanced at the plate a few days later, what he saw changed the course of human history. He noticed that the growing *Penicillium* mould appeared to dissolve the colonies of staphylococci.

> *Serendipity, finding something valuable or delightful when you are not looking for it, plays a role in chemistry research and learning, as in all human activity.*

Fleming realized that the mould must be producing a compound that killed bacteria, and he spent several years trying to isolate the substance. Finally, in 1939, the Australian pathologist Howard Florey and the German refugee Ernst Chain managed to isolate the active substance, called *penicillin*. Penicillin's dramatic ability to cure infections in mice was soon demonstrated, and successful tests in humans followed shortly thereafter. By 1943, penicillin was produced on a large scale for military use, and by 1944, it was used for civilians. Since then, many millions of lives have been saved. In 1945, Fleming, Florey, and Chain shared the Nobel Prize in Medicine.

Now called benzylpenicillin, or penicillin G, molecules of the substance discovered by Fleming contain a carboxylic acid and two amine-containing carboxylic acid derivatives, called amides. The key site of reactivity in the molecule is the cyclic amide (a **lactam**) in the four-membered ring (the "β" designation, two carbon atoms outside the amide group in the ring). Members of this class of drugs are called *β-lactam antibiotics*.

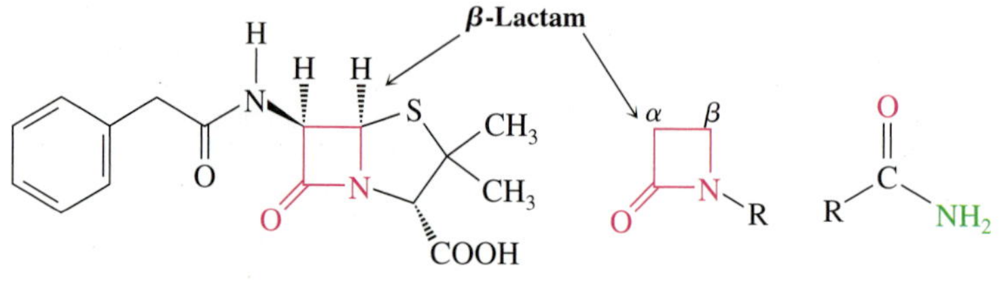

Benzylpenicillin
(penicillin G)

Amide

You will learn in this chapter that amides are relatively unreactive compared with other carboxylic acid derivatives, such as acyl chlorides or esters. Hydrolysis (reaction with water) of amides to produce carboxylic acids requires harsh conditions. From these observations, we would conclude that nucleophilic substitution at the amide carbon doesn't occur readily.

Should it come as a surprise, then, to find out that the β-lactam family of antibiotics fights bacteria by making use of the reactivity of an amide? Perhaps not. Ring strain makes this β-lactam much more reactive than most amides. We know that bond angles are typically 109.5° for single-bonded carbon atoms and 120° for double-bonded carbon atoms. When atoms are confined into a four-membered ring, the bond angles are squeezed closer to 90°. This *ring strain* [<<Section 9.7] makes the amide reactive toward some nucleophiles.

β-Lactam antibiotics react with a nucleophile on an active site of the enzyme transpeptidase. Transpeptidase is responsible for building the cell walls of bacteria. A nucleophilic hydroxyl group of the enzyme reacts with a molecule of the β-lactam antibiotic to open up the four-membered ring and form a stable ester, another class of carboxylic acid derivatives introduced in this chapter. This acyclic ester prevents transpeptidase from doing any additional chemistry. The bacterial cell wall is never finished, and the bacteria ultimately die.

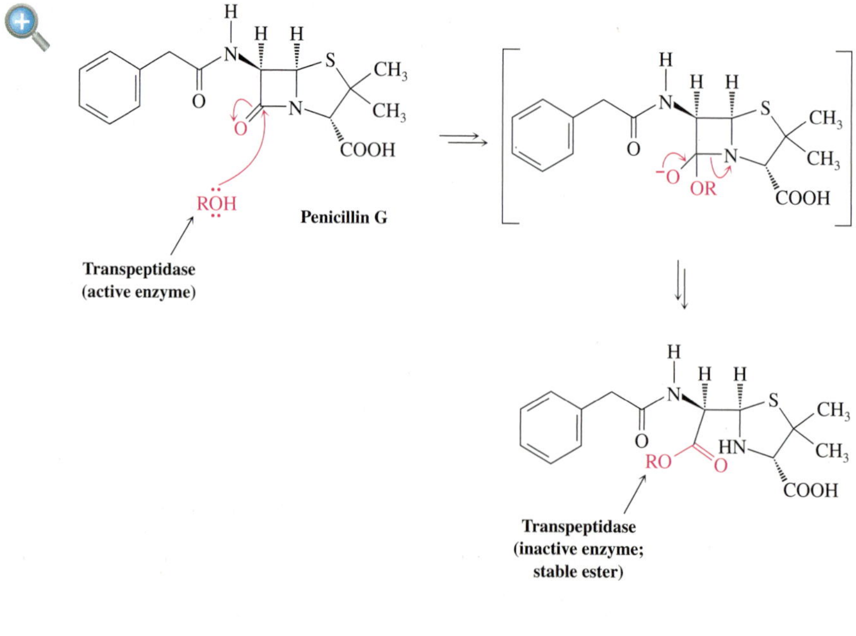

Molecular Modelling (Odyssey)

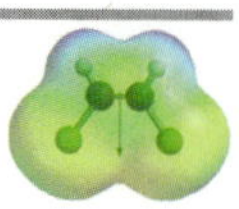

e24.1 Compare the amide group in penicillin G molecules with other amides.

Penicillin G

Transpeptidase (active enzyme)

Transpeptidase (inactive enzyme; stable ester)

Unfortunately, bacteria have become resistant to penicillin by evolving enzymes similar to transpeptidase, such as β-lactamase, which more effectively attack penicillin and hydrolyze the amide. The hydrolyzed penicillin has no antibacterial activity.

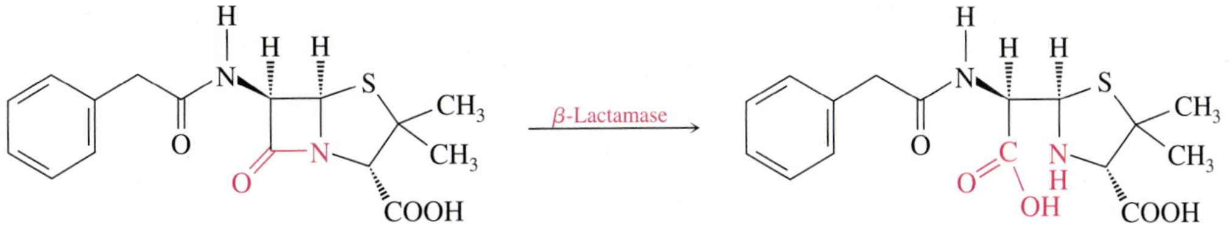

Active drug
(penicillin G)

Inactive product

Chemistry, however, gives humans tools to try to outwit evolving bacteria—at least for a while. One way is to change the structure of the drug molecules slightly, with the expectation that the β-lactamase will no longer recognize it. The acyl-amino side chain, shown on the left side of the penicillin G structure above, can be varied in the laboratory to provide literally hundreds of penicillin analogues with different biological activity profiles. Ampicillin, for instance, has an extra amine group in that side chain. Closely related to the penicillins are the cephalosporins, a group of β-lactam antibiotics that contain an unsaturated six-membered sulfur-containing ring. Cephalexin, marketed under the trade name Keflex, is one example. Cephalosporins generally have much greater antibacterial activity than do penicillins, particularly against resistant strains of bacteria.

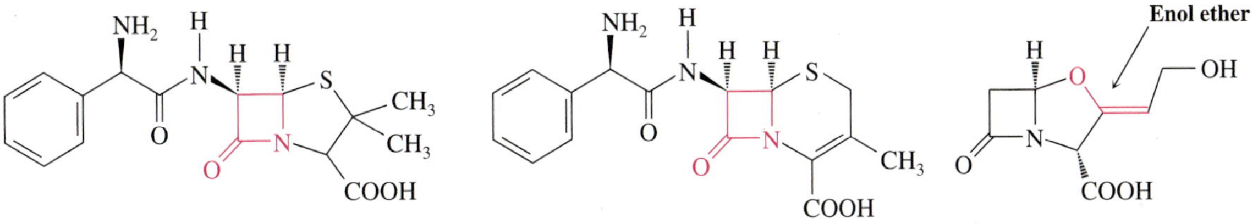

Ampicillin

Cephalexin
(a cephalosporin)

Clavulanic acid

Another strategy is to protect the drug by adding a second drug that inhibits the β-lactamase enzyme. Molecules of clavulanic acid have a β-lactam ring and a shape that is recognized by β-lactamase; however, the sulfur atom of the penicillins is replaced with an oxygen, and this enol ether functional group leads to different reactivity. A series of reactions leads to the formation of a complex between clavulanic acid and the β-lactamase, trapping the β-lactamase and preventing it from binding and hydrolyzing the β-lactam antibiotic that is the other part of the mixture. Augmentin is one of the most widely used two-drug mixtures.

It's hard enough to envision living comfortably to a ripe old age without cyclic peptides like penicillin. But you wouldn't even be here without the peptide oxytocin that stimulated labour to help your mother bring you into the world. Now try to imagine life without hemoglobin to carry oxygen through your blood, α-keratin proteins to give structure to your skin, polyethylene terphthalate polymers to contain your satisfying carbonated soft drink, or hexanoic acid (caproic acid) to give goats as well as sweaty soccer shoes their unmistakable aromas. These are just a few examples of carboxylic acids and their derivatives that are the focus of this chapter, and that make up some of the most abundant and important organic compounds in living systems.

24.2 Naming Carboxylic Acids and Derivatives

You can recognize a **carboxylic acid** molecule as having an —OH group directly attached to a carbonyl carbon atom —C=O. Examples include acetic acid (CH_3COOH), the principal organic component of vinegar; butanoic acid ($CH_3CH_2CH_2COOH$), responsible for the rancid odour of sour butter; and lactic acid ($CH_3CHOHCOOH$), produced during metabolism and exercise in animals.

In **carboxylic acid derivatives**, an electronegative substituent such as an $-X$ (halogen), $-OR$, $-SR$, or $-NH_2$ group replaces the $-OH$ group attached to the carbonyl carbon atom of a carboxylic acid. Although there are many different kinds of carboxylic acid derivatives, we'll be concerned only with some of the most common ones: **acyl halides, acid anhydrides, esters, amides**, and related compounds called **nitriles**. In addition, *acyl phosphates* and *thioesters* are acid derivatives of great importance in numerous biological processes [>>Chapter 29]. By contrast, aldehydes and ketones [<<Chapter 23] have an alkyl group or hydrogen atoms on both sides of the C=O group.

Naming Carboxylic Acids: RCOOH

Simple open-chain carboxylic acids are named by replacing the terminal –e of the alkane name with –oic acid. The $-COOH$ carbon (the carbonyl group carbon atom) is always numbered C1.

Alternatively, compounds that have a $-COOH$ group bonded to a ring are named by using the suffix –carboxylic acid. In this alternative system, the carboxylic acid carbon atom is *attached to* C1 on the ring but is not itself numbered.

Because many carboxylic acids were among the first organic compounds to be isolated and purified, there are a large number of acids with trivial (also called common) names. We use systematic names in this book, with the exception of formic (methanoic) acid (HCOOH), acetic (ethanoic) acid (CH_3COOH), and lactic acid (above) whose names are so well known that it makes little sense to refer to them in any other way. Also listed in Table 24.1 are the names for acyl derivatives of the parent carboxylic acids.

CARBOXYLIC ACID		ACYL GROUP		**TABLE 24.1** Some Trivial (Common) Names of Carboxylic Acids and Acyl Groups
Structure	**Name**	**Name**	**Structure**	
HCOOH	Formic	Formyl	HCO—	
CH_3COOH	Acetic	Acetyl	CH_3CO—	
CH_3CH_2COOH	Propionic	Propionyl	CH_3CH_2CO—	
$CH_3CH_2CH_2COOH$	Butyric	Butyryl	$CH_3(CH_2)_2CO$—	
HOOCCOOH	Oxalic	Oxalyl	—OCCO—	
$HOOCCH_2COOH$	Malonic	Malonyl	—$OCCH_2CO$—	
$HOOCCH_2CH_2COOH$	Succinic	Succinyl	—$OC(CH_2)_2CO$—	
H_2C=CHCOOH	Acrylic	Acryloyl	H_2C=CHCO—	
[benzene ring with CO₂H]	Benzoic	Benzoyl	[benzene ring with C=O]	

Naming Acyl Halides: RCOX

Acyl halides are named by identifying first the acyl group and then the halide. The acyl group name is derived from the acid name by replacing the *–ic acid* ending with *–yl*, or the *–carboxylic acid* ending with *–carbonyl*. For example,

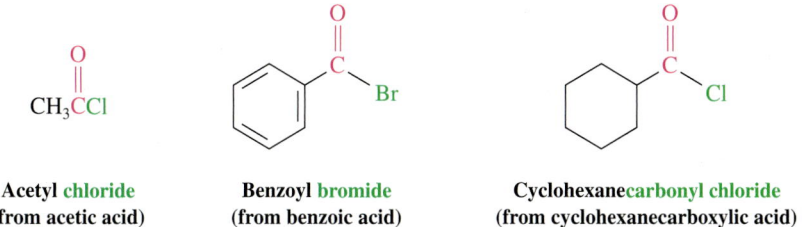

Acetyl chloride	**Benzoyl bromide**	**Cyclohexanecarbonyl chloride**
(from acetic acid)	(from benzoic acid)	(from cyclohexanecarboxylic acid)

Naming Acid Anhydrides: RCOOCOR′

Anhydrides from simple carboxylic acids and cyclic anhydrides from dicarboxylic acids are named by replacing the word *acid* with *anhydride:*

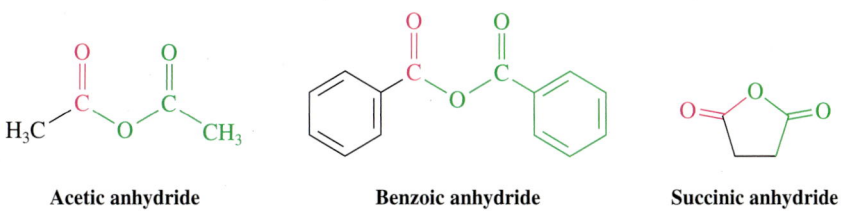

Acetic anhydride	**Benzoic anhydride**	**Succinic anhydride**

Naming Amides: RCONH₂

Amides with an unsubstituted —NH_2 group are named by replacing the *–oic acid* or *–ic acid* ending with *–amide*, or by replacing the *–carboxylic acid* ending with *–carboxamide:*

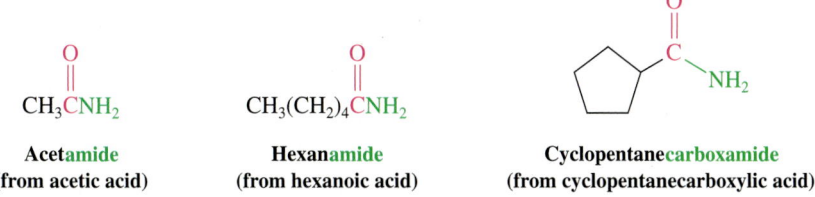

Acetamide	**Hexanamide**	**Cyclopentanecarboxamide**
(from acetic acid)	(from hexanoic acid)	(from cyclopentanecarboxylic acid)

If the nitrogen atom is substituted, the amide is named by first identifying the substituent group and then the parent. The substituents are preceded by the letter *N* to identify them as being directly attached to nitrogen.

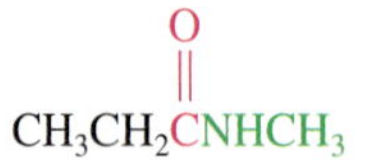

N-**Methyl**propanamide

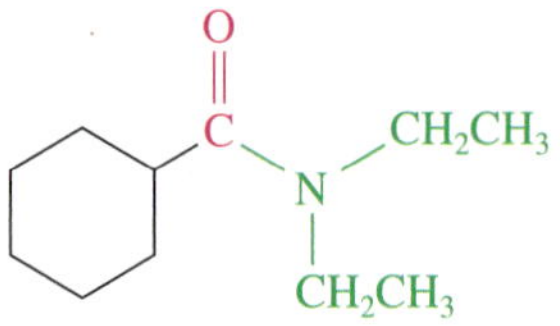

N,N-**Diethyl**cyclohexanecarboxamide

Naming Esters: RCOOR′

Systematic names for esters are derived by first giving the name of the group attached to oxygen and then identifying the carboxylic acid. In so doing, the *–ic acid* ending is replaced by *–ate:*

Ethyl acetate
(the ethyl ester of
acetic acid)

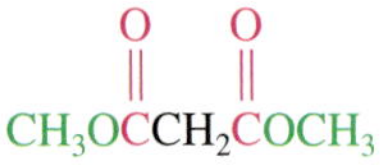

Dimethyl malonate
(the dimethyl ester of
malonic acid)

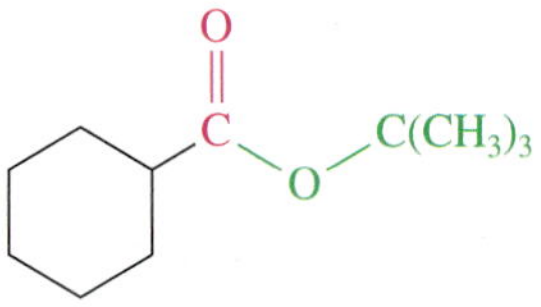

tert-**Butyl** cyclohexanecarboxylate
(the *tert*-butyl ester of
cyclohexanecarboxylic acid)

**Interactive Exercise
24.1**

SUBMIT

Draw structures of
molecules from IUPAC names.

EXERCISE 24.2—NAMING CARBOXYLIC ACIDS

Draw structures of molecules corresponding to the following names:
(a) 2,3-dimethylhexanoic acid
(b) 4-methylpentanoic acid
(c) 2-hydroxybenzoic acid
(d) *trans*-1,2-cyclobutanedicarboxylic acid

EXERCISE 24.3—NAMING CARBOXYLIC ACID DERIVATIVES

Draw structures of molecules corresponding to the following names:
(a) 2,2-dimethylpropanoyl chloride
(b) *N*-methylbenzamide
(c) ethyl *p*-nitrobenzoate
(d) *tert*-butyl butanoate
(e) *trans*-2-methylcyclohexanecarboxamide
(f) 4-methylbenzoic anhydride

24.3 Spectroscopy of Carboxylic Acids and Derivatives

Infrared Spectroscopy

Two features are particularly useful in identifying solution (or liquid) and solid phase samples of *carboxylic acids:* an intense C=O stretching vibration and a very broad hydrogen-bonded O—H stretching vibration (Figure 24.1). The O—H bonds of carboxylic acids typically give an exceptionally broad signal between 2400 and 3300 cm^{-1} due to the many species with a wide variation in the extent of hydrogen bonding in solution [<<Section 22.3]. The C=O stretching absorption is an intense peak between 1710 and 1760 cm^{-1}.

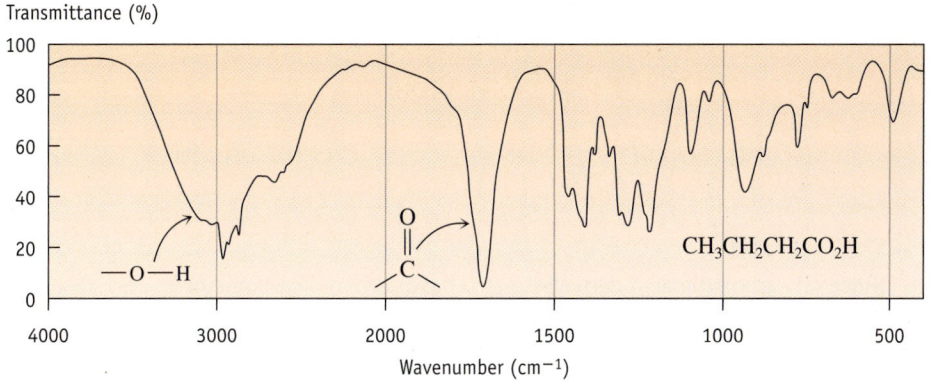

FIGURE 24.1 Infrared spectrum of a liquid phase sample of butanoic acid.

Carboxylic acid derivatives all possess a carbonyl group, and so show intense IR absorptions for C=O stretching vibrations in the range 1650–1850 cm^{-1}. The exact wavenumber depends on the bond strength of the C=O group, which changes depending on the amount of electron donation or withdrawal to the C=O bond in each derivative. This can be used as a valuable means of identification (Table 24.2).

Carbonyl Type	Absorption (cm^{-1})
Saturated acyl chloride	1810
Aromatic acyl chloride	1770
Saturated acid anhydride	1820, 1760
Saturated ester	1735
Aromatic ester	1720
Saturated amide	1690
N-Substituted amide	1680
N,N-Disubstituted amide	1650
Aromatic amide	1675

TABLE 24.2 Characteristic Wavenumber for C=O Stretching of Carboxylic Acid Derivatives

As expected, substituents such as a —Cl, —OR, —SR, or —NH$_2$ group directly attached to the carbonyl carbon atom of a carboxylic acid derivative also show characteristic stretching and bending absorptions.

Think about It

e24.2 Compare animated IR spectra of carboxylic acids and their derivatives.

EXERCISE 24.4—IR SPECTROSCOPY

What functional groups might compounds have if they show the following IR absorptions?
(a) 1735 cm^{-1}
(b) 1810 cm^{-1}
(c) 2400–3300 cm^{-1} and 1710 cm^{-1}
(d) 1715 cm^{-1}

EXERCISE 24.5—IR SPECTROSCOPY

Propose structures for compounds with the following formulas and IR absorptions:
(a) C$_6$H$_{12}$O$_2$, 1735 cm^{-1} (b) C$_4$H$_9$NO, 1650 cm^{-1} (c) C$_4$H$_5$ClO, 1780 cm^{-1}

Nuclear Magnetic Resonance Spectroscopy

In the ^{13}C NMR spectrum, carbonyl carbon atoms of both carboxylic acid molecules and their derivatives absorb in the range 160–185 δ. Aromatic and conjugated carboxylic acids absorb near the upfield end (165 δ) of this region and saturated acids near the downfield end (~185 δ).

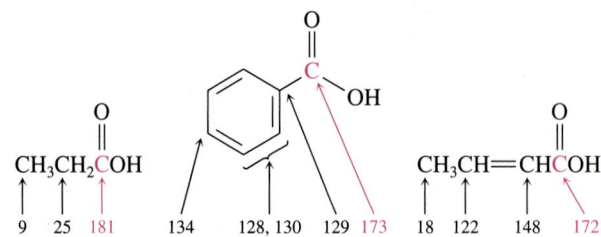

Since all carboxylic acid derivatives also have carbonyl carbon atoms, they absorb in this downfield range also, and ^{1}H NMR spectroscopy is required to easily distinguish a carboxylic acid from one of its derivatives.

In the ^{1}H NMR spectrum, the acidic —COOH proton normally absorbs as a singlet with a very unusual chemical shift near 12 δ. As with alcohols, the —COOH proton can undergo rapid exchange with deuterium atoms when D$_2$O is added to the sample tube, causing this absorption to disappear from the ^{1}H NMR spectrum. Figure 24.2 shows the ^{1}H NMR spectrum of phenylacetic acid.

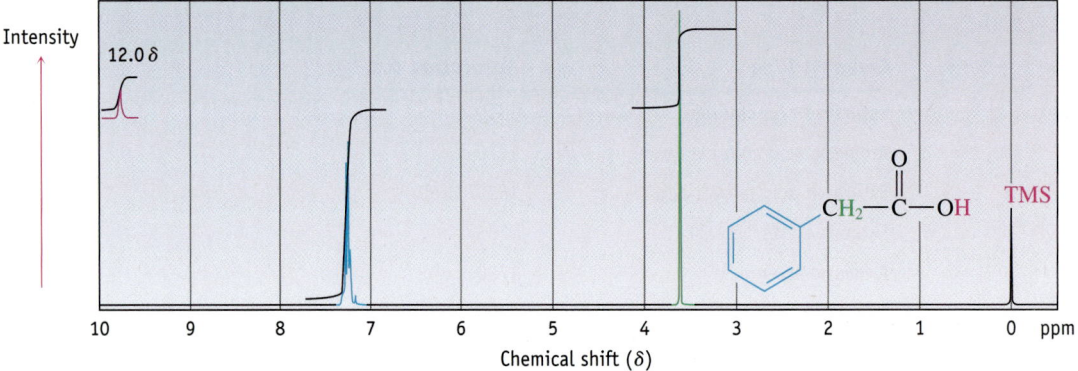

FIGURE 24.2 The ^{1}H NMR spectrum of phenylacetic acid.

Hydrogen atoms on the carbon atom next to a carbonyl group are slightly deshielded and absorb near 2 δ in the ^{1}H NMR spectrum. The identity of the carbonyl group can't be determined by ^{1}H NMR however, because all acid derivatives with hydrogen atoms in this position absorb in a similar range. Figure 24.3 shows the ^{1}H NMR spectrum of ethyl acetate.

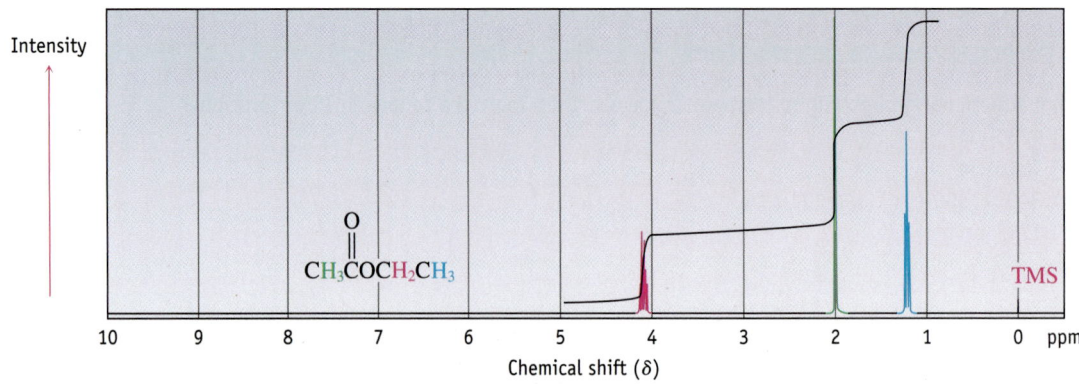

FIGURE 24.3 ^{1}H NMR spectrum of ethyl acetate.

Interactive Exercise 24.7

Compare the IR and NMR spectra of carboxylic acids and derivatives.

SUBMIT

EXERCISE 24.6—NMR SPECTROSCOPY

Predict the key features in the ^{1}H and ^{13}C NMR spectra of the isomers cyclopentanecarboxylic acid and 4-hydroxycyclohexanone, and explain how you would differentiate between them.

Carboxylic acids can be identified by infrared spectroscopy, through their characteristic C=O and O—H stretching absorptions; by C=O carbon atom chemical shifts at about 200 δ in their ^{13}C NMR spectra; and by carboxylic acid O—H hydrogen atoms chemical shifts at about 12 δ in their ^1H NMR spectra.

24.4 Electronic Structure and Reactivity: Carboxylic Acids

Consider two experimental observations about carboxylic acids. First, carboxylic acids boil at much higher temperatures than alkanes or alkyl halides of similar molecular weight. Acetic acid, for example, boils at 118 °C, whereas chloropropane boils at 46.6 °C. We saw a similar surprising increase in the boiling points of alcohols relative to less polar compounds. Second, look more closely at the infrared spectra of two samples of acetic acid (Figure 24.4). In (a) the sample of acetic acid is dissolved in a solvent, and in (b) the spectrum of a gas-phase sample is recorded. What might be the cause of the dramatic change in the sharpness and position of the O—H stretching vibration near 3000 cm^{-1}? In the gas-phase spectrum (Figure 24.4b), a sharp band at 3600 cm^{-1} appears. In the solution spectrum (Figure 24.4a), the O—H absorption is at lower wavenumber, and much broader—extending from 2600 to 3500 cm^{-1}. We saw a similar change in the infrared spectra of ethanol in Section 22.3.

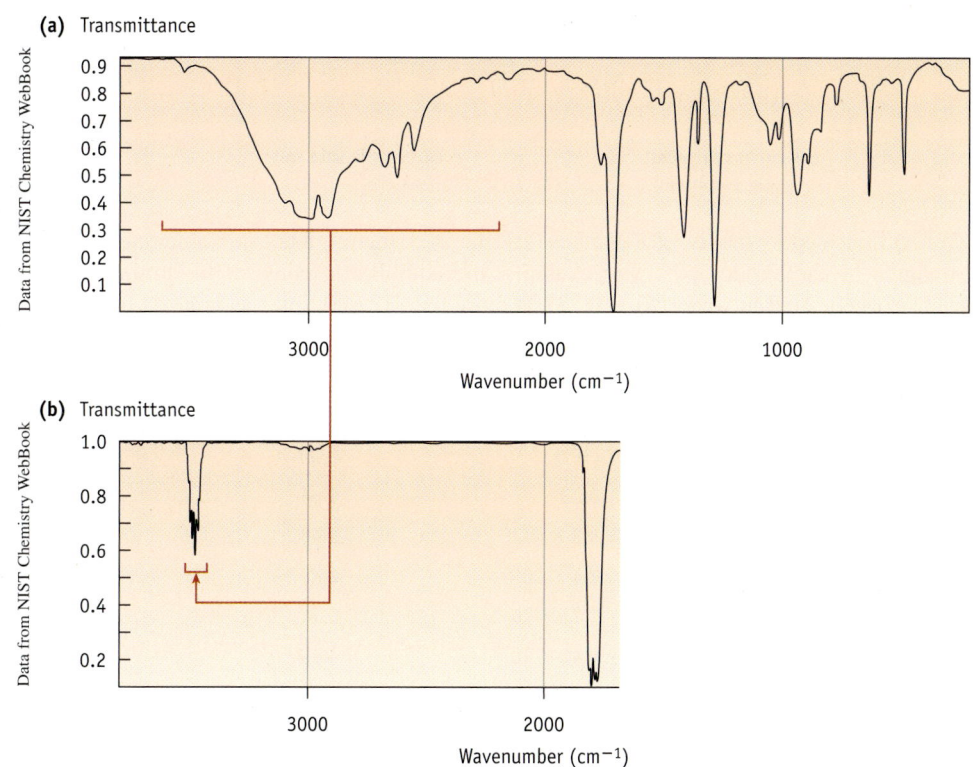

(a) Transmittance

Data from NIST Chemistry WebBook

Wavenumber (cm^{-1})

(b) Transmittance

Data from NIST Chemistry WebBook

Wavenumber (cm^{-1})

FIGURE 24.4 (a) Infrared spectrum of a solution of acetic acid, and (b) the relevant portion of the infrared spectrum of a gas-phase sample. Note the dramatic change in sharpness and position of the O—H stretching vibration above 3000 cm^{-1}.

A reasonable molecular-level explanation for both of these observations is that strong hydrogen bonding forces attract carboxylic acid molecules to each other in solution (or in the liquid phase), increasing the energy input required to separate them when boiling sends them into the vapour phase. The change in the two infrared spectra is an elegant direct experimental observation of hydrogen bonding, since the degree of hydrogen bonding affects the vibrational energy of the O—H stretch. Hydrogen bonding between acetic acid molecules is not important in the gas phase, as molecules are much further apart than in a solution. In solution, however, hydrogen bonding is much stronger, and constantly in flux, producing a lower energy, much broader signal. This explanation is reinforced by noting the strong polarity of the O—H bond in the electrostatic potential map for a carboxylic acid molecule. This leads to very strong *intermolecular* forces of attraction between carboxylic acid molecules. These forces of attraction are so significant that most carboxylic acids exist as dimers – pairs of molecules in solution are held together by hydrogen bonds [<<Section 6.3]:

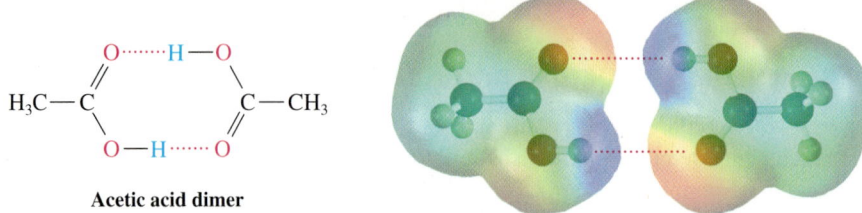

Acetic acid dimer

The acidity implied by the name carboxylic acid is also consistent with a molecular-level look at the high polarity of the O—H bonds in their molecules. Although much weaker than mineral acids like HCl, carboxylic acids are nevertheless much stronger acids than alcohols; acetic acid, for example, has $K_a = 1.76 \times 10^{-5}$ ($pK_a = 4.75$), while ethanol has $K_a = 10^{-16}$ ($pK_a = 16$). In practical terms, a K_a value near 10^{-5} means that only about 1% of the molecules in a 0.1 mol L^{-1} aqueous solution are ionized. Because of their acidity, carboxylic acids react with bases such as NaOH to give water-soluble metal carboxylates ($RCOO^- Na^+$).

A carboxylic acid   **A carboxylic acid salt**
(water-insoluble)   **(water-soluble)**

As indicated by the list of K_a values in Table 24.3, there is a considerable range in the strengths of carboxylic acids. For example, the equilibrium constant for ionization of trichloroacetic acid in water ($K_a = 0.23$), is more than 12 000 times as large as for acetic acid ($K_a = 1.76 \times 10^{-5}$).

TABLE 24.3 Relative Acidity of Some Carboxylic Acids and Reference Compounds

Name	K_a	pK_a	
HCl (hydrochloric acid)*	(10^7)	(-7)	Stronger acid
CCl_3COOH	0.23	0.64	
$CHCl_2COOH$	3.3×10^{-2}	1.48	
$CH_2ClCOOH$	1.4×10^{-3}	2.85	
HCOOH	1.77×10^{-4}	3.75	
C_6H_5COOH	6.46×10^{-5}	4.19	
$H_2C{=}CHCOOH$	5.6×10^{-5}	4.25	
CH_3COOH	1.76×10^{-5}	4.75	
CH_3CH_2OH (ethanol)*	(10^{-16})	(16)	Weaker acid

* Values for HCl and ethanol are shown for reference.

How can we account for the large differences in pK_a between carboxylic acids and alcohols, and among carboxylic acids in Table 24.3? Because proton dissociation is an equilibrium process, anything that stabilizes (lowers the free energy of) product species (the aquated conjugate base anions) relative to reactants (the aquated acid molecules) favours increased ionization and increased acidity. Following this line of thinking, we might start by asking why carboxylic acids are so much more acidic than alcohols even though molecules of both contain polar O—H groups.

It is tempting to give an intuitive explanation that focuses only on the relative stabilities of individual carboxylate anions versus individual alkoxide anions (Figure 24.5). Our reasoning would be as follows: in an alkoxide ion, the negative charge is localized on the only oxygen atom. In a carboxylate ion, however, the negative charge is shared by both oxygen atoms. One way to represent this delocalization is to show a single carboxylate anion as a stabilized resonance hybrid of two equivalent structures. Since a carboxylate ion is more stable (lower in free energy) relative to its conjugate carboxylic acid than an alkoxide ion is compared to its alcohol conjugate acid, the carboxylate ion is present in greater amount at equilibrium.

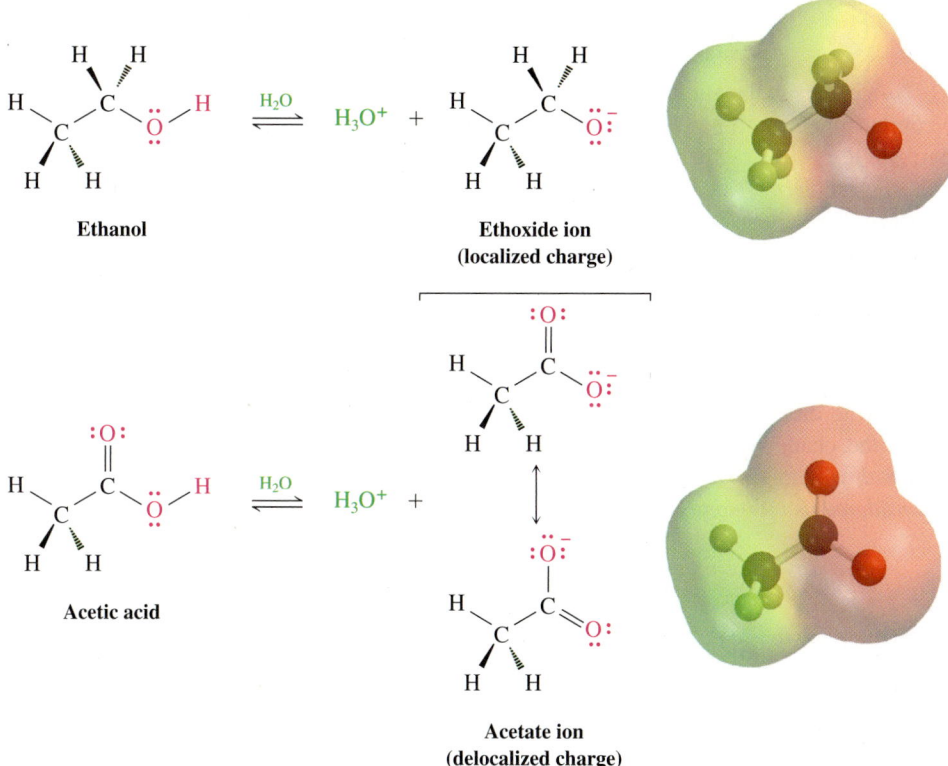

Ethanol

Ethoxide ion
(localized charge)

Acetic acid

Acetate ion
(delocalized charge)

FIGURE 24.5 Charge delocalization. An alkoxide ion has its charge localized on one oxygen atom and is less stable relative to its conjugate acid, while a carboxylate ion has the charge spread equally (delocalized) over both oxygen atoms, as shown by the two resonance structures above and is therefore more stable relative to its conjugate acid.

Following this same line of thinking, can we account for the large differences in pK_a listed in Table 24.3? Because acid ionization is an equilibrium process, anything that stabilizes (lowers the free energy of) the carboxylate anion favours increased ionization into ions and increased acidity. So the electron-withdrawing chlorine atom delocalizes the negative charge on the anion even more and makes chloroacetic acid stronger than acetic acid. With two electronegative chlorine atoms, the K_a for dichloroacetic acid is 3000 times that of acetic acid, and with three chlorine atoms, the K_a for trichloroacetic acid is 12 000 times greater.

$pK_a = 4.75$ $\qquad$ $pK_a = 2.85$ $\qquad$ $pK_a = 1.48$ $\qquad$ $pK_a = 0.64$

Weaker acid $\longrightarrow$ Stronger acid

This satisfying and intuitive explanation, however, is incomplete and misleading. It neglects the fact that these acid and conjugate base species do not exist as individual entities in solution. Each is constantly surrounded by, and interacting with, water molecules, and it is the total free energy difference between the water-solvated carboxylic acid molecules and water-solvated conjugate base anions that provides the real explanation for the observed changes in acid strength. Recall from Section 17.7 that the equilibrium constant K is related to the *standard free energy change* between reactants and products ($\Delta_r G°$). The free energy difference involves two terms, one which depends on the enthalpy difference between reactants and products, and one which depends on the entropy difference.

$$\Delta_r G° = -RT \ln K \quad \text{and} \quad \Delta_r G° = \Delta_r H° - T\Delta_r S°$$

A better explanation for the increased acidity of chlorinated acetic acid molecules comes from examining the relative magnitude of the enthalpy and entropy terms that contribute to $\Delta_r G°$. In experimental data for the series from acetic acid to chloroacetic acid, the *entropy term* ($T\Delta_r S°$) is clearly the dominant factor contributing to the change in free energy. This suggests that differences in aquation may be the most important factor. The ordering of water molecules around a negatively charged anion leads to a reduction in entropy. Delocalization of the negative charge on each of the carboxylate anions that have chlorine atoms leads to less tight interaction with the water solvent molecules surrounding the anions. This gives a smaller reduction in entropy for acid ionization. So a better explanation for why $\Delta_r G°$ becomes less positive with increasing chlorine substitution has more to do with differences in the amount of order in the solvent molecules (an *entropy* factor) than the changing strength of the O—H bond or the stability of the conjugate base (*enthalpy* factors). *This is another reminder that reactants and products should never be considered as individual, isolated species when they are in solution.*

Taking It Further

e14.15 Read about how acid strength is related to molecular structure.

Molecular Modelling (Odyssey)

e24.5 Examine electrostatic potential maps of carboxylic acid molecules and their conjugate bases.

WORKED EXAMPLE 24.1—RELATIVE ACIDITY

Which would you expect to be the stronger acid: benzoic acid or *p*-nitrobenzoic acid?

Solution

Using only an enthalpy argument, the more stabilized the conjugate base carboxylate anion, the stronger the acid. We know from its effect on aromatic substitution [<<Section 20.8] that a nitro group is electron-withdrawing and can stabilize a negative charge. A *p*-nitrobenzoate ion is more stable than a benzoate ion, and *p*-nitrobenzoic acid is stronger than benzoic acid.

Nitro group withdraws electrons from ring and stabilizes negative charge.

Remembering that all of these species are aquated, the more delocalized *p*-nitrobenzoate anion has less tightly associated water molecules, leading to a smaller reduction in entropy for the ionization of its conjugate acid. $\Delta_r G°$ is less positive for *p*-nitrobenzoic acid ionization.

EXERCISE 24.8—REACTIONS OF CARBOXYLIC ACIDS

Draw structures of molecules of the products of the following reactions:

(a)

$$\text{C}_6\text{H}_5\text{COOH} \xrightarrow{\text{NaOCH}_3} \text{?}$$

(b)

$$\underset{\underset{\text{CH}_3}{|}}{\overset{\overset{\text{CH}_3}{|}}{\text{CH}_3\text{CCOOH}}} \xrightarrow{\text{KOH}} \text{?}$$

EXERCISE 24.9—RELATIVE ACIDITY

Rank the following compounds in order of increasing acidity: sulfuric acid, methanol, phenol, *p*-nitrophenol, acetic acid.

EXERCISE 24.10—RELATIVE ACIDITY

Rank the following compounds in order of increasing acidity:
(a) CH_3CH_2COOH, $BrCH_2COOH$, $BrCH_2CH_2COOH$
(b) benzoic acid, ethanol, *p*-cyanobenzoic acid

Reactions of carboxylic acids can be understood by recognizing that they are weak acids. This can be predicted from looking at electrostatic potential maps of their molecules. Although weaker than mineral acids like HCl, carboxylic acids are much more acidic than alcohols because of electron delocalization in their conjugate base carboxylate ions. Most carboxylic acids have pK_a values near 5, but the exact acidity of an acid depends on its structure. Electron-withdrawing substituents make carboxylic acids more acidic.

24.5 Electronic Structure and Reactivity: Derivatives

While some carboxylic acids are foul-smelling compounds, many simple esters found in nature are pleasant-smelling liquids responsible for the fragrant odours of fruits and flowers. Methyl butanoate, for example, has been isolated from pineapple oil, and isopentyl acetate has been found in banana oil. The ester linkage is also present in animal fats and other biologically important molecules. The lack of strong intermolecular forces, like the hydrogen bonds of carboxylic acid molecules, gives esters lower boiling points and makes them more volatile at room temperature. So not only do they interact differently with receptors in your nose than carboxylic acids (due to different shapes and polarity), but also more ester molecules make it to your nose due to their volatility.

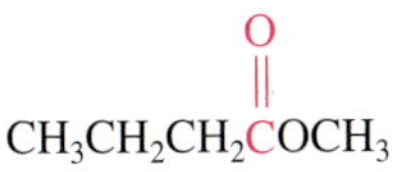

Methyl butanoate
(from pineapples)

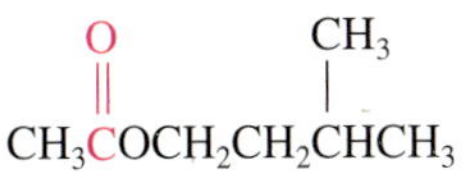

Isopentyl acetate
(from bananas)

A fat
(R = C_{11-17} chains)

As seen in Section 24.1, amides are less reactive than esters; this stability makes amides ideal linkages in peptides and proteins. Hydrogen bonding between amide molecules increases their boiling points relative to other carboxylic acid derivatives. A diverse range of biological events—from protein folding to the action of drugs like the antibiotic vancomycin—depend on hydrogen bonding between amides. Nylon, a synthetic polymer, also is composed of molecules with amide links between its repeating units, and owes its strength to hydrogen bonding between adjacent molecules:

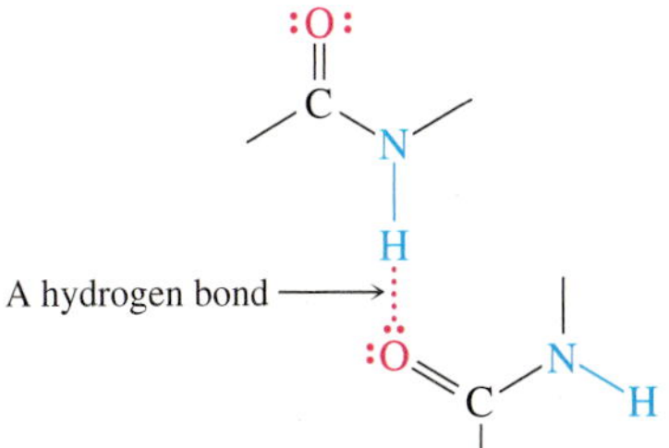

Acyl chlorides and anhydrides are commonly used as synthetic starting materials in the chemical and pharmaceutical industries, as they are very reactive, and can be easily converted into other carboxylic acid derivatives such as esters or amides. With few exceptions, these two functional groups are not found in molecules in nature due to their reactivity. We examine some uses of these synthetic molecules in Section 24.7.

A hydrogen bond →

Nucleophilic Substitution Reactions of Carboxylic Acid Derivatives

What reactions would we predict for carboxylic acid derivatives? Electrostatic potential maps (Figure 24.8) show that their carbonyl carbon atoms are electron-deficient. We would predict that the most important chemistry of carboxylic acid derivatives would involve electron-rich nucleophiles reacting with their molecules' electron-deficient carbonyl carbon atoms, just as we saw with the reactions of carbonyl groups of aldehydes and ketones.

However, unlike aldehydes and ketones, which undergo nucleophilic addition reactions [<<Section 23.5], carboxylic acid derivatives undergo *nucleophilic substitution reactions*. The reason is evident from remembering the general mechanism that was postulated for nucleophilic addition reactions. Molecules of aldehydes and ketones have either two alkyl groups (ketones) or one alkyl group and one hydrogen atom (aldehydes) attached to the carbonyl carbon atom. Carboxylic acid derivative molecules have one alkyl group and one electronegative substituent such as an —X (halogen), —OR, —SR, or —NH$_2$ group. Each of these substituents of carboxylic acid derivatives has one important reactive feature in common. Under the right conditions, they can act as leaving groups in substitution reactions because they are much weaker bases than the groups on aldehyde and ketone molecules. By looking in more detail at the proposed general reaction mechanism for these substitution reactions, we can make sense of why this difference in structure leads to such an important difference in observed patterns of reactivity.

24.6 Nucleophilic Acyl Substitution: Reaction Mechanism

Look again at the general reaction mechanism (Figure 23.2) proposed to explain patterns of *addition* of nucleophiles to the polar C=O bonds of aldehyde and ketone molecules. Zooming in at the molecular level to consider only one reactant pair and ignoring the solvent and the many other productive and non-productive collisions, we visualize a nucleophile first adding to the planar carbonyl carbon atom of an aldehyde or ketone molecule. In most cases, the only feasible next step, besides the simple reversal of this addition step, is for the resulting alkoxide ions (called *tetrahedral intermediates* because of their change in geometry at the carbon atoms) to undergo protonation to yield alcohol molecules. This leads to net *addition* reactions.

The different overall reaction pattern seen when nucleophiles react with carboxylic acid derivatives is a consequence of structure. Unlike aldehydes and ketones, carboxylic acid derivative molecules have a carbonyl carbon atom bonded to a group that can leave as a *weak base*.

Zooming in on a reacting pair, we can imagine that following *addition* of a nucleophile to the carbonyl carbon atom of a carboxylic acid derivative molecule, the same type of tetrahedral intermediate is formed as in the case of aldehydes/ketones. But following the addition step, a new possibility emerges. The leaving group can be *eliminated* to re-form the exceptionally strong C=O bond of a carbonyl group. This reaction involves the replacement of —Y with —Nu, an overall *nucleophilic acyl substitution reaction* (Figure 24.6), but *the proposed pathway results from first an addition step, followed by an elimination step*.

Recall that leaving groups in substitution reactions must be weak bases. [<<Section 21.5].

Since aldehydes and ketones have no such good leaving group, they are unable to undergo substitution reactions and, with most nucleophiles, undergo simple protonation to give addition products.

Ketone or aldehyde: nucleophilic addition

Carboxylic acid derivatives: nucleophilic acyl substitution

FIGURE 24.6 **Accepted general mechanisms** for nucleophilic addition reactions to aldehyde and ketone molecules and nucleophilic substitution reactions of carboxylic acid derivatives. Both reaction mechanisms are thought to begin with a productive collision leading to the addition of a nucleophile to a polar C=O bond to give a tetrahedral alkoxide ion intermediate. The intermediate formed from an aldehyde or ketone is protonated to give an alcohol, but the intermediate formed from a carboxylic acid derivative has a good leaving group (a weak base), which it can expel to give a new, substituted carbonyl group.

A good example of nucleophilic acyl substitution is the conversion of carboxylic acids into esters by reaction with an alcohol, the substitution of —OH by —OR. Called the Fischer esterification reaction, the simplest method involves heating the carboxylic acid with an acid catalyst in an alcohol solvent.

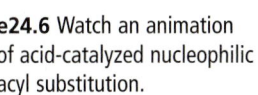

Benzoic acid   **Propan-1-ol**   **Propyl benzoate**

In the proposed mechanism for this reaction (Figure 24.7), the acid catalyst first protonates the C=O oxygen atom of the —COOH group, which makes the carbonyl carbon atom more electron-deficient and more reactive toward nucleophiles. An alcohol molecule then adds to the protonated carboxylic acid, and subsequent loss of water yields the ester product. Under these acid-catalyzed conditions, an essential step is the transfer of a proton to the —OH group in the third step, so water (a weak base) can be the leaving group rather than OH⁻ (a strong base).

All steps in the proposed mechanism for the Fischer esterification reaction are reversible, and we observe that the relative amounts of acid and ester at equilibrium can be increased or decreased depending on the reaction conditions. Ester formation is favoured when alcohol is used as solvent, but carboxylic acid is favoured when water is used as solvent.

Evidence shows the mechanism postulated for this overall nucleophilic substitution reaction of carboxylic acid derivatives shown in Figure 24.6 is different than the S_N1 and

Think about It

e24.6 Watch an animation of acid-catalyzed nucleophilic acyl substitution.

FIGURE 24.7 Accepted reaction mechanism of the esterification reaction of a carboxylic acid. The reaction is an acid-catalyzed nucleophilic acyl substitution.

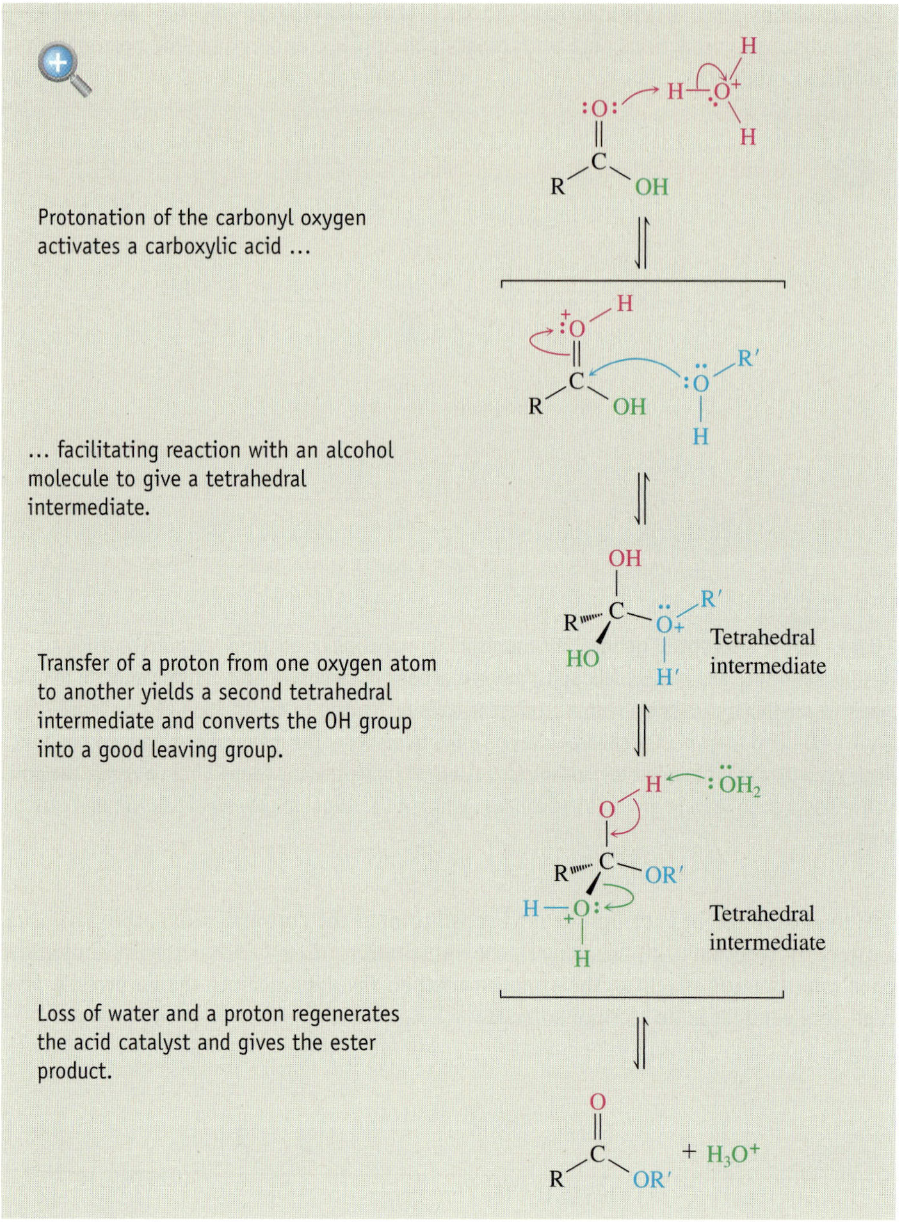

Protonation of the carbonyl oxygen activates a carboxylic acid ...

... facilitating reaction with an alcohol molecule to give a tetrahedral intermediate.

Transfer of a proton from one oxygen atom to another yields a second tetrahedral intermediate and converts the OH group into a good leaving group.

Loss of water and a proton regenerates the acid catalyst and gives the ester product.

S_N2 pathways proposed in Sections 21.6–21.7 for reactions of nucleophiles with alkyl halides and other substituted hydrocarbons.

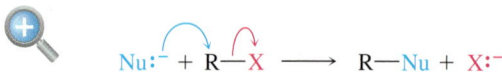

The presence of the carbonyl group in molecules of carboxylic acid derivative substrates leads to substitution by a two-step addition-elimination reaction mechanism.

Comparing Reactivity of Different Carboxylic Acid Derivatives

In comparing the reactivity of different acyl derivatives, we observe that the more electron-poor the molecule's C=O carbon atom, the more readily the compound reacts with nucleophiles. Acyl chlorides are the most reactive compounds because their electronegative chlorine atoms strongly withdraw electrons from their carbonyl carbon atoms, whereas amides are the least reactive compounds. Electrostatic potential maps (Figure 24.8) of amides (like dimethylacetamide), esters (like methyl acetate), acid anhydrides (like acetic anhydride), and acyl chlorides (like acetyl chloride) indicate these differences among various carboxylic acid derivatives by the small but significant changes in colour on the C=O carbon atoms.

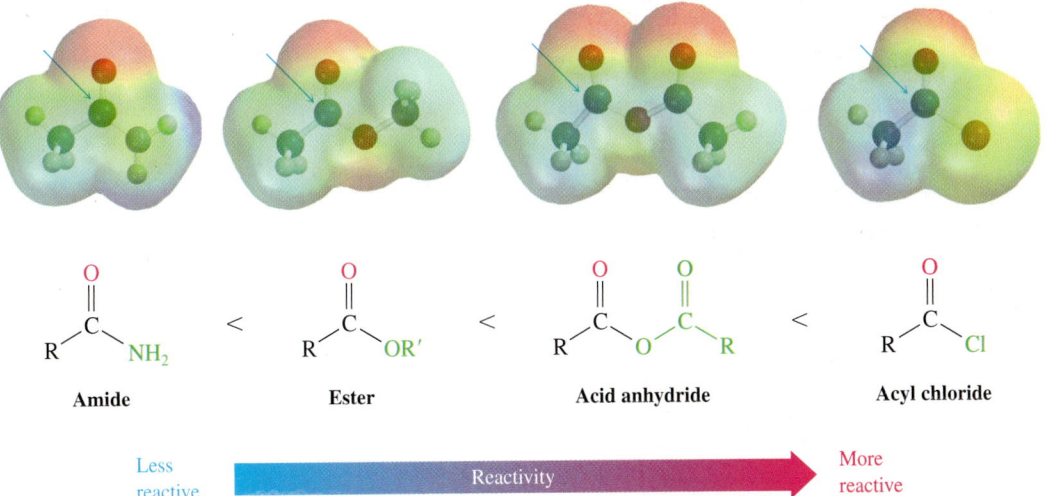

FIGURE 24.8 Electrostatic potential maps of an amide, ester, acid anhydride, and acyl chloride.

A consequence of these reactivity differences is that it is usually possible to convert a more reactive acid derivative into a less reactive one. Acyl chlorides and anhydrides, for example, can be converted into esters and amides, and the reactions are so fast that neither an acid or base catalyst is usually needed. But amides and esters can't be converted easily into acyl chlorides. Remembering the reactivity order is therefore a useful way to keep track of a large number of reactions (Figure 24.9).

Molecular Modelling (Odyssey)

e24.7 Compare electrostatic potential maps of molecules of carboxylic acid derivatives.

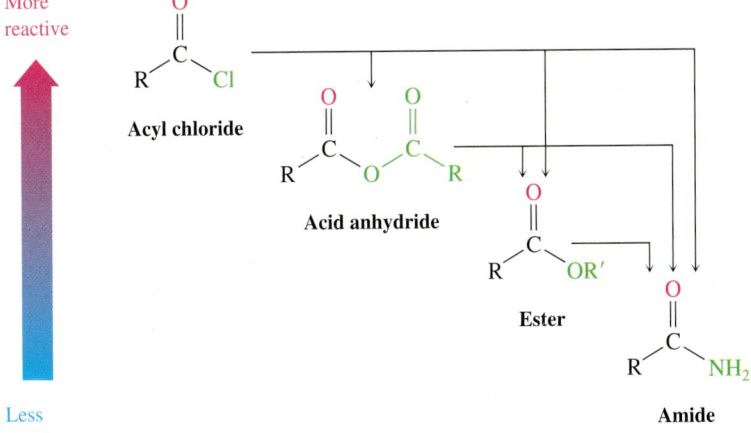

FIGURE 24.9 Observed order of reactivity of carboxylic acid derivatives. More reactive compounds can be readily converted into less reactive ones.

Molecular Modelling (Odyssey)

e24.8 Use models to understand when reaction of carboxylic acid derivatives requires acid catalysis.

This order of reactivity is consistent with what we see in nature. Organisms contain many amides and esters, but not acyl chlorides and anhydrides, which would react rapidly in aqueous environments of living systems. On the other hand, the most useful synthetic starting materials in laboratory settings are the very reactive acyl chlorides and anhydrides.

WORKED EXAMPLE 24.2—REACTIVITY OF CARBOXYLIC ACID DERIVATIVES

Which is more reactive in a nucleophilic acyl substitution reaction with hydroxide ion: CH_3CONH_2 or CH_3COCl?

Strategy

Identify the two potential leaving groups, $-NH_2$ and $-Cl$, and predict which will increase more strongly the polarity of the neighbouring carbonyl group.

Solution

Since Cl atoms are more electronegative than N atoms, the carbonyl group of an acyl chloride is more polar than the carbonyl group of an amide, and the acyl chloride CH_3COCl is more reactive than the amide CH_3CONH_2. Cl^- is also a better leaving group than NH_2^-.

EXERCISE 24.11—REACTIVITY OF CARBOXYLIC ACID DERIVATIVES

Which compound in each of the following sets is more reactive in nucleophilic acyl substitution reactions?
(a) CH_3COCl or CH_3COOCH_3
(b) $(CH_3)_2CHCONH_2$ or $CH_3CH_2COOCH_3$
(c) CH_3COOCH_3 or $CH_3COOCOCH_3$
(d) CH_3COOCH_3 or CH_3CHO

EXERCISE 24.12—REACTIVITY OF CARBOXYLIC ACID DERIVATIVES

Methyl trifluoroacetate (CF_3COOCH_3) is more reactive than methyl acetate (CH_3COOCH_3) in nucleophilic acyl substitution reactions. Explain.

> Acyl chlorides and anhydrides are the most reactive carboxylic acid derivatives, while esters and amides are much less reactive toward nucleophilic substitution.

24.7   Reactions of Carboxylic Acids and Derivatives

Since acyl halides and acid anhydrides are much more reactive toward nucleophilic substitution than other derivatives, we start with methods of synthesizing them from carboxylic acids, and then show how they, in turn, can be readily converted into other derivatives.

Synthesis and Reactions of Acyl Halides

Acyl chlorides are prepared from carboxylic acids by reaction with thionyl chloride ($SOCl_2$):

For example:

2,4,6-Trimethylbenzoic acid   2,4,6-Trimethylbenzoyl chloride (90%)

Molecular Modelling (Odyssey)

e24.9 Use an electrostatic potential map of a thionyl chloride molecule to identify the electrophilic centre.

Synthesis and Reactions of Acid Anhydrides

The best method for preparing acid anhydrides is by a nucleophilic acyl substitution reaction of an acyl chloride with a carboxylate anion. Both symmetrical and unsymmetrical acid anhydrides can be prepared in this way.

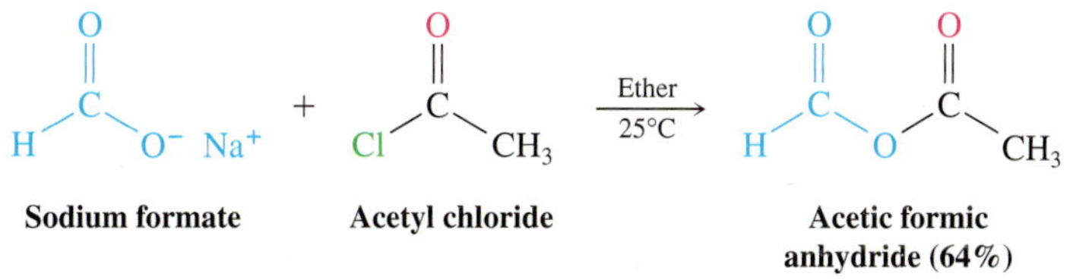

Sodium formate	+	Acetyl chloride		Acetic formic anhydride (64%)

The chemistry of acid anhydrides is similar to that of acyl chlorides. Acid anhydrides react with water to form acids, with alcohols to form esters, and with amines to form amides:

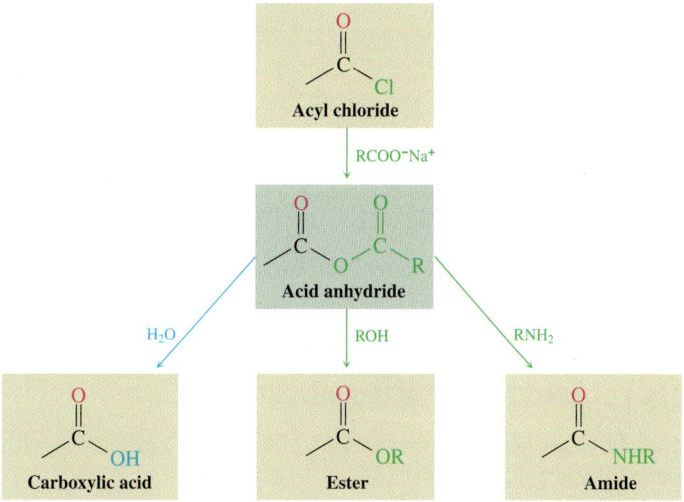

Acetic anhydride is often used to prepare acetate esters of complex alcohols and to prepare substituted acetamides from amines. For example, aspirin (an ester) is prepared by reaction of acetic anhydride with *o*-hydroxybenzoic acid. Similarly, acetaminophen (an amide; the active ingredient in Tylenol) is prepared by reaction of acetic anhydride with *p*-hydroxyaniline.

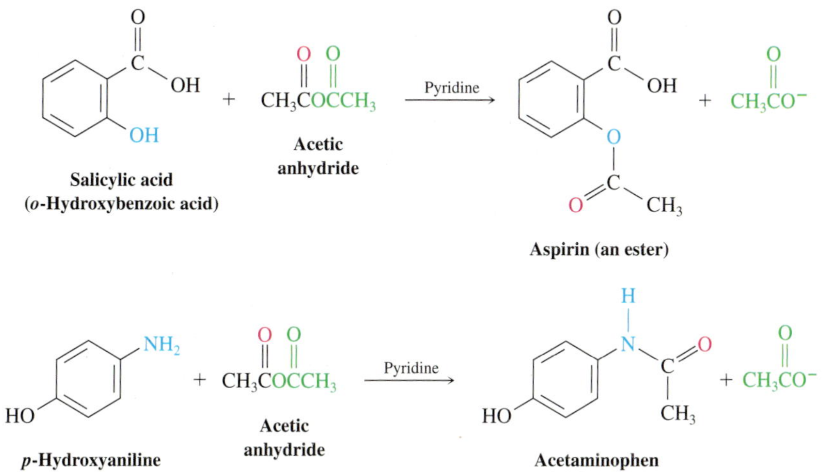

Notice in both of these examples that only "half" of the anhydride molecule is used; the other half acts as the leaving group during the nucleophilic acyl substitution step and produces carboxylate anion as a by-product. For this reason, anhydrides are less efficient to use from the perspective of *atom economy* [<<Section 5.10], and acyl chlorides are normally used instead.

Molecular Modelling (Odyssey)

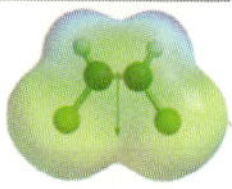

e24.12 Model why anhydrides are more effective (but less efficient) than esters for acylation reactions.

WORKED EXAMPLE 24.5—REACTIONS OF ACID ANHYDRIDES

What is the product of the following reaction?

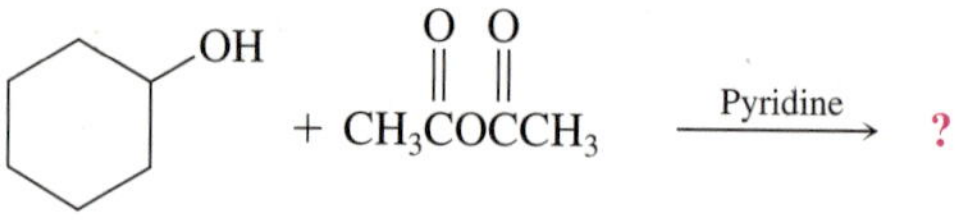

Solution

Remember that acid anhydrides undergo nucleophilic acyl substitution reaction with alcohols to give esters. Reaction of cyclohexanol with acetic anhydride yields cyclohexyl acetate by nucleophilic acyl substitution of the —$OCOCH_3$ group of the anhydride by the —OR group of the alcohol.

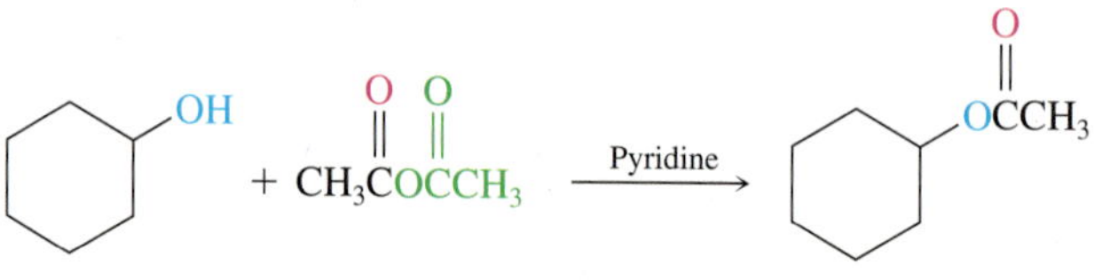

Cyclohexanol **Cyclohexyl acetate**

EXERCISE 24.16—MECHANISM FOR ACID ANHYDRIDE REACTIONS

Write the steps in the mechanism of the reaction between *p*-hydroxyaniline and acetic anhydride to prepare acetaminophen.

EXERCISE 24.17—REACTIONS OF ACID ANHYDRIDES

What product would you expect from the reaction of equimolar amounts of methanol and the cyclic anhydride phthalic anhydride?

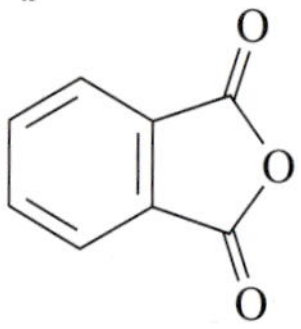

Phthalic anhydride

Synthesis and Reactions of Esters

Esters are usually prepared by nucleophilic substitution of an alcohol for chloride ion of an acyl chloride, as discussed earlier. They can also be made from carboxylic acids directly by esterification with an alcohol and an acid catalyst.

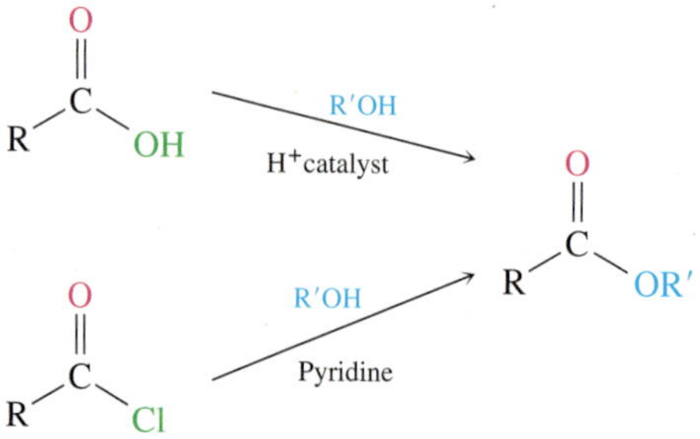

Esters undergo the same reactions as for other acyl derivatives, but they are less reactive toward nucleophiles than acyl chlorides or anhydrides.

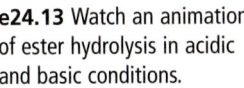

Conversion of Esters into Acids (RCOOR' → RCOOH) Esters are hydrolyzed either by aqueous base or by aqueous acid to yield a carboxylic acid plus an alcohol:

Ester   →   **Acid**   + R'OH

Hydrolysis of an ester in basic solution is called *saponification*, after the Latin word *sapo*, meaning "soap." Soap is made by the base-induced hydrolysis of animal fat, whose fatty acid molecules are formed through ester linkages. Ester hydrolysis is thought to occur by a typical nucleophilic acyl substitution pathway in which OH⁻ nucleophile adds to the ester carbonyl group, yielding a tetrahedral intermediate. Loss of —OR'⁻ then gives a carboxylic acid, which is deprotonated under reaction conditions to give the acid carboxylate. Acid must be added at the end of a saponification reaction to regenerate the carboxylic acid:

Ester   **Tetrahedral intermediate**   **Acid**   **Acid salt**

Think about It

e24.13 Watch an animation of ester hydrolysis in acidic and basic conditions.

Molecular Modelling (Odyssey)

e24.14 Model saturated and unsaturated fatty acids and see the difference between soaps and detergents.

WORKED EXAMPLE 24.6—REACTIONS OF ESTERS

Write the products of the following saponification reaction:

$$CH_3CHCH_2COCH_2CH_3 \xrightarrow[\text{2. H}_3O^+]{\text{1. NaOH, H}_2O} \ ?$$

Ethyl 3-methylbutanoate

Solution

Treatment of an ester with aqueous base cleaves the ester into its acid and alcohol components by breaking the bond between the carbonyl carbon and alcohol oxygen atoms.

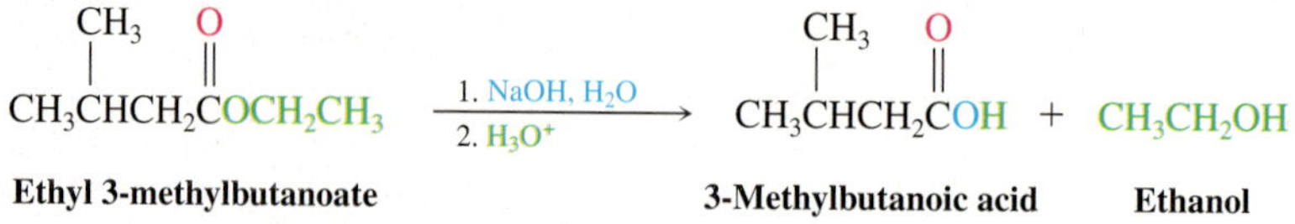

Ethyl 3-methylbutanoate   →   **3-Methylbutanoic acid**   **Ethanol**

reaction conditions: 1. NaOH, H_2O 2. H_3O^+

EXERCISE 24.18—REACTIONS OF ESTERS

Show the products of hydrolysis of the following esters:

(a)

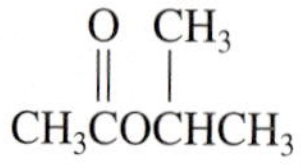

(b)

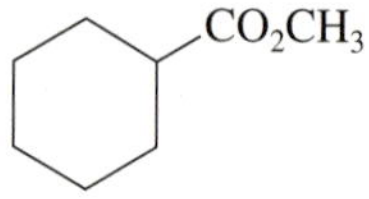

Hydrolysis of Esters with Enzyme Catalysts Enzymes, complex proteins produced as catalysts in living cells, can also be used to catalyze laboratory reactions [<<Section 18.9]. The result is often a great advantage in the stereoselectivity of reactions, relative to using a drop of strong acid or base as a catalyst.

Enzyme molecules have complex three-dimensional shapes and possess remarkable abilities to recognize and differentiate reactants based on stereochemical differences. One example of the use of enzymes to do organic chemistry in reaction flasks is the hydrolysis of esters. Often these esters are of great importance to the pharmaceutical and chemical industries. The partial hydrolysis of the diester shown using NaOH produces a *racemate* [<<Section 9.9] because the two monoester products are enantiomers. The separation of these enantiomers is difficult because, like all pairs of enantiomers, their compounds have identical physical properties in a symmetric environment, except for the sign of their rotation of plane-polarized light.

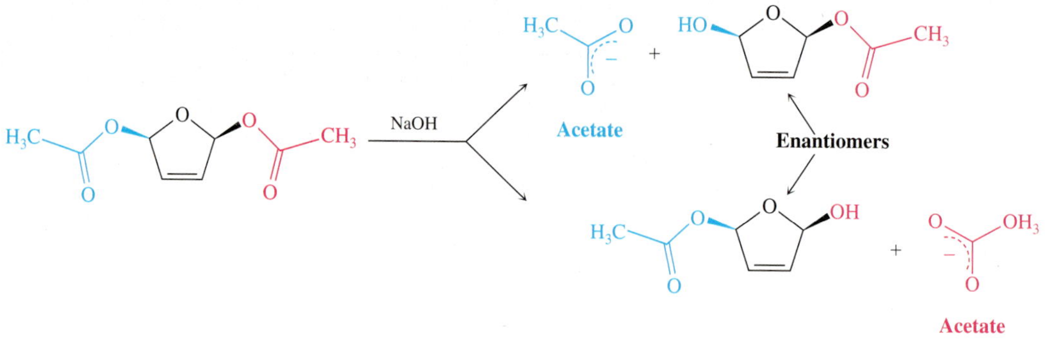

What if only one of these two enantiomers is useful? A strategy that produces only the desired monoester would be ideal. One solution to this problem is to use a chiral hydrolysis catalyst. Enzymes called *esterases* (because they do chemistry on esters) provide the necessary *chiral environment* to allow selective hydrolysis. Molecules of esterases from different organisms have different three-dimensional shapes and, as a result, perform different chemistry. It so happens that acetylcholine esterase from an electric eel selectively cleaves one ester, while the lipase (a lipid esterase) from a pig's pancreas cleaves the other.

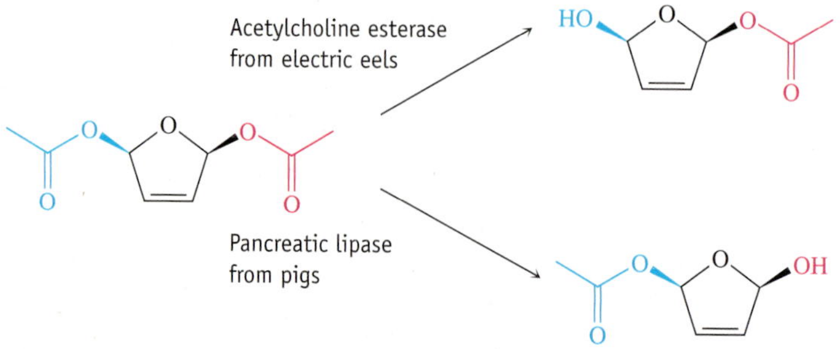

Enzymes can be used as laboratory catalysts, providing chiral environments for selective reactions of chiral carboxylic acid derivatives.

Conversion of Esters into Alcohols by Reduction (RCOOR' → RCH$_2$OH) Esters are reduced by treatment with LiAlH$_4$ to yield primary alcohols [<<Section 22.5]

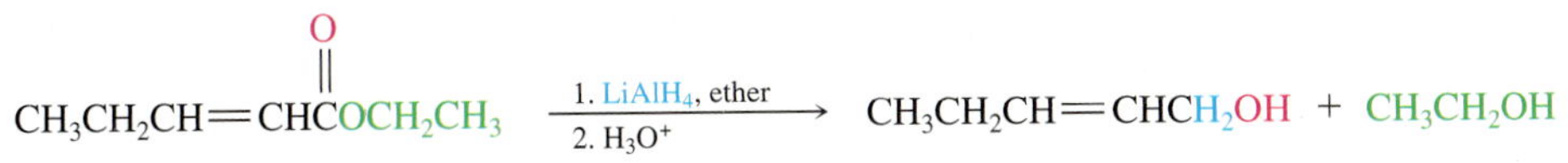

$$CH_3CH_2CH=CHCOCH_2CH_3 \xrightarrow[\text{2. H}_3O^+]{\text{1. LiAlH}_4,\ \text{ether}} CH_3CH_2CH=CHCH_2OH\ +\ CH_3CH_2OH$$

Ethyl pent-2-enoate   **Pent-2-en-1-ol (91%)**

A simplified reaction mechanism can be pictured as involving an initial nucleophilic acyl substitution reaction, as hydride ion first adds to the carbonyl group of an ester molecule, followed by elimination of an alkoxide ion to yield an aldehyde intermediate. Further reduction of the aldehyde gives the primary alcohol.

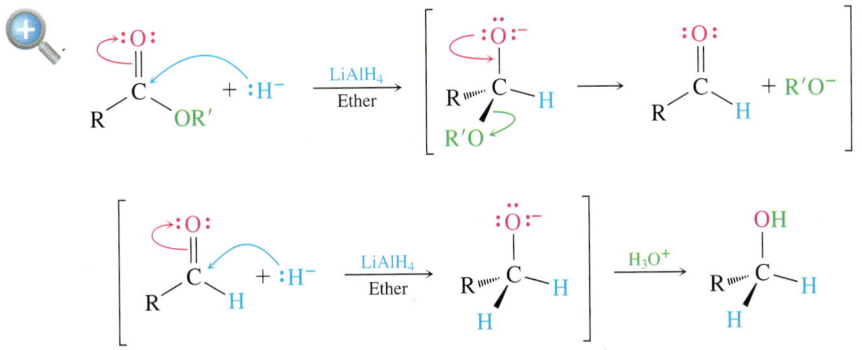

WORKED EXAMPLE 24.7—REACTIONS OF ESTERS

What products would you obtain by reduction of propyl benzoate with LiAlH$_4$?

Solution

Reduction of an ester with LiAlH$_4$ yields two molecules of alcohol product, one from the acyl part of the ester and one from the alkoxy part. Reduction of propyl benzoate yields benzyl alcohol (from the acyl group) and propan-1-ol (from the alkoxy group).

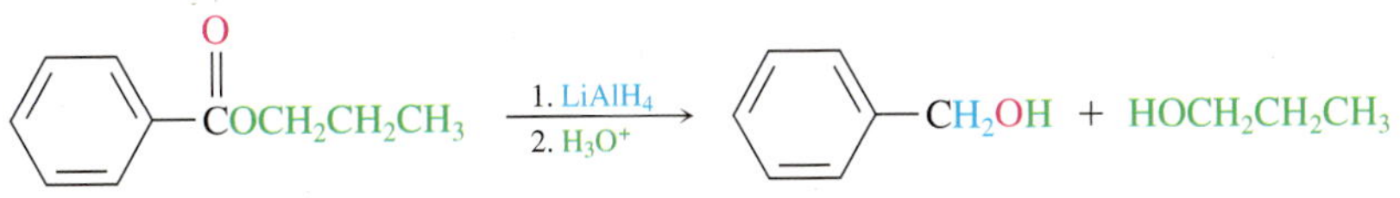

Propyl benzoate   **Benzyl alcohol**   **Propan-1-ol**

EXERCISE 24.19—REACTIONS OF ESTERS

What products would you obtain by reduction of the following esters with LiAlH$_4$?

(a)

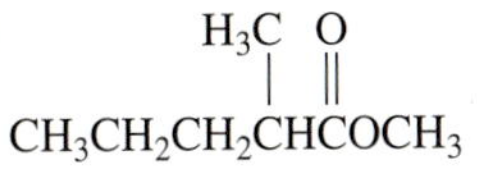

$$\underset{\text{CH}_3\text{CH}_2\text{CH}_2\text{CHCOCH}_3}{\overset{\overset{\text{H}_3\text{C}}{|}\ \overset{\text{O}}{\|}}{}}$$

(b)

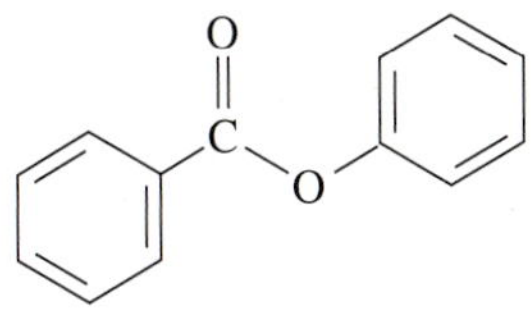

Reaction of Esters with Grignard Reagents Grignard reagents react with esters to yield tertiary alcohols in which two of the substituents on the hydroxyl-bearing carbon are identical. This is a particularly important reaction, because it involves formation of C—C bonds, essential in synthetic strategies. For example, each mole of methyl benzoate reacts with 2 mol

of CH_3MgBr to yield 2-phenylpropan-2-ol. The reaction mechanism is thought to occur by addition of a Grignard reagent to an ester, elimination of alkoxide ion to give an intermediate ketone, and further addition to the ketone to yield a tertiary alcohol (Figure 24.11).

FIGURE 24.11 Reactive intermediate in the reaction of an ester with a Grignard reagent.

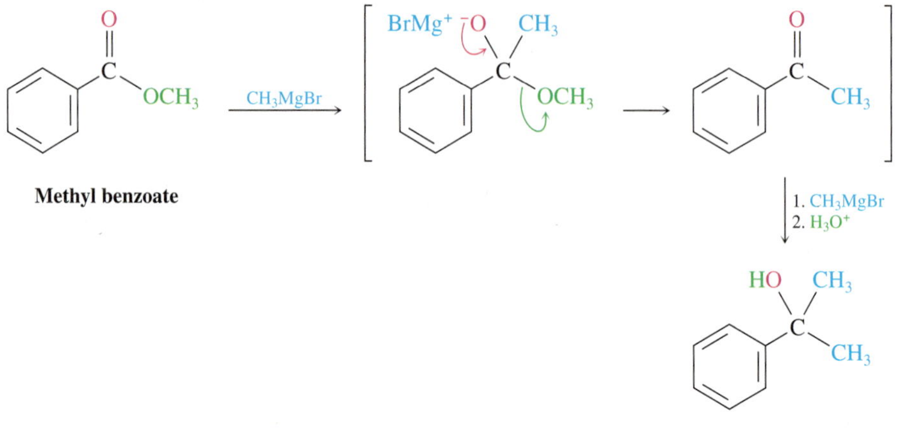

Methyl benzoate

2-Phenylpropan-2-ol
(95%)

WORKED EXAMPLE 24.8—GRIGNARD REACTIONS OF ESTERS

How could you use the reaction of a Grignard reagent with an ester to synthesize 1,1-diphenylpropan-1-ol?

Solution

The product of the reaction between a Grignard reagent and an ester is a tertiary alcohol in which the alcohol carbon atom and one of the attached groups have come from the ester, and the remaining two groups bonded to the alcohol carbon atom have come from the Grignard reagent. Since 1,1-diphenylpropan-1-ol has two phenyl groups and one ethyl group bonded to the alcohol carbon atom, it must be made from reaction of a phenylmagnesium halide with an ester of propanoic acid.

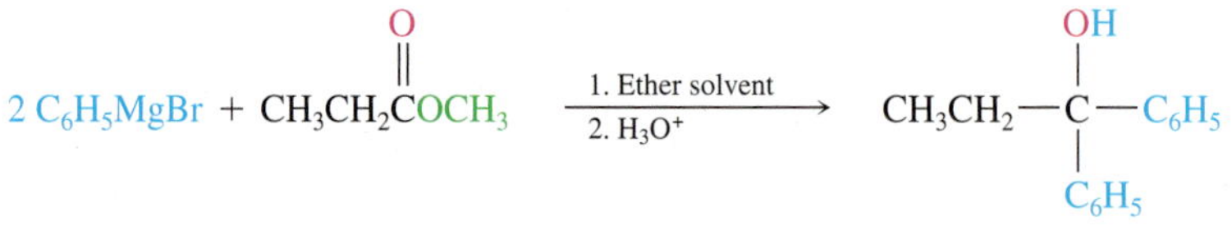

1,1-Diphenylpropan-1-ol

EXERCISE 24.20—GRIGNARD REACTIONS OF ESTERS

What ester and what Grignard reagent might you use to synthesize the following alcohols?

(a)   **(b)** 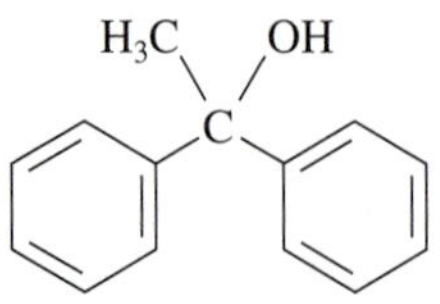   **(c)**

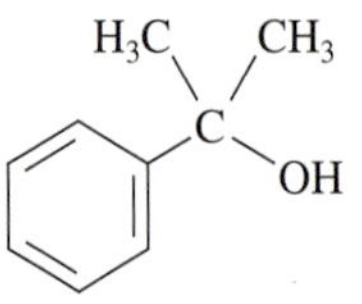

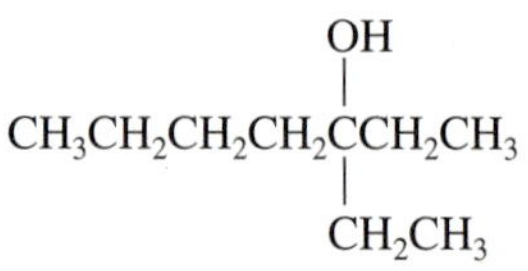

Synthesis and Reactions of Amides

Amides are usually synthesized by reaction of an acyl chloride with an amine, as we saw earlier in this section. Ammonia, monosubstituted amines, and disubstituted amines all undergo this reaction.

Acyl chloride   **Acid anhydride**   **Ester**

RNH$_2$ RNH$_2$ RNH$_2$

Amide

H$_2$O   H$^-$

Carboxylic acid   **Amine**

Amides are much less reactive than acyl chlorides, acid anhydrides, and esters. We'll see in Chapter 29 that the amide linkage is stable enough to serve as the fundamental unit connecting amino acids to form peptides and proteins.

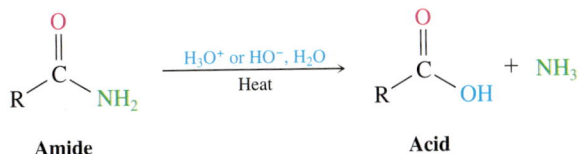

Amino acids   **A protein (polyamide)**

Conversion of Amides into Acids ($RCONH_2 \rightarrow RCOOH$) Amides undergo hydrolysis to yield carboxylic acids plus amines on heating in either aqueous acid or base. Although the reaction is slow and requires prolonged heating, the overall transformation is a typical nucleophilic acyl substitution of —OH for —NH$_2$. Although the amide is more stable than the acid, the aqueous environment shifts the equilibrium toward formation of the acid product.

Think about It

e24.15 Watch an animation of amide hydrolysis in acidic and basic conditions.

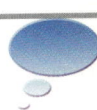

Amide   **Acid**

Now zoom in at the molecular level and think about the mechanism for this reaction. Recall from the general mechanism proposed for nucleophilic acyl substitution that a reactive intermediate, called a *tetrahedral intermediate* [<<Section 23.5] is thought to form after water adds to the carbonyl carbon atom. The change in geometry of the carbonyl carbon atom from a trigonal-planar to a tetrahedral shape during this reaction has profound implications for **molecular recognition** [<<Section 9.1]. Pharmaceutical chemists can use this change in shape to design similar yet slightly different molecules that will inhibit these reactions in living organisms. For example, a critical step of a human immunodeficiency virus (HIV) infection involves an amide hydrolysis reaction that is catalyzed by an enzyme called HIV protease. Drugs such as Amprenavir are successful in treating HIV because the secondary alcohol (red —OH) of the drug molecules mimics the shape of the tetrahedral intermediate. The stereochemistry and other structural elements (blue) of Amprenavir match that of the amide reagent.

The ability of HIV protease to propagate the infection is greatly reduced because Amprenavir binds tightly to this enzyme catalyst. This interferes with the function of the enzyme in the replication of the HIV virus, and results in the formation of immature noninfectious viral particles.

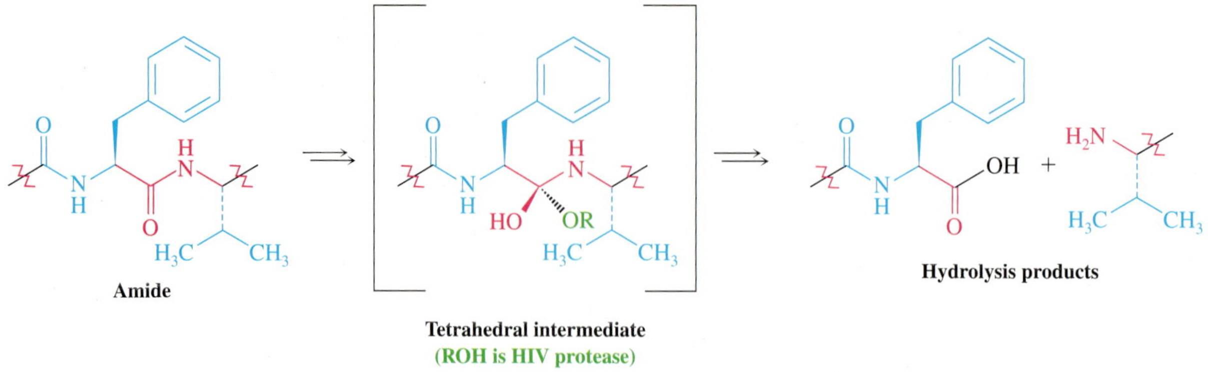

Amide

Tetrahedral intermediate
(ROH is HIV protease)

Hydrolysis products

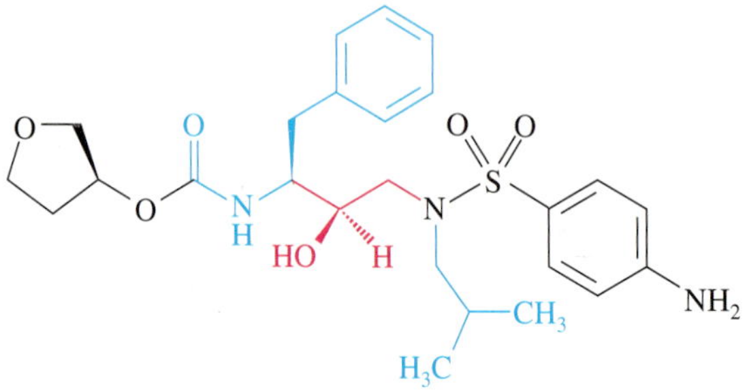

Amprenavir
(mimic of the tetrahedral intermediate)

Reduction of Amides to Amines ($RCONH_2 \rightarrow RCH_2NH_2$) Like other carboxylic acid derivatives, amides are reduced by $LiAlH_4$. The product of this reduction, however, is an *amine* rather than an alcohol. Recall that one way to recognize reduction in organic chemistry is to check for the addition of hydrogen atoms or the loss of oxygen atoms [<<Section 22.5]:

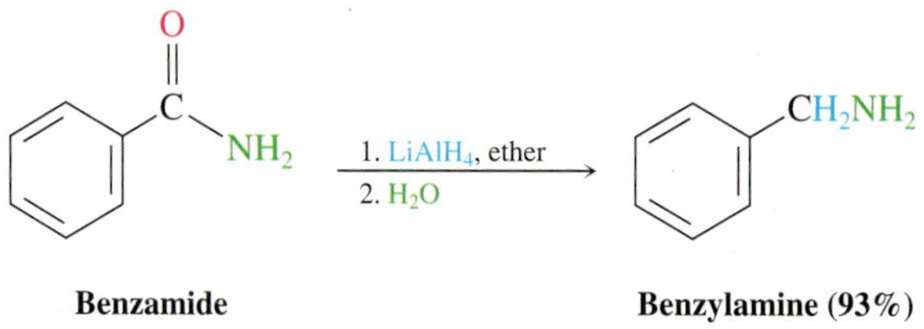

Benzamide

Benzylamine (93%)

The effect of amide reduction is to convert the amide carbonyl group of a molecule into a methylene group ($C{=}O \longrightarrow CH_2$).

WORKED EXAMPLE 24.9—REACTIONS OF AMIDES

How could you synthesize *N*-ethylaniline by reduction of an amide with $LiAlH_4$?

Solution

Reduction of an amide with $LiAlH_4$ yields an amine. To find the starting material for synthesis of *N*-ethylaniline, look for a CH_2 position next to the nitrogen atom of the amide molecule and replace that CH_2 by $C{=}O$. In this case, the amide is *N*-phenylacetamide.

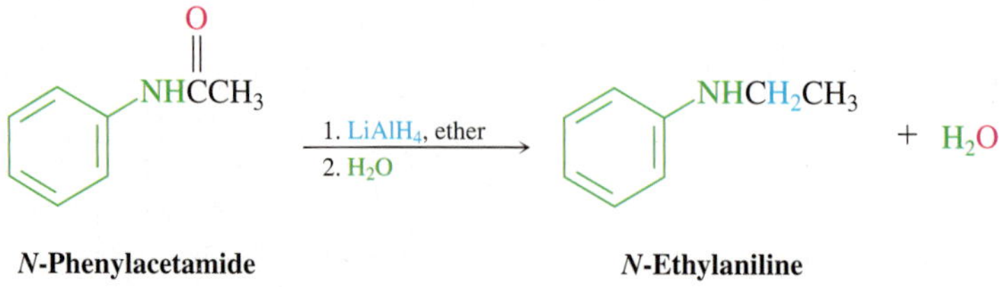

N-Phenylacetamide

N-Ethylaniline

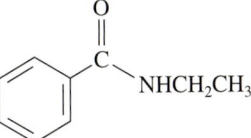

N-Ethylbenzamide

EXERCISE 24.21—REACTIONS OF AMIDES

How would you convert *N*-ethylbenzamide into the following substances?
(a) benzoic acid
(b) benzyl alcohol
(c) *N*-ethylbenzylamine (C$_6$H$_5$CH$_2$NHCH$_2$CH$_3$)

Polymers from Carbonyl Compounds: Polyamides and Polyesters

You are now equipped to better understand the structures of the important carboxylic acid derivative polymers, classified as *polyamides* and *polyesters*.

The best-known *condensation* polymers [<<Section 4.6] are the polyamides, or nylons, first prepared by Wallace Carothers at the DuPont Company. Nylon polymers are usually made by reaction between a dicarboxylic acid and a diamine. For example, nylon 66 is made by heating the six-carbon adipic acid (hexanedioic acid) with the six-carbon hexamethylenediamine (hexane-1,6-diamine) at 280 °C:

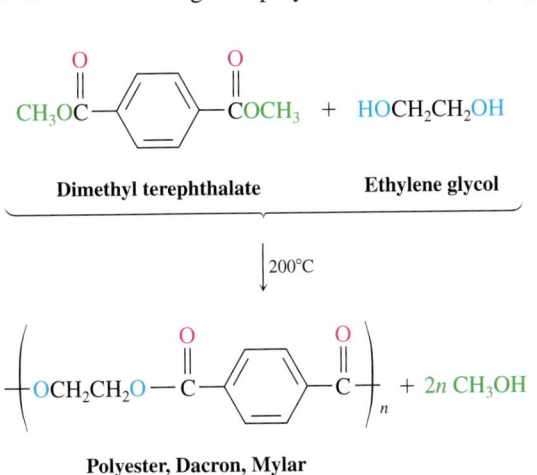

Nylon 66

Nylons are used both in engineering applications and in making fibres. A combination of high-impact strength and abrasion resistance makes nylon an excellent metal substitute for bearings and gears. As fibres, nylon is used in a variety of applications, from clothing to tire cord to mountaineering ropes.

Just as a **polyamide** is made by reaction between a diacid and a diamine, a **polyester** is made by reaction between a dicarboxylic acid or its derivative and a diol. The most generally useful polyester is polyethylene terephthalate (PET), made by a nucleophilic acyl substitution reaction between a diester (dimethyl terephthalate) and a diol (ethylene glycol). The product is used under the trade name Dacron to make clothing fibre and tire cord and under the name Mylar to make plastic film, diaphragms of loudspeakers, and layers of space suits. The tensile strength of polyester film is nearly equal to that of steel.

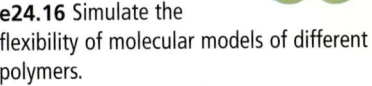

Molecular Modelling (Odyssey)

e24.16 Simulate the flexibility of molecular models of different polymers.

Polyester, Dacron, Mylar

WORKED EXAMPLE 24.10—POLYAMIDES

Draw the structure of Qiana, a polyamide made by high-temperature reaction of hexanedioic acid with cyclohexane-1,4-diamine.

Solution

$$HOC(CH_2)_4COH$$

Hexanedioic acid

+

$$H_2N \quad NH_2$$

Cyclohexane-1,4-diamine

$$\left(HN \quad NH - C(CH_2)_4C \right)_n$$

Qiana

EXERCISE 24.22—POLYAMIDES

Kevlar, a nylon polymer used in bulletproof vests, is made by reaction of benzene-1,4-dicarboxylic acid with benzene-1,4-diamine. Show the structure of Kevlar.

Well-known condensation polymers—polyamides and polyesters—are carboxylic acid derivatives, and undergo polymerization through nucleophilic acyl substitution reactions.

24.8 α-Carbon Atoms: A Second Reactive Site in Carbonyl Compounds

Taking It Further

e24.17 Read about additional reactions of carboxylic acids to produce alcohols and amides.

In our treatment of carbonyl compounds, we have focused so far on reactions that involve nucleophiles adding to the carbonyl carbon atoms of aldehyde and ketone molecules [<<Chapter 23] and substituting for a leaving group attached to the carbonyl carbon atom of carboxylic acid derivative molecules (this chapter).

A second reactive site besides the carbonyl carbon atom in aldehydes, ketones, and carboxylic acid derivatives is the carbon atom adjacent to the carbonyl carbon. This carbon atom is referred to as the α-carbon atom and can be made into a nucleophilic site by removing hydrogen atoms attached to it, which are referred to as α-hydrogen atoms. Chemistry that takes place at the α-carbon atom can be used in synthesis both in the laboratory and living systems to make carbon–carbon bonds in *alkylation* and *acylation* reactions.

We start exploring this chemistry by returning to aldehyde and ketone molecules and looking at experimental evidence that the hydrogen atoms on a carbon atom that is α to (between) two carbonyl groups in pentane-2,4-dione are quite unusual.

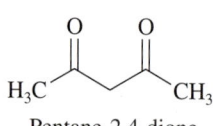
Pentane-2,4-dione

How many peaks would you predict in its ^{13}C NMR spectrum? Its ^1H NMR spectrum? In the Lewis structure shown, the two —CH$_3$ groups are equivalent to each other by symmetry, as are the two C=O carbon atoms. You would expect the ^{13}C spectrum to show three absorption peaks for the unique carbon atoms, one downfield at about 200 δ where carbonyl carbon atoms absorb, and two much further upfield, where saturated carbon atoms absorb. The ^1H NMR spectrum should show only two absorption peaks: one for the —CH$_2$ hydrogen atoms and one for the two equivalent —CH$_3$ hydrogen atoms.

But that's not what the experimental spectra (Figure 24.12) show! Twice the number of peaks as expected, in each case, are apparent: six in the ^{13}C NMR spectrum and four in the ^1H NMR spectrum.

If impurities in the sample are ruled out, the most reasonable explanation for the unexpected appearance of exactly twice as many NMR signals as expected is that molecules of pentane-2,4-dione can exist in two different forms, both of which are found in solution in

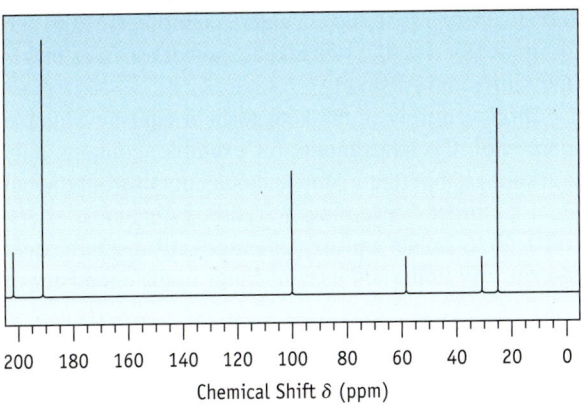

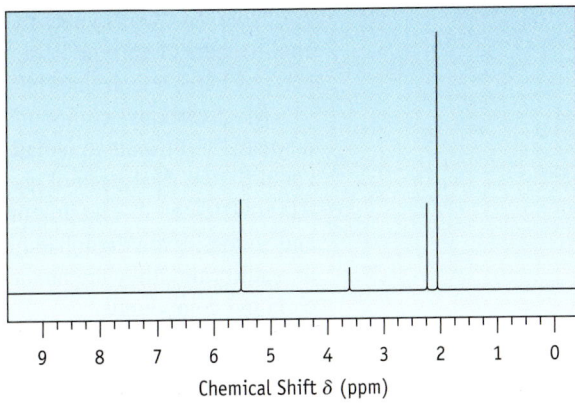

FIGURE 24.12 NMR spectra of pentane-2,4-dione, (a) ^{13}C NMR spectrum showing six peaks, and (b) ^{1}H NMR spectrum showing four peaks absorptions.

an NMR tube. Let's see if we can explain this by referring back to acid-base concepts, and then return to analyze the NMR spectra.

A carbonyl compound whose molecules have a hydrogen atom on the carbon atom right next to the carbonyl group (its alpha- [*α*-]carbon atom) is more acidic than an unsubstituted alkane. Ketones have pK_a values of about 20, and esters about 24. While they are still very weak acids, their acid dissociation constants are 35–40 orders of magnitude greater than an alkane like ethane, whose pK_a is about 60. This difference in acidity becomes even more dramatic when a hydrogen atom is on a carbon atom that is between two carbonyl carbon atoms (alpha to both) in a molecule. The pK_a of pentane-2,4-dione is about 9!

Molecular Modelling (Odyssey)

e24.18 Use electrostatic potential maps to explain differences in acid strength of ketones.

Acetone
(pK_a = 19.3)

Ethane
(pK_a ≈ 60)

Keto–Enol Tautomerism

This NMR spectral evidence is consistent with other observations that show that as a result of this unusual acidity of *α*-hydrogen atoms, ketones, esters, and other carbonyl compounds rapidly interconvert between two isomers. A chemical reaction takes place in which this *α*-hydrogen atom of a carbonyl-containing molecule is transferred to the oxygen atom of the carbonyl group, with rearrangement of *π* electrons, creating a vinyl alcohol called an *enol* (*ene* + *ol;* unsaturated alcohol). This special kind of isomerism between what are called the *keto* and *enol* forms is called *tautomerism* (Greek *tauto,* meaning "the same," and *meros,* meaning "part"). The individual isomers are called *tautomers.* The interconversion of tautomers involves the movement of atoms in a chemical reaction, so the process must be shown differently than when we visualize resonance form representations of a molecule of the same substance. We use equilibrium arrows to indicate the chemical reaction that interconverts tautomers.

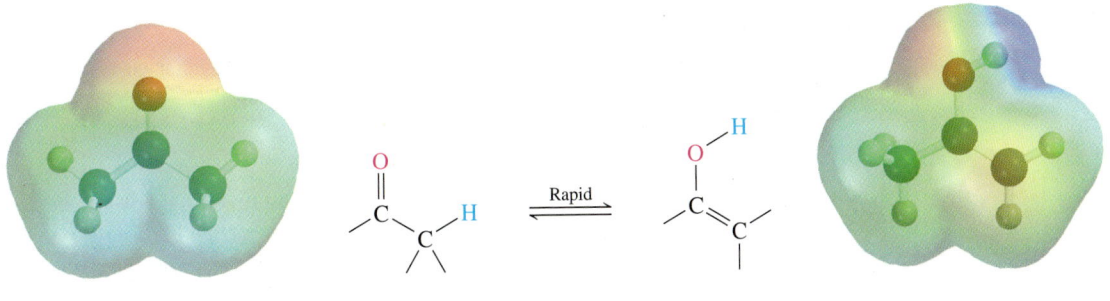

Keto tautomer   **Enol tautomer**

Note that two isomers must interconvert *rapidly* to be considered tautomers. Keto and enol isomers of carbonyl compounds are tautomers, but two alkene isomers such as but-1-ene and but-2-ene are not because they interconvert slowly.

Most carbonyl compounds exist almost entirely in the keto form at equilibrium, and it's usually difficult to isolate the pure enol. Cyclohexanone, for example, contains only about 0.0001% of its enol tautomer at room temperature, and acetone contains only about 0.000 001% enol. The amount of enol tautomer is even less for carboxylic acids, esters, and amides. Even though enols are difficult to isolate and are present to only a small extent at equilibrium, they are nevertheless critically important intermediates in the chemistry of carbonyl compounds.

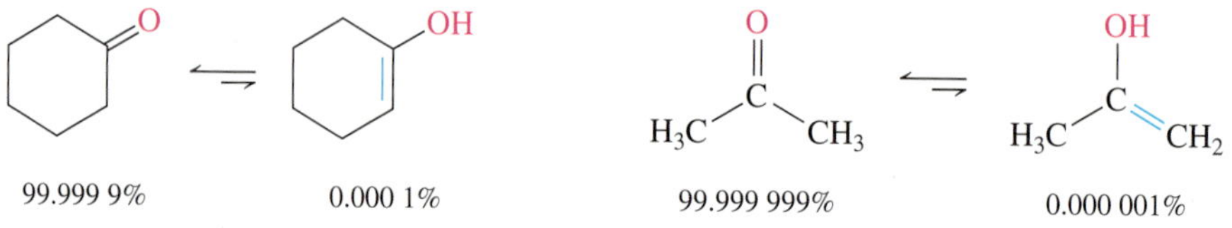

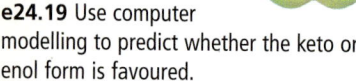

99.999 9% 　　　　　0.000 1% 　　　　　 99.999 999% 　　　　　 0.000 001%

Cyclohexanone 　　　　　　　　　　　　　　　**Acetone**

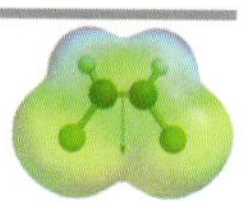

Most aldehydes, ketones, or esters don't have a high enough concentration of the enol form for this isomer to be seen in their NMR spectra. When a hydrogen atom is on a carbon atom that is between two carbonyl groups, however, this equilibrium starts to favour the enol form, and both tautomers are present in sufficient concentrations to be easily detected. In the structure of the enol form of a diketone, the —OH of an enol can form intramolecular hydrogen bonds with the carbonyl oxygen atom in the same molecule (Figure 24.13), stabilizing the enol form. By integrating the peaks in the ^{1}H NMR spectrum it is possible to tell exactly how much of each isomer is present.

With this in mind, let's return to the ^{13}C NMR and ^{1}H NMR spectra of pentane-2,4-dione. Closer analysis of the ^{13}C NMR spectrum shows three peaks with the right chemical shifts for the keto tautomer at 202, 58, and 31 δ. But the two possible enol tautomers give only three, not six absorbances. We've seen this before [<<Section 9.5] as resulting from two equivalent structures that interconvert more rapidly than the time scale for the NMR experiment. For example, the methyl carbon atoms, even though they are different in each of the two enol structures, give only a single time-averaged signal. Analysis of the ^{1}H NMR spectrum is consistent with this same interpretation, only we note that the enol —OH hydrogen atom is not shown on our spectrum. It is in an unusual electronic environment and absorbs even further downfield at 14.5 δ—shown as a broad single peak only if the scale is expanded.

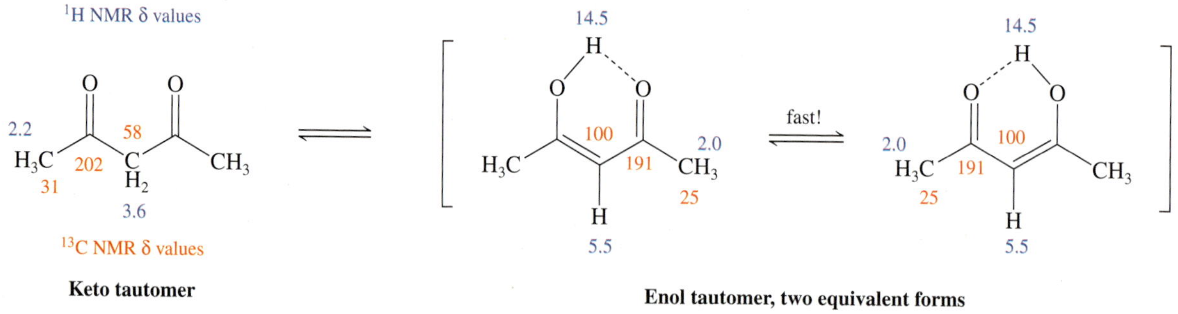

FIGURE 24.13　The keto and two rapidly interconverting enol forms of pentane-2,4-dione, showing the intramolecular hydrogen bonding in enol tautomer. Observed ^{13}C NMR chemical shifts are shown in red and ^{1}H NMR shifts in blue, with their assignments.

Keto–enol tautomerism of carbonyl compounds is catalyzed by both acids and bases. The proposed mechanism for acid catalysis involves protonation of the carbonyl oxygen atom (a Lewis base) to give an intermediate cation that then loses H$^+$ from the α-carbon atom to yield the enol (Figure 24.14). This proton loss from the positively charged intermediate is analogous to what is proposed for an E1 mechanism when a carbocation loses an H^+ from the neighbouring carbon atom to form an alkene [<<Section 21.10].

Spectroscopic and other evidence shows that molecules containing carbonyl groups with a hydrogen on the alpha- [α-]carbon next to a carbonyl group are more acidic than other alkanes, and equilibrate between two isomeric forms: keto and enol tautomers. This reaction can be catalyzed by acid or base.

Think about It

e24.20 Watch an animation of tautomerization in acidic conditions.

Protonation of the carbonyl oxygen atom by an acid catalyst HA yields a cation that can be represented by two resonance structures.

Keto tautomer

Loss of H^+ from the α position by reaction with a base A^- then yields the enol tautomer and regenerates HA catalyst.

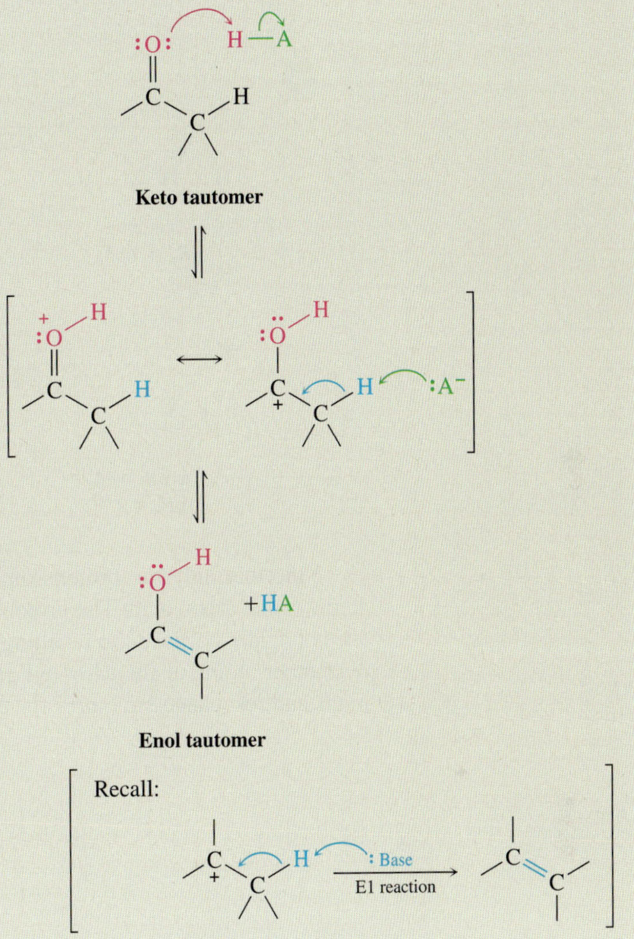

$+HA$

Enol tautomer

Recall:

$\xrightarrow[\text{E1 reaction}]{: \text{Base}}$

FIGURE 24.14   **Accepted reaction mechanism** for acid-catalyzed enol formation.

24.9 **The Enolate Anion: A Carbon Nucleophile**

We've seen that keto–enol tautomerism takes place and can be catalyzed by acids. What happens under basic conditions? Since carbonyl compounds are weak acids, a strong base should be able to abstract an α proton from a molecule of a carbonyl compound to produce an **enolate anion** [<<Section 23.11].

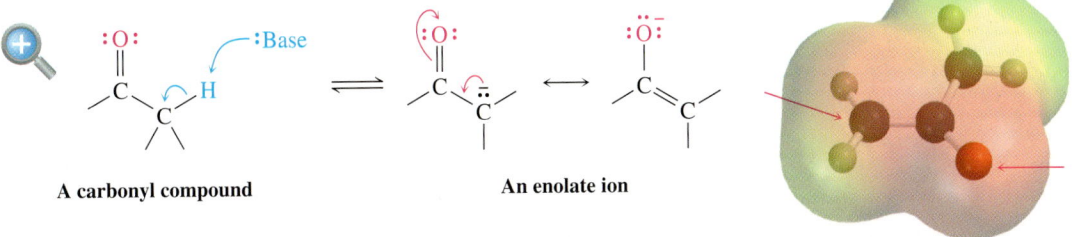

A carbonyl compound   **An enolate ion**

As seen in Section 24.4, both enthalpy changes between acids and their conjugate bases and entropy changes due to their interaction with water solvent molecules must be considered for a complete explanation.

Carbonyl compounds are more acidic than alkanes for the same reason that carboxylic acids are more acidic than alcohols [<<Section 24.4]. In both cases, the anions are stabilized because the negative charge can be spread out over a larger molecular volume, over several atoms of the same molecule. Focusing only on the enthalpy term, electrostatic potential maps and resonance structures show that in the case of the enolate anion, the negative charge can be distributed between the α-carbon atom and the electronegative oxygen atom of the same ion. In the acetate ion charge is delocalized primarily over two oxygen atoms.

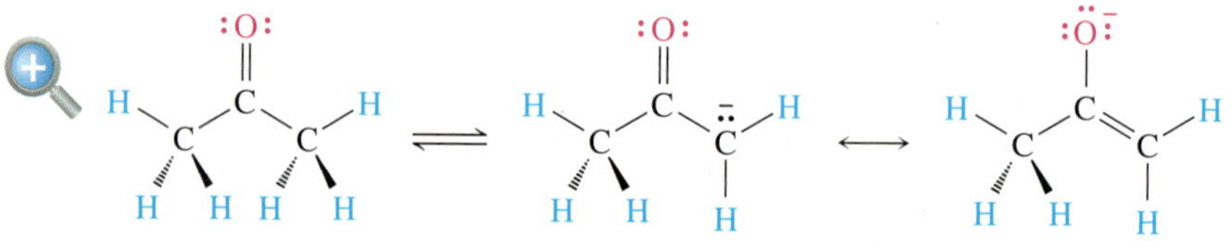

Acetone
(pK$_a$ = 19.3)

Non-equivalent resonance forms

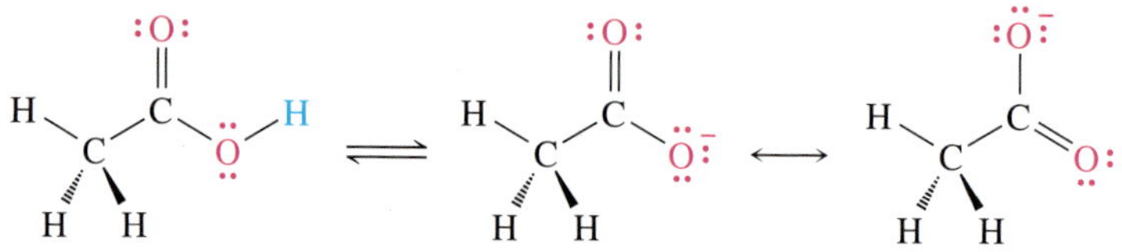

Acetic acid
(pK$_a$ = 4.7)

Equivalent resonance forms

Note that only the protons on the α position of carbonyl compounds are unusually acidic for a carbon acid. The protons at beta (β), gamma (γ), delta (δ), and other positions are not acidic because the resulting anions cannot have their negative charge delocalized over other atoms of the carbonyl group. This is evident from their electrostatic potential maps and resonance structures for their conjugate bases.

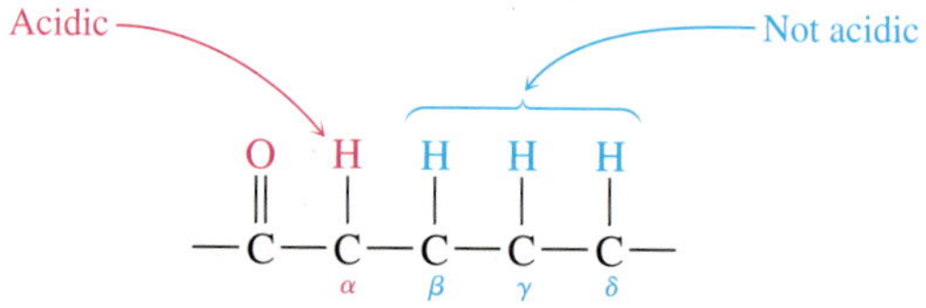

Acidic — Not acidic

Because α-hydrogen atoms of carbonyl compounds are only weakly acidic, strong bases are needed to form enolate ions. If an alkoxide ion, such as sodium ethoxide is used, ionization of acetone takes place only to the extent of about 0.1% because acetone (pK$_a$ = 19.3) is a weaker acid than ethanol (pK$_a$ = 16). If, however, a more powerful base such as sodium amide (NaNH$_2$, the sodium salt of ammonia) or sodium hydride (NaH, the sodium salt of H$_2$) is used, then a carbonyl compound is completely converted into its enolate ion.

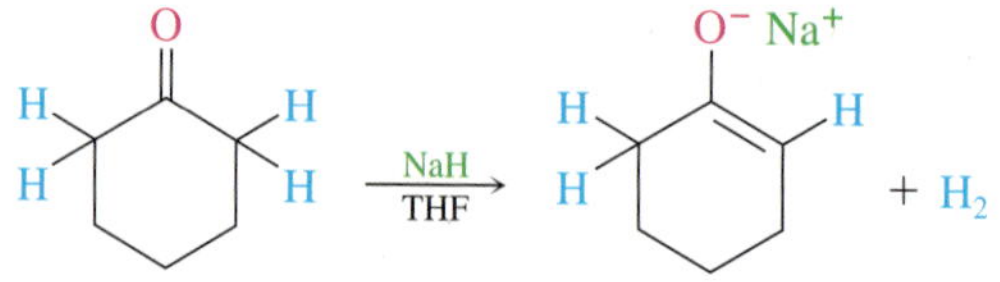

Cyclohexanone

Cyclohexanone enolate ion (100%)

$\xrightarrow[\text{THF}]{\text{NaH}}$ $+ \text{H}_2$

Table 24.4 lists the approximate pK$_a$ values of different kinds of carbonyl compounds and shows how these values compare with other common acids.

TABLE 24.4 Acid Dissociation Constants for Carbonyl-Containing and Other Reference Compounds. A more complete listing is found in Appendix A.

Compound Type	Example	pKₐ	
Carboxylic acid	CH_3COOH	5	
1,3-Diketone	$CH_2(COCH_3)_2$	9	Stronger acid
3-Keto ester	$CH_3COCH_2COOCH_3$	11	
1,3-Diester	$CH_2(COOCH_3)_2$	13	
Water	HOH	15.74	
Primary alcohol	CH_3CH_2OH	16	
Acyl chloride	CH_3COCl	16	
Aldehyde	CH_2CHO	17	
Ketone	CH_3COCH_3	19	
Ester	CH_3COOCH_3	25	
Nitrile	CH_3CN	25	
Dialkylamide	$CH_3CON(CH_3)_2$	30	Weaker acid
Ammonia	NH_3	35	

As we've seen in the example of pentane-2,4-dione, when a C—H bond is flanked by *two* carbonyl groups, its acidity is enhanced even more. Table 24.4 shows that 1,3-diketones (called *β-diketones*), 3-keto esters (*β-keto esters*), and 1,3-diesters are more acidic than water. The enolate ions derived from these β-dicarbonyl compounds are stabilized by sharing of the negative charge onto *two* neighbouring carbonyl oxygen atoms. For example, there are three resonance forms for the enolate ion from pentane-2,4-dione, showing the delocalization of negative charge:

Molecular Modelling (Odyssey)
e24.21 Use electrostatic potential maps to compare the stability of the dione and the enolate.
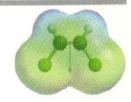

Pentane-2,4-dione (pKₐ = 9)

Base

Pentane-2,4-dione also has the historical (trivial) name acetylacetone, and the ease with which it forms an enolate ion is exploited in coordination chemistry [>>Section 27.4]. Its conjugate base, the acetylacetonate ion (commonly called acac⁻), is used as a ligand (Lewis base) to coordinate to transition metals. One example is Mn(acac)₃, where three acac ligands bind to a Mn³⁺ ion, with each acac⁻ ion binding in two places at the enolate ion oxygen atoms. Note that Mn(acac)₃ is a chiral molecule.

Our molecular-level picture of enolate ion formation shows the base abstracting a proton from the α-carbon atom of a carbonyl-containing molecule to form a resonance-stabilized enolate ion, which is then protonated to yield a neutral molecule. If protonation of the enolate ion takes place on the α-carbon atom, the keto tautomer is regenerated and no net change occurs. If, however, protonation takes place on the oxygen atom, then an enol tautomer is formed (Figure 24.15).

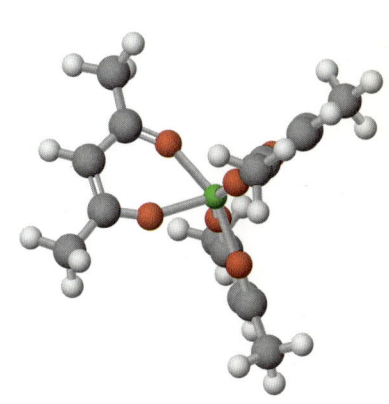

Think about It

e24.22 Watch an animation of tautomerization in basic conditions.

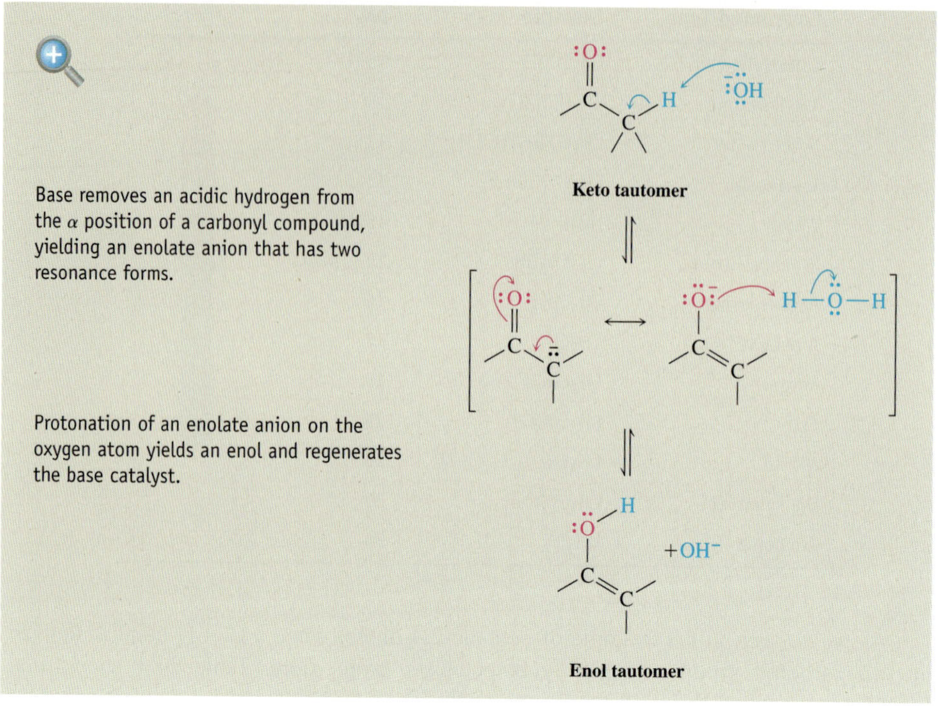

Base removes an acidic hydrogen from the α position of a carbonyl compound, yielding an enolate anion that has two resonance forms.

Protonation of an enolate anion on the oxygen atom yields an enol and regenerates the base catalyst.

Keto tautomer

Enol tautomer

FIGURE 24.15 **Accepted reaction mechanism** for base-catalyzed enol formation.

WORKED EXAMPLE 24.11—ENOLATE ANIONS

Draw structures of the two enolate ions you could obtain by deprotonation of 3-methylcyclohexanone.

Solution

Locate the acidic hydrogen atoms, and then remove them one at a time to generate the possible enolate ions. In this case, 3-methylcyclohexanone can be deprotonated either at C2 or at C6.

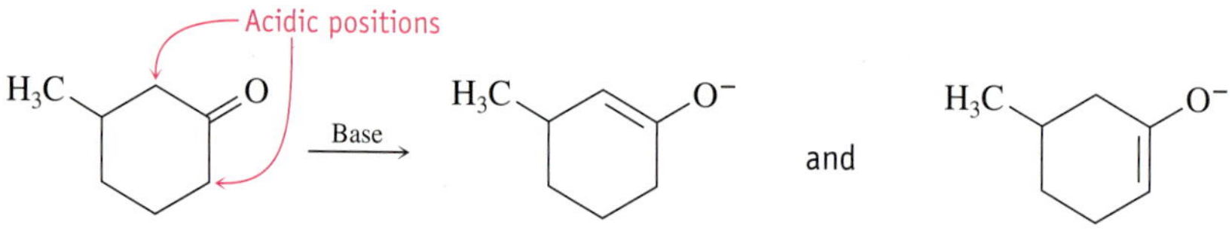

EXERCISE 24.23—ACIDIC HYDROGEN ATOMS

Identify all acidic hydrogen atoms in the following molecules:
(a) CH_3CH_2CHO
(b) $(CH_3)_3CCOCH_3$
(c) CH_3COOH
(d) 1,3-cyclohexanedione

EXERCISE 24.24—ENOLATE ANIONS

Show the enolate ions you would obtain by deprotonation of the following carbonyl compounds. Include any important resonance structures:

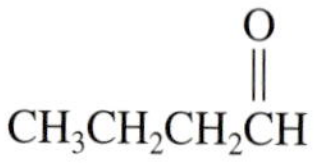

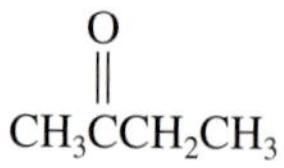

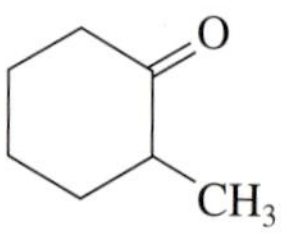

EXERCISE 24.25—ENOLATE ANIONS

Draw three resonance forms for the most stable enolate ion you would obtain by deprotonation of methyl 3-oxobutanoate.

$$CH_3CCH_2COCH_3$$

Methyl 3-oxobutanoate

24.10 Reactions of Enolate Ions: Alkylation

What kind of chemistry would we expect from compounds containing an enolate ion? Because enolate ions are the conjugate bases of weak acids, we would predict that they would be good Lewis bases or nucleophiles, and should react with electrophiles. Using the resonance hybrid model, enolate ions can be thought of either as α-keto carbanions ($-C-C=O$) or as vinylic alkoxides ($C=C-O-$). Thus, enolate ions can react with electrophiles either on oxygen or on carbon atoms. Reaction on an oxygen atom yields an enol derivative, while reaction on a carbon atom yields an α-substituted carbonyl compound (Figure 24.16). Both kinds of reactivity are known, but reaction on the α-carbon atom is more common, and synthetically very useful.

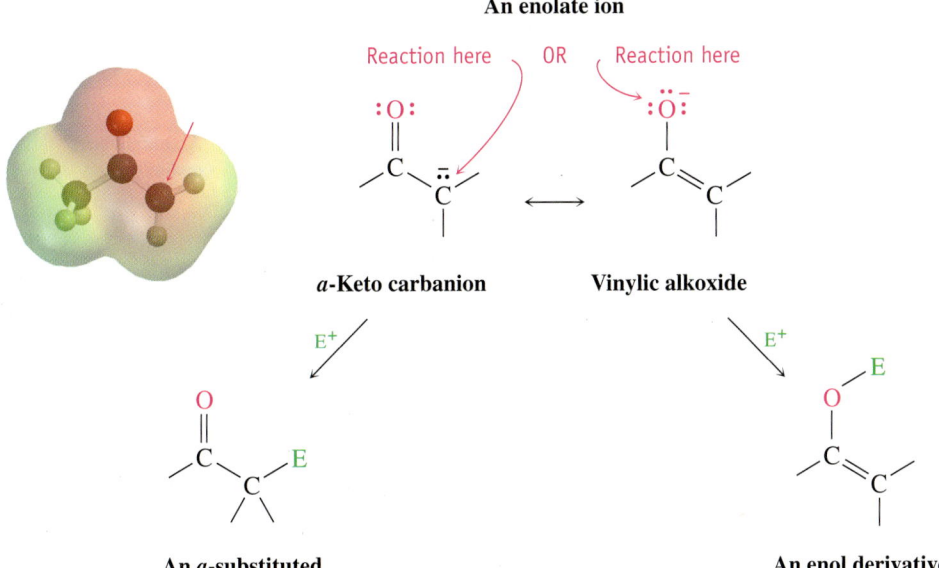

An enolate ion

a-Keto carbanion

Vinylic alkoxide

An *a*-substituted carbonyl compound

An enol derivative

FIGURE 24.16 Two modes of enolate ion reactivity. Reaction on the α-carbon atom is more common, and yields an α-substituted carbonyl compound.

While many reactions of enolate ions as good nucleophiles are possible, we will focus on one of the most important—their *alkylation* by reaction with an alkyl halide. The reaction is an essential feature in the toolbox of an organic chemist, as it forms a new C—C bond, joining two smaller pieces into one larger molecule. Alkylation occurs when the nucleophilic enolate ion reacts with an electrophilic alkyl halide in an S_N2 reaction, displacing the halide ion in the usual way.

Enolate ion   **Alkyl halide** S_N2 reaction

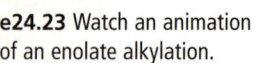

Think about It

e24.23 Watch an animation
of an enolate alkylation.

Like all S_N2 reactions [<<Section 21.6], alkylations are successful only when a primary alkyl halide (RCH_2—) or methyl halide (CH_3—) is used, because a competing E2 elimination occurs if a secondary or tertiary halide is used. The leaving group X can be chloride, bromide, or iodide.

One of the best-known carbonyl alkylation reactions, a process called the *malonic ester synthesis*, is an excellent method for synthesizing a substituted acetic acid in several steps from an alkyl halide:

$$R-X \xrightarrow[\text{ester synthesis}]{\text{Via malonic}} R-CH_2\overset{\overset{\displaystyle O}{\|}}{C}OH$$

Alkyl halide   **α-Substituted**
                           acetic acid

Diethyl propanedioate, commonly called diethyl malonate or *malonic ester*, is relatively acidic ($pK_a = 13$) because its α-hydrogen atoms are flanked by two carbonyl groups. Malonic ester is easily converted into its enolate ion by reaction with sodium ethoxide in ethanol. The enolate ion, in turn, is readily alkylated by treatment with an alkyl halide, yielding an α-substituted malonic ester. In the following examples, the abbreviation "Et" is used for an ethyl group ($-CH_2CH_3$).

$$\underset{\substack{\text{Diethyl propanedioate}\\ \text{(malonic ester)}}}{H-\overset{\overset{\displaystyle CO_2Et}{|}}{\underset{\underset{\displaystyle H}{|}}{C}}-CO_2Et} \xrightarrow[\text{EtOH}]{\text{Na}^+ \ ^-\text{OEt}} \left[\underset{\substack{\text{Sodio malonic ester}}}{Na^+ \ ^-\!:\overset{\overset{\displaystyle CO_2Et}{|}}{\underset{\underset{\displaystyle H}{|}}{C}}-CO_2Et} \right] \xrightarrow{\text{RX}} \underset{\substack{\text{An alkylated}\\ \text{malonic ester}}}{R-\overset{\overset{\displaystyle CO_2Et}{|}}{\underset{\underset{\displaystyle H}{|}}{C}}-CO_2Et}$$

The product of a malonic ester alkylation has molecules with one acidic α-hydrogen atom remaining, so the alkylation process can be repeated a second time to yield a dialkylated malonic ester:

$$\underset{\substack{\text{An alkylated}\\ \text{malonic ester}}}{R-\overset{\overset{\displaystyle CO_2Et}{|}}{\underset{\underset{\displaystyle H}{|}}{C}}-CO_2Et} \xrightarrow[\text{2. R'X}]{\text{1. Na}^+ \ ^-\text{OEt}} \underset{\substack{\text{A dialkylated}\\ \text{malonic ester}}}{R-\overset{\overset{\displaystyle CO_2Et}{|}}{\underset{\underset{\displaystyle R'}{|}}{C}}-CO_2Et}$$

On heating with hydrochloric acid solution, the alkylated (or dialkylated) malonic ester undergoes hydrolysis of the ester and *decarboxylation* (loss of a very stable CO_2 molecule) to yield a substituted monocarboxylic acid. Decarboxylation is not a general reaction of carboxylic acids but is a unique feature of compounds like malonic acids that have a second carbonyl group two atoms away from the —COOH.

$$\underset{}{R-\overset{\overset{\displaystyle CO_2Et}{|}}{\underset{\underset{\displaystyle H}{|}}{C}}-CO_2Et} \xrightarrow[\text{Heat}]{\text{H}_3\text{O}^+} R-\overset{\overset{\displaystyle H}{|}}{\underset{\underset{\displaystyle H}{|}}{C}}-COOH \ + \ CO_2 \ + \ 2 \ EtOH$$

The overall result of the malonic ester synthesis is to convert an alkyl halide into a carboxylic acid and to lengthen the carbon chain by two atoms ($RX \rightarrow RCH_2COOH$).

$$CH_3CH_2CH_2CH_2Br + Na^+ \; ^-:CH(CO_2Et)_2 \longrightarrow CH_3CH_2CH_2CH_2CH(CO_2Et)_2$$

1-Bromobutane   **Sodio diethylmalonate**   **(84%)**

$$\downarrow H_3O^+$$

$$\overset{\overset{\textstyle O}{\textstyle \|}}{CH_3CH_2CH_2CH_2CH_2COH}$$

Hexanoic acid (75%)

WORKED EXAMPLE 24.12—MALONIC ESTER SYNTHESIS

How would you make heptanoic acid by a malonic ester synthesis?

Strategy

The malonic ester synthesis converts an alkyl halide molecule into a carboxylic acid with two more carbon atoms. Thus, a seven-carbon acid chain must be derived from a five-carbon alkyl halide such as 1-bromopentane.

Solution

$$CH_3CH_2CH_2CH_2CH_2Br + CH_2(CO_2Et)_2 \xrightarrow[\text{2. H}_3\text{O}^+,\text{ heat}]{\text{1. Na}^+ \text{ }^-\text{OEt}} \overset{\overset{\textstyle O}{\textstyle \|}}{CH_3CH_2CH_2CH_2CH_2CH_2COH}$$

EXERCISE 24.26—MALONIC ESTER SYNTHESIS

What alkyl halide would you use to make the following compounds by a malonic ester synthesis?

(a)

$$\overset{\overset{\textstyle O}{\textstyle \|}}{CH_3CH_2CH_2COH}$$

(b)

$$\overset{\overset{\textstyle O}{\textstyle \|}}{CH_2CH_2COH}$$ (attached to a benzene ring)

(c)

$$\overset{\overset{\textstyle CH_3}{\textstyle |}}{CH_3CHCH_2CH_2CH_2}\overset{\overset{\textstyle O}{\textstyle \|}}{COH}$$

Alpha-hydrogen atoms in carbonyl compounds are acidic and can be removed by bases to yield enolate ions. The most important reaction of enolate ions is their S_N2 alkylation by reaction with alkyl halides. The mechanism is thought to involve nucleophilic enolate ion reaction with an alkyl halide, displacing the leaving halide group and yielding an α-alkylated product. The malonic ester synthesis, which involves alkylation of diethyl malonate with an alkyl halide, is a good method for preparing a monoalkylated or dialkylated acetic acid.

Enolate Alkylation and the Dawn of Modern Medicine

While herbal remedies have been used to treat illness and disease for thousands of years, the prescription of substances prepared in the laboratory has a much shorter history. The barbiturates, a class of potent sedatives, constitute an early example of medicinal chemistry. The synthesis and medical use of these compounds traces back to the early 1900s and relies on reactions that are now quite familiar—enolate alkylations and the chemistry of esters. Starting with diethyl malonate, enolate alkylations with simple alkyl halides provide a wealth of potential derivatives (Figure 24.17). Conversion to the barbiturate occurs on heating a diester with urea in the presence of a base. While the —NH_2 groups of urea molecules are not strong nucleophiles, deprotonation generates a more reactive anion, —NH^-.

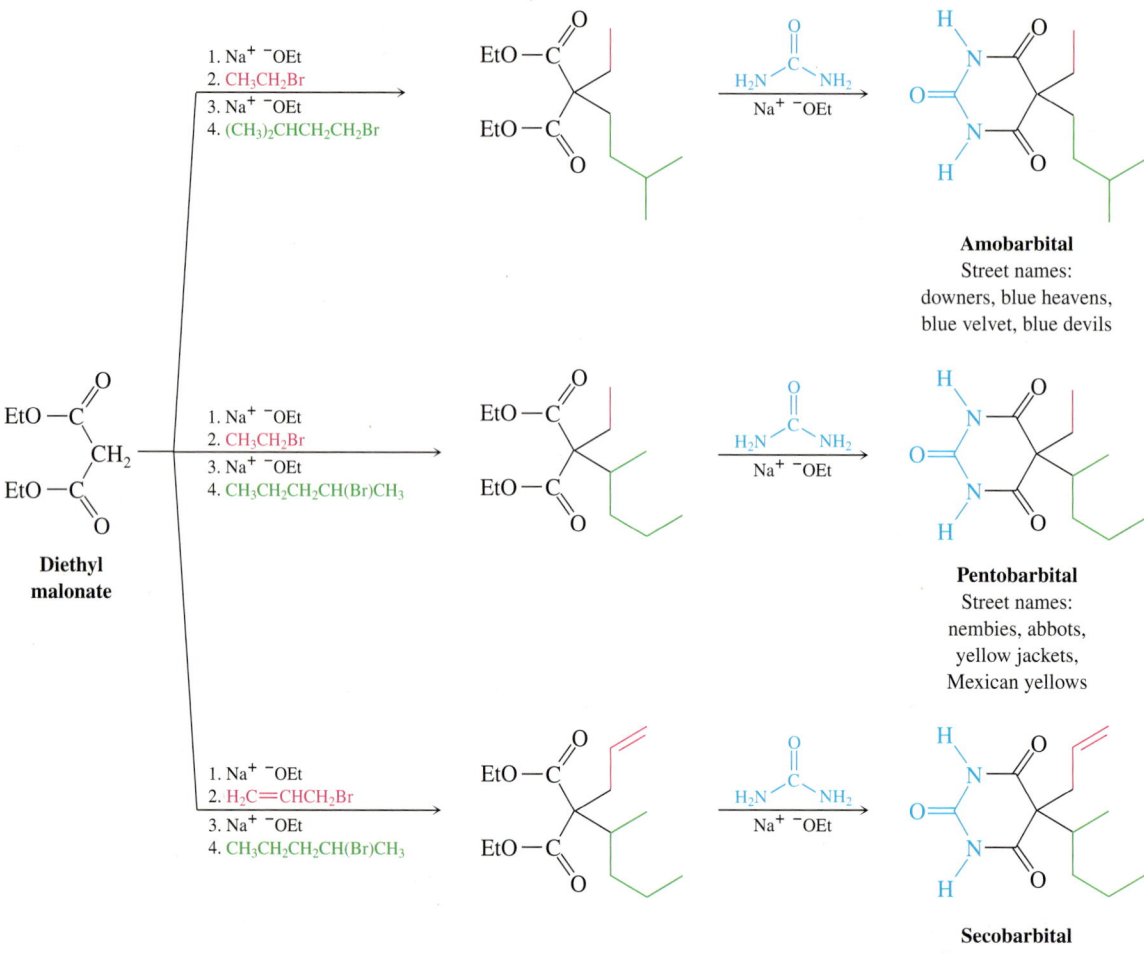

FIGURE 24.17   The synthesis of barbiturates relies on enolate alkylation and nucleophilic acyl substitution reactions of carbonyl compounds.

Each barbiturate comes as a tablet of regulated size, shape, and colour. The street names used when the drugs are trafficked illegally are equally colourful and are often derived from the colour of the pill. Although still used today, most of these drugs have been replaced with a safer class of sedative-hypnotics, including diazepam (Valium), compounds with markedly different structures.

24.11 Carbonyl Condensation Reactions

We've seen now that carbonyl compounds can behave as either electrophiles or nucleophiles. Our molecular-level picture shows that in both nucleophilic addition reactions and nucleophilic acyl substitution reactions, the carbonyl group behaves as an electrophile by accepting electrons from a nucleophile that serves as a reaction partner. In an α-substitution reaction, however, the carbonyl compound behaves as a nucleophile after conversion into an enolate ion.

Electrophilic carbonyl group reacts with nucleophiles.

Nucleophilic enolate ion reacts with electrophiles.

Carbonyl condensation reactions, in which two molecules combine to form one larger one, involve *both* kinds of reactivity. These reactions take place between two carbonyl partners and involve a combination of nucleophilic addition or substitution and α-substitution steps. One partner (the nucleophilic *donor*) is converted into an enolate ion and undergoes an α-substitution reaction, while the other partner (the electrophilic *acceptor*) undergoes a nucleophilic addition or substitution reaction. There are numerous variations of carbonyl condensation reactions, depending on the two carbonyl partners, but the general mechanism remains the same (Figure 24.18).

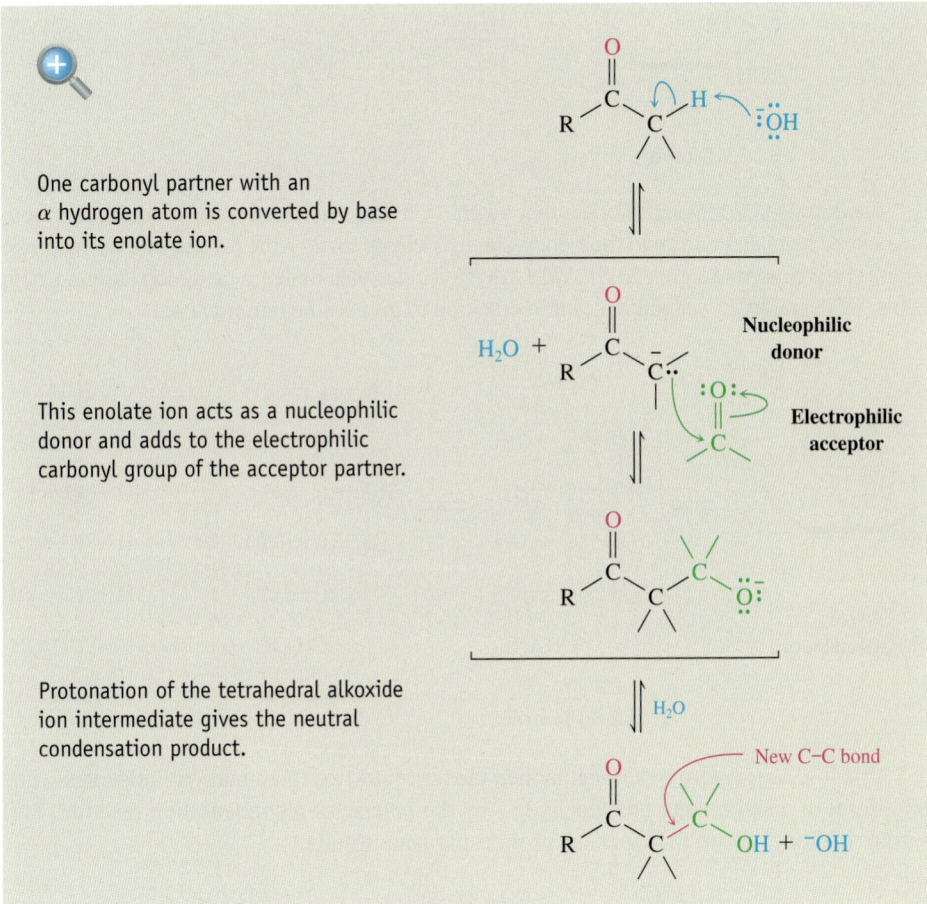

One carbonyl partner with an α hydrogen atom is converted by base into its enolate ion.

This enolate ion acts as a nucleophilic donor and adds to the electrophilic carbonyl group of the acceptor partner.

Protonation of the tetrahedral alkoxide ion intermediate gives the neutral condensation product.

FIGURE 24.18 **Accepted general mechanism** of a carbonyl condensation reaction. Our molecular-level view suggests that one partner (the donor) acts as a nucleophile, while the other partner (the acceptor) acts as an electrophile. In this case, since the two carbonyl compounds are ketones, nucleophilic addition of the enolate ion takes place. If the carbonyl compound were a carboxylic acid derivative, nucleophilic substitution would occur.

Two examples of the synthetic importance of carbonyl condensation reactions are the Aldol and Claisen Condensations. For illustrations and the mechanisms of these reactions, see e24.24 and e24.25.

A carbonyl condensation reaction takes place between two carbonyl components and the mechanism is thought to involve a combination of nucleophilic addition and α-substitution steps. One carbonyl partner (the donor) is converted into its enolate ion, which then adds to the carbonyl group of the second partner (the acceptor).

Think about It

e24.24 Watch an animation of the Aldol Condensation.

Think about It

e24.25 Watch an animation of the Claisen Condensation.

24.12 Biological Carbonyl Reactions

While carbonyl condensation reactions are also enormously useful in the laboratory to make C—C bonds as key parts of synthetic sequences of larger molecules, we will focus on an example of their importance in living systems. *Much of biochemistry is carbonyl chemistry.*

Almost all metabolic processes used by living organisms involve one or more of the carbonyl group reactions we've seen in this chapter and Chapter 23. The digestion and metabolic breakdown of all major classes of food molecules—fats, carbohydrates, and proteins—take place by nucleophilic addition reactions, nucleophilic acyl substitutions, α-substitutions, and carbonyl condensations. Similarly, hormones and other crucial biological molecules are built up from smaller precursors by these same carbonyl group reactions.

Take *glycolysis,* for example—the metabolic pathway by which organisms convert glucose to pyruvate as the first step in extracting energy from carbohydrates:

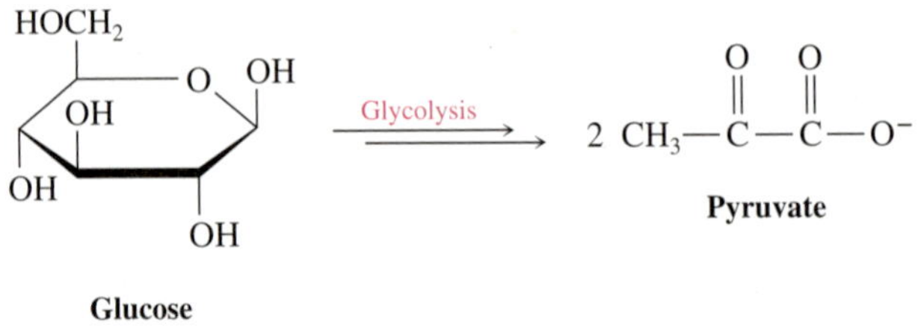

Glycolysis is a ten-step process that begins with conversion of glucose molecules from their cyclic hemiacetal form to their open-chain aldehyde form— the reverse of a nucleophilic addition reaction [<<Section 23.5]. Aldehyde molecules then undergo tautomerization to yield enols, which undergo yet another tautomerization to give the ketone fructose.

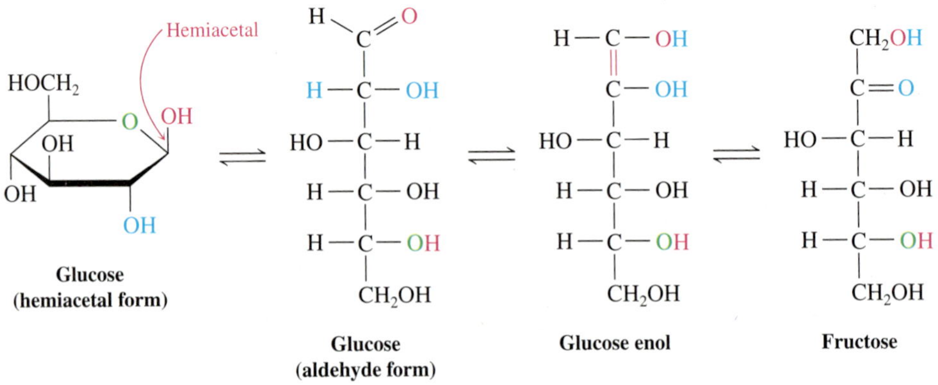

Fructose, a β-hydroxy ketone, is then cleaved into two three-carbon molecules—one enol form of a ketone and one aldehyde—by the reverse of a condensation reaction. Still further carbonyl group reactions occur until pyruvate results.

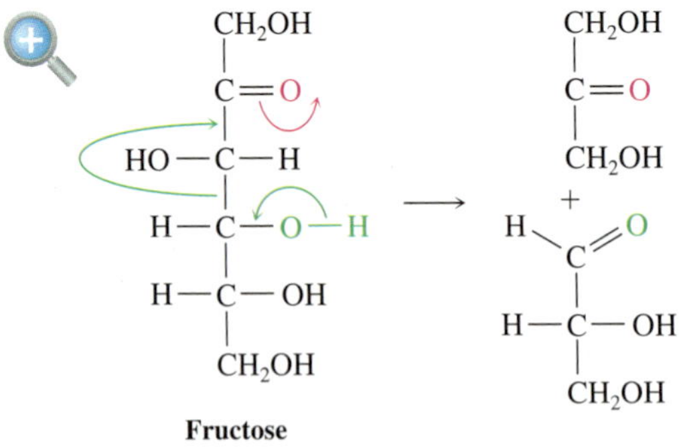

Another example of a biological carbonyl reaction, nature uses the two-carbon acetate fragment of acetyl CoA as the major building block for synthesis. Acetyl CoA can act not only as an electrophilic acceptor, reacting with nucleophiles at the carbonyl group, but also as a nucleophilic donor by loss of its molecule's acidic α-hydrogen atom. Once formed, the enolate ion of acetyl CoA can add to another carbonyl group in a condensation reaction. For example, citric acid is biosynthesized by nucleophilic addition of acetyl CoA to the ketone carbonyl group of oxaloacetic acid (2-oxobutanedioic acid) molecules in a condensation reaction involving two different carboxylic acid derivatives.

$$CH_3CSCoA \longrightarrow \left[\overset{..}{:}CH_2CSCoA \right] + O=C \longrightarrow HO-C-CH_2COOH$$

Acetyl CoA
(a thioester)

Oxaloacetic acid   **Citric acid**

The few examples just given are only an introduction; we'll look at several of the major metabolic pathways and see many more carbonyl group reactions in Chapter 29. A good grasp of carbonyl chemistry is crucial to understanding biochemistry.

SUMMARY

Key Concepts

- The characteristic reactions of **carboxylic acids** are as weak acids. This can be predicted from looking at electrostatic potential maps. Although weaker than mineral acids like HCl, carboxylic acids are much more acidic than alcohols because of electron delocalization in their conjugate base carboxylate ions. Most carboxylic acids have pK_a values near 5, but the exact acidity of an acid depends on its structure. Electron-withdrawing substituents make carboxylic acids more acidic. (Section 24.4)
- Carboxylic acids can be identified by infrared spectroscopy, through their characteristic C=O and O—H stretching absorptions; by C=O carbon atom chemical shifts at about 200 δ in their ^{13}C NMR spectra; and by carboxyl O—H hydrogen atoms chemical shifts at about 12 δ in their ^1H NMR spectra. (Section 24.3)
- Carboxylic acids can be transformed into a variety of carboxylic acid derivatives in which the acid —OH group has been replaced by another substituent. **Acyl halides**, **acid anhydrides**, **esters**, and **amides** are the most common. (Section 24.2)
- The chemistry of all these carboxylic acid derivatives is similar and is dominated by a single general reaction type: the *nucleophilic acyl substitution reaction*. The accepted mechanism for these substitutions is thought to take place by addition of a nucleophile to the polar carbonyl group of an acid derivative, followed by expulsion of the leaving group. Carboxylic acid derivatives can undergo reaction with many different nucleophiles. Among the most important are reaction with water, alcohols, amines, hydride ions, and Grignard reagents.

where Y = Cl, Br, I (acid halide); OR (ester); OCOR (anhydride); or NH₂ (amide)

Acyl chlorides and anhydrides are the most reactive **carboxylic acid derivatives**, while esters and amides are much less reactive toward nucleophilic substitution. (Sections 24.5–24.7)
- *Enzymes* can be used as laboratory catalysts, providing *chiral environments* for selective reactions of chiral carboxylic acid derivatives. (Section 24.7)
- Well-known *condensation polymers*, **polyamides** and **polyesters**, are carboxylic acid derivatives, and undergo polymerization through nucleophilic acyl substitution reactions. (Section 24.7)

- Spectroscopic and other evidence shows that molecules containing carbonyl groups with a hydrogen on the α-carbon atom next to a carbonyl group are more acidic than other alkanes, and equilibrate between two isomeric forms: *keto and enol tautomers*. This reaction can be catalyzed by acid or base. Enol tautomers are normally present to only a small extent, and pure enols usually cannot be isolated. (Section 24.8)

- α-Hydrogen atoms in carbonyl compounds are acidic and can be abstracted by bases to yield **enolate anions**. Ketones, aldehydes, esters, and amides can all be deprotonated. The most important reaction of enolate ions is their S$_N$2 *alkylation* by reaction with alkyl halides. The mechanism is thought to involve nucleophilic enolate ion reaction with an alkyl halide, displacing the leaving halide group and yielding an α-alkylated product. The *malonic ester synthesis*, which involves alkylation of diethyl malonate with an alkyl halide, is a good method for preparing a monoalkylated or dialkylated acetic acid. (Sections 24.9–24.10)

- A **carbonyl condensation reaction** takes place between two carbonyl components and the mechanism is thought to involve a combination of nucleophilic addition and α-substitution steps. One carbonyl partner (the donor) is converted into its enolate ion, which then adds to the carbonyl group of the second partner (the acceptor). (Section 24.11)

Key Reactions

1. Reactions of carboxylic acids
   (a) Reactions as weak acids (Section 24.4)

      A carboxylic acid   **A carboxylic acid salt**
      (water-insoluble)   **(water-soluble)**

   (b) Conversion into acyl chlorides (Section 24.6)

   (c) Conversion into esters (Fischer esterification) (Section 24.6)

2. Reactions of acyl halides
   (a) Conversion into carboxylic acids (Section 24.7)

   (b) Conversion into esters (Section 24.7)

(c) Conversion into amides (Section 24.7)

$$
\underset{R}{\overset{O}{\parallel}}\!\!C\!\!-\!\!Cl + NH_3 \longrightarrow \underset{R}{\overset{O}{\parallel}}\!\!C\!\!-\!\!NH_2
$$

3 Reactions of acid anhydrides
   (a) Conversion into esters (Section 24.7)

$$
\underset{R}{\overset{O}{\parallel}}\!\!C\!\!-\!\!O\!\!-\!\!\overset{O}{\underset{R}{\parallel}}\!\!C + R'OH \xrightarrow{\text{Pyridine}} \underset{R}{\overset{O}{\parallel}}\!\!C\!\!-\!\!OR'
$$

   (b) Conversion into amides (Section 24.7)

$$
\underset{R}{\overset{O}{\parallel}}\!\!C\!\!-\!\!O\!\!-\!\!\overset{O}{\underset{R}{\parallel}}\!\!C + NH_3 \xrightarrow{\text{Pyridine}} \underset{R}{\overset{O}{\parallel}}\!\!C\!\!-\!\!NH_2
$$

4. Reactions of esters
   (a) Conversion into acids (Section 24.7)

$$
\underset{R}{\overset{O}{\parallel}}\!\!C\!\!-\!\!OR' + H_2O \xrightarrow[\text{NaOH}]{\text{H}^+ \text{ or}} \underset{R}{\overset{O}{\parallel}}\!\!C\!\!-\!\!OH
$$

   (b) Conversion into primary alcohols by reduction (Section 24.7)

$$
\underset{R}{\overset{O}{\parallel}}\!\!C\!\!-\!\!OR' \xrightarrow[\text{2. H}_3\text{O}^+]{\text{1. LiAlH}_4} RCH_2OH
$$

   (c) Conversion into tertiary alcohols by Grignard reaction (Section 24.7)

$$
\underset{R}{\overset{O}{\parallel}}\!\!C\!\!-\!\!OR' \xrightarrow[\text{2. H}_3\text{O}^+]{\text{1. R''MgX}} R\!-\!\underset{R''}{\overset{OH}{\underset{|}{\overset{|}{C}}}}\!-\!R''
$$

5. Reactions of amides
   (a) Conversion into carboxylic acids (Section 24.7)

$$
\underset{R}{\overset{O}{\parallel}}\!\!C\!\!-\!\!NH_2 + H_2O \xrightarrow[\text{NaOH}]{\text{H}^+ \text{ or}} \underset{R}{\overset{O}{\parallel}}\!\!C\!\!-\!\!OH
$$

   (b) Conversion into amines by reduction (Section 24.7)

$$
\underset{R}{\overset{O}{\parallel}}\!\!C\!\!-\!\!NH_2 \xrightarrow[\text{2. H}_2\text{O}]{\text{1. LiAlH}_4} RCH_2NH_2
$$

6. Formation of enolate ions (Section 24.9)

A carbonyl compound   **An enolate ion**

7. Alkylation of enolate anions (Section 24.10)

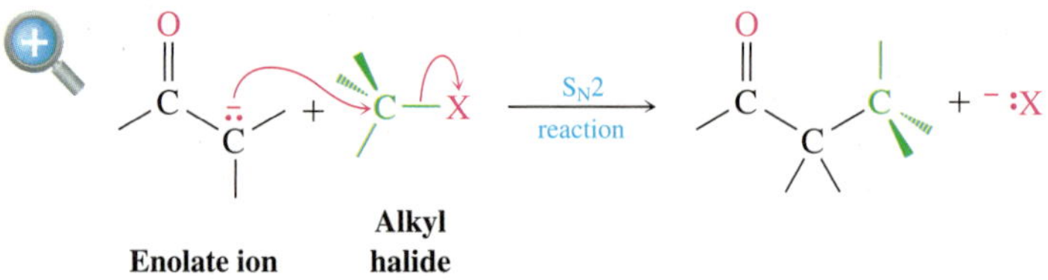

Enolate ion   **Alkyl halide**

8. Malonic ester synthesis (Section 24.10)

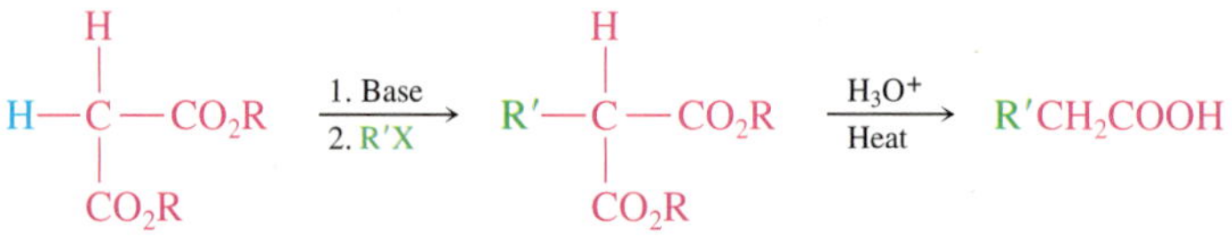

REVIEW QUESTIONS

Section 24.2: Naming Carboxylic Acids and Derivatives

24.27 Draw and name the eight carboxylic acids with formula $C_6H_{12}O_2$. Which are chiral?

24.28 Give IUPAC names for the carboxylic acids whose molecules are shown:

(a)

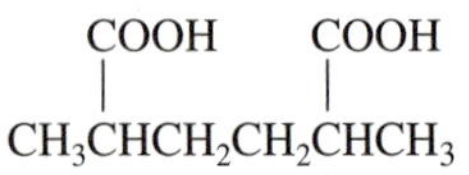

(b) $(CH_3)_3CCOOH$

(c)

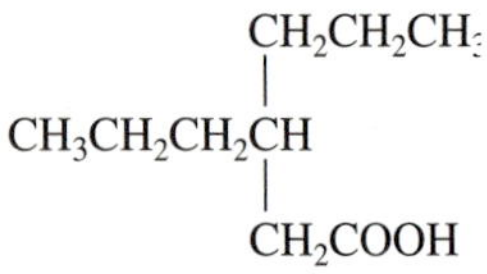

(d)

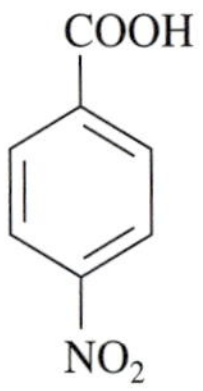

(e)

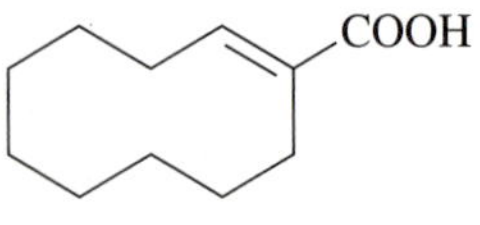

(f) $BrCH_2CH(Br)CH_2CH_2COOH$

24.29 Give IUPAC names for the carboxylic acid derivatives whose molecules are shown:

(a)

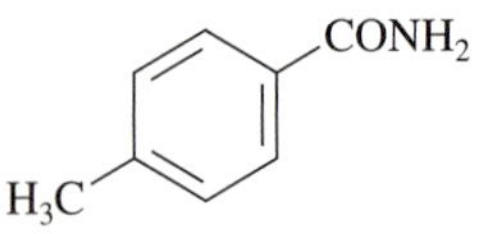

(b) $(CH_3CH_2)_2CHCH=CHCN$

(c) $CH_3O_2CCH_2CH_2CO_2CH_3$

(d)

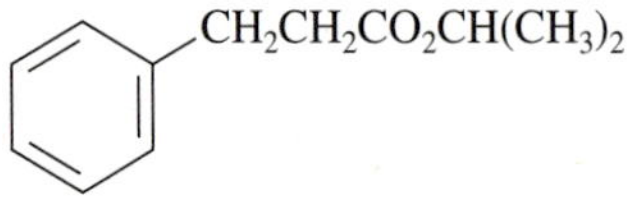

(e)

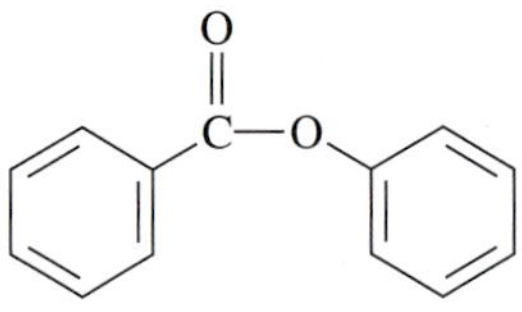

(f) $CH_3CH(Br)CH_2CONHCH_3$

(g)

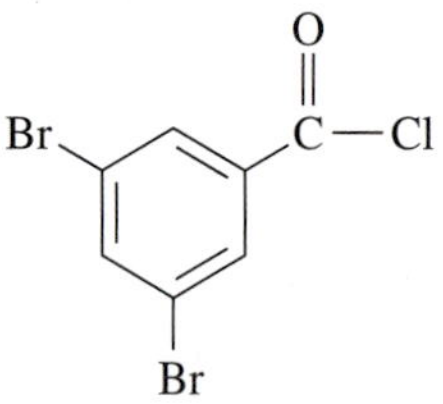

(h)

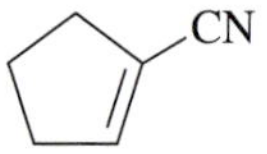

24.30 Draw structures of molecules corresponding to the following IUPAC names:
(a) 4,5-dimethylheptanoic acid
(b) *cis*-cyclohexane-1,2-dicarboxylic acid
(c) heptanedioic acid
(d) triphenylacetic acid
(e) 2,2-dimethylhexanamide
(f) phenylacetamide
(g) dimethyl malonate
(h) ethyl cyclohexanecarboxylate

24.31 Draw structures of their molecules, and then name the following compounds:
(a) three acyl chlorides, C_6H_9ClO
(b) three amides, $C_7H_{11}NO$
(c) three anhydrides, $C_6H_{10}O_3$
(d) three esters, $C_5H_8O_2$

Section 24.3: Spectroscopy of Carboxylic Acids and Derivatives

24.32 Compound A ($C_4H_8O_3$) has infrared absorptions at 1710 and 2500–3100 cm^{-1} and the ^{1}H NMR spectrum shown. Propose a structure for A.

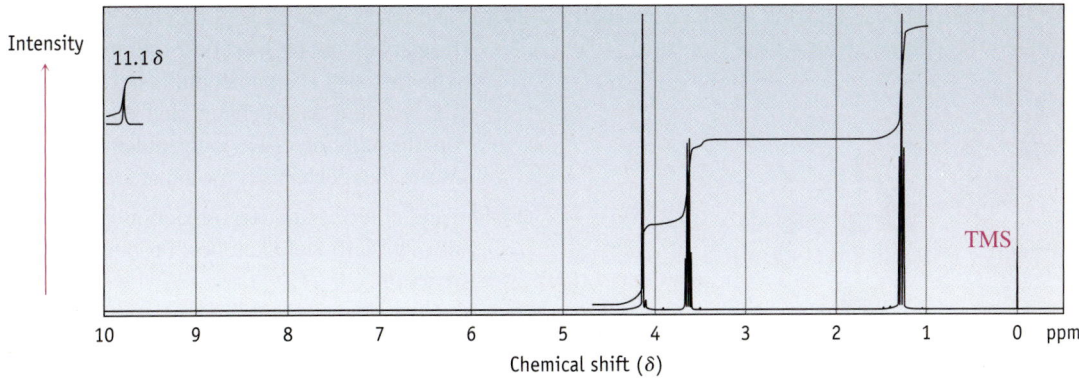

24.33 Propose the structure of molecules of the compounds with the following ^{1}H NMR spectra:
(a) C_4H_7ClO; IR: 1810 cm^{-1}

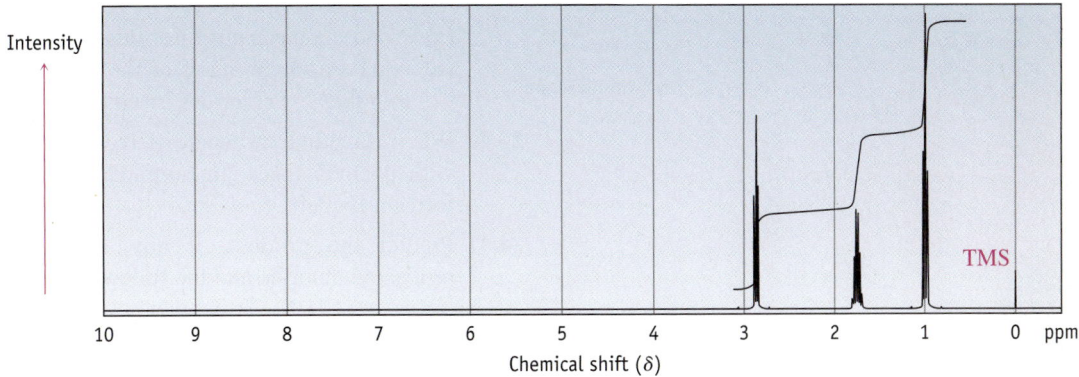

(b) $C_5H_{10}O_2$; IR: 1735 cm^{-1}

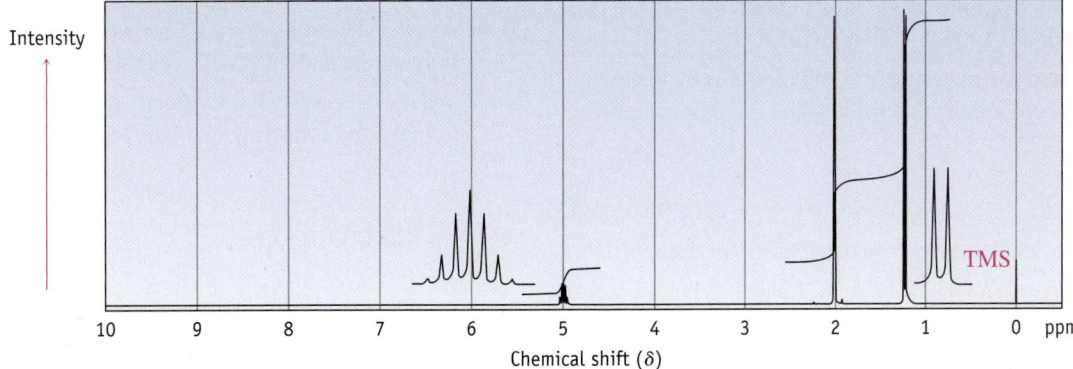

Section 24.4: Electronic Structure and Reactivity: Carboxylic Acids

24.34 Citric acid has $pK_a = 3.14$, and tartaric acid has $pK_a = 2.98$. Which acid is stronger?

24.35 Rank the compounds in each of the following sets with respect to increasing acidity:
(a) acetic acid, chloroacetic acid, trifluoroacetic acid
(b) benzoic acid, *p*-bromobenzoic acid, *p*-nitrobenzoic acid
(c) acetic acid, phenol, cyclohexanol

24.36 How can you explain the fact that 2-chlorobutanoic acid has $pK_a = 2.86$, 3-chlorobutanoic acid has $pK_a = 4.05$, 4-chlorobutanoic acid has $pK_a = 4.52$, and butanoic acid itself has $pK_a = 4.82$?

Sections 24.5–24.7: Reactions of Carboxylic Acids and Derivatives

24.37 The following structure represents a tetrahedral alkoxide ion intermediate formed by addition of a nucleophile to a carboxylic acid derivative. Identify the nucleophile, the leaving group, the reactant, and the ultimate product (red = O, blue = N, yellow-green = Cl).

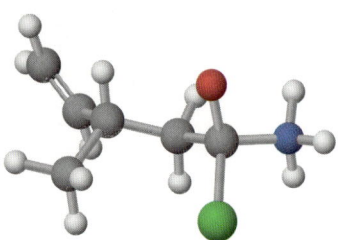

24.38 Electrostatic potential maps of methyl thioacetate and methyl acetate molecules are shown. Which of the two do you think is more reactive in nucleophilic acyl substitution reactions? Explain.

Methyl thioacetate

Methyl acetate

24.39 Rank the following compounds in order of their reactivity toward nucleophilic acyl substitution: CH_3COOCH_3, CH_3COCl, CH_3CONH_2, $CH_3COOCOCH_3$

24.40 One method for preparing acid anhydrides is by treatment of an acyl chloride with a carboxylate ion. For example:

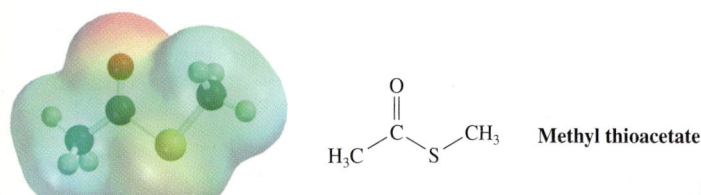

Propose a detailed, step-wise mechanism for this reaction.

24.41 If 5-hydroxypentanoic acid is treated with an acid catalyst, an intramolecular (within the same molecule) esterification reaction occurs. What is the structure of the product?

$$HOCH_2CH_2CH_2CH_2CO_2H$$ **5-Hydroxypentanoic acid**

24.42 Acyl chlorides undergo reduction with $LiAlH_4$ in the same way that esters do to yield primary alcohols. What are the products of the following reactions?

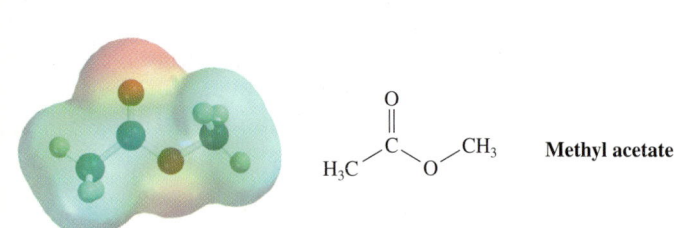

24.43 The reaction of an acyl chloride with $LiAlH_4$ to yield a primary alcohol (Question 24.42) takes place in two steps. The first step is a nucleophilic acyl substitution of H^- for Cl^- to yield an aldehyde, and the second step is nucleophilic addition of H^- to the aldehyde to yield an alcohol. Write the mechanism for the reduction of CH_3COCl.

24.44 Acyl chlorides undergo reaction with Grignard reagents at −78 °C to yield ketones. Propose a mechanism for the reaction.

$$\underset{R}{\overset{O}{\underset{\|}{C}}}\underset{Cl}{} + R'MgX \longrightarrow \underset{R}{\overset{O}{\underset{\|}{C}}}\underset{R'}{}$$

24.45 If the reaction of an acyl chloride with a Grignard reagent (Question 24.44) is carried out at room temperature, a tertiary alcohol is formed.
(a) Propose a mechanism for this reaction.
(b) What are the products of the reaction of CH_3MgBr with the acyl chlorides given in Question 24.42?

24.46 When dimethyl carbonate ($CH_3OCOOCH_3$) is treated with phenylmagnesium bromide, triphenylmethanol is formed. Explain.

24.47 Predict the product, if any, of reaction between propanoyl chloride and the following reagents:
(a) excess CH_3MgBr in ether
(b) NaOH in H_2O
(c) methylamine
(d) $LiAlH_4$
(e) cyclohexanol
(f) sodium acetate

24.48 Answer Question 24.47 for reaction between methyl propanoate and the listed reagents.

24.49 What esters and what Grignard reagents would you use to make the following alcohols? Show all possibilities.

(a)
$$\underset{CH_3CH_2CH_2}{\overset{\overset{OH}{|}}{C}}\underset{\underset{CH_3}{|}}{CH_3}$$

(b)

24.50 What products would you obtain on saponification of the following esters?

(a)

(b) Cyclohexyl propanoate

24.51 What product would you expect from the reaction of a cyclic ester such as butyrolactone with $LiAlH_4$?

Butyrolactone

Section 24.8: α-Carbon Atoms: A Second Reactive Site in Carbonyl Compounds

24.52 Indicate all acidic hydrogen atoms in the following molecules:

(a) $\underset{\displaystyle \ \ \ \ \ \ \ \overset{\textstyle O}{\underset{\textstyle \|}{}}}{}$ HOCH$_2$CCH$_3$

(c) Cyclopentane-1,3-dione

(b) HOCH$_2$CH$_2$CC(CH$_3$)$_3$

(d) CH$_3$CH=CHCHO

24.53 Draw structures for the monoenol tautomers of cyclohexane-1,3-dione. How many enol forms are possible, and which would you expect to be most stable? Explain.

24.54 Why do you suppose pentane-2,4-dione is 76% enolized at equilibrium although acetone is enolized only to the extent of about 0.0001%?

24.55 Write resonance structures for the following anions:

(a) CH$_3$CCHCCH$_3$

(b) :CH$_2$C≡N

(d) N≡CCHCO$_2$C$_2$H$_5$

24.56 How do the mechanisms of base- and acid-catalyzed enolization differ?

24.57 Why is an enolate ion generally more reactive than a neutral enol?

Sections 24.9–24.11: Reactions of Enolate Ions and Condensation Reactions

24.58 Show how you could use a malonic ester synthesis to prepare the following compounds:
(a) 4-methylpentanoic acid   (b) 2-methylpentanoic acid

24.59 Show how you could use a malonic ester synthesis to prepare the following compound:

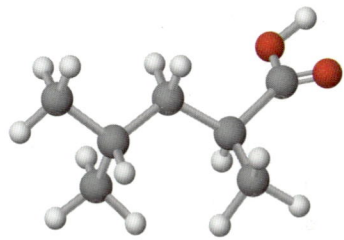

24.60 Monoalkylated acetic acid (RCH$_2$COOH) and dialkylated acetic acid molecules (R$_2$CHCOOH) can be made by malonic ester synthesis, but trialkylated acetic acids (R$_3$CHCOOH) cannot. Explain.

SUMMARY AND CONCEPTUAL QUESTIONS

24.61 Show how you could synthesize each of the following compounds starting with an appropriate carboxylic acid and any other reagents needed (red = O, blue = N, reddish-brown = Br):

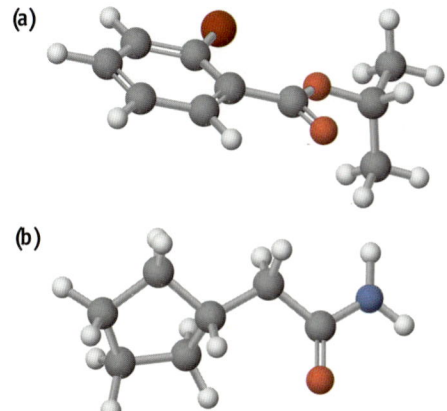

(a)

(b)

24.62 Acetic acid boils at 118 °C, but its ethyl ester boils at 77 °C. Why is the boiling point of the acid so much higher, even though it has a lower molecular weight?

24.63 The following reactivity order has been found for the saponification of alkyl acetates by aqueous NaOH:

CH$_3$COOCH$_3$ > CH$_3$COOCH$_2$CH$_3$ >
CH$_3$COOCH(CH$_3$)$_2$ > CH$_3$COOC(CH$_3$)$_3$

How can you explain this reactivity order?

24.64 Predict the product(s) of the following reactions:

(a)

$$\underset{\text{2. H}_3\text{O}^+}{\overset{\text{1. CH}_3\text{CH}_2\text{MgBr}}{\xrightarrow{\hspace{1.5cm}}}}\ ?$$

with CO$_2$CH$_2$CH$_3$

(b)

$$\overset{\displaystyle \text{CH}_3}{\underset{\displaystyle |}{}}$$
CH$_3$CHCH$_2$CH$_2$C≡N $\underset{\text{2. H}_2\text{O}}{\overset{\text{1. LiAlH}_4}{\xrightarrow{\hspace{1.5cm}}}}$?

(c)

COCl $\xrightarrow{\text{CH}_3\text{NH}_2}$?

$$\xrightarrow[\text{H}_2\text{SO}_4]{\text{CH}_3\text{OH}} \quad ?$$

(e)

$$\text{H}_2\text{C}=\text{CHCHCH}_2\text{CO}_2\text{CH}_3 \xrightarrow[\text{2. H}_3\text{O}^+]{\text{1. LiAlH}_4} \quad ?$$

with CH$_3$ substituent

(f)

cyclohexanol with OH $\xrightarrow[\text{Pyridine}]{\text{CH}_3\text{CO}_2\text{COCH}_3} \quad ?$

(g)

benzene ring with CONH$_2$ and CH$_3$ $\xrightarrow[\text{2. H}_2\text{O}]{\text{1. LiAlH}_4} \quad ?$

(h)

benzene ring with CH$_2$CO$_2$H and Br $\xrightarrow{\text{SOCl}_2} \quad ?$

24.65 How can you prepare acetophenone (methyl phenyl ketone) from the following starting materials? (More than one step may be required.)
(a) bromobenzene (b) methyl benzoate (c) benzene

24.66 How might you prepare the following products from butanoic acid? (More than one step may be required.)
(a) $\text{CH}_3\text{CH}_2\text{CH}_2\text{CH}_2\text{OH}$
(b) $\text{CH}_3\text{CH}_2\text{CH}_2\text{CHO}$
(c) $\text{CH}_3\text{CH}_2\text{CH}_2\text{CH}_2\text{Br}$
(d) $\text{CH}_3\text{CH}_2\text{CH}_2\text{COOCH}_3$
(e) $\text{CH}_3\text{CH}_2\text{CH}=\text{CH}_2$
(f) $\text{CH}_3\text{CH}_2\text{CH}_2\text{CH}_2\text{NH}_2$

24.67 Predict the product of the reaction of *p*-methylbenzoic acid with each of the following reagents:
(a) LiAlH_4 (b) CH_3OH, HCl
(c) SOCl_2 (d) NaOH, then CH_3I

24.68 Show how you might synthesize the anti-inflammatory agent ibuprofen starting from isobutylbenzene. More than one step is needed.

Isobutylbenzene   **Ibuprofen**

24.69 How would you prepare the following substances from the indicated starting materials? More than one step is required in each case.
(a) $(\text{CH}_3)_2\text{CHCH}_2\text{CH}_2\text{NH}_2$ from $(\text{CH}_3)_2\text{CHCH}_2\text{I}$
(b) 1-phenylbutan-2-one from benzyl bromide, $\text{C}_6\text{H}_5\text{CH}_2\text{Br}$

24.70 When *methyl* acetate is heated in pure ethanol containing a small amount of HCl catalyst, *ethyl* acetate results. Explain.

24.71 *tert*-Butoxycarbonyl azide, an important reagent used in protein synthesis, is made by treating *tert*-butoxycarbonyl chloride with sodium azide. Propose a mechanism for this reaction.

24.72 What product would you expect to obtain on treatment of the cyclic ester butyrolactone with excess phenylmagnesium bromide?

Butyrolactone

24.73 *N,N*-Diethyl-*m*-toluamide (DEET) is the active ingredient in many insect repellents. How might you synthesize DEET from *m*-bromotoluene?

N, N-**Diethyl-*m*-toluamide**

24.74 In the *iodoform reaction,* a triiodomethyl ketone reacts with aqueous NaOH to yield a carboxylate ion and iodoform (triiodomethane). Propose a mechanism for this reaction.

$$\text{R}-\overset{\overset{\text{O}}{\|}}{\text{C}}-\text{CI}_3 \xrightarrow{\text{NaOH, H}_2\text{O}} \text{R}-\overset{\overset{\text{O}}{\|}}{\text{C}}-\text{O}^- + \text{CHI}_3$$

24.75 The step-growth polymer called nylon 6 is prepared from caprolactam. The reaction involves initial reaction of caprolactam with water to give an intermediate amino acid, followed by heating to form the polymer. Propose mechanisms for both steps, and show the structure of nylon 6.

Caprolactam

24.76 Draw a representative segment of the polyester that would result from reaction of pentanedioic acid $(\text{HO}_2\text{CCH}_2\text{CH}_2\text{CH}_2\text{COOH})$ and pentane-1,5-diol.

24.77 Rank the following compounds in order of increasing acidity:

$$\text{CH}_3\text{CH}_2\overset{\overset{\text{O}}{\|}}{\text{C}}\text{OH}, \quad \text{CH}_3\overset{\overset{\text{O}}{\|}}{\text{C}}\text{CH}_3, \quad \text{CH}_3\text{CH}_2\text{OH}, \quad \text{CH}_3\overset{\overset{\text{O}}{\|}}{\text{C}}\text{CH}_2\overset{\overset{\text{O}}{\|}}{\text{C}}\text{CH}_3$$

24.78 The marine natural product and potential chemotherapeutic agent spongistatin is rich in functional groups. How many of the 15 oxygen-containing functional groups are alcohols? Ethers? Aldehydes? Ketones? Carboxylic acids? Hemiacetals? Acetals?

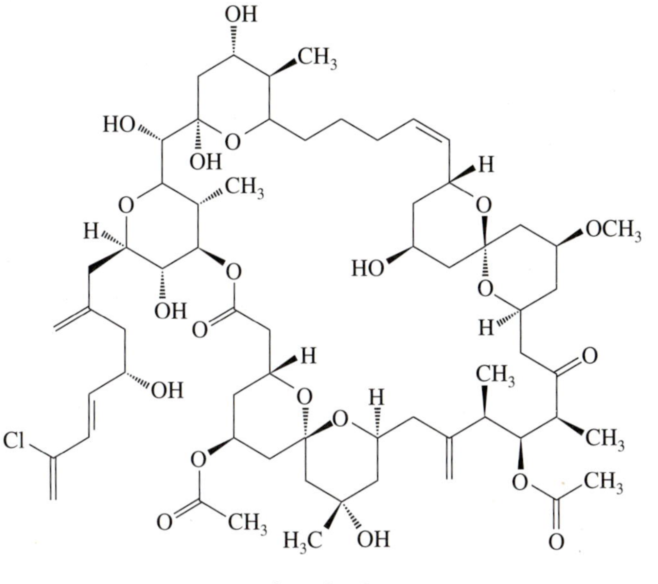

Spongistatin

24.79 Yesterday's drug is often today's poison. Cocaine enjoyed a much better reputation 100 years ago, when it was used as a stimulant in products such as "Force March," which helped Shackleton explore Antarctica, as well as in drops to treat toothaches and depression. What three molecules are produced on hydrolysis of a cocaine molecule?

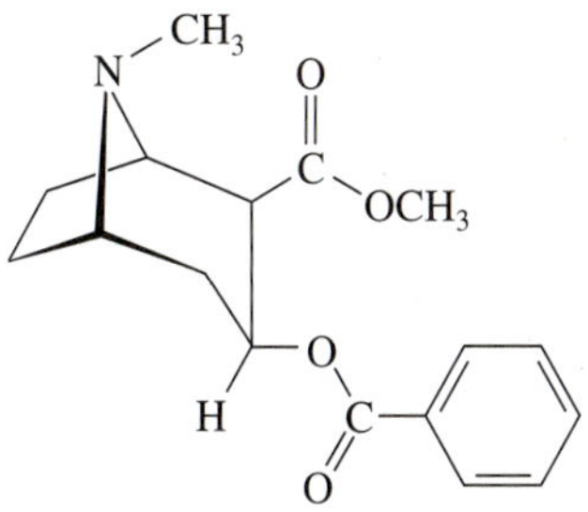

Cocaine

24.80 Because of cocaine's addictive properties, researchers have looked for less addictive alternatives to relieve pain. Lidocaine and Novocain preserve many of the common structural features of cocaine (red) but do not have the same risk factors.

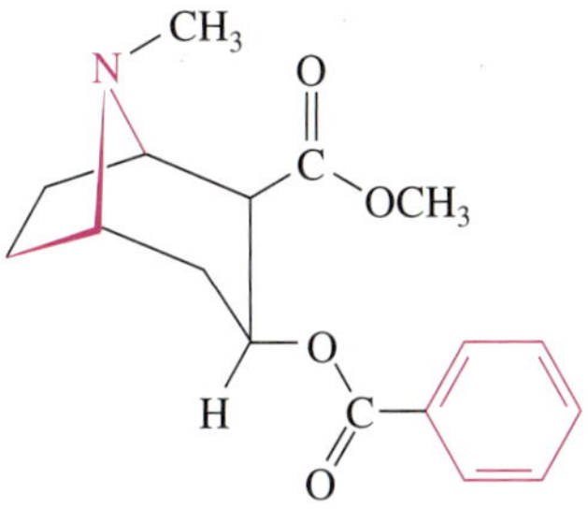

Cocaine 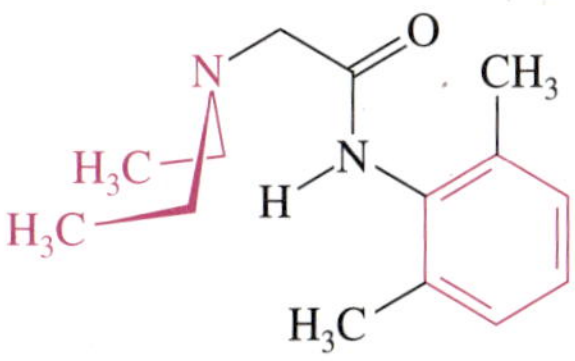   **Lidocaine** 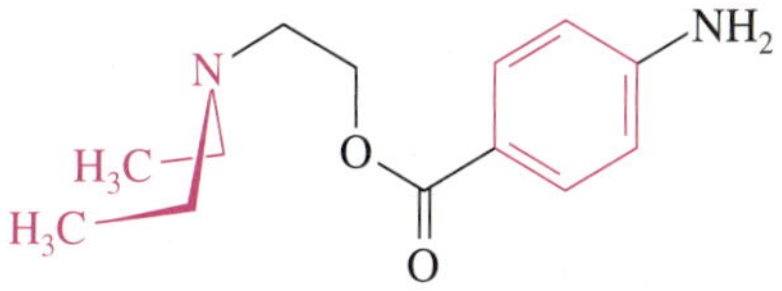   **Novocain**

Lidocaine is prepared by the reaction sequence shown. For each step indicate the type of reaction and draw a mechanism.

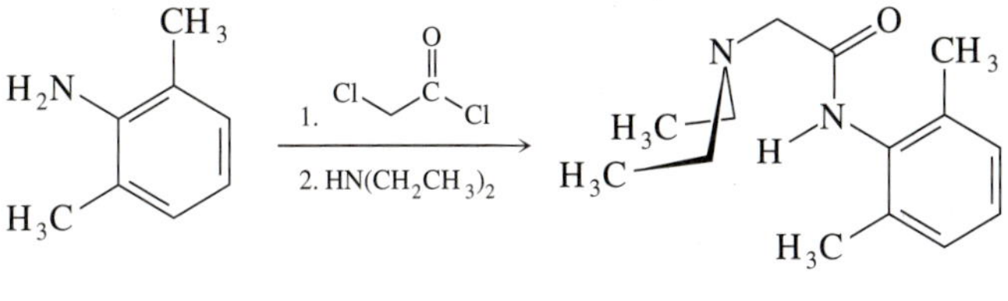

Lidocaine

24.81 Consumption of wheat contaminated with certain fungi (that appear like brown kernels) produces hallucinations and a condition called St. Anthony's fire. Long-term exposure causes gangrene due to the potent vasoconstrictive action of a certain molecule present—a natu-

rally occurring derivative of lysergic acid. LSD, or lysergic acid diethylamide, is synthesized by the hydrolysis of the amide bond indicated and reaction with diethylamine.

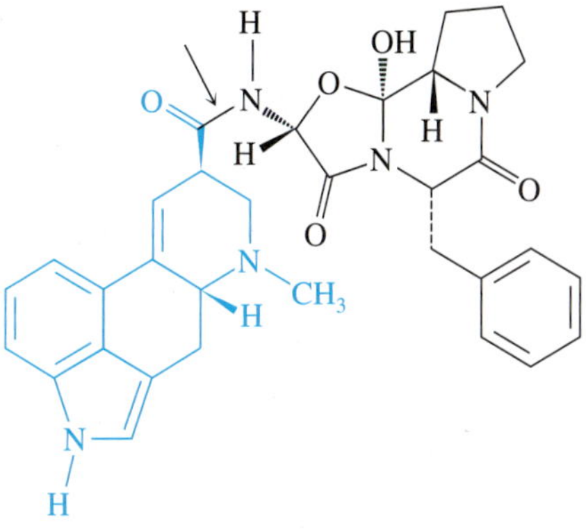

Tom Volk at TomVolkFungi.net

(a) Draw the product of the hydrolysis of the indicated bond.

(b) Mixing lysergic acid with diethylamine does not produce LSD. What is the product of this acid-base reaction?

(c) Reaction of lysergic acid with NaOH, followed by acetyl chloride, provides a mixed anhydride. What is the structure of the anhydride?

(d) Reaction of this anhydride with a two-fold excess of diethylamine provides LSD. Why is a two-fold excess required?

24.82 Derivatives of lysergic acid (Question 24.81) cause vasoconstriction. By mimicking LSD, a number of vasoconstrictors have been designed and marketed to treat migraines, caused by pressure exerted by dilated blood vessels in the head. Sumatriptan is one example. The structural features common to lysergic acid and the triptan family of vasoconstrictors are highlighted in blue.

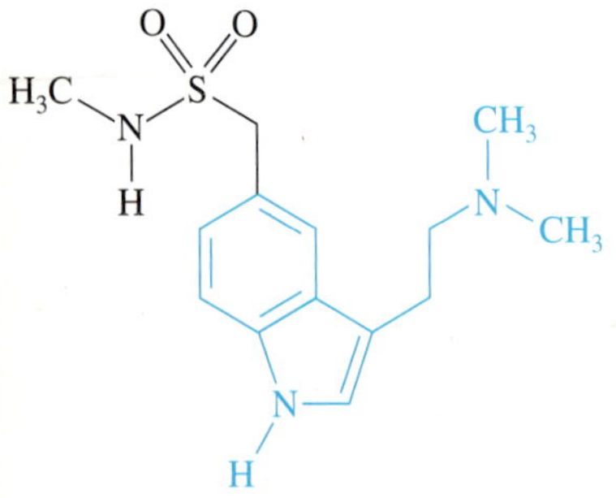

Sumatriptan

Sumatriptan contains a sulfonamide group. These groups can be prepared from amines and sulfonyl chlorides. Draw a mechanism for the reaction of CH_3NH_2 with $ArCH_2SO_2Cl$ (Ar signifies an *ar*omatic group.)

24.83 Acetylcholine is a neurotransmitter that is constantly synthesized and hydrolyzed at nerve endings. Poisoning the enzyme acetylcholine esterase, which is responsible for hydrolysis of acetylcholine, kills organisms. A nucleophilic hydroxyl group on the enzyme (abbreviated ROH) reacts with acetylcholine to produce choline and an acetylated enzyme intermediate ($ROCOCH_3$). In a

subsequent step, a water molecule hydrolyzes the acetyl group from the enzyme to regenerate ROH and produce acetic acid. Show a mechanism for both steps of this reaction.

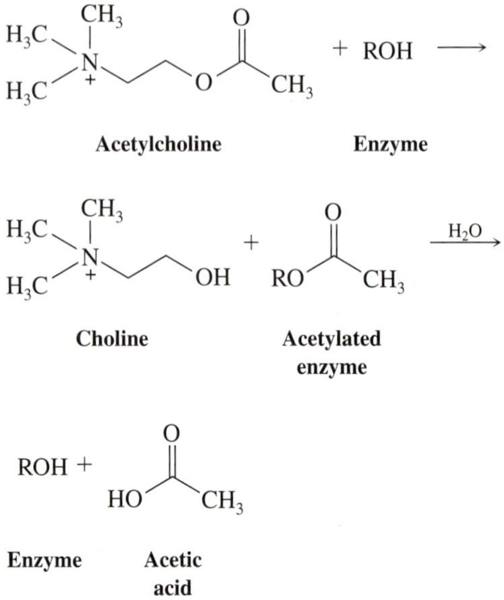

Acetylcholine **Enzyme**

Choline **Acetylated enzyme**

Enzyme **Acetic acid**

24.84 Carbaryl is made by reacting an aromatic alcohol nucleophile with methyl isocyanate, $H_3C-N=C=O$. Propose a mechanism for this reaction. Methyl isocyanate is the poisonous gas responsible for the accident that led to the 1984 tragedy in Bhopal, India.

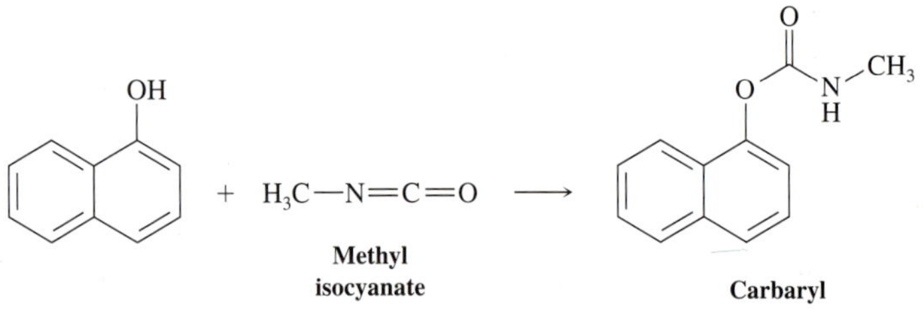

Methyl isocyanate **Carbaryl**

24.85 The *acetoacetic ester synthesis* is closely related to the malonic ester synthesis but involves alkylation with the anion of ethyl acetoacetate rather than diethyl malonate. Treatment of the ethyl acetoacetate anion with an alkyl halide, followed by decarboxylation, yields a ketone product:

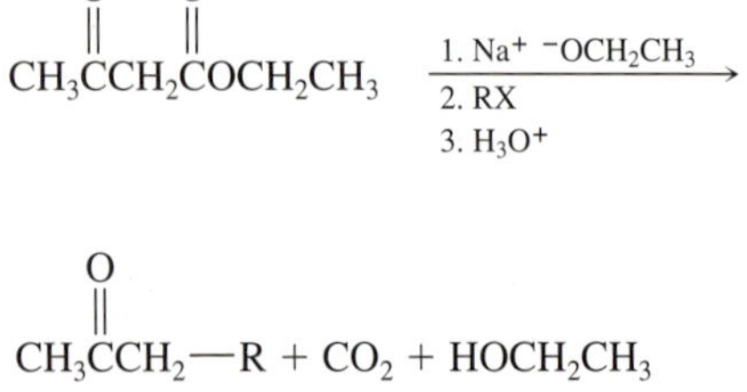

$$CH_3CCH_2COCH_2CH_3 \xrightarrow[\text{2. RX}]{\text{1. Na}^+ \ ^-OCH_2CH_3} \xrightarrow{\text{3. H}_3O^+}$$

$$CH_3CCH_2-R + CO_2 + HOCH_2CH_3$$

How would you prepare the following compounds using an acetoacetic ester synthesis?

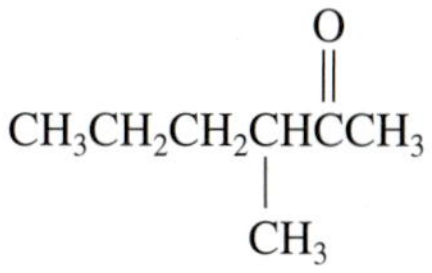

24.86 Which of the following compounds cannot be prepared by an acetoacetic ester synthesis (see Question 24.85)? Explain.

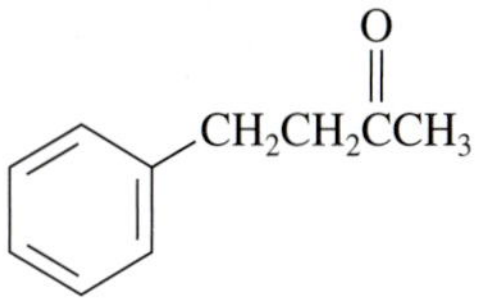

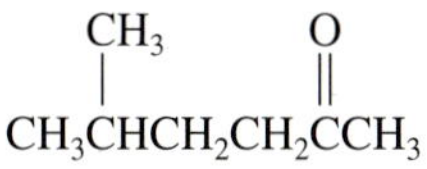

(a)

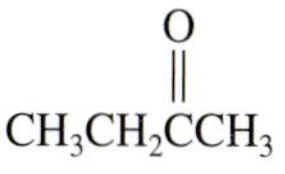

$$CH_2CH_2CCH_3$$ with O double bond, attached to benzene ring

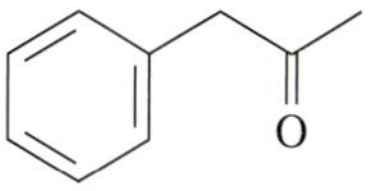

(c)

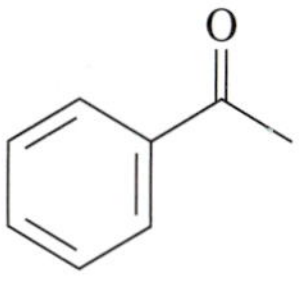

$$CH_3CH_2CH_2CHCCH_3$$ with O double bond and CH_3 substituent

(b)

$$CH_3CHCH_2CH_2CCH_3$$ with CH_3 substituent and O double bond

24.86 (a)

$$CH_3CH_2CCH_3$$ with O double bond

(b) benzene ring—CH_2—C(=O)—CH_3

(c) benzene ring—C(=O)—CH_3

(d)

$$CH_3 - \overset{H_3C}{\underset{H_3C}{C}} - CCH_3$$ with O double bond

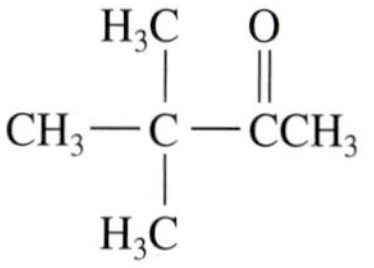

With the **eCHACR** single-sign-on access card bundled with your text, log on (http://login.cengage.com/sso) and access the e-book and click on any in-margin icons for dynamic molecular-level animations and simulations, videos of laboratory reactions, interactive exercises, reviews of background concepts, and other online supplementary materials as noted by the icons in the text margins.

Also go to www.chemistry.nelson.com <http://www.chemistry.nelson.com> for Answers to in-chapter exercises and selected Review Questions, Test Yourself questions, weblinks, crossword puzzles, flashcards, glossary of key terms, and other student resources.

25 Amines and Nitrogen Heterocycles

Outline

© Biosphoto / Thiriet Claudius / Peter Arnold Inc.

25.1 Case Study: The Amino World: "An Organic Chemist's Playground"

The importance of nitrogen fixation, which converts unreactive dinitrogen (N_2) gas into forms that are usable by organisms, was shown in Chapter 14. We discussed estimates that half of the nitrogen atoms in your body have passed at some point through the Haber-Bosch process for the synthesis of ammonia (NH_3) from hydrogen (H_2) and nitrogen (N_2) gases. While this industrial synthesis requires high temperature and pressure, and the right catalyst, *biological nitrogen fixation* takes place at ambient pressure and temperature in certain legumes and bacteria through nitrogenase bacterial enzymes. These enzymes convert N_2 into ammonia, nitrites (NO_2^-) or nitrates (NO_3^-), which plants then convert into proteins, nucleic acids, and many other organic nitrogen-containing compounds.

Many of these organic nitrogen compounds are **amines**, molecules formally derived from ammonia by replacing one or more hydrogen atoms with alkyl groups. The many amines that are produced by plants, animals, and fungi are called *alkaloids* [>>Section 25.5]. Taking cues from what happens in nature, chemists have learned to synthesize complex nitrogen-containing organic compounds using simpler starting materials—an area of research seen as so important by 1924 Nobel Laureate Robert Robinson that he described the chemistry of alkaloids as becoming "an organic chemist's playground."

Three nitrogen-containing organic compounds in Figure 25.1 illustrate the evolution of the sophisticated and powerful organic chemistry playground we know today.

Urea: Creating the Playing Field

The laboratory synthesis of urea (chapter opening photo) by Friederich Wöhler in 1828 marks the formal beginning of organic chemistry. By the mid-1700s, alchemists noticed unexplainable differences between compounds derived from living sources and those derived from minerals. Compounds from plants and animals were often difficult to isolate

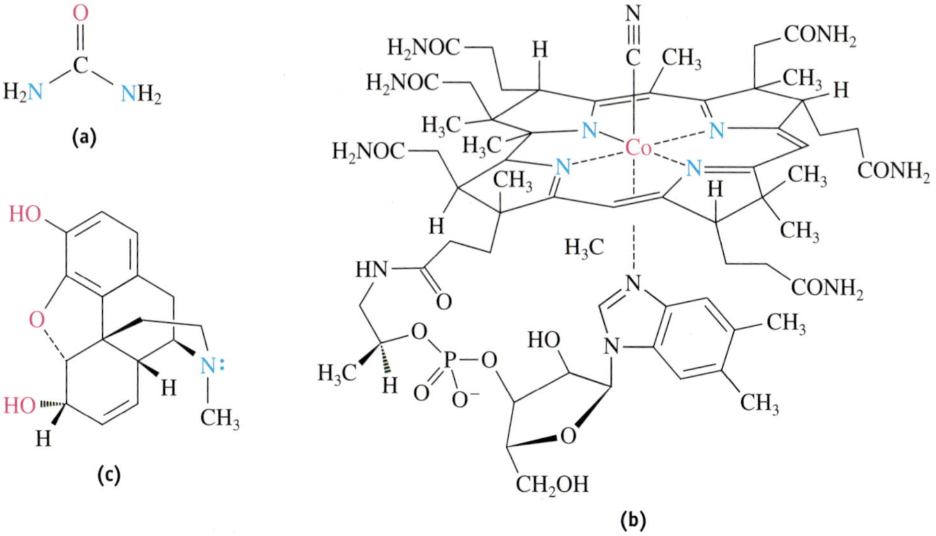

FIGURE 25.1 Molecules of three important nitrogen-containing organic compounds: (a) urea—a diamide; (b) a form of Vitamin B-12—with nitrogen bases bound to a cobalt ion; and (c) morphine—an alkaloid used as a powerful analgesic.

and purify. Even when pure, they were difficult to work with and tended to decompose more easily than compounds from minerals. The Swedish chemist Torbern Bergman was the first to express this difference between "organic" and "inorganic" substances in 1770, and the term *organic chemistry* soon came to mean the chemistry of compounds from living organisms.

To many chemists of the time, the only explanation for the difference in behaviour between organic and inorganic compounds was that organic compounds contained an indefinable "vital force" as a result of their origin from living sources. Wöhler's discovery that it was possible to convert the "inorganic" salt ammonium cyanate into urea, an "organic" compound isolated from urine, showed that chemistry is chemistry, whether carried out by plants or by humans in laboratories. This discovery created the playing field for future work in synthetic organic chemistry.

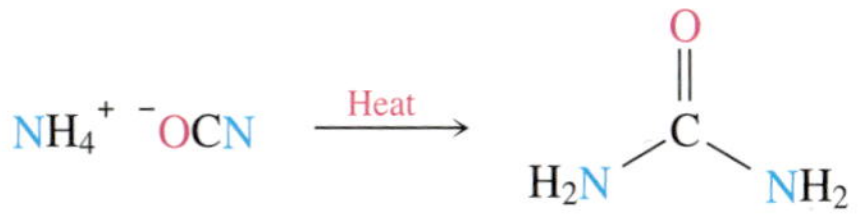

Compounds with functional groups often behave differently when several different ones are present in molecules of a substance and urea is not a simple amine, but a di*amide* of carbonic acid [<<Section 24.2]. Its molecule's two bonds between nitrogen and carbonyl carbon atoms are peptide bonds like those connecting amino groups from one amino acid molecule to the acyl group of a second amino acid (Figure 25.2) in proteins and peptides. It's not surprising, then, that urea is a monomeric building block. It finds use both as

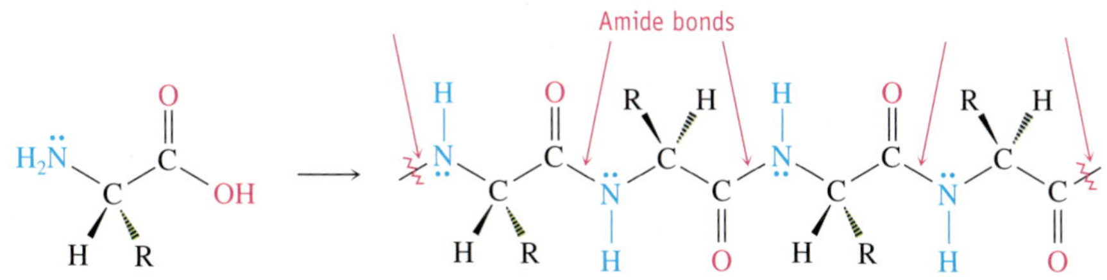

FIGURE 25.2 Amino acid monomers link together to form peptides and proteins through amide (peptide) bonds.

a fertilizer, which plants can use to synthesize protein polymers, and in industry where chemists can co-polymerize it with formaldehyde to form resins with applications such as adhesives, finishes, and insulation.

Amines are Lewis bases, and this explains their reactivity with compounds containing electron-deficient carbonyl carbon atoms (acylation reactions). Other important reactions of amines in this chapter involve nitrogen atoms in amine molecules reacting as Lewis bases to donate electron pairs that form bonds to protons or electron-deficient carbon atoms of alkyl halide molecules (alkylation reactions).

Vitamin B-12: Building the Playground Equipment

Vitamin B-12, a common form of which is called *cyanocobalamin*, plays an essential role in the functioning of the brain and nervous system and the formation of red blood cells. Since humans can't biosynthesize it, trace amounts (about $1\mu g$ per day) must be included in our diet to prevent anemia. Discovery of the function and structure of Vitamin B-12, followed by its laboratory synthesis, played a transforming role in the 20th century development of theoretical and experimental tools for chemists. It equipped chemists to take a quantum leap in their ability to understand structures, predict reactivity, and synthesize complex substances like natural products. Four Nobel prizes resulted!

The first, in medicine, went to U.S. chemists George Whipple, George Minot, and William Murphy in 1934 for isolating Vitamin B-12 from raw liver, which proved the key to treating pernicious anemia, a disease resulting from a deficiency of red blood cells.

Next, Dorothy Hodgkin at Oxford University in the U.K. helped biochemistry emerge as a mature and relevant discipline by successfully characterizing B-12's three-dimensional molecular structure using X-ray crystallography [<<Section 9.3]. At the time she started, all that was known was Vitamin B-12's approximate empirical formula. It took eight years of work by her research group on a few red crystals to understand the exact connectivity and structure. In the initial stages, they carried out thousands of calculations on a cast-off punched card computer. For determining the three-dimensional structures of penicillin and Vitamin B-12 molecules, Hodgkin was awarded the 1964 Nobel Prize in Chemistry.

Examine the structure of a molecule of a synthetic form of Vitamin B-12 in Figure 25.1. You can see that the amine functional groups are in the rings of the molecule that look like pyrrole rings [<<Section 20.3], except they have fewer double bonds. These four reduced pyrrole rings are connected together with a cobalt metal ion bound to the middle of four nitrogen atoms. The bonding between the amine nitrogen atoms and the metal ion are good examples of Lewis base–Lewis acid interactions, resulting from each nitrogen atom donating an electron pair to form a bond to the metal ion. The CN^- group is picked up by the cobalt ion during a purification step of the vitamin. In the physiologically active form of B-12, the CN^- group is replaced with a deoxyribonucleoside.

Once the structure was solved by Hodgkin, the brightest minds in organic chemistry set out to try to synthesize Vitamin B-12, whose molecules are enormously complex. Each molecule is about three times larger than all other vitamins, and contains nine stereocentres in the four reduced pyrrole rings alone (can you find them in Figure 25.1b?). In principle, this means a maximum of 2^9 or 512 possible stereoisomers might exist, even if smaller molecules could be assembled with just the right connectivity to give the overall framework of the vitamin. The challenge was taken on by Robert Woodward (at Harvard University) and Albert Eschenmoser (at the Swiss federal Institute of Technology). Eleven years, over one hundred chemists, and almost a hundred reaction steps later, the synthesis was complete. Along the way, Woodward turned to theoretical chemist Roald Hoffmann at Cornell University for an understanding of the complex stereochemical constraints of some of these key reactions. Woodward was awarded the 1965 Nobel Prize for his landmark synthesis, which convinced chemists that even very complex natural products could

"I should not like to leave an impression that all structural problems can be settled by X-ray analysis or that all crystal structures are easy to solve. I seem to have spent much more of my life not solving structures than solving them."

Dorothy Crowfoot Hodgkin, 1964 Nobel Prize in Chemistry. Hodgkin was deeply committed to humanitarian and peace causes and to mentoring many women in her research group at Oxford.

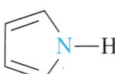

Pyrrole

be rationally synthesized. Hoffmann shared the 1981 Nobel Prize for his theoretical studies and new rules for the control of reactions that arose from the challenges of this synthesis.

One hundred and fifty years after Wöhler's synthesis of urea, the organic chemist's playground had some very impressive equipment!

Altering Alkaloids: Playing by the Rules

Medical uses of the poppy, *Papaver somniferum,* have been known at least since the 17th century, when crude extracts—called *opium*—were used for the relief of pain. Morphine was the first pure alkaloid to be isolated from opium, but its close relative, codeine, also occurs naturally (Figure 25.3). It makes sense that morphine should be used for pain relief: very recently, it has been shown through ^{18}O labelling studies that morphine is formed *endogenously*—it can be synthesized in very low concentrations by human cells. Codeine, which is simply the methyl ether of morphine, is used in prescription cough medicines and as an analgesic.

Molecular Modelling (Odyssey)

e25.1 Compare molecular models of morphine and cocaine with their structural formulas in Figure 25.3.

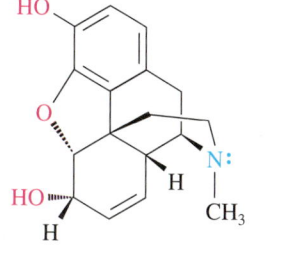
Morphine

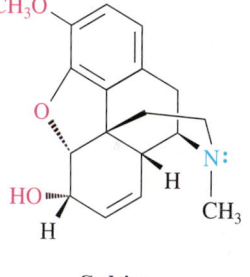
Codeine

FIGURE 25.3 Morphine, codeine, and related compounds can be isolated from opium poppy *Papaver somniferum*.

Chemical investigations into the structure of morphine molecules occupied many chemists of the 19th and early 20th centuries, and in 1924 the puzzle was finally solved by Robert Robinson, who penned the metaphor of an organic chemist's playground.

The characterization and modification of these alkaloids has led to the development of several different kinds of rules for the organic playground. The first kind is rules of analysis to guide *structure–activity relationships*. Morphine and its relatives are extremely useful pharmaceutical agents, yet they also pose an enormous social problem because of their addictive properties. Much effort has therefore gone into understanding how morphine works and into developing modified morphine analogs that retain the analgesic activity but don't cause physical dependence. Our present understanding is that morphine functions by non-covalent binding to so-called μ-opiod receptor sites both in the spinal cord, where it interferes with the transmission of pain signals, and in brain neurons, where it changes the brain's reception of the signal.

Hundreds of compounds with morphine-like molecules have been synthesized and tested for their analgesic properties. Research has shown that only part of the complex framework of morphine molecules is necessary for biological activity. According to the "morphine rule," biological activity requires (1) an aromatic ring attached to (2) a quaternary carbon atom and (3) a tertiary amine situated (4) two carbon atoms farther away. Meperidine (Demerol), a widely used analgesic, and methadone, a substance used in the treatment of heroin addiction, are two of many compounds that fit the morphine rule.

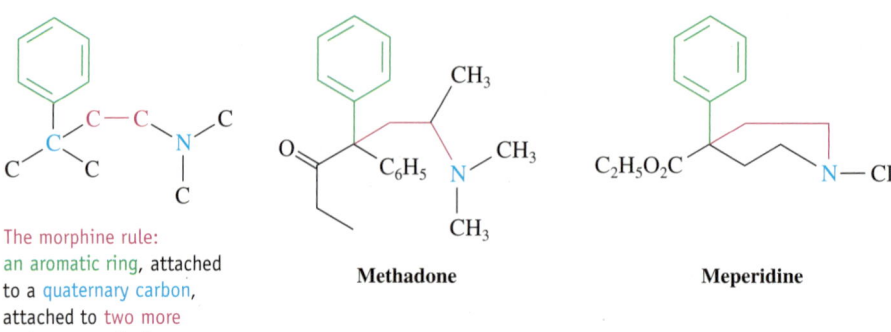

Heroin

The morphine rule:
an aromatic ring, attached
to a quaternary carbon,
attached to two more
carbons, attached to a
tertiary amine

Methadone

Meperidine

Creative Eye-images Photography

"Molecules are molecules. Chemists and engineers make new ones, transform old ones. Still others in the economic chain sell them, and we all want them and use them. Each of us has a role in the use and misuse of chemicals … There is no way to close one's eyes to creation or discovery. If you don't find that molecule, someone else will. At the same time, I believe that scientists have absolute responsibility for thinking about the uses of their creation, even the abuses by others. And they must do everything possible to bring those dangers and abuses before the public. If not I, then who?"

Roald Hoffmann, 1981 Nobel Laureate in Chemistry for his development of the Woodward-Hoffmann Rules.

IUPAC, which formulates rules for nomenclature, also raises awareness about the need for ethical guidelines for the multiple uses of chemicals. See www.iupac.org/multiple-uses-of-chemicals.

As synthetic methodologies become ever easier to carry out, a completely different kinds of rules—ethical ones—have become necessary to monitor and control the organic playground. Heroin, another close relative of morphine, does not occur naturally but can be easily synthesized by diacetylation of morphine. A university chemistry student such as yourself could design such a synthesis, having learned the reactions of carboxylic acid derivatives and alcohols to produce esters in Chapter 24.

> *Ethical choices about the beneficial use, misuse, and abuse of chemicals lie in the hands of each researcher and student entering the playground of organic chemistry.*

In need of similar guidance by societal rules is the widespread transformation of the alkaloid pseudoephedrine into methamphetamine ("crystal meth"). Pseudoephedrine is extracted from one of the world's oldest medicinal plants, Ma-Huang, an *Ephedra* shrub native to Northern China. It is used in pharmaceutical decongestants and anti-asthmatic medicines as a nasal decongestant and bronchodilator to open air passages to the lungs. Yet pseudoephedrine, which can also be synthesized in the laboratory, has become infamous as the readily available precursor to one of the world's fastest growing drugs of abuse, methamphetamine (Figure 25.4). The simple reduction of an alcohol functional group is all that is required.

Pseudoephedrine
(cough suppressant)

Methamphetamine
("crystal meth")

FIGURE 25.4 The reduction of pseudoephedrine to form methamphetamine.

Like many chemical substances, pseudoephedrine has multiple uses—it can be a beneficial medicine or converted in clandestine basement laboratories into crystal meth. Ethical choices about the beneficial use, misuse, and abuse of chemicals lie in the hands of each researcher and student entering the playground of organic chemistry.

25.2 Naming Amines

Amine molecules are classed as *primary (RNH₂), secondary (R₂NH),* or *tertiary (R₃N),* depending on the number of organic substituents attached to nitrogen atoms. For example, methylamine (CH_3NH_2) is a primary amine, and trimethylamine [($CH_3)_3N$] is a tertiary amine. Note that this usage of the terms *primary, secondary,* and *tertiary* is different from our previous usage. When we speak of a tertiary alcohol [<<Section 22.2] or alkyl halide, we refer to the degree of substitution at a molecule's alkyl *carbon* atom, but when we speak of a tertiary amine, we refer to the degree of substitution at the *nitrogen* atom.

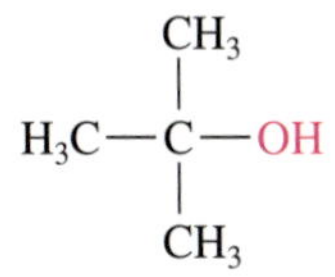

tert-Butyl alcohol
(a tertiary alcohol)

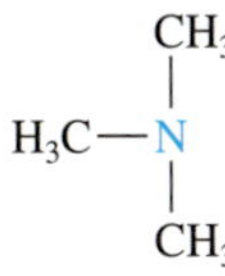

Trimethylamine
(a tertiary amine)

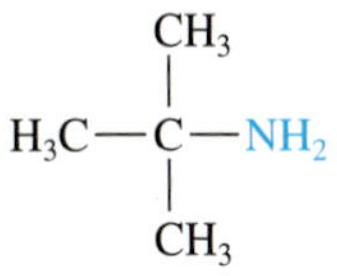

tert-Butylamine
(a primary amine)

Compounds with four groups attached to nitrogen atoms are also known, in which the nitrogen atom in the molecule must carry a positive charge. Such compounds are called *quaternary ammonium salts.*

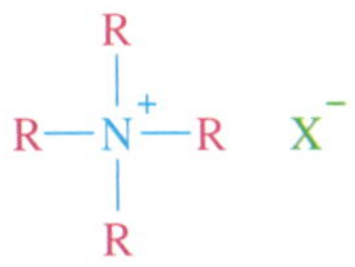

A quaternary ammonium salt

Amine molecules can be either alkyl-substituted (*alkylamines*) or aryl-substituted (*arylamines*). Although much of the chemistry of the two classes is similar, we'll see that there are also important differences.

Ethylamine
(an alkylamine)

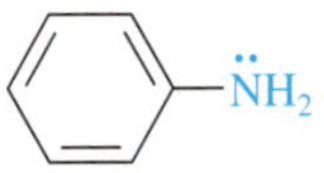

Aniline
(an arylamine)

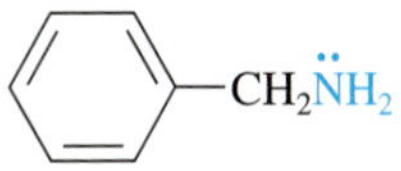

Benzylamine
(an alkylamine)

Primary amine molecules (RNH_2) are named in the IUPAC system by adding the suffix *–amine* to the name of the organic substituent:

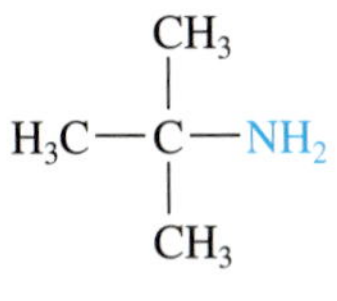

tert-Butylamine

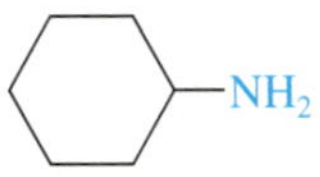

Cyclohexylamine

Butane-1,4-diamine

Amines that have additional functional groups are named by considering the —NH_2 as an *amino* substituent on the parent molecule:

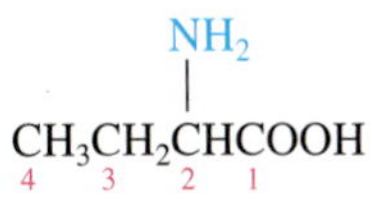

2-Aminobutanoic acid

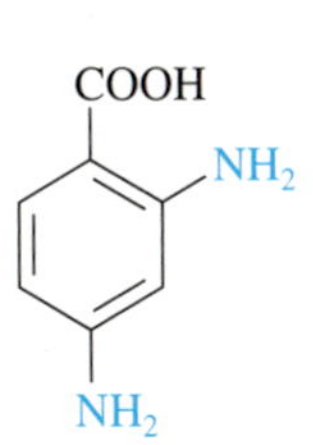

2,4-Diaminobenzoic acid

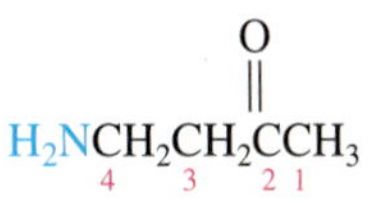

4-Aminobutan-2-one

Symmetrical secondary and tertiary amines are named by adding the prefix *di–* or *tri–* to the alkyl group:

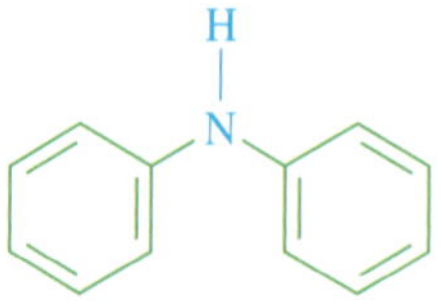

Diphenylamine 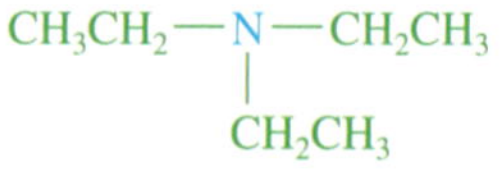   **Triethylamine**

Unsymmetrically substituted secondary and tertiary amines are named as *N*-substituted primary amines. The largest organic group is chosen as the parent, and the other groups are considered as *N*-substituents on the parent (*N* because they are attached to a nitrogen atom).

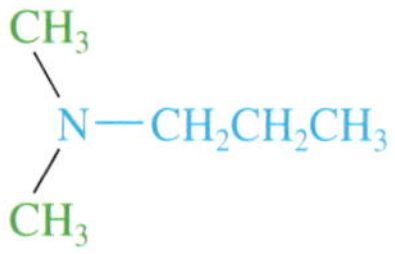

N,N-**Dimethylpropylamine**
(propylamine is the parent name; the two
methyl groups are substituents on nitrogen)

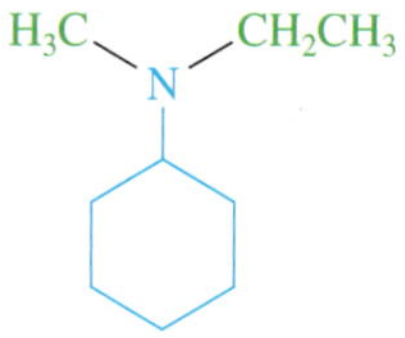

N-Ethyl-*N*-methylcyclohexylamine
(cyclohexylamine is the parent name;
methyl and ethyl are *N*-substituents)

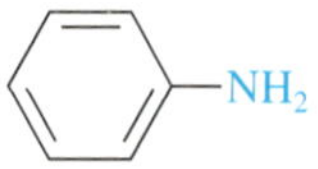

Aniline

There are few common names for simple amines, although phenylamine is usually called *aniline*.

Heterocyclic amines—whose molecules have nitrogen atoms as part of a ring—are also common, and each different heterocyclic ring system has its own parent name. In all cases, the nitrogen atom is numbered as position 1.

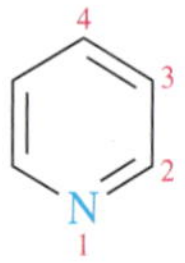

Pyridine 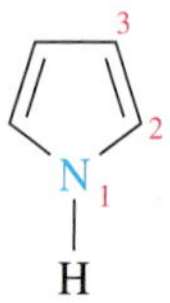   **Pyrrole** 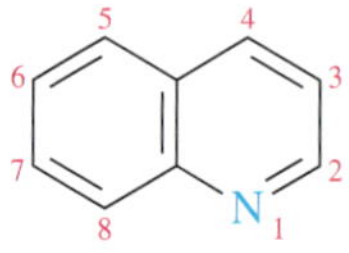   **Quinoline** 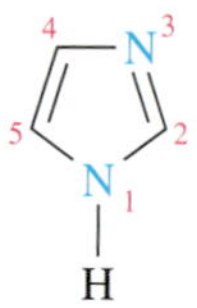   **Imidazole** 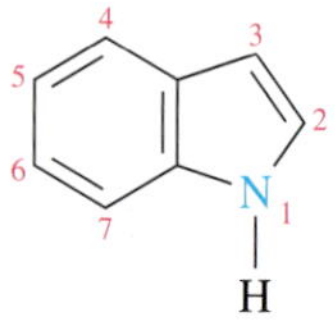   **Indole** 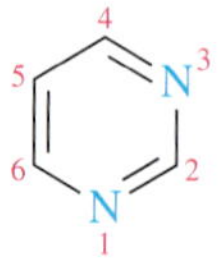   **Pyrimidine**

**Interactive
Exercise 25.1**

Draw structural formulas
for amine molecules from their
IUPAC names.

SUBMIT

WORKED EXAMPLE 25.1—NOMENCLATURE

Classify the following amines as primary, secondary, or tertiary:

(a) 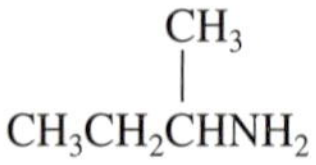
$$CH_3$$
$$CH_3CH_2CHNH_2$$

(b) 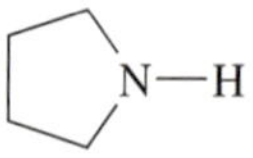
N—H

(c)

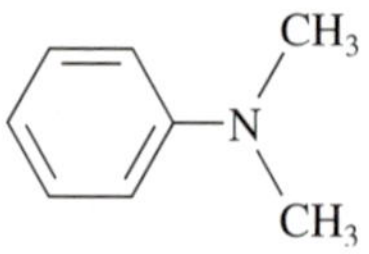

Solution

A molecule of amine (a) has one organic group attached to its nitrogen atom and is primary; (b) has two organic groups attached to its nitrogen atom and is secondary; and (c) is tertiary.

EXERCISE 25.2—NOMENCLATURE

Classify each of the following compounds as either a primary, secondary, or tertiary amine, or as a quaternary ammonium salt:

(a) $(CH_3)_2CHNH_2$

(c)
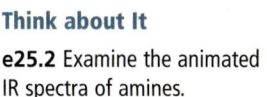

(b) $(CH_3CH_2)_2NH$

(d)

Amines are formally organic derivatives of ammonia. They are named in the IUPAC system either by adding the suffix *–amine* to the name of the alkyl substituent or by considering the amino group as a substituent on a more complex parent molecule.

25.3 Identifying Amines: Spectroscopy

Infrared Spectroscopy

Primary and secondary amines can be identified by a characteristic N—H stretching absorbance in the region between 3300 and 3500 cm^{-1} (Figure 25.5). This can be difficult to distinguish from O—H stretching bands of alcohols, but amine bands are often sharper and less intense than hydroxyl bands. Primary amines show two bands for symmetric and antisymmetric stretching of their molecules' two N—H bonds at about 3350 and 3450 cm^{-1}. Secondary amines usually show a single absorption at about 3350 cm^{-1}. Tertiary amines have no N—H bonds and so don't absorb in this region. N—H "scissor" bending vibrations typically absorb in the 1650 and 1580 cm^{-1} region of the spectrum and N—H "wagging" bends between 910 and 660 cm^{-1}. A weak C—N stretching band is sometimes seen between 1020 and 1250 cm^{-1}.

Think about It

e25.2 Examine the animated IR spectra of amines.

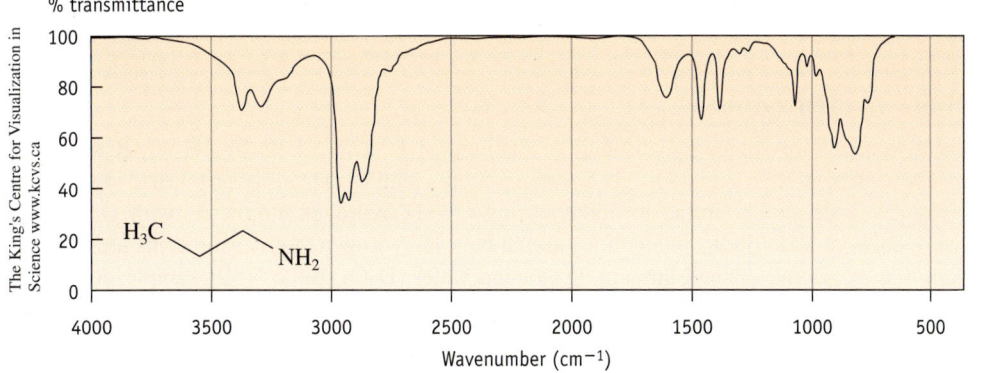

FIGURE 25.5 Infrared spectrum of propanamine.

Mass Spectrometry

An important feature in the mass spectrum of amines is the appearance of a molecular ion peak, which corresponds to the molecular weight of the compound, at an odd mass/charge ratio. The *nitrogen rule* of mass spectrometry states that molecules with an odd number of nitrogen atoms (i.e., one nitrogen atom) will have an odd-numbered molecular weight. This is because nitrogen atoms form three bonds, instead of four like carbon atoms, and require an odd number of hydrogen atoms or substituents. For example, morphine has the formula $C_{17}H_{19}NO_3$ and a small molecular ion peak appears at an *m/z* value of 285.

The mass spectrum of *N*-ethylpropylamine (Figure 25.6) shows an odd-numbered molecular ion at *m/z* = 87, and fragment ions at *m/z* = 72 and *m/z* = 58, which result from loss of methyl and ethyl radicals, respectively. Breaking the C—C bond immediately adjacent to the nitrogen atom (bond to the α-carbon atom) in the molecular ion is commonly seen, as this leads to a nitrogen-stabilized carbocation. This fragmentation is referred to as α-cleavage.

FIGURE 25.6 Mass spectrum of N-ethylpropylamine, showing an odd-numbered molecular ion, and two peaks resulting from α-cleavage.

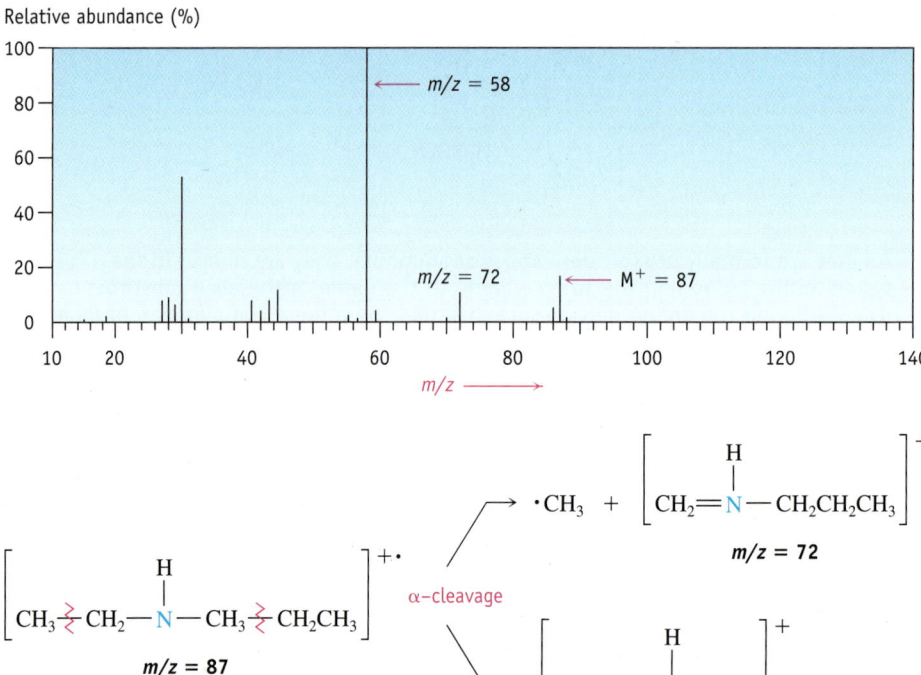

In practice, identification of unknown amines or other organic compounds by mass spectrometry is routinely done through comparison with computer libraries consisting of spectra of tens of thousands of known structures.

NMR Spectroscopy

Again, amines are somewhat difficult to distinguish from alcohols by ^{1}H NMR spectroscopy, as both N—H and O—H hydrogen atoms tend to appear as very broad signals without well-defined coupling to neighbouring C—H hydrogen atoms. As with O—H hydrogen atom absorptions in alcohols, amine N—H hydrogen atom absorptions are best identified by adding a small amount of labelled water (D_2O) to the NMR sample. Rapid exchange of N—D for N—H takes place, and the N—H hydrogen atom peak disappears from the NMR spectrum.

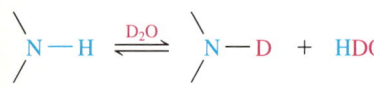

In p-ethoxyaniline, for example, the N-methyl hydrogen atoms are distinctive in the ^{1}H NMR spectrum, where they appear as a singlet at 3.3 δ (Figure 25.7).

EXERCISE 25.3—^{1}H NMR SPECTRA

Identify all of the peaks in the ^{1}H NMR spectrum of p-ethoxyaniline in Figure 25.7, and explain their chemical shift, splitting, and integration.

Carbon atoms next to amine nitrogen atoms, like those next to oxygen atoms, are slightly deshielded in the ^{13}C NMR spectrum and absorb about 20 ppm downfield from where they would absorb in a similar alkane.

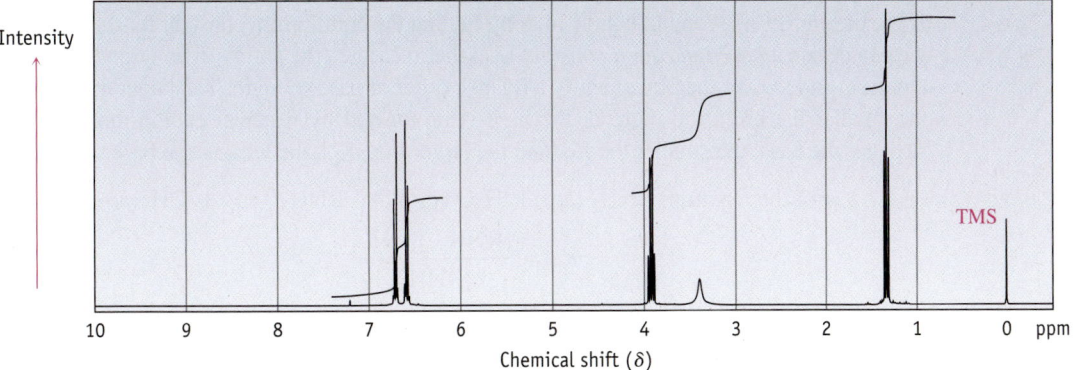

***p*-Ethoxyaniline**

Intensity

10   9   8   7   6   5   4   3   2   1   0   ppm

Chemical shift (δ)

TMS

FIGURE 25.7 ^{1}H NMR spectrum of *p*-ethoxyaniline.

EXERCISE 25.4—^{13}C NMR SPECTRA

Predict the number of peaks in the ^{13}C NMR spectrum of *p*-ethoxyaniline, a molecule of which is shown in Figure 25.7, and the general regions for each chemical shift

---

Using IR spectroscopy, amines can be detected by peaks corresponding to stretching of N—H bonds (strong bands at 3300–3500 cm^{-1}, two bands for primary amines) and C—N bonds (weak band at 1020–1250 cm^{-1}), and weak bands for N—H scissor (1650–1580 cm^{-1}) and wagging (910–660 cm^{-1}) bending vibrations. In the MS, the key feature is due to the "nitrogen rule"—a molecular ion has an odd-numbered mass/charge ratio for compounds with an odd number of nitrogen atoms. In the ^{13}C NMR spectrum, carbon atoms alpha to nitrogen atoms are deshielded and absorb about 20 ppm downfield from normal values. In the ^{1}H NMR spectrum, N—H hydrogen atoms show broad signals that can be confirmed by adding D$_2$O and seeing if the signals disappear due to exchange of deuterium atoms for hydrogen atoms.

25.4 Electronic Structure and Reactivity

The bonding in amine molecules is similar to the bonding in ammonia. The shape about an amine nitrogen atom is approximately tetrahedral, with the three bonded atoms occupying three corners of a regular tetrahedron and the lone pair of electrons occupying the fourth. As expected, the C—N—C bond angles are very close to the 109° tetrahedral value. For trimethylamine, the C—N—C angle is 108° and the C—N bond length is 147 pm. Using the hybrid orbital model, the nitrogen atom is *sp*^{3} hybridized, and a lone pair of electrons is in an *sp*^{3} orbital, where it can readily be shared with electrophiles.

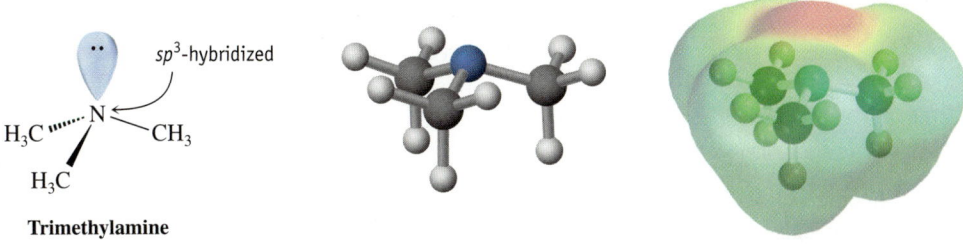

sp^3-hybridized

Trimethylamine

Basicity

The physical properties and chemical reactivity of amines can be understood with the help of an electrostatic potential map. The red region of an amine molecule shows the availability of the lone pair of electrons on its nitrogen atom for interaction with electrophilic sites on other molecules. Amines are much stronger bases than alcohols, ethers, or water. When an amine is dissolved in water, an equilibrium is established in which water acts as an acid and donates H^+ to the amine. As seen in Section 14.4, by finding the equilibrium constant for the reaction, we can define a *base ionization constant* (K_b) that measures the ability of an amine to accept a proton, and we can thereby quantify a relative order of base strength. The larger the K_b (and the smaller the pK_b), the more of the base is protonated by water at equilibrium, and the stronger the base; the smaller the K_b (and the larger the pK_b), the weaker the base.

$$\text{For the reaction: } RNH_2 \text{ (aq)} + H_2O \text{ (ℓ)} \rightleftharpoons RNH_3^+\text{(aq)} + OH^-\text{(aq)}$$

$$K_b = \frac{[RNH_3^+][OH^-]}{[RNH_2]}$$

$$pK_b = -\log K_b$$

Table 25.1 gives the pK_b values of some common amines. As indicated, substitution has relatively little effect on alkylamine basicity. Most simple alkylamines have pK_b values in the narrow range 3–4, regardless of their exact structure.

TABLE 25.1 Basicity of Some Common Amines

Name	Structure		pK_b
Triethylamine	$(CH_3CH_2)_3N$	More basic	2.99
Ethylamine	$CH_3CH_2NH_2$		3.19
Dimethylamine	$(CH_3)_2NH$		3.27
Methylamine	CH_3NH_2		3.34
Diethylamine	$(CH_3CH_2)_2NH$		3.51
Trimethylamine	$(CH_3)_3N$		4.19
Ammonia	NH_3		4.74
Pyridine	(pyridine structure with N)		8.75
Aniline	(benzene ring with $-NH_2$)	Less basic	9.37

Molecular Modelling (Odyssey)

e25.3 Compare electrostatic potential maps of amines and amide molecules to understand their basicities.

As seen from Table 25.1, *arylamines,* such as aniline, are much weaker bases than alkylamines, with K_b values lower by a factor of about 1 000 000. The nitrogen lone-pair electrons in an arylamine molecule are shared by orbital overlap with the π orbital of the aromatic ring, and they are less available for bonding to an acid. The electrostatic potential map in Figure 25.8 shows this electron delocalization, as do the five resonance structures that can be drawn for a single aniline molecule:

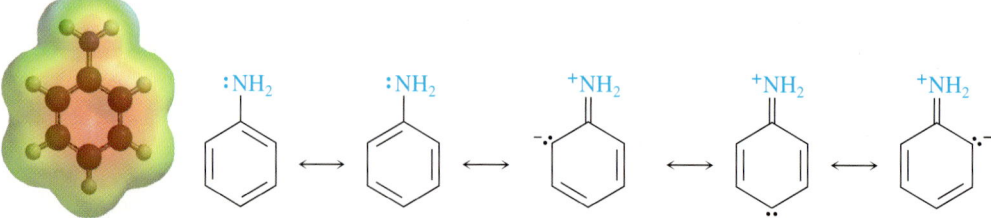

FIGURE 25.8 Electrostatic potential map and resonance structures, showing the delocalization of the electron pair in a single aniline molecule.

In contrast to amines, *amides* ($RCONH_2$) are much weaker bases. Amides don't usually react with acids except under harsh conditions, and their aqueous solutions are neutral. The main reason for the decreased basicity of amides relative to amines is that their molecules' nitrogen lone-pair electrons are shared by orbital overlap with the neighbouring carbonyl group's π orbital. The electrons are therefore much less available for bonding to an acid. Using a delocalization model, amides are less reactive as bases than amines because their molecules' electron density is delocalized over both the oxygen and nitrogen atoms, whereas they are localized on the nitrogen atom in amine molecules. Electrostatic potential maps show clearly this decreased electron density on the amide nitrogen atom, and two resonance structures show how the electrons are distributed within one amide molecule.

Think about It

e25.4 Watch an animation of the electron delocalization over the amide group.

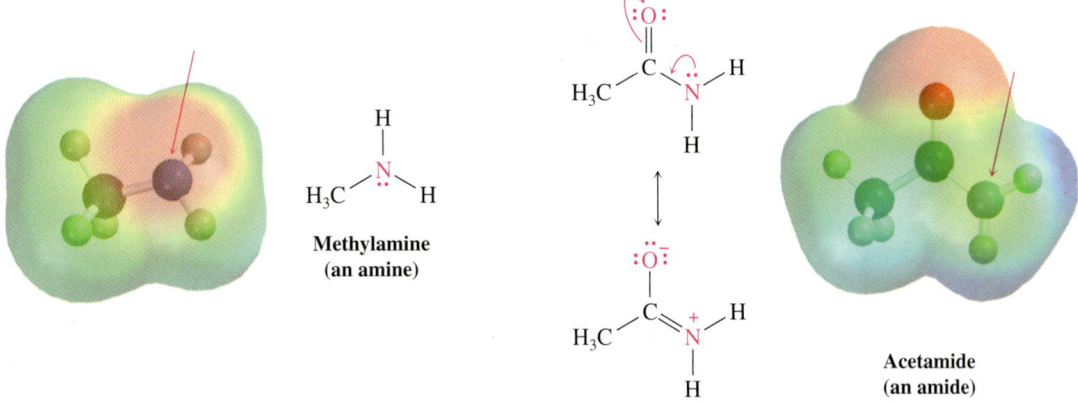

Methylamine (an amine)

Acetamide (an amide)

As a result of their basicity, the lone pair of electrons on the nitrogen atom of amines can interact with electrophilic sites in other molecules. This helps explain both the physical properties and chemical reactivity of amines.

EXERCISE 25.5—ACID-BASE PROPERTIES

Which compound in each of the following pairs is more basic?
(a) $CH_3CH_2NH_2$ or $CH_3CH_2CONH_2$
(b) $NaOH$ or $C_6H_5NH_2$
(c) CH_3NHCH_3 or $CH_3NHC_6H_5$
(d) CH_3OCH_3 or $(CH_3)_3N$

The three atoms bonded to the nitrogen atom of amine molecules are directed to three corners of a regular tetrahedron, and the lone pair of electrons occupies the fourth corner of the tetrahedron. Using the hybrid orbital model the nitrogen atom is considered sp^3 hybridized. The chemistry of amines is dominated by the presence of the lone-pair electrons on nitrogen atoms in amine molecules, which make amines both basic and nucleophilic. Arylamines and amides are generally weaker bases than alkylamines because their nitrogen atom lone-pair electrons are delocalized by orbital overlap with the aromatic π electron system of an arylamine or the carbonyl group of an amide.

Physical Properties

Like alcohols, molecules of primary and secondary amines form hydrogen bonds, giving amines higher boiling points than alkanes of similar molecular weight. Also like alcohols, amines have high dipole moments due to their molecular polarity, and those with fewer than five carbon atoms are generally water-soluble.

Molecular Modelling (Odyssey)

e25.5 Use electrostatic potential maps of DNA base pairs to see H-bonding interactions.

The importance of non-covalent hydrogen-bonding interactions involving amines can play a major role in determining overall shape of macromolecules. For example, the two strands of the DNA double helix are held together by complementary pairing of nitrogen or oxygen bases on one strand with N—H hydrogen atoms from amine or amide functional groups on the other strand (Figure 25.9).

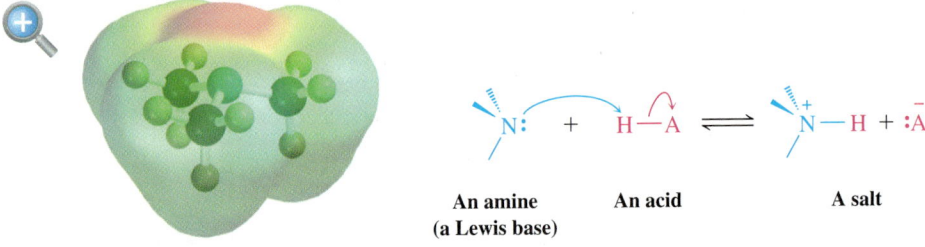

FIGURE 25.9 Hydrogen bonding between complementary base pairs in DNA. The centres of molecules of the bases are relatively neutral (green) while the edges have positive (blue) and negative (red) regions. A guanosine base (G) on one strand is shown here aligned with a cytidine base (C) on the second strand.

Reactions of Amines as Bases

The chemistry of amines is also dominated by the lone pair of electrons on their nitrogen atoms. Amines can react with acids to form acid-base salts, and with electrophiles in many of the polar reactions seen in previous chapters.

It's often possible to take advantage of this basicity to purify an amine. For example, if a mixture of an amine (basic) and a ketone (neutral) is dissolved in an organic solvent and aqueous HCl solution is added, the basic amine dissolves in the acidic water solution as its ammonium ion, while the ketone remains in the organic solvent. Separation of the water layer and neutralization of the ammonium ion by addition of NaOH then provides the pure amine (Figure 25.10).

FIGURE 25.10 Separation and purification of an amine from a mixture.

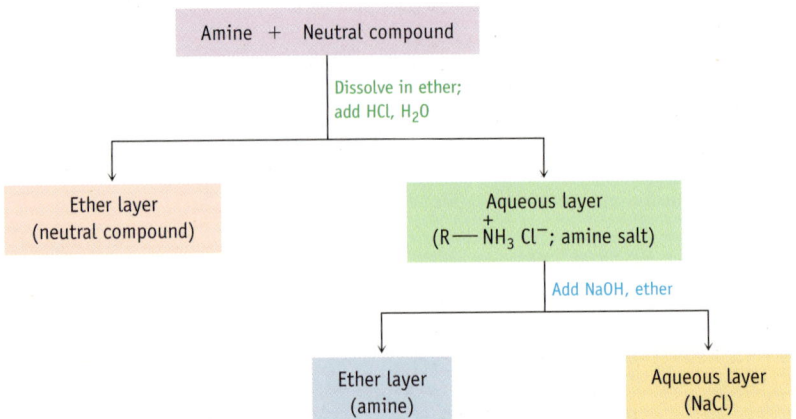

WORKED EXAMPLE 25.2—ACID-BASE REACTIONS OF AMINES

Predict the product of the following reaction:

$$CH_3CH_2NHCH_3(aq) + H_3O^+(aq) \longrightarrow ?$$

Strategy

Amines are protonated by strong acids to yield ammonium ions.

Solution

$$CH_3CH_2NHCH_3(aq) + H_3O^+(aq) \longrightarrow [CH_3CH_2NH_2CH_3]^+(aq)$$

EXERCISE 25.6—ACID-BASE REACTIONS OF AMINES

Predict the product of the following reaction:

$$\text{(cyclopentyl)}-N\begin{smallmatrix}CH_3\\H\end{smallmatrix} + HBr \longrightarrow ?$$

Alkylation Reactions of Amines

Ammonia or alkylamines react with alkyl halides in S_N2 alkylation reactions [<<Section 21.6]. In principle, this can be used to synthesize more highly substituted amines from simple ones. If ammonia is used, a primary amine results; if a primary amine is used, a secondary amine results; and so on. Even tertiary amines react with alkyl halides to yield quaternary ammonium salts, $R_4N^+ X^-$. Alkyl halides are referred to as **alkylating agents** in these reactions because they add alkyl groups to the nitrogen atoms of amine molecules.

S_N2 reaction

Ammonia	$\ddot{N}H_3 + R-X$	$\longrightarrow$	$RNH_3^+ X^-$	$\xrightarrow{NaOH} RNH_2$	Primary
Primary	$R\ddot{N}H_2 + R-X$	$\longrightarrow$	$R_2NH_2^+ X^-$	$\xrightarrow{NaOH} R_2NH$	Secondary
Secondary	$R_2\ddot{N}H + R-X$	$\longrightarrow$	$R_3NH^+ X^-$	$\xrightarrow{NaOH} R_3N$	Tertiary
Tertiary	$R_3\ddot{N} + R-X$	$\longrightarrow$	$R_4N^+ X^-$	Quaternary ammonium salt	

Unfortunately, none of these reactions stops cleanly after a single alkylation has occurred. Because primary, secondary, and tertiary amines all have similar reactivity, the initially formed monoalkylated amine often undergoes further reaction to yield a mixture of products. For example, treatment of 1-bromooctane with a two-fold excess of ammonia leads to a mixture containing only a 45% yield of octylamine. A nearly equal amount of dioctylamine is produced by double alkylation, along with smaller amounts of trioctylamine and tetraoctylammonium bromide.

$$CH_3(CH_2)_6CH_2Br + :NH_3 \longrightarrow CH_3(CH_2)_6CH_2\ddot{N}H_2 + [CH_3(CH_2)_6CH_2]_2\ddot{N}H$$

1-Bromooctane   **Octylamine (45%)**   **Dioctylamine (43%)**

$$+ [CH_3(CH_2)_6CH_2]_3N: + [CH_3(CH_2)_6CH_2]_4\overset{+}{N} \overset{-}{Br}$$

Trace   Trace

The ability of amines to act as nucleophiles to initiate alkylation reactions is also the reason why many alkylating agents are considered carcinogens (cancer-promoting substances). We saw one example of this with the nucleophilic ring opening of epoxides of polycyclic aromatic hydrocarbons in Section 22.9. Nitrogen atoms (and oxygen atoms) in molecules of guanine bases in DNA nucleotides can donate electron pairs for bond formation to electrophilic carbon atoms on alkylating agent molecules, with loss of a suitable leaving group (Figure 25.11).

Think about It

e25.6 Watch an animation of the S_N2 mechanism for these reactions.

Molecular Modelling (Odyssey)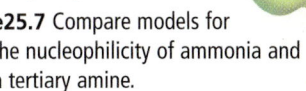

e25.7 Compare models for the nucleophilicity of ammonia and a tertiary amine.

FIGURE 25.11 Alkylation of a guanine base nitrogen atom in DNA by a carcinogen alkylating agent. This changes the interactions between base pairs in DNA strands and results in mutations that disrupt DNA function. The carbonyl oxygen atom can also be alkylated.

Anti-neoplastic alkylating agents inhibit or control the development of neoplasms—abnormal tissue growth—by alkylating DNA and disrupting DNA function.

Once a nitrogen atom in a molecule has been alkylated, its lone pair of electrons is no longer available for being a Lewis base or for hydrogen-bonding interactions with base pairs on another strand of DNA. This can lead to events, such as the mis-pairing of nucleotides, that disrupt DNA function in several different ways and ultimately lead to cell death. Ironically, this ability of alkylating agents to bind to DNA can be used as an advantage when the goal is to develop pharmaceutical products [*chemotherapeutic drugs*, <<Section 1.2] that will kill rapidly growing cells. An example is anti-neoplastic alkylating agents, drugs important in cancer chemotherapy, which work by first alkylating DNA.

A good example is *cis*-diamminedichloroplatinum(II) [cisplatin, <<Section 9.6], which has an electrophilic Pt(II) ion and *two* coordination sites occupied by Cl⁻ leaving groups *cis* to each other. One molecule of cisplatin binds to nitrogen atoms of two different guanine and/or adenine bases in DNA (Figure 25.12). This creates covalent bonds between atoms in two different places on one DNA strand and also creates cross-bridges between two strands. The result? DNA replication and DNA repair mechanisms are inhibited.

FIGURE 25.12 The chemotherapeutic agent cisplatin is (a) injected into the blood and converted to its active form inside the cell. (b) Here, it can bind to nitrogen atoms on adjacent guanine (G) bases on the same strand, or different strands, of DNA. (c) This distorts the DNA structure and interferes with DNA repair mechanisms, leading to cell death.

Molecular Modelling (Odyssey)

e25.8 Model the Pt binding sites on DNA, and see how binding affects DNA structure.

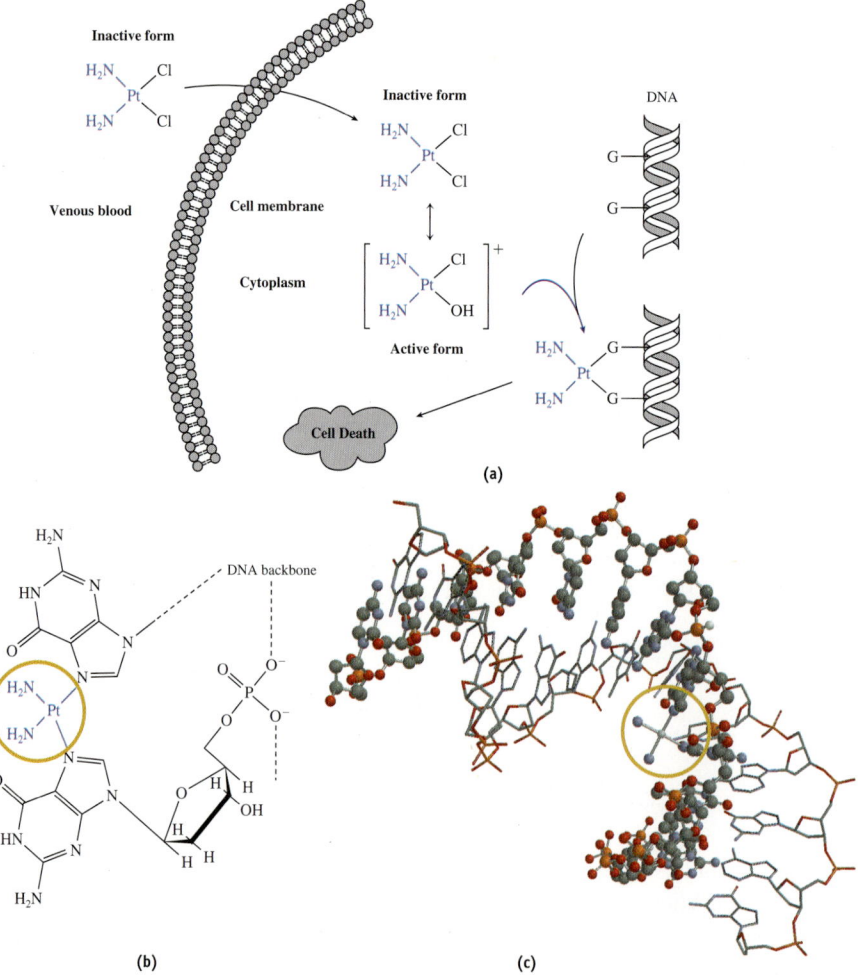

How could you synthesize diethylamine from ammonia and an alkyl halide?

Strategy

Look at the starting material (NH_3) and the product ($(CH_3CH_2)_2NH$, and note the difference. Since two ethyl groups have become bonded to the nitrogen atom, the reaction must involve 2 mol of ethyl halide for each mol of ammonia.

Solution

$$2\ CH_3CH_2Br + NH_3 \longrightarrow (CH_3CH_2)_2NH$$

How could you synthesize the following amines from ammonia and appropriate alkyl halides?
(a) triethylamine
(b) tetramethylammonium bromide

A substance containing the following molecules (blue = nitrogen atom) can be made by reaction between a primary amine and a *dihalide*. Identify the two reactants, and write the reaction equation.

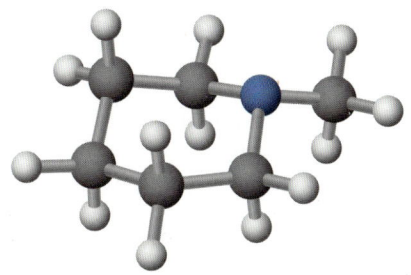

Acylation of Amines

Primary and secondary (but not tertiary) amines can also be acylated by nucleophilic acyl substitution reactions with acyl chlorides or acid anhydrides to give amides [<<Section 24.7]. This same reaction between amino groups of amino acid molecules and carboxyl groups of other amino acids results in the amide bonds that hold together peptides and proteins.

Think about It

e25.9 Watch an animation of amide formation.

Amines react with alkyl halides in S$_N$2 reactions and with acyl chlorides in nucleophilic acyl substitution reactions. S$_N$2 alkylation of ammonia yields a primary amine; alkylation of a primary amine yields a secondary amine; and so on. This is both an important reaction of amines and a useful method for synthesizing substituted amines. Alkylation by amines in DNA is the mechanism by which both carcinogenic alkylating agents and some types of chemotherapeutic drugs work to disrupt DNA replication. Amines also react as nucleophiles with carbonyl compounds in acylation reactions, giving amides in reaction with carboxylic acid derivatives.

25.5 More about Alkaloids: Naturally Occurring Amines

Naturally occurring amines derived from plant sources were once known as "vegetable alkali" because their aqueous solutions are slightly basic, but they are now referred to as **alkaloids**. Alkaloids vary widely in structure, from the simple to the enormously complex. The odour of rotting fish, for example, is caused by the simplest alkaloid, methylamine. (The common use of acidic lemon juice to mask fish odours is an acid-base reaction that forms the non-volatile methyl-ammonium salt.) Diamines such as putrescine (butane-1, 4-diamine) and cadaverine (pentane-1,5-diamine)—found in rotting flesh—smell just like you would predict from their names!

Many alkaloids have pronounced biological properties, and many of the pharmaceutical agents used today are alkaloids from natural sources. You were introduced to morphine and pseudoephedrine in Section 25.1. A few others of the many thousand examples:

- Atropine, an anti-spasmodic agent used for the treatment of colitis, is obtained from the flowering plant *Atropa belladonna,* commonly called the deadly nightshade.
- Cocaine, both an anesthetic and a central nervous system stimulant, is obtained from the coca bush *Erythroxylon coca,* endemic to upland rainforest areas of Colombia, Ecuador, Peru, Bolivia, and western Brazil.
- Reserpine, a tranquilizer and anti-hypertensive, comes from powdered roots of the semi-tropical plant *Rauwolfia serpentina.*

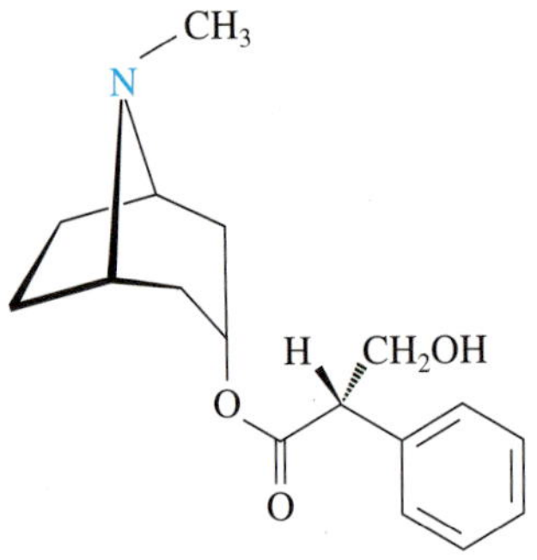

Atropine

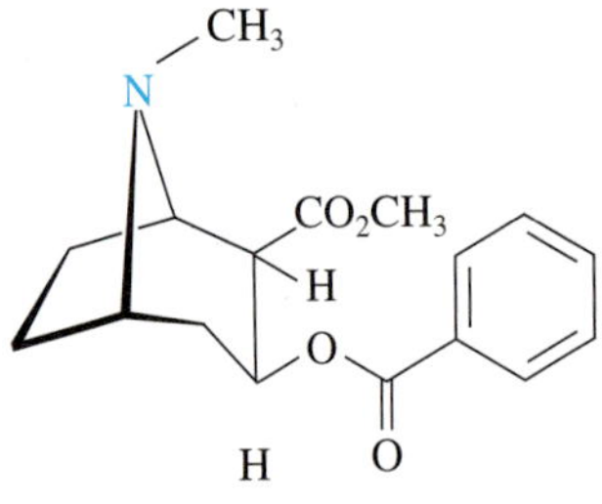

Cocaine

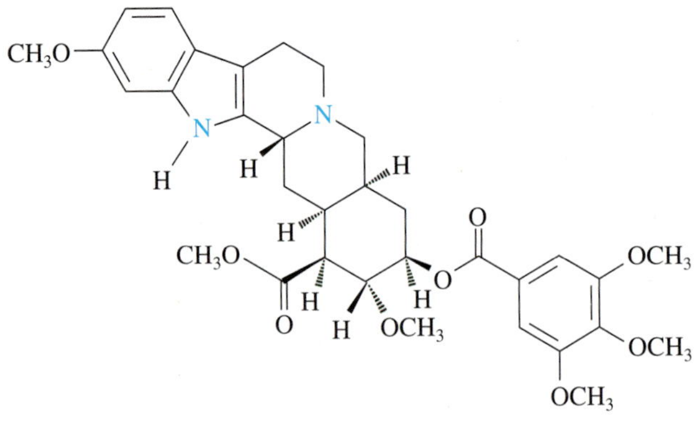

Reserpine

25.6 Heterocyclic Aromatic Amines

Recall [<<Sections 9.7 and 20.3] that **heterocyclic compounds** contain one or more different atoms in addition to carbon atoms in their rings. We've already seen as examples alkaloids such as cocaine and morphine and reduced pyrrole heterocycles in Vitamin B-12. Particularly important in biochemistry are the heterocyclic nitrogen bases in the

nucleotides of DNA. Other examples of heterocyclic amines include the anti-ulcer agent cimetidine and the sedative phenobarbital:

Cimetidine (an antiulcer agent),
with multiple nitrogen-containing functional groups

Phenobarbital (a sedative),
a heterocycle

For the most part, heterocyclic amines have the same chemistry as their open-chain counterparts. In certain cases, though, particularly when their rings are unsaturated, heterocycles have unique and interesting properties. Let's look at several examples.

Pyrrole, a Five-Membered Aromatic Heterocycle

Molecules of pyrrole are made of five-membered heterocyclic amines with two double bonds and one nitrogen atom. Although pyrrole is both an amine and a conjugated diene, its chemistry is not consistent with either of these structural features. Unlike most amines, pyrrole is not basic; unlike most conjugated dienes, pyrrole doesn't undergo electrophilic addition reactions. How can we explain these observations?

Pyrrole is *aromatic*. Even though its molecules have five-membered rings, those rings have six π electrons in a cyclic, conjugated π orbital system, just as benzene does, fitting the $4n + 2$ rule [<<Section 20.2]. Using the hybrid orbital model, each of the four carbon atoms contributes one π electron, and the sp^2 hybridized nitrogen atom contributes two more (its lone pair). The six π electrons occupy π orbitals with lobes above and below the plane of the flat ring, as shown in Figure 25.13. Because the lone-pair electrons on the nitrogen atom are delocalized around the aromatic ring, they are less available for donation to an acid and pyrrole is not very basic. In the electrostatic potential map of a pyrrole molecule, you can see that the nitrogen atom is neutral (green) rather than electron-rich (red).

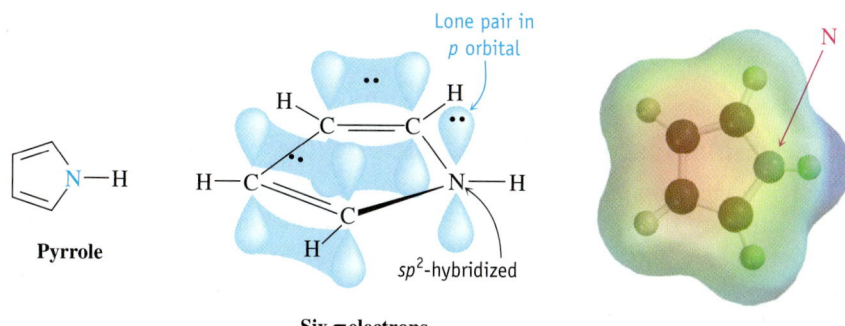

Pyrrole

Six π electrons

FIGURE 25.13 Pyrrole, an aromatic heterocycle, has molecules with a π electron structure similar to that of benzene. The nitrogen atom is much less basic than other amines because the electron pair is delocalized around the aromatic ring.

Like benzene, pyrrole molecules undergo substitution of a ring hydrogen atom on reaction with an electrophile, rather than reaction at the nitrogen atom. Substitution normally occurs on the carbon atom next to the nitrogen atom, as the following nitration shows. This orientation can be understood by comparing the relative stability of the carbocation intermediates for substitution at the carbon atom next to the nitrogen atom and the carbon atom adjacent to this one (see Exercise 25.9) [<<Section 20.8].

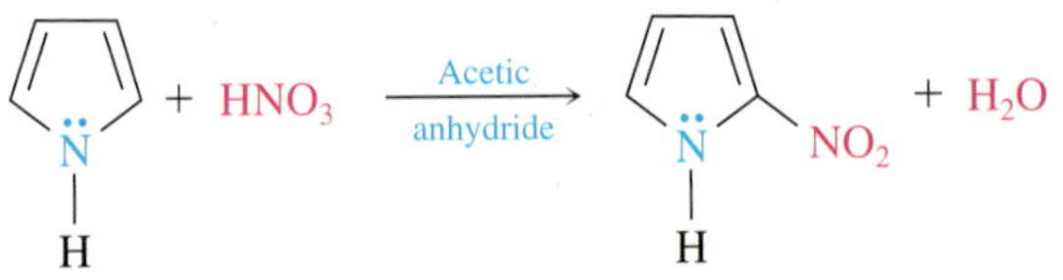

Pyrrole

2-Nitropyrrole
(83)%

Substituted pyrrole rings form the basic building blocks from which many important plant and animal pigments are constructed. Among these are the porphyrins used in photodynamic therapy and *heme,* an iron-containing tetrapyrrole found in blood [<<Section 1.2]. Note that the pyrrole nitrogen atoms complexed to the metal ion in heme molecules do not have a proton attached to them. A deprotonation step before complexation makes the pyrrole nitrogen atoms good Lewis bases.

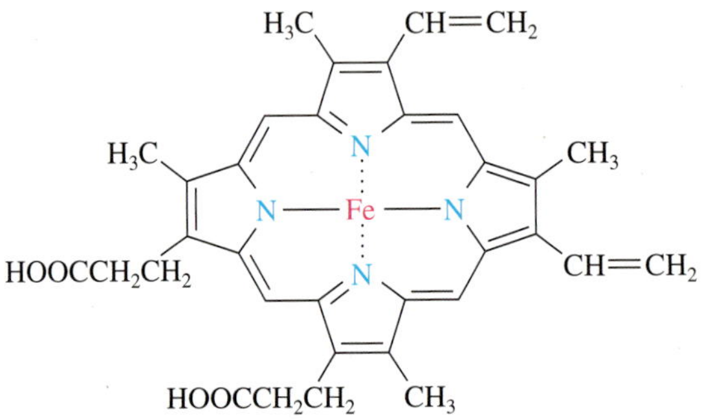

Heme

Reduced pyrrole rings also form the core skeleton of Vitamin B-12.

EXERCISE 25.9—ELECTROPHILIC AROMATIC SUBSTITUTION

Pyrrole undergoes other typical electrophilic substitution reactions in addition to nitration. What products would you expect to obtain from reaction of *N*-methylpyrrole with the following reagents?

(a) Br_2 (b) CH_3Cl, $AlCl_3$ (c) CH_3COCl, $AlCl_3$

EXERCISE 25.10—ELECTROPHILIC AROMATIC SUBSTITUTION

Review the mechanism of the nitration of benzene [<<Section 20.6], and then propose a mechanism for the nitration of pyrrole.

Pyridine, a Six-Membered Aromatic Heterocycle

Pyridine is a nitrogen-containing heterocyclic analog of benzene. Like benzene, pyridine molecules are flat with bond angles of approximately 120° and with C—C bond lengths of 139 pm, intermediate between normal single and double bonds. Also like benzene, pyridine is aromatic with six π electrons in a cyclic, conjugated π orbital system. Counting electrons in the π system using the hybrid orbital model, we can think of the sp^2 hybridized nitrogen atom and the five carbon atoms each contributing one π electron to the cyclic, conjugated π orbitals of the ring. Unlike the situation in pyrrole, however, the lone-pair electrons on the pyridine nitrogen atom are not part of the π orbital system but instead occupy an sp^2 orbital in the plane of the ring (Figure 25.14). As a result, the pyridine lone-pair electrons are available for donation to an acid and pyridine is therefore a reasonable base. Compare the electrostatic potential maps of molecules of pyrrole (Figure 25.13) and pyridine (Figure 25.14) to see this difference in basicity.

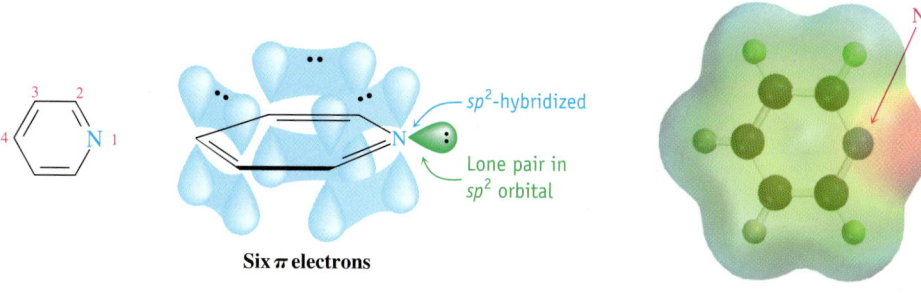

Six π electrons

FIGURE 25.14 Electronic structure of pyridine, a nitrogen-containing analog of benzene. Although less basic than typical alkylamines, pyridine ($pK_b = 8.75$) is nevertheless used in a variety of organic reactions when a base catalyst is required. Recall, for instance, that the reaction of an acyl chloride with an alcohol to yield an ester is commonly done in the presence of pyridine [<<Section 24.7].

Substituted pyridines, such as the B-6 complex vitamins pyridoxal and pyridoxine, are important biologically. Present in yeast, cereal, and other foodstuffs, the B-6 vitamins are necessary for the synthesis of some amino acids.

EXERCISE 25.11—HETEROCYCLIC AMINES

Molecules of the five-membered heterocycle imidazole contains two nitrogen atoms, one "pyrrole-like" and one "pyridine-like." Draw an orbital picture of imidazole, and indicate the orbital in which each nitrogen atom has its electron lone pair.

Imidazole

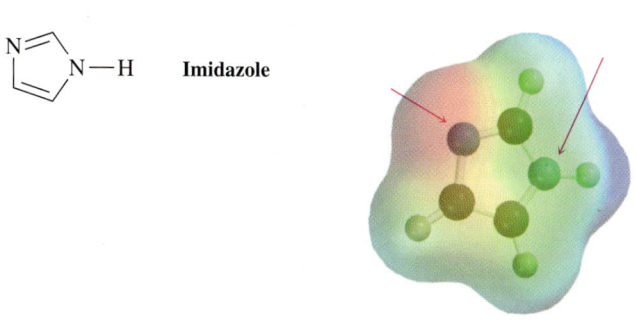

Imidazole

Pyridoxal

Pyridoxine

EXERCISE 25.12—HETEROCYCLIC AMINES

Which nitrogen atom in an imidazole molecule (Exercise 25.11) is more basic according to the electrostatic potential map above? Why?

Fused-Ring Aromatic Heterocyclic Amines

Fused-ring heterocycles like quinoline, isoquinoline, and indole are more complex than simple monocyclic compounds. Molecules of all three have a benzene ring and a heterocyclic ring sharing a common bond. These and many other fused-ring systems occur widely in nature, and many members of the class have useful biological properties. Quinine, a quinoline derivative found in the bark of the South American cinchona tree, was one of the first important anti-malarial drugs. *N,N*-Dimethyltryptamine, which contains an indole ring, is a powerful hallucinogen.

Heterocyclic amines, whose molecules all have a nitrogen atom in a ring, have a great diversity in their structures and properties. Pyrrole, pyridine, indole, and quinoline all show aromatic properties.

Quinoline

Isoquinoline

Indole

Quinine, an antimalarial drug
(a quinoline alkaloid)

N, *N*-Dimethyltryptamine, a hallucinogen
(an indole alkaloid)

Taking It Further

e25.10 Read about how amines are synthesized.

We have discussed the synthesis of amines from ammonia and simpler amines in Section 25.4. See e25.10 to learn other ways to synthesize amines: by reduction of nitriles and amides, the reductive amination of aldehydes and ketones, and the reduction of nitrobenzenes.

SUMMARY

Key Concepts

- Nitrogen-containing compounds such as **amines**, amides, and **heterocyclic compounds**, play crucial roles in the chemistry of living systems, and in the history of organic chemistry. (Section 25.1)
- Amines are formally organic derivatives of ammonia. They are named in the IUPAC system either by adding the suffix –*amine* to the name of the alkyl substituent or by considering the amino group as a substituent on a more complex parent molecule. (Section 25.2)
- Amines can be identified by characteristic spectral features. Using IR spectroscopy, amines can be detected by peaks corresponding to stretching of N—H bonds (strong bands at 3300–3500 cm^{-1}, two bands for primary amines) and C—N bonds (weak band at 1020–1250 cm^{-1}), and weak bands for N—H scissor (1650–1580 cm^{-1}) and wagging (910–660 cm^{-1}) bending vibrations. In the MS, the key feature is due to the "nitrogen rule"—a molecular ion has an odd-numbered mass/charge ratio for compounds with an odd number of nitrogen atoms. In the ^{13}C NMR spectrum, carbon atoms alpha to nitrogen atoms are deshielded and absorb about 20 ppm downfield from normal values. In the ^{1}H NMR spectrum, N—H hydrogen atoms show broad signals that can be confirmed by adding D$_2$O and seeing if the signals disappear due to exchange of deuterium atoms for hydrogen atoms. (Section 25.3)
- The three atoms bonded to the nitrogen atom of amine molecules are directed to three corners of a regular tetrahedron, and the lone pair of electrons occupies the fourth corner of the tetrahedron. Using the hybrid orbital model, the nitrogen atom is considered sp^3 hybridized. The chemistry of amines is dominated by the presence of the lone-pair

electrons on nitrogen atoms in amine molecules, which make amines both basic and nucleophilic. Arylamines and amides are generally weaker bases than alkylamines because their nitrogen atom lone-pair electrons are delocalized by orbital overlap with the aromatic π electron system of an arylamine or the carbonyl group of an amide. (Section 25.4)

- Many of the reactions of amines are familiar from previous chapters. Amines react with alkyl halides in S_N2 reactions and with acyl chlorides in nucleophilic acyl substitution reactions. S_N2 alkylation of ammonia yields a primary amine; alkylation of a primary amine yields a secondary amine; and so on. This is both an important reaction of amines and a useful method for synthesizing substituted amines. Alkylation by amines in DNA is the mechanism by which both carcinogenic **alkylating agents** and some types of chemotherapeutic agents work to disrupt DNA replication. Amines also react as nucleophiles with carbonyl compounds in acylation reactions, giving amides in reaction with carboxylic acid derivatives. (Section 25.4)
- Heterocyclic amines, whose molecules all have a nitrogen atom in a ring, have a great diversity in their structures and properties. Pyrrole, pyridine, indole, and quinoline all show aromatic properties. (Section 25.6)
- Amines can be synthesized from S_N2 reaction of ammonia or less-substituted amines with alkylating agents (Section 25.4); from amides and nitriles by reduction with $LiAlH_4$; and from aldehydes and ketones by reductive amination with ammonia and a reducing agent. Arylamines are synthesized by nitration of an aromatic ring followed by reduction of the nitro group. (e25.10)

Key Reactions

1. Reactions of amines
   (a) With acids (Section 25.4)

An amine   **An acid**   **A salt**
(a Lewis base)

   (b) In S_N2 reactions with alkylating agents such as alkyl halides (Section 25.4)

S_N2 reaction

Ammonia	$NH_3 + R-X \longrightarrow RNH_3^+ X^- \xrightarrow{NaOH} RNH_2$	Primary
Primary	$RNH_2 + R-X \longrightarrow R_2NH_2^+ X^- \xrightarrow{NaOH} R_2NH$	Secondary
Secondary	$R_2NH + R-X \longrightarrow R_3NH^+ X^- \xrightarrow{NaOH} R_3N$	Tertiary
Tertiary	$R_3N + R-X \longrightarrow R_4N^+ X^-$	Quaternary ammonium salt

   (c) With carbonyl compounds to produce amides (Section 25.4)

   (d) Heterocyclic aromatic amines undergo electrophilic aromatic substitution (Section 25.6)

Pyrrole   **2-Nitropyrrole**
(83)%

2. Synthesis of amines
   (a) By alkylation of ammonia and less substituted amines (see 1(b) above)
   (b) By reductive amination of aldehydes and ketones (e25.10)

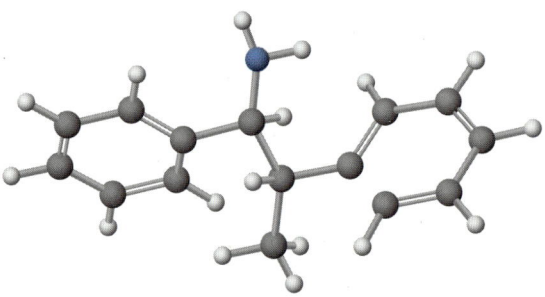

   (c) Synthesis of arylamines by reduction of nitrobenzenes (e25.10)

REVIEW QUESTIONS

Section 25.2: Naming Amines

25.13 Name the amines whose molecules are shown below, and identify each as primary, secondary, or tertiary:

(a)

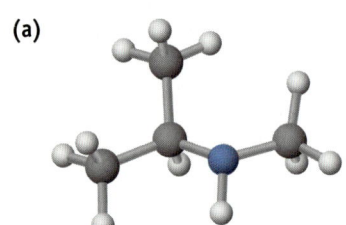

(b)

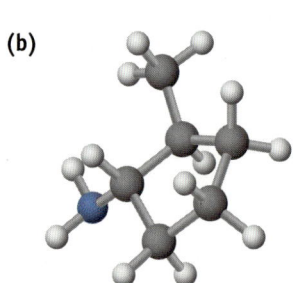

(c)

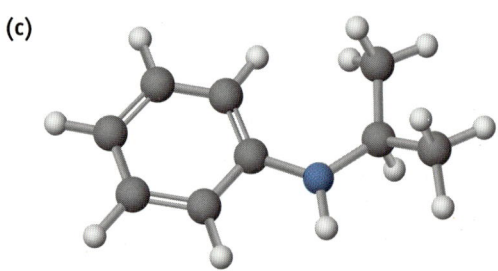

25.14 Name the amine whose molecule is shown, including *R, S* stereochemistry, and draw the structure of the product molecule of its reaction with (a) CH₃CH₂Br, and (b) CH₃COCl.

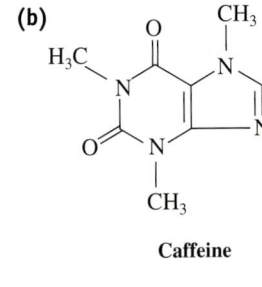

25.15 Classify each of the amine (not amide) nitrogen atoms in the following molecules as primary, secondary, or tertiary:

(a)

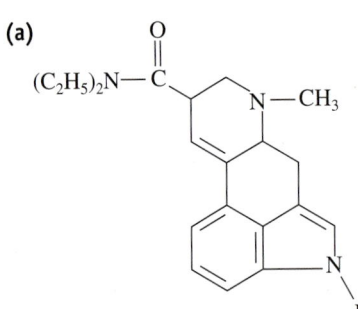

Lysergic acid diethylamide

(b)

Caffeine

25.16 Draw structures of molecules that meet the following descriptions:
   (a) a secondary amine with one isopropyl group
   (b) a tertiary amine with one phenyl group and one ethyl group
   (c) a quaternary ammonium salt with four different groups bonded to nitrogen

25.17 Draw structures of molecules corresponding to the following IUPAC names:
   (a) triethylamine
   (b) *N*-methylaniline

(c) tetraethylammonium bromide
(d) *p*-bromoaniline
(e) *N*-ethyl-*N*-methylcyclopentylamine

25.18 Draw structures of molecules corresponding to the following IUPAC names:
(a) *N,N*-dimethylaniline
(b) *N*-methylcyclohexylamine
(c) (cyclohexylmethyl)amine
(d) (2-methylcyclohexyl)amine
(e) 3-(*N,N*-dimethylamino)propanoic acid

25.19 Mescaline, a powerful hallucinogen derived from the peyote cactus, has the systematic name 2-(3,4, 5-trimethoxyphenyl)ethylamine. Draw the structure of a mescaline molecule.

25.20 There are eight isomeric amines with the formula $C_4H_{11}N$. Draw structures of their molecules, name them, and classify each as primary, secondary, or tertiary.

25.21 Propose structures for amine molecules that fit the following descriptions:
(a) a secondary arylamine
(b) a 1,3,5-trisubstituted arylamine
(c) an achiral quaternary ammonium salt
(d) a five-membered heterocyclic amine

Section 25.3: Identifying Amines: Spectroscopy

25.22 Propose structures for the molecules of amines with the following ^1H NMR spectra:

(a) C_3H_9NO

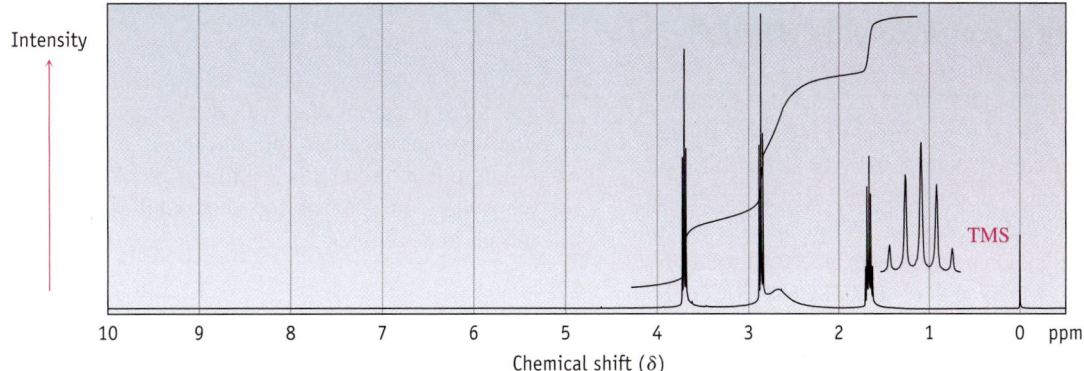

(b) $C_4H_{11}NO_2$

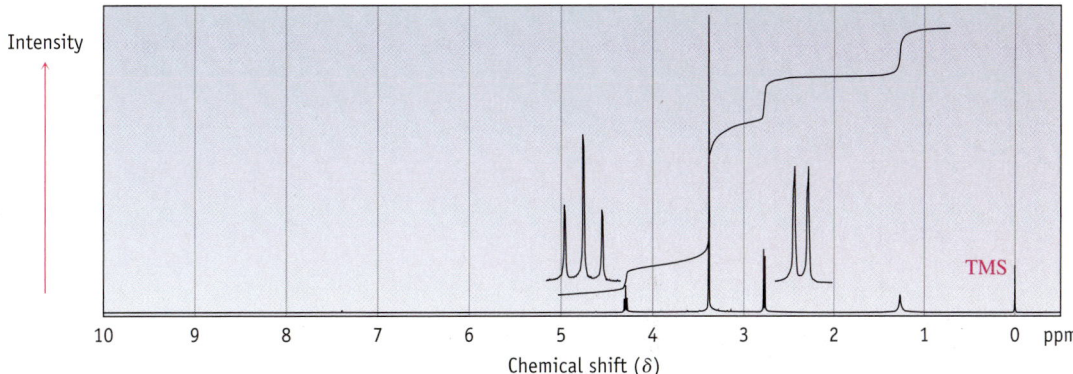

Section 25.4: Electronic Structure and Reactivity

25.23 Molecules of the following amine contain three nitrogen atoms. Rank them in order of expected increasing basicity.

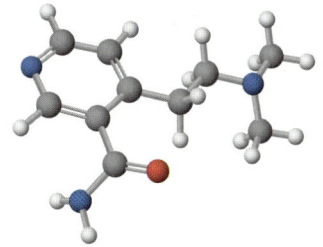

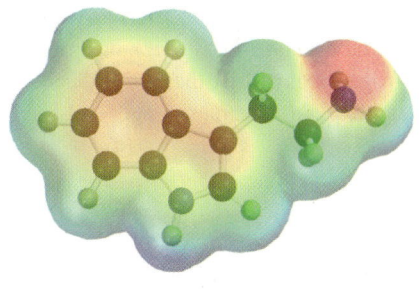

25.24 Which nitrogen atom in a molecule of the alkaloid tryptamine is more basic? Explain.

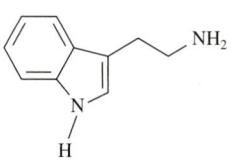

Tryptamine

25.25 How can you explain the fact that trimethylamine (boiling point: 3 °C) has a lower boiling point than dimethylamine (boiling point: 7 °C) even though it has a higher molecular weight?

25.26 How might you synthesize the following amines from ammonia and any alkyl halides needed?

(a) $CH_3CH_2CH_2CH_2CH_2CH_2NH_2$

(b) $(CH_3)_4N^+ \ I^-$

(c) [structure: benzene ring with CH₂NH₂ substituent]

(d) [structure: cyclohexane ring with NHCH₃ substituent]

Section 25.5: More about Alkaloids: Naturally Occurring Amines

25.27 Atropine ($C_{17}H_{23}NO_3$) is a poisonous alkaloid isolated from the leaves and roots of the deadly nightshade, *Atropa belladonna*. In small doses, atropine acts as a muscle relaxant: 0.5 ng (1 nanogram = 10^{-9} g) is sufficient to cause pupil dilation. On reaction with an NaOH solution, atropine yields tropic acid ($C_6H_5CH(CH_2OH)COOH$) and tropine ($C_8H_{15}NO$). Tropine, an optically inactive alcohol, yields tropidene on dehydration. Propose a structure for an atropine molecule.

[structure: Tropidene — bicyclic ring system with N–CH₃ bridge]
Tropidene

Section 25.6: Heterocyclic Aromatic Amines

25.28 Draw structures for molecules of the following amines:
(a) 2-ethylpyrrole
(b) 2,3-dimethylaniline
(c) 3-methylindole

25.29 Furan, the oxygen-containing analog of pyrrole, is aromatic in the same way that pyrrole is. Draw an orbital picture of a furan molecule, and show how it has six electrons in its cyclic conjugated π orbitals.

[structure: furan ring with O] **Furan**

25.30 By analogy with the chemistry of pyrrole, what product would you expect from the reaction of furan with Br_2 (see Question 25.29)?

25.31 Write mechanisms for the nitration of pyrrole at the 2- and 3- positions and, with reference to the stability of the two carbocation intermediates, explain why the predominant product is 2-nitropyrrole. You may wish to consult the mechanism for electrophilic aromatic substitution in Section 20.5.

[reaction scheme: Pyrrole + HNO₃ →(Acetic anhydride) 2-Nitropyrrole + H₂O]

Pyrrole   **2-Nitropyrrole**
                         (83)%

SUMMARY AND CONCEPTUAL QUESTIONS

25.32 To help understand which of the guanine nitrogen atoms in Figure 25.11 preferentially undergoes alkylation, consider a model compound. Histamine, whose release in the body triggers nasal secretions and constricts airways (this is why you might take an *anti-histamine*), has molecules with three nitrogen atoms. List them in order of increasing basicity and explain your reasoning.

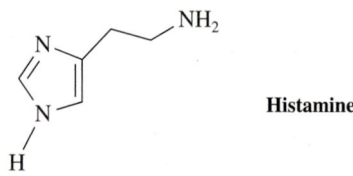
Histamine

25.33 How might you synthesize the following amines from 1-bromobutane?
(a) butylamine
(b) dibutylamine
(c) pentylamine

25.34 How might you synthesize each of the amines in Question 25.33 from butan-1-ol?

25.35 Which compound is more basic: $CH_3CH_2NH_2$ or $CF_3CH_2NH_2$? Explain.

25.36 Which compound is more basic: *p*-aminobenzaldehyde or aniline? Explain.

and as a vapour in the atmosphere. Many common minerals also contain these elements, including limestone ($CaCO_3$) and quartz (SiO_2). Aluminum and silicon occur together in many minerals; among the more common ones are feldspar, granite, and clay.

In Section 8.2, we saw the periodic variation of observable chemical and atomic properties of the main group elements, giving rise to trends in these properties across the periods and down the groups of the periodic table. The elements lower down groups 1 and 2 are the most metallic: they have the smallest first ionization energies, are the most powerful reducing agents, have ionic compounds with the non-metals, and their oxides are strong bases. They are the least electronegative (most electropositive) elements. Putting aside the noble gases of group 18, the elements high up on the right side of the periodic table (groups 16 and 17) are the most non-metallic.

The quantum mechanical model of electrons in atoms [<<Section 8.4] leads to periodic patterns of the electron configurations of atoms of the elements [<<Section 8.5]. Taken in conjunction with the concept of shielding and effective nuclear charge [<<Section 8.6], the electron configurations of the elements can be used to rationalize the periodic patterns of behaviour of the elements [<<Section 8.7]. For example, all of the elements of group 14 have s^2p^2 valence electron configurations.

Despite the commonalities among properties of the elements in each of the groups, each element is unique. In this chapter, we consider the specific chemical behaviour of selected elements and their compounds from each group, how their chemical properties influence where they are found, how they are obtained from their natural sources, as well as their uses. We also look at how the physical properties of some elements or their compounds can be interpreted in terms of their structures.

In this chapter, you will see how the simple concept of charge density (or charge-to-radius ratio) can help to make sense of differences among the chemical properties of compounds of the main group elements.

> The main group elements are those in groups 1, 2, and 13–18 of the periodic table. Eight of the ten most abundant elements on the earth are in this group, and the top ten chemicals produced by the chemical industry are all main group elements or their compounds. Although there are commonalities among properties of the elements in each of the groups, each element is unique.

26.3 Charge Density of Cations: An Explanatory Concept

All of chemistry, except nuclear chemistry [>>Chapter 30], is related to the extent and intensity of electrostatic interactions between species (atoms, ions, or molecules). This is how we interpret, for example, differences between melting points of substances and differences between solubilities in water, and why some substances, when mixed together, undergo a redistribution of electrons so that new substances are formed—that is, a chemical reaction.

In this section, we try to visualize the strength of electrostatic interaction between cations and other atoms, molecules, or ions in their environment. We do this from the perspective of how *charge density* of a cation influences the intensity of these interactions and affects the chemical properties of its compounds. This concept helps to explain the differences in the general chemical behaviours of compounds of the elements from group to group of the periodic table, as well as the particular differences between the chemistries of the elements within each group.

Although this chapter is concerned with the chemistry of the elements of the main block of the periodic table, the concept is equally applicable to cations of the transition elements [>>Chapter 27].

The Concept of Charge Density

We should expect that the strength of electrostatic interaction between a cation and negatively charged ions, atoms, or parts of molecules to be greater (a) the higher the charge on the cation, and (b) the smaller the cation and the closer it can get to the adjacent species.

This is consistent with the simple relationship for electrostatic force between two point charges q_1 and q_2 a distance d apart:

$$F = \frac{q_1 \times q_2}{d^2}$$

This relationship tells us that (a) a cation with double the charge of another will attract a negatively charged species with twice the force (at the same distance of separation), and (b) if the distance between oppositely charged species is halved, the force between them increases four-fold.

A simple way of bringing these two factors into one single quantitative measure is through *the ratio of charge/radius*, which is called **charge density**, implying some measure of concentration of positive charge. An alternative name is **charge-to-radius ratio**. Values of charge, radius, and charge-to-radius ratio for some common cations are listed in Table 26.2.

The concept of *charge density* of a cation is an entirely different concept from *effective nuclear charge* (which is an estimate of the positive charge that valence electrons of an atom or ion "feel" is on its own nucleus—and which we can use to rationalize properties like atom size, ionic radius, electronegativity, and ionization potential). Charge density refers to interactions external to the ion, whereas effective nuclear charge refers to interactions within an atom or ion.

In general, highly charged cations are small, so they have high charge density. Some chemical properties that can be attributed to high charge densities of cations are discussed in the following sections.

> The charge density, or charge-to-radius ratio, of a cation is a combined measure of magnitude of charge and size of the ion. The higher the charge density of a cation, the greater its ability is to attract negatively charged species near it.

TABLE 26.2 Charge-to-Radius Ratios of Some Cations

Ion	Charge (q)	Radius (r, in pm)	q/r
Be^{2+}	+2	34	0.059
Al^{3+}	+3	57	0.053
Cr^{3+}	+3	63	0.048
Fe^{3+}	+3	64	0.047
Cu^{2+}	+2	72	0.028
Fe^{2+}	+2	74	0.027
Mg^{2+}	+2	79	0.025
Ba^{2+}	+2	143	0.014
Li^{+}	+1	78	0.013
Ag^{+}	+1	126	0.008
K^{+}	+1	133	0.008

Covalent-Ionic Bond Character

There are correlations between covalent character of salts and oxides and (a) high charge density of the cation, and (b) large size of the anion. For example, KCl(s) has the characteristic properties that we classify as ionic compounds, while AlBr$_3$(s) has typical covalent characteristics. An explanation that chemists use is that the higher the charge density of cations, the more they can distort the electron distribution in a nearby atom, molecule, or ion toward itself. This distortion is called *polarization* (Figure 26.3). The greater the attraction of the anion's electrons toward the cation and into the bonding region, the more covalent the bond is.

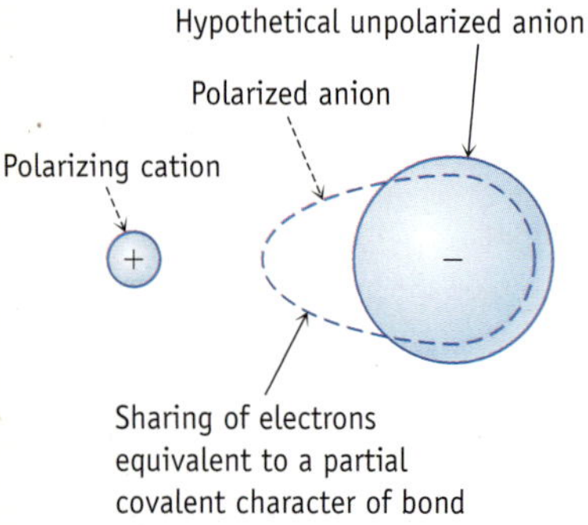

The following terms are used:

- **Polarization** refers to the distortion of the electron cloud of an anion by the attracting influence of the positive charge of an adjacent cation.
- **Polarizing power** of a cation refers to the degree to which it can induce polarization of the electron clouds of neighbouring species—and, according to the model we have been using, correlates with the cation's charge density.
- **Polarizability** of an anion refers to the ease with which its electron cloud can be distorted. As you might expect, the bigger an anion is, the more polarizable it is (for a given charge on the anion).

This model helps to rationalize the observations that the chemical behaviours of halides such as AlCl$_3$, FeCl$_3$, SnI$_4$, PbCl$_4$, BeCl$_2$, and BCl$_3$ are typical of covalent compounds—as evidenced by the melting points of chlorides and bromides of selected metals in Table 26.3. For a given anion, the higher the charge density of the cation, the lower the melting point is.

FIGURE 26.3 Polarization of an electron cloud. The closer the cation can approach, and the larger the charge, the greater the attraction for, and distortion of, electrons on an anion. The greater the electron density between cation and anion, the more covalent character the compound has.

TABLE 26.3 Melting Points of Metal Chlorides and Bromides

Chlorides			Bromides	
BeCl$_2$(s)	405 °C	Increasing charge density	AlBr$_3$(s)	98 °C
CaCl$_2$(s)	772 °C	of cation, increasing	MgBr$_2$(s)	700 °C
		covalent character	NaBr(s)	755 °C

The terms *ionic* and *covalent* are categories that chemists assign to materials to rationalize observed properties. Two simple observations for this purpose are melting point and solubility in non-polar solvents: lower melting point and higher solubility in non-polar solvents suggest more covalent character. The distinction is not clear-cut: the properties of a substance may place it anywhere on a continuum between pure ionic and pure covalent character. Here we are concerned less with categorization of substances as "ionic" or "covalent" in an absolute sense, and more with comparison of the degree of covalent (or ionic) character of different substances.

A similar correlation between high charge density of the cation and covalent character is shown in the melting points of the chlorides of elements of the third period of the periodic table (Table 26.4):

$SiCl_4(\ell)$, like $CCl_4(\ell)$, is a typically covalent compound, so we assume that there is no such species as the Si^{4+} ion. In cases like this, when we use the concept of cationic charge density to rationalize properties, we are really considering what the charge density would be if the substance were ionic.

If we compare two compounds with the same anion, in which a metal has different oxidation states, the one in which the metal has the higher oxidation state has more covalent character. Another perspective of this comparison is that the higher the oxidation state of a metal in a compound (or the higher the charge density of its ion, if we regard it as being ionic), then the less metallic the metal is. In other words, how metallic a metal is in its compounds depends on its oxidation state.

For a given cation, the larger the *anion* is, the more polarizable (more easily distorted) its electron cloud, and the more covalent the substance—as evidenced by the trend in melting points of halides of lithium (Table 26.5).

TABLE 26.4 Melting Points of Chlorides of the Third Period

NaCl(s)	MgCl$_2$(s)	AlCl$_3$(s)	SiCl$_4$(ℓ)
808 °C	714 °C	192 °C	−68 °C

Increasing polarizing power of cation :

Increasing covalent character of compound :

LiF(s)	870 °C
LiCl(s)	613 °C
LiBr(s)	547 °C
LiI(s)	446 °C

Increasing size (*and polarizability*) of anion, more covalent character.

TABLE 26.5 Melting Points of Lithium Halides

By now you may have realized that the common generalization that compounds formed between a metal and a non-metal are ionic is less true as we consider compounds of metals in which the cations have high charge density.

Degree of Ionic-Covalent Character of Bonds

It is perhaps easy to visualize a lattice of ions in an ionic solid, and to visualize a substance composed of covalent molecules. The properties of a substance like beryllium chloride ($BeCl_2$) are intermediate between those of ionic substances and those of covalent molecular substances, but what is the nature of such a substance that we describe as "somewhat covalent" or "partially ionic"? What happens when such a substance melts? These are questions about electron distribution at the intramolecular level [<<Section 10.1].

It's good to think about the concepts that are used in chemistry, but it is even more important is to realize that all concepts are, in some sense, attempts to model the behaviour of substances at the invisible, molecular level, and they do not represent the "truth" (not even for exam purposes). You should realize that questions such as those asked above cannot be answered even by professional chemists—they can't see bonds; they can only observe properties and ponder. The chemistry research community continuously tries to understand better by devising better models that help us to rationalize behaviour.

Oxides—A Special Case The generalization that oxides and hydroxides of metals are basic is less true as we go to compounds of metals whose ions have higher charge density. There is a correspondence between high charge density of metal ions and covalent molecular character of their oxides and hydroxides, as well as with acidity of the oxides and hydroxides.

Calcium oxide, CaO(s), is typical of compounds that we describe as ionic, and when we dissolve calcium hydroxide, $Ca(OH)_2(s)$, in water the ions separate to form a basic solution:

$$Ca(OH)_2(s) \xrightarrow{H_2O(\ell)} Ca^{2+}(aq) + 2\,OH^-(aq)$$

On the other hand, chromium(VI) oxide, $CrO_3(s)$, is typical of substances that we call *covalent molecular*, and the hydroxide formed when it is dissolved in water—$CrO_2(OH)_2$ or H_2CrO_4—is acidic, as a result of breaking of an O—H bond. This is typical of covalent molecular hydroxides.

In a similar way to H_2SO_4, H_2CrO_4 is strong in the first step of acid ionization, and weak in the second step. In H_2CrO_4 molecules, Cr is in the +6 oxidation state, as is S in H_2SO_4. The structure of a molecule of H_2CrO_4 corresponds with that of a molecule of H_2SO_4.

Aluminum hydroxide, $Al(OH)_3(s)$, and beryllium hydroxide, $Be(OH)_2(s)$, are amphoteric.

> For salts with a particular anion, the higher the charge density of the cation, the more covalent character the salt has. This can be explained by polarization (distortion) of the electron cloud of the anion into the bonding region. The higher the charge density of the cation is, the greater its polarizing power. The larger the anion is, the more easily its electron cloud is distorted—that is, the greater its polarizability. In the case of oxides, increased covalent character correlates with higher acidity.

Waters of Crystallization of Solid Salts

When their solutions are evaporated to dryness, many salts with cations of high charge density crystallize with several molecules of water of crystallization. X-ray crystallography shows that at least some of the H_2O molecules are bound directly to the cation. Analysis of X-ray crystallographic data from a blue crystal of copper sulfate pentahydrate ($CuSO_4 \cdot 5H_2O$) gives the structure represented in Figure 6.2.

The formula $CuSO_4 \cdot 5H_2O$ tells us about the composition of copper sulfate pentahydrate, but nothing about the structure. The X-ray evidence about the structure suggests the existence of $[Cu(OH_2)_4]^{2+}$ complex ions. Perhaps a more informative formula that gives some information about the structural arrangement of species in the crystal is $[Cu(OH_2)_4]^{2+} \cdot SO_4^{2-} \cdot H_2O$.

Other examples (there are many) of salts that crystallize with water of crystallization are

- $Al(NO_3)_3 \cdot 9H_2O$ or $[Al(OH_2)_6]^{3+} \cdot (NO_3^-)_3 \cdot 3H_2O$
- $CrCl_3 \cdot 6H_2O$ or $[Cr(OH_2)_6]^{3+} \cdot (Cl^-)_3$

The solids NaCl, KNO_3, and $AgNO_3$, in which the cations have low charge density, recrystallize from solution without waters of crystallization.

> Many salts with cations of high charge density crystallize with several molecules of water of crystallization.

Strength of Aquation of Cations

All ions, cations, and anions are aquated in aqueous solution. For cations, aquation may be regarded as formation of complex ions—a Lewis acid-base reaction [<<Section 6.7].

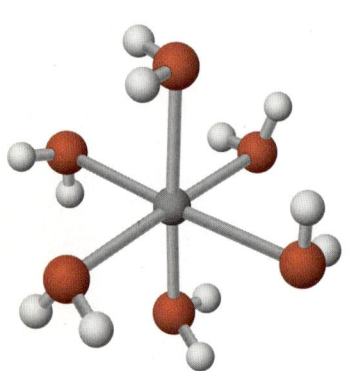

The $[Fe(OH_2)_6]^{3+}$ ion. All the bonds in this polyatomic complex ion can be classified as covalent bonds.

The strengths of the M—O bonds in various aquated metal ions, $M^{n+}(aq)$, can be compared using the energy released when 1 mol of the gaseous ions is put into water (not something we can measure by direct experimental measurement). This is the enthalpy change of aquation (or *hydration*) of the metal ion ($\Delta_{hyd}H$), which is negative for all ions:

$$M^{n+}(g) \xrightarrow{H_2O(\ell)} M^{n+}(aq) \qquad \Delta_rH = \Delta_{hyd}H$$

The larger the charge density of a cation is, the more exothermic aquation is. For example, across a period of the main block, the strength of hydration increases in the following order:

$$Na^+ - OH_2 < Mg^{2+} - OH_2 < Al^{3+} - OH_2$$

Down a column, the M—O bond strengths decrease in the following order:

$$Be^{2+} - OH_2 > Ca^{2+} - OH_2 > Sr^{2+} - OH_2 > Ba^{2+} - OH_2$$

The larger the charge density of a cation is, the more exothermic aquation is.

Acidity of Aqueous Solutions of Salts

In pure water, only a very small fraction of water molecules are ionized to $H^+(aq)$ and $OH^-(aq)$. But solutions made by dissolving $Al(NO_3)_3(s)$, $Fe(NO_3)_3(s)$, and $Cr(NO_3)_3(s)$ in water are acidic [<<Section 14.4]. This can be explained by assuming that the O—H bonds in water molecules bound to cations with high charge density are more easily broken than the O—H bonds in bulk water. This can in turn be accounted for by strong attraction for electrons by the high charge density cation—withdrawing electron density from the water molecule and weakening the O—H bond (Figure 26.4).

Because of the weakened O—H bonds, nearby water molecules can more easily remove H^+ ions from the bound water molecules (with both electrons of the O—H bond remaining on the O atom). The net result is an equilibrium condition represented by the following equation:

$$\left[Al(OH_2)_6\right]^{3+}(aq) + H_2O(\ell) \rightleftharpoons \left[Al(OH_2)_5(OH)\right]^{2+}(aq) + H_3O^+(aq)$$

The $[Al(H_2O)_6]^{3+}$ ion is a weak acid. The acid ionization constants [<<Table 14.5] at 25 °C for a few aquated cations with high charge density are listed here:

$$[Fe(H_2O)_6]^{3+} \qquad K_a = 6 \times 10^{-3}$$
$$[Cr(H_2O)_6]^{3+} \qquad K_a = 1.3 \times 10^{-4}$$
$$[Al(H_2O)_6]^{3+} \qquad K_a = 1.3 \times 10^{-5}$$

For comparison, K_a(acetic acid) $= 1.8 \times 10^{-5}$, so we can see that the aquated metal(III) ions listed have acid strengths similar to that of acetic acid. On the other hand, KNO_3 and $Ca(NO_3)$ solutions, in which the cations have low charge density, have negligible acidity. For $Ca^{2+}(aq)$, K_a is approximately 10^{-13}.

The higher the charge density of a cation is, the more acidic its aqueous solutions are. This can be explained by withdrawal of the electrons from complexed water molecules, weakening the O—H bonds in the water molecules.

Lattice Enthalpies

Lattice enthalpy (often called *lattice energy*) is the energy evolved when ions in the gas phase come together to form 1 mol of a solid crystal:

$$M^{n+}(g) + X^{n-}(g) \longrightarrow MX(s) \qquad \Delta_rH = \Delta_{lattice}H$$

Molecular Modelling (Odyssey)

e26.1 Examine simulations showing the effect of charge density on aquation of cations and anions.

Molecular Modelling (Odyssey)

e26.2 Compare the aquation around cations of different charge density.

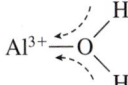

FIGURE 26.4 Electron-withdrawal from an H_2O molecule in a $[Al(OH_2)_6]^{3+}$ ion. The curved arrows indicate withdrawal of electrons from the water molecule toward the Al^{3+} ion, weakening the O—H bonds in the water molecule.

The yellow colour of iron(III) solutions is due to the $[Fe(H_2O)_5(OH)]^{2+}$ complex ion rather than the $[Fe(H_2O)_6]^{3+}$ ion. You can demonstrate this by adding concentrated nitric acid solution to a solution of iron(III) nitrate, to convert all of the $[Fe(H_2O)_5(OH)]^{2+}$ to the $[Fe(H_2O)_6]^{3+}$ ion. The solution becomes colourless.

It may be easier to understand the concept of lattice energy by considering the energy needed to separate all of the ions in 1 mol of a salt to infinite distance from each other:

$$MX(s) \longrightarrow M^{n+}(g) + X^{n-}(g) \qquad \Delta_r H = -\Delta_{\text{lattice}} H$$

The higher the charge density of metal cations, the higher the lattice energies of their salts are. You might expect this because the higher the charge density of the cation is, the higher the force of attraction to a specified anion (at a specified distance). For example, the lattice energy of $LiF(s)$ is 1050 kJ mol^{-1}, and that of $KF(s)$ is 825 kJ mol^{-1}.

> The higher the charge density of metal cations, the higher the lattice energies of their salts are.

Mobility of Ions in Water

In the field of an applied electric potential, ions move through water—cations in one direction and anions in the other. The speed with which they move is called their *mobility*.

Given that the radius of a Li^+ ion is 76 pm, and the radius of a K^+ ion is 138 pm, you might expect that Li^+ ions would have a higher mobility than K^+ ions. In fact, Li^+ ions are much slower. The rationale for this is that Li^+ ions hold on to H_2O molecules so strongly that the mobile species is a complex ion such as $[Li(OH_2)_6]^+$—rather like a family of seven holding hands as they try to pass through a crowd. On the other hand, the K^+—OH_2 bond is more easily broken and re-formed with other H_2O molecules, so that K^+ ions exchange their solvating water molecules as they move along.

> Aquated lithium ions move through water slower than we might expect, and this can be explained by their strong attraction to the water molecules: instead of exchanging them as they move (as other ions do), lithium ions carry their "aquation spheres" with them.

Element	$E°$/V
$Li^+(aq) + e^- \longrightarrow Li$	−3.045
$Na^+(aq) + e^- \longrightarrow Na$	−2.714
$K^+(aq) + e^- \longrightarrow K$	−2.925
$Rb^+(aq) + e^- \longrightarrow Rb$	−2.925
$Cs^+(aq) + e^- \longrightarrow Cs$	−2.92

The Reducing Ability of Metals

The values of standard reduction potentials reveal that in aqueous media, Li is the most powerful reducing agent in group 1, whereas Na is the poorest, and the remainder of these metals have roughly comparable reducing ability.

This order might seem in contradiction to the trend to lower ionization energies of the elements as we go down groups of the periodic table [<<Section 8.2]. Since lithium has the highest ionization energy of these elements, we might expect it to be the poorest reducing agent. We must remember, however, that ionization energies are a measure of the energy required to remove an electron from an isolated atom (approximated by atoms in the gas phase), whereas reduction potentials are measured in aqueous solution, and the interaction of ions with solvent becomes an important factor governing stability.

We can reconcile the difference between the trend of ionization energies and the trend of $E°$ values using our thermodynamics knowledge. Reducing agents are oxidized when they perform their task, and we can understand better the relative reduction potentials in aqueous solution if we break the process of metal oxidation, $M(s) \longrightarrow M^+(aq) + e^-$, into a series of imagined steps: (a) the metal sublimes to vapour, (b) an electron is removed to form the gaseous cation, and (c) the cation is aquated.

The first two steps require energy, but the last is exothermic. From Hess's law [<<Section 7.7], we deduce that the net energy change is given by

$$\Delta H_{\text{net}} = \Delta_{\text{sub}} H + IE + \Delta_{\text{hyd}} H$$

From an enthalpy consideration, and ignoring any entropy factor, the element with the most negative value of ΔH_{net} should be the best reducing agent. For the group 1 metals, enthalpies of aquated range from −506 kJ mol^{-1} for Li^+ to −180 kJ mol^{-1} for Cs^+. The fact that aquation of Li^+ ions is so much more exothermic than for Cs^+ ions largely accounts for the difference in reducing ability. In other words, aquation of the ions formed after the metal atoms have provided an electron to the species being reduced is a significant "driving force" of reaction.

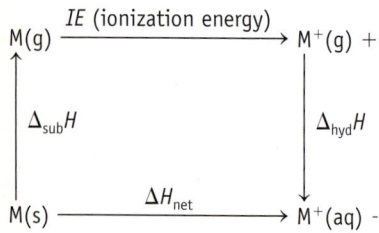

The reducing ability of lithium in aqueous solutions is enhanced by the strong interaction of Li^+ ions with water, because of their high charge density.

EXERCISE 26.1—INFLUENCE OF CHARGE DENSITIES OF CATIONS ON PROPERTIES

Which do you predict is the more polarizing cation: Be^{2+} or Ba^{2+}? Why?

EXERCISE 26.2—INFLUENCE OF CHARGE DENSITIES OF CATIONS ON PROPERTIES

The melting point of chromium(III) oxide is 2435 °C and that of chromium(VI) oxide is 196 °C.
(a) Rationalize the difference between the melting points.
(b) Predict the relative acid-base properties of the two oxides when dissolved in water.

26.4 Hydrogen

Most (73%) of the atoms in the universe are hydrogen atoms, although there are negligible amounts in the earth's atmosphere because the speeds of the molecules are greater than "escape velocity."

There are three isotopes of the element with atomic number 1, two of which (protium and deuterium) are stable, while tritium is radioactive (Table 26.6).

Of the isotopes, only H and D are found in nature in significant quantities. In contrast, tritium, which is produced by cosmic ray bombardment of nitrogen in the atmosphere, is found to the extent of 1 atom per 10^{18} atoms of ^1H. This is because tritium is radioactive, with a half-life [<<Section 18.5] of 12.26 years.

Most of the atoms in the universe are hydrogen atoms, of which there are three isotopes.

TABLE 26.6 Isotopes of Hydrogen

Isotope Mass (u)	Symbol	Name
1.0078	^1_1H or H	Hydrogen (protium)
2.0141	^2_1H or D	Deuterium
3.0160	^3_1H or T	Tritium

Properties and Reactions of the Substance Hydrogen

Under ambient conditions, hydrogen is a gas. Its very low boiling point, 20.7 K, reflects its non-polar character and low molar mass. As the least dense gas known, it is ideal for filling lighter-than-air craft. In 1783, Jacques Charles first used hydrogen to fill a balloon large enough to float above the French countryside, a method used in World War I to float observation balloons. The *Graf Zeppelin*, a passenger-carrying dirigible built in Germany in 1928, was also filled with hydrogen. It carried more than 13 000 people between Germany and the United States until 1937, when it was replaced by the *Hindenburg*.

The *Hindenburg* was designed to be filled with helium. World War II was approaching, and the United States, which has much of the world's supply of helium, would not sell the gas to Germany. As a consequence, hydrogen was used. The airship burned when landing in Lakehurst, New Jersey, in May 1937 (Figure 26.5). Of the 62 people on board, only about half escaped uninjured. As a result of this disaster, hydrogen has acquired a reputation as being a very dangerous substance. Actually, it is as safe to handle as any other fuel, as evidenced by the large quantities used today, even in public transportation. Indeed hydrogen is regarded as one of the main future portable energy sources [<<Section 7.1].

Deuterium compounds have been thoroughly studied. Since D atoms have twice the mass of H atoms, reactions involving D atom transfer are slightly slower than those involving H atoms. This knowledge leads to a way to produce D_2O or "heavy water." Hydrogen can be produced by electrolysis of water (Figure 16.13).

$$H_2O(\ell) \xrightarrow{\text{electrical energy}} H_2(g) + \tfrac{1}{2} O_2(g)$$

Any sample of natural water contains a tiny concentration of D_2O. In electrolysis cells, H_2O is electrolyzed more rapidly than D_2O, so the water remaining during electrolysis is enriched in D_2O. This "heavy water" is valuable as a moderator of some nuclear reactions used in power generation.

Mary Evans Picture Library/Photo Researchers, Inc.

FIGURE 26.5 The Hindenburg. With the hydrogen burning, this dirigible crashed in Lakehurst, New Jersey, in May 1937. Some have speculated that a spark from the aluminum paint coating the skin of the craft initiated the fire.

Hydrogen combines chemically with virtually every other element except the noble gases. There are three categories of binary hydrogen-containing compounds. *Ionic hydrides* are formed in the reaction of $H_2(g)$ with group 1 and 2 metals. These compounds contain the hydride ion (H^-), in which hydrogen is in oxidation state -1.

$$2\,Na(s) + H_2(g) \longrightarrow 2\,NaH(s)$$

$$Ca(s) + H_2(g) \longrightarrow CaH_2(s)$$

The formulas of these ionic hydrides can be represented as, for example, Li^+H^- and $Ca^{2+}(H^-)_2$. They are white crystalline solids with high melting points. During electrolysis of molten ionic hydrides, H_2 gas is produced by oxidation of H^- ions at the anode (in contrast to electrolysis of dilute acid solutions in which $H_2(g)$ is formed at the cathode):

$$2\,H^- \longrightarrow H_2(g) + 2\,e^-$$

Ionic hydrides are good reducing agents, capable of reducing ketones to alcohols and nitrates to ammonia, with oxidation of H^- ions to $H_2(g)$. They undergo vigorous reaction with water to produce hydrogen and to form a basic solution (Figure 26.7).

$$Ca^{2+}(H^-)_2(s) + 2\,H_2O(\ell) \longrightarrow Ca^{2+}(aq) + 2\,OH^-(aq) + 2\,H_2(g)$$

Covalent molecular hydrides are formed between hydrogen and the more electronegative elements.

$$N_2(g) + 3\,H_2(g) \longrightarrow 2\,NH_3(g)$$

$$F_2(g) + H_2(g) \longrightarrow 2\,HF(g)$$

Examples are $H_2O(\ell)$, $H_2S(g)$, $HCl(g)$, $HBr(g)$, $NH_3(g)$, $PH_3(g)$, and $CH_4(g)$. The oxidation state of H in these compounds is $+1$. Typically, they are non-conducting gases, liquids, and soft solids and, except for the more polar ones, are soluble in organic solvents. Covalent molecular hydrides with highly electronegative elements E have polar E—H bonds. These are acidic in aqueous solution because of ionization to form $H_3O^+(aq)$ ions:

$$HCl(g) \xrightarrow{H_2O(\ell)} H_3O^+(aq) + Cl^-(aq)$$

$$H_2S(aq) + H_2O(\ell) \rightleftharpoons H_3O^+(aq) + HS^-(aq)$$

With less electronegative elements like N and P, the E—H bond is less polar and is not broken in aqueous solution. If the E atom has a lone pair of electrons, it can accept H^+ from water, resulting in basic solution:

$$NH_3(aq) + H_2O(\ell) \rightleftharpoons NH_4^+(aq) + OH^-(aq)$$

Hydrogen is absorbed by many metals, particularly the transition metals, to form *metallic hydrides*. Hydrides of this type are metallic in appearance and conduct electricity. Their formulas are non-stoichiometric—that is, the ratio of metal and hydrogen is variable and usually not a whole number. They are best regarded as alloys, such as we get by mixing two or more metals. When palladium metal is used as an electrode for the electrolysis of water, the metal can soak up 1000 times its volume of hydrogen. When heated, H_2 gas is driven off. This phenomenon allows these materials to store hydrogen, just as a sponge can store water, and presents one way to store hydrogen for use as a fuel in motor vehicles [<<Section 7.1].

Elemental hydrogen is a gas composed of diatomic molecules, H_2. Hydrogen combines chemically with all of other element except the noble gases. With the most electropositive elements, it forms ionic hydrides in which H is in oxidation state -1, and these are powerful reducing agents. With the most electronegative elements, it forms covalent molecular hydrides with H in oxidation state $+1$, and with the many metals it forms metallic hydrides of variable composition, which are rather like alloys.

Making Hydrogen

About 300 billion litres (at SATP) of hydrogen gas is produced annually worldwide, and virtually all is used in the manufacture of ammonia [<<Section 13.1], methanol [<<Section 18.6], or other chemicals.

In the past, hydrogen has been made from coal and steam, by injecting water into a bed of red-hot coke:

$$C(s) + H_2O(g) \longrightarrow H_2(g) + CO(g)$$

The mixture of gases produced, called "water gas" or "town gas" was reticulated to houses until about 1950 as a fuel for cooking, heating, and lighting. However, its use for this purpose has serious drawbacks: it produces only about half as much heat as an equal amount of methane, and its flame produces very little light. Moreover, because it contains carbon monoxide, water gas is toxic.

Most hydrogen is now produced by the *catalytic steam re-formation* of hydrocarbons such as methane, obtained from natural gas (Figure 26.6). Methane reacts with steam at high temperature to give $H_2(g)$ and $CO(g)$:

$$CH_4(g) + H_2O(g) \longrightarrow 3 H_2(g) + CO(g) \quad \Delta_r H^\circ = +206 \text{ kJ mol}^{-1}$$

The reaction is rapid at 900–1000 °C and goes nearly to completion. Mixtures of hydrogen and carbon monoxide are also called *synthesis gas* or *syngas* because they are used, in conjunction with specific catalysts, to produce liquid fuels, methanol or ammonia [<<Section 18.6].

Perhaps the cleanest way to make hydrogen on a relatively large scale is the electrolysis of water (Figure 16.13). Because electricity is quite expensive, this method is not generally used commercially. As the cost of electricity from solar panels comes down, the prospects for this process will be enhanced.

Table 26.7 and Figure 26.7 give examples of reactions used to produce H_2 gas in the laboratory. The most common is the reaction of a metal with an acid. Alternatively, the reaction of aluminum with NaOH solution generates hydrogen. In the past, this reaction has been used to obtain hydrogen to inflate small balloons for weather observation and to raise radio antennas. Aluminum is plentiful.

FIGURE 26.6 Production of synthesis gas. Synthesis gas, or syngas, is a mixture of $H_2(g)$ and $CO(g)$. It is produced by reaction of coal, coke, or a hydrocarbon with steam at high temperatures in plants such as that pictured here. Methane has the advantage that it gives more $H_2(g)$ per gram than other hydrocarbons, and the ratio of the by-product $CO_2(g)$ to $H_2(g)$ is lower.

TABLE 26.7 Methods for Making $H_2(g)$ in the Laboratory

1. Metal + acid $\longrightarrow$ metal ions(aq) and $H_2(g)$

   $Mg(s) + 2 H^+(aq) \longrightarrow Mg^{2+}(aq) + H_2(g)$

2. Reactive metal + $H_2O(\ell) \longrightarrow$ metal ions(aq), $OH^-(aq)$ and $H_2(g)$

   $2 Na(s) + 2 H_2O(\ell) \longrightarrow 2 Na^+(aq) + 2 OH^-(aq) + H_2(g)$

3. Aluminum + NaOH solution $\longrightarrow H_2(g)$ and aluminate ions(aq)

   $2 Al(s) + 2 OH^-(aq) + 6 H_2O(\ell) \longrightarrow 2 [Al(OH)_4^-](aq) + 3 H_2(g)$

4. Metal hydride + $H_2O \longrightarrow$ metal hydroxide and $H_2(g)$

   $CaH_2(s) + 2 H_2O(\ell) \longrightarrow Ca(OH)_2(s) + 2 H_2(g)$

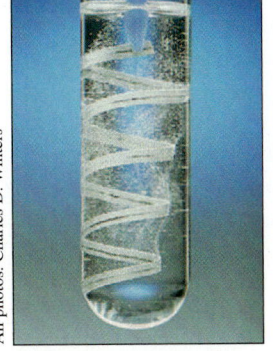

(a) The reaction of magnesium and acid. The products are hydrogen gas and a magnesium salt.

(b) The reaction of aluminum and NaOH solution. The products of this reaction are hydrogen gas and $[Al(OH)_4]^-(aq)$ ions.

(c) The reaction of $CaH_2(s)$ and water. The products are hydrogen gas and $Ca(OH)_2(s)$.

FIGURE 26.7 Producing hydrogen gas in the laboratory. (a) The reaction of magnesium with acidic solution. (b) The reaction of aluminum with NaOH solution. (c) The vigorous reaction of calcium hydride, $CaH_2(s)$, an ionic hydride, with water.

The reaction of an ionic metal hydride and water is an efficient but expensive way to make $H_2(g)$ in the laboratory. The reaction is more commonly used in laboratories to dry organic solvents because the metal hydride reacts with traces of water in a solvent.

The main mode of industrial production of hydrogen is by reaction of steam with methane from natural gas: the product mixture of hydrogen and carbon monoxide is synthesis gas. Production of hydrogen by electrolysis of water may become more important.

Hydrogen: Group 1, Group 17, or a Group on Its Own?

Hydrogen has some properties in common with the elements of group 1 of the periodic table. These include the formation of ions with +1 charge, and its ability to alloy with many metals. On the other hand, there are some properties that suggest that H is better classified as a group 17 element. Examples of such properties include its existence as a substance in the gaseous state at SATP, as well as being composed of diatomic molecules. Furthermore, like the group 17 elements, it forms covalent compounds with non-metallic elements, and ionic compounds in which it forms negatively charged ions (ionic metal hydrides) with group 1 metals and some of the group 2 metals.

Because of the ambiguity of its classification, perhaps H should be classified as neither a group 1 nor a group 17 element, and is best placed in a group on its own, as indicated in the periodic table on the inside front cover. This is supported by estimates of its electronegativity, which is very close to that of phosphorus, and not like those of the group 1 nor group 2 elements.

Hydrogen has some properties that are common to the group 1 elements, some in common with the group 17 elements, and some that suggest it is better classified in a group on its own.

EXERCISE 26.3—HYDROGEN AND ITS COMPOUNDS

Write balanced chemical equations for the reaction of hydrogen gas with oxygen, chlorine, and nitrogen.

EXERCISE 26.4—HYDROGEN AND ITS COMPOUNDS

Write an equation for the reaction of potassium and hydrogen. Name the product. Is it ionic or covalent? Predict one physical property and one chemical property of this compound.

26.5 The Alkali Metals, Group 1

Sodium and potassium are the sixth and eighth most abundant elements in the earth's crust by mass. In contrast, lithium, rubidium, and cesium are relatively rare (Figure 26.2). Only traces of francium occur in nature, because all of its isotopes are radioactive. Its longest-lived isotope (^{223}Fr) has a half-life of only 22 minutes.

All of the group 1 elements are metals, and all are highly reactive with oxygen, water, and other oxidizing agents. In all cases, compounds of the group 1 metals contain the element as an ion with +1 charge. None is ever found in nature as the uncombined element.

Most sodium and potassium compounds are water-soluble, so it is not surprising that sodium and potassium compounds are found in the oceans and in underground deposits that are the residue of ancient seas. To a much smaller extent, these elements are also found in minerals, such as Chilean saltpetre ($NaNO_3$).

Although sodium is only slightly more abundant than potassium on the earth, sea water contains significantly more sodium than potassium (2.8% by mass NaCl compared with 0.8% KCl). Why the great difference? Both are water-soluble, so why didn't the rain dissolve

Group 1
Alkali metals

Lithium
3
Li
20 ppm

Sodium
11
Na
23 600 ppm

Potassium
19
K
21 000 ppm

Rubidium
37
Rb
90 ppm

Cesium
55
Cs
0.0003 ppm

Francium
87
Fr
trace

Element abundances are in parts-per-million in the earth's crust.

Na- and K-containing minerals over the centuries and carry them down to the sea, so that they are at the same concentration in the oceans as on land? The answer lies in the fact that potassium is an important factor in plant growth. Most plants contain four to six times as much combined potassium as sodium. So most of the potassium ions in groundwater from dissolved minerals are taken up preferentially by plants, whereas sodium salts continue on to the oceans.

Some NaCl is essential in the diet of humans and other animals because many biological functions are controlled by the concentrations of Na^+ and Cl^- ions (Figure 26.8). The fact that salt has long been recognized as important is evident in surprising ways. We are paid a "salary" for work done. This word is derived from the Latin *salarium*, meaning "salt money" because Roman soldiers were paid in salt.

> All of the group 1 elements are metals, and all are highly reactive with oxygen, water, and other oxidizing agents. None is ever found in nature as the uncombined element. In all cases, compounds of the group 1 metals contain the element as an ion with +1 charge. Most sodium and potassium compounds are water-soluble.

Production of Sodium and Potassium

The methods used to produce sodium and potassium are governed by the ease of oxidation of the metals—and therefore by the difficulty of reducing their ions to the metal. Sodium is produced by reducing the sodium ions in sodium salts, but because common chemical reducing agents (such as carbon and hydrogen) are not sufficiently powerful reducers to reduce sodium ions, an electrolytic method is used to "force" the reduction to occur.

The English chemist Sir Humphry Davy first isolated sodium in 1807 by the electrolysis of molten sodium carbonate. The element remained a laboratory curiosity until 1824, when it was found that sodium could be used to reduce aluminum chloride to aluminum metal. Because aluminum was so valuable, this discovery inspired considerable interest in manufacturing sodium. By 1886, a practical method of sodium production had been devised (the reduction of NaOH by carbon). Unfortunately for sodium producers, in this same year Hall and Heroult invented the electrolytic method for aluminum production, thereby eliminating this market for sodium.

Sodium is currently produced by the electrolysis of molten NaCl [<<Section 16.7]. The Downs cell for the electrolysis of molten NaCl (Figure 26.9) operates at 7–8 V with currents of 25 000–40 000 amps. The cell contains a molten mixture of NaCl, $CaCl_2$, and $BaCl_2$. Adding other salts to NaCl lowers the melting point from that of pure NaCl (800.7 °C) to about 600 °C. Sodium is produced at an iron cathode that surrounds a

Because the biology of plants depends on the presence of potassium ions, fertilizers often contain a significant concentration of potassium salts. One of the major benefits to plants is increased flowering, and in the case of fruits, strengthening of the attachment of fruit.

Charles D. Winters

FIGURE 26.8 The importance of salt. All animals, including humans, need salt in their diet. Sodium ions are important in maintaining electrolyte balance and in regulating osmotic pressure. For an interesting account of the importance of salt in society, culture, history, and economy, see *Salt, A World History*, by M. Kurlansky (2002), New York: Penguin Books.

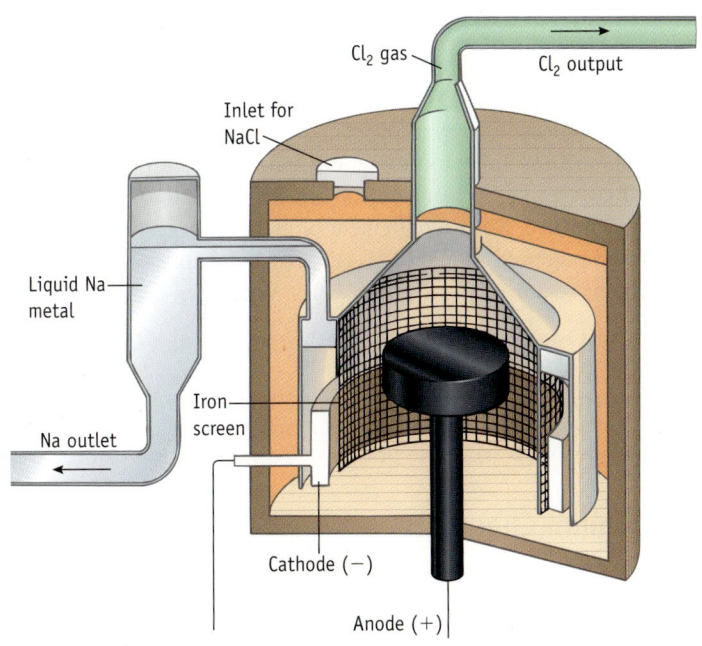

FIGURE 26.9 A Downs cell for producing sodium. An iron screen separates the circular iron cathode from the graphite anode. At the temperature of electrolysis, about 600 °C, sodium produced at the cathode is a liquid. It floats to the top and is drawn off periodically. Chlorine gas is produced at the anode and collected inside the inverted cone in the centre of the cell. If the products are not kept apart, they will react to form NaCl(s).

circular graphite anode. Directly over the cathode is an inverted trough in which the low-density, molten sodium (melting point: 97.8 °C) collects. The valuable by-product, Cl_2 gas, is collected at the anode.

Potassium can also be made by electrolysis. However, since molten potassium is soluble in molten KCl, separation of the metal is difficult. The preferred method for preparation of potassium uses the reaction of sodium vapour with molten KCl, with potassium being continually removed from the equilibrium mixture.

$$Na(g) + KCl(\ell) \longrightarrow K(g) + NaCl(\ell)$$

Because sodium and potassium are so reactive, their ions cannot be reduced by common reducing agents. Sodium is obtained by electrolysis of molten NaCl.

Properties of Sodium and Potassium

Sodium and potassium are silvery metals that are soft and easily cut with a knife. Their densities are a little less than the density of water, and their melting points are quite low (97.8 °C for sodium and 63.7 °C for potassium).

All of the alkali metals (the elemental substances) are highly reactive as powerful reducing agents, and their main uses are due to their reducing abilities. When exposed to moist air, the metal surface quickly becomes coated with a film of oxide or hydroxide, so the metals must be stored in a way that avoids contact with air, typically by putting them under kerosene or mineral oil.

The high reactivity of group 1 metals is exemplified by the reaction of potassium with water, which generates hydrogen gas and aquated potassium ions (Figure 26.10),

$$2 K(s) + 2 H_2O(l) \longrightarrow 2 K^+(aq) + 2 OH^-(aq) + H_2(g)$$

and their reaction with any of the halogens to yield metal halides (Figure 1.7):

$$2 Na(s) + Cl_2(g) \longrightarrow 2 NaCl(s)$$
$$2 K(s) + Br_2(\ell) \longrightarrow 2 KBr(s)$$

Chemistry sometimes produces surprises. Group 1 metal oxides (M_2O) are known, but they are not the main products of reactions between the group 1 elements and oxygen: the main product of reaction between sodium and oxygen is sodium *peroxide*, $Na_2O_2(s)$, while the reaction of potassium and oxygen gives mainly potassium *superoxide*, $KO_2(s)$.

$$2 Na(s) + O_2(g) \longrightarrow Na_2O_2(s)$$
$$K(s) + O_2(g) \longrightarrow KO_2(s)$$

Both $Na_2O_2(s)$ and $KO_2(s)$ are ionic compounds: the lattice contains cations of the group 1 metal and either the peroxide ion (O_2^{2-}) or the superoxide ion (O_2^-). These compounds are not merely laboratory curiosities. They are used in oxygen-generation devices in places where people are confined, such as submarines, aircraft, and spacecraft, or when an emergency supply is needed (Figure 26.11). When a person breathes, 0.82 mol of $CO_2(g)$ is exhaled for every 1 mol of $O_2(g)$ inhaled. So, a requirement of an oxygen-generation system is that the amount (mol) of $O_2(g)$ produced should be greater than the amount of $CO_2(g)$ absorbed from the breath. This requirement is met with superoxides. With KO_2, the reaction is

$$4 KO_2(s) + 2 CO_2(g) \longrightarrow 2 K_2CO_3(s) + 3 O_2(g)$$

The alkali metals are powerful reducing agents. They react with oxygen, water, and the halogens, forming ionic compounds. In reactions with oxygen, sodium forms mainly the peroxide, $Na_2O_2(s)$, while potassium forms the superoxide, $KO_2(s)$.

Non-Typical Lithium Chemistry

Besides its reactions that are typical of group 1 elements, lithium displays a degree of non-metallic behaviour that is not typical of the other members. This is consistent with the general trend to less metallic behaviour as we go to elements toward the top of each group,

FIGURE 26.10 Reaction of potassium with water. Potassium metal is a powerful reducing agent and reacts vigorously with water.

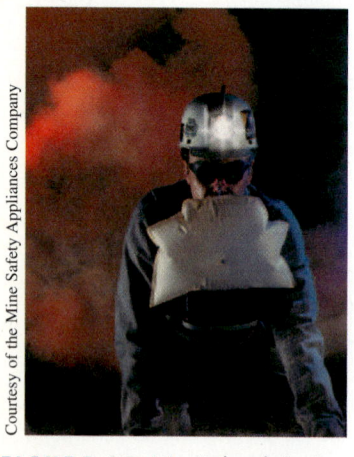

FIGURE 26.11 A closed-circuit breathing apparatus that generates oxygen. One source of oxygen is potassium superoxide (KO_2). Both carbon dioxide and moisture exhaled by the wearer into the breathing tube react with the KO_2 to generate oxygen. Because the rate of the chemical reaction is determined by the quantity of moisture and carbon dioxide exhaled, the production of oxygen is regulated automatically. With each exhalation, more oxygen is produced than the user requires.

although lithium is a little more non-metallic than we might expect in comparison with the elements below it. This is a feature that lithium has in common with the first members of other groups. This non-typical behaviour can be attributed to the high charge density of the Li^+ cation, compared with those of the Na^+ and K^+ ions, as discussed in Section 26.3.

Lithium forms covalent molecular compounds with hydrocarbon groups by reaction with organic halides.

$$2\,Li(s) + CH_3CH_2Cl(g) \longrightarrow CH_3CH_2Li(s) + LiCl(s)$$

This reaction is similar to that of magnesium to form Grignard reagents [<<Section 21.12]. These "organolithium" compounds are widely used as reducing agents in the synthesis of organic compounds. They are liquids or low-melting solids, and dissolve in non-polar solvents such as hexane. The lithium halides $LiCl(s)$, $LiBr(s)$, and $LiI(s)$ have much more covalent character than their sodium and potassium counterparts, being more soluble in polar solvents such as acetone and ethanol.

The high lattice energy of the fluoride, hydroxide, carbonate, and phosphate salts of lithium causes these compounds to be less soluble in water than their sodium and potassium counterparts—even though dissolving is favoured by the high enthalpy change of hydration of Li^+ ions. Of the group 1 metals, only lithium reacts with $O_2(g)$ and $N_2(g)$ in the air to form the oxide $Li_2O(s)$ and the nitride $Li_3N(s)$.

Some of these non-typical properties of lithium are similar to properties of magnesium, and the charge densities of the Li^+ and Mg^{2+} ions are similar. This comparison is the start of the *diagonal relationship* between second-row elements with third-row elements of the next group, discussed again in Sections 26.6 and 26.7.

The chemical properties of lithium are those of an element less metallic than the other members of group 1. As one example of the diagonal relationship, many of its properties are more like those of magnesium.

Important Lithium, Sodium, and Potassium Compounds

Electrolysis of aqueous sodium chloride solution to produce chlorine [<<Section 16.7] is the basis of a huge worldwide "chlor-alkali" chemical industry.

$$2\,Cl^-(aq) + 2\,H_2O(\ell) \longrightarrow Cl_2(g) + 2\,OH^-(aq) + H_2(g)$$

Because of environmental problems associated with chlorine gas and chlorine compounds, considerable interest has arisen in the possibility of manufacturing sodium hydroxide by methods other than brine electrolysis. This has led to a revival of the old "soda-lime process," which produces $NaOH(s)$ from inexpensive lime (CaO) and sodium carbonate (Na_2CO_3).

$$CO_3{}^{2-}(aq) + CaO(s) + H_2O(\ell) \longrightarrow 2\,OH^-(aq) + CaCO_3(s)$$

After filtering off the calcium carbonate by-product, the residual solution is evaporated. The calcium carbonate is recycled into the process by heating it (*calcining*) to form lime again:

$$CaCO_3(s) \longrightarrow CaO(s) + CO_2(g)$$

Sodium carbonate, called *soda ash*, is used in glass manufacture, as well as for making sodium silicate, sodium phosphate, and sodium cyanide. In addition, it is used in the pulp and paper industry, in water treatment, and in detergents. Depending upon available resources, in some countries it is manufactured by combining $NaCl(s)$ and $CO_2(s)$ by a method known as the Solvay process, while in others it is obtained from naturally occurring deposits of the mineral *trona*, $Na_2CO_3\ NaHCO_3.2H_2O$.

Sodium hydrogencarbonate, $NaHCO_3(s)$, known as *baking soda*, is used in cooking, and is also added in small amounts to table salt. Table salt is often contaminated with small amounts of $MgCl_2(s)$, which is hygroscopic (it absorbs water from air) and causes the NaCl to clump. Adding $NaHCO_3(s)$ converts $MgCl_2(s)$ to magnesium carbonate, which is not hygroscopic:

$$MgCl_2(s) + 2\,NaHCO_3(s) \longrightarrow MgCO_3(s) + 2\,NaCl(s) + H_2O(\ell) + CO_2(g)$$

The slight non-metallic behaviour of lithium is often referred to as the "anomalous" behaviour of lithium.

Large deposits of sodium nitrate, $NaNO_3(s)$, are found in Chile, which explains its common name of "Chile saltpetre." These deposits are thought to have formed by bacterial action on organisms in shallow seas. The initial product was ammonia, which was subsequently oxidized to nitrate ions; then combination with sea salt led to sodium nitrate. Because nitrates in general, and alkali metal nitrates in particular, are highly water-soluble, deposits of $NaNO_3(s)$ are found only in areas of very little rainfall [<<Section 13.1].

Sodium nitrate is important because potassium nitrate can be formed by mixing a solution of it with potassium chloride solution:

$$Na^+(aq) + NO_3^-(aq) + K^+(aq) + Cl^-(aq) \longrightarrow K^+(aq) + NO_3^-(aq) + NaCl(s)$$

Of the four rather soluble salts that might form from this mixture, NaCl is the least soluble in hot water. As the solution is concentrated by evaporation, NaCl(s) precipitates. When it has all precipitated and separated off, $KNO_3(s)$ can be obtained by evaporation.

Potassium nitrate has been used for centuries as the oxidizing agent in gunpowder. A mixture of $KNO_3(s)$, charcoal, and sulfur reacts very rapidly when ignited, with the formation of relatively large volumes of gases, and it is these that propel a bullet from a gun or cause a firecracker to explode:

$$2\,KNO_3(s) + 4\,C(s) \longrightarrow K_2CO_3(s) + 3\,CO(g) + N_2(g)$$
$$8\,KNO_3(s) + S_8(s) \longrightarrow 4\,K_2SO_4(s) + 4\,SO_2(g) + 4\,N_2(g)$$

Lithium carbonate, $Li_2CO_3(s)$, has been used for more than 40 years as a treatment for manic depression, an illness that involves alternating periods of depression and mania or over-excitement that can extend over a few weeks to a year or more. The mechanism of action of the alkali metal salt in controlling the symptoms of manic depression is not understood.

EXERCISE 26.5—THE ALKALI METALS, GROUP 1

Select one of the alkali metals and write a balanced chemical equation for its reaction with chlorine. Is the reaction likely to be exothermic or endothermic? Is the product ionic or molecular? For the product of the reaction, predict the following physical properties: colour, state of matter (s, l, or g), and solubility in water.

EXERCISE 26.6—THE ALKALI METALS, GROUP 1

(a) Write equations for the half-reactions that occur at the cathode and the anode when an aqueous solution of KCl is electrolyzed. Which chemical species is oxidized, and which chemical species is reduced in this reaction?

(b) Predict the products formed when an aqueous CsI solution is electrolyzed.

26.6 The Alkaline Earth Elements, Group 2

The "earth" part of the name *alkaline earth* dates back to the days of medieval alchemy. To alchemists, any solid that did not melt and was not changed by fire into another substance was called an "earth". Compounds of the group 2 elements, such as CaO(s), were alkaline according to experimental tests conducted by the alchemists: they had a bitter taste and neutralized acids. These compounds have very high melting points and are unaffected by heat.

Calcium and magnesium, always in compounds with other elements, rank fifth and eighth in abundance on the earth. Both elements form many commercially important compounds.

The group 2 elements, like those of group 1, are very reactive, so they are found in nature only in compounds. Unlike group 1 metals, however, many compounds of the group

---

Group 2
Alkaline earths

Beryllium 4	Be	2.6 ppm
Magnesium 12	Mg	23 000 ppm
Calcium 20	Ca	41 000 ppm
Strontium 38	Sr	370 ppm
Barium 56	Ba	500 ppm
Radium 88	Ra	6×10^{-7} ppm

Element abundances are in parts-per-million in the earth's crust.

2 elements have low water solubility, which explains their occurrence as minerals. Common calcium minerals include limestone ($CaCO_3$), gypsum ($CaSO_4.2H_2O$), and fluorite (CaF_2) (Figure 26.12). Magnesite ($MgCO_3$), talc ($3MgO.4SiO_2.H_2O$), and asbestos ($3MgO.4SiO_2.2H_2O$) are common magnesium-containing minerals. The mineral dolomite, $MgCa(CO_3)_2$, contains both magnesium and calcium.

Limestone, a sedimentary rock, is found widely on the earth's surface (Figure 26.12). Many of these deposits contain the fossilized remains of marine life. Other forms of calcium carbonate include marble and Iceland spar, which forms large, clear crystals (Figure 26.12).

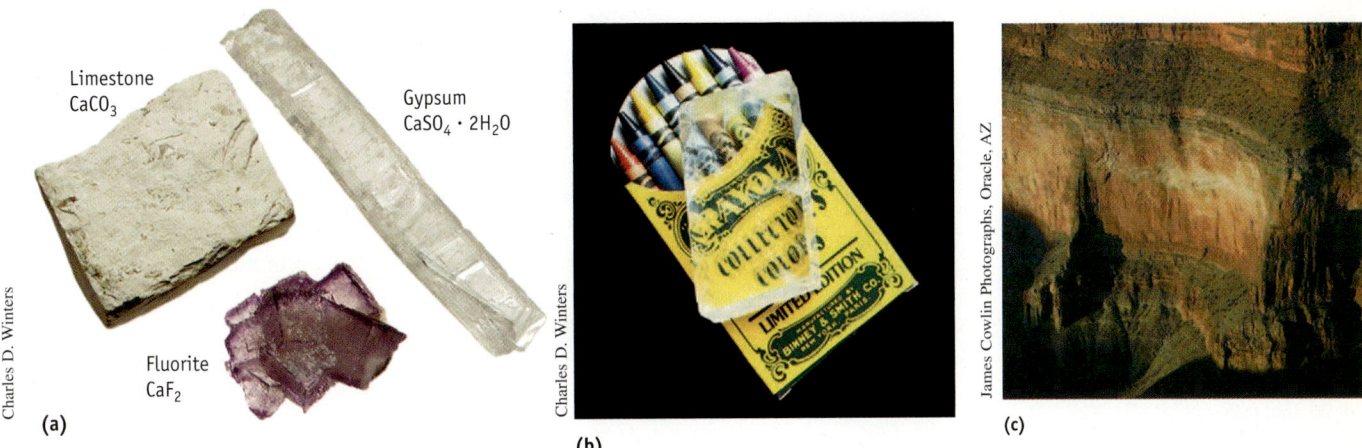

FIGURE 26.12 Various minerals containing calcium. (a) Limestone, gypsum, and fluorite. (b) A crystal of Iceland spar, one of the forms of $CaCO_3$, exhibits birefringence: light passing through the crystal forms a double image. (c) The walls of the Grand Canyon in Arizona are largely limestone or dolomite.

> The group 2 elements are very reactive and are found in nature only in compounds. Unlike group 1 metals, many compounds of the group 2 elements have low water solubility.

Properties of Calcium and Magnesium

Calcium and magnesium are fairly high-melting, silvery metals. They are oxidized by a wide range of oxidizing agents to form ionic compounds that contain the M^{2+} ion. For example, these elements combine with halogens to form halides, with oxygen or sulfur to form oxides or sulfides, and with water to form hydrogen and the metal hydroxide, $M(OH)_2(s)$ (Figure 26.13). With acids, hydrogen is evolved (see Figure 26.7 and Table 26.7), and a salt of the metal cation and the anion of the acid results.

> Calcium and magnesium are reducing agents that undergo most of the reactions that sodium and potassium do, but are less reactive than the group 1 metals.

Non-Typical Behaviour of Beryllium Compounds

Beryllium's behaviour is more non-metallic than the other members of its group—much more different from the other members of its group than lithium is different from other group 1 elements. This is directly attributable to the very high charge density of the Be^{2+} cation. The properties of beryllium are similar in many respects to those of aluminum, and it is no coincidence that the Be^{2+} and Al^{3+} ions have similar charge densities. This similarity is an example of the diagonal relationship referred to in the discussion of the non-typical behaviour of lithium in Section 26.5.

All beryllium compounds exhibit covalent bonding: even the most ionic, $BeF_2(s)$, has a relatively low melting point, and the molten compound is a poor conductor of electricity.

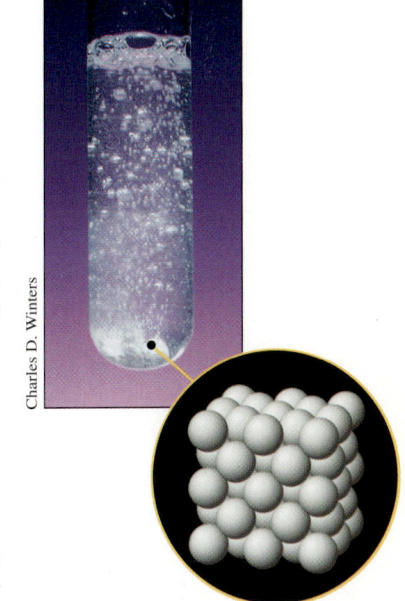

FIGURE 26.13 The reaction of calcium and warm water. Hydrogen bubbles are seen rising from the metal surface, and after a while a precipitate of $Ca(OH)_2(s)$ forms. The inset is a model of the lattice of calcium ions in the metal.

Beryllium hydroxide is amphoteric, like aluminum hydroxide but unlike the hydroxides of the other group 2 elements, which are basic:

$$Be(OH)_2(s) + 2\,H^+(aq) \longrightarrow Be^{2+}(aq) + 2\,H_2O(\ell)$$

$$Be(OH)_2(s) + 2\,OH^-(aq) \longrightarrow Be(OH)_4^{2-}(aq)$$

Like solutions of aluminum salts, solutions of beryllium salts are acidic—much more so than solutions of magnesium salts, while solutions of salts of the other members are neutral. The rationale for this is related to the high charge density of the Be^{2+} ion, and is described for Al^{3+} solutions in Section 26.3. An over-simplified equation is shown below:

> In the equation describing the acidity of solutions of beryllium salts, n may be taken to be 4 or 6.

$$[Be(OH_2)_n]^{2+} + H_2O(\ell) \rightleftharpoons [Be(OH_2)_{n-1}(OH)]^{2+} + H_3O^+(aq)$$

The coordination complexes of Be^{2+} are more stable than those of ions of other group 2 elements, and this again can be attributed to its high charge density. At high temperatures in the gas phase, $BeCl_2$ consists of linear covalent molecules with only four electrons in the Be valence shell.

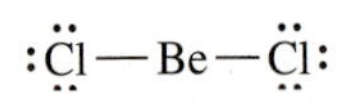

$$:\ddot{C}l - Be - \ddot{C}l:$$

A $BeCl_2$ molecule, with only four electrons in the valence shell of the Be atom.

In the solid phase, $BeCl_2$ is polymeric—and we can represent this (Figure 26.14) with 8 electrons in the valence shell of each Be atom through Lewis acid-base interactions [<<Section 6.7]. In this representation, the molecules are aligned, and each Be atom accepts a pair of electrons from each of two Cl atoms on different molecules.

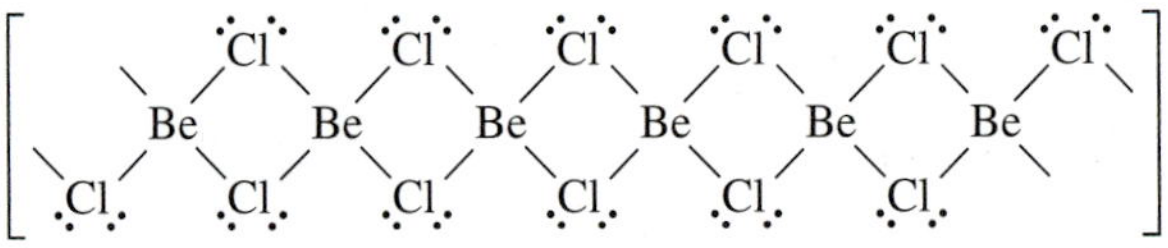

FIGURE 26.14 The polymeric structure of solid $BeCl_2$.

> **Beryllium is significantly less metallic than the other group 2 elements. For example, its hydroxide is amphoteric.**

Metallurgy of Magnesium

Several hundred thousand tonnes of magnesium metal are produced annually, largely for use in lightweight alloys. (The density of magnesium, $1.74\ \text{g cm}^{-3}$, is about 4.5 times less than that of iron.) In fact, most aluminum used today contains about 5% magnesium to improve its mechanical properties and to make it more resistant to corrosion. Other alloys having more magnesium than aluminum are used when a high strength-to-weight ratio is needed and when corrosion resistance is important, such as in aircraft and automotive parts and lightweight tools.

Minerals are not the source of magnesium: most magnesium is obtained from sea water, in which Mg^{2+} ions are present at a concentration of about $0.05\ \text{mol L}^{-1}$. To obtain the metal, magnesium ions are first precipitated from sea water as the relatively insoluble hydroxide, by addition of slaked lime (a slurry of calcium hydroxide formed by adding water to ground lime). $Mg(OH)_2(s)$ is more insoluble than $Ca(OH)_2(s)$.

$$Mg^{2+}(aq) + Ca(OH)_2(s) \longrightarrow Mg(OH)_2(s) + Ca^{2+}(aq)$$

Magnesium hydroxide is separated by filtration and then neutralized with hydrochloric acid.

$$Mg(OH)_2(s) + 2\,H^+(aq) \longrightarrow Mg^{2+}(aq) + 2\,H_2O(\ell)$$

After evaporating the water, magnesium chloride remains. Solid $MgCl_2$ melts at 714 °C, and the molten salt is electrolyzed in a process similar to that of the Downs cell used to produce sodium metal (Figure 26.10), to give magnesium metal (at the cathode) and chlorine.

$$\textit{Cathode}:\ Mg^{2+} + 2\,e^- \longrightarrow Mg(\ell)$$

$$\textit{Anode}:\ 2\,Cl^- \longrightarrow Cl_2(g) + 2\,e^-$$

> **Most Magnesium is obtained from sea water by a method that depends on the lower solubility of $Mg(OH)_2$ than $Ca(OH)_2$.**

Calcium Minerals and Their Applications

The most common calcium minerals are the fluoride, phosphate, and carbonate salts. Fluorite, $CaF_2(s)$, and fluorapatite, $Ca_5F(PO_4)_3(s)$, are important as commercial sources of fluorine. Almost half of the $CaF_2(s)$ mined is used in the steel industry, where it is added to the mixture of materials that is melted to make crude iron. The $CaF_2(s)$ acts to remove some impurities and improves the separation of molten metal from silicate impurities and other by-products resulting from the reduction of iron ore to the metal [>>Section 27.3.] A second major application of fluorite is in the manufacture of hydrofluoric acid by reaction with concentrated sulfuric acid:

$$CaF_2(s) + H_2SO_4(\ell) \longrightarrow 2\,HF(g) + CaSO_4(s)$$

Hydrofluoric acid is used to make cryolite (Na_3AlF_6), a material needed in aluminum production [>>Section 26.7] and in the manufacture of fluorocarbons such as tetrafluoroethylene, the precursor to Teflon [<<Section 19.13].

Apatites (Figure 26.15) have the general formula $Ca_5X(PO_4)_3$, in which X = F, Cl, or OH. More than 100 million tonnes of apatite is mined annually. Most of this is converted to phosphoric acid by reaction with sulfuric acid.

Calcium carbonate ($CaCO_3$) and calcium oxide or lime (CaO) are of special interest. Lime is one of the top ten industrial chemicals produced, with about 20 billion kilograms produced annually. The thermal decomposition of $CaCO_3(s)$ to lime is one of the oldest chemical reactions utilized by humans:

$$CaCO_3(s) \xrightarrow{\text{heat}} CaO(s) + CO_2(g)$$

Limestone, which consists mostly of calcium carbonate, has been used in agriculture for centuries. Apart from its use in lime production, it is spread on fields to neutralize acidic compounds in the soil and to supply the essential nutrient Ca^{2+}. Because magnesium carbonate ($MgCO_3$) is often present in limestone, "liming" a field also supplies Mg^{2+}, another important nutrient for plants.

For several thousand years, lime has been used in mortar (a slurry made from lime, sand, and water) to hold stones to one another in houses, walls, and roads. The Chinese used it to set stones in the Great Wall. The Romans perfected its use, and the fact that many of their constructions still stand today is testament both to their skill and to the usefulness of lime. The famous Appian Way used lime mortar between several layers of its stones. It is now used in buildings to hold bricks in place.

The utility of mortar depends on some simple chemistry. Mortar consists of one part lime to three parts sand, with water added to make a thick paste. The first reaction that occurs, referred to as *slaking*, produces a slurry of calcium hydroxide, known as *slaked lime*.

$$CaO(s) + H_2O(\ell) \longrightarrow Ca(OH)_2(s)$$

When the mortar is placed between bricks or stone blocks, it slowly absorbs and reacts with $CO_2(g)$ from the air, and the slaked lime reverts to calcium carbonate. The sand grains are bound together by the particles of calcium carbonate.

$$Ca(OH)_2(s) + CO_2(g) \longrightarrow CaCO_3(s) + H_2O(\ell)$$

"Hard water" contains dissolved metal ions, mainly $Ca^{2+}(aq)$ and $Mg^{2+}(aq)$ ions. These ions are found in water due to the reaction of limestone or the related mineral dolomite, $CaMg(CO_3)_2$, with water containing dissolved CO_2.

$$CaCO_3(s) + H_2O(\ell) + CO_2(g) \rightleftharpoons Ca^{2+}(aq) + 2\,HCO_3^-(aq)$$

This reaction can be reversed. When hard water is heated, the solubility of CO_2 decreases and that gas is driven from solution. Then net reaction happens in the direction that forms solid $CaCO_3$. If this happens in a heating system or a steam-generating plant, the walls of the hot-water pipes can become coated or even blocked with solid $CaCO_3$. You may be able to see such a coating of calcium carbonate on the inside of your kitchen kettle.

Charles D. Winters

FIGURE 26.15 Apatite. The mineral has the general formula of $Ca_5X(PO_4)_3$ (X = F, Cl, OH). The apatite is the elongated crystal in a matrix of other rock.

The previous equations also describe the chemistry occurring inside limestone caves. Water rendered slightly acidic by dissolved CO_2 and vegetable matter can dissolve limestone, and over hundreds of years this can lead to formation of underground caves. As water containing $Ca^{2+}(aq)$ and $HCO_3^-(aq)$ ions seeps into caves from above, if the air in the cave is deficient in CO_2, deposition of white, solid $CaCO_3$ leads to slow formation of stalactites and stalagmites [<<Section 13.2].

There are many important calcium compounds that occur naturally as minerals, the most useful of which is limestone, a form of calcium carbonate. Cave stalactites and stalagmites are composed of calcium carbonate.

Alkaline Earth Metals and Biology

Plants and animals derive energy from the oxidation of glucose with oxygen. Plants are unique, however, in being able to synthesize glucose from carbon dioxide and water by using sunlight as an energy source. This process is initiated by chlorophyll, with very large, magnesium-based molecules.

In our bodies, the metal ions Na^+, K^+, Mg^{2+}, and Ca^{2+} serve regulatory functions. Although Mg^{2+} and Ca^{2+} ions are required by living systems, ions of the other group 2 elements are toxic. Beryllium compounds are carcinogenic, and soluble barium salts are poisons. You may be concerned if your physician asks you to drink a "barium cocktail" to check the condition of your digestive tract. Barium sulfate is opaque to x-rays, so its path through your organs appears on the developed X-ray [<<Figure 15.3]. There is no need for concern, because the "cocktail" contains very insoluble $BaSO_4$ (K_{sp} = 1.1×10^{-10}), so the concentration of $Ba^{2+}(aq)$ ions in solution is negligible.

The calcium-containing compound hydroxyapatite is the main component of tooth enamel. Cavities in your teeth form when acids decompose the weakly basic apatite coating.

$$Ca_5(OH)(PO_4)_3(s) + 4\ H^+(aq) \longrightarrow 5\ Ca^{2+}(aq) + 3\ HPO_4^{2-}(aq) + H_2O(\ell)$$

This reaction can be prevented by converting hydroxyapatite to the much more acid-resistant coating of fluoroapatite.

$$Ca_5(OH)(PO_4)_3(s) + F^-(aq) \longrightarrow Ca_5F(PO_4)_3(s) + OH^-(aq)$$

Chlorophyll, one of the most important biological compounds, is a complex of magnesium. Tooth enamel is composed mainly of hydroxyapatite, a calcium phosphate.

A molecule of one of the forms of chlorophyll, a coordination compound of magnesium.

The source of the fluoride ion in toothpastes may be sodium fluoride or sodium monofluorophosphate, $Na_2(F)(PO_3)$. In some parts of the world, public water supplies are "fluoridated"—most often by addition of sodium hexafluorosilicate (Na_2SiF_6).

EXERCISE 26.7—THE ALKALINE EARTH ELEMENTS, GROUP 2

Select one of the alkaline earth metals and write a balanced chemical equation for its reaction with oxygen. Is the reaction likely to be exothermic or endothermic? Is the product ionic or molecular? For the product of the reaction, predict the following physical properties: colour, state of matter (s, l, or g), and solubility in water.

EXERCISE 26.8—THE ALKALINE EARTH ELEMENTS, GROUP 2

Would you expect to find calcium occurring naturally in the earth's crust as a free element? Why or why not?

26.7 Boron, Aluminum, and the Group 13 Elements

The elements of group 13 vary widely in their relative abundances on the earth. Aluminum is the third most abundant element in the earth's crust (82 000 ppm). In contrast, the other elements of the group are all relatively rare and, except for boron compounds, have limited commercial uses.

The General Chemistry of the Group 13 Elements

With group 13 we see a marked change from metallic behaviour of the elements at the left side of the periodic table to non-metal behaviour on the right side. Boron is a metalloid, whereas all the other elements of group 13 are metals.

Group 13 elements are characterized by the valence electron configuration ns^2np^1. In their compounds, they are commonly in the +3 oxidation state as a result of removal of three electrons from their atoms during reactions, although the heavier elements, especially thallium, also form compounds in oxidation state +1.

> Boron is a metalloid. As in the other groups, going down group 13, the elements become more metallic. The most common oxidation state is +3, although thallium has some +1 compounds.

Boron Chemistry and the Diagonal Relationship

It is generally recognized that a chemical similarity exists between some elements diagonally situated in the periodic table. As well as the previously mentioned diagonal relationship between Li and Mg, and Be and Al, boron and silicon share some chemical properties, of which the following are examples:

- Boric oxide, B_2O_3(s), and boric acid, $B(OH)_3$(s) or H_3BO_3(s), are weakly acidic, as are SiO_2(s) and its acid, orthosilic acid, $Si(OH)_4$(s) or H_4SiO_4(s).
- The boron-oxygen compounds, called borates, are in many ways chemically similar to the silicon-oxygen compounds called silicates.
- Chlorides, bromides, and iodides of boron and silicon (such as BCl_3 and $SiCl_4$) react vigorously with water.
- The hydrides of boron and silicon are simple molecular species that are volatile and flammable, and react readily with water.

The pairs of elements whose properties exhibit the diagonal relationship can be portrayed as follows:

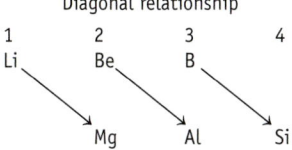

This relationship should not be unexpected since going down a column from a given element takes us to a more metallic element, and if we then go to the right along a row takes us to a less metallic element—rather like the one from which we started. The common cations of the elements of each diagonally related pair have similar charge densities.

> Consistent with the diagonal relationship, boric acid is a weak acid like orthosilicic acid, borates have similarities with silicates, and the hydrides of both boron and silicon are volatile, flammable, covalent molecular compounds.

Boron Minerals and Production of the Element

Although boron has a low abundance on the earth, its minerals are found in concentrated deposits. Large deposits of borax, $Na_2B_4O_7 \cdot 10H_2O$(s), are currently mined near the town of Boron in California. A crystal of this mineral is shown in Figure 26.16.

Isolation of pure, elemental boron is extremely difficult and is done in small quantities. Like most metals and metalloids, boron can be obtained by chemically or electrolytically reducing an oxide or halide. Magnesium has often been used for chemical reduction of boron compounds, but the boron produced in this way is non-crystalline and of low purity.

$$B_2O_3(s) + 3\,Mg(s) \longrightarrow 2\,B(s) + 3\,MgO(s)$$

Boron has several allotropes, all characterized by having the icosahedron of B atoms as a structural element (Figure 26.17). Partly as a result of extended covalent bonding, elemental

Group 13

Boron
5
B
10 ppm
Aluminum
13
Al
82 000 ppm
Gallium
31
Ga
18 ppm
Indium
49
In
0.05 ppm
Thallium
81
Tl
0.6 ppm

Element abundances are in parts-per-million in the earth's crust.

Charles D. Winters

FIGURE 26.16 Crystalline borax, $Na_2B_4O_7 \cdot 10H_2O$.

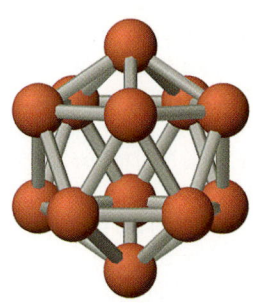

FIGURE 26.17 Icosahedron of boron atoms. All allotropes of elemental boron have an icosahedron of 12 covalently linked boron atoms as a structural unit.

boron is very hard, refractory (resistant to heat), and a semiconductor. In this regard, it differs from the other group 13 elements; Al, Ga, In, and Tl are all relatively low-melting (Figure 26.18), rather soft metals with high electrical conductivity.

FIGURE 26.18 Gallium is one of the few metals that is a liquid at or near room temperature. Others are mercury and cesium. Gallium has a melting point of 29.8 °C.

Unlike the low-melting other members of group 13, boron is a network solid whose allotropes all include in their structure an icosahedron of boron atoms.

Metallic Aluminum and Its Production

The low cost of aluminum and the excellent characteristics of its alloys with other metals (low density, strength, ease of handling in fabrication, and inertness toward corrosion, among others) have led to its widespread use. We know it best in the form of aluminum foil, aluminum cans, and parts of aircraft.

Pure aluminum is soft and weak; moreover, it loses strength rapidly at temperatures higher than 300 °C. Usually aluminum is alloyed with small amounts of other elements to strengthen the metal and improve its properties. A typical alloy may contain about 4% copper with smaller amounts of silicon, magnesium, and manganese. Softer, more corrosion-resistant alloys for window frames, furniture, highway signs, and cooking utensils may include only manganese.

The low standard reduction potential of aquated aluminum ions tells us that aluminum is easily oxidized. From this, we might expect aluminum to be highly susceptible to corrosion but, in fact, it is quite resistant, due to the formation of a thin, tough, and transparent layer of $Al_2O_3(s)$ that adheres to the metal surface. An important feature of the protective oxide layer is that it rapidly self-repairs. If you penetrate the surface coating by scratching it or using some chemical agent, the exposed metal surface immediately reacts with oxygen to form a new layer of oxide over the damaged area (Figure 26.19).

FIGURE 26.19 Corrosion of aluminum. (a) A ball of aluminum foil is added to a solution made by dissolving copper(II) nitrate and sodium chloride in water. Normally, the coating of chemically inert $Al_2O_3(s)$ on the surface of aluminum protects the metal from further oxidation. (b) In the presence of the Cl^- ion, the coating of $Al_2O_3(s)$ is breached, and aluminum reduces copper(II) ions to copper metal. The reaction is rapid and so exothermic that the water can boil on the surface of the foil. The blue colour of aqueous copper(II) ions will fade as these ions react.

(a)

(b)

Aluminum was first produced by reducing $AlCl_3(s)$ using sodium or potassium. This was a costly process and, in the 19th century, aluminum was a precious metal. At the 1855 Paris Exposition, in fact, a sample of aluminum was exhibited along with the crown jewels of France. In an interesting coincidence, in 1886, two men, Frenchman Paul Heroult (1863–1914) and American Charles Hall (1863–1914), simultaneously and independently conceived of the electrochemical method used today (see below). The Hall-Heroult method bears the names of the two discoverers.

Aluminum is found in nature as aluminosilicates, minerals such as clay that are based on aluminum, silicon, and oxygen. As these minerals weather, they break down to various forms of hydrated aluminum oxide, $Al_2O_3 \cdot nH_2O(s)$, called *bauxite*. Mined in huge quantities, bauxite is the raw material from which aluminum is obtained. The first step is to purify the ore, separating alumina (Al_2O_3) from iron and silicon oxides. This is done by the *Bayer process*, which relies on the amphoteric nature of $Al_2O_3(s)$ to separate it from acidic silica, $SiO_2(s)$, and basic $Fe_2O_3(s)$. Silica and alumina dissolve in a hot concentrated solution of caustic soda (NaOH), leaving insoluble $Fe_2O_3(s)$ to be filtered off:

$$Al_2O_3(s) + 2\,OH^-(aq) + 3\,H_2O(\ell) \longrightarrow 2\,[Al(OH)_4]^-(aq)$$
$$SiO_2(s) + 2\,OH^-(aq) + 2\,H_2O(\ell) \longrightarrow [Si(OH)_6]^{2-}(aq)$$

By treating the solution containing aluminate and silicate anions with $CO_2(g)$, $Al_2O_3(s)$ precipitates and the silicate ion remains in solution. CO_2 is an acidic oxide that forms the weak acid $H_2CO_3(aq)$ in water, so the $Al_2O_3(s)$ precipitation is an acid-base reaction.

$$H_2CO_3(aq) + 2\,[Al(OH)]_4^-(aq) \longrightarrow CO_3^{2-}(aq) + Al_2O_3(s) + 5\,H_2O(\ell)$$

Metallic aluminum is obtained from alumina by electrolysis (Figure 26.20). Bauxite is first mixed with cryolite (Na_3AlF_6) to form a lower-melting mixture (melting temperature = 980 °C) that is electrolyzed in a cell with graphite electrodes. The cell operates at a relatively low voltage (4.0–5.5 V) but, in order to obtain large amounts of aluminum in short times, with a very high current (50 000–150 000 amps). Aluminum is produced at the cathode and oxygen at the anode. To produce 1 kg of aluminum requires 13–16 kWh of energy, as well as the energy required to keep the electrolyte molten.

Charles Martin Hall (1863–1914). Hall was only 22 years old when he worked out the electrolytic process for extracting aluminum from Al_2O_3 in a woodshed behind the family home in Oberlin, Ohio. He went on to found a company that eventually became ALCOA, the Aluminum Corporation of America.

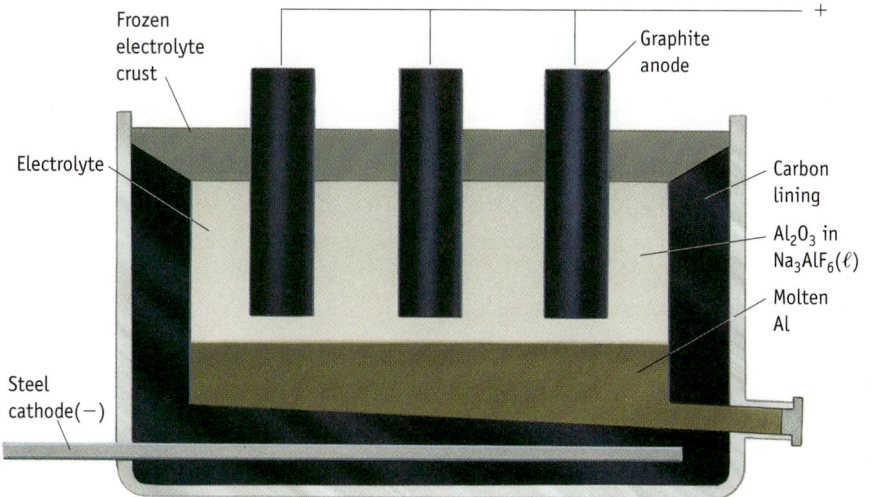

(a) Electrolysis of aluminum oxide to produce aluminum metal.

(b) Molten aluminum from recycled metal.

FIGURE 26.20 Industrial production of aluminum. (a) Alumina (Al_2O_3) is mixed with cryolite (Na_3AlF_6) to give a mixture that melts at a lower temperature than pure Al_2O_3. The molten mixture is the electrolyte of the cell, and alumina is reduced at the steel cathode to form molten aluminum. Oxygen is produced at the carbon anode, and the gas reacts slowly with the carbon to give CO_2, leading to eventual destruction of the electrode. (b) Molten aluminum alloy, produced from recycled metal, at 760 °C, in crucibles of 16 tonnes capacity.

Aluminum is very reactive, but its corrosion in air is slow because of a protective oxide layer. Pure alumina is obtained from bauxite ores by dissolving in hot sodium hydroxide—a reaction that occurs because alumina is amphoteric—and filtering off insoluble iron oxides. Alumina cannot be reduced by common reducing agents, so aluminum metal is obtained by electrolysis of alumina.

Boron Compounds

Borax, $Na_2B_4O_7.10H_2O(s)$, is the most important boron–oxygen compound and is the form of the element most often found in nature. It has been used for centuries as a low-melting flux in metallurgy, because of the ability of molten borax to dissolve other metal oxides. This cleans the surfaces of metals to be joined and permits a good metal-to-metal contact.

The formula $Na_2B_4O_7.10H_2O$ for borax is misleading. The salt contains an ion better described by the formula $B_4O_5(OH)_4^{2-}$ (Figure 26.21). This ion illustrates two commonly observed structural features in inorganic chemistry. First, many minerals consist of MO_n groups that share O atoms. Second, the sharing of O atoms between atoms of two metals or metalloids often leads to rings with $—M—O—M—O—$ sequences.

After refinement, borax can be treated with sulfuric acid to produce boric acid, $B(OH)_3$:

$$Na_2B_4O_7.10H_2O(s) + 2\,H^+(aq) \longrightarrow 4\,B(OH)_3(aq) + 2\,Na^+(aq) + 5\,H_2O(\ell)$$

The chemistry of boric acid incorporates both Lewis and Brønsted-Lowry acid behaviour. Hydronium ions are produced by a Lewis acid-base reaction between boric acid and water. Boric acid is a monoprotic acid: 1 mol of it will react with only 1 mol of $OH^-(aq)$ ions.

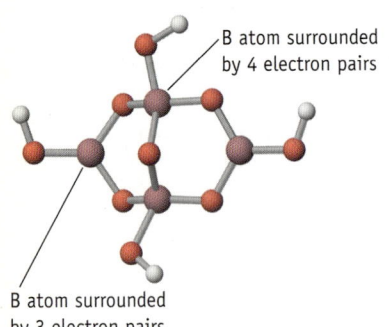

B atom surrounded by 4 electron pairs

B atom surrounded by 3 electron pairs

FIGURE 26.21 The $B_4O_5(OH)_4^{2-}$ ion in the structure of borax.

$$K_a = 7.3 \times 10^{-10}$$

Because of its weak acid properties and slight biological activity, boric acid has been used for many years as an antiseptic. Furthermore, because the acid is so weak, salts of borate ions, such as the $B_4O_5(OH)_4^{2-}$ ion in borax, are hydrolyzed in water to give a basic solution.

Boric acid is dehydrated to boric oxide when strongly heated:

$$2\,B(OH)_3(s) \longrightarrow B_2O_3(s) + 3\,H_2O(\ell)$$

By far the largest use for the oxide is in the manufacture of borosilicate glass. This type of glass is composed of 76% SiO_2, 13% B_2O_3, and much smaller amounts of Al_2O_3 and Na_2O. The presence of boric oxide gives the glass a higher softening temperature, imparts a better resistance to attack by acids, and makes the glass expand less on heating.

Like its metalloid neighbour silicon, boron forms a series of molecular compounds with hydrogen. Because boron is slightly less electronegative than hydrogen, these compounds are best described as hydrides, in which the H atoms bear a slight negative charge. More than 20 neutral boron hydrides, or *boranes*, with the general formula B_xH_y are known. The simplest of these is diborane, $B_2H_6(g)$. This colourless, gaseous compound has a boiling point of $-92.6\ °C$.

Molecules of diborane are unusual. You might have expected that the simplest boron hydride would have the formula BH_3 and a planar, trigonal geometry like that of the boron trihalides. Diborane seems even more curious if you examine its molecular structure (Figure 26.22). Two of the H atoms are bonded not to a single boron atom, but rather each is bonded to two boron atoms. Furthermore, there appears to be a shortage of electrons for all the bonds. Two boron atoms and six hydrogen atoms bring a total of 12 valence electrons to bind the molecule together. If you take each of the eight lines in the structural diagram as a two-electron bond, 16 electrons would be required. Because there appears not to be enough electrons for all the bonds in the molecule (and the same is true in other boron hydrides), these compounds came to be called *electron-deficient* molecules.

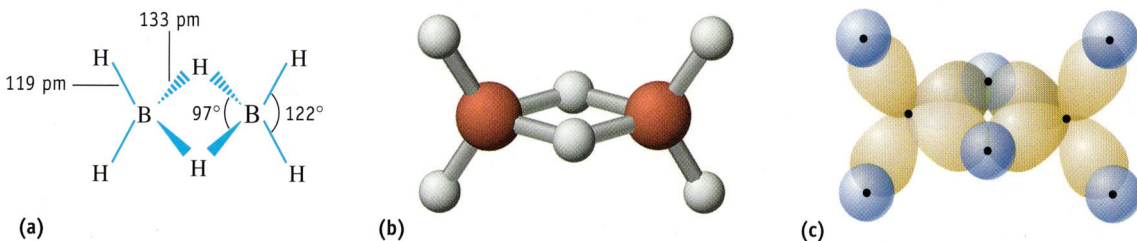

FIGURE 26.22 Bonding in diborane molecules. (a, b) The structure of diborane (B_2H_6) molecules. (c) After accounting for bonding to two terminal H atoms, two sp^3 hybrid orbitals remain on each B atom. One such orbital from each boron may overlap a hydrogen $1s$ orbital in the bridge to give a "three-centre bond" (involving three atoms), to which two electrons are assigned.

There are several ways to solve the diborane-bonding dilemma, but just one model is outlined here. According to this model, since the boron atoms are surrounded more or less tetrahedrally by H atoms, we assume the B atoms are sp^3 hybridized. The "outside" or terminal B—H bonds are assumed to be normal, two-electron bonds formed by the overlap of an H atom's $1s$ orbital with a B atom's sp^3 orbital. The four bonds of this type require 8 of the 12 electrons available. Each boron atom has two additional sp^3 hybrid orbitals, which extend into the bridging region (Figure 26.22c). Here a spherical hydrogen $1s$ orbital can overlap with one sp^3 hybrid orbital from each boron, creating a *three-centre bond*. This three-centre bond can accommodate two electrons, so the two bridges account for the four remaining valence electrons.

Diborane has a very endothermic enthalpy change of formation: $\Delta_f H° = +41.0$ kJ mol^{-1}. It is not surprising, then, that it and other boron hydrides have potential use as rocket fuels. They burn in air to give boric oxide and water vapour in an extremely exothermic reaction.

$$B_2H_6(g) + 3\ O_2(g) \longrightarrow B_2O_3(s) + 3\ H_2O(g) \qquad \Delta_r H° = -2038\ \text{kJ mol}^{-1}$$

Diborane can be synthesized from sodium borohydride, $NaBH_4(s)$, the only B—H compound produced in tonne quantities.

$$2\ NaBH_4(s) + I_2(s) \longrightarrow B_2H_6(g) + 2\ NaI(s) + H_2(g)$$

Sodium borohydride, $NaBH_4(s)$, a white, crystalline, water-soluble solid, is made from $NaH(s)$ and a borate.

$$4\ NaH(s) + B(OCH_3)_3(g) \longrightarrow NaBH_4(s) + 3\ NaOCH_3(s)$$

One of the main uses of sodium borohydride is as a reducing agent in organic synthesis, to reduce aldehydes, carboxylic acids, and ketones [<<Section 23.9].

> Borates commonly have B—O—B sequences in their structures. Boric acid is a monoprotic weak acid. There is a series of boranes that release a lot of energy on combustion. Diborane, the simplest, has a structure that can be explained by the presence of three-centre, two-electron B—H—B bonds.

Aluminum Compounds

Aluminum is an excellent reducing agent, so it reacts readily with hydrochloric acid. In contrast, it does not react with nitric acid. The latter rapidly oxidizes the surface of aluminum, and the resulting film of Al_2O_3 protects the metal from further reaction (Figure 26.23). This protection allows nitric acid to be shipped in aluminum tanks.

Various salts of aluminum dissolve in water, forming the hydrated Al^{3+}(aq) ion. As described in Section 26.3, these solutions are acidic because the hydrated ion is a weak Brønsted-Lowry acid.

$$[Al(OH_2)]_6]^{3+}(aq) + H_2O(\ell) \rightleftharpoons [Al(OH_2)_5(OH)]^{2+}(aq) + H_3O^+(aq)$$

Addition of base causes more and more H_2O molecules bound to the Al^{3+} ion to be replaced by OH$^-$ ions, and eventually addition of sufficient hydroxide ions results in precipitation of the hydrated oxide $[Al(OH_2)_3(OH)_3](s)$ or $Al_2O_3 \cdot 3H_2O(s)$.

Think about It

e26.3 Work through a tutorial on boron hydrides.

Molecular Modelling (Odyssey)

e26.4 Examine the electron density isosurface of diborane.

Charles D. Winters

FIGURE 26.23 Aluminum does not react with nitric acid. Nitric acid, a strongly oxidizing acid, reacts vigorously with copper (left) but aluminum (right) is untouched.

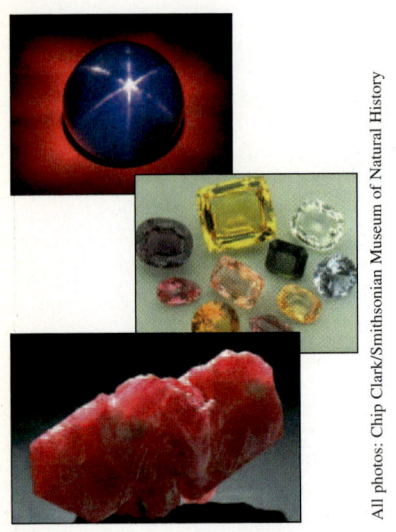

FIGURE 26.24 Sapphires and rubies. Both minerals are based on Al_2O_3 in which a few Al^{3+} ions have been replaced by ions such as Cr^{3+}, Fe^{2+}, or Ti^{4+}. (*Top*) The Star of Asia sapphire; (*middle*) various sapphires; (*bottom*) uncut corundum.

All photos: Chip Clark/Smithsonian Museum of Natural History

FIGURE 26.25 The Al_2Br_6 molecule.

Think about It

e26.5 Work through a tutorial on aluminium compounds.

Aluminum oxide (Al_2O_3), formed by dehydrating the hydrated oxide, is quite insoluble in water and generally resistant to chemical attack. In the crystalline form, aluminum oxide is known as *corundum*. This material is extraordinarily hard, a property that leads to its use as an abrasive in grinding wheels, "sandpaper," and toothpaste.

Some gems are aluminum oxide containing traces of impurities. Rubies, beautiful red crystals prized for jewellery and used in some lasers, are $Al_2O_3(s)$ contaminated with a small amount of Cr^{3+} (Figure 26.24). The Cr^{3+} ions replace some of the Al^{3+} ions in the crystal lattice. Synthetic rubies were first made in 1902, and the worldwide annual production is now about 200 000 kg; much of this production is used for jewel bearings in watches and instruments. Blue sapphires consist of $Al_2O_3(s)$ with Fe^{2+} and Ti^{4+} impurities in place of Al^{3+} ions.

Boron forms gaseous halides such as $BF_3(g)$ and $BCl_3(g)$ that have the expected trigonal planar shape of a boron atom surrounded by three halogen atoms. In contrast, the aluminum halides are all solids with interesting structures. Aluminum bromide, which is made by the very exothermic reaction of aluminum metal and bromine, is composed of two units of $AlBr_3$.

$$2\ Al(s) + 3\ Br_2(\ell) \longrightarrow Al_2Br_6(s)$$

That is, Al_2Br_6 is a *dimer* of $AlBr_3$ units (Figure 26.25). The structure resembles that of diborane (Figure 26.22) in that there are bridging atoms between the two Al atoms. However, Al_2Br_6 is not electron-deficient; the bridge is formed when a Br atom on one $AlBr_3$ unit uses a lone pair to form a coordinate covalent bond to a neighbouring aluminum atom. Each Al atom has four Br atoms tetrahedrally arranged around it.

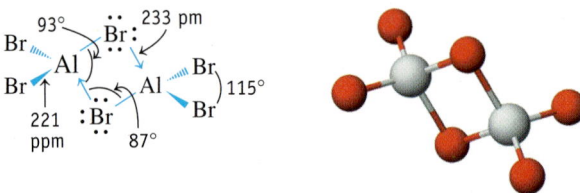

Both aluminum bromide and aluminum iodide have this structure, whereas aluminum chloride exists as a dimer only in the vapour state.

Aluminum chloride can react with chloride ions to form the simple ion $AlCl_4^-$. Aluminum fluoride, in contrast, can accommodate three additional F^- ions to form an octahedral AlF_6^{3-} ion. This form of aluminum is found in cryolite, $Na_3AlF_6(s)$. Apparently, the Al^{3+} ion can bind to six of the smaller F^- ions, whereas only four of the larger Cl^-, Br^-, or I^- ions can surround an Al^{3+} ion.

Aluminum metal reacts with dilute hydrochloric acid solutions, but not with nitric acid solutions, because of a protective oxide layer. Solutions of aluminum salts are acidic. Aluminum metal reacts with bromine to form $Al_2Br_6(s)$.

EXERCISE 26.9—THE GROUP 13 ELEMENTS

(a) Gallium hydroxide, like aluminum hydroxide, is amphoteric. Write balanced equations to represent how this hydroxide reacts in both HCl solution and NaOH solution.

(b) The aquated gallium ion, $Ga^{3+}(aq)$, has a K_a value of 1.2×10^{-3}. Is this ion a stronger or a weaker acid than the $Al^{3+}(aq)$ ion?

(c) What volume of 0.0112 M HCl solution is needed to react completely with 1.25 g of $Ga(OH)_3(s)$?

EXERCISE 26.10—THE GROUP 13 ELEMENTS

Write equations for the reactions of aluminum metal with HCl solution, $Cl_2(g)$, and $O_2(g)$.

26.8 Silicon and the Group 14 Elements

The elements of group 14 have the broadest range of chemical behaviours of any group in the periodic table. Carbon is distinctly non-metallic in its chemistry, but silicon and germanium are classed as metalloids, while tin and lead are metals.

Atoms of all of the group 14 elements are characterized by half-filled valence shells with two electrons in the *ns* orbital and two electrons in *np* orbitals (where *n* is the period in which the element is placed). The bonding in carbon and silicon compounds is largely covalent and involves sharing of four electron pairs with neighbouring atoms. In germanium compounds, the +4 oxidation state is common (GeO_2 and $GeCl_4$), but some +2 oxidation state compounds exist (GeI_2). An oxidation state of +2, as well as +4, is even more common for tin and lead (such as $SnCl_2$ and PbO). The increasing importance of the +2 oxidation state for heavier elements in the group illustrates a trend seen in elements of groups 15, 16, and 17: lower oxidation states (such as those of Tl^+, Pb^{2+}, and Bi^{3+}) are more stable (therefore, more common) for the heavier members of the groups.

Silicon

Silicon is second after oxygen in abundance in the earth's crust, so it is not surprising that we are surrounded by silicon-containing materials: bricks, pottery, porcelain, lubricants, sealants, computer chips, and solar cells. The computer revolution is based on the semiconducting properties of silicon.

Reasonably pure silicon can be made in large quantities by heating pure silica sand with purified coke to approximately 3000 °C in an electric furnace.

$$SiO_2(s) + 2\,C(s) \longrightarrow Si(\ell) + 2\,CO(g)$$

The molten silicon is drawn off the bottom of the furnace and allowed to cool to a shiny blue-grey solid. Because extremely high-purity silicon is needed for the electronics industry, purifying raw silicon requires several steps. First, the silicon in the impure sample is allowed to react with chlorine to convert the silicon to liquid silicon tetrachloride.

$$Si(s) + 2\,Cl_2(g) \longrightarrow SiCl_4(\ell)$$

Silicon tetrachloride (boiling point: 57.6 °C) is carefully purified by distillation and then reduced to silicon using magnesium.

$$SiCl_4(\ell) + 2\,Mg(s) \longrightarrow 2\,MgCl_2(s) + Si(s)$$

The magnesium chloride is washed out with water, and the silicon is remelted and cast into bars. A final purification is carried out by *zone refining*, a process in which a special heating device is used to melt a narrow segment of the silicon rod. The heater is moved slowly down the rod. Impurities contained in the silicon tend to remain in the liquid phase because the melting point of a mixture is lower than that of the pure element. The silicon that crystallizes above the heated zone is therefore of a higher purity (Figure 26.26).

> Silicon is made by reduction of silica sand with coke in an electric furnace. Purification is carried out by zone refining.

Silicon Dioxide

The simplest oxide of silicon is, $SiO_2(s)$, commonly called *silica*, a constituent of many rocks such as granite and sandstone. Quartz is a pure crystalline form of silica, but impurities in quartz produce gemstones such as amethyst (Figure 26.27).

Silicon dioxide and carbon dioxide are both oxides of group 14 elements, but they are very different: SiO_2 is a high-melting solid (quartz melts at 1610 °C), whereas CO_2 is a gas at SATP. This great disparity arises from their different structures. Carbon dioxide is a

Group 14

Carbon 6 **C** 480 ppm	
Silicon 14 **Si** 277 100 ppm	
Germanium 32 **Ge** 1.8 ppm	
Tin 50 **Sn** 2.2 ppm	
Lead 82 **Pb** 14 ppm	

Element abundances are in parts-per-million in the earth's crust.

© Science VU/Visuals Unlimited

FIGURE 26.26 Pure silicon. The manufacture of very pure silicon begins with producing the volatile liquid silanes $SiCl_4$ or $SiHCl_3$. After carefully purifying these by distillation, they are reduced to elemental silicon with extremely pure Mg or Zn. The resulting spongy silicon is heated to produce molten silicon, which is then purified by zone refining. The end result is a cylindrical rod of ultrapure silicon such as those seen in this photograph. Finally, thin wafers of silicon are cut from the bars and are the basis for the semiconducting chips in computers and other devices.

Molecular Modelling (Odyssey)

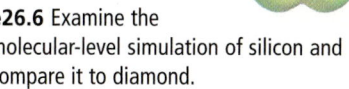

e26.6 Examine the molecular-level simulation of silicon and compare it to diamond.

FIGURE 26.27 Forms of quartz. (a) Pure quartz is colourless, but the presence of small amounts of impurities adds colour. Purple amethyst and brown citrine crystals are quartz with iron impurities. (b) Quartz is a network solid in which each Si atom is bonded in tetrahedral directions to four O atoms, each O atom linked to another Si atom.

Amethyst

Citrine

Quartz

Photos: Charles D. Winters

Synthetic quartz. These crystals were grown from silica in sodium hydroxide. The colours come from added Co^{2+} ions (blue) or Fe^{2+} ions (brown).

Charles D. Winters

Charles D. Winters

Silica gel. Silica gel is solid, non-crystalline SiO_2. Packages of the material are often used to keep electronic equipment dry when stored. Silica gel is also used to clarify beer; passing beer through a bed of silica gel removes minute particles that would otherwise make the brew cloudy. Yet another use is in kitty litter.

Molecular Modelling (Odyssey)

e26.7 Examine the molecular-level simulation of silicon dioxide and compare it to diamond and silicon.

molecular compound, and its melting point is determined by the relatively weak dispersion forces between the molecules. In contrast, SiO_2 is a solid covalent network solid. The contrast between $SiO_2(s)$ and $CO_2(g)$ exemplifies a more general phenomenon: multiple bonds, often encountered between second-period elements, are rare among elements in the third and higher periods. There are many compounds with multiple bonds to carbon atoms, but very few compounds featuring multiple bonds to silicon atoms.

Quartz crystals are used to control the frequency of radio and television transmissions. Because these and related applications use so much quartz, there is not enough natural quartz to fulfill demand, and quartz is therefore made artificially. Non-crystalline, or vitreous, quartz, made by melting pure silica sand, is placed in a steel "bomb" and dilute aqueous NaOH solution is added. A "seed" crystal is placed in the mixture. When the mixture is heated above the critical temperature of water (400 °C and 1700 bar) over a period of days, pure quartz crystallizes.

Silica is resistant to attack by all acids except HF, with which it reacts to give $SiF_4(g)$ and water. It also dissolves slowly in hot, molten NaOH or Na_2CO_3 to give sodium silicate, $Na_4SiO_4(s)$.

$$SiO_2(s) + 4\,HF(aq) \longrightarrow SiF_4(g) + 2\,H_2O(\ell)$$
$$SiO_2(s) + 2\,Na_2CO_3(\ell) \longrightarrow Na_4SiO_4(s) + 2\,CO_2(g)$$

After the molten mixture has cooled, hot water under pressure is added. This partially dissolves the material to give a solution of sodium silicate. After filtering off insoluble sand or glass, the solvent is evaporated to leave sodium silicate, called *water glass*. The biggest single use of this material is in household and industrial detergents, in which it is included because a sodium silicate solution maintains pH by its buffering ability. Additionally, sodium silicate is used in various adhesives and binders, especially for gluing corrugated cardboard boxes.

If sodium silicate is treated with acid, a gelatinous precipitate of SiO_2 called *silica gel* is obtained. Washed and dried, silica gel is a highly porous material with dozens of uses. It is a drying agent, readily absorbing up to 40% of its own weight of water. Small packets of silica gel are often placed in packing boxes of merchandise during storage. The material is frequently coated with $(NH_4)_2CoCl_4(s)$, a humidity detector that is pink when hydrated and blue when dry. The $[CoCl_4]^{2-}$ complex ion is blue. When its environment becomes moist, the Lewis base chloride ions are replaced by water molecules, forming pink $[Co(OH_2)_6]^{2+}$ complex ions. On heating, water is driven off and the reaction is reversed.

$$[CoCl_4]^{2-} + 6\,H_2O(\ell) \rightleftharpoons [Co(OH_2)_6]^{2+} + 4\,Cl^-(aq)$$

Quartz is a pure crystalline form of silicon dioxide, called silica; impurities produce gemstones. Silica is a high-melting network solid, resistant to attack by acids except HF. Silica gel is a gelatinous form of silica that absorbs large quantities of water.

Silicate Minerals with Chain and Ribbon Structures

Silicate minerals are a world in themselves. All silicate structures have tetrahedral SiO_4 units, but they have different properties because of the way these tetrahedral SiO_4 units link together.

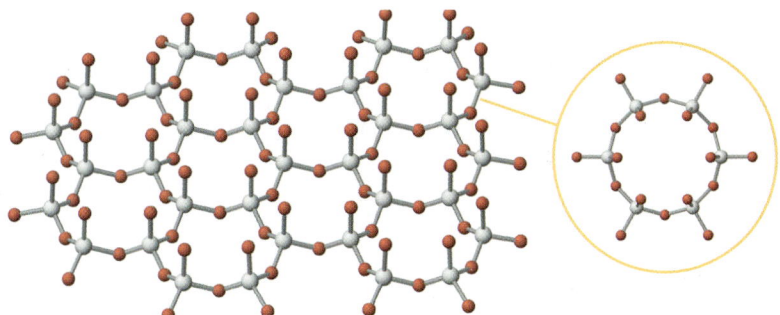

The simplest silicates, *orthosilicates*, contain SiO_4^{4-} anions. The −4 charge of the anion is balanced by four M^+ ions, two M^{2+} ions, or a combination of ions. Calcium orthosilicate (Ca_2SiO_4) is a component of Portland cement. Olivine, an important mineral in the earth's mantle, contains Mg^{2+} and Fe^{2+} ions, with the Fe^{2+} ion giving the mineral its characteristic olive colour.

A group of minerals called *pyroxenes* have as their basic structural unit a chain of SiO_4 tetrahedra.

If two such chains are linked together by sharing oxygen atoms, the result is an *amphibole*, of which the asbestos minerals are one example. The molecular chain results in asbestos being a fibrous material.

> Othosilicates contain SiO_4^{4-} anions in a lattice with metal cations. Pyroxenes have structures based on chains of SiO_4 tetrahedra.

Silicates with Sheet Structures and Aluminosilicates

Sheets of SiO_4 tetrahedra, which we can imagine to be formed by linking many silicate chains together (Figure 26.28) are the basic structural feature of some of the earth's most important minerals, particularly the clay minerals, mica, and the chrysotile form of asbestos.

FIGURE 26.28 Mica, a sheet silicate. The sheet-like structure of mica explains its physical appearance. As in the pyroxenes, each silicon is bonded to four oxygen atoms, but the Si and O atoms form a sheet of six-membered rings of Si atoms with O atoms in each edge. The ratio of Si to O is 1 to 2.5. (A formula of $SiO_{2.5}$ requires a positive ion, such as Na^+, to counterbalance the charge. Thus, mica and other sheet silicates, and aluminosilicates such as talc and many clays, have positive ions between the sheets.)

The sheet structure leads to the characteristic feature of mica, which is often found as "books" of thin, silicate sheets that easily cleave apart. Mica is used in furnace windows and as insulation, and flecks of mica give the glitter to "metallic" paints.

Mica, a large family of clays, and asbestos are actually *aluminosilicates*, substances containing both aluminum and silicon. In kaolinite clay, for example, the sheet of SiO_4 tetrahedra is bonded to a sheet of AlO_6 octahedra (Figure 26.29). In addition, some Si atoms can be replaced by Al atoms. Because Si^{4+} ions are being replaced by Al^{3+} ions with a smaller charge, nature adds positive ions such as Na^+ and Mg^{2+} for every aluminum ion in the lattice. This feature leads to some interesting uses of clays, such as in medicine (Figure 26.30a). In certain cultures, clay is eaten for medicinal purposes. Several remedies for the relief of upset stomach contain highly purified clays that absorb excess stomach acid as well as potentially harmful bacteria and their toxins by exchanging the intersheet cations in the clays for the toxins, which are often organic cations.

Other aluminosilicates include the feldspars, common minerals that make up about 60% of the earth's crust, and zeolites (Figure 26.30b). Both materials are composed of SiO_4 tetrahedra in which some of the Si atoms have been replaced by Al atoms, along with alkali and alkaline earth metal ions for charge balance. The main feature of zeolite structures is their regularly shaped tunnels and cavities. Hole diameters are between 300 and 1000 pm, and small molecules such as water can fit into the cavities of the zeolite. As a result, zeolites can be used as drying agents to selectively absorb water from air or a solvent. Small amounts of zeolites are often sealed into multipane windows to keep the air dry between the panes.

FIGURE 26.29 A model of kaolinite clay. The basic structural feature of many clays, and kaolinite in particular, is a sheet of SiO_4 tetrahedra (black and red spheres) bonded to a sheet of AlO_6 octahedra (grey and green spheres).

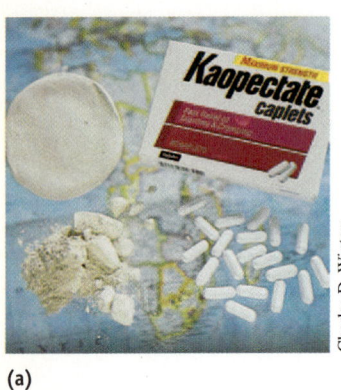
(a)

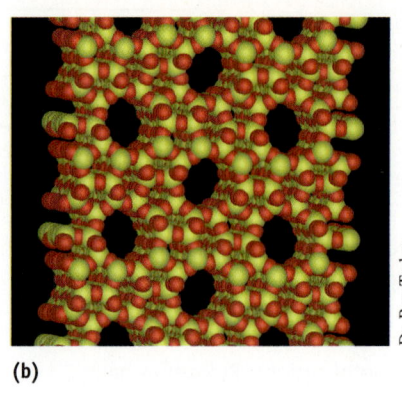
(b)

(c)

(d)

FIGURE 26.30 Aluminosilicates. (a) Remedies for stomach upset. One of the ingredients in these products is kaolin, a clay. The off-white objects are from Ghana in Africa where it is eaten as a remedy for stomach upsets—a practice widespread among the world's different cultures. (b) The structure of a zeolite. Zeolites, which have Si, Al, and O linked in a polyhedral framework, are often portrayed in diagrams like this. Each edge is a Si—O—Si, Al—O—Si, or Al—O—Al bond. The channels in the structure can selectively capture small molecules or ions, or act as catalytic sites. (c) Apophyllite, a crystalline zeolite. (d) Consumer products that remove odour-causing molecules from the air often contain zeolites.

Zeolites are also used as catalysts. ExxonMobil, for example, has patented a process in which methanol (CH_3OH) is converted to gasoline in the presence of specially designed zeolites. In addition, zeolites are added to detergents, where they function as water-softening agents because the sodium ions of the zeolite can be exchanged for Ca^{2+} ions in hard water, effectively removing Ca^{2+} ions from the water.

Many aluminosilicate minerals and clays have structures that are based on parallel sheets of silicates.

Silicone Polymers

Silicones are covalent molecular polymers in which the molecules have a "backbone" with alternate Si and O atoms (Figure 26.31).

FIGURE 26.31 "Backbone" of a silicone polymer molecule. Silicones can be more simply represented as

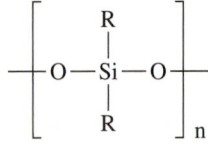

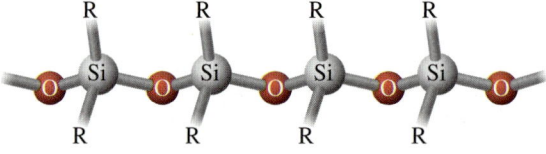

Silicone polymers are non-toxic and have good stability to heat, light, and oxygen; they are chemically inert and have valuable anti-stick and anti-foam properties. These materials are used in a wide variety of products: lubricants, peel-off labels, lipstick, suntan lotion, car polish, and building caulk.

Silicones are produced by polymerization of compounds called *silanols*. For example, dimethylsilanol, $(CH_3)_2Si(OH)_2$, is produced in two stages from silicon and methyl chloride:

$$Si(s) + 2\ CH_3Cl(g) \xrightarrow[300\ °C]{Cu\ catalyst} (CH_3)_2SiCl_2(\ell)$$

$$(CH_3)_2SiCl_2(\ell) + 2\ H_2O(\ell) \xrightarrow{hydrolysis} (CH_3)_2Si(OH)_2 + 2\ HCl(g)$$

Polymerization of dimethylsilanol with elimination of water molecules produces a silicone called *polydimethylsiloxane*.

$$n\ (CH_3)_2Si(OH)_2 \longrightarrow [-(CH_3)_2SiO-]_n + n\ H_2O$$

In the more general case, polymerization of silanols with the formula $R_2Si(OH)_2$ is portrayed in Figure 26.32.

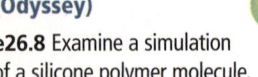

Molecular Modelling (Odyssey)

e26.8 Examine a simulation of a silicone polymer molecule.

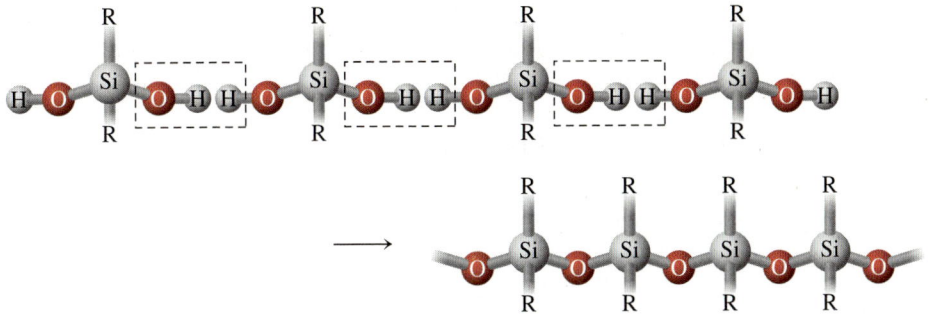

FIGURE 26.32 Silicone formation. Polymerization of silanol molecules by elimination of water forms a silicone.

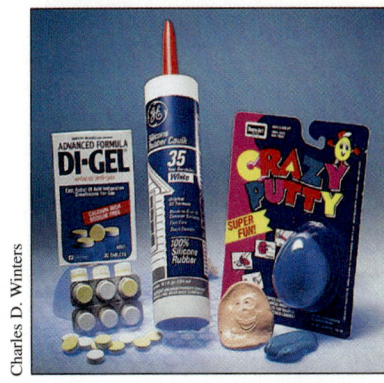

Charles D. Winters

Silicones. Some examples of polymers with repeating —Si—O—Si— units.

Silicones with relatively short polymer chains are oils, and longer-chain polymers are greases. When a silanol of the general formula $RSi(OH)_3$ is polymerized, Si—O—Si links are formed between polymer chains, and the resulting cross-linked polymers are rigid materials called *resins*.

When the cross-links between silicone chains are alkyl groups (achieved by polymerization in the presence of peroxide initiators), the polymer is a rubber, more correctly called an *elastomer* (Figure 26.33).

By varying the conditions of manufacture, chemists can achieve different degrees of cross-linking, and different lengths of the alkyl groups, to produce elastomers with particular properties for particular purposes.

> Silicones are covalent molecular polymers in which the molecules have a "backbone" with alternate Si and O atoms. The properties of silicones depend on the degree of interlinking between chains and the nature of the links (silicone or alkyl).

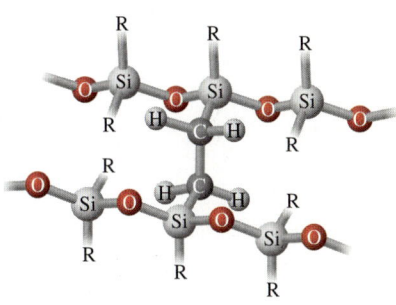

FIGURE 26.33 Portion of an elastomer molecule, with alkyl group cross-links.

EXERCISE 26.11—THE GROUP 14 ELEMENTS

Describe the structures of $SiO_2(s)$ and $CO_2(g)$. Explain why silicon dioxide has a very high melting point, whereas carbon dioxide is a gas.

EXERCISE 26.12—THE GROUP 14 ELEMENTS

Bearing in mind the influence of charge density of cations, predict which chloride of lead ($PbCl_2$ or $PbCl_4$) has the lower melting point.

26.9 Group 15 Elements and Their Compounds

Atoms of the group 15 elements are characterized by the ns^2np^3 valence electron configuration, with a half-filled p sub-shell. In compounds of these elements, the most common oxidation states are $+3$ and $+5$, although in a set of common nitrogen compounds a range of oxidation states from -3 to $+3$ and $+5$ is seen. Once again, as in groups 13 and 14, the compounds of the heavier elements with the highest oxidation state are less stable. In many arsenic and bismuth compounds, the element is in oxidation state $+3$. Compounds of these elements with oxidation state $+5$ are powerful oxidizing agents.

We will concentrate on the chemistries of nitrogen and phosphorus. Nitrogen is found primarily as $N_2(g)$ in the atmosphere, where it constitutes 78.1% of the amount, in moles (75.5% by weight). In contrast, phosphorus occurs in the earth's crust in solid compounds. More than 200 different phosphorus-containing minerals are known; all contain the tetrahedral phosphate ion (PO_4^{3-}) or a derivative of it. By far the most abundant minerals are apatites (Figure 26.15).

Nitrogen and its compounds play a key role in our economy, with ammonia making particularly notable contributions. Phosphoric acid is an important industrial chemical whose major use is in the manufacture of fertilizers.

Group 15

| Nitrogen |
| 7 |
| **N** |
| 25 ppm |

| Phosphorus |
| 15 |
| **P** |
| 1000 ppm |

| Arsenic |
| 33 |
| **As** |
| 1.5 ppm |

| Antimony |
| 51 |
| **Sb** |
| 0.2 ppm |

| Bismuth |
| 83 |
| **Bi** |
| 0.048 ppm |

Element abundances are in parts-per-million in the earth's crust.

Both phosphorus and nitrogen are part of every living organism. Phosphorus is contained in biochemicals called nucleic acids and phospholipids, and nitrogen occurs in proteins and nucleic acids. Indeed, phosphorus was first derived from human urine.

Properties of Nitrogen and Phosphorus

Nitrogen (N_2) is a colourless gas that liquefies at 77 K (-196 °C). Although there are very many nitrogen compounds, the most notable feature of the elemental substance (sometimes called dinitrogen, N_2) is its lack of reactivity with other elements or compounds. This can be attributed to the strength of the N≡N triple bond (dissociation energy: 945 kJ mol^{-1}), and a resultant high energy of activation. Nitrogen does, however, react with hydrogen to give ammonia in the presence of a catalyst [<<Sections 13.1 and 13.7] and with a few metals to give metal nitrides, compounds containing the N^{3-} ion.

Elemental nitrogen is a very useful material. Because of its lack of reactivity, it is used to provide a non-oxidizing atmosphere for packaged foods and wine and to pressurize electric cables and telephone wires. Liquid nitrogen is valuable as a coolant in freezing biological samples such as blood and semen, in freeze-drying food, and for other applications that require extremely low temperatures.

Liquid nitrogen. Biological samples—such as embryos or semen from animals and humans—can be stored in liquid nitrogen for long periods of time.

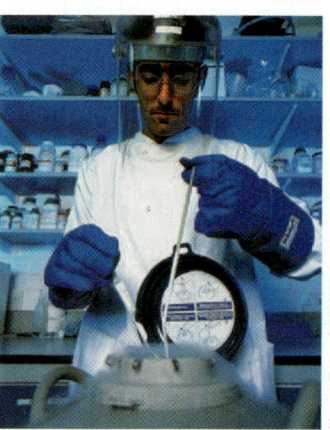

Simon Fraser/MRC Unit, Newcastle General Hospital/Science Photo Library/Photo Researchers, Inc.

The most stable allotrope of phosphorus is white phosphorus. Rather than occurring as a diatomic molecule with a triple bond like its second-period relative nitrogen (N_2), white phosphorus is made up of tetrahedral P_4 molecules, in which each P atom is joined to three others via single bonds (Figure 26.34). Red phosphorus is a polymer of P_4 units.

White phosphorus, P_4

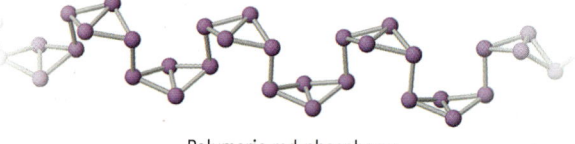

Polymeric red phosphorus

FIGURE 26.34 Molecules of the allotropes of phosphorus.

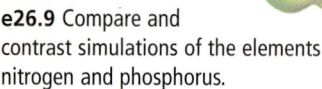

The red and white allotropes of phosphorus.

Charles D. Winters

Elemental white phosphorus is produced by the reduction of phosphate minerals in an electric furnace.

$$2\ Ca_3(PO_4)_2(s) + 10\ C(s) + 6\ SiO_2(s) \longrightarrow P_4(g) + 6\ CaSiO_3(s) + 10\ CO(g)$$

Nitrogen as $N_2(g)$ is not very reactive—even though there are many nitrogen-containing compounds—because of the strength of the N≡N bonds. Allotropes of phosphorus are white phosphorus, with P_4 molecules, and red phosphorus, with polymer chains of P_4 units.

Molecular Modelling (Odyssey)

e26.9 Compare and contrast simulations of the elements nitrogen and phosphorus.

Nitrogen Compounds

A notable feature of the chemistry of nitrogen is the wide diversity of its compounds. Compounds with nitrogen in all oxidation states between −3 and +5 are known, and molecules of some of these compounds are shown in Figure 26.35.

> Compounds with nitrogen in all oxidation states between −3 and +5 are known.

Hydrogen Compounds of Nitrogen: Ammonia and Hydrazine

Ammonia is a gas at ambient temperatures and pressures. It has a very penetrating odour and condenses to a liquid at −33 °C at 1 bar. In solutions in water, it is a weak base [<<Section 14.2]. Ammonia is a major industrial chemical and is manufactured by the Haber-Bosch process [<<Sections 13.1 and 13.7], largely for use as a fertilizer.

Hydrazine (N_2H_4) is a colourless, fuming liquid with an ammonia-like odour (melting point: 2.0 °C; boiling point: 113.5 °C). Almost 1 million kg of hydrazine is produced annually by the Raschig process—the oxidation of ammonia with sodium hypochlorite in alkaline solution.

$$2\ NH_3(aq) + ClO^-(aq) \rightleftharpoons N_2H_4(aq) + Cl^-(aq) + H_2O(\ell)$$

Hydrazine, like ammonia, is a base in both the Brønsted-Lowry and Lewis senses:

$$N_2H_4(aq) + H_2O(\ell) \rightleftharpoons N_2H_5^+(aq) + OH^-(aq) \qquad K_b = 8.5 \times 10^{-7}$$

It is also a powerful reducing agent, as reflected in the standard reduction potential of the N_2/N_2H_4 half-reaction in basic solution:

$$N_2(g) + 4\ H_2O(\ell) + 4\ e^- \longrightarrow N_2H_4(aq) + 4\ OH^-(aq) \quad E° = -1.15\ V$$

Hydrazine's reducing ability is exploited in its use in wastewater treatment for chemical plants. It removes ions such as CrO_4^{2-} by reducing them and thus prevents them from entering the environment. A related use is the treatment of water boilers in large electric-generating plants. Oxygen dissolved in the water presents a serious problem in these plants because the dissolved gas can oxidize the metal of the boiler and pipes and lead to corrosion. Hydrazine reduces the dissolved oxygen to water.

$$N_2H_4(aq) + O_2(g) \longrightarrow N_2(g) + 2\ H_2O(\ell)$$

> Ammonia is important industrially, especially in manufacture of fertilizers. Hydrazine is a weak base in aqueous solution as well as a powerful reducing agent.

Oxides and Oxoacids of Nitrogen Nitrogen is unique among elements in the number of binary oxides it forms (Table 26.8). All are thermodynamically unstable with respect to decomposition to $N_2(g)$ and $O_2(g)$; that is, all have positive $\Delta_f G°$ values. Most are slow to decompose, however, and so are described as kinetically inert.

Dinitrogen monoxide, $N_2O(g)$, commonly called nitrous oxide, is a non-toxic, odourless, and tasteless gas in which nitrogen has the lowest oxidation state (+1) among nitrogen oxides. It can be made by the careful decomposition of ammonium nitrate at 250 °C.

$$NH_4NO_3(s) \longrightarrow N_2O(g) + 2\ H_2O(g)$$

$N_2O(g)$ is used as an anesthetic in minor surgery and has been called "laughing gas" because of its euphoriant effects. Because it is soluble in vegetable fats, the largest commercial use of $N_2O(g)$ is as a propellant and aerating agent in cans of whipped cream. $N_2O(g)$ is released into the atmosphere by nitrification and denitrification processes in soils. Because it is quite inert, it has a long tropospheric lifetime and is a significant 'greenhouse gas' [<<Section 4.3]

Nitrogen monoxide, or nitric oxide, NO(g), has odd-electron molecules: each has 11 valence electrons, giving it one unpaired electron and making it a free radical. The compound has recently been the subject of intense research because it has been found to be important in a number of biochemical processes.

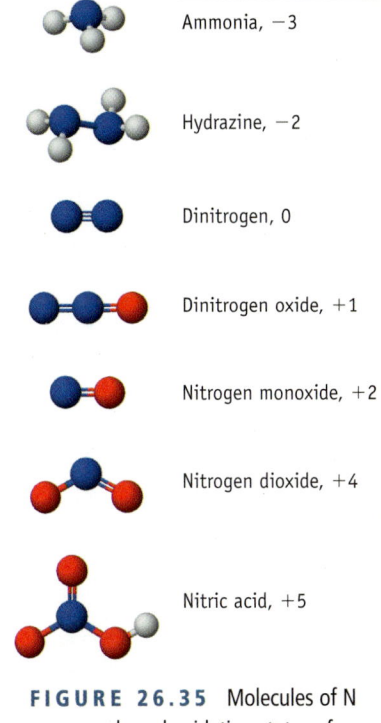

Name and Oxidation State
Ammonia, −3
Hydrazine, −2
Dinitrogen, 0
Dinitrogen oxide, +1
Nitrogen monoxide, +2
Nitrogen dioxide, +4
Nitric acid, +5

FIGURE 26.35 Molecules of N compounds and oxidation states of nitrogen. In its compounds, nitrogen can be in oxidation states from −3 to +5.

Nitrogen dioxide, $NO_2(g)$, is the brown gas you see when a bottle of nitric acid is allowed to stand in the light.

$$2 H^+(aq) + 2 NO_3^-(aq) \longrightarrow 2 NO_2(g) + H_2O(\ell) + \tfrac{1}{2}O_2(g)$$

Nitrogen dioxide is also a culprit in air pollution. Nitrogen monoxide is often present in urban polluted air and forms when atmospheric nitrogen and oxygen are heated in internal combustion engines. In the presence of excess oxygen, NO rapidly forms NO_2, which, if not removed by a catalytic exhaust system in an automobile, enters the atmosphere and can contribute, under the right conditions, to formation of photochemical smogs [<<Sections 17.1 and 17.10].

TABLE 26.8 Some Oxides of Nitrogen

Formula	Name	Structure	Nitrogen Oxidation Number	State Description
N_2O	Dinitrogen monoxide (nitrous oxide)	$:N{\equiv}N{-}\ddot{O}:$ linear	+1	Colourless gas (laughing gas)
NO	Nitrogen monoxide (nitric oxide)	*	+2	Colourless gas; odd-electron molecule (paramagnetic)
N_2O_3	Dinitrogen trioxide	planar	+3	Blue solid (mp: −107 °C); reversibly dissociates to NO and NO_2
NO_2	Nitrogen dioxide		+4	Brown, paramagnetic gas; odd-electron molecule
N_2O_4	Dinitrogen tetraoxide	planar	+4	Colourless liquid/gas; dissociates to NO_2 (see Figure 13.8)
N_2O_5	Dinitrogen pentaoxide		+5	Colourless solid

* There is no satisfactory Lewis structure that accurately represents the electronic distribution in NO. See Chapter 10.

Molecular Modelling (Odyssey)

e26.10 Compare and contrast molecular level simulations of these nitrogen oxides.

$$2 NO(g) + O_2(g) \longrightarrow 2 NO_2(g)$$

Molecules of $NO_2(g)$ are also free radicals, having 17 valence electrons. Because the odd electron largely resides on the N atom, pairs of NO_2 molecules readily combine, forming an N—N bond and producing N_2O_4 molecules.

Dinitrogen tetraoxide, $N_2O_4(g)$, is a colourless gas that always exists in equilibrium with nitrogen dioxide, the relative amounts of each depending upon the temperature (and, therefore, the equilibrium constant).

$$2 NO_2(g) \rightleftharpoons N_2O_4(g)$$

Solid N_2O_4 (melting point: −11.2 °C) is colourless and consists of N_2O_4 molecules. However, as the solid melts and the temperature increases to the boiling point, the colour darkens as dissociation of $N_2O_4(g)$ to form brown $NO_2(g)$ increases. At the normal boiling point (21.5 °C), the distinctly brown gas mixture consists of 15.9% $NO_2(g)$ and 84.1% $N_2O_4(g)$.

When $NO_2(g)$ is bubbled into water, nitric acid and nitrous acid form.

$$2 NO_2(g) + H_2O(\ell) \longrightarrow H^+(aq) + NO_3^-(aq) + HNO_2(aq)$$

nitric acid solution   nitrous acid

Nitric acid has been known for centuries and has become an important compound in our modern economy. The oldest way to make nitric acid is to treat sodium nitrate with concentrated sulfuric acid (Figure 26.36).

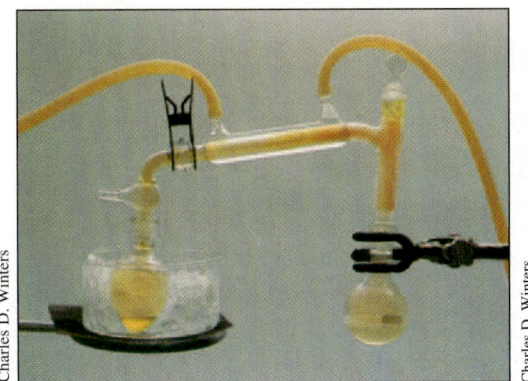

(a) Making nitric acid.

(b) Reaction of HNO₃ solution with copper.

FIGURE 26.36 The production and properties of nitric acid. (a) Nitric acid is made by the reaction of concentrated sulfuric acid and sodium nitrate. Pure HNO₃ is colourless, but some acid decomposes to give brown NO₂(g). This gas fills the apparatus and colours the liquid in the distillation flask. (b) When concentrated nitric acid solution reacts with copper, the metal is oxidized to copper(II) ions, and NO₂ gas is a reaction product.

$$2\,\mathrm{NaNO_3(s)} + \mathrm{H_2SO_4(\ell)} \longrightarrow 2\,\mathrm{HNO_3(\ell)} + \mathrm{Na_2SO_4(s)}$$

Enormous quantities of nitric acid are now produced industrially by the oxidation of ammonia (in the multi-step Ostwald process). The acid has many applications, but by far the greatest amount is turned into ammonium nitrate, for use as a fertilizer, by the reaction of nitric acid and ammonia.

Nitric acid (actually, the nitrate ion in acidic conditions) is a powerful oxidizing agent, as the large, positive $E°$ values for the following half-reactions illustrate:

$$\mathrm{NO_3^-(aq)} + 4\,\mathrm{H^+(aq)} + 3\,\mathrm{e^-} \longrightarrow \mathrm{NO(g)} + 2\,\mathrm{H_2O(\ell)} \quad E° = +0.96\ \mathrm{V}$$
$$\mathrm{NO_3^-(aq)} + 2\,\mathrm{H^+(aq)} + \mathrm{e^-} \longrightarrow \mathrm{NO_2(g)} + \mathrm{H_2O(\ell)} \qquad E° = +0.80\ \mathrm{V}$$

Concentrated nitric acid solution attacks and oxidizes most metals, although aluminum is an exception [<<Section 26.7]. In this process, the nitrate ion is reduced to one of the nitrogen oxides. The oxide formed depends on the metal and on reaction conditions. In the case of copper, for example, either NO(g) or NO₂(g) is produced, depending on the concentration of the acid in solution (Figure 26.36b).

In dilute nitric acid solution:

$$3\,\mathrm{Cu(s)} + 8\,\mathrm{H^+(aq)} + 2\,\mathrm{NO_3^-(aq)} \longrightarrow 3\,\mathrm{Cu^{2+}(aq)} + 4\,\mathrm{H_2O(\ell)} + 2\,\mathrm{NO(g)}$$

In concentrated nitric acid solution:

$$\mathrm{Cu(s)} + 4\,\mathrm{H^+(aq)} + 2\,\mathrm{NO_3^-(aq)} \longrightarrow \mathrm{Cu^{2+}(aq)} + 2\,\mathrm{H_2O(\ell)} + 2\,\mathrm{NO_2(g)}$$

Four metals (Au, Pt, Rh, and Ir) that do not react with nitric acid are often described as the "noble metals." The alchemists of the 14th century, however, knew that if they mixed HNO₃ with HCl in a ratio of about 1:3, this *aqua regia*, or "kingly water," would attack even gold, the noblest of metals.

$$10\,\mathrm{Au(s)} + 6\,\mathrm{NO_3^-(aq)} + 40\,\mathrm{Cl^-(aq)} + 36\,\mathrm{H^+(aq)} \longrightarrow 10\,[\mathrm{AuCl_4}]^- + 3\,\mathrm{N_2(g)} + 18\,\mathrm{H_2O(\ell)}$$

Nitrogen forms a wide variety of oxides, with N in a range of oxidation states. NO(g) and NO₂(g) are associated with photochemical smog formation in particular circumstances. As well as being an acid, nitric acid is a powerful oxidizing agent.

Hydrogen Compounds of Phosphorus and Other Group 15 Elements

The phosphorus analog of ammonia, phosphine (PH₃), is a poisonous, highly reactive gas with a faint garlic-like odour. Industrially, it is made by the alkaline hydrolysis of white phosphorus.

$$\mathrm{P_4(s)} + 3\,\mathrm{OH^-(aq)} + 3\,\mathrm{H_2O(\ell)} \longrightarrow \mathrm{PH_3(g)} + 3\,\mathrm{H_2PO_2^-(aq)}$$

The other hydrides of the heavier group 15 elements are exceedingly toxic and become more unstable as the atomic number of the element increases. Nonetheless, arsine (AsH_3) is used in the semiconductor industry as a starting material in the preparation of gallium arsenide (GaAs) semiconductors.

Phosphine, $PH_3(g)$, is a gaseous analog of ammonia.

Phosphorus Oxides and Sulfides

The most important compounds of phosphorus are those with oxygen, and there are at least six simple binary compounds containing just phosphorus and oxygen. All of them can be thought of as being derived structurally from the P_4 tetrahedron of white phosphorus. For example, if $P_4(s)$ is carefully oxidized, $P_4O_6(s)$ is formed; an O atom is inserted into each P—P bond in the tetrahedron (Figure 26.37).

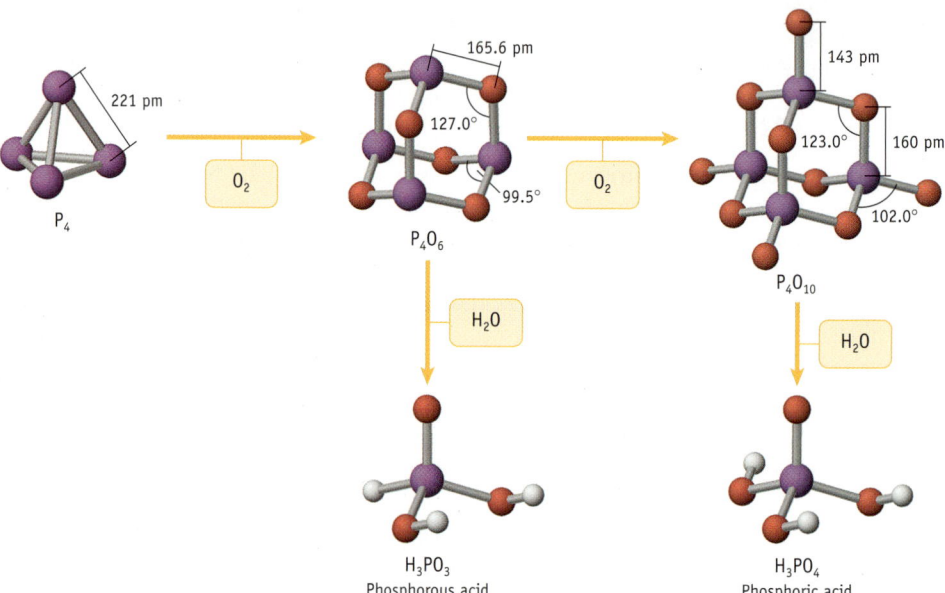

FIGURE 26.37 Phosphorus oxides. Other binary P—O compounds have formulas between P_4O_6 and P_4O_{10}. Their formation can be imagined to start with P_4O_6, adding O atoms successively to the P atom vertices.

Matches. The head of a "strike anywhere" match contains P_4S_3 and the oxidizing agent $KClO_3$. (Other components are ground glass, Fe_2O_3, ZnO, and glue.) Safety matches have sulfur (3–5%) and $KClO_3$ (45–55%) in the match head and red phosphorus in the striking strip.

The most common and important phosphorus oxide is $P_4O_{10}(s)$, a fine white powder commonly called "phosphorus pentaoxide" because its empirical formula is P_2O_5. In molecules of P_4O_{10}, each P atom is surrounded tetrahedrally by O atoms.

Unlike nitrogen, phosphorus also forms a series of compounds with sulfur. Of these, the most important is P_4S_3.

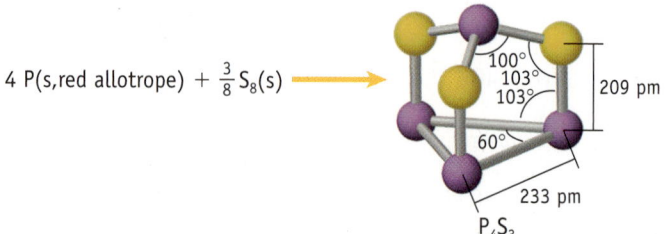

$$4\ P(s, \text{red allotrope}) + \tfrac{3}{8}\ S_8(s) \longrightarrow$$

In P_4S_3 molecules, S atoms are inserted into only three of the P—P bonds. The principal use of P_4S_3 is in "strike anywhere" matches—the kind that light when you rub the head against a rough object. The active ingredients are P_4S_3 and the powerful oxidizing agent potassium chlorate ($KClO_3$). The "safety" match is now more common than the "strike anywhere" variety. In safety matches, the head consists mainly of $KClO_3$, and the material on the matchbox is red phosphorus (about 50%), Sb_2S_3, Fe_2O_3, and glue.

The two most important phosphorus oxides, P_4O_6(s) and P_4O_{10}(s), are solids whose structures contain expanded P_4 tetrahedra. P_4S_3(s) is used in the heads of safety matches.

Phosphorus Oxoacids and Their Salts

The structures of molecules of a few of the many known phosphorus oxoacids are illustrated in Table 26.9. There are so many acids and their salts in this category that structural principles have been developed to organize and understand them:

- All P atoms in the oxoacids and their anions (conjugate bases) are four-coordinate and tetrahedral.
- All the P atoms in the acids are bonded to at least one —OH group. In every case, the H is ionizable as H^+.
- A few oxoacids have one or more P—H bonds. This H atom is not ionizable as H^+.
- Polymerization can occur by P—O—P linkages to give both linear and cyclic species. Two P atoms are never joined by more than one P—O—P bridge.
- When a P atom is surrounded only by O atoms (as in H_3PO_4), its oxidation state is +5. For each P—OH that is replaced by P—H, the oxidation state drops by 2 (because P is considered more electronegative than H). For example, the oxidation state of P in H_3PO_2 is +1.

TABLE 26.9 Phosphorus Oxoacids

Formula	Name	Structure
H_3PO_4	Orthophosphoric acid	
$H_4P_2O_7$	Pyrophosphoric acid (diphosphoric acid)	
$(HPO_3)_3$	Metaphosphoric acid	
H_3PO_3	Phosphorous acid	
H_3PO_2	(phosphonic acid) Hypophosphorous acid (phosphinic acid)	

Orthophosphoric acid (H_3PO_4, usually called simply phosphoric acid) and its salts are far more important commercially than other phosphorus oxoacids. Millions of tonnes of phosphoric acid are made annually, some using white phosphorus as the starting material. The phosphorus is burned in oxygen to give P_4O_{10}(s), and the oxide reacts with water to produce the acid (Figure 26.38).

$$P_4O_{10}(s) + 6\ H_2O(\ell) \longrightarrow 4\ H_3PO_4(aq)$$

This approach gives a pure product, so it is employed to make phosphoric acid for use in food products in particular. When the pure acid is dilute, it is non-toxic and gives the

FIGURE 26.38 Reaction of P_4O_{10}(s) and water. The white solid oxide reacts vigorously with water to give orthophosphoric acid (H_3PO_4). The heat generated vaporizes the water, so steam is visible.

Charles D. Winters

tart or sour taste to carbonated soft drinks, such as various colas (about 0.05% H_3PO_4) or root beer (about 0.01% H_3PO_4).

As illustrated in Figure 26.39, a major use for phosphoric acid is to impart corrosion resistance to metal objects such as nuts and bolts, tools, and car-engine parts by plunging the object into a hot acid bath. Car bodies are similarly treated with phosphoric acid containing metal ions such as Zn^{2+}(aq) ions, and aluminum trim is "polished" by treating it with the acid.

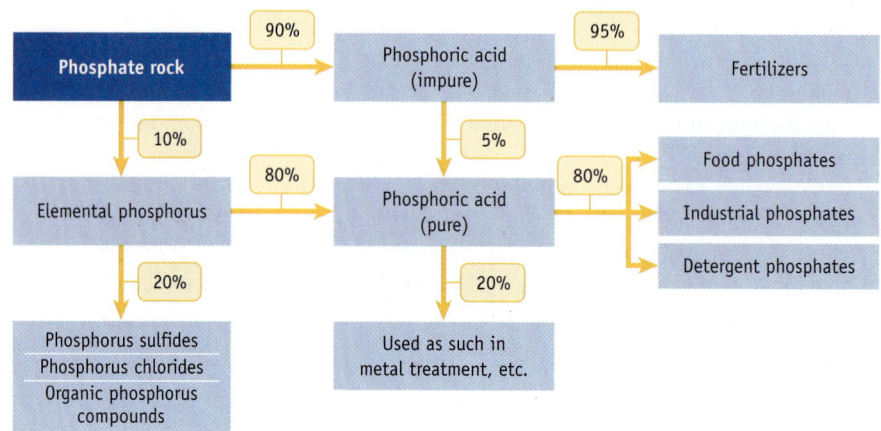

FIGURE 26.39 Uses of phosphate rock, phosphorus, and phosphoric acid.

Step-wise neutralization of aqueous H_3PO_4 solutions [<<Section 14.6] can be used to produce salts such as NaH_2PO_4(s), Na_2HPO_4(s), and Na_3PO_4(s). In industry, $H_2PO_4^-$(aq) ions and HPO_4^{2-}(aq) ions are produced using Na_2CO_3 in solution as the base, but an excess of the stronger base NaOH is required to give PO_4^{3-}(aq) ions.

Sodium phosphate, Na_3PO_4(s), is used in scouring powders and paint strippers because the anion PO_4^{3-} is a reasonably strong (weak) base in water ($K_b = 2.8 \times 10^{-2}$). Sodium monohydrogenphosphate, Na_2HPO_4(s), which has a less basic anion than PO_4^{3-}, is widely used in food products. Kraft has patented a process using the salt in the manufacture of pasteurized cheese, for example. Thousands of tonnes of Na_2HPO_4(s) are still used for this purpose, even though the function of the salt in this process is not completely understood. In addition, a small amount of Na_2HPO_4(s) in pudding mixes enables the mix to gel in cold water, and this basic anion raises the pH of cereals to provide quick-cooking breakfast cereal. Apparently, the OH^-(aq) ions from hydrolysis of HPO_4^{2-}(aq) ions accelerate the breakdown of the cellulose material in the cereal.

Calcium phosphates are used in many products. For example, the weak acid $Ca(H_2PO_4)_2.H_2O$(s) is used as the acid leavening agent in baking powder. A typical baking powder contains 28% $NaHCO_3$, 10.7% $Ca(H_2PO_4)_2.H_2O$, and 21.4% $NaAl(SO_4)_2$. The weak acids react with sodium bicarbonate to produce CO_2 gas. For example,

$$Ca(H_2PO_4)_2.H_2O(s) + 2\ HCO_3^-(aq)]$$
$$\longrightarrow 2\ CO_2(g) + 3\ H_2O(\ell) + HPO_4^{2-}(aq) + CaHPO_4(aq)$$

Finally, calcium monophosphate, $CaHPO_4$(s), is used as an abrasive and polishing agent in toothpaste.

> The ionizable hydrogen ions in the various phosphorus oxoacids are produced by breaking of an O—H bond. Large quantities of orthophosphoric acid (H_3PO_4) are used in manufacture of fertilizers. Other uses are in corrosion protection and in foods.

Arsenic in Drinking Water

In the early 1990s, hundreds of people in the very poor country of Bangladesh became ill with what some thought was leprosy. Soon, however, the problem was recognized as arsenic poisoning. An attempt by world health organizations to alleviate the problem of

made in 1774 by the Swedish chemist Karl Wilhelm Scheele (1742–1786), who combined sodium chloride with an oxidizing agent in an acidic solution (Figure 26.43).

Industrially, chlorine is made by electrolysis of brine (concentrated aqueous NaCl solution). The other product of the electrolysis, NaOH, is also a valuable industrial chemical. About 80% of the chlorine produced is made using an electrochemical cell similar to the one depicted in Figure 26.44. Oxidation of aquated chloride ions to Cl_2 gas occurs at the anode and reduction of water occurs at the cathode.

$$\text{Anode reaction (oxidation):} \quad 2\ Cl^-(aq) \longrightarrow Cl_2(g) + 2\ e^-$$

$$\text{Cathode reaction (reduction):} \quad 2\ H_2O(\ell) + 2\ e^- \longrightarrow H_2(g) + 2\ OH^-(aq)$$

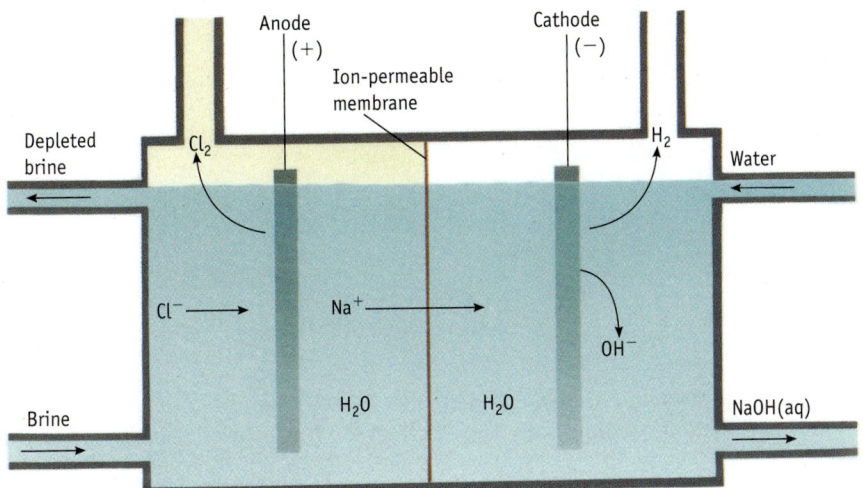

FIGURE 26.44 A membrane cell for the production of NaOH and Cl_2 gas from a saturated, aqueous solution of NaCl (brine). Here the anode and cathode compartments are separated by a water-impermeable, but ion-conducting, membrane. A widely used membrane is made of Nafion, a fluorine-containing polymer that is a relative of polytetrafluoroethylene (Teflon). Brine is fed into the anode compartment and dilute sodium hydroxide or water into the cathode compartment. Overflow pipes carry the evolved gases and NaOH solution away from the chambers of the electrolysis cell.

Activated titanium is used for the anode and stainless steel or nickel is preferred for the cathode. The membrane separating the anode and cathode compartments is not permeable to water, but it does allow sodium ions to pass so as to maintain the charge balance. Thus, the membrane functions as a salt bridge between the anode and cathode compartments. The energy consumption of these cells is in the range of 2000–2500 kWh per tonne of NaOH produced.

Bromine All halogens are oxidizing agents, but that ability declines as we go to elements lower in the group (Table 26.10).

As a result, $Cl_2(g)$ will oxidize $Br^-(aq)$ ions in aqueous solution.

$$Cl_2(aq) + Br^-(aq) \longrightarrow 2\ Cl^-(aq) + Br_2(aq)$$

$$E^\circ_{net} = E^\circ_{cathode} - E^\circ_{anode} = 1.36\ V - (1.08\ V) = +0.28\ V$$

In fact, this is the commercial method of preparing bromine when sodium bromide is obtained from natural brine wells in various parts of the world.

Iodine Iodine is a lustrous, purple-black solid that sublimes at ambient temperatures and pressures. The element was first isolated in 1811 from seaweed and kelp, extracts of which had long been used for treatment of goitre, the enlargement of the thyroid gland. It is now known that the thyroid gland produces a growth-regulating hormone (thyroxine) that contains iodine. Iodized table salt has 0.01% NaI added to provide the necessary iodine in the diet.

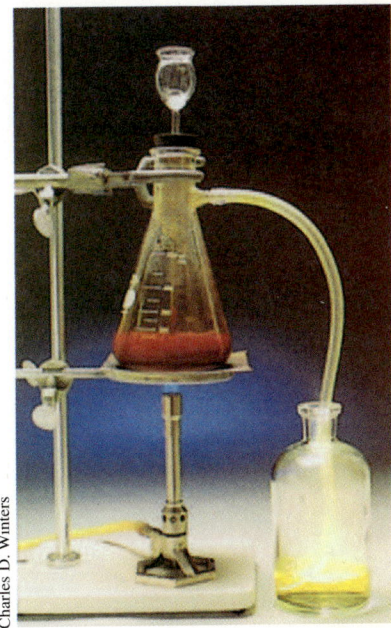

FIGURE 26.43 Chlorine production. Chlorine gas is made by oxidation of aquated chloride ions using a strong oxidizing agent. Here, oxidation of $Cl^-(aq)$ in NaCl solution is accomplished by $Cr_2O_7^{2-}$ (aq) ions in acidic solution. The Cl_2 gas is bubbled into water in a receiving flask.

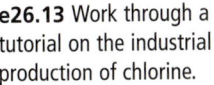

Think about It

e26.13 Work through a tutorial on the industrial production of chlorine.

TABLE 26.10 Standard Reduction Potentials of Halogens at 25 °C

Half-Reaction	E° at 25 °C
$F_2(g) + 2\ e^- \longrightarrow 2\ F^-(aq)$	2.87 V
$Cl_2(g) + 2\ e^- \longrightarrow 2\ Cl^-(aq)$	1.36 V
$Br_2(\ell) + 2\ e^- \longrightarrow 2\ Br^-(aq)$	1.08 V
$I_2(s) + 2\ e^- \longrightarrow 2\ I^-(aq)$	0.535 V

Charles D. Winters

A laboratory method for making solid I_2 (Figure 26.45) is the oxidation of sodium iodide by $MnO_2(s)$ in acidic conditions.

$$2\,NaI(s) + 4\,H^+(aq) + MnO_2(s) \longrightarrow 2\,Na^+(aq) + Mn^{2+}(aq) + 2\,H_2O(\ell) + I_2(s)$$

FIGURE 26.45 Making iodine. A mixture of sodium iodide and manganese(IV) oxide was placed in the flask (*left*). On adding concentrated sulfuric acid (*right*), brown iodine vapour is evolved. This condenses directly to the solid.

Charles D. Winters

Commercial production, however, depends on the source of I^- and its concentration. In one method, iodide ions are first precipitated with silver ions to give "insoluble" $AgI(s)$.

$$I^-(aq) + Ag^+(aq) \longrightarrow AgI(s)$$

This is reduced by clean scrap iron to release aquated iodide ions by reduction of the Ag^+ ions in $AgI(s)$ to metallic silver.

$$2\,AgI(s) + Fe(s) \longrightarrow Fe^{2+}(aq) + 2\,I^-(aq) + 2\,Ag(s)$$

Finally, the aquated iodide ions are oxidized to iodine by chlorine.

$$6\,I^-(aq) + 3\,Cl_2(aq) \longrightarrow 3\,I_2(s) + 6\,Cl^-(aq)$$

Fluorine is produced by electrolysis of KF in anhydrous HF solution, and chlorine by electrolysis of saturated NaCl solution. Bromine is made by oxidation of bromide ions by chlorine. In the laboratory, iodine is made by oxidation of iodides.

Fluorine Compounds

Fluorine is the most reactive of all of the elements, forming compounds with every element except He and Ne. In most cases, the elements combine directly, and some reactions can be so vigorous as to be explosive. This reactivity can be explained by at least two features of fluorine chemistry: the relatively weak F—F bond compared with the other halogens, and the relatively strong bonds formed by fluorine to other elements (Table 26.11).

In addition to its oxidizing ability, another notable characteristic of fluorine is its small size. These properties lead to the formation of compounds where a number of F atoms can be bonded to a central element in a high oxidation state. Examples include PtF_6, AgF_2, UF_6, IF_7, and XeF_4.

All the halogens form hydrogen halides with the formula HX. Hydrogen fluoride and hydrogen chloride are especially important industrial chemicals. More than 1 million tonnes of hydrogen fluoride is produced annually worldwide, almost all by reaction of concentrated sulfuric acid with fluorspar.

$$CaF_2(s) + H_2SO_4(\ell) \longrightarrow CaSO_4(s) + 2\,HF(g)$$

TABLE 26.11 Bond Energies of Some Halogen Compounds (kJ mol^{-1})

X	X—X	H—X	C—X (in CX$_4$)
F	155	565	485
Cl	242	432	339
Br	193	366	285
I	151	299	213

HF(g) is used in a broad range of industries: in the production of refrigerants, herbicides, pharmaceuticals, high-octane gasoline, aluminum, plastics, electrical components, and fluorescent light bulbs.

The fluorspar used to produce HF(g) must be very pure and free of SiO_2 because HF(g) reacts readily with silicon dioxide.

$$SiO_2(s) + 4\ HF(g) \longrightarrow SiF_4(g) + 2\ H_2O(\ell)$$

$$SiF_4(g) + 2\ HF(aq) \longrightarrow H_2SiF_6(aq)$$

This series of reactions explains why HF(g) can be used to etch or frost glass (such as the inside of fluorescent light bulbs). It also explains why HF(g) is not shipped in glass containers—unlike HCl, for example.

The aluminum industry consumes about 10–40 kg of cryolite (Na_3AlF_6) per tonne of aluminum produced. The reason is that cryolite is added to aluminum oxide to produce a lower-melting mixture for electrolysis, thus saving energy costs. Cryolite is found in only small quantities in nature, so it is made in various ways, among them the following:

$$6\ HF(aq) + Al(OH)_3(s) + 3\ Na^+(aq) + 3\ OH^-(aq) \longrightarrow Na_3AlF_6(s) + 6\ H_2O(\ell)$$

About 3% of the hydrogen fluoride produced is used in uranium fuel production. Naturally occurring uranium is processed to give UO_2. To separate uranium isotopes in a gas centrifuge, the uranium must be in the form of a gaseous compound. Therefore, uranium oxide is treated with hydrogen fluoride to give UF_4, which is then reacted with F_2 to produce the volatile solid UF_6.

$$UO_2(s) + 4\ HF(aq) \longrightarrow UF_4(s) + 2\ H_2O(\ell)$$

$$UF_4(s) + F_2(g) \longrightarrow UF_6(s)$$

Fluorine compounds that have found a dramatic increase in usage as aerosol propellants in recent times are the hydrofluorocarbons (HFCs), of which a common example has the formula $CH_2F{-}CF_3$ (named HFC-134a). These compounds are being used as replacements for CFCs, which have been identified as contributors to the depletion of the ozone layer as a result of production of Cl atoms in the stratosphere [<<Section 21.1].

> Fluorine is the most reactive of all of the elements, forming compounds with every element except He and Ne, usually by direct reaction. Important compounds are hydrogen fluoride, cryolite, and hydrofluorocarbons used as aerosol propellants (in place of CFCs).

Chlorine Compounds

Hydrogen Chloride Hydrogen chloride is a valuable industrial chemical. Hydrogen chloride gas is mostly made by the rapid exothermic reaction of hydrogen and chlorine gases. The classical method of making HCl(g) in the laboratory is by reaction of sodium chloride and concentrated sulfuric acid, a procedure that takes advantage of the facts that HCl is a gas and that H_2SO_4 will not oxidize chloride ions.

$$2\ NaCl(s) + H_2SO_4(\ell) \longrightarrow Na_2SO_4(s) + 2\ HCl(g)$$

Hydrogen chloride gas has a sharp, irritating odour. One of its most important uses is as "hydrochloric acid" in aqueous solution. Both gas-phase HCl and aqueous HCl solutions react with metals and metal oxides to give metal chlorides and, in the case of aqueous solutions, hydrogen or water. In these reactions, $H^+(aq)$ ions bring about oxidation.

$$Mg(s) + 2\ H^+(aq) \longrightarrow Mg^{2+}(aq) + H_2(g)$$

$$ZnO(s) + 2\ H^+(aq) \longrightarrow Zn^{2+}(aq) + H_2O(g)$$

> Hydrogen chloride is made by reaction between hydrogen gas and chlorine gas. Its aqueous solutions are called hydrochloric acid.

Oxoacids of Chlorine Oxoacids of chlorine range from HOCl (or HClO), in which chlorine has an oxidation state of $+1$, to HClO$_4$, in which the oxidation state is $+7$ (Table 26.12). All are strong oxidizing agents.

TABLE 26.12 Oxoacids of Chlorine

Acid	Name	Anion	Name
HOCl or ClOH	Hypochlorous acid	ClO$^-$	Hypochlorite
HClO$_2$ or ClO(OH)	Chlorous acid	ClO$_2^-$	Chlorite
HClO$_3$ or ClO$_2$(OH)	Chloric acid	ClO$_3^-$	Chlorate
HClO$_4$ or ClO$_3$(OH)	Perchloric acid	ClO$_4^-$	Perchlorate

Hypochlorous acid (HOCl) forms when chlorine dissolves in water. In this reaction, half of the chlorine is oxidized to hypochlorite ions and half is reduced to chloride ions in a *disproportionation* reaction—a reaction in which an element or compound is simultaneously oxidized and reduced. In this case, Cl$_2$(g) is oxidized to ClO$^-$(aq) ions and reduced to Cl$^-$(aq) ions. Some Cl$_2$ molecules oxidize others and are reduced in doing so.

$$Cl_2(g) + 2H_2O(\ell) \rightleftharpoons H_3O^+(aq) + HClO(aq) + Cl^-(aq)$$

In aqueous solution, chlorine, chloride ion, and hypochlorous acid exist in equilibrium. A low pH increases the relative concentration of Cl$_2$(g), whereas a high pH increases the relative amounts of HOCl(aq) and Cl$^-$(aq). So, if Cl$_2$(g) is dissolved in cold aqueous NaOH solution, rather than in pure water, aquated hypochlorite ions and chloride ions are produced.

$$Cl_2(aq) + 2OH^-(aq) \rightleftharpoons ClO^-(aq) + Cl^-(aq) + H_2O(\ell)$$

The resulting alkaline solution is the "liquid bleach" used in home laundries. The bleaching action of this solution is a result of the oxidizing ability of ClO$^-$(aq) ions. Most dyes and stains are coloured organic compounds, and hypochlorite ion oxidizes them to colourless products.

When calcium hydroxide reacts with Cl$_2$(g), solid calcium hypochlorite, Ca(ClO)$_2$, is the product. This compound is easily handled and is the "chlorine" that is sold for swimming pool disinfection.

When a basic solution containing hypochlorite ions is heated, disproportionation occurs, forming aquated chlorate and chloride ions.

$$3ClO^-(aq) \longrightarrow ClO_3^-(aq) + 2Cl^-(aq)$$

Sodium chlorate (NaClO$_3$) and potassium chlorate (KClO$_3$) are made in large quantities via this reaction. The sodium salt can be reduced to ClO$_2$(g), a compound used for bleaching paper pulp. KClO$_3$ is the preferred oxidizing agent in fireworks and is a component of safety matches.

Perchlorates, salts containing ClO$_4^-$ ions, are powerful oxidants. Pure perchloric acid (HClO$_4$) is a colourless liquid that explodes if shocked. It explosively oxidizes organic materials and rapidly oxidizes silver and gold. Dilute aqueous solutions of the acid are safe to handle, however.

Perchlorate salts of most metals are usually relatively stable, although somewhat unpredictable. Great care should be used when handling any perchlorate salt. Ammonium perchlorate, for example, bursts into flame if heated above 200 °C.

$$2NH_4ClO_4(s) \longrightarrow N_2(g) + Cl_2(g) + 2O_2(g) + 4H_2O(g)$$

The strong oxidizing ability of the ammonium salt accounts for its use as the oxidizer in the solid booster rockets for the Space Shuttle. The solid propellant in these rockets is largely NH$_4$ClO$_4$(s), the remainder being powdered aluminum, the reducing agent (Figure 26.46). Each launch requires about 750 tonnes of ammonium perchlorate. More than half of the sodium perchlorate currently manufactured is converted to the ammonium salt.

FIGURE 26.46 Use of a perchlorate in rocket fuel. The solid-fuel booster rockets of the Space Shuttle utilize a mixture of NH$_4$ClO$_4$ (oxidizing agent) and Al powder (reducing agent).

Oxoacids of chlorine range from HOCl, in which Cl is in oxidation state +1, to $HClO_4$, in which Cl is in oxidation state +7. With increasing oxidation state, the oxoacids are stronger acids and the oxoanions are more powerful oxidizing agents.

EXERCISE 26.16—THE GROUP 17 ELEMENTS

Oxidation of aquated halide ions by another halogen can be represented as

$$X_2(aq) + 2Y^-(aq) \longrightarrow 2X^-(aq) + Y_2(aq)$$

(a) List a pair of halogens that could be X and Y in this reaction.
(b) List a pair of halogens that cannot be X and Y in this reaction.

26.12 Group 18, the Noble Gases

All of the noble gases have the s^2p^6 electron configuration. They have very low melting points and boiling points [<<Table 8.1], which can be attributed to their existence as monoatomic particles, so that there is no opportunity for intermolecular forces of attraction other than dispersion forces. Going down the group from helium to radon, the melting and boiling points increase. Dispersion forces are attributed to attractions between instantaneous fluctuating dipoles [<<Section 6.3], so this trend makes sense if we accept that the electron clouds of the bigger atoms such as krypton and xenon are more polarizable.

In the past, elements in group 18 have been called the "rare gases." This is because their lack of reactivity rendered them essentially "invisible" chemically. The label is inappropriate, however, because helium constitute 23% of atoms in the universe (hardly rare!), some natural gas deposits contain up to 10% helium, and the amount of argon in air is 30 times that of CO_2. There is an insignificant amount of helium in air because the average speed of helium atoms at ordinary temperatures (about 1500 m s^{-1}) is greater than that needed to escape the earth's gravitational field.

The first evidence of helium, in 1868, was from a yellow line in the spectrum from the sun's chromosphere that did not correspond with an emission from any known substance. So helium was discovered extra-terrestrially before the existence of any of the noble gases was known on Earth! These elements were unknown at the time of Mendeleev's first periodic table: a whole column of the periodic table as we now know it was missing.

The existence of argon in air was demonstrated by the extraordinarily meticulous work of Lord Rayleigh in the 1890s. He produced nitrogen by two methods:

- "Atmospheric nitrogen," by removal of oxygen from air, by reaction with red hot copper:

$$2\,Cu(s) + O_2(g) \xrightarrow{\text{heat}} 2\,CuO(s)$$

- "Synthetic nitrogen" by heating ammonium nitrite:

$$NH_4NO_2(s) \xrightarrow{\text{heat}} N_2(g) + 2\,H_2O(g)$$

Over two years, he made dozens of measurements of the densities of the nitrogen formed in the two ways, continually eliminating errors from his measurements. Collecting gases quantitatively and measuring their density is not easy even today. Eventually, statistical analysis convinced him that samples of "atmospheric nitrogen" were about 0.5% more dense that samples of "synthetic nitrogen." He then removed nitrogen from his atmospheric sample by burning magnesium in it (forming magnesium nitride, Mg_3N_2), and found that there was a small amount of a residual heavier gas. He called it *argon* (Greek for "lazy").

The group 18 elements have also been called the "inert gases" because they were thought, until the 1960s, to be totally unreactive. Not any more (see below). Because they

Group 18
Noble gases

| Helium |
| 2 |
| **He** |
| 8000 ppm |

| Neon |
| 10 |
| **Ne** |
| 0.005 ppm |

| Argon |
| 18 |
| **Ar** |
| 3.5 ppm |

| Krypton |
| 36 |
| **Kr** |
| 0.0001 ppm |

| Xenon |
| 54 |
| **Xe** |
| 0.00005 ppm |

| Radon |
| 86 |
| **Rn** |
| trace |

Elemental abundances are in ppm in the earth's crust.

are unreactive in most common situations, they are now called *noble gases*: "noble" means relatively unreactive (like the noble metals gold, silver, and platinum).

The way that the periodic table is most commonly represented, the noble gases appear at the "end" of each period. But if we arrange the elements in a line in order of atomic number, each noble gas is situated between elements from the two most reactive groups—group 17 and group 1. This begs the question of why they are noble. One level of explanation is as follows:

- They have very high ionization energies, so they are not able to participate in bonding either by having an electron removed (to form cations in ionic compounds) or by donating an electron pair as a ligand in a complex compound, or even to share their electrons in a covalent bond.
- They also have very low electron affinities (Table 8.7). As a result, they have little tendency to participate in bonding either by taking electrons from other atoms or molecules (to form anions in ionic compounds), or by accepting an electron pair from a ligand to form complex compounds, or even to share the electrons of other substances in a covalent bond.

Perhaps that makes sense, but it begs the further questions of why they have high ionization energies, and why they have low affinity to accept additional electrons. Answers are to be had in calculation of the effective nuclear charges on the atoms or their negative ions [<<Section 8.6].

EXERCISE 26.17—NOBLE GASES

(a) Account for the high ionization energy of argon by calculating the effective nuclear charge experienced by a valence electron in an argon atom, and comparing this with the effective nuclear charges in atoms of elements near argon in the periodic table.

(b) Account for the low electron affinity of argon by calculating the effective nuclear charge experienced by a valence electron in an Ar^- ion, and comparing this with the effective nuclear charges on negative ions of elements near argon in the periodic table.

Compounds of Higher Members

Although group 18 elements have higher first ionization energies than those of adjacent elements, going down the group, ionization energies decrease. For example, $IE_1(\text{He}) = 2372$ kJ mol^{-1} and $IE_1(\text{Rn}) = 1037$ kJ mol^{-1}. An alert chemist, such as Neill Bartlett at the University of California, would recognize that it is increasingly likely that heavier members might react with highly electronegative reagents. Demonstrating the power of factual knowledge, in 1962, he realized the similarity of $IE_1(\text{Xe}) = 1170$ kJ mol^{-1} and $IE_1(\text{O}_2) = 1170$ kJ mol^{-1}. And he thought that if oxygen (O_2) can undergo electron transfer to highly electronegative species, why shouldn't xenon?

Bartlett knew well the electron transfer reaction represented as

$$O_2(g) + PtF_6(s) \longrightarrow [O_2^+][PtF_6^-](s)$$

and he was very quickly able to tip the chemistry beliefs of the time upside down by carrying out a relatively simple corresponding reaction of xenon, to form an orange-coloured solid:

$$Xe(g) + PtF_6(s) \longrightarrow [Xe^+][PtF_6^-](s)$$

Soon after that he made, by direct reaction between xenon and fluorine, and under not very harsh conditions, XeF_2, XeF_4, and XeF_6. And then, by reaction of some of these species with water, compounds such as $XeOF_4$, XeO_2F_2, and the very explosive XeO_3. There are now many compounds of xenon, krypton, and even argon with fluorine and oxygen and a whole range of electrophilic compounds.

Compounds of radon are not stable—probably because of decomposition by radiation emitted by radioactive radon atoms.

Uses of the Noble Gases

The noble gases have many uses, including the following:

- To provide an inert atmosphere to avoid oxidation of metals at high temperatures during welding, metallurgical processes, and in electric light bulbs.
- Low temperature coolants, particularly for scientific experiments carried out at a temperature of 4 K—the boiling point of liquid helium.
- $Ne(g)$, $Ar(g)$, $Kr(g)$, and $Xe(g)$ are used in coloured electrical discharge tubes, called "neon lights." Each emits characteristically coloured emission lines.
- Helium is used in balloons, including those used as people transporters, because its density is much less than that of air.
- In "artificial air" mixtures (80% helium, 20% O_2) for deep-sea diving [<<Section 12.4]. Because helium is less soluble in blood than nitrogen, there is less chance of gas bubbles being released into the bloodstream (inducing the "bends") as pressure decreases during ascent to the surface. But you develop a squeaky voice!
- Radon is used in radiological treatments because it is radioactive.

"Neon" lights.

EXERCISE 26.18—NOBLE GASES

From the molecular weights of $N_2(g)$ and $He(g)$, calculate the ratio of densities of these two gases at the same temperature and pressure. Account for the buoyancy of helium-filled vessels in air.

EXERCISE 26.19—NOBLE GASES

Explain why there are compounds of xenon and krypton, but none are known of neon.

SUMMARY

Key Concepts

- The main group elements are those in groups 1, 2, and 13–18 of the periodic table. Eight of the ten most abundant elements on the earth are in this group, and the top ten industrial chemicals are all main group elements or their compounds. Although there are commonalities among properties of the elements in each of the groups, each element is unique. (Section 26.2)
- The **charge density,** or **charge-to-radius ratio,** of a cation is a combined measure of magnitude of charge and size of the ion. The higher the charge density of a cation is, the greater its ability to attract negatively charged species near it. (Section 26.3)
- For salts with a particular anion, the higher the charge density of the cation is, the more covalent character the salt has. This can be explained by **polarization** (distortion) of the electron cloud of the anion into the bonding region. The higher the charge density of the cation is, the greater its **polarizing power**. The larger the anion is, the more easily its electron cloud is distorted—that is, the greater its **polarizability**. In the case of oxides, an increased covalent character correlates with higher acidity. (Section 26.3)
- Salts whose cations have high charge density are more likely to crystallize with several molecules of water of crystallization and to have relatively high lattice energies. (Section 26.3)

- The larger the charge density of a cation is, the more exothermic its aquation is, and the more acidic its aqueous solutions are. (Section 26.3)
- The relatively slow movement of aquated lithium ions through solution is explained by their high charge density, resulting in them carrying their water ligands with them as they move. (Section 26.3)
- The reducing ability of lithium in aqueous solutions is enhanced by the strong interaction of Li^+ ions with water, because of their high charge density. (Section 26.3)
- Most of the atoms in the universe are hydrogen atoms, of which there are three isotopes. (Section 26.4)
- Elemental hydrogen is a gas composed of diatomic molecules (H_2). Hydrogen combines chemically with all of other element except the noble gases. With the most electropositive elements, it forms ionic hydrides in which H is in oxidation state -1, and these are powerful reducing agents. With the most electronegative elements, it forms covalent molecular hydrides with H in oxidation state $+1$, and with the many metals it forms metallic hydrides of variable composition, which are rather like alloys. (Section 26.4)
- The main mode of industrial production of hydrogen is by reaction of steam with methane from natural gas: the product mixture of hydrogen and carbon monoxide is synthesis gas. Production of hydrogen by electrolysis of water may become more important. (Section 26.4)
- Hydrogen has some properties that are common to the group 1 elements, some in common with the group 17 elements, and some that suggest it should be classified in a group on its own. (Section 26.4)
- All of the group 1 elements are metals, and all are highly reactive with oxygen, water, and other oxidizing agents. None is ever found in nature as the uncombined element. In all cases, compounds of the group 1 metals contain the element as an ion with $+1$ charge. Most sodium and potassium compounds are water-soluble. (Section 26.5)
- Because sodium and potassium are so reactive, their ions cannot be reduced by common reducing agents. Sodium is obtained by electrolysis of molten NaCl. (Section 26.5)
- The alkali metals are powerful reducing agents. They react with oxygen, water, and the halogens, forming ionic compounds. In reactions with oxygen, sodium forms mainly the peroxide, $Na_2O_2(s)$, while potassium forms the superoxide, $KO_2(s)$. (Section 26.5)
- The chemical properties of lithium are those of an element less metallic than the other members of group 1. As one example of the diagonal relationship, many of its properties are more like those of magnesium. (Section 26.5)
- The group 2 elements are very reactive, and are found in nature only in compounds. Unlike group 1 metals, many compounds of the group 2 elements have low water solubility. (Section 26.6)
- Calcium and magnesium are reducing agents that undergo most of the reactions that sodium and potassium do, but are less reactive than the group 1 metals. (Section 26.6)
- Beryllium is significantly less metallic than the other group 2 elements. For example, its hydroxide is amphoteric. (Section 26.6)
- Most magnesium metal is obtained from sea water by a method that depends on the lower solubility of $Mg(OH)_2$ than $Ca(OH)_2$. (Section 26.6)
- There are many important calcium compounds that occur naturally as minerals, the most useful of which is limestone, a form of calcium carbonate. Cave stalactites and stalagmites are composed of calcium carbonate. (Section 26.6)
- Chlorophyll, one of the most important biological compounds, is a complex of magnesium. Tooth enamel is composed mainly of hydroxyapatite, a calcium phosphate. (Section 26.6)
- Boron is a metalloid. As in the other groups, going down group 13, the elements become more metallic. The most common oxidation state is $+3$, although thallium has some $+1$ compounds. (Section 26.7)
- Consistent with the diagonal relationship, boric acid is a weak acid like orthosilicic acid, borates have similarities with silicates, and the hydrides of both boron and silicon are volatile, flammable, covalent molecular compounds. (Section 26.7)

- Unlike the low-melting other members of group 13, boron is a network solid whose allotropes all include in their structure an icosahedron of boron atoms. (Section 26.7)
- Aluminum is very reactive, but its corrosion in air is slow because of a protective oxide layer. Pure alumina is obtained from bauxite ores by dissolving in hot sodium hydroxide—a reaction that occurs because alumina is amphoteric—and filtering off insoluble iron oxides. Alumina cannot be reduced by common reducing agents, so aluminum metal is obtained by electrolysis of alumina. (Section 26.7)
- Borates commonly have B—O—B sequences in their structures. Boric acid is a monoprotic weak acid. A series of boron–hydrogen compounds called boranes release a lot of energy on combustion with oxygen. Diborane, the simplest, has a structure which can be explained by the presence of a three-centre, two-electron B—H—B bond. (Section 26.7)
- Aluminum reacts with dilute hydrochloric acid solutions, but not with nitric acid solutions, because of a protective oxide layer. Solutions of aluminum salts are acidic. Aluminum reacts with bromine to form $Al_2Br_6(s)$. (Section 26.7)
- Silicon is made by reduction of silica sand with coke in an electric furnace. Purification is carried out by zone refining. (Section 26.8)
- Quartz is a pure crystalline form of silicon dioxide, called silica; impurities produce gemstones. Silica is a high-melting network solid, resistant to attack by acids except HF. Silica gel is a gelatinous form of silica that absorbs large quantities of water. (Section 26.8)
- Othosilicates contain SiO_4^{4-} anions in a lattice with metal cations. Pyroxenes have structures based on chains of SiO_4 tetrahedra. (Section 26.8)
- Many aluminosilicate minerals and clays have structures that are based on parallel sheets of silicates. (Section 26.8)
- Silicones are covalent molecular polymers in which the molecules have a "backbone" with alternate S and O atoms. The properties of silicones depend on the degree of interlinking between chains, and the nature of the links (silicone or alkyl). (Section 26.8)
- Nitrogen as $N_2(g)$ is not very reactive, even though there are many nitrogen-containing compounds, because of the strength of the N≡N bonds. Allotropes of phosphorus are white phosphorus, with P_4 molecules, and red phosphorus, with polymer chains of P_4 units. (Section 26.9)
- Compounds with nitrogen in all oxidation states between -3 and $+5$ are known. (Section 26.9)
- Ammonia is important industrially, especially in manufacture of fertilizers. Hydrazine is a weak base in aqueous solution as well as a powerful reducing agent. (Section 26.9)
- Nitrogen forms a wide variety of oxides, with N in a range of oxidation states. NO(g) and $NO_2(g)$ are associated with photochemical smog formation in particular circumstances. As well as being an acid, nitric acid is a powerful oxidizing agent. (Section 26.9)
- Phosphine, $PH_3(g)$, is a gaseous analogue of ammonia. (Section 26.9)
- The two most important phosphorus oxides, $P_4O_6(s)$ and $P_4O_{10}(s)$, are solids whose structures contain expanded P_4 tetrahedra. $P_4S_3(s)$ is used in the heads of safety matches. (Section 26.9)
- The ionizable hydrogen ions in the various phosphorus oxoacids are produced by breaking of an O—H bond. Large quantities of orthophosphoric acid (H_3PO_4) are used in manufacture of fertilizers. Other uses are in corrosion protection and in foods. (Section 26.9)
- In its various forms, oxygen is the most abundant element in the earth's crust. It is present as the element, $O_2(g)$, in the atmosphere and is combined with other elements in water and in many minerals. Sulfur is found in nature in its elemental form, as well as in compounds natural gas and oil. In minerals, sulfur occurs as sulfide, disulfide, and sulfate ions. Sulfur oxides occur in nature, mostly from volcanic activity. (Section 26.10)
- Allotropes of the element oxygen are diatomic oxygen, $O_2(g)$, and ozone, $O_3(g)$. Ozone is important as an absorber of high-energy ultraviolet radiation in the upper atmosphere. The most common forms of sulfur have S_8 molecules. Selenium and tellurium are relatively rare, and are often found as selenides and tellurides associated with sulfide minerals. (Section 26.10)

- By far the most important sulfur compound industrially is sulfuric acid, most of which is used in the manufacture of superphosphate fertilizer from phosphate rock. (Section 26.10)
- Fluorine is produced by electrolysis of KF in anhydrous HF solution, and chlorine by electrolysis of saturated NaCl solution. Bromine is made by oxidation of bromide ions by chlorine. In the laboratory, iodine is made by oxidation of iodides. (Section 26.11)
- Fluorine is the most reactive of all of the elements, forming compounds with every element except He and Ne, usually by direct reaction. Important compounds are hydrogen fluoride, cryolite, and hydrofluorocarbons used as aerosol propellants (in place of CFCs). (Section 26.11)
- Hydrogen chloride is made industrially by reaction between hydrogen gas and chlorine gas. Its aqueous solutions are called hydrochloric acid. (Section 26.11)
- Oxoacids of chlorine range from HOCl, in which Cl is in oxidation state +1, to $HClO_4$, in which Cl is in oxidation state is +7. With increasing oxidation state, the oxoacids are stronger acids, and the oxoanions are more powerful oxidizing agents. (Section 26.11)

REVIEW QUESTIONS

Section 26.3: Charge Density of Cations: An Explanatory Concept

26.20 Which do you predict is the more polarizable anion: Cl^- or I^-? Why?

26.21 Rationalize the following:
(a) The properties of $BeCl_2$, BCl_3, $AlCl_3$, $TiCl_4$, and $FeCl_3$ are typical of covalent compounds, but those of KCl and AgCl are typical of ionic compounds.
(b) The melting points of AlF_3, $AlCl_3$, and $AlBr_3$ are 1200 °C, 192 °C, and 98 °C respectively.
(c) KCl, $AgNO_3$ and NH_4NO_3 crystallize from evaporated solutions as anhydrous salts, while salts of Mg^{2+}, Fe^{3+}, and Cr^{3+} usually crystallize as hydrated compounds.
(d) The ionic radii of Li^+ and Rb^+ ions are 76 pm and 152 pm, respectively. Because it is smaller, we might expect the Li^+ ion to move through water under the influence of an electrical field more quickly than Rb^+ does. In fact, the opposite is the case.

26.22 The compositional formula $CuSO_4.5H_2O$ can be written as $[Cu(H_2O)_4]^{2+}.SO_4^{2-}.H_2O$ to indicate better the structure of the salt. Write sensible corresponding formulas for the following:
(a) $Cr(NO_3)_3.9H_2O$ (b) $Al_2(SO_4)_3.18H_2O$

26.23 For each of the sets of cations listed, account for the comparative values of the molar enthalpy change of hydration, $\Delta_{hyd}H$:
(a) Be^{2+} (2455 kJ mol^{-1}) and Ba^{2+} (1275 kJ mol^{-1})
(b) Fe^{2+} (1890 kJ mol^{-1}) and Fe^{3+} (4340 kJ mol^{-1})

26.24 An example of the so-called "diagonal relationship" between properties of a second-row element and the third-row element in the next group is the similar chemistry, in some respects, of beryllium and aluminum. This can be attributed to the similar charge-to-radius ratios of the Be^{2+} and Al^{3+} ions. Explain how these two ions can have similar charge-to-radius ratios.

26.25 Comparing cations of the same metal with different oxidation states, we find that lead(IV) chloride has a lower melting point and is more soluble in non-polar solvents (that is, it is more characteristic of covalent molecular compounds) than lead(II) chloride. Rationalize this observation.

Section 26.4: Hydrogen

26.26 One of the pieces of evidence relating to the hydride ion in ionic metal hydrides comes from electrochemistry. Predict the reactions that occur at each electrode when molten LiH is electrolyzed.

26.27 Use bond energies (Table 7.3) to calculate the enthalpy change for the reaction of methane and water to give hydrogen and carbon monoxide (with all compounds in the gas phase). Considering the bond energies of reactants and products, suggest why the reaction is endothermic.

26.28 Write a balanced chemical equation for the preparation of $H_2(g)$ (and $CO(g)$) by the reaction of $CH_4(g)$ and water. Using data in Appendix C, calculate $\Delta_r H°$, $\Delta_r G°$, and $\Delta_r S°$ for this reaction.

26.29 Using values in Appendix C, calculate $\Delta_r H°$, $\Delta_r G°$, and $\Delta_r S°$ for the reaction of carbon and water to form $CO(g)$ and $H_2(g)$.

26.30 A method recently suggested for the preparation of hydrogen (and oxygen) from water proceeds as follows:
(a) Sulfuric acid and hydrogen iodide are formed from sulfur dioxide, water, and iodine.
(b) The sulfuric acid from the first step is decomposed by heat to water, sulfur dioxide, and oxygen.
(c) The hydrogen iodide from the first step is decomposed with heat to hydrogen and iodine.

Write a balanced equation for each of these steps and show that their sum is the decomposition of water to form hydrogen and oxygen.

Section 26.5: The Alkali Metals, Group 1

26.31 Write equations for the reaction of sodium with each of the halogens. Predict at least two physical properties that are common to all of the alkali metal halides.

26.32 Write balanced equations for the reaction of lithium, sodium, and potassium with O_2. Specify which metal forms an oxide, which forms a peroxide, and which forms a superoxide.

Section 26.6: The Alkaline Earth Elements, Group 2

26.33 When magnesium burns in air, it forms both an oxide and a nitride. Write balanced equations for the formation of both compounds.

26.34 Calcium reacts with hydrogen gas at 300–400 °C to form a hydride. This compound reacts readily with water, so it is an excellent drying agent for organic solvents.
(a) Write a balanced equation showing the formation of calcium hydride from $Ca(s)$ and $H_2(g)$.
(b) Write a balanced equation for the reaction of calcium hydride with water.

26.35 Name three uses of limestone. Write a balanced equation for the reaction of limestone with $CO2(aq)$ in water.

26.36 Explain what is meant by "hard water." What causes hard water, and what problems are associated with it?

26.37 Calcium oxide, $CaO(g)$, is used to remove $SO_2(g)$ from powerplant exhaust. These two compounds react to give solid $CaSO_3$. What mass of $SO_2(g)$ can be removed using 1.2×10^3 kg of $CaO(s)$?

26.38 The solubility product (K_{sp}) for $Ca(OH)_2$ is 5.5×10^{-5}, whereas that for $Mg(OH)_2$ is 5.6×10^{-12}. Calculate the equilibrium constant for the reaction:

$$Ca(OH)_2(s) + Mg^{2+}(aq) \rightleftharpoons Ca^{2+}(aq) + Mg(OH)_2(s)$$

26.39 Explain why the reaction in Question 26.38 can be used in the commercial isolation of magnesium from sea water.

26.40 Using data in Appendix C, calculate $\Delta_r G°$ values for the decomposition of $MCO_3(s)$ to $MO(s)$ and $CO_2(g)$ where M = Mg, Ca, and Ba. What is the relative tendency of these carbonates to decompose?

26.41 Beryllium, the lightest element in group 2, has some important industrial applications, but exposure (by breathing) to some of its compounds can cause berylliosis. Search the Internet for the uses of the element and the causes and symptoms of berylliosis.

Section 26.7: Boron, Aluminum, and the Group 13 Elements

26.42 (a) Write an equation for the reaction of $Al(s)$ and $H_2O(\ell)$ to produce $H_2(g)$ and $Al_2O_3(s)$.
(b) Using thermodynamic data in Appendix C, calculate $\Delta_r H°$, $\Delta_r S°$, and $\Delta_r G°$ for this reaction. Do these data indicate that the reaction should favour formation of the products?
(c) Why is aluminum metal unaffected by water?

26.43 Diborane can be made by the reaction of $NaBH_4$ and I_2. Which substance is oxidized and which is reduced?

26.44 Alumina, $Al_2O_3(s)$, is amphoteric. Among examples of its amphoteric character are the reactions that occur when it is heated strongly or "fused" with acidic oxides and basic oxides.
(a) Write a balanced equation for the reaction of alumina with silica, an acidic oxide, to give aluminum metasilicate, $Al_2(SiO_3)_3(s)$.
(b) Write a balanced equation for the reaction of alumina with the basic oxide CaO to give calcium aluminate, $Ca(AlO_2)_2(s)$.

26.45 Halides of the group 13 elements are excellent Lewis acids. When a Lewis base such as the Cl^- ion interacts with $AlCl_3$, the $AlCl_4^-$ ion is formed. Draw a Lewis electron dot structure for this ion. What structure do you predict for the $AlCl_4^-$ ion?

26.46 Aluminum dissolves readily in hot aqueous NaOH solution to form the aluminate ion, $Al(OH)_4^-(aq)$, and $H_2(g)$. Write a balanced equation for this reaction. If you begin with 13.2 g of $Al(s)$, what volume of H_2 gas is produced when the gas is measured at 1.0 bar pressure and 25 °C?

26.47 In water, the boron trihalides (except BF_3) hydrolyze completely to boric acid and the acid HX.
(a) Write a balanced equation for the reaction of BCl_3 with water.
(b) Calculate $\Delta_r H°$ for the hydrolysis of BCl_3 using data in Appendix C and the following information: $\Delta_f H°[BCl_3(g)] = -403$ kJ mol^{-1} and $\Delta_f H°[B(OH)_3(s)] = -1094$ kJ mol^{-1}.

Section 26.8: Silicon and the Group 14 Elements

26.48 Silicon–oxygen rings are a common structural feature in silicate chemistry. Draw the structure for the anion $Si_3O_9^{6-}$, which is found in minerals such as bentonite. The ring has three Si atoms and three O atoms, and there are two other O atoms on each Si atom.

26.49 Describe how ultra-pure silicon can be produced from sand.

26.50 Describe the structure of pyroxenes. What is the ratio of silicon to oxygen in this type of silicate?

26.51 One material needed to make silicones is dichlorodimethylsilane, $(CH_3)_2SiCl_2$. It is made by treating silicon powder at about 300 °C with CH_3Cl in the presence of a copper-containing catalyst.
   (a) Write a balanced equation for the reaction.
   (b) Assume you carry out the reaction on a small scale with 2.65 g of silicon. To measure the CH_3Cl gas, you fill a 5.60 L flask at 24.5 °C. What pressure of CH_3Cl gas must you have in the flask to have the stoichiometrically correct amount of the compound?
   (c) What mass of $(CH_3)_2SiCl_2$ is produced from 2.65 g of Si?

Section 26.9: Group 15 Elements and Their Compounds

26.52 Review the structure of phosphorous acid in Table 26.9.
   (a) What is the oxidation state of the phosphorus atom in this acid?
   (b) Predict and draw the structure of diphosphorous acid $(H_4P_2O_5)$. What is the maximum number of protons this acid can dissociate in water?

26.53 $CaHPO_4(s)$ is used as an abrasive in toothpaste. Write a balanced equation showing a possible preparation for this compound.

26.54 Consult the data in Appendix C. Are any of the nitrogen oxides listed there stable with respect to decomposition to $N_2(g)$ and $O_2(g)$?

26.55 Use data in Appendix C to calculate $\Delta_r H°$ and $\Delta_r G°$ at 25 °C for the reaction

$$2\ NO_2(g) \longrightarrow N_2O_4(g)$$

Is this reaction exothermic or endothermic? Is the reaction product- or reactant-favoured?

26.56 A major use of hydrazine (N_2H_4) is in steam boilers in powerplants.
   (a) The reaction of hydrazine with O_2 dissolved in water gives $N_2(g)$ and water. Write a balanced equation for this reaction.
   (b) O_2 dissolves in water to the extent of 3.08 mL (gas measured at 1 bar and 0 °C) in 100.0 mL of water at 20 °C. To react with all of the dissolved O_2 in 3.00 $\times$ 10^4 L of water (enough to fill a small swimming pool), what mass of N_2H_4 is needed?

26.57 A common analytical method for hydrazine in aqueous solution involves its oxidation with aquated iodate ions (IO_3^-) in acidic conditions. In the process, hydrazine acts as a four-electron reducing agent.

$$N_2(g) + 5\ H_3O(aq) + 4\ e^- \longrightarrow N_2H_5(aq) + 5\ H_2O(\ell)$$
$$E° = -0.23\ V\ \text{at}\ 25\ °C$$

Write the balanced equation for the reaction of hydrazine in acid solution $(N_2H_5^+)$ with IO_3^-(aq) ions to give $N_2(g)$ and $I_2(s)$. Calculate $E°$ at 25 °C for this reaction.

26.58 The steering rockets in the Space Shuttle use N_2O_4 and a derivative of hydrazine, 1,1-dimethylhydrazine. This mixture is called a *hypergolic fuel* because it ignites when the reactants come into contact:

$H_2NN(CH_3)_2(\ell) + 2\ N_2O_4(\ell)$
$$\longrightarrow 3\ N_2(g) + 4\ H_2O(g) + 2\ CO_2(g)$$

   (a) Identify the oxidizing agent and the reducing agent in this reaction.
   (b) The same propulsion system was used by the Lunar Lander on moon missions in the 1970s. If the Lander used 4100 kg of $H_2NN(CH_3)_2$, what mass of N_2O_4 was required to react with it? What mass of each of the reaction products was generated?

26.59 Dinitrogen trioxide (N_2O_3) has the structure shown here.

The oxide is unstable, decomposing to NO and NO_2 in the gas phase at 25 °C.

$$N_2O_3(g) \longrightarrow NO(g) + NO_2(g)$$

   (a) Explain why one bond distance in N_2O_3 is 114.2 pm, whereas the other two bonds are longer (121 pm) and nearly equal to each other.
   (b) For the decomposition reaction, $\Delta_r H° = +40.5$ kJ mol^{-1} and $\Delta_r G° = -1.59$ kJ mol^{-1}. Calculate $\Delta_r S°$ and K for the reaction at 298 K.
   (c) Calculate $\Delta_f H°$ for $N_2O_3(g)$.

26.60 Use $\Delta_f H°$ data in Appendix C to calculate the enthalpy change of the reaction

$$2\ N_2(g) + 5\ O_2(g) + 2\ H_2O(\ell) \longrightarrow 4\ HNO_3(aq)$$

Speculate on whether such a reaction could be used to "fix" nitrogen. Would research to find ways to accomplish this reaction be a useful endeavour?

Section 26.10: Group 16 Elements and Their Compounds

26.61 Metal sulfides roasted in air produce metal oxides.

$$2\ ZnS(s) + 3\ O_2(g) \longrightarrow 2\ ZnO(s) + 2\ SO_2(g)$$

Use thermodynamics to decide if the reaction is product-favoured at 298 K. Will the reaction be more or less product-favoured at a high temperature?

26.62 In the "contact process" for making sulfuric acid, sulfur is first burned in air to form $SO_2(g)$. Environmental restrictions allow no more than 0.30% of this SO_2 to be vented to the atmosphere.

(a) If enough sulfur is burned in a plant to produce 1.80×10^6 kg of pure, anhydrous H_2SO_4 per day, what is the maximum amount of $SO_2(g)$ that is allowed to be exhausted to the atmosphere?

(b) One way to prevent any SO_2 from reaching the atmosphere is to "scrub" the exhaust gases with slaked lime, $Ca(OH)_2$:

$$Ca(OH)_2(s) + SO_2(g) \longrightarrow CaSO_3(s) + H_2O(\ell)$$
$$2\, CaSO_3(s) + O_2(g) \longrightarrow 2\, CaSO_4(s)$$

What mass of $Ca(OH)_2$ is needed to remove the $SO_2(g)$ calculated in part (a)?

26.63 Sulfur forms a range of compounds with fluorine. Draw Lewis electron dot structures for S_2F_2 (connectivity is FSSF), SF_2, SF_4, SF_6, and S_2F_{10}. What is the formal oxidation state of sulfur in each of these compounds?

26.64 A sulfuric acid plant produces an enormous amount of heat. To keep costs as low as possible, much of this heat is used to make steam to generate electricity. Some of the electricity is used to run the plant, and the excess is sold to the local electrical utility. Three reactions are important in sulfuric acid production: (a) burning $S_8(s)$ to $SO_2(g)$; (b) oxidation of $SO_2(g)$ to $SO_3(g)$; and (c) reaction of $SO_3(g)$ with H_2O:

$$SO_3(g) + H_2O(\text{in 98\% } H_2SO_4) \longrightarrow H_2SO_4(\ell)$$

The enthalpy change of the third reaction is -130 kJ mol^{-1}. Estimate the total heat produced when 1.00 mol of sulfur is used to produce 1.00 mol of $H_2SO_4(\ell)$. How much heat is produced per tonne of $H_2SO_4(\ell)$?

Section 26.11: The Halogens, Group 17

26.65 To make chlorine from chloride ions, a strong oxidizing agent is required. The aquated permanganate ion, $MnO_4^-(aq)$, is one example. Consult the table of standard reduction potentials (Appendix E) and identify several other oxidizing agents that may be suitable. Write balanced equations for the reactions of these species with aquated chloride ions.

26.66 Metals generally react with hydrogen halides such as HCl to give the metal halide and hydrogen.

$$Ag(s) + HCl(g) \longrightarrow AgCl(s) + \tfrac{1}{2} H_2(g)$$

The reaction is thermodynamically product-favoured if $\Delta_r G°$ is negative. Is this true for all of the hydrogen halides reacting with silver? Relevant free energy changes of formation are listed.

$\Delta_f G°$ of HX (kJ mol^{-1})	$\Delta_f G°$ of AgX(s) (kJ mol^{-1})
HF, -273.2	AgF, -193.8
HCl, -95.09	AgCl, -109.76
HBr, -53.45	AgBr, -96.90
HI, $+1.56$	AgI, -66.19

26.67 The halogen oxides and oxoanions are good oxidizing agents. For example, the reduction of bromate ion has an $E°$ value of 1.44 V in acid solution at 25 °C.

$$2\, BrO_3^-(aq) + 12\, H^+(aq) + 10\, e^- \longrightarrow Br_2(aq) + 6\, H_2O(\ell)$$

Is it possible to oxidize $Mn^{2+}(aq)$ ions in 1.0 mol L^{-1} aqueous solution to $MnO_4^-(aq)$ ions (also at 1.0 mol L^{-1}) by reaction with 1.0 mol L^{-1} $BrO_3^-(aq)$ ions at 25 °C?

26.68 Bromine is obtained from brine wells. The process involves treating water containing bromide ions with $Cl_2(g)$ and extracting the product Br_2 from the solution using an organic solvent. Write a balanced equation for the reaction of $Cl_2(g)$ and $Br^-(aq)$ ions. What are the oxidizing and reducing agents in this reaction? Using the table of standard reduction potentials (Appendix E), verify that this is a product-favoured reaction.

26.69 Halogens combine with one another to produce *interhalogens* such as $BrF_3(g)$. Sketch a possible structure for a molecule of bromine trifluoride.

26.70 The standard enthalpy of formation of OF_2 gas is $+24.5$ kJ mol^{-1}. Calculate the average O—F bond energy.

26.71 Calcium fluoride can be used in the fluoridation of municipal water supplies. If you want to achieve a fluoride ion concentration of 2.0×10^{-5} mol L^{-1}, what mass of CaF_2 must you use in 1.0×10^6 L of water?

SUMMARY AND CONCEPTUAL QUESTIONS

26.72 Give examples of two basic oxides. Write equations illustrating the formation of each oxide from its component elements. Write another chemical equation that illustrates the basic character of each oxide.

26.73 Give examples of two acidic oxides. Write equations illustrating the formation of each oxide from its component elements. Write another chemical equation that illustrates the acidic character of each oxide.

26.74 Give the name and symbol of each element having the valence configuration [noble gas]ns^2np^1.

26.75 Give symbols and names for four monatomic ions that have the same electron configuration as argon.

26.76 Which of the first ten elements in the periodic table are found as free elements in the earth's crust? Which elements in this group occur in the earth's crust only as part of a chemical compound?

26.77 Place the following oxides in order of increasing basicity: CO_2, SiO_2, SnO_2.

26.78 Place the following oxides in order of increasing basicity: Na_2O, Al_2O_3, SiO_2, SO_3.

26.79 Complete and balance the equations for the following reactions. Assume an excess of oxygen for (d).
(a) $Na(s) + Br_2(\ell) \longrightarrow$ (c) $Al(s) + F_2(g) \longrightarrow$
(b) $Mg(s) + O_2(g) \longrightarrow$ (d) $C(s) + O_2(g) \longrightarrow$

26.80 Complete and balance the equations for the following reactions:
(a) $K(s) + I_2(g) \longrightarrow$ (b) $Al(s) + S_8(s) \longrightarrow$
(c) $Ba(s) + O_2(g) \longrightarrow$ (d) $Si(s) + Cl_2(g) \longrightarrow$

26.81 The electrolysis of aqueous NaCl solution is used to make $NaOH(s)$, $Cl_2(g)$, and $H_2(g)$.
(a) Write a balanced equation for the process.
(b) Is the electrolysis of aqueous NaCl solution the only source of these chemicals?

26.82 For each of the third-period elements (Na through Ar), identify the following:
(a) whether the element is a metal, non-metal, or metalloid
(b) the colour and appearance of the element
(c) the state of the element (s, ℓ, or g) under ambient conditions

26.83 For each of the second-period elements (Li through Ne), identify the following:
(a) whether the element is a metal, non-metal, or metalloid
(b) the colour and appearance of the element
(c) the state of the element (s, ℓ, or g) under ambient conditions

26.84 Consider the chemistries of the elements sodium, magnesium, aluminum, silicon, and phosphorus.
(a) Write a balanced chemical equation for the reaction of each element with elemental chlorine.
(b) Describe the bonding in each of the products of the reactions with chlorine as ionic or covalent.

(c) Draw Lewis electron dot structures for the products of the reactions of silicon and phosphorus with chlorine. What are their electron-pair and molecular geometries?

26.85 Complete and balance the following equations.
(a) $KClO_3(s) \xrightarrow{\text{heat}}$
(b) $H_2S(g) + O_2(g) \longrightarrow$
(c) $Na(s) + O_2(g) \longrightarrow$
(d) $P_4(s) + KOH(aq) + H_2O(\ell) \longrightarrow$
(e) $NH_4NO_3(s) \xrightarrow{\text{heat}}$
(f) $In(s) + Br_2(\ell) \longrightarrow$
(g) $SnCl_4(\ell) + H_2O(\ell) \longrightarrow$

26.86 You have a 1.0 L flask that contains a mixture of argon and hydrogen. The pressure inside the flask is 99.3 kPa and the temperature is 22 °C. Describe an experiment that you could use to determine the percentage of hydrogen in this mixture.

26.87 How would you extinguish a sodium fire in the laboratory? What is the worst thing you could do?

26.88 You are given a stoppered flask that contains either hydrogen, nitrogen, or oxygen. Suggest an experiment to identify the gas.

26.89 When 1.00 g of a white solid A is strongly heated, you obtain another white solid, B, and a gas. An experiment is carried out on the gas, showing that it exerts a pressure of 27.86 kPa in a 450 mL flask at 25 °C. Bubbling the gas into a $Ca(OH)_2$ solution gives another white solid, C. If the white solid B is added to water, the resulting solution turns red litmus paper blue. Addition of aqueous HCl solution to the solution of B and evaporation of the resulting solution to dryness yields 1.055 g of a white solid D. When D is placed in a Bunsen burner flame, it colours the flame green (you may need to do some research about this property). Finally, if the aqueous solution of B is treated with sulfuric acid, a white precipitate, E, forms. Identify the lettered compounds in the reaction scheme.

With the **eCHACR** single-sign-on access card bundled with your text, log on (http://login.cengage.com/sso) and access the e-book and click on any in-margin icons for dynamic molecular-level animations and simulations, videos of laboratory reactions, interactive exercises, reviews of background concepts, and other online supplementary materials as noted by the icons in the text margins.

Also go to www.chemistry.nelson.com <http://www.chemistry.nelson.com> for Answers to in-chapter exercises and selected Review Questions, Test Yourself questions, weblinks, crossword puzzles, flashcards, glossary of key terms, and other student resources.

Transition Elements and Their Compounds

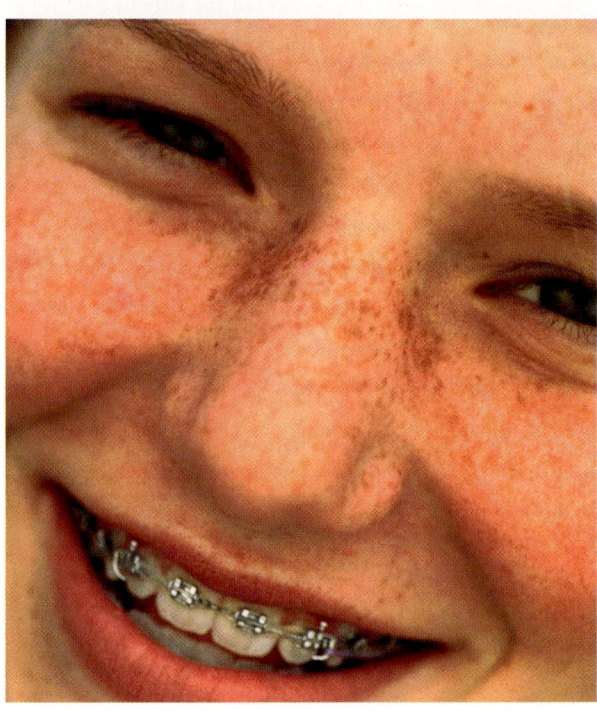

© Francoise Gervais/Corbis

27.1 Case Study: **Metals with Memory**

In the early 1960s, metallurgical engineer William J. Buehler, a researcher at the Naval Ordnance Laboratory in White Oak, Maryland, U.S., was experimenting with alloys made of two metals. He was looking for an impact- and heat-resistant material, intended for use in the nose cone of a missile. The material also needed to be fatigue-resistant so that it would not lose these properties if it were often bent and shaped. An alloy of two transition metals, nickel and titanium, appeared to fit the requirements. Buehler prepared long, thin strips of this alloy to demonstrate that it could be folded and unfolded many times without breaking. At a meeting to discuss this alloy, one of his associates decided to see what would happen when the strip was heated. He held a cigarette lighter to a folded-up piece of metal and was amazed to observe that the metal strip immediately unfolded and assumed its original shape. Memory metal was discovered. This unusual alloy is now called *Nitinol*, a name constructed out of **ni**ckel, **ti**tanium, and **N**aval **O**rdnance **L**aboratory.

> *He held a cigarette lighter to a folded-up piece of metal and was amazed to observe that the metal strip immediately unfolded and assumed its original shape. Memory metal was discovered.*

Memory metal is an alloy with approximately an equal number of Ni and Ti atoms. It "remembers" its shape because of the arrangement of these atoms in two solid phases,

The word *transition* in "phase transition temperature" is unrelated to the fact that the metals in nitinol are transition metals. This is a purely coincidental use of the same word in two contexts with different meanings.

corresponding with the way that water and silicon have polymorphic forms [<<Section 11.15]. In the form known as austenite, the atoms are in a highly symmetrical arrangement. When it is in this phase, the metal can be twisted or bent into a desired shape that it will "remember." When the alloy is cooled below a temperature called its *phase transition temperature*, it changes to a phase called martensite, which has a less ordered arrangement of the atoms. The martensite form of nitinol is fairly soft and may be bent and twisted out of shape. When warmed, it changes again to the austenite phase, which is more stable at the higher temperature and regains its original shape. Application of mechanical stress also causes change of austenite to martensite, with reversion of the structure on relief of the stress.

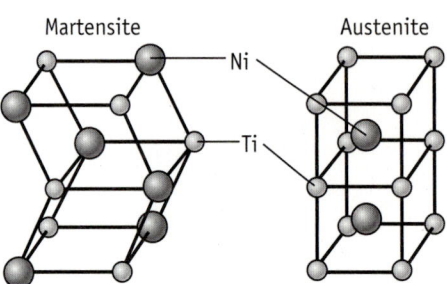

The two solid phases of nitinol, with different crystalline arrangements of the Ni and Ti atoms.

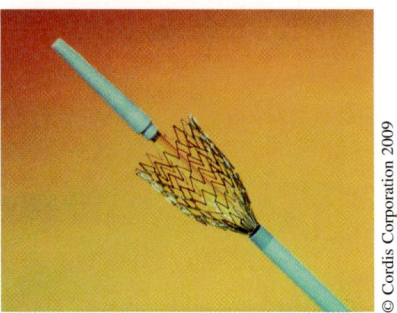

A nitinol stent. A stent made of nitinol is used to reinforce blood vessels.

Nitinol frames for glasses. These sunglass frames are made of nitinol that has a transition temperature below ambient temperatures, so that it is in the austenite form. Under pressure (as well as with cooling), austenite changes to the more flexible martensite, and its shape can be deformed. With relaxation of the pressure, it reverts to austenite and regains its original shape.

Nitinol is made with a range of compositions of Ni and Ti, each with a characteristic transition temperature in the range from −125 °C to 70 °C. Because of this, nitinol alloys with different compositions have different applications.

Memory metal did not make it into missile nose cones, but it has found a wide variety of other applications. Some of the most interesting are in the practice of medicine. One application involves the fabrication of vascular stents to reinforce blood vessels. The stent is crushed and inserted into a blood vessel through a very fine needle. When the nitinol stent warms to body temperature, it returns to its "memorized" shape and reinforces the walls of the blood vessel.

Nitinol is also used in orthodontics. Made-to-measure dental braces are cooled to the more flexible martensite phase and applied. As they come to body temperature, they revert to the austenite phase, contracting as they do so to apply a steady pressure on the teeth. The pressure is maintained as the teeth move, so that tightening is not necessary.

27.2 The *d*-Block Elements and Compounds

The transition elements are the block of metallic elements in the central portion of the periodic table, bridging the *s*-block elements at the left and the *p*-block elements on the right (Figure 27.1). They are often divided into two groups: the ***d*-block elements,** which differ in the number of electrons in *d* orbitals; and the ***f*-block elements,** characterized by unfilled *f* orbitals. Contained within the latter group are two sub-groups of elements: the *lanthanides* from lanthanum (La, $Z = 57$) to hafnium (Hf, $Z = 72$), and the *actinides* from actinium (Ac, $Z = 89$) to rutherfordium (Rf, $Z = 104$).

This chapter primarily focuses on the *d*-block elements, and within this group we concentrate mainly on the elements in the fourth period— that is, the elements of the first transition series, scandium to zinc.

The *d*-block metals include elements with a wide range of properties. They encompass the most common metal used in construction and manufacturing (iron), metals that are valued for their beauty and chemical inertness (gold, silver, and platinum), and metals used in coins (nickel, copper, and zinc). There are metals used in modern technology (titanium) and metals known and used in early civilizations (copper, silver, gold, and iron). The *d*-block includes the most dense elements (osmium, $d = 22.49$ g cm^{-3}, and iridium, $d = 22.41$ g cm^{-3}), the metals with the highest and lowest melting points (tungsten, $T_m = 3410$ °C, and mercury, $T_m = -38.9$ °C), and one of the only two elements with atomic

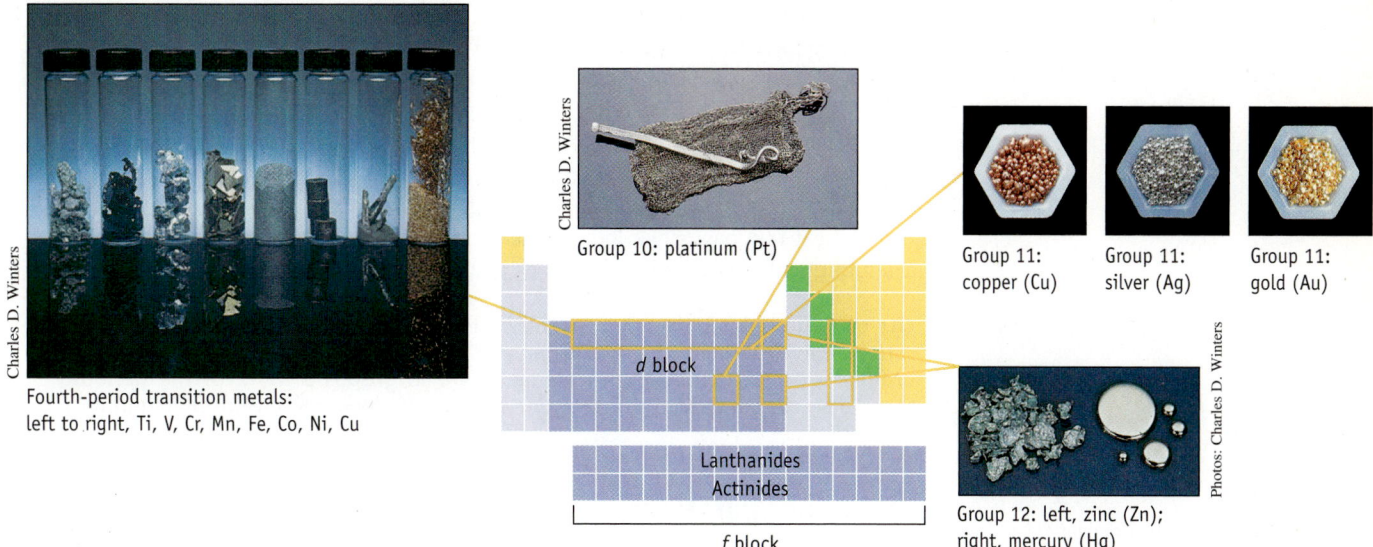

FIGURE 27.1 The transition metals. The *d*-block and *f*-block elements are highlighted in a darker shade of purple.

numbers less than 83 that are radioactive (technetium, Tc, $Z = 43$; the other is promethium, Pm, $Z = 61$, in the *f*-block).

With the exception of mercury, the transition elements are solids, mostly with high melting and boiling points. They have a metallic sheen and conduct electricity and heat well. They react with a range of oxidizing agents to produce ionic compounds, although there is considerable variation in such reactions among the elements. Silver, gold, and platinum, for example, resist oxidation and do not readily tarnish.

Compounds of several *d*-block elements are particularly important in the biochemistry of living organisms. Cobalt is the crucial element in Vitamin B-12, which is part of a catalyst essential for several biochemical reactions. As discussed in Section 27.4, the oxygen-carrying and -storage proteins hemoglobin and myoglobin contain iron. Molybdenum and iron, together with sulfur, form the reactive portion of nitrogenase, a biological catalyst used by nitrogen-fixing organisms to convert atmospheric nitrogen into ammonia.

Many transition metal compounds are highly coloured and are used as pigments in paints and dyes (Figure 27.2). Prussian blue, $Fe_4[Fe(CN)_6]_3 \cdot 14H_2O$, has been used as a "bluing agent" in engineering blueprints. Cadmium yellow, a pigment often used by artists, contains cadmium sulfide (CdS), and the pigment in most white paints is titanium(IV) oxide (TiO_2).

(a) Paint pigments: yellow, CdS; green, Cr_2O_3; white, TiO_2 and ZnO; purple, $Mn_3(PO_4)_2$; blue, Co_2O_3 and Al_2O_3; and ochre, Fe_2O_3.

(b) Small amounts of transition metal compounds are used to colour glass: blue, Co_2O_3; green, Cu and Cr oxides; purple, Ni and Co oxides; red, Cu oxide; and iridescent green, U oxide.

(c) Traces of transition metal ions are responsible for the colours of green jade (Fe), red corundum (Cr), blue azurite (Cu), blue-green turquoise (Cu), and purple amethyst (Fe).

FIGURE 27.2 Colourful chemistry. Many transition metal compounds are coloured, a property that leads to specific uses.

The presence of transition metal ions in crystalline silicates or alumina transforms these common materials into gemstones. Iron(II) ions cause the yellow color in citrine, and chromium(III) ions produce the red colour of ruby. Transition metal complexes in small quantities add colour to glass. For example, blue glass contains small amounts of a cobalt(III) oxide, and addition of chromium(III) oxide to glass gives a green colour. Old window panes sometimes have a purple tinge because of oxidation of traces of manganese(II) ions (Mn^{2+}) in the glass to permanganate ions (MnO_4^-).

In the next few pages, we look at the properties of the *d*-block transition elements, concentrating on the underlying principles that govern these properties.

The transition elements comprise the *d*-block elements, whose electron configurations have unfilled *d* orbitals, and the *f*-block elements, characterized by unfilled *f* orbitals. The *d*-block elements are in groups 3 to 12 of the periodic table. The elements of the first series of *d*-block elements have atomic numbers 21 to 30.

Electron Configurations

The chemical behaviour of elemental substances is related to their electron configurations. The configurations of the *d*-block elements are listed in Table 27.1 and those of the other transition elements are listed in Table 8.10. The valence electrons for the fourth-period transition elements are the 4*s* and 3*d* sub-shells.

The electron configurations of chromium and copper have only one electron in the 4*s* orbital, and are not what you would predict by use of the Aufbau principle and the order of assigning electrons shown in Figure 8.24. This is evidence that electrons in the 3*d* and 4*s* orbitals have nearly the same energy, and it is difficult to predict which orbitals are "occupied" in ground-state atoms. The "discrepancies" have little or no effect on the chemical behaviour of these elements and their compounds.

The elements in the first *d*-block series (in the fourth period of the periodic table) have valence electrons in the 3*d* and 4*s* orbitals.

Oxidation-Reduction Chemistry

A characteristic chemical property of metals is that they undergo oxidation by a wide range of oxidizing agents such as oxygen, the halogens, and aqueous solutions of acids (Table 27.2). All of the transition metals react with these reagents, except that vanadium and copper are not oxidized in aqueous HCl solution. Sometimes, and especially in the case of corrosion of iron [<<Section 16.8], this reactivity is undesirable.

TABLE 27.1 Electron Configurations of the Fourth-Period Transition Elements

	spdf Configuration	Box Notation 3*d*	4*s*
Sc	[Ar]3$d^1$4s^2	↑ □ □ □ □	↑↓
Ti	[Ar]3$d^2$4s^2	↑ ↑ □ □ □	↑↓
V	[Ar]3$d^3$4s^2	↑ ↑ ↑ □ □	↑↓
Cr	[Ar]3$d^5$4s^1	↑ ↑ ↑ ↑ ↑	↑
Mn	[Ar]3$d^5$4s^2	↑ ↑ ↑ ↑ ↑	↑↓
Fe	[Ar]3$d^6$4s^2	↑↓ ↑ ↑ ↑ ↑	↑↓
Co	[Ar]3$d^7$4s^2	↑↓ ↑↓ ↑ ↑ ↑	↑↓
Ni	[Ar]3$d^8$4s^2	↑↓ ↑↓ ↑↓ ↑ ↑	↑↓
Cu	[Ar]3d^{10}4s^1	↑↓ ↑↓ ↑↓ ↑↓ ↑↓	↑
Zn	[Ar]3d^{10}4s^2	↑↓ ↑↓ ↑↓ ↑↓ ↑↓	↑↓

TABLE 27.2 Products of Reactions of the Fourth-Row Transition Metals with O_2(g), Cl_2(g), and Aqueous HCl solution

Element	Reaction with O_2(g) *	Reaction with Cl_2(g)	Reaction with Aqueous HCl Solution
Scandium	Sc_2O_3	$ScCl_3$	Sc^{3+}(aq)
Titanium	TiO_2	$TiCl_4$	Ti^{3+}(aq)
Vanadium	V_2O_5	VCl_4	NR[†]
Chromium	Cr_2O_3	$CrCl_3$	Cr^{2+}(aq)
Manganese	MnO_2	$MnCl_2$	Mn^{2+}(aq)
Iron	Fe_2O_3	$FeCl_3$	Fe^{2+}(aq)
Cobalt	Co_2O_3	$CoCl_2$	Co^{2+}(aq)
Nickel	NiO	$NiCl_2$	Ni^{2+}(aq)
Copper	CuO	$CuCl_2$	NR[†]
Zinc	ZnO	$ZnCl_2$	Zn^{2+}(aq)

* Not all possible compounds are listed

[†] NR = no reaction

In the most common compounds of the *d*-block elements, the metal is in oxidation states +2 or +3 (Table 27.2). For example, iron reacts with oxygen to form $Fe_2O_3(s)$, with chlorine gas in aqueous solution to form $FeCl_3(s)$, and with acidic solutions to produce $Fe^{2+}(aq)$ ions and H_2 gas (Figure 27.3).

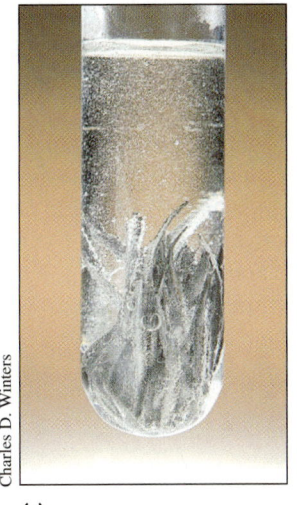

(a)   (b)   (c)

FIGURE 27.3 Reactions of iron typical of the transition metals. These metals react with oxygen, with halogens, and with acidic solutions. (a) Steel wool reacts with $O_2(g)$, (b) steel wool reacts with chlorine gas, $Cl_2(g)$, and (c) iron filings react with aqueous HCl solution.

Although there is a preponderance of the +2 and +3 oxidation states in compounds of the first transition metal series, the range of oxidation states in these compounds is broad (Figure 27.4). For example, in chromate ions (CrO_4^{2-}) and dichromate ions ($Cr_2O_7^{2-}$), chromium is in oxidation state +6, manganese has oxidation states up to +7 (in the permanganate ion, MnO_4^-), silver and copper form +1 cations, and vanadium oxidation states range from +2 to +5.

Think about It

e27.1 Work through a tutorial on assigning oxidation states to elements in compounds.

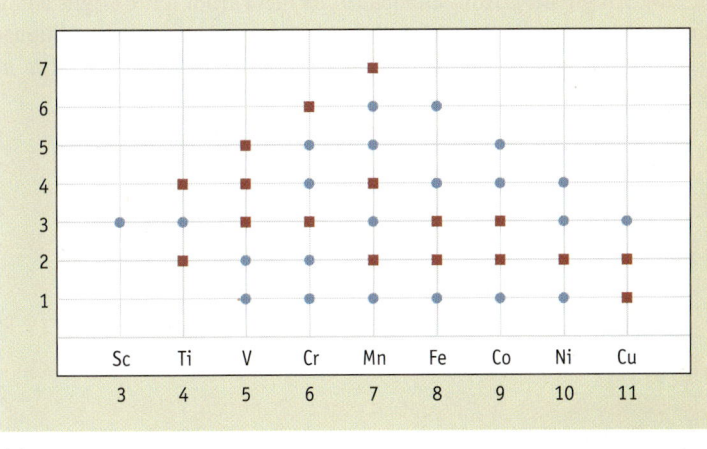

(a)

(b)

FIGURE 27.4 Oxidation states of the transition elements in the first transition series. (a) The most common oxidation states are indicated with red squares; less common oxidation states are indicated with blue dots. (b) Aqueous solutions of chromium compounds with two different oxidation states: +3 in $Cr(NO_3)_3$ solution (violet) and $CrCl_3$ solution (green), and +6 in K_2CrO_4 solution (yellow) and $K_2Cr_2O_7$ solution (orange).

The oxidation-reduction chemistry of the transition metals is due to the existence (in most cases) of more than one relatively stable oxidation state (Figure 27.4). An oxidizing agent may bring about a change of a transition metal species to another with higher oxidation state, and a reducing agent may cause a change to a species with a lower oxidation state. When a transition metal is oxidized, the highest-energy *s* electrons are removed, perhaps along with one or more *d* electrons. During oxidation of iron, for example, Fe atoms with electron configuration $[Ar]3d^64s^2$ are converted to Fe^{2+} ions with configuration $[Ar]3d^6$ or Fe^{3+} ions, $[Ar]3d^5$. In contrast to ions formed by main group elements, most transition metal cations are paramagnetic, a property attributable to unpaired electrons. Many of these ions are coloured, due to the absorption of light in the visible region of the electromagnetic spectrum. A rationalization of the magnetism and colours of the transition metal ions from the perspective of *crystal-field theory* is provided in Section 27.8.

Higher oxidation states are more common in compounds of the elements in the second and third transition series. For example, the naturally occurring sources of molybdenum and tungsten are the ores molybdenite (MoS_2) and wolframite (WO_3). This general trend is carried over in the *f*-block. The lanthanides form ions primarily in the +3 oxidation state. In contrast, actinide elements usually have higher oxidation states in their compounds; +4 and even +6 are typical. For example, UO_3 is a common oxide of uranium, and UF_6 is a compound important in processing uranium fuel for nuclear reactors [>>Chapter 30].

> Almost all of the *d*-block elements have multiple oxidation states, and this gives rise to their oxidation-reduction chemistry.

Periodic Trends: Size, Density, and Melting Point

The periodic table organizes the elements with respect to their chemical and physical properties. Let's look at three properties of the transition elements that vary periodically with increasing atomic number: atomic radius, density, and melting point.

Atomic Radius The variation in atomic radii for the transition elements in the fourth, fifth, and sixth periods is illustrated in Figure 27.5. Unlike the trends of atomic radii of the main block elements [<<Figure 8.4], the radii of the transition elements are quite similar, with a small decrease to a minimum around the middle members of the groups. This similarity of radii can be understood by reference to the effective nuclear charge experienced by the outer *s* electrons [<<Section 8.6]. Progressing from element to element from left to right along fourth period, for example, the increased attraction of the 4*s* electrons due to an additional proton in the nucleus is almost cancelled out by repulsion from the additional electron in the "inner" 3*d* orbitals.

> Atomic radius is an atomic property that can be inferred from experimental measurements, while density and melting point are macroscopic, observable physical properties of elemental substances.

FIGURE 27.5 Atomic radii of the transition elements of the fourth, fifth, and sixth periods, compared with those in groups 1 and 2.

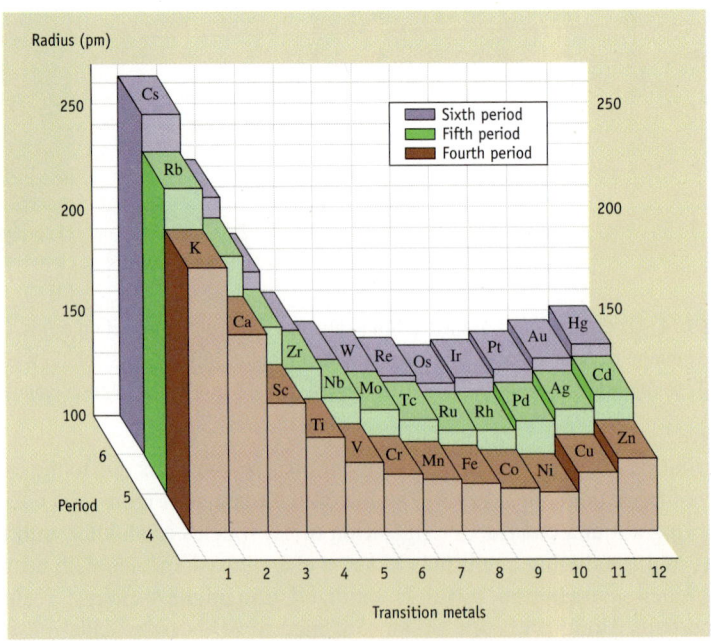

The radii of the *d*-block elements in the fifth and sixth periods in each group are almost identical. The reason is that the lanthanide elements immediately precede the third series of *d*-block elements. Going from element to element with an additional electron in the 4*f* orbitals, there is a steady contraction in size, consistent with the general trend of decreasing size from left to right in the periodic table. By the time we come to hafnium ($Z = 72$) and the third *d*-block elements, the radii have decreased to a size similar to that of elements in the previous period. The cumulative decrease in size across the elements with unfilled 4*f* orbitals is called the *lanthanide contraction*.

The similar sizes of the second- and third-period *d*-block elements have significant consequences for their chemistry. For example, the "platinum group metals" (Ru, Os, Rh, Ir, Pd, and Pt) form similar compounds, so it is not surprising that minerals containing these metals are found in the same geological zones on Earth. Nor is it surprising that it is difficult to separate compounds containing these elements from one another by chemical means because they have similar chemical reactivities.

Density The density of a solid depends upon the number of atoms per unit volume, and the mass of its atoms. Across a period, the densities of the transition elements first increase and then decrease, consistent with the variation in their atomic radii (Figure 27.6a). Although the overall differences in radii among these elements is small, the effect is magnified because the volume changes with the cube of the radius ($V = (4/3)\pi r^3$).

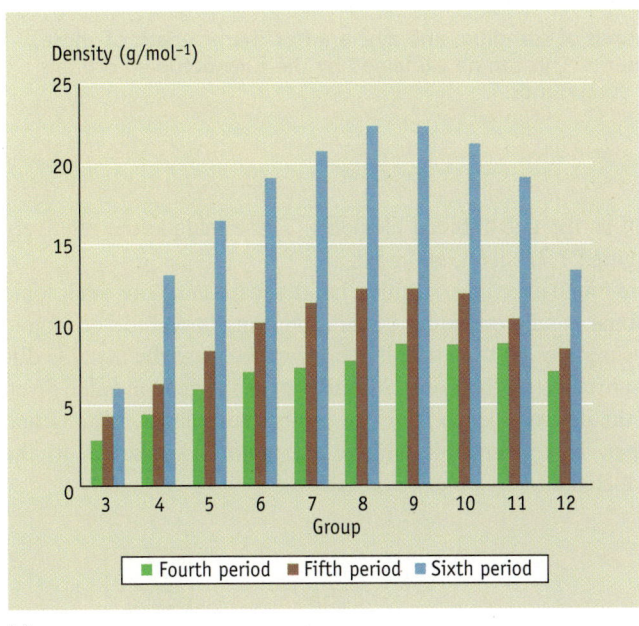

(a)

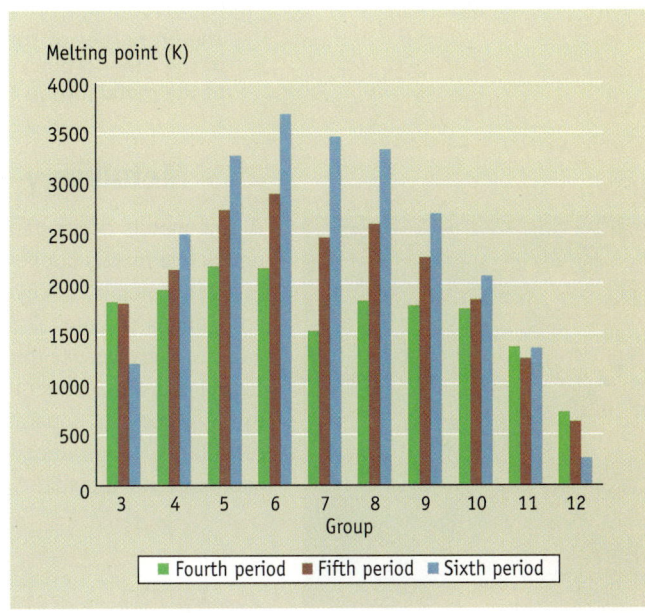

(b)

FIGURE 27.6 Periodic properties in the transition series. (a) Density and (b) melting point of the transition series metals.

The lanthanide contraction explains why elements in the sixth period have the highest density. The relatively small radii of sixth-period transition metals means that more atoms pack into a given volume. In addition, their atomic masses are larger than their counterparts in the fifth period.

Melting Point The melting point of any substance reflects the forces of attraction between the atoms, molecules, or ions that compose the solid. With transition elements, the melting points rise to a maximum around the middle of the series (Figure 27.6b), then decrease. Again, these elements' electron configurations provide us with an explanation. The variation in melting point indicates that the strongest metallic bonds occur when the *d* sub-shell is about half-filled. And these elements have the largest number of electrons in the bonding molecular orbitals in the metal (see the discussion of bonding in metals in Section 28.2).

> The atomic radii of any of the *d*-block series of elements vary little, with a minimum near the middle of the series. Within each series, there are maxima of density and melting point at elements near the middle.

Non-typical Scandium and Zinc

The properties of compounds of scandium and zinc are not characteristic of those of the other *d*-block elements, and these two elements are not generally regarded as transition elements.

Scandium compounds are neither coloured nor magnetic. The Sc^{3+} ion forms a rather limited number of coordination compounds. This lack of transition-element characteristics is consistent with the negligible evidence of multiple oxidation states, almost all compounds and the aquated ion being in the +3 oxidation state. In formation of the +3 oxidation state, all three valence electrons are removed from the 3*d* and 4*s* orbitals, so that there are no partially filled shells, and there is no opportunity for unpaired electrons to give rise to magnetism (as discussed in Section 27.8).

Although there are numerous coordination compounds of zinc, almost all of its ions and compounds are in the +2 oxidations state, and these are colourless and not magnetic. These properties can be attributed to the full 3*d* and 4*s* shells of zinc in the +2 oxidation state.

> The properties of compounds of scandium and zinc are not characteristic of those of the other *d*-block elements. This can be explained by the fully occupied orbitals of the metal ions in their compounds.

27.3 Metallurgy

A few metals occur naturally as the uncombined elements. These include the relatively unreactive metals copper (Figure 27.7), silver, and gold.

Most metals, however, are found as oxides, sulfides, halides, carbonates, or other ionic compounds (Figure 27.8). Some metal-containing mineral deposits have little economic value, either because the concentration of the metal is too low or because the metal is difficult to separate from other substances in their ores. **Metallurgy** is the general name given to the process of separating the desired metals from these other substances. Mining and metallurgy to produce copper, nickel, and aluminium are major contributors to the economies of Canada, Australia, and many other countries.

Charles D. Winters

FIGURE 27.7 Naturally occurring forms of copper. Copper occurs as the metal ("native copper") and as minerals such as blue azurite [$2CuCO_3.Cu(OH)_2$] and green malachite [$CuCO_3.Cu(OH)_2$].

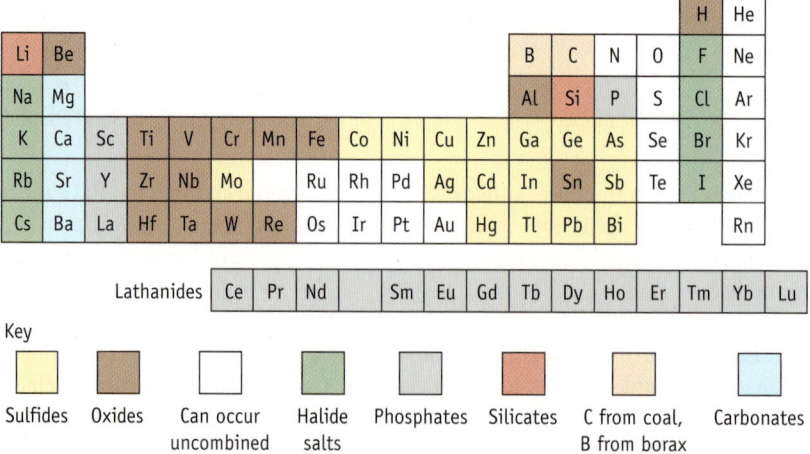

FIGURE 27.8 Types of minerals in which elements are found in ores. Most transition elements are found naturally as oxides or sulfides.

Ores usually consist of one or more minerals (compounds of the desired metals) mixed with large quantities of impurities such as sand and clay, called *gangue* (pronounced "gang"). Generally, the first step in a metallurgical process is to separate the mineral from the gangue. Then the ore is converted, by a reduction process, to the metal. Pyrometallurgy and hydrometallurgy are two methods of recovering metals from their ores. As the names imply, **pyrometallurgy** involves high temperatures and **hydrometallurgy** uses aqueous solutions (and so is limited to the relatively low temperatures at which water is a liquid). Iron and copper metallurgy are examples of these two methods of metal extraction.

Iron Extraction from Ores: Pyrometallurgy

The production of iron from its ores is carried out in a blast furnace (Figure 27.9). The furnace is loaded from the top with a mixture of ore (usually hematite, Fe_2O_3), coke (which is mainly carbon), and limestone ($CaCO_3$). A blast of hot air forced in at the bottom of the furnace causes the coke to burn with such an intense heat that the temperature at the bottom is almost 1500 °C. The quantity of air input is controlled so that carbon monoxide is the main product of combustion. Both carbon and carbon monoxide participate in the reduction of iron(III) oxide to give impure metal.

$$Fe_2O_3(s) + 3\,C(s) \longrightarrow 2\,Fe(\ell) + 3\,CO(g)$$
$$Fe_2O_3(s) + 3\,CO(g) \longrightarrow 2\,Fe(\ell) + 3\,CO_2(g)$$

Coke is made by heating coal in a large oven that is sealed to keep out oxygen. Heating drives off volatile compounds, including benzene and ammonia. What remains (coke) is mostly carbon.

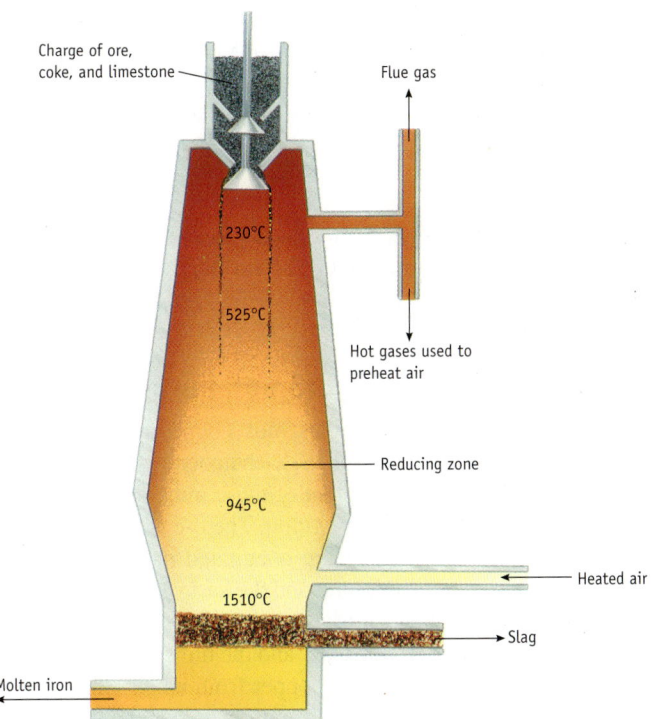

Charge of ore, coke, and limestone

Flue gas

230°C

525°C

Hot gases used to preheat air

Reducing zone

945°C

Heated air

1510°C

Slag

Molten iron

FIGURE 27.9 A blast furnace. The largest modern furnaces have hearths 14 metres in diameter. They can produce as much as 10 000 tonnes of iron per day.

Much of the carbon dioxide formed during the reduction of hematite (as well as from decomposition of the limestone) is reduced again to carbon monoxide on contact with unburned coke.

$$CO_2(g) + C(s) \longrightarrow 2\,CO(g)$$

Iron ores generally contain silicate minerals and silicon dioxide. Lime (CaO), formed when limestone is heated, reacts with these materials to produce calcium silicate.

$$SiO_2(s) + CaO(s) \longrightarrow CaSiO_3(\ell)$$

This is an acid-base reaction: CaO is a basic oxide and SiO_2 is an acidic oxide. The calcium silicate, molten at the temperature of the blast furnace and less dense than molten

Think about It

e27.2 Work through a tutorial on the thermodynamics of the extraction of metals from their ores.

iron, floats on the iron. Other non-metal oxides dissolve in this layer and the mixture, called *slag*, is removed.

The molten iron flows down through the furnace and collects at the bottom, where it is tapped off through an opening in the side. This impure iron is called *pig iron*. Usually, pig iron is either brittle or soft (undesirable properties for most uses) due to the presence of non-metal impurities, and may contain as much as 4.5% elemental carbon, 0.3% phosphorus, 0.04 % sulfur, 1.5 % silicon, as well as some other elements. A small amount of pig iron is used to make *cast iron* by removal of most of the impurities, but most is used to make steel.

To make steel, the non-metal impurities in pig iron must be removed. This is most commonly achieved by the *basic oxygen furnace* (Figure 27.10), which removes most of the carbon and all of the phosphorus, sulfur, and silicon. Pure oxygen is blown into the molten pig iron and oxidizes phosphorus to P_4O_{10}, sulfur to SO_2, and carbon to CO_2. These non-metal oxides either escape as gases or react with basic oxides such as CaO that are added to the molten mixture or are used to line the furnace. For example:

$$P_4O_{10}(g) + 6\,CaO(s) \longrightarrow 2\,Ca_3(PO_4)_2(\ell)$$

FIGURE 27.10 Molten iron being poured into a basic oxygen furnace for steel production.

Courtesy of ArcelorMittal

The result is *carbon steel*. Wide ranges of flexibility, hardness, strength, and malleability can be achieved in carbon steel by reheating and cooling in a process called *tempering*. The resulting materials can then be used in a variety of applications. The major disadvantages of carbon steel are that it corrodes easily and that it loses its properties when heated strongly.

Other transition metals, such as chromium, manganese, and nickel, can be added during the steel-making process, forming *alloys* (solid solutions of two or more metals) that have specific physical, chemical, and mechanical properties. One well-known alloy is stainless steel, which contains 18% to 20% Cr and 8% to 12% Ni. Stainless steel is much more resistant to corrosion than carbon steel. Another alloy of iron is Alnico V. Used in loudspeaker magnets because of its permanent magnetism, it contains five elements: Al (8%), Ni (14%), Co (24%), Cu (3%), and Fe (51%).

In the blast furnace, iron oxides are reduced to iron by carbon and carbon monoxide at high temperatures. Many of the impurities are removed in the slag. In preparation for making steel, non-metal impurities are removed by oxidation with oxygen. Other metals can be added to the melt to make alloys such as stainless steels.

Copper Extraction from Ores: Hydrometallurgy

In contrast to iron ores, which are mostly oxides, most copper minerals are sulfides. Copper-bearing minerals include chalcopyrite ($CuFeS_2$), chalcocite (Cu_2S), and covellite (CuS). Because the ores generally have a low content of these minerals, enrichment is necessary. This is carried out by a process known as *flotation*. The ore is finely powdered, a specially designed surfactant is added, and the mixture is agitated in a large tank (Figure 27.11). Compressed air is forced through the mixture, and the surfactant causes the fine copper sulfide particles to attach to air bubbles and be carried to the top as a frothy mixture. The heavier gangue settles to the bottom of the tank, and the mineral-laden froth is skimmed off.

FIGURE 27.11 Enriching copper ore by the flotation process. The fine particles of copper sulfides attach to the air bubbles and rise to the surface. The denser gangue settles to the bottom.

GS International/GetStock.com

Hydrometallurgy can be used to obtain copper from an enriched ore. In one method, enriched chalcopyrite ore is treated with a solution of copper(II) chloride. A reaction ensues that forms solid, insoluble copper(I) chloride (CuCl), which is easily separated from the iron that remains in solution as $Fe^{2+}(aq)$ ions.

$$CuFeS_2(s) + 3\,Cu^{2+}(aq) + 6\,Cl^-(aq) \longrightarrow 4\,CuCl(s) + Fe^{2+}(aq) + 2\,Cl^-(aq) + 2\,S(s)$$

Concentrated aqueous NaCl solution is then added and CuCl dissolves because of the formation of the complex ion $[CuCl_2]^-$.

$$CuCl(s) + Cl^-(aq) \longrightarrow [CuCl_2]^-(aq)$$

Copper(I) compounds are unstable with respect to Cu(0) and Cu(II), and $[CuCl_2]^-$ ions disproportionate to the metal and $Cu^{2+}(aq)$ ions, and the latter are used again to treat more ore.

$$2\,[CuCl_2]^-(aq) \longrightarrow Cu(s) + Cu^{2+}(aq) + 4\,Cl^-(aq)$$

An alternative process called *bio-extraction*, or *bio-leaching* of copper ores is becoming more common is some situations. An acidic solution is sprayed onto heaps of the ore. As the solution trickles down through the crushed rock, the bacterium *Thiobacillus ferrooxidans* breaks down the iron sulfides in the rock and converts iron(II) to iron(III). Iron(III) ions oxidize the sulfide ions of copper sulfide to sulfate ions, leaving copper(II) ions in solution. Then, the copper(II) ions are reduced to metallic copper by reaction with iron.

$$Cu^{2+}(aq) + Fe(s) \longrightarrow Cu(s) + Fe^{2+}(aq)$$

The purity of the copper obtained via these metallurgical processes is about 99%, but this is not acceptable because even traces of impurities greatly diminish the electrical conductivity of the metal. A further purification step involves electrolysis [<<Section 16.7]. Slabs of the impure copper and thin sheets of pure copper metal and are immersed in aqueous sulfuric acid solution. The impure slabs are made the anode of an electrolysis cell, and the pure copper sheets the cathode (Figure 27.12). Copper in the impure sample is oxidized to copper(II) ions at the anode, and copper(II) ions in solution are reduced to pure copper at the cathode. If the applied voltage is carefully controlled, metals more noble than copper in the impure anode (gold, silver) are not oxidized and fall to bottom, while ions of more reactive metals (zinc, iron, lead) are oxidized at the anode, but not reduced at the cathode, and so remain in solution.

Thiobacillus ferrooxidans is an extremophile bacterium that speeds up oxidation of iron sulfides in natural situations, causing acidification of soils and water [<<Section 26.1].

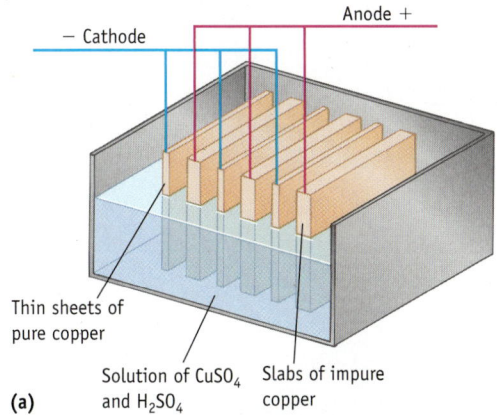

(a) Thin sheets of pure copper — Solution of CuSO₄ and H₂SO₄ — Slabs of impure copper

(b)

FIGURE 27.12 Electrolytic refining of copper. (a) Slabs of impure copper, called "blister copper," form the anode and pure copper is deposited at the cathode. (b) The electrolysis cells at a copper refinery.

Copper sulfide ores are enriched by froth flotation. In one hydrometallurgical process, copper minerals are reduced to the +1 oxidation state as the $[CuCl_2]^-$ ion, which disproportionates to Cu(s) and Cu^{2+}(aq) ions. Copper is also extracted from ores as Cu^{2+}(aq) ions by bioleaching. Small amounts of impurities, both more reactive and less reactive than copper, are removed by electrolysis with control of the applied emf.

27.4 Coordination Compounds

The nature of coordination compounds (complexes) has been introduced previously [<<Section 6.7]. The transition metals and their ions form a wide range of coordination compounds, often with beautiful colours, magnetic properties, and interesting structures. One purpose of this chapter is to explore some commonly observed structures and explain their magnetic properties and the origin of their colours.

Complexes and Ligands

A green solution formed by dissolving nickel(II) chloride in water contains Ni^{2+}(aq) and Cl^-(aq) ions (Figure 27.13). If the solvent water is evaporated off, a green crystalline solid is obtained, whose composition can be represented by the formula $NiCl_2 \cdot 6H_2O$. Addition of ammonia to the aqueous nickel(II) chloride solution gives a lilac-coloured solution from which we can obtain another compound whose composition, $NiCl_2 \cdot 6NH_3$, corresponds with substitution of the water molecules in the hydrate by ammonia molecules.

Molecular Modelling (Odyssey)

e27.3 Examine simulations of some metal complexes on their own, and in solution.

Add NH$_3$

[Ni(H$_2$O)$_6$]$^{2+}$

Add NaOH

[Ni(NH$_3$)$_6$]$^{2+}$

Insoluble Ni(OH)$_2$

Add
ethylenediamine
NH$_2$CH$_2$CH$_2$NH$_2$

Add
dimethylglyoxime
(dmg)

[Ni(NH$_2$CH$_2$CH$_2$NH$_2$)$_3$]$^{2+}$

Ni(dmg)$_2$

Photos: Charles D. Winters

FIGURE 27.13 Coordination complexes of Ni^{2+} ions. Complex ions of the same metal ion with different ligands have different colours.

What are these two nickel compounds? The formulas specify their compositions, but do not give information about their structures. Because properties of compounds derive from their structures, we need to evaluate the structures in more detail. Typically, metal compounds are ionic, and solid ionic compounds have structures with cations and anions arranged in a regular array. The structure of hydrated nickel(II) chloride contains cations with the formula $[Ni(OH_2)_6]^{2+}$ and Cl^- anions. The ammonia-containing compound has $[Ni(NH_3)_6]^{2+}$ cations and Cl^- anions.

Ions such as $[Ni(H_2O)_6]^{2+}$ and $[Ni(NH_3)_6]^{2+}$, in which a metal ion and either water or ammonia molecules comprise a single structural unit (Figure 27.14), are examples of *coordination complexes*, also known as *complex ions* [<<Sections 6.7 and 9.7]. Compounds containing a coordination complex as part of the structure are called *coordination compounds*, and their chemistry is known as *coordination chemistry*. The conventional method of writing the formula for coordination compounds places the metal atom or ion and the ligand molecules or anions directly bonded to it within brackets to show that it is a single entity. Thus, a formula for the hydrated nickel chloride that indicates its structure is $[Ni(OH_2)_6]Cl_2$, and it is named hexaaquanickel(II) chloride. The structural formula of the nickel(II)-ammonia compound (hexamminenickel(II) chloride) is $[Ni(NH_3)_6]Cl_2$.

All coordination complexes contain a metal atom or ion, bonded to which are molecules or ions called *ligands* or *Lewis bases* [<<Section 6.7]. In the preceding examples, water and ammonia are ligands. Ligands are Lewis bases because they furnish the electron pair; the metal ion is a Lewis acid because it accepts electron pairs.

Ligands can be either neutral molecules or anions (or, in rare instances, cations). The characteristic feature of a ligand is that it contains a lone pair of electrons. In the classic description of bonding in a coordination complex, the lone pair of electrons on a ligand is shared with the metal ion. The attachment is sometimes called a *coordinate covalent bond*, although it is indistinguishable from any other covalent bond.

The net charge on a coordination complex is the sum of the charges on the metal and its attached groups. Complexes can be cations (as in the two nickel complexes used as examples above), anions, or uncharged.

The number of donor atoms to which a metal ion is bonded is the **coordination number** of the metal ion. The **coordination geometry** of a complex ion refers to the distribution in space of ligands around the metal atom or ion. In the $[Ni(NH_3)_6]^{2+}$ ion, the six donor N atoms are arranged in a regular octahedral geometry around the central metal ion (Figure 27.14). The structures of some coordination compounds give rise to isomerism corresponding with the types of isomerism in organic compounds [<<Chapter 9], discussed in Section 27.7.

Ligands such as H_2O and NH_3, which coordinate to the metal via one donor atom, are termed **monodentate ligands**. Some ligands attach to the metal with more than one donor atom, and are called **polydentate ligands**. Ethylenediamine (ethane-1,2-diamine), $H_2NCH_2CH_2NH_2$, abbreviated as en; oxalate ion, $C_2O_4^{2-}$ (ox^{2-}); and phenanthroline, $C_{12}H_8N_2$ (phen), are examples of the wide variety of bidentate ligands (Figure 27.15).

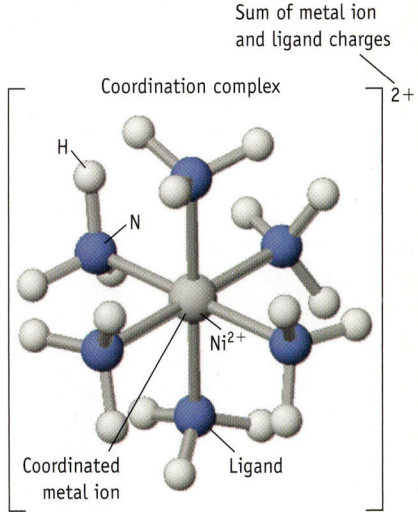

Coordination complex

Sum of metal ion and ligand charges

2+

H

N

Ni^{2+}

Coordinated metal ion

Ligand

$[Ni(NH_3)_6]^{2+}$

FIGURE 27.14 A $[Ni(NH_3)_6]^{2+}$ complex ion. In the formation of this complex ion from a Ni^{2+} ion and NH_3 molecules, the Ni^{2+} ion is a Lewis acid, and each NH_3 molecule is a Lewis base. Because the NH_3 molecules have zero charge, the charge on the complex ion is +2.

Taking It Further
e27.4 Read and apply the nomenclature rules for coordination compounds.

The word *dentate* comes from the Latin *dentis*, meaning "tooth," so an ammonia molecule is a "one-toothed" ligand, while an ethylenediamine molecule is "two-toothed."

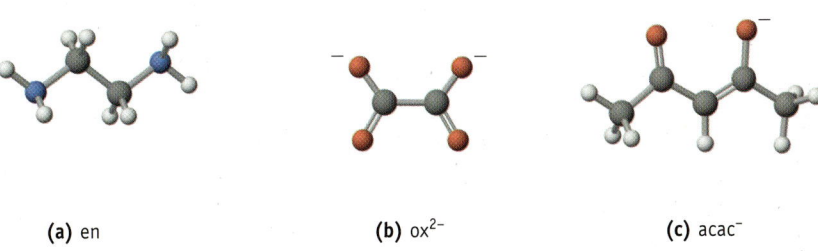

(a) en   **(b)** ox^{2-} **(c)** $acac^-$ **(d)** phen

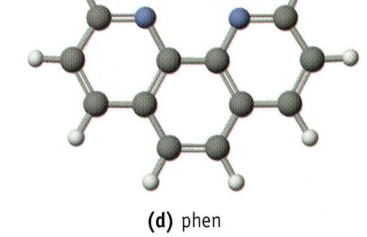

FIGURE 27.15 Common bidentate ligands. (a) Ethylenediamine, $H_2NCH_2CH_2NH_2$; (b) oxalate ion, $C_2O_4^{2-}$ (c) acetylacetonate ion, $CH_3COCHCOCH_3^-$; (d) phenanthroline, $C_{12}H_8N_2$. Coordination of these bidentate ligands to a transition metal ion results in five- or six-membered rings containing the metal ion.

Examples of some coordination compounds whose complex ions have bidentate ligands are shown, along with the structures of the complex ions, in Figure 27.16.

[Fe(C$_2$O$_4$)$_3$]$^{3-}$ [Co(en)$_3$]$^{3+}$ Cr(acac)$_3$

FIGURE 27.16 Complex ions with bidentate ligands. See Figure 27.15 for abbreviations of ligand formulas.

Polydentate ligands are also called **chelating ligands,** or **chelates** (pronounced *key-lates*), deriving from the Greek *chele*, meaning "claw." Complexes with chelating ligands have greater stability than those with monodentate ligands with the same donor atoms, and this is discussed in more detail in Section 27.5.

Chelated complexes are important in everyday life. One way to clean the rust out of water-cooled automobile engines and steam boilers is to add a solution of oxalic acid. Iron oxide reacts with oxalic acid to give a water-soluble iron oxalate complex ion:

$$Fe_2O_3(s) + 6\,H_2C_2O_4(aq) + 6\,H_2O(\ell) \longrightarrow 2\,[Fe(C_2O_4)_3]^{3-}(aq) + 6\,H_3O^+(aq) + 3\,H_2O(\ell)$$

Ethylenediaminetetraacetate ion (EDTA^{4-}), a hexadentate ligand, is an excellent chelating ligand (Figure 27.17). It can wrap around a metal ion, encapsulating it. Because of this property, EDTA is used in treatment of people with metal poisoning, since it binds with a wide range of metal ions and eliminates them from body fluids. Salts of this anion are also often added to commercial salad dressings to remove traces of free metal ions from solution; otherwise, these metal ions act as catalysts for the oxidation of the oils in the dressings, and the dressings quickly becomes rancid. Another use is in bathroom cleansers: EDTA^{4-} ions remove deposits of CaCO$_3$(s) and MgCO$_3$(s) residues from hard water by coordinating to Ca^{2+} or Mg^{2+} ions to create soluble complex ions.

FIGURE 27.17
(a) Ethylenediaminetetraacetate (EDTA^{4-}), a hexadentate ligand. (b) [Co(EDTA)]$^-$. The complex has five- and six-member rings that contain the metal ion.

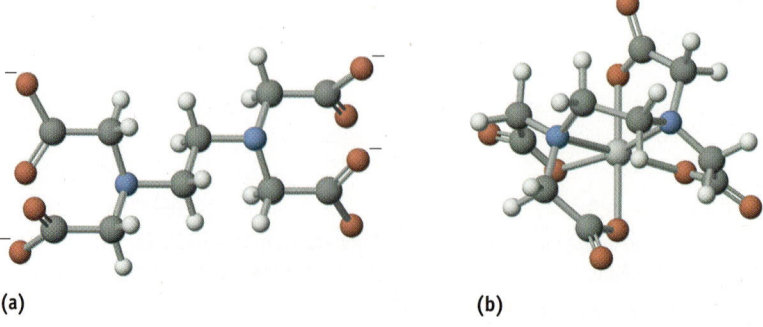

(a)   (b)

Which of the following ligands do you expect to be monodentate and which might be polydentate?
(a) CH_3NH_2 (d) en
(b) CH_3CN (e) Br^-
(c) N_3^- (f) phen

EXERCISE 27.2—LIGANDS

One of the following nitrogen compounds or ions cannot be a ligand: NH_4^+, NH_3, NH_2^-. Identify this species and explain your answer.

In coordination complexes, ligands behave as Lewis bases in providing an electron pair to bond to a metal ion. The bond formed in this way is indistinguishable from any other covalent bond. Ligands are monodentate or polydentate. Polydentate ligands are called chelating ligands. The coordination number of the metal ion is the number of electron pair donor atoms that are bound to it. The coordination geometry refers to the spatial arrangement of ligands around the metal ion.

Hemoglobin

Metal-containing complexes with polydentate ligands play particularly important roles in many biochemical reactions. Perhaps the best-known example is hemoglobin, the chemical in the blood responsible for O_2 transport. It is also one of the most thoroughly studied bioinorganic compounds.

Hemoglobin (Hb) is a large iron-containing protein [<<Section 1.2, >>Section 29.3]. It includes four polypeptide segments, each containing an iron(II) ion locked inside a porphyrin ring system (Figure 27.18) and coordinated to a nitrogen atom from another part of the protein. A sixth site is available to attach to an O atom of O_2 molecules.

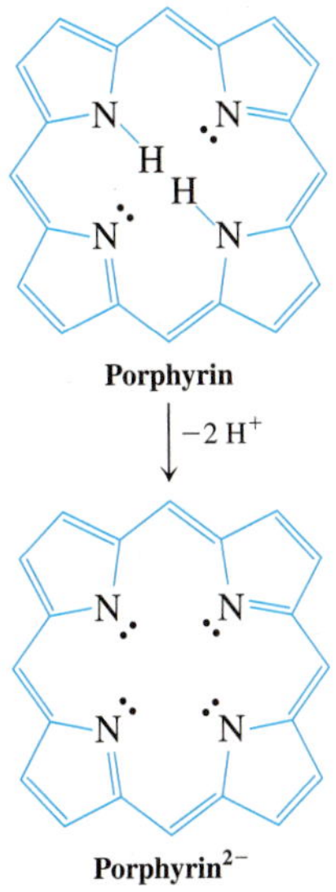

One segment of the hemoglobin molecule resembles the myoglobin structure (Figure 27.19). In this case, the iron-containing heme group is enclosed with a polypeptide chain. The first and sixth coordination positions are taken up by N atoms from amino acids of the polypeptide chain.

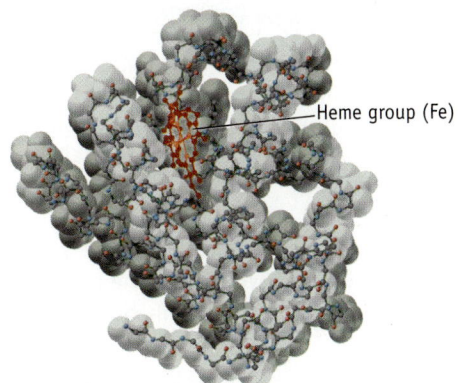

FIGURE 27.19 The heme group in myoglobin. Myoglobin, the oxygen-storage protein in muscle, has only one polypeptide chain with an enclosed heme group. This protein is a close relative of hemoglobin.

FIGURE 27.18 Porphyrin ring of the heme group. The tetradentate ligand surrounding the iron(II) ion in hemoglobin is an anion of a porphyrin. Because of the double bonds in this structure, all of the carbon and nitrogen atoms in the anion of the porphyrin lie in a plane. The molecular dimensions are such that a metal ion fits nicely into the cavity.

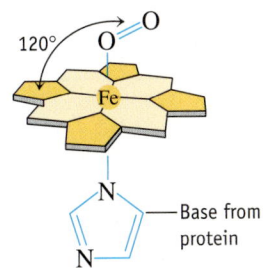

FIGURE 27.20 Oxygen binding. Oxygen binds to the iron of the heme group in hemoglobin (and in myoglobin). Interestingly, the O_2 molecule sits at an angle to the Fe–O bond.

Hemoglobin functions by reversibly adding oxygen to the sixth coordination position of each iron, giving a complex called oxyhemoglobin (Figure 27.20).

Because hemoglobin has four iron centres, a maximum of four O_2 molecules can bind to the molecule. The binding to O_2 molecules is cooperative; that is, binding one molecule enhances the tendency to bind the second, third, and fourth molecules. The equilibrium constant for the reaction binding O_2 molecules to hemoglobin

$$Hb(aq) + O_2(g) \rightleftharpoons HbO_2(aq)$$

is reasonably large. If it were too large, however, the equilibrium constant for displacement of O_2 molecules by CO_2 molecules

$$HbO_2(aq) + CO_2(aq) \rightleftharpoons HbCO_2(aq) + O_2(g)$$

would be too small to allow transport of enough CO_2 back to the lungs.

An increase in acidity (decrease in pH) leads to a decrease in the stability of the oxygenated complex, because of protonation of the Lewis base sites, as discussed further in Section 27.5. As a consequence, release of oxygen in tissues is facilitated by an increase in acidity that results from the presence of CO_2 formed by metabolism.

Among the notable properties of hemoglobin is its ability to form a very stable complex with carbon monoxide.

$$HbO_2(aq) + CO(g) \rightleftharpoons HbCO(aq) + O_2(g) \qquad K = 200$$

When CO complexes with iron, the oxygen-carrying capacity of hemoglobin is lost. Consequently, CO is highly toxic to humans. Exposure to even small amounts greatly reduces the capacity of the blood to transport oxygen. See also the discussion of "blood gas" measurement in Section 16.1.

> Hemoglobin is a protein with four polypeptide segments, each containing an iron(II) ion coordinated to four N atoms in a porphyrin ring system and to a N atom from another part of the protein. Through its ability to coordinate to O_2 molecules at the sixth site on each Fe atom, hemoglobin in blood is able to transport oxygen to tissues.

Formulas of Coordination Compounds

It is useful to be able to deduce the formula of a coordination complex, given the metal ion and ligands, and to derive the oxidation state and coordination number of the coordinated metal ion, given the formula. Worked Examples 27.1 and 27.2 illustrate the procedures.

WORKED EXAMPLE 27.1—FORMULAS OF COORDINATION COMPLEXES

Write the formulas of the following coordination complexes:
(a) a complex with a Ni^{2+} ion bound to two water molecules and two bidentate oxalate ions
(b) a complex with a Co^{3+} ion bound to one Cl^- ion, one ammonia molecule, and two bidentate ethylenediamine (en) molecules

Strategy

The task requires determining the net charge, which equals the sum of the charges of the various component parts of the complex ion. Symbols for the metal and ligands can be assembled in the formula, which is placed in brackets, and the net charge shown outside of the bracket.

Solution

(a) This complex ion is constructed from two neutral H_2O molecules, two $C_2O_4^{2-}$ ions, and one Ni^{2+} ion, so the net charge on the complex is -2. The formula for the complex ion is $[Ni(C_2O_4)_2(H_2O)_2]^{2-}$.

(b) This cobalt(III) complex combines two neutral en molecules and one neutral NH_3 molecule, as well as one Cl^- ion and a Co^{3+} ion. The net charge is $+2$. The formula for this complex is $[Co(H_2NCH_2CH_2NH_2)_2(NH_3)Cl]^{2+}$.

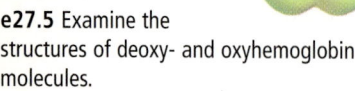

Molecular Modelling (Odyssey)

e27.5 Examine the structures of deoxy- and oxyhemoglobin molecules.

Molecular Modelling (Odyssey)

e27.6 Examine the speciation in solution of different complexes with the same composition, but different ligands and counter ions.

WORKED EXAMPLE 27.2—FORMULAS OF COORDINATION COMPLEXES

From the formula of each of the following complexes, deduce the metal's oxidation state and the coordination number.

(a) $[Co(en)_2(NO_2)_2]Cl$ (c) $[Pt(NH_3)_2Cl_4]$

(b) $[Pt(NH_3)_2(C_2O_4)]$ (d) $[Co(NH_3)_5Cl]SO_4$

Strategy

Each formula consists of a complex ion or molecule made up of the metal ion, neutral and/or anionic ligands (inside the square brackets), and a counterion (outside the brackets). Recall that the algebraic sum of the oxidation states of the atoms in a polyatomic ion is equal to the charge on the ion [<<Section 16.2]. In complex ions, it is easier to deduce the oxidation state of the metal as the charge that it must have to be consistent with the overall charge on the complex ion, recognizing the charge on each ligand molecule or ion.

The coordination number is the number of donor atoms in the ligands that are bonded to the metal. Remember that bidentate ligands (en, oxalate ion) attach to the metal at two sites. The counterion is not part of the complex ion; that is, it is not a Lewis base in this compound.

Solution

(a) The chloride counterion has a −1 charge, so the charge on the complex ion must be +1. There are two nitrite ions (NO_2^-), each with charge −1 and two neutral bidentate ethylenediamine ligands in the complex.

$$\text{Oxidation state of Co} + 2(0) + 2(-1) = +1$$

So the cobalt is in oxidation state +3. Each en ligand forms bonds through lone pairs on each of two N atoms, and each nitrite ion forms a bond from one atom, so the coordination number of the metal is 6.

(b) There is an oxalate ion ($C_2O_4^{2-}$) and two neutral ammonia ligands.

$$\text{Oxidation state of Pt} + 2(0) + 1(-2) = 0$$

The platinum is in oxidation state of +2. The coordination number is 4, with an oxalate ligand having two Lewis base sites, and each ammonia molecule one.

(c) The charge on the complex is zero. There are four chloride ions (Cl^-) and two neutral ammonia ligands.

$$\text{Oxidation state of Pt} + 2(0) + 4(-1) = 0$$

In this complex, the oxidation state of the metal is +4 and, since each of the ligands is monodentate, the coordination number is 6.

(d) The counter ion is sulfate with a −2 charge, so the net charge on the complex ion is +2. There is one chloride ion (Cl^-) and five neutral ammonia ligands.

$$\text{Oxidation state of Co} + 5(0) + 1(-1) = +2$$

The oxidation state of the metal is +3 and the coordination number is 6 (the sulfate ion is not coordinated to the metal).

EXERCISE 27.3—FORMULAS OF COORDINATION COMPLEXES

(a) What is the formula of a complex ion composed of one Co^{3+} ion, three ammonia molecules, and three Cl^- ions?

(b) Deduce the metal's oxidation state and coordination number in (i) $K_3[Co(NO_2)_6]$, and (ii) $[Mn(NH_3)_4Cl_2]$.

The formula of a coordination complex displays the metal and the number and identity of the ligand(s) comprising the complex, and allows us to deduce the oxidation state of the metal and coordination number.

27.5 Complexation Equilibria, Stability of Complexes

So far, we have talked about coordination complexes with the implication that they are stable and form readily. We should, however, realize that complexes differ in their stabilities: some complexes for which we can write a formula may not form, and some may need very high concentrations of metal ions or ligands for appreciable amounts to form. In the case of a given metal ion in aqueous solution along with two different ligands, one of the possible complex ions may be formed to the almost complete exclusion of the other. A similar situation may apply in the case of one type of ligand in solution with two (or more) metal ions. Complexation may be in competition (successful or unsuccessful, depending upon the identities of species and their concentrations) with precipitation reactions [<<Section 15.3], as well as with protonation of ligands that are bases [<<Section 15.5]. Speciation [<<Section 14.7] varies with concentrations of reagents, and the application of mathematics to calculate which species is dominant under specified conditions depends on treating complexation reactions as equilibrium reactions [<<Section 5.5 and Chapter 13], for which equilibrium constants can be measured.

The formation of $[Cu(NH_3)_4]^{2+}$ complex ions in aqueous solution containing aquated $Cu^{2+}(aq)$ ions and ammonia can be represented by the equilibrium equation

$$Cu^{2+}(aq) + 4\,NH_3(aq) \rightleftharpoons [Cu(NH_3)_4]^{2+}$$

or, better still because it shows the substitution of one Lewis base (water molecules) by another (ammonia molecules),

$$[Cu(OH_2)_6]^{2+}(aq) + 4\,NH_3(aq) \rightleftharpoons [Cu(NH_3)_4]^{2+} + 6\,H_2O(\ell)$$

As for any other equilibrium reaction, we can quantitatively express the extent to which this reaction proceeds in specified conditions through an equilibrium constant.

Formation Constants of Complexes

The equilibrium constant for a reaction in which a metal ion complex is formed from the aqua complex of the metal ion is called the **formation constant (β),** or the *stability constant*. In the case of the above reaction forming tetraamminecopper(II) ions from hexaaquacopper(II) ions, at 25 °C the formation constant is 3.3×10^{12}:

$$Q = \frac{[Cu(NH_3)_4{}^{2+}]}{[Cu(OH_2)_6{}^{2+}][NH_3]^4} = \beta = 3.3 \times 10^{12} \text{ at } 25\,^{\circ}C$$

One way to interpret this formation constant is to show, by transposing one term in the reaction quotient, that the ratio of concentrations of the ammine complex and the aqua complex is dependent on the concentration of ammonia. At 25 °C,

$$\frac{[Cu(NH_3)_4{}^{2+}]}{[Cu(OH_2)_6{}^{2+}]} = \beta \times [NH_3]^4 = (3.3 \times 10^{12}) \times [NH_3]^4$$

Applying this relationship to a solution at 25 °C in which the reaction producing tetraamminecopper(II) ions from hexaaquacopper(II) ions is at equilibrium when $[NH_3] = 1.0 \times 10^{-3}$ mol L^{-1}:

$$\frac{[Cu(NH_3)_4{}^{2+}]}{[Cu(OH_2)_6{}^{2+}]} = \beta \times [NH_3]^4 = (3.3 \times 10^{12}) \times (1.0 \times 10^{-3})^4 = 3.3$$

We can see that under these conditions more of the copper is present in the ammine complex ion than in the aqua complex ion.

We should expect that at higher $NH_3(aq)$ concentrations, more of the water ligands are replaced by ammonia ligands. This is demonstrated by the higher ratio calculated at, for example, $[NH_3] = 1.0$ mol L^{-1}:

$$\frac{[Cu(NH_3)_4{}^{2+}]}{[Cu(OH_2)_6{}^{2+}]} = \beta \times [NH_3]^4 = (3.3 \times 10^{12}) \times (1.0)^4 = 3.3 \times 10^{12}$$

Square brackets are usually used in formulas to identify the bound species in coordination complexes, with the charge on the complex indicated outside the brackets. However, square brackets are also used as the symbol for concentration of species in solution. In this and the following equations, to avoid confusion the square brackets are used only to mean concentration, and they have been eliminated from the formulas of the complex ions.

Table 27.3 lists the formation constants at 25 °C of selected metal ion complexes with monodentate ligands. The larger the formation constant is, the more stable the complex is compared with the aqua complex of the metal ion. In practical terms, this means that the larger the formation constant is, the greater the ratio of concentrations of the complex ion to the aqua complex is at any specified concentration of the ligand.

From the data in Table 27.3, we can make the following generalizations:

- The stabilities of complex ions (compared with the aqua complex of the same metal ion) differ considerably from metal to metal—even for complexes with the same number of the same ligand, such as $[Ag(CN)_2]^-$ and $[Au(CN)_2]^-$, or $[Cu(NH_3)_4]^{2+}$ and $[Zn(NH_3)_4]^{2+}$.
- For complexes of a given metal with a given ligand, the higher the charge on the metal ion, the more stable the complex is. This can be seen in a comparison of the formation constants of $[Fe(CN)_6]^{4-}$ and $[Fe(CN)_6]^{3-}$, $[Co(en)_3]^{2+}$ and $[Co(en)_3]^{3+}$, and $[Fe(ox)_3]^{4-}$ and $[Fe(ox)_3]^{3-}$. This can be rationalized in terms of the higher charge density [<<Section 26.3] of the Fe^{3+} ion compared with that of the Fe^{2+} ion, and a consequent higher electrostatic force between the metal ion and the ligands.
- Although the cyanide ligand (CN^-) is sometimes called a pseudo-halide, cyano complexes are more stable than corresponding halide complexes of the same metal. This is illustrated by comparisons of, for example, the formation constants of $[Ag(CN)_2]^-$ and $[AgCl_2]^-$, or those of $[Cd(CN)_4]^{2-}$ and $[CdCl_4]^{2-}$.

> The equilibrium constant for a reaction in which a metal ion complex is formed from the aqua complex of the metal ion is called the formation constant. The larger the formation constant, the more stable the complex is. Complexes of a given metal are generally more stable the higher the oxidation state of the metal ion is.

Speciation among Complex Ions

We have seen that polyprotic acids ionize in a series of overlapping steps, and that the equilibrium constants allow us to calculate the relative amounts of the various species as $[H^+]$ is changed, and so produce a speciation plot [<<Section 14.7]. In a corresponding way, complexation of metal ions by substitution of ligands for water occurs to form a variety of species whose relative concentrations are such that the equilibrium constants for a series of reactions are simultaneously satisfied. The reaction forming tetraamminecopper(II) ions from hexaaquacopper(II) ions, for example, can be imagined to occur in the following series of overlapping steps:

$$[Cu(OH_2)_6]^{2+}(aq) + NH_3(aq) \rightleftharpoons [Cu(OH_2)_5(NH_3)]^{2+} + H_2O(\ell) \qquad K_{f1} = 1.5 \times 10^4$$

$$[Cu(OH_2)_5(NH_3)]^{2+} + NH_3(aq) \rightleftharpoons [Cu(OH_2)_4(NH_3)_2]^{2+} + H_2O(\ell) \qquad K_{f2} = 3.3 \times 10^3$$

$$[Cu(OH_2)_4(NH_3)_2]^{2+} + NH_3(aq) \rightleftharpoons [Cu(OH_2)_3(NH_3)_3]^{2+} + H_2O(\ell) \qquad K_{f3} = 5.8 \times 10^2$$

$$[Cu(OH_2)_3(NH_3)_3]^{2+} + NH_3(aq) \rightleftharpoons [Cu(NH_3)_4]^{2+} + 3\,H_2O(\ell) \qquad K_{f4} = 1.1 \times 10^2$$

The sum of these step-wise equations is that for the overall equation

$$[Cu(OH_2)_6]^{2+}(aq) + 4\,NH_3(aq) \rightleftharpoons [Cu(NH_3)_4]^{2+} + 6\,H_2O(\ell)$$

and the product of the step-wise formation constants is, allowing for rounding off errors, the same as the overall formation constant, $\beta = K_{f1} \times K_{f2} \times K_{f3} \times K_{f4} = 3.3 \times 10^{12}$ (at 25 °C).

Using the separate step-wise equilibrium constants, we can calculate the relative concentrations of any two of the complex species in a solution in which $[NH_3]$ is known. For example, using the first step, we can calculate the ratio of $[Cu(OH_2)_6{}^{2+}]$ to $[Cu(OH_2)_5(NH_3)^{2+}]$ in any solution in which $[NH_3] = 1.0 \times 10^{-2}$ mol L^{-1}:

$$K_{f1} = \frac{[Cu(OH_2)_5(NH_3)^{2+}]}{[Cu(OH_2)_6{}^{2+}][NH_3]} = \frac{[Cu(OH_2)_5(NH_3)^{2+}]}{[Cu(OH_2)_6{}^{2+}](1.0 \times 10^{-2})} = 1.5 \times 10^4$$

$$\therefore \quad \frac{[Cu(OH_2)_5(NH_3)^{2+}]}{[Cu(OH_2)_6{}^{2+}]} = (1.5 \times 10^4)(1.0 \times 10^{-2}) = 1.5 \times 10^2$$

TABLE 27.3 Formation Constants of Some Metal Ion Complexes at 25 °C

Complex	β
$[Ag(NH_3)_2]^+$	1.6×10^7
$[Cd(NH_3)_4]^{2+}$	1.0×10^7
$[Cu(NH_3)_4]^{2+}$	1.0×10^{13}
$[Ni(NH_3)_6]^{2+}$	1.0×10^9
$[Zn(NH_3)_4]^{2+}$	2.9×10^9
$[AgCl_2]^-$	1.1×10^5
$[CdCl_4]^{2-}$	1.1×10^2
$[PdCl_4]^{2-}$	1.7×10^{13}
$[CdI_4]^{2-}$	1.3×10^6
$[Ag(CN)_2]^-$	1.0×10^{22}
$[Cd(CN)_4]^{2-}$	1.2×10^{17}
$[Au(CN)_2]^-$	2.0×10^{38}
$[Fe(CN)_6]^{4-}$	7.7×10^{36}
$[Fe(CN)_6]^{3-}$	7.7×10^{43}
$[Zn(OH)_4]^{2-}$	2.0×10^{20}
$[Cd(en)_2]^{2+}$	3.8×10^{10}
$[Co(en)_3]^{2+}$	6.6×10^{13}
$[Co(en)_3]^{3+}$	4.9×10^{48}
$[Fe(ox)_3]^{4-}$	1.7×10^5
$[Fe(ox)_3]^{3-}$	3.3×10^{20}

We carry out such calculations for all species, over a range of values of [NH₃] to produce a speciation plot. Figure 27.21 is such a plot, shown in the format commonly used by chemists who work in this area. Just as acid-base speciation plots [<<Section 14.7] show the relative amounts of species with change of pH, where $pH = -\log_{10}[H^+]$, speciation of complex ions is graphed against $p(ligand) = -\log_{10}[ligand]$. And usually the values of p(ligand) on the horizontal axis increase from left to right; that is, the concentration of the ligand decreases from left to right. Figure 27.21 shows how changes of p(NH₃) brings about changing relative concentrations of species in solutions at pH 12, at which essentially all of the ammonia is present as NH₃ (and a negligible fraction of it is protonated to NH₄⁺).

FIGURE 27.21 Distribution among Cu²⁺-ammine species in a solution at pH 12 as [NH₃] is increased from 1×10^{-4} mol L⁻¹ (p(NH₃) = 4, on right side) to 1×10^{-2} mol L⁻¹ (p(NH₃) = 2, on left side). As [NH₃] is increased, the relative amounts of the more highly substituted complexes increase.

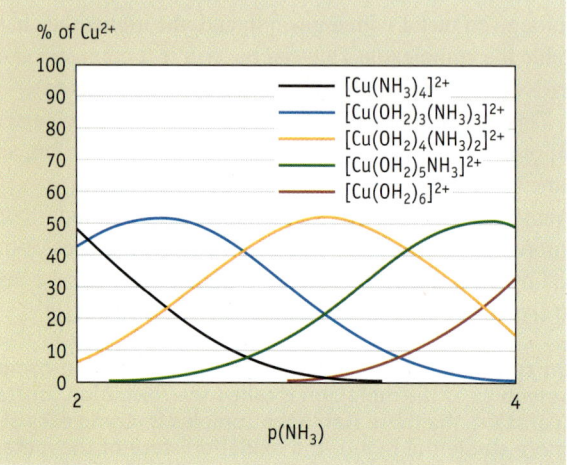

Many of the ligands that form complex ions with metals are weak bases. In these cases, as the pH of solution is decreased, in a particular pH range (near pK_a of the conjugate acid) the ligands are protonated, their ability to act as Lewis bases is eliminated, and they are displaced by water molecules to form the aqua complex. The speciation plot in Figure 27.22 illustrates this for copper-ammine complex ions.

FIGURE 27.22 pH dependence of the relative distribution amongst Cu²⁺-ammine species. Modelling has been carried out here for solutions containing 1×10^{-3} mol L⁻¹ Cu²⁺ ions and 5×10^{-3} mol L⁻¹ NH₃.

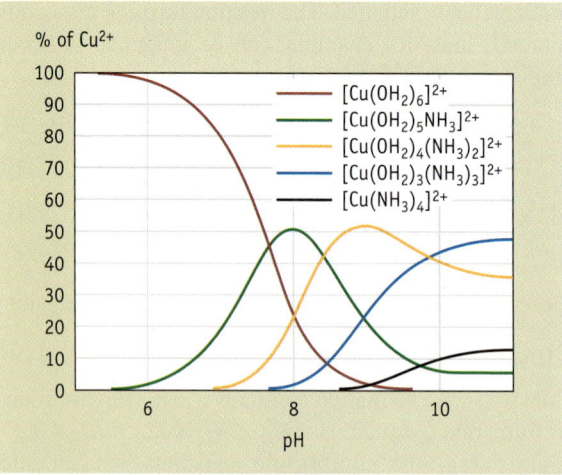

The ammonium ion (NH₄⁺), the conjugate acid of the Lewis base NH₃, has $pK_a = 9.47$. Consequently, at pH = 9.47, [NH₄⁺] = [NH₃], and at pH = 7.47, [NH₄⁺] = 100 × [NH₃]. Since only about 1% is present as NH₃ at pH 7.5, you might expect the concentration of all Cu²⁺-ammine species to be negligible at this, and lower, pH values. To expect this would be to ignore that Cu²⁺ ions "compete" with the H⁺ ions to bond to the Lewis base molecules. The H⁺ ions are not able to protonate the NH₃ molecules until the pH is lower than it needs to be in the absence of the metal ions.

EXERCISE 27.4—SPECIATION AMONG COMPLEX IONS

The first two steps of complexation of Zn^{2+} ions with cyanide ions in aqueous solution can be represented as follows:

$$[Zn(OH_2)_6]^{2+} + CN^-(aq) \rightleftharpoons [Zn(OH_2)_5(CN)]^+ + H_2O(\ell) \qquad K_{f1} = 5.0 \times 10^5$$

$$[Zn(OH_2)_5(CN)]^+ + CN^-(aq) \rightleftharpoons [Zn(OH_2)_4(CN)_2] + H_2O(\ell) \qquad K_{f2} = 2.5 \times 10^5$$

(a) At what $[CN^-]$ is the concentration of $[Zn(OH_2)_5(CN)]^+$ the same as the concentration of $[Zn(OH_2)_6]^{2+}$?

(b) At what $[CN^-]$ is the concentration of $[Zn(OH_2)_4(CN)_2]$ the same as the concentration of $[Zn(OH_2)_5(CN)]^+$?

(c) What is the ratio of concentrations of $[Zn(OH_2)_6]^{2+}$ and $[Zn(OH_2)_5(CN)]^+$ when $[CN^-] = 0.10$ mol L^{-1}?

The distribution of a metal ion among various complexes formed with a ligand depends upon the equilibrium constants of each of a series of step-wise reactions. Using the step-wise equilibrium constants, we can calculate the relative concentrations of any two of the complex species in a solution in which the ligand concentration is known. In this way, we can calculate speciation plots that indicate the relative concentrations of the various species as the ligand concentration is changed at a defined pH. An alternative speciation plot shows the pH-dependence of the distribution among species as pH is changed in a solution of given concentration of metal ion and ligand. Ligands that are weak bases lose their Lewis base ability when they are protonated.

Chelate Effect

A generally observed phenomenon, called the **chelate effect**, is that chelates (complex ions with polydentate ligands) are more stable than complexes that have the same number and type of donor atoms in monodentate ligands. For example, the $[Cd(en)_2]^{2+}$ and $[Cd(NH_3)_4]^{2+}$ complex ions both have a Cd^{2+} ion bound to four N atoms—in one case in two bidentate ethylenediamine ligand molecules, and in the other in four ammonia ligand molecules. We see from Table 27.3 that the $[Cd(en)_2]^{2+}$ complex ion is more stable than the $[Cd(NH_3)_4]^{2+}$ ion.

Let's compare the reactions to which the formation constants apply:

$$[Cd(OH_2)_4]^{2+} + 4\,NH_3(aq) \rightleftharpoons [Cd(NH_3)_4]^{2+} + 4\,H_2O(\ell) \qquad \beta = 1.0 \times 10^7$$

$$[Cd(OH_2)_4]^{2+} + 2\,en(aq) \rightleftharpoons [Cd(en)_2]^{2+} + 4\,H_2O(\ell) \qquad \beta = 3.8 \times 10^{10}$$

The difference in stabilities is mainly due to a difference in entropy change accompanying these two reactions, as can be seen by inspection of the thermodynamic data at 25 °C.

	$\Delta_r H°$ (kJ mol^{-1})	$\Delta_r S°$ (J K^{-1} mol^{-1})	$\Delta_r G°$ (kJ mol^{-1})
Formation of $[Cd(NH_3)_4]^{2+}$	−53.1	−35.5	−42.5
Formation of $[Cd(en)_2]^{2+}$	−56.5	+13.8	−60.7

The standard enthalpy change of reaction ($\Delta_r H°$) for these two reactions is similar, as we might expect since each involves the breaking of four Cd—O bonds, and the formation of four Cd—N bonds. The more negative value of $\Delta_r G°$ for the reaction forming $[Cd(en)_2]^{2+}$, and the consequent larger equilibrium constant, can be attributed almost entirely to the greater standard entropy change of reaction. This seems reasonable when we see from the chemical equations that during formation of $[Cd(NH_3)_4]^{2+}$ ions, five particles react to form five particles of products, while in the formation of the chelate, five particles of products are formed from only three particles of reactants.

A simplified view of the chelate effect, using the examples above, is to view the formation of the complexes as competition for the metal ions between the donor atoms of the ligands and water molecules. Remember that the equilibrium condition involves a dynamic condition with reactions in both directions. If a $Cd-N$ bond in a $[Cd(NH_3)_4]^{2+}$ ion breaks, and an NH_3 molecule is replaced by a water molecule to form a $[Cd(NH_3)_3(OH_2)]^{2+}$ ion, the likelihood of re-formation of the $[Cd(NH_3)_4]^{2+}$ complex ion depends partly on the probability of a collision of an NH_3 molecule in solution with particular orientation and sufficient energy. If a $Cd-N$ bond breaks in a $[Cd(en)_2]^{2+}$ ion, one N atom of the $H_2NCH_2CH_2NH_2$ ligand is still attached, and the "free" N atom is more likely to find its way back to the Cd^{2+} ion.

The chelating effect is increased further if the chelating ligand molecules have three (or four or six) donor atoms. For example, the triamine $H_2NCH_2CH_2(NH)CH_2CH_2NH_2$ (abbreviation *trien*) with three donor atoms forms even more stable complexes than the en ligand. The formation constants of some complexes of the Co^{3+} ion with six Co–N bonds are in the following order:

$$[Co(trien)_2]^{3+} > [Co(en)_3]^{3+} > [Co(NH_3)_6]^{3+}$$

EXERCISE 27.5—THE CHELATE EFFECT

The chelate effect is illustrated by the thermodynamics of the following reaction involving substitution of ligands in aqueous solution at 25 °C:

$$[Ni(NH_3)_6]^{2+} + 3\ en(aq) \rightleftharpoons [Ni(en)_3]^{2+} + 6\ NH_3(aq)$$

For this reaction, $\Delta_r H° = -12.1$ kJ mol^{-1} and $\Delta_r S° = +185$ J K^{-1} mol^{-1}. Calculate (a) the value of $\Delta_r G°$, and (b) K for this reaction. Comment on the contribution of the entropy change of this substitution reaction to the magnitude of $\Delta_r G°$.

The chelate effect refers to the observation that complex ions of a given metal ion with polydentate ligands are more stable than complexes that have the same number and type of donor atoms in monodentate ligands. This is largely due to differences in entropy of formation of the complexes.

Stability and Lability of Complexes

The point has previously been made [<<Section 17.1] that thermodynamics and kinetics are independent of each other, and the same necessarily applies to formation of complexes. The thermodynamic stability of a complex refers to the equilibrium constant for its formation from an aqua complex, but this is unrelated to the rate of substitution of ligands. Complexes that undergo rapid substitution of ligand species are said to be **labile**, and those in which substitution is slow are **inert**. These concepts are well illustrated by reactions of some cyano complexes.

The formation constant for $[Ni(CN)_4]^{2-}$ ion is 1×10^{30}. This is large, and tells us that at equilibrium in a solution with $[CN^-] = 1$ mol L^{-1}, the concentration of this cyano complex ion is 10^{30} times greater than the concentration of aqua complex ions. Even so, there is rapid exchange between the CN^- ions in the complex ions and those in solution: half of the CN^- ions in the complexes are exchanged about every 30 s. Evidence of this *lability* has been obtained by using cyanide ions with a radioactive carbon atom, and tracking the distribution of the radioactivity between the complex ions and the solution.

$$[Ni(CN)_4]^{2-} + 4\ C^*N^-(aq) \underset{t_{1/2} = 30\,s}{\rightleftharpoons} [Ni(C^*N)_4]^{2-} + 4\ CN^-(aq)$$

By contrast, an inert complex has a slow rate of ligand substitution:

$$[Cr(CN)_6]^{3-} + 6\ C^*N^-(aq) \underset{t_{1/2} = 24\,d}{\rightleftharpoons} [Cr(C^*N)_6]^{3-} + 6\ CN^-(aq)$$

Both of the isotope substitution reactions above have equilibrium constants of 1, although one is fast, and one is slow. There are many reactions in which formation of a more stable complex as a result of ligand substitution (spontaneous reactions) is very slow. Consider the substitution of NH_3 ligand molecules in the cobalt(III) complex ion $[Co(NH_3)_6]^{3+}$ by water molecules in acidic aqueous solution:

$$[Co(NH_3)_6]^{3+} + 6 \, H_3O^+(aq) \rightleftharpoons [Co(OH_2)_6]^{3+} + 6 \, NH_4^+(aq)$$

This reaction has a very large equilibrium constant of about 1×10^{64} (so when all species are at 1 mol L^{-1} the reaction is spontaneous in the direction written), and yet it is so slow that the $[Co(NH_3)_6]^{3+}$ ions remain in solution for weeks. The $[Co(NH_3)_6]^{3+}$ ions are thermodynamically unstable (in acidic solution), but kinetically inert.

On the other hand, although the corresponding complex of Co^{2+} ions is also thermodynamically unstable in acidic solution, it is kinetically labile: ligand substitution happens in a matter of seconds.

$$[Co(NH_3)_6]^{2+} + 6 \, H_3O^+(aq) \rightleftharpoons [Co(OH_2)_6]^{2+} + 6 \, NH_4^+(aq)$$

An explanation of the instability of $[Co(NH_3)_6]^{3+}$ ions in acidic solution is that in acidic solutions NH_3 molecules are essentially all protonated to form NH_4^+ ions, which, because they do not have a "lone pair" of electrons, cannot bind to the Co^{3+} ions. Those that are bound before the solution is acidified may be slow to detach, but once detached they are quickly protonated.

> The equilibrium constant for formation of a metal ion complex from the aqua complex is a measure of its thermodynamic stability. This is unrelated to its lability, or rate of substitution of ligands. Complexes that undergo rapid substitution of ligand species are said to be labile, and those in which substitution is slow are inert.

27.6 Structures of Coordination Complexes

The three-dimensional shape of a coordination complex is defined by the spatial arrangement of bonds from the donor atoms of the ligands around the central metal ion. Metal ions in coordination compounds can have coordination numbers ranging from 2 to 12. Only complexes with coordination numbers of 2, 4, and 6 are common, however, and we will concentrate on these.

Common Three-Dimensional Shapes of Complexes

Among the complex species with coordination numbers of 2, 4, or 6, the following shapes are encountered (Figure 27.23):

- All complexes with monodentate ligands and coordination number 2 are linear. The two ligands are on opposite sides of the metal. Common examples include $[Ag(NH_3)_2]^+$ and $[CuCl_2]^-$.
- Many complexes with coordination number 4 are tetrahedral. Examples with monodentate ligands include $TiCl_4$, $[CoCl_4]^{2-}$, $[NiCl_4]^{2-}$, and $[Zn(NH_3)_4]^{2+}$.
- Some complexes with coordination number 4 are square-planar. This shape is most often the case with metal ions that have eight d electrons. Examples include $Pt(NH_3)_2Cl_2$, $[Ni(CN)_4]^{2-}$, and the nickel complex with the dimethylglyoximate (dmg^-) ligand in Figure 27.13.
- Complexes with coordination number 6 are octahedral.

> Common spatial orientations of donor atoms around a metal ion include linear in complexes with two donor atoms, square-planar or tetrahedral in complexes with four donor atoms, and octahedral in complexes with six donor atoms.

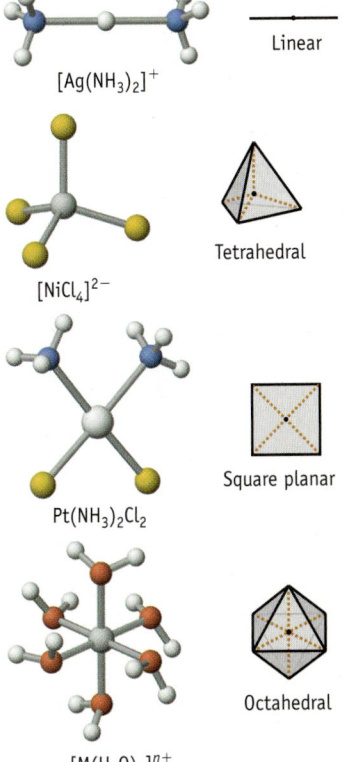

$[Ag(NH_3)_2]^+$ Linear

$[NiCl_4]^{2-}$ Tetrahedral

$Pt(NH_3)_2Cl_2$ Square planar

$[M(H_2O)_6]^{n+}$ Octahedral

FIGURE 27.23 Common three-dimensional geometries of coordination complexes. For each geometry type, an example of a complex with this geometry is shown.

Think about It

e27.7 Work through a tutorial on the shape of complexes.

27.7 Isomerism in Coordination Complexes

Isomers are compounds whose molecules or ions have the same number of atoms of each element but have some difference of arrangement that gives rise to different properties of the isomers. Most of the types of isomerism already referred to in organic chemistry [<<Sections 3.8, 4.5, and 9.6] apply to coordination complexes. As in organic chemistry, isomerism can be broadly classified here into two types:

- *Constitutional isomerism* (sometimes called *structural isomerism*) occurs when molecules of the isomers have different connectivities—that is, a different sequence of atoms that are bonded to each other.
- *Stereoisomerism* is when molecules of the isomers have the same atom-to-atom connectivity, but the atoms differ in their spatial arrangement in space.

Constitutional Isomerism

The most important types of constitutional isomerism among coordination compounds are ionization isomerism and linkage isomerism.

Ionization isomerism occurs when it is possible to exchange a coordinated ligand and an un-coordinated counterion. For example, dark violet $[Co(NH_3)_5Br]SO_4$ and red $[Co(NH_3)_5SO_4]Br$ are ionization isomers. In the violet compound, bromide ion is a ligand and sulfate is the counterion, while in the other, sulfate is a ligand and bromide is the counterion. A diagnostic test for ionization isomers can be based on chemical reactions of ions in solutions of the isomers. For example, addition of $Ag^+(aq)$ ions to a solution of $[Co(NH_3)_5SO_4]Br$ forms a precipitate of $AgBr(s)$, indicating the presence of bromide ions in solution. In contrast, no precipitate forms if $Ag^+(aq)$ ions are added at low concentration to a solution of $[Co(NH_3)_5Br]SO_4$. In this latter complex, bromide ions are bound to the Co^{3+} ion and are not free ions in solution.

$$[Co(NH_3)_5SO_4]Br + Ag^+(aq) \longrightarrow AgBr(s) + [Co(NH_3)_5SO_4]^{2+}(aq)$$

$$[Co(NH_3)_5Br]SO_4 + Ag^+(aq) \longrightarrow \text{no reaction}$$

On the other hand, a $BaSO_4(s)$ precipitate forms on addition of a barium chloride solution to an aqueous solution of $[Co(NH_3)_5Br]SO_4$, but not on addition to a solution of $[Co(NH_3)_5SO_4]Br$.

These tests demonstrate that these coordination compounds, like simple salts, separate into aquated cations and anions when they dissolve in water.

$$[Co(NH_3)_5SO_4]Br \xrightarrow{H_2O(\ell)} [Co(NH_3)_5SO_4]^+(aq) + Br^-(aq)$$

$$[Co(NH_3)_5Br]SO_4 \xrightarrow{H_2O(\ell)} [Co(NH_3)_5Br]^{2+}(aq) + SO_4^{2-}(aq)$$

A particular type of ionization isomerism is *hydration isomerism*, in which the isomers differ as a result of interchange of water molecules between the coordinating ligands and waters of crystallization external to the coordination sphere. For example, compounds with the formulas shown below are hydration isomers.

$$[CrCl(OH_2)_5]Cl_2 . H_2O(s) \qquad \text{blue-green}$$

$$[CrCl_2(OH_2)_4]Cl . 2H_2O(s) \qquad \text{green}$$

Linkage isomerism occurs when a ligand can bond to the metal through different types of donor atoms. The two most common ligands with which linkage isomerism arises are thiocyanate (SCN^-) and nitrite (NO_2^-). The Lewis structure of the thiocyanate ion shows that there are lone pairs of electrons on sulfur and nitrogen. A thiocyanate ion can attach to a metal either through a sulfur atom (S-bonded thiocyanate) or through a nitrogen atom (N-bonded thiocyanate). A nitrite ion can bond by lone-pair donation either from an oxygen atom or from its nitrogen atom. The former are called nitrito complexes, and the latter are nitro complexes (Figure 27.24).

Ligands forming linkage isomers

Bind to metal ion
using either lone pair

Bind to metal ion
using either lone
pair

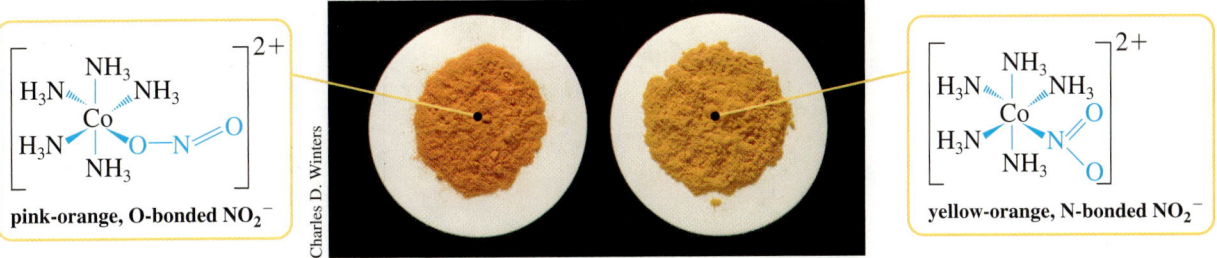

pink-orange, O-bonded NO_2^-

yellow-orange, N-bonded NO_2^-

Charles D. Winters

FIGURE 27.24 Linkage isomers, $[Co(NH_3)_5ONO]^{2+}$ and $[Co(NH_3)_5NO_2]^{2+}$. These complexes, whose systematic names are the pentaamminenitritocobalt(III) ion and the pentaamminenitrocobalt(III) ion, were the first known examples of this type of isomerism.

EXERCISE 27.6—CONSTITUTIONAL ISOMERISM IN TRANSITION METAL COMPLEXES

Excess silver nitrate solution is added to a solution containing 1.0 mol of $[Co(NH_3)_4Cl_2]Cl$. What amount of AgCl (in mol) will precipitate?

EXERCISE 27.7—CONSTITUTIONAL ISOMERISM IN TRANSITION METAL COMPLEXES

Two different coordination compounds containing one cobalt(III) ion, five ammonia molecules, one bromide ion, and one sulfate ion exist. The dark violet form (A) gives a precipitate upon addition of aqueous $BaCl_2$ solution. No reaction is seen upon addition of aqueous $BaCl_2$ solution to the violet-red form (B). Suggest structures for these two compounds, and write a chemical equation for the reaction of (A) with aqueous $BaCl_2$ solution.

In ionization isomers, ligand(s) and counterion(s) are interchanged. Linkage isomerism is possible when a ligand can bond to a metal ion through lone pairs on different atoms.

Stereoisomerism

Complex ions that are stereoisomers have the same ligands bonded directly to the same metal ion, but with different spatial arrangements around the metal ion. By and large, the types and characteristics of stereoisomerism are similar to those in organic compounds, in which there are different spatial arrangements of atoms or groups around carbon atoms [<<Section 9.6]. Because the formulas of coordination complexes indicate connectivity, the formulas of constitutional isomers are different, while those of stereoisomers are the same. The main classes of stereoisomers in complexes are *cis–trans* isomers, *fac–mer* isomers, and enantiomers.

Cis–trans **isomerism** in complexes corresponds with that in organic chemistry [<<Section 9.6], in the sense that two identical ligands are adjacent to each other in *cis* isomers, and on opposite sides of the metal ion in *trans* isomers. This class of isomerism occurs in both square-planar and octahedral complexes, but cannot occur in a tetrahedral complex because each metal–ligand bond is adjacent to all of the other three metal–ligand bonds.

An example of *cis–trans* isomerism in a square-planar complex occurs with the compounds whose formula is $[Pt(NH_3)_2Cl_2]$, of which the *cis* isomer is called *cis-platin*. In the *cis* isomer, the two Cl^- ion ligands are adjacent, and in the *trans* isomer, they are on opposite sides of the metal ion (Figure 27.25a). The *cis* isomer is effective in the treatment of testicular, ovarian, bladder, and osteogenic sarcoma cancers, but the *trans* isomer has no effect on these diseases.

Cis–trans isomerism in an octahedral complex is illustrated by $[Co(H_2NCH_2CH_2NH_2)_2Cl_2]^+$, a complex ion with two bidentate ethylenediamine ligands and two chloride ligands. In the purple *cis* isomer, the two Cl^- ion ligands are adjacent, while in the green *trans* isomer they are opposite (Figure 27.25b).

Molecular Modelling (Odyssey)

e27.8 Examine models illustrating different types of stereoisomers of metal complexes.

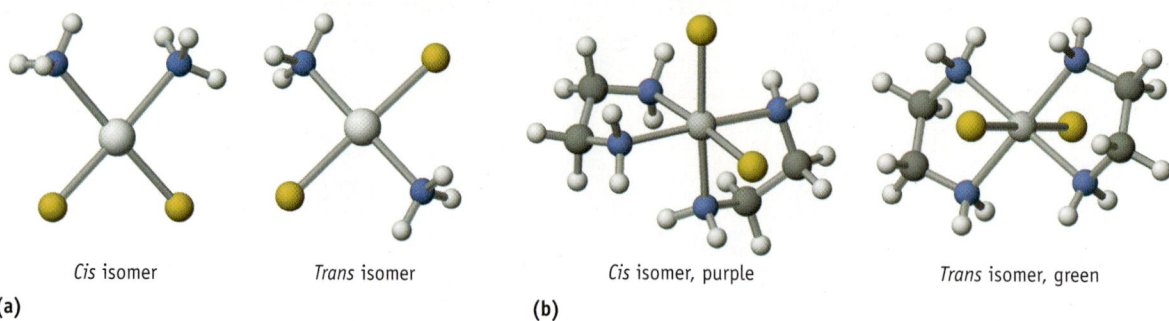

FIGURE 27.25 *Cis–trans* isomers. (a) There are *cis* and *trans* isomers of the square-planar complex [Pt(NH₃)₂Cl₂]. (b) Similarly, there are *cis* and *trans* isomers of the octahedral [Co(en)₂Cl₂]⁺ complex ion.

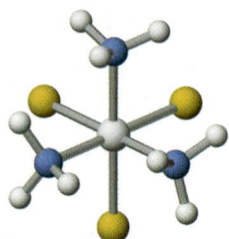

fac isomer

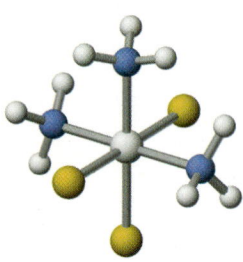

mer isomer

FIGURE 27.26 *Fac* and *mer* isomers of [Cr(NH₃)₃Cl₃]. In the *fac* isomer, the three chloride ligands (and the three ammonia ligands) are arranged at the corners of a triangular face. In the *mer* isomer, the three similar ligands are at sites on a meridian.

***Fac–mer* isomerism** occurs in octahedral complexes with the general formula MX_3Y_3, where X and Y are ligands. A *fac* isomer has three identical ligands at the corners of a triangular face of an octahedron defined by the donor atoms (*fac* = facial), whereas the ligands are situated on a meridian in the *mer* isomer (*mer* = meridional). *Fac* and *mer* isomers of the [Cr(NH₃)₃Cl₃] complex are shown in Figure 27.26.

EXERCISE 27.8—STEREOISOMERS OF COORDINATION COMPLEXES

Draw all possible stereoisomers of the following complexes:
(a) [Fe(NH₃)₄Cl₂]
(b) [Pt(NH₃)₂(SCN)(Br)] in which SCN⁻ is bonded to Pt²⁺ through S
(c) [Co(NH₃)₃(NO₂)₃] in which NO₂⁻ is bonded to Co³⁺ through N
(d) [Co(en)Cl₄]⁻

Stereoisomers have the same ligands bonded directly to the same metal ion, but with different spatial arrangements around the metal ion. *Cis–trans* isomers differ in that in the *cis* isomer, identical ligands are adjacent, while they are on opposite sides of the metal ion in the *trans* isomer. Octahedral complexes with the general formula MX_3Y_3 have a *fac* isomer with three identical ligands at the corners of a triangular face of an octahedron defined by the donor atoms, and a *mer* isomer with identical ligands on a meridian.

Chirality of molecules and ions [<<Section 9.8], occurs for octahedral complexes when the metal ion is bound to three bidentate ligands, or to two bidentate ligands and two monodentate ligands in a *cis* position. [Co(en)₃]³⁺ complex ions are chiral. Each [Co(en)₃]³⁺ ion is either one of an *enantiomeric pair*: ions whose structures are mirror images which are not superimposable (Figure 27.27). Solutions of each of the enantiomers rotate plane-polarized light in opposite directions, and are said to be *optically active*. Similarly, there are enantiomers of the chiral *cis*-[Co(en)₂Cl₂]⁺ complex, also illustrated in Figure 27.27.

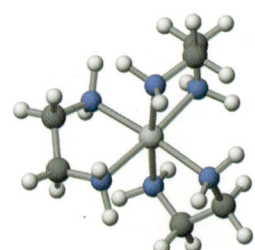
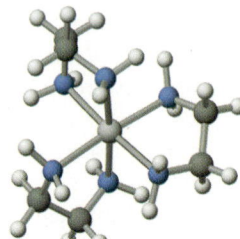
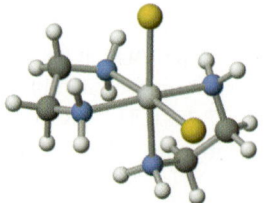
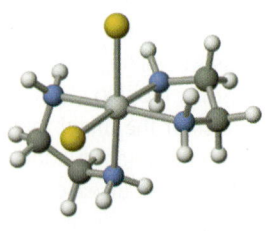

FIGURE 27.27 Chiral complexes. Both [Co(en)₃]³⁺ and *cis*-[Co(en)₂Cl₂]⁺ are chiral. There is a pair of non-superimposable mirror-image enantiomers of each complex.

Unless the ligands themselves are chiral, square-planar complexes cannot be chiral; the mirror images are superimposable. Chirality in tetrahedral complexes is possible, but examples of such complexes with a metal ion bound to four different monodentate ligands are rare.

WORKED EXAMPLE 27.3—ISOMERISM IN COORDINATION COMPLEXES

For which of the complexes whose formulas are shown do isomers exist? If isomers are possible, identify the type of isomerism (constitutional or stereoisomerism). Decide whether there are enantiomers of this complex.

(a) $[Co(NH_3)_4Cl_2]^+$ (b) $[Pt(NH_3)_2(CN)_2]$ (square-planar)
(c) $[Co(NH_3)_3Cl_3]$ (d) $[Zn(NH_3)_2Cl_2]$ (tetrahedral)
(e) $K_3[Fe(C_2O_4)_3]$ (f) $[Co(NH_3)_5SCN]^{2+}$

Think about It

e27.9 Work through a tutorial on isomerism in coordination complexes.

Strategy

Determine the number of ligands attached to the metal and decide whether the ligands are monodentate or bidentate. Knowing how many donor atoms are coordinated to the metal (the coordination number) will allow you to establish the three-dimensional shape of the complex. Building models, or drawing pictures, of the complexes will help you visualize the possibilities of isomers.

Solution

(a) There are *cis–trans* isomers of octahedral complexes with a formula of MA_4B_2. The *cis* isomer has the two Cl$^-$ ions adjacent, and the *trans* isomer has the Cl$^-$ ligands with 180° angle between the Co–Cl bonds. Neither of these isomers are chiral.

(b) In this square-planar complex, the two NH_3 ligands (and the two CN$^-$ ligands) can be either *cis* or *trans*. Neither the *cis* isomer nor the *trans* isomer is chiral.

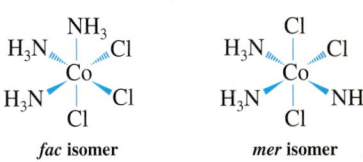

cis isomer   *trans* isomer

cis isomer   *trans* isomer

(c) There are *fac* and *mer* isomers of this octahedral complex. In the *fac* isomer, the three Co–Cl bonds are all at 90°. In the *mer* isomer, two Cl$^-$ ligands are at 180°, and the third is 90° from the other two. Neither of the isomers is chiral.

(d) Only a single structure is possible for tetrahedral complexes such as $[Zn(NH_3)_2Cl_2]$. No matter where you put the NH_3 ligands (or the Cl$^-$ ligands) on a model, you will find that they are adjacent.

fac isomer   *mer* isomer

(e) Ignore the counterions, K$^+$. The anion is an octahedral complex. The bidentate oxalate ion occupies two coordination sites of the metal, and that three oxalate ligands means that the metal has a coordination number of 6. Mirror images of complexes of the stoichiometry M(bidentate)_3 are not superimposable, so two chiral enantiomers exist. (Here the ligands, $C_2O_4{}^{2-}$, are drawn abbreviated as O–O.)

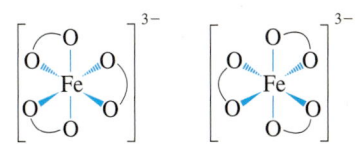

Non-superimposable mirror images of $[Fe(ox)_3]^{3-}$

(f) Only linkage isomerism is possible for this octahedral cobalt complex. In one isomer, the S atom of the SCN$^-$ ligand is the donor atom, while in the other, the N atom is the donor atom.

S-bonded SCN$^-$ **N-bonded SCN**$^-$

EXERCISE 27.9—ISOMERISM IN COORDINATION COMPLEXES

What isomers are possible for the complexes whose formulas are shown?

(a) $K[Co(NH_3)_2Cl_4]$

(b) square-planar $[Pt(en)Cl_2]$

(c) $[Co(NH_3)_5Cl]^{2+}$

(d) $[Ru(phen)_3]Cl_3$

(e) tetrahedral $Na_2[MnCl_4]$

(f) $[Co(NH_3)_5NO_2]^{2+}$

EXERCISE 27.10—ISOMERISM IN COORDINATION COMPLEXES

Determine whether the metal ion in the following complexes is a stereocentre.

(a) $[Fe(en)_3]^{2+}$

(b) *trans*-$[Co(en)_2Br_2]^+$

(c) *fac*-$[Co(en)(H_2O)Cl_3]$

(d) square-planar $[Pt(NH_3)(H_2O)(Cl)(NO_2)]$

> Complexes whose mirror images are not superimposable are chiral, with two enantiomers. There are enantiomers of octahedral complexes with the metal ion bound to three bidentate ligands, or to two bidentate ligands and two monodentate ligands in a *cis* position.

27.8 Bonding in Coordination Complexes

Metal–ligand bonding in a coordination complex was described earlier in this chapter as being covalent, resulting from the sharing of the electron pair of the donor atom between the metal and the ligand. Although frequently used, this model is not capable of explaining the colour and magnetic behaviour of complexes. As a consequence, the covalent bonding picture has now largely been superseded by two other bonding models: molecular orbital theory and crystal-field theory.

Molecular orbital theory [<<Section 10.9] assumes that the electrons belong to the whole complex, and quantum theory is applied to calculate the electron energies and the probabilistic distribution of the electrons in their various orbitals.

The *crystal-field theory*, in contrast, focuses on the electrons in the *d* orbitals of the transition metal ion, and the effect of ligands on their energies. The crystal-field theory assumes that the metal–ligand bond is due to electrostatic attraction between the positively charged metal ion and the donor atom. For the most part, the molecular orbital and crystal-field theories predict similar qualitative results regarding colour and magnetic behaviour. Here, we will focus on the simpler crystal-field approach because it lends itself better to visualization of the electron energies, and it is sufficient to explain the colours and magnetic properties of transition metal complexes. First, we look at the observations that we require our model to explain.

Colours of Coordination Complexes

One of the most interesting features of coordination complexes of the transition elements is their colours. For example, the aqua complexes of some fourth-period transition elements are displayed in Figure 27.28.

Chemists' models need to be able to explain why these solutions are coloured, and why they have different colours. Further, they need to explain why different complexes of the same element may have different colours; for example, why the $[CoF_6]^{3-}$ ion is green, while the $[Co(NH_3)_6]^{3+}$ ion is yellow-orange.

The colours of solutions, such as those shown in Figure 27.28, are the colours of light transmitted through the solutions. If we accept that the colour of transmitted light

> There is some debate among chemists about whether the aqua complexes of Cu^{2+} and Zn^{2+} are better represented as $[Cu(OH_2)_4]^{2+}$ and $[Zn(OH_2)_4]^{2+}$ than $[Cu(OH_2)_6]^{2+}$ and $[Zn(OH_2)_6]^{2+}$.

Charles D. Winters

FIGURE 27.28 Aqueous solutions of some transition metal ions. Compounds of transition metal elements are often coloured, whereas those of main group metals are usually colourless. Pictured here, from left to right, are solutions of Fe^{3+}, Co^{2+}, Ni^{2+}, Cu^{2+}, and Zn^{2+} nitrates. The colours are due to the $[Fe(OH_2)_6]^{3+}$, $[Co(OH_2)_6]^{2+}$, $[Ni(OH_2)_6]^{2+}$, $[Cu(OH_2)_6]^{2+}$, and $[Zn(OH_2)_6]^{2+}$ ions.

is due to absorption of complementary colours from incident white light (Figure 27.29), we can infer the following:

- Solutions with $[CoF_6]^{3-}$ ions are green because wavelengths corresponding with red light are absorbed by the solution.
- Solutions with $[Co(NH_3)_6]^{3+}$ ions are yellow-orange because wavelengths corresponding with blue light are absorbed.

Atomic spectra are obtained when electrons are excited from one energy level to another [<<Section 8.3]. The energy of the light absorbed or emitted is the same as the difference between energy levels of the electrons. The concept that light is absorbed when electrons move from lower to higher energy states applies to all substances, not just atoms. It is the basic premise for the absorption of light for transition metal coordination complexes. On this basis, we can now further infer that the differences between energy levels of electrons in coordination complexes is the same as the energy of photons in the visible light range—red in the case of $[CoF_6]^{3-}$ ions, and blue in the case of $[Co(NH_3)_6]^{3+}$ ions.

Since photons of blue light have more energy than those of red light, we can also infer that the energy difference ΔE in the case of $[Co(NH_3)_6]^{3+}$ ions (which absorb blue light) is greater than ΔE for $[CoF_6]^{3-}$ ions (which absorb red light). The crystal-field model gives us a sensible way of rationalizing this.

Solutions of most transition metal compounds are coloured. We can infer that the differences between energy levels of electrons in complexes is the same as the energy of photons in the visible light range.

Charles D. Winters

FIGURE 27.29 Light absorption and colour. The colour of a solution is due to the colour of the light *not* absorbed by the solution. Here an aqueous solution containing $Ni^{2+}(aq)$ ions absorbs red and blue light and so we see the residual green light.

Magnetic Properties of Coordination Complexes

Paramagnetism, the property of being attracted into a magnetic field, is attributed to the presence of unpaired electrons; the more unpaired electrons in a species, the more magnetic it is [<<Section 10.9]. Diamagnetic substances have no unpaired electrons. A model of bonding in coordination complexes needs to be able to provide explanations for phenomena such as the following:

- Although they both have cobalt in the +3 oxidation state, the $[Co(NH_3)_6]^{3+}$ ion is diamagnetic, while the $[CoF_6]^{3-}$ ion is paramagnetic and its degree of magnetism suggests the presence of four unpaired electrons.
- Although they both have iron in the +2 oxidation state, the $[Fe(CN)_6]^{4-}$ ion is diamagnetic, while the $[Fe(H_2O)_6]^{2+}$ is paramagnetic to a degree corresponding with four unpaired electrons.
- Although they both have manganese in the +2 oxidation state, the $[Mn(CN)_6]^{4-}$ ion is weakly paramagnetic, corresponding with one unpaired electron, while the stronger magnetism of the $[Mn(OH_2)_6]^{4-}$ ion corresponds with five unpaired electrons.

The complexes of a given metal ion may be diamagnetic or paramagnetic. Paramagnetism is attributed to the presence of unpaired electrons, and the degree of paramagnetism to the number of unpaired electrons in the complex.

Crystal-Field Theory: *d*-Orbital Energy Splitting

To use **crystal-field theory** as a model to account for the colours and magnetic properties of transition metal complexes, we consider the energies of electrons in the *d* orbitals of the metal ion, particularly with regard to orientation of its *d* orbitals relative to the positions of ligands around the metal ion. We look first at octahedral complexes.

Before engaging in bonding, the five 3*d* orbitals on the metal ion are presumed to be of equal energy (degenerate), with distribution of electron densities with the following characteristics, as illustrated in Figure 27.30:

- The $d_{x^2-y^2}$ orbital has electron density concentrated along the *x*- and *y*- axes.
- The d_{z^2} orbital has electron density concentrated along the *z*-axis.
- The d_{xy}, d_{yz}, and d_{xz} orbitals have electron density concentrated *between* the *x*-, *y*- and *z*- axes.

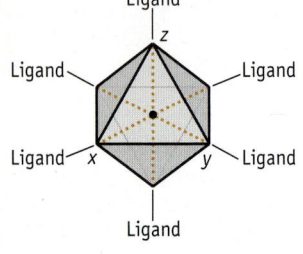

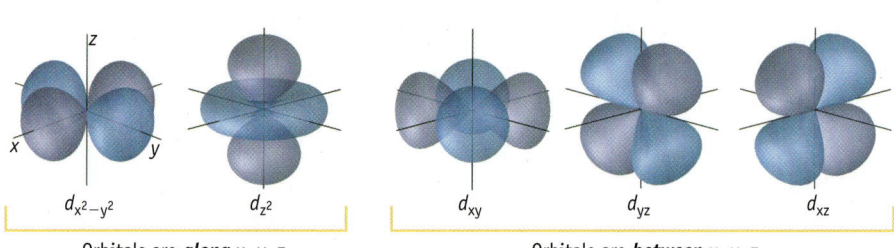

FIGURE 27.30 The concentration of electron density in the *d* orbitals of a transition metal ion. The five *d* orbitals, and the direction of their regions of concentration of electron density in relation to the spatial orientation of the ligands on the arbitrarily defined *x*-, *y*-, and *z*-axes.

We can imagine the six ligands in an octahedral complex as point negative charges on the arbitrarily defined *x*-, *y*-, and *z*- axes. According to the crystal-field model of bonding in coordination complexes, there is a repulsive interaction between electrons in the orbitals of the ligands and those in the *d* orbitals of the metal ion. This results in the electrons of the metal ion having higher energy in the complex than in the non-bonded ion—but not equally high for all of the electrons. The energy increase depends on which orbitals they "occupy." Since the concentrations of electron density in the $d_{x^2-y^2}$ and d_{z^2} orbitals are in directions pointing toward the ligands, the interaction between each of these orbitals with the ligand orbitals is stronger than the interaction between each of the d_{xy}, d_{yz}, and d_{xz} orbitals with the ligand orbitals. Consequently, the energy of the electrons in the $d_{x^2-y^2}$ and d_{z^2} orbitals is higher than the energy of those in the other three *d* orbitals (Figure 27.31). The difference

FIGURE 27.31 Ligand-field splitting in an octahedral complex. In the crystal-field model, the energies of the metal ion electrons are higher in the complex than in the non-bonded ion as a result of interaction with the ligands imagined as point charges on the *x*-, *y*-, and *z*-axes. The energy of electrons in the $d_{x^2-y^2}$ and d_{z^2} orbitals, which point directly toward the ligands, is higher than the energy of those in the d_{xy}, d_{yz}, and d_{xz} orbitals. The resultant difference is the ligand-field splitting (Δ_0).

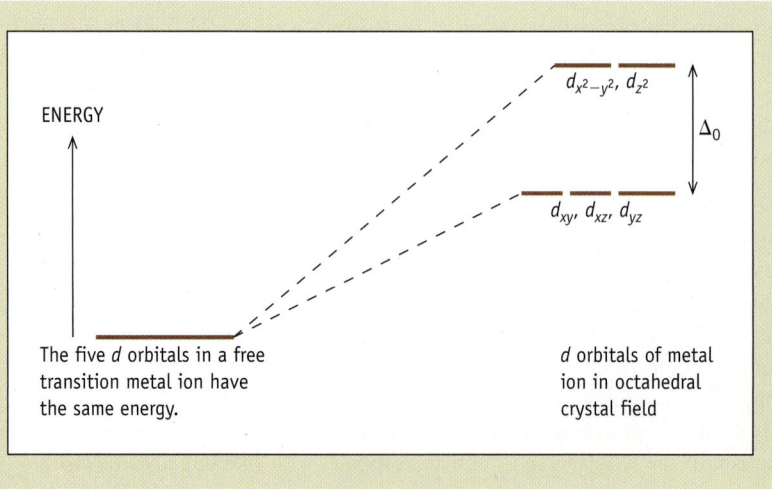

between the energy of the electrons in the $d_{x^2-y^2}$ and d_{z^2} orbitals and the energy of those in the d_{xy}, d_{yz}, and d_{xz} orbitals is called the **ligand-field splitting**, denoted Δ_o. The magnitude of Δ_o depends on which metal ion and which ligands comprise the complex, and varies predictably from one complex to another.

Whether Δ_o is large or small depends on which ligands are bound to the metal ion. Ligands that interact strongly with the d orbitals of the metal ion are called **strong-field ligands**: they cause larger increases in energy of the metal ion orbitals and larger "splitting" between the two sets of levels (i.e., larger Δ_o). Ligands that have less effect on the energies of the metal ion orbitals and cause smaller ligand-field splitting are called **weak-field ligands**.

Ligands can be approximately ordered according to their ability to split the d-orbital energies. This rank order is called the **spectrochemical series**. A selection of the more common ligands from this series is shown in Figure 27.32.

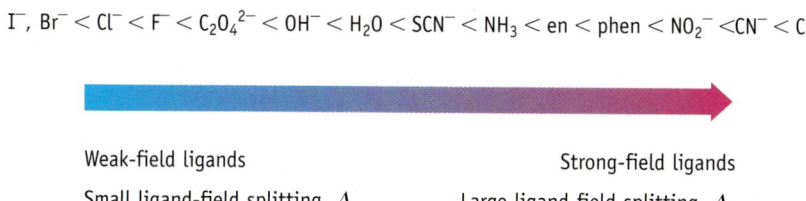

I^-, $Br^- < Cl^- < F^- < C_2O_4^{2-} < OH^- < H_2O < SCN^- < NH_3 < en < phen < NO_2^- < CN^- < CO$

Weak-field ligands

Small ligand-field splitting, Δ_o

Strong-field ligands

Large ligand-field splitting, Δ_o

FIGURE 27.32 The spectrochemical series. Ligands are listed in general order of their increasing ability to interact with metal d orbitals and bring about ligand-field splitting. The order depends somewhat on the metal ion to which they are bonded.

A different order of energies of the metal ion electrons is postulated for square-planar complexes (Figure 27.33). Assume that the four ligands are along the x- and y-axes. The $d_{x^2-y^2}$ orbital also points along these axes, so it is raised to the highest energy. The d_{xy} orbital, which also lies in the xy-plane but does not point at the ligands, is next highest in energy, followed by the d_{z^2} orbital. The d_{xz} and d_{yz} orbitals, both of which partially point in the z-direction, interact with the ligands the least, and so have the lowest energy.

The d-orbital splitting pattern for a tetrahedral complex is the reverse of the pattern observed for octahedral complexes. Three orbitals (d_{xz}, d_{xy}, d_{yz}) are higher in energy, whereas the $d_{x^2-y^2}$ and d_{z^2} orbitals are below them in energy (Figure 27.33).

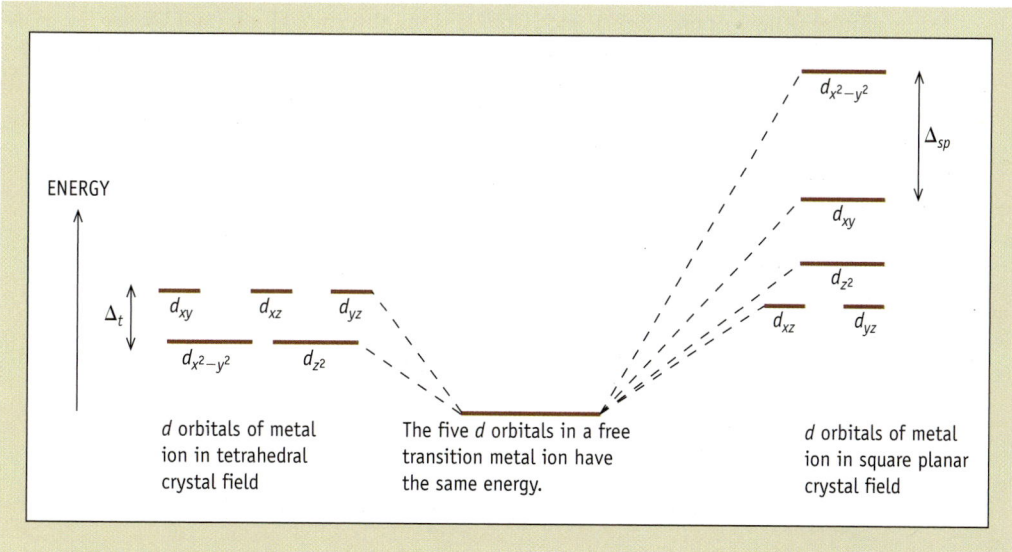

FIGURE 27.33 Splitting of the d orbitals in tetrahedral (*left*) and square-planar (*right*) complexes.

Crystal-field theory assumes a "splitting" of the energies of electrons in the d orbitals of the metal ion due to differences in the degree of interaction of the d orbitals with orbitals of the ligands. The outcomes of splitting is different in octahedral, tetrahedral, and square-planar complexes. Strong-field ligands give rise to large ligand-field splitting, and weak-field ligands to small splitting of energies. The spectrochemical series is a rank order of ability of ligands to split the d-orbital energies.

Crystal-Field Theory and Colours

The application of crystal-field theory to account for colours of transition metal complexes is indicated here by reference to octahedral complexes. Among other observations, our model needs to account for the fact that both $[Co(NH_3)_6]^{3+}$ ions and $[CoF_6]^{3-}$ ions are coloured, the former being yellow-orange, and the latter green (page 1081).

In coordination complexes, the splitting between d orbitals often corresponds with the energy of visible light, so light in the visible region of the spectrum is absorbed when electrons are excited from a lower-energy d orbital to a higher-energy d orbital. This excitation is called a *d-to-d transition*. Qualitatively, such a transition for $[Co(NH_3)_6]^{3+}$ might be represented by an energy-level diagram such as that in Figure 27.34.

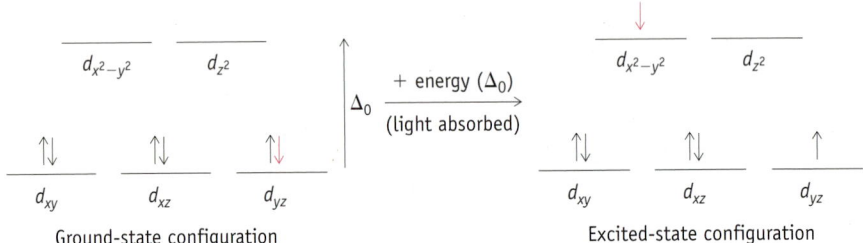

Ground-state configuration   Excited-state configuration

FIGURE 27.34 A *d-to-d* electron transition in the $[Co(NH_3)_6]^{3+}$ complex ion. From white light passing through the sample, photons with energy Δ_0 are absorbed as an electron is excited from the d_{yz} (or d_{xy} or d_{xz}) orbital to the $d_{x^2-y^2}$ (or d_{z^2}) orbital, and the other photons are transmitted. We would expect a smaller energy gap Δ_0 in the case of weak-field F^- ions, with absorption of light of longer wavelengths. Discussion of why there are six electrons in the three lower-energy orbitals is presented below.

If Δ_0 is relatively large, blue light is absorbed by electron excitation and, in contrast, a small Δ_0 can result in less energetic red light being absorbed. The NH_3 molecule is a "strong-field ligand" and induces a larger gap Δ_0 than the "weak-field ligand" F^- ion. So we should expect the $[Co(NH_3)_6]^{3+}$ ion to absorb higher energy light (shorter wavelengths) than does the $[CoF_6]^{3-}$ ion. Indeed, $[Co(NH_3)_6]^{3+}$ ions absorb blue light (and we see the transmitted yellow-orange light), and $[CoF_6]^{3-}$ ions absorb red light (and we see the transmitted green light).

By way of further illustration, some spectroscopic data for several cobalt(III) complexes is shown in Table 27.4.

TABLE 27.4 The Colours of Some Co^{3+} Complexes

Complex Ion	Wavelength of Light Absorbed (mm)	Colour of Light Absorbed	Colour of Complex
$[CoF_6]^{3-}$	700	Red	Green
$[Co(C_2O_4)_3]^{3-}$	600, 420	Yellow, violet	Dark green
$[Co(H_2O)_6]^{3+}$	600, 400	Yellow, violet	Blue-green
$[Co(NH_3)_6]^{3+}$	475, 340	Blue, ultraviolet	Yellow-orange
$[Co(en)_3]^{3+}$	470, 340	Blue, ultraviolet	Yellow-orange
$[Co(CN)_6]^{3-}$	310	Ultraviolet	Pale yellow

The absorption maxima of the complexes listed range from 700 nm for $[CoF_6]^{3-}$ to 310 nm for $[Co(CN)_6]^{3-}$. The ligands are different from member to member of this series, and we can conclude that the energy of the light absorbed by the complex is related to the different ligand-field splittings (Δ_0), caused by the different ligands. Fluoride ion causes the smallest splitting of the d orbitals among the complexes listed in Table 27.4, whereas cyanide causes the largest splitting.

We can make sense of some of the data in Table 27.4 as follows:

- Both $[Co(NH_3)_6]^{3+}$ and $[Co(en)_3]^{3+}$ are yellow-orange because they absorb light in the blue portion of the visible spectrum. These compounds have very similar spectra, to be expected because both have six amine-type donor atoms ($H-NH_2$ or $R-NH_2$).
- Although $[Co(CN)_6]^{3-}$ does not have an absorption band in the visible region, it is pale yellow. Light absorption occurs in the ultraviolet region, but the absorption is broad and extends at least minimally into the visible (blue) region.
- $[Co(C_2O_4)_3]^{3-}$ and $[Co(H_2O)_6]^{3+}$ have similar absorptions, in the yellow and violet regions. Their colours are shades of green with a small difference due to the different amount of light of each wavelength that is absorbed.

Think about It

e27.10 Work through a tutorial on the spectroscopy of transition metal complexes.

WORKED EXAMPLE 27.4—COLOURS OF COORDINATION COMPLEXES

An aqueous solution containing $[Fe(OH_2)_6]^{2+}$ ion is light blue-green. Is this observation consistent with the water molecule being a strong-field ligand or a weak-field ligand?

Strategy

The colour of the complex, blue-green, tells us what kind of light is transmitted (blue and green), from which we learn what kind of light has been absorbed.

Solution

The complex must absorb red light, which is at the low-energy end of the visible spectrum. The low energy of the light absorbed suggests that the ligand-field splitting (Δ_o) is small, so the water molecule is a weak-field ligand.

EXERCISE 27.11—COLOURS OF COORDINATION COMPLEXES

Rank the complexes $[Ti(NH_3)_6]^{3+}$, $[Ti(CN)_6]^{3-}$, and $[Ti(OH_2)_6]^{3+}$ in order of the magnitudes of the ligand-field splitting and the energy of visible light absorbed.

The colours of solutions of transition metal complexes can be explained by ligand-field splitting. Complexes of strong-field ligands absorb higher energy radiation from white light than those of weak-field ligands.

Crystal-Field Theory and Magnetic Properties

To account for the magnetic properties of coordination complexes listed on page 1081, we need to determine the electron configuration of the complexes using the crystal-field model, so that we can see if there are unpaired electrons in the complex. Some complexes of Cr^{2+} ions are paramagnetic and some are diamagnetic. Crystal-field theory can account for this in the following way.

An isolated Cr^{2+} ion (in the absence of interaction with any other species) has the electron configuration $[Ar]3d^4$, and the five $3d$ orbitals have the same energy. The four electrons occupy different d orbitals, according to Hund's rule, and the Cr^{2+} ion has four unpaired electrons.

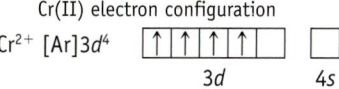

Cr(II) electron configuration

Cr^{2+} $[Ar]3d^4$ 3d   4s

In an octahedral complex, according to the crystal-field model, the d electrons do not have identical energies: there are two sets of degenerate orbitals, with electrons in the d_{xy}, d_{xz}, and d_{yz} orbitals having lower energy than those in the $d_{x^2-y^2}$ and d_{z^2} orbitals (Figure 27.31). This means that two different electron configurations are possible (Figure 27.35). Three of the four d electrons in the Cr^{2+} ion can be expected to occupy the lower-energy d_{xy}, d_{xz}, and d_{yz} orbitals. There are two possibilities for the fourth electron:

FIGURE 27.35 High- and low-spin cases for an octahedral chromium(II) complex. (*Left, high spin*) If the ligand-field splitting (Δ_o) is smaller than the pairing energy (P), all the electrons occupy different orbitals and the complex has four unpaired electrons. (*Right, low spin*) If the splitting is larger than the pairing energy, all four electrons will be in the lower-energy orbital set. This requires pairing two electrons in one of the orbitals, so the complex will have two unpaired electrons.

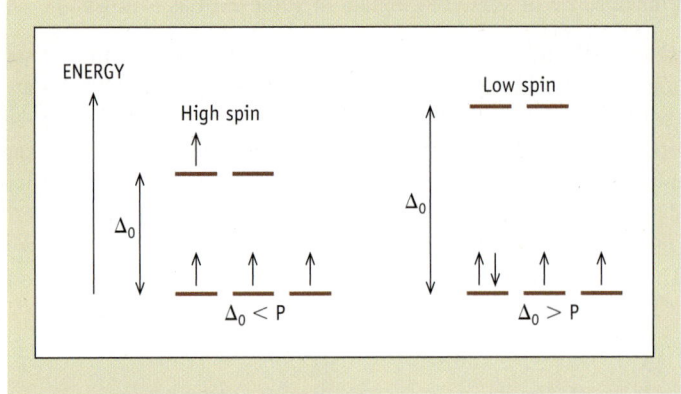

- If the complex has strong-field ligands, the energy gap Δ_o is large, and the fourth electron pairs up with another in the lower-energy d_{xy}, d_{xz}, and d_{yz} orbitals. This configuration is called a **low-spin configuration** because it has the minimum possible number of unpaired electrons (or the maximum number of paired electrons). Low-spin chromium(II) complexes, are paramagnetic, with a degree of magnetism corresponding with two unpaired electrons.

- If the complex has weak-field ligands, with a low energy gap, the complex may be more stable with the fourth electron in the higher energy $d_{x^2-y^2}$ or d_{z^2} orbitals (compared with pairing up with another electron in one of the lower-energy orbitals). This results in a **high-spin configuration**, because it has the maximum possible number of unpaired electrons—four in the case of an octahedral complex of Cr^{2+} ions with weak-field ligands.

A high-spin configuration may seem counterintuitive. It may seem logical that the most stable configuration would have all four electrons in the lowest-energy orbitals. A second factor intervenes, however. Because electrons are negatively charged, they repel each other when they occupy the same region of space (same orbital). This destabilizing effect is called *pairing energy*. The preference for an electron to be in the lowest-energy orbital and the pairing energy have opposing effects.

Low-spin complexes arise when the ligand-field splitting of the *d* orbitals is large—that is, when Δ_o has a large value. The energy gained by putting all of the electrons in the lowest-energy level is the dominant effect. In contrast, high-spin complexes occur if the value of Δ_o is small.

For octahedral complexes, high- and low-spin complexes can occur only for configurations d^4 through d^7 (Figure 27.36).

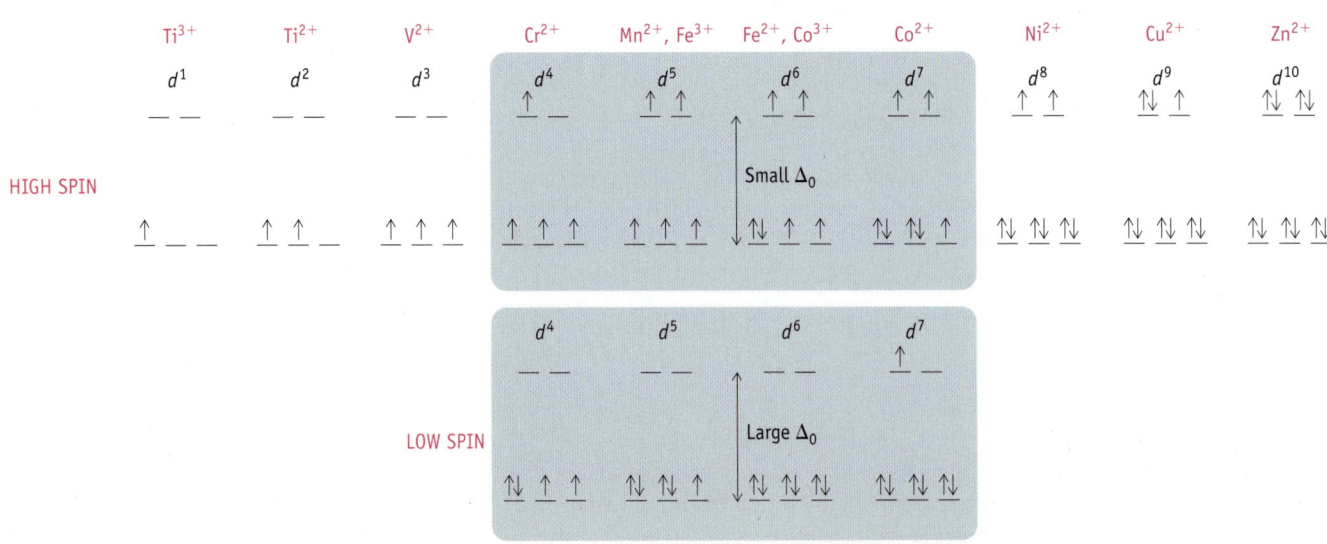

FIGURE 27.36 High- and low-spin octahedral complexes. *d*-Orbital occupancy for octahedral complexes of metal ions. Only the d^4 through d^7 cases have both high- and low-spin configurations.

Complexes of the d^6 metal ion Fe^{2+}, for example, can have either high spin or low spin. The complex with weak-field water ligands, $[Fe(OH_2)_6]^{2+}$, has low-energy splitting, and is high spin, with paramagnetism corresponding to the presence of four unpaired electrons. On the other hand, the $[Fe(CN)_6]^{4-}$ complex ion with strong-field ligands has a large Δ_o, and is low spin, being diamagnetic (no unpaired electrons).

Electron configuration for Fe^{2+} in an octahedral complex

$\uparrow\ \uparrow$ $d_{x^2-y^2}, d_{z^2}$	$\Delta_o(H_2O)$	$\overline{}\ \overline{}$ $d_{x^2-y^2}, d_{z^2}$	$\Delta_o(CN^-)$
$\uparrow\downarrow\ \uparrow\ \uparrow$ d_{xy}, d_{xz}, d_{yz}		$\uparrow\downarrow\ \uparrow\downarrow\ \uparrow\downarrow$ d_{xy}, d_{xz}, d_{yz}	
high spin $[Fe(H_2O)_6]^{2+}$		low spin $[Fe(CN)_6]^{4-}$	

Most complexes of Pd^{2+} and Pt^{2+} ions are square-planar, with eight valence d electrons. In a square-planar complex, there are four sets of orbitals (Figure 27.33). All except the highest-energy orbital are filled, and all electrons are paired, resulting in diamagnetic (low-spin) complexes.

On the other hand, nickel, which is above palladium in the periodic table, forms both square-planar and tetrahedral complexes. For example, the complex ion $[Ni(CN)_4]^{2-}$ is square-planar, whereas the $[NiCl_4]^{2-}$ ion is tetrahedral. Magnetism allows us to differentiate between these two geometries. Paramagnetism of the chloride complex, consistent with two unpaired electrons, matches predictions of the ligand-field theory applied to tetrahedral complexes. The cyanide complex is diamagnetic, consistent with ligand-field theory predictions for a square-planar complex (Figure 27.37).

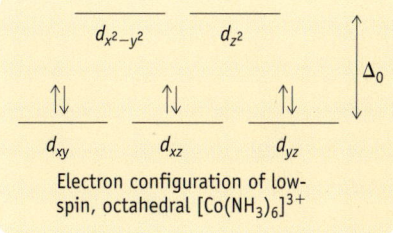

FIGURE 27.37 The $[NiCl_4]^{2-}$ anion is a paramagnetic tetrahedral complex. In contrast, $[Ni(CN)_4]^{2-}$ is a diamagnetic square-planar complex.

WORKED EXAMPLE 27.5—MAGNETISM OF TRANSITION METAL COMPLEXES

Give the electron configuration for each of the following complexes. How many unpaired electrons are present in each complex? Are the complexes paramagnetic or diamagnetic?
(a) low-spin $[Co(NH_3)_6]^{3+}$ (b) high-spin $[CoF_6]^{3-}$

Strategy

These ions are complexes of Co^{3+}, which has a d^6 valence electron configuration. Set up an energy-level diagram for an octahedral complex. In low-spin complexes, the electrons are added preferentially to the lower-energy set of orbitals. In high-spin complexes, the first five electrons are added singly to each of the five orbitals, then additional electrons are paired with electrons in orbitals in the lower-energy set.

Solution

(a) The six electrons of the Co^{3+} ion fill the lower-energy set of orbitals entirely. This d^6 complex ion has no unpaired electrons and is diamagnetic.
(b) To obtain the electron configuration in high-spin $[CoF_6]^{3-}$, place one electron in each of the five d orbitals, and then place the sixth electron in one of the lower-energy orbitals. The complex has four unpaired electrons and is paramagnetic.

EXERCISE 27.12—MAGNETISM OF TRANSITION METAL COMPLEXES

For each of the following complex ions, give the oxidation state of the metal, depict possible low- and high-spin configurations, give the number of unpaired electrons in each configuration, and deduce whether it is paramagnetic or diamagnetic.
(a) $[Ru(H_2O)_6]^{2+}$ (b) $[Ni(NH_3)_6]^{2+}$

Think about It

e27.11 Work through a tutorial on the electronic structures of transition metal complexes.

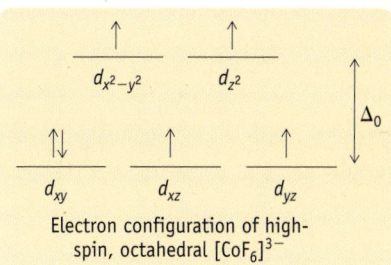

Electron configuration of low-spin, octahedral $[Co(NH_3)_6]^{3+}$

Electron configuration of high-spin, octahedral $[CoF_6]^{3-}$

The magnitude of the d-orbital splitting, and the consequent magnetic character of complexes, often has a correspondence with the colours of the complexes that can be interpreted from the spectrochemical series (Figure 27.32). Recall that d^4, d^5, d^6, and d^7 complexes can be high or low spin, depending on the ligand-field splitting (Δ_o). Complexes formed with weak-field ligands near the left of the spectrochemical series (such as $[CoF_6]^{3-}$ ions) can be expected to have small Δ_o values and, therefore, are likely to be high spin, and also to absorb low-energy radiations and have colours at the blue end of the spectrum. In contrast, complexes with strong-field ligands near the right end of the series, such as the $[Co(NH_3)_6]^{3+}$ ion, have large Δ_o values and low-spin configurations, and absorb high-energy radiation and appear yellow to red.

EXERCISE 27.13—MAGNETISM OF TRANSITION METAL COMPLEXES

The following are low-spin complexes. Use the crystal-field theory to find the electron configuration of each ion. Determine which are diamagnetic. Give the number of unpaired electrons for the paramagnetic complexes.

(a) $[Mn(CN)_6]^{4-}$ (c) $[Fe(H_2O)_6]^{3+}$

(b) $[Co(NH_3)_6]Cl_3$ (d) $[Cr(en)_3]SO_4$

The magnetism of transition metal complexes can be explained by the crystal-field model if we presume that the energy splitting due to strong-field ligands gives rise to low-spin electron configurations of the metal ion because electrons pair up in the lower-energy set of orbitals. With weak-field ligands, the energy gap is not sufficient to prevent occupation of the higher-energy orbitals, so unpaired electrons are maximized and high-spin configurations result.

SUMMARY

Key Concepts

- The transition elements comprise the **d-block elements**, whose electron configurations have unfilled *d* orbitals, and the **f-block elements**, characterized by unfilled *f* orbitals. The *d*-block elements are in groups 3 to 12 of the periodic table. The elements of the first series of *d*-block elements, in the fourth period of the periodic table, with atomic numbers 21 to 30, have valence electrons in the 3*d* and 4*s* orbitals. (Section 27.2)
- Almost all of the *d*-block elements have multiple oxidation states, and this gives rise to their oxidation-reduction chemistry. (Section 27.2)
- The properties of compounds of scandium and zinc are not characteristic of those of the other *d*-block elements. This can be explained by the fully occupied orbitals of the metal ions in their compounds. (Section 27.2)
- Transition metals can be obtained from their ores either by **pyrometallurgy** or **hydrometallurgy**. (Section 27.2)
- In coordination complexes, ligands behave as Lewis bases in providing an electron pair to bond to a metal ion. The bond is indistinguishable from any other covalent bond. Ligands are **monodentate** or **polydentate**. Polydentate ligands are called **chelating ligands**. The **coordination number** of the metal ion is the number of electron pair donor atoms to which it is bonded. The **coordination geometry** refers to the spatial arrangement of ligands around the metal ion. (Section 27.4)
- Hemoglobin is a coordination complex of particular biological importance. (Section 27.4)
- The formula of a coordination complex displays the metal and the number and identity of the ligand(s) comprising the complex, and allows us to deduce the oxidation state of the metal and coordination number. (Section 27.4)
- The equilibrium constant for a reaction in which a metal ion complex is formed from the aqua complex of the metal ion is called the **formation constant**. The larger the

formation constant is, the more stable the complex. Complexes of a given metal are generally more stable the higher the oxidation state of the metal ion. (Section 27.5)

- The distribution of a metal ion among various complexes formed with a ligand depends upon the equilibrium constants of each of a series of step-wise reactions. Using the step-wise equilibrium constants, we can calculate speciation plots that indicate the relative concentrations of the various species as the ligand concentration is changed at a defined pH. An alternative speciation plot shows the pH-dependence of the distribution among species as pH is changed in a solution of given concentration of metal ion and ligand. Ligands that are weak bases lose their Lewis base ability when they are protonated. (Section 27.5)

- Complex ions of a given metal ion with polydentate ligands are more stable than complexes that have the same number and type of donor atoms in monodentate ligands. This **chelate effect** is largely due to differences in entropy of formation of the complexes. (Section 27.5)

- The thermodynamic stability of a complex is unrelated to its lability, or rate of substitution of ligand species. Complexes that undergo rapid substitution of ligand species are said to be **labile**, and those in which substitution is slow are **inert**. (Section 27.5)

- Common spatial orientations of donor atoms around a metal ion include linear in complexes with two donor atoms, square-planar or tetrahedral in complexes with four donor atoms, and octahedral in complexes with six donor atoms. (Section 27.6)

- Constitutional isomers of complexes include **ionization isomers** in which ligand(s) and counterion(s) are interchanged, and **linkage isomers** in which a ligand is bonded to a metal ion through lone pairs on different atoms. (Section 27.7)

- Stereoisomers of complexes have the same ligands bonded directly to the same metal ion, but with different spatial arrangements around the metal ion. These include *cis–trans* isomers, *fac–mer* isomers, and enantiomers. (Section 27.7)

- **Crystal-field theory** assumes a "splitting" of the energies of electrons in the *d* orbitals of the metal ion due to differences in the degree of interaction of the *d* orbitals with orbitals of the ligands. The outcomes of this **ligand-field splitting** is different in octahedral, tetrahedral, and square-planar complexes. **Strong-field ligands** give rise to large ligand-field splitting, and **weak-field ligands** to small splitting of energies. The **spectrochemical series** is a rank order of ability of ligands to split the *d*-orbital energies. (Section 27.8)

- The colours of solutions of transition metal complexes can be explained by ligand-field splitting. Complexes of strong-field ligands absorb higher energy visible radiation from white light than those of weak-field ligands. (Section 27.8)

- Paramagnetism of complexes is attributed to the presence of unpaired electrons, and the degree of paramagnetism to the number of unpaired electrons in the complex. The degree of magnetism of metal complexes can be explained by the crystal-field model if we presume that the energy splitting due to strong-field ligands gives rise to **low-spin configurations** of the metal ion because electrons pair up in the lower-energy set of orbitals. With weak-field ligands, the energy gap is not sufficient to prevent occupation of the higher-energy orbitals, so unpaired electrons are maximized and **high-spin configurations** result. (Section 27.8)

REVIEW QUESTIONS

Section 27.2: The *d*-Block Elements and Compounds

27.14 Give the electron configuration for each of the following ions, and state whether each is paramagnetic or diamagnetic:
(a) Cr^{3+} (b) V^{2+} (c) Ni^{2+} (d) Cu^+

27.15 Identify two transition metal ions with the following electron configurations:
(a) $[Ar]3d^6$ (b) $[Ar]3d^{10}$ (c) $[Ar]3d^5$ (d) $[Ar]3d^8$

27.16 Identify an ion of a first series transition metal that is isoelectronic with each of the following:
(a) Fe^{3+} (b) Zn^{2+} (c) Fe^{2+} (d) Cr^{3+}

Section 27.3: Metallurgy

27.17 The following equations represent various ways of obtaining transition metals from their compounds. Balance each equation.

(a) $Cr_2O_3(s) + Al(s) \longrightarrow Al_2O_3(s) + Cr(s)$
(b) $TiCl_4(\ell) + Mg(s) \longrightarrow Ti(s) + MgCl_2(s)$
(c) $[Ag(CN)_2]^-(aq) + Zn(s)$
$\longrightarrow Ag(s) + [Zn(CN)_4]^{2-}(aq)$
(d) $Mn_3O_4(s) + Al(s) \longrightarrow Mn(s) + Al_2O_3(s)$

Section 27.4: Coordination Compounds

27.18 Deduce the oxidation state of the metal ion in each of the following compounds:

(a) $[Mn(NH_3)_6]SO_4$ (c) $[Co(NH_3)_4Cl_2]Cl$
(b) $K_3[Co(CN)_6]$ (d) $Cr(en)_2Cl_2$

27.19 Give the formula of a complex constructed from one Cr^{3+} ion, two ethylenediamine ligands, and two ammonia molecules. Is the complex neutral or is it charged? If charged, give the charge.

27.20 Give the formula of the complex formed from one Co^{3+} ion, two ethylenediamine molecules, one water molecule, and one chloride ion. Is the complex neutral or charged? If charged, give the net charge on the ion.

27.21 When $CrCl_3$ dissolves in water, three different species can be obtained.

- $[Cr(H_2O)_6]Cl_3$, violet
- $[Cr(H_2O)_5Cl]Cl_2$, pale green
- $[Cr(H_2O)_4Cl_2]Cl$, dark green

If diethylether is added, a fourth complex can be obtained: $Cr(H_2O)_3Cl_3$ (brown). Describe an experiment that will allow you to differentiate these complexes.

27.22 Three different compounds of chromium(III) with water and chloride ion have the same composition: 19.51% Cr, 39.92% Cl, and 40.57% H_2O. One of the compounds is violet and dissolves in water to give a complex ion with a +3 charge and three chloride ions. All three chloride ions precipitate immediately as AgCl on adding $AgNO_3$. Draw the structure of the complex ion and name the compound. Write a net ionic equation for the reaction of this compound with silver nitrate.

Section 27.5: Complexation Equilibria, Stability of Complexes

27.23 The step-wise complexation of Cd^{2+} ions with the uncharged bidentate ligand ethylenediamine (en) can be represented as follows:

$[Cd(OH_2)_6]^{2+} + en \rightleftharpoons [Cd(OH_2)_4(en)]^{2+}$
$+ 2 H_2O(\ell) \quad K_{f1} = 2.5 \times 10^5$

$[Cd(OH_2)_4(en)]^{2+} + en \rightleftharpoons [Cd(OH_2)_2(en)_2]^{2+}$
$+ 2 H_2O(\ell) \quad K_{f2} = 3.2 \times 10^4$

$[Cd(OH_2)_2(en)_2]^{2+} + en \rightleftharpoons [Cd(en)_3]^{2+}$
$+ 2 H_2O(\ell) \quad K_{f3} = 63$

(a) Calculate the equilibrium constant for the overall complexation reaction.

$[Cd(OH_2)_6]^{2+} + 3 en \rightleftharpoons [Cd(en)_3]^{2+} + 6 H_2O(\ell)$

(b) What is the ratio of concentrations of the species $[Cd(OH_2)_6]^{2+}$ and $[Cd(OH_2)_4(en)]^{2+}$ when [en] $= 1 \times 10^{-3}$ mol L^{-1}?

(c) At what concentration of ethylenediamine is the concentration of $[Cd(OH_2)_4(en)]^{2+}$ equal to the concentration of $[Cd(OH_2)_2(en)_2]^{2+}$?

27.24 Consider the formation constants (β) for the complexes formed in the following reactions:

$[Co(OH_2)_6]^{2+} + 6 NH_3(aq) \rightleftharpoons [Co(NH_3)_6]^{2+}$
$+ 6 H_2O(\ell) \quad \beta = 1.5 \times 10^5$

$[Co(OH_2)_6]^{2+} + 3 en(aq) \rightleftharpoons [Co(en)_3]^{2+}$
$+ 6 H_2O(\ell) \quad \beta = 8 \times 10^{13}$

(a) Are these equilibrium constants consistent with the chelate effect? Explain.

(b) Calculate the equilibrium constant for the reaction

$[Co(NH_3)_6]^{2+} + 3 en(aq) \rightleftharpoons [Co(en)_3]^{2+} + 6 NH_3(aq)$

27.25 The stability of analogous complexes $[ML_6]^{n+}$ (relative to ligand dissociation) is in the general order Mn^{2+}, Fe^{2+}, Co^{2+}, Ni^{2+}, Cu^{2+}, Zn^{2+}. This order of ions is called the Irving-Williams series. Look up the values of the formation constants for the ammonia complexes of Co^{2+}, Ni^{2+}, Cu^{2+}, and Zn^{2+} in Appendix D and verify this statement.

Section 27.6: Structures of Coordination Complexes

27.26 Describe an experiment that would determine whether nickel in $K_2[NiCl_4]$ is square-planar or tetrahedral.

Section 27.7: Isomerism in Coordination Complexes

27.27 How many stereoisomers are possible for the square-planar complex $[Pt(NH_3)(CN)Cl_2]^-$?

27.28 In which of the following complexes are stereoisomers possible? If isomers are possible, draw their structures and label them as *cis* or *trans*, or as *fac* or *mer*.

(a) $[Co(H_2O)_4Cl_2]^+$ (c) $[Pt(NH_3)Br_3]^-$
(b) $[Co(NH_3)_3F_3]$ (d) $[Co(en)_2(NH_3)Cl]^{2+}$

27.29 Four isomers are possible for $[Co(en)(NH_3)_2(H_2O)Cl]^+$. Draw the structures of all four (two of the isomers are chiral).

27.30 Which of the following complexes containing the oxalate ion are chiral?
(a) $[Fe(C_2O_4)Cl_4]^{2-}$
(b) cis-$[Fe(C_2O_4)_2Cl_2]^{2-}$
(c) $trans$-$[Fe(C_2O_4)_2Cl_2]^{2-}$

27.31 Diethylenetriamine (dien) is capable of serving as a tridentate ligand.

$$H_2\ddot{N}CH_2CH_2-\overset{|}{\underset{H}{\ddot{N}}}-CH_2CH_2\ddot{N}H_2$$

(a) Draw the structures of fac-$[Cr(dien)Cl_3]$ and mer-$[Cr(dien)Cl_3]$.
(b) Two different cis–$trans$ isomers of mer-$[Cr(dien)Cl_2Br]$ are possible. Draw the structure for each.
(c) Three different isomers are possible for the $[Cr(dien)_2]^{3+}$ complex ion. Two have the dien ligand in a fac configuration, and one has the ligand in a mer orientation. Draw the structure of each isomer.

27.32 Three cis–$trans$ isomers are possible for $[Co(en)(NH_3)_2(H_2O)_2]^{3+}$. One of the three is chiral—that is, it has non-superimposable mirror-image enantiomers. Draw the structures of the three isomers. Which one is chiral?

27.33 The square-planar complex $[Pt(en)Cl_2]$ has chloride ligands in a cis configuration. No $trans$ isomer is known. Based on the bond lengths and bond angles of carbon and nitrogen in the ethylenediamine ligand, explain why the $trans$ compound is not possible.

Section 27.8: Bonding in Coordination Complexes

27.34 Which of the complexes, $[V(NH_3)_6]^{3+}$ and $[V(OH_2)_6]^{3+}$, would you expect to have a colour nearer the red end of the visible spectrum?

27.35 For a tetrahedral complex of a metal in the first transition series, which of the following statements concerning energies of the $3d$ orbitals is correct?
(a) The five d orbitals have the same energy.
(b) The $d_{x^2-y^2}$ and d_{z^2} orbitals are higher in energy than the d_{xz}, d_{yz}, and d_{xy} orbitals.
(c) The d_{xz}, d_{yz}, and d_{xy} orbitals are higher in energy than the $d_{x^2-y^2}$ and d_{z^2} orbitals.
(d) The d orbitals all have different energies.

27.36 The complex $[Mn(H_2O)_6]^{2+}$ has five unpaired electrons, whereas $[Mn(CN)_6]^{4-}$ has only one. Using the crystal-field model, depict the electron configuration for each ion. What can you conclude about the effects of the different ligands on the magnitude of Δ_o?

27.37 Experiments show that $K_4[Cr(CN)_6]$ is paramagnetic and has two unpaired electrons. The related complex $K_4[Cr(SCN)_6]$ is paramagnetic and has four unpaired elec-

trons. Account for the magnetism of each compound using the crystal-field model. Predict where the SCN^- ion occurs in the spectrochemical series relative to CN^-.

27.38 The following are high-spin complexes. Use the crystal-field model to find the electron configuration of each ion and determine the number of unpaired electrons in each.
(a) $K_4[FeF_6]$
(c) $[Cr(H_2O)_6]^{2+}$
(b) $[MnF_6]^{4-}$
(d) $(NH_4)_3[FeF_6]$

27.39 Determine the number of unpaired electrons in the following tetrahedral complexes. All tetrahedral complexes are high spin.
(a) $[FeCl_4]^{2-}$
(c) $[MnCl_4]^{2-}$
(b) $Na_2[CoCl_4]$
(d) $(NH_4)_2[ZnCl_4]$

27.40 For the high-spin complex $[Fe(H_2O)_6]SO_4$, identify the following:
(a) the coordination number of iron
(b) the coordination geometry for iron
(c) the oxidation state of iron
(d) the number of unpaired electrons
(e) whether the complex is diamagnetic or paramagnetic

27.41 For the low-spin complex $[Co(en)(NH_3)_2Cl_2]ClO_4$, identify the following:
(a) the coordination number of cobalt
(b) the coordination geometry for cobalt
(c) the oxidation state of cobalt
(d) the number of unpaired electrons
(e) whether the complex is diamagnetic or paramagnetic
Draw any geometric isomers.

27.42 The coordination compound formed in an aqueous solution of cobalt(III) sulfate is diamagnetic. If NaF is added, the complex in solution becomes paramagnetic. Why does the magnetism change?

27.43 An aqueous solution of iron(II) sulfate is paramagnetic. If NH_3 is added, the solution becomes diamagnetic. Why does the magnetism change?

27.44 Which of the following low-spin complexes has the greatest number of unpaired electrons?
(a) $[Cr(H_2O)_6]^{3+}$
(b) $[Mn(H_2O)_6]^{2+}$
(c) $[Fe(H_2O)_6]^{2+}$
(d) $[Ni(H_2O)_6]^{2+}$

27.45 How many unpaired electrons are expected for high-spin and low-spin complexes of Fe^{2+}?

27.46 The complex ion $[Co(CO_3)_3]^{3-}$, an octahedral complex with bidentate carbonate ions as ligands, has one absorption in the visible region of the spectrum at 640 nm. From this information:
(a) predict the colour of this complex, and explain your reasoning
(b) decide whether the carbonate ion is a weak- or strong-field ligand
(c) predict whether $[Co(CO_3)_3]^{3-}$ will be paramagnetic or diamagnetic

SUMMARY AND CONCEPTUAL QUESTIONS

27.47 Early in the 20th century, complexes sometimes were given names based on their colours. Two compounds with the formula $CoCl_3.4NH_3$ were named praseo-cobalt chloride (*praseo* = green) and violio-cobalt chloride (violet). We now know that these compounds are octahedral cobalt complexes and that they are *cis* and *trans* isomers. Draw the structures of these two compounds and name them using systematic nomenclature.

27.48 Give the formula and name of a square-planar complex of Pt^{2+} with one nitrite ion (NO_2^-, which binds to Pt^{2+} through the N atom), one chloride ion, and two ammonia molecules as ligands. Are isomers possible? If so, draw the structure of each isomer, and tell what type of isomerism is observed.

27.49 The glycinate ion ($H_2NCH_2COO^-$), formed by deprotonation of the amino acid glycine, can function as a bidentate ligand, coordinating to a metal through the nitrogen of the amino group and one of the oxygen atoms.

Glycinate ion, a bidentate ligand

Site of bonding to transition metal ion

A copper complex of this ligand has the formula $[Cu(H_2NCH_2COO)_2(H_2O)_2]$. For this complex, determine the following:

(a) the oxidation state of copper

(b) the coordination number of copper

(c) the number of unpaired electrons

(d) whether the complex is diamagnetic or paramagnetic

27.50 Draw structures for the five possible stereoisomers of $[Cu(H_2NCH_2COO)_2(H_2O)_2]$. Are any of these species chiral? (See the structure of the ligand in Question 27.49.)

27.51 The transition metals form a class of compounds called metal carbonyls, an example of which is the tetrahedral complex $[Ni(CO)_4]$. Given the following thermodynamic data:

	$\Delta_f H°$ (kJ mol^{-1})	S° (J K^{-1} mol^{-1})
Ni	0	29.87
CO(g)	−110.525	+197.674
Ni(CO)$_4$(g)	−602.9	+410.6

(a) Calculate the equilibrium constant for the formation of $Ni(CO)_4(g)$ from nickel metal and CO gas.

(b) Is the reaction of Ni(s) and CO(g) product-favoured or reactant-favoured?

(c) Is the reaction more or less product-favoured at higher temperatures? How could this reaction be used in the purification of nickel metal?

27.52 In this question, we explore the differences between metal coordination by monodentate and bidentate ligands. Formation constants, K_f, for $[Ni(NH_3)_6]^{2+}$(aq) and $[Ni(en)_3]^{2+}$(aq) are as follows:

$$Ni^{2+}(aq) + 6\ NH_3(aq) \longrightarrow [Ni(NH_3)_6]^{2+}(aq) \quad \beta = 10^8$$

$$Ni^{2+}(aq) + 3\ en(aq) \longrightarrow [Ni(en)_3]^{2+}(aq) \quad \beta = 10^{18}$$

The difference in K_f between these complexes indicates a higher thermodynamic stability for the chelated complex, caused by the *chelate effect*. Recall that β (like all equilibrium constants) is related to the standard free energy of the reaction by $\Delta_r G° = -RT \ln \beta$, and $\Delta_r G° = \Delta_r H° - T\Delta_r S°$. We know from experiment that $\Delta_r H°$ for the reaction with NH_3 reaction is -109 kJ mol^{-1}, and $\Delta_r H°$ for the reaction with ethylenediamine is -117 kJ mol^{-1}. Is the difference in $\Delta_r H°$ sufficient to account for the 10^{10}-fold difference in β? Comment on the role of entropy in the second reaction.

The Chemistry of Modern Materials

Peter Titmuss/Getstock.com

28.1 Case Study: Materials: Ancient and Modern Building Blocks

If they could speak, the white cliffs of Dover would tell many stories of their importance in British history, including their vital role in building defences dating as far back as the time of the Roman invasion during Julius Caesar's reign in 55 B.C. But the pure white chalky substance of which they are composed, streaked with flecks of black flint, would also tell much older chemical stories. You might learn of their origin through the uptake of atmospheric carbon dioxide by tiny algae, coral, sponges, and other sea creatures, followed by the deposition of their calcium carbonate skeletons more than 70 million years ago when a shallow sea covered this part of Britain. Then, after being compressed under thick glaciers, catastrophic floods several hundred thousand years ago from bursting glacial lakes carved the cliffs of the English Channel through these massive silica-speckled calcium carbonate deposits.

Silica and calcium carbonate—these and other substances with a similarly rich natural history now form the basis for the construction of 21st century structural materials. The properties of materials depend on the structural arrangement of their molecules, atoms, or ions. By studying the composition and structure of naturally occurring materials, chemists and engineers are able to gain insight into what gives each material its properties. They can then use synthetic techniques to create materials that have predictable behaviours, making them suitable for particular applications. Chemists can also extract and purify naturally occurring substances that have properties desirable for specific uses. Ironically, they've learned that it is often advantageous to intentionally add small amounts of impurities to give a material particular electrical or thermal properties. The study and synthesis of structural substances is the general domain of **materials science.** While chemistry serves as the foundation of materials science, the field crosses into physics, biology, and engineering.

The properties of materials depend on the structural arrangement of their molecules, atoms, or ions. By studying the composition and structure of naturally occurring materials, chemists and engineers are able to gain insight into what gives each material its properties.

FIGURE 28.1 Forms of calcium carbonate. *(Clockwise from top)* The shell of an abalone, a limestone paving block from Europe, crystalline aragonite, common blackboard chalk ($CaCO_3$ and a binder), and transparent Iceland spar.

In Chapter 11, we focused on the properties that many gases, regardless of their identity, have in common under certain conditions. Unlike gases, the solids that we focus on in this chapter are usually characterized by their remarkable differences, as illustrated by the forms of calcium carbonate.

The white chalk of Dover cliffs consists of the mineral calcite, one form of calcium carbonate. The minerals calcite, aragonite, and Iceland spar are all composed mostly of calcium carbonate. So are limestone, eggshells, and sea shells. These materials have distinctly different physical characteristics (Figure 28.1), yet they are composed primarily of the same particles, Ca^{2+} and CO_3^{2-} ions. The differences in the macroscopic characteristics result from small differences in composition (due to the presence of impurities) and differences in the ways the ions are arranged in the substance.

The flecks of flint impurities in the white chalk of the cliffs are a form of the mineral quartz, which is a crystal lattice of silicon dioxide (SiO_2, also called silica). Silicon dioxide is another important building block of modern materials. Often carefully controlled amounts of impurities are added, leading to remarkable changes in properties of the material.

This chapter explores a variety of ancient and modern materials and examines the connection between composition, atomic arrangements, and bulk properties. Organic polymers were described in Sections 4.6 and 24.7.

28.2 Metals

Bonding in Metals

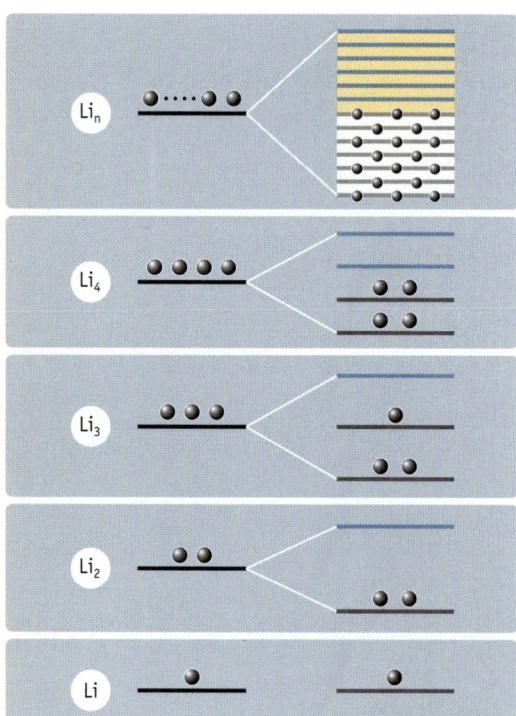

FIGURE 28.2 Bands of molecular orbitals in a metal crystal. In this case, the 2s valence orbitals of Li atoms are combined in a thought exercise to form molecular orbitals. As more and more atoms are added, the number of MOs increases until the orbitals are so close in energy that they merge into a band of molecular orbitals. If the 2s and 2p valence atomic orbitals of 1 mol of Li atoms are combined, 24×10^{23} molecular orbitals are formed. There are only 1 mol (6×10^{23}) of valence electrons, so only a small fraction of the MOs is "occupied" by electron pairs.

Molecular orbital (MO) theory was introduced in Section 10.8 as a complementary model to rationalize covalent bonding. Recall the basic outline used in MO theory: In a thought exercise, atomic orbitals from different atoms in a molecule are mathematically combined to form molecular orbitals spanning two or more atoms, with the number of MOs equal to the number of atomic orbitals combined. Electrons "occupy" the lower-energy, bonding MOs, so the molecule is more stable than the separated atoms from which it is made.

MO theory can also be used to describe metallic bonding. A metal is a kind of "supermolecule," and to describe the bonding in a metal, we have to consider all the atoms in a sample.

Even the tiniest piece of a metal contains a very large number of atoms and an even larger number of valence orbitals. As an example, let's look at 1 mol of lithium atoms (6×10^{23} atoms). Considering only the 2s and 2p valence orbitals of lithium, there are $4 \times 6 \times 10^{23}$ atomic orbitals, from which 24×10^{23} molecular orbitals can be created. The molecular orbitals that we envision in lithium will span all the atoms in a crystal. A mole of lithium has 1 mol of valence electrons, and these electrons occupy the lower-energy bonding orbitals. The bonding is described as delocalized; that is, the valence electrons are associated with all the atoms in the crystal and not with particular bonds between pairs of atoms.

The theory of metallic bonding is called **band theory.** An energy-level diagram would show the molecular orbitals blending together into a band of molecular orbitals (Figure 28.2), with the MOs being so close together in energy that they are indistinguishable. The band is composed of as many molecular orbitals as there are contributing atomic orbitals, and each molecular orbital can accommodate two electrons of opposite spin.

In metals, there are not enough electrons to fill all of the molecular orbitals. In 1 mol of Li atoms, for example, 6×10^{23} valence electrons are sufficient to fill only 1/8 of the 24×10^{23} MOs—remembering that the orbitals are "occupied" by pairs of electrons. The lowest energy for a system occurs with all electrons in orbitals with the lowest possible energy, but this can be reached only at 0 K. At 0 K, the highest filled level is called the **Fermi level** (Figure 28.3).

At temperatures above 0 K, the thermal energy causes some of the electrons to occupy higher-energy orbitals. Even a small input of energy (for example, raising the temperature a few degrees above 0 K) will cause electrons to move from filled orbitals to higher-energy orbitals. For each electron promoted, two singly occupied levels result: a negative electron in an orbital above the Fermi level and a positive "hole"—from the absence of an electron—below the Fermi level.

The electrical conductivity of a metal arises from the movement of electrons and holes in singly occupied states in the presence of an applied electric field. When an electric field is applied, negative electrons move toward the positive side, and the positive "holes" move to the negative side. (Positive holes "move" because an electron from a filled orbital can move into the hole, thereby leaving behind a new "hole.")

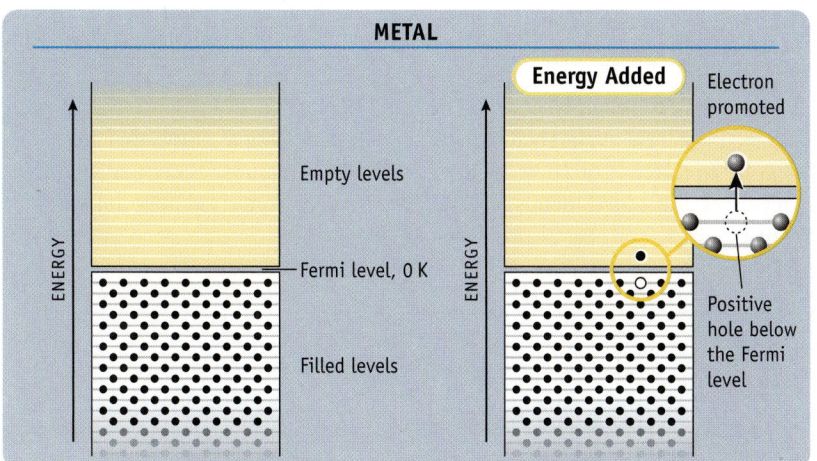
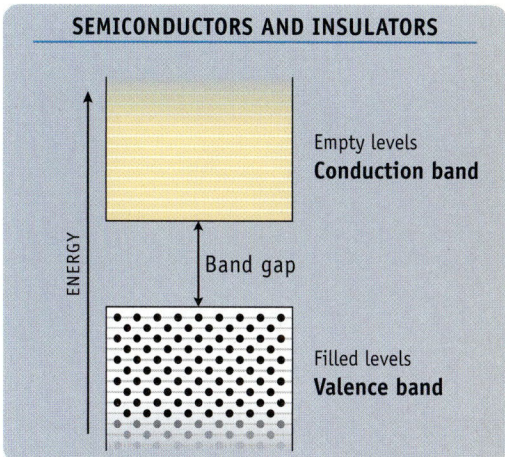

FIGURE 28.3 Band theory applied to metals, semiconductors, and insulators. The bonding in metals and semiconductors can be described using molecular orbital theory. MOs, which can be considered to be created by mathematical combination of the valence atomic orbitals, are delocalized over all the atoms. *(Metal, left and centre)* The highest filled level at 0 K is referred to as the Fermi level. *(Semiconductors and insulators, right)* In contrast to metals, the band of filled levels (the valence band) is separated from the band of empty levels (the conduction band) by a significant energy difference, called the "band gap." In insulators, the size of the band gap is large.

Because the band of unfilled energy levels in a metal is essentially continuous—that is, because the energy gaps between levels are extremely small—a metal can absorb energy of nearly any wavelength. When light is absorbed, causing an electron in a metal to have a higher energy state, the now-excited system can immediately emit a photon of the same energy, and the electron returns to the original energy level. This rapid and efficient absorption and re-emission of light makes polished metal surfaces reflective, appearing lustrous (shiny).

The MO picture for metallic bonding provides an interpretation for other physical characteristics of metals. For example, most metals are malleable and ductile, meaning they can be rolled into sheets and drawn into wires. In these processes, the metal atoms must be able to move reasonably freely with respect to their nearest neighbours. This is possible because metallic bonding is delocalized and non-directional. The layers of atoms can slide past one another relatively easily, as if the delocalized electrons were ball bearings that allow this motion, while at the same time keeping the layers bonded through coulombic attractions between the nuclei and the electrons.

In contrast to metals, rigid covalent network solids such as diamond, silicon, and silica (SiO_2) have localized bonding between pairs of atoms, which anchors the component atoms or ions in fixed positions [<<Section 3.3]. Movement of atoms in these structures relative to neighbouring atoms requires breaking of chemical bonds. As a result, such substances are typically hard and brittle. They will not deform under stress as metals do, but instead tend to cleave along crystal planes.

Band theory can be used to describe the bonding in metals and semiconductors. Molecular orbitals are constructed from the valence orbitals on each atom and are delocalized over all the atoms. The highest filled level at 0 K is referred to as the Fermi level. In contrast to metals, the band of filled levels (the valence band) is separated from the band of empty levels (the conduction band) by a band gap. In insulators, the energy of the band gap is large.

Alloys: Mixtures of Metals

Pure metals often do not have the ideal properties suited to particular uses. The properties can sometimes be improved, however, by adding one or more other elements to the metal to form an alloy (Table 28.1). In fact, most metallic objects we use are made of **alloys**, mixtures of a metal with one or more other metals or even with a non-metal such as carbon (as in carbon steel). For example, sterling silver, commonly used for jewellery, is an alloy composed of 92.5% Ag and 7.5% Cu. Pure silver is soft and easily damaged, and the addition of copper makes the metal more rigid. You can confirm that an article of jewellery is sterling silver by looking for the stamp that says "925."

Gold used in jewellery is rarely pure (24 karat, 24 carat, or 24 kt) gold. More often, you will find 18 kt, 14 kt, or 9 kt stamped in a gold object, referring to alloys that are 18/24, 14/24, or 9/24 gold. In the case of 18 kt "yellow" gold, the remainder is copper and silver. As with sterling silver, the added metals lead to a harder and more rigid material (and one that is less costly).

TABLE 28.1 Some Common Alloys

Sterling silver	92.5% Ag, 7.5% Cu
18 kt "yellow" gold	75% Au, 12.5% Ag, 12.5% Cu
Pewter	91% Sn, 7.5% Sb, 1.5% Cu
Low-alloy steel	98.6% Fe, 1.0% Mn, 0.4% C
Carbon steels	Approximately 99% Fe, 0.2–1.5% C
Stainless steel	72.8% Fe, 17.0% Cr, 7.1% Ni, and approximately 1% each of Al and Mn
Alnico magnets	10% Al, 19% Ni, 12% Co, 6% Cu, remainder Fe
Brass	95–60% Cu, 5–40% Zn
Bronze	90% Cu, 10% Sn

Alloys are categorized into three classes: solid solutions, which are homogeneous mixtures of two or more elements; heterogeneous mixtures; and intermetallic compounds.

In solid solutions, one element is usually considered the "solute" and the other the "solvent." As with solutions in liquids, the solute atoms are randomly and evenly dispersed throughout the solid such that the bulk structure is *homogeneous*. Unlike liquid solutions, however, there are limitations on the sizes of solvent and solute atoms. For a solid solution to form, the solute atoms must be incorporated into the structure in such a way that the crystal structure of the solvent metal is essentially preserved. Solid solutions can be achieved in two ways: with solute atoms as *interstitial* atoms, or as *substitutional* atoms in the crystalline lattice. In interstitial alloys, the solute atoms occupy the interstices, the small spaces between solvent atoms (Figure 28.4a). The solute atoms must be substantially smaller than the metal atoms making up the lattice to fit into these positions. In substitutional alloys, the solute atoms replace solvent atoms in the original crystal structure (Figure 28.4b). For this to occur, the solute and solvent atoms must be very similar in size.

If the size constraints are not met, then the alloy is more likely to form a heterogeneous mixture: when viewed under a microscope, regions of different composition and crystal structure can be seen (Figure 28.5).

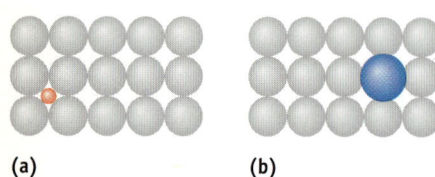

(a)   (b)

FIGURE 28.4 Alloys. (a) The solute atoms may be interstitial atoms, fitting into holes in the crystal lattice. (b) The solute atoms can also substitute for lattice atoms.

For a solid solution to form, the electronegativities of the alloy components must also be similar. When the two metals have different electronegativities, the possibility exists for *intermetallic compounds*—substances with a definite stoichiometry and formula. Examples of intermetallic compounds include those with formulas $CuAl_2$, Mg_2Pb, and $AuCu_3$. In general, intermetallic compounds are likely when one element is relatively electronegative and the other is more electropositive. In Mg_2Pb, for example, χ for Pb = 2.3 and χ for Mg = 1.3 ($\Delta\chi = 1.0$).

The macroscopic properties of an alloy vary depending on the ratio of the elements in the mixture. For example, "stainless" steel is highly resistant to corrosion and is roughly five times stronger than carbon and low-alloy steels. Melting point, electrical resistance, thermal conductivity, ductility, and other properties can also be adjusted by changing the compositions of alloys.

Metals and their alloys are good examples of how small changes in the atomic composition and structure of a crystalline substance can have profound effects on its bulk chemical and physical characteristics. The same is true of semiconductors, the next class of materials that we will examine.

Most metallic objects we use are alloys, mixtures of a metal with one or more other metals or a non-metal such as carbon. Alloys can be categorized into three classes: solid solutions, heterogeneous mixtures, and intermetallic compounds.

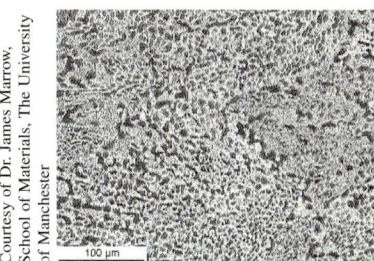

FIGURE 28.5 Photomicrograph of the surface of a heterogeneous alloy of lead and tin.

Courtesy of Dr. James Marrow, School of Materials, The University of Manchester

28.3 Semiconductors

Semiconducting materials are at the heart of all solid-state electronic devices, including such well-known devices as computer chips and diode lasers. **Semiconductors** will not conduct electricity easily but must be encouraged to do so by the input of energy. This property allows devices made from semiconductors to essentially have "on" and "off" states, which form the basis of the binary logic used in computers. We can understand how semiconductors function by looking at their electronic structure, following the molecular orbital model for bonding used for metals.

Bonding in Semiconductors: The Band Gap

The group 14 elements carbon (in the diamond form), silicon, and germanium have similar structures. Each atom is surrounded by four other atoms at the corners of a tetrahedron (Figure 28.6).

Using the band model of bonding, we can carry out another thought experiment in which we imagine combining the orbitals of each carbon atom (or each silicon or germanium atom) to form molecular orbitals that are delocalized over the solid. Unlike with metals, however, the result for group 14 elements is two bands. The partially filled band of molecular orbitals typical of a metal is split into a lower-energy *valence band* and a higher-energy *conduction band*. In metals, there is no energy barrier for an electron to go from the filled molecular orbitals to empty molecular orbitals, and electricity can flow easily. In electrical insulators, such as diamond, and in semiconductors, such as silicon and germanium, the valence and conduction bands are separated from each other by an energy barrier, known as the *band gap* (Figure 28.3). In the group 14 elements, the orbitals of the valence band are filled, but the conduction band is empty.

The band gap in diamond is 580 kJ mol^{-1}—so large that electrons are trapped in the filled valence band and cannot make the transition to the conduction band, even at elevated temperatures. So, it is not possible to create positive "holes," and diamond is an insulator, or a non-conductor. Semiconductors, in contrast, have a smaller band gap. For common semiconducting materials, this band gap is usually in the range of 10–240 kJ mol^{-1}. The band gap is 106 kJ mol^{-1} in silicon, and 68 kJ mol^{-1} in germanium. The relatively small band gap in semiconductors is such that these substances are able to pass small currents, but, as their name implies, they are much poorer conductors than metals.

FIGURE 28.6 The structure of diamond. The structures of silicon and germanium are similar to that of diamond in that each atom is bound tetrahedrally to four others.

Charles D. Winters

Semiconductors can conduct a current because thermal energy is sufficient to promote some of the electrons from the valence band to the conduction band (Figure 28.7). Conduction then occurs when the electrons in the conduction band migrate in one direction and the positive holes in the valence band migrate in the opposite direction.

Pure silicon and germanium are called *intrinsic semiconductors*: their conducting ability is an intrinsic property of the pure material. In these, the number of electrons in the conduction band is governed by the temperature and the magnitude of the band gap. The smaller the band gap is, the smaller the energy required to promote a significant number of electrons. As the temperature increases, more electrons are promoted into the conduction band and a higher conductivity results.

In contrast to intrinsic semiconductors are materials known as *extrinsic semiconductors*. The conductivity of these materials is controlled by adding small numbers of atoms (typically 1 in 10^6 to 1 in 10^8) of different elements called *dopants*. That is, the characteristics of semiconductors can be changed by altering their chemical makeup, just as the properties of alloys differ from the properties of pure metals.

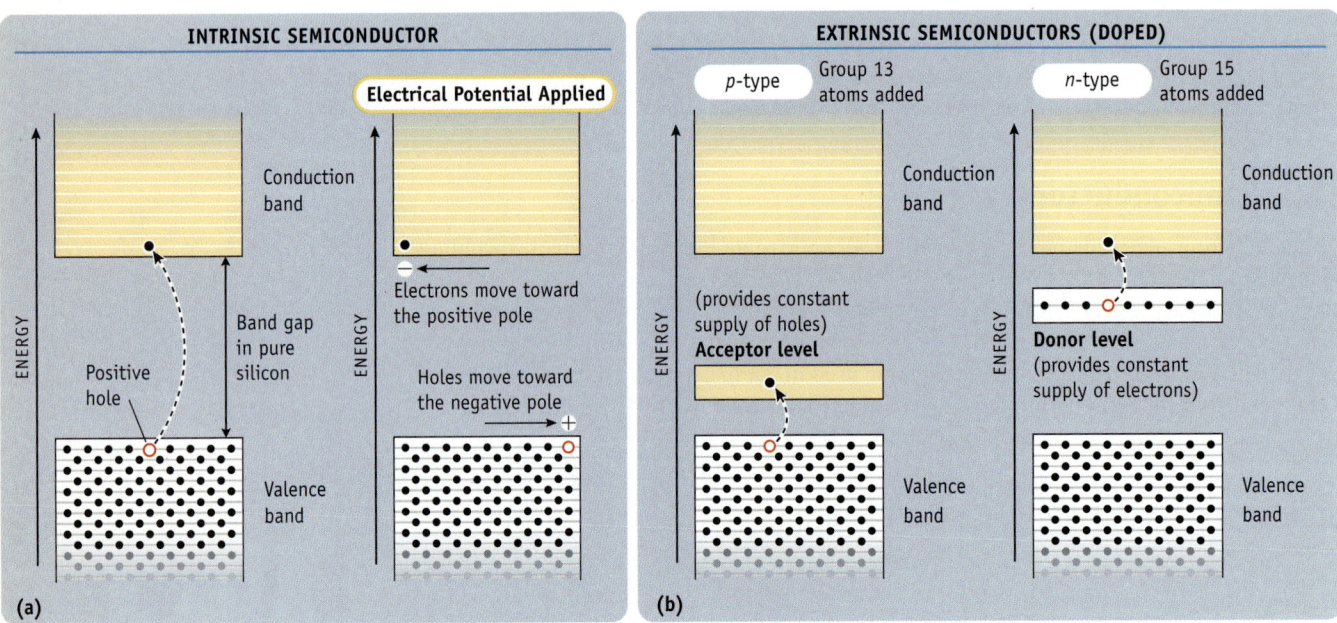

FIGURE 28.7 Intrinsic and extrinsic semiconductors.

Suppose a few silicon atoms in the silicon lattice are replaced by aluminum atoms (or atoms of some other group 13 element). Aluminum has only three valence electrons, whereas silicon has four. Four Si–Al bonds are created per aluminum atom in the lattice, but these bonds must be deficient in electrons. According to band theory, the Si–Al bonds form a discrete band at an energy level higher than the valence band. This level is referred to as an *acceptor level* because it can accept electrons. The gap between the valence band and the acceptor level is usually quite small, so electrons can be promoted readily to the acceptor level. The positive holes created in the valence band are able to move about under the influence of an electric potential, so electrical conduction results from the hole mobility. Because positive holes are created in an aluminum-doped semiconductor, this is called a *p*-type semiconductor (Figure 28.7b, left).

Now suppose phosphorus atoms (or atoms of some other group 15 element such as arsenic) are incorporated into the silicon lattice instead of aluminum atoms. The material is still a semiconductor, but it now has extra electrons because phosphorus has one more valence electron than silicon. Semiconductors doped in this manner have a discrete, partially filled *donor level* that resides just lower in energy than the conduction band. Electrons are promoted readily to the conduction band from this donor band, and electrons in the conduction band carry the charge. Such a material, consisting of negative charge carriers, is called an *n-type semiconductor* (Figure 28.7b, right).

One group of materials that have desirable semiconducting properties are the 13–15 semiconductors, so called because they are formed by combining elements from group 13 (such as Ga and In) with elements from group 15 (such as As or Sb). Replacing Si atoms in pure silicon with equal numbers of Ga and As atoms, for example, does not change the number of valence electrons present. (Two Si atoms, for example, have eight valence electrons, as does the combination of a Ga atom with an As atom.)

Gallium arsenide (GaAs) is a common semiconducting material that has electrical conductivity properties that are sometimes preferable to those of pure silicon or germanium. The crystal structure of GaAs is similar to that of diamond; each Ga atom is tetrahedrally coordinated to four As atoms, and vice versa. This structure is often referred to as the *zinc blende* structure.

It is also possible for group 12 and 16 elements to form semiconducting compounds, such as CdS. The farther apart the elements are found in the periodic table, however, the more ionic the bonding becomes. As the ionic character of the bonding increases, the band gap increases and the material becomes an insulator rather than a semiconductor. For example, the band gap in GaAs(s) is 140 kJ mol^{-1}, whereas it is 232 kJ mol^{-1} in CdS(s).

These materials can be modified further by substituting other atoms into the structure. For example, in one widely used semiconductor, aluminum atoms are substituted for gallium atoms in GaAs(s), giving materials with a range of compositions (Ga$_{1-x}$Al$_x$As). The importance of this modification is that the band gap depends on the relative proportions of the elements, so it is possible to control the size of the band gap by adjusting the stoichiometry. As Al atoms are substituted for Ga atoms, for example, the band gap energy increases. This consideration is important for the specific uses of these materials in devices such as LEDs.

Applications of Semiconductors: Diodes, LEDs, and Transistors

The combination of *p*- and *n*-type semiconducting materials in a single electronic device launched the microelectronics and computer industries. When a semiconductor is created such that it is *p*-type on one half and *n*-type on the other, a marvellous device known as the *p-n rectifying junction,* or *diode,* results. It allows current to flow easily in only one direction when a voltage is applied. The *p-n* junction is the fundamental building block of solid-state electronic devices. It is used for many circuitry applications, such as switching and converting between electromagnetic radiation and electric current.

The lights you see on the dashboard of your car and in its rear warning lights, in traffic lights, Christmas trees, and children's shoes are *LEDs,* or *light-emitting diodes* (Figure 28.8). These semiconducting devices are made by combining elements such as gallium, phosphorus, arsenic, and aluminum. When attached to a low-voltage (say 6–12 V) source, they emit light with a wavelength that depends on their composition. Furthermore, they emit light with a brightness that rivals standard incandescent lights, and the light can be focused using a tiny plastic lens. Because their energy consumption is low, LED lights hold a great deal of promise for household lighting.

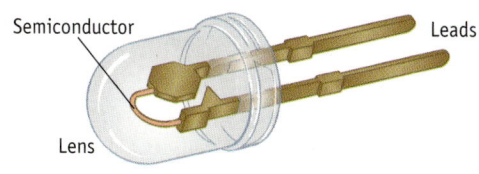

Semiconductor · Leads · Lens

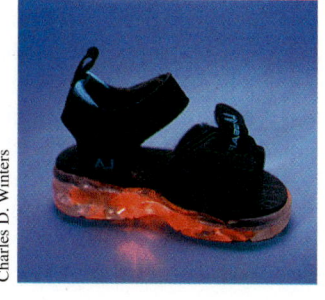

Charles D. Winters

FIGURE 28.8 (*Left*) A schematic drawing of a typical light emitting diode (LED). (*Right*) A child's shoe with red LEDs.

An LED has a simple construction. It consists of a *p*-type semiconductor joined to an *n*-type semiconductor (Figure 28.9). A voltage is applied to the material, perhaps by hooking the positive terminal of a battery to the *p*-type semiconductor and the negative terminal to the *n*-type semiconductor. Negative electrons move from the *n*-type to the *p*-type, and positive holes move from the *p*-type to the *n*-type. When electrons move across the *p-n* junction, they can drop from the conduction band into a hole in the valence band of the *p*-type semiconductor, and energy is released. (The mechanism of light emission by an LED is similar to that described for excited atoms in Section 8.3.) If the band gap energy is equivalent to the energy of light in the visible region, light can be observed. Because the band gap energy can be adjusted by changing the composition of the doped semiconductor, the wavelength of the light can also be altered.

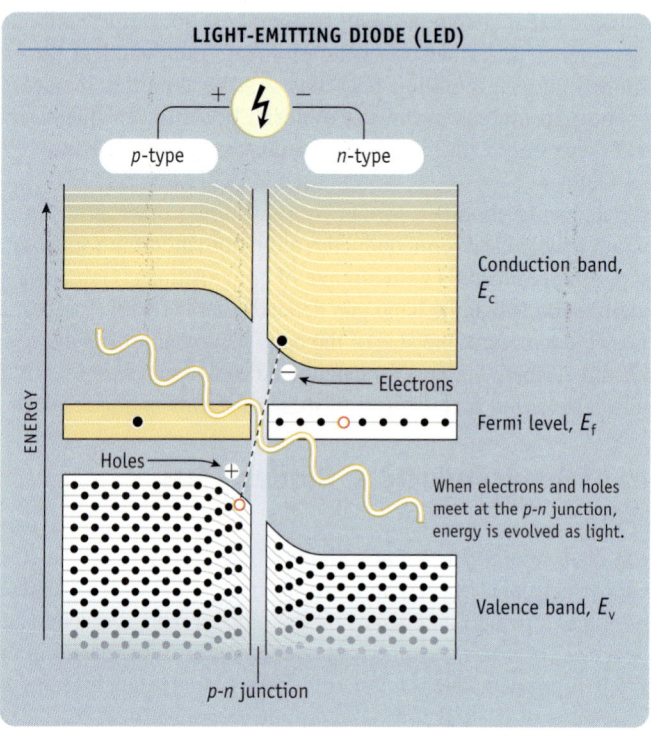

LIGHT-EMITTING DIODE (LED)

p-type   *n*-type

ENERGY

Conduction band, E_c

Electrons

Fermi level, E_f

Holes

When electrons and holes meet at the *p-n* junction, energy is evolved as light.

Valence band, E_v

p-n junction

FIGURE 28.9 Mechanism for the emission of light from an LED constructed from *n*- and *p*-type semiconductors. When *p*- and *n*-type semiconductors are joined, the energy levels adjust so that the Fermi levels (E_F) are equal. This causes the energy levels of the conduction (E_c) and valence (E_v) bands to "bend." Also, holes flow from the *p* side to the *n* side, and electrons flow from *n* to *p* until equilibrium is reached. No more charge will flow until a voltage is applied. When an electric field is applied, occasionally electrons in the conduction band will move across the band gap and combine with holes in the valence band. Energy is then evolved as light. The energy of the emitted light is approximately equal to the band gap. Therefore, by adjusting the band gap, the colour of the emitted light can be altered. (See S.M. Condren, et al., (2001), "LEDs: New Lamps for Old and a Paradigm for Ongoing Curriculum Modernization," *Journal of Chemical Education*, 78: 1033–40.)

NASA JPL

FIGURE 28.10 Gallium arsenide (GaAs) solar panel. This panel was built for NASA's Deep Space 1 probe. The array uses 3600 solar cells, which convert light to electricity to power an ion propulsion system. (DS1 was launched on October 24, 1998, and sent back images of Comet Borrelly in deep space. The spacecraft was retired on December 18, 2001.)

The same device that forms the LED can be run in reverse to convert light that falls on it into an electrical signal. Solar panel cells work in this manner (Figure 28.10). They are generally GaAs-based *p-n* junction materials that have a band gap corresponding to the energy of visible light. When sunlight falls on these devices, a current is induced. That current can be used either immediately or stored in batteries for later use. A similar technology is used in simpler devices referred to as photodiode detectors. They have an abundance of applications, ranging from the light-sensitive switches on elevator doors to sensitive detection equipment for spectrometers and other scientific instruments.

The *p*- and *n*-type semiconductor materials can also be constructed into a sandwich structure of either *p-n-p* or *n-p-n* composition. This arrangement forms a device known as a *transistor*. A transistor amplifies an electrical signal, making it ideal for powering loudspeakers, for example. Transistors can also be used for processing and storing information, a critical function for computer chips. By combining thousands of these transistors and diodes, an integrated circuit can be made that is the basis of what we commonly refer to as computer chips, devices for controlling and storing information (Figure 28.11).

FIGURE 28.11 Integrated circuits. (*Left*) A wafer on which a large number of integrated circuits has been printed. (*Right*) A close-up of a semiconductor chip showing the complex layering of circuits that is now possible.

Diodes, LEDs, and transistors are semiconductors, materials that can be encouraged to conduct electricity by the input of energy. The properties of semiconductor materials can be changed by adjusting the band gap.

Microfabrication Techniques Using Semiconductor Materials

The technology for semiconductor fabrication has become so sophisticated that scientists are reaching the point where they can make features as small as 50–90 nm across. The techniques have been perfected to the level where new devices called microelectromechanical systems (MEMS) are being developed. These machines have moving parts so small that a spider mite, about 0.4 mm (400 000 nm) across, appears colossal in comparison (Figure 28.12). MEMS devices are being used in "smart" materials applications, where microsensors will detect signals from the environment around them and cause certain

 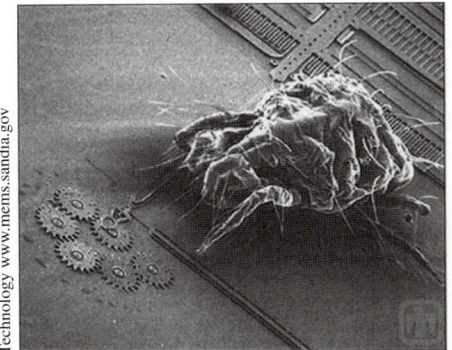 

FIGURE 28.12 Microelectromechanical systems (MEMS). These tiny systems are made from polycrystalline silicon and have promising applications in many fields. In the top photograph, a spider mite is crawling on the device. A spider mite is about 0.8 mm long and 0.4 mm across.

responses in the MEMS devices. An example of an application is in the deployment of airbags, where the MEMS device acts as an accelerometer and triggers the release of the bag when the vehicle comes to an abrupt stop.

28.4   Ceramics

Having looked at metals and semiconductors, let's return to our original examples of various forms of calcium carbonate ($CaCO_3$), including sea shells and chalk. Chalk is so soft that it will rub off on the rough surface of a blackboard. In contrast, sea shells are inherently tough, and they can protect their soft and vulnerable inhabitants from the powerful jaws of sea-borne predators or rough conditions underwater. Chalk, sea shells, and the spines of sea urchins (Figure 28.13) are all ceramics, but they are obviously different from one another. Clearly, there is a great deal of variability in the single class of materials known as ceramics.

FIGURE 28.13 A sea urchin. The spines of the urchin are composed primarily of $CaCO_3$ and also contain $MgCO_3$.

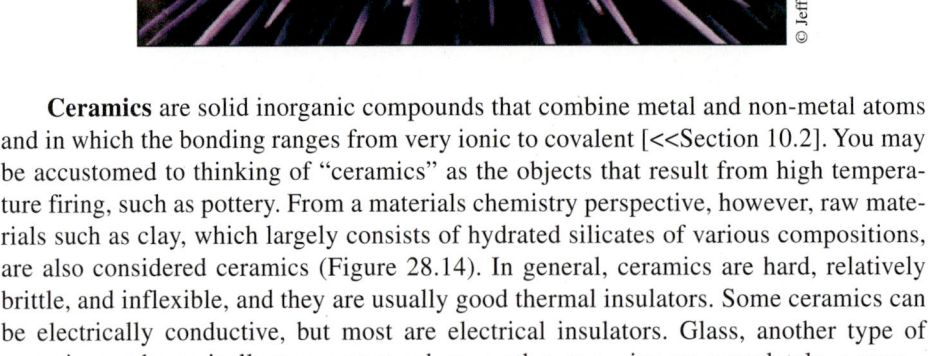

Ceramics are solid inorganic compounds that combine metal and non-metal atoms and in which the bonding ranges from very ionic to covalent [<<Section 10.2]. You may be accustomed to thinking of "ceramics" as the objects that result from high temperature firing, such as pottery. From a materials chemistry perspective, however, raw materials such as clay, which largely consists of hydrated silicates of various compositions, are also considered ceramics (Figure 28.14). In general, ceramics are hard, relatively brittle, and inflexible, and they are usually good thermal insulators. Some ceramics can be electrically conductive, but most are electrical insulators. Glass, another type of ceramic, can be optically transparent, whereas other ceramics are completely opaque.

It is also possible to have ceramics in which impurity atoms are included in the composition. As we saw with metals and semiconductors, impurity atoms can have dramatic effects on the characteristics of a material.

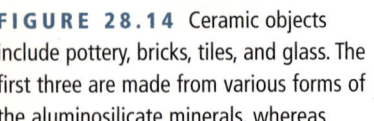

FIGURE 28.14 Ceramic objects include pottery, bricks, tiles, and glass. The first three are made from various forms of the aluminosilicate minerals, whereas glass is primarily silicon dioxide.

Glass: A Disordered Ceramic

An amorphous, or non-crystalline, solid structure is generally referred to as a glass. Glasses are formed by melting the raw ceramic material and then cooling it from the liquid state rapidly so that the component atoms do not have time to crystallize into a regular lattice structure. A wide range of materials, including metals and organic polymers, can be coaxed into a glassy form. However, the best-known glasses are silicate glasses. These are derived from SiO_2, which is plentiful, inexpensive, and chemically unreactive. Each silicon atom is linked to four oxygen atoms in the solid structure, with a tetrahedral arrangement around each silicon atom. The SiO_2 units are linked together to form a large network of atoms (Figure 28.15). Over a longer distance, however, the network has no discernable order or pattern.

Glasses can be modified by the presence of alkali metal oxides (such as Na_2O and K_2O) or other metal or non-metal oxides (such as CaO, B_2O_3, and Al_2O_3). The added impurities change the silicate network and alter the properties of the material. The oxide ions are incorporated into the silicate network structure, and the resultant negative charge is balanced by the interstitial metal cations (Figure 28.15c). Because the network is changed by such an addition, these network modifiers can dramatically alter the physical characteristics of the material such as melting point, colour, opacity, and strength. Soda-lime glass—made from SiO_2, Na_2O (soda), and CaO (lime)—is a common glass used in windows and for containers. The metal oxides lower this glass's melting temperature by about a thousand degrees from that of pure silica. Pyrex glass, also called borosilicate glass, incorporates boric oxide. The boric oxide raises the softening temperature and minimizes the coefficient of thermal expansion, enabling the glass to better withstand temperature changes. Because of its excellent thermal properties, this type of glass is used for beakers and flasks in chemistry laboratories and for ovenware for the kitchen.

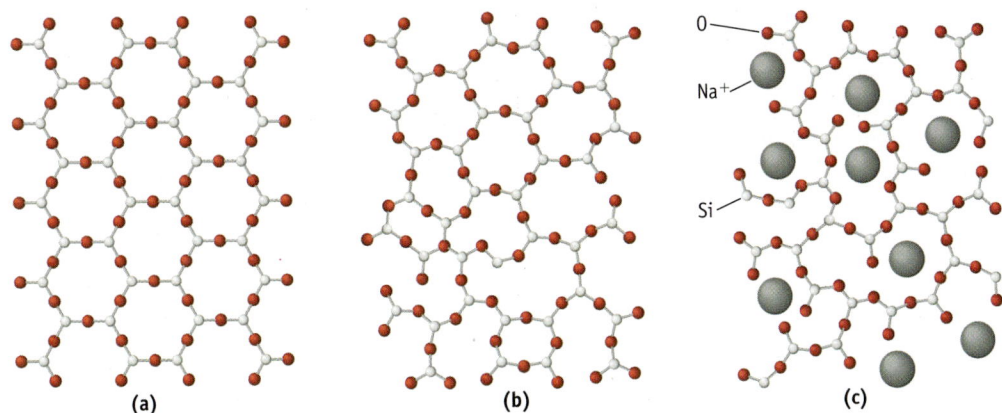

(a)   **(b)**   **(c)**

FIGURE 28.15 Molecular-level representation of glass structure. (a) Silica glass (SiO_2) may have some order over a short distance but much less order over a larger portion of the solid (b). (c) The SiO_2 structure can be modified by adding metal oxides, which leads to a lower melting temperature and other desirable properties. (In this simple representation, the Si atoms are shown at the centre of a planar triangle of O atoms; in reality, each Si atom is surrounded tetrahedrally by O atoms.)

An important characteristic of glasses is their optical transparency, which allows them to be used as windows and lenses. Glasses can also be reflective. The combination of transparency and reflectivity of a material is controlled by its *refractive index* (*n*): the ratio of the velocity of light in a vacuum to the velocity of light in the material. The refractive index of water is 1.333 (the velocity of light in water is about 75% of that in a vacuum), and typical values for silicate glasses are in the range of 1.5 to 1.9. Police authorities use the different refractive indices of glass to identify fragments found at crime scenes.

When a beam of light in air strikes the surface of a material, some of the light is reflected, with the angle of reflection to a line perpendicular to the surface the same as the angle of incidence, and some enters the material. The change in the velocity of the light as it enters the material causes the beam to change direction (Figure 28.16a). You can observe this effect by putting a long object in a glass of water and you will see an apparent bend in the object at the air-water interface (Figure 28.16b). The fraction of light that is reflected and the refracted angle (the change in direction of the beam as it enters the material) depend on both the incident angle and refractive index of the material.

FIGURE 28.16 Refraction of light. (a) When light enters a different medium, its velocity changes. This causes the path of a photon to change direction at the interface. (b) Observing an object in a glass of water illustrates the effect of light refraction.

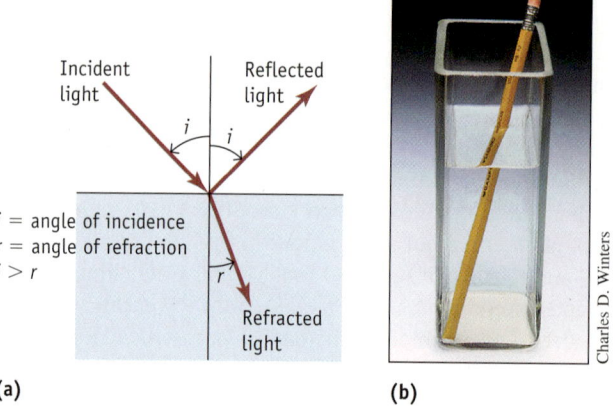

Incident light

Reflected light

i = angle of incidence
r = angle of refraction
$i > r$

Refracted light

(a)

(b)

Charles D. Winters

This combination of the transmission and reflection characteristics of glass has allowed scientists and engineers to develop optical fibres (Figure 28.17). Optical fibres are designed to have a property called *total internal reflection,* whereby all the light that enters at one end of the fibre stays within the fibre because of reflections at the interior surface as the light travels from one end of the fibre to the other. Total internal reflection in these fibres is achieved by controlling the ratio of the refractive indices between the fibre's core and its outside surface. Chemically, the refraction index is controlled by adjusting the quantity and type of cationic network modifiers that are added to the glass. The refraction index of a glass fibre can be controlled so that it has one value at the core of the fibre but changes gradually across the radius of the fibre to a different value at the surface. This is accomplished by an ion-exchange process during fibre production in which, for example, K^+ ions are replaced by Tl^+ ions.

Optical fibres are being used to transform the communications industry in an amazing fashion. Instead of transmitting information using electrons travelling through wires, optical fibres allow communication to occur by transmitting photons through glass fibre bundles. Signal transmission by optical fibres, known as *photonics,* is much faster and more economical than using copper wires and cables. For example, the quantity of copper required to carry the equivalent amount of information transmitted by optical fibre would weigh 300 000 times more than the optical fibre material!

FIGURE 28.17 Optical fibres. (*Left*) Glass fibres transmit light along the axis of the fibre. (*Right*) Recent discoveries show that a deep-sea sponge, made chiefly of silica (SiO_2), has a framework that has the characteristics of optical fibres.

Simon Fraser/Photo Researchers, Inc.

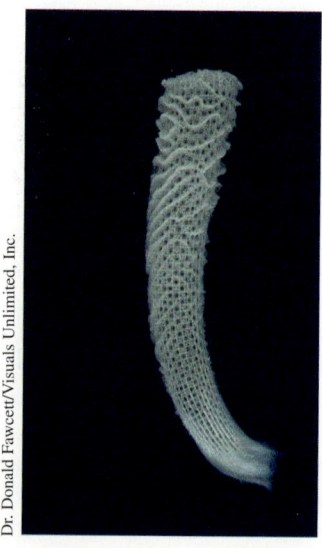

Dr. Donald Fawcett/Visuals Unlimited, Inc.

Fired Ceramics for Special Purposes: Cements, Clays, and Refractories

Other classes of ceramics include cements, clays, and refractories. Unlike glasses, these ceramics are processed by shaping, drying, and then firing, without ever melting the solid.

Cements are extremely strong and are commonly used as structural materials. They can be formed into almost any shape. When mixed with water, they produce a paste that can be poured into moulds and allowed to dry and harden.

Clays are generally mixtures of hydrated alumina (Al_2O_3) and silica (SiO_2), but may also contain other ingredients, such as tricalcium silicate, ($3CaO.SiO_2$), dicalcium silicate, ($2CaO.SiO_2$), and magnesium oxide (MgO). Their composition is irregular and, because they are powders, their crystallinity extends for only short distances.

Clays have the useful property of becoming very plastic when water is added, a characteristic referred to as *hydroplasticity*. This plasticity, and clay's ability to hold its shape during firing, are very important for the forming processes used to create various objects.

The layered molecular structure of clays results in microscopic platelets that can slide over each other easily when wet. The layers consist of SiO_4 tetrahedra joined with AlO_6 octahedra. In addition to the basic silicon- and aluminum-based structures, different cations can be substituted into the framework to change the properties of the clay. Common substituents include Ca^{2+}, Fe^{2+}, and Mg^{2+} ions. Different clay materials can then be created by varying the combinations of layers and the substituent cations.

Refractories constitute a class of ceramics that are capable of withstanding very high temperatures without deforming, in some cases up to 1650 °C, and that are thermally insulating. For these reasons, refractory bricks are used in applications such as furnace linings and in metallurgical operations. These materials are thermally insulating largely because of the porosity of their structure; that is, holes (or pores) are dispersed evenly within the solid. However, while porosity will make a material a thermal insulator, it will also weaken it. As a consequence, refractories are not as strong as cements.

An amazing example of the use of porosity to increase the insulating capacities of a ceramic is found in a material developed at NASA called *aerogel* (Figure 28.18). Aerogel is more than 99% air, with the remainder consisting of a networked matrix of SiO_2. This makes aerogel about 1000 times less dense than glass but gives the material extraordinary thermal-insulating abilities. NASA used aerogel on a mission in which a spacecraft flew through the tail of the comet Wild 2 and returned to Earth with space particles embedded in the aerogel.

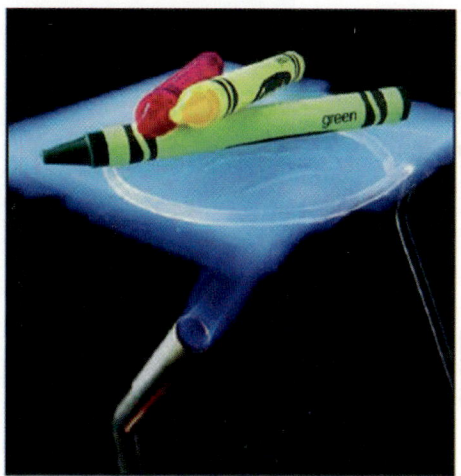

NASA JPL

FIGURE 28.18 Aerogel, a networked matrix of SiO_2. (*Left*) NASA's Peter Tsou holds a piece of aerogel. It is 99.8% air, 39 times more insulating than the best fibreglass insulation, and 1000 times less dense than glass. (*Centre*) A demonstration of the insulating properties of aerogel. (*Right*) Aerogel was used on a NASA mission to collect the particles in comet dust. The particles entered the gel at a very high velocity, but were slowed gradually. Scientists studied the tracks made by the particles and later retrieved the particles and studied their composition.

Modern Ceramics with Exceptional Properties

In 1880, Pierre Curie and his brother Jacques worked in a small laboratory in Paris to examine the electrical properties of certain crystalline substances. Using nothing more than tin foil, glue, wire, and magnets, they were able to confirm the presence of surface charges on samples of materials such as tourmaline, quartz, and topaz when they were subjected to mechanical stresses. This phenomenon, now called **piezoelectricity,** is the property that allows a mechanical distortion (such as a slight bending) to induce an electrical current and, conversely, an electrical current to cause a distortion in the material.

Not all crystalline ceramics exhibit piezoelectricity. Those that do have a specific unit cell structure that can loosely trap an impurity cation. The ion's position shifts when the unit cell is deformed by mechanical stress. This shift causes an induced dipole and, therefore, a potential difference across the material, detected as an electrical signal.

In addition to the minerals originally tested by the Curie brothers, materials known to exhibit the piezoelectric effect include titanium compounds of barium and lead, lead zirconate ($PbZrO_3$), and ammonium dihydrogen phosphate ($NH_4H_2PO_4$).

Materials that exhibit piezoelectricity have a great many applications, ranging from home gadgets to sophisticated medical and scientific applications. One use with which you may be familiar is the automatic ignition systems on some barbecue grills and lighters (Figure 28.19). All digital watch beepers are based on piezoceramics, as are smoke detector alarms. A less familiar application is found in the sensing lever of some atomic force microscopes (AFMs), and scanning-tunnelling microscopes (STMs), which convert mechanical vibrations to electrical signals.

Scientists and engineers are always searching for ceramic materials with new and useful properties. A great deal of effort has gone into developing new ceramics that are **superconductors** at relatively high temperatures. Superconductivity is a phenomenon in which the electrical resistivity of a material drops to nearly zero at a particular temperature referred to as the *critical temperature* (T_c) (Figure 28.20). Most metals have resistivities that decrease with temperature in a constant manner but still have significant resistivity even at temperatures near 0 K.

FIGURE 28.19 Devices that depend on the piezoelectric effect. These devices work by using a mechanical stress to produce an electric current. Piezoelectric devices are widely used in ignitors and in devices that convert electric impulses to vibrations, such as in the timing circuit of a wristwatch.

Charles D. Winters

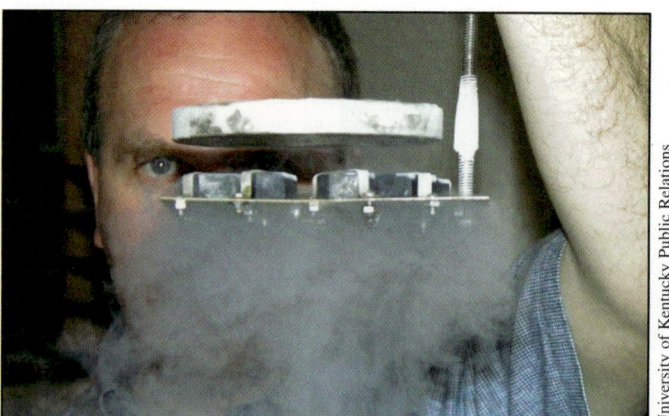

University of Kentucky Public Relations

FIGURE 28.20 Superconductivity. Superconducting material is cooled to a low temperature (in liquid nitrogen, for example), creating conditions in which induced magnetization causes a strong enough repelling force on a magnet to counter the force of gravity. In this photo a 2 kg magnet is levitated in the field created by the cooled superconductor. The electromagnets for the ultra high field NMR instruments used to characterize structures of large molecules are made of many km of superconducting wire, cooled in a bath of liquid helium to 4 K, where the resistance of the wire drops almost to zero. Current continues to flow in the coil as long as the temperature is maintained.

A few metals and metal alloys have been found to exhibit superconductivity. For metals such as those used in NMR magnets, however, the critical temperatures are extremely low, between 0 and 20 K. These temperatures are costly to achieve and difficult to maintain. Recent scientific attention has, therefore, focused on a class of ceramics with superconductive critical temperatures near 100 K. These materials include $YBa_2Cu_3O_7$, with $T_c = 92$ K (Figure 28.21), and $HgBa_2Ca_2Cu_2O_3$, with $T_c = 153$ K.

Once again, we see that combining atoms into sometimes complex chemical compositions allows scientists to develop materials with particular properties. In ceramics, which are normally electrically insulating, this includes even the ability to conduct electricity.

Ceramics are solid inorganic compounds that combine metal and non-metal atoms and in which the bonding ranges from very ionic to covalent. Ceramics are generally hard, relatively brittle, and inflexible, and they are usually good thermal insulators. Some ceramics can be electrically conductive, but most are electrical insulators. Other ceramics include glass, clays, cements, and refractory and superconducting materials.

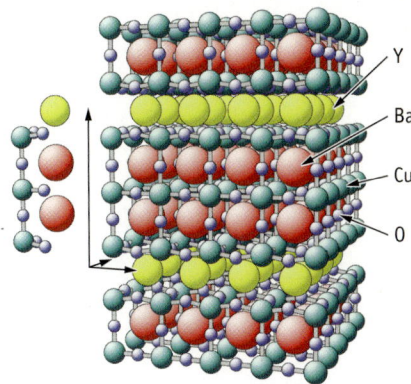

FIGURE 28.21 The lattice of $YBa_2Cu_3O_7$, a superconductor. Yttrium atoms are yellow, barium atoms are red, copper atoms are green, and oxygen atoms are blue.

28.5 Biomaterials: Learning from Nature

Most of the materials described so far in this chapter come from non-living sources and, in many cases, are the result of laboratory syntheses. However, an important branch of materials research deals with examining, understanding, and even copying materials produced by living systems.

The study of naturally occurring materials has led to the development of synthetic materials that can mimic important properties. A good example is rubber [<<Section 4.5]. The basic polymer we know as rubber is produced by certain trees, but chemical modification is needed to convert it to a useful material. Natural rubber was found to be so useful that chemists eventually achieved the synthesis of a structurally identical material. This work, which spanned more than 200 years, has had important consequences for humans as evidenced by the myriad applications of rubber today.

Today, scientists continue to look to nature to provide new materials and to provide clues to improve the materials we already use. The sea urchin and its ceramic spines (Figure 28.13) and the sponge whose skeleton has the characteristics of optical fibres (Figure 28.17) are just two examples of biomaterials research that has focused on sea life in a search for new materials. Scientists have also examined conch shells to understand their incredible fracture strength. They used scanning electron microscopy (SEM) to scrutinize the structure of the shell when it was fractured. What they discovered was a crisscrossed, layered structure that is the equivalent of a "ceramic plywood" (Figure 28.22). This microarchitecture prevents fractures that occur on the outside surface of the shell from being transferred into the inner layers. The discovery has inspired materials engineers to create materials that are significantly strengthened by incorporating a fibrous ceramic matrix, such as SiC (silicon carbide) whiskers.

FIGURE 28.22 A scanning electron microscope picture of the shell of the conch.

Photos: Courtesy of Dr. Arthur Heuer

In another area of research focusing on sea creatures, the connective tissues of sea cucumbers and other echinoderms (marine invertebrates with tube feet and calcite-covered, radially symmetrical bodies) have been studied in an attempt to discover how these animals can reversibly control the stiffness of their outer skin. The connective tissues of these animals include the protein *collagen* in a cross-linked fibre structure, similar to the dermis of many mammals. At the same time, other proteins and soluble molecules in the echinoderm system allow the animals to change the characteristics of the connective tissue through their nervous system. As a result, creatures such as sea cucumbers can move about and, in some cases, can defend themselves by hardening their skin to an almost shell-like consistency. The ensuing laboratory research has focused on the formulation of a synthetic collagen-based polymer composite material in which the stiffness can be changed repeatedly through a series of oxidation and reduction reactions. Scientists are now developing models for synthetic skin and muscle based on their findings.

Research on adhesive materials represents another area in which sea creatures can provide some clues. Getting things to stick together is important in a multitude of applications. The loss of the U.S. space shuttle *Columbia* in early 2003 as a result of a piece of insulating foam that fell off during launch offered a sobering lesson in adhesive failure under extreme conditions of temperature and humidity. If you look around, you will probably find something with an adhesive label, something with an attached plastic part, something with a rubber seal, or perhaps something taped together. For every type of sticking application, different properties are needed for the adhesive material.

Nature provides numerous examples of adhesion. Geckos and flies that can walk on glass while upside down hold clues to the kind of biologically based adhesion that could be the basis of synthetic analogues [<<Section 9.1]. Marine mussels, which can stick equally well to wood, metal, and rock, also hold great interest for scientists studying adhesion (Figure 28.23). Such adhesives have proven useful for medical applications, where specialized glues help doctors seal tissues within the human body.

FIGURE 28.23 Strong mussels. (*Left*) A common blue mussel can cling to almost any surface like this Teflon sheet, even underwater. (*Right*) The adhesive precursor is a protein interlinked with iron(III) ions. Side chains on the protein are dihydroxyphenylalanine (DOPA), and an iron(III) ion binds to the hydroxy groups (—OH) in three side chains. The iron–DOPA complex reacts with oxygen, forming very reactive free radicals. These radicals may act as polymerization agents, forming the adhesive.

Jonathan Wilker, Purdue University

Scientists who have researched mussel adhesives have been able to determine that the amino acid 3,4-dihydroxyphenylalanine (DOPA) is the agent primarily responsible for the strength of the adhesion. But DOPA alone cannot explain the incredible strength of the mussel glues. The secret lies in the combination of an Fe^{3+} ion with DOPA to form a cross-linked matrix of the mussel's protein (Figure 28.23). The curing process, or hardening of the natural proteinacious liquid produced by the mussel, is a result of the iron–protein interaction that occurs to form $Fe(DOPA)_3$ cross-links.

Biomaterials research looks to natural materials, including sea life, in a search for substances, approaches, and processes that can be mimicked and modified in the laboratory. Examples include the development of collagen-based polymer composite materials and new adhesives.

28.6 The Future of Materials

The modern tools and techniques of chemistry are making it possible for scientists not only to develop novel materials, but also to proceed in new and unforeseen directions. The field of *nanotechnology* [<<Section 20.12] is an example. In nanotechnology, structures with dimensions on the order of nanometres are used to carry out specific functions. Instead of building a ball bearing by polishing a piece of metal until it is very smooth, scientists can now create a tube of carbon atoms embedded in a slightly larger carbon tube to act as a ball bearing at the molecular level (Figure 28.24).

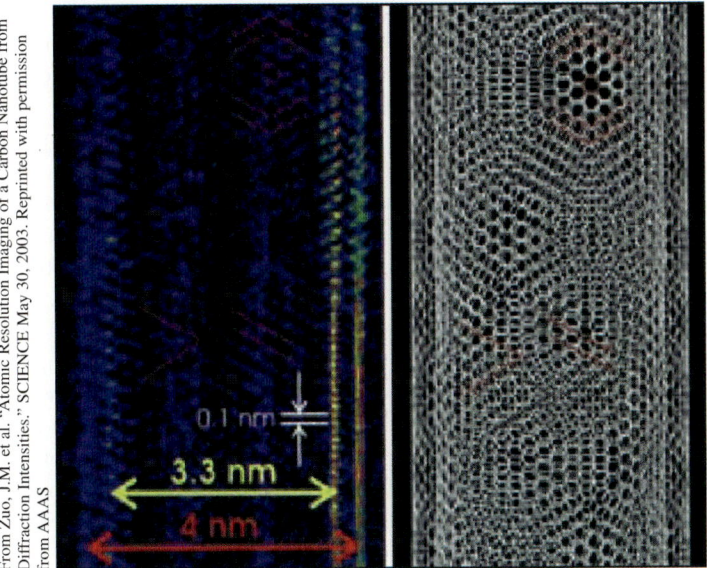

From Zuo, J.M. et al. "Atomic Resolution Imaging of a Carbon Nanotube from Diffraction Intensities." SCIENCE May 30, 2003. Reprinted with permission from AAAS

FIGURE 28.24 A molecular bearing: double-walled carbon nanotubes. (*Left*) An electron diffraction image of a double-walled carbon nanotube. (*Right*) A model of the material.

Nanoscience has provided profoundly important applications for medicine, computing, and energy consumption. For example, scientists have developed *quantum dots* (Figure 28.25), nanometre-scale crystals of different materials that can emit light and can even be made to function as lasers. Quantum dots have been used as biological markers by attaching them to various cells. By shining light on them, the quantum dots will fluoresce in different colours, allowing the cells to be imaged.

Other scientists are examining nanoscale drug delivery technologies that can inject medicinal agents directly into target cells. One way that scientists have been able to achieve these breakthroughs is by studying the structures of materials that are already known to us. They have been learning to manipulate atoms and molecules so that the atoms and molecules will arrange themselves in specific ways to achieve desired shapes and functions. In a process referred to as *self-assembly* [<<Section 9.1], molecules or atoms will arrange themselves based on their shapes, the intermolecular forces between them, and the interactions with their environment.

Chemistry is the key to understanding and developing materials. As we have seen, the atomic compositions and long-term atomic arrangements of different materials determine their properties and characteristics. Chemists can use analytical instruments to characterize these structures. They can then exploit this knowledge to manipulate or develop materials to achieve different properties for special functions. In many cases, we can look to nature to provide answers and suggestions on how to proceed.

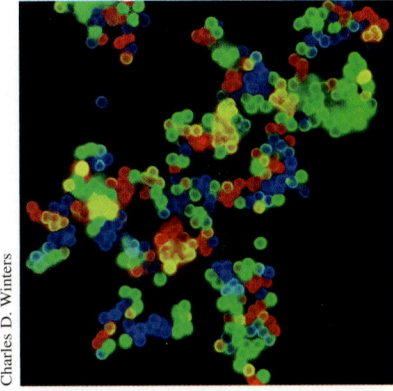

Charles D. Winters

FIGURE 28.25 Quantum dots. Polymer beads embedded with quantum dots fluoresce in five different colours.

In nanotechnology, structures with dimensions on the order of nanometres are used to carry out specific functions. Chemists have learned to manipulate atoms and molecules so that they will undergo self-assembly—that is, arrange themselves in specific ways to achieve desired shapes and functions.

SUMMARY

Key Concepts

- **Band theory** can be used to describe the bonding in metals and semiconductors. Molecular orbitals are constructed from the valence orbitals on each atom and are delocalized over all the atoms. The highest filled level at 0 K is referred to as the **Fermi level.** In contrast to metals, the band of filled levels (the valence band) is separated from the band of empty levels (the conduction band) by a band gap. In insulators, the energy of the band gap is large. (Sections 28.2, 28.3)
- Most metallic objects we use are **alloys,** mixtures of a metal with one or more other metals or a non-metal such as carbon. Alloys fall in three general classes: solid solutions, which are homogeneous mixtures of two or more elements; heterogeneous mixtures; and intermetallic compounds. (Section 28.2)
- Diodes, LEDs, and transistors are **semiconductors**, materials that can be encouraged to conduct electricity by the input of energy. The properties of semiconductor materials can be changed by adjusting the band gap. (Section 28.3)
- **Ceramics** are solid inorganic compounds that combine metal and non-metal atoms and in which the bonding ranges from very ionic to covalent. Ceramics are generally hard, relatively brittle, and inflexible, and they are usually good thermal insulators. Some ceramics can be electrically conductive, but most are electrical insulators. Other ceramics include glass, clays, cements, and refractory and superconducting materials. (Section 28.4)
- Biomaterials research looks to natural materials, including sea life, in a search for substances, approaches, and processes that can be mimicked and modified in the laboratory. Examples include the development of collagen-based polymer composite materials and new adhesives. (Section 28.5)
- In **nanotechnology**, structures with dimensions on the order of nanometres are used to carry out specific functions. Chemists have learned to manipulate atoms and molecules so that they will undergo **self-assembly**—that is, arrange themselves in specific ways to achieve desired shapes and functions. (Section 28.6)

REVIEW QUESTIONS

Section 28.2: Metals

28.1 In magnesium, what percentage of the molecular orbitals that can be formed from s and p atomic orbitals can be filled by valence electrons?

28.2 Which of the following would be good substitutional impurities for an aluminum alloy?
(a) Sn   (b) P   (c) K   (d) Pb

28.3 Suggest what "solvent" metals might be used as (a) interstitial atoms, and (b) substitutional atoms in solid solutions in iron.

Section 28.3: Semiconductors

28.4 Compare (a) the electrical properties, and (b) the band gap in electrical conductors and semiconductors.

28.5 What is the maximum wavelength of light that can excite an electron transition across the band gap of GaAs? To which region of the electromagnetic spectrum does this correspond?

Section 28.4: Ceramics

28.6 Calculate an approximate value for the density of aerogel using the fact that it is 99% air, by volume, and the remainder is SiO_2. What is the mass of a 1.0 cm^3 piece of aerogel?

28.7 Why is Pyrex glass superior to soda glass for cookware?

28.8 Find the refractive index of (a) water, (b) benzene, (c) Pyrex glass, and (d) diamond. In which does light travel with highest velocity?

28.9 Describe what is meant by (a) hydroplasticity and (b) refractory.

SUMMARY AND CONCEPTUAL QUESTIONS

28.10 The amount of sunlight striking the surface of the earth (when the sun is directly overhead on a clear day) is approximately 925 watts per square metre ($W\ m^{-2}$). The area of a typical solar cell is approximately 1.0 cm^2. If the cell is running at 25% efficiency, how much energy will be produced in one minute?

28.11 Using the result of the calculation in Question 28.10, estimate the number of solar cells that would be needed to power a 700 W microwave oven. If the solar cells were assembled into a panel, what would be the approximate area of the panel?

28.12 Use the information in Figure 28.12 to estimate the diameter of the gears in the photo. How does this compare with the diameter of a typical human red blood cell?

28.13 Using the data in Table 28.1, calculate the density of pewter. (Look up the densities of the constituent elements on a website such as www.webelements.com. Click on the periodic tool, and then click on the symbol of each of the elements in pewter. A table of atomic properties includes the element's density.)

28.14 Explain, in terms of structure and bonding, the differences between the properties of graphite and diamond.

28.15 Is it possible to make a generalization about the strength of hydrogen bonding compared with the strength of dispersion forces? Is your answer consistent with the observations that at ambient temperatures water is a liquid and iodine is a solid? Rationalize.

28.16 Distinguish between the meanings of critical temperature used in the contexts of (a) phase changes, and (b) superconductivity.

With the **eCHACR** single-sign-on access card bundled with your text, log on (http://login.cengage.com/sso) and access the e-book and click on any in-margin icons for dynamic molecular-level animations and simulations, videos of laboratory reactions, interactive exercises, reviews of background concepts, and other online supplementary materials as noted by the icons in the text margins.

Also go to www.chemistry.nelson.com <http://www.chemistry.nelson.com> for Answers to in-chapter exercises and selected Review Questions, Test Yourself questions, weblinks, crossword puzzles, flashcards, glossary of key terms, and other student resources.

29 Biomolecules

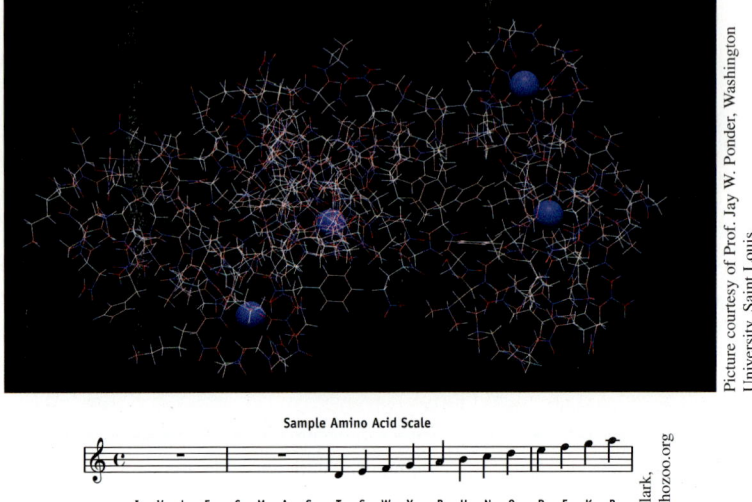

Picture courtesy of Prof. Jay W. Ponder, Washington University, Saint Louis

Sample Amino Acid Scale

I V L F C M A G T S W Y P H N Q D E K R

M.A. Clark, www.whozoo.org

29.1 Case Study: Molecules and Melodies of Life

Along with many of your classmates, you might usually settle in with your MP3 player before reading your chemistry textbook. Perhaps today, you will find the soothing strains of a Latin American sounding rhythm will help you process what you're reading. Follow the margin link in e29.1 to listen to a piece we thought might set an appropriate mood for learning about the molecules of life.

The melody we've provided is much more than just background music. And it, like the molecules of life, has been synthesized for you. Biochemists and artists have collaborated to create several musical scales that emphasize important features of each of the amino acids that make up peptides and proteins. You can hear the melody of a connectivity pattern as the musical score travels along the amino acid units that link together with peptide bonds to make polymeric chains of a protein molecule. More subtle differences can also be heard. Water-loving (hydrophilic) and water-avoiding (hydrophobic) amino acids can be assigned notes on a different scale or given different voices by appropriate choice of instruments.

> *Biochemists and artists have collaborated to create several musical scales that allow you to hear the melody of a connectivity pattern as the musical score travels along the amino acid units that link together with peptide bonds to make polymeric chains of a protein molecule.*

The structure of a complex protein molecule has features that go beyond the simple connectivity of amino acid units (primary structure). Data from X-ray crystallography and NMR spectroscopy can reveal the orientation patterns of segments of a protein backbone (secondary structure), as well as the overall coiling of a large molecule into a three-dimensional shape (tertiary structure). These features, too, can be represented well by subtle variations in a musical composition. Your mental picture of the structures of biological polymers can be enriched by hearing, as well seeing, and because our brains are good at recognizing differences in speech, the ear can often pick up subtle differences in structure more easily than the eye.

The composition you've been listening to sets to music a chain of 148 amino acids that comprise the primary structure of a molecule of calmodulin (CaM), an important protein that controls levels of calcium, the most plentiful mineral in your body. The protein binds to calcium ions and then to other proteins, affecting their activity. Appropriately, the calcium bound form of the substance calmodulin plays a role in controlling both your thinking about chemistry and your foot-tapping. It helps to regulate proteins that play a role in many different processes, including muscle contraction and nerve signalling. It would be impossible to either study chemistry or dance without it!

If you could "play" CaM proteins that had been extracted from the cells of different organisms, you would find that they sound almost exactly the same. The sequence of amino acids in CaM from plants, animals, fungi, and protozoans differ by only a few dozen amino acid units. Human CaM is also not very different from that of other animals.

An intriguing feature of CaM is that the overall protein changes shape dramatically after it binds to Ca^{2+} ions. The calcium-free protein molecule is shaped something like a dumbbell, with two calcium-binding regions (domains) joined by a flexible alpha helix connector.

Three different representations of a molecule of this protein are shown in Figure 29.1. The first two show different representations of the calcium-free protein molecule. In (a), the amino acid chains are shown as ribbons, and you can see the two calcium binding domains on the top and bottom, connected by an alpha helix. In (b), a space-filling representation more accurately shows electron density and the overall shape of the calcium-free form. Representation (c) shows the protein molecule after it has bound to four calcium ions, two in each of the calcium-binding domains.

The *primary structure* of a protein molecule refers to the number and sequence of amino acids that are linked together.

"Music is not a mere linear sequence of notes. Our minds perceive pieces of music on a level far higher than that. We chunk notes into phrases, phrases into melodies, melodies into movements, and movements into full pieces. Similarly proteins only make sense when they act as chunked units. Although a primary structure carries all the information for the tertiary structure to be created, it still 'feels' like less, for its potential is only realized when the tertiary structure is actually physically created."

Douglas Hofstadter, Professor of Cognitive Science, Indiana University, U.S.

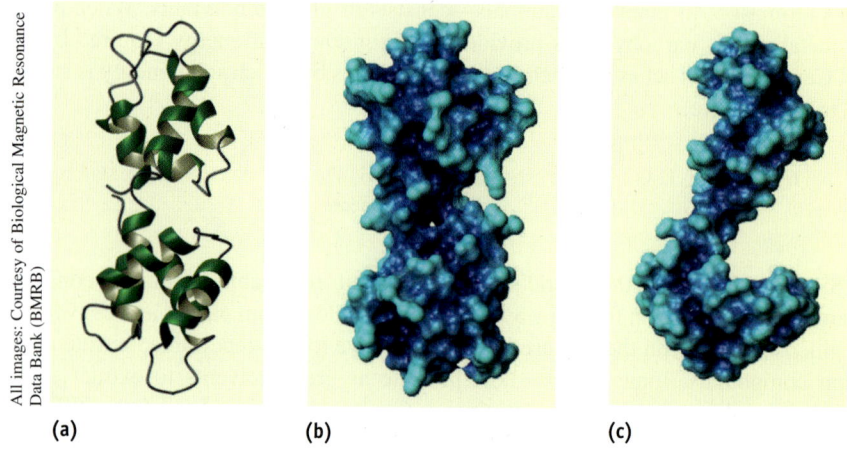

All images: Courtesy of Biological Magnetic Resonance Data Bank (BMRB)

(a) (b) (c)

FIGURE 29.1 Three representations of calmodulin (CaM). (a) and (b) show the calcium-free form of the protein molecule; (c) shows the conformational change that occurs following binding to four Ca^{2+} ions.

Of great interest to chemists is the conformational change that a CaM molecule undergoes following binding to calcium ions. In the unbound form (b), most of the hydrophobic regions are located on the inside of the protein molecule. When the molecule uncoils after binding to calcium ions, these hydrophobic regions move to the surface, creating sites where they can interact with the target hydrophobic protein molecules. Since the alpha helix that connects the two domains of CaM is quite flexible, these two domains of the protein can wrap themselves around target protein molecules of different sizes and shapes.

As the calmodulin protein example demonstrates, a marvellously intricate set of interactions at the molecular level controls processes in your body, and in every other organism on the earth. And proteins are only one of several important classes of biomolecules controlling living organisms. What are these substances that are present in you, and can we make sense of how their molecules determine their properties? How is genetic information passed from generation to generation? How does your body carry out the numerous reactions that are needed for life?

In this chapter, we give an overview of three classes of substances that are particularly important in controlling the chemistry of life: carbohydrates, proteins like CaM, and nucleic acids. We take a molecular-level look at the structures of molecules of these substance, and identify patterns of reactivity that are consistent with the chemistry of their functional groups. Many of the laboratory reactions of organic compounds in Chapters 19–25 are seen again in living systems, and you will be able to recognize both similarities and differences from the organic reactions you've already learned. The biggest differences are that multiple functional groups are usually present within each large biomolecule, and the most efficient catalysts we know, enzymes, control the selectivity and rates of competing reactions along biochemical pathways.

Let the music begin.

29.2 Carbohydrates

Carbohydrates, which are polyhydroxylated aldehydes and ketones commonly called *sugars,* occur in every living organism. The sugar and starch in food and the cellulose in wood, paper, and cotton are nearly pure carbohydrate. Modified carbohydrates form part of the coating around living cells; other carbohydrates are found in the DNA that carries genetic information, and still others are used as medicines.

Carbohydrates are made by green plants during photosynthesis, a complex process in which sunlight provides the energy to convert CO_2 and H_2O into glucose. Many molecules of glucose are then chemically linked for storage by the plant to form either the substances' cellulose or starch. It has been estimated that more than 50% of the dry weight of the earth's biomass—all plants and animals—consists of glucose polymers. When eaten and then metabolized, carbohydrates provide the major source of energy required by organisms. Carbohydrates act as the chemical intermediaries by which solar energy is stored and used to support life.

$$6\,CO_2 \;+\; 6\,H_2O \;\xrightarrow{\text{Sunlight}}\; 6\,O_2 \;+\; \underset{\textbf{Glucose}}{C_6H_{12}O_6} \;\longrightarrow\; \text{Cellulose, starch}$$

Since cellulose is the world's most abundant renewable resource, considerable research is being done on finding ways to extract cellulose from plants, and convert it back into glucose, which can then be used as an alternative to non-renewable, petroleum-based carbon compounds. Ionic liquids show promise as green solvents to extract cellulose, which is insoluble in many common solvents. Glucose can be used to synthesize a variety of useful substances, such as Vitamin C and sorbitol, which is a sugar substitute and laxative. More exotic substances can also be made from glucose with the aid of enzymes (Figure 29.2). These include monomers for polymer synthesis, aromatic compounds, and specialty chemicals that serve as the basis for more elaborate syntheses, like the anti-viral drug Tamiflu (oseltamivir). Because microbes commonly use glucose, chemists are not limited to traditional, one-step reactions carried out in laboratory glassware. Microbes can be put to work to carry out syntheses in highly selective and efficient reactions.

You've already seen one important example of small macrocyclic polymers of glucose, called cyclodextrins [<<Section 22.1], which are particularly important in drug delivery. Cyclodextrins form host–guest complexes when relatively non-polar molecules (guests) of the right size and shape are held in place by non-covalent forces in the cavity of a cyclodextrin host. The alcohol functional groups on the exterior of the cavity of a cyclodextrin molecule make it very polar and hydrophilic, and the interior cavity is relatively non-polar. Such non-covalent interactions are important to the functioning of other carbohydrates, as well as proteins and nucleic acids.

Carbohydrates are generally classed into two groups: *simple* and *complex. Simple sugars,* or *monosaccharides,* are carbohydrates like glucose and fructose whose molecules can't be converted into smaller sugar molecules by hydrolysis reactions. The molecules of *complex carbohydrates* are composed of two or more simple sugar molecules linked together.

Carbohydrate molecules contain both R—O—H (hydroxyl) and R—O—R (ether) functional groups, classifying them as both alcohols and ethers. In the aqueous solutions found in living organisms, some of the alcohol and ether functional groups in certain cyclic carbohydrate molecules can be interconverted with open-chain forms that have aldehyde and ketone functional groups.

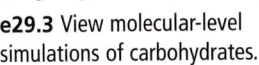

Molecular Modelling (Odyssey)

e29.3 View molecular-level simulations of carbohydrates.

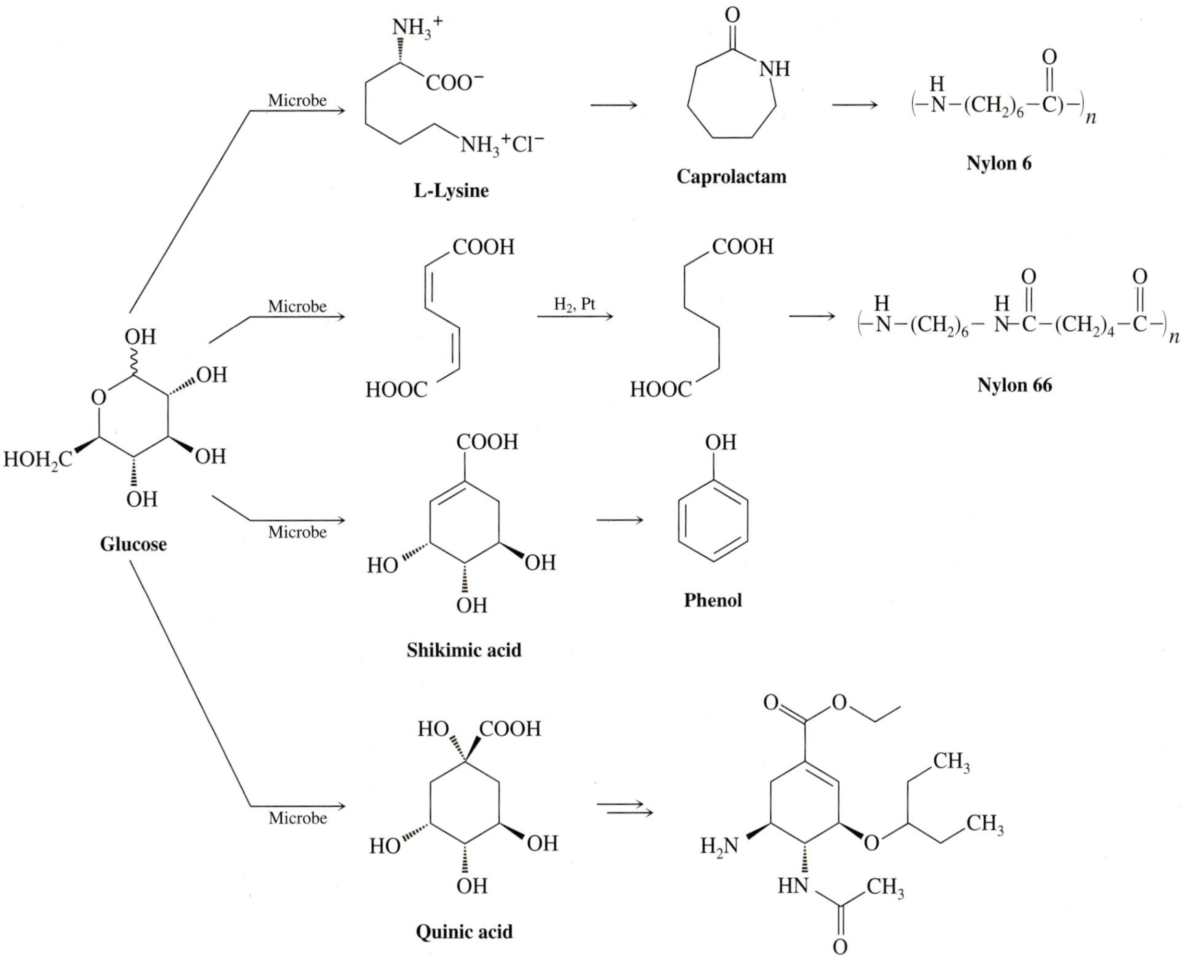

FIGURE 29.2 Using plants as a renewable alternative to petroleum for feedstocks. Cellulose can be extracted from plants such as corn, rice, or wheat using green solvents, such as ionic liquids, and converted to glucose. Glucose can then be converted into a wide range of useful substances.

A sucrose molecule (from table sugar), for example, is a *disaccharide* made up of one glucose molecule linked to one fructose molecule. Similarly, cellulose is a *polysaccharide* made up of several thousand glucose molecules linked together. Hydrolysis of polysaccharides breaks their macromolecules down into their constituent monosaccharide units.

$$1 \text{ Sucrose} \xrightarrow{\text{H}_3\text{O}^+} 1 \text{ Glucose } + 1 \text{ Fructose}$$

$$\text{Cellulose} \xrightarrow{\text{H}_3\text{O}^+} \sim 3000 \text{ Glucose}$$

Monosaccharide molecules are further classified as either *aldoses or ketoses*. The —*ose* suffix is used as the family name ending for carbohydrates, and the *aldo–* and *keto–* prefixes identify the nature of the carbonyl group in molecules of the open-chain form, whether aldehyde or ketone. The number of carbon atoms in a monosaccharide molecule is indicated by using *tri–, tetr–, pent–, hex–,* and so forth, in the name. For example, a molecule of glucose is an *aldohexose,* a six-carbon aldehyde sugar; fructose is a *ketohexose,* a six-carbon ketone sugar; and ribose is an *aldopentose,* a five-carbon aldehyde sugar. Most of the commonly occurring simple sugars are either aldopentoses or aldohexoses.

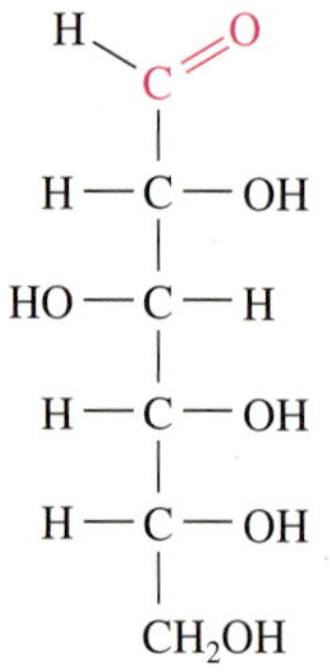

Glucose
(an **aldo**hexose)

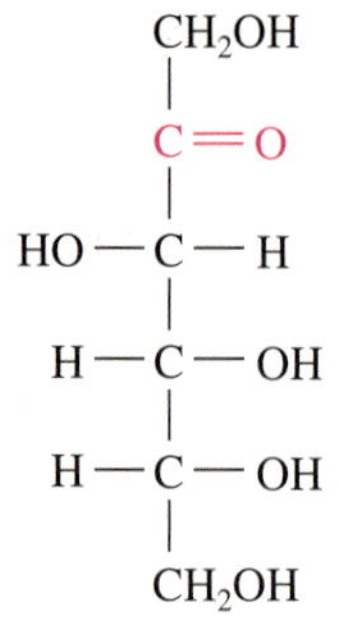

Fructose
(a **keto**hexose)

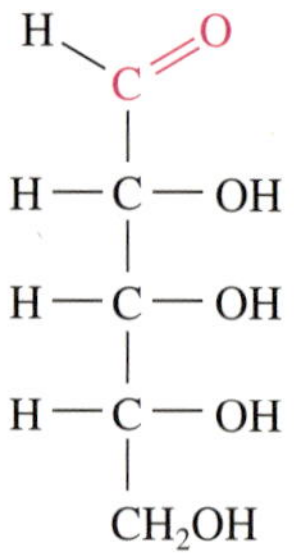

Ribose
(an **aldo**pentose)

WORKED EXAMPLE 29.1—NOMENCLATURE

Classify the monosaccharide allose:

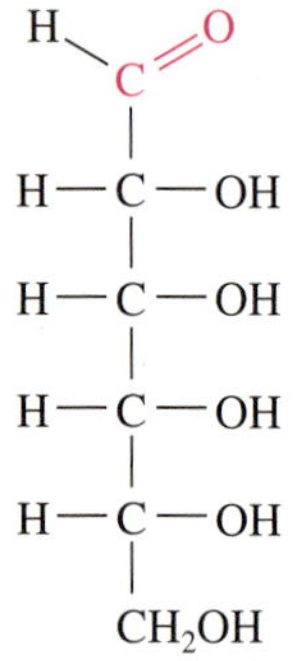

Solution

Since each allose molecule has six carbon atoms and an aldehyde carbonyl group, it's an aldo-hexose.

Allose

EXERCISE 29.1—NOMENCLATURE

Classify each of the following monosaccharides:

(a)

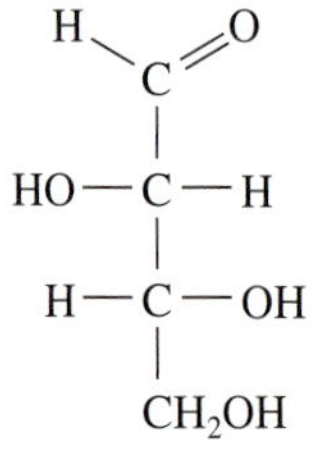

Threose

(b)

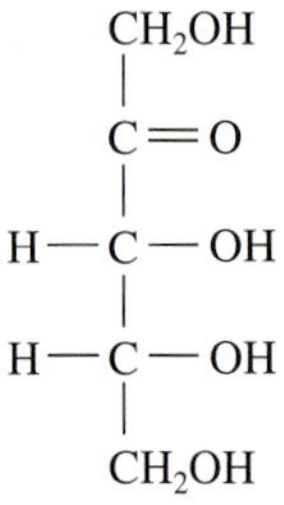

Ribulose

(c)

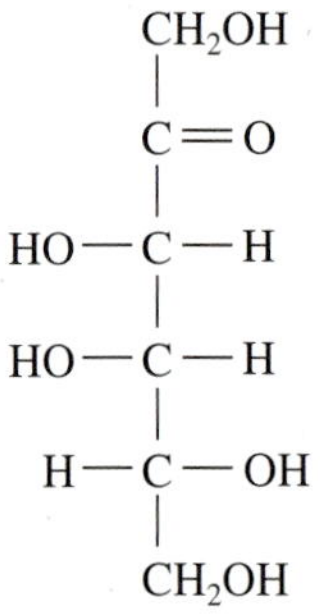

Tagatose

(d)

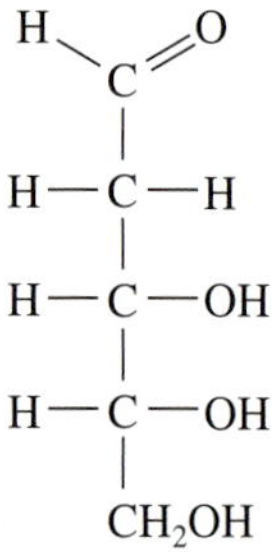

2-Deoxyribose

Configurations of Monosaccharides: Fischer Projections

Since all carbohydrate molecules are chiral and have stereocentres [<<Section 9.8], it was recognized long ago that a standard method of representation is needed to describe the stereocentres in three-dimensional carbohydrate molecules in the two-dimensional world of paper. In 1891, Emil Fischer suggested a method based on the projection of a tetrahedral carbon atom onto a flat surface. These *Fischer projections* were soon adopted and are now a standard means of depicting stereochemistry.

A tetrahedral carbon atom is represented in a Fischer projection by two perpendicular lines. The horizontal lines are a simple way to indicate wedges—bonds connecting to atoms in front of the plane of the paper, and the vertical lines are a simple way to indicate dashes—bonds connecting to atoms behind the plane of the paper:

By convention, the carbonyl carbon atom is placed at or near the top in Fischer projections of carbohydrate molecules. A molecule of (*R*)-glyceraldehyde, the simplest

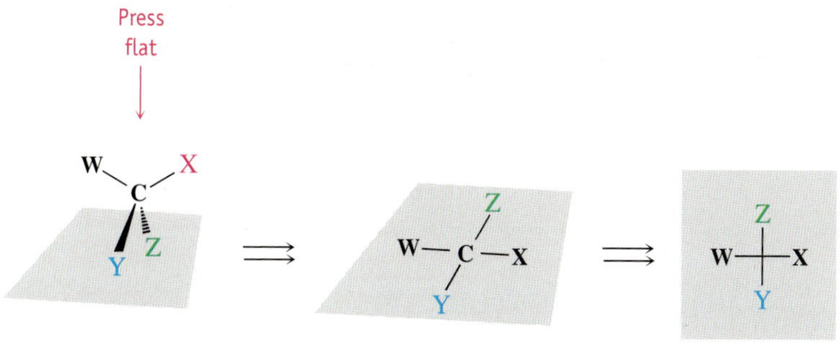

Press
flat

Fischer
projection

monosaccharide, is represented as shown in Figure 29.3. To review the *R/S* system for specifying configuration, see the rules in Section 9.10.

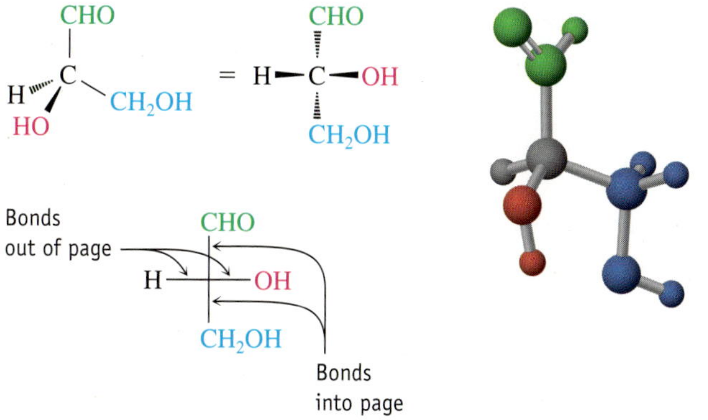

**Fischer projection of
(R)-glyceraldehyde**

FIGURE 29.3 Fischer projection of an (*R*)-glyceraldehyde molecule.

Carbohydrates whose molecules have more than one stereocentre are shown by drawing each stereocentre, one above the other, with the carbonyl carbon atom at or near the top of the Fischer projection. Glucose molecules, for example, have four stereocentres, arranged vertically in a Fischer projection:

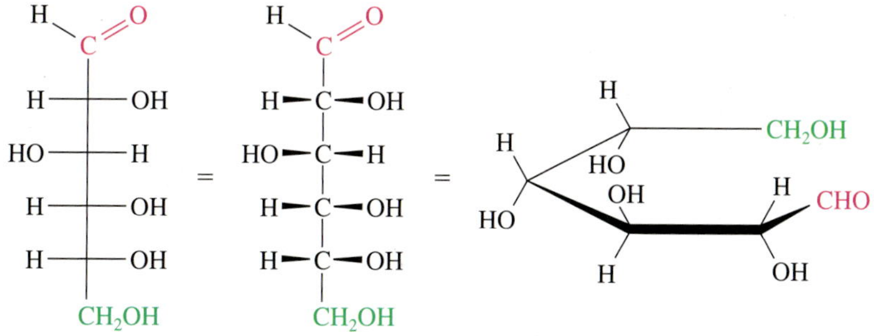

**Glucose
(carbonyl group at top)**

WORKED EXAMPLE 29.2—STEREOCHEMISTRY

Convert the following tetrahedral representation of an (*R*)-butan-2-ol molecule into a Fischer projection.

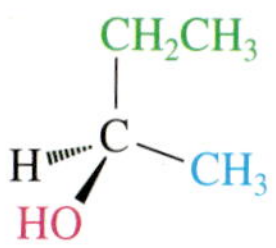

(*R*)-Butan-2-ol

Solution

Orient the molecule so that there are two horizontal bonds coming toward you and two vertical bonds receding away from you. Then press the molecule flat into the paper, indicating the stereocentre as the intersection of two crossed lines.

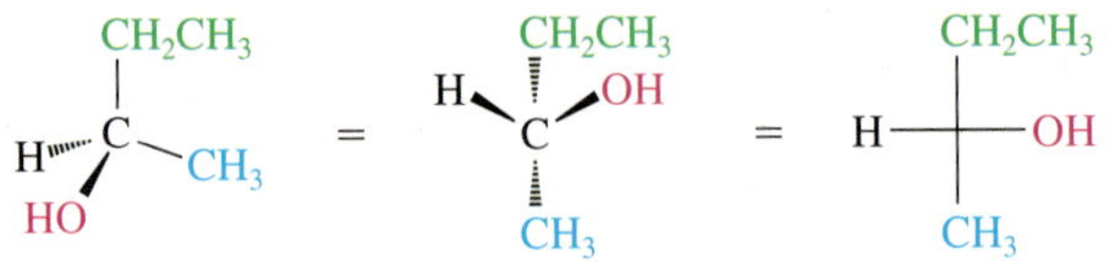

$$CH_2CH_3 - C(H^{\text{''''}})(CH_3)(OH) = H(CH_2CH_3)C(OH)(CH_3) = H - \overset{CH_2CH_3}{\underset{CH_3}{|}} - OH$$

(*R*)-**Butan-2-ol**

COOH
H———OH   **Lactic acid**
CH₃

WORKED EXAMPLE 29.3—STEREOCHEMISTRY

Convert the following Fischer projection of a lactic acid molecule into a tetrahedral representation, and indicate whether the molecule is (*R*) or (*S*):

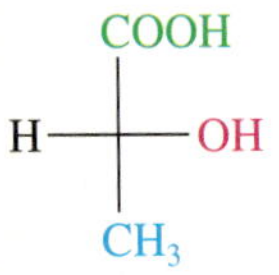

Solution

Place a carbon atom at the intersection of the two crossed lines, and imagine that the two horizontal bonds are coming toward you and the two vertical bonds are receding away from you. The projection represents an (R)-lactic acid molecule.

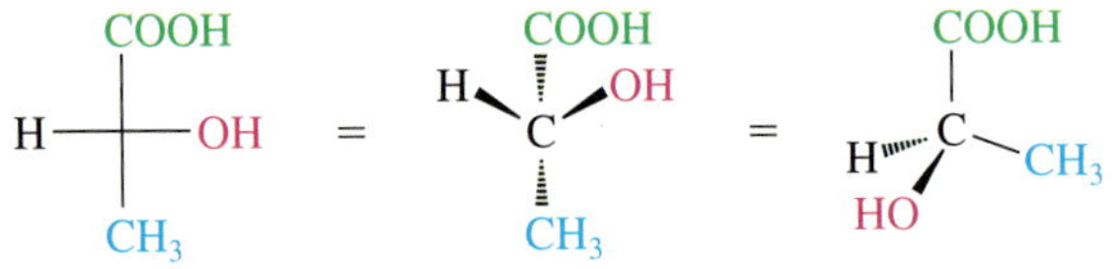

$$H - \overset{COOH}{\underset{CH_3}{|}} - OH = H(COOH)C(OH)(CH_3) = H^{\text{''''}}(COOH)C(CH_3)(HO)$$

(*R*)-**Lactic acid**

EXERCISE 29.2—FISCHER PROJECTIONS

Convert the following tetrahedral representation of an (*S*)-glyceraldehyde molecule into a Fischer projection:

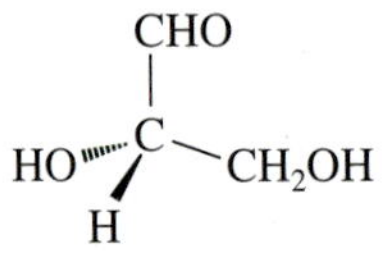

$$HO^{\text{''''}}\overset{CHO}{\underset{H}{C}} - CH_2OH$$ (*S*)-**Glyceraldehyde**

EXERCISE 29.3—FISCHER PROJECTIONS

Draw Fischer projections of molecules of both (*R*)-2-chlorobutane and (*S*)-2-chlorobutane.

EXERCISE 29.4—STEREOCHEMISTRY

Convert the following Fischer projections into tetrahedral representations, and assign *R* or *S* stereochemistry to each:

(a)
COOH
H₂N———H
CH₃

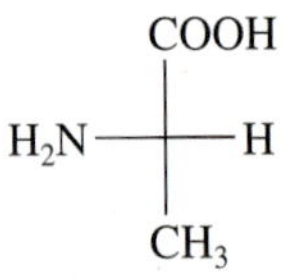

(b)
CHO
H———OH
CH₃

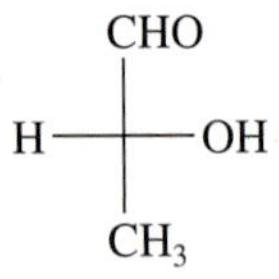

(c)
CH₃
H———CHO
CH₂CH₃

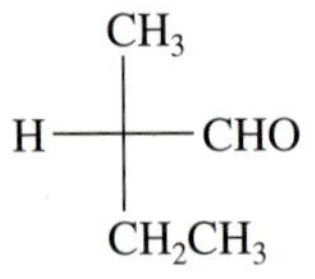

D, L Sugars

Glyceraldehyde molecules have only one stereocentre and therefore can exist in two enantiomeric (mirror-image) forms. Only the dextrorotatory form of the substance glyceraldehyde occurs naturally, however. That is, a sample of naturally occurring glyceraldehyde placed in a polarimeter [<<Section 9.9] rotates plane-polarized light in a clockwise direction, denoted (+). Since molecules making up the substance (+)-glyceraldehyde are

known to have the *R* configuration at C2, they can be represented as in Figure 29.4. For historical reasons dating from long before the adoption of the *R,S* system, (*R*)-(+)-glyceraldehyde is also referred to as *D-glyceraldehyde* (D for dextrorotatory). The other enantiomer, (*S*)-(−)-glyceraldehyde, is known as *L-glyceraldehyde* (L for levorotatory).

H C=O	H C=O	H C=O	CH₂OH
H——OH	H——OH	H——OH	C=O
CH₂OH	H——OH	HO——H	HO——H
	H——OH	H——OH	H——OH
D-Glyceraldehyde	CH₂OH	H——OH	H——OH
[(*R*)-(+)-glyceraldehyde]	**D-Ribose**	CH₂OH	CH₂OH
		D-Glucose	**D-Fructose**

FIGURE 29.4 Molecules of some naturally occurring D sugars. The —OH group at the bottom stereocentre of each molecule (the one farthest from the carbonyl group) is on the right in Fischer projections.

The D and L notations for the configuration of individual molecules of sugars come from historical practice and have no relation to the direction in which a physical sample of a given sugar rotates plane-polarized light. A sample of a sugar made of D molecules may have the physical property of being either dextrorotatory or levorotatory, and this can't be predicted without carrying out an experimental measurement. The prefix D indicates only that in molecules of this sugar, the stereochemistry of the stereocentre farthest from the carbonyl group is the same as that of D-glyceraldehyde and is to the right in a Fischer projection when the molecule is drawn in the standard way with the carbonyl group at or near the top. The D, L system of carbohydrate nomenclature describes the configuration at only *one* stereocentre of each molecule and says nothing about the configuration of other stereocentres that may be present.

Because of the way that monosaccharides are synthesized in nature, molecules of glucose, fructose, ribose, and most other naturally occurring monosaccharides have the same *R* stereochemical configuration as D-glyceraldehyde at the stereocentre farthest from the carbonyl group. In Fischer projections, therefore, most naturally occurring sugars have the —OH group at the bottom stereocentre pointing to the *right* (Figure 29.4). Such compounds are referred to as D sugars. Fischer projections of molecules of the four-, five-, and six-carbon aldoses are shown in Figure 29.5 for the D-series.

In Figure 29.5, a total of eight molecules of D-aldohexoses are shown. Since each of these aldohexose molecules has four stereocentres, you might predict [<<Section 9.8] that the maximum number of possible stereoisomeric aldohexose molecules should be $2^4 = 16$. Besides the eight D sugars, shown above, another eight L aldohexose molecules are known, each of which has an *S* configuration at the stereocentre furthest away from the carbonyl group, with the bottom —OH group in Fischer projections pointing to the *left*. A molecule of an L sugar is the mirror image (enantiomer) of the corresponding D sugar and has the opposite configuration at all stereocentres.

WORKED EXAMPLE 29.4—STEREOCHEMISTRY

Look at the Fischer projection of a D-fructose molecule in Figure 29.4, and draw a Fischer projection of L-fructose.

Solution

Since L-fructose molecules are the enantiomers (mirror images) of D-fructose molecules, we simply take the structure of a D-fructose molecule and reverse the configuration at each stereocentre.

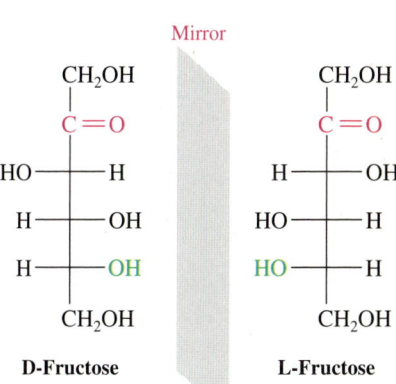

Mirror

CH₂OH		CH₂OH
C=O		C=O
HO——H		H——OH
H——OH		HO——H
H——OH		HO——H
CH₂OH		CH₂OH
D-Fructose		**L-Fructose**

EXERCISE 29.5—STEREOCHEMISTRY

Which of the following are molecules of L sugars and which are D sugars?

(a) CHO	**(b)** CHO	**(c)** CH₂OH
HO——H	H——OH	C=O
HO——H	HO——H	HO——H
CH₂OH	H——OH	H——OH
	CH₂OH	CH₂OH

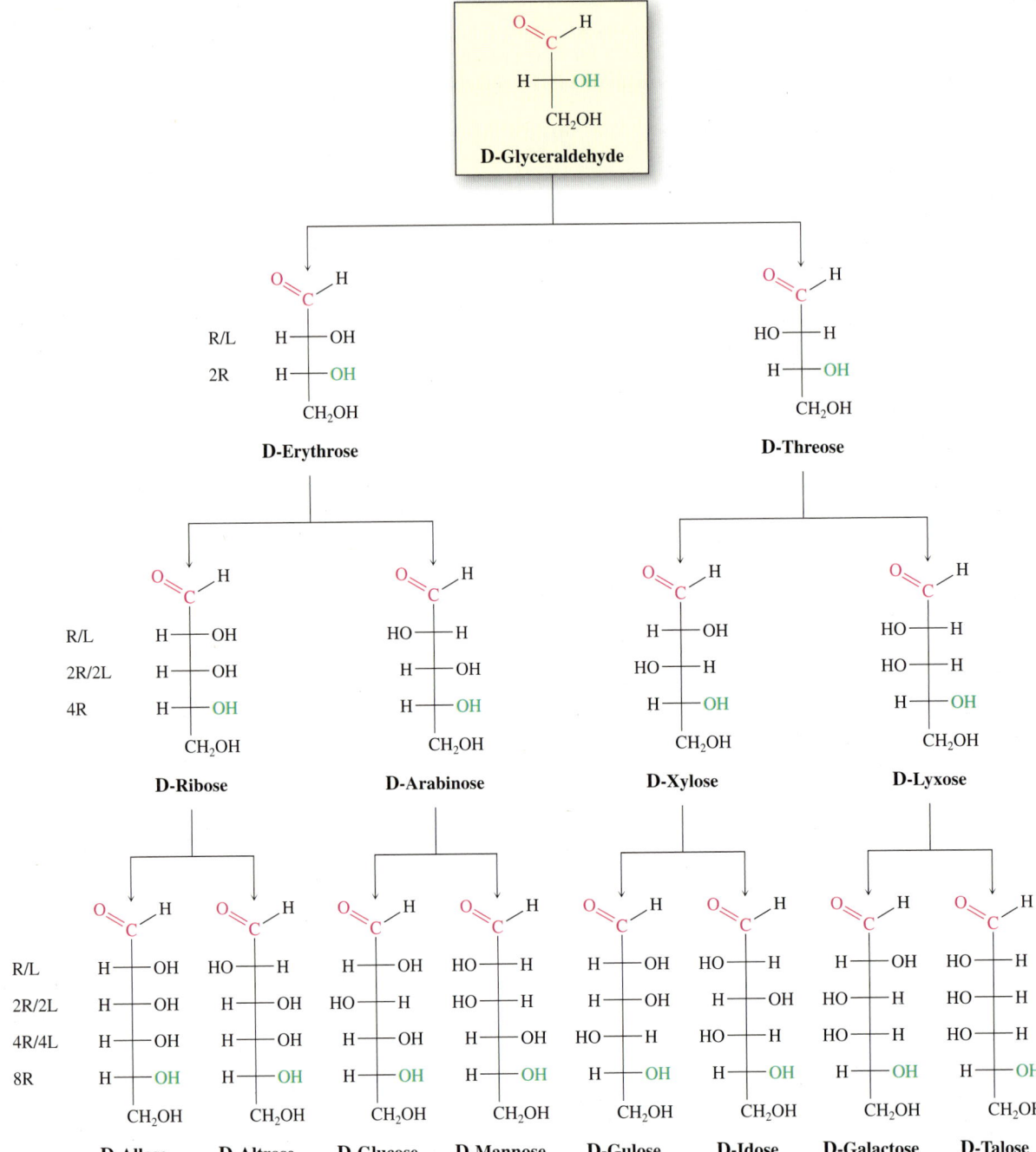

FIGURE 29.5 Configurations of molecules of D-aldoses. The structures are arranged from left to right so that the —OH groups on C2 of each molecule alternate from right to left (R/L) going across a series. Similarly, the —OH groups at C3 alternate two right/two left (2R/2L), the —OH groups at C_4 alternate 4R/4L, and the —OH groups at C_5 are to the right in all eight (8R). The symbols R and L in this figure refer only to right and left orientation and should not be confused with *R* and *S*.

EXERCISE 29.6—STEREOCHEMISTRY

Draw the enantiomers (mirror images) of the carbohydrate molecules shown in Exercise 29.5, and identify each as a D sugar or an L sugar.

Carbohydrates are polyhydroxy aldehydes and ketones. They are classified by the number of carbon atoms and the kind of carbonyl group in their molecules. Monosaccharides are further classified as either D or L sugars, depending on the stereochemistry of the stereocentre farthest from the carbonyl group. Most naturally occurring sugars are in the D series.

Cyclic Structures of Monosaccharides: Hemiacetal Formation

During the discussion of carbonyl-group chemistry in Section 23.7, you learned that alcohols undergo a rapid and reversible nucleophilic addition reaction with aldehydes and ketones to form hemiacetals:

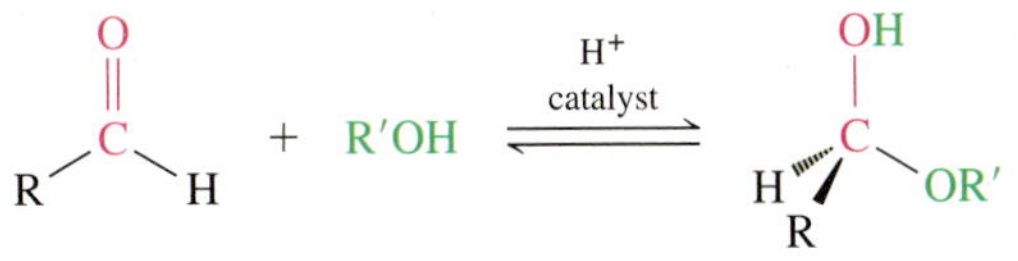

An aldehyde   **A hemiacetal**

> A hemiacetal carbon atom in a molecule can be recognized as it has two oxygen atoms directly attached to it—one is part of an alcohol group, the other part of an ether.

In carbohydrates, this same reaction takes place, although both the hydroxyl and the carbonyl functional groups are in the same molecule, and an *intramolecular* nucleophilic addition can take place, leading to the formation of a molecule with a *cyclic* hemiacetal functional group. Because the rings are relatively strain free, five- and six-membered cyclic hemiacetal molecules form particularly easily, and many carbohydrate molecules exist in an equilibrium between open-chain and cyclic hemiacetal forms. For example, glucose exists in aqueous solution primarily in the form of six-membered *pyranose* molecules resulting from intramolecular nucleophilic addition of the —OH group at C5 to the C1 aldehyde group. Fructose molecules, on the other hand, exist to the extent of about 72% in the pyranose form and about 28% in the five-membered *furanose* form resulting from addition of the —OH group at C5 to the C2 ketone. (The names *pyranose* for a molecule with a six-membered ring and *furanose* for one with a five-membered ring are derived from the names of the simple cyclic ethers pyran and furan.) The cyclic forms of molecules of glucose and fructose are shown in Figure 29.6.

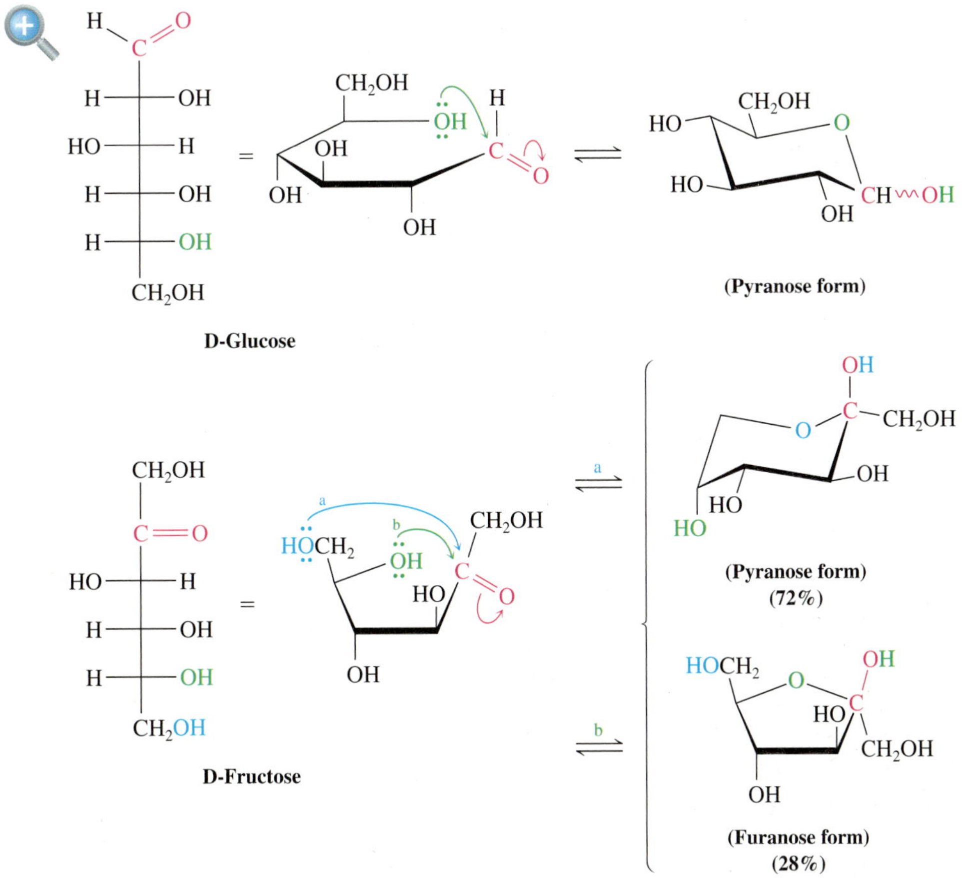

FIGURE 29.6 Molecules of glucose and fructose in their cyclic pyranose and furanose forms.

Like cyclohexane rings [<<Section 9.7], pyranose rings have a chair-like geometry with axial and equatorial substituents. By convention, the rings are usually drawn by placing the hemiacetal oxygen atom at the right rear, as shown in Figure 29.6. Note that an —OH group on the *right* in a Fischer projection is on the *bottom* face of the pyranose ring, and an —OH group on the *left* in a Fischer projection is on the *top* face of the ring. For D sugars, the terminal —CH₂OH group is on the top of the ring, whereas for L sugars, the —CH₂OH group is on the bottom.

WORKED EXAMPLE 29.5—CARBOHYDRATE MOLECULAR STRUCTURE

A molecule of D-mannose differs from one of D-glucose in its stereochemistry at C2. Draw a D-mannose molecule in its pyranose form.

Solution

First draw a Fischer projection of a D-mannose molecule. Then lay it on its side, and curl it around so that the —CHO group (C1) is on the right front and the —CH₂OH group (C6) is toward the left rear. Now, connect the —OH at C5 to the C1 carbonyl group to form the pyranose ring. In drawing the chair form, raise the leftmost carbon atom (C4) up and drop the rightmost carbon atom (C1) down.

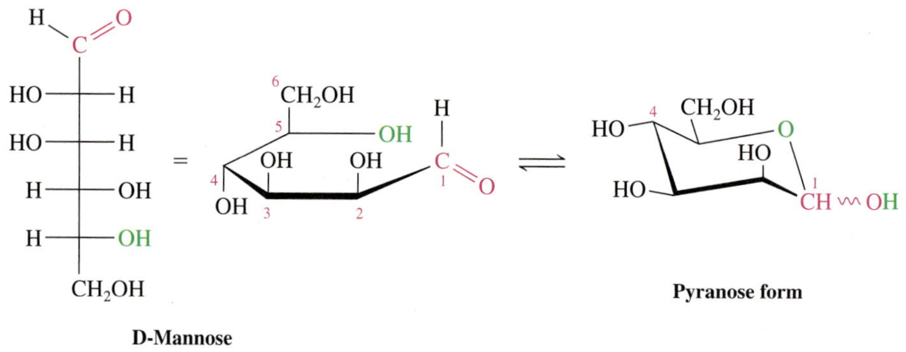

D-Mannose

Pyranose form

EXERCISE 29.7—CARBOHYDRATE MOLECULAR STRUCTURE

D-Galactose molecules differ from those of D-glucose in their stereochemistry at C4. Draw a D-galactose molecule in its pyranose form.

EXERCISE 29.8—CARBOHYDRATE MOLECULAR STRUCTURE

Ribose molecules exist largely in their furanose form, produced by addition of the C4 —OH group to the C1 aldehyde. Find the structure of a D-ribose molecule in Figure 29.5, and draw it in its furanose form.

Much of the chemistry of monosaccharides is the familiar chemistry of alcohol and carbonyl functional groups. Monosaccharides normally exist in solution as cyclic hemiacetals rather than as open-chain aldehydes or ketones. The hemiacetal linkage results from reaction of the carbonyl group with an —OH group three or four carbon atoms away. A five-membered ring hemiacetal is a furanose, and a six-membered ring hemiacetal is a pyranose.

Monosaccharide Anomers: Mutarotation

When an open-chain monosaccharide molecule cyclizes to a pyranose or furanose form, a new stereocentre forms at the former carbonyl carbon atom. Two diastereomers, called *anomers*, are produced, with the hemiacetal carbon atom referred to as the **anomeric**

centre. For example, glucose molecules cyclize reversibly in aqueous solution to yield a 37:63 mixture of two anomers (Figure 29.7). The minor anomer, which has the C1 —OH group trans to the CH$_2$OH substituent at C5, is called the *alpha* (*α*) *anomer*; its full name is *α*-D-glucopyranose. The major anomer, which has the C1 —OH group *cis* to the —CH$_2$OH substituent at C5, is called the *beta* (*β*) *anomer*; its full name is *β*-D-glucopyranose.

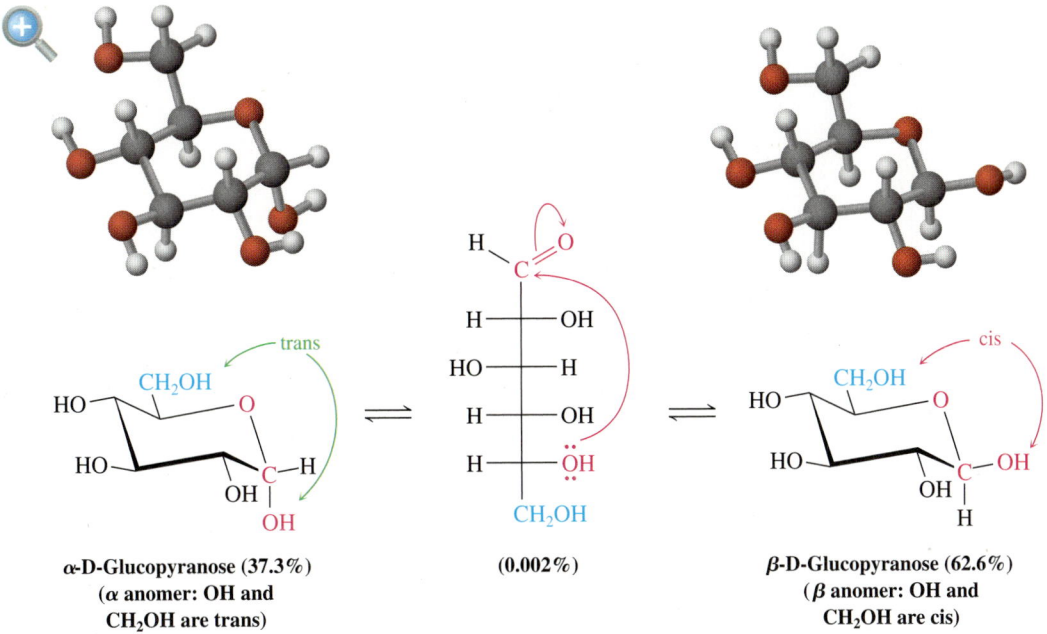

FIGURE 29.7 Structures of the alpha and beta anomers of molecules of glucose.

Substances made of both anomers of D-glucopyranose can be crystallized and purified. Pure *α*-D-glucopyranose has a melting point of 146 °C and a specific rotation [*α*]$_D$ = 112.2°; pure *β*-D-glucopyranose has a melting point of 148–155 °C and a specific rotation [*α*]$_D$ = 18.7°. When a sample of either pure *α*-D-glucopyranose or pure *β*-D-glucopyranose is dissolved in water, however, the specific rotation slowly changes and ultimately reaches a constant value of 52.6°. The specific rotation of the *α* anomer solution decreases from 112.2° to 52.6°, and the specific rotation of the *β* anomer solution increases from 18.7° to 52.6°. Called **mutarotation**, this spontaneous change in optical rotation is caused by the slow conversion of the pure *α* and *β* enantiomers into the 37:63 equilibrium mixture.

At the molecular level, you can think of mutarotation as occurring by a reversible ring opening of each hemiacetal anomer to the open-chain aldehyde form, followed by reclosure. Although equilibration is slow at neutral pH, it is catalyzed by either acid or base.

WORKED EXAMPLE 29.6—CARBOHYDRATE MOLECULAR STRUCTURE

Draw the two pyranose anomers of D-galactose, and identify each as α or β.

Solution

The α anomer has the —OH group at C1 pointing down, trans to the CH$_2$OH, and the β anomer has the —OH group at C1 pointing up, cis to the CH$_2$OH.

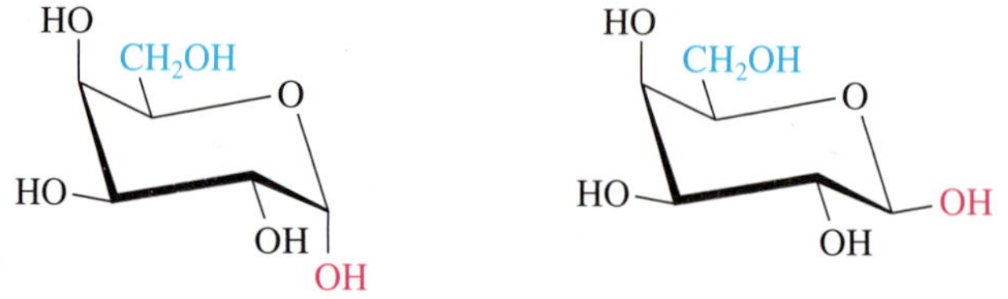

α-D-Galactopyranose β-D-Galactopyranose

EXERCISE 29.9—CARBOHYDRATE MOLECULAR STRUCTURE

At equilibrium in aqueous solution, D-fructose consists of 70% β-pyranose, 2% α-pyranose, 23% β-furanose, and 5% α-furanose forms. Draw molecules of all four.

EXERCISE 29.10—CARBOHYDRATE MOLECULAR STRUCTURE

Draw a β-D-mannopyranose molecule in its chair conformation, and label all substituents as axial or equatorial. Which would you expect to be more stable: mannose or galactose (Worked Example 29.6)?

Cyclization of an open-chain sugar molecule leads to the formation of a new stereocentre (the anomeric centre) and the production of two diastereomeric hemiacetals called alpha (α) and beta (β) anomers. These two anomers can be interconverted.

Glycoside Formation

We saw in Section 23.7 that treatment of a hemiacetal with an alcohol and an acid catalyst yields an acetal:

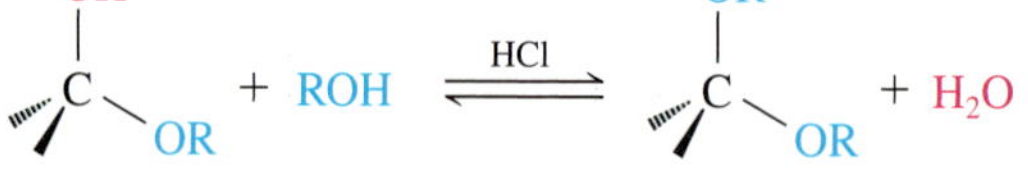

In the same way, treatment of a substance made of monosaccharide hemiacetal molecules with an alcohol and an acid catalyst yields a substance with acetal molecules in which each anomeric —OH group has been replaced by an —OR group. For example, reaction of glucose with methanol gives a mixture of α and β methyl D-glucopyranoside molecules:

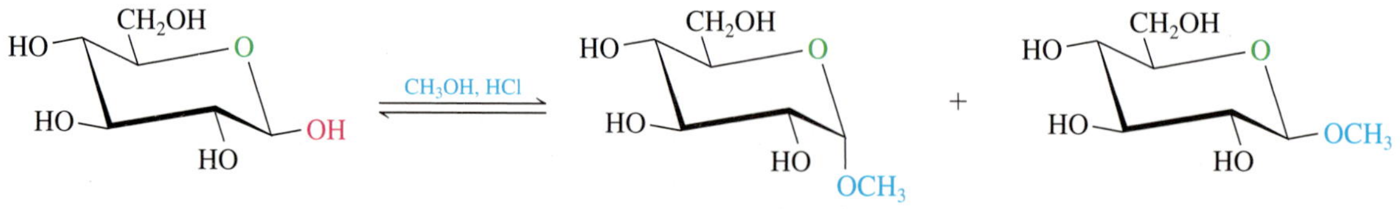

β-D-Glucopyranose
(a cyclic hemiacetal)   Methyl α-D-glucopyranoside
(66%)   Methyl β-D-glucopyranoside
(33%)

Called *glycosides,* carbohydrate acetal molecules are named by first citing the alkyl group and then replacing the *–ose* ending of the sugar with *–oside.* Like all acetals, glycosides are stable to water. They aren't in equilibrium with an open-chain form, and they don't show mutarotation. They can, however, be converted back to the free monosaccharide by hydrolysis with aqueous acid.

Glycosides are widespread in nature, and many biologically active molecules contain glycosidic linkages. For example, digitoxin, the active component of the digitalis preparations used for treatment of heart disease, has glycoside molecules consisting of a complex steroid alcohol linked to a trisaccharide. Note that the three sugars are also linked by glycoside bonds.

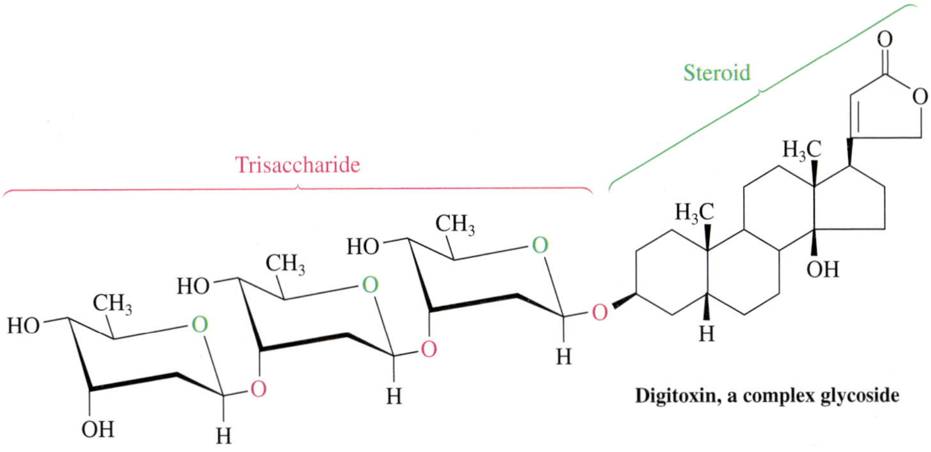

Digitoxin, a complex glycoside

WORKED EXAMPLE 29.7—GLYCOSIDES

What product would you expect from the acid-catalyzed reaction of β-D-ribofuranose with methanol?

Solution

The acid-catalyzed reaction of a monosaccharide with an alcohol yields a glycoside in which the anomeric —OH group is replaced by the —OR group of the alcohol:

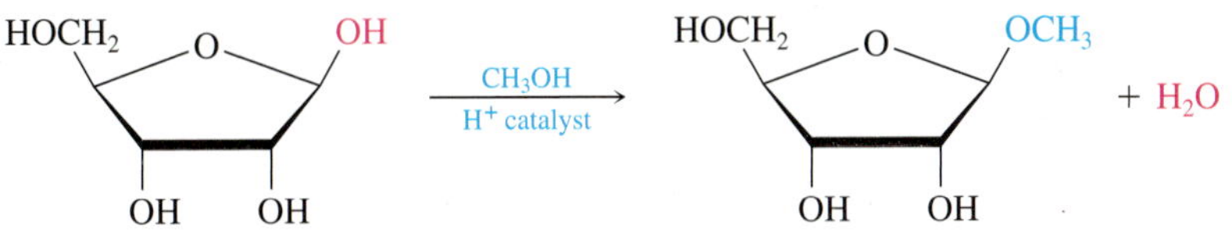

β-D-Ribofuranose   **Methyl β-D-ribofuranoside**

EXERCISE 29.11—GLYCOSIDES

Draw the product you would obtain from the acid-catalyzed reaction of β-D-galactopyranose with ethanol.

Reducing Sugars

Like other aldehydes, an aldose is easily oxidized to yield the corresponding carboxylic acid, called an *aldonic acid.* Aldoses react with Tollens' reagent (with Ag^+ ions in aqueous ammonia solution), Fehling's reagent (with Cu^{2+} ions in aqueous sodium tartrate solution), and Benedict's reagent (with Cu^{2+} ions in aqueous sodium citrate solution) to yield the oxidized sugar and a reduced metallic species. All three reactions serve as simple chemical tests for what are called **reducing sugars** (*reducing* because the sugar reduces the metal ion oxidizing agent).

β-D-Galactose

D-Galactonic acid
(an aldonic acid)

If Tollens' reagent is used, metallic silver is produced as a shiny mirror on the walls of the reaction flask or test tube [<<Section 23.12]. If Fehling's or Benedict's reagent is used, a reddish precipitate of Cu_2O signals a positive result. Some simple diabetes self-test kits sold in drugstores for home use employ Benedict's test. As little as 0.1% glucose in urine gives a positive test.

The importance of these tests is that they allow us to easily identify sugars whose molecules have only glycoside groups. All aldoses are reducing sugars because they contain aldehyde carbonyl groups, but glycosides are non-reducing. Glycosides don't react with Tollens' or Fehling's reagents because the acetal group can't open to an aldehyde under basic conditions [<<Section 23.12]. Any sugar whose molecules have all their anomeric carbon atoms in the form of glycosides (acetals or ketals) will be non-reducing sugars. If the carbohydrate contains one or more anomeric carbon atoms in the form of a hemiacetal, they are reducing sugars.

> The reaction of monosaccharide hemiacetal molecules with an alcohol and an acid catalyst yields acetal molecules in which each anomeric —OH group has been replaced by an —OR group, called a glycoside. Glycosides are non-reducing sugars, because their acetal groups can't open to an open-chain aldehyde form required to reduce metallic oxidizing agents under basic conditions.

Disaccharides

We saw in the previous section that reaction of a monosaccharide with a hemiacetal functional group with an alcohol yields a glycoside in which the anomeric —OH group of the sugar is replaced by an —OR substituent. If the alcohol is a sugar, the glycoside product is a disaccharide.

Maltose and Cellobiose Disaccharide molecules can contain a glycosidic acetal bond between the anomeric carbon atom (the carbonyl carbon) of one sugar molecule and an —OH group at *any* position on the other sugar molecule. A glycosidic link between C1 of the first sugar and C4 of the second sugar is particularly common. Such a bond is called a *1,4′ link*, where the "prime" superscript indicates that the 4′ position is on a different sugar molecule than the 1 position.

A glycosidic bond can be either α or β. Maltose, the disaccharide obtained by partial hydrolysis of starch, consists of two D-glucopyranose molecules joined by a 1,4′-α-glycoside bond. Cellobiose, the disaccharide obtained by partial hydrolysis of cellulose, consists of two D-glucopyranose molecules joined by a 1,4′-β-glycoside bond.

Maltose and cellobiose are both reducing sugars because the right-hand saccharide unit in each has a hemiacetal group. Both are therefore in equilibrium with aldehyde forms, which can reduce Tollens' or Fehling's reagent. For a similar reason, both maltose and cellobiose undergo mutarotation.

Despite the similarities of their structures, maltose and cellobiose are dramatically different in biochemical reactions. Cellobiose can't be digested by humans and can't be fermented by yeast. Maltose, however, is digested without difficulty and is readily fermented. This striking difference in reactivity is due to the presence of enzymes in

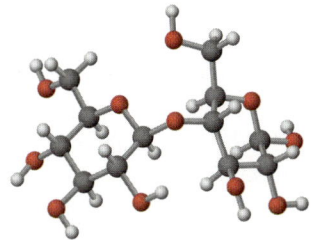

Maltose, a 1,4′-α-glycoside
[4-*O*-(α-D-glucopyranosyl)-
α-D-glucopyranose]

Cellobiose, a 1,4′-β-glycoside
[4-*O*-(β-D-glucopyranosyl)-
β-D-glucopyranose]

humans that selectively break apart only α-glycoside linkages to produce smaller sugars, and not beta ones.

Sucrose *Sucrose,* or ordinary table sugar, is probably the most abundant organic substance in the world. Whether from sugar cane (20% by weight) or from sugar beets (15% by weight), and whether raw or refined, all table sugar is sucrose.

Sucrose is a disaccharide that yields equal amounts (mol) of glucose and fructose on hydrolysis. This 1:1 mixture of glucose and fructose is often referred to as *invert sugar* because the sign of optical rotation changes (inverts) during the hydrolysis from sucrose, $[\alpha]_D = 66.5°$, to a glucose-fructose mixture, $[\alpha]_D = -22°$. Insects such as honeybees have enzymes called *invertases* that catalyze the hydrolysis of sucrose to glucose-fructose. Honey is primarily a mixture of glucose, fructose, and sucrose.

Unlike most other disaccharides, sucrose is not a reducing sugar and does not exhibit mutarotation. These observations imply that sucrose molecules have no hemiacetal groups and that the glucose and fructose units must *both* be glycosides. This can happen only if the two sugars are joined by a glycoside link between the anomeric carbon atoms of both sugars—C1 of glucose and C2 of fructose.

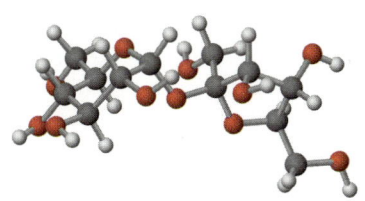

Sucrose, a 1,2′-glycoside
[2-*O*-(α-D-glucopyranosyl)-β-D-fructofuranoside]

Polysaccharides

Polysaccharides are carbohydrates whose molecules have tens, hundreds, or even thousands of simple sugars linked by glycoside bonds. Cellulose and starch are the two most widely occurring polysaccharides.

Cellulose Cellulose molecules consist of several thousand D-glucose units linked by 1,4′-β-glycoside bonds like those in cellobiose. Different cellulose molecules can then interact to form a large aggregate structure held together by hydrogen bonds:

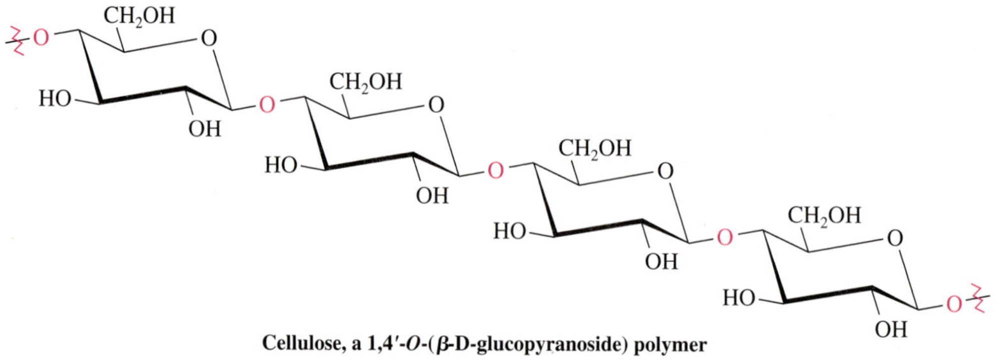

Cellulose, a 1,4′-O-(β-D-glucopyranoside) polymer

Nature uses cellulose primarily as a structural material to impart strength and rigidity to plants. Wood, leaves, grasses, and cotton are primarily cellulose. Cellulose also serves as a raw material for the manufacture of cellulose acetate, known commercially as *rayon,* and cellulose nitrate, known as *guncotton.* Guncotton is the major ingredient in smokeless powder, the explosive propellant used in artillery shells and in ammunition for firearms.

Starch and Glycogen Potatoes, corn, and cereal grains contain large amounts of *starch,* a polymer of glucose in which the monosaccharide units are linked by 1,4′-α-glycoside bonds like those in maltose. Starch can be separated into two fractions: *amylose,* which is insoluble in cold water, and *amylopectin,* which is soluble in cold water. Amylose accounts for about 20% by weight of starch and consists of several hundred glucose molecules linked together by 1,4′-α-glycoside bonds.

Amylopectin, which accounts for the remaining 80% of starch, is more complex in structure than amylose. Unlike cellulose or amylose, which are linear polymers, amylopectin contains 1,6′-α-glycoside *branches* approximately every 25 glucose units. As a result, amylopectin has an exceedingly complex three-dimensional structure.

Nature uses starch as the medium by which plants store energy for later use. When eaten, starch is digested in the mouth and stomach by enzymes called *glycosidases,* which catalyze the hydrolysis of glycoside bonds and release individual molecules of glucose. Like most enzymes, glycosidases are highly selective in their action. They hydrolyze only the α-glycoside links in starch and leave the β-glycoside links in cellulose untouched. As a result, humans can digest potatoes and grains but not grass.

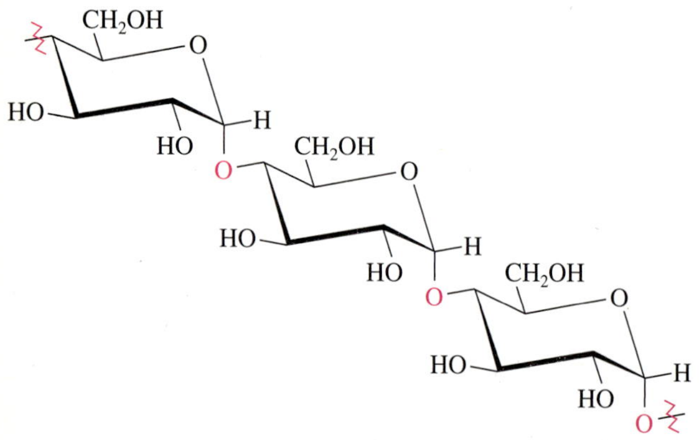

Amylose, a 1,4′-O-(α-D-glucopyranoside) polymer

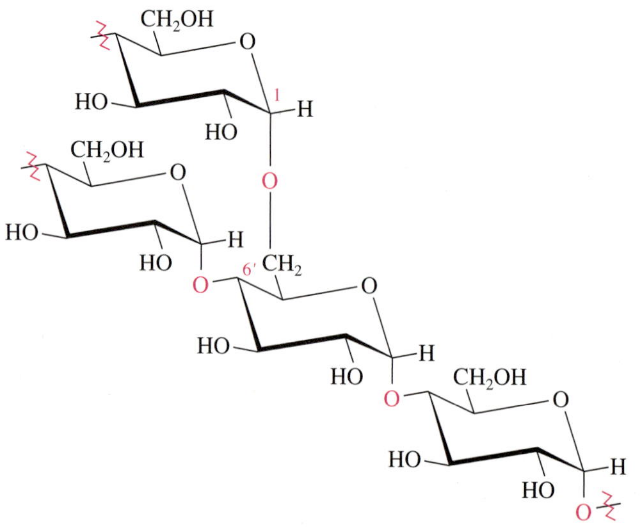

Amylopectin

Glycogen is a polysaccharide that serves the same energy-storage function in animals that starch serves in plants. Dietary carbohydrate not needed for immediate energy is converted by the body to glycogen for long-term storage. Like the amylopectin found in starch, glycogen molecules have a complex three-dimensional structure with both 1,4′ and 1,6′ links (Figure 29.8). Glycogen molecules are larger than those of amylopectin—up to 100 000 glucose units—and contain even more branches.

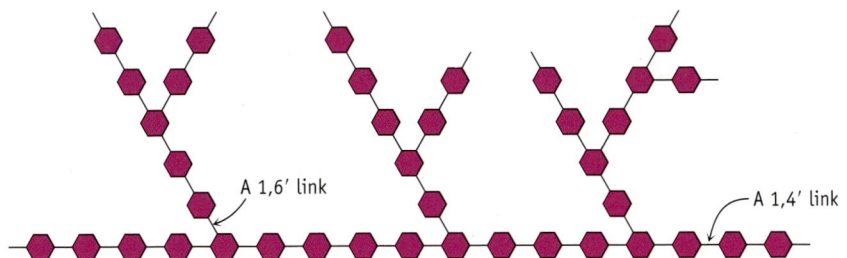

A 1,6′ link

A 1,4′ link

FIGURE 29.8 A representation of the connectivity of monosaccharide units in a portion of a glycogen molecule. The hexagons represent glucose units linked by 1,4' and 1,6' acetal bonds.

Disaccharides are complex carbohydrates in which two simple sugar molecules are linked by a glycoside bond between the anomeric carbon atom of one unit and an —OH of the second unit. The two sugars can be the same, as in maltose and cellobiose, or different, as in sucrose. The glycoside bond can be either α (maltose) or β (cellobiose) and can involve any —OH of the second sugar molecule. A 1,4′ link is most common (cellobiose, maltose), but other links, such as 1,2′ (sucrose), also occur. Polysaccharides, such as cellulose, starch, and glycogen, are used in nature both as structural materials and for long-term energy storage.

Other Important Carbohydrates

In addition to the common carbohydrates mentioned in previous sections, a variety of important carbohydrate-derived materials have structures that have been chemically modified. Their structural resemblance to sugars is clear, but they aren't simple aldoses or ketoses.

Deoxy sugars have molecules with one of their oxygen atoms "missing." That is, an —OH group is replaced by an —H. The most common deoxy sugar is 2-deoxyribose, a monosaccharide found in DNA [*deoxyribonucleic acid*, >>Section 29.4]. 2-deoxyribose exists in water solution as a complex equilibrium mixture of both furanose and pyranose forms.

Oxygen missing

CHO

H——H
H——OH
H——OH
CH_2OH

HOCH₂

α-D-2-Deoxyribopyranose (40%)
(+ 35% β anomer)

(0.7%)

α-D-2-Deoxyribofuranose (13%)
(+ 12% β anomer)

Amino sugars, such as D-glucosamine, have molecules with one of their —OH groups replaced by an —NH_2. The *N*-acetyl amide derived from D-glucosamine is the monosaccharide unit from which *chitin,* the hard crust around insects and shellfish, is made. Antibiotics such as streptomycin and gentamicin contain still other amino sugars.

CH_2OH

HO

HO

NH_2

OH

β-D-Glucosamine

Cell-Surface Carbohydrates and Carbohydrate Vaccines

It was once thought that carbohydrates were useful in nature only as structural materials and energy sources. Although carbohydrates do indeed serve these purposes, they also have many other important biochemical functions. For example, polysaccharides are centrally involved in cell–cell recognition, the critical process by which one type of cell distinguishes another. Small polysaccharide chains, covalently bound by glycosidic links to hydroxyl groups on proteins (*glycoproteins*), act as biochemical markers on cell surfaces, as illustrated by the human blood-group antigens.

It has been known for more than a century that human blood can be classified into four blood-group types (A, B, AB, and O) and that blood from a donor of one type can't be transfused into a recipient with another type unless the two types are compatible (Table 29.1). Should an incompatible mix be made, the red blood cells clump together, or *agglutinate.*

The agglutination of incompatible red blood cells, which indicates that the body's immune system has recognized the presence of foreign cells in the body and has formed antibodies against them, results from the presence of polysaccharide markers on the surface of the cells. Types A, B, and O red blood cells each have characteristic markers, called *antigenic determinants*; type AB cells have both type A and type B markers. The structures of all three blood-group determinants are shown below:

TABLE 29.1 Human Blood-Group Compatibilities

DONOR BLOOD TYPE	ACCEPTOR BLOOD TYPE			
	A	B	AB	O
A	o	x	o	x
B	x	o	o	x
AB	x	x	o	x
O	o	o	o	o

o = compatible; x = incompatible.

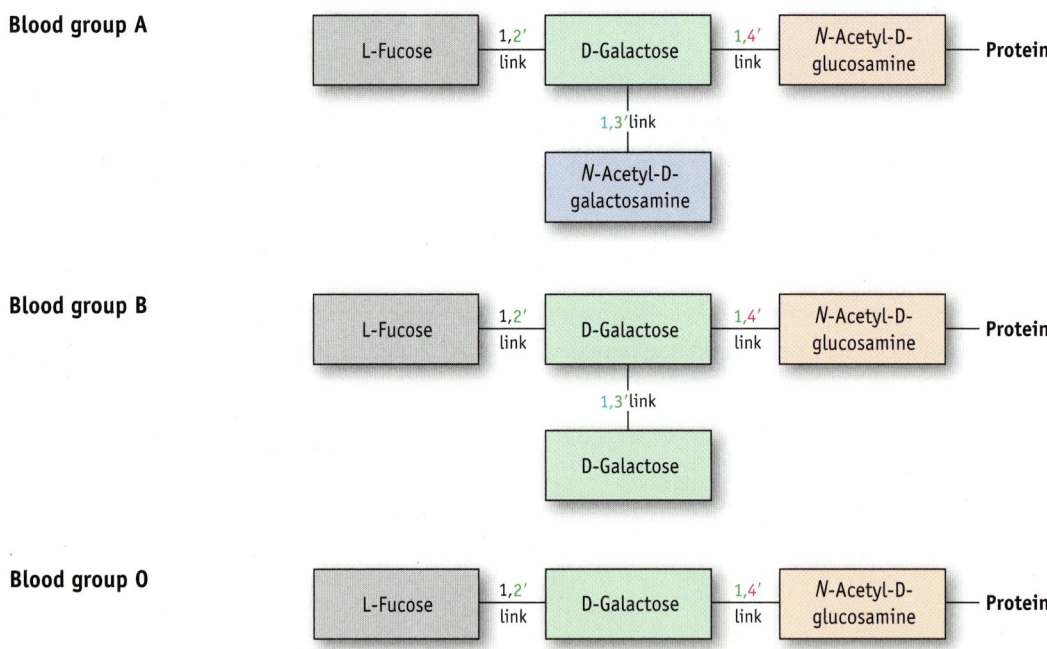

Blood group A

L-Fucose — 1,2′ link — D-Galactose — 1,4′ link — N-Acetyl-D-glucosamine — Protein

1,3′ link — N-Acetyl-D-galactosamine

Blood group B

L-Fucose — 1,2′ link — D-Galactose — 1,4′ link — N-Acetyl-D-glucosamine — Protein

1,3′ link — D-Galactose

Blood group O

L-Fucose — 1,2′ link — D-Galactose — 1,4′ link — N-Acetyl-D-glucosamine — Protein

Note that some unusual carbohydrates are involved. All three blood-group antigenic determinants contain N-acetyl amino sugars as well as the unusual monosaccharide L-fucose.

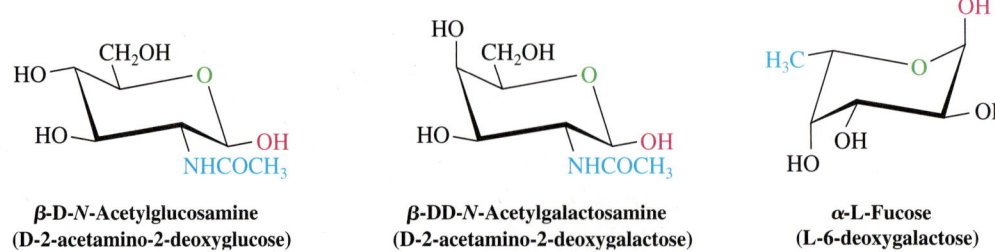

β-D-*N*-Acetylglucosamine
(D-2-acetamino-2-deoxyglucose)

β-DD-*N*-Acetylgalactosamine
(D-2-acetamino-2-deoxygalactose)

α-L-Fucose
(L-6-deoxygalactose)

Understanding the role of carbohydrates in cell recognition is a vigorous area of current research that offers hope of breakthroughs in the understanding of a wide range of diseases from bacterial infections to cancer. Particularly exciting is the possibility of developing useful anti-cancer vaccines to help mobilize the body's immune system against tumour cells. Recent advances along these lines have included a laboratory synthesis of the so-called globo H antigen, found on the surface of human breast, prostate, colon, and pancreatic cancer cells. Mice treated with the synthetic globo H hexasaccharide linked to a carrier protein developed large amounts of antibodies, which then recognized tumour cells.

Globo H antigen

29.3 Amino Acids, Peptides, and Proteins

Proteins are large biomolecules found in every living organism. Many types of proteins exist, with many different biological functions. The keratin of skin and fingernails, the insulin that regulates glucose metabolism in the body, calmodulin that binds Ca^{2+} ions [<<Section 29.1], and the DNA polymerase that catalyzes the synthesis of DNA in cells are all proteins. Regardless of their appearance or function, all proteins are chemically similar. All of their molecules are made up of many *amino acid* monomer units linked together by amide bonds in a long polymer chain.

Molecular Modelling (Odyssey)

e29.4 View molecular-level simulations of amino acids.

Amino Acids

Amino acid molecules [<<Section 14.8], as their name implies, have two functional groups. They contain both a basic amino group and an acidic carboxylic acid group. However, they almost never exist as the uncharged molecules shown on the left below, because the carboxylic acid and amino functional groups can each undergo proton transfer reactions:

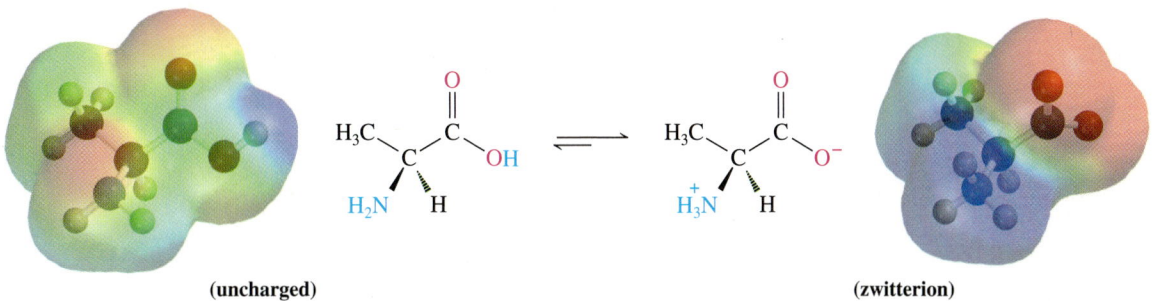

(uncharged) (zwitterion)

Alanine

At low pH, both functional groups are protonated, and the species has a positive charge. At high pH, both functional groups are deprotonated, and the species has a negative charge. Between these two pH values, the species exists primarily in the form of a dipolar ion, or *zwitterions* (German *zwitter*, meaning "hybrid.")

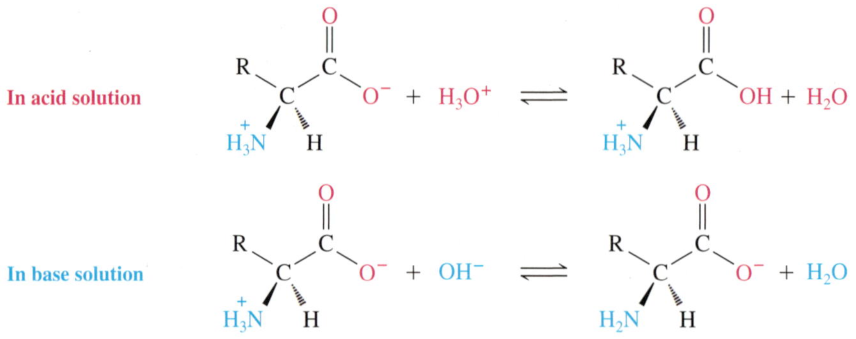

The relative amount of cations, anions, and zwitterions of an amino acid solution depends on the effect of the different —R groups on the acidity and basicity of an amino acid [<<Section 14.8]. Varying the pH of an amino acid solution is an easy way to regulate the balance of cations and anions in solution and to characterize the acid/base properties of an amino acid. *Electrophoresis* experiments [<<e14.25] take advantage of this property. When a solution of an amino acid is placed in an electric field, the migration of anions and cations toward the two electrodes depends on the **isoelectric point** (given the abbreviation pI) of the amino acid and on the pH of the solution. Recall that the isoelectric point (pI) is defined as the pH at which the amino acid is exactly balanced between anionic and cationic forms and exists primarily as the neutral, dipolar, zwitterion. [<<Section 14.6]. A mixture of amino acids can be separated by buffer control of the solution pH in an electric field (Figure 29.9).

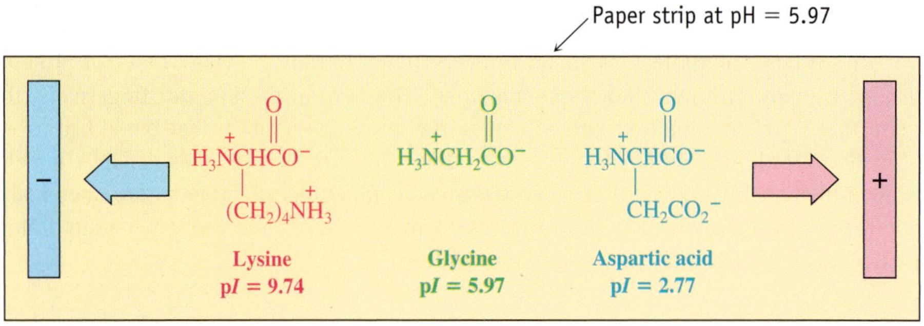

FIGURE 29.9 Separation of an amino acid mixture by electrophoresis. At pH 5.97, glycine exists mostly as neutral zwitterions and do not migrate, lysine molecules are protonated and migrate toward the cathode, and aspartic acid molecules are deprotonated and migrate toward the anode.

The structures, abbreviations, and pK_a values of the 20 amino acids commonly found in proteins are shown in Table 14.10 in the form that predominates within cells at a physiological pH of 7.4. All 20 are α-amino acids, meaning that the amino group is a substituent on the α-carbon atom—the one next to the carbonyl group in the molecule. Nineteen of the twenty are primary amines (—NH$_2$) and differ only in the identity of the *side chain*—the substituent attached to the α-carbon. Proline, however, is a secondary amine whose nitrogen and α-carbon atoms are part of a five-membered pyrrolidine ring.

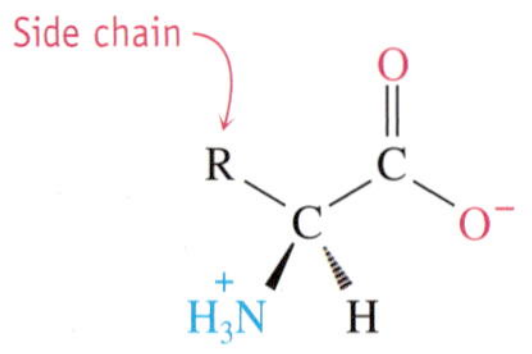

A primary α-amino acid

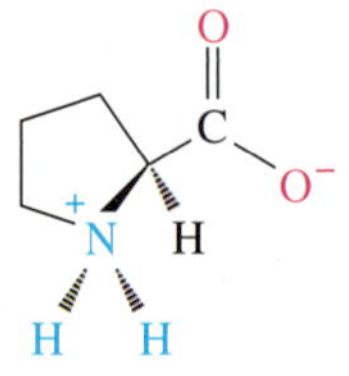

Proline, a secondary α-amino acid

In addition to the 20 amino acids found in proteins, there are a number of other biologically important non-protein amino acids. γ-Aminobutyric acid (GABA), for instance, is found in the brain and acts as a neurotransmitter; homocysteine is found in blood and is linked to coronary heart disease; and thyroxine is found in the thyroid gland, where it acts as a hormone.

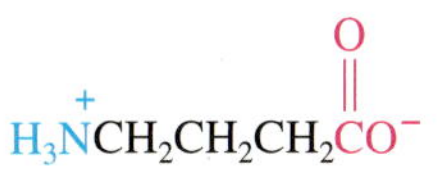

$$\overset{+}{H_3N}CH_2CH_2CH_2\overset{O}{\overset{\|}{C}}O^-$$

**γ-Amino-
butyric acid**

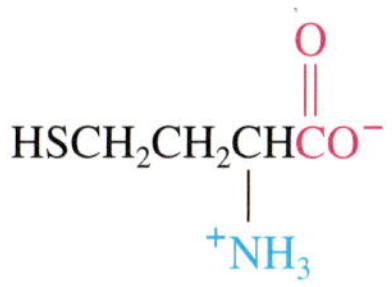

$$HSCH_2CH_2\overset{O}{\overset{\|}{\underset{\overset{|}{\overset{+}{N}H_3}}{C}HC}}O^-$$

Homocysteine

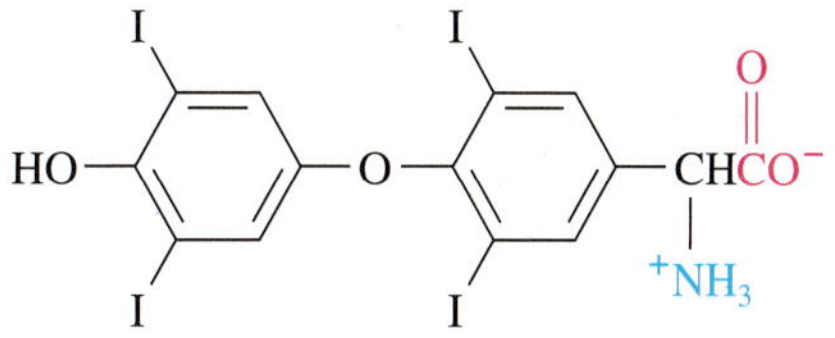

Thyroxine

With the exception of glycine ($H_2NCH_2CO_2H$), the α-carbon atoms of the 20 amino acid molecules are stereocentres. Two enantiomeric forms are therefore possible, but nature uses only a single enantiomer to build proteins. In Fischer projections, naturally occurring amino acids are represented by placing the carboxyl group at the top as if drawing a carbohydrate [<<Section 29.2] and then placing the amino group on the left. Because of their stereochemical similarity to L sugars, the naturally occurring *a*-amino acids are often referred to as L-amino acids.

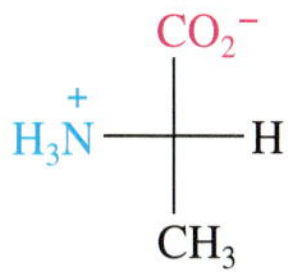

L-Alanine
(*S*)-Alanine

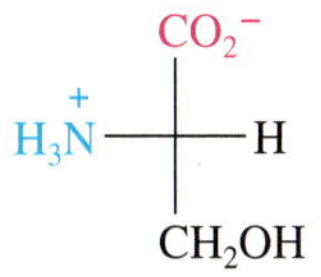

L-Serine
(*S*)-Serine

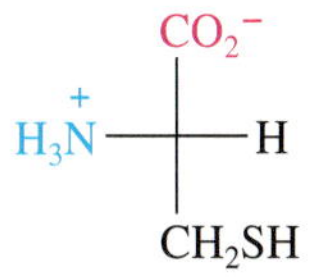

L-Cysteine
(*R*)-Cysteine

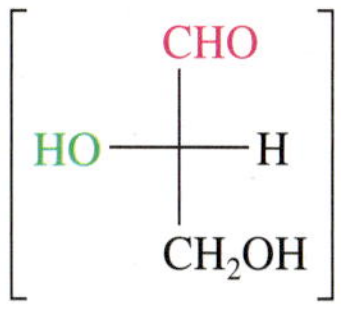

L-Glyceraldehyde

The 20 common amino acids can be further classified as either neutral, acidic, or basic, depending on the structure of their molecule's side chain. Of the 20 amino acids, 15 have neutral side chains, 2 (aspartic acid and glutamic acid) have an extra carboxylic acid functional group in their side chains, and 3 (lysine, arginine, and histidine) have basic amino groups in their side chains. Note, however, that both cysteine (a thiol) and tyrosine (a phenol), although classified as neutral amino acids, have weakly acidic side chains and can be deprotonated in strongly basic solution.

At the pH of 7.4 found within cells, the side-chain carboxyl groups of aspartic acid and glutamic acid are ionized and exist as carboxylate ions, $-CO_2^-$. Similarly, the basic side-chain nitrogen atoms of lysine and arginine are protonated at pH 7.4 and exist as alkyl ammonium ions, $R-NH_3^+$. Histidine, however, which contains a heterocyclic imidazole ring in its molecules' side chain, is not quite basic enough to be protonated at pH 7.4. Note that only the pyridine-like doubly bonded nitrogen atom in histidine is basic. The pyrrole-like singly bonded nitrogen atom is not as basic because its lone pair of electrons is delocalized as part of the aromatic imidazole ring [<<Section 20.3].

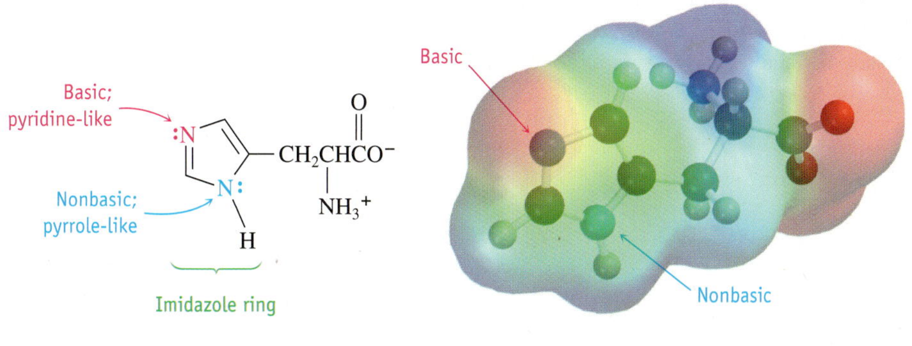

Histidine

All 20 of the amino acids are necessary for protein synthesis, but the human body can synthesize only 10 of the 20. The other 10 (shown in red in Table 14.10) are called *essential amino acids* because they must be obtained from diet. Failure to include an adequate dietary supply of any of these essential amino acids leads to poor growth and general failure to thrive.

Peptides and Proteins

As seen in Section 29.1, the value of amino acids as building blocks for proteins stems from the fact that amino acid molecules can link together into long polymeric chains. Linkages between amino acid monomers are comprised of amide (peptide) bonds between the —NH$_2$ of one amino acid and the —COOH of another [<<Section 24.7]. For classification purposes, chains with fewer than 50 amino acids are usually called **peptides**, while the term **protein** is used for longer chains. The amine end of the peptide is referred to as the *N-terminal* end and the carboxylic acid end the *C-terminal* end.

Molecular Modelling (Odyssey)

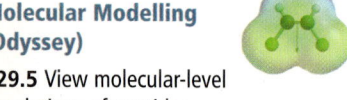

e29.5 View molecular-level simulations of peptides.

In considering the primary structure of protein molecules, we focus on the properties of the linking amide bonds [<<Section 25.4]. Amide nitrogen atoms are much less basic than amine nitrogen atoms because their non-bonding electron pair is delocalized by interaction with the carbonyl group. This overlap of the nitrogen *p* orbital with the *p* orbitals of the carbonyl group imparts a certain amount of double-bond character to the C—N bond and restricts rotation around it. The atoms attached to the C and N of the amide bond are planar, and the O—C—N—H torsional angle is 180°.

A second kind of covalent bonding in peptides occurs when a disulfide linkage, RS—SR, is formed between two cysteine residues. As we saw in Section 22.10, a disulfide bond is easily formed by mild oxidation of a thiol, RSH, and is easily cleaved by mild reduction.

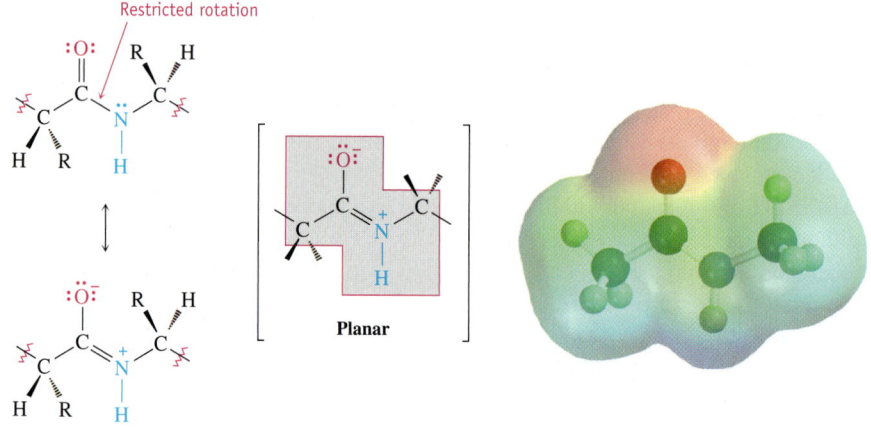

Two cysteines

A disulfide bond between cysteine residues in different peptide chains links the otherwise separate chains together, while a disulfide bond between cysteine residues in the same chain forms a loop in the chain, giving a cyclic peptide. Such is the case with vasopressin, an anti-diuretic hormone found in the pituitary gland. Note that the C-terminal end of vasopressin occurs as the primary amide, $—CONH_2$, rather than as the free acid.

$$
\begin{array}{c}
\text{Disulfide bridge} \\
\text{S} \longrightarrow \text{S} \\
| \qquad\qquad | \\
\text{Cys-Tyr-Phe-Glu-Asn-Cys-Pro-Arg-Gly-NH}_2
\end{array}
$$

Vasopressin

Besides the amide and disulfide covalent bonds that link amino acids to each other, non-covalent interactions such as dipole-dipole forces are also important in determining the overall shape of a protein.

Proteins and peptides are biomolecules made of amino acid residues linked together by amide bonds. Twenty amino acids are commonly found in proteins; all are α-amino acids, and all except glycine have stereochemistry similar to that of L sugars.

Classification of Proteins

Proteins are classified into two major types by their composition. Simple proteins, such as blood serum albumin, are those that yield only amino acids and no other compounds on hydrolysis. Conjugated proteins, which are much more common than simple proteins, yield other compounds such as carbohydrates, fats, or nucleic acids in addition to amino acids on hydrolysis. As shown in Table 29.2, conjugated proteins can be further classified by the chemical nature of the non-amino acid portion.

The term *conjugated* is used in several different senses in chemistry. In Section 19.1, it was used to refer to molecules with alternating single and double or triple bonds. In the context of proteins, it means something completely different—conjugated proteins produce compounds besides amino acids upon hydrolysis.

TABLE 29.2 Some Conjugated Proteins

Name	Composition
Glycoproteins	Proteins bonded to a carbohydrate; cell membranes have a glycoprotein coating.
Lipoproteins	Proteins bonded to fats and oils (lipids); these proteins transport cholesterol and other fats through the body.
Metalloproteins	Proteins bonded to a metal ion; the enzyme cytochrome oxidase, necessary for biological energy production, is an example.
Nucleoproteins	Proteins bonded to RNA (ribonucleic acid); these are found in cell ribosomes.
Phosphoproteins	Proteins bonded to a phosphate group; milk casein, which stores nutrients for growing embryos, is an example.

Another way to classify proteins is as either *fibrous* or *globular*, by their three-dimensional shape. *Fibrous proteins*, such as collagen and keratin, consist of polypeptide chains arranged side by side in long filaments. Because these proteins

TABLE 29.3 Some Common Fibrous and Globular Proteins

Name	Occurrence and Use
Fibrous Proteins (insoluble)	
Collagens	Animal hide, tendons, connective tissues
Elastins	Blood vessels, ligaments
Fibrinogen	Necessary for blood clotting
Keratins	Skin, wool, feathers, hooves, silk, fingernails
Myosins	Muscle tissue
Globular Proteins (soluble)	
Hemoglobin	Involved in oxygen transport
Immunoglobulins	Involved in immune response
Insulin	Hormone for controlling glucose metabolism
Ribonuclease	Enzyme controlling RNA synthesis

are tough and insoluble in water, they are used in nature for structural materials such as tendons, hooves, horns, and muscles. *Globular proteins,* by contrast, are usually coiled into compact, nearly spherical shapes. These proteins are generally soluble in water and are mobile within cells, and often play regulatory functions. Most of the 2000 or so known enzymes are globular. Table 29.3 lists some common examples of both kinds.

Yet a third way to classify proteins is by their function. As shown in Table 29.4, there is an extraordinary diversity to the biological roles of proteins.

TABLE 29.4 Some Biological Functions of Proteins

Type	Function and Example
Enzymes	Proteins such as chymotrypsin that act as biological catalysts
Hormones	Proteins such as insulin that regulate body processes
Protective proteins	Proteins such as antibodies that fight infection
Storage proteins	Proteins such as casein that store nutrients
Structural proteins	Proteins such as keratin, elastin, and collagen that form the structure of an organism
Transport proteins	Proteins such as hemoglobin that transport oxygen and other substances through the body

Protein Structure

Protein molecules are so large that the word *structure* takes on a broader meaning than it does with simpler organic compounds. Chemists speak of four different levels of structure when describing proteins:

- The *primary structure* of a protein is simply the amino acid sequence within a protein molecule.
- The *secondary structure* of a protein describes how *segments* of peptide backbones orient into a regular pattern.
- The *tertiary structure* describes how the *entire* protein molecule coils into an overall three-dimensional shape.
- The *quaternary structure* describes how different protein molecules come together to yield large aggregate structures.

Molecular Modelling (Odyssey)

e29.6 View molecular-level simulations of proteins.

Let's look at three examples—α-keratin (fibrous), fibroin (fibrous), and myoglobin (globular)—to see how higher structure affects a protein's properties.

α-Keratin α-Keratin is the fibrous structural protein found in wool, hair, nails, and feathers. Studies show that segments of the *a*-keratin chain are coiled into a right-handed helical secondary structure like that of a telephone cord. Illustrated in Figure 29.10, this so-called *α-helix* is stabilized by hydrogen bonding between amide N—H groups and C=O groups four residues away. Each coil of the helix (the *repeat distance*) contains 3.6 amino acid residues, and the distance between coils is 540 pm, or 5.40 Å.

Evidence also shows that the α-keratins of wool and hair have a quaternary structure. The individual helical strands are themselves coiled about one another in stiff bundles to form a *superhelix* that accounts for the threadlike properties and strength of these proteins. Although α-keratin is the best example of an almost entirely helical protein, most globular proteins also contain α-helical segments.

Fibroin Fibroin, the fibrous protein found in silk, has a secondary structure called a *β-pleated sheet* in which neighbouring polypeptide chains line up in a parallel arrangement held together by hydrogen bonds between chains (Figure 29.11). The

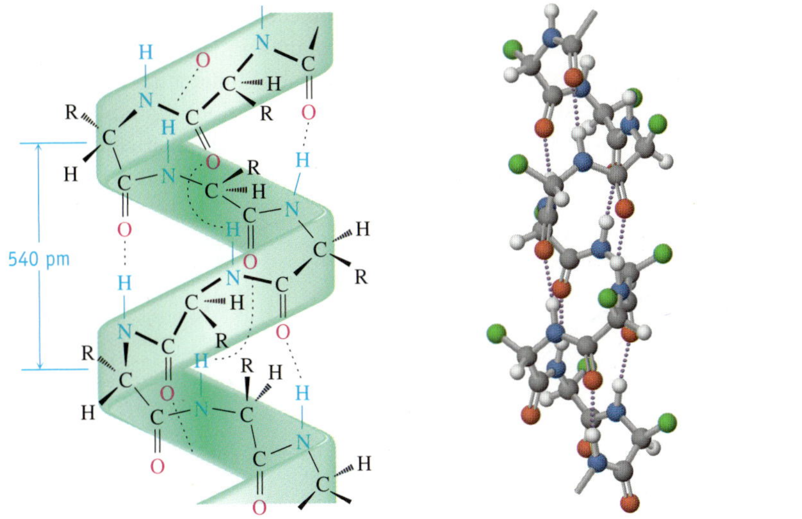

FIGURE 29.10 The helical secondary structure present in α-keratin.

540 pm

neighbouring chains can run either in the same direction (parallel) or in opposite directions (anti-parallel), although the anti-parallel arrangement is more common and energetically somewhat more favourable. While not as common as the *a*-helix, small β-pleated-sheet regions occur frequently in globular proteins.

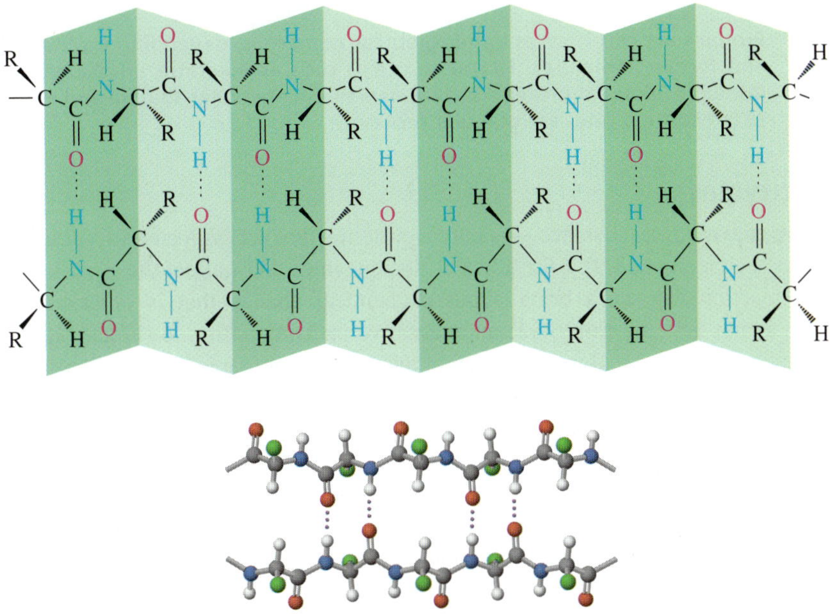

Chain 1

Chain 2

FIGURE 29.11 The β-pleated sheet structure in silk fibroin.

Myoglobin Myoglobin is a small globular protein containing 153 amino acid residues in a single chain. A relative of hemoglobin, myoglobin is found in the skeletal muscles of sea mammals, where it stores oxygen needed to sustain the animals during long dives. Myoglobin consists of eight helical segments connected by bends to form a compact, nearly spherical, tertiary structure (Figure 29.12).

Why does myoglobin adopt the shape it does? The forces that determine the tertiary structure of myoglobin and other globular protein molecules are the same simple forces that act on all molecules, regardless of size, to provide maximum stability in the environment that the protein finds itself in. Particularly important are the hydrophobic (water-repelling) interactions of hydrocarbon side chains on neutral amino acids. Those amino acids with neutral, non-polar side chains have a strong tendency to congregate on the hydrocarbon-like interior of a protein molecule, away from the aqueous medium. Those acidic or basic amino acids with charged side chains, by contrast, tend to congregate on the exterior of the protein where they can be solvated by water.

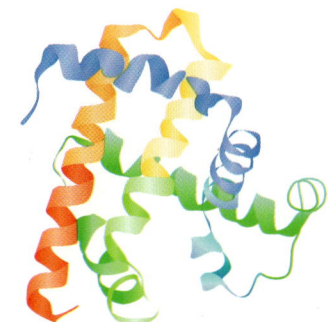

FIGURE 29.12 Secondary and tertiary structure of myoglobin, a small protein with extensive helical sections, shown here as ribbons.

Think about It

e29.7 "Listen" to several features of beta-globulin protein represented by different musical voices.

Also important for stabilizing a protein molecule's tertiary structure are the formation of disulfide bridges between cysteine residues, the formation of hydrogen bonds between nearby amino acid residues, and the development of ionic attractions, called *salt bridges,* between positively and negatively charged sites on various amino acid side chains within the protein. The various kinds of stabilizing forces are summarized in Figure 29.13.

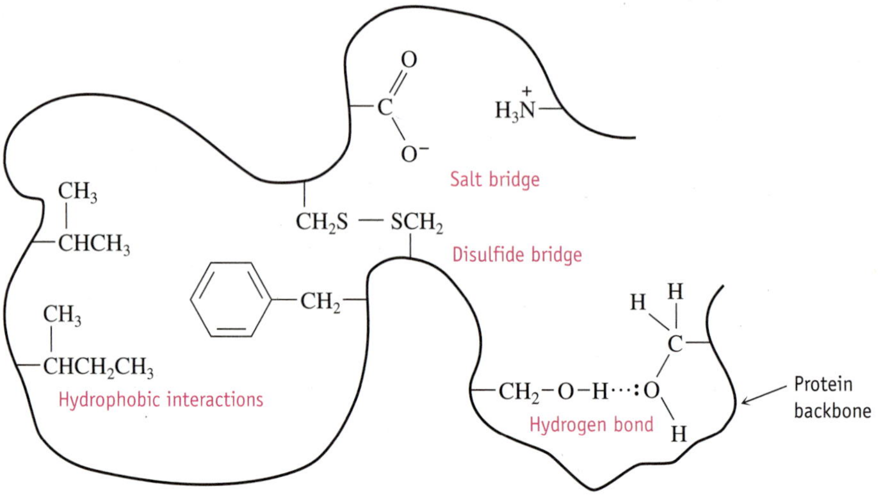

FIGURE 29.13 Kinds of interactions among amino acid side chains that stabilize a protein molecule's tertiary structure.

Proteins are classified as either globular or fibrous, depending on their secondary and tertiary structures. Fibrous proteins such as α-keratin are tough and water-insoluble; globular proteins such as myoglobin are water-soluble and mobile within cells. Most enzymes are globular proteins.

Enzymes

An **enzyme** is a substance—usually a protein—that acts as a catalyst for a biological reaction [<<Section 18.9]. Like all catalysts, enzymes don't affect the equilibrium constant of a reaction and can't bring about a chemical change that is otherwise unfavourable. Enzymes act only to provide an alternative pathway with a lower activation energy for a reaction, thereby making the reaction take place more rapidly. Sometimes the rate acceleration brought about by enzymes is extraordinary. The glycosidase enzymes that hydrolyze polysaccharides, for example, increase the reaction rate by a factor of more than 10^{17}, changing the time required for the reaction from millions of years to milliseconds.

Unlike many of the catalysts that chemists use in the laboratory, enzymes are usually specific in their action. Often an enzyme will catalyze only a single reaction of a single compound, called the enzyme's *substrate*. For example, the enzyme amylase found in the human digestive tract catalyzes only the hydrolysis of starch to yield glucose; cellulose and other polysaccharides are untouched by amylase.

Different enzymes have different specificities. Some, such as amylase, are specific for a single substrate, but others operate on a range of substrates. Papain, for instance, a globular protein of 212 amino acids isolated from papaya fruit, catalyzes the hydrolysis of many kinds of peptide bonds. This ability to hydrolyze peptide bonds makes papain useful as a meat tenderizer and a cleaner for contact lenses.

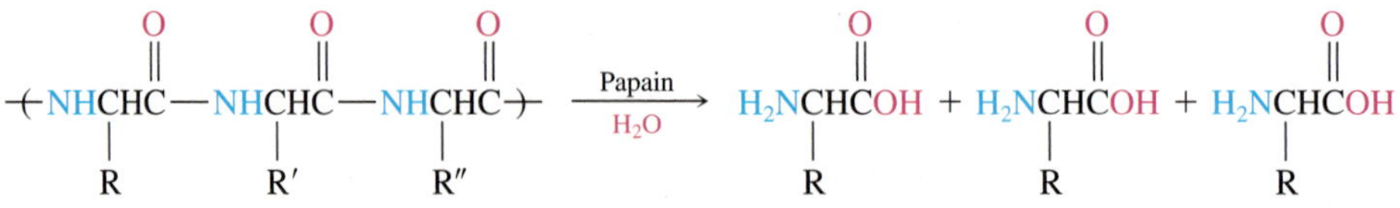

Most of the more than 2000 known enzymes are globular proteins. In addition to the protein part, most enzymes also have a small non-protein part called a *co-factor*. The protein part in such an enzyme is called an *apoenzyme*, and the combination of apoenzyme plus cofactor is called a *holoenzyme*. Only holoenzymes have biological activity; neither co-factor nor apoenzyme can catalyze reactions by themselves.

$$\text{Holoenzyme} = \text{co-factor} + \text{apoenzyme}$$

A co-factor can be either an inorganic ion, such as Zn^{2+}, or a small organic molecule, called a *co-enzyme*. The requirement of many enzymes for inorganic co-factors is the main reason for our dietary need of trace minerals. Iron, zinc, copper, manganese, and numerous other metal ions are all essential minerals that act as enzyme co-factors, although their exact biological role is not known in all cases.

A variety of organic molecules act as co-enzymes. Many, though not all, co-enzymes are *vitamins*, small organic molecules that must be obtained in the diet and are required in trace amounts for proper growth. Table 29.5 lists the 13 known vitamins required in the human diet and their enzyme functions.

TABLE 29.5 Vitamins and Their Enzyme Functions

Vitamin	Enzyme Function	Deficiency Symptoms
Water-Soluble Vitamins		
Ascorbic acid (Vitamin C)	Hydrolases	Bleeding gums, bruising
Thiamin (Vitamin B-1)	Reductases	Fatigue, depression
Riboflavin (Vitamin B-2)	Reductases	Cracked lips, scaly skin
Pyridoxine (Vitamin B-6)	Transaminases	Anemia, irritability
Niacin	Reductases	Dermatitis, dementia
Folic acid (vitamin M)	Methyltransferases	Megaloblastic anemia
Vitamin B-12	Isomerases	Megaloblastic anemia, neurodegeneration
Pantothenic acid	Acyltransferases	Weight loss, irritability
Biotin (Vitamin H)	Carboxylases	Dermatitis, anorexia, depression
Fat-Soluble Vitamins		
Vitamin A	Visual systems	Night blindness, dry skin
Vitamin D	Calcium metabolism	Rickets, osteomalacia
Vitamin E	Antioxidant	Hemolysis of red blood cells
Vitamin K	Blood clotting	Hemorrhage, delayed blood clotting

Enzymes are grouped into six classes by the kind of reaction they catalyze (Table 29.6). *Hydrolases* catalyze hydrolysis reactions; *isomerases* catalyze isomerizations; *ligases* catalyze the bonding together of two molecules with participation of

TABLE 29.6 Classification of Enzymes

Main Class	Some Subclasses	Type of Reaction Catalyzed
Hydrolases	Lipases	Hydrolysis of an ester group
	Nucleases	Hydrolysis of a phosphate group
	Proteases	Hydrolysis of an amide group
Isomerases	Epimerases	Isomerization of a stereocentre
Ligases	Carboxylases	Addition of CO_2
	Dehydrases	Loss of H_2O
Oxidoreductases	Dehydrogenases	Introduction of double bond by removal of H_2
	Oxidases	Oxidation
	Reductases	Reduction
Transferases	Kinases	Transfer of a phosphate group
	Transaminases	Transfer of an amino group

adenosine triphosphate (ATP); *lyases* catalyze the breaking away of a small molecule such as H_2O from a substrate; *oxidoreductases* catalyze oxidations and reduction reactions; and *transferases* catalyze the transfer of a group from one substrate to another.

Although some enzymes, like papain and trypsin, have uninformative common names, the systematic name of an enzyme has two parts, ending with *-ase*. The first part identifies the enzyme's substrate, and the second part identifies its class. For example, *hexose kinase* is an enzyme that catalyzes the transfer of a phosphate group from adenosine triphosphate to glucose, a six-carbon atom (hexose) sugar molecule.

EXERCISE 29.12—ENZYMES

To what classes do the following enzymes belong?
(a) pyruvate decarboxylase   (b) chymotrypsin   (c) alcohol dehydrogenase

---

Enzymes are globular proteins that act as biological catalysts. They are classified into six groups by the kind of reaction they catalyze.

How Do Enzymes Work? Citrate Synthase

Enzymes exert their catalytic activity by bringing reactant molecules together, holding them in the orientation necessary for reaction, and providing any necessary acidic or basic sites to catalyze specific steps. Let's look, for example, at *citrate synthase,* an enzyme that catalyzes the aldol-like addition of acetyl CoA to oxaloacetate to give citrate [<<Section 24.12]. This reaction is the first step in the so-called *citric acid cycle,* in which acetyl groups produced by degradation of food molecules are metabolically "burned" to yield CO_2 and H_2O.

$$\underset{\textbf{Oxaloacetate}}{^-O_2CCH_2\overset{\overset{\textstyle O}{\|}}{C}CO_2^-} + \underset{\textbf{Acetyl CoA}}{CH_3\overset{\overset{\textstyle O}{\|}}{C}SCoA} \xrightarrow[\text{synthase}]{\text{Citrate}} \underset{\textbf{Citrate}}{^-O_2CCH_2\underset{\underset{\textstyle CO_2^-}{|}}{\overset{\overset{\textstyle OH}{|}}{C}}CH_2CO_2^-} + HSCoA$$

Citrate synthase is a globular protein molecule with a deep cleft lined by an array of functional groups that can bind to oxaloacetate. Upon binding oxaloacetate, the original cleft closes and another opens up to bind acetyl CoA. This second cleft is also lined by appropriate functional groups, including a histidine at position 274 and an aspartic acid at position 375. The two reactants are now held by the enzyme in close proximity and with a suitable orientation for reaction. Figure 29.14 shows the structure of citrate synthase as determined by X-ray crystallography.

FIGURE 29.14 Models of citrate synthase. Part (a) is a space-filling model, which shows the deep clefts in the enzyme. Part (b) is a ribbon model, which emphasizes the α-helical segments of the protein chain and indicates that the enzyme is dimeric; that is, it consists of two identical chains held together by hydrogen bonds and other intermolecular attractions.

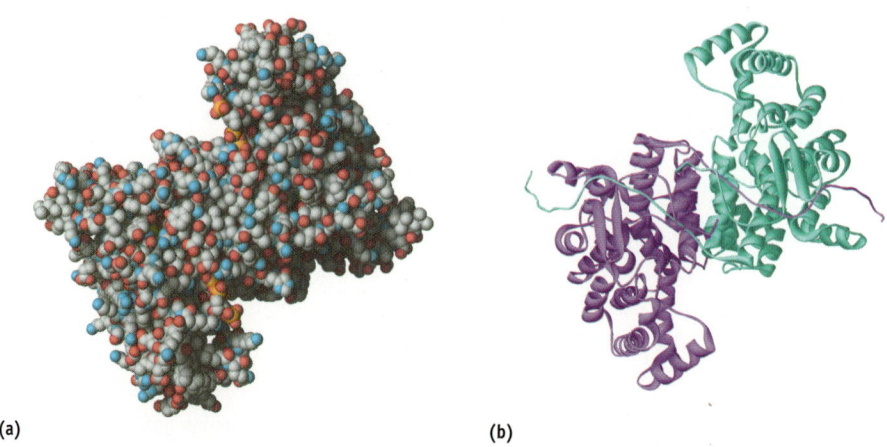

(a)   (b)

The first step in the aldol reaction is generation of the enol of acetyl CoA. The side-chain carboxyl group of Asp-375 acts as base to abstract an acidic α-proton, while at the same time the side-chain imidazole ring of His-274 donates H$^+$ to the carbonyl oxygen atom. The enol thus produced then does a nucleophilic addition to the ketone carbonyl group of oxaloacetate. The His-274 acts as a base to remove the -OH hydrogen from the enol, while another histidine residue at position 320 simultaneously donates an H$^+$ to the oxaloacetate carbonyl group, giving citryl CoA. Water then hydrolyzes the thioester group in citryl CoA, releasing citrate and co-enzyme A as the final products. An accepted mechanism is shown in Figure 29.15.

Acetyl CoA is held in the cleft of the citrate synthase enzyme with His-274 and Asp-375 nearby. The side-chain carboxylate group of Asp-375 acts as a base and removes an acidic α proton, while an N–H group on the side chain of His-274 acts as an acid and donates a proton to the carbonyl oxygen. The net result is formation of an enol.

A nitrogen atom on the His-274 side chain acts as a base to deprotonate the acetyl CoA enol, which adds to the ketone carbonyl group of oxaloacetate in an aldol-like reaction. Simultaneously, an acidic N–H proton on the side chain of His-320 protonates the carbonyl oxygen, producing citryl CoA.

The thioester group of citryl CoA is hydrolyzed in a typical nucleophilic acyl substitution step, breaking the C–S bond and producing citrate plus coenzyme A.

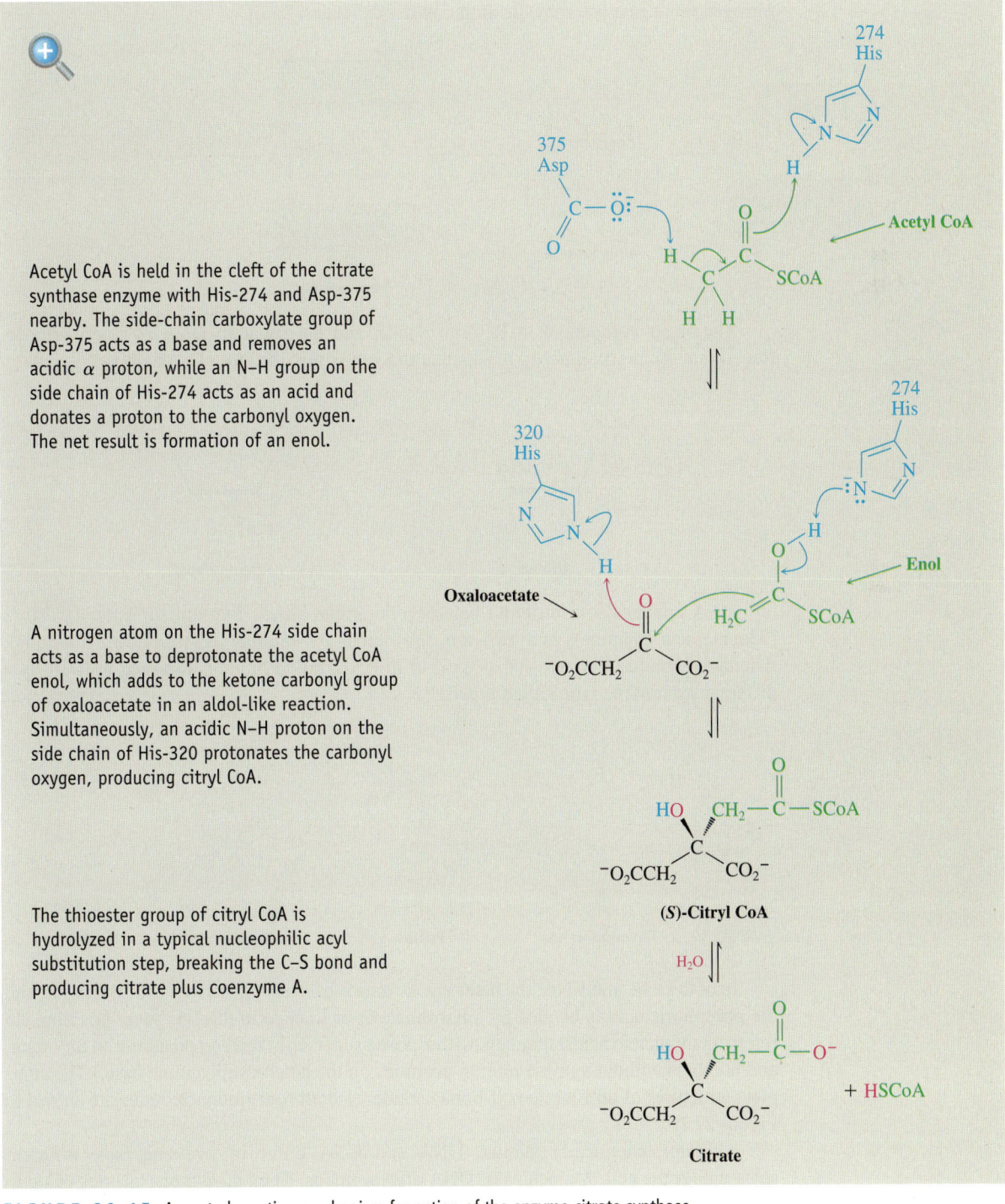

FIGURE 29.15 Accepted reaction mechanism for action of the enzyme citrate synthase.

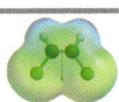
29.4 Nucleic Acids and Nucleotides

The **nucleic acids**, *deoxyribonucleic acid (DNA) and ribonucleic acid (RNA),* are the carriers and processors of a cell's genetic information. Coded in a cell's DNA is all the information that determines the nature of the cell, controls cell growth and division, and directs biosynthesis of the enzymes and other proteins required for all cellular functions.

Just as proteins are polymers of amino acid units, nucleic acids are polymers of building blocks called *nucleotides* linked together to form a long chain. Each nucleotide molecule is composed of a *nucleoside* bonded to a phosphate group, and each nucleoside is composed of an aldopentose sugar joined through its anomeric carbon atom to the nitrogen atom of a heterocyclic amine base [<<Section 25.6].

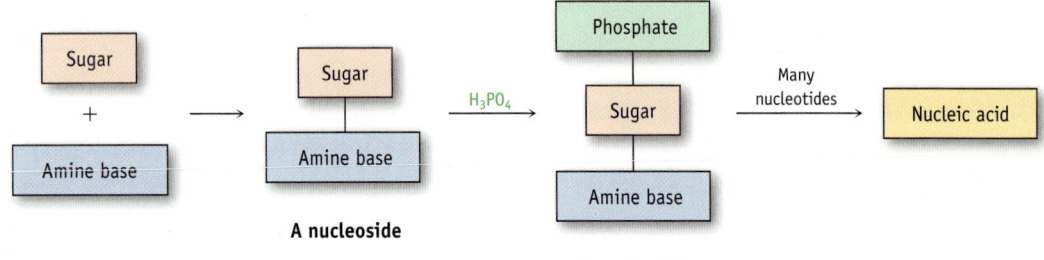

The sugar component in RNA is ribose, and the sugar in DNA is 2-deoxyribose. Recall that the prefix *2-deoxy* means that oxygen is missing from C2 of ribose.

Ribose   **2-Deoxyribose**

DNA contains four different heterocyclic amine bases. Two are substituted *purines* (adenine and guanine), and two are substituted *pyrimidines* (cytosine and thymine). Adenine, guanine, and cytosine also occur in RNA, but thymine is replaced in RNA by a different pyrimidine base called uracil.

Pyrimidine   Cytosine   Uracil (RNA)   Thymine (DNA)   Purine   Adenine   Guanine

In both DNA and RNA, the heterocyclic amine base is bonded to C1′ of the sugar, and the phosphoric acid is bonded by a phosphate ester linkage to the C5′ sugar position. (In referring to nucleic acids, numbers with a prime superscript refer to positions on the sugar, and numbers without a prime refer to positions on the heterocyclic amine base.) The complete structures of all four deoxyribonucleotides and all four ribonucleotides are shown in Figure 29.16.

Although chemically similar, DNA and RNA differ in size and have different roles within the cell. Molecules of DNA are enormous. They have molar masses of up to 150 *billion* and lengths of up to 12 cm when stretched out, and they are found mostly in

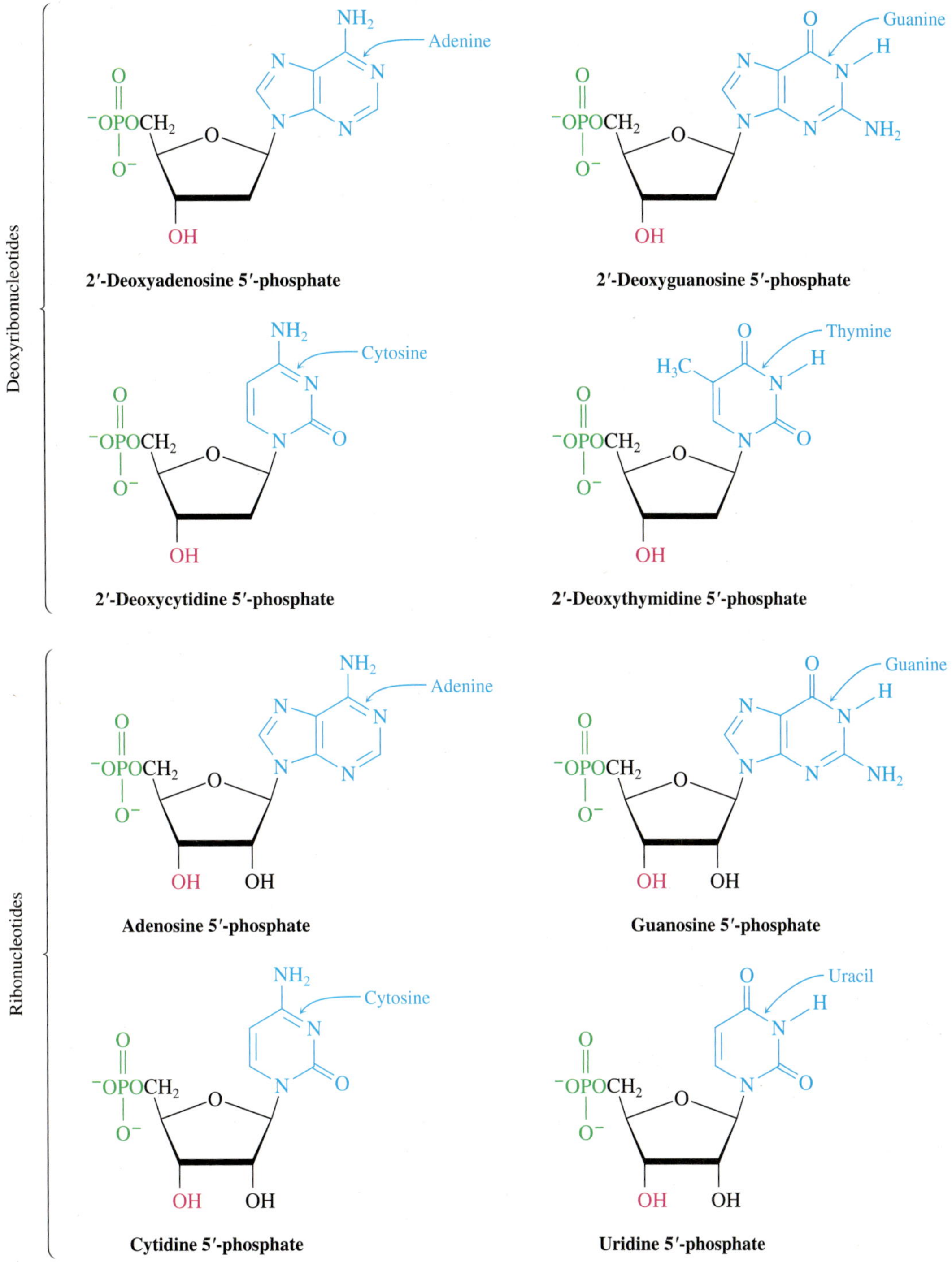

FIGURE 29.16 Structures of the four deoxyribonucleotides and the four ribonucleotides.

the nucleus of cells. Molecules of RNA, by contrast, are much smaller (as low as 35 000 in molar mass) and are found mostly outside the cell nucleus. We'll consider the two kinds of nucleic acids separately, beginning with DNA.

The nucleic acids, DNA (deoxyribonucleic acid) and RNA (ribonucleic acid), are biological polymers that act as chemical carriers of an organism's genetic information. Nucleic acids are made of nucleotides, which consist of a purine or pyrimidine heterocyclic amine base linked to C1' of a pentose sugar (ribose in RNA and 2-deoxyribose in DNA), with the sugar in turn linked through its C5' hydroxyl to a phosphate group.

Molecular Modelling (Odyssey)

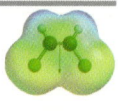

e29.9 Watch molecular-level simulations of DNA.

FIGURE 29.17 Generalized structure of several nucleotides that form DNA.

Structure of DNA

Nucleotides are joined together in DNA by a phosphate ester bond between the 5′-phosphate group on one nucleotide and the 3′-hydroxyl group on the sugar of another nucleotide (Figure 29.17). One end of the nucleic acid polymer thus has a free hydroxyl group at C3′ (the *3′ end*), and the other end has a phosphate group at C5′ (the *5′ end*).

Just as the structure of a protein molecule depends on the sequence in which individual amino acids are connected, the structure of a nucleic acid polymer chain depends on the sequence of individual nucleotides. To carry the analogy further, just as a protein has a polyamide backbone with different side chains attached to it, a nucleic acid has an alternating sugar–phosphate backbone with different amine bases attached.

The sequence of nucleotides in a chain is described by starting at the 5′ end and identifying the bases in order of occurrence, using the abbreviations A for adenosine, G for guanosine, C for cytidine, and T for thymine (or U for uracil in RNA). A typical sequence might be written as TAGGCT.

WORKED EXAMPLE 29.8—NUCLEOTIDES

Draw the full structure of the DNA dinucleotide CT.

Solution

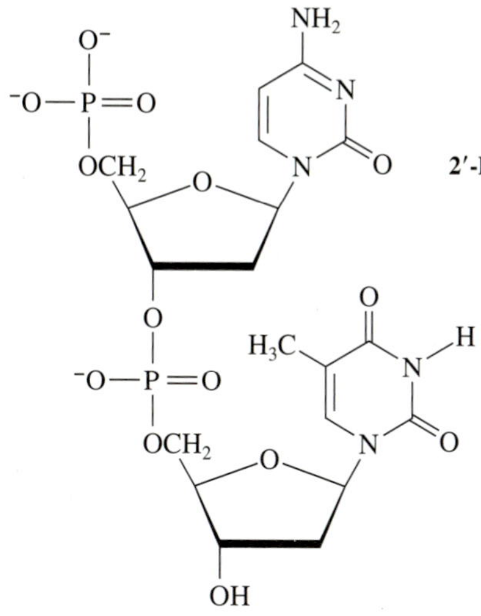

2′-Deoxycytidine (C)

2′-Deoxythymidine (T)

EXERCISE 29.13—NUCLEOTIDES

Draw the full structure of the DNA dinucleotide AG.

EXERCISE 29.14—NUCLEOTIDES

Draw the full structure of the RNA dinucleotide UA.

Base Pairing in DNA: The Watson–Crick Model

Samples of DNA isolated from different tissues of the same species have the same proportions of heterocyclic bases, but samples from different species can have greatly different proportions of bases. Human DNA, for example, contains about 30% each of A and T and about 20% each of G and C. The bacterium *Clostridium perfringens,* however, contains about 37% each of A and T and only 13% each of G and C. Note that in both examples, the bases occur in pairs; A and T are usually present in equal amounts, as are G and C. Why should this be?

In 1953, James Watson and Francis Crick made their now classic proposal for the secondary structure of DNA. According to the Watson–Crick model, DNA consists of two polynucleotide strands, running in opposite directions and coiled around each other in a *double helix* like the handrails on a spiral staircase. The strands run in opposite directions and are held together by hydrogen bonds between specific pairs of bases. Adenine (A) forms a strong hydrogen bond to thymine (T) but not to G or C. Similarly, G and C form strong hydrogen bonds to each other but not to A or T. The nature of this hydrogen bonding is particularly apparent in electrostatic potential maps, which show the alignment of electron-rich and electron-poor regions along the edges of the bases (Figure 29.18).

The two strands of the DNA double helix are not identical; rather, they're complementary because of hydrogen bonding. Whenever a G occurs in one strand, a C occurs opposite it in the other strand. When an A occurs in one strand, a T occurs in the other strand. This complementary pairing of bases explains why A and T are always found in equal

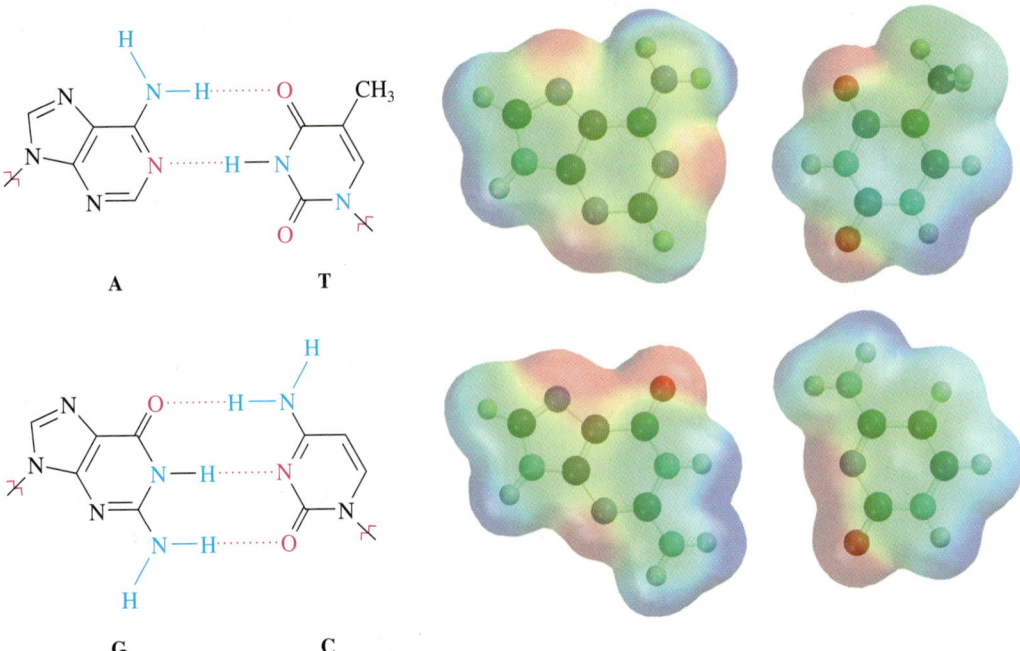

FIGURE 29.18 Hydrogen bonding between complementary base pairs in DNA. The faces of the bases are relatively neutral (green), while the edges have positive (blue) and negative (red) regions. Base A is aligned for hydrogen bonding with T, and G is aligned with C.

amounts, as are G and C. A full turn of the DNA double helix is shown in Figure 29.19. The helix is 2.0 nm (20 Å) wide, there are 10 base pairs per turn, and each turn is 3.4 nm (34 Å) in chain length.

The two strands of the double helix coil in such a way that two kinds of "grooves" result, a *major groove* 1200 pm (12 Å) wide and a *minor groove* 600 pm (6 Å) wide. The major groove is slightly deeper than the minor groove, and both are lined by

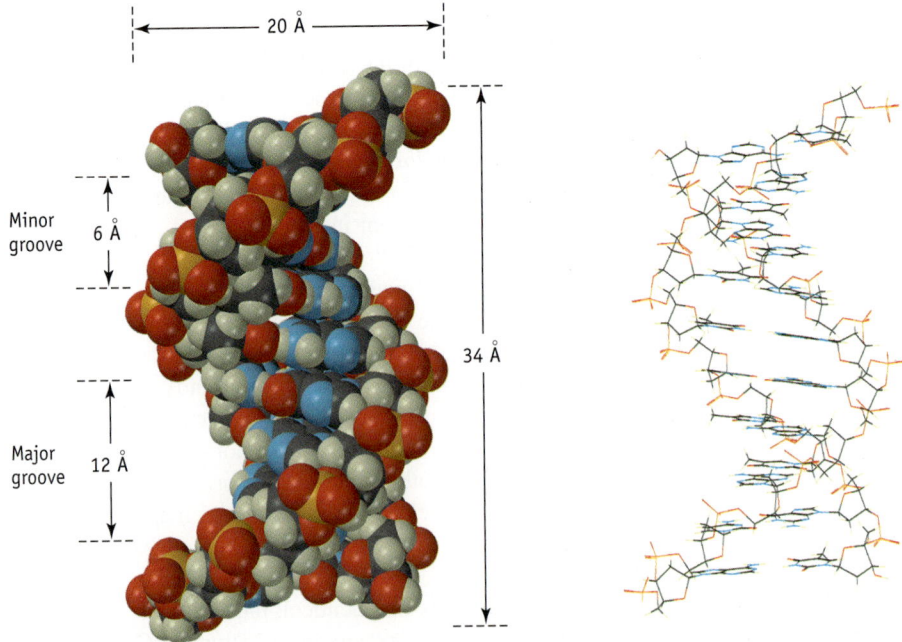

FIGURE 29.19 A turn of the DNA double helix in both space-filling and wire-frame formats. The sugar-phosphate backbone runs along the outside of the helix, and the amine bases hydrogen bond to one another on the inside.

potential hydrogen bond donors and acceptors. As a result, a variety of flat, polycyclic aromatic molecules are able to insert sideways, or *intercalate,* between the stacked bases [<<Section 20.11]. Many cancer-causing and cancer-preventing agents function by interacting with DNA in this way.

WORKED EXAMPLE 29.9—DNA

What sequence of bases on one strand of DNA is complementary to the sequence TATGCAT on another strand?

Solution

Remember that A and G form complementary pairs with T and C, respectively, and then go through the sequence replacing A by T, G by C, T by A, and C by G.
     Original: TATGCAT
     Complement: ATACGTA

EXERCISE 29.15—NUCLEOTIDES

What sequence of bases on one strand of DNA is complementary to the following sequence on another strand?
     GGCTAATCCGT

Molecules of DNA consist of two complementary strands held together by hydrogen bonds between heterocyclic bases on the different strands and coiled into a double helix. Adenine (A) and thymine (T) form hydrogen bonds to each other, as do cytosine (C) and guanine (G).

Nucleic Acids and Heredity

The genetic information of an organism is stored as a sequence of deoxyribonucleotides strung together in the DNA chain. For the information to be preserved and passed on to future generations, a mechanism must exist for copying DNA. For the information to be used, a mechanism must exist for decoding the DNA message and implementing the instructions it contains.

What Crick called the "central dogma of molecular genetics" says that the function of DNA is to store information and pass it on to RNA. The function of RNA, in turn, is to read, decode, and use the information received from DNA to make proteins. By decoding the right bit of DNA at the right time, an organism uses genetic information to synthesize the thousands of proteins necessary for functioning.

Three fundamental processes take place in the transfer of genetic information:

1. *Replication* is the process by which identical copies of DNA are made so that genetic information can be preserved and handed down to succeeding generations.
2. *Transcription* is the process by which the genetic messages are read and carried out of the cell nucleus to ribosomes, where protein synthesis occurs.
3. *Translation* is the process by which the genetic messages are decoded and used to synthesize proteins.

Replication of DNA

DNA *replication* is an enzyme-catalyzed process that begins by a partial unwinding of the double helix. As the strands separate and bases are exposed, new nucleotides line up on each strand in a complementary manner, A to T and C to G, and two new strands begin to grow. Each new strand is complementary to its old template strand, and two new DNA double helices are produced (Figure 29.20). Since each of the new DNA molecules contains one old strand and one new strand, the process is described as *semi-conservative replication*.

FIGURE 29.20 Schematic representation of DNA replication. The original double-stranded DNA partially unwinds, bases are exposed, nucleotides line up on each strand in a complementary manner, and two new strands begin to grow.

The process by which the individual nucleotides are joined to create new DNA strands involves many steps and many different enzymes. Addition of new nucleotide units to the growing chain takes place in the $5' \rightarrow 3'$ direction and is catalyzed by the enzyme *DNA polymerase*. The key step is the addition of a 5'-mononucleoside triphosphate to the free 3'-hydroxyl group of the growing chain as the 3'-hydroxyl attacks the triphosphate and expels a diphosphate leaving group.

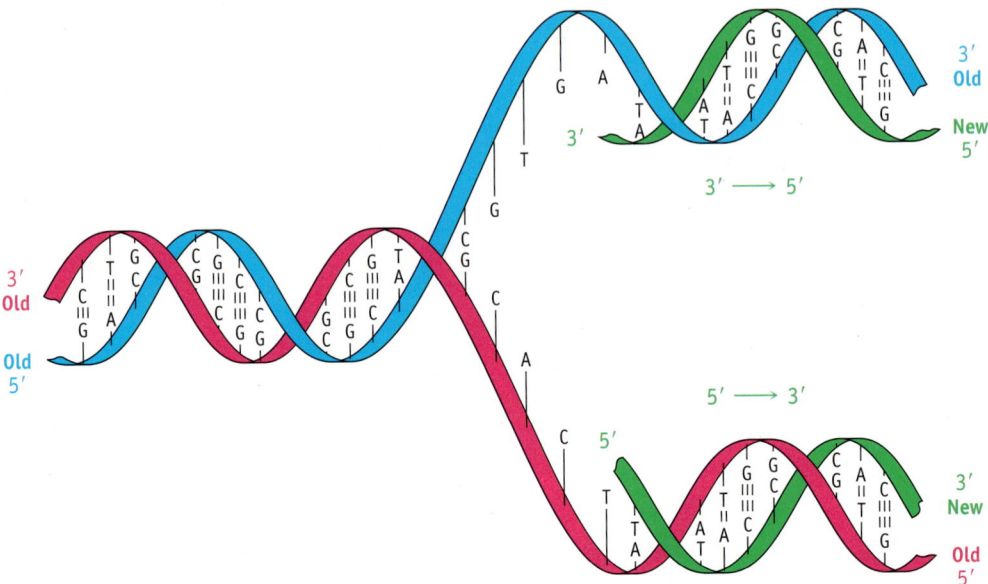

The magnitude of the replication process is staggering. The nucleus of a human cell contains 46 chromosomes (23 pairs), each of which consists of one very large DNA molecule. Each chromosome, in turn, is made up of several thousand DNA segments called *genes,* and the sum of all genes in a human cell (the *genome*) is estimated to be approximately 3 billion base pairs. Despite the size of these massive molecules, the base sequence is faithfully copied during replication, with an error occurring only about once each 10 to 100 billion bases.

Structure and Synthesis of RNA: Transcription

RNA is structurally similar to DNA. Both are sugar–phosphate polymers, and both have heterocyclic bases attached. The only differences are that RNA contains ribose rather than 2-deoxyribose and uracil rather than thymine. Uracil in RNA forms strong hydrogen bonds to its complementary base, adenine, just as thymine does in DNA. In addition, RNA molecules are much smaller than DNA, and RNA remains single-stranded rather than double-stranded.

There are three major kinds of ribonucleic acid, each of which serves a specific function:

1. *Messenger RNA (mRNA)* carries genetic messages from DNA to ribosomes, where protein synthesis occurs.
2. *Ribosomal RNA (rRNA)* provides the physical makeup of ribosomes.
3. *Transfer RNA (tRNA)* transports specific amino acids to the ribosomes, where they are joined together to make proteins.

The conversion of the information in DNA into proteins begins in the nucleus of cells with the synthesis of mRNA by the process of *transcription.* In the case of double-stranded DNA, several turns of the DNA double helix unwind, forming a "bubble" and exposing the bases of the two strands. Ribonucleotides line up in the proper order by hydrogen bonding to their complementary bases on DNA, bond formation occurs in the $5' \rightarrow 3'$ direction, and the growing RNA molecule unwinds from DNA (Figure 29.21).

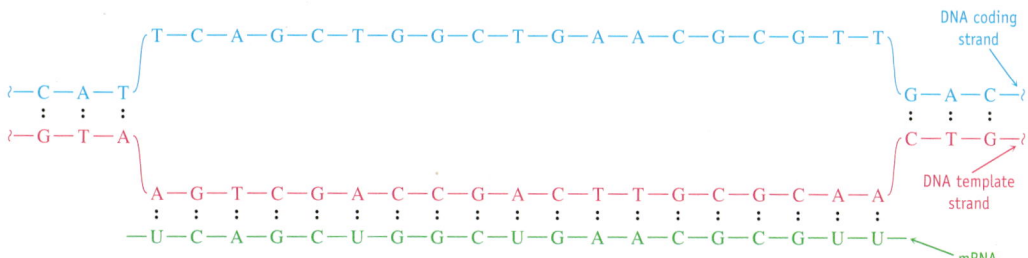

FIGURE 29.21 Synthesis of RNA using a DNA segment as a template.

Unlike what happens in DNA replication, where both strands are copied, only one of the two DNA strands is transcribed into mRNA. The strand that contains the gene is called the *coding strand,* or *sense strand,* and the strand that gets transcribed is called *the template strand,* or *antisense strand.* Since the template strand and the coding strand are complementary, and since the template strand and the RNA molecule are also complementary, *the RNA molecule produced during transcription is a copy of the coding strand.* The only difference is that the RNA molecule has a U everywhere the DNA coding strand has a T.

Transcription of DNA by the process just discussed raises many questions. How does the DNA know where to unwind? Where along the chain does one gene stop and the next one start? How do the ribonucleotides know the right place along the template strand to begin lining up and the right place to stop? The picture that has emerged is that a DNA chain contains specific base sequences called *promoter sites* that lie at positions 10 base pairs and 35 base pairs upstream from the coding region and signal the beginning of a gene. Similarly, there are other base sequences near the end of the gene that signal a stop.

WORKED EXAMPLE 29.10—RNA

What RNA base sequence is complementary to the following DNA base sequence?
    (5′)TAAGCCGTG(3′)

Solution

Go through the sequence replacing A by U, G by C, T by A, and C by G.
    Original DNA: (5′)TAAGCCGTG(3′)
    Complementary RNA: (3′)AUUCGGCAC(5′)

EXERCISE 29.16—RNA

Show how uracil can form strong hydrogen bonds to adenine, just as thymine can.

EXERCISE 29.17—RNA

What RNA base sequence is complementary to the following DNA base sequence?
    (5′)GATTACCGTA(3′)

EXERCISE 29.18—RNA

From what DNA base sequence was the following RNA sequence transcribed?
    (5′)UUCGCAGAGU(3′)

TABLE 29.7 Codon Assignment of Base Triplets

FIRST BASE (5′ END)	SECOND BASE	THIRD BASE (3′ END)			
		U	C	A	G
U	U	Phe	Phe	Leu	Leu
	C	Ser	Ser	Ser	Ser
	A	Tyr	Tyr	Stop	Stop
	G	Cys	Cys	Stop	Trp
C	U	Leu	Leu	Leu	Leu
	C	Pro	Pro	Pro	Pro
	A	His	His	Gln	Gln
	G	Arg	Arg	Arg	Arg
A	U	Ile	Ile	Ile	Met
	C	Thr	Thr	Thr	Thr
	A	Asn	Asn	Lys	Lys
	G	Ser	Ser	Arg	Arg
G	U	Val	Val	Val	Val
	C	Ala	Ala	Ala	Ala
	A	Asp	Asp	Glu	Glu
	G	Gly	Gly	Gly	Gly

DNA and Protein Biosynthesis: Translation

The primary cellular function of RNA is to direct biosynthesis of the thousands of diverse peptides and proteins required by an organism. The mechanics of protein biosynthesis are directed by mRNA and take place on *ribosomes,* small granular particles in the cytoplasm of a cell that consist of about 60% rRNA and 40% protein. On the ribosome, mRNA serves as a template to pass on the genetic information it has transcribed from DNA.

The specific ribonucleotide sequence in mRNA forms a message that determines the order in which different amino acid residues are to be joined. Each "word," or *codon,* along the mRNA chain consists of a sequence of three ribonucleotides that is specific for a given amino acid. For example, the series UUC on mRNA is a codon directing incorporation of the amino acid phenylalanine into the growing protein. Of the $4^3 = 64$ possible triplets of the four bases in RNA, 61 code for specific amino acids (most amino acids are specified by more than one codon) and 3 code for chain termination. Table 29.7 shows the meaning of each codon.

The message carried by mRNA is read by tRNA in a process called *translation.* There are 61 different tRNAs, one for each of the 61 codons in Table 29.7 that specifies an amino acid. A typical tRNA molecule is roughly the shape of a cloverleaf, as shown in Figure 29.22. It consists of about 70–100 ribonucleotides and is bonded to a specific amino acid by an ester linkage through the 3′-hydroxyl on ribose at the end of the tRNA. Each tRNA also contains in its chain a segment called an *anticodon,* a sequence of three ribonucleotides that is complementary to the codon sequence. For example, the codon sequence UUC present on mRNA is read by a phenylalanine-bearing tRNA having the complementary anticodon sequence AAG. (Remember that nucleotide sequences are written in the 5′ → 3′ direction, so the sequence in an anticodon must be reversed. That is, the complement to (5′)-UUC-(3′) is (3′)-AAG-(5′), which is written as (5′)-GAA-(3′).)

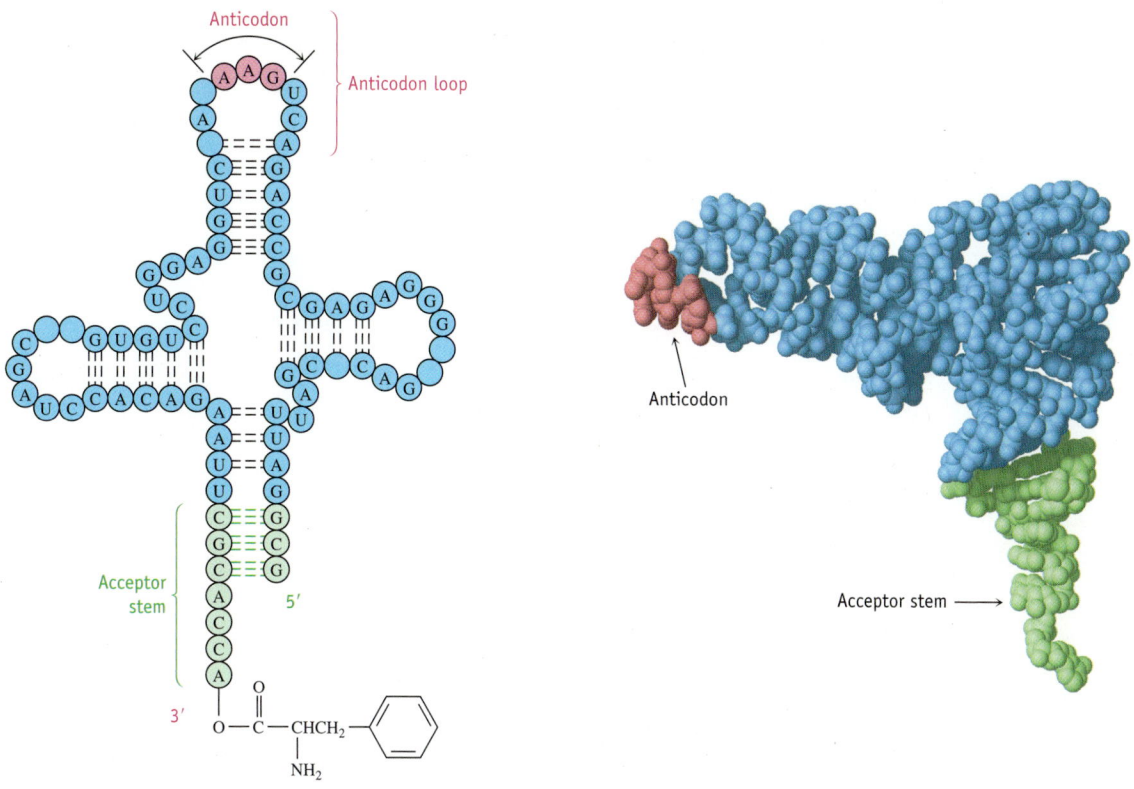

FIGURE 29.22 Structure of a tRNA molecule. The tRNA is a roughly cloverleaf-shaped molecule containing an anticodon triplet on one "leaf" and a covalently attached amino acid residue at its 3' end. The example shown is a yeast tRNA that codes for phenylalanine. The nucleotides not specifically identified are chemically modified analogues of the four usual nucleotides.

As each successive codon on mRNA is read, appropriate tRNAs bring the correct amino acids into position for enzyme-mediated transfer to the growing peptide. When synthesis of the proper protein is completed, a "stop" codon signals the end, and a polypeptide is released from the ribosome, and undergoes post-translational modification to produce the protein. The entire process of protein biosynthesis is illustrated schematically in Figure 29.23.

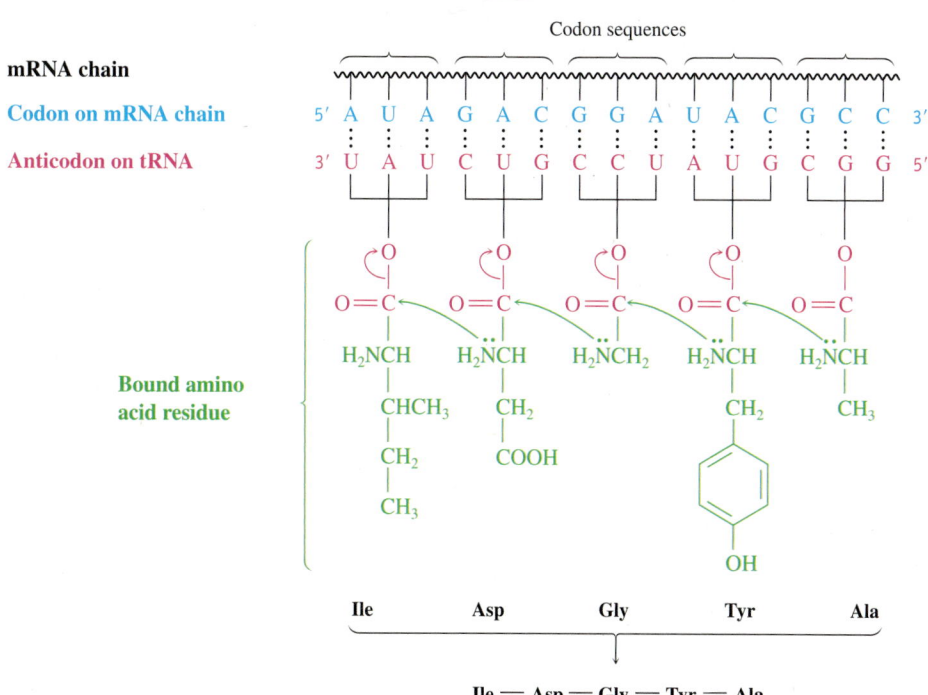

FIGURE 29.23 A schematic representation of protein biosynthesis. The mRNA containing codon base sequences is read by tRNA containing complementary anticodon base sequences. Transfer RNAs assemble the proper amino acids into position for incorporating into the peptide.

WORKED EXAMPLE 29.11—RNA

Give a codon sequence for valine.

Solution

According to Table 29.7, there are four codons for valine: GUU, GUC, GUA, GUG.

WORKED EXAMPLE 29.12—RNA

What amino acid sequence is coded by the mRNA base sequence AUC-GGU?

Solution

Table 29.7 indicates that AUC codes for isoleucine and GGU codes for glycine. AUC-GGU codes for Ile-Gly.

EXERCISE 29.19—RNA

List codon sequences for the following amino acids:
(a) Ala (b) Phe (c) Leu (d) Tyr

EXERCISE 29.20—RNA

What amino acid sequence is coded by the following mRNA base sequence?
CUU-AUG-GCU-UGG-CCC-UAA

EXERCISE 29.21—RNA

What anticodon sequences of tRNAs are coded by the mRNA in Exercise 29.20?

EXERCISE 29.22—RNA

What was the base sequence in the original DNA strand on which the mRNA sequence in Exercise 29.20 was made?

Three main processes take place in deciphering the genetic information in DNA. (a) Replication of DNA is the process by which identical copies are made. In the case of double-stranded DNA, the helix unwinds, complementary deoxyribonucleotides line up, and two new DNA molecules are produced. (b) Transcription is the process by which RNA is produced. This occurs when a segment of the DNA double helix unwinds and complementary ribonucleotides line up to produce messenger RNA (mRNA). (c) Translation is the process by which mRNA directs protein synthesis. Each mRNA has a three-base segment called a codon along its chain. Codons are recognized by small molecules of transfer RNA (tRNA), which carry the appropriate amino acids needed for protein synthesis.

Sequencing DNA

One of the greatest scientific revolutions in history is now occurring in molecular biology as scientists are learning how to manipulate and harness the genetic machinery of organisms. None of the extraordinary advances of the past decade would have been possible, however, were it not for the discovery in 1977 of methods for sequencing immense DNA chains.

The first step in DNA sequencing is to cleave the enormous chain at predictable points to produce smaller, more manageable pieces, a task accomplished by the use of enzymes called *restriction endonucleases.* Each different restriction enzyme, of which more than 200 are available, cleaves a DNA molecule at a well-defined point in the chain wherever a specific base sequence occurs. For example, the restriction enzyme *Alu*I cleaves between G and C in the four-base sequence AG-CT. Note that the sequence is a *palindrome,* meaning that it reads the same from left to right and right to left; that is, the *sequence* (5′)-AG-CT-(3′) is identical to its *complement,* (3′)-TC-GA-(5′). The same is true for other restriction endonucleases.

If the original DNA molecule is cut with another restriction enzyme having a different specificity for cleavage, still other segments are produced whose sequences partially overlap those produced by the first enzyme. Sequencing of all the segments, followed by identification of the overlapping regions, then allows complete DNA sequencing.

Two methods of DNA sequencing are in general use. Both operate along similar lines, but the *Maxam–Gilbert method* uses chemical techniques, while the *Sanger dideoxy method* uses enzymatic reactions. The Maxam–Gilbert method is used in specialized instances, but it is the Sanger method that has allowed the sequencing of the entire human genome of 3 billion base pairs. The dideoxy method used in commercial sequencing instruments begins with a mixture of the following:

- the restriction fragment to be sequenced
- a small piece of DNA called a *primer,* whose sequence is complementary to that on the 3′ end of the restriction fragment
- the four 2′-deoxyribonucleoside triphosphates (dNTPs)
- very small amounts of the four 2′, 3′-dideoxyribonucleoside triphosphates (ddNTPs), each of which is labelled with a fluorescent dye of a different colour (a 2′, 3′-*dideoxy*ribonucleoside triphosphate is one in which both 2′ and 3′ —OH groups are missing from ribose)

A 2′-deoxyribonucleoside triphosphate

A labelled 2′,3′-dideoxyribonucleoside triphosphate

DNA polymerase enzyme is then added to this mix, and a strand of DNA complementary to the restriction fragment begins to grow from the end of the primer. Most of the time, only normal deoxyribonucleotides are incorporated into the growing chain, but every so often, a *dideoxy*ribonucleotide is incorporated. When that happens, DNA synthesis stops because the chain end no longer has a 3′-hydroxyl group for adding further nucleotides.

After reaction is complete, the product consists of a mixture of DNA fragments of all possible lengths, each terminated by one of the four dye-labelled dideoxyribonucleotides. When this product mixture is then submitted to electrophoresis [<<e14.25], each fragment migrates at a rate that depends on the number of negatively charged phosphate groups (the number of nucleotides) it contains. Smaller pieces move rapidly, and larger pieces move more slowly. The technique is so sensitive that up to 1100 DNA fragments, differing in size by only one nucleotide, can be separated.

After separation by electrophoresis by size, the identity of the terminal dideoxyribonucleotide in each piece—and thus the sequence of the restriction fragment—is identified simply by noting the colour with which it fluoresces. Figure 29.24 shows a typical result.

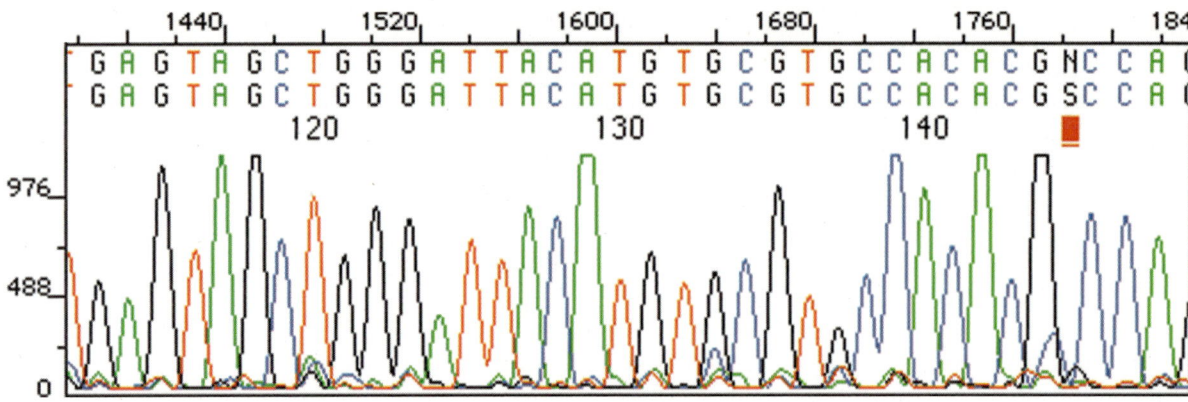

FIGURE 29.24 The sequence of a restriction fragment determined by the Sanger dideoxy method can be read simply by noting the colours of the dye attached to each of the various terminal nucleotides.

So efficient is the automated dideoxy method that sequences up to 1100 nucleotides in length can be rapidly sequenced with 98% accuracy. After a decade of work, preliminary sequence information for the entire human genome of 3 billion base pairs was announced early in 2001. Remarkably, our genome appears to contain only 20 000 to 30 000 genes, about one-third the generally predicted number and only twice the number found in much simpler organisms such as the common roundworm and the fruitfly.

The Polymerase Chain Reaction

Once a gene sequence is known, obtaining an amount of DNA large enough for study is often the next step. The method used is the **polymerase chain reaction (PCR)**, which has been described as being to genes what Gutenberg's invention of the printing press was to the written word. Just as the printing press produces multiple copies of a book, PCR produces multiple copies of a given DNA sequence. Starting from less than 1 *picogram* of DNA with a chain length of 10 000 nucleotides (1 pg = 10^{-12} g; about 100 000 molecules), PCR makes it possible to obtain several micrograms (1 μg = 10^{-6} g; about 10^{11} molecules) in just a few hours.

The key to the polymerase chain reaction is the discovery of *Taq* DNA polymerase, a heat-stable enzyme isolated from the thermophilic bacterium *Thermus aquaticus* found in a hot spring in the U.S. Yellowstone National Park [<<Section 26.1]. *Taq* polymerase is able to take a single strand of DNA and, starting from a short "primer" piece that is complementary to one end of the chain, finish constructing the entire complementary strand. The overall process takes three steps, as shown schematically in Figure 29.25.

Step 1   The double-stranded DNA to be amplified is heated in the presence of *Taq* polymerase, Mg^{2+} ion, the four deoxyribonucleotide triphosphate monomers (dNTPs), and a large excess of two short DNA primer pieces of about 20 bases each. Each primer is complementary to the sequence at the end of one of the target DNA segments. At a temperature of 95 °C, double-stranded DNA breaks apart into two single strands.

Step 2   The temperature is lowered to between 37 °C and 50 °C, allowing the primers, because of their relatively high concentration, to anneal to a complementary sequence at the end of each target strand.

Step 3   The temperature is then raised to 72 °C, and *Taq* polymerase catalyzes the addition of further nucleotides to the two primed DNA strands. When replication of each strand is finished, *two* copies of the original DNA now exist. Repeating the denature–anneal–synthesize cycle a second time yields four DNA copies, repeating a third time yields eight copies, and so on, in an exponential series.

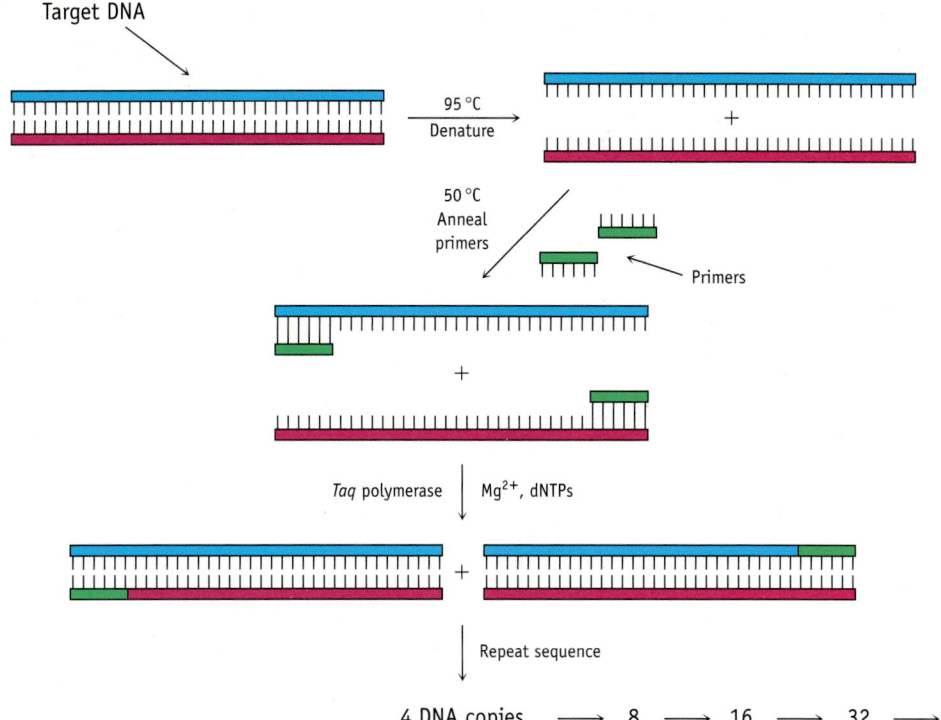

FIGURE 29.25 The polymerase chain reaction. Double-stranded DNA is heated to 95 °C in the presence of two short primer sequences, each of which is complementary to the end of one of the strands. After the DNA denatures, the temperature is lowered and the primer sequences anneal to the strand ends. Raising the temperature in the presence of *Taq* polymerase, Mg^{2+} and a mixture of the four deoxynucleotide triphosphates (dNTPs) effects strand replication, producing two DNA copies. Each further repetition of the sequence again doubles the number of copies.

PCR has been automated, and 30 or so cycles can be carried out in an hour, resulting in a theoretical amplification factor of 2^{30} ($\sim 10^9$). In practice, however, the efficiency of each cycle is less than 100%, and an experimental amplification of about 10^6 to 10^8 is routinely achieved for 30 cycles.

Sequencing of DNA fragments is done by the Sanger dideoxy method. Small amounts of DNA can be amplified using the polymerase chain reaction (PCR).

RNA: A Paradigm Breaker

For a long time, it was generally believed that DNA functioned as a permanent repository of information, that RNA functioned primarily as a transient communicator of information, and that proteins did chemistry. This belief, often referred to as the *central dogma*, has slowly been refined, and these paradigms have been broken. Viruses, such as the AIDS virus, that use RNA as a repository of genetic information offer one example. The discovery of viral proteins called *reverse transcriptases* that synthesize DNA from a viral RNA template overturned what was classically believed to be a one-way flow of information from DNA to RNA.

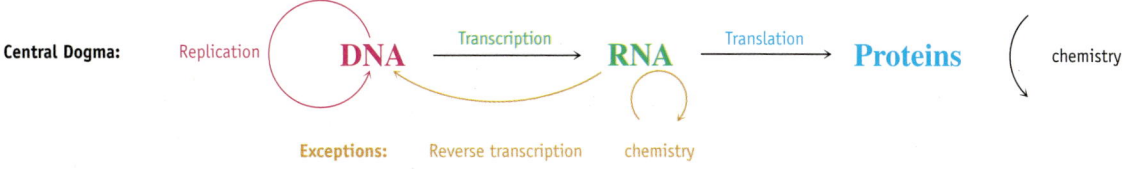

In the last few years, scientists have asked the question, "Can RNA do catalysis like proteins?" The answer is clearly yes: RNA can do chemistry. RNA enzymes, called *ribozymes*, have been found in nature and created in the laboratory. In Figure 29.26, one strand of a ribozyme (shown in red) catalyzes the cleavage of the phosphodiester bond of another RNA (the yellow strand) at the position indicated by the green residue. Groups on the ribozyme act as acids (HA^+) and bases (B:).

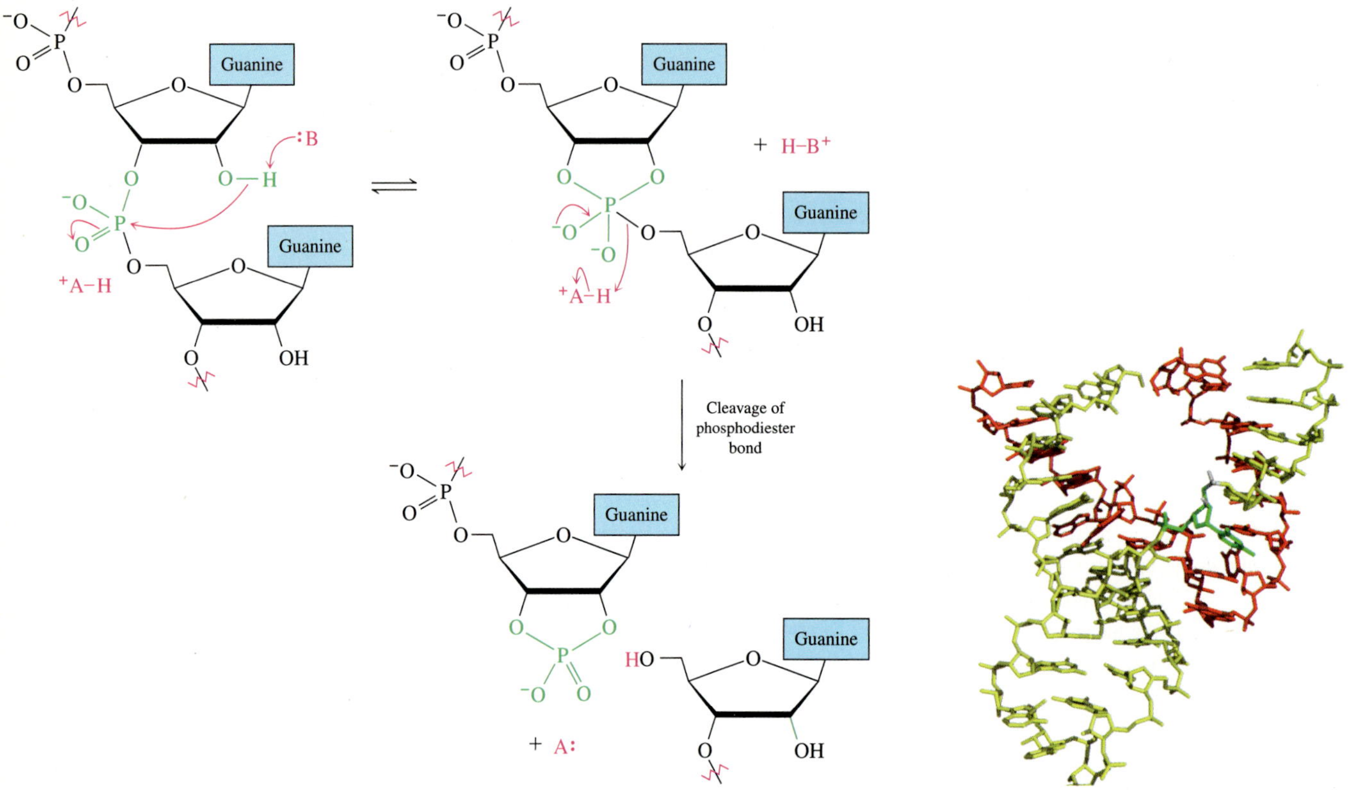

FIGURE 29.26 A ribozyme (red) is an RNA that catalyzes a chemical reaction. In this case, the robozyme cleaves the phosphodiester bond of another RNA (yellow) at the position denoted in green.

SUMMARY

Key Concepts

- **Carbohydrates** are polyhydroxy aldehydes and ketones. They are classified by the number of carbon atoms and the kind of carbonyl group their molecules contain. For example, glucose is an *aldohexose,* a six-carbon aldehydo sugar. *Monosaccharides* are further classified as either *D or L sugars*, depending on the stereochemistry of the stereocentre farthest from the carbonyl group. Most naturally occurring sugars are in the D series. (Section 29.2)

- Much of the chemistry of monosaccharides is the familiar chemistry of substances whose molecules have alcohol and carbonyl functional groups. Monosaccharides normally exist in solution as cyclic hemiacetals rather than as open-chain aldehydes or ketones. The hemiacetal linkage results from reaction of the carbonyl group with an —OH group three or four carbon atoms away. A five-membered ring hemiacetal is a furanose, and a six-membered ring hemiacetal is a pyranose. Cyclization leads to the formation of a new stereocentre (the **anomeric centre**) and the production of two diastereomeric hemiacetals called alpha (α) and beta (β) anomers. (Section 29.2)

- The reaction of monosaccharide hemiacetal molecules with an alcohol and an acid catalyst yields acetal molecules in which each anomeric —OH group has been replaced by an —OR group, called a *glycoside*. Glycosides are non-reducing sugars, because their acetal groups can't open to an open-chain aldehyde form required to reduce metallic-oxidizing agents under basic conditions. (Section 29.2)
- Disaccharides are complex carbohydrates in which two simple sugars are linked by a glycoside bond between the anomeric carbon of one unit and an —OH of the second unit. The two sugars can be the same, as in maltose and cellobiose, or different, as in sucrose. The glycoside bond can be either α (maltose) or β (cellobiose) and can involve any —OH of the second sugar. A 1,4′ link is most common (cellobiose, maltose), but other links, such as 1,2′ (sucrose), also occur. **Polysaccharides**, such as cellulose, starch, and glycogen, are used in nature both as structural materials and for long-term energy storage. (Section 29.2)
- **Proteins** and **peptides** are biomolecules made of **amino acid** residues linked together by amide bonds. Twenty amino acids are commonly found in proteins; all are α-*amino acids,* and all except glycine have stereochemistry similar to that of L sugars. (Section 29.3)
- Proteins are classified as either *globular* or *fibrous*, depending on their *secondary* and *tertiary structures*. Fibrous proteins such as α-keratin are tough and water-insoluble; globular proteins such as myoglobin are water-soluble and mobile within cells. Most of the 2000 or so known enzymes are globular proteins. (Section 29.3)
- **Enzymes** are globular proteins that act as biological catalysts. They are classified into six groups by the kind of reaction they catalyze: *oxidoreductases* catalyze oxidations and reductions; *transferases* catalyze transfers of groups; *hydrolases* catalyze hydrolysis; *isomerases* catalyze isomerizations; *lyases* catalyze bond breakages; and *ligases* catalyze bond formations. (Section 29.3)
- In addition to their protein part, many enzymes contain *co-factors*, which can be either metal ions or small organic molecules. If the co-factor is an organic molecule, it is called a *co-enzyme*. The combination of protein (*apoenzyme*) plus co-enzyme is called a *holoenzyme*. Often, the co-enzyme is a *vitamin*, a small molecule that must be obtained in the diet and is required in trace amounts for proper growth and functioning. (Section 29.3)
- The **nucleic acids**, *DNA (deoxyribonucleic acid)* and *RNA (ribonucleic acid),* are biological polymers that act as chemical carriers of an organism's genetic information. Nucleic acids are made of *nucleotides*, which consist of a purine or pyrimidine heterocyclic amine base linked to C1′ of a pentose sugar (ribose in RNA and 2-deoxyribose in DNA), with the sugar in turn linked through its C5′ hydroxyl to a phosphate group. (Section 29.4)
- Molecules of DNA consist of two complementary strands held together by hydrogen bonds between heterocyclic bases on the different strands and coiled into a *double helix*. Adenine (A) and thymine (T) form hydrogen bonds to each other, as do cytosine (C) and guanine (G). (Section 29.4)
- Three main processes take place in deciphering the genetic information in DNA. (a) *Replication* of DNA is the process by which identical copies are made. The DNA double helix unwinds, complementary deoxyribonucleotides line up, and two new DNA molecules are produced. (b) *Transcription* is the process by which RNA is produced. This occurs when a segment of the DNA double helix unwinds and complementary ribonucleotides line up to produce *messenger RNA (mRNA)*. (c) *Translation* is the process by which mRNA directs protein synthesis. Each mRNA has a three-base segment called a *codon* along its chain. Codons are recognized by small molecules of *transfer RNA (tRNA)*, which carry the appropriate amino acids needed for protein synthesis. (Section 29.4)
- Sequencing of DNA fragments is done by the Sanger dideoxy method. Small amounts of DNA can be amplified using the **polymerase chain reaction (PCR)**. (Section 29.4)

REVIEW QUESTIONS

Section 29.2: Carbohydrates

29.23 The following molecular model is that of an aldo-hexose:

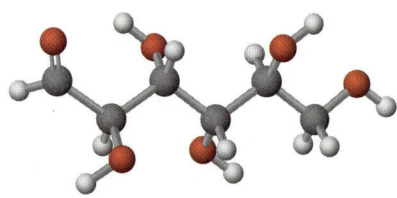

(a) Draw Fischer projections of the sugar molecule, its enantiomer, and a diastereomer.
(b) Is this a D sugar or an L sugar? Explain.
(c) Draw the β anomer of the sugar in its furanose form.

29.24 Write open-chain structures for a ketotetrose and a ketopentose molecule.

29.25 Write an open-chain structure for a deoxyaldo-hexose molecule.

29.26 Write an open-chain structure for a molecule of a five-carbon amino sugar.

29.27 The structure of an ascorbic acid (Vitamin C) molecule is shown. Does it have a D or L configuration?

Ascorbic acid

29.28 Assign *R* or *S* stereochemistry to each stereocentre in an ascorbic acid molecule (Question 29.27).

29.29 Define the following terms, and give an example of each:
(a) monosaccharide   (e) reducing sugar
(b) anomeric centre   (f) pyranose form
(c) Fischer projection   (g) 1,4′ link
(d) glycoside   (h) D-sugar

29.30 The following cyclic structure is that of a gulose molecule. Is this a furanose or pyranose form? Is it an α or β anomer? Is it a D sugar or L sugar?

Gulose

29.31 What is the stereochemical relationship of molecules of D-allose to L-allose? What generalizations can you make about the following properties of the two sugar substances?
(a) melting point   (c) specific rotation
(b) solubility in water   (d) density

29.32 What is the stereochemical relationship of molecules of D-ribose to L-xylose? What generalizations can you make about the following properties of the two sugar substances?
(a) melting point   (c) specific rotation
(b) solubility in water   (d) density

29.33 Convert the following Fischer projections into tetrahedral representations:

(a)

$$H \underset{CH_3}{\overset{Br}{\rule{2cm}{0.4pt}}} OCH_3$$

(b)

$$H \underset{CH_2CH_3}{\overset{CH_3}{\rule{2cm}{0.4pt}}} NH_2$$

29.34 Gentiobiose is a rare disaccharide found in saffron and gentian. It is a reducing sugar and forms only glucose on hydrolysis with aqueous acid. If gentiobiose contains a 1,6′-β-glycoside link, what is its molecular structure?

29.35 Many other sugars besides glucose exhibit mutarotation. For example, α-D-galactopyranose has $[\alpha]_D = +150.7°$, and β-D-galactopyranose has $[\alpha]_D = +52.8°$. If either anomer is dissolved in water and allowed to reach equilibrium, the specific rotation of the solution is +80.2°. What are the percentages of each anomer at equilibrium?

Section 29.3: Amino Acids, Peptides, and Proteins

29.36 (a) Draw the Lewis structure for the amino acid valine, showing the amino group and the carboxylic acid group in their unionized forms.
(b) Draw the Lewis structure for the zwitterionic form of valine.
(c) Which of these structures will be the predominant form at physiological pH?

29.37 Consider the amino acids alanine, leucine, serine, phenylalanine, lysine, and aspartic acid. Which have polar R groups, and which have non-polar R groups?

29.38 Using Lewis structures, show two different ways that alanine and glycine may be combined in a peptide bond.

29.39 When listing the sequence of amino acids in a polypeptide or protein, the sequence always begins with the amino acid that has the free amino group and ends with the amino acid that has the free carboxylic acid group. Draw the Lewis structure for the tripeptide: serine-leucine-valine.

29.40 Draw two Lewis structures for the dipeptide alanine-isoleucine that show the resonance structures of the amide linkage.

29.41 Identify the type of structure (primary, secondary, tertiary, or quaternary) that corresponds to the following statements:
(a) This type of structure is the amino acid sequence in the protein molecule.
(b) This type of structure indicates how different peptide chains in the overall protein are arranged with respect to one another.
(c) This type of structure refers to how the polypeptide chain is folded, including how amino acids that are far apart in the sequence end up in the overall molecule.
(d) This type of structure deals with how amino acids near one another in the sequence within a molecule arrange themselves.

Section 29.4: Nucleic Acids and Nucleotides

29.42 What type of information would be given in a description of the quaternary structure of reverse transcriptase (page 1136)?

29.43 (a) Draw the Lewis structure for the sugar ribose.
(b) Draw the Lewis structure for the nucleoside adenosine (it consists of ribose and adenine).
(c) Draw the Lewis structure for the nucleotide adenosine 5′-monophosphate.

29.44 DNA or RNA sequence is usually written from the end with a free 5′-OH to the end with a free 3′-OH. Draw the Lewis structure for the tetranucleotide AUGC.

29.45 Do the DNA sequences ATGC and CGTA represent the same molecule?

29.46 Which base pairs were proposed by Watson and Crick in their structure of DNA?

29.47 (a) Describe what occurs in the process of transcription.
(b) Describe what occurs in the process of translation.

SUMMARY AND CONCEPTUAL QUESTIONS

29.48 Raffinose, a trisaccharide found in sugar beets, is formed by a 1,6′ α linkage of D-galactose to the glucose unit of sucrose. Draw the molecular structure of raffinose.

29.49 Is raffinose (see Question 29.48) a reducing sugar? Explain.

29.50 Glucose and fructose can be interconverted by treatment with dilute aqueous NaOH. Propose a mechanism (see Sections 24.8–24.9).

29.51 The accumulation of glycolipids called *glucocerebrosides* in different organs can lead to severe neurological problems and a life-threatening condition called Gaucher's disease.

A glucocerebroside

(a) Identify the functional groups in this glucocerebroside molecule.
(b) Hydrolysis of glucocerebrosides in acidic water produces glucose, sphingosine, and a fatty acid. Draw mechanisms for these hydrolysis reactions.
(c) One treatment for Gaucher's disease is enzyme replacement therapy. Cerezyme is a protein that catalyzes the hydrolysis of glycosidic bonds. What are the molecular structures of the products?

29.52 (a) What type of interaction holds DNA's double-helical strands together?
(b) Why would it not be good for DNA's double-helical strands to be held together by covalent bonds?

29.53 Complementary strands of nucleic acids run in opposite directions. That is, the 5′ end of one strand will be lined up with the 3′ end of the other. Given the following nucleotide sequence in DNA:
5′—ACGCGATTC—3′:
(a) Determine the sequence of the complementary strand of DNA. Report this sequence by writing it from its 5′ end to its 3′ end (the usual way of reporting nucleic acid sequences).
(b) Write the sequence (5′ → 3′) for the strand of mRNA that would be complementary to the original strand of DNA.

(c) Assuming that this sequence is part of the coding sequence for a protein and that it is properly lined up so that the first codon of this sequence begins with the 5′ nucleotide of the mRNA, write the sequences for the three anticodons that would be complementary to this strand of mRNA in this region.

(d) What sequence of amino acids is coded for by this mRNA?

29.54 (a) According to the genetic code in Table 29.7, which amino acid is coded for by the mRNA codon GAA?

(b) What is the sequence in the original DNA that led to this codon being present in the mRNA?

(c) If a mutation occurs in the DNA in which a G is substituted for the nucleotide at the second position of this coding region in the DNA, which amino acid will now be selected?

With the **eCHACR** single-sign-on access card bundled with your text, log on (http://login.cengage.com/sso) and access the e-book and click on any in-margin icons for dynamic molecular-level animations and simulations, videos of laboratory reactions, interactive exercises, reviews of background concepts, and other online supplementary materials as noted by the icons in the text margins.

Also go to www.chemistry.nelson.com <http://www.chemistry.nelson.com> for Answers to in-chapter exercises and selected Review Questions, Test Yourself questions, weblinks, crossword puzzles, flashcards, glossary of key terms, and other student resources.

Nuclear Chemistry

30.1 Case Study: Human Activity, Chemical Reactivity

As evidence mounts that the scale of human activity over the past 150 years is causing major changes to the earth's climate, attention is increasingly drawn from temperature changes themselves to the implications of temperature and climate changes for other processes, such as the provision of fresh water supplies. Effects are becoming apparent in places far removed from those countries with the greatest per capita emissions of greenhouse gases.

Nuclear chemistry has recently been used to provide a radioisotope timepiece to help monitor the rate and scale at which glacial sources of fresh water resources are being lost. The disappearance of distinctive radioisotopic signals in some of the planet's ice fields gives strong evidence that glaciers have shrunk substantially since those radioactive isotopes from atmospheric nuclear testing were deposited a half-century ago. Lonnie Thompson, one of the world's leading glaciologists, suggests that the warning this should give us about diminishing water resources is as important to hear as that given by canaries that stopped singing in coal mines. In the 19th and early 20th centuries, canaries were used as an early warning system for the presence of lethal concentrations of gases such as methane and carbon monoxide in coal mines.

> *"In the past, when the canaries stopped singing and died, the miners knew to get out of the mine. Our problem is, we live in the mine."*
>
> **U.S. glaciologist Lonnie Thompson**

Melt water from some of the 15 000 glaciers in the Himalayan Mountains feeds the Ganges, Brahmaputra, Indus, and other major rivers that provide fresh water to almost one-sixth of the world's population on the Indian sub-continent. The scale of glacial ice in this region is so large that it is often referred to as the Third Pole. About 12 000 km^3 of fresh water is stored in these glaciers, and the rate at which the glaciers melt is

U.S. glaciologist Lonnie Thompson (chapter opening photo) has shown that β-decay signals from atmospheric nuclear testing a half-century ago are missing from ice core samples taken from Mount Naimona'nyi on the Tibetan plateau (*right*), suggesting the glaciers have shrunk significantly in the past 60 years.

Jake Norton/Aurora Photos/Getstock.com

critically important for daily life in communities fed by this source of surface water. Melting that is too rapid causes flooding, and the surprisingly rapid loss of ice from these ice fields raises serious concerns about long-term availability of water in the region.

Ironically, the experimental evidence for this loss of glacial ice results from an even more obviously planet-threatening human activity: the extensive atmospheric testing of nuclear weapons during the Cold War period of the 1950s and 60s. Atmospheric weapons testing sent material containing radioactive isotopes into the atmosphere, where it dispersed throughout the troposphere, and left a measurable trace in the layers of glacial ice deposited during that time period. The β-decay signals from residual radioactive isotopes such as ^{90}Sr, ^{137}Cs, ^{3}H, and ^{36}Cl have now been measured in ice core samples from polar and tropical glaciers.

Thompson, working with Chinese collaborators, reported in December 2007 that 150 m long core samples drilled from the summit of Mount Naimona'nyi on the Tibetan plateau (chapter-opening photo) are missing the distinctive β-decay signals found in the top layers of almost every other ice core sampled earlier from around the world. The team's findings suggest that the Tibetan glaciers have begun to shrink since large-scale atmospheric weapons testing 60 years ago.

The detail of climate records from analysis of core samples taken from glaciers and ice caps is so fine, that annual ice layers deposited over each year for the past 550 years can be distinguished. The chemical composition of those layers gives a record of both human activity and natural phenomena such as volcanic eruptions. The top several metres of cores would be expected to contain evidence of nuclear weapons testing. The fact that this is missing in Mount Naimona'nyi ice suggests that these layers have melted. Much deeper in the cores, the ratio of different isotopes of hydrogen and oxygen gives data from which we can infer atmospheric temperature over much longer periods of time.

Besides providing information about ice deposition and major climatic events, substances containing radioactive isotopes from atmospheric nuclear weapons in the 1950s and '60s can help solve other timing puzzles, such as in forensic analysis. Atmospheric testing of nuclear weapons significantly increased the fraction of ^{14}C atoms in carbon-containing species in our atmosphere. Evidence for this was obtained by measurement of the ^{14}C/^{12}C isotope ratio in CO_2, which peaked in 1963 before a steady decline (Figure 30.1).

Increased concentrations of ^{14}C in human tissues as a result of nuclear weapons testing lag behind increased atmospheric levels by several months to a year or two, due to the time that it takes for this radioactive isotope to be taken up by plants and then

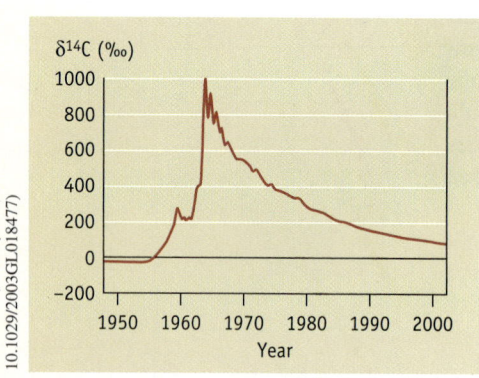

Levin et al., 2003 (GRL, doi. 10.1029/2003GL018477)

FIGURE 30.1 Nuclear explosions in the atmosphere caused the ^{14}C/^{12}C isotope ratio in atmospheric CO_2 in the 1960s to be much higher than previous levels. δ^{14}C is the difference in the ^{14}C/^{12}C isotope ratio (measured in parts per thousand, symbol ‰) from an internationally accepted reference value.

animals in the human food chain. Forensic scientists have been able to use the levels of these radioisotopes in human skeletons as a clock to determine whether a person died before or after the mid-1950s when atmospheric ^{14}C levels increased substantially. The bomb curve data in Figure 30.1 has been used in finer-grain analyses of gallstones removed from patients, to determine how many years prior to surgery the gall stones began to form.

We have given the first section of this last chapter the same title as the book. As you read this chapter, reflect on "Chemistry: Human activity, chemical reactivity." For better, and sometimes for worse, in this nuclear chemistry case study, the link between human activity and chemical reactivity is inseparable. Radioactive isotopes deposited in the ice on pristine glaciers were produced as a result of harnessing the massive amount of energy from nuclear reactions [>>Section 30.7] to produce nuclear bombs. More than 500 of these nuclear weapons were detonated by the United States, Soviet Union, United Kingdom, France, and China in the open atmosphere between 1945 and 1980. β-decay from those residual radioactive isotopes now provides a helpful way of monitoring the rapid loss of glacial ice due to abrupt climate change, which in turn results at least in part from the large-scale human activity of fossil fuel combustion to produce energy. Those fossil fuels themselves are long-term storage devices for the nuclear fusion reactions taking place in the sun. And humans can now carry out controlled nuclear fission to produce large amounts of power without combustion of fossil fuels, an energy technology (not without drawbacks) that has taken on increasing importance as one key strategy to limit the production of greenhouse gases. Finally, nuclear chemistry has become an essential tool for other areas of human activity—including both diagnosis and treatment in medicine and in the preservation of food. The dependence of modern society on these beneficial medical uses of nuclear chemistry became very apparent in late 2007, when a Canadian nuclear reactor that produces a significant percentage of the medical isotopes used globally was temporarily shut down due to safety concerns, leaving thousands of people in several countries with limited access to radioisotopes for diagnosis and treatment.

> The terms *atomic* and *nuclear* are often commonly and incorrectly used interchangeably in phrases such as "atomic/nuclear energy" and "atomic/nuclear bombs." Nuclear chemistry refers to the part of chemistry that studies nuclei and nuclear reactions. It must be carefully distinguished from other processes that involve the rearrangement or transfer of electrons as chemical species interconvert.

30.2 Natural Radioactivity

The discovery of missing radioactive signatures in the top layers of Third Pole ice cores was an unexpected result that has given us insight into the rate of melting ice caps. Other unexpected discoveries about radiation from natural ore samples in the 19th century played a crucial role in our understanding the basics of nuclear chemistry. French scientist Antoine Henri Becquerel was responsible for one of the most important serendipitous discoveries in science in 1896 when he developed a photographic plate that had been stored in the dark next to a sample of uranium ore. He found a completely unexpected bright image on the film caused by radiation from the crystal. The photographic plate had been placed in a drawer because cloudy Parisian skies had prevented him from continuing an experiment in which he thought sunlight shining on the uranium crystal would induce the emission of penetrating radiation. The uranium was producing radiation in the dark, not as a result of emission following exposure to sunlight.

A key scientist following up on Becquerel's discovery was Ernest Rutherford (1871–1937). Rutherford was born in New Zealand and carried out a remarkable program of research in Canada and the U.K. that involved ten future recipients of the Nobel Prize in Chemistry. While studying radiation emanating from uranium and thorium, he concluded that "there are present at least two distinct types of radiation—one that is readily absorbed, which will be termed for convenience $\boldsymbol{\alpha}$ **(alpha) radiation**, and the other of a more penetrative character, which will be termed $\boldsymbol{\beta}$ **(beta) radiation**." Subsequently, charge-to-mass ratio measurements showed that α radiation is composed of helium nuclei (He^{2+} ions) and β radiation is composed of electrons (e^-) (Table 30.1).

Following Rutherford's work, a third type of radiation was discovered by the French scientist Paul Villard (1860–1934); he named it $\boldsymbol{\gamma}$ **(gamma) radiation**, using the third letter in the Greek alphabet in keeping with Rutherford's scheme. Unlike α

TABLE 30.1 Characteristics of α, β, and γ Radiation. The symbol $_{-1}^{0}\beta$, does not mean that a β particle (electron) has atomic number −1 (which would be meaningless). Rather, it means that when an atom emits a β particle during radioactive decay, the atomic number of the atom increases by one because of conversion of a neutron to a proton (see below)

Name	Symbols	Charge	Mass (g/particle)
Alpha	$_{2}^{4}\text{He}$, $_{2}^{4}\alpha$	2+	6.65×10^{-24}
Beta	$_{-1}^{0}\text{e}$, $_{-1}^{0}\beta$	1−	9.11×10^{-28}
Gamma	γ	0	0

and β radiation, γ radiation is not affected by electric and magnetic fields. Rather, it is a form of electromagnetic radiation like x-rays but even more energetic.

Early studies measured the penetrating power of the three types of radiation (Figure 30.2). Alpha radiation is the least penetrating; it can be stopped by several sheets of ordinary paper or clothing. Aluminum that is at least 0.5 cm thick is needed to stop β particles; they can penetrate several millimetres of living bone or tissue. Gamma radiation is the most penetrating. Thick layers of lead or concrete are required to shield the body from this radiation, and γ-rays can pass completely through the human body. The energy associated with each type of radiation is transferred to any material used to stop the particle or absorb the radiation. This fact is important because the damage caused by radiation is related to the energy absorbed.

FIGURE 30.2 The relative penetrating ability of α, β, and γ radiation. Highly charged alpha particles interact strongly with matter and are stopped by a piece of paper. Beta particles, with less mass and a lower charge, interact to a lesser extent with matter and thus can penetrate farther. Gamma radiation is the most penetrating.

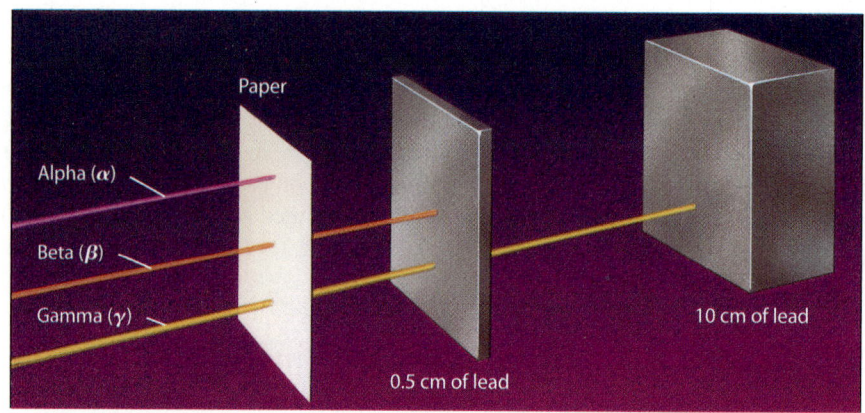

Three distinct types of radiation are given off by natural radioactive sources. α and β particles and γ radiation all possess high kinetic energies, but have different abilities to penetrate matter.

30.3 Nuclear Reactions and Radioactive Decay

Equations for Nuclear Reactions

In 1903, Rutherford and Frederick Soddy (1877–1956) proposed that radioactivity is the result of a natural change of a substance consisting of isotopes of one element into a substance consisting of isotopes of a different element. Such processes are called **nuclear reactions**. In a nuclear reaction, nuclei are changed into other nuclei with different atomic number and often different mass number as well.

Consider a reaction in which radium-226 nuclei (of the isotope of radium with mass number 226) emit an α particle to form radon-222 nuclei. The equation for this reaction is

$$^{226}_{88}\text{Ra} \longrightarrow {}^{4}_{2}\alpha + {}^{226}_{86}\text{Rn}$$

Symbols used in equations for nuclear reactions. To help balance equations for nuclear reactions, the mass number of each nucleus is included as a superscript and the atomic number is included as a subscript preceding the symbols for nuclei on the reactant and product sides of the equation.

In writing an equation for a nuclear reaction, the sum of the mass numbers of reacting particles must equal the sum of the mass numbers of product particles. Furthermore, to maintain nuclear charge balance, the sum of the atomic numbers of the product nuclei must equal the sum of the atomic numbers of the reactant nuclei. These principles are illustrated using the preceding nuclear equation:

	$^{226}_{88}\text{Ra}$	$\longrightarrow$	$^{4}_{2}\alpha$	+	$^{222}_{86}\text{Rn}$
	radium-226	$\longrightarrow$	α particle	+	radon-222
Mass number (protons + neutrons)	226	=	4	+	222
Atomic number (protons)	88	=	2	+	86

Alpha particle emission must produce product nuclei with atomic number two less than that of the reacting nuclei, and with mass number four less. Similarly, nuclear mass and nuclear charge balance accompany β particle emission by uranium-239:

	$^{239}_{92}\text{U}$	$\longrightarrow$	$^{0}_{-1}\beta$	+	$^{239}_{93}\text{Np}$
	uranium-239	$\longrightarrow$	β particle	+	neptunium-239
Mass number (protons + neutrons)	239	=	0	+	239
Atomic number (protons)	92	=	-1	+	93

The β particle has a charge of -1. Charge balance requires that the atomic number of the product atoms be one unit greater than the atomic number of the reacting nuclei. The mass number of the product nuclei is the same as that of the reactant nuclei.

How does a nucleus, composed of protons and neutrons, eject an electron? It is a complex process, but the net result is the conversion *within the nucleus* of a neutron to a proton and an electron.

$$\underset{\text{neutron}}{^{1}_{0}\text{n}} \longrightarrow \underset{\text{electron}}{^{0}_{-1}\text{e}} + \underset{\text{proton}}{^{1}_{1}\text{p}}$$

The term *extra-nuclear electrons* is sometimes used for the electrons surrounding nuclei to distinguish them from the electrons described here, which are produced by a reaction *within the nucleus* of an atom.

Notice that the mass and charge numbers balance in this equation.

What is the origin of the gamma radiation that accompanies most nuclear reactions? Recall that a photon of visible light is emitted when electrons in an atom undergo a transition from an excited *electronic* state to a lower-energy state [<<Section 8.3]. Gamma radiation originates from transitions between *nuclear* energy levels. Nuclear reactions often result in the formation of products in which nuclei are in an excited nuclear state. One option is to return to the ground state by emitting the excess energy as a photon. The high energy of γ radiation is a measure of the large energy difference between the energy levels in the nucleus.

EXERCISE 30.1—EQUATIONS FOR NUCLEAR REACTIONS

Write equations for the following nuclear reactions and confirm that they are balanced with respect to nuclear mass and nuclear charge:
(a) the emission of an α particle by radon-222 to form polonium-218
(b) the emission of a β particle by polonium-218 to form astatine-218

EXERCISE 30.2—ENERGY OF PHOTONS

Calculate the energy per photon, and the energy per mole of photons, of γ radiation with a wavelength of 2.0×10^{-12} m. (*Hint*: Review similar calculations on the energy of photons of visible light, Section 8.3.)

Gamma ray energies are often reported with the unit *MeV*, which stands for 1 million electron volts. One electron volt (1 eV) is the energy of an electron that has been accelerated by a potential of 1 volt. The conversion factor between electron volts and joules is 1 eV = 1.60218×10^{-19} J.

In writing an equation for a nuclear reaction, the sum of the mass numbers of reacting particles must equal the sum of the mass numbers of product particles. To maintain nuclear charge balance, the sum of the atomic numbers of the product nuclei must equal the sum of the atomic numbers of the reactant nuclei.

Radioactive Decay Series

Naturally occurring radioactive isotopes sometimes decay to form products that are also radioactive. When this happens, the initial nuclear reaction is followed by a second nuclear reaction; if the situation is repeated, a third and a fourth nuclear reaction occur; and so on. Eventually, a non-radioactive isotope is formed to end the series. Such a sequence of nuclear reactions is called a *radioactive decay series*. In each step of this nuclear reaction sequence, the reactant nucleus is called the *parent* and the product called the *daughter*.

Uranium-238, the most abundant of three naturally occurring uranium isotopes, heads one of four radioactive decay series. This series begins with the loss of an α particle from the parent $^{238}_{92}\text{U}$ to form the daughter radioactive $^{234}_{90}\text{Th}$. Thorium-234 then decomposes by β emission to $^{234}_{91}\text{Pa}$, which emits a β particle to give $^{234}_{92}\text{U}$. Uranium-234 is an α emitter, forming $^{230}_{90}\text{Th}$. Further α and β emissions follow, until the series ends with formation of the stable, non-radioactive isotope, $^{206}_{82}\text{Pb}$. In all, this radioactive decay series converting $^{238}_{92}\text{U}$ to $^{206}_{82}\text{Pb}$ is made up of 14 reactions, with eight α and six β particles being emitted. The series is portrayed graphically by plotting atomic number versus mass number (Figure 30.3). An equation can be written for each step in the sequence. Equations for the first four steps in the uranium-238 radioactive decay series are as follows:

$$\text{Step 1} \quad {}^{238}_{92}\text{U} \longrightarrow {}^{234}_{90}\text{Th} + {}^{4}_{2}\alpha$$

$$\text{Step 2} \quad {}^{234}_{90}\text{Th} \longrightarrow {}^{234}_{91}\text{Pa} + {}^{0}_{-1}\beta$$

$$\text{Step 3} \quad {}^{234}_{91}\text{Pa} \longrightarrow {}^{234}_{92}\text{U} + {}^{0}_{-1}\beta$$

$$\text{Step 4} \quad {}^{234}_{92}\text{U} \longrightarrow {}^{230}_{90}\text{Th} + {}^{4}_{2}\alpha$$

Uranium ore has trace quantities of various substances containing the radioactive elements formed in the radioactive decay series. A significant development in nuclear chemistry was Marie Curie's discovery in 1898 of radium and polonium as trace components

In 1891, at the age of 23, Marie Sklodovska sat on her own stool in a fourth class train compartment as she travelled from her home in Warsaw, Poland, to Paris, to obtain an education. Twenty years later she was known as Marie Curie, and had won the first of two Nobel Prizes. The 1911 Nobel Prize in Chemistry recognized her isolation of the elements radium and polonium from pitchblende, a uranium ore. Curie's contributions to the development of nuclear chemistry are highlighted in many activities in 2011, the International Year of Chemistry, and the 100th anniversary of her first Nobel Prize. Curie died of leukemia in 1934, almost certainly the result of her exposure to radiation in her research, the effects of which were not fully appreciated at the time.

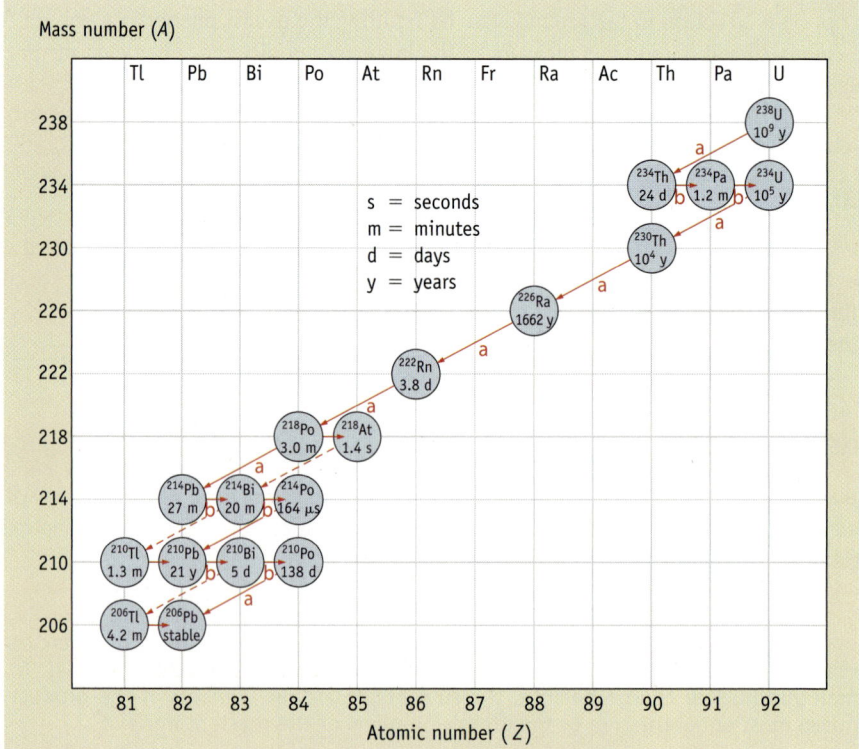

FIGURE 30.3 The uranium-238 radioactive decay series. The steps in this radioactive decay series are shown graphically in this plot of mass number versus atomic number. Each α decay step lowers the atomic number by two units and the mass number by four units. Beta particle emission does not change the mass number but raises the atomic number by one unit. Half-lives of the isotopes are included on the chart.

of pitchblende, a uranium ore. The amount of each of these elements is small because the isotopes of these elements have short half-lives. It is reported that Curie isolated only 1 g of radium from 7.1 tonnes of ore. It is a credit to her skills as a chemist that she extracted sufficient amounts of radium and polonium from uranium ore to identify these elements.

The uranium-238 radioactive decay series is also the source of the environmental hazard radon. Studies of uranium miners who developed lung cancer have shown a correlation between radon exposure and deaths from lung cancer. Lung cancer risks in the home can also increase as a result of trace quantities of uranium that are often present naturally in the soil and rocks in which radon-222 is being continuously formed. Because radon is chemically inert, it is not trapped by chemical processes occurring in soil or water and is free to seep into mines or into homes through pores in cement block walls, through cracks in the basement floor or walls, or around pipes. Because it is more dense than air, radon tends to collect in low spots, so its concentration can build up in a basement if measures are not taken to remove it.

The major health hazard from radon, when it is inhaled by humans, arises not from radon itself but from its decomposition product, polonium.

$$^{222}_{86}\text{Rn} \longrightarrow {}^{218}_{84}\text{Po} + {}^{4}_{2}\alpha \qquad t_{1/2} = 3.82 \text{ days}$$

$$^{218}_{84}\text{Po} \longrightarrow {}^{214}_{82}\text{Pb} + {}^{4}_{2}\alpha \qquad t_{1/2} = 3.04 \text{ minutes}$$

Radon does not undergo chemical reactions to form compounds that can be taken up in the body. Polonium, however, can undergo nuclear reactions. Polonium-218 can lodge in body tissues, where it undergoes α decay to give lead-214, another radioactive isotope. The penetration of an α particle through body tissue is quite small, perhaps 0.7 mm. This is approximately the thickness of the epithelial cells of the lungs, however, so α particle radiation can cause serious damage to lung tissues.

Mean levels of radon in homes around the world vary greatly—from 4 Bq m^{-3} in New Zealand and 18 Bq m^{-3} in Australia to 30 Bq m^{-3} in Canada and 270 Bq m^{-3} in the U.S.A. In certain regions in these countries where many homes have some level of radon, kits can be purchased to test for the presence of this gas (Figure 30.4). If radon gas is detected in your home, you should take corrective actions such as sealing cracks around the foundation and in the basement. It may be reassuring to know that the health risks associated with radon are low. The likelihood of getting lung cancer from exposure to radon is about the same as the likelihood of dying in an accident in your home.

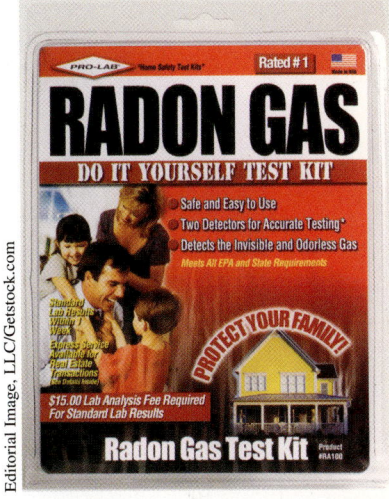

FIGURE 30.4 Radon detector. This kit is intended for use in the home to detect radon gas. Popular radon kits expose charcoal canisters or alpha track detectors to household air (usually in a basement) for a given time period. The exposed material is then sent to a laboratory to measure the amount of radon that might be present.

The SI unit for radioactivity is the Becquerel (Bq), which is the number of disintegrations from a sample in one second.

WORKED EXAMPLE 30.1—RADIOACTIVE DECAY

A second radioactive decay series begins with $^{235}_{92}\text{U}$ and ends with $^{207}_{82}\text{Pb}$.

(a) For each $^{235}_{92}\text{U}$ nucleus that undergoes decay, how many α and β particles are emitted in this series?

(b) The first three steps of this series are (in order) α, β and α emission. Write an equation for each of these steps.

Strategy

First, find the total change in atomic number and mass number. A combination of α and β particles is required that will decrease the total nuclear mass by 28 (235 − 207) and at the same time decrease the atomic number by 10 (92 − 82). Each equation must give symbols for the parent and daughter nuclei and the emitted particle. In the equations, the sums of the atomic numbers and mass numbers for reactants and products must be equal.

Solution

(a) Mass declines by 28 mass units (235 − 207). Because a decrease of 4 mass units occurs with each α emission, 7 α particles must be emitted. Also, for each α emission, the atomic number decreases by 2. Emission of 7 α particles would cause the atomic number to decrease by 14, but the actual decrease in atomic number is 10 (92 − 82). This means that 4 β particles must also have been emitted because each β emission *increases* the atomic number of the product by one unit. The radioactive decay sequence involves emission of 7 α and 4 β particles.

(b) Step 1 $^{235}_{92}\text{U} \longrightarrow ^{231}_{90}\text{Th} + ^{4}_{2}\alpha$

Step 2 $^{231}_{90}\text{Th} \longrightarrow ^{231}_{91}\text{Pa} + ^{0}_{-1}\beta$

Step 3 $^{231}_{91}\text{Pa} \longrightarrow ^{227}_{89}\text{Ac} + ^{4}_{2}\alpha$

Comment

Notice in Figure 30.3 that all daughter nuclei for the series beginning with $^{238}_{92}\text{U}$ have mass numbers differing by four units: 238, 234, 230, . . . , 206. This series is sometimes called the *4n + 2 series* because each mass number (*M*) fits the equation $4n + 2 = M$, where *n* is an integer (*n* is 59 for the first member of this series). For the series headed by $^{235}_{92}\text{U}$, the mass numbers are 235, 231, 227, . . . , 207; this is the *4n + 3 series*.

Two other decay series are possible. One, called the *4n series* and beginning with ^{232}Th, is found in nature; the other, the *4n + 1 series*, is not. No member of this series has a very long half-life. During the 5 billion years since this planet was formed, all members of this series have completely decayed.

EXERCISE 30.3—RADIOACTIVE DECAY

(a) Six α particles and four β particles are emitted in the thorium-232 radioactive decay series before a stable isotope is reached. What is the final product in this series?

(b) The first three steps in the thorium-232 decay series (in order) give rise to α, β, and β emission. Write an equation for each step.

Other Types of Radioactive Decay

> Positrons were discovered by Carl Anderson (1905–1991) in 1932. The positron is one of a group of particles that are known as *antimatter*. If matter and antimatter particles collide, mutual annihilation occurs, with energy being emitted.

Most naturally occurring radioactive elements decay by emission of α, β, and γ radiation. Other nuclear decay processes became known, however, when new radioactive elements were synthesized by artificial means. These include **positron ($^{0}_{+1}\beta$) emission** and **electron capture**.

Positrons ($^{0}_{+1}\beta$) and electrons have the same mass and equal charge, but of the opposite sign. The positron is the antimatter analogue to an electron. Positron emission by polonium-207, for example, results in the formation of bismuth-207.

> Emission of a positron leads to reduction of atomic number by 1, not because the positron has a proton, but because its emission is accompanied by conversion of a proton to a neutron in the nucleus.

	$^{207}_{84}\text{Po}$	$\longrightarrow$	$^{0}_{+1}\beta$	+	$^{207}_{83}\text{Bi}$
	polonium-207	$\longrightarrow$	positron	+	bismuth-207
Mass number (protons + neutrons)	207	=	0	+	207
Atomic number (protons)	84	=	+1	+	83

To retain charge balance, positron decay results in a decrease in the atomic number.

In electron capture, an extra-nuclear electron is captured by the nucleus. The mass number is unchanged and the atomic number is reduced by one. (In an old nomenclature, the innermost electron shell was called the K shell, and electron capture was called *K capture*.)

> Neutrinos and antineutrinos. Beta particles with a wide range of energies are emitted. To balance the energy associated with β decay, it is necessary to postulate the concurrent emission of another particle, the *antineutrino*. Similarly, neutrino emission accompanies positron emission. Much study has gone into detecting neutrinos and antineutrinos. These massless, chargeless particles are not included when writing nuclear equations.

	$^{7}_{4}\text{Be}$	+	$^{0}_{-1}\text{e}$	$\longrightarrow$	$^{7}_{3}\text{Li}$
	beryllium-7	+	electron	$\longrightarrow$	lithium-7
Mass number (protons + neutrons)	7	+	0	=	7
Atomic number (protons)	4	+	-1	=	3

In summary, most unstable nuclei decay by one of four paths: α or β decay, positron emission, or electron capture. Gamma radiation often accompanies these processes. Section 30.7 introduces a fifth way that nuclei decompose, *fission*.

WORKED EXAMPLE 30.2—BALANCED NUCLEAR REACTION EQUATIONS

Complete the following equations. Give the symbol, mass number, and atomic number of the product species.

(a) $^{37}_{18}\text{Ar} + ^{0}_{-1}\text{e} \longrightarrow ?$

(b) $^{11}_{6}\text{C} \longrightarrow ^{11}_{5}\text{B} + ?$

(c) $^{35}_{16}\text{S} \longrightarrow ^{35}_{17}\text{Cl} + ?$

(d) $^{30}_{15}\text{P} \longrightarrow ^{0}_{+1}\beta + ?$

Strategy

The missing product in each nuclear reaction can be determined by recognizing that the sums of mass numbers and atomic numbers for products and reactants must be equal. When you know the nuclear mass and nuclear charge of the product, you can identify it with the appropriate symbol.

Solution

(a) This is an electron capture reaction. The product has a mass number of $37 + 0 = 37$ and an atomic number of $18 - 1 = 17$. Therefore, the symbol for the product is $^{37}_{17}\text{Cl}$.

(b) This missing particle has a mass of zero and a charge of $+1$; these are the characteristics of a positron, $(^{0}_{+1}\beta)$. If this particle is included in the equation, the sums of the atomic numbers $(6 = 5 + 1)$ and the mass numbers (11) on either side of the equation are equal.

(c) A beta particle $(^{0}_{-1}\beta)$ is required to balance the mass numbers (35) and atomic numbers $(16 = 17 - 1)$ in the equation.

(d) The product nucleus has mass number 30 and atomic number 14. This identifies the unknown as $^{30}_{14}\text{Si}$.

EXERCISE 30.4—BALANCED NUCLEAR REACTION EQUATIONS

Indicate the symbol, the mass number, and the atomic number of the missing product in each of the following nuclear reactions:

(a) $^{13}_{7}\text{N} \longrightarrow ^{13}_{6}\text{C} + ?$

(b) $^{41}_{20}\text{Ca} + ^{0}_{-1}\text{e} \longrightarrow ?$

(c) $^{90}_{16}\text{Sr} \longrightarrow ^{90}_{39}\text{Y} + ?$

(d) $^{22}_{11}\text{Na} \longrightarrow ? + ^{0}_{-1}\beta$

Naturally occurring radioactive elements decay by emission of α, β, and γ radiation. Artificially induced nuclear decay processes also include positron $(^{0}_{+1}\beta)$ emission and electron capture.

30.4 **Stability of Atomic Nuclei**

We can learn something about nuclear stability from Figure 30.5. In this plot, the horizontal axis represents the number of protons, and the vertical axis gives the number of neutrons for known isotopes. Each circle represents an isotope identified by the number of neutrons and protons contained in its nucleus. The black circles represent stable (non-radioactive) isotopes, some 300 in number, and the red circles represent some of the known radioactive isotopes. For example, the three isotopes of hydrogen are $^{1}_{1}\text{H}$ and $^{2}_{1}\text{H}$ (neither is radioactive) and $^{3}_{1}\text{H}$ (tritium, radioactive). For lithium, the third element, isotopes with mass numbers 4, 5, 6, and 7 are known. The isotopes with masses of 6 and 7 (shown in black) are stable, whereas the other two isotopes (in red) are radioactive.

Figure 30.5 contains the following information about nuclear stability:

- Stable isotopes fall in a very narrow range called the *band of stability*. It is remarkable how few isotopes are stable.
- Only two stable isotopes ($^{1}_{1}\text{H}$ and $^{3}_{2}\text{He}$) have more protons than neutrons.
- Up to calcium ($Z = 20$), stable isotopes often have equal numbers of protons and neutrons or only one or two more neutrons than protons.

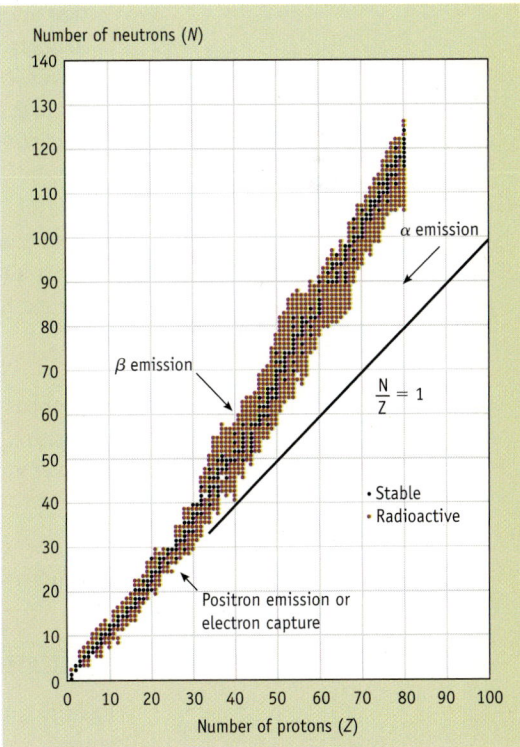

FIGURE 30.5 Stable and unstable isotopes. A graph of the number of neutrons (N) versus the number of protons (Z) for stable (black circles) and radioactive (red circles) isotopes from hydrogen to bismuth. This graph is used to assess criteria for nuclear stability and to predict modes of decay for unstable nuclei.

- Beyond calcium, the neutron/proton ratio is always greater than 1. As the mass increases, the band of stable isotopes deviates more and more from a line in which $N = Z$.
- Beyond lead-208 (82 protons and 126 neutrons), all isotopes are unstable and radioactive. There is apparently no nuclear "superglue" strong enough to hold heavy nuclei together.
- The lifetimes of unstable nuclei are shorter for the heaviest nuclei. For example, half of a sample of $^{238}_{92}\text{U}$ disintegrates in 4.5 billion years, whereas half of a sample of $^{257}_{103}\text{Lr}$ is gone in only 0.65 second. Isotopes that fall farther from the band of stability tend to have shorter half-lives than do unstable isotopes nearer to the band of stability.
- Elements of even atomic number have more stable isotopes than do those of odd atomic number. More stable isotopes have an even number of neutrons than have an odd number. Roughly 200 isotopes have an even number of neutrons and an even number of protons. Only about 120 isotopes have an odd number of either protons or neutrons. Only five stable isotopes ($^{1}_{1}\text{H}$, $^{6}_{3}\text{Li}$, $^{10}_{5}\text{B}$, $^{14}_{7}\text{N}$, and $^{180}_{73}\text{Ta}$) have odd numbers of both protons and neutrons.

The Band of Stability and Radioactive Decay

Besides being a criterion for stability, the neutron/proton ratio can assist in predicting what type of radioactive decay will be observed. Unstable nuclei decay in a manner that brings them toward a stable neutron/proton ratio—that is, toward the band of stability.

- All elements beyond lead-208 ($Z = 82$) are unstable. To eventually form isotopes that are in the band of stability starting with these elements, a process that decreases the atomic number takes place. Alpha emission is an effective way to lower Z, the atomic number, because each emission decreases the atomic number by two. For example, americium, the radioactive element used in smoke detectors, decays by α emission:

$$^{243}_{95}\text{Am} \longrightarrow {}^{4}_{2}\alpha + {}^{239}_{93}\text{Np}$$

- Beta emission occurs for isotopes that have a high neutron/proton ratio—that is, isotopes above the band of stability. With β decay, the atomic number increases by 1 and the mass number remains constant, resulting in a lower n/p ratio.

$$^{60}_{27}\text{Co} \longrightarrow {}^{0}_{-1}\beta + {}^{60}_{28}\text{Ni}$$

- Isotopes with a low neutron/proton ratio, below the band of stability, decay by positron emission or by electron capture. Both processes lead to product nuclei with a lower atomic number and the same mass number:

$$^{13}_{7}\text{N} \longrightarrow {}^{0}_{+1}\beta + {}^{13}_{6}\text{C}$$

$$^{41}_{20}\text{Ca} + {}^{0}_{-1}\text{e} \longrightarrow {}^{41}_{19}\text{K}$$

WORKED EXAMPLE 30.3—RADIOACTIVE DECAY

Identify probable mode(s) of decay for each isotope and write an equation for the decay process.

(a) oxygen-15, $^{15}_{8}\text{O}$

(b) uranium-234, $^{234}_{92}\text{U}$

(c) fluorine-20, $^{20}_{9}\text{F}$

(d) manganese-56, $^{56}_{25}\text{Mn}$

Strategy

In parts (a), (c), and (d), compare the mass number with the atomic weight. If the mass number of the isotope is higher than the atomic weight, then there are too many neutrons and

β emission is likely. If the mass number is lower than the atomic weight, then there are too few neutrons and positron emission or electron capture is the more likely process. It is not possible to choose between the latter two modes of decay without further information. For part (b), note that isotopes with atomic number greater than 83 are likely to be α emitters.

Solution

(a) Oxygen-15 has 7 neutrons and 8 protons so the n/p ratio is less than 1—too low for ^{15}O to be stable. Nuclei with too few neutrons are expected to decay by either positron emission or electron capture. In this instance, the process is $_{+1}^{0}\beta$ emission and the equation is $^{15}_{8}O \longrightarrow {}^{0}_{+1}\beta + {}^{15}_{7}N$.

(b) Alpha decay is common for isotopes of elements with atomic numbers higher than 83. The decay of uranium-234 is one example:

$$^{234}_{92}U \longrightarrow {}^{230}_{90}Th + {}^{4}_{2}\alpha$$

(c) Fluorine-20 has 11 neutrons and 9 protons, a high n/p ratio. The ratio is lowered by β emission:

$$^{20}_{9}F \longrightarrow {}^{0}_{-1}\beta + {}^{20}_{10}Ne$$

(d) The atomic weight of manganese is 54.85. The higher mass number, 56, suggests that this radioactive isotope has an excess of neutrons, in which case it would be expected to decay by β emission:

$$^{56}_{25}Mn \longrightarrow {}^{0}_{-1}\beta + {}^{56}_{26}Fe$$

Be aware that predictions made in this manner will be right much of the time, but exceptions will sometimes occur.

EXERCISE 30.5—RADIOACTIVE DECAY

Write an equation for the probable mode of decay for each of the following unstable isotopes, and write an equation for that nuclear reaction.
(a) silicon-32, $^{32}_{14}Si$
(b) titanium-45, $^{45}_{22}Ti$
(c) plutonium-239, $^{239}_{94}Pu$
(d) potassium-42, $^{42}_{19}K$

Nuclear Binding Energy

Harnessing the energy released in exothermic nuclear processes had been dreamed of for decades, prior to the first successful operation of an artificial nuclear reactor under the University of Chicago football stadium in December 1942. The Chicago tests broke ground for the development of the first nuclear weapons dropped on Japan during World War II. The motivation for the huge scientific effort by some 130 000 people working on the Manhattan Project was the realization that potentially the scale of energy that could be produced in nuclear processes was much greater than in even the most exothermic chemical reactions. The immediate application was to tip the balance of global power through military use, but there was great potential to make use of nuclear processes for the production of energy. For example, a modern 1000 MW nuclear reactor requires 75 kg per day of uranium fuel, while to produce the same amount of electrical energy in a coal-fired power plant would require burning 8 600 000 kg of coal. To understand why the scale of energy released in nuclear processes is so much greater, we must look more closely at the forces holding together the nuclei of atoms.

A nucleus can contain as many as 82 protons and still be stable. For stability, nuclear binding (attractive) forces must be greater than the electrostatic repulsive forces between the closely packed protons in the nucleus. **Nuclear binding energy** (E_b) is defined as the energy required to separate the nucleus of an atom into protons and neutrons. For example, the nuclear binding energy for deuterium is the energy required to convert 1 mol of deuterium ($^2_1$H) nuclei into 1 mol of protons and 1 mol of neutrons.

$$^2_1\text{H} \longrightarrow ^1_1\text{p} + ^1_0\text{n} \qquad E_b = 2.15 \times 10^8 \text{ kJ mol}^{-1}$$

The positive sign for E_b indicates that energy is required for this process. A deuterium nucleus is more stable than an isolated proton and an isolated neutron, just as the H_2 molecule is more stable than two isolated H atoms. Recall, however, that the H—H bond energy is only 436 kJ mol^{-1}. The energy holding a proton and a neutron together in a deuterium nucleus, 2.15×10^8 kJ mol^{-1}, is about 500 000 times larger than the typical covalent bond energies.

To further understand nuclear binding energy, we turn to an experimental observation and a theory. The experimental observation is that the mass of a nucleus is always less than the sum of the masses of its constituent protons and neutrons. The theory is that the "missing mass," called the *mass defect*, is equated with energy that holds the nuclear particles together.

The mass defect for deuterium is the difference between the mass of a deuterium nucleus and the sum of the masses of a proton and a neutron. Mass spectrometric measurements [<<Section 2.9] give the accurate masses of these particles to a high level of precision, providing the numbers needed to carry out calculations of mass defects.

Masses of atomic nuclei are not generally listed in reference tables, but masses of atoms are. Calculation of the mass defect can be carried out using masses of atoms instead of masses of nuclei. By using atomic masses, we include in this calculation the masses of extra-nuclear electrons in the reactants and the products. Because the same number of extra-nuclear electrons appears in products and reactants, this does not affect the result. For 1 mol of deuterium nuclei, the mass defect is found as follows:

$$^2_1\text{H} \longrightarrow ^1_1\text{H} + ^1_0\text{n}$$

$$2.01410 \text{ g mol}^{-1} \qquad 1.007825 \text{ g mol}^{-1} \qquad 1.008665 \text{ g mol}^{-1}$$

Mass defect = Δm = mass of products − mass of reactants

$$= [1.007825 \text{ g mol}^{-1} + 1.008665 \text{ g mol}^{-1}] - 2.01410 \text{ g mol}^{-1}$$

$$= 0.00239 \text{ g mol}^{-1}$$

The relationship between mass and energy is contained in Albert Einstein's 1905 theory of special relativity, which claims that mass and energy are different manifestations of the same quantity. Einstein defined the energy–mass relationship: energy is equivalent to mass times the square of the speed of light; that is, $E = mc^2$. In the case of atomic nuclei, it is assumed that the missing mass (the mass defect, Δm) is equated with the binding energy holding the nucleus together.

$$E_b = (\Delta m)c^2 \qquad (30.1)$$

If Δm is given in kilograms and the speed of light is given in metres per second, E_b will have units of joules (because 1 J = 1 kg × m^2 s^{-2}). For the decomposition of 1 mol of deuterium nuclei to 1 mol of protons and 1 mol of neutrons, we have

$$E_b = (2.39 \times 10^{-6} \text{ kg mol}^{-1})(2.988 \times 10^8 \text{ m s}^{-1})^2$$

$$= 2.15 \times 10^{11} \text{ J mol}^{-1} \text{ of } ^2_1\text{H nuclei } (= 2.15 \times 10^8 \text{ kJ mol}^{-1} \text{ of } ^2_1\text{H nuclei})$$

The nuclear stabilities of different elements are compared using the *binding energy per mol of nucleons*. (**Nucleon** is the general name given to nuclear particles—that is, protons and neutrons.) A deuterium nucleus contains two nucleons, so the binding energy per mol of nucleons, E_b/n, is 2.15×10^8 kJ mol^{-1} divided by 2, or 1.08×10^8 kJ per mol of nucleons.

$$E_b/n = \left(\frac{2.15 \times 10^8 \text{ kJ}}{\text{mol } {}_1^2\text{H nuclei}} \right) \left(\frac{1 \text{ mol of } {}_1^2\text{H nuclei}}{2 \text{ mol nucleons}} \right)$$

$$E_b/n = 1.08 \times 10^8 \text{ kJ mol}^{-1} \text{ nucleons}$$

The binding energy per nucleon can be calculated for any atom whose mass is known. Then, to compare nuclear stabilities, binding energies per nucleon are plotted as a function of mass number (Figure 30.6). The greater the binding energy per nucleon is, the greater the stability of the nucleus. From the graph in Figure 30.6, the point of maximum nuclear stability occurs at a mass of 56 (that is, at iron in the periodic table).

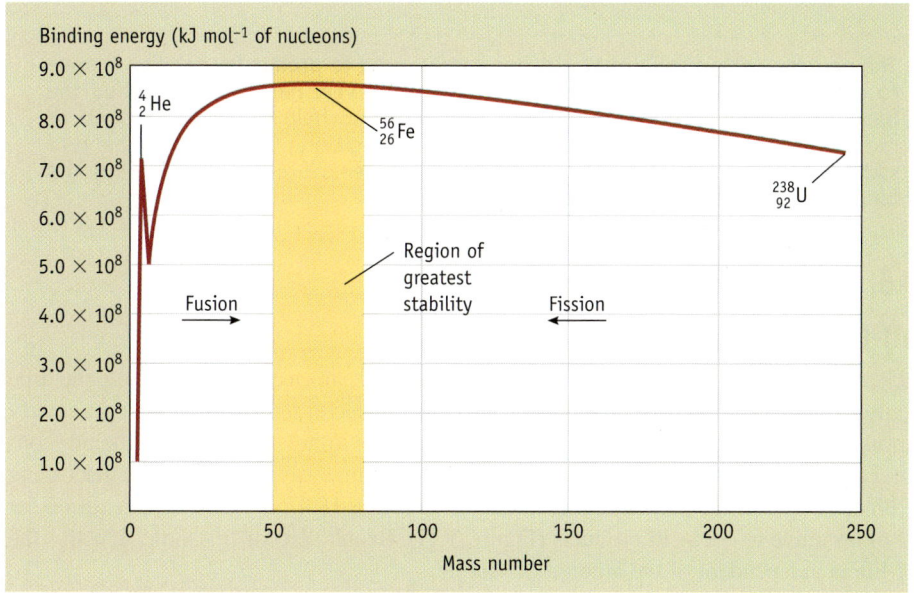

FIGURE 30.6 Relative stability of nuclei. Binding energy per nucleon for the most stable isotope of elements between hydrogen and uranium is plotted as a function of mass number. (Fission and fusion are discussed in Sections 30.7 and 8.)

WORKED EXAMPLE 30.4—BINDING ENERGY

Calculate the binding energy, E_b (in kJ mol^{-1}), and the binding energy per nucleon, E_b/n (in kJ mol^{-1} nucleon), for carbon-12.

Strategy

First determine the mass defect, then use equation 30.1 to determine the binding energy. There are 12 nuclear particles in carbon-12, so dividing the nuclear binding energy by 12 will give the binding energy per nucleon.

Solution

The mass of ${}_1^1$H is 1.007825 g mol^{-1}, and the mass of ${}_0^1$n is 1.008665 g mol^{-1}. Carbon-12 (${}_6^{12}$C) is the standard for the atomic masses in the periodic table, and its mass is defined as exactly 12 g mol^{-1} (12.000000 g mol^{-1}).

$$\Delta m = [(6 \times \text{mass } {}_1^1\text{H}) + (6 \times \text{mass } {}_0^1\text{n})] - \text{mass } {}_6^{12}\text{C}$$

$$= [(6 \times 1.007825) + (6 \times 1.008665)] - 12.000000$$

$$= 9.8940 \times 10^{-2} \text{ g mol}^{-1} \text{ nuclei}$$

The binding energy is calculated using equation 30.1. Using the mass in kilograms and the speed of light in metres per second gives the binding energy in joules:

$$E_b = (\Delta m)c^2$$

$$= (9.8940 \times 10^{-5} \text{ kg mol}^{-1})(2.99792 \times 10^8 \text{ m s}^{-1})^2$$

$$= 8.89 \times 10^{12} \text{ J mol}^{-1} \text{ nuclei } (= 8.89 \times 10^9 \text{ kJ mol}^{-1} \text{ nuclei})$$

The binding energy per nucleon (E_b/n) is determined by dividing the binding energy by 12 (the number of nucleons):

$$\frac{E_b}{n} = \frac{8.89 \times 10^9 \text{ kJ mol}^{-1} \text{ nuclei}}{12 \text{ mol nucleons mol}^{-1} \text{ nuclei}}$$

$$= 7.41 \times 10^8 \text{ kJ mol}^{-1} \text{ nucleons}$$

EXERCISE 30.6—BINDING ENERGY

Calculate the binding energy per nucleon (in kJ mol^{-1}) for the formation of lithium-6. The molar mass of $^6_3$Li is 6.015125 g mol^{-1}.

The neutron/proton ratio of an isotope can be used to predict its stability relative to other isotopes, with stable isotopes falling in a very narrow range called the band of stability. The neutron/proton ratio can also help predict what kind of radioactive decay an unstable isotope will undergo to bring it closer to the band of stability. The nuclear stabilities of different elements are compared using the binding energy per mole of nucleons.

30.5 Rates of Nuclear Decay

Half-Life

Nuclear fission is a first-order process. Half-life provides an easy way to estimate the time required for the amount of a radioactive element to decrease. This can be particularly helpful in predicting how long a radioisotope in the environment will pose a health concern. Half-life ($t_{1/2}$) is used in nuclear chemistry in the same way it is used when discussing the kinetics of first-order chemical reactions [<<Section 18.4]: It is the time required for half of a sample to decay to products (Figure 30.7). Recall that for first-order kinetics the half-life is independent of the amount of sample.

Unlike what is observed in chemical kinetics, temperature does not affect the rate of nuclear decay.

FIGURE 30.7 Decay of 20.0 mg of oxygen-15. After each half-life period of 2.0 min, the mass of oxygen-15 decreases by one half.

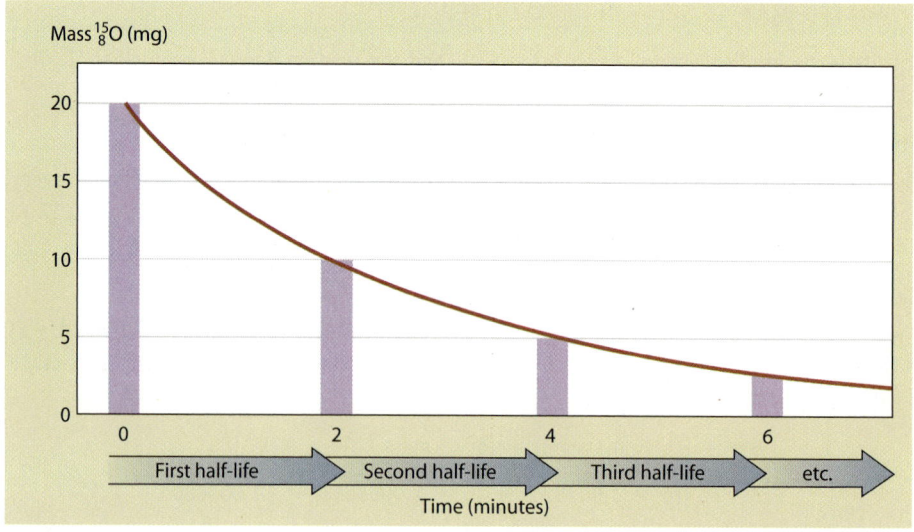

Significant quantities of strontium-90, a β emitter with a half-life of 29.1 years, were dispersed into the environment in atmospheric nuclear weapons tests described in Section 30.1. From the half-life, we know that a quarter of what was dispersed into the environment is still around. The health problems associated with strontium-90 arise because calcium and strontium have similar chemical properties. Strontium-90 is taken into the body and deposited in bone, taking the place of calcium. Radiation damage by strontium-90 in bone has been directly linked to bone-related cancers.

Half-lives for radioactive isotopes cover a wide range of values. Uranium-238 has one of the longer half-lives, 4.47×10^9 years, a length of time close to the age of the earth (estimated at 4.5–4.6×10^9 years). Roughly half of the uranium-238 present when the planet was formed is still around. At the other end of the range of half-lives are isotopes such as element 112, whose 277 isotope has a half-life of 240 microseconds ($1 \; \mu s = 1 \times 10^{-6}$ s).

WORKED EXAMPLE 30.5—HALF-LIFE

Radioactive iodine-131, used to treat hyperthyroidism, has a half-life of 8.04 days.
(a) If you have 8.8 μg of this isotope, what mass remains after 32.2 days?
(b) How long will it take for a sample of iodine-131 to decay to one-eighth of its activity?
(c) Estimate the length of time necessary for the sample to decay to 10% of its original activity.

Strategy

This problem asks you to use half-life to qualitatively assess the rate of decay. After one half-life, half of the sample remains. After another half-life, the amount of sample is again decreased by half to one-fourth of its original value (this situation is illustrated in Figure 30.7). To answer these questions, assess the number of half-lives that have elapsed and use this information to determine the amount of sample remaining.

Solution

(a) The time elapsed, 32.2 days, is 4 half-lives ($32.2/8.04 = 4$). The amount of iodine-131 has decreased to 1/16 of the original amount [$1/2 \times 1/2 \times 1/2 \times 1/2 = (1/2)^4 = 1/16$]. The amount of iodine remaining is 8. 8 μg $\times (1/2)^4$ or 0.55 μg.
(b) After 3 half-lives, the amount of iodine-131 remaining is 1/8 ($= (1/2)^3$) of the original amount. The amount remaining is 8.8 μg $\times (1/2)^3 = 1.1 \; \mu$g.
(c) After 3 half-lives, 1/8 (12.5%) of the sample remains; after 4 half-lives, 1/16 (6.25%) remains. It will take between 3 and 4 half-lives, between 24.15 and 32.2 days, to decrease the amount of sample to 10% of its original value.

Comment

You will find it useful to make approximations as has been done in (c). An exact time can be calculated from the first-order rate law [<<Section 18.4].

EXERCISE 30.7—HALF-LIFE

Tritium ($^{3}_{1}$H), a radioactive isotope of hydrogen, has a half-life of 12.3 years.
(a) Starting with 1.5 mg of this isotope, how many milligrams remain after 49.2 years?
(b) How long will it take for a sample of tritium to decay to one-eighth of its activity?
(c) Estimate the length of time necessary for the sample to decay to 1% of its original activity.

Kinetics of Nuclear Decay

The rate of nuclear decay is determined from measurements of the **activity** (A) of a sample. Activity refers to the number of disintegrations observed per unit time, a quantity that can be measured readily with devices such as a Geiger-Müller counter (Figure 30.8). *Activity is proportional to the number of radioactive atoms present (N), and is expressed by the SI unit the Becquerel (Bq), which is the number of disintegrations per second.*

$$A \propto N$$

If the number of radioactive nuclei N is reduced by half, the activity of the sample will be half as large. Doubling N will double the activity. This evidence indicates that the rate of

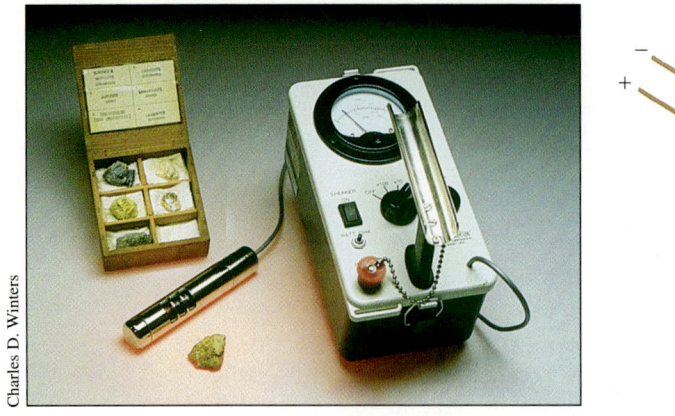
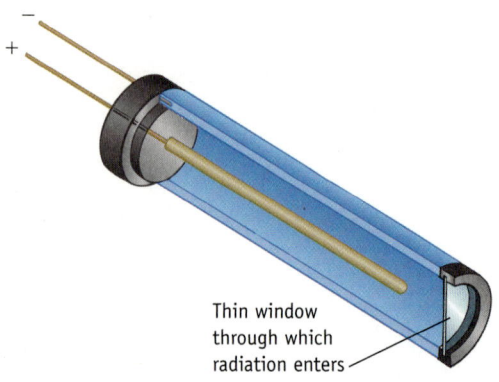

Charles D. Winters

Thin window
through which
radiation enters

FIGURE 30.8 A Geiger-Müller counter. A charged particle (an α or β particle) enters the gas-filled tube (*diagram at the right*) and ionizes the gas. The gaseous ions migrate to electrically charged plates and are recorded as a pulse of electric current. The current is amplified and used to operate a counter. A sample of carnotite, a mineral containing uranium oxide, is also shown in the photograph.

decomposition is first order with respect to N. Consequently, the equations describing rates of radioactive decay are the same as those used to describe first-order chemical reactions; the change in the number of radioactive atoms N per unit of time is proportional to N:

$$\frac{\Delta N}{\Delta t} = -kN$$

The integrated rate equation can be written in two ways depending on the data used:

$$\ln\left(\frac{N}{N_0}\right) = -kt \tag{30.2}$$

or

$$\ln\left(\frac{A}{A_0}\right) = -kt \tag{30.3}$$

Here, N_0 and A_0 are the number of atoms and the activity of the sample initially, respectively, and N and A are the number of atoms and the activity of the sample after time t, respectively. Thus, N/N_0 is the fraction of atoms remaining after a given time (t), and A/A_0 is the fraction of the activity remaining after the same period. In these equations, k is the rate constant (decay constant) for the isotope in question. The relationship between half-life and the first-order rate constant is the same as seen with chemical kinetics [<<Section 18.5]:

$$t_{1/2} = \frac{0.693}{k} \tag{30.4}$$

Equations 30.2 -30.4 are useful in several ways:

- If the activity (A) or the number of radioactive nuclei (N) is measured in the laboratory over some period (t), then k can be calculated. The decay constant k can then be used to determine the half-life of the sample.
- If k is known, the fraction of a radioactive sample (N/N_0) still present after some time (t) has elapsed can be calculated.
- If k is known, the time required for that isotope to decay to a fraction of the original activity (A/A_0) can be calculated.

WORKED EXAMPLE 30.6—KINETICS OF NUCLEAR DECAY

A sample of radon-222 has an initial α particle activity (A_0) of 7.0×10^4 Bq (disintegrations per second). After 6.6 days, its activity (A) is 2.1×10^4 Bq. What is the half-life of radon-222?

Strategy

Values for A, A_0, and t are given. Solve the problem using equation 30.3 with k as the unknown. Once k is found, calculate the half-life using equation 30.4.

Solution

$$\ln \frac{2.1 \times 10^4 \ \cancel{Bq}}{7.0 \times 10^4 \ \cancel{Bq}} = -k(6.6 \ d)$$

$$\ln(0.30) = -k(6.6 \ d)$$

$$k = 0.18 \ d^{-1}$$

From k, we obtain $t_{1/2}$:

$$t_{1/2} = 0.693/0.18 \ d^{-1} = 3.8 \ d$$

Comment

Notice that the activity decreased to between one-half and one-fourth of its original value. The 6.6 days of elapsed time represents one full half-life and part of another half-life.

WORKED EXAMPLE 30.7—KINETICS OF NUCLEAR DECAY

Gallium citrate, containing the radioactive isotope gallium-67, is used medically as a tumour-seeking agent. It has a half-life of 78.2 h. How long will it take for a sample of gallium citrate to decay to 10.0% of its original activity?

Strategy

Use equations 30.3–4 to solve this problem. In this case, the unknown is the time t. The rate constant k is calculated from the half-life using equation 30.4. Although we do not have specific values of activity, the value of A/A_0 is known. Because A is 10.0% of A_0, the value of A/A_0 is 0.100.

Solution

First determine k:

$$k = 0.693/t_{1/2} = 0.693/78.2 \ h$$

$$k = 8.86 \times 10^{-3} \ h^{-1}$$

Then substitute the given values of A/A_0 and k into equation 30.3:

$$\ln(A/A_0 = kt)$$

$$\ln(0.100) = (8.86 \times 10^{-3} \ h^{-1}) \ t$$

$$t = 2.60 \times 10^2 \ h$$

Comment

The time required is between three half-lives ($3 \times 78.2 \ h = 235 \ h$) and four half-lives ($4 \times 78.2 \ h = 313 \ h$).

EXERCISE 30.8—KINETICS OF NUCLEAR DECAY

A sample of $Ca_3(PO_4)_2$ containing phosphorus-32 has an activity of 55.8 Bq. Two days later, the activity is 53.0 Bq. Calculate the half-life of phosphorus-32.

EXERCISE 30.9—KINETICS OF NUCLEAR DECAY

A highly radioactive sample of nuclear waste products with a half-life ($t_{1/2}$) of 200 years is stored in an underground tank. How long will it take for the activity to diminish from an initial activity of 1.08×10^{11} Bq to a fairly harmless activity of 5.00×10^1 Bq?

Radiocarbon Dating

In Section 2.8, you were introduced to Ötzi, the mummy discovered in 1991 by two hikers in the melting ice of a glacier high above the Alps. The isotopic ratios of strontium, lead, and oxygen in samples from the "ice man" were compared with environmental sources in the region to determine where Ötzi lived. PCR techniques [<<Section 29.4] were used to amplify DNA extracted from the contents of Ötzi's intestines, from which it was determined that his last meal likely consisted of red deer meat and cereals. A nuclear chemistry technique based on the rate of decay of carbon-14 isotopes was used on mg-amounts of bone and tissue by researchers in Zürich and Oxford to determine that he lived about 5300 years ago (Figure 30.9).

FIGURE 30.9 Ötzi, the Ice Man, with mountaineer discoverers Hans Kammerlander (*left*) and Reinhold Messner (*right*). Carbon-14 dating techniques were used to determine that Ötzi lived about 5300 years ago; making this accidental find the world's oldest preserved human remains.

Picture taken by K. Fritz (Photo Paul Hanny)

A 70 kg human body contains about 16 kg of carbon in various forms. Almost all of this carbon is present as carbon-12 and carbon-13 with isotopic abundances of 98.9% and 1.1%, respectively. In addition, traces of a third isotope, carbon-14, are present to the extent of about 1 in 10^{12} atoms in atmospheric CO_2 and in living materials. Carbon-14 nuclei are β emitters with a half-life of 5730 years. A 1 g sample of carbon from living material will show about 14 disintegrations per minute (0.23 Bq), not a lot of radioactivity but nevertheless detectable by modern methods.

Carbon-14 is formed in the upper atmosphere by nuclear reactions initiated by neutrons in cosmic radiation:

$$^{14}_{7}N + ^{1}_{0}n \longrightarrow ^{14}_{6}C + ^{1}_{1}H$$

Once formed, carbon-14 is oxidized to $^{14}CO_2$. This product enters the carbon cycle, circulating through the atmosphere, oceans, and biosphere.

The usefulness of carbon-14 for dating comes about in the following way. Plants absorb CO_2 and convert it to organic compounds, thereby incorporating carbon-14 into living tissue. As long as a plant remains alive, this process will continue and the percentage of carbon that is carbon-14 in the plant will equal the percentage in the atmosphere. When the plant dies, carbon-14 will no longer be taken up. Radioactive decay continues, however, with the carbon-14 activity decreasing over time. After 5730 years, the activity will be 0.12 Bq g^{-1}; after 11 460 years, it will be 0.06 Bq g^{-1}; and so on. By measuring the activity of a sample, and knowing the half-life of carbon-14, it is possible to calculate when a plant (or an animal that was eating plants) died.

As with all experimental procedures, carbon-14 dating has limitations. The procedure assumes that the amount of carbon-14 in the atmosphere hundreds or thousands of years ago is the same as it is now. We know that this isn't exactly true; the percentage has varied by as much as 10% (Figure 30.10). Furthermore, it is not possible to use carbon-14 to date an object that is less than about 100 years old; the radiation level from carbon-14 will not change enough in this short time period to permit accurate detection of a difference from the initial value. Finally, it is not possible to determine ages of objects much older than about 40 000 years. By then, after nearly seven half-lives, the radioactivity will have decreased virtually to zero. But for the span of time between 100 and 40 000 years, this technique has provided important information (Figure 30.10) with a typical accuracy of several hundred years.

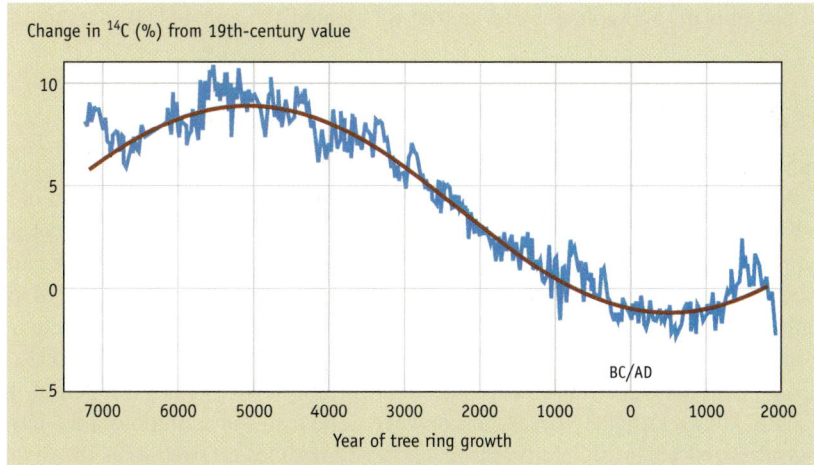

Change in ^{14}C (%) from 19th-century value

Year of tree ring growth

Oesper Collection in the History of Chemistry, University of Cincinnati

FIGURE 30.10 Variation of atmospheric carbon-14 activity. The amount of carbon-14 has varied with variation in cosmic ray activity. To obtain the data for the pre-1990 part of the curve shown in this graph, scientists carried out carbon-14 dating of artefacts for which the age was accurately known (often through written records). Similar results can be obtained using carbon-14 dating of tree rings.

WORKED EXAMPLE 30.8—CARBON DATING

To test the concept of carbon-14 dating, J.R. Arnold and W.F. Libby applied this technique to analyze samples of acacia and cyprus wood whose ages were already known. (The acacia wood, which was supplied by the Metropolitan Museum of Art in New York, came from the tomb of Zoser, the first Egyptian pharaoh to be entombed in a pyramid. The cyprus wood was from the tomb of Sneferu.) The average activity based on five determinations was 0.117 Bq per gram of carbon. Assume (as Arnold and Libby did) that the activity of carbon-14 (A_0) is 0.21 Bq per gram of carbon. Calculate the approximate age of the sample.

Strategy

First, determine the rate constant for the decay of carbon-14 from its half-life ($t_{1/2}$ for ^{14}C is 5.73×10^3 a). Then, use equation 30.3.

Solution

$$k = 0.693/t_{1/2} = 0.693/5730 \text{ a} = 1.21 \times 10^{-4} \text{ a}^{-1}$$

$$\ln (A/A_0) = -kt$$

$$\ln \left(\frac{0.117 \text{ Bq g}^{-1}}{0.21 \text{ Bq g}^{-1}} \right) = (-1.21 \times 10^{-4} \text{ a}^{-1})t$$

$$t = 4.8 \times 10^3 \text{ a}$$

The wood is about 4800 years old.

Comment

This problem uses historical data from an early research paper in which the carbon-14 dating method was being tested. The age of the wood was known to be 4750 ± 250 years. (See J.R. Arnold and W.F. Libby, (1949), "Age Determinations by Radiocarbon Content: Checks with Samples of Known Age," *Science*, 110: 678.)

Willard Libby (1908–1980) received the 1960 Nobel Prize in Chemistry for developing carbon-14 dating techniques.

EXERCISE 30.10—CARBON DATING

A sample of the inner part of a redwood tree felled in 1874 was shown to have ^{14}C activity of 0.155 Bq g^{-1}. Calculate the approximate age of the tree when it was cut down. Compare this age with that obtained from tree ring data, which estimated that the tree began to grow in 979 ± 52 B.C. Use 0.223 Bq g^{-1} for the value of A_0.

Radioactive decay processes follow first-order kinetics, but are independent of temperature. Half-life ($t_{1/2}$) is used to estimate the time for the amount of a radioactive isotope to decrease to a certain value or to change over a particular period of time. Radiocarbon dating, the rate of decay of carbon-14 isotopes, can be used to estimate the age of plant and animal material that is between 100 and 40 000 years old.

30.6 Artificial Nuclear Reactions

How many different isotopes are found on the earth? Most naturally occurring isotopes are stable. A few unstable (radioactive) isotopes that have long half-lives are found in nature; the best-known examples are uranium-235, uranium-238, thorium-232, and potassium-40. Trace quantities of other radioactive isotopes with short half-lives are present because they are being formed continuously by nuclear reactions. They include isotopes of radium, polonium, and radon, along with other elements produced in various radioactive decay series, and carbon-14, formed in a nuclear reaction initiated by cosmic radiation.

Naturally occurring isotopes account for only a very small fraction of the currently known radioactive isotopes, however. The rest—several thousand—have been synthesized via artificial nuclear reactions, sometimes referred to as *transmutation*.

The first artificial nuclear reaction was identified by Rutherford about 90 years ago. Recall the classic experiment that led to the nuclear model of the atom [<<Section 2.7] in which gold foil was bombarded with α particles. In the years following that experiment, Rutherford and his co-workers bombarded many other elements with α particles. In 1919, one of these experiments led to an unexpected result: when nitrogen atoms were bombarded with α particles, protons were detected among the products. Rutherford correctly concluded that a nuclear reaction had occurred. Nitrogen atoms had undergone a *transmutation* to oxygen atoms:

$$\ce{^4_2He + ^{14}_7N -> ^{17}_8O + ^1_1H}$$

During the next decade, other nuclear reactions were discovered by bombarding other elements with α particles. Progress was slow, however, because in most cases α particles are simply scattered by target nuclei. The bombarding particles cannot get close enough to the nucleus to react because of the strong repulsive forces between the positively charged α particle and the positively charged atomic nucleus.

In 1932, two advances were made that greatly extended nuclear reaction chemistry. The first involved the use of particle accelerators to create high-energy particles as projectiles. The second was the use of neutrons as the bombarding particles.

The α particles used in the early studies on nuclear reactions came from naturally radioactive materials such as uranium and had relatively low energies, at least by today's standards. Particles with higher energy were needed, so J.D. Cockcroft (1897–1967) and E.T.S. Walton (1903–1995), working in Rutherford's laboratory in Cambridge, England, turned to protons. Protons are formed when hydrogen atoms ionize in a cathode-ray tube, and it was known that they could be accelerated to higher energy by applying a high voltage. Cockcroft and Walton found that when energetic protons struck a lithium target, the following reaction occurs:

$$\ce{^7_3Li + ^1_1p -> 2 ^4_2He}$$

This was the first example of a reaction initiated by a particle that had been artificially accelerated to high energy. Since this experiment was conducted, the technique has been developed much further, and the use of particle accelerators in nuclear chemistry is now commonplace. Particle accelerators operate on the principle that a charged particle placed between charged plates will be accelerated to a high speed and high energy. Modern examples of this process are seen in the synthesis of the transuranium elements, several of which are described in more detail in e30.1.

Experiments using neutrons as bombarding particles were first carried out in both the United States and Great Britain in 1932. Nitrogen, oxygen, fluorine, and neon were bombarded with energetic neutrons, and α particles were detected among the products. Using neutrons made sense: because neutrons have no charge, it was reasoned that these particles

A coffee mug created in honour of Glenn T. Seaborg (1912–1999), who figured out that thorium and the elements that followed it fit under the lanthanides in the periodic table. For this insight, he and Edwin McMillan shared the 1951 Nobel Prize in Chemistry. Over a 21-year period, Seaborg and his colleagues synthesized 10 new transuranium elements (Pu through Lr). To honour Seaborg's scientific contributions, the name "seaborgium" was assigned to element 106. It marked the first time an element was named for a living person.

Courtesy of ACT2. Photo by Dr. Diana Mason

Taking It Further

e30.1 Read about the search for new elements.

would not be repelled by the positively charged nucleus particles. Thus, neutrons did not need high energies to react.

In 1934, Enrico Fermi (1901–1954) and his co-workers showed that nuclear reactions using neutrons are more favourable if the neutrons have low energy. A low-energy neutron is simply captured by the nucleus, giving a product in which the mass number is increased by one unit. Because of the low energy of the bombarding particle, the product nucleus does not have sufficient energy to fragment in these reactions. The new nucleus is produced in an excited state, however; when the nucleus returns to the ground state, a γ-ray is emitted. Reactions in which a neutron is captured and a γ-ray is emitted are called *(n, γ) reactions*.

The (n, γ) reactions are the source of many of the radioisotopes used in medicine and chemistry. An example is radioactive phosphorus ($^{32}_{15}P$), which is used in chemical studies such as tracing the uptake of phosphorus in the body.

$$^{31}_{15}P + {}^{1}_{0}n \longrightarrow {}^{32}_{15}P + \gamma$$

Transuranium elements, elements with an atomic number greater than 92, were first made in a nuclear reaction sequence beginning with an (n, γ) reaction. Scientists at the University of California at Berkeley bombarded uranium-238 with neutrons. Among the products identified were neptunium-239 and plutonium-239. These new elements were formed when ^{239}U decayed by β radiation.

$$^{238}_{92}U + {}^{1}_{0}n \longrightarrow {}^{239}_{92}U$$

$$^{239}_{92}U \longrightarrow {}^{239}_{93}Np + {}^{0}_{-1}\beta$$

$$^{239}_{93}Np \longrightarrow {}^{239}_{94}Pu + {}^{0}_{-1}\beta$$

Four years later, a similar reaction sequence was used to make americium-241. Plutonium-239 was found to add two neutrons to form plutonium-241, which decays by β emission to give americium-241.

WORKED EXAMPLE 30.9—ARTIFICIAL NUCLEAR REACTIONS

Write equations for the nuclear reactions described below.
(a) Fluorine-19 undergoes an (n, γ) reaction to give a radioactive product that decays by $^{0}_{-1}\beta$ emission. (Write equations for both nuclear reactions.)
(b) A common neutron source is a plutonium-beryllium alloy. Plutonium-239 is an α emitter. When beryllium-9 (the only stable isotope of beryllium) reacts with α particles emitted by plutonium, neutrons are ejected. (Write equations for both reactions.)

Strategy
The equations are written so that both mass and charge are balanced.

Solution
(a) $^{19}_{9}F + {}^{1}_{0}n \longrightarrow {}^{20}_{9}F + \gamma$

$^{20}_{9}F \longrightarrow {}^{20}_{10}Ne + {}^{0}_{-1}\beta$

(b) $^{239}_{94}Pu \longrightarrow {}^{235}_{92}U + {}^{4}_{2}\alpha$

$^{4}_{2}\alpha + {}^{9}_{4}Be \longrightarrow {}^{12}_{6}C + {}^{1}_{0}n$

Bombardment of nuclei with high-energy alpha particles or neutrons in particle accelerators can be used to artificially create different nuclei, in a process called transmutation.

30.7 Nuclear Fission

The radioactive isotopes ^{90}Sr, ^{137}Cs, ^{3}H, and ^{36}Cl found near the top of most ice core samples from around the world were produced in the 1950s and '60s by nuclear reactions that used Uranium-235 as the source of energy for nuclear bombs. A key step in understanding how

Neutrons had been predicted to exist for more than a decade before they were identified in 1932 by James Chadwick (1891–1974). Chadwick produced neutrons in a nuclear reaction between α particles and beryllium:
$^{4}_{2}\alpha + {}^{9}_{4}Be \longrightarrow {}^{12}_{6}C + {}^{1}_{0}n$.

Neptunium and plutonium were unknown prior to their preparation via these nuclear reactions. Later, these elements were found to be present in trace quantities in uranium ores.

In fission reactions, a heavy nucleus splits into two or more fragments. Uranium-236 undergoes fission to form a large number of different fission products (elements). Barium was the element first identified, and its identification provided the key that led to recognition that fission had occurred.

Lise Meitner (1878–1968). Meitner's greatest contribution to 20th-century science was her explanation of the process of nuclear fission. She and her nephew, Otto Frisch, also a physicist, published a paper in 1939 that was the first to use the term *nuclear fission*. Element number 109 is named meitnerium to honour Meitner's contributions. The leader of the team that discovered this element said that "She should be honoured as the most significant woman scientist of [the 20th] century."

uranium can be converted into isotopes of other elements in nuclear reactions came in 1938, when two chemists, Otto Hahn (1879–1968) and Fritz Strassman (1902–1980), isolated and identified barium in a sample of uranium that had been bombarded with neutrons. How was barium formed? The answer to that question explained one of the most significant scientific discoveries of the 20th century. The uranium nucleus had split into smaller pieces in the process we now call **nuclear fission**.

The details of nuclear fission were unravelled through the work of a number of scientists. They determined that a uranium-235 nucleus initially captured a neutron to form uranium-236. This isotope underwent nuclear fission to produce two new nuclei, one with a mass number around 140 and the other with a mass around 90, along with several neutrons (Figure 30.11). The nuclear reactions that led to formation of barium when a sample of ^{235}U was bombarded with neutrons are

$$^{235}_{92}U + {}^{1}_{0}n \longrightarrow {}^{236}_{92}U$$

$$^{236}_{92}U \longrightarrow {}^{141}_{56}Ba + {}^{92}_{36}Kr + 3\,{}^{1}_{0}n$$

An important aspect of fission reactions is that they produce more neutrons than are used to initiate the process. Under the right circumstances, these neutrons then serve to continue the reaction. If one or more of these neutrons are captured by another ^{235}U nucleus, then a further reaction can occur, releasing still more neutrons. This sequence repeats over and over. Such a mechanism, in which each step generates a reactant to continue the reaction, is called a *chain reaction*.

FIGURE 30.11 Nuclear fission. Neutron capture by $^{235}_{92}U$ produces $^{236}_{92}U$. This isotope undergoes fission, which yields several fragments along with several neutrons. These neutrons initiate further nuclear reactions by adding to other $^{235}_{92}U$ nuclei. The process is highly exothermic, producing about 2×10^{10} kJ mol^{-1}.

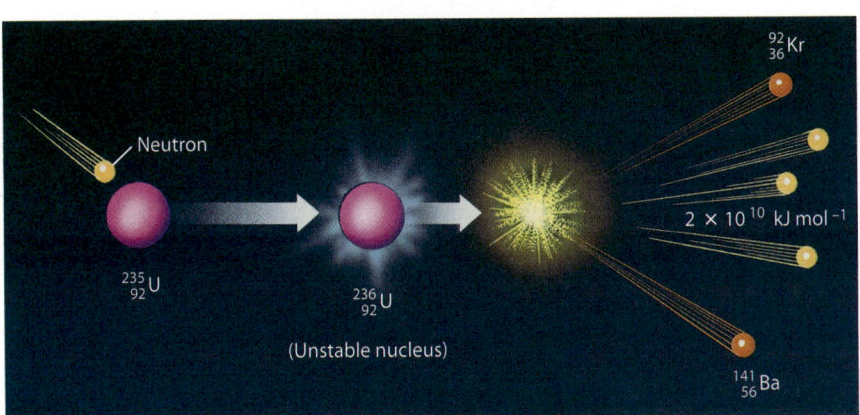

A nuclear fission chain reaction has three general steps:

1. *Initiation.* The reaction of a single atom is needed to start the chain. Fission of ^{235}U is initiated by the absorption of a neutron.
2. *Propagation.* This part of the process repeats itself over and over, with each step yielding more product. The fission of ^{236}U releases neutrons that initiate the fission of other uranium atoms.
3. *Termination.* Eventually the chain will end. Termination could occur if the reactant (^{235}U) is used up, or if the neutrons that continue the chain escape from the sample without being captured by ^{235}U.

In a nuclear bomb, each nuclear fission step produces 3 neutrons, which leads to about 3 more fissions and 9 more neutrons, which leads to 9 more fission steps and 27 more neutrons, and so on. The rate depends on the number of neutrons, so the nuclear reaction occurs faster and faster as more and more neutrons are formed, leading to an enormous output of energy in a short time span.

To harness the energy produced in a nuclear reaction, it is necessary to control the rate at which a fission reaction occurs. This is managed by balancing the propagation and termination steps by limiting the number of neutrons available. In a nuclear reactor, this balance is accomplished by using cadmium rods to absorb neutrons. By withdrawing or inserting the rods, the number of neutrons available to propagate the chain can be changed, and the rate of the fission reaction (and the rate of energy production) can be increased or decreased.

Uranium-235 and plutonium-239 are the fissionable isotopes most commonly used in nuclear power reactors. Natural uranium contains only 0.72% of uranium-235; more than 99% of the natural element is uranium-238. The percentage of uranium-235 in natural uranium is too small to sustain a chain reaction, however, so the uranium used for nuclear fuel must be enriched in this isotope. One way to do so is by gaseous diffusion [<<Section 11.8]. Plutonium, which occurs naturally in only trace quantities, must be made via a nuclear reaction. The raw material for this nuclear synthesis is the more abundant uranium isotope, ^{238}U. Addition of a neutron to ^{238}U gives ^{239}U, which, as noted earlier, undergoes two β emissions to form ^{239}Pu.

At the time this is written, there are 438 operating nuclear power plants in 31 countries around the world, and 34 new ones under construction worldwide. About 17% of the world's energy comes from nuclear power, although huge variation exists. Nuclear energy provides less than 2% of China's electricity, but over 70% of the needs of Lithuania and almost 80% for France. With the world's attention drawn to climate disruption resulting from fossil fuel combustion, nuclear energy may be called upon to meet the ever-increasing energy needs of society. But among other things, the disaster at Chernobyl (in the former Soviet Union) and incident at Three Mile Island (in Pennsylvania) have raised questions in the public about the risks associated with nuclear powerplants. The cost to construct a nuclear powerplant (measured in terms of dollars per kilowatt-hour of energy) is considerably more than the cost for a natural gas–powered facility, and the regulatory restrictions for nuclear power are challenging. Disposal of highly radioactive nuclear waste is another thorny problem, with 20 tonnes of waste being generated per year at each reactor.

In addition to technical problems, nuclear energy production brings with it significant geopolitical security concerns. The process for enriching uranium for use in a reactor is the same process used for generating weapons-grade uranium. Also, some nuclear reactors are designed so that one by-product of their operation is the isotope plutonium-239, which can be removed and used in a nuclear weapon. Despite these problems, nuclear fission must be considered as one piece of the energy pie to meet the needs of an increasingly energy-hungry world.

> In a nuclear fission reaction, a heavy nucleus splits into two or more fragments. Nuclear energy, released from the self-sustaining chain reactions of uranium-235, can be converted to electrical energy in nuclear powerplants. It is also released, but in a much less controlled way, in nuclear weapons.

30.8 Nuclear Fusion

The ultimate source of energy for the fossil fuels we burn is **nuclear fusion** reactions in the sun. The energy has been stored by chemical reactions in plants that have then become fossilized over geological time scales. In a nuclear fusion reaction, several small nuclei react to form a larger nucleus. Enormous amounts of energy can be generated by such reactions. An example is the fusion of deuterium and tritium nuclei to form $^{4}_{2}$He and a neutron:

$$^{2}_{1}\text{H} + ^{3}_{1}\text{H} \longrightarrow ^{4}_{2}\text{He} + ^{1}_{0}\text{n} \qquad \Delta E = -1.7 \times 10^{9} \text{ kJ mol}^{-1}$$

Fusion reactions provide the energy of our sun and other stars. Scientists have long dreamed of being able to harness fusion to provide power. To do so, a temperature of 10^{6} to 10^{7} K, like that in the interior of the sun, would be required to bring the positively charged nuclei together with enough energy to overcome nuclear repulsions. At the very high temperatures needed for a fusion reaction, matter does not exist as atoms or molecules; instead, matter is in the form of a *plasma* made up of unbound nuclei and electrons [<<Section 11.1].

Three critical requirements must be met before nuclear fusion could represent a viable energy source. First, the temperature must be high enough for fusion to occur. The fusion of deuterium and tritium, for example, requires a temperature of 10^{8} K or more. Second, the plasma must be confined long enough to release a net output of energy. Third, the energy must be recovered in some usable form.

Harnessing a nuclear fusion reaction for a peaceful use has not yet been achieved. Nevertheless, many attractive features encourage continuing research in this field. The hydrogen isotopes used as "fuel" are cheap and available in almost unlimited amounts. As a further benefit, most radioisotopes produced by fusion have short half-lives, so they remain a radiation hazard for only a short time.

> In a nuclear fusion reaction, several small nuclei react to form a larger nucleus.

30.9   Radiation Health and Safety

Units for Measuring Radiation

We have already introduced the basic unit for radiation activity, the becquerel (Bq), which tells us the activity of a substance—that is, how many disintegrations of a radioactive substance take place per second. A Geiger counter can be used to measure activity.

By itself, the activity of a radioactive source isn't enough to give us information about the harm that may be caused by different forms of radiation acting on different types of body tissue. In addition, we need to determine *exposure*, which is the effect of that radiation on tissues that absorb it. An understanding of exposure requires knowing how much energy is absorbed, and the effectiveness of the particular kind of radiation in causing damage to a particular kind of living tissue.

The SI unit that measures the amount of energy absorbed by living tissue is the *gray* (Gy); 1 Gy denotes the absorption of 1 joule of energy per kilogram of tissue.

Different forms of radiation cause different amounts of biological damage. The amount of damage depends on how strongly a form of radiation interacts with matter. Alpha particles cannot penetrate the body any deeper than the outer layer of skin. If α particles are emitted within the body, however, they will do between 10 and 20 times the amount of damage done by γ-rays, which can go entirely through a human body without being stopped. In determining the amount of biological damage to living tissue, differences in damaging power are accounted for using a radiation weighting factor. This weighting factor has been set at 1 for β and γ radiation absorbed by human tissue, 5 for low-energy protons and neutrons, and 20 for α particles or high-energy protons and neutrons.

Biological damage is measured by a quantity called *equivalent dose*, which relates the absorbed dose in human tissue to the damage that radiation will cause to human tissue. Equivalent dose is quantified in an SI unit called the *sievert* (Sv), which is determined by multiplying the dose in grays by the radiation weighting factor, which is different for each type of radiation.

Radiation Doses and Effects

Exposure to a small amount of radiation is unavoidable. The earth is constantly being bombarded with radioactive particles from outer space. There is also some exposure to radioactive elements that occur naturally on the earth, including ^{14}C, ^{40}K (a radioactive isotope that occurs naturally in 0.0117% abundance), ^{238}U, and ^{232}Th. Radioactive elements in the environment that were created artificially (in the fallout from nuclear bomb tests, for example) also contribute to this exposure. For some people, medical procedures using radioisotopes are a major contributor.

The average dose of background radioactivity in Australia is about 2 millisieverts per annum (Table 30.2). Well over half of that amount comes from natural sources over which we have no control. Of the exposure that comes from artificial sources, nearly 90% is delivered in medical procedures such as x-ray examinations and radiation therapy. Considering the controversy surrounding nuclear power, it is interesting to note that less than 0.5% of the total annual background dose of radiation that the average person receives can be attributed to the nuclear power industry.

What Is a Safe Exposure?

Is the exposure to natural background radiation totally without effect? Can you equate the effect of a single dose and the effect of cumulative, smaller doses that are spread out over a long period of time? The assumption generally made is that no "safe maximum dose," or level below which absolutely no damage will occur, exists. However, the accuracy of this assumption has come into question. These issues are not testable with human subjects, and

	mSv yr^{-1}	Percentage
Natural Sources		
Cosmic radiation	0.500	25.8
The earth	0.470	24.2
Building materials	0.030	1.5
Inhaled from the air	0.050	2.6
Elements found naturally in human tissue	0.210	10.8
Subtotal	**1.26**	**64.9**
Medical Sources		
Diagnostic x-rays	0.500	25.8
Radiotherapy	0.100	5.2
Internal diagnosis	0.0100	0.5
Subtotal	**0.610**	**31.5**
Other Artificial Sources		
Nuclear power industry	0.0085	0.4
Luminous watch dials, TV tubes	0.020	1.0
Fallout from nuclear tests	0.040	2.1
Subtotal	**0.069**	**3.5**
Total	**1.939**	**99.9**

TABLE 30.2 Annual Radiation Exposure for an Individual from Natural and Artificial Sources

tests based on animal studies are not completely reliable because of the uncertainty of species-to-species variations.

The model used by government regulators to set exposure limits assumes that the relationship between exposure to radiation and incidence of radiation-induced health effects, such as cancer, anemia, and immune system problems, is linear. Under this assumption, if a dose of $2x$ mSv causes damage in 20% of the population, then a dose of x mSv will cause damage in 10% of the population. But is this true? Cells do possess mechanisms for repairing damage. Many scientists believe that this self-repair mechanism renders the human body less susceptible to damage from smaller doses of radiation, because the damage will be repaired as part of the normal course of events. They argue that, at extremely low doses of radiation, the self-repair response results in less damage.

However, much still remains to be learned in this area. And the stakes are significant.

Describing the biological effects of a dose of radiation precisely is not a simple matter. The amount of damage done depends not only on the kind of radiation and the amount of energy absorbed, but also on the particular tissues exposed and the rate at which the dose builds up. A great deal has been learned about the effects of radiation on the human body by studying the survivors of the bombs dropped over Japan in World War II and the workers exposed to radiation from the reactor disaster at Chernobyl. From studies of the health of these survivors, we have learned that the effects of radiation are not generally observable below a single dose of 0.25 Sv. At the other extreme, a single dose of >2 Sv will be fatal to about half the population (Table 30.3).

Dose (Sv)	Effect
0–0.25	No effect observed
0.26–0.50	Small decrease in white blood cell count
0.51–1.00	Significant decrease in white blood cell count, lesions
1.01–2.00	Loss of hair, nausea
2.01–5.00	Hemorrhaging, ulcers, death in 50% of population
5.00	Death

TABLE 30.3 Effects of a Single Dose of Radiation

Our information is more accurate when dealing with single, large doses than it is for the effects of chronic, smaller doses of radiation. One current issue of debate in the scientific community is how to judge the effects of multiple smaller doses or long-term exposure.

> Activity, measured in becquerels (Bq), is a measure of how many disintegrations take place in a second by a radioactive sample. Exposure is the effect of that radiation on tissues that absorb it. The amount of energy absorbed by living tissue is measured in gray (Gy) units; 1 Gy denotes the absorption of 1 joule of energy per kilogram of tissue. Biological damage is measured by a quantity called equivalent dose, which relates the absorbed dose in human tissue to the damage that radiation will cause to human tissue. It is measured in sievert (Sv) units, determined by multiplying the dose in grays by the radiation weighting factor, which is different for each type of radiation.

30.10 Applications of Nuclear Chemistry

We tend to think about nuclear chemistry first as having applications in powerplants and bombs. In truth, radioactive elements are now used in all areas of science and medicine, and they are of ever-increasing importance to our lives. Because describing all of their uses would take several books, we have selected just a few examples to illustrate the diversity of applications of radioactivity.

Nuclear Medicine: Medical Imaging

Diagnostic procedures using nuclear chemistry are essential in medical imaging, which entails the creation of images of specific parts of the body. There are three principal components to constructing a radioisotope-based image:

1. a radioactive isotope, administered as the element or incorporated into a compound, that concentrates the radioactive isotope in the tissue to be imaged
2. a method to detect the type of radiation involved
3. a computer to assemble the information from the detector into a meaningful visual image

The choice of a radioisotope and the manner in which it is administered are determined by the tissue in question. A compound containing the isotope must be absorbed more by the diseased tissue than by the rest of the body. Table 30.4 lists radioisotopes that are commonly used in nuclear imaging processes, their half-lives, and the tissues they are used to image. All of the isotopes in Table 30.4 are γ emitters; γ radiation is preferred for imaging because it is less damaging to the body in small doses than either α or β radiation.

Technetium-99m is used in more than 85% of the diagnostic scans done in hospitals each year (see e30.2). The "m" stands for *metastable*, a term used to identify an unstable state that exists for a finite period of time. Recall that atoms in excited electronic states emit visible, infrared, and ultraviolet radiation [<<Section 8.3]. Similarly, a nucleus in an excited state gives up its excess energy, but in this case a much higher energy is

Taking It Further

e30.2 Take a closer look at medical applications of technetium-99m.

TABLE 30.4 Radioisotopes Used in Medical Diagnostic Procedures

Radioisotope	Half-Life	Imaging
^{99m}Tc	6.0	Thyroid, brain, kidneys
^{201}Tl	73.0	Heart
^{123}I	13.2	Thyroid
^{67}Ga	78.2	Various tumours and abscesses
^{18}F	1.8	Brain, sites of metabolic activity

involved and the emission occurs as γ radiation. The γ-rays given off by ^{99m}Tc are detected to produce the image (Figure 30.12).

Another medical imaging technique based on nuclear chemistry is positron emission tomography (PET). In PET, an isotope that decays by positron emission is incorporated into a carrier compound and given to the patient. When emitted, the positron travels no more than a few millimetres before undergoing matter–antimatter annihilation.

$$^{0}_{+1}\beta + ^{0}_{-1}e \longrightarrow 2\,\gamma$$

The two emitted γ-rays travel in opposite directions. By determining where high numbers of γ-rays are being emitted, one can construct a map showing where the positron emitter is located in the body.

An isotope often used in PET is ^{15}O. A patient is given gaseous O_2 that contains some ^{15}O atoms. As O_2 travels throughout the body in the bloodstream, radiation emitted by the ^{15}O atoms allows images of the brain and bloodstream (Figure 30.13) to be obtained. Because positron emitters are typically very short-lived, PET facilities must be located near a cyclotron where the radioactive nuclei are prepared and then immediately incorporated into a carrier compound.

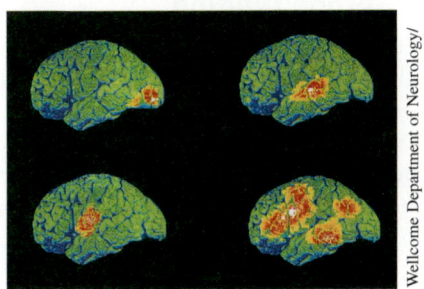

FIGURE 30.13 PET scans of the brain. These scans show the left side of the brain; red indicates an area of highest activity. (*Upper left*) Sight activates the visual area in the occipital cortex at the back of the brain. (*Upper right*) Hearing activates the auditory area in the superior temporal cortex of the brain. (*Lower left*) Speaking activates the speech centres in the insula and motor cortex. (*Lower right*) Thinking about verbs, and speaking them, generates high activity, including in the hearing, speaking, temporal, and parietal areas.

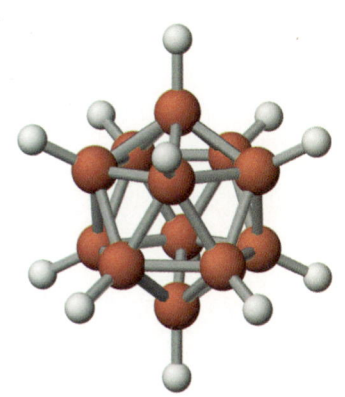

David Parker/Photo Researchers, Inc.

FIGURE 30.12 Heart imaging with technetium-99m. The radioactive element technetium-99m, a gamma emitter, is injected into a patient's vein in the form of the pertechnetate ion (TcO_4^-) or as a complex ion with an organic ligand. A series of scans of the gamma emissions of the isotope are made while the patient is resting and then again after strenuous exercise. Bright areas in the scans indicate that the isotope is binding to the tissue in that area. The scans in this figure show a normal heart function.

Wellcome Department of Neurology/Science Photo Library/Photo Researchers, Inc.

Nuclear Medicine: Radiation Therapy

To treat many cancers, it is necessary to use radiation that can penetrate the body to the location of the tumour. Gamma radiation from a cobalt-60 source is commonly used. Unfortunately, the penetrating ability of γ-rays makes it virtually impossible to destroy diseased tissue without also damaging healthy tissue in the process. Nevertheless, this technique is a regularly sanctioned procedure and its successes are well known.

To avoid the side effects associated with more traditional forms of radiation therapy, a new form of treatment has been explored in the last 10 to 15 years, called *boron neutron capture therapy* (BNCT). BNCT is unusual in that boron-10, the isotope of boron used as part of the treatment, is not radioactive. Boron-10 is highly effective in capturing neutrons, however—2500 times better than boron-11, and 8 times better than uranium-235. When the nucleus of a boron-10 atom captures a neutron, the resulting boron-11 nucleus has so much energy that it fragments to form an α particle and a lithium-7 atom. Although the α particles do a great deal of damage, because their penetrating power is so low, the damage remains confined to an area not much larger than one or two cells in diameter.

In a typical BNCT treatment, a solution of a boron compound is injected into the tumour. After a few hours, the tumour is bombarded with neutrons. The α particles are produced only at the site of the tumour, and the production stops when the neutron bombardment ends.

One of the compounds used in BNCT is $Na_2[B_{12}H_{12}]$. The structure of the $B_{12}H_{12}^{2-}$ anion is a regular polyhedron with 20 sides, called an icosahedron.

Analytical Methods: The Use of Radioactive Isotopes as Tracers

Radioactive isotopes can be used to help determine the fate of compounds in the body or in the environment. These studies begin with a compound that contains a radioactive isotope of one of its component elements. In biology, for example, scientists can use radioactive isotopes to measure the uptake of nutrients. Plants take up phosphorus-containing compounds from the soil through their roots. By adding a small amount of radioactive ^{32}P (a β emitter with a half-life of 14.3 days) to fertilizer and then measuring the rate at which the radioactivity appears in the leaves, plant biologists can determine the rate at which phosphorus is taken up. The outcome can assist scientists in identifying hybrid strains of plants that can absorb phosphorus quickly, resulting in faster-maturing crops, better yields per hectare, and more food or fibre at less expense.

To measure pesticide levels, a pesticide can be tagged with a radioisotope and then applied to a test field. By counting the disintegrations of the radioactive tracer, information can be obtained about how much pesticide accumulates in the soil, is taken up by the plant, and is carried off in runoff surface water. After these tests are completed, the radioactive isotope decays to harmless levels in a few days or a few weeks because of the short half-lives of the isotopes used.

Analytical Methods: Isotope Dilution

Imagine, for the moment, that you wanted to estimate the volume of blood in an animal subject. How might you do this? Obviously, draining the blood and measuring its volume in volumetric glassware is not a desirable option.

One technique uses a method called *isotope dilution*. In this process, a small amount of radioactive isotope is injected into the bloodstream. After a period of time to allow the isotope to become distributed throughout the body, a blood sample is taken and its radioactivity measured. The calculation used to determine the total blood volume is illustrated in the next example.

WORKED EXAMPLE 30.10—ISOTOPE DILUTION

A 1.00 mL solution containing 8.8×10^3 Bq of tritium is injected into a dog's bloodstream. After a period of time to allow the isotope to be dispersed, a 1.00 mL sample of blood is drawn. The radioactivity of this sample is found to be 16 Bq mL^{-1}. What is the total volume of blood in the dog?

Strategy

For any solution, concentration equals the amount of solute divided by volume of solution. In this problem, we relate the activity of the sample (in Ci) to concentration. The total amount of solute is 8.8×10^3 Bq, and the concentration (measured on the small sample of blood) is 16 Bq mL^{-1}. The unknown is the volume.

Solution

The blood contains a total of 8.8×10^3 Bq of the radioactive material. We can represent its concentration as 8.8×10^3 Bq $/x$, where x is the total blood volume. After dilution in the bloodstream, 1.00 mL of blood is found to have an activity of 16 Bq mL^{-1}.

$$8.8 \times 10^3 \text{ Bq } /x = 16 \text{ Bq mL}^{-1}$$

$$x = 560 \text{ mL}$$

EXERCISE 30.11—ISOTOPE DILUTION

Suppose you hydrolyze a 10.00 g sample of a protein. Next, you add to it a 3.00 mg sample of ^{14}C-labelled threonine, an amino acid with a specific activity of 50 Bq. After mixing, part of the threonine (60.0 mg) is separated and isolated from the mixture. The activity of the isolated sample is 20 Bq. How much threonine was present in the original sample?

Space Science: Neutron Activation Analysis and the Moon Rocks

The first manned space mission to the moon brought back a number of samples of soil and rock—a treasure trove for scientists. One of their first tasks was to analyze these samples to determine their identity and composition. Most analytical methods require chemical reactions using at least a small amount of material; however, this was not a desirable option, considering that the moon rocks were at the time the most valuable rocks in the world.

A few fortunate scientists got a chance to work on this unique project, and one of the analytical tools they used was *neutron activation analysis*. In this non-destructive process, a sample is irradiated with neutrons. Most isotopes add a neutron to form a new isotope that is one mass unit higher in an excited nuclear state. When the nucleus decays to its ground state, it emits a γ-ray. The energy of the γ-ray identifies the element, and the number of γ-rays can be counted to determine the amount of the element in the sample. Using neutron activation analysis, it is possible to analyze for a number of elements in a single experiment (Table 30.5).

Neutron activation analysis has many other uses. This analytical procedure yields a kind of fingerprint that can be used to identify a substance. For example, this technique has been applied in determining whether an artwork is real or fraudulent. Analysis of the pigments in paints on a painting can be carried out without damaging the painting to determine whether the composition resembles modern paints or paints used hundreds of years ago.

Food Science: Food Irradiation

Refrigeration, canning, and chemical additives provide significant protection from food spoilage, but in some parts of the world these procedures are unavailable and stored-food spoilage may claim as much as 50% of the food crop. Irradiation with γ-rays from sources such as ^{60}Co and ^{137}Cs is an option for prolonging the shelf life of foods. Relatively low levels of radiation may retard the growth of organisms, such as bacteria, moulds, and yeasts, which can cause food spoilage. After irradiation, milk in a sealed container has a minimum shelf life of three months without refrigeration. Chicken flesh normally has a three-day refrigerated shelf life; after irradiation, it may have a three-week refrigerated shelf life.

Higher levels of radiation, in the $1–5 \times 10^4$ Gy range, will kill every living organism. Foods irradiated at these levels will keep indefinitely when sealed in plastic or aluminum-foil packages. Ham, beef, turkey, and corned beef sterilized by radiation have been used on many Space Shuttle flights, for example. An astronaut said, "The beautiful thing was that it didn't disturb the taste, which made the meals much better than the freeze-dried and other types of foods we had."

These procedures are not without their opponents, and the public has not fully embraced irradiation of foods yet. An interesting argument favouring this technique is that radiation is less harmful than other methodologies for food preservation. This type of sterilization offers greater safety to food workers because it lessens chances of exposure to harmful chemicals, and it protects the environment by avoiding contamination of water supplies with toxic chemicals.

Food irradiation is commonly used in European countries, Canada, and Mexico. In Australia and New Zealand, only herbs and spices, herbal teas, and some tropical fruits have been approved to be irradiated.

In addition to supplying energy, numerous applications of nuclear chemistry are now found in modern life, including applications in medicine, for imaging and radiation therapy, and food preservation. Important analytical methods also make use of radioisotopes as tracers and isotope dilution to determine the volume of aqueous solutions such as blood in living organisms. Neutron activation analysis is a non-destructive application to determine elemental composition.

TABLE 30.5 Rare Earth Analysis of Moon Rock Sample 10022 (a fine-grain igneous rock)

Element	Concentration (ppm)
La	26.4
Ce	68
Nd	66
Sm	21.2
Eu	2.04
Gd	25
Tb	4.7
Dy	31.2
Ho	5.5
Er	16
Yb	17.7
Lu	2.55

L.A. Haskin, P.A. Helmke, and R.O. Allen, (1970),"Rare Earth Elements in Returned Lunar Samples," *Science*, 167: 487. The concentrations of rare earths in moon rocks were quite similar to the values in terrestrial rocks except that the europium concentration is much depleted

SUMMARY

Key Concepts

- Nuclear chemistry refers to the part of chemistry that studies nuclei and nuclear reactions. Directly and indirectly, nuclear fusion reactions taking place in the sun provide humans with the energy that fuels modern life, and other nuclear processes have become centrally important in the provision of energy and tools for diagnosis and treatment in medicine. (Section 30.1)

- Three distinct types of radiation are given off by natural radioactive sources. **α, β**, and **γ radiation** all possess high kinetic energies, but have different abilities to penetrate matter. The energy associated with this radiation is transferred to any material used to stop the particle or absorb the radiation. This fact is important because the damage caused by radiation is related to the energy absorbed. (Section 30.2)

- In writing an equation for a **nuclear reaction**, the sum of the mass numbers of reacting particles must equal the sum of the mass numbers of product particles. Furthermore, to maintain nuclear charge balance, the sum of the atomic numbers of the product nuclei must equal the sum of the atomic numbers of the reactant nuclei. (Section 30.3)

- Naturally occurring radioactive elements decay by emission of α and β particles, and γ radiation. Artificially induced nuclear decay processes also include **positron ($_{+1}^{\ 0}\beta$) emission** and **electron capture**. (Section 30.3)

- The neutron/proton ratio of an isotope can be used to predict its stability relative to other isotopes, with stable isotopes falling in a very narrow range called the *band of stability*. The neutron/proton ratio can also help predict what kind of radioactive decay an unstable isotope will undergo to bring it closer to the band of stability. The nuclear stabilities of different elements are compared using the **nuclear binding energy** per mole of **nucleons**. (Section 30.4)

- Radioactive decay processes follow first-order kinetics, but are independent of temperature. Half-life ($t_{1/2}$) is used to estimate the time for the amount of a radioactive isotope to decrease. Radiocarbon dating, the rate of decay of carbon-14 isotopes, can be used to estimate the age of plant and animal material that is between 100 and 40 000 years old. (Section 30.5)

- Bombardment of nuclei with high-energy alpha particles or neutrons in particle accelerators can be used to artificially create different nuclei, in a process called *transmutation*. (Section 30.6)

- In a **nuclear fission** reaction, a heavy nucleus splits into two or more fragments. Nuclear energy, released from the self-sustaining chain reactions of uranium-235, can be converted to electrical energy in nuclear powerplants. It is also released, but in a much less controlled way, in nuclear weapons. (Section 30.7)

- In a **nuclear fusion** reaction, several small nuclei react to form a larger nucleus. (Section 30.8)

- **Activity**, measured in Becquerels (Bq), is a measure of how many disintegrations take place in a second by a radioactive sample. *Exposure* is the effect of that radiation on tissues that absorb it. The amount of energy absorbed by living tissue is measured in *gray* (Gy) units; 1 Gy denotes the absorption of 1 joule of energy per kilogram of tissue. Biological damage is measured by a quantity called *equivalent dose*, which relates the absorbed dose in human tissue to the damage that radiation will cause to human tissue. It is measured in *sievert* (Sv) units, determined by multiplying the dose in grays by the radiation weighting factor, which is different for each type of radiation. (Section 30.9)

- In addition to supplying energy, numerous applications of nuclear chemistry are now found in modern life, including applications in medicine, for imaging and radiation therapy, and food preservation. Important analytical methods also make use of radioisotopes as tracers and isotope dilution to determine the volume of aqueous solutions such as blood in living organisms. Neutron activation analysis is a non-destructive application to determine elemental composition. (Section 30.10)

Key Equations

$E_b = (\Delta m)c^2$ (30.1, Section 30.4)

$\ln\left(\dfrac{N}{N_0}\right) = -kt$ (30.2, Section 30.5)

$\ln\left(\dfrac{A}{A_0}\right) = -kt$ (30.3, Section 30.5)

$t_{1/2} = \dfrac{0.693}{k}$ (30.4, Section 30.5)

REVIEW QUESTIONS

Section 30.1: Human Activity, Chemical Reactivity

30.12 Of what significance to our understanding of anthropogenic climate change is the absence of β-decay signals from the top layers of glacial ice on the Tibetan plateau?

Sections 30.2–30.3: Natural Radioactivity and Nuclear Reactions and Radioactive Decay

30.13 Complete the following nuclear equations. Write the mass number and atomic number for the remaining particle, as well as its symbol.
(a) $^{54}_{26}\text{Fe} + ^{4}_{2}\text{He} \longrightarrow 2\,^{1}_{0}\text{n} + ?$
(b) $^{27}_{13}\text{Al} + ^{4}_{2}\text{He} \longrightarrow ^{30}_{15}\text{P} + ?$
(c) $^{32}_{16}\text{S} + ^{1}_{0}\text{n} \longrightarrow ^{1}_{1}\text{H} + ?$
(d) $^{96}_{42}\text{Mo} + ^{2}_{1}\text{H} \longrightarrow ^{1}_{0}\text{n} + ?$
(e) $^{98}_{42}\text{Mo} + ^{1}_{0}\text{n} \longrightarrow ^{99}_{43}\text{Tc} + ?$
(f) $^{18}_{9}\text{F} \longrightarrow ^{18}_{8}\text{O} + ?$

30.14 Complete the following nuclear equations. Write the mass number, atomic number, and symbol for the remaining particle.
(a) $^{9}_{4}\text{Be} + ? \longrightarrow ^{6}_{3}\text{Li} + ^{4}_{2}\text{He}$
(b) $? + ^{1}_{0}\text{n} \longrightarrow ^{24}_{11}\text{Na} + ^{4}_{2}\text{He}$
(c) $^{40}_{20}\text{Ca} + ? \longrightarrow ^{40}_{19}\text{K} + ^{1}_{1}\text{H}$
(d) $^{241}_{95}\text{Am} + ^{4}_{2}\text{He} \longrightarrow ^{243}_{97}\text{Bk} + ?$
(e) $^{246}_{96}\text{Cm} + ^{12}_{6}\text{C} \longrightarrow 4\,^{1}_{0}\text{n} + ?$
(f) $^{238}_{92}\text{U} + ? \longrightarrow ^{249}_{100}\text{Fm} + 5\,^{1}_{0}\text{n}$

30.15 Complete the following nuclear equations. Write the mass number, atomic number, and symbol for the remaining particle.
(a) $^{111}_{47}\text{Ag} \longrightarrow ^{111}_{48}\text{Cd} + ?$
(b) $^{87}_{36}\text{Kr} \longrightarrow ^{0}_{-1}\beta + ?$
(c) $^{231}_{91}\text{Pa} \longrightarrow ^{227}_{89}\text{Ac} + ?$
(d) $^{230}_{90}\text{Th} \longrightarrow ^{4}_{2}\text{He} + ?$
(e) $^{82}_{35}\text{Br} \longrightarrow ^{82}_{36}\text{Kr} + ?$
(f) $? \longrightarrow ^{24}_{12}\text{Mg} + ^{0}_{-1}\beta$

30.16 Complete the following nuclear equations. Write the mass number, atomic number, and symbol for the remaining particle.
(a) $^{19}_{10}\text{Ne} \longrightarrow ^{0}_{+1}\beta + ?$
(b) $^{59}_{26}\text{Fe} \longrightarrow ^{0}_{+1}\beta + ?$
(c) $^{40}_{19}\text{K} \longrightarrow ^{0}_{-1}\beta + ?$
(d) $^{37}_{18}\text{Ar} + ^{0}_{-1}\text{e}$ (electron capture) $\longrightarrow ?$
(e) $^{55}_{26}\text{Fe} + ^{0}_{-1}\text{e}$ (electron capture) $\longrightarrow ?$
(f) $^{26}_{13}\text{Al} \longrightarrow ^{25}_{12}\text{Mg} + ?$

30.17 The uranium-235 radioactive decay series, beginning with $^{235}_{92}\text{U}$ and ending with $^{207}_{82}\text{Pb}$, occurs in the following sequence: α, β, α, β, α, α, α, α, β, β, α. Write an equation for each step in this series.

30.18 The thorium-232 radioactive decay series, beginning with $^{232}_{90}\text{Th}$ and ending with $^{208}_{82}\text{Pb}$, occurs in the following sequence: α, β, α, β, α, α, α, β, β, α. Write an equation for each step in this series.

Section 30.4: Stability of Atomic Nuclei

30.19 What particle is emitted in the following nuclear reactions? Write an equation for each reaction.

(a) Gold-198 decays to mercury-198.
(b) Radon-222 decays to polonium-218.
(c) Cesium-137 decays to barium-137.
(d) Indium-110 decays to cadmium-110.

30.20 What is the product of the following nuclear decay processes? Write an equation for each process.
(a) Gallium-67 decays by electron capture.
(b) Potassium-38 decays with positron emission.
(c) Technetium-99m decays with γ emission.
(d) Manganese-56 decays by β emission.

30.21 Predict the probable mode of decay for each of the following radioactive isotopes, and write an equation to show the products of decay.
(a) bromine-80m   (c) cobalt-61
(b) californium-240 (d) carbon-11

30.22 Predict the probable mode of decay for each of the following radioactive isotopes, and write an equation to show the products of decay.
(a) manganese-54   (c) silver-110
(b) americium-241   (d) mercury-197m

30.23 (a) Which of the following nuclei decay by $_{-1}^{0}\beta$ decay?

$$^{3}\text{H} \quad ^{16}\text{O} \quad ^{20}\text{F} \quad ^{13}\text{N}$$

(b) Which of the following nuclei decays by $_{+1}^{0}\beta$ decay?

$$^{238}\text{U} \quad ^{19}\text{F} \quad ^{22}\text{Na} \quad ^{24}\text{Na}$$

30.24 (a) Which of the following nuclei decay by $_{-1}^{0}\beta$ decay?

$$^{1}\text{H} \quad ^{23}\text{Mg} \quad ^{32}\text{P} \quad ^{20}\text{Ne}$$

(b) Which of the following nuclei decay by $_{+1}^{0}\beta$ decay?

$$^{235}\text{U} \quad ^{35}\text{Cl} \quad ^{38}\text{K} \quad ^{24}\text{Na}$$

30.25 Boron has two stable isotopes, ^{10}B and ^{11}B. Calculate the binding energies per nucleon of these two nuclei. The required masses (in grams per mole) are $_{1}^{1}\text{H} = 1.00783$, $_{0}^{1}\text{n} = 1.00867$, $_{5}^{10}\text{B} = 10.01294$, and $_{5}^{11}\text{B} = 11.00931$.

30.26 Calculate the binding energy (in kJ mol^{-1}) of nucleons of P for the formation of ^{30}P and ^{31}P. The required masses (in g mol^{-1}) are $_{1}^{1}\text{H} = 1.00783$, $_{0}^{1}\text{n} = 1.00867$, $_{15}^{30}\text{P} = 29.97832$, and $_{15}^{31}\text{P} = 30.97376$.

30.27 Calculate the binding energy per nucleon for calcium-40, and compare your result with the value for calcium-40 in Figure 30.6. Masses needed for this calculation are $_{1}^{1}\text{H} = 1.00783$, $_{0}^{1}\text{n} = 1.00867$, and $_{20}^{40}\text{Ca} = 39.96259$.

30.28 Calculate the binding energy per nucleon for iron-56. Masses needed for this calculation are $_{1}^{1}\text{H} = 1.00783$, $_{0}^{1}\text{n} = 1.00867$ and $_{26}^{56}\text{Fe} = 55.9349$. Compare the result of your calculation to the value for iron-56 in the graph in Figure 30.6.

30.29 Calculate the binding energy per mole of nucleons for $_{8}^{16}\text{O}$. Masses needed for this calculations are $_{1}^{1}\text{H} = 1.00783$, $_{0}^{1}\text{n} = 1.00867$, and $_{8}^{16}\text{O} = 15.99492$.

30.30 Calculate the binding energy per nucleon for nitrogen-14. The mass of nitrogen-14 is 14.003074.

Section 30.5: Rates of Nuclear Decay

30.31 Copper acetate containing ^{64}Cu is used to study brain tumours. This isotope has a half-life of 12.7 h. If you begin with 25.0 μg of ^{64}Cu, what mass (in μg) remains after 64 h?

30.32 Gold-198 is used in the diagnosis of liver problems. The half-life of ^{198}Au is 2.69 days. If you begin with 2.8 μg of this gold isotope, what mass remains after 10.8 days?

30.33 Iodine-131 is used to treat thyroid cancer.
(a) The isotope decays by β particle emission. Write a balanced equation for this process.
(b) Iodine-131 has a half-life of 8.04 days. If you begin with 2.4 μg of radioactive ^{131}I, what mass remains after 40.2 days?

30.34 Phosphorus-32 is used in the form of Na_2HPO_4 in the treatment of chronic myeloid leukemia, among other things.
(a) The isotope decays by β particle emission. Write a balanced equation for this process.
(b) The half-life of ^{32}P is 14.3 days. If you begin with 4.8 μg of radioactive ^{32}P in the form of Na_2HPO_4, what mass remains after 28.6 days (about one month)?

30.35 Gallium-67 ($t_{1/2} = 78.25$ h) is used in the medical diagnosis of certain kinds of tumours. If you ingest a compound containing 0.015 mg of this isotope, what mass (in mg) remains in your body after 13 days? (Assume none is excreted.)

30.36 Iodine-131 ($t_{1/2} = 8.04$ days), a β emitter, is used to treat thyroid cancer.
(a) Write an equation for the decomposition of ^{131}I.
(b) If you ingest a sample of NaI containing ^{131}I, how much time is required for the activity to decrease to 35.0% of its original value?

30.37 Radon has been the focus of much attention recently because it is often found in homes. Radon-222 emits α particles and has a half-life of 3.82 days.
(a) Write a balanced equation to show this process.
(b) How long does it take for a sample of ^{222}Rn to decrease to 20.0% of its original activity?

30.38 A sample of wood from a Thracian chariot found in an excavation in Bulgaria has a ^{14}C activity of 0.187 Bq g^{-1}. Estimate the age of the chariot and the year it was made ($t_{1/2}$ for ^{14}C is 5.73×10^3 years, and the activity of ^{14}C in living material is 0.21 Bq g^{-1}).

30.39 A piece of charred bone found in the ruins of a Native American village has a $^{14}C:^{12}C$ ratio that is 72% of the radio found in living organisms. Calculate the age of the bone fragment.

30.40 Strontium-90 is a hazardous radioactive isotope that resulted from atmospheric nuclear testing. A sample of strontium carbonate containing ^{90}Sr is found to have an activity of 16.67 Bq. One year later the activity of this sample is 16.25 Bq.
(a) Calculate the half-life of strontium-90 from this information.
(b) How long will it take for the activity of this sample to drop to 1.0% of the initial value?

30.41 Radioactive cobalt-60 is used extensively in nuclear medicine as a γ-ray source. It is made by a neutron capture reaction from cobalt-59, and it is a β emitter; β emission is accompanied by strong γ radiation. The half-life of cobalt-60 is 5.27 years.
(a) How long will it take for a cobalt-60 source to decrease to one-eighth of its original activity?
(b) What fraction of the activity of a cobalt-60 source remains after 1.0 year?

30.42 Scandium occurs in nature as a single isotope, scandium-45. Neutron irradiation produces scandium-46, a β emitter with a half-life of 83.8 days. If the initial activity is 1.16×10^3 Bq, draw a graph showing Bq as a function of time during a period of one year.

30.43 Sodium-23 (in a sample of NaCl) is subjected to neutron bombardment in a nuclear reactor to produce ^{24}Na. When removed from the reactor, the sample is radioactive, with β activity of 423 Bq. The decrease in radioactivity over time was studied, producing the following data:

Activity (Bq)	Time (h)
423	0
403	1
385	2
333	5
267	10
168	20

(a) Write equations for the neutron capture reaction and for the reaction in which the product of this reaction decays by β emission.
(b) Determine the half-life of sodium-24.

30.44 The isotope of polonium that was most likely isolated by Marie Curie in her pioneering studies is polonium-210. A sample of this element was prepared in a nuclear reaction. Initially its activity (α emission) was 131 Bq. Measuring radioactivity over time produced the following data:

Activity (Bq)	Time (days)
131	0
126	7
122	14
99	56
91	72

Determine the half-life of polonium-210.

Section 30.6: Artificial Nuclear Reactions

30.45 There are two isotopes of americium, both with half-lives sufficiently long to allow the handling of large quantities. Americium-241, with a half-life of 432 years, is an α emitter. It is used in smoke detectors. The isotope is formed from ^{239}Pu by absorption of two neutrons followed by emission of a β particle. Write a balanced equation for this process.

30.46 Americium-240 is made by bombarding plutonium-239 with α particles. In addition to ^{240}Am, the products are a proton and two neutrons. Write a balanced equation for this process.

30.47 To synthesize the heavier transuranium elements, a nucleus must be bombarded with a relatively large particle. If you know the products are californium-246 and four neutrons, with what particle would you bombard uranium-238 atoms?

30.48 Element $^{287}114$ was made by firing a beam of ^{48}Ca ions at ^{242}Pu. Three neutrons were ejected in the reaction. Write a balanced nuclear equation for the synthesis of this super-heavy element.

30.49 Deuterium nuclei (^2_1H) are particularly effective as bombarding particles to carry out nuclear reactions. Complete the following equations:
(a) $^{114}_{48}Cd + {}^2_1H \longrightarrow ? + {}^1_1H$
(b) $^6_3Li + {}^2_1H \longrightarrow ? + {}^1_0n$
(c) $^{40}_{20}Ca + {}^2_1H \longrightarrow {}^{38}_{19}K + ?$
(d) $? + {}^2_1H \longrightarrow {}^{65}_{30}Zn + \gamma$

30.50 Some of the reactions explored by Rutherford and others are listed below. Identify the unknown species in each reaction.
(a) $^{14}_7N + {}^4_2He \longrightarrow {}^{17}_8O + ?$
(b) $^9_4Be + {}^4_2He \longrightarrow ? + {}^1_0n$
(c) $? + {}^4_2He \longrightarrow {}^{30}_{15}P + {}^1_0n$
(d) $^{239}_{94}Pu + {}^4_2He \longrightarrow ? + {}^1_0n$

30.51 Boron is an effective absorber of neutrons. When boron-10 adds a neutron, an α particle is emitted. Write an equation for this nuclear reaction.

Section 30.7: Nuclear Fission

30.52   If a shortage in worldwide supplies of fissionable uranium arose, it would be possible to use other fissionable nuclei. Plutonium, one such fuel, can be made in "breeder" reactors that manufacture more fuel than they consume. The sequence of reactions by which plutonium is made is as follows:

(a) A ^{238}U nucleus undergoes an (n, γ) to produce ^{239}U.

(b) ^{239}U decays by β emission ($t_{1/2} = 23.5$ min) to give an isotope of neptunium.

(c) This neptunium isotope decays by β emission to give a plutonium isotope.

(d) The plutonium isotope is fissionable. On collision of one of these plutonium isotopes with a neutron, fission occurs, with at least two neutrons and two other nuclei as products.

Write an equation for each of the nuclear reactions.

Section 30.8: Nuclear Fusion

30.53   Tritium ($^{3}_{1}H$) is one of the nuclei used in fusion reactions. This isotope is radioactive, with a half-life of 12.3 years. Like carbon-14, tritium is formed in the upper atmosphere from cosmic radiation, and it is found in trace amounts on the earth. To obtain the amounts required for a fusion reaction, however, it must be made via a nuclear reaction. The reaction of $^{6}_{3}Li$ with a neutron produces tritium and an α particle. Write an equation for this nuclear reaction.

SUMMARY AND CONCEPTUAL QUESTIONS

30.54   A graph of binding energy per nucleon is shown in Figure 30.6. Explain how the data used to construct this graph were obtained.

30.55   Explain how carbon-14 is used to estimate the ages of archaeological artefacts. What are the limitations for use of this technique?

30.56   What is a radioactive decay series? Explain why radium and polonium are found in uranium ores.

30.57   The interaction of radiation with matter has both positive and negative consequences. Discuss briefly the hazards of radiation and the way that radiation can be used in medicine.

30.58   A technique to date geological samples uses rubidium-87, a long-lived radioactive isotope of rubidium ($t_{1/2} = 4.8 \times 10^{10}$ years). Rubidium-87 decays by β emission to strontium-87. If the rubidium-87 is part of a rock or mineral, then strontium-87 will remain trapped within the crystalline structure of the rock. The age of the rock dates back to the time when the rock solidified. Chemical analysis of the rock gives the amounts of ^{87}Rb and ^{87}Sr. From these data, the fraction of ^{87}Rb that remains can be calculated.

Analysis of a stony meteorite determined that 3.2 mmol of ^{87}Rb and 0.2 mmol of ^{87}Sr were present. Estimate the age of the meteorite. (*Hint*: The amount of ^{87}Rb at t_0 is moles ^{87}Rb + moles ^{87}Sr.)

30.59   The oldest-known fossil found in South Africa has been dated based on the decay of Rb-87.

$$^{87}Rb \longrightarrow \, ^{87}Sr + \, ^{0}_{-1}\beta \qquad t_{1/2} = 4.8 \times 10^{10} \text{ years}$$

If the ratio of the present quantity of ^{87}Rb to the original quantity is 0.951, calculate the age of the fossil.

30.60   The age of minerals can sometimes be determined by measuring the amounts of ^{206}Pb and ^{238}U in a sample. This determination assumes that all of the ^{206}Pb in the sample comes from the decay of ^{238}U. The date obtained identifies when the rock solidified. Assume that the ratio of ^{206}Pb to ^{238}U in an igneous rock sample is 0.33. Calculate the age of the rock. ($t_{1/2}$ for ^{238}U is 4.5×10^9 years.)

30.61   In June 1972, natural fission reactors, which operated billions of years ago, were discovered in Oklo, Gabon. At present, natural uranium contains 0.72% ^{235}U. How many years ago did natural uranium contain 3.0% ^{235}U, the amount needed to sustain a natural reactor? ($t_{1/2}$ for ^{235}U is 7.04×10^8 years.)

30.62   When a neutron is captured by an atomic nucleus, energy is released as γ radiation. This energy can be calculated based on the change in mass in converting reactants to products. For the nuclear reaction, $^{6}_{3}Li + \, ^{1}_{0}n \longrightarrow \, ^{7}_{3}Li + \gamma$.

(a) Calculate the energy evolved in this reaction (per atom). Masses needed for this calculation are $^{6}_{3}Li$ = 6.01512, $^{1}_{0}n$ = 1.00867, and $^{6}_{3}Li$ = 7.01600.

(b) Use the answer in part (a) to calculate the wavelength of the γ-rays emitted in the reaction.

30.63   The average energy output of a good grade of coal is 2.9×10^7 kJ tonne^{-1}. Fission of 1 mol of ^{235}U releases 2.1×10^{10} kJ. Find the number of tonnes of coal needed to produce the same energy as 1 kg of ^{235}U.

30.64 Collision of an electron and a positron results in formation of two γ-rays. In the process, their masses are converted completely into energy.

(a) Calculate the energy evolved (in kJ mol^{-1}) from the annihilation of an electron and a positron.

(b) Using Planck's equation [<<Section 8.3], determine the frequency of the γ-rays emitted in this process.

30.65 To measure the volume of the blood system of an animal, the following experiment was done. A 1.0 mL sample of an aqueous solution containing tritium, with an activity of 2.0×10^6 Bq, was injected into the animal's bloodstream. After time was allowed for complete circulatory mixing, a 1.0 mL blood sample was withdrawn and found to have an activity of 1.5×10^4 Bq. What was the volume of the circulatory system? (The half-life of tritium is 12.3 years, so this experiment assumes that only a negligible amount of tritium has decayed in the time of the experiment.)

30.66 The principle underlying the isotope dilution method can be applied to many kinds of problems. Suppose that you, a marine biologist, want to estimate the number of fish in a lake. You release 1000 tagged fish, and after allowing an adequate amount of time for the fish to disperse evenly in the lake, you catch 5250 fish and find that 27 of them have tags. How many fish are in the lake?

30.67 Radioactive isotopes are often used as "tracers" to follow an atom through a chemical reaction. The following is an example of this process: acetic acid reacts with methanol (CH_3OH) by eliminating a molecule of H_2O to form methyl acetate ($CH_3CO_2CH_3$). Explain how you would use the radioactive isotope ^{15}O to show whether the oxygen atom in the water product comes from the —OH of the acid or the —OH of the alcohol molecule.

30.68 Radioactive decay series begin with a very long-lived isotope. For example, the half-life of ^{238}U is 4.5×10^9 years. Each series is identified by the name of the long-lived parent isotope of highest mass.

(a) The uranium-238 radioactive decay series is sometimes referred to as the $4n + 2$ series because the masses of all 13 members of this series can be expressed by the equation $m = 4n + 2$, where m is the mass number and n is an integer. Explain why the masses are correlated in this way.

(b) Two other radioactive decay series identified in minerals in the earth's crust are the thorium-232 series and the uranium-235 series. Do the masses of the isotopes in these series conform to a simple mathematical equation? If so, identify the equation.

(c) Identify the radioactive decay series to which each of the following isotopes belongs: $^{226}_{88}Ra$, $^{215}_{86}At$, $^{228}_{90}Th$, $^{210}_{83}Bi$.

(d) Evaluation reveals that one series of elements, the $4n + 1$ series, is not present in the earth's crust. Speculate why.

30.69 You might wonder how it is possible to determine the half-life of long-lived radioactive isotopes such as ^{238}U. With a half-life of more than 10^9 years, the radioactivity of a sample of uranium will not measurably change in your lifetime. You can calculate the half-life using the mathematics governing first-order reactions. It can be shown that a 1.0-mg sample of ^{238}U decays at the rate of 12 α emissions per second. Set up a mathematical equation for the rate of decay, $\Delta N / \Delta t = -kN$, where N is the number of nuclei in the 1.0 mg sample and $\Delta N / \Delta t$ is 12 Bq. Solve this equation for the rate constant for this process, and then relate the rate constant to the half-life of the reaction. Carry out this calculation, and compare your result with the literature value, 4.5×10^9 years.

30.70 The last unknown element between bismuth and uranium was discovered by Lise Meitner (1878–1968) and Otto Hahn (1879–1968) in 1918. They obtained ^{231}Pa by chemical extraction of pitchblende, in which its concentration is about 1 ppm. This isotope, an α emitter, has a half-life of 3.27×10^4 years.

(a) Which radioactive decay series (the uranium-235, uranium-238, or thorium-232 series) contains ^{231}Pa as a member?

(b) Suggest a possible sequence of nuclear reactions starting with the long-lived isotope that eventually forms this isotope.

(c) What quantity of ore would be required to isolate 1.0 g of ^{231}Pa, assuming 100% yield?

(d) Write an equation for the radioactive decay process for ^{231}Pa.

With the **eCHACR** single-sign-on access card bundled with your text, log on (http://login.cengage.com/sso) and access the e-book and click on any in-margin icons for dynamic molecular-level animations and simulations, videos of laboratory reactions, interactive exercises, reviews of background concepts, and other online supplementary materials as noted by the icons in the text margins.

Also go to www.chemistry.nelson.com <http://www.chemistry.nelson.com> for Answers to in-chapter exercises and selected Review Questions, Test Yourself questions, weblinks, crossword puzzles, flashcards, glossary of key terms, and other student resources.

A P P E N D I X   A

pK$_a$ Values for Acids in Aqueous Solution at 25 °C*

Molecular Structure	Name, Formula	pK$_a$
HI	Hydroiodic acid	−10
	Perchloric acid (HClO$_4$)	−10
HBr	Hydrobromic acid	−9
HCl	Hydrochloric acid	−7
	Sulfuric acid (H$_2$SO$_4$)	−3, 1.92
	Nitric acid (HNO$_3$)	−1.3
CH$_3$SO$_3$H	Methanesulfonic acid	−1.8
CH(NO$_2$)$_3$	Trinitromethane	0.1
	Picric acid 2,4,6-Trinitrophenol	0.3
CCl$_3$COOH	Trichloroacetic acid	0.5
CF$_3$COOH	Trifluoroacetic acid	0.5
CBr$_3$COOH	Tribromoacetic acid	0.7
HOOCCOOH	Oxalic acid	1.23, 4.19
CHCl$_2$COOH	Dichloroacetic acid	1.3
	Phosphoric acid (H$_3$PO$_4$)	2.12, 7.21, 12.44
	Citric acid	2.13, 4.77, 6.40
	Arsenic acid (H$_3$AsO$_4$)	2.25, 6.77, 11.60
	2-Nitrobenzoic acid	2.4

Molecular Structure	Name, Formula	pK$_a$
CH$_2$FCOOH	Monofluoroacetic acid	2.7
CH$_2$ClCOOH	Monochloroacetic acid	2.8
HOOCCH$_2$COOH	Malonic acid Propanedioic acid	2.8, 5.6
CH$_2$BrCOOH	Monobromoacetic acid	2.9
	2-Chlorobenzoic acid	3.0
	Salicylic acid 2-Hydroxybenzoic acid	3.0
HF	Hydrofluoric acid	3.2
	Nitrous acid (HNO$_2$)	3.2
	4-Nitrobenzoic acid	3.4
CH$_2$(NO$_2$)$_2$	Dinitromethane	3.6
CH$_3$COCH$_2$COOH	Acetoacetic acid	3.6
HCOOH	Formic acid Methanoic acid	3.74
	4-Chlorobenzoic acid	4.0
CH$_2$BrCH$_2$COOH	3-Bromopropanoic acid	4.0
	Dinitrophenol	4.1
	Benzoic acid (C$_6$H$_5$COOH)	4.2
HOOCCH$_2$CH$_2$COOH	Succinic acid Butanedioic acid	4.2, 5.7
HOOCCH$_2$CH$_2$CH$_2$COOH	Glutaric acid Pentanedioic acid	4.3, 5.4
	Pentachlorophenol	4.5

(continued)

Molecular Structure	Name, Formula	pK_a
—NH$_3$$^+$	Anilinium ion ($C_6H_5NH_3^+$)	4.60
CH$_3$COOH	Acetic acid Ethanoic acid	4.74
CH$_3$COCH$_2$NH$_2$	Aminoacetone	5.1
NH$^+$	Pyridinium ion	5.19
	Cyclohexane-1,3-dione	5.3
O$_2$NCH$_2$COOCH$_3$	3-Nitromethyl acetate	5.8
	Carbonic acid (H$_2$CO$_3$)	6.38, 10.32
—SH	Benzenethiol	6.6
	Hydrogen sulfide (H$_2$S)	7.00, 17
	2-Nitrophenol	7.2
HOCl	Hypochlorous acid	7.46
	1-Chloro-2-nitrobenzene	8.5
HOBr	Hypobromous acid	8.60
F$_3$C— —OH	4-(Trifluoromethyl)phenol	8.7
CH$_3$COCH$_2$COCH$_3$	Acetylacetone Pentane-2,4-dione	9.0
	Boric acid (H$_3$BO$_3$)	9.14
NH$_4$$^+$	Ammonium ion	9.26
HCN	Hydrocyanic acid Hydrogen cyanide	9.21
—CH$_2$SH	Phenylmethanethiol	9.4
HO— —OH	Hydroxyquinone	9.9, 11.5
—OH	Phenol (C$_6$H$_5$OH)	9.89

Molecular Structure	Name, Formula	pK_a
CH$_3$NO$_2$	Nitromethane	10.3
CH$_3$SH	Methanethiol	10.3
CH$_3$COCH$_2$COOCH$_3$	Ethylacetoacetate	10.6
CH$_3$CH$_2$NH$_2$$^+$	Ethylaminium ion	10.63
CH$_3$NH$_3$$^+$	Methylaminium ion	10.70
HOOH	Hydrogen peroxide	11.62
CCl$_3$CH$_2$OH	2,2,2-Trichloroethanol	12.2
C$_6$H$_{12}$O$_6$	Glucose	12.3
CH$_2$(COOCH$_3$)$_2$	Dimethyl malonate	12.9
CH$_2$ClCH$_2$OH	2-Chlorethanol	14.3
	Cyclopentadiene	15.0
—CH$_2$OH	Phenylmethanol	15.4
CH$_3$OH	Methanol	15.5
CH$_3$CH$_2$OH	Ethanol	16.0
CH$_3$COCH$_2$Br	Bromoacetone	16.1
=O	Cyclohexanone	16.7
CH$_3$CHO	Acetaldehyde Ethanal	17
(CH$_3$)$_2$CHOH	iso-Propyl alcohol Propan-2-ol	17.1
(CH$_3$)$_3$COH	tert-Butyl alcohol	18.0
CH$_3$COCH$_3$	Acetone Propan-2-one	19.3
CH$_3$COOCH$_2$CH$_3$	Ethyl acetate	25
HC≡CH	Acetylene Ethyne	25
CH$_3$CN	Acetonitrile	25
CH$_3$SO$_2$CH$_3$	Methylsulfonylmethane	28
CH$_3$SOCH$_3$	Dimethyl sulfoxide	35
NH$_3$	Ammonia	36
CH$_3$CH$_2$NH$_2$	Ethylamine	36
—CH$_3$	Toluene (C$_6$H$_5$CH$_3$)	41
	Benzene (C$_6$H$_6$)	43
H$_2$C=CH$_2$	Ethylene Ethene	44
CH$_4$	Methane	~ 60

* pK_b values of the conjugate bases can be calculated using $pK_a + pK_b = 14$.

APPENDIX B

Solubility Products of Slightly Soluble Salts in Aqueous Solution at 25 °C

Cation	Compound	K_{sp}	Cation	Compound	K_{sp}
Ba^{2+}	*$BaCrO_4$	1.2×10^{-10}	Mg^{2+}	$MgCO_3$	6.8×10^{-6}
	$BaCO_3$	2.6×10^{-9}		MgF_2	5.2×10^{-11}
	BaF_2	1.8×10^{-7}		$Mg(OH)_2$	5.6×10^{-12}
	*$BaSO_4$	1.1×10^{-10}	Mn^{2+}	$MnCO_3$	2.3×10^{-11}
Ca^{2+}	$CaCO_3$ (calcite)	3.4×10^{-9}		*$Mn(OH)_2$	1.9×10^{-13}
	*CaF_2	5.3×10^{-11}	Hg_2^{2+}	*Hg_2Br_2	6.4×10^{-23}
	*$Ca(OH)_2$	5.5×10^{-5}		Hg_2Cl_2	1.4×10^{-18}
	$CaSO_4$	4.9×10^{-5}		*Hg_2I_2	2.9×10^{-29}
$Cu^{+,2+}$	$CuBr$	6.3×10^{-9}		Hg_2SO_4	6.5×10^{-7}
	CuI	1.3×10^{-12}	Ni^{2+}	$NiCO_3$	1.4×10^{-7}
	$Cu(OH)_2$	2.2×10^{-20}		$Ni(OH)_2$	5.5×10^{-16}
	$CuSCN$	1.8×10^{-13}	Ag^+	*$AgBr$	5.4×10^{-13}
Au^+	$AuCl$	2.0×10^{-13}		*$AgBrO_3$	5.4×10^{-5}
Fe^{2+}	$FeCO_3$	3.1×10^{-11}		$AgCH_3CO_2$	1.9×10^{-3}
	$Fe(OH)_2$	4.9×10^{-17}		$AgCN$	6.0×10^{-17}
Pb^{2+}	$PbBr_2$	6.6×10^{-6}		Ag_2CO_3	8.5×10^{-12}
	$PbCO_3$	7.4×10^{-14}		*$Ag_2C_2O_4$	5.4×10^{-12}
	$PbCl_2$	1.7×10^{-5}		*$AgCl$	1.8×10^{-10}
	$PbCrO_4$	2.8×10^{-13}		Ag_2CrO_4	1.1×10^{-12}
	PbF_2	3.3×10^{-8}		*AgI	8.5×10^{-17}
	PbI_2	9.8×10^{-9}		$AgSCN$	1.0×10^{-12}
	$Pb(OH)_2$	1.4×10^{-15}		*Ag_2SO_4	1.2×10^{-5}
	$PbSO_4$	2.5×10^{-8}			

(continued)

Solubility products at 25 °C (*continued*)

Cation	Compound	K_{sp}	Cation	Compound	K_{sp}
Sr^{2+}	$SrCO_3$	5.6×10^{-10}	Zn^{2+}	$Zn(OH)_2$	3×10^{-17}
	SrF_2	4.3×10^{-9}		$Zn(CN)_2$	8.0×10^{-12}
	$SrSO_4$	3.4×10^{-7}			
Tl^+	$TlBr$	3.7×10^{-6}			
	$TlCl$	1.9×10^{-4}			
	TlI	5.5×10^{-8}			

The values reported in this table were taken from J. A. Dean: *Lange's Handbook of Chemistry*, 15th Edition. New York, McGraw Hill Publishers, 1999. Values have been rounded off to two significant figures.

* Calculated solubility from these K_{sp} values will match experimental solubility for this compound within a factor of 2. Experimental values for solubilities are given in R. W. Clark and J. M. Bonicamp: *Journal of Chemical Education*, Vol. 75, p. 1182, 1998.

Equilibrium Constants* for Dissolution of Some Metal Sulfides at 25 °C

Substance	K
HgS (red)	4×10^{-54}
HgS (black)	2×10^{-53}
Ag_2S	6×10^{-51}
CuS	6×10^{-37}
PbS	3×10^{-28}
CdS	8×10^{-28}
SnS	1×10^{-26}
FeS	6×10^{-19}

* The equilibrium constant K for metal sulfides refers to the equilibrium
$MS(s) + H_2O(\ell) \rightleftharpoons M^{2+}(aq) + OH^-(aq) + HS^-(aq)$;
see R. J. Myers, *Journal of Chemical Education*, Vol. 63, p. 687, 1986.

Selected Thermodynamic Data* at 25 °C

Species	$\Delta_f H°$ (kJ mol^{-1})	$S°$ (J K^{-1} mol^{-1})	$\Delta_f G°$ (kJ mol^{-1})
Aluminum			
$Al(s)$	0	28.3	0
$AlCl_3(s)$	−705.63	109.29	−630.0
$Al_2O_3(s)$	−1675.7	50.92	−1582.3
Barium			
$BaCl_2(s)$	−858.6	123.68	−810.4
$BaCO_3(s)$	−1213	112.1	−1134.41
$BaO(s)$	−548.1	72.05	−520.38
$BaSO_4(s)$	−1473.2	132.2	−1362.2
Beryllium			
$Be(s)$	0	9.5	0
$Be(OH)_2(s)$	−902.5	51.9	−815.0
Boron			
$BCl_3(g)$	−402.96	290.17	−387.95
Bromine			
$Br(g)$	111.884	175.022	82.396
$Br_2(\ell)$	0	152.2	0
$Br_2(g)$	30.91	245.47	3.12
$BrF_3(g)$	−255.60	292.53	−229.43
$HBr(g)$	−36.29	198.70	−53.45
Calcium			
$Ca(s)$	0	41.59	0
$Ca(g)$	178.2	158.884	144.3
$Ca^{2+}(g)$	1925.90	—	—
$CaC_2(s)$	−59.8	70	−64.93
$CaCO_3(s, \text{calcite})$	−1207.6	91.7	−1129.16
$CaCl_2(s)$	−795.8	104.6	−748.1
$CaF_2(s)$	−1219.6	68.87	−1167.3
$CaH_2(s)$	−186.2	42	−147.2
$CaO(s)$	−635.09	38.2	−603.42
$CaS(s)$	−482.4	56.5	−477.4
$Ca(OH)_2(s)$	−986.09	83.39	−898.43
$Ca(OH)_2(aq)$	−1002.82		−868.07
$CaSO_4(s)$	−1434.52	106.5	−1322.02

(continued)

Selected Thermodynamic Data* at 25 °C *(continued)*

Species	$\Delta_f H°$ (kJ mol^{-1})	$S°$ (J K^{-1} mol^{-1})	$\Delta_f G°$ (kJ mol^{-1})
Carbon			
C(s, graphite)	0	5.6	0
C(s, diamond)	1.8	2.377	2.900
C(g)	716.67	158.1	671.2
CCl$_4$(ℓ)	−128.4	214.39	−57.63
CCl$_4$(g)	−95.98	309.65	−53.61
CHCl$_3$(ℓ)	−134.47	201.7	−73.66
CHCl$_3$(g)	−103.18	295.61	−70.4
CH$_4$(g, methane)	−74.87	186.26	−50.8
C$_2$H$_2$(g, ethyne)	226.73	200.94	209.20
C$_2$H$_4$(g, ethene)	52.47	219.36	68.35
C$_2$H$_6$(g, ethane)	−83.85	229.2	−31.89
C$_3$H$_8$(g, propane)	−104.7	270.3	−24.4
C$_6$H$_6$(ℓ, benzene)	48.95	173.26	124.21
CH$_3$OH(ℓ, methanol)	−238.4	127.19	−166.14
CH$_3$OH(g, methanol)	−201.0	239.7	−162.5
C$_2$H$_5$OH(ℓ, ethanol)	−277.0	160.7	−174.7
C$_2$H$_5$OH(g, ethanol)	−235.3	282.70	−168.49
CO(g)	−110.525	197.674	−137.168
CO$_2$(g)	−393.509	213.74	−394.359
CS$_2$(ℓ)	89.41	151	65.2
CS$_2$(g)	116.7	237.8	66.61
COCl$_2$(g)	−218.8	283.53	−204.6
Cesium			
Cs(s)	0	85.23	0
Cs$^+$(g)	457.964	—	—
CsCl(s)	−443.04	101.17	−414.53
Chlorine			
Cl(g)	121.3	165.19	105.3
Cl$^-$(g)	−233.13	—	—
Cl$_2$(g)	0	223.08	0
HCl(g)	−92.31	186.2	−95.09
HCl(aq)	−167.159	56.5	−131.26
Chromium			
Cr(s)	0	23.62	0
Cr$_2$O$_3$(s)	−1134.7	80.65	−1052.95
CrCl$_3$(s)	−556.5	123.0	−486.1
Copper			
Cu(s)	0	33.17	0
CuO(s)	−156.06	42.59	−128.3
CuCl$_2$(s)	−220.1	108.07	−175.7
CuSO$_4$(s)	−769.98	109.05	−660.75

(continued)

Selected Thermodynamic Data* at 25 °C *(continued)*

Species	$\Delta_f H°$ (kJ mol^{-1})	$S°$ (J K^{-1} mol^{-1})	$\Delta_f G°$ (kJ mol^{-1})
Fluorine			
$F_2(g)$	0	202.8	0
$F(g)$	78.99	158.754	61.91
$F^-(g)$	−255.39	—	—
$F^-(aq)$	−332.63		−278.79
$HF(g)$	−273.3	173.779	−273.2
$HF(aq)$	−332.63	88.7	−278.79
Hydrogen			
$H_2(g)$	0	130.7	0
$H(g)$	217.965	114.713	203.247
$H^+(g)$	1536.202	—	—
$H_2O(\ell)$	−285.83	69.95	−237.15
$H_2O(g)$	−241.83	188.84	−228.59
$H_2O_2(\ell)$	−187.78	109.6	−120.35
Iodine			
$I_2(s)$	0	116.135	0
$I_2(g)$	62.438	260.69	19.327
$I(g)$	106.838	180.791	70.250
$I^-(g)$	−197	—	—
$ICl(g)$	17.51	247.56	−5.73
Iron			
$Fe(s)$	0	27.78	0
$FeO(s)$	−272	—	—
$Fe_2O_3(s, \text{hematite})$	−825.5	87.40	−742.2
$Fe_3O_4(s, \text{magnetite})$	−1118.4	146.4	−1015.4
$FeCl_2(s)$	−341.79	117.95	−302.30
$FeCl_3(s)$	−399.49	142.3	−344.00
$FeS_2(s, \text{pyrite})$	−178.2	52.93	−166.9
$Fe(CO)_5(\ell)$	−774.0	338.1	−705.3
Lead			
$Pb(s)$	0	64.81	0
$PbCl_2(s)$	−359.41	136.0	−314.10
$PbO(s, \text{yellow})$	−219	66.5	−196
$PbO_2(s)$	−277.4	68.6	−217.39
$PbS(s)$	−100.4	91.2	−98.7
Lithium			
$Li(s)$	0	29.12	0
$Li^+(g)$	685.783	—	—
$LiOH(s)$	−484.93	42.81	−438.96
$LiOH(aq)$	−508.48	2.80	−450.58
$LiCl(s)$	−408.701	59.33	−384.37
Magnesium			
$Mg(s)$	0	32.67	0

(continued)

Selected Thermodynamic Data* at 25 °C *(continued)*

Species	$\Delta_f H°$ (kJ mol^{-1})	$S°$ (J K^{-1} mol^{-1})	$\Delta_f G°$ (kJ mol^{-1})
$MgCl_2(s)$	−641.62	89.62	−592.09
$MgCO_3(s)$	−1111.69	65.84	−1028.2
$MgO(s)$	−601.24	26.85	−568.93
$Mg(OH)_2(s)$	−924.54	63.18	−833.51
$MgS(s)$	−346.0	50.33	−341.8
Mercury			
$Hg(\ell)$	0	76.02	0
$HgCl_2(s)$	−224.3	146.0	−178.6
$HgO(s, red)$	−90.83	70.29	−58.539
$HgS(s, red)$	−58.2	82.4	−50.6
Nickel			
$Ni(s)$	0	29.87	0
$NiO(s)$	−239.7	37.99	−211.7
$NiCl_2(s)$	−305.332	97.65	−259.032
Nitrogen			
$N_2(g)$	0	191.56	0
$N(g)$	472.704	153.298	455.563
$NH_3(g)$	−45.90	192.77	−16.37
$N_2H_4(\ell)$	50.63	121.52	149.45
$NH_4Cl(s)$	−314.55	94.85	−203.08
$NH_4Cl(aq)$	−299.66	169.9	−210.57
$NH_4NO_3(s)$	−365.56	151.08	−183.84
$NH_4NO_3(aq)$	−339.87	259.8	−190.57
$NO(g)$	90.29	210.76	86.58
$NO_2(g)$	33.1	240.04	51.23
$N_2O(g)$	82.05	219.85	104.20
$N_2O_4(g)$	9.08	304.38	97.73
$NOCl(g)$	51.71	261.8	66.08
$HNO_3(\ell)$	−174.10	155.60	−80.71
$HNO_3(g)$	−135.06	266.38	−74.72
$HNO_3(aq)$	−207.36	146.4	−111.25
Oxygen			
$O_2(g)$	0	205.07	0
$O(g)$	249.170	161.055	231.731
$O_3(g)$	142.67	238.92	163.2
Phosphorus			
$P_4(s, white)$	0	41.1	0
$P_4(s, red)$	−17.6	22.80	−12.1
$P(g)$	314.64	163.193	278.25
$PH_3(g)$	22.89	210.24	30.91
$PCl_3(g)$	−287.0	311.78	−267.8
$P_4O_{10}(s)$	−2984.0	228.86	−2697.7
$H_3PO_4(\ell)$	−1279.0	110.5	−1119.1

(continued)

Selected Thermodynamic Data* at 25 °C *(continued)*

Species	$\Delta_f H°$ (kJ mol^{-1})	$S°$ (J K^{-1} mol^{-1})	$\Delta_f G°$ (kJ mol^{-1})
Potassium			
K(s)	0	64.63	0
KCl(s)	−436.68	82.56	−408.77
KClO$_3$(s)	−397.73	143.1	−296.25
KI(s)	−327.90	106.32	−324.892
KOH(s)	−424.72	78.9	−378.92
KOH(aq)	−482.37	91.6	−440.50
Silicon			
Si(s)	0	18.82	0
SiBr$_4$(ℓ)	−457.3	277.8	−443.9
SiC(s)	−65.3	16.61	−62.8
SiCl$_4$(g)	−662.75	330.86	−622.76
SiH$_4$(g)	34.31	204.65	56.84
SiF$_4$(g)	−1614.94	282.49	−1572.65
SiO$_2$(s, quartz)	−910.86	41.46	−856.97
Silver			
Ag(s)	0	42.55	0
Ag$_2$O(s)	−31.1	121.3	−11.32
AgCl(s)	−127.01	96.25	−109.76
AgNO$_3$(s)	−124.39	140.92	−33.41
Sodium			
Na(s)	0	51.21	0
Na(g)	107.3	153.765	76.83
Na$^+$(g)	609.358	—	—
NaBr(s)	−361.02	86.82	−348.983
NaCl(s)	−411.12	72.11	−384.04
NaCl(g)	−181.42	229.79	−201.33
NaCl(aq)	−407.27	115.5	−393.133
NaOH(s)	−425.93	64.46	−379.75
NaOH(aq)	−469.15	48.1	−418.09
Na$_2$CO$_3$(s)	−1130.77	134.79	−1048.08
Sulfur			
S(s, rhombic)	0	32.1	0
S(g)	278.98	167.83	236.51
S$_2$Cl$_2$(g)	−18.4	331.5	−31.8
SF$_6$(g)	−1209	291.82	−1105.3
H$_2$S(g)	−20.63	205.79	−33.56
SO$_2$(g)	−296.84	248.21	−300.13
SO$_3$(g)	−395.77	256.77	−371.04
SOCl$_2$(g)	−212.5	309.77	−198.3
H$_2$SO$_4$(ℓ)	−814	156.9	−689.96
H$_2$SO$_4$(aq)	−909.27	20.1	−744.53

(continued)

Selected Thermodynamic Data* at 25 °C *(continued)*

Species	$\Delta_f H °$ (kJ mol^{-1})	$S°$ (J K^{-1} mol^{-1})	$\Delta_f G°$ (kJ mol^{-1})
Tin			
Sn(s, white)	0	51.08	0
Sn(s, gray)	−2.09	44.14	0.13
SnCl$_4$(ℓ)	−511.3	258.6	−440.15
SnCl$_4$(g)	−471.5	365.8	−432.31
SnO$_2$(s)	−577.63	49.04	−515.88
Titanium			
Ti(s)	0	30.72	0
TiCl$_4$(ℓ)	−804.2	252.34	−737.2
TiCl$_4$(g)	−763.16	354.84	−726.7
TiO$_2$(s)	−939.7	49.92	−884.5
Zinc			
Zn(s)	0	41.63	0
ZnCl$_2$(s)	−415.05	111.46	−369.398
ZnO(s)	−348.28	43.64	−318.30
ZnS(s, sphalerite)	−205.98	57.7	−201.29

* Most thermodynamic data are taken from the NIST Webbook at **http://webbook.nist.gov.**

Formation Constants of Complex Ions in Aqueous Solution at 25 °C

(See Section 27.5.)

Formation Equilibrium Reaction*	β
$[Ag(OH_2)_2]^+ + 2\ Br^-(aq) \rightleftharpoons [AgBr_2]^- + 2\ H_2O(\ell)$	1.3×10^7
$[Ag(OH_2)_2]^+ + 2\ Cl^-(aq) \rightleftharpoons [AgCl_2]^- + 2\ H_2O(\ell)$	1.1×10^5
$[Ag(OH_2)_2]^+ + 2\ CN^-(aq) \rightleftharpoons [Ag(CN)_2]^- + 2\ H_2O(\ell)$	1.0×10^{22}
$[Ag(OH_2)_2]^+ + 2\ S_2O_3{}^{2-}(aq) \rightleftharpoons [Ag(S_2O_3)_2]^{3-} + 2\ H_2O(\ell)$	2.0×10^{13}
$[Ag(OH_2)_2]^+ + 2\ NH_3(aq) \rightleftharpoons [Ag(NH_3)_2]^+ + 2\ H_2O(\ell)$	1.6×10^7
$[Al(OH_2)_6]^+ + 6\ F^-(aq) \rightleftharpoons [AlF_6]^{3-} + 6\ H_2O(\ell)$	5.0×10^{23}
$[Al(OH_2)_6]^+ + 4\ OH^-(aq) \rightleftharpoons [Al(OH)_4(OH_2)_2]^- + 4\ H_2O(\ell)$	7.7×10^{33}
$[Au(OH_2)_2]^+ + 2\ CN^-(aq) \rightleftharpoons [Au(CN)_2]^- + 2\ H_2O(\ell)$	2.0×10^{38}
$[Cd(OH_2)_4]^{2+} + 4\ CN^-(aq) \rightleftharpoons [Cd(CN)_4]^{2-} + 4\ H_2O(\ell)$	1.2×10^{17}
$[Cd(OH_2)_4]^{2+} + 4\ Cl^-(aq) \rightleftharpoons [CdCl_4]^{2-} + 4\ H_2O(\ell)$	1.0×10^2
$[Cd(OH_2)_4]^{2+} + 4\ NH_3(aq) \rightleftharpoons [Cd(NH_3)_4]^{2+} + 4\ H_2O(\ell)$	1.0×10^7
$[Co(OH_2)_6]^{2+} + 6\ NH_3(aq) \rightleftharpoons [Co(NH_3)_6]^{2+} + 6\ H_2O(\ell)$	7.7×10^4
$[Cu(OH_2)_4]^{2+} + 4\ NH_3(aq) \rightleftharpoons [Cu(NH_3)_4]^{2+} + 4\ H_2O(\ell)$	1×10^{13}
$[Fe(OH_2)_6]^{2+} + 6\ CN^-(aq) \rightleftharpoons [Fe(CN)_6]^{4-} + 6\ H_2O(\ell)$	7.7×10^{36}
$[Hg(OH_2)_4]^{2+} + 4\ Cl^-(aq) \rightleftharpoons [HgCl_4]^{2-} + 4\ H_2O(\ell)$	1.2×10^{15}
$[Ni(OH_2)_6]^{2+} + 4\ CN^-(aq) \rightleftharpoons [Ni(CN)_4]^{2-} + 6\ H_2O(\ell)$	1.0×10^{31}
$[Ni(OH_2)_6]^{2+} + 6\ NH_3(aq) \rightleftharpoons [Ni(NH_3)_6]^{2+} + 6\ H_2O(\ell)$	1×10^9
$[Zn(OH_2)_4]^{2+} + 4\ OH^-(aq) \rightleftharpoons [Zn(OH)_4]^{2-} + 4\ H_2O(\ell)$	2.0×10^{20}
$[Zn(OH_2)_4]^{2+} + 4\ NH_3(aq) \rightleftharpoons [Zn(NH_3)_4]^{2+} + 4\ H_2O(\ell)$	2.9×10^9

* There is uncertainty about the number of water molecules in the aqua complexes of the metal ions. This does not affect the value of the formation constants.

Standard Reduction Potentials in Aqueous Solution at 25 °C

Acidic Solution	Standard Reduction Potential, $E°$ (volts)
$F_2(g) + 2 e^- \longrightarrow 2 F^-(aq)$	2.87
$Co^{3+}(aq) + e^- \longrightarrow Co^{2+}(aq)$	1.82
$Pb^{4+}(aq) + 2 e^- \longrightarrow Pb^{2+}(aq)$	1.8
$H_2O_2(aq) + 2 H^+(aq) + 2 e^- \longrightarrow 2 H_2O$	1.77
$NiO_2(s) + 4 H^+(aq) + 2 e^- \longrightarrow Ni^{2+}(aq) + 2 H_2O$	1.7
$PbO_2(s) + SO_4{}^{2-}(aq) + 4 H^+(aq) + 2 e^- \longrightarrow PbSO_4(s) + 2 H_2O$	1.685
$Au^+(aq) + e^- \longrightarrow Au(s)$	1.68
$2 HClO(aq) + 2 H^+(aq) + 2 e^- \longrightarrow Cl_2(g) + 2 H_2O$	1.63
$Ce^{4+}(aq) + e^- \longrightarrow Ce^{3+}(aq)$	1.61
$NaBiO_3(s) + 6 H^+(aq) + 2 e^- \longrightarrow Bi^{3+}(aq) + Na^+(aq) + 3 H_2O$	≈ 1.6
$MnO_4{}^-(aq) + 8 H^+(aq) + 5 e^- \longrightarrow Mn^{2+}(aq) + 4 H_2O$	1.51
$Au^{3+}(aq) + 3 e^- \longrightarrow Au(s)$	1.50
$ClO_3{}^-(aq) + 6 H^+(aq) + 5 e^- \longrightarrow \frac{1}{2} Cl_2(g) + 3 H_2O$	1.47
$BrO_3{}^-(aq) + 6 H^+(aq) + 6 e^- \longrightarrow Br^-(aq) + 3 H_2O$	1.44
$Cl_2(g) + 2 e^- \longrightarrow 2 Cl^-(aq)$	1.36
$Cr_2O_7{}^{2-}(aq) + 14 H^+(aq) + 6 e^- \longrightarrow 2 Cr^{3+}(aq) + 7 H_2O$	1.33
$N_2H_5{}^+(aq) + 3 H^+(aq) + 2 e^- \longrightarrow 2 NH_4{}^+(aq)$	1.24
$MnO_2(s) + 4 H^+(aq) + 2 e^- \longrightarrow Mn^{2+}(aq) + 2 H_2O$	1.23
$O_2(g) + 4 H^+(aq) + 4 e^- \longrightarrow 2 H_2O$	1.229
$Pt^{2+}(aq) + 2 e^- \longrightarrow Pt(s)$	1.2
$IO_3{}^-(aq) + 6 H^+(aq) + 5 e^- \longrightarrow \frac{1}{2} I_2(aq) + 3 H_2O$	1.195
$ClO_4{}^-(aq) + 2 H^+(aq) + 2 e^- \longrightarrow ClO_3{}^-(aq) + H_2O$	1.19
$Br_2(\ell) + 2 e^- \longrightarrow 2 Br^-(aq)$	1.08
$AuCl_4{}^-(aq) + 3 e^- \longrightarrow Au(s) + 4 Cl^-(aq)$	1.00
$Pd^{2+}(aq) + 2 e^- \longrightarrow Pd(s)$	0.987
$NO_3{}^-(aq) + 4 H^+(aq) + 3 e^- \longrightarrow NO(g) + 2 H_2O$	0.96
$NO_3{}^-(aq) + 3 H^+(aq) + 2 e^- \longrightarrow HNO_2(aq) + H_2O$	0.94
$2 Hg^+(aq) + 2 e^- \longrightarrow Hg_2{}^{2+}(aq)$	0.920
$Hg^{2+}(aq) + 2 e^- \longrightarrow Hg(\ell)$	0.855
$Ag^+(aq) + e^- \longrightarrow Ag(s)$	0.7994
$Hg_2{}^{2+}(aq) + 2 e^- \longrightarrow 2 Hg(\ell)$	0.789
$Fe^{3+}(aq) + e^- \longrightarrow Fe^{2+}(aq)$	0.771
$SbCl_6{}^-(aq) + 2 e^- \longrightarrow SbCl_4{}^-(aq) + 2 Cl^-(aq)$	0.75
$[PtCl_4]^{2-}(aq) + 2 e^- \longrightarrow Pt(s) + 4 Cl^-(aq)$	0.73

(continued)

Standard Reduction Potentials (continued)

Acidic Solution	Standard Reduction Potential, $E°$ (volts)
$O_2(g) + 2\ H^+(aq) + 2\ e^- \longrightarrow H_2O_2(aq)$	0.682
$[PtCl_6]^{2-}(aq) + 2\ e^- \longrightarrow [PtCl_4]^{2-}(aq) + 2\ Cl^-(aq)$	0.68
$H_3AsO_4(aq) + 2\ H^+(aq) + 2\ e^- \longrightarrow H_3AsO_3(aq) + H_2O$	0.58
$I_2(s) + 2\ e^- \longrightarrow 2\ I^-(aq)$	0.535
$TeO_2(s) + 4\ H^+(aq) + 4\ e^- \longrightarrow Te(s) + 2\ H_2O$	0.529
$Cu^+(aq) + e^- \longrightarrow Cu(s)$	0.521
$[RhCl_6]^{3-}(aq) + 3\ e^- \longrightarrow Rh(s) + 6\ Cl^-(aq)$	0.44
$Cu^{2+}(aq) + 2\ e^- \longrightarrow Cu(s)$	0.337
$Hg_2Cl_2(s) + 2\ e^- \longrightarrow 2\ Hg(\ell) + 2\ Cl^-(aq)$	0.27
$AgCl(s) + e^- \longrightarrow Ag(s) + Cl^-(aq)$	0.222
$SO_4{}^{2-}(aq) + 4\ H^+(aq) + 2\ e^- \longrightarrow SO_2(g) + 2\ H_2O$	0.20
$SO_4{}^{2-}(aq) + 4\ H^+(aq) + 2\ e^- \longrightarrow H_2SO_3(aq) + H_2O$	0.17
$Cu^{2+}(aq) + e^- \longrightarrow Cu^+(aq)$	0.153
$Sn^{4+}(aq) + 2\ e^- \longrightarrow Sn^{2+}(aq)$	0.15
$S(s) + 2\ H^+ + 2\ e^- \longrightarrow H_2S(aq)$	0.14
$AgBr(s) + e^- \longrightarrow Ag(s) + Br^-(aq)$	0.0713
$2\ H^+(aq) + 2\ e^- \longrightarrow H_2(g)$ (reference electrode)	0.0000
$N_2O(g) + 6\ H^+(aq) + H_2O + 4\ e^- \longrightarrow 2\ NH_3OH^+(aq)$	−0.05
$Pb^{2+}(aq) + 2\ e^- \longrightarrow Pb(s)$	−0.126
$Sn^{2+}(aq) + 2\ e^- \longrightarrow Sn(s)$	−0.14
$AgI(s) + e^- \longrightarrow Ag(s) + I^-(aq)$	−0.15
$[SnF_6]^{2-}(aq) + 4\ e^- \longrightarrow Sn(s) + 6\ F^-(aq)$	−0.25
$Ni^{2+}(aq) + 2\ e^- \longrightarrow Ni(s)$	−0.25
$Co^{2+}(aq) + 2\ e^- \longrightarrow Co(s)$	−0.28
$Tl^+(aq) + e^- \longrightarrow Tl(s)$	−0.34
$PbSO_4(s) + 2\ e^- \longrightarrow Pb(s) + SO_4{}^{2-}(aq)$	−0.356
$Se(s) + 2\ H^+(aq) + 2\ e^- \longrightarrow H_2Se(aq)$	−0.40
$Cd^{2+}(aq) + 2\ e^- \longrightarrow Cd(s)$	−0.403
$Cr^{3+}(aq) + e^- \longrightarrow Cr^{2+}(aq)$	−0.41
$Fe^{2+}(aq) + 2\ e^- \longrightarrow Fe(s)$	−0.44
$2\ CO_2(g) + 2\ H^+(aq) + 2\ e^- \longrightarrow H_2C_2O_4(aq)$	−0.49
$Ga^{3+}(aq) + 3\ e^- \longrightarrow Ga(s)$	−0.53
$HgS(s) + 2\ H^+(aq) + 2\ e^- \longrightarrow Hg(\ell) + H_2S(g)$	−0.72
$Cr^{3+}(aq) + 3\ e^- \longrightarrow Cr(s)$	−0.74
$Zn^{2+}(aq) + 2\ e^- \longrightarrow Zn(s)$	−0.763
$Cr^{2+}(aq) + 2\ e^- \longrightarrow Cr(s)$	−0.91
$FeS(s) + 2\ e^- \longrightarrow Fe(s) + S^{2-}(aq)$	−1.01
$Mn^{2+}(aq) + 2\ e^- \longrightarrow Mn(s)$	−1.18
$V^{2+}(aq) + 2\ e^- \longrightarrow V(s)$	−1.18
$CdS(s) + 2\ e^- \longrightarrow Cd(s) + S^{2-}(aq)$	−1.21
$ZnS(s) + 2\ e^- \longrightarrow Zn(s) + S^{2-}(aq)$	−1.44

(continued)

Standard Reduction Potentials (continued)

Acidic Solution	Standard Reduction Potential, $E°$ (volts)
$Zr^{4+}(aq) + 4\ e^- \longrightarrow Zr(s)$	−1.53
$Al^{3+}(aq) + 3\ e^- \longrightarrow Al(s)$	−1.66
$Mg^{2+}(aq) + 2\ e^- \longrightarrow Mg(s)$	−2.37
$Na^+(aq) + e^- \longrightarrow Na(s)$	−2.714
$Ca^{2+}(aq) + 2\ e^- \longrightarrow Ca(s)$	−2.87
$Sr^{2+}(aq) + 2\ e^- \longrightarrow Sr(s)$	−2.89
$Ba^{2+}(aq) + 2\ e^- \longrightarrow Ba(s)$	−2.90
$Rb^+(aq) + e^- \longrightarrow Rb(s)$	−2.925
$K^+(aq) + e^- \longrightarrow K(s)$	−2.925
$Li^+(aq) + e^- \longrightarrow Li(s)$	−3.045

Basic Solution*	
$ClO^-(aq) + H_2O + 2\ e^- \longrightarrow Cl^-(aq) + 2\ OH^-(aq)$	0.89
$OOH^-(aq) + H_2O + 2\ e^- \longrightarrow 3\ OH^-(aq)$	0.88
$2\ NH_2OH(aq) + 2\ e^- \longrightarrow N_2H_4(aq) + 2\ OH^-(aq)$	0.74
$ClO_3^-(aq) + 3\ H_2O + 6\ e^- \longrightarrow Cl^-(aq) + 6\ OH^-(aq)$	0.62
$MnO_4^-(aq) + 2\ H_2O + 3\ e^- \longrightarrow MnO_2(s) + 4\ OH^-(aq)$	0.588
$MnO_4^-(aq) + e^- \longrightarrow MnO_4^{2-}(aq)$	0.564
$NiO_2(s) + 2\ H_2O + 2\ e^- \longrightarrow Ni(OH)_2(s) + 2\ OH^-(aq)$	0.49
$Ag_2CrO_4(s) + 2\ e^- \longrightarrow 2\ Ag(s) + CrO_4^{2-}(aq)$	0.446
$O_2(g) + 2\ H_2O + 4\ e^- \longrightarrow 4\ OH^-(aq)$	0.40
$ClO_4^-(aq) + H_2O + 2\ e^- \longrightarrow ClO_3^-(aq) + 2\ OH^-(aq)$	0.36
$Ag_2O(s) + H_2O + 2\ e^- \longrightarrow 2\ Ag(s) + 2\ OH^-(aq)$	0.34
$2\ NO_2^-(aq) + 3\ H_2O + 4\ e^- \longrightarrow N_2O(g) + 6\ OH^-(aq)$	0.15
$N_2H_4(aq) + 2\ H_2O + 2\ e^- \longrightarrow 2\ NH_3(aq) + 2\ OH^-(aq)$	0.10
$[Co(NH_3)_6]^{3+}(aq) + e^- \longrightarrow [Co(NH_3)_6]^{2+}(aq)$	0.10
$HgO(s) + H_2O + 2\ e^- \longrightarrow Hg(\ell) + 2\ OH^-(aq)$	0.0984
$O_2(g) + H_2O + 2\ e^- \longrightarrow OOH^-(aq) + OH^-(aq)$	0.076
$NO_3^-(aq) + H_2O + 2\ e^- \longrightarrow NO_2^-(aq) + 2\ OH^-(aq)$	0.01
$MnO_2(s) + 2\ H_2O + 2\ e^- \longrightarrow Mn(OH)_2(s) + 2\ OH^-(aq)$	−0.05
$CrO_4^{2-}(aq) + 4\ H_2O + 3\ e^- \longrightarrow Cr(OH)_3(s) + 5\ OH^-(aq)$	−0.12
$Cu(OH)_2(s) + 2\ e^- \longrightarrow Cu(s) + 2\ OH^-(aq)$	−0.36
$S(s) + 2\ e^- \longrightarrow S^{2-}(aq)$	−0.48
$Fe(OH)_3(s) + e^- \longrightarrow Fe(OH)_2(s) + OH^-(aq)$	−0.56
$2\ H_2O + 2\ e^- \longrightarrow H_2(g) + 2\ OH^-(aq)$	−0.8277
$2\ NO_3^-(aq) + 2\ H_2O + 2\ e^- \longrightarrow N_2O_4(g) + 4\ OH^-(aq)$	−0.85
$Fe(OH)_2(s) + 2\ e^- \longrightarrow Fe(s) + 2\ OH^-(aq)$	−0.877
$SO_4^{2-}(aq) + H_2O + 2\ e^- \longrightarrow SO_3^{2-}(aq) + 2\ OH^-(aq)$	−0.93
$N_2(g) + 4\ H_2O + 4\ e^- \longrightarrow N_2H_4(aq) + 4\ OH^-(aq)$	−1.15
$[Zn(OH)_4]^{2-}(aq) + 2\ e^- \longrightarrow Zn(s) + 4\ OH^-(aq)$	−1.22

(continued)

Standard Reduction Potentials *(continued)*

Basic Solution*

$Zn(OH)_2(s) + 2\ e^- \longrightarrow Zn(s) + 2\ OH^-(aq)$	-1.245
$[Zn(CN)_4]^{2-}(aq) + 2\ e^- \longrightarrow Zn(s) + 4\ CN^-(aq)$	-1.26
$Cr(OH)_3(s) + 3\ e^- \longrightarrow Cr(s) + 3\ OH^-(aq)$	-1.30
$SiO_3^{2-}(aq) + 3\ H_2O + 4\ e^- \longrightarrow Si(s) + 6\ OH^-(aq)$	-1.70

*Unless insoluble species are formed at $[OH^-] = 1.0$ mol L^{-1}, $E°$ values in basic solution can be calculated from the corresponding $E°$ values in acidic solution by application of the Nernst Equation using $[H^+] = 1 \times 10^{-14}$ mol L^{-1}.

Physical Quantities and Their Units of Measurement

All measurements that we make are of one of the seven fundamental quantities—*length*, *mass*, *time*, *temperature*, *amount of substance*, *electric current*, and *luminous intensity*—or of quantities that are mathematically derived from these fundamental ones. For example:

$$\text{Volume} = \text{length} \times \text{length} \times \text{length} = (\text{length})^3$$

$$\text{Density} = \frac{\text{mass}}{(\text{length})^3} = \text{mass} \times (\text{length})^{-3}$$

$$\text{Energy} = \text{mass} \times \text{length} \times \text{acceleration} = \text{mass} \times (\text{length})^2 \times (\text{time})^{-2}$$

Units of Measurement of Quantities in the SI System

At various times, in various geographical situations and various disciplines, each of the fundamental quantities has been measured in different units. For example, an object measured as 2.00 metres in one system is been measured as 78.74 inches in another. This has caused considerable nuisance in communication between people in different places or disciplines. As a result, in 1960, the 11th General Conference on Weights and Measures adopted the International System of Units, or Système International (SI). In this SI system, now recommended by the International Union of Pure and Applied Chemistry (IUPAC) and the parent bodies of the other disciplines, the units used for measurement of the seven basic quantities are those listed in Table F.1.

TABLE F.1 Physical Quantities and Their SI Units

Physical Quantity	Symbol for the Quantity	SI Unit of Measurement	Symbol for the Unit
Length	l	metre	m
Mass	m	kilogram	kg
Time	t	second	s
Temperature	T	kelvin	K
Amount of Substance	n	mole	mol
Electric Current	I	ampere	A
Luminous Intensity	I_ν	candela	cd

The convention for recording a time duration of 6.354 seconds, for example, is as follows:

$$t = 6.354 \text{ s}$$

The first five units listed in Table F.1 are particularly useful in chemistry, and are defined as follows:

- A **metre** is the length of path travelled by light in vacuum during a time interval of 1/299 792 458 of a second.
- A **kilogram** is equal to the mass of the international prototype of the kilogram: a platinum-iridium block kept at the International Bureau of Weights and Measures at Sevrès, France.
- A **second** is the duration of 9 192 631 770 periods of the radiation corresponding to the transition between two defined electron energy levels in the ground-state cesium-133 atom.

- A *kelvin* is 1/273.16 of the interval between the absolute zero of temperature and the temperature of the triple point of water.
- A *mole* is the amount of substance in a system that contains as many defined entities as there are atoms in 0.012 kilogram of carbon-12.

Prefixes Used with SI Units

When quantities have very large or very small values when expressed in SI units, it is common to use decimal multiples or fractions of the SI units so that the value is between 1 and 1000. These multiples of fractions are indicated by a prefix to the unit, indicated in the unit symbol. For example, one kilometre (1 km) = one thousand metres (1000 m), and one millimole (1 mmol) = one-thousandth of a mole (0.001 mol). The prefixes used to indicate multiples or fractions of units, and the symbols used for them, are listed in Table F.2.

TABLE F.2 Prefixes Used for Multiples and Fractions of SI Units

Multiple	Prefix	Symbol	Fraction	Prefix	Symbol
10^{12}	tera	T	10^{-1}	deci	d
10^{9}	giga	G	10^{-2}	centi	c
10^{6}	mega	M	10^{-3}	milli	m
10^{3}	kilo	k	10^{-6}	micro	μ
10^{2}	hecto	h	10^{-9}	nano	n
10^{1}	deka	da	10^{-12}	pico	p
			10^{-15}	femto	f
			10^{-18}	atto	a

Derived SI Quantities and Their Units

In the International System of Units, all physical quantities are mathematical combinations of the base quantities listed in Table F.1, and the units are the corresponding mathematical combinations of the units of the base quantities. A list of derived quantities, and their units, that are frequently used in chemistry are shown in Table F.3.

TABLE F.3 Derived SI Quantities and Their Units

Physical Quantity	Symbol	Name of Unit	Symbol of Unit	Combination of SI Units
Area	A	square metre	m^2	m^2
Volume	V	cubic metre	m^3	m^3
Density	ρ	kilogram per cubic metre	$kg\ m^{-3}$	$kg\ m^{-3}$
Force	F	Newton	N	$kg\ m\ s^{-2}$
Pressure	p	Pascal	Pa	$N\ m^{-2} = kg\ m^{-1}\ s^{-2}$
Energy	E	Joule	J	$kg\ m^2\ s^{-2}$
Electric charge	q	Coulomb	C	$A\ s$
Electric potential difference	E	Volt	V	$J\ A^{-1}\ s^{-1} = kg\ m^2\ A^{-1}\ s^{-3}$

Alternative SI Units

Some exact decimal multiples or fractions of the SI units are accepted as alternative units that can be used, and often are. These include the *gram* (1 g = 0.001 kg), the *litre* (1 L = 0.001 m³), the *tonne* (1 t = 1000 kg), and the *bar* (1 bar = 1 × 10⁵ Pa). Indeed, it is more common for prefixes to be appended to these units than to the basic SI units. For example, 0.006 g is written as 6 mg, rather than the very awkward 6 × 10⁻³ mkg. In this resource, the litre is always used as the unit of measurement of volume, rather than cubic metres.

Conversion Factors between Values Expressed in Different Units

Occasionally, you will come across values of quantities expressed in units other than those of the SI system. For selected quantities, Table F.4 lists conversion factors that will enable you to convert such values into the appropriate SI values.

For example, the list shows that the factor for conversion of a value of energy measured in calories to a value in joules is 4.184 J cal^{-1} (which means that an energy of 1 cal is equivalent to 4.184 J). So if we wish to express a value of 1.45×10^6 cal in the SI unit of energy measurement, we use:

$$E = 6.84 \times 10^6 \text{ cal} = (1.45 \times 10^6 \text{ cal})(4.184 \text{ J cal}^{-1}) = 6.07 \times 10^6 \text{ J}$$

TABLE F.4   Factors for Conversion to Values in SI Units

Non-SI Unit	Conversion Factor
Length	
inch	2.54 cm inch^{-1}
mile	1.6093 km mile^{-1}
ångstrom (Å)	1.00×10^{-10} m Å^{-1}
Volume	
cubic foot (ft^3)	28.316 L ft^{-3}
U.K. gallon (gal)	4.55 L gal^{-1}
U.S. gallon (gal)	3.79 L gal^{-1}
Mass	
pound (lb)	453.59237 g lb^{-1}
ounce	28.35 g ounce^{-1}
ton (= 2240 lb)	1.016 t ton^{-1} or 1016 kg ton^{-1}
Temperature	
celsius degree (°C)	1 K °C^{-1} (exactly)
	$T(\text{K}) = T(°\text{C}) + 273.15$
Pressure	
atmosphere (atm)	101.325 kPa atm^{-1} or 1.01325 bar atm^{-1}
1 torr (= 1 mm Hg)	133.3 Pa torr^{-1}
Energy	
calorie (cal)	4.184 J cal^{-1}
kilocalorie (kcal)	4.184 kJ kcal^{-1} or 4.184×10^3 J kcal^{-1}
kilowatt hour (kWh)	3.6×10^6 J kWh^{-1}
British thermal unit (Btu)	1.054×10^3 J Btu^{-1}

APPENDIX G

Making Measurements: Precision, Accuracy, Error, and Significant Figures

Precision and Accuracy

The *precision* of a measurement indicates how well several determinations of the same quantity agree. This is illustrated by the results of throwing darts at a target. In Figure G.1a, the dart thrower was apparently not skillful as the precision of the dart's placement on the target is low. In Figures G.1b and G.1c, the darts are clustered together, indicating much better consistency on the part of the thrower—that is, greater precision.

FIGURE G.1 Precision and accuracy.

(a) Poor precision and poor accuracy   (b) Good precision and poor accuracy   (c) Good precision and good accuracy

Accuracy is the agreement of a measurement with the accepted value of the quantity. Figure G.1c shows that our thrower was accurate as well as precise—the average of all shots is close to the targeted position, the bull's eye.

Figure G.1b shows that it is possible to be precise without being accurate—the thrower has consistently missed the bull's eye, although all the darts are clustered precisely around one point on the target. This is analogous to an experiment with some consistent flaws of measurement (due to faulty instruments, faulty experimental design, or human errors such as incorrect record keeping). These are called *systematic errors*, a consequence of which is that all results differ from the true value by about the same amount.

The degree of precision of a series of measurements of a quantity is due to so-called "indeterminate" errors arising from uncertainties in measurement, the cause of which is not known and cannot be eliminated. The most common way to evaluate the extent of indeterminate error in measurements of a quantity is to calculate the *standard deviation*. The statistical significance of this measure is that for a large number of measurements, 68% of the values can be expected to be within one standard deviation of the average value. The smaller is the standard deviation, the more tightly bunched are the measurements around the average value.

The standard deviation of a series of measurements is equal to the square root of the sum of the squares of the deviations (differences) of each measurement from the average value, divided by the number of measurements. For n measurements giving values denoted x_i, with average value $\bar{x}$, the standard deviation (σ) is expressed mathematically as

$$\sigma = \sqrt{\frac{(\sum (x_i - \bar{x})^2)}{n}}$$

Consider a simple example. Suppose you carefully measured the mass of water delivered by a 10 mL pipette. For five attempts at the measurement (shown in the table below, column 2), the standard deviation is found as follows. First, the average of the measurements is calculated (here, 9.984). Next, the deviation of each individual measurement from this value is determined (column 3). These values are squared, giving the values in column 4, and then the sum of these values is determined. The standard deviation is then calculated by dividing this sum by 5 (the number of measurements) and taking the square root of the result.

Determination	Measured Mass, (g)	Difference between Average and Measurement (g)	Square of Difference
1	9.990	0.006	4×10^{-5}
2	9.993	0.009	8×10^{-5}
3	9.973	0.011	12×10^{-5}
4	9.980	0.004	2×10^{-5}
5	9.982	0.002	0.4×10^{-5}

Average mass = 9.984 g

Sum of squares of differences = 26×10^{-5}

$$\text{Standard deviation, } \sigma = \sqrt{\frac{26 \times 10^{-5}}{5}} = 0.007$$

Based on this calculation it is appropriate to represent the measured mass as 9.984 ± 0.007 g. This tells a reader that if this measurement were repeated a large number of times, 68% of the values would fall in the range of 9.977 g to 9.991 g. It is common for the standard deviation (in this case ± 0.007) to be referred to as the *uncertainty* of the measurement.

Experimental Error

If you are measuring a quantity in the laboratory, you may be required to report the *error* in the result—that is, the difference between your result and the accepted value:

Error = experimentally determined value− accepted value

or the *percentage error*.

$$\text{Percentage error} = \frac{\text{error}}{\text{accepted value}} \times 100\%$$

The National Institute for Standards and Technology (NIST) is the most important resource for the standards used in science. Comparison with the NIST data is the best test of the accuracy of the measurement. See www.nist.gov.

WORKED EXAMPLE G.1—PRECISION AND ACCURACY

A coin has an "accepted" diameter of 28.054 mm. In an experiment, two students measure this diameter. Student A makes four measurements using a precision tool called a micrometer. Student B uses a simple plastic ruler. They report the following results:

Student A	Student B
28.246 mm	27.9 mm
28.244 mm	28.0 mm
28.246 mm	27.8 mm
28.248 mm	28.1 mm

What is the average diameter and percentage error obtained in each case? Which student's data are more accurate? Which are more precise?

Strategy

For each set of values, calculate the average of the results and then compare this average with 28.054 mm.

Solution

The average for each set of data is obtained by summing the four values and dividing by 4. For student A's measurements, the average value is 28.246 mm, while student B's average value is 28.0 mm.

Student A's data are all very close to the average value, so they are quite precise: the standard deviation is low. Student B's data, in contrast, have a wider range and are less precise. However, student A's result is less accurate than that of student B. The average diameter for student A differs from the "accepted" value by 0.192 mm and has a percent error of 0.684%:

$$\text{Percentage error} = \frac{28.246\,\text{mm} - 28.054\,\text{mm}}{28.054\,\text{mm}} \times 100\% = 0.684\%$$

Student B's average value has an error of only about 0.2%.

Comment

Possible reasons for the error in student A's result are incorrect use of the micrometer or a flaw in the instrument.

EXERCISE G.1—ERROR, PRECISION, AND ACCURACY

Two students measured the freezing point of an unknown liquid. Student A used an ordinary laboratory thermometer calibrated in 0.1 °C units. Student B used a thermometer certified by NIST and calibrated in 0.01 °C units. Their results were as follows:

Student A: 0.3 °C; 0.2 °C; 0.0 °C; –0.3 °C

Student B: 273.13 K; 273.17 K; 273.15 K; 273.19 K

Calculate the average value and, knowing that the liquid was water, calculate the percentage error for each student. Which student has the more precise values? Which has the smaller error?

Significant Figures

The value 1.1200 g does not mean the same as 1.12 g. The former implies a measurement known with greater precision than the latter: the value 1.1200 g suggests that there is uncertainty in the fourth decimal place, while 1.12 g suggests uncertainty in the second decimal place. In expressing measured values, we imply the level of precision through the number of *significant figures*: the number of digits that are known with some certainty. For example, the value 1.1200 g has five significant figures, with some uncertainty in the fifth (that is, the fourth decimal place), while the value 1.12 g has only three significant figures, with some uncertainty in the third.

In the above discussion about standard deviation, we used the example of a measurement that was known to be 9.984 g with an uncertainty of ±0.007 g. That is, the last digit of our measurement (0.004 g) is uncertain to some degree. So the measured value has four significant figures, the last of which is uncertain to some extent. It would not be sensible to express a measurement of, say, 9.984 ±0.07 g, because there is so much uncertainty in the third digit (0.08 g) that we really have no idea what the fourth digit should be. In this case, we should use only three significant figures and write 9.98 ±0.07 g.

It is common sense that a result calculated from experimental data can be no more precise than the least precise piece of information that went into the calculation. Suppose that we want to calculate the density of a piece of metal from its mass and dimensions, determined by standard laboratory techniques, listed in the following table.

Measurement	Data Collected	Significant Figures
Mass of metal	13.56 g	4
Length	6.45 cm	3
Width	2.50 cm	3
Thickness	3.1 mm	2

Although most of these measured values have two digits to the right of the decimal, they differ in the number of significant figures. The value 3.1 mm has two significant figures: the 3 is exactly known, but the 0.1 is not. Unless stated otherwise, it is common practice to assign an uncertainty of ± 1 to the last significant digit. This means the thickness of the metal piece may have been as small as 3.0 mm or as large as 3.2 mm.

We could derive the density from the measured data as follows:

$$\text{Density} = \frac{\text{mass}}{\text{volume}} = \frac{13.56\,\text{g}}{6.45\,\text{cm} \times 2.50\,\text{cm} \times 0.31\,\text{cm}} = 2.712678\,\text{g cm}^{-3}$$

But since one piece of input data is only known to two significant figures, the calculated value cannot be known more precisely than that, and we should report the density as $2.7\,\text{g cm}^{-3}$.

When doing calculations using measured values of quantities, we follow some basic rules so that the results reflect the precision of all the data that go into the calculations. The rules used for significant figures in this resource are as follows:

Rule 1 To determine the number of significant figures in a value, count all digits from left to right, starting with the first digit that is not zero. The examples in the following table illustrate this.

Example	Number of Significant Figures
1.23	3; all non-zero digits are significant.
0.00123 g	3; the zeros to the left of the 1 (which is the first significant digit) simply locate the decimal point. To avoid confusion, write numbers of this type in scientific notation; in this case, as 1.23×10^{-3}.
2.040 g	4; when a number is greater than 1, all zeros to the right of the decimal point are significant.
0.02040 g	4; for a number less than 1, only zeros to the right of the first non-zero digit are significant.
100 g	1; in numbers that do not contain a decimal point, "trailing" zeros may or may not be significant. To make the precision clear, it is common to write such numbers, in which all digits are significant. For example, 1.00×10^{2} has three significant digits, whereas 1×10^{2} has only one significant digit.
100 cm m^{-1}	This is a *defined quantity*. Defined quantities can be regarded as having an infinite number of significant digits, and do not limit the number of significant figures in a calculated result.
$\pi = 3.1415926$	The value of certain constants such as π is known to a greater number of significant figures than you will ever use in a calculation.

Rule 2 When adding or subtracting numbers, the number of decimal places in the answer is equal to the number of decimal places in the number with the fewest digits after the decimal.

0.12	2 decimal places	2 significant figures
+ 1.9	1 decimal place	2 significant figures
+10.925	3 decimal places	5 significant figures
12.945	3 decimal places	

The sum should be reported as 12.9, with one decimal place, because 1.9 has only one decimal place.

Rule 3 In multiplication or division, the number of significant figures in the answer should be the same as that in the quantity with the fewest significant figures.

$$\frac{0.01208}{0.0236} = 0.512 \text{ or, in scientific notation, } 5.12 \times 10^{-1}$$

Because 0.0236 has only three significant digits and 0.01208 has four, the answer should have three significant digits.

Rule 4 When a number is rounded off, the last digit to be retained is increased by one only if the following digit is 5 or greater.

Full Number	Number Rounded to Three Significant Digits
12.696	12.7
16.349	16.3
18.35	18.4
18.351	18.4

You should recognize that different outcomes are possible from one input value depending on the type of mathematical operation. For example, for operations on the number 4.68:

- 4.68 + 4.68 + 4.68 = 14.04 (four significant figures)
- 4.68 × 3 = 14.0 (three significant figures)

When working problems, you should do the calculation with all the digits allowed by your calculator and round off only at the end of the calculation. Rounding off in the middle can introduce errors. If your answer to a problem in this resource does not quite agree with the answers provided, the discrepancy may be the result of rounding the answer after each step and then using that rounded answer in the next step.

WORKED EXAMPLE G.2—USING SIGNIFICANT FIGURES

Consider the following calculation, in which we will ignore the units of the input data for the purpose of the exercise:

$$\text{Volume of gas (L)} = \frac{(0.120)(0.08206)(273.15 + 23)}{(230/760.0)}$$

Calculate the final answer to the correct number of significant figures.

Strategy

First decide on the number of significant figures represented by each number (rule 1), and then apply rules 2 and 3.

Solution

Number	Number of Significant Figures	Comments
0.120	3	The trailing 0 is significant. See rule 1.
0.08206	4	The first 0 to the immediate right of the decimal is not significant. See rule 1.
273.15 + 23 = 296	3	23 has no decimal places, so the sum can have none. See rule 2.
230/760.0 = 0.30	2	230 has two significant figures because the last 0 is not significant. In contrast, there is a decimal point in 760.0, so there are four significant digits. The quotient may have only two significant digits. See rules 1 and 3.

One of the pieces of information is known to only two significant figures. Therefore, the volume of gas should be expressed to two significant figures, as 9.6 L.

EXERCISE G.2—USING SIGNIFICANT FIGURES

(a) How many significant figures are indicated by the numbers (i) 2.33×10^7, (ii) 50.5, and (iii) 200?

(b) What are (i) the sum and (ii) the product of 10.26 and 0.063?

(c) What is the result of the following calculation?

$$x = \frac{(110.7 - 64)}{(0.056)(0.00216)}$$

Mathematics for Chemistry: Exponential Notation, Logarithms, Graphing, and Quadratic Equations

Chemistry is a quantitative science. Chemists make measurements of, among other things, size, mass, volume, time, and temperature. Scientists then manipulate that quantitative numerical information to search for relationships among properties and to provide insight into the molecular basis of matter.

This section reviews some of the mathematical skills you will need in chemical calculations. It also describes ways to perform calculations and ways to handle quantitative information. The background you should have to be successful includes the following skills:

- the ability to express and use numbers in exponential or scientific notation
- the ability to use logarithms in calculations
- the ability to read information from graphs
- the ability to make a graph of numerical information and, if the graph produces a straight line, to find the slope and equation of the line
- the ability to express quantitative information in an algebraic equation and solve that equation

Exponential or Scientific Notation

Consider the following quantitative information obtained by extensive environmental studies on a lake:

Lake Characteristics	Measurement Data
Area	2.33×10^7 m^2
Maximum depth	505 m
Dissolved solids in lake water	2×10^2 mg L^{-1}
Average rainfall in the lake basin	1.02×10^2 cm year^{-1}
Average snowfall in the lake basin	198 cm year^{-1}

Some data are expressed in *fixed notation* (505 m, 198 cm year^{-1}), whereas other data are expressed in *exponential*, or *scientific*, *notation* (2.33×10^7 m^2). Scientific notation is a way of presenting very large or very small numbers in a compact and consistent form that simplifies calculations. Because of its convenience, it is widely used in sciences such as chemistry, physics, engineering, and astronomy. In addition, the exponential format is often the best way of indicating the precision of a value through the number of significant figures [<<Appendix G].

In scientific notation, the number is expressed as a product of two numbers: $N \times 10^n$. N is the *digit* term and is a number between 1 and 9.9999 .... The second number, 10^n, the *exponential* term, is some integer power of 10. For example, 1234 is written in scientific notation as 1.234×10^3, or 1.234 multiplied by 10 three times:

$$1.234 = 1.234 \times 10^1 \times 10^1 \times 10^1 = 1.234 \times 10^3$$

Conversely, a number less than 1, such as 0.01234, is written as 1.234×10^{-2}. This notation tells us that 1.234 should be divided twice by 10 to obtain 0.01234:

$$0.01234 = \frac{1.234}{10^1 \times 10^1} = 1.234 \times 10^{-1} \times 10^{-1} = 1.234 \times 10^{-2}$$

Some other examples of scientific notation follow:

$$10000 = 1 \times 10^4 \qquad\qquad 12345 = 1.2345 \times 10^4$$
$$1000 = 1 \times 10^3 \qquad\qquad 1234.5 = 1.2345 \times 10^3$$
$$100 = 1 \times 10^2 \qquad\qquad 123.45 = 1.2345 \times 10^2$$
$$10 = 1 \times 10^1 \qquad\qquad 12.345 = 1.2345 \times 10^1$$
$$1 = 1 \times 10^0 \qquad \text{(any number to the zero power} = 1)$$
$$1/10 = 1 \times 10^{-1} \qquad\qquad 0.12 = 1.2 \times 10^{-1}$$
$$1/100 = 1 \times 10^{-2} \qquad\qquad 0.012 = 1.2 \times 10^{-2}$$
$$1/1000 = 1 \times 10^{-3} \qquad\qquad 0.0012 = 1.2 \times 10^{-3}$$
$$1/10000 = 1 \times 10^{-4} \qquad\qquad 0.00012 = 1.2 \times 10^{-4}$$

When converting a number to scientific notation, the exponent n is positive if the number is greater than 1 and negative if the number is less than 1. The value of n is the number of places by which the decimal is shifted to obtain the number in scientific notation, as indicated by the following examples:

$$1\ 2\ 3\ 4\ 5. = 1.2345 \times 10^4$$

Decimal point shifted four places to the left. So n is +4.

$$0.0\ 0\ 1\ 2 = 1.2 \times 10^{-3}$$

Decimal point shifted three places to the right. So n is −3.

If you wish to convert a number in scientific notation to one using fixed notation (that is, not using powers of 10), the procedure is reversed:

$$0.0\ 0\ 1\ 2 = 1.2 \times 10^{-3}$$

Decimal point shifted two places to the right, because n is +2.

$$0\ 0\ 6.273 \times 10^{-3} = 0.006273$$

Decimal point shifted three places to the left, because n is −3.

Be aware that calculators and computers often express a number such as 1.23×10^3 as 1.23E3 or 6.45×10^{-5} as 6.45E−5. Some electronic calculators can convert numbers in fixed notation to scientific notation. If you have such a calculator, you may be able to do this by pressing the EE or EXP key and then the "=" key (but check your calculator manual to learn how your device operates).

In chemistry, you will often have to use numbers in exponential notation in mathematical operations. The following five operations are important.

1. *Adding and Subtracting Numbers Expressed in Scientific Notation*
   When adding or subtracting two numbers, first convert them to the same powers of 10. The digit terms are then added or subtracted as appropriate:

$$(1.234 \times 10^{-3}) + (5.623 \times 10^{-2}) = (0.1234 \times 10^{-2}) + (5.623 \times 10^{-2})$$
$$= 5.746 \times 10^{-2}$$

2. *Multiplication of Numbers Expressed in Scientific Notation*
   The digit terms are multiplied in the usual manner, and the exponents are added algebraically. The result is expressed with a digit term with only one non-zero digit to the left of the decimal:

$$(6.0 \times 10^{23})(2.0 \times 10^{-2}) = (6.0)(2.0) \times 10^{23-2} = 12 \times 10^{21} = 1.2 \times 10^{22}$$

3. *Division of Numbers Expressed in Scientific Notation*
The digit terms are divided in the usual manner, and the exponents are subtracted algebraically. The quotient is written with one non-zero digit to the left of the decimal in the digit term:

$$\frac{7.60 \times 10^3}{1.23 \times 10^2} = \frac{7.60}{1.23} \times 10^{3-2} = 6.18 \times 10^1$$

4. *Powers of Numbers Expressed in Scientific Notation*
When raising a number in exponential notation to a power, treat the digit term in the usual manner. The exponent is then multiplied by the number indicating the power:

$$(5.28 \times 10^3)^2 = (5.28)^2 \times 10^{3 \times 2} = 27.9 \times 10^6 = 2.79 \times 10^7$$

5. *Roots of Numbers Expressed in Scientific Notation*
Unless you use an electronic calculator, the number must first be put into a form in which the exponent is exactly divisible by the root. For example, for a square root, the exponent should be divisible by 2. The root of the digit term is found in the usual way, and the exponent is divided by the desired root:

$$\sqrt{3.6 \times 10^7} = \sqrt{36 \times 10^6} = \sqrt{3.6} \times \sqrt{10^6} = 6.0 \times 10^3$$

Logarithms

There are two main types of logarithms: (a) common logarithms (abbreviated *log*) whose base is 10, and (b) natural logarithms (abbreviated *ln*) whose base is e (= 2.71828):

$$\log x = n, \text{ where } x = 10^n$$
$$\ln x = m, \text{ where } x = e^m$$

Most equations in chemistry and physics were developed in natural, or base e, logarithms, and we follow this practice in this course. The relation between log and ln is

$$\ln x = 2.303 \log x$$

Despite the different bases of the two logarithms, their operation is similar in principle. What follows is largely a description of the use of common logarithms.

A common logarithm is the power to which you must raise 10 to obtain the number. For example, the log of 100 is 2, since you must raise 10 to the second power to obtain 100. Other examples are

$$\log 1000 = \log(10^3) = 3$$
$$\log 10 = \log(10^1) = 3$$
$$\log 1 = \log(10^0) = 0$$
$$\log 0.1 = \log(10^{-1}) = -1$$
$$\log 0.0001 = \log(10^{-4}) = -4$$

To obtain the common logarithm of a number other than a simple power of 10, you must resort to a calculator or a log table. On your calculator, enter the number, and then press the "log" key. For example:

$$\log 2.10 = 0.3222, \text{ which means that } 10^{0.3222} = 2.10$$
$$\log 5.16 = 0.7126, \text{ which means that } 10^{0.7126} = 5.16$$
$$\log 3.125 = 0.49485, \text{ which means that } 10^{0.49485} = 3.125$$

The log of any number between 10 and 100 is between 1 and 2, and the log any number between 100 and 1000 is between 2 and 3. And, for example, the log of any number between 0.001 and 0.01 is between -3 and -2.

To obtain the natural logarithm (ln) of the numbers shown here, enter each number and press "ln:"

$$\ln 2.10 = 0.7419, \text{ which means that } e^{0.7419} = 2.10$$
$$\ln 5.16 = 1.6409, \text{ which means that } e^{1.6409} = 5.16$$

To find the common logarithm of a number greater than 10 or less than 1 with a log table, first express the number in scientific notation. Then find the log of each part of the number and add the logs. For example:

$$\log 241 = \log (2.41 \times 10^2) = \log 2.41 + \log 10^2$$
$$= 0.382 + 2 = 2.382$$
$$\log 0.00573 = \log (5.73 \times 10^{-3}) = \log 5.73 + \log 10^{-3}$$
$$= 0.758 + (-3) = -2.242$$

Significant Figures and Logarithms

The number to the left to the decimal in a logarithm is called the *characteristic*, and the number to the right of the decimal is the *mantissa*. The mantissa should have as many significant figures as the number whose log was found. (This rule was not strictly followed until the last two examples so that you could more clearly see the result obtained with a calculator.)

Obtaining Antilogarithms

If you are given the logarithm of a number, and find the number from it, you have obtained the *antilogarithm* or *antilog* of the number. Consider the following examples:

1. Find the number whose log is 5.234:
    Recall that $\log x = n$, where $x = 10^n$. In this case $n = 5.234$, so we want to find the number x such that $x = 10^{5.234}$. Enter 5.234 in your calculator, and find the value of $10^{5.234}$, the antilog. In this case:

$$10^{5.234} = 10^{0.234} \times 10^5 = 1.71 \times 10^5$$

Notice that the characteristic (5) sets the decimal point; it is the power of 10 in the exponential form. The mantissa (0.234) gives the value 1.71. Check that $10^{4.234} = 1.71 \times 10^4$, and that $10^{11.234} = 1.71 \times 10^{11}$.

2. Find the number whose log is −3.456:

$$10^{-3.456} = 10^{0.544} \times 10^{-4} = 3.50 \times 10^{-4}$$

Notice here that −3.456 must be expressed as the sum of −4 and + 0.544.

Mathematical Operations Using Logarithms

Because logarithms are exponents, operations involving them follow the same rules used for exponents. Thus, multiplying two numbers can be done by adding their logarithms because

$$\log xy = \log x + \log y$$

For example, we multiply 563 by 125 by adding their logarithms and finding the antilogarithm of the result:

$$\begin{aligned} \log 563 &= 2.751 \\ \log 125 &= \underline{2.097} \\ \log (563 \times 125) &= 4.848 \\ 563 \times 125 &= 10^{4.848} = 10^4 \times 10^{0.848} = 7.05 \times 10^4 \end{aligned}$$

One number (x) can be divided by another (y) by subtraction of their logarithms:

$$\log \frac{x}{y} = \log x - \log y$$

For example, to divide 125 by 742:

$$\log 125 = 2.097$$
$$-\log 742 = \underline{2.870}$$
$$\log \frac{125}{742} = -0.773$$

$$\frac{125}{742} = 10^{-0.773} = 10^{0.227} \times 10^{-1} = 1.68 \times 10^{-1}$$

Similarly, powers and roots of numbers can be found using logarithms.

$$\log x^y = y(\log x)$$

$$\log \sqrt[y]{x} = \log x^{\frac{1}{y}} = \tfrac{1}{y}\log x$$

As an example, find the fourth power of 5.23; that is $(5.23)^4$. We first find the log of 5.23 and then multiply it by 4. The result, 2.874, is the log of the answer, so we find the antilog of 2.874:

$$\log (5.23)^4 = 4(\log 5.23) = 4(0.719) = 2.874$$
$$(5.23)^4 = 10^{2.874} = 748$$

As another example, find the fifth root of 1.89×10^{-9}:

$$\sqrt[5]{1.89 \times 10^{-9}} = (1.89 \times 10^{-9})^{\frac{1}{5}} = \ ?$$
$$\log (1.89 \times 10^{-9})^{\frac{1}{5}} = \tfrac{1}{5}\log (1.80 \times 10^{-9} = \tfrac{1}{5}(-8.724) = -1.745$$
$$(1.89 \times 10^{-9})^{\frac{1}{5}} = 10^{-1.745} = 1.80 \times 10^{-2}$$

Graphing

In a number of instances in this resource, graphs are used when analyzing experimental data with a goal of obtaining a mathematical equation. The procedure used sometimes results in a straight line, which has the equation

$$y = mx + b$$

In this equation, y is usually referred to as the dependent variable; its value is determined from (that is, is dependent on) the values of x, m, and b. In this equation, x is called the independent variable and m is the slope of the line. The parameter b is the y-intercept—that is, the value of y when $x = 0$. Let's use an example to investigate two things: (a) how to construct a graph from a set of data points, and (b) how to derive an equation for the line generated by the data.

A set of data points to be graphed is presented in Figure H.1. We first mark off each axis in increments of the values of x and y. Here our x-data range from –2 to 4, so the x-axis is marked off in increments of 1 unit. The y-data range from 0 to 2.5, so we mark off the y-axis in increments of 0.5. Each data set is marked as a circle on the graph.

After plotting the points on the graph (round circles), we draw a *straight line* that comes as close as possible to representing the trend in the data. (Do not connect the dots!) Because there is always some inaccuracy in experimental data, this line may not pass exactly through every point.

In principle, to identify the specific equation corresponding to our data, we must determine the y-intercept (b) and slope (m) for the equation $y = mx + b$. The y-intercept is the point at which $x = 0$. (In Figure H.1, $y = 1.87$ when $x = 0$). The slope is determined by selecting two points on the line (marked with squares in Figure H.1) and calculating the difference in values of y ($\Delta y = y_2 - y_1$) and x ($\Delta x = x_2 - x_1$). The slope of the line is then the ratio of these differences, $m = y/x$. Here, the slope has the value –0.525. With the slope and intercept now known, we can write the equation for the line

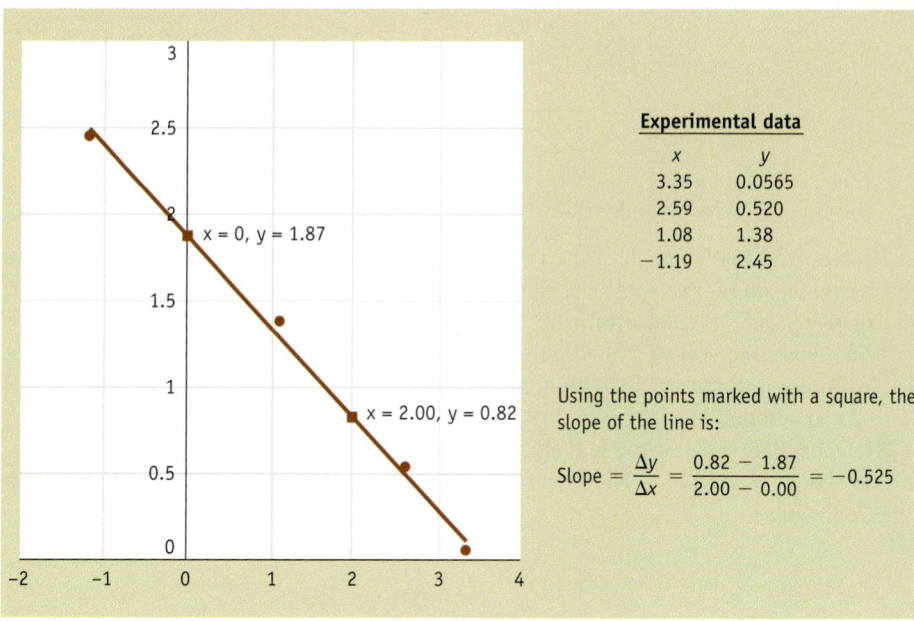

Experimental data

x	y
3.35	0.0565
2.59	0.520
1.08	1.38
−1.19	2.45

Using the points marked with a square, the slope of the line is:

$$\text{Slope} = \frac{\Delta y}{\Delta x} = \frac{0.82 - 1.87}{2.00 - 0.00} = -0.525$$

FIGURE H.1 Data for the variable x are plotted along the horizontal axis (abscissa), and data for y are plotted along the vertical axis (ordinate). The slope of the line, m in the equation $y = mx + b$, is given by $\Delta y/\Delta x$. The intercept of the line with the y-axis (when $x = 0$) is b in the equation. Using a plotting program with these data, and doing a linear regression (or least-squares) analysis, we find $y = 0.525x + 1.87$.

$$y = -0.525x + 1.87$$

More likely, you will obtain this equation by inputting the data into a computer graphing program, and performing a "least squares" or "linear regression" analysis to find the equation for a straight line that best fits the data.

The equation for the best-fit straight line can now be used to calculate y-values corresponding to values of x that are not part of our original set of x-y data. For example, when $x = 1.50$, $y = 1.08$.

EXERCISE H.1—GRAPHING

To find the mass of 50 jelly beans, some students weighed several samples of beans.

Number of Beans	Mass
5	12.82 g
11	27.14 g
16	39.30 g
24	59.04 g

Plot these data with the number of beans on the horizontal or x-axis, and the mass of beans on the vertical or y-axis. What is the slope of the line? Use your equation of a straight line to calculate the mass of 50 jelly beans.

Quadratic Equations

Algebraic equations of the form $ax^2 + bx + c = 0$ are called *quadratic equations*. The coefficients a, b, and c may be either positive or negative. The two roots of the equation may be found using the *quadratic formula*:

$$x = \frac{-b \pm \sqrt{b^2 - 4ac}}{2a}$$

As an example, solve the equation $5x^2 - 3x - 2 = 0$. Here $a = 5$, $b = -3$, and $c = -2$. Therefore,

$$x = \frac{3 \pm \sqrt{(-3)^2 - 4(5)(-2)}}{2(5)}$$

$$= \frac{3 \pm \left\{\sqrt{9 - (-40)}\right\}}{10} = \frac{3 \pm \sqrt{49}}{10} = \frac{3 \pm 7}{10}$$

$$= 1 \text{ and } -0.4$$

How do you know which of the two roots is the correct answer? You have to decide in each case which root has physical significance. It is *usually* true in this course, however, that negative values are not significant.

When you have solved a quadratic expression, you should always check your values by substitution into the original equation. In the previous example, we find that $5(1)^2 - 3(1) - 2 = 0$ and that $5(-0.4)^2 - 3(-0.4) - 2 = 0$.

The most likely places you will encounter quadratic equations are in solutions to problems on chemical equilibria. Here, you will often be faced with solving an equation such as the following:

$$1.8 \times 10^{-4} = \frac{x^2}{0.0010 - x}$$

This equation can certainly be solved using the quadratic formula (to give $x = 3.4 \times 10^{-4}$). You may find the *method of successive approximations* to be convenient, however. Here, we begin by making a reasonable approximation of x. This approximate value is substituted into the original equation, which is then solved to give what is hoped to be a more correct value of x. This process is repeated until the answer converges on a particular value of x—that is, until the value of x derived from two successive approximations is the same.

Step 1 First assume that x is so small that $(0.0010 - x) \approx 0.0010$. This means that

$$x^2 = 1.8 \times 10^{-4}(0.0010)$$

$$x = 4.2 \times 10^{-4} \text{ (to 2 significant figures)}$$

Step 2 Substitute the value of x from step 1 into the denominator of the original equation, and again solve for x:

$$x^2 = 1.8 \times 10^{-4}(0.0010 - 0.00042)$$

$$x = 3.2 \times 10^{-4}$$

Step 3 Repeat step 2 using the value of x found in that step:

$$x = \sqrt{1.8 \times 10^{-4}(0.0010 - 0.0032)} = 3.5 \times 10^{-4}$$

Step 4 Continue repeating the calculation, using the value of x found in the previous step:

$$x = \sqrt{1.8 \times 10^{-4}(0.0010 - 0.0035)} = 3.4 \times 10^{-4}$$

Step 5 $x = \sqrt{1.8 \times 10^{-4}(0.0010 - 0.0034)} = 3.4 \times 10^{-4}$

Here, we find that iterations after the fourth step give the same value for x, indicating that we have arrived at a valid answer (and the same one obtained from the quadratic formula).

If your calculator has a memory function, successive approximations can be carried out easily and rapidly.

The method of approximations works as long as $K < 4C$, which is always the case for weak acids and bases in aqueous solution. In cases where K is very large, as may be the situation for reactions in the gas phase, for example, the method does not work because successive steps may give answers that are random or that diverge from the correct value.

INDEX/GLOSSARY

Page references with "f" indicate figures; those with "t" indicate tables. Glossary terms are printed in blue.

abiogenic (without life) methane, 85

absolute zero The lowest possible temperature, equivalent to –273.15 °C, used as the zero point of the Kelvin scale, 204

absorption spectrum A plot of the percentage of radiation transmitted over a range of incident radiation energies (or wavelengths), 67

acceptor level, 1098

acetal Functional group having the structure $R_2C(OR')_2$, 895

acetylacetonate ion ($acac^-$), 951

achiral Molecules or objects that are superimposable on their mirror images and therefore not chiral, 306

acid A proton donor or electron-pair acceptor, 177, 489–490, 503

acid anhydride Functional group with two acyl groups bonded to the same O atom, $RC(=O)OC(=O)R'$, 920

acid anhydrides, 921, 937–938

acid-base adduct, 181

acid-base equilibria in aqueous solution

　acid-base properties of amino acids and proteins, 521–525

　　pH-dependent speciation of amino acids, 521–525, 523–524t

　acid-base titrations. *See* titration

　Brønsted-Lowry model, 489–494

　　characteristics of acids, bases, and amphoteric species, 490–494

　buffer solutions

　　buffer capacity, 533–535

　　change of pH, 532–533

　　composition and mode of operation, 526–527

　　design of, 530–532

　　quantitative calculations of pH, 527–530

　Lewis model of acids and bases, 503–507

　　electron pair transfer, 503–504

　　visualization of reactive sites of organic acids and base, 504–507

　speciation, relative concentration of species and, 516–521

　　acid-base speciation and complexation with metal ions, 519–520

　　distribution among species from polyprotic acids, 520–521

　　distribution between acid and base species as pH is changed, 516–519

　water and the pH scale, 494–497

　　logarithmic scale of hydronium ion concentrations, 496–497

　　water self-ionization and the water ionization constant (K_w), 494–496

　of weak acids and bases, 507–516

　　aqueous solutions of weak bases, 511–513

　　dependence of percentage ionization on magnitude of K_a, 509–510

　　dependence of percentage ionization on solution concentration, 510

　　effect of common ions on percentage ionization, 510–511

　　equilibrium concentrations, pH, and percentage ionization from K_a, 508–509

　　estimating K_a from solute concentration and measured pH, 507–508

　　relative strengths of, 498–503

　　solutions of polyprotic acids or their bases, 513–514

　　two measures of acidity of a solution, 515–516

acid-base neutralization Reaction due to transfer of protons, with formation of water, 179

acid-base properties of amino acids and proteins, 521–525

　pH-dependent speciation of amino acids, 521–525, 523–524t

acid-base reaction An exchange reaction between an acid and a base producing a salt and water, 116, 172

　proton transfer and, 177–179

　　acids in aqueous solution, 177–178, 177f

　　bases in aqueous solution, 178

　　common acids and bases, 178t

　　neutralization, reactions of acids with bases and, 178–179

acid-base titrations. *See* titration

acidic amino acids, 524n

acidic solution An aqueous solution in which $[H_3O^+] > [OH^-]$, 495

acid ionization constant (K_a) The equilibrium constant for ionization of a weak acid in aqueous solution, 498–499

acidity, 515–516

actinides The series of elements between actinium and rutherfordium in the periodic table, 1054

activating groups, 793

activation energy (E_a) The minimum combined kinetic energy that a pair of colliding particles must have in excess of the average for their collision to result in reaction, 687–688, 737

activity (A) (nuclear chemistry) Number of disintegrations observed per unit time, measured in becquerel (Bq), 1175–1176

activity (a) The effective concentration of a solute species, 443

acyl halide Functional group with acyl group bonded to a halogen atom, $RC(=O)X$, 920, 921

　synthesis and reactions of, 934–937, 935f

addition reaction Reaction of two molecules to form molecules of a new one with no atoms "left over," 729

adduct, 503–504

adrenaline, 490

aerogel, 1105, 1105f

aerosol Fine droplets of liquid or dust suspended in the atmosphere, 90

age-related macular degeneration (AMD), 3–5

albedo The fraction of sunlight incident on the earth that is reflected, 89

alcohol Functional group having the structure $R—OH$, 57, 70

　naming, 845–846

　reactions of, 857–863

　　conversions into ethers as, 858–859

　　dehydration of, 859–861, 860f

　　oxidation of, 861–863

　　as weak acids and bases, 857–858, 857t, 858f

　spectroscopy of, 849–853

　　infrared, 849–850, 849f–850f

　　mass, 852–853, 852f

　　nuclear magnetic resonance, 850–851, 850f

　structure and reactivity of, 847–849, 848f–849f

　synthesis of, 853–856

　　aldehydes and ketones, reduction of, 854–855

　　esters and carboxylic acids, reduction of, 855–856

　　organic chemistry, oxidation/reduction in, 853–854, 854f

aldehyde Functional group having the structure $RC(=O)H$, with a carbonyl group bonded to one hydrogen atom, 883–888

　naming, 883–885, 884t

　oxidation of, 907–908

　reduction of, 901–903, 902f

　spectroscopy of, 886–887

　synthesis of, 888

aldonic acid, 1125

alicyclic (*aliphatic cyclic*) compounds, 297

Correlation of ¹³C Chemical Shift with Environment

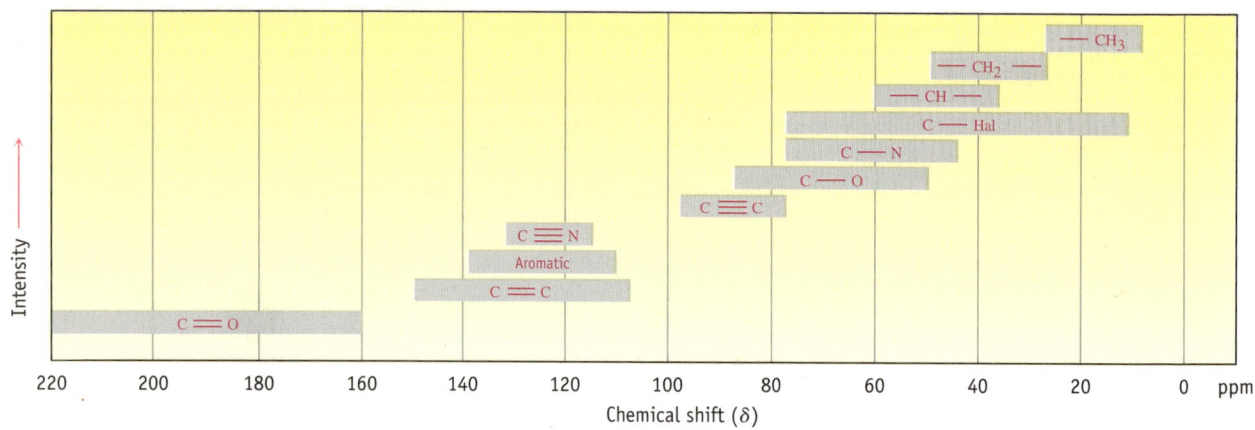

Correlation of ¹H Chemical Shift with Environment

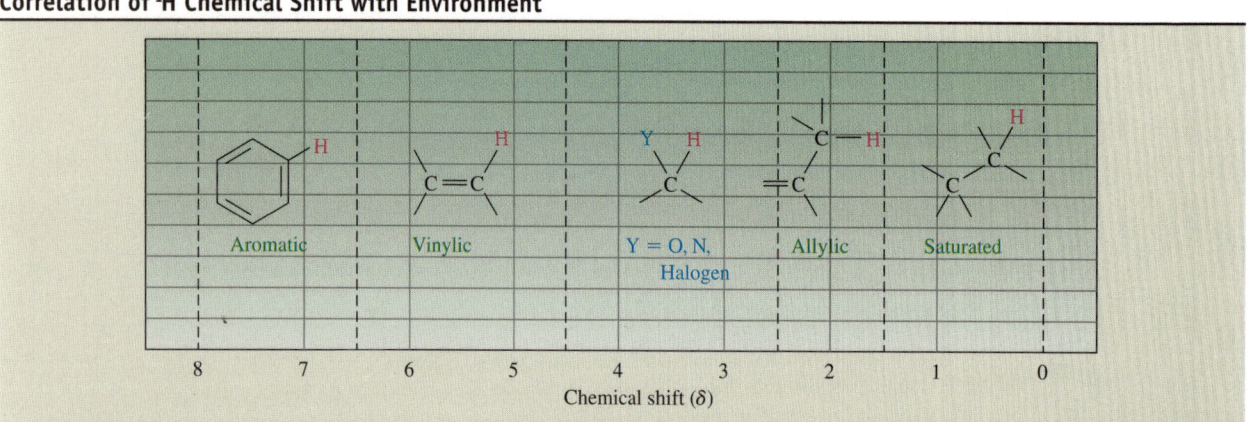

Type of Hydrogen		Chemical Shift (δ)	Type of Hydrogen		Chemical Shift (δ)
Reference	$(CH_3)_4Si$	0			
Saturated primary	$-CH_3$	0.7–1.3	Alkyl halide X = Cl, Br, I		2.5–4.0
Saturated secondary	$-CH_2-$	1.2–1.6	Alcohol		2.5–5.0 (Variable)
Saturated tertiary		1.4–1.8	Alcohol, ether		3.3–4.5
Allylic		1.6–2.2	Vinylic		4.5–6.5
			Aromatic	Ar—H	6.5–8.0
Methyl ketone		2.0–2.4	Aldehyde		9.7–10.0
Aromatic methyl	Ar—CH₃	2.4–2.7			
Alkynyl	$-C{\equiv}C-H$	2.5–3.0	Carboxylic acid		11.0–12.0